MATERIALS HANDLING HANDBOOK

MATERIALS HANDLING HANDBOOK

2nd Edition

Editor-in-Chief

RAYMOND A. KULWIEC

Executive Editor
MODERN MATERIALS HANDLING *Magazine*

Formerly Senior Editor, Material Handling
PLANT ENGINEERING *Magazine*

Sponsored by

THE AMERICAN SOCIETY OF MECHANICAL ENGINEERS

and the

INTERNATIONAL MATERIAL MANAGEMENT SOCIETY

A Wiley-Interscience Publication

JOHN WILEY & SONS

New York • Chichester • Brisbane • Toronto • Singapore

Library of Congress Cataloging in Publication Data:

Main entry under title:

Materials handling handbook.

"A Wiley-Interscience publication."
Includes index.

1. Materials handling—Handbooks, manuals, etc.
I. Kulwiec, Raymond A. II. American Society of Mechanical
Engineers. III. International Material Management
Society.

TS180.M315 1984 621.8'6 84-10443
ISBN 0-471-09782-9

Printed in the United States of America

10 9 8 7 6 5 4

BOARD OF REVIEWERS

CONTRIBUTING AUTHORS

Kenneth B. Ackerman
The K. B. Ackerman Company
Columbus, Ohio

Edmund P. Andersson
Computer Identics Corporation
Canton, Massachusetts

Sterling Anthony, Jr.
Sterling Anthony, Inc.
Detroit, Michigan

Thomas C. Aude
Pipeline Systems Incorporated
Orinda, California

Kishan Bagadia
Unik Associates
Milwaukee, Wisconsin

S. Gene Balaban
Telemotive Division
Dynascan Corporation
Chicago, Illinois

Ben Bayer
Aero-Go, Inc.
Seattle, Washington

Bruce Boldrin
Eaton-Kenway
Salt Lake City, Utah

James M. Cahill
Rapistan Division
Lear Siegler, Inc.
Oak Brook, Illinois

Robert J. Cammack
Taylor-Dunn Manufacturing Company
Anaheim, California

M. R. Carstens, Ph.D.
Atlanta, Georgia

A. J. Cason
Economy Engineering Company
Bensenville, Illinois

John Castaldi
Supreme Equipment & Systems Corporation
Brooklyn, New York

Guy A. Castleberry
Conco-Tellus Inc.
Mendota, Illinois

Conveyor Equipment Manufacturers Association
Vibrating Equipment Section
Washington, DC

Ralph M. Cox
Industrial Handling Engineers, Inc.
Houston, Texas

Thomas P. Cullinane, Ph.D.
Northeastern University
Boston, Massachusetts

Byrl Curry
Jervis B. Webb Company
Farmington Hills, Michigan

Robert E. DeCrane
K. L. Cook Associates Inc.
Sugarcreek, Ohio

William Devaney
Stanley-Vidmar
Allentown, Pennsylvania

John G. Dorrance
SI Handling Systems, Inc.
Easton, Pennsylvania

Lawrence Feit
Crane Consultant
Hebron, Indiana

Daniel T. Fitzpatrick
Prab Conveyors, Inc.
Kalamazoo, Michigan

Irving M. Footlik, P.E.
Footlik & Associates
Evanston, Illinois

Robert B. Footlik, P.E.
Footlik & Associates
Evanston, Illinois

W. Scott Fowler
Lyon Metal Products, Inc.
Aurora, Illinois

David L. Fowlston
The Raymond Corporation
Greene, New York

David R. Freeman, Ph.D.
Northeastern University
Boston, Massachusetts

Frank J. Gerchow, P.E.
Koppers Co., Inc.
Muncy, Pennsylvania

Saul B. Green
Green Associates, Inc.
Pittsburgh, Pennsylvania

B. J. Hinterlong
Continental Screw Conveyor Corporation
St. Joseph, Missouri

Clifton H. Hubbell
Sauerman Bros. Inc.
Bellwood, Illinois

John R. Huffman, Ph.D.
Semco, Sweet & Mayers, Inc.
Los Angeles, California

Charles A. Isenberger
Grove Manufacturing Company
Shady Grove, Pennsylvania

Jerry R. Johanson
Jenike & Johanson, Inc.
No. Billerica, Massachusetts

Herbert H. Klein
Unarco Materials Storage Division
Unarco Industries, Inc.
Chicago, Illinois

Garry A. Koff
Barrett Electronics Corporation
Subsidary of Mannesmann Demag Corporation
Northbrook, Illinois

Robert Kolatac
West Milford, New Jersey

Raymond A. Kulwiec
Modern Materials Handling Magazine
Boston, Massachusetts

Karl E. Lanker, P.E., C.M.C.
The Sims Consulting Group, Inc.
Lancaster, Ohio

William S. Lewis, P.E.
Jaros, Baum & Bolles
New York, New York

Gary E. Lovested, C.S.P.
Deere and Company
Moline, Illinois

Daniel Mahr, P.E.
Orba Corporation, a unit of Amca International Corporation
Fairfield, New Jersey

Charles E. Manley
The Raymond Corporation
Greene, New York

Lori A. May
Taylor-Dunn Manufacturing Company
Anaheim, California

Ambrose Mazzola, P.E.
Materials Handling Consultants, Inc.
Simpsonville, South Carolina

Frank J. Meiners
Carlisle Engineering Management, Inc.
Carlisle, Massachusetts

W. R. Midgley
Midgley, Clauer & Associates
Youngstown, Ohio

J. D. Mitchell, P.E.
Mar Hook & Equipment, Inc.
Sandy, Oregon

Edward Moon
Acco Babcock, Inc., Conveyor Div.
Detroit, Michigan

Richard Muther, P.E., C.M.C.
Richard Muther & Associates, Inc.
Kansas City, Missouri

Alex J. Nagy
Materials Handling & Warehousing Consultant
Farmington, Michigan

James Nolan
Stubbs, Overbeck and Associates, Inc.
Houston, Texas

William P. O'Connell
Yale Materials Handling Corporation
Flemington, New Jersey

John F. Oyler
Dravo Engineers and Constructors
Pittsburgh, Pennsylvania

Alexander Pomerantsev
California State University at Fullerton
Fullerton, California

Robert Promisel
H & P Associates
Edison, New Jersey

Daniel J. Quinn
Southworth, Inc.
Portland, Maine

S. H. Raskin
Tri-Cell Company
Rockwall, Texas

Robert R. Reisinger, P.E.
Acco Babcock, Inc. Hoist & Crane Division
York, Pennsylvania

Charles Rose
Coal Technology Consultants, Inc.
Birmingham, Alabama

Robert Roth
Webb-Norfolk Div.,
Jervis B. Webb Company
Cohasset, Massachusetts

John William Russell
Liberty Mutual Insurance Company
Glastonbury, Connecticut

William E. Sabina, P.E.
William E. Sabina, Inc.
Denver, Colorado

Brian W. Sanford
Litton Unit Handling Systems
Florence, Kentucky

George A. Schultz
Epstein Process Engineering, Inc.
Chicago, Illinois

Clark C. Simpson
Clark Equipment Company
Industrial Truck Division
Battle Creek, Michigan

E. Ralph Sims, Jr., P.E., C.M.C.
The Sims Consulting Group, Inc.
Lancaster, Ohio

Ted P. Smyre, P.E.
Jeffrey Div., Dresser Industries, Inc.
Woodruff, South Carolina

James M. Snyder, P.E., C.M.C.
The Sims Consulting Group, Inc.
Lancaster, Ohio

Ivan D. Stankovich, P.E.
Consulting Engineer
Piedmont, California

Terry Strombeck
Spacesaver Corporation
Fort Atkinson, Wisconsin

William H. Tanner
Productivity Systems, Inc.
Farmington, Michigan

Terry L. Thompson
Pipeline Systems Incorporated
Orinda, California

James A. Tompkins, Ph.D.
Tompkins Associates, Inc.
Raleigh, North Carolina

Harold VanAsselt
Rapistan Div., Lear Siegler, Inc.
Grand Rapids, Michigan

Richard C. Wahl
Vibra Screw Inc.
Totowa, New Jersey

Dennis B. Webster, Ph.D., J.D., P.E.
Auburn University
Auburn, Alabama

Donald J. Weiss
White Storage & Retrieval Systems, Inc.
Kenilworth, New Jersey

Lyle F. Yerges, P.E.
Consulting Engineer
Downers Grove, Illinois

Dr. A. T. Yu, P.E.
ORBA Corp., a unit of Amca International Corp.
Fairfield, New Jersey

Fred Zacharias
Buckhorn, Inc.
Cincinnati, Ohio

Frederick A. Zenz, P.E.
Frederick A. Zenz, Inc.
Garrison, New York

FOREWORD

Publication of the first edition of the *Materials Handling Handbook* of 1958 was a major step forward in the synthesis of the many aspects of materials handling engineering into a single comprehensive source of reference. In recent years, however, it has become increasingly clear that the time had come to prepare a new edition of the *Handbook* that would reflect the changes and developments that have taken place in the industry in the quarter century since the first publication.

To carry out this project, the sponsoring societies, The American Society of Mechanical Engineers and the International Material Management Society (successor to the American Material Handling Society), appointed a Joint Handbook Committee, and charged it with the responsibility of establishing an operating budget, preparing a draft outline, seeking qualified contributing authors and reviewers, and selecting the Editor-in-Chief.

Members of the ASME and IMMS who served on the Joint Committee include:

Ambrose Mazzola, P.E. (ASME)—Chairman
Materials Handling Consultants, Inc.
Simpsonville, South Carolina

Fred Sudhoff, P.E. (IMMS)—Vice Chairman
Procter & Gamble
Cincinnati, Ohio

Ralph M. Cox, P.E. (IMMS)
Industrial Handling Engineers, Inc.
Houston, Texas

Irving M. Footlik, P.E. (IMMS)
Footlik & Associates
Evanston, Illinois

Harry R. Mack, P.E. (ASME)
Smith, Hinchman & Gryll Associates
Detroit, Michigan

John F. Oyler, P.E. (ASME)
Dravo Corp.
Pittsburgh, Pennsylvania

William E. Sabina, P.E. (ASME)
William E. Sabina, Inc.
Denver, Colorado

David L. Schaefer, (IMMS)
David L. Schaefer, Inc.
Elkins Park, Pennsylvania

George Van Hare, Jr., P.E. (IMMS)
Fenco Engineers, Inc.
Toronto, Ontario, Canada

George H. Weaver (ASME)
Eaton Corp.
Philadelphia, Pennsylvania

Jack D. Wetzel, P.E. (ASME)
FMC Corp.
Colmar, Pennsylvania

Most of the major disciplines comprising the overall field of materials handling were represented on the Committee, whose members gave selflessly of their time and talent in numerous meetings over a three-year period in order to complete their assigned task.

The fruition of this project, through the publication of the second edition of the *Materials Handling Handbook,* was made possible by the strong moral and financial support of the sponsoring societies; by the dedicated contributions of the Joint Committee; by the tireless efforts of the Editor-in-Chief; and by the professional competence of the many contributing authors and reviewers.

And although all those involved take pride in its completion, it is ultimately up to you, the reader, to judge the extent to which we have succeeded in meeting our oft-stated objective of providing in this volume, ". . . a useful and objectively written working tool for users and designers of materials handling systems and equipment."

AMBROSE MAZZOLA, P.E.
Chairman, Joint ASME/IMMS Handbook Committee

PREFACE

The history of materials handling is as broad and fascinating as that of industry itself. A variety of manual handling techniques and mechanical aids have been described and illustrated throughout the records of mankind's development. Early records show many uses of simple levers and slings. Handling was already well advanced in ancient Egypt, and further developed in Greek and Roman civilizations. Uses of treadmills, cranes, pulleys, and other devices are well documented in medieval literature.

Further materials handling developments took place during the Industrial Revolution, and giant strides were made in the early twentieth century, when the fledgling automotive industry began to use conveyors and mechanical assembly lines.

The developments in materials handling technology that played such a pivotal role in World War II have been well documented. Personnel shortages caused by the war effort led to the use of mechanical methods for handling unit loads, and one of the major symbols of materials handling—the lift truck—came into its own during that period.

Another major development at that time was the extra-long-span, multiple-runway crane. Some of the World War II aircraft plants had clear spans of up to 300 feet. Cranes having up to eleven runways were developed during this period to provide overhead hoist coverage of these areas.

The developments in materials handling technology accelerated during the war were subsequently turned to use in industry in post-war years.

Today, the successful operation of every process plant, manufacturing facility, and warehouse depends to a large extent on the efficient handling, storage, and flow of materials. Depending on the industry and type of operation, materials handling may account for 30–75% or more of the cost of making a product.

Yet inefficient materials handling does not always surface as the root cause of a problem, because its role is often obscured by other functions. What is in reality a materials handling problem may be perceived as a production problem, or an inventory or quality control problem.

In the past, this identity problem was compounded by the lack of a central reference source on materials handling for engineers in industry. All too often, working engineers had to consult numerous industry sources, handbooks, and texts, in order to accumulate the basic information they needed to solve materials handling problems in their facilities.

A major breakthrough took place in 1958, when the 1st Edition of the *Materials Handling Handbook* was published by The Ronald Press Company (subsequently acquired by John Wiley & Sons, Inc.). The *Handbook* provided, for the first time, a single-source reference work for engineers and other technical professionals and managers engaged in work in unit and bulk materials handling.

Today industry stands on the threshold of what many have termed the Second Industrial Revolution. Where mechanization was viewed as the major advance of the World War II era and the post-war period, the 1980s and beyond will be seeing the increasing advance of the systems concept, culminating in the automatic plant and automatic warehouse. Rapid developments in computer technology have helped spur this trend. Predictably, materials handling represents the "glue" that brings and holds together all the elements of the modern manufacturing and processing plant and distribution facility.

This second edition is designed to meet today's needs for updated technical information on materials handling. Its emphasis is on presenting data that the working engineer can use to select, size, or apply equipment and systems. Tables, charts, graphs, calculations, and illustrations are presented wherever appropriate.

The American Society of Mechanical Engineers and International Material Management Society are the sponsoring societies for this *Handbook*. Although they have provided funding and human resources necessary to support development of the *Handbook,* the societies, and the Editor-In-Chief, disclaim responsibility for any information provided by individual contributing authors. At the same time, they encourage the reader to seek qualified professional assistance where needed in implementing new or improved materials handling systems.

The success of an undertaking of this type depends heavily on the contributing authors and reviewers. Over 100 professionals in various aspects of materials handling have participated in this venture. These distinguished contributing authors and reviewers are listed on the preceding pages. This *Handbook* would not have been possible without their dedicated efforts. Appreciation is also extended to the organizations with which these individuals are associated, for the support that was provided in many cases.

The Editor would like to extend personal thanks to many friends in industry who have provided assistance and advice during this venture. They are too numerous to mention here, but their support has been invaluable. Special appreciation is expressed to Leo Spector, Editor, and Robert C. Baldwin, Managing Editor, *Plant Engineering* Magazine, for their cooperation. Finally, I would like to thank Mary Claire, Ray, and Alexia Kulwiec for their patience, understanding, and support.

RAYMOND A. KULWIEC

Arlington Heights, Illinois
September 1984

CONTENTS

MATERIALS HANDLING HANDBOOK

PART I
INTRODUCTION

CHAPTER 1
BASIC MATERIALS HANDLING CONCEPTS

RAYMOND A. KULWIEC

Modern Materials Handling **Magazine**
Boston, Massachusetts

1.1 DEFINITION AND SCOPE OF MATERIALS HANDLING

Materials handling has often been called "the art and science of moving, packaging, and storing of substances in any form." However, in recent years it has taken on broader connotations.

To begin with, the concept of time and place utility must be incorporated into any definition of materials handling. Products are of no value to the retailer, for example, unless they are available for display on the shelf when the customer is in the store. In a similar vein, a part or tool is of little value in an assembly or manufacturing operation unless it is available at a specific workplace, when needed at a specific time in the work cycle. Thus, properly applied, materials handling not only moves, packages, and stores, but performs these functions within specified parameters of time and space.

Materials handling also should be considered within a system context. Rarely, if ever, are activities performed in one area or department of a facility without having an impact on other operations. Examples: The efficiency of a storeroom will affect the efficiency with which production operations are performed out on the shop floor. The positioning of a conveyor line in a plant might improve materials flow through the facility, or it could present a hindrance to plant traffic. A significant improvement in the efficiency of one operation, without a corresponding improvement in a subsequent step in the work sequence, may only result in a piling up of materials down the line.

These simple examples illustrate the point that to maximize overall productivity of the plant or warehouse, the materials handling steps that support production, order assembly, and other operations must be integrated into a system of activities, rather than being viewed as a number of isolated, independent procedures.

In addition to considering time and place utility and the systems approach, a thorough definition of materials handling must also include the human aspect. People are always a part of materials handling whether the operation is a simple one, involving only a few items of equipment, or a large, complex, automated system. Operators must carry out required procedures. Maintenance personnel keep the equipment working properly and keep downtime to a minimum. Foremen and supervisors oversee overall operations, making sure they meet the objectives of the department or plant. Training in operating procedures, and in safety practices, is usually required to make handling operations pay off as expected.

The facility or space in which operations are housed should be considered part of the system. Building configuration, levelness tolerances, fire protection measures, and energy requirements all impact the manner in which the system functions.

Finally, the definition of materials handling must contain an economic consideration. Certainly the delivery of parts and materials to a specific point, and at a specific time, it not completely meaningful unless accomplished at an acceptable cost so that an adequate return is realized.

Considering all the factors, a more complete definition might be the following:

Materials handling is a system or combination of methods, facilities, labor, and equipment for moving, packaging, and storing of materials to meet specific objectives.

It is important to note the factors that are not part of the definition, as well as those that are. For example, size is not a part of the definition, nor is degree of mechanization. A materials handling operation can be simple and small, and involve only a few pieces of basic equipment. Or, it may be large, complex, or automated.

1.2 THE SYSTEMS CONCEPT IN MATERIALS HANDLING

Originally a methodology used in military and aerospace planning, the systems point of view has been applied in recent years in industry, and in particular industrial materials handling, with considerable success. Most recently, the concept has been extended with the broad application of solid-state electronic controls, and wide proliferation of computer technology.

Scanners and other control devices now provide on-line or "real time" data on the status of equipment and materials in an operation or entire plant. This information in turn forms the basis for actuating controls or for making operating decisions. The linking of materials handling activities with controls and computers has produced a potential for attaining levels of productivity that were unheard of a few years ago, and has made the automated factory and warehouse a real possibility.

A materials handling system can encompass an entire plant and, in some cases, even the facilities of suppliers and customers. In a manufacturing plant, for example, it may begin at the receiving dock, and continue through inspection, storage, processing, packaging, and shipping. It can also include packaging and shipping operations at the supplier's plant, as well as unloading and handling at the customer's site. Within the overall system, smaller systems or subsystems operate in various areas of the plant. Ideally they should be integrated into the entire operation, so as to optimize materials handling efficiency throughout the facility.

Regardless of size and complexity, a materials handling system should contain two parallel flows—the physical flow of materials, and a corresponding flow of information (Fig. 1.2.1). It is necessary to know why an item is moving past a particular point at a given time, where it is going, and what is to be done with it next in order to fulfill operating objectives. The flow of information, therefore, provides the basis for controlling the operation.

Depending on the type of system, the control mode may be manual, mechanized, or automatic. The basic manual mode is typified by individuals observing operations, making records, and taking corrective actions when needed. Job tickets, move orders, and control documents may accompany materials traveling their routes in containers or on pallets.

Mechanized control systems include the use of limit switches, photoelectric cells, and other devices to actuate stopping, starting, diverting, accumulating, and other functions.

Automated modes include the use of code-reading scanners (including laser scanners) that either record operating information or cause actions to take place, or both. The system may be under control of a local computer, which in turn may be linked to a higher-level computer. Small subsystems may in turn be directed by programmable controllers or microcomputers.

A high degree of automation or even mechanization may not be required—or even desired—for many operations. However, as the trend toward integrating individual operations into systems continues, the engineer should keep possible future information needs of the system or the organization in mind while designing an operation to meet current needs.

The following benefits can be achieved by applying the systems approach to materials handling:

Better adaptability to control
Better coordination with suppliers and customers
Continuous flow of materials and information
Fewer delays between operations and departments
Higher levels of equipment utilization
Improved scheduling
Less product damage
Lower labor costs
Optimum return on investment
Reduced inventories
Reduced space needs
Safer, more systematic work procedures

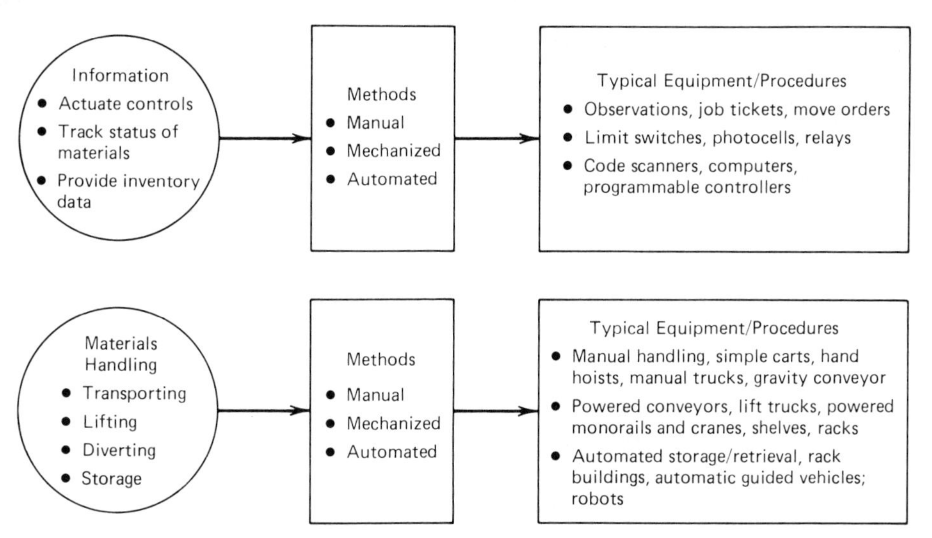

Fig. 1.2.1 In a materials handling system, the physical flow of materials, bottom, is accompanied by a parallel flow of information, top. The information flow provides the basis for decision-making and for controlling the physical materials flow.

1.3 MAJOR CLASSIFICATIONS—UNIT AND BULK

The "material" in materials handling can be any substance from which a product is assembled or packaged for further processing or sale. Or, it is any item that is assembled, packaged, and distributed for further use.

There are two broad categories of materials—unit and bulk. Units are separate, discrete items, ranging in size from (1) nuts and bolts to (2) pallet loads of bags, (3) car bodies, and (4) airplane wings. They are characterized by the fact that they can be distinguished as separate entities.

Common examples of unit handling include automotive assembly lines, order picking in warehouses, can conveying in food plants, and stacking materials in storage racks with lift trucks.

Bulk materials, on the other hand, are stored and handled in volume, often in unpackaged form. Examples include dry powders, granules, flakes, and resins. Coal, fertilizer, sulfur, and salt are other common types.

Bulk materials handling is characterized by continuous-flow operations, involving materials in an aggregate form. In many cases the bulk materials assume flow characteristics similar to those of fluids.

Elements of a typical bulk materials handling system are: (1) bins, silos, or hoppers for storing the material, (2) discharge devices or feeders, (3) conveyors, and (4) flow-aid devices when needed.

Common elements of a unit handling system include lift trucks, conveyors, storage racks, and overhead cranes and hoists.

This handbook covers the two basic types of materials handling in separate, major parts. Special attention is paid to the important role of material characteristics in bulk solids handling, and to the various types of equipment options possible in unit handling applications.

1.4 MATERIALS HANDLING AND PRODUCTIVITY

Materials handling is increasingly being recognized as a primary tool for improving productivity. Thus, any evaluation of alternative materials handling plans must consider how each approach will affect the productivity of the facility or operation it is intended to serve.

The basic measure of productivity is the ratio of output to input. The ratio can be expressed in terms such as number of damaged loads per total number of loads, cases packed per labor hour, items stored per square foot of space, and so on. Such ratios are used to show how efficiently resources are being used to generate work, products, or levels of service. They provide a measuring stick of relative performance. In general they are not used to measure absolute values against a standard. At present there are few industry standards available relative to materials handling productivity.

The primary value, then, of these ratios lies in their use for monitoring performance over time. Comparisons can be made against ratios achieved during past periods. Or, if data are available, a plant's performance can be compared with that achieved by other plants in the same industry, or in factories or warehouses where similar types of handling take place. Trends or changes in productivity measures can be used to evaluate performance of a system, and point to the need for corrective action where appropriate. Some common productivity ratios related to materials handling are as follows:

Manufacturing Cycle Efficiency (MCE)

$$\text{MCE} = \frac{\text{Actual production machine time}}{\text{Time spent in production department}}$$

Among other things, this ratio can be an indicator of the efficiency with which materials handling is conducted at the workplace. In many instances, machines are idle because of delays caused by inefficient scheduling of parts and tools, ineffective storage and retrieval operations, and slow machine loading and unloading.

Materials Handling Labor (MHL) Ratio

$$\text{MHL} = \frac{\text{Personnel assigned to materials handling duties}}{\text{Total operating personnel}}$$

This ratio can be interpreted in various ways, depending on conditions prevailing in a particular organization. For example, it can reflect a company's commitment to providing full-time staffing to materials handling functions. On the other hand, it can also be indicative of a large proportion of tasks in the plant being performed manually, rather than being mechanized.

Storage Space Utilization (SSU) Ratio

$$\text{SSU} = \frac{\text{Storage space occupied}}{\text{Total available storage space}}$$

This ratio, which should be based on cubic space rather than on floor space, is a measure of the efficiency of storage. It also bears a relationship in warehouses to the following measure.

Aisle Space Percentage (ASP)

$$\text{ASP} = \frac{\text{Space occupied by aisles}}{\text{Total space}}$$

The following are some other common ratios.

$$\text{Equipment Utilization (EU)} = \frac{\text{Actual output}}{\text{Theoretical output}}$$

$$\text{Damaged Loads (DL) Ratio} = \frac{\text{Number of damaged loads}}{\text{Total number of loads}}$$

$$\text{Inventory Turnover (IT)} = \frac{\text{Annual sales (or cost of goods sold)}}{\text{Average annual inventory investment}}$$

$$\begin{array}{c}\text{Receiving (shipping)}\\ \text{Productivity Ratio}\end{array} = \frac{\text{Weight received (shipped) per day}}{\text{Labor hours per day}}$$

$$\begin{array}{c}\text{Throughput Performance}\\ \text{Index (TPI)}\end{array} = \frac{\text{Actual throughput per day}}{\text{Daily throughput capacity}}$$

$$\begin{array}{c}\text{Output per}\\ \text{labor-hour}\end{array} = \frac{\text{Dollar shipment/price deflator}}{\text{Total labor-hours}}$$

$$\begin{array}{c}\text{Unit Labor}\\ \text{Cost}\end{array} = \frac{\text{Payroll including fringe benefit costs, in dollars}}{\text{Shipments, in dollars/price deflator}}$$

$$\begin{array}{c}\text{Labor Cost as}\\ \text{\% of Sales}\end{array} = \frac{\text{Payroll including fringe benefit costs, in dollars}}{\text{Sales, in dollars}}$$

1.5 MATERIALS HANDLING AND SAFETY

Over one-fifth of all reported industrial accidents can be attributed to handling activities, according to industry studies. Such accidents particularly involve lifting and related types of manual effort. Thus, both mechanized and automated materials handling equipment can play a key role in saving life and limb in the industrial plant or warehouse. The use of even simple, manually operated types of trucks, hoists, and ramps can go a long way toward providing a safe workplace.

Obviously equipment alone cannot do the job, however. Achieving safe working conditions requires the commitment of top management in supporting a plant-wide safety program. It also requires a thorough training program applied to workers, supervisors, and other personnel in contact with plant operations and equipment.

A primary reason for having a safety program is a humanitarian one—ensuring the well-being of workers. In addition, many safety programs and practices are now mandated by federal, state, and local governments. But, the need for compliance aside, safety has also been proven to be good business. Insurance statistics indicate that companies with thorough, well-planned safety programs generally show better bottom-line results than those that do not.

Usually, an injury to a worker does not affect the productivity of that worker alone. Often a work-related injury involves downtime, and the idling of many other operators. And, it often also involves damage to equipment and products. Insurance data show that companies not having a well-planned and enforced safety program typically pay twice—once in the cost of downtime and the direct costs of the incident, and secondly in the form of increased insurance premiums.

Although materials handling equipment can improve the safety of operations, it is incumbent upon engineers and managers to make sure that the new equipment does not also introduce new hazards. Safety designs should be checked carefully before equipment is purchased. Typical types of equipment safety features that should be considered include the following:

Hoist Braking Systems. Typically, electrical hoists are equipped with two braking systems, a motor brake and a supplementary load or control brake.

Lift Truck Safeguards. Standards have been developed for use of certain types of trucks in specific kinds of operating environments. Overhead guards and other safeguards are prescribed by law, and units using attachments must be derated. Overload protection systems also are available.

Cranes. Control systems are used to prevent excessive swinging of loads. Runways are equipped with bumper stops, and anticollision controls are available.

Conveyors. Guards are used to cover nip and pinch points. Proper location of on–off controls is essential, and warning labels are provided at critical areas.

Storage Racks. Ability to withstand seismic forces is designed into some units. Fire protection systems are an important part of rack installations, particularly high-rise configurations.

Industrial Batteries. Charging rooms must be properly ventilated, and battery lifting equipment provided. Special procedures are followed to avoid accidental spills and injuries.

Periodic inspections are an important part of a materials handling safety program. Detailed inspection procedures and recommended frequencies are available from most equipment dealers and manufacturers.

1.6 MATERIALS MANAGEMENT

Properly applied materials handling is, as has been noted, a system of interrelated handling activities. The materials handling function is, itself, part of a larger system of interrelated plant or corporate functions. In some companies, this system, or grouping of disciplines, has been formally organized under the heading of "Materials Management." This function coordinates and directs all those activities concerned with the control of materials. These activities or responsibilities include (1) purchasing, (2) materials handling, (3) packaging, (4) production and inventory control, (5) receiving and shipping, (6) distribution, and (7) transportation.

Increasingly frequent tools or methodologies for directing materials management are Material Requirements Planning (MRP I) and Manufacturing Requirements Planning (MRP II). These are time-phased systems of procuring and scheduling materials tied to specific operating plans or manufacturing schedules.

In organizations having a materials management structure, materials handling obviously plays a key role in controlling materials, whether they are flowing through the plant, or being stored, counted, or inspected. Functional relationships can vary, depending on the organization. Perhaps more important than any particular organization chart, however, is making sure that materials handling is identified as a separate function, and recognizing its full contribution to overall operating efficiency and the profitability of the enterprise as a whole.

1.7 MATERIALS HANDLING AT WORK

It is difficult to conceive of a situation, in either private life, industry, or commerce, where some form of handling does not come into play. Common, everyday examples of facilities where materials handling is practiced include the grocery supermarket, the bank, the library, and the airport luggage area. Some typical examples of the key role played by materials handling in industry, commerce, and institutions are examined briefly in this section.

1.7.1 Manufacturing

The factory conveyor belt and the progressive automotive assembly line are common, long-standing examples of materials handling in industrial manufacturing. Others can be found in any facility. In fact, depending on the industry, materials handling may account for 30–75%, or more, of the cost of making a product. Handling always has to be performed, the question is how efficiently it can be done. When materials handling is inefficient, telltale signs begin to appear, such as those in the listing shown in Fig. 1.8.1.

Materials handling equipment often is selected to support the type of manufacturing layout used. A job-shop or batch processing line, for example, usually has a relatively random pattern of activities or moves. The materials handling equipment used to support such operations tends to be of a general-purpose type. Industrial trucks, overhead cranes, and air-film devices are typical examples.

A line or flow pattern of manufacturing (such as an assembly line) is more structured and sequenced, and thus has little or no backtracking. More specialized, fixed-path types of equipment are commonly used, such as towlines or overhead trolleys or power-and-free conveyors.

Modern flexible manufacturing systems provide a highly productive, generally automated means of batch processing, under computer control. Usually, handling equipment also is under the system control, and includes such devices as automated guided vehicles that bring parts and tools to the workplace, and robots that feed materials to machines, or directly perform processing steps such as welding and painting.

1.7.2 Processing

In process industries, such as those that produce chemicals, plastics, fertilizers, pharmaceuticals, and food, the handling of dry bulk solids materials is an important part of the entire operation. A classic and familiar type of solids handling equipment is the belt conveyor, which consists of an endless belt, supported by idlers, that forms a trough for holding and moving materials along long inclines. Such units are commonly seen operating at quarries, power stations, and large processing plants.

A typical sequence of equipment and operations in many bulk handling systems includes (1) a conveyor system (pneumatic, belt, screw, other) that carries incoming materials from receiving and deposits them into (2) a storage bin or silo. When called for, materials are withdrawn from the bin, often passing through (3) a hopper section, sometimes with the aid of (4) a flow-promoting device (pneumatic, mechanical, vibratory). From the discharge opening, solids are removed by (5) a feeder that charges them at a controlled rate to (6) a takeaway conveyor, which moves the materials to processing operations. Often the success or failure of a bulk handling installation depends on whether material characteristics have been carefully evaluated.

1.7.3 Coal Handling

A type of bulk handling that is common to both processing plants and manufacturing facilities is the movement of coal to boilers for producing both environmental and process heat. The heart of most coal handling systems is, again, a conveyor; but other associated types of equipment such as trippers, chutes, silos, and grates also are highly important.

Coal handling is also an extremely important function in power generating plants. The type of equipment and systems design used depends on (1) the grade and quality of coal, (2) the volume of the operation, and (3) whether a new or existing installation is involved.

1.7.4 Warehousing

One of the classic applications of materials handling can be found in the warehousing function. Warehousing can be defined as the activity concerned with the orderly storage and issuing of goods or products, either within the plant proper, or at remote locations. The remote facilities may be operated by a manufacturer, or by various agents in the distribution process.

Materials handling activities that parallel the physical flow of materials include (1) receiving at the dock, (2) identification and sorting, (3) inspection, (4) storage, (5) order picking, (6) order assembly, (7) packing, (8) loading, (9) shipping, and (10) record-keeping. The end goal of materials handling in a warehouse is the efficient distribution of goods and products.

1.7.5 Hospitals

A full-service hospital actually functions as both a distribution center and a processing facility, with materials handling activities paralleling those found in industry.

Elements common to most hospitals include (1) central stores and distribution, (2) mail handling, (3) central processing, (4) pharmacy operations, (5) dietary services, and (6) laundry and clothing services.

As in a plant, incoming materials are first received, checked, unpacked, and sent to stores, where an inventory is maintained. The materials are then distributed to using departments as needed.

Meanwhile, the pharmacy department is typically responsible for its own purchasing, receiving, storing, and packaging of items such as drugs and intravenous solutions. And materials handling goes on continually in service departments such as the mailroom, laundry, and kitchen.

1.7.6 Office Buildings

A unique category of handling requirements are found in modern office buildings, incorporating vertical and horizontal transport of people, equipment, furniture, supplies, mail, and waste materials. Typical types of equipment include docks and dock levelers, lift trucks, elevators, mail carts and automated mail carriers, pneumatic tube conveyors, containers, and chutes.

1.7.7 Stores

In department stores, numerous examples of transportation and materials handling can be found, all targeted toward getting merchandise on display at appropriate locations for customer viewing and selection. Typical types of handling equipment include garment conveyors and racks, merchandise freight elevators, and cantilever racks for storing furniture.

The grocery supermarket provides an excellent example of materials handling at work. Typical kinds of equipment that can be readily seen include shelves, containers, pallets, pallet trucks, carts, checkout counter conveyors, and bar-code scanners. Dock lifts and waste handling equipment are at work in the rear areas of the stores not normally seen by customers.

Besides the examples cited, other common facilities that are obvious users of materials handling equipment include mail and parcel services, airports, restaurants, commercial kitchens, libraries, and military bases.

1.8 SOLVING MATERIALS HANDLING PROBLEMS

Usually there are various alternative methods of handling materials in any given facility. The question is, how does one go about selecting the right approach? The following sequence of steps is a recommended approach for solving materials handling problems:

Identify and define the problem.

Collect relevant data.

Analyze the data.

Evaluate alternative approaches.

Select the preferred approach.

Develop a plan.

Implement the solution.

1.8.1 Identifying the Problem

Identifying a materials handling problem is not always an easy task. Often problems in a plant are attributed to other factors, such as production or quality control, when the underlying cause actually stems from the handling approach being used.

The checklist in Fig. 1.8.1 can be used as a starting point and aid in spotting handling problems in an existing plant. Checklists of this type should not be relied on in and of themselves; sometimes they can be misleading. However, in general they do help to spot symptoms that more often than not are associated with poor materials handling practices.

- ☐ Crowded operating conditions
- ☐ Cluttered aisles
- ☐ Cluttered docks
- ☐ Poor housekeeping
- ☐ Jam-ups in service departments
- ☐ Backtracking in materials flow
- ☐ Obstacles in materials flow
- ☐ Manual loading and unloading
- ☐ Manual handling of loads weighing more than 50 lb
- ☐ Two-man lifting jobs
- ☐ Excess temporary storage
- ☐ Excess time spent retrieving stored goods
- ☐ Unused building cube space (air rights)
- ☐ Excessive rehandling (too much picking up, setting down)
- ☐ Single pieces handled instead of unit loads
- ☐ Production delays
- ☐ Idle equipment and machines
- ☐ High damage rate
- ☐ High demurrage charges
- ☐ High indirect labor costs
- ☐ Skilled employees waste time handling materials
- ☐ Materials handling equipment more than 10 years old

Fig. 1.8.1 Materials handling problem checklist.

A good starting point in evaluating the quality of materials handling in an existing facility is to take a plant tour, armed with a clipboard and checklist. While trying to spot telltale signs of inefficient materials handling, the observer should also try to see if relationships can be noted between the different problems. First, the engineer or analyst should look for relationships between problems *within* an operating section or department. The next step is to look for relationships among problems in different departments. It is important that handling efficiency be optimized throughout the facility, not just within one isolated department.

1.8.2 Defining the Problem

Once a problem has been identified, the next important step is to define it fully. The problem definition must include its scope.

For example, suppose that considerable clutter and confusion exist at a work area on the shop floor of a plant. What is the scope of the problem? Is it limited to the work area itself, and attributable to a lack of on-site storage facilities, or poor workplace handling practices?

Or, might the problem also encompass the way materials are being delivered from the receiving area, or from the adjacent department? Perhaps the delivery of materials should be in an even flow throughout the day, rather than in large, staggered increments. Possibly the difficulty is caused by a poor layout of the production area.

The problem definition should, wherever possible, contain quantitative information. How many feet or yards is the adjacent department away from the work site? How many square feet of floor space, or cubic feet of space, have been allocated to storage? How many different parts and tools are involved, and how have they been organized before delivery to production?

1.8.3 Collecting Data

Answers to some of the necessary questions may not be immediately available. Rather, some data collection and analysis may be necessary in order to uncover the desired information. Often information must be developed regarding the flow of materials through the facility, along with the types of moves that take place. Various graphical and analytical techniques can be used. They are discussed in Chapter 2.

Care must be taken to ensure that the data being generated and collected are reliable. The solution that is eventually developed can be only as good as the data on which it is based.

1.8.4 Evaluating Alternatives

Once appropriate data have been assembled and analyzed, the engineer can begin developing and evaluating alternative solution plans. A good first step in building a solution is to consider the 20 Principles of Materials Handling (Fig. 1.8.2). These principles are a distillation of accumulated experience and knowledge on the part of many practitioners and students of materials handling. They have been compiled by the College–Industry Council on Material Handling Education [1]. As with any such listings, they should be viewed as general principles that can be used as a starting point in developing a solution. However, they do not represent absolute rules in any sense. Rather, they should be combined with other factors before arriving at a solution. For example, although the use of gravity should be encouraged whenever practical, in certain applications powered conveyors are clearly the preferred solution when compared with gravity chutes.

The elements of a materials handling solution include people, equipment, facilities, dollars, and time. Thus formulation of a solution involves questions of the following type:

How many operators will be involved?

What kind of training will they require?

How many supervisors will be needed?

How large a maintenance staff will be needed, and what types of skills should they have?

What types of equipment will be used?

What are the power requirements?

Will a new building addition be required?

How soon can we get on stream?

How much will it cost?

What is the expected return on investment?

To answer these and related questions, both technical and economic factors must be considered. Usually the primary technical factor is a thorough knowledge of the types of handling equipment available, their advantages and disadvantages for specific applications, their purchase, installation, and

The Twenty Principles of Material Handling

1. **Orientation Principle:** Study the system relationships thoroughly prior to preliminary planning in order to identify existing methods and problems, physical and economic constraints, and to establish future requirements and goals.
2. **Planning Principle:** Establish a plan to include basic requirements, desirable options, and the consideration of contingencies for all material handling and storage activities.
3. **Systems Principle:** Integrate those handling and storage activities which are economically viable into a coordinated system of operation including receiving, inspection, storage, production, assembly, packaging, warehousing, shipping and transportation.
4. **Unit Load Principle:** Handle product in as large a unit load as practical.
5. **Space Utilization Principle:** Make effective utilization of all cubic space.
6. **Standardization Principle:** Standardize handling methods and equipment wherever possible.
7. **Ergonomic Principle:** Recognize human capabilities and limitations by designing material handling equipment and procedures for effective interaction with the people using the system.
8. **Energy Principle:** Include energy consumption of the material handling systems and material handling procedures when making comparisons or preparing economic justifications.
9. **Ecology Principle:** Minimize adverse effects on the environment when selecting material handling equipment and procedures.
10. **Mechanization Principle:** Mechanize the handling process where feasible to increase efficiency and economy in the handling of materials.
11. **Flexibility Principle:** Use methods and equipment which can perform a variety of tasks under a variety of operating conditions.
12. **Simplification Principle:** Simplify handling by eliminating, reducing, or combining unnecessary movements and/or equipment.
13. **Gravity Principle:** Utilize gravity to move material wherever possible, while respecting limitations concerning safety, product damage and loss.
14. **Safety Principle:** Provide safe material handling equipment and methods which follow existing safety codes and regulations in addition to accrued experience.
15. **Computerization Principle:** Consider computerization in material handling and storage systems, when circumstances warrant, for improved material and information control.
16. **System Flow Principle:** Integrate data flow with the physical material flow in handling and storage.
17. **Layout Principle:** Prepare an operational sequence and equipment layout for all viable system solutions, then select the alternative system which best integrates efficiency and effectiveness.
18. **Cost Principle:** Compare the economic justification of alternate solutions in equipment and methods on the basis of economic effectiveness as measured by expense per unit handled.
19. **Maintenance Principle:** Prepare a plan for preventive maintenance and scheduled repairs on all material handling equipment.
20. **Obsolescence Principle:** Prepare a long range and economically sound policy for replacement of obsolete equipment and methods with special consideration to after-tax life cycle costs.

Fig. 1.8.2 The 20 principles of materials handling.

operating costs, and their adaptability to different situations. Plant visits, seminars, short courses, trade shows, and business publications are all good sources of information.

1.8.5 Choosing the Solution

Whenever possible, tests should be applied to various alternative approaches. In some cases, alternative schemes might be tested with simulation models or other quantitative techniques. In other cases, particularly those involving bulk solids, laboratory or pilot-plant runs may provide required data.

The various proposed solutions should also be tested against economic criteria. Factors such as cash flow, investment tax credit, and income tax must always be taken into account.

Technical and economic factors involved in analyzing potential materials handling solutions are covered in Chapters 2 and 3. The expertise of consultants and suppliers can be utilized in developing and evaluating solution plans.

After technical and economic factors have been considered, however, another set of factors that must be dealt with are the intangibles. Often, these items can tip the scales in one direction or another. Typical intangibles include the following:

Increase in morale

Job enrichment

Improved customer service

Compatability with company philosophy

Operating feasibility (considering availability of labor and skills)

Operator comfort

Ability to cope with changing conditions

Adaptability to future changes in technology

Adaptability for expansion (or reduction)

Quality of service

Reputation with customers and vendors

Durability of equipment

Simulation and other methods of computer analysis can be used to help evaluate feasibility of alternatives. It is important, however, that the assumptions underlying any model, and the relationships represented in it, are thoroughly understood.

The final hurdle in selecting a preferred solution is obtaining approval of top management to expend funds. An entire chapter, or even perhaps an entire book, can be written on the subject of making presentations to management. However, the basic principles generally reduce to economics, after technical feasibility has been established. An understanding of engineering economy principles, as discussed in Chapter 3, is essential in this regard.

1.8.6 Applying the Solution

Once a preferred solution has been identified, the major challenge is developing the implementation plan. Obviously a different degree of effort and expertise is required for obtaining a hoist, a section of conveyor, or a shelving section, as is the case for planning an engineered materials handling system. Depending on the complexity of the job, assistance may be required from equipment manufacturers, distributors, consultants, and systems contractors.

Generally the following steps are involved in implementing a materials handling systems project:

Develop specifications.

Solicit bids.

Evaluate bids.

Award contracts.

Manage the project.

Train operating and maintenance personnel.

Proceed with startup.

Perform post audit.

It is most important that the bid specification be written, be well organized, and spell out clearly and precisely what various vendors will be bidding against. Even in a small job, involving only a few pieces of equipment, competitive bids can vary widely in price if vendors do not understand clearly what the specifications are to which they must bid. As a minimum, requests for bids should be accompanied by scaled drawings whenever appropriate. If larger systems are involved, consultants are often brought in to assist with bid preparation.

1.8.7 Evaluating Suppliers

For a materials handling system of any significant size, qualifying those suppliers that will be invited to submit bids is an important part of the bid preparation process. Evaluation of supplier capabilities is also a factor when subsequent bids are being evaluated.

Depending on the scope of the project, it might be a good idea to tour the prospective supplier's facilities, with the following questions in mind:

What is the condition of the supplier's plant? How well organized are plant operations? What is the condition and age of equipment?

How good is the quality control?

How busy is the supplier? Will his workload permit him to give proper attention to my project?

What full-time skills are available in mechanical, electrical, and structural crafts? How about data processing capabilities?

If possible, customers of the supplier also should be visited to see how their systems are working out. If possible, operators and mechanics should be interviewed, as well as supervisory and managerial personnel, in order to get a balanced viewpoint. Information about the following matters should be sought:

How smooth was the system installation and startup?

Did the supplier assist in setting up a training program for the user's personnel?

How helpful was the supplier in setting up a maintenance program and spare parts inventory?

Was the supplier effective in solving problems that surfaced during installation and startup?

Did supplier personnel have adequate skills, and were the same people available throughout the duration of the project?

Is the system delivering what was promised, in terms of performance and uptime?

1.8.8 Evaluating the Bid

An important part of evaluating competitive bids is making sure that all vendors are bidding to the same specifications. This job is made easier when well-written, precise specifications are prepared in advance. Otherwise, prices quoted may have no relation to duty classification or construction grade of equipment quoted.

In the case of a large system or facility, a performance specification may be the basis on which prospective suppliers bid. This type of specification spells out the type of performance required (pallets per hour handled, number of picks per hour, etc.), but does not necessarily restrict the supplier to the type of approach to be used. Rather, the burden is on the bidder to suggest the approach to be taken.

1.8.9 Selecting the Supplier

When such a performance specification is evaluated, a meticulous cost analysis is an important, if not the major, part of the evaluation. Operating costs as well as initial costs must be evaluated carefully for every alternative approach. Anticipated maintenance costs and spare parts inventory costs are among the items to be included in the operating cost category. In a surprisingly large number of cases, the initial cost of a system does not represent the overriding consideration when compared against total life-cycle costs.

Of course, absolute cost figures should be balanced against perceptions of overall supplier capabilities. A supplier presenting the lowest overall cost bid may also carry the lowest overall confidence level. Qualitative judgments must be made at this point. In some cases, numerical point rating systems might be applied to help factor such issues into the evaluation.

1.8.10 Awarding the Contract

Once the supplier has been selected and the winning bid chosen, the contract must be awarded. Often the success of the system falls down at this point. Practicality must be kept in mind when the contract is drawn up. The important thing is that the system meets performance specifications, on time, and within budget. The supplier should not be bogged down with an overly restrictive contract that limits his abilities to perform. On the other hand, the needs of the customer, as identified and approved by top management, must not be compromised. Basically, the contract should be viewed as a tool for helping both parties—the customer and the supplier—in managing the project. A good materials handling contract generally contains the following elements:

Objectives of the system

Modes of operation

Environmental factors (temperature, atmosphere, seasonal factors)

Description of loads to be handled, along with volume and throughputs

Target date for system to be operational at specified performance level

Designated responsibility on the part of supplier and user for insurance, safety, scheduling, and fire protection

Warranty details

Supplier and user share of project management responsibilities

Acceptance criteria

Terms of payment

Procedures for handling system changes and new requirements

Spare parts stocking

Supplier support activities and materials (training, manuals, monitoring)

1.8.11 Implementing the Project

Typically, project implementation involves design, manufacturing, testing, installation, checkout, and acceptance phases. Depending on the size and complexity of the project, design can incorporate layout, engineering analysis of racks and other structures, and a definition of control functions. As the design work progresses, engineering drawings and bills of material are prepared for fabrication of equipment.

Manufacturing and testing are performed by the supplier and any subcontractors the supplier may use. As much in-shop testing of integrated operations as possible is performed, to minimize delays at the installation site.

The installation phase marks the beginning of the work at the customer's site. It is an ideal time to begin having the user's operating and maintenance personnel become familiarized with the various parts of the system.

Another critical phase is checkout, which involves testing each major piece of equipment when it is installed. This is an acid test, because equipment is no longer being tested under laboratory or simulation conditions, but under actual operating conditions.

The acceptance stage involves an agreement by the user that the system that has been supplied operates properly and meets agreed-upon criteria of performance. In a larger system, often this phase is reached only after a significant number of tests, changes, corrections, and debugging procedures have been made. The importance to all parties concerned of a well-developed contract will become visible at this point.

Acceptance occurs either simultaneously with, or is closely followed by, final startup of the system. Typically startup is begun at a low throughput rate and light loading, and gradually increased to full-volume operation. Simulations should be run carefully if startup at relatively high volumes is required.

After the system has been operating for a number of months, its performance levels should be audited to ensure that it is delivering the performance upon which the investment was justified. This procedure is called post-startup audit, or just "post audit."

In a number of cases, the post audit will reveal that further refinement is still necessary to make the system live up to its potential for cost savings and productivity improvement. Additional engineering work and installation modifications may be required.

It is wise to include a provision for post-audit procedures in the original bid package upon which suppliers will compete. The criteria upon which a post audit is based can include uptime, throughput, operating speeds, meeting of safety standards, and energy efficiency.

1.9 PROJECT MANAGEMENT

As with many other activities, a successful materials handling system is the result of the combined efforts of a number of people. A simple equipment transaction may of course involve only an engineer and a vendor salesperson. However, implementing a larger installation may require the talents of various individuals with the user and supplier organization, as well as distributors and independent consultants in some cases.

For a fairly large or complex system, the user company normally will provide a team to see the project through to completion. Typical members of the team represent such functions as manufacturing, plant engineering, quality control, purchasing, and data processing. A project leader is chosen from one of the represented disciplines. The leader directs the activities of the team, and also provides the basic interface with the supplier organization, which also may develop a parallel project team of its own.

The mission of the project team is managing the design, manufacturing, testing, installation, checkout, and acceptance activities detailed previously, through to the completion of the system. The duties of the team include the following items:

1. Identify system requirements and establish performance standards.
2. Perform economic justification.
3. Control project finances.
4. Make supplier and equipment decisions.
5. Establish and control schedules.
6. Monitor project status.
7. Monitor supplier and subcontractor performance.
8. Generate and/or approve changes and modifications.
9. Oversee acceptance and audit procedures.

The project leader or manager has significant responsibility with limited authority. Typically the project manager directs members of a multidisciplined team over whom he or she has no line authority. And, the work must be accomplished within the context of short time frames, changing requirements, and unanticipated problems in hardware and software.

Obviously the project leader and the team can use as many tools and aids that are available in order to successfully complete the assignment. Various scheduling techniques are in the forefront of such tools, and are typically used to allocate finite resources within established time frames. Typical scheduling techniques include the following:

Gannt or Bar Charts. These commonly used graphical tools use lines or bars whose lengths are proportional to time spent in activities (Fig. 1.9.1).

Network Diagrams. Two common types are critical-path method (CPM) and project evaluation and control technique (PERT). Each of these related methods uses lines terminated by an arrow to indicate duration and direction of individual activities. Interrelationships among activity lines are analyzed to identify control points and determine schedules (Fig. 1.9.2).

Load-Leveling Techniques. Various methods have been developed for evening out the use of people or equipment to avoid wasted or idle time.

Least-Cost Analysis. Trade-offs between cost of an activity and its duration are evaluated in this approach. For example, a critical path that threatens the meeting of a deadline can be shortened by adding more people and equipment. However, the cost impact on the overall project of such an action also must be analyzed.

1.10 THE PEOPLE FACTOR

In the final analysis, the success of a materials handling system more often than not hinges on how successfully people have been integrated into the operation. Even highly automated systems depend heavily upon people, at the infeed and discharge points to the system, and in the context of a carefully planned and executed maintenance program that will ensure uptime.

One of the people factors to be considered in any new system—or even a modification to an existing one—is resistance to change, or "fear of the unknown." A considerable body of knowledge has been built up concerning transition management—the art of guiding an organization from where it has

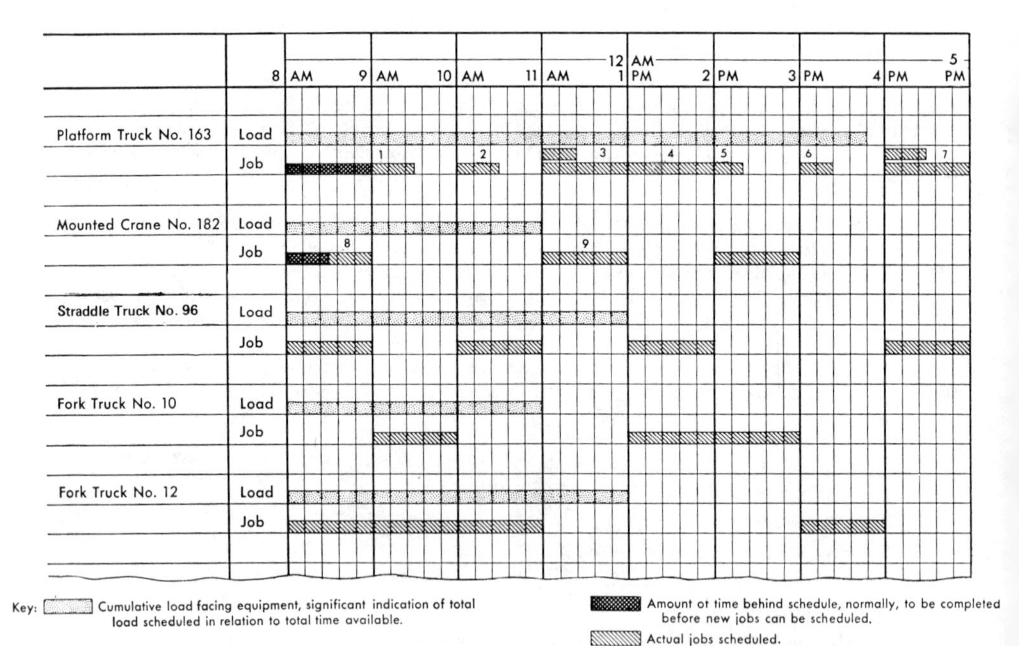

Fig. 1.9.1 Scheduling materials handling equipment by means of a Gantt chart (based on an 8-hr day).

been to where it must go in order to remain healthy and competitive. However, one of the basic tenets of transition management, as of any activity involving the planning of equipment and systems, is getting people involved early.

Involving supervisors, operators, and maintenance personnel in the early stages of the project, well before startup, is the best way to ensure ultimate success. There are two basic reasons for this factor: (1) operators and others will feel it is "their" system rather than an unknown quantity forced upon them, and (2) knowledgeable operating and maintenance personnel can provide valuable, practical suggestions regarding system details that planners and engineers may have overlooked.

The following are among the factors that can make personnel feel favorable toward a system:

A feeling of safety

An understanding of how the system works

Adequate access for maintenance and servicing

Sufficient manpower assigned to the operation

Confidence in support and service groups assigned to the system

Information and instructions presented in a manner that makes them easy to comprehend and use

Training must be an important part of a materials handling system plan if human resources at all levels are to be utilized effectively. Both classroom and hands-on equipment training are desirable. Often the system supplier can provide initial training, as well as supplying necessary operator and maintenance manuals. Sharing of training responsibilities can be incorporated into the system contract.

Ultimately, training of personnel is the long-term responsibility of the user. Management commitment must be secured for the concept that training is an ongoing activity throughout the life of the system

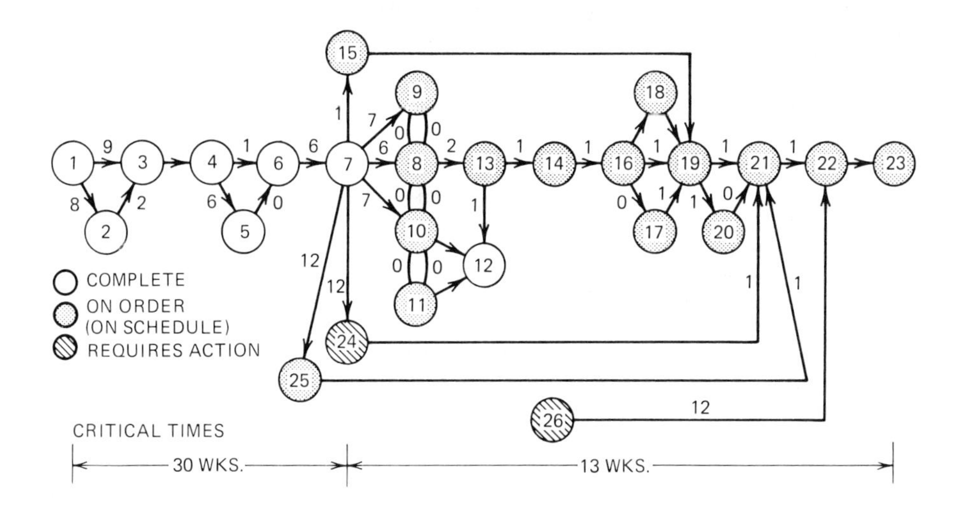

Place Order

1 Design
2 Specifications
3 Request quotes
4 Evaluate
5 Approve funds
6 Place order

Fabricate, Install

7 Approve shop drawings
8 Modify floor
9 Window, door
10 Power
11 Phones
12 Terminal
13 Ship carriages
15 Ship cabinets
16 Preparation for tracks
14 Install tracks
17 Carpets
18 Carriages, power
19 Install racks
20 Install end panels
21 Prove-in
22 Stock, I.D.
24 Move maintenance
25 Lighting
26 Scale

Fig. 1.9.2 Critical-path schedule is used for managing project from start to finish. Thick lines form critical path. Numbers beside lines represent time, in weeks. If critical-path schedule times are not met, entire project is subject to delay. From *Plant Engineering* Magazine, 3/3/83, p. 196. Courtesy Technical Publishing, a company of the Dun & Bradstreet Corporation.

operation. An adequate training program can have a significant impact on safety as well as on productivity of an operation. Outside sources of assistance for developing a materials handling training program include equipment manufacturers and distributors, industry associations and professional societies, and junior and community colleges and trade schools.

REFERENCE

1. College–Industry Council on Material Handling Education is sponsored by The Material Handling Institute, Inc., 1326 Freeport Rd., Pittsburgh, PA 15238.

CHAPTER 2
PLANT LAYOUT AND MATERIALS HANDLING

RICHARD MUTHER

Richard Muther & Associates
Kansas City, Missouri

DENNIS B. WEBSTER

Auburn University
Auburn, Alabama

2.1 RELATIONSHIP BETWEEN MATERIALS HANDLING AND PLANT LAYOUT

Richard Muther

2.1.1 Essential Interrelationship

Materials handling and plant layout are tied together. Indeed, they are interdependent. One can hardly function effectively without the other.

This interrelationship applies both in the planning or analysis of materials handling and plant layout *and* in the physical facilities as installed and operating. How well they tie together is directly dependent on how well they are planned and how well they are installed and operated. One simple example of the interdependence is shown in Fig. 2.1.1.

One cannot move materials at all if there is no space in which to move them. Thus, the amount of space, the type of space, and its shape or configuration influence the kinds and types of materials handling equipment that can be used. The amount, kind, and configuration of space for the activities or the equipment they represent form the layout.

In shipbuilding, if only one large ship is to be built, the keel is laid down in one location and all material is brought to it, where it is constructed. The finished hull is then launched. However, if small boats of a standardized design are to be made in large quantity, the keel structure is put in place as part of the midsection or hull, then moved to other stations for additional hull assembly. Obviously, the movement of all materials to one fixed location is quite a different problem than moving different materials to various locations. Thus, a change in products and quantities affects both the layout and the handling methods.

Figure 2.1.2 shows the movement to one work area. By breaking down the total job so that the work content is more specialized—to gain the economies of specialized equipment and skills of labor—more handling is introduced. However, if the total net cost of the operations and handling is lowered by increasing the number of moves, the stretched-out layout is economically better. In either case, two things are readily apparent:

1. There has to be movement to and from each point of work.
2. The number of operations and their layout will affect the method of handling the materials to and from the points of work and between them.

2.1.2 Fundamentals of Handling and Layout

The fundamentals underlying every materials handling problem are:

MATERIALS, MOVES, and METHODS

That is, the various materials in their varying quantities have to be moved over various routes by certain kinds of materials handling equipment and containers. Thus, the problem is to establish the *methods* to *move* the *materials.*

The basic fundamentals of every layout planning project are:

RELATIONSHIPS, SPACE, and ADJUSTMENT

That is, the relationships indicate the relative closeness desired between things; the space establishes the amount, kind, and configuration of space for each thing; and the adjustment involves the arrangement of this space to satisfy the closeness relationships and space requirements. Thus, the problem is to *adjust* the *space* to honor the *relationships.*

To establish the relationships (or closeness desired between various machines or departments or buildings on a site), one needs to establish what kind of material and how much of it has to be moved among these activity-areas. As a result, both the materials and the moves need to be determined in order to establish the relationships. However, the methods of moving the materials are, in part, a function of the distances involved, which are determined by the layout. Therefore, in terms of the basic fundamentals, the layout depends on the materials and moves and the selection of handling methods depends upon the layout.

2.1.3 Various Kinds of Projects

There are typically three kinds of materials handling projects insofar as layout is concerned:

1. Layout is fixed; project is to improve the handling methods.
2. Handling methods are fixed; improve the layout.

3. Neither are fixed; improve both the handling methods and the layout.

In 1, the analyst solves the handling problem with the location of activity-areas established and, therefore, the distances involved serve as identifiable constraints.

In 2, the handling methods are already established, so the analyst has the problem of rearranging the space with minor modifications only of the handling methods.

In 3, both can be improved or established anew. Here, the analyst must integrate both, usually by (a) analyzing flow, (b) using it as a basis for closeness desired among activity-areas, and (c) tying handling methods to the resulting layout.

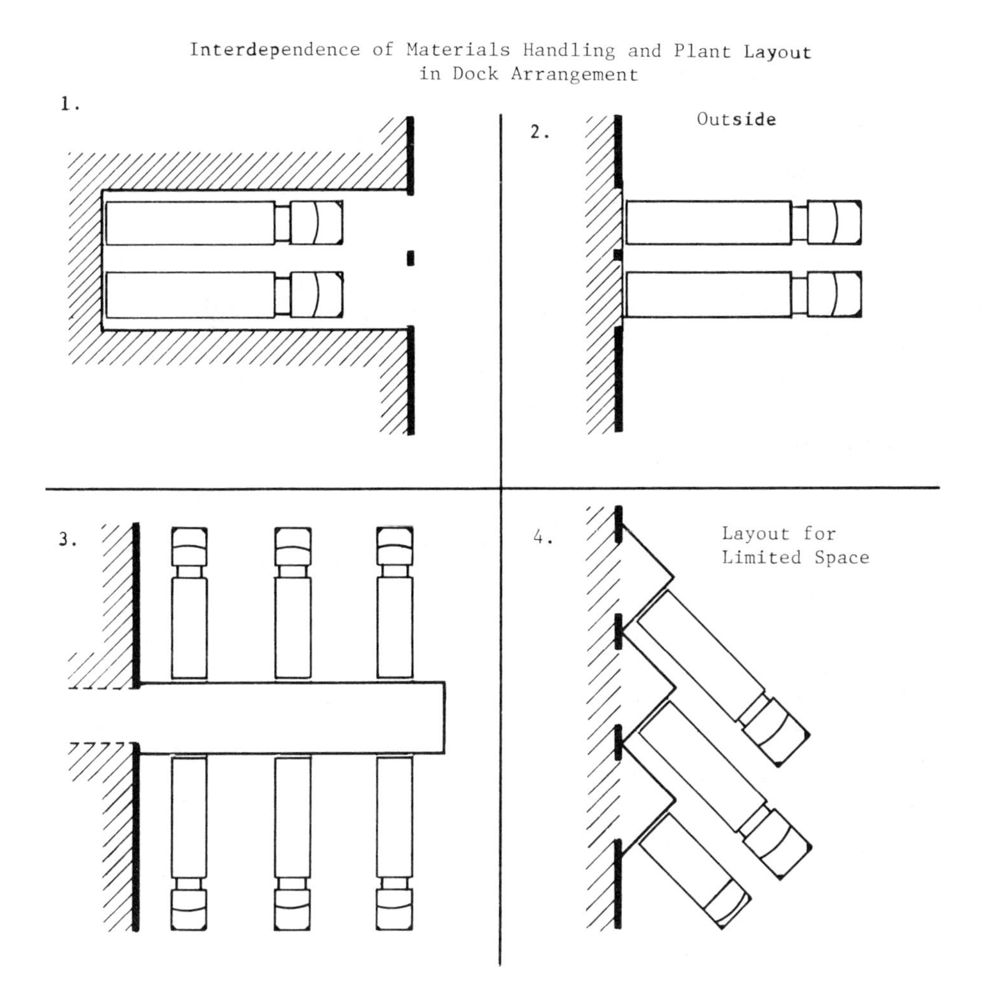

1. One way of arranging the receiving or shipping function is to back highway trucks inside the building. Tailgate and side loading by mobile industrial truck, as well as by overhead crane or monorail hoist, are possible.

2. If there is no room in the building layout, trucks may backup to doors in the building wall.

3. If layout permits only one door, outside dock finger may be used for all handling.

4. If the space for truck turning is limited, the dock layout may have to be sawtooth in design.

Fig. 2.1.1 Interdependence of materials handling and plant layout in dock arrangement.

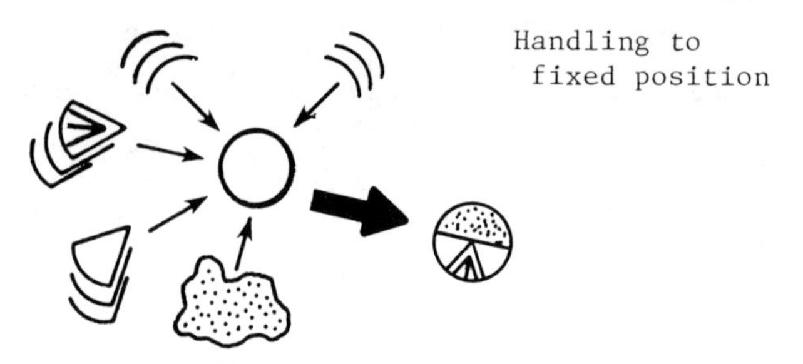

Handling to divided operations

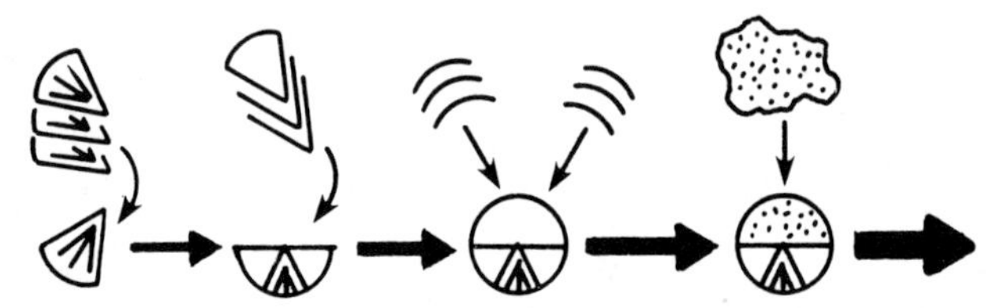

Fig. 2.1.2 Handling to fixed position and to divided operations.

2.1.4 Basic Handling Cost Elements Affected by Distance

The methods of handling depend on the materials and moves. The materials are determined by the physical characteristics of the products, items, varieties, formulations, mixes, and so on, and by the quantity of each required or desired per period of time. The moves involve the quantity or intensity of items moving over each route and the distance from origin to destination of that route. As a result, the theoretical cost of material handling varies with:

1. The physical characteristics of the materials
2. The quantity of each product or material
3. The intensity (or quantity per period of time) of each item on each route
4. The distance the material has to move on each route

This transforms into the formula:

$$\text{Cost of handling} \simeq TW$$

where TW is the transport work.

$$TW = I \times D$$

where I is the intensity of material moving, and D is the distance moved.

$$I = (n \times p) \times D$$

where n = the number of units per period of time, and p = the measure of one unit, piece, part, or item.

Additionally, several other factors affect "distance." These include:

1. Establishing a meaningful common unit of measure for the various items to be moved.
2. The changing physical or chemical characteristics of each material as it is moved through individual operations or storage areas.
3. The physical condition or congestion of the route.
4. The operating situation at the points of pickup and setdown.
5. Timing considerations such as number of shifts available for work, seasonal peaks, and urgency of move.

So distance D is not the true measure of material handling cost even if pickup and setdown (or loading and off-loading) are considered. See Fig. 2.1.3.

Every Move involves:
1. Pick up or load the vehicle
2. Move or transport loaded
3. Set down or unload the vehicle.

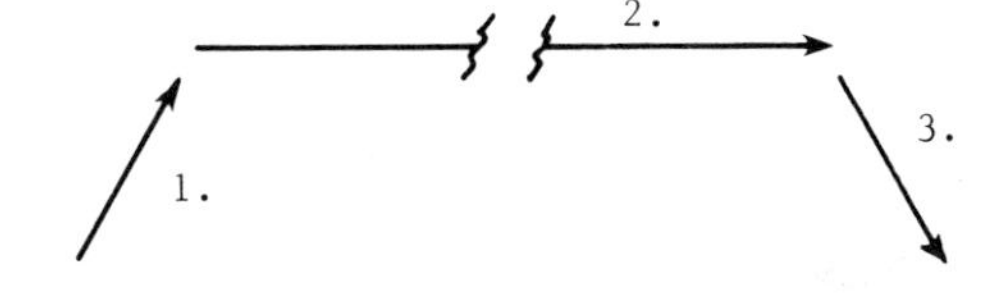

But 1 and 3 are not related to distance moved. This divides handling cost into Terminal Cost and Travel Cost.

And when the distance moved is short, most of the cost is in the terminal cost.

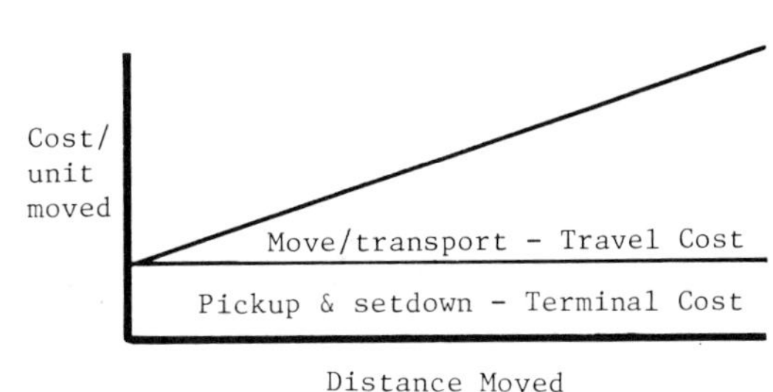

Fig. 2.1.3 Pickup and setdown—terminal and travel cost.

2.1.5 Phases of Planning

Almost every project of layout planning is different. There are large projects in which the "layout" becomes the *location* of various plants and warehouses throughout the world. At the other extreme are small projects where the handling methods relate to the layout within an individual workplace. But, there is a general sequence of planning phases that is always involved:

Phase I. *Location* of the area to be laid out and the *external* movements of material to and from that location.

Phase II. An overall plan for the *blocked-out layout* and an *overall handling plan* for the total area.

Phase III. *Detailed layouts* of each specific piece of machinery, equipment or storage rack, and the *specific handling methods* to and from each specific place.

Phase IV. *Installation* including *approval* of plans, *procurement* and/or *rearrangement.*

These phases come in sequence and, for best results, the planner should try to schedule the project so that each phase overlaps the one following it. See Fig. 2.1.4.

All projects, except those for very small areas, involve Phase II and Phase III. And each phase involves integration of layout and handling. Moreover, integration is required between an overall plan and the detailed plans, both for layout and handling. That is, if you select overhead cranes for movement *between* departments, the moves *within* the department must be compatible with the overhead cranes. Similarly, external handling equipment must be integrated with the handling equipment within the plant. Conversely, the overall handling methods can be reversely influenced by detailed layouts in each departmental area.

2.1.6 Flow of Material in Determining Layouts

Flow of material is a basic determinant for layout planning. Where the intensity of flow is high, the distance should be short; where the intensity of flow is low, the distance can be long.

The terms *flow of material* and *movement summary* are similar. Flow implies a sequence of operations and therefore tends to involve the entire sweep of movement throughout the area involved. It involves sequence and all materials to be moved. Movement summary, on the other hand, involves the movements of each item or class of material on each individual route. As a result, flow of material relates more closely to layout planning; movement summary relates to analysis and selection of materials handling methods. This distinction is noted in Fig. 2.1.5. The figure shows the relation of layout planning and materials handling analysis.

Flow of materials (the total move of all items on each route) helps determine the closeness required between each origin and destination. That is, those activity-areas having a high flow of material between them should be close to each other; those pairs of activities having a low or negligible flow of material between them can be located farther away. Moreover, the sequence of flow between several operations

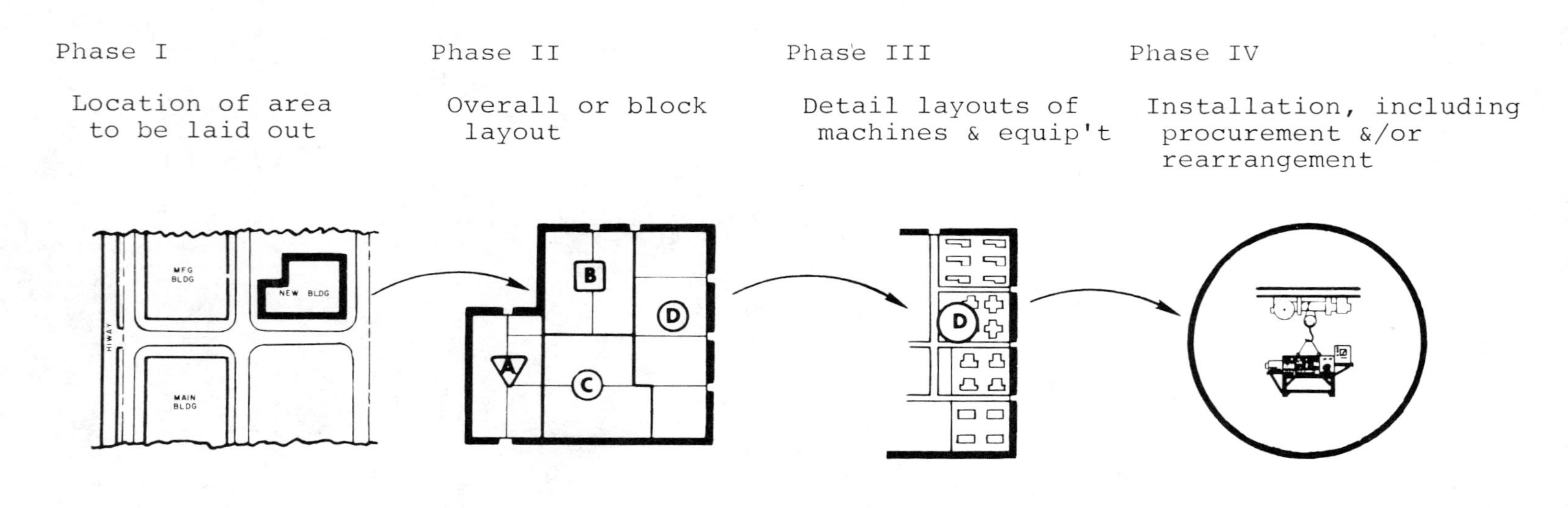

Fig. 2.1.4 Phases of layout planning.

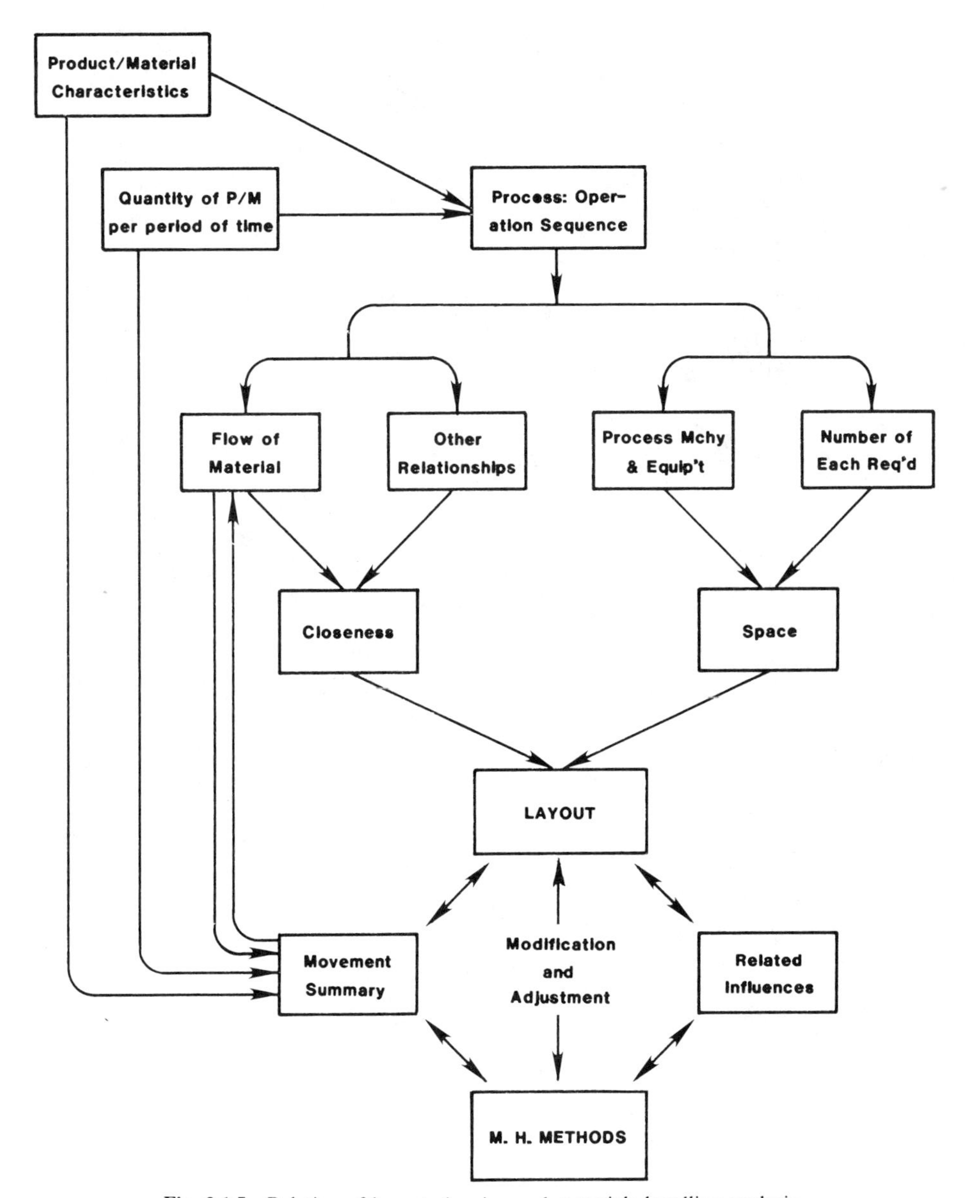

Fig. 2.1.5 Relation of layout planning and materials handling analysis.

establishes a pattern of flow. And, where quantities of materials to be moved are high, the flow pattern often has a major influence on determining the configuration of the layout.

The movement summary shows, for each item or class of material on each route (pick-up point/origin to set-down point/destination), the intensity of material moving. Thus, if the items or classes of materials are arranged across the top of a worksheet and the routes are arranged down the side of the sheet, then each column represents a class of material and each line represents a specific route. By recording, at the intersection of each class and each route, the intensity of material moved and the situation at each end of the move, much of the information needed to select the handling methods is indicated.

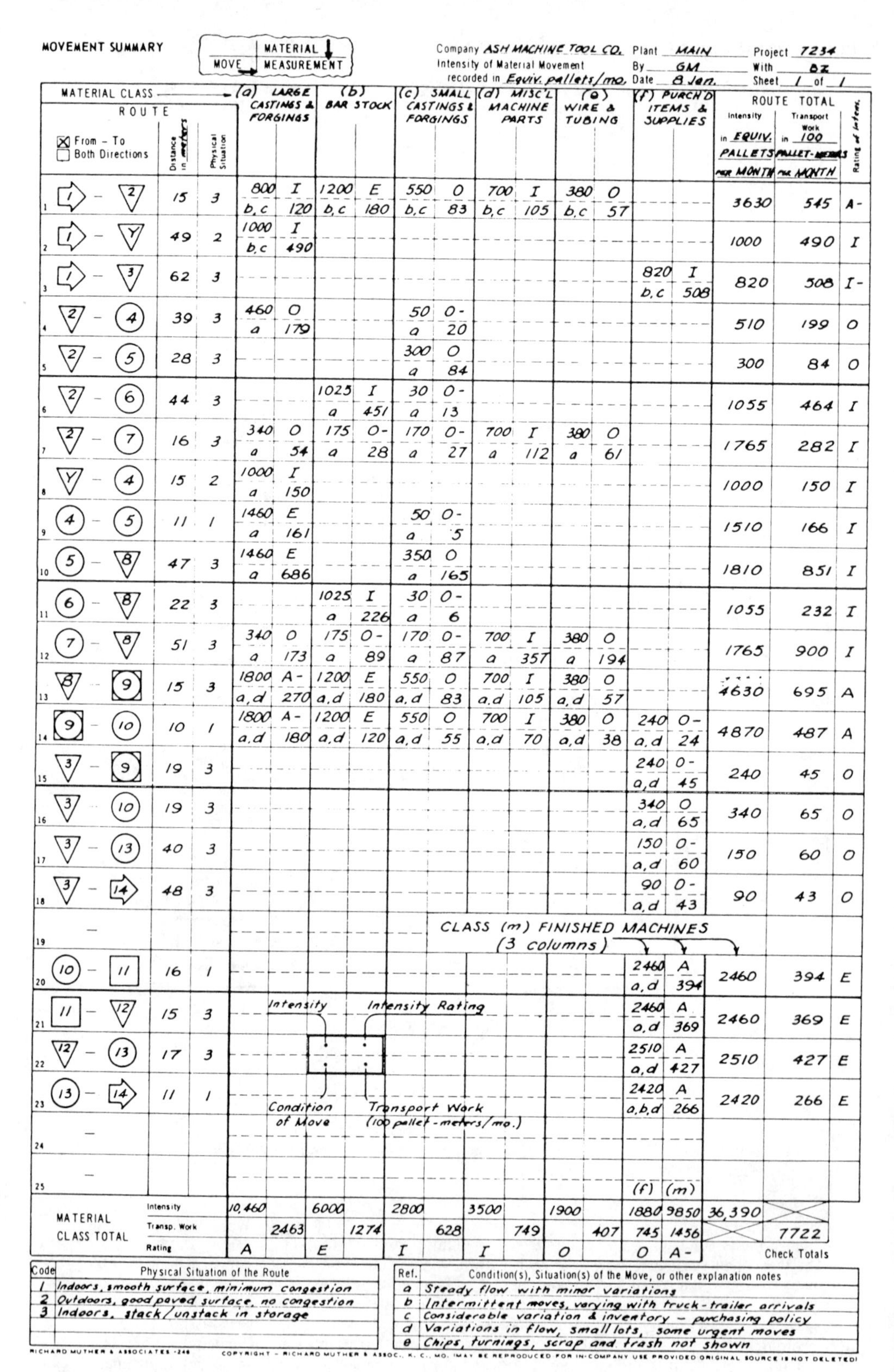

MOVEMENT SUMMARY — MOVE → / MATERIAL ↓ MEASUREMENT

Company *ASH MACHINE TOOL CO.* Plant *MAIN* Project *7234*
Intensity of Material Movement recorded in *Equiv. pallets/mo.* By *GM* With *BZ*
Date *8 Jan.* Sheet *1* of *1*

Route: ☒ From – To ☐ Both Directions. Each material-class cell shows: Intensity, Intensity Rating / Condition of Move, Transport Work (100 pallet-meters/mo.)

No.	Route	Distance in meters	Physical Situation	(a) Large Castings & Forgings	(b) Bar Stock	(c) Small Castings & Forgings	(d) Misc'l Machine Parts	(e) Wire & Tubing	(f) Purch'd Items & Supplies	Route Total Intensity in Equiv. Pallets per Month	Route Total Transport Work in 100 Pallet-Meters per Month	Rating of Intens.
1	1 – 2	15	3	800 I; b,c 120	1200 E; b,c 180	550 O; b,c 83	700 I; b,c 105	380 O; b,c 57		3630	545	A-
2	1 – Y	49	2	1000 I; b,c 490						1000	490	I
3	1 – 3	62	3						820 I; b,c 508	820	508	I-
4	2 – 4	39	3	460 O; a 179		50 O-; a 20				510	199	O
5	2 – 5	28	3			300 O; a 84				300	84	O
6	2 – 6	44	3		1025 I; a 451	30 O-; a 13				1055	464	I
7	2 – 7	16	3	340 O; a 54	175 O-; a 28	170 O-; a 27	700 I; a 112	380 O; a 61		1765	282	I
8	Y – 4	15	2	1000 I; a 150						1000	150	I
9	4 – 5	11	1	1460 E; a 161		50 O-; a 5				1510	166	I
10	5 – 8	47	3	1460 E; a 686		350 O; a 165				1810	851	I
11	6 – 8	22	3		1025 I; a 226	30 O-; a 6				1055	232	I
12	7 – 8	51	3	340 O; a 173	175 O-; a 89	170 O-; a 87	700 I; a 357	380 O; a 194		1765	900	I
13	8 – 9	15	3	1800 A-; a,d 270	1200 E; a,d 180	550 O; a,d 83	700 I; a,d 105	380 O; a,d 57		4630	695	A
14	9 – 10	10	1	1800 A-; a,d 180	1200 E; a,d 120	550 O; a,d 55	700 I; a,d 70	380 O; a,d 38	240 O-; a,d 24	4870	487	A
15	3 – 9	19	3						240 O-; a,d 45	240	45	O
16	3 – 10	19	3						340 O; a,d 65	340	65	O
17	3 – 13	40	3						150 O-; a,d 60	150	60	O
18	3 – 14	48	3						90 O-; a,d 43	90	43	O
19	—			CLASS (m) FINISHED MACHINES (3 columns)								
20	10 – 11	16	1						(m) 2460 A; a,d 394	2460	394	E
21	11 – 12	15	3						(m) 2460 A; a,d 369	2460	369	E
22	12 – 13	17	3						(m) 2510 A; a,d 427	2510	427	E
23	13 – 14	11	1						(m) 2420 A; a,b,d 266	2420	266	E
24	—											
25	—											

Material Class Total	(a)	(b)	(c)	(d)	(e)	(f)	(m)	Check Totals
Intensity	10,460	6000	2800	3500	1900	1880	9850	36,390
Transp. Work	2463	1274	628	749	407	745	1456	7722
Rating	A	E	I	I	O	O	A-	

Code	Physical Situation of the Route
1	Indoors, smooth surface, minimum congestion
2	Outdoors, good paved surface, no congestion
3	Indoors, stack/unstack in storage

Ref.	Condition(s), Situation(s) of the Move, or other explanation notes
a	Steady flow with minor variations
b	Intermittent moves, varying with truck-trailer arrivals
c	Considerable variation & inventory – purchasing policy
d	Variations in flow, small lots, some urgent moves
e	Chips, turnings, scrap and trash not shown

RICHARD MUTHER & ASSOCIATES -246

Fig. 2.1.6 Movement summary

By cross-totalling for each route, the total movement of all materials on that route is obtained. By down-totalling each column, the total movement for each class of material on all routes is attained. In the final analysis, the material handling methods selected must provide a method of moving material for *each* class on *each* route. See Fig. 2.1.6.

The movement summary may or may not indicate the distance to be moved. But, with the movement summary available and the layout established, the materials handling methods can be selected, subject, of course, to the other factors that may affect the choice. These include the building features, the distribution of utilities, the communications and controls required, personnel available, safety requirements, potential hazards involved, and regulations of external agencies such as OSHA, EPA, city building code, railroad agent, and highway engineer.

2.1.7 Relationships Other Than Material Flow

To determine the closeness desired between each activity-area and every other activity-area, consideration of flow of material alone is not sufficient. Even in situations involving high intensities of flow and large or awkward materials to transport, the layout must include such non-flow-related areas as maintenance department, tool storage, boiler room, air compressor, offices, cafeteria or lunch room, and toilet areas. These are part of the layout and therefore must be accommodated in planning the layout.

Another example is the relationship between furnaces in a glass plant. If two furnaces are involved, each melting a different color glass, there is no flow of material between the two furnaces. However, they both require the same kind of utilities, heat exhausting equipment, maintenance, fire-brick supplies, and so on. So, it is logical to have them close to each other.

Figure 2.1.7 shows how flow-of-material and other relationships are involved in establishing the net or combined closeness. This equates each vowel-letter closeness rating (see Section 2.2.7) to a number value, adding the values and striking a new, combined vowel-letter rating.

Cases involving negative relationships (X = keep the activities away from each other) need special examination. For example, shot-blast cleaning often immediately precedes painting of heavy weldments.

FLOW OF MATERIALS

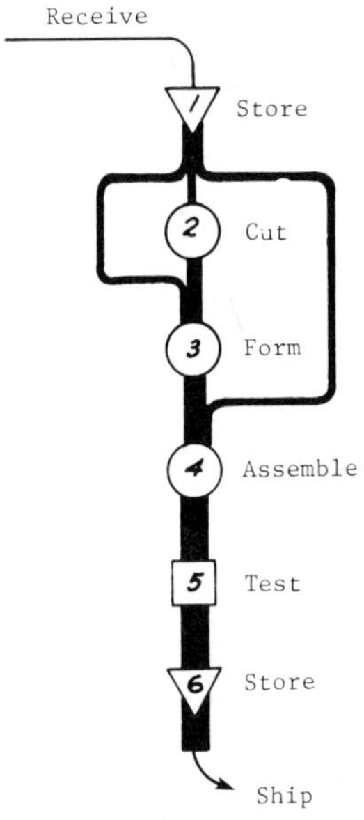

Receive
1 Store
2 Cut
3 Form
4 Assemble
5 Test
6 Store
Ship

OTHER-THAN-FLOW RELATIONSHIPS

Activity Pairs	Rating	Reasons for Closeness
1 Store - 9 Office	0	Communications/controls
2 Cut - 3 Form	I	Same Supervision
2 Cut - 5 Test	X	Dirt and Noise
2 Cut - 7 Maintenance	I	Lubrication & Repairs
2 Cut - 8 Personnel	0	Servicing
3 Form - 7 Maintenance	E	Lubrication & Repairs

COMBINED CLOSENESS RELATIONSHIPS

Activity Pairs	Flow		Other		Combined	
1 - 2	0	1	-	0	1	0-
1 - 3	0	1	-	0	1	0-
1 - 4	0	1	-	0	1	0-
1 - 9	-	0	0	1	1	0-
2 - 3	I	2	I	2	4	I
2 - 5	-	0	X	X	X	X
2 - 7	-	0	I	2	2	0
2 - 8	-	0	0	1	1	0-
3 - 4	E	3	-	0	3	I-
3 - 7	-	0	E	3	3	I-

Closeness rating

Equivalent number value (4 to zero)

Combined number value (8 to 0) and closeness rating

A E I O U X order-of-magnitude rating, originally adopted in Systematic Layout Planning (SLP) and used generally today, indicates closeness desired. Vowels convert to numbers: A = 4.. X is negative. Half a degree of closeness is rated with minus sign.

Fig. 2.1.7 Combining closeness—flow and other-than-flow relationships.

Blasting is a dirty operation; it should be kept away from painting. This can be accommodated by separating the two activities some distance from each other, or by constructing the building so there is no contamination between the two operations.

2.1.8 Classical Types of Layouts

There are three general types of layouts:

1. Layout by fixed position
2. Layout by process
3. Layout by product

Layout by Fixed Position (by Fixed Location). Here all operations are performed with the material (in the case of forming or treating) or the major component (in the case of assembly or packing) remaining in one fixed location. That is, the chief material is held at a fixed place. Examples are specialty-tool making and large-ship construction.

This type of layout is best when the product or material is physically large, when the quantity required is low, and when the process operations are fairly simple.

The material handling is usually characterized as large or sturdy, as mobile or flexible for assembly parts, and as relatively occasional.

Layout by Process (by Function). Here all operations of the same type are performed in the same area; similar assembly operations or like machines are grouped together. That is, the material is moved through process departments or functional areas. Examples are: machine shop divided into lathes, grinders, and milling machines; and assembly areas divided into weld, rivet, clean, paint, and pack.

This type of layout is used when the product or material is relatively diversified, when the quantity is moderate or small, and when the process equipment itself is dominant, complex, or highly expensive.

The handling in this type of layout is usually characterized as mobile or flexible (if fixed equipment), as versatile or adaptable, and relatively intermittent.

Layout by Product (Line Production). Here the machines or assembly work stations are arranged in the sequence of operations, successive work stations being located immediately adjacent to each other. That is, the material is moved from one operation directly to the next. Examples are a typical assembly line, an automatic car-washing line, and assembling a tray of food in a cafeteria.

This type of layout is generally used when the product or material is relatively standardized, the quantity is relatively high, and the process operations are relatively simple.

The handling for this type of layout is usually characterized as fixed, straight or direct, and relatively continuous.

2.1.9 Basic Patterns of Material Flow

The primary patterns of material flow in industry are:

1. Straight or straight through
2. U-Shape or circular
3. L-Shape or right angle

Most other flow patterns such as zig-zag or spine are combinations or modifications of these.

The *straight-through flow* (in one end and out the other) allows the cleanest layout and simplest handling. It is also the easiest to expand. But, it requires separated receiving and shipping areas, with resultant costs in extra doors in the building, extra roadways on the site, additional storage space for inbound and outbound items, and little opportunity to combine receiving, shipping, and storage personnel.

The *U-shape layout* (in and out the same end or side) almost by definition leads to combining receiving and shipping. This provides an opportunity for savings in fewer doors in the building, less pavement and land for roads on the outside, an opportunity to combine storage space, and to combine receiving, shipping, and storage personnel. Additionally, if a well-coordinated dispatching of mobile-handling equipment is in operation, the utilization of handling equipment will be higher. The U-shape flow pattern has the disadvantage of being less easy to expand, of having prospective problems of dock-and-yard congestion for trucks, and of mixing inbound and outbound materials.

The *L-shape layout* (in one end and out the side, or in one side and out the end) is usually caused by the need to get multiple access and egress to and from a building by various means of external transport (rail, truck, barge, pipeline, aircraft, etc.). It does have the advantage of being able to segregate receiving of heavy materials from the receiving of purchased parts and small components. Thus, in appliance manufacture, coils of sheet metal are received, stored, and passed through forming presses, welding operations, and on into a clean-and-paint area. It is nice to have Clean and Paint in the corner because it has a greater opportunity to expand, which, as integrated process equipment, is relatively costly and disruptive to expand and/or relocate. From painting, the assembly operations change direction, moving along a leg perpendicular to the sheet-metal cabinet-making flow. The storage of purchased parts and components and also of finished goods and/or packing materials are located inside the "L," thus squaring up the building.

Figure 2.1.8 indicates the benefits of primary types of flow patterns.

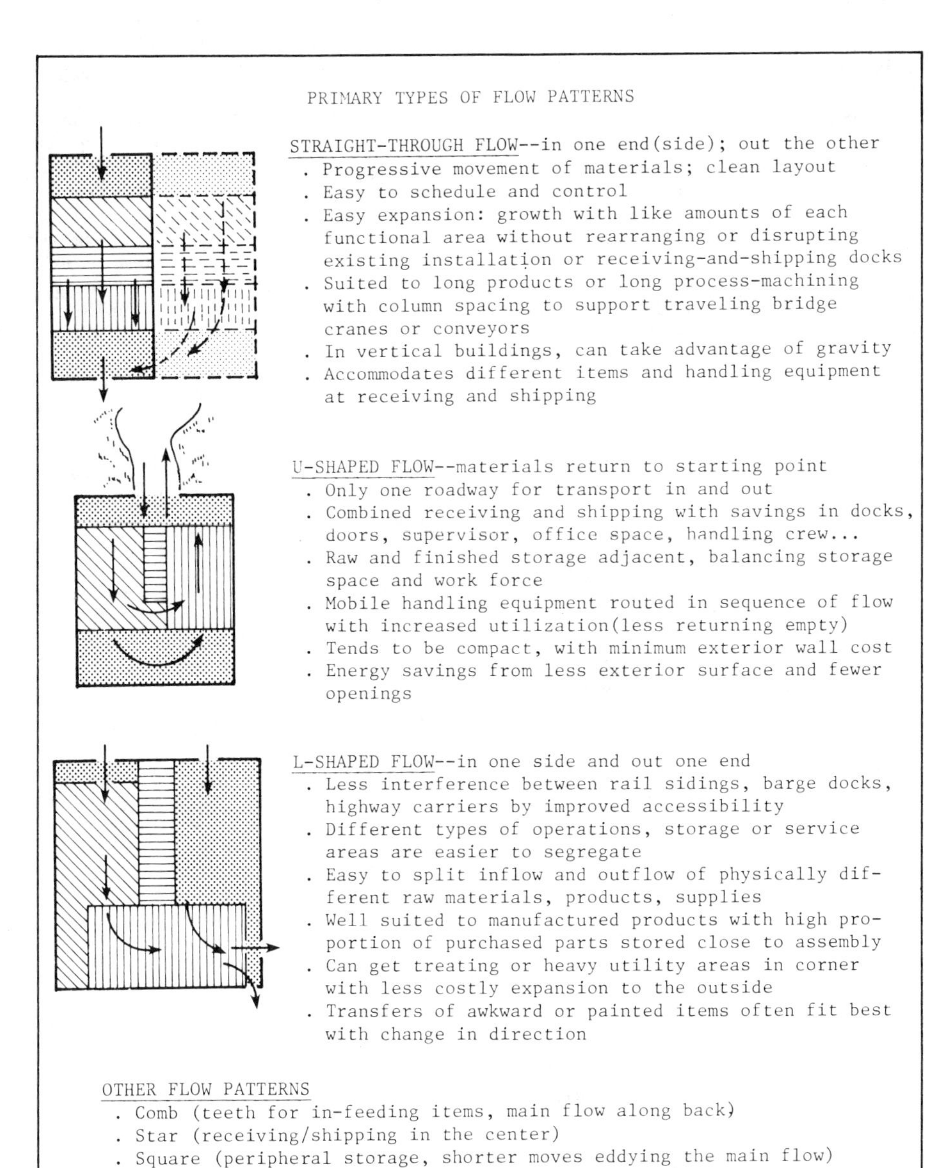

Fig. 2.1.8 Primary types of flow patterns.

In all cases, where the layout is planned around the flow of material, locating the offices and supporting services becomes a problem for they tend to get in the way of a clean flow pattern and to block easy expansion. And, where the layout is planned for ease of quick expansion with modular building "blocks," locating the handling corridor(s) and placement of receiving and shipping activities become critical.

2.1.10 Measuring Flow of Materials

When measuring flow of materials in basic steel, tons is a realistic measure. In an oil refinery, barrels is logical. In a warehouse of packaged food or toiletries, number of cartons (or cases) measures the quantity, throughput, or volume. But in manufacturing, selection of a meaningful unit becomes a problem. One spark plug is not the same as one engine block or one automobile body. Indeed, an auto body has a greater handling effort when removing it from the paint conveyor than when hanging it on the conveyor, for its painted surface can be damaged.

To treat this handling inequality, planners typically convert all items to an equivalent unit of measure—equivalent tons, equivalent pallet loads, or equivalent containers, for instance. Moreover, the material changes its characteristics in almost every operation; so for flow measures that realistically compare the relative difficulty of handling various items, a series of factors should be considered. The most comprehensive way of doing this is the Mag Count. It quantifies the size, density, shape, risk of damage, and condition (including value) into a score for any item. Because the size of one mag (10 cubic inches or one handful) is small, the unit of macromag (one cubic meter, or approximately 40 inches cube) is often used as an easier unit of size to work with. The size unit is then modified by factors B, C, D, E, and F (if F is not included in E). Figure 2.1.9 shows an application of macromag determination for a plant producing a variety of metal office furniture. Figure 2.1.10 indicates (for the same plant) how meaningful this common unit is in locating its departments by overlaying the handling effort required.

2.1.11 Materials Handling and Long-range Site Layout

A recent study of British industry concluded that the single most important cause of high materials handling costs is ad hoc expansion of plant facilities performed without a strategic site-development plan. Inasmuch as plant facilities are most often integrated around a conceptual layout(s), long-range facilities, including handling equipment, rest on macro-layout planning. Long-range site planning should certainly accommodate, among its major considerations, basic flows of materials.

A summary of site-planning principles set forth by Muther and Hales in *Systematic Planning of Industrial Facilities* (SPIF) includes the following:

1. Group similar activities or functions together.
2. Develop a basic plan of growth for the site.
3. Establish a basic pattern(s) of material flow and/or product–process (or other primary) relationships.
4. Orient or align the proposed facilities with the property lines or existing dominant features.
5. Take advantage of the natural features of the site.
6. Develop a basic infrastructure for the site.
7. Establish a pattern of internal transport and/or circulation.
8. Establish a pattern and dedicate corridors for primary distribution of utilities.
9. Keep the planned facilities flexible.
10. Plan facilities for ease of expansion.
11. Plan for implementation of plans and the site's development.
12. Protect the resale value of the property(ies).
13. Stay in compliance with all regulations.
14. Conserve energy through orientation, alignment, short distances, and minimum openings.
15. Provide for appearance, beautification, company image, and natural environment.
16. Provide for safety and convenience of employees.
17. Aid security considerations—fire, theft, espionage.
18. Avoid overcrowding the site. Keep some uncommitted space.

2.1.12 Systematic Layout Planning (SLP)

A basic approach to planning layouts is available. It consists of the four phases of planning; a series of conventions for recording, rating, and visualizing; and a specific pattern of planning steps. These are shown graphically in Fig. 2.1.11. A pictorialized sketch of SLP in application is shown in Fig. 2.1.12.

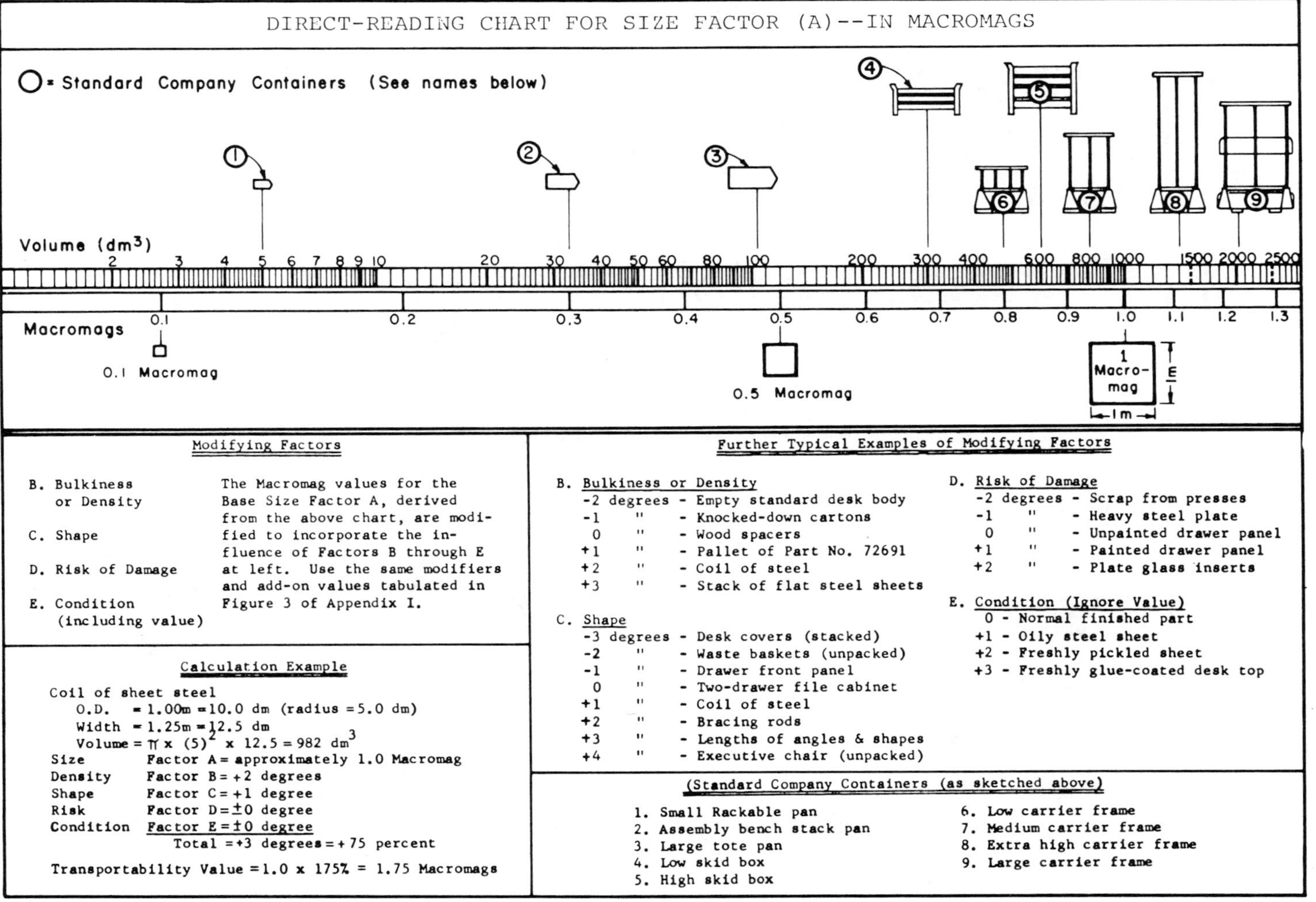

Fig. 2.1.9 Macromag figures. From R. Muther, *Systematic Layout Planning* (SLP), Copyright 1973, Richard Muther.

MATERIAL FLOW INTENSITY (IN MACROMAGS) PER YEAR
SUPERIMPOSED ON PROPOSED LAYOUT PLAN

Note the high intensity (high effort required) from Painting through Assembly

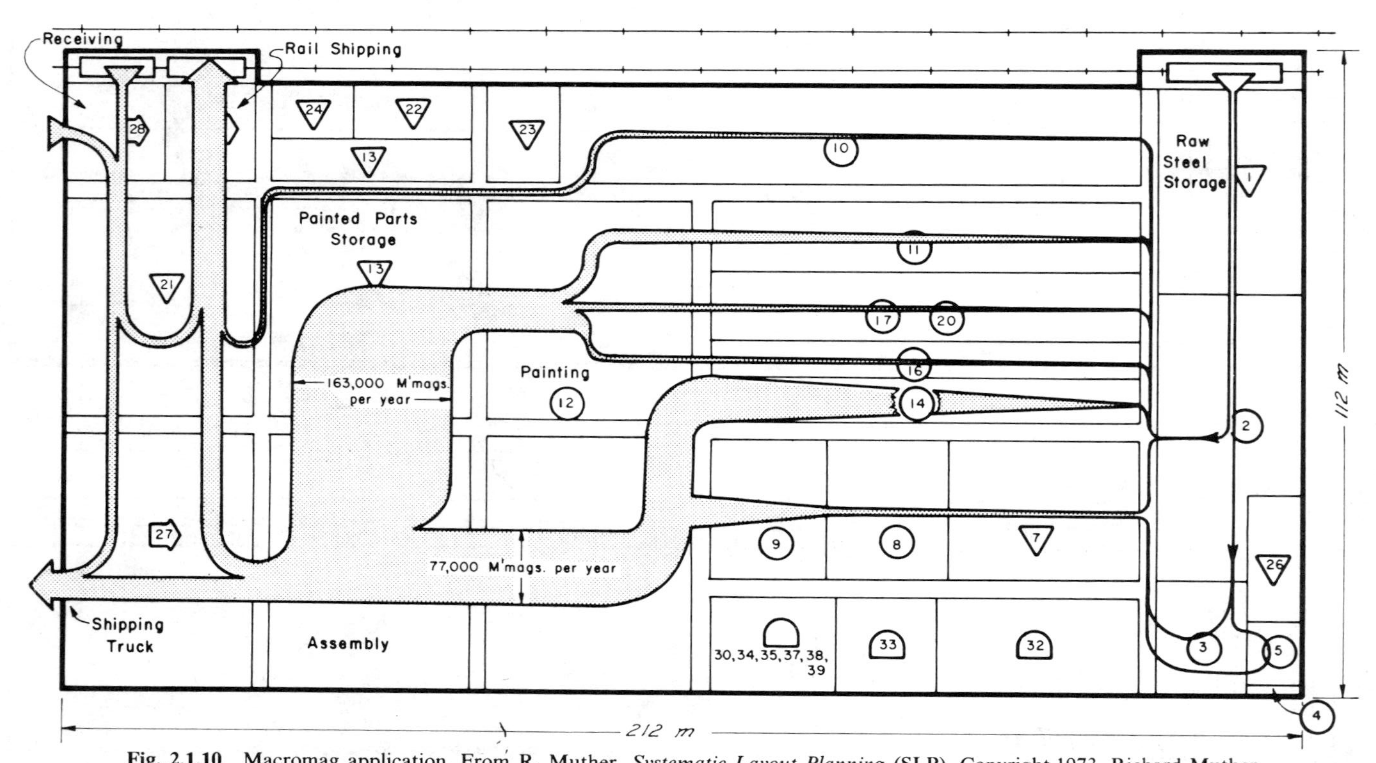

Fig. 2.1.10 Macromag application. From R. Muther, *Systematic Layout Planning* (SLP). Copyright 1973, Richard Muther.

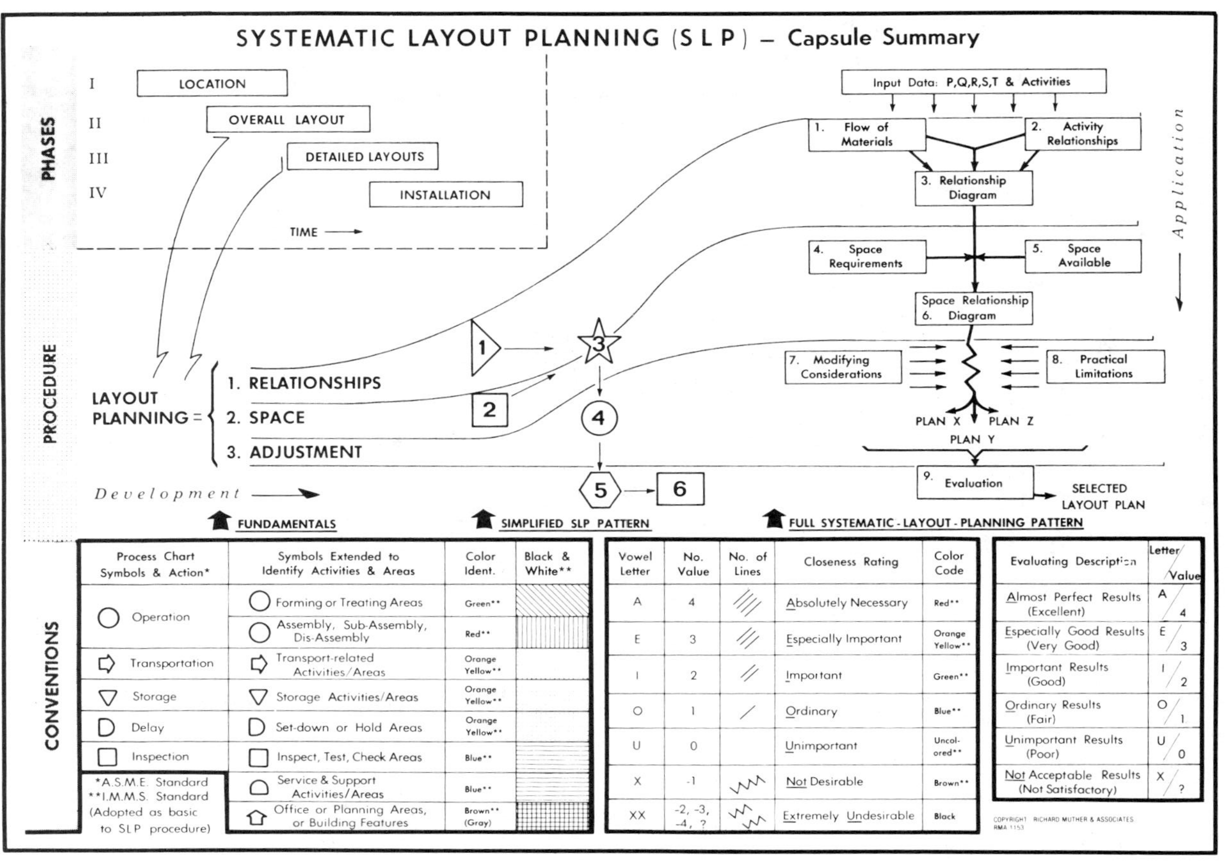

Process Chart Symbols & Action*	Symbols Extended to Identify Activities & Areas	Color Ident.	Black & White**
Operation	Forming or Treating Areas	Green**	
	Assembly, Sub-Assembly, Dis-Assembly	Red**	
Transportation	Transport-related Activities/Areas	Orange Yellow**	
Storage	Storage Activities/Areas	Orange Yellow**	
Delay	Set-down or Hold Areas	Orange Yellow**	
Inspection	Inspect, Test, Check Areas	Blue**	
*A.S.M.E. Standard **I.M.M.S. Standard (Adopted as basic to SLP procedure)	Service & Support Activities/Areas	Blue**	
	Office or Planning Areas, or Building Features	Brown** (Gray)	

Vowel Letter	No. Value	No. of Lines	Closeness Rating	Color Code
A	4	////	Absolutely Necessary	Red**
E	3	///	Especially Important	Orange Yellow**
I	2	//	Important	Green**
O	1	/	Ordinary	Blue**
U	0		Unimportant	Uncolored**
X	-1	zigzag	Not Desirable	Brown**
XX	-2, -3, -4, ?	zigzag	Extremely Undesirable	Black

Evaluating Description	Letter / Value
Almost Perfect Results (Excellent)	A / 4
Especially Good Results (Very Good)	E / 3
Important Results (Good)	I / 2
Ordinary Results (Fair)	O / 1
Unimportant Results (Poor)	U / 0
Not Acceptable Results (Not Satisfactory)	X / ?

Fig. 2.1.11 Capsule summary of systematic layout planning.

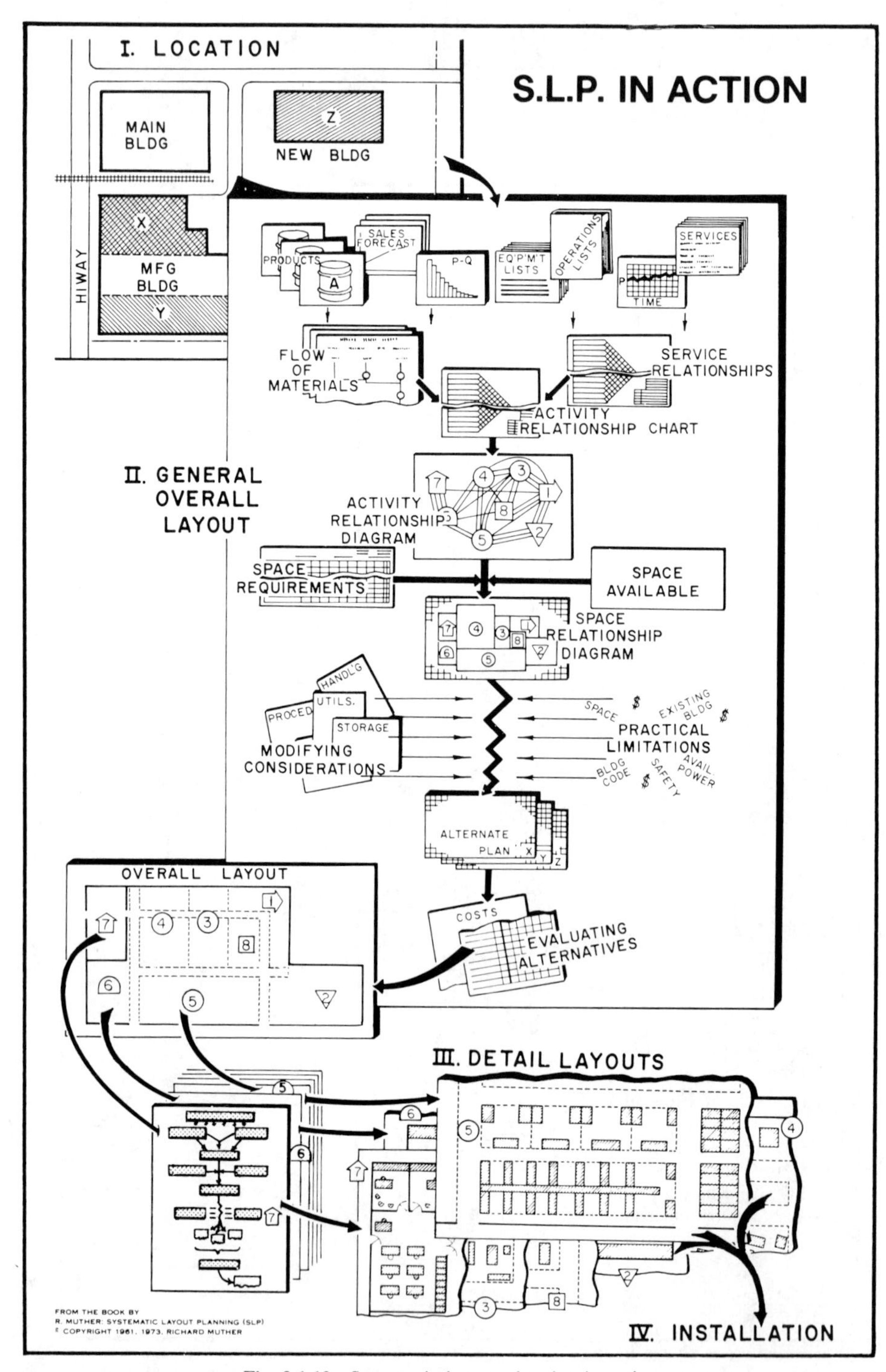

Fig. 2.1.12 Systematic layout planning in action.

2.2 GRAPHICAL TECHNIQUES OF LAYOUT ANALYSIS

Richard Muther

2.2.1 Approach to Analysis

Layout planning starts with: *What materials* in *what quantities* go over *what routes?*

Materials (products, parts, items, etc.) involve the physical or chemical characteristics of the material (solid, liquid, gas, bulk, contained, individual item, size, weight, shape, risk of damage, condition, value, etc.).

Quantities involve the number of pieces or units (kilograms, pounds, tons, gallons, barrels, linear meters, board feet, square inches, cubic yards, etc.).

Routes include the paths between the points of origin and destination (the pick-up place and the set-down place). A series of operations (where material is physically or chemically changed in its characteristics) connected by routes defines the process.

Flow of materials is then the quantity of material per period of time going over a sequence of routes. From a layout standpoint, this flow of materials establishes the closeness of various activity-areas to each other. On routes where the intensity of flow is high, for reduced handling cost, the distance should be short. That is, closeness *relationships are based on flow of materials.*

In addition to relationships based on flow of materials, there are *service and nonflow relationships.* These are always involved, especially in job shops, laboratories, and offices, where flow of material is less significant. Additionally, the *space* of each activity-area is fundamental—in amount, kind, and shape or configuration.

Finally, the *adjustment* of spaces into a layout involves the other physical components or elements of a total facility—controls and communications, utilities and auxiliary conductors, and the building or structure. Certainly the layout of the process-or-storage equipment and the physical material handling methods must integrate with all of these.

2.2.2 Types of Graphical Techniques

Having recognized the larger context of layout planning, let's return to the graphical techniques of analysis.

There are many techniques for planning layouts. Generally, those used for planning new layouts may also be applied to improving existing ones, perhaps with some modifications.

The graphical techniques involved with layout planning break down into four general classes:

1. Data analysis—data gathering and charting.
2. Data visualization—data plotting and diagraming.
3. Layout development—arranging and adjusting.
4. Layout verification—checking and evaluating.

This section concentrates on the first two, dividing its treatment into overall layouts and detail layout plans.

2.2.3 Analysis of Flow of One Material

The analysis of material flow depends largely on its complexity: one material only, few or several materials, many materials.

To understand how a product is made, an assembly or "gozinto" chart is helpful. See Fig. 2.2.1.

The best way of understanding the flow of material is to prepare a process chart. This uses a standard set of symbols and a standard procedure. See Fig. 2.2.2.

Process charts may be drawn on a preprinted form as shown in Fig. 2.2.3. Or, they may be made on a blank sheet of paper, as shown in Fig. 2.2.4.

Additionally, the chart may show the major operations or activities only; or it may show every activity that takes place in actually producing the item. The former is called an *operation process chart.* The latter, as shown in Fig. 2.2.5, is termed a *flow process chart.* Here again, the chart may be made on a preprinted form or blank sheet of paper, as best suits the situation.

2.2.4 Analysis of Material Flow for Several Items

If there are two or three materials, the planner will probably make a process chart for each of them. However, when there are as many as five or ten items, it becomes too time-consuming to repeat

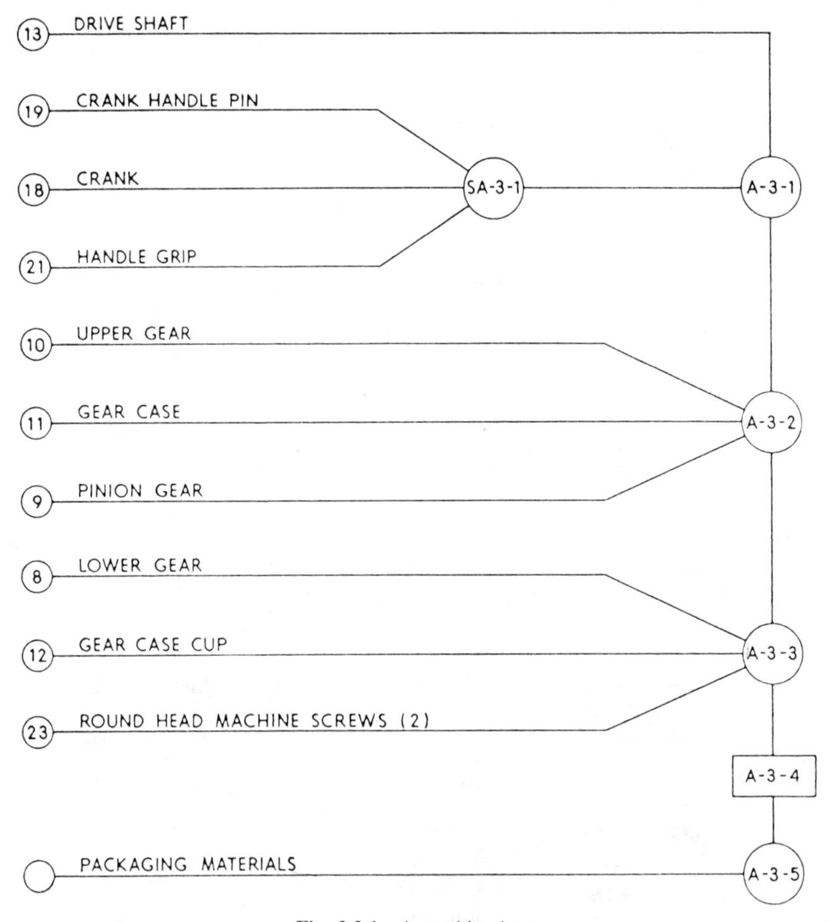

Fig. 2.2.1 Assembly chart.

much of the same information. Rather, the planner should move to a multiproduct or multiple-material process chart. This chart arrays each material in columns vertically and prelists each operation on one horizontal line. See Fig. 2.2.6. Here, if the operations are listed in the best sequence—or the existing squence—then each diagonal line represents backtracking or counterflow.

2.2.5 Group-of-Items Flow

When the items involved become very numerous, they can probably be put into groups or families of items. In simple terms, if the items look alike, they are probably made by the same or similar operations. Therefore, these items can be segregated for analysis and production. The classifying of materials and grouping them into their own layout "cell" is termed *group technology.* Although the classifying and sorting is typically done by data-processing equipment, the flow analysis can be done by means of a multiproduct process chart (Fig. 2.2.6), by a from–to chart (see Fig. 2.2.8), or by other forms of flow analysis and machine loading typically involving a computer. Figure 2.2.7 shows a classical example of grouping items.

If grouping is not possible, the planner may select a sample of representative items, the representative or typical items being basically the same as all others. Or, the planner may pick the worst-condition items, such as the heaviest, most fragile, or most awkward. Their flow can then be treated by operation process chart or multiproduct process chart as appropriate.

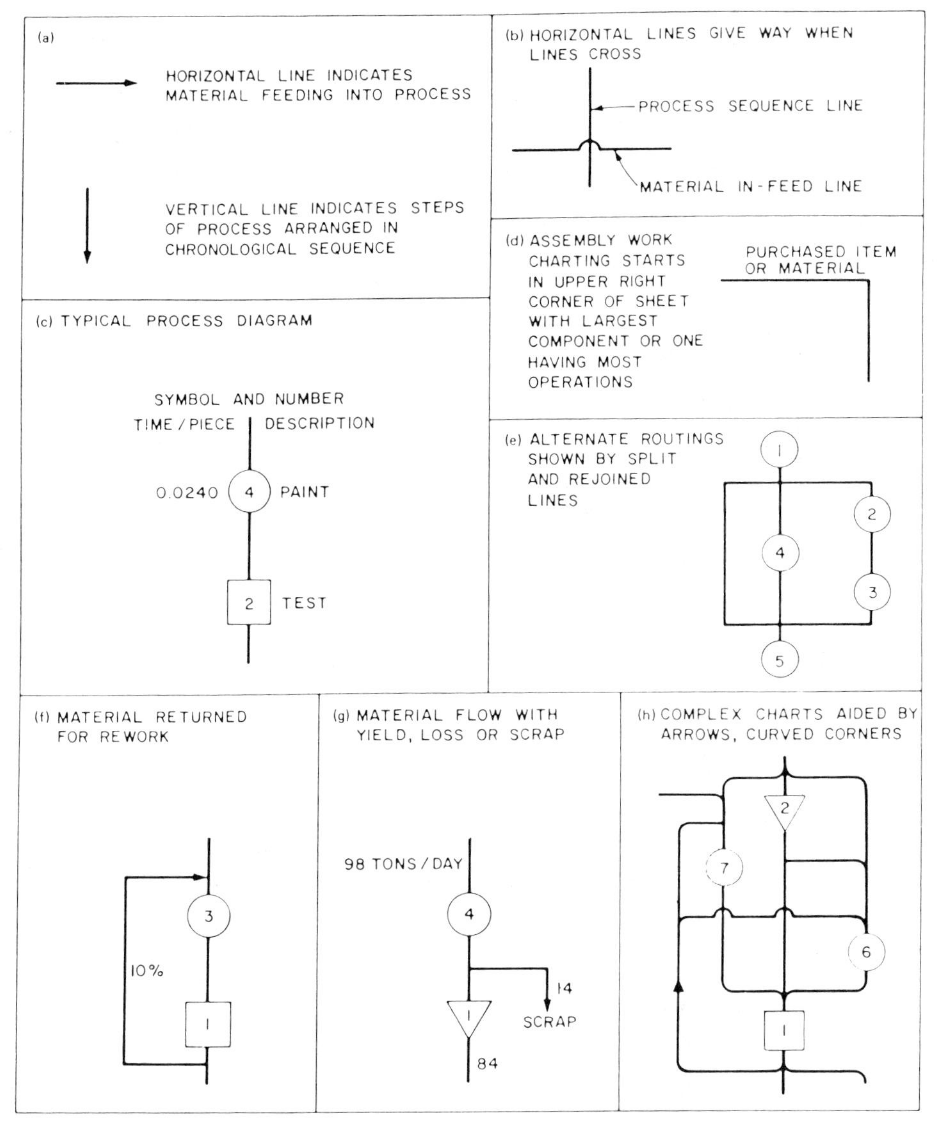

Fig. 2.2.2 Standards for process construction. From American National Standard, Process Charts, ASME, Subcommittee Y15.3 Process Charts.

2.2.6 Analysis of Flow for Many Items

When a great many items are involved, it becomes highly desirable to classify the materials. The classification is made around the physical characteristics of the materials, plus usually the quantity, timing considerations, and any special control problems, such as with drugs in pharmaceutical plants or radioactive items.

Basically, each class defines items with similar characteristics such that any item in that class could be handled by the same handling equipment and the same or similar type of container. Data for this kind of situation is usually compiled, manually or by data-processing equipment, on one of the following bases:

1. Tracking of all materials on each route (use a Route Chart).

PROCESS CHART

Conversions for Charted Unit to End Unit		
Charted Unit	Size or Weight	Quantity per End Unit
Protector	2 lbs.	(End unit)
Carton (filled)	4½ lbs.	1 per 2 protectors
Formed piece	2 lbs.	1 per 1 protector
Blank	2 lbs.	1 per 1 protector
Steel Plate	65 lbs.	1 per 2.3 protector

Process Charted: Make door protectors from steel plates.

Plant ABC Project 34-7
By LFT With REX
Date 6/8 Sheet 1 of 1

Starting Point Steel storage
Ending Point Finished Goods Storage

☒ Present or ☐ Proposed (Alt. #)
Description of Alternative

Quantity of End Unit per (time)
1380 protectors per day

Charted Unit and Units per Load	Activity Symbol	Description of Action	Weight of Load in lbs.	Number of Trips per Day	Distance in feet	Notes
1. Steel Plates	▽1	Stacked on floor	–	–	–	
2. Steel Plates – 12	⇨1	To Cutting on 4-wheel push-truck	780	5	280	
3. Steel Plate – One	○1	Cut to size	–	–	–	Offal
4. Blank – One	⇨2	To form press on conveyor	2	1380	20	
5. Blank – One	○2	Formed	–	–	–	
6. Formed Piece – 400	⇨3	To In-process Storage, pallet & lift truck	800	3.5	320	Hand-Powered low-lift truck
7. Formed Piece – 400	▽2	On pallet on floor	–	–	–	Sometimes moved to pallet rack with hi-lift truck
8. Formed Piece – 400	⇨4	To Grinding on pallet & lift truck	800	3.5	80	Hand-powered low-lift truck
9. Formed Piece – One	○3	Ground to break all edges	–	–	–	
10. Protectors – 260	⇨5	To Packing by lift truck and pallet box	520	5.3	370	260 protectors per pallet box.
11. Protectors – Two	○4	Packed in carton	–	–	–	Packing materials supplied in advance
12. Cartons – 140	⇨6	To F.G. Storage on pallet & lift truck	630	5	210	
13. Cartons – 140	▽3	Pallets with cartons stacked on floor	–	–	–	
14.						
15.						
Totals:	○4 ⇨6 ▽3			Total	1280	

Fig. 2.2.3 Process chart—form. From American National Standard, Process Charts, ASME, Subcommittee Y15.3 Process Charts.

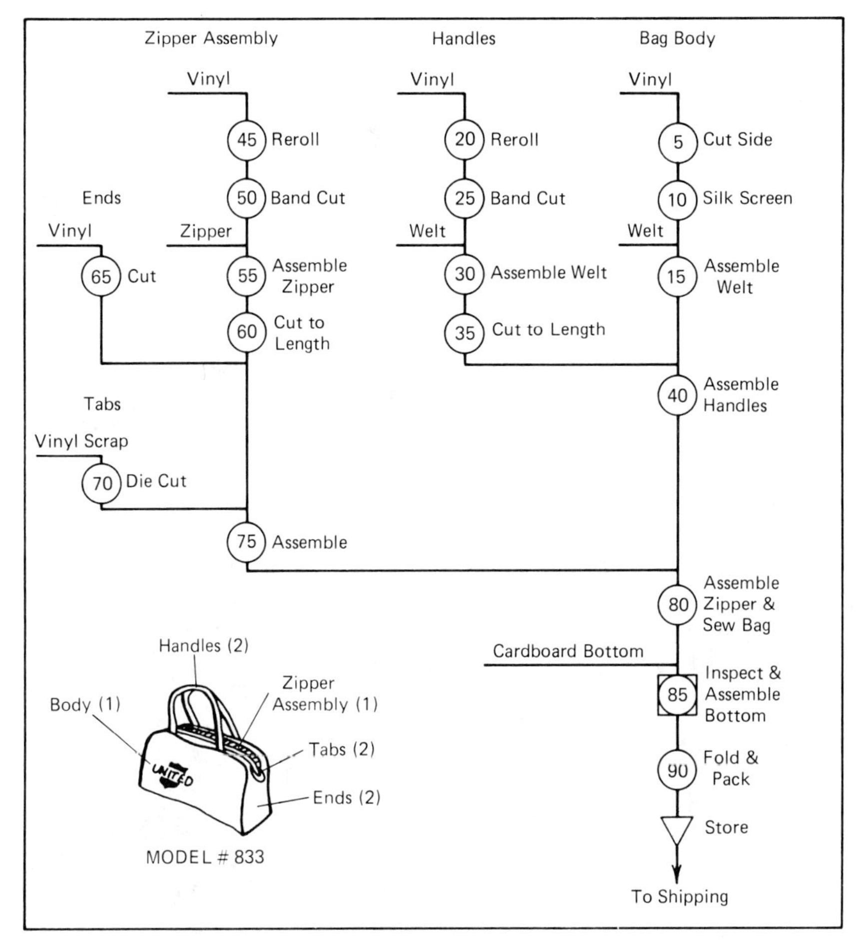

Fig. 2.2.4 Process chart—plain paper.

2. Tracking of all materials in and out of each activity-area or department (use a Flow-in, Flow-out Chart).
3. Tracking of each class of material on all routes (use From–To Charts, one for each material class). See Fig. 2.2.8.

Regardless of how the data is compiled and/or processed, it is usually arrayed in the form of a Movement Summary. Such a summary can be simple or complex, depending on how much information the planner wants to show. Basically, it is a matrix showing across the sheet each class of material in one column and down the sheet each route on one line. The intensity of flow is recorded, in the intersecting boxes, for each class of material on each route. When the data is cross-totalled, it shows the total intensity of flow on each route. When the columns are down-totalled, they show the total intensity of flow for each class of material. Refer back to Fig. 2.1.6.

2.2.7 Combining Flow and Other-Than-Flow Relationships

The need to recognize other-than-flow relationships was discussed previously. (Refer to Fig. 2.1.7.) The method for doing this involves essentially the following sequence:

1. Establish the intensity of flow between each pair of activity-areas.
2. Array these in decreasing order of magnitude.
3. Divide the magnitude of the intensities into four grades or ratings, using the following two guidelines:

FLOW PROCESS CHART

Process charted: Drums all through storage, filling and warehouse

☐ Man or ☒ Material
Starting point
Ending point

Conversions for Charted Unit to End Unit		
Charted Unit	Size/Weight	Quantity/End Unit
Freight car,	in coming	1/600
Empty drum	35 lbs.	1/1
Drum, filled	320 lbs.	1/1
Freight car,	outgoing	1/200 *
Over-the-road truck		1/100 *
Filled drum	regarded as	"End Unit"

Plant Nordic Mixtures Ltd. Project Packing Department
By CL With MRC
Date Sheet 1 of 1

Quantity of End Unit (unit of end item) per (time) 200 Drums per day

☒ Present ☐ Proposed (Alternative #)
Description of Alternative:

	CHARTED UNIT (Unit of Product or Material Charted)	UNITS PER LOAD	Operation	Handling	Transport	Inspection	Delay	Storage	DESCRIPTION OF ACTION	Weight or Size of Load in lbs	Number of trips per day	Distance in feet	Time in manhrs per day	Cost in ___ per ___	NOTES Verify: Product-Quantity-Route-Support-Time Analyse: Why-What-Where-When-Who-How Eliminate-Combine-Rearrange
1	Freight car	-		1					Open door; place slide	-	-				
2	Empty drum	1 drum			1				From freight car to storage	35	200	10-50	8		Step 1
3	Empty drum	1 drum		2					Stacked manually 2-4 high	35					
4	Empty drums							1	Max 1200 drums in storage						
5	Empty drum	1 drum		3					Unstacked by hand	35			5		Step 2
6	Empty drum	1 drum			2				Rolled and lifted to packing	35	200	10-50			
7	Empty drums						1		Max 20 drums in bank						
8	Empty drum	1 drum			3				Moved to scale	35	200	5-10			
9	Empty drum	1 drum				1			Tare Weighed				5½		Step 3
10	Empty drum	1 drum			4				Moved to bank by machine	35		10-20			
11	Empty drum	1 drum	1	4					Up-ended - lid removed						
12	Empty drums						2		Max 20 drums in bank						
13	Empty drum	1 drum		5					Placed in packing machine						
14	Empty drum	1 drum	2						Filled in packing machine				8		Step 4
15	Drum	1 drum		6					Moved out of packing machine	320	200				
16	Drum	1 drum	3						Lid and destination tag fastened						
17	Drum	1 drum			5				Moved to scale with handtruck	320		10-20			
18	Drum	1 drum				2			Weighed				6½		Step 5
19	Drum	1 drum	4						Lid secured with a bolt						
20	Drum	1 drum			6				Moved to stencil area	320	200	5-10			
21	Drum	1 drum	5						Stenciled (spray paint)				5		Step 6
22	Drum	1 drum			7				Moved to pick-up area	320	200	10-20			
23	Drums						3		Max 10 drums in pick-up area						
24	Drum	4 drums			8				With fork-lift to warehouse	1280	50	100-200			Step 7
25	Drums							2	Max 7000 drums in warehouse				7		
26	Drums	4 drums			9				With fork-lift to shipping	1280	50	30-150			Step 8
27	Car or truck					3			Shipment checked by foreman						
28	-n-			7					Lock and secure						* 90% of production is shipped by rail, 10% by road.
29															
30															
		Totals	5	7	9	3	3	2			Totals	300-500	45		

Fig. 2.2.5 Flow process chart.

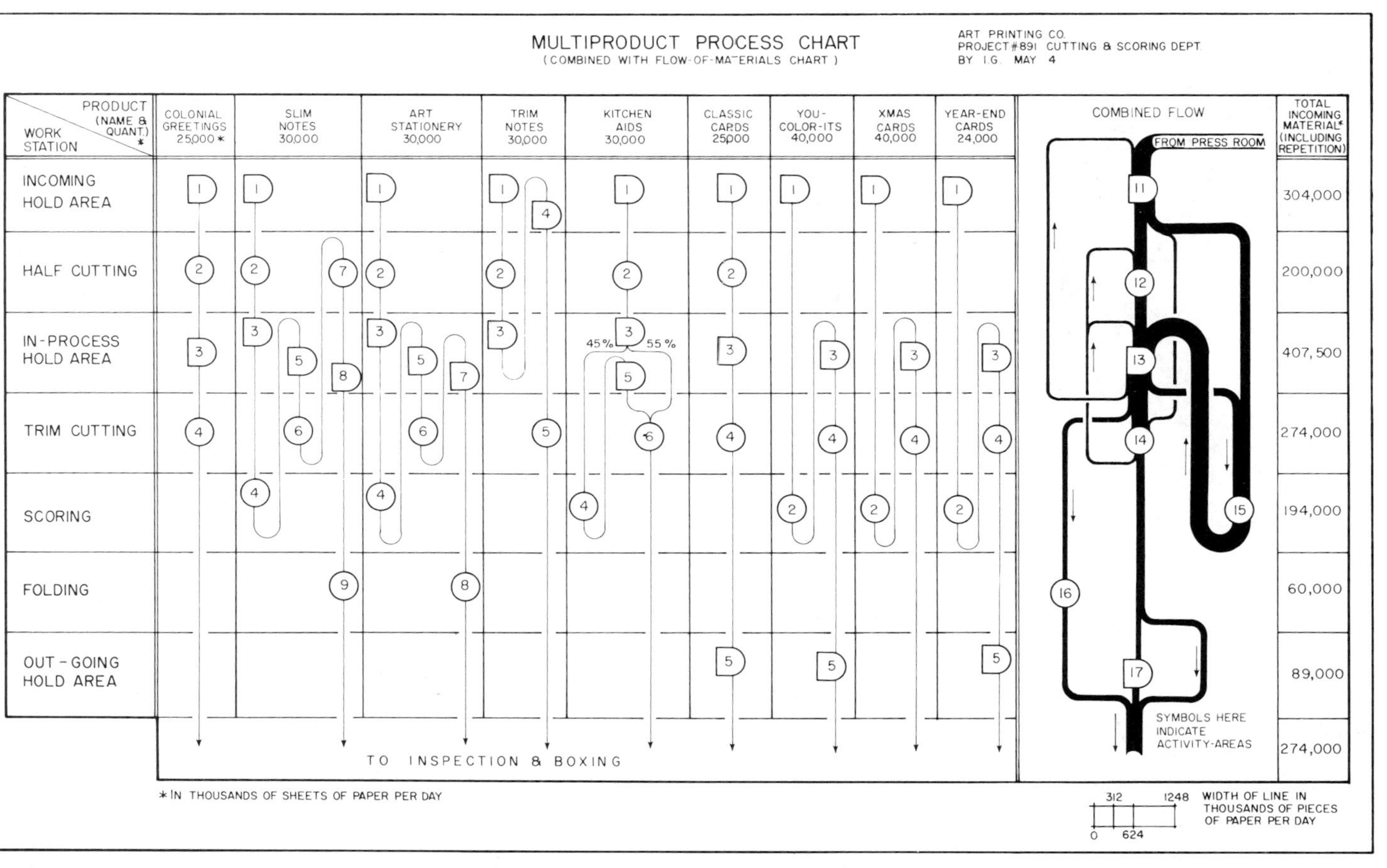

Fig. 2.2.6 Multiple material process chart.

a. Divide the intensity ratings for each route where there are major differences in the intensities.

b. Divide the ratings approximately into the highest 10% of the routes, next highest 20% of the routes, third highest 30% of the routes, and the remaining 40% of the routes. This will almost automatically lead to 10% of the routes encompassing routes with intensities over 60% of the highest intensity on any route.

4. Assign an order-of-magnitude vowel letter to the rating of intensity for each route:

A—*A*bnormally high intensity of flow

E—*E*specially high intensity of flow

I—*I*mportant intensity

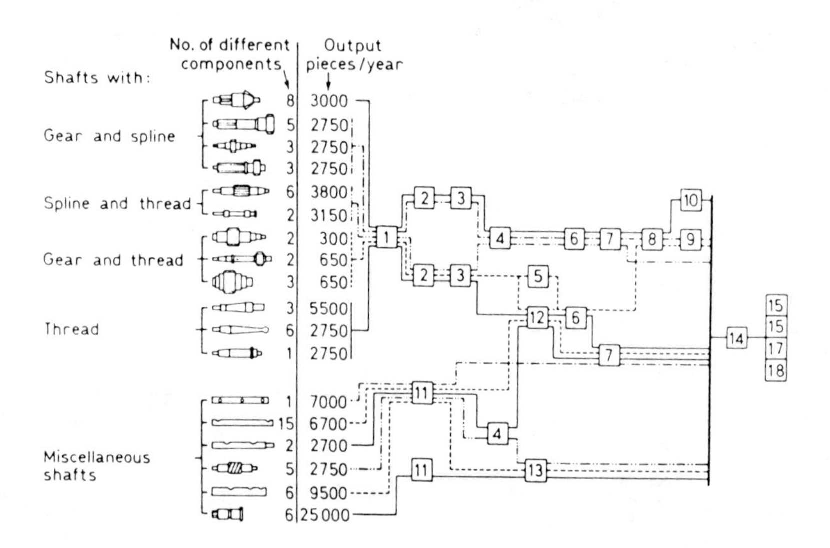

1. Centring lathe
2. Hydraulic contour lathe
3. Finish engine lathe
4. Spline milling machine
5. Key milling machine
6. Thread milling machine
7. Three-speed drilling press
8. Cylinder grinder
9. Gear milling machine
10. Gear cutting machine
11. Turret lathe
12. Universal milling machine
13. Hydraulic plain milling machine
14. Inspection
15. Hardening department
16. Grinding department
17. Final inspection
18. Assembly stock

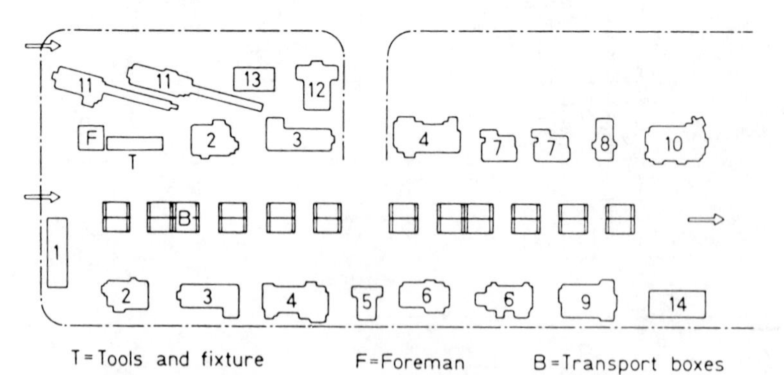

Fig. 2.2.7 Group production.

FROM – TO CHART

Item(s) Charted: All Items
Basis of Values: Equiv. Skids/Year (in 000's)
Plant: White Lighting Co.
By: K. W. M.
Date: 22 June
Project: –
With:
Page: 1 of 1

FROM \ TO (Activity or Operation)	1 Receiving	2 Material Storage	3 Machining	4 Wire Stringing	5 Small Parts Sub-Assembly	6 Fluorescent Assembly	7 Mercury Vapor Assembly	8 Facade Light Assembly	9 Finished Fixture Storage	10 Pipe Receiving & Storage	11 Pipe Bending	12 Welding	13 Painting	14 Outside Pole Storage ⓑ	15 Shipping	16	17	18	19	20	TOTALS
Receiving 1		60	–	–	–	–	–	–	–	–	–	–	–	–	–						60
Material Storage 2	–		9	4	–	22	1	30	–	–	–	–	–	–	–						66
Machining 3	–	1		–	9	–	–	–	–	–	–	–	–	1	–						11
Wire Stringing 4	–	–	–		–	2	1	3	–	–	–	–	–	–	–						6
Small Parts Sub-Assembly 5	–	–	–	–		6	1	–	–	–	–	–	–	–	–						7
Fluorescent Assembly 6	–	2	–	–	–		–	–	24	–	–	–	–	–	–						26
Mercury Vapor Assembly 7	–	–	–	–	–	–		–	2	–	–	–	–	–	–						2
Facade Light Assembly 8	–	3	–	–	–	–	–		25	–	–	–	–	–	–						28
Finished Fixture Storage 9	–	–	–	–	–	4	½	5		–	–	–	–	–	50						59½
Pipe Receiving & Storage 10	–	–	–	–	–	–	–	–	–		60	40	–	–	–						100
Pipe Bending 11	–	–	–	–	–	–	–	–	–	–		80	–	–	–						80
Welding 12	–	–	–	–	–	–	–	–	–	–	–		120	–	–						120
Painting 13	–	–	–	–	–	–	–	–	–	–	–	–		140	–						140
Outside Pole Storage ⓑ 14	–	–	–	–	–	–	–	–	–	–	–	–	–		140						140
Shipping 15	–	–	–	–	–	–	–	–	10ⓐ	–	–	–	–	–							10
16																					
17																					
18																					
19																					
20																					
TOTALS	–	66	9	4	9	34	3½	38	61	–	60	120	120	141	190						855½

NOTES: ⓐ Packing materials are best stored in the finished fixture storage area.
ⓑ Including scrap-accumulation depot.

Fig. 2.2.8 From–to chart.

O—*O*rdinary intensity

U—*U*nimportant or negligible intensity of flow

5. Establish for each pair of activity-areas a closeness-desired rating for reasons other than flow. This may include activity pairs already identified as having flow, the pairs involved in process-areas but not having any movement of material between them, and the service or non-process activity-areas as they relate to both the activity-areas involving processing of materials and to those activity-areas which are strictly service or non-process areas. In rating other than flow, order-of-magnitude vowel letters are typically used also, but with closeness-desired/required meanings:

   A—*A*bsolutely necessary closeness

   E—*E*specially important closeness

   I—*I*mportant closeness

   O—*O*rdinary closeness

   U—*U*nimportant closeness

   X—*Not* desirable closeness

   Note that there are nonflow relationships that involve separating activity-areas. It may even be as important sometimes to be sure certain activities do not get close to each other as to place others close.

6. Combine the flow and other-than-flow ratings, weighting the relative importance of flow to other-than-flow. If there is a one-to-one relative importance, then let A for flow = 4 and A for other-than-flow = 4, making a total of 8. The new or combined rating then would simply be divided by two to get a net, resultant or combined rating for the vowel-letter sequence ($8 \div 2 = 4 = A$). If flow is twice as important as other-than-flow, then A for flow would represent 8, and A for other-than-flow would represent 4, and the total would start at 12. Back this down into 4 divisions, with anything above 9 now becoming equal to a net, resultant, or combined rating of A. A series of matrix tables for ready working of these data is shown in Fig. 2.2.9.
7. Now post the resultant relationship ratings on a combined relationship chart. See Fig. 2.2.10.

The result of this combining now assures the planner that:

1. Both flow and other-than-flow considerations have been recognized and simply scored for closeness.
2. The relationship of each activity-area to every other activity-area has indeed been considered.
3. The supporting reasons are recorded for each rating.
4. All the relationships are on one convenient piece of paper.
5. The opportunity exists to have these reviewed and agreed to before the planner goes farther with the layout planning.

2.2.8 Diagraming Flow and Closeness Relationships

Diagraming of materials flow can be done several ways. One form is shown in the top half of Fig. 2.2.7. Others are shown in Figs. 2.1.10 and 2.2.11. The most frequently followed variations are indicated in Fig. 2.2.12.

When flow and other-than-flow have been combined, a number-of-lines code is typically used. Each line can be likened to an elastic band with four lines pulling things up close and lesser lines allowing activity-areas to be farther away. See standard symbols as indicated in Fig. 2.1.11.

Generally speaking, in developing diagrams of relationships, several steps should be kept in mind:

1. Diagram the A's, using four lines. If improvement is practical, redraw the diagram for best fit before adding the E's.

2. Redraw the diagram for best fit after each set of closeness lines (each vowel letter) is recorded. Do not develop the whole diagram for all ratings. Note that the wiggly line is likened to a compressed spring, pushing activities away from each other.

3. When diagraming, place three-line relationships (E) about twice as far apart as activities with four lines connecting them. Then, place two lines twice as far again. Another way to consider this is that the area inside each band of connecting lines (except X) should be kept approximately equal, thus equalizing the transport work, or Intensity × Distance.

4. Work with black-and-white paper and pencil on a clear sheet. Redraw rather than erase.

5. When the final diagram is completed, color in or shade the band of connecting lines, using the closeness-desired color code as shown in the standard established by the International Material Management Society. See Fig. 2.1.11.

COMBINING FLOW AND OTHER-THAN-FLOW BY MATRIX METHOD

IN VARIOUS SITUATIONS, "FLOW OF MATERIALS" CARRIES DIFFERENT DEGREES OF IMPORTANCE RELATIVE TO FACTORS "OTHER THAN FLOW". ON EACH PROJECT, THE PLANNER OR ANALYST SHOULD DETERMINE THIS DEGREE OF RELATIVE IMPORTANCE, AND THEN ASSIGN A COMBINED CLOSENESS RATING FOR EACH PAIR OF ACTIVITY-AREAS.

THIS CAN BE DONE BY BUILDING A MATRIX OF FLOW AND OTHER-THAN-FLOW; ESTABLISHING A RELATIVE WEIGHT BETWEEN THE TWO; MULTIPLYING THE WEIGHT VALUE BY THE EQUIVALENT CLOSENESS-DESIRED NUMBER VALUE; AND THEN DIVIDING THE TOTALS AT CONSISTENT BREAK-OFF PLACES.

THE DIVISIONS WITHIN ANY MATRIX NEED NOT BE THE SAME FOR DIFFERENT PROJECTS, AS THE PLANNER SHOULD HAVE THE OPPORTUNITY TO END UP WITH A REALISTIC DISTRIBUTION OF NET RATINGS: SAY 1 TO 3% A's; 2 TO 5% E's; 3 TO 8% I's; 5 TO 15% O's; 20 TO 85% U's; 0 TO 10% X's.

FLOW AND OTHER-THAN-FLOW AT RELATIVE IMPORTANCE OF 1 TO 1

FLOW		OTHER-THAN-FLOW 4	3½	3	2½	2	1½	1	½	0	-1	-2
		A	A-	E	E-	I	I-	O	O-	U	X	XX
4	A	8	7½	7	6½	6	5½	5	4½	4	3	2
3½	A-	7½	7	6½	6	5½	5	4½	4	3½	2½	1½
3	E	7	6½	6	5½	5	4½	4	3½	3	2	1
2½	E-	6½	6	5½	5	4½	4	3½	3	2½	1½	½
2	I	6	5½	5	4½	4	3½	3	2½	2	1	0
1½	I-	5½	5	4½	4	3½	3	2½	2	1½	½	-½
1	O	5	4½	4	3½	3	2½	2	1½	1	0	-1
½	O-	4½	4	3½	3	2½	2	1½	1	½	-½	-1½
0	U	4	3½	3	2½	2	1½	1	½	0	-1	-2

FLOW CONSIDERED TWICE AS IMPORTANT AS OTHER-THAN-FLOW.

FLOW		OTHER-THAN-FLOW 4	3½	3	2½	2	1½	1	½	0	-1	-2
		A	A-	E	E-	I	I-	O	O-	U	X	XX
8	A	12	11½	11	10½	10	9½	9	8½	8	7	6
7	A-	11	10½	10	9½	9	8½	8	7½	7	6	5
6	E	10	9½	9	8½	8	7½	7	6½	6	5	4
5	E-	9	8½	8	7½	7	6½	6	5½	5	4	3
4	I	8	7½	7	6½	6	5½	5	4½	4	3	2
3	I-	7	6½	6	5½	5	4½	4	3½	3	2	1
2	O	6	5½	5	4½	4	3½	3	2½	2	1	0
1	O-	5	4½	4	3½	3	2½	2	1½	1	0	-1
0	U	4	3½	3	2½	2	1½	1	½	0	-1	-2

OTHER-THAN-FLOW TWICE AS IMPORTANT AS FLOW

FLOW		OTHER-THAN-FLOW 8	7	6	5	4	3	2	1	9	-2	-4
		A	A-	E	E-	I	I-	O	O-	U	X	XX
4	A	12	11	10	9	8	7	6	5	4	2	0
3½	A-	11½	10½	9½	8½	7½	6½	5½	4½	3½	1½	-½
3	E	11	10	9	8	7	6	5	4	3	1	-1
2½	E-	10½	9½	8½	7½	6½	5½	4½	3½	2½	½	-1½
2	I	10	9	8	7	6	5	4	3	2	0	-2
1½	I-	9½	8½	7½	6½	5½	4½	3½	2½	1½	-½	-2½
1	O	9	8	7	6	5	4	3	2	1	-1	-3
½	O-	8½	7½	6½	5½	4½	3½	2½	1½	½	-1½	-3½
0	U	8	7	6	5	4	3	2	1	0	-2	-4

FLOW CONSIDERED 1½ TIMES AS IMPORTANT AS OTHER-THAN-FLOW.
(FLOW 3; OTHER THAN FLOW 2)

FLOW		OTHER-THAN-FLOW 8	7	6	5	4	3	2	1	0	-2	-4
		A	A-	E	E-	I	I-	O	O-	U	X	XX
12	A	20	19	18	17	16	15	14	13	12	10	8
10½	A-	18½	17½	16½	15½	14½	13½	12½	11½	10½	8½	6½
9	E	17	16	15	14	13	12	11	10	9	7	5
7½	E-	15½	14½	13½	12½	11½	10½	9½	8½	7½	5½	3½
6	I	14	13	12	11	10	9	8	7	6	4	2
4½	I-	12½	11½	10½	9½	8½	7½	6½	5½	4½	2½	½
3	O	11	10	9	8	7	6	5	4	3	1	-1
1½	O-	9½	8½	7½	6½	5½	4½	3½	2½	1½	-½	-2½
0	U	8	7	6	5	4	3	2	1	0	-2	-4

OTHER-THAN-FLOW CONSIDERED 1½ TIMES AS IMPORTANT AS FLOW.
(FLOW 2; OTHER THAN FLOW 3)

FLOW		OTHER-THAN-FLOW 12	10½	9	7½	6	4½	3	1½	0	-3	-6
		A	A-	E	E-	I	I-	O	O-	U	X	XX
8	A	20	18½	17	15½	14	12½	11	9½	8	5	2
7	A-	19	17½	16	14½	13	11½	10	8½	7	4	1
6	E	18	16½	15	13½	12	10½	9	7½	6	3	0
5	E-	17	15½	14	12½	11	9½	8	6½	5	2	-1
4	I	16	14½	13	11½	10	8½	7	5½	4	1	-2
3	I-	15	13½	12	10½	9	7½	6	4½	3	0	-3
2	O	14	12½	11	9½	8	6½	5	3½	2	-1	-4
1	O-	13	11½	10	8½	7	5½	4	2½	1	-2	-5
0	U	12	10½	9	7½	6	4½	3	1½	0	-3	-6

Fig. 2.2.9 Matrix tables for flow and other than flow.

2.2.9 Space Requirements

In determining space, five methods are most frequently used:

Calculating the space for equipment required
Converting from what you happen to have to what you will require
Roughing out a layout by scale sketch or by placing templets
Employing space standards for repetitive areas
Plotting a projection of a trend for a ratio that includes space

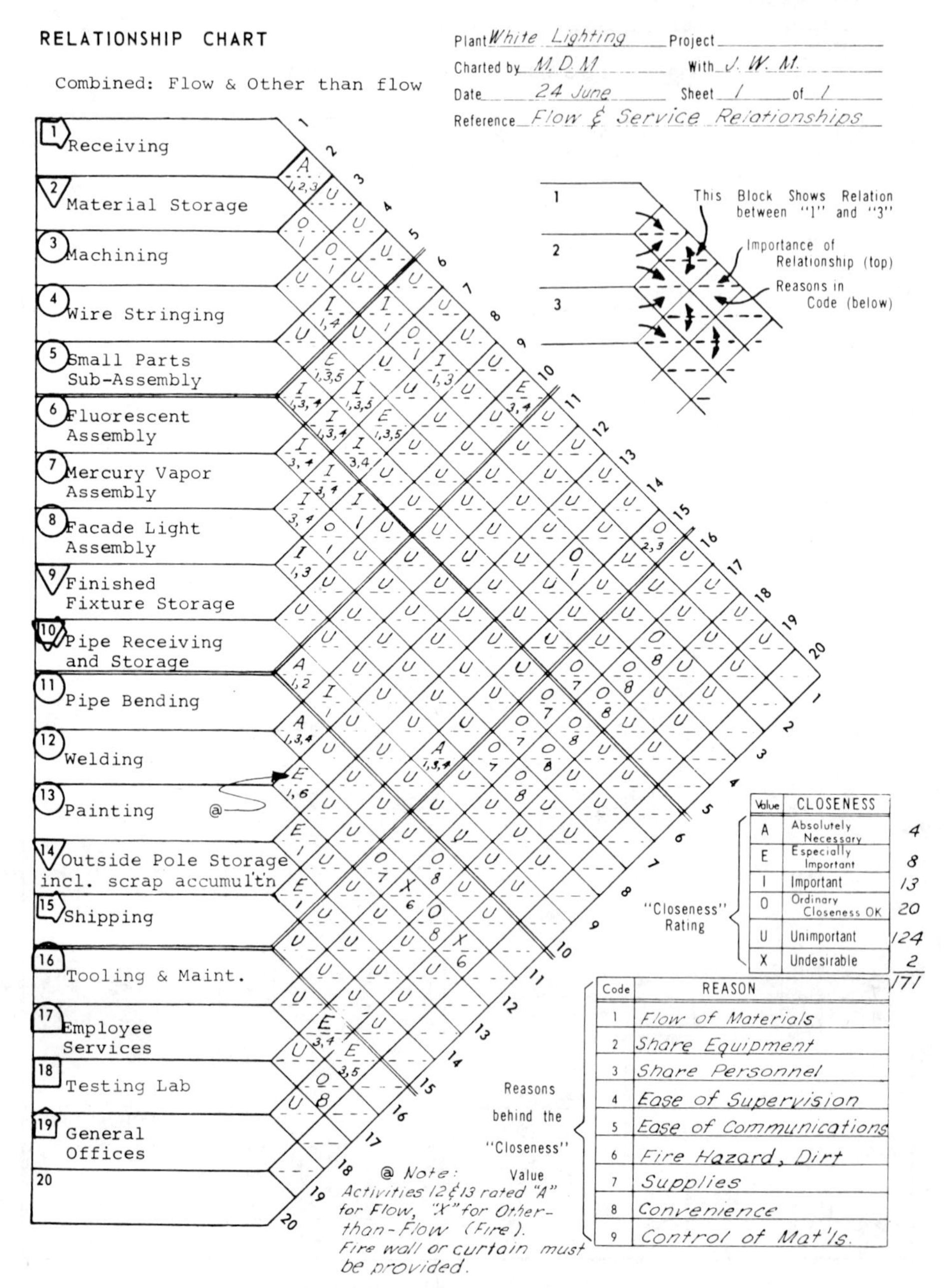

Fig. 2.2.10 Combined relationship chart.

In actual practice, the space for different activity-areas may require different methods of space determination on the same project. And not infrequently, more than one method for the same area is used, lending credibility to the figures. Thus, all five methods could be used on the same project.

On many projects, these methods are combined, especially when a few dominant pieces of equipment are involved. The space required for that equipment can be calculated; the secondary and support areas converted by ratio of their areas to the area for the dominant equipment; and the storage areas established by rough layout of various kinds of storage racks based on the anticipated quantities of each type required. In any case, some form of space specifications is needed. See Fig. 2.2.13.

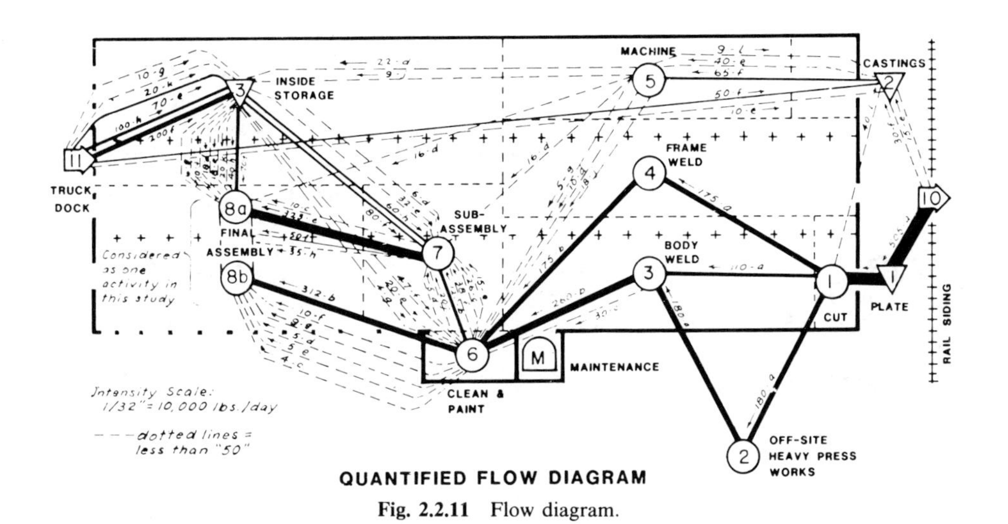

Fig. 2.2.11 Flow diagram.

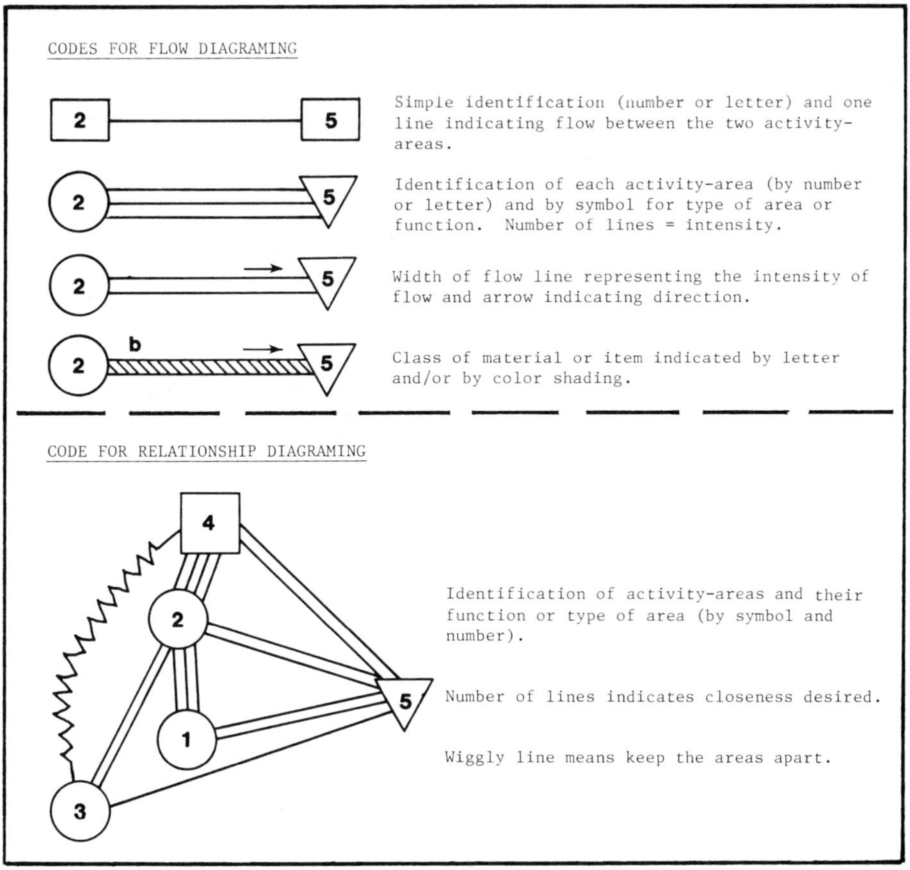

CODES FOR FLOW DIAGRAMING

2 — 5

Simple identification (number or letter) and one line indicating flow between the two activity-areas.

2 — 5

Identification of each activity-area (by number or letter) and by symbol for type of area or function. Number of lines = intensity.

2 → 5

Width of flow line representing the intensity of flow and arrow indicating direction.

2 b → 5

Class of material or item indicated by letter and/or by color shading.

CODE FOR RELATIONSHIP DIAGRAMING

4 2 1 3 5

Identification of activity-areas and their function or type of area (by symbol and number).

Number of lines indicates closeness desired.

Wiggly line means keep the areas apart.

Fig. 2.2.12 Variations for activity–relationship diagram.

ACTIVITIES AREA & FEATURES SHEET

Plant Novelty Luggage
Project #621
By CHC With JBS
Date Mar. 20 Page ___ of ___

Activity: No.	Activity: Name	Area in sq.ft.	Minimum O'Head Clearance	Max. Overhead Supported Load	Max. Floor Loading	Min. Column Spacing	Water & Drains	Steam	Compressed Air	Foundations -or Pits	Fire or Explosion Hazard	Special Ventilation	Special Electrification	Air Condition	Requirements for Shape or Configuration of Area (Space)
		Total: 11,200	Enter Unit and Required Amount under each: ft.	lbs.			Relative Importance of Features: A – Absolutely Necessary; E – Especially Important; I – Important; O – Ordinary Importance; — Not Required								Enter Requirements for Shape or Configuration and Reasons therefore
1.	Cutting	900	12	(a) 700			–	–	–	–	–	–		O	
2.	Silk Screen	1,100				(b)	E	–	–	–	A	A		(c) I	Min. length dimension is 45 ft. (for dryer).
3.	Sub-assembly	300					–	–	–	–	–	–		O	
4.	Final Assembly	2,300					–	–	–	–	–	–		O	
5.	Packaging	400					–	–	–	–	–	–		O	
6.	Rec'g.& Ship'g.	1,500					–	–	–	–	–	–		O	
7.	Mat'l. Storage	1,200					–	–	–	–	–	–		O	
8.	Darkroom	250	Same	Normal	150 lbs./sq.ft.		A	–	–	–	–	E	220/110 Volt	A	
9.	Art Rm.& Design	200					–	–	–	–	–	–		A	
10.	Office	1,700					–	–	–	–	–	–		A	
11.	Maintenance	300					E	–	–	–	–	E		O	
12.	Rest Rooms	200					A	–	–	–	–	–		O	
13.	Lunch Room	850					E	–	–	–	–	–		A	
14.															
15.															

Sub-Activities or Areas

Notation References:

a. Possible use of overhead hoist, light,--500 lb. maximum load.
b. Should accept automated dryer, 35 ft. long and 6 ft. wide.
c. Humidity controlled room is desirable
d.

Fig. 2.2.13 Space requirements/space specifications.

2.2.10 Space Relationship Diagram

Combining space and relationships takes one of two forms:

Flow of material and space
Relationships (flow and other than flow) and space

When materials are large, heavy, awkward, or require special handling or protection, the layout is usually developed around the flow of material desired or required. A flow diagram with space shown can be directly helpful in developing the layout.

When the service and support activities are at all significant, the layout is best developed around a diagram of combined relationships. See Fig. 2.2.14. Here, the space required for each activity-area is superimposed on an activity-relationship diagram. The resulting diagram shows the space in its best-fit arrangement, ready for adjustment.

As shown by the shaded areas in Fig. 2.2.14, the space is usually accentuated by color, shading, or other meaningful indicators. The standard colors or shadings for type of space are indicated in Fig. 2.1.11.

Numerous computer-based algorithms have been developed to improve or construct layouts. Most of these print out a rough block layout. That is, they automate part of the adjustment process, and the diagraming itself is seldom printed. The basic characteristic of these algorithms is that they can generate a large number of alternative arrangements, evaluate them against some preestablished objective, and print out the best. In spite of some claims by protagonists, none of these layout algorithms guarantee

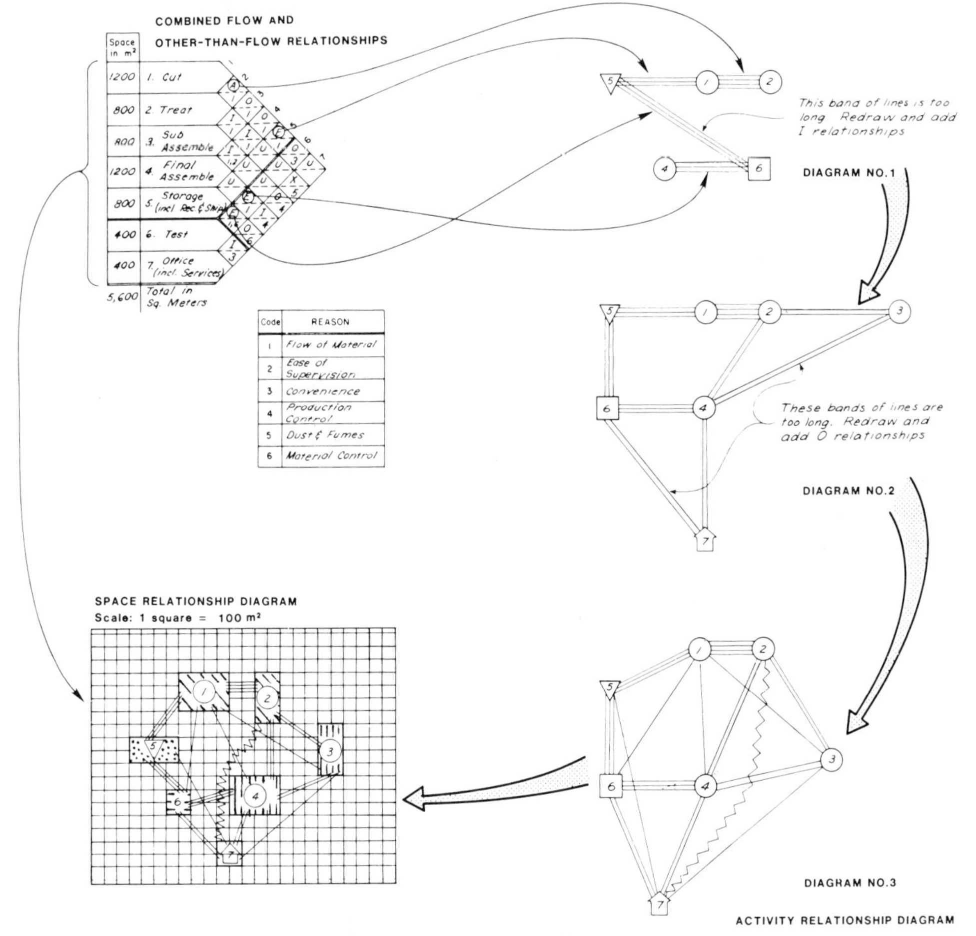

Fig. 2.2.14 Space relationship diagram.

a best or optimal solution. So, there is always further adjustment of the output. And, although they have been available for some 15 years, planners who have relied on them for developing realistic layouts have largely been disappointed. Interested planners can find more on this in *Computer-Aided Layout Planning: A Users Guide* by Tompkins and Moore, available from the Institute of Industrial Engineers.

Most users experienced with algorithms feel they should be viewed as adjuncts to a sound graphic approach. See Fig. 2.2.15.

2.2.11 Adjusting the Space into a Layout

Adjusting the diagram (i.e., arranging the space) may involve any or all of the following:

Rotating the diagram (in the plane of the paper); taking the mirror image (turning it over); or a combination of the two.

Adjusting the diagram by fitting the spaces together, and by splitting, dividing, or shaping the space—keeping in mind the actual function of the operations it represents.

Integrating with the diagram the method of handling (its equipment, transport unit or container, and its pattern or system of moves) and any adjustments required for utilities, building features, production dispatching or inventory control, safety, security, building code, budget limits, and so on.

In practice, when adjusting the diagram manually, it is best to:

1. Convert the space for each activity-area to a number-of-squares count.
2. Draw on a sheet of grid (graph) paper to show the space for each activity-area.
3. Use a piece of tracing paper with fade-out grid printed thereon. Count squares to save calculations.
4. Use a scale that gets the whole layout on one sheet of paper. You can change the scale later for more detail.
5. Work with a straightedge and black pencil. You can ink, color, or shade the areas later.
6. Lay the grid sheet over a print of the site, building, or floor plan in question as you integrate the layout with the existing facilities or physical features.
7. Develop several alternatives, incorporating various concepts of operation and not being afraid to violate some closeness relationships if you or others get a better idea as you see the layout plan develop.

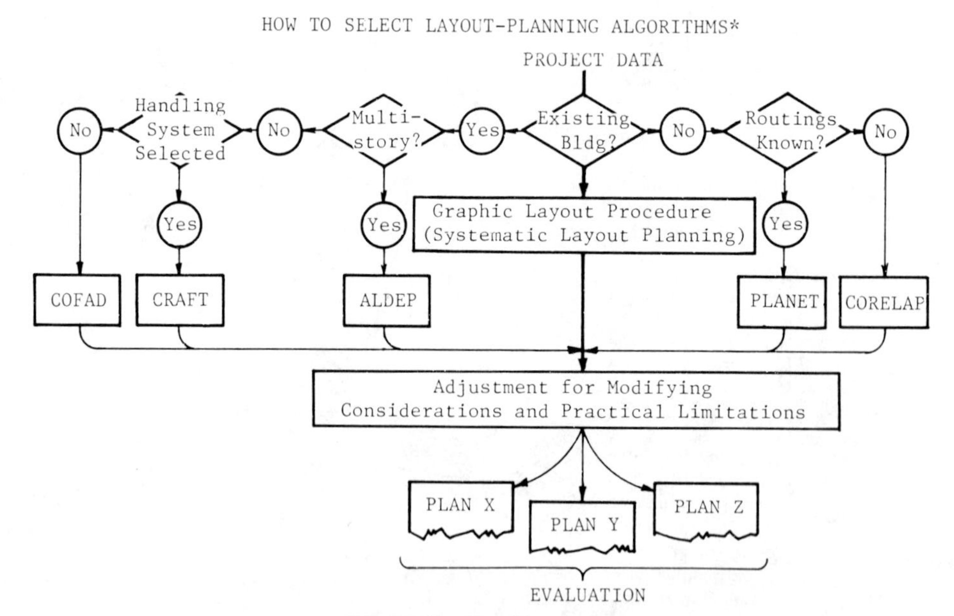

Fig. 2.2.15 Algorithm tie-in.

To visualize the overall or block layouts, planners generally use the following methods:

1. *Drawing* is probably the easiest and fastest way to visualize layouts. By use of basic symbols, colors, and shadings, the spaces can be accentuated. When grid sheets, especially tracing paper with a fade-out grid, are used, layouts can be generated quickly. Pre-copied blank floor sheets—with grid, columns, and walls indicated—are similarly very practical.

When integrating handling equipment and building features, elevation drawings are helpful. And, multistory buildings that have to integrate vertically can use orthographic drawings and orthographic grid paper.

2. *Templets* are *not* generally recommended for block layout planning. They imply a size and shape that becomes fixed in the planner's mind, and they are not easy to separate or re-shape. Unit-area templets, developed to get around this inflexibility of shape, have not proved practical.

In detail layout, templets are fine because the machine or equipment they represent has a finite size and shape.

3. Three-dimensional *models* can show the layout more clearly than drawings. They are time-consuming to make and often costly. Most planners do not have time, or really the need, to develop such models for block layouts. However, when presenting an overall plan to approvers, boards of directors, planning commissions, regulatory agencies, or bidders and suppliers of major equipment, models make the layout more understandable.

4. A modification is the *three-dimensional drawing.* This is usually an orthographic or perspective drawing of the facilities involved, so it tends to be more than just a visualization of the layout. But, in the final presentation or when seeking approvals, it can be colored and dressed up to real advantage. So also can a colored photograph of the 3-D model.

5. *Overlay sheets,* usually of clear plastic material, are truly helpful. Use them to diagram the flow of material, over the existing or a proposed layout. Similarly, storage or material staging areas, utility runs (overhead or underfloor), building features, and communications/control equipment can be integrated with the proposed layout, each on its own overlay sheet as appropriate.

Overlay sheets are especially helpful in showing materials handling equipment, where it is to be located and how it will operate. And, when comparing alternative layouts, such overlays can show how well each layout meets the anticipated flow of material or honors the desired relationships.

6. *Computers* to make drawings are becoming more widely used every day. Most Computer-Aided Design equipment can produce drawings at a high rate of speed. Depending on the details of each "layer" of data, various features can be called out for inclusion on the display screen. Working interactively, the planner can make changes directly and call for an adjusted view immediately, either in two or three dimensions. Depending on primary and output equipment, the planner can enlarge or reduce plans or sections thereof, and can get copies of finished drawings at the press of a key. Figure 2.2.16 shows a layout made by CAD equipment. Preparing layouts this way is especially helpful in generating many different alternatives. It is, of course, very fast, once the system is in place, in final adjusting and making changes.

2.2.12 Evaluating Alternative Layouts

Seldom is only one layout plan developed. There are too many ways space can be arranged. And there are always trade-offs in costs, intangible considerations, short-versus-long-range questions and personal preferences. So a number of layout plans will be developed. These will be refined and screened, usually down to the last two to five alternatives. Evaluating alternative layouts requires several considerations: cost analysis, nonfinancial considerations, and hidden factors.

Economic or Cost Analysis

Estimated or quoted cost for providing the layout versus the costs of operating the layout must be evaluated. There are too many ways to justify capital investments to discuss here (see Chapter 3 for discussion).

Intangible or Nonfinancial Considerations

Intangible factors may be evaluated using a simple listing of pros and cons (advantages and disadvantages) or a weighted-factor comparison of alternative layouts. See Fig. 2.2.17. The planner should try to get upper managers or approvers of the layout to select the factors and/or assign or approve the weight values for them. The planner should try to get operating managers or supervisors to participate in the rating of plan-against-plan on each factor. This helps others understand the plans; it presells the solution; it keeps the planner from missing any oversights.

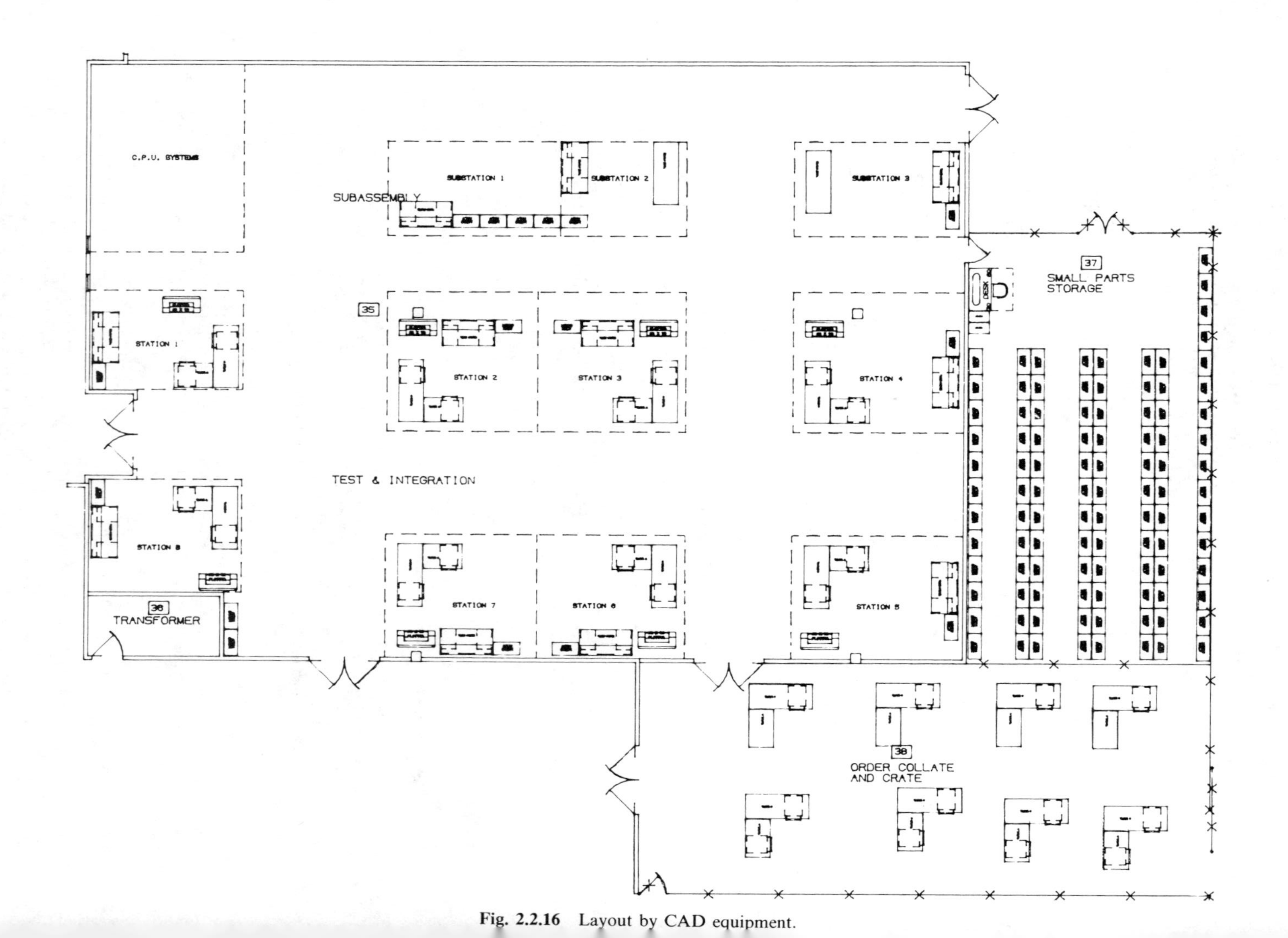

Fig. 2.2.16 Layout by CAD equipment.

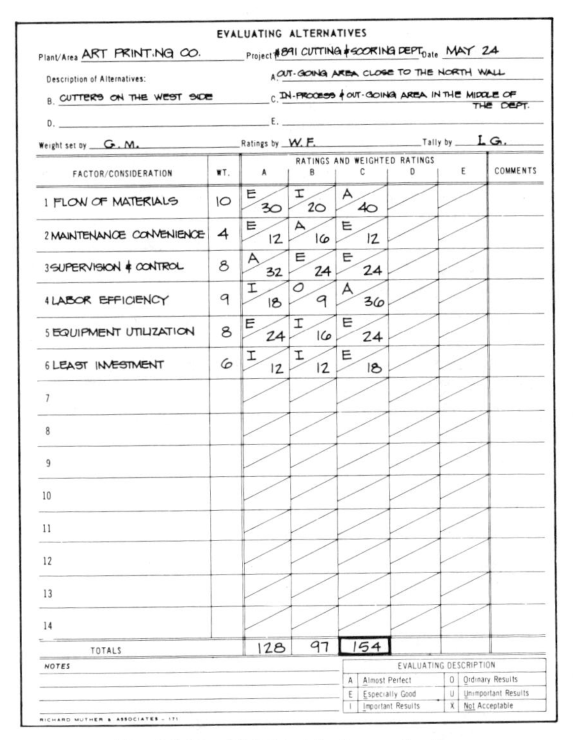
EVALUATING ALTERNATIVES

Plant/Area ART PRINTING CO. Project #891 CUTTING & SCORING DEPT. Date MAY 24

Description of Alternatives:

A. OUT-GOING AREA CLOSE TO THE NORTH WALL

B. CUTTERS ON THE WEST SIDE

C. IN-PROCESS & OUT-GOING AREA IN THE MIDDLE OF THE DEPT.

D.

E.

Weight set by G. M. Ratings by W. F. Tally by L. G.

FACTOR/CONSIDERATION	WT.	RATINGS AND WEIGHTED RATINGS A	B	C	D	E	COMMENTS
1 FLOW OF MATERIALS	10	E 30	I 20	A 40			
2 MAINTENANCE CONVENIENCE	4	E 12	A 16	E 12			
3 SUPERVISION & CONTROL	8	A 32	E 24	E 24			
4 LABOR EFFICIENCY	9	I 18	O 9	A 36			
5 EQUIPMENT UTILIZATION	8	E 24	I 16	E 24			
6 LEAST INVESTMENT	6	I 12	I 12	E 18			
7							
8							
9							
10							
11							
12							
13							
14							
TOTALS		128	97	154			

NOTES

EVALUATING DESCRIPTION			
A	Almost Perfect	O	Ordinary Results
E	Especially Good	U	Unimportant Results
I	Important Results	X	Not Acceptable

RICHARD MUTHER & ASSOCIATES - 171

Fig. 2.2.17 Weighted-factor evaluation.

Hidden Factors

There always seem to be some hidden factor(s) influencing the choice of alternatives. Things move so fast in modern industry that the layout planner cannot keep informed of all of management's prospective or potential shifts; and often it is inadvisable for planners to be made privy to many business opportunities. These conditions are especially true in high-technology companies.

So the planner should anticipate new factors coming into the evaluation-and-approval step of the layout-planning project. Hopefully, these latent considerations can be "smoked out" by involving the approvers of the plan when selecting the factors and weights to be used for the planner's evaluation.

2.2.13 Detail Layout Planning

Detail layouts involve the location or placement of each machine or piece of equipment. Detail layouts show more detail; they show it at a larger scale than overall plans; they work with spaces that have finite shapes or configuration. These differences cause detail plans to require somewhat different methods of diagraming and visualizing. But essentially, their planning follows the same fundamentals and logic as for block layouts (discussed previously).

Relationships based on flow and/or other-than-flow factors

Space in amount, kind, and finite shape

Adjustment to meet all modifications and limitations

A sequence of steps to develop layout alternatives and to evaluate them. (Now the handling analysis that supports the detail layouts encompasses movement to and from each workplace.)

Detail relationships can be based on flow of materials, other-than-flow considerations only, or a combination of the two. Space, in detail layout planning, represents each machine or piece of equipment, so a finite templet or model can be used. Thus, a rough arrangement of templets in detail layouts is comparable to a space relationship diagram in overall planning.

Detail diagraming and visualizing are now more realistic. And, the "shape" being specific means that prerecording and filing of it for ready use is often done before a planning project begins. As a result, templets, models, and computer-filed outlines of equipment can be used to advantage.

Templets are typically made of plastic. They are reproduced photographically or by copy machine. They are then fastened to a transparent plastic grid sheet along with adhesive-backed tapes for aisles, building walls, utility lines, and so on. The composite layout sheet is then run through a print or copying machine to record the particular arrangement.

Models are used when the layout involves complex three-dimensional considerations, like overhead conveyors, underfloor utilities, or interlaced piping lines; new or changed processes, products, or manufacturing systems; very substantial capital investments; checking and reviewing of the layout by others not frequently exposed to visualizing spacial relationships; new employees or supervisors; and suppliers unfamiliar with the type of facilities being planned.

Computer-Aided Design (*CAD*) is beginning to replace much of the foregoing. Its costs are reducing rapidly. Once the make-ready data are in place, this equipment can reduce tremendously the time for recording, visualizing, and redrawing. Thus, it is most helpful in detail layout planning and in preparation of installation drawings. This is particularly so for avoiding 3-D interferences and when walls, columns, toilet facilities, multiple rows of storage racks, and so on, are involved.

With all of these methods of analysis and visualization, work areas or banks of machines can be predesigned and reproduced as one discreet unit of space. Thus, standardized toilet facilities, dock space, multiple rows of racks, office work stations, assembly areas, and the like, can all be put into place as one unit. See Fig. 2.2.18.

There are many applications of computer-based data in layout, handling, and facilities planning generally. These include the following, roughly in the order of currently most-frequent use.

Gathering and processing input data for planning
Scheduling and/or managing planning projects
Analyzing processes and/or machinery requirements
Documenting space requirements
Analyzing the flow of materials between areas
Documenting existing/historical space allocations
Developing alternative layout plans
Scheduling/managing equipment installations
Listing plant or personnel services
Making building/equipment drawings
Evaluating alternative layout plans

In the early phases of most planning projects, computer aids are primarily computational, relieving the planner of tedious calculations and performing complex analysis that would not be possible with manual methods. Detail layouts are used chiefly in developing, visualizing, and preparing drawings of the equipment planned, including the building and handling features. During construction and installation, they aid project management, drastically reducing the clerical work involved in scheduling, reporting, and control.

Many detail modifications and limitations come into the adjustment steps. Adequate set-down space for in-process materials, balancing of operating times for optimum machine utilization, placement of equipment so one worker can conveniently do several tasks, are typical of the kinds of detail analysis that the planner may have to make. Many of these will be addressed in the following section.

2.3 ANALYTICAL TECHNIQUES OF ANALYSIS

Dennis B. Webster

A new era appears to be dawning in the analysis phase of defining and resolving plant layout and materials handling problems. Quantitative methods and modeling approaches that were readily dismissed

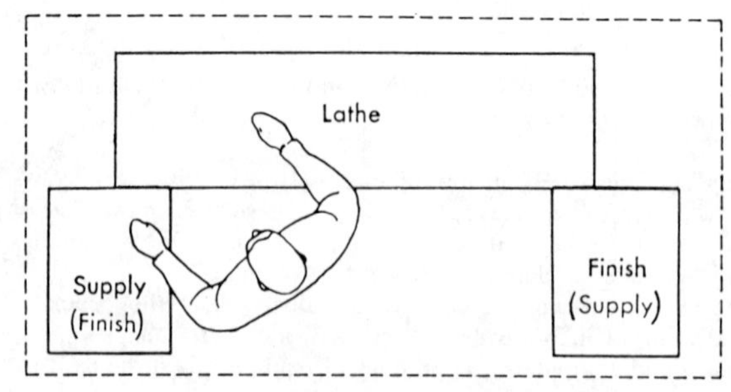

Fig. 2.2.18 Workplace templet.

a few years ago as being unsuitable within this complex problem solving arena are now finding stalwart advocates. This change is taking place for many reasons; the principal ones include the following:

1. Advances in the quantitative techniques themselves which allow more realistic modeling approaches
2. The widespread use and availability of the digital computer by analysts in systems planning
3. Increased sophistication and complexity of production and materials handling systems
4. Increased sophistication and capability of the individuals who are planning and designing these systems

A reader only has to look at the earlier edition of the *Materials Handling Handbook*, published in 1958, and see that there is not even a section on analytical techniques of analysis, to realize the significance and depth of the changes taking place.

The purpose of this section is to describe and discuss the major types of quantitative and computer-based techniques that have been found to be of benefit in modeling and resolving a wide variety of plant layout and materials handling problems. The techniques are discussed under the four major headings of (1) simulation analysis, (2) waiting-line analysis, (3) routing analysis, and (4) location and layout analysis. The descriptions and discussions are not exhaustive and references are provided at the end of the chapter for those wishing to further pursue any of the topics.

2.3.1 Simulation

Of all analytical techniques that currently exist to assist in the design and analysis of facilities and materials handling related problems, simulation is by far the most versatile and widely used. There are many types of simulation, but as most commonly encountered, and the only type discussed here, simulation is a process by which a mathematical and/or logical model of a system is developed and then the model is operated or exercised in order to draw inferences about the system. This type of simulation often goes under the name of monte carlo, discrete event, event oriented, or computer simulation.

The Simulation Process

Regardless of the name it goes by, one common element of this type of simulation is that it involves the drawing of samples from distributions as part of operating the model, generally in a random fashion. For example, suppose the Industrial Engineering Department has collected data on the time it takes a fork truck to perform each move request it receives from the Milling Machine Department and it appears as in Fig. 2.3.1. Further suppose that as part of a simulation model of the Milling Machine Department it is necessary to draw random samples of times from this distribution. A common procedure to do this would consist of the following steps.

1. Translate the data distribution (here a probability distribution) into a cumulative distribution. See Fig. 2.3.2. Since the data was collected in 5-min increments, an average value for each interval is assigned, that is, 7.5 min for the interval between 5 and 10 min, 12.5 min for the interval between 10 and 15 min, and so on. Thus if a sample is found to lie in the 25–30-min interval, a value of 27.5 min would be assumed for that request. The cumulative distribution accumulates the probability or relative frequencies of a random sample being less than or equal to a specific value in the collected data.

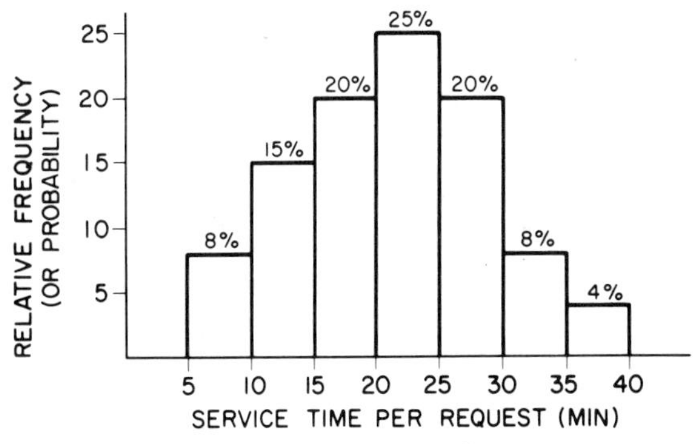

Fig. 2.3.1 A histogram of service times per request.

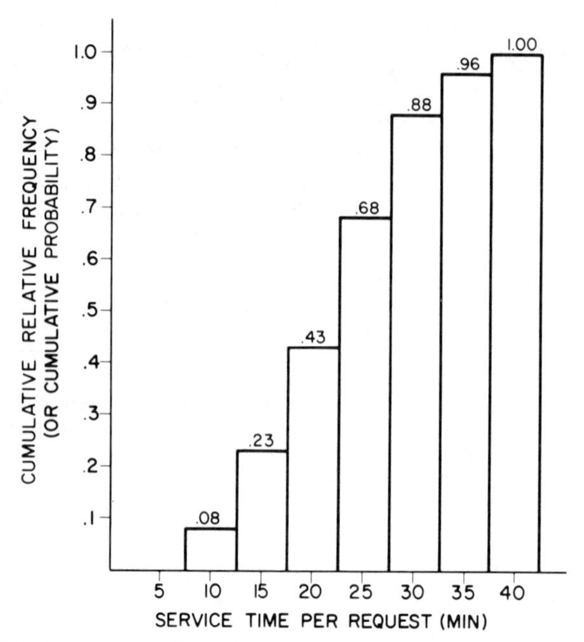

Fig. 2.3.2 A cumulative distribution of services times per request.

2. Draw a random number between 0 and 1. This can be performed by using a table of random numbers or from a computerized random number generator. Mathematically this is the same as randomly choosing a number from a uniform distribution on the (0,1) interval.

3. Set the cumulative relative frequency or probability equal to the random number, and find which service time includes this value in its range of cumulative relative frequencies or probabilities. This service time will be a random sample from the data distribution of service times.

4. For additional random samples from this distribution, steps 2 and 3 are repeated for each additional sample desired.

As an example of this procedure, drawing a random number of 0.500 would result in a sample time of 22.5 min being drawn from the distribution given in Fig. 2.3.2.

The process of simulation modeling, which has the preceding random sample generation procedure embedded within it, involves a number of interrelated steps similar to other modeling efforts.

1. Define the problem by stating the objectives or questions that answers are being sought for and the criteria by which one can measure the effectiveness of obtaining these goals. This step involves specifying the system boundaries, restrictions, and assumptions that can be used to guide the model's development.

2. Define the model to obtain answers to the objectives. This will involve an abstraction process whereby

   a. The model's static form is patterned after the system that forms its basis.
   b. The components that made up the model are then specified as being deterministic or probabilistic.
   c. Sample generation procedures are designed for each probabilistic component.
   d. Mathematical or logical linkages which are patterned after the system are designed to tie the model's components together as a representation of the system.
   e. Statistical collection and reporting procedures are built into the model for each system variable of interest.

3. Translate the abstracted model into a computer model. This step involves the selection of an appropriate computer language and developing the resultant computer model.

4. Collect any necessary data for the required input distributions or other system data. For those system components defined as probabilistic, data is required for each sampling distribution.

5. Verify that the model will operate as intended by a preliminary set of test runs of the computer model. Computer models often tend to be complex. This step should be used to investigate the input and output of model components to ensure each is performing as designed.

6. Validate the model by performing a preliminary set of test runs of the computer model. This step is to ensure that the abstraction process in the model definition step does indeed duplicate the actual system to the desired degree. Generally this involves comparing the model's results to actual system's results if an actual process is being modeled, or it involves "tests of reasonableness" in looking at the model's results in a step-wise manner through the model to ensure that no unexpected occurrences appear to be happening for a nonexistent system.

7. Establish the experimental design necessary to provide the answers desired by the objective definition. This may be as simple as comparing the performance of two model configurations or as complicated as formalizing a statistical design of experiments, depending on the objectives of the model.

8. Determine initial, startup conditions for the model. These should duplicate as much as possible those that the actual system experiences. In general most simulation models are interested in steady-state conditions, so typical conditions should be used to start the model. Where typical conditions may not be initially obtainable, it may be necessary to operate the model for a period of time to allow the model's usual state to be determined and then discard the first part of the run so the results will not be biased. The reader is referred to Fishman [1] for more detail on this problem area.

9. Determine the number of samples or the model run length necessary for appropriate inferences to be drawn from the model. Simulation models are used generally in comparing systems based on statistical results that the model generates. Therefore the number of samples required to answer an objective consideration is most often a statistical question. The reader is referred to Shannon [2] for current methods used for developing run size estimates.

10. Operate the model for the required number of runs and type of configuration and analyze the results. This step is the execution step where the desired data is collected, the sensitivity analyses are performed, and the discovered conclusions are drawn.

11. Make appropriate decisions and plan for implementation of results.

12. Document the model. This step is very important although one of the least likely to be done in most instances. Many simulation models start out being designed for one purpose and end up being used for purposes other than that for which they were designed. Documentation is necessary if there is any chance that it may be used in the future or by someone else. As there is often the cursed philosophy of "publish or perish" in universities, there should be a "document or die" philosophy in private industries' use of simulation models.

Example of Simulation Modeling

A typical, small-scale problem for which simulation modeling would be a suitable solution procedure is the following.

A manufacturing plant that is organized as a general flow shop produces a family of similar products. As a consequence, palletized in-process goods are moved from various locations in the Drill and Tap Department for their next processing operation. Data has been gathered on the time between requests for moves and this is given in Fig. 2.3.3. Currently two walkie battery-powered lift trucks are being

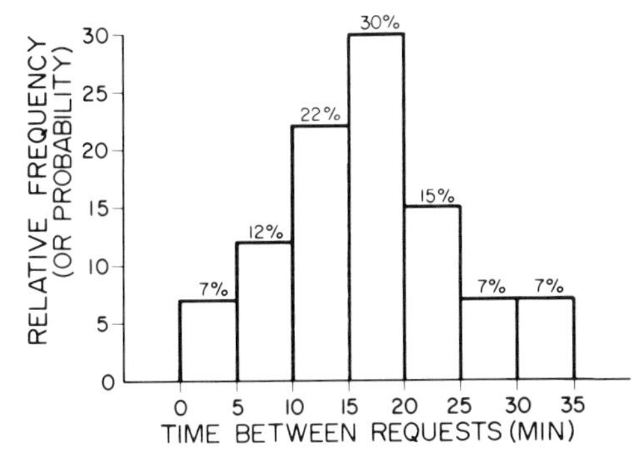

Fig. 2.3.3 A histogram of the time between move requests.

used to perform these moves. Data has been collected on the time an operation takes to service each request and are shown in Fig. 2.3.1. Earlier studies have shown the two operators and their equipment to be sufficiently similar so that their times were lumped together. Due to the costs of performing these moves, the relatively long distances required in transport, and their being no apparent safety problems, the manufacturing manager has instructed industrial engineering to investigate the possibility of using one battery-powered, rider fork truck to perform these moves.

Using this brief description, the following is a commentary of the steps used to examine this system through the process of simulation modeling.

1. Questions the model can provide answers to are those which relate to the times associated with the performance of each of the methods of performing this particular move operation. The boundaries of the process being studied begin with the move requests and end when the requests have been satisfied by performance of the move. Among the questions of interest might be:

   a. What is the utilization of each of the walkie lift trucks versus the utilization of the rider lift truck?
   b. How long does a move request wait with the current handling procedures versus using the rider lift truck?
   c. What are the associated material handling costs which relate to each of these procedures?

2. a. The system under study can be characterized as in Fig. 2.3.4. The model consists of generating move requests which are to be performed by one of the two alternatives under consideration, that is, by one of the walkie lift trucks or the rider lift truck.
   b. Deterministic quantities in this model are: (i) Only two walkie trucks are being considered and (ii) only one rider truck is being considered. Probabilistic processes in this model are: (i) The time between occurrences of move request, (ii) the time it takes a walkie truck to service a move request, and (iii) the time it takes the rider truck to service a move request.
   c. Sample generation procedures, such as those discussed previously, would now be designed to generate random samples for each of the probabilistic processes.
   d. Linkages that are used to tie the various components of this system together would include providing a procedure whereby the move requests are "stacked" in a request file if all equipment is currently busy with other move requests. After completing one move request, the operator of the equipment would determine if any other move requests are waiting to be processed. If not, he goes into an idle state. If there is, he proceeds to take care of the request that has been waiting the longest, that is, first-come, first-served basis.

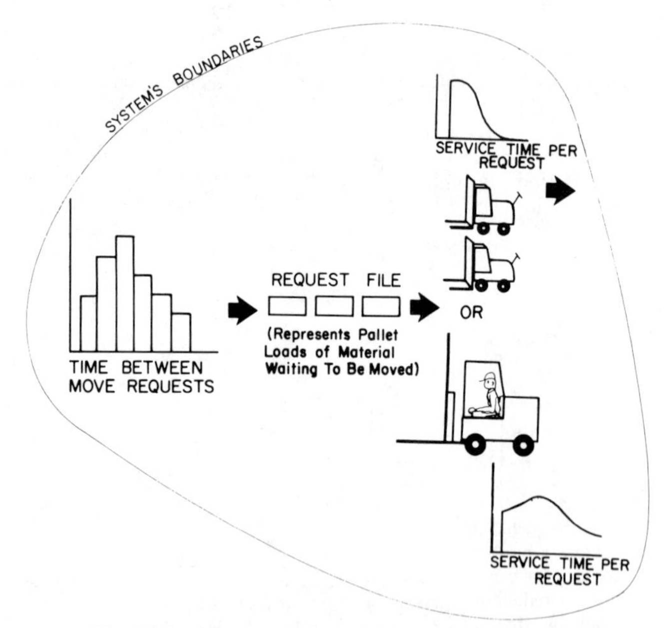

Fig. 2.3.4 A representation of the system under study.

e. Based on the question of interest, statistical collecting procedures would be developed to collect (i) the time that each piece of equipment is being used to perform a move request, and (ii) a distribution of times that each move request waits before being processed.

3. The model's elements and logic as defined are now ready to be translated into a computer model representation. In this instance only one model need be developed. The only difference between the alternatives is the rapidity of performing a move request, which can be modeled by altering the service sampling distribution. The choice of computer languages available to the user is discussed in the following section.

4. Data required to simulate this system which has already been collected include (i) a distribution of the time between move requests (Fig. 2.3.3), and (ii) a distribution of the times to service a moverequest by a walkie-type lift truck (Fig. 2.3.1). Additional data necessary to provide for simulation of this problem and to answer the questions of interest include (i) a distribution of times required for a rider fork truck to service the move requests, and (ii) equipment and labor costs so the alternative's material handling costs may be calculated. The distribution of move times for the rider truck may be estimated using standard times for a set of representative moves between the two departments. Equipment and labor costs likewise can be estimated once the equipment is specified by combining these costs with the labor rates for the specified time requirements.

5, 6. The model should be relatively easy to verify and validate, since model runs using the walkie-type lift trucks can be compared to actual system performance. Any seeming discrepancies should be explained or corrected at this stage.

7. The experimental design in this problem situation would consist simply of comparing the results of utilization for each piece of equipment, the waiting times for move requests, and the associated costs of each alternative to determine which system has the most desirable traits. Because the alternatives are to be compared to each other, it would be most desirable to subject each alternative to the same set of move requests, that is, use the same sequences of random numbers to generate the time between move requests and the times required to perform the moves for each alternative.

8. Initial, startup conditions for this problem can be easily obtained from observing the actual system. Typical conditions should be chosen.

9–12. These steps require that the model be run, samples collected, and sensitivity analyses be performed. Other than calculation of the number of samples that should be collected (for which the reader has already been referred to additional reading, if desired), the remaining steps are straightforward and are felt to require no further explanation.

Choice of Computer Language in Simulation Modeling

As discovered in the preceding discussion, a computer model is generally developed as a result of progressing through the simulation modeling process. But what are the choices and how do you select which computer language to use? Many authors have discussed these questions from a number of perspectives; no true consensus has ever developed. At the risk of oversimplifying the attributes most often found to be determinative, the following comments are offered.

The types of languages commonly available include general purpose languages like BASIC, FORTRAN, PL/1, and PASCAL, and special purpose simulation languages like GPSS, SIMSCRIPT, GASP, SLAM, and GEMS. Reasons for choosing one over the other are presented in Table 2.3.1.

Table 2.3.1 Reasons for Selecting General or Special Purpose Computer Language for Simulation Modeling

General Purpose Languages

1. The user already knows the language.
2. The language is currently available at computer installation.
3. Excellent documentation is available for the language.
4. Little restrictions on form of input, model design, or output is imposed by the language.
5. The user anticipates infrequent, and/or relatively small, simulation modeling requirements.

Special Purpose Languages

1. The user anticipates frequent, and/or relatively large, simulation modeling requirements.
2. The language provides a built-in pattern for framework of modeling design efforts.
3. The language provides for ease of statistical collection efforts and display of model's results.
4. The language allows a reduction in time required to develop model, once the language is known.
5. Aids for error checking often are built into the language.

The primary purposes in selecting a special purpose language are, generally, (1) to reduce the time to develop the computer model, once the language has been learned, and (2) to provide automatic mechanisms by means of the language to collect and report on statistics of interest. The language requires the user to frame or structure the problem within the modeling framework provided by the language, however, and there is a tradeoff in ease of learning and the restrictiveness of the language structure. As a general rule, the more flexibility desired and provided in the language, the more difficult the language is to learn.

Experience has shown that answers to the following questions most often determine the selection between the categories of special versus general purpose languages as well as selection of the specific language within the category.

1. Who is going to build and translate the model? What current computer skills does that person possess? What languages does he or she already know?
2. What languages are currently available at the computer installation where the model will be developed?
3. Are manuals and technical assistance for the language readily available? Is the documentation readable?
4. What are the costs associated with obtaining, learning, and maintaining additional languages?
5. How much time and resources are to be devoted to simulation modeling? Is this a one-time effort or part of a long, continuing development project?
6. What computer resources are available to support the project?
7. How large is the model going to be? How many uses is the model being developed to serve?

Pitfalls or Areas to Watch

Although simulation modeling is one of the easiest of the analytical techniques to use, it is potentially one of the most hazardous when drawing conclusions based on the results generated from a simulation modeling effort. Of all the pitfalls that may trap the unwary, the following represent the areas that should be of most concern.

1. *Generation of Random Numbers.* All computer simulation models are based on using random number streams for generating samples from distributions. Most simulation languages have their own built-in generators. It is essential that the random number streams used in the modeling process be sufficiently tested to ensure their performances are statistically adequate. As a practical matter, most modelers often develop a set of random number streams that they have tested to their satisfaction as to lack of correlation among samples and with proper distributional characteristics and they then use only these in any subsequent modeling effort.

2. *Relative Versus Absolute Results.* Simulation models are more efficient at developing relative rather than absolute results. In the modeling example given previously, the performance of the two alternatives were compared. This is the best use for simulation analysis. If on the other hand the model was used just to develop an absolute figure, for example, the number of move requests which may be waiting at any time, larger sample sizes (i.e., run lengths) for the model would be required. Also, the model necessarily includes less variability than experienced by the actual system. This will result in output distributions demonstrating less variability than experienced by the actual system, which may greatly affect the significance of any conclusions drawn.

3. *Using the Model Correctly.* The model can only answer questions for which it was designed. This should be obvious to most users, but is often overlooked in simulation modeling. Simulation models are often designed for one purpose in mind and then used in an attempt to answer a variety of other questions. One of the major advantages to developing simulation models is their ease of use. But the user should be cautioned that when the objectives of the modeling effort change, the validity of the model to provide answers to the new situation may be cast in doubt. To prevent this may require changing the model structure or sampling procedure.

4. *Hidden Assumptions.* As in any modeling effort, the simulation process that results in the development of a computer model requires the developer to make certain assumptions. As opposed to more mathematical procedures, however, simulation model assumptions are often hidden. For example, in the model developed previously, the assumptions that (i) operators work at a constant pace, (ii) the equipment never breaks down, and (iii) perfect information is received in terms of which move requests have waited the longest for service, are generally made. Although these assumptions could be relaxed by making the model more complex, any model has its built-in assumptions. In simulation these assumptions are more likely to be hidden, thus it is important for the user to look for them to ensure they are valid for the system under study.

2.3.2 Waiting Line Analysis

Waiting line analysis is a set of probabilistic models based on queueing theory which describe how a system will operate assuming certain arrival and service conditions. Although there may be many dissimilarities that exist among these models all are concerned with a population of items each requiring service of some sort. Typically the analysis is intended to answer such questions as: (1) What is the utilization of the service facility? (2) What is the average number of items waiting to be serviced? (3) What percentage of time is the service facility empty? (4) What is the average amount of time an item spends waiting? (5) What is the average amount of time an item spends waiting and being serviced? The primary purpose of waiting line analysis is to provide performance measures of systems where due to the lack of synchronization in the demands for service from when the service facility may be available, a waiting line or queue may be formed.

Definition of Terms

Frequently encountered terms which we used to describe waiting line problems include the following:

Service Facility(ies). There may only be one service facility or there may be multiple service facilities.

Service. The action required by an arriving item.

Item Requiring Service. The person or thing that requires the service facility to act upon it; an arrival to the system that requires some action upon it.

Arrival Process. The probability distribution that describes the manner in which the items arrive and are serviced or added to the waiting line.

Service Process. The probability distribution that describes the length and variability of time required to provide service for each arriving item.

Service Discipline. The order in which waiting items are processed, that is, first-come, first-served; last-in, first-out; random selection, or some priority discipline.

Queue Length. The number of items which are in the waiting line, that is, waiting for service.

To describe a system so that it may be amenable for waiting line analysis several factors have to be defined. These include:

1. The arrival distribution
2. The service distribution
3. The number of service facilities
4. The service discipline

Arrival Distribution

The arrival distribution is for describing the probabilistic behavior of demand that is being placed on the service facilities by the arriving items. The term arrival is used in a figurative sense, and is really concerned with the occurrence of demands that require the use of a service facility. For example, if a lift truck is used in a manufacturing department to move pallets of in-process materials from one work station to another as required, the move requests themselves are the arrival items. If a move request occurs while the truck is being used to perform another move, the request in effect joins a queue or waiting line awaiting the services of the lift truck.

In order to use waiting line analysis, the most common arrival pattern assumed is the Poisson distribution. This distribution is appropriate in many material handling situations where an occurrence happens over a period of time, like the arrival of a move request. When it is as likely that the occurrence will happen in one interval as in any other and one occurrence has no effect on whether or not another happens, the number of occurrences in a fixed period of time is often assumed to have a Poisson distribution. It is important to note that a Poisson distribution is an arrival rate distribution.

Because of this, the time between arrivals will follow an exponential distribution. For example, if λ is the mean arrival rate of occurrences, then $1/\lambda$ is the mean length of the time between consecutive arrivals.

Besides the pattern itself, a second attribute of importance is the number of items making up the arrival distribution. In much queueing-oriented development, the assumption is often made that the distribution of arriving items is made up of an infinite number of items. This is often acceptable if the items arriving to the service facility are part of a continuous process or if the proportion of items requiring service at any one time is very small when compared to the total number that might require service.

But if the population of items that might require service is small or large when compared to the total number that may require service, a finite population should be assumed. This would be the case where a lift truck is used to service more requests in a department that has relatively few machines. Or it might also be applicable where the number of move requests at any time is a significant proportion of the total number of machines.

Service Distribution

The service distribution is for describing the length and variability in time that each arriving item requires for service. The most commonly assumed service distributions in facilities design/materials handling are constant and exponential distributions. Constant distributions are generally encountered in machine-controlled situations where the output from a machine is being considered as the service mechanism, or is closely controlled automatic equipment. Exponential distributions, however, are often encountered in other service situations where there is built-in variability due to the type of service provided, the use of human operators, or other variable circumstances. Of all the service distributions that might be encountered, the exponential is the most widely used because it allows relatively simple formulas to be developed for most of the variables of interest in simple waiting lines when used with a Poisson arrival pattern.

Number of Service Facilities and Service Discipline

The number of service facilities, of course, will affect the values of performance variables used in describing and analyzing a waiting line system. It may be that the number is fixed in a particular system and the analysis used to describe the behavior of the operation, or the analysis may be used to determine the number of service facilities that should be provided for a certain level of service.

Likewise in describing a system for study, the order in which arriving items are processed is of importance. The most widely used discipline in most material handling situations is first-come, first-served, either in the single or multiple service facility system. This is also the service discipline that is generally the easiest to analyze in waiting line analysis. A closely allied service discipline is random selection; that is, if there is more than one item waiting for service, each item has an equally likely chance of being selected as the next item to be serviced. The statistical behavior of random selection is identical to that of first-come, first-served, so the formulas developed for performance measures of one system can be used in the other.

Some situations require the use of other service disciplines, such as priority selection based on scheduling priorities. Or it may be necessary to expedite jobs through the handling system. In these instances the basic formulas given in Table 2.3.3 may no longer be applicable, and it may be very difficult to mathematically define the situation.

Example Using Waiting Line Analysis

Examples of situations in plant layout and materials handling where waiting line analysis may be of use are frequently encountered wherever there is the opportunity for congestion or interference. Table 2.3.2 lists some such examples. Table 2.3.3 provides mathematical relationships which may be used for the analysis when certain conditions hold. Using these as the basis for an example, the following problem demonstrates the type of information waiting line analysis can provide.

The problem: One lift truck is currently being used in a manufacturing department to move pallets of in-process materials from one work station to another. It appears to the department head that the time between when a move request is initiated and when it was actually started is perhaps too long and he is interested in exploring the possibility of acquiring the use of a second truck. What insight might waiting line analysis provide?

More requests (arrivals) occur at an average rate of 5 per hour, and can be assumed as coming from a Poisson distribution. Service times have been collected from the existing lift truck and can be assumed as being exponentially distributed with the truck being able to perform, on the average, one move every 10 min.

Selecting Models 1 and 2 as being representative of this situation with one and two lift trucks, respectively, the results are presented in Table 2.3.4. With respect to the results of the model representing the present, one-lift-truck alternative, these can be checked against measures of the current system to see if the model is valid.

Assuming the model, and therefore queueing, is a valid approach, the proposed two-lift-truck alternative results can then be evaluated as to whether it is desirable. The question is then purely economic.

Table 2.3.2 Commonly Encountered Plant Layout/Material Handling Situations That May Be Amenable to Waiting Line Analysis

Type of Problem	Arriving Items	Service Facilities	Type of Service
1. Number of milling machines to provide in a department in a process-oriented layout	Orders that require milling operation	Milling machines	Manufacturing operation on each order
2. Number of lift trucks to service move requests between two or more manufacturing departments	Move requests requiring a lift truck	Lift trucks	Performing the necessary move of in-process materials between the departments
3. Length of accumulating conveyor storing discrete units of product	Product to be temporarily stored on the conveyor	Next processing operation, which removes product from the conveyor	Manufacturing operation on each unit of product
4. Number of dock facilities to provide in receiving/inspection	Over-the-road trucks	Rider and walkie lift trucks	Loading and unloading of over-the-road trucks
5. Amount of in-process storage at input end of transfer machine line	Product to be processed on transfer line	Transfer machine line	Production operations on items

Table 2.3.3 Mathematical Relationships for Calculating Performance Variables for Four Common Waiting Line Models[a]

Performance Variable	1 Server Model 1	2 Servers Model 2	1 Server Model 3	2 Servers Model 4
Utilization of server(s), ρ	$\rho = \frac{\lambda}{\mu}$	$\rho = \frac{\lambda}{\mu s}$	$\rho = \frac{\bar{\lambda}}{\mu}$	$\rho = \frac{\bar{\lambda}}{\mu s}$
Average number of items in the queue, L_q	$L_q = \frac{\lambda^2}{\mu(\mu - \lambda)}$	$L_q = \frac{P_0 (\lambda/\mu)^s \rho}{s!\,(1-\rho)^2}$	$L_q = \sum_{n=1}^{M} (n-1) P_n$	$L_q = \sum_{n=s}^{M} (n-s) P_n$
Average number of items in the system, L	$L = \frac{\lambda}{\mu - \lambda}$	$L = L_q + \frac{\lambda}{\mu}$	$= M - \frac{\lambda + \mu}{\lambda}(1 - P_0)$	
Percentage of time system is empty, P_0	$P_0 = \left(1 - \frac{\lambda}{\mu}\right)$	$P_0 = 1 \Big/ \left[\sum_{n=0}^{s-1} (\lambda/\mu)^n + \frac{(\lambda/\mu)^s}{s!} \frac{1}{1 - (\lambda/s\mu)} \right]$	$L = M - \frac{\mu}{\lambda}(1 - P_0)$ $P_0 = 1 \Big/ \sum_{n=0}^{M} \left[\frac{M!}{(M-n)!} \left(\frac{\lambda}{n}\right)^n \right]$	$L = \sum_{n=0}^{s-1} n p_n + L_q + s\left(1 - \sum_{n=0}^{s-1} P_n\right)$ $P_0 = 1 \Big/ \left[\sum_{n=0}^{s-1} \frac{M!}{(M-n)!\,n!} \left(\frac{\lambda}{m}\right)^n + \sum_{n=s}^{m} \frac{M!}{(M-n)!\,s!\,s^{n-s}} \left(\frac{\lambda}{\mu}\right)^n \right]$
Average amount of time that item spends in system, W	$W = \frac{1}{\mu - \lambda}$	$W = W_q + \frac{1}{\mu}$	$W = \frac{L}{\bar{\lambda}}$	$W = \frac{L}{\bar{\lambda}}$
Average amount of time that item spends in the queue, W_q	$W_q = \frac{\mu}{\mu(\mu - \lambda)}$	$W_q = \frac{L_q}{\lambda}$	$W_q = \frac{L_q}{\bar{\lambda}}$	$W_q = \frac{L_q}{\bar{\lambda}}$
Probability of exactly n items in the system, P_n	$P_n = \left(\frac{\lambda}{\mu}\right)^n \left(1 - \frac{\lambda}{\mu}\right)$	$P_n =$ $\frac{(\lambda/\mu)^n}{n!} P_0$ if $0 \le n \le s$ $\frac{(\lambda/\mu)^n}{s!\,s^{n-s}} P_0$ if $n \ge s$	$P_n = \frac{M!}{(M-n)!} \left(\frac{\lambda}{n}\right)^n P_0$	$P_n =$ $P_0 \frac{M!}{(M-n)!\,n!} \left(\frac{\lambda}{\mu}\right)^n$ if $0 \le n \le s$ $P_0 \frac{M!}{(M-n)!\,s!\,s^{n-s}} \left(\frac{\lambda}{\mu}\right)^n$ if $s \le n \le M$ 0 if $n > M$

[a] Models 1 and 2, Infinite Arrival Population; Models 3 and 4, Finite Arrival Population (M = number of items in arrival population). Poisson-distributed arrivals (λ = arrival rate), where $\bar{\lambda} = (M - L)$; exponentially distributed service times (μ = service rate).

Table 2.3.4 Results of the Two Alternatives

Parameter	Alternative 1 One Lift Truck (Present Method)	Alternative 2 Two Lift Trucks (Proposed Method)
Utilization of trucks	$\frac{5}{6} = 0.83$ or 83%	0.42 or 42%
Average number of move requests waiting for service	4.17	0.175
Average number of move requests at any one time in the system	5	1.01
Percentage of time the truck(s) is(are) idle	$\frac{1}{16} = 0.16$ or 17%	0.412 or 41.2%
Percentage of time that at least one of the two trucks is idle $(P_0 + P_1)$	N.A.	0.754 or 75.4%
Average amount of time a move request spends waiting and being serviced	1 hr	0.202 hr or 12.10 min
Average amount of time a move request spends waiting	0.83 hr or 49.8 min	0.035 hr or 2.1 min

Is it more costly to provide a second lift truck where the utilization of both trucks is low (i.e., 42%), or to accept the amount of time that a move request has to wait before being serviced by a single truck (i.e., 49.8 min)?

Additional Remarks

Waiting line analysis can provide a very useful modeling tool for use in many congestion situations. Its primary uses have occurred where the analyst has been interested in obtaining good answers to queueing situations without having to resort to the use of a computer. Even though all assumptions may not be satisfied, waiting line analysis often is used to provide a means of checking the sensitivity of any answers obtained to determine if further refinements are necessary or cost can be justified.

It is, of course, important that the assumptions made in applying a particular model apply reasonably well to the system under study. There have been many waiting line models developed that relax one or more of the assumptions concerning arrival distributions, service distributions, and service discipline, and references have been provided to assist the reader in further pursuit of this topic. The mathematics can often become complex. If the system under study is overly complicated, the user might be well advised to investigate the use of simulation as an alternate modeling approach.

2.3.3 Routing Analysis

Routing analysis is concerned with selecting a path between two or more locations within a move system where the objective is to minimize the time or costs associated with traversing the path. In general, routing problems assume deterministic rather than probabilistic values and the analysis seeks to provide the best or optimum path based on these values. This type of analysis has been used for many years in physical distribution problems outside the plant where efficient delivery routes for supplying customers from a distribution center or for deciding which distribution centers should be supplied from which plant facilities were the major concerns. While routing analysis may have much use outside the plant, increasing use is also being made of it for inside the plant and material handling functions.

Almost all realistic size routing problems require the use of a computer and efficient algorithms. The wider use of computers thus has increased interest in this modeling technique. Unfortunately, algorithms available to solve many of the formulations are not as widespread, and it may be a considerable burden to find and obtain efficient algorithms and computer code for solving a problem once it is formulated. Fortunately new codes are in almost continuous development and may soon be commercially available to a large body of users.

Of all routing techniques and problem formulations available for use in plant layout and material handling applications, the most common and useful can be grouped under the headings of (1) transportation problems, (2) assignment problems, (3) shortest route problems, and (4) traveling salesman problems. Each of these is discussed next by describing a typical problem of that class, mathematically formulating the problem, followed by a description of solution approaches.

Transportation Problems

An example of a transportation problem would be that of evaluation of alternate locations in a finished goods warehouse to store finished product. Suppose we have M different locations in the plant where pallet loads of finished product are generated that are to be stored in a finished-goods warehouse. Each of the locations generate a_i number of pallet loads per week, where $i = 1, 2, 3, \ldots, M$ (for each of the locations). The finished-goods warehouse has N different bay locations for storage of these pallet loads and each location has a capacity of b_j number of pallet loads per week, where $j = 1, 2, 3, \ldots, N$. Industrial engineering, through time study and cost analysis, has estimated the cost of transporting a pallet load from each of the M locations in the plant to each of the N locations in the finished-goods warehouse and then removal from the warehouse and loading on a truck at the shipping dock. This cost estimate is designated as c_{ij}, that is, the transportation costs for one pallet load if the pallet from location i in the plant is placed in location j in the warehouse. With the objective of minimizing the sum of transportation costs, and thus determine the storage locations (and also the paths) for each of the pallet loads from each of the plant locations, the problem could be formulated as:

$$\text{Minimize: } Z = \sum_{i=1}^{M} \sum_{j=1}^{N} c_{ij} x_{ij}$$

$$\text{Subject to: } \sum_{j=1}^{N} x_{ij} = a_i \quad \text{for } i = 1, 2, 3, \ldots, M$$

$$\sum_{i=1}^{M} x_{ij} \leq b_j \quad \text{for } j = 1, 2, 3, \ldots, N$$

$$x_{ij} \geq 0 \quad \text{for all } i \text{ and } j$$

where x_{ij} = the number of pallet loads per week moved from plant location i to finished-goods storage location j.

This formulation is in the form of a linear programming model and can be solved by the simplex or revised simplex procedure found in any text discussing linear programming. Although these problem formulations are manually solvable, the usual case is that a computer is required due to the number of calculations for a realistic sized problem. Standard, commercially available packages are readily available, the best known of which is IBM's Mathematical Programming Systems Extended (MPSX) [3]. Newer techniques, based upon network analysis have also been developed which can solve this problem, such as the out-of-kilter algorithm developed by Ford and Fulkerson (see Bazaraa et al. [4]), but suitable computer routines are not readily obtainable.

Assignment Problems

Assignment problems are a special variation of transportation problems. These problems can generally be formulated when the number of units (or locations) available exactly equals the number of units (or locations) required, and one unit (or location) of supply can be assigned to only one unit (or location) of the demand.

For example, in the problem discussed previously, suppose M locations were available in the finished-goods warehouse (rather than N) and we wanted to segregate the products coming from each of the M plant locations into one of each of the M finished-goods locations in the warehouse for control purposes. Of course each of the warehouse locations would have to have the capacity to store each of the plant location's production. The cost estimate c_{ij} would now be the transportation costs for moving total production per week from location i in the plant to location j in the warehouse, plus removal from the warehouse and loading onto trucks for shipping. The objective is still to minimize the sum of transportation costs by determining the finished-goods storage locations (and also the paths) for each of the location's production, and can be formulated as:

$$\text{Minimize: } Z = \sum_{i=1}^{M} \sum_{j=1}^{M} c_{ij} x_{ij}$$

$$\text{Subject to: } \sum_{i=1}^{M} x_{ij} = 1 \quad \text{for } j = 1, 2, 3, \ldots, M$$

$$\sum_{j=1}^{M} x_{ij} = 1 \quad \text{for } i = 1, 2, 3, \ldots, M$$

$$\text{where } x_{ij} = \begin{cases} 1 \text{ if production from plant location } i \text{ is assigned to storage location } j \\ 0 \text{ otherwise} \end{cases}$$

As with transportation problems, commercially available computer routines can be easily obtained to solve these types of problems. Any routine that can solve the more general transportation problem can also provide solutions to the more specialized assignment problem. Because of the special structure that exists in assignment problems, the solution procedure is also generally simpler and faster.

Shortest Route Problems

Shortest route problems are often encountered in plant layout and material handling situations but often receive little analysis because few alternatives or external constraints force a particular solution. It is widely used in physical distribution. The common problem formulation consists of starting and ending locations. The objective is to find the route between the locations that will minimize the time or costs associated with traversing the path. The various alternate locations and paths that may be taken are often displayed as nodes and arcs in a network representation.

An example where this modeling approach might be used would be where lift trucks out of a centralized material handling department are radio dispatched and perform all move requests of palletized in-process materials among the manufacturing departments. After completing a move request that ends up at location A in the plant, Operator 1 calls in for another task assignment from the dispatcher in centralized materials handling. The dispatcher has a request for in-process materials to be moved from Location B to Location C. In assigning this job to Operator 1, the objective would be to specify the path that he should follow to minimize either the time or the distance in going from A to B to pick up the material and then on to C in delivering the material. There are, therefore, two shortest route problems, that is, the first in going from A to B; the second in going from B to C.

One way of formulating this type of problem is as a linear programming problem by interpreting it as a transportation problem in which one unit of product is shipped from Location A to B (and then B to C) where a unit "cost" of shipping from Location i to j (all locations which might be on the path) is c_{ij} (which could either be transportation times or distances between Locations i and j). Although this can be done, with the resultant benefit of being able to use the easily obtainable linear programming computer packages, there are better methods available for solving this problem.

One of the most efficient and often referenced algorithms is that which was developed by Dijkstra (see Phillips et al. [5]). The steps are quite straightforward, but can be very tedious for realistic sized problems. No commercially available computer routine that implements this algorithm is known to exist, but the steps are easily computerized and should represent no real difficulty in developing a programmed routine.

Traveling Salesman Problems

Traveling salesman problems are shortest routing problems cast in a little different light. A common description of this class of problems would be where there are a number of different locations that must be visited once and only once in a routing scheme that starts and ends at some origin location. The objective is generally to minimize the total distance (or time) required to complete the route.

Many examples are encountered in the plant layout material handling context. A common demonstrative situation is in determining routing sequences for order pickers in warehouses or distribution centers. In many order picking situations, each order picker is given a set of orders he is to fill in a given period of time. To fill these orders, it is necessary for the worker to travel to many different locations in the facility before returning the completed orders to an assembly area for shipment. From the set of orders, the individual locations can be determined, from which a route can be generated that is used to minimize either the distance or time required for each worker to complete his or her set of picking orders. Many distribution centers use variations of the traveling salesman problem to computer generate order picking lists for their workers.

The traveling salesman problem can be formulated as a linear programming model but it is a complex formulation and there appears to be no computationally feasible optimal algorithm available for all except relatively small problems. Many heuristics do exist, however. One simple heuristic often used is known as the Closest Unvisited City algorithm, in which a route is constructed by starting from the origin and always selecting the closest location not yet visited. This algorithm does not require the use of a computer routine, and by selecting different locations as the origin, alternate routes can be developed from which the best of the set can be chosen.

Other general purpose suboptimal procedures such as branch and bound and neighborhood search techniques are also used to develop feasible routes, but the effectiveness of these depend heavily on the ingenuity of the person developing the algorithm being properly matched to the particular characteristics of the problem. All of these generally require the use of a computer routine for implementation and are not readily commercially available.

2.3.4 Location and Layout Analysis

Location and layout analysis is another major area in plant layout and material handling situations where modeling techniques can provide assistance. In fact the prime objective in plant arrangement is to locate plant facilities, departments, and other physical resources so as to facilitate the manufacturing process. Just as in any problem area there are degrees of complexity in the type of layout problem which might be encountered and there are degrees of complexity of the tools available to assist in their resolution.

Location and layout problems can be classified in many ways, but based on the complexity of the problem, a suitable classification consists of the following categories:

1. Additions to an existing area
2. Rearrangement of an existing area or layout
3. Development of a new area or layout

Additions or replacement of equipment within an existing manufacturing area often require reevaluation of flow paths and other objectives being sought within the plant area. A new machine, or the area where a manufacturing department's centralized in-process storage should be situated may be the object of location. The number of facilities or areas to be located may be either single or multiple, but there are generally physical or monetary constraints that limit the location problem to this classification. Rearrangement of an existing area or layout, while less constrained in terms of the extensiveness of alternatives that may be considered, is still constrained by the physical limitations of the existing area. Models and tools that can profitably be used must be capable of taking these constraints into account in specification of objectives and criteria. Development of a new area or layout in location analysis is the category most free of physical restrictions, but is the most complex in terms of the number of alternatives and objectives which must be considered. Location and layout analysis, therefore, include more than just plant layout.

One of the difficult but critical aspects of location and layout analysis is the selection of suitable objectives and criteria. Objectives are those goals one is seeking to achieve by a particular arrangement, and the criteria are the measures of effectiveness used to evaluate the degree that an alternative achieves each objective. Usual types of objectives cited are:

1. To provide for effective materials handling
2. To avoid unnecessary capital investment
3. To provide for ease of future expansion
4. To provide for flexibility in the layout
5. To provide for effective space utilization
6. To provide for cost of supervision and control

Usual criteria for objectives often include:

1. Minimizing the cost of material handling
2. Minimizing the cost of capital investment
3. Minimizing the number of cross product path flows
4. Minimizing the distance that products travel
5. Minimizing the volume times distance measure that products travel
6. Maximizing the use of cube volume in inventory areas
7. Minimizing the time that products require to be produced in the facility

Objectives and criteria must be clearly delineated before attempts are made to develop alternatives. This is often the most difficult aspect of location analysis because of two interrelated problems. First there is often more than one decision maker who will eventually determine which alternative will be chosen. Often, depending on the scope of the problem, there are department heads of the areas affected, manufacturing managers, and plant managers, each with his own sense of what a solution to the problem should achieve. Rarely if ever will the individuals involved initially agree upon the set of objectives and criteria that should be used. The second reason is that multiple objectives and criteria that represent both quantitative and qualitative data are generally used to evaluate alternatives. When multiple objectives and criteria are used, therefore, there must also be some weighting scheme used to indicate the importance of each to the overall decision process. The task of resolving these two problems often falls upon the layout analyst, because he recognizes that unless he can get some agreement there is no hope of providing satisfactory alternatives for the defined location or layout problem.

Unfortunately most location and layout analysis models allow only one type of objective and criterion to be considered. Of the three types of modeling tools discussed in this section (i.e., assignment models, level curves, and computer aided layout techniques), only certain computer-aided layout techniques generally allow the use of multiple objectives and criteria. As with all models, however, the model should be considered as a design aid, and the results will provide a benchmark against which other solutions can be compared. No model can contain every consideration that may influence the resultant proposed solution, so every solution requires modification and adjustments.

Additions to an Existing Area

One of the most common and least complex problems is the location of a new machine, or inventory area among currently used machining or inventory areas. This may come about in replacing old equipment with new, higher production equipment, or as a result of a decision to move a machine or inventory area because of congestion it is creating at its present location.

A graphical procedure available to solve this type of problem is the use of Level Curves. Level curves are contour lines like those found on topographical maps but represent iso-costs and/or distances between the facility to be added and the existing facilities. The procedure allows an optimal location or locations to be found based upon the sum of costs (generally materials handling costs) and/or distances for the new facility, and if that location is not available for one reason or the other, the iso-cost and/or distance curves allow the best nonoptimal location to be found.

It is assumed in the use of this technique that there is some required relationship, generally movement of products, which exists between the new facility to be added and the existing ones. (If not, the facility should not be located in the area.) Also assumed are: (1) the existing equipment and/or areas is to remain fixed in the existing area, (2) the location of each piece of equipment or area is sufficiently described by defining a point on the floor, and (3) if costs are used as the measure of effectiveness, then they are directly proportional to the distance between the facilities.

Two methods of model formulation typically are encountered in the use of level curves: the straight movement case and the rectilinear movement case.

Level Curves—Straight Movement Case. The straight movement case is used when the paths over which the traffic or product travels between the new facility and the existing ones are straight lines. This may happen for instance when lengths of conveyors would be used to connect the facilities together. For m existing facilities, (a_i, b_i) is used to denote the rectangular coordinates of the ith existing facility where the origin is arbitrarily located. The problem is stated as determining the coordinates (x,y) of the new facility so that the measure M is minimized, where

$$M = \sum_{i=1}^{m} c_i[(x - a_i)^2 + (y - b_i)^2]^{1/2}$$

The c_i's are costs or weights that can be applied to differentiate between the importance of travel between the new facility and each of the existing facilities. For example, M may be seen as a measure of total transportation costs between the new and existing facilities. Then the c_i's could be expressed as (dollars per foot traveled) × (trips per year), where the quantity inside the brackets in the preceding equation is the number of feet per trip. M would then be the total cost per year for locating the new facility at location (x,y). In many applications the costs of moving each product a unit distance among the facilities is a constant. This might be true, for example, where the same type of conveyor system is being used for each move. If this is the case, the problem reduces to determining the location of the new facility that minimizes the sum of the distances from the new facility to each of the existing facilities.

For the cases where there is just one existing machine, the curves are concentric circles around the existing machine. See Fig. 2.3.5. Where the c_i's are constant and there are two existing machines,

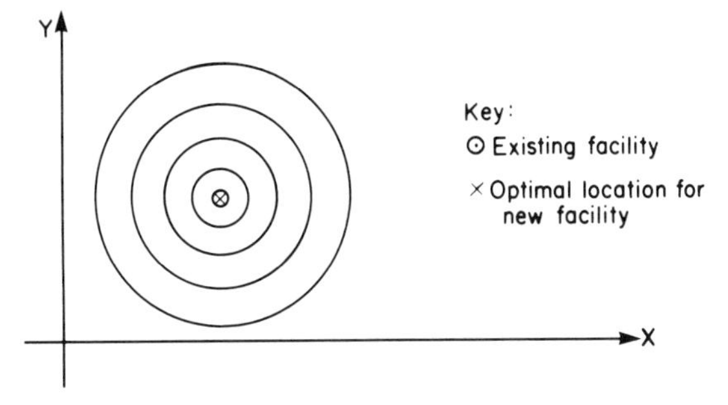

Fig. 2.3.5 Level curves for the one existing facility case.

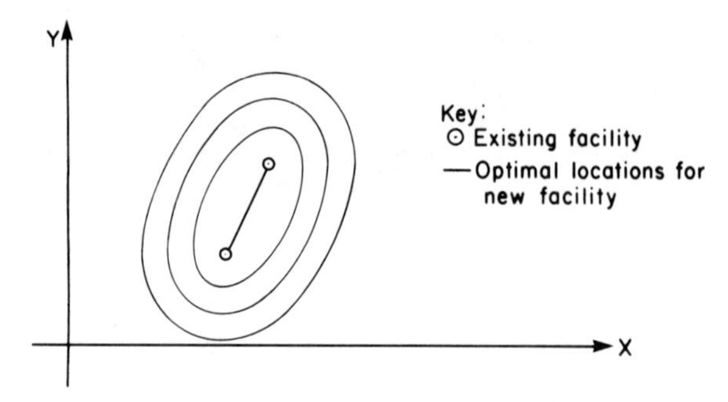

Fig. 2.3.6 Level curves for the two existing facilities case where the c_i's are constant.

the level curves are concentric elipsices around the two machines. See Fig. 2.3.6. It should be noted that for the one existing facility case, the optimal location for the new facility is on top of the existing one. Since this of course is not possible, the iso-cost curves provide a way to evaluate alternate feasible locations. For two existing facilities, the optimal locations for the new facility are anywhere on the line between the existing facilities.

As the problem becomes more realistic by considering more existing facilities, the iso-cost curves become much more difficult to sketch. Fortunately, in a particular sense, there also is less reason to do so. Plant conditions often reduce the number of potential locations for adding any new facility to a relatively few, so the solution procedure can reduce to substituting each of the available locations in turn into the foregoing equation and solving M. The location that minimizes M would be selected. If large areas are available for placement of the new facility, a second, alternate approach is to calculate iso-cost curve segments for various locations in these areas. This would provide information as to how the value of M changes within the potential areas.

To determine the optimal location for the new facility, the partial derivations of M with respect to both x and y can be taken and set equal to zero. This will result in the following two equations:

$$\frac{\partial M}{\partial x} = \sum_{i=1}^{m} \frac{c_i(x - a_i)}{[(x - a_i)^2 + (y - b_i)^2]^{1/2}}$$

$$\frac{\partial M}{\partial y} = \sum_{i=1}^{m} \frac{c_i(y - b_i)}{[(x - a_i)^2 + (y - b_i)^2]^{1/2}}$$

Although this results in two equations and two unknowns (i.e., the x,y coordinates of the new facility), it is messy to solve. It should also be noted that if the optimal location turns out being the location of an existing facility, then these two equations are undefined. The two equations may be quickly solved by computer, using an iterative routine and successively substituting different values for the x,y coordinates. This is often characterized as a two-dimensional search problem, where values of x,y are sought that result in the two partial derivatives being equal to zero. For the user who is acquainted with search procedures, such as the method of steepest ascent or pattern search, this problem is readily programmable and solvable.

A second iterative approach, attributable to Kuhn [6] and found in Francis and White [7], can also be used to find an optimal location.

Let

$$g_i(x,y) = \frac{c_i}{[(x - a_i)^2 + (y - b_i)^2]^{1/2}} \quad \text{for } i = 1, 2, \ldots, m$$

$$x = \frac{\sum_{i=1}^{m} a_i g_i(x,y)}{\sum_{i=1}^{m} g_i(x,y)}$$

$$y = \frac{\sum_{i=1}^{m} b_i g_i(x,y)}{\sum_{i=1}^{m} g_i(x,y)}$$

Then

$$x^{(k)} = \frac{\sum_{i=1}^{m} a_i g_i(x^{(k-1)},y^{(k-1)})}{\sum_{i=1}^{m} g_i(x^{(k-1)},y^{(k-1)})}$$

$$y^{(k)} = \frac{\sum_{i=1}^{m} b_i g_i(x^{(k-1)},y^{(k-1)})}{\sum_{i=1}^{m} g_i(x^{(k-1)},y^{(k-1)})}$$

where the superscripts denote the iteration number. Therefore a starting value $(x^{(0)},y^{(0)})$ is required to calculate $(x^{(1)},y^{(1)})$, and $(x^{(1)},y^{(1)})$ is used to calculate $(x^{(2)},y^{(2)})$, and so on. The iterative procedure is guaranteed to converge to the optimal location and starts at $(x^{(0)},y^{(0)})$ and continues until no appreciable change in the estimate of the optimal location for the new facility occurs.

Kuhn has shown that a location (x^*,y^*) is optimal if $R(x^*,y^*) = (0,0)$, where

$$R(x,y) = \begin{cases} \dfrac{U_k - c_k}{U_k} s_k, \dfrac{U_k - c_k}{U_k} t_k, & \text{if } U_k > c_k \\ (0,0) & \text{, if } U_k \leq c_k \end{cases}$$

and

$$s_k = \sum_{\substack{i=1 \\ \neq k}}^{m} \frac{c_i(a_k - a_i)}{[(a_i - a_k)^2 + (b_i - b_k)^2]^{1/2}}$$

$$t_k = \sum_{\substack{i=1 \\ \neq k}}^{m} \frac{c_i(b_k - b_i)}{[(a_i - a_k)^2 + (b_i - b_k)^2]^{1/2}}$$

$$U_k = (S_k^2 + t_k^2)^{1/2}$$

Typically, the iterative approach is used until little change is noted in both the x and y coordinates. Then the resultant coordinates are checked using Kuhn's [6] optimality conditions to ensure that an optimal location has been found.

Both of the iterative procedures discussed can be manually manipulated, but due to the extensiveness of the required calculations, a computer should be used.

Level Curves—Rectilinear Movement Case. The rectilinear movement case is used when the distances over which the traffic or product travels between the new facility and existing ones must be measured along existing aisles, assumed to be set up in rectangular grid patterns. This often is true where traffic aisles are marked off and the material handling among the facilities is by use of lift trucks. For m existing facilities, (a_i,b_i) is used to denote the rectangular coordinates of the ith existing facility, where the aisles are aligned with the x,y axes. Using the c_i's as defined in the straight movement case, the problem is again formulated as finding the coordinates (x,y) of the new facility such that M is minimized, where

$$M = \sum_{i=1}^{m} c_i[|x - a_i| + |y - b_i|]$$

Fortunately this formulation not only is often more representative of most single facility location problems that are encountered, but it is also much easier to treat analytically, since the foregoing expression is equivalent to

$$M = \sum_{i=1}^{m} c_i|x - a_i| + \sum_{i=1}^{m} c_i|y - b_i|$$

This allows the optimal x coordinate to be found independently of the optimal y coordinate for the new facility, and vice versa.

It can be shown that the optimal location for the new facility has the same x and y coordinates as those of some of the existing facilities; that is, the optimal x coordinate is the same as the x coordinate for one of the existing facilities, and the optimal y coordinate is the same as the y coordinate for one of the existing facilities, but not necessarily the same one. The optimal x and y coordinates are determined by the c_i values, and are such that no more than half of the cost (or weight) of the movement is on either side of the optimal location in either the x or y directions. This optimal location is generally referred to as a median location.

Depending on the number of existing facilities and the c_i values, the optimal location may be a point, line, or area. If the optimal location is a line or area, this means that there are alternative optima, that is, alternative locations that result in the same value for M. When this occurs, the alternative locations are often referred to as areas or lines of indifference.

To sketch the iso-cost or level curves the steps are as follows:

1. Delimit all rectangular areas by tracing the lines where $x = a_i$ and $y = b_i$ for $i = 1, 2, 3, \ldots, m$. (Within each of these rectangular areas, all iso-cost curves are straight lines and have the same slope.)
2. Determine the median locations for x and y. This is accomplished in the following manner:

   a. For each of the x and y coordinates, sum the total costs (or weights) and then divide this sum by two, that is, $\sum_{i=1}^{m} c_i/2$. The median location will correspond to this result.
   b. Arrange each of the x and y coordinates in ascending order.
   c. Starting with the lowest a_i value, sum the weights associated with each a_i. When this cumulative sum is equal to $\sum_{i=1}^{m} c_i/2$, the value of a_i is the median location, that is, the optimal x coordinate for the new facility.
   d. Step c is repeated for the b_i's to find the optimal y coordinate.

3. Determine representative slopes in each of the rectangular areas by finding first a value of M that falls within each rectangular area, then secondly, two points in the area that have the same value of M. The line drawn between these two points in the area has the same slope as all other iso-cost or level curve line segments passing through the rectangle.
4. Once the set of slopes has been determined, each iso-cost or level curve can be constructed by starting at any arbitrary point and successively connecting the line segments with the slopes as determined in the previous step.

Example Using Rectilinear Movement. As an example of developing iso-cost or level curves, consider the following problem. The machining department currently has seven different machine tools which are used in manufacturing a wide variety of parts. Due to the milling machine's age, it is going to be replaced by a newer machine. The department head is interested in determining the best location for the new machine, rather than just placing it in the same location as the old one. Because the newer machine will have additional versatility as well as higher productive capabilities, estimates of the number of trips of pallet loads of in-process inventory between the new machine and the remaining six machines were made and are given in Table 2.3.5. Previous costs calculated for the machining department for the lift trucks being used averaged \$2.45 per 100 feet traveled. By locating the x and y axes along two sides of the machining department's boundaries, the locations for the six remaining machines are designated as being located at (10,10), (20,50), (40,90), (80,50), (100,80), and (100,20). Movement between the machines is restricted to rectilinear flow paths for the lift trucks, since the aisles had to be marked off and kept free of obstructions due to OSHA regulations.

By following the steps just delineated, the optimal location of the new machine is found to be at (40,50) with a minimum cost of \$2254 per year incurred for materials handling. The optimal location as well as some iso-cost curves are shown in Fig. 2.3.7. It should again be emphasized that although this technique can allow optimum placement of a new facility based on the required movement and/or costs among the machines, other factors such as utilities requirements, locations of in-process inventory, and floor loading may influence its ultimate location.

These procedures for both straight line and rectangular movement cases can be extended to locating multiple new facilities in an existing area, but the procedures are correspondingly more complex. In

Table 2.3.5 Estimated Number of Trips per Year and the Resultant Coefficients between the New Facility and Existing Facilities

Existing Machines	Trips/yr	c_i (\$/ft × trips/yr)
1	300	7.35
2	100	2.45
3	500	12.25
4	100	2.45
5	200	4.90
6	300	7.35

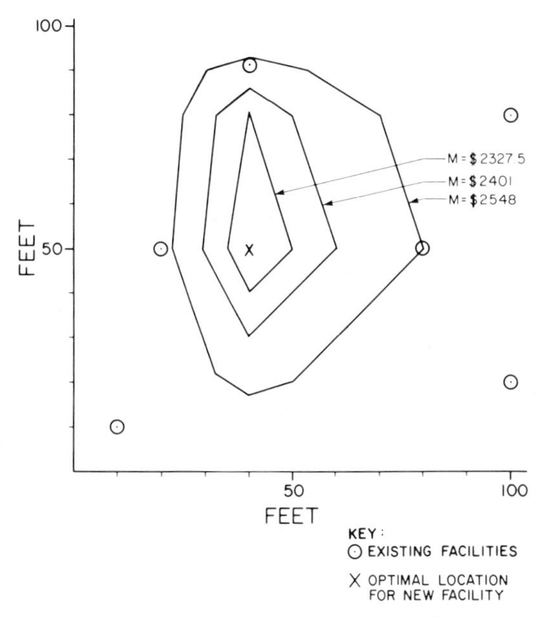

Fig. 2.3.7 Optimal location for a new machine and contour curves for an example problem with six existing machines.

general, it is no longer possible to construct contour lines except for certain special cases where only the new facilities are to be added. The primary emphasis on the solution procedures is therefore to find the optimal locations for each of the new facilities. As before, if there are relatively few locations in which to locate the new facilities this may present no real practical difficulty, since costs for each alternate location configuration may be evaluated for comparison purposes. The reader is referred to Francis and White [7] for description of solution procedures for the multiple new facility location problems.

Rearrangement of an Existing Area

This classification of location problems is likewise commonly encountered and may be modeled in a variety of ways. Depending on the physical constraints surrounding the problem, the time and resources available to explore alternate solutions, the number of alternate solutions that may be of interest, and/or perhaps other factors, this type of problem is often approached in one of two ways.

If there are discrete locations available for each facility or area, a linear program assignment problem formulation might be used where the objective uses some measure of material handling costs or distances. The example given in the Section on routing analysis using an assignment problem format is a good example of a problem that might also be classed as a re-layout problem. Here the objective is to assign a separate facility or area to each location so as to minimize the material handling costs associated with moving product through this area. Constraints such as infeasible locations can be handled for facilities that may be too small or too large for a candidate area, by either subdividing the areas and/or assigning very high costs associated with a particular assignment. A less mathematically rigorous, but more flexible, approach employs one or more of the computerized layout techniques. Some of these procedures, such as CRAFT, allow the analyst to specify the size and/or the perimeter of the area to be rearranged as well as fixing certain facilities in the area that are not capable of being moved, such as stairwells or elevators. Several of the available algorithms are discussed later in more detail. As this section is being read, the reader is encouraged to relate the usefulness of each to the rearrangement location problem.

Development of a New Area or Layout

Developing a new layout represents the most complicated location problem. It requires the use of tools that are capable of dealing with this complexity and yet be flexible enough to resolve a variety of problems. From a modeling perspective, the result has been the development of computer routines that have algorithmic procedures embedded in them to develop block diagram layouts. Although the problem that the algorithm is attempting to solve generally has a mathematical formulation, the solution procedures are heuristic.

Computerized arrangement procedures capable of assisting in the development of a new layout are generally classified as being quantitative or qualitative. Quantitative routines use criteria based on materials handling relationships such as minimizing the volume times distance traveled or the materials handling systems costs. Therefore the input data to these models must also be quantitative. Qualitative models use criteria associated with activity relationships represented by the letters A, E, I, O, U, and X. These reflect the need of departments or areas to be close together and range from absolutely necessary (A), essential (E), important (I), ordinary closeness (O), unimportant (U), to undesirable (X). Proponents of the quantitative routines cite the importance of using mathematically sound procedures to evaluate alternate layouts such as those contained in the quantitative models, whereas proponents of qualitative routines cite the importance of being able to consider multiple objectives when assigning the activity relationships to the various departments.

Computerized arrangement procedures may also be classified according to the way the final layout is generated. Some routines develop the layout from the data input by successively selecting departments and in turn placing them in a layout. This "construction" process continues until all areas or departments have been added to obtain the final layout. A second type, called "improvement" routines, requires the user to input an initial layout. The computer routine then attempts to improve on this initial layout by interchanging the areas or departments in specified ways. As long as improvements can be found (as measured by the evaluation scheme), the procedure iterates. When no further improvements can be made, the procedure generates the resultant layout and then stops.

Many different computerized arrangement procedures have been developed and many more are likely to be available in the future. Currently some 30 or more different algorithms are known to exist with probably 20 additional ones only available to internal consultant proprietary groups. It is not feasible to attempt to compare features of all of these, but five of the most commonly encountered have been selected and are presented in Tables 2.3.6 and 2.3.7.

CRAFT and COFAD are representative of the improvement type of layout routine and use quantitative data as input. CRAFT uses two matrices for material flow input data expressed in from–to chart format: (1) a cost matrix for movement per unit distance among all departments, and (2) a volume matrix for expressing the volume (in number of trips, weight, number of pieces, etc.) among all departments. The algorithm then calculates a third matrix's values, distances, where the distances are the centroid-to-centroid distances among all pairs of departments in the current layout. These three matrices are multiplied together and the objective is to rearrange the departments so as to minimize the sum of the resultant cost–volume–distance products. COFAD additionally allocates the material handling costs to each of the moves among available alternate methods in an attempt to select the lowest cost material handling system.

To develop improved layouts CRAFT considers interchanging departments in the layout that have equal areas or common borders in an effort to reduce the distances among the department centroids. The following alternate interchange options are available:

1. Two department interchanges
2. Three department interchanges
3. Two department interchanges followed by three department interchanges
4. Three department interchanges followed by two department interchanges
5. The best of two and three department interchanges

The objective function is approximated for each proposed interchange by assuming the department centroids were exchanged. The interchange that offers the greatest reduction is made and the affected departments are interchanged. The new department centroid locations are calculated, a new distance matrix is found by CRAFT, and the objective function for the improved layout is calculated. This process continues until no further reduction in the objective function can be found.

Alternatively, CORELAP, ALDEP, and PLANET, as representatives of the construction type of layout routine, generally use qualitative data as input. Each requires quantitative departmental data, however, that specify the area and number of the departments. The primary departmental flow data is usually given in a relationship chart format specifying the A, E, I, O, U, and X closeness ratings. Of these three, CORELAP is generally acknowledged as requiring the least amount of detailed data. PLANET is the most flexible, in that input data can be specified either as quantitative or qualitative, depending on the one of the three available forms selected for the data. The objective of each of these routines is to construct a layout that maximizes the sum of closeness ratings of adjacent departments.

To construct a layout, CORELAP calculates a closeness rating for each department by summing the relationship values each department has with every other department. It then selects the first department for placement as the department with the highest closeness rating. This is placed in the center of the layout. Then the relationship chart is scanned to find departments with high closeness

Table 2.3.6 Features of Five Common Computerized Layout Programs

Name of Routine	Classification	Objective	Maximum Number of Departments in Layout	Shape of Departments in Final Layouts (Regular/Irregular)
Computerized Relative Allocation of Facilities Technique (CRAFT)	Quantitative; Improvement	Minimizes the sum of the cost–volume–distance product	40	Regular
Computerized Facilities Design (COFAD)	Quantitative; Improvement	Minimizes the cost of material handling systems	40	Regular
Computerized Relationship Layout Planning (CORELAP)	Qualitative; Construction	Maximizes the sum of closeness relationships of adjacent departments	70	Irregular
Automated Layout Design Program (ALDEP)	Qualitative; Construction	Maximizes the sum of the closeness relationship of adjacent departments	64	Regular
Plant Layout Analysis and Evaluation Technique (PLANET)	Qualitative; Construction	Maximizes the sum of the closeness relationship of adjacent departments	98	Irregular

Table 2.3.7 Data Requirements and Special Features of Five Common Computerized Layout Programs

Name of Routine	Minimum Input Data Requirements	Special Features
CRAFT	1. Cost matrix for movement per unit distance 2. Volume matrix expressing flow along all departments 3. Initial layout (with sizes of all departments specified)	1. Departments can be fixed in layout 2. Alternate methods are available in the algorithmic interchange of departments
COFAD	Same as CRAFT, except the costs of performing all moves in the layout must be given for each of the alternate handling equipment types under consideration	Embeds the CRAFT algorithm in a procedure seeking to select the best layout in combination with the best handling system configuration
CORELAP	1. Departmental data, primarily sizes 2. Relationship chart	1. Requires least amount of detail—data as input, among the programs listed in this table 2. Can fix location of department, but only to a quadrant of the layout
ALDEP	1. Departmental data, primarily sizes 2. Relationship chart	1. Departments can be fixed in layout 2. Has multiple- (up to three) story layout capability
PLANET	1. Department data (size and placement priority) 2. Material flow data in one of three forms: a. Extended parts list (frequency, sequence, and cost per move), b. From–to chart, or c. Penalty chart (very much like a relationship chart)	1. Alternate methods are available input of data 2. Alternate methods are available for selecting the order in which departments are placed in the layout

ratings with the one already selected. These departments are placed in turn around the perimeters of the department first located. After all departments that have high closeness ratings with the first department have been placed in the layout, the relationship chart is scanned for departments that have high closeness ratings with the second department that was placed. If none exists that are not yet in the layout, then the procedure is repeated for the third department that was placed. This process continues until all departments are placed, and the layout grows in a spiral fashion as each successive department enters the layout.

PLANET has three alternate selection methods to determine the order in which departments enter the layout. All three methods are based on the importance of the relationships specified among the departments and each selects the first pair of departments to enter the layout that appear to have the greatest importance ratings with the remaining departments. Each method in turn selects from the remaining departments in order of importance to the departments already placed, and the PLANET layout grows in spiral fashion, much like a CORELAP layout.

ALDEP approaches the order in which departments enter the layout in a different fashion. The first department is chosen randomly. Then the relationship chart is scanned to find departments having high closeness ratings with the first department. If more than one exists, a random selection is made among them and placed in the layout. Once the second department has been entered, the relationship chart is scanned for all departments that have high closeness ratings with it, and the procedure is repeated. If no departments are found with high closeness ratings to that of the most recent added to the layout, the next department is randomly selected. This process continues until all departments have been placed in the layout. Rather than placing departments in a spiral fashion like CORELAP and PLANET, ALDEP places departments by utilizing a vertical sweep procedure as shown graphically in Fig. 2.3.8. Since ALDEP has a random aspect in its layout generation routine, it is often used to provide several layouts from which one or more may be selected for further study and adjustment.

The layouts generated from any of the computerized routines often require manual adjustment and consideration of other factors before final layouts may be designed. The goal of the use of these algorithms is to provide the analyst with an objective look at alternatives and a method by which a rather large amount of data can be condensed and summarized into a resultant layout. Often these

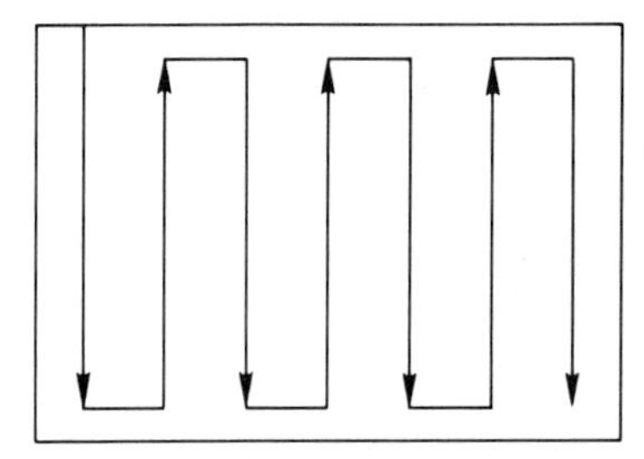

Fig. 2.3.8 Vertical sweep procedure used by ALDEP.

routines may be combined in a search phase, such as using a construction routine to generate an initial layout, followed by the use of an improvement routine to see if improvements can be found. As with all modeling efforts, the analyst must understand that these computerized layouts are aids to, and not substitutes for, the design process.

2.3.5 Summary

The four major areas discussed demonstrate the variety of applications possible in the use of quantitative methods and modeling procedures. Additional areas and tools will be added as future advancements in methodologies take place. Before using any of these techniques, however, it is imperative that the user understand the assumptions behind them, and the sensitivity of any decisions on these assumptions. Often analytical tools are used to provide base line results against which other solutions can be compared. The analyst realizes that rarely if ever will a modeling approach provide the answer to a problem he or she is trying to solve. But it can provide an excellent means to approach a problem and insight into the major variables and interactions that may affect any solution developed. Use of these techniques will not by themselves insure that better analysis and thus better alternatives will be developed by material handling and layout analysts. The alternatives themselves still must be scoped out, developed, and at least generally designed by a human mind; that is still essential. Careful and understanding use of the techniques discussed, however, will allow the analyst to develop a more rational approach in the development phase and a firmer basis in the justification phase for any solution developed.

2.3.6 REFERENCES

1. Fishman, George S., *Concepts and Methods in Discrete Event Digital Simulation,* Wiley, New York, 1973.
2. Shannon, Robert E., *Systems Simulation: The Art and Science,* Prentice-Hall, Englewood Cliffs, N.J., 1975.
3. IBM Mathematical Programming System Extended/370, Primer Manual, GH19–1091–1, IBM Corporation, Data Processing Division, White Plains, N.Y.
4. Bazaraa, Mokhtar S. and John J. Jarvis, *Linear Programming and Network Flows,* Wiley, New York, 1977.
5. Phillips, Don T., A. Ravindran, and James J. Solberg, *Operations Research: Principles and Practice,* Wiley, New York, 1976.
6. Kuhn, H. W., "On a Pair of Dual Non-Linear Problems," *Non-Linear Programming,* Chapter 3, J. Abadic, Ed., Wiley, New York, 1967.
7. Francis, Richard L. and John A. White, *Facility Layout and Location: An Analytical Approach,* Prentice-Hall, Englewood Cliffs, N.J., 1974.

2.3.7 BIBLIOGRAPHY

Apple, James M., *Plant Layout and Material Handling,* 3rd ed., Ronald Press, New York, 1977.

Hillier, Frederick S. and Gerald J. Lieberman, *Operations Research,* 2nd ed., Holden-Day, San Francisco, California, 1974.

Kleinrock, Leonard, *Queueing Systems, Volume I: Theory,* Wiley, New York, 1975.

Mihram, G. Arthur, *Simulation: Statistical Foundations and Methodology,* Academic Press, New York, 1972.

Tompkins, James A. and James M. Moore, *Computer Aided Layout: A User's Guide,* FP & D Monograph Series No. 1, Institute of Industrial Engineers, Norcross, Georgia, 1977.

CHAPTER 3
EVALUATING AND JUSTIFYING MATERIALS HANDLING PROJECTS

THOMAS CULLINANE
DAVID FREEMAN

Northeastern University
Boston, Massachusetts

3.1 PLANNING AND BUDGETING

3.1.1 Capital Budgeting Factors

Firms typically use some form of capital budgeting in the selection of major projects requiring investment capital. Capital budgeting might be described as the process of analyzing and determining the optimum mix of projects that consume a scarce resource, capital.

Materials handling projects usually require capital expenditure and thus become a part of the budget process. The amount of capital investment involved in some major systems can be substantial. For example, a large automated warehouse system could involve several million dollars of investment capital.

The details of capital budgeting will vary from firm to firm, but a typical budget process for a multidivision firm can be described. The process occurs over time with capital need projections made for periods of three to five years in the future. Typically those budget projections farthest into future time are made in overall dollars, with little or no individual project identification.

Budgets become more specific as to division or plant *and* more specific in terms of actual projects as the projection approaches present time. Thus the capital budget for the next fiscal year might identify amounts allotted to each division and projects for which funds have been budgeted. The same process occurs within a specific plant but the planning sequence would be department to plant to division.

3.1.2 Project Inclusion in a Capital Budget

In virtually all firms or government agencies, capital is a scarce resource with more opportunities to consume capital than there is capital available. The whole process then becomes one of deciding which of the many projects should be included at a specific time.

Usually budgets are built "from the bottom up." That is, input is requested from the plants, accumulated by division, and finally compiled into an overall budget for the firm. Some compilations may come from gross projections with little formal determination of projects. They may very well become extrapolation of past trends in capital expenditure. This process of extrapolation can be dangerous. If a division anticipates need for a new plant to increase capacity or get into a new product venture, a major capital expenditure will arise, one that far exceeds the normal needs. Firms may very well recognize this, and include special needs in addition to those that could be handled by funds extrapolated.

As the next fiscal year approaches, the budgets become more explicit as to project. The usual means of inclusion of projects is by engineering economic analysis. Projects are included that show a satisfactory return on the capital required. Details of this analysis process are described later. At this point it suffices to say that projects are included if the return expected meets the firm's economic acceptance level.

One engineering economy analysis tool frequently used at the budget phase is payback analysis. This analytical technique is rejected later as a sound economic analysis technique, but has some value as a device at budget preparation time, particularly early budget considerations.

In payback analysis, the payback period is calculated as follows:

Let I = dollars investment in project
S_t = difference between revenues in year t and costs in year t for $t = 1, 2, \ldots$

To find the payback, sum the S_t until the sum equals I. The number of years (or fraction thereof) is the payback period. Example: A miniload system is contemplated for small parts stores. It is estimated that the installation in total will require $800,000 capital investment. Savings of the following are expected:

Year	Savings ($)	Cumulative ($)
1	100,000	100,000
2	175,000	275,000
3	225,000	500,000
4	225,000	725,000
5	225,000	950,000
6	225,000	

The payback period is between 4 and 5 years ($4\frac{1}{3}$ years, or 4 years, 4 months if we interpolate).

A firm that required the payback period to be 4 years or less would exclude the proposed project, whereas a firm with a 5-year cutoff would include it. The defense of this procedure is that it's very easy to apply and may effectively screen the many proposals down to those sensible ones that can then undergo more accurate analysis.

The following is a summary of a typical 3-year cycle for a firm.

Budget Period	Type	How Prepared
3 years hence	Overall dollars, No project I.D.	Extrapolation of capital expenditures
2 years hence	Broken down by division or plant	Requests for division
1 year hence	Plant, division breakdown; Specific projects	Rough economic analysis

3.1.3 Implementation of a Capital Budget

It needs to be stressed that a capital budget is not approval to spend. Rather, it is a plan. Generally, in the fiscal year, the current capital budget becomes equivalent to an authorization to expend funds. If the capital budget for a plant has included within it $20,000 for a forklift truck for the receiving area, can the truck be purchased? No simple answer is possible. Most firms require a formal "request for authorization" to expend capital funds. This usually requires a formal economic evaluation. The financial officers will make a decision based on the evaluation and corporate factors such as capital availability and business trends.

In many firms, rules exist defining the spending authority of a given management level. For example, a firm might have:

Level	Spending Authority, Single Project
Department Manager	maximum $10,000
Plant Manager	maximum $30,000
Division Head	maximum $150,000
Projects over $150,000	Budget Committee

3.2 MATERIALS HANDLING NEEDS

3.2.1 Overall Needs

Nearly every activity in manufacturing involves some form of material handling. Materials handling is clearly of major importance because it impacts on the manufacturing function in at least five different ways. They are: (1) the cost of manufacturing a product, (2) the safety and health of workers, (3) the damage done to products, (4) the volume of products and materials lost or stolen, and (5) work-in-process inventory levels.

It has been documented that the cost of materials handling can account for between 15 and 50%, or more, of the cost of manufacturing a product. In most companies materials handling is the controlling element of operations such as receiving, inspections, storage, packaging, packing, shipping, and distribution. Research reports from private sources and government agencies point to the fact that a large percentage of recorded accidents take place while a person is performing a task that can be classified as materials handling. Many back injuries, lacerated fingers, and crushed toes are the result of a poorly planned or poorly performed handling task. The improper use of certain types of handling equipment often leads to worker injuries. Regardless of how a materials handling accident takes place, it can be translated into increased production costs.

During the time period that a product is being transported between two points within the manufacturing process, it is usually fairly secure and safe from damage. A majority of the damage sustained by a product takes place at the point where the unit is picked up, put down, or packaged. Although equipment manufacturers have attempted to provide specialized fixtures for use in reducing product damage during materials handling operations, very few companies have been able to reduce product damage costs to acceptable levels. It is a very rare firm that does not have a stack of crushed cartons and damaged products that are a direct result of poor handling practices.

The cost of pilferage is growing at astounding rates in many companies. The methods for handling materials and the storage configuration of goods can have significant influence on the ease with which goods can be stolen. In the case of heavy items, if handling equipment is not readily available for use in stealing a product, chances are reduced that it will be taken. When work-in-process inventories are under control and the location of the product, as well as the location of the materials handling equipment, is always known, pilferage can be reduced. To achieve this goal in a large facility, investments in automatic identification systems, bar coding equipment, and computers may have to be made.

Product residence time in a manufacturing system can have a significant bearing on overall production cost. When interest rates hover in the double digits, the cost of financing work-in-process inventory is significant enough to warrant a close investigation. If a materials handling system has been properly planned and the equipment selected according to the needs of the system, work-in-process inventories and the raw materials inventories can be minimized. The speed and the efficiency with which a product is handled throughout the manufacturing process have to be maintained.

A brief review of the trade literature for the past few years would reveal several articles in which very substantial cost reductions and savings have been realized through improvements in a handling system. Improvements in operations and policies, as well as improvements in equipment, can result in a more profitable circumstance for a manufacturing firm. Seeking out and solving materials handling problems can result in high returns on the time and capital devoted to the study.

3.2.2 Project Identification

In the previous section it was illustrated that it is economically wise to study problem areas in materials handling and to remedy any unacceptable conditions. Project identification is not always an easy task. Correct problem identification is extremely important. Without a well-documented problem statement an analyst always runs the risk of investing time and capital to solve the wrong problem. The process of identifying and defining materials handling problems is covered in Chapter 1.

3.3 DATA SOURCES

3.3.1 Introduction

In economic evaluation of materials handling projects, it becomes imperative to obtain good cost data on all aspects of a project's behavior. Sources of this cost data are many and varied, and are both internal and external to the organization.

The typical materials handling system for evaluation might involve (1) purchasing and installing some equipment, (2) operating and maintaining it over a reasonable period of years, and (3) disposing of it (Fig. 3.3.1).

3.3.2 Data Source Components

Purchase of the Equipment

The usual data source in this case stems from requested vendor quotes, and information from other users of similar equipment and consultants. Several vendors should be contacted to obtain comparison.

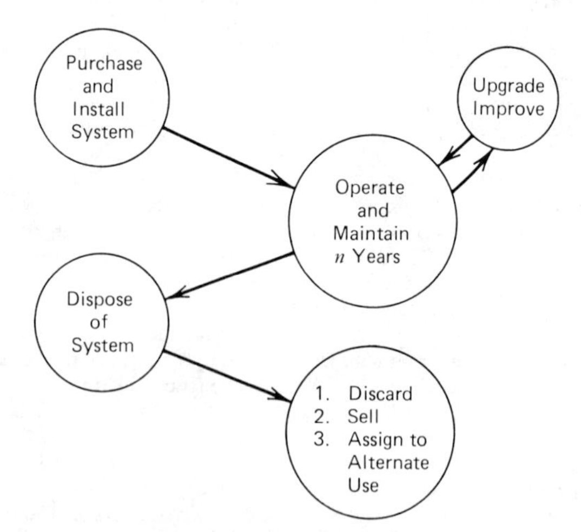

Fig. 3.3.1 A typical materials handling system for evaluation.

Most quotes are applicable for a fixed period of time and are honored for that period of time. If actual implementation is well in the future, provision must be made for price increases. Trends in the Wholesale Price Index (WPI) or any historical data on price trends in the industry may be used.

The investment in a materials handling system for analysis purposes is the "installed cost." Thus, a vendor quote needs to be examined and adjusted to accommodate all costs. Of particular interest are shipping costs, site preparation costs, and hidden costs.

Shipping Costs. The freight or traffic department will have data on shipping costs by the various modes of transportation considered feasible. If equipment is purchased from "off-shore" suppliers, there can be additional costs for freight forwarding, duties, entry fees, and so on.

Site Preparation Costs. Most materials handling systems require site preparation, including equipment relocation, construction activity, mechanical and electrical modifications, and some HVAC work. Here the best source of cost data is the Plant Engineering Department.

Hidden Costs. There can be costs that are easy to forget but are quite significant because they can be a substantial part of the total cost. In some major materials handling systems, installation requires substantial renovation and modification, changes that disrupt related activity in the area. The added cost involved in maintaining necessary output levels in these areas of production must be ascertained and assessed to the project cost. In developing costs it is necessary to do so with the *plan* of how the work is to be done.

Some possibilities are (1) shut down for a sufficient period of time to complete the work, (2) do the work on off-shifts such as second and third shift, or (3) attempt to do the renovation while maintaining activity in related areas.

Other hidden costs are often neglected but should be included. As materials handling systems become more complex these costs become more significant. They could arise as training costs, software development, debugging the system, and so on. A computer-controlled system behaves differently than more traditional approaches. Thus, not only must those operating the system be properly trained, but also those who interface with it must receive training.

Software development may be required. This can be true even when the supplier includes it as part of the installation. In major systems such as automated warehouses, a start-up phase always exists. One never simply "throws a switch." In fact, many installations involve a planned period of break-in during which operations are maintained by the old system while the new installation emulates the action in a test environment. Costs are involved in running such a dual system.

A final set of costs often overlooked are costs of such items as sprinkler systems, controls, guards, safety devices, feeders, and auxiliary equipment. Planning for these costs initially makes economic evaluation realistic.

Costs of Operating and Maintaining a System

The operating costs of a system include labor, energy, and various miscellaneous costs. Labor costs can generally be obtained from accounting records of the firm. They must be estimated over the life of the project so that past trends of wage increases must be evaluated for extrapolation into the future. If new technology for the firm is involved, it may be necessary to seek labor data from outside. Government statistics, trade associations, and other companies are the best sources.

Many firms have time standards for work performance. They can be invaluable sources of data for operating costs, and can be used to synthesize the amount of various types of labor needed to operate a system.

This source of time data may also exist for the maintenance crafts. If so, synthesis of the installation costs discussed earlier may be possible.

Maintenance costs also come from both internal accounting records and external sources. In a simple materials handling project such as replacement of a forklift truck, past history will provide clear insight. However, in complex systems, particularly if the firm has had little experience, the internal records will be of little value. Vendor data are helpful but must be examined carefully. The following questions should always be asked. Do quoted figures come from experience comparable to your installation? Is the operational environment similar?

One crucial concern of highly automated systems is downtime. What can be reasonably anticipated? What is the implication of "going down"? When a forklift truck fails, consequences are relatively minor. Another is pressed into service or alternative handling techniques may be feasible. However, when an extensive automated system goes down, it may impact an entire production area. Parts cannot be fed. Alternatives may simply not exist.

One example is a complex delivery system for feeding parts to a large, high-volume assembly area. Failure of the materials handling system fundamentally causes production to stop. Manual feeding is not viable, because the handling is an integrated part of the whole manufacturing system.

To alleviate this impact, drive motors exist in spares with rapid and easy changeover possible.

Frequently, extensive PM is done at all key points. Such measures have associated costs that must be incorporated into the system plan.

Also, some complex systems require maintenance skills not currently available in the firm. A computer-controlled system requires special maintenance skills new to the firm on the first installation. It is costly initially. Consideration might be given to contracts for maintenance. These are more costly as the demanded response time is shortened. The need versus the cost must be assessed.

A final area of operating costs is energy. Energy consumption by materials handling systems is relatively easy to predict. The cost of providing this energy in future periods is difficult. The past decade has seen energy costs escalate at high rates. This trend could continue. If past costs are to be used, recent past trends are likely to be the most reliable.

Disposal Costs and/or Revenue

The terminal value of the system needs to be considered. Accuracy here is most difficult. First, when will the system terminate? Second, what can be salvaged and for how much money? Past history with similar assets only provides a clue.

In analysis of highly specialized systems, a conservative attitude is advocated, which assumes relatively short life and low value of the assets downstream. Highly specialized systems can become obsolete as conditions that made them attractive change. They also tend to be useless in other applications, and thus are nearly valueless.

For example, a forklift truck is a general piece of equipment. It could easily have a long period of utility. Chances are it can be sold or relegated to an alternative function. However, a complex computer-controlled system of conveyors, towlines, and so on, has little or no value except to the specific application.

Another example is the difference in a standard warehouse building with 15–20 ft ceilings and column spacing of 25–40 ft, versus a high bay, rack-supported building for automated storage and retrieval. If need for a warehouse in the area ceases, the former building is convertible to manufacturing or office space, or is resaleable. The rack-supported building is useless except as a warehouse.

Solid engineering work and investigation are essential for establishing proper estimates of costs and revenues. Accounting records and accountants are useful sources of data, but cannot provide all the data needed. Full understanding of all aspects of the proposal is essential. Only then can reasonable estimates be developed. When performing a cost analysis on a piece of equipment, the purchase price is only the beginning. Maintenance requirements, personnel requirements, cost of spare parts, service life, and operating costs should also be considered. One recently documented case involved a lift truck having an initial purchase price of $16,000. Fuel costs over a 5-yr period amounted to $45,000, parts and maintenance were $12,000, and the operator costs were $72,000. In total, the original purchase price accounted for only 5% of the total cost of owning the vehicle for 5 yr.

3.4 ECONOMIC ANALYSIS

3.4.1 Introduction

Analysis of proposed materials handling systems to determine their economic merits is no different than the analysis done for any capital expenditure made by the firm.

A partial bibliography of books on engineering economy is contained at the end of the chapter. Material in this chapter is consistent with the approach taken in the various texts.

3.4.2 Assumptions of Economic Analysis

This discussion focuses on how one makes the decision as to what materials handling system should be recommended for implementation. The entire focus is toward *the decision.* Analysis techniques to arrive at the decision will be explained.

In this context it is assumed that materials handling systems have been suggested for implementation. The proposed systems accomplish a particular desired function and are technologically feasible. Furthermore, each proposed system accomplishes the specific desired objective. The analysis will reflect how one system accomplishes the objective differently from another.

For example, suppose the concern is to purchase appropriate materials handling equipment to remove 20 pallets/hr from the discharge end of an automated storage and retrieval system (AS/RS), and deliver these pallets to an adjacent manufacturing area. Two systems have been proposed: (1) use of forklift trucks, or (2) a conveyorized system.

Suppose the lift truck can handle the 20 pallets/hr but not more than this. The conveyorized system can also handle the 20 pallets/hr but, in fact, has sufficient capacity to handle up to 50 pallets/hr.

Indeed, the added capacity of the second alternative may be attractive but the difference in performance should be properly integrated into the analysis. If 20 pallets/hr is an appropriate volume and will so remain for a reasonable period of time (perhaps it is the capacity of the AS/RS), then the extra 30-pallets/hr capability of the conveyor system is useless and the *analysis* and *decision* should not be influenced by its presence.

On the other hand, if 20 pallets/hr is current demand on the system and demand is expected to steadily increase to 50 pallets/hr in 4 yr, then another approach is required. Proper alternatives would be to compare the conveyorized system (full capacity immediately) with lift trucks phased in over a 5-yr period.

The important point is that we must be careful to identify clearly just what problem we are solving. Failure to do so can lead to bad decisions.

3.4.3 A Systematic Analysis Approach

In any economic evaluation of proposed capital expenditure, a systematic approach to the entire decision process is essential. Although the step-by-step approach advocated by various engineering economy writers may differ somewhat, all bear a certain similarity.

One such approach that will be found useful in this discussion is called "Systematic Economic Analysis Technique" or SEAT [2]. This procedure uses an eight-step process for performing the analysis:

1. Define the set of feasible mutually exclusive alternatives to be compared.
2. Define the planning horizon to be used.
3. Develop cash flow profiles for each alternative.
4. Specify the time value of money (investment rate).
5. Specify the measure of merit.
6. Compare the alternatives using the measure of effectiveness.
7. Perform supplementary analysis.
8. Select the preferred alternative.

Defining Alternatives

This step involves clearly identifying the methods of accomplishing the function. In a sense, it is the "engineering" portion of engineering economic analysis. Alternatives must be identified clearly so that all aspects of the alternatives are specified. When the analysis is performed, the economic comparison is for accomplishing the desired function.

To illustrate, consider the example of (1) a conveyorized system; or, (2) a forklift truck as a means of handling pallets from an AS/RS to a manufacturing area. If these are two proposed alternatives, the key question to be answered is: do they each accomplish the objective of moving the desired volume of pallets through the system? As pointed out earlier, if the firm currently requires that 20 pallets/hr be handled but this volume will increase over the next few years to 40 pallets/hr, the lift truck alternative must be specified in such a way that it accomplishes the goal. This may mean this alternative really involves purchase of one truck immediately and a second truck 2 yr from now.

Note: It is imperative that *all* alternatives be specified. No amount of elegant economic analysis can overcome a deficiency in the engineering aspect of proposed expenditures. Overlooking or ignoring a good alternative forces the analyst to select the best alternative from a set of poor choices.

Defining the Planning Horizon

When investment alternatives are compared, a decision is being sought that has an impact on cash flows that will occur in the future. Precisely what period of time should be used for comparison purposes is not a trivial problem. No simple answer applicable to all cases exists to this problem.

In a situation where the alternative being analyzed is which of two proposed pieces of equipment to purchase to perform a material handling task, the planning horizon problem *might* be trivial. A "life cycle" might be chosen, and would be appropriate if each piece of equipment had a natural life that was the same.

However, it is common to encounter a situation in which the natural life of the alternatives is quite different. A commonly encountered analysis situation in which alternatives have very different lives is that of replacement. In the typical replacement problem the alternatives are often (1) purchase a new piece of equipment to perform a task, or (2) continue to utilize the existing piece of equipment for one or more years. Here we find one alternative with a long life versus one with a relatively short life.

Proposed bases for choosing a planning horizon that exist in the literature are:

1. The common life of cash alternative if they are the same.
2. The least-common multiple of the lives of the alternatives.

3. An arbitrary life—perhaps the life of the shortest, the longest, or some arbitrary number of years.

All engineering economy authorities seem to agree that basis 1 is valid when the situation arises and basis 2 is valid although the analysis can be cumbersome. For example, comparing a 7-yr life and a 9-yr life alternative becomes analysis over a 63-yr planning horizon. The implicit assumption in such an approach is seven life cycles of a 9-yr alternative equals nine life cycles of a 7-yr alternative.

Basis 3, an arbitrary life, is not generally supported. It clearly biases toward one of the alternatives. Nonetheless, it is found in practice. At one time the U.S. Postal Service used a 10-yr horizon for all economic evaluation. This horizon was used for vehicle replacement (with lives typically 5–7 yr) or new buildings (30–50-yr life).

Developing Cash Flow Profiles

Essential to economic evaluation is the development for each alternative of the proper, real cash flows that will arise if the alternative is implemented. Two methods of presenting cash flow profiles are encountered. The first, a cash flow diagram, is useful as a visual reminder and learning device. The second, a cash flow table, is probably more useful as a problem solving tool. A simple example is used to illustrate each.

A lift truck is to be purchased for $30,000. It will be used for 7 yr and retired at that time. It is anticipated that it will be possible to sell it for $3,000. Yearly costs for operation and maintenance are expected to be $20,000.

Cash Flow Diagram

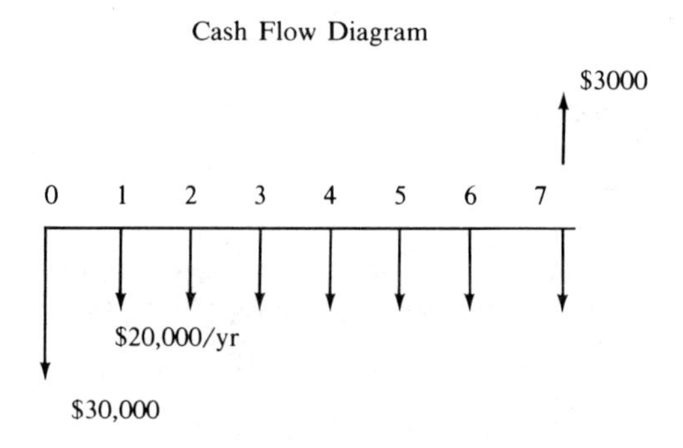

Cash Flow Table

End of Year	Cash Flow ($)
0	−30,000
1	−20,000
2	−20,000
3	−20,000
4	−20,000
5	−20,000
6	−20,000
7	−20,000
7	+ 3,000

In each case the following generally accepted conventions have been followed:

1. Money flow *to* the firm is indicated by ↑ (diagram) or + (table).
2. Money flow *out,* such as costs and expenses, are shown as ↓ (diagram) or − (table).
3. All flows are displayed as end-of-period flows.
4. Periods for our purposes are assumed to be years.

Specifying the Time Value of Money

In the next section, the means of establishing equivalence between money flows at different points in time will be reviewed. In this development an interest rate i, the time value of money, is used. In economic analysis it is essential to have an interest rate specified that, if viewed as a return on invested capital, could be considered acceptable to the firm. Firms typically have such a desired return, frequently called a "minimum attractive rate of return" or MARR. This convention will be followed.

How a firm arrives at such a figure is beyond the scope of this handbook. Discussion can be found in books on corporate finance or managerial economics. A few generalities can be made that might be useful. To illustrate the point a few examples are useful.

Example 1. A firm can buy a monorail device to move material from workplace to storage at a cost of $20,000. It will save $5000/yr for 5 yr. Alternatively it could purchase transfer tables and conveyors to accomplish the task. This scheme would cost $20,000 and save $6,000 over a 5-yr horizon. Which is preferred?

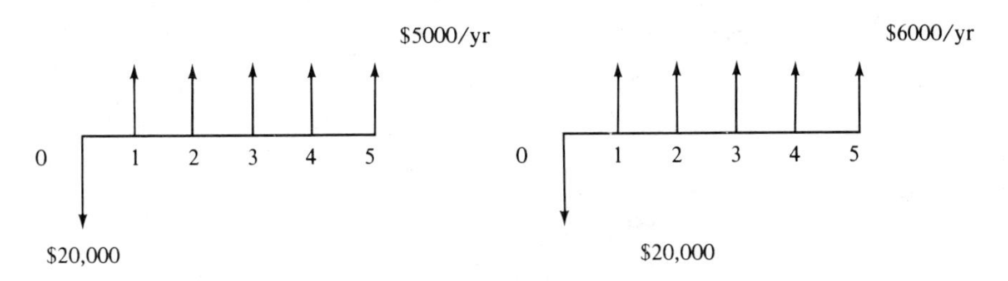

Clearly the second alternative is preferred in this example. Timing is the same and only differences in *amount* occur with the cash flows.

Example 2. For this example the same alternatives will be used as in example 1. However, changes in the relevant cash flow will be made. Assume the lift truck cash flow is unchanged. For the second alternative, the first cost is $25,000 and the life that can be expected is 6 yr.

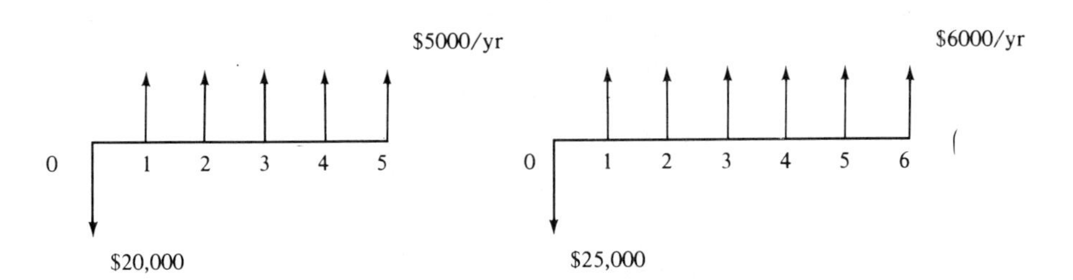

A preferred alternative is no longer obvious. The second alternative requires an extra $5,000 investment at $t = 0$. It does yield an extra positive cash flow of $1,000 for $t = 1, 2, \ldots, 5$ plus a positive $6,000 at year 6. These differences make the decision of the preferred alternative difficult.

The procedure used is to convert the money flows for each alternative to some single measure. Common measures are:

1. Present worth
2. Future worth
3. Annual worth
4. Benefit-cost ratio
5. Rate of return

Comparing Alternatives Using Measure of Effectiveness

The alternatives under investigation are compared by one of the methods mentioned. This requires calculating the value of the measure of effectiveness chosen for each alternative.

It is assumed that money flowing to the firm has been given a positive value and money expended is treated as a negative flow.

In the case where present worth, future worth, or annual worth are chosen, the analyst calculates for each alternative the equivalent value at time 0, time n, or flowing annually from years 1 to n, respectively. The equivalence is taken at $i =$ MARR. Establishing equivalence relationships is discussed in Section 3.4.4.

Problems arise if alternatives have different lives. An approach was discussed earlier for such a situation.

In benefit-cost ratio (B/C) measures, the measure is the ratio of the present worth of the alternative's benefit stream divided by the present worth of the cost to achieve the benefits (investment) where present worth is calculated at i = MARR.

In rate of return (ROR) analysis, the return on invested capital is calculated for alternatives and compared to the MARR. If the return is greater than the MARR, the alternative is acceptable.

A caution is necessary for application of the last two techniques—B/C and ROR. Proper analysis leading to correct decisions requires that this analysis be done incrementally. To perform incremental analysis, the following steps must be taken:

1. Order alternatives from lowest investment to highest.
2. Obtain the incremental cash flow between adjacent cash flows (higher of the two minus the lower of the two).
3. Calculate the incremental rate of return i^*.
4. If i^* is greater than MARR, retain the higher investment alternatives, reject the lower.
5. Go to step 2.
6. If i^* is less than MARR, reject the higher and continue to step 2.

The same incremental procedure is required in B/C analysis.

Performing Supplemental Analysis

This step is discussed later.

Selecting the Preferred Alternative

The preferred alternative from an economic viewpoint can now be established. In the case where the measure of merit is present worth, future worth, or annual worth, the alternative having the maximum value is preferred.

Benefit-cost ratio techniques and rate of return techniques require the incremental approach. The preferred alternative is that which remains after the analysis has been performed on all alternatives. The process requires that each increment of investment have sufficient economic return (at least the MARR).

A simple example can be used to illustrate. The cash flow is taken deliberately to reflect a principle and is unlikely to represent any true materials handling alternatives.

	Alternative					
	Case 1			Case 2		
Year	A	B	C	A	B	C
0	−10,000	−20,000	−30,000	−10,000	−20,000	−30,000
1–5	+3,000	+5,000	+6,000	+2,000	5,000	6,000
5	+10,000	+20,000	+30,000	+10,000	+20,000	+30,000
ROR	30%	25%	20%	20%	20%	25%

(Note: Cash flows with 100% salvage values have been selected deliberately since then the ROR is A, the yearly flow, divided by P, the initial investment.)

In the foregoing example it is tempting to conclude that A in Case 1 and C in case 2 are preferred *because* ROR is maximized. This is not a valid conclusion. An incremental approach is essential.

Let us assume a MARR of 20%. For case 1 the analysis is:

1. ROR on alternative A = 30%
2. Incremental cash flow = alternative B − A

Year	B − A
0	−10,000
1–5	+2,000
5	+10,000

3. ROR = 20%, B is acceptable
4. Incremental Flow C − B

Year	Flow
0	−10,000
1–5	+1,000
5	+10,000

5. ROR = 10%, reject C

Therefore alternative B is selected. In Case 2 the analysis is:

1. Alternative A: ROR = 20%; accept
2. Incremental Flow B − A

Year	Flow
0	−10,000
1–5	+3,000
5	+10,000

3. ROR = 30%, accept
4. Incremental Flow C − B

Year	Flow
0	−10,000
1–5	1,000
5	10,000

5. ROR = 10%, reject C

The result for both cases is alternative B.

It is interesting to note the importance of including all feasible alternatives. Suppose in each case the middle alternative had not been identified and included as a consideration.

	Case 1			Case 2		
Year	A	C	C − A	A	C	C − A
0	−10,000	−30,000	−20,000	−10,000	−30,000	−20,000
1–5	+3,000	+6,000	+3,000	+2,000	+6,000	+4,000
5	+10,000	+30,000	+20,000	+10,000	+30,000	+20,000
ROR	30%	20%	15%	20%	20%	20%

Case 1: Reject C, ROR < 20%

Select Alternative A

Case 2: Accept C, ROR = 20% = MARR

Note: In Case 1, alternative A is very attractive (i = 30%), yet the increment from A to C is not (i = 15%). In Case 2, A is just at the MARR as is C, thus C is accepted. Yet when we had alternative B, it was attractive as an increment over A (i = 30%). But the increment from B to C is poor (i = 10%). Ignoring the existence of B makes C acceptable.

3.4.4 Equivalence—Time Value of Money

A fundamental concept in engineering economy is that of equivalence. The concept is that to individuals or to the firm, a fixed sum of money flowing at two different points in time is not considered comparable.

If you have a choice of receiving \$100 now or \$100 in five years you would clearly take the \$100 now. There is some sum of money S that would make you indifferent to the choice of \$100 now or S five years from now. S will be greater than \$100. How much greater depends on many factors.

This concept is that of money having value over time. An equivalence relationship can be formalized by identifying this time value of money as interest.

For simplicity, consider 10% as an appropriate interest rate. Then \$100 now and \$110 in one year would be equivalent.

$$\$100 + 0.10(\$100) = \$110$$

In 2 years we have

$$\$110 + 0.10(110) = \$121$$

You will observe that we have taken the original sum and multiplied by $1 + i$ ($i = 0.10$). In year 2 we took the starting value of dollars times $1 + i$. We can get the second year sum of money as:

$$[100(1 + i)]\,(1 + i) = 100(1 + i)^2$$

The foregoing is generalized easily. Some standard nomenclature is in order:

P = a present sum of money at time 0.
F = the future sum of money at the end of n interest periods.
i = the interest rate
n = the number of interest periods.
A = the sum of money flowing at the end of each period commencing one period hence and ending n periods hence.

From the point of view of a cash flow diagram we have:

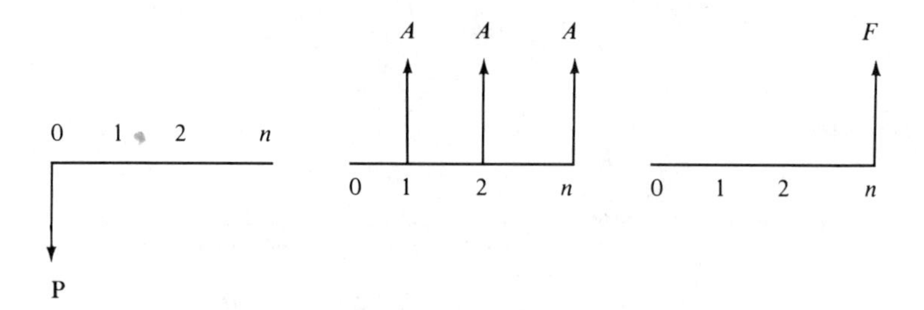

It is relatively easy to establish equivalence between P, F, or A. The simple example given illustrates that

$$F = (1 + i)^n P$$

By simple algebraic manipulation we can obtain

$$P = (1 + i)^{-n} F$$

Establishing equivalence between any of the flows P, F, and A is always a function of i, the interest rate and n, the number of periods. The equation is shown for P and F; the equations involving A are a bit more complex but easily developed.

A number of years ago, the Engineering Economy Division of the American Society for Engineering Education standardized nomenclature used to define the mathematical factors in making money conversions to establish equivalence. Using the example given, to convert a present sum to a future sum:

$$F = (1 + i)^n P$$

The term $(1 + i)^n$ is known as the single payment compound amount factor. The standard nomenclature is:

$$F/P, i, n$$

To convert a future sum to a present sum:

$$P/F, i, n = (1 + i)^{-n}$$

If the vertical line is read "given," the factors become easy to remember and use. Thus, $F/P, i, n$ can be read as the factor to find F, a future sum, given P, a present sum; i, an interest rate; and n, the number of periods. Table 3.4.1 is a summary of all the common factors.

Table 3.4.1 A Summary of Common Factors

To Find	Given	Factor	Expression	Name of Factor
F	P	$F/P,i,n$	$(1+i)^n$	Single payment compound amount
P	F	$P/F,i,n$	$(1+i)^{-n}$	Single payment present worth
P	A	$P/A,i,n$	$\frac{(1+i)^n - 1}{n}$	Series present worth
A	P	$A/P,i,n$	$\frac{i(1+i)^n}{(1+i)^{n-1}}$	Capital recovery factor
F	A	$F/A,i,n$	$\frac{(1+i)^n - 1}{i}$	Series compound amount
A	F	$A/F,i,n$	$\frac{i}{(1+i)^{n-1}}$	Sinking fund factor

Other factors have been developed that handle more complex flows that are encountered fairly frequently. For example, most texts have a factor to convert an annual flow of money in which each year's flow is greater than its predecessor by a fixed amount (called a uniform gradient). White et al. [2] have factors to convert flows where each year's flow is greater than its predecessor by some given percentage. These factors are useful in handling a cash flow such as labor. Labor to operate handling equipment might be described as being X dollars now but expected to be 10% more costly each year due to general increases. Factors that handle this can be useful.

These factors are not included here because they are readily available in other texts. Also, it should be pointed out that all factors are straightforward functions of interest rate i and number of periods n. Programs can and are easily developed on computers, even microcomputers. Readers interested in such programs might view the appendixes in the text by Stevens [3]; they provide a FORTRAN listing of several useful programs. Also, the Institute of Industrial Engineers markets several microcomputer libraries for economic analysis. They are written in Basic language.

Some Examples of Equivalence

In the examples that follow, $i = 10\%$ is used throughout for simplicity.

Example 1. What is the equivalent future sum 10 yr hence, F, for \$1,000 now?

$$F = 1000\ (F/P,\ 10\%,\ 10) = \$1000\ (2.5937) = \$2593.70$$

Example 2. What present sum P is equivalent to \$5000 5 yr from now?

$$P = 5000\ (P/F,\ 10\%,\ 5) = (\$5000)(0.6209) = \$3104.50$$

Example 3. How much money must be placed in a fund now such that \$1000 per year can be withdrawn for 20 yr and leave the fund with a zero balance after the last withdrawal?

$$P = 1000\ (P/A,\ 10\%,\ 20) = (\$1000)(8.5136) = \$8513.60$$

Example 4. This is a more complicated situation. A machine will cost \$100,000 to purchase and operating costs will be \$20,000 per year for the next 10 yr. In addition, at the end of the fifth year, \$3000 must be spent for a major overhaul. Finally, the machine will be sold for \$10,000 in 10 yr.

What sum of money A paid annually to acquire the service would be equivalent?

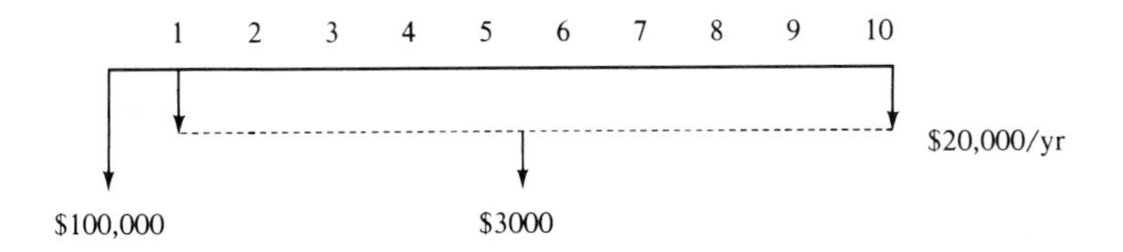

Solution: It is perhaps easier to first calculate a present sum P.

$$\begin{aligned} P &= 100{,}000 + 20{,}000\,(P/A, 10{,}10) + 3{,}000(P/F, 10{,}5) - 10{,}000(P/A, 10{,}10) \\ &= 100{,}000 + 20{,}000\,(6.1446) + 3{,}000\,(0.7513) - 10{,}000(0.3855) \\ &= 100{,}000 + 122{,}932 + 2{,}254 - 3{,}855 \\ &= \$191{,}331 \end{aligned}$$

Finally

$$\begin{aligned} A &= 191{,}331(A/P,\ 10,\ 10) = 191{,}331\ (0.1627) \\ &= \$31{,}130 \end{aligned}$$

The last example is more typical of the situation that arises in economic evaluation of handling systems. As shall be seen later, if the example had represented two choices of moving finished goods from the dock to a warehouse, the first cash flow might represent purchase of requisite equipment to do the job, plus the operating costs (labor, maintenance, etc.). The second might be a contract hauler. If the hauler would charge $31,130/yr and money is valued at 10%, either alternative is acceptable. Should the hauler charge more than $31,130/yr, owning your own equipment is preferred.

3.5 INTANGIBLE FACTORS

3.5.1 Introduction

Generally speaking, intangible factors are those factors that could have some influence on the decision yet cannot readily be translated into a cash flow. Some factors in this category really cannot be reflected in a cash flow, while others may be such that a cash flow impact could perhaps be estimated with such great difficulty and/or possible inaccuracy that it is not deemed practical to do so.

3.5.2 Quality

Some materials handling projects may have as a side benefit (or liability) the fact that quality of manufactured goods would be affected. The level of change may be difficult or impossible to measure or translate into the cash flow. If it can be predicted and translated into dollars, this should be done. Examples might be as follows:

Comparison is under consideration of (1) a traditional warehouse with low-bay pallet racks and handling by forklift versus (2) an automated high-bay AS/RS. One advantage anticipated for alternative 2 is reduced damage to material on pallets. Damage can arise because forklift truck operators occasionally inadvertently spear pallets in their routine movements. In this case it may be possible to predict the impact in dollar terms, provided data have been kept. Frequently no information exists, leaving the analyst the alternative of treating the issue as an intangible. If the economic decision is very close, this factor would favor the automated system. Alternatively, the analyst might make an estimate of the cost impact and include it in the analysis.

Quality "dis-benefit" might arise also. For example, a proposal for an automated system has included pallet unloaders. There is great variety in carton content on pallets, particularly with respect to fragility of carton content. It is believed that some damage could arise to goods during the unloading cycle. The extent is not known but the fact, an intangible, mitigates against the unloader and favors manual unloading.

3.5.3 Environmental, Safety, Health Hazards

If an ingredient of a project, or the project itself, is proposed to correct an environmental problem, overcome safety hazards, or rectify a health hazard, the cash flow impact *may* be difficult to measure accurately. Perhaps some data exist to permit predicting the reduction in lost-time injuries, reduction in indemnification costs, and/or change in insurance premiums. If so, it should clearly be included.

One note on projects in this area is relevant. Some capital investments in this area are simply required. In this case elaborate analysis is unnecessary.

For example, OSHA may require that an existing overhead handling system be equipped with protection beneath the system to protect workers from falling items. Elaborate analysis is not essential, because the investment may be required as a condition of continued operation.

A note of caution is in order. The investment above is necessary to correct a safety hazard. It should not be assumed that there is only one way to deal with the hazard. An example might be rerouting the system so that its path is never over areas occupied by workers. In this case, economic evaluation is both desired and essential. When human exposure to hazard opportunity is reduced, an intangible gain is realized.

3.5.4 Development of Materials Handling Projects

Materials handling projects may be undertaken on a developmental basis. Elaborate economic analysis is probably not wise. The nature of such development is that formal analysis against traditional criteria will not work. Such projects should be handled the same as experimental projects in new manufacturing techniques. They should be approved if the tangible benefits that might be realized are substantial enough to make the major investment worthwhile. For example, a firm might make a modest investment to permit testing the feasibility of affixing bar code ID labels to bins of manufactured parts for routing through the manufacturing floor. Envisioned is a system of conveyors and transfers throughout the shop for controlling material flow. The trial would be conducted to assess the ability to affix, the reliability of the read, and the ability to encode the read and establish feasible routing. The total system envisioned could involve a total equipment investment of many thousands of dollars. Prior to such a commitment, for a small investment the trial can detect problems that could arise. Full implementation may be performed later. And, experiments have the added advantage of being a source of cost data, for analysis of full implementation.

3.5.5 Inflation

Including the topic of inflation with intangibles is perhaps not accurate. However, the impact inflation has in economic analysis needs to be discussed.

Including inflation in economic analysis is complex, and is handled in various ways.

Obviously inflation cannot be entirely ignored. A common treatment by firms is to incorporate inflationary consequences in the MARR. That is, where inflationary effects are heavy and anticipated future inflation rates are expected to be high, a high MARR is used relative to that used when rates are low. A variation on this would be to have two MARR's, one for use in cash flow profiles insensitive to inflation, and another for those in which inflation would impact heavily.

An analysis can be conducted as follows: Recall that A dollars flowing n years hence is equivalent to $P = A(1 + i)^{-n}$ dollars at present, given interest at i. If we further assume inflation at j fraction per year, then the A dollars has equivalent purchase power to $A(1 + j)^{-n}$. Discounting this to present dollars by our interest i would give us

$$P = A(1 + i)^{-n}(1 + j)^{-n}$$

This process could be used in the analysis. Actually, any cash flow can be translated to an equivalent present sum by discounting each cash flow A_t in year t by $1/(1 + i)^t$ and summing. If the equivalence is established as a multiple of $1/(1 + i)^t(1 + j)^t$, it has explicitly handled inflation at a rate j. Incorporating it into a single interest rate, say i_f, and discounting year t by $1/(1 + i_f)^t$ is really saying that

$$\frac{1}{(1 + i_f)^t} \approx \frac{1}{(1 + i)^t(1 + j)^t}$$

Mathematically speaking, no such indentity exists for a given i and j (>0) that would hold for all t. However, an approximation can yield acceptable results for the decision maker.

The reason the approximation is valid is that precision in the treatment on i and j, although a good idea, may be ignoring the fact that cash flows A_t are estimates prone to errors as severe as approximations on interest. Furthermore, analysis is usually comparison of differences in cash flows of competing alternatives, and the inflationary impact on the difference may be insignificant in affecting the decision.

The reader seeking a more thorough analysis of inflation consequences might obtain a copy of "Inflation, Taxes, and Other Economic Frustrations in the Justification of Material Handling Investments" by James Riggs [5]. Also, an entire issue of *Industrial Engineering* was devoted to the issue of inflation. The reader seeking more detail should look at Vol. 12, No. 3, March 1980 of that journal.

3.6 SPECIAL MATERIALS HANDLING CONSIDERATIONS AND TAXES

3.6.1 Taxes

Some firms use a "before-tax" cash flow for analysis purposes. The rationale for doing this is usually simplicity, but it can lead to bad decisions. Taxes are significant cash flows, and can impact alternative means of accomplishing an objective. Because of the dynamic nature of the tax laws, a tax specialist should be consulted prior to any decision making.

Taxes can be treated as any other cash flow. The secret to after-tax analysis is obtaining the proper tax flow consequence of each alternative.

Every firm is confronted with many taxes. Relevant to a material handling project are:

1. Federal income tax:
   Regular income
   Capital gains and losses
   Investment tax credit
2. State income taxes
3. Property taxes
4. Other, including local taxes

Federal taxes are of the greatest importance. They have far more impact on the decision process because of their magnitude due to depreciation and other considerations, and because substantial differences can arise from one alternative to another.

The general behavior of state income taxes is very similar to behavior of federal taxes, except that the rate is substantially lower. In fact, many states have tax codes that follow the pattern of the IRS but at lower rates. Property taxes are paid to the local community and are based on asset value of real and personal property. The rate is usually in the order of magnitude of 1–3%.

Income tax impact is the most significant with respect to decisions among alternative investments. Major changes in the tax laws occurred with the Economic Recovery Tax Act of 1981. This act established several fundamental changes that were to be implemented over several years, essentially stabilizing by 1983.

A few provisions of this act are important and will be briefly described. Then a discussion through examples will be presented to show how taxes can influence the decision process.

The above act had within it a major provision known as the Accelerated Cost Recovery (ACR) system. This system accelerated depreciation schedules and, in turn, the recovery of capital expenditures for projects, including materials handling projects, by:

Eliminating the "useful life" concept and replacing it with a shorter recovery period
Allowing more cost recovery (depreciation) in early years (accelerated recovery)

Prior to 1981, new equipment and buildings were given a useful life and then depreciated by one of the allowed methods over that useful life. The useful life was supposed to be a reasonable period in comparison to actual service life. Thus, equipment such as trucks, tractors, and cranes, would usually have life of 7–10 yr. Conventional buildings would have life of 30–50 year.

Under ACR, capital expenditures are recovered over a "recovery period" of 3, 5, 10, or 15 yr. The appropriate period depends on the type of equipment, *not* any measure of useful life. The periods and type of investment are as follows:

Three-Year Life. Property with a depreciation life of 4 yr or less. It includes equipment such as automobiles and lightweight vehicles, tools, and property used in conjunction with research.

Five-Year Life. Generally machinery and equipment. This would include cranes, monorails, industrial trucks, and so on.

Ten-Year Life. This class generally defined life for certain public utility property and property that, under the Asset Depreciation Rate (ADR) system in effect before 1981, had been assigned lives of $18\frac{1}{2}$–25 yr. Generally speaking, materials handling investments do not fall in this life category.

Fifteen-Year Life. Used essentially for real depreciable property such as conventional buildings.

Depreciation (called recovery rate) is established for property acquired and assigned a period above according to a specific rate for each year. These rates are given in Table 3.6.1.

Two points should be made regarding differences from depreciation methods used prior to ACRS. First, no salvage value is used. Rather than depreciating the initial installed cost less the salvage value, the installed cost is depreciated. Second, the "half-year" convention is used in all categories except 15-yr life. In effect, one-half of a year's depreciation is taken in the year the asset is acquired.

There is a provision for "expensing" some capital investments. By 1986, a firm will be allowed to expense (write off in the year acquired) $10,000 of capital investment and still claim Investment Tax Credit. This will have no real impact on decisions for large firms, but could be significant for the small, emerging firm.

The 1981 act retained the Investment Tax Credit but modified it slightly, so that a tax credit is allowed as follows:

Property Class	ITC Rate
3-year	6%
5-year	10%

Table 3.6.1 Accelerated Recovery Tables

Recovery Year	Property Class %		
	3-year	5-year	10-year
Property Placed in Service After 1980 and Before 1985			
1	25	15	8
2	38	22	14
3	37	21	12
4		21	10
5		21	10
6			10
7			9
8			9
9			9
10			9
Property Placed in Service During 1985			
1	29	18	9
2	47	33	19
3	24	25	16
4		16	14
5		8	12
6			10
7			8
8			6
9			4
10			2
Property Placed in Service After 1985			
1	33	20	10
2	45	32	18
3	22	24	16
4		16	14
5		8	12
6			10
7			8
8			6
9			4
10			2

Provisions for ITC should be checked. The primary purpose of inclusion of ITC was to induce capital expenditure for overall economic reasons. The provision might be labeled the "ping-pong" tax, since it bounces in and out of tax acts with some regularity.

Tax on capital gains and capital losses represents a final area of consequence to the decision maker. Gains and losses are easily defined. They are the difference between the sale price of an asset and the book value of the asset—a gain if sold for more than book value, a loss if sold for less. The law is very complex with respect to whether the gain or loss will have tax impact equal at the regular tax rate, or at the lower rate of tax on capital gains and losses, 30%.

Before leaving the descriptive aspects of taxes, the rate structure should be given. Federal income tax consists of two parts, a normal tax and a surtax. Currently, exemptions exist for each, but for a firm earning over $100,000 the combined taxes will amount to 46%. This incremental rate will be accurate unless earnings are low.

Since state taxes are a deductible business expense that reduces taxable income for calculation of federal taxes, a combined incremental rate can be calculated as follows:

r = combined incremental rate
s = incremental state tax rate
f = incremental federal tax rate

then

$$r = s + (1 - s)f$$

For example, with $f = 0.46$, $s = 0.04$, the net combined rate is 0.4816, or just over 48%. For this reason, engineering economy analysis is often done using an assumed tax rate of 50%, an excellent approximation for a firm with substantial net income.

3.6.2 Tax Implications of Materials Handling Projects

If the capital investment for either of two feasible alternatives is quite similar, taxes will not have major effects on one and thus be a dominant influence. For example, suppose a firm has narrowed alternatives to consideration of a narrow aisle, high-reach stacker for use in a warehouse. The alternatives are a vehicle supplied by company A versus one by company B. Costs differ somewhat for each. However, tax impact would be similar for both alternatives. In both cases, the purchase would place the vehicle in the "5-year" category. Depreciation would be according to the same schedule, with dollar amounts differing only by the difference in first cost.

However, an example that illustrates the influence and differing impact on alternatives might be given by a slight change in the alternatives just illustrated:

Example: Assume one alternative is purchase of a suitable stacker truck for $25,000. It is assumed the truck will be used 8 yr and will be valueless at the end of 8 yr.

Alternatively, the firm could lease a comparable vehicle for $6000/year. Annual lease payments are to be made at the start of each year.

Which alternative is preferred if the firm has an after-tax MARR of 15% and taxes are assumed to be 50%?

It will be assumed that all other costs and revenues are comparable in each alternative and thus will not be included in the analysis. An 8-yr study period is considered valid. A tax rate of 50% will be used.

The truck would go into the 5-yr property class. Appropriate rates for years 1–5 are 15%, 22%, 21%, 21%, 21%. Further, the purchase would allow an investment tax credit of 10%. Table 3.6.2 is a cash flow table for the purchase alternative. For the lease alternative the cash flow table would be as shown in Table 3.6.3. Because the present worth (PW) purchase is lower, it is the preferred alternative. Note that absence of the ITC would make the lease alternative the preferred choice. This illustrates that taxes do have an impact on decisions.

Two other facts are illustrated in the example.

First, depreciation charges *are not* cash flows. No cash flows when a depreciation cost is incurred. The year-2 charge of $5500 for depreciation involves *no* cash flow. The cash flow occurs at the time of purchase. It does result in a tax reduction that year of $2750, an amount that would be paid had no depreciation charge occurred.

Table 3.6.2 Cash Flow for Purchase Alternative[a]

Year A	Before-Tax Cash Flow B	Depreciation C	Taxable Income D (B − C)	Tax E	After-Tax Cash Flow F
0	−25,000			+2,500[b]	−22,500
1	0	3,750	−3,750	+1,875	1,875
2	0	5,500	−5,500	+2,750	2,750
3	0	5,250	−5,250	+2,625	2,625
4	0	5,250	−5,250	+2,625	2,625
5	0	5,250	−5,250	+2,625	2,625
6	0	0	0	0	0
7	0	0	0	0	0
8	0	0	0	0	0

[a] PW(15%) = 14,258.
[b] Investment Tax Credit; could assume actual impact, end of year one.

Table 3.6.3 Cash Flow for Lease Alternative[a]

Year A	Before-Tax Cash Flow B	Depreciation C	Taxable Income D (B − C)	Tax E	After-Tax Cash Flow F
0	−6,000	0	+6,000	−3,000	−3,000
1	−6,000	0	+6,000	−3,000	−3,000
2	−6,000	0	+6,000	−3,000	−3,000
3	−6,000	0	+6,000	−3,000	−3,000
4	−6,000	0	+6,000	−3,000	−3,000
5	−6,000	0	+6,000	−3,000	−3,000
6	−6,000	0	+6,000	−3,000	−3,000
7	−6,000	0	+6,000	−3,000	−3,000
8	0	0	0	0	0

[a] PW(15%) = −15,480.

A second point illustrated by the example is that lease/buy situations require no new or special analytical tools. Leasing is merely an alternative to which analysis can be applied.

Assume that 2 years ago an industrial truck was purchased for moving pallets in a receiving area. It was given a 10-yr life. The initial cost was $22,000, with an estimated salvage value of $2000. Straight-line depreciation has been used. Maintenance has been about $3000/yr and is expected to continue. Operating costs have been $30,000/yr, which includes regular shift operation and some overtime to keep pace with handling requirements in the area.

A replacement vehicle is under consideration. It will cost $30,000 and will be used for the next 8 yr. The principal advantages will be operating costs of only $24,000/yr, since it has greater capacity and improved characteristics precluding need for overtime. Furthermore, maintenance is expected to be only $2500/yr. The supplier has offered a $15,000 trade-in value of the original vehicle against the $30,000 purchase price. This is believed to be a reasonable value for the vehicle. After-tax MARR is assumed to be 15%. A tax rate of 50% is assumed on both regular income and gains and losses. Neither vehicle is expected to have any significant value at the end of the 8-yr period.

Again it is useful to focus on a cash flow for each alternative. Table 3.6.4 is a relevant flow for retention of the old. In replacement analysis, authors differ on their approach to the investment value of the existing asset. Some take the new asset less trade-in of the old as a net investment for the new. The approach here will be to assign the old asset an investment equivalent to those funds tied up by the decision to retain the asset. The decision under either approach is the same.

The reader may verify that at 15% for an MARR:

$$\text{PW(old)} = -85{,}726$$
$$\text{PW(new)} = -76{,}566$$

Table 3.6.4 Cash Flow for Retaining Existing Equipment

Year	Before-Tax Cash Flow	Depreciation	Effect on Tax Inc.	Tax	After-Tax Cash Flow
0	−16,500[a]				−16,500
1	−33,000	2,000	−35,000	−17,500	−15,500
2	−33,000	2,000	−35,000	−17,500	−15,500
3	−33,000	2,000	−35,000	−17,500	−15,500
4	−33,000	2,000	−35,000	−17,500	−15,500
5	−33,000	2,000	−35,000	−17,500	−15,500
6	−33,000	2,000	−35,000	−17,500	−15,500
7	−33,000	2,000	−35,000	−17,500	−15,500
8	−33,000	2,000	−35,000	−17,500	−15,500
8	+ 1,000[b]				+ 1,000

[a] The asset was acquired for $22,000. Straight line depreciation of $2000/yr [(22,000 − 2,000)/10] has been taken for 2 years leaving a book value of $18,000. The capital loss of $3000 @ 50% tax results in a tax saving of $1500. Thus a net investment of $16,500.

[b] At the end of 8 years (10 years for vehicle) a book value of $2000 will cause a capital loss taxed at 50% yielding a tax reduction of $1000.

The preferred alternative is to purchase the new, replacing the old. It is worth examining the results and doing some post-optimality analysis as suggested in earlier sections.

The cost history of the current vehicle is fairly well established but not that of the new vehicle. Suppose projected operating savings do not materialize as anticipated. How high would annual operating costs have to go before the advantage of the new machine is eliminated? There is an advantage of $9160 in PW over an 8-year period at 10%. For this period *P/A*, 15%, 8 yr = 4.487. Thus, an increase in ATCF of $2041 for 8 years would eliminate the advantage. Since taxes are at a 50% level, this means an increase is assumed before tax cost of $4082 must prevail to remove the advantage. Thus, any combination of annual cost for operation and maintenance of $4082 more than projected eliminates the advantage.

Before leaving this example it might be useful to focus on ingredients that make attractive investing an extra $10,500 (net) to retire a piece of equipment. First, the advent of an operating and maintenance cost saving of $6500/yr for 8 yr is attractive even if it is only $2500/yr. Of course, a $2500/yr saving for 8 yr would not normally justify investing $10,500 at any reasonable interest rate. It is made attractive because the old vehicle is constrained to a depreciation schedule far less attractive than currently allowed rapid writeoff at high rates in early years—a tax flow gain for the challenging new asset.

3.6.3 Other Special Considerations

The previous discussion has briefly covered the impact of taxes on economic analysis. There are some materials handling capital expenditures that arise having special features that require attention. These will be briefly discussed.

Major Systems—Computer Controlled

In major systems projects such as automated warehouses, miniload systems, or complex manufacturing systems with integrated handling and control, the total investment is the composite of many different kinds of system costs. There is a tendency to enter into a "turn-key" commitment wherein one supplier handles the entire system. The actual system may consist of conveyors, racks, transfer devices, pallet loaders, computers, software, construction, installation, and training. Conceivably the parts could have been acquired from a host of different sources by the customer and combined to form the system. Major systems suppliers have the advantage in the turn-key approach because their components fit together properly to form an integrated whole.

Turn-key quotes should always be "unbundled." That is, the total system cost should be broken down into the separate parts—building costs, equipment costs, labor, software development, training, and so on. Then, those items that can be expensed will be identified, along with those that must be capitalized. This approach takes full advantage of the tax laws.

For example, a multi-million dollar system will have:

Expense items—installation labor, training, tools, software

5-year equipment—cranes, racks, conveyors, and so on

15-year items—building

Failure to properly identify the different categories will result in a poor economic picture for the system.

Rack-Supported Warehouse

The rack-supported automated warehouse has come into common usage over the last few decades. In such warehouses, the building (or a major portion of the building) is merely a skin hung on the sides of the rack system, with the roof also supported by the racks. Although this practice will increase the loading of the racks above that are required to support the stored material, the increase is usually modest, in the neighborhood of 5–10%. Thus, it has only modest impact on rack cost.

The added cost is offset by some cost advantages. The design eliminates the cost of the redundant steel structure for sidewalls and roof. Erection cost savings exist for walls and roof, since they are quickly and easily added to the rack structure. The cost saving is typically in the range of $5 to $10 per square foot. This figure should be used judiciously since, in truth, cubic feet of space represents a more significant variable than square feet.

The biggest advantage comes not from the gains just cited, but from a tax advantage. Rack-supported structures can be treated as equipment rather than buildings for depreciation purposes. With ACRS, this means 5-yr property versus 15-yr property, since the bulk of what would be treated as a building with 15-yr life can be treated as equipment in a 5-yr property class. The early depreciation and consequent tax reduction is a significant advantage from a cash flow perspective.

3.7 POST-AUDIT PROCEDURES

Authors disagree as to the need for post-audit, the extent of post-audit, and the timing of post-audits. Those arguing against post-audit quite correctly point out that once a project has been approved and implemented, the calendar cannot be reversed and the decision negated if things are not working out.

While this is true, the argument for post-audit is not to specifically check on a given project but rather to glean as much information as possible about actual behavior versus predicted behavior, so that mistakes or bad assumptions can be corrected that might impact future decisions.

The focus should be an examination of all actual costs versus all predicted costs. In doing so, the emphasis should be on the form of costs that show the greatest disparity between actual and predicted. Are labor costs, maintenance costs, and operating costs close to predicted? Frequent errors in a given area should be scrutinized and necessary adjustment made.

Occasionally a post-audit will detect a problem that can be corrected. Perhaps a cost is far out of line, and investigation uncovers the fact that the system is being used in a different manner than intended, thereby causing unusual costs. For example, an approved stacker system was designed to handle pallets with gross weight under 2000 lb. The stackers have had high maintenance costs. Post-audit investigation uncovers the fact that product mix shift has created a situation where material receipts are frequently for pallets over 2000 lb. Consideration can now be given to alternatives for handling the problem such as stacker replacement, changes in order practice, or possibly, accepting the high maintenance cost.

The timing for performing post-audits is also important. Most material handling systems installed are expected to be in service for many years. However, waiting too long before performing a post-audit virtually guarantees a difference between actual and predicted behavior. So many changes occur over time that it is virtually certain that some aspect is different. Waiting too long will make determination of why cost variations occur impossible to establish.

On the other hand, early post-audit will also be frustrating. In complex systems, "steady state" results will not be immediate. A reasonable period of maturation is needed. For most material handling systems, post-audit is the most valuable when performed about 2 years after implementation.

Finally, on the topic of post-audit, the question arises as to the need to audit all projects. Again, views differ, but most firms either audit all or none. If only certain projects are to be audited, the decision as to which becomes difficult. Furthermore, there is a tendency to ignore those known to be misbehaving. Of course, these are precisely the ones that should be audited.

Some firms audit all projects annually for a reasonable period beyond implementation, perhaps 2–5 yr. This can be very effective and useful and may go beyond post-audit. It may, in fact, be a form of feedback control in the sense that deviations from expectation are examined and corrected if possible.

One final caution is in order with respect to post-audit. The firm should examine the time devoted to post-audit. Post-audits do look back in time. The principal thrust of key executives should be forward in time, not backward. Only when the backward look is constructive with regard to the future is anything positive going to arise. A systematic method of performing post-audit is helpful.

3.8 POST-OPTIMALITY ANALYSIS—SENSITIVITY

Most economic analysis is performed by a prescribed, step-by-step procedure. One crucial step prior to the decision and implementation is sensitivity analysis. It is the process of seeking answers to many "what if" questions.

These "what if" questions are posed against cash flows, directly or indirectly. They are of the form:

What if labor rates rise dramatically?

What if system performance is below expectation?

What if volume of activity rises (falls) substantially above (below) the predicted level?

What if material costs rise?

Asking such questions requires some formal procedure or mechanism to permit proper analysis. One fairly simple, yet very practical, approach will be described. For each cash flow, rather than assigning a specific value as is often done, three estimates, a pessimistic, a "most likely," and an optimistic, are assigned.

The most likely is probably that which was used in the initial analysis—that which was assumed to apply. The pessimistic is that which would arise when things go badly—the worst case, initial cost of installation or worst case operational behavior. Optimistic estimates are those which would arise if everything were to go right.

Performing analysis on the basis of comparing the outcome with various combinations of cash flow estimates can provide insight as to what combinations make the proposal attractive.

It should be pointed out that even this simplistic approach is not without problems. Given four cash flows—an initial cost, an annual operating cost, an annual maintenance cost, and a salvage value—there are 81 (3^4) combinations yielding a different cash flow profile.

However, some cash flows may be known with a high degree of certainty. Perhaps initial cost is based primarily upon a quote that is contractual. Thus it becomes fixed. Cash flows far in the future may be relatively small. Salvage, for example, may be a small fraction of cash flows in total.

Cash flows in the future, when discounted to present value over many years, have only a slight impact on the decision. Thus, judiciously focusing on only those factors with major impact and the greatest potential for error reduces the amount of analysis required to manageable levels.

In general, it will be wise to focus on those parameters that most heavily influence the acceptability of a project. Generally the decision is heavily influenced by:

1. Initial installed cost
2. Operating costs
3. Maintenance costs (in complex systems)

Generally the following parameters have only modest influence on acceptability:

1. Salvage values
2. Life (if minimum expected life is over 4 yr)

REFERENCES

1. Riggs, James, Inflation, Taxes, and Other Economic Frustrations in the Justification of Materials Handling Investments, *Proceedings, AIIE/MHI Seminar,* 1980, The Material Handling Institute, Inc., Pittsburgh, 1979.
2. White, J., K. Case, and M. Agee, *Principles of Engineering Economic Analysis,* Wiley, New York, 1984.
3. Stevens, G. T., *Economic and Financial Analysis of Capital Investments,* Wiley, New York, 1979.

BIBLIOGRAPHY

Apple, James A., *Material Handling Systems Design,* Ronald Press, New York, 1972.

Apple, James A., *Plant Layout and Material Handling,* 3rd Ed., Ronald Press, New York, 1977.

Basics of Material Handling, The Material Handling Institute, Inc., Pittsburgh, PA, 1981.

Canada, J., and J. White, *Capital Investment Decision Analyses for Management & Engineering,* Prentice-Hall, Englewood Cliffs, N.J., 1980.

Collier, C., and W. Ledbetter, *Engineering Costs Analysis,* Harper & Row, New York, 1982.

Grant, E., W. G. Ireson, and R. Leavenworth, *Principles of Engineering Economy,* 7th Ed., Wiley, New York, 1982.

Lesson Guide Outline, The Material Handling Institute, Inc., Pittsburgh, PA, 1983.

Riggs, J., *Engineering Economics,* 2nd Ed., McGraw Hill, New York, 1982.

Riggs, J., *Essentials of Engineering Economics,* McGraw Hill, New York, 1977.

Tompkins, James, How to Solve Material Handling Problems, *Proceedings, AIIE/MHI Seminar,* The Material Handling Institute, Inc., Pittsburgh, PA, 1978.

White, John A., How to Identify Material Handling Problems, *Proceedings, AIIE/MHI Seminar,* The Material Handling Institute, Inc., Pittsburgh, PA, 1980.

PART II
UNIT MATERIALS HANDLING

CHAPTER 4
DEFINITIONS AND CLASSIFICATIONS

KISHAN BAGADIA

Unik Associates
Milwaukee, Wisconsin

4.1 UNIT LOAD CONCEPT

The unit load principle implies that materials should be handled in the most efficient, maximum size unit, using mechanical means to reduce the number of moves needed for a given amount of material.

The many facets of the broad field of unit materials handling are covered in this section, which comprises Chapters 4 through 18. Each of the chapters and its subsections deal with a major subject within this field.

Unit materials handling is concerned with the handling of discrete, individual items or loads, large or small. Thus, it involves the handling of such diverse items as car bodies, pallet loads of bags, cartons, packages, bottles, bricks, shoes, nuts and bolts, subassemblies, and airplane wings. This type of handling is distinguished from that involving bulk materials such as powders, granules, and liquids, which are stored and handled in volume, often in unpackaged form.

Although the unit materials just mentioned may be quite different from one another, their handling often follows certain basic principles that are applicable in many situations. The following chapters discuss these principles, and the various categories of equipment used to put them into practice.

This introductory chapter provides a capsule overview of the entire field of unit materials handling, and a classification of various types of handling equipment and systems. As such, it serves as a basic outline of the contents of Chapters 5 through 18, and follows the sequence in which various topics are treated in subsequent chapters.

Types of unit loads are discussed next.

4.1.1 Platform

Skid

The skid is usually a nonstackable, single-faced device with only two horizontal runners or stringers serving as supports for the deck surface, and elevated above the floor to allow lifting by a platform truck. A skid may be corrugated metal or a combination of wood and metal. A skid box may be a metal panel, corrugated metal, or wood.

Pallet

A pallet is a horizontal platform device used as a base for assembly, storing, and handling materials in a unit load. There are several types of pallets:

Corrugated paper—expendable, low cost. Usually used in shipping.

Wood—an economical, reusable pallet. Often can be repaired readily, at about two-thirds the cost of new pallet.

Plywood—when continuous, uniform support is desired under the load. Relatively light for ease of handling.

Plastic—when product protection and uniform tare weight are desired. Can be steam cleaned for sanitary applications.

Metal—for heavy-duty applications and fixturing of work, and when precise weighing is important.

4.1.2 Sheet

The slipsheet is ideal for use in distribution systems, when it is desirable to keep pallets captive to the plant because of their higher cost. Uses space efficiently in trailers. Little cube space is required for storage when not in use. There are various types of slipsheets:

Corrugated paperboard—primarily for one-way trips. Degraded by moisture or temperature and humidity variations.

Fiberboard—more durable than corrugated, can last for several trips.

Plastic—very durable, resistant to temperature and humidity changes. Can be used for up to 20 trips, recyclable.

Molded—specially formed, leaving hollow spaces for fork entry.

Flexible—used as a "sling" between the forks, particularly for materials packed in bags.

4.1.3 Rack

A rack is usually specially designed to hold parts in a desired position, often equipped with inserts, pegs, or holes to orient parts.

4.1.4 Container

Parts are unitized with a container of specific configuration.

4.1.5 Self-contained Unit Load

Stretch Wrapping. Tight, strong securement of loads, usually on pallets or slipsheets. Application is easy and relatively low in cost. Load is completely covered. Various stretch films used, depending on application:

Hand held—for small loads or light-duty, infrequent operations.

Walk around—a low-cost, basic system for low-volume wrapping.

Mechanized walk around (robot)—eliminates operator effort for somewhat higher-volume jobs.

Mechanized platform—higher-volume applications and large loads. Depending on type of system and degree of mechanization and automation, volume can be from 20 to 100 loads per hour.

Full web—for relatively uniform, constant-height loads. Equipment ranges from simple mechanized platforms to automated systems. Film width covers entire load surface with each revolution.

Spiral—for handling various sized loads. Roll of film is shorter than load height, and wrap is applied spiral fashion. High wrap tension is obtained.

Overwrap—wrapping is applied over top of load, instead of around side as with foregoing systems. Often used on rolled goods, bags, sheets. Full-web, spiral, and multiple-banding methods can be used.

Shrink Wrapping. Provides total load encapsulation for high product protection, suited for outdoor storage. Handles heavy and irregular shifting loads. Shrink heating process required. Applications:

Hand-held gun—small loads, low volumes.

System with pedestal, dolly, or conveyor base—heavy or large loads, irregular configurations. Range of throughput rates available.

Strapping. Securing of pallet loads and other unit loads to prevent shifting during handling and shipment. Strapping materials can be steel (high strength, no-stretch applications), polypropylene (low cost, for light, medium duty), nylon (high tension), polyester (low elongation, good for stable or expanding loads). Applications:

Hand tools—for low volume, or when worker must take tool to the load.

Two-piece (manual)—tensioning and sealing operations performed separately.

Combination (manual)—tensioning and sealing performed by single unit.

Combination (pneumatic)—faster operation, less consistent tension than with preceding.

Hand tools with predrape equipment—various methods for positioning or feeding strapping around load. Tensioning and sealing still done by hand.

Power strapping machines (semiautomatic, automatic, operatorless)—high productivity, more uniformly and securely strapped loads, and reduced labor needs. For carton closing, bundling, and unitizing. Some machines can apply 60 straps per minute.

4.1.6 Palletless Handling

Another development in unit load handling includes no pallet or other support in the final unit load. This method commonly makes use of a clamp-type attachment on a forklift truck. Unit loads of this type are most commonly made up of single items, such as rolls, bales, or cartons, bags, and so on, arranged in an optimum pattern so they can be picked up, moved, and stacked using the squeezing action of the clamp attachment.

4.2 INDUSTRIAL HAND TRUCKS

Applications for hand trucks can be found in virtually any environment, commercial and even industrial. Hand trucks are suitable if:

Travel distance is relatively short.

Volume and frequency of moves are low.

Physical limitations prohibit use of other types of equipment (e.g., narrow doors, stairs, ramps).

Minimum maintenance requirements and low cost are desirable.

Used in warehousing, manufacturing, shipping, and distribution. Types of industrial hand trucks:

Dolly. An inexpensive, small, low platform-type load carrier without handle and with [illegible] or more rollers, casters, or wheels. Applications: Usually 2000–3000 lb capacity, short travel distances, low volume.

Two-Wheel Hand Truck. A simple definition would be a lever with wheels used for moving material too heavy or awkward to be carried. Most commonly and widely used. Capacity up to 1000 lb.

Four-Wheel Hand Truck. A rectangular load-carrying platform with four or six wheels, for manual pushing, usually by means of a handle or a rack at one or both ends. Applications: Can handle up to 4000 lb, low volume, short travel distances.

Hand Lift Truck. A platform-type or fork-type wheeled platform equipped with a lifting device to raise loads just high enough to clear the floor to enable moving the load. Hand-propelled, mechanically or hydraulically lifted. Platform type is used for handling skids, and fork type for handling pallets. Applications: 2500–5000 lb capacity, short travel distance, low volume.

Power Lift Stacker. Light to intermediate duty, short travel distance. Capacity 1000–2000 lb.

4.3 POWERED INDUSTRIAL TRUCKS

4.3.1 Walkie Trucks

The term *walkie trucks* is applied to power-operated trucks but with operator walking and operating the truck by means of controls on the handle. There are several types:

Low Lift. This type is similar to the nonpowered hydraulic hand pallet truck. The tips of the forks are tapered for easy pallet entry. Applications: 3000–5000 lb. If no stacking is required, the low lift is best suited for horizontal moves.

Straddle Walkies. Straddle walkies have structural extensions (or straddles) mounted parallel to and outside of the lift forks. Applications: 2000–4000 lb. Best suited for narrow-aisle operations.

The straddle–reach truck is a variation of straddle truck in which the forks reach out for the load. Forks travel forward to engage the load, lift it, and then retract it to the mast for traveling.

Counterbalanced. These trucks balance load and carriage by adding length and weight to the truck body. Application: Where various pallet sizes are encountered.

4.3.2 Rider Trucks

Electric Counterbalanced. Clean, quiet operation indoors and in some outdoor service when equipped with pneumatic tires.

Three-Wheel Sitdown. (Capacity 2000–4000 lb.) Combines maneuverability in tight quarters with flexibility of counterbalanced operation.

Four-Wheel Standup. (Capacity 2000–4000 lb.) For continuous, demanding duty cycles when ability to ride increases operator productivity and comfort. Standup design reduces necessary truck length.

Four-Wheel Sitdown. (Capacity 200–12,000 lb.) For heavy-duty, continuous work cycles when sitdown mode reduces operator fatigue and promotes efficient operation. Suited for loading and unloading trailer trucks and rail cars.

Electric Narrow Aisle. Permits reduction of costly aisle space and more efficient use of building cube space. Many versions and models are available. Basic categories are listed.

Straddle. (Capacity 2000–4000 lb.) Quick handling of single-size pallet loads. Base legs straddle the bottom load. Loads usually stacked directly above one another.

Reach. (Capacity 2000–6000 lb.) Has pantograph mechanism that extends forks beyond base legs; various pallet sizes and rack openings can be accommodated. Deep-reach versions to permit two-deep stacking and eliminate aisles are available.

Order Selector. (Capacity 2500 lb.) Quick selection of items to fill orders. Operator rides platform to stock height.

Heavy-duty Order Selector. (Capacity 2000 lb.) For bulky items such as furniture. Operator rides platform.

Sideloader. (Capacity 6000 lb.) For long, difficult-to-handle items. Truck operates in aisles only $2\frac{1}{2}$ ft wider than load. Load length is parallel to aisle length during travel.

Heavy-duty Sideloader. (Capacity 10,000 lb.) Heavy-duty truck for transporting heavy materials, such as mill loads, and for stacking from the side.

IC Cushion Tire. (Capacity 2,000–13,500 lb.) General-purpose handling for indoor, dockside, and lighter outdoor jobs. Suited for loading and unloading trailer trucks and rail cars.

IC Pneumatic Tire. (Capacity 2,000–10,000 lb.) For both inplant and yard operations, including travel on semi-improved surfaces, ramps, and long-distance runs.

IC Pneumatic Tire, Heavy Load. (Capacity 11,000–30,000 lb.) Heavy-duty operations including outdoor handling of lumber, steel, and concrete products. Suited for high-duty cycles on uneven yard surfaces.

IC Rough-Terrain Truck. (Capacity 4000–8000 lb.) Rugged operation at construction sites and rough yards where extra traction is needed. Some models are towed to the worksite.

Rider Type. For handling individual wheeled loads or load trains over long distances.

Electric Cushion Tire. (Capacity 1000–4000 lb.) Transport over relatively smooth surfaces when quiet, fume-free operation is desired.

Electric Pneumatic Tire. (Capacity 3000 lb.) When electric operation is desired on rougher surfaces requiring pneumatic tires.

IC Pneumatic Tire. (Capacity 3,000–10,000 lb.) Variety of towing jobs over various quality surfaces.

4.3.3 Tractor-Trailer Train

The tractor-trailer train is a handling system consisting of a three- or four-wheeled, self-propelled vehicle designed for pulling loaded carts or trailers. Common versions are the rider type, walkie type, and the electronically guided type. Applications: Low-cost movement of large quantities. Warehousing, receiving and shipping, order picking, collecting, and delivering loads to a number of specific locations.

4.3.4 Personnel/Burden Carriers

These carriers are used for efficient movement of personnel and goods in the plant. Types:

One or two riders—load capacity up to 300 lb. Primarily used for personnel such as supervisors or stock chasers, and simple loads.

Multiple riders or cargo—load capacity 1000–6000 lb. Used for transporting several riders or supplementing lift truck operation by providing load transportation over long runs.

4.3.5 Industrial Crane Trucks

The crane truck is a self-propelled, wheeled vehicle designed to accommodate a crane for lifting and carrying objects. It can be gas, diesel, LP, or battery powered. Applications: Loads up to 10,000 lb. Lifting, loading positioning, maneuvering loads. Used for maintenance and yard work, short moves, and handling awkward shapes with slings, chains, and so on.

4.4 AUTOMATIC GUIDED VEHICLE

The automatic guided vehicle is a driverless battery-powered vehicle that follows guide wires imbedded in the floor. The guided path system can be easily modified and expanded. It has programming capabilities for path selection and positioning. Applications: Material movement from one point to another. Types of automated guided vehicle system:

1. Automatic horizontal transportation
   a. Towing vehicles (driverless tractors)
   b. Unit load carriers
   c. Guided pallet trucks
   d. Light load transporters
2. Automatically positioned stock selectors

4.5 CONVEYORS

4.5.1 Package Handling Conveyors

Gravity Conveyors. The nonpowered conveyor is the most basic and least costly method of moving objects between points. Material can be moved forward at a straight decline, down a spiral path, or, with slight manual assistance, horizontally. Applications: Individual items, packages, containers, cloth and paper bags, wood or cardboard cases, slat bottom, wire-bound, or shaped hardwood bundles, metal tote boxes. Load capacity up to several tons. Types of gravity conveyors:

Spiral chutes—can handle individual items up to 250 lb. Double or triple runway spirals can be used for sorting and delivering products to different levels, as well as conveying.

Gravity roller—constructed of metal tubing with bearing pressed in each end with the roller mounted on a hex axle supported in either a channel or angle frame.

Gravity wheel—conveyor is constructed of a series of wheels mounted on a common axle supported in a channel frame.

Powered Package Handling. Conveyors are used for transporting over long distances and along inclines, and for moving heavy materials that cannot be handled on gravity equipment without substantial manual assistance. Powered conveyors also are used for paced, controlled operations and are suited for mechanized and automated handling systems, both on the plant floor and in storage areas. Types of powered conveyors:

Live roller conveyor—moves packages horizontally and up to 5–7° slopes. It can be used for transporting heavy unit loads and items with sharp corners and irregular shapes, and for intermittent loading and unloading operations.

Belt conveyor—provides complete support under the package being moved. Normally used for transportation of light and medium weight loads. Particularly useful when an incline or decline is required in the conveyor path. It can move items of unusual shapes and configurations that cannot be handled on wheel or roller conveyors.

Applications of powered conveyors:

Chain conveyor—primarily used for transportation of heavy products.

Single-strand chain conveyor—for conveying dollies or pallets. Loads are supported on rollers, wheels, or slide plates.

Two-strand conveyor—used in process operations, such as dipping of loads into tanks. Loads are supported on crossbars or in chain pockets.

Multiple-strand conveyor—handles heavy unit loads or parallel lanes of chain.

Slat conveyor—handles heavy loads with abrasive surfaces. Has load-supporting slats attached to the chain.

Roller flight conveyor—uses rollers instead of slats. Used for accumulation of heavy loads and in rough service applications.

Sorter conveyor—uses tilting slats for diverting loads along flow path.

Pusher bar type chain—conveyor does not support load, but propels it along slider or roller bed.

4.5.2 Overhead Trolley Conveyors

These conveyors generally consist of chain or cable suspended from trolleys running in a structural I-beam or other specially designed track. The chain has a fixed path, but can negotiate horizontal and vertical changes easily. A wide range of capacities is available to handle loads up to several thousand pounds. Unit loads are usually carried on trays, hooks, baskets, or specially designed suspension means. Applications: Overhead trolley systems are generally used for fixed path, paced flow, operation for transportation, assembly work, processing, live storage, and automatic load/unload systems.

4.5.3 Power-and-Free System

The power-and-free system consists of two major construction features. The upper power portion is similar to the overhead conveyor. Chain section is equipped with dogs which engage and push the free trolley below the chain. The lower portion consists of a four-wheeled trolley, which carries the load. This runs on two structural angles, channels, or other specially designed free trolley track. There is no positive fixed, pinned, or bolted connection between the power pusher and the free trolley. Horizontal and vertical changes in path are negotiated easily. The free trolley can be switched into alternate paths and drop or lift sections can be used. Controls permit dispatching to various destinations. Variable load spacing and accumulation are easily accomplished.

Applications: This type is most frequently applied to production line work, assembly, process operations, segregated storage and dispatching systems, temporary accumulation, and variable load spacing. It is also used in applications requiring hand-pushed travel, powered travel, stops on line, and lift sections for vertical elevation changes.

4.5.4 Tow Conveyor or In-Floor Towline

The unit consists of a drop forged chain suspended by wheeled trolleys running on flanges of two opposing structural members. Cover bars, which are flush with the surrounding floor, form a slot $1-1\frac{1}{2}$ in. wide. Entire track is imbedded in concrete floor. Pusher dogs on 10–20 ft centers engage tow pin on cart.

Applications: Tow conveyors are generally used to convey four-wheeled carts through warehouse or freight terminals. There are special applications for production line work. Loads up to 3000 lb are easily handled, with heavier loads possible. Speeds range from 80 to 120 ft/min resulting in movement of tremendous tonnages per hour at a paced flow over long distances. Switching system can be furnished to add flexibility. Control device can be added for dispatching to spurs, or to transfer any accumulation areas.

4.5.5 Overhead Tow Conveyors

These units are basically the same as overhead trolley conveyors except that special attachments are used to tow four-wheeled carts running on the floor. They are flexible in horizontal path. However, caution and special provisions are required when inclined paths are involved. Applications: Uses are similar to those for a tow conveyor. Switching systems can be used when cart has a tow mast.

4.5.6 Unit Load and Engineered Systems

These are similar in design to packaged roller or chain conveyors, but are capable of handling heavy loads. The construction is heavy-duty throughout.

4.5.7 Vertical Conveyor

The vertical conveyor transfers packages, totes, boxes, or unit loads between various levels in operations with intermittent material flow. The unit consists of an upright column structure within which a series of powered, guided carriers travel up and down. Products can be received at one level and discharged at another. Operation may be manual, push button, or fully automatic.

4.5.8 Scrap and Chip Handling Conveyors

Scrap handling is basically removal of chips, curls, and flakes from the production area. Types of metal scrap:

Bushy steel scrap
Broken steel chips
Bushy aluminum chips
Broken aluminum chips
Bushy brass scrap
Fine brass scrap
Cast iron scrap
Stamping scrap
Die-casting scrap

Types of metal scrap handling conveyors:

- Harpoon
- Drag
- Tubular
- Oscillating
- Hydraulic/sluice
- Screw
- Hinge belt
- Compacveyor
- Magnetic
- Pneumatic
- Hydraulic grab
- Troughing belt
- Dumper/lifter
- Bucket elevator

4.6 CRANES, HOISTS, AND MONORAILS

4.6.1 Bridge Crane

A bridge crane is a lifting device mounted on a bridge consisting of one or two horizontal girders, which are supported at each end by trucks riding on runways installed at right angles to the bridge. Runways are installed on building columns, overhead trusses, or frames. Lifting device moves along bridge while bridge moves along runway. Provides full coverage of working area or bay.

Types: Top running (end trucks ride on top of runway tracks) and underhung (end trucks are suspended from lower flanges of runway tracks).

Applications: Machine shops, foundries, steel mills, heavy assembly, and repair shops; intermittent moves; warehousing and yard storage; can handle wide range of loads with attachments.

4.6.2 Gantry Crane

The gantry crane is commonly of bridge type, rigidly fixed at one or both ends to supporting columns or legs. Wheels on end trucks roll on track. It can be built to handle very large loads. Types of gantry cranes:

- Single leg, single girder
- Single leg, double girder
- Double leg, single girder
- Double leg, double girder

4.6.3 Jib Crane

A jib crane is a lifting device traveling on a horizontal boom which is mounted on a column or mast, which is fastened to the floor, the floor and a top support, or a wall bracket or rails. It is relatively inexpensive. Generally, used for serving individual work areas in machine shops, and so on, within its radius.

4.6.4 Monorail

A monorail system is one on which loads are suspended from wheeled carriers or trolleys that are readily rolled along an overhead track. The carrier wheels usually roll along the top surface of the lower flange of the rail forming the track or in a similar fashion with other track shapes.

A monorail is used for fixed-path, point-to-point handling; low-volume moves; handling through processes (paint, bake, dry, test, etc.); intermittent handling tasks, and so on.

4.6.5 Stacker Crane

A device with a rigid upright mast or supports, suspended from a carriage, mounted on an overhead traveling crane (or equivalent) and fitted with forks or a platform to permit it to take place in or retrieve items from racks on either side of the aisle it traverses.

This crane is used for handling unit or containerized bulk loads, storage and warehousing operations, adaptable to "automatic" warehousing operations, with attachments can handle a wide variety of loads, excellent for long loads (metal bars, shapes, sheets, pipes, tubes, etc.). Types of stacker cranes:

Top running single girder
Underhung single girder
Underhung double girder

4.6.6 Hoist

A hoist is a device for lifting or lowering objects suspended from a hook on the end of retractable chains or cables. It is usually supported from overhead by a hook or it travels on a track. Applications: Serves a machine or work area. Types of hoists:

Hand hoist (chain and ratchet)
Power hoist (electric wire rope, electric chain, air wire rope, and air chain)

Types of below-the-hook lifting devices:

Nylon and polyester slings
Wire rope slings
Chain slings
Wire mesh slings
Coil hook
Pallet lifter
Sheet lifter
Battery lifting beams
Cargo tie-downs
Special lifting devices

4.7 BASIC STORAGE EQUIPMENT AND SYSTEMS

Storage of a unit load involves holding the material efficiently until it is needed in the plant or is to be shipped to another facility.

4.7.1 Storage Racks

The pallet rack is the most familiar type of storage rack. In its form it is a frame structure designed to allow individual pallet loads to be stored and retrieved. It can be modified to handle nonpalletized loads. There are several types of pallet racks.

Standard. Load readily accessible from either of two directions.

Two-Deep. Aisle space is conserved by placing two standard sections back to back. Two-deep retrieval can be effected by a truck with reach mechanism. Continuous pallet support front to back.

Drive-in. Pallets densely loaded in front-to-back or side-to-side rack lanes. Lift trucks drive in for last-in, first-out retrieval. Good for identical nonspoiling goods.

Cantilever. For storing long, narrow items such as pipe and tubing, or larger long loads such as furniture. Usually not economical for unit loads.

Pallet Flow Rack. First-in, first-out stock rotation in high-density storage. Pallets move on rollers in lanes, from rear to front. Lanes sloped for gravity flow. Motor power can be supplied if necessary.

4.7.2 Stacking Frames

Upright frames can be fitted to pallets permitting unit-load stacking without use of separate racks.

Steel Pallet. Heavy-duty applications can be used as part of distribution system, as well as storage.

Wood Pallet. Same function as steel pallet system, typically for lighter-duty jobs.

4.7.3 Parts and Package Storage

The most common type of shelving is the familiar sections consisting of four steel uprights and a number of adjustable steel shelves. Hand loaded. Accessories include shelf boxes, inserts, and dividers.

Flow Shelving. First-in, first-out control of totes and case goods. Typically for fast-moving items.

Modular Drawers. Modular drawer storage is a relatively new concept, based on the use of standard drawer modules that can be altered or adapted as required to concentrate the storage of tools and nonbulk inventories systematically and safely within human reach and sight. Only the size of the items to be stored dictates the size of the drawer and cabinet. This flexibility is the key to the usefulness of the drawer-storage concept.

Mezzanine. Makes use of two or three tiers of shelving separated by a specially constructed floor. Its main advantage is doubling or tripling the amount of standard height steel shelving that can be placed on a given amount of floor space.

4.8 STORAGE SYSTEMS

A high-rise storage system is served by:

Guide Truck. Stacking heights to 40 ft. Narrow-aisle, high-lift or turret truck can be under rail or wire guidance. Somewhat lighter tolerances than on conventional selective pallet racks.

Storage/Retrieval (S/R) Machine. High degree of inventory control in large-volume, high-throughput applications. Operations frequently automated under computer control. Exacting tolerances on racks and floors. Storage heights can reach 75 ft or more.

High-Density. Loads confined to specific lanes under automated control. Traveling carrier or retrieval mechanism removes loads on first-in, first-out basis. High use of cube space through dense storage of like items. High-rise configurations can be used.

Operator-to-Part. Operator goes to where materials are stored to retrieve them. Mechanized or automated equipment may be used.

For High Rise

Manual Equipment. From 10 to 15 ft high storage served by manually pushed rolling ladders or stock-picking trucks. Relatively low number of line items and low throughput.

Powered Order-Picker (Operator Steered). From 10 to 20 ft. Generally feasible when 500 items are stocked and inventory turns five or more times a year. About 50–70 picks per hour.

Aisle-to-Aisle Vehicle. Storage heights to 40 ft. Operator-aboard vehicle usually under rail or wire guidance. About 60–80 picks per hour.

Captive-Aisle Vehicle. Storage heights to 40 ft. Vehicle dedicated to one aisle for high-speed, high-volume applications. From 90 to 125 picks per hour, or more, can be achieved.

Part-to-Operator. Materials are brought to operator in mechanized fashion.

Carousel. Parts are contained in baskets, tubs, or shelves suspended in revolving carousels. Systems save aisle space, provide rapid access to parts.

For Maintenance and General Stores

Miniload. Enclosed structure houses coded parts bins that are retrieved by automated S/R machine and brought to operator at work station. High throughput, high security, suited to automation.

Light Duty. Individual bin capacities range to 200 lb. Operation typically directed by operator through keyboard entry terminal.

Minimum Duty. Bin capacities to 500 lb. S/R machine typically equipped with on-board microprocessor. Operation often tied to minicomputer.

Heavy Duty. Bin capacities range to 700 lb.

Rack-Supported Building. Due to the fact that S/R systems achieve a density that generally does not permit other operations functioning in the same area, the natural progression was to take the rack structure and utilize it to support the building, thus eliminating all of the building's heavy structural steel. Rack-supported buildings are gaining widespread acceptance.

4.9 CONTAINERS

4.9.1 In-Plant Containers

These are used for handling parts and unit loads within the plant. The containers are used over and over again. Types:

Metal shop pans—for handling heavy small parts. Expanded metal or perforated metal designs available for dipping or draining applications.

Hopper-front storage bins—organizing and storing small- to medium-sized items, maintenance parts. Can be mounted on floors, shelves, or racks.

Modular containers (plastic)—handling small parts, assemblies, and components. Good for organizing into families or groupings.

General-purpose nesting, stacking containers—general carrying of various products. Some designed for product protection, others to save space in storage.

Tote boxes—industrial-grade box for small castings, stampings, large quantities of small parts with high total weight.

Collapsible wire containers—can contain large, heavy, or irregularly shaped items that are awkward to handle, such as foundry castings. High product visibility. Compact storage. Maximum stack of 3 or 4 high.

Rigid wire containers—heavy-duty units for high stacking, use with order-picking vehicles. Loads to 6000 lb.

Corrugated steel containers heavy-duty handling of castings, forgings, stampings, fasteners, and other items. Can be used with tilting stands or dumping attachments.

Wood boxes—commonly used for textiles or soft goods, or when high impact resistance of wood is desirable for parts handling. Readily repaired.

Wire-bound boxes—handling and shipping of components, assemblies, and implements. Can be readily built or repaired in the field.

4.9.2 Shipping Containers

Shipping containers are used to handle parts and unit loads from one facility to another. These containers may or may not be used more than once. Types:

Cardboard boxes
Corrugated containers
Metal containers
Plastic containers
Solid fiber containers
Wire-bound containers
Wire-rod containers
Wooden containers
Wooden crates
Fiber drums
Plastic bags
Multiwall paper bags

4.10 FREIGHT ELEVATORS

Elevators are used for moving products or personnel from one level to another. They are used in multilevel plants to fill vertical handling needs. There are several types of freight elevators: Light duty, general purpose, and power truck.

Light Duty. Capacity up to 2500 lb, rises to 35 ft. May have drum, traction, or plunger electric driving machines.

General Purpose. Capacity 2,500–10,000 lb. Traction or plunger driving machines.

Power Trucks. Capacity 10,000–20,000 lb. Facilitates fork truck handling.

Systems with automated freight elevators interfacing with powered conveyors also have applications.

4.11 POSITIONING EQUIPMENT

4.11.1 Lift Tables

The lift table is a valuable tool as a positioning device. Although its primary function is the vertical move, it can perform other operations in conjunction with fork trucks, conveyors, hoists, or other handling equipment. Types of lift tables:

Screw-jack lift
Scissor lift
Cylinder (vertical post) lift

4.11.2 Force-Balance Manipulators

A manipulator allows the operator to handle heavy, awkward, or fragile objects easily, efficiently, and quickly within a limited space. Grabbing devices are usually custom made.

Applications: In heavy-duty manufacturing (castings, sand cores, machined parts, engines, bumpers, etc.); components; finished goods such as heavy steel coils, carpeting, and insulation. Types of manipulators:

Balancing hoists
Articulated jib cranes
Triaxial manipulators

4.11.3 Power Dumpers

These are used for lifting and dumping of various containers such as drums and boxes to make the operation fast and efficient. They can be portable or stationary; automatic, semiautomatic, or manually controlled; and can be modified or specially designed to handle a wide variety of functions.

4.11.4 Die Handling

Live storage rack is equipped with a series of easy-moving wheels
Overhead lifting device
Retractable mast truck used as an elevator in the storing of heavy dies
Track-mounted elevating travel vehicle (cart)

4.11.5 Industrial Robots

An industrial robot is a programmable manipulator capable of performing a variety of work tasks such as spotwelding, tool handling, painting, or parts transfer. It is versatile and flexible and can be retooled and reprogrammed easily to perform a new task when the old task changes or goes away. Robots come in a variety of sizes and shapes with physical capabilities covering a wide range.

Applications: Materials handling, assembly, welding, machine loading and unloading, and painting.

4.12 DOCK OPERATIONS AND EQUIPMENT

Dock Board. A specially designed platform device to bridge the gap between the edge of the dock and carrier floor and to compensate for variations in truck bed heights. They are also known as bridge plates. Types: Pit mounted, self-standing, self-forming, and step-down.

Ramp. A portable device for placement at the door of a carrier or building and for bridging the vertical distance to the ground level with a sloping runway. Used for providing easy access to carrier without a raised dock and providing access to a dock from ground level.

Dockleveler. A platformlike device built into the dock surface with capability of raising and lowering to accommodate truck height when bridging the gap between dock and truck floor. Types:

Mechanical–Manual Lift—(Capacity 20,000–30,000 lb.) Lighter-duty operations. Counterbalanced by springs that offset most of dockleveler weight. Operator lifts unit, then walks ramp down; lip extends automatically.

Activated Release—(Capacity 20,000–60,000 lb.) Commonly used type, counterbalanced by powerful spring. Unit rises to highest position when operator pulls release lever; operator then walks ramp down.

Hydraulic—(Capacity 20,000–100,000 lb.) Dockleveler is raised by a hydraulic cylinder activated by pushbutton. It automatically stops upon making contact with truck, then returns to flush, dock-level position—with lip retracted—when truck leaves.

Truck Actuated—(Capacity 25,000–100,000 lb.) Automatic actuation without operator. For heavy-duty operation, high-traffic volume, low number of dock attendants.

Rail.—(Capacity to 60,000 lb.) For rail-car loading, unloading. Operated hydraulically.

Dock Lifts. An elevatorlike device designed to lift and lower loads between ground level and carrier floor level when a dock is not convenient or available. May be portable or built-in at door, rail siding, and so on.

Door Protection. Strategically located cement-filled posts, constructed behind overhead door rails, will help eliminate much of the damage done to the rails by forklifts.

Dock Lights. Wall-mounted lights that swing into position are relatively inexpensive and enable forklift operators to see more clearly when entering and leaving the truck, thus reducing safety and damage hazards. For open docks and inside docks, lights can be mounted from the ceiling or canopy if a structural member is not close by.

Weatherseal. Flush docks, equipped with docklevelers, should have weatherseals applied to the full length and rear width of the dockleveler. This will minimize the amount of cold or heat entering the building through the gaps between the dockleveler and the dock. Foal insulation beneath the dockleveler ramp is effective in reducing compensation in frozen food operations.

Dock Bumpers. Used to protect docks from backing trucks.

Strip Doors. Transparent curtains of overlapping PVC strips which hang straight, close snugly to block heat losses, and shut out noise, dust, and fumes.

Dock Seal. The truck seals itself against the dock seal as it backs against the dock. The seal also:

Absorbs dynamic force for the backing truck.

Prevents freight-damaging rain runoff on loading end.

Dock shelters serve similar purpose as the dock seal.

BIBLIOGRAPHY

Apple, J. M., *Material Handling Systems Design,* Ronald Press, New York, 1972.

Basic Conveyor Selection Guidelines, Plant Engineering Library, Technical Publishing, a company of The Dun & Bradstreet Corp., Barrington, IL, 1976.

Considerations for Planning and Installing Automatic Guided Vehicle Systems, The Material Handling Institute Inc., Pittsburgh, PA, 1980.

Kulwiec, R., Conveyors, *Plant Engineering,* Sept. 3, 1981.

Lift Trucks, Plant Engineering Library, Technical Publishing, a company of The Dun & Bradstreet Corp., Barrington, IL, 1978.

Material Handling Engineering Handbook and Directory, Penton/IPC, Inc., Cleveland, OH, 1979/80.

Modern Materials Handling, Casebook Directory, Cahners Pub. Co., Boston, MA, 1981.

Modern Materials Handling, Casebook Directory, Cahners Pub. Co., Boston, MA, 1982.

Overhead Cranes and Hoists, Plant Engineering Library, Technical Publishing, a company of The Dun & Bradstreet Corp., Barrington, IL, 1972.

Plant Engineering Directory and Specifications Catalog, Technical Publishing, a company of The Dun & Bradstreet Corp., Barrington, IL, 1979.

Specialized Handling Systems, Plant Engineering Library, Technical Publishing, a company of The Dun & Bradstreet Corp., Barrington, IL, 1978.

Storage Systems, Plant Engineering Library, Technical Publishing, a company of The Dun & Bradstreet Corp., Barrington, IL, 1970.

Treer, K. R., *Automated Assembly,* Society of Manufacturing Engineers, 1978.

Unit Load Conveyors, Plant Engineering Library, Technical Publishing, a company of The Dun & Bradstreet Corp., Barrington, IL, 1971.

CHAPTER 5
UNIT LOAD CONCEPTS

ROBERT PROMISEL

Stubbs, Overbeck & Associates, Inc.
Houston, Texas

Unitizing can be accomplished on a pallet or slipsheet. And, bin boxes can be used on pallets or slipsheets.

Modern technology has brought advancements in research and development on pallet construction, slipsheet construction, and on ways to handle these concepts efficiently and economically.

Pallets and slipsheets can be loaded and unloaded with a push–pull unit, loaded into a 10-, 20-, or 40-ft container, and loaded on trucks that go on railroad tracks. Also, they can be loaded on dock conveyors and fed to a trailer truck in less than 3 min; then transported to another location and unloaded on dock conveyors, again in less than 3 min.

In reality, these concepts are here today, and this chapter defines the type of systems available and how the reader may develop the entire system or parts of the system to suit his or her particular need.

5.1 SLIPSHEETS

Slipsheets are most commonly manufactured from solid fiber, corrugated, and plastic materials.

Solid Fiber

Solid fiber slipsheets are plys of Kraft or Kraft linerboard and cylinder board laminated together to provide adequate tensile strength or other requirements of a slipsheet. The thickness of a solid fiber slipsheet is referred to as *caliper of board* and is measured in *points.* They are usually 3- or 4-ply. See Figs. 5.1.1 and 5.1.2.

Fiberboard slipsheets are flat sheets of *corrugated* board or *solid fiber* possessing one or more edge tabs. They are used as a base on which product/materials can be assembled, stored, handled, and transported as a unit load. Slipsheets are used as an alternative to wooden pallets.

Solid fiber slipsheet calipers range from 0.035 in. (0.9 mm) to 0.150 in. (3.8 mm). Slipsheets with calipers at the lower end of the scale are generally used for lighter-weight boxed products. The higher-caliper slipsheets are often used for heavy products, or bagged products and multiple use applications.

Folded tabs are more manageable in lighter-caliper sheet. Products should be kept behind the scoreline on a slipsheet so that tabs will fold properly into a 90° position when backed up with another load. Improper stacking will damage the flaps on reuse.

It is recommended that, when moving into a slipsheet configuration, the load be grabbed by the tabs and not scooped up, because this can fracture the slipsheet. (Note: as indicated in other parts of this chapter, this should not be performed until the lift truck operator is totally familiar with the equipment, slipsheets, and the working area.)

In instances where scoring of slipsheets can assist the handling of loads, the male scoring head will be positioned on the bottom shaft, and the matching female head on the top shaft. This will result in the scoring being *up* rather than down.

Score sheets will then be stacked in this relative position (with score being up) when shipped to a customer. To help operation, all slipsheets should be printed "THIS SIDE UP" on the top side to facilitate proper folding operation.

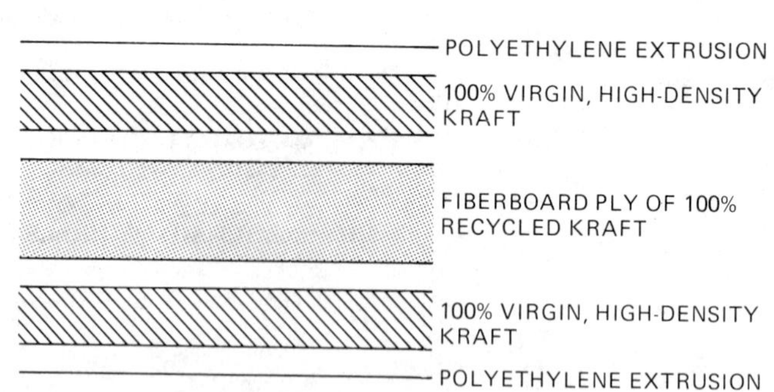

Fig. 5.1.1 Laminate diagram of 0.050-in. (1.27-mm) sheet. Also available in 0.095-in. (2.41-mm) sheets.

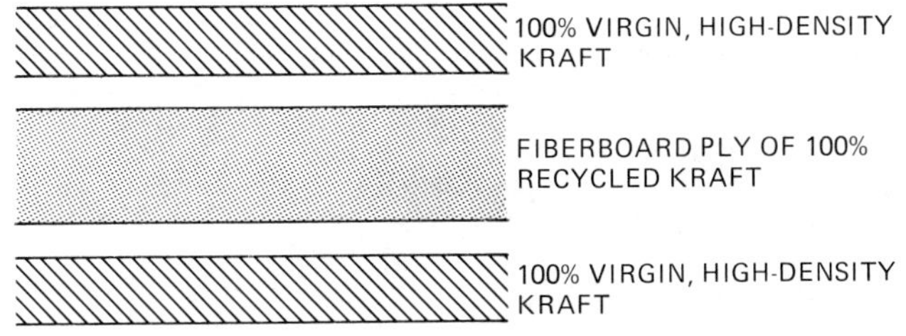

Fig. 5.1.2 Laminate diagram of 0.050-in. (1.27-mm) sheet. Also available in a four-ply, heavier-caliper construction.

Corrugated

Corrugated slipsheets are Kraft linerboard facings with corrugated medium bonded together to provide adequate tensile strength. Corrugated slipsheet is referred to in terms of flute size and board test [e.g., B flute, 250 lb test (43,781.7 N/m)].

Plastic

Plastic slipsheets include any kind of combination of polymerized materials, including but not limited to polyethylene or polypropylene, that provides adequate tensile strength and other requirements of the slipsheet. Thickness of a plastic slipsheet is referred to in mils [1 mil equals 0.001 in. (0.025 mm)].

5.1.1 Specifications for Solid Fiber Slipsheets

Performance for a solid fiber slipsheet with push–pull equipment can be improved considerably with proper attention to several important factors during slipsheet manufacturing process. These factors are:

Tensile strength
Tear resistance
Moisture content
Quality of scoring
Mullen-bursting or puncture resistance
Stiffness
Moisture resistance

Tensile Strength

Because the slipsheet tab will be engaged by the gripper bar of the push–pull attachment and used to pull the unit load onto the push–pull platens, the slipsheet must contain sufficient tensile strength to avoid tab failure during this pulling operation. The tensile strength can be measured easily in a laboratory with an electrohydraulic tensile tester.

Testing has not determined whether slipsheet tabs are stronger across or with the grain. In the past it was assumed that tabs across the grain were stronger because pulling of these tabs is in the grain direction which has 2 to 3 times greater tensile strength than does the cross direction. Failure comes primarily from flexing at the scoreline. Primary causes of cracking are *inadequate scoring and low moisture content.*

As the tensile strength increases, so does the cost. Polyethylene liners also increase slipsheet cost. Important considerations in deciding on minimum slipsheet specification for a specific materials handling situation are:

1. Weight of the unit load to be handled
2. The type product being unitized
3. Stability of the unit load
4. How the slipsheet and unit load will be warehoused (with or without wood pallets)
5. How the unit load will be shipped (by truck and railcar as opposed to truck or railcar only)

Tensile strength of the slipsheet must be adequate for the specific application. In general, higher tensile strengths are required for heavier unit loads or loads subjected to repeated use with push–pull equipment.

Depending on the test of board used, tensile values (lb/in.) for corrugated slipsheet are as follows:

Cross direction: 90–155 lb (15,761.4–27,144.7 N/m)
Machine direction: 170–300 lb (29,771.6–52,538.0 N/m)

Tensile values for the solid fiber slipsheets are:

Cross direction: 130–350 lb (22,766.5–61,294.4 N/m)
Machine direction: 300–680 lb (29,771.6–119,086.2 N/m)

Tensile testing procedures utilize TAPPI 494 or ASTM D828. Samples should be tested through the scoreline and after the scoreline has been folded a full 90°.

Tear Resistance

During normal operations of the push–pull attachment, tears may start in the tab areas of the slipsheet after repeated pulls on the tab. Such tearing could also be caused by crumbling of the tab as one load is pushed against another in warehousing or in loading operations.

Moisture Content

Quality of any solid fiber slipsheet will be adversely affected whenever the board contains insufficient moisture. Even using the very best scoring equipment available, quality scores cannot be obtained if moisture content in the slipsheet is too low. If the sheet is too dry, outer liners will crack when the tab is folded at the scoreline, resulting in premature tab failure. As with other important factors, exact moisture content can be accurately determined in the laboratory.

Quality of Scoring

As noted under Moisture Content, the score is a very important part of any slipsheet and can often be the cause of tab failure. Proper moisture content and certain guidelines should be followed in matching the scoring dies to the slipsheet caliper. The objective of quality slipsheet scoring should be to permit a 90° fold at the scoreline with minimum disturbance to the fibers.

Mullen-Bursting or Puncture Resistance

Overall strength of slipsheet can be measured to a certain degree with Mullen and/or Beach Puncture Test. Testing methods are described in TAPPI Publications T810 and T803. There is a pulling action on the slipsheet when handled by push–pull attachment. The tab areas should be constructed to withstand punctures and other crushing that may be caused by repeated clamping of the gripper bar or by placing unit loads in tight storage areas.

Stiffness

Stiffness of the slipsheet may be extremely important in certain materials handling situations. These involve storage of bagged or odd-shaped materials, the need for an especially wide unit load, and so on.

If the slipsheet is not designed for adequate stiffness, pressure points may cause excessive sagging, to the degree that the tab cannot be properly engaged by the attachment's gripper bar.

Stresses caused by excessive pressure points within the unit load may result in slipsheet failure beneath the products. A heavier caliper sheet is normally required for these conditions. Stiffness laboratory test method is described in ASTM D747.

Moisture Resistance

There are many materials handling situations that involve relatively wet conditions. These could include frozen food storage and shipment, refrigerated shipments, export shipments, high humidity conditions, wet product such as produce, wet floors, and so on.

A slipsheet used under these or similar conditions may fail unless properly treated to resist moisture. One such treatment is the extrusion of polyethylene to outer liners. Laboratory tests can be made to determine effectiveness of the moisture barrier. A moisture resistance specification should not be established if handling and shipping conditions are relatively dry because slipsheet costs would be increased unnecessarily.

5.1.2 Slipsheet Specification Guidelines

Several factors must be taken into consideration in selecting slipsheets for specific applications. These are:

The number of times the unitized load is handled by push–pull equipment (i.e., 5 pulls, 10 pulls, etc.)
Unit load weight, composition, and stability
Moisture exposure
Operator experience
Receiver requirements
Case size, product characteristics, and packaging (type of container)
Type of materials handling equipment to be used

Slipsheet Sizes and Configurations

Slipsheet sizes are nonstandard. The size is determined by the user and is based on the length and width dimensions of the unit load.

Slipsheet dimensional tolerances are $\pm \frac{1}{4}$ in. (6.4 mm). Edge tabs should have a minimum depth of 3 in. (76.2 mm) and a maximum depth of 4 in. (101 mm). Slipsheets are supplied with tabs on one or more sides as specified by the user.

The purpose of the edge tabs is to facilitate use with materials handling equipment generally of the push–pull type (e.g., lift trucks equipped with gripper bars).

Slipsheet specifications should be developed between consultation of the supplier and user, considering all the factors listed.

Slipsheet end-user applications vary. Therefore, different qualities and types are sometimes required to obtain optimum results. Several of the more important considerations are shown in Table 5.1.1.

5.1.3 Slipsheet Tabs

In some instances where the sheets have less tensile strength, it is likely that the tabs may be reinforced to sustain the pull with the push–pull attachment. Thus, a 0.06-in. (1.5-mm) caliper sheet may not be 0.06 in. (1.5 mm) caliper across the entire face of the sheet; the tabs are built up. Tabs can also be scored.

Tensile strength in the machine direction (MD) and in the cross direction (CD) are not the same

Table 5.1.1 Recommended Slipsheets for Various Unit Load Weights

Unit Load Weights	Slipsheet Caliper	Tensile Strength (lb)	
		MD[a]	CD[b]
Less than 1000 lb	3-Ply, 0.035 in. sheet (0.9 mm)	319 (55,865.4 N/m)	122 (21,365.5 N/m)
1000 lb (453.6 kg)	3-Ply, 0.046 in. sheet (1.2 mm)	356 (62,345.1 N/m)	147 (25,743.6 N/m)
1500 lb (680.4 kg)	4-Ply, 0.046 in. sheet (1.2 mm)	420 (73,553.3 N/m)	157 (27,494.9 N/m)
1800 lb (816.5 kg)	4-Ply, 0.058 in. sheet (1.5 mm)	457 (80,032.9 N/m)	181 (31,698.0 N/m)
2200 lb (997.9 kg)	4-Ply, 0.063 in. sheet (1.6 mm)	463 (81,083.7 N/m)	184 (32,223.3 N/m)
2500 lb (1134.0 kg)	4-Ply, 0.070 in. sheet (1.8 mm)	494 (86,512.6 N/m)	206 (36,076.1 N/m)
3000 lb (1360.8 kg)	4-Ply, 0.080 in. sheet (2.0 mm)	506 (88,614.2 N/m)	210 (36,776.6 N/m)

[a] MD = machine direction.
[b] CD = cross direction.

because the direction of pull is most affected by moisture or humidity. This factor can affect plastic and corrugated sheets exposed for any extended length of time.

5.1.4 Building Unit Loads

Use an interlocking pattern wherever possible when using common slipsheet size for multiple case sizes. Always stack cases to edges of slipsheet, leaving necessary voids between cases. Keep cases $\frac{1}{4}$ in. (6.4 mm) to $\frac{1}{2}$ in. (12.7 mm) away from all scorelines (tabs must fall freely on scores).

In ordering a slipsheet for a specific case size, determine overall dimensions of cases, outside to outside, after stacking at best interlocking pattern. Add $\frac{1}{2}$ in. (12.7 mm) in each direction to calculate stacking surface for the slipsheet. Dimensions for necessary tabs must then be added. Unless unusual conditions exist, a 3-in. (76.2-mm) tab will be adequate for any style push–pull attachment. [Note: In early development stages, when in training mode, it is recommended that 4-in. (101.6-mm) flaps be used. When in daily operational mode after break-in period, chiseling under slipsheets with platens may be done. This reduces the access time to measure the tab distance and approach it squarely.]

Use palletizing-type adhesive (high shear, low tensile) whenever adhesive on cases is not objectional. Caution should be used to prevent freezing of this type of adhesive. Adhesives may be applied with air-pressure spray cans, hoses, or plastic squeeze bottles.

Apply the adhesive generously to the slipsheet before stacking the first layer of cases. Apply in a close "S" design so that several $\frac{1}{8}$-in. (3.2-mm) wide adhesive strips contact the bottom of each case.

Tie a string or tape around the top layer of cases, especially if cases are very light.

5.1.5 Loading of Trailer or Railcar

Avoid the need for decking unit loads within the trailer or railcar whenever possible. An improved vertical alignment of deck loads can be obtained in less time by performing the decking operation at the dock against the solid surface (wall, column, etc.).

Good vertical alignment inside trailers is often difficult because the area is often dark and space for maneuvering of the attachment is very limited. (Note: Side-shifting mechanism on attachment is vital to alignment.) The nose area of many trailers is rounded so that no solid surface is available to push the second tier load when backing the push–pull truck away. This resulting offset condition of the second tier load may cause breakdown of first tier containers because the corner stacking strength of the containers is not properly utilized.

Avoid placement of unit loads flush against side walls of trailer or railcar whenever possible. It is difficult to approach the unit load squarely with the conventional push–pull attachment because the side of the gripper bar will bind against the side walls. If the load consists of a mixture of slipsheets and unit loads and loose cases, position unit loads to the *center* of the trailer or railcar, filling in side voids with loose cases. Leave at least a 12-in. (304.8-mm) void between the top-most case and the trailer or railroad car roof.

5.1.6 Unloading and Undecking of Unit Loads

Avoid use of the gripper bar and in undecking operation inside the trailer or railcar, whenever possible. With the pantograph fully retracted, platen tips resting on floor, and platen tilted forward slightly, scoop under the bottom-most unit load. If load is tiered two or three high, remove the entire stack to the dark area in one trip (this assumes capacity of lift truck is adequate).

If unit loads are to be positioned on captive wood pallets for warehousing, position the empty wood pallet against the wall. Bring the stack of unit loads completely over the empty pallet, lower to approximately 1 in. (25.4 mm) above the pallet deck, and push the load tightly against the wall while backing the lift truck away. This will result in good vertical alignment with the wood pallet in minimum time.

When undecking the second tier load, keep the platen tips pressed firmly against the top tier cases [approximately 1 in. (25.4 mm) from the case top] while pulling the second tier load onto the platen. Be certain the lift truck brake is set.

This will prevent upsetting top tier cases of the first tier load while undecking. Pull load completely onto platen with truck in stationary position rather than attempting to drive the truck forward under tiered load.

Repeat the earlier procedure to position the second tier load on its wood pallet.

5.1.7 Side Tabs on Slipsheet

Only one tab is needed if unloading only in trailer or only in railcar, provided the approach to load and warehouse is always from same direction. A side and end tab are needed if loading both railcars and trailers. A 40 in. × 42 in. (1016 mm × 1066.8 mm) wide slipsheet is ideal for trailer width and

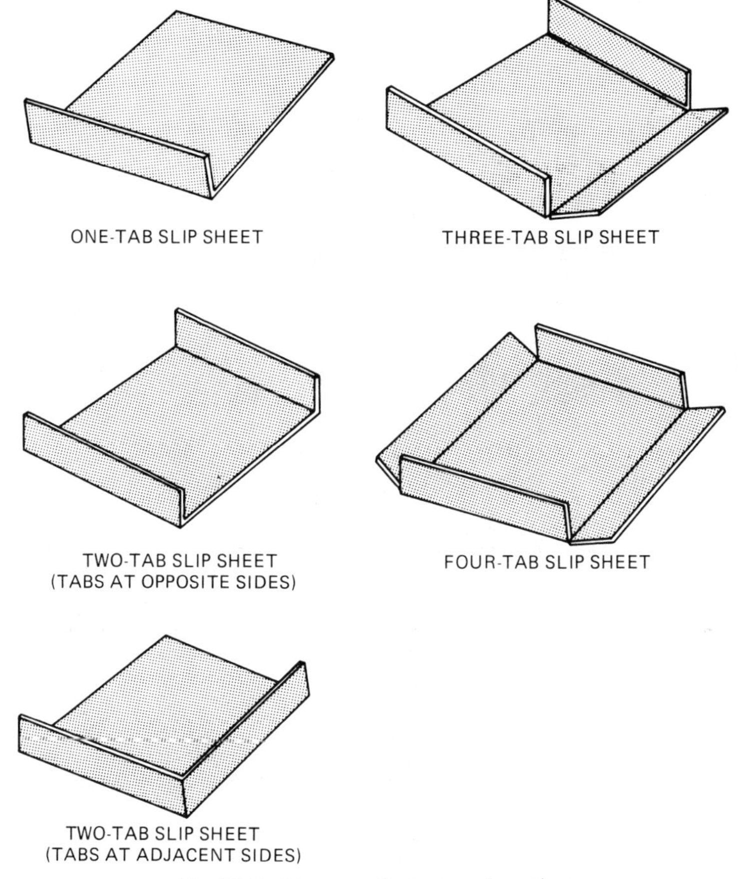

Fig. 5.1.3 Common slipsheet configurations.

a 48-in. × 52-in. (1219.2 mm × 1320.8 mm) width slipsheet is ideal for railcar loading. These specifications assume a two-load width in either truck or railcar; if the case size lends itself to three-wide car loading, a 35-in. (889-mm) slipsheet width may be ideal.

More than one end and one side tab may be useful only in conjunction with stretch or shrink film. Third and fourth tabs tend to interfere with necessary close placement of unit loads in trailers or railcars. See Fig. 5.1.3.

5.1.8 Slipsheet Box

A practical, economical, and sound response to the pressures imposed by rising transportation costs, materials costs, and by the fluctuating price of wooden pallets is the slipsheet box. The containair slipsheet box is an idea whose time has come. No new equipment or facilities are required.

Container slipsheet boxes fit smoothly and easily into existing handling procedures, on both the shipping and receiving ends. The slipsheet box is designed to allow the forklift to slip under the box without using pallets.

The containair slipsheet box measures 51 in. × 41 in. × 40 in. (1295.4 mm × 1041.4 mm × 1016 mm) and contains 50 ft^3 of usable volume. It is built of heavy-duty triple-wall fiberboard, with integral solid fiberboard slipsheet base. It trims tare weight to 38 lb (17.2 kg) per box with a 12% increase in usable volume. This extra-low tare weight makes it ideal for air shipping.

5.2 PALLETS

Pallet standards in the United States began in 1953 under the auspices of the organization now known as The American National Standards Institute (ANSI). The MH1 Standards Committee has responsibil-

Table 5.2.1 Standard Pallet Dimensions

Imperial (in.)	Hard Metric (mm)	Hard Metric (in.)	Soft Metric (mm)
24 × 32	600 × 800	23.64 × 31.52	609 × 812
32 × 40	800 × 1000	31.52 × 39.40	812 × 1016
32 × 48	800 × 1200	31.52 × 47.28	812 × 1219
36 × 42	900 × 1060	35.46 × 41.75	914 × 1066
36 × 48	1060 × 1200	35.46 × 47.28	1066 × 1219
40 × 48	1000 × 1200	39.40 × 47.28	1016 × 1219
42 × 54	1060 × 1370	41.75 × 53.96	1066 × 1371
48 × 60	1200 × 1500	47.28 × 59.10	1219 × 1523
48 × 72	1200 × 1800	47.28 × 70.90	1219 × 1828
36 × 36	900 × 900	35.46 × 35.46	914 × 914
42 × 42	1060 × 1060	41.75 × 41.75	1066 × 1066
48 × 48	1200 × 1200	47.28 × 47.28	1219 × 1219

ity for developing standards for "pallets, slip sheets and other bases for unit loads." The first national standard was published in 1959, revised in 1965, and later subdivided into three standards. The present status of work is:

MH1.1.2-1978 Pallet Definitions and Terminology. Published revision of 1972 standard.

MH1.4.1977 Procedures for Testing Pallets. First comprehensive national standard on testing.

MH1.2.2-1975 Pallet Sizes. Presently have 12 standard sizes; 9 oblong, 3 square. Dimensions are stated in imperial inches and comparable hard and soft metric dimensions (Table 5.2.1).

5.2.1 Planning for Palletization

An efficient palletization program is one tailor-made to a particular application. Detailed discussions of all factors involved in planning an individual pallet program are beyond the scope of this chapter. There are, however, three basic principles relating to the science of materials handling that apply to all handling programs. These principles are:

Handle as large units as possible within practical limits.

Handle materials as few times as possible.

Utilize mechanical equipment rather than manual labor.

In addition, three factors involving type of movement, nature of material to be palletized, and characteristics of planned storage areas must be considered carefully in the course of arriving at an efficient system. Consideration of these factors must precede the selection of the pallet.

Movement of goods is one of the governing factors in determining the efficiency of palletization as relates to the individual operation. Within this field, four elements are important in the movement of goods:

Fixed paths

Variable paths

Distance

Frequency

If an operation involves movement of small quantities of lightweight goods over extremely short distances, or material and packages of limited sizes from fixed origin to fixed destination, then use of pallets may *not* be justified. The need for pallets and other forms of handling equipment, however, increases in direct proportion to the increase and quantity of goods, distance, and frequency of movement, and to variability of flow paths.

5.2.2 Types of Pallets

Pallets fall into three general groups:

Expendable

General purpose
Special purpose

Because expendable pallets are usually a one-trip affair, cost is the essential factor and their design and construction becomes a matter for negotiation between the pallet manufacturer and the customer.

Special purpose pallets may be of any design suitable for the product to be carried. Their design and construction are also a matter for agreement between the pallet manufacturer and the customer.

Styles, Designs, and Constructions

The most common designs of wooden pallets are:

Two-way pallets, which permit the entry of forks or hand pallet trucks from two sides, and in opposite directions.

Four-way pallets, which permit entry on all four sides: Notched-stringer design, which has four-way entry only with forklift trucks and two-way entry with hand pallet trucks; and block design, which has four-way entry with both forklift and hand pallet trucks.

There are two styles of wooden pallets:

Single-face pallets, which have only one deck as the top surface.

Nonreversible, in which top and bottom decks have different openings and upon which goods may be stacked only on the top deck.

Wooden pallet constructions are as follows:

Flush Stringer. A pallet in which the outside stringers or blocks are flush with the end of the deckboards.

Single Wing. A pallet in which the outside stringers are set inboard of the top deck and the stringers are flush with the ends of the bottom of the deckboards.

Double Wing. A pallet in which the outside stringers are set inboard of both top and bottom deckboards to accommodate bar slings or other devices for handling pallets.

Designation of Pallet Types

For the mutual convenience of pallet users and manufacturers, the following numerical designations have been adapted to express combinations of styles and construction:

Type 1. Single-face, nonreversible pallet
Type 2. Double-face, flush stringer or block, nonreversible pallet
Type 3. Double-face, flush stringer or block, reversible pallet
Type 4. Double-face, single-wing, nonreversible pallet
Type 5. Double-face, double-wing, nonreversible pallet
Type 6. Double-face, double-wing, reversible pallet

Each of these types exists in more than one design, therefore the customer must specify a choice of design.

Nomenclature

Deckboards are the boards that make up the faces of a pallet and which either carry or rest upon the goods packed thereon.

Stringers are the wooden runners to which the deckboards are fastened, and which serve as a spacer between the top and bottom decks to permit the entry of mechanical handling devices.

Stringer boards are boards used over blocks below the deckboards on the four-way block-type pallet.

Blocks are square or rectangular wooden parts employed on some four-way pallets in place of stringers, and which serve the same purpose.

See Fig. 5.2.1 for other features of pallets.

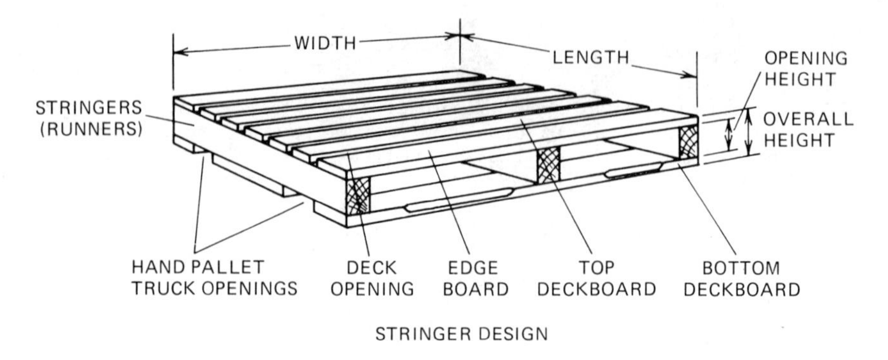

STRINGER DESIGN

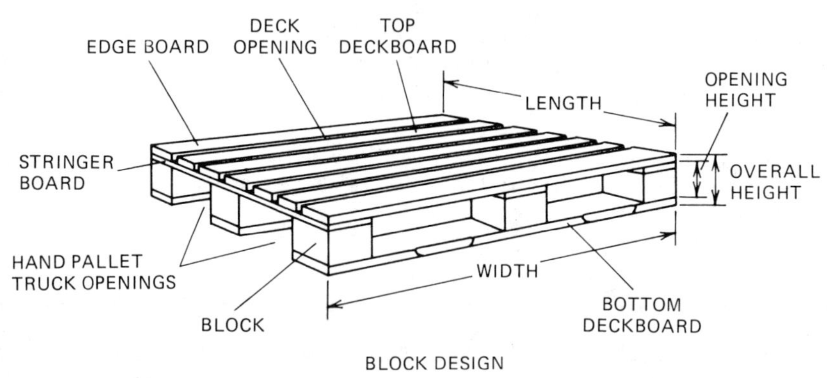

BLOCK DESIGN

Fig. 5.2.1 Principal parts and commonly used construction features of stronger and block design pallets.

Definitions

In speaking of pallet sizes, dimensions should be stated in inches, and LENGTH should be designated before WIDTH. The WIDTH should always be the dimension parallel to the top of the deckboard.

Species

The following specification applies only to hardwood pallets. The National Wooden Pallet and Container Association (NWPCA) has separate specifications for other species of wood. The following classification groups woods according to physical properties.

Class A. The softer woods of the broad-leaved species. They are relatively free from splitting when being nailed, have moderate nail-holding power, moderate strength as a beam, and moderate shock-resisting capacity. They are soft, lightweight, easy to work, hold their shape well after manufacture, and are easy to dry.

Aspen
Basswood
Buckeye
Cottonwood
Willow

Class B. The medium-density hardwood species. They are more inclined to split when being nailed than the wood of Class A, but have greater nail-holding power and greater strength as a beam. They are softer, lighter, easier to work, and easier to dry than woods in Class C.

Ash (except white)
Soft elm

Tupelo
Butternut
Soft maple
Yellow poplar
Chestnut
Sweet gum
Sycamore
Walnut
Magnolia

Class C. The heaviest hardwood species. They have the greatest nail-holding power, greater strength as a beam, and the greatest shock-resisting capacity. They are difficult to drive nails into, have the greatest tendency to split at the nails, and are difficult to dry. They are the heaviest and hardest domestic woods and therefore are difficult to work with.

Beech
Birch
Hackberry
Hard maple
Hickory
Oak
Pecan
Rock elm
White ash

Deckboards

The edge boards of the top and bottom decks shall not be less than a nominal 6-in. (152.4-mm) board. The balance of the deckboards may be random width stock 4–8 in. (101.6–203.2 mm) wide. However, no deckboards shall be less than a nominal 4-in. (101.6-mm) board, and not more than a nominal 8-in. (203.2-mm) board.

The type of merchandise to be palletized determines the amount of lumber coverage and spaces between deckboards. Therefore, the number of inches of top and/or bottom deckboards, and the maximum and minimum individual and total spacing shall be specified in all orders by the purchaser.

The amount of lumber required for the deckboards in the pallet faces shall be as per Table 5.2.2.

Table 5.2.2 Minimum Lumber Requirements for Construction of Pallet Decks [a]

Stringer Length		Maximum Deckboard Spacing				
in.	mm	1 in. (25.4 mm)	$1\frac{1}{2}$ in. (38.1 mm)	2 in. (50.8 mm)	$2\frac{1}{2}$ in. (63.5 mm)	3 in. (76.2 mm)
32	(812.8)	28	26	24	24	23
34	(863.6)	29	28	26	24	24
36	(914.4)	31	29	28	26	25
38	(965.2)	33	31	30	28	26
40	(1016.0)	34	33	30	30	28
42	(1066.8)	36	34	32	30	30
44	(1117.6)	38	36	34	32	30
46	(1168.4)	39	37	34	34	31
48	(1219.2)	41	39	36	34	33
50	(1270.0)	43	41	38	36	34
52	(1320.8)	44	42	40	38	36
54	(1371.6)	47	44	41	39	36
56	(1422.4)	50	47	43	39	38
58	(1473.2)	52	48	45	41	39
60	(1524.0)	52	49	46	44	40

[a] Single-face, reversible pallet decks, and top face of nonreversible pallets (expressed in accumulated inches of nominal width boards before sizing).

When pallets are intended for use with hand pallet trucks, the space between the edge boards of the bottom deck and their adjacent deckboards shall be specified by the customer as to the minimum and maximum openings allowed.

Stringers and Blocks

All pallets shall have *not less* than the following number of stringers or blocks:

Deckboard Length	Stringers	Blocks
Not exceeding 24 in. (609.6 mm)	2	6
25–48 in. inclusive (635.0–1219.2 mm)	3	9
Over 48 in. (1219.2 mm)	3	9

On deckboard lengths over 48 in. it is recommended that additional stringers or blocks be used.

Chamfering

Unless otherwise specified by the purchaser, only the outside edges of the bottom edge deckboard shall be chamfered. These chamfers shall be at least 12 in. (304.8 mm) long and shall be cut on a 35° angle to the face so as to leave an edge adjacent to the chamfer not less than $\frac{1}{4}$ in. (6.4 mm) from the outer edge of the deckboard.

Fastenings

There are a number of acceptable fasteners for wooden pallets. These include drive screws (helically threaded nails), annularly threaded nails, helically deformed auto nails, staples, and bolts and nuts.

5.2.3 Pallet Performance

A pilot study was performed by Virginia Polytechnic Institute and State University Wood Research & Wood Construction Laboratory, Pallet & Container Research Center in Blacksburg, Virginia. The study tested the hypothesis that the bending stiffness of a pallet, with a concentrated load applied along the pallet centerline parallel to the stringers (Figs. 5.2.2 and 5.2.3), may be predicted from the modulus of elasticity of the deck wood lumber and the modulus of the deckboard–stringer joints. The validity of this prediction was proved by the results of static tests on pallet sections assembled with two common American wood species and two sizes of helically threaded pallet nails.

Standardized Methods for Testing Pallets

The American Society for Testing and Materials provides an ASTM Standard D1085-73: A Method Using Quota Point Loading for Determining the Bending Stiffness and Strength of Pallets. In another approach, the pallet is supported at its corners and the load is applied at the pallet center over a 12 in. × 14 in. (304.8 mm × 355.6 mm) area. Both methods provide useful comparative information on the design of pallets but do not advance a procedure for the engineering design of pallets.

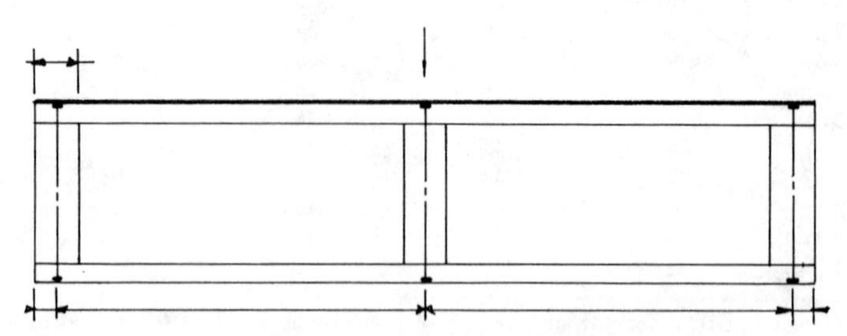

Fig. 5.2.2 Pallet dimensions and loading conditions.

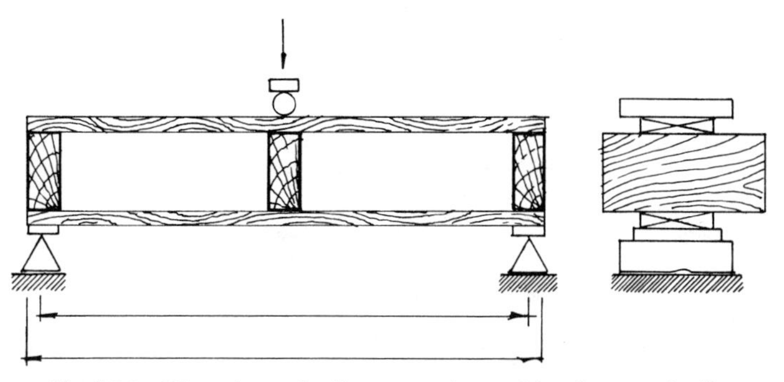

Fig. 5.2.3 Dimensions of pallet test section and bending test details.

The work described in the ASTM D1085-73 report was undertaken to examine the possibility of predicting the bending stiffness of pallets using basic properties of the materials and joints, which are easily determined.

Deflection of a Pallet Section Under Central Concentrated Load

Prior to attempting a theoretical analysis and testing, a preliminary examination was made of the bending behavior of a section of a wooden pallet. The section consisted of single top and bottom deckboards nailed to three stringers and loaded with a central concentrated load over a span parallel to the deckboards.

During this examination, it was observed that there was essentially no shear deformation at the deckboard stringer joints when the normal nail fastening was used, probably because of the relatively high lateral (shear) stiffness of the joints and the friction between the members.

If the pallet is loaded with a concentrated line load along the center stringer and is supported along the centerlines of the outer stringers, and if it is assumed that the shear deformation at the deckboard–stringer joints is negligible, it may be deduced by the computations presented that the deflection, in inches, at the stringer and mid-span is given by the following equation.

$$D = \frac{PK(L-b)^3L^2}{192} \times \frac{L(32S_B + nkb^2L) + 48bS_B}{L^3S_T(32S_B + nkb^2L) + 96bLS_BS_T(L+b) + K(L-b)^3\,S_B(8S_B + nkb^2L)}$$

where: $K = \dfrac{32S_T + nkb^2(L-b)}{8S_T + nkb^2(L-b)}$

D = deflection of stringer at mid-span (inches)
P = central concentrated load (pounds)
L = span = distance between centerlines of outer stringers (inches)
b = thickness of stringers (inches)
S_TS_B = EI product for top and bottom deckboards, respectively, where E = modulus of elasticity and I = total moment of inertia of the deckboards (pound-inches squared)
n = number of fasteners joining deckboards to stringers at each end of each deckboard
k = modulus of separation of deckboard from stringer per single fastener (pounds per inch)

The term K is a property that varies with the species and the fastener and is the ratio of the force applied in the direction of the fastener to the amount of separation at that force. It is dependent on the withdrawal resistance of the fastener in the stringer and the head pull-through resistance in the deckboard.

To test the validity of this equation, static bending tests were performed on pallet sections consisting of single top and bottom deckboards, each fastened with two nails to each of three short stringer sections. The pallet sections were assembled with (1) green red oak, (2) green, and (3) dry southern hardened-steel, with 3 in. × 0.120 in. (76.2 mm × 3.0 mm), pointless, umbrella-headed, helically threaded, hardened-steel pallet nails or 2.25 in. × 0.113 in. (57.2 mm × 2.9 mm), helically threaded, medium-carbon-steel pallet nails with flat heads and diamond points. Top and bottom deckboards were sawn from different planks. However, the top deckboards, and similarly the bottom deckboards, were sawn from the same plank for the assemblies with the two nail types. Two replications of each species and each nail type were tested.

The mid-span deflection was measured with a dial gauge deflectometer with reference to the neutral axis of the assembly. The loading rate was 0.20 in. (5.1 mm) per minute of the testing machine cross-head travel. A load–deflection diagram was plotted up to a deflection of approximately $\frac{3}{4}$ in. (19.1 mm). See Fig. 5.2.4.

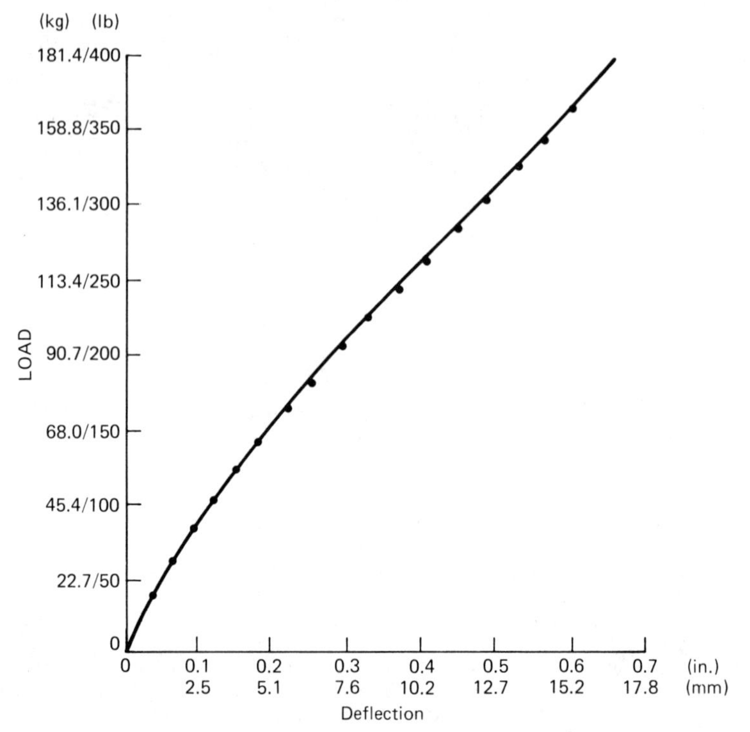

Fig. 5.2.4 Load versus deflection curve for pallet sections computed from average basic data. Average test data for all species and both nail types.

5.2.4 Pallet Containers

Pallet containers, in basic terms, are containers used to apply mechanized handling and storage methods to solid materials of irregular shapes and sizes to granular materials and to liquids.

Pallet containers are literally the instruments of a revolution in materials handling, a revolution that has brought with it reductions in processing costs, handling costs, storage costs, and product distribution costs.

For example, in agriculture, containerization is approaching the stage in which fruits and vegetables may be picked, loaded into pallet containers, and then washed, transported, and placed in local supermarkets, still in the same containers.

In industry, the pallet container into which parts or small units are to be loaded for storage or distribution may become part of the assembly line process, further reducing handling costs.

Types of Pallet Containers

As in most rapidly expanding and developing industries, there is considerable confusion about types of pallet containers and their designations. Lacking any established standard, manufacturers of pallet containers have developed their own terminology as well as their own trade names for pallet containers. The purpose of this section is to help pallet container users eliminate this confusion, and provide standard terminology which can be widely understood.

There are only three basic types of pallet containers (Fig. 5.2.5) and all special purpose variations that NWPCA knows about can be properly placed under one of the following types.

Pallet Bins. Pallet containers with pallet bases and closed sides and ends.

Pallet Boxes. Pallet containers with pallet bases, closed sides and ends and top.

Pallet Crates. Pallet containers with pallet bases, open or slotted sides and ends.

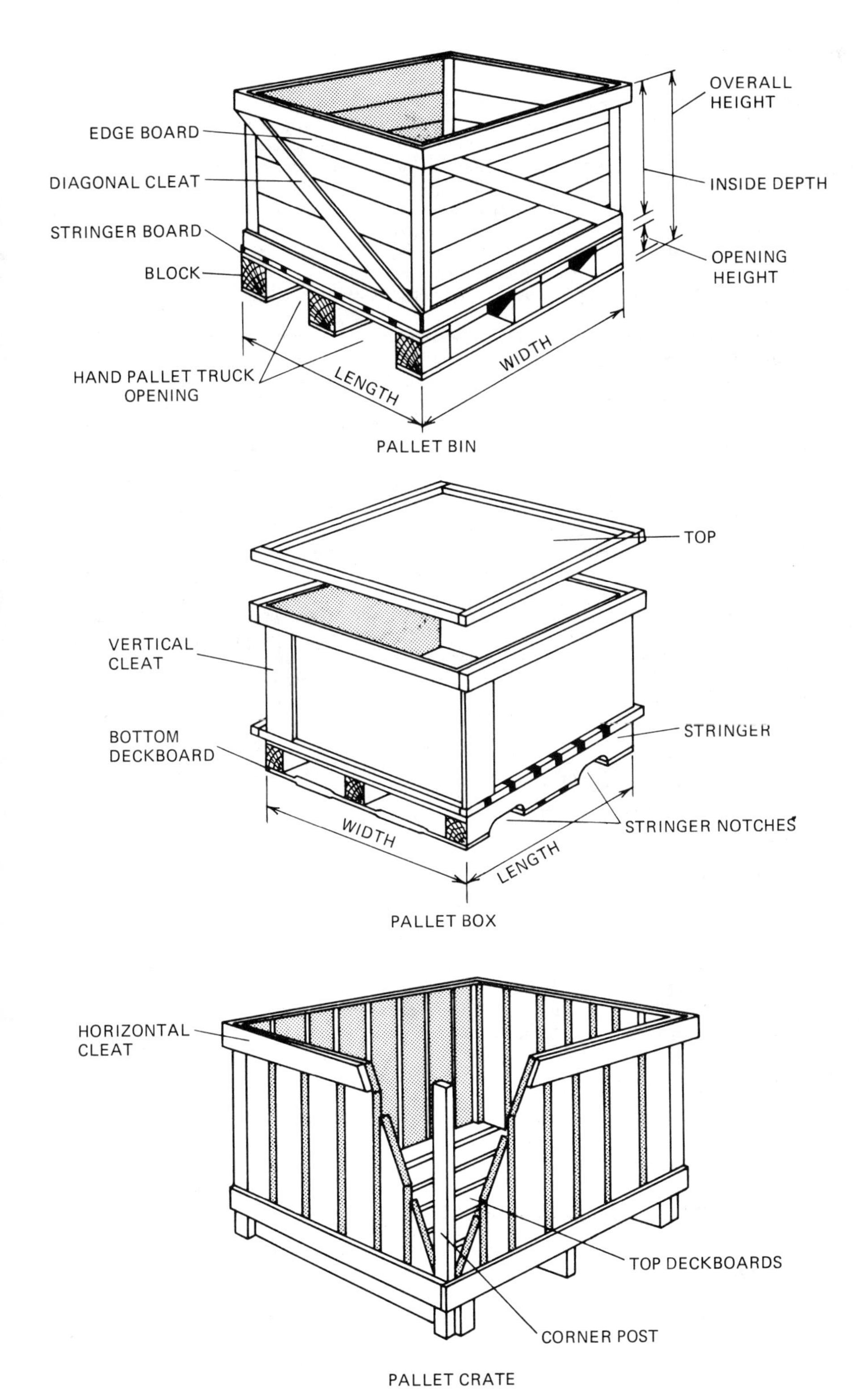

Fig. 5.2.5 Principal parts of typical pallet containers and bases.

Basic Forms of Pallet Container Design

Expendable and Noncollapsible. This type of container is considered primarily a shipping unit designed to pay its way in one shipment. Its construction is lightweight and economical within the limits necessary to withstand shipping stresses and to protect the merchandise it contains.

As a practical method, this type of container is often used in secondary shipping, and further used internally at the receiver's operation. The expendable, noncollapsible container is particularly useful in shipping parts or increments from a manufacturer to an assembly plant, or for shipping merchandise to a wholesaler for further distribution. In some cases, the container may become a merchandising tool carrying the imprint of the *receiver* of goods, who may then reload it with his own finished products for shipment to the customer. The expendable, noncollapsible container is used chiefly in industry, since its light, open construction limits its usefulness in agricultural produce handling.

Expendable and Collapsible. This container has the same general properties and uses as the expendable noncollapsible container. In addition, it has the advantage of being shipped and stored, knocked down prior to use. It permits the user to take greater advantage of the economies of mass production without creating problems of additional storage space.

Reusable and Noncollapsible. The reusable, noncollapsible container serves all or any of the uses for which pallet containers are designed, including harvesting, toting, processing, storing, shipping, agricultural commodities, and the in-process handling, shipping, and storing of industrial materials and products. Its durable construction makes it particularly effective where continuous handling is required, as in movement of crops between fields and processing plant, where high density storage is required, placing great load stress on the container. It is readily preparable. By its nature it can be considered a capital investment subject to amortization.

Reusable and Collapsible. This container possesses many of the characteristics of the reusable, noncollapsible container, and shares with it many of the same uses. It has the additional advantage of being shipped and stored in reduced space, when it is empty. It may be collapsed and returned either to its source or another point for reuse, thus adding flexibility to operations where shipping and storing requirements fluctuate. The reusable, collapsible container offers a wide variety of styles and fastenings, a number of them patented. There are, however, three basic types that the user should know:

The fully collapsible, demountable pallet container

The fully collapsible, non-demountable pallet container

The partially collapsible, non-demountable pallet container

In the fully collapsible, demountable pallet container, the unitized side and end panels are completely separated from each other and from the pallet, and are stored or stacked on the pallet for shipment. This type has the advantage of easy, economical repair or replacement of a single damaged panel. Its disadvantage is the necessity to match parts in sets for use, a requirement that may complicate storage and shipping procedures, and add to setup labor costs.

The fully collapsible, non-demountable pallet container is of hinged construction and folds flat for storage within the dimensions of its pallet. At the same time it permits the frame to stay together by the nature of its hardware for ease of reassembly. This is the most versatile type.

In the partially collapsible, non-demountable pallet container, the container frame folds flat for storage, but *does not* fold within the dimensions of its pallet. This is slightly less versatile than the fully collapsible non-demountable container because the pallet bases and container frames have to be bundled or piled separately for reshipment or storage when empty.

Pallet Container Ends and Sides

There are two principal types of end panel and side panel constructions in pallet containers: solid construction and horizontal slat construction. The type selected depends on the use to which the container is to be put, the material to be contained, and the strength and rigidity desired.

Solid construction is normally of plywood, but may also be of lumber, with vertical, horizontal, or diagonal lumber cleats, or a combination thereof. Plywood, because of its alternating laminated grain, provides great strength, as well as smooth interior surface. It is particularly suited for handling produce or other materials that might easily be bruised or damaged by board edges and slat construction.

Horizontal slat construction is normally used for shallower containers than those of vertical, slat construction. Again, end use of the container controls spacing between slats, if any. This type of container may also be constructed with inside or outside vertical or diagonal cleats for resistance to racking.

Connectors and Interlocks for Stacking

In general, pallet containers properly designed for rigidity and load bearing, and using standard base designs, may be stacked on each other with reasonable care for proper vertical placement of the upper containers on the lower. However, there are also various special container designs that provide interlocking features and complete security of the stack if desired. These include:

Extended Bottom Deckboards. Bottom deckboards of a stringer-type pallet are extended beyond the stringer to give greater bearing surface. This provides some latitude for misalignment and stacking and is ideal for in-plant storage.

Inset Bottom Deckboards. Bottom deckboards of a stringer-type pallet are inset from the stringer edge, so that the bottom deck of the pallet container base fits inside the open top of the unit below.

Extended Side Design. When a single-wing pallet base is used to facilitate straddle truck handling, two sides of the container may be extended vertically to receive the unit.

Extended Cleat Design. Pallet containers may also be designed with extended vertical cleats for nesting a single-wing pallet base. The base on top fits inside the upward-projecting cleats of the container beneath.

Cover Interlock Design. A pallet box lid or cover with cleats framing the top interlocks with the pallet base of the unit stacked.

Outset Design. The bottom of the pallet base fits into the open top of the container on which it is stacked, and interlocks. An additional horizontal cleat on the bottom of the container's side panels and projecting stringer ends create the equivalent of an inset bottom deck.

Special Types of Pallet Containers

The diversity of construction methods possible with the variety of wood members and fastening materials available makes any comprehensive catalog of special types of pallet containers impossible in this brief text. Pallet container users who require special designs and construction features will find NWPCA's staff expert and helpful in supplying technical information and suggesting manufacturers well equipped to meet a particular need.

However, a number of special pallet container construction features and applications are rather widely used and are noted here as a matter of general interest.

Wood and Wire Construction. Pallet boxes of wood and wire construction generally provide the economy of lightweight construction with the added strength of wire binding. They may be either expendable or reusable types. They are of two kinds: wire bound and woven wood and wire.

In wire-bound containers, premanufactured panels are reinforced with vertical and horizontal wire binding stapled to the panels and twist-closed at one corner when the container is assembled for use.

In woven wood and wire containers, end and side panels are premanufactured of woven wood and wire, a method that can produce great strengths, depending on methods of reinforcement and corner closure used.

Corner Interlock Construction. Pallet containers are available with a number of patented corner interlock features for side and end panels. These offer extremely rigid corner construction, and very fast assembly and disassembly.

One version, in either noncollapsible or collapsible models, is manufactured with double scotch mitered panels which are fitted together and further reinforced with steel strapping recessed into the horizontal cleats.

A combination container/pallet, whose side and end panels are equipped with a series of heavy-duty metal hinges, comes in two variations, one partially collapsible and one completely collapsible. The completely collapsible version may be stacked and stored in piles on a pallet base, thus saving space when not in use. Both models permit a buildup of this container to any desired depth.

A convertible container offers other variation of corner design featuring pallet container sides constructed around a welded steel-rod frame hinged at the corners with steel helices. This type of container may be locked to any pallet for shipment or storage without disassembly.

The strong box container is a collapsible container with metal corners of a patented design. These corners may be fitted together to provide additional height. Grooved panels also slide in and out of the corners for easy assembly and disassembly, but are held in place by a flange at the metal corners.

Another version has side and end panels fitted with extruded aluminum, interlocking corners that permit assembly and disassembly simply by sliding matching panels together.

5.2.5 Household Moving and Storage

Pallet Containers

These containers have brought the economies of containerization to long-distance moving and storage of household goods and similarly bulky items of moderate to light weight. Normally they are of plywood construction, and may be built for regular duty or heavy duty, or to government specifications, depending on distance to be shipped and ultimate use. They are normally dustproof and may also be made waterproof for overseas shipment. A typical size is 7 ft × 5 ft × 7 ft (2.1 m × 1.5 m × 1.1 m).

Air Cargo Containers

Air cargo pallet containers are representative of a great many special-purpose containers built in a variety of standard and nonstandard shapes. The air cargo pallet container, for example, has one beveled side to fit the aircraft fuselage contour. Preferential air cargo rates are granted to materials shipped in approved types of this pallet container.

5.2.6 Automatic Storage and Retrieval Systems

Wooden pallets are often used in automatic storage and retrieval systems (AS/RS). In the design of such systems it is important to take into account the standard manufacturing practice of the pallet industry.

While close tolerances such as maximum deflection of $\frac{1}{4}$ in. (6.4 mm) are achievable in wooden pallet design, this type of restriction often unnecessarily doubles the cost of the pallet.

On the other hand, average deflections of up to 1 in. (25.4 mm) across the open span of the pallet are common and, if designed into AS/RS equipment, would reduce the pallet's cost. Similar considerations with the pallets out of square deviations are recommended ± 1 in. (25.4 mm) and overall length and width deviations, of up to $\pm\frac{1}{2}$ in. (12.7 mm), should be designed into the system. Actual standard manufacturing tolerances are tighter than these. However, in normal handling environments, the pallet wears and requires slightly more operating tolerances than the handling equipment.

5.3 PALLET VERSUS SLIPSHEET COST COMPARISON

5.3.1 Pallet System—Data Required

1. Number of units shipped ________
2. Cost per pallet ________
3. Number of pallets owned ________
4. Average pallet life ________
5. Cost of money in percent ________
6. Average weight of outbound freight ________
7. Weight of pallets ________
8. Cost of pallet return freight ________
9. Total pallet repair cost annually ________
10. Time required to store and remove loads from storage ________
11. Time required to stage and load with pallets ________
12. Labor rates ________
13. Time to unload and return pallets, inspect, sort, and administrate ________

Formula (Store and Ship on Pallets)

A. Cost per pallet ________
× Number of pallets owned ________
÷ Average pallet life ________
= Annual replacement pallet expense ________

B. Cost per pallet ________
× Number of pallets owned ________
÷ Cost of money, percent ________
= TOTAL interest expense ________

C. Total pallet repair cost annually ________
TOTAL pallet repair costs ________

D. Number of units shipped ________
× Weight of pallets ________
× Average of outbound freight ________
= TOTAL freight costs to ship pallets ________

E. Number of units shipped ________
× Weight of pallets ________
× Average cost of inbound returned freight ________
= TOTAL freight costs to return pallets ________

F. Time to unload and return pallets, inspect, sort, and administrate ________
× Number of loads ________
× Labor rates ________
= TOTAL labor cost to stage and load ________

G. Time required to store and remove load from storage ________
× Number of loads ________
× Labor rates ________
= TOTAL labor cost to store and remove from storage ________

H. Time required to stage and load with pallets ________
× Number of loads ________
× Labor rates ________
= TOTAL labor cost to stage and load ________

TOTAL COST TO SHIP ON PALLETS ________

TOTAL COST TO SHIP ON PALLETS ________
÷ NUMBER OF LOADS ________
= COST PER LOAD ________

5.3.2 Slipsheet System—Data Required

1. Number of loads ________
2. Cost of slipsheets ________
3. Cost of slipsheet dispenser ________
4. Slipsheet dispenser average life ________
5. Slipsheet dispenser number of units required ________
6. Cost of load push–pull unit ________
7. Load push–pull average life ________
8. Load push–pull number of units required ________
9. Time to store and remove with slipsheets ________
10. Time to stage and remove load from storage on captive pallet ________
11. Labor rate ________
12. Cost of captive pallet ________
13. Number of captive pallets ________
14. Average life of captive pallets ________
15. Cost of money ________
16. Pallet repair cost of captive pallets ________

Formula (Store on Captive Pallets—Ship on Slipsheet)

A. Number of loads shipped ________
× Cost of slipsheets ________
= TOTAL slipsheet expense ________

B. Cost of slipsheets ________
× Number of units required ________
÷ Average life ________
= Annual capital expense for slipsheet dispenser ________

C. Cost of load push–pull unit ________
× Number of units required ________
÷ Average life ________
= Annual capital expense for load push–pull units ________

D. Time to store and remove load from storage on captive pallet ________
× Number of loads ________
× Labor rate ________
= TOTAL labor cost to store and remove loads from storage ________

E. Time to stage and load with slipsheets ________
× Number of loads ________
× Labor rate ________
= TOTAL labor cost to store and remove from storage ________

F. Cost of captive pallets ________
× Number of captive pallets ________
÷ Average life of captive pallets ________
= Annual captive pallet replacement cost ________

G. Cost of captive pallets ________
× Number of captive pallets ________
× Cost of money ________
= TOTAL interest expense ________

H. Pallet repair cost of captive pallets ________
= TOTAL repair cost of captive pallets ________

TOTAL COST TO STORE ON CAPTIVE PALLETS AND SHIP WITH SLIPSHEETS ________

TOTAL COST TO STORE ON CAPTIVE PALLETS AND SHIP WITH SLIPSHEETS ________
÷ *NUMBER OF LOADS* ________
= *COST PER LOAD* ________

5.4 PUSH–PULL EQUIPMENT

Following are some of the most important factors in selecting the appropriate attachments.

Weight of the Attachment

Weights vary depending on overall size of attachment required. Some newer attachments weigh a little more than 1000 lb (453.6 kg) and others range up to 1600 or 1700 lb (725.8 or 771.1 kg). The weight of the attachment reduces the lift truck's capacity. For example, if the capacity of an existing piece of equipment is 5000 lb (2268 kg) at a 24-in. (609.6-mm) load center with mast fully extended and the slipsheet unit load maximum weight is 3000 lb (1360.8 kg), weight of the attachment will not be a factor. However, a lift truck with a capacity of 5000 lb (2268 kg) carrying a push–pull attachment weighing 1600 lb (725.8 kg) *will not lift 5000 lb (2268 kg)*. A formula used to determine the proper lift truck attachment is shown in Fig. 5.4.1.

Size of the Attachment

The recommended load overhang on the attachment platen or forks is 2 in. (50.8 mm) on each side and front of the load. Low-profile and narrow-aisle attachments are available, if slipsheeted unit loads

$$\text{Net capacity} = \frac{A(B + C) - D(E + F)}{E + G + H} \text{ pounds}$$

where: A = truck basic capacity, lb
B = inches from front wheel centerline to fork face
C = inches from fork face to truck rating point (usually 24 in., 609.6 mm)
D = weight of attachment, lb
E = inches from front wheel centerline to carriage face
F = inches from carriage face to attachment center of gravity (CG)
G = inches from carriage face to rear face of load (Attachment Effective Thickness)
H = inches from rear face of load to load center

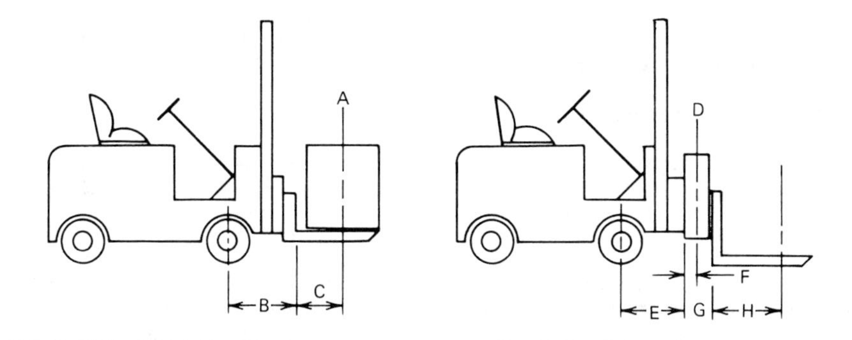

Note: This formula provides an estimated downrated capacity for the combination fork truck with attachment. The fork truck manufacturer should be contacted to determine the actual capacities to be shown on the truck nameplate.

Assume: Rotating fork clamp with 42 in. (1066.8 mm) long forks. Attachment mounted on a 5500 lb (2494.8 kg) truck handling a load 48 in. (1219.2 mm) long.

(U.S.) $A = 5500$ lb $\quad C = 24$ in. $\quad E = 11\frac{1}{2}$ in. $\quad G = 13\frac{1}{2}$ in.
$B = 13\frac{1}{4}$ in. $\quad D = 1475$ lb $\quad F = 9\frac{3}{8}$ in. $\quad H = 24$ in.

(U.S.)
$$\text{Net capacity} = \frac{5500(13\frac{1}{4} + 24) - 1475(11\frac{1}{2} + 9\frac{3}{8})}{11\frac{1}{2} + 13\frac{1}{2} + 24}$$

$$\text{Net capacity} = \frac{204{,}875 - 30{,}791}{49} = 3553 \text{ lb}$$

(metric) $A = 2494.8$ kg $\quad C = 609.6$ mm $\quad E = 292.1$ mm $\quad G = 342.9$ mm
$B = 336.6$ mm $\quad D = 669.1$ kg $\quad F = 238.1$ mm $\quad H = 609.6$ mm

(metric)
$$\text{Net capacity} = \frac{2494.8(336.6 + 609.6) - 669.1(292.1 + 238.1)}{292.1 + 342.9 + 609.6}$$

$$\text{Net capacity} = \frac{2{,}360{,}579.7 - 354{,}756.8}{1244.6} = 1611.6 \text{ kg}$$

Fig. 5.4.1 Speedy capacity formula for forklift truck attachments.

must be positioned side by side in a truck trailer or double tiered in a freight car. Larger attachments of more rugged design are available for warehousing operations.

Height

Height of the attachment's face plate and the need for a tilting push plate must be considered.

Side-shifting Requirements

Side-shifting of unit load only versus the entire assembly will have a bearing on selection.

Quick-change Assembly

If attachment must be removed periodically for the use of conventional forks, a quick-change assembly may be useful.

Costs

Generally there is little difference between manufacturers' cost for comparable equipment. Total cost will depend on how many additional features (as just discussed) are required. The price pages you may request from various organizations will give you a good idea of basic costs and the added costs of the various options. In addition to the equipment price, there is an additional mounting charge from the dealer for preparing the truck for the necessary valves and the plumbing and for installing the attachment. This usually varies between $700 and $1000 per truck.

5.4.1 Optional Equipment to Push–Pull Attachments

Several attachments are available that can clamp, rotate, side shift, side stabilize, and push, allowing users to retrieve costly pallets for loading out. This arrangement enables users to warehouse on in-house captive pallets and to ship on slipsheets, customer pallets, boxcars, or trailers.

The unit operates in the following manner:

1. Bottom forks are inserted under load. Load is raised to clamp against stationary upper forks. Side stabilizer automatically engages load from side to secure it prior to rotation.
2. As truck moves to load-out point, the attachment is rotated 180° so that pallet is on top of load. Slipsheet or customer pallet (if used) is now on bottom.
3. Forks lift pallet clear of load. Operator pushes off and backs away simultaneously. Attachment is then rotated to original position as operator returns to deposit empty pallet and pick up next load.

There are load transfers and pallet stager units that have a capacity to transfer 4000 lb of product. It can transfer loads from pallets to platens or multiple forks of lift trucks. It can also transfer product from pallet to slipsheet, from pallet to pallet. The entire operation takes only seconds and it can transfer any type of commodity normally warehoused on pallets. Figure 5.4.2 shows a push-pull transfer station.

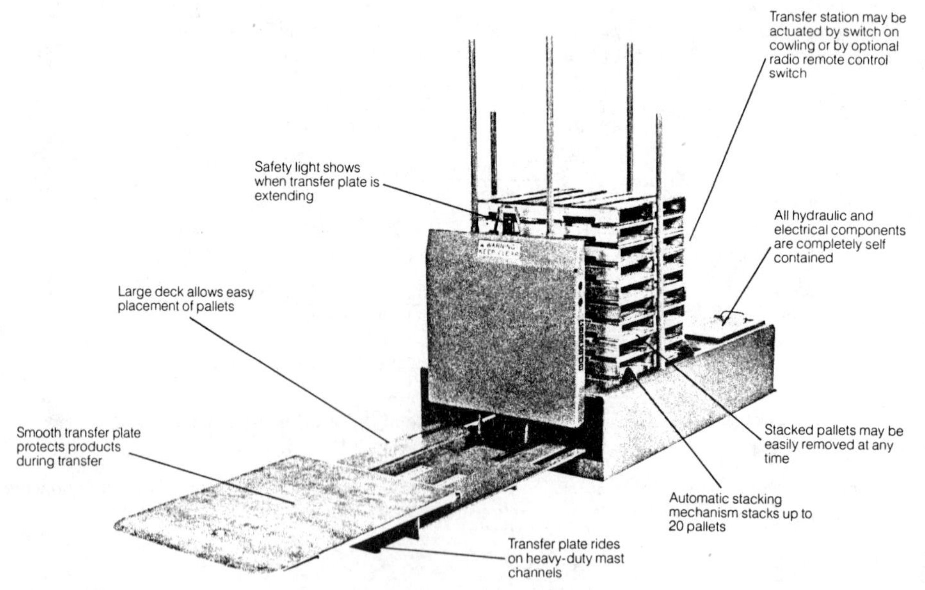

Fig. 5.4.2 Push-pull transfer station.

Load Transfer Cycle

Figure 5.4.3 shows the transfer cycle. The major steps are as follows.

Load and Position for Transfer. Lift truck driver has placed a pallet load of product in position for transfer. Note that lift truck is equipped with a push–pull attachment and multiple forks.

1. THE PALLET LOAD IS PLACED IN THE STATION AND THE FORKS WITHDRAWN FROM THE PALLET.

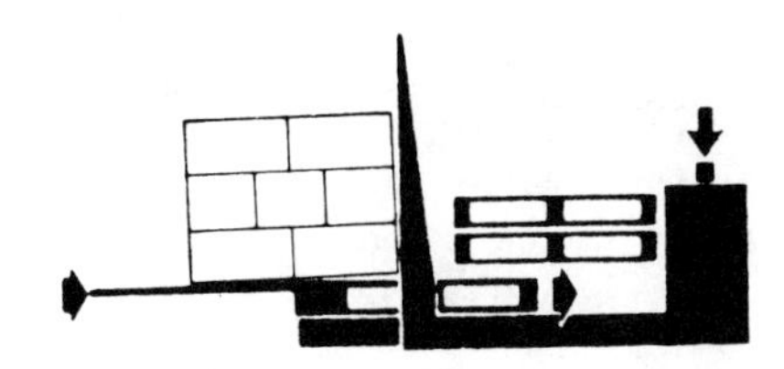

2. THE UNIT CYCLE SWITCH, OR RADIO CONTROLLED CYCLE SWITCH, IS ACTUATED AND THE STATION AUTOMATICALLY REMOVES AND STACKS THE PALLETS.

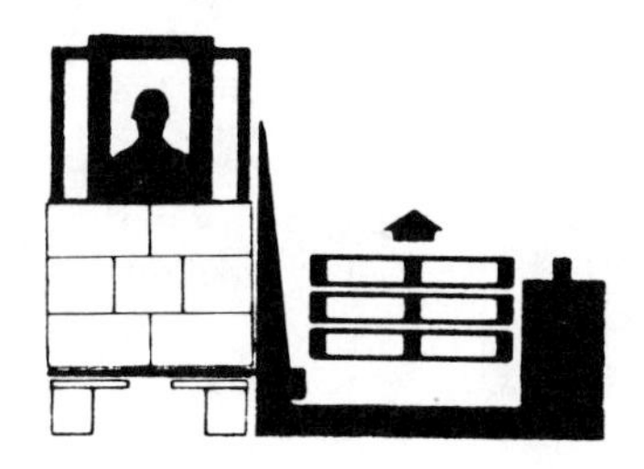

3. THE SHIPPING PALLET, SLIPSHEET OR BARE PLATENS ARE PLACED DIRECTLY UNDER THE TRANSFER PLATE FOR RECEIVING THE LOAD ON THE NARROW FACING.

3a. THE SHIPPING PALLET, SLIPSHEET OR PLATENS MAY ALSO BE POSITIONED FOR RECEIVING THE LOAD ON THE WIDE FACING.

4. THE UNIT CYCLE SWITCH OR RADIO CONTROLLED CYCLE SWITCH IS AGAIN ACTUATED. THE TRANSFER PLATE SLIDES OUT FROM UNDER THE LOAD, TRANSFERRING IT TO THE RECEIVING LIFT TRUCK.

5. ONCE THE TRANSFER PLATE IS COMPLETELY WITHDRAWN, THE TRUCK IS FREE TO PROCEED WITH THE LOAD.

Fig. 5.4.3 Load transfer cycles.

Load Transfers from Pallet. Automatic cycle has been actuated by lift truck driver (by pulling rope suspended from ceiling or by otherwise pushing start button). Hydraulically powered transfer plate pushes load onto receiving plate, which has moved to the full up position.

Multiple Forks Engage Loads. The load is now resting on support bars ready for pickup by lift truck with six forks. At this point receiving plate has automatically lowered. As the transfer plate retracts, the empty pallet has moved into position for automatic staging.

Pallet Automatically Staged. The pallet has been staged, the lift truck driver elevates his multiple forks to clear the load support bars. Load is now in position for transporting to load railcar, truck, or in warehouse area.

Another attachment has the following characteristics:

Load transfer capacity: 4000 lb (1814.4 kg)

Pallets stacking at storage capacity: 15 pallets

Range of pallet sizes: Any combination to 48 in. × 48 in. (1219.2 mm × 1219.2 mm)

Weight (with receiving deck): Approx. 7500 lb (3402 kg)

Length: 15 ft 0 in. (4.6 m)

Width: 6 ft 0 in. (1.8m)

Height: 4 ft 8 in. (1.4 m)

Electric motor size: 5 Hp

Power required: 220–440 v, 3-phase

Size hydraulic pump: 5 gpm

Total cycle time: 23 sec

5.4.2 Forklift Tine Specifications

Full tapered tines are desirable. If these are not available, regular tines can be modified as shown below. Tines must be tapered for a dull knife edge, with all sharp edges rounded, and ends of tines must be even with each other.

Tines should be polished and kept lubricated with grease, oil, or wax (preferably wax) to facilitate sliding beneath the slipsheet. Bottoms of tines should be polished and lubricated for double tiering use. Following are recommended tine specifications:

Tilt: 3–5° forward necessary

Length of tine: 36 in. (914.4 mm) minimum

Width of tine: 5 in. (127 mm) preferred, 4 in. (101.6 mm) can be used.

Alignment of tines: When lowered to floor, each tine should rest even on level floor

Taper of tines: See diagram of tine tip in Fig. 5.4.4

5.4.3 Miscellaneous Push–Pull Equipment

For organizations that do not have sufficient capacity to use push–pull equipment attachments of heavy capacity, there is a low-cost unit that makes slipsheet load handling faster and easy. The system transports bulky slipsheet loads quickly onto the pallet, and motorized hand pallet trucks operate the unit, which is powered by a 24-v system. A pallet truck with the unit is the simple, fast, affordable approach to moving slipsheet and pallet carloads in-house.

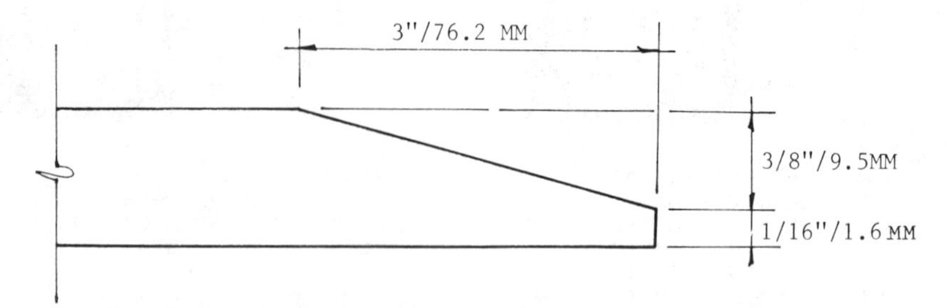

Fig. 5.4.4 Recommended taper for last 3 in. of fork tine.

Specifications for Hand Pallet Truck

Maximum load capacity: 4500 lb (2041.2 kg)
Minimum fork height: 2.9 in. (69.9 mm)
Maximum fork height: $7\frac{3}{4}$ in. (196.9 mm)
Width overall of forks: 27 in. (685.8 mm)
Width of fork tines: 6 in. (152.4 mm)
Distance between fork tines: 15 in. (381.0 mm)
Fork length: 48 in. (1219.2 mm)
Wheel dimensions, steering: 7 in. × $2\frac{1}{2}$ in.
Fork, single urethane: 2.9 in. × 4 in. (69.9 mm × 101.6 mm)
Weight with truck: 528 lb (239.5 kg)

Specifications for Motorized Pallet Truck

Maximum load capacity: 2650 lb (1202.0 kg)
Minimum fork height: 3.3 in. (82.3 mm)
Maximum fork height: 7.9 in. (196.9 mm)
Fork length: 48 in. (1219.2 mm)
Travel speed with load: 2.8 mph
Travel speed without load: 3.7 mph
Batteries and motor: Battery, 24-v system; 2 12-v batteries
Maximum charge in current: 10 A
Drive motor: 0.07 Hp
Speed control: 2-Speed
Weight with truck: 725 lb (328.9 kg)
Winch specifications, 24-v DC Motor 450: 1 gear ratio
Capacity: 2650 lb (1202.0 kg) 7° incline
Speed: 16 ft (4.9 m)/min with load
Cable: 7 × 19 aircraft type $\frac{7}{32}$ in. × 11 ft long
Hand actuator: Remote control
Hand actuator clutch: Equipped with pallet stop safety switch

Specifications for Gripping Bar

18 in. gripping width (457.2 mm)
4000 lb clamp force (1814.4 kg)
Weight: 23 lb (10.4 kg)

Specifications for Ramp

40 in. × 40 in. (1016 mm × 1016 mm) 7° incline
Hard coat anodyzed aluminum with Teflon-impregnated surface
Weight: 46 lb (20.9 kg)

5.5 PACKAGING METHODS

This section covers the three most widely used methods of unitizing loads with packaging methods. They are:

Stretch film
Shrink wrapping
Banding (strapping)

5.5.1 Stretch Film

To evaluate the type of film to be used for each particular need, a simple analysis must be made to evaluate the application. The following factors should be considered:

What product or products are being loaded on the pallet or slipsheet?

Is load configuration regular or irregular?

What is the average load dimension, in inches (length × width × height)?

What is the pallet weight per pound (this is product weight)?

How many pallets per day and slipsheet loads per day?

Describe the service conditions the film must withstand: outdoor exposure; how long (how many days, weeks, months, years), maximum temperature (Fahrenheit or Celsius), normal temperature, minimum temperature, and abrasives.

Are you currently using or evaluating stretch film?

What type of film is now being used [0.8 mil PVC, 1.0 mil LDPE, 1.0 mil EVA (below 12% VA content)]?

Stretch Wrap Material

Stretch wrapping has been established as an important unitizing alternative for approximately one decade now. A major factor in the growth and acceptance of this technology is the cost reduction it offers (Figs. 5.5.1 and 5.5.2). Most users feel that a cost saving of .20¢ to .30¢ per unit load is possible, compared to other unitizing methods, often coupled with greatly improved load protection.

Stretch wrap systems would perform best with some attainable magic film that offered near-infinite stretch, total resistance to rips and tears, extremely high strength, and bargain basement costs. Unfortunately, no such film exists. The resins most commonly used in today's stretch films are:

1. Linear low-density polyethylene (LLDPE)
2. Low-density polyethylene (LDPE)
3. Ethyl vinyl acetate copolymer (EVA)
4. Polyvinyl chloride (PVC)

Each resin has specific properties which it brings to the film and thus determines the film's primary performance characteristics. Blends of more than one resin are possible when specific characteristics are required. Tables 5.5.1 and 5.5.2 compare pallet wraps.

Strength of the film is typically quoted by its supplier in terms of tensile ultimate stress (in units of psi). An average film such as PVC develops strength of approximately 3700 psi; a high-strength film like LLDPE is more likely to be around 6000 psi. These figures are readily available from film suppliers as are the thicknesses of the film. When film tensile stress is multiplied by film thickness and width, force resistance of the film (in pounds) is determined.

Stretchability is a measure on how much a given film can be stretched over a given load profile. You can classify the three major load profiles:

A—Regular loads with no puncture hazards

B—Irregular loads with puncture hazards up to 3 in. (76.2 mm)

C—Random shape loads with puncture over 3 in. (76.2 mm)

Some types of film are much more stretchable than others, and such is indeed the truth. On ordinary equipment, the linear low-density polyethylene (type 1) can be stretched up to 55% over a profile A load. In contrast, a type 3 low-density polyethylene film can be stretched only 15% over the same type of load. These are "production" stretch figures attainable with good management, and they are far less than figures obtained under ideal conditions.

Stretchability is an extremely important point of comparison between the different types of films because it directly affects installed costs. It is stressed as a point of competitive advantage among manufacturers of different films. It is useful and significant only if you are actually getting as much stretch as you think you are.

Prices differ among stretch films in relation to their performance characteristics. As you might expect, films with the most desirable strength and stretchability properties are the most expensive. However, films that cost more may actually be less expensive to use if they can be stretched as much in real applications as their properties seem to imply. Generally, type 1 films stretch 50–100% farther than type 2 films. Type 2 films stretch 33–50% farther than type 3 films.

Stretch Film Test Procedure

STEP 1. Determine the Proper Stretch Percentage for the Film. Wrap the load at a given brake setting. After wrapping, place a 10-in. (254-mm) vertical slit in the film which has been applied to the load. If the slit does propagate, decrease the brake setting slightly and rewrap.

NET SAVINGS CALCULATIONS

Number of Loads/Hour	× Ⓩ Net Savings/Load	= Net Savings/Hour
Number of Working Hours ×	Net Savings/Hour	= Net Savings/Day
Number of Days/Week ×	Net Savings/Day	= Net Savings/Week
Number of Weeks/Year ×	Net Savings/Week	= Net Savings/Year

EQUIPMENT COST AMORTIZATION CHART

The chart below is an aid to help you determine the length of time it will take for the stretch wrap savings to pay for the equipment.
By intersecting the **YEARLY $ SAVINGS COLUMN** with the **CAPITAL EXPENDITURE COLUMN** you will determine the number of months for payback. It's straight savings from then on.
Example: If you saved $20,000 in your first year and purchased an $8,000 unit, you would pay for the machine in 4.8 months. Just follow the two dark lines on the chart.

SAVINGS $ THOUSANDS/YEAR											
40	.30	.60	.90	1.2	1.5	1.8	2.1	2.4	3	4.5	7.5
30	.40	.80	1.2	1.6	2	2.4	2.8	3.2	4	6	10
25	.48	.96	1.4	1.9	2.4	2.9	3.4	3.8	4.8	7.2	12
20	.60	1.2	1.8	2.4	3	3.6	4.2	4.8	6	9	15
15	.80	1.6	2.4	3.2	4	4.8	5.6	6.4	8	12	20
10	1.2	2.4	3.6	4.8	5	7.2	8.4	9.6	12	18	30
8	1.5	3	4.5	6	7.5	9	10.5	12	15	22.5	37.5
5	2.4	4.8	7.2	9.6	12	14.4	16.8	19.2	24	36	60
	1	2	3	4	5	6	7	8	10	15	25

CAPITAL EQUIPMENT EXPENDITURE $ THOUSANDS

Fig. 5.5.1 Summary calculations.

Continue this process until the brake setting is achieved that will stretch the film to its maximum extent without propagation of the 10-in. (254-mm) slit. (Rationale: The maximum stretch possible is desirable, but not so much that a tear in the film during pallet shipping or storage would result in zippering.) The recorded tension setting corresponds to the maximum tension that can be applied to the film.

STEP 2. Record the Percentage of Stretch. After completing Step 1, remove film from the pallet and record its weight. Lower the tension setting to zero and record the weight of film for the same pallet. Stretch, defined as increase in length over original length, may then be determined as follows:

Select the appropriate or closest looking pallet style from the examples given, utilizing the suggested material value range—or your own. Fill in the blanks and complete the calculations.

Refer to the summary section and compute the net savings—the difference between your existing unitization costs and the proposed stretch wrap methods—for the various duration periods.

The cost amortization chart (Fig. 5.5.2) will determine the equipment write-off, in months. See instructions.

EXAMPLE A
STRAP AND ANGLEBOARD

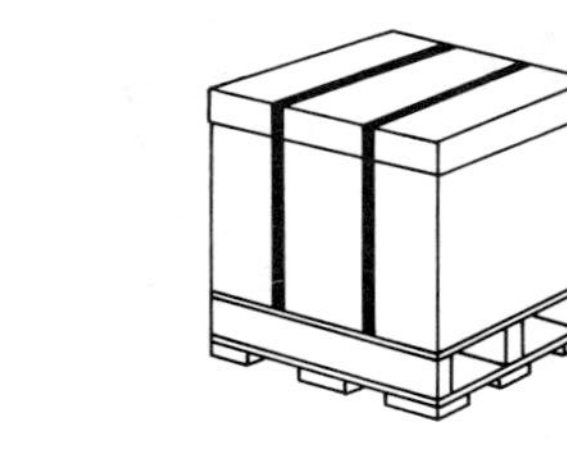

EXAMPLE B
STRAP AND MASTER CARTON

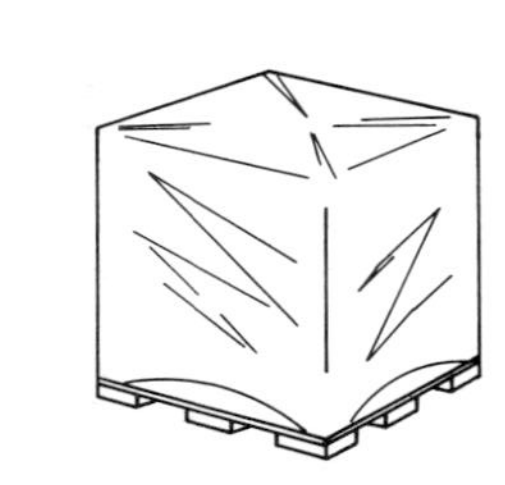

EXAMPLE C
SHRINKWRAP

MATERIAL COSTING—EXAMPLE A

$\frac{3}{8}$	$\frac{1}{2}$	1	¢/ft Strapping	
Straps ×			ft × ¢/ft =	
3	5	7	¢/ft Angleboard	
Ft			× ¢/ft =	
$\frac{3}{8}$	$\frac{1}{2}$	1	Cents clips	
Pieces ×			¢ each =	
			Subtotal a	

MATERIAL COSTING—EXAMPLE B

$\frac{3}{8}$	$\frac{1}{2}$	1	¢ ft Strapping	
Straps ×			ft × ¢/ft =	
$\frac{3}{8}$	$\frac{1}{2}$	1	Cents clips	
Pieces			× ¢ each =	
\$3	\$5	\$8	Master carton	
Unit cost ÷ trips =				
			Subtotal b	

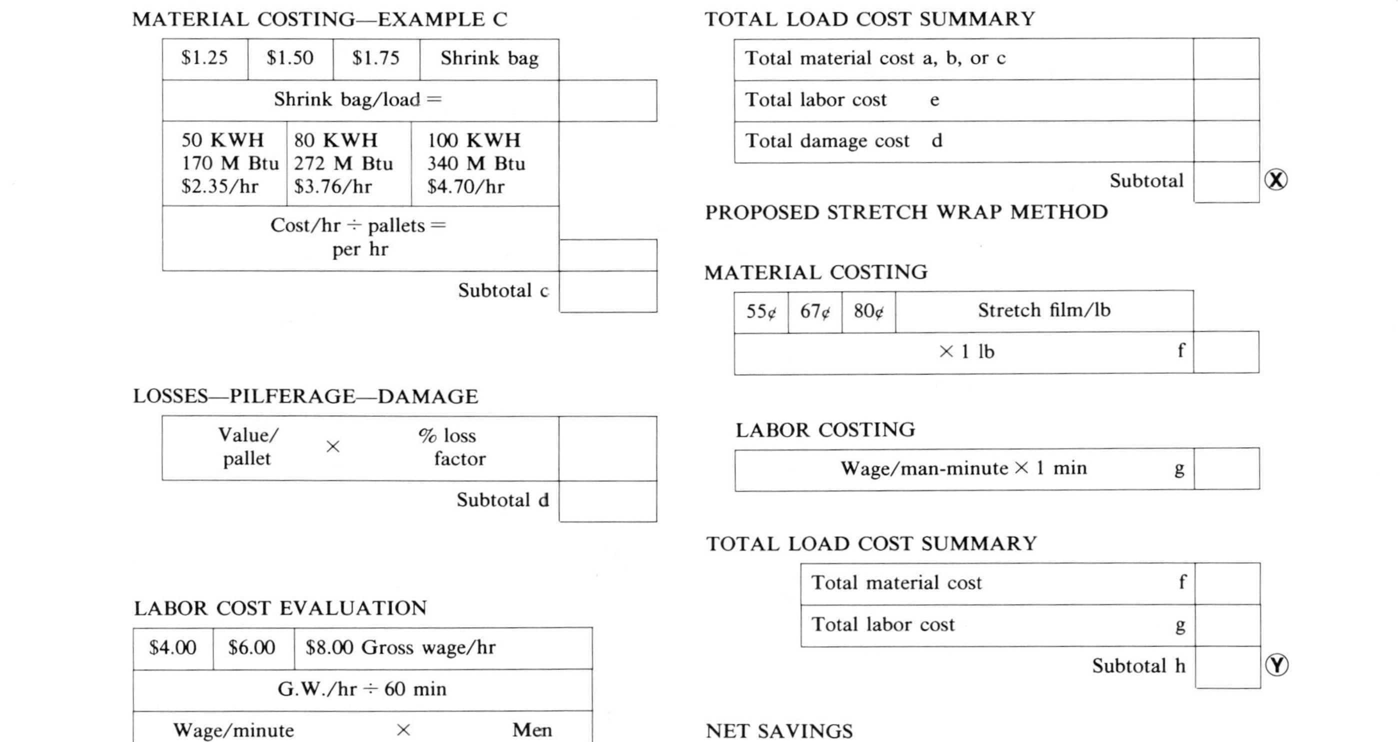

Fig. 5.5.2 Evaluation procedure.

Table 5.5.1 Comparison of Pallet Wraps by Typical Values

Physical Properties	LDPE[a]	PVC[b]	EVA[c]	Coextruded Polyethylene	Modified EVA
Gauge in mils	1.00	0.80	0.80	0.90	0.80
Yield per pound (in.^{2}/lb)	30,000 (19354 80 mm)	28,000 (18064 48 mm)	36,500 (23548 34 mm)	33,600 (21677 38 mm)	37,200 (23999 95 mm)
Tensile strength (psi)	1,500 (680.4 kg)	3,500 (1587.6 kg)	5,500 (2494.8 kg)	5,300 (2404.1 kg)	6,000 (2721.6 kg)
Ultimate tensile strength is measured on the Instron Tensile Tester in pounds per square inch. The tester stretches a 1-in. wide piece of film at a rate of 20 in./min. The force and amount of elongation when the film rips apart is shown as ultimate tensile strength and elongation.					
Elongation (% MD)	200	250	450	580	700
Ultimate elongation is also measured on the Instron Tensile Tester in percent.					
Cling (lb/in.)	0.6 (0.27 kg)	1.5 (0.68 kg)	1.2 (0.54 kg)	1.6 (0.73 kg)	1.3 (0.59 kg)
Lap shear is the test used to measure the cling of the film. Lap shear is the force needed to pull apart two 1-in. (25.4-mm) wide film strips pressed together over 1 in.^{2} of contact area.					
Tear resistance (g/mil)	150	125	120	550	500
Tear resistance can be measured using the Elmendorf Tear Test. In this test a rapidly moving blade continues the tear. The results are measured in grams of pressure needed to sustain the tear. The higher the number, the more resistant the film is to this kind of damage.					
Stress retention (% of original) after 16 hr:					
30% Stretch	60	30	55	65	62
60% Stretch	N/A	N/A	55	62	65

Property					
This is strength retained 16 hr after wrap had been stretched 30% and 60%. Higher numbers mean better long-term strength.					
Spencer puncture resistance (psi)	5,500	16,700	21,000	11,900	19,000
Spencer puncture resistance is measured using a modification of the Elmendorf equipment. The units are the pounds per square inch needed to puncture the film being tested. The higher the number the tougher the film.	(2494.8 kg)	(7575.1 kg)	(9525.6 kg)	(5397.8 kg)	(8618.4 kg)
Dart impact: 0% failure (g)	60	300	280	90	230
The dart impact test measures the weight in grams of a dart dropped 26 in. to puncture unstretched film held by a fixture. These tests were done at 0% failure.					
Clarity: Gloss (%)	77	89	74	84	74
Haze (%)	1.4	1.3	1.6	1.3	3
In the gloss measurement, a higher percentage indicates a more shiny surface. In the haze measurement, the higher percentage indicates a greater degree of cloudiness. However, it should be noted that all values are relatively equal, and there is little or no loss of product identification when using any of the films.					
Temperature range (°F)	0/120	20/120	−20/120	0/120	−20/120
All five films perform acceptably up to 120°F. However, the lowest application temperature for PVC is around 20°F, when it becomes brittle. LDPE's lose their stretch around 0°F. Certain films stretch well at low temperatures and can be used down to −20°F.	(0/49°C)	(7°C/ 49°C)	(−29°C/ 49°C)	(0/49°C)	(−29°C/ 49°C)

[a] Typical cling low-density polyethylene.
[b] Typical polyvinyl chloride.
[c] High vinyl acetate content.

Table 5.5.2 Comparison of Pallet Wraps by Typical Wrapping Properties

Wrapping Properties[a]	LDPE[b]	PVC[c]	EVA[d]	Coextruded Polyethylene	Modified EVA
Normal stretch (%)[e] This value is the percent stretch obtained using a flat-sided load at approximately the same tension.	15–20	15–45	20–45	30–55	40–70
Yield at maximum stretch (in.^{2}/lb) This demonstrates how much area a pound of each film will cover at the optimum machine tension.	36,000 (23225 76 mm)	41,000 (26451 56 mm)	53,000 (34193 48 mm)	52,000 (33548 32 mm)	63,000 (40645 08 mm)
Conformation This indicates how well a film will conform to a highly irregular-shaped load.	Low	Medium	High	High	High
Tear resistance This indicates if the film can be wrapped over an irregular load using appropriate tension without tearing and also passes the immediate 10 in. (254 mm) cut test.	Seldom	Occasionally	Occasionally	Nearly every case	Nearly every case
Long-term elasticity This relates to the effect of vibration on wrapped loads during shipping. The more elastic films hold the load better.	Medium	Low	High	Medium	High

[a] These values were obtained in laboratory tests and have been verified in numerous field tests. Due to the variety of conditions found in the field, results may differ.
[b] Typical cling low-density polyethylene.
[c] Typical polyvinyl chloride.
[d] High vinyl acetate content.
[e] The maximum stretch values are for narrower widths.

$$\% \text{ Stretch} = \frac{\text{Film weight at zero tension} - \text{Film weight at maximum tension}}{\text{Film weight at maximum tension}} \times 100$$

STEP 3. Determine the Cost of the Film Applied to the Load. After completing Step 1, wrap the same load several times, cutting the film completely from the load after each wrapping. Weigh each bundle of film, and divide the total weight by the per pound cost of the film. (Since the weight of each bundle of film may vary slightly, an average weight for several bundles is used to improve the validity of the test results.)

STEP 4. Determine If the Load Will, in Fact, Reach Its Destination. Since the "acid test" of a film's suitability is its performance in an actual storage and shipping environment, identical pallets wrapped with competitive films should be stored and shipped under identical conditions (shipped on the same truck if possible) and the condition of the film and the load examined at its destination. This test is also important to determine if a film's cling level is adequate, since there is not another reliable method for testing cling.

Stretch Film Methods

With conventional stretch wrappers, the rotating load on the turntables exerts the stretching force that pulls the film from the roll. Film stretch is accomplished by restricting the film on wind motion with some type of frictional braking device. See Fig. 5.5.3.

The roller-stretch method, on the other hand, isolates the stretching process between two rollers. There is no film break; the exit roller rotates faster than the entry roller to apply the film stretch. The rotating load on the turntable simply maintains the tension on a film as it is applied to the load.

With the conventional stretch wrappers, all of the force to stretch the film is exerted between the film roll and the unit load. Force application is affected by the load profile, the film itself, and the braking system. The variation on even the best braking device is up to 16%. When you add the fact that virtually all of the film stretch force is concentrated on the corners—the worst puncture hazards on the low profile—it is little wonder that operators are prompted to turn back the tension control to prevent premature film failure. This loosely controlled stretching process simply cannot tap the additional stretch in every film. By isolating the stretch film between the two rollers, as in the roller-stretch method, the undesirable variables have been eliminated. That brings a critical stretching process off the load and under control. And because the film stretch is so precisely controlled, roller-stretch systems provide at least twice the stretch possible with any conventional system, no matter what type of film is used.

5.5.2 Shrink Wrapping

Shrink wrapping of pallet load insures the highest possible protection against distortion and stress during transportation. Shrink wrapping can follow immediately after the palletizing process. A number of organizations have pioneered shrink wrapping techniques as well as developing the tubular shrink wrapping system. The gusseted tubular film is drawn from the supply reel, expanded, and pulled down over the load without touching the sides. When the length of the tube corresponds to the height of the load, the downward movement is interrupted and the tube is separated, formed into a hood by a heat sealer, and then drawn below the base of the pallet.

Following this, the load is transferred into the shrink oven where the film is exposed to the action of warm air and thus shrinks. On cooling, the film forms a skin-tight wrapping around the pallet and the load, resulting in one compact unit.

Pallet shrink wrapping protection can be used for all kinds of palletized goods, even for products sensitive to heat and pressure. Shrink wrapping is an additional protection against dampness and dust and obviates pilferage. It also enables the contents of the load to be clearly seen, and the use of printed shrink film provides an extensive advertising medium. Polyethylene film is not detrimental to the environment—disposal of used film has no adverse effects.

The early innovation of shrink wrap packaging came at a time when the high cost of energy was not a significant factor. The palletized load is completely encased in a layer of film, or bag which is seamed, then subjected to heat; this causes the film to shrink around the load and accept its shape. Various thicknesses of film must be used to allow for variations in load strength requirements. Shrink wrapping is best applied when standard-size items can be automated, will not be affected by heat, and require complete weather protection.

5.5.3 Banding

Metal banding is one of the earliest methods of securing materials loaded on pallets. It has been modernized by substituting other types of materials in place of the metal bands. However, it offers

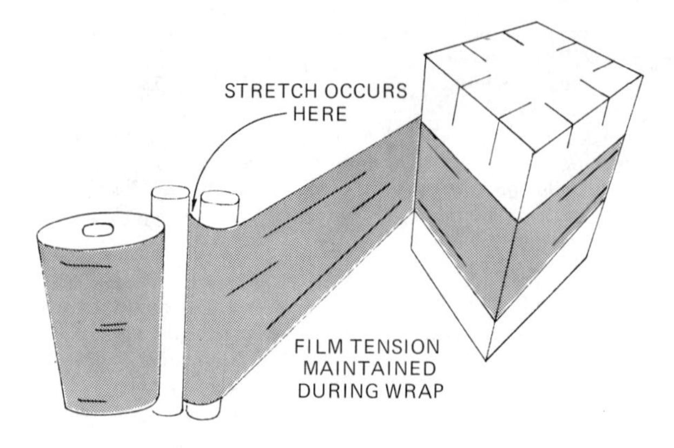

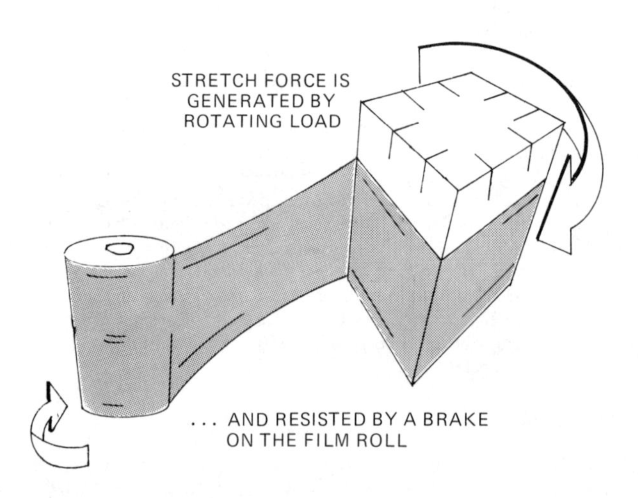

Fig. 5.5.3 Stretch wrap methods.

no weather protection and you can often secure certain loads without having them actually damaged by the straps cutting into the packaging during handling and load shifts while in transit. Banding is best suited for extremely heavy loads of pipe, steel bars, or heavy objects that cannot be damaged by banding.

5.5.4 Automatic Wrapping Concepts

Stretch Bagger

The stretch bagger uses a specialized form of stretch film, usually 3–4 mil thickness, and provides only 12% maximum stretch to the film material, limiting its application to uniform pallet loads such as cartons or bagged products. The system is capable of processing 40–50 loads per hour including all materials handling, but requires an extensive amount of floor area with an excessive ceiling height clearance for the stretch tower.

Hood Shrink Wrapping System

The reverse hood-wrapping system is one of the latest developments in load unitizing. A palletless stack load is shrink wrapped by the inner hood and reserve hood unit. The load can be handled in the same way as a normal pallet load.

This system has the advantage of eliminating pallets and offers a completely weatherproof unit load. Palletless shrink wrap loads can contain sensitive products, and can be transported on open vehicles, loaded and unloaded a number of times, and stored outdoors. Standard machines of palletizing and shrink wrapping techniques are main parts of the palletless packaging line.

Pass-Through Systems

The pass-through systems are manufactured in the United States by two companies and provide anywhere from a 6–10% elongation of the film utilized. The system incorporates a wall of film 2–3 mil thick and is extremely fast, boasting production speeds of over 100 loads per hour. It is an ideal solution for the handful of major corporations with that scope of productivity, providing that the degree of elastic stability required is not demanding. The seals are the key to the success of this concept, and a high maintenance and adjustment factor of the complicated sealing mechanism has somewhat tarnished its history. The application of this machine for light, unstable, or random-shaped loads makes it inappropriate for general industry.

Full Web Rotary System

This system concept, currently manufactured only by one firm, grips the bottom 3 in. (76.2 mm) of the film and leads it around almost the entire perimeter of the pallet load under no tension before the stretch effect is applied. The stretch elongation applied, in practical measurements, is often less than 15%. The full web rotary system is fast and can produce in excess of 60 loads per hour.

Spiral Rotary Systems

The spiral rotary systems are manufactured by at least three major corporations and are the most economical in terms of film consumption because the elongation of the narrow web of the film can be maximized by the braking system. The spiral rotary systems are operated in a convolute (cross-spiral) function which, unless the entire system is speeded up, holds the maximum production rate to the 40–55 load per hour range.

Dual-Roll Rotary System

This system, produced only by one manufacturer, operates with a double spiral action, wrapping both up and down simultaneously to shorten the duration on the machine. It has all of the advantages of the standard spiral system. Speeds are quoted at 100 loads per hour.

Overhead Rotary System

This system was designed primarily for shrink tunnel replacement applications where an existing tower conveyor could be utilized. Its concept is similar to the rotary system, but the pallet remains motionless on the conveyor and the film tower is driven around it, stretching the film as it travels. This system has several benefits in that very tall or unstable pallets of bulk-pack materials are not disturbed and no compression or steadying platen is required. Throughput on this system averages 40–55 loads per hour.

Portable Platform Rotary System

This system has excellent application where a highly organized operation exists that restricts the activities of the forklift driver. Operating in the same manner as the spiral rotary system, the lift truck driver places the load on the turntable, backs away, and activates the cycle of the machine through a remote pendent switch. This system may become the automatic approach of the future due to potential mass production.

There are, of course, minor deviations to all of these system concepts, but these eight comprise a wide scope of fully automatic operatorless stretch wrap systems.

5.5.5 Horizontal Stretch Wrappers or Stretch Bundling

Horizontal bundles utilize a web of stretch film from 10 in. to 40 in. (254–1016 mm). A product or unit load is put into the ring, the film is attached (manually or automatically), and the roll carriage is rotated around and over the product. The film is then cut and sealed with or without an operator.

The horizontal stretch wrapping application is effective in the following industries:

Textiles (bolt cloth, carpet)
Sheeted goods (styrofoam, paneling, plywood)

Extruded products (aluminum, plastic)
Carton products (bicycles)
Bag products (dog food, agriculture feed)
Electronic products (picture tubes)

5.6 DISTRIBUTION PACKAGING

In the broadest sense, any product that is distributed has to be physically contained. The physical process of containing the product in distribution can be called distribution packaging. Distribution packaging is the function that is fundamentally concerned with the economical preparation of protection of merchandise for shipment and distribution. Distribution packaging includes considerations of transportation, warehousing, and materials handling with primary emphasis on performance and economy. If the basic function of packaging is to provide for performance in the distribution process, it is distribution packaging; if sales acceptance of a product is influenced by packaging, it is consumer packaging.

An example of the compounding effect of packaging inefficiencies follows. Although the results will vary considerably from item to item, the example shown is typical of many products.

5.6.1 Compounding Packaging Inefficiencies

The following hypothetical data will create an example of the compounding of packaging inefficiencies:

Product: Desk-top tape dispenser
Product dimensions: 6 in. × 2 in. × 2.5 in. (152.4 mm × 50.8 mm × 63.5 mm)
Product cube: 30 in.^{3}
Dispensers per case: 12
Case dimensions (outside): 16 in. × 10 in. × 7.125 in. (406.4 mm × 254.0 mm × 181.0 mm)
Case cube (outside): 1140 in.^{3}
Cases per pallet load: 60
Dispensers per pallet load: 720
Pallet load dimensions (maximum): 48 in. × 45 in. (including 3 in. of overhang) × 42 in. (height) [1219.2 mm × 1143.0 mm (including 76.2 mm of overhang) × 1066.8 mm (height)]
Pallet load cube (maximum): 90,720 in.^{3} (52.5 ft^3)
Pallets per 40-ft (12.2-m) trailer (double decked): 36
Trailer dimensions: 40 ft × 8 ft × 8 ft (12.2 m × 2.4 m × 2.4 m)
Trailer cube (maximum): 2560 ft^3 (70.27M^3)
Warehouse storage area per pallet load: 10 ft^2 (0.93 m^2)
[Assumes typical warehouse with four-high pallet storage in 8.25 ft, (2.5 m), narrow aisle racks, allowing 10% of space for staging.]
Warehouse storage cube per pallet load: 200 ft^3 (5.67 m^3) [10 ft^3/pallet (0.93 m^3) × 4 pallets × 20 ft (6.1 m) ceiling ÷ 4 pallets high = 200 ft^3 (5.67 m^3)]

Four types of packaging efficiency can be calculated using the foregoing data: the amount of product per case, the number of cases per pallet load (unit shipping unit module), the number of pallet loads per trailer, and the amount of warehouse storage space required per pallet load.

Case Efficiency

$$\text{Case efficiency} = \frac{\text{Product cube/case}}{\text{Case cube}}$$

$$= \frac{30 \text{ in.}^3/\text{roll} \times 12 \text{ rolls/case}}{1140 \text{ in.}^3/\text{case}}$$

$$= 31.6\%$$

In other words, 31.6% of the cube of the case is actual product in this example.

Case efficiency (31.6%) affects material costs, case forming, packing, and sealing costs and all forms of case handling costs (manual and automated) as well as pallet load efficiency (storage and transportation) explained later.

Pallet Load Efficiency

$$\text{Pallet load efficiency} = \frac{\text{Case cube/pallet load}}{\text{Maximum pallet load cube}}$$

$$= \frac{1140 \text{ in.}^3\text{/case} \times 60 \text{ cases/pallet load}}{90{,}720 \text{ in.}^3\text{/pallet load}}$$

$$= 75.4\%$$

Only 75.4% of the maximum cube of the pallet load is utilized with cases that contain 31.6% product. Therefore, 23.8% (75.4% × 31.6%) of the maximum pallet load cube is actually product in this example. Pallet load efficiency affects materials costs (pallets, slipsheets, stretch wrapping, etc.), pallet load materials handling costs, pallet load storage costs, and transportation costs when shipping by pallet load.

Trailer Load Efficiency

$$\text{Trailer load efficiency} = \frac{\text{Pallet load cube/trailer}}{\text{Maximum trailer cube}}$$

$$= \frac{52.5 \text{ ft}^3\text{/pallet load} \times 36 \text{ pallet loads/40-ft trailer (12.2-m)}}{2560 \text{ ft}^3\text{/40-ft trailer (12.2-m)}}$$

$$= 73.8\%$$

With that, 73.8% of the maximum cube of the trailer is utilized with pallet loads that utilize 75.4% of their maximum cube with cases that contain 31.6% product. Therefore, 17.6% (73.8% × 75.4% × 31.6%) of the maximum trailer cube is actually product in this example.

Trailer utilization with regard to cases is 55.6% (73.8% × 75.4%) of maximum. This figure is substantiated by samples of "fully loaded" trailers that contained approximately 1300 ft^3 of cases [(1300 ft^3 ÷ 2560 ft^3 × 100 = 50.8%)].

Warehouse Storage Efficiency

$$\text{Warehouse storage efficiency} = \frac{\text{Maximum pallet load cube}}{\text{Warehouse cube to store one pallet}}$$

$$= \frac{52.5 \text{ ft}^3}{200 \text{ ft}^3}$$

$$= 26.2\%$$

As a result, only 26.2% of the typical warehouse cube is utilized with pallet loads that utilize 75.4% of their maximum cube, with cases that contain 31.6% product. Therefore, 19.8% (26.2% × 75.4%) of the warehouse storage cube is occupied with cases and 6.3% (19.8% × 31.6%) of the warehouse storage cube is occupied with actual product (assuming 100% slot occupancy in both cases).

This hypothetical (but typical) example is shown solely to focus attention on how inefficiencies in packaging and distribution tend to compound. See Tables 5.6.1 and 5.6.2 for a summary of the example showing this compounding effect.

Obviously, 100% efficiency in distribution packaging is impossible. However, significant improvement over these figures is possible. Remember that these efficiencies, particularly case efficiency, vary considerably depending on type of product.

Also, since these efficiencies are fairly easy to calculate, they can be used to determine which products or types of products have the greatest potential for improvements in distribution packaging.

Table 5.6.1 Trailer Utilization Example

Product/Case = 32%	
times	= 24% Product/Pallet
Cases/Pallet = 75%	
times	= 56% Cases/Trailer
Pallets/Trailer = 74%	
equals	
Product/Trailer = 18%	

Table 5.6.2 Warehouse Utilization Example

Product/Case = 32%	
times	= 24% Product/Pallet
Cases/Pallet = 75%	
times	= 20% Cases/Warehouse
Pallets/Warehouse = 26%	
equals	
Product/Warehouse = 6%	

5.7 COMPUTERIZED PACKAGING SPECIFICATIONS

The information that follows is a capsule of the changes and improvement in modern computer program technology. It is a combination of computer-assisted services and techniques for packaging-oriented programs available today.

Data entry features are needed to effectively automate specifications.

5.7.1 Computerized Packaging

Hardware

Additional terminals, printers, memory, and storage space can easily be added. The system can cope with organizational changes and evolve with organizational needs. The system can be expanded from one terminal and 10 million characters of information storage, to dozens of terminals and billions of characters of storage, if needed.

Package Design Programs

New package design software is available that enables the user to better coordinate package design with other departments. This coordination results in substantial savings. New pallet patterns, case arrangements, and package sizes can be found at reduced materials cost while improving distribution and transportation efficiencies.

Project Management Control

Software can also be added to control and adequately track major projects, their sequence of tests, and relay due dates. Any slip in project schedules will be immediately identifiable.

Computing Ability

Additional computer programs can be written or purchased for installation on your computer to solve other information needs such as standard word processing, packaging line simulation, engineering calculations, and so on.

Software Enhancements and New Releases

Purchase of subscription entitles the users to get their own copy of the newest specification system enhancements and software tools for packaging engineers and distribution groups.

User-Designed Questions

Menus for creating specifications and package component data bases are built by the user using simple English commands. Anyone subsequently creating a specification or adding information to the data base must answer these predefined questions and must provide complete information; responses are checked for accuracy.

Standardization

Specifications are always of a standard format, containing all the needed information, regardless of who created them. Training new employees is streamlined.

The system itself enters most of the general data. The user enters a product code, the computer validates that product code, looks up and enters its description, finds the plant manufacturing the product, and enters the plant's name and address. The computer will even ask different sets of questions depending on the type of code product entered, and will list all possible answers to each question to help the user to select the best answer.

Automatic Calculations and Conversions

The engineer enters inside dimension data, and the computer calculates outside dimensions and blank sizes. Fractional dimensions can be entered and decimal dimensions will be calculated. English and metric conversions are readily available. Unit load and case weights can be computed automatically.

Extensive Multichannel Editing

Far more than just numeric range checking, proper data formatting, and field length verification are provided. Tables of allowable responses for specified questions are built. Responses to previous questions are then used in limiting the acceptable table of responses even further. For example, allowable board grade responses are 125 lb B flute, 150 lb B flute, 200 lb C flute, and 275 lb C flute.

However, if the user indicates a very heavy case load, the computer automatically eliminates 125 lb B flute and 150 lb B flute from the list of acceptable responses. When a question "enter board grade" is asked, the only permitted responses are 200 lb C flute and 275 lb C flute.

Unique Reports for Each Department

Each department can lay out its own easy-to-read report format in minutes without requiring any special programming. Only the information the department is interested in is printed.

Written and Packaged Terminology

Reports are easy to request and understand since the computer converses using packaging phrases and terms (e.g., RSC, folding carton, flute, manufacturing joint, linerboard, tare weight, bursting strength, facings, compression, vibration, and corrosiveness).

Complete Indexing and Cross-Referencing

Almost any specification-related question can be answered accurately and quickly. For example:

1. What products use that particular specification?
2. What package specifications are required by that product?
3. Which specifications have not been revised the last year?
4. Which products are effected by the label change?

Conditional Reporting

Conditional data searching (i.e., "and/or" questioning) allows each user to extract information based on his specialized request. For example:

1. Print a report showing all the types of plastic closures that house products containing high levels of corrosive ingredients A, B, or C.
2. Search existing shipping case specifications to find one that may be used for a new or redesigned product.

Graphical Output

Technical specifications and ideas that are tough to verbalize are easily expressed through drawings and diagrams. Pallet patterns are clearly illustrated. Scaled engineering and mechanical drawings as well as color graphics capabilities are enhancements planned for the near future.

Multiplant Inquiry

Avoid delays associated with the typewriter, copier, and mail by harnessing the speed of the computer and the power of phone-line data transmission. Each plant can have a remote computer terminal connected to the central specification data base via phone. Current specification information is instantly available to authorized users.

Individual Departments Have Access

Portable lightweight computer terminals (smaller and lighter than many portable typewriters) can be placed in key department managers' offices. If one of these managers needs quick specification information for a supplier or a customer, he or she simply sits down at the desk, asks the computer to find and print the needed data, and seconds later the needed information is in hand.

No Reliance on the Central Computer

All the specification information is stored on a stand-alone minicomputer. The user does not have to wait until accounting has finished payroll, or marketing has completed their sales forecasting. The user has access to the data at any time, any working day, or at odd hours, because the computer is designed specifically to be responsive to his needs.

Built-In Security Controls

Each user must know the initial time-sharing sign-on procedure as well as the software password. In addition, each user is assigned a security category and each category permits access to only certain pieces of the data base (e.g., an engineer developing a request for quotation that would be sent to a supplier, would not be permitted access to sensitive information items).

Restricted Authorization of New Specifications

In addition to password security, three different permission levels and an authorization code exist to protect against unapproved releases or unauthorized changes of new specifications.

Storage of Historical Specifications

Avoid losing valuable information or violating legal requirements by losing out-of-date specifications. Up to 10 versions of each specification can be stored in the master data base with an unlimited number of older versions stored in the history data base.

5.7.2 Packaging Design and Palletization

The computer can be used to increase pallet load unitization; reduce materials, shipping, and distribution costs; calculate load and stacking strengths; select board grades and flute type; and to design cartons and their arrangements in shippers and on the pallets.

Computerized packaging specifications can create, modify, and reproduce package specifications; provide rapid multiplant communications; automatically file and index specifications; and provide flexible and easy information retrieval.

5.7.3 Cube Utilization Audit

A cube utilization audit can evaluate unit shipping modularization and case consolidation; perform slipsheet analysis for the entire product line; compile pallet pattern reference manuals; and increase cube utilization of transportation vehicles and warehouse space.

Cube utilization audit provides an in-depth analysis of the distribution system space utilization. Packaging professionals meet with packaging and distribution personnel to gather information and understand the operations.

The first step in a cube utilization audit is to identify the shipping unit module. The shipping unit module is the cubic space that the products must be shipped within. For each company the shipping unit module may be different. The unique operational and handling requirements determine the dimensions of the shipping unit module. Pallet load dimensions, truck/railcar dimensions, rack dimensions, and warehouse configurations all contribute to defining the shipping module.

Product lines usually evolve through time with little consideration given for distribution efficiencies. Thus, a cube utilization audit is probably needed to improve space utilization in your distribution network.

Because capital evaluation will produce a shipping unit module that best suits your operations, the efficient utilization of the shipping unit module is the goal. A cube utilization audit includes a complete review of the palletization efficiencies of all the major products. If possible, alternate pallet patterns will be determined for those products with poor efficiencies.

For the remaining products, all possible rearrangements or product within the shipping case will be evaluated to find the most cost-effective configuration. Results should yield an increased shipping unit module cubic efficiency, and may bring about a decrease in packaging material costs. Additional areas that will be looked into are:

Clampability evaluation
Stacking strength calculations
Customized pallet pattern manuals
New product design packaging
Customized computer programs for special application
Slipsheet analysis

5.8 ANTISKID METHODS OF RESTRAINING LOADS

5.8.1 Introduction

There are several types of treatments on the market for application to the outer surface of fiberboard shipping containers for the intended purpose of reducing slipping or sliding. The containers usually slip or slide during movement in transit over the road or by lift trucks in-house. The treatments may be classified as:

Colloidal silicas
Aluminum oxides
Nonskid inks
Nonskid varnishes
Mechanical embossing
Resins or latices
Adhesives

5.8.2 Testing for Antiskid Effectiveness

There are two methods commonly used for testing antiskid: (1) Determining the "angle of slide" or coefficient of friction of the linerboard, applicable to all types; and (2) spraying an indicator onto the surface of the box, applicable when the container has been treated with colloidal silica or aluminum oxide.

The angle of slide test is preferred by many as giving a measure of the resistance to slide. In brief, this test consists of a weighted specimen being placed face-to-face on a similar specimen mounted on an adjustable inclined plane. The angle of the plane is increased until the specimen slides. The coefficient of friction is the tangent of the angle of slide.

Another type of tester determines the coefficient of friction by pulling one specimen, mounted under a small weight or sled, over the other on a horizontal plane and measuring the force required to pull it. The force divided by the weight of the sled is the coefficient of friction. This may be stated as the static friction (the force required to *start* the specimen sliding) or the kinetic friction (the force at which the specimen continues to slide). (Reference: TAPPI Suggested Methods, T-815 and T-816; and ASTM Standard Methods, D-3248 and D-3247.)

In the indicator method, a chemical (lactone) is sprayed on the treated surface from an aerosol spray can. The treated area turns color. Response is due to reaction with moisture in the silica or alumina of the colloid treatment. The intensity of color (a measure of concentration of application) will be affected by the amount of moisture in the material at the time of checking with the indicator. Also, the treatments show less response with aging, color response varies with different materials (the original blue indicator did not show up on some samples under certain conditions), and blue tends to fade. Therefore, a red indicator has been introduced since the advent of alumina type antiskid material.

This method, as the name implies, indicates the presence of the silica or alumina treatment and is used to monitor production, but does not measure the antiskid effect.

5.8.3 Comments on Antiskid Treatments

The degree of effectiveness of antiskid treatments is inversely proportional to the original slickness of the substrate (i.e., the rougher the surface of the linerboard, the less effective is the antiskid treatment).

Present-day regular Kraft facings are not finished highly, purposely, and only a few degrees increase in the angle of slide can be achieved by treatment and this may not result in a noticeable improvement in service. Another consideration is that there is an optimum amount of antiskid material that can be applied; a greater amount does not improve effectiveness.

5.8.4 Packaging Methods

Following are some treatments explaining the various methods:

Colloidal Silica

The dominant factor in colloidal silica treatment is stated to be not the roughness provided by the particles, but rather the stiffening of the paper fibers so that the irregular fiber network provides the frictionizing effect.

From a boxmaker's point of view, colloidal silica presents a cleanup problem because of buildup on the applicator equipment and also the abrasive effect on converting equipment. A modified colloidal silica has been introduced to provide easier cleanup. However, the cost of the solids is about twice that of the original formulation.

Aluminum Oxide

The characteristic of this type, which is emphasized to the boxmaker, is ease of cleanup. If excess of aluminum oxide is deposited on equipment, it forms a loose powder which easily wipes away with a wet rag. On the other hand, some users report it is worse in pitting equipment because it is acidic (pH 4.3).

Cost of aluminum oxide application, based on twice the dilution as the silica application, is claimed to be equal and performance is said to be the same.

Nonskid Inks

Inks with antiskid materials added are purposely made abrasive. This results in poor rub qualities and smearing can result. Furthermore, printability is poor. Such materials have been proposed as tints only. These would create fewer rub problems, but printing in colors cannot be matched readily.

Their antiskid effectiveness is dependent on how much of the normal printing qualities can be sacrificed. Cost is greater than with other types of antiskid treatment.

Nonskid Varnishes

These are similar to nonskid inks except that the color problems are not present.

Mechanical Embossing

This consists of punching a pattern of small projections on the outer facing and flap areas of a box. It is done on the corrugator just before joining the outer facing to the single-face board by means of a tool with a series of needlelike points. This was one of the original attempts to decrease sliding, but it is not used to an appreciable extent today.

The projections themselves do not improve the coefficient of friction, but are effective only if they happen to catch on similar projections or some other very rough surface. The weight of a filled box tends to mash the projections.

Resins or Latices

These are intended for application at the end of the packing line. They can be applied to give high antiskid properties.

Adhesives

These also are intended for application at the end of the packing line as boxes are stacked. They are special types of adhesive known as break-away or soft-seal type. They will not shear, as in a sliding action, but can be broken away by a pulling force such as tapping the side with the hand when unloading.

5.8.5 Protective Packaging

The following are some of the products for which protective packaging is desirable: Acoustical products, adhesives and sealants, agricultural supplies, appliances, automotive accessories, bag products (concrete, chemicals, salt and flour), barrel, drum, and pail goods, bottle and can products (beer, liquor, soft drinks), biological products, building materials and hardware, castings and metalwork, chemical and allied products, clay and brick products, dairy products, detergents and soaps, drugs, electrical apparatus, electronic products, farm supply products, feeds and fertilizers, fixtures and furnishings, food products, fresh fruits and vegetables, garden and lawn equipment, glassware and china products, greenhouse, nursery, and florist supplies, laboratory and hospital equipment, masonry, stonework and bricks, millwork and lumber products, paper products, plastic products, screw machine products, sheet metal products, varnish and paint supplies, just to name a few.

5.9 AUTOMATIC PALLET LOADERS

5.9.1 Purpose and Distinguishing Features

The purpose of automatic pallet loaders is to accept cartons at productionline speeds and stack them in a predetermined pattern. This machine has great flexibility which enables it to stack a variety of boxes in various patterns and load sizes. It offers many variables and accessories, which enables it to handle bags as well as boxes, either on pallets or without pallets as a unitizer.

There are many desirable features (explained later), but what distinguishes this machine from other pallet loaders is the following:

Boxes are oriented by a right-angle turning device which selectively turns the axis of the package by 90° horizontally to form a pattern.

It has an articulated apron (stripper) to permit it to turn from a horizontal to a vertical plane and thus save space.

It is hydraulically operated for positive and flexible control and long wear.

Hoist table is raised and lowered by hydraulic ramp and layers of packages are added on top.

Maintenance is simplified by plug-in control boxes.

Controls are completely automatic and numerous auxiliary electrical controls are available, such as automatic control of a multiline installation feeding to one palletizer.

Things to consider in dealing with the automatic pallet loader are as follows:

Unit load
Pallet height maximum
Pallet height minimum
Number of different pallets—not mixed
Number of pallets in magazine
Package size maximum
Package size minimum
Package sizes, number handled, layer lengths
Package speeds
Package types
Load height
Load height selections
Layer-stacking patterns
Pallet size versus load size

5.9.2 Pallet Loader Accessories

Package and Load Selection

1. Pattern control boxes
2. Special pattern control boxes
3. Additional package sizes
4. Additional layer height selections
5. Special pattern box changing next to last layer

Pattern-forming Auxiliary Equipment

1. Pattern-forming stops
2. Extra pattern-forming stops
3. Two pattern-forming stops in the same row
4. Overhead stops

Overall Pallet Loader Changes and Those Affecting Layout

1. Combination pallet loader and unitizer
2. Side-loading pallet magazine
3. Piers
4. Sway-bracing hoist cylinder
5. Larger-size feeder and turning device
6. Special provisions for sanitation
7. Heavier loads
8. Machine size

Pallet Magazine and Dispenser

1. Adapters and magazine for handling two or three pallets
2. Adjustable finger frame
3. Low pallet level warning
4. Low pallet level control
5. Longer pallet magazine
6. Additional pallet magazine (tandem or external)

Sheet Dispensers

1. Sheet dispenser for bottom of load
2. Sheet dispenser used between layers
3. Sheet dispenser, side-loading

Hydraulic System Accessories

1. Shut-off valves outside and close to tank
2. External oil filters
3. Boot for hoist cylinder
4. Tank heater

Machine Speeds

1. Increased box speeds (not cfpm)
2. Two-speed pallet loader

Equipment today, in the pallet loader and unitizer models, is capable of handling slipsheets only, slipsheets on a pallet, or pallets only. Any of these conditions can be met by adapting a unit already in-house or by developing a new unit to be installed for the operations indicated.

5.10 AUTOMATED TRUCK-LOADING SYSTEMS

Modern technology has improved truck loading to the point that we are able to rapidly load trailers with up to 55,000 lb (24,948 kg) of freight utilizing a cable floor. With push-button ease, the floor moves only when power unit is activated. A portable switch station can be mounted on a swing-arm (for fork truck operator's easy reach) to control load or unload movements. The cargo rests on cables at all times. Cables and bulkhead move only when loading or unloading. There are no rollers. The cable floor system consists of a series of optionally spaced cables which slide along the vehicle floor in shallow tracks. The cargo rests directly on the cables, which when moved act much like a belt conveyor.

A bulkhead is attached to the cables to keep freight from falling forward. A driving mechanism powers the cables toward the front to load or toward the rear to unload (Fig. 5.10.1). Optional limit switches, to start/stop the action, control the unit for completely automatic interface with forklift or other systems.

The entire system in a 40-ft (12.2-m) trailer weighs less than 900 lb (408.2 kg). In a 45-ft (13.7-m) trailer, it weighs less than 1000 lb (453.6 kg).

The sidewall load-securing system is based on the principle of a parallelogram (Fig. 5.10.2). It is mounted on the inside trailer walls. A series of hinges and channels supports the plywood walls of

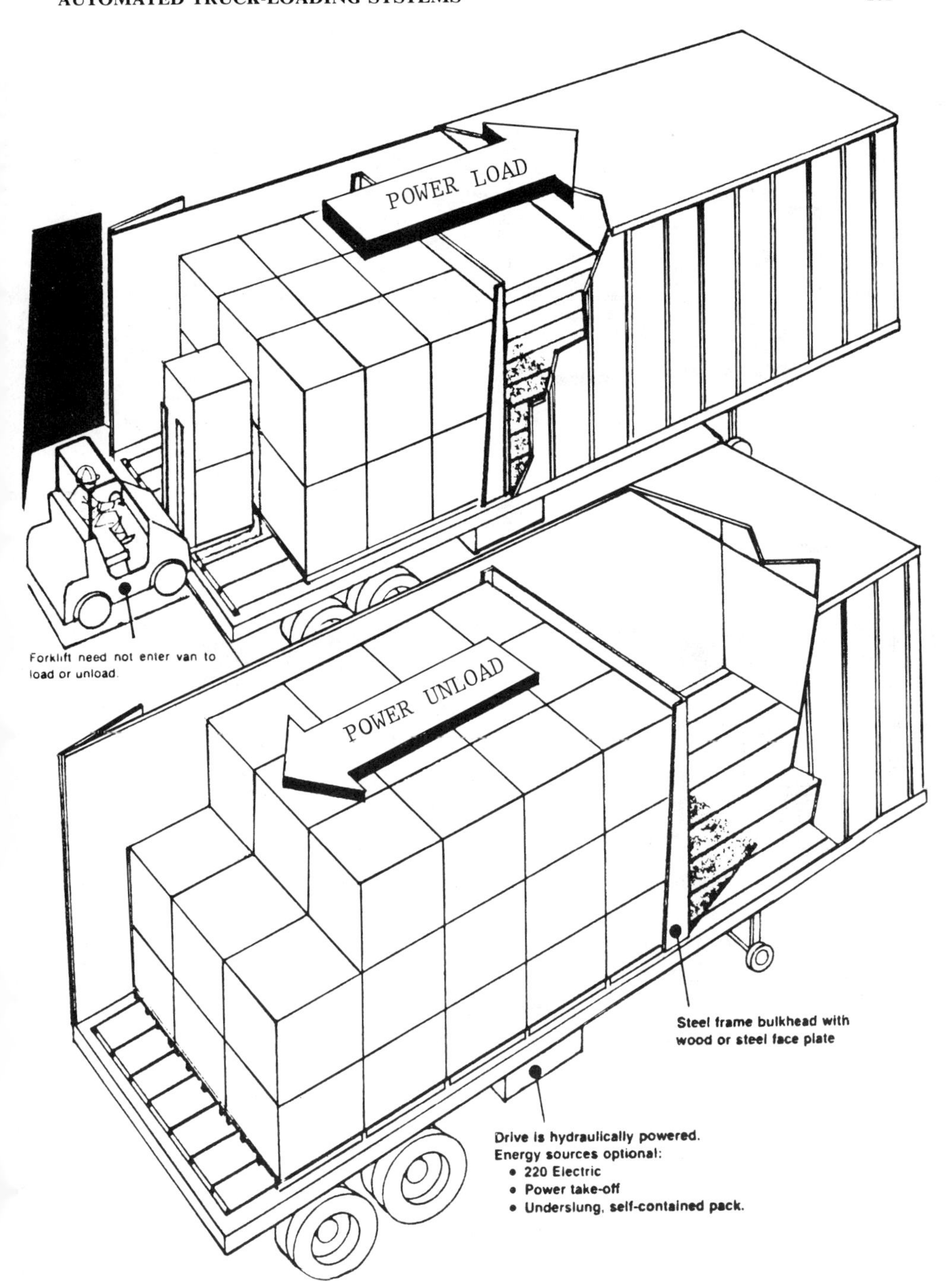

Fig. 5.10.1 Cable conveyor system for trucks.

the system. But expanded, the rear hinges are locked, and the entire load is held securely for travel. Support for this system is a dock-mounted loader. As the truck backs up to the dock facility, the vehicle locks into place, and hooks up with the dock-mounted loader; up to 50,000 lb (22,680 kg) of pre-stage freight can be loaded in approximately 5 min or better. The advantage of this type of system is that one can load pallets, slipsheets, boxes, and mixtures of any type of loads which may include

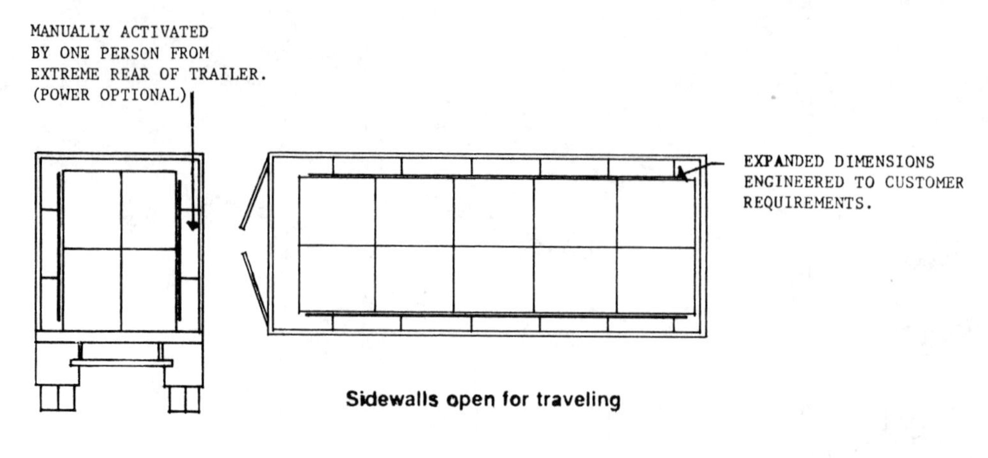

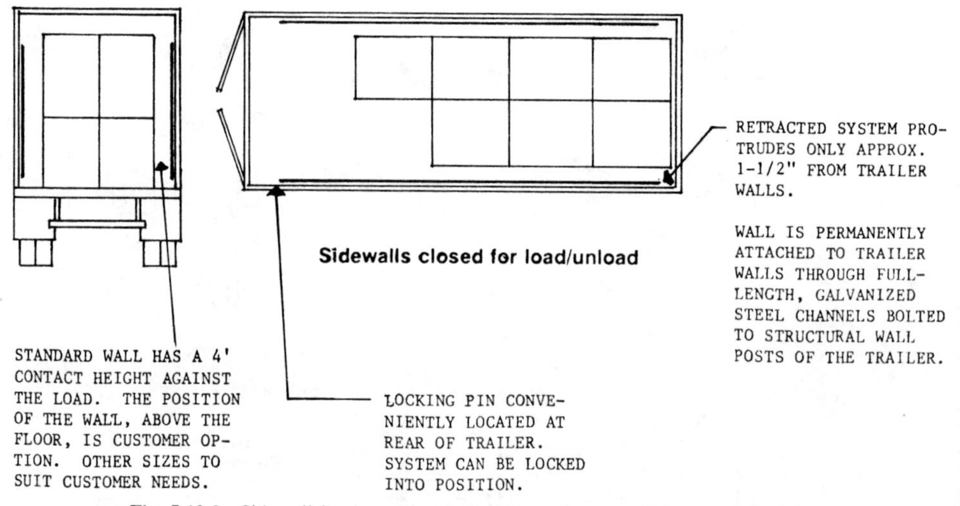

Fig. 5.10.2 Sidewall load-securing systems based on parallelogram principle.

barrels, bags, or any other items. There are other variations which include a nonpower conveyor system—a roller conveyor bed. However, depending on the manner and ability to load rapidly, they are not as quick and require manpower to do the loading.

5.11 ROAD RAIL VEHICLES

Road railers are bimodal freight vehicles that can be pulled by truck tractors over highways or pulled by locomotives on special railroad trains. Each Road railer has two sets of running gear; a set of tandem axle rubber tires for highway operations and a single axle with steel wheels for rail operations (Fig. 5.11.1).

Each set of wheels is locked into running or stored position, depending on the desired mode of operation. Each running gear is mounted with its own braking system and an air suspension. These

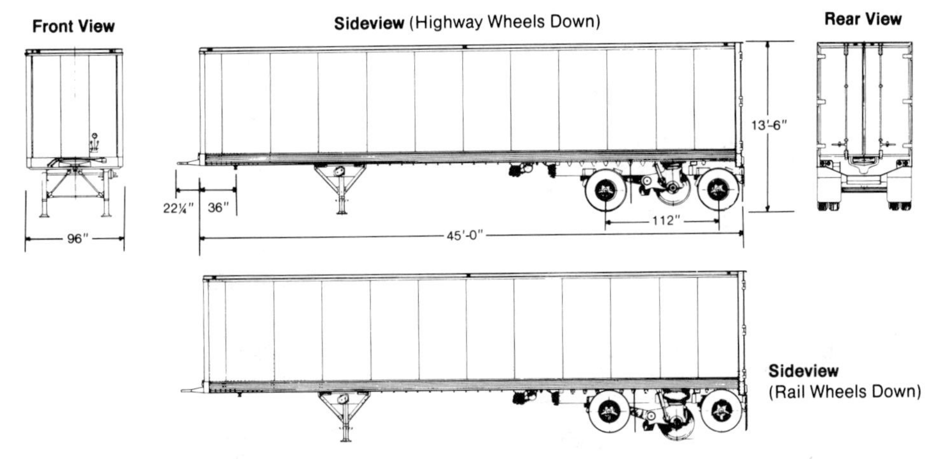

Fig. 5.11.1 Road railer specifications.

air suspensions provide a smooth ride in each mode, as well as accomplishing the mode-to-mode transfer operation.

5.12 TECHNICAL SOURCES

The following people and organizations supplied valuable technical material which aided in the development of this chapter. Without their assistance, this chapter would not have been completed.

Mr. Jonathan A. Hix
American Paper Institute, Inc.
260 Madison Avenue
New York, NY

Mr. David W. Lutz
Automatic Truck Loading Systems, Inc.
P.O. Box 810
Carlisle, PA 17013

Mr. J. J. Mack
Virginia Polytechnic Institute
and State University
Wood Research & Wood Constr. Lab.
Pallet & Container Research Center
Blacksburg, VA

Mr. Henry A. Fahl
RoadRailer Bi-Modal Corporation
200 Railroad Avenue
P.O. Box 767
Greenwich, CT 06836

Mr. Tom Dickman
Lantech, Inc.
Blue Grass Industrial Park
11000 Blue Grass Parkway
Louisville, KY 40299

Mr. Ken Frees
Hoover Universal
537 East Highway 54
Camdenton, MO 65020

Mr. William C. Baldwin
National Wooden Pallet & Container Assoc.
1619 Massachusetts Avenue, N.W.
Washington, D.C. 20036

Mr. Don Kueser
Alvey, Inc.
9301 Olive Blvd.
St. Louis, MO 63132

Mr. Roy W. Verstraete
Little Giant Products,
1600 N. E. Adams Street
Peoria, IL 61601

Mr. Julius B. Kupersmit
Containair Systems Corp.
145-80 228 Street
Springfield Gardens, NY

Mr. David K. Spencer
Cascade Corp.
Sylvan Westgate Bldg.
P.O. Box 25240
Portland, OR 93225

Mr. William J. Rehring
Compucon
13749 Neutron Road
P.O. Box 401229
Dallas, TX 75240

Mr. Lloyd C. Dick
Mead Paperboard Products
4400 Marburg Avenue
Cincinnati, OH 45209

Mr. Robert D. Carter
Longview Fibre Company
3832 North 3rd Street
P.O. Box 2008
Milwaukee, WI 53201

CHAPTER 6
INDUSTRIAL HAND TRUCKS

IRVING M. FOOTLIK

Footlik & Associates
Evanston, Illinois

Hand trucks, or floor trucks as they are sometimes called, are grouped into two divisions, two-wheel hand trucks and multiple-wheel hand trucks. Multiple-wheel trucks are those having more than two wheels and are used mostly for horizontal movement. These units are often employed in towline (tow conveyor) operations, where they are pulled by a power unit driving an overhead or underfloor conveyor chain or trailer-train operations.

Under these two major divisions there are a great many subdivisions, depending on the use to which the respective types of trucks are put. Many of the trucks are of general application in manufacturing plants, warehouses, terminals, department stores, delivery services, and numerous other industries. Other types of trucks are of a specialized nature for handling particular kinds of materials and commodities where certain special requirements exist.

The following factors should be considered in selecting the specific class or type of truck for a particular use:

1. *Load.* Bulk of the load, weight in relation to floor load, susceptibility to damage in transit, and possible operator abuse.
2. *Floor Conditions.* Trucks are cheaper to replace than floors. Therefore, truck wheels should be selected to minimize wear of the floor over which the trucks will operate. Use of steel wheels on concrete floors is not recommended. Rubber-tired wheels should be used.
3. *Truck Wheels.* The type and size of wheel that should be used can best be determined by a consideration of the kind and condition of the floors over which the truck will run, and the weights of trucks and their loads. Generally, a roller-bearing wheel is considered best.
4. *Overall Conditions.* After analyzing the foregoing elements of the problem, the prospective truck user should select the correct type of truck for each kind of handling operation. Before ordering trucks, the user should check the widths of doorways through which the trucks will have to pass. In older plants there may not be sufficient clearance for the wider types of new trucks.
5. *Supplemental Equipment.* Industrial hand trucks were the original backbone of the materials handling industry. At one time, it was thought that the way to increase productivity was to eliminate hand trucks and replace them with fork trucks or powered industrial equipment. Subsequently, it has been proven that the ideal situation is a combination of both. For this reason, we now use the lift truck to lift the hand truck to the elevations desired and finish the movement on the hand truck by moving the product thereon directly into the transporting vehicle or to the point of processing. The hand truck has also become extremely popular by grouping into trains which are moved by tugs. Some of these tugs are electrified, and guided by wires in the floor that allow the automatic hitching and unhitching and switching at strategic points. The four-wheel trucks are also used as part of a handling system where they are tugged by underground or overhead chains as they proceed in a continuous flow pattern through designated areas.
6. *Multiple Use.* Trucks that are designed for multiple use in conjunction with a fork truck, such as those used to pick orders in narrow aisles, provide significant handling efficiencies.

6.1 TWO-WHEEL HAND TRUCKS

6.1.1 Nomenclature

The common nomenclature within this group should be recognized for its import in explaining the features of the several subdivisions. Figure 6.1.1 will help in making this nomenclature clear.

6.1.2 Basic Types

The two basic subdivisions of two-wheel hand trucks are:

1. Eastern type
2. Western type

The eastern type of truck (Fig. 6.1.2) has a tapered frame with wheels located outside the frame and either curved or flat cross members. It usually has straight handles, although it can be equipped with curved handles if this is desired. The height ranges from 48 to 60 in. This type of truck is useful for handling mixed freight such as boxes, barrels, cartons, bags, and other bulky objects.

The western type of truck (Fig. 6.1.3) has a parallel frame with wheels inside the frame and curved or flat cross members. Usually equipped with curved handles, it may also be supplied with

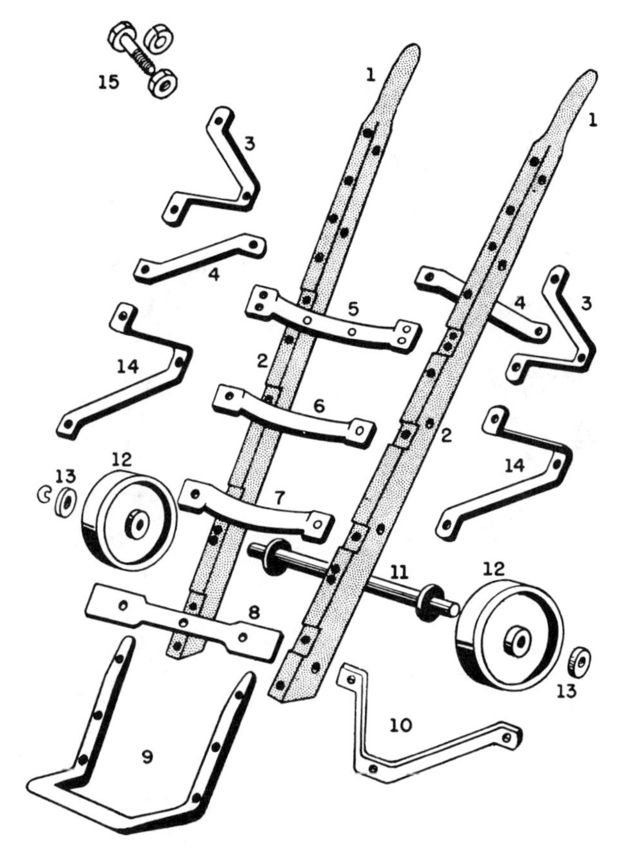

1. Handle
2. Side rail
3. Leg
4. Leg brace
5. Top crossbar
6. 3d crossbar
7. 2d crossbar
8. 1st crossbar
9. Nose
10. Axle brace
11. Axle
12. Pressed steel wheel
13. Retaining ring
14. Axle bracket
15. Nut, bolt, and lock washer

Fig. 6.1.1 Parts of a common two-wheel hand truck.

straight handles, if desired. The height ranges from 48 to 60 in. and this type is used in heavy handling around motor truck and railroad terminals with reinforcing usually in the form of two strengthening members running the length of the truck as extra bracing for the cross members.

There are basic design deviations that should be noted:

1. Barrel trucks
2. Bag trucks
3. Beverage trucks
4. Appliance trucks
5. Utility trucks
6. Special purpose trucks

Barrel Trucks

Barrel trucks are designed for handling heavy barrels and drums. Equipped with short nose prongs and a special hook mechanism, it usually stands vertically at right angles to the floor. Many of these units are equipped with floating axle construction. All these features, with some variations in wheel equipment, facilitate easy loading, handling, and unloading (see Figs. 6.1.4 and 6.1.5).

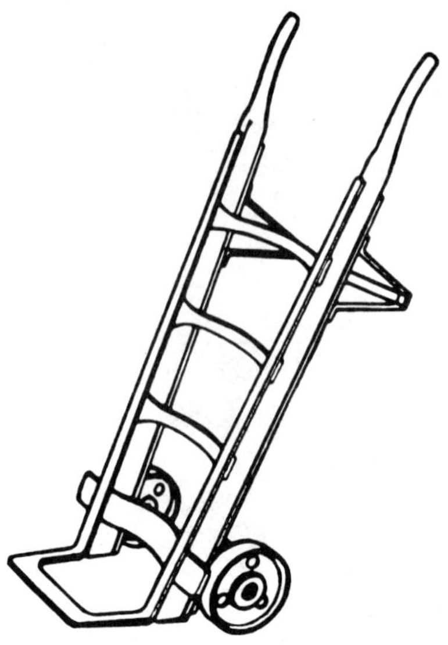

Fig. 6.1.2 Hand truck, eastern type, with curved cross members.

Fig. 6.1.3 Hand truck, western type, with flat cross members.

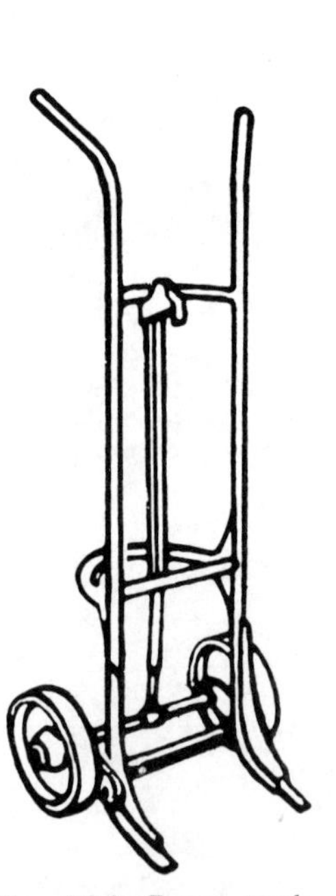

Fig. 6.1.4 Drum truck for heavy barrels and drums.

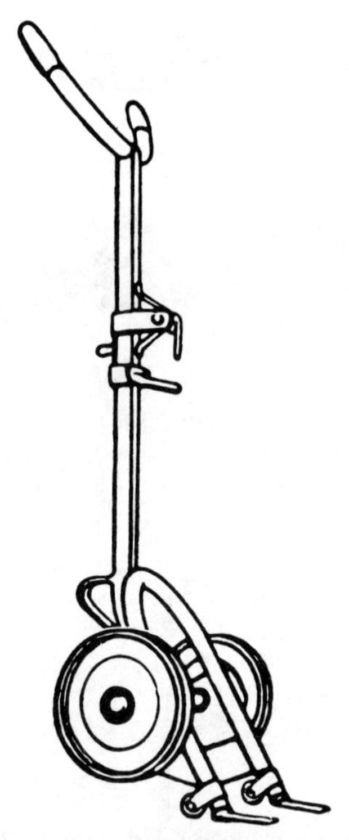

Fig. 6.1.5 Drum truck with large and small rubber-tired wheels.

Bag Trucks

These trucks are of either eastern or western design. The difference from the basic truck design is in length or type of nose, for ease of handling bagged materials (see Fig. 6.1.6).

Beverage Trucks

Recently, a standard pattern of truck design has evolved for case and beverage handling that makes use of any one of three handle designs; tipped-top bar handle (Fig. 6.1.7), single-grip handle (Fig. 6.1.8), and the usual double-grip handle. All these trucks have parallel side rails and a plate or open nose, and when empty, stand upright at right angles to the floor. Some unique features have been developed, one of which is axle interchangeability so that several sizes of wheels can be interchanged

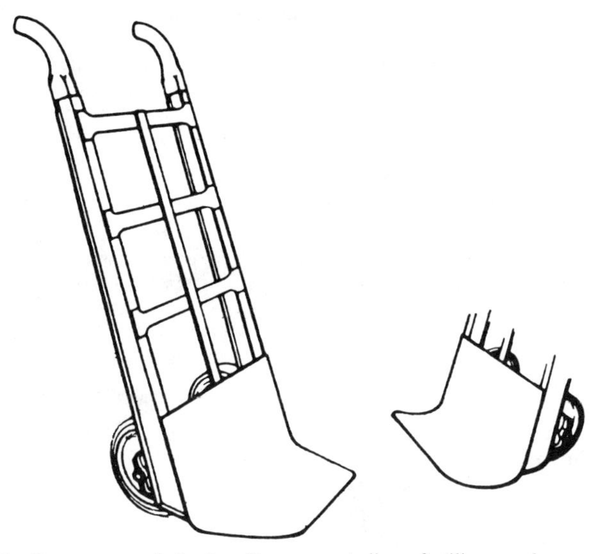

Fig. 6.1.6 Bent-nose truck for handling cement, lime, fertilizer, and so on, in bags.

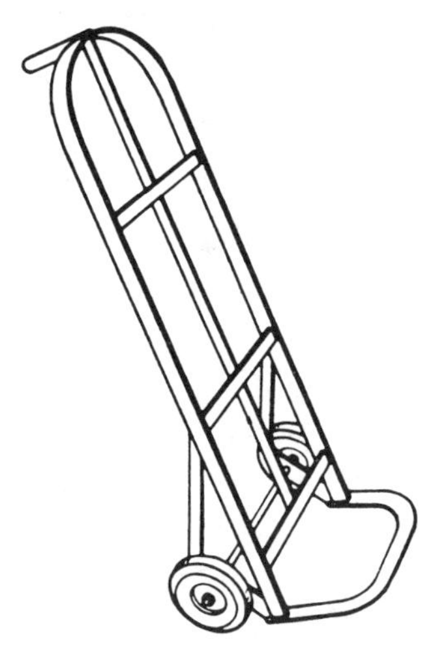

Fig. 6.1.7 Tipped-top-bar-handle truck.

Fig. 6.1.8 Single-grip-handle truck.

quickly without affecting the balance of the truck. These trucks are of light, but rigid, construction to withstand shock loads such as those brought on by running over door sills and curbings.

Appliance Trucks

These are designed for handling refrigerators, stoves, freezers, and other items of household equipment. Some appliance trucks are equipped with endless-belt mechanisms, which enable the trucks to operate on stairs, literally making a ramp of the stairway (Fig. 6.1.9). Others are caster equipped for right-angle movement in such restricted areas as stairwells. Patented locking devices are incorporated on web strapping for fastening units securely to the truck. There are also units that are adjustable in width and height to fit various sizes of appliances and furniture (Fig. 6.1.10).

Utility Trucks

There are various types of light, inexpensive, general purpose trucks that are used for handling materials of varying bulk, shape, and weight. Two such trucks are shown in Figs. 6.1.11 and 6.1.12. These trucks have different kinds of one- and two-handle construction, with frames also of varied design. The construction is usually determined by the individual decision of the manufacturer, depending on trade demand. There is no fixed pattern, and the manufacturer may produce several different models. These are used mostly on delivery trucks serving beverage distributors, supermarkets, and so on, to enable the driver to handle deliveries more readily. The prime requisite is light weight and sturdiness.

Special Purpose Trucks

These trucks differ from those of specific classifications previously covered in that they have certain characteristics that adapt them to specific handling jobs. The more common types, which serve important industrial and transportation, are included here.

Cannery Trucks. Shown in Fig. 6.1.13, this type of truck has a fork-type nose and two pairs of handles located on different planes, the larger pair forming long side rails. The nose of the truck is

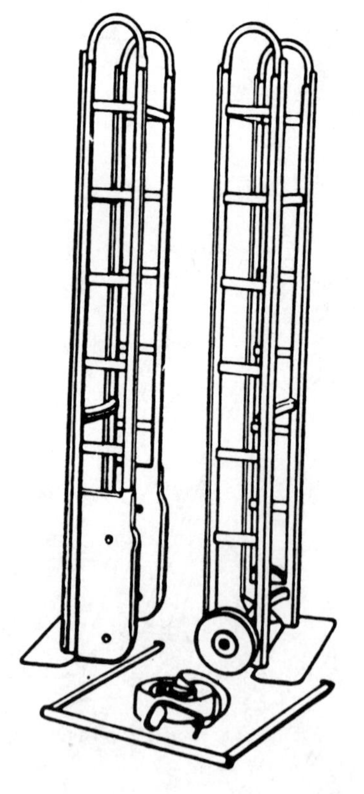

Fig. 6.1.9 Trucks for handling large applicances. Model on left has endless belt in place of wheels.

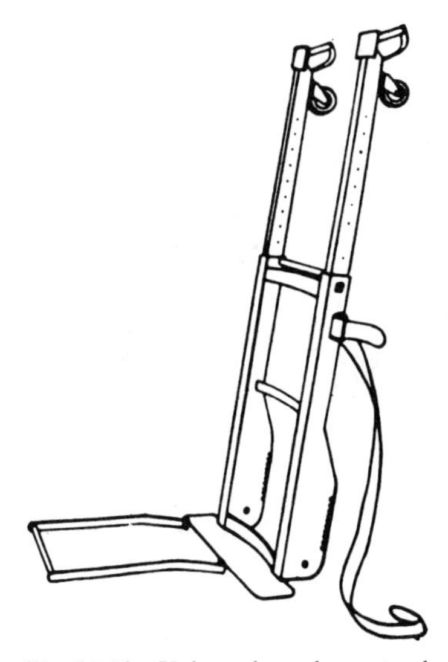

Fig. 6.1.10 Universal warehouse truck adjustable for width and height.

Fig. 6.1.11 Heavy-duty truck for heavy cases or boxes.

run under a small pallet, where cases are piled. The second set of handles is used to lower the load to easy trucking position.

Grain Trucks. The grain truck is a standard eastern-type truck, except for a longer nose and wheel guards, which protect the load of grain in burlap bags in handling. Many models are equipped with hub caps to prevent tearing of bags. This type of truck is widely used for handling general cargo, also, in many sections of the United States.

Cotton Trucks. These are of conventional basic design with the exception that they are equipped with short, tapered, prong-type noses that are engaged into the bale instead of under it, thus providing better balance of the load. They can be either eastern (more common and illustrated in Fig. 6.1.14), or western in type.

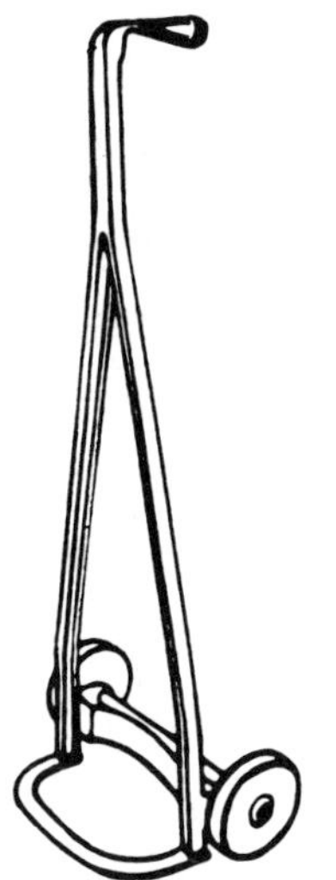

Fig. 6.1.12 Single-grip truck for transporting long, lightweight cases.

Fig. 6.1.13 Cannery or bottling truck for handling goods piled high on pallets.

Fig. 6.1.14 Truck for handling baled or bundled material (eastern type).

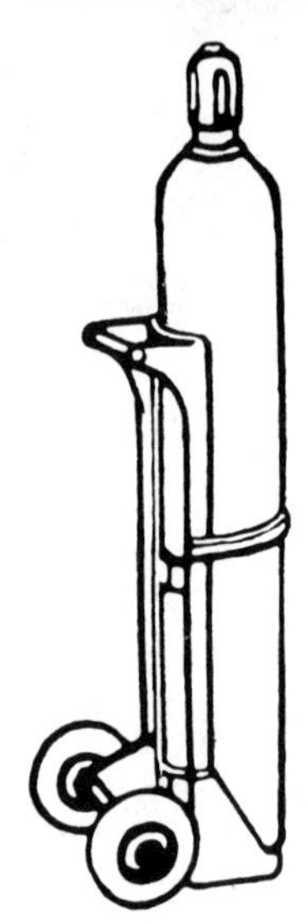

Fig. 6.1.15 Single-cylinder acetylene truck.

Cylinder Trucks. Trucks for handling cylinders depart completely from the basic designs previously mentioned. They have large wheels, a right-angle plate nose, and a chain-locking device for holding the cylinders on the trucks. Cylinder trucks are often equipped with a toolbox to hold welding tips, torches, and other small tools. Figure 6.1.15 shows a single-cylinder handling truck and Fig. 6.1.16 shows a multiple-cylinder truck with a toolbox.

Pry Trucks. Trucks in this class (Fig. 6.1.17) pry up the load on a crowbar nose and roll it away. These are used for loads too heavy or bulky for ordinary two-wheel trucks, and are well adapted for use in motor trucks, freight cars, warehouses, and so on, where space is limited. In many cases, they are used in pairs by two people.

6.1.3 Other Types of Two-wheel Trucks

Additional types of special purpose trucks include those for handling carboys, paper rolls, radiators, furniture, ash cans, fish, and so on. Roll and cylinder trucks for heavy and light duty are also available

Fig. 6.1.16 Multicylinder acetylene tank truck.

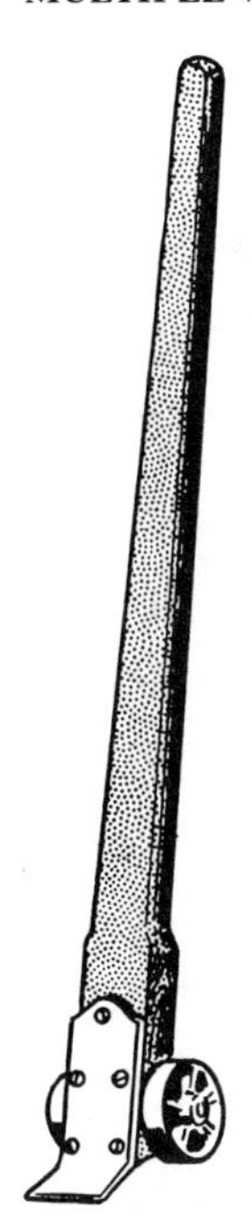

Fig. 6.1.17 5000-lb capacity pry truck.

for handling paper rolls of 30–54 in. in diameter, and in capacity ranges from 500 to 2000 lb. Certain heavy-duty models not only transport, but also raise objects to machine height. There are several special two-wheel hand trucks designed and built for one purpose or for a specific customer, and their use to date has been limited. Thus, they are not mentioned here.

6.2 MULTIPLE-WHEEL FLOOR TRUCKS

6.2.1 Standards Defining Running Gear

There is no clearly defined classification for multiple-wheel floor trucks, as there is for the two-wheel types, but certain standards have been developed by the Caster and Floor Truck Manufacturers' Association that defines the basic running-gear arrangements. These are illustrated in Fig. 6.2.1. Attention should be given to the six factors outlined earlier when determining the specific type of truck required.

Dollies

Wood or metal is used in construction, in various sizes, depending upon the load to be handled. There are four basic designs: four or six rigid wheels for short, straight movement; three or four swivel casters for short, straight, angle, or turning movement; two rigid and two swivel casters also for straight, angle, or turning movement; and six-wheel tilt-type where center wheels are either of larger diameter or center axles are on a different level for straight, angle, or turning movement (see Fig. 6.2.1). The selection is based on loads to be handled and the customer's preference.

Dolly frames may be rectangular, triangular, or circular. Solid or open decks are optional, or they may be specifically designed to hold specific loads. Typical dollies are shown in Figs. 6.2.2 and 6.2.3. The triangular dolly in Fig. 6.2.4 is designed to facilitate its being pushed under barrels and crates that have been tilted. Rollers are sometimes used in place of or in combination with wheels. A stevedoring dolly has a row of four rollers mounted in its steel frame, on either side, permitting movement in a straight line. A timber dolly consists of a single large roller mounted across the middle of the frame at right angles to the axis of the load and to the line of movement. Some dollies, Fig. 6.2.5, combine rollers and swivel casters to allow a 360° lateral revolution.

Platform Trucks

Platform trucks are actually a larger edition of the dolly and have two basic chassis constructions: tilt or balance type, and the nontilt type. They are produced in many sizes and in light-, medium-, and heavy-duty construction.

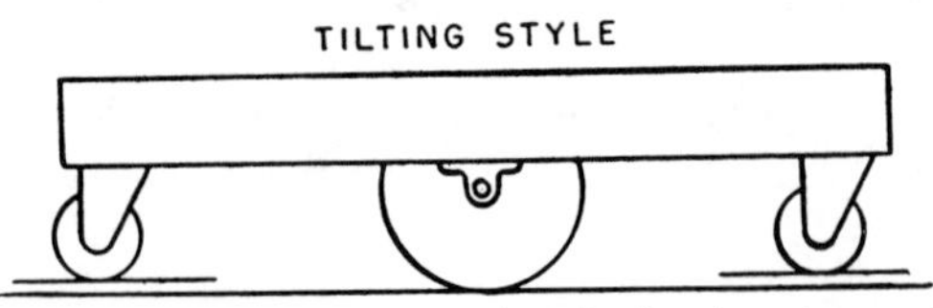

No. 1 Type. Four-wheel tilting style, two main wheels at center, and one swivel caster at each end; for light and medium duty.

No. 2 Type. Six-wheel tilting style, two main wheels at center and two swivel casters at each end; for medium and heavy duty.

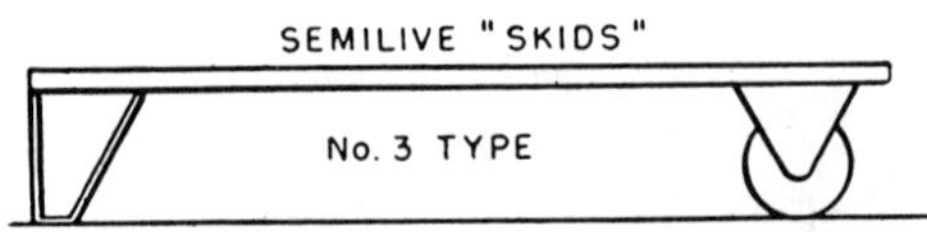

No. 3 Type. Semi-live skid with two main wheels or rigid casters at one end and rigid legs at the other end; to be used with a lift jack.

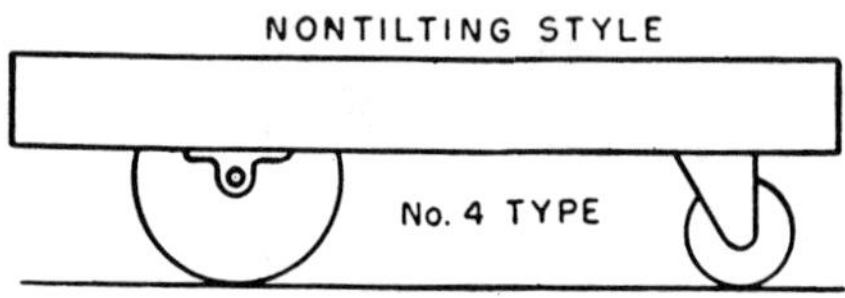

No. 4 Type. Four-wheel nontilting style, two main wheels near one end and two swivel casters at the other end; for medium and heavy duty.

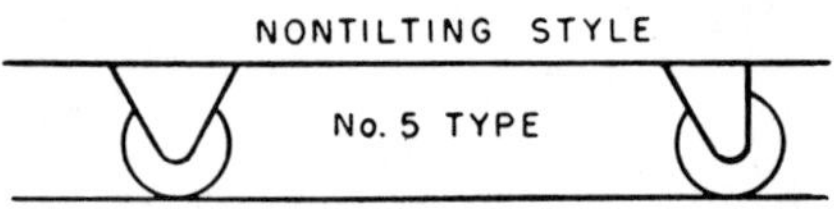

No. 5 Type. Caster type, two rigid casters at one end and two swivel casters at the other end; desirable for low platform requirements; for light and medium duty.

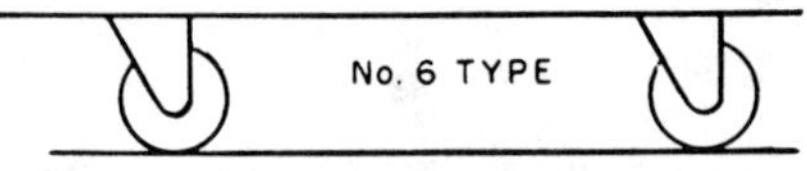

No. 6 Type. Caster type, one swivel caster at each corner; for light and medium duty.

No. 7 Type. Caster type, 6-wheel tilting style with two rigid casters at center and two swivel casters at each end; for regular- and medium-heavy-duty low-platform requirements.

No. 8 Type. Caster type, 4-wheel tilting style with two rigid casters at center and one swivel caster at each end; for light- and medium-duty low-platform requirements.

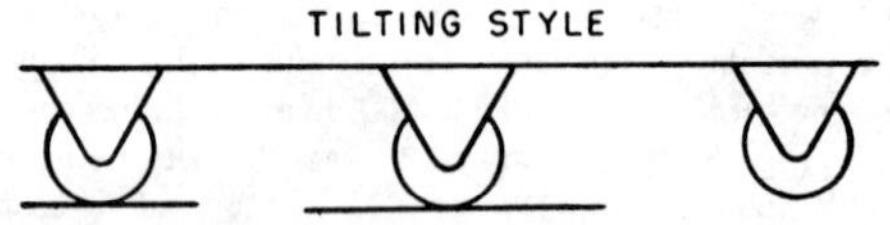

No. 9 Type. Caster type, 4-wheel tilting style with two rigid casters at center and one rigid caster at each end; for light- and medium-duty low-platform requirements.

Fig. 6.2.1 Caster and wheel diagrams of standard running-gear arrangements.

Tilt or Balance Type of Truck. The chassis on this type of truck has rigid load wheels or rigid casters located in the center and one or two swivel casters, usually smaller in diameter, located at or near each end of the platform, permitting maneuverability, with load shift between load wheels and swivel casters at each end by a slight tilting action.

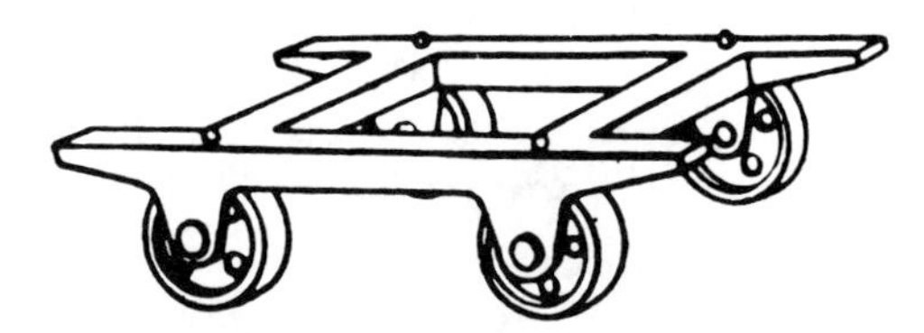

Fig. 6.2.2 Four-rigid-wheel, nontilt dolly.

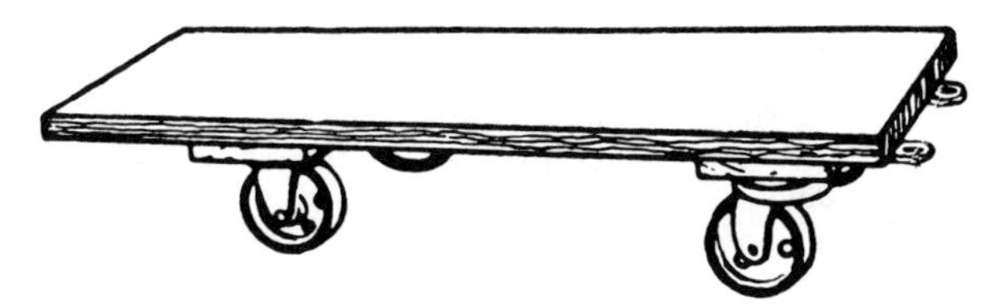

Fig. 6.2.3 Nontilt dolly with two rigid, two swivel casters.

Fig. 6.2.4 Rubber-tired, 24-in. triangular dolly.

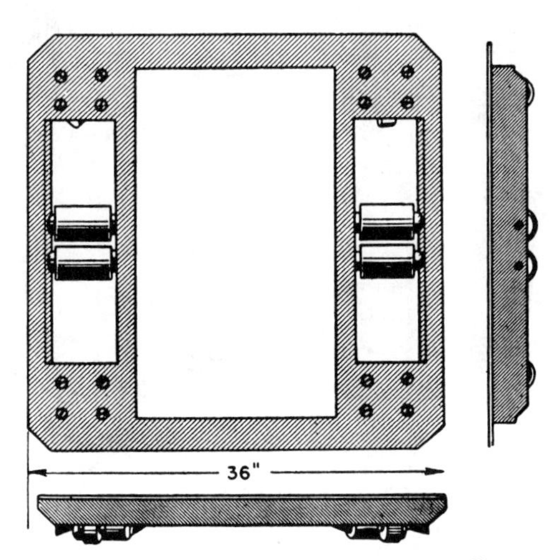

Fig. 6.2.5 Swivel dolly with caster and rollers.

Nontilt Type. The chassis on this type of truck has load wheels at or near one end and swivel casters, usually of smaller diameter, located at the other end so that the load is distributed at all times and all wheels function at all times. Corner post holders attached to these trucks enable high loads to be braced against spilling. They are also made with steel slat-racks and steel frames.

For very heavy work, trucks of this class have heavier chassis construction. To meet the load requirements, the running gear is also heavier and made in tilt or balance types with the same maneuverability. It is not recommended that they be operated on steep ramps. By means of couplers, and with extra reinforcement, such trucks may be used for light-duty trailer service or towline conveyor systems.

Wagon-type Trucks

The difference between the wagon-type truck and the standard nontilt platform truck is that the swivel casters are replaced by either a knuckle-steer arrangement or a fifth-wheel assembly to which a tongue-type handle is fitted for pulling the truck. Handling long, overhanging loads that are difficult to push is the general use for this type of truck. Knuckle-steer arrangement is preferred for maximum stability because it maintains a wide four-point wheel support which prevents tipping or spilling of the load.

Superstructures for Trucks

Superstructures are of many varieties and range from four corner stakes to completely closed containers. The design is based entirely on the function to be performed and the requirements of the truck involved (Fig. 6.2.6). Materials of both platform trucks and their superstructures may be all wood, all metal, or a combination of both. Certain superstructures also have combinations of fiber or canvas with wood or metal, or both. Wheels and casters vary in size in the proportion to unit load ratings and required truck platform heights.

Semilive Skid Platforms

These are platforms with two load wheels at one end and two permanent legs at the other end. The skid platform is activated by a lift jack, which consists of a long handle mounted on a pair of wheels,

Fig. 6.2.6 Various types of rack bodies used on trucks.

Fig. 6.2.7 Semilive skid platform.

with a hook arrangement over the wheels to engage the coupling on the dead end of the skid and to exert a jacking or prying action. The unit formed is equivalent to a three-point suspended-platform truck (see Fig. 6.2.7).

6.3 MAINTENANCE OF INDUSTRIAL HAND TRUCKS

An adequate maintenance program and proper procedures to keep trucks in good operating condition are of vital importance. The average truck will give many years of satisfactory service when given regular inspection, adequate lubrication, and repairs and careful operation. In practically all cases, the truck manufacturer has supplied his trucks with suitable lubricating fittings or has provided oil holes that lead to the various bearings. Periodic lubrication is imperative for satisfactory operating service, and the dates when inspections are made and the required adjustments taken care of should be noted in the repair department's records. It is also advisable to maintain a reasonable stock of maintenance parts so that repairs can be made promptly and the trucks restored to service without any costly delays. Typical safety procedures for hand trucks are shown in Fig. 6.3.1.

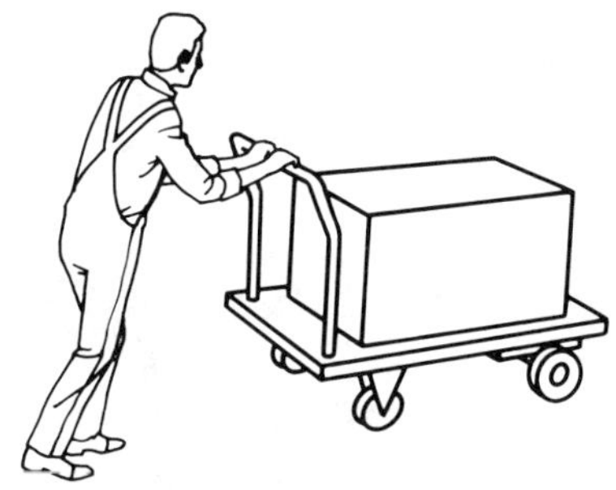

1. CHECK THE ROLLABILITY and capacity rating of trucks, for your conditions.

2. PLACE LOADS PROPERLY on four-wheeled trucks, to avoid tipping, ensure steering ease.

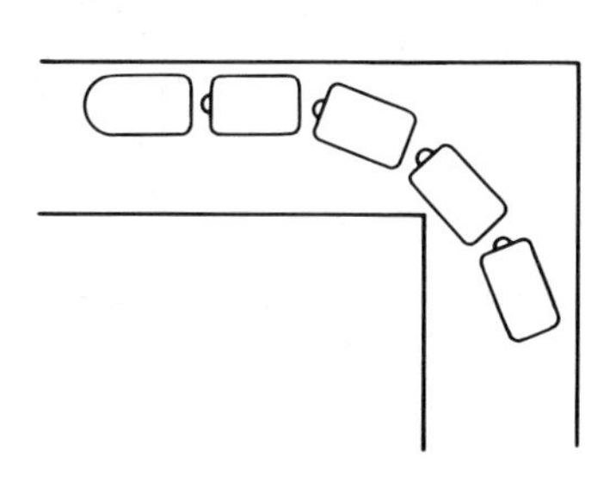

3. USE MANAGEABLE TRAINLENGTHS, that can make the turns when several trucks are towed.

4. TILT LOAD FORWARD to insert nose of a two-wheeled truck. Avoid unsafe ramming action.

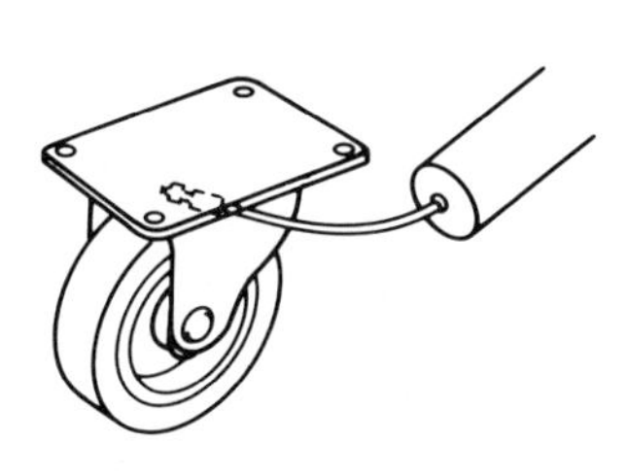

5. REGULARLY LUBRICATE wheels and casters to minimize friction, improve rollability.

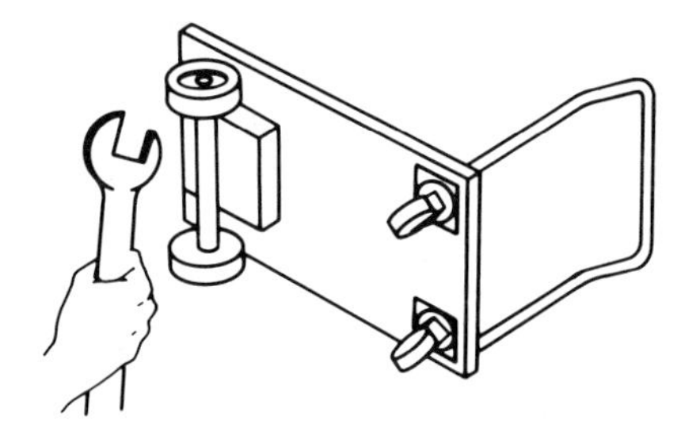

6. REGULARLY INSPECT your trucks, to avoid gradual deterioration and increased hazards.

Fig. 6.3.1 Six ways to boost floor truck safety.

6.4 APPLICABILITY OF HAND LIFT TRUCKS

Hydraulic and mechanically operated hand lift trucks are especially adaptable to handling operations where space is at a premium, and it is necessary to operate within narrow aisles. Because of its light weight it also is suitable for use on floors of low capacity and on elevators. Improvements in the mechanical features of these trucks and in the design of wheels and bearings have made it possible for truck operators to handle much larger and heavier loads with less effort.

The hand truck's lifting mechanism is operated by the actuation of a series of levers. The lifting action, with a hydraulic hand lift truck, is started mechanically by the operator and is transmitted through the hydraulic system to levers that raise and lower the loads. Hand lift trucks are usually classified as follows:

1. Trucks: Hand lift hydraulic
   a. Pallet, low-lift
   b. Platform, low-lift
   c. Special
2. Trucks: Hand lift mechanical
   a. Pallet, low-lift
   b. Platform, low-lift
   c. Special

6.4.1 Analyzing Truck Requirements

The selection of the right truck for particular needs is largely a matter of studying the pertinent factors that enter into a plant's handling operations. Key factors are listed in Fig. 6.4.1. The special use trucks, such as those designed for handling such items as appliances, produce, bricks, chairs, and odd order picking have brought on an entirely new field of development. To attempt to illustrate all these trucks would be very difficult. It is, therefore, recommended that users of the trucks contact the manufacturer and indicate the pertinent information as illustrated on the specification sheet and illustrations that follow. The manufacturer then can make a specialty truck for the purpose desired.

The art and science of designing these specialty trucks has now become a full-time occupation, and there are not many industries left for which specialty trucks are not available. When working with these designs, care should be taken to try to obtain as many multiple uses as possible. Example: There are carts on the market that are used in hotels that can be handled as two-wheel or four-wheel carts, depending on the type of baggage being handled. Similarly, there are manufacturers who make appliance carts with small motors that allow the cart to climb stairs. To locate these manufacturers, consult the Yellow Pages in your local telephone directory.

Size and Capacity of Truck

This depends on the loads to be carried—their length, width, height, weight, and shape. Work-flow conditions and locations of inclines, or ramps, loading docks, elevators, aisles, doorways, and so on, also affect truck selection, and the size and capacity of trucks presently in service must also be taken into account when deciding on trucks to be used for a particular job.

Loads to Be Handled

Two factors must be considered in analyzing the loads which must be handled:

1. Maximum weight of the loads to be moved. The determination of the capacity of the truck to be selected is based on the greatest weight to be handled in any one load. Sample loads should be weighed. Loads should be limited in weight so they are not excessively heavy for the particular trucks to handle. The truck that can handle the heaviest loads likely to be encountered should be the one selected.

2. Maximum size of the load. Load volume or size should be limited so that loaded trucks can be operated easily in aisles, in and out of elevators, through doorways, on loading docks, and elsewhere. For handling a variety of loads, skid-platform sizes should be selected that will accommodate any and all of the pallets in use, with a minimum of wasted space.

6.4.2 Truck Selection

Assume that a truck is to be selected and used for certain handling operations. The following factors should be taken into consideration.

	Name of Manufacturer				
		Technical Data			
1.	Rated capacity				
2.	Maximum lift				
3.	Lowered height				
4.	Number of strokes required				
5.	Lb. of effort required to lift load				
6.	Service weight				
7.	Over-all length				
8.	Over-all width				
9.	Size of front wheel or wheels				
10.	Type of wheel				
11.	Underclearance in lowered position				
12.	Type and size of bearings				
13.	Method of lifting				
	a. Handles (type)				
	b. Actuating lever				
	c. Foot pedal				
14.	Turning radius				
15.	Percent of grade				
16.	Wheel base				
17.	Safety features				
18. Straddle Trucks	a. Distance between outriggers				
	b. Width of outriggers				
	c. Height of outriggers				
	d. Length of outriggers				
	e. Number of wheels in outriggers				
	f. Size of outrigger wheels or casters				
	g. Type of fork or platform				
	h. Length of fork or platform				
	i. Width of fork or platform				
	j. Forks adjustable or fixed				
	k. Outriggers adjustable or fixed				
19. Pallet Trucks	a. Length of pallet arms				
	b. Width of pallet arms				
	c. Distance between pallet arms				
	d. Underclearance at center in lowered position				
	e. Size of wheel—single or double				
	f. Type of lead edge—(R) Roller or (S) Slide				
	g. Adaptable to platform truck				
20. Platform Trucks	a. Length of platform				
	b. Width of platform				
	c. Type of wheels				
	d. Size of wheels				
21.	F.O.B. Price				

Fig. 6.4.1 General specifications for hand lift trucks, hydraulic and mechanical.

Truck Capacity

Loads carried by trucks should not exceed 90% of their rated load-carrying capacity. For easy lifting, a multiple-stroke truck should be used instead of a single-stroke truck. Single-stroke trucks are recommended for the lighter loads, 3000 lb or less, and in cases where single-lift operation is desired for some particular reason. Single-stroke trucks are built with a compound lifting mechanism to provide the easiest possible elevation with one full stroke of the handle. The one-stroke lift is a little faster, but it is not as easy to operate.

Multiple-stroke trucks are recommended for loads over 4000 lb. The multiple-stroke action, either mechanical or hydraulic, makes it easier for the operator to elevate the load. Loads are lifted by making several short strokes with the truck handle.

Truck Width Compared to Width of Load

The load carried on a truck should not overhang the truck frame more than 8 in. on each side. The best practice is to select the widest appropriate truck, thereby increasing the stability of the loads and permitting speedier handling. A good rule to follow, to obtain proper truck width, is to select a narrow model of truck for skid platforms up to 34 in. wide and a wide model of truck for skid platforms over 34 in. in width. On extremely wide loads, guide rails can be installed on the underside of the skid platform (Fig. 6.4.2) to insure proper centering of the truck.

Clearance between truck frame and skid platform legs should be at least 1 in. on narrow-model trucks and 2 in. on wide-model trucks (see Fig. 6.4.3). The truck should be centered under the load for stability and better operation.

Truck Length Compared to Length of Load

Relative dimensions have been set for length of loads versus length of trucks. Where skid platforms are 54 in. long, or longer, truck frames 12 in. shorter are considered best (Fig. 6.4.4). Where extremely long skids are necessary, a greater overhang is permissible. However, when skids overhang the truck excessively, there is danger of the load's "hanging up" where ramp or floor levels change.

Truck frames should not be extended beyond the length of the skids to be handled. If the truck frame is too long, skids cannot be spotted close to a wall or close to other spotted skids, and valuable storage space will be wasted. But, if the skid overhangs the truck too much, the load will "lift" at the rear when traveling up or down inclines.

Heights of Truck Platforms

This depends on the wheel diameter of the trucks. The United States Department of Commerce recommends 7-in. and 11-in. high truck platforms (Fig. 6.4.5). A truck with 7-in. wheels, measuring 7 in. from top of frame to floor; one with 11-in. wheels, 11 in. to the floor, for example.

Wheel Sizes

Trucks should have wheels of large diameter when handling heavy loads. This facilitates easier starting and rolling, and when floors are rough or uneven, it is especially advisable, so as to reduce the possibility of jarring boxes off the pallets.

If power lift trucks are also used, the hand lift trucks should be of the standard 11-in. type. Any

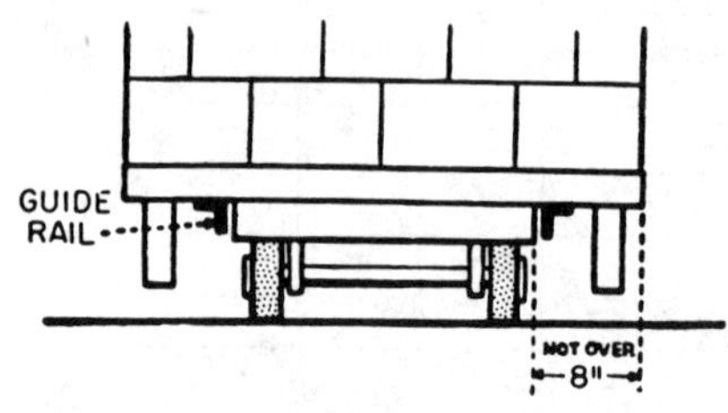

Fig. 6.4.2 Guide rails to position wider skid on standard lift truck.

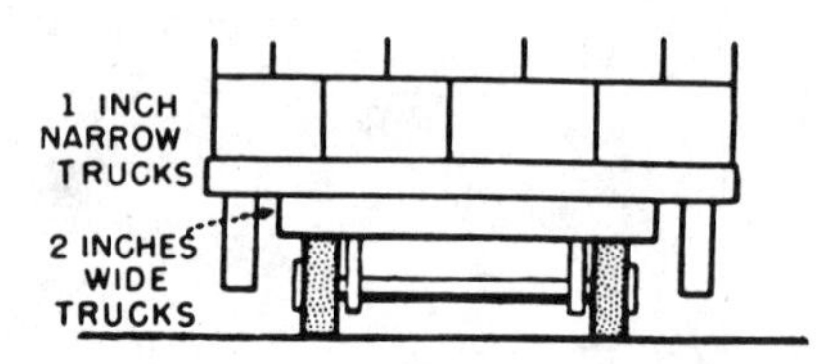

Fig. 6.4.3 Side clearance of standard skid platform on standard lift truck.

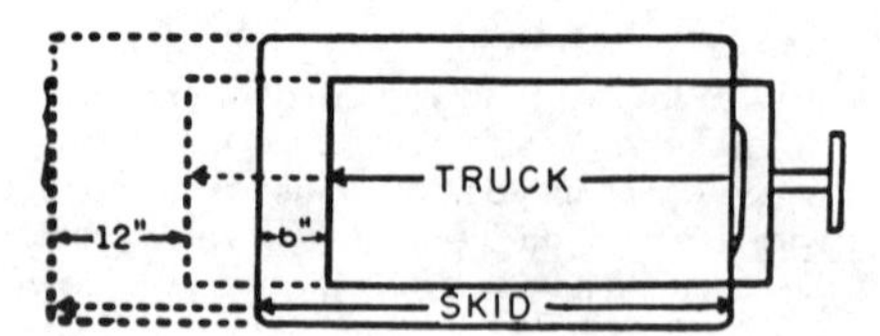

Fig. 6.4.4 Normal versus extreme permissible lengths of loads on standard lift trucks.

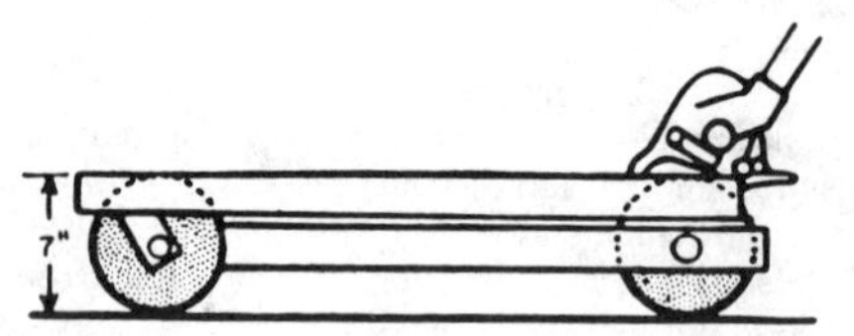

Fig. 6.4.5 Standard height of lift truck platform.

Fig. 6.4.6 Effects of insufficient clearance in truck movement over an uneven surface.

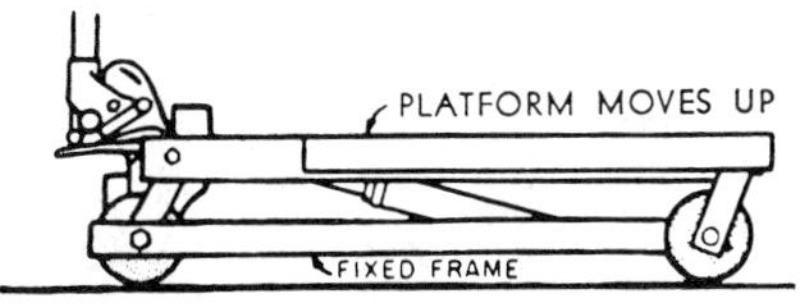

Fig. 6.4.7 Fixed-frame truck is hampered in moving over uneven surface.

further additions to the system should be of the same size and type for the sake of uniformity and maintenance convenience.

Clearance should be $\frac{1}{2}$ in. between the truck frame and the underside of the skid deck. However, on trucks with a higher lift, a 1-in. clearance between the truck platform and the underside of the skid deck provides a greater allowance for possible sagging of the skid boards under heavy loads.

Ground Clearance

Attention must be given to problems that may arise when trucks are to be used on ramps, in and out of freight cars, on motor trucks, or on and off elevators. Trucks may "hang up," as shown in Fig. 6.4.6, through lack of proper ground clearance. Trucks with high lift or greater ground clearance should be used in these areas. Note that although the hydraulic lift truck in Fig. 6.4.7 has a fixed frame and low ground clearance, the entire frame of the truck shown in Fig. 6.4.8 will move upward, giving better ground clearance.

Skid Platform Leg Clearance

The height of lift of skid-handling trucks is in definite relationship to the clearance between the bottom of the skid legs and the floor, after the skid has been picked up by the lift truck. A truck with a higher lift gives greater leg clearance, which is often a decided advantage when the truck is operating over rough or uneven floors. Multiple-stroke hand lift trucks provide higher lifts and get skid legs higher above the floor.

Selecting Wheels to Suit Floors

Either semisteel or forged steel wheels are usually equipped on lift trucks. Floor protective wheels are also available. Rubber-tired wheels are resilient and protect floors from damage. When of suitable size and capacity, they require a slightly higher starting effort than metal wheels, but roll over rough floors as easily and reduce vibration.

6.4.3 Checking Operating Conditions

This is done to ascertain whether existing operations are satisfactory.

Check aisles over which trucks must operate. Unless aisles are sufficiently wide, larger trucks and skids may not be usable.

Check the floor and aisle layout. Narrow passageways or short turns may prevent trucks from traversing certain aisles or departments, as in Fig. 6.4.9.

Check door openings. Entrances to elevators and doorways between departments may not be wide enough to allow loaded hand lift trucks to pass through. The narrowest points on the routes over which the trucks travel determine the widths of the skids that can be used in these areas.

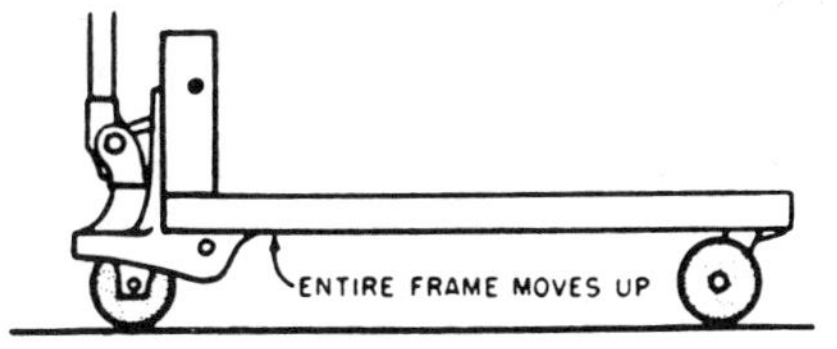

Fig. 6.4.8 Entire frame of truck will rise, giving ample ground clearance.

Fig. 6.4.9 Narrow aisles may prevent turns.

6.5 HYDRAULIC HAND LIFT TRUCKS

6.5.1 Lift Mechanism of Hydraulic Trucks

There is a greater mechanical advantage in hydraulic hand lift trucks because they operate on the principle of the simple lever and require less effort to lift the load. The longer the operating arm, the less effort required. Such lifts are of two types: the foot-operated and the hand-operated. In both, the lifting mechanism is actuated by moving the handle arm up and down like a pump handle for a well. The hand-operated type requires less effort because of the greater leverage. The lowering and lifting are done with the same pump unit in the hydraulic hand lift truck. When unstable loads are being handled or a truck is used in a multistory building with low-capacity floors, the lowering valve should be controllable so that lowering speed can be regulated.

Important factors are simplification of mechanism and ease of removing the operating unit. Handles must be balanced so as to prevent them from falling over and causing injury to the operator. The grip and design should be comfortable. Trucks should have a minimum underclearance of 2 in. or there is a tendency for the wheels to be raised off the ground while the truck is going up ramps or dock boards. Steel wheels with ball bearings are usually recommended and where light loads are being used, tapered bearings are suitable. Capacities normally range from 1000 to 20,000 lb. The single-stroke lift truck is used for light loads, and the multiple-stroke truck is for heavier loads. Figure 6.5.1 illustrates the hydraulic mechanism of a hydraulic hand lift truck, with oil reserves, pump mechanisms, and the lowering and release valve.

6.5.2 Low-lift Pallet Trucks

Figure 6.5.2 shows a totally hand-operated lift truck, on which the load is raised and pulled by hand power, the truck frame being arranged so that it can enter between the top and bottom boards of a pallet. The rear wheels are lowered through 8-in. minimum openings between the boards in the bottom of the pallet.

These trucks are constructed with large forward wheels connected to a steering handle attached to the main frame, and connected through a linkage or hydraulic system to the rear wheels, which are usually depressed in the main frame. At the end of the main frame, near the back, additional small wheels or slides are mounted in the frame to assist the operator in propelling the forks into the pallet. The pallets have openings in the bottom deck, both front and rear, so that the truck can enter from either end. After the truck has been run into the pallet, the lifting mechanism is operated and the small wheels in the rear of the truck are lowered through the pallet openings to raise the pallet 2 or 3 in. above the floor for transport.

Applications of Low-lift Trucks

The hand pallet truck is used in conjunction with pallets constructed with bottom openings. The purpose of the hand pallet truck system is to reduce the number of handlings. A change of location

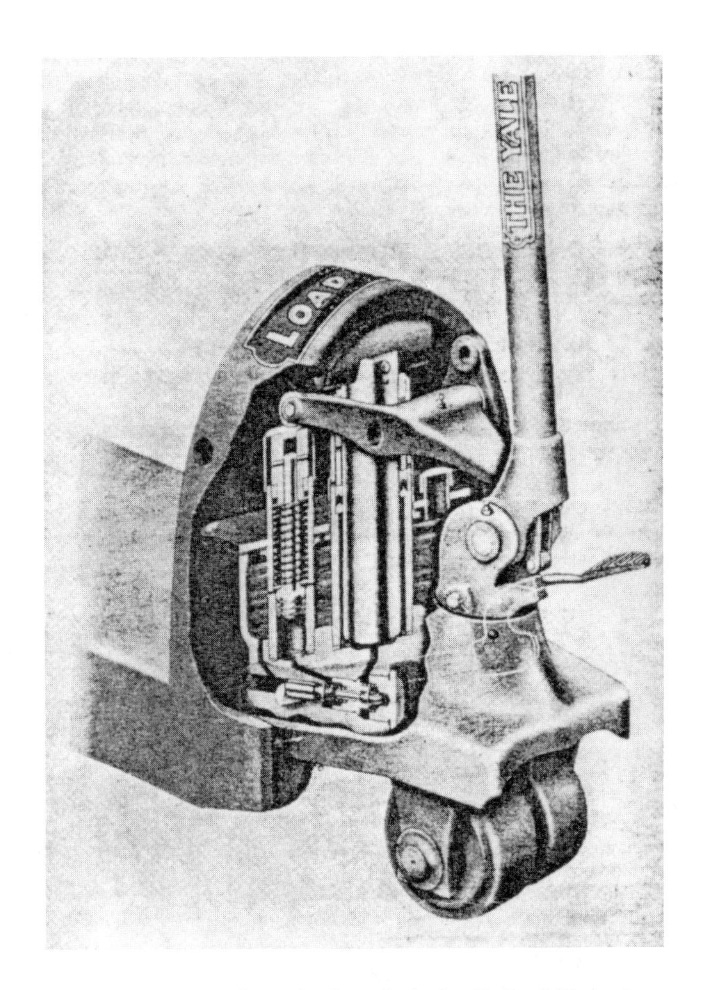

Fig. 6.5.1 Hydraulic mechanism of a hydraulic hand lift truck.

Fig. 6.5.2 Hand-operated low-lift pallet truck with hand-operated pump.

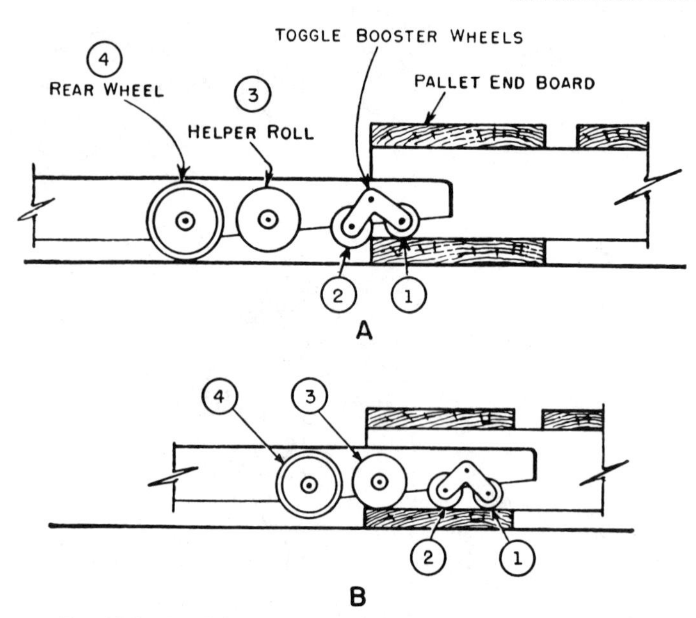

Fig. 6.5.3 Toggle-booster wheel system used on pallet lift trucks.

or handling between operations may be accomplished quickly by having materials loaded on pallets. The hand pallet truck is for horizontal movements only and is not suitable for stacking. It works well with power-elevating fork trucks and can be used where power trucks cannot be operated, such as in packing rooms or freight terminals, or for the movement of palleted loads in boxcars, trailers, and so on, and where aisle space or floor load is limited. The small rear wheels limit its use to short hauls (50–75 ft).

It has been widely accepted throughout the industry that standard truck models be 27 in. wide and the length should be equal to the length of pallet. The trucks are designed for loads up to 60 in. long and load capacities range from 1000 to 6000 lb, with a 4-in. lift to provide greater underclearance.

Pallets used with these trucks should have 8-in. minimum openings on the bottom side and 6-in. chamfered end boards and center boards for truck entry.

Figure 6.5.3 shows the toggle-booster wheel system and is typical of the arrangements used to facilitate insertion of the truck into the pallet. As the truck is inserted into the pallet, Roll 2 strikes the edge of the end board, causing Roll 1 to engage with the face of the end board. The inertia of the truck causes Roll 1 to raise the truck's rear wheels (4) from the floor as shown in part B of Fig. 6.5.3. The helper roll (3) and the rear wheel (4) are now in position to roll through the pallet, placing the truck in lifting position. The pallet should be positioned on the pallet truck so that the edge of the pallet rides against the inside end of the left platform. Forks are inserted immediately under the deck with the pallet in elevated position and the rear wheels of the truck are dropped through the space between the lower deckboards, contacting the floor surface.

Hand Low-lift Platform Trucks

These are constructed with the forward wheels connected to a steering handle attached to the main frame. The rear wheels, which are attached to the same frame, are actuated up and down by the hydraulic system. Platforms are solid or open and the trucks are generally used with skids.

The capacities on standard trucks of this kind are from 1000 to 6000 lb, widths of 18–27 in., lengths of 30–72 in. in 6-in. increments, and lift heights of 6, 7, 9, and 11 in. Depending on the use to which they will be put, they have either wood or steel platforms. Special trucks are built in capacities up to 20,000 lb and with platforms wider and longer than the standard.

6.6 MECHANICAL HAND LIFT TRUCKS

6.6.1 Mechanical Lift Mechanism

Similar to hydraulic hand lift trucks, these are operated by a system of levers rather than by hydraulic cylinders. This type of action can be compared to that of an automobile hand jack. The platform is

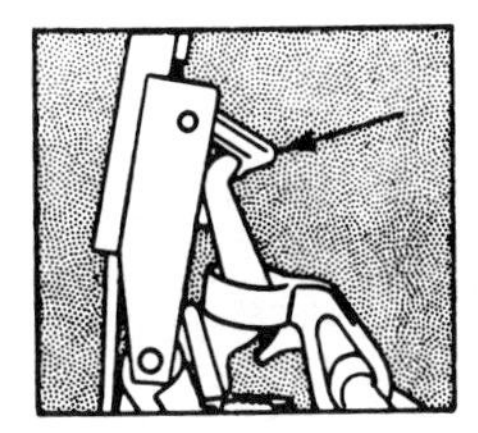
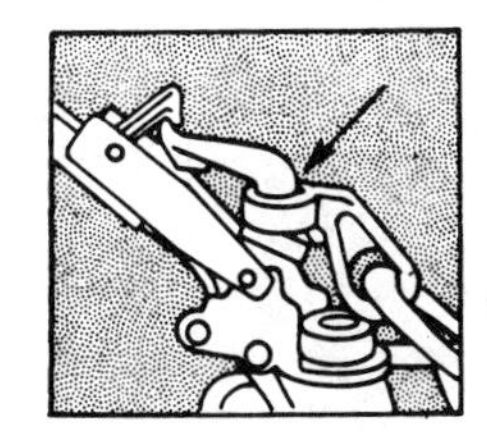
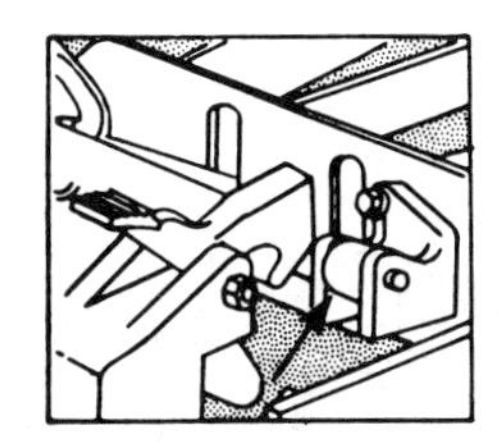

Fig. 6.6.1 One-stroke lever-type lift mechanism.

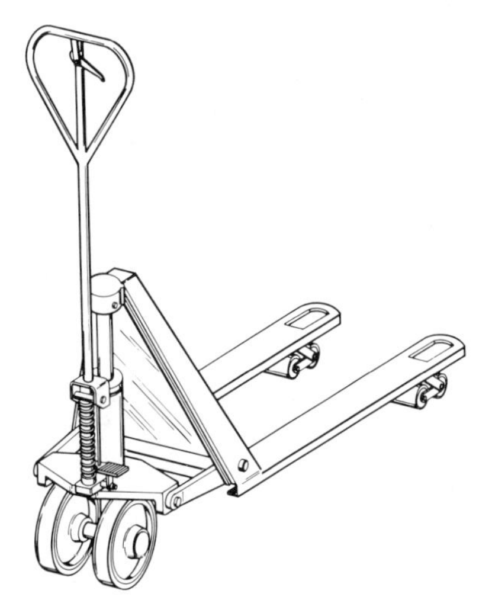

Fig. 6.6.2 Multiple-stroke, mechanically operated pallet lift truck.

raised by actuating a handle which raises a pawl that falls into a slot or groove. By releasing the pawl, lowering is accomplished. There are many modifications of this unit. In a single-stroke, low-lift unit, a latch on the tow handle engages a lifting lever which is coupled to the platform. The platform is secured into a raised position by another latch, which is released for lowering. The tow handle may be equipped with a spring that relieves the operator of the handle weight and keeps the handle off the floor when the unit is not in use. Figure 6.6.1 shows certain features of this single-stroke lever-type mechanism.

When relatively light loads are to be handled, low initial cost is an important factor, and frequency of use is not a problem, mechanical hand lift trucks are used. However, they do require more effort than the multiple-stroke type truck to operate.

6.6.2 Multiple-stroke Truck

Figure 6.6.2 shows a 2000-lb capacity multiple-stroke pallet lift truck. It requires five strokes for raising and it lifts 4 in. Operation is achieved with double-faced pallets having a $3\frac{5}{8}$-in. minimum opening for truck entry. Rear wheels are available in steel, rubber, aluminum, and plastic and are 3 in. × 5 in. in diameter. The weight of these trucks is normally around 300 lb each.

CHAPTER 7
POWERED INDUSTRIAL TRUCKS

WILLIAM O'CONNELL

Yale Materials Handling Corp.
Flemington, New Jersey

KARL E. LANKER
JAMES M. SNYDER

The Sims Consulting Group, Inc.
Lancaster, Ohio

CLARK C. SIMPSON

Clark Equipment Company
Industrial Truck Division
Battle Creek, Michigan

ROBERT CAMMACK
LORI MAY

Taylor-Dunn Manufacturing Company
Anaheim, California

CHARLES A. ISENBERGER

Grove Manufacturing Company
Shady Grove, Pennsylvania

A. J. CASON

Economy Engineering Company
Bensenville, Illinois

7.1 LIFT TRUCKS

7.1.1 Walkie Trucks

William O'Connell

Low-lift Pallet Trucks

Low-lift pallet trucks are the most widely known walkie trucks that are self-loading and equipped with wheeled forks of dimensions to permit the forks to go between the top and bottom boards of a double-faced pallet. The wheels are capable of lowering into spaces between the bottom boards so as to raise the pallet off the floor for transporting.

Nonpowered low-lift pallet trucks are available in 1000–2500-kg capacity. They have fork spreads that permit them to be used with 800-, 1000-, and 1200-mm wide pallets. Fork lengths must correspond to pallet length when used with double-faced pallets to insure that the wheels project through the bottom boards of the pallet. Nonpowered low-lift pallet trucks are generally used in applications where travel distance and tonnage of material to be transported are moderate.

Powered low-lift pallet trucks are available in capacities through 6000 kg. These trucks handle double-faced pallets through 3000 kg. The width and length of pallets compatible with trucks are combinations of 800, 1000, and 1200 mm. The trucks with capacities of 4000–6000 kg are capable of handling single-faced pallets only. The powered low-lift pallet trucks are economical alternatives to a riding-type lift truck where horizontal transporting over relatively short distance is required. See Fig. 7.1.1.

For applications that require not only short-distance transporting, but also require longer hauls, a walkie/rider truck is available. This unit allows the operator to ride on the truck, yet also permits off-truck maneuvering by the operator. When riding, the travel speed is approximately 10 km/hr, whereas the walkie speed is approximately 5 km/hr. The operator's position is either on the front of the truck or in the center of the truck. The latter position is frequently used in order-picking applications to position the operator closer to the load. Because of the higher travel speed that is available when the operator rides on this truck, it is also equipped with caster wheels to stabilize the truck for operator and load. See Fig. 7.1.2.

With the advent of palletless or slipsheet handling, low-lift pallet trucks have been developed that are compatible with this technique of unitizing loads. The trucks are designed in such a way that the platens used for slipsheet loads are "split" to enable them to handle standard pallets also. This versatility permits a slipsheet load to be retrieved from the floor and then be palletized. The palletized load can then be transported. The reverse handling process from pallet to floor can also be accomplished. See Fig. 7.1.3.

A very popular derivative of the pallet truck is a truck equipped with fork assemblies that will handle two pallets. Double pallet handling is very common in grocery order-selecting applications. These trucks handle two 1000 mm wide × 1200 mm long pallets with the wheels projected through either of the openings in the bottom boards of the pallet on the rear end of the truck.

Low-lift Platform Trucks

A low-lift platform truck is a self-loading truck equipped with a load platform intended primarily for transporting skids and bins. A wide variety of platform lengths and widths are available to insure compatibility with the skid. The lowered height of the platform is generally a minimum of 150 mm. Other lowered heights are also available. The raising of the platform is accomplished through a linkage mechanism. However, contrary to a pallet truck, the wheels are not necessarily part of the linkage. Generally they are affixed to the truck frame. The wheels under the platform are larger in diameter than those used on a pallet truck because they do not have to go into a double-faced pallet. See Fig. 7.1.4.

Low-lift platform trucks have a capacity of 2000–6000 kg; however, special trucks are available with capacities of 10,000 kg. These trucks demonstrate their versatility in many special applications, wherein the platform is modified to accept customized unit loads such as steel coils, wire and cable reels, bolts of yarn, and so on.

Low-lift platform trucks are also available in a configuration in which the platform is created by what is termed a "skid adapter" option that is available. A truck equipped with this option permits double-faced pallets to be handled with the adapter raised and locked in a vertical position or single-faced pallets or skids when the skid adapter is resting on the forks. See Fig. 7.1.5.

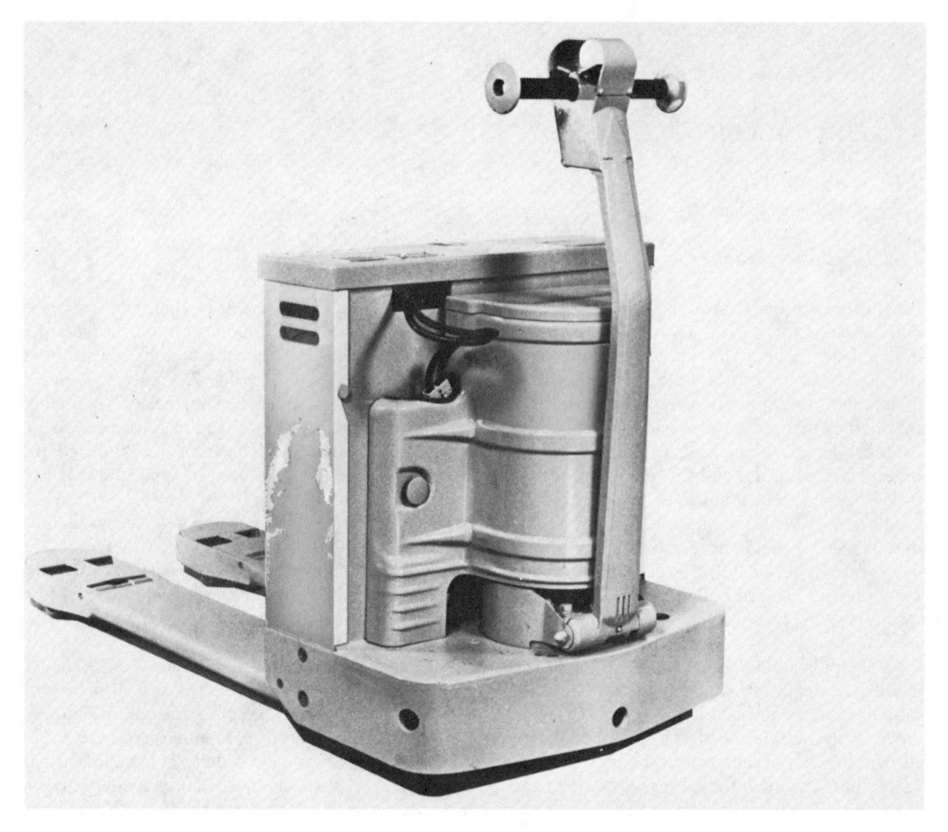

Fig. 7.1.1 Low-lift pallet truck.

High-lift Pallet Trucks

High-lift pallet trucks are capable of self-loading. They are equipped with an elevating mechanism designed to permit tiering. Again, they fall into two areas: powered and nonpowered. For the sake of our discussion in this chapter we discuss primarily powered trucks of the various types used in industry.

One type of high-lift pallet truck is a counterbalanced truck. This is a truck equipped with a load-engaging means wherein the load is suspended forward of the front wheels. With this design the truck offers all of the high-lift capabilities of a rider forklift truck. It can handle single- and double-faced pallets, skids, and bins, and it can be equipped with many of the popular attachments offered on rider type trucks. See Fig. 7.1.6.

Walkie high-lift pallet trucks are also available as straddle trucks. Straddle trucks may be defined as trucks with a horizontal structural member supporting wheels extending forward from the main body of the trucks. These trucks are designed to operate in narrow aisles as compared to a counterbalanced truck. The load is carried between the outrigger arms of the truck. This type of truck is utilized where uniformly sized pallet or skid loads are encountered. See Fig. 7.1.7.

Another type of high-lift pallet truck is that which is called a "fork over arm" truck. This truck is a non-counterbalanced truck. The forks are placed directly over the load wheels. This truck is limited to handling single-faced pallets, skids, or bins.

The walkie reach truck is a self-loading truck having a load-engaging means mounted so it can be extended forward under control to permit a load to be picked up and deposited in the extended position and transported in the retracted position. This truck has an advantage over the straddle truck in that pallets of various widths can be handled and stored close together. Because the base arms do not enter the rack sections, the truck can be used with any existing rack installation. See Fig. 7.1.8.

Fig. 7.1.2 Walkie/rider low-lift pallet truck. Note the lift/lower control on bar and operator riding the platform.

High-lift Platform Trucks

High-lift platform trucks are self-loading and are equipped with a load platform. They are designed primarily for transporting and tiering skids. In contrast to the "fork over arm" truck, these trucks do not handle single-faced pallets.

Walkie Tractors

Walkie tow tractors are powered trucks designed primarily to draw one or more nonpowered trucks, trailers, or other mobile loads. These tractors are generally capable of pulling loads of 4000 kg. See Fig. 7.1.9.

Fig. 7.1.3 Walkie/rider pallet truck handling slipsheet load.

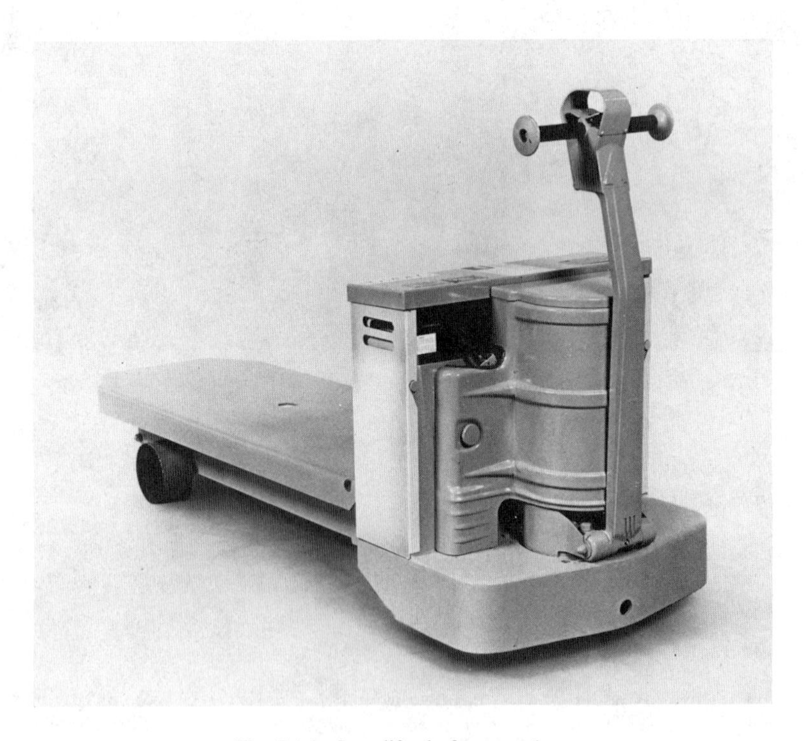

Fig. 7.1.4 Low-lift platform truck.

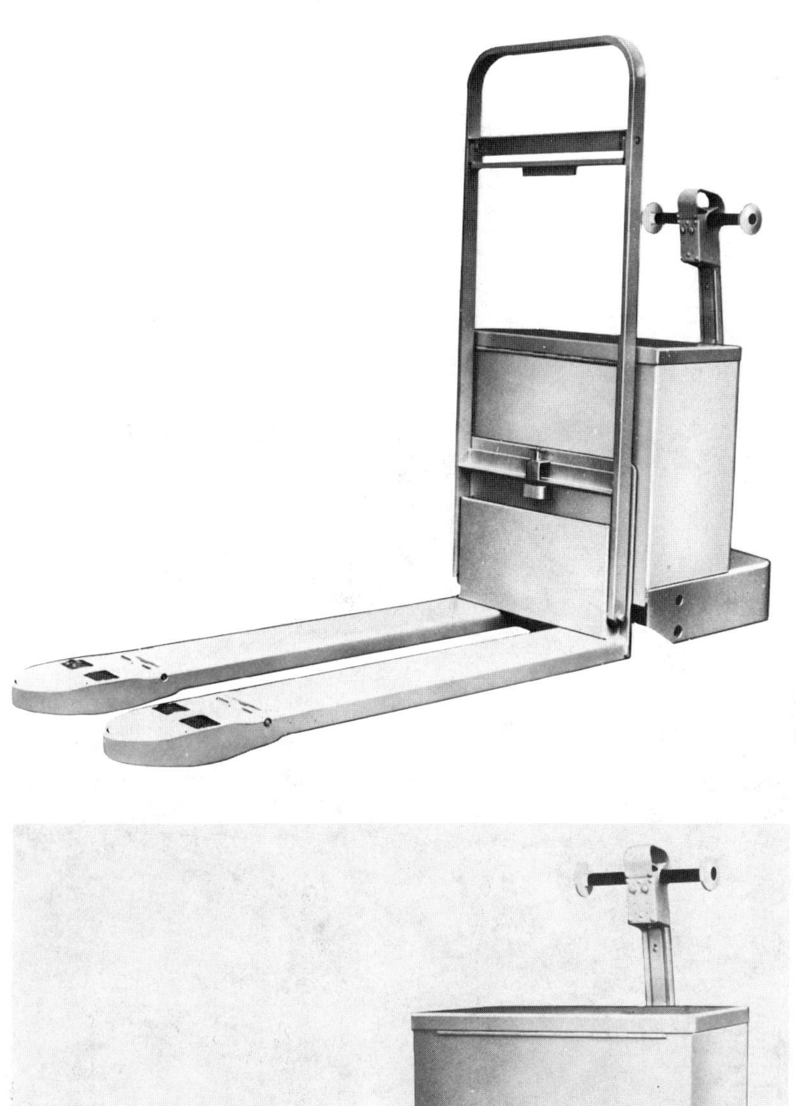

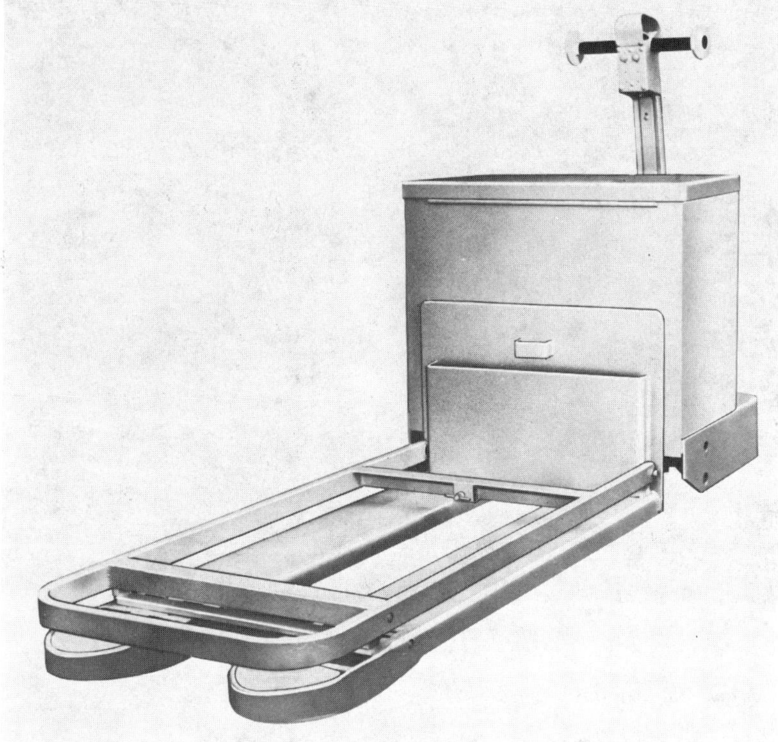

Fig. 7.1.5 Low-lift pallet truck equipped with skid adapter. *Top,* skid adapter raised to handle double-faced pallets; *Bottom,* adapter lowered to handle skids or bins.

Fig. 7.1.6 High-lift counterbalanced truck.

Fig. 7.1.7 Walkie straddle truck.

Walkie Truck Applications

Now that we have described the various types of walkie trucks and the typical type of pallet or skid used with each vehicle, it is important to identify the additional criteria you must be aware of when making a selection of this type of truck for your materials handling needs.

Fig. 7.1.8 Walkie reach truck.

Horizontal Travel Distance. Because the operating speed is limited to 5 km/hr, the travel distance for transporting unitized loads should be limited to maximum distances of 100 m. The frequency of the need to travel this distance will also determine whether a rider truck versus a walkie truck would be more effective.

Operating Aisles. The operating aisle required for a walkie truck can be determined by referring to Tables 7.1.1 and 7.1.2. For a description of right-angle aisle (RAA) and intersecting aisle refer to Fig. 7.1.10.

The importance of determining the aisle requirements is that it permits us to ensure we have optimized equipment consistent with space utilization.

Stacking Heights. High-lift pallet trucks of the types defined are generally limited to a maximum stacking height of 5 m. More commonly they are used to stack three-high unit loads to a height of 3 m. The time to stack or lift to 3 m is 12 sec without a load and 20 sec with a load. One should consult manufacturers, specifications because this specification is greatly dependent on battery voltage, type of mast, and load weight to be carried.

The basic information identified here will permit you to determine the general application of a walkie truck.

Vehicle Specifications

We can combine the basic information just provided with the specifications of our unit load. We must know its size in height, width, and length and its weight. Also we should know if this load is uniformly distributed and homogeneous. The load information is then combined with our selection of the type of pallet or skid we will employ consistent with our transporting and/or stacking requirements, considering optimum storage requirements defined by type of racks, block storage, and so on.

Additional facility-related environmental specifications (such as ramps, grades to be negotiated) must be considered and must be compatible with: vehicle grade capability and underclearance; floor loading distribution with vehicle weight and weight distribution; and elevation limitations with truck overall height. Atmospheric conditions as defined in NFPA 505 should be evaluated and the "type" designation of the equipment should be specified.

Fig. 7.1.9 Walkie tow tractor.

Facility support requirements to consider for battery charging are:

Power requirements
Location within facility
Ventilation
Water: washing and drainage
Battery-charging means, if necessary

Equipment maintenance, if done internally or by an outside service, may dictate tools, shop equipment, and parts inventory, which will be considered with facility needs.

Table 7.1.1 Required Dimensions of Right-angle and Intersecting Aisles for Low-lift Trucks[a]

Low-lift Truck	Right-Angle Aisle	Intersecting Aisle
Walkie pallet	1500–2000 mm	1500–2000 mm
Walkie/rider pallet	2000–2500 mm	1500–2000 mm
Walkie platform	1500–2000 mm	1500–2000 mm

[a] Load dimension, 1200 mm × 1200 mm (maximum); load weight, 3000 kg (maximum).

Table 7.1.2 Aisle and Height Dimensions for High-lift Trucks

High-lift Truck	Right-angle Aisle	Intersecting Aisle	Collapsed Height	Lift Height
Straddle[a]	1500–2000 mm	1500–2000 mm	2000–2500 mm	4000 mm
Reach[a]	1500–2000 mm	1500–2000 mm	2000–2500 mm	4000 mm
Counterbalanced[b]	3000–3500 mm	1500–2500 mm	2000–2500 mm	3000 mm

[a] Load dimension, 1200 mm × 1200 mm (maximum); load weight, 2000 kg (maximum).
[b] Load dimension, 1200 mm × 1200 mm (maximum); load weight, 1500 kg (maximum).

Walkie Truck Features

In addition to determining the type of truck based on specifications that will match your application needs, there are several features that will affect the total operating efficiency of your handling methods.

The horizontal traction control is activated from the end of the control handle. Trucks are available with either "stepped" 2 or 3 speeds in forward and reverse direction, or with infinitely variable speed; these are equipped with a transistor or silicon controlled rectifier (SCR) electronic control. The choice of speed control is dependent on the amount of close-quarter maneuvering to be encountered and the fragility of product to be handled. Direction is generally controlled simultaneously with speed.

Lift and lower controls are either electrohydraulically or manual-hydraulically operated through a lever. The former is activated by electrical switches in the immediate proximity to speed and direction controls.

The brakes of walkies are activated by the steering handle being placed in the vertical and/or horizontal position. When not in either of these positions, horizontal travel is possible. The steering handle automatically returns to a vertical position, applying the brakes, when released by the operator. See Figs. 7.1.11 and 7.1.12.

Many vehicles have an automatic reversing feature available. This feature incorporates an activating device at the end of the steering handle which reverses the truck when the handle encounters a resisting force. The objective of this feature is additional safety in inhibiting the operator from being trapped between the truck and stationary objects. See Fig. 7.1.13.

As is the case with rider trucks, many other features are available such as state-of-charge gauges indicating condition of the battery, hour meters indicating elapsed time of traction motor use, and dynamic braking to obtain braking on grades electrically while maintaining the steering handle in the operating position.

Safety

The walkie truck is designed with the safety of the operator and the control of vehicle in mind. However, as with all lift trucks, as well as mobile equipment in general, operator training is mandatory.

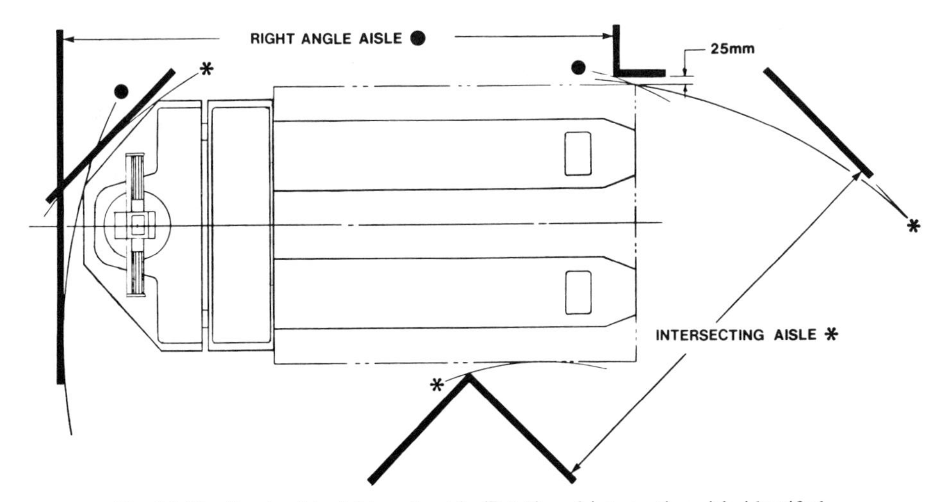

Fig. 7.1.10 Truck with right-angle aisle (RAA) and intersecting aisle identified.

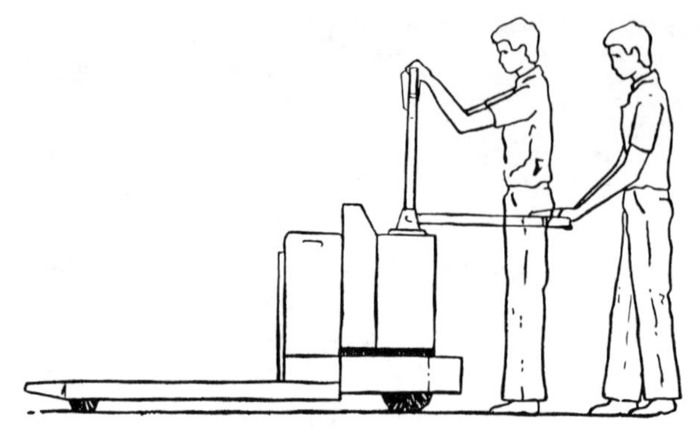

Fig. 7.1.11 Braking positions of handle.

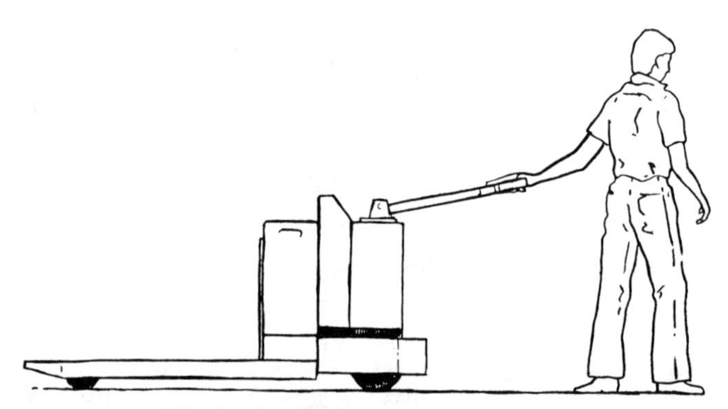

Fig. 7.1.12 Traveling position of handle.

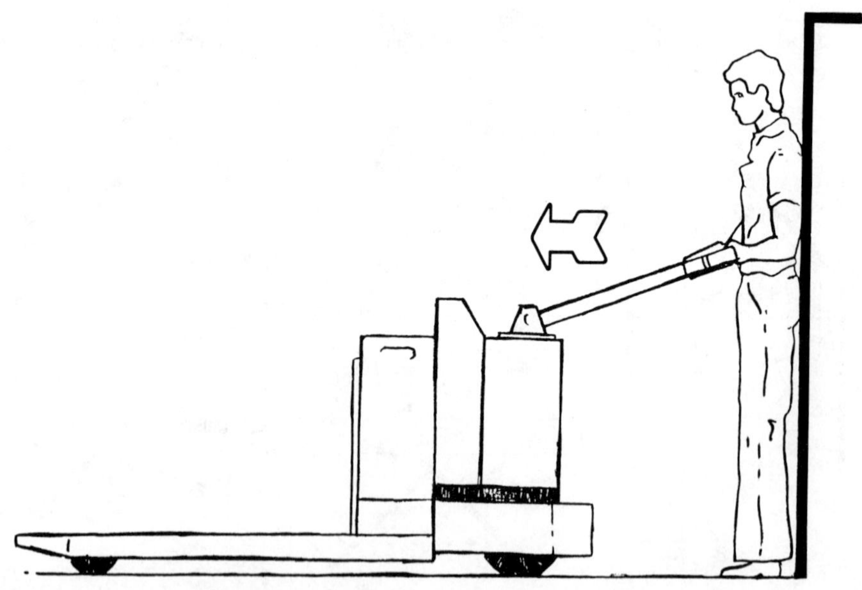

Fig. 7.1.13 Reversing feature incorported in handle of pallet truck.

The relative simplicity of the walkie truck in comparison to other types of equipment makes the training of production workers as part-time operators practical.

7.1.2 Rider Trucks and General Considerations

Karl E. Lanker and James M. Snyder

The development of powered industrial materials handling trucks has seen great advances since the first crude vehicles began to achieve widespread use after World War II. These early vehicle designs have been refined and improved to lift heavier and larger loads higher while at the same time requiring less floor space for both handling and storage. Power sources have diversified from strictly gasoline-powered units to many models with gasoline, liquified petroleum gas (LPG), diesel, and electric motors. Many manufacturers now make a variety of vehicles. Some models have similar characteristics while other versions have unique capabilities for specific applications.

Because of the diversity of modern forklift trucks, it is important for the user to develop a set of operating characteristics for the job to be accomplished and then have the supplier demonstrate that the product conforms to those characteristics. The selection of a specific supplier should be made based on compliance with the desired operating characteristics and with consideration given to product design, reliability, and serviceability within the specific plant location. Although the pricing of vehicles within general capacity ranges between suppliers is computable, total ownership costs may not be computable and an analysis of operating cost, parts pricing, and location of service facility should be included in any purchase price analysis.

Types of Trucks

Powered industrial trucks can be broadly categorized by load-lifting capacity and aisle width requirements as illustrated in Fig. 7.1.14. Additional subclassifications can be made based on the power source and method of control (operator walks or rides).

In general, forklift trucks are produced in "families" based upon chassis capacity. These family groupings are generally divided into load capacity ranges of 1000–2000 lb (450–900 kg), 2000–5000 lb (900–2300 kg), 5000–8000 lb (2300–3600 kg), 8000–10,000 lb (3600–4500 kg), and vehicles with a load capacity in excess of 10,000 lb (4500 kg). Although capacity ratings are certified by the manufacturers, the ability of a truck to handle a particular load depends on both the truck capacity and the relationship of that capacity to the family group chassis capacity. Expected operating and repair costs may be accelerated if a light-duty chassis is used or if the vehicle is used in continuous or heavy-duty operation. The use of a chassis at the high end of its load range will be more expensive when compared to using a higher-rated chassis at the low end of its load range. Therefore, it is important to know the chassis grouping and capacity being considered and the means by which the purchaser can select proper vehicle capacity to obtain long life and low operating cost.

Vehicles with capacities under 10,000 lb (4500 kg) can be further subdivided by their relative aisle width requirements and the method of control, that is, whether the operator rides on the vehicle or leads the vehicle by means of a hand or remote control. These latter vehicles, the walkie trucks, have specific applications in relatively low lift height or low-speed operations and are covered in a previous section of this handbook. The rider-type vehicles are most commonly used in warehousing and materials handling application where rapid load movement is required. When divided by aisle width requirements, the terms *conventional, narrow,* and *very narrow* aisle are usually used.

Conventional Aisle. The first industrial fork trucks consisted of a counterbalance design in which the overturning moment of the load was offset by the weight of the vehicle behind the load center of gravity. The aisle space required for these vehicles was, and is today, approximately 10–15 ft (3.1–4.6 m) for a 48 in. × 40 in. (1200 mm × 1000 mm) pallet. These vehicles require what is now known as a conventional aisle because these were the standard for material handling for many years. The counterbalanced fork truck is the primary type of conventional aisle vehicle. These vehicles are available with either internal combustion or electric power sources. Internal combustion engines are further available in gasoline, diesel, or LPG versions.

Narrow Aisle. Engineering efforts to reduce the amount of aisle space required to handle loads resulted in design changes which reduced aisle requirements to approximately 7–10 ft (2.1–3.1 m). Trucks requiring this aisle size became known as narrow aisle vehicles because the aisle required was more narrow than that of the conventional fork trucks. Almost all narrow-aisle vehicles currently available are electrically powered.

The basic classification for narrow-aisle vehicles is a straddle-type truck. These vehicles use outrigger arms to straddle the load and thus reduce the requirement for the amount of counterbalancing weight needed to prevent overturning. The true straddle truck obtains its high stability and low vehicle weight from the fact that the vehicle has a low center of gravity and the load is carried entirely within the footprint of the vehicle.

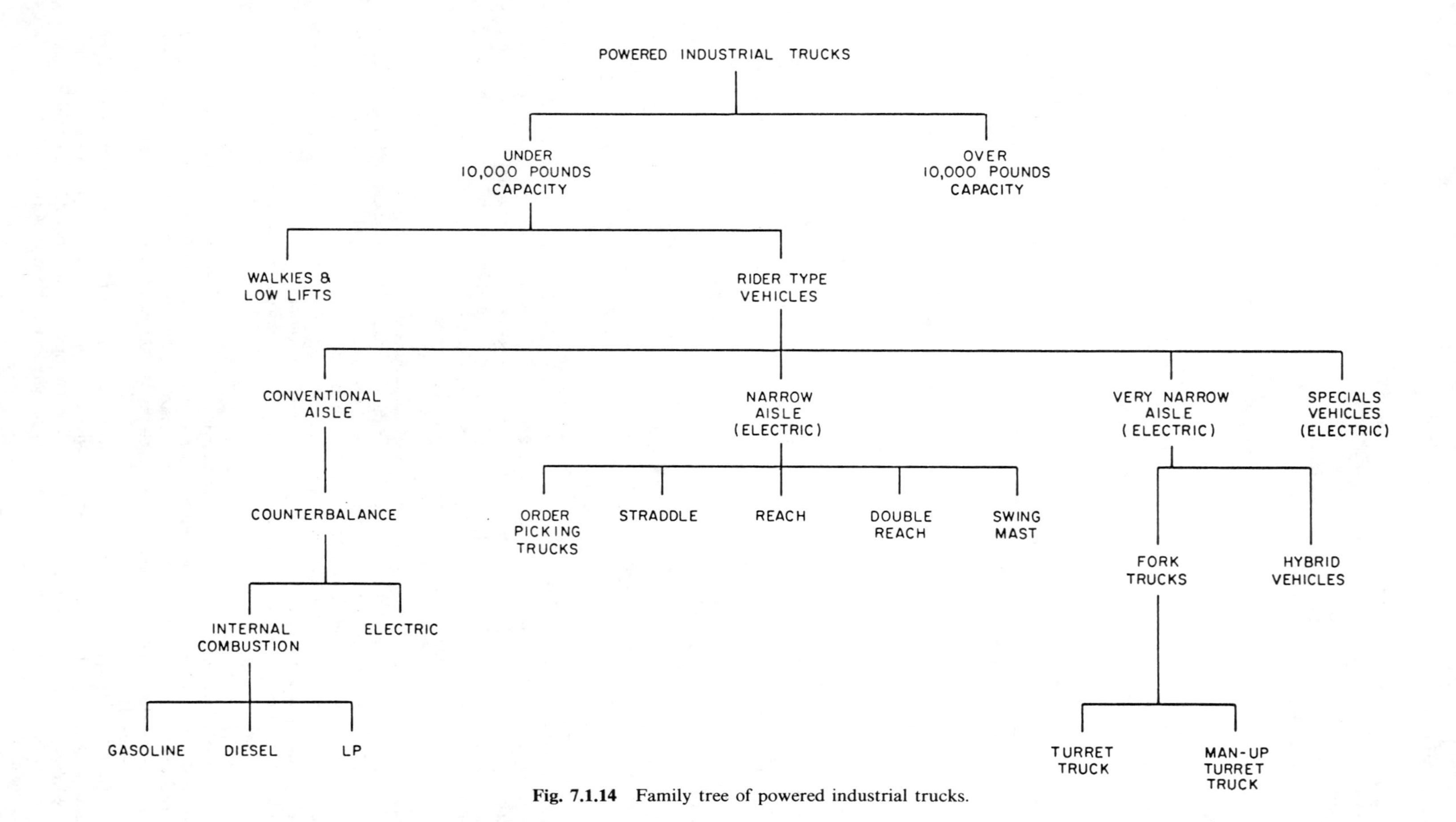

Fig. 7.1.14 Family tree of powered industrial trucks.

Variations of the straddle truck concept include the reach truck and double reach truck. Reach trucks do not straddle the entire length of the load but rather use a pantograph-type mechanism to extend the forks beyond shortened outrigger arms. These trucks do require a small amount of counterbalancing to provide stability when the load is in the extended position. The double reach version of the reach truck has an extra-long pantograph assembly which permits storing loads two pallets deep, thus obtaining greater storage density with no increase in aisle requirements.

A third form of narrow-aisle vehicle is a variation on the counterbalance truck and is known as a swing mast vehicle. This vehicle uses a rotating mast assembly to place the load at a right angle to the direction of travel, thus permitting storage within an aisle narrower than that required for a conventional counterbalance truck. This type of vehicle is also suited for use with long loads such as pipe and lumber. The appearance of the vehicle and its load stability are, however, very similar to that of a counterbalance truck.

Very Narrow Aisle. In the quest to obtain greater storage efficiency for materials handling vehicles, aisle requirements were further reduced by placing the load at a right angle to the direction of travel and side loading the load from the vehicle into the storage position. These side-loading vehicles are generally referred to as turret trucks due to the use of a rotating fork assembly or turret assembly. This rotating fork mechanism permits the forks to be positioned to face either the right or left side of the aisle and store and retrieve a pallet load. These vehicles are available in configurations in which the operator remains at floor level or in which the operator can ride with the load in a mast-mounted cab. The man-up configuration provides the operator with greater visibility of the storage and retrieval operations and also permits the vehicle to be used for order picking.

Another technique used to produce a very-narrow-aisle vehicle is the adaptation of automated storage and retrieval (AS/R) machine technology to produce a hybrid vehicle. Hybrid vehicles combine the tall, top-guided mast technology of the AS/R machines with the battery-powered wheeled vehicle technology of forklift trucks. When transferring between aisles, the vehicle is driven by battery power. When operating within the aisle, the vehicle runs on AC power collected by top-mounted collectors. This in-aisle AC power source is also used to recharge the batteries. This marriage of technologies results in the aisle transfer capability associated with fork trucks and the high speed, high lift capability of the AS/R machines.

Special Vehicles. In addition to the conventional-aisle, narrow-aisle, and very-narrow-aisle fork trucks, a number of special vehicles are also available in capacities under 10,000 lb (4500 kg). These special vehicles consist of both side-loading trucks typically used to handle long loads and operate in narrow aisles, and specially designed straddle trucks used for order picking. Aisle requirements for the side-loading trucks are similar to those for the narrow-aisle vehicles but large maneuvering areas at the ends of the storage aisles are generally required due to the length of the load being handled. These vehicles are usually equipped with traveling masts to retract the load to a transport position over the bed of the truck. Other units have reach-type fork assemblies to accomplish the same purpose. The order-picking trucks are equipped with a platform that permits the operator to pick cases from a rack storage area and place them on a pallet held on forks located in front of the operator platform.

Four directional trucks are capable of traveling as either conventional units or as side-loading narrow-aisle machines. These special application vehicles are generally used for narrow aisle side loading of long loads and where multiple deep stacking of these loads is advantageous to increase cube utilization.

Typical Fork Truck Construction

An industrial fork truck, such as the typical sit-down rider counterbalanced truck illustrated in Fig. 7.1.15, consists of a number of component systems essential to the operation or safety of the vehicle. These basic components consist of:

The Overhead Guard. Overhead guards are required by the U.S. Occupational Safety and Health Administration (OSHA) to protect the driver from falling objects unless operating conditions do not permit. Examples of conditions that do not permit falling objects to strike the driver include trucks designed to prevent the load from being lifted above the driver or situations where a vehicle is used in low-headroom truck trailers for loading and unloading and where high lift is not used. Some fork trucks can be equipped with retractable overhead guards which permit the truck to be used in both high-lift and low-overhead conditions.

The Mast Assembly. The mast assembly of the fork truck consists of a steel upright assembly and hydraulic cylinders to provide the lifting capability of the fork truck. The mast consists of stationary outer channels and telescoping inner channels which provide for high lifting capabilities while permitting a low collapsed height. Various configurations of lift cylinders, mast geometry, and chains are used to provide equalized reaction to loads, increase stability, and decrease uncontrollable mast deflection. Multiple-stage masts increase mast complexity and require more energy than a low freelift design.

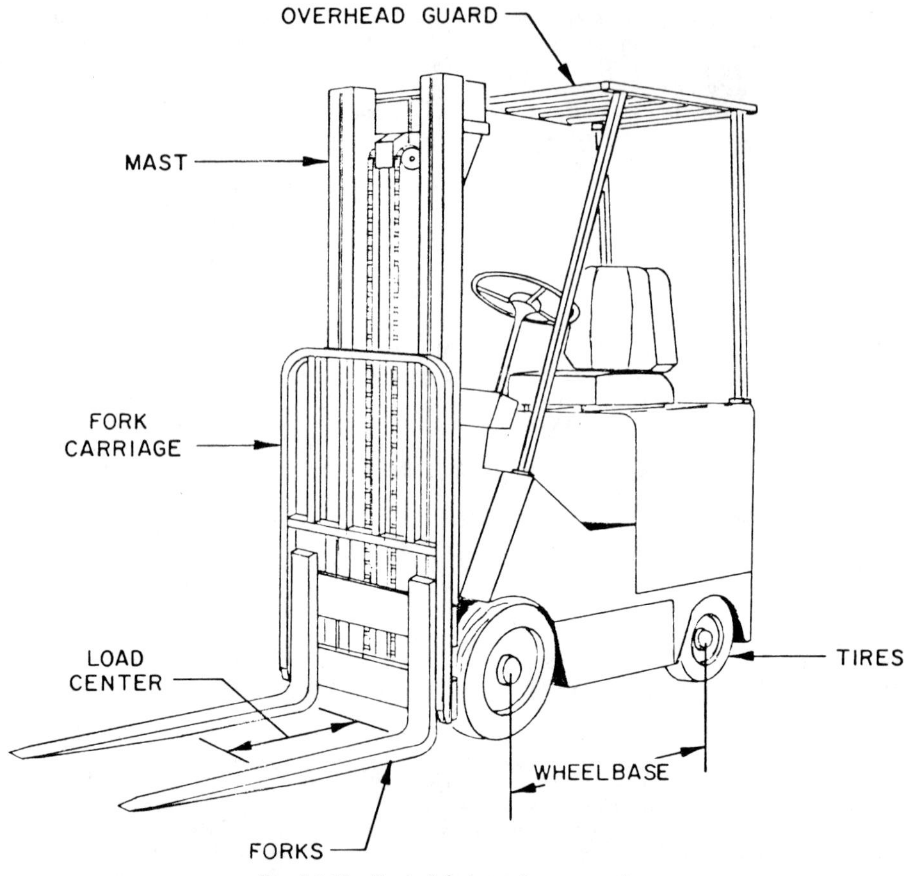

Fig. 7.1.15 Typical fork truck construction.

The most economical mast for general purpose storage and in-plant transportation is a two-stage mast with low freelift. The most common type of mast assemblies are discussed next.

Single-Stage Mast. The single-stage mast consists of two upright channels in which the fork carriage travels at twice the speed of lifting. This type of mast assembly results in the highest overall lowered height in comparison to the available lifting height. This mast is typically used in low-lift or high-lift applications where the overall lowered height is not critical. Because it has the least number of moving or sliding parts, this type of mast provides the greatest stability in terms of load movement at a given lift height.

Two-Stage or Three-Stage Mast. Duplex and triplex mast assemblies provide greater lift heights while retaining low overall lowered heights. Generally these masts are equipped with double-acting cylinders and chain reeving to achieve the higher lift heights. Fork travel speed varies according to the stage in which the upright assembly is lifting.

Four-Stage Mast. Four-stage or quad masts operate in a manner similar to the two- and three-stage assemblies but provide even greater lifting height in relationship to the overall lowered height. Because of the large number of moving parts and the clearances required for movement in each of the telescoping assemblies, this type of mast has the greatest of uncontrolled load movement of all mast configurations. This uncontrolled movement requires that additional clearances be allowed at high lift heights to permit maneuvering space to compensate for this movement.

An important operational aspect of mast assemblies is freelift. Freelift is defined as the vertical distance forks can raise above the floor before the mast begins to extend from its collapsed height. The amount of freelift available can be critical in low-headroom applications such as truck loading or unloading where high-lift capacity is required within the plant but low headroom is available within the truck. The amount of freelift available is dependent on the mast and truck design. Freelift may range from only a few inches on some models to over 60 in. (1500 mm) on the man-up turret truck. Most manufacturers offer trucks with either standard or "high" freelift available.

Another aspect of mast design is fore and aft tilt. Mast tilt is controlled by a hydraulic cylinder and is adjustable from approximately 5° to 6° forward to up to 10° backward. Mast tilt is useful for providing additional load stability and security during transport (backward tilt), and assists in positioning the forks under a load or removing the forks from a load (forward tilt).

Fork Carriage. The fork carriage is the assembly to which forks or other attachments are mounted. The Industrial Truck Association (ITA) has developed classifications for fork carriages based on the capacity of the fork truck. These categories are identified as Class 1, Class 2, and Class 3 carriage assemblies. A load backrest is usually provided on the fork carriage to prevent loads from falling back when the mast is tilted backward. In addition to providing the mounting and lifting base for the forks, the carriage assembly is also the mounting base for various attachments. The location of mounting holes has been standardized by the ITA and helps assure the interchange and versatility of attachments by providing a standard hole pattern for mounting attachments.

Forks. Forks are the actual loading engagement device most commonly used on forklift trucks. Forks are generally 4–6 in. (100–150 mm) wide, 42–48 in. (1070–1200 mm) long, and are typically $1\frac{1}{2}$ in. (40 mm) thick at the heel. Forks are available in a number of variations with chiseled ends, full bottom taper, and spark-resistant materials.

Two options available on many trucks are side-shifters and fork spreaders. The side-shifter permits the forks to retain a given spacing but be moved side to side on the fork carriage thus permitting a load to be picked up even if it is not perfectly aligned with the truck. Fork spreaders permit the spacing of the forks to be varied to accommodate varying load sizes.

Wheelbase. The wheelbase of a truck is defined as the centerline distance between the front and rear wheels. The wheelbase of a truck determines, to a great extent, the operating and handling characteristics of the truck. These characteristics include the overall load capacity (especially when applied to counterbalance trucks), turning radii, aisle space required for right-angle load stacking, and the gradeability or maximum grade that can be negotiated by the truck without scraping bottom.

Load Center. The truck load center (LC) is the distance from the front face of the forks to the center of gravity of the load. Most forklift trucks in capacity ranges under 10,000 lb (4500 kg) have a 24-in. (600-mm) LC specified as the standard. The LC is especially critical for counterbalance vehicles in which the wheelbase and counterbalancing weight determine the ultimate capability of a truck to support a load at a given load center and lift height. If oversized loads have load centers greater than 24 in. (600 mm), the truck capacity may be substantially decreased or derated.

Tires. Fork truck tires are most commonly available in either a pneumatic or cushion tire type. Trucks are generally designed to use only one type of tire. Specifics on tires are further described in the Equipment Selection and Engineering Data sections.

Power Source. With the exception of the electrically powered narrow- and very-narrow-aisle vehicles, most forklift trucks are available in either internal combustion (IC) or electric models. Internal combustion fork trucks may be powered by standard gasoline engines, gasoline engines converted to burn liquified petroleum gas (LPG), or diesel engines. Details on the various power sources are described in the Equipment Selection and Engineering Data sections.

Fork Truck Characteristics

Each of the various categories of fork trucks has evolved to serve a specific purpose or group of objectives such as high load capacity, high lift, narrow aisle size, high speed, double-depth stacking capability, ability to stack to either side of an aisle without turning the truck, and so on. Before defining variations in performance characteristics, it is important to recognize the characteristics of, and differences between, each of the major types of forklift truck.

Counterbalance Trucks. The counterbalance truck is available in two basic designs, either the sitdown rider (Fig. 7.1.16) or standup rider (Fig. 7.1.17). The sitdown rider version is better adapted to long-distance applications in which the operator is not required to frequently mount and dismount the vehicle. Due to its relatively long wheelbase, the sitdown rider has a greater load capacity than the standup version. Because of its shorter wheelbase, the maximum capacity of the standup rider counterbalance truck is less than that of the sitdown version. The standup rider can operate in a smaller aisle due to its shorter wheelbase. Because of the standing operator position, the standup

Fig. 7.1.16 Typical counterbalanced truck.

rider is better suited to short-distance operations requiring the operator to mount and dismount the vehicle frequently.

The load handling capacity of a counterbalance truck is determined by the relationship of the load center to the center of gravity of the truck. Because the load is cantilevered in front of the front load axle, the overturning moment of the load must be counterbalanced by the truck weight. Sitdown rider trucks are available in load capacities from 2000 to 10,000 lb (900–4500 kg) and higher. The standup rider versions are available in load capacities from 2000–5000 lb (900–2300 kg).

Straddle Trucks. The straddle truck (Fig. 7.1.18) can be considered a true non-counterbalanced vehicle because the load capacity and stability of this vehicle are obtained from outrigger arms that straddle the load, hence the name straddle truck. By maintaining the load within the wheelbase of the truck, a very small and stable vehicle is possible. Because the load must be contained between the straddles, this vehicle is not suited to extremely wide loads since the width across the straddles becomes excessive and increases the aisle requirement of the machine. Straddle trucks are available in both standup rider and walkie versions. The length of the straddles can vary depending on the application and ranges up to 5–6 ft (1.5–1.8 m) for order-picking use.

Since the straddles are only as long as the load, the stability of the truck and the working aisle size are determined by the overall width over the straddles. If small, heavy loads must be lifted to high elevated heights, the width over the straddles necessary to achieve adequate stability may result in an excessively wide aisle. In this case, another type of truck may be better suited to the application.

Reach Trucks. The reach truck (Fig. 7.1.19) is a variation of the straddle truck in which the outrigger arms are approximately one-half of the load length and do not have to be spaced wide enough to straddle the load. When a typical 40–48-in. (1000–1200-mm) wide load is used, the minimum straddle width required to assure adequate load stability will usually permit the pallet load to be retracted within the outriggers.

Loads are stored and retrieved using the pantograph-mounted fork mechanism which has a fork extension several inches longer than the length of the outrigger arms. Two types of handling operations are possible with this truck, depending on the size of the load relative to the outrigger spacing. If the outrigger spacing is wider than the load, direct load withdrawal and transport is possible. When

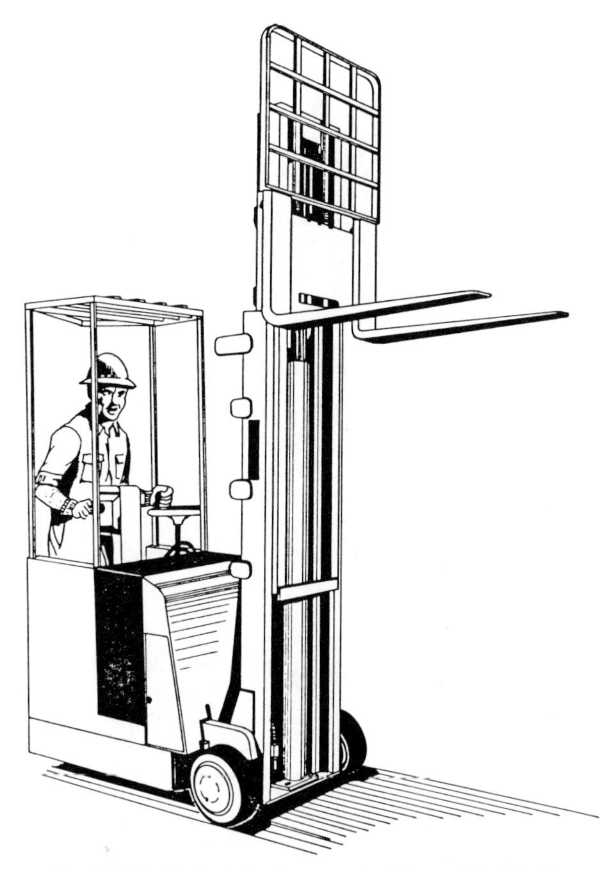

Fig. 7.1.17 Typical stand-up counterbalanced fork truck.

the load is wider than the outrigger opening, the forks are extended to the front of the truck, the load is picked up, raised, and withdrawn to a position directly over the outriggers. This type of load retraction is known as "up and over." The "up and over" method requires additional maneuvering space and/or first-level pallet rack opening height to perform the fork movement.

The reach truck is a semi-counterbalanced vehicle because the load is not fully contained within the wheelbase. Depending on the load size, the load center is positioned approximately over the centerline of the outrigger wheels. When the load is extended, vehicle stability is derived from the counterbalance capability of the truck and the width of the outriggers.

The reach truck relies on a partial counterbalance effect to retain load stability when traveling. With the load in a fully extended position, the load is entirely in front of the outrigger wheels. Because of the relatively long wheelbase and the rear location of the operator and battery, the low weight of the battery and truck still provides a significant counterbalancing force to maintain truck stability with loads up to 4000–6000 lb (1800–2700 kg).

Two types of outrigger wheels are available on reach trucks. One type consists of dual 4 in. × 5 in. (100 mm × 130 mm) casters in each outrigger arm. The second type of wheel consists of a single 5 in. × 12 in. (130 mm × 300 mm) wheel in each outrigger arm. The large wheels can negotiate uneven floors and floor cracks better than the small wheels and are suitable for use where the pallet load can be retracted within the outriggers. If loads will be handled that are larger than the outrigger spacing, thus requiring the "up and over" retraction technique, the small dual wheels should be used to minimize the lift height required to retract a load over the outriggers to the travel position.

Double Reach Trucks. The double reach truck (Fig. 7.1.20) is similar in most characteristics to the conventional reach truck. The pantograph mechanism, however, is designed to provide additional extension of approximately one-half pallet load depth beyond the front of the outriggers. The double reach truck is used to store loads two pallets deep. The front load is stored in the normal manner in which the truck is driven up to the pallet rack (Fig. 7.1.21) and the load is placed in the first storage

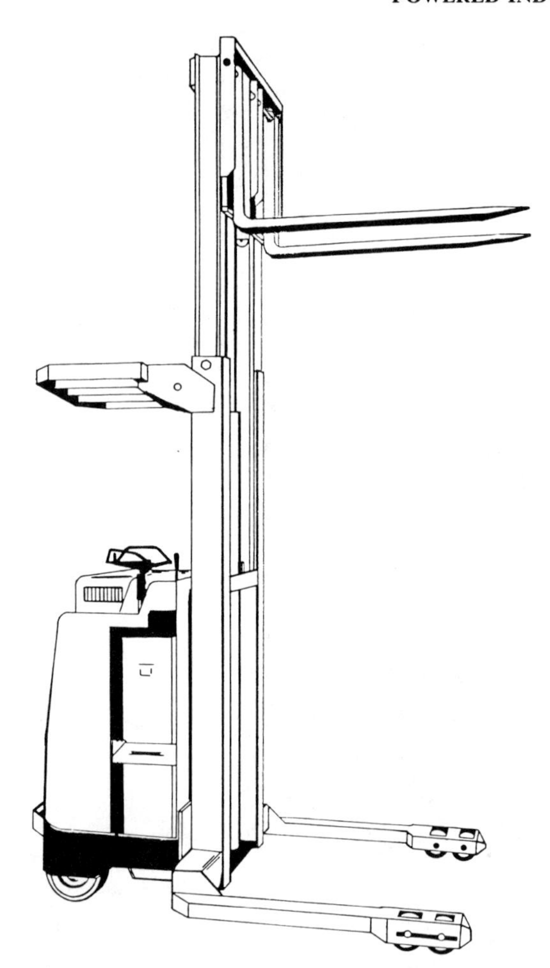

Fig. 7.1.18 Typical straddle truck.

position. If a load is to be placed in the second position, the double reach truck must be driven up to the rack with the mast assembly close to the load support beam. When the forks are extended, the additional extension capability of the double reach truck permits the pallet load to be placed in the second storage position. Provisions must be made to have the straddles either straddle the bottom pallet on the floor or fit under it by setting the pallet on a rack beam.

Two critical operating dimensions in this double-depth application are the maximum throat dimension for the pantograph mechanism or load backrest, and the reach distance of the pantograph mechanism. The pallet rack storage window must have a clearance sufficient to permit the pantograph mechanism to extend to the rear position. If the pallet loads being stored are not at least as high as the greater of the load backrest or pantograph throat dimension, storage space will be wasted because additional height will be required for equipment access but is not needed for material storage.

If loads are especially deep or there is insufficient reach capability, the second load cannot be directly placed in position. Instead, a "double bite" procedure must be used. In this procedure, the rear load is set as far back as possible, the forks are retracted slightly, and a "second bite" is used to place the load in the final position. This double movement technique not only slows load handling but can also result in inconsistent placement of the second load, causing irregular storage patterns.

Turret Trucks. The turret truck (Fig. 7.1.22) combines characteristics of the side-loading and counterbalance trucks. Turret trucks are typically designed with a long wheelbase for stability. The batteries and operator positioned at the rear of the machine provide a sufficient counterbalancing weight to ensure load stability at the high lift height these machines are designed for. In addition to the turret fork mechanism, turret trucks also differ from conventional counterbalance trucks in the

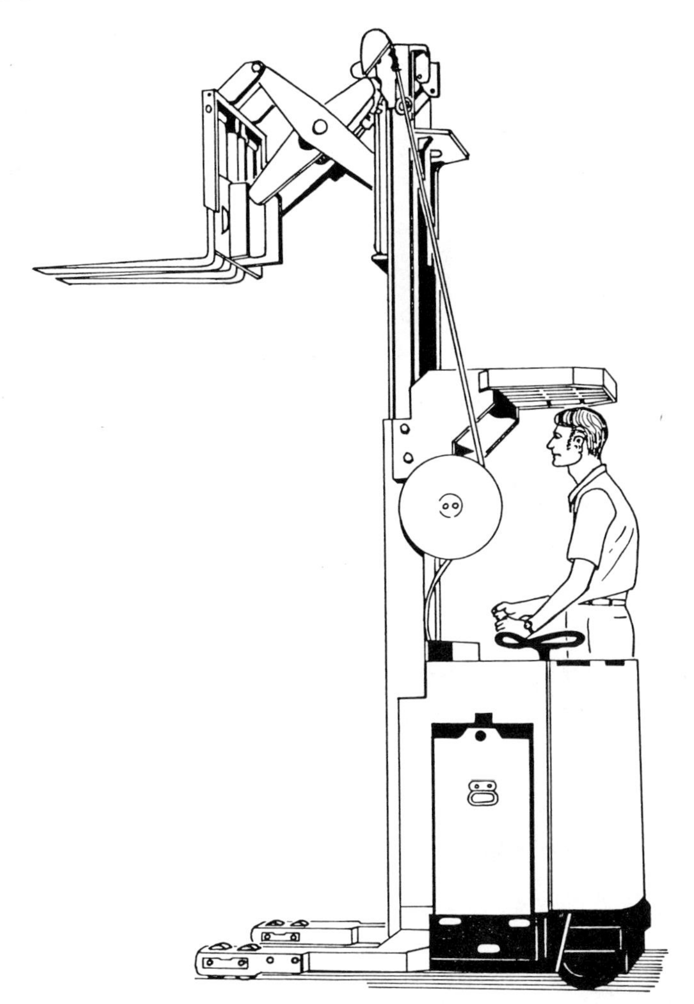

Fig. 7.1.19 Typical reach truck.

mast width, which is extended to almost the full width of the truck, providing the increased stability needed for the higher lifts.

The ability of the turret truck to operate in a very narrow aisle is due to the side-loading nature of the rotating (turret) fork mechanism, which permits the forks and load to be rotated to a position perpendicular to the direction of aisle travel and then be side loaded to a storage position.

The turret truck is capable of performing two basic load movements in addition to lift. Rotation turns the load to face either right or left of the direction of travel in the aisle. Traverse is the side-to-side movement of the load and is used for the storage and retrieval operation to pick and place loads in the pallet rack.

The turret head mechanism is available in either a J head configuration (Fig. 7.1.23) or an L head configuration (Fig. 7.1.24). The J head mechanism was originally designed to facilitate turning a load in the working aisle. Because of the overhead support mechanism, the ultimate forklift height is restricted in this configuration. The throat dimension of the head limits the load height that can be handled by this system. In addition, the more massive support structure increases the amount of suspended weight, thus decreasing the load capacity of the truck when compared to other turret fork mechanisms. The L head mechanism, which has achieved more widespread acceptance, has no throat restriction. It is therefore more suitable for variable-height loads. The lower overall height of the L head mechanism also permits the vehicle to store loads higher than is possible with the J head mechanism.

Rotation of a load within the aisle requires synchronization of the fork rotation and traverse mechanisms. Controls on most trucks make this synchronization relatively easy to accomplish. Rotation of

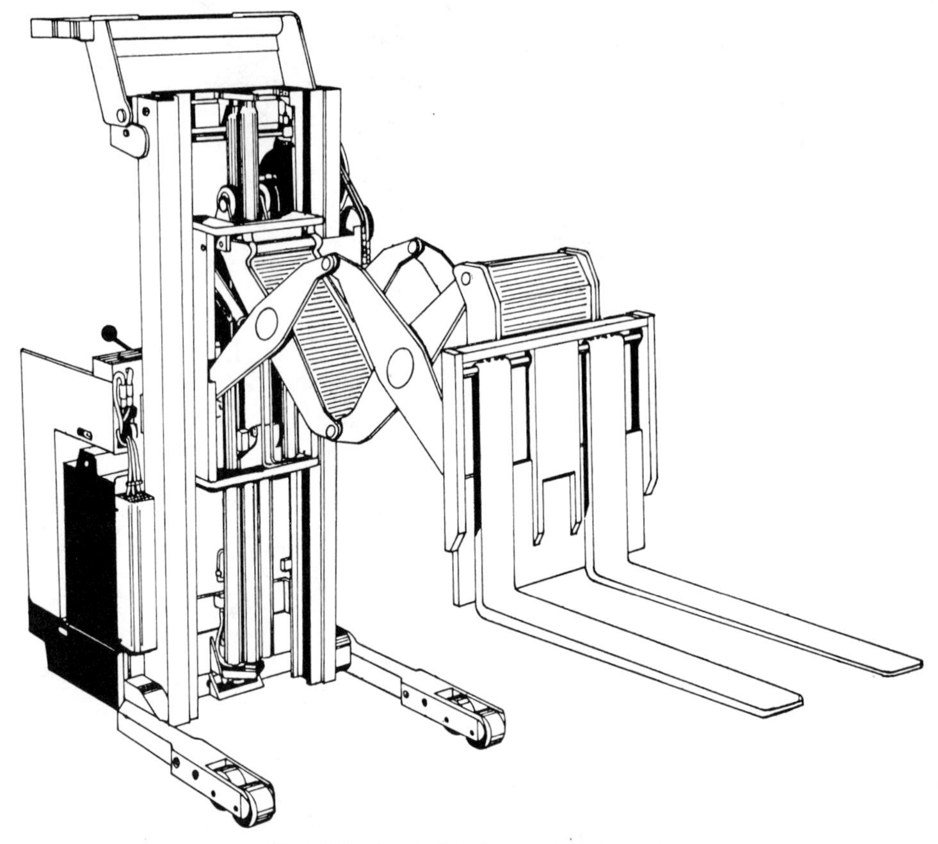

Fig. 7.1.20 Typical double reach fork truck.

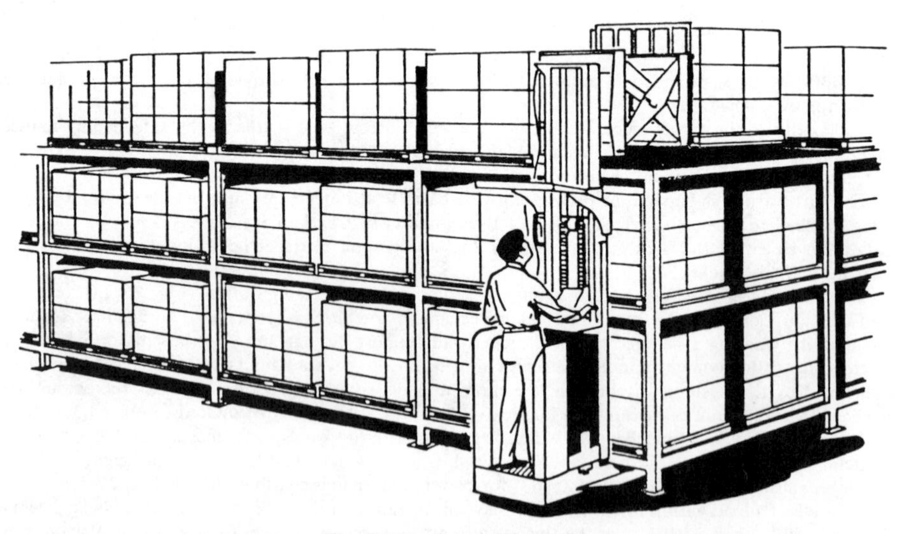

Fig. 7.1.21 Reach lift truck.

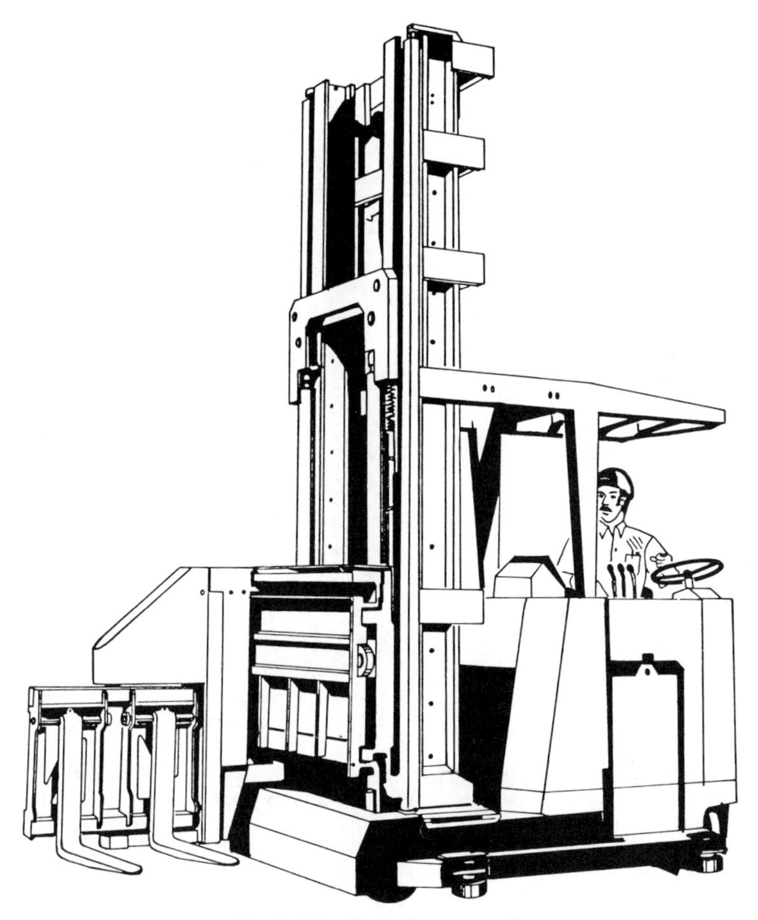

Fig. 7.1.22 Typical turret truck.

the load while the truck is in motion is not recommended by most manufacturers. Generally, a wider aisle is required if load rotation is performed within the aisle. The actual aisle size depends on the design of the fork rotation and traverse mechanisms, the diagonal dimension of the load, and the relative width and length of the load.

A variation of the conventional turret truck is a man-up version (Fig. 7.1.25), which provides an operator platform at the front of the truck. This up-front operator position permits the use of the man-up vehicle for order picking as well as conventional pallet storage and retrieval. Keeping the operator near the load permits the operator to observe the storage and retrieval operations without relying on supplemental devices such as guide marks or shelf height selectors to determine when the pallet load is positioned properly.

Because of the increased load center caused by the operator platform, as well as the need for greater stability to ensure operator confidence, the man-up machines employ additional mast bracing designs to increase stability and decrease uncontrolled deflections. These designs include mast supports, balanced reaction chains, and changes in mast geometry and construction.

Swing-Mast Trucks. The swing-mast truck (Fig. 7.1.26) is another side-loading variation of the counterbalance truck. The swing-mast truck is equipped with a rotating mast assembly instead of the rotating fork assembly used on turret trucks. This design limits these vehicles to only right-hand load rotation. Because the load can only be rotated to one side of the aisle, the swing mast truck must leave the aisle and turn around in order to store or retrieve loads from the opposite side of the aisle. These vehicles closely resemble the conventional counterbalance truck and can operate within narrow aisles of approximately 6–8 ft (1.8–2.4 m). The additional weight added by the rotating mast mechanism requires a substantially higher truck capacity to achieve the same load carrying capacity when compared to a counterbalance vehicle.

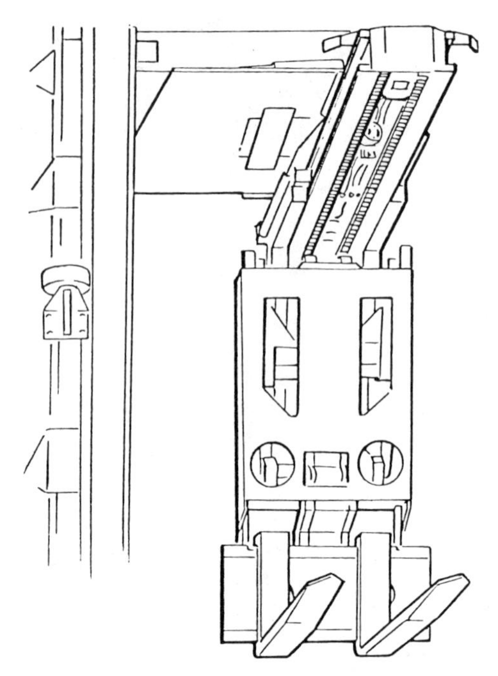

Fig. 7.1.23 J-type turret head.

Hybrid Vehicles. Hybrid vehicles (Fig. 7.1.27) represent a combination of fork truck and automated storage and retrieval (AS/R) machine technology. These units operate on AC power in the storage aisle and on self-contained batteries when moving outside the aisle. The operator travels up with the load, providing direct visibility and control of storage and retrieval operations. The man-up design also permits order picking. The guided-aisle arrangement and AC power operation permit these hybrid vehicles to operate at speeds in excess of 400 ft/min (2.1 m/sec) when traveling in the storage aisle. Storage heights of up to 60 ft (18.3 m) are available, which contrasts with a maximum height of approximately 40 ft (12.2 m) for high-lift turret trucks. Hybrid vehicles have the additional advantage of permitting simultaneous aisle travel and lifting or lowering, which reduces cycle time and increases

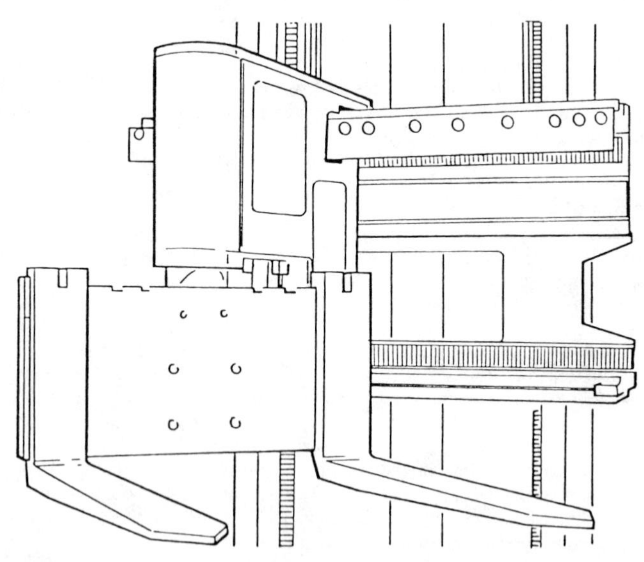

Fig. 7.1.24 L-type turret head.

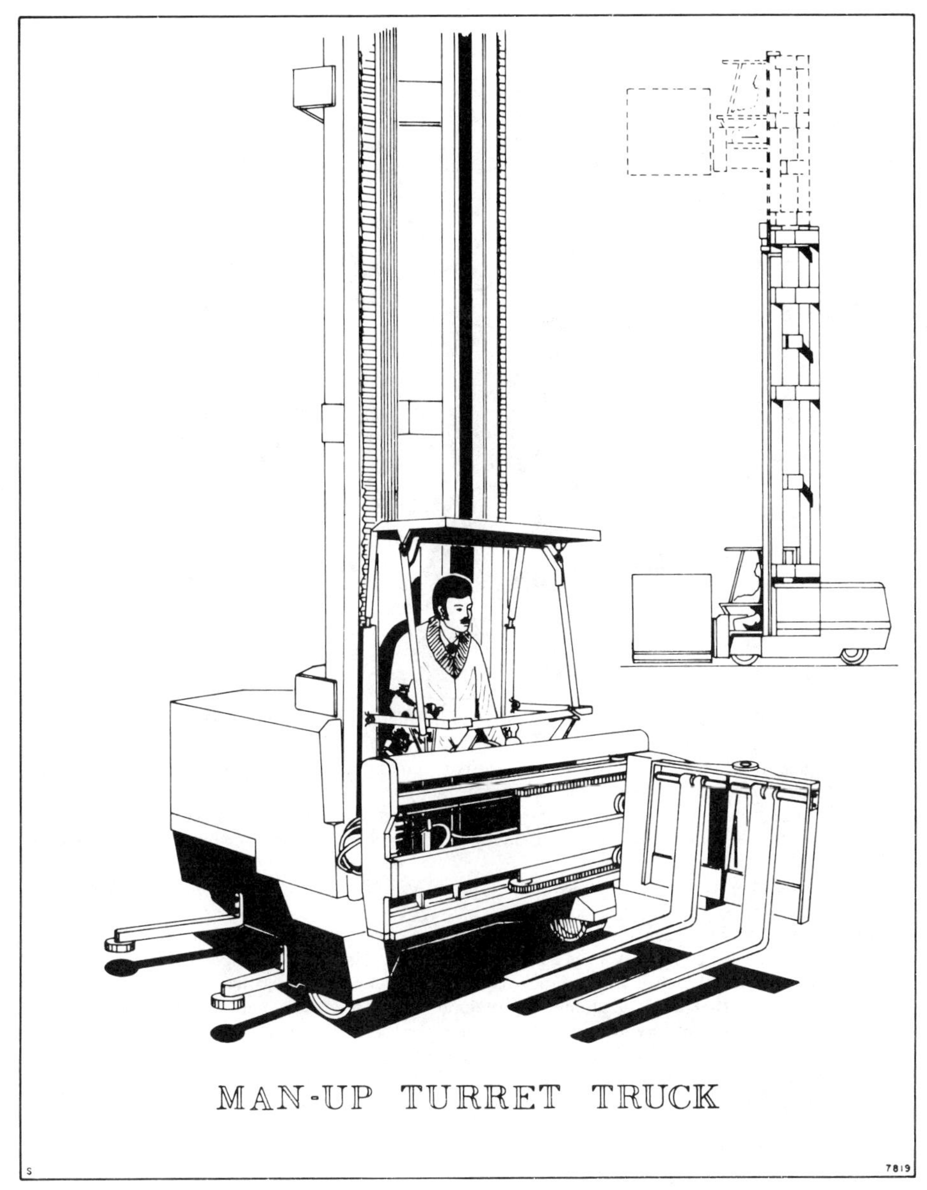

Fig. 7.1.25 Man-up turret truck.

productivity. This type of simultaneous movement is generally not available in turret trucks unless they are operated at fork elevated heights of 6 ft (1.8 m) or less.

Hybrid vehicles are equipped with shuttle table mechanisms which lift pallet loads from below, as opposed to fork mechanisms which lift loads from within fork pockets. The shuttle mechanism simplifies the mechanics of the storage and retrieval system and permits direct movement of the pallet load across the aisle without requiring the rotational movement or additional aisle width required by turret forks. Hybrid vehicles provide increased versatility when compared to dedicated aisle AS/R machines.

Performance Characteristics

Regardless of the specific fork truck design, be it counterbalance, straddle, reach, turret, or hybrid, there are five basic performance characteristics which define the capability of a truck to perform a

Fig. 7.1.26 Swing-mast lift truck (rack storage, nonguided aisles).

given job. These characteristics include load capacity, lift height, travel and lift speed, maneuverability, and ramp climbing capability.

The design of any fork truck represents a compromise among these five factors. Performance emphasis is placed on the factor or factors considered most critical by the manufacturer to a specific application. The purchaser of a fork truck must evaluate the relative importance of each of the five major performance characteristics to determine whether an individual vehicle will satisfy the intended operational requirements.

Performance characteristics for the various types of forklift trucks are summarized in Table 7.1.3.

Load Capacity. Load capacity is generally the first performance characteristic considered since the fork truck must be capable of lifting the intended load. The load capacity of a truck is specified at a given load center, which is the distance from the front face of the forks to the center of gravity of the load. Factors that affect load capacity include variations in the load center and the intended lift height. The presence of any attachments can increase the load center and/or add to the apparent weight of the load itself, resulting in a reduction of truck capacity.

There are five definitions of load capacity (Fig. 7.1.28). Each definition is determined by a specific set of operating conditions. The five definitions of capacity are:

Load capacity. Load capacity is specified at a given load center and is the maximum weight that a truck, with forks or specified attachments, can lift to a stated maximum elevation. At capacity, the truck must be able to carry and stack loads while maintaining structural strength and stability. Load capacity is shown on the truck nameplate.

Alternate capacity. Alternate capacity is established at less than the maximum elevation and for the same load center as the load capacity rating. Generally, the alternate capacity is greater than the load capacity due to the lower elevation which increases the stability of the truck.

Rated capacity. Rated capacity is the maximum weight, at a given load center, that a truck can carry and stack at an elevation specified by the manufacturer. This capacity is based on structural strength, stability tests, or calculations. The load used to determine rated capacity is considered to be a homogeneous cube with overall dimensions of twice the 24-in. (610-mm) load center dimension.

Alternate rated capacity. The alternate rated capacity is the maximum weight that a truck can carry and stack to an elevation other than that originally specified by the manufacturer. The load center and load size used to determine alternate rated capacity can vary from the 24 in. (600 mm) dimension used to establish the rated capacity.

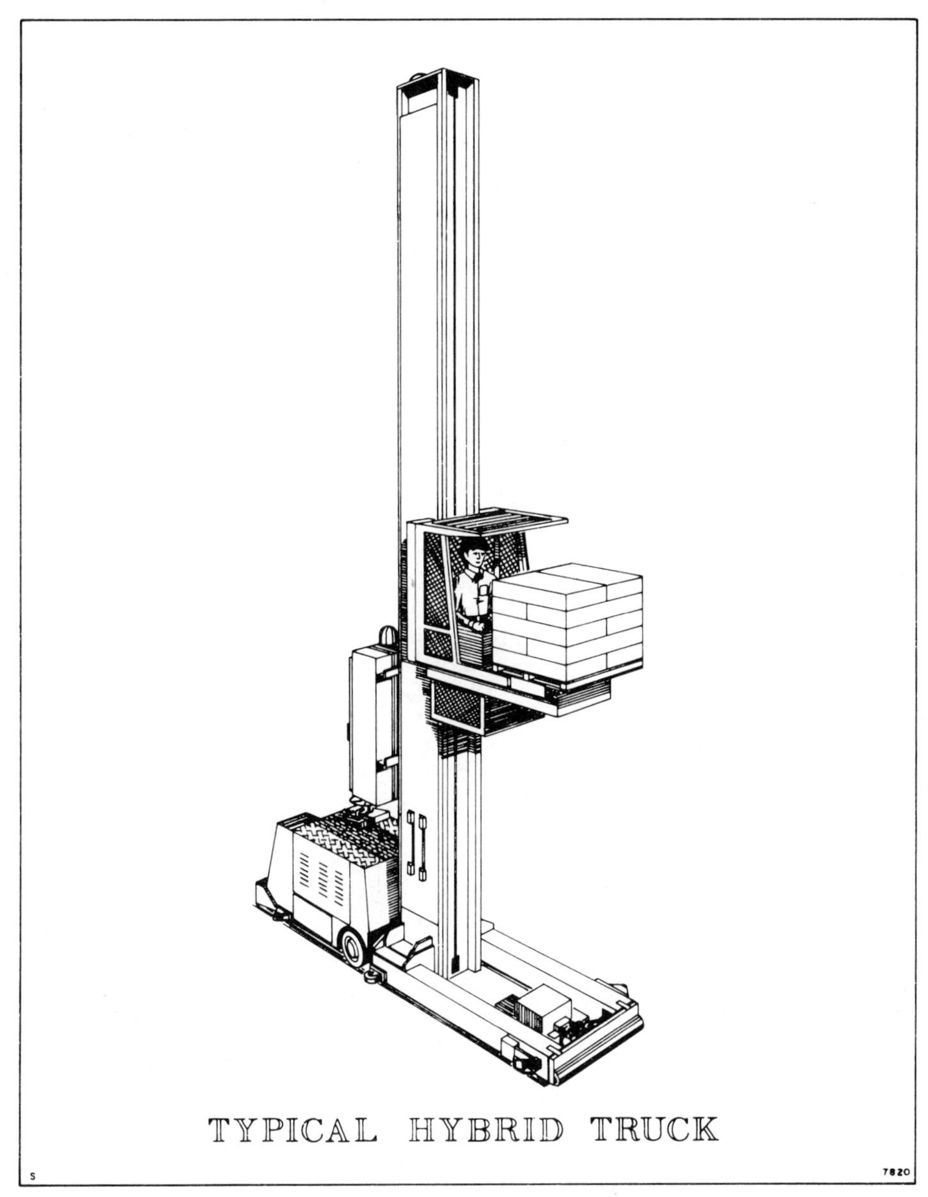

Fig. 7.1.27 Typical hybrid truck.

Rated capacity with attachments. The rated capacity with attachments is established at a load center and elevation specified by the manufacturer.

Load capacity ratings are subject to derating by the manufacturer due to increases in the maximum elevation or changes in the load center. Derating is a reduction of rated capacity to a lesser capacity appropriate for increased lift height or changes in load center. Derating results primarily from the decrease in truck stability exhibited at extended lift heights. Stability can be increased by using dual-load wheels, increasing the mast width, and increasing the footprint of the machine by making the truck wider.

Because of the numerous ways in which truck capacity can be specified, it is important for a purchaser to verify the actual capacity with the manufacturer by describing the load size, weight,

Table 7.1.3 Typical Powered Industrial Truck Performance Characteristics

Vehicle	Lift Capacity Range	Storage Aisle Width[a]	Lift Height	Lift Speed	Travel Speed	Grade Clearance
Sit-down counterbalanced	2,000–10,000 lb (900–4,500 kg)	12–15 ft (3.6–4.6 m)	264 in. (6,700 mm)	80 ft/min (0.41 m/sec)	6.0 mi/hr (9.7 km/hr)	35%
Stand-up counterbalanced	2,000–6,000 lb (900–2,700 kg)	10–12 ft (3.0–3.6 m)	240 in. (6,100 mm)	65 ft/min (0.33 m/sec)	5.0 mi/hr (8.0 km/hr)	35%
Straddle	2,000–6,000 lb (900–2,700 kg)	$6\frac{1}{2}$–9 ft (2.0–2.7 m)	252 in. (6,400 mm)	60 ft/min (0.30 m/sec)	5.3 mi/hr (8.5 km/hr)	15%
Reach	2,000–5,000 lb (900–2,300 kg)	6–8 ft (1.8–2.4 m)	360 in. (9,150 mm)	50 ft/min (0.25 m/sec)	5.5 mi/hr (8.9 km/hr)	15%
Turret	3,000–4,000 lb (1,360–1,800 kg)	5–7 ft (1.5–2.1 m)	480 in.) (12,200 mm)	75 ft/min (0.38 m/sec)	5.5 mi/hr (8.9 km/hr)	[b]
Man-up turret	3,000–4,000 lb (1,360–1,800 kg)	5–7 ft (1.5–2.1 m)	480 in. (12,200 mm)	50 ft/min (0.25 m/sec)	5.5 mi/hr (8.9 km/hr)	[b]
Sideloader	2,000–10,000 lb (900–4,500 kg)	5–7 ft (1.5–2.1 m)	360 in. (9,150 mm)	50 ft/min (0.25 m/sec)	5.0 mi/hr (8.0 km/hr)	20%
Swing mast	2,000–8,000 lb (900–3,600 kg)	5–6 ft (1.5–1.8 m)	360 in. (9,250 mm)	50 ft/min (0.25 m/sec)	6.4 mi/hr (10.3 km/hr)	15%
4-directional	2,000–3,000 lb (900–1,360 kg)	6 ft (1.8 m)	204 in. (5,180 mm)	40 ft/min (0.20 m/sec)	4.5 mi/hr (7.2 km/hr)	12%
Hybrid	2,000–4,000 lb (900–1,800 kg)	5 ft (1.5 m)	600 in. (15,240 mm)	60 ft/min (0.30 m/sec)	5.5 mi/hr (8.9 km/hr)	[b]
Order picker	2,000–4,000 lb (900–1,800 kg)	4 ft (1.2 m)	264 in. (6,700 mm)	30 ft/min (0.15 m/sec)	2.9 mi/hr (4.7 km/hr	15%

[a] Aisle width determined by specific load size and handling method.
[b] Not designed for use in applications on other than flat floors.

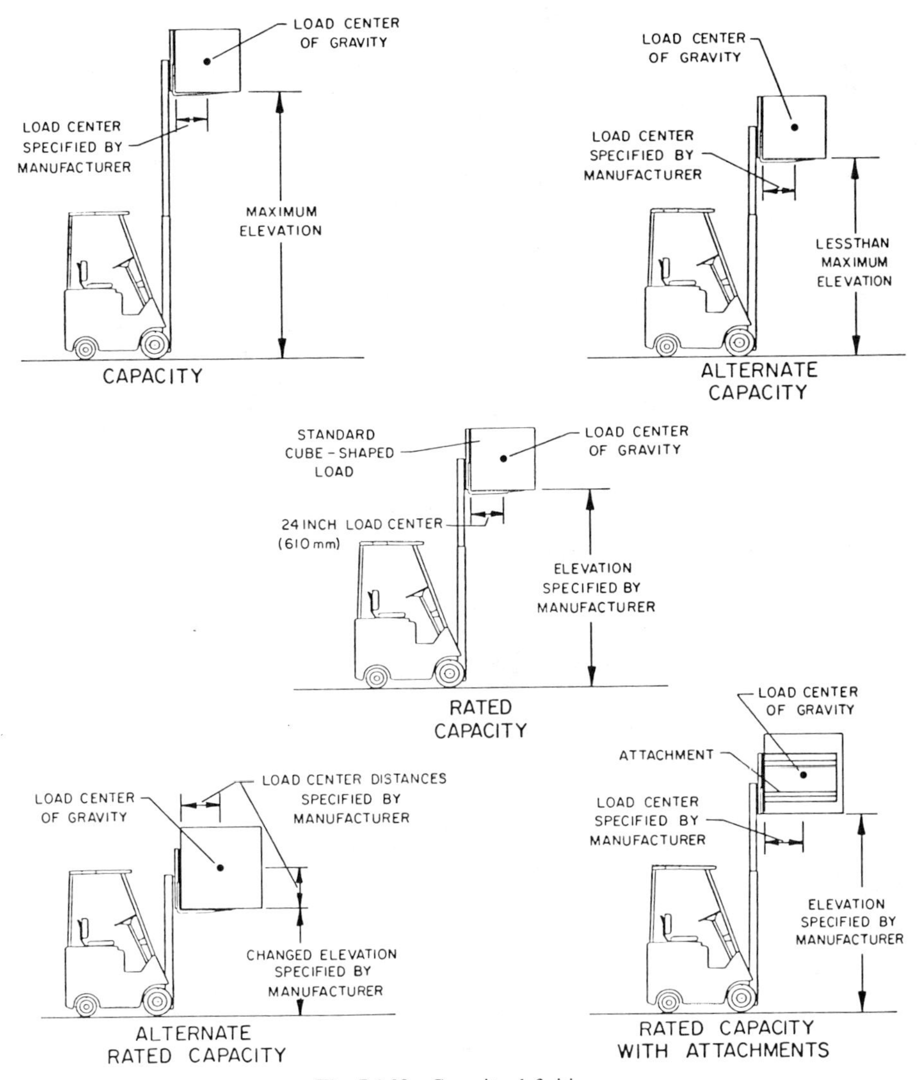

Fig. 7.1.28 Capacity definitions.

and intended lift height. Often, capacity ratings are quoted without including the derating information necessary to determine the actual load handling capacity at extended elevations.

Lift. The second most important performance factor, after load capacity, is lift. Lift specifications include a variety of dimensions that determine the overhead clearance dimensions of the truck, the ability of the truck to stack loads in low-overhead situations, the maximum lift height available, and the overall stability of the vehicle. Lift performance can be measured by the following characteristics, illustrated in Fig. 7.1.29.

Elevated height. Elevated height is the maximum lift height of the vehicle. Depending on the stacking height desired, the elevated height should be approximately 6–12 in. (150–300 mm) more than the pallet level storage height of the highest load to be stored. This additional lift capability ensures that loads stacked in the highest storage positions can be lifted sufficiently to clear the storage rack for pickup or putaway. The load capacity available at the desired lift height should equal or exceed the greatest weight intended for storage.

Extended height. Extended height, sometimes referred to as overall elevated height (OAEH), is the maximum elevation attained by the top of the mast, fork carriage, load backrest, or operator cage,

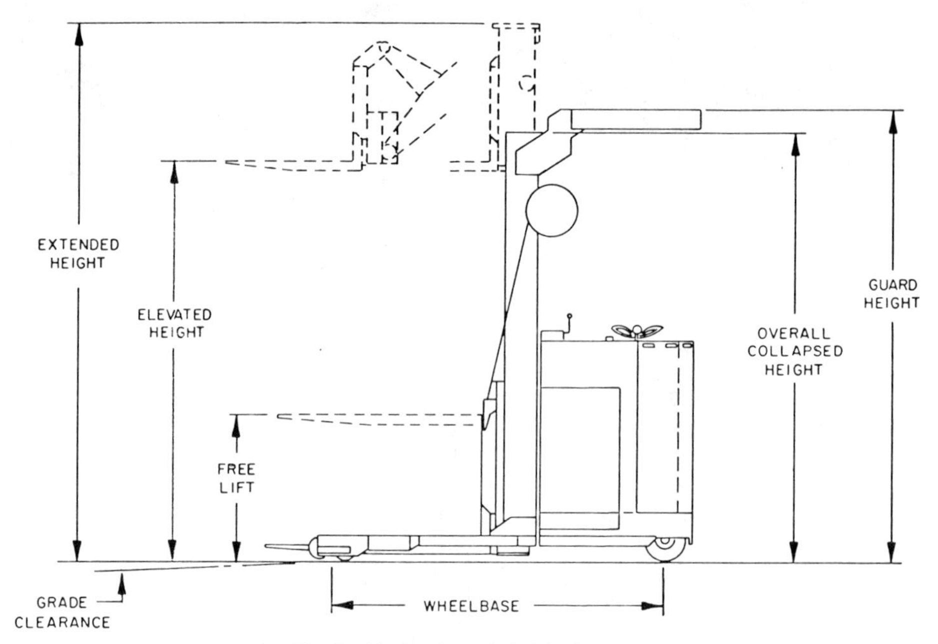

Fig. 7.1.29 Fork truck height factors.

when the mast is extended to the maximum elevated height. The maximum extended height will determine the minimum building clearance required to accommodate the truck when operating at the maximum elevated height. At least 12 in. (300 mm) of clearance should be allowed between the extended height and the lowest overhead building obstructions. This allowance may vary depending on sprinkler and lighting requirements and load height. High loads may require greater clearances.

The difference between the elevated height and extended height provides a measure of the height of the load backrest, mast, or other equipment items protruding above the load-carrying surface. If this difference exceeds the height of the highest load, the building clearance dimensions will be determined by the equipment. If the highest load exceeds this difference, the load will contact building obstructions before the truck does. In man-up turret trucks and order-picking vehicles, the maximum elevated height may be significantly less than the maximum extended height, due to the height of the operator cab. Therefore, it is possible, if a truck is specified for a given elevated height without considering the extended height, that the operator cab could contact overhead obstructions before the forks reached the desired elevation.

Overall collapsed height. Overall collapsed height (OACH) is the height from the floor to the top of the mast, load backrest, or operator cab, when the forks are in their lowest position and the mast is fully collapsed. Overall collapsed height is generally the determining factor in the overall lowered clearance requirements of a fork truck used for truck dock operations. Overall collapsed heights of 83–84 in. (2110–2135 mm) are common to permit loading and unloading of semitrailers.

Overall collapsed height, however, is not the only determining factor in the overall lowered height of a fork truck. In many standup rider machines, the overhead guard may be higher than the collapsed mast height. Therefore, the overhead guard determines the minimum clearance requirements for the truck. On trucks in which the mast determines the overall collapsed height, more compact dimensions can be achieved by using multiple-stage masts to permit a high elevated height while retaining a low overall collapsed height.

Freelift. Freelift is defined as the amount of fork elevation obtainable before the mast moves from its fully collapsed position. The amount of freelift available is determined by the mast design. Trucks are available with a variety of two-, three-, and four-stage masts with either low freelift or high freelift. The term *low freelift* is generally applied to freelifts of 24 in. (600 mm) or less. High freelifts are often available to slightly less than the height of the lowered mast, or approximately 60 in. (1520 mm). High freelift mast designs permit loads to be stacked in low-headroom conditions. Such masts are often used to stack loads inside boxcars and highway trucks.

Travel and Lift Speed. Because the travel and lift functions are basic to every handling cycle, the travel and lift speed performance of a fork truck has a direct effect on the productivity of the fork truck operator. A wide variety of travel and lift speed combinations are available from the manufacturers but, in any case, the combination of speed plus maneuvering ability must not add up to more than an operator can safely control. The basic power source of the vehicle generally determines the ultimate speed capabilities for travel and lift. Improvements in batteries, motors, and controls have enabled electric cushion-tire trucks to achieve travel and lift speeds comparable to internal combustion powered fork trucks. Common travel speeds are 6–8 mph (10–13 km/hr) maximum.

An important specification in any lift truck purchase is that of lift speed. Lift speed is determined by the design of the truck hydraulic system and is affected most by the pump motor horsepower and pump volume rating. One area for economizing in truck design is undersizing the lift pump capacity to reduce costs. With an undersized pump, the truck can still achieve the desired load capacity but loaded lifting speed will be decreased to the point where operator efficiency will suffer because the truck takes too long to elevate a load. Typical loaded lift speeds for modern electric trucks range from 60 to 90 ft/min (0.3–0.5 m/sec) when operating at the design load.

Maneuverability. The maneuverability of a fork truck is defined in terms of its capability to operate within a given aisle width. The required aisle is determined by a combination of load length, load spacing, truck size, and outside and inside turning radii. The basic maneuverability index for most forklift trucks is the minimum right-angle stacking aisle without clearance. Fork truck dimensions that affect maneuverability include the width, length, and wheelbase. Details on aisle width and means of calculating the width are given in the Equipment Selection and Engineering Data sections.

Controls

Electric-powered fork trucks have a variety of controls available, both standard and optional, to enhance the performance, maneuverability, safety, and service life of the truck. These controls can be divided into two major categories consisting of drive controls and guidance controls. Drive controls are associated with control of vehicle functions such as forward and reverse travel, braking, and raising and lowering of the forks. Guidance controls are associated with guidance of the vehicle, usually when operating within the storage aisle.

Drive Controls. Two types of drive controls are commonly available on modern fork trucks. They consist of the mechanical stepped-resistance control and the electronic, silicon-controlled rectifier (SCR) control. The stepped-resistance control consists of a bank of resistors which are mechanically switched in response to depression of an accelerator pedal. The resistors in this type of system consume a great deal of power, especially at the lower speed settings. At full speed, the battery is connected directly to the motor and the resistors consume no power. Power is delivered to the drive motor in discrete steps which are determined by the relative values of the resistors and the number of contacts provided by the control. This type of control is relatively inexpensive and simple to maintain but does require periodic maintenance of the contact assemblies. Only the most inexpensive vehicles, or those having only occasional use, use mechanical controls.

Solid-state electronic drive and lift controls, based on SCR technology, have largely replaced mechanical controls on most fork trucks. SCR controls are available in a variety of designs which provide smooth, step-free power delivery to the drive motor. These controls are efficient and consume very little power themselves, thus providing longer working shifts and reducing power requirements. SCR controls operate on a switching principle in which full power is delivered to the drive motor in short bursts. The spacing of these bursts of power is controlled by the accelerator pedal, thereby directly varying the effective power delivered to the drive motor. Because full battery power is delivered to the drive motor with each burst of energy, these controls combine a high-torque low-speed capability with smooth speed control.

Modern electronic controls are protected against reverse polarity, surge currents, and transients generated by reversing direction and plugging. The newest form of control is an enhancement of the SCR technology and employs regenerative energy-recovery techniques. Regenerative systems recover some of the energy stored in the motion of the vehicle and direct this energy to the battery, thereby replacing some of the energy used during acceleration or lifting operations. The use of regenerative systems increases the amount of working time available during a shift by restoring to the battery some of the energy consumed. Since the hydraulic system and load-lifting function consume the greatest power during most material handling cycles, the most significant increase in service time is obtained when these systems are applied to the lift pump motor.

Guidance Controls. As aisle sizes have been continually reduced in an effort to obtain maximum storage efficiency, working aisle clearances have been reduced to the point where an operator cannot safely steer the vehicle in the narrow storage aisles. To solve this problem, automatic guidance systems, both mechanical and electronic, have been developed to free the operator from steering while in the storage aisle.

The mechanical systems consist of guide rails bolted to the floor on both sides of the storage aisle (Fig. 7.1.30). Guide rollers attached to the sides of the fork truck contact these guide rails and passively steer the truck within the aisle. This rail guide method is relatively economical if the total length of guided aisle is low and there are few vehicles. The system is easy to install and requires almost no maintenance. When the length of guided aisle and/or number of vehicles becomes large, it usually becomes more economical to use an electronic guidance system.

Electronic guidance controls provide active electronic control of the truck steering when operating within the storage aisle. These systems consist of a small guide wire buried in a shallow slot in the floor. The slot is filled with an epoxy filler material to provide a smooth floor surface. The wire, activated by a centrally located driver unit, transmits a low frequency signal which is picked up by an antenna and receiver located on the fork truck. The signal is decoded and is used to steer the truck. Some systems utilize two vehicle antennas with one installed at each end of the truck, while

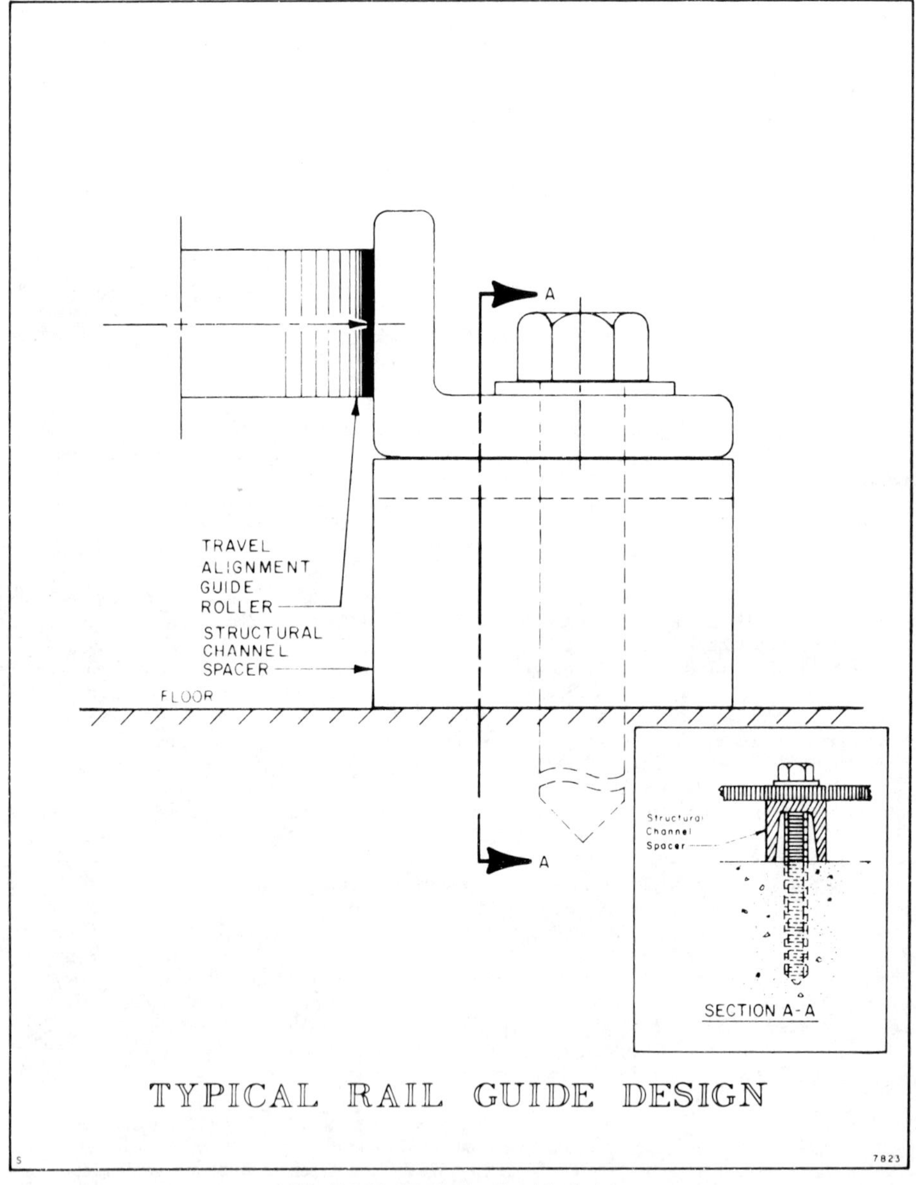

Fig. 7.1.30 Typical rail guide design.

other systems use only one antenna. Systems employing two antennas can provide steering guidance when the truck is traveling either forward or backward. The single-antenna systems provide guidance in one direction only. The receiver antennas are located on the bottom of the truck just above the floor. The in-floor wire is driven by a loop driver unit which plugs into a standard 120-v outlet and provides the proper drive signal. Modifications can be made to the wire guide system by installing new wire at the desired locations, disconnecting the old wire where it is no longer needed, and splicing the new wire into the loop. Wire guide systems can also be used for travel guidance in areas other than the storage aisle.

The electronic guidance systems provide hands-free steering control and permit the operator to perform storage, retrieval, or order-picking functions without having to steer the vehicle. These guidance systems are equipped with a variety of safety interlocks which disable vehicle motion in the event of malfunction or loss of guidance signal.

Electronic guidance systems, when compared to the rail guide method, are economically justifiable if a large number of vehicles are involved or if there is a large quantity of guided aisle. The economic evaluation must be made on a case-by-case basis because the factors involved include the number of vehicles (and associated receivers), the total length of aisle to be guided, and the number of system driver units or boosters required for the system.

Also affecting the comparison of rail and wire guidance systems is the fact that wire guide systems can have an optional capability added that permits either one-way or two-way transmission of data on the guide wire. This technique enables a unit located on the fork truck to receive instructions from a central computer. Two-way system capability permits the fork truck operator to report to a central computer and note any abnormal situations such as incorrect stock or insufficient stock. These systems can form the base for a completely mechanized and computer-controlled order-picking and inventory reporting system.

Equipment Selection and Engineering Data

Lifting Capacity. The lift capacity (load lift) is a weight–distance relationship between the load being lifted and the rear counterbalance weight of the truck on opposite sides of a fulcrum. The term is the moment and is expressed in inch-pounds. The fulcrum is centerline of front axle about which the load moment resists the forward tipping of the truck.

A manufacturer's rating of lift truck lifting capacity is based on the load weight and its center of gravity relative to the lifting mechanism of the lift truck and not necessarily based on the fulcrum. The industrial truck industry has standardized a load capacity rating as previously stated on the load center (LC). This is the distance from the face of the forks to the load's center of gravity. The accepted LC standard is 24 in. (600 mm), which is half of the load length for a 48-in. (1200-mm) pallet. The load configuration has a major impact on truck manufacturers' design specifications and should be investigated and compared before selecting a specific truck for an assigned service.

The issues of investigation are load weight, load size, truck footprint, lift height, mast fore and aft tilt, and front-end attachments. Manufacturers' specifications state the lift capacity at a standard LC for the basic truck without extra attachments and extended high lifts. As an example, a truck with a capacity of 3000 lb (1364 kg) and a 24-in. (600-mm) LC is rated to safely lift an evenly distributed weight of 3000 lb (1364 kg) at a 48-in. (1200-mm) load length. If the load configuration is increased in weight and size, the capacity should be recalculated to assure that the rated stability has not been exceeded, creating an operational hazard. Unless otherwise specified, the lifting capacity is based for a single-mast lift of forks elevated to a height (FEH) of 12–15 ft (3.6–4.6 m). Higher lifts downrate the lifting capacity. This aspect is covered in a subsequent section.

The calculation of the lift capacity and the effects of load changes on a rated truck is based on the moment of the load as related to the truck fulcrum and is illustrated in Fig. 7.1.31 and subsequent equations.

$$M = PD \tag{7.1.1}$$

where: M = load moment in inch-pounds (N-m)
P = weight of load in pounds (kg)
D = distance from the fulcrum (centerline of the front axle) to the center of gravity of the load on the fork

Assume a 300-lb (1364-kg) lift truck with a 24-in. (600-mm) load center. The distance from the face of the load against the forklift truck carriage is 24 in. (600 mm) and from there to the center of the front axles (fulcrum) is an additional 12 in. (300 mm).

$$D = 24 \text{ in. } (600 \text{ mm}) + 12 \text{ in. } (300 \text{ mm}) = 36 \text{ in. } (900 \text{ mm})$$
$$M = PD = 36 \text{ in. } (900 \text{ mm}) \times 3000 \text{ lb } (1364 \text{ kg})$$
$$= 108{,}000 \text{ in.-lb } (12{,}200 \text{ N-m})$$

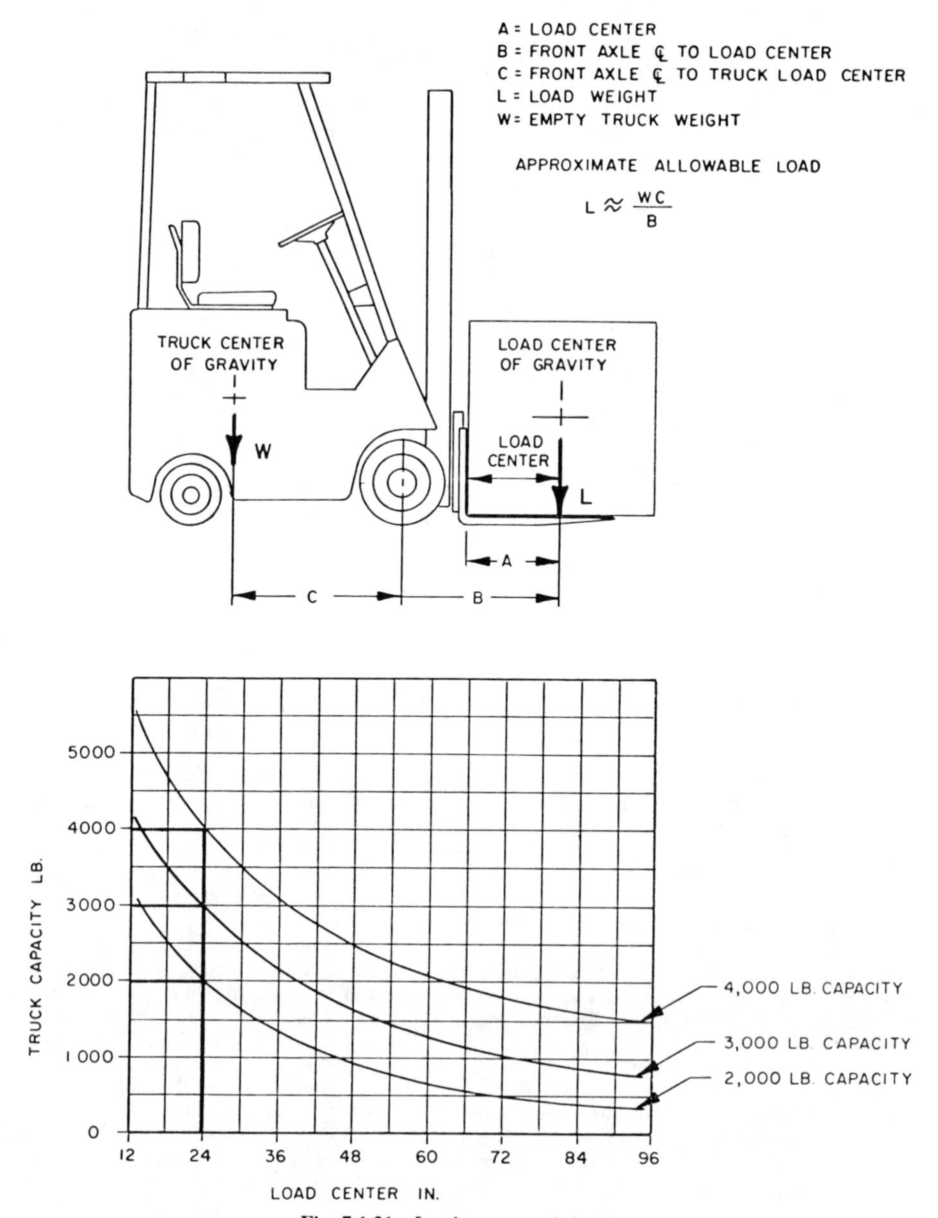

Fig. 7.1.31 Load center and derating.

If the pallet length is increased from 48 in. (1200 mm) to 60 in. (1524 mm), the question is: What is the safe lifting capacity for the extended load with an LC of 30 in. (760 mm)?

$$D' = 30 \text{ in. } (760 \text{ mm}) + 12 \text{ in. } (300 \text{ mm}) = 42 \text{ in. } (1067 \text{ mm})$$

The allowable load is $M = PD = 108{,}000$ in.-lb (12,200 N-m). The adjusted load moment equation is $M' = P'D'$ in which M' must equal M with no basic change to the lift truck. Therefore, if $P'D' = 108{,}000$ in.-lb (12,200 N-m) and $D' = 42$ in. (1070 mm), then $P' = 2571$ lb (1170 kg). The rated 3000-lb (1364-kg) truck capacity lift is now downrated to 2571 lb (1170 kg).

The moment calculation is an acceptable method of approximating lift capacity with load changes but does not compensate for inclusion of such other factors as load tilting and extended lift heights as related to the truck stability. The effects of these issues on stability must be verified with the manufacturer. Many truck manufacturers include other stability factors in determining the safety rating of trucks which include the longitudinal and lateral running and stacking factors. These are defined as:

1. Longitudinal running—to safely stop while traveling empty and with a load 12 in. (300 mm) above the floor.
2. Longitudinal stacking—to safely handle elevated loads while making sudden stops at low speeds.
3. Lateral running—to safely negotiate quick directional changes while traveling empty and with a load 12 in. (300 mm) above the floor.
4. Lateral stacking—to safely stack high loads and the effects on stability with slightly sloped floors.

All possible safety precautions cannot be built into industrial lift trucks to prevent mishaps. The operator is still responsible for the safe operation of the trucks. Therefore it is extremely important that proper operating instructions be given initially and on a continuing basis to all operators.

In addition to load weights and center of gravity, other safety considerations are wheelbase, tire tread, size and type of mast and frame, and fork deflection. All of these factors must be considered in the truck design, and if the margin of variation in maneuverability, stability, and capacity is exceeded, the vehicle will be unsafe. The graph in Fig. 7.1.31 illustrates the downrating effects on 2000-lb (910-kg), 3000-lb (1364-kg), and 4000-lb (1820-kg) lift trucks as the load center is increased.

Lift Derating for Attachments. In addition to standard tapered forks and a low carriage backrest, any additional device attached to or forward of the mast becomes an accessory or attachment for special or improved load handling, placing, or positioning. Attachments directly affect the load center and reduce the rated capacity. In any case, the effect of attachments can be calculated and must be part of the downrating specifications of the equipment or a larger rated capacity truck specified.

As a general rule in calculating the effects of attachment, the following equation can be used as a planning guide in calculating the derated lift capacity of a specific truck:

$$P' = \frac{M - D_a P_a}{D'} \tag{7.1.2}$$

where: P' = load weight that can be safely handled with the attachment on the lift truck in pounds (kg)
D_a = the distance from the centerline on the front axle to the attachment center of gravity in inches (mm)
P_a = the weight of the attachment in pounds (kg)
D' = distance from the centerline of the front axle to the center gravity of the load held by the attachment in inches (mm)

Assume the same 3000-lb (1364-kg) lift truck examined in (7.1.1) with a side-shift attachment of an additional 4 in. (100 mm) and 500 lb (227 kg) to the truck carriage.

From (7.1.1), M = 108,000 in.-lb (12,200 N-m).

$$D_a P_a = 2000 \text{ in.-lb } (226 \text{ N-m})$$
$$D' = 40 \text{ in. } (1016 \text{ mm})$$
$$P' = \frac{108{,}000 - 2{,}000}{40} = \frac{106{,}000}{40} = 2650 \text{ lb } (1205 \text{ kg})$$

The attachment reduced the truck capacity from 3000 lb (1364 kg) to 2650 lb (1205 kg).

Some of the attachments that are generally considered standard equipment on trucks are: side shifter, fork gripper, ram, revolving carriage, put-and-take device, push–pull device and carton, drum, C-clamp, paper roll, and bale clamp and bale device.

Even though they may be considered standard equipment, each attachment will have a downrating effect on any type truck and its effects should be checked and verified before specifying a specific truck.

Derating for Lift Heights. Extended lift heights are accomplished through the mast and hydraulic lifting mechanism previously identified as dual, triple, or quad mast. A single mast is standard equipment with all lift trucks. Multilift masts have a downrating effect on the lift capacity of a truck and will derate the originally designed specifications. The extended lift downrating factors have a nonlinear effect as heights are extended, which affects not only the lifting moment, but also the stability of the truck. Therefore, extended lifts must be verified by the truck manufacturer. As an initial planning guide in determining the effects of extended lift downrating, an acceptable procedure is a decrease of 100 lb for each additional foot of lift above 15 ft to 20 ft and 200 lb per foot above 20 ft (45 kg per 0.3 m of lift above 4.6 m to 6.1 m and 90 kg per 0.3 m above 6.1 m) in addition to the weight of the larger mast. Furthermore, the effects of the longitudinal and lateral stability of the truck must be reviewed with manufacturer's representative or specifications.

Most truck manufacturers use the same truck frame within a limited series or a family of lift capacities. Such a series of models may be within a basic group or configuration such as a 2500–3500-lb (1136–1590-kg) capacity lift truck may have the same frame and footprint. However, the counterbalance weight will be increased proportionately according to the rated lifting capacity. If the specified truck is at the low lift range of a family group of say 2500 lb (1136 kg) lift, then the lift height may be extended with a multilift mast without materially affecting the stability of the truck as long as it is operated within the design safety parameters of the changes. However, if the upper range in the family group is used, then the next larger family group or frame size should be used.

Right-Angle Stacking. The maneuvering space required to effectively operate lift trucks is critical in the initial planning and design of materials handling operations in warehouse, assembly, and manufacturing facilities. The constraints should be known early. It may be necessary to compromise equipment in order to gain the optimum space utilization and equipment performance for an effective operation. One of the major issues of concern is right-angle stacking and cross-aisle requirements. In both requirements, a space and performance compromise becomes necessary to maintain minimum aisle width while providing adequate clearance to effectively operate the truck within the designed aisle width without creating barriers and damaging product and equipment. As an effective planning guide, the calculation of right-angle stacking and cross-aisle maneuvering is based on the following three dimensional factors:

1. Truck turning radii
2. Truck frame configuration
3. Unit load length and width

Depending on the type of truck and its steering mechanism, all trucks have two significant turning radii, the outer and inner turning radii. For a two-rear-wheel steering truck, the pivot point of the inside radius is about 3–4 in. (75–100 mm) outside the truck drive wheels. These trucks generally cannot pivot within their own frame footprint. They maneuver slightly around an outside circle. The outside radius is the overall swing of the truck frame, relative to the pivot point of the inside radius, to the farthermost part of the rear frame or extended backcouple attachment. Both radii should be obtained from the manufacturer when calculating or sizing aisle requirements. The outside radius should be specified with the truck loaded to its rated capacity and with the load about 6 in. (150 mm) off the floor. The reason for using a loaded truck is that the rear steering wheels have less weight and have a tendency to slip or creep slightly sideways when maneuvering in right-angle turns. Slippage occurs because the rear steering wheels are in a near-balanced state. With an empty lift truck, the steering wheels are carrying the full counterbalance dead-load weight which is only a true operating condition for about 50% of the time. Slippage is a true operating condition that should be recognized in right-angle aisle width requirements.

The calculation of right-angle aisle stacking is influenced by several maneuvering factors and by the size and space relationship of the area. To assure effective results, tests should be made to verify the calculated design width.

For a double rear-steer-wheel counterbalance truck, the calculation shown in Fig. 7.1.32 can be used. The equation also applies to a one-wheel rear-steer counterbalance truck that pivots within its own frame footprint. In that case, the inside radius (R_1) is zero and B is half the width of the truck. Since the inside radius only impacts right-angle stacking requirements by a few inches, the single rear-steer wheel has only a marginal effect on right-angle aisle requirements. For narrow-aisle lift trucks, such as reach or straddle trucks, the right-angle stacking aisle is smaller than for counterbalance trucks. The aisle width may vary from 3 to 5 ft (0.9–1.5 m) less for narrow-aisle trucks. The difference depends on the truck design, battery compartment, load size, and operating requirements. Most manufacturers of narrow-aisle trucks specify the right-angle stacking clearance. However, these are usually rub clearances without the operating clearance being specifically stated.

For operating clearance, 12–18 in. (300–450 mm) should be added to the calculated aisle rub width or stated width. Clearances greater than this are not generally needed; however, there are tradeoffs between the aisle width and truck productivity. A wider aisle generally improves productivity and helps to minimize damages. Frequently, it is necessary to review operational requirements and tailor aisle allowances accordingly.

For intersecting (cross) aisles, the dimensions are frequently given by the manufacturer and are usually stated based on a truck without a load. However, this can be misleading since provisions should be made to safely maneuver loaded trucks among aisles. A general rule is for cross aisles to be slightly wider than right-angle aisles for conventional lift trucks. For turret, swing-mast, and hybrid trucks, the cross-aisle width will be directly proportional to the truck length plus load length if the aisle is guided. The manufacturer should be consulted to assure that proper clearances are provided.

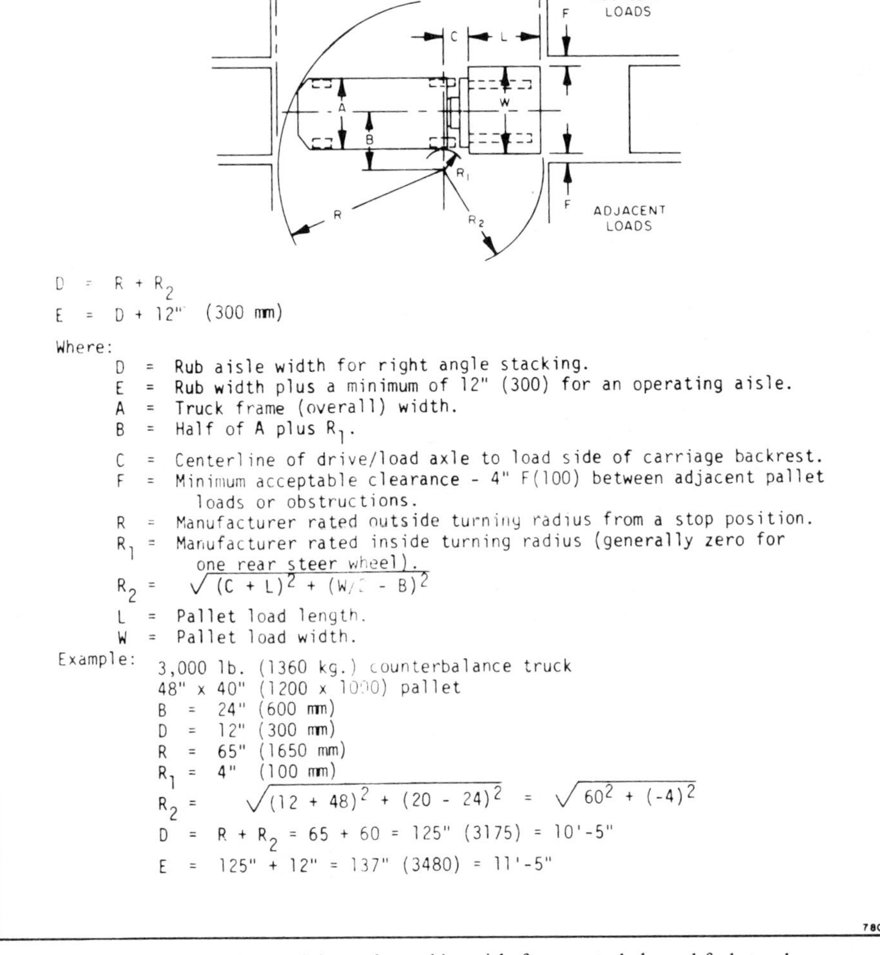

Fig. 7.1.32 Minimum right-angle stacking aisle for counterbalanced fork trucks.

Additionally, cross aisles and main service aisles should be wide enough to permit two-way traffic without slowdowns or delays. These aisles should be double the width of the load or truck, whichever is greater, plus a minimum of 18 in. (450 mm) for passing clearance for conventional trucks, or larger for special purpose trucks.

Guided Aisle Width. As previously mentioned, storage aisles can be guided by either mechanical systems, using floor-mounted guide rails and truck-mounted rollers, or by electronic systems, using an in-floor wire and a truck-mounted receiver. The systems are not interchangeable without physically modifying the equipment. In some systems, the required storage aisle width may differ between the mechanical and electronic methods.

The required clear aisle width and suitability of either system are influenced by the following factors:

1. The physical characteristics of the lift truck, including length, width, and height.
2. The type of load-handling mechanism used. Common mechanisms are turret forks, pass-through forks, and articulating forks or carriages.
3. Rigidity of the elevated mast assembly and the presence or lack of upper-level mast guides.
4. The type and size of the loads to be stored.
5. The storage height.
6. The type of guidance system used (mechanical or electronic).
7. Compatibility of the vehicle and guidance system voltages.
8. Possible sources of electrical interference to an electronic guidance system.
9. Flatness characteristics of the warehouse floor.

These factors should be reviewed with the truck manufacturer and the guidance system supplier before committing to a system or storage layout. It is extremely important that all of the foregoing factors be incorporated into the design of the system. After the guidance system has been installed, the operating performance is locked into the system. If the aisles are too narrow, or if the floor is too uneven, the vehicle cannot operate at its design speed. The result may be a performance reduction of 30–60%. In the initial design stage of a facility, only a few of the listed factors need be considered after the type of guidance system, load size, and type of vehicle have been determined.

For a mechanical guidance system, the aisle width will be determined by the location of the guide rails and their relationship to the storage rack. The rail width will depend on the type of truck selected and the physical location of the truck guide wheels. If the rails are very large, the base of the pallet rack may have to be recessed to provide the necessary rail clearance while retaining a narrow load-handling aisle. The need for, or amount of, rack recess depends on the following:

1. Load size
2. Load and mast sway clearance required while traveling at the allowable elevated height
3. The rack upright width and pallet load overhang
4. Load traverse or articulation requirements

A rule of thumb for developing a first estimate of aisle width states that 4–6 in. (100–150 mm) of clearance between the load (or truck) and rack (or load overhang) be provided on each side of the aisle.

For electronic guidance systems, the same factors apply and must be evaluated when selecting an aisle width. However, the clearance provided on each side of the load (or truck) should not be less than 4 in. (100 mm) and should preferably be 6–8 in. (150–200 mm). Slightly greater clearance is preferred for electronic guidance systems because, unlike the mechanical systems, they are not rigidly restrained within the aisle.

Gradeability. Gradeability is defined as the greatest slope or incline a lift truck can have and still have the ability to negotiate without a considerable loss of forward speed or considerable drain on the source of energy. The angle of incline is stated as a percent and is the tangent of the slope, which is generally between 7% and 25% depending on the truck design and power source. For counterbalance electric trucks, the accepted grade is 10–12%. For electric reach trucks, the grade is 10% or less. Internal combustion (IC) engine trucks can negotiate steeper grades because of their multiple-speed transmissions and more powerful engines. IC trucks can negotiate inclines between 15% and 25%.

Grade Clearance. Grade clearance is defined as the underframe clearance in which a truck can negotiate a change in slope without a "hang up" or a rub contact with the grade apex and the underside of the truck frame. This is defined as the clearance between the fore and aft wheels at the lowest part of the frame.

Figure 7.1.33 shows the general configuration of the angle of grade a truck will clear. This dimenison should be verified by the equipment manufacturer. However, the lowest clearance of the truck is directly beneath the mast, but it is not generally as critical as the underframe clearance between the wheelbase.

Drawbar Pull. Vehicle drawbar pull is the amount of effort with which a vehicle can pull loads to overcome friction and sustain motion with a series of carts in tow. Generally, manufacturers do not recommend that lift trucks be used for pulling carts, but towing functions are performed effectively in a number of warehousing operations. Drawbar rating is expressed in both nominal and capacity loads ranging from 200 to 700 lb (90–300 kg) for regular electric industrial trucks. The conversion of the nominal weight to rolling weight is determined by dividing the drawbar pull by the friction factor, which ranges from 2% to 3%. The rolling weight, therefore, ranges from 10,000 to 35,000 lb (4,500–15,900 kg).

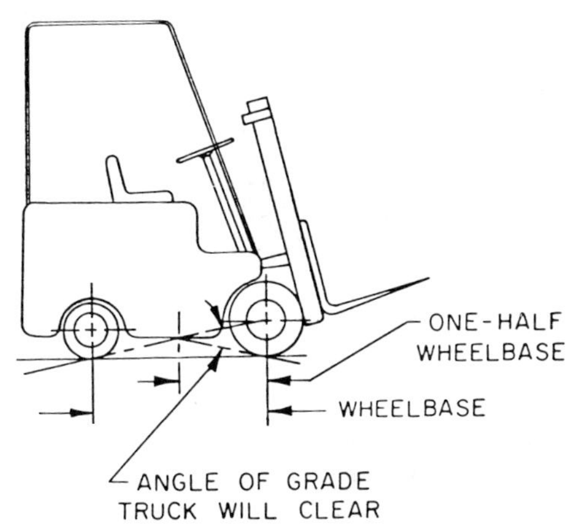

Fig. 7.1.33 Grade clearance.

Tires. Industrial truck tires are available in two basic types, either pneumatic or solid. Pneumatic tires are inflated with air or filled with a resilient composition to provide an operating casing profile for softer rides and greater traction than solid or cushion tires. They are similar to those used on highway trucks and some passenger cars. Resilient materials are used to resist deflation from punctures and still provide cushion on rough terrain. Pneumatic types are used almost exclusively for outside operations and some inside operations. The inside operations are primarily those where higher speeds are permitted, where cushioning of fragile loads is desired, or when driver fatigue is encountered due to long continuous operation.

Solid tires are of the same basic material composition but with a different structural profile and tread design. Each type and design serve a specific purpose.

Solid tires are available in two general types known as solid-cushion and solid tires. Both are low profile and are used where small diameter tires and high-stacking truck stability is required. The solid tires are mounted on a steel ring, rim, or hub by one of two common processes, either press-on or cured-on. The press-on tire is bonded to a steel ring which is pressed onto the truck wheel. The cured-on tire is molded or vulcanized directly to the wheel casing, forming an integral unit which is attached to the axle. When a cured-on tire wears out, the entire wheel must be replaced, whereas press-on tires and rings are replaceable when worn out. Press-on tires and rings are generally used on the drive, steering, and braking wheels and normally have tread. Cured-on tires are used for the free-rolling wheels and normally do not have tread. A glossary of industrial truck tire terms is summarized in Fig. 7.1.34.

The solid tire compound is generally identified by the following nomenclature:

1. Universal service—for general-purpose usage and resistance to chipping and cutting.
2. Low-powered consumption with low rolling resistance—for electric lift trucks to increase battery life.
3. Oil resistance—for use where separation, swelling, and excessive cracking may be caused by constant contact with oil, grease, or solvent.
4. Static conductors—to reduce spark hazard in highly explosive environments.
5. Nonmarking—to eliminate marks or smudges on floors in special environments such as showrooms, bakeries, pharmaceutical plants, or hospitals.
6. Polyurethane—a resilient synthetic compound for high loading applications and high resistance to abrasion.
7. Shredded wire—a mixture of short wire lengths blended with the rubber tread material as a protection against cutting, chipping, and separation caused by floor hazards.
8. Metal studs—to improve traction on inclines and ramps that are wet, oily, or icy.

Hazardous Operating Conditions

Before selecting the type of power source for an industrial truck, the environmental conditions in which the truck is to operate should be analyzed to determine the nature, quantity, and concentration of combustible materials that may occupy the operating environment. Hazardous environments have been defined by Underwriters Laboratories, Inc. (UL) and the National Fire Protection Association

GLOSSARY OF LIFT TRUCK TIRE TERMS

PNEUMATIC TIRES

Bead—that part of the tire made of steel wires, wrapped or reinforced by ply cords, that is shaped to fit the rim.

Carcass—the tire structure, except tread and sidewall rubber.

Load Rating (LR)—the maximum load a tire is rated to carry for a given inflation pressure.

Maximum Load Rating—the maximum load and inflation for a given ply rating (and/or LR) at a specified speed for given operating conditions.

Maximum Permissible Inflation Pressure—maximum pressure to which cold tire may be inflated.

Overall Diameter (OD)—diameter measured from the tread centerline of an inflated and unloaded tire.

Overall Width—the section width, adjusted to include added width from protective side ribs, bars, or decorations.

Percent Deflection—tire deflection, divided by the unloaded section height above the top of the rim flange, multiplied by 100.

Ply Rating (PR)—an index of tire strength expressed as an equivalent, but not necessarily the actual number, of cord plies in a tire.

Revolutions Per Mile (rpm)—number of revolutions deflected tire travels in 1 mile.

$$\text{RPM} = \frac{5280 \times 12}{2\pi\ \text{SLR}} = \frac{10084}{\text{SLR}}$$

where static load radius, SLR, (see definition farther on in this glossary) is in inches.

Rim—a metal support for a tire, or a tire and tube assembly, upon which the tire beads are seated.

Section Height (SH)—radial distance between base of tire at nominal rim diameter radially to a line tangent to outside tread arc at highest point.

$$\text{SH} = \frac{\text{OD} - \text{Nominal Rim Diameter}}{2}$$

Section Diameter or Section Width (SD)—the calipered width of a new inflated tire, including 24-hr inflation growth and normal sidewalls, but not including protective side ribs, bars, or decorations as mounted on a standard rim.

Size Identification—a coded sequence of numbers and letters:

a. Conventional—the first number designates approximate section width (SD) and the second the nominal rim diameter (that is, 6.50-10 indicates a tire of approximately 6½-in. cross section and mounted on a 10-in. nominal diameter rim).

b. Low Profile (or Low SH)—the first number designates the approximate OD, the second the approximate SD, and the third the nominal rim diameter (that is, 27x10-12 indicates a tire with 27-in. OD and 10 in. SD, used on nominal 12-in dia rim).

c. NHS Designation—means *N*ot for *H*ighway *S*ervice (that is, 6.90-9 NHS).

Static Loaded Radius (SLR)—the deflected radius of a tire measured from its wheel axis to contact surface, with the tire at rated load and inflation

$$\text{SLR} = \frac{\text{OD}}{2} - \text{Radial Deflection}$$

Tire Deflection—the difference between unloaded and loaded section heights.

Tread—that portion of a tire that comes into contact with the road.

Tread Radius (TR)—the transverse radius of tread curvature (or arc) across tread width, determined by measuring-templates of known radii fitted to the convex tread profile.

Tread Width (TW)—the arc width distance of an inflated tire, as measured across the tread between the extreme tread edges.

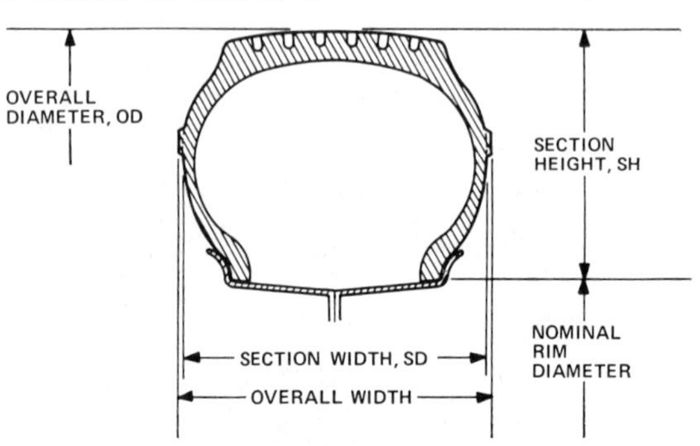

SOLID TIRES

Base Width (W)—base band width (also equal to tire base width).

Hi-Load Solid—tire made of special high-load-carrying material, such as polyurethane.

Multiple Tire Rating—when multiple tires are used as a unit on a single wheel, the maximum load rating is equal to the sum of the individual tire load ratings.

Overall Tire Diameter (OD)—a dimension equal to the sum of wheel diameter and twice the section height (including base band thickness).

Rubber Section Height (H_2)—does not include base band thickness.

Section Height (H_1)—includes base band thickness.

$$H_1 = \frac{\text{OD} - \text{actual wheel diameter}}{2}$$

Size Identification—a coded sequence of numbers:

a. Pressed-on—the first number designates overall tire diameter, the second number is the base width, and the third, the nominal diameter of the wheel or hub on which the tire is to be applied (example: 9x5x5).

b. Cured-on—the first number designates overall tire diameter, the second number is the tire width at the metal wheel diameter (example: 8x3).

Tread Design Types—the three basic forms are: smooth, grooved, nondirectional traction.

Tread Width (TW)—tread surface width contacting the ground.

Wheel Inner Diameter (ID)—the diameter which corresponds to the inside diameter of the metal base band.

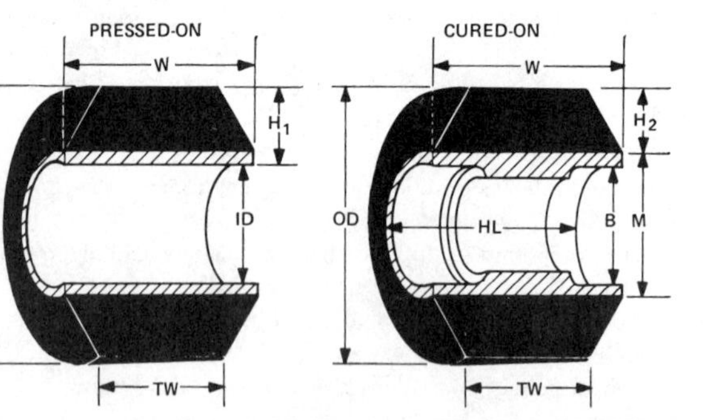

Hi-Load Exception for Intermittent Service—permissible tire loadings for industrial counterbalanced fork lift trucks in intermittent service are higher than those for the same tire in other types of service. This intermittent service exception is in contrast to other load classifications, in which continuous operating conditions are used to establish tire ratings.

Fig. 7.1.34 Glossary of lift truck tire terms.

(NFPA) and are published in the *Federal Register* as part of OSHA regulations. The three principal classes of hazardous environments are:

Class 1—Combustible gases or liquids
Class 2—Combustible dust
Class 3—Combustible fibers

There are 11 different industrial truck designations that have been identified by type of power that can be operated in nonclassified or classified hazardous environments. UL and NFPA have developed specific fire protection ratings for electric-, gas-, and diesel-powered trucks that can be operated safely within a known environment. The safety requirements apply to the fuel source and the exhaust or electrical supply systems. Therefore, a safety rating according to the environment should be established and designated in order to operate equipment in accordance with safety codes and within insurance underwriter requirements for a facility.

Vehicle Safety Classification. The type and power designation of industrial trucks are classified as follows:

1. Electrically powered lift trucks
   a. Type E units are electrically powered units having minimum combustible safeguards against inherent fire and electrical shock hazards.
   b. Type ES units are electrically powered units that, in addition to all the requirements of the Type E units, are provided with additional safeguards to the electrical system to prevent emission of hazardous sparks and to limit surface temperatures.
   c. Type EE units are electrically powered units that have, in addition to all the requirements for the Types E and ES units, electric motors and all other electrical equipment completely enclosed.
   d. Type EX units are electrically powered units that differ from Types E, ES, and EE in that the electrical fittings and equipment are so designed, constructed, and assembled that the units may be used in atmospheres containing specifically flammable vapors, dust, and fibers.
2. Gasoline-powered lift trucks
   a. Type G units are gasoline-powered units having minimum acceptable safeguards against inherent fire hazards.
   b. Type GS units are gas-operated units having, in addition to the requirements for Type G units, additional safeguards to the exhaust, fuel, and electrical systems.
3. Liquified petroleum gas-powered lift trucks
   a. Type LP units are liquefied petroleum gas-powered units having minimum acceptable safeguards against inherent fire hazards.
   b. Type LPS units are liquefied petroleum gas-powered units that, in addition to the requirements for the Type LP units, are provided with additional safeguards to the exhaust, fuel, and electrical systems.
4. Dual fuel-powered lift trucks
   a. Type G/LP units operate on either gasoline or liquefied petroleum gas and have minimum acceptable safeguards against inherent fire hazards.
   b. Type GS/LPS units operate on either gasoline or liquefied petroleum gas and, in addition to the requirements for the Type G/LP units, have additional safeguards to the exhaust, fuel, and electrical systems.
5. Diesel-powered lift trucks
   a. Type D units are diesel-powered units having minimal safeguards against inherent fire hazards.
   b. Type DS units are diesel-powered units that, in addition to all requirements for the Type D units, have additional safeguards to the exhaust, fuel, and electrical systems.
   c. Type DY units are diesel-powered units that have all the safeguards of Type DS units and, in addition, do not have any electrical equipment, including ignition.

Operating Environment. Table 7.1.4 is a summary of the allowable-use conditions for the various types of powered lift trucks. These guidelines are extracted from the NFPA 505 Industrial Truck Standards and describe the operating environment by class and division by groups within a class in which certain types of vehicles can be operated. Approval should be given by an authorized safety engineer before a truck is selected for use in a hazardous location.

Table 7.1.4 Powered Industrial Trucks Allowable-Use Summary[a]

Locations	Diesel Powered			Electric Powered				Gasoline Powered		LP-Gas Powered		Dual Fuel	
	D	DS	DY	E	ES	EE	EX	G	GS	LP	LPS	G/LP	GS/LPS
Class I													
Division 1													
Group A													
Group B													
Group C													
Group D							A						
Class I													
Division 2													
Group A		X	X		X	X	X		X		X		X
Group B		X	X		X	X	X		X		X		X
Group C		X	X		X	X	X		X		X		X
Group D		*	A		*	A	A		*		*		*
Class II													
Division 1													
Group E							*						
Group F							*						
Group G							A						
Class II													
Division 2													
Group E		X	X		X	X	X		X		X		X
Group F		X	X		X	X	X		X		X		X
Group G		*	A		*	A	A		*		*		*
Class III													
Division 1			A			A	A						
Class III													
Division 2		A	A	*	A	A	A		A		A		A

[a] A = type truck authorized in location described; * = Type truck authorized in location described with approval of the authority having jurisdiction; X = type truck authorized to be determined by the authority having jurisdiction; blank spaces = type truck not authorized in location described.

Some flammable gases or vapors where powered lift trucks should not be used in Class I, Division 1, Groups A, B, and C are: acetylene, allyl alcohol, butane, ethyl sulfide, hydrogen, hydrogen cyanide, and propylene oxide.

Some flammable liquids and gases that exist in a normal operating condition of Class I, Division 1, Group D are: acetic acid, acetone, benzene, butane, ethane, gasoline, isoprene, methylethyl ketone, octanes, styrene, toluene, and vinyl chloride.

Class I, Division 1 is an environment in which the gases and vapors are transferred from one container to another or involve open operations in which the solutions are used in processing operations. In Division 2, the various materials are used but are confined within a container or a closed system in which they possibly could escape but only in case of an accident, rupture, or breakdown in the containers of the system.

Class II, Division 1 is represented by a contaminated atmosphere of metal dust, carbon black, coke, or coal dust in which some of the following metals are found: aluminum, magnesium, and other commercial alloys and other metals similar in characteristics to coal and coke dust.

In Class II, Division 1, Group G, combustible dust is or may be suspended under normal operating conditions in quantities sufficient to produce an explosive or ignitable mixture. Examples of this type of working environment are handling and storage plants for starch, sugar, malting, wood, flour, hay grinding, and similar combustible items. Room must be provided for such equipment as grinders, pulverizers, cleaners, graders, scalpers, open conveyors or spouts, open bins or hoppers, mixers and blenders.

Class II, Division 2, Group G contains the same materials as Class II, Division 1, Group G, but these materials are contained within a closed system and are not normally dispersed into suspension by normal operation of the equipment, thus minimizing the explosive and ignitable mixtures in the atmosphere.

Class III, Division 1 is an atmosphere in which combustible materials are suspended in quantities sufficient to produce an ignitable mixture. Locations containing such fibers are operations involving nylon, cotton, and textile mills, combustible fiber manufacturing and processing plants, cotton gins, and wood-working operations. Class III, Division 2 is a similar environment to Division 1, but one in which the fibers are stored or handled in closed containers and are not normally suspended in the air by processing or manufacturing operations.

In Class I, Division 2, Groups A, B, and C and Class II, Division 2, Groups E and F, hazardous areas are difficult to classify. Before operating powered industrial trucks in these areas, an engineering survey of the property and an evaluation of the fire and explosion hazards should be made.

Operational Safety

Industrial trucks are the backbone of materials handling operations within a plant. Frequently, they are considered a typical vehicle to which an operator can be assigned and operate with a minimum of instruction in the fundamental operation of the machine. Although the industrial truck appears to be a simple and foolproof machine, it is instead a complex piece of machinery that requires a skilled and proficient operator.

A careful application screening process should be used in the selection of candidates to operate lift trucks. Operators should be given a physical examination that includes steadiness, field of vision, hearing, reaction time, and distance judgment tests.

In the operating training program, a systematic procedure of teaching should be combined with classroom instruction and hands-on training in operation factors such as maneuverability and sensitivity. A program should include the following elements:

1. Instruction on driving techniques, including types of hazardous environments and ways to avoid them.
2. Familiarization with the use of the various types of trucks a driver will operate, including a complete explanation of the operation and control of each.
3. Obstacle-course driving instruction on courses set up expressly for this purpose.
4. Driving sessions devoted to situations arising under actual driving conditions.
5. Verbal coaching during driving sessions.
6. Written or oral tests given during and at the end of the course to provide a measure of training and learning.
7. Emphasis on safety rules, driver responsibility, and a discussion on maintenance problems.

General safety rules that should be followed by industrial truck operators include the following:

General conditions

Truck condition. Inspect the condition of the truck at the start of each shift. Check operational and mechanical parts as well as safety features and controls such as dead-man brake, lights, and horn.

Truck cleaning. Remove dirt and rubbish that would make footing uncertain and remove obstructions lodged in the mechanism.

Truck operation. Report faulty truck operation to the maintenance department and do not operate the truck until the problem is corrected.

Report accidents. Report all accidents immediately.

External conditions

Clearances. Check overhead and doorway clearances when in doubt.

Load inspection. Inspect the load for stability, obstructions, and damaged skids or pallets before picking up the load.

Debris. Report the presence of rubbish, chips, and so on, in aisles or storage areas.

Obstructions. Never drive over obstructions or in pedestrian lanes.

Elevators. Check elevator capacity before entering.

Lights. Never drive in the dark or in poorly lighted areas.

Congestion. Restrict driving in congested areas or during periods of heavy traffic.

Operating Rules. Apply prudent and courteous operating practices and operate the vehicle with concern for the safety of others, the facility, and yourself. Be alert, drive defensively, and observe all safety procedures. Do not harm the equipment, the facility, the operator, or associates.

Motive Power Systems

Internal combustion (IC) engines and electric motors are the two basic types of motive power for industrial lift trucks in the 2,000–10,000-lb (900–4,500-kg) lift capacity range. The following four factors should be considered when evaluating feasible power systems for industrial trucks:

1. Economy (purchase and operating)
2. Performance
3. Environment
4. Type of truck generally manufactured

Internal combustion (IC) engine trucks are generally supplied with an engine that is fueled by gasoline, diesel, or liquefied petroleum gas. The engine is connected to the drive wheels through either a manual transmission or fluid torque converter (automatic transmission). Gasoline engines generally have four or six cylinders and are water cooled. Diesel engines are similar and are equipped with fuel injection systems. The LPG-powered trucks are usually equipped with gasoline engines converted to operate on liquefied petroleum gas.

Electric trucks are powered by a lead–acid industrial traction storage battery. The drive power is provided by an electric motor the speed of which is controlled by a motor controller circuit. Separate motors may be used to provide auxiliary power to other functions such as the hydraulic, tilting, lifting, steering, and attachment systems. The battery is rated by both voltage and ampere-hours (A-hr). In place of ampere-hours, kilowatt hours (kW-hr) are sometimes used. Operating voltages are typically 12, 24, 36, 48, and 72 v. The 72-v systems are powered by two 36-v batteries in series. The A-hr capacity ranges from 200 to 1800 A-hr over a 6-hr period of continuous use.

Selection of the type of power used in a lift truck will depend on the environment and the operating characteristics in the facility. For long periods of operation inside a well-ventilated facility, or outside, diesel engines will be the most economical source of power. In the same general environmental surroundings, a gas- or LPG-powered truck could perform the same functions. Internal combustion engines are generally used where long runs and high speeds are required. IC engine trucks perform more effectively in rough terrain than electric trucks. In general, when grades greater than 10% must be climbed regularly, IC trucks should be used.

When selecting an engine for an industrial truck it is first necessary to determine the requirements of the truck. After the maximum gross weight load has been determined, that weight should be matched to a truck that has the lift capacity to safely meet the needs. The second step in matching needs is to determine the power required to accelerate and maintain safe speed and maneuverability within the operating environment and comply with any hazardous operating conditions that may exist. Although engines are usually rated in horsepower, it is the torque output that is the prime factor in selecting IC industrial trucks. Torque is the amount of force an engine can exert at the flywheel, whereas horsepower is the amount of work an engine can do in a given period of time. Torque is expressed in units of weight and distance such as foot-pounds (N-m). As an example, an engine having a torque rating of 100 ft-lb (13.5 N-m) exerts 100 lb (445 N) of rotational force at the distance of 1 ft (300 mm) from the center of the flywheel. Torque indicates the true amount of energy available to drive a unit after all the energy losses from friction and cooling have been deducted from the operating characteristics.

Internal Combustion Powered Trucks. The liquefied petroleum (LP) gas used in IC engines is commonly a mixture of propane and butane. The chief source of the fuel is natural gas. LP gas weighs about 4.46 lb/gal (0.5 kg/liter) and is normally colorless and odorless. For industrial use, odor additives are introduced so that escaping gas can be recognized. LP tanks normally used on industrial trucks are either the ICC detachable type or the permanently attached ASME type. For the detachable tanks, refueling is accomplished by exchanging the empty tank for a full tank by means of quick-connect fittings. Full tanks can be delivered to the truck operating areas in quantity. Trucks with permanently mounted tanks must be driven to a bulk fueling station. Safety requirements stipulate that a fueling station must be maintained a safe distance from all permanent facilities, usually a minimum of 50 ft (15.2 m). The need for this separation causes travel problems when refueling trucks since they must be driven to the refueling station, generally once per shift. The driver and truck are nonproductive while the truck is being refueled, and additional lost time may occur because of queueing problems at the refueling station. If a truck runs out of fuel while in service, it must be towed to the fueling station.

Engines originally designed for LP gas use have a higher compression ratio than converted gasoline engines. The advantages of an LP gas engine result from the air fuel manifold receiving vaporized gas, whereas the gas-fueled engine must vaporize the fuel. The LP system includes a simple carburetion system and a solenoid valve which cuts off all fuel when the engine stops. The combined vaporizer–pressure regulator assures vaporization and control of the pressure at the carburetor. The fuel system is designed for either liquid or vapor withdrawal for operation. For operation over a wide ambient temperature range, the liquid withdrawal method is favored since vaporization is aided by the use of engine heat instead of depending on air temperature.

Some of the advantages of LP gas are:

1. More complete combustion because the fuel enters the engine in a gaseous state
2. Reduction of crankcase oil dilution since no liquid enters the engine
3. Increased engine life and a reduction of fuel-caused engine deposits
4. Greater engine efficiency because of the higher octane rating of the fuel and more complete combustion
5. Reduced fuel costs in some areas

Some of the disadvantages of LP gas-powered industrial trucks are:

1. Increased fuel costs in some areas
2. Greater initial cost for equipment
3. Handling and storage of empty and refueled tanks

Battery-Powered Trucks. The electric motor in a battery-powered truck is a low-voltage, direct-current (DC) device using a storage battery as the power source. Motors are not rated by horsepower output; rather, they are chosen on the basis of the torque they produce for a given voltage and current draw. Compared to IC engines, the horsepower output is relatively low. A 4000-lb (1800-kg) capacity truck may have a drive motor rating of approximately $4\frac{1}{2}$ Hp (3.4 kW) compared to the approximately 40 Hp (29.8 kW) in a similarly rated IC truck.

The motors used in industrial trucks are heavy-duty industrial designs having an overload capacity rating of approximately 500%. The torque available for momentary surges is usually more than 10 times greater than that required to move the truck on a level grade with full load. A continuously variable silicon-controlled rectifier (SCR) control system varies the effective power delivered to the motor. This controls the torque to provide a much smoother change in vehicle speed when compared to any form of gear shifting.

Electric Truck Battery Selection. The selection of batteries depends mostly on voltage and ampere-hour (A-hr) rating and should be tailored to the characteristics of the equipment and the operating environment. The voltage required is directly related to the speed characteristics of the truck and the ampere-hour(s) rating is related to the service period over which the truck is to be operated. Some manufacturers also rate batteries in kilowatt-hours (kW-hr). This is the relationship of volts and amperes for a sustained period of rated use. The calculation is the volts times amps times hours of available use divided by one thousand and expressed as kilowatt-hour(s). Although kilowatt-hour(s) are used to rate batteries, the most commonly used measure is ampere-hour(s).

Voltage is the unit of electrical potential or the pressure from a complete circuit from the battery to the load and back to the battery. A typical lead–acid traction battery is made up of cells that are nominally rated at 2 v per cell. Therefore, an 18-cell battery is rated at 36 v. However, there are several schools of thought as to what is the practical voltage in the cell. The operating range is generally from 1.6 to 1.7 v per fully charged cell. In a discharged state, the voltage can drop to as low as 1.2 v per cell.

The transfer of current from a battery through an electrical motor is measured in amperes at a steady draw. The rate at which the draw can be sustained depends on the actual activity of truck and, therefore, the battery ampere-hour(s) usage rating is stated as one ampere flow for one hour. For a battery that can be discharged at 125 A over a sustained operating period of 6 hr, a 750-A-hr usable capacity is required. The effective operating capacity of the battery is generally measured to a discharged state which should not be less than 20% of capacity. Therefore, in order to have a 750-A-hr usable battery capacity, the rated battery capacity should be at least 900 A-hr, and preferably 950 A-hr.

Generally, lift truck manufacturers do not recommend that batteries be operated below the 80% discharge level. Below this point, further use puts an extra strain on the electrical system and can damage the circuits. There are two methods of preventing over-discharging batteries. One method is an ampere-hour meter (battery discharge meter) which measures the power consumed and indicates the amount of power available before reaching the 80% discharge level. At this point, a light will come on indicating that the truck must be taken back to the battery charging area for recharge or a battery change. The meter operation is similar to a fuel gauge. The second method involves installing a low-voltage cutout in the lifting circuit. This control will cut off the truck's lifting mechanism when the battery reaches the 80% discharge level and will not permit the operator to raise the forks. Therefore, the operator knows that the truck must be taken back to the charging area while there is still enough power to travel before the battery is completely discharged.

Usually, trucks are required to perform more work than they were originally designed or specified to do at the time of purchase. Incentive-paced specifications usually result in a truck being worked to its maximum capacity rather than the defined duty cycle. More recently, various state-of-the-art information systems have improved the effectiveness of operators. In addition to improving operations, the systems can also monitor the duty cycle of the battery, the truck, and the operator and also monitor equipment utilization to assist in preventive or predictive maintenance programs.

Battery Chargers. Battery chargers are available in motor-generator (MG), ferroresonant, and pulsed units. The MG type consists of a drive motor turning a generator which supplies the required charging voltage and current. These units are generally for special conditions and are not frequently used. The ferroresonant charger is the most widely used charger for traction batteries. It provides a tapered charge in which the initial charge is at a high rate. As the battery is charged, the voltage is gradually reduced until the battery is fully charged and thereafter maintains a trickle charge. Pulse-type chargers supply maximum voltage until the battery is at full charge, at which time the charge is cut off until the battery charge drops to a preset level. The charger will start again for a short burst of charge, constantly repeating the charging cycle. The charger ratings must be compatible with the battery being charged. Specific chargers or ratings are generally provided by the truck or battery suppliers.

Battery Connectors. There are numerous battery connectors of various types. Care should be taken to make sure the battery charger and truck are equipped with the same type of connectors.

In the selection of connectors, caution should be practiced to assure that there is no sparking or gapping of any electrical circuits that may create a hazard in the charging area.

Battery Charging Area

Generally, the least amount of planning and design arrangements are made in selecting and equipping an effective charging area. The following two basic requirements should be emphasized when designing a charging area:

1. What are the equipment requirements?
2. What are the safety requirements?

For equipment requirements, it is necessary to determine the size or projected size of the truck fleet, shifts of operation, and locations. The issues that should be considered include:

1. Number of truck positions to park at each charger.
2. Number of shifts the trucks will operate.
3. The position of battery charge equipment to effectively remove and reinsert battery by carts, cranes, or roll-out/roll-in conveyors, if there is more than one shift.
4. Special treatment of floors to resist acid spills, with controlled floor drains.
5. Utilities required for cleaning the area and battery maintenance.

For the safety aspects, the following must be provided:

1. Proper ventilation to reduce hazardous gas buildup
2. Deluge shower and eye wash
3. No-smoking area marked
4. Safety clothing and eye protection

7.2 INDUSTRIAL TOWING TRACTORS

Clark C. Simpson

The variety of machinery designed and built for the purpose of towing or pulling is immense. For this section, we limit ourselves to wheeled vehicles that generally operate on improved surfaces, moving material by means of trailers, or moving trailer-like equipment, commonly known as tow tractors. Although the mathematical methods used may apply to other types of tractors, they are specifically aimed at tow tractors.

The purpose of this discussion is to familiarize the reader with the more common types of tow tractors, and to provide information on the application of this equipment. It should be emphasized that the application information container herein is of a general nature, and specific application of a specific manufacturer's equipment requires consultation with the manufacturer or his representative.

7.2.1 Types and Vehicle Nomenclature

Tow tractors can be broadly clsssified as electrical (storage battery powered) or internal combustion engine powered, see Fig. 7.2.1.

Electric tow tractors have evolved into various types of vehicles with specific designations. A walkie is a small vehicle that has no provision for the operator to ride, and is guided by a control handle of

Fig. 7.2.1 Typical stand-up tugger. (Courtesy Crown Controls Corp.)

some type, while the operator walks along side the vehicle. A walkie rider is usually a small vehicle with an operator station to allow either sitting or standing on the vehicle while driving. The control is still a handle, which also allows the vehicle to be guided by walking along side. A stand-up has an operator compartment with no seat and commonly uses a tiller-type steering control. Sit-down rider usually indicates a larger-size machine with a more conventional operator compartment with seat and steering wheel, and the rider must be seated while operating, see Fig. 7.2.2.

Typical power trains used in electric tow tractors include some type of DC traction motor, usually series wound and connected by a controller to a storage battery. The design of controllers is a very active area with electric tow tractor manufacturers and results in a variety of performance parameters. Control designs vary from resistor types to SCR controls. SCR types are the most costly and are more efficient. Applications requiring higher horsepower, such as climbing grades, lend themselves to SCR control. The most popular sizes of electric tow tractors range from drawbar pulls of 175–2500 lb (778–11,121 N), although there are electric tow tractors in the 85,000-lb (378,810-N) range, see Fig. 7.2.3.

Internal combustion engine-powered tow tractors are almost exclusively the sit-down rider type, with minimums of approximately 2500 lb (11,121 N) drawbar pull. Typical power trains include friction clutch and sliding gear transmissions in smaller models, as well as hydrostatic and hydrokinetic (torque convertor) drives. Medium-size tractors tend to share power trains with on-highway automotive equipment. Automatic shift, planetary type multispeed light highway truck transmissions are plentiful in the 3000–6000-lb (13,340–16,680-N) drawbar pull sizes. Sizes above 6000 lb drawbar pull lean more toward off-highway powershift transmissions; and in very large sizes, diesel electric drives have been seen. Engines include gasoline and LP gas-fueled versions in the smaller sizes with a majority of larger sizes opting for diesel-type power plants, see Fig. 7.2.4.

7.2.2 Typical Applications

The most common, and also varied, use for tow tractors is in in-plant material transfer by means of trailers or trailer systems. Tow tractors commonly move material from one marshalling area to another by means of multitrailer trains. Order picking is also done with tow tractors, see Fig. 7.2.5. Tow tractors are a viable alternative when fairly long runs or congested traveling areas are encountered.

Tow tractors have also found widespread use in the air transportation industry. Typical uses include moving passenger baggage and air freight, up to and including air freight containers. Medium to large tow tractors have also been popularly used in aircraft push-out applications at major airports.

7.2.3 Definition of Terms

Any discussion of the selection of a particular size and style of tow tractor will necessitate the definition of some terms commonly used in discussion of the vehicle's performance.

Fig. 7.2.2 Typical sit-down rider electric tow tractor. (Courtesy Lansing Bagnall Corp.)

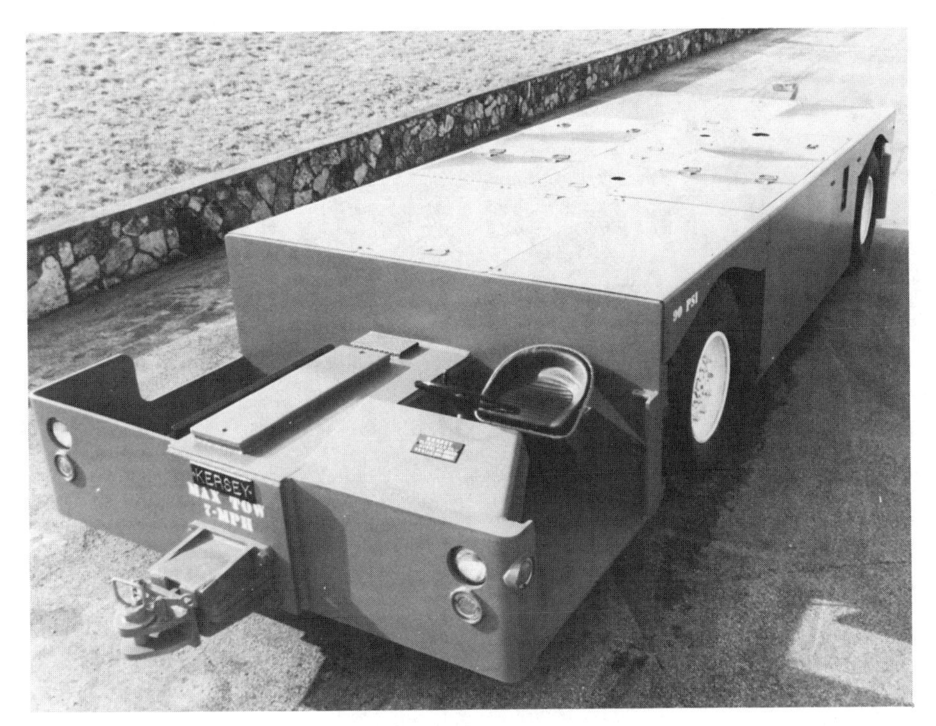

Fig. 7.2.3 A 60,000 lb (266,893 N) drawbar pull battery-powered electric tow tractor. (Courtesy Kersey Mfg. Co.)

Fig. 7.2.4 Typical internal combustion-powered 5000 lb (22, 241 N) drawbar pull tow tractor. (Courtesy Clark Equipment Co.)

Fig. 7.2.5 A walkie-rider at work in an order-picking application. (Courtesy Crown Controls Corp.)

***Acceleration resistance* (R_A).** The force required to overcome the inertia of the tractor and towed load.

This force is actually the difference between the drawbar pull the tractor is exerting and the sum of the rolling and grade resistance. The drawbar pull of most tow tractors decreases as vehicle speed increases, due to gear ratio changes. Less force is available, then, to accelerate the tow tractor and towed load. As a result, the rate of acceleration slows as the vehicle picks up speed.

The calculation of the speed a tow tractor and load reaches in a given time, or the distance required to achieve a certain top speed, requires solution of the equation:

$$d = \frac{WV^2}{2gR_A}$$

where: d = distance
W = weight of tow tractor and load
V = velocity of tow tractor and load
$g = 32.2$ ft/sec²

It can be seen that the mathematically correct solution requires integration of R_A from initial start to top speed. Two approximations are commonly done in calculations involving acceleration. The first is to merely assume a constant force for R_A and calculate d. The second is to assume 30–40 ft/min·sec (0.046–0.062 m/sec²) as a requirement for nominal acceleration. To accelerate at a rate of 96.6 ft/min·sec or 1.61 ft/sec² (0.150 m/sec²), requires 100 lb (444.8 N) per 2000 lb (907 kg) of gross train weight as follows:

$$F = \frac{Wa}{g}$$

$$F = \frac{2000 \text{ lb} \times 1.61 \text{ ft/sec}^2}{32.2 \text{ ft/sec}^2} = 100 \text{ lb (444.8 N)}$$

Rounding 96.6 off to 100 ft/min·sec results in 10 lb/2000 lb for each 10 ft/min of acceleration per second. Our initial assumption of 30–40 ft/min·sec results in R_A of 30–40 lb/2000 lb. This can be expressed in the English system as a percentage of gross train weight, 0.015–0.020.

This value, or factor, can safely be used for most general calculations. Consideration of acceleration is important for short travel periods with much starting and stopping. It is usually not considered for long runs or calculations of tow tractor operations on grades.

Coefficient of friction. The relationship of the force required to cause tire slippage at the tire–ground interface to the normal or vertical load on the tire.

***Drawbar pull* (D_p).** The force a towing vehicle is able to exert on a moving towed load.

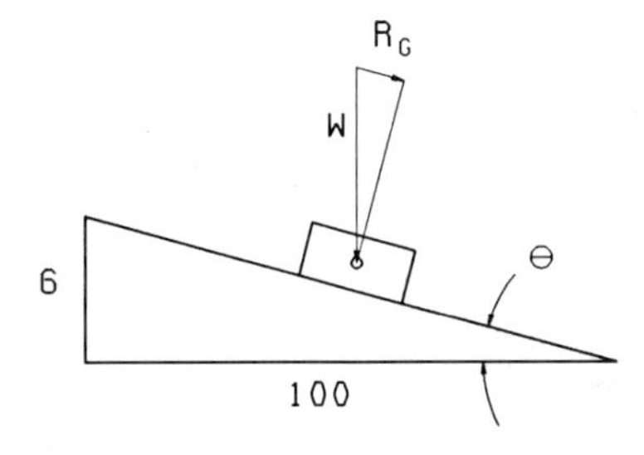

Fig. 7.2.6 Grade resistance factors.

Grade. A measure of the slope of the working surface. It reflects the amount of vertical rise in a specific horizontal distance.

Grade resistance **(R_G).** The force required to move a frictionless load up a grade, see Fig. 7.2.6. It is calculated by the following equation:

$$R_G = \sin\theta\ W$$

where: W = weight of load
θ = grade angle in degrees

It can be seen that for slopes expressed in grade or gradient,

Tangent θ = percent grade (expressed as a percent of 100; 6% = 0.06)

Therefore,

θ = inverse tangent of grade

Gross train weight **(W_{GT}).** The combined weight of the tow vehicle and its towed load.

Load weight **(W_L).** The weight of the towed load.

Rolling resistance **(R_R).** The force required to overcome a vehicle's resistance to rolling on a flat surface as a result of tire and surface interaction.

Rolling resistance is expressed in pounds of force (Newtons) per unit of vehicle weight, such as 20 lb/1000 lb (89 N/454 kg). The following table represents typical rolling resistances of bias-ply tires on various surfaces, assuming antifriction bearings at all wheels:

Surface	Rolling Resistance lb/1000 lb (N/454 kg)
Smooth concrete	15–25 (66.7–111.2)
Wood blocks	17–25 (75.6–111.2)
Smooth asphalt	15–25 (66.7–111.2)
Paving blocks, poor brick	22–38 (97.9–169.0)
Hard-packed gravel	30–40 (133.4–178.0)

An average figure of 20 lb/1000 lb (89 N/454 kg) is commonly used for well-maintained improved surfaces, and is generally the value used for calculation of drawbar pull by manufacturers.

Service weight **(W_S).** Weight of tow tractor.

Tractive effort **(T_E).** The force the tow vehicle's powertrain is able to develop at the ground–tire interface. Tractive effort does not include the rolling resistance of the tow vehicle.

$$T_E = D_P + R_R$$

Weight transfer **(W_R).** A change in the static weight of the tow tractor's drive and steer axles that occurs dynamically when towing because of the height of the drawbar. Weight transfer also results from operation on grades, see Fig. 7.2.7. Weight transfer on the level can be calculated by:

$$W_R = \frac{D_P h}{WB} \tag{7.2.1}$$

where: D_P = drawbar pull
WB = wheelbase
h = height of towbar from ground
W_R = weight added to the static weight of the rear axle

Weight transfer due to operation on grades requires knowledge of the vertical center of gravity of the tow tractor. It is commonly assumed that the center of gravity of the tow tractor is at towbar height, in which case R_G can be substituted in (7.2.1) for D_P. A simpler method is to simply add D_P and R_G and solve for W_R.

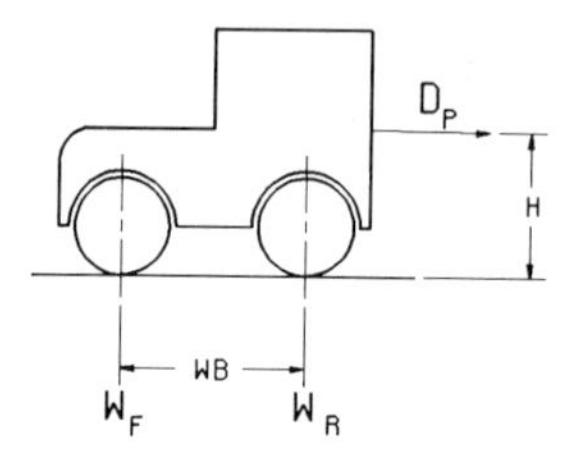

Fig. 7.2.7 Weight transfer factors.

7.2.4 Drawbar Pull Requirement (IC)

The theoretical calculation of the tractive effort to tow a load is represented by the equation:

$$T_E = R_G + R_A + R_R$$

Example. Determine the tractive effort required to tow four trailers weighing 10,000 lb (4537 kg) each, over a course that includes a 6% grade. Assume a tow tractor weight of 6500 lb (2949 kg).

$R_G = \sin\theta \ W_{GT}$
$\tan\theta = 0.06,\ \theta = 3.433°$
$W_{GT} = 4\ (10{,}000) + 6500 = 40{,}000 + 6500 = 46{,}500$ lb
$R_G = \sin(3.433°)\ 46{,}500 = 2785$ lb (12,388 N)

For R_A, assume 0.015 (30 lb/1000 lb).

$$R_A = 46{,}500\ (0.015) = 697.5 \text{ lb (3102 N)}$$

For R_R, assume 0.02 (40 lb/1000 lb).

$$R_R = 46{,}500\ (0.02) = 930 \text{ lb (4136 N)}$$

To climb the 6% grade requires:

$$T_E = R_R + R_G = 930 + 2785 = 3715 \text{ lb (16,525 N)}$$

To accelerate and move the load on the level would require:

$$T_E = R_R + R_G = 647.5 + 930 = 1627.5 \text{ lb (7239 N)}$$

Since level operation only requires T_E of 1627.5 lb, our tractive effort is determined by the grade requirement. Thus, a tow tractor with a tractive effort of 3715 lb (16,525 N) is required. You will notice that the calculation was made for tractive effort rather than drawbar pull. This is a conservative solution which assumes the data from a manufacturer is tractive effort, and will relate to static values empirically arrived at through test.

A common practice in internal combustion tow tractors is to use R_R of 20 lb (89 N) per 1000 lb (454 kg) for rolling calculations and use 40 lb (178 N) per 1000 lb (454 kg) for breakaway or start-up calculations. The following charts on drawbar pull versus speed and towing capacity on grade are typical and reflect breakaway or startup conditions. The weight of the load is gross train weight W_{GT}, see Figs. 7.2.8 and 7.2.9.

7.2.5 Drawbar Pull Considerations (Electric Tow Tractors)

Electric tow tractors are commonly given two ratings: one that is continuous, and one that is considered maximum. The continuous rating relates to the rating of the drive motor and reflects energy level at

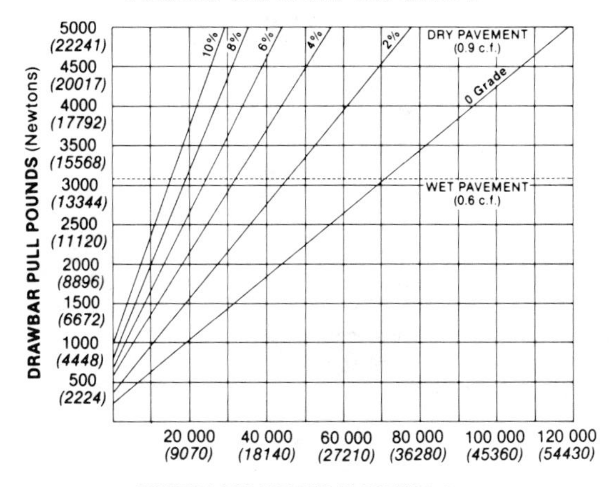

Fig. 7.2.8 Towing capacity on grade.

which it can operate for extended periods of time. Manufacturers of electric tow tractors have arrived at empirical factors for sizing their vehicles, and one commonly used factor relates required D_P to towed load. It is generally safe to assume a requirement of 25 lb (111.2 N) of D_P for every 1000 lb (454 kg) of towed load. This is based on level operation and trailers with antifriction bearings and tires, which results in rolling resistances of about 20 lb (89 N) per 1000 lb (454 kg). The diameter and composition of trailer tires has a marked effect on drawbar pull requirement.

Smaller-size electric tow tractors are generally designed for operation on level surfaces. The inevitable requirement for operation on grades and the need for extra D_P for starting loads generate a need for ratings that reflect maximum torque available from the drive motor. While there is widespread use of a maximum D_P rating, there is considerable difference among manufacturers as to how the maximum

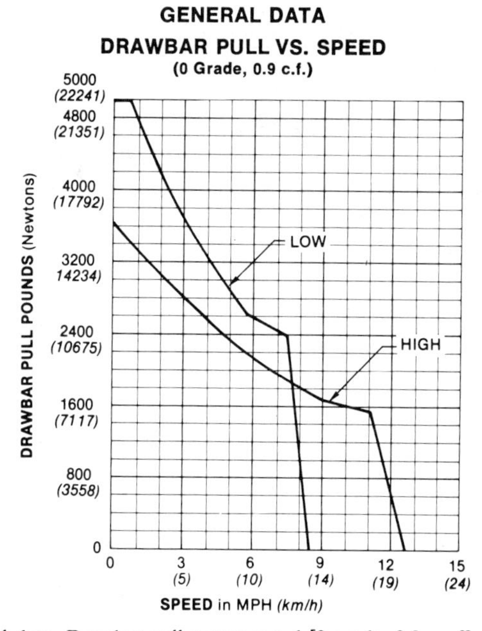

Fig. 7.2.9 General data. Drawbar pull versus speed [0 grade, 0.9 coefficient of friction (c.f.)].

rating should be applied. The area of electric tow tractor performance at maximum motor torques is one of high activity in the application of new technology, and results in much design variation. Battery design and controller design can have a significant effect on the time between charges. The ability of a specific drawbar pull to perform successfully on grades depends on how long the grade and how long the train. The individual manufacturer should be consulted in applications requiring use of the maximum rating.

7.2.6 Battery Life

An obvious concern in the application of electric tow tractors is the length of time between charges, or more importantly, how large a battery will be required. Since a battery is an energy storage device, its size will relate to the amount of work to be done. An approximation of the power requirement can be made using the following illustrations. Fig. 7.2.10 reflects the power requirement for overcoming rolling resistance R_R and acceleration resistance R_A and Fig. 7.2.11 indicates a method for predicting the power requirement for overcoming grade resistance R_G.

It must be understood that the methods in the figures are approximations, and must be used as such. They do not reflect such things as the many stops and starts during a cycle, or the effects of coasting. Operators may extend battery life by applying a burst of power and then coasting.

7.2.7 Braking Requirements

In addition to being able to tow the load, stopping is also a necessity. Most tow tractor applications do not have trailers with separate braking systems, and rely on the tow tractors to supply the retarding force. Stopping distance is the parameter we are most concerned about, and it can be determined from the equation:

$$d = \frac{WV^2}{2gF}$$

where: d = stopping distance
W = weight of the train
V = velocity of the train
F = retarding force
g = 32.2 ft/sec^2
F = brake drag + R_R + R_G

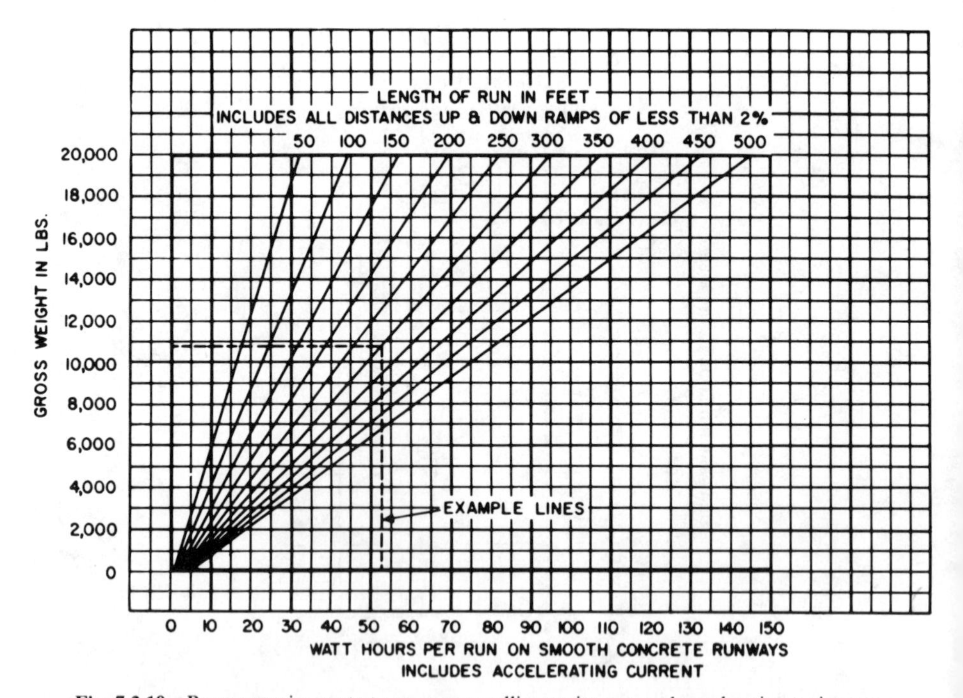

Fig. 7.2.10 Power requirements to overcome rolling resistance and acceleration resistance.

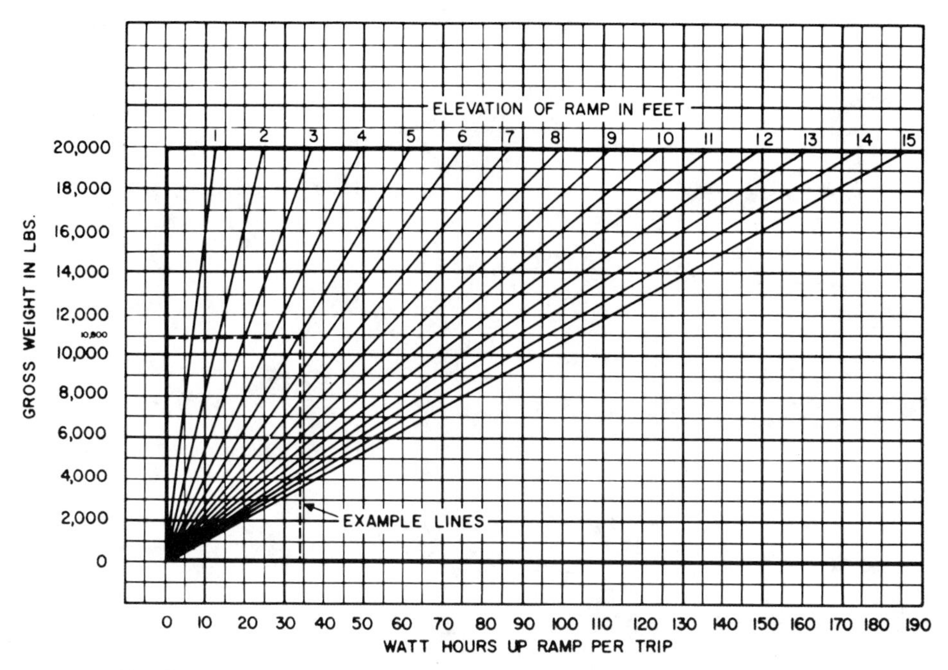

Fig. 7.2.11 Determining power requirements for overcoming grade resistance.

It should be nóted that R_G is positive in sign when traveling up grades, but negative when stopping on a downhill grade.

Values for brake drag are available from manufacturers, and are significantly different if two-wheel or four-wheel brakes are employed, due to the effects of weight transfer. The force of the towed load against the tow tractor transfers weight off the rear axle.

7.3 PERSONNEL AND BURDEN CARRIERS

Robert Cammack and Lori May

7.3.1 Types of Vehicles

The category of personnel and burden carriers encompasses a broad spectrum of vehicles, ranging from one-man scooters to 3200-kg heavy haulers. They can be of either three- or four-wheel type with propulsion generated by internal combustion or electric motors. Various types are illustrated in Fig. 7.3.1.

The personnel carriers are characterized by a driver (compartment) area and, depending on transportation needs, additional seating configurations for up to 16 passengers. The burden carriers also have a driver (compartment) area and a cargo bed area for carrying loads.

Various construction materials are utilized, with steel being the most prevalent. The fiberglass/plastic and aluminum bodies and parts are most common to lighter duty units.

Vehicle applications are virtually limitless. Table 7.3.1 lists users by industry and some of the applications in which the vehicles are being used.

The primary purpose of the personnel and burden carriers is to transport people and/or materials. As material movers, the carriers are most often selected because they are the most economical alternatives available. When compared to a forklift, the burden carrier is less expensive, just as maneuverable, has its own cargo platform for large and small loads, is faster, and is both easier and less expensive to operate. When compared to tow tractors and trailers, the burden carrier is again less expensive, more maneuverable, has its own cargo bed for small loads, is faster, and can also tow up to 12,500 kg. When utilized for personnel transportation, the vehicle saves steps, thereby saving money. Typically, an individual walks at a speed of 4 km/hr whereas a vehicle travels at 16 km/hr. Using the equation,

$$\text{Time spent walking per day} \times 0.75 \times 240 \times \text{Hourly rate of pay} = \text{Annual savings}$$

it can be shown that for an individual who walks more than 2 hours per day, time and cost savings greater than a vehicle's cost will be realized in one year if the individual rides instead. Cost justification

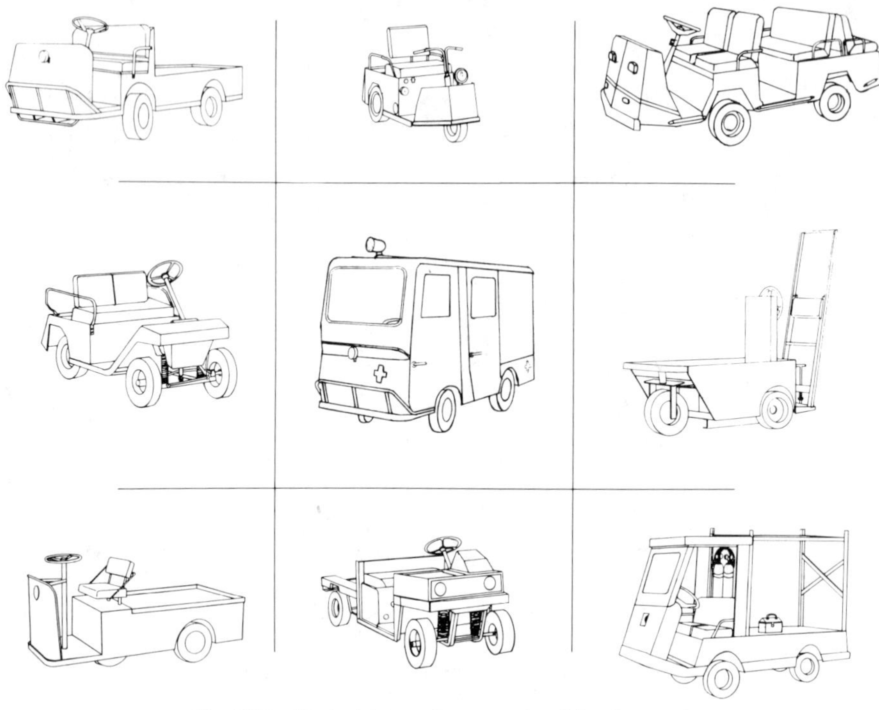

Fig. 7.3.1 Typical types of personnel and burden carriers.

is simplified where there are no other practical alternatives available or where the savings in physical effort is obvious.

One of the earliest applications for burden carriers was in nurseries for transportation of heavy and cumbersome plants. The existing alternatives at the time included physically dragging, hauling, or pushing carts or trailers; the savings with a vehicle were obvious. Although there are a variety of vehicles available today, the basic types have been summarized on Table 7.3.2. The table provides a preliminary guide to vehicle types and applications.

7.3.2 Standards

The design, specification, construction, and operation of personnel and burden carriers are governed primarily by two national standards: ANSI B 56.8-1981 and OSHA 1910.178. All vehicle manufacturers

Table 7.3.1 Users and Applications of Personnel and Burden Carriers

Typical User	Typical Application
Manufacturing	Trash truck
Warehouse	Parts moving
Airport	Order picking
Nursery	Grounds keeping
Hospital	Security service
Sports complex	Visitor transport
School, university	Executive transport
Zoo, amusement park	Maintenance truck
Hotel, resort	Emergency vehicle
Government agency	Tow tractor
Parking garage	Message/mail service
Shopping center	Equipment hauling

Table 7.3.2 Types of Vehicles

Wheels	Driver Position	Steering	Capacity (kg)	Bed (cm)	Drawbar (N)	Purpose (User)
3	Rear-stand	Tiller	50–150	25 × 50	—	Supervisor, expeditor, security
3	Mid-sit	Tiller	90–230	45 × 70	400	Supervisor, security, warehouse
3	Rear-stand	Wheel	230–460	75 × 120	450	Stock picking, warehouse, nursery
4	Rear-stand	Wheel	460–1365	80 × 110	1200	Warehouse, manufacturing
3 or 4	Mid-sit	Wheel	230–460	60 × 110	550	Security, visitor, car replacement
3	Front-sit	Wheel	460–820	64 × 75	700	Narrow aisles, manufacturing, towing
4	Front-sit	Wheel	460–2725	104 × 191	1100	Manufacturing, heavy haul, towing

Table 7.3.3 Personnel/Burden Carrier Standards

ANSI B 56.8-1981	Personnel and Burden Carriers
OSHA 1910.178	Powered Industrial Trucks
ANSI/NFPA 505-1978	Powered Industrial Trucks
ANSI/UL 583-1977	Electric-Battery-Powered Industrial Trucks
ANSI/UL 558-1977	Industrial Trucks, Internal Combustion Engine-Powered
ANSI/NFPA 30-1976	Flammable and Combustible Liquids Code
ANSI/NFPA 58-1976	Storage and Handling Liquefied Petroleum Gases

must comply with these standards and so state on product labeling. Additionally, ANSI B 56.8-1981 lists the standards to which the owner/operator must conform. Copies of this standard are available from American National Standards Institute, Inc., 1430 Broadway, New York, NY 10018. Additional standards that have been incorporated by reference into ANSI B 56.8-1981 are listed in Table 7.3.3.

Each manufacturer must self-test to insure that the vehicle design conforms to the standards. Although not required, some manufacturers have also contracted for outside test agency confirmation of design. An additional indicator of a manufacturer's adherence to standards is its membership in the Personnel/Burden Carrier Manufacturers Association, A Product Division of The Material Handling Institute, Inc.

7.3.3 Maintenance

The first consideration in owning or operating a group of vehicles is facilities availability. Whether the units are securely stored in one location or in various locations throughout the plant, adequate space with good ventilation must be provided. In addition, for battery-powered vehicles, electrical service must be provided to allow charging.

A servicing area will be required or arrangement made with an outside mobile servicing agency. For inside service, any forklift or auto/truck repair facility will be sufficient. If not available, allow a 3.5 m × 7 m area to include a workbench for servicing motors and chargers. Also required would be a hoist with a minimum 500-kg capacity to lift vehicles.

An additional cost of ownership is supplying fuel/power. For gasoline-powered units, a fuel storage area is required. The governing safety and design standards are ANSI/NFPA 505-1978 and ANSI/NFPA 30-1976. With electric units, 20-A AC service minimum can be expected for each vehicle. If the units have built-in chargers, this will require that AC outlets be strategically located throughout the facility.

Vehicle maintenance is another item requiring planning. Figure 7.3.2 shows a typical lubrication schematic for a vehicle and lists frequency of various service times.

Some service can be performed by the driver, such as checking gas, oil, or water level. However, general estimates allow 2 hr/mo in repairs for gas vehicles and 1 hr/mo in repairs for electric vehicles. Advances in design, such as lube-free bushings, and the installation of hour meters and battery charge indicators will lower these figures.

7.3.4 Performance and Specifications

The primary use of gasoline vehicles is outdoors. Speeds above 30 km/hr are also better suited to gasoline propulsion, primarily because gasoline can provide greater range through easier refueling. Electric vehicles are ideally suited for indoor applications where exhaust fumes would be objectionable. Where daily travel requirements are under 45 km and an 8-hr time period is available for recharging, electric vehicles are preferred, indoor or outdoor, because operating costs for electric are up to 70% less than gasoline propelled vehicles.

In specifying a vehicle, first analyze the facility layout. Aisle widths and turnaround areas should be recorded. Manufacturers will specify vehicle characteristics according to Fig. 7.3.3.

The second consideration should be the grade of terrain over which the vehicle will be operating; a rise of 3–4° can greatly alter the performance of a fully loaded vehicle. Specifications and calculations use percent of grade rather than degrees. As illustrated in Fig. 7.3.4, the percent is merely the tangent of the slope angle. Conversion for calculations will be required.

The third consideration is payload and the vehicle's load capacity. To safely operate a vehicle, manufacturer specifications are listed as limits that should not be exceeded. Additionally, payloads with load centers above the deck greater than one-half the deck width or length should be noted. Extra capacity may be required to maintain vehicle stability (see ANSI B 56.8-1981, Section 703i).

Grease		Check
1	Ball joints	Monthly
2	Wheel hubs/bearings	3 Months
3	Axle spindles	3 Months
4	Brake linkage	Monthly
5	Accelerator linkage	Monthly
6	Steering gear	Yearly
7	Rheostat	Weekly

Oil		Check
8	Engine	Daily
9	Transmission	Daily
10	Drive assembly	Yearly
11	Master cylinder	Monthly

Servicing		Check
12	Tune-up	Monthly
13	Air tires	Monthly
14	Adjust brakes	Monthly
15	Replace lining	2 Years
16	Battery water	Weekly

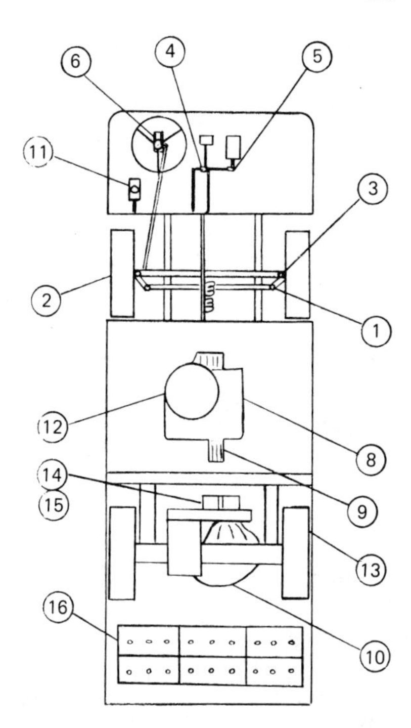

Fig. 7.3.2 Servicing requirements.

Towing applications will require determination of drawbar pull needed. Flat ground applications, such as inside a warehouse, are calculated as:

Trailing load × Coefficient of friction = Normal drawbar pull

When grades are involved, the effect is significant and the equation becomes:

(Trailing load × Coefficient) + (Vehicle + Trailing load) × (Grade) = Normal drawbar pull

Typical coefficients are listed in Table 7.3.4. Compare the results to the manufacturer's stated "normal drawbar pull." Peak or ultimate drawbar pull figures provide power available for acceleration or transgressing obstacles, and should not be used for duty application calculations.

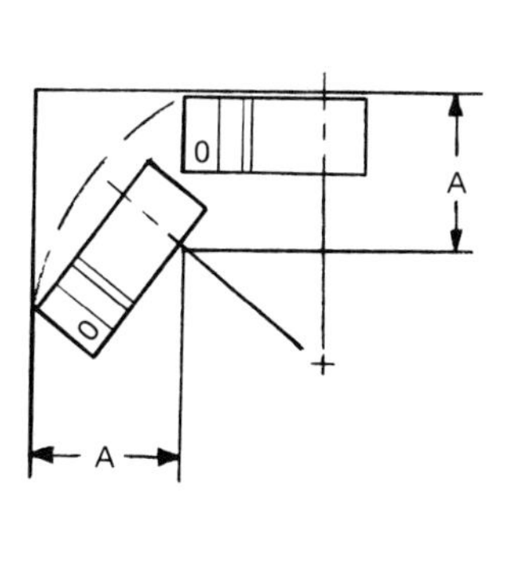

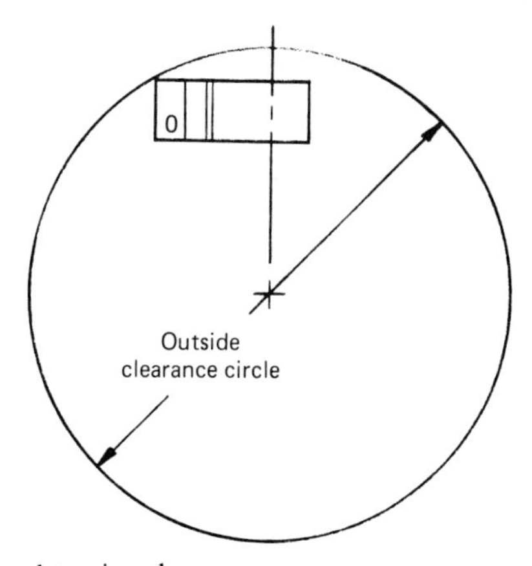

Fig. 7.3.3 Aisle and turning clearance.

PERCENT OF GRADE IS THE RATIO
OF THE VERTICLE RISE TO THE
HORIZONTAL DISTANCE
EXAMPLE: IN SLOPE BELOW, RISE IS
3 m, HORIZONTAL DISTANCE IS 10 m

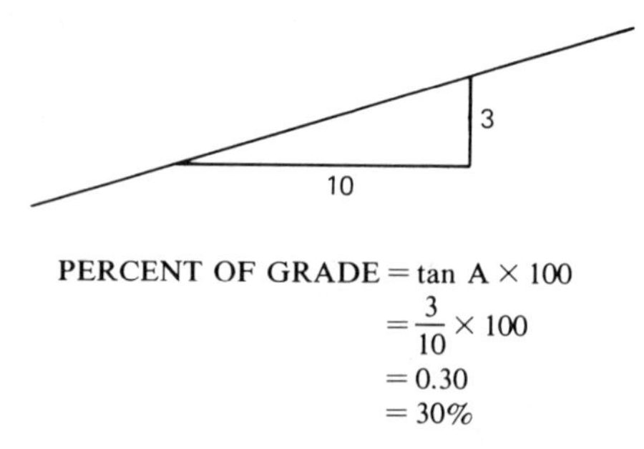

$$\text{PERCENT OF GRADE} = \tan A \times 100 = \frac{3}{10} \times 100 = 0.30 = 30\%$$

Fig. 7.3.4 Percent of grade.

In determining propulsion system and quantity of vehicles required, the normal operating time and distance must be determined. Generally, a duty factor of 35% is used. Allowing for stops, starts, loading, unloading, and other activities, a vehicle is in physical operation approximately $2\frac{3}{4}$ hr during any 8-hr shift. There are exceptions, however, such as a parking patrol service or a continual tram service, which should be considered.

Range is just as important as operating time. For electric vehicles, estimate a 50-km range between charges. For gasoline, 4 km/liter to 6 km/liter is the normal consumption rate, increasing with unit size. Ranges for electric and gas vehicles can be extended by adding larger batteries or fuel tanks.

Also, safety should be considered. Speed of operation must be specified. For indoor applications where manufacturing is being conducted and pedestrian traffic is predominant, speed should be in the 10–15 km/hr range. In airport baggage handling, a 20 km/hr speed is appropriate; however, on the flight line, where the vehicle is competing with full-size trucks and automobiles, 40 km/hr would be more appropriate.

Environmental conditions must also be evaluated. Table 7.3.5 summarizes the vehicle types from OSHA 1910.178. Gasoline vehicles are built standard to Type G rating, whereas electric vehicles are built to comply with Type E rating.

7.3.5 Gasoline versus Electric Propulsion

For economic reasons, electric propulsion is normally preferred. However, as outlined in specifications, a variety of factors must be analyzed, such as fumes, range, operating time, and speed. Another factor to consider is the availability of gasoline storage or electric charging facilities. Ease of operation is another consideration; electric vehicles below 25 km/hr require no gear-shifting transmission.

Gasoline engines vary in size between 4.5 kW (6 Hp) for single-driver three-wheel units and 13.5 kW (18.2 Hp) for high-speed or heavy-load-carrying units. Four-cycle, air-cooled, gasoline engines are standard with diesel offerings being minimal. LPG conversions are available as options. Tank sizes range from 8 to 25 liters, allowing driving ranges of 50–100 km. Transmissions, normally three-speed, are standard on larger units, and electric start is standard on all.

Table 7.3.4 Friction Coefficients

Road Surface	Coefficient
Flat block floor	.02
Concrete	.02
Asphalt	.02
Rolled gravel	.025
Firm soil or grass	.035
Firm soil or grass, with turf tires	.030
Wheels on rail	.002

Table 7.3.5 Vehicle Ratings[a]

Rating	Use
G	Gasoline; for all areas not containing hazardous atmosphere
GS	Gasoline; fitted with safeguards to reduce sparks, heat
E	Electric; for all areas not containing hazardous atmosphere
ES	Electric; meets E, plus fitted with safeguards to reduce sparks, heat
EE	Electric; meets E and ES, plus electric equipment totally enclosed
EX	Electric; meets EE, plus designed to operate in atmosphere containing flammable vapors or dust

[a] Consult OSHA 1910.178

Electric motors are more varied, with size matched to voltage and vehicle requirements. Table 7.3.6 lists typical sizes, characteristics and normal current draw, which are useful in determining vehicle range. Some motors under 0.7 kW (one-hour rating) are permanent magnet type. At 0.7 kW and above, motors are series wound with copper or aluminum field coils. Above 2.5 kW (one-hour rating) some have fans to assist in cooling. A direct power comparison between gasoline and electric motors is not possible because of their very different operating characteristics. Most manufacturers give a 5-min rating for electric motors to indicate power available for climbing and heavy hauling.

Electric vehicles are usually in the under 20 km/hr speed range, therefore no transmission is used. Some higher-speed applications use continuously variable transmissions (CVT's) to utilize high-torque characteristics of electric motors and still allow adequate acceleration. CVT efficiency is low at low speeds; however, the savings of a transmission and ease of operation can justify the application.

7.3.6 Batteries

Common batteries today are the lead–acid type. Gasoline starting batteries fall in the BCI Group 21–27 range and are 6- or 12-v size classified at 20-hr discharge rate. Electric vehicle batteries of 6-v size are BCI Group GC2 and are classified in amp-hours at a 20-hr discharge rate and in minutes at a 75-A continuous draw. By connecting batteries in series, voltage packs of 12 v, 24 v, 36 v, 48 v, and 72 v in size can be constructed.

Industrial batteries are also available in 12 v, 24 v, 36 v, 48 v, and 72 v sizes. These batteries are single units, complete in steel tray with cable leads. Ratings are based on vehicle rating system of OSHA 1910.178 (see Table 7.3.5). Type E is standard, with both Types EE and EX available. Ampere-hour ratings are based on 6- or 8-hr discharge rates.

Electric vehicle batteries of BCI Group GC2 will last approximately 300–400 cycles, or 2 yr.

Table 7.3.6 Motor Sizes

Voltage	1-Hour Rate	5-Minute Rate	Normal Current Draw (A)
24	1.0 Hp/0.75 kW 1600 rpm	3.0 Hp/2.2 kW 900 rpm	35
24	1.5 Hp/1.1 kW 1600 rpm	4.5 Hp/3.4 kW 900 rpm	40
24	2.25 Hp/1.7 kW 1900 rpm	6.5 Hp/4.8 kW 1000 rpm	60
24	3.5 Hp/2.6 kW 1485 rpm	10 Hp/7.5 kW 650 rpm	90
36	2.0 Hp/1.5 kW 2800 rpm	6 Hp/4.5 kW 1400 rpm	50
36	3.5 Hp/2.6 kW 2800 rpm	10 Hp/7.5 kW 1400 rpm	75
36	5 Hp/3.7 kW 2300 rpm	15 Hp/11.2 kW 1000 rpm	100

Industrial battery life is approximately 1000 cycles or 5–7 yr. An additional consideration with industrial batteries is their 500–1500-kg weight and the need for lifting devices for battery removal.

Electric vehicle operating range is based on ampere-hour rating (adjusted for current draw rating of motors) and average motor current (see Table 7.3.6). Electric vehicle batteries with 20-hr discharge rate must be adjusted with a factor of 0.75 to calculate running time:

$$\frac{\text{Ampere-hour rating} \times 0.75}{\text{Avg. motor current}} = \text{Running time}$$

For an 8-hr rated industrial battery the factor is 0.95:

$$\frac{\text{Ampere-hour rating} \times 0.95}{\text{Avg. motor current}} = \text{Running time}$$

For a 6-hr rated industrial battery the factor is 0.98:

$$\frac{\text{Ampere-hour rating} \times 0.98}{\text{Avg. motor current}} = \text{Running time}$$

These equations are estimates only and are subject to variations of motor size, terrain, driving conditions, and specific vehicle applications. Running time estimates will be decidedly different for forklifts. Driving range is calculated by multiplying the running time by the average driving speed.

7.3.7 Chargers

Chargers are available in two versions: portable and built-in. Portable chargers can be either bench top (up to 40 A output) or industrial case stationary (wheels optional) of the 50–150-A size. Bench-top units are 115 v or 230 v (48 v, 40 A output), 60-Hz type. Fan cooling is not ordinarily used and chargers may be stacked or mounted in rows in a well-ventilated area. Output plugs are two-blade type, with or without shield, conforming to angular position of NEMA 10-30R receptable. Industrial chargers use SB-type plugs and are often fan cooled. Due to the higher current output, AC input is 230 or 460 v.

Built-in chargers are similar to bench top units. Some manufacturers install portable chargers under vehicle deck or seat and hard wire to vehicle. Built-in chargers require extra protection from water damage and greater allowance for cooling because they are mounted in the vehicle. Units consist of a control console, transformer cabinet (often located below console, under dash), and an AC cord. Built-in chargers allow charging wherever AC power is available, permitting greater mobility and operating flexibility.

Chargers are of three design types: standard, semiautomatic, and fully automatic. Standard (timed charge) units are the most common, least expensive, and easiest to service. Line is compensated and equipped with 12- or 24-hr timers. The operator simply selects a time for the charger to run. Since operator selection is independent of battery requirements, irregular charging can result and battery life is affected.

Semiautomatic (transistorized) chargers have a 4- to 8-hr timer. These units supply power until a preset charge level is sensed, at which time the timer starts, completing the charge cycle. These units provide greater uniformity of charge, yet still rely on a timer.

Fully automatic (electronic) chargers automatically apply charge to the battery when connected and continue charging until a preset current and voltage level is reached. The unit stops at that point, but will restart if for any reason the battery voltage drops below a certain level. Although these units are more expensive, the extended battery life from their use outweighs the additional cost. Fully automatic chargers are extremely effective in two applications: where vehicles are operated almost continuously, thus having only short periods to recharge; or where vehicles with batteries installed are stored for extended periods.

Chargers are rated according to peak output. Actual output is a function of battery voltage and will drop as the battery voltage rises. For sizing estimations, use an average output of 66% of rated output. The goal in battery charger sizing is to sustain the recharge time under 10 hr using the smallest practical charger. If the battery is recharged too quickly (under 6 hr), battery life will be reduced. For batteries rated using a 20-hr discharge rate, an 80% factor must be included:

$$\frac{\text{(20-hr discharge rate) Ampere-hour rating} \times 0.80}{0.66 \times \text{Charger current rating}} = \text{Avg. charge time}$$

For industrial-type batteries:

$$\frac{\text{(6–8-hr discharge rate) Ampere-hour rating}}{0.66 \times \text{Charger current rating}} = \text{Avg. charge time}$$

Check local building codes when determining proper AC requirements. Allowance should be made for charger to operate at full rated output at 90% efficiency:

$$\text{Minimum AC current} = \frac{\text{DC charge voltage} \times \text{DC charger current}}{\text{AC line voltage} \times 0.90}$$

7.3.8 Speed Controls

Internal combustion engines are controlled by mechanical linkage from the accelerator pedal of engine carburetor. Servicing and adjustment is easily accomplished during normal tune-up. Electric vehicles are controlled by two basic types of controls: resistor and solid state.

Resistor-based controls utilize either a wiper assembly or solenoid/contactor assembly to direct the current flow. Using three to five resistor steps, voltage at the motor is changed. Motor current, however, is still allowed to rise to whatever level the motor demands. Wiper type controls are less expensive to purchase, but do require continuous maintenance. Solenoids/contactors do not require the weekly cleaning and lubrication that wipers do, but still require periodic repair and/or replacement.

Solid-state controls are of the SCR or transistor type. These devices control motor voltage and current without the power loss produced by resistors. In normal use, efficiency increases of 30% over resistor type controls are common. Although initial cost is higher, boosted efficiency plus elimination of maintenance requirements provide great savings. Additional features, such as plug braking and positive current limit on transistors to protect against motor overloading, make solid-state controls preferable.

7.3.9 Drive Systems

Internal combustion engines are usually coupled to automotive type differential drive through three-speed transmission and direct gearing. V-belts and pulleys are used on lighter-duty units.

Electric vehicles use V-belts, chains, and direct gearing. Belt drives are the least expensive and most versatile in providing speed ranges. Also, belt drives are easy to service and adjust. However, efficiencies tend to be about 95% and torque-transmitting capabilities are limited. Chain-driven systems also provide a broad range of speed selections. These drives, open and closed, use three types of chain. In ascending order of strength, they are single roller chain, silent chain, and double roller chain. Open-chain systems are noisier and experience wear much faster than oil-bath-enclosed chain drivers. Open-chain drives should include case or shield to keep outside contaminants off chain and sprockets.

Gear drives are found in the least expensive as well as the most expensive vehicles. The least expensive drives are light-duty (up to 700 kg) adaptations of golf cart drives. The most costly drives are geared for payloads over 2200 kg and towing applications.

Gear drives are not as flexible in speed and drawbar pull selections. Gear drives are most efficient, but are subject to greater shock loading wear and are noisier in operation.

Differential/axle assemblies coupled to the drive are of the helical or spiral/bevel gear type. The smallest, one-man units, do not use a differential. On these, the chain or belt runs directly from the motor to a sprocket or pulley on a solid drive axle. Two-man vehicles under 700 kg capacity use both aluminum and steel drive housings; adaptations of garden tractor and golf cart drives. Larger units use automotive or truck-sized assemblies.

7.3.10 Brakes

Brakes are mounted either on the drive line or at the wheels. Drive-line brakes mounted at motor shaft or differential pinion can be of either the band or disc type. These brakes utilize the gear advantage of the drive in braking. Axles must be sized large enough to handle the extra braking torque transmitted. These brakes perform well as service brakes up to 2000 kg vehicle rating.

Wheel brakes can be either mechanical or hydraulic, drum or disc type. At the low speed rating of the vehicles, mechanical brakes are very effective and less expensive.

Mechanical wheel brakes are best suited for vehicles up to 1000 kg capacity. Above that rating, the hydraulic wheel brakes are preferred.

7.3.11 Tires

Pneumatic (tube or tubeless) tires are standard. Pneumatic tires allow for greater operator comfort and a smoother ride. The three primary types of pneumatics include:

Terra-tires: used for off-road applications, have large footprint for low turf loading, and are typical of golf carts, park maintenance vehicles, and nursery vehicles.

4.80 Series: used primarily on single-passenger vehicles and payload capacities under 700 kg.

5.70 Series: used on burden and multipassenger units, providing good stability, wider tread, and broad capacity range and selection.

Tire capacities are shown in Table 7.3.7. It is important to match tire to vehicle load at operating speed. For estimation, the rule of thumb is that unit loading is 65% on the drive axle and 35% on the steering axle. Required capacity per tire is determined by the following equation:

$$\frac{\text{Vehicle weight} + \text{Capacity}}{2} \times 0.65 = \text{Required capacity per driving tire}$$

Pneumatic tires can be made more puncture resistant by using foam-filling urethane liners, or vinyl sealants. Capacity of foam-filled tires should be computed at 95% of capacity in Table 7.3.7.

Extra-cushion tires are also available, yet are more costly. They are a cross between pneumatic and solid rubber tires, have the consistency of a very dense sponge, and do not go flat. If the environment in which the vehicles operate tends to cause flats, these tires could be cost-effective; however, rider discomfort will be greater as will wear and tear on the vehicle.

Solid cushion tires are essentially solid rubber bonded to a steel ring and pressed onto a cast-iron wheel. They are for flat, level ground applications only. Producing the roughest ride and heaviest shock loading, they are intended for heavy haulers (over 2000 kg) and tow tractors. Equipment must be extremely rugged to handle these tires.

7.3.12 Suspension and Steering

Suspensions are of automotive type design. Front wheels may have independent suspension or beam axle. Both leaf and coil springs are used. At low vehicle speeds, shock absorbers provide little benefit. Shock absorbers are sometimes used on drive assemblies to dampen shock from high-torque motor startup. Some slow-speed vehicles have no suspension at all, relying upon pneumatic tires, seat cushions, and level operating surfaces for driver comfort.

Steering systems are also of automotive design with worm or rack and pinion gearing. With wheel effort in the 80–175 N range, power steering is not required. Tiller steering is inexpensive and used on small three-wheel units. Tiller steering is unacceptable at speeds over 16 km/hr or capacities above 300 kg.

7.3.13 Options

Full value of the truck will be realized only if the vehicle is completely equipped for the application. Proper consideration should be given to the variety of optional equipment available, such as:

Lighting. Headlights, taillights, turn indicators, and warning lights may have to be added for night or hazardous area operation.

Indicators. Hour meters, discharge indicators, and speedometers will aid in servicing and operating.

Enclosures. Cabs, tops, and cargo boxes should be considered if weather is inclement. Cabs should have removable doors for cooler summer operation.

Hitches. Pintle or automatic coupling will allow double duty of truck in towing and hauling.

Battery packs. Roll-out or lift-out battery trays can add in servicing as well as increase operating range of vehicle.

Identification. Custom colors of vehicle or seat cushions and/or vehicle numbering will provide easier identification of ownership, operator responsibility, or servicing schedule.

7.3.14 Purchasing

Personnel and burden carriers are available from a variety of manufacturers. Before purchasing, consider again:

1. Is manufacturer a member of an organization such as The Material Handling Institute, Inc.? Most major manufacturers have met their standards and become members.
2. Are parts and service readily available? Downtime can be expensive. Look for a sound manufacturer with a dealer network.
3. Are vehicles manufactured and labeled to comply with ANSI B 56.8-1981 and OSHA 1910.178? This should be used as a minimum requirement for any purchase. Additional assistance can be obtained by contacting:

    Personnel and Burden Carrier
    Manufacturers Association,

Table 7.3.7 Tire Capacities

Size	Type	Load Range	Ply Rating	Cold Inflation (psi/pascal)	5-mph Capacity (kg)	10-mph Capacity (kg)	15-mph Capacity (kg)
18 × 8.50 × 8	Terra-tire	B	4	22/1.5 E5	—	370	—
4.80 × 8	Highway tread	A	2	35/2.4 E5	290	229	213
4.80 × 8	Highway tread	B	4	70/4.8 E5	435	345	322
4.80 × 8	Steel guard	C	6	100/6.9 E5	553	435	406
5.70 × 8	Highway tread	B	4	60/4.1 E5	562	444	415
5.70 × 8	Highway tread	C	6	90/6.2 E5	689	562	526
5.70 × 8	Steel guard	D	8	100/6.9 E5	844	667	621

A Product Section of
The Material Handling Institute, Inc.
1326 Freeport Road
Pittsburgh, PA 15238

Acknowledgements

The authors would like to acknowledge the following groups providing general information used:

American Lead Association
Eagle Vehicles, Inc.
Kalamazoo Manufacturing
Nordskog Electric Vehicles
Taylor Dunn Manufacturing Company
The Material Handling Institute, Inc.
The Tire and Rim Association, Inc.
Trojan Battery

7.4 MOBILE HYDRAULIC CRANES

Charles A. Isenberger

7.4.1 Definition

The typical modern mobile hydraulic crane, designed for industrial applications (Fig. 7.4.1) is characterized by three principal features:

1. A multisection power telescoping boom, used to lift objects for materials transfer in a pick-and-carry operation.
2. A basic front-wheel-drive power train combined with rear-wheel steer, viz-a-viz the customary 4-wheel drive and steer configuration found in hydraulic cranes for rough terrain operation.
3. Compact size for performing maintenance in tight quarters.

Work assignments given to telescoping-boom industrial cranes cover a broad range of materials handling and plant maintenance requirements. Jobs performed by hydraulic cranes for industry tend to fall into three general categories: routine pick-and-carry work in which objects such as product components or stacks of material are lifted and transferred from one production operation to another; lifting and placing material in a repetitive work-cycle operation, for example, in loading and off-loading trucks or railcars; and in filling a variety of typical plant maintenance needs—installation, disassembly, replacement, repair of large machinery, handling overhead ductwork, air conditioning systems, heating units, and so on.

Fig. 7.4.1 Typical configuration of a modern mobile hydraulic industrial crane.

In general, materials handling assignments for mobile hydraulic cranes involve the lifting, carrying, and placing of items or components which cannot be handled effectively by any other materials handling system.

For example, in cases where the size, shape, or weight of a component or object restricts its being palletized for forklift truck handling, the optimum approach is to handle it with a pick-and-carry industrial yard crane. There are also occasions when suppliers deliver materials that were palletized at the source to expedite handling, and where the receiver does not have or cannot justify a fork truck; and therefore the load is handled by an industrial crane.

To enhance lifting versatility, special crane hook attachments are available such as counterbalanced forks for handling loaded pallets, "C" hooks, and drum tongs.

Other instances might involve the need to lift the load over tight-quarter obstructions. Then a mobile telescoping-boom crane is the optimum solution to the handling problem.

Materials and odd-shaped objects that do not lend themselves to handling either by forklift, roller dollies, belt conveyors, or overhead monorail are therefore handled best by a mobile crane. These might include long lengths of pipe or conduit, I-beams, stacks of lumber, fabricated components in irregular shapes and sizes, large, bulky, and unwieldy foundry castings—loads that can be readily lifted and placed by cranes using riggings of chains, slings, spreader bars, or lifted by built-in eye-rings for crane hoisting.

Table 7.4.1 is an alphabetical listing of some of the types of industries currently using mobile hydraulic front-wheel-drive cranes for day-to-day handling requirements.

Table 7.4.1 Applications for Industrial Telescoping-Boom Cranes

Air cargo services
Air conditioning manufacturers/contractors
Air pollution device mfrs.
Air product suppliers
Aircraft operations manufacturing/maintenance
Alloy metal plants
Aluminum fabricators
Anodizing operations
Antennae manufacturers
Art metal works
Asphalt mixing plants
Auto engine rebuilders
Automobile machine shops
Automobile manufacturers

Bakery equipment plants
Bank equipment manufacturers
Barge lines/barge builders
Barrel and drum manufacturers
Bleacher/grandstand rentals
Blower system manufacturers
Boat builders and dealers
Boiler manufacturers
Brass foundries
Brewery maintenance
Building supply houses
Buildings, prefabricated
Burial vault mfrs/dealers
Bus body manufacturers

Cement companies
Chemical manufacturers
Chime and bell factories
Clay products plants
Coal handling equipment manufacturing/maintenance
Coal mining companies
Compressor manufacturers
Concrete block plants
Concrete (ready mix) plants
Condenser/heat exchangers
Conduit suppliers
Construction equipment manufacturers/dealers
Contractors equipment rentals
Conveyor manufacturers
Culvert manufacturers
Curtain wall manufacturers
Cylinder manufacturers (air and hydraulic)

Die block repairers
Die casting plants
Display production shops
Distilleries
Drill manufacturers
Drilling contractors

Electric motor manufacturers
Electric power utilities
Electrical contractors
Electrical equipment builders
Electroplating plants
Elevator manufacturers
Engine manufacturers/dealers
Engine rebuilding companies
Excavating contractors
Exposition erection services

Fan and propeller manufacturers
Farm equipment firms/dealers
Fastener firms (bolts, etc.)
Fence manufacturers/dealers
Fire escape manufacturers
Food product equipment firms
Forge equipment manufacturers
Forge operations
Foundries
Furnace manufacturers/dealers
Furniture manufacturers

Table 7.4.1 Continued

Galvanizing plants
Gas companies
Glazing operations

Heat exchanger manufacturers
Heat treating plants
Heating installation firms
Hydraulic equipment producers and distributors

Iron works (ornamental)

Laundry equipment companies
Light and power plants
Lumber mills, lumber yards

Machine shops
Machine tool distributors
Machinery movers/installers
Marble quarries and dealers
Marinas/boat storage yards
Marine equipment suppliers
Materials handling equipment manufacturers and dealers
Mechanical contractors
Metal cleaners, finishers
Metal rolling, forming firms
Metal stamping operations
Meter manufacturers
Millwrights
Mining companies
Mold makers
Monorail sales and service
Monument manufacturers
Motion picture productions
Motor freight lines
Municipal maintenance depts.

Nylon manufacturers

Oil well suppliers
Oxygen suppliers

Paneling firms/suppliers
Paper manufacturers
Pile driving companies
Piling manufacturers/dealers
Pipe bending/fabrication firms
Pipe coating companies
Pipe fitting manufacturers
Pipe line contractors
Pipe yards (steel and concrete)
Plumbing contractors
Port authorities
Power transmission gear firms
Precast concrete plants
Pump manufacturers/dealers

Railroads
Refractories
Refrigeration equipment firms
Rental equipment storage yards
Restaurant equipment sales
Riggers and haulers
River freight terminals
Road building equipment/sales
Rolling mill machinery firms
Roof deck manufacturers
Roof truss manufacturers
Roofing installers

Safe and vault companies
Salt mines
Salvage operations
Sand and gravel suppliers
Scaffolding builders/dealers
Scales sales/repair firms
Scrap metal dealers
Salvage operations
Septic tank manufacturers
Sewage treatment equipment firms
Sewer pipe manufacturers (clay, concrete, and iron)
Shipyards and drydocks
Sign builders
Slag producers and processors
Smelters and refineries
Smoke abatement equipment firms
Snow removal equipment firms
Stack manufacturers
Stair builders
Steam fitters
Steel brokers
Steel distributors/warehouses
Steel fabricators
Steel processing operations
Steel production mills

Tank manufacturers (metal and fiberglass)
Telephone companies
Tire dealers
Tool makers
Tractor (crawler) dealers
Trailer manufacturers
Truck body builders

Wire and cable manufacturers
Woodworking equipment companies

7.4.2 Capacities and Performance Characteristics

Mobile hydraulic cranes for industrial applications are available in a fairly broad range of load-lifting capacities ranging from 2 tons (1814 kg) to a maximum of 35 tons (31,752 kg), although the selection of makes and models is relatively limited.

Available units tend to have generally similar configurations in design and componentry—in much the same way as with forklift trucks. Major differences are mostly in overall size, lifting capacity, selection of componentry and power—plus, as is expected, some innovative or exclusive feature found in one make or another.

An almost universally common feature is two-wheel (rear-axle) steering combined with single- or dual-wheel front-axle drive. All makes have hydraulically extendable and retractable multiple-section telescoping booms to facilitate reaching under or over obstacles to position loads. The telescoping-boom characteristic is the single most important advantage of the hydraulic yard crane over the conventional fixed boom (cable operated) "wagon crane" still occasionally seen operating in a scrap-metal yard.

Figure 7.4.2 illustrates and describes principal componentry found in a more or less typical mobile hydraulic yard crane.

For the most part, modern industrial mobile cranes are designed to have low profiles, permitting easy access through plant and shop doorways for handling of production materials. In the case of some models, they are low enough to make them readily adaptable for use in performing turnaround work in hydrocarbon and petrochemical refinery maintenance; space is tight and required lifts often mean projecting loads between or over/under obstacles using telescoping boom action.

7.4.3 Maneuverability

The modern front-wheel drive industrial hydraulic crane can be maneuvered into a lift location practically anywhere in areas wide enough for it to travel, both inside and outside industrial facilities—plants, warehouses, inventory storage areas, shipyards, and freight and dockside operations.

In general, the turning radii of industrial hydraulic cranes are short enough to negotiate 90° turns in most yard or plant aisleways (Fig. 7.4.3). Mobility and traction are excellent on typical industrial paved traffic lanes, and most units are equipped with mud and snow tires on drive axles, delivering adequate traction for most unpaved storage yard areas. The radius of the largest unit's turning circle is 26 ft 2 in. (8.00 m).

Industrial travel areas are generally located on firm, level ground with a surface of asphalt, concrete or brick paving, or compacted earth and gravel. Given this type of travel surface firmness, it would be a misapplication of overall capability to acquire or assign a large lug-tired four-wheel drive and steer rough-terrain hydraulic crane to perform routine hard-surface pick-and-carry materials handling. Optimum use for the rough-terrain telescoping-boom crane is on the unimproved ground of construction jobsites, mining operations, cross-country powerline tower erection jobs, and so on.

7.4.4 Speed and Gradeability

In addition to overall mobility and maneuverability, gradeability is an important factor in selecting an industrial crane suitable for one's handling requirements. If the crane is expected to carry loads

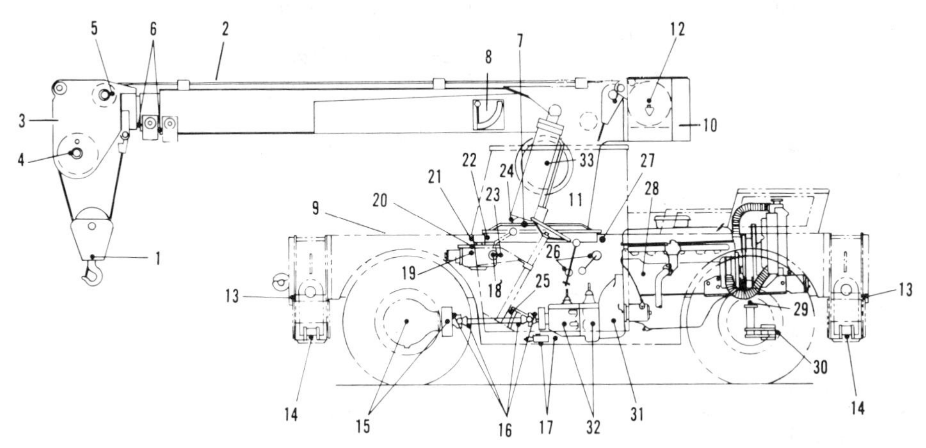

Fig. 7.4.2 Typical componentry of U.S.-sourced mobile telescoping-boom cranes. 1, Hook block; 2, telescoping boom; 3, boom nose; 4, nose sheave; 5, idler sheave; 6, boom wear pads; 7, boom elevation cylinder; 8, boom angle indicator; 9, carry deck; 10, counterweight; 11, turntable; 12, hydraulic hoist; 13, outrigger assembly; 14, outrigger pad; 15, differential; 16, universal joints; 17, clutch/brake cylinders; 18, hydraulic control shaft; 19, swing bearing; 20, gearbox bearing; 21, swing gearbox; 22, drive pinion; 23, control levers; 24, hydraulic oil reservoir; 25, foot control shafts; 26, shift linkage; 27, hydraulic oil filter; 28, engine crankcase; 29, axle pivots; 30, steering axle; 31, clutch; 32, transmission; 33, hydraulic hose reel.

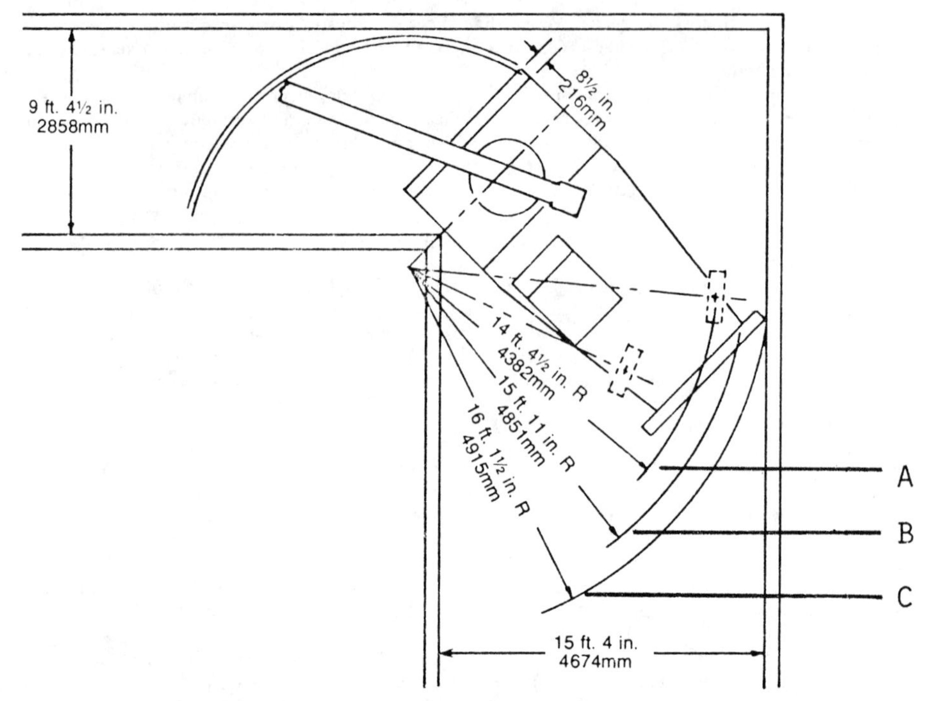

Fig. 7.4.3 Turning diagram for a rear-axle-steer materials handling hydraulic crane showing 3 radii: A, turning radius arc; b, curb radius arc; c, overall tailswing clearance.

up loading ramps or other inclined surfaces, specifications should be checked for maximum drawbar pull and gradeability. If expressed in the manufacturer's specifications, these factors will probably be shown with speed ranges relative to the unit's gear ratio. Refer to Fig. 7.4.4.

Several manufacturers offer a selection of power trains (options that should be considered governed by total crane requirements), engines, transmissions, and types of fuel. Among available units there is a wide variety of transmissions involving both manual and foot pedal gear changing: shuttle-shift, power-shift, automatic and clutchless with manual selection. Each of these factors influences the choice of a given unit as it relates to work requirements of speed and gradeability.

PERFORMANCE—STANDARD ENGINE & TRANSMISSION

Axle ratio: 6.167:1 Auxiliary Gear Box: 2.41:1

Gear Range	Ratio	Speed (mph)	Drawbar Pull (lb)	% Gradeability
1st	6.324:1	3.5 (5.6 km/hr)	6950 (3152 kg)	26.0
2nd	3.092:1	7.2 (11.6 km/hr)	4802 (2178 kg)	17.8
3rd	1.686:1	13.2 (21.2 km/hr)	2413 (1094 kg)	8.8
4th	1.000:1	22.2 (35.7 km/hr)	1248 (566 kg)	4.4

Note: Speed and gradeability performance measured on concrete. Maximum drawbar pull and gradeability limited by torque required to slip wheels.

Fig. 7.4.4 Example of exerpt from mobile industrial crane specification showing relationship of drive functions to power ratios.

7.4.5 Boom Rotation

Most available industrial hydraulic cranes have turntable swing gears that rotate the boom-and-hoist superstructure with a continuous 360° swing capability. A few models have standard swings restricted to perhaps 270° with continuous turntable rotation optional.

A smooth boom swing is a vital characteristic in the selection of any hydraulic materials handling or maintenance type crane. The function should be free of jerking or surging motion. Any sudden, unpredictable surge in the boom swing could be dangerous if it should result in loss of load control.

7.4.6 Stabilizing Outriggers

Of the approximately 13 models of U.S.-sourced mobile industrial hydraulic cranes available, all but four offer as either standard or optional equipment, extendable and retractable outriggers (refer to Item 14 in Fig. 7.4.2) for increasing stability when lifting with the boom swung through 360° rotation.

Therefore, the factor of outriggers becomes another consideration to be resolved in the final selection of an industrial hydraulic yard crane.

A key to the decision will be how the crane is operated. If the boom is centered over the front end of the crane in pick-and-carry operation and no lifting over the side is involved, outriggers are not necessary.

If, however, in routine materials handling assignments the crane will be expected to swing from front-centered travel position 90° to either side to pick loads from inventory rows or warehouse storage racks, for the sake of stability and overall safety, the unit should be equipped with outriggers, both front and rear. Specifications should also be checked for the amount of *outrigger spread*—a dimension that becomes important if the crane is to be operated in restricted-space areas, see Fig. 7.4.5.

Optional on at least two of the available cranes are dual-position hydraulic outriggers which can be extended either vertically where space is limited, or obliquely for an extended stance where tight-quarter lifting is not a factor.

7.4.7 Carry Decks

Aside from capacities for carrying loads suspended from the crane hook, a number of available industrial type hydraulic cranes also feature carry decks—front and/or side chassis areas designed to be used

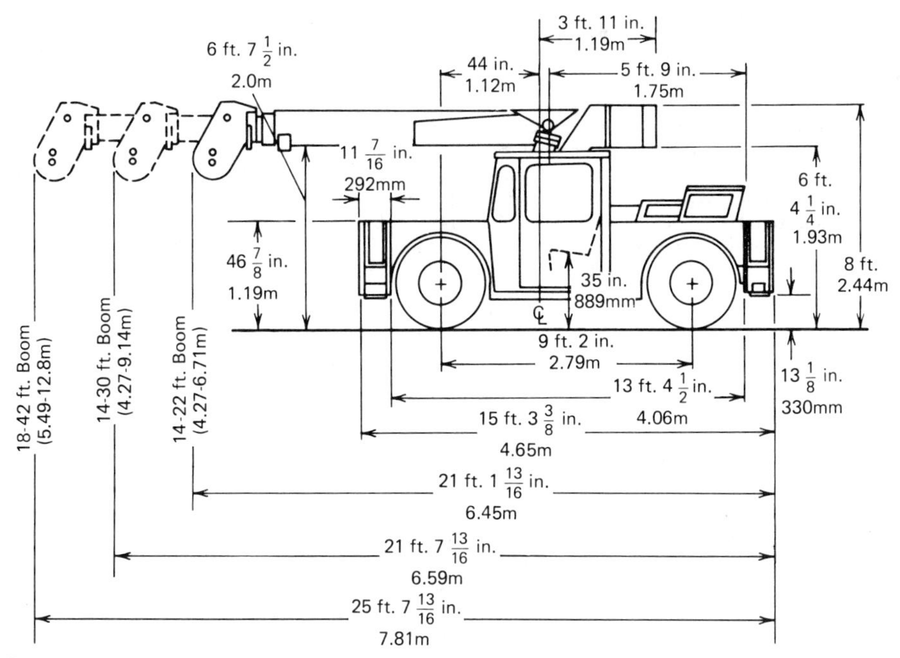

Fig. 7.4.5 Dimensions of a typical mid-size mobile hydraulic industrial crane showing lengths of optional booms in retracted attitude.

for transporting materials, typically lengths of pipe, bundles of lumber, and so on. These carry-deck areas vary of course with the size and capacity of the mobile crane, ranging from just a few square feet at the small end of the selection to some 130 ft^2 (12.09 m^2) on the largest unit available.

In some cases, the capacity of the carry deck is equivalent to that of the crane when used for pick-and-carry work. Example: in the case of the 35-tonner just mentioned, the deck can accommodate an evenly distributed load of 50,000 lb (22,680 kg), which is the same as the crane's maximum travel hook-load capacity during pick-and-carry handling—25 tons (22.6 mt). The 35-ton (31.75 mt) rating is with outriggers extended.

7.4.8 Hook and Boom Attachments

Industrial crane manufacturers and other sources make available a variety of attachments to enhance the mobile yard crane's versatility and profitability.

Hook attachments include such items as counterbalanced lifting forks for handling palletized materials; "C" hooks for concrete pipe sections, and so on; drum tongs, grapples, clamshell buckets, and even magnets, powered by an on-board electric generator kit, for steel plate handling or for operating in a scrap-metal area.

Aside from the expected availability of jib extensions for augmenting boom reach capability, some makes of cranes can accommodate boom attachments such as aerial work platforms (without remote crane-operating controls, however) and fiberglass man-baskets for elevating personnel to overhead work locations. These devices are generally self-leveling relative to the raised boom and are stabilized by a locking lever.

7.4.9 Hydraulically Powered Hoists

For materials handling operations, in which experience has been mainly with forklifts, introduction of hydraulic cranes with the added functions of hydraulic hoists means getting used to job requirement considerations such as hoist drum (spool) capacity, line pull, line speed, how many parts of line, and the related decision on the use of hookblocks and headache balls.

For example, a crane to be used for materials handling during construction on a plant expansion project might be hoisted from floor to floor for pick-and-carry handling as building erection proceeds. Working on an upper floor level, the crane is capable of hoisting materials from the ground *provided* it is equipped with sufficient wire rope on a sufficiently large hoist drum.

7.4.10 Operator Cab Controls

Different makes of industrial mobile hydraulic cranes obviously have different arrangements of operator cab gauges and controls, but Fig. 7.4.6 provides identification of those found in a typical, mid-size yard-type hydraulic crane.

One sees that the operator has access to console gauges showing engine oil pressure, engine coolant temperature, engine hour meter (if not in the control cab, then in the engine compartment), fuel level, and voltmeter. These are all essential to monitoring "vital functions" of the engine for optimum performance. Other gauges, not shown, might include transmission oil temperature, brake air pressure (if the unit is equipped with air brakes), and tachometer.

For controlling the standard functions of a mobile hydraulic crane, the operator has before him levers for operating hydraulic directional control values for these crane functions: boom telescope (out/in), boom elevation (up/down), turntable rotation (boom swing), hoist drum rotation, and (if the crane is so equipped) hydraulic outriggers (out/in).

7.4.11 Devices to Aid Operator

Operator aids that, depending on the make of crane, are either standard equipment or offered as extra-cost options include the following:

Anti-two-block warning system: A device designed to alert the operator to an impending contact between the load being lifted and the crane boom nose during hoisting. Such a condition, if not interrupted, could cause the hook to separate from the hoist line (or the line itself could break) releasing the load. Alerted by the device, the operator can prevent the impending contact by reverse action: lowering the load. The warning is an audible and visual signal (buzzer and flashing light) in the control cab.

Load-moment indicator system: An electronic sensing, monitoring, and reporting device that provides both a console-mounted visual (dial read-out) and audible (buzzer) warning to the operator should the combined factors of hook load, boom angle, and working radius create an impending overload

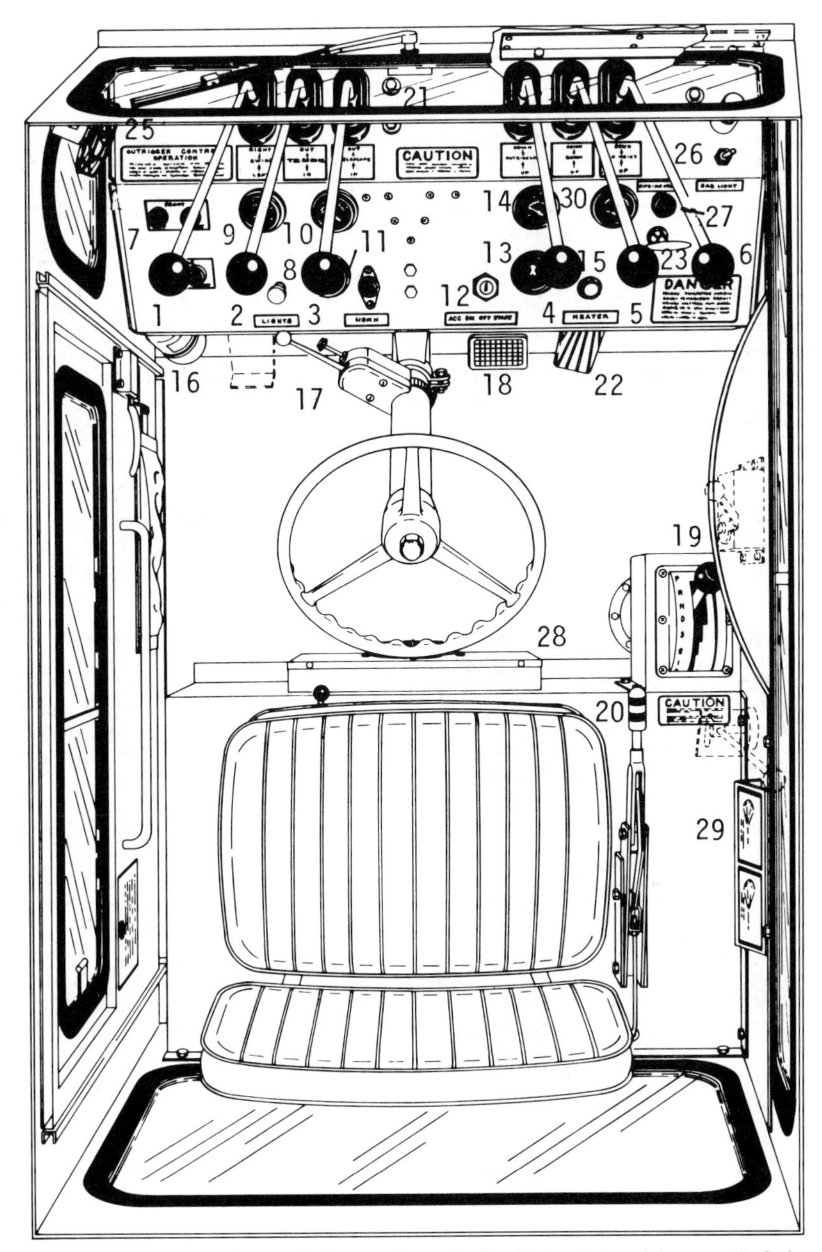

Fig. 7.4.6 Cab controls and indicators for typical mid-size industrial crane: 1, Swing control lever; 2, Mid telescope control lever; 3, Fly telescope control lever; 4, Outrigger control lever; 5, Boom lift control lever; 6, Hoist control lever; 7, Outrigger control panel; 8, Lights switch; 9, Voltmeter; 10, Engine oil pressure gauge; 11, Engine coolant temperature gauge; 12, Ignition switch; 13, Hourmeter gauge; 14, Fuel quantity gauge; 15, Heater switch; 16, Fire extinguisher; 17, Turn signal switch; 18, Foot brake; 19, Transmission 4-speed lever; 20, Park brake lever; 21, Windshield wiper switch; 22, Accelerator pedal; 23, Choke switch; 24, Horn switch button; 25, Defroster switch; 26, Cab light switch; 27, Engine stop switch; 28, Heater box; 29, Load chart; 30, Converter oil indication switch.

condition. Some LMI systems are available that also automatically neutralize the pertinent crane controls, thus preventing the impending overload condition from developing.

Back-up alarm: Electronic pulsating-signal alarm as well as rear-mounted warning lights which are activated automatically when the crane is shifted into reverse.

7.4.12 Operator Training Aids

Most crane manufacturers supply with the crane some sort of operator's handbook which also generally contains information and advice on safe operating practices.

In the interests of protecting lives, property, and the owner's investment in the equipment, this information should be thoroughly absorbed by anyone involved in the operation and maintenance of the mobile industrial crane.

The power of a modern hydraulic system is nothing less than awesome in its potential force. With that in mind (as is only prudent with any hydraulically powered equipment but especially mobile cranes), strict orders should be issued to all personnel and administered without exception:

All manuals, operating instructions, and so on, provided with this equipment shall be read and understood by anyone assigned to operate it, BEFORE attempting to start it!

Operator training programs should be instituted using the manufacturer's materials as guides. Everyone should be impressed with the fact that a mobile hydraulic crane is in no way as simple to operate as an automobile, and that basic operation training is a MUST. It follows that any piece of mobile equipment is only as safe as the person who takes over the controls. A mobile crane cannot think and use reason; it responds only to the directions given it by the human operator manipulating the controls.

That is why equipment supervisors should emphasize the importance of having any potential operator of any mobile hydraulic crane read and understand the training material provided before entering the crane control compartment.

Prospective operators should also be able to interpret and comply with the manufacturer's published lifting capacity chart and range diagram (see Fig. 7.4.7) for any designated mobile hydraulic crane assigned to a lifting job. They should become familiar with the individual components of each machine and be able to discern during operation any abnormal functional characteristics.

7.4.13 Hoist Line Reeving

Manufacturer's recommendations for hoist line reeving should also be understood and applied. Depending on the gross weight of the load and hook attachments, the lift may require reeving the line on a hook block using multiple parts of line rather than through a single-line "headache ball." In all cases the lifting capacity chart for the crane should be consulted to determine both capabilities and restrictions. Lower-capacity cranes do not normally come equipped with enough boomhead sheaves to accommodate more than two parts of line. It follows that for major lifts, higher-capacity cranes will be used with a hoist line hook block reeved with multiple parts of line. Figure 7.4.8 indicates the proper method for reeving a hook block with four parts of line.

7.4.14 Boom Assembly Nomenclature

Personnel involved with the scheduling and operation of industrial cranes should make a practice of describing components by the terms generally used throughout the crane manufacturing field. For example, in identifying sections of power-telescoping booms, when there are merely two boom sections, the lower section is the "base" and the outer is called the "fly" section. If there are three sections, the center section is the "mid," and with a four-section boom, the two center sections are in "outer mid" (next to the fly) and "inner mid" (next to the base section). In the construction industry, giant hydraulic cranes may have as many as five sections, but those designed for industrial applications are seldom equipped with more than three.

7.4.15 Hand Signal Coordination

Personnel assigned to assist mobile crane operators in both pick-and-carry and load-lifting and placing assignments must be able to convey instructions to the operator by using universal hand signals for desired crane functions such as lower boom, hoist hook, extend boom, and swing.

To avoid confusion these signals should be familiar to everyone working with the materials handling operation. They should be conveyed clearly so that the operator has no doubt about what load movement is required. If loads are handled in a pick-and-carry operation, they often need to be controlled by a tagline—a rope attached to one end of the load to minimize swinging during transit and placement.

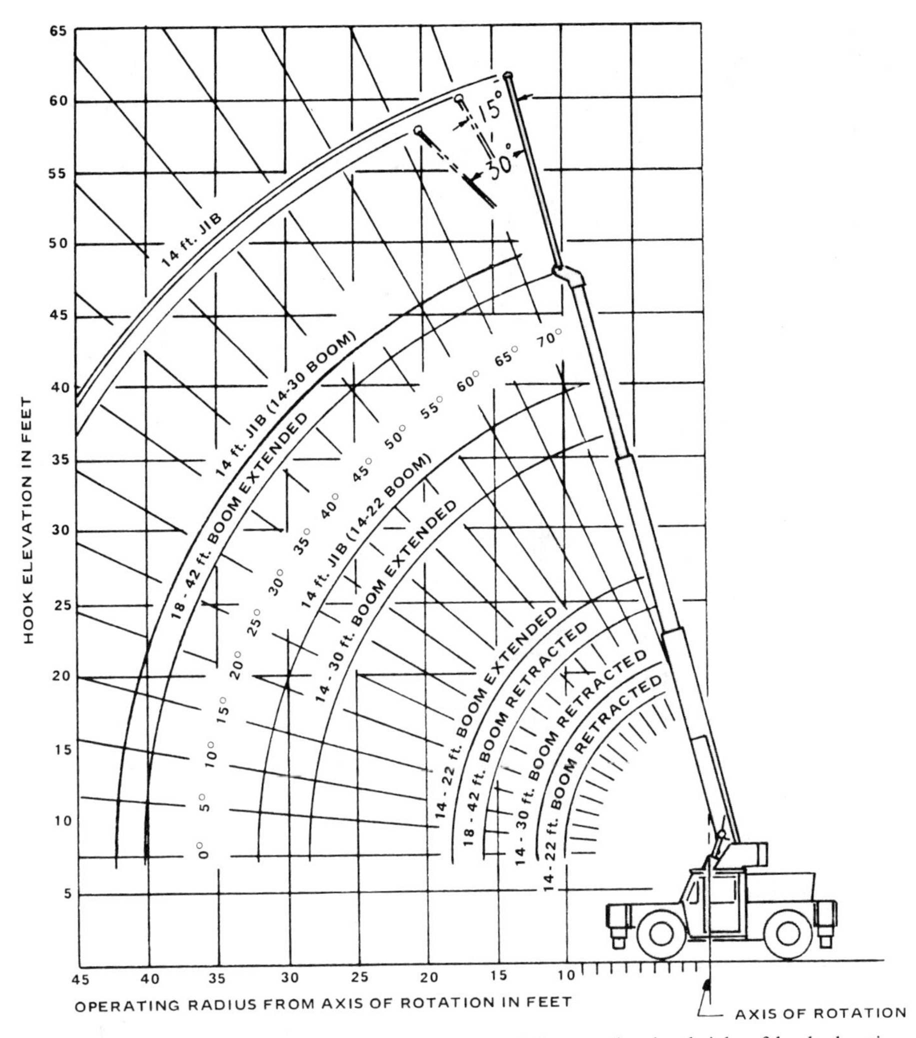

Fig. 7.4.7 Typical range diagram for a mobile industrial crane showing height of hook elevation at various boom lengths and radii (measured from center line of rotation on the crane superstructure).

Signals have been standardized by the American Society of Mechanical Engineers and are reproduced for reference in Fig. 7.4.9. A qualified signalman should be available at all times, but especially under the following conditions:

Work is in the vicinity of overhead power lines.

The operator cannot clearly see the load at all times.

The operator cannot clearly see travel path for crane.

Ground personnel are working in the area of lift.

Common sense is very often the most valuable ingredient in the safe, sound, and profitable operation of mobile hydraulic cranes.

Materials handling engineers would do well to explore the potential for time-saving, labor-saving work inherent in mobile telescoping-boom cranes. Considering their capability of handling such a wide variety of material handling assignments, these workhorses deserve to be seen and evaluated on the basis of their maneuverability and all-around load-handling versatility.

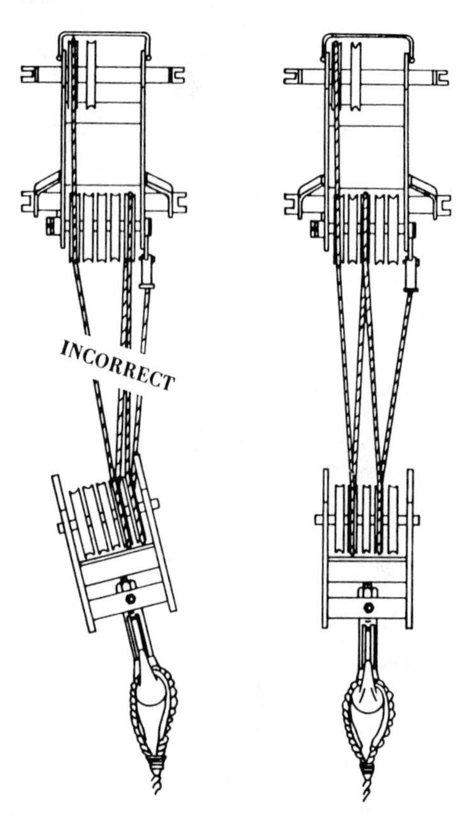

Fig. 7.4.8 Hook block must be reeved with wire rope so that device is balanced below boom nose. Illustration shows right and wrong routes for reeving four parts of line.

7.5 MAINTENANCE VEHICLES

A. J. Cason

7.5.1 Basic Considerations

Selecting the proper high-reach maintenance equipment for industrial applications is a matter of asking questions of oneself, co-workers, and vendors. A good starting point is to conduct a maintenance audit of all overhead lift needs. Answers should be sought to the following questions:

How high?
How often?
How critical? (How much is it worth?)

Start by listing all the overhead jobs maintenance crews will be expected to handle, including the following:

Cleaning
Welding
Crane maintenance
Electrical work
Painting
Ductwork
Plumbing
Lamp replacement
Production
Conveyor maintenance

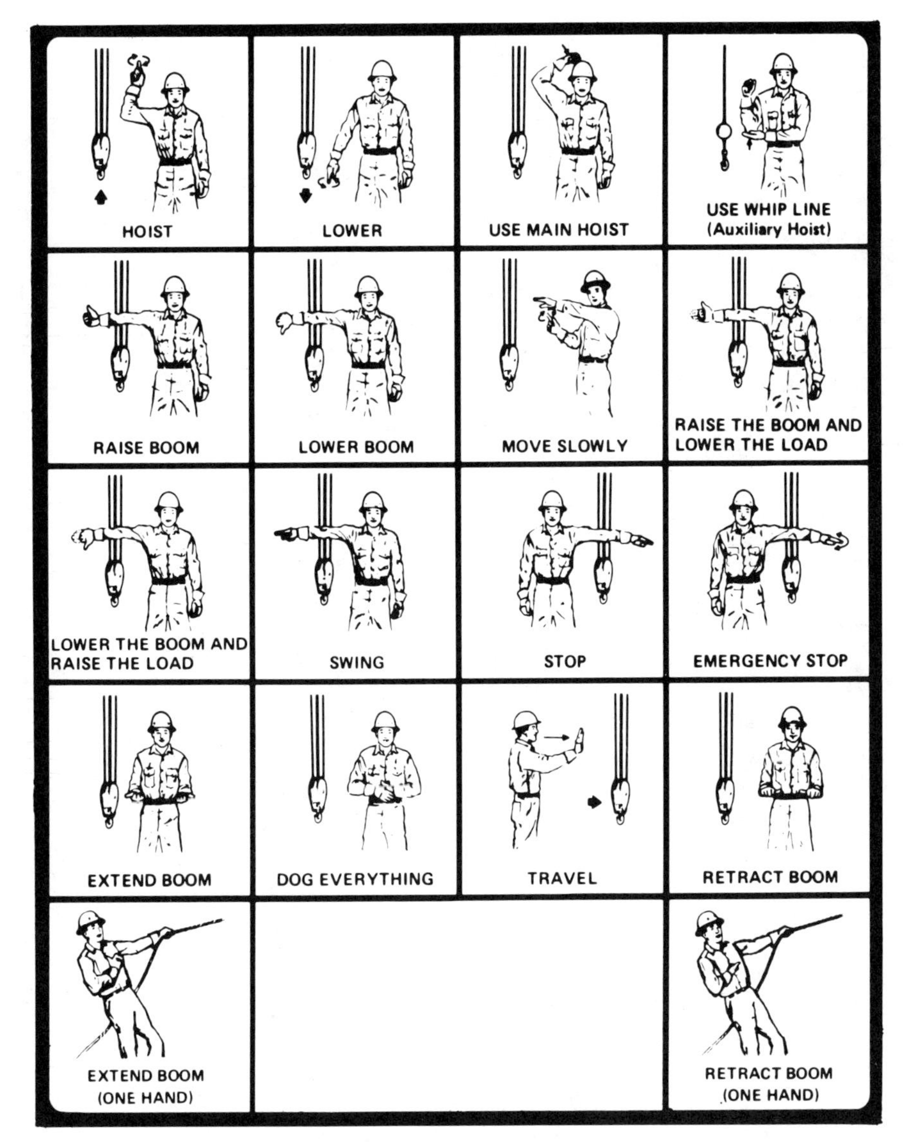

Fig. 7.4.9 Standardized hand signals for conveying to crane operator desired hook-load movements.

Application survey forms are available from manufacturers to assist in the audit procedure. All specific tasks should be listed on such a form (Fig. 7.5.1). Frequencies of different jobs should be estimated. Will they be performed routinely on a regularly scheduled basis? On an emergency, or on an "as-the-need-arises" basis?

What other departments or services will be able to use new high-reach equipment? The answer here can be of great significance when the need and cost can be spread among several departments.

Where will proposed new equipment be used? Inside? Outside? What kind of travel surfaces are available? Answers to these questions will help determine the kind of running gear that will be required.

How high must the equipment go? What is the highest point in the plant? Will the equipment be used to service this area? How frequent is this job? A careful study of height requirements helps prevent overbuying. It may be economical to use short-term rentals for infrequent maximum-height job assignments.

WORK PLATFORM REQUIREMENT SURVEY

Date ________

Manufacturer ________

Individual Contacted ________ Phone ________

Address ________

City ________ State ________ Zip ________

A. Maintenance applications (inside)
   ( ) Lighting fixtures
   ( ) Overhead electrical/mechanical systems
   ( ) Installation of new equipment
   ( ) Heating, ventilation, air conditioning
   ( ) General building/structure
   ( ) Painting
   ( ) Piping
   ( ) Ceiling
   ( ) Cranes, conveyors, other material handling equipment
   ( ) Process/production equipment
   ( ) Sprinkler system
   ( ) Other

B. Maintenance application (outside)
   ( ) Building exterior
   ( ) Roof
   ( ) Cooling tower
   ( ) Erection projects
   ( ) Tree trimming
   ( ) Parking-lot lights
   ( ) Other

C. Other possible applications
   ( ) Order picking
   ( ) Inventory stocking
   ( ) Storage area
   ( ) Other

D. Platform/bucket size desired
   Width ________ Length ________ Diameter ________

E. Working height required
   Minimum ________ Maximum ________

F. Lifting capacity desired
   ________ lb

G. Types most suitable for in-plant applications
   Scissors ______ Boom ______ Mast ______ Articulated arm ______

H. Size of units that will fit majority of plant doorways
   Height ________ Width ________ Length ________

I. Power source preferred
   Electric ______ Gasoline ______ Propane ______ Diesel ______

J. Other features

Fig. 7.5.1 Application survey form.

What size platform is needed? Analysis of specific plant jobs can determine the answer to this question. How many people will the job require? What kinds of tools and materials will they have to take with them? Again, beware of overbuying. Too large a platform can mean a base unit that presents difficulties in moving it around the plant.

If there are frequent jobs requiring only one operator, plant needs may be better served by employing several smaller units strategically located throughout the facility. This is often the case in installations with high concentrations of conveyor lines.

How much platform capacity is needed? Estimate the weight of crew members and their tools. Calculate what added materials they may need. The most popular capacity is between 750 and 1000 lb for two-person-sized platforms. One-person platforms generally have a 300–500 lb capacity.

Minimum dimensions and clearances can have an important bearing on any kind of unit ultimately selected. Check all door openings, conveyor heights, aisle widths, and turning radii at cross aisles and in storage areas. These dimensions will dictate the size of device that can be used within the physical limitations of the plant.

Special safety requirements. Are explosion-proof units needed for volatile environments? Or is heavy-duty equipment required for prolonged or extensive duty cycles? If so, the parameters covering particular job requirements must be established.

How about mobility? What method of travel is right for the installation? There are three possible options, as follows:

Manual or push-around. This equipment is generally lowest in price and designed for one- or two-person operation. Look for large, easy-rolling casters or wheels to best cope with particular floor surface conditions.

Towable. These units are designed to be towed behind trucks, forklift trucks, or prime movers, at up to highway speeds. Most units for intraplant use fall into the 2½ to 15 mph speed range. Since these units do not have a sprung chassis, they work best on relatively smooth, level surfaces.

Self-propelled. A power-propelled chassis that permits this unit to be driven to and from job sites is a major feature. This is the highest-cost option, and one that raises several questions relative to size and type of power.

DC electric (battery) power is an increasingly preferred power source for inside use. It is quiet and safe, and emits no noxious exhaust fumes. There are no flammable, hazardous, or volatile fuels to worry about. For propulsion of medium- to smaller-size units, it is an ideal power source.

One of the major reasons favoring self-propelled units lies with the distance between maintenance headquarters and job sites within the plant. When distances are great or frequency of use is a factor, the self-propelled unit justifies itself in time saved. Urgency of use is another big reason in favor of being able to drive to the job.

Once at the job, the ability to control movement from the platform can be desirable. A self-propelled unit provides faster access to most overhead jobs.

On the other hand, self-propelled units can be difficult to maneuver in confined areas. Negotiating narrow aisles becomes a time-consuming and frustrating experience for all but the most expert operators. Areas congested by machinery and people can be doubly hazardous. Such conditions present an operating hindrance and a high potential for accidents. Make doubly sure of dimensions and the operator's ability to maneuver before investing in self-propelled units.

Many plant facilities include ramps and grades which maintenance vehicles must be able to climb. Make sure the unit being considered has the power needed to contend with physical conditions of the facility, such as a 15% grade.

Sometimes going up means going out. Usually, overhead jobs are located in hard-to-reach places—above a new press brake, over shelving, over the loading dock, or in other nearly inaccessible locations. When this happens, servicing calls for going up and reaching out. Most vertical lift units have limited overreach capability.

When overreach becomes a major consideration, there are alternatives. Scissors manufacturers offer platform extensions that can provide up to 7½ ft of outreach. Boom-type devices have masts that reach out to distances of 80 ft or more. These devices are heavy, so that load-bearing capacity of plant floors becomes a serious consideration.

7.5.2 Economic Justification

Justifying an equipment decision calls for considering many intangibles. Some examples are:

Employee comfort
Employee safety
Reduced operator fatigue
Improved morale

Improved work quality
Increased productivity from older, skilled workers
Better overall housekeeping
Improved ability to handle emergency problems
New efficiency; lower energy costs
Lower workmen's compensation rates

Methods are available for calculating man-hour savings by comparing one type of equipment to another. Some manufacturers provide an economic survey form to help users determine the time differential between the use of self-propelled lifts and ladders or scaffolds.

A survey chart of this type (Fig. 7.5.2) calculates time factors for doing jobs at various distances

Positioning and Returning Time

Steps involved: Going to location, positioning, raising platform to proper height, extending the platform fully, retracting the platform, lowering platform to travel position, clearing the area with the Lift-A-Loft and going out into the aisle, and returning to the maintenance office.

Distance from Maintenance Office (ft)	Working Height				
	20 ft (min: sec)	25 ft (min: sec)	30 ft (min: sec)	40 ft (min: sec)	50 ft (min: sec)
100	1:38	1:51	2:04	2:30	2:56
150	1:48	2:01	2:14	2:40	3:06
200	1:58	2:11	2:24	2:50	3:16
250	2:08	2:21	2:34	3:00	3:26
300	2:18	2:31	2:44	3:10	3:36
350	2:28	2:41	2:54	3:20	3:46
400	2:38	2:51	3:04	3:30	3:56
450	2:48	3:01	3:14	3:40	4:06
500	2:58	3:11	3:24	3:50	4:16
550	3:08	3:21	3:34	4:00	4:26
600	3:18	3:31	3:44	4:10	4:36
650	3:28	3:41	3:54	4:20	4:46
700	3:38	3:51	4:04	4:30	4:56
750	3:48	4:01	4:14	4:40	5:06
800	3:58	4:11	4:24	4:50	5:16
850	4:08	4:21	4:34	5:00	5:26
900	4:18	4:31	4:44	5:10	5:36
950	4:28	4:41	4:54	5:20	5:46
1000	4:38	4:51	5:04	5:30	5:56
1100	4:58	5:11	5:24	5:50	6:16
1200	5:18	5:31	5:44	6:10	6:36
1300	5:38	5:51	6:04	6:30	6:56
1400	5:58	6:11	6:24	6:50	7:16
1500	6:18	6:31	6:44	7:10	7:36
1600	6:38	6:51	7:04	7:30	7:56
1700	6:58	7:11	7:24	7:50	8:16
1800	7:18	7:31	7:44	8:10	8:36
1900	7:38	7:51	8:04	8:30	8:56
2000	7:58	8:11	8:24	8:50	9:16
2100	8:18	8:31	8:44	9:10	9:36
2200	8:38	8:51	9:04	9:30	9:56
2300	8:58	9:11	9:24	9:50	10:16
2400	9:18	9:31	9:44	10:10	10:36
2500	9:38	9:51	10:04	10:30	10:56
2600	9:58	10:11	10:24	10:50	11:16

Industry-wide standards of time required to operate various types of access equipment have been used in the table. This table has been tabulated assuming the best possible conditions, in open accessible areas.

Fig. 7.5.2 Typical working-time factors chart.

from the maintenance office. Norms are established for going to the job, setting up ladders, going up to the job, coming down, securing equipment, and returning to the maintenance office.

7.5.3 Types of Equipment

Mobile work platforms come in various shapes, sizes, and price categories, including lightweight, one-man push-around models; manually propelled or towable scissors-type units; and large, self-propelled booms that reach out 80 ft or more. Each has a specific application here.

Sliding-Frame Elevating Platform. The sliding-frame elevating platform (Fig. 7.5.3) is the simplest, most economical version of a one-man push-around towable unit. It consists of a railed platform affixed to a frame that slides up or down on a vertical mast set on a mobile base. The platform can be raised or lowered mechanically by a hand crank winch and cable suspension system. Or, the units can be powered up and down hydraulically or by compressed air. Power sources can be AC or DC electric motors or plant air.

The mechanical hand crank version is usually the lowest cost and most dependable choice. It requires little maintenance other than an occasional greasing.

The unit must be rolled into place and cranked up to the desired working height. The operator then climbs its ladderlike structure to enter the railed platform. These units have a holding capacity of 300 lb.

Powered versions offer the operator the advantage of riding the platform up to the required work height. They can also serve as a material lift, with a lifting capacity of 500 lb.

AC-powered units offer the advantage of electrical power to the platform for use with electric tools or accessories. This feature is a very definite plus for maintenance crews. A possible disadvantage exists when power failure is the maintenance problem the crew must solve.

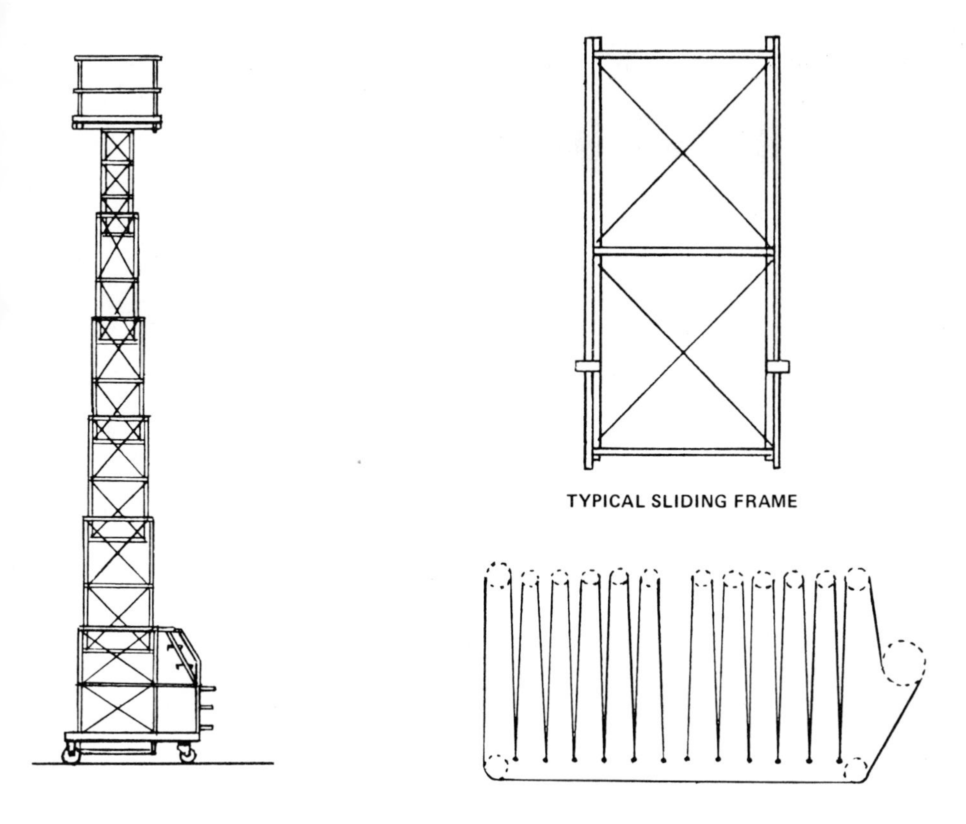

Fig. 7.5.3 Sliding-frame elevating platform.

DC-powered units can be considered ideal emergency standby units. Their only disadvantage is the inconvenience of maintaining and charging the battery.

All these units are highly mobile. They retract to pass through standard doorways. They are compact, relatively lightweight, and easily stored. Yet they can be moved and set up in minutes.

Sliding-frame units are available with working heights from 7 ft to 31 ft, and lifting capacities up to 500 lb.

Vertical Telescoping Tower Units. The vertical telescoping tower unit consists of a mast comprised of telescoping sections nested on a mobile base frame. The mast goes up or down as supporting sections are raised or lowered by an AC electric-powered drum and cable suspension system.

Vertical telescopers (Fig. 7.5.4) are designed to reduce overhead maintenance costs and manpower requirements. For working at heights to 60 ft they are safe, dependable units that can be operated by one person. No crews are required. Their use eliminates the costly erection of scaffolding ordinarily required at these heights. One person can roll the unit into place.

The operator controls height adjustment from the platform. Floor locks and outriggers provide stability when the unit is fully extended. These mobile units operate from any 120-v electrical power source.

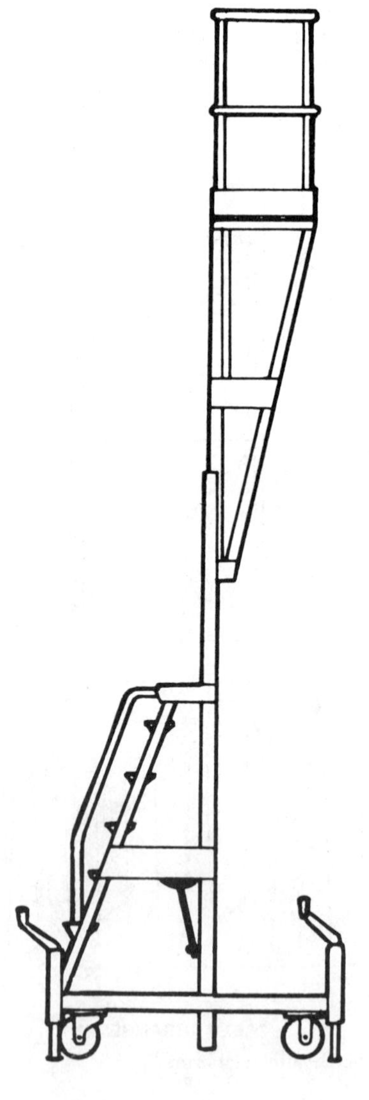

Fig. 7.5.4 Vertical telescoping tower unit.

Constructed of aluminum or steel, the vehicles have dual operating controls, cable brake systems to hold the platform safely in place whenever current is interrupted, and slack cable and secondary safety systems.

Standard bases generally have two fixed and two swivel casters. Thus, the units are highly maneuverable. Some models retract to pass through standard doorways. They are available from heights of 25–60 ft with capacities from 850 lb to 400 lb.

High-reach telescopers simplify indoor and outdoor maintenance and installation projects; save time and effort in replacing lights, electrical repair, painting, window washing, wall and fixture cleaning, and heating and ventilating service; and are useful in countless other routine or emergency assignments.

Telescopers are the most economical means of providing straight vertical placement of men and materials up to work heights of 40 ft or more.

Scissor Lifts. Over the past years the use and popularity of the scissor lift has proliferated immensely. As a design concept the scissor lift idea goes back to the days of the ancient Egyptians.

All scissor units consist of a roomy, two-man-size platform surrounded by a sturdy railing with at least 4-in. high toe boards or kick plates (Fig. 7.5.5). The platform is raised or lowered by an elevating assembly made up of two or more sets of scissor arms. The platform and lifting assembly are mounted on a mobile carrier or base frame.

Scissor lifts are rectangular-shaped machines with a four-wheeled, steerable chassis. They can be self-propelled or manually pushed around on casters.

Scissor lifts are usually powered hydraulically. Pumps are driven by gasoline, diesel, or LPG engines, or by battery power.

Such units have sufficient size and power to lift men, tools, and materials and to support them for extended periods while overhead jobs are being performed. Scissor lifts are available in a range of sizes—from 9-ft to 42-ft platform heights and lift capacities to 4000 lb.

Because scissor lifts go straight up or down they are ideally suited for direct overhead jobs. Most manufacturers, however, offer some type of accessory to extend the deck overhang or increase operator outreach with movable extension sections. Crank-out platform sections provide up to 7½ ft of added reach over the front base frames. These extension sections have weight limits of 250–300 lb.

Self-propelled scissor lifts are designed to be driven from the platform. They are highly maneuverable. The operator can drive in the lowered position (usually at speeds of 3–5 mph), fully extended (at 0.75 mph), and anywhere in between. Most are designed for operation on slabs. They operate best over smooth, level, and stable terrain.

Some manufacturers offer rough terrain units. These are generally contractor-operated machines. Most industrial plant applications are well served by DC-powered slab-type units.

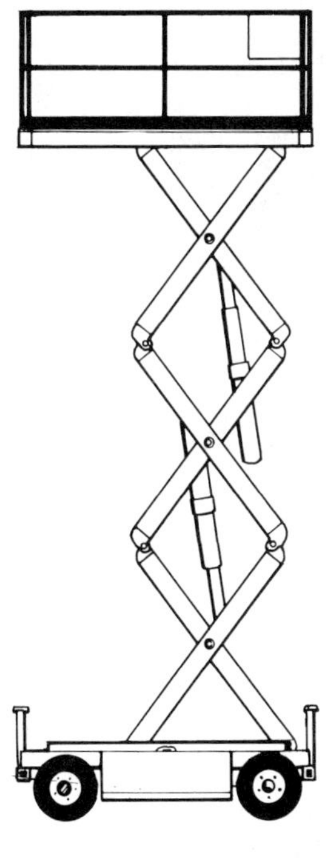

Fig. 7.5.5 Scissors lift.

Self-propelled Booms. Boom-type aerial work platforms (Fig. 7.5.6) are usually four-wheel machines with a telescoping boom that can be operated from −20° to +75° above horizontal, or they can be telescoped horizontally to a full extension at 0°. Booms can be driven from the platform and can be sustained in any position for prolonged periods of time while aerial work is performed. These units can rotate a full 360° and are ideal for maneuvering in and around obstacles.

Self-propelled aerial work platforms are available with maximum platform heights that extend 20–110 ft. Platform load capacities range from 500 to 1750 lb.

Some manufacturers of self-propelled aerial work platforms use a multiple-platform load rating system, whereby load capacities vary with the angle and extension of the boom. Other manufacturers use a load system that specifies a maximum load in any boom position. It is important that the difference between unrestricted load ratings and multiple ratings is understood before a final selection decision is made.

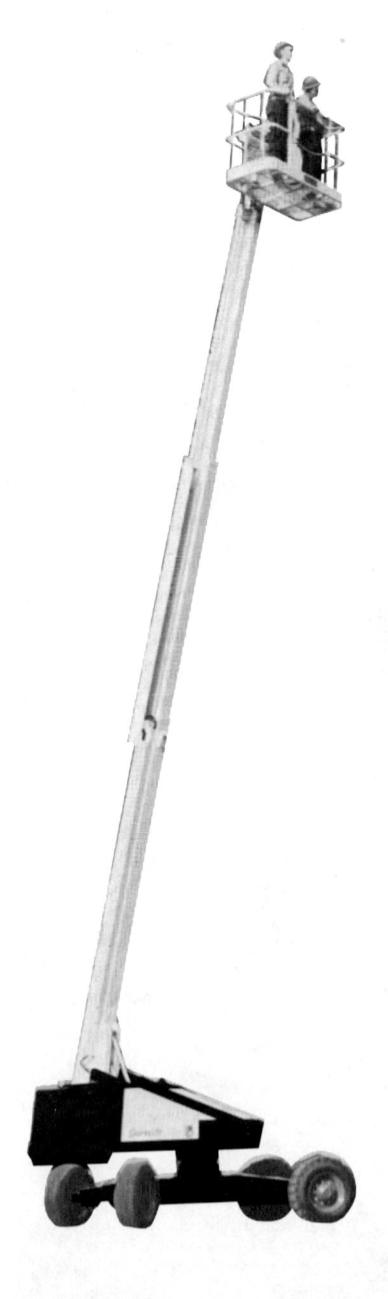

Fig. 7.5.6 Self-propelled boom.

There are three basic types of self-propelled boom machines. Most models have a counterweight that swings out beyond the width of the chassis to offset the weight of the extending boom. There are narrow-aisle models that can travel and rotate in aisles as narrow as 6 ft. They have a counterweight with a zero tail swing. Models with zero tail swing can operate with the wheels or tires directly up against a wall or other obstruction and still rotate in a 360° circle. Some models require outrigger or extendable axles to attain adequate stability at heights ranging from 60 to 110 ft. Most self-propelled aerial work platforms can be driven at full vertical extension except those models that utilize outriggers.

Like scissor lifts, self-propelled units are available with gasoline, diesel, propane, and electric power systems. Most electric self-propelled units have sufficient battery capacity to provide 8 hr of continuous operation with an 8-hr recharging period.

Self-propelled, boom-type aerial work platforms have high ground clearance, so they can operate over rough terrain more easily than most scissor lifts. Many manufacturers offer rough terrain models with large flotation tires and a drive ratio that provides high torque. A few manufacturers offer four-wheel-drive units for operation on extremely difficult terrain such as mud, sand, and loose rock. Although such units are designed as rough terrain models, their use is restricted to operation on slopes of 5° or less, in order to comply with stability requirements.

When considering boom-type units, particular attention should be given to aisle widths, turning radii, door openings, and overhead clearances. Their overall larger sizes can present operating difficulties. Indeed, storage of the unit when not in use can pose a problem where space is restricted.

All self-propelled lifts require careful driver supervision and training. Improperly used, these devices can cause serious or even fatal injuries.

7.5.4 Selecting the Equipment

At the beginning of the selection process, some of the most frequent and time-consuming overhead projects should be analyzed. This sort of analysis helps in making an accurate estimate of overall savings and benefits to be derived from the addition of lift equipment to the plant's pool of labor-saving devices.

The lowest-priced unit may not be the best buy. Crews must be willing to use the equipment and be comfortable with it.

Any equipment considered should comply with current safety standards and tests. Ask for a plant demonstration. Drive or move the equipment around and about the facilities exactly as if it were being put on the job. How does it perform? This is the opportunity to double check critical dimensions and tolerances.

Have plant crews go up on the unit. Is it to their liking? Do they feel safe and comfortable working on it? They are really the ones who have to be satisfied.

Ask about service. Is the product backed by factory warranty? Is the local dealer capable of keeping the equipment on the job and *at work?*

Does the product operate dependably? Does the manufacturer furnish operating instructions, parts, and service manuals to teach plant personnel how the unit works, and how to troubleshoot it for minor repairs or problems?

Renting should be considered as an alternative to ownership. It provides the benefits of using the equipment and none of the penalties of owning it. And rental provides an opportunity to try different types of units. Furthermore, rental fees can be written off for tax purposes.

CHAPTER 8
AUTOMATED GUIDED VEHICLES

GARRY A. KOFF

Barrett Electronics Corporation
A Subsidiary of Mannesmann Demag Corp.
Northbrook, Illinois

BRUCE BOLDRIN

Eaton-Kenway
Salt Lake City, Utah

8.1 DRIVERLESS TRAINS AND WIRE-GUIDED PALLET TRUCKS

Garry A. Koff

As recently as five years ago, perhaps only three out of 10 people involved in materials handling even knew what an automatic guided vehicle (AGV) was. Today, perhaps eight or nine people out of ten not only know what an AGV system (AGVS) is, but have either seen or been involved in the installation of one. The 1980s are going to be the decade for explosive growth in AGV systems, just as the 1970s were the period of rapid growth and maturity for automatic storage and retrieval (AS/R) systems.

It may be a surprise to some that AGV systems are not a new development; they have been around for 25 years. In the early years, there was a great deal of pioneering and only those who were innovative and willing to risk an investment on a new concept installed AGV systems. The systems were then called driverless vehicle systems, and in many respects they did much of what AGV systems do today. To understand what these systems are doing now and will do in the future, one must look at the past to see how this concept developed and how it was applied.

8.1.1 History

The earliest wire-guided AGV systems were developed in 1954. A. M. Barrett, Jr., then a young electronics engineer out of college, saw a way to apply electronic technology to industrial materials handling vehicles being manufactured by his father's company. Thus came the first generation of driverless vehicles capable of following a predetermined path and making simple route selection decisions (see Fig. 8.1.1).

Electronics Developments

AGVS technology has benefitted greatly from developments made in the field of electronics during the last 20 years. The rapid introduction of transistors in the late 1950s was readily incorporated into AGVS control system technology, helping to expand the range and improve the reliability of AGVS applications. During the 1960s the first remotely controlled AGV systems were designed and installed using discrete solid-state electronics. In one system, an onboard vehicle computer using discrete electronic components was designed to permit remote data communication with the driverless vehicle, which functioned in a system involving completely automated load transfer and remote control.

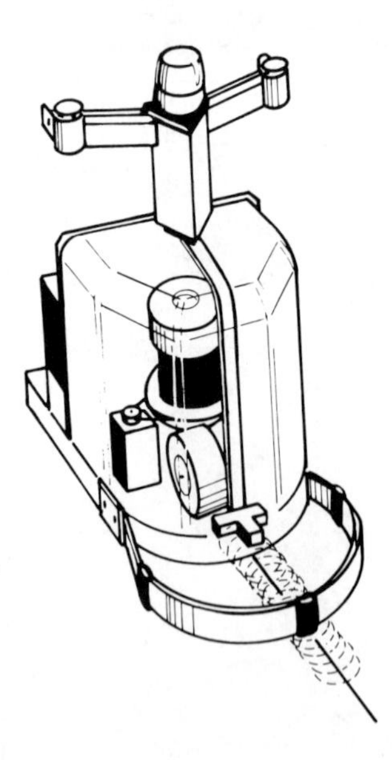

Fig. 8.1.1 Wire-guided AGVS tractor senses electromagnetic field generated around wire buried in floor.

During the 1960s and the early 1970s, integrated circuit (IC) electronic technology was adapted to AGV control systems. The IC is a small electronic chip that contains many thousands of transistor circuits. This development furthered the automation and control aspects of AGVS vehicles by permitting an increasingly sophisticated control system design. One system using such IC technology was capable of fully automatic "hands-off" control, including multiple logic sequences. The vehicles in this system were capable of determining where loads were at various plant locations, picking up available loads, and delivering them to predetermined drop-off points throughout the system.

During the 1970s, with the emergence of IC electronic technology, computer control of AGV systems became increasingly popular. Minicomputers were used to remotely communicate with vehicles in a system and to direct their movements in performing specific automatic functions. It became routine with the use of a minicomputer to design and install systems that responded to calls for service. Remote locations in a system could call for a vehicle to come to their location. The central minicomputer would dispatch an available vehicle to that location without the need for human intervention. Integration with other automated systems such as conveyor systems or AS/RS systems became more popular because the central computer technologies were similar and could readily be interfaced.

In the late 1970s microprocessors became increasingly popular in fixed equipment and onboard AGVS vehicles. Whereas the central minicomputer used software to control the overall AGV system operation, the microprocessor on each vehicle permitted it to be controlled by individual vehicle software. In systems using onboard vehicle microprocessor technology, software has eliminated a great deal of hardware within the system and onboard the vehicles. Microprocessor software contained in small chips gives vehicles sophisticated control capabilities without the previously required hardware complexity. In addition, advanced control capabilities such as real-time communications with all vehicles, continuous location and status monitoring of all vehicles, and enhanced system control options are now easily available. The use of microprocessors in AGV system design is an increasingly popular trend which is bound to continue the AGV technology momentum.

Technical support organizations were formed to facilitate interchange of information and to promote the growth of AGV systems within the United States. In 1978, the AGVS product section of The Material Handling Institute, Inc. was formed by five AGVS manufacturers.

Current Systems

In general, current systems can be divided into three basic categories: (1) the standard system with basic control electronics, (2) the advanced system with microprocessor or sophisticated control electronics, and (3) the microprocessor or sophisticated system with interface to other systems for fully automated materials handling.

Standard System with Basic Controls. This class of system is simple in concept and operation. It may involve as few as three of four "stop" stations and only one or two vehicles plus associated trailers. The total length of the guidepath is likely to be in the range of 1000–2000 ft and certainly under 5000 ft. If the guidepath is longer, the system will probably require sophisticated blocking and other controls to operate at optimum efficiency.

Advanced System with Microprocessor or Sophisticated Controls. Where a number of vehicles are necessary, and where guidepaths are longer and routings more complex, this class of system may be required. Its greater sophistication permits the use of automatic loading/unloading stations and remote programming of individual vehicles. Since there are more controls aboard the vehicles, both the control costs and the vehicle costs are generally higher. The increased number of blocking beacons in the floor, the complexity of the guidepath, and use of auxiliary guidepath wiring boost the installed cost per foot of this class of system.

Microprocessor or Sophisticated System with Interface to Other Systems. Interfacing an AGV system with other automated systems within a plant offers a number of advantages. But the increased complexity requires more engineering, more software, and additional system planning expenses, plus more sophisticated controls. However, vehicle and installation costs do not increase.

Automation Trends

The increasing need for greater productivity in manufacturing contributed to the increased acceptance of AGV systems. Labor costs, new plant construction, and increased emphasis on automation jumped AGV systems into the forefront of materials handling technology.

These AGVs have a tremendous range of automated capabilities. Many systems are called upon to go up and down ramps, open and close automatic doors, cross drawbridges, run outside, use elevators, and perform advanced automatic loading and unloading functions (see Figs. 8.1.2 and 8.1.3). These capabilities give a materials handling engineer a powerful tool to use in solving horizontal materials transportation problems.

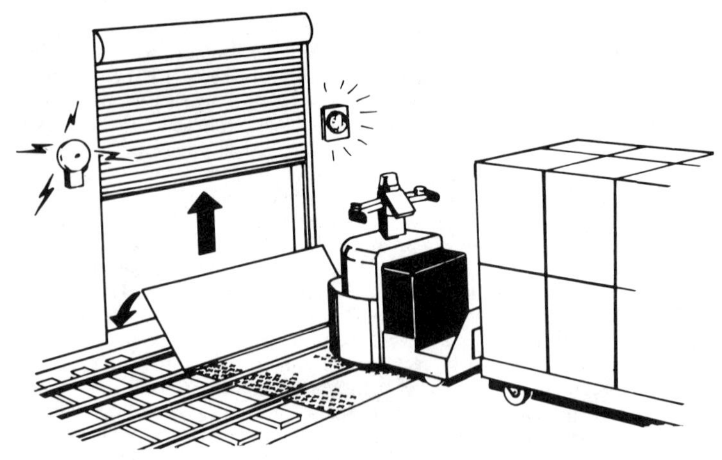

Fig. 8.1.2 Automatic controls on AGV can actuate draw bridge, door, and appropriate warning signals.

Fig. 8.1.3 AGV can pick up or drop off pallets at special floor stand.

8.1.2 Benefits

Flexibility

An advantage of the AGV system is its adaptability and flexibility. In fact, most AGV systems are installed in existing facilities without major changes in facility layout or material process flow. AGVS vehicles use existing aisles to transport materials throughout a facility. Anywhere fork trucks were previously transporting loads an AGV can work, although it is recommended that AGVS equipment operate in dedicated lanes not normally used by other vehicles or humans. Although many systems start out with a simple plan in mind, there are always changes that are made before, during, and after the system becomes functional. These changes usually are not expensive because the AGV system is relatively easy to change. Guidepaths can be rerouted easily and "stop" stations in a guidepath system can be added, deleted, or moved. Vehicles can be added to increase throughput and new system functions can be added.

Ease of Installation

The installation of an AGV system is relatively easy compared with conveyors or in-floor towlines. A guidepath can be established by simply putting a $\frac{1}{2}$-in. deep by $\frac{1}{8}$-in. wide slot in the floor. A flexible guidepath wire is then placed in the slot, which is sealed. The wire carries the guidance signal which the vehicle senses as it follows the route. A guidance sensor on each vehicle detects the small electromagnetic field radiating from the guidepath wire and provides the vehicle with steering instructions.

In most systems, magnets or small metal plates are placed in the floor at key points in the system to provide positional information to the AGV. As the vehicle passes over these codes in the floor, an onboard sensor tells the vehicle where it is in the system and what functions it should perform. These floor codes are used to designate "stop" stations within the system, decision points (places where the path splits into two paths), and other functions such as activating the vehicle's horn or telling the vehicle to change speed.

Installation time for an average 2000-ft system is generally two to three weeks. During that time, plant operations are rarely disrupted to any great extent because the saw cutting and wire laying do not require that sections of the plant be shut down. Once the saw cutting is complete in an area, it is available for other traffic until the wire is placed in the floor, an operation that takes only a few hours to accomplish.

Portability

One often overlooked advantage of the AGV system is its portability. Many plants change operations and transfer equipment to other plant sites. If an AGVS was installed in the original plant, it can be reinstalled in the new plant site.

AGV systems do not require concrete floors, although concrete is the best surface. Systems have been installed without problems on wood block, asphalt surfaces, and tile floors. Special installation precautions are required for these surfaces, however.

8.1.3 Determining the Proper Vehicle

AGV systems include a variety of wire-guided vehicles. By far the most numerous are the driverless train and wire-guided pallet truck systems.

Driverless Trains

The driverless train consists of a towing tractor that pulls trailers on which loads are carried (see Fig. 8.1.4). Driverless trains are the most efficient when:

1. A heavy flow of material must be moved between two points.
2. Several pallet loads (or equivalents) must be dropped off or picked up at intermediate locations along a route at least 500 ft long.
3. Finished work on pallets is to be transferred to storage or another manufacturing operation.

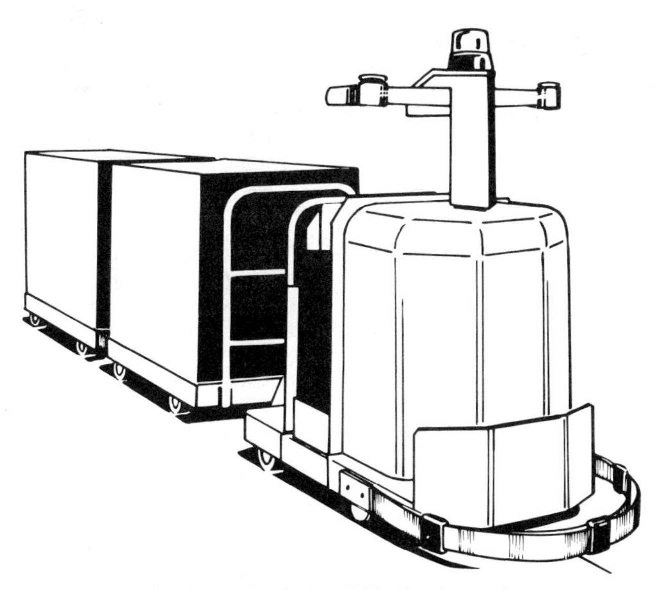

Fig. 8.1.4 Typical AGVS driverless train.

Application. Driverless trains have been workhorses in a broad range of industries for many years. In textiles, driverless train systems move spools of yarn on A-frame trailers, rolls of fabric, and boxed materials. In the paper industry, driverless trains tow trailers loaded with rolls of paper or flat stock. In the rubber industry, green and finished tires are moved with driverless trains. In general manufacturing, driverless trains are used to move incoming material to warehouse locations, and from warehouse locations to shipping docks. In many other industries driverless trains are used to move material in some phase of their operation.

Use of Driverless Trains for Bulk or Chain Moves. Besides handling bulk movements, driverless trains are highly effective for chain movements. A chain movement occurs when material is loaded onto a train for dispersal throughout a system. In this case, the driverless train can make several stops during one cycle through the system. At each stop, loads are either removed from the train or added to the train. Operators in these areas can direct the train to specific stops corresponding to the load movement requirements. Trains used in chain-type movements typically circulate a system picking up loads and delivering them with a flexible and changeable pattern.

Sizing Tractors for Train Operation. The driverless tractor must be sized for the application. Capacities in driverless tractors range from 8000 to 50,000 lb of rolling load. This assumes that the rolling load resistance of the towed trailers is 2%, so that the drawbar pulls of the tractors range from 160 to 1000 lb. To convert tractor drawbar pull to rolling load capacity, merely divide the drawbar pull by 0.02.

$$\frac{\text{Tractor pull}}{0.02} = \text{Rolling load capacity}$$

The driverless train is used mainly in systems where the load movements are at least 500 ft and the volumes are moderate. The load movements are either batch movements where the trailers are loaded with product all destined for one discharge point or distributed where the train makes several stops on a cycle through the system picking up and delivering loads. Labor savings are usually sizeable in these systems due to the elimination of fork truck travel. The fork trucks now can be used in local areas where they travel less than 200 ft and are used for stacking or loading/unloading.

Trailers for Train Operation. Driverless tractors can tow a wide variety of trailers. These range from simple caster steer trailers to automated trailers with capabilities to load and unload automatically. Next to the standard caster steer trailer, the most popular trailer is the four-wheel steer trailer with powered conveyor top. Driverless tractors towing these types of trailers are used to interface with fixed conveyor systems where the load transfer is completely automatic.

Other types of trailers towed by driverless tractors include hand pallet trucks with special towing attachments, roll-handling trailers, and trailers with special designs such as shuttle mechanisms or tilt decks (see Fig. 8.1.5).

Space requirements for driverless trains vary, depending upon the length of train (Fig. 8.1.6), the type of steering on the trailers (Fig. 8.1.7), and the width of the aisles it will be operating in (Fig. 8.1.8). Caster steer trailers do not trail well, and have a tendency to cut corners when the train makes turns. This requires wide aisles or a short train. Four-wheel or fifth-wheel steer trailers steer better and are almost universally used where the loading and unloading of the trailers is done automatically. However, where trailers are intended to be uncoupled from tractors and moved about manually, four-wheel or fifth-wheel steer trailers are not a good choice because they are hard to handle manually.

Other Considerations. Most driverless train system applications require side tracks or spurs to permit moving trains to pass trains parked at "stop" stations. These sidings require space which must be carefully planned for, see Fig. 8.1.9.

Most driverless train systems are controlled through the use of the onboard station selector panel. Once the train has been loaded or unloaded at a "stop" station, the operator uses the onboard panel to program the vehicle to its next "stop" station. This method of onboard control requires that all operators be trained in the use of these controls. This is a very simple procedure and usually requires about a 15-min operator training session.

Wire-guided Pallet Trucks

Wire-guided pallet trucks (Fig. 8.1.10) are ideal in applications where load volumes are moderate and distances are moderate to long. Whereas driverless trains excel in chain load movements (multiple, distributed stops on each train cycle), wire-guided pallet trucks excel where the load movement requirements are basically from point A to point B and back to point A again.

Wire-guided Pallet Truck Application. The wire-guided pallet truck has a unique capability to automatically unload itself when it arrives at a programmed "stop" station. The vehicle is loaded

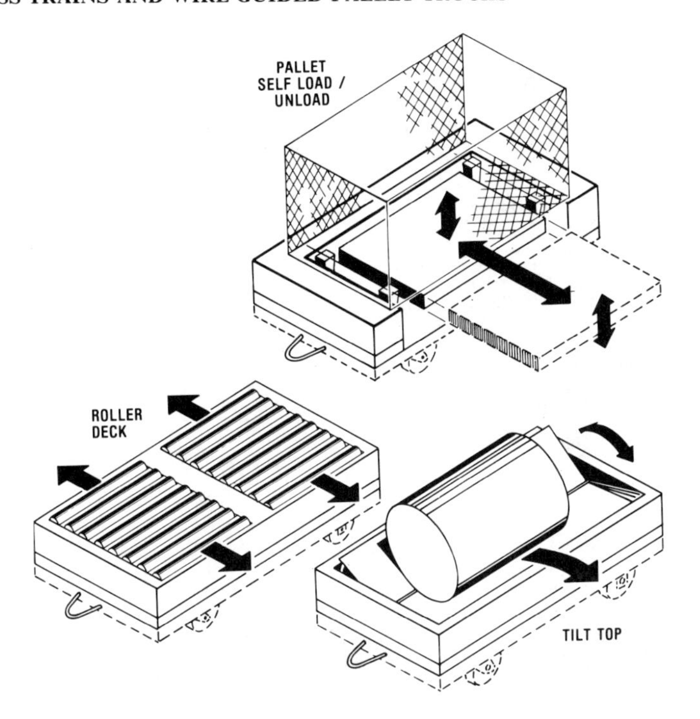

Fig. 8.1.5 Some automated trailers for use with driverless trains. Tilt top, roller deck, self-load/unload features are shown.

manually and then dispatched to a drop location situated on a side track. When the vehicle arrives at the drop location, it pulls off onto the side track and lowers its forks, which automatically places the pallets on the floor. The vehicle then restarts, pulling out from the pallets and returns to the main path. The vehicle now proceeds to the next "stop" station assigned to it. At the new loading point, a man will board the vehicle and back it underneath new pallet loads, return it to the guidepath, dispatch it to an automatic drop location where the process repeats (Fig. 8.1.11). In this manner, pallet loads can automatically be delivered and left on the floor at a drop location without any manual intervention. Fork trucks are normally used to pick the pallets up off the floor and store them into racks or move them to the next process.

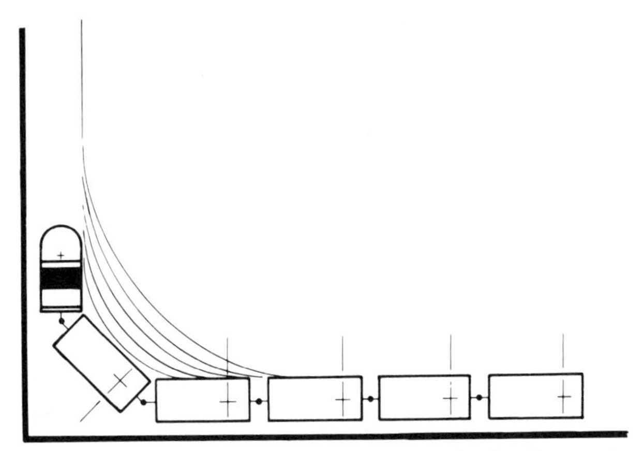

Fig. 8.1.6 Turning radius of driverless train depends on the type of trailer steering and the number of trailers being pulled.

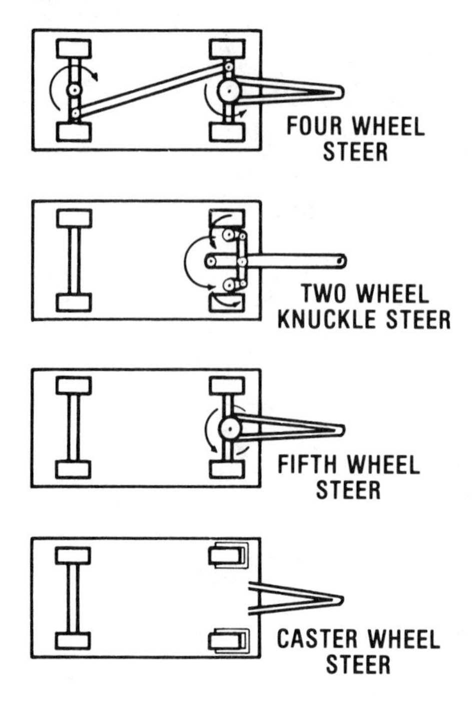

Fig. 8.1.7 Typical steering arrangements for driverless train trailers.

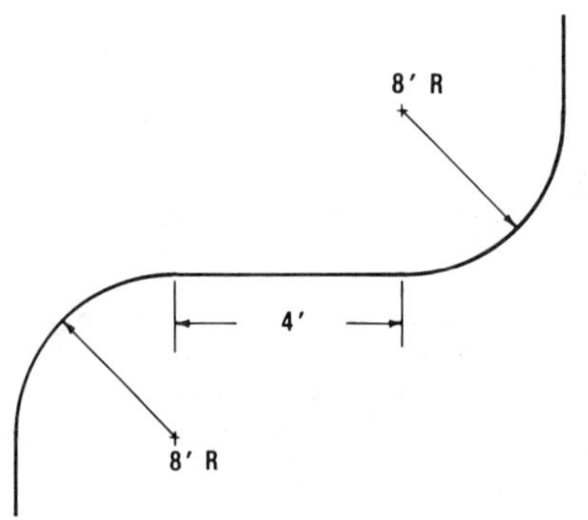

Fig. 8.1.8 Minimum guidepath turning dimensions for automated guided vehicles and trailers.

This use of the wire-guided pallet truck is extremely efficient, and has gained much popularity in recent years. Under manual control, these trucks normally travel twice the speed they travel in automatic.

Sizes Available. Wire-guided pallet trucks range in load capacities from 4000 to 6000 lb. Models capable of carrying either one or two pallets are available. The load forks on the rear of the pallet truck are designed to slip into double-faced pallets or under load skids. Proper care must be exercised when specifying the vehicle pallet fork lengths and rear pallet fork wheel locations so that the pallet forks properly slide into double-faced pallets. Since the pallet forks are supported by small retractable wheels, the rear pallet fork wheel spacing must be such that when the pallet forks are raised, the bottom boards of the pallet do not interfere with the extension of the pallet fork wheels. Skids can be picked up automatically because there are no bottom boards to obstruct the load wheels.

Special Requirements. In most cases, wire-guided pallet trucks require that the loads be staged on the floor in pick-up areas where operators can easily back the vehicle into the loads. They also require that the drop-off locations be on side spurs where the automatically delivered loads will not interfere with the traffic flow on the main path.

Aisle space requirements for guided pallet trucks are highly predictable. The space required depends on the length of the guided pallet truck, which dictates its turning radius. Most guided pallet trucks

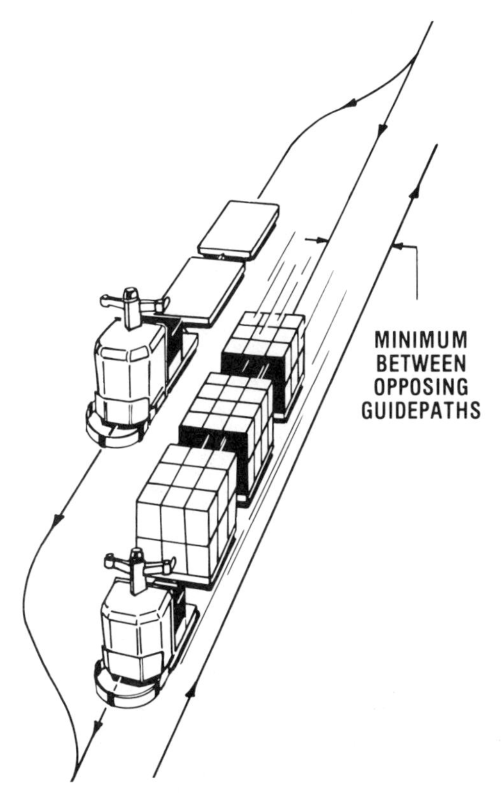

Fig. 8.1.9 Opposing guidepaths must be separated by the greatest load width plus 18 in.

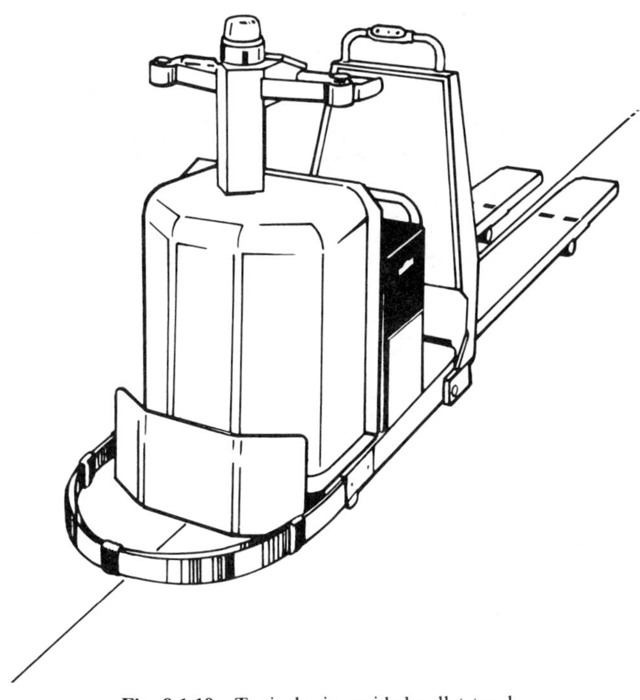

Fig. 8.1.10 Typical wire-guided pallet truck.

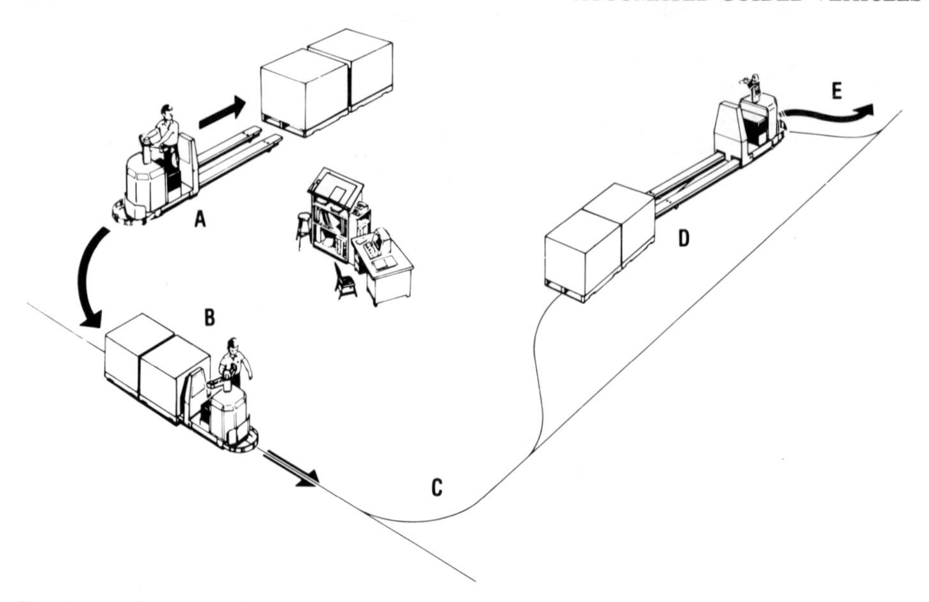

Fig. 8.1.11 Typical application of wire-guided pallet truck for automatic delivery. Operator backs truck into load (A), places truck on guidepath (B), which truck follows (C) to unloading side track (D). Here, it automatically drops load and proceeds back onto main guidepath (E).

are a double-pallet type and require at least twice their length to straighten out after a turn. This is an important consideration when laying out the guidepath and the side tracks for the system operation, see Fig. 8.1.12.

Automatic-delivery pallet truck systems are extremely efficient because, like the driverless train, these vehicles can be used manually whenever required. This is an added benefit for nonsystem functions such as using the vehicle manually to move material to areas of a plant where there is no guidepath to follow.

8.1.4 Manual Operation

Driverless tractors and wire-guided pallet trucks are in most instances manual vehicles that have been converted for automatic guided vehicle operation. They can be used manually, as well as under automatic control. This feature offers a number of advantages.

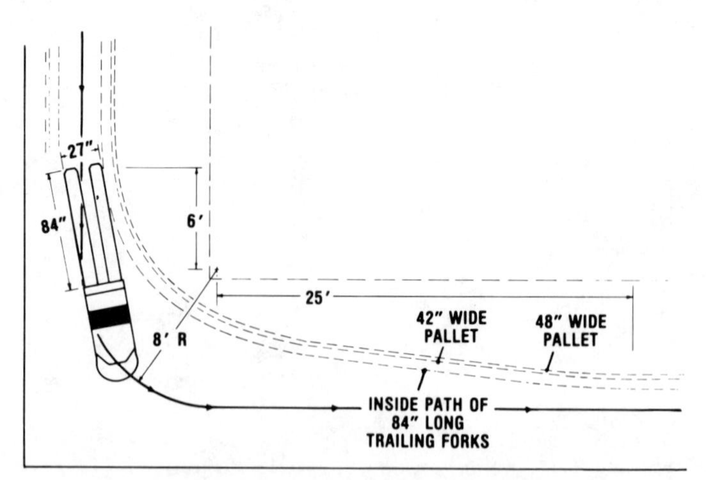

Fig. 8.1.12 Trailing characteristics of wire-guided double pallet truck.

Operating these vehicles manually is really quite simple. An operator merely boards the vehicle, turns the key switch to manual operation, and drives the vehicle manually off the guidepath. With this kind of flexibility, driverless tractors can be used to pull trailers to areas of the system where there is no guidepath or position loads in areas adjacent to the guidepath. Many driverless train systems start out with short or medium-length guidepaths and are often expanded. Prior to the expansion, driverless tractors are used manually in those areas where there is no guidepath to help determine an efficient flow for the future guidepath.

Wire-guided pallet trucks are often used manually. They must be manually backed up under a load at which time the rear pallet forks are elevated several inches so that the load is now off the floor and fully onboard the vehicle. The vehicle is then manually returned to the guidepath for automatic operation or manually driven to a destination in the system where the pallet is to be delivered.

Manual operation of AGVS equipment is important if the automatic controls fail. When a driverless tractor or wire-guided pallet truck fails in the automatic mode, it can be driven manually. This is an important feature which allows the maintenance department to drive the vehicle to the maintenance area for proper trouble shooting. It also allows regular plant personnel to remove vehicles quickly from a guidepath if they have malfunctioned.

8.1.5 Controls

The controls onboard the vehicle and within the AGV system are one of the most important selection considerations. The more suitable the controls are to your plant requirements, the better the system will operate, in terms of efficiency, throughput, reliability, and flexibility.

Control Functions

What are the areas of control in an AGV system? The most fundamental area of control is onboard the vehicle. The vehicle has to be able to guide itself (i.e., follow the buried wire). In addition, the controls must start and stop the vehicle and make routing decisions without an operator's assistance. They must also enable the vehicle to avoid other AGV traffic in the system through the use of anticollision (blocking) controls. Other optional control capabilities include remote communication with a central computer or automatic loading and unloading in systems that interface with other materials handling equipment, such as conveyors or AS/R systems.

Vehicle Routing Control. In a simple AGV system, there may be only one pathway in the form of an unbroken loop. There may be a number of stopping points along the loop where material is to be dropped off or picked up. The controls must tell the AGV train to make the required stop; skipping some stops where there is no load dropoff or pickup. The controls can be preprogrammed by manual means or can be set to accept new commands as the train progresses on its guidepath and as load requirements change after the initial programming is completed.

Control Programming

Manual. Manual programming is usually accomplished in one of three ways:

1. Toggle switches
2. Thumbwheel switches
3. Push-button numeric pad

Onboard the vehicle will be a "stop" station selector panel which an operator uses to direct the vehicle to specific stations in the system. This station selector panel can consist of toggle switches (one for each "stop" station in the system), thumbwheel switches, or a push-button numeric pad with a digital display window. Any of these three methods allow an operator to select a "stop" station in the system and dispatch the vehicle to it. The toggle switch dispatch panel is the simplest of the three and is used in most system applications. The push-button key pad allows greater sophistication in systems and is required when there are many program "stops" in the system or if priority station selection routing is required. It allows the operator to dispatch a vehicle to any "stops" within the system in any specific order required. This is not permitted with a toggle switch programmer which is limited to specifying "stops" but cannot establish an order of priority. For example, if two or three "stop" stations are selected, the vehicle will arrive at them in a random order, depending upon which "stop" station it physically comes to first. Priority "stop" selection routing means that the vehicle will service those stations selected in the order in which they were selected. In certain controlled material process movements, this is a necessary requirement.

Automatic Guidepath Selection. Complex guidepaths require sophisticated control systems. For example, a guidepath loop may have a number of side spurs or branches. Other systems may have a number of alternate routes or multiple loops through a plant or warehouse. At each point where the guidepath splits, the controls must make a decision based on instructions received as to which pathway to follow.

When a vehicle is dispatched to a "stop" station, it proceeds along the guidepath until a decision point is reached. Then reading a floor code (magnet code or floor plate code), the vehicle makes a decision whether to go straight or to turn. This decision is made dependent upon whether the "stop" station it wishes to reach is off the main line and on the branch or somewhere further down the main line. The decision is made instantaneously.

Usually, the vehicle is selecting a guidance frequency in the floor to follow when it makes its routing decision. At the decision point one guidance frequency goes straight and the other frequency follows the branch line. When the vehicle selects a frequency to follow after reading the floor code, it continues to follow that frequency until the instructions are changed.

A less sophisticated method which has been used in the past is for the guidepath to be switched "on" by local controllers. In this method, the vehicle signals a local control unit in the floor which turns "off" one path, leaving the other path "on." Since the vehicle only sees the path which is "on," it continues along that path. This also accomplishes the routine decision, but requires the use of a control package, either in the floor or on the wall, near each decision point.

System Anticollision Blocking Controls. Vehicle anticollision controls are commonly referred to as the blocking system. Driverless trains and pallet trucks on the same path avoid hitting one another by use of these blocking controls. They can either be onboard the vehicle or along the guidepath.

The simplest form of blocking is to place *optical controls* on each vehicle that sense reflective targets on the vehicles in front of them. This requires optical reflective tape to be placed on the trailers of the leading train. Optical blocking is not used on guided pallet truck systems because of the difficulty in locating optical reflectors on the rear of pallet trucks in such a way so that other vehicles will "see" the reflections in all operating modes, both loaded and unloaded.

Optical blocking is not the most foolproof method of blocking and is used mostly where there are long straightaways. Turns in a system make optical blocking impossible because the targets are at an angle to the trailing vehicle's optical receiver unit. In those areas, the optical blocking is supplemented with other fixed blocking controls.

Most multiple vehicle systems rely on some form of *zone blocking.* In this method, the guidepath is divided into zones that are large enough to contain one train or vehicle, plus some safety margin. The control algorithm is that only one vehicle is permitted in a zone at a time. If the zone is occupied, then any vehicle wishing to enter that zone will wait in a zone prior to it until the leading zone is cleared. Only then will the following vehicle be allowed to move into the new zone. As long as zones are clear, vehicles proceed smoothly along the path without hesitation. Once a zone in front of a vehicle is occupied, the trailing vehicle must go into a "hold" and wait until that zone is cleared before proceeding (Fig. 8.1.13). This method of blocking is accomplished by three basic methods. The selection choice depends upon the size of the system, the sophistication of traffic control required, and cost.

Point-to-point zone blocking uses individual zone control boxes mounted on columns, walls, or in the floor adjacent to each zone. When a vehicle passes from one zone to another, it triggers the zone control package to activate the zone behind it. This prevents other vehicles from colliding into the rear of the lead vehicle. As the lead vehicle passes into the next zone, two things happen: (1) the previously blocked zone is deactivated, and (2) the zone just vacated is now activated. In this method, the zones are turned "on" and "off" by the vehicle in a sequential fashion.

Continuous blocking is a method that does not use fixed equipment on walls or in the floor. Each vehicle in the system has an onboard blocking signal transmitter and receiver. The vehicles transmit a high-frequency signal into an auxiliary wire buried in the floor. This auxiliary or blocking frequency wire carries the signal to the zone behind the vehicle. There, the signal appears in a loop of wire in the floor which trailing vehicles pass over. When a trailing vehicle senses a signal in the blocking wire beneath it, it immediately goes into a blocking hold. Remember that this signal is caused by the leading vehicle transmitting its blocking signal into the blocking wire. When the leading vehicle moves into the next zone in the system, it will be transmitting into a new blocking wire and the previous blocking wire signal will disappear, allowing the trailing vehicle to move forward. The trailing vehicle moves forward into the next zone until it passes over another blocking wire in the floor and stops there since a blocking signal is now present at this new location.

In *computer zone blocking,* the vehicles communicate with other vehicles or to a central computer controller, to identify which zones they are in. The vehicles carry an onboard microprocessor system which decodes blocking information picked up from the guidepath wire in the floor. This information causes the vehicle to take the proper blocking actions.

In this system, each vehicle is equipped with a transmitting and receiving antenna, and communicates either directly into the guidepath wire or through an FM radio link to other vehicles or a central computer controller. The information transmitted identifies the zone that the vehicle is in. The information is obtained by the vehicle which "reads" magnets or plates in the floor for each zone. When a

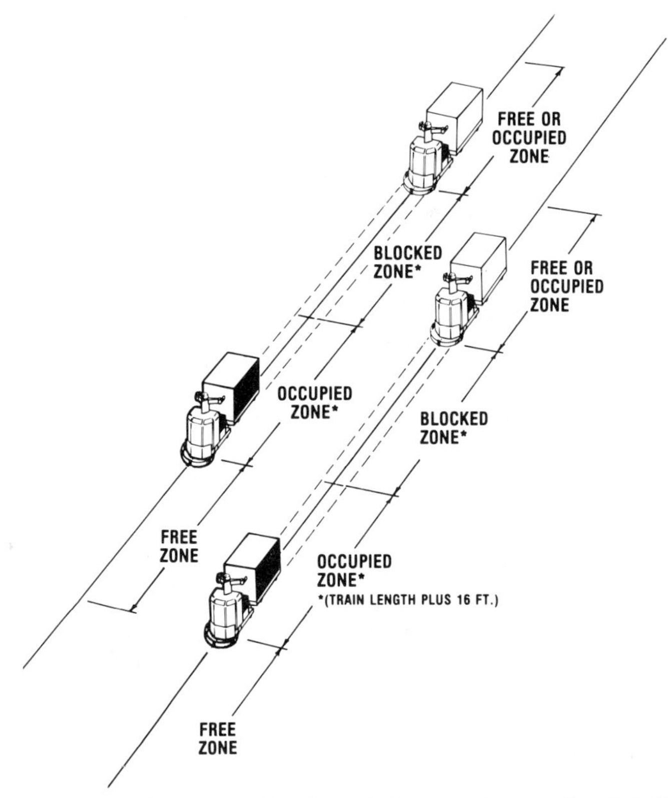

Fig. 8.1.13 Blocking controls prevent a driverless train from entering zone directly behind an occupied zone.

central computer controller is used, it receives the zone information and keeps track of each vehicle location. If the zone about to be entered is already occupied, a holding signal will be transmitted.

The anticollision blocking system used in multivehicle driverless train and guided pallet truck systems is one of the most important areas in the planning and implementation of a system. The anticollision blocking system design will influence system throughput, system reliability, flexibility of operation, and the ease with which systems may be modified or expanded. The wrong blocking system for a given application could unnecessarily limit a system's capability and thereby affect efficiency and operating cost.

Advantages and Disadvantages of Different Blocking Methods. Optical blocking is a more passive method of blocking and is generally not as foolproof as the positive zone blocking methods. The operating range of optical blocking transceivers on the vehicles does have a tendency to change at times and optical targets can fall off trailers. This would cause a blocking failure to occur. Systems with many turns do not lend themselves to the application of these controls.

Optical blocking controls do have certain advantages, however. They do not require fixed-length block zones and, therefore, do allow more traffic flow flexibility. In some applications this is very advantageous. In addition, vehicles using these controls can be removed from the guidepath or inserted onto the guidepath anywhere within the system without affecting the flow of the remaining vehicles. Another advantage is that these controls used in systems with long straightaways reduce the overall cost of the blocking controls for the system.

Point-to-point zone blocking is used predominately in small- to medium-size systems (less than 3000 ft in length and four to five vehicles) and provides positive blocking control which keeps vehicles from catching up and colliding with one another. A disadvantage is encountered when a vehicle is removed from the guidepath. In this situation, the block zone control unit is still activated in the floor or on the wall, and following vehicles will not proceed through automatically. To deactivate blocking, the vehicle that has been removed must be returned to the path exactly where it was removed and restarted to proceed into the next zone. Otherwise a reset button must be provided for each

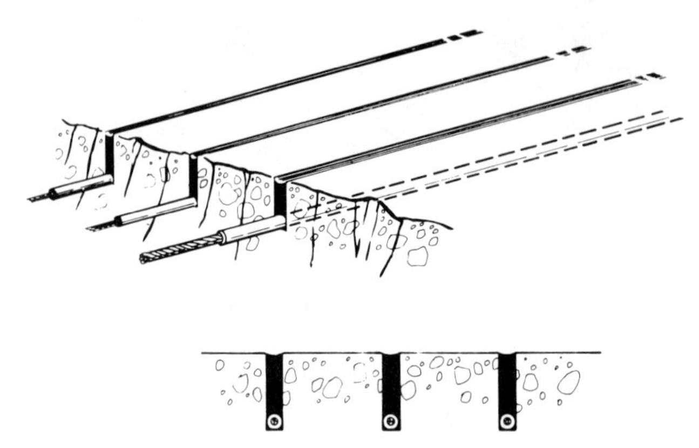

Fig. 8.1.14 Continuous blocking requires two extra wires embedded in the floor in addition to the guidepath wire.

zone control box that will allow the removal of a vehicle and the resetting of the zone so that other traffic may continue through.

In systems where vehicles will be removed at random, this blocking method is not well suited unless strict control is maintained over the system operation. The installed cost of each block zone varies from $1000 to $1500 each. Therefore, a number of more wires must be installed (see Fig. 8.1.14). This usually makes continuous blocking more expensive initially than point-to-point zone blocking.

Since there are no control units involved in continuous blocking, there is practically no maintenance required in a continuous blocking system as compared to a point-to-point blocking system. Furthermore, it is economical to increase the number of zones in order to allow more throughput and traffic flow. This is due to the fact that the price of a continuous blocking system is directly related to the length of the guidepath. While two extra slots must be cut for the auxiliary blocking signal wire, the cost of a separate control panel at each block zone point is eliminated.

Computer zone blocking has similar advantages to continuous blocking whereby vehicles may be removed or inserted into the system at random without resetting block zones. In addition, there are no control boxes on the walls or in the floor that need periodic maintenance. Unique traffic flow problems are easily solved because the vehicle is "smart" and can make routing decision based on traffic situations in its immediate vicinity. Point-to-point and continuous blocking methods are fixed and usually allow only one blocking action to occur, regardless of traffic conditions. In computer blocking systems, more complicated guidepath layouts can be installed and handled more efficiently in contrast to point-to-point or continuous blocking. One of the biggest advantages of computer blocking is that it readily permits real-time vehicle tracking and monitoring.

8.1.6 Safety and Warning Devices

AGVS vehicles have a very good safety record. The standard safety and warning devices used on driverless trains and wire-guided pallet trucks include emergency bumpers, warning lights, guidepath monitor circuits, and audible warning devices.

Emergency Bumper

The emergency bumper protrudes from the front of the vehicle and upon contact with an object causes the vehicle to emergency stop. Driverless trains and wire-guided pallet trucks are usually designed so that they apply hard braking whenever the emergency bumper is activated. Depending on the load the vehicle is carrying or towing, the floor surface conditions, vehicle speed, and so on, the vehicle stopping distance will vary. Nominally, this distance is between 2 and 5 ft. Once the emergency bumper has been activated, the vehicle must be manually restarted by an operator. This is an added safety precaution in case the object that deflected the bumper in the first place is still in front of the vehicle.

Other passive forms of stopping the vehicle are optionally available. These include sonic and optical object detection which "look" for objects in front of the vehicle at a given range. Once an object is detected by these types of sensors, the vehicle can be designed to either stop or continue at reduced speed, allowing the bumper to make gentle contact with the object, thus completely stopping the vehicle. The latter technique is frequently used because the sonic or optical devices may detect an object ahead of the vehicle which is not actually in the vehicle's path. In this case, the vehicle should not stop. It does, however, slow down and proceed at slow speed until the object is no longer detected by the sensors. This happens frequently at turns where an object ahead of the vehicle, such as a column or rack, will not be in the vehicle's path because the vehicle is making a turn shortly before the object would be encountered. In this case, the object detectors may "see" the column or rack and cause the vehicle to proceed slowly around the curve until the object is no longer detected. Stopping the vehicle would hold up traffic unnecessarily behind the vehicle.

Lights

Warning devices on driverless trains and wire-guided pallet trucks are generally rotating red lights. A rotating red light produces a beam of light which is highly noticeable even to people operating in aisles perpendicular to the direction of travel of the vehicle. The rotating red light not only alerts those directly in front of the vehicle of its presence, but also those off to the side of the guidepath. Strobe lights can also be used with a similar effect.

Guidepath Monitor

The guidepath monitor circuit onboard the vehicle shuts the vehicle down should it ever deviate from the guidepath. Normally this circuit is set to activate when the vehicle strays more than 2 in. from the guidepath. This circuit prevents the vehicle from continuing to travel with no guidepath to guide it. In the event there is a power failure in a plant, the guidepath energizer would cease operation. Since the guidepath energizer produces the guidance frequency in the wire in the floor, this signal would then disappear. With a power loss, all vehicles operating on the path in the system would then automatically shut down.

Audible Signal

The audible warning signal usually is a mechanical warning bell mounted to a wheel on the vehicle. Sometimes, electronic warning horns or other types of audible signaling devices are used. Depending on the application, the correct audible warning device can be provided. Very noisy applications oftentimes require a loud intermittent warning horn. Quiet applications usually employ a simple warning bell or electronic tone.

Supplementary Warning Signals

It is always useful in systems to employ overhead warning signs and flashing lights in particularly congested areas. Occasionally the route of the driverless vehicle takes it past a pedestrian area and a warning sign is employed overhead with flashing lights to signal pedestrians that a driverless vehicle is approaching.

Painting the floor to clearly distinguish guidepath right-of-way for the driverless vehicles is another safety measure. This deters fork truck drivers or others from putting material in the driverless vehicle's right-of-way.

Good safety practices should be followed when determining the right-of-way for the driverless vehicles to follow. At least 18 in. should be maintained between the driverless train or wire-guided pallet truck and any columns or obstructions. This is particularly critical on turns or through doorways. No part of the driverless vehicle or any trailers (including overhanging loads) it may be towing should encroach closer than 18 in. to a fixed obstruction.

In systems employing ramps, two vehicles should never be allowed in either direction on the ramp at the same time. The proper system controls can be designed to prevent this from occurring. Also, distinct right-of-way on ramps is necessary so that there is room for the driverless vehicle and other traffic on the ramp at the same time.

8.1.7 Advanced Capabilities

Driverless trains and pallet trucks have been solving many basic materials handling needs for many years. However, recently these systems have taken on more sophistication to meet growing demands for greater automation and control, particularly in manufacturing operations and in integrated systems where automatic storage and retrieval systems (AS/RS) and powered conveyors may be computer-linked to AGV systems. These vehicles can be equipped with advanced controls such as onboard microprocessors and/or controlled by a central dispatch computer which allow these AGV systems to perform advanced functions. The trend toward integrated warehousing and manufacturing has brought AGVS vehicles into close interaction with conveyor and AS/R systems. A greater proportion of these integrated systems are now being installed. Driverless tractors pulling automated trailers routinely interface with conveyors or AS/R systems for completely automatic load transfer in a reliable and efficient manner. This is possible because of the advent of microprocessors and remote communication capabilities that now exist in guided tractor and pallet truck technology.

Automatic Summons with Accountability

The key to a fully integrated system is control and monitoring. When an AS/R system retrieves a load for delivery to a dock location or manufacturing area, the AGVS vehicle can be automatically summoned to pick that load up and carry it to the proper destination. The vehicles can be constantly monitored to insure that they reach their proper destinations and that the loads they are carrying have been delivered. The concept of accountability is keyed to this advanced system capability. Many plants and warehouses are moving hundreds of pallets every day with advanced AGV systems. With the monitoring available through these systems individual loads can be tracked to insure accountability.

Central Computer Control. Two basic methods of advanced system controls are available. Both involve a central computer that remotely manages the vehicles in the system.

The first method employs the central computer to remotely dispatch the vehicles to "stop" stations where they are needed *and* controls the movement of the vehicles until they reach those locations. This sometimes involves switching paths on and off at decision points, so that the vehicle can follow the energized path to its next stop station.

Onboard Microprocessor. In the second approach, the central computer again remotely dispatches the vehicle to a "stop" location in the system. However, the vehicle carries its own onboard intelligence (usually a microprocessor) that allows it to make its own decision as it travels along the guidepath toward its "stop" location (Fig. 8.1.15). The central computer in this case is only responsible for remotely dispatching the vehicle to its next "stop" station.

With both methods, the vehicles can be tracked and monitored for complete system control.

Vehicles with onboard microprocessors have generally been called "smart" vehicles. They have the capability, once they have been told where to go, to proceed there totally on their own without any additional central computer control. Under this operating scheme, if one of these vehicles fail, the others continue to operate. The central computer is not essential to the vehicle's operation. In fact, if the central computer is not functioning, the vehicles may still carry out their functions under semiautomatic onboard control or via a backup central control. Vehicles that require a central computer for control of vehicle routing cannot be used in this semiautomatic control mode if the central computer is not functioning.

Color Graphics

One of the new advances in AGV system technology is real-time color graphics monitoring of the vehicles in the system, see Fig. 8.1.16.

A color TV monitor is used to graphically display the system guidepath and location of all vehicles. As vehicles move along the path, the graphic depiction of the vehicle moves on the graphic guidepath. Word messages can also be displayed, which give the status of each vehicle in the system. This status information can include the vehicle's location, its next destination, whether it is moving, stopped, blocked, or malfunctioned, and whether its battery is low, plus other vital information. Plant supervisors can quickly spot system slowdowns with this real-time monitoring tool. Systems with multiple vehicles and/or long complicated guidepaths are natural applications for a graphics monitor display. System efficiency is kept at a high level because a greater measure of supervisory control can be provided with this new tool.

8.1.8 Remote Control Refinements

AGV systems permit an operator at a work station to "call" for a vehicle to pick up a load. This is usually accomplished by pushing a button on a control panel that is interconnected into the AGV system.

Several refinements have been added through the use of individual control stations with programming capabilities. Using thumbwheel switches or a push-button key pad, an operator at a work station can "call" a vehicle and also specify where it is to take a load. Specific "stop" stations are simply dialed in. When the vehicle reaches its new station, it again can receive further instructions from the control panel at the station on where to proceed next.

An application of this concept is an automatic load transfer operation which "calls" the vehicle when it senses a load is ready for transfer, then instructs the vehicle where to take the load. Note that no operator intervenes at any time during the loading–transfer–unload sequence.

As a further refinement, more than one vehicle can be remotely dispatched at a time.

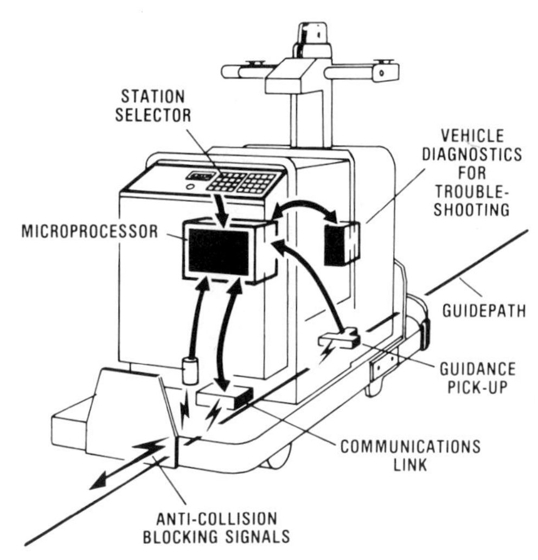

Fig. 8.1.15 Onboard microprocessor adds independent intelligence mode to AGVS operation.

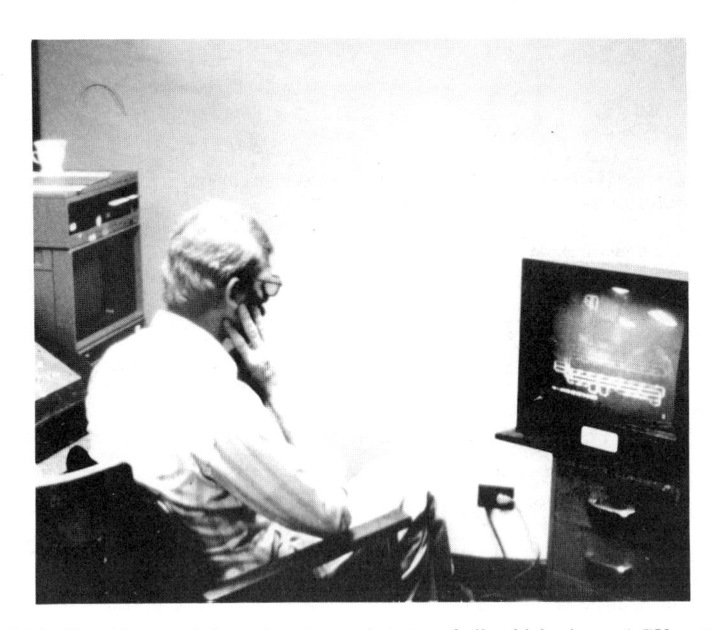

Fig. 8.1.16 Graphics panel shows location and status of all vehicles in an AGV system.

In some advanced systems, the call and routing instructions are input into the AGV system by means of a code-reading device. It can read, for example, a label on a pallet to initiate a vehicle "call" or to assign a destination station. Operationally, the product code is read, and the data sent to a control computer. The computer determines the destination for the load and communicates this information to the AGVS vehicle when it arrives to pick up the load. Upon receiving its instructions, the vehicle proceeds to the designated drop-off station. Two examples of call stations are shown in Fig. 8.1.17.

8.1.9 Battery Requirements

Driverless tractors and pallet trucks usually operate with 24-v electrical systems, and require, for most applications, a minimum 450 A-hr capacity battery. Maximum battery capacity required in most

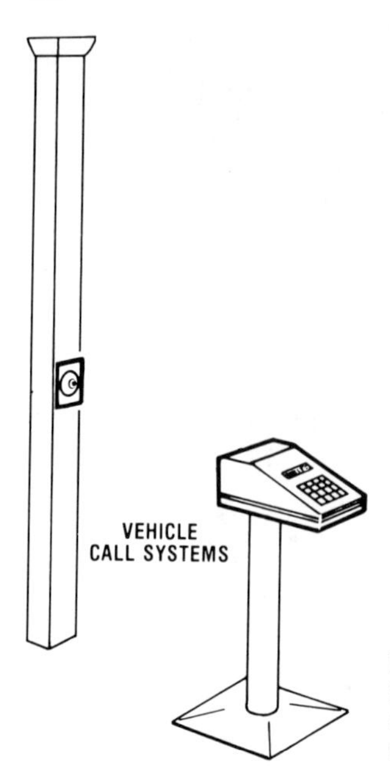

Fig. 8.1.17 Two examples of call stations. Left, simple call button is panel-mounted near a work station. Right, pushbutton keypad and readout permit calling vehicle *and* selection of next "stop" stations.

systems never exceeds 800 A-hr. The approximate battery capacity required for a given system application can be arrived at as follows:

STEP 1. Determine average distance vehicle is physically moving per shift, in miles. (All numbers are per shift.)

$$\frac{\text{Total load moves req'd.}}{\text{No. of loads carried/vehicle/trip}} = \text{No. of vehicle trips/shift}$$

$$\frac{\text{No. of vehicle trips/shift}}{\text{No. of vehicles}} = \text{No. of trips/shift/vehicle}$$

Determine distance traveled per shift per vehicle.

$$\text{No. of trips/shift/vehicle} \times \text{Avg. distance (ft)/trip} = \text{Distance/shift}$$

$$\frac{\text{Distance (ft)/shift}}{\text{5280 ft/mi}} = \text{Distance (mi) traveled/vehicle/shift}$$

STEP 2. Determine average load in tons moved per vehicle cycle.
(This is an estimate of the average load carried. It is an average of loaded and empty conditions weighted as required.)

STEP 3. Determine vehicle ton-miles (t-mi) per shift.

$$\text{Tons} \times \text{Distance} = \text{Ton-miles}$$

STEP 4. Determine battery watt-hour requirement (assuming for AGVS applications vehicle uses 120 W-hr/t-mi).

$$\text{120 W-hr/t-mi} \times \text{Ton-miles} = \text{Watt-hours}$$

STEP 5. Divide watt-hours by battery voltage (usually 24 v) to get battery size (ampere-hours).

$$\frac{\text{Watt-hours}}{\text{24 v}} = \text{Battery ampere-hours}$$

This figure represents actual battery consumption and should be divided by 0.8 to avoid total battery discharge. Since these vehicles do not require drivers, the battery usage is more controlled and constant than on manually operated vehicles. Battery life of 8 yr is not unusual.

8.1.10 Ramp Capacities

Both driverless tractors and guided pallet trucks can negotiate ramps, see Fig. 8.1.18. Ramp applications require special controls on these vehicles which regulate the speed of the vehicle while it is descending the ramp. Normal rolling load capacities are derated for vehicles operating on ramps. AGVS vehicles can negotiate ramps of up to 10% with restricted capacity under normal circumstances. Above 10%, special design requirements are necessary. An approximate method for determining ramp capacity of a vehicle is as follows:

STEP 1. Determine percent of grade of ramp.

$$\frac{\text{Rise}}{\text{Length (horizontal)}} \times 100 = \text{Percent of grade}$$

STEP 2. Determine trailer drawbar pull required to move trailers at a constant speed on a level surface.

$$\text{Trailer weight loaded} \times 2\% = \text{Trailer drawbar pull}$$

STEP 3. Determine ramp drawbar pull required to move at constant speed up ramp.

$$(\text{Vehicle wt.} + \text{Trailer wt. loaded}) \times \text{Percent of grade} = \text{Ramp drawbar pull}$$

STEP 4. Determine total drawbar pull required for ramp.

$$\text{Trailer drawbar pull} + \text{Ramp drawbar pull} = \text{Total drawbar pull}$$

If the drawbar pull determined exceeds the maximum drawbar pull rating of the vehicle, then the capacity assumed in the calculation is not feasible. Lower the capacity and refigure the drawbar pull to determine an approximate feasible ramp capacity.

The length of the ramp is also a determining factor. Long ramps cause further derating of the capacity due to possible overheating of the vehicle's drive motor. In addition, the vehicle may be required to stop while it is on the ramp. In this case, the manufacturer should be consulted to determine if the vehicle has sufficient initial starting drawbar pull to restart while on the ramp at a given load capacity. Do not overlook this requirement.

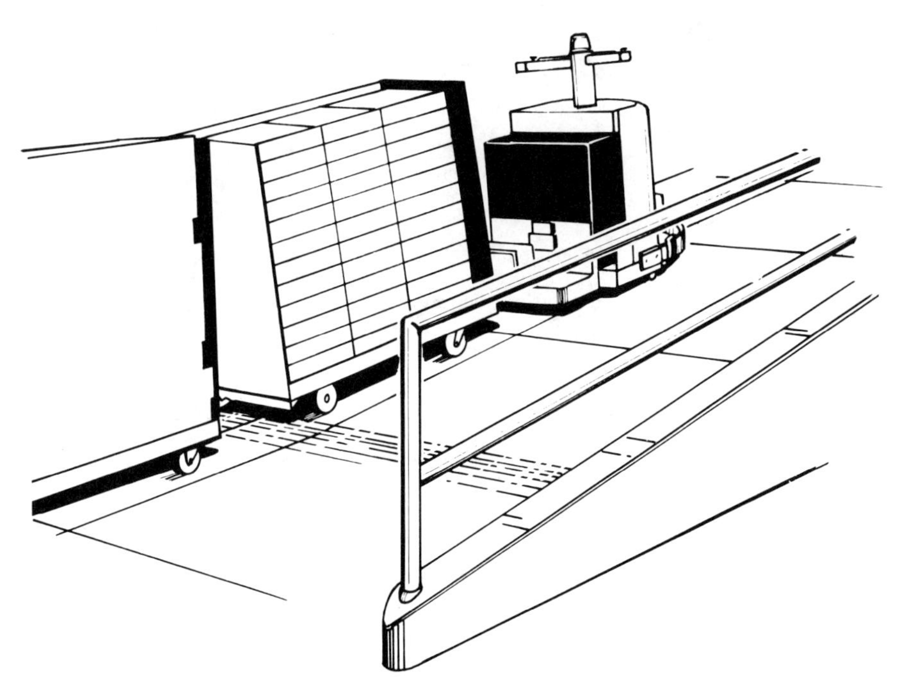

Fig. 8.1.18 Driverless tractors operate successfully on ramps, but require accurate calculation of drawbar pull and braking capacity.

Braking on Ramps

One of the most important concerns on ramps is the ability of the vehicle to stop while descending the ramp. Maximum control is achieved through the use of individual trailer brakes. These brakes are automatically applied as the train descends the ramp and greatly enhance the safety margins required for ramp operations.

8.1.11 Calculating Number of Vehicles Required

Before one can determine the number of driverless trains or guided pallet trucks required in a system, the parameters must be decided. Where is the guidepath going to be located? How will the system be used? What type of blocking system will be utilized? Such questions have considerable bearing on the calculation of the number of vehicles.

When these questions have been answered, there are several methods that can be used to determine the number of vehicles required. There are various mathematical techniques from operations research, such as shortest-route theory, queue line theory, and network theory. In addition, the growing use of various computer simulation models can yield very good results. In controlled applications (i.e., where little human intervention is involved) and where there are specific nonvarying defined system tasks, these mathematical models yield very accurate results.

However, for most typical applications the approximate number of vehicles can be quickly estimated at little cost by the average individual. The following is a simple procedure.

Handling Capacity Calculations

STEP 1. Determine the average distance the AGVS vehicle travels per trip, in feet.

(Estimate average travel distance based on the logical trip routes the vehicle negotiates in normal operation.)

STEP 2. Determine the average trip time.

$$\frac{\text{Distance}}{\text{Vehicle speed (ft/min)}} = \text{Running time (RT) in minutes}$$

$$\text{RT} \times \text{BTF*} = \text{Blocking time}$$

$$\text{RT} \times \text{ITF} = \text{Idle time}$$

$$\text{No. of load transfers} \times \text{LTT} = \text{Load/unload time}$$

$$\text{Theoretical trip time (minutes)} = \text{Running time} + \text{Blocking time} + \text{Idle time} + \text{Load/unload time}$$

$$\text{Actual trip time} = \frac{\text{Theoretical trip time}}{0.85\ \text{(efficiency factor)}}$$

STEP 3. Determine how many trips one vehicle can make per hour.

$$\text{60 min/trip time} = \text{Trips/hour/vehicle}$$

STEP 4. Determine how many vehicle trips/hour are required.

$$\frac{\text{Required load movements/hour}}{\text{Loads/vehicle/trip}} = \text{Vehicle trips/hour req'd.}$$

STEP 5. Determine how many vehicles are required.

$$\frac{\text{Vehicle trips/hour required}}{\text{Trips/hour/vehicle}} = \text{No. of vehicles req'd.}$$

8.1.12 Training

When a driverless train or wire-guided pallet truck system is installed, the customer usually receives several types of training programs.

---

* BTF = blocking time factor (10–15%); ITF = idle time factor (10–15%); and LTT = load transfer time (2 min/load if manual loading, 0.5 min/load if automatic loading).

Basic Operation School

The first is the basic operation school which is conducted by the AGV vendor at the customer's site. This school teaches plant personnel how to operate the vehicles both manually and automatically. Basic fundamentals such as putting the vehicle on the path and starting it or removing it from the path are covered. Operators are taught how to use the onboard dispatch controls for routing of the vehicles and what various indicator lights on the vehicle mean. Operators are familiarized with how to operate the vehicles on a day-to-day basis within their job function.

The operating school generally takes less than an hour to conduct, but it is a good idea to repeat this after a week of system operation. Plant personnel can develop bad habits using equipment and this can be corrected by refamiliarizing them with proper equipment and system operation. During the first session, the plant personnel merely pick up the basic operating highlights of the equipment.

It is vitally important that plant personnel thoroughly understand the system operation, its purpose, and how it should be operated to achieve the desired results. They should be fully aware of the anticollision blocking feature of the system. In addition, it is essential that fork truck drivers quickly become aware of the fact that the vehicles must have a clear right-of-way.

Maintenance Training

As another aspect of the training program, plant maintenance personnel receive in-depth training on maintenance and troubleshooting. This is an on-site school which is generally two to three days in length. It covers the basic operating principles of the mobile equipment, the operation of the fixed equipment, preventive maintenance procedures, troubleshooting procedures, familiarization with equipment wiring diagrams, and an in-depth study of the key mechanical and electronic subsystems onboard the vehicles and fixed equipment. The maintenance training school is conducted by a qualified technician or engineer. Usually several customer maintenance personnel are involved. If the system is to operate on a multishift basis, maintenance people from each shift should be present. They are given maintenance manuals on the system and are taught how the equipment operates through in-class training sessions combined with hands-on experience. It is a good idea to follow up this training with a second session, either at the customer's site or preferably at the vendor's plant.

Factory Schools

Schools held at the vendor's plant are an effective way to establish a high level of competence among key maintenance personnel about their new AGV system. These schools are held several times a

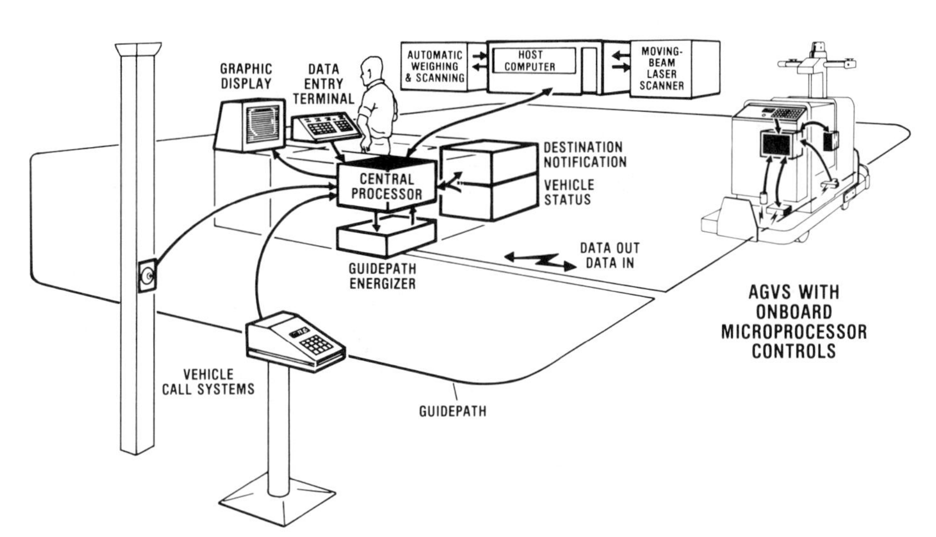

Fig. 8.1.19 Advanced distributed control system links many auxiliary functions into an integrated, automatic materials handling operation with onboard vehicle microprocessor controls as the central interacting link.

year and permit the new AGVS user to send maintenance personnel to a dedicated maintenance school which covers all aspects of the system in great detail. Generally these schools last from three to four days, 8 hr/day. Often, users send new plant maintenance personnel through the vendor-conducted maintenance schools. This insures that the user has properly trained personnel at all times.

8.1.13 Future Considerations

Highly advanced AGV systems have already been designed and are in operation.These systems interface with automatic storage systems on one hand and with manufacturing operations on the other. They are coordinated through a centralized computer to provide virtually automatic flow of materials through a manufacturing plant, taking raw materials to machines, and carrying finished products into storage or shipping. Such totally automatic systems are bound to become more prevalent in the future. An advanced distribution control system is shown in Fig. 8.1.19.

Another trend gaining momentum is the great sophistication of controls that permits a vehicle to operate more intelligently and thus fit more efficiently into complex handling patterns. For example, blocking controls will allow the departure and rapid reentry of independently programmed vehicles into a fast-moving guidepath, which maximizes material flow. Shortcuts will be built into the guidepath for faster movement between random loading and unloading locations.

Guidance controls on other materials handling equipment such as lift trucks will permit further reduction of human drivers as materials handling becomes more controllable.

8.2 UNIT LOAD CARRIERS

Bruce Boldrin

Originally, unit load carriers were defined as a class of automatic guided vehicle systems (AGVS) that moved unit loads between stations. Since then, there has been a substantial proliferation of models due to technological improvements in mechanisms and controls, and marketplace demand.

All unit load carriers have the defined characteristics of AGVS:

They are battery powered.
They are driverless.
They are guided on wire, chemical, and optical paths.
They automatically route and position.

Beyond these characteristics are many variations, including the following capabilities:

Automatic loading and unloading
Bidirectional travel
Towing
One, two, and four directions of travel

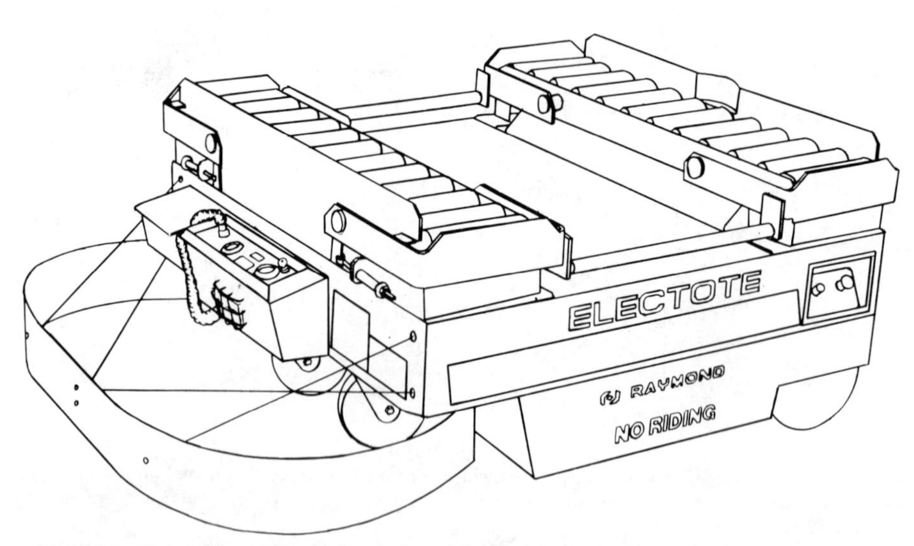

Fig. 8.2.1 Unit load vehicle, passive deck, unidirectional travel. (Courtesy of Raymond Corp.)

Computer control of vehicles and stations

Loads are moved between floor, rack, and pickup and delivery stations by bidirectional forked vehicles

This section defines the major categories and applications of unit load carriers. Considerations and guidelines for system design follows, including vehicle subsystems, pickup and deposit stations, system controls, determination of vehicle requirements, traffic control strategies, system planning, and justification criteria.

8.2.1 Categories and Applications

To set perspective, major vehicle categories are identified, along with typical loads and applications.

Unit Load Vehicles

Figures 8.2.1 through 8.2.4 show four representative models of unit load vehicles: two individual unit load transporters, a unit load vehicle that tows, and a roller deck model.

Fig. 8.2.2 Unit load vehicle, lift–lower deck, bidirectional travel. (Courtesy of Eaton-Kenway)

Fig. 8.2.3 Unit load vehicle with towing capability. (Courtesy of Teleco)

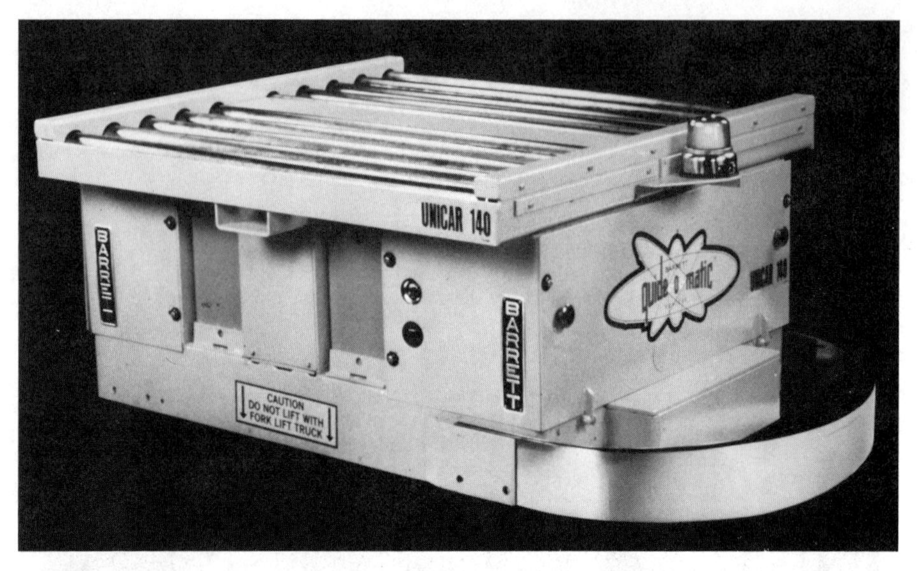

Fig. 8.2.4 Unit load vehicles with roller deck for side load/unload. (Courtesy of Barrett Electronics Corp.)

Module Transporters

Figure 8.2.5 shows a low-profile unit load carrier. The basic method of moving a load is to pick the castered module up slightly off the floor and transport it to a destination, where it is automatically off-loaded.

Fig. 8.2.5 Module transporter vehicle, carrying four-caster module on lift-deck. (Courtesy of Amsco)

Light-Load Transporters

Figures 8.2.6 and 8.2.7 show two versions of light-load transporters. One is a mail delivery vehicle. The second is a small unit load vehicle of 1000-lb capacity.

High-lift Unit Load Trucks

Figure 8.2.8 shows a bidirectional unit load truck that can pick up loads on the floor and deliver them to the floor, rack, or to elevated stations, and perform the reverse operation. An option is a highly accurate measuring wheel that allows the vehicle to leave the guidewire, perform programmed functions, and return to the guidewire—all under computer control.

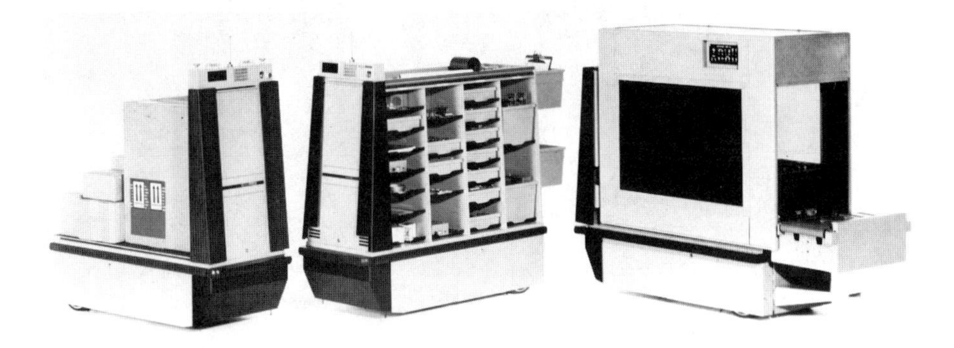

Fig. 8.2.6 Light-load transporters, used for mail, small parts, and tote pan delivery. (Courtesy of Litton)

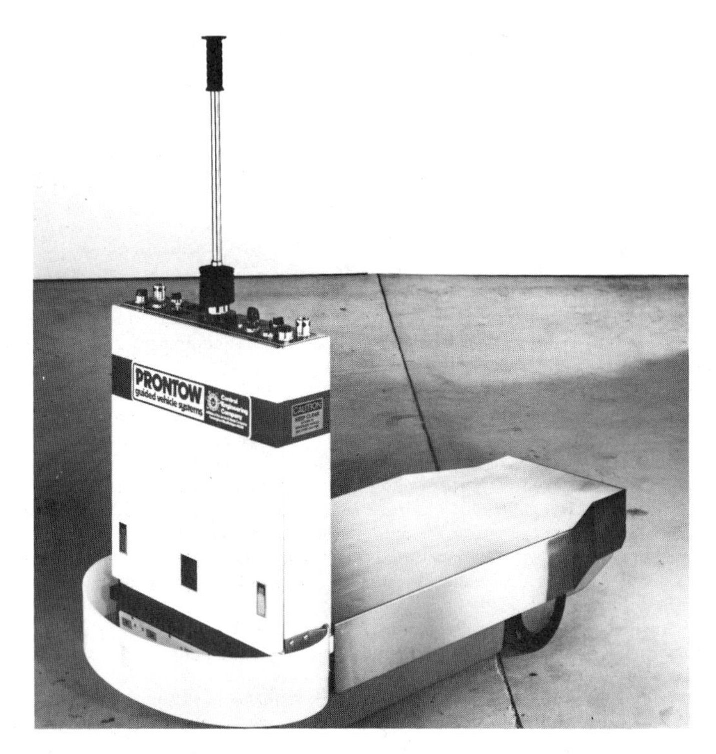

Fig. 8.2.7 Light-load transporter, used for unit loads within 2 ft × 4 ft envelope. (Courtesy of Control Engineering Co., Affiliate of Jervis B. Webb Co.)

Fig. 8.2.8 High-lift unit load truck moves loads between floor and elevated stations. (Courtesy of Conco-Tellus Inc.)

Types of Loads Handled

Unit loads can include productive material, nonproductive material, fixtures, and tooling. Unit loads

Types of Loads Handled

Unit loads can include productive material, nonproductive material, fixtures, and tooling. Unit loads
range in weight from 20 to 20,000 lb; in size from 14 in. × 20 in. × 6 in. height to 6 ft × 16 ft × 6 ft. Typical unit loads are 4 ft × 4 ft × 4 ft pallets or containers. In general, smaller loads consist of either cartons or tote trays. Larger loads may be palletized or directly handled (e.g., large paper rolls). Other types of unit loads include wheeled modules, trash and metal chip containers, and assemblies-in-process.

Typical Applications

Unit load carriers are being used in distribution and manufacturing industries, the office, and in hospitals. They are being used from the receiving dock throughout the manufacturing process to the shipping dock. These vehicles interface to storerooms using conventional equipment and to automated storage and retrieval systems (AS/RS).

Bidirectional vehicles are used to deliver material to picking platforms for order picking and parts kitting. In other applications individual loads of line stock are delivered to assembly stations by guided vehicles. To accumulate line stock, the vehicles off-load onto a queueing conveyor.

Unit load carriers are increasingly used in flexible manufacturing and assembly. Computer numerically controlled machining centers have tooling, fixtures, and parts automatically delivered on steel pallets on guided vehicles (Fig. 8.2.9). Figures 8.2.10 and 8.2.11, respectively, show a small vehicle for engine detailing, and a large carrier for automobile chassis assembly.

From manufacturing and assembly operations, unit loads can be taken to in-line inspection stations. At the station, visual checks are made by an inspector, who can address the vehicle to a specific destination premised on the inspection results.

Vehicles with precise positioning ability (±0.25 in. in x and y axes) can deliver loads not only to flexible manufacturing systems, but also to automatic testing equipment—including coordinate measuring machines and dynamometer test cells.

The light-load transporter with a rack on its deck can be used in a pickup/delivery route for mail and light packages. Multiple tote handling is possible with a side-loading mechanism.

Loads can be picked up directly on slipsheets or pallets and taken through a pass-through stretch-wrapping machine. This facility is useful to secure and stabilize loads that are going to be delivered by over-the-road tractor trailers, or subjected to rough handling before the point of use.

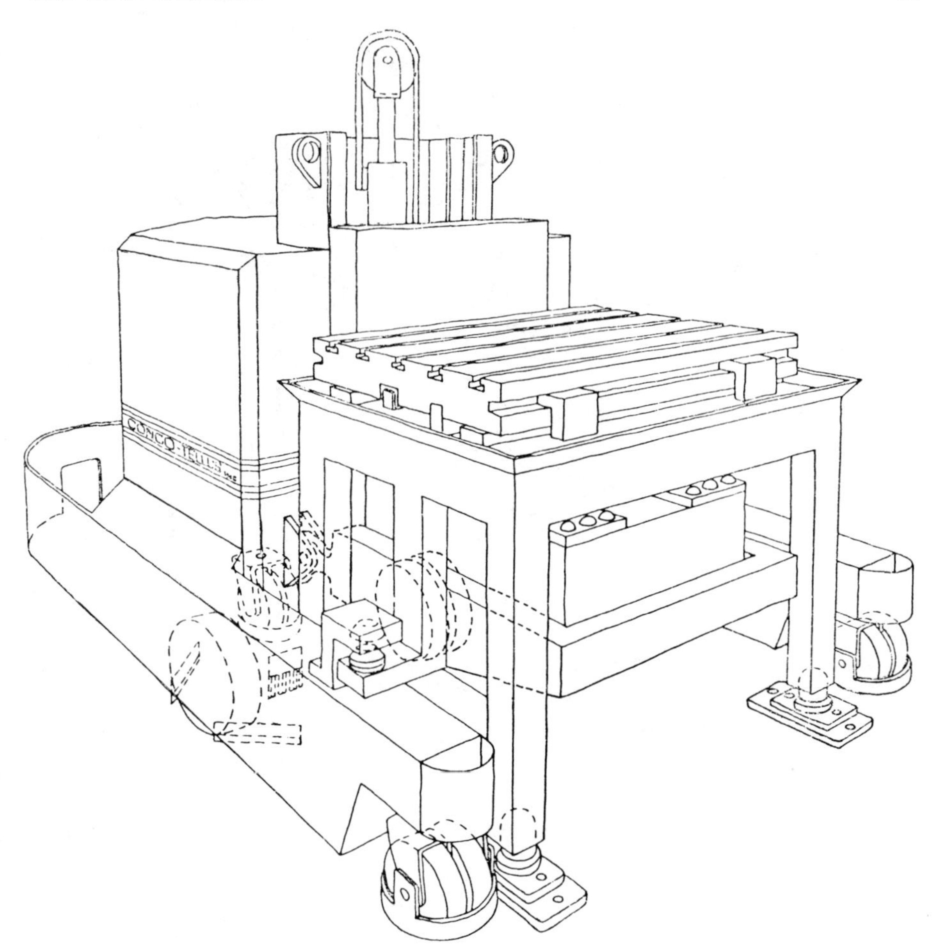

Fig. 8.2.9 Carrier brings parts and tools to machining center under computer control. (Courtesy of Conco-Tellus Inc.)

Module transporters are used in hospital applications. In one case, a bidirectional unit load vehicle transports modules of linens, surgical supplies, and food to multiple floors of multiple towers. All towers are connected through a single service level. Call and control of the lifts is automatic.

8.2.2 Subsystem Components

There are three subsystem components in a unit carrier AGV system: vehicles, controls, and pickup and deposit (P&D) stations. The description of controls in the earlier part of this chapter, on driverless tractors and pallet trucks, has substantial application to unit load carriers. It will be cross-referenced in this section with comments on those aspects unique to unit load carriers.

Vehicle Subsystems

Guidance. The major guidance techniques include optical, chemical, and inductive tracking.

Optical and chemical guidepaths are the easiest to install and change. Paths can be taped down, painted, or sprayed on the floor. However, these guidepaths must be clear and clean. If the guidepath is faded or covered up, the vehicle sensor will not properly sense the path, and the vehicle will automatically stop.

Chemical path tracking uses an ultraviolet light source on the vehicle that stimulates fluorescent particles in the guidepath to emit light. This light is then "read" and tracked by the vehicle.

Wire or inductive guidance is the main technique used in industry. The embedded guidepath carries a signal with low current (less than 400 mA), low voltage (less than 40 v), and low frequency (1–15

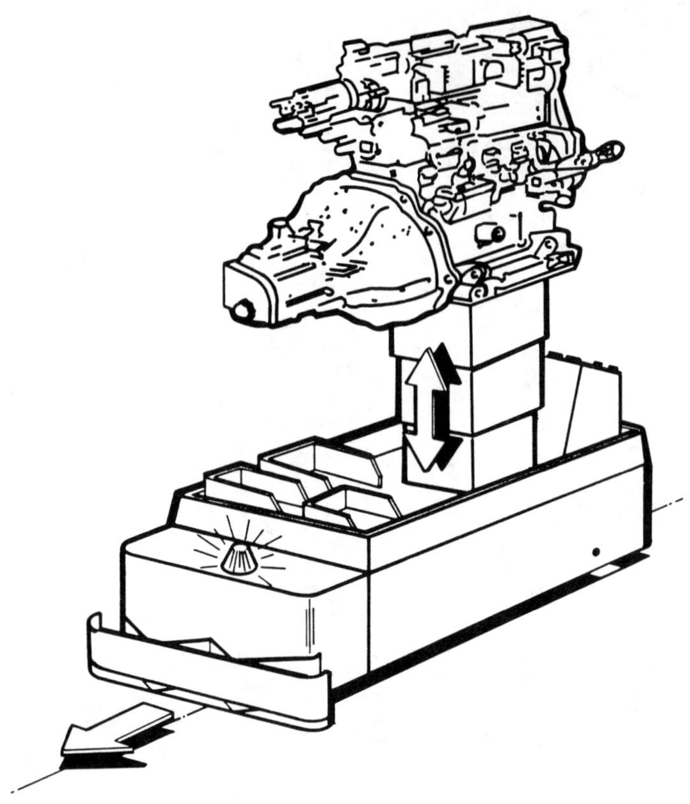

Fig. 8.2.10 Engine dress-up is completed in flexible assembly application. (Courtesy of Eaton-Kenway)

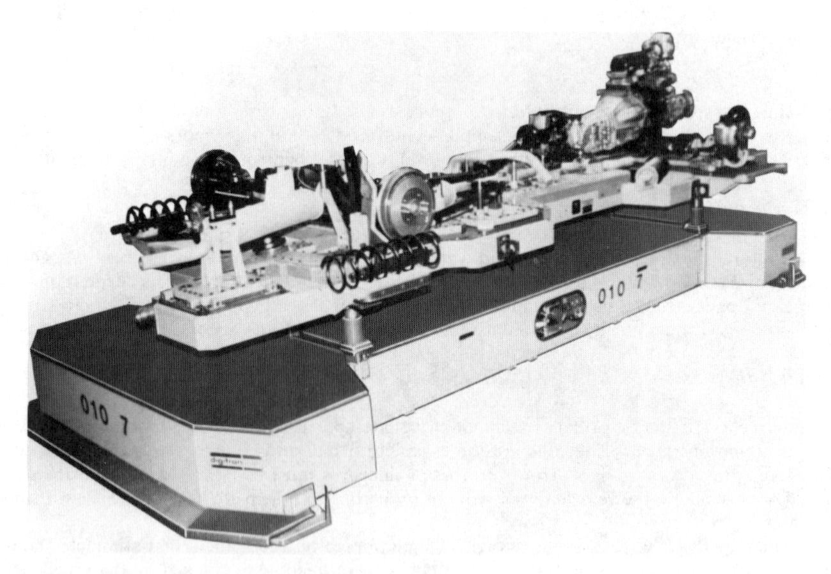

Fig. 8.2.11 Complete automobile chassis assembly is performed on fixture carried by vehicle. (Courtesy of Eaton-Kenway)

kHz). Stable frequency sources and high discrimination on the vehicle against signals outside the tracking frequency contribute to high noise rejection and reliable guidance. CAUTION: Close proximity of the wire guidepath to an intense broad-spectrum noise generator, such as arc welders, should be avoided. To date, no standards on setback distance have been set. Noise avoidance is established empirically—by moving welder, guidepath, or both. A guidepath setback of 10 ft from the welder tends to be a safe distance.

To place vehicles on the guidepath, acquisition distance to obtain lock-on to the guidepath is between 2 and 4 in. for most AGVs. Reciprocally, departure of the vehicle from the guidepath by 2–4 in. will cause automatic shutdown.

Drive and Steering. Drive and steering are achieved with the same drive unit(s). In general, one of four mechanisms is used: (1) single-stacked motor-transmission-drive (M-T-D) wheel, with steering servosystem; (2) two fixed M-T-Ds located at one end of vehicle for single-directional travel; (3) two fixed independently driven M-T-Ds, located in center of vehicle, for bidirectional travel; and (4) independently driven and steered wheels for four-directional travel. Figure 8.2.12 shows a plan view of mechanisms 1, 3, and 4.

Some vehicles require flat floors to maintain wheel traction over all parts of the guidepath. The reason for this levelness ($\frac{1}{4}$ in. in 5 ft) requirement is the rigid vehicle frame construction where the supporting wheels could place the driving wheel over a depression and lose traction. Pivoted drive axles, three-wheel vehicles, and articulated frames have been used to improve operation over rough and unlevel floors.

The advent of microprocessor controls has provided the ability on some unit load vehicles to "dead reckon" in-path travel and turns. This feature enables inductively guided vehicles to cross steel plate, by controlling travel between departure and reacquisition of the guidewire path.

Speed Control. Speed control selection is available on some AGVs. Vehicle speed is generally set by codes on the floor where the codes are magnets or plates. Speed control zones are easy to move or modify.

Vehicle Addressing. Vehicle addressing can be performed directly or remotely. The devices for direct programming include onboard thumbwheel switches, rotary switches, and touch pad keyboards.

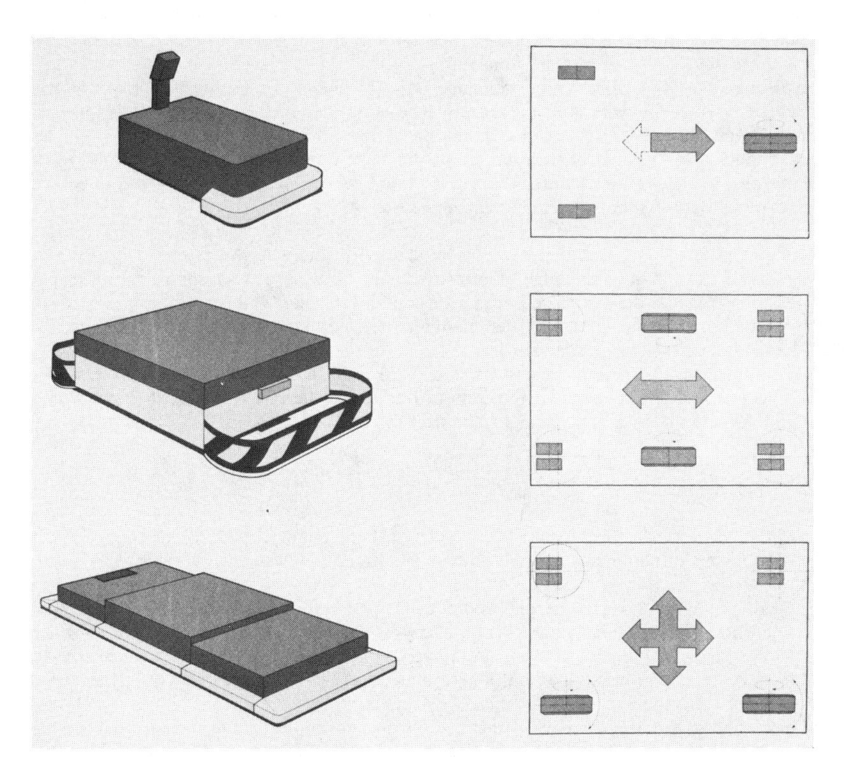

Fig. 8.2.12 Three drive-steer geometries. (Courtesy of Eaton-Kenway)

Off-board devices include remote addressing panels and automatic addressing through radio frequency (RF) transmission or direct commands by a higher-level controller (computer).

To maintain material tracking integrity some automatic systems do not readily allow manual intervention. Vehicles in this category generally interface with other automatic processes or storage systems, and require load tracking information to be transferred from one computer to another.

Additional methods and further details are provided earlier in this chapter, as is the subject of automatic route control.

Positioning Accuracy is important for unit load carriers. Vehicle stopping accuracy ranges from ±3 in. to ±0.010 in. Manual load and unload can tolerate the ±3 in. repetition. Greater accuracy is required for the automatic transfer onto and off P&D stands. Examples of acceptable tolerances follow:

Interface Method	Tolerance
Manual load and unload	±3 in.
Conveyor interface	±1 in.
AS/RS P&D station	±0.25 in.
Machine tool interface	±0.010 in.

Collision Control is covered earlier. These techniques apply in general to unit load carriers. Safety and warning devices also are covered earlier.

Manual Controllers. Manual controllers are either built in or can be plugged into the vehicle. When vehicles are designed to operate in both manual and automatic modes, the controller (for vehicle movement functions) is built in. For dedicated automatic operation, the plug-in controller avoids unauthorized use of the vehicle and potential disruption of the load tracking system.

Battery Requirements. Battery power requirements are a function of the duty cycle. The design parameters for most AGVs are to support 8 or 16 operational hours with 85% vehicle utilization and either charge onboard or change batteries when depleted. Battery capacity calculations are provided earlier in this chapter.

Status monitors of battery charge can be displayed onboard the vehicle for manual monitoring, or transmitted to a central controller for some AGVs. On these vehicles, the transmitted battery status can then trigger a subroutine where the vehicle is commanded to complete its current task and be routed to the battery change area. A message identifying these actions is displayed on the system operator's console for immediate followup.

In a typical battery charge area, the exchange on one vehicle is typically performed in less than 5 min. Eyewash, deluge showers, and floor drain to a sump are included in a well-designed battery charge area. Space requirement is a function of the battery change technique for the vehicles, and the battery quantity per vehicle. A rule of thumb for side removal of batteries is to allow for the vehicle width plus maneuvering room for the battery handler. AGV manufacturers can provide details of space and power requirements in the battery charge area.

Maintenance Aids. Maintenance aids are available with many AGVs. Some diagnostic aids are built in, such as LED lights at the edge of printed circuit (PC) cards and control panel displays on the vehicle. Some plug-in diagnostic aids can exercise the vehicle while monitoring and displaying the status of components and circuits. Other maintenance aids include vehicle jack stands, PC card extenders, and manual controllers.

Test tracks, adjacent to maintenance bays, are often used in AGV systems. These tracks enable off-line checking of a repaired vehicle before reinserting it into the system. Typical tracks would have left and right turns with speed zones (e.g., a "figure-8").

Guidepath Layout

Three elements of guidepath layout must be considered: floor quality, AGV envelope, and equipment interface. A summary of comments and design considerations follows:

Floor Quality. What levelness is required for the types of AGVs being considered? What are the ramp conditions—slope and length? Is there any steel plate on the floor surface? Surface steel with lengths greater than 1 in. could affect inductively guided vehicles. Is there sufficient traction for the AGV to start, turn, and brake under all operating conditions? If not, slower speeds or resurfacing may be required in the zone.

Wood block floors tend to become problematic with time, due to the flexing and eventual replacement. A guidepath will not out-survive the floor; replacement of path must occur if the floor is replaced.

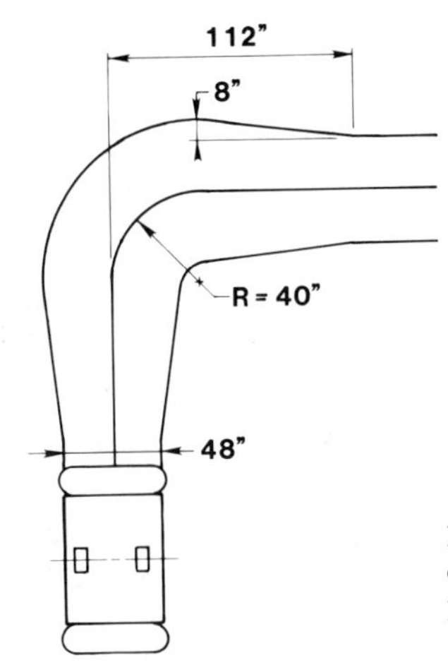

Fig. 8.2.13 Travel envelope of center-steered vehicle with uncompensated, tangential turns.

AGV Envelope. Every AGV sweeps out a repetitive space envelope during travel. Normally, 6 in. of clearance is added to all sides of the physical envelope to assure spatial separation between the AGV and nearby objects. There are two common exceptions to the 6 in. rule: (1) where safety considerations dictate greater space for personnel, and (2) where the physical interface to another piece of equipment requires closer spacing (e.g., P&D stands).

Characteristics of the space envelope, in turn, depend on the physical shape of the vehicle and steering geometry. For example, nose overswing of rear-wheel-drive vehicles is greater than for vehicles with front-wheel- or center-wheel-drive; however, tail swing is negligible.

Figure 8.2.13 shows the travel envelope of a center-steered vehicle for 90° or 180° turns. The actual envelope shape is defined by both nose and tail swing. Characteristic of most unit load carriers, peak overswing is approximately opposite the center of the radius turn.

Vehicle alignment at a specific station depth after a turn is information needed for placing P&D stands and avoiding obstructions. Stand location can be inferred from this figure. P&D stands need to be placed so that load transfer is within acceptable alignment tolerances.

If the distance to a P&D stand, or an obstruction, is too short for a 90° tangential turn, a compensated turn can be made. Compensated turns overshoot the 90° path and require the vehicle to turn more than 90° before the radius turn ends. This technique is comparable to a truck driver overturning a 90° corner with a semitrailer rig to avoid the corner post. Figure 8.2.14 shows compensated 90° turns. By comparison with Fig. 8.2.13, the vehicle is aligned with the guidepath 2 ft closer to the main path using a compensated turn.

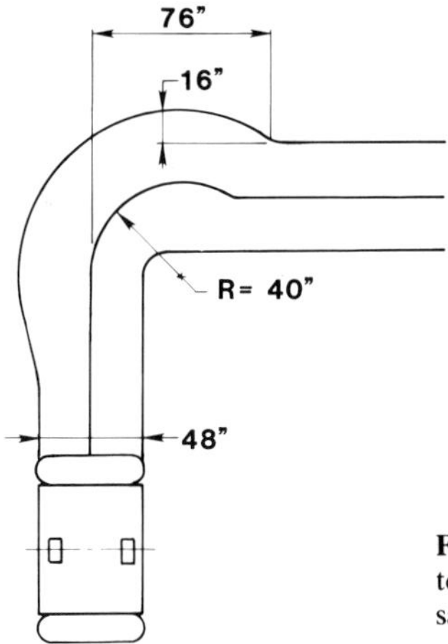

Fig. 8.2.14 Travel envelope of center-steered vehicle with compensated turns.

The newer class of bidirectional vehicles can make a turn about their own axis, producing the minimum spatial envelope for a change in direction.

Four-directional AGVs move laterally into a P&D station. This rectilinear motion allows stations to be at the edge of AGV aisles, and minimize the production area required for P&Ds.

Several models of AGV use a mitered 90° turn (vs. a radius turn). The benefit of mitered turns is in the guidepath installation: Only straight-cutting concrete saws are required to cut the guidepath. Typically, these AGV models operate at lower travel speeds than models using radius turns.

Equipment Interface. There are two types of AGV interface to be evaluated: mechanical and electrical. The mechanical interface consists of specific positioning of vehicle and P&D station—in order that the appropriate load transfer can be performed. The AGV envelope covers a portion of this interface. Levelness of vehicle deck and stopping accuracy are two other considerations. For example, proper right-angle load transfer between a roller deck vehicle and a roller P&D stand requires both to be level. This requires a level floor and rigid vehicle structure. Due to the vertical deck travel, lift–lower decks compensate for greater unlevelness of floor and P&D stands.

Additionally there are electrical and control requirements to assure automatic transfer. Two such devices are a load present sensor and alignment photocells on stands—used for final fine positioning by some AGVs.

A steady-state signal source is required for all inductively guided vehicles. These sources require low power, typically a 115-v, 15-A circuit.

Optical and chemical guidepaths are passive, requiring no electrical power.

Pickup and Deposit (P&D) Stands

P&D stand or station is a general name given to any position where loads are picked up or deposited by an AGV. The stands are typically elevated, although with the module transporter and high-lift unit load transporter, P&D stands can be stations on the floor. A summary of typical P&D stands follows:

Station	Comments
Manual	Uses other equipment or hand power to load and unload.
Simple stand	Automatic load and unload off and onto a mechanical stand.
Conveyor stand	Chain or roller conveyor. Can be powered or nonpowered roller. Nonpowered roller may be gravity or slave driven.
Slave drive stand	The P&D is powered by a mechanical (Fig. 8.2.15) or electrical drive mechanism on vehicle. Both mechanisms use AGV power to transfer a load.
Floor station	Usually has mechanical load alignment devices (guide angles for pallets, detents for module wheels) (Fig. 8.2.15).

Changing Load Elevation. Scissor lifts and multifloor lifts are used to change elevation of individual loads. Lifts are used to move vehicles between floors. Interactive AGV and lift controls are required to call, dispatch, and accurately move and stop the lift. Lift capacity and floor space must be adequate for an AGV with full load. Threshold gaps of $\frac{1}{4}$–$\frac{1}{2}$ in. are typical. Some AGVs require $\frac{1}{4}$ in. and levelness within $\frac{1}{4}$ in.; other AGVs can "dead reckon" across the metal threshold.

P&D Design Rules. The following items must be considered in the design and placement of a P&D stand:

1. Is any load *accumulation* required? If so, conveyor or drive-through and drop (set-down) techniques must be used. Single load positions can use simple stands.
2. *Human engineering:* If personnel access to a load is required, there should be sufficient room on three sides. Containers should be low enough to allow operators to reach over the top. Guided vehicles with lower deck heights or lowering P&Ds could be selected to provide improved top access.
3. *Fork truck* servicing requires that the stand be reinforced or mechanical guarding be provided.
4. *AGV alignment* is critical to several types of stands. The guidepath and stands must be positioned to assure this alignment and the necessary mechanical clearances.

Fig. 8.2.15 Slave driven pickup and deposit conveyor (Courtesy of Control Engineering Co., Affiliate of Jervis B. Webb Co.)

5. *Load sensors* monitor load presence/absence. They provide a control signal to assure load transfer at a station.
6. *Electrical power* for conveyor drive motors and air pressure for accumulation devices may be required. Specify applicable codes; for example, National Electric Code.

Size and Weight Profiling Stations

In applications involving an AS/RS, inbound loads must be within a size profile and weight limit to be automatically handled by equipment beyond the AGV system. Loads outside acceptable limits would be rejected for subsequent corrections.

Weighing and sizing can be done with the AGV and load parked at a size/weight profile station, or "on-the-fly." On-the-fly sizing can be performed when the vehicle crosses a floor-flush weight table and passes through the (stacked) beams of a photoelectric device. Size/weight profilers can achieve size accuracies within $\pm\frac{1}{4}$ in. and weight within 1%.

Increasing the accuracy of weighing requires the load be set down on a weigh station and tare-out of the container weight.

System Controls

Manual Control Panels. Manual control panels allow station addressing. Figure 8.2.16 shows a panel at a P&D station. The station panel calls an empty vehicle to pick up a load, transfers the load address to the vehicle, and dispatches the vehicle when load transfer is complete.

Move-Request Stations. There are several types of move-request stations. Selection depends on the number of addresses to which a load may be dispatched.

Address	Mechanism
Single-dedicated destination	Push button or switch
3 or less dedicated destinations	Push-button panel
4–99 destinations	Move-request console (Fig. 8.2.16)
4–500 destinations	CRT terminal in centrally controlled systems

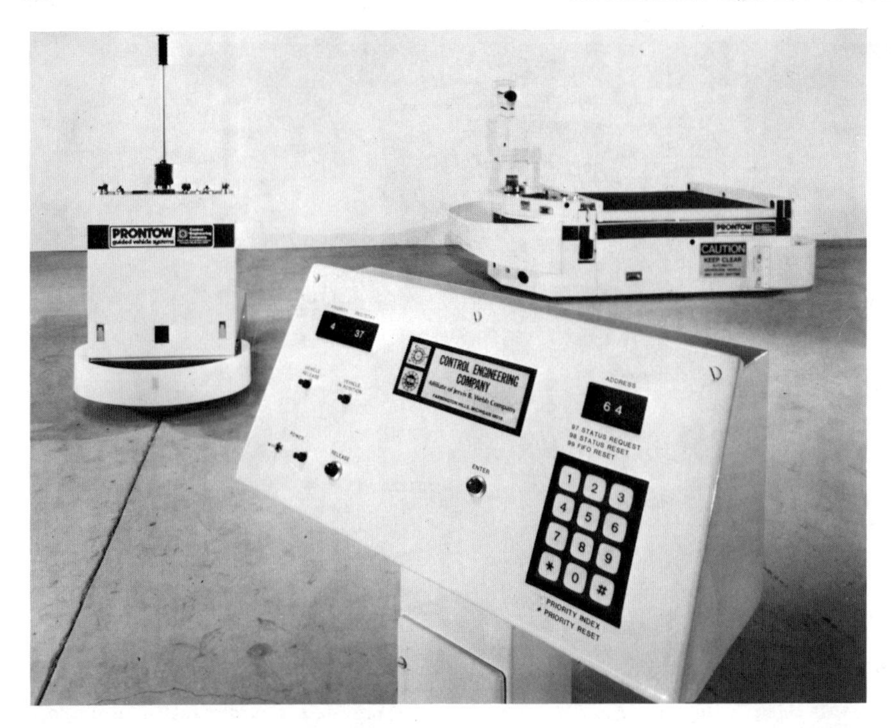

Fig. 8.2.16 Manual entry panel performs unattended load dispatch. (Courtesy of Control Engineering Co., Affiliate of Jervis B. Webb Co.)

CAUTION: Do not use "load present" switches on manually loaded stations to automatically request load movement. The rationale for this restriction is that a person may have second thoughts about load alignment and could be attempting to correct positioning when a requested vehicle automatically arrives.

Door Controls. For centrally (computer) controlled systems, the interface for door control is a discrete "open–close" signal from the computer with a limit switch to assure the door is open for vehicle passage. Distributed systems use signals from the local controller, traffic detector loops, or RF transmission from the vehicle to open and close doors.

8.2.3 Control Architecture

Simple guidance and control systems use a line driver and control codes in the floor to achieve vehicle speed control and station addressing. They do not track material. More sophisticated systems directly tie into factory production and inventory control systems, providing on-line feedback of material information status.

In an on-line system, the corporate Materials Resource Planning (MRP) and production control systems are directly linked to an AS/R system that uses an AGV system for material transportation. Material is tracked directly to the using department by the real-time controller (RTC) and notification of its delivery is passed to the corporate computer system. For critical operations, the host computer is backed up with a redundant computer and often, an uninterruptible power supply (UPS) to allow controlled shutdown and recovery in the event of a computer failure.

8.2.4 Backup Strategies

A series of backup operations must be planned into the system during design phases. Some typical levels of backup follow:

System Situation	Backup
Loss of guidance or addressing	Operate vehicle manually.
In system with direct command by remote programming station—loss of remote addressing	Command input directly to vehicle.
Stand-alone computer-controlled systems—loss of computer for	
1. Minutes	1. Suspend operations, recover computer, and proceed with automatic operation.
2. Hours	2. Operate vehicles semiautomatically without load tracking. Recover the data base after computer is repaired.
For systems using computer control with tasking by a host computer—loss of host computer	Operate with AGV computer control using terminal input for commands or work off buffer of stored commands. Maintain audit trail of completed tasks to update host computer when it returns to an operational status.

8.2.5 Calculating Vehicle Requirements

There are three practical methods to develop vehicle requirements. Two methods are computational, that is, deterministic. The third involves simulation, is probabilistic, and requires a computer. These methods and applicability are:

Basic computation. Applies to single-directional vehicles and bidirectional vehicles without central dispatching control.

Advanced computation. Applies to bidirectional vehicles with central, automatic dispatching control.

Simulation. Applies to the more complex systems, especially those involving direct interface to other automated equipment or processes and having substantial amounts of activity and/or complex travel paths.

Simulation techniques, such as general purpose system simulation (GPSS), have broad validity. But due to the computer requirement, these techniques are not valid subjects for a handbook; hence, they are treated elsewhere.

Basic Computation

The assumptions for this method are:

1. Each vehicle completes a loop of travel, returning to its starting point.
2. There is minimal vehicle contention allowing superposition of vehicle requirements for individual loops.
3. The equipment utilization and acceleration/deceleration/contention factor is 80%, equating to a planning window of 0.80 × 60 min/hr = 48 min/hr.

The vehicle requirement for travel loop L is computed as follows:

$$C_i = \frac{L}{V} + P + D \qquad (8.2.1)$$

$$N_i = \frac{TC}{48} \qquad (8.2.2)$$

where: C_i = unimpeded, full-speed travel time over loop L_i (min/cycle)
L_i = length of travel, returning to the starting point (ft; m)
V = maximum vehicle speed (ft/min; m/min)
P_i = pickup time to load the vehicle (min)
D_i = deposit time to unload the vehicle (min)
N_i = number of required vehicles
T = throughput of loop L_i (loads/hr)

For multiple loops $L_1, L_2, \ldots, L_n$, perform the calculations for each loop and add the results. If the start and end points of all loops do not coincide, provide additional vehicles to allow for interloop travel, using (8.2.1) and (8.2.2) with L = travel distance between starting positions and $P + D$ = dispatch and coordination time. Using (8.2.3), add vehicle requirements for all loops to obtain the total.

$$N = N_1 + N_2 + \ldots + N_n \tag{8.2.3}$$

Example. Two loops with the same starting points:

Loop 1: $L = 600$ ft, $V = 150$ ft/min, $P = D = 1$ min, $T = 15$ loads/hr

Loop 2: $L = 450$ ft, $V = 150$ ft/min, $P = D = 2$ min, $T = 20$ loads/hr

(8.2.1) $C_1 = \dfrac{600}{150} + 1 + 1 = 6$ min

(8.2.2) $N_1 = \dfrac{15 \times 6}{48} = 1.79$ vehicles

(8.2.1) $C_2 = \dfrac{450}{150} + 2 + 2 = 7$ min

(8.2.2) $N_2 = \dfrac{20 \times 7}{48} = 2.92$ vehicles

(8.2.3) $N = N_1 + N_2 = 1.79 + 2.92 = 4.71$ vehicles

Round up to 5 vehicles.

This method of subdividing the problem is a good first-order approximation of vehicle requirements. In applying this analysis, test the results for sensitivity to different factors, by inserting values for the best and worst cases. For example, manual (fork truck) loading and unloading of vehicles may require substantially different times than were planned.

Peak loading can be evaluated by assigning a "peaking factor" to throughput, that is, multiplying throughput by the peaking factor before using (8.2.2). Factors of 15–30% are typical in distribution operations, 10–20% in manufacturing. Different loops can have different peaking factors, for example, 30% on the receiving dock (1.30 T) and 15% in manufacturing (1.15 T). Be confident in using these factors as they apply to a specific operation, because they more accurately reflect actual conditions.

Advanced Computation

The concept embodied in this technique is optimization of vehicle travel based on current vehicle locations. The equipment represented is typically computer controlled. The computer has travel algorithms that can rapidly analyze current tasks and traffic conditions and assign the best resource (vehicle) to move a load.

The base for these calculations is actual load movement by a vehicle; unloaded travel is the variable factor. The amount of unloaded travel depends on the control algorithms, guidepath layout, and vehicle capabilities. These factors can be netted out with a multiplier that depends on system configuration. In effect, the multiplier accounts for additional travel and strategic parking—for vehicles to be prepositioned for the probable appearance of downstream loads.

To apply this technique, the methodology requires a load movement table (Table 8.2.1) showing peak throughput rates. All moves are entered in the table. The form is a matrix with all pickup locations on the ordinate and all deposit locations on the abcissa. All columns and rows are totalled.

Next, develop a table of required move distances in the form of Table 8.2.2.

The calculations for throughput include an optimization factor K, and utilization factor of 80% (0.80 × 60 min/hr = 48 min/hr):

$$N = \frac{1}{48}\left[(91 + K)(\Sigma L_i T_i) + \Sigma(\text{P\&D})_i T_i\right] \tag{8.2.4}$$

Total vehicles = The sum of all N_i

where:
N = number of vehicles
K = optimization factor (0.45–0.65) for small, highly optimized computer-controlled systems; (0.60–1.0) for larger, optimized systems
L_i = length of travel for throughput T over segment I
T_i = throughput over segment i
$(\text{P\&D})_i$ = pickup and deposit time for segment i

Table 8.2.1 Load Movement Table[a]

	Deposit						
Pickup	Rcvg.	Dept. 10	Dept. 20	Dept. 30	. . .	Ship.	Total Pickup
Receiving	X						
Dept. 10		X					
Dept. 20			X				
Dept.30				X			
⋮					X		
Shipping						X	
Total deposit							

[a] The second step is to lay out the guidepath to accomplish all required flow. Use a "best guess" and expect to modify the layout as the analysis reveals potential improvements. The process is iterative.

The *optimization factor K* represents several elements: "deadhead" travel distance to pick up the next load, parking between moves, and traffic contention. Cross-checking can be accomplished through relating the physical flow of loads to loaded and unloaded vehicle movement. In reality, all travel legs must be included to obtain vehicle requirements. For example, if the vehicle always travels with a load and has no deadheading, $K = 0$ for travel, with some finite value for parking and contention. Alternatively, if the vehicle deadheads the same distance back as it travels loaded out to a P&D, $K = 1.0$ (plus parking and contention). Excessive deadhead travel means K will exceed 1.0.

8.2.6 Calculating Battery Requirements

Proper battery sizing in an application is important to assure proper duration and magnitude of performance. Battery requirements relate to the move cycle dynamics. Normally, they are calculated for one typical cycle and extrapolated to the full shift or full day requirements.

Usage factors to include are stated in ampere-seconds of ampere-hours (3600 A-sec = 1 A-hr). Perform separate calculations for the ramp and level portions of travel. Add together for total requirements. A summary of usage factors follows:

Travel, loaded (A-sec/ft)
Travel, unloaded (A-sec/ft)
Acceleration (A-sec/each)

Table 8.2.2 Required Movement and P&D Times

Load Movement						
Pickup	Deposit	Travel Distance L_i	Throughput T_i	$L_i \times T_i$	P&D Time $(\text{P\&D})_i$	$(\text{P\&D}) \times T_i$
Rcvg	Dept. 10					
Rcvg	Dept. 20					
.		(From layout)	(From Table 8.2.1)	Units are $\frac{\text{Load-ft}}{\text{Vehicle-hr}}$		
.						
.						
Dept. 10	Dept. 20					
Dept. 10	Dept. 30					
.						
.						
.						
Dept. 10	Shipping					
.						
.						
.						
Total				$\Sigma L_i T_i$		$\Sigma(\text{P\&D})_i T_i$

Deceleration (A-sec/each)
Picking up a load (A-sec/each)
Depositing a load (A-sec/each)
Idle (not traveling) (A-sec/min)

After adding usage factors for an average cycle, multiply by the number of cycles per day and restrict the total usage to less than 80% of the rated battery capacity. Obtain the battery ampere-hour capacity and usage factors from AGVS suppliers.

For a good first-order approximation, use the loaded and unloaded *travel factors* and 70% of total capacity.

8.2.7 Flexible Assembly Systems

Flexible-path assembly systems provide several distinct operational benefits compared to conventional linear-path assembly systems. Flexible assembly systems are applied when the number of models exceeds 100, the time per assembly has exceeded a 2:1 ratio for different models, and production rates are 800–1600 units per day.

There are two flexible path assembly concepts using driverless unit load carriers. The typical vehicles and concepts are identified by the type of work station in the system:

1. *Mobile Work Platform.* Units to be assembled are transported by and remain on the unit load vehicle through all *work stations.*
2. *Fixed Work Stand.* Units to be assembled are picked up, transported, and deposited on fixed *work stands* by the unit load vehicles.

Both flexible path concepts have a commonality of modular design:

Work Grouping. Every station in a work group is in parallel and has the same work scope. Typically, a station has 4 or more minutes of elemental assembly time. In most systems, work content per station group varies from 4 to 10 min. As model mix increases, the work group efficiency increases relative to linear lines.

Parts Provisioning. The unit to be assembled stops at a parts provisioning station between work groups. The parts provisioner places major parts on the unit or vehicle, often prepositioning parts to assist the assembler. This factor requires a readjustment of the assembler's work standards to reduce materials handling functions.

Work Station Tools and Parts. Expense stock, such as bolts, washers, and fasteners, are stocked and replaced at the work station. Thus, these parts are not the concern of the parts provisioner. Tooling is duplicated at each station in the work group, a factor that must be included in system cost.

Bypass Capability. If assembly is not required in a work group, the group can be bypassed. A second aspect of bypass is used for problem solution or rework. When a problem is identified that cannot, or by policy will not, be repaired at a work station, the assembly unit can be moved by vehicle to a grief/repair/rework station.

Intergroup Buffers. For throughput leveling, there is inherent intergroup buffering. These buffers allow decoupling of the work groups, so that problems or events within one work group do not immediately affect the next, downstream work group; nor do they immediately affect the upstream group. This characteristic can be used to smooth line flow.

Universal Vehicle. The unit load vehicle is a universal carrier for all models to be assembled. Thus, there is a standard interface to the product, such as a pallet or fixture, whose mating mechanism to the vehicle is constant. For new models, the pallet or fixture top is modified.

8.2.8 System Specifications

Define System Mission and Objectives

The first step in planning an automatic guided vehicle system is to define the system mission and develop quantized objectives. The next step is to set system performance criteria.

System Performance Criteria

1. Specify *loads* to be handled by the AGVS, including length, width, height, weight, and stability. Shape and material of the load conveying surface are often important specifications.

2. Define the *material flow* rates in and out of stations and through the system. Flow rates between load stations can be specified in a load-movement matrix. Peak rates and time frame are critical to develop accurate system sizing.

3. Specify *layout* requirements, including floor material and levelness, location and grade of ramps, fire doors, elevators, and available electrical service.

4. Specify all *machine interfaces.* Quantify layout position and dimensions, operator's location, and safety constraints.

5. Specify *system information requirements,* including format and data to command load movement, on-line inquiry for vehicle position and destination, and management reports. Identify required locations and activity rate of centralized dispatch versus distributed command stations. Centralized dispatch may be achieved with a host computer, which requires specification of the host computer and model, line protocol, and information content to be transmitted and received. Specify system controller requirements, if your plant standards require programmable controllers or computers from specific manufacturers.

6. *Maximize availability.* System availability is the total time a system may be used to sustain its throughput rate. Failure of individual components of subsystems should not affect the measurement of system availability unless the specified throughput rate cannot be met for critical operation. Specify a backup computer and uninterruptible power supply (UPS) system for maximum system availability and orderly shutdown of the computer in the event of power failure.

7. *Minimize downtime.* Downtime is that percentage of the total available time when a system is not capable of performing its intended function due to failure of system components or subsystems. It is difficult to accurately predict downtime, since downtime is directly related to many factors beyond the control of the system supplier, including proper maintenance, operators' skills, compliance with operating procedures, and availability of spare parts and repair facilities. With proper maintenance, spares, and training, downtime has been held below 2–5%.

8. *Plan utilization.* Utilization is a measure of the percentage of time that the AGVS is actually used during the time it is available. Allowance has to be made for errors, operational interruptions, and so on, to permit peak rate processing and perform preventive maintenance. It is common to plan for system utilization at something less than 90%; however, this varies from application to application.

9. Specify *service, parts, documentation, and training* requirements.

Typical Scope of Work

The division of work in a project can be conveniently divided into vendor's and customer's scope of work. This specific statement of scope assists both parties in identifying their respective responsibilities. Scopes of work are quantified when applicable. In a typical AGV system under computer control the scope would appear as follows:

Vendor Scope of Work:

(Qty) Vehicles
(Qty) Batteries
(Qty) Chargers
(Qty) Charge stands or charge stations
(Length) Guidepath (traveled path)
(Qty) Frequency generator (line driver, signal source)
(Qty) Zone controllers
(Qty) Controlled fire doors
(Qty) Pickup and deposit stands (stations)
(Qty) Computer system
(Lot) Design services
(Lot) Installation
(Lot) Documentation
(Lot) Training
(Lot) Spare parts
(Lot) Warranty
(Lot) Freight onboard (location)

Customer Scope of Work

(Lot) Floor quality and levelness
(Lot) Electrical service to specified service panels or locations
(Lot) Hours (shifts) available for installation
(Lot) Water for floor cutting
(Lot) Type of saws and splash protection
(Lot) Personnel for acceptance test
(Lot) Maintenance after acceptance
(Lot) Containers and loads to be used in operational system
(Lot) Integrity and quality of pallets or other containers

System Installation Sequence

All AGVS vendors offer vehicles and guidepath installed, with service manuals and warranty. In addition, several AGVS vendors offer feasibility studies, simulation, design services, manuals, training, and maintenance contracts—to the extent of assuming turnkey responsibility.

Installation normally occurs in the following sequence:

1. Layout of the guidepath
2. Cutting the guidepath
3. Placement and connection of the guide wire, frequency generator, and floor controllers
4. Physical and electrical continuity check of the guidepath
5. Covering the guide wires with grout
6. Installation of P&D stations with load sensors
7. Installation of the computer
8. Completion of controls wiring
9. Subsystem checkout (guidepath, vehicles, controls, computer)
10. System checkout
11. Training of customer's maintenance and operational personnel
12. Acceptance tests
13. System commissioning and start of beneficial use by customer

8.2.9 System Justification

The following criteria are typically the major elements of system justification. The list is not total, nor is it prioritized for all applications. However, it is a checklist of elements to consider. Payback, return on investment, or other methods used by your company are based on the amount of investment and potential savings. Thus, it is important to thoroughly analyze each alternative system and quantize the savings.

The evaluation should compare all vendor-provided machinery, controls, and services. Expect some justification criteria to be more important than another, possibly more important than labor savings. In an AGVS, the savings often extend beyond materials handling labor.

Item	Contributors
Labor savings	Direct: material handlers and production personnel Indirect: clerks, dispatcher, and expediters Increased productivity by reduction of search and handling of materials by production personnel Improved equipment utilization, because AGVS can work while the labor force rests, or is away from the plant
Space recovery	Reduction or elimination of line stock in production areas Narrower aisles, due to tight tracking and two-way travel of AGVS
Real-time inventory control	Planned delivery schedules Reduction or elimination of time lag in information updating

Item	Contributors
	Transaction audit trails
	Reduced paperwork
	Reduced lost material
	Reduced inventory carrying costs
	Timely, accurate information to make the best management decision
Better handling	Reduced damage to material
	Less equipment damage, thus reduced maintenance and unplanned downtime
Greater availability	Extra production time
Improved customer service	Increased share of market in service-oriented society
	Reduced production line stoppage
Energy and environment considerations	Less electric power is required by transporters than lifting vehicles
	Clean operation
	Off-shift operation in reduced light or total darkness
System expandability vs. alternative systems	Cost
	Schedule
	Install controls and minimum number of vehicles for startup; add vehicles, as throughput is increased
Automatic interface to automated systems	Improved timeliness
	Performance is synchronized to other automated systems
	Reduced operating costs
	On-line information tracking
	Accurate positioning
Job enrichment	Dull, demanding, dangerous, and routine jobs cause personnel turnover. Reduce recruitment and retraining costs
	Increase in job scope and prestige as a "system operator"
Tax incentives	Investment tax credit
	Equipment depreciation schedules
	Interest expense deduction

8.2.10 System Acceptance Procedure

1. Before signing a contract for procurement of an automated guided vehicle system, it is preferable that an acceptance procedure be agreed upon with the system supplier. The scope of an acceptance procedure is dependent on the system complexity. A recommended procedure for computer-controlled systems is discussed here. Without specific agreement it is customary to consider beneficial use of the system by the owner as constituting system acceptance.
2. A demonstration of the system is appropriate. The following guidelines are suggested for this demonstration.
   a. The system should demonstrate all physical and control functions as outlined in the system performance description.
   b. The system should demonstrate capability of meeting the throughput rates for a specified period of time (which is recommended not to exceed 4 hr). This is done by operating the system under actual production conditions for a fixed period of time. It is customary for the system user (owner) to supply test loads and for operators to use the system under the direction of the system supplier.
   c. The demonstration may extend up to one or two days, or more if necessary, it being understood that all systems require an initial shake-down period. Problems encountered, if relatively minor and easily corrected, are customarily considered to be under a supplier's warranty.
   d. Should a demonstration of uptime or system availability be required, the demonstration may be conducted, for example, for a period of up to 8 hr. Downtime due to load faults, operator error, and waiting for maintenance personnel are not generally charged against system uptime or availability.
   e. Should a system demonstration fail after being partially completed, upon resuming the

demonstration begin at the point of failure. Successfully completed portions of the demonstration need not be repeated.

f. The owner normally has a responsibility to schedule any demonstration within a reasonable time after having been made aware that the system is ready to run. The notification as to system readiness and the time for scheduling a demonstration may be specified in the contract.

CHAPTER 9
CONVEYORS

JAMES M. CAHILL, PCMH

Rexnord Material Handling Division
Lombard, Illinois

JOHN G. DORRANCE

SI Handling Systems, Inc.
Easton, Pennsylvania

HAROLD VANASSELT

Rapistan Division of Lear Siegler, Inc.
Grand Rapids, Michigan

EDWARD MOON

Acco Babcock, Inc., Conveyor Division
Detroit, Michigan

BYRL CURRY

Jervis B. Webb Company
Farmington Hills, Michigan

ROBERT H. ROTH

Norfolk Conveyor Division
Jervis B. Webb Company
Cohasset, Massachusetts

9.1 PACKAGE-HANDLING CONVEYORS

James M. Cahill

9.1.1 Introduction

Package-handling conveyors are a broad classification of materials handling equipment sometimes referred to as "unit-handling conveyors" or "on-floor conveyors." They may be described as equipment for conveying a definable shape such as a carton, pallet, or casting. The load is usually placed directly on the conveyor, which is why the relationship between the load and the conveying surface is very important.

In this section, a basic understanding of the types of package-handling conveyors is covered, including how to select the proper type of conveyor, how to apply the conveyor, and how to maintain the conveyor. Emphasis is given to the information necessary to make these evaluations and to the information that is critical to success.

Probably the single most important step in proper application of package-handling conveyors is to determine what is to be conveyed. Some of the required information includes overall dimensions, weight, type of container and its condition, material, center of gravity, overhang, whether it is fragile or soft, conveying rate, and all other characteristics of the load. All products to be conveyed must be considered: smallest and largest, heaviest and lightest, and sealed and unsealed.

Consideration should be given to future requirements. Since most package-handling conveyors can be rearranged, lengthened, shortened, or modified, they are often used for many applications beyond the initial requirement. A little planning for these future applications can pay off in large savings.

Although most conveyors are designed with safety a prime concern, the proper application and selection of equipment is also necessary for a safe conveyor installation. Compliance with safety standards, including OSHA and other federal, state, and local codes or regulations, is usually the responsibility of the owner of the conveyor equipment. Placement of guards and other safety equipment in accordance with these safety standards is determined by the location of the conveyor and the use of the equipment.

Energy conservation should also be considered in conveyor design and system layout. Drives should be designed using the proper horsepower and should not be run except when required.

Noise should also be considered in selecting conveyor equipment. A conveyor that may be acceptable in a machine shop or other noisy atmosphere, will usually not be acceptable in an office or hospital application. Although there are government regulations in regard to allowable noise limits, the surrounding building features and other equipment will effect the final combined noise level of the installation.

9.1.2 Types of Conveyors

Gravity

In selecting package-handling conveyors, the College Industry Council on Material Handling Education recommends that you "utilize gravity to move material wherever possible, while respecting limitations concerning safety, product damage and loss" (*Twenty Principles of Material Handling*). Many times all nonpowered conveyors are referred to as gravity conveyors. In considering nonpowered conveyors, we should be specific if we truly mean gravity, level roller or wheel conveyor, or a "helper grade." This is discussed further under roller conveyor applications.

Skatewheel

In addition to the obvious economic advantage, other reasons for considering skatewheel conveyors are: portability (lightweight sections), differential action of wheels in curves, low inertia of wheels for lightweight products, and quick and easy assembly or disassembly. Also, some products that may not convey properly on roller conveyors, such as heavy wall bags or sacks, may move more readily on wheel conveyor. For bags or sacks, an in-line pattern of wheel conveyor is recommended since the material will form tracks and not be displaced as when passing over rollers.

In considering capacity, both the frame and the wheels should be investigated. The frame load capacity is determined by the deflection of the frame and the distance between supports. Excessive deflection causes shock loading on the wheels, unequal load distribution on the wheels, and varying grades for the moving loads due to sag between the supports. Good conveyor practice limits normal deflection to $\frac{1}{360}$ of the unsupported span. Table 9.1.1 gives properties for typical $2\frac{1}{2}$ in. × 1 in. × 12-gauge formed steel channel frames and $2\frac{1}{2}$ in. × 1 in. × 11-gauge extruded aluminum channel frames.

Table 9.1.1 Properties of Channel Frames

Length between Supports	Deflection/100 lb of Load		Recommended Maximum Deflection
	Steel	Aluminum	
5 ft	0.0128 in.	0.0236 in.	0.167 in.
10 ft	0.095 in.	0.208 in.	0.333 in.

Standard wheels are $1\frac{15}{16}$ in. diameter with $\frac{1}{4}$ in. diameter axles. Spacers are used between the wheels and the frames to form the various patterns. Steel wheels have 60 lb capacity each and aluminum wheels have 40 lb capacity each. Nylon wheels and tires of rubber and other composition are available for special applications. These wheels may require additional pitch because of the increased friction and deflection of the wheel outer surface.

Only experience can determine the proper number of wheels per foot. In general, the smaller the package, the more wheels are required per foot. Typical patterns are shown in Fig. 9.1.1.

Grades for conveying cartons vary with the weight of the carton and the condition of its conveying surface. Full cartons and cases require a grade of approximately 3 in. per 10-ft section and 3 in. per 90° curve. Empty cartons require 4 in. per 10-ft section. Loosely tied bundles, soft-bottomed articles, and bagged material require steeper grades. Caution should be used in determining the grade since the steep grade adequate to start movement of a product at rest may allow excessive speed to develop

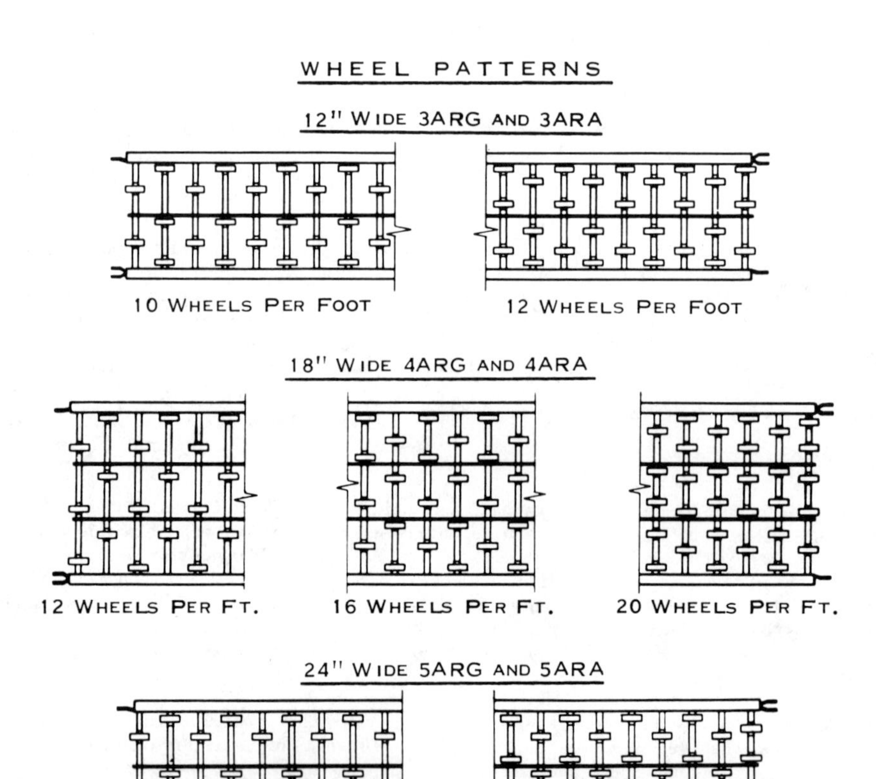

Fig. 9.1.1 Standard patterns for gravity wheel conveyors.

once the load is moving. If excessive speed develops, one solution is to use retarding rollers or plates in the line to slow the packages. If there is doubt as to the proper grade, the product should be tested before the final grades are determined.

When assembling sections of wheel conveyor and curves, couplings that will be compatible with the adjoining sections must be specified. This is especially true when wheel conveyor must be joined to roller or powered conveyor.

Roller Conveyor

As mentioned previously, use gravity conveyor wherever possible. There is the obvious savings of initial cost, ease of assembly, low maintenance cost, and adaptability for future layout revisions. Roller conveyor can be used in a much broader range of applications than skatewheel, but the cost is correspondingly greater with the heavier roller conveyor sections. Where skatewheel is normally considered for portable applications or with lightweight articles, roller conveyor is usually not as portable and can be designed for loads of several thousand pounds.

Again, it is of primary importance to define what is to be conveyed now and in future applications of the conveyor equipment. All available information must be considered, largest and smallest, lightest and heaviest, shapes, conveying surface, and so on.

Because of the wide range of applications of roller conveyor, there are numerous combinations of rollers, bearings, axles, frame rails, and supports. The correct combination of these components is a judgment based on the product to be conveyed, the environment of the installation, and the cost. Each of these conveyor components must be considered in developing the conveyor specification.

Most roller conveyors use nonprecision bearings. These bearings are an unground, commercial grade sometimes referred to as "conveyor grade" bearings. The races, the inner and outer surfaces that the balls ride on, are surface-hardened and unground. The balls usually fill the spaces between the races and are in contact with each other. This full complement of balls assures maximum static and low-speed carrying capacity along with shock-loading resistance. Conveyor grade bearings are in most cases advantageous in gravity applications because their wide manufacturing tolerances make them more forgiving of variations and mounting inaccuracies. Also, these added clearances allow foreign substances to fall through the bearings and not jam the balls.

For higher speeds, continuous operation at maximum load, or where noise is a problem, precision bearings may be required. These bearings have a cage or a retainer to maintain equal space between the balls and have hardened, ground races. Because of their tight tolerances and precision fit, friction at the start of rotation is normally higher and they are therefore generally not suitable for gravity applications. Conveyor grade bearings are available as open, shielded (or sealed), grease packed, or regreasable. In most gravity applications, the open or steel-shielded bearing is preferable to avoid increased friction in the bearing. In powered applications, the sealed, grease-packed bearing is recommended for the continuous duty cycle. For higher temperatures (above 200°F) or higher speeds (over 100 rpm), the regreasable bearings should be considered.

Conveyor grade bearings can be used up to 350°F with proper lubrication. However, felt-type seals should not be used above 180°F since the felt will char and contaminate the bearing. Steel seals are used for applications where the temperature is over 180°F but not over 225°F.

If the rollers are carrying high-temperature loads such as castings or forgings, the same temperature limits apply to the roller temperature. By moving the loads fast and not allowing them to stop on the rollers or by spacing the loads, the rollers will not "soak up" the heat from the loads. Also, there should be good circulation of air around the rollers to dissipate the heat.

Low temperatures require consideration mainly of the proper lubricant or no lubricant and not of the bearing capacity or construction.

Rollers are available in diameters from under 1 in. O.D. to over $7\frac{1}{2}$ in. O.D. and with various wall thicknesses. The bearings are inserted in the smaller rollers by a straight press fit, swaged, or curled end (see Fig. 9.1.2).

In the medium-size rollers, adapter cups may also be used to fit the smaller bearings and axles (see Fig. 9.1.3). In the larger rollers, often the ends are counterbored to receive the bearings. Although the method of assembly is usually determined by the manufacturer's equipment, the user should be aware of the type of assembly because some of these assemblies do not allow replacement of the bearings. Also, if not properly assembled or if overloaded, the adapter cups or bearings may become loose or collapse.

The selection of the diameter of the roller is usually determined by the capacity required, but other considerations are lower rpm of larger-diameter rollers, frame clearance, and available wall thickness of the roller.

Most conveyor rollers have hexagon axles to prevent the axles from rotating in the frame rails and to prevent the inner bearing cone from rotating on the axle. Light- and medium-duty package-conveyor rollers have spring-loaded axles. These rollers can be installed or removed from one side of the conveyor and basically within the overall dimension of the conveyor. Other axle arrangements are an up-set on one end and cotter pin or hog ring on the other end. These types of axle require

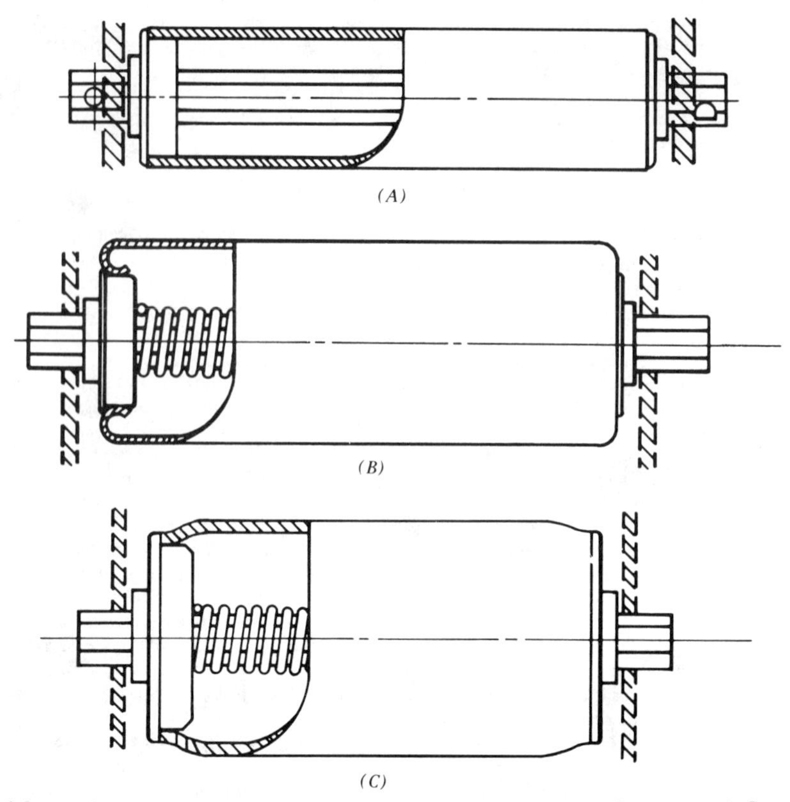

Fig. 9.1.2 Bearings may be inserted in rollers by straight press fit (*A*), curled (*B*), or swaged (*C*).

that the axle be completely removed from the roller for assembly or disassembly from the frame. This requires clearance next to the conveyor to remove the axle and access on the other side of the conveyor to remove the cotter pin or the hog ring. When assembling longer rolls in the frames with this type of construction, it is difficult to feed the axle through the roller and the far bearing and frame rail. When installing rollers with removable axles next to a wall or some other fixed object, it is important to install the conveyor so that the axle can be removed on the side away from the fixed obstacle.

Other axle types are flush axle and drop-in axle construction (see Fig. 9.1.4). The flush axle construction has clearance between the bearings and the frame rail for the cotter pin or other retainer. The axle is cut to the same length as the outside overall dimension of the frame. Flush construction prevents conveyed articles or personnel from coming in contact with the ends of the axles. Drop-in construction has flats milled on the axles and the frames are slotted from the top edge. This construction is often used in foundries where rollers are replaced frequently. Frame capacity must be reduced for this type of construction because of the slots.

Because of the many combinations of bearings, axles, and tubing, most manufacturers stock only

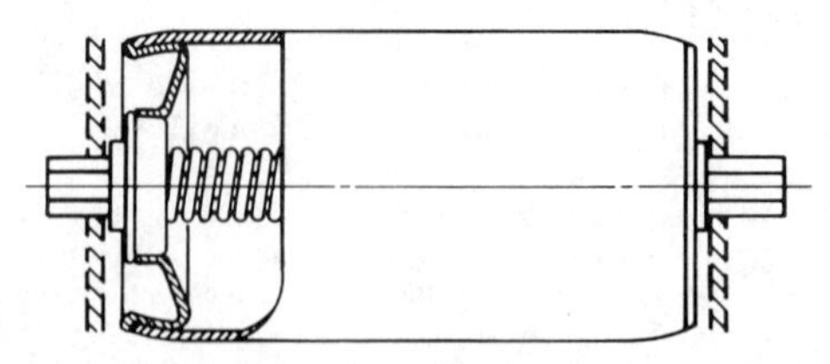

Fig. 9.1.3 Adapter cups may be used to fit smaller bearings and axles in medium-size rollers.

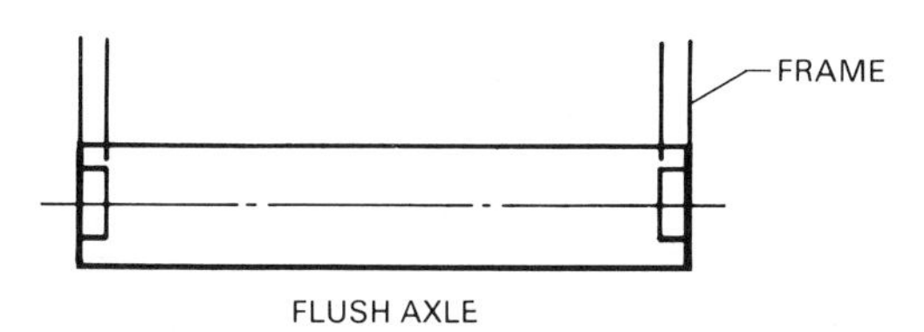

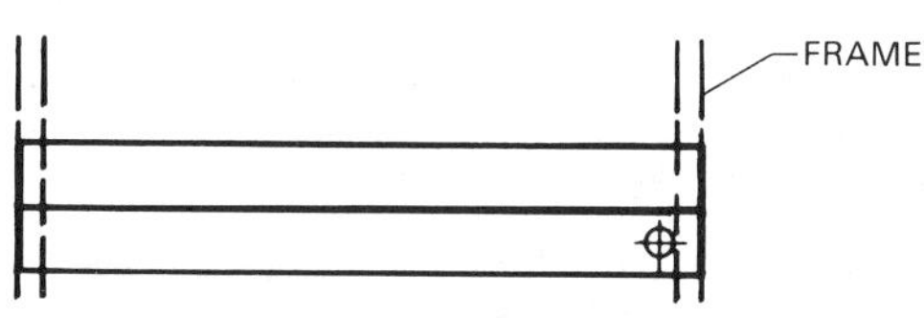

Fig. 9.1.4 Axles for conveyor rollers.

a few diameters and widths of rollers assembled. However, they also stock the tubing, bearings, and axle stock to assemble rollers to specific requirements. This custom assembly usually increases the price and delivery time.

Roller capacity is determined by both the basic bearing capacity and the width of the roller. The basic conveyor bearing load rating is defined as the constant, radial bearing load that a group of apparently identical bearings can endure for one million revolutions with 90% of the bearings surviving (Conveyor Equipment Manufacturers Standard No. 401-1978). This can be calculated by the equation:

$$C = fz^{2/3}D^{1.8} \tag{9.1.1}$$

where C = basic conveyor load rating in pounds; f = factor depending on geometry of bearing components, accuracy of parts, and material; z = number of balls; and D = ball diameter in inches.

The rating life of conveyor grade bearings operating at loads other than the basic conveyor bearing load rating can be calculated by the equation:

$$L_{10} = \frac{C}{Pe} \times 10^6 \text{ revolutions} \tag{9.1.2}$$

where L_{10} = rating life in revolutions, C = basic conveyor bearing load rating in pounds, and Pe = applied radial load on bearings in pounds. This load/life relationship is based on continuous running and constant load with laboratory conditions. The service life resulting from reduced load and speeds can be calculated from the equation:

$$L_{10}S = N(L_{10}) \tag{9.1.3}$$

where $L_{10}S$ = conveyor bearing service life, N = approximately 30, and L_{10} = rating life. Life in hours of operation can be calculated by:

$$t = \frac{L\pi D}{12(60)S} \tag{9.1.4}$$

where t = hours of life, L = service life $L_{10}S$ in revolutions, D = roller diameter in inches, and S = roller surface speed per minute.

Equations 9.1.1 through 9.1.4 are from the Conveyor Equipment Manufacturers Standard No. 401-1978.

In addition to the basic bearing capacity, the roller length must be considered. Axle deflection is directly proportional to roller span for a given load and therefore roller capacity decreases with conveyor width. The clearance built into the conveyor-grade bearings allows approximately 1° axle deflection before the rating is affected (see Fig. 9.1.5).

Just as with rollers, there are many frame combinations available. The most common components

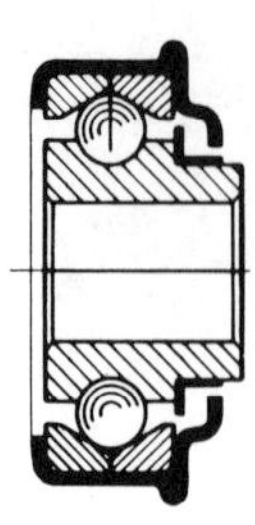

Fig. 9.1.5 Bearing clearance allows 1° deflection before rating is affected.

are roll formed channels, structural channels, structural angles, and flats. Usually the conveyor frames are the same on both sides with welded cross braces between. However, in some applications, the side frame members may not be the same on both sides. Also, angle frame rails may be assembled with the toe in or toe out (see Fig. 9.1.6).

The rollers may be set high or low in the frame rails depending on the axle hole punching. If the rollers are set low, the frame rails will contain the product and guard rails will not be necessary. However, wider loads cannot be accommodated and the conveyor is therefore not as adaptable for future applications.

In loading areas, a wear bar may be welded to the frames under the axle hole to prevent the axles from enlarging the holes due to impact on the rollers (see Fig. 9.1.7).

Most formed frame rails have standard patterns of axle holes such as $1\frac{1}{2}$ in. or 2 in. centers. Rollers can be installed on centers that are multiples of the punching patterns. With structural frames, the axle holes are usually punched to order for the application. Since most manufacturers do not charge for this additional punching in structural frames, it may be good insurance to punch additional holes in case additional rollers need be added at a future date.

Although three-rail construction is used most often in curves, it is also used in straight sections to provide additional rollers for better conveyability and capacity. Another application of three-rail conveyor is the herringbone section. This conveyor is used for centering loads such as before a case sealer.

A roller conveyor designed similar to the herringbone is the skewed roller section. This design is used for moving loads to one side of the conveyor. Usually wheel guides are included on this type of conveyor to guide the load with the minimum amount of friction. Typical application for this style of conveyor would be before a code reader or a powered pusher.

Gravity curves require special consideration in two areas: the width between frames or guards, and tracking of the conveyed article. For all items except circular ones, such as drums, the product will require additional conveyor width in the curves because of the cording action of the product. As with the straight sections, it is usually desirable to have 2–3 in. total clearance between the product

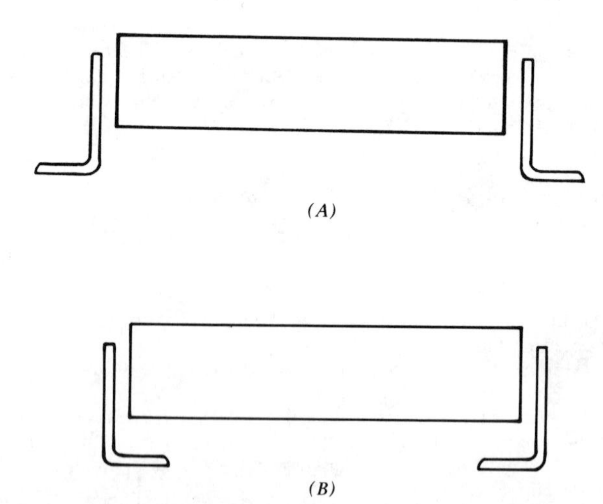

Fig. 9.1.6 Angle frame rail assemblies: (*A*), toe out; (*B*), toe in.

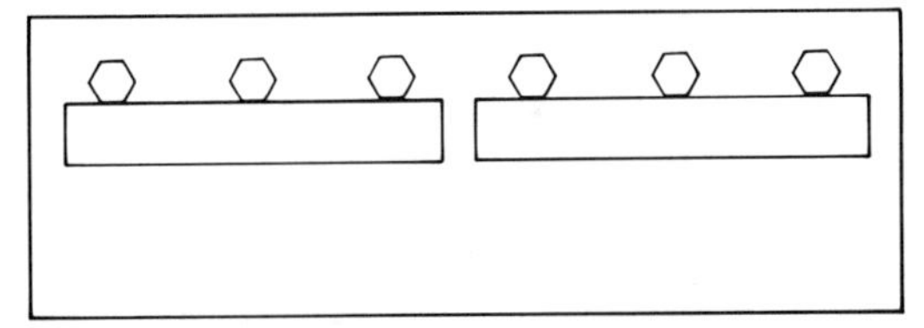

Fig. 9.1.7 Wear bar welded to frame.

and the frames or guards. An equation for determining the distance between the frames or guard rails is:

$$G = \sqrt{(\text{Radius} + \text{Package width})^2 + \left(\frac{\text{Package length}}{2}\right)^2} - (\text{Radius} - 2\text{ in.}) \qquad (9.1.5)$$

where G = the distance between frame or guard rails (see Fig. 9.1.8).

The inside radius of roller curves should be greater than the length of the longest item to be conveyed. Curves are usually not used for accumulation since products may lock together or hang up on the guard rail and not proceed when released.

For the proper tracking of all but circular products, the selection of the type of curve is very important. Since the distance around the outside of the curve is farther than the distance around the inside, the outside conveying surface must turn faster than the inside surface or the product will rotate on the conveyor. In the case of round articles like drums, this rotation should not affect the operation. For most other products such as cartons or pallets, skewing of the product could be a problem. The best solution is usually a tapered roller curve. The taper on the roller forms a cone with the focal point at the centerline of the radius. Any point on a tapered roller will rotate proportionately faster as the distance from the focal point increases and the roller diameter increases. In using tapered rollers, however, the curve radius selection is limited by the taper manufacturing equipment. In other words, when the conveyor width is selected, this predetermines the radius of the curve because of the taper on the rollers.

Other types of curves that assist proper tracking of the product are wheel curves and three-rail curves. Both allow differential action of the outer conveying surface.

Curves should not be connected directly to the end of a belt conveyor because the greater friction of the belt on the product will force the product across the roller into the outside guard rail. A section of straight roller conveyor at least one-half as long as the longest product should be used between the curve and the belt conveyor.

A curve should also not be connected directly to the discharge end of a chute. A section of straight conveyor at least three times the length of the longest product should be used between the curve and the chute.

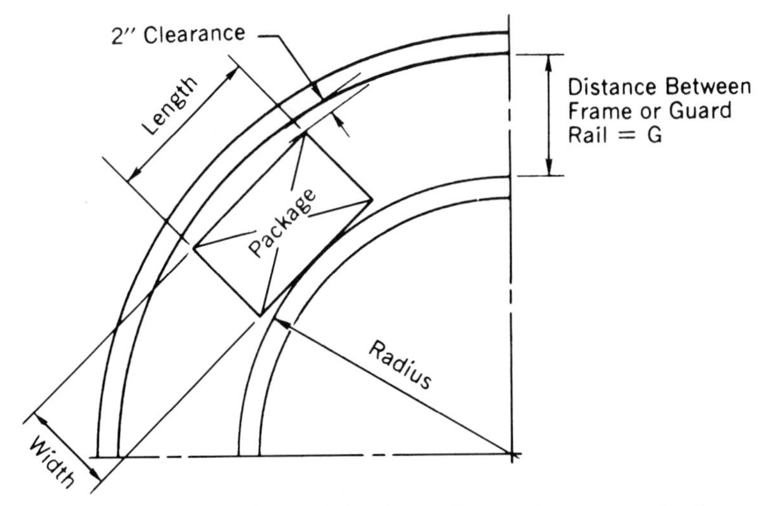

Fig. 9.1.8 Method of determining distance between frame or guard rails.

Since there are many variations of roller lengths and bearing projections from the roller, the conveyor width is commonly referred to as the "between frames" dimension. This also avoids confusion over various frame widths and types of frame flanges (see Fig. 9.1.9).

Supports. Supports are assembled to suit the type of frame, the between frames dimension, and the height. Except for the heavy-duty type, most supports have adjustable height. This adjustment allows the conveyor to be leveled in spite of the condition of the floor or to change the pitch of gravity conveyor (within the limits of the height adjustment). Supports over 30 in. elevation should be braced off the conveyor frame with ties referred to as knee braces. Support height should not be more than three times the width without external bracing into the building or a fixed machine.

Unless otherwise specified, manufacturers usually supply the hardware for bolting the conveyor section together and to the supports. However, the owner of the equipment must usually provide the floor anchors or overhead stringer steel.

When selecting the roller, the frame rails, and the supports for an application, the capacity of these components and the characteristics of the load must be evaluated.

To determine the roller capacity required, all items to be conveyed must be considered. The weight of the loads should be divided by the number of rollers under the load. A good practice is to figure only two-thirds of the rollers under the load as actually carrying the weight because of uneven load and conveyor surfaces.

Another good practice is to have a minimum of three rollers under the load at all times. To determine the roller spacing, subtract one inch from the load length and divide the remainder by three. The next smaller available roller spacing should then be used. After checking the roller loading as described previously, it may be necessary to use closer roller centers if the loading is greater than the roller capacity, or use a larger capacity roller. Roller centers for hard-surface items such as castings or steel tubs may be as much as 12–18 in. or greater. However, roller centers should be no greater than 6–8 in. for wooden pallets or other softer-surface products.

Conveyor loading areas require more rollers, heavier wall rollers, and increased capacity because of possible impact. Load characteristics, speed, and many other factors determine the amount of impact, but a general rule is to double the required capacity of the components in these areas. For severe applications, spring-mounted rollers should be used to absorb the impact and to distribute the load over all the rollers.

After determining the roller size and spacing, the frame can be selected. The live load (the total weight of the product on the section), the total weight of the rollers, and the frame weight should be compared with the frame capacity (see Table 9.1.2). Since the standard length of most conveyor sections is 10 ft, the sections are supported on 10-ft centers at section splices whenever possible. For additional capacity, it may be necessary to support on 5-ft centers.

When determining the pitch for gravity roller conveyor, experience and caution are important. An average pitch for roller conveyor is approximately $\frac{1}{2}$ in. per foot of conveyor length. This pitch must be increased for lighter loads or soft-surface products and decreased for heavier loads or hard-surface items. Additional pitch is also required when the loads must start from rest. The use of gravity roller conveyor for handling heavy loads must be limited to short runs, or brakes or other speed control devices must be added. When both heavy and light loads are mixed on the conveyor, it is difficult to select a pitch for satisfactory, safe gravity movement. Most loads reach their average maximum speed of approximately 180 ft/min within three times their length. Roller brakes are usually spaced

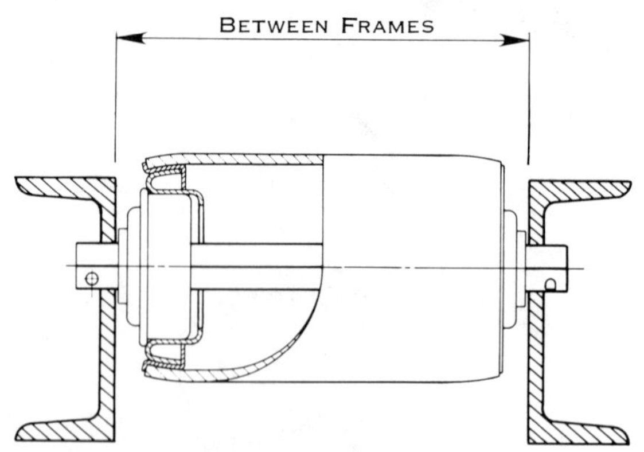

Fig. 9.1.9 "Between frames" dimension defines conveyor width.

Table 9.1.2 Frame Capacities[a]

Frame Members	Type of Forming	5 Feet between Supports		10 Feet between Supports	
		Level	Gravity	Level	Gravity
2.5 × 1 × 12 ga. [	Cold	1,300 lb	1,000 lb	350 lb	260 lb
2.5 × 1.5 × $\frac{3}{16}$ L	Hot	1,400	1,100	410	300
2.5 × 1.5 × $\frac{1}{4}$ L	Hot	1,900	1,400	500	440
3 × 2 × $\frac{1}{4}$ L	Hot	2,800	2,100	960	720
3.5 × 1.5 × $\frac{1}{8}$ [	Cold	3,300	2,400	1,200	910
3 @ 4.1 [	Hot	5,800	4,400	1,400	1,100
5 × 1.5 × $\frac{3}{16}$ [	Cold	6,800	5,100	3,400	2,500
4 @ 5.4 [	Hot	10,000	7,600	3,300	2,500
5 × 3.5 × $\frac{5}{16}$ L	Hot	10,000	7,600	5,000	3,800
5 @ 6.7 [	Hot	16,000	12,000	6,600	5,000
6 @ 8.2 [	Hot	22,000	17,000	11,000	8,600
6 @ 13.0 [	Hot	30,000	23,000	15,000	11,000
7 × 4 × $\frac{1}{2}$ L	Hot	30,000	23,000	15,000	11,000
7 @ 9.8 [	Hot	32,000	24,000	16,000	12,000
7 × 4 × $\frac{3}{4}$ L	Hot	44,000	33,000	22,000	16,000
10 @ 25 [	Hot	96,000	72,000	48,000	36,000
15 @ 40 [	Hot	240,000	180,000	120,000	92,000
Aluminum					
2.5 × 1 × $\frac{1}{8}$ [	—	710	530	160	120
3.5 × 1.5 × $\frac{1}{8}$ [	—	1,500	1,200	360	280

[a] Based on two-rail construction with allowance for axle and bolt holes.

at approximately the length of the product. This can obviously run into considerable additional cost. Long unbroken runs of gravity roller conveyors should be avoided where it is possible for one package to catch up with another and form a train. As mentioned previously, additional pitch is necessary for a load to start from rest because of the friction within the roller bearings and the mass of the roller. However, once the load is moving and the roller is rotating, this friction decreases and the load may reach an excessive speed.

Listed in Table 9.1.3 are some suggested pitches for various types of loads on gravity roller conveyor. These recommendations are based on average conditions and products. Final determination of the slope should be made after experimenting with the actual product and installation conditions.

Table 9.1.3 Suggested Grades for Roller Conveyors

Item Being Conveyed	Weight (lb)	Slope (in./ft)	Slope (in./10 ft)
Wood cases	150–200		
Steel drums	50–100		
Steel tote, boxes with smooth-riding surface	15– 50	$\frac{3}{8}$	$3\frac{3}{4}$
Wood cases	50–100		
Steel and wood beer cases (full)	40– 60	$\frac{7}{16}$	$4\frac{3}{8}$
Wood cases	20– 50		
Baskets with wood runners	50– 70		
Lumber (standard boards)	Any	$\frac{1}{2}$	5
Brick (standard smooth)	8– 12		
Cartons	50– 75		
Cartons	15– 50	$\frac{5}{8}$	$6\frac{1}{4}$
Cartons	5– 15	$\frac{3}{4}$	$7\frac{1}{2}$
Cartons	1– 5	$\frac{7}{8}$	$8\frac{3}{4}$

Ball Transfers

Ball transfers consist of a single large-diameter steel ball resting on a quantity of smaller-diameter balls contained in a hardened steel cup with a cap and a base for mounting. The large ball is free to rotate in any direction with a minimum of resistance. Ball transfers are intended to facilitate the movement of smooth, hard-surface commodities without lifting, for precise positioning, and for transferring from conveyor lines. Although primarily designed for level mounting with the large ball on top, ball transfers can be tilted up to approximately 30° from horizontal without significant capacity loss. However, they should not be used inverted because the principle of the flow of the small balls is upset and the small balls will settle down into the clearance area around the cap and the large ball will rest directly on the base of the cup.

Ball transfers should be operated without lubrication. When lubricated, they tend to pick up dust and other foreign matter which, when mixed with oil, prevents the balls from rotating freely. Ball transfers should not be applied in dusty conditions or where waste products from the process retard the balls and prevent them from rotating freely. Foundries and woodworking plants are usually bad applications for ball transfers. Sluggish operation of ball transfers due to glue or carton dust can sometimes be relieved by rubbing paraffin over the large ball each day.

In operation, ball transfer will mark the riding surface of some materials such as brass, soft lumber, or highly finished steel sheets. For lighter applications, if this marking is objectionable, a nylon ball should be used.

Ball transfers should not be used for soft-bottomed articles such as soggy cartons or bags, pallets, drums with chimes, baskets, totes with runners, or wire crates.

Since they are conveying a hard surface which is usually uneven, good practice suggests that capacity calculations should be based on only three ball transfers carrying the load. The use of spring-mounted ball transfers allows the total number of ball transfers under the load to be used in calculating capacity. Maximum spacing can be determined by dividing the minimum conveying surface dimension by 2.5.

The pushing force required varies with weight and the condition of the conveying surface. Hard-surface loads will be easier to push than softer ones. The force required is usually between 5 and 15% of the weight of the load to be moved. Since ball transfers are a multidirectional device, the load will move in the direction it is pushed and may rotate or skew depending on the skill of the operator.

Belt Conveyors

In addition to being an economical means of horizontal transportation, belt conveyors are also used for inclined and declined movement of loads. The belt is supported on rollers or on a metal bed called a slider bed (see Fig. 9.1.10). Both types of belt conveyors have channel frame rails that also support the end rollers. Roller bed is usually preferred because of lower power requirements and longer belt life. Slider bed is not recommended for heavy loads or high speeds (above 100 ft/min). Slider bed is used for loads with small, irregular bases, conveyor loading areas, and next to operators such as at assembly stations.

The between frames dimension is usually 3 in. wider than the belt. Since the belt carries the load

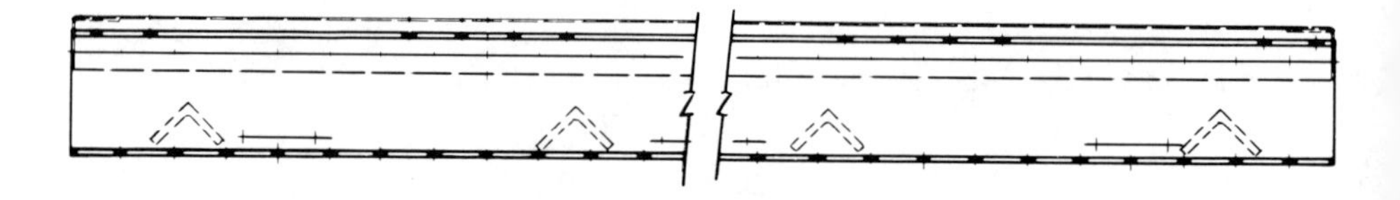

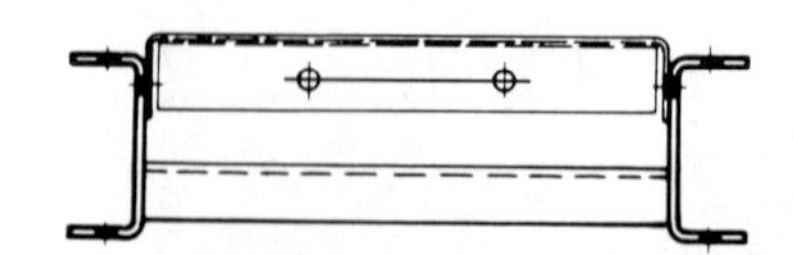

Fig. 9.1.10 Belt conveyor with slider-bed construction.

above the conveyor frame, sometimes the loads do overhang the conveyor. The belt-carrying rollers are spaced for a minimum of two rollers under the product at all times, but tall packages should have three rollers under them. If the roller spacing is too great, flexing of the belt passing over the rollers will cause the package to rock and may cause tumbling. The return belt idler rollers are spaced on 10 ft 0 in. maximum centers.

If the conveyor direction of travel is reversible, the drive is usually located at the center or toward the more heavily loaded end of the conveyor. If the conveyor is nonreversible, the drive is located toward the discharge end. Reversible belt conveyors require more care in manufacturing and assembly since the belt is more difficult to track. Also, long (over 50 ft) or short reversible belt conveyors are difficult to track. As a general rule, belt conveyors should not be shorter than three times the belt width to prevent tracking problems.

There are many combinations of belt types, rollers, frames, and drives depending on the product to be conveyed and the application. The belt type is determined by the belt pull (tension), the environment (oil, water, temperature, etc.), incline or decline (friction surface, rough top, or cleated), and the belt back surface (bare, friction, or slick). Because of the many types of belt available, a belting supplier should be consulted when special conditions exist. The belt section also affects end roller and drive design since thicker belts and heavier belting materials require larger pulley diameters.

The selection of rollers, frames, and supports is the same as that discussed for gravity roller conveyors except the rollers usually have grease-packed or regreasable bearings. If there are to be high speeds, precision bearings (over 150 ft/min) and balanced rollers (over 250 ft/min) may be required.

End drives are usually used for short conveyors, one-direction travel, and lighter loads. Center drives are used for reversible travel and medium or heavy applications. Another drive consideration is the conveyor speed. The speed should be as high as practical so that loads are spaced farther apart, reducing the tension in the belt. If the delivery rate or speed cannot be predetermined or must be relative to another conveyor or piece of equipment, a variable-speed drive may be used. Mechanical variable-speed drives have up to a 10:1 speed ratio. An electrical variable-speed drive may have as much as a 50:1 ratio. When using belt conveyors in a system, each succeeding belt should increase speed approximately 5 ft/min to insure a safer transfer between the conveyors. When a conveyor may start and stop more than eight times per minute (as for an indexing conveyor) the motor and reducer duty cycle may require special consideration.

Incline belt conveyors are used to raise or lower material at various degrees of slope. Where carton size permits, roller beds are preferable although a short slider bed is generally provided at the transition between the feeder section and the incline portion.

The maximum angle of elevation is governed by the configuration and surface of the conveyed product, the type of belt used, and the method of feeding the incline. The relationship of the height of the package to its base length is important in determining the maximum slope. A general rule to follow for a uniform load is to make the slope such that a perpendicular line down through the center of gravity of the package falls within the middle one-third of the package's base length. Caution should be used in locating the center of gravity since not all packages are uniformly loaded. Special consideration must also be given to load stability where an incline belt conveyor might be required to start and stop. In addition to checking the load on the incline, it should also be checked at transfer points such as at the end rollers or three-pulley device (see Fig. 9.1.11).

As mentioned previously, selection of the belt's surface requires experience and knowledge of the product's surface. Loads like smooth-bottom plastic trays require a special belt in order to be handled even at a 15° slope. A slope of 25° has been found to be the maximum for conveying most carton sizes. Some loads may be handled at slopes greater than 30° but these are special applications and must be given careful consideration.

Up to a 10° and, sometimes, a 15° slope, loads up to 24 in. long can be transferred from roller or wheel conveyor to an incline belt without a feeder section. The feeder section may be a short belt conveyor chain driven from the tail shaft of the incline conveyor or a two- or three-pulley device where the incline conveyor belt is snubbed down over an arrangement of pulleys. Slave-driven feeder sections make it possible to vary the conveyor speed for carton separation. However, the rules governing conveyor length versus belt width must be considered for slave-driven feeders to avoid belt tracking problems.

Belt conveyors must be checked to see that smaller packages transferring from the end roller land safely on the end roller or first roller of the next conveyor. Packages making a number of transfers in a system can change orientation on the conveyor, so always use the smallest dimension as the length of the package when making this check unless the package orientation can be assured.

When an incline conveyor with a three-pulley device for a tail feeder receives cartons from a roller conveyor section, there is often a problem with the cartons bridging across the three-pulley device. Since the speed of the cartons on roller conveyor may be two to three times the speed of the cartons on a belt conveyor, the cartons may form a train on the roller conveyor and enter the belt conveyor end to end. When this train reaches the three-pulley device, the resulting bridging action can damage the cartons or even cause them to fall off the conveyor. This problem can be particularly serious when using rough top belting because of the high coefficient of friction of the belt on the

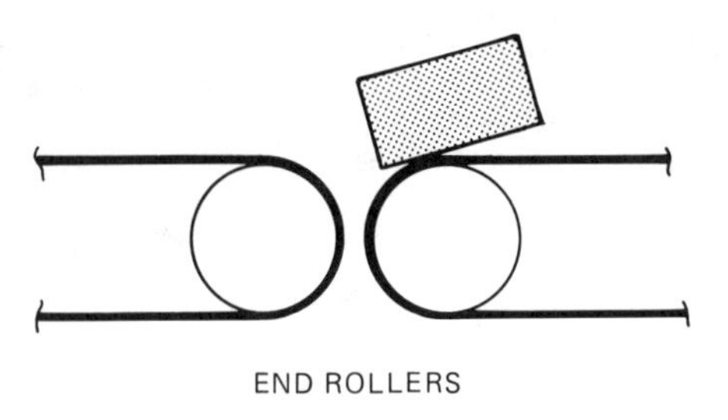

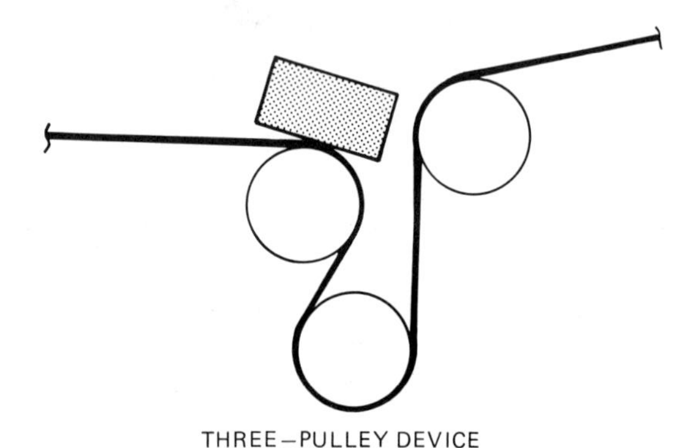

Fig. 9.1.11 Load should be checked at these points besides at the incline.

product. Also, when a long line of roller conveyor is feeding the belt conveyor, the line pressure will force the cartons together onto the belt. As mentioned previously, one solution to this bridging problem is to use a slave-driven conveyor for the tail feeder and use the sprocket ratio between the two conveyors to effect a slower speed in the tail section to achieve carton separation at the incline.

A motor brake should be used on all inclined or declined belt conveyors over 10°. This will prevent the conveyor from drifting when the electric power is shut off.

Because of the resulting force on the rollers from the belt tension and the weight of the live load, vertical bends with heavier-capacity rollers are recommended for inclines over 10°. This section is installed at the transition between the incline tread sections and the horizontal tread section at the top.

See Fig. 9.1.12 for a quick calculation of conveyor horizontal and incline lengths at slopes of 10–30°.

When calculating the effective belt pull, use the equation:

$$BP = F \times (L + B + R + 0.05T) + \sin\theta(I) + 0.3D \tag{9.1.6}$$

where BP = belt pull; F = friction (5% for roller tread and 30% for slider tread); L = live load (total weight of conveyed product); B = weight of belt; R = weight of rollers (tread rollers and return rollers); T = weight of load on tail feed section; θ = angle of incline; I = weight of live load on incline; and D = weight of heaviest load to be deflected.

Effective belt pull = belt pull × 1.25. This additional 25% is for belt flexing and bearing losses.

For motor horsepower use:

$$\text{Horsepower} = \frac{\text{Effective belt pull} \times \text{Speed in ft/min}}{33{,}000} \tag{9.1.7}$$

and use the efficiency of chain drive (95%) and reducer (manufacturer's recommendation).

Take-ups are required on all belt conveyors to compensate for changes in belt length and to maintain belt tension. Take-up devices may be located at any point along the return belt after the drive or at

BELT LIFT GUIDE

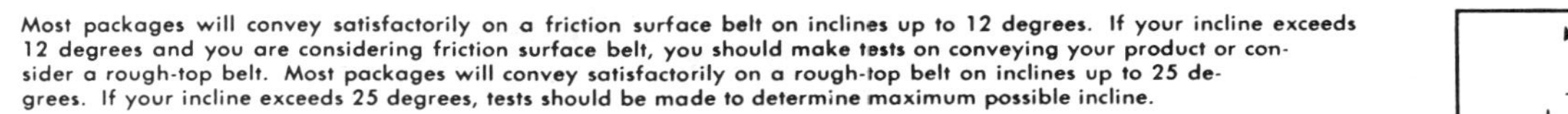

Most packages will convey satisfactorily on a friction surface belt on inclines up to 12 degrees. If your incline exceeds 12 degrees and you are considering friction surface belt, you should make tests on conveying your product or consider a rough-top belt. Most packages will convey satisfactorily on a rough-top belt on inclines up to 25 degrees. If your incline exceeds 25 degrees, tests should be made to determine maximum possible incline.

MAXIMUM INCLINE IS DEPENDENT ON:

1. Package Size
2. Package Material
3. Center of Gravity of Package
4. Loading Conditions
5. Belt Material
6. Belt Surface
7. Belt Speed
8. Age of Belt (New Belts Will Convey on Greater Inclines)

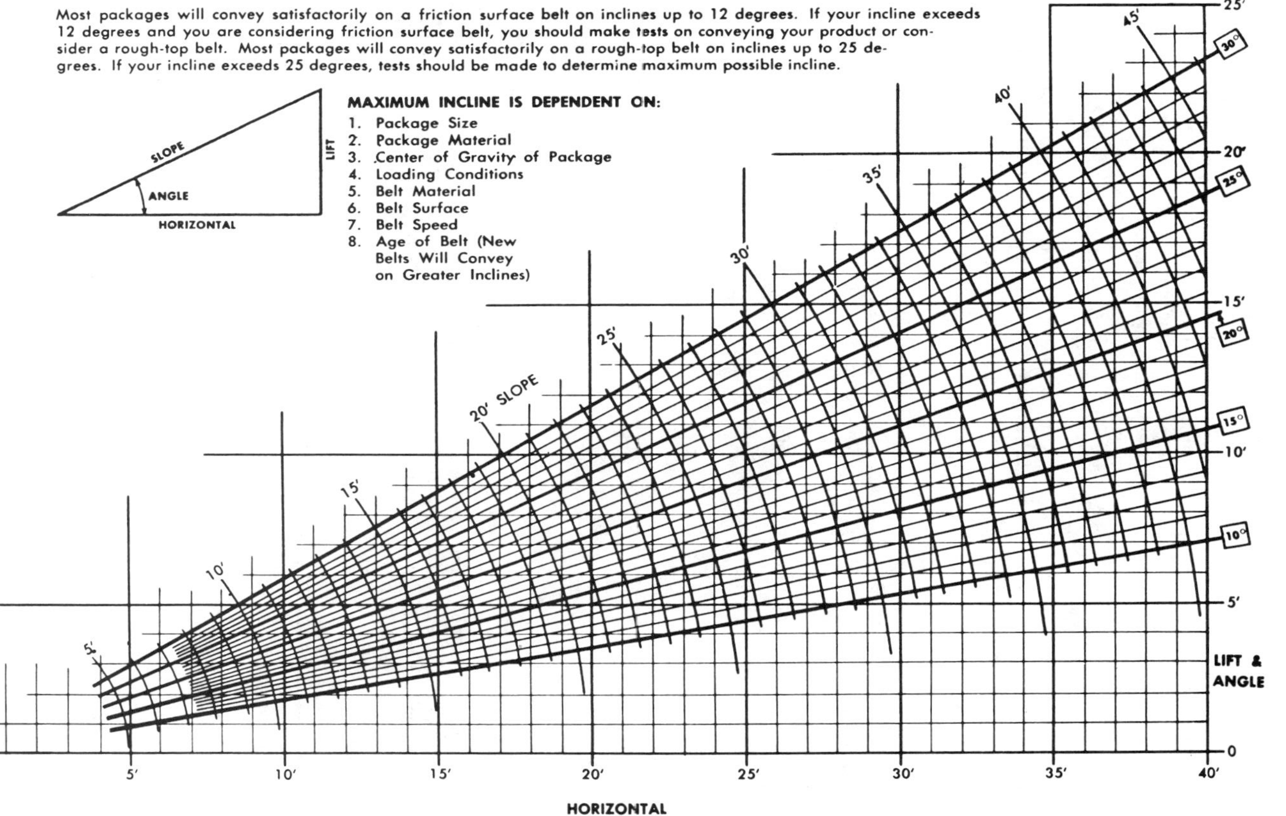

Fig. 9.1.12 Quick method for calculating conveyor horizontal and incline lengths at slopes of 10–30°.

the tail pulley. The most desirable location is immediately following the drive, on the slack side of the belt. Take-ups should be designed to provide a minimum of movement of 1% of the conveyor length. Belts that operate at higher stresses require 2% take-up movement. These figures apply to rubber-filled belts. For stitched canvas or solid woven belts, these figures should be doubled. Gravity-type take-ups should be used on all belt conveyors when conveyor lengths exceed the following: 150 ft for rubber- and neoprene-impregnated belting; 70 ft for stitched canvas belting; and 50 ft for woven cotton belting.

Gravity take-ups must be used regardless of length on inclined conveyors where the drive must be located at the bottom, when slack-side belt tension is critical for drive purposes as with heavily loaded conveyors, and in installations where there is fluctuating humidity or large changes in temperature.

On conveyors operating in one direction only, the gravity take-up weight must equal twice the required slack-side tension. Also, the take-up must be located on the slack side of the drive unit. On reversible conveyors, the take-up weight must equal twice the tension that would normally occur on the tight side of the belt. Another consideration is to use two gravity take-ups for reversible conveyors, one located on each side of the drive unit with the weight for each take-up being twice the slack-side tension. These take-ups must be provided with stops to limit the movement of the take-up roller when it is on the tight side of the belt.

Live Roller

As previously discussed, belt conveyors are used mainly for economical transportation. Live roller conveyors are used for a much wider range of applications such as accumulation; diverting on or off the conveyor; heavy loads; oily, dirty, or wet conditions; to turn, skew, or move load on the conveyor; high or low temperatures; or to eliminate drives. This wide range of applications requires different types of live rollers.

Live rollers are categorized by the type of drive used to turn the rollers. The common types of live roller are:

1. Flat belt
2. V-belt
3. Cable
4. Line shaft
5. Chain
   a. Continuous
   b. Roll to roll
6. Patented accumulation

The components for flat-belt-driven live rollers are very similar to belt conveyors except the belt is below the carrying rollers and is supported by snub rollers (see Fig. 9.1.13).

The selection and spacing of the tread rollers is the same as with gravity roller conveyor. There should be a minimum of three rollers under the load at all times. Snub rollers are located between

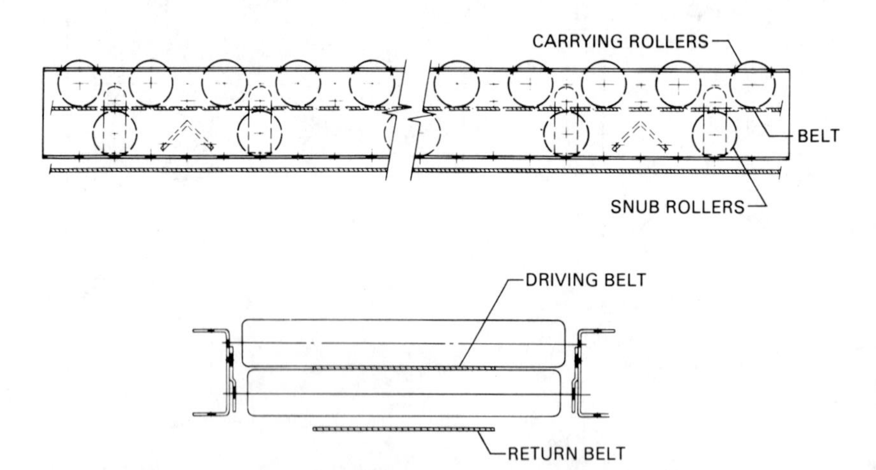

Fig. 9.1.13 In flat-belt-driven, live-roller conveyor, belt is below the carrying rollers and is supported by snub rollers.

carrying rollers and are adjustable up or down for increasing or decreasing the belt contact with the carrying rollers. At a point on the conveyor where the load may be stopped while the conveyor continues to run, the snub roller should be lowered for less belt contact with the carrying roller and therefore less driving force on the load. A more complete discussion of types of accumulation conveyor will follow later in this chapter.

In an area where a carton must be diverted from the conveyor, the snub rollers should be raised to provide increased driving force on the load. As a general rule, adjust for the minimum belt pressure required to keep the load moving. This will keep belt tension and drive loading to a minimum.

Since the belt width is not related to the actual conveying surface as with a belt conveyor, narrower belts can be employed on belt-driven live-roller conveyors. The belt width is selected by determining the effective belt pull and relating that number to the belt capacity per inch of width. The maximum belt width is determined by clearances at the end rollers and in the drive unit as specified by the conveyor manufacturer. The minimum belt width is usually also based on the end roller and drive construction. Most manufacturers use "crowned face" pulleys to help maintain the belt tracking. The manufacturer will specify a minimum belt width to be used in conjunction with the crowned face pulleys. Friction surface belting is usually used on belt-driven live roller.

The carrying rollers on belt-driven live-roller end rollers and on end drives will be as close as possible to the end of the conveyor. However, the pulleys in these devices will be set in from the end of the conveyor so that the belt does not travel past the end of the conveyor and possibly interfere with the next conveyor. These end-carrying rollers therefore may not be powered and caution should be used when conveying small articles (see Fig. 9.1.14).

V-belt-driven live roller is the same as the described flat-belt-driven live roller except the flat belt is replaced with a V-belt and the snub rollers are replaced with snub wheels mounted on one side frame. The drive and end pulleys employ V-belt sheaves.

V-belt-driven live roller is used for light loads and short conveyors. Along with the limitation on belt capacity, there is the concern for the strength and life of the splice in this type of belt and it is therefore preferable to use endless belts when possible.

V-belt-driven live roller is ideally suited for curves because the V-belt can flex around the curve radius. Also, the same drive and V-belt can power straight sections along with the curve.

Cable-driven live roller is very similar to the V-belt driven. The cable is snubbed under the carrying roller using wheels mounted on the conveyor frame.

Different types of cables are employed by various manufacturers. Some cables have a steel inner core with a soft covering for a drive surface on the roller and to prevent marking and wearing on the roller. Splicing this type of cable can be a problem because of the flexibility required in the steel core and the soft outer covering.

Other types of cables use synthetic materials without the steel core. These are usually easier to splice but stretching may be a problem. Since the stretching is greater than for other types of conveyors, larger take-up devices must be provided to maintain tension through the drive unit. In extreme conditions, the continued stretching will reduce the diameter of the cable to a point where it loses its driving contact with the carrying rollers.

In addition to the obviously lower equipment costs of the components, cable conveyors can be used to eliminate drive units because the cable can drive through corners and straight sections with the same drive unit. This results in equipment savings not only in the drive units but also electrical controls, and in mechanical and electrical installation.

As with the cable conveyor, the main purpose of line shaft conveyors is to eliminate drive units. The line shaft is mounted on one side of the conveyor frame and can be mounted around curves using universal joints. Spurs can also be powered using jackshafts and couplings. The rollers are driven from the line shaft using O-rings over the roller and spools on the line shaft. The O-rings ride in grooves in the roller to maintain proper alignment and tension. The spool is made from a

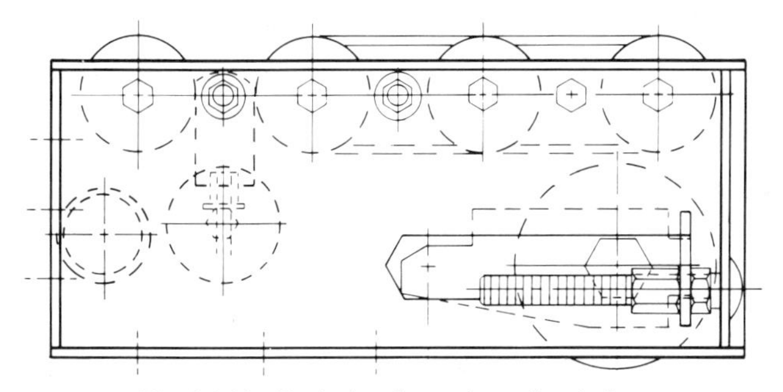

Fig. 9.1.14 Typical end-carrying roller design.

low-friction material which will slip on the line shaft when the roller rotation is stopped thus allowing minimum pressure accumulation.

Line shaft conveyors are used for light- to medium-duty applications. They should not be used in oily or wet applications where the friction coefficient between the spool and the drive shaft will be affected. The O-rings should not be exposed to continuous direct sunlight or radiation which can cause deterioration. As with the V-belt and cable-type live-roller conveyors, splicing O-rings in the field requires the proper equipment and materials. In some cases, additional rings are placed on the line shaft so that they can be quickly used to replace a broken ring. The rings are also limited to a temperature range of approximately 32–150°F.

Chain-driven live roller is used in more severe applications such as heavy loads; oily, dirty, or wet conditions; and high or low temperatures.

There are two types of chain-driven live rollers: continuous and roller to roller. The continuous-chain type has the lower initial cost but is more limited in application.

Continuous, chain-driven live roller consists of a single strand of chain traveling over a single welded rack tooth sprocket on each roller (see Fig. 9.1.15). Contact with the sprocket teeth is maintained by a continuous cover plate and hold-down. The chain is returned in a channel guide or over sprocket idlers. Since the chain is only in contact with the sprocket teeth on the top of the sprocket, this design should not be used for start–stop applications, and only with moderate-weight loads. However, by using side bow chain, it can also be used to power curves. Roller spacing is therefore influenced by sprocket diameter. When close center rollers are required, a double-width chain with staggered sprockets can be used, or an idler roller between each pair of driven rollers. The drive unit should be located at the discharge end of the conveyor.

In roller-to-roller construction, two sprockets are welded side by side on one end of each roller. Individual loops of chain connect pairs of rollers in a staggered pattern along the length of the conveyor. This driving arrangement is used for heavier loads or for frequent stopping or reversing service because of a greater arc of chain and sprocket contact. The conveyor length is limited because of cumulative chain pull and slack chain. The maximum number of consecutive chain loops should not exceed 80. By locating the drive in the center of the conveyor, there can be 80 chain loops on each side of the drive for a total of 160 loops. For moderate applications, the chain can be manually lubricated periodically. For speed over 150 ft/min, the chains can be automatically lubricated.

Belt-driven live-roller drive and belt pull calculations are similar to those for belt conveyors.

$$BP = F \times (L + B + R) + \sin\theta(I) + 0.3D \tag{9.1.8}$$

where BP = belt pull; F = friction (10%); L = live load; B = weight of belt; R = weight of rollers (tread + snub + idlers); θ = angle of incline; I = weight of live load on incline; D = weight if heaviest load to be deflected; and effective belt pull = belt pull × 1.25.

$$\text{Motor horsepower} = \frac{\text{Effective belt pull} \times \text{Speed in ft/min}}{33{,}000} \times \text{Efficiency of the chain drive (95\%)} \times \text{Reducer efficiency (manufacturer's recommendations)} \tag{9.1.9}$$

Belt take-ups are required as described in the belt conveyor section.

The maximum recommended incline on belt-driven live roller is 5°. Greater inclines may be possible but the product must be tested and the rollers should be on close centers.

Declines up to 4° can be considered level in calculating the pull. Over 4°, the belt pull should be calculated two ways, completely empty and completely loaded. The completely empty calculation will be the belt pull and horsepower required to run the conveyor. The fully loaded calculation will be the belt pull and drive capacity necessary to hold back the load.

Caution should be used when matching incline or decline conveyor with level live roller. Solid

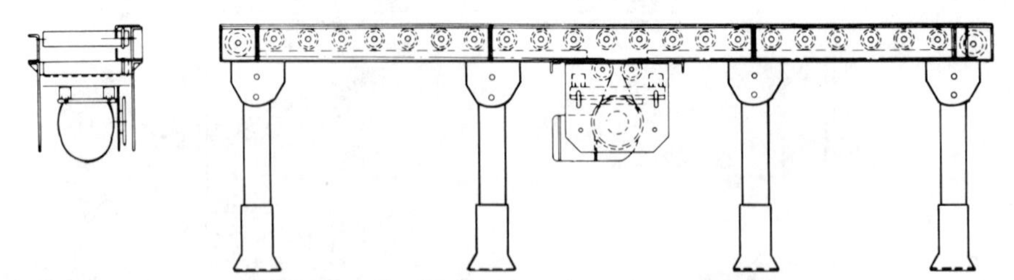

Fig. 9.1.15 Continuous, chain-driven live-roller conveyor consists of single strand of chain traveling over a single welded rack tooth sprocket on each roller.

bottom loads such as pallets or castings, will cord across the transition and may hang up on the rollers (see Fig. 9.1.16). To calculate the drive and chain size for chain-driven live roller, use:

$$CP = F \times (L + R + S + C) + \sin \theta(I) + 0.3D \tag{9.1.10}$$

where CP = effective chain pull; F = friction (6% for continuous chain and 5% for roller to roller); L = live load; R = weight of rollers; S = weight of sprockets; C = weight of chain; θ = angle of incline; I = weight of live load on incline; D = weight of heaviest load to be deflected.

$$\text{Horsepower} = \frac{\text{Effective chain pull} \times \text{Speed in ft/min}}{33{,}000 \times \text{Efficiency of chain drive (95\%)} \times \text{Reducer efficiency (manufacturer's recommendation)}} \tag{9.1.11}$$

To determine the size chain required, the effective chain pull must be multiplied by one or more of the following service factors:

Condition	Service Factor	
	8–10 hours	16–24 hours
Continuous single strand	1.0	1.0
Roller-to-roller drive	1.2	1.2
Intermittent load	0.8	1.0
Even load	1.0	1.2
Any of the following:	1.2	1.4
Stopping or reversing		
Heavy unit loads		
No lubrication		
Dirty conditions		
Moisture		

The allowable chain pull is in Table 9.1.4.

Chain Conveyors

Chain conveyors are another general classification of unit-handling conveyors. There are four types of chain conveyors: sliding, rolling, pusher, and vertical. The type is determined by the application.

Chain conveyors are used to transport loads that must be carried because of their configuration such as tubs, pallets, or other articles with the runners perpendicular to travel. Chain conveyors are also used for heavy loads such as large castings, stacks of steel sheets, or steel coils. In many cases, chain conveyors are more economical than other types of conveyors for transporting long distances.

The load carrying surface for sliding or rolling chain conveyors is the chain side bars or an attachment bolted onto a matched pitch pair of chains. Pallets, tubs, or stacks of steel sheet are usually conveyed directly on the chain with the number of chains determined by the total live load to be conveyed and the capacity of the chains. Small castings or parts, irregular shapes, and fixtures are conveyed on an attachment such as a steel or wooden slat attached to a pair of chains. Inclines on slat conveyors should be limited to 10° for wood and 8° for steel slats when receiving directly from roller conveyor, or 14° for wood and 12° for steel when a level section of the slat conveyor is provided. Cleats should be added to the pallets for any greater inclines. All articles negotiating inclines should be checked for stability using the same procedure as previously described for inclined belt conveyors. Radii of bottom bends on slat conveyors should be checked against the following equation:

$$R = \frac{L^2}{8} \tag{9.1.12}$$

where R = the radius to top of slats and L = the length of the article.

Roller Flight Conveyors

Roller flight conveyors are chain conveyors with rollers between the matched pitch chain. The load is conveyed and accumulated on the rollers. When a mechanism stops the load, the chains continue to move and the rollers turn under the load. The only line pressure is that resulting from the friction of the bearings in the rollers. When the stop mechanism is released, the bearing friction causes the roller to stop turning and the load is conveyed along with the rollers and chain. Separation or speeding up of the product can also be accomplished on this type of a conveyor by moving a section of friction

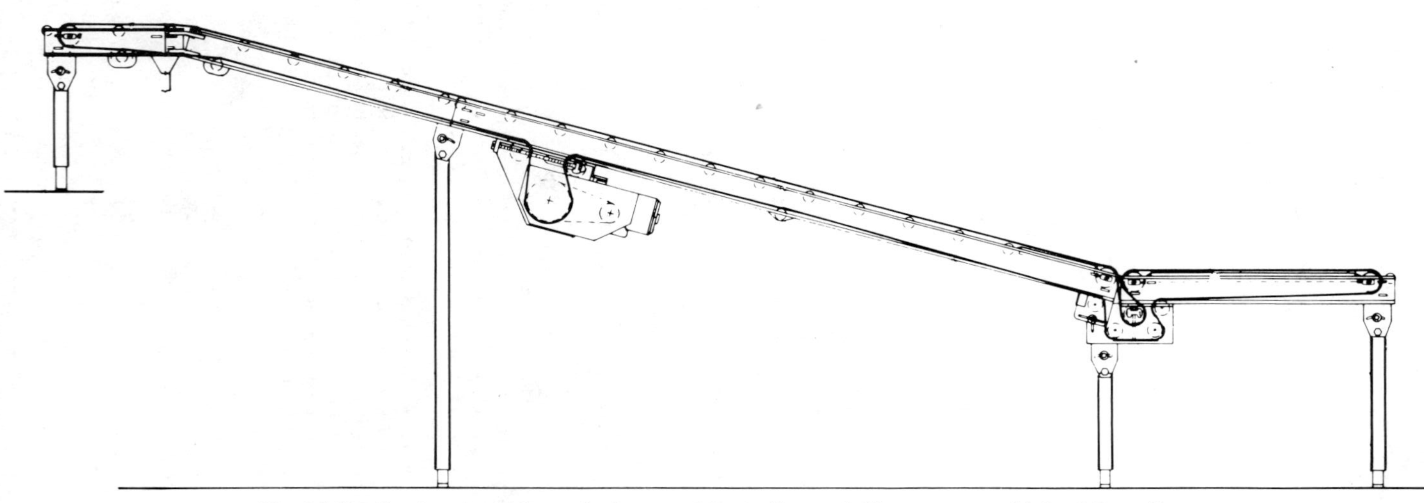

Fig. 9.1.16 Caution should be used when matching incline or decline conveyor with level live roller.

Table 9.1.4 Allowable Chain Pull, lb for

Chain No.	Conveyor Speed in ft/min 0–60	80[a]	100	150[b]	200	250	300
40	560 lb	500 lb	450 lb	325 lb	—	—	—
50	875 lb	800 lb	750 lb	650 lb	—	—	—
60	1200 lb	1075 lb	1000 lb	850 lb	775 lb	—	—
80	2100 lb	1950 lb	1800 lb	1550 lb	1350 lb	1225 lb	—
100	3550 lb	3250 lb	3000 lb	2600 lb	2300 lb	2100 lb	1900 lb

[a] Suggested maximum speed for continuous-strand chain.
[b] Suggested maximum speed for roller-to-roller conveyor without automatic lubrication.

surface material up against the lower surface of the conveying rollers. This results in the load being conveyed at double the speed of the chains.

Sliding Chain Conveyors

Sliding chain conveyors carry the load directly on the chains and the chain side plates ride directly on the track. The track may be abrasion resistant or be a low coefficient-of-friction material. Because of the higher friction of this design, it is used for relatively light loads and shorter distances.

Rolling Chain Conveyors

Rolling chain conveyors have lower horsepower requirements and lower wear rate. This design is used for the heavier loads and longer conveyors. There are two types of rolling chain conveyors: the chain roller on a track, or the chain side bars on a roller track. With the latter design, the chain side bar length and the roller spacing is important to successful operation. The side bars should always be on at least two rollers.

The Pusher-type Chain Conveyor

The pusher-type chain conveyor does not support the load but merely propels it either on a slider bed or on a roller tread. The pusher assembly may be either a bar between two matched pitch chains on both sides of the carrying surface or a chain running between two conveyor surfaces. With the pusher bar, care must be taken that the bar does not damage fragile loads and that the load is placed on the conveyor ahead of the pusher bar. When used on an incline, the pusher and pusher bar height should be checked against the following equation in order to assure adequate grip on the articles when passing over the top bend.

$$G = \frac{R - \sqrt{R^2 - 2gR - L^2}}{4} \tag{9.1.13}$$

where G = the minimum pusher height; R = radius to top of pushers or centerline of pusher bar; L = length of longest article; g = grip on article at top of bend, 1 in. minimum; and r = radius to top of slat or top of tread plate.

With the direct pusher chain arrangement, the bottom surface of the load must be flat and compatible with the chain passing beneath it.

Vertical Chain Conveyors

Vertical chain conveyors are either reciprocating or continuous type. Reciprocating conveyors have limited applications because of a lower throughput capacity. The continuous chain conveyors can handle a higher rate but usually require more expensive automatic loading and unloading stations. Local codes must be checked before applying either type.

Drive and chain pull calculations for chain conveyors are similar to those previously discussed.

$$CP = F \times (L + C + A) + \sin \theta(I) + 0.3D \tag{9.1.14}$$

where CP = effective chain pull; F = friction (see Table 9.1.5); L = live load; C = weight of chains; A = weight of attachments and slats; θ = angle of incline; I = weight of live load on incline; and D = weight of heaviest load to be deflected.

Table 9.1.5 Surface Friction versus Roller Diameter

Roller Diameter, in.	Friction, % Dry	Friction, % Slight Lubrication
$1\frac{1}{2}$	25	$20\frac{3}{4}$
2	20	15
$2\frac{1}{2}$	15	12
3	12	9
4	11	9
5	10	7
6	9	6
Rolls frozen	30	20

On vertical conveyors, figure the up-side full and the down-side empty unless there is a positive means of assuring otherwise. The blocked line friction for roller flight conveyors is 5% in place of 30% for all other chain conveyors.

To determine the size and/or quantity of the chain, the allowable chain pull must be compared with the maximum chain pull. The allowable chain pull will usually be reduced by one or more of the factors in Table 9.1.6. To determine the actual allowable chain pull, divide the rated capacity (manufacturer's recommendation) by the appropriate factors.

The horsepower can be calculated using the equation:

$$\text{Hp} = \frac{\text{Effective chain pull} \times \text{Speed in ft/min}}{33{,}000 \times \text{Reducer efficiency} \times \text{Chain efficiency (95\%)}} \tag{9.1.15}$$

(manufacturer's recommendation)

Accumulation Conveyors

Accumulation conveyors may be roller, live roller, slat, or belt type. In reality, control of product movement is the key, since we can convey on basically any type of conveyor equipment. Controls can be mechanical, electrical, pneumatic, or a combination.

Before deciding on the type of accumulation conveyor, it is first necessary to determine that accumulation is actually necessary. A few common examples of accumulation are as follows: a manufacturing operation that may require a buffer zone between machine operations that operate at different rates; a staging area for similar products such as at a packing station or prior to a palletizing operation; a high-rate conveyor system where products are merged as slugs rather than randomly; or in a cooling or curing process such as a foundry. There are many other examples of accumulation requirements but the application must be defined prior to selecting the equipment.

There are three general types of accumulation conveyors: flexible accumulation, slug accumulation, and fixed accumulation. Flexible accumulation is where the input and output are independent of each other. Examples of this type of accumulation are gravity roller and skatewheel, roller flight, belt-driven live roller, segmented chain-driven live roller, and line shaft conveyors.

Slug accumulation is where groups of products are moved simultaneously to the farthest open down-conveyor position. On these conveyors, a slug is built by indexing products at the receiving

Table 9.1.6 Conveyor Service Factors

Condition	Service Factor
Steady load	1.0
Intermittent use	0.8
8–10-hr service	1.0
24-hr service	1.2
Moderate shock	1.1
Heavy shock	1.3
Dusty atmosphere, moderate temperature	1.2
Abrasive, corrosive atmosphere	1.4
High temperatures	1.4

end of the first conveyor. Examples of this type of accumulation are belt conveyor and chain-driven live-roller conveyors.

Fixed accumulation is where there is one load in and one load out. Examples of this are chain conveyors, reciprocating beam, belt conveyor, and walking beam.

Other factors that must be considered in selecting the type of accumulation conveyor are product size, product weight, product rate, operating conditions, and product characteristics. As previously discussed for the various types of conveyors, these factors must be considered in determining the roller size, spacing, bearing construction, and so on.

In addition to the previously listed types of accumulation, many manufacturers have their own patented design of accumulation conveyor. Most of these conveyors have linkages, clutches, and so on, for neutralizing zones of the conveyor. In applying these types of conveyors, the conveyor operation must be reviewed in relation to the application. In some instances, such as feeding a palletizer, the entire length of loads should be released at one time. In other applications, such as before a vertical lift or a case sealer, the loads should be singularly released. In any case, the system operation should be reviewed in detail with the manufacturer for the proper selection of equipment.

9.1.3 Miscellaneous Equipment

In designing a unit-handling conveyor system, there are many miscellaneous devices that will be used with the previously described conveyors. Most manufacturers have their own standards for this type of equipment, or it may be necessary to custom design it for a particular application.

Turntables

Turntables are basically a conveyor tread section mounted on a bearing surface. Both the conveyor tread and rotation can be either powered or manual depending on the application and the product to be handled. The product is conveyed onto the turntable and the turntable rotated into alignment with the take-away conveyor. In a powered application, control devices are necessary to insure that the load is completely on the turntable prior to rotation and that the receiving conveyor has room to receive the load. Also, when the load has been discharged, it must be completely clear of the turntable before the turntable returns to its original position. Turntables are also used in a conveyor line to reverse the product on the conveyor. The product is conveyed onto the turntable, rotated 180°, and discharged. When handling pallets or other loads with runners, turntables are required at transfer points to properly orient the runners with the direction of travel on the conveyor.

Transfer Cars

Transfer cars are made up of a conveyor tread section mounted on a frame with wheels. Transfer car treads and travel can be either manual or powered depending on the application and the product. Transfer cars are usually used where the throughput is low and there are several transfer points.

There are several methods used for powering the movement of the transfer cars. The car wheels may be powered, the car may pull itself along on a chain, or a cable mechanism may pull the car forward and back. For short travel, a pneumatic or hydraulic cylinder can be used. Again, the selection of the method usually depends on the system application. In designing for a transfer car, there are several types of rails for the car to ride on. For heavier cars, industrial rails are recessed into the floor. The lighter applications use an inverted angle on one side and a flat steel strip on the other side. Consideration should also be given to cross traffic in the area of the transfer car operation. This is particularly important when the car is automatic and may move at any time. The car path should be fenced in and interlocked with the crossing area to eliminate possible collisions with pedestrians or other mechanized traffic.

For higher-rate systems where the product is conveyable in both directions, pop-up chain transfers are used at conveyor intersections. If the product is not conveyable in both directions, a "slave" pallet consisting of a piece of plywood or other hard, flat surface material can be used to carry the load. There are several designs for pop-up chain transfers depending on the load size, space requirements, and the manufacturer. For heavier loads, equalizer arms should be used to insure that the loads are raised and lowered uniformly.

Some type of stop device is required at the end of accumulation conveyors. A fixed-end stop is used where all loads are removed from the conveyor and a manual or powered stop is used where loads proceed on the conveyor when the stop is retracted. Another type of stop is a brake belt which can also be used to meter loads out of the accumulation line. In selecting the stop arrangement, care must be used to make sure that the stop can properly engage between loads on the conveyor. When a train of loads is advancing, it may be difficult to raise the stop in front of a particular load in the train. Another device that is used in combination with a stop in this type of application is an escapement device. The escapement holds back the train of loads allowing one load at a time to be released. In systems where loads must be diverted from a conveyor, there are a variety of devices available. No

Table 9.1.7 Comparison of Diverters

COMPARISON OF DIVERTERS

	Manual	Pop-Up Rollers	Air Pusher	Diverter Arm	Puller	Cross Chain Belt or Live Roller	Rotating Paddle Pusher	Swivel Wheel	Chain Tilt Device (Tray or Slat)
Rate: Units/Min.	15-25	15-20	25-30	30-35	30-40	50-70	50-70	120	65-300
Load Range/Lbs.	1-75	10-200	1-100	1-75	10-100	1-75	1-75	3-100	1-300
Impact on Load	Gentle	Gentle	Medium to Rough	Medium to Rough	Medium to Rough	Medium to Rough	Medium	Gentle	Medium to Rough
Type Recognition	Visual	Keyboard or Scanning	Keyboard or Scanning	Keyboard or Scanning	Keyboard or Scanning	Keyboard or Scanning	Keyboard or Scanning	2-3 Keyboard or Scanning	3 or more Keyboard or Scanning
Spur Spacing	Touching	Almost Touching	Almost Touching	2′-7′	2′-3′	Touching	9′	4′-5′	7′
Accumulation on Spurs	Excellent	Excellent	Excellent	Good	Excellent	Poor	Good	Good	Poor
Load Orientation	As Desired	Sideways or Lengthwise	Sideways	Lengthwise	Sideways	Random	Lengthwise	Lengthwise	Random
Maintenance Cost	Lowest	Low	Low	Low	Medium	High	High	Medium	High

single diverter will handle the full range of applications. Rates, impact on load, load size, and range are some of the important criteria used in selecting the proper diverter. The diverter rate can go up or down based on the carton size and the "clear time" required by the diverter. This clear time relates directly to the gap needed between cartons and is a function of conveyor speed.

Table 9.1.7 compares features of several types of diverters. Since a thorough knowledge of the system is required in selecting a diverter, this table should be used as a guideline and not the final solution.

When conveyors intersect pedestrian and vehicle traffic, it is necessary to include a means of crossing the conveyor. Some typical arrangements are a removable conveyor section if the traffic is infrequent, a gate section which can be raised manually or automatically, or a pedestrian stile for walking across the conveyor. In making the selection, it is necessary to know the frequency of traffic, pedestrian or vehicular, and the effect of interrupting the conveyor traffic. It is also important to consider emergency exits from the conveyor area in the case of fires or other calamities.

9.2 IN-FLOOR TOWLINE CONVEYORS

John G. Dorrance

Towline transport systems provide a simple and dependable, yet versatile, method of transporting raw materials, in-process work, and finished products. Manufacturing, distribution, warehousing, and storage operations can all benefit from the application of this type of transport system. Towlines, when used in these operations, can be successfully integrated with manual or automatic processes, depending on the needs of the operation. Towline systems have effectively solved many of the problems associated with the horizontal transport of unit loads.

Towlines create an orderly, uncongested flow of traffic through the facility by following a predetermined path designed specifically for each application. In manufacturing, carts carry in-process work in an orderly progression from one machine tool to the next. In the distribution center or warehouse facility, separate orders can be placed on each cart by manual or automatic selection systems. Orders can then proceed directly to shipping areas. In-process or finished products are transported to and from designated storage areas, thereby optimizing all available building space. In all of these cases, towlines have enabled their users to achieve greater efficiency and improved organization in the workplace.

Rising energy and labor costs have made it necessary for industry to find alternative handling and transportation methods. Fork trucks, tractor trains, and overhead trolleys are now being replaced by the more energy-efficient towline method of transporting products and materials. In addition to conserving energy, towline systems also enable companies to make better use of manpower skills. Employees can be taken out of hazardous environments such as freezers, ovens, and dangerous chemical areas and transferred to safer areas of the facility where they can be more productive.

Companies that have installed towline systems are now realizing significant increases in productivity rates, while at the same time cutting their overall operating costs. Because towline systems provide the benefits of paced production, inventory control, and increased operating efficiency, industry is better able to meet the needs of a competitive market.

Towline systems were first designed as deep-channel in-floor conveyors. In the 1960s, however, advancements in technology brought about the low-profile or shallow-trench design. With the introduction of this new conveyor design, towlines could now be installed in existing facilities. This new installation capability stimulated a rapid growth of low-profile towline system sales. The emphasis of this discussion is on the low-profile (shallow-trench) system.

9.2.1 Operation

Towlines are available in various sizes and configurations, but the general operation of the low-profile system is basically the same. Depending on a facility's size and requirements, the system may be comprised of a single or multiple loop of track. The track is recessed in a specially designed trench that is placed within the floor of the plant. The track houses a powered, sliding chain which in turn moves the carts through the system layout (see Fig. 9.2.1).

The most efficient type of low-profile drive is a vertical sprocket-drive assembly. With this arrangement the chain is engaged by the drive sprocket, passes over the sprocket, under an idler wheel, and is returned to track-level elevation. The drive sprocket maintains a full 180° chain engagement throughout the drive area, thus eliminating surge and binding of the chain. Pusher dogs on the chain pick up carts and transport them through the system.

Shallow versus Deep Trench Systems

As mentioned, towlines were first introduced as deep-channel in-floor conveyors. As engineering technology progressed, new companies brought out systems that utilized a shallow-trench (channel) design. This new design concept was based on the theory that the ideal theoretical place to apply a horizontal

Fig. 9.2.1 Chain runs along recessed track.

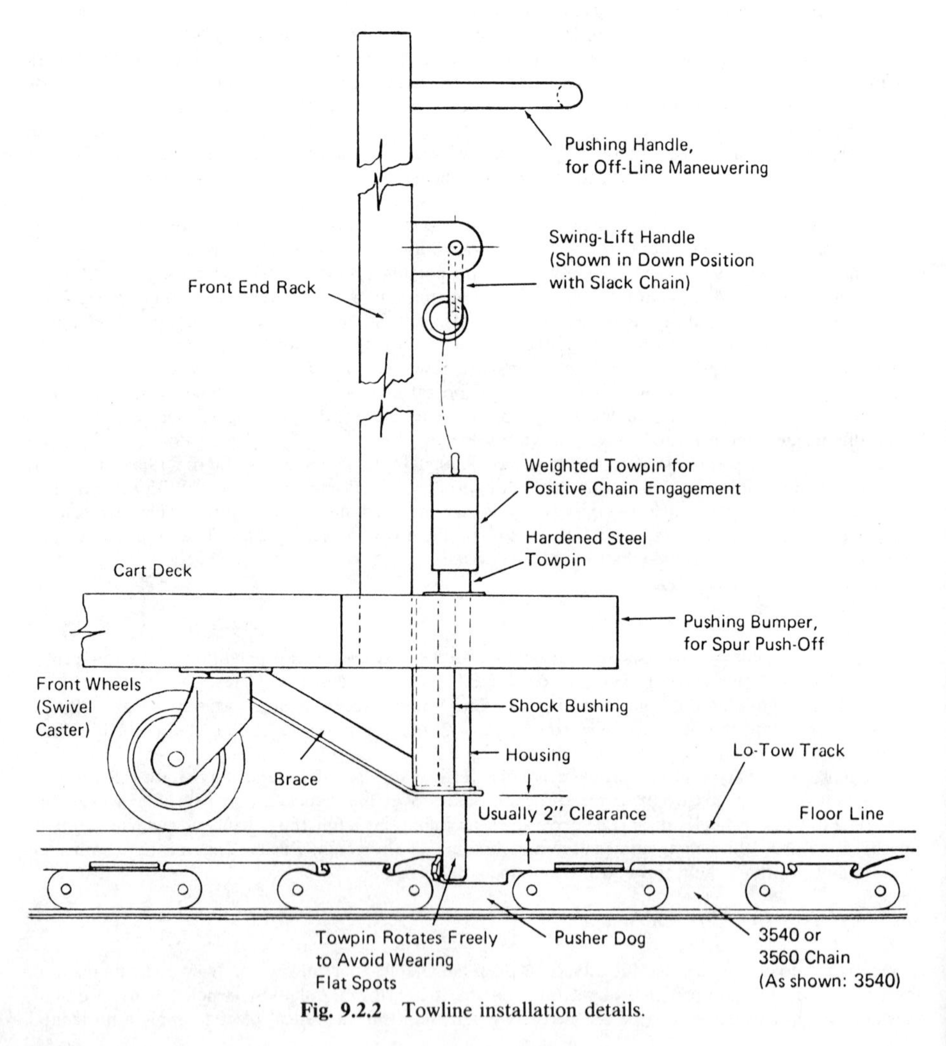

Fig. 9.2.2 Towline installation details.

towing force would be at the cart deck height above the floor. (This would eliminate the overturning moment and therefore, minimize the force necessary to move the cart.) This theoretical idea being impractical, the best solution was to shorten the length of the moment arm, as much as possible, thereby reducing the resulting overturning moment (see Fig. 9.2.2).

9.2.2 Capabilities

Other developments in materials handling technology permit some towline systems to be installed in multilevel structures. Lifts or ramps, used expressly for this purpose, are specified according to the desired elevation. This multilevel capability is a major factor in the expanded use of towline systems.

Towline systems have also benefited from advancements in computerization. Systems now interface with other manufacturing and warehousing functions such as assembly, inspection, and order selection. Advanced control systems permit independent monitoring of individual carts.

Towlines operate in hazardous environmental conditions such as dirt, dust, chemicals, freezers, and ovens. This capability enables companies to get their employees out of potentially dangerous situations and place them in more productive areas.

9.2.3 Components

Track

The towline track is a U-shaped receptacle on which three critically spaced steel wear bars are solidly plug-weld mounted. The wear bars insure a smooth and quiet gliding motion of the chain through the layout. The track top opening is narrow enough for foot and wheeled traffic to cross safely (see Fig. 9.2.3).

Chain

The towline chain is constructed of steel and is comprised of alternate links of side bars and center blocks. Connecting links are attached to bars and blocks by horizontally placed, high-strength alloy steel rivets. Pusher links are factory inserted at predetermined uniform locations throughout the chain length. Chain pitch is a varying characteristic that is dependent on the maximum cart size and payload capacity of the system.

The chain in most shallow-trench system designs is self-cleaning. The top surfaces of the links carry debris to frequently spaced clean-out boxes. In the drive pit, the overturning motion of the

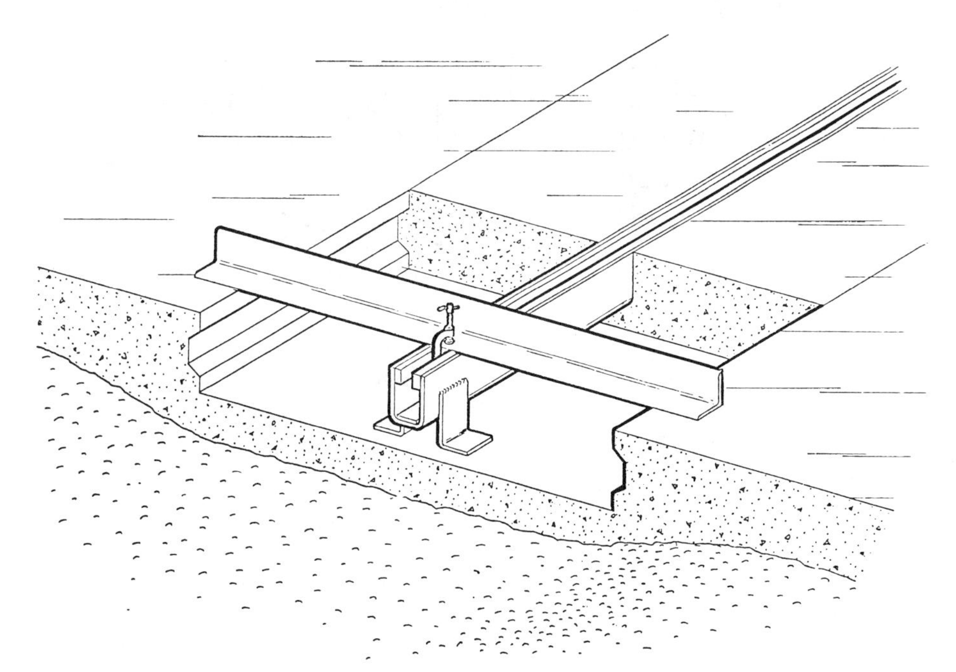

Fig. 9.2.3 Track top opening details.

chain releases any debris and allows it to drop to the bottom of the pit. A system designed in this manner requires less maintenance and therefore affords fewer operating problems. Automatic lubrication of links and pins occurs at zero chain tension when the chain is in the inverted position passing through the drive.

Drives

A towline system can be powered by a constant- or variable-speed drive, depending on the user's requirements. Variable-speed drives are frequently used to pace production.

A low-profile conveyor drive system consists of a vertically mounted sprocket on a speed reducer and an adjustable idler wheel. The eccentric rolling-gear type reducer is driven through V-belt by an electric motor. This drive arrangement has fewer moving parts than conventional reduction systems. It is proven to be a more efficient system, and also requires less maintenance. Motor horsepower and electrical characteristics are dependent on the system design and installation site requirements.

A soft-start controller is also used to provide a gradual accelerated start-up of the conveyor. Thus, every time the system is started, the conveyor gradually is accelerated to its designed speed. Variable-speed drives have been successfully employed in driving the low-profile towline system.

As the chain enters the drive pit it is engaged by the sprocket, passes over the sprocket, and turns over. It then passes under the idler sprocket, is wrapped around it, and exits the drive pit at track elevation (see Fig. 9.2.4).

Each drive unit contains a solenoid-operated automatic lubricator. This device applies oil directly to the sliding surfaces of the bar and block links of the chain. When the chain is in the overturned position, between the drive and idler sprockets, there is virtually zero tension except for a slight catenary. Oil is applied on the chain-connecting pins and sprocket teeth at this point. The automatic lubrication feature is important, in that it reduces chain and sprocket wear and increases the system's life.

Turns

Roller-chain turns are used to negotiate a horizontal change of direction with no sliding friction. This auxiliary chain operates at one-half of the normal conveyor speed. Resulting forces from the main conveyor chain tension are transmitted through the auxiliary chain and absorbed by a steel back-up bar. Two boxes, one at each end of the turn, provide a dual function: accumulation of dirt and debris, and a point where the auxiliary chain can reverse its direction and return to the starting position. In effect, the turn performs like a large horizontal roller bearing.

The standard roller chain has a 6-ft radius, although specially designed radii are available. Standard curves include 15°, 30°, 45°, 60°, 90°, and 180°.

Spur Diverters

Carts can be either mechanically or electrically diverted to separate sections of track. In mechanical switching, a bump (spherical projection) protrudes approximately one inch above the floor level. A cart selector pin trips the bump as the cart passes by. When the bump is tripped, a tab that is connected

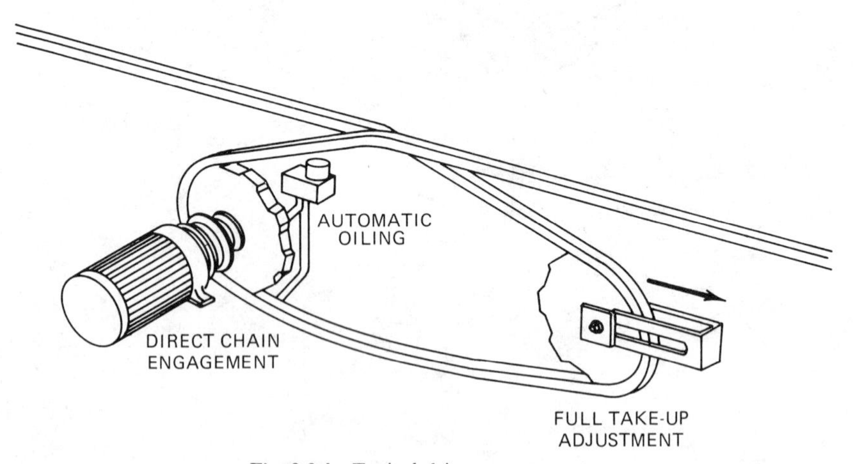

Fig. 9.2.4 Typical drive arrangement.

to a cable below floor level mechanically opens a diverter. The diverter has a spring-loaded blade which is held in the closed position by a latch mechanism. This mechanism is connected to the bump tab by the cable and opens when the bump is tripped. The cart tow pin is cammed from the conveyor track onto the spur track. After the tow pin passes over the diverter blade, the blade returns to the closed position and is held in place by the latch mechanism. The distance between the bump centerline and the conveyor centerline determines the station or selector pin to which the spur responds.

Electrical switching is activated by a magnetic cart selector probe. When the probe passes over a reed switch at a diverter station, the magnet completes the reed circuit. The now-energized reed circuit activates a solenoid, which in turn releases the diverter blade. The blade springs open across the conveyor track and cams the cart tow pin onto the spur track. The rear of the diverter blade cams the blade closed and it is retained in the closed position by a latching mechanism. The distance between the reed circuit board centerline and the conveyor centerline determines which station will be activated.

The circuit may contain either single or dual reed switches. In a dual reed circuit, the passing selector probe magnet must close two switches simultaneously. The advantage of this system over the single-reed system is its greater ability to discriminate between adjacent codes.

Electrical switching has distinct advantages over mechanical switching. There are no protruding bumps within the system which may cause safety hazards. Spur locations and codes are easily revised to adapt to new traffic patterns. Electrical control of switching adapts to varying degrees of automation, including system monitoring, cart counting, and computer control.

Spur Configurations. After the cart has been diverted from the main conveyor, it may follow one of three spur configurations. Spur flow can be perpendicular, parallel, or at a 45° angle off the main conveyor flow. In all of these cases, change of direction should be accomplished in no greater than 45° increments. Spur motion can be nonpowered, gravity, or powered.

Most nonpowered spurs are designed for only level or horizontal operation. Conveyor speed, cart load, and wheel variables are determining factors for the level spur capacity and geometry. Soon after the cart has been diverted into the spur, it comes to rest. In this manner carts are accumulated and stored without affecting system operation.

Partial gravity spurs are also available for use when the slope of the downward grade is not as steep as that of the full gravity spur. This method is less costly to design and execute and can also handle a higher load capacity.

A lock-out mechanism is included in all of the preceding types of spur designs. The mechanism includes a limit switch which is deactivated when the spur is full and thus prevents additional carts from entering the spur. Spur cleanout boxes are also available for collecting loose debris.

Electrical or mechanical diverters can be used to transfer carts onto a transfer or a powered spur. The powered transfer or powered spur utilizes a specially designed side-finger chain to move the carts. This chain has a vertical connecting pin and roller design which enables the chain to more easily negotiate radius turns. A pusher extends out and away from the conveyor centerline and tows the cart tow pin alongside. This offsetting pushing finger allows the cart to be transferred positively from the towline main conveyor to the transfer or powered spur. A wide steel channel liner houses a narrower steel channel. A steel bar divides the channels in two equal parts and the whole assembly is enclosed with a $\frac{1}{2}$-in. steel cover plate (see Fig. 9.2.5). A side-finger track rides within each half of the narrow channel, one side for the active portion, the other side for the return portion of the chain. The chain side fingers protrude out into the two channels. The track is driven by an electric motor coupled to

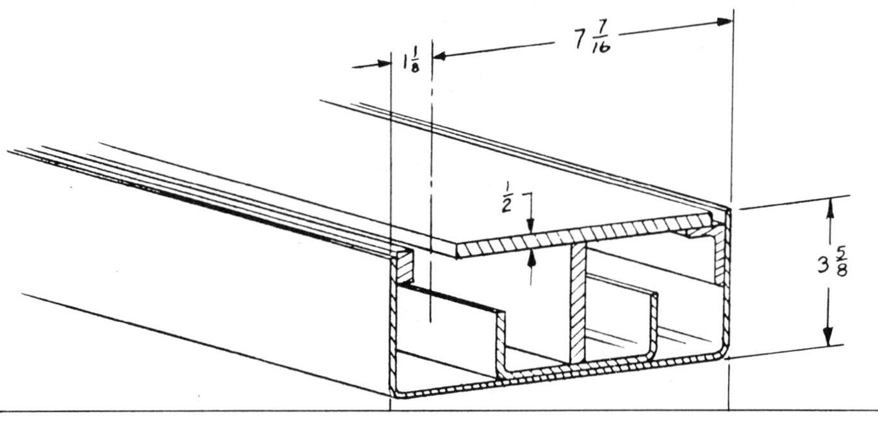

Fig. 9.2.5 Channel assembly.

a speed reducer on which the drive sprocket is direct-mounted to the vertical output shaft. This drive assembly can be below the floor or if necessary above the floor.

Carts accumulate by the use of side action stops in transfer and powered spur conveyors. Main line accumulation is possible on systems where carts are equipped with accumulation bumpers and vertical blade accumulation stops. On these applications an approaching cart signals an electric motor to raise the vertical blade above floor level which in turn activates the accumulation bumper and stops the cart.

In loop-type systems, carts are able to change elevations through the use of ramps and vertical lifts. Track sections on a ramp incorporate vertical radius turns. To hold the tow pin down in the pusher dog when the cart travels on a ramp, a hold-down rail of angle iron is bolted to the side of the track. Vertical lifts consist of a powered entrance spur, a powered load/unload mechanism in the elevator bed, and a powered take-away spur at each destination floor. When the lift is in position, the load mechanism automatically loads a cart onto the lift bed as it simultaneously off-loads a cart at the back end of the lift cage.

Carts

Towline carts are available in a variety of sizes, shapes, and materials, depending on the manufacturer and the function the carts will serve. The most common construction is the all-steel cart deck. A formed steel or angle-iron support is located at the front and rear of the cart. Rigid casters are mounted to the rear support, and swivel casters are located on the front support. A steel deck folds over the side sills of the cart and covers the entire deck. This is the most economical deck design. Decks can also be constructed of wood or a combination of steel and wood. A wood deck can provide a more stable and quieter surface.

Rubber or polyurethane treaded wheels are usually supplied on the carts. Steel or plastic wheels should only be used for special environmental conditions when rubber or polyurethane wheels would prove unreliable. Steel or plastic wheels can quickly wear grooves in a concrete floor. When they are used, a steel floor plate is mounted throughout the wheel path to prevent concrete surface deterioration.

Controls

As discussed previously, towline carts can be mechanically or electrically diverted to follow a predetermined traffic pattern. In addition to these basic types of switching, new methods are being designed as technological advancements are developed. Photoelectric switching employs an electronic reader (photoelectric cell) mounted below the floor surface, which monitors the passing of each cart. Reflective disk codes located on the carts are read by the photoelectric cells (located in the floor) as the cart passes over them. Each specific cart is identified and electronically mapped through the system layout. In optical switching, the towline system is equipped with optical scanners strategically placed along the cart path. The scanners read coded identification labels which have been affixed to the top side of each cart and guide the cart along a predetermined path.

When electrical or electronic switching is employed in a system, additional electronic monitoring and control functions are easily integrated to advance the system's capabilities. Through the use of computers, programmable logic controllers, and microprocessors, many production functions can be coordinated to achieve the desired degree of automation. In the manufacturing plant, products or materials can be automatically loaded or unloaded from the towline carts and placed on computer-controlled processing equipment. Towline systems in the warehouse or distribution center can be interfaced, by computer, with automatic order selection systems to improve productivity rates. Both manufacturing and warehousing operations benefit when towline systems are integrated with automatic storage and retrieval systems. Corporate management functions such as inventory control, receiving, invoicing, and shipping can be linked by computer and interfaced with the towline transport system and provide daily management reports. Computer hardware and software can either be supplied by the towline manufacturer or purchased separately from one of many reputable computer firms. Many towline systems can be controlled by adding new programs to existing corporate computer systems. Whichever method of controlling the system is selected, provisions should be made for expanding these control capabilities to meet the growing needs of the business.

9.2.4 Applications and Installations

Towline systems offer warehousing and distribution centers as an economical means of transporting merchandise in an organized pattern throughout the facility. Any type of product can be handled, including groceries, paper products, newsprint rolls, in-process machinery, furniture, and heavy appliances. Towlines are especially efficient when used in conjunction with automated order selection systems. The order selection machine automatically deposits each store order on a towline cart. Computer control systems then guide each cart through the system layout, continually monitoring its location, until it reaches a preprogrammed destination. With the aid of a towline system, store orders are quickly filled and delivered to specified shipping areas. In this manner, more timely, accurate deliveries are achieved.

Towlines also benefit warehousing and distribution operations by integrating receiving, bulk storage, and order-picking functions. This integration can result in reduced congestion within the workplace and a better utilization of available building space. Product damages are minimized because there is less manual contact. All in all, the system guarantees increased productivity rates through a better utilization of manpower skills and production performance.

Recently, a towline system was installed in a midwestern warehousing and distribution center. The system consists of one main loop, six powered spurs, and two bypass loop transfers. An operator enters a cart identification number and transfer spur code number into a computer terminal keyboard station. Thus, when the cart reaches an activated in-floor photocell scanner (located before the spur's divert point), the cart is then switched into the spur.

Various merge and stage stops are located throughout the system. Because the cart identification codes have been entered into the towline control system, carts proceed automatically to predetermined stop points where they are identified and loaded or unloaded.

An increasingly popular application for towline systems is in newsprint roll handling operations. At one recent installation, precoded dollies (carts) circulate the loop and, depending on the type of roll, are automatically switched onto designated spurs. This system also reads product information and identifies empty carts for rerouting.

In the manufacturing plant, towline systems transport anything from soft goods to heavy-duty industrial products weighing up to 10,000 lb. The system provides a physical link between all process functions, resulting in a unified, smooth-flowing operation. Towline carts can be specially designed to function as mobile workbenches for assembly and inspection procedures. In this case, carts travel from one work station to another. Employees at each station assemble the product—from start to finish—and inspect the finished product without ever having to remove the work from the cart. Finished products can then be transported directly to shipping docks or storage areas. Because products are never removed from the carts, production can be paced according to desired rates. Towline systems can also be integrated with automated assembly and manufacturing processes. Carts are accurately positioned automatically at each work station. Automatic welding, machining or assembly machines can then perform their functions, using the towline cart as a workbench. Applications of this kind are now employed throughout the world to increase productivity and reduce operating costs.

One of the first applications of a towline system was in large-volume transportation terminals. Today, towlines are used throughout bulk mail centers and truck, rail, and air freight terminals to expedite large-volume dispatching. The system offers the benefits of rapid unloading and sorting of incoming and outgoing shipments. Integration with automatic loading and unloading equipment is also possible through the use of electronic control systems. The use of advanced electronics also facilitates auxiliary functions such as receiving and shipping records-keeping.

In this furniture manufacturing facility, a towline system transports furniture through the finishing process. The system can only be started from the main control panel, but can be stopped at any remote stop location. Whenever the system is started, warning horns sound for 10 sec prior to system movement. Each powered spur operates whenever the loop is running. Specially designed rework index power spurs operate only when a code (on a cart) is read by the spur's read switch assembly.

Carts in the system are equipped with coded devices which enable them to automatically travel to predetermined spur locations. Once the cart has entered a spur it is either manually or automatically stopped and loaded or unloaded.

An industrial manufacturing facility in the eastern United States uses a sophisticated towline system in the production of metal parts. The system incorporates a main control panel for starting or stopping towline operation and remote stop and reset controls along the system path. A mimic outline board located with the main control panel in the supervisor's office gives a graphic representation of the towline system coordinated with the building outline and column grid locations. With the outline board the supervisor can monitor cart location and travel throughout the facility.

Electrical switching is employed to divert coded carts onto transfers. Carts then proceed to nonpowered spurs located along the transfer where they are loaded or unloaded according to their scheduled operation. Carts pass through an inbound and intermediate section before returning to the towline loop.

9.2.5 System Planning and Installation Guidelines

The first step in planning the installation of a towline system is to analyze the specific transport requirements to a particular application. This is accomplished by following the guidelines discussed next.

Organize a logical traffic flow pattern through the facility by developing a step-by-step sequence of the operations to be performed. Paths should be as simple as possible, with limited turns and changes in elevation.

Integrate handling activities and operations by combining steps where possible and eliminating unnecessary stops. Processes such as inspection can be integrated with machining and assembly operations. In-process storage may be eliminated in cases where work is stored on the towline carts.

Standard equipment should be selected whenever possible to assure quick delivery and easy maintenance. This standard equipment varies among suppliers, so a supplier should be chosen who best suits your particular needs.

This system must interface with any existing or new equipment for operational efficiency. For example, in the distribution center, the towline system can be interfaced with new or existing order-selection processes. By providing for this interface in the design stages, costly system revisions are eliminated.

The company that plans for the future growth and expansion of its business will profit most by the installation of a towline system. This planning, when realized in the design stage of a towline system, can actually add to the company's growth potential by allowing for the flexibility necessary to adapt to changing world markets.

Maintenance and component accessibility are major factors to consider when planning for a towline system. A supplier should be chosen who has a reputation for providing timely service in addition to quality products. With proper maintenance, the system should prove to be both efficient and reliable.

Evaluate any special needs or operating conditions that are unique to your application. This includes any hazardous environments or safety requirements to which the system must conform.

By taking into consideration all of these selection guidelines, the system will prove to be a valuable asset to your operation. Once the preceding guidelines have been evaluated, then the specific design factors should be considered.

Anticipated maximum load weights and dimensions must be known in order to design a system that can adequately handle the actual load capacity and throughput. At this time, special carts can also be designed to meet your specific needs. The proposed installation site must also be reviewed to insure adequate area for trench installation and cart travel.

If you have followed the preceding guidelines and recommendations, a towline system will be a profitable investment.

9.3 SPINNING-TUBE CONVEYORS

John G. Dorrance

Spinning-tube transport systems provide a precise, efficient means of conveying products to aid in manufacturing and distribution center processes. These systems have been proven effective in increasing productivity, while at the same time reducing overall operating costs and product damages. The modularity of the systems facilitates installation in new or existing buildings and provides the flexibility necessary to adapt to changing industrial requirements.

The spinning-tube method of transporting products and materials is operating successfully in facilities throughout the world. These systems benefit manufacturing plants by providing an interface with machining equipment and assembly operations. The systems are easily integrated with manual or newly developed automated manufacturing processes. In the warehouse or distribution center, spinning-tube transport systems safely convey products integrating with order selection and automated storage and retrieval functions. These spinning-tube transport systems are especially useful with products which are bulky, delicate, or otherwise difficult to convey, and where environmental conditions are detrimental to employee health. In all of these applications, spinning-tube transport systems are used to improve productivity and to effectively control flow in the workplace.

9.3.1 Operation

Operation of the spinning-tube transport system is simple, with only one moving subassembly. That subassembly is the spinning drive tube which is mounted between the rails of a two-rail section of track. Sections of track are joined together according to the system layout. Carriers are mounted on and guided by this track and propelled by the spinning tube within.

A spring-loaded drive wheel, mounted on the underside of the carrier, is compressed against the spinning tube. Contact between this drive wheel and spinning tube creates a forward motion of the carrier. When the drive wheel centerline is parallel with the spinning drive tube (Fig. 9.3.1) there is no driving force and the wheel merely rotates in place. As the angle between the drive tube and drive wheel centerline increases, a thrust is generated which propels the carrier forward. Maximum carrier speed is obtained when this angle reaches 45°. Carrier speed can be decreased by decreasing the angle and therefore reducing the thrust.

A pivoting housing contains the drive wheel and limits its movement from 0° to 45°. This pivoting housing also provides for independent control of carrier and tube speeds. Specially designed queue station control bars are located on the track throughout the system layout to control the position of the pivoting housing. These control bars provide for the precise positioning and controlled acceleration or deceleration of each carrier. For accumulation capabilities, a cam follower arm on the front of each carrier is linked to the pivoting housing. A cam is also located at the rear of the carrier. Contact

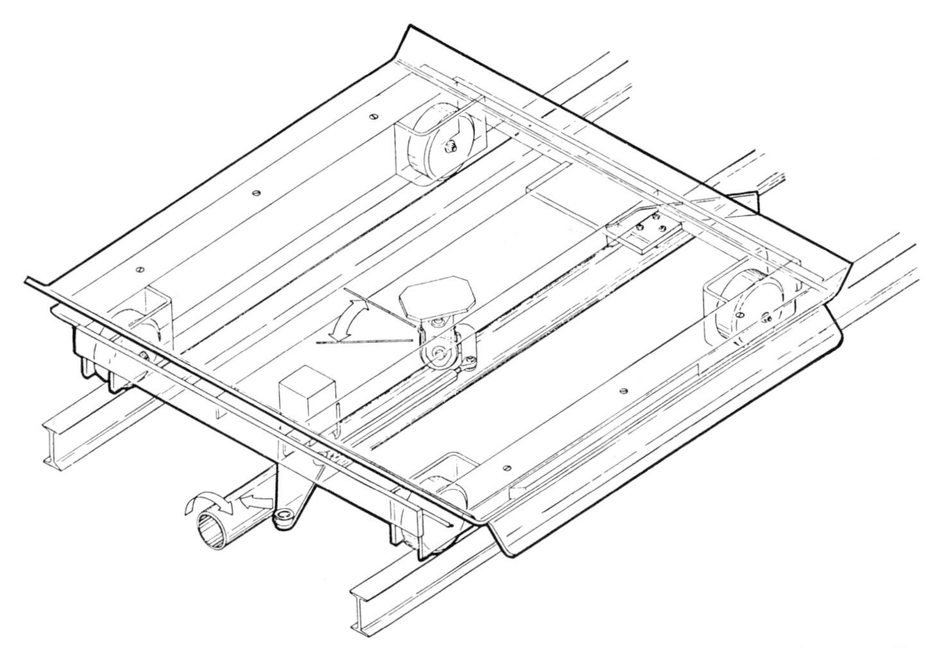

Fig. 9.3.1 Wheel rotates in place, with no driving force being applied, when drive wheel centerline is parallel with spinning drive tube.

between the cam follower arm of one carrier and the cam of another provides controlled acceleration or deceleration in the accumulation of carriers.

9.3.2 Capabilities

The capabilities of this unique transportation system have solved many difficult in-process production and distribution problems.

Nonsynchronous, independent movement permits carriers to be individually indexed throughout the system. Carriers move from one station to another without affecting the rest of the system operation. This feature is unique to the spinning-tube transport system for it provides versatility and flexibility within the workplace, which is not the case with other transportation methods.

Controlled acceleration and deceleration is especially useful in applications where delicate or unstable loads must be transported. This smooth, gentle movement prevents the product damage often incurred by fork truck or manual transport. Controlled acceleration/deceleration also enables carriers to be precisely positioned for interface with other production equipment. Positioning can be as precise as ±0.005 in. in three planes. This precise positioning is a necessity when the transport system is integrated with precise automated machining and assembly operations.

Different speeds may be used simultaneously in separate sections of a system, according to the user's requirements. Speeds can be varied from 0.5 to 600 or more feet per minute. In addition, this capability permits individual sections of a system to be shut down in order to conserve energy or perform routine maintenance procedures.

Spinning-tube transport systems have been proven to be operationally effective in environmentally harsh industrial and warehousing applications. These systems can be used in freezers with temperatures as low as −40°F and in ovens where temperatures reach 200°F. They function in the presence of dirt, dust, and chemicals and in environments that would otherwise be unsafe for employees.

In addition to these capabilities, the spinning-tube transport system increases efficiency through better use of manpower and space, resulting in greater system throughputs. The system's modular design allows for easy installation in new or existing facilities and can be totally relocated if necessary.

9.3.3 Components

Track Modules

The function of the track module is to support, guide, and propel the carriers through the system (see Fig. 9.3.2). Tube drives, queue stations, track supports, and limit switches are all mounted on

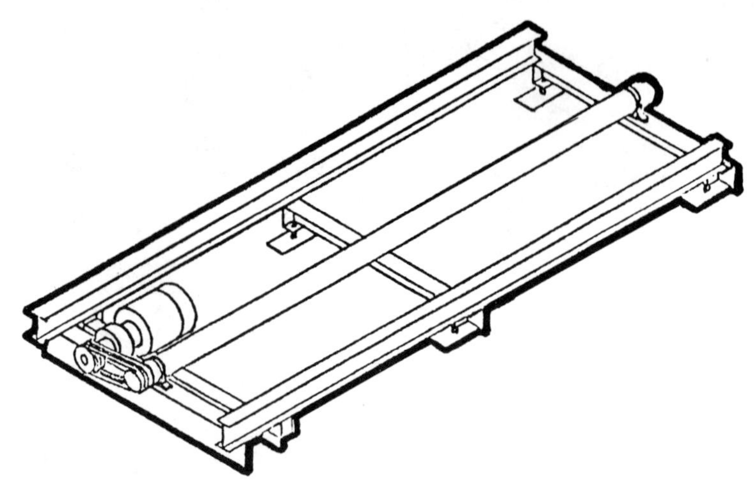

Fig. 9.3.2 The track module supports, guides, and propels carriers.

the track module. Each track module is comprised of three basic assemblies: the track rails, crossmembers, and the drive tube.

The track rails are formed from I-beams and are designed for the structural strength and rigidity necessary to support both the carriers and the maximum load capacity of the system. The track frame is constructed by welding these rails to steel angle crossmembers. Holes in the crossmembers are provided for mounting the drive tube, track supports, queuing devices, and tube drives.

The drive tube is constructed of mechanical steel tubing with hex-shaft plugs capping off each end. Hex-bore ball bearing pillow blocks, sealed and permanently lubricated, are clamped rigidly at the tube ends. This hex-shaft/hex-bore bearing combination assures positive bearing rotation, greater reliability, and longer life. When a tube drive is to be located at a track junction, the driven end will require a sheaved tube end to cap off the drive tube. If a tube drive is not required at a track junction, the tubes are joined with a male/female coupling.

Track Supports

The track supports provide a rigid base for supporting the track modules at a predetermined elevation above the floor level. A support consists of a top pan and two feet bolted together and anchored to the floor. Track crossmembers are bolted to the top pan.

Tube Drive

The spinning-tube drive provides the power necessary to propel the carriers throughout the system. This is accomplished by rotating the drive tubes within the track modules. A tube drive is capable of rotating one or more drive tubes, depending on the horsepower requirements of the system.

The drive package consists of a motor, speed reducer, sheave and bushing, V-belts, and mounting frame. The horsepower and electrical characteristics of the motor are dependent on the system and installation site requirements. A flexible coupling connects the motor shaft to the high-speed side of a speed reducer. A grooved sheave is attached to the output shaft of the speed reducer. V-belts transmit power from the reducer to the sheaved-tube end of the track module drive tube.

Turntables

Turntables are used to change the carrier's direction of travel by 90°. In performing this function, the turntable also supports, guides, and propels the carrier through the system. The two main components of the turntable are the deck and the supporting frame (Fig. 9.3.3).

The deck is the rotating portion of the turntable. It is supported by a large-diameter ball bearing, and includes a deck plate, track rails, drive tube, queue station, and limit switch. As a carrier approaches the turntable, the queue station stops the carrier's motion precisely on top of the turntable. The limit switch detects the carrier's presence and activates the rotational drive unit. This rotational drive unit, which is mounted on the supporting frame, rotates the carrier 90°. A second limit switch senses that the carrier has completed its rotation, pivots the queue station control bar, and the carrier accelerates forward. This is the most basic operation of a turntable.

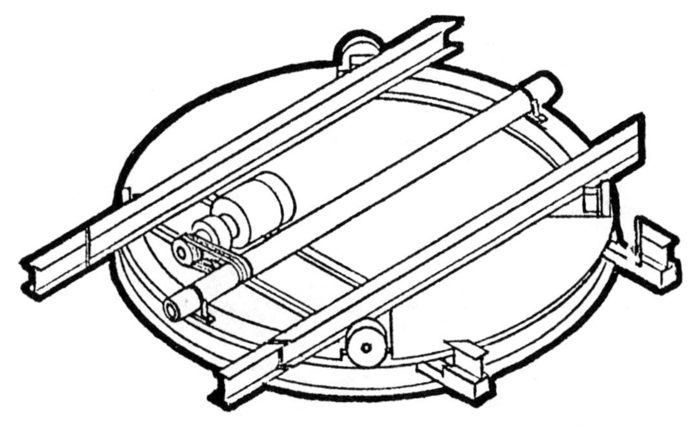

Fig. 9.3.3 Turntable components are deck and support frame.

A more complex type of turntable is also available which, in addition to changing the carrier direction by 90°, also serves as a merge or divert location. Operation is basically the same as previously described; however, this turntable may accept carriers from two locations and merge these carriers into a third output location. This "T" type turntable may also accept a carrier from one input location and divert it into two possible output locations.

Transfers

Transfers are used to transport a carrier perpendicularly from one path to a parallel path. In doing so, the transfer is capable of discharging the carrier in the original or reverse direction of travel.

The two major components of a transfer are the transfer car and transfer track. The transfer car is similar to a carrier, in that it utilizes a chassis with integral drive, guide, and load-carrying wheels and a limit switch cam. The fixture mounted atop the transfer car is a standard spinning-tube track section. It incorporates a tube drive, queue station, and limit switch. The transfer track propels the transfer car from one end of the transfer to the other. Its construction is similar to that of a standard track module. It utilizes track rails, crossmembers, drive pan, and drive tube. Tube drive, queue stations, and limit switches are also included.

As a carrier approaches a transfer, its wheels move onto the adjacent track rails of the transfer car. A queue station on the transfer car brings the carrier to a controlled stop. The carrier's presence activates the transfer car limit switch which initiates transfer car travel. The transfer car travels to its predetermined destination where it is stopped by a queue station on the transfer track. The presence of the transfer car activates a limit switch which in turn causes the carrier to be released from the transfer car. The carrier accelerates off of the transfer car and onto the adjacent track rails. Once the carrier clears the transfer car, a limit switch is activated which initiates transfer car movement to the next selected position.

Radius Turns

Radius turns are also used to change carrier's direction of travel within the spinning-tube transport system. The turns utilize the basic concept of the track module, incorporating track rails, drive tubes, universal joints, and tube drive units. A radius turn also serves to support, guide, and propel a carrier through the system (see Fig. 9.3.4).

Turns are formed by rolling standard track rails to a desired radius. The curved track rails are then welded to steel crossmembers to form a rigid frame. Holes in the crossmembers are provided for mounting and supporting the drive tubes and for attaching the track supports. Two types of drive tubes are used: fixed tubes with pillow blocks and floating tubes without pillow blocks. Fixed and floating tubes are placed alternately throughout the turn and joined by universal joints. The same hex-shaft/hex-bore bearing combination found on standard track modules is utilized here. The standard tube drive unit powers the drive tubes of the radius turn.

As in the straight track modules, carriers are propelled through the turn by the interaction of the carrier drive wheel against the drive tube. Guide rollers on the swivel wheels of the carrier follow the path of the curved rails, effecting a smooth, uninterrupted change of direction. Supports are bolted to the turn structure in the same manner as on the straight track modules.

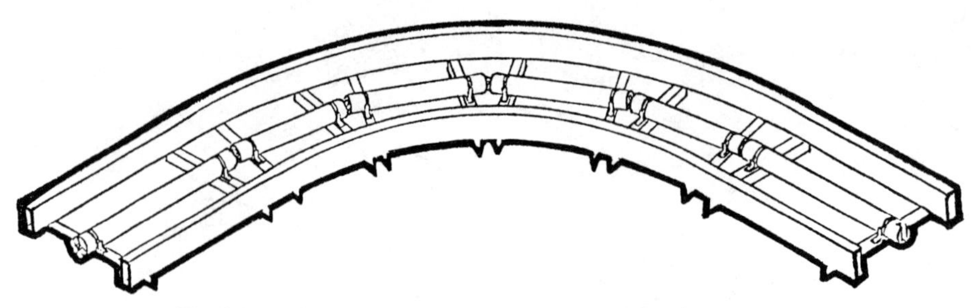

Fig. 9.3.4 The radius turn supports, guides, and propels carrier.

Carriers

Carriers are the transport vehicles of the spinning tube system. They are propelled through the track layout in a smooth, quiet, and safe manner, stopping as required at the predetermined locations. A typical carrier is constructed of these subassemblies: chassis, drive wheel, guide wheels, traveling wheels, limit switch cam, accumulation nose, and accumulation tail (see Fig. 9.3.5).

The chassis is a rigidly formed steel frame, comprised of deck plate, end members, and a center channel. Specially designed fixtures for carrying products or materials are mounted on top of the chassis. The drive wheel is a vertically spring-loaded urethane wheel which provides positive contact with the spinning drive tube. Sealed and permanently lubricated ball bearings on the drive wheel assure smooth, dependable operation. The top plate of the drive wheel is bolted to the center channel

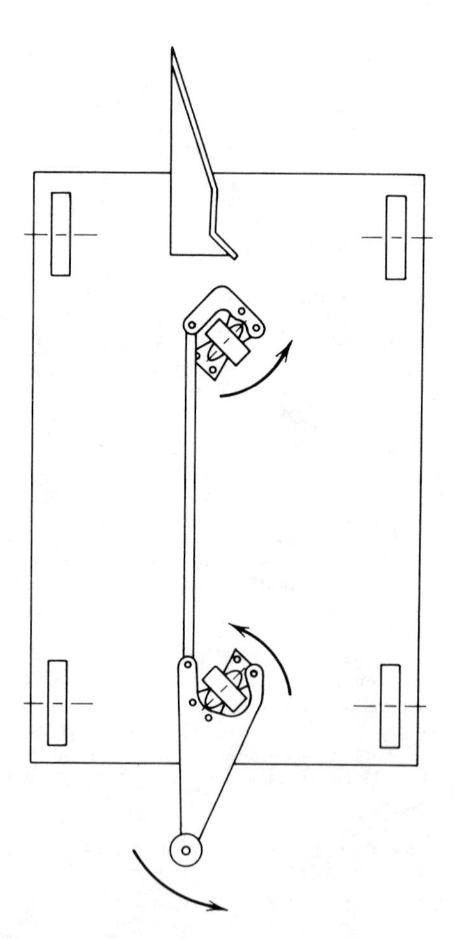

Fig. 9.3.5 Carrier is composed of chassis, drive wheel, guide wheels, traveling wheels, limit switch cam, accumulation nose, and accumulation tail.

of the chassis. Two guide-wheel subassemblies are bolted to the deck plate on one side of the carrier. A traveling, load-carrying wheel supports the carrier, while guide rollers are used to negotiate the system layout. Traveling wheels on the opposite side also support the carrier.

Automatic accumulation is made possible through the use of an accumulation nose cam follower (mounted on the front of the carrier) and accumulation tail cam (mounted at the rear of the carrier). The accumulation nose mechanism consists of a cam roller mounted to a pivoting nose plate, which is linked to the drive wheel by a connecting rod. The accumulation tail is a plate with a cam surface.

The carrier operates when the spring-loaded drive wheel is compressed against the drive tube. Front and rear guide wheels straddle one side of the track. At a queue station (stop location), the accumulation nose contacts the queue station control bar (cam surface) mounted on the track module. The carrier decelerates at a smooth, controlled rate and is stopped completely when the carrier stop bar contacts the roller on the queue station. Acceleration can only begin again after the queue station control bar has been manually or electronically pivoted to one side, clear of the stop bar on the carrier.

Accumulation may be accomplished on line when a moving carrier comes in contact with a carrier that has stopped. The roller on the front of the approaching carrier contacts the cam plate on the rear of the stopped carrier. The cam plate moves the roller and lever arm, which in turn moves the drive wheel assembly to the neutral position. As a result, the approaching car slows and stops. Any number of carriers may be accumulated in this manner on a straight section of track.

Carrier location is detected throughout the system by the use of limit or proximity switches. The cam bolted to the carrier chassis operates the roller lever on the limit switch thus relaying an electrical signal.

A typical system layout uses all of these components to create an orderly flow of traffic throughout the facility. Through the use of limit switches and advanced electronic sensing devices, the system is able to interface with modern manufacturing and warehousing processes.

Controls

Controls for the spinning-tube transport system can be simple or sophisticated, depending on individual requirements and capital available to invest in the system. As with the spinning-tube system layout, control systems can be updated to meet future needs and expanded to incorporate additional operating functions.

The most basic control system is one that functions only to regulate carrier traffic flow. Once power is supplied to the system, the operator has no individual control over the different sections of the layout. Carriers travel in a predetermined path, with no remote identification capabilities.

In advanced control systems, operators monitor the movement of each carrier in relation to the total system layout. Addressing capabilities can determine where a carrier must go, what path it should take and what it must do once it reaches that destination. The operator has a means of identifying the location of each carrier at any time. These types of systems can use escort memory, nonsynchronous memory, or computer control. Advanced control systems are imperative if the system is to be interfaced with other warehousing or manufacturing operations.

These systems are easily integrated with advanced manufacturing processes such as industrial robots and computer-controlled machine tools. In the modern distribution center, computerized order-selection systems are interfaced with spinning-tube transport systems to guarantee the quick, efficient delivery of customer orders.

Controls for the spinning-tube transport system can also be linked to corporate management functions such as inventory control, invoicing, shipping and receiving, and up-to-the-minute status reports. This unique data highway is possible through the use of programmable logic controllers, microprocessors, and large-scale process computers. Hardware and software packages are often available from the spinning-tube transport system manufacturer.

9.3.4 Applications and Installations

The particular type of in-process material transport system can easily be applied for use in any manufacturing or distribution center. Its versatility and flexibility permits its use with products as diverse as fruits and vegetables to large 12-cylinder diesel truck engines. What follows is a brief discussion of some of the most common applications in service today.

A rapidly growing function of spinning-tube transport systems involves the movement of parts within a manufacturing facility from one machine tool operation to another. This application eliminates the need for batch handling of parts between machines and leaves the area uncluttered of in-process work. Parts are delivered, as required, by the carriers which then move on to the next operation after work has been completed.

Mechanization of manufacturing operations is easily accomplished. Because of its accurate positioning capabilities, the spinning-tube transport system can be effectively integrated with robots or automatic cycling devices. Preprogrammed routing indexes parts directly from one operation to the next. Production rates can be predetermined by adjusting the transport system's rate of travel.

This type of system is used in the production of printing press rolls. The rolls, in various stages of completion, are transported from one machining operation to another. In one area, robots remove the rolls from the carriers and place them on numerically controlled lathes. In other areas of the factory, the system interfaces with manual operations. This application has proven that a spinning-tube transport system can be successfully integrated with both automated and manual manufacturing processes.

Spinning-tube transport systems can also be applied to assembly operations. Many types of materials and parts can be transported with ease by specially designed fixtures mounted on the carriers. By maintaining a fixed, constant speed, the system assures that predetermined production rates will be met. System layout can be designed so that one main system is used for final assembly, while subsystems are employed to carry parts to the main system. With advanced computer capabilities, carriers can be randomly selected to move materials from one station to another in any desired sequence.

A U.S. company that assembles automobile parts has installed this type of system in four of their facilities. An over/under spinning-tube system transports the parts through a multistation robot spot-welding operation. Carriers are positioned precisely (±0.035 in.) and travel rapidly from station to station. Parts are transferred to and from the system by a crane.

The spinning-tube system is also used to link products to high-rise automated storage and retrieval systems (AS/RS). Parts are carried to and from production areas by the transport system and are manually or automatically placed in storage or production vehicles. This transport method optimizes the available space within a warehouse and provides a constant control of inventory. With this system, the cost of fork truck maintenance and fuel is significantly reduced. A facility may employ one spinning-tube system for in-process work and another for AS/RS, or the functions can be combined to form one large system.

One foreign automobile manufacturer utilizes a spinning-tube transport system to interface with an AS/RS. This particular storage system is capable of housing 5,000 automobiles at one time. Cars are driven onto the system carriers and then transferred to the required storage aisle.

Order picking is a distribution center function which lends itself well to the use of this type of transport system. The system can be employed in conjunction with manual or automatic product selection to achieve higher productivity rates and reduce product damages. The order-picking transport system is designed as a closed-loop circuit with designated stations established for selection and replenishment operations. The improved product flow feature maintains an uncongested traffic pattern and optimizes space utilization of the facility.

One such installation utilizes a closed-loop spinning-tube transport system for the selection of produce in a distribution center. This particular system employs over 400 ft of track and 80 carriers. The carriers automatically stop at two manually operated selector stations. Here, orders are filled according to store requirements listed on computer printouts. Through the integration of advanced computer capabilities and innovative material transport systems, the company has succeeded in increasing productivity and improving service to their stores.

9.3.5 System Planning and Installation Guidelines

The most effective material transport systems are the result of careful planning and an in-depth evaluation of present and future material management requirements. When considering the installation of any materials handling system, the supplier chosen should be capable of performing this evaluation. The following factors should be considered when evaluating the total material handling requirements of any operation, and these should be examined before purchasing any spinning-tube transport system.

Whether the system is to be used in a manufacturing or distribution facility, it is important to integrate as many handling activities as possible. Each activity (i.e., storage, production, inspection, etc.) is an integral part of the whole system and the user will benefit most by the coordination of these operations.

To optimize material flow and reduce traffic congestion, a progressive sequence of operations should be determined. In new or existing buildings this sequence is dictated by the logical progression of work to be done.

The spinning-tube transport system layout should be simplified by reducing, combining, or eliminating unnecessary movement. In manufacturing, for example, a machining operation and inspection station may be combined at one location. In this instance, an unnecessary queue station is eliminated, thereby reducing the cost of the system.

Once operations have been combined or eliminated and there is no longer a need for in-process storage, vacant spaces will be available within the facility. This means that equipment and operations can be moved closer together. The spinning-tube transport system now requires fewer track modules and therefore will be less expensive.

Standardized equipment and components should be selected whenever possible. This reduces maintenance and replacement costs and provides for faster installation.

Probably the most important factor to be considered is flexibility and modularity within the system. As companies grow and products change, the spinning-tube transport system must also be able to adapt. Many systems can even be relocated if necessary. By planning for expansion when a system is first installed, these growing pains will be kept to a minimum.

Maintenance time and costs for a spinning-tube transport system can significantly affect a company's production capabilities. In most cases, the maintenance of this type of system is less costly and time-consuming than fork lift trucks or manual transportation. If the system does require repair, however, parts, components, and labor should be readily accessible from the manufacturer.

Lastly, any special needs or operating conditions must be considered which may affect the system layout or operation.

Before a spinning-tube transport system design can be initiated the following operating conditions must be established:

Maximum live load weight
Load dimensions
Mounting accessibilities (floor, wall, or ceiling)
Environmental characteristics (i.e., temperature, chemicals, dirt, etc.)

9.4 VERTICAL LIFT CONVEYORS

Harold VanAsselt

9.4.1 Introduction

Vertical conveyors, or powered vertical lifts, provide a convenient and efficient means of moving materials between floors and/or between different levels on a single floor. They are used in manufacturing plants, wholesale and retail stores, food processing plants, hotels, restaurants, hospitals, and so on. There are two inherent characteristics of vertical conveyors that distinguish them from other methods of vertical transport such as forklifts, freight elevators, or incline conveyors. Vertical conveyors occupy a minimum of valuable floor space and they provide continuous or immediate availability.

Types of Vertical Lift Conveyors

Vertical conveyors are classified into two basic types: reciprocating and continuous. Although reciprocating conveyors share some characteristics in common with conventional elevators, a closer look at both their design and operation clearly reveals a number of significant differences between the two. The major distinction is that vertical lift conveyors are not intended to carry people and are thus usually designed to preclude a person "boarding" the lift. Generally, reciprocating vertical lift conveyors are specified when operation is intermittent yet availability is a critical requirement. As their name implies, continuous vertical lifts are used to handle a continuous flow of materials or components as might be encountered in manufacturing operations.

Vertical Conveyor Speeds

Reciprocating vertical lifts are able to operate over a wide range of speeds; speeds greater than 200 ft/min (61 m/min) are not uncommon. More important than velocity, however, is the transfer rate associated with a reciprocating lift. This in turn is a function of distance traveled and the fact that the lift usually makes its return trip unloaded.

As a rule of thumb, reciprocating lifts can be considered for application where the transfer rates don't exceed three items per minute; for higher transfer rates, continuous vertical lifts should be considered. The upper limit for light-duty continuous service is on the order of 30 loads per minute while for heavy-duty service, rates of 600 loads per hour are possible.

Levels

Vertical conveyors may be used to transport items between different levels on the same floor, or between two or more floors. Traditionally, these devices were specified where available space didn't permit the installation of an incline conveyor, or where it wasn't feasible for incline conveyor to handle the designated loads (e.g., pallets).

More recently, the use of vertical conveyors for intrafloor movement has accelerated with the trend toward installing horizontal conveyor overhead to conserve valuable floor space. Thus, a part might exit a production line, be loaded onto a vertical conveyor, lifted and discharged to an overhead horizontal conveyor which transports it to another production line or inventory storage area. This type application could be served by either a reciprocating or continuous vertical conveyor, depending on production rates and the physical characteristics of the items being conveyed.

Another intralevel application especially suited to continuous vertical conveyor is in bridging obstacles at floor level such as other horizontal conveyor, towline, or rail tracks. In these instances, material is received at say floor level, lifted and transported horizontally above the obstruction at or near overhead level, lowered on the opposite side and discharged.

9.4.2 Standards, Safety Codes, and Precautions

The ASME published standard, AMSI/ASME B20.1, which is distributed by both the ASME and ANSI, serves as a guideline for continuous and reciprocating conveyor installations. As with horizontal conveyor, most manufacturers have developed a variety of standard designs which in turn are tailored to meet individual requirements.

A vertical conveyor should never used to carry people. The fundamental difference between a vertical conveyor and an elevator is that a conveyor carries material only while an elevator carries people and material.

One of the first steps when considering a conveyor installation is to review ANSI/ASME B20.1 for operating and guarding safety requirements. Where operation to multilevels is involved, requiring building modifications, applicable fire and building codes should be reviewed.

Rarely is a permit required for the installation of a conveyor. The conveyor manufacturer or supplier should be conversant with the applicable codes. Your contractual agreement with the conveyor supplier should spell out clearly the areas of responsibility, which include:

Compliance with applicable codes
Responsibility for installation of mechanical and electrical components
Responsibility for providing required guarding and gates
Operational training

Protective Devices

Protective devices associated with vertical lift conveyors fall into three categories: personnel safety, equipment protection, and load protection. Some safety devices provide protection in two or all three of these areas.

Basically, the same safety features outlined in ANSI B.20.1 for horizontal conveyor should be incorporated into vertical conveyor designs (e.g., guarding of nip and pinch points, electromechanical interlocks, etc.). There are in addition, however, a number of safety features and precautions that apply specifically to vertical conveyors which should be kept in mind when considering their installation and operation.

Reciprocating Lifts. The primary personnel safety feature unique to reciprocating lifts is a method of preventing access to a dangerous area such as the platform guideway. In cases where a reciprocating vertical conveyor automatically receives or discharges loads, one of two safety devices should be present: either (1) station doors or (2) entrances and exits guarded by a suitable enclosure, barrier, or guard extending on all sides.

Similar to the guarding of horizontal conveyor, effective guarding of vertical guideways on reciprocating lifts is an important consideration. This requirement is usually met by installing guard tunnels at lift entrances and exits. The tunnels should be a minimum of 30 in. (0.76 m) long and have cross-sectional dimensions that allow material or product to enter the lift, but prevent a person from boarding the unit as well as preventing arms or hands to come near the moving parts of the lift. An alternative is a set of doors that must be shut before the reciprocating lift can operate and that can be opened only when the carriage has stopped and is properly positioned at the door's level. Naturally, an override provision must be made to permit access to the unit for repair or maintenance.

For protection of equipment and loads, the first step is to provide an adequate safety factor in the design. Beyond this there are a number of fairly standard safety devices which should provide a completely adequate margin of safety. These include a slack chain limit switch, appropriately positioned over travel limit switches, and mechanically operated stops on associated (horizontal) conveyor which prevent loads from being conveyed into a void.

Continuous Vertical Lifts. As with reciprocating lifts, ANSI B.20.1 contains the basic safety guidelines for continuous vertical lifts. In addition, appropriate controls should be provided to stop the lift in the event an oversize load appears at an entrance. Provision should also be made to prevent premature or late entry of a load onto the lift platform, to stop the lift if there is inadequate room at the exit side to accept the discharged load, or in the event a load shifts and jams the platform. Especially important from the standpoint of personnel safety is a manual control to restart the lift after having stopped for overloading or jamming; under no circumstances should a lift be able to restart automatically under these conditions.

9.4.3 Reciprocating Vertical Lift Conveyors

There are five basic attributes by which these devices are classified:

1. *Frame and Guiding Means.* The most common types are the two-mast (or post) and four-mast design. Although some single-mast units have been installed, their load capacity is a definite limitation.

Two-mast reciprocating vertical lifts (Fig. 9.4.1) have a lower intrinsic maximum load capacity than four-mast designs. Because carriage platforms on the two-mast designs are cantilevered, loads must be considered in terms of their general shape, their position on the platform, and their resulting moment loads. (When considering load-carrying capacity, the weight of the carriage arms and platform must be subtracted from total load capacity to arrive at net capacity.)

Four-mast reciprocating lifts (Fig. 9.4.2) are generally specified for loads from 1000 lb (454 kg) up to 25,000 lb (11,340 kg) and beyond. Because the lifting means are centrally located on opposite sides of the carriage, positioning of loads is not as critical a concern as with the cantilevered two-mast designs.

Four-mast lifts can utilize either two-chain or four-chain suspensions. Two-chain lifts can be used for external loads up to 6000–8000 lb (2722–3629 kg); for heavier loads a four-chain model is recommended.

2. *Carriages and Platforms.* The typical carriage on a reciprocating lift consists of a frame, to which guide rollers are attached, and a carriage on which a platform is mounted. The platform can consist of horizontal or tilting gravity conveyor or various types of powered (horizontal) conveyor. Powered conveyor segments can be belt, live roller, or chain which serve as the basic induction and discharge mechanisms for the lift. In general, some mechanical device is used to bring loads to and remove them from a vertical lift conveyor, because most state and local codes forbid manual loading and unloading.

3. *Drive Location.* With reciprocating vertical lift conveyors, users have the option of locating drives either at the top or bottom of the lift unit. Depending on available space, top-mounted drives can be installed either vertically or horizontally; bottom-mounted drives can be located at the rear, to the side or under the conveyor carriage.

The decision between top-mounting and bottom-mounting is further tempered by the following considerations. Mounting the drive at the top of the conveyor is preferable from a purely mechanical standpoint (there is less chain in high tension). This approach will also result in comparatively lower initial installation costs. From the standpoint of accessibility for maintenance or repair, bottom-mounting would be the preferred choice. It should also be pointed out, however, that bottom-mounting requires valuable floor space which might be put to better use for other purposes.

4. *Counterweights.* As with conventional elevators, counterweights are commonly incorporated into reciprocating vertical lift conveyors, albeit usually only with four-mast designs. Although it is possible that two-mast designs might be counterweighted, their lighter load capacities usually make this unnecessary. With four-mast designs counterweights provide a decided assist in overcoming inertia during acceleration and deceleration. As is the practice with conventional elevator design, the counterweight mass should approximately equal the weight of the carriage plus one-half the weight of the maximum external load. This then permits selection of a smaller motor for a given application; in effect, with counterweighting only half the load need be powered. (It might be noted that the major portion of the inertial load in a vertical conveyor is contributed by the internal inertia of the motor itself rather than the external load.) With the four-mast designs, the reduction in inertial load and therefore drive motor horsepower requirements resulting from counterweighting are significant.

5. *Multilevel Reciprocating Lifts.* Frequently, reciprocating lifts are used to convey loads between two or more levels (floors) of a building. In these cases, there are several factors that must be studied carefully in order to assure a successful installation. First, of course, are the floor structures through which the lift will pass. When linking one level with another, particular attention should be directed to overall material flow patterns; in effect, the vertical lift is meshing a material flow pattern on one floor with the flow pattern on another floor. Related to this overall flow pattern are the questions of distances between charge and discharge points, speed of travel, and required transfer rates.

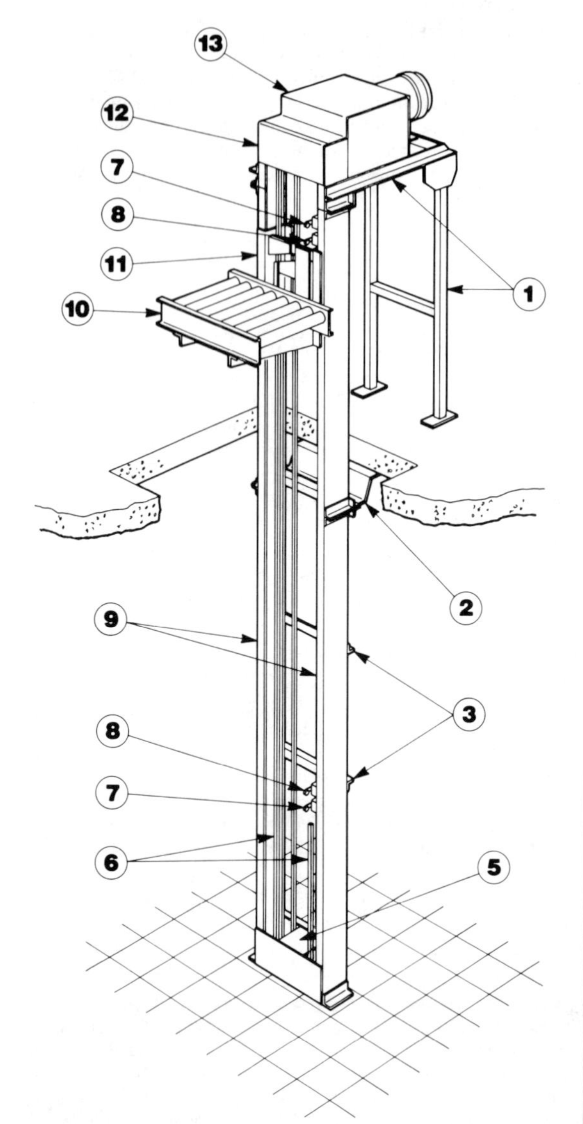

1. DRIVE UNIT SUPPORT FRAME
2. STRUCTURAL BRACING (BY USER OR INSTALLER
3. CROSS BRACING
4. DRIVE UNIT (BOTTOM MOUNTED)
5. IDLER SECTION
6. LIFT CHAIN
7. STATION OVER-TRAVEL LIMIT SWITCH
8. STATION LEVEL LIMIT SWITCH
9. FRAME COLUMNS
10. PLATFORM
11. CARRIAGE
12. DRIVE CAP
13. DRIVE UNIT (TOP MOUNTED)

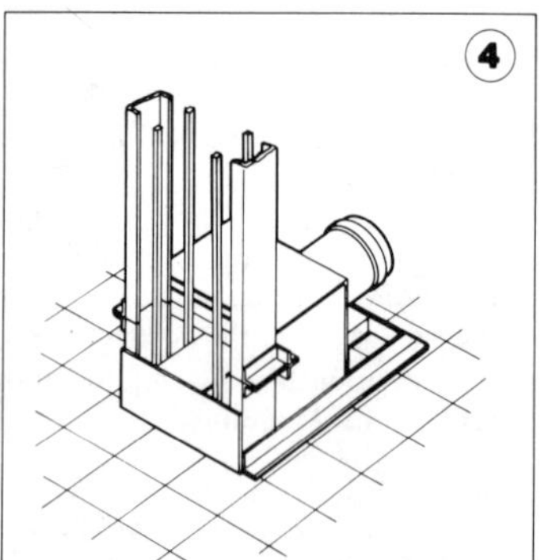

Fig. 9.4.1 Two-masted, reciprocating vertical lift conveyor (equipment features common to most designs).

Reciprocating lifts can be controlled either manually, automatically, or by a combination of the two. Here, two factors should be considered: the frequency at which loads arrive at and leave the lift (transfer rate) and the time required for one complete up-down excursion of the lift (lift duty cycle). With relatively low or intermittent load transfer rates, reciprocating vertical lifts can readily be controlled by manual means, that is, stop-start buttons. Conversely, when the load transfer rate is constant and approaches the lift duty cycle, it might be worthwhile to consider automatic control of the lift. Basically the same types of control devices are required as are used for automatic control of continuous vertical conveyors.

9.4.4 Continuous Vertical Lifts

The most common type platform used with continuous vertical lifts is the flexible (or articulated) platform design. The platforms are made up of groups of narrow bars or slats which are permanently attached to continuous-loop drive chains on opposite sides of the lift frame (see Fig. 9.4.3). The chain

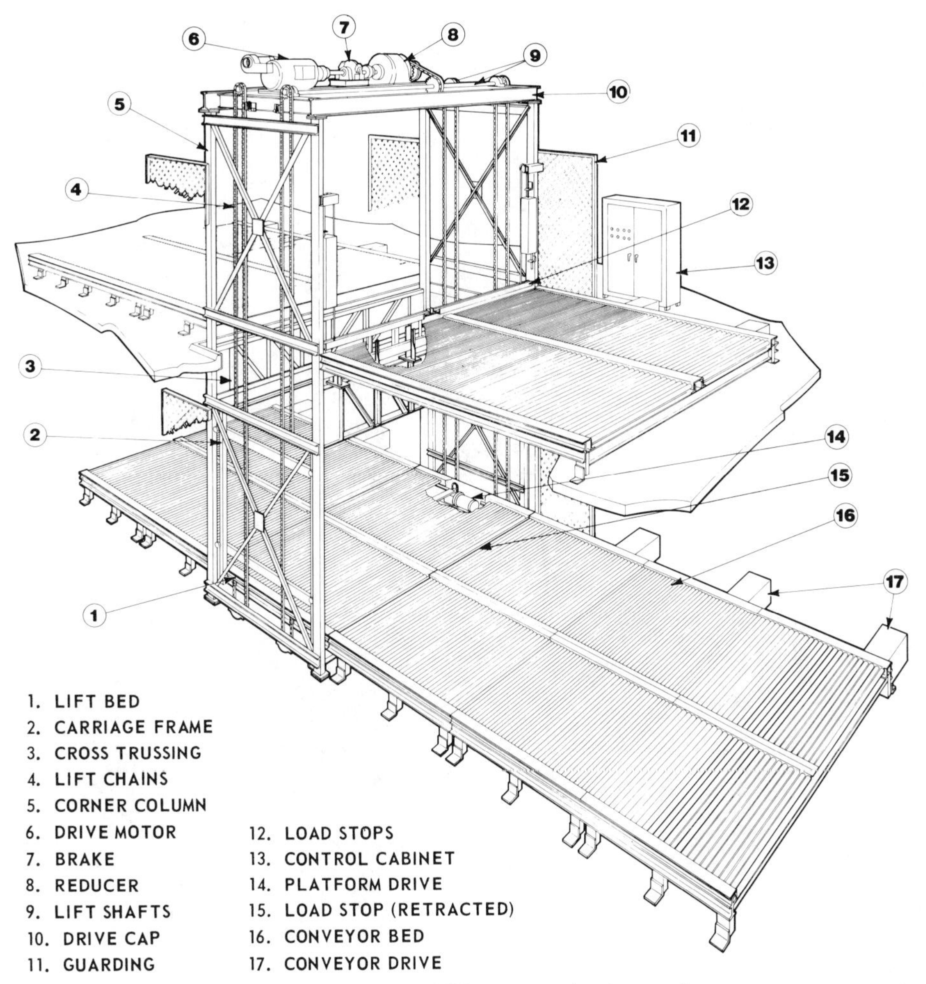

Fig. 9.4.2 Four-masted, reciprocating vertical lift conveyor (equipment features common to most designs).

paths are defined by a series of guides and sprockets, which cause the platform elements to deploy horizontally between specified receiving and discharge points. For the remainder of their travel, the groups of elements follow the chain loop pattern, returning to the top or bottom of the lift and resuming a horizontal deployment.

The two other platform configurations used with continuous vertical lifts are the swinging carriage and opposing paddle designs. These types have more limited applications and are generally considered special purpose designs. The opposing paddle lift, for example, is used with baggage-handling containers at some airports.

Load Considerations

When considering continuous vertical lifts, the main concern about loads is an accurate estimate of load dimensions—especially height. Load heights have a direct impact on platform spacing and consequently the lift speed necessary to attain the desired transfer rate. On the other hand, weight of the load isn't as critical as with reciprocating lifts.

Rigidity of loads, however, is a factor that should also be considered. A relatively rigid load is recommended in order to assure stability on the platform during transport.

The term "continuous" refers not only to the path traversed by the lift platform but to its mode of operation as well. Thus, continuous vertical conveyor suppliers recommend that a system be turned

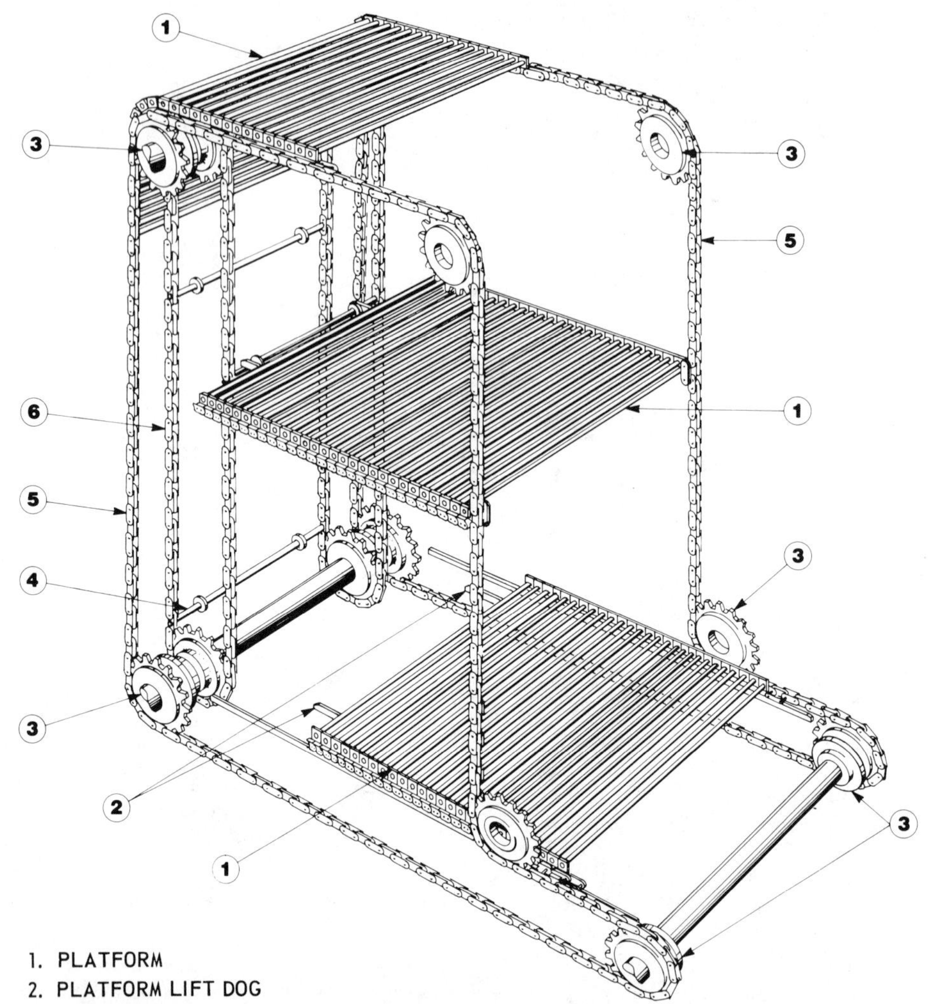

1. PLATFORM
2. PLATFORM LIFT DOG
3. DRIVE OR IDLER SHAFT & SPROCKET (DRIVE UNIT CAN BE COUPLED TO EITHER TOP OR BOTTOM SHAFT WITH DUAL SPROCKET FOR PRIMARY AND SECONDARY CHAINS. REMAINING SHAFTS AND SPROCKETS ARE IDLER ASSEMBLIES.)
4. LIFT BARS
5. PRIMARY CHAIN (ONE LOOP ON EACH SIDE OF LIFT.)
6. SECONDARY CHAIN (ONE LOOP ON EACH SIDE OF LIFT.)

Fig. 9.4.3 Typical mechanical features of continuous vertical lift conveyor (lift frame, bed, guarding, etc., omitted; these are similar to corresponding components for reciprocating lifts).

on and operated continuously as opposed to start-stop sequencing. This characteristic in turn implies automated control for the conveyor.

Typically, continuous vertical conveyors are interfaced with horizontal conveyor at both the receiving and discharge sides. Thus, the final section of horizontal conveyor feeding the vertical unit is usually an appropriate length of metering conveyor interlocked with the vertical conveyor through a sensing switch. In this way, the metering conveyor will hold any article on the metering section until the vertical unit is ready to accept it. As a lift platform moves into its takeover position it overrides the stop signal at the metering conveyor. The metering conveyor resumes its motion, delivering the article to the moving lift platform. At the delivery point the load is automatically discharged from the lift.

In some instances, continuous vertical conveyors must operate bidirectionally. This requires that metering conveyor and interlocked sensing switches be provided at both ends of the conveyor. If, for example, flow is manually reversed (by pushing a button) the system should "remember" to first empty any load present in the original direction and then automatically reverse to accept loads destined in the opposite direction.

Oversize-load sensors also should be incorporated as a part of the control system for continuous vertical lift conveyors. The purpose of these controls is to prevent oversized loads from entering and jamming the conveyor. The sensor might simply shut down the continuous lift until the unwanted load is removed or it might be designed to actuate a diverter to automatically shunt the load out of the way.

As an essential safety precaution, continuous lifts should *never* incorporate devices that automatically restart the conveyor after it has been stopped for overloading or jamming. Operating and maintenance personnel should be thoroughly acquainted with procedures for checking out and restarting the conveyor under these circumstances.

9.4.5 Material Flow Patterns

The use of vertical lift conveyor implies a three-dimensional flow pattern. The flow pattern of material being brought to and discharged from a lift usually can be described in two dimensions (i.e., in the horizontal plane). The introduction of a vertical lift adds the third dimension to that pattern. When planning the installation of the vertical lift conveyor, it is essential that one of the earliest steps be a thorough mapping of flow patterns. A vertical lift can either change the direction of flow (in the horizontal plane) of its load upon discharge or it can allow the load to continue in the original direction.

The two most frequently encountered receive/discharge patterns for vertical lifts are the so-called "Z-load" and "C-load" configurations. With a Z-load pattern, the load exits the lift from the side opposite the charging side, and proceeds in the same horizontal direction. With a C-load pattern, the load exits the same side from which it was charged, thus reversing the horizontal direction of travel. Four-mast reciprocating lifts can readily accommodate either pattern. Two-mast reciprocating lifts on the other hand, can be charged and discharged from the front or from either side of the carriage platform.

For continuous vertical lifts, the question is somewhat more complex. The Z-load pattern represents the more straightforward approach from the standpoint of lift design and is thus the more common configuration. A C-load pattern introduces additional mechanical complexity to the continuous lift design. As an alternative, a Z-load pattern is often specified for the vertical lift, with travel direction (in the horizontal plane) being changed with an appropriate turn on a horizontal conveyor at the discharge (or receiving) end of the lift. Flow patterns are shown in Fig. 9.4.4.

9.4.6 Drives

Drives for vertical lift conveyors can be either electrical (electromechanical) or hydrostatic (hydraulic) systems. In practice, the largest percentage are electrical. The reasons for this are obvious: the technology is well established and widely understood; drive designs can be based on standard, off-the-shelf components; electrical drives perform with a high degree of reliability and with an adequate degree of accuracy; and maintenance people are generally familiar with these kinds of systems.

Hydrostatic (hydraulic) drive technology, however, can provide solutions to certain demanding applications. One example is a requirement for extremely high positioning accuracy or instances where space or environmental conditions (temperature) preclude installation of an electrical drive. In such instances, the additional cost and special maintenance requirements might be justified.

9.4.7 Operating Temperatures

The normal operating temperature range for continuous vertical lifts is 32–150°F (0–65.5°C). Extended temperature ranges can be accommodated, however, through special design. With such special designs, particular attention is paid to items such as lubricants, electrical insulation, seals, equipment enclosures, and location of controls. Freezer and cold-room applications, for example, represent areas where field-proven design and installation information already exist.

9.4.8 Building Structure

Almost without exception, vertical lift conveyors are freestanding structures that place no unusual demands on the structural system of the buildings in which they are used. Occasionally, circumstances dictate installation of the conveyor immediately adjacent to an exterior wall of the building they serve. Whether erected internally or externally, the vertical lift requires only a solid footing or foundation [usually a concrete base rated at 300 psi (1465 kg/m^2)] and a measure of horizontal bracing when height of the lift structure exceeds about 10 ft (3 m). Usually, such bracing is recommended at 10-ft intervals to assure that the lift structure remains plumb.

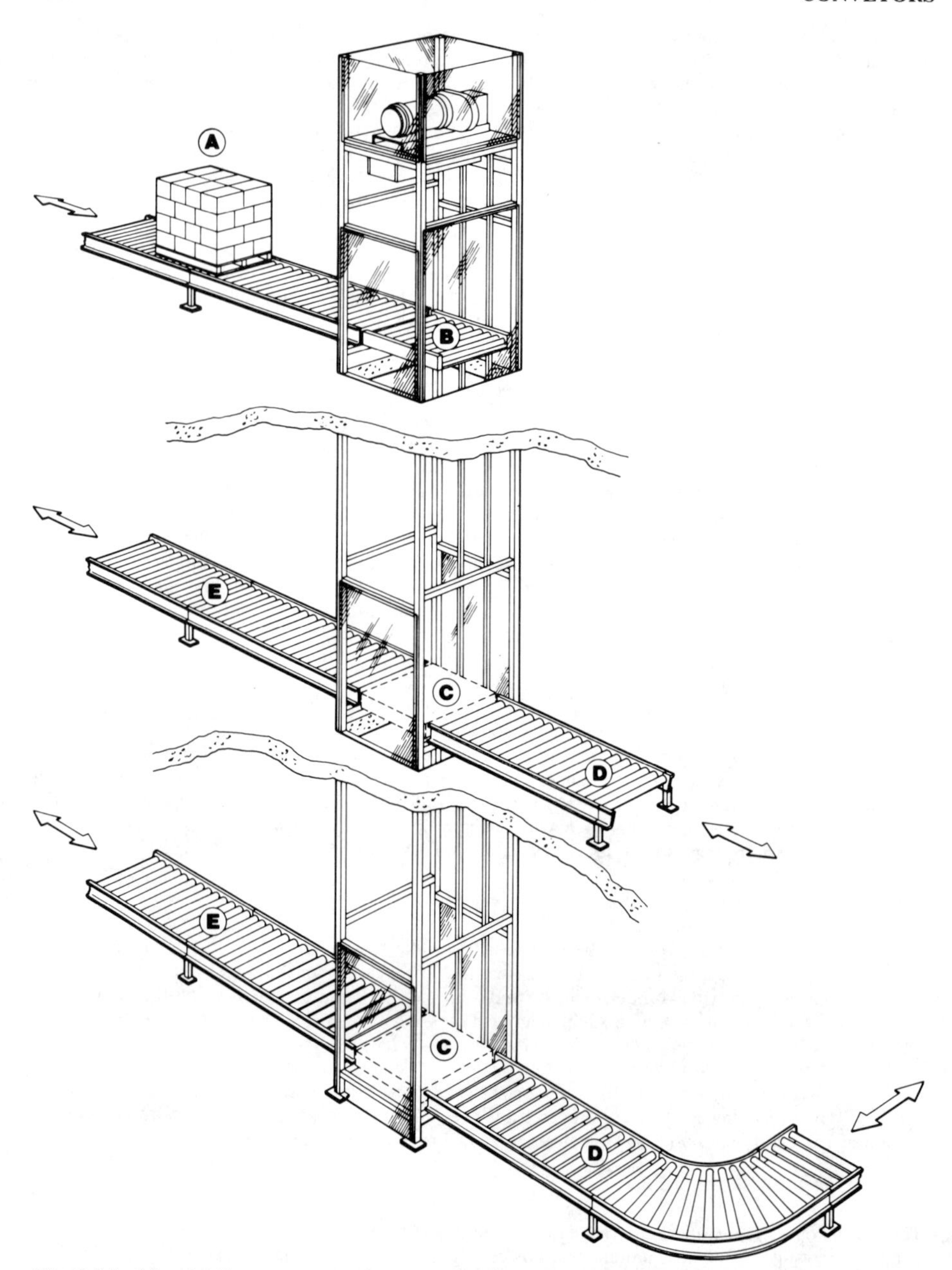

Fig. 9.4.4 Material flow patterns common to all types of vertical lift conveyor. Movement through points A, B, C, and D defines basic Z-load pattern; movement through points A, B, C, and E defines C-load pattern. Vertical movement can be either up or down or both.

Although their freestanding nature usually makes it possible to install a vertical lift outside the building it is intended to serve, the practice is generally discouraged because of the additional cost of a protective enclosure as well as the cost of installing openings in the building's walls for lift entrances and exits.

When multilevel lifts are to link two or more floors together, adequate steps must be taken to assure that the floor opening through which they pass is structurally sound. With new construction, this provision would be made in the original architectural plans. With existing buildings, a careful review should be made to insure that floors will remain structurally sound after openings are made to accommodate the lift.

9.4.9 Installation Time

Because the major share of vertical lift conveyors is based on essentially standardized designs, their installation is usually completed in a relatively short time span. A typical schedule for reciprocating lifts would call for installation of mechanical components in one or two weeks; wiring, and debugging, from one to two weeks, depending on the complexity of associated systems. Similarly with continuous vertical lifts, installation times can vary from three days to two weeks, again depending on the complexity of the control system.

Human Factors

The two major considerations in terms of human engineering are accessibility of components for maintenance and guarding for personnel safety (both discussed elsewhere in this section). Other considerations might include, in the case of reciprocating vertical lifts, the location and method of actuating safety doors, and the location of start-stop controls.

Maintenance

In normal operation (within design specifications) vertical conveyors require only routine maintenance. The key items for either reciprocating or continuous conveyors are regular chain lubrication and regular inspection and lubrication of bearings and speed reducers. The one category of critical items that requires frequent (monthly) inspection is limit switches in order to assure they are providing their intended protection for the system.

9.4.10 Specifications

Although there is a variety of standard designs from which to choose, vertical lift conveyors are not considered off-the-shelf items and must be specified in detail. The items requiring definition include:

1. *Net Lift.* The effective distance that loads are to be conveyed. (An additional space allowance must be made to accommodate components such as lift drives and idler assemblies, tension sheaves, or carriage buffers.)
2. *Net Capacity.* The effective external load carrying capacity of the conveyor. Additional allowance must be made for the weight of the entire carriage assembly.
3. *Overall Headroom Limitations.* Take into account vertical space (under the roof of the building) occupied by structural beams, ductwork, piping, electrical and air supply lines, and so on.
4. *Maximum Acceptable Lower Conveying Height.* The required elevation of the carriage assembly (top surface of platform) is a key datum for establishing clearance dimensions for associated system components (drives, etc.).
5. *Provision for Pit.* It is sometimes necessary to install a pit beneath a vertical lift for any number of reasons, for example, to accommodate a particularly low elevation of the bottom extremity of a carriage excursion. Whenever possible, however, it is preferable to avoid the need for a pit for the obvious reasons—additional construction and maintenance costs.
6. *Material Flow and Control System.* Development of control system requirements is one of the most demanding parts of the specification process. Because vertical conveyors invariably interface with materials handling operations on two or more levels, a precise statement of charge/discharge requirements must be provided.

   With two-level lifts, the task is relatively simple and straightforward, whether the control system is to be manual or automatic. With multilevel installations, however, the task becomes significantly more complex because the number of departure–destination combinations increases geometrically with the number of levels served by the conveyor. Similarly, as the number of these combinations served by the control system increases, so does the cost. Therefore, it is essential that material flow patterns as well as charge/discharge requirements be closely defined and that control system capabilities be accurately matched to those requirements.
7. *Definition of Cycle Times (Rates).* In order to be integrated smoothly with overall material flow, cycle times for vertical conveyors must be carefully specified. Where vertical conveyors are interfaced with horizontal conveyors, for example, conveying speeds of the latter should be taken into account in determining vertical cycle times.
8. *Load Size.* Maximum dimensions (i.e., length, width, and height) of loads should be clearly defined so that carriage and platforms can be properly sized. The height dimension is especially important for continuous lifts.

9. *Service Factors.* Normal service factors for vertical lift conveyors take into account variables such as projected number of hours per day the unit will operate and anticipated shock loading; normal design recommendations usually reflect a service factor of 1.0.

   The components that determine the service factor for a vertical lift conveyor include the drive motor, speed reducer, chain, and sprockets. When specifying service factors, whether for the entire lift conveyor or for individual components, it is important that ratings for all components be consistent with one another and with accepted power transmission design practices.

   When evaluating the installation of vertical conveyor, however, it is recommended that additional conditions be considered in arriving at required service factors. An obvious example is the impact on associated equipment and overall operations resulting from unscheduled downtime of the vertical lift. Service factors ranging from 1.5 to 3.0 might be specified in cases where uninterrupted operation of the lift is essential.

10. *Space Available for Conveyor Frame.* An accurate description of available space as well as permanent obstructions is necessary in order to size the conveyor frame as well as to provide guidance in locating auxiliary components such as control cabinets.

11. *Applicable State and Local Codes.* Because of variations from one locale to another, a review of specific state and local codes is required before any design or installation of a vertical lift conveyor can be finalized.

9.5 OVERHEAD TROLLEY CONVEYORS

Edward Moon

9.5.1 Definitions and Applications

Definitions

A *trolley* is an assembly of wheels, bearings, and brackets used for supporting suspended loads or load-carrying elements. A *trolley conveyor* is a series of trolleys supported from or within an overhead track and connected by an endless chain, cable, or other linkage. The loads are usually supported from the trolleys, although they may be supported by the connecting element.

Typical Trolley Conveyor

In its most widely used form, a trolley conveyor consists of an endless drop-forged chain suspended from two-wheel trolleys that roll on the lower flange of an I-beam track. A typical arrangement of rail, trolley chain, and load-carrying attachments is shown in Fig. 9.5.1. Loads are suspended from attachments bolted to the trolley brackets. If the loads are light, they may be suspended from attachments bolted to the chain between trolleys. The loads are supported by a great variety of carriers; hooks and trays are the commonest.

The path of the conveyor consists of straight runs which may be horizontal or inclined. The straight

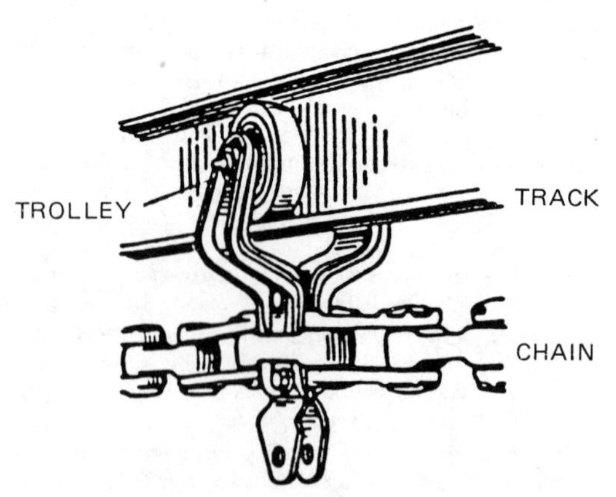

Fig. 9.5.1 Arrangement of chain track and trolley.

runs are joined together by horizontal turns and/or vertical curves to form an endless loop. The trolley conveyor may be powered by one or more drive units, or may take power from some other conveyor or machine.

The basic components of a trolley conveyor are:

1. Chain
2. Trolleys
3. Track and supports
4. Horizontal turns
5. Vertical curves
6. Drives
7. Take-ups
8. Carriers

There are many other types of trolley conveyors. In some, the drop-forged rivetless chain is replaced by other chain types such as link or so-called universal chains. This latter chain has both vertical and horizontal wheels. The chain may also be replaced by cable or sections of rod. The I-beam may also be replaced by other shapes such as T's, angles, channels, or formed rectangles. In the case of universal chain, the track is usually a rectangular or round formed shape with a slot at or near the bottom through which the load connection is made. Universal chain also permits horizontal turns without auxiliary devices. See Figs. 9.5.2 and 9.5.3 for various available arrangements.

Each of the special types of conveyor has its own characteristics and limitations as defined by the manufacturer. They provide a useful source for conveyors in what is called the light-to-medium range. Note that although the variations are too numerous to cover in detail here, the basic principles of use are the same for conventional systems using drop-forged chain.

Uses of Trolley Conveyors

The trolley conveyor is one of the most versatile types of conveyor. This is primarily because it functions in three dimensions, whereas almost all of the other fundamental types of conveyors operate in a single plane. Its three-dimensional capability gives the trolley conveyor an enormous advantage. Horizontal and inclined runs, vertical curves and horizontal turns can be put together in many combinations to follow complicated routes. There is no need for separate power units at transitions, yet with the

Fig. 9.5.2 Conveyor using link chain.

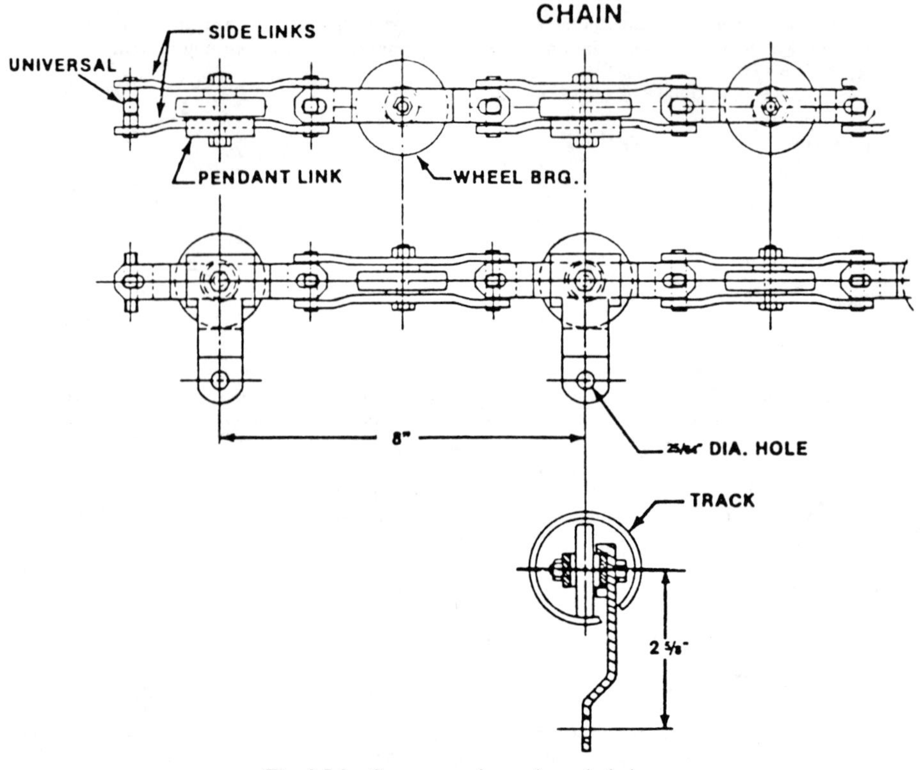

Fig. 9.5.3 Conveyor using universal chain.

use of multiple drives there is practically no limitation on length. The restrictions on the path of travel are more often those of the building or process than those caused by the capability of the conveyor.

The trolley conveyor can be used to carry almost any type of material through most kinds of operations or processes. Loads can be very large and heavy, or very light and small. Different sizes and weights of materials can be handled at the same time on the same conveyor. Materials can be hot or cold, and can be carried through processes such as cleaning and washing, bonderizing, painting, drying and baking, degreasing, sand blasting, and many others. The trolley conveyor can be, and usually is, used as a storage conveyor at the same time it functions as a processing and delivery conveyor. It can be a pusher or a towing conveyor. The same conveyor can be used for a combination of these functions. Except for carousels, other conveyors have a fixed head and tail end and an empty return run. Trolley conveyors are endless, and the entire length may generally be used. The recirculation of materials is often of tremendous advantage in manufacturing.

A major use of trolley conveyors in recent years is as the power source for power-and-free conveyors. Almost all of the trolley conveyor hardware has been adapted without change into this product. The limits on design and hardware discussed here, therefore, apply to the same elements used in power-and-free conveyors, discussed in another section.

Trolley conveyors are used for reasons other than their ease of application. Additional advantages of trolley conveyors are low initial unit cost, long wear, low maintenance, and inexpensive repair and replacement. Salvage value is very high. Conveyors can be inexpensively taken down from one location and reinstalled in another. They can easily be altered in path, shortened, lengthened, combined, and divided into almost any desired combination. Spacing and types of loads can be changed easily. Another advantage is the small space taken up by the trolley conveyor.

9.5.2 Trolley Conveyor Components

Rivetless Chain

This drop-forged chain is ideal for trolley conveyors. It combines light weight with high strength at the lowest cost per pound of working load of any comparable chain. When used with the pin joints vertical, it will negotiate small horizontal turns, yet it can be flexed transversely on a comparatively short radius. Its open center link makes it easy to attach trolleys without the use of special attachments. The chain is assembled from three parts: (1) side links, (2) center links, and (3) chain pins. It may be easily assembled or disassembled without tools.

Chain Sizes. Four sizes of chain are in common use. They are designated X-228, X-348, X-458, and X-678 (see Fig. 9.5.4). There are other sizes of rivetless chain such as 478, 658, and 9118. They have other uses such as in towing conveyors. There are also chains without the X prefix. These are little used in trolley conveyors since the X prefix designates an improved design which has better proportioned links for twisting strength and transverse flexibility. These chains are so widely used that they are often identified by number only, without other specifications. Some manufacturers have other prefix letters to identify their own chain design. Chains of different manufacturers are, however, in general, interchangeable.

Chain Options. All drop-forged chain should be sized after the forging operation. This ensures uniform initial size. It should also be purchased in the heat-treated condition. It is not considered desirable to have both unheated and heat-treated chain in the same plant, since trolley conveyors are frequently altered and chain taken from one conveyor to be used in another. Some manufacturers can provide alloyed steel chain with a higher tensile strength and added wear capabilities. The added cost is often warranted if the conveyor is an essential conveyor located in a difficult environment.

Allowable Chain Pull. The working load allowed on a trolley conveyor chain is less than for the same chain used in some other types of conveyors. The factors that limit the maximum chain pull are:

1. The pressure per square inch on the bearing area
2. The track life at vertical curves due to trolley load plus chain reaction load
3. Pin bending and/or pin bearing pressure at vertical curves

There is no uniform agreement between trolley-conveyor manufacturers on the proper values to use for maximum allowable chain pull. Ratios between ultimate strengths and allowable loads that are in general use are 16:1 for multiplane conveyors and 12:1 for monoplane conveyors. Larger values than the allowable loads given in Table 9.5.1 are sometimes used under very favorable conditions.

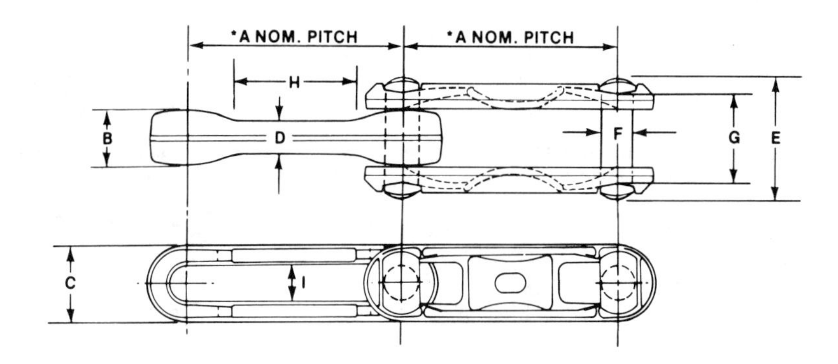

Chain	A *	B	C max	D	E max	F	G min	H min	I min
X-228 X-50-6	2 50	15/32 11.90	11/16 17.46	3/8 9.53	1-1/8 28.57	1/4 6.35	53/64 21.03	1-1/16 26.98	5/16 7.94
X-348 X-75-13	3 75	3/4 19.05	1-3/32 27.78	1/2 12.70	1-27/32 46.83	1/2 12.70	1-9/32 32.54	1-5/8 41.27	9/16 14.28
X-458 X-100-16	4 100	1 25.4	1-13/32 35.71	5/8 15.87	2-1/4 57.15	5/8 15.87	1-5/8 41.27	2-1/4 57.15	11/16 17.46
X-678 X-150-22	6 150	1-9/32 32.50	2 50.80	13/16 20.63	3-1/8 79.37	7/8 22.22	2-1/4 57.15	3-3/8 85.72	31/32 24.61

Fig. 9.5.4 Dimensions of rivetless chain.

Table 9.5.1 Load Ratings for Chain

Chain No.	Approximate Pitch	Pin Diameter	Ultimate Strength (lb)	Maximum Allowable Load (lb)	
				Multiplane[a]	Monoplane[b]
X-228	2 in.	$\frac{1}{4}$ in.	6,000	400	600
X-348	3 in.	$\frac{1}{2}$ in.	24,000	1000–1400	2000
X-458	4 in.	$\frac{5}{8}$ in.	48,000	2000–3000	4000
X-678	6 in.	$\frac{7}{8}$ in.	85,000	4000–5000	6000

[a] Multiplane indicates conveyors which have vertical (transverse) curves.
[b] Monoplane indicates conveyors operating in a horizontal plane only.

The actual rating used is a judgment decision based on acceptable rates of wear, environment, maintenance, and predictability of load. Most large users try to use conservative ratings.

Attachments. Loads are usually hung from trolleys. However, when light loads must be closely spaced, they may be hung from the chain. Attachments can be furnished as extensions to chain pins in styles similar to the attachments used with trolleys (see Fig. 9.5.5).

Radii for Transverse Bending. All chains used in trolley conveyors have limitations on the minimum radius for vertical curves. For drop-forged chain this is a function of the capability of the chain and the spacing of the trolleys. Table 9.5.2 shows the values recommended by most manufacturers on systems with average loadings. Heavy loads and adverse operating conditions may dictate larger values. Short-radius curves and steep inclines can cause operational limitations but are very desirable for plant layout purposes. They save valuable floor space. In the case of paint-dip and plating tanks and similar dipping operations, they allow the use of smaller tanks. The use of short-radius curves also makes it easier to run the conveyors through floors. Another advantage of short-radius curves is that they facilitate the conveyors' entrance into and exit from ovens.

Trolleys

There are many different types of trolleys, with different kinds of wheels, brackets or side arms, attachments, and methods of connecting brackets and wheels. The most popular type of trolley has heavy, machined, ball-bearing wheels, drop-forged brackets, and steel attachments, which may be fabricated, drop forged, or cast.

Wheels. The wheel is the most important part of the trolley. The foremost manufacturers generally make trolley wheels to the following specifications:

1. The wheel tread is one piece, heavy, and machined to run true and smoothly. The tread is also the outer race of the ball bearing.
2. The inner race is one piece.
3. The ball bearing has large diameter balls. If the wheel is of the Conrad bearing design, the balls are accurately spaced by a metal separator. If they are of the full ball type, the wheel is filled with properly sized balls through a loading slot.
4. The balls and ball races compare favorably with precision ball bearings in quality of metal and heat treatment. Ball races are ground.
5. Fit of bearing balls and races is purposely made loose so that there is a noticeable "rock" in the bearing. It prevents binding from changes in temperature and allows the wheels to roll when lubricant cakes up or dirt gets in the bearings. Close tolerance is held on the fit.
6. Seals, when used, are selected for long life. Steel labyrinth seals (by far the most widely used) are installed with very small clearances.
7. The side of the trolley wheel facing the I-beam is closed when seals are used, making a dust- and grease-tight enclosure.
8. Wheels are cambered and the tread specially contoured to provide good tracking and wear characteristics with respect to the I-beam.

Trolley wheels that meet these specifications are used in 3 in., 4 in., and 6 in. systems. See Fig. 9.5.6.

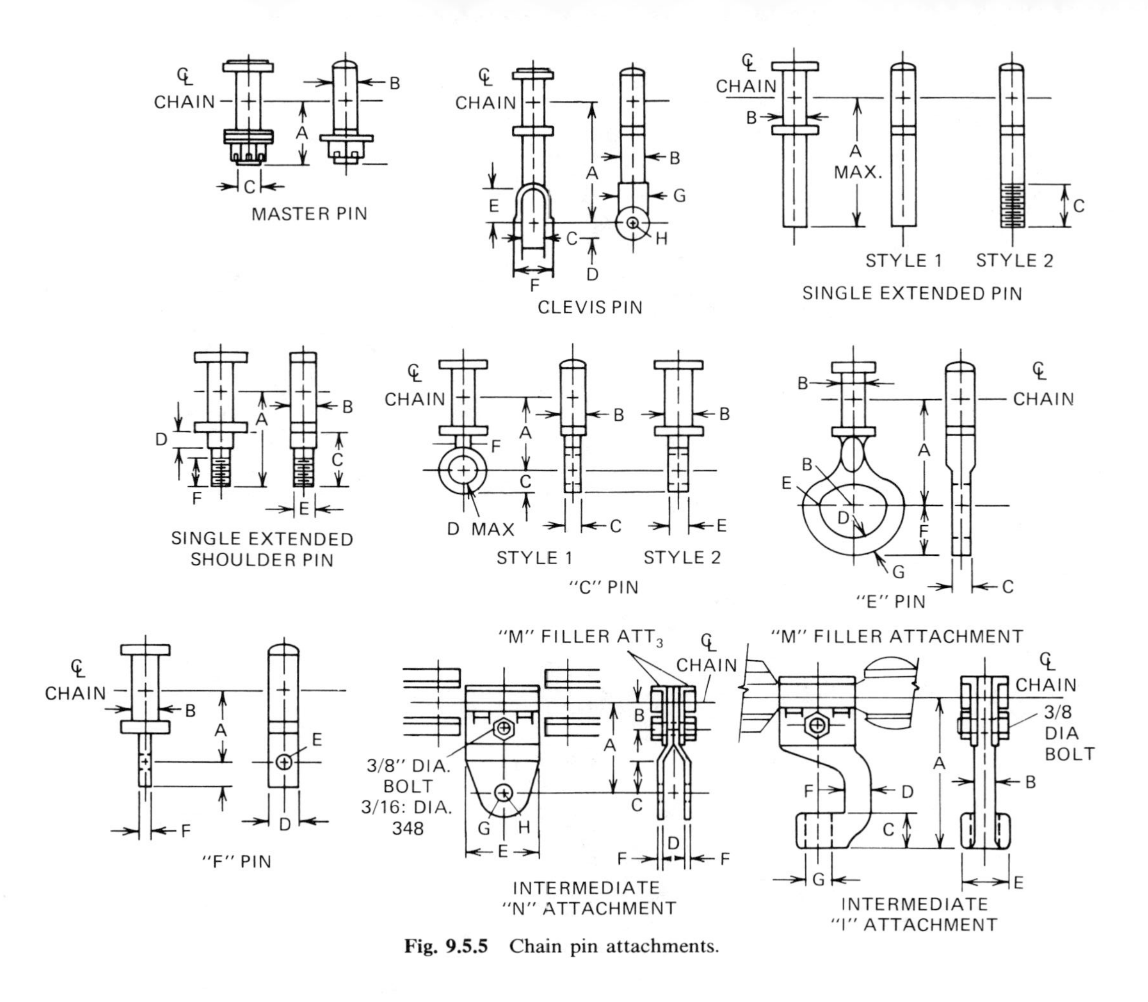

Fig. 9.5.5 Chain pin attachments.

Table 9.5.2 Minimum and Recommended Radius of Vertical Curve Track for Overhead Conveyors as Related to Trolley Spacing

	Chain Number							
	X-228 X-50-6		X-348 X-75-13		X-458 X-100-16		X-678 X-150-22	
Trolley Spacing	min.[a]	rec.[b]	min.	rec.	min.	rec.	min.	rec.
8″(203.2)[c]	2′–0 (609.6)	4′–0 (1219.2)			3′–0 (1086.8)	6′–0 (1828.8)		
12″(304.8)	3′–0 (914.4)	4′–0 (1219.2)	4′–0 (1219.2)	5′–0 (1524.0)			6′–0 (1828.8)	12′–0 (3657.6)
16″(406.4)	4′–0 (1219.2)	4′–0 (1219.2)			5′–6 (1676.4)	8′–0 (2438.4)		
18″(457.2)			5′–0 (1524.0)	6′–8 (1961.2)				
20″(508.0)	6′–0 (1828.8)	6′–0 (1828.8)						
24″(609.6)	8′–0 (2438.4)	8′–0 (2438.4)	6′–6 (1961.2)	8′–0 (2438.4)	7′–0 (2133.6)	10′–0 (3048.0)	11′–0 (3352.8)	15′–0 (4572.0)
30″(762.0)			7′–8 (2336.8)	10′–0 (3048.0)				
32″(812.8)					9′–0 (2743.2)	12′–0 (3657.6)		
36″(914.4)			9′–0 (2743.2)	12′–0 (3657.6)			16′–0 (4876.8)	20′–0 (6096.0)

[a] Minimum radii to be used only when absolutely required and only after considering chain pull, imposed load on trolley, beam wear, and possibility of surge.
[b] rec. = recommended.
[c] Dimensions in parentheses are in mm.

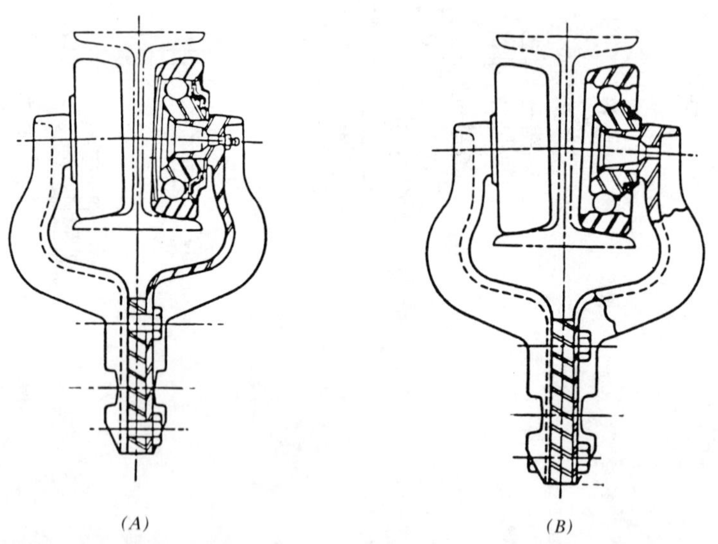

Fig. 9.5.6 Types of trolley wheels. (*A*), Labrinth sealed conrad type with ball retainers and grease fitting; (*B*), open type full ball complement—no seals or grease fitting.

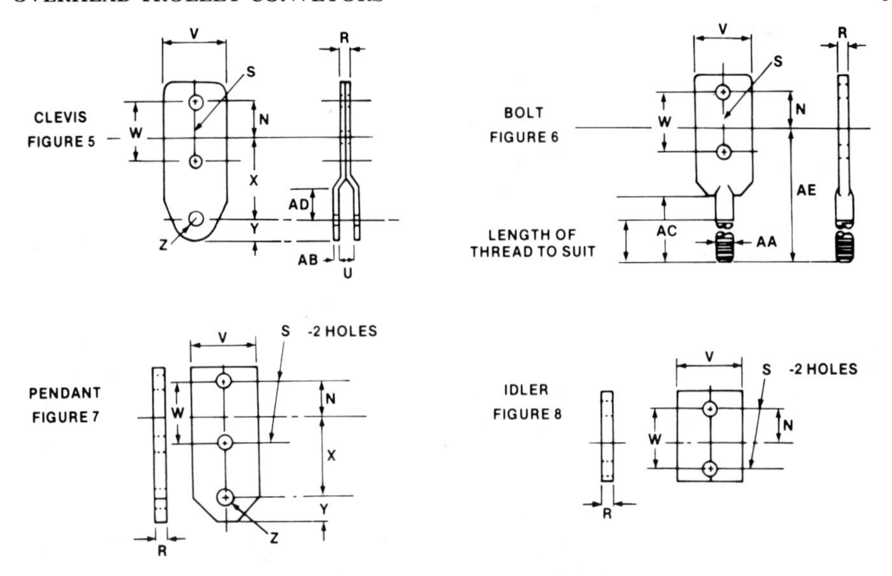

Beam	Chain	R	S Dia.	U	V	W	X	Y	Z	N	AA	AB	AC min.	AD min.	AE min.
3" I @ 5.7# per foot	X-348	1/4	5/16	9/16	1-1/2	1-7/8	3-1/8	5/8	1/2	1	1/2	1/8	1-13/16	1	4-1/4
76.2 I @ 8.46 kg/m	X-75-13	6.35	7.93	14.28	38.10	47.60	79.38	15.87	12.70	25.40	12.70	3.17	46.03	25.40	184.15
4" I @ 7.7# per foot	X-458	3/8	3/8	11/16	2-1/8	2-1/8	2-7/8	3/4	1/2	1-5/16	5/8	3/16	2-1/8	1	4-15/16
101.60 I @ 11.45 kg/m	X-100-60	9.52	9.52	17.46	53.97	53.97	73.03	19.05	12.70	33.33	15.87	4.76	53.97	25.40	125.41
6" I @ 12.5 #/ft	X-678	1/2	1/2	13/16	3	2-3/4	3-5/8	1-1/8	3/4	1-5/8	7/8	1/4	3	1-1/8	6-1/8
152.4 I @ 18.60 kg/m	X-150-22	12.70	12.70	20.64	76.20	69.85	92.07	28.58	19.05	41.27	22.23	6.35	76.20	28.58	155.57

Fig. 9.5.7 Trolley attachments.

Wheel Types. The Conrad-type wheel remains the industry standard. It is the most widely used, most efficient, and usually the most economical of its type. The full ball complement wheel theoretically has a greater capacity when it is used without seals in adverse environments. It tends to be self-cleaning. These wheels are available with a variety of seals and with or without lubrication fitting. Some users operate without seals and lubricate with an oil spray. This method is usually used when a high-temperature oven is in the system. Other users make no attempt to relubricate and run the trolleys until they wear out.

Wheels with precision sealed-for-life bearings are available for special application. They are, however, limited to use at moderate temperatures. A better application is to obtain a standard wheel with a felt or Teflon insert. At the other end of the cost range there are wheels constructed of pressed steel with two-piece ball races or wheels with a commercial unground ball bearing inserted. These wheels are limited in life unless used at moderate speeds and loadings.

Trolley wheels for use with 228 chain are pressed steel wheels, such as used in roller skates. A wide variation in quality, load ratings, and performance may be found in different makes of these wheels. Special types of conveyors are made with tapered roller-bearing flanged wheels for very heavy duty. For some types of service, trolleys with bronze bushings, Textolite treads, and other variations are used. It is advisable to use standard types whenever possible.

Brackets. These are sometimes called side arms. They connect wheels to the chain and load attachments. Brackets are usually drop-forged steel with wide, heavy ribs. Brackets are bolted or riveted to trolley wheels. Stamped steel brackets are used for some makes of trolleys. However, better distribution of metal and greater strength are available in drop-forged brackets at little extra cost.

Attachments. The third important part of the trolley is the attachment. Some representative attachments are shown in Fig. 9.5.7. Commonly used types include:

1. I, or idler attachment—used when the trolley is connected to the chain only and does not carry a load.
2. H, or clevis attachment—used more than all others, for suspended load carriers.
3. B, or rod attachment—used mostly for connection to load bars, and for oven-sealing plates.
4. C, or pendant tongue attachment—used in the same manner as the H attachment, except that the clevis connection is on the load carrier.
5. P attachment—similar to the C attachment, but designed for rigid connection of load carrier to trolley.
6. J attachment—for use with rod-connected carriers, rotating loads, oven-sealing plates, and so on.
7. T attachment—used almost exclusively for carrying load bars.

Numerous modifications of these basic types are available.

For most chains, except the drop-forged chains, the attachments and the trolley brackets are bolted to the top and bottom of the chain. This requires special chain attachments or special chain pins. Thus, the chain supports must be strong enough to carry the loads. The open center links of the drop-forged chains allow the loads to be carried directly through the links from the attachments to the trolley brackets. The chain pulls the loads, but does not have to carry them.

Drop. The distance from the I-beam track to the centerline of the chain is called the "drop." This distance is sometimes measured from the top of the I-beam and sometimes from the bottom. It is an important dimension. It varies with different manufacturers but trolleys can generally be obtained with commonly used drops. It is possible, and very desirable, to have a uniform drop throughout a plant. Conveyors can then be made interchangeable, even though obtained from several sources. It is also desirable that distances from the centerline of the chain to the attachment bolt holes conform to a common standard. Typical dimensions for the most commonly used trolleys are shown in Fig. 9.5.8.

Trolley Loads. Trolleys must support two loads: (1) the suspended weight of the carrier and its live load, and (2) the load imposed on the trolley by chain pull at vertical curves where there is a change in elevation. When chain pull is large, or radius of vertical curves is small, the imposed load can become excessive. It should be calculated to be sure that the maximum trolley ratings are not exceeded (see Fig. 9.5.9). Use the following equation:

$$L = \frac{PS}{R}$$

where: L = imposed load on a trolley (lb)
P = chain pull (lb)
S = trolley's spacing (in.)
R = curve radius at centerline of chain (in.)

If W = suspended weight of carrier (lb), then $W + L$ = radial load on the trolley (vector addition). In practice, arithmetical addition of $W + L$ is usually close enough.

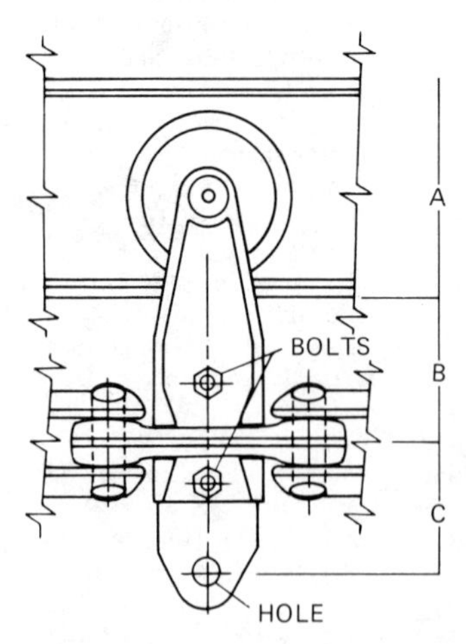

Chain Size	A	B	C
228	$2\frac{5}{8}$ in.	$1\frac{7}{8}$ in.	2 in.
348	3	$2\frac{1}{2}$	$3\frac{7}{8}$
458	4	$3\frac{3}{16}$	$2\frac{7}{8}$
458	4	4	$2\frac{7}{8}$
678	6	4	$3\frac{5}{8}$

Fig. 9.5.8 Trolley dimensions and "drop."

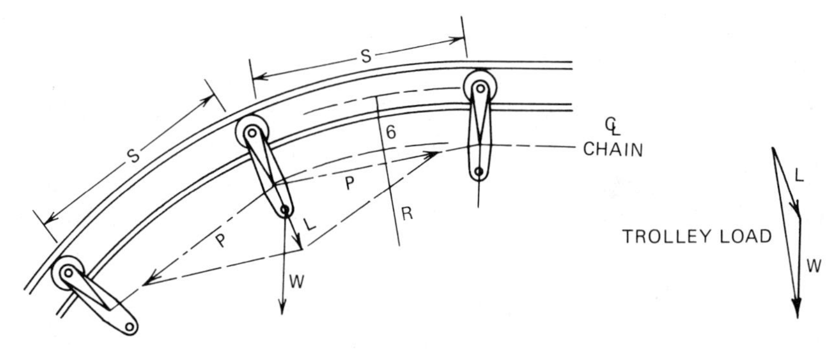

Fig. 9.5.9 Diagram of imposed load.

This equation is for uniform trolley spacing. When trolley spacing is not uniform, the average of the trolley spacing on each side of the loaded trolley should be used for S.

The modern type of trolley wheel is a large ball bearing with a very high load capacity. The track on which it runs cannot support nearly as great a load as the wheel can carry. Thus, the I-beam wear is the limiting factor in setting trolley load limits. Load ratings generally preferred are:

Size of I-Beam	Maximum Allowable Trolley Load	
	$W + L$	W
$2\frac{5}{8}$ in.	100 lb	75 lb
3 in.	400 lb	250 lb
4 in.	800 lb	400 lb
6 in.	1600 lb	1000 lb

Load Bars. When a trolley load exceeds the allowable rating of a single trolley, it can be hung from a load bar that is connected to two trolleys. Sometimes two load bars are connected by a third one so that the load is suspended from four trolleys. Load bars are necessary on steep inclines and vertical runs to hold the suspended loads away from the chain and to prevent the chain from kinking. A semirigid load bar attachment is shown in Fig. 9.5.10.

Track

Many different shapes have been used for overhead trolley conveyor track. These include I-beams, double angles, structural tees, double T-rails, steel bars, rods and pipe, and formed track similar to sliding door track. I-beam track is the most frequently used. Standard sizes of I-beams are:

$2\frac{5}{8}$ in., 3.74 lb
3 in., 5.7 lb
4 in., 7.7 lb
6 in., 12.5 lb

Standard low-carbon I-beam is not usually adequate as a track material. Generally the wear on vertical bends of a system limits the loading. Most manufacturers, therefore, offer specially rolled sections in the medium-carbon range and which have small amounts of alloys to improve the wear quality. The best analysis provides for good weldability as well as wear.

I-beams should be selected for straightness, with webs both central and square with the flanges. Track must be installed plumb; otherwise the trolleys will ride to one side, which cuts the web of the track and causes excessive wear on the trolleys due to eccentric loading.

Joints. Track joints are usually welded and ground smooth where trolley wheels make contact or must clear. Top flange joints are reinforced with angles. Sometimes track must be installed in places where there are fire hazards and where welding, therefore, is not permitted, such as paint-spray rooms, oil and paint storage, oil washing, and other flammable locations. In such locations

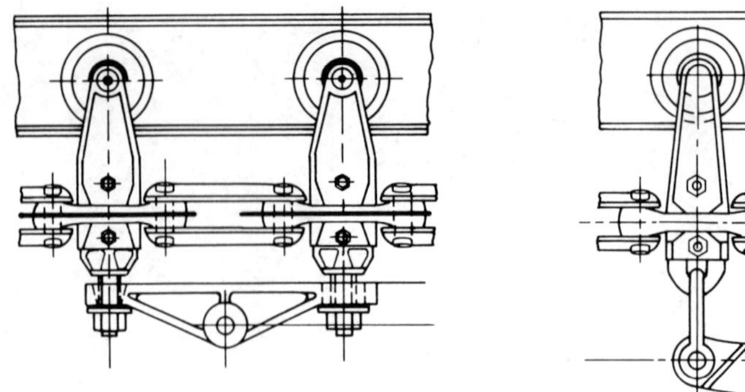
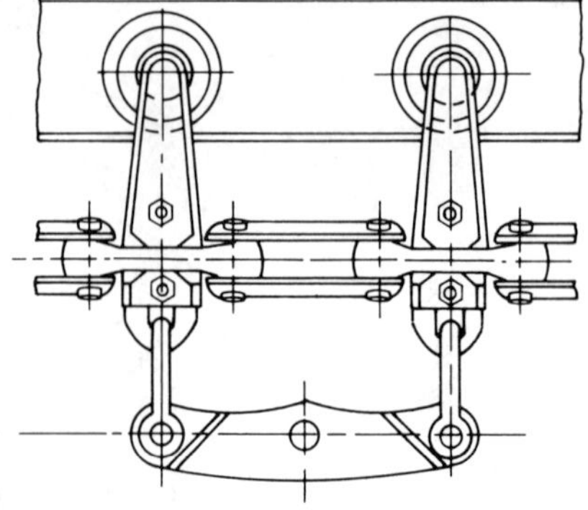

Fig. 9.5.10 Load bars.

track joints should be bolted. The joints should occur at or close to supports and the track ends should be doweled together for alignment. This is usually a more expensive construction than welded joints.

Expansion Joints. Expansion joints in the track are used when a conveyor operates in an oven or where the temperature varies between wide limits or at building expansion joints. Long oven lines may require several of these joints. They prevent distortion, warping, and twisting in the track. The change in length of the track is determined by the following equation:

$$E = \frac{0.078\,TL}{1000}$$

where: E = expansion of track (in.)
T = range in temperature (°F)
L = length of track affected (ft)

Spans. I-beams are furnished in 20-ft lengths, and, on occasion, supports may be as much as 20 ft apart. When sag is excessive, intermediate hangers are added. In most cases, shorter spans are required for heavier loads. The simplicity and low cost of the single I-beam track are so desirable that any deviation from this construction should usually be avoided. There may, however, be times when long spans are compulsory. In such cases, cap channels or extra supporting beams are used in combination with the I-beam track.

Supports

The selection of the supporting method depends on the track loading and construction of the building. The building load includes the track loads and support load.

Track loads include the weight of track, trolleys, chain, carriers, and live loads. In addition to these loads, when guards are used, their weight is usually carried by the track beam.

Loads. Track, trolleys, and chain can be considered uniform loads. Carriers and live loads, if light and closely and evenly spaced, can also be considered uniform loads. When carriers and live loads are heavy or widely spaced (i.e., more than 24 in. centers), they should be considered concentrated loads. Hangers for guards are widely spaced and transfer the guard weight to the track as concentrated loads.

Hangers. Most trolley conveyors are suspended from ceilings, trusses, floor beams and slabs, and overhead structures. Hangers consist of light angles, rods, or steel bars, with clip angles for connection to the track, and clamps for connection to roof trusses and beams. Inserts or expansion bolts are used for connection to concrete ceilings and lag bolts or clamps for connection to wood.

Hangers over 12 in. long are periodically side or sway braced. One leg should be vertical to simplify field work in aligning the track, both horizontally and vertically. Horizontal bends, drives, and take-ups are heavily braced; when they are long, the hanger frames are usually tower structures. The final location and arrangement of the hangers is usually determined in the field by erection superintendents who know by experience how to hang and brace a conveyor to get the best results. Manufacturers do, however, establish standards for supporting their conveyors. Table 9.5.3 illustrates recommended spans for various loadings while Fig. 9.5.11 illustrates a typical support for a horizontal turn.

Table 9.5.3 Recommended Track Spans and Loads[a]

I-Beam Size (in.)	Span, ft, for Maximum Load, lb, between Supports of: 6 ft	8 ft	10 ft	12 ft	15 ft	20 ft
$2\frac{5}{8}$	1500	1800	600	—	—	—
3	—	1600	1000	700	500	—
4	—	3000	2000	1600	1100	500
6	—	9000	5600	4000	2600	1400

[a] Total load = weight of carriers plus load on carrier × carriers in span. Guard weight must be added if it is attached between supports.

Floor Supports. Where the ceiling structure will not support the conveyor load, or in very high bays where hangers would be too long, or where overhead cranes or other conditions prevent the use of hangers, column floor supports are used. When it is necessary to use columns for floor supports, they can be selected by structural design methods for cantilever columns with eccentric loads. However, it should be noted that the length l, used in determining allowable loads for columns, is equal to twice the height A of such a column (i.e., $l = 2A$).

As floor support columns must be very stiff, bracing is seldom used. One column may even be used to support a corner turn, since the columns in the adjacent straight lines are usually sufficiently stiff to take the conveyor chain pull. A good anchorage is required, such as a wide base with large anchor bolts.

Superstructure. Any extra framing that must be installed at points where hangers cannot be directly connected to the building structure is called superstructure. This framing is usually clamped to the building steel so that it can be removed with the conveyor when changes are made, leaving the building in its original condition. It is customary to consider the superstructure as part of the conveyor, to be furnished and built on the job by the conveyor contractor.

Horizontal Turns

Horizontal turns are made at roller turns, at sprocket turns such as are used with sprocket drives, or at traction wheel turns. Horizontal turns can be of short radius (compared to long-radius vertical curves), since the chain articulates at the joints. The radius is determined by the clearance between

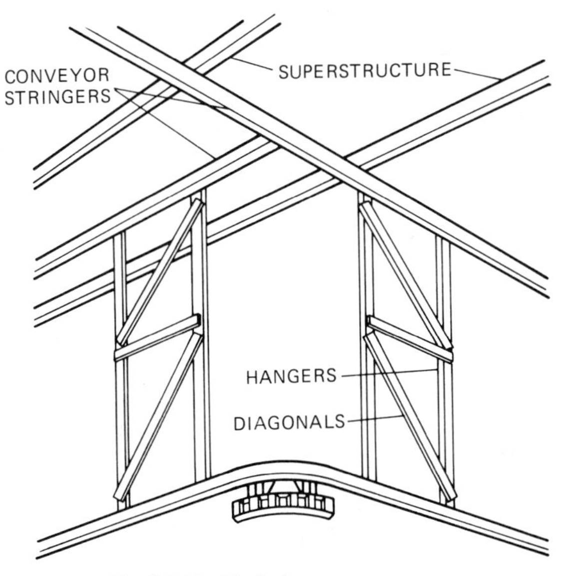

Fig. 9.5.11 Typical turn support.

Table 9.5.4 Minimum Recommended Radius and Diameter Turns for Various Trolley Spacings

Chain Size	Trolley Spacing	Traction Wheel Diameter	Roller Turn Radius
X-348,	up to 18 in. (457.2 mm)	24 in. (609.6 mm)	18 in. (457.2 mm)
X-75-13	24 in. (609.6 mm)	30 in. (762.0 mm)	
	30 in. (762.0 mm)	36 in. (914.4 mm)	
X-458,	up to 24 in. (609.6 mm)	30 in. (762.0 mm)	24 in. (609.6 mm)
X-100-16	32 in. (812.8 mm)	36 in. (914.4 mm)	
X-678,	12 in. (304.8 mm)	36 in. (914.4 mm)	36 in. (914.4 mm)
X-150-22	24 in. (609.6 mm)	42 in. (1066.8 mm)	
	36 in. (914.4 mm)	48 in. (1219.2 mm)	

adjacent carriers or loads. Radii should be as large as possible, especially for heavy chain pulls, and near the points where chain pull approaches the maximum. However, it is better practice to use the same radius for all horizontal turns.

The minimum radius for standard roller turns (depending on pitch of the chain) is 15 in. Minimum diameter for traction wheels is 12 in. Most manufacturers have individual standards, and the radii or diameters may vary slightly. Standard angles for turns are 15 (kinker), 30, 45, 60, 90, and 180° and these are used whenever possible. When required by special conditions, turns can be made any angle or radius. A guide for minimum turn radius for various trolley spacings is shown in Table 9.5.4.

Roller Turns. For angles of 90° or less, roller turns generally cost less and occupy less space than wheel turns of the same radius. A large-radius roller turn can be used around a corner or column where only a small-diameter wheel turn can be used. There is no limit on the maximum radius of a roller turn except cost. Where very large radii are required, they are usually constructed in a segmented fashion using a number of small-angle and radius turns. Figure 9.5.12 illustrates a typical 90° roller turn.

Rollers. It is generally considered good practice to use double-row ball bearing, roller turn rolls, with hardened treads concentric with the ball races. The rolls should be spaced close enough to prevent bumping or pulsation of the conveyor chain. A typical roller turn mounting is shown in Fig. 9.5.13. Rollers are available in a number of constructions using different methods of sealing and lubrication. The choice is based on environment and degree of use.

Wheel Turns. Wheel turns are made with flat-face traction wheels and sprocket wheels. The power loss at these turns is less than at roller turns of the same radius. For angles of 90° or less their cost is usually higher than for corresponding roller turns. A 180° traction wheel is shown in Fig. 9.5.14.

Traction wheels are preferred in ovens, washers, degreasers, sandblast operation, foundries, and similar severe-duty locations where lubrication, maintenance, and wear must be reduced to a minimum. Traction wheels may be cast iron with finished or chilled treads, or have steel rims with welded pipe or rod spokes, or steel rims with rod spokes up-set and riveted to the rims or plates with welded rims and either welded or removable hubs. Hubs may have either bronze bushings or roller bearings. In some cases, such as inside an oven, traction wheels have been fastened to rotating shafts with bearings located outside the oven.

Alignment. The centerline of the chain should be slightly above the center of roller turn rollers or traction wheel treads to allow for chain sag between trolleys. Traction wheel treads must be concentric with the bores, and both roller turns and traction wheels should be in proper alignment with the

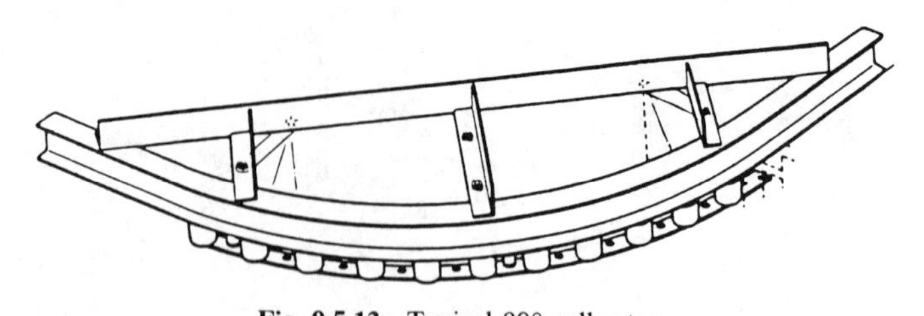

Fig. 9.5.12 Typical 90° roller turn.

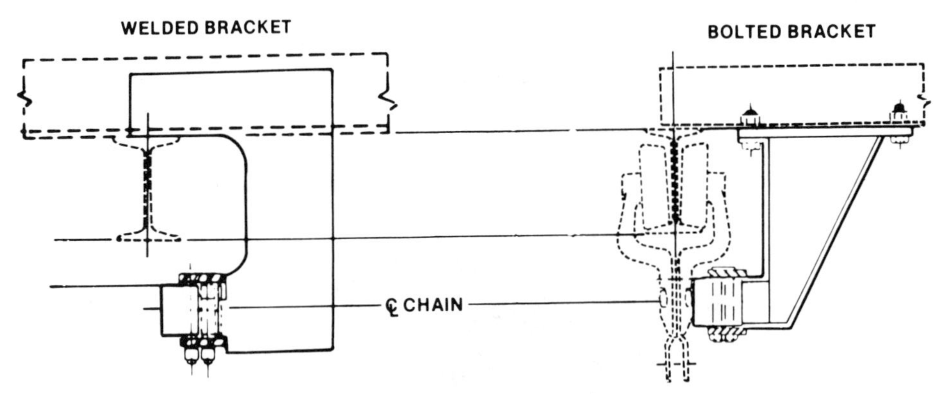

Fig. 9.5.13 Roller turn mounting.

curved track. Otherwise, trolleys will be pushed or pulled to one side, resulting in excessive wear both on them and on the track.

Vertical Curves

Changes in elevation are made with vertical curves. For each change, often called a "dip," one reverse curve or compound vertical curve is required. To go from a higher to a lower elevation requires a dip down, and the opposite requires a dip up. When the change in elevation is large, the top and bottom sections of the reverse curve are connected by an inclined straight section of track.

The maximum inclination from the horizontal used in standard practice is 45°. Steeper inclines can be used, and conveyors have been built with vertical runs, but these require special consideration. Heavy suspended loads on steep inclines tend to cock the trolleys and to kink the chain. Some users do not permit inclines exceeding 30°, but this prohibition, if used dogmatically, sacrifices valuable space.

For layout purposes it is desirable to use the shortest possible radius and steep inclines for making vertical curves and changes in elevation. For optimum conveyor operation it is best to use the longest possible radius and gradual inclines. Therefore, it is advisable to check carefully the radii of vertical curves, the trolley spacings that determine those radii, and the angle of incline, for both new and existing conveyors. With floor space at a premium, changes in existing standards may pay big dividends. Equations and values useful in curve computation are given in Figs. 9.5.15, 9.5.16, and 9.5.17.

Reinforcement. The heaviest trolley loads occur at vertical curves because of imposed loads due to chain pull. Thus, the maximum wear and peening of track beam flanges will usually occur at these points. Careful location of drives can often reduce this load to a minimum. Sometimes, however, very heavy loads require reinforcement of the curve flanges. This can be done by stitch welding thin plates to the flanges of the I-beam. Special heat treating will also improve the wear capabilities.

Drives

The standards for drives are those of individual manufacturers. However, the designs are all similar and the selection of components can be tailored to the needs of the user. Two types of drives predominate: sprocket drives and caterpillar drives. Examples of each are shown in Figs. 9.5.18 and 9.5.19.

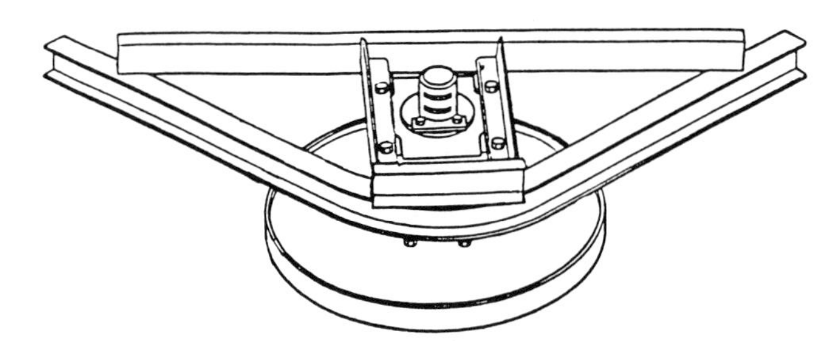

Fig. 9.5.14 Typical 180° wheel turn.

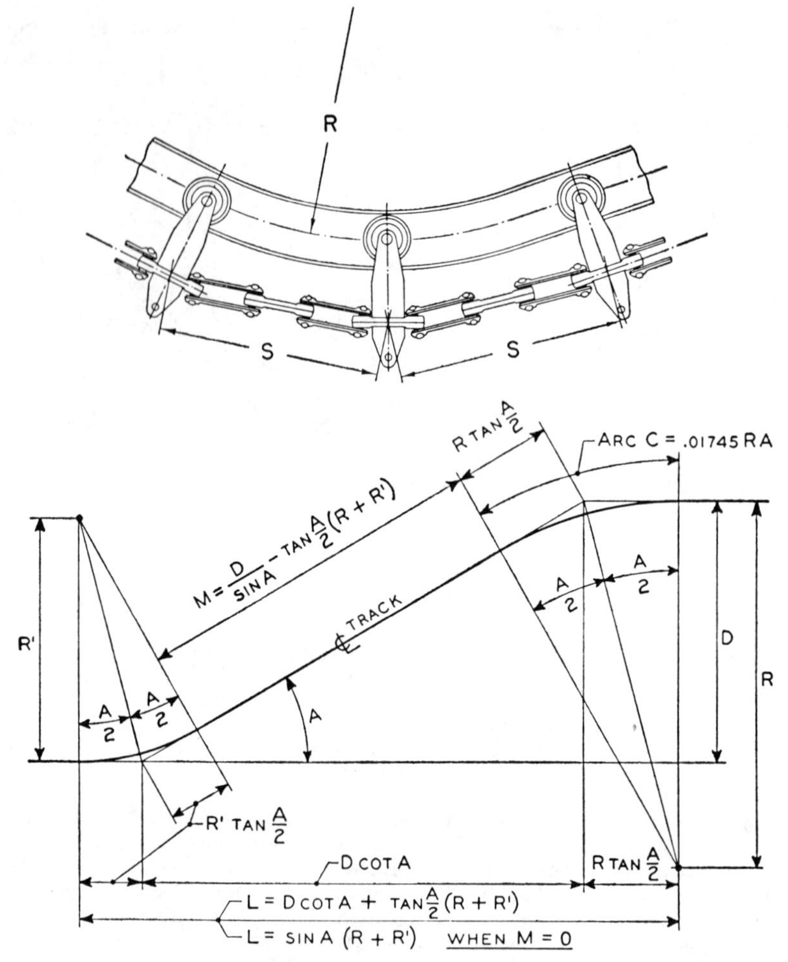

Fig. 9.5.15 Diagram and formula for vertical curves.

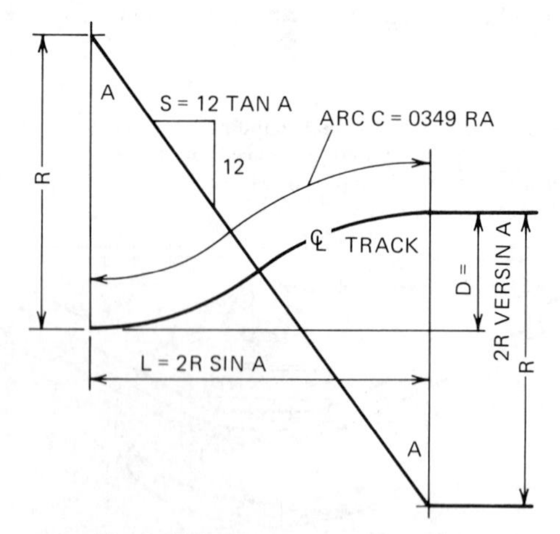

Fig. 9.5.16 Reverse curve formulas.

		Radius R									
		2′ 0″		4′ 0″		6′ 0″		12′ 0″		25′ 0″	
A	S	D	L	D	L	D	L	D	L	D	L
5°	$1\frac{1}{16}''$							$0'\ 1\frac{1}{8}''$	$2'\ 1\frac{1}{8}''$	$0'\ 2\frac{1}{4}''$	$4'\ \frac{5}{16}''$
10°	$2\frac{1}{8}''$	$0'\ \frac{3}{4}''$	$0'\ 8\frac{5}{16}''$	$0'\ 1\frac{1}{2}''$	$1'\ 4\frac{5}{8}''$	$0'\ 2\frac{3}{16}''$	$2'\ 1''$	$0'\ 4\frac{3}{8}''$	$4'\ 2''$	$0'\ 9\frac{1}{8}''$	$8'\ 8\frac{3}{16}''$
15°	$3\frac{7}{32}''$	$0'\ 1\frac{5}{8}''$	$1'\ \frac{7}{16}''$	$0'\ 3\frac{1}{4}''$	$2'\ \frac{7}{8}''$	$0'\ 4\frac{7}{8}''$	$3'\ 1\frac{1}{4}''$	$0'\ 9\frac{13}{16}''$	$6'\ 2\frac{9}{16}''$	$1'\ 8\frac{7}{16}''$	$12'\ 11\frac{5}{16}$
20°	$4\frac{3}{8}''$	$0'\ 2\frac{7}{8}''$	$1'\ 4\frac{7}{16}''$	$0'\ 5\frac{3}{4}''$	$2'\ 8\frac{7}{8}''$	$0'\ 8\frac{11}{16}''$	$4'\ 1\frac{1}{4}''$	$1'\ 5\frac{3}{8}''$	$8'\ 2\frac{1}{2}''$	$3'\ 0\frac{3}{16}''$	$17'\ 1\frac{3}{16}''$
25°	$5\frac{19}{32}''$	$0'\ 4\frac{1}{2}''$	$1'\ 8\frac{5}{16}''$	$0'\ 9''$	$3'\ 4\frac{9}{16}''$	$1'\ 1\frac{1}{2}''$	$5'\ \frac{7}{8}''$	$2'\ 3''$	$10'\ 1\frac{11}{16}''$	$4'\ 8\frac{3}{16}''$	$21'\ 1\frac{9}{16}''$
30°	$6\frac{15}{16}''$	$0'\ 6\frac{7}{16}''$	$2'\ 0''$	$1'\ \frac{7}{8}''$	$4'\ 0''$	$1'\ 7\frac{5}{16}''$	$6'\ 0''$	$3'\ 2\frac{9}{16}''$	$12'\ 0''$	$6'\ 8\frac{3}{8}''$	$25'\ 0''$
35°	$8\frac{13}{32}''$	$0'\ 8\frac{11}{16}''$	$2'\ 3\frac{9}{16}''$	$1'\ 5\frac{3}{8}''$	$4'\ 7\frac{1}{16}''$	$2'\ 2\frac{1}{16}''$	$6'\ 10\frac{5}{8}''$	$4'\ 4\frac{1}{16}''$	$13'\ 9\frac{3}{16}''$	$9'\ \frac{1}{2}''$	$28'\ 8\frac{1}{8}''$
40°	$10\frac{1}{16}''$	$0'\ 11\frac{1}{4}''$	$2'\ 6\frac{7}{8}''$	$1'\ 10\frac{1}{2}''$	$5'\ 1\frac{11}{16}''$	$2'\ 9\frac{11}{16}''$	$7'\ 8\frac{9}{16}''$	$5'\ 7\frac{3}{8}''$	$15'\ 5\frac{1}{8}''$	$11'\ 8\frac{3}{8}''$	$32'\ 1\frac{11}{16}''$
45°	$12''$	$1'\ 2\frac{1}{16}''$	$2'\ 9\frac{15}{16}''$	$2'\ 4\frac{1}{8}''$	$5'\ 7\frac{7}{8}''$	$3'\ 6\frac{3}{16}''$	$8'\ 5\frac{13}{16}''$	$7'\ \frac{3}{8}''$	$16'\ 11\frac{5}{8}''$	$14'\ 7\frac{3}{4}''$	$35'\ 4\frac{1}{4}''$

Fig. 9.5.17 Vertical curve dimensions.

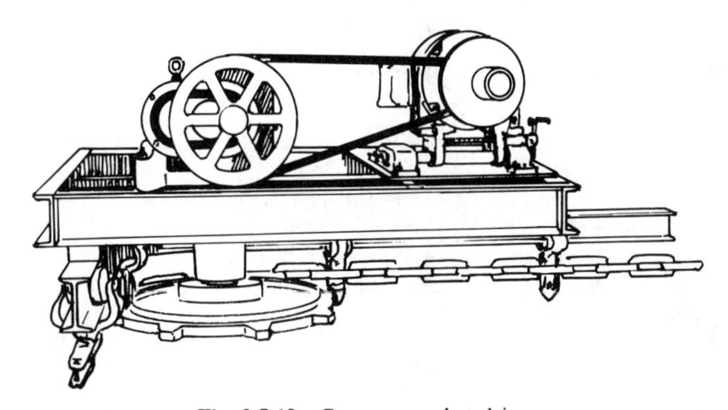

Fig. 9.5.18 Corner sprocket drive.

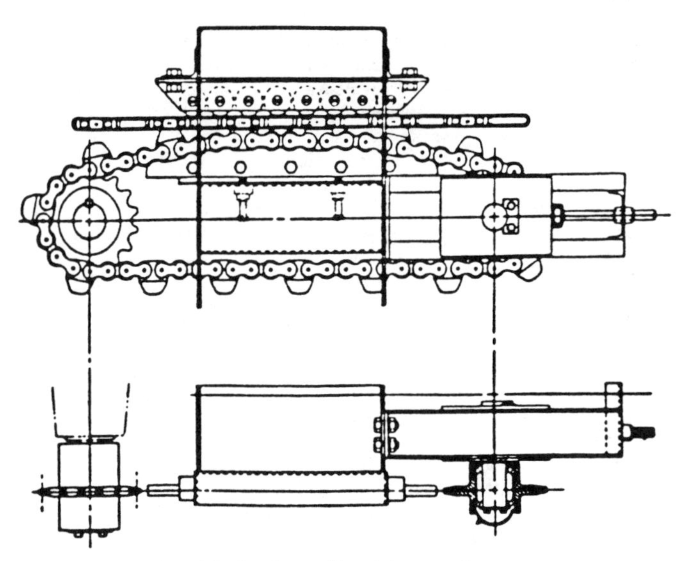

Fig. 9.5.19 Caterpillar drive attachment.

General Design. Trolley conveyors are very often installed with close ceiling and side clearances. Therefore, the drives are usually constructed with the components (except the drive element) arranged in a frame just above the track. The parts are arranged horizontally in a line. The drive is long but compact in width and height. The arrangement of the drive machinery (motor, sheaves, reducer, etc.) is the same for both sprocket and caterpillar drives. They differ only in the manner in which they engage the conveyor chain.

Sprocket Drives. This type of conveyor drive requires that the conveyor chain be wrapped around and driven by the sprocket, usually attached to the output shaft of the drive-gear reducer. A sprocket drive is thus always located at the corner turn, usually 90° or 180°, but, if necessary, at turns of other angles. The sprocket drive is rated for chain pull at the head sprocket. The size of the sprocket is related to the minimum radius of corner permitted in the system. It is obvious that large corners are a deterent to the use of sprocket drives. Sprockets with a 36-in. pitch diameter are commonly used with drop-forged chains. The largest size sprocket is limited by output torque and speed ratios of available reducers. The smallest size sprocket is limited by product clearance and the standards of the manufacturer.

Caterpillar Drives. The caterpillar drive employs a short loop of auxiliary drive chain alongside and parallel to the conveyor chain. Driving dogs on the caterpillar chain engage and drive the conveyor chain. Back-up bars on the caterpillar side and rollers on the driven chain side confine the chains and prevent them from separating. A caterpillar drive can be located on any straight run of conveyor. The ratings of caterpillar drives are based on chain pull capacity.

Advantages of Caterpillar Drives. Although sprocket drives are frequently used, caterpillar drives have many advantages:

1. The caterpillar drive can be located in the best place to take care of chain pull, such as the middle of a straight run.
2. If load conditions change, or if the conveyor is altered, a caterpillar drive is easier to move and relocate without changing the existing conveyor.
3. When chain pull is moderate to heavy, a caterpillar drive is usually less expensive than a sprocket drive, owing to the much lower torque requirements.
4. A caterpillar drive can be reused on any conveyor, regardless of the requirements for clearance or of the radius or angle of horizontal turns. With a sprocket drive, the load clearance will be limited by the diameter of the drive sprocket.

Constant and Variable Speeds. Both sprocket and caterpillar drives employ the same basic drive mechanism. These come in three different sytles: (1) constant speed, (2) varipulley with infinite speed variation up to 3:1, and (3) full-variable with infinite speed variation up to 10:1 or even more.

Constant-speed drives use a V-belt drive from an electric motor to a vertical gear reducer. The output shaft of the reducer carries the conveyor sprocket. This sprocket can be either the one that engages the conveyor chain, making a sprocket drive, or the drive sprocket of a caterpillar chain drive. This drive can also be made variable speed by modulating the motor rpm.

A varipulley drive unit is arranged the same as the constant speed type except that a variable pitch pulley unit is used instead of a V-belt drive between the motor and reducer. The ratio between maximum and minimum speeds is approximately 3:1.

In the third style, a full variable transmission is introduced into the drive between the motor and reducer. Connection from the transmission to the reducer is by V-belt or silent chain drives and to the motor by V-belt drives.

Speeds can be obtained from an imperceptible movement of a fraction of an inch per minute up to 150 ft/min. Drives for very low or very high speeds and for very large chain pulls must be specially designed. Manufacturers' standard drives can usually meet most ordinary requirements. Drive ratings should have an allowance for short-period overloads without damage to the drive machinery.

Construction of the Drive. The use of trolley conveyors as part of vital material handling systems in manufacturing plants makes it mandatory that drives be of the utmost dependability. An essential part of the drive is the speed reducer. There are a number of gear reducers available that have been developed especially for trolley conveyors. These have an extended lower bearing support which provides excellent overhung load capacity, continuous gear lubrication, and a deep well which prevents the gear lubricant from leaking through the lower bearing.

The basic power source for trolley conveyors is an electric motor. In most large manufacturing plants, where all operations are interdependent, only standard stock motors are used in trolley conveyor drives. They are preferred because a stock motor can be replaced quickly in case of a motor failure. This prevents a long, costly shutdown while a damaged motor is being repaired. Motors with special windings, shafts, or other features, gear motors, and motors with built-in mechanical or electrical adjustable speed changers are generally not used because replacements cannot be made in a hurry.

The AC constant-speed squirrel cage motor is frequently used. Both motor and control are inexpensive and dependable. Under normal conditions they are easily obtained from stock. Across-the-line starting requires only simple wiring, which is inexpensive and easy to maintain.

Where small speed variations will suffice, an AC motor with a variable-pitch pulley on the output shaft provides the most economical arrangement. This type of unit can be adjusted manually or automatically but is generally limited to single drives.

An important requirement for variable-speed drives is that they hold to a set speed regardless of variations of loading. Where this variation exceeds about 3:1, but less than 10:1, a mechanical speed changer is interspaced between the motor and reducer. Combination motor/speed changer units are readily available but are seldom used.

A frequent method of speed variation now being used is the combination of a standard DC motor with an adjustable voltage rectified AC power supply. This method provides useful speed variations of up to 20:1 and with tachometer feedback, speed regulation within 1% of the set speed. A generated DC power to supply may also be used.

Another variable-speed method is the use of an eddy current clutch in conjunction with an AC motor. The motor runs at a constant speed while the clutch has a variable-speed output speed and an eddy current clutch with a variable-speed output shaft. There is no mechanical connection between the input and output rotors of the clutch, the connection being made through a magnetic field at an air gap. The slip at the air gap can be varied by means of an electrical control to obtain any desired speed. The drive has stable speed characteristics.

V-belts or adjustable pulley belts are usually used to connect the motor to the reducer and mechanical speed changer. These belts should be conservatively rated and provided with a suitable guard. The connection between a mechanical speed changer and the reducer is usually silent chain. It also must be guarded.

Motor horsepower for a trolley conveyor is calculated from the following equation:

$$\text{Motor horsepower} = \frac{\text{Chain pull (lb)} \times \text{Speed (ft/min)}}{33{,}000 \times \text{Total drive efficiency}}$$

The horsepower to drive an average trolley conveyor is quite modest. For example, for a heavy drive of 3000 lb used to drive a conveyor operating at 50 ft/min and having an efficiency of 75%, the net horsepower of the motor is:

$$\frac{3{,}000 \times 50}{33{,}000 \times 0.75} = 6.1 \text{ Hp}$$

Multiple Drives. When the chain pull of a conveyor is too great to be handled by a single drive, two or more drives are used. Additional drives can be added when a conveyor is lengthened or when the chain pull is increased. Theoretically there is no limit to the number of drives or length of conveyor that can be used. However, long conveyors invariably have many inclines and load variations so that expert advice is usually required.

The problem of matching drive speeds is complicated, even at constant speed. Long strands of chain of equal length will not have exactly the same number of pitches. Thus, a 1000-ft strand of 458 chain may have 2970 pitches and the next 1000 ft only 2950 pitches. The same length (measured in feet, not pitches) must pass each drive during the same time interval. But the drive sprockets (whether for sprocket or caterpillar drives), which have equal numbers of teeth and advance equal numbers of chain pitches per revolution, must therefore rotate at different rates. Thus, each drive turns a different number of revolutions for each phase and section of chain, and during any time interval the drives are revolving at different, varying rates. This requires a continuous "hunting" action of the drives so that they will keep in proper relationship and will not tend to tighten or loosen the sections of chain they pull.

In some cases there are wide variations in chain pull at any one drive, without corresponding changes in drive loading at the other drives. Then the problem of controlling the speed of the multiple drives on a conveyor becomes so complicated that electrical controls must be used to get workable installations.

Constant-Speed Multiple Drives. For constant-speed drives, high-slip, squirrel-cage motors are often used. These slow down under heavy loads or tight chains and regain speed under light loads or loose chains. Slip couplings of the fluid type are also frequently used for balancing of individual drives on a multiple drive system. If more than two or three are required, it is advisable to keep chain pull per drive low. Wound-rotor and synchronous-motor drives are not satisfactory.

With the advent of economical DC power supplies, it is becoming more usual to solve the problem of multiple drives and load sharing through use of master and slave drives. One drive is selected as a master drive, the others are adjusted to be slave. A tachometer on the master drive provides the speed reference which the other drives endeavor to match through feedback circuits.

Variable-Speed Multiple Drives. Most multiple-speed, multiple-drive systems are constructed using either DC drives or eddy current clutches. These systems are not generally limited as to the number of drives. These drives are almost a necessity when the load on an individual drive varies greatly and independently of the other drives. In addition, both systems contain provision for balancing loads when physical considerations preclude accurate balancing by means of location of the drive. Both systems offer remote control of the speed as a result of the control wiring to each drive.

Drive Location. A drive should be located at the top of a drop or just ahead of a take-up to eliminate slack. Location just before a rise might allow the chain to bunch up and jam in a caterpillar drive or come off the sprocket in a sprocket drive. The portion of the building to which the drive is attached must be strong enough to sustain the chain pull as well as the conveyor weight. Location of multiple drives requires expert consideration and should be determined by the point-to-point or progressive calculation method. Location of drives for reversing service and automatic loading and unloading are difficult problems. Ovens are poor locations for drives, but sometimes must be used.

Overload Devices. All drives require some form of mechanical device for protection of the drive and conveyor in case of an overload or jam in the line. The overload device is usually placed to directly oppose the chain pull. It can also be mounted on the output shaft of the reducer, speed changer, or motor. However, a mechanical overload device can be used at a motor, or on the drive from the motor to the reducer, only when the speed is constant. If the speed is varied, an overload device at the motor will furnish protection only at the high speed, and little or none at lower speeds.

Floating Drive. The best overload device is one that will stop the motor (or motors, when more than one drive is used) when an overload occurs and will allow the motor to be started again when the cause of the overload is removed. An overload device that does not require changing shear pins, or any other alteration to the drive, is desirable. Also, the overload device should be located in the drive as close to the load as possible so that it will respond to the load quickly and will not require a superfine adjustment or calibration. A drive that meets these requirements is the floating drive. It is made in both caterpillar and sprocket types.

The floating caterpillar drive is made up of a drive frame free to move in a fixed secondary frame to which the conveyor track is attached. The driving effort is resisted by compression-coil springs of suitable capacity and so arranged that the sliding frame will contact a limit switch to shut off the motor in case of an overload or jam on the line. The shut-off point can be adjusted to allow for different chain pulls. A shear pin is not required, and after the cause of shutdown has been corrected, the unit is at once ready to go back into service. The floating drive measures only the load between the conveyor and caterpillar chains. It does not protect the drive against misalignment or extraneous material in the caterpillar base. Sprocket drives can also be made the floating (pivoting) type, arranged so that the springs resist rotation.

Shear Sprocket. These sprockets consist of two halves: the sprocket half, not keyseated and free to turn on the driving shaft; and the driving flange half, keyed and set screwed to the shaft (see Fig. 9.5.20). The driving force is transmitted to the sprocket from the driving flange by a single cold-rolled steel-necked shear pin carried in a pair of hardened steel bushings. The shear sprocket has the advantages of simplicity, economy, instantaneous reaction, and a location that tends to protect all of the drive machinery. However, there are factors that can limit its application. Replacement of pins is

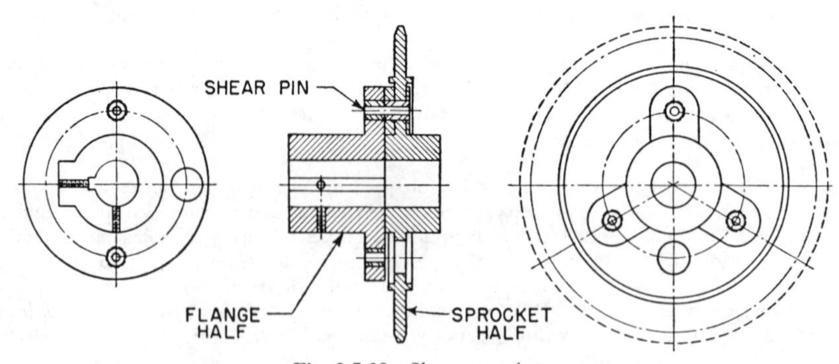

Fig. 9.5.20 Shear sprocket.

sometimes inconvenient and time consuming. In addition there is a tendency of maintenance personnel to ignore the cause of repeated pin failure and replace the pins with an un-necked pin, thereby limiting protection. Finally, the failure of a pin does not normally shut off the motor and this can result in seizure between the stationary sprocket and the rotating member. This difficulty can be overcome by the use of an electric shear sprocket which turns off the drive motor or all motors on a multiple-drive conveyor.

Overload Cutout. Drive sheaves can be equipped with a torque limiting device with limit switch cutout. These are usually spring-loaded devices that are adjustable for the desired drive effort.

Take-ups

Take-ups are used on almost all trolley conveyors. Sometimes a dip following a drive will take care of some slack in the chain and in some plants, where excellent maintenance facilities exist, the preference is to omit take-ups. Generally, ideal conditions for omitting take-ups do not occur and most conveyors are installed with take-ups.

Level conveyors should all have take-ups. Oven conveyors and outdoor installations with wide variations in temperature require counterweighted or air-operated take-ups, unless the expected change in length is small, in which case spring take-ups may be used.

Types of Take-ups. There are five types of take-ups: track type, screw take-ups, spring take-ups, air cylinder take-ups, and counterweighted take-ups. Track take-ups are used only on very inexpensive installations. The track is provided either with inserted filler pieces or expansion-type sliding joints. When taking up slack in the chain, the hangers are adjusted or moved. This type of take-up is used at 90° and 180° turns. The screw take-up is used in conveyor installations where slack must be taken up but the turn must not move or float. These conditions are required in some multiple-drive installations and for the exact positioning necessary in automatic loading and unloading.

Spring take-ups have the same construction as the screw type, except that springs automatically adjust the take-up to a tension on the chain. Screws may or may not be used for initial adjustment or to release the take-up when removing slack in the chain. The counterweighted and air cylinder take-ups are also similar to the screw type, except that they maintain a constant tension on the chain. All take-ups require attention. Otherwise, when excessive chain slack develops, they have lost their function and they do not become useful until the slack is removed.

Unless it is absolutely impossible, 180° take-ups should be used. They work better and adjust for much more slack than 90° take-ups. Whenever there is no place in the layout for a 180° turn, a 90° type is used. This has a single sliding joint and the leg at right angles has a hinge or the track is sprung.

When the distance between the parallel lines at a take-up is greater than the diameter of the 180° turn, the take-up can be made of two 90° turns with a straight section between them. Thus a turn with two lines 6 ft apart using 24-in. radius roller turns would be a "24-in. radius roller turn 180° take-up with a 6-ft spread."

Take-ups can be very useful in the control of multiple drives. If there are two drives on a conveyor, one can be the master and the second the slave, controlled by the movement of the adjacent take-up. In some cases when drives are electronically controlled, the take-up following one drive is used to control other drives.

Location. Take-ups are located close to drives, on the slack side of the conveyor chain. A straight horizontal run or a short curve upward should follow a take-up, especially when it is the automatic (counterweighted or spring) type. The take-up should not be followed by a downward curve which will put a pull on it through the weight of the conveyor line. Sometimes a take-up must be located where there is a heavy chain pull so that a drive pulls through the take-up. This requires a very heavy counterweight, or a screw take-up, and should be avoided.

Guards

The ideal guard not only should prevent a load from dropping on someone below but should allow enough room so that moving loads will clear a fallen load resting on the floor of the guard. In practice, there is seldom enough headroom to provide that much clearance, and the cost of the larger guard would be excessive. Clearances must be enough to keep swaying or tilted loads from hitting or dragging on the sides or bottom of the guard. Guards must be carefully laid out to provide these clearances. This is especially important at turns and dips.

Types of Guards. Falling-part guards are usually made of woven wire mesh, welded wire mesh, or expanded metal so as to offer the least obstruction to light and to keep the guards light in weight. Drip pan type guards are used to keep oil or dirt from dropping on workers or equipment. They are

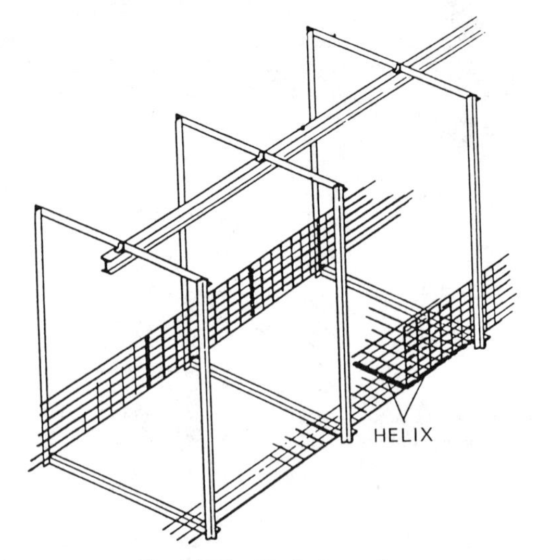

Fig. 9.5.21 Typical guard.

also used following paint or other dip tanks to catch drip from loads and, if possible, to allow it to drain back to the tanks. See Fig. 9.5.21 for a typical guard construction.

When a conveyor drops down to a work station, the guards follow the dips down from the high lines until the bottom clearance line of the guard framing is approximately 7 ft above the floor. This minimum height is in general use, as it allows walking headroom under the conveyor. In some plants, where this is not considered sufficiently safe, additional protection is furnished by installing handrails around the lower curves of these dips. Then if any load falls on an inclined guard, it will roll down into a guarded space on the floor. Handrails will also keep workers from walking into moving conveyor loads at the inclines. Finger guards are also used in some plants at low runs of conveyors. These cover the chain, trolleys, track, and turns to prevent hands or fingers from getting injured.

Cost. Safety guards are quite expensive and have been known to equal the cost of the balance of the conveyor. The guard structure and mesh should be designed only for protection and not for structural rigidity. It is much cheaper to replace, infrequently, a piece of guard damaged by a falling load than to build it strong enough to resist damage which will seldom occur, and then only in a

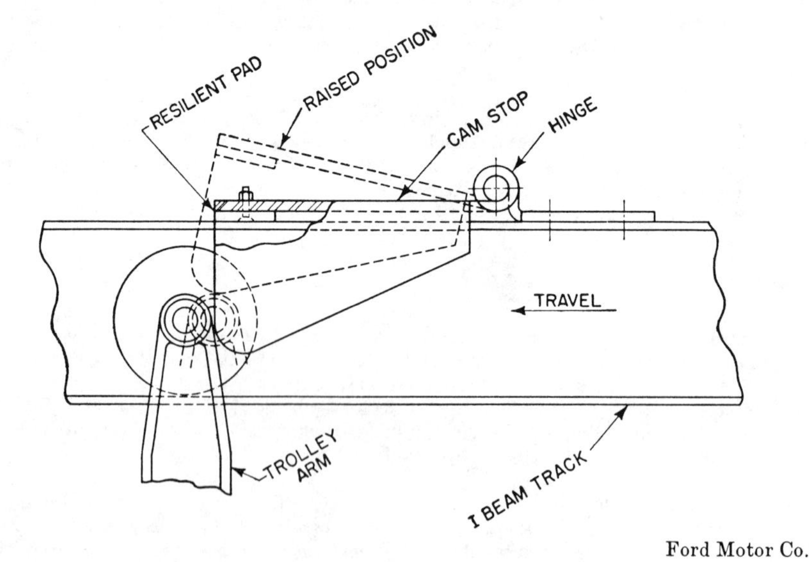

Ford Motor Co.

Fig. 9.5.22 Uphill safety stop.

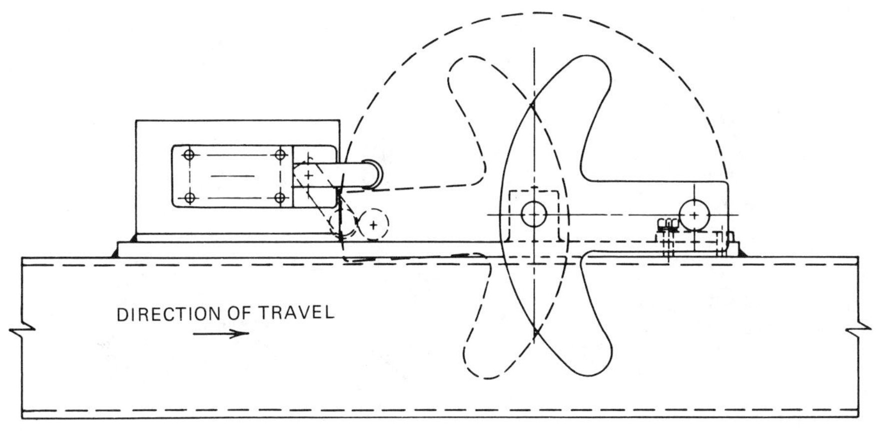

Fig. 9.5.23 Downhill safety stop.

few spots. It is advisable to make the construction as simple as possible and to see that guards are specified only where they are necessary.

Safety Stops

For conveyors with inclines, uphill and downhill stops are sometimes used to prevent a runaway of trolleys and loads if a chain breaks. An uphill safety stop may be seen in Fig. 9.5.22; Fig. 9.5.23 shows a downhill stop. To be effective, at least the downhill units should be equipped with limit switches wired into the drive circuitry.

Lubricators

Like all other types of conveyors, the life is related to the quality of maintenance. The life of trolley conveyors is greatly enhanced by regular periodic applications of oil to the chain pin bearing areas, the replenishing of the grease in sealed trolleys, or applications of oil to open trolley wheels. This maintenance can be done manually but the operation is cumbersome and, when done safely, often interferes with the use of the conveyor. Essential and heavily used conveyors should, therefore, be equipped with automatic lubricators, preferably installed so that they cycle on and off without human attention. These units are available through the conveyor manufacturer or from the numerous suppliers of lubrication equipment. Lubrication should be according to the conveyor manufacturer's recommendation or at least to proven plant standards. One type of trolley lubricator is shown in Fig. 9.5.24.

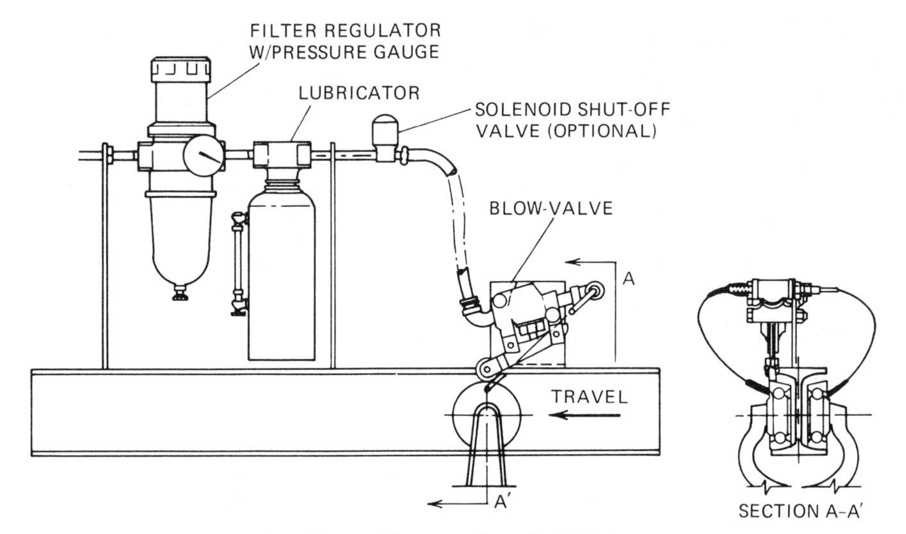

Fig. 9.5.24 Oil-type trolley wheel lubricator.

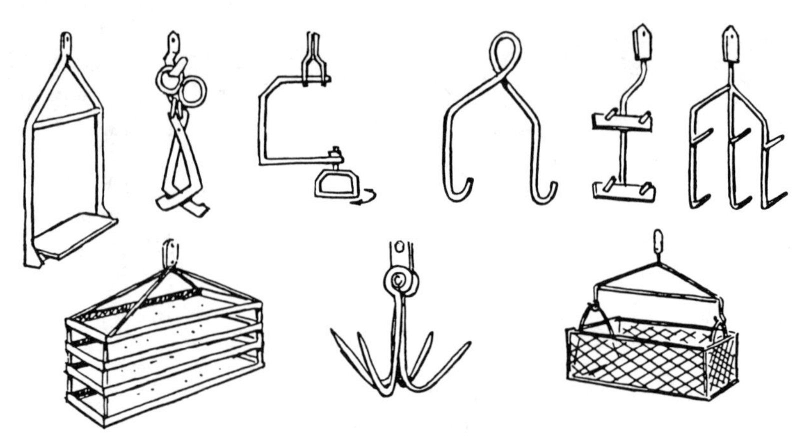

Fig. 9.5.25 Typical carriers.

Carriers

There is almost no limit on the ingenuity that has gone into the design of carriers. Every major supplier and user of trolley conveyor has devised hundreds. The essential design requirement, other than the ability to carry the load safely, is to keep the weight and size to a minimum. The carriers are a dead load that the conveyor has to move constantly. For manually loaded carriers, it is important to keep the size and design within the capabilities of human beings to load and unload. Automatically loaded carriers almost always require a guiding system to assure that the carrier is correctly positioned for loading. Figure 9.5.25 shows a few of the more common types of carriers.

9.5.3 Laying Out the Conveyor

Procedure

Laying out a trolley conveyor requires the use of the same plant engineering techniques as other conveyors. Good layouts, however, make full use of the great flexibility of the trolley conveyor. Some conveyors are easy to lay out. Others, because the route is long or irregular, require the exercise of considerable ingenuity. The usual practice is to make the conveyor fit the layout of buildings and machinery, rather than vice versa.

Typical Example

A manufacturing plant has the departmental layout shown in Fig. 9.5.26*A*. Raw materials must go from Receiving Storage to Departments 1, 2, 3, and 4 and must be further processed in Department 5. The output of Departments 5, 6, and 7 must go to Department 8 for inspection and packing, and then to Shipping Storage. The problem is to lay out a single conveyor to handle materials in the proper order.

STEP 1. From detailed department layouts, locate on the plan (Fig. 9.5.26) the points X where materials will be unloaded from the conveyor to enter each department, and the points O where they will be loaded back on the conveyor when they leave the department. Conveyor dips (vertical curves) will be required at each of these points. The direction of the conveyor line at the vertical curves must suit the department layout and is now marked at each point.

STEP 2. Join the points marked X in Departments 1, 2, 3, 4, 6, and 7. Also join the warehouse loading and unloading points and connect the point O in Department 8 (Fig. 9.5.26*B*).

STEP 3. Now join points O in Departments 1, 2, 3, and 4, and point X in Department 5. Continue to O in Departments 5, 6, and 7 and to X and then O in Department 8 (Fig. 9.5.26*C*).

STEP 4. Combine the layouts in *B* and *C* of Fig. 9.5.26, as in Fig. 9.5.26*D*. Join O in Department 1 with X in Department 4, and O in the receiving warehouse to X in Department 6, taking care to make one run high and the other one low where the conveyor lines cross.

If more than one path can be drawn, try all possible arrangements. Select the one that is the

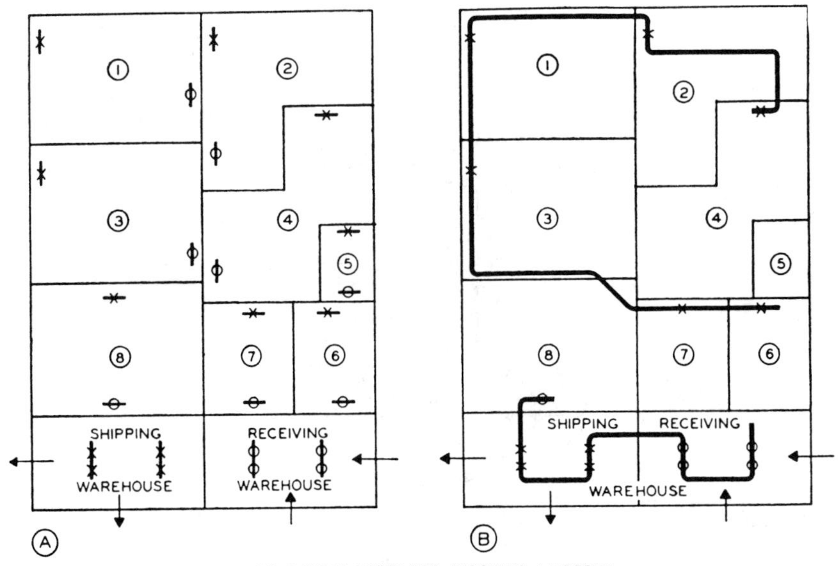

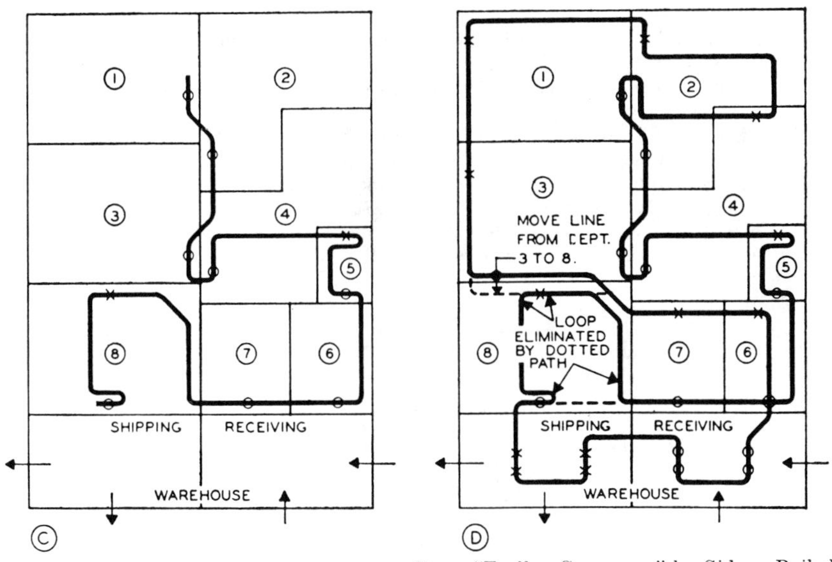

From "Trolley Conveyors" by Sidney Reibel

Fig. 9.5.26 Laying out the conveyor.

shortest and has the least number of turns, provided it is satisfactory from the operating viewpoint. Try to see whether backtracking can be eliminated. A slight change in conditions may improve the layout. For instance, if raw, processed, and semiprocessed materials can be loaded on the conveyor at the same time (either on the same or different carriers), and if the capacity is ample, a change can be made that will effect a considerable saving. In this example, the dotted path in Fig. 9.5.26*D* will eliminate the backtracking loops in Department 8.

Clearance Diagrams

The path of the conveyor must allow for external clearance, such as headroom over aisles, workers, and machines, and under trusses, beams, piping, ducts, wiring, and lights. Horizontally, the conveyor

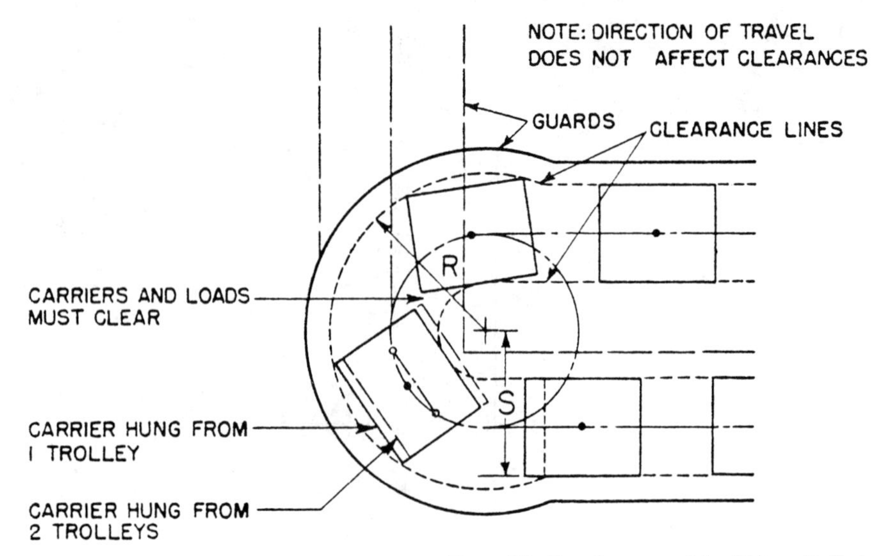

From "Trolley Conveyors" by Sidney Reibel

Fig. 9.5.27 Carrier and guard clearance at horizontal turns.

must clear the same obstructions, as well as columns, walls, and other items. Guards for high conveyor lines add to the clearance requirements. Internal clearance must be provided so that carriers, loads, guards, and chain do not foul or interfere with each other. Figure 9.5.27 shows the method to use in laying out clearances at vertical curves. To lay out the bottom clearance line, determine first points D and H at the top and bottom of the curves, where the loads or carriers start to change elevation. Then with radius $CD = AB$, and radius $GH = EF$, lay out arcs DJ and HK. Join tangent points J and K, and the clearance line $DJKH$ is established. (Note that radius EF is not equal to AB.)

Figure 9.5.28 shows the clearances that must be determined at horizontal turns. Clearance radius R is greater than dimension S. The clearance lines shown are for loads and carriers suspended from one trolley. When carriers are suspended from two trolleys, less clearance is needed at the outside,

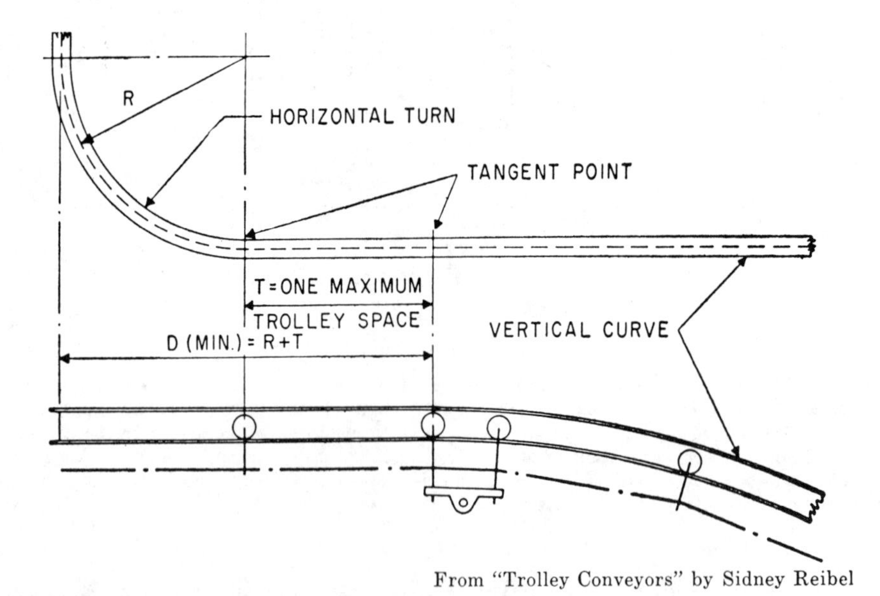

From "Trolley Conveyors" by Sidney Reibel

Fig. 9.5.28 Clearance between horizontal turn and vertical curve.

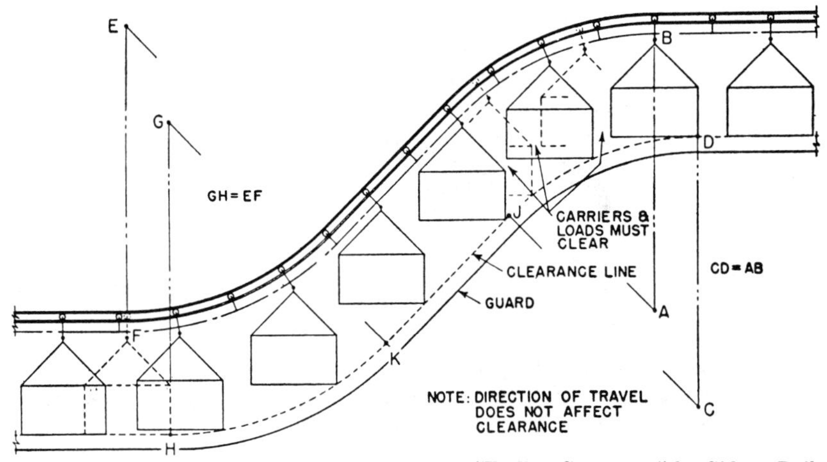

From "Trolley Conveyors" by Sidney Reibel

Fig. 9.5.29 Clearance on vertical curves.

but more is required on the inside curve, and the spacing between loads may have to be increased so that they will clear each other.

Figure 9.5.29 shows the clearance required between adjacent horizontal and vertical curves. The distance T between tangent points of the two curves should not be less than one trolley space. When trolley spacing is not uniform, use T equal to the largest trolley space. The space between horizontal and vertical curves prevents twisting of the chain, which would cause excessive wear on all the chain and trolleys, and on the track at this point.

9.5.4 Calculating Chain Pull

Basis of Calculations

Calculations of the chain pull of trolley conveyors are based on determining the losses due to friction of the moving conveyor parts and loads, and the forces necessary for elevating loads. The method of calculation is the same whether the parts are carried or towed. The calculation must also include any special load. This type of load occurs in washers, dip tanks, paint-spray and sand-blast cabinets, and places where the parts are revolved. The power requirements for such cases, except sealing plates, are small, and the excess capacity of drives will usually be enough to take care of the extra loads.

Moving conveyor parts include trolleys, chain, carriers, and suspended live loads. The rolling friction is due to the weight of these parts, which is carried by the trolley wheels. Additional friction is due to losses at horizontal turns and to imposed loads and bending losses at vertical curves. There is also the pull required when live loads, elevated at vertical curves, are not balanced by corresponding loads lowered at other vertical curves; such as when the conveyor is loaded going up and empty or partly empty going down. When loads are towed by mast attachments on trucks, the horizontal draw-bar pull at the trucks is added to the chain pull due to the weight of the conveyor chain, trolleys, and attachments.

Variable Factors. The factors that are the most difficult to determine are the amount of live load and the friction losses. The variations in loading to consider are:

1. Weight of loads:
   a. Maximum (for trolley capacity).
   b. Average (for chain pull calculation).
2. Percentage of loading of carriers:
   a. With maximum loads.
   b. With average loads.
   c. Partly loaded.
   d. Empty.

3. Distribution of loads on the conveyor when:
   a. Conveyor starts empty and is gradually loaded.
   b. Variations in loading occur at different sections of the conveyor during normal operation.
   c. The conveyor is unloaded for shutdown.
   d. Loads are not removed and are carried again around the circuit.

Friction Factors. These vary considerably at different times, even on the same conveyor. They can be affected by the following conditions:

1. Trolley construction:
   a. Smoothness and roundness of treads—whether or not the threads are concentric with the bearings.
   b. Construction of trolley wheel bearings—accuracy, finish, and fit of balls and races.
   c. Ratio of tread diameter to mean diameter of bearing.
   d. Accuracy of alignment of trolley wheels and brackets and proper suspension of loads—whether wheels run true and carry equal loads.
   e. Condition of trolley wheels—new, partly worn, badly worn.
2. Lubrication:
   a. Type and efficiency of lubricant.
   b. Whether and how long the lubricant remains in the trolley bearing and on the chain. (Conveyors running through washers, degreasers, ovens, etc., may operate under widely different conditions varying from well lubricated to dry.)
   c. Whether bearings are clean or dirty. (Contamination may be dust, dirt, water, etc., which enters from the outside, or burned lubricant inside a well-sealed bearing.)
3. Alignment and condition of the track:
   a. On straight runs.
   b. At vertical curves.
   c. At horizontal turns.
4. Conditions at roller and traction wheel turns, similar to those listed for trolley construction and lubricant.

The preceding list of variables indicates it is virtually impossible to make truly accurate calculations of chain pull for any trolley conveyor, no matter how good the design and layout, how well lubricated it is, how carefully loaded and operated, or how well maintained. The best way to handle this situation is to be sure that, under the worst possible conditions, the chain and trolleys will not be overloaded and the drive unit and motor will have ample capacity.

Approximate Method for Calculating Chain Pull

1. Add the total weight of conveyor chain, trolleys, carriers, and live loads. In most cases, live loads are figured for the entire length of the conveyor. Even when part of the conveyor is expected to be empty, if there is any possibility at all that it may be loaded, it should be figured 100% loaded. Sometimes, part of the conveyor, or all of the carriers, definitely will not be loaded. This is taken into account when figuring the loading.

Multiply the total moving weight by:

For $2\frac{5}{8}$-in. I-228 Chain Conveyor	4.00%
For 3-in. I-348 Chain Conveyor	3.00%
For 4-in. I-458 Chain Conveyor	2.50%
For 6-in. I-678 Chain Conveyor	2.25%

When operating conditions are good, and there are only a few turns and dips, as in a monoplane conveyor, slightly lower factors can be used. When operating conditions are bad (accumulated dirt, poor lubrication, etc.), or when there are a great number of turns and dips, the factors should be increased.

2. The chain pull due to change in elevation is ignored for ordinary dips on the same floor. For considerable changes in elevation (over 8 ft or from one floor to another) and where loads are elevated without balancing descending loads, the chain pull equals the maximum live load per foot of conveyor multiplied by the difference in elevations (expressed in feet).

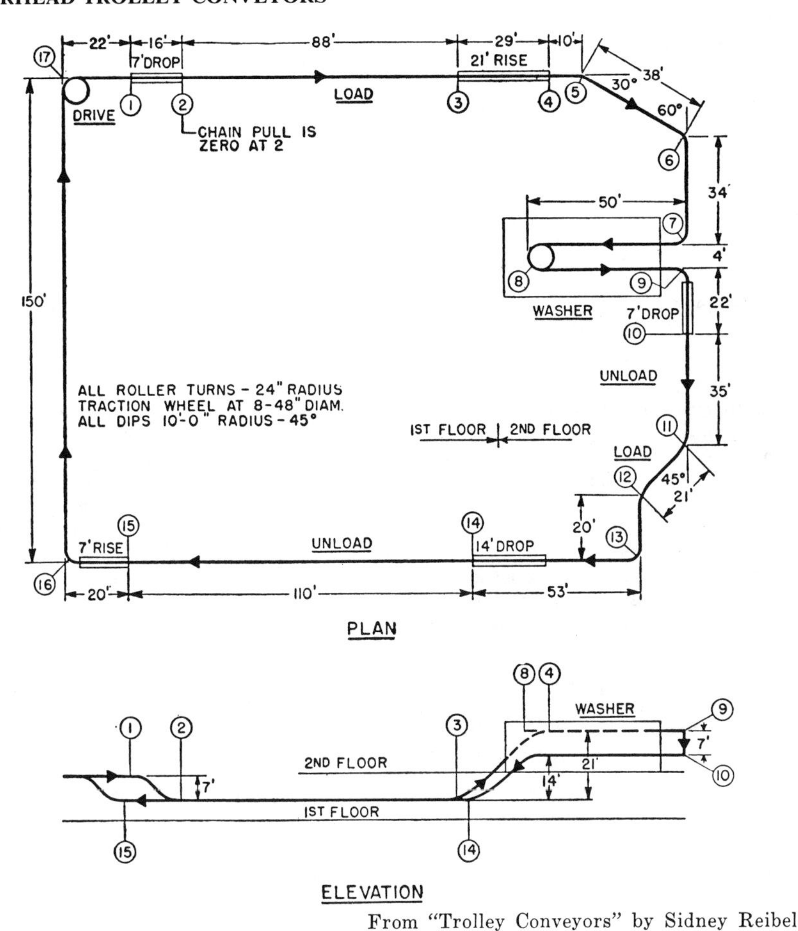

From "Trolley Conveyors" by Sidney Reibel

Fig. 9.5.30 Typical conveyor, for chain pull calculations.

3. Add the values of 1 and 2 to get the maximum speed (ft/min).
4. Find the motor horsepower using the following equation.

$$\text{Motor Hp} = \frac{\text{Chain pull (lb)} \times \text{Maximum speed (ft/min)}}{33{,}000 \times \text{Drive efficiency}}$$

For example, a conveyor follows the path shown in Fig. 9.5.30, and has carriers spaced at 6-ft centers. The maximum load on the carriers is 200 lb, but the average load is 135 lb. Not more than 80% of the carriers will be loaded. The carriers weigh 132 lb. Chain is × 458, track is 4-in. I, trolleys are ball bearing and spaced 24 in. center to center. Conveyor speed is 10–30 ft/min. The chain and trolleys are well protected from the sprays in the washer. Determine the chain pull and the sizes of the drive unit and motor.

The data are as follows:

Total length of conveyor: 772 ft (approx.)
Maximum lift: 21 ft
Weight of chain (772 × 3): 2,300 lb
Weight of trolleys (772 ÷ 2 × 8): 3,100 lb
Weight of carriers (772 ÷ 6 × 132): 17,000 lb
Weight of loads (772 ÷ 6 × 135 × 0.80): 13,900 lb
Total moving load: 36,300 lb

Calculations as follows:

Chain pull from rolling friction (0.025 × 36,300): 910 lb

Chain pull from change in elevation (200 ÷ 6 × 21): 700 lb

Total chain pull: 1,610 lb

Use a 2000-lb drive unit.

Assuming a drive efficiency of 75%:

$$\text{Motor Hp} = \frac{1{,}610 \times 30}{33{,}000 \times 0.75} = 1.95$$

Use a 2 Hp motor.

Progressive Method

Quite often, the general method of calculating chain pull is not accurate enough. This is usually the case with long, heavily loaded conveyors that have high lifts and a great many horizontal turns and changes in elevation. Calculations must then be made progressively from one point to the next, adding percentages by which the chain pull is increased at horizontal turns and vertical curves. An analysis can also be made by drawing a profile of the conveyor, which is very helpful in locating drives. Major trolley conveyor manufacturers have developed computer programs for investigating the varying loadings on conveyors. Anyone contemplating the purchase of a complex multidrive conveyor should utilize their services. Given enough time, anyone can duplicate the computerized results by longhand methods. However, if a supplier is given the loading conditions, they will usually verify the drive(s) size and location at no charge. The computerized calculation is a refinement of the following example of progressive calculations.

Establishing the Factors. With this type of calculation, friction or loading factors are established for each element of the conveyor. The manufacturer will supply recommended factors for his equipment. To facilitate calculations, constants are also calculated for each type of loading. For example:

Weight of carrier (132 ÷ 6): 22 lb/ft

Weight of trolleys (8 ÷ 2): 4 lb/ft

Weight of chain: 3 lb/ft

Conveyor weight (carrier empty): 29 lb/ft

Average carrier load [(135/6) × 0.80]: 18 lb/ft

Average loaded conveyor weight: 47 lb/ft

Average half-loaded conveyor weight: 38 lb/ft

In addition, on the assumption that the conveyor is in average condition and is lubricated, a friction factor of 1.5% would be adequate for straight runs of chain. This results in the following constants:

Unloaded chain pull (0.15 × 29): 0.5 lb/ft

Half-loaded chain pull (0.15 × 38): 0.6 lb/ft

Loaded chain pull (0.15 × 47): 0.7 lb/ft

For the other conveyor elements the following would be typical percentages to add.

Vertical bends 2%

Corner drive 5%

24-in. radius 30° roller turn 1.0%

24-in. radius 45° roller turn 1.5%

24-in. radius 60° roller turn 2.0%

24-in. radius 90° roller turn 3.0%

48-in. diameter traction wheel, bronze bushed 4%

Performing the Calculation. Using the foregoing factors, progressive calculation is made in chart form as follows:

Conveyor Section	Horizontal Pull	Rise or Drop		Factor for Turn or Curve		Chain Pull (lb)
2–3	0 + (88 × 0.6)					53
3–4	53 + (29 × 0.7)	+(21 × 47)	=	1060 × 1.02	=	1081
4–5	1081 + (10 × 0.7)		=	1088 × 1.01	=	1099
5–6	1099 + (38 × 0.7)		=	1126 × 1.02	=	1149
6–7	1149 + (34 × 0.7)		=	1173 × 1.03	=	1208
7–8	1208 + (52 × 0.7)		=	1244 × 1.04	=	1294
8–9	1294 + (52 × 0.7)		=	1330 × 1.03	=	1370
9–10	1370 + (22 × 0.7)	−(7 × 47)	=	1056 × 1.02	=	1077
10–11	1077 + (35 × 0.6)		=	1098 × 1.015	=	1153
11–12	1153 + (21 × 0.6)		=	1166 × 1.015	=	1184
12–13	1184 + (20 × 0.7)		=	1198 × 1.03	=	1234
13–14	1234 + (53 × 0.7)	−(14 × 47)	=	613 × 1.02	=	625
14–15	625 + (110 × 0.6)		=	691 × 1.0	=	691
15–16	691 + (20 × 0.6)	−(7 × 38)	=	969 × 1.05	=	1010*
16–17	1010 + (150 × 0.6)		=	1100 × 1.05	=	1155

On the other side of the drive corner, there will be a negative pull:

17–1–2	0 + (38 × 0.6)	−(7 × 29)	=	180 ÷ 1.02	=	177

Net pull at drive = 1155 − 184 = 971 lb.

In the foregoing calculation it was assumed that the declines were favorably loaded. If the reverse were true, then the effort at the drive would have increased 400–500 lb, making the drive effort comparable to the result obtained with the approximate method. This illustrates the need for thorough examination of the loading conditions of long and/or complicated conveyors and for the conservative application of drives.

Operating conditions determine the size of the drive to use. If the worst condition will last only 15–20 min, and does not occur too often, a drive should tolerate overload if the motor is large enough. Drives are then selected on the basis of the chain pull resulting from normal operating conditions. However, such decisions should be made only when reliable operating information is available.

It is good practice to be liberal in the selection of a drive unit after determining chain pull. Quite often, conditions change, or the conveyor may be lengthened, increasing the original chain pull. Some users have minimum requirements of 2000 lb for the chain pull capacity of a drive unit and $1\frac{1}{2}$ Hp for the size of a motor. They have found that oversize drives are insurance against loss of production due to breakdown, or costly drive changes when a conveyor is altered. Also, when conveyors are taken down to be used elsewhere, the salvage possibilities of the larger-capacity drive units are much better.

The chain pull at the drive unit may not be the maximum. This is especially the case where the difference in elevations is large. The maximum chain pull in Fig. 9.5.26 is more than the net pull at the drive. The difference in pull on some conveyors may be much greater. It is not necessary to determine the maximum chain pull, except when it may equal or exceed the maximum allowable values for the chain and trolleys.

Profile Diagram. Heavily loaded conveyors with very high lifts or drops can be analyzed by making a profile diagram. Vertical lines can be used to indicate each horizontal turn and other symbols to identify special features. A similar plot can be made of the chain pull at each location. Analysis of the two usually indicates approximate locations of drives and take-ups, making it possible to resolve relatively complex problems.

9.6 POWER-AND-FREE CONVEYORS

Byrl Curry

The term "power and free" describes a conveying system that is comprised of unpowered or "free," trolley-supported carriers which move along a track usually made up of double channel, mounted

* 1.02 for base × 1.03 for 90° turn = 1.05.

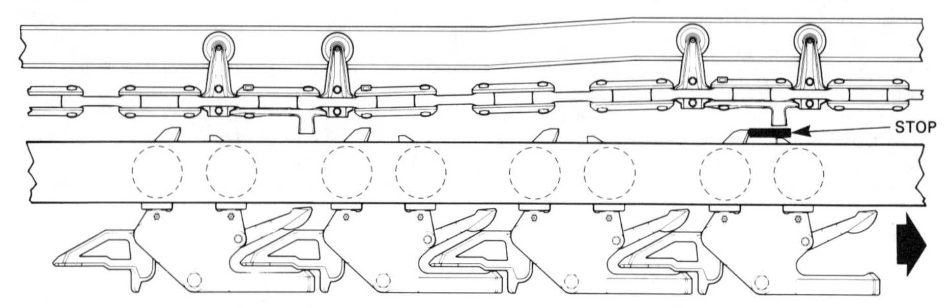

Fig. 9.6.1 Power-and-free trolleys queued at a stop.

below or adjacent to an I-beam track from which a powered conveyor chain is suspended. Pusher dogs on the chain contact the free trolleys and propel them along the free track.

Power-and-free systems convey unit loads of the widest range of shapes and sizes with weights from fractions of a pound to many tons.

9.6.1 Applications

The power-and-free conveyor can meet a wider range of materials handling applications than any other single conveying method. Although it is best known for its use in large, complex manufacturing facilities, there are many simpler applications that utilize the unique functions made possible by its design.

Power-and-free conveying should be considered for the following applications:

Buffer Banks. In any operation where an individual work station must be kept supplied with product for processing, the power-and-free conveyor is an appropriate choice. A stop mechanism mounted on the conveyor track will halt load-bearing carriers at the work station. The accumulating feature of the free trolley allows carriers to queue at the work station without interrupting carrier movement on other sections of the conveyor (Fig. 9.6.1). An operator-controlled push button can be used to release the carrier from the stop when work has been completed. All carriers in the bank will automatically advance to replace the released load.

Surge between Processes. Power-and-free conveyors are used when there is a need to accommodate differences in production rates among varying manufacturing processes. Surges between the processes can be accommodated to maintain a steady throughput.

On-Line Storage and Segregation. When there is a need to hold work-in-process or completed units in temporary storage awaiting shipment or further production needs, the power-and-free conveyor should be considered. Carriers can be switched from the main line of traffic flow into a recirculating storage loop (Fig. 9.6.2) or into dedicated spurs for batched storage (Fig. 9.6.3). Carriers may be released from storage in a random mix or batched in any appropriate order.

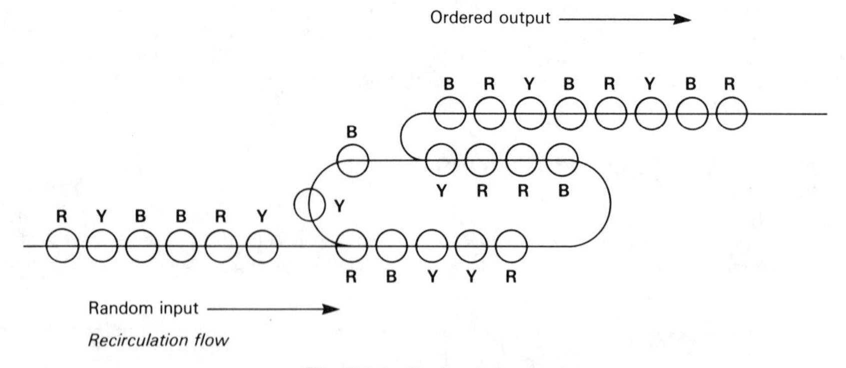

Fig. 9.6.2 Recirculation.

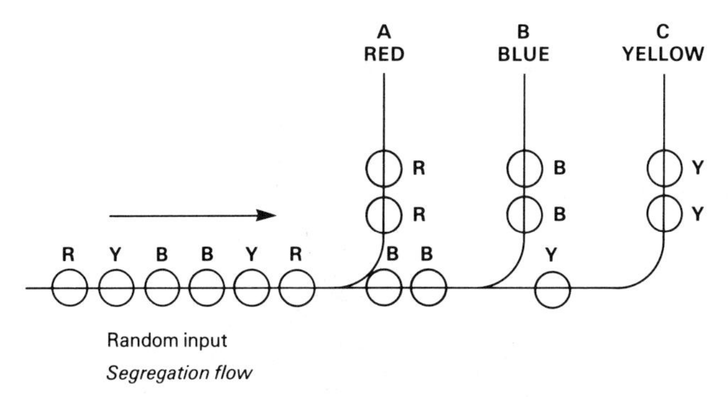

Fig. 9.6.3 Segregation.

Varying Load Centers. Different spacing between carriers is often a requirement and one that can be easily provided by the power-and-free conveyor. For example, parts should be well separated as they go through the paint spray to permit room for painting. Later, the parts should be brought close together as they go through the bake oven for reasons of energy conservation and economy. The power-and-free design makes it possible to switch the carriers to a power chain with the appropriate wide dog spacings as they enter the paint booth. Chain speed can also be changed to suit the process requirements. By transferring the carriers to a second chain with closer dog spacing, the needs of the bake oven can be met.

When loads of varying length are to be conveyed with fixed or minimum gap between loads, the chain is provided with pusher dogs on close centers. The load length is measured and following carriers put a minimum distance or fixed distance from the rear of the lead carrier (Fig. 9.6.4).

Differing Load Speeds. The difference in speed requirements shown in the example of the paint booth followed by a bake oven represents another production variable easily handled through the use of the power-and-free conveyor. If the number of units per hour that must be moved past production stations varies in different parts of the plant, this too can be accomplished with the power-and-free system. Those sections of the system that require greater production rates are serviced by power chains operating at higher speeds. The transfer between chains of different speeds is accomplished with established transfer techniques that will be discussed later.

Changes in Elevation. The power-and-free conveyor lends itself readily to changes in elevation. When transporting loads over distances, from one work location to another, the ability of the power-and-free conveyor to travel up sloped track to levels well above the work areas frees valuable floor space for productive activity. Loads can also be lifted or lowered directly from one level to another by means of a lift (Fig. 9.6.5). A common application is to lower a carrier so that the load is immersed in a paint bath, or some other similar process, then returned to the track level for further travel.

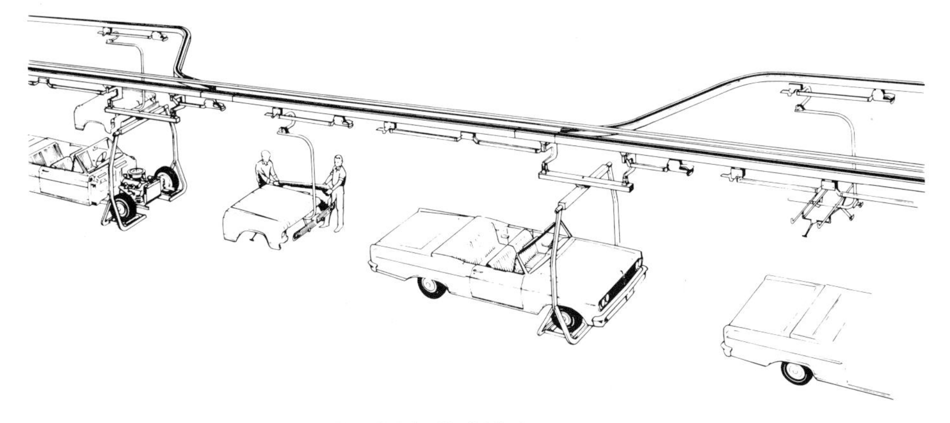

Fig. 9.6.4 Variable load centers.

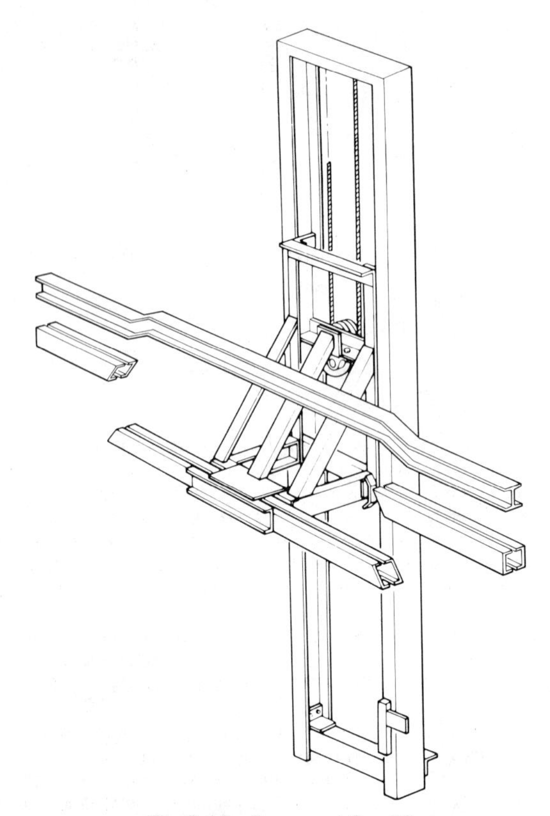
Fig. 9.6.5 Power-and-free lift.

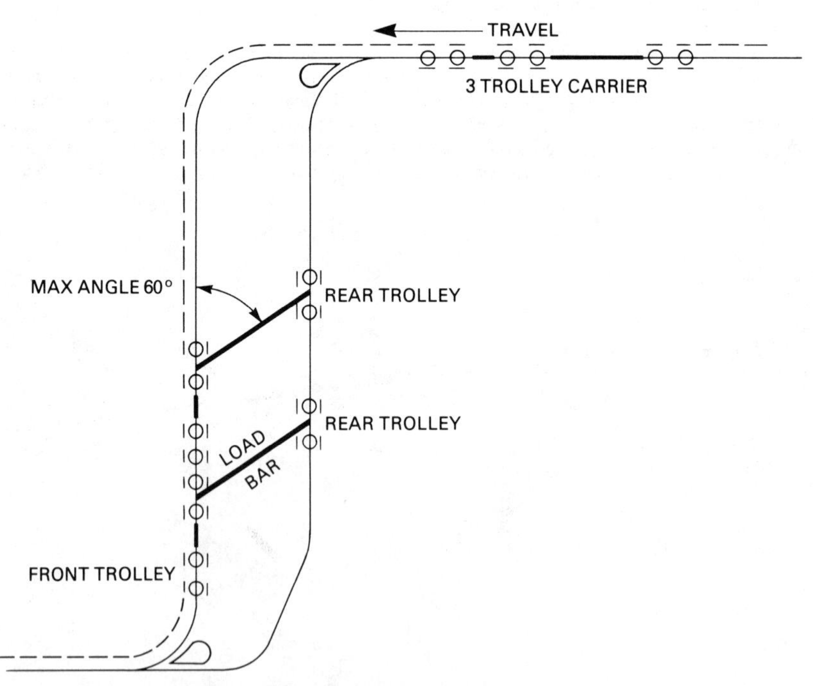

Fig. 9.6.6 Typical bias banking layout.

Fig. 9.6.7 Bias banking carrier trolley configuration.

Another application moves the carrier with its trolley from one track level to another. Proper lift section design includes:

Assurance that the moveable and stationary track ends are properly aligned.

Mechanical end stops to close off open track ends. These are in addition to the functional automatic control stops.

Safety circuits for emergency stopping of the lift and transferring conveyor if there is a malfunction during the carrier interchange cycle.

Bias Banking. When you wish to bank long loads, it is often desirable to reduce the track length used. This can be accomplished by bias banking. To bank on a bias, the track is split, the front trolley taking one track, the trailing trolley the second (Fig. 9.6.6).

To provide pusher disengagement from the chain when banking either on straight track or bias track, an actuating cam is provided on both the center and rear trolley of a three-trolley load bar (Fig. 9.6.7).

The angle between the bias-banked load bar and the line of travel of the rear trolley should not exceed 60°.

Indexing. The intermittent motion to index a carrier through a washer, a shot blast, or through automatic load or unload stations can be provided by:

A separately driven indexing chain

A cylinder/pusher beam, air or hydraulically driven

A conventional stop-to-stop control with a continuously driven chain

9.6.2 Conveyor Configurations

Track Cross Section, Vertical

The track cross section of a typical power-and-free conveyor consists of an I-beam or power rail over a double-channel free track (Fig. 9.6.8). Yokes welded to the tracks at regular intervals provide structural integrity and vertical spacing. The power chain provides the driving force for the load trolley. It is supported by a trolley which rides on the lower flanges of the I-beam. See Fig. 9.6.8 and Table 9.6.1 for typical cross-sectional dimensions.

Because 3-in. and 4-in. I-beam power track and trolleys have adequate strength to handle loads requiring 3-, 4-, or 6-in. chain, it is common to interchange chain size, power track and trolleys, and free track. Table 9.6.2 shows typical combinations for open and closed track systems.

Track Cross Section, Light to Medium Duty

Power-and-free systems whose load requirements do not exceed 250 lb use a cross section commonly called enclosed track. In this configuration (Fig. 9.6.9), both the power and free tracks are constructed of roll-formed steel. The two track sections are mated by connecting yokes for stability and proper spacing. In these assemblies, the power chain is constructed of some form of universal link chain with the driving power transmitted to the free trolley by pusher dogs that form an integral part of the chain. A significant number of light- to medium-duty installations utilize a side-by-side track section (Fig. 9.6.10).

Track Cross Section, Three Rail

In certain applications of power and free, it is important to keep the load level through changes of elevation or to provide added load stability. The three-rail system is used for this purpose. These systems use a conventional vertical cross section (I-beam over channel) with additional free rails running parallel, but at a different elevation.

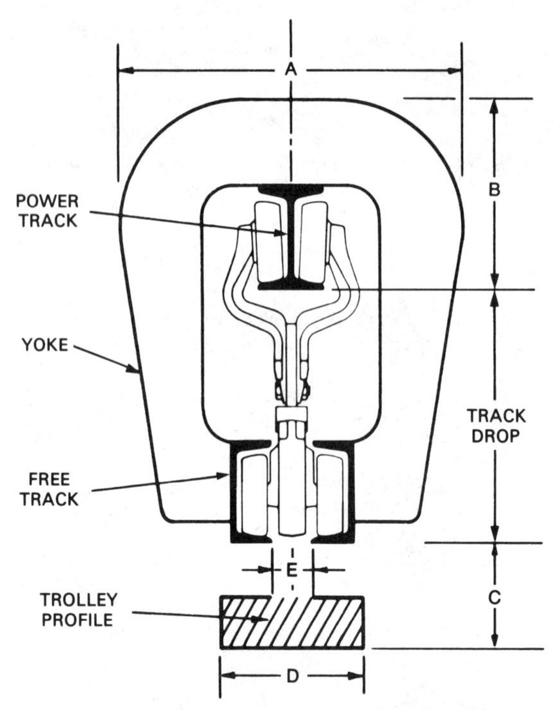

Fig. 9.6.8 Typical power-and-free cross section.

9.6.3 Components

Power Trolleys

Power trolleys for medium- to heavy-duty applications consist of paired half-trolley assemblies. A half-trolley consists of one trolley wheel mounted on a drop-forged bracket. After the paired brackets and wheels are inserted into the chain link, an "I" attachment is placed between the brackets and the assembly is bolted together (Fig. 9.6.11). The trolley wheels are cambered to match the angle of the power rail flange on which the wheel rides. In medium- to light-duty applications, the power wheels are incorporated into the universal link chain design.

Table 9.6.1 Cross-Sectional Dimensions of Power-and-Free Conveyor Track

Free Track (in.)	Power Track (in.)	Power Chain	Normal Track Drop (in.)	Dimensions (in.) from Fig. 9.6.8				
				A	B	C	D	E
3	3	348	$9\frac{1}{8}$	$14\frac{1}{2}$	6	$4\frac{5}{8}$	$5\frac{1}{2}$	$1\frac{3}{4}$
	4	458	$10\frac{1}{2}$	$14\frac{1}{2}$	7	$4\frac{5}{8}$	$5\frac{1}{2}$	$1\frac{3}{4}$
4	3	348	$9\frac{1}{2}$	$15\frac{1}{2}$	$6\frac{1}{4}$	8	7	$1\frac{3}{4}$
	4	458	$10\frac{7}{8}$	$15\frac{1}{2}$	$7\frac{1}{4}$	8	7	$1\frac{3}{4}$
	4	678	$11\frac{5}{8}$	$15\frac{1}{2}$	$7\frac{1}{4}$	8	7	$1\frac{3}{4}$
6	4	458	13	19	9	$8\frac{3}{4}$	7	$3\frac{1}{16}$
	4	678	$13\frac{3}{4}$	19	9	$8\frac{3}{4}$	7	$3\frac{1}{16}$

Table 9.6.2 Track/Chain Combinations

Power Track	Chain	Free Track
Enclosed	Enclosed track type	Formed channel
Enclosed	Enclosed track type	3 in.
Enclosed	Enclosed track type	4 in.
3 in.	348	3 in.
3 in.	348	4 in.
4 in.	458	3 in.
4 in.	458	4 in.
4 in.	458	6 in.
4 in.	678	4 in.
4 in.	678	6 in.

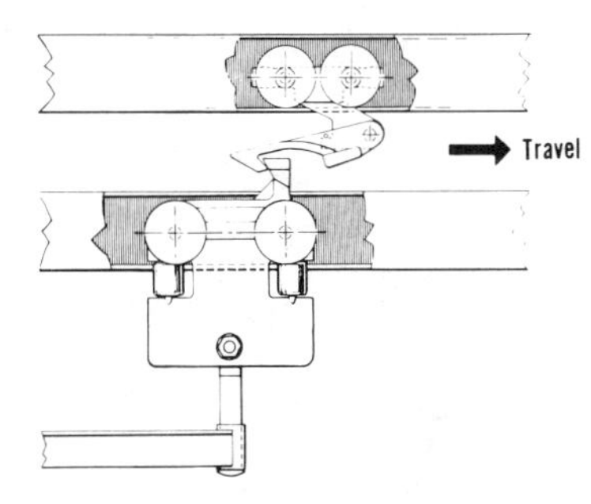

Fig. 9.6.9 Enclosed track power and free—vertical configuration.

Free (or Load) Trolleys

The power-and-free concept normally uses multiple free trolleys per carrier (Fig. 9.6.12), although there are many installations in which the load is suspended from a single trolley. The front, rear (and, if used, intermediate) trolleys travel inside the double-channel free track on four wheels with horizontal guide rollers. The front, or lead, trolley is normally equipped with a pusher and a hold-back dog that engages and entraps the chain pusher for positive carrier control.

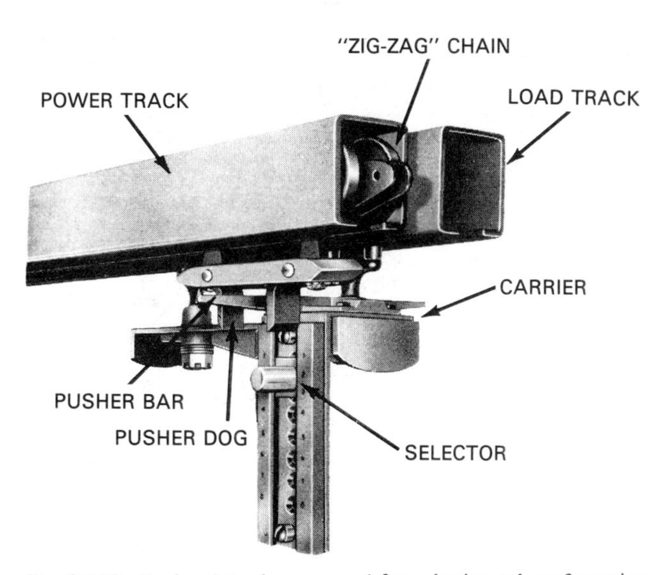

Fig. 9.6.10 Enclosed track power and free—horizontal configuration.

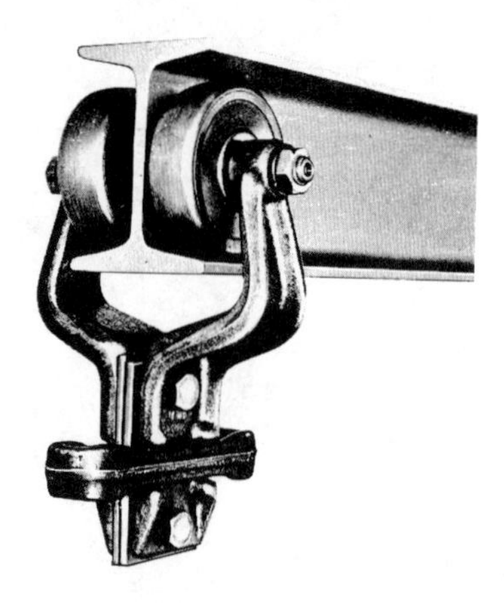

Fig. 9.6.11 Power trolley assembly.

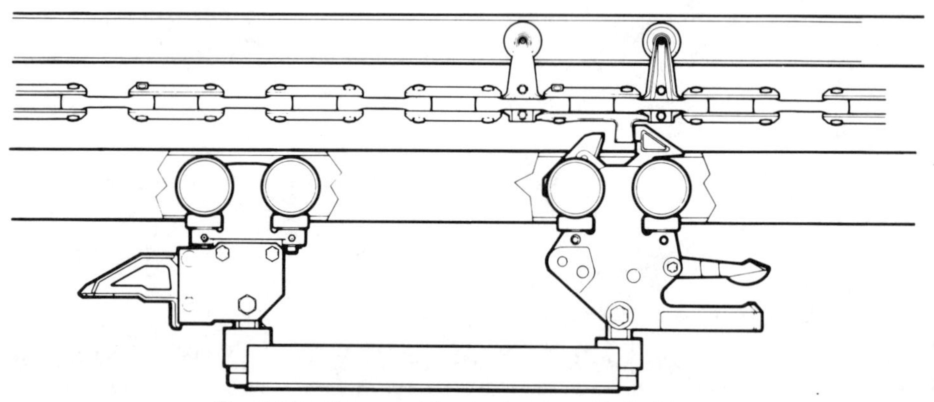

Fig. 9.6.12 Power-and-free load trolley with load bar.

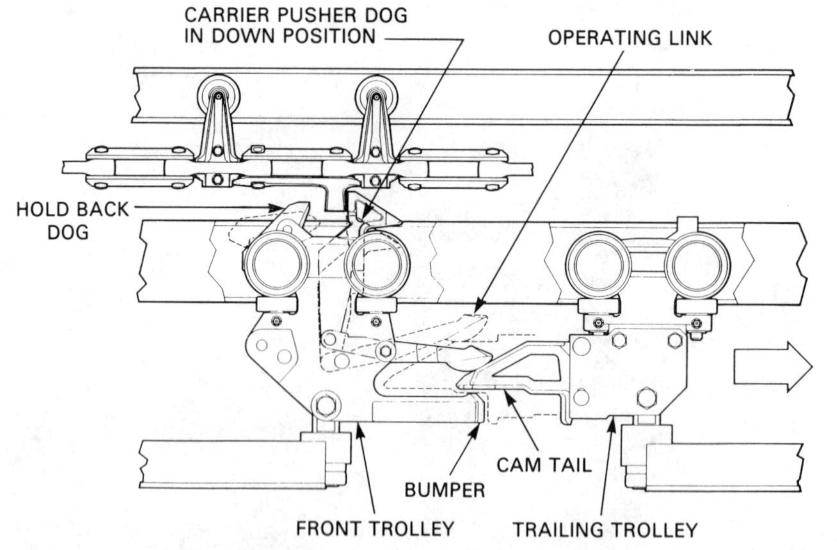

Fig. 9.6.13 Queued power-and-free carriers (detail).

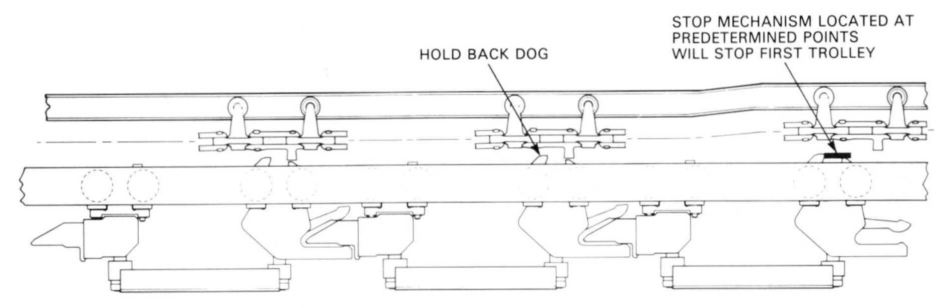

Fig. 9.6.14 Power-and-free accumulation.

The trailing trolley of a carrier has a rear cam. This raises the operating link of a succeeding trolley to disengage the carrier pusher dog from the power chain. In Fig. 9.6.13 the front trolley operating link is raised over the cam tail of the trolley in front of it. This lowers the pusher dog. When this happens, the carrier accumulates, or queues, on the cam tail of the preceding carrier while the power chain continues to move. In a similar manner, following carriers disengage from the power chain and accumulate behind the carriers ahead of them (Fig. 9.6.14).

Trolleys of multiple-trolley carriers are connected by an articulating load bar. In the single trolley assembly, the operating link, rear cam, hold-back, and pusher dog form one complete assembly.

Load trolleys are available for enclosed track, 3-in., 4-in., and 6-in. channels. Tables 9.6.3 and 9.6.4 summarize their characteristics.

Power Chain

Rivetless chain for open-track power-and-free conveyors is available in three common sizes—3 in., 4 in., and 6 in. pitch. The purpose of the chain is to provide the motive force needed to move the load through the system. It must be able to withstand continuous wear and tear of daily use and not be damaged under jam conditions that subject it to tensions far in excess of normal. Chain design for enclosed track systems is unique to each manufacturer and is rated by each. One typical design has alternating pairs of vertical support wheels and a single lateral guide wheel. Universal joints at each connection provide needed flexibility (Fig. 9.6.15).

Rivetless chain has three basic parts: center link, side links, and chain pins (Fig. 9.6.16). It has four basic ratings, that is, light-, normal-, heavy-duty, and ultimate tension capability. Table 9.6.5 summarizes these characteristics by chain size.

Chain Pull Calculations

Calculation of chain pull is the determining of the losses due to friction of the moving conveyor parts and loads, and the forces necessary for elevating loads. Moving conveyor parts include trolleys, chain, carriers, and suspended loads. Rolling friction loss is due to the weight carried on the trolleys. Additional friction losses result from horizontal turns and chain-articulating vertical curves.

Table 9.6.3 Load Ratings and Operating Characteristics for Load Trolleys[a]

Size (in.)	Nominal Load Rating per Four-Wheel Trolley (lb)	Minimum Radius Accumulation (in.)	Minimum Accumulation Length Single Trolley (in.)	Maximum Accumulation Length Single Trolley (in.)
3	500	15	12	18
4	1000	18	$17\frac{3}{4}$	20
6	5000	24	$18\frac{3}{8}$	22

[a] These values are for estimating purposes only. Actual values may be higher or lower, depending on application and/or manufacturer.

Table 9.6.4 Typical Free-and-Power System Capacities

System Size (Free Track)	Service Class	Trolley System[a]	Allowable Carrier Load (lb) Level System	Vertical Curves 15°	30°
Enclosed track	Light	2	400	350	300
	Light	3	400	400	400
	Normal	2	300	250	200
	Normal	3	300	300	300
	Heavy	2	250	225	200
	Heavy	3	250	250	250
3″	Light	2	1,500	1,200	1,000
	Light	3	1,500	1,500	1,500
	Normal	2	1,200	950	800
	Normal	3	1,200	1,200	1,200
	Heavy	2	1,000	800	700
	Heavy	3	1,000	1,000	1,000
4″	Light	2	2,500	2,000	1,650
	Light	3	2,500	2,500	2,500
	Normal	2	2,000	1,600	1,350
	Normal	3	2,000	2,000	2,000
	Heavy	2	1,500	1,200	1,000
	Heavy	3	1,500	1,500	1,500
6″	Light	2	15,000	12,500	10,000
	Light	3	15,000	15,000	15,000
	Normal	2	10,000	9,000	8,000
	Normal	3	10,000	10,000	10,000
	Heavy	2	8,000	7,000	6,000
	Heavy	3	8,000	8,000	8,000

[a] Two-trolley load bar system with load equally distributed on each trolley. On three-trolley load bar system the load is equally divided between the two rear trolleys, no load on the front trolley.

Chain Pushers

Chain pushers are available in varying designs and constructions. The one shown in Fig. 9.6.17 is a forging that combines the pusher, one side link, and one chain pin. The side link is slotted at the rear to accept a standard chain pin. The pusher is immediately below the rigid chain pin. This assembly, when supported between two adjacent power trolleys, is stable, minimizes objectionable reactions and internal stresses, and provides long wear life.

Drives

There are no general standards for conveyor drives except those which individual manufacturers specify. The most common drive used in power-and-free systems is the caterpillar drive (Fig. 9.6.18). This type of drive uses a short loop of auxiliary drive chain alongside and parallel to the conveyor chain

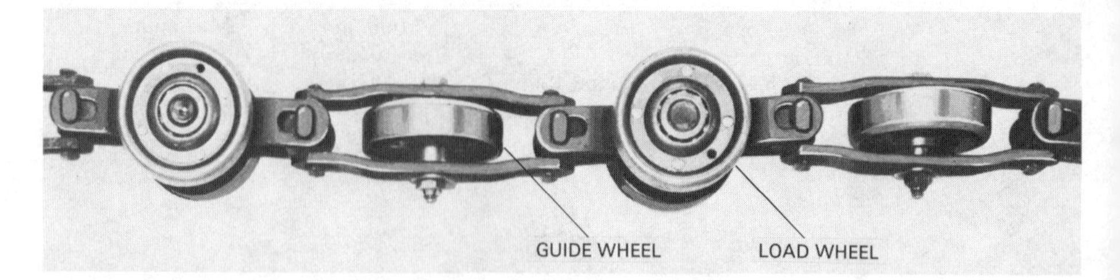

Fig. 9.6.15 Enclosed track chain.

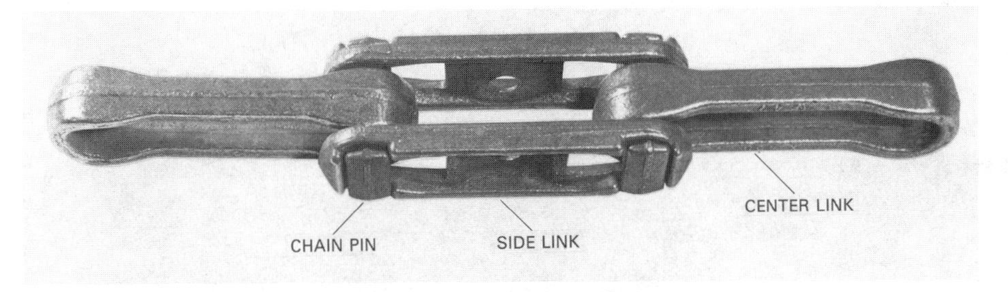

Fig. 9.6.16 Rivetless chain.

Table 9.6.5 Rivetless Chain Characteristics[a]

Chain Size	Pitch	Tension Rating/Duty, lb Light	Normal	Heavy	Ultimate Strength, lb
Enclosed track	4″	700	600	450	10,000
348	3″	2000	1500	1200	24,000
458	4″	3300	2500	2000	48,000
678	6″	6000	5000	4000	85,000

[a] These values are for estimating purposes only. Actual values may be higher or lower, depending on application and/or manufacturer.

(Fig. 9.6.19). A back-up bar on the caterpillar side and rollers on the driven-chain side confine the conveyor and drive chains. Drives are usually located on accessible, straight, level sections of conveyor track.

Drive Speed. Drives for transportation and accumulation are usually constant speed. Production rate or synchronized power-and-free conveyors may use variable-speed drives, such as direct current,

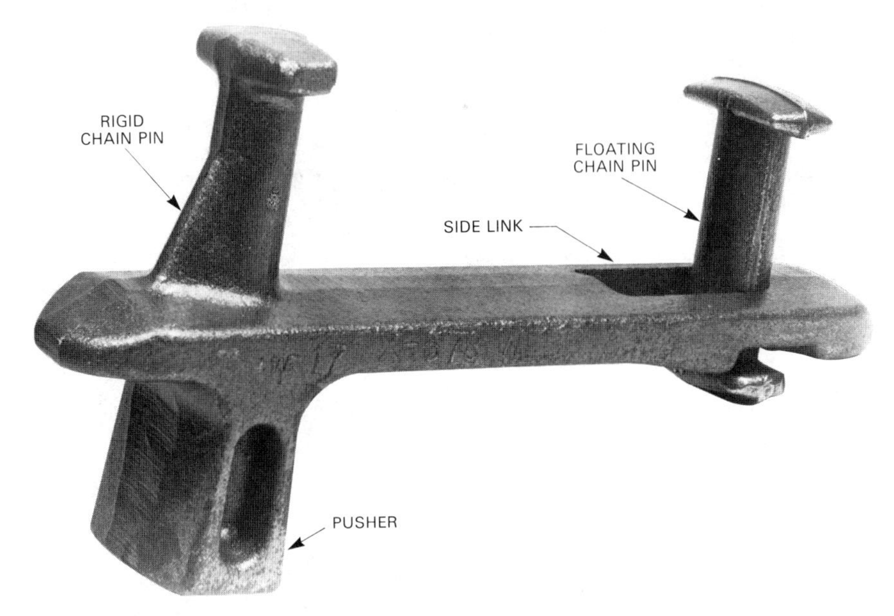

Fig. 9.6.17 Chain pusher.

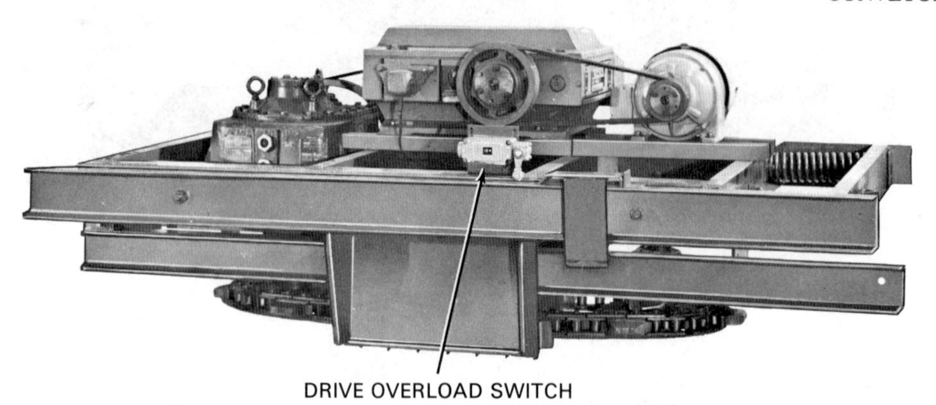

Fig. 9.6.18 Caterpillar drive.

variable frequency, or eddy current couplings. If the conveyor is multiplane, the drive should be provided with a brake to prevent unbalanced vertical curve loads from causing the conveyor to drift or coast after the drive is de-energized.

Drive Horsepower. Drive motor horsepower is calculated as follows:

$$\text{Motor horsepower} = \frac{\text{Chain pull (lb)} \times \text{Speed (ft/min)}}{33{,}000 \times \text{Drive efficiency (decimal)}}$$

Chain pull is calculated as previously discussed. Speed is as required. The 33,000 constant is the number of foot-pounds per minute in one horsepower.

Drive Overload Protection. The most common form of overload protection is the mounting of the drive on a floating frame. The drive is held in operating position by springs. If the force transferred from the floating frame to the conveyor chain exceeds the spring force, the frame moves. The movement compresses the restraining springs. If the motion continues to a predetermined position, a limit switch is operated, shutting down the drive (see Fig. 9.6.18).

Other forms of overload detection include torque hub cutouts and electrical drive overcurrent detection. Shear pin protection is not recommended.

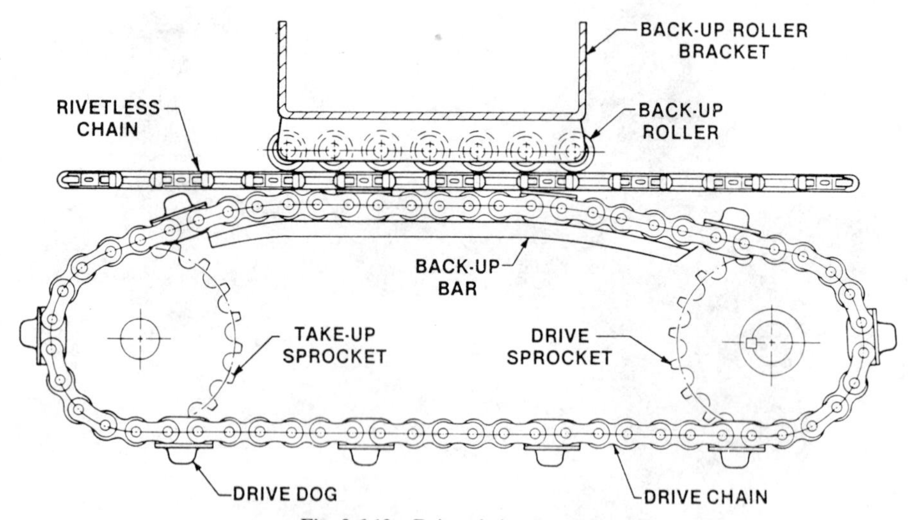

Fig. 9.6.19 Drive chain assembly.

Take-ups

Chain conveyors, whether power and free or monorail, open or enclosed track, normally use a take-up device to remove slack chain. There are five types of take-ups in common use:

Screw
Spring
Air cylinder
Counterweight
Air/hydraulic

These can be supplied with either roller turns or traction wheels (Fig. 9.6.20). *Screw take-up* frames are made to roll or slide in a fixed frame. The tracks have expansion-type sliding joints. Adjustment is made by the use of a long threaded rod. *Spring take-ups* are of the same construction as a screw take-up, except that springs automatically adjust for any slack chain by maintaining pressure on the floating frame. *Counterweight* and *air cylinder take-ups* are similar to the spring take-up, except that they maintain a constant, preset tension on the chain. A typical take-up force is 400 lb, providing 200 lb chain tension. *Air/hydraulic take-ups* are used where conditions are such that heavy loads negotiating vertical declines create excessive tensions that would collapse a normal take-up. This device has a hydraulic cylinder with its oil supply under air pressure providing normal tension forces. If slack chain develops, it is removed just as it would be with an air cylinder take-up. If excessive tension develops, threatening the collapse of the take-up, the noncompressible hydraulic oil holds the cylinder in position, preventing collapse. After the intermittent excessive force reduces to normal, the take-up again functions as an air take-up. A hydraulic pressure relief valve limits the tension to a safe maximum.

Switches

Most power-and-free track switches use a deflector (pivoting tongue) mechanism. A *converging switch* uses a free pivoting tongue. As the carrier reaches the switch point, the free trolley guide roller deflects the tongue to the merge position (Fig. 9.6.21). A diverging switch tongue is operated by means of a pull chain, by air cylinder, electric actuator, or cammed by a tripper on the carrier.

Stops

Stops are mechanical devices used to halt a power-and-free carrier at some predetermined point and hold it at that location until released. This device is affixed to the track at the chosen location. When the stop is "closed," a steel blade is positioned over the opening of the free track between the free track and the power rail. The stop blade forces the retractable dog on the lead free trolley down and comes into contact with the hold-back dog (Fig. 9.6.22). As long as the stop is closed, the retractable dog is held down, allowing the pusher on the power chain to pass over. When the stop is "opened," the carrier pusher dog is allowed to rise and engage the next available chain pusher and resume forward movement. Stops are used for traffic control at switch points, to establish carrier positioning at work stations, and for queueing and banking purposes. They can be operated manually, by air cylinder, or electric actuators.

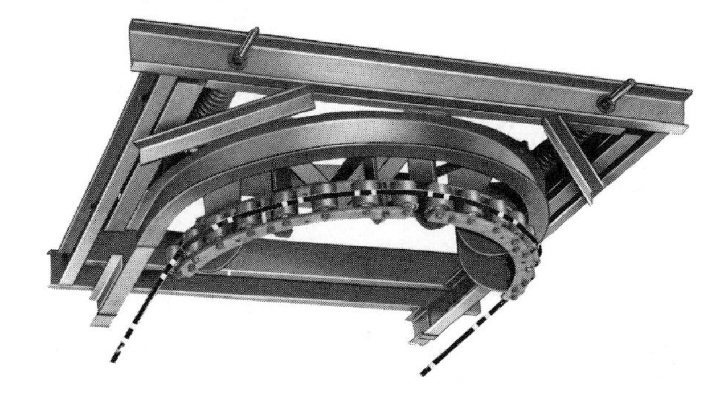

Fig. 9.6.20 Spring take-up on 180° roller turn.

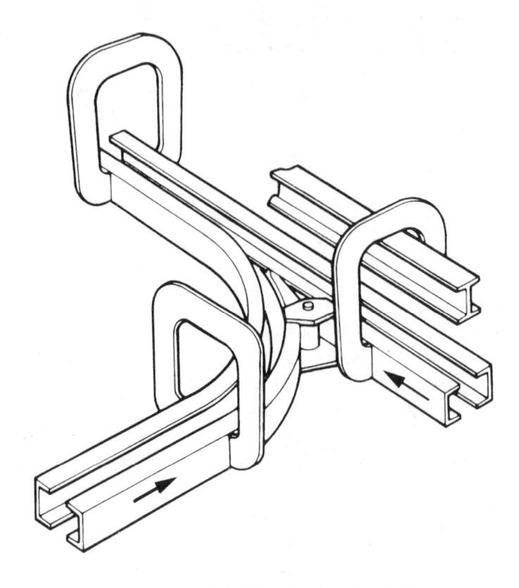

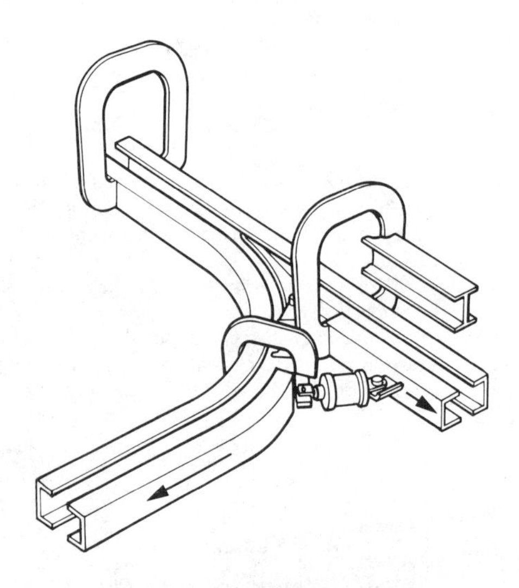

Fig. 9.6.21 Track switches.

Carrier Transfer, Chain to Chain and Switching

Since most power-and-free systems utilize more than one power chain and frequent switching, effective transfer of carriers from one chain to another is of major concern to the system designer. Carrier movement through a chain transfer must take into account chain speed, pusher centers, and the size of the load. Consideration must be given to the result of stopping either the delivery or receiving

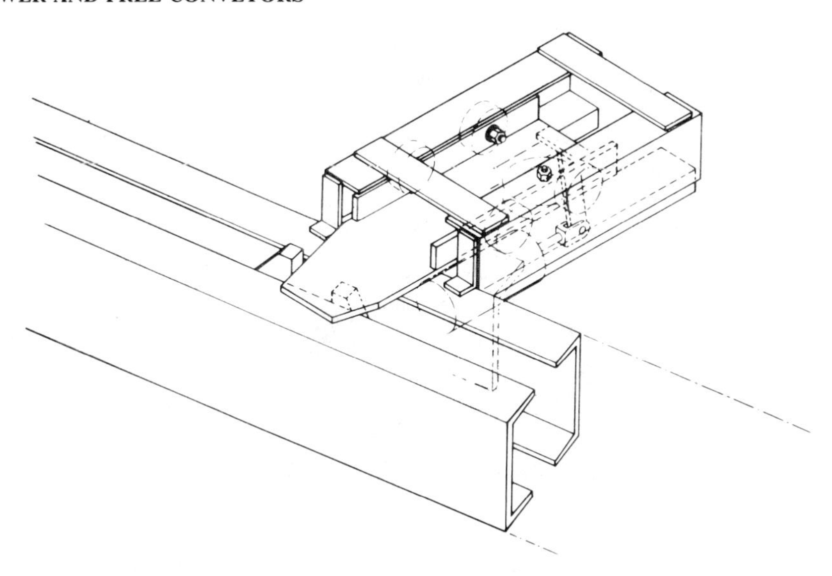

Fig. 9.6.22 Power-and-free stop.

chain. Once a carrier begins its transfer, forward progress must be assured. Most manufacturers use positive means to transfer the carrier from one chain to another, such as:

Push-across
Air-cylinder pushers
Dog-to-dog
Feeder chains
Paddle wheel

Gravity is rarely a desirable option.

A frequently used method is the *push-across transfer.* In this transfer, the power rail is lowered in relation to the free channel track a short distance ahead of the transfer area. This makes it possible for the power chain pushers to engage a shorter push-across dog on the trailing free trolley. As the chain wipes off the front retractable dog, the next chain pusher engages the push-across dog on the trailing trolley and pushes the carrier across the gap. Once across, the take-away chain engages the retractable dog and takes the carrier away (Fig. 9.6.23). Rather than a shorter push-across dog, some manufacturers use a movable dog that is cammed into position.

Another method of transfer uses a *paddle wheel.* In this transfer, the delivery chain wraps around

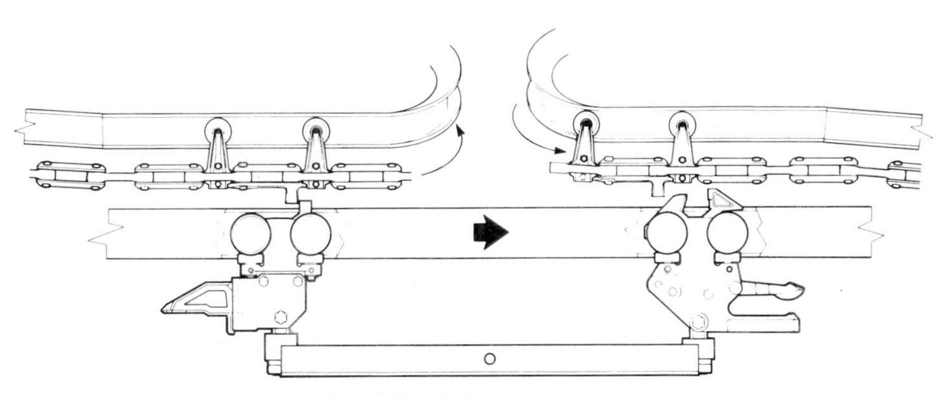

Fig. 9.6.23 Push-across transfer.

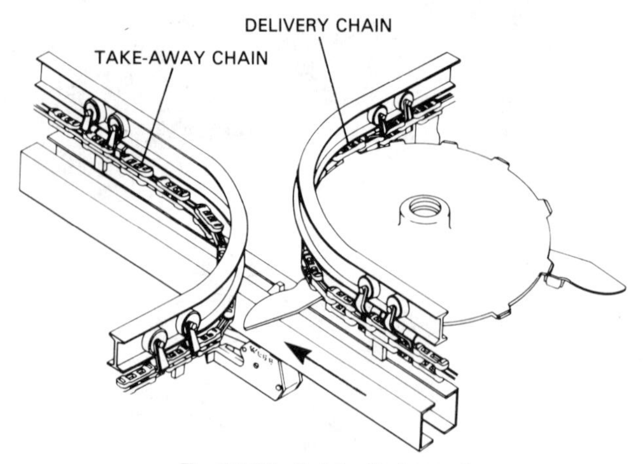

Fig. 9.6.24 Paddlewheel transfer.

a sprocket which has paddle-arms extending from the under side. These paddles act as wipers which push the carrier across the gap between the two chains (Fig. 9.6.24).

It is important that all chains in an integrated system operate independently so that stoppages may be kept to a minimum. For this reason, chains are operationally interlocked only during a carrier chain-to-chain transfer, and then only when absolutely necessary.

Horizontal turns are constructed so that the power chain and trolleys travel smoothly around the turn. One such construction is the *roller turn* (Fig. 9.6.25). Another is the *traction wheel* (Fig. 9.6.26). Horizontal turns are available in standard 30, 45, 60, 90, and 180° arcs. Horizontal turns for power-and-free track combine a power-only turn with the double channel for the free trolleys (Fig. 9.6.27).

The radius and angle of arc of the horizontal turn are designed with the requirements of the load bar in mind. Care must be taken when two or more trolley load bars are used. The angle of the load bar in relation to a trailing trolley line of travel must not be so great that forward force is less than resistive force. The resistive force is a function of the friction between the trailing trolley and the rail. To accommodate adverse conditions, it is assumed that the guide rollers are frozen, that is, they

Fig. 9.6.25 Roller turn.

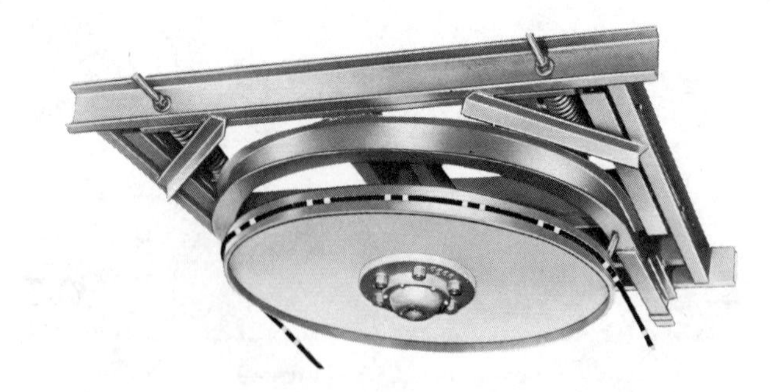

Fig. 9.6.26 Traction wheel turn.

Fig. 9.6.27 Power and free 180° roller turn.

do not turn. Under these conditions, a sliding coefficient of friction for steel on steel is used. For purposes of analysis, a conservative value of 30 percent (0.30) is used. With a 0.30 coefficient of friction, the maximum angle is calculated to be 73°, as follows:

Problem Definition. Calculate the angle θ (Fig. 9.6.28) at which the resistive force ($R1 + R2$) equals the forward force.

Solve for resistive force $R1 + R2$.

$$R1 + R2 = uF1 + uF2$$
$$R1 + R2 = u(F1 + F2)$$
$$F1 + F2 = (P \sin \theta)$$

Therefore:

$$R1 + R2 = u(P \sin \theta)$$

Solve for forward force. Forward force (FF) component equals the carrier chain-pull force (P) times the cosine of the angle θ.

$$FF = P \times \cos \theta$$

Solve for angle θ when the resistive force equals the forward force.

$$R1 + R2 = FF$$
$$u(P \sin \theta) = P \cos \theta$$
$$u = \frac{\text{P} \cos \theta}{\text{P} \sin \theta}$$
$$u = \frac{\cos \theta}{\sin \theta}$$

For simplicity of calculation, this expression is inverted.

$$\frac{1}{u} = \frac{\sin \theta}{\cos \theta}$$
$$\frac{1}{u} = \tan \theta$$

Therefore: When the tangent of the angle θ equals the reciprocal of the coefficient of friction, the resistive force equals the motive force.

In our example,

$$u = 0.3$$

Therefore:

$$\tan \theta = \frac{1}{0.3}$$
$$\theta = 73.3°$$

With a margin of safety, a maximum angle of 60° is recommended.

The 60° limit on the angle of the load bar to trolley line of travel limits the ratio of the load bar

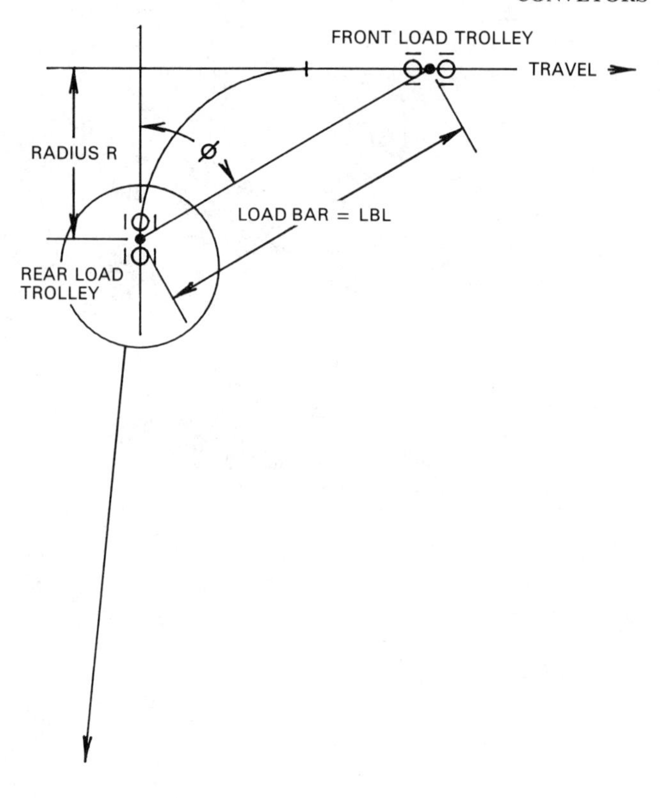

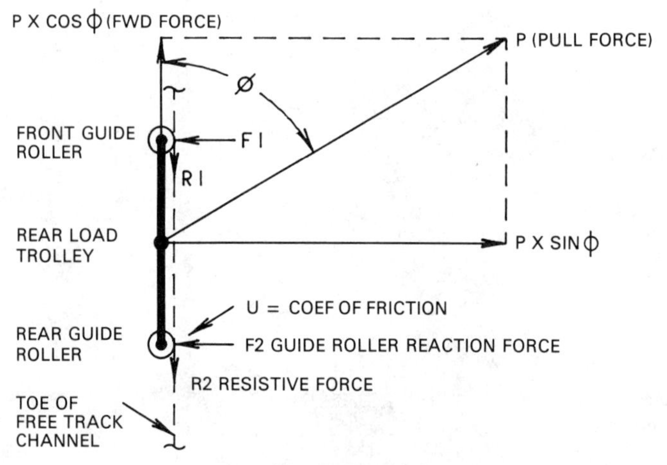

Fig. 9.6.28 Force vectors for 90° turn.

length to the radius of the turn. For example, in a 90° turn, the trolley center-to-center dimension cannot be greater than twice the radius.

This is proved as follows:

What is the ratio ($\cos\theta$) of the radius (R) to the load bar length (LBL) when the angle θ equals 60°?

$$\text{Ratio} = \frac{R}{LBL} = \cos\theta$$

Therefore:

$$\frac{R}{LBL} = \cos 60° = 0.5$$

$$\frac{LBL}{R} = 2$$

Safety Considerations for Inclines and Declines

Although it is important that the entire power-and-free system be "designed for safety," this is especially true where changes of elevation are involved. On inclines and declines, the free carrier should be properly "dogged up"; that is, there should be a positive "bite." To ensure that the free carrier, when properly dogged up, maintains this positive bite, the gap between the power and free tracks is decreased. This increases the bite and prevents disengagement of the carrier from the power chain.

Antibackups

These are pivoted arms that move freely out of position in the forward direction, but lock in a restraining position in the reverse direction. They can be actuated by the power chain or the free carrier trolley. When actuated by the chain, they provide a safety against chain breakage. When actuated by the trolley, they provide a safety against possible disengagement of the carrier from the chain pusher (Fig. 9.6.29). These devices are also used on inclines.

Antirunaways

Antirunaway devices are used on track declines. They are designed to restrain a runaway load or stop runaway chain. If a conveyor chain were to break, the free end on a sloping track would run downhill at an uncontrolled speed and stack up on the horizontal track below. To prevent this, chain antirunaway devices are used. One design uses a pivoting rocker arm, shown in Fig. 9.6.30. At normal speeds, the trolley lifts the stop just enough to pass by. The stop then falls back to its normal position. At runaway speeds, however, the trolley will impart sufficient energy to rotate the stop into a reverse position, locking or catching the next descending trolley (Fig. 9.6.31). If a position detector is used in conjunction with the antirunaway, the stop will trip the detector and hold it in an actuated position, shutting off the chain drive.

Antirunaway Conveyor

This is a separate, closed-loop chain, mounted beside and parallel to a vertical curve. As each carrier descends, it engages the antirunaway conveyor and drives it in tandem (Fig. 9.6.32). An overspeed switch on the antirunaway conveyor will detect a runaway load and engage a brake to bring the carrier to a stop. Electrical interlocks shut off the primary conveyor.

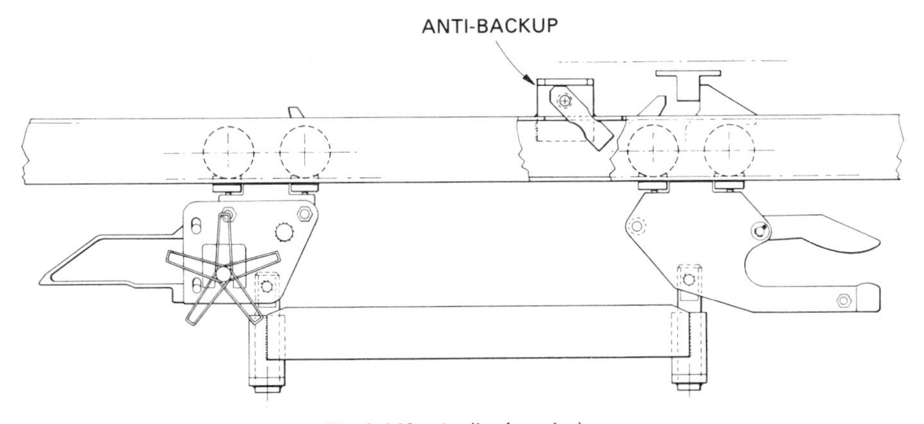

Fig. 9.6.29 Antibackup device.

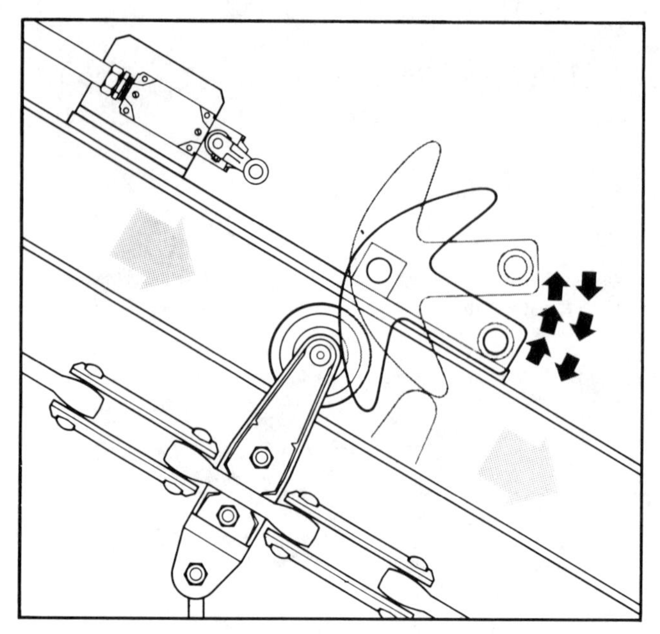

Fig. 9.6.30 Antirunaway device, normal position.

9.6.4 Power-and-Free Control

Introduction

The design of a power-and-free control system must take into consideration system complexity; the method used to identify the load; whether variable, fixed, or none; and whether central, distributed, or centrally directed distributed control is to be used. A selection of control hardware must be made between relays, programmable controllers, microcomputers, computers, or some combination of these.

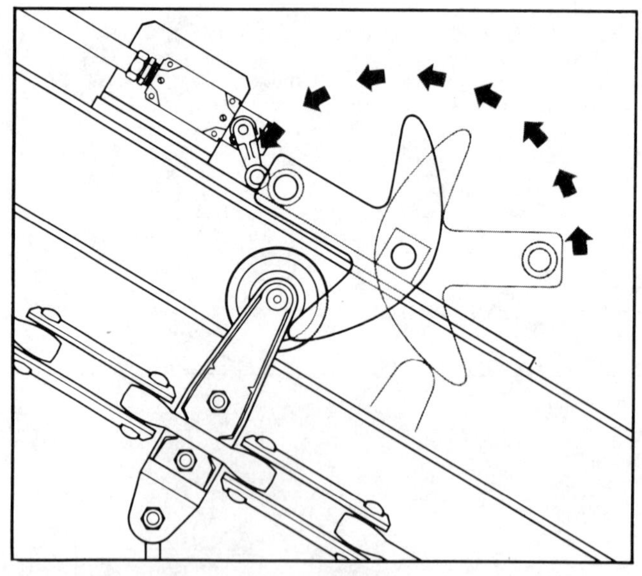

Fig. 9.6.31 Antirunaway device, loaded position.

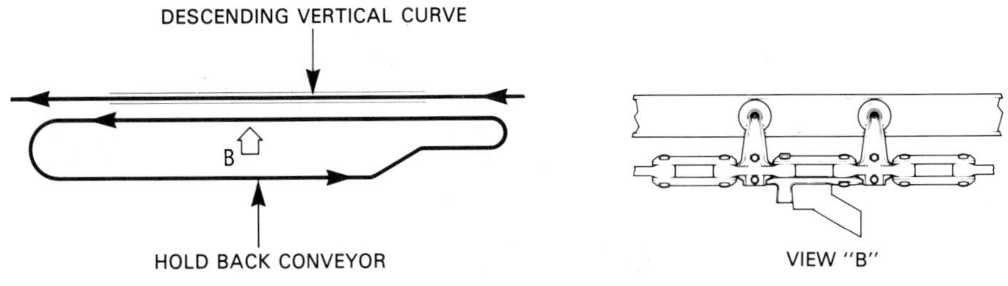

Fig. 9.6.32 Antirunaway conveyor.

Distributed control is the most common technique used with power-and-free conveyors. If relays are used, individual area control panels are provided to segment the control. If programmable controllers are used, remote I/O panels are provided in each area, with limited-wire communication to the processor.

Distributed control divides large, complex systems into manageable local modules. Distributed control also provides local access to the control for an immediate area. Segmentation simplifies servicing. The impact of a control failure is reduced because only one segment is affected by a failure.

Basic Motions

In power-and-free systems, three basic modes of material movement are available:

1. Point-to-point
2. Segregation
3. Recirculation

These three principles can be combined, duplicated, mirrored, paralleled, or put in series to solve the most complex material handling problem.

Point-to-point flow is similar to traffic on a one-way street. It is limited to start and stop.

Segregation divides randomly delivered parts or loads into independent groups (Fig. 9.6.3).

Recirculation is the movement of randomly delivered loads around a closed loop where they are tested at a control point for possible withdrawal (Fig. 9.6.2).

Mechanical Control Hardware

The application of these three principles uses only two items of mechanical control: stops and track switches. *Track switches,* as their name implies, direct carriers from one conveyor line to another. There are two types: converging and diverging.

A converging track switch merges two lines into one (Fig. 9.6.33). When deflector switches are used, no actuator is needed for a converging track switch. The tongue is moved by the carrier.

A diverging track switch is used to divert a carrier into one of two lines (Fig. 9.6.34). In this configuration, air, electrical, or manual actuators move the deflector.

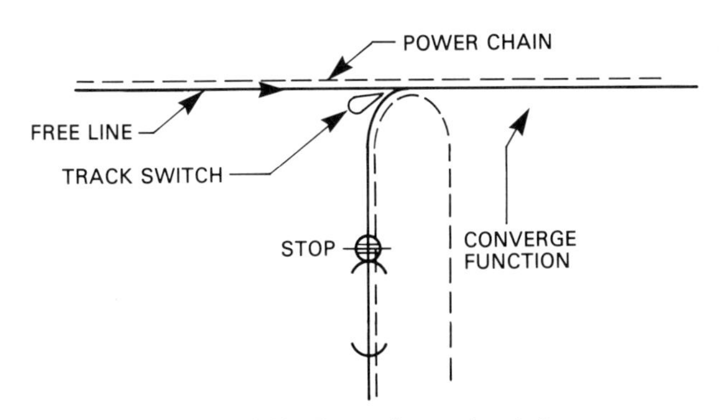

Fig. 9.6.33 Converging track switch.

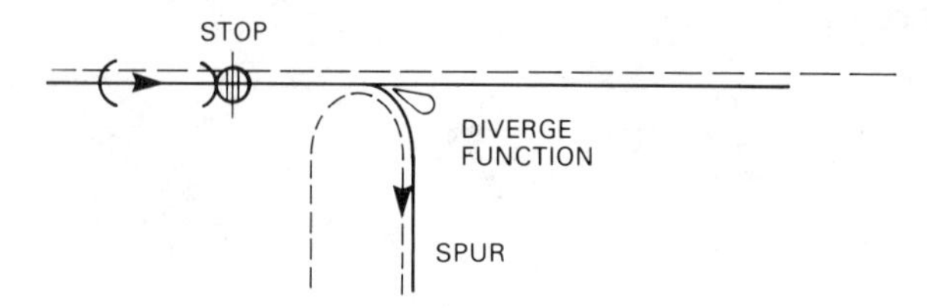

Fig. 9.6.34 Diverging track switch.

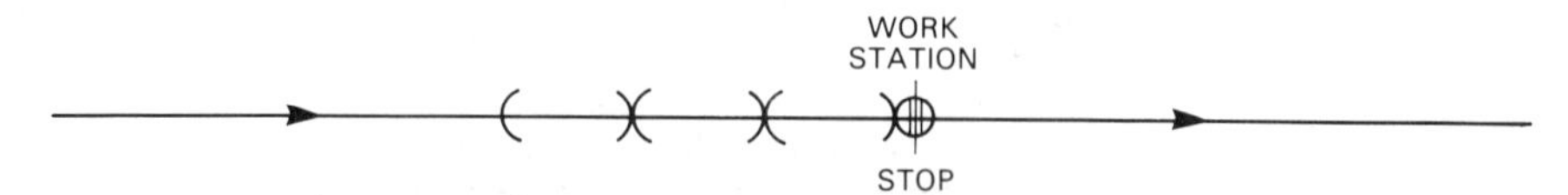

Fig. 9.6.35 Carriers queued at a work station stop.

Stops are used to control the forward movement of a carrier and to control carrier traffic through track switches. The following examples illustrate these and other uses of stops:

1. Stops are used to hold a carrier at a work station until released automatically or manually (Fig. 9.6.35).
2. Carriers cannot queue on a vertical change in elevation. Stops are used in these situations to limit the number of carriers on the vertical curve (Fig. 9.6.36).
3. In some cases, carriers cannot queue around a horizontal turn. Properly placed stops can be used to limit the number of carriers at these locations (Fig. 9.6.37).
4. Stops are placed on each leg of converging lines to control traffic through the switch (Fig. 9.6.38).

Electrical Control Hardware

To operate a track switch, a control decision must be made based on logic associated with the carrier being switched. In the segregation example given earlier (Fig. 9.6.3), red parts are directed into Line A, blue into B, and yellow into Line C. In the recirculation example, the carriers are withdrawn in

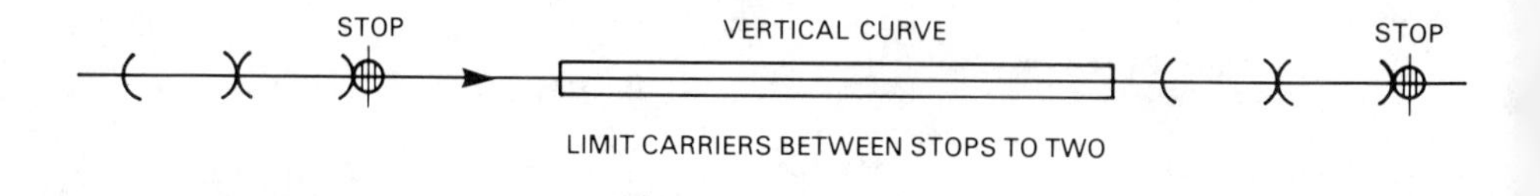

Fig. 9.6.36 Use of stops at vertical curve.

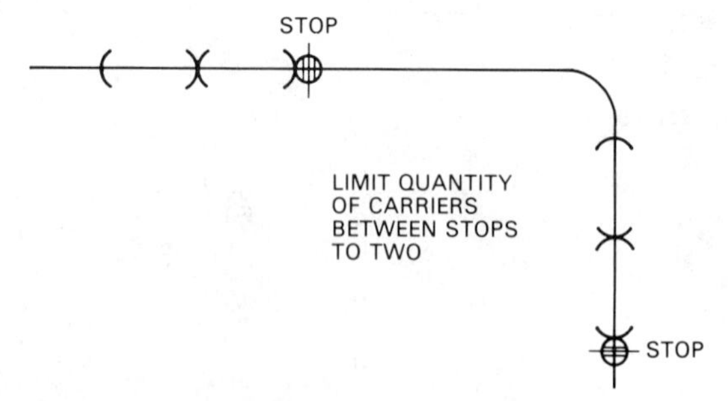

Fig. 9.6.37 Use of stops at horizontal turn.

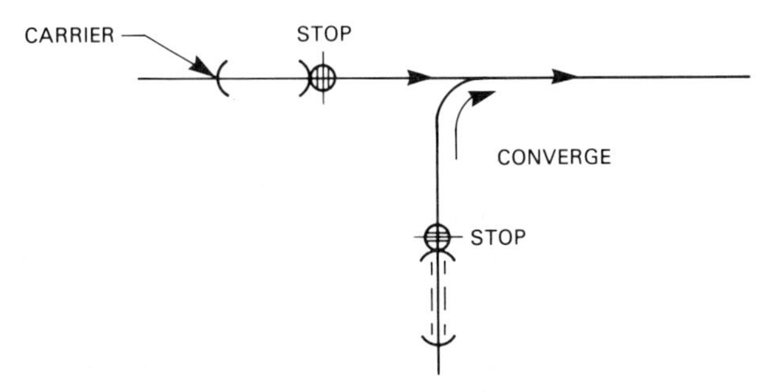

Fig. 9.6.38 Use of stops at converging track switch.

sequence: red, blue, yellow; R, B, Y; and so on. The logic needed to inform control that a carrier has red, blue, or yellow parts can be obtained in a number of ways.

Code Cards

A common method used is an escort memory (a coding system that escorts the carrier). Escort memory can be obtained by using an aluminum plaque or code card on which a code representing the red, blue, or yellow characteristic is magnetically inscribed on magnetizable inserts. These are variable load-identity code cards; that is, they can be changed for each different load they carry (Fig. 9.6.39). The code card is mounted on the lead free trolley. The information carried on the card is "read" by code readers mounted on the conveyor track at points of decision throughout the system

An alternate to the variable identity code cards is one that is mechanically similar, but has permanent magnets. With this device, each carrier is given a unique, sequential, permanent number. The red, blue, or yellow characteristic is married to the carrier number at the time of loading. After loading, the carrier/characteristic is determined by table look-up. Fixed load identity is used when the amount of data is too great to put on a variable code card.

The part characteristic (R, B, Y) can also be married to the carrier without a code card. Here, the characteristic is entered into a logic table at the moment of carrier loading. After loading, the characteristic for each carrier is maintained in logic shift registers that represent the physical system.

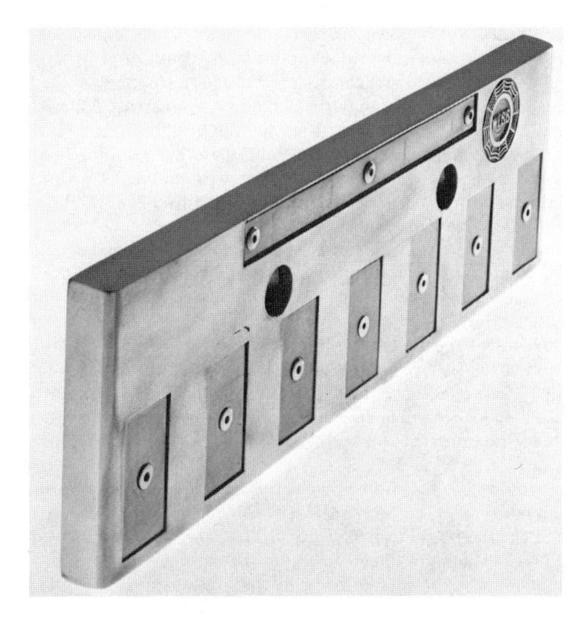

Fig. 9.6.39 Variable identity code card.

Control Hardware Selection

Hardware selection is based on the answers to such questions as: Do carriers use fixed, variable, or no code identity? Is central or distributed control desired? Is the control simple (e.g., point-to-point) or complex? With all the facts understood, a selection is made between relays, programmable controllers, microcomputer, computer, or some combination.

Relays are used on simple systems having variable or nonidentified loads.

Programmable controllers are an option on simple or complex algorithm systems using any form of load identity and distributed or central control.

The *microcomputer* is appropriate for fixed-function programs using variable or noncoded loads on distributed or centrally directed distributed control systems.

The *computer* meets the needs of systems having complex algorithms or large data files using any form of coding and central or centrally directed distributed control.

There are several subordinate factors that also must be given close scrutiny, since they play a subtle but critical role in hardware selection. One is field wiring, a major cost item. This factor is greatly affected by the type of interelement communication used: fixed wire for relays, matrix for computers, or bit serial for programmable controllers.

Consideration of local or remote start-up and servicing must address the following questions: Is start-up, debug, and manual control to be done locally or remotely? Who performs these services? Keep in mind that computers require programmers and technicians, whereas relays and PCs require electricians.

Care must be taken to be sure that control cycle rates do not exceed acceptable limits. A relay is essentially a fixed-time device. Registration of a change of state is direct, not a function of internal clocks or scan rates, as used in PCs or computers.

When physical material flow requirements, project specifications, and control principles are examined in sequence, the solution to control becomes evident.

9.6.5 History

The power-and-free conveyor made its first appearance in the late 1930s when demand arose for an overhead conveying method that could be designed as a truly integrated system. Prior to this time, product or work-in-process could only be moved overhead on carriers hung from the "captive trolleys" of a monorail conveyor chain. The overhead monorail is still an important conveying method, of course, with many useful applications, but it lacked the flexibility needed to meet the growing demands of integrated production techniques. This flexibility was achieved by mounting a second, "free" rail either beside or below the monorail I-beam. Load-bearing trolleys ran on this track. They were not fixed or interconnected to the power chain in any way, but were moved along by pusher dogs that projected from the power chain and engaged similar dogs on the free trolleys.

With the newly developed power-and-free conveyor, it became possible to switch loaded carriers from one chain to another, thereby allowing the connecting of all the operations within a plant into an integrated system. This new conveyor design could perform many of the functions that had previously required several different types of material handling equipment. Especially important was its ability to halt individual loads for processing without halting all the carriers on that section of track. Carriers could be accumulated in static storage, one behind the other awaiting processing, or moved at different production rates by assigning different speeds to individual chains within the system.

The early power-and-free track configuration went through several design changes before industry standardized with the present I-beam over double-channel cross section. Techniques used to disengage the free load-bearing trolley from the power chain have also changed since the conveyor was introduced. One of the earliest methods was to open the spacing between the power-and-free tracks so that the chain dogs could pass freely over the pusher dogs on the free trolleys. Gravity sloped track was used to accumulate carriers and to move them through track switches.

The next important step in the evolution of the power-and-free trolleys as they are seen today was the development of the spring pusher. This device formed part of the pusher dog assembly on the power chain. The pusher was held in operating position by a spring with sufficient rigidity to push a loaded carrier to a manual or automatic stop or against the last carrier in a bank. There the increased resistance compressed the spring, allowing the pusher to slide over the pusher dog on the free trolley.

Use of the spring pusher in power-and-free systems was extensive during the late 1940s and through the 1950s. Although most have been replaced, some spring pusher systems are still operating today.

Refinement of the methods of moving free trolleys through switches, chain transfers, and disengagement from the power chain continued until, by the early 1960s the last major development emerged.

This development was the introduction of the "releasing dog" trolley. This design uses a pusher dog on the free trolley that can be cammed down by a stop blade, or by the cam tail of a preceding trolley. This trolley design has replaced the spring pusher, as well as the use of gravity, for switching and transfers and provides positive control of carriers at all times, eliminating the undesirable elements found on the earlier designs.

9.7 UNIT LOAD CONVEYORS AND ENGINEERED SYSTEMS

Robert H. Roth

9.7.1 Types of Unit Loads Typically Conveyed

For the purpose of this section we will consider hea'y-unit-load conveyors only, which, according to guidelines established by the Conveyor Equipment Manufacturer's Association (CEMA), are those which handle unit loads exceeding 500 lb each. These unit loads are most often conveyed on pallets, or occasionally their bottoms if the bottom construction has one of the configurations found in pallets. These load configurations are as follows:

Type A. Slave pallets and other solid bottom loads.
Type B. Double-faced pallets with spaced bottom boards.
Type B-1. Traveling with bottom boards parallel to direction of travel.
Type B-2. Traveling with bottom boards perpendicular to direction of travel.
Type C. Skids or objects with two, sometimes more, bottom runners.
Type C-1. Traveling with runners parallel to direction of travel.
Type C-2. Traveling with runners perpendicular to direction of travel.

The orientation of the pallet with respect to the direction of travel is important because it dictates the type of conveyor on which the unit load can be handled.

9.7.2 Most Frequently Used Types of Conveyors

Chain-driven live roller is the most commonly selected conveyor for heavy unit loads (Fig. 9.7.1). It can be constructed ruggedly enough to withstand a wide range of weights, and, with roll-to-roll chains, provides positive drive that is not destructive to the bottom of the pallet. For this reason, it is the conveyor of choice where loads must be squared against end stops, or diverted against a guide rail for accurate alignment. With proper design, it can provide gaps between the rollers to accommodate multiple-strand chains for transfer of loads.

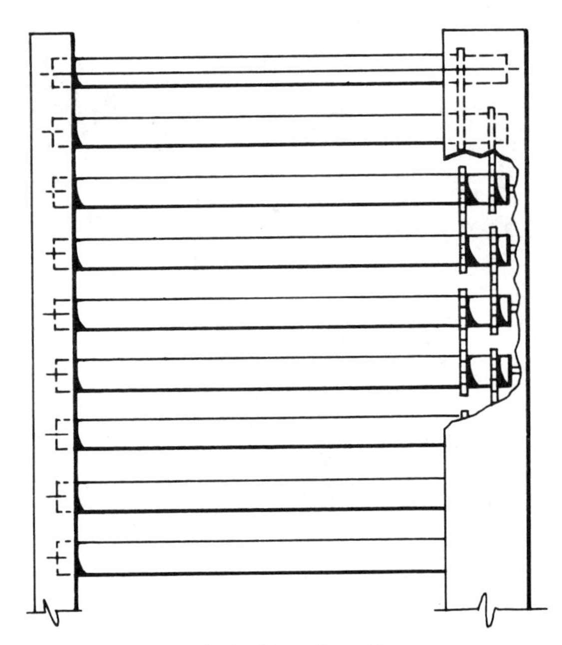

Fig. 9.7.1 Chain-driven live-roll conveyor.

Chain-driven live roller is ideal for handling loads with solid bottoms or bottom boards running in the direction of travel (Types A, B-1, and C-1). Roll spacing can be selected to suit the rigidity of the surface being conveyed and the weight of the load. In selecting the capacity of the rollers to be used, and the number of those rollers, it is important to remember that it is always better to have a small number of heavy-duty rolls than a large number of lighter rolls, in order to better withstand shock loading and uneven distribution of the weight.

Chain-driven live rollers can also be used to handle Types B-2 and C-2 loads if the rolls can be spaced closely enough to be slightly closer than the width of the bottom boards or runners. In this case, it is preferred that every roll be driven; and at close centers, this will require chain drives on each side of the conveyor. An alternative for Type B-2 only is to drive every other roll, with free-turning rolls between powered rolls. This is acceptable only if there are enough bottom boards to assure that at least half the rolls supporting the load, with a minimum of two, are driven. It is also important that care be taken to keep the tops of the rolls accurately aligned and that the bottom of the load be of uniform quality.

Two-strand chain conveyor is suited for handling loads of Types A, B-2, or C-2 (Fig. 9.7.2). Although this conveyor is sometimes used with bottom boards in the direction of travel, care must be taken to accurately position the narrow surface above the chains. Since this is often difficult, it is usually avoided.

Two-strand chains are typically used either as long transport conveyors, or as short transfer devices to be raised and lowered between rollers. From a cost standpoint, transportation conveyors must be long enough to absorb the relatively high cost of the drive and take-up terminals.

Gravity roller conveyor (Fig. 9.7.3) is used both in storage lines, and where conditions permit, on short runs of transportation. In storage lines and in long, flow-through racks using gravity conveyor, it is usually required that devices be included to retard the flow, limiting the speed to that which will permit the pallet to stop abruptly against an end stop or another pallet without causing the product to shift or be damaged. Rollers or wheels with built-in centrifugal braking devices, or wheels made of an energy-absorbing plastic material, are typically used to control the speed.

The pitch in a gravity line can best be determined by testing with the actual product. This is especially important when adjustment is not practical or when a large number of runs is involved, as in a flow rack. For preliminary calculations, use $\frac{3}{8}$ in./ft. The following factors tend to decrease the pitch required:

Exceptionally hard, flat surfaces
Heavier than normal loads
Slug loads, creating a train effect

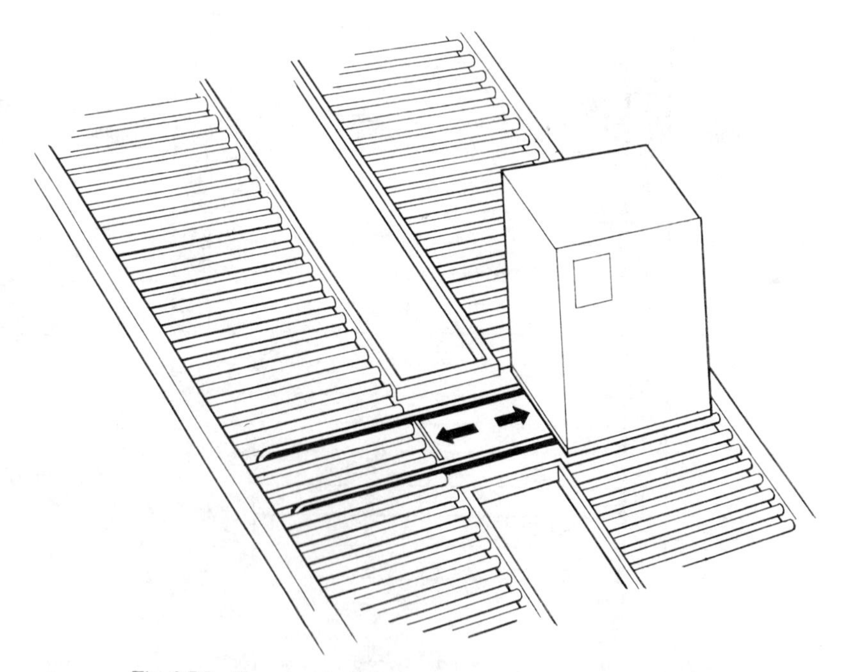

Fig. 9.7.2 Two-strand chain conveyor used as a transfer device.

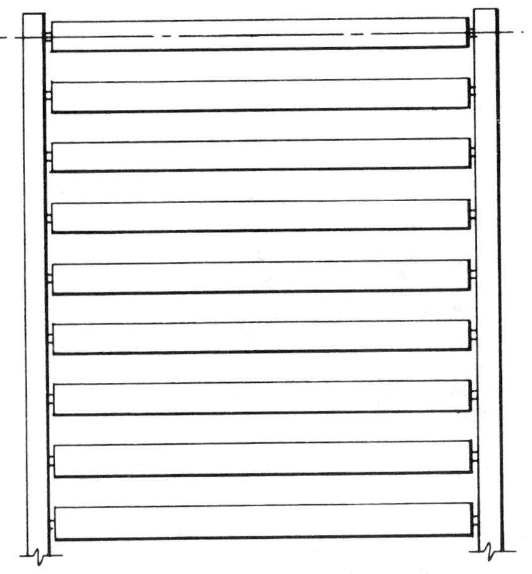

Fig. 9.7.3 Gravity roller conveyor.

Factors that tend to increase the pitch required are:

Soft or irregular surfaces
Poor quality pallets
Light loads
Guard rail contact

When all these have been considered, the final pitch must be tested with the actual range of products, and adjusted to the figure that best represents a compromise of all factors.

9.7.3 Other Heavy-Unit-Load Conveyors

Belt conveyors are sometimes used but care must be taken to recognize their limitations. They are best limited to loads in the 500–1000-lb range, with good quality bottom surfaces that will not damage the belt. Because traditional methods of removing items from the conveyor at intermediate points, such as diverters and pushers, are not applicable to heavy unit loads, the belt conveyor is generally limited to straight transportation with end loading and unloading.

Belt-driven live-roller conveyors may be used to handle unit loads in the 500–2000-lb range. Because they provide a less positive drive than chain-driven live roller, the quality of the conveyed surface is very important. Warped or split boards and protruding nails are more likely to present a problem, and guide rail friction can cause a hang-up.

Slave pallets and those with bottom boards or runners in the direction of travel (Types A, B-1, and C-1) are easily handled on belt-driven live-roller conveyors. Type B-2 pallets can be handled with rolls at close centers, and with special placement of the belt pressure rolls to drive every roll.

Because belt-driven live roller is less positive than chain-driven live roller, it offers a less abrupt start, an advantage to be considered in handling somewhat unstable loads.

Unpowered rollers with a pusher are available in several configurations for handling heavy pallets, with the roll centers selected to suit the type of bottom involved. Typically, two narrow runs of rollers support the pallet on its outer edges, and the center is clear for the pusher.

One version of this conveyor utilizes a series of dogs on a continuous chain, with the dogs engaging the load and moving it the full length of the conveyor. Another utilizes a continuous chain supported by air bags or an inflatable hose, to drive frictionally against the bottom of the load. On a third version, pusher dogs, which are hinged in one direction, are spaced on a bar which oscillates a distance equal to the pallet length plus about six inches.

Slat conveyors, because they have a rugged and relatively solid surface, can handle any unit load (Fig. 9.7.4). They are well suited to loading and unloading by fork truck. Their rugged construction absorbs impact well. Feeding to and from the ends of slat conveyors requires special consideration to

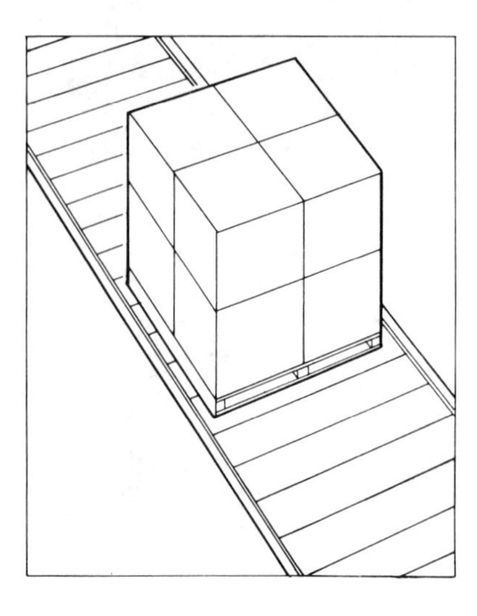

Fig. 9.7.4 Slat conveyor.

bridge the gaps that result from the need for large sprockets for the conveyor chain and the necessity of an end take-up. This is most often accomplished with fill-in rollers, chain driven where necessary. Because of its relatively high cost, slat conveyor is used only where its special capabilities are needed.

Walking beam conveyors (Fig. 9.7.5) are most frequently used in hostile environments that would adversely effect other conveyors. High temperature; dirt, dust, sand, and other gritty substances; water and other liquids are all within the design capabilities of this conveyor. Since a walking beam conveyor uses horizontal beams with reciprocating motion to lift and move the loads on supporting rails, it is best suited to pallets or loads with solid bottoms, or crosswise slats or runners (Types A, B-2, and C-2). Because its cost is high, its use is limited to those applications which require its special capabilities.

9.7.4 Accumulating Conveyors for Heavy Unit Loads

Chain-driven live roller is adapted for accumulation by dividing its length into short segments, or zones, equal in length to the load plus an average 6-in. clearance. In the earliest versions, each zone was provided with a motor, and limit-switch actuated controls were provided to move loads forward in a cascade manner to the last available downstream zone. Although this is an effective way of accumulating with no contact between loads, it is expensive from the standpoint of field wiring and adjustment.

Pallet accumulators, which replace the motors and controls with mechanical clutches and interlocking mechanical linkages, have recently become available. A single motor is capable of driving 30 zones, and loads of up to 8000 lb are possible. Where electrical control of a given zone is desirable, a solenoid can replace or supplement the mechanical actuators. Mechanical time delay devices can be incorporated in selected zones to facilitate loading and unloading by fork truck.

Other variations of the pallet accumulator replace the mechanical linkages with air or electrical controls, to achieve the features of no-contact accumulation with loads moving forward to fill the last available downstream zone.

Unpowered rollers with a pusher chain can be used for accumulation in several ways. When the chain is supported by air bags or an inflatable hose, the pressure can be controlled to provide enough

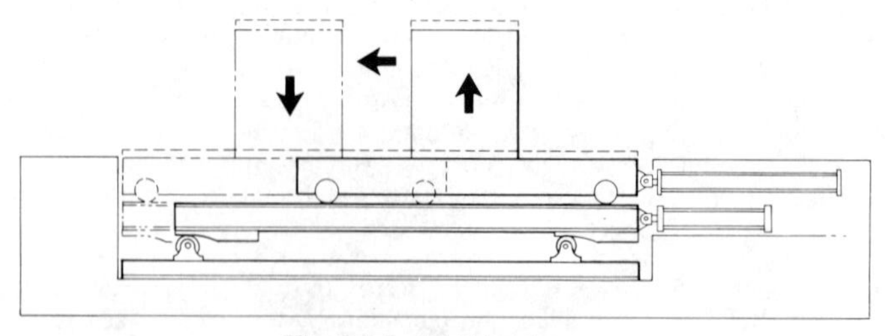

Fig. 9.7.5 Walking beam.

force to move an unobstructed load, but still permit the chain to slip under a blocked load. This does result in a certain amount of wear on the bottom of the load under which the chain is slipping—a factor to be considered in selecting this type of equipment. If this is objectionable, a version is available that segments the chain support bars in zones which, by means of air controls, selectively drops the chain to provide cascade accumulation similar to that described for chain-driven live rolls.

Belt-driven live roller, within the limitations of capacity established previously, can provide accumulation in a wide variety of forms. By providing adjustment to the pressure rolls, the contact between the belt and the carrying rollers can be adjusted to obtain driving force high enough to move the load, but low enough that the belt will slip under the carrying rolls when there is an obstruction to forward motion. This type of accumulation has several features that might not be desirable. The forward pressure increases with the number of loads accumulated; and it is more than a straight line increase because, as the belt tension becomes greater, it becomes increasingly more difficult to flex under the rollers. It is often difficult or impossible to develop enough driving force to move heavy loads or loads with bottom irregularities without frequent hang-ups.

The accumulated forward pressure can be reduced by modifying the design to include a ripple belt. In this concept, the belt is designed with thickened sections or pads at selected intervals throughout its length such that the padded portion provides driving force to the carrying rollers and the remainder of the belt does not. In addition to reduced forward pressure, the use of ripple belt results in lower horsepower requirements, and because the belt is moving opposite to the direction of unit load flow, it offers singulating capabilities. If the lowering of the end stop is synchronized with the approach of a pad, loads will start forward one at a time, with space between them in which to raise the stop as desired. On the negative side is the fact that loads in transit do not move forward continuously, but start and stop based on the length and spacing of pads. Also, discharge from the end is not always immediate, but must await the approach of a pad. This requires careful consideration of throughput requirements in selecting the pad ratio and the belt speed.

Belt-driven live-roller conveyor can be equipped with zones to achieve controlled accumulation. The belt is selectively dropped in areas to reduce or eliminate the driving force. This can be accomplished with mechanical linkage, air-controlled, or electrically controlled devices.

Two-strand chain conveyor accomplishes accumulation in two basic ways. In one version, the accumulator conveyor consists of a series of short conveyors, each with their own drive. As described earlier, these can be individual motors, or clutches that are interlocked mechanically, pneumatically, or electrically. In every case, the objective is to provide driving force to the individual conveyors in a manner that will move loads forward one at a time and always fill the last downstream conveyor.

The other version utilizes a long two-strand chain conveyor. Air-operated lifting devices are placed between the chains, spaced at a distance equal to the load length plus clearance. Pneumatic or electrical sensors control the lifting devices, raising the loads above the chain when there is an obstruction immediately downstream, and lowering them when there is not.

Roller slat conveyor is constructed of two strands of conveyor chain, with free-turning rollers suspended between them to provide a support surface for the unit loads. The rollers carry the load in

Table 9.7.1 Conveyor Selection Guide

		TYPE OF UNIT LOAD				
		A	B-1	B-2	C-1	C-2
	Chain Driven Live Roller	X	X	0	X	0
C	Two Strand Chain	X	–	X	–	X
O	Gravity Roller	X	X	0	X	0
N	Belt Conveyor	X	X	X	X	X
V	Belt Drive Live Roller	X	X	0	X	0
E	Rollers With Pusher	X	–	0	–	0
Y	Slat Conveyor	X	X	X	X	X
O	Walking Beam	X	–	X	–	X
R	Roller Slat	X	X	0	X	0

X = Well suited.

– = Requires accurate positioning of longtitudinal members above support beams.

0 = Suited only if roll centers can be close enough to support the load.

the same manner as slats in a conventional slat conveyor when there is no obstruction to the forward motion of the load. However, when the load is obstructed, either by a stop or by another load, the rollers turn under the load as the chains continue to move, providing the ability to accumulate. It should be noted that there is a cumulative buildup of pressure against the stop or last load, which is, at a minimum, equal to the total load accumulated times the factor of friction of the rollers; and it can be much higher if the loads are soft or irregular.

When metal unit loads such as baskets, hoppers, or bins are handled on a roller slat conveyor, the sound level often becomes objectionable, because the rollers are constantly moving forward under the load during accumulation.

A unique positive feature of this equipment is its ability to release loads one at a time from an accumulated line. This is accomplished by providing a pressure bar with a high friction surface to be raised against the bottom of the rolls under the load to be discharged. The rollers turn, and the load moves forward at twice the speed of the chain, creating a gap in which the end stop can be raised.

Gravity roller conveyor used for accumulation is the most economical equipment available, but seldom gives ideal results. Because the pitch must be kept to a minimum to avoid excessive accumulated forward pressure, occasional hang-ups must be accepted. It is difficult to handle a wide range of weights satisfactorily.

It should be noted that, although retarding devices control the speed at which unit loads move on a gravity roller accumulating conveyor, they do not significantly reduce the resultant forward force inherent in a load on a decline. For this reason, escapements are generally required at the discharge end to hold back all but the end load, thus facilitating removal by fork truck or other means.

Table 9.7.1 provides a conveyor selection guide for handling different types of unit loads.

9.7.5 Interfacing with Other Equipment

Automatic Guided Vehicles

There are two basic types of conveyor used in conjunction with AGVs. The first is a split conveyor with two narrow runs of chain-drive live roller or two-strand chain conveyor. This is used when a lift-and-carry vehicle must discharge to or receive from a conveyor that is two or more unit loads in length. In most applications, the conveyor accumulates a number of loads after set-down or before pick-up. The narrow runs of conveyor are open on the end from which the AGV approaches. The loaded vehicle drives between the legs of the conveyor, lowers the load onto the rollers which are stopped and then reverses itself to withdraw, while the conveyor starts and indexes the load forward on, in most applications, zoned pallet accumulator. Where a load is being picked up, it is usually necessary to guide it on the incoming conveyor for accurate final positioning. Where a load is being deposited, the width must be great enough to accommodate the positioning tolerances of the vehicle. Since this accuracy varies with the type and manufacture of the vehicle, it is a factor that must be determined for each application.

The other conveyor used with AGVs is a snubber-driven stand (Fig. 9.7.6) designed to interface with those vehicles that support the load on a section of conveyor during transport, and load or unload from the side. The snubber drive is most practical when a small number of vehicles must serve a large number of positions. By providing a wheel on the vehicle that is driven with the load-carrying conveyor and that is pivoted to swing into engagement with a mating wheel on the stands, any number of stands can be driven from the vehicle. This can drive in either direction to load to or from the vehicle.

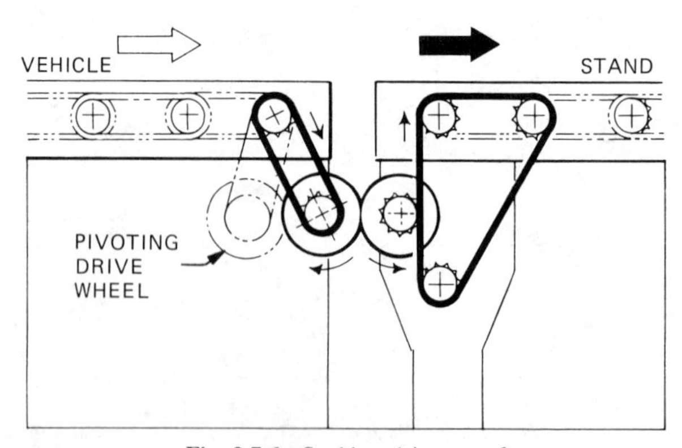

Fig. 9.7.6 Snubber drive transfer.

With snubber-driven stands, accuracy of load positioning for pick-up by the vehicle, and width of clearance for receiving from the vehicle are both dependent on its stopping accuracy. In those applications where a load is discharged to a stand and is later picked up for transport to yet another stand, the locating errors are compounded. In these cases, it is common to add a clamping device to the vehicle which will center the load after it is on the vehicle conveyor. This clamp, as well as the snubber wheel pivot, can be operated by means of an onboard hydraulic pump.

The most common means of transferring power in a snubber drive is frictionally through two rubber wheels, connected by roller chain on each conveyor. The drive reversal necessary to keep both conveyors moving in the same direction is incorporated in one of the chain drives. In a similar concept, the friction wheels are replaced by a pivoted sprocket which meshes directly with the roller chain on the driven stand. This provides a more positive drive, but is more expensive because an auxiliary centering device is required to obtain the lateral accuracy needed to mesh the sprocket with the chain.

Automatic Storage and Retrieval Systems

There are basically three unique requirements when heavy-unit-load conveyors must interface with AS/RS. They are: creating a window for the crane shuttle, squaring the load for accuracy at pick- up, and sizing and weighing the load to limit those that will not fit in the racks or will overload the crane.

There are many techniques for providing a clear area under the load to accommodate the shuttle device with which the crane picks up the load. One method provides a lifting device with two blades which come up between the rollers or chains to raise the load. Or the conveyor can be lowered, by pivoting one end or by dropping it vertically, to leave the load supported on shelf angles and the area beneath it clear. Another method is to feed the load perpendicular to the crane aisle on two strands of chain, spaced to provide the window.

It should be noted that not all cranes require windows for a shuttle. Some are provided with forks that engage the pockets in a pallet or require cutouts in the conveyor under the load. Chain-driven live roller will easily accommodate cutouts as long as the unit load will bridge the resultant gaps. This should be no problem with Types A, B-1, or C-1 loads. With B-2 and C-2 loads, special attention must be given to the design of the pockets to properly support the load.

Squaring the load can most often be accomplished by taking advantage of the transfer points common to most layouts. By driving against a squaring stop on the cross aisle conveyor prior to transfer and then against an end stop at the pick-up station prior to creating the window, adequate squaring is usually obtained. If the cross aisle transfer is on the fly, it is possible to provide skewed rolls in the aisle conveyor which will shift the load to an edge guide, before contacting the end stop. Final accurate positioning can be accomplished at the shuttle pick-up point by providing sloped sides on the shelves or corner pockets which raise the load or onto which the load is lowered.

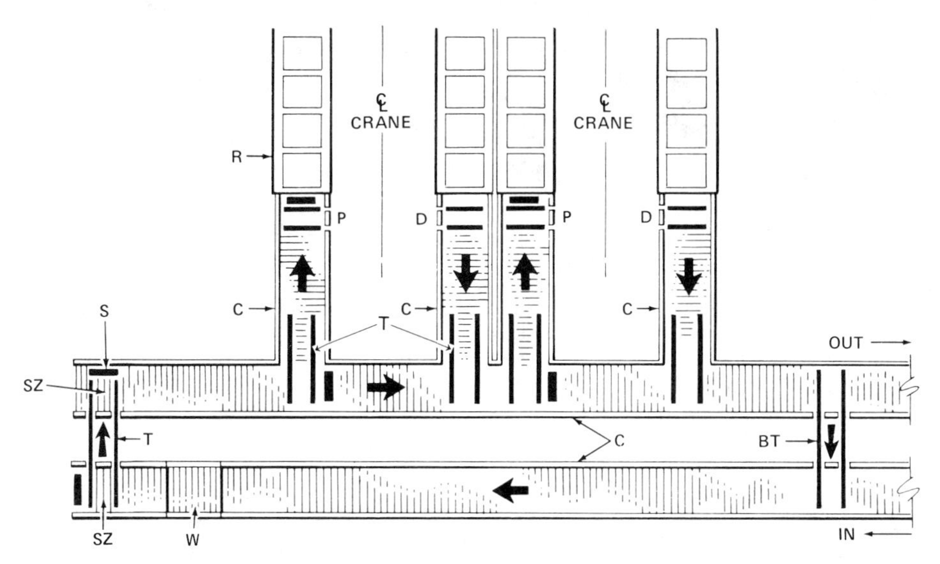

Fig. 9.7.7 Typical AS/RS front end conveyor. P = pickup station; D = deposit station; R = rack opening; S = squaring stop; W = weighing device; C = chain-driven live roller; T = two-strand chain transfer; BT = bypass transfer for missed or rejected loads; SZ = sizing stations.

Sizing and weighing of the unit load is usually accomplished on the conveyors that feed the cross aisle conveyor. Where the layout permits, the load is driven against a squaring stop and the length is read, then against another stop at 90° and the width is read. Photocells are most often used to detect dimensions that exceed an established limit. If the flow pattern does not permit this method of sizing, skewed rollers can be used to drive the load to a side guide, and then a retractable stop can be contacted. With two edges thus in known positions, the load can be scanned for overhanging dimensions. Weighing is on a short section of chain-driven live roller equipped with load cells. Readings can be taken with the load stopped or on the fly.

After sizing and weighing, provision must be made to correct or by-pass the rejected loads. As a general rule, a rejected load should be removed or corrected as close as possible to the sizing and weighing point. Where throughput permits, the load can be stopped at the sizing or weighing station to signal for manual correction. On a front-end system which is input only, a gravity runout can be provided at the first transfer after sizing or at the discharge end of the cross aisle conveyor where fork trucks can pick up the load for repair and reintroduction to the system. On a system with high throughput, a recirculation loop is often provided, with repair being accomplished during a pause on the by-pass transfer. In this case, it is important that the conveyors in the loop have a high load-carrying capacity and that adequate clearance be provided for oversized loads.

A typical AS/RS front-end conveyor is shown in Fig. 9.7.7.

9.7.6 Special Devices

The transfer of heavy unit loads from chain-driven live-roller conveyor is usually accomplished with two or more strands of chain that are raised between the rollers. Occasionally, when lighter loads of approximately 500 lb are involved, a right-angle pusher may be used to slide the load across the surface of the rollers.

The chain transfers may be hinged at one end at a point approximately one load length from the edge of the conveyor. The other end is supported by an air bag or a cylinder to raise and lower the chain between the rollers. In an alternate arrangement, the chain conveyor is confined within the width of the live roller and is raised vertically, with stabilizing torque arms provided to keep the chains level. For both types of chain transfers, air bags are used most frequently, whereas air cyclinders are used where high-speed operation requires that air power be provided for both up and down travel.

Whenever possible, the unit load is stopped before the transfer is begun. This permits driving against a fixed or movable end stop for squaring. The chains are started after they have raised the load, to maintain its square position.

When transfer on the fly is required, it is most effective when transferring from the chains onto the live roller. In the other direction, the chain comes up under a moving load and this is destructive to the bottom of the load. In any case, guide attachments should be provided on the chain or the track to keep the chain positioned.

Transfer from long two-strand chain transport conveyors is accomplished by raising a short section of chain-driven live rollers between the chains. The options of hinged or vertical travel and air bags or cylinders are similar to those detailed previously. In this case, the lift-up section of rollers can transfer the unit load to or from the chain on the fly.

90° corner transfers, of the type described earlier, are the most common and economical way of negotiating a 90° corner with a heavy unit load. However, it should be noted that this does change the orientation of the load with respect to the direction of travel. While this might not be a concern with slave pallets or other solid bottoms, alternate means should be considered when handling skids or other loads with bottom runners.

A 90° chain-driven live-roller curve does not change the orientation of the load. While these curves are available with a choice of straight or tapered rollers, the tapered rollers are preferred since they keep the load reasonably square. Tapered rollers are mandatory when handling skids or loads with only two bottom runners, because severe skewing cannot be avoided on straight rollers.

A turntable is another device that does not reorient the pallet with respect to the direction of travel. It is usually equipped with a chain-driven live-roller top, and is powered for rotation by a chain, gear, or friction motor drive, or by air or hydraulic cylinders. In addition to accommodating a 90° corner, it can be used in a "T" or a four-way intersection by making the roller top reversible.

CHAPTER 10
OVERHEAD LIFTING: CRANES, HOISTS, AND MONORAILS

LARRY FEIT

Crane Consultant
Hebron, Indiana

AMBROSE MAZZOLA

Materials Handling Consultants, Inc.
Simpsonville, South Carolina

ROBERT R. REISINGER

Acco Babcock Inc.
Hoist and Crane Division
York, Pennsylvania

J. D. MITCHELL

MAR Hook and Equipment, Inc.
Sandy, Oregon

10.1 TOP-RUNNING BRIDGE CRANES, GANTRY CRANES, AND JIB CRANES

Larry Feit

10.1.1 Single-girder Top-running Cranes

In the single-girder top-running crane, the bridge travels on rails and runway beams that are supported by stools on the building columns (fig. 10.1.1). This arrangement brings the bridge close up to the underside of the roof trusses. Thus, the top-running crane is adaptable to many applications where headroom is limited.

The top-running crane has its specific place in light industry and warehousing. It offers cubic operation, with rectangular orientation and a lifting range that meets the requirements of a wide range of crane users at a reasonable initial cost.

It is usually limited to 5-ton capacity, and 60-ft span. It can be obtained for both manual or electrical operation of all motions. This type of crane lends itself readily to use for maintenance in machine shops, pumping stations and similar operations where its use is intermittent.

10.1.2 Double-girder Cranes

The double-girder crane is distinguished from the single-girder top-running crane principally by the bridge construction, which utilizes two girders or beams. The hoist trolley runs on top of these two beams with the hook and rope usually dropped between them. There are various categories of these units, as follows:

Intermittent Industrial Cranes

This double-girder construction provides increased capacity, up to 20 tons, and span up to 60 ft.

The only lost space under the roof with this type of crane is that from the top of the hoist to the hook in its highest position (Fig. 10.1.2).

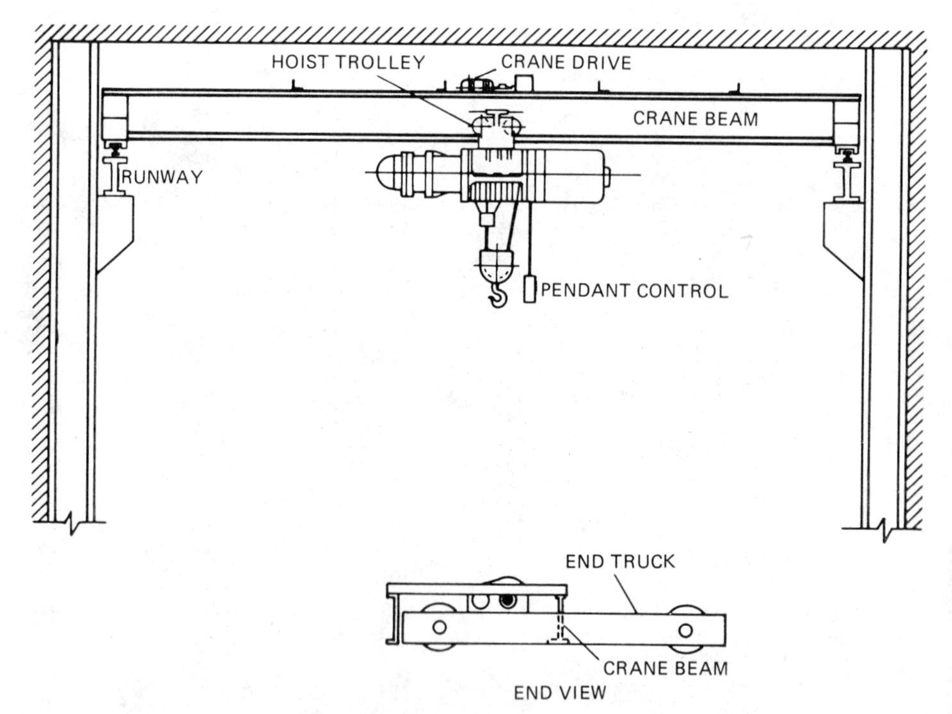

Fig. 10.1.1 Single-girder top-running crane.

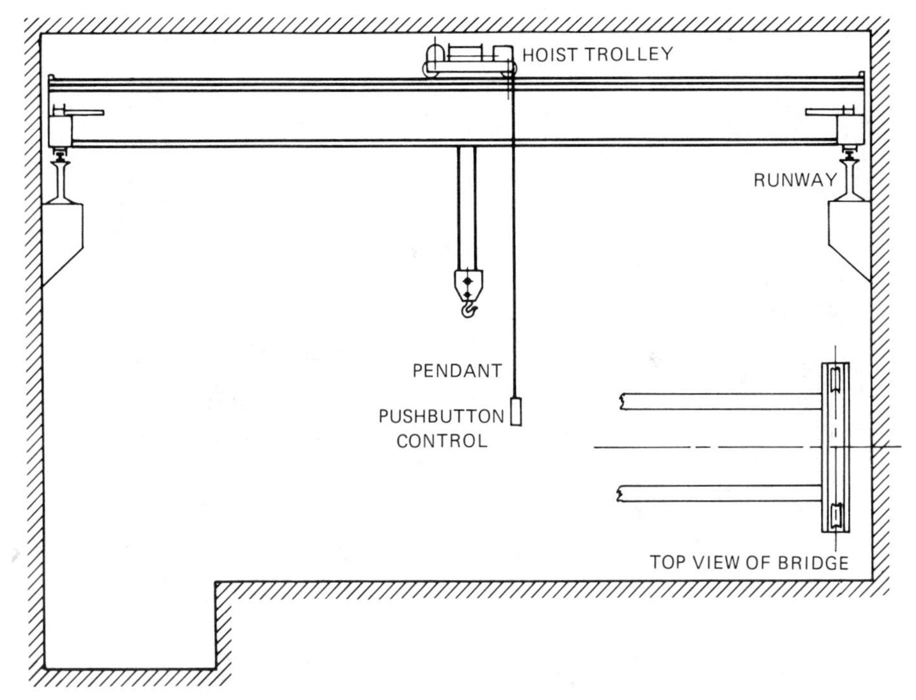

Fig. 10.1.2 Intermittent industrial crane.

Many manufacturers utilize a maximum of outside-purchased standard components for hoists, bridge, and trolley drives. These are usually more economical than the units found on completely custom-built cranes.

The span limitation is due to the fact that the bridge beams are constructed from open rolled sections rather than the box construction used for cranes with longer spans or capacities higher than 20 tons. The usual safety factor is 5/1 on the mechanics of this type of crane.

Most crane manufacturers guarantee this type of crane for one year. It provides a very serviceable unit with a much lower initial cost than its custom-built counterpart.

Powerhouse and Standby Cranes

Powerhouse and standby cranes, as the name implies, are for very intermittent service. Because of the operational demands, however, capacity runs from 40 to 200 tons, and spans up to 150 ft (Fig. 10.1.3).

They are invariably cab controlled. Speeds of all motions are relatively slow. Usually they operate from AC current, driven by wound-rotor motors with 3- or 5-step control.

Cranes of this type are usually found in hydroelectric and thermal power stations, big pumping stations, and heavy machinery warehouses.

The higher-capacity and long-span cranes of this type are frequently fitted with 8 to 16 wheels on the bridge.

General Service, Regular Industrial Cranes

General service, regular industrial cranes are custom built. This type is in widest use today. They are available in capacities ranging from 5 to 100 tons, and spans up to 120 ft (Fig. 10.1.4). Invariably these units are cab controlled.

These cranes are used in machine shops, foundries, fabricating shops, railroad workshops, and similar operations where the average cycle is about 15 lifts per hour. Usually designed with a safety factor of 5/1.

All wheels, axles, and shafts are fitted with antifriction bearings. Powered by wound-rotor motors with selective speed controls.

Economy designs are available using pin-type axles providing about 7000 hr. B10 bearing life.

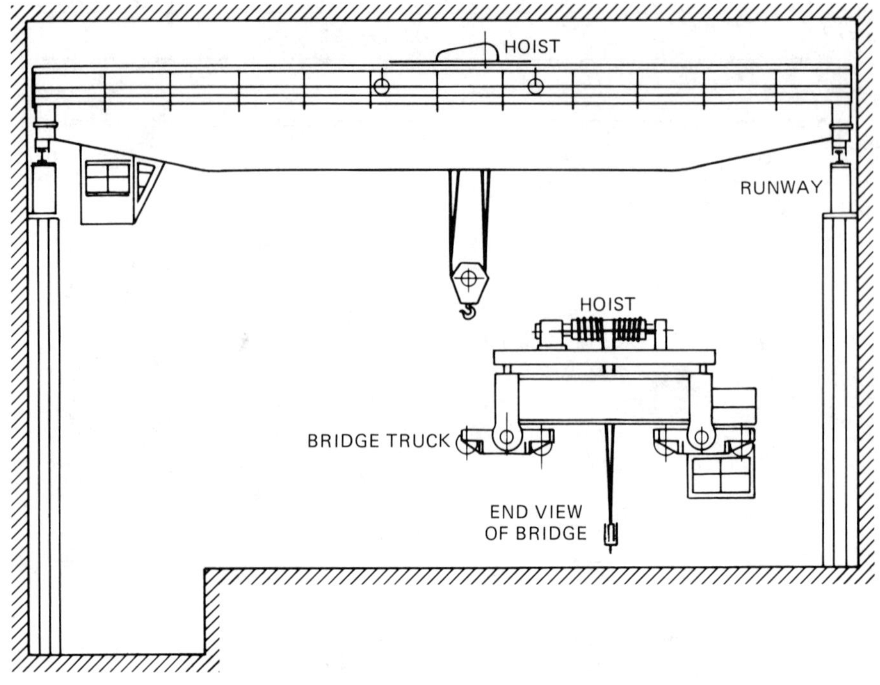

Fig. 10.1.3 Powerhouse and standby crane.

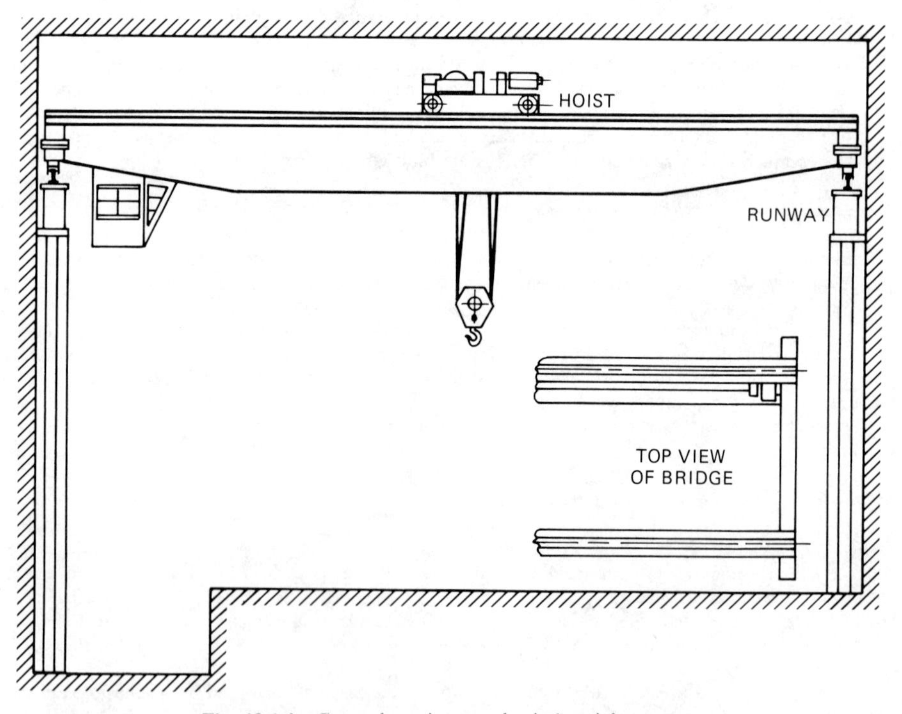

Fig. 10.1.4 General service, regular industrial crane.

These cranes usually have open trolley frames and no deck plate. Machined pads for motor and gear box support are replaced by shims for leveling the machinery.

The majority of catalogs issued by manufacturers are applicable to this group.

Besides the solenoid-operated electromagnetic brake on the hoist, cranes of this type use a mechanical load brake in the gear box that is capable of arresting the rated load in case of electrical or mechanical failure. All cranes and hoists are required to have a supplementary safety braking device. The extra cost of this feature is insignificant when included in the initial purchase.

Since this type of crane is custom built, many variations can be supplied with the off-the-shelf equipment available. Electrical control systems alone would fill several volumes.

Alternative braking systems used on this type of crane usually operate by feeding DC current back into the AC driving motor to act as a suppressor. This provides a cushion effect when lowering the load and provides an approximation of the precise control obtainable with DC power.

These cranes are also available for operating on DC current. However, since few if any power stations distribute DC power, the AC normally available must be converted to DC in order to take advantage of the superior characteristics of the DC mode. Controller manufacturers have endeavored with marked success to reproduce DC characteristics for such applications.

Rapid-handling, Continuous-duty Cranes

Rapid-handling, continuous-duty custom-built cranes are more sophisticated than the general-service type. These are built in accordance with AISE specification No. 6.

The significant characteristic is that all motions are performed at higher speeds than the fast industrial cranes. They are usually required to accommodate a duty cycle.

Cranes in this classification are usually bucket, pipe-handling, magnet, storage yard, and double hoist types (Fig. 10.1.5). They are used principally in mills, yards, docks, and foundries. Their function is the handling of scrap, coal, cement, stone, lumber, sand, fertilizer, and similar heavy loads.

These cranes are frequently employed as auxiliary mill cranes. In this application they have MCB bearings and safety factors of the order of 7/1. They are essentially a mill crane without the additional refinements such as tool steel gears, wheels, and similar components.

This type of crane can be powered by either AC or DC. When operating from AC the motors are either 60-min rated or continuous-rated, wound rotor, totally enclosed, and fan-cooled. Controls are designed for high duty cycles.

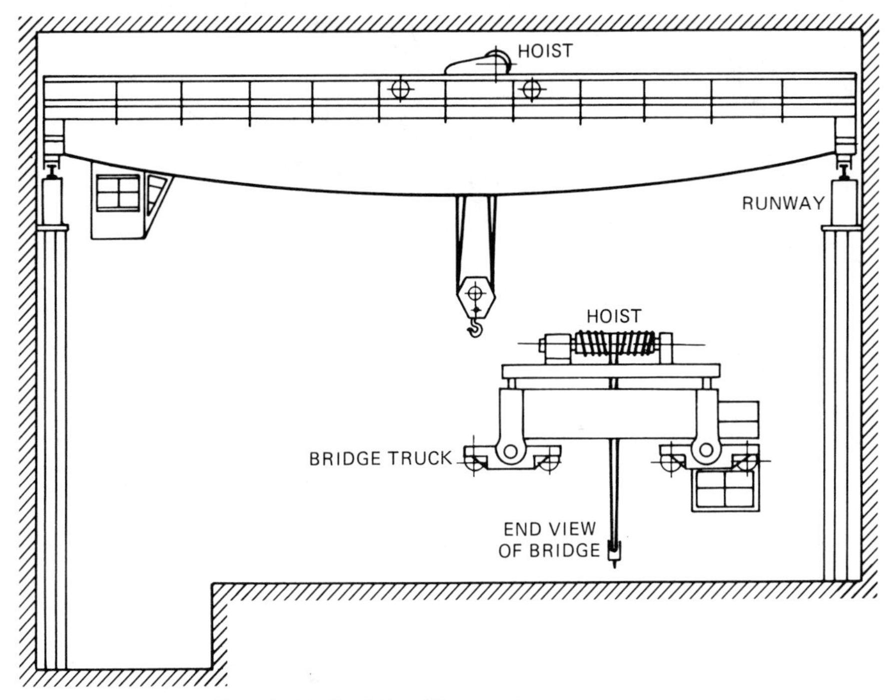

Fig. 10.1.5 Rapid-handling, continuous-duty crane.

10.1.3 Gantry Cranes

The full range of gantry cranes covers a wide field. They include unloading towers, dockside cranes, powerhouse gantries, and similar very large and heavy-duty installations.

There are three types of gantry configuration used in industrial installation. These are used primarily in manufacturing and powerhouses.

Gantries are built in a wide range of sizes, up to 150 ft span and as high as 60 ft. The capacity in such cranes is up to 150 tons. Larger ones have been constructed for special applications.

The heavy-capacity gantries are usually required on hydroelectric installations where head works and tail-race gantries provide high lifting capacity, but with very infrequent lifts. They are used primarily for installation of heavy equipment and subsequent intermittent maintenance.

The principal feature of the gantry crane which distinguishes it from other bridge types is that it incorporates its own vertical supports (Fig. 10.1.6). Gantry cranes fall into three categories: standard double leg, cantilever, and single leg.

In the standard gantry the bridge traversed by the hoist trolley is rigidly fixed at each end to supporting columns or legs. Wheels mounted on the bottom of the legs allow the crane to travel along a track of any reasonable length. Some small standard gantries dispense with the track and travel on any hard surface of suitable strength to support the rolling load.

The cantilever gantry is similar to the standard type, except that part of the bridge is cantilevered beyond the uprights at either or both ends.

The single-leg gantry is a special type or "half-gantry." The runway for one end of the bridge is mounted on the wall of a building or columns supporting some existing structure. The other support is normal gantry configuration.

The gantry crane offers advantages for many difficult applications. For instance, when a crane must be installed in an existing building with neither footings nor structure adapted for cranes, rails to carry a gantry can be laid on the floor. Current for its operation can be collected from the underside of roof trusses or from collectors mounted on the building columns.

Gantries serve their most useful purpose in industry for shipping and receiving. They have a distinct

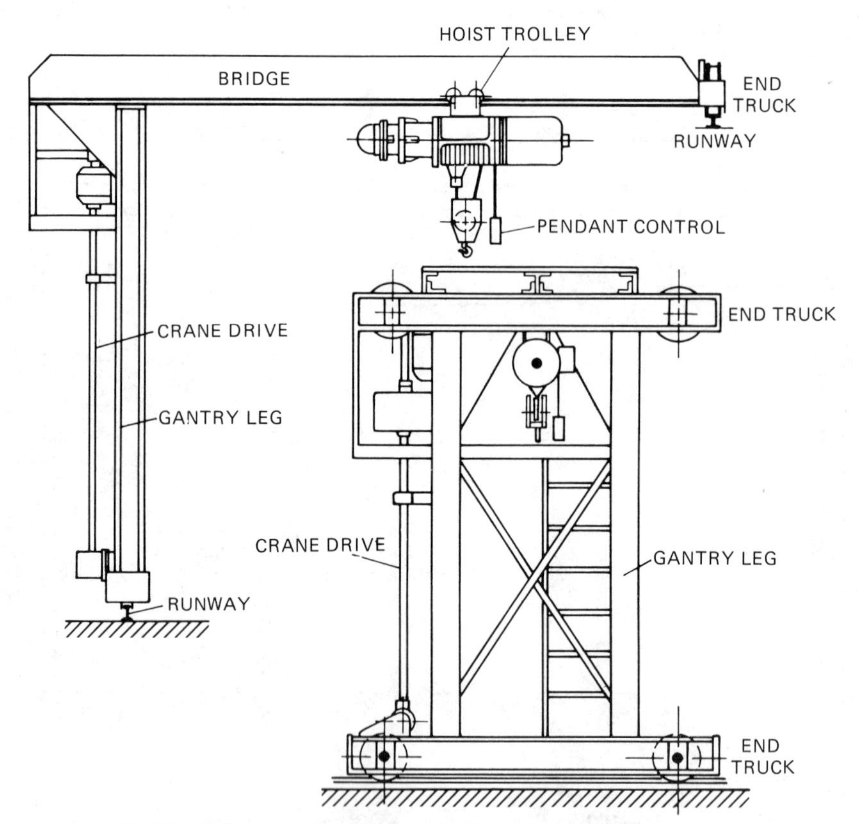

Fig. 10.1.6 Gantry cranes: single leg, top; double-leg, bottom.

advantage when runway extension is necessary. All that is required is additional concrete foundation and rail and extension to the electrical collector system.

Single-leg gantries are usually used in conjunction with internal overhead cranes. Such installations provide an integrated materials handling system of unusual versatility. They are employed widely in heavy industrial plants.

Many unusual features have been incorporated into gantries, such as self-contained diesel or gasoline generator sets to provide current for operation. Some installations operate on tracks that deviate from the usual straight line.

10.1.4 Jib Cranes

A jib crane is in effect a monorail that is cantilevered from its supporting members and pivoted at one end. The horizontal beam provides the track for the hoist trolley.

Jib cranes have three degrees of freedom: vertical, radial, and rotary. However, they cannot reach into corners. They are usually used where activity is localized, such as in machine shops. The following are basic configurations (Figs. 10.1.7–10.1.10).

Wall-mounted jib cranes usually are supported from the existing structure of the building. They have a swing of approximately 180°.

Floor or pit-mounted jib cranes also usually have 180° of movement. When the pivots are offset from the column, it is possible to swing the beam as much as 270°.

Floor-mounted jib cranes (having a column that is free to rotate) can swing 360°.

The walking-wall type is the most sophisticated and versatile jib crane. Its beam is cantilevered from two horizontal tracks mounted one above the other on building columns.

Many types of jib cranes have been designed for specific applications, but they are all basically variations of the types mentioned. Jib cranes may be manually or power operated. They are used frequently as auxiliary equipment with electric overhead cranes.

Standard jib-crane capacities range to 5 tons, and reaches range to 30 ft. Special units can be built with far greater capabilities.

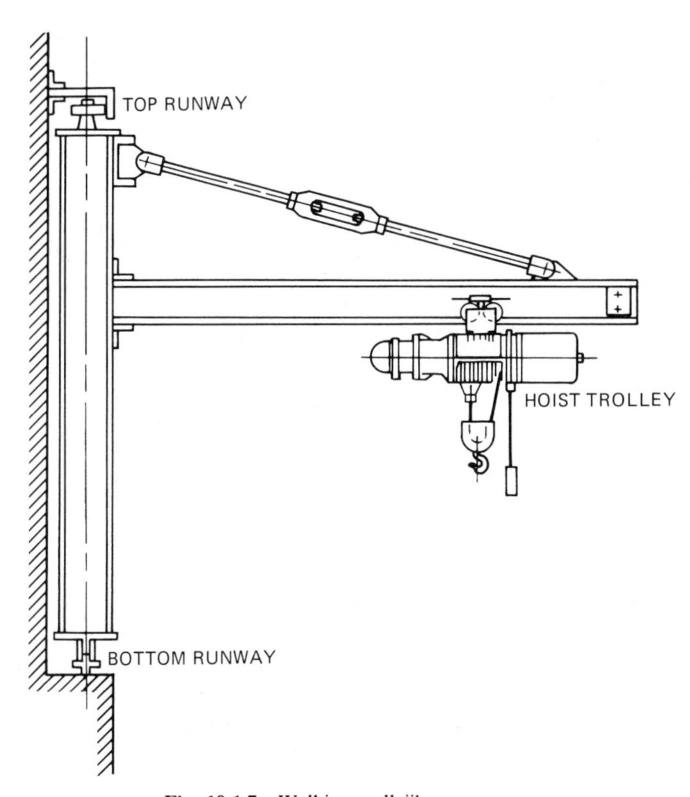

Fig. 10.1.7 Walking wall jib crane.

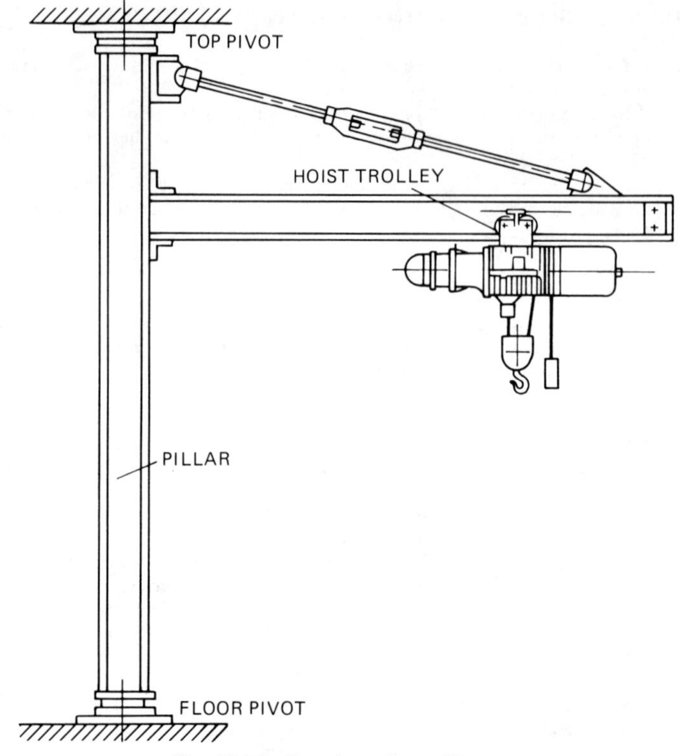

Fig. 10.1.8 Rotating column jib crane.

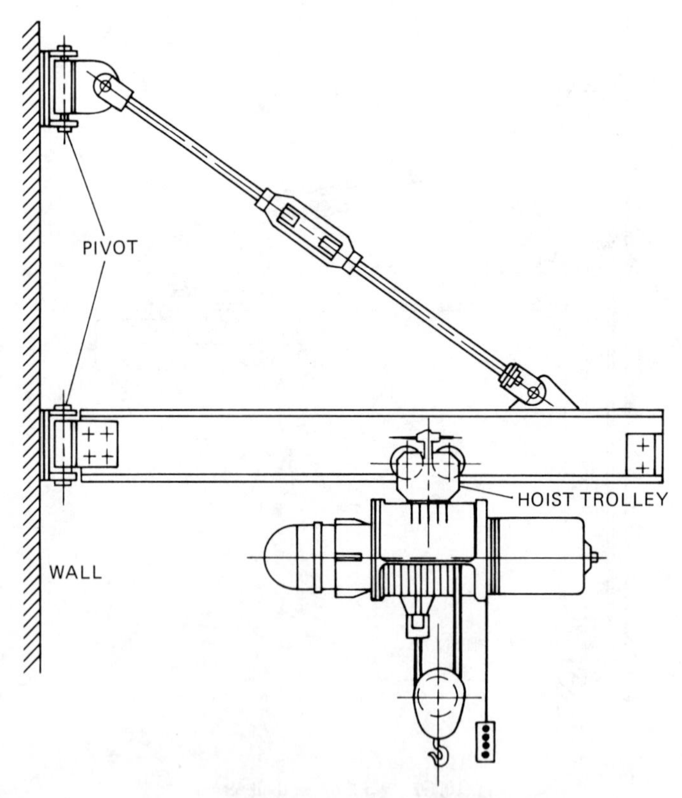

Fig. 10.1.9 Wall-mounted jib crane.

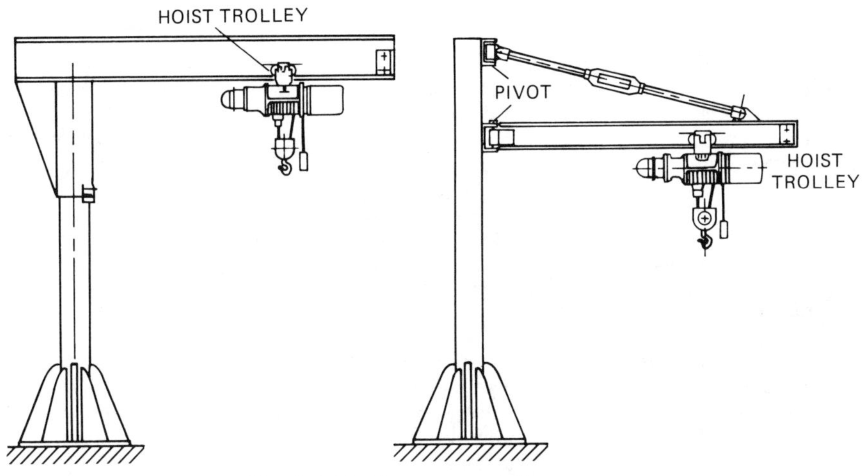

Fig. 10.1.10 Floor-mounted jib cranes.

10.1.5 Controls

Motor control systems are provided for all three crane motions. Controlling motor operation in DC systems is relatively easy. However, motor controls typically must be provided for AC motors.

Wound Rotor Motors

Maximum starting torque is obtained with the wound rotor motor. In a simple reversing wound rotor induction motor, the motor winding is connected directly to the power source through contacts operated magnetically. The secondary resistance governs motor speed and torque output. As more resistance is shorted out, motor speed and torque increase. Manual drum controllers can also be used to select control points on the resistors for wound rotor motors. However, advances in magnetic controller design have made this method of control obsolete.

On electric cranes, control accuracy is more important for hoisting and lowering motions than for travel motion. Few problems arise with bridge or trolley drive units.

Bridge and trolley reverse plugging and directional control employs simple resistance control of a wound rotor motor secondary. The principal disadvantage of the system is speed variation. A change in the weight of the live load can cause changes in speed. This approach is therefore limited to general hook service, where loads are heavy and similar in nature. It is usually employed for bridge and trolley electric crane motion in industrial applications.

Squirrel-cage Motors

Single-speed, squirrel-cage motors have high starting torque. Controls are used to provide smooth starting and gradual acceleration, thus preventing excessive swinging of the load. The following are different approaches that can be taken to control torque and acceleration.

Ballast Resistance Control

Resistance is permanently connected in the motor circuit to reduce load swing. There is no choice of torque steps once initial adjustments are made; generally used on low-capacity, light-duty cranes.

Multiple-speed Motors

A two-speed motor with dual windings operating at a 2:1 or 3:1 ratio provides a controlled approach to full-speed operation; usually it is used on hoists. Multiple-speed motors with up to five points of speed regulation are used on heavy-duty cranes. Resistors cushion the approach between different speed levels.

Fluid Clutch

No-load starting and gradual acceleration can be obtained with this device, generally for bridge and trolley of low-capacity, light-duty cranes. Seals must be maintained.

Autotransformer

Autotransformer circuits can provide soft starting for bridge and trolley drives. Taps on the autotransformer provide different voltages that can be applied to the drive motor to obtain different torque levels. Because of heat buildup, the practical crane capacity limit is about 20 tons.

Eddy-current Controls

In this method, an electromagnet is used to exert torque on the motor. An eddy-current load brake provides speed control for hoists. This system is particularly suited for hoists on large cranes, when heat generation may limit the utility of some other control systems.

Static Controls

Modular circuitry with saturable reactors or silicon-controlled rectifiers (SCR's) are used for stepped or stepless speed regulation of bridge, trolley, and hoist travel. Precise spotting and inching of loads can be achieved with these solid-state systems. In some instances they may be combined with other control approaches such as eddy-current controls.

10.1.6 Control Centers

Although manually operated drum controllers are still found on some old cranes, most crane controls are activated through magnetic contactors by the use of switches in an operator cab, push buttons on a floor-operated pendant, or remote radio controls.

Cabs are used when a heavy-duty application is involved, operating speeds are high, the crane is operated continuously, the operator must be removed from the floor because of obstacles or because an elevated view is needed, or hydraulic braking available with cabs is preferred to electric braking.

A cab usually requires a two-man crew. The cab operator stays at his station full time and communicates with a hook man on the floor through hand signals.

A push-button pendant permits one-man operation from the floor. And, when a pendant is used, a full-time crane operator may not be required. Various individuals can operate the crane with a pendant. The pendant station is suspended about 4 ft from the floor. When a pendant is suspended with festooned cables riding on the trolley, the operator can be positioned some distance from the load for safety or flexibility.

If it is desirable to keep the operator off the floor, a skeleton cab can be provided. The empty cab is not equipped with any controls, but can house the operator working a pendant.

Remote radio control is another way of keeping the operator off the floor or away from the immediate operating area. A radio control system consists of a portable transmitter worn or carried by the operator, an antenna and receiver on the bridge, an intermediate relay panel on the bridge to amplify the signals for crane contactors, and, in the case of DC cranes, a solid-state inverter to change DC to AC.

Radio controls permit one-man operation and a high degree of operator mobility on the floor. The operator can also be positioned in a skeleton cab, on a walkway, or at a stationary pulpit. Radio controls are generally justified on the basis of the high degree of operating flexibility and economy of operation they provide.

10.1.7 Crane Specifications

The types of classifications of cranes are many and varied. The dividing line between the groups is very thin. With such a wide range available, only detailed analysis will ensure that the crane you obtain is appropriate to the job. The Crane Manufacturers Association of America (CMAA) defines the following service classifications for overhead traveling cranes:

Class A (Standby or Infrequent Service)

Precise handling of equipment at slow speeds with long, idle periods between lifts. Typical installations are motor rooms, transformer stations, turbine rooms, and powerhouses. Rated-capacity loads may be handled for initial installation of equipment and for infrequent maintenance.

Class B (Light Service)

Lifting in repair shops, light assembly operations, service buildings, and light warehousing, where service requirements are light and speed is slow. Loads may vary from no load to occasional full-rated load, with two to five lifts/hr, averaging 10 ft per lift.

Class C (Moderate Service)

Applications in machine shops or other facilities where service is moderate. Crane handles loads that average 50% of rated capacity. About 5 to 10 lifts are handled per hour, averaging 15 ft, with no more than 50% of the lifts being at rated capacity.

Class D (Heavy Service)

Lifting in heavy machine shops, foundries, fabrication plants, steel warehouses, and standard-duty bucket and magnet operations where heavy-duty production is required. Loads approaching 50% of the rated capacity will be handled constantly during the working period. High speeds are involved, with 10 to 20 lifts/hr, averaging 15 ft, with no more than 65% of the lifts being at rated capacity.

Class E (Severe Service)

Loads approach rated capacity throughout the life of the crane. Applications include magnet, bucket, and magnet/bucket combination cranes for scrap yards, cement mills, lumber mills, and fertilizer plants. Application involves 20 or more lifts/hr, at or near rated capacity.

Class F (Continuous Severe Service)

Loads approach rated capacity continuously, under severe service conditions, throughout the life of the crane. Applications include custom-designed specialty cranes essential to performing critical work tasks affecting the total production facility. Cranes typically require high reliability, together with ease of maintenance.

Table 10.1.1 provides service classifications as defined by the Crane Manufacturers Association of America.

A crane is a capital equipment investment that should serve well for the next 30 to 50 years. Its versatility is such that any type of load can be slung, lifted, and moved within the limitations of its rated capacity.

The request to the manufacturers should give all information available, for example, location of job site, erection required, field test required, taxes extra, firm delivery, terms of payment. Submit an abbreviated specification with the inquiry, based on information available.

Frequently a reputable crane manufacturer will suggest a more economical unit, with specifications that vary from those requested. These are worthy of investigation, as, due to partial standardization of certain components, some manufacturers may find it more expensive to fabricate to a detailed specification.

Table 10.1.1 Load Classes and Load Cycles for Various Crane Service Classifications

	Crane Service Classification (A to F) for Various Load Cycle Ranges				
Load Class[a]	20,000 to 200,000 Cycles	200,000 to 600,000 Cycles	600,000 to 2 million Cycles	Over 2 million Cycles	K = Mean Effective Load Factor
L_1	A	B	C	D	0.35–0.53
L_2	B	C	D	E	0.531–0.67
L_3	C	D	E	F	0.671–0.85
L_4	D	E	F	F	0.851–1.00
	Application				
	Irregular occasional use followed by long idle periods	Regular use in intermittent operation	Regular use in continuous operation	Regular use in severe continuous operation	

[a] L_1 = Cranes that hoist the rated load exceptionally and, normally, very light loads.
L_2 = Cranes that rarely hoist the rated load, and normal loads of about $\frac{1}{3}$ of the rated load.
L_3 = Cranes that hoist the rated load fairly frequently and normally, loads between $\frac{1}{3}$ and $\frac{2}{3}$ of the rated load.
L_4 = Cranes which are regularly close to the rated load.

Source: Crane Manufacturers Association of America (CMAA) Specification No. 70-1983. "Specifications for Electric Overhead Traveling Cranes."

Data Required for Cranes

The following information should be supplied to the requested bidders:

1. Service. State the intended use of the crane.
2. Capacity. What is the maximum intended weight to be lifted?
3. Span. Center to center of runway rails.
4. Power supply. Volts, number of phases, and frequency.
5. Lift.
6. Dimension of top of runway rail to floor.
7. Dimension of top of runway rail to underside of roof truss.
8. Dimension center of runway rail to nearest side obstruction.
9. Length of runway and size of runway rail.
10. Runway wiring. Down shop conductors.
11. Bridge bumpers. Type required.
12. Control location. State preference.

In the event the crane is to be installed in an existing building or on an existing outside runway, a line sketch showing principal clearances should be included.

Additional equipment is frequently required on certain types of cranes. This should be specified on the inquiry. Typical items are: air conditioned cabs, sirens, alarm bells, motor generator for DC magnets on AC cranes, selenium rectifiers, take-up reels, extra walkways, emergency ladders, and floodlights.

Crane Proposal Analysis

An evaluation of crane proposals should be based on the analysis of the various components as follows:

Capacity	Bridge trucks	Main hoist block	Holding brake
Span hook lift	Bridge wheels	Main hoist details	Trolley frame
Service motors	Bridge drive	Main hoist gear box	Trolley wheels
Speeds	Bridge brake	Lower brake	Trolley drive
Controls	Foot walk	Holding brake	Trolley brake
Operator controls	Runway collectors	Aux. hoist block	Trolley collectors
Resistors	Operator's cab	Aux. hoist details	Cross bridge conductors
Protection	Bridge bumpers	Aux. hoist gear box	Trolley bumpers
Girders	Bridge rails	Lowering brake	Clearance dimension
			Loads and weights
			Additional specifications

Anyone contemplating the installation of an electric overhead crane should obtain and be guided by the following specifications:

AISE Standard No. 6-1949. Specifications for Electric Overhead Traveling Cranes for Steel Mill Service. Published by the Association of Iron and Steel Engineers, Suite 2350, Three Gateway Center, Pittsburgh, PA 15222.

CMAA. Specification No. 70. (Revised 1983.) Specifications for Electric Overhead Traveling Cranes. Published by the Crane Manufacturers Association of America, Inc., 1326 Freeport Road, Pittsburgh, PA 15238.

U.S.A. Standard Safety Code for Cranes, Derricks, Hoists, Jacks, and Slings.

U.S.A. B30.2.0-1967 Overhead and Gantry Cranes, published by the American Society of Mechanical Engineers, United Engineering Center, 345 East 47th Street, New York, NY 10017.

10.2 UNDERHUNG CRANES AND MONORAILS

Ambrose Mazzola

10.2.1 Introduction

Monorails and underhung cranes are grouped together for purposes of discussion because their development is tied together historically; because they use many of the same components such as rails, rail suspensions, trolleys, drives, electrical conductors, and controls; and because they are frequently used in conjunction with each other in integrated overhead materials handling systems.

Underhung cranes and monorail systems, including curves, switches, and vertical transfer sections, may be interlocked together so as to permit the transfer of loads between cranes, and between cranes and interlocking sections of monorail. Because of this unique characteristic, systems may vary in complexity from a simple monorail comprising nothing more than a short straight section of rail supporting a hand-pushed trolley, to systems of great complexity covering an entire building, or perhaps several buildings, all tied together with an interconnecting system of monorails, switches, and cranes, which are not necessarily limited to operation in a single horizontal plane.

Because of space limitations, many aspects of monorail and underhung crane design and application are treated briefly. Where further detailed information is required, the reader is referred to the various manufacturers' catalogs, many of which contain valuable engineering data, in addition to the specifications and detail dimensions necessary for proper application of the equipment under consideration.

10.2.2 Monorails and Accessories

Industrial monorails are available in five basic types, each of which has characteristics that should be considered in selecting the monorail system best suited to a specific application.

The basic types include the following:

Meat Track

The "meat track," so called because of its widespread use in the meat-packing industry, consists of a flat steel bar on edge, supported from one side, and employing trolleys with flanged wheels that ride along the top edge of the bar as shown in Fig. 10.2.1.

Meat track in common usage is limited to loads of about 500 lb or less, depending on bar size and hanger spacing.

Tubular Track

Tubular track is generally rectangular in shape, formed of sheet or plate steel, and provided with a continuous slot along the under side to permit passage of the load-carrying member, which is usually a hook or eye-bolt suspended from a two-wheel or four-wheel trolley riding *within* the tubular track (see Fig. 10.2.2).

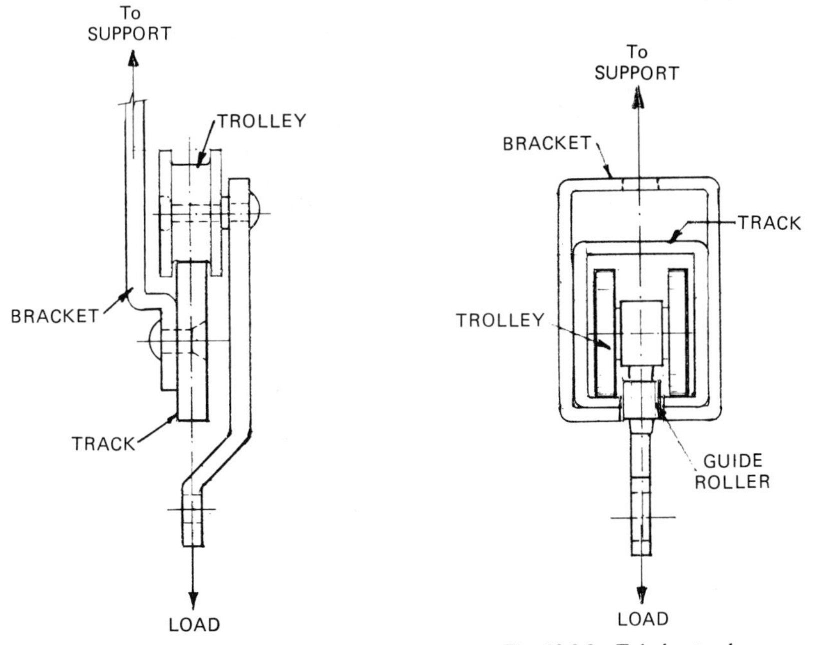

Fig. 10.2.1 Meat track.

Fig. 10.2.2 Tubular track.

Tubular track monorail systems are generally employed for loads of 1000 lb per trolley, or less. The actual capacity is determined by the size of the tube, the gauge (thickness) of the steel from which the tube is formed, and the distance between supports.

This type of monorail is available in several sizes, along with a variety of mating trolleys and suspension fittings. Load rating tables are published by the manufacturers of this type of equipment, and the actual loads imposed should be held within the manufacturer's recommended ratings.

Horizontal travel of the load-carrying trolleys for tubular-track monorails may be either manual or motorized.

In manual or "hand-propelled" systems, horizontal motion is generally obtained by the operator pushing on the suspended load.

Motorization of the horizontal travel, when used, is most commonly provided by means of a pusher-conveyor, comprising a motor-driven chain conveyor running parallel to, and a few inches distant from, the tubular-track monorail.

The trolleys to be propelled are engaged by a pusher dog, or pawl, which is attached to the driving chain. This method may be employed to propel either individual trolleys, or a group of trolleys simultaneously.

"Patented" Track

The patented track designation includes a group of monorail track sections of varying configurations, as shown in Fig. 10.2.3. Although these members vary in cross section, they share certain characteristics that are common to this class of equipment.

The generic term, "patented-track," is derived from the fact that over the years during which this class of equipment was developed, several patents were granted covering certain innovative features of these special track sections.

The tread width of the commercially available sections varies from 2 to 4 in., when used "as-rolled"; and the width must remain constant in any continuous system so that the trolleys may roll freely from one part of the system to another.

These types of monorail sections are rolled specifically for industrial monorail applications, from high-carbon steel, generally equivalent to an SAE 1060, in order to increase surface hardness to a minimum of 195 Brinnell, thereby extending the wear-life of the tread surface on which the trolley wheels bear.

The rolling tread surfaces are usually flat, rather than tapered; and some sections incorporate a raised tread wearing strip. When a raised wear-strip is provided, this portion of the track section, typically, is not included when calculating section properties, so that an allowance for wear is provided before encroaching on the strength of the monorail section.

The patented truck monorail sections, when used as-rolled, that is, without being fabricated into beams of greater depth, are generally employed on systems ranging in capacity from a few hundred

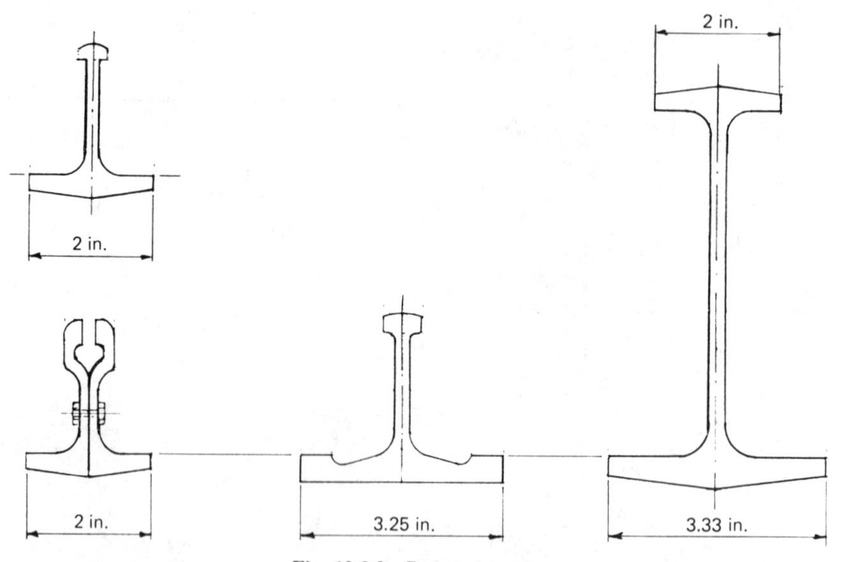

Fig. 10.2.3 Patented track.

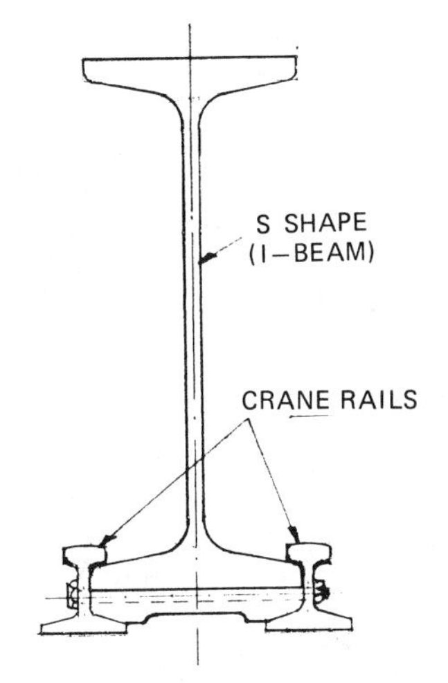

Fig. 10.2.4 Combination track.

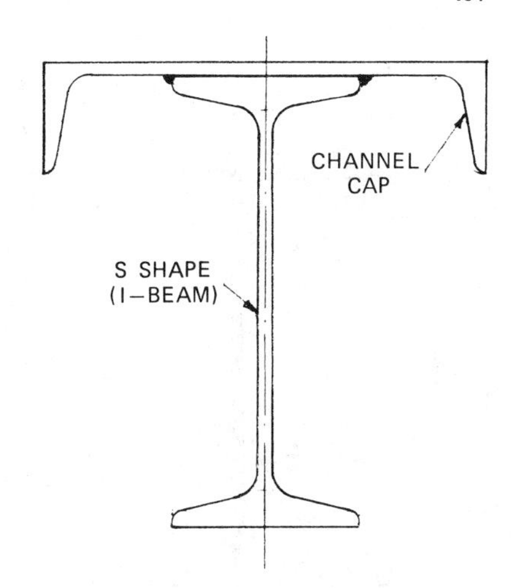

Fig. 10.2.5 Capped I-beam track.

pounds, up to about 2 or 3 tons. Actual capacity depends on the specific section under consideration, distribution of the load or loads, and the spacing between supports.

Here again, each manufacturer publishes load tables, and the actual loads imposed should not exceed the manufacturer's recommendations.

Loads are suspended from trolleys engaging the lower flange, and may be either hand-pushed or motor propelled.

I-Beam

Standard structural I-beam* sections have been used as the rail component of monorail systems for many years, in part because of their universal availability. However, except in applications where the duty cycle is low, or the service is intermittent, such as in maintenance or standby service, the standard mild steel, or A-36 steel, does not have sufficient surface hardness to provide long wear when used as a rail.

It is therefore recommended that when an I-beam system is employed, in normal service, that a higher-carbon steel, such as an SAE 1050 or SAE 1060 be specified for the rail, in lieu of the standard ASTM-A36, in order to increase the wear-life of the system.

Because of the wide range of sizes commercially available, the capacity of I-beam systems normally ranges from a few hundred pounds up to a maximum of about 12–15 tons. Heavier systems may be designed, but they are not common. Trolleys may be hand-pushed, or motor propelled.

For certain applications requiring heavier loads and a severe duty cycle, such as furnace-charging, combination rail has been used, in which two high-carbon ASCE rails are clamped to the edges of the lower flange of a standard I-beam in the manner shown in Fig. 10.2.4, in order to provide greater resistance to wear.

I-beams used for monorail service may be, and frequently are, capped with structural steel channels or angles in order to increase their load-carrying capacity, and in some applications, in order to provide greater lateral stiffness (Fig. 10.2.5).

Built-up Patented Track Sections

In order to obtain greater carrying capacity, most of the monorail track sections illustrated in Fig. 10.2.3, or slight modifications of them, are also used as the lower members of built-up, or capped, monorail tracks as shown in Fig. 10.2.6.

Built-up monorail sections range in depth from about 7 in. to 40 in. or more, with tread widths

---

* Current AISC practice is to refer to I-beam sections (old designation) as "S-shapes" (new designation). However, the old nomenclature is still current, *as a classification,* in the monorail industry when referring to I-beam monorail systems, as distinguished from patented-track monorail systems. Accordingly, and in that context, the term I-beam systems is used herein. However, it is to be understood that the call-out of such sections on engineering drawings should conform to the current AISC practice.

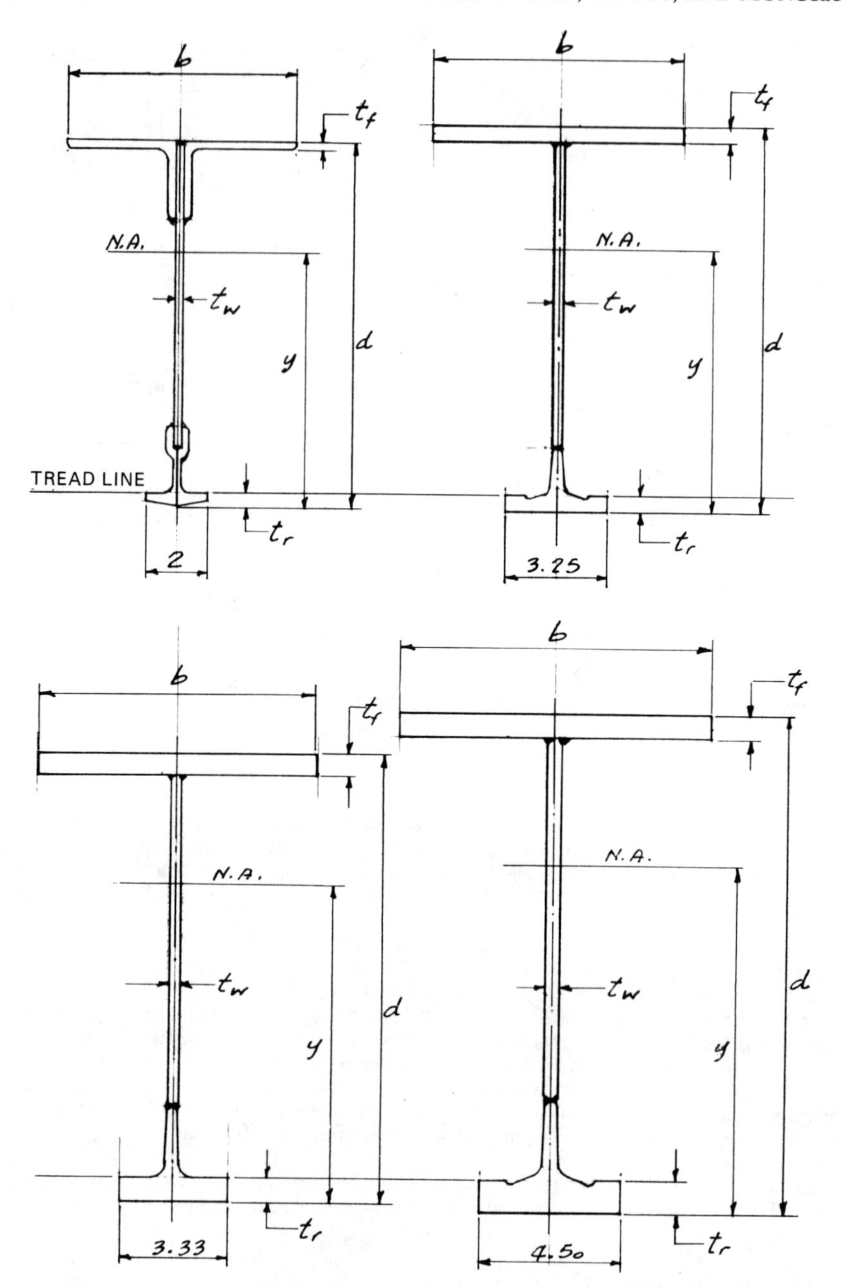

Fig. 10.2.6 Capped patented track.

of 2–4½ in. They are used in systems ranging in capacity from a few hundred pounds to about 25 tons.

The built-up rail incorporates the high-carbon flat-tread track section as its lower member, on which the trolley wheels ride; but the web and top flange, which form the cap, are generally made of ASTM-A36 structural steel. Some manufacturers split a wide flange beam to form two T sections. The web of the resulting T-shaped cap is then welded to the stem of the lower high-carbon rail member. Others employ a capping T made of two plates. In the latter construction, the web is welded to the top flange at the top, and to the high-carbon rail member at the bottom, to form the complete built-up rail section.

By using this method of construction, it is possible to assemble a track system employing monorail sections of varying depths, while retaining a uniform tread width so that the trolleys may pass freely from one part of the monorail system to another.

Although there is no theoretical limit to the number of variations in depth and top flange width that may be employed for capped monorail sections, in practice most monorail manufacturers have standardized on a relatively small number of sizes which cover the usual range of spans and capacities encountered in industrial applications.

It is possible to design an optimum rail for any specific application, that is, a rail designed to the absolute limit of the allowable stresses and deflection, so that in theory no steel is wasted. From a practical production standpoint, however, unless the application is one requiring a minimum of several thousand feet of nonstandard, custom-built rail, it is doubtful that any cost savings would be realized by doing so.

Accordingly, the rail sections in the standard sizes listed by the several monorail manufacturers are those most likely to be employed in the majority of industrial applications.

Monorail systems employing patented track rail as a bottom member may be used for hand-pushed systems in the lighter capacities; however, in the higher-capacity systems, the trolley travel is generally motorized. A variety of powered drives are commercially available, and their characteristics and application are covered in Section 10.2.6.

Curves and Switches

Industrial monorails are generally used to transport a suspended load from point to point along a predetermined path.

In their simplest applications, the points lie in a straight line and in a horizontal plane. Consequently, a length of straight monorail will suffice. The straight path may be very short, in which case a single piece of rail may be used; or it may be quite long, requiring several lengths of rail, spliced together at the joints, in order to cover the distance.

In most applications the points to be serviced by the monorail systems do *not* lie in a straight line. It is therefore necessary to provide curves in the monorail system; and where multiple destination points are involved, it becomes necessary to provide the monorail system with switching devices which will permit the traveling load to be switched from one section of monorail to another.

Thus, the scope of industrial monorail applications can range from a very simple short straight run, to long and very complex systems incorporating switches, turntables, and perhaps interlocking cranes, all interconnected so as to provide an integrated monorail system servicing numerous destination points in a plant.

If all of the area to be serviced is at one elevation, and if the supporting structure permits, then the entire monorail system, regardless of its complexity, may lie in a single horizontal plane.

However, if the loads are to be transported from floor to floor, or other circumstances require a change in rail elevation, then the system may be further equipped with devices for raising and lowering a section of rail, so as to permit the moving load to be transferred from one rail elevation to another. Such devices are generally called vertical transfer sections, because they provide a switching function in the vertical plane.

If the process requires either the lowering or raising of the load, *without* a subsequent transfer to a connecting monorail at a different elevation, as, for example, in a dipping operation, the device is generally designated a "drop section" or "lift section," and such sections do *not* serve a true switching function.

All types of switching devices are available without electrical conductor bars, for use on systems in which the load is lifted and transported manually through the use of hand-operated hoists and hand-pushed trolleys. Most are also available equipped with electrical conductor bars, so as to permit the use of electric hoists, and motorized trolleys. When so equipped, the conductor bars must generally be of the rigid type, and must be of the same, or compatible, manufacture, and match the spacing of the conductor bars on the incoming and outgoing monorails.

Most monorail manufacturers' catalogs provide sufficient dimensional data on their standard curves and switches to permit the layout of a monorail system. Such catalog data will also include, on electrified switches, the type of conductors employed and their exact location with respect to the monorail. It is important that this information be verified *prior* to attempting the layout of a monorail system.

The large majority of monorail switches are operated by hand, generally by pulling on one of a pair of hand-chains located at, or close to, the switch. However, for some systems, particularly those that are automatic in operation, it is necessary or desirable to motorize the switch-throw so that the switch can be operated from a push button, or other control device, located at a distance from the switch.

There are numerous switching devices commercially available, and although they differ in detail, certain general principles apply to all such devices. Generally, in the diagrams which follow, dimensions are omitted, except as they may be useful in illustrating a particular point; but such dimensions as may be required to lay out a specific monorail system are available from the various manufacturers. For the more common devices, such layout dimensions as may be required to design a system are published in their equipment catalogs.

Monorail Curves

Monorail curves are sections of monorail track, bent to a specified radius, and with the ends of the rail curve finished in the manner appropriate to the specific use for which the curve is intended.

The angle of arc subtended by the monorail curve may vary from a few degrees to 90° or more; although a single piece of rail is seldom bent through an angle in excess of 180°. For ease of handling, curves in excess of 90° are generally made in two or more sections which are then spliced together during installation.

Although there is no theoretical limit to the radius of a monorail curve, because of practical considerations, for normal industrial applications, monorail curve radii will lie in the range of 24–96 in.

The *minimum* practical radius on any particular monorail system is generally controlled by the wheelbase of the longest trolley operating on the system. For ease of operation, it is recommended that the minimum radius be not less than $1\frac{1}{2}$ times the trolley wheelbase.

The *maximum* radius is limited only by practical considerations of space required and the availability of an adequate supporting structure.

The end detail of a monorail curve varies, depending on its intended use. Where the curve is used simply to splice two rails together, the ends of the curve usually include a short tangent section, and each end of the curve is cut off square, thus providing for a simple splice to the connecting straight section.

On the other hand, if either end of the curve is connected to a switch, then that end must be cut in a manner compatible with the switch to which it is to be attached.

In general, on switch curves the following will apply:

	Angle of Curve End	
Type of Switch	Inlet	Outlet
Tongue	90°	90°
Glide	90°	30–45°

It is common practice to provide a short tangent section of track integral with the outlet curve, but because switch details vary among manufacturers, and even among switch models produced by the same company, the length of the tangent portion, and the exact angle at which the curve end is cut, should be coordinated with the details of the monorail switch to be employed. Here again, the dimensions shown in published catalog literature may serve as a guide, but exact dimensions for a specific application should be certified by the equipment manufacturer.

Monorail Switches

Monorail switches of the tongue or glide type are devices for switching a load traveling along a monorail, in a common horizontal plane, from the rail entering the switch, at its "heel" side, to any one of two or three rails leaving the "toe" side of the switch; or conversely, permitting the switching of a load entering the switch on the toe side, from any one of two or three entering rails, to a common outlet rail on the heel side of the switch. This basic concept is shown graphically in Figs. 10.2.7 and 10.2.8.

Cross track switches and turn-table switches are rotary switch devices, also designed for use in applications where all connecting tracks are in a common horizontal plane.

Vertical transfer sections are generally custom designed, and built to suit a particular application. Where they are shown and described in the several manufacturers' catalogs, they are usually included under the heading of track devices, or some similar designation. They are included here in the switch section because, functionally, they share much in common with other types of monorail switches, except of course, that they operate in the vertical plane rather than the horizontal.

An important distinction among the several types of monorail switches available is whether or not they are operable with a load on the switch. It should be noted that tongue switches, glide switches, and cross-track switches are designed to be operated *without* a load on the switch. In other words, the switch, *under no load,* must first be shifted or rotated to its desired position, and *then* the load may enter and pass through the switch to the connecting track on the other side.

On the other hand, turn-table switches and vertical transfer sections are designed to operate with a loaded trolley *on* the rail segment forming part of the switch. In operation, a loaded trolley from the entering rail is first rolled onto the turn-table or transfer section; the switching device is then rotated, raised, or lowered, as the case may be, so as to be locked in alignment with the outgoing track; and then the load is rolled off the track segment forming part of the switch, and onto the outgoing track with which it was aligned.

Tongue Switches. A tongue switch comprises a short section of monorail track, pivoted at one end (the "heel"), and movably supported at its other end (the "toe"), so that the movable, or toe, end can be aligned, and locked into position, with either of two or three connecting rails.

The swinging section of the track (the "tongue"), its pivot, and the support at the toe end are all mounted in, and supported from, a suitable frame. Typically, the frame includes a method of support, and means for attaching the ends of all connecting tracks, suitable latching devices for holding the toe end in proper alignment with the outlet track, and manual or powered means for throwing the switch from one position to another.

The tongue switch may be designated as a two-way, right hand or left hand; a two-way wye switch; or a three-way switch. The meaning of these designations is illustrated in Fig. 10.2.7.

Tongue switches are available with or without conductor bars to match the monorail system on which they are to be used. However, when employed with certain types of conductor bars, which require wide spacing between bars, both the throw and the length of the switch may become excessive, limiting the compactness of the monorail system in which they are to be used. For that reason, although tongue switches are usually somewhat less costly, glide switches may be preferred in electrified systems if the monorail spurs must be grouped closely together.

Glide Switches. A glide switch comprises a generally rectangular, stationary supporting frame, with means for attaching the incoming rails to the stationary frame; and from which is suspended a laterally movable subframe incorporating either two or three sections of track, together with locking devices for holding the switch rails in proper alignment with the incoming and outgoing rails; and means, either manual or motorized, for throwing the switch into one or another of its operating modes.

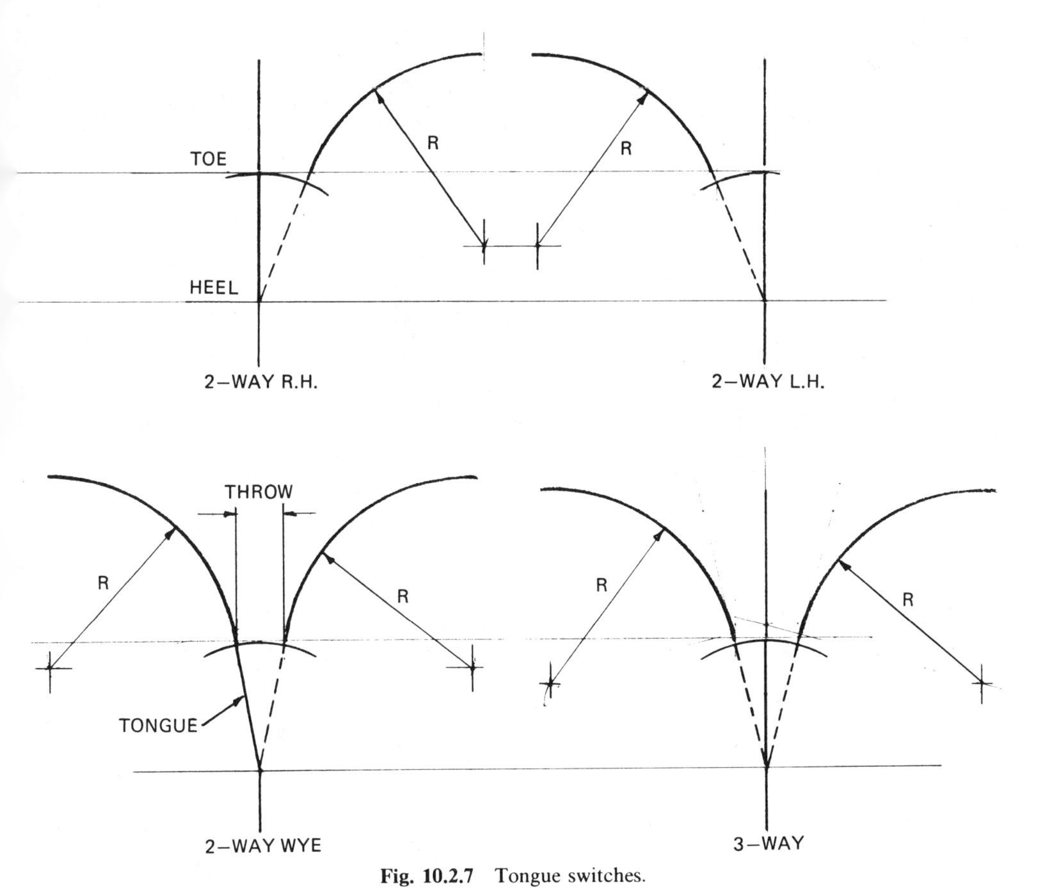

Fig. 10.2.7 Tongue switches.

As with the tongue switch, glide switches are available as two-way right or left hand, two-way wye, or three-way. Although the configuration is different from that of a tongue switch, as a matter of convenience, the heel and toe designations are retained, referring respectively to the side of the switch attached to the single incoming rail, and the side attached to the two or three outgoing rails.

The several types of glide switches are illustrated diagrammatically in Fig. 10.2.8.

Glide switches are available either with or without conductor bars. As with the tongue switches, when electrified, the minimum throw, and hence the *width*, of the switch may be determined by the spacing of the conductor bars. However, in a glide switch, the *length* of the switch is primarily a function of the radius of the switch curve, and is not necessarily increased when the throw and width are increased. So the influence of the conductor-bar spacing is not so pronounced in a glide switch, and for that reason, in an electrified monorail system, glide switches, as a general rule, are inclined to provide more flexibility in the layout of the track system.

Cross-Track Switches. A cross-track switch is a device that may be used at the intersection of two monorail tracks running at right angles to each other. When set in one position, it permits through travel on one of the two intersecting monorails; and when rotated 90°, it permits through travel on the other track, but not on the first.

For example, as shown in the illustration in Fig. 10.2.9, when the switch is in Position I (shown), loads can be transported through the switch from B to D; in Position II, loads can travel through the switch from A to C.

In either case, as noted previously, the empty switch must be thrown and set in one position or the other, *before* attempting to roll a loaded trolley across it.

Turn-table Switches. A turn-table switch is similar to a cross-track switch only in the fact that in both instances the switching is achieved by means of a rotary motion. There the similarity ends.

A turn-table is designed to accept the loaded trolley from the entering track, to rotate and lock into its new position *under full load,* and then allow the loaded trolley to enter the discharge track and proceed to its destination. A typical example is illustrated in Fig. 10.2.10.

Rotary Switches. Closely related to the turn-table switch, but somewhat more complex, is the rotary switch.

Like the turn-table switch, the rotating member is equipped with a segment of straight rail along one diameter. This center rail may be aligned with either of two monorail tracks disposed at right

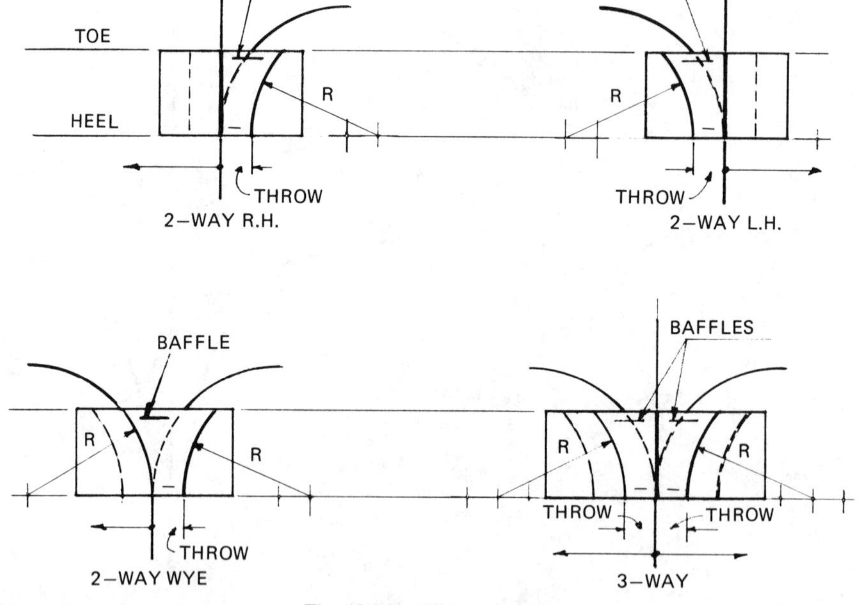

Fig. 10.2.8 Glide switches.

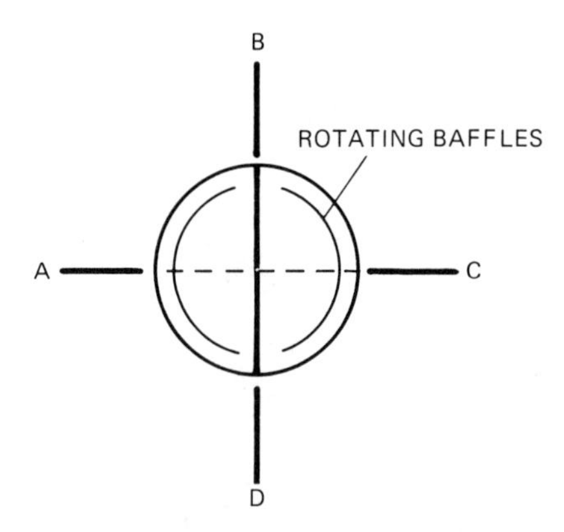

Fig. 10.2.9 Cross-track switch. Position I—shown; position II—90°.

angles to each other, accept a load, rotate 90°, under load, and permit the trolley to be discharged on to the intersecting track. To this extent it is similar to a turn-table switch.

However, the rotary switch is provided with two additional track elements, greatly increasing its versatility. As shown in Fig. 10.2.11, in addition to the straight segment of rail, the rotating member is equipped with two 90° curves, one on each side of the straight track.

With the switch in Position I (shown in Fig. 10.2.11), travel is permitted along the B–D axis. Rotation of the switch 90° (Position II) permits travel along the A–C axis. In the intermediate position (Position III, 45°) trolleys may negotiate two 90° curves simultaneously, permitting travel in either direction between points A and B, and between points C and D.

This switch is not designed to be rotated with a load on either of the curved segments; but may be operated with a load on the straight center track segment, in the same manner as a true turn-table switch.

In the illustration shown, if the switch is *nonelectrified,* it may be rotated so as to also permit travel between points A and D, and between points C and B.

However, if the switch is electrified in the normal manner, then this might lead to an undesirable condition known as "phase reversal," a phenomenon described in greater detail in Section 10.2.7. Accordingly, the switch may not be safely used in this manner without special provisions to compensate for the phase-reversal condition.

Vertical Transfer Sections. A vertical transfer section may be considered to be a special type of monorail switch, designed to be operated in the vertical plane, thus permitting the transfer of a loaded trolley from one monorail level to another at a higher or lower elevation. Units of this type are generally custom designed to suit specific applications.

The operating member comprises a short section of monorail track, generally straight (but not necessarily so), supported from a vertically movable frame, capable of limited up and down travel along stationary vertical guides.

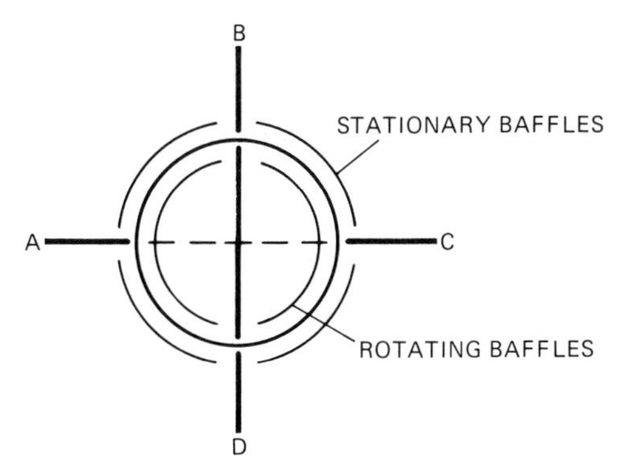

Fig. 10.2.10 Turn-table switch. Position I—shown; position II—90°.

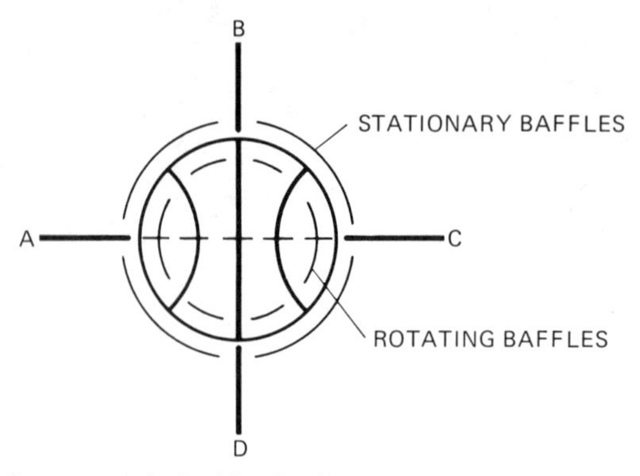

Fig. 10.2.11 Rotary switch. Position I—shown; position II—90°; position III—45°.

In operation, the movable section is aligned with the monorail at one level and locked into position; the load trolley is then rolled from the entering monorail onto the transfer section, and the unit is raised or lowered until it is aligned with the discharge rail. At this point, it is again locked in position, and the load may then travel off the transfer section and on to the discharge rail.

Vertical travel may be manual, as by means of a hand chain, or motorized by any one of several methods. Those most commonly employed include electric hoists, pneumatic or hydraulic cylinders, and screw jacks. The preferred method of powering the unit will vary according to the specific application under consideration.

Switch Safety Devices

Switches of all types must be provided with safety devices to prevent a trolley from rolling through an open end of monorail. The open end may occur, when in mid-throw, on one of the entering or discharging rails; or, as in the case of a turn-table, on the segment of rail that is part of the switch.

The safety devices may be mechanical, electrical, or a combination of both. Mechanical safety stops are usually included as an integral part of all standard switches; whereas electrical safety devices are generally custom designed to meet specific application requirements.

Mechanical. The most common mechanical safety device is called a switch baffle. A baffle is simply a section of steel plate built into the *movable portion* of the switch, which effectively blocks off any entering or discharge rail, unless and until the switch is accurately aligned and locked in one position or another. In other words, when the switch is in its mid-throw position, the baffles prevent a trolley waiting on any connecting rail from entering, or attempting to enter, the switch.

Switching devices that move under load (e.g., turn-tables) require additional baffles attached to the *stationary frame* to prevent a trolley on the switch rail segment from rolling off the switch, unless and until it is properly aligned with the discharge rail.

Mechanical switch baffles, as they are commonly applied, are shown diagrammatically in Figs. 10.2.8 through 10.2.11.

It is important to note that the effectiveness of a mechanical switch baffle, essentially an interference device, is closely related to the relative dimensions of the switch, the switch throw, the baffle, and the width and height of the trolleys normally traversing the switch.

When the switches and trolleys are the product of the same manufacturer, these dimensions are usually well coordinated. However, if additions are made to an existing system, or if for any reason components of the monorail system are from two or more sources, then it is important that specific dimensions and clearances be checked in order to be assured of the proper degree of safety.

Electrical. In addition to the mechanical baffles, some monorail systems may require electrical safety devices of varying degrees of sophistication.

In its simplest form, an electrical safety device might comprise a signal light, electrically interlocked to the switch, and color-coded to provide visual indication of the switch position with respect to the connecting rails.

Where the trolleys are motor-driven, a limit switch, or other electrical device, on the monorail switch may also be used to cut off the power on one or more of the connecting rails, for a limited distance back from the switch itself. By this method, further travel of the motor-driven trolley toward the switch is prevented, unless and until the switch has been moved into proper position to restore power, and thus to permit the trolley to travel through it.

When used in this manner, the segments of conductor bar adjacent the switch, on which power is automatically cut off when the switch is in certain positions, are sometimes called electric baffles because of the similarity of their function to that of mechanical baffles.

Other types of electrical interlocking may be, and frequently are, used on the more complex monorail systems. This occurs most often on automatic or semiautomatic monorail systems, in which case the electrical reading of the switch position may serve an operational as well as a safety function.

Switch Electrification

When switches are used as part of a monorail system that is equipped with conductor bars, then the segments of rail incorporated in the switch must be equipped with compatible mating bars, mounted in the same position with respect to the rail, so as to permit uninterrupted passage of the current collectors through the switch.

Generally, power must be maintained as the trolley travels through the switch, but since the switch rail is movable, special provisions are required to feed line current to the conductor bars on the switch rails. This is usually accomplished by means of a "switch harness."

A switch harness, when supplied, becomes an integral part of the electrified monorail switch. The harness comprises a pattern of rigid conduit, through which are run jumper wires, which are in turn flexibly connected to the moving conductor-bar segments on the switch, and solidly connected to the monorail conductors at each of the several entering rails. When these connections are properly phased, electrical power continuity is maintained as the load trolley travels through the switch in any direction.

10.2.3 Underhung Cranes

An underhung crane is one in which the crane bridge is suspended from, and rides below, its supporting runway rails (Fig. 10.2.12). Underhung cranes evolved from monorail systems, and are frequently used as an integral part of such systems through the use of rail interlocking devices which permit trolleys to travel from the crane bridge on to an interlocking monorail, and vice versa. Bridges on adjacent parallel runways may also be interlocked to each other, so as to permit transfer of the load from bridge to bridge.

The runways for an underhung crane comprise two, or more, parallel monorails running below, and suspended from, an overhead supporting structure. For indoor installations, the underhung runways may be, and most frequently are, suspended from the roof structure. For outdoor installations, or because of other special considerations, the runway support may be a freestanding structure designed and built for the purpose.

In either case, the runways are connected to their overhead supports by clamping or bolting, or by means of flexible suspension fittings. Because proper operation of the crane requires that the runway spacing and elevation be held to closer tolerances than those commonly employed in steel building construction, the runway support fittings must include means for both lateral and vertical adjustment.

While most underhung cranes are suspended from a single pair of parallel runway rails, a unique and important characteristic of underhung cranes is that they *may* be suspended from three or more

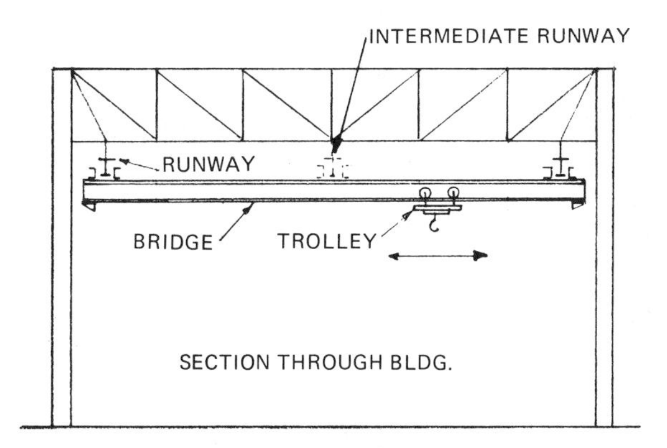

Fig. 10.2.12—Underhung crane.

runways. This feature permits a light bridge member to span relatively long distances because of the intermediate support, or supports, provided by the multiple runways (Fig. 10.2.13).

Three runway cranes are quite common, and underhung cranes with as many as 11 runways have been built in order to bridge clear spans of as much as 300 ft, such as may be encountered in an aircraft assembly plant, for example.

As in top-riding equipment, underhung cranes are commercially available in either the single-girder or double-girder configuration.

The most commonly used types of underhung cranes are shown schematically in Fig. 10.2.13, and are as follows:

Two-runway, single-girder
Two-runway, double-girder
Three-runway, single-girder
Three-runway, double-girder

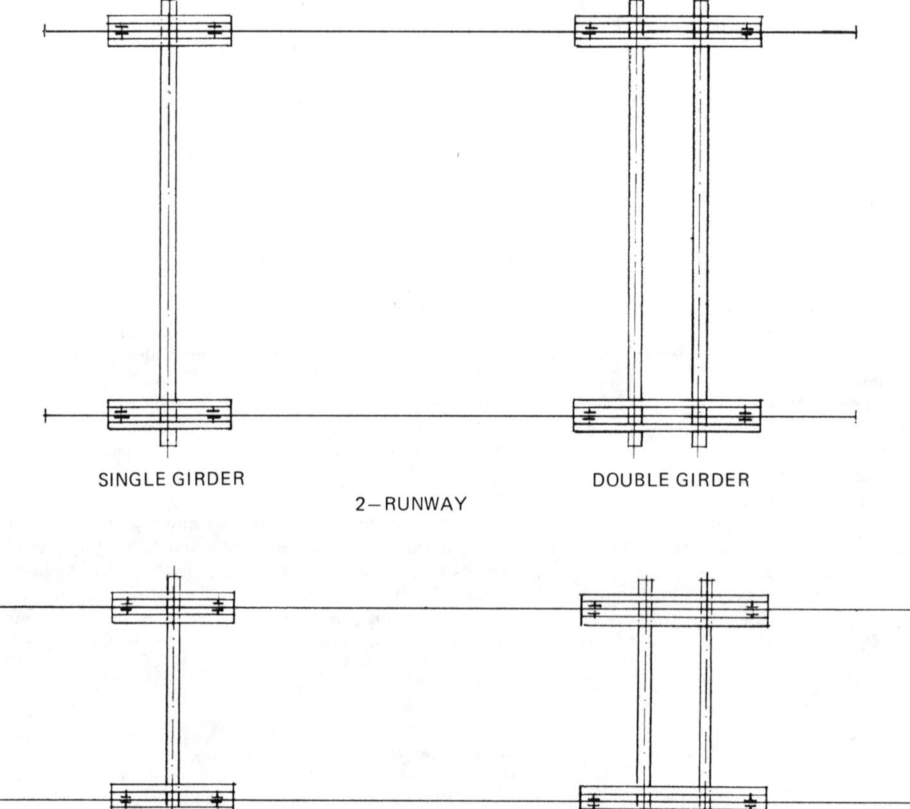

Fig. 10.2.13 Underhung crane types.

The type best suited to any particular application must be determined by an analysis of several pertinent factors including: the capacity of the crane; the width and length of the building; the type of overhead support available; the need to transfer loads from bridge to bridge, or from bridge to interlocking monorail; and, of course, once the functional requirements have been met, the relative costs.

For a given capacity and span, the two-runway single-girder type is generally the most economical; and if it meets the functional criteria, may be the best selection. However, the operating characteristics required may dictate other choices.

For example, a two-runway double-girder unit, which allows the hoist trolley to be located *between* the bridge girders, and well above the tread of the girders, provides greater headroom. Therefore, where headroom is critical, a double-girder unit may be the best choice. Or, the crane may require two bridge girders to properly support the weight of the hoist trolley and live load simply because of the span involved.

A three-runway single-girder crane should be considered when the span is too great to be handled by a two-runway unit, and interlocking with a monorail system, or other considerations preclude the use of a double-girder crane bridge. The third runway down the center of the building cuts the effective bridge span in two, thereby reducing both the depth and weight of the bridge girder required for a given capacity crane.

A three-runway double-girder crane combines the advantages of a lighter bridge for a given span, with the greater headroom obtainable by locating the hoist trolley up between the two bridge girders.

In general, the double-girder units provide better headroom; however, when used as part of an overall system, their flexibility is more limited than that of a single-girder bridge. This is because a single-girder bridge may be interlocked to a monorail system, permitting the transfer of the hoist trolley onto a connecting monorail system which may include curves and switches.

The double-girder cranes may also be interlocked, but because of the configuration of the double-girder hoist trolley, transferring is usually limited to a bridge-to-bridge type operation, or from the bridge to a relatively short, straight double-rail spur, in which the gauge of the double-rail spur matches that of the bridge girders. Because it requires double rails, it is generally not feasible to operate a double-girder hoist trolley through a complex monorail system which may include curves, switches, and other track devices.

For a given capacity and span, the three-runway units, whether the bridge is single-girder or double-girder, are generally lighter in weight than their two-runway counterparts. However, because of the center runway, their application may be limited by building considerations.

The center runway carries virtually the entire weight of the live load, plus the weight of the hoist and trolley, and a portion of the bridge weight. It is therefore essential that the building trusses be designed to support this load at, or near, the center of the span.

This usually requires a truss having a proper depth-to-span ratio, and designed to support the stipulated crane load, *in addition* to the statutory roof loads, and with the truss deflection held within very narrow limits. A rigid-frame type building is seldom capable of providing proper support for a center runway; and even trusses must generally be designed in advance to sustain the loads imposed by the crane system at midspan.

Regardless of the type of monorail or crane selected, or the method of suspension employed, the live loads imposed on the monorail or crane runways, the rail suspension fittings, and the building or supporting structure must be accurately determined. Some of the methods employed for calculating these loads are discussed in Section 10.2.4 which follows.

In order to travel along the runway smoothly, the wheelbase of the end trucks must be properly proportioned with respect to the span of the crane. If the wheelbase is too short, the crane may have a tendency to skew as it travels, especially when the hoist trolley is moved to a position close to either end of the bridge.

Skewing attributable to improper proportions may be kept under control, and smooth travel along the runway obtained, by adhering to the following ratios:

For two-runway cranes,

$$WB \geqslant \frac{S}{7} \tag{10.2.1}$$

For multiple-runway cranes,

$$WB \geqslant \frac{S_{mx}}{7} \tag{10.2.2}$$

where: WB = *minimum* end-truck wheelbase
S = span of two-runway crane
S_{mx} = maximum span between any two adjacent end trucks in a multiple-runway crane

Under certain special conditions, the span-to-wheelbase ratio may exceed these criteria slightly, but for best results in normal applications they should be adhered to. It is especially important that the end-truck wheelbase be generously proportioned with respect to the bridge span, when the crane is to be employed in an interlocking system.

Interlocking

As previously noted, a unique characteristic of underhung cranes is that the bridge may be interlocked to a monorail or another bridge. When used in interlocking applications, provision must be made to assure a smooth transfer of the trolley from one unit to the other.

Interlocking may be employed so as to provide for three separate and distinct functions, each of which requires special preparation at the points of transfer in order to assure proper operation. The three interlocking functions include:

1. Bridge-to-monorail transfer
2. Bridge-to-bridge transfer at preestablished fixed transfer points
3. Bridge-to-bridge interlocking and transfer at any point along the runway

CASE 1. BRIDGE-TO-MONORAIL. When interlocking and transferring a load from a crane bridge to a monorail, the end of the interlocking monorail must be supported *from* the runway rail, and *not* independently of the runway rail, so as to assure that both members, at the point of transfer, deflect equally and thus retain their vertical relationship to each other as the trolley travels across the narrow gap. This is accomplished by means of a special support commonly called a "gooseneck." The gap between rails is controlled by a horizontal guide roller and flared guide. See Fig. 10.2.14.

CASE 2. BRIDGE-TO-BRIDGE, FIXED LOCATION. In those applications where the load is transferred from bridge to bridge at a fixed point along the runway, a transfer rail between cranes is usually employed. Both ends of the transfer rail are supported from the adjacent runways, in a manner resembling a double gooseneck, and, in effect, the transfer rail becomes a short section of monorail, each end of which may be interlocked to one of the cranes. See Fig. 10.2.15.

CASE 3. BRIDGE-TO-BRIDGE, ANY POINT ALONG RUNWAY. Direct interlocking of one bridge to another at any point along the runway is employed primarily to permit transfer of the trolley from one bridge to the other. However, in certain applications, the two bridges may be operated and controlled from a single point when interlocked together so that, functionally and operationally, while in the interlocked mode they become, in effect, a single multiple-runway crane.

When used in this manner, the gooseneck cannot be employed. Instead, the differential deflection between the mating ends of the interlocking bridges is kept to a minimum by means of an additional set of vertical guide rollers. As in the previous cases, the horizontal gap is kept within operational limits by means of horizontal guide rollers (see Fig. 10.2.16).

When it is necessary to travel in the interlocked mode, and both cranes are to be controlled from a common point, then the interlocking mechanism must also include a series of sliding electrical contacts, so as to permit the transfer of the electrical control circuits from one crane to the other.

Hoist Considerations

A complete crane system includes the hoist and its supporting trolley, which provide the means for lifting the load and propelling it along the length of the bridge, while the bridge provides longitudinal

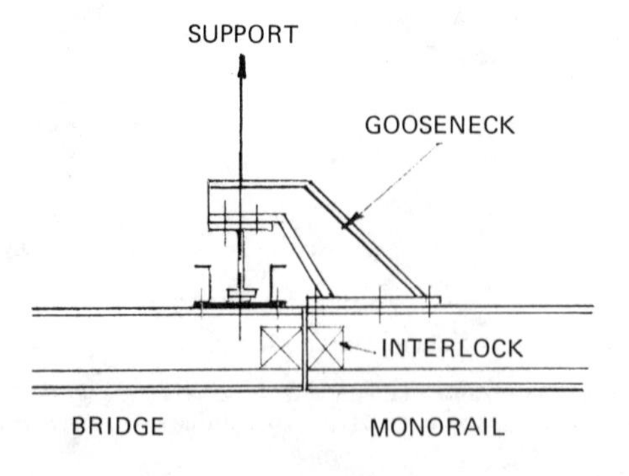

Fig. 10.2.14 Interlock, gooseneck. Case 1: Bridge to monorail.

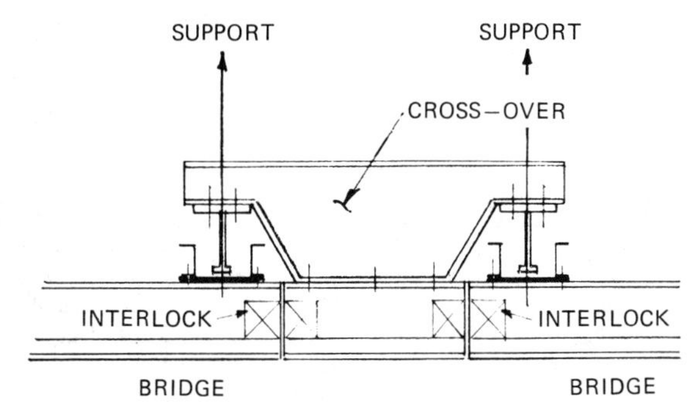

Fig. 10.2.15 Interlock, transfer section. Case 2: Bridge to bridge (fixed location).

travel along the runway. Because the bridge, unlike a stationary monorail, can move at right angles to, and simultaneously with, the trolley travel, the type of hoist selected for use on a crane requires special consideration.

As outlined in Section 10.3, wire rope hoists of the type commonly employed on monorails and cranes may be single-reeved or double-reeved. In a single-reeved hoist, the drum end of the wire rope travels from one end of the drum to the other as the hook is raised from its lowest to its highest position. As a result, as the hook is raised and lowered, it travels parallel to the axis of the drum a distance equal to approximately one-half the active length of the drum.

Therefore, in order to avoid an unbalanced condition when suspended from a monorail or single-girder crane, a single-reeved hoist must be hung with its drum parallel to the rail or bridge girder on which it travels.

On the other hand, in a double-reeved hoist, the wire rope is anchored to both ends of the drum, and reeved in such a manner that the drum ends of the rope travel simultaneously, and at equal speeds, from the ends of the drum in toward its center as the hook is raised; and from the center back out toward the ends of the drum as the hook is lowered. Thus, the hook is always centered with respect to the drum length, and the hoist may be balanced.

Because the hook remains centered, and the hoist balanced, a double-reeved, or "low-headroom," hoist may in theory be cross-mounted on either a monorail or a single-girder bridge. However, although this arrangement is generally satisfactory on a monorail, when it is used on a single-girder crane, the travel of the bridge along the runway, especially when starting or stopping, may induce swinging in a cross-mounted hoist.

If severe enough, the swinging could cause the wire rope to leave its proper groove. This would not only unbalance the hoist, causing it to tilt, but could lead to overwrapping and possibly serious damage. Although there are devices available that detect overwrapping and stop the hoist in the event it occurs, it is generally safer, on a single-girder crane, to employ a single-reeved hoist with the drum parallel to the bridge.

If the need for greater headroom requires the use of a low-headroom, cross-mounted hoist, then it may be well to consider the use of a double-girder bridge in which the hoist trolley may be mounted

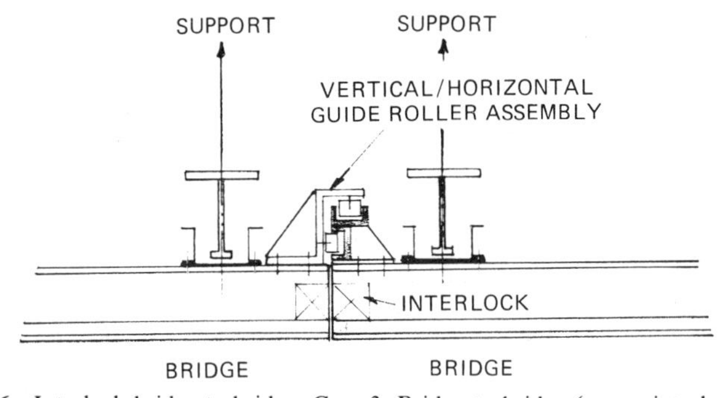

Fig. 10.2.16 Interlock bridge-to-bridge. Case 3: Bridge to bridge (any point along runway).

in a manner that prevents swinging. Or, as another but less desirable alternative, a single-girder crane may be equipped with stabilizer bars designed to limit the swinging of the hoist to very narrow, acceptable limits.

10.2.4 Rail Calculations and Suspensions

As pointed out in Section 10.2.2, rails used for monorails and underhung crane runways are manufactured in a wide range of sizes. Rail depths may vary in commercially available sections from as little as 2 or 3 in., up to 40 in. or more. Rail tread widths range from 2 to $4\frac{1}{2}$ in. in patented track, and from $2\frac{3}{8}$ to 8 in. in I-beam sections (S-shapes). Top-flange widths run the gamut from about $\frac{1}{2}$ in. in T-shaped light monorails, up to 18 in. or more in the heaviest built-up sections. Accordingly, section properties, which in turn control the load-carrying capacities of the monorail members, cover a very broad spectrum.

As a result, the lightest monorails may require supports on as little as 18–24-in. centers in order to support loads of a few hundred pounds, while the heavier members may safely support loads up to 10 tons or more on free spans of 60–80 ft.

The various manufacturers of monorail equipment publish load tables listing the maximum concentrated load that may be applied to the center of each size of rail they manufacture. The tabulated maximum loads vary inversely with the free span between supports. Additionally, the AISC Manual of Steel Construction tabulates the section properties of I-beam sections, from which the load carrying capacity may be calculated.

In order to determine the *rated* capacity, the capacity of the monorail section is first calculated using three different criteria, each of which imposes a maximum load limit, and the *minimum* of these three calculated values, less an allowance for the weight of the rail, becomes the tabular value appearing in the published load tables.

The criteria employed to determine the beam capacity are:

1. The maximum allowable tensile stress
2. The maximum allowable compression stress
3. The maximum allowable deflection

The maximum allowable tensile stress varies with the type of steel employed in the monorail section, but should not exceed 20% of the ultimate tensile strength of the material used in the lower flange.

Typically, in an I-beam section rolled from A-36 structural steel (F_u = 58–80 ksi), the tensile stress in the lower flange is held to 12 ksi, or less. In patented track, the lower flange is generally rolled from an SAE 1060 steel, or equivalent, having a carbon content of 0.55–0.65%, and an ultimate tensile strength (F_u) of 115–125 ksi. Accordingly, in this type rail the tensile stress in the lower flange is limited to 25 ksi, as a maximum.

In the top flange, the maximum allowable compression stress is governed in part by the type of steel employed, and in part by the ratio of the flange area with respect to both the depth of the beam and the length of the unsupported span.

Thus the maximum allowable compression stress is determined by applying the following equation:

$$F_b \leqslant \frac{12 \times 10^3}{ld/A_f} \text{ ksi} \tag{10.2.3}$$

but not more than 0.6 F_y

where: F_b = maximum *allowable* compression stress
F_y = minimum *specified* yield stress
l = unsupported span in inches (mm)
d = depth of track in inches (mm)
A_f = area of top flange in square inches (mm)

On longer spans, with I-beam systems, where compression may be the limiting factor, it is not unusual to cap the I-beam with a channel or pair of angles in order to increase both the area of the top flange and the section modulus of the section in compression (S_c), thus increasing its load-carrying capacity (see Fig. 10.2.5).

The *maximum* allowable deflection, in good practice, is limited to $\frac{1}{450}$ of the span, but not more than $1\frac{1}{4}$ in. (32 mm). For some applications, the allowable deflection may be limited to $\frac{1}{600}$, $\frac{1}{800}$, or even $\frac{1}{1000}$ of the span. When handling loads that might be adversely affected by vibration, it is advisable to specify the more stringent deflection criteria.

The actual maximum stresses allowable will, of course, vary with the type of steel comprising the monorail section; but the values given are typical for many of the monorail sections commercially available at this time.

Since there is considerable variation in the detail dimensions of monorail sections available, each manufacturer publishes tables giving the dimensions, properties, and load capacities for the several

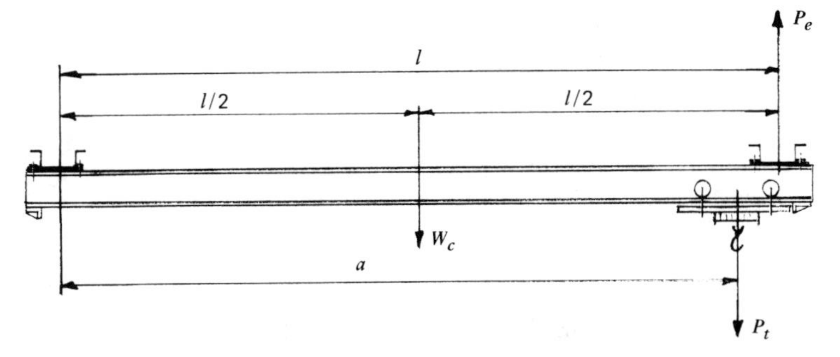

Fig. 10.2.17 Maximum end-truck load.

sizes of rail it produces. Separate load tables are frequently published for each of the commonly adhered to deflection ratios, that is, $\frac{1}{450}$, $\frac{1}{600}$, $\frac{1}{800}$, and $\frac{1}{1000}$ of the span.

Since the tabular values include only the *least* of the three calculated capacities for each rail size and span, the tables usually indicate whether the listed capacity is limited by tension stress, compression stress, or deflection. The published values should be, and generally are, *net* capacities with an allowance for the weight of the monorail having been deducted.

As stated earlier, the published values indicate the maximum single concentrated load that may be applied at the center of the span. In practice, the actual load is usually suspended from one or more monorail trolleys or end trucks, and is therefore transmitted to the supporting monorail, or crane runway, at two or more points. It therefore becomes necessary, in order to use the published load tables, first to convert the actual distributed load into an equivalent center load, or ECL, as it is commonly designated in the industry.

Determination of ECL

The first step in determining the ECL is to calculate the total design load P imposed on the monorail, bridge, or crane runway by the loaded trolley or end truck.

For a monorail system, or a crane bridge, the total load imposed on the monorail, by *each* trolley in the system, is determined as follows:

$$P_t = W_h + W_t + LL + I \tag{10.2.4}$$

where: P_t = maximum design load imposed by trolley
W_h = weight of hoist
W_t = weight of trolley
LL = live load, including grabs, or other handling devices, that may be suspended from, and lifted by, the hoist
I = impact allowance

For an underhung crane system, the total load imposed by *each* end truck on each runway in turn, is determined as follows:

$$P_e = \frac{a}{l} P_t + \frac{W_c}{2} \tag{10.2.5}$$

where: P_e = maximum design load imposed by end truck
P_t = maximum design load imposed by trolley
a = distance from one end truck to center of gravity of P_t, with trolley positioned so as to impose maximum reaction on end truck (Fig. 10.2.17)
l = span of bridge, center-to-center of end trucks
W_c = weight of bridge

The foregoing is based on a symmetrical two-runway bridge.

For a three-runway bridge, position the hoist c.g. directly under the center end truck, and use *half* the weight of the bridge ($W_c/2$) at the *center* runway. Use *one-quarter* of the total bridge weight ($W_c/4$) at each of the *outside* runways, again positioning the trolley so as to impose the maximum reaction on the end truck under consideration.

The impact allowance I varies with the type of equipment and the hoisting speed. For general service applications the following guidelines conform with current good practice, and meet MMA specifications:

For manually operated hoists: $I = 0\text{–}0.05\ LL$

For powered hoists: $I = \frac{1}{2}\%$ of LL per each foot per minute of lifting speed, but not less than $0.15\ LL$

For powered hoists, in bucket and magnet applications: $I = 0.50\ LL$

Once the total trolley or end-truck load has been established, the ECL may then be accurately determined by methods in the following section.

Calculation of ECL

In the design of a monorail or underhung crane system, the determination of the ECL, which takes into account the effect of the distribution of the total design load P, is a frequently recurring, and important, calculation.

Accordingly, certain design methods have evolved in the industry which save engineering time, yet provide generally acceptable results. These methods are presented on the following pages, and for convenience have been grouped into three categories, as follows:

1. Equal wheel loading, four-wheel trolley
2. Equal wheel loading, eight-wheel trolley
3. Unequal wheel loading, general solution

CASE 1. EQUAL WHEEL LOADING, FOUR-WHEEL TROLLEY OR END TRUCK.

$$P_1 = P_2 = \frac{P}{2}$$

In Fig. 10.2.18, the four-wheel trolley or end truck is shown in the position that will produce the maximum bending moment in the rail. In the figure:

P = maximum design load (equally distributed)
P_1, P_2 = trolley loads
l = rail span between supports
a = wheelbase of trolley or end truck

To determine the ECL, the design load P is multiplied by the coefficient C, so that

$$ECL = C \times P \tag{10.2.6}$$

in which C = coefficient due to load distribution.

When the wheelbase is relatively short with respect to the span, that is, $a \leqslant l/4$, then the approximate value of C may be calculated by the use of a very simple equation. Although the result is an approximation, the error is on the order of 2% or less, and because of its simplicity the following equation is widely used within the monorail industry.

$$C \cong \frac{l-a}{l} \tag{10.2.7}$$

Where greater accuracy is necessary, or desired, then the exact value of C may be determined by the following:

$$C = \frac{\left(l - \frac{a}{2}\right)^2}{l^2} \quad \text{but not less than } 0.50 \tag{10.2.8}$$

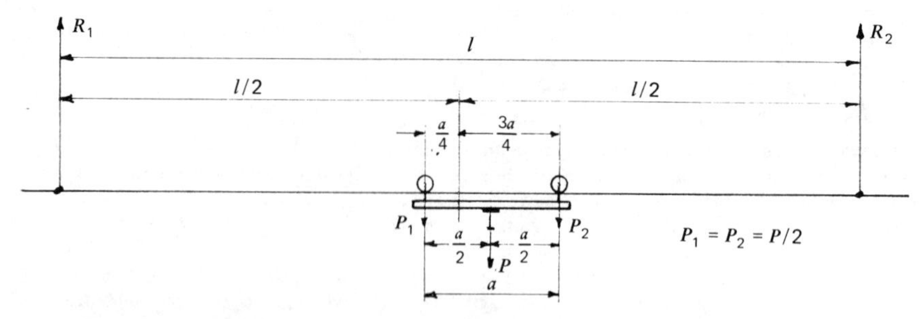

Fig. 10.2.18 Maximum bending moment with four-wheel trolley or end truck.

The above equation applies for all values of a up to and including $a = 0.586\ l$. For values of a greater than $0.586\ l$, $C = 0.50$.

CASE 2. EQUAL WHEEL LOADING, EIGHT-WHEEL TROLLEY OR END TRUCK.

$$P_1 = P_2 = P_3 = P_4 = \frac{P}{4}$$

In Fig. 10.2.19, the eight-wheel trolley or end truck is shown in the position that will produce the maximum bending moment in the rail. In the figure:

P = maximum design load (equally distributed)
P_1, P_2, P_3, P_4 = trolley loads
l = rail span between supports
a = principal wheelbase of end truck or trolley
t = wheelbase of auxiliary four-wheel trolleys

As in Case 1, C may be calculated by use of (10.2.7),

$$C \cong \frac{l - a}{l}$$

when $a \leqslant l/4$, and $t \leqslant a/4$, after which, by (10.2.6),

$$ECL \cong C \times P$$

This takes into account the fact that when $t \leqslant a/4$, the influence of the spread of the auxiliary trolley wheels is minimal, and for ease of calculation each pair of auxiliary trolley loads, $P_1 + P_2$ and $P_3 + P_4$, may be assumed to be acting at a single point.

As in the previous case, this results in an approximation in which the magnitude of the error is on the order of 2% or less; and because of its simplicity this method is commonly used to determine ECL for the purpose of selecting a rail from the published load tables.

However, when $t \geqslant a/4$, or where greater accuracy is required, for any reason, the exact value of C may be determined by the equation:

$$C = \frac{0.25(2l - a + t)^2 - lt}{l^2}, \text{ but not less than } 0.50 \qquad (10.2.9)$$

When a is greater than $0.586\ l$, the maximum bending moment is obtained by shifting the end truck along the rail so that only one of the auxiliary four-wheel trolleys is within, and near the center of, the span under consideration. That will place the auxiliary trolley at the opposite end of the endtruck, well beyond the nearest support and into the adjacent runway span, where it will have no effect on the ECL calculation.

When this situation occurs, the auxiliary four-wheel trolley is next positioned so as to produce the maximum bending moment in the span being analyzed, as shown in Case 1 (Fig. 10.2.18) and the calculation may then proceed in the same manner as in the case of a four-wheel trolley or end truck (Case 1, Eq. 10.2.6).

CASE 3. UNEQUAL WHEEL LOADING, GENERAL SOLUTION. The preceding illustrations, Cases 1 and 2, cover a very special set of conditions in which the design load P is equally distributed among

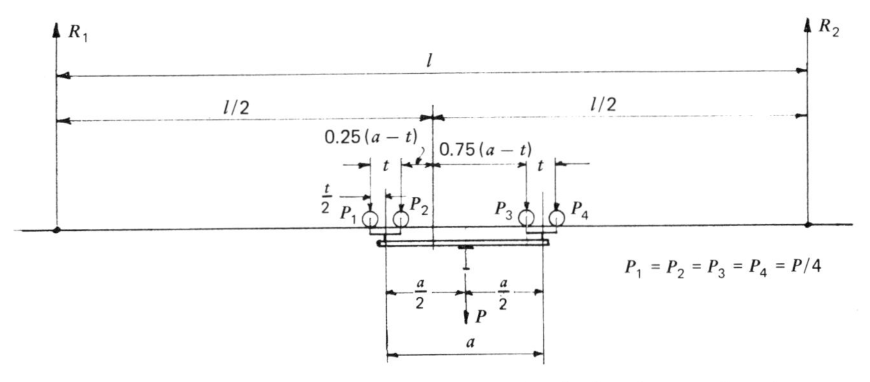

Fig. 10.2.19 Maximum bending moment with eight-wheel trolley or end truck.

the supporting trolley wheels. While this condition occurs quite frequently in monorail and underhung crane applications, especially when there is only one trolley on a monorail system, or one crane on a runway, there are many other applications in which the total design load is *not* borne equally by the trolleys and wheels that can accumulate in a single span.

So while the methods previously described find widespread usage where they are applicable, primarily because of their simplicity, it becomes necessary in many instances to employ a general solution that is valid regardless of the distribution of the design loads.

The method outlined in the following paragraphs is applicable to any combination of wheel loads, including those previously described in Cases 1 and 2, but is not limited by the stipulations set forth in the previous cases. In using the general solution, the ECL is determined directly, and accordingly the calculation of a distribution coefficient C is redundant and unnecessary.

The general solution requires five basic calculations following, and in addition to, the determination of the design load per two-wheel trolley by the methods previously described.

The five-step solution includes the following:

1. Calculate c.g. of all loads in span.
2. Position loads (trolleys) along beam so as to produce maximum bending moment.
3. Calculate end reactions at rail supports.
4. Calculate maximum bending moment.
5. From maximum bending moment, calculate the ECL.

STEP 1. Calculate the c.g. of all loads in span.

Definition. The location X of the center of gravity (c.g.) of any group of parallel downward forces acting in the same plane, with respect to any assueed point of rotation, such as the point o, is that distance x from the point o at which the sum of the forces (ΣP) acting together produces a moment equal to the sum of the moments (ΣM) produced by the individual forces acting alone.

This may be expressed by the equation:

$$X = \frac{\Sigma M}{\Sigma P} \tag{10.2.10}$$

Figure 10.2.20 illustrates this principle. Since,

$$M_1 = P_1 l_1 \quad M_2 = P_2 l_2 \quad M_3 = P_3 l_3 \quad M_4 = P_4 l_4$$

then

$$X = \frac{\Sigma M}{\Sigma P} = \frac{P_1 l_1 + P_2 l_2 + P_3 l_3 + P_4 l_4}{P_1 + P_2 + P_3 + P_4} \tag{10.2.11}$$

$$e = X - l_2 \text{ (distance between c.g. and nearest wheel)}$$

As stated in the definition, the point o may be assumed to be anywhere along the line x–y. However, as a practical matter, it is generally most convenient to locate it coincident with force P_1. By doing so, $P_1 l_1 = 0$, and the calculation becomes somewhat shorter.

STEP 2. Position loads along beam so as to produce the maximum bending moment.

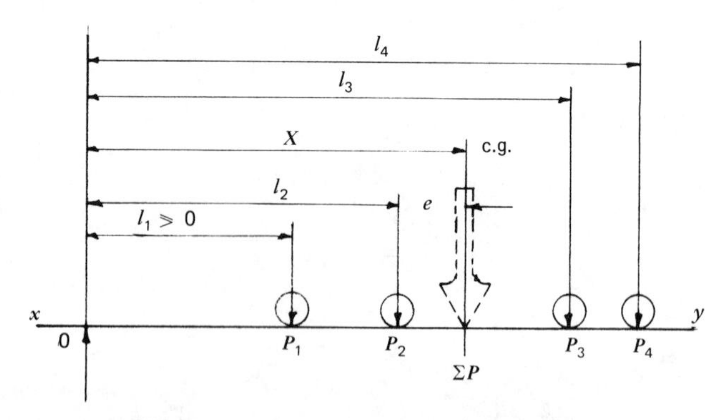

Fig. 10.2.20 Determining the center of gravity.

Definition. When there are two or more rolling loads producing downward forces within a single span, the maximum bending moment within that span occurs when the rolling loads are positioned so that the centerline of the span falls midway between the center of gravity of all the loads within that span and the wheel (or trolley) nearest the c.g.

This condition is illustrated in Fig. 10.2.21, in which the group of trolleys has been positioned so as to produce the maximum bending moment in the beam.

STEP 3. Calculate end reactions at rail supports.

From the actual dimensions of the equipment under consideration, calculate and insert the values of moment arms a, b, c, d, and the span l (see Fig. 10.2.21). Then,

$$R_2 = \frac{P_1 a + P_2 b + P_3 c + P_4 d}{l} \tag{10.2.12}$$

$$R_1 = \Sigma P - R_2 \tag{10.2.13}$$

STEP 4. Calculate maximum bending moment.

Having determined the end reactions R_1 and R_2, draw a shear diagram (Fig. 10.2.22).

From the foregoing,

$$M_{mx} = R_1 b - P_1(b - a) \tag{10.2.14}$$

STEP 5. Calculate the ECL.

Since for a single concentrated load P in a simple span,

$$M = \frac{Pl}{4} = \frac{ECL \times l}{4} \tag{10.2.15}$$

it follows that:

$$ECL = \frac{4M}{l} \tag{10.2.16}$$

or

$$ECL = \frac{4[R_1 b - P_1(b - a)]}{l} \tag{10.2.17}$$

The general solution outlined may be employed to determine the ECL for any combination of wheel or trolley loads not covered by the conditions stated in Cases 1 and 2.

Calculation of Hanger Loads

After having determined the ECL, a rail of the proper depth and weight may be selected to suit the span and specified deflection limits. It then becomes necessary to calculate the maximum hanger loads (MHL). The accurate determination of the maximum hanger loads will, in turn, permit the selection of the proper hanger rods, clamps, or other suspension fittings; and provide the loading data necessary for designing, or checking the design of, the overhead supporting structure.

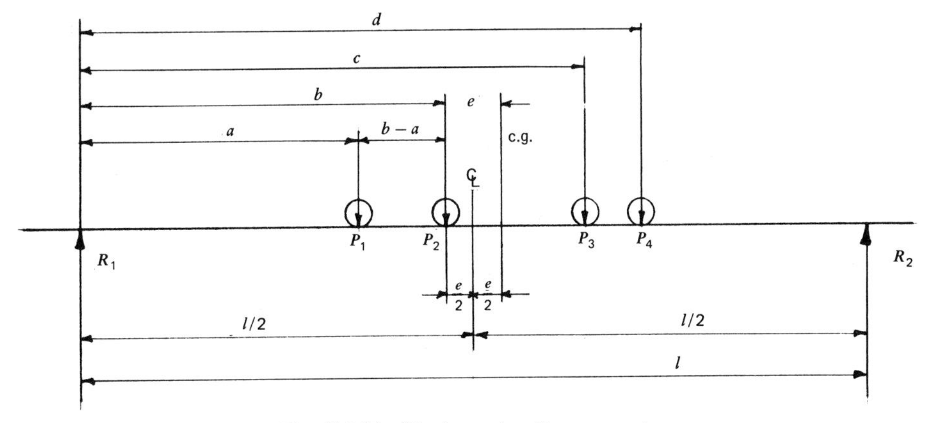

Fig. 10.2.21 Maximum bending moment.

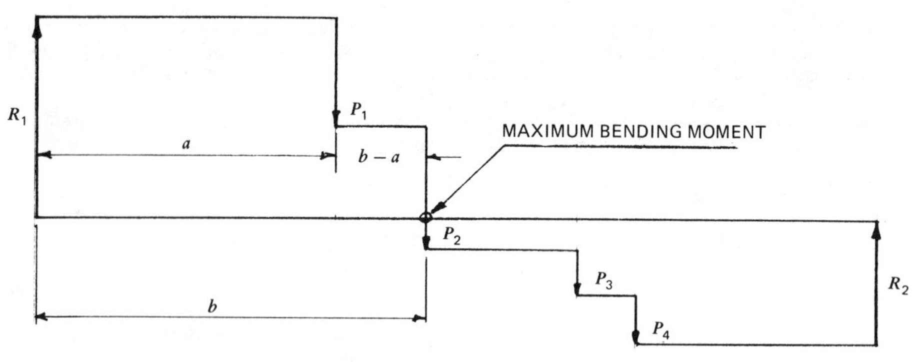

Fig. 10.2.22 Shear diagram.

As in the calculation of the ECL, the methods employed for determining the MHL differ according to the distribution of the total load P applied to the monorail or crane runway. For convenience, they are grouped into two categories:

1. Equal wheel loading, equal spans
2. Unequal wheel loading, or unsymmetrical distribution

In the following discussion, the term *maximum hanger load* is defined as the maximum reaction occurring at the hanger being analyzed due to the maximum live load P imposed by the trolley or end truck, plus the weight of the rail. Although in most instances the weight of the hanger itself is relatively minor, it should, nevertheless, be added to the MHL in order to arrive at a total load to be used in the design of any overhead supporting structure.

CASE 1. EQUAL WHEEL LOADING, EQUAL SPANS. Both Figs. 10.2.23 and 10.2.24 meet the stipulated conditions, that is, wheel loads are equal, spans are equal, and, in Fig. 10.2.24, the auxiliary wheelbases t are equal.

In order to determine the MHL, the trolley or end truck is first centered under the hanger being analyzed (H_2), and the diagram should include the hangers immediately adjacent, each side of H_2. In the figures the adjacent hangers are designated H_1 and H_3, and as stipulated, $l = l$ and $t = t$. Under these conditions, the load at hanger H_2, or the MHL, may be determined by the use of the simple equation:

$$MHL = H_2 = KP + \text{Rail weight} \tag{10.2.18}$$

where: K = distribution factor
P = total design load

$$K = \frac{l - \frac{a}{2}}{l} \tag{10.2.19}$$

The factor K is applicable to either four-wheel or eight-wheel trolleys or end trucks, so long as all of the stipulated conditions have been met.

When the wheel loads are *not* equal, or the conditions of symmetrical distribution are not present, then the foregoing equation does not apply. See Case 2.

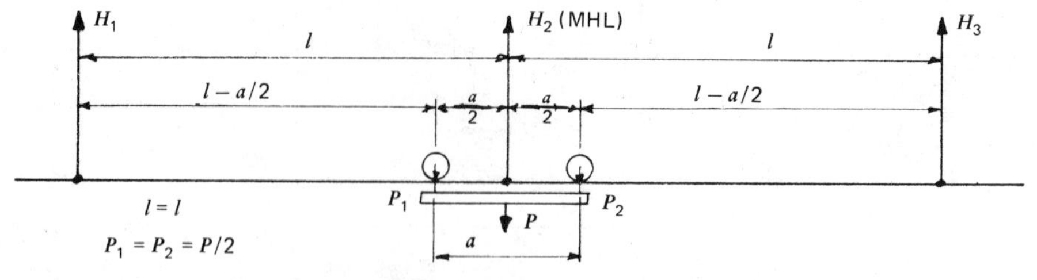

Fig. 10.2.23 Maximum hanger load, four-wheel trolley.

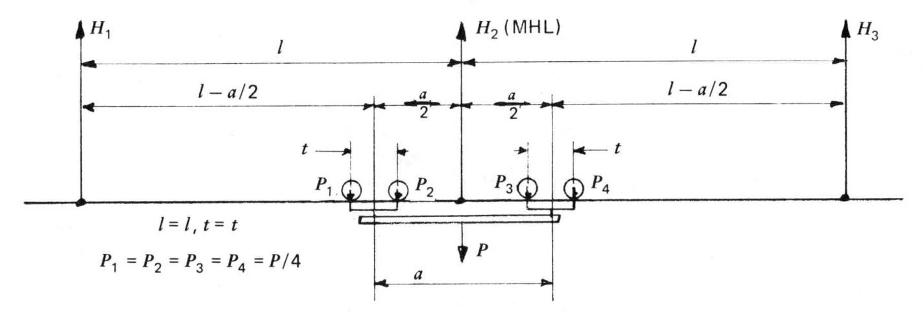

Fig. 10.2.24 Maximum hanger load, eight-wheel trolley.

Case 2. Unequal Wheel Loads, or Unsymmetrical Distribution. The general solution that follows covers any combination of wheel loads and spans, including, but not limited to, those covered in Case 1. However, because the method previously described is so simple, the general solution is usually employed only in those cases where the wheel loads are unequal, or their distribution is not symmetrical about the center hanger H_2.

As in the previous case, the first step is to construct a loading diagram with the trolley or end truck positioned to produce the maximum reaction at the hanger being analyzed. In general, when the loads are unequal, this is most likely to occur (but not always) with the heaviest wheel load placed directly under the hanger; and with the remaining wheels located so as to produce the greatest reaction at the hanger under analysis.

Referring to Fig. 10.2.25, $l_2 > l_1$, $P_2 \geqslant P_1$, and $P_2 > P_3$ or P_4. Then,

$$\text{MHL} = H_2 = P_2 + \frac{P_1 a}{l_1} + \frac{P_3 b + P_4 c}{l_2} + \text{Rail weight} \tag{10.2.20}$$

For most combinations of loads and spacing encountered in practice, the positioning of the wheels in the manner shown will produce the maximum reaction at H_2. However, this is not always the case; and in some instances it may be necessary to shift the wheels to other positions and make two or more trial calculations in order to arrive at the maximum loading condition.

A visual inspection of the loads with respect to the trolley wheel spacing, and the relative length of the rail spans either side of H_2, generally suffice to position the trolley so as to obtain the maximum total reaction at H_2. However, in those cases where the best position is not obvious from a visual inspection, then a second or third trial position and calculation may be necessary.

In the example shown, conditions that might suggest a shift of the trolley assembly to the right in the figure are the following: If P_1 and P_2 are very nearly equal, and both are much larger than either P_3 or P_4, and t_1 is a substantial percentage of l_1, which is much smaller than l_2, then the influence of P_1 is increased by moving it closer to, and perhaps directly under, H_2. The exact effect of this shift must be determined by a trial calculation.

In any case, the method employed is the same. After the wheels have been positioned, the total

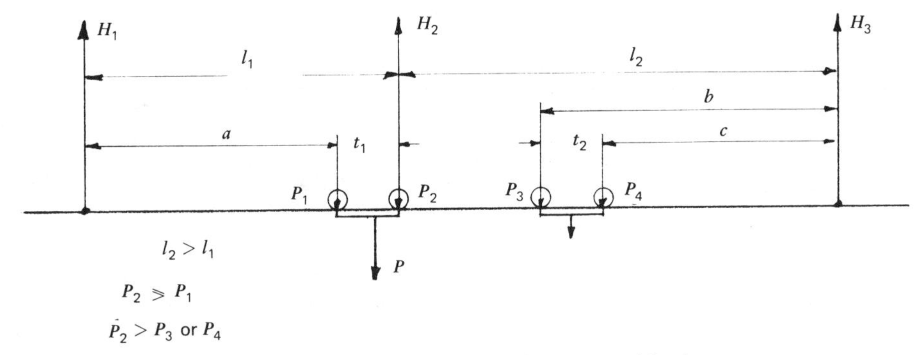

Fig. 10.2.25 Maximum hanger load, unequal loads.

reaction at H_2 is: The sum of the moments about point H_1 produced by all of the wheels within span l_1, divided by l_1; plus the sum of the moments about point H_3 produced by all of the wheels within span l_2, divided by l_2; plus the load imposed by any wheel lying directly below and in line with H_2; plus the weight of the rail.

When there is any doubt as to the trolley position that will result in the greatest reaction at H_2, the load should be calculated as outlined for each position of the trolley that places one wheel under H_2, and the maximum result thus obtained should then be used for design of the hanger and supporting structure.

10.2.5 Rolling Stock

To suspend and transport a load along a monorail track, it is necessary to provide a *trolley,* sometimes called a *carrier,* equipped with wheels for engaging the monorail track and a yoke or load bar from which the load is suspended.

When the rolling member is used to support one end of an underhung crane, it is commonly called an *end truck.*

While there are, or may be, substantial differences in the appearances and function of monorail trolleys when compared to underhung end trucks, they share certain basic characteristics, and are therefore grouped together for discussion.

Monorail Trolleys

Monorail trolleys are commercially available in capacities ranging from 100 lb or less, up to a maximum of about 25 tons. All monorail trolleys are capable of traversing a straight track, but if curves are included in the monorail system, as they frequently are, then certain characteristics must be designed into the trolleys to permit them to negotiate curves of various radii.

A monorail trolley must obviously be suitable for operation on the track with which it is intended to be used, and since the monorail tracks vary in configuration, monorail trolleys also vary considerably, both in appearance and in the details of construction.

Trolleys for Meat Track. Because of their light capacity, ranging from 50 to 250 lb, meat-track trolleys usually employ a single wheel, or at most two wheels, mounted on a single trolley frame which is disposed to the side of the track on which it operates (see Fig. 10.2.1).

Because the trolley wheels ride along the top edge of the track, they are double-flanged, with the flanges straddling the edges of the rail so as to keep the wheel centered over the track section.

These trolleys are nearly always hand-pushed. In some systems, the track is installed with a slight pitch downward from the loading point, so that the trolley is either gravity operated, or if hand-pushed, gravity assisted.

Meat-track trolleys have one operating characteristic that is unique, and not shared by monorails of other configurations. Because the trolley operates outside the rail and is suspended from one side only, the trolley, with or without its load attached, may be lifted up and off the rail, and removed at any point in the system. This characteristic is generally taken into consideration, and may be incorporated in the system planning, when a meat-track monorail system is in the design stage.

Trolleys for Tubular Track. Trolleys for tubular-type monorail tracks must operate entirely *within* the track, except for the hook or eyebolt from which the load is suspended.

The load-carrying hook or eyebolt portion of the trolley protrudes through the longitudinal slot which runs full length of the track along its underside. The track is supported from its topside, and, accordingly, the trolley with its suspended load may travel along any portion of the monorail system, including any curves or switches in the installation (see Fig. 10.2.2).

Standard individual trolleys are generally of the two-wheel or four-wheel type, equipped with small flangeless wheels that roll within the tubular housing, and bear on the inner surfaces of the lower in-turned flanges.

Some trolleys also incorporate small guide rollers that bear against the inner surfaces of the longitudinal slot, or the sidewalls, thus reducing rolling friction, especially when negotiating curves.

Standard two-wheel and four-wheel trolleys range in capacity from about 100 to 300 lb per trolley. Greater loads may be handled by coupling two or more four-wheel trolleys together through the use of equalizing auxiliary load bars. By this method, the load may be distributed to 8 or 16 wheels, increasing the maximum capacity accordingly.

The practical limit to the number of auxiliary load bars that may be added is usually dictated by the geometry at the curves. The chordal length of the main load bar should not exceed $\frac{2}{3}$ of the minimum radius curve that the trolley is to traverse. If this ratio is exceeded, the trolley may have difficulty in negotiating the curve smoothly.

I-Beam and Patented Track Trolleys. The basic configuration on a trolley designed to operate along the bottom flange of a monorail track section is markedly different from one designed to operate along the top edge of a steel bar or within a tube.

So although trolleys intended for operation on I-beam and patented track incorporate differences in detail, and are generally available from different sources, some basic concepts are similar, and the fundamental principles to be discussed are applicable to both.

As pointed out in Section 10.2.2, patented tracks generally have a flat rolling tread, whereas I-beam tracks have a rolling tread incorporating a taper or slope of about $9\frac{1}{2}°$. Accordingly, there are some differences in the wheel profiles, and at times in the manner in which the wheels are mounted.

The accommodation of the variable tread widths of I-beams, as they increase in depth or weight, is another factor reflected in the details of construction of trolleys designed to be operated on this type track.

The influence of these variations on the design and construction of the various types and capacities of trolleys will become apparent in the discussion that follows.

Wheels. The basic element of the trolley, and perhaps the most critical component in a monorail system, is the wheel that rides on the monorail track.

Wheels for use on monorails may be of either the *flanged* or *flangeless* type, depending on the configuration of the monorail section on which it rolls, and sometimes on the type of service.

Wheels for use on meat track must be double-flanged, while those for use in tubular track must be flangeless. For I-beam and patented-track systems, the wheels may be either single flanged, or flangeless with side guide rollers incorporated in the trolley frame. In either type, the design of the wheel has considerable influence on the rolling resistance of the trolley, and to a certain extent on the wear life of the rail.

In order to induce minimum wear, and hence maximum life, in both the wheel and the monorail track, the wheel tread must have a true rolling motion with respect to the track flange, and not a combination of rolling and sliding.

An analysis of the geometry involved shows that to insure true rolling, the wheel tread must be cylindrical, rather than conical, and the wheel axis must be parallel to the surface of the track tread on which it rolls.

As can be observed in Fig. 10.2.26, Case I and Case II permit true rolling. Because the wheel

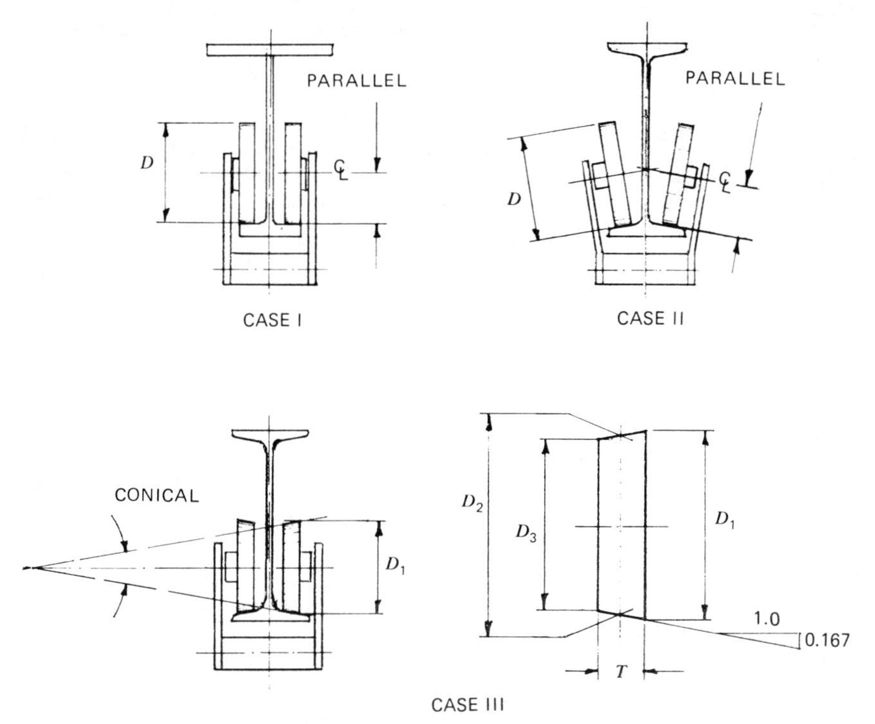

Fig. 10.2.26 Wheel profiles.

treads are cylindrical, wheel circumference is uniform wherever measured along that surface. As a result, for each revolution of the wheel, all elements of the wheel tread surface traverse equal distances along the track, and, by definition, true rolling has occurred.

In Case III, the wheel tread is conical, and Diameters D_1, D_2, and D_3 get progressively smaller. Accordingly, the *circumference* of the wheel will vary from one plane to another across the contact surface of the wheel tread.

As a result, the distance traversed by each element of the wheel tread tends to become greater as we progress from the inner (web) side of the wheel to the outer side. However, since the wheel is a solid member, not divided into planes that can rotate with respect to each other, the actual length of track traversed by one revolution of the wheel must be exactly the same, *wherever* measured along the width of the contact surfaces.

This can occur if, and only if, portions of the wheel tread each side of a neutral plane are, in fact, sliding on the rail.

As an example, a wheel with a major diameter of 6 in., and a 1-in. tread width, is analyzed. Here we see that the differential between the major circumference and the minor circumference is a little over 1 in.

If we assume the neutral plane to be at or near the center of the tread width D_2, then for each revolution of the wheel, the outer circumference tends to lead, and the inner circumference tends to lag by approximately $\frac{1}{2}$ in. per revolution. Assuming true rolling at the neutral plane, a wheel 6 in. in diameter will make approximately 66 revolutions per 100 ft of travel. Therefore, at the major and minor diameters, slippage is on the order of 34 in. plus, or nearly 3 ft of sliding, for each 100 ft of travel.

The exact figures shown apply only to the wheel profile illustrated in Fig. 10.2.26. The actual percentage of slippage varies with the angle of slope, the wheel diameter, and the tread width, but the principle remains the same.

The slope shown in the calculations ($16\frac{2}{3}\%$) is the standard slope for S-shapes, or I-beams, rolled in the United States, but may vary slightly depending on the source. The flatter the slope, the less the slippage.

The wheel dimensions affect the slippage in two ways: For any given slope and travel distance, the relative slippage *increases* as the tread gets wider, and *decreases* as the tread width (and hence the differential in diameters) narrows, and *increases* as the wheel diameter gets smaller.

A design sometimes employed to reduce differential slippage between the bearing surface of the wheel tread and a tapered rail is to crown the wheel tread as shown in Fig. 10.2.27. The actual radius of the convex wheel tread surface is exaggerated in the illustration for clarity.

In this arrangement, under no-load conditions, the point of contact lies in the neutral plane x–x', and since other portions of the wheel tread, lying either side of x–x', are not in contact with the rail, true rolling can occur.

However, since a point contact is incapable of supporting any load, when the wheel is loaded, slight deformation takes place at the point of contact until equilibrium is reached. The amount of deformation is a function of the load, the wheel diameter, the crown radius, and the relative hardness of the wheel and track. By means of rational mathematical analysis outside the scope of this discussion, the true shape and area of the deformed support surface may be calculated for any given set of conditions.

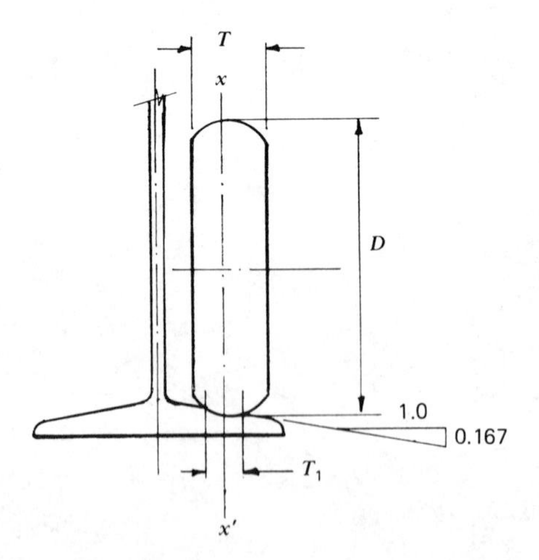

Fig. 10.2.27 Crowned-tread wheel.

For our present purposes, however, it is sufficient to note that under load, in order to provide a bearing area, the contact surfaces deform slightly, so that after deformation the width of the wheel tread in contact with the rail will be dimension T_1, lying somewhere between zero (true point contact) and T, the full tread width which would be in contact when the wheel is cylindrical or conical.

Hence, to the extent that the contact width T_1 embraces differential wheel diameters lying within the width of contact, relative slippage will occur each side of the neutral plane x–x'. However, with a convex wheel tread, the variation in wheel contact diameters is considerably less than that occurring in the case of a conical wheel.

Therefore, from a wear standpoint, the use of a crowned (or convex) tread lies somewhere between a cylindrical and a conical tread in function. Unlike a cylindrical wheel, some differential slippage does occur, the amount depending on the load and other factors involved; but it is generally substantially less than that occurring with a conical wheel.

Obviously, the practical effects of the differential slippage is largely dependent on the type of service for which the monorail is intended, the load per unit area, and such controllable factors as the hardness of the rolling surfaces of both the wheel and the rail.

Since the wheel, whatever its profile, incurs many more wear cycles per 100 ft of travel than the rail, one might reason that the higher the wheel hardness, the longer the wheel life, and therefore the harder the better. While this may be true in theory, from a practical standpoint, a wheel is generally a relatively inexpensive part to replace, while the replacement of the rail in a monorail system, or an underhung crane runway, is a relatively expensive procedure.

Accordingly, in practice, the relative hardness of wheel and track is a compromise based on experience. In general, good practice dictates that the wheel be harder than the monorail track on which it rolls, but not so much so that it damages the rail, and that the rail itself be harder than standard structural steel.

While certain conditions outside the normal range of industrial service might require special consideration, the following hardnesses recommended by the Monorail Manufacturers Association (MMA), and adhered to by a number of independent manufacturers as well, are considered by many to be the best current practice for industrial monorails:

1. For monorail track, "The minimum hardness of the lower carrying (tension) flange shall be 195 Brinell." (MMA Sec. 3.2.)
2. For wheels, "Wheels . . . shall have a minimum tread hardness of 425 Brinell." (MMA Sec. 5.1.1.)

Wheel Load Ratings. The load rating of a monorail wheel assembly, must take into account (1) the rating of the wheel itself, (2) the rating of wheel bearing, and (3) the strength of the wheel pin. For any given wheel assembly, the rated capacity of the unit must not exceed the lowest of these three values.

The first of these values, that is, the basic rating of the wheel itself, is a function of the wheel diameter, the wheel tread, and, in current practice, a series of arbitrary constants based on the cumulative experience of the monorail crane industry.

Although much research remains to be done to establish a more rational method of selecting the arbitrary constants, there are a number of guidelines in current usage that have general acceptance within the industry.

For monorails and underhung cranes, the maximum wheel loads permitted by CMAA Specification No. 74 are as follows:

1. For contour tread, that is, one in which the wheel tread matches the rolling surface on which it rides,
$$P = 1000\ W\ D \text{ (Table 4.7.1.2-1)}$$
2. For convex tread,
$$P = 600\ W\ D \text{ (Table 4.7.1.2-1)}$$

In both cases, P is the maximum recommended load, in pounds; and W and D are the wheel tread width and diameter, respectively, in inches. Figures given are for steel wheels.

By way of comparison, CMAA Specification No. 70, covering top-riding cranes, allows wheel loadings ranging from 1200 $W\ D$ to 1600 $W\ D$, depending on the class of service. However, it should be noted that these ratings are for wheels riding on ASCE-type steel crane rails, generally complying with specification ASTM-A1, in which the carbon content of the steel rail ranges from 0.55 to 0.82%.

Since on patented-track systems the rail tip loading is usually the limiting factor, the current MMA specifications for underhung cranes and monorail systems do not address this aspect of wheel design directly. They do, however, spell out wheel and track hardness, and minimum bearing life for various classes of service (MMA-ANSI MH27.1-1981).

Since the carbon content in patented-track monorail sections seldom exceeds 0.60%, and may be considerably lower in some of the I-beam sections, it is considered good practice, based on operating experience, to stay *within* the recommendations of CMAA Specification No. 74. That is, wheel loads should not exceed 1000 $W\ D$ for contoured wheels, and should be held at or below 600 $W\ D$ for wheels with convex treads.

In addition to the wheel itself, the bearing rating must be taken in to consideration in rating the assembly. In this area, the recommended guidelines are more precise, and are directly related to the class of service.

The MMA specifications for underhung cranes and monorails recommend the following for antifriction ball or roller bearings (Sec. 5.1.2.1):

Class of Service	Minimum B-10 Life
A1, A2, and B (standby to light service)	3,000 hr
C (moderate service)	5,000 hr
D (heavy duty)	10,000 hr
E (severe duty)	15,000 hr

Additionally (Sec. 5.1.2.2), it is recommended that design bearing life be based on 75% of the maximum rated wheel load, an assumed travel speed of 150 ft/min for hand-pushed trolleys, and the actual maximum travel speed of motor-driven equipment.

In general, these standards appear to meet the operating requirements of the vast majority of industrial monorail applications. However, in any specific installation, consideration should be given to service conditions, or other factors that might suggest a modification of these guidelines.

In any case, the purchaser and manufacturer should reach agreement, prior to purchase, as to the exact criteria and specifications that shall govern the design and manufacture of the wheels, and all other elements of the monorail system under consideration.

Two-wheel Trolleys. Wheels for monorail service are usually assembled into a basic two-wheel or four-wheel trolley, which may in turn be combined with other such trolley assemblies to obtain a multiple-wheel unit of higher capacity.

The two-wheel trolley most often consists of a U-shaped yoke, to which the wheel assemblies are attached, forming a typical unit as shown in Fig. 10.2.28. The yoke may be forged, cast, or fabricated out of structural steel bars or channels, depending on its capacity and usage.

For trolleys that operate on I-beams or patented track, it is desirable, from a servicing standpoint, to employ a design that permits removal and replacement of a wheel anywhere along the monorail system. This may be accomplished, in a one-piece yoke, by proper proportioning of the yoke, the wheel, and the wheel pin. Or it may be done by employing a two-piece yoke, in which case the two halves of the yoke are bolted together when in normal operation, but may be separated when necessary to remove and replace a wheel.

Two-wheel trolleys are usually equipped with a suspension fitting attached to the lower cross member of the yoke. The suspension fitting may consist of an eyebolt, a hook, or any one of a number of special devices. The best choice in any particular case is usually determined by the load to be handled. In most cases the suspension fitting may be rotated, but in some applications this may not be necessary

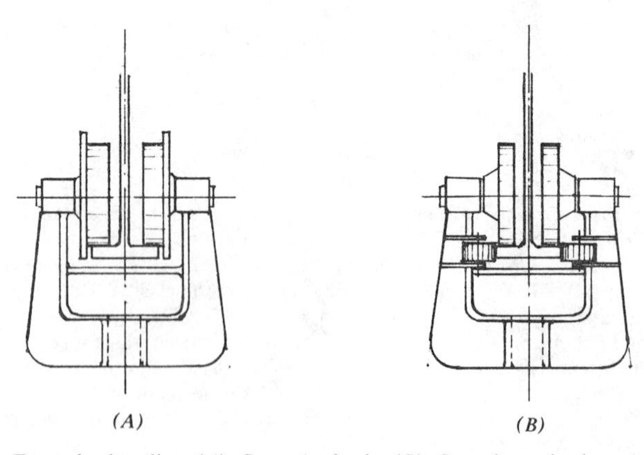

Fig. 10.2.28 Two-wheel trolley. (*A*), flanged wheels; (*B*), flangeless wheels and guide rollers.

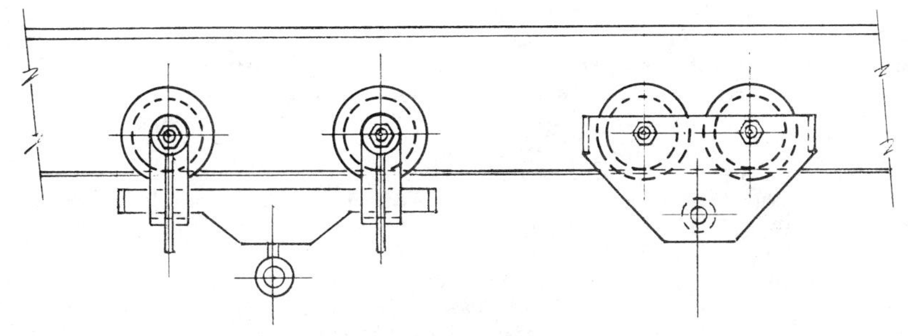

Fig. 10.2.29 Four-wheel trolley.

or desirable. The catalogs published by the several manufacturers indicate the wide variety of suspension fittings available.

The two-wheel trolley assembly may include a pair of flanged wheels, in which the wheel flanges straddling the lower flange of the monorail track serve to keep the trolley centered on the track, or flangeless wheels and side guide rollers. The latter construction is somewhat more costly than its flanged-wheel equivalent, but reduces rolling resistance and is generally to be preferred where service is heavy, or in systems including numerous curves.

Four-wheel Trolleys. Four-wheel trolleys are manufactured in two basic configurations, swiveling and nonswiveling. The first type employs a pair of individual two-wheel trolleys, of the type previously described, joined together by a load bar. The nonswiveling type comprises two sideplates, bolted together below the rail, each of the sideplates being equipped with a pair of wheels, the complete assembly thus forming a single four-wheel trolley. The two basic types are illustrated in Fig. 10.2.29.

It should be noted that both types may employ either flanged wheels, or flangeless wheels and guide rollers, the choice depending on the specific application.

Swiveling Trolleys. As the name implies, in a swiveling trolley, each of the two-wheel trolleys is connected to the load bar by means of a vertical pin that permits the trolleys to swivel with respect to the load bar. The limits of swiveling, or the maximum number of degrees through which each trolley may rotate, is determined by the relative dimensions of the trolley yoke and the load bar. See Fig. 10.2.30.

This type trolley may be used on either straight or curved track, but lends itself particularly well to operation on monorail systems incorporating curves and switches.

The minimum radius curve that can be negotiated is a function of the wheelbase and the degree of rotational freedom of the trolleys. In general, it is good practice to keep the curve radius somewhat greater than the theoretical minimum, so as to avoid cramping the yoke tightly against the load bar. And, as pointed out previously, for the best operation, wherever possible the wheelbase should not exceed $\frac{2}{3}$ of the minimum radius curve.

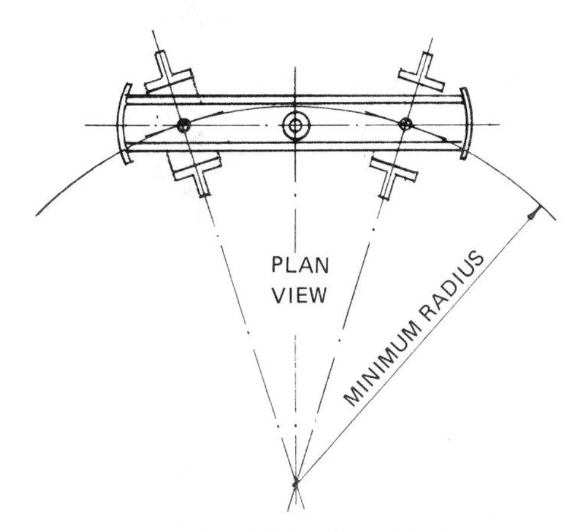

Fig. 10.2.30 Swivel limits.

Nonswiveling Trolleys. The sideplates of four-wheel nonswiveling trolleys may be forged or cast, or they may be fabricated from steel plates. All three types are in common usage, and the manufacturers' catalogs usually specify the design employed, which may vary depending on the capacity of the trolley.

When the sideplates of these trolleys are bolted together, the load-carrying suspension member is frequently provided with spacers so that the width of the trolley can be adjusted to accommodate several flange widths.

While this type trolley is used in many applications, it is particularly common with hoist manufacturers. In a typical hoist application, the sideplates are bolted to each side of a lug on the hoist frame, thus providing rolling support for the hoist. Here again, through the use of proper spacers, the trolley may be adjusted to fit a range of variable flange widths.

Four-wheel, nonswiveling trolleys may be hand propelled by pushing on the load, or two of the wheels may be provided with geared flanges that engage a driving pinion. The pinion is then rotated by means of a hand-chain and pocket sheave attached to the drive pinion shaft, in which case it is called a hand-racked trolley; or the pinion shaft may be driven by a suitable motor, in which case it is called a motor-driven trolley.

Although four-wheel nonswiveling trolleys do not permit rotation about a vertical axis by individual pairs of wheels, they are generally built with relatively short wheelbases so that they may negotiate curves.

The minimum radius curve that a nonswiveling trolley may traverse is a function of the wheelbase, and the clearance between the wheel flanges and the edges of the monorail track on which it rolls. For any specific trolley, the manufacturer's catalog will generally indicate the minimum radius curve that it will negotiate.

When nonswiveling trolleys are employed on monorail systems that incorporate switches, it is important that the trolley clearance through the switch and switch baffles be carefully coordinated in order to avoid unplanned interferences. If the trolleys are hand-pushed, the necessary clearances may generally be provided.

However, when this type trolley is hand-racked, or motor-driven from one side (the so-called "sidewinder" drive), the lateral projection of the chain sprocket or the drive motor is such that the trolley, when so equipped, will *not* clear any commercially available standard switch. So when used in this type service, it is usually necessary to employ a different type drive, or to use swiveling trolleys, in order to obtain the necessary operating clearances. See Section 10.2.6 for types of drives.

Eight-wheel Trolleys. Eight-wheel trolleys generally comprise a pair of four-wheel trolleys, rotatably connected to a load bar, to which is attached a suspension fitting suitable for the load to be handled.

By this method, the capacity of a four-wheel trolley assembly may be doubled, without increasing either the individual wheel loads or the load imposed on the rail tip.

The basic four-wheel trolleys employed in this type multiple-wheel trolley may themselves comprise a subassembly of two-wheel trolleys and load bar, as previously described, or they may be of the nonswiveling type. See Fig. 10.2.31.

Either way, the resulting eight-wheel trolley assembly is capable of negotiating curves, the minimum radius of which is governed by the major wheelbase of the trolley.

In theory, this system of doubling and redoubling the capacity of multiple-wheel trolleys by the introduction of additional levels of load bars has no limit; however, because of practical considerations, it is seldom employed to assemble trolleys having more than sixteen wheels (three levels of load bars), and in most cases this configuration is used for trolleys having eight wheels (two levels of load bars) or less.

End Trucks for Underhung Cranes

The basic components of monorail systems also serve as some of the essential elements of underhung crane systems. Some of the components, such as the rail for example, are adaptable to crane service

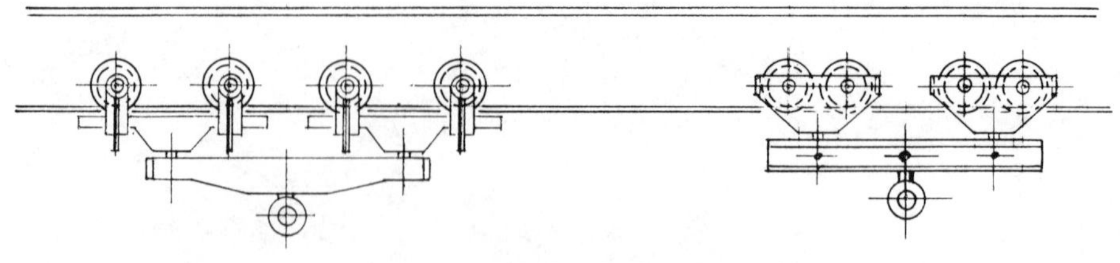

Fig. 10.2.31 Eight-wheel trolley.

without modification; others, such as the trolleys, require special adaptation for use as crane end trucks while still retaining some of the characteristics of the monorail trolleys from which they evolved.

Functionally, underhung end trucks differ from comparable monorail trolleys in two basic respects. First, they are intended almost exclusively for operation on straight track only, rather than curves. Second, because proper crane operation requires that the length of the end-truck wheelbase be not less than that required to comply with certain minimum ratios with respect to the bridge span, end trucks are usually longer than monorail trolleys of similar capacity (see Section 10.2.3).

In most cases, many of the wheel assemblies employed in underhung end trucks are identical to those incorporated in monorail trolleys produced by the same manufacturer. The wheel *mounting* may differ, as may be required to suit the end-truck function, but the essential elements remain the same.

Although underhung end trucks for industrial use vary in detail, they all fall into one of two basic categories. One type employs rigidly mounted wheels, and these end trucks are generally equipped with four wheels. The other basic type is the articulating end truck, in which the wheels are mounted in the end-truck frame with a certain degree of freedom or articulation. Articulating trucks may be equipped with either four or eight wheels, or even more if required.

Underhung end trucks, like monorail trolleys, may be designed to be hand-propelled, or motor-driven. The types of motor drive vary considerably, and are discussed in detail in Section 10.2.6. As will be observed, the type of drive employed has some bearing on the design and configuration of the end truck.

Rigid End Trucks. A rigid end-truck frame comprises a pair of parallel structural members, usually channels, spaced far enough apart to straddle the lower flange of the track on which it rides, and tied together at each end with steel plates bolted or welded to the side members. There may be, and usually are, additional ties across the bottom of the frame at the points of attachment of the crane girder or girders.

The complete end-truck assembly for a single-girder crane includes the structural frame, the four wheels mounted in pairs opposite each other near each end of the frame, a connection for the bridge girder at the center of the frame, end bumpers, and safety lugs. See Fig. 10.2.32.

End trucks for double-girder cranes are similar, but have two points of attachment for the bridge girders, usually, but not necessarily, equidistant from the wheel centers.

As in the case with monorail trolleys, the wheels employed on rigid end trucks may be either flanged or flangeless.

When flanged wheels are used, a provision is usually made for adjusting the space between the flanges of each opposing pair of wheels, so as to match the flange width of the runway track on which they ride. This may be done either by sliding the wheel axles in or out, so as to decrease or increase the gap, or in some models by adjusting the space between the side channels.

In end trucks incorporating flangeless wheels, the truck is kept in alignment with the runway track through the use of side guide rollers, usually four per end truck, rotating on vertical axles.

The side guide rollers may be set low enough in the end-truck frame to engage the edges of the lower flange of the runway track, thus replacing the function of the wheel flanges. In that type of construction, the position of the guide rollers must be laterally adjustable in order to accommodate a specified range of flange widths. This may be accomplished by changing roller diameters, or by adding or removing spacers between the guide-roller brackets and the truck frame so as to obtain the proper spacing, face to face, of the guide rollers.

An alternate method employs guide rollers set at an elevation *above* the lower flange, and set in close to the centerline of the end truck, so that the rollers engage the *web* of the runway track instead of the lower flange. This type of construction has the advantage of being suitable for operation on any number of flange widths within a given range, without requiring adjustment to a specific width.

When this type truck is employed, however, it is important that the lower edges of the runway web splices be kept high enough above the tread line to avoid interference with the guide rollers.

Rigid trucks of the two general types described are available in standard models in a wide range

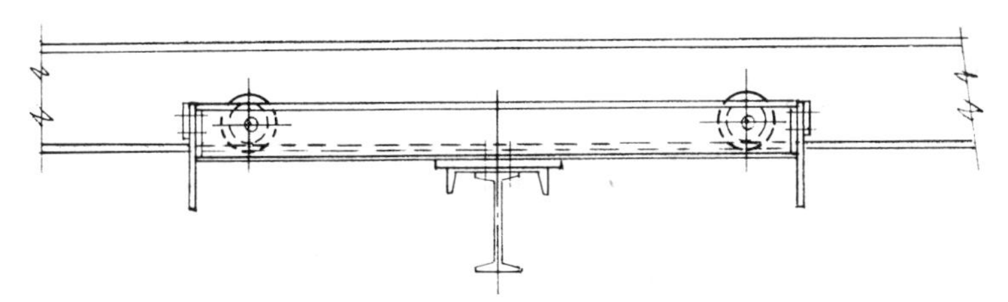

Fig. 10.2.32 Rigid four-wheel end truck.

of capacities starting at about 1000 lb, and going up to a maximum of about 25 tons. Wheel diameters range from a minimum of about 4 in., to about 12 in. maximum; and the wheelbase may range from about 24 in., to 10 ft or more.

Exact details vary, of course, from one manufacturer to another, but the manufacturers' catalogs usually include capacity ratings and sufficient dimensional data for the proper application of the end trucks.

Unfortunately, at this writing, there remains a lack of uniformity in standards for rating the load-carrying capacity of the many end trucks that are commercially available. Some are rated according to the actual *net capacity* of the end truck itself, while others are rated according to the *live-load* capacity of the cranes on which they are used.

The net capacity rating is preferred, and should be acertained whenever possible. The reason for this is that the live load method of rating the end trucks includes certain assumed average weights for the bridge, hoist, and hoist trolley, which may or may not reflect actual conditions in any given application. As a result, end trucks rated in this manner may be underrated or overrated in any specific situation, depending on how accurately the assumed average deadweights reflect the actual conditions.

In any event, it is important to determine clearly which of the two methods is employed when selecting end trucks for any application. If the cataloged data is unclear on this point, the information may usually be obtained from the manufacturer.

End-truck net-capacity ratings should also reflect the class of service for which the equipment is intended to be used. As previously pointed out, the class of service is directly related to minimum bearing life, and hence to wheel-bearing selection. Although consideration of this aspect may not affect the basic strength of the structural components of the end truck, it has a direct influence on the wear-life that may be expected of the equipment selected, and in good practice it must be taken into account.

Articulating End Trucks. An articulating end truck comprises an end-truck frame, two or more sets of monorail-type trolleys, safety lugs, end bumpers, and means for connecting the end truck to the bridge girder.

In a four-wheel end truck, the ends of the truck frame resemble, in some respects, the ends of a four-wheel swiveling monorail trolley. The similarity lies in the fact that, like the trolley load bar, each end of the truck is supported by, and is pivotally connected to, a two-wheel trolley yoke (Fig. 10.2.33).

An eight-wheel articulating end truck includes an end-truck frame and the same basic elements as the four-wheel end truck, except that the rolling components at each end are four-wheel, rather than two-wheel, trolleys. The two four-wheel trolleys employed may be of either the swiveling or nonswiveling type (Fig. 10.2.34).

Four-wheel Articulating End Trucks. Four-wheel articulating end trucks in the lighter capacities, and having a relatively short wheelbase of 18–24 in., may be nothing more than a monorail trolley with clips added to the load bar for attaching and supporting the crane girder. For short-span, single-girder cranes of light capacity, this is frequently adequate.

However, for cranes of longer span and rated capacities in excess of 1 ton, the design and construction of the end truck differs from that of a monorail trolley in several important respects.

These differences in design are the result of functional requirements. In a monorail trolley having four wheels, and operating on a single rail, the two-wheel trolleys that are the basic component are connected to the load bar by means of vertical swivel pins. These connections are proportioned so as to provide the maximum degree of rotational freedom about a vertical axis in order that the trolley may negotiate small-radius curves.

If the monorail trolley has eight wheels, then the auxiliary load bars of a pair of four-wheel trolleys are, in turn, connected to the main load bar in the same manner, that is, by means of vertical swivel pins, and for the same purpose: to provide the capacity to negotiate curves freely. As additional

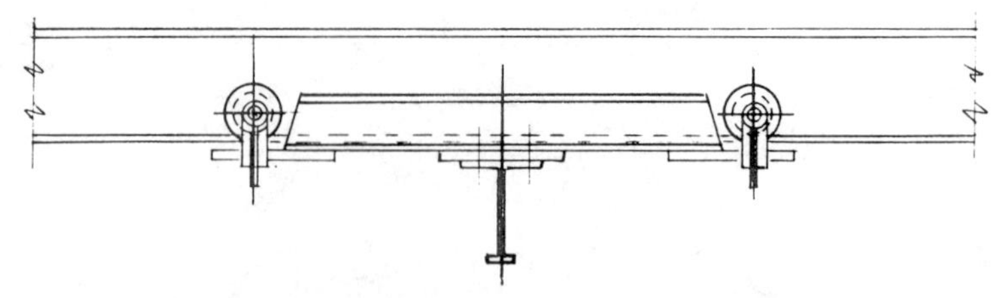

Fig. 10.2.33 Articulating four-wheel end truck.

levels of load bars are added to increase capacity, this detail is repeated at each level so that maximum flexibility in the horizontal plane is maintained.

An end truck, however, designed for operation on a straight runway rail need not negotiate curves and, accordingly, the swiveling or articulation of the two-wheel trolleys within the end-truck frame serves an entirely different purpose, and may be more limited.

While good practice and standard specifications widely accepted within the crane industry provide that "runway rails shall be straight, parallel, level and at the same elevation," it must be understood that, in the real world, these terms are relative. An underhung crane runway, suspended from a building structure, however carefully manufactured and installed, is not a planer bed. While specifications generally allow a variance of $\pm \frac{1}{8}$ in. in these criteria, the specified tolerances represent ideal conditions which, in actual installations, are rarely maintained over a long period of time.

Therefore, in order to accommodate commonly encountered field conditions, the articulation designed into an end truck is intended to equalize end-truck wheel loadings even when runway rail spacing, levelness, and straightness fall somewhat short of the ideal conditions specified.

If the connection between each of the two-wheel trolleys and the end-truck frame allows a few degrees of freedom about a horizontal axis, then the end truck has the principal characteristic required to compensate for inaccuracies in the underhung runway.

Commercially available articulating end trucks achieve this basic purpose in a number of different ways. The trolley to end-truck frame connection may include horizontal swivel pins, double-axis trunnions, spherical seats, or a combination of a vertical pin and spherical washer. All of these methods are in current use, and although they vary in detail, their function remains the same: to provide the limited articulation required to equalize end-truck wheel loads when rolling on runway rails that deviate, within reasonable limits, from ideal conditions.

Eight-wheel Articulating End Trucks. As capacity requirements increase, and it becomes necessary to step up to an eight-wheel end truck, the problem of compensating for runway inaccuracies is solved in a manner that differs in detail from that employed in four-wheel trucks, but which incorporates the same basic principles. The individual four-wheel trolleys within the overall end-truck assembly must be mounted in a manner that permits the wheel loads to equalize.

Here again, the details vary among manufacturers, but the essential components include a pair of four wheel trolleys, each of which has its load bar connected to one end of the end-truck frame in a manner that permits articulation (Fig. 10.2.34).

Among commercial models currently available, the articulating connection of the end-truck frame to the auxiliary load bars is made in two basic manners. In one type of connection, the ends of the main frame are suspended from swiveling eyebolts in the auxiliary load bars. In the other type, a cross member located at or near each end of the main frame rests on top of, and is connected to, each of the auxiliary load bars through a suitable load-bearing device, such as a ball or spherical bearing, that is shaped to permit the degree of freedom required for equalizing the wheel loads.

A safety-related detail in the construction of an eight-wheel end truck that should be considered is the manner in which the runway end-stops impact the truck. If the ends of the auxiliary load bars are allowed to project beyond the ends of the main frame, then the impact of stopping the crane is first taken on the end of the auxiliary load bar, and then through the articulating bearing device, to the main frame. This is a condition to be avoided.

The preferred construction is to provide an extension on each end of the main frame, so that the impact of striking the end-stops may be transmitted directly into the main frame, thus protecting the articulating bearings from damage and possible failure.

The load-carrying, rolling components are among the most critical elements in any monorail or underhung crane system. It is important, therefore, that the operating requirements, the class of service, and the conditions surrounding a planned installation be carefully analyzed. This will help to assure that the equipment selected is best suited to the application under consideration.

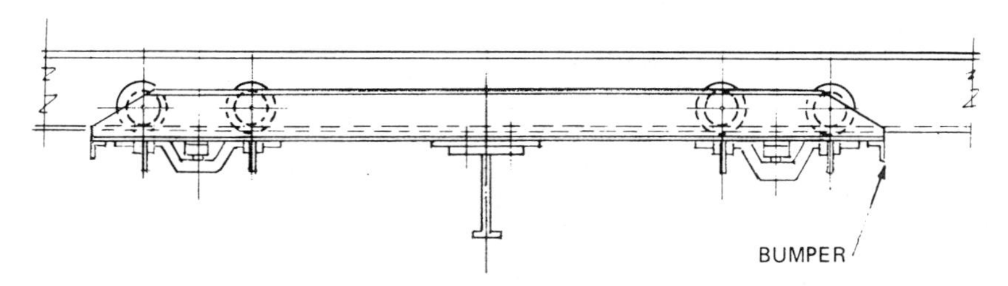

Fig. 10.2.34 Articulating eight-wheel end truck.

10.2.6 Drives

Underhung cranes and trolleys on monorail systems, may be either hand-propelled or motor-driven. Hand-propelled units may be either *hand-pushed* or *hand-racked;* whereas motor-driven units may be powered by electric motors or, more infrequently, by pneumatic or hydraulic motors.

Hand-propelled Units

Hand-pushed units, as the name implies, are propelled along the supporting rail by the operator pushing on the suspended load or, in some cases, on a "pusher-arm" provided for that purpose. Hand-pushed systems are limited to the lighter capacities, generally about 2–3 tons as a practical maximum, and to those applications where the distance to be traveled is short, and where the travel speeds required are not faster than a slow walk. A typical example would be a trolley operating along the boom of a jib crane over a lathe or other machine tool. The force required to push a load on the level will vary from 20 to 30 lb per ton, depending on the condition of the rail and trolley.

Where the loads are heavier, but the duty cycle or other considerations do not warrant the use of powered equipment, a hand-racked unit may be the most practical choice.

In a *hand-racked unit,* one pair of the trolley or end-truck wheels is provided with gear teeth. The geared load-bearing wheels are then driven through a pinion shaft connected at one end to a pocket sheave. An endless hand-chain engages the pocket sheave, and drops to a position within reach of the operator, who can then drive the unit in either direction by pulling on one side or the other of the hand chain. Thus, while the unit is still hand propelled, the mechanical advantage obtained through use of the proper gear reduction ratio and pocket-sheave diameter makes it possible to propel loads as great as 25 tons at slow speeds. Again, however, this type of manual propulsion, especially in the heavier capacities, is limited to those applications where travel distances are relatively short, slow travel speeds are acceptable, and the duty cycle is low.

Motor-driven Units

For those applications requiring a motorized drive, aside from the fact that the motive power may be either electric, pneumatic, or hydraulic, there are two basic methods in general use.

These include the geared-wheel type of drive, in which tractive effort is applied through the load-bearing wheels; and the so-called "underdrive" type, in which the tractive effort is applied independently of the load-bearing wheels by a separate drive wheel forced into frictional engagement with the underside, or soffit, of the rail by springs or other means of applying pressure.

Geared-Wheel Drives. In underhung cranes and monorail trolleys, the basic design and construction details of geared-wheel drives vary, depending upon whether the end truck or trolley is of the swiveling or nonswiveling rigid type (see Section 10.2.5).

If a geared-wheel drive is applied to a close-coupled nonswiveling trolley having a wheelbase only slightly longer than the wheel diameters, then in many cases a side-winder type of design is employed. In the side-winder drive, two wheels on the same side of the rail are provided with gear teeth. A common driving pinion fits between, and meshes with, the two geared wheels; the driving pinion is in turn connected through suitable reduction gearing to the drive motor. See Fig. 10.2.35.

Although the side-winder drive has the advantage of being economical, the configuration is such that it will not clear most standard monorail switches; and because in this type of drive the tractive effort is applied to one side of the rail only, the load may have a tendency to skew. Where clearance is not a consideration and the side-winder drive may be safely employed, flangeless wheels and side guide rollers will help to reduce rail flange wear.

When the rigid wheel mounting, or nonswiveling design, is employed in a long-wheelbase trolley, or an end truck, or in any application where the side-winder is not suitable, then the method of applying power to the driving wheels is somewhat different.

In these applications, a pair of opposing wheels at one end of the end truck or trolley is provided with geared flanges. The pinion shaft then crosses under the end truck or trolley frame (and hence under the runway rail as well), and a pair of pinions keyed to a common shaft engage the geared flanges of the load-bearing drive wheels.

The pinion shaft may be driven by either of two methods which are in current usage.

On trolleys and on cranes using the individual-type drive, each end truck is provided with a motor that is connected to its pinion shaft through suitable reduction gearing. The two or more motors comprising the crane drive are operated from a common control, so as to act together.

A method somewhat more common at this time, in this type of crane, is to drive both pinion shafts from a single motor and reducer located at or near the midpoint of the bridge span. The output shaft of the center drive unit is in turn connected to the drive-pinion shafts, mounted in the end trucks, through suitable intermediate shafting and shaft couplings. This design is commonly referred to as a "squaring-shaft drive" (Fig. 10.2.36).

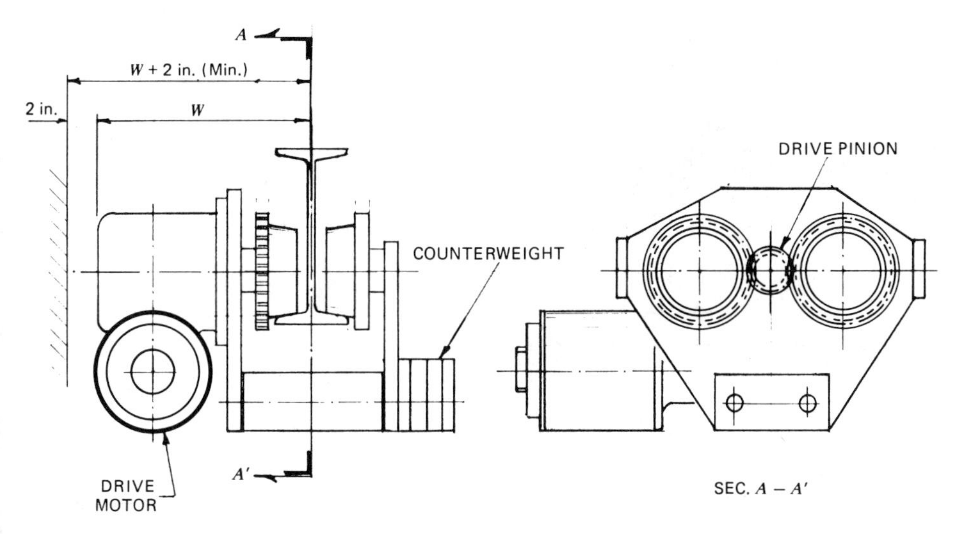

Fig. 10.2.35 Side-winder drive.

Articulated Trolleys. When geared-wheel drives are applied to swiveling trolleys or articulated end trucks the design is, of necessity, somewhat different than that employed when the wheels are mounted rigidly.

It is noted in Section 10.2.5 that the basic component of the swiveling trolley or articulated end truck is the two-wheel trolley, which in combination with the proper load bars and other two-wheel trolleys may be assembled into four-wheel or eight-wheel units.

The yoke of the two-wheel trolley comprises a U-shaped steel frame, in which the upper ends of the vertical members provide support for the wheels, and in which the lower horizontal cross member provides articulating support for one end of the load bar or end-truck frame that is connected to it.

Since the ability to swivel, or articulate, with respect to the mating load bar is an important function of an articulating trolley, the capacity for relative motion between mating components must not be impaired when this type unit is motorized.

Accordingly, when articulating trolleys are motor-driven, the motor and gear-reduction assembly are built in the configuration of a modified yoke, in which the driving wheels are also mounted, thus forming a compact integrated unit. This design permits the end of the load bar or end-truck frame to be movably supported by the motor-driven trolley in essentially the same manner as when the unit is hand-propelled (Fig. 10.2.37).

On underhung cranes using articulating end trucks and geared-wheel drives, it is not feasible to use a center drive and squaring-shaft, as previously described, since the coupling of the drive shaft to the geared trolleys would necessarily inhibit their freedom to articulate. Because of this, motor-driven cranes with articulating end trucks in which the propelling force is applied through the load-bearing wheels must employ individual drives as described previously with a minimum of one drive per end truck.

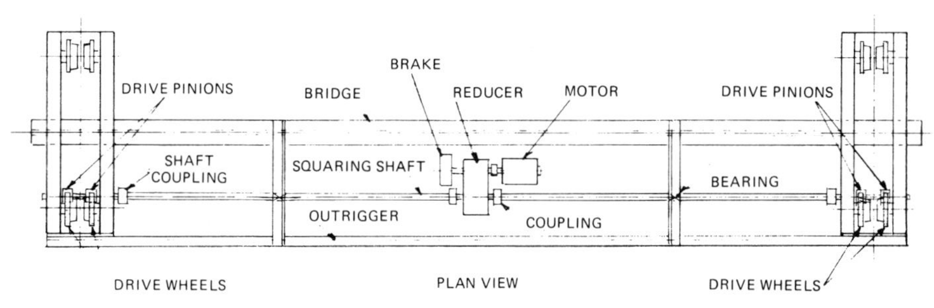

Fig. 10.2.36 Squaring-shaft drive.

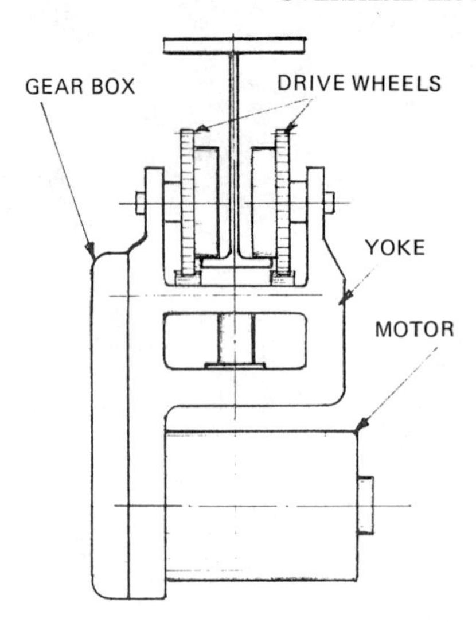

Fig. 10.2.37 Geared articulating drive.

Underdrives. An alternative to the geared-wheel drive widely used in the monorail industry is the underdrive. As the name implies, in an underdrive the driving wheel is located below the rail, and tractive effort is developed by applying pressure at the point of contact between the rotating drive wheel and the underside of the lower flange of the monorail.

Underdrives may be used for propelling a load along a monorail, or for driving an underhung crane. They may be applied to either rigid or articulating trolleys or end trucks; and because the driving element is independent of, and not rigidly connected to, the load-bearing wheels, either individual units, one per end-truck, or a squaring-shaft construction may be employed.

An underdrive may be built into the trolley or end truck which it is intended to propel, or it may be independently supported from its own wheels, and connected to the moving load by means of a tow bar. When used in the latter configuration, it is commonly called a tractor drive (Fig. 10.2.38).

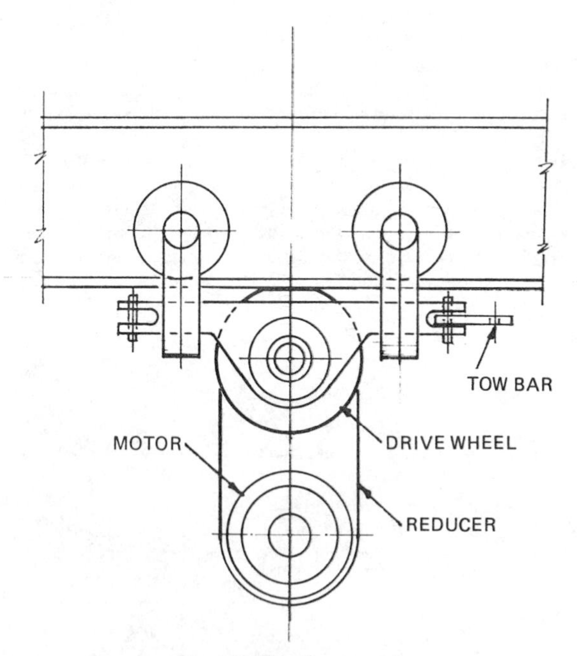

Fig. 10.2.38 Tractor drive.

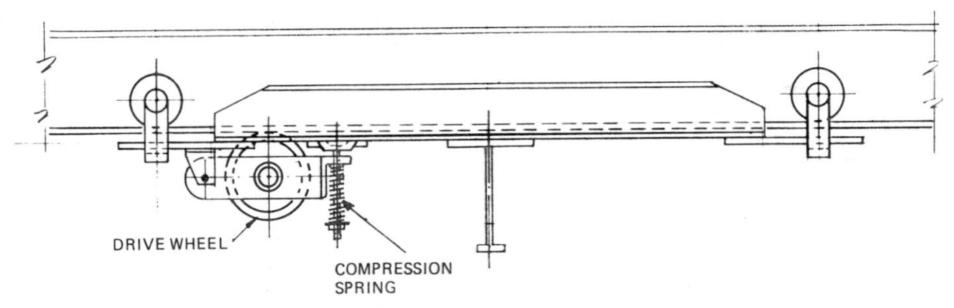

Fig. 10.2.39 Underdrive, spring biased.

The driving element of underdrive units commercially available may be a steel wheel, a rubber-tired wheel, or a steel hub with a solid tire made of rubber, neoprene, polyurethane, or other compressible plastic material suited to the application.

The pressure required to develop the rated tractive capacity of any particular design varies with the type of material used for the tread of the drive wheel and its coefficient of friction with respect to the rail. Typically, a drive wheel with a steel tread requires greater pressure than a rubber-tired drive wheel in order to develop the same traction when operating on a steel rail. Since the pressure on the drive wheel imposes an equal and opposite force on the load-supporting wheels, this load must be taken into account when calculating the net rating of a trolley or end truck driven by an underdrive.

The pressure between the drive wheel and the rail may be applied by the use of adjustable compression springs, or by any one of a number of alternative methods. When employing compression springs, the frame supporting the drive wheel axle and bearings must be pivotally, or otherwise movably, supported from the end truck or trolley so that the drive wheel is free to move up and down, within limits, in order to maintain the predetermined pressure as the rail flange thickness and the drive wheel diameter vary slightly due to wear, manufacturing tolerances, or other reasons (Fig. 10.2.39).

The use of compression springs to apply pressure to the drive wheel is usually employed when the drive wheel is made of steel, or when the drive wheel tire is solid, relatively thin, and subject to relatively minor deformation when under full load.

One method of applying pressure without the use of springs is through the use of a pneumatic tire on the drive wheel. In this arrangement, developed in the early days of the industry, the axle of the drive wheel is held at a fixed distance from the soffit of the rail, the fixed distance being slightly less than the outside radius of the fully inflated tire. Consequently, the surface of the tire is slightly flattened at its point of contact with the rail, and as the air pressure within the tire is increased, the surface pressure between the tire tread and the rail, and thus the traction, is increased proportionately (Fig. 10.2.40).

In recent years, a method similar to the foregoing but employing a so-called "airless" tire has largely replaced the pneumatic tire. The airless tire resembles the pneumatic tire in appearance, but because the sidewalls are stiff enough to maintain reasonable pressure against the rail when deformed, without the introduction of air under pressure, field maintenance is simplified.

Yet another method that has found limited usage, is to employ the weight of the driven unit itself to apply pressure to the drive wheels through a system of leverage. In this construction, typically, the drive-wheel supporting frame, carrying the wheel axle and its bearings, is pivoted to, and supported by, the end-truck trolley. The drive wheel is then located on the *outboard* side of the trolley, and

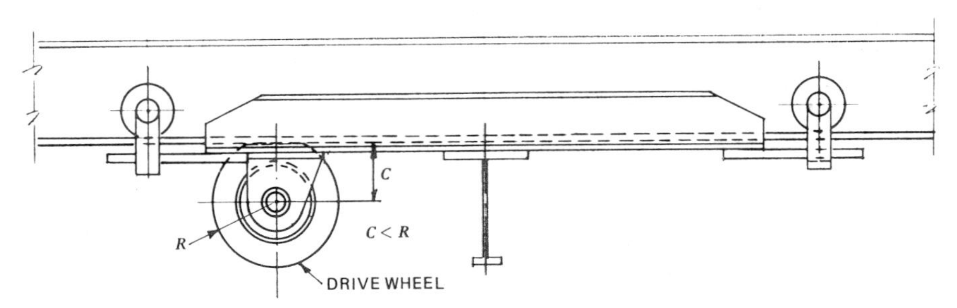

Fig. 10.2.40 Underdrive with pneumatic tire.

the end truck is pivotally connected to the drive wheel frame on the *inboard* side of the trolley. Thus the trolley lies in a pivotal position between the drive wheel on one side and the end-truck frame on the other. Consequently, as the load on the end truck is increased, the pressure on the drive wheel, and hence the traction, is increased proportionately as required to propel the heavier load (Fig. 10.2.41).

Power Requirements

To calculate the horsepower required in a drive, it is first necessary to estimate the force required to propel the load on the level. The force F, or tractive effort, must be sufficient to overcome the rolling resistance R_r with the monorail or runway in a horizontal plane.

Expressed mathematically,

$$F \geqslant R_r \qquad (10.2.21)$$

where: F = propelling force
R_r = rolling resistance

Having determined the force F and the desired speed of the moving load, the basic horsepower required may be calculated. If the monorail is on an incline, an additional allowance to raise the load must be added to the basic tractive horsepower in order to determine the total power requirement.

In a monorail or underhung crane system, the rolling resistance R_r will vary considerably depending on several factors. These factors include the basic design of the unit, such as wheel size and configuration (see Section 10.2.5); the material and relative hardness of the wheels and the rail tread; the type of bearings employed, and the method of rail suspension; and the conditions pertaining to a particular installation, including the general condition of the equipment (e.g., new or excessively worn), and atmospheric conditions (e.g., dirty or dusty surroundings, excessive moisture, or extreme temperatures).

Because these conditions may vary widely, the rolling resistance in any specific application is, of necessity, an estimate based on experience and good judgment.

For most industrial installations employing steel wheels on steel rails, under average conditions, the rolling resistance falls in the range of 1–3% of the total weight of the moving load, or

$$R_r \approxeq 0.01\ W \text{ to } 0.03\ W \qquad (10.2.22)$$

where: R_r = rolling resistance
W = total weight of moving load

For general industrial applications, manufacturers can usually provide this information, as it applies to their equipment, provided all pertinent conditions are made known. However, in very unusual situations, where special conditions prevail, it is sometimes necessary or advisable to conduct a pull-test in order to make an accurate determination of the rolling resistance.

Having determined R_r and the minimum force F required to propel the load, it then becomes necessary to consider the method by which the propelling force is applied to the equipment being analyzed.

As previously stated, there are two basic types of drives available for use on monorails and underhung cranes: the geared-wheel drive and the underdrive. In either type, the propelling force F that can be generated is the product of the coefficient of friction f_o between the rail and the drive wheel, and the pressure p applied to the drive wheel, or

$$F = p \times f_o \qquad (10.2.23)$$

As with the rolling resistance, the coefficient of friction f_o may vary depending on the specific conditions. Included among the variable conditions affecting f_o are the materials comprising the drive

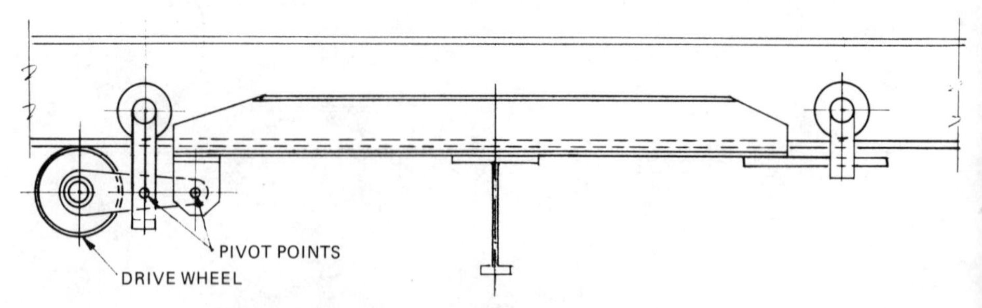

Fig. 10.2.41 Underdrive, weight actuated.

wheel and the rail (e.g., steel on steel, bronze on steel, or rubber on steel), and the condition of the mating surfaces (e.g., new or worn, painted or unpainted, clean or dusty or oily).

For average industrial applications, f_o generally falls within the following ranges:

Steel on steel,	$f_o \approx 0.1$ to 0.2
Bronze on steel,	$f_o \approx 0.075$ to 0.15
Rubber on steel,	$f_o \approx 0.3$ to 0.5

Unlike the factor f_o, the factor p, or the pressure between the drive wheel and the rail, may be accurately determined. As previously noted, pressure p may be applied directly by the weight of the rolling equipment when it is driven through the load-bearing wheels, or indirectly by means of spring pressure when employing an underdrive.

It is important, to minimize slippage and provide adequate propelling force, that the proper pressure be applied to the drive wheels. When an underdrive is employed, the pressure between the drive wheel and the rail is fairly constant regardless of the position of the load. It may be quite accurately controlled, in the case of a steel or solid-tired drive wheel, by proper adjustment of the springs, or by adjusting the position of the drive wheel so as to obtain the proper degree of deformation when a pneumatic or airless rubber tire is the driving element.

When the drive is through the load-bearing wheels, then the pressure p is a function of the weight of the rolling load. This being the case, p varies from a minimum value under a no-load condition to a maximum value when operating under full load. In the case of an underhung crane, p also varies with the position of load along the length of the bridge.

In the determination of the pressure p on the drive wheels of a monorail trolley supporting a hoist, due consideration should be given to the relative position of the load hook with respect to the trolley wheels, when the hook is in both the upper and lower limits of travel. On hoists with single line reeving, the hook travels longitudinally along the axis of the drum as it is raised and lowered. This changes the relative distribution of the wheel loads, and hence the pressure p on the driving wheels. If possible, it is usually preferable to select as the drive wheels those bearing the heaviest load when the hook is in the up position, since that is usually the position of the load when traveling, and hence the arrangement providing the greatest traction.

It is also very important, on underhung cranes, to make certain that the drive wheels in a geared-wheel drive are bearing a large enough percentage of the dead load to provide traction without slippage.

For example, on a crane equipped with four-wheel end trucks, if one end of each truck is equipped with a geared-wheel drive, then 50% of the dead load is borne by the driving wheels and helping to generate traction. This is generally more than adequate.

If, on the other hand, a similar crane of the same weight is equipped with eight-wheel end trucks, then only 25% of the dead load is borne by the driving wheels. As a result, other things being equal, the total tractive force F is one-half that obtainable with the four-wheel truck, and may be insufficient to propel the crane.

Accordingly, when the end truck incorporates eight or more wheels, the tractive force obtainable should be carefully checked against the minimum required to overcome the rolling resistance. This is particularly important where the dead weight of the crane is relatively low as compared to its load-carrying capacity.

For example, in a multiple-runway single-girder crane of moderate span employing eight-wheel end trucks, the factor p may be so low that the total tractive force (10.2.23),

$$F = p \times f_o$$

may be inadequate to overcome the rolling resistance R_r, since as previously noted the tractive force must equal or exceed the rolling resistance.

When this is the case, it may be necessary either to replace the eight-wheel end trucks with four-wheel trucks of equal capacity, or to provide two drives per end truck, or to replace the geared-wheel drives with underdrives in which the tractive force is independent of the end-truck wheel loads.

Horsepower Calculations

Once the minimum required total tractive force F has been determined, the calculation of the horsepower required for any given travel speed may proceed.

Let the total horsepower required be designated by HP. Then,

$$HP = \frac{F \times FPM}{33{,}000 \times e} \qquad (10.2.24)$$

where: F = tractive force required, in pounds
FPM = velocity in feet per minute
e = efficiency of the total drive system

The efficiency e will vary considerably, depending on the gear reduction ratio, the type of gearing employed, drive bearings and other details of construction, and the losses at the drive wheel which vary with the material used in the drive wheel tire, and the amount of deformation when under full pressure. The efficiency e generally lies somewhere in the range of 60–90%.

Most manufacturers of monorail equipment publish rating tables on their drives, in which the efficiency has been taken into account. On custom-built units employing purchased gear reducers, or combination motor-reducer units, it will be found that the manufacturer's rating tables also give net output capacities, with the internal efficiency factor provided for.

When the total horsepower required has been determined, if the crane is using a single drive unit and squaring shaft, power units having the next higher commercially available horsepower ratings should be selected.

If the crane is to be driven by two or more unit drives, then the *total* horsepower required should be divided by the total number of drive units. This figure should then be multiplied by a factor of 1.1, and motors of the next higher standard horsepower should be selected for each drive.

In each case, the output torque to be delivered by each unit should be calculated, and the drive components selected should have an output torque capacity equal to, or slightly in excess of, this figure.

In general, the total torque T required is equal to the total tractive force F, multiplied by the drive wheel radius, or

$$T = F \times r \qquad (10.2.25)$$

When there is more than one drive wheel, the torque required may be divided equally among them. In the determination of the torque capacity required of a squaring shaft, due allowance should be made for the total number of drive wheels connected to each segment of the shaft, and for the ratio of the gear or chin-and-sprocket reduction, if any, between the squaring shaft and the drive wheels to which it is connected. Torsional shaft deflection should be held to a minimum.

The preceding calculations outline the method of determining the total horsepower HP for propelling any given load along a *level* track, and will cover the vast majority of industrial applications. However, in those cases where it is necessary to propel a load up an incline, the horsepower required to lift the load must be added to the total.

For any given load, the incremental horsepower HP_i required will vary with the travel speed and the slope of the incline, thus:

$$HP_i = \frac{W \times FPM \times \sin\theta}{33{,}000 \times e} \qquad (10.2.26)$$

where: HP_i = incremental HP due to incline
θ = angle of incline
W, FPM, and e retain their previous definitions

As in other aspects of crane design and selection, the best results are obtained when all known factors bearing on a particular application are taken into consideration, and made known to any potential supplier, prior to making a final selection.

10.2.7 Electrification

In a monorail or crane system, if any part of the moving equipment is motorized, either the hoist, bridge, or trolley, then it becomes necessary to provide electrical current to one or more motors that are moving with respect to a stationary power source.

If the travel path is straight, or nearly so, and there are no monorail switches in the system, then power may be supplied to the moving equipment by means of a multiple-conductor flexible cable.

Flexible Conductors

When the path is very short, on the order of 25 ft or less, a cable reel may be employed, provided there are no interferences with the cable when fully extended. However, if the travel path is of such length that the power cable requires intermediate supports, then a tagline or festoon system, in which the power cable is looped and movably supported at frequent intervals, is the better choice.

Tagline and festoon systems are somewhat similar in appearance, but the method of supporting the cable differs and has a direct bearing on the maximum practical length of such systems. In either case, the power is fed to the flexible cable at one end of the rail to be served, and the other end of the cable is connected to the moving load.

In a tagline system, the flexible cable is attached to steel rings at regular intervals, depending on the maximum allowable length per loop. The steel rings are in turn movably supported from, and are free to slide along, a taut wire which is threaded through them, and which runs parallel to, and the full length of, the monorail, bridge, or crane runway that it serves (Fig. 10.2.42).

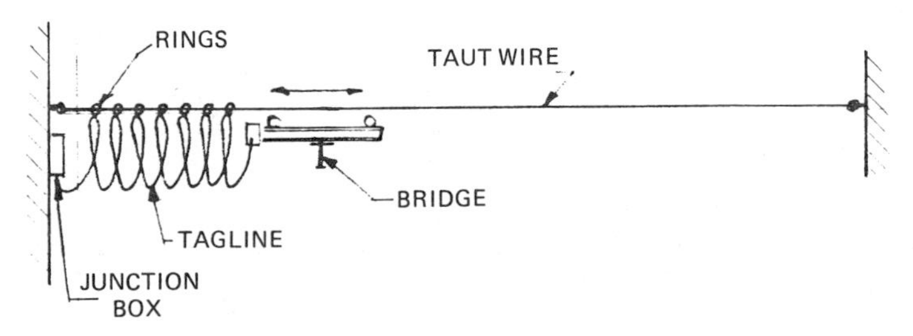

Fig. 10.2.42 Tagline conductor system.

Since the supporting taut wire may be anchored only at its ends, the maximum length of a tagline system is generally limited by the sag in the catenary formed by the supporting taut wire, and in good practice seldom exceeds 100 ft.

A festoon system is similar to a tagline system to the extent that the power cable is looped and supported at frequent intervals. The cable supporting system, however, comprises a series of small trolleys supported by, and are free to travel along, a light-duty monorail. The light supporting monorail may consist of a T section, a light I-beam, or a tubular track built for the purpose (see Section 10.2.2).

As in the case of a taut wire, the light monorail supporting the festoon system must be mounted parallel to, and the full length of, the monorail, bridge, or runway that it serves. However, since the supporting rail may itself be supported at frequent intervals, its length is not limited and it may therefore be used on much longer systems (Fig. 10.2.43).

In the design of either a tagline or festoon system, consideration must be given to the longitudinal distance required to store the loops of power cable, when the crane or trolley is at its extreme limit of travel nearest the power source. The length of rail required for cable storage will depend on the number of loops in the power cable, and, in the case of a festoon system, on the cumulative length of the trolleys supporting each loop. On long systems, this may limit the end approach of the crane or trolley, and should be taken into account and provided for as may be required.

Rigid Conductors

Rigid conductors running parallel to, and supported from, the rail they serve can be used in the same types of systems that employ flexible conductors. But they are more widely applicable because they are not subject to the same limitations in track layout, and may therefore be installed on systems incorporating curves, switches, interlocking transfer cranes, and such other track devices as may be included in the system.

Rigid conductors are available from all manufacturers of monorail equipment and other nonrelated sources as well. And although there are many variations in detail, some proprietary, as a class of equipment they share many characteristics that should be taken into account in the design of a monorail or underhung crane system.

While there may remain a few existing monorail systems that have been in place for many years, still employing open conductor bars, the majority of systems installed since the middle 1950s incorporate enclosed safety-type conductors. Since the use of bare conductor bars no longer complies with existing safety codes, our discussion is limited to enclosed conductors which are now used exclusively on new installations.

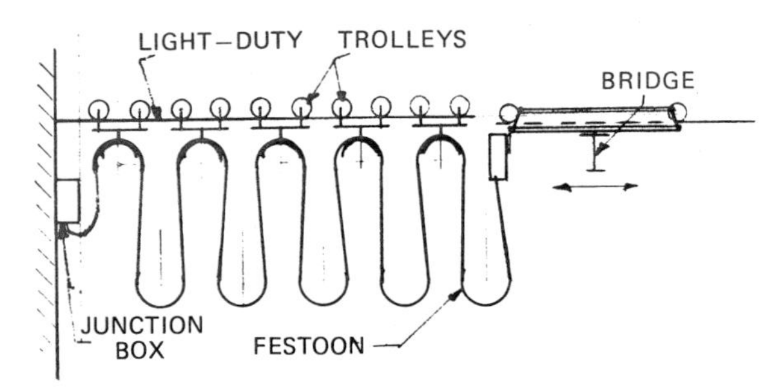

Fig. 10.2.43 Festoon conductor system.

Enclosed conductors are made in two basic types. In one configuration, two, three, or more conductor bars are enclosed in a common enclosure, designed to support a traveling current collector within itself, and provided with a narrow longitudinal slot along its entire length, through which the collector leads may travel. This type of enclosed conductor is usually available in standard straight lengths of about ten feet, but may also be obtained formed into curves of various radii and lengths.

The other basic type of enclosed conductor bar widely used on monorail systems comprises a single steel bar for each phase, enclosed in an insulating sheath, which is also provided with a narrow longitudinal aperture along its entire length through which the spring-biased current-collector shoe can extend to make contact with the power bar.

The rigid conductor bar is usually made of steel instead of copper, which may be a better conductor, in order to avoid differential thermal expansion between the conductor bar and the steel monorail to which it is bracketed. Where temperature fluctuations are likely, the differential in the coefficient of expansion between dissimilar metals may cause the conductor bar to bulge or bow between supports, a condition to be avoided.

Individual conductor bars are manufactured in several shapes, including light T bars, channel bars, angles, figure-8 bars, and various special shapes rolled for the purpose. All are provided with an extruded insulating sheath, and current collectors to match the specific bar employed are generally available from the same manufacturer.

The current-carrying capacities of the various conductor bars available for industrial monorail use range from about 70 to 200 A or more, and the matching current collectors are also manufactured in several capacities. The electrical ratings for the components of any particular system are obtainable from the manufacturer, and this information is generally included in the published literature.

Each system includes, in addition to the conductor bar and current collectors, such necessary auxiliary components as splices, support brackets, power-feed connectors, end closures, and the special end fittings required to provide smooth passage of the current collector as it traverses the narrow gap between the ends of mating conductor bars that are movable with respect to each other. This condition exists at all transfer points such as those that occur at switches, interlocking cranes, and other similar applications.

Conductor bars are usually mounted on the monorails to which they are attached in one of two patterns. In one arrangement they are "out-bracketed," and as the name implies, lie alongside the rail, but spaced out several inches from the rail centerline (Fig. 10.2.44). The other method commonly employed is known as "web-mounted." In this arrangement, the conductor bars are mounted by means of suitable insulating brackets directly, and in close proximity, to the web of the monorail or crane runway.

If the conductor bars are to be used on a crane runway, a bridge girder, or a monorail system without tongue switches, either the out-bracketed or the web-mounted method of supporting the conductor bars may be employed.

However, wherever tongue switches are included in an electrified monorail system, the web-mounted conductor bars may be preferred because they reduce both the required throw and length of the switch, and the reduced switch dimensions provide greater flexibility in the monorail layout (see Section 10.2.2).

In a three-phase electrified system, the moving equipment must be grounded, and a fourth conductor

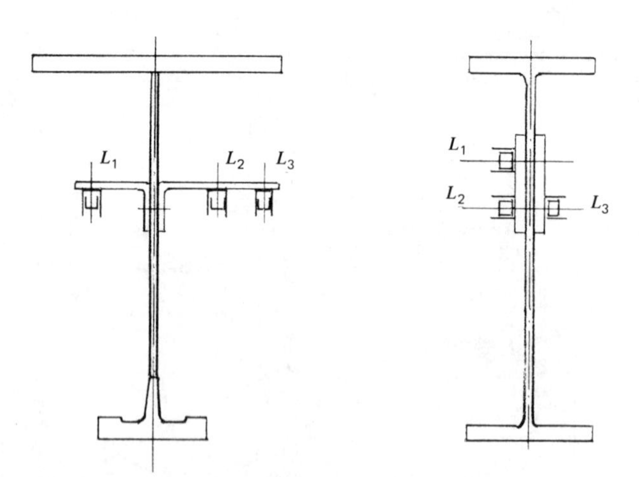

Fig. 10.2.44 Rigid conductor system.

and grounding collector are frequently provided to serve this function. However, at the present time, partly because it is somewhat less costly, the rail itself is more commonly employed for this purpose. When the system is grounded through the monorail track, it is important that all paint be removed from the treads of the rail and the grounding wheels, and that at some point the rail be electrically bonded to a positive ground. A separate grounding brush on the rolling equipment is also preferable to grounding through the wheels and bearings.

It is common practice to provide one current collector per bar on motorized equipment, and, if properly sized, this is usually sufficient. However, there are some applications where even an instantaneous interruption of current is undesirable, or even dangerous. Such a condition sometimes exists with electronic control systems where control voltages may be very low. In such situations it is advantageous to use two collectors per phase, spaced a few inches to a foot or two apart, with each pair connected in parallel. This insures continuity of current even if one collector of a paralleled pair should momentarily lose contact with its conductor bar.

A condition to be scrupulously avoided in the layout of an electrified monorail system is known as "phase reversal." Phase reversal can occur when the rail switching arrangement is such that an electrified trolley, equipped with current collectors, can be turned end-for-end on the monorail track on which it operates (see Fig. 10.2.45).

If this is allowed to occur, the phases of the power supply to the trolley are reversed. This in turn reverses the functions of the control push buttons, so that pressing the "forward" button causes the trolley to move in reverse, and pressing the "down" button on the hoist causes it to raise. The latter is particularly dangerous because, when in this mode, the upper limit switch on the hoist is no longer operative, and serious damage can result.

The best remedy for this situation is to rearrange the track layout so that it cannot occur. However, if the functional requirements of the system are such that the switch arrangement cannot be altered, so that a trolley can in fact be turned end-for-end, then there are two ways to compensate for this condition. Both require prior planning.

The first method involves increasing the number of conductor bars to six, and a rearrangement of the conductor bars and current collectors in such a manner that they are mechanically and electrically symmetrical about the centerline of the rail. If this is done, then the trolley may face in either direction on the rail, without changing the phasing of the power to the trolley control panel. See Fig. 10.2.46. Although this requires additional conductor bars and collectors, it does avoid the dangerous situation created by an unexpected phase reversal.

Another method requires the use of four conductor bars instead of six, plus a device known as a "phase reversal relay."

In this system, the bars are again arranged so that they are mechanically symmetrical about the rail, but only one phase is duplicated electrically (see Fig. 10.2.47). The phase reversal relay is a device placed in the trolley control panel, *between* the power leads from the current collectors and the reversing contactors in the panel that control the moving equipment. Its function, when it senses a phase reversal in the incoming current, is to actuate a separate contactor *ahead* of the reversing contactors, and which redirects the current to the reversing contactors in such a manner that it is properly phased with respect to the internal wiring of the control panel and the push-button markings.

The method best suited to a particular application must be evaluated on the basis of a careful analysis of all pertinent factors. For example, on a long system incorporating many switches and only one powered trolley, the cost of a phase reversal relay may be much less than the cost of supplying six conductor bars along the entire system. On the other hand, if there are several motorized trolleys operating on a relatively short track system, the economic advantage might revert to the other alternative. Cost, of course, is only one of the many factors that should be considered in arriving at a decision. The important thing with phase reversal is to avoid it, or provide for it.

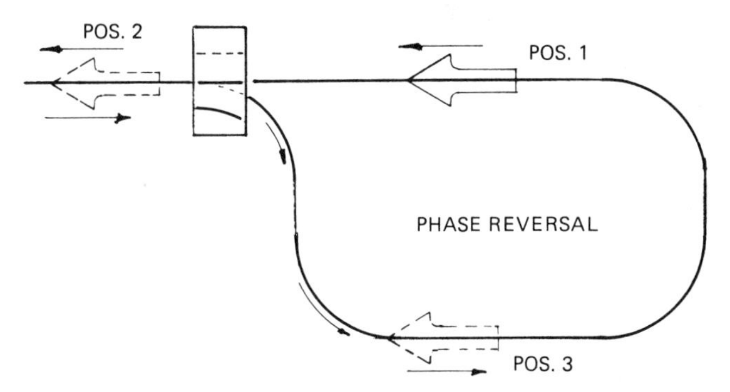

Fig. 10.2.45 Phase reversal, track layout.

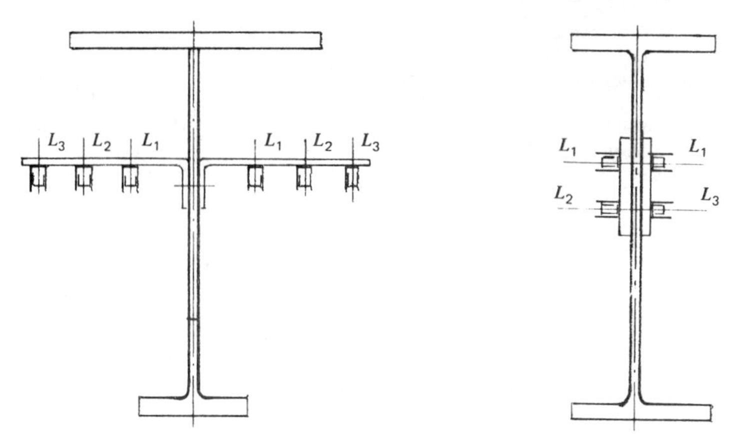

Fig. 10.2.46 Phase reversal, six-bar system. **Fig. 10.2.47** Phase reversal, four-bar system.

10.3 HOISTS

Robert R. Reisinger

10.3.1 Introduction

Industrial hoists constitute one of the most basic materials handling devices used in industry. In fact, the history of hoisting devices predates the very inception of any industrial society.

Today's industrial hoist is commonly referred to as an overhead hoist. Types of hoists are hand-chain manually operated chain hoists that utilize load chain as the lifting medium, and electric or air-powered hoists that utilize either load chain or wire rope as the lifting medium.

Industrial hoists, whether manually or power-operated, are designed and intended for vertical lifting of freely suspended materials handling loads. Industrial hoists should never be used to lift, transport, or suspend personnel. Although the general design of an overhead hoist is based on vertical lifting, the use of a hoist is not strictly limited to only vertical-type lifts; however, for such applications, a detailed analysis of the application must be made to verify that all types of load forces transmitted into the hoist unit can be handled by the hoist and its suspension, or to determine if a special hoist design is necessary.

Various suspension means are available for hoists. Manually operated chain hoists are normally hook, clevis, lug, or trolley suspended. Hook-, clevis-, or lug-suspension units can be mounted in a fixed location or suspended from a trolley unit that will accept one of these types of suspensions. Trolley-hoist units can, therefore, be a combination of a separate hoist and trolley, or a trolley-hoist design wherein the hoist unit is integrally designed with the trolley. Trolley travel for manually operated trolley hoists can be manually operated by pulling a trolley that has plain wheels, or manually operated by use of a hand chain that transfers motion to the geared wheels of the trolley.

Electric or air-powered chain hoists can have the same suspension types as manually operated chain hoists. In addition, trolleys for these types of hoists can be powered by either electric or air means.

Electric or air-powered wire rope hoists can have many types of suspensions for either fixed or trolley mounting. Deck-type mountings are provided for mounting on trolley units of double-girder cranes, and base-mounted units allow mounting in typical winch-type applications.

10.3.2 Manually Operated Chain Hoists

Manually operated chain hoists are powered by the manual effort of the operator. The operator's effort is transferred into hoist motion by the operator pulling on the hoist hand chain. The hand chain turns a pocketed hand-chain wheel that actuates the hoist mechanism.

Hoist Mechanism

The hoist mechanism consists of a gear train and mechanical load brake. The mechanical load brake is an automatic type of brake that will stop and hold the load when hand-chain pull is released, and permits controlled lowering of the load when hand-chain pull is applied in the lowering direction. Although this unidirectional mechanical device requires an applied hand-chain pulling force to lower a load, it does not require any additional hand-chain pulling force when lifting a load other than that normally required through the gear reduction.

The output shaft of the hoist gear reduction drives a load wheel which is a component that transmits motion to the hoist load chain. The load chain is the lifting medium used to lift or lower the load. The load wheel can also be called load sheave, load sprocket, pocket wheel, chain wheel, or lift wheel.

Load Chain and Hand Chain

The pockets of hand-chain wheels and load wheels are designed to allow proper fit and engagement for the specific hand chain or load chain used on the hoist. Hand chain and load chain are both accurately pitched within specified tolerances to fit the wheel pockets. Hoist load chain is manufactured and heat treated for hoist service, and is not similar to standard chains made for slings. This is sufficient reason why hand or load chains cannot be interchanged between hoist products, or why standard type chain cannot be installed on hoists as replacement chain.

Hand Hoist Efficiency

Two items of information normally published in manual hoist catalogs are hand-chain pull to lift rated hoist load, and hand-chain overhaul to lift the load 1 ft (meter). Manually operated chain-hoist efficiency can be calculated from these published figures or figures collected by actual test.

$$\text{Efficiency} = \frac{\text{Output}}{\text{Input}}$$

$$= \frac{W}{PH} \times 100$$

where: W = load, pounds (kg)
P = hand-chain pull, pounds (kg)
H = hand-chain overhaul in feet (meters) to lift load 1 ft (meter)

Hand Hoist Terminology

Standard terminology applicable to hook-suspended hand-chain manually operated chain hoists is shown in Fig. 10.3.1, and is defined below.

Headroom is the distance from the saddle of the top hook to the saddle of the load hook with the load hook at its upper limit of travel. For other types of suspensions, headroom is measured from the saddle of the load hook when it is at its upper limit of travel to the suspension-hole centerline of clevis- or lug-suspended hoists, or to the bottom of the beam on trolley-suspended hoists.

Lift is the maximum vertical distance that the load hook can travel between its lower limit of travel and upper limit of travel.

Reach is the sum of headroom and lift, and for hook-suspended hoists, is the distance from the saddle of the top hook to the saddle of the load hook at its lower limit of travel. For other types of suspensions, reach is measured from the suspension points discussed under headroom.

Hand-chain drop permits location of the hand chain in relation to the hoist unit and the limits of load-hook travel. Hand-chain drop is the measured distance from the saddle of the load hook with the load hook at its upper limit of travel to the lowest point of the hand chain.

Suspensions and Trolleys

Hook-suspended hoists are available in capacities from $\frac{1}{2}$ ton to 60 tons. Accurate spotting of load level can be obtained and many high-capacity units are used in maintenance and overhaul applications. High capacities are obtained by multiple reeving of the load chain. Some capacities use two basic hoist units and, therefore, have two hand chains. These units are intended to have one operator pull on each chain.

Trolley-suspended units offer a wide range of flexible selections, especially in regard to headroom. One type of trolley unit is a hook-suspended hoist attached to a separate trolley. This arrangement is the least expensive type of trolley-suspension unit; however, it does result in the largest headroom dimension. Two types of integral trolley-hoist units are available, commonly identified as an army-type trolley hoist and a close headroom trolley hoist. The army-type has a smaller headroom dimension than the combination hook-suspended hoist and separate trolley. The close headroom type offers the minimum headroom dimension. Most close headroom trolley-hoist designs have double load chains which allow the load hook to be raised up against the monorail beam on which the trolley operates. The difference in these trolley units is shown in Fig. 10.3.2. Integral trolley-hoist units can usually operate on smaller-radius curves than the separate hoist and trolley arrangement. Integral trolley-hoist units are available in capacities from $\frac{1}{2}$ ton to 24 tons.

Plain trolleys are recommended where motion of the trolley is infrequent or for a relatively short travel distance. Hand-chain-operated trolleys are recommended for capacities of 3 tons (3000 kg) and greater, locations where the elevation of the monorail beam is greater than 20 ft (6 m) above the operator's position, and applications requiring accurate spotting of the load.

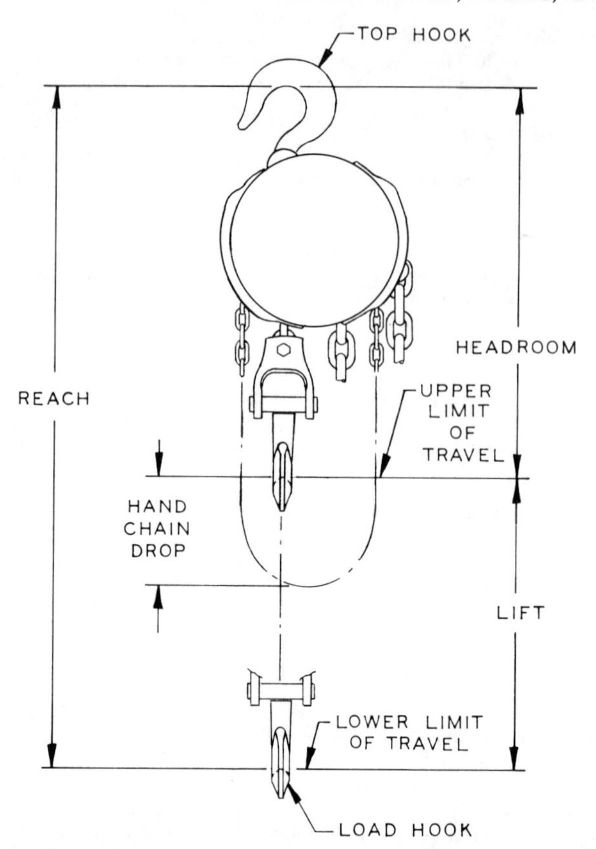

Fig. 10.3.1 Hook-suspended hoist terminology. Used with the permission of Acco Babcock, Inc.

Design Factors

Load-suspension parts of a manually operated hand-chain hoist are the parts that are subject to direct load forces from the load attached to the load hook. Load-suspension parts are the means of suspension, including the trolley, suspension hook, clevis or lug, the hoist frame which supports the load wheel, the load wheel, load chain, and the load block. Load-suspension parts are normally designed so that

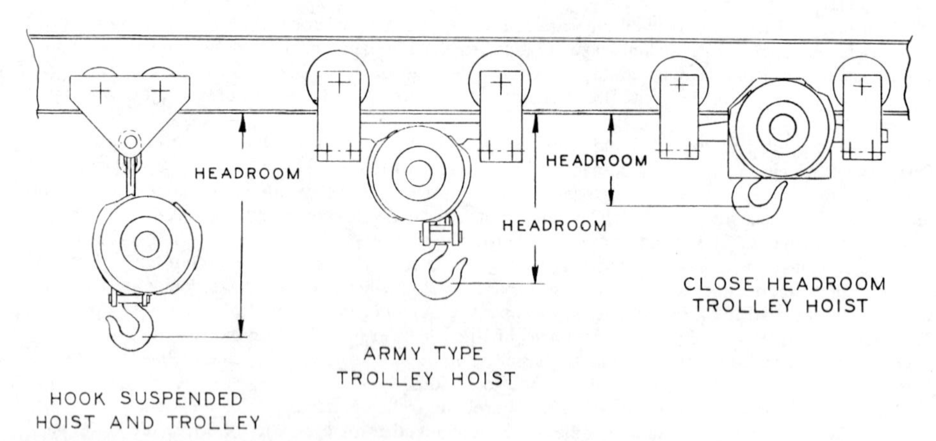

Fig. 10.3.2 Headroom dimensions of various trolley-suspended hand hoists. Used with the permission of Acco Babcock, Inc.

the static stress calculated for the rated capacity load does not exceed 25% of the average ultimate strength of the material used for the part. The resultant design factor is 4.

Standard models of manually operated hand-chain hoists are intended for general industrial use in ambient temperatures from 0°F (−18°C) to 130°F (54°C). Environmental conditions such as hazardous locations, excessively low or high ambient temperatures, corrosive-fume atmospheres, dust-laden atmospheres, moisture-laden atmospheres, and exposure to outdoor-weather conditions that could be detrimental to a manually operated hand-chain hoist are considered abnormal operating conditions. Special features are available to permit operation in such conditions. Specific details regarding the environmental conditions that could be encountered should be referred to a hoist manufacturer in order to obtain a unit that will perform satisfactorily under such conditions.

Manually Operated Hand-Chain Hoist Standards

Standards that apply to overhead hand hoists are HMI (Hoist Manufacturers Institute) 200, Standard Specification for Hand Operated Chain Hoists [2] and ANSI (American National Standards Institute) B30.16–1981, Safety Standard for Overhead Hoists (Underhung) [4]. Other standards, applicable to hand hoists when they are used in conjunction with trolleys or manually operated cranes, are referenced in Section 10.3.4. An ASME (American Society of Mechanical Engineers) Standards Committee on Overhead Hoists was organized in 1979 for the purpose of developing performance standards for various types of overhead hoists [13]. As standards developed by this committee are adopted, they will replace the HMI standards.

10.3.3 Power-operated Hoists

Power-operated overhead hoists are either electric or air powered, and use either load chain or wire rope as the lifting medium. Other power means, such as hydraulic motors or internal-combustion engines, can be used to operate these hoists; however, these types of powered-hoist units are not commercially available and would require specialized designs.

The basic design and function of a powered overhead industrial hoist is the same regardless of whether it is electric powered or air powered. The only substantial difference is the motor and controls necessary to operate the motor and other devices supplied as part of the hoist unit.

Air-powered hoists have an inherent variable-speed characteristic and are used in hazardous and spark-danger atmospheres because of the absence of electrical control arcing. Electric-powered units are also used in these locations when supplied with the proper motors, controls, and wiring for use in hazardous locations. Location of the hoist may present problems in furnishing an air supply to air-powered hoists. The air supply must include accessories that will ensure that clean, dry air is being supplied to the air motor.

10.3.4 Electric-powered Wire Rope Hoists

Electric-powered wire rope hoists are available in many variations of reeving, suspension, and control, and are therefore extremely flexible in being adapted to a specific application requirement.

Wire Rope

The majority of wire rope used on overhead hoists is of a 6 × 37 class construction because of its flexible properties in wrapping around the hoist drum and sheaves. As shown in Fig. 10.3.3, the second number of this designation is the nominal number of preformed wires to form a strand, and the first number of this designation is the number of strands wrapped around the core of the wire rope. The core may be fiber, plastic, or an independent wire-rope core designated IWRC. Independent wire-rope cores are wire ropes in themselves. Wire rope materials are plow steel, improved plow steel, extra improved plow steel, or other materials, such as stainless steel, for use in abnormal operating environments. These variables result in different strength limits for the same size diameter wire rope and readily indicate why wire rope replacement on an overhead hoist must be of the same size, construction, core, and material as that originally supplied on the hoist. The diameter of the wire rope is the diameter of a circle that would enclose the wire-rope strands. Correct and incorrect methods of measuring wire-rope diameter are shown in Fig. 10.3.4. Construction classes other than 6 × 37 are used on overhead hoists depending on design and application requirements.

Reeving

Type of wire-rope reeving will affect the capacity, lifting speed, and available lifting distance of a specific hoist model. Hoist-reeving terminology includes *single* or *double,* and *parts. Single* or *double* refers to the number of wire ropes wrapping or winding on the hoist drum. *Parts* refers to the mechanical advantage achieved by multiple reeving of the wire rope.

A single-reeved hoist has one rope attached to one end of the drum. The single rope winds around the drum until the drum is full. Hoist lifting distance is related to the amount of rope that can be wound on the drum and, therefore, is dependent on the drum length. Single reeving will create a

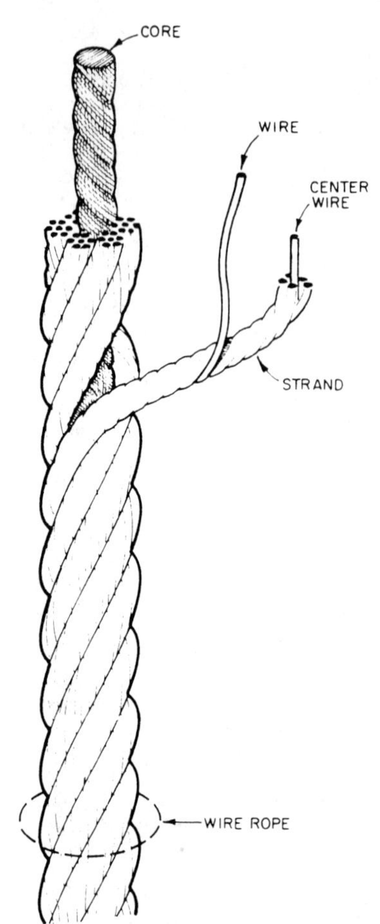

Fig. 10.3.3 Wire rope construction and terminology. From the *Wire Rope User's Manual,* 2nd Ed. Copyright 1981 by the Committee of Wire Rope Producers, American Iron and Steel Institute. Used with the permission of the American Iron and Steel Institute.

shifting hook position. The load hook, either attached to the rope directly or attached to a load block with sheaves operating over multiple rope parts, will shift or travel horizontally as the rope wraps on the drum from one drum end to the other drum end.

A double-reeved hoist has two ropes winding on the drum with one rope attached to each end of the drum. The two ropes wind toward the center of the drum until the load hook reaches its high-hook position. As in single reeving, lifting distance is related to the amount of double ropes that can be wound on the drum and, therefore, is dependent on the drum length. For a given hoist model, lifting distance of a double-reeved unit is approximately one-half the lifting distance of a single-reeved unit with identical drum lengths. Unlike a single-reeved hoist, the load hook of a double-reeved hoist will remain centered under the hoist as the double ropes are wound onto the drum. This feature,

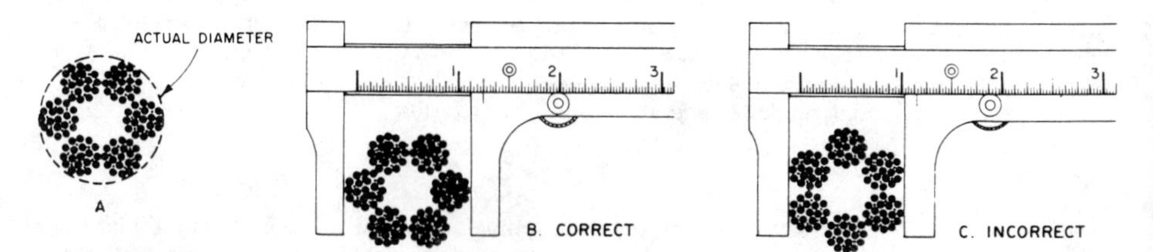

Fig. 10.3.4 How to measure (or caliper) a wire rope correctly. Since the "true" diameter (A) lies within the circumscribed circle, always measure the larger dimension (B). From the *Wire Rope User's Manual,* 2nd Ed. Copyright 1981 by The Committee of Wire Rope Producers, American Iron and Steel Institute. Used with the permission of the American Iron and Steel Institute.

termed *true vertical lift,* is of particular importance when horizontal hook movement must be avoided during the lifting or lowering operation.

Single and double reeving with different part designations are shown in Fig. 10.3.5. One-part single reeving (1PS) has the load or load hook supported by only one part of a single rope. The speed of the load hook is equal to the speed of the rope at the drum. Reeving a single rope into two parts results in two-part single reeving (2PS) and distributes the load into the two rope parts. Mechanical advantage achieved by the two parts doubles the capacity of one-part single reeving; however, it reduces the speed of the load hook to one-half the speed of the rope at the drum. In the same manner, reeving a single rope into four parts results in four-part single reeving (4PS), quadruples the capacity of one-part single reeving, and reduces the speed of the load hook to one-fourth the speed of the rope at the drum. Designation of parts for double reeving refers to the parts of each of the two ropes winding from the drum. With all types of reeving, variations in mechanical advantage or hoist capacity and load-hook speed occur, but one is obtained at a sacrifice of the other.

Hoist Types

Trolley-suspended wire rope hoists using single reeving are usually mounted with the drum centerline parallel to the beam on which they are suspended. This is necessary because of the horizontal hook movement that is characteristic of single reeving. Double-reeved units are usually mounted with the drum centerline perpendicular to the beam on which they are suspended. Since double reeving permits the use of smaller-diameter wire rope and, therefore, smaller-diameter sheaves as compared to the same capacity for single reeving, and double reeving usually allows the hoist to be mounted closer to the beam than single reeving, double reeving normally allows the load hook to be raised to a higher position than single reeving. For this reason, single-reeved hoists are often termed standard headroom hoists, and double-reeved hoists are often termed close headroom hoists. Double-reeved hoists having the drum parallel to the beam are offered for special applications.

Rigid mountings are available in many configurations to permit ceiling mounting, wall mounting, and deck or base mounting. A ceiling-mounted hoist is a type of mounting allowing the hoist to be rigidly mounted to the underside of a horizontal supporting surface. A wall-mounted hoist is a type

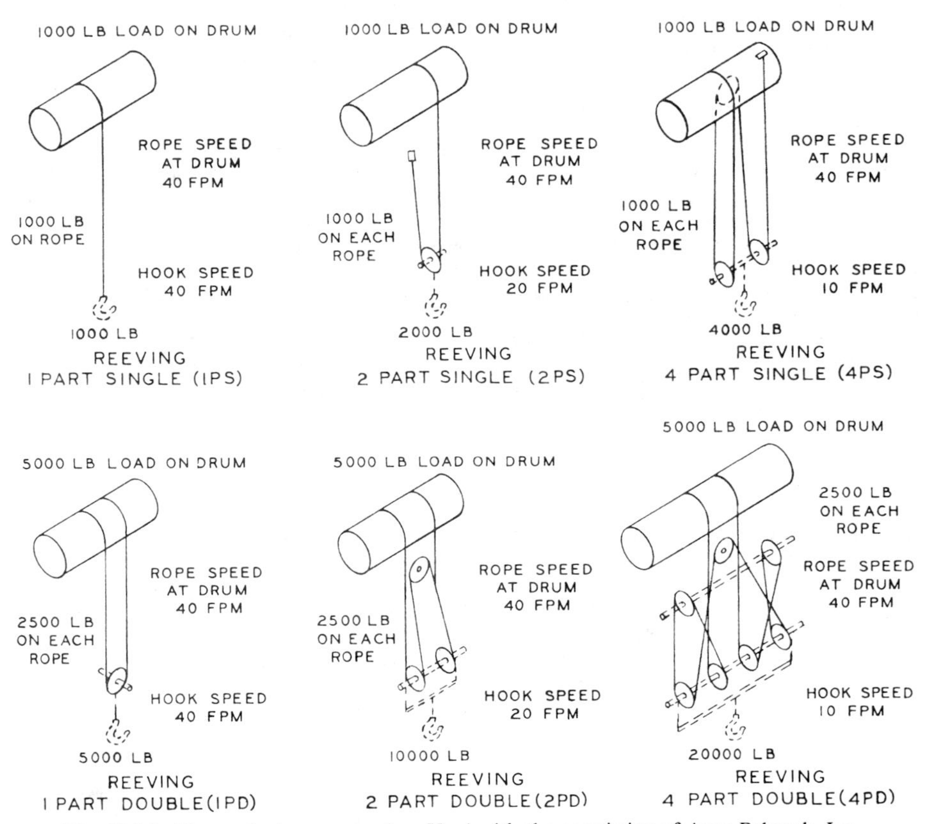

Fig. 10.3.5 Types of wire rope reeving. Used with the permission of Acco Babcock, Inc.

of mounting allowing the hoist to be rigidly mounted to a vertical supporting surface. Ceiling- and wall-mounted suspensions can be one of two types. One type is where the suspension is fabricated as an integral part of the hoist or hoist drum hanger. The other type is more common and consists of a ceiling- or wall-mounting suspension frame to which is attached by means of suspension pins, a standard lug-suspended hoist.

Deck- or base-mounted hoists are types of mountings that allow the hoist to be rigidly mounted to the top side of a horizontal supporting surface. While either the words deck or base can be applied to this type of mounting, base mounting is usually when the horizontal supporting surface is below the hoist unit, and the base-mounting support means is attached to a standard lug-suspended hoist. The term deck-mounted hoist usually refers to a horizontal mounting surface that is not limited to a below-the-hoist unit location. Typical deck-mounted units are used in top-running trolleys that operate on double-girder top-running overhead and gantry cranes. When a deck-mounted hoist is used in this type of trolley mounting, it is then called a top-running trolley hoist.

Base-mounted hoists are often used in configurations commonly called winches. Typical locations of the base-mounted hoist unit in relation to the load hook and rope reeving are shown in Fig. 10.3.6. Arrangements of this type offer advantages where the hoist unit cannot be located where the load hook is required, or to obtain an improved headroom dimension by moving the hoist unit away from directly above the load hook travel path. In arrangements of this type, care must be exercised in locating the lead sheave that guides the wire rope onto the drum. The distance between the lead sheave and the drum, identified as (b) in Fig. 10.3.6, must be of sufficient length to limit the fleet angle of the wire rope within permissible limits.

Powered Wire Rope Hoist Standards

Several standards are published that apply to overhead powered wire rope hoists, as well as standards that apply to some overhead cranes that utilize overhead powered wire rope hoists for the hoist lifting mechanism. All hoist users or potential hoist users should become familiar with the requirements of these standards. The standards that apply directly to overhead powered wire rope hoists are HMI (Hoist Manufacturers Institute) 100–74, Standard Specifications for Electric Wire Rope Hoists [1], and ANSI (American National Standards Institute) B30.16–1981, Safety Standard for Overhead Hoists (Underhung) [4]. An ASME (American Society of Mechanical Engineers) Standards Committee on Overhead Hoists was organized in 1979 for the purpose of developing performance standards for various types of overhead hoists [14, 16]. As standards developed by this committee are adopted, they will replace the HMI standards.

Standards that apply to overhead cranes that make reference to and utilize wire rope hoists covered by HMI 100–74 and ANSI B30.16–1981, are ANSI B30.11–1980, Safety Standard for Monorails and Underhung Cranes [5]; ANSI B30.17–1980, Safety Standard for Overhead and Gantry Cranes (Top-Running Bridge, Single Girder, Underhung Hoists [6]; CMAA (Crane Manufacturers Association of America) Specification No. 74, Specification for Top Running and Under Running Single Girder Electric Overhead Traveling Cranes [8]; and ANSI MH27.1–1981, Specifications for Underhung Cranes and Monorail Systems [10].

When an overhead powered hoist is mounted in a top running trolley for operation in a double-girder overhead crane, it should also comply with the requirements of CMAA Specification No. 70, Revised 1975, Specification for Electric Overhead Traveling Cranes [9], and ANSI B30.2.0–1976, Safety Standard for Overhead and Gantry Cranes (Top-Running Bridge, Multiple Girder) [7].

Hoists used in the handling of hot and molten material require additional considerations to operate in such an environment, and should comply with the applicable provisions of ANSI Z241.2–1981, Safety Requirements for Melting and Pouring of Metals in the Metalcasting Industry [11].

Hoist Duty Service Classification

Major causes of downtime and maintenance costs associated with electric overhead hoisting equipment can usually be attributed to misapplication or inadequate maintenance. In many instances, these two possible causes are closely related. Failure to fully analyze an application and select a hoist capable of performing the work required can result in excessive maintenance costs on a relatively new piece of hoisting equipment and reduce the useful life of the equipment. A specified capacity hoist is available in many types of reevings and suspensions, and also in physical sizes. The reason for this is the ability of each to perform different levels of work. Hoist selection for a particular application requires a review of the work that is to be performed in some unit of time. Selection of a hoist capable of performing this amount of work in the required unit of time will greatly influence the resultant performance of the hoist and the degree of maintenance necessary throughout the expected operating life of the unit. Hoist performance is defined in HMI 100–74, Standard Specification for Electric Wire Rope Hoists [1], by five hoist-duty service classifications identified as H1, H2, H3, H4, and H5. Table 10.3.1 is reprinted from HMI 100–74, and lists the five HMI duty classes that have been established, along with a description of typical application areas where each class can normally be used.

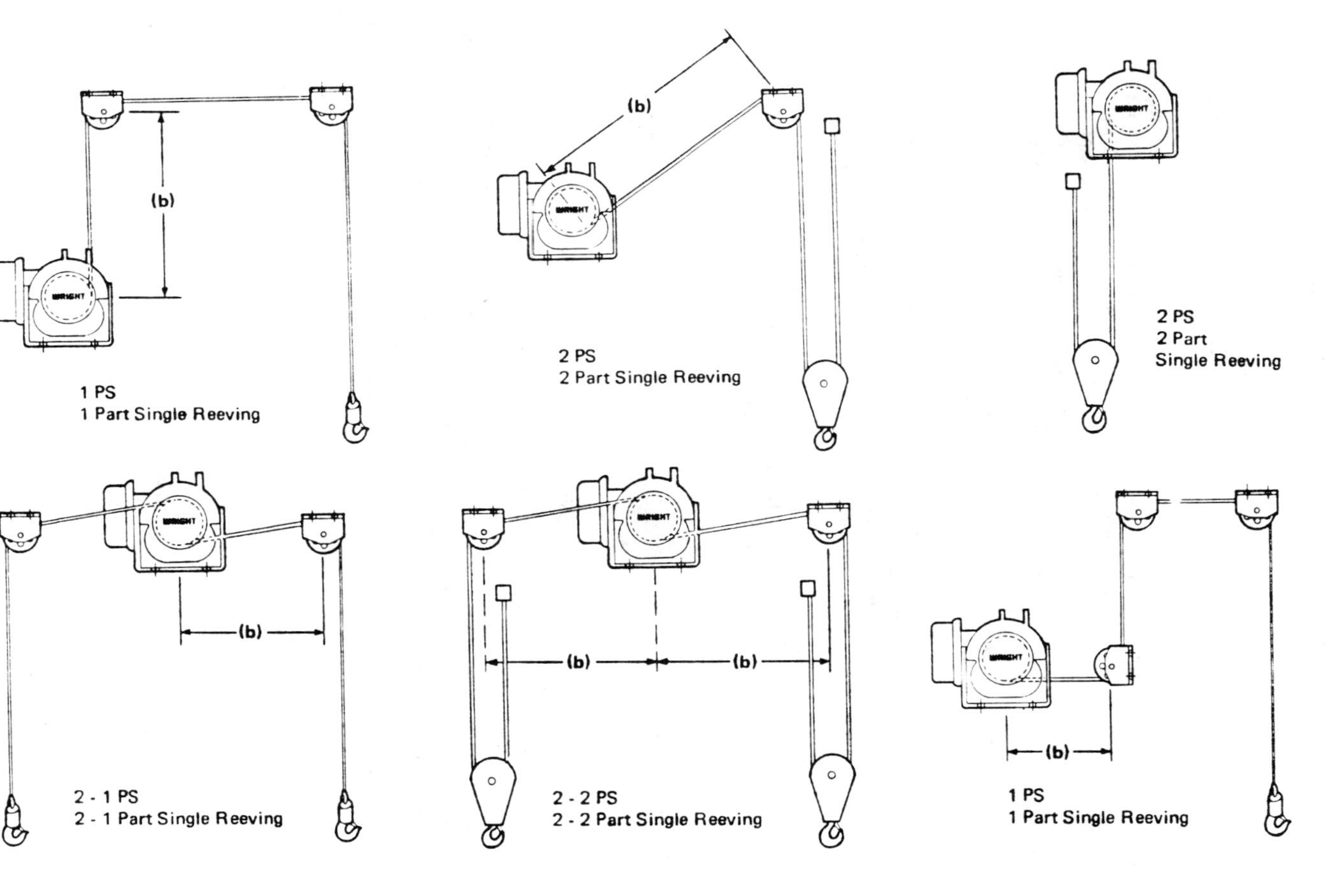

Fig. 10.3.6 Typical reevings of base-mounted hoists used in winch configurations. Used with the permission of Acco Babcock, Inc.

Table 10.3.1 Hoist Duty Service Classification

Hoist Class	Service Classification	Typical Areas of Application
H1	Infrequent or standby	Powerhouse and utilities, infrequent handling. Hoists used primarily to install and service heavy equipment, where loads frequently approach hoist capacity, with periods of utilization being infrequent and widely scattered.
H2	Light	Light machine shop and fabricating industries and service and maintenance work, where loads and utilization are randomly distributed with capacity loads infrequently handled, and where total running time of equipment does not exceed 10–15% of the work period.
H3	Standard	General machine shop, fabricating, assembly, storage, and warehousing, where loads and utilization are randomly distributed, with total running time of equipment not exceeding 15–25% of the work period.
H4	Heavy	High-volume handling in steel warehousing, machine shops, fabricating plants, mills, and foundries. Manual or automatic cycling operations in heat treating and plating operations. Total running time of equipment normally approaches 25–50% of work period, with loads at or near rated capacity frequently handled.
H5	Severe	Bulk handling of material in combination with buckets, magnets, or other heavy attachments. Equipment often cab operated. Duty cycles approaching continuous operation are frequently necessary. User must specify exact details of operation, including weight of attachments.

Source: From HMI 100–74, Standard Specifications for Electric Wire Rope Hoists. Copyright 1974, Hoist Manufacturers Institute. Used with the permission of the Hoist Manufacturers Institute.

Many factors must be considered in the selection of the proper size hoist to perform under given requirements. These include load distribution, operational time, work distribution, number of starts and stops, repeated long-lowering operations that generate heat in control braking means, and environmental conditions. Because hoist-duty service classification is a function of many factors, the actual performance of any particular hoist within a given class will be based on the extent that the actual loads and running times conform to those used in the evaluation. Table 10.3.2 is reprinted from HMI 100–74, and represents a more detailed description of the five HMI duty classes in performance figures.

The majority of hoist applications fall into classes H1, H2, and H3. Many industrial applications handling randomly distributed loads periodically during the work period or uniform loads that do not exceed 65% of the hoist rated capacity during the work period, can be generalized and selected

Table 10.3.2 Performance Specifications for Standard Electric Wire Rope Hoists

Hoist Class	Hoist B10 Bearing Life at K = 0.65 (hr)	Operational Time Ratings @ K = 0.65: Uniformly Distributed Work Periods		Infrequent Heavy Work Periods	
		Max. On Time (min/hr)	Max. Number of Starts per Hour	Minutes	Starts
H1 Infrequent standby	1,250	8	75	30	100
H2 Light	2,500	8	75	30	100
H3 Standard	5,000	15	150	60	200
H4 Heavy	10,000	30	300	N/A	N/A
H5 Severe	20,000	Up to continuous	600	N/A	N/A

Source: From HMI 100–74, Standard Specifications for Electric Wire Rope Hoists. Copyright 1974, Hoist Manufacturers Institute. Used with the permission of the Hoist Manufacturers Institute.

by the application descriptions of Table 10.3.1. A detailed analysis should be made of the duty cycle and duty class selected in accordance with the requirements of Table 10.3.2 if the application cannot be generalized as stated above and by the descriptions of Table 10.3.1. Classes H4 and H5 always require an analysis of the duty cycle.

The operational time ratings and hoist bearing B10 life listed in Table 10.3.2 are based on a mean effective load factor of 0.65. Mean effective load is a theoretical single load value that will have the same effect on the hoist mechanism as various randomly distributed loads that are actually applied to the hoist in some specified period of time. The letter K is used to denote the mean effective load factor and is expressed as:

$$K = \sqrt[3]{(W_1^3 P_1) + (W_2^3 P_2) + (W_3^3 P_3) + \ldots + (W_n^3 P_n)}$$

where: K = mean effective load factor
W = load magnitude
P = load probability

Mean effective load factor is a ratio of the theoretical mean effective load to the hoist rated capacity load.

Load magnitude is the ratio of the hoist actual operating load to the hoist rated capacity load expressed as a decimal. Operation with no load must be included, along with the weight of any dead load such as lifting attachments or devices suspended from the hoist hook.

$$W = \frac{\text{Hoist actual operating load}}{\text{Hoist rated capacity load}}$$

Load probability is the ratio of running time under each load magnitude condition to the hoist total running time expressed as a decimal. A standard total running time used in duty cycle analysis is one hour. The sum of all load probabilities must equal 1.0.

$$P_1 + P_2 + P_3 + \ldots + P_n = 1.0$$

Since hoist bearing B10 life, specified in Table 10.3.2, is based on a mean effective load factor of 0.65, the actual hoist bearing B10 life for other mean effective load factor values can be determined from the equation:

$$\text{Actual B10 hours} = \text{Rated B10 hours} \left(\frac{0.65}{K}\right)^3$$

Hoist design criteria normally includes design of all rotating elements for life expectancy equivalent to the bearing B10 life hours of Table 10.3.2 based on normal operation and proper maintenance.

Columns three and four of Table 10.3.2 give maximum running time in minutes per hour and maximum starts per hour based on the hoist use being uniformly distributed over the specified period of one hour.

Columns five and six of Table 10.3.2 are based on maximum elapsed time from a cold start at ambient temperature and that during this elapsed time, the hoist is only actually running 50% of the listed time. These values apply to infrequent periods of heavy service and cannot be repeated until the hoist unit is permitted to cool down to the ambient temperature.

The question most likely to be faced is when to use the generalized approach based on the descriptions of Table 10.3.1, and when to use the detailed analysis of Table 10.3.2. The generalized selection is based on handling randomly distributed loads during the work period. "Randomly distributed" implies that loads applied to the hoist, other than loads at rated load capacity, are assumed to be evenly scattered over the hoist capacity range in decreasing steps of 80% of the previous load value. Random loads are, therefore, assumed to be 80, 64, 51, 41, 26%, and so on, of rated capacity load. Operation with randomly distributed loads is assumed to occur on an equal time basis for the permissible operating time remaining after accounting for the time the hoist is operating at rated load and no-load conditions.

Detailed analysis of randomly distributed loads will show that under certain specified conditions, the mean effective load factor will be 0.65 or less [17]. The conditions are:

1. The hoist is operating without load during one-half of its total operating time. (Load magnitude = 0, and load probability = 0.5)
2. The hoist is operating with rated capacity load for a period of time that does not exceed 20% of its total operating time. (Load magnitude = 1.0, and load probability = 0.2 or less)
3. Loads applied to the hoist during the remainder of its permissible operating time are randomly distributed.

If all of the three conditions can be met, the generalized approach can be used. If actual operation is expected to exceed any one of these conditions, the detailed analysis should be used.

Design Factors

Hoist components and parts are divided into two classifications as to design in relation to the stresses that will be induced during operation. These are load-suspension parts and power-transmission parts. Load-suspension parts are the parts that are subject to direct load forces from the load attached to the load hook. A direct line of suspension can be visualized through these parts from the load hook to the hoist means of suspension. Load-suspension parts are the means of suspension, including trolley, suspension hook, lug or other mounting brackets, drum housing or other structure supporting the drum, the drum, wire rope, sheaves, load block, and load hook.

Load-suspension parts are commonly designed so that the static stress calculated for the rated capacity load does not exceed 20% of the average ultimate strength of the material used for the part. The resultant design factor is 5. Although the standard design factor for wire rope is 5, ANSI Z241.2 requires a wire rope design factor of 8 when the hoist is handling hot-molten materials.

Power-transmission parts are the machinery components of the hoist and include gears, shafts, clutches, couplings, bearings, motors, and brakes. Power-transmission parts are designed so that the dynamic stress calculated for the rated capacity load does not exceed the fatigue and endurance limits of the material used for the part or as established by the component manufacturer.

Headroom, Lift, and Reach

Headroom is the distance from the saddle of the load hook to the suspension-hole centerline of lug-suspended hoists as shown in Fig. 10.3.7. For other types of suspensions, headroom is measured from the saddle of the load hook to the saddle of the top hook for hook-suspended hoists, to the bottom of the beam for trolley-suspended hoists, and to the mounting surface for deck or base-mounted hoists. *Lift* is the maximum vertical distance that the load hook can travel between its lower limit of travel and upper limit of travel. *Reach* is the sum of headroom and lift.

Braking System

The motor brake shown in Fig. 10.3.7 is part of the hoist braking system. The motor brake may be located at many points within the hoist unit but is most commonly located as shown in Fig. 10.3.7 because of its convenient accessibility for maintenance and adjustment.

The hoist braking system consists of a holding brake and a control braking means to control the lowering speed. The performance of the braking system under normal operating conditions of capacity load, and under test conditions with test loads up to 125% of rated capacity load, should stop and hold the load when controls are released to the off position; stop and hold the load hook in the event of a complete power failure; and control and limit the speed of the load hook during lowering to a maximum speed of 120% of rated lowering speed for the load being handled.

The motor brake serves as the holding brake in the hoist braking system. The motor brake is

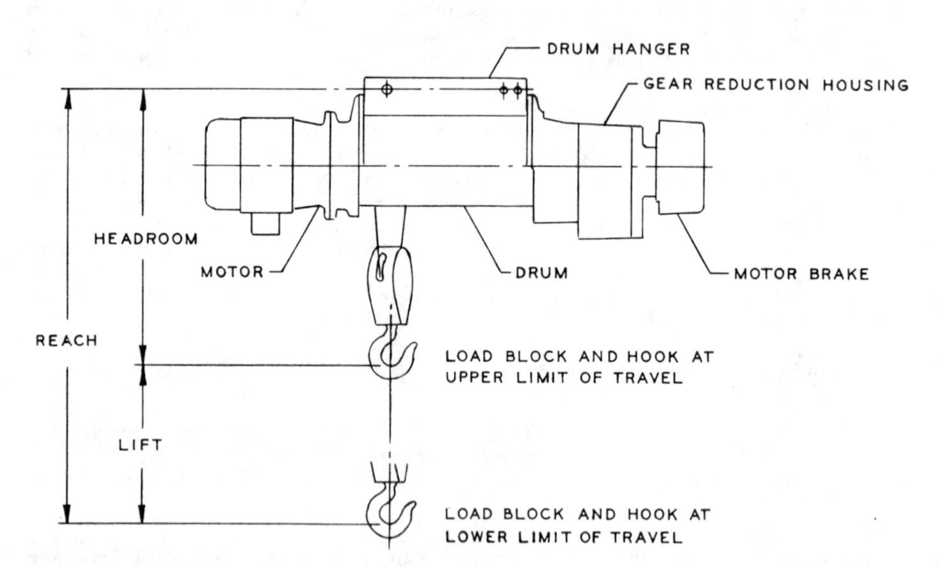

Fig. 10.3.7 Electric wire rope hoist terminology. Used with the permission of Acco Babcock, Inc.

normally in a closed position, and when the controls are activated to supply power to the motor, the brake is released to the open position allowing the hoist to operate. The brake should be applied or closed automatically when power is removed from the hoist motor. Some braking systems are designed to prevent the hoist from operating if the motor brake is unable to open because of adjustment or other causes. This feature prevents the hoist from operating against a closed brake that could cause subsequent damage to other components. Holding brakes are rated in foot-pounds of torque for stopping and holding a load and must have thermal capacity for the frequency of operation required by the hoist duty cycle classification. Holding-brake minimum torque requirements are stated as a percentage of the hoist motor rated load hoisting torque at the point where the brake is applied, and are:

1. 125% when used with a control braking means other than mechanical
2. 100% when used with a mechanical braking means
3. 100% per brake when two holding brakes are provided

Hoists used to handle hot-molten materials are usually required to have two holding brakes in addition to the control braking means.

Control braking means is a method of controlling the lowering speed whereby energy is removed from the moving body or energy is imparted in the opposite direction, and can be mechanical, hydraulic, or electrical. Electrical control braking means include dynamic, eddy current, countertorque, and regenative. *Dynamic* uses the motor as a generator and dissipates the energy as resistance. *Eddy current* reduces the speed by means of an electrical induction load brake. *Countertorque* applies a variable motor torque in a direction opposite to the direction the motor is being rotated by overhaul of the load. *Regenative* uses the motor as a generator and the electrical energy generated is fed back into the power system. Hydraulic braking controls, powers, or brakes by means of displacement of a liquid. Mechanical braking controls, reduces speed, or brakes by frictional means.

Worm-geared hoists having worm gears with worm angles that prevent the speed of the load from accelerating in the lowering direction do not normally require an additional control braking means. The self-locking feature of the worm angle serves as a control braking means and only a holding brake is required.

Gear Reduction

The hoist gear reduction serves the basic purpose of any gear reduction: to reduce speed and increase torque between the motor and output shaft driving the drum. Reduction steps, size, and type of gearing are dependent upon the design criteria established by the hoist manufacturer. The design must consider the design stresses for power-transmission parts and life expectancy related to hoist-duty cycle classifications. Mechanical control braking means are designed within the gear reduction and utilize the gear oil bath and housing to dissipate the heat generated by removing energy through friction means of the mechanical load brake. Housing design and oil bath housing contact area are important as the ability to dissipate this heat is often the limiting factor in establishing the duty-cycle classification of a hoist unit.

Although hoists are designed for normal operation in an ambient temperature range of 0–100°F (−18–40°C), the oil supplied with the hoist may not provide optimum performance throughout this entire range. Instructions provided by the manufacturer regarding lubrication should be consulted when the actual operating ambient temperatures are near either end of this range, or if significant changes occur in the ambient temperature due to seasonal change. The design characteristics of the mechanical load brake in regard to lubrication are of major importance in using only lubricants that are recommended by the manufacturer. Lubricants having additives for increased lubricity could completely alter the operational design of a mechanical load brake, and should never be used unless approved by the hoist manufacturer.

Drum

The drum provides motion to the wire rope thereby raising or lowering the load block. The drum also is a storage device for spooling the wire rope on as the load block is raised through the length of hoist lift. The drum should be grooved to achieve smooth and guided spooling across the drum and to reduce rope wear and wire breaks. The groove and radius at the bottom of the groove should result in forming a close-fitting saddle for the size of rope used to prevent crushing-type damage to the rope. Special applications may use drums without grooves or multiple-lay spooling of the rope.

The ratio of drum pitch diameter to rope diameter is one factor in determining service life of the rope. Hoist industry experience has established a recommendation that this ratio be not less than 18. This does not preclude the use of a smaller ratio as design application limitations may require the use of a smaller ratio. Smaller ratios may require more frequent rope inspection and rope replacement as determined by replacement criteria.

Hoists are not normally provided with a lower limit of travel limit switch unless specifically required. Therefore, the hoist must be installed and operated in locations where no less than two wraps of rope remain on each anchorage of the drum when the load hook is at its lower limit of travel. If a lower limit of travel limit switch is provided, the hoist can be used where no less than one wrap of rope remains on each anchorage of the drum.

Drum Hanger

The drum hanger is a structural member that supports the drum and transmits the forces induced by the load into the means used to suspend the hoist. Several other components are often incorporated into the drum hanger and include limit switches, overload limit device, rope sheaves for the reeving system, and control enclosures, among others.

Rope Sheaves

Rope sheaves provide a function similar to the drum except that they do not serve as a storage device. Rope sheaves are either running sheaves or equalizing sheaves. Running sheaves rotate as the rope passes over them in the process of lifting or lowering the load block. An equalizing sheave is a nonrotating sheave used to equalize the tension in opposite nonrunning parts of a double-reeved rope.

Sheaves, whether installed in a structural member or in the load block, could be subjected to conditions where the rope can be momentarily unloaded and attempt to jump the sheave groove. Provisions should be made to guide the rope back into the groove when the load is reapplied and to prevent the ropes from completely leaving the sheave.

Ratio of sheave pitch diameter to rope diameter has an effect on service life of the rope. Hoist industry experience has established recommendations that this ratio be not less than 16 for running sheaves and not less than 12 for equalizing sheaves. This does not preclude the use of smaller ratios as was discussed under drums, and therefore the comment about more frequent rope inspection or rope replacement also applies to sheaves.

Limit Switches

Limit switches are electromechanical devices that limit the travel of the load block. Most hoists have a gravity-operated limit switch that limits only the upper limit of travel. A geared limit switch can be installed on most hoists and offers limit protection at both the upper and lower limits of travel.

Gravity limit switches can be weight operated or lever operated. Either a weight or the weight of the lever keeps the switch, which is connected in the hoist control circuit, in the closed position. The switch is activated by the load block coming into contact with and lifting the weight or lever, causing the switch to trip to the open position. This opens the raising control circuit causing the hoist contactors to open and break the power being supplied to the hoist motor. The majority of gravity limit switches are provided with a plugging feature wherein if the load block drifts upward an excessive distance, the switch will close the lowering control circuit and lower the load block before it comes in contact with the drum or other structural member. The plugging feature should be connected in the circuit to enable it to function even if the operator's controls are released.

Geared limit switches are driven by the drum or gear reduction. This results in some known ratio between drum revolutions and geared limit switch revolutions. This device is connected into the hoist control circuit. It not only provides both upper and lower limits of travel, but because of its adjustability allows either the upper or lower limit to be activated at any predetermined point throughout the hoist lift range. Since geared limit switches operate with the definite ratio of drum revolutions, if the rope overwraps or jumps grooves on the drum, due to misapplication, the limit switch is no longer synchronized and will not be activated when the load hook reaches its upper limit of travel. A geared limit switch should never be used without a gravity limit switch, except in single-line winch applications which may require special consideration. The geared limit switch should always be adjusted to trip before the gravity limit switch. This makes the geared limit switch the normal means to limit the upper limit of travel, and the gravity-operated limit switch as a backup device.

Gravity and geared limit switches are connected in the hoist control circuit because their electrical ratings can only handle the current encountered in the control circuit. Because they are connected in the control circuit, they are rendered inoperative in limiting travel if the hoist motor contactor becomes frozen in a closed position. *Power-circuit limit switches* can be provided to prevent this possibility from occurring. Power-circuit limit switches are sized to handle the currents of the power supply provided to the hoist motor and are connected directly in the power supply leads that feed the hoist motor. Activation of a power-circuit limit switch breaks all power to the hoist. If a power-circuit limit switch is used, it should be used in conjunction with control circuit limit switches set to trip before the power-circuit limit switch is activated.

Limit switches normally supplied on a hoist are intended to serve as an emergency device and should not be used to repeatedly stop load-hook travel as a production means. If a limit switch is to be used for repeated stopping of the load hook at a set limit of travel, a second limit switch should be used to serve as the emergency device. A typical example is the use of a geared and gravity limit switch for the upper limit of travel. The geared limit is adjusted for the planned limit of travel on a repeated basis, and the gravity switch is set at a higher level of travel to serve as the emergency device in case a malfunction occurs in the geared switch. For automatic production purposes, multiple-circuit geared limit switches can be used to stop the load hook at a specific level of travel, and then initiate some other function, such as trolley travel.

Overload Limit Device

Overload limit devices are supplied as standard equipment on some hoists. Hoists that do not contain overload limit devices as standard equipment can usually be furnished with one on an optional basis when required. An overload limit device is intended to serve as an emergency device only, and is not to be used as a weighing device in picking up loads.

Many variables are experienced within a hoist system that make it impossible to adjust an overload limit device that would prevent the lifting of any overload or load in excess of rated capacity load. These include acceleration of the load, dynamics of the total system, type and length of wire rope, and operator experience, among others. An overload device is designed to permit hoist operation within its rated load range and to limit the *amount* of overload that can be lifted. The device will, therefore, allow lifting of an overload of such magnitude that will not cause damage to the hoist, and prevent lifting of an overload of such magnitude that could cause damage to the hoist.

An overload limit device is activated only by loads and forces incurred when lifting a freely-suspended load on the load hook. It cannot be relied upon to render the hoist mechanism inoperative if other sources, such as but not limited to, snagging of the load, driving the load block into the hoist frame, or attempting to lift a fixed load, induce forces into the hoisting system. Since the overload limit device is connected into the hoist control circuit, it will not prevent damage to the hoist, trolley, or other means of suspension if excessive overloads are induced into the hoisting system when the hoist unit is in a nonoperating or static condition.

Electric Motors

Hoist electric motors are available in various voltages, speeds, and classes of insulation. Typical hoist motors are class B insulation with 55°C temperature rise in a totally enclosed nonventilated frame, having an intermittent 30-min duty rating. The 30-min duty rating means that the motor can operate at full load for at least 30 min, and that the temperature rise experienced by the motor during this time will not exceed the allowable temperature rise for the class of insulation. The 30-min motor rating is sufficient for intermittent operation experienced in the majority of hoist applications. Motors having 60-min or continuous duty rating are available for high-cycle and severe-duty applications. The power supply voltage must be known in order to select the proper motor for the hoist. The motor nameplate voltage will normally be less than the nominal power supply system voltage. Recommended standards for both power supply system voltage and motor nameplate voltage are shown in Table 10.3.3.

Motor insulations are classified by their ability to withstand a specified temperature for a specified

Table 10.3.3 Standard Power Supply Voltages

Power Supply	Nominal Power System Voltage	NEMA Standard Motor Nameplate Voltage	Permissible Motor Operating Range Volts
AC, single phase, 60 Hz	120	115	104–126
	240	230	207–253
AC, polyphase, 60 Hz	208	200	180–220
	240	230	207–253
	480	460	414–506
	600	575	518–632
AC, polyphase, 50 Hz	208	200	180–220
	230	220	198–242
	400	380	342–418
DC	125	115	104–126
	125	120	108–132
	250	230	207–253
	250	240	216–264

Table 10.3.4 Temperature Ratings Generally Applying to TENV Hoist and Crane Motors

Insulation Class	Maximum Allowable Ambient Temperature	Allowable Motor Nameplate Temperature Rise	Total Allowable Motor Temperature
B	60°C	55°C	115°C
F	80°C	55°C	135°C
H	100°C	55°C	155°C

length of time without deteriorating. The total allowable temperature is the sum of the ambient temperature in which the motor operates and the heat generated by the motor, in allowable temperature rise. Some typical temperature ratings that generally apply to totally enclosed nonventilated (TENV) hoist and crane duty motors are shown on Table 10.3.4. Special environmental motor insulations are available and should be specified for hoists that will be operating in abnormal or adverse operating environments. Typical conditions are:

High humidity and fungus (tropical protection)

Steam and excess moisture from vapor, splashing, or dripping water (such as a packing house)

Excessive amounts of acid or alkali vapor, fumes, or dust (such as a chemical plant)

Conducting or abrasive dusts (such as cast iron dust, carbon, graphite, coke)

Some optional motor features are motor thermostats and motor space heaters. The motor thermostat is a bimetallic automatic thermostat built into the motor windings. It will deenergize the motor control if the winding temperature of the motor exceeds its allowable limit. The motor thermostat provides running protection against overheating caused by plugging, increased ambient temperature, obstruction of normal ventilation, variation in line voltage, and any other overheating in which the temperature rise is gradual. Motor space heaters prevent moisture condensation on the motor windings after shutdown or during prolonged periods of idleness. They are recommended on all motors using tropical protection insulation.

Hoist motors are either single speed, two speed, or variable speed. Single-speed hoists have only one lifting speed and will accelerate to their rated speed when the control is energized. Two-speed units have the low speed as some ratio of the high speed. Common ratios are 4 to 1, 3 to 1, and 2 to 1. The selection of lifting speed is optional by the operator and is normally achieved by two distinct speed-step depressions in the pendant push-button station. The first step provides low speed and the second step provides high speed. Installing a time delay in the hoist control circuit will prevent going into high speed without first going into low speed.

Variable-speed control, as applied to hoists, means a stepped variable speed. Normal steps are either three or five. With variable-speed control, speed is usually regulated by adding resistance in the circuit to obtain the varying speed steps. Generally, a variation in lifting speed is obtained when lifting various loads that are less than rated hoist capacity, and is shown in Fig. 10.3.8. For example, if a 5-ton hoist has its resistance values based on a 10,000-lb (4536-kg) rated capacity load, and has five speed steps, the hoist would not be able to lift a 6000-lb (2722-kg) load in step 1. At step 2, 20% of the full load speed would be obtained. At step 3, approximately 50% of full load speed would be obtained. At step 4, approximately 68% of full load speed would be obtained. At step 5, 100% of full load speed would be obtained. A time delay can also be added in the steps of a variable-speed hoist control circuit to prevent inadvertent starting operation of the hoist in the higher speeds. More sophisticated speed controls, including infinitely variable speed, are available for applications where extremely accurate speed selection is desirable or mandatory. Required specifications and availability should be discussed with a hoist manufacturer. Two-speed hoists are often a better selection over variable-speed hoists. Two-speed motors have definite speed steps that are not affected by varying load conditions, and are more economical because of the motor itself and the controls required.

Control Enclosures

The control enclosure is a means of mounting the control and protective devices in a group for assembly, inspection, and maintenance. It protects personnel from inadvertently contacting the enclosed electrical devices, and protects the enclosed internal devices against specified external conditions. Enclosures are classified according to the protection they are designed to offer to the enclosed devices, and are referred to as NEMA (National Electrical Manufacturers Association) types. The types most frequently used on hoists are listed in Table 10.3.5.

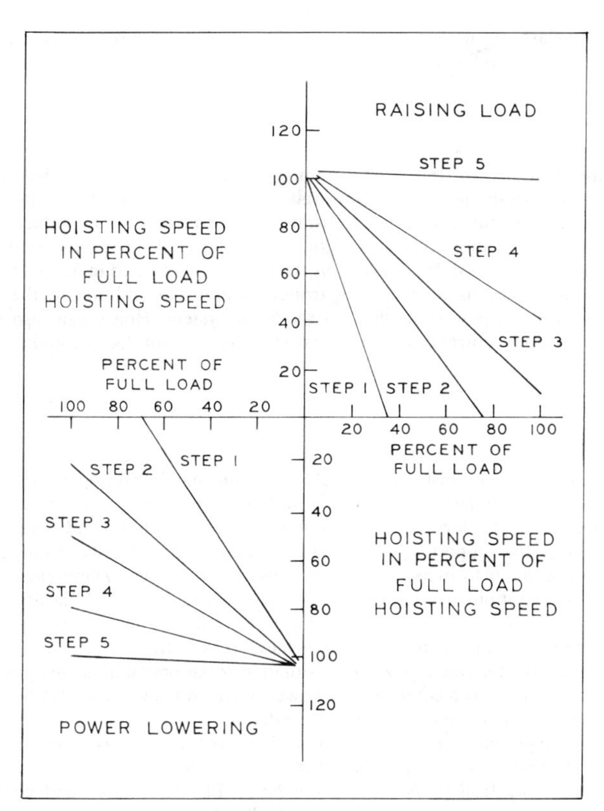

Fig. 10.3.8 Typical speed vs. load curves for a five-step variable-speed electric hoist. Used with the permission of Acco Babcock, Inc.

Enclosures for hazardous locations are specially designed to meet applicable requirements of the National Electric Code, and in accordance with specifications of the Underwriter's Laboratories, Inc. Such enclosures are NEMA Type 7 and Type 9. Hazardous or classified locations are locations where fire or explosive hazards may exist. These locations are classified depending on the properties of the flammable vapors, liquids, or gases, or combustible dusts or fibers which may be present, and the likelihood that a flammable or combustible concentration or quantity is present. Hazardous locations

Table 10.3.5 NEMA Enclosure Classifications Generally Applying to Overhead Hoists and Cranes

Type	Purpose	Description
NEMA Type 1	General	Indoor use only, where surrounding conditions are fairly clean and dry.
NEMA Type 3R	Weather resistant	Suitable for all outdoor applications except in hazardous locations. Ventilated types do not protect against windblown dust. Most indoor applications except hazardous locations or where cleaning requires hosedown.
NEMA Type 4	Water tight and dust resistant, indoor and outdoor	Suitable for all indoor or outdoor applications except hazardous locations. Offers protection where cleaning requires hosing down.
NEMA Types 7 & 9	Hazardous locations	Further defined by class, group, and division.
NEMA Type 12	Industrial dust tight and drip tight	Suitable for indoor use where fibers, dust, splashing, oil seepage and dripping may be present.

are further defined by class, group, and division; therefore, specific requirements and availability should be discussed with a hoist manufacturer.

Pendant Control

Pendant push-button stations are the most common device used to operate hoists. These pendant stations are suspended from the hoist by a strain-relief chain or rope, or other mechanically supported means. This is to prevent strain from being transmitted into the electrical conductors as an operator moves or pulls the pendant station during operation of the hoist. Pendant push-button enclosures are classified in the same manner as control enclosures by NEMA types listed in Table 10.3.5. The pendant station can include push buttons to operate a trolley or crane in addition to the hoist, as well as other buttons to control the power supplied to the hoist system. Hoists can also be controlled by wall-mounted controls, radio controls, or lever-operated master controls located in the cab of an overhead crane.

Trolleys

Trolleys for powered wire rope hoists are available in plain, hand-chain-operated, and power-driven types. Plain trolleys are recommended only where motion of the trolley is infrequent or for relatively short travel distances. Plain trolleys are not recommended for handling loads over 3 tons (3000 kg) or where the elevation of the supporting beam is greater than 20 ft (6 m) above the operator's position because of the forces required to manually operate this type of trolley. Hand-chain-operated trolleys are recommended for operations similar to plain trolleys where loads and elevations of the support beam exceed the recommended limits for plain trolleys. The operator must be positioned almost directly under a plain or hand-chain-operated trolley to achieve trolley travel motion. The physical size of the load being handled could create a hazardous situation to an operator in this position and in such cases, a power-operated trolley is recommended. Power-operated trolleys are also recommended where frequent operation or long travel distances are encountered.

Power-operated trolleys can be either electric or air powered. The basic design and function of a powered trolley is the same regardless of whether it is electric or air powered. The comments of Section 10.3.3 apply to the trolley, as well as the hoist. Electric motors and controls for electric power-operated trolleys are the same as for the hoist. Standard trolley designs are usually based on operation on straight beam track sections. Special design considerations may be required for operation on curved beams or beam systems containing switches. Requirements of this type should be discussed with a hoist manufacturer. Two-speed or variable-speed motors, or a cushioned start is recommended for trolley travel speeds in excess of 100 ft/min (30 m/min).

Power-operated trolleys are not required to have a motor brake. This is because the deceleration by mechanical losses, especially if a worm-gear reduction is used, is usually sufficient to stop trolley travel within required distances. Also, abrupt stopping of trolley travel can result in severe swinging of the load being handled. Travel of the trolley, with or without a brake, should be stopped within a distance in feet or meters equal to 10% of the rated travel speed in ft/min (m/min) when the trolley is traveling at rated travel speed with rated capacity load.

Hoist and Trolley Speeds

Hoists and trolleys are available in a wide range of lifting speeds and travel speeds. Common or readily available speeds are shown in Table 10.3.6. The figures of Table 10.3.6 are not intended to imply that other capacities or speeds are not available. Specific requirements should be discussed with a hoist manufacturer.

Table 10.3.6 Typical Powered Wire Rope Hoist Capacity and Speed Characteristics

Rated Capacity Load [tons (kg)]	Hoist Lifting Speed [ft/min (m/min)]	Powered Trolley Travel Speed [ft/min (m/min)]
1–2 (900–2,000)	10–60 (3–18)	30–100 (9–30)
3–5 (2,700–5,000)	10–40 (3–12)	30–100 (9–30)
$7\frac{1}{2}$–10 (6,800–10,000)	10–30 (3–9)	30–100 (9–30)
15–20 (13,500–20,000)	10–25 (3–7.5)	30–100 (9–30)
25–30 (22,500–30,000)	10–20 (3–6)	30–100 (9–30)

10.3.5 Electric-powered Chain Hoists

Electric-powered chain hoists are available with various suspensions and controls. Capacities available were for many years limited to 2-ton capacity. The available capacity range has increased to about 12 tons over the last several years. Combining multiple hoists with multiple reeving of the load chain has resulted in capacities as high as 50 tons in Europe and the Far East.

Electric chain hoists are basically similar to electric wire rope hoists except that the lifting medium is load chain rather than wire rope. Since load chain can only flex at and between links, the pitch diameter of the load wheel varies slightly between links of load chain. The result is an almost unnoticeable slight difference in smoothness of operation between wire rope and chain.

Load Chain

Two types of load chain are used in electric chain hoists. These are welded-link chain and roller-link chain. Welded-link chain consists of a series of interwoven links that are formed and welded. While the majority of welded-link chain is made of alloy steel, other materials, such as stainless steel, are available for special applications. Welded-link load chain is manufactured and heat treated for powered-hoist service and is completely different from standard link chain used in the manufacture of slings. In addition to the material and properties of load chain, it is dimensionally designed and accurately pitched during manufacture within specified tolerance limits to operate in conjunction with pockets of a load wheel for a specific hoist. Welded-link load chain cannot be interchanged between hoist types and manufacturers. Standard type welded-link chain cannot be used in a hoist as replacement for the welded-link load chain used in the design of that hoist.

Roller-link chain consists of a series of alternately assembled roller links and pin links in which the pins articulate within bushings and the rollers are free to turn on bushings. Pins and bushings are press-fitted into their respective link plates. Roller-link load chain is normally alloy steel; however, other materials are available for specific applications. Hoist roller-link load chain looks very similar to standard roller-link chain used in power transmission applications. This similarity applies to appearance only and major differences exist between hoist roller-link chain and power transmission roller-link chain in material, heat treatment, and method of assembly. Any replacement of a hoist roller-link chain should only be made by the use of roller-link chain designed and manufactured for hoist service.

Load Wheels

The load wheel provides motion to the load chain thereby raising or lowering the load block. The load wheel may also be called load sheave, load sprocket, pocket wheel, chain wheel, or lift wheel. Regardless of the term used by any individual or manufacturer, they all provide the same basic function in the electric chain hoist unit. The pockets of the load wheel are designed to allow proper fit of the load chain and engagement of the load chain links as they enter and leave the load-wheel pockets during operation. Since welded-link load chain has adjacent links at 90° to each other, the load wheel must provide for proper seating of the link that is flat or horizontal to the wheel diameter, as well as allow for clearance or support of the vertical length. Some welded-link chain load wheels are designed to support the vertical link whereas others support only horizontal links and, therefore, need clearance between the wheel and vertical link. Load-wheel design is critical in relation to the type and size of load chain with which it is designed to operate. Load chain and load wheels are not interchangeable. Replacement load chain must be the exact type and size originally intended for operation with a specific load wheel.

Overall design of the load wheel area of the hoist should include provisions to guide the load chain into proper position with relation to the load wheel, guard against jamming of the load chain within the hoisting mechanism, and disengage the chain from the load wheel on the unloaded side of the load wheel. These design provisions apply to normal operating conditions only. Powered chain hoists are intended for vertical operation of the load chain only. Attempting to use a standard powered chain hoist in horizontal or upside-down positions is abnormal operation that is abusive to the equipment and could create a hazardous situation. This could result in property damage to the load or equipment, or injury to personnel in the vicinity of the hoist. Special designs can be made to allow other than vertical operation and should be discussed with a hoist manufacturer if such operation is required.

Environmental atmospheres containing an excessive dust content, such as cement or coal, also create an abnormal condition of the load chain and load wheel. If conditions exist wherein this dust can enter the load wheel portion of the hoist and be deposited on the load-wheel pockets, it is possible that the load chain will pack the dust into the pocket as the load chain operates over the load wheel. This action could result in alteration of the pocket shape to such a degree that proper operation of the load chain and load-wheel pocket is seriously affected. These types of environmental atmospheres should be considered before installing a hoist, and discussed with a hoist manufacturer.

As the load chain operates over the load wheel during lifting, the load chain on the unloaded side of the load wheel is normally allowed to hang in a free position. If free suspension of the unloaded, or dead side, chain could cause interference with other equipment, a chain container can be used for storage of this load chain. The chain container must be of sufficient size to permit storage of the chain as it feeds into the container in a random manner, and must be positioned in relation to the hoist unit to allow alignment of the load chain and load wheel as the load chain feeds out of the container during lowering operations. Chain containers, when used, may reduce the amount of lift available or cause interference with the load being lifted. Only chain containers designed for a particular hoist unit should be used with that hoist unit. *Homemade* chain containers should not be used.

Other Components

Many components or areas of concern for powered chain hoists are identical with those of powered wire rope hoists. Specific areas include hoist duty service classifications, design factors, headroom, lift, reach, braking system, gear reduction, limit switches, overload limit devices, electric motors, control enclosures, pendant push-button stations, and trolleys. These items are discussed in Section 10.3.4.

Powered Chain Hoist Standards

Several standards are published that apply to overhead powered chain hoists, as well as standards that utilize overhead powered chain hoists for the hoist lifting mechanism. All hoist users should become familiar with the requirements of these standards.

The standards that apply directly to overhead powered chain hoists are HMI (Hoist Manufacturers Institute) 400, Standard Specification for Electric Chain Hoists [3], and ANSI (American National Standards Institute) B30.16, Safety Standards for Hoists (Overhead) [4]. An ASME (American Society of Mechanical Engineers) Standards Committee on Overhead Hoists was organized in 1979 for the purpose of developing performance standards for various types of overhead hoists [12, 15]. As standards developed by this committee are adopted, they will replace the HMI standards.

Other standards applicable to powered chain hoists, when powered chain hoists are used with equipment covered by these standards, are CMAA Specification No. 74, Specification for Top Running and Under Running Single Girder Electric Overhead Traveling Cranes [8]; ANSI B30.11–1980, Safety Standard for Overhead and Underhung Cranes [5]; ANSI B30.17, Safety Standard for Overhead and Gantry Cranes (Top-Running Bridge, Single Girder, Underhung Hoists) [6]; and ANSI MH27.1–1981, Specifications for Underhung Cranes and Monorail Systems [10]. Hoists used in the handling of hot-molten material require additional considerations to operate in such an environment and should comply with the applicable provisions of ANSI Z241.2–1981, Safety Requirements for Melting and Pouring of Metals in the Metalcasting Industry [11].

Hoist and Trolley Speeds

Hoists and trolleys are available in a wide range of lifting speeds and travel speeds. Common or readily available speeds are shown in Table 10.3.7. The figures of Table 10.3.7 are not intended to imply that other capacities or speeds are not available. Specific requirements should be discussed with a hoist manufacturer.

10.3.6 Hoist Installation

Proper installation of overhead hoists is essential in achieving the expected performance and providing a safe environment to the operators of the equipment.

Support Structure

All structures involved in supporting an overhead hoist must be designed to handle all of the loads and forces that will be imposed on the structure by operation of the hoist. Supporting structures

Table 10.3.7 Typical Powered Chain Hoist Capacity and Speed Characteristics

Rated Capacity Load [tons (kg)]	Hoist Lifting Speed [ft/min (m/min)]	Powered Trolley Travel Speed [ft/min (m/min)]
$\frac{1}{4}$–1 (220–1,000)	7–64 (2–20)	30–100 (9–30)
$1\frac{1}{2}$–3 (1,360–3,000)	4–40 (1.2–12)	30–100 (9–30)
4–12 (3,630–12,000)	4–25 (1.2–7.5)	30–100 (9–30)

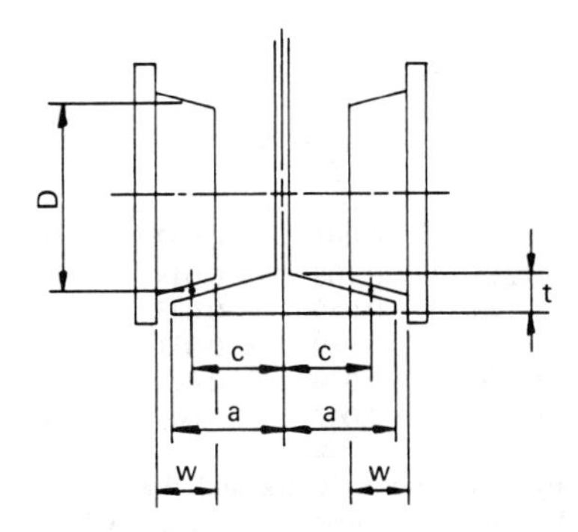

Fig. 10.3.9 Trolley wheel on beam flange showing location of dimensions c and a. From CMAA Specification No. 74, Specification for Top Running and Under Running Single Girder Electric Overhead Traveling Cranes. Copyright 1974 by the Crane Manufacturers Association of America. Used with the permission of the Crane Manufacturers Association of America.

include trolleys, monorails, and cranes, and therefore these pieces of equipment must have a load capacity rating at least equal to that of the hoist.

Support beams used for trolley travel must be sized for the trolley to fit on the beam flange. One area that cannot be neglected is analysis to determine that the beam flange can support the loads imposed by the trolley wheels. It is possible to select a beam that can adequately support the hoist and trolley load under typical beam loading analysis and yet experience failure of the beam flange because of localized bending stresses induced into the flange by the trolley wheel loads.

Localized bending due to trolley wheel loads on beam flanges is determined by considering each wheel load acting as a concentrated load applied at the contact point between the wheel and flange. The resultant flange bending stress is a function of the c/a ratio shown in Figs. 10.3.9 and 10.3.10. The bending stress at any point on the flange can be determined by the equation

$$S = K_m \frac{6P}{t^2}$$

where: S = flange bending stress, psi
P = trolley wheel load, pounds
t = flange thickness at beam web, inches
K_m = dimensionless coefficient selected from Table 10.3.8, depending on the location of the load

Table 10.3.8 and Fig. 10.3.10 are from *Formulas for Stress and Strain,* 5th Ed., by Raymond J. Roark and Warren C. Young, copyright 1975, McGraw-Hill Book Company, and are used with the permission of McGraw-Hill Book Company. A single trolley wheel produces the maximum localized beam flange bending stress at the point $x = 0$ and $z = 0$. If the trolley has more than one wheel on each flange extending from the beam centerline, the trolley wheel centers should be equal to or greater than dimension c. If the trolley wheel centers are less than dimension c, the localized bending stress must be determined by the combined effort of each wheel. Localized flange bending stress is combined with the beam bending stress by means of Mohr's circle to determine the maximum stress.

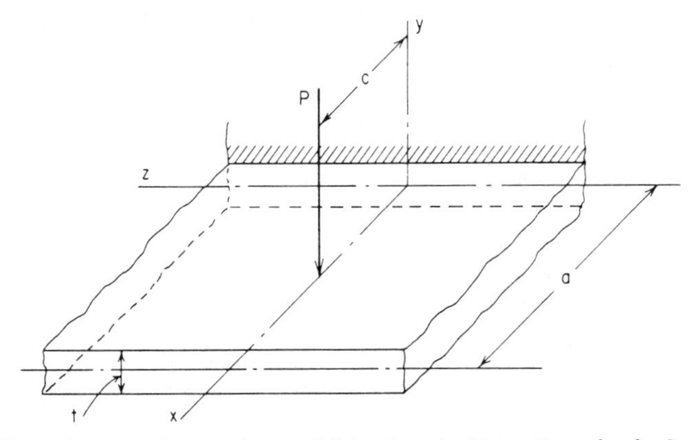

Fig. 10.3.10 Dimensions c and a, used to establish c/a ratio. From *Formulas for Stress and Strain,* 5th Ed., by Raymond J. Roark and Warren C. Young. Copyright 1975, McGraw-Hill Book Company. Used with the permission of McGraw-Hill Book Company.

Table 10.3.8 Flange Bending Stress

		z/a						
c/a	Coefficient	0	0.25	0.50	1.0	1.5	2	∞
1.0	K_m	0.509	0.474	0.390	0.205	0.091	0.037	0
	K_y	0.524	0.470	0.380	0.215	0.108	0.049	0
0.75	K_m	0.428	0.387	0.284	0.140	0.059	0.023	0
	K_y	0.318	0.294	0.243	0.138	0.069	0.031	0
0.50	K_m	0.370	0.302	0.196	0.076	0.029	0.011	0
0.25	K_m	0.332	0.172	0.073	0.022	0.007	0.003	0

Source: From *Formulas for Stress and Strain,* 5th Ed., by Raymond J. Roark and Warren C. Young. Copyright 1975, McGraw-Hill Book Company. Used with the permission of McGraw-Hill Book Company.

Location

Hoists should be installed in locations that enable the operator to stand free of the load being lifted during operation. Hand-chain or pendant controls should be positioned at a level convenient to the operator.

Procedures

Before installing any new hoist, the installation and general instructions furnished by the hoist manufacturer should be read by the people directly responsible for performing the installation. Installation and inspection steps must be followed in the proper order as recommended by the manufacturer. Step-by-step procedures for installing, inspecting, and checking a new hoist are as follows:

Prior to Installation.

1. Check for any damage during shipment. Do not install a damaged hoist.
2. Check all lubricant levels and install vent plugs where instructed.
3. Check wire rope for damage if hoist is wire rope type. Be sure wire rope is properly seated in drum and sheave grooves.
4. Check chain for damage if hoist is chain type. Be sure chain properly enters sprockets and chain guiding points.
5. Check to be sure that power supply shown on the serial plate of an electric-powered hoist is the same as the power supply to be connected to the hoist.

Installation. Install stationary mounting or trolley mounting to monorail beam exactly as instructed by the instructions furnished with the hoist. Check that clearance between trolley wheel flanges and toe of monorail beam flange is within the limits specified in the instructions.

Power Supply Connection.

1. Make sure all electrical connections are made in accordance with the manufacturer's wiring diagram.
2. Make sure electrical supply system is in accordance with the National Electrical Code [18].

Check Motor Phasing.

1. Depress the "up" button on the pendant control to determine direction of hook travel. If hook travel is upward, the hoist motor is properly phased. If hook travel is downward, discontinue operation until hoist motor is properly phased.
2. If a three-phase hoist motor is improperly phased, correction is made by interchanging any two power line leads to the hoist. Never change internal wiring connections in the hoist or pendant control.
3. If a single-phase hoist hook travel does not agree with the directional markings, consult the hoist manufacturer.

Check Upper Limit Switch.

1. Check upper limit switch only after first checking for proper motor phasing.
2. Raise the unloaded load block by using the pendant push-button station. Manually activate the gravity limit switch to verify that it is operative and will stop load-block travel.
3. If upper limit switch does not operate or trip point is too close to hoist, disconnect power supply and check all electrical connections or make necessary adjustments.
4. Reconnect power supply and recheck upper limit switch.

Check Lower Limit Switch. Check operation of a hoist having a lower limit switch in the same manner used to check upper limit switch. Never adjust the lower limit switch to a point where less than one full wrap of wire rope remains on each anchorage of the drum of wire rope hoist.

Check Lower Limit of Travel.

1. When hoist does not have a lower limit switch, lower the unloaded hook block to its lowest possible operating point or, for wire rope hoists, until two full wraps of wire rope remain on each anchorage of the drum.
2. If it appears that less than two full wraps of wire rope will remain on each anchorage of the drum at the lowest possible operating point, the hoist cannot be used at that location unless it is equipped with a lower limit switch.

Trolley Operation. Operate a trolley-mounted hoist over its entire travel distance while the hoist is unloaded to check all clearances and verify that no interference occurs.

Braking system.

1. Raise and lower load block, without load, stopping the motion at several points to test operation of the braking system.
2. Raise load block with rated capacity load several inches and stop to check that brake system holds and that the load block does not drift downward. If drift does not occur, raise and lower load block with rated capacity load, stopping the load at several points to test the operation of the brake system.

Load Test. Load test the hoist or system with a test load and procedure in accordance with standards applicable to hoist, or system utilizing the hoist.

Report. Report should be written outlining the installation procedures, any problems encountered, and results of all checks and tests conducted. This report should indicate the approval or certification for production use, and should be signed by the responsible individual and filed in the equipment record folder.

Operating Instructions.

1. All warning tags or labels furnished on the hoist must remain in place to provide this information to operators.
2. Issue instructions to all personnel who will operate the hoist in accordance with the manufacturer's manual and applicable standards.

10.3.7 Operation

Proper operating procedures are essential in using any overhead hoist. All operators of hoists must be instructed by the employer in the fundamentals of correct hoist operation, to prevent abuse of the equipment, and safe operating procedures, to assure a safe working environment for the operator and operator's coworkers. Operator training must include instruction on proper procedures to follow in hooking up and handling loads. Operators must be advised to exercise intelligence and common sense in order to anticipate the consequencial motions of the load that will occur as a result of their operation of the hoist controls.

Instruction of operators is a continuing requirement. It includes updating present operators and instituting programs to train and inform new employees about correct hoist operation. The ANSI B30 series of standards extensively outline safe operating procedures for the type of equipment each standard of the series covers. The hoist manufacturer will provide detailed operating instructions and warnings in the manual furnished with each piece of hoisting equipment. Operating information from

OPERATION

Any hoist can be dangerous in the hands of a careless operator. The operator should read and observe the following recommendations to attain optimum performance under the best possible conditions.

NEVER ALLOW
UNTRAINED OR PHYSICALLY UNFIT PERSONS
TO OPERATE HOIST.

Before initial operation of hoist:

OPERATOR SHALL

... Become familiar with equipment and its proper care.

... Read and heed all instruction and warning information on or attached to hoist or controls

... Check lubricant (See LUBRICATION, Page 4)

... Check phasing (See Page 3, Paragraph C-2)

... Check brakes (See Page 4, Paragraph C-5)

... Determine that
Rope is well seated in drum grooves and sheaves
Rope is not twisted, kinked, or damaged

Before each shift:

OPERATOR SHALL

... Test controls and limit switch. Test brake.

... Inspect all running ropes in continuous service.

... Inspect hooks for nicks; gouges; cracks; signs of deforming, pulling apart, or twisting.

... Replace WARNING label if lost or illegible.

Before operating hoist:

OPERATOR SHALL

... Be certain all personnel are clear

... Make sure loads will clear stock piles, machinery, or other obstructions when raising, lowering, or traveling.

... Center hoist over load

... Inspect ropes (on hoist idle for a month or more).

... Be sure load attachment is properly seated in saddle of hook. Balance load properly. Avoid tip loading.

While operating hoist:

OPERATOR SHALL

... Avoid swinging load or hook when moving hoist.

... Take up slack slowly.

... Avoid plugging, inching, and quick reversals.

... Avoid sharp contact between two hoists, hoist and en post, or hook and hoist body.

... Disconnect hoist from power supply before beginnir any maintenance.

OPERATOR SHALL NOT - AT ANY TIME

... Operate hoist IF ....
1. It is damaged or malfunctioning.
2. It is not protected by a properly functioni limit switch.
3. Rope is twisted, kinked, damaged, or imprope spooled on drum or sheave.

... Change wiring leads of limit switch or push-butt station to correct improper phasing

... Use hoist rope as a sling or as a ground for weldi (NEVER touch a live welding electrode to the rop

... Overload hoist.

... Divert his attention while operating hoist.

... Use hoist for side loading.

... Rotate drum in lowering direction beyond a po where two wraps of rope remain on drum.

EXCEPT
When hoist is equipped with geared limit switch, one wrap remaining is sufficient.

... Transport loads over the heads of workmen, or perm anyone to stand under loaded hook.

... Use hoist to lift humans

... Leave a load suspended and unattended.

... Use limit switch as a means of stopping hoist. **This i emergency device only.**

... Exceed fuse rating recommended by the Nati Electric Code or recommended duty cycle of h

... Remove or obliterate WARNING label.

Fig. 10.3.11 Typical operation instructions from a hoist manufacturer's manual. Used with the permission of Acco Babcock, Inc.

these two sources should be the basis for the training of operators. A typical operating summary from a manufacturer's manual is shown in Fig. 10.3.11.

Tags or labels containing warning information on hoist operation will be supplied as part of the equipment by the hoist manufacturer. A typical warning label furnished with a powered hoist is shown in Fig. 10.3.12. On powered hoists, this label is usually attached to the pendant push-button station. It should not be removed, and if it does become inadvertently detached, it should be replaced immediately. Figure 10.3.12 is not intended to imply that a warning label cannot contain additional or less information than shown. The extent of warnings listed is dependent on the specific equipment and manufacturer.

10.3.8 Inspection

Regularly scheduled inspections are important in maintaining an overhead hoist for optimum performance and as a safe piece of equipment. Failure to inspect hoists can lead not only to serious production delays, it may result in unsafe operation. The majority of hoist malfunctions can be attributed to components that are related to and perform safety functions. Items to inspect and frequency for performing inspections are listed in the Hoist Manufacturers Manual and ANSI B30.16 [4]. Tables 10.3.9 and 10.3.10, reprinted from ANSI B30.16, summarize inspection items and frequency for manual and power-operated hoists. Certain items that require constant attention for hoist operation are discussed in this section. This does not imply that other areas addressed in the hoist manual or ANSI B30.16 can be neglected.

Hooks

Load hooks on hoists in continuous service should be visually inspected daily. Hoists operating around-the-clock should be visually inspected at the start of each shift. The visual inspection may indicate the need for a more thorough inspection. Scheduled inspections with written reports should be made in accordance with Tables 10.3.9 and 10.3.10, and ANSI B30.16. Specific inspection procedures and conditions requiring load hook replacement are:

1. Measure hook throat opening, dimension E of Fig. 10.3.13. Throat opening for a new hook is usually specified in the manual furnished with the hoist. If this figure is not provided, measure the throat opening when the hook is new. The hook must be replaced when the throat opening has increased by 15%. For example, a hook with an initial 2-in. throat opening should be replaced when the throat opening becomes 2.30 in., or approximately $2\frac{5}{16}$ in.
2. Check load bearing point of hook. When hook thickness at the load bearing point, Fig. 10.3.13, is worn by 10%, the hook must be replaced.
3. Inspect hook tip. If it is twisted 10° or more from the plane of the unbent hook, the hook must be replaced.
4. Check for excessive damage from chemicals, and for deformation and cracks. If grinding to remove this type of damage reduces the original thickness by more than 10%, the damage is excessive and the hook must be replaced.
5. Check for and replace damaged, inoperative, or missing hook latches. Investigate to determine if excessive throat opening or hook twist damage has been caused by abuse or overloading of the hoist. Hook damage caused by abuse or overloading is an indication that other load-bearing parts should be inspected to determine if any other damage to the hoist has occurred. Actions should be taken immediately to eliminate the reoccurrence of abuse or overloading of the hoist.

Wire Rope

Wire rope on hoists in continuous service should be visually inspected daily. Hoists operating around-the-clock should be visually inspected at the start of each shift. The visual inspection may indicate the need for a more thorough inspection. Scheduled inspections with written reports should be made in accordance with Table 10.3.10 and ANSI B30.16. Strength of used wire rope must be evaluated carefully because safety of operation depends upon this remaining strength. Signs of wire rope deterioration and guidelines for wire rope replacement are shown in Fig. 10.3.14. The term *one rope lay* refers to the axial wire-rope length for one wire-rope strand to completely wrap around the wire-rope assembly. When a wire-rope inspection is made, particular attention should be given to sections of wire rope subjected to reverse bends or operation over drums and all sheaves, including equalizers.

Replacement wire rope should be of the same size, grade, and construction as the original furnished by the hoist manufacturer, unless otherwise recommended by the hoist or wire-rope manufacturer because of actual operating conditions. It is recommended that replacement wire ropes be stocked

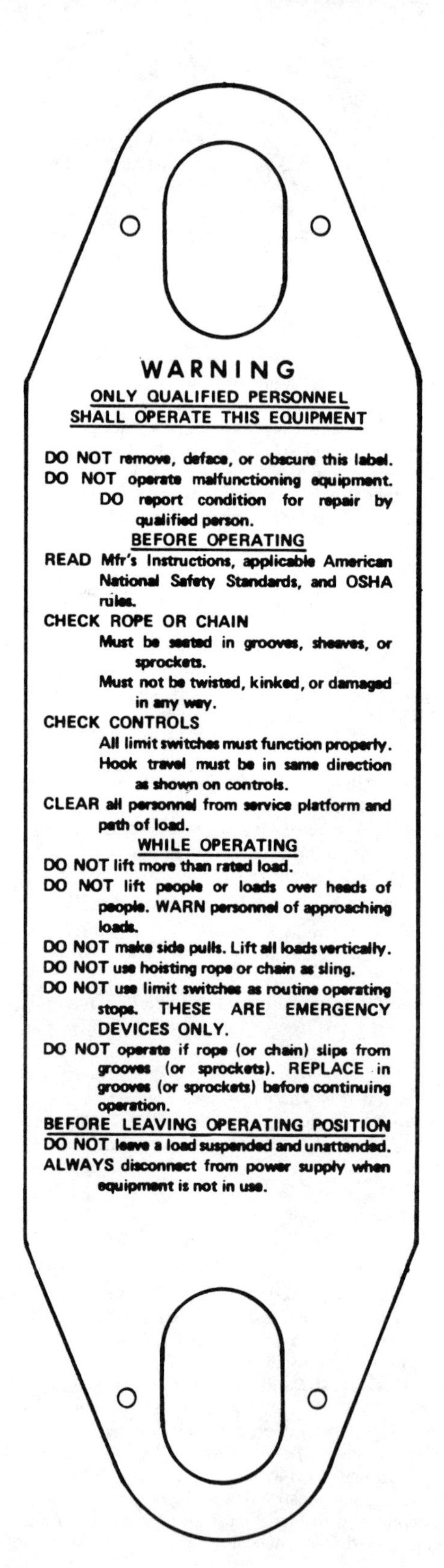

Fig. 10.3.12 Typical warning label attached to electric-powered hoists. Used with the permission of Acco Babcock, Inc.

Table 10.3.9 Minimum Inspection for Hand Chain Operated Hoists

Item	Normal Service		Heavy Service		Severe Service	
	Visual[a] Monthly	Record[b] Yearly	Visual[a] Weekly to Monthly	Record[c] Semi-annually	Visual[a] Daily to Weekly	Record[c] Quarterly
Frequent Inspection (Refer to 16–2.1.2)						
All functional operating mechanisms for maladjustment interfering with proper operation	X		X		X	
Hooks for damage, cracks, or excessive throat opening [Refer to 16–2.3.3(c)(5)]	X		X		X	
Hook latch operation, if used	X		X		X	
Load chain in accordance with 16–2.5.1 or 16–2.6.1	X		X		X	
Load chain reeving for compliance with hoist manufacturer's recommendations	X		X		X	
Periodic Inspection (Refer to 16–2.1.3)						
Requirements of frequent inspection		X		X		X
External evidence of loose bolts, nuts, or rivets		X		X		X
External evidence of worn, corroded, cracked, or distorted parts such as load blocks, suspension housing, hand chain wheels, chain attachments, clevises, yokes, suspension bolts, shafts, gears, bearings, pins, rollers, locking and clamping devices		X		X		X
External evidence of damage to hook retaining nuts or collars and pins and welds or rivets used to secure the retaining members		X		X		X
External evidence of damage or excessive wear of load sprockets, idler sprockets, or hand chain wheel		X		X		X
External evidence of worn, glazed, or oil-contaminated friction discs; worn pawls, cams, or ratchet; corroded, stretched, or broken pawl springs in brake mechanism		X		X		X
External evidence of damage of supporting structure or trolley, if used		X		X		X
Warning label required by 16–1.1.4, except as provided in 16–2.3.3(c)(8)		X		X		X
End connections of load chain		X		X		X

[a] By operator or other designated personnel with records not required.

[b] Visual inspection by appointed person making records of apparent external conditions to provide the basis for a continuing evaluation.

[c] As in *b* unless external conditions indicate that disassembly should be done to permit detailed inspection.

Source: From ANSI B30.16, Safety Standard for Overhead Hoists (Underhung). Copyright 1981, the American Society of Mechanical Engineers. Used with the permission of The American Society of Mechanical Engineers.

Table 10.3.10 Minimum Inspection for Electric or Air-powered Hoists

Item	Normal Service		Heavy Service		Severe Service	
	Visual[a] Monthly	Record[b] Yearly	Visual[a] Weekly to Monthly	Record[c] Semi-annually	Visual[a] Daily to Weekly	Record[c] Quarterly
Frequent Inspection (Refer to 16–2.1.2)						
All functional operating mechanisms for maladjustment interfering with proper operation	X		X		X	
Limit devices for operation	X		X		X	
Air lines, valves, and other parts for leakage	X		X		X	
Hooks for damage, cracks, or excessive throat opening [Refer to 16–2.3.3(c)(5)]	X		X		X	
Hook latch operation, if used	X		X		X	
Hoist rope in accordance with 16–2.4.1(a)	X		X		X	
Load chain in accordance with 16–2.5.1 or 16–2.6.1	X		X		X	
Rope or load-chain reeving for compliance with hoist manufacturer's recommendations	X		X		X	
Periodic Inspection (Refer to 16–2.1.3)						
Requirements of frequent inspection		X		X		X
Hoist rope in accordance with 16–2.4.1(b)		X		X		X
External evidence of loose bolts, nuts, or rivets		X		X		X
External evidence of worn, corroded, cracked, or distorted parts such as load blocks, suspension housing, chain attachments, clevises, yokes, suspension bolts, shafts, gears, bearings, pins, rollers, locking and clamping devices		X		X		X
External evidence of damage to hook retaining nuts or collars and pins and welds or rivets used to secure the retaining members		X		X		X
External evidence of damage or excessive wear of load sprockets, idler sprockets, and drums or sheaves		X		X		X
External evidence of excessive wear on motor or load brake		X		X		X
Electrical apparatus for signs of pitting or any deterioration of visible controller contacts		X		X		X
External evidence of damage of supporting structure or trolley, if used		X		X		X
Warning label required by 16–1.1.4, except as provided in 16–2.3.3(c)(8)		X		X		X
End connections of rope or load chain		X		X		X

[a] By operator or other designated personnel with records not required.

[b] Visual inspection by appointed person making records of apparent external conditions to provide the basis for a continuing evaluation.

[c] As in *b* unless external conditions indicate that disassembly should be done to permit detailed inspection.

Source: From ANSI B30.16, Safety Standard for Overhead Hoists (Underhung). Copyright 1981, the American Society of Mechanical Engineers. Used with the permission of The American Society of Mechanical Engineers.

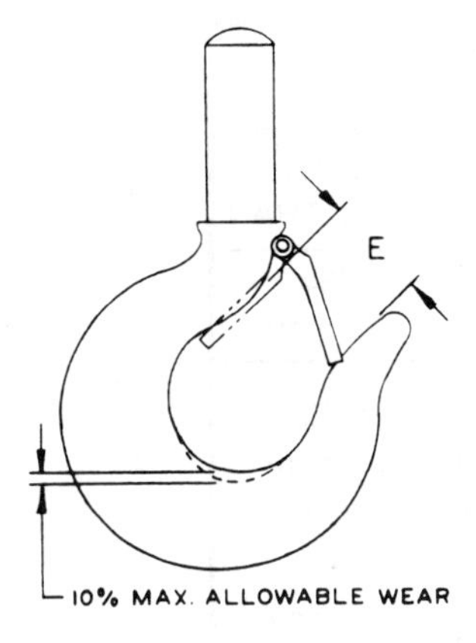

Fig. 10.3.13 Hoist hook showing throat opening measurement and point of load bearing wear. Used with the permission of Acco Babcock, Inc.

for hoists in continuous service. Care must be exercised in storing the replacement wire ropes to prevent damage and deterioration. Wire rope should be installed as recommended by the hoist manufacturer and with extreme care to avoid kinking and twisting.

Load Chain

Load-chain inspection procedures differ, depending on whether welded-link or roller-link chain is used as the lifting medium. In either case, the chain should be properly cleaned and lubricated, and the hoist tested, with load, in the lifting and lowering directions to observe operation of the chain and sprockets. The chain should feed smoothly into and out of the sprockets. If the chain binds, jumps, or is noisy, the chain and mating parts must be inspected. Chain elongation or stretch must be checked. Replacement chain must be the same size, type, grade, and construction as the original furnished by the manufacturer. Replacement chain should be installed only in the manner recommended by the hoist manufacturer.

Welded-link Chain

Wear in welded-link chain occurs on the inner contact points at each end of the chain link. In most cases, wear is most severe on the links that rest horizontally in the load wheel pockets. Chain must be slackened to check and measure this inner-link wear. Any link that has been worn by an amount equivalent to 10% or more of its original wire diameter (Fig. 10.3.15) requires replacement of the chain. Elongation or stretch should be measured in accordance with the hoist manufacturer's instructions, and the chain should be replaced when it exceeds the length recommended by the manufacturer. If instructions or recommendations are not available, a procedure outlined in ANSI B30.16 should be followed. Welded-link load chain should never be repaired by welding, or by attempting to replace or add sections by welding them to old chain.

Roller-link Chain

While the hoist is in its normal operating position, apply a light load of approximately 50 lb and check the chain for twist and side-bow or camber. The chain should be replaced if the twist in any 5-ft-long section exceeds 15°, or if side-bow in any 5-ft-long section exceeds $\frac{1}{4}$ in. Elongation or stretch should be measured in accordance with the hoist manufacturer's instructions. Chain should be replaced when it exceeds the length recommended by the manufacturer. If instructions or recommendations are not available, procedure outlined in ANSI B30.16 should be followed.

The chain must be inspected to determine the degree of impairment and necessity for replacement resulting from any deficiencies such as the following:

Pins turned from their original position.

Rollers that do not turn freely with light finger pressure.

Joints that cannot be flexed by light hand pressure.

Link plates that are spread open.

Corrosion, pitting, or discoloration.

Gouges, nicks, or weld splatter.

SIGNS OF DETERIORATION	CAUSE FOR REPLACEMENT	
	FOR ROPE DIAMETERS	REDUCTION MORE THAN
Reduction of rope diameter because of: Loss of core support Corrosion Worn outside wires	Up to 5/16 (8.0 mm)	1/64 (0.4 mm)
	3/8 (9.5 mm) to 1/2 (13.0 mm)	1/32 (0.8 mm)
	9/16 (14.5 mm) to 3/4 (19.0 mm)	3/64 (1.2 mm)
Broken outside wires	Twelve randomly distributed broken wires in one rope lay. Four broken wires in one strand in one rope lay.	
Worn outside wires	Wear of one-third of the original diameter of outside individual wires.	
Corroded or broken wires at end connections. Corroded, cracked, bent or worn end connections.	Any of these conditions indicates need for replacement.	
Severe kinking, crushing, cutting, or unstranding.	When such kinking, etc. results in distortion of rope structure	

NOTE

Evidence of any heat damage from any cause, or weld splatter on rope, are sufficient reasons for questioning safety and considering replacement.

Fig. 10.3.14 Table of deterioration causes for wire rope replacement. Used with the permission of Acco Babcock, Inc.

Braking System

Hoist braking systems should be checked daily, or at the start of each shift on around-the-clock operations. The load hook should be raised and lowered, without load, and stopped at several points to test the brake operation. Before raising any load the entire distance to be lifted, the operator should stop the hoist to check the brake system operation after raising the load several inches. The motor brake should be inspected, and a report prepared and maintained on a scheduled basis in accordance with Tables 10.3.9 and 10.3.10, and ANSI B30.16. Each type of control brake has specific operating characteristics. The hoist manufacturer should be consulted for procedures to be used in checking the control brake used in a particular hoist.

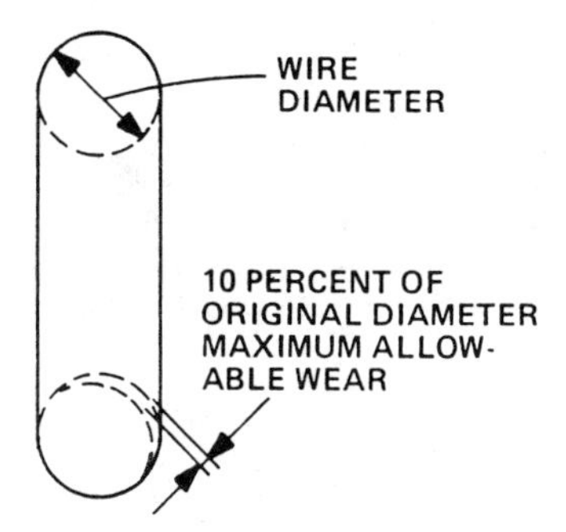

Fig. 10.3.15 Point of measurement for inter-link wear of welded-link load chain. Used with the permission of Acco Babcock, Inc.

Reports

Assignment of equipment numbers to hoists, inspection reports, and maintenance reports are part of a good hoist inspection and maintenance program. Reports describing inspections and maintenance work performed on a hoist should be maintained as part of the hoist equipment file. Repairs resulting from an inspection should be included in the inspection report. Maintenance forms and reports should provide information indicating that the hoist was checked and tested, if necessary, after the maintenance work was completed, and that the hoist was recertified for production use. A history of parts replacement on a particular hoist should be maintained. This record should include dates and causes for the replacement. The replacement-parts record can be used in establishing preventive maintenance procedures and in justification to obtain new hoisting equipment.

REFERENCES

1. HMI 100–74, Standard Specifications for Electric Wire Rope Hoists, Hoist Manufacturers Institute, 1326 Freeport Road, Pittsburgh, PA 15238, 1974.
2. HMI 200–74, Standard Specifications for Hand Operated Chain Hoists, Hoist Manufacturers Institute, 1326 Freeport Road, Pittsburgh, PA 15238, 1974.
3. HMI 400–71, Standard Specifications for Electric Chain Hoists, Hoist Manufacturers Institute, 1326 Freeport Road, Pittsburgh, PA 15238, 1971.
4. ANSI B30.16–1981, Safety Standard for Overhead Hoists (Underhung), American Society of Mechanical Engineers, 345 East 47th Street, New York, NY 10017, 1981.
5. ANSI B30.11–1980, Safety Standard for Monorails and Underhung Cranes, American Society of Mechanical Engineers, 345 East 47th Street, New York, NY 10017, 1980.
6. ANSI B30.17–1980, Safety Standard for Overhead and Gantry Cranes (Top-Running Bridge, Single Girder, Underhung Hoists), American Society of Mechanical Engineers, 345 East 47th Street, New York, NY 10017, 1980.
7. ANSI B30.2.0–1976, Safety Standard for Overhead and Gantry Cranes (Top-Running Bridge, Multiple Girder), American Society of Mechanical Engineers, 345 East 47th Street, New York, NY 10017, 1976.
8. CMAA Specification No. 74, Specification for Top Running and Under Running Single Girder Electric Overhead Traveling Cranes, Crane Manufacturers Association of America, 1326 Freeport Road, Pittsburgh, PA 15238, 1974.
9. CMAA Specification No. 70, Revised 1975, Specification for Electric Overhead Traveling Cranes, Crane Manufacturers Association of America, 1326 Freeport Road, Pittsburgh, PA 15238, 1975.
10. ANSI MH27.1–1981, Specifications for Underhung Cranes and Monorail Systems, Monorail Manufacturers Association, 1326 Freeport Road, Pittsburgh, PA 15238, 1981.
11. ANSI Z241.2–1981, Safety Requirements for Melting and Pouring of Metals in the Metalcasting Industry, American Foundrymen's Society, Golf and Wolf Roads, Des Plaines, IL 60016, 1981.
12. *Proposed* ANSI/ASME HST-1M, Performance Standard for Electric Chain Hoists, American Society of Mechanical Engineers, 345 East 47th Street, New York, NY 10017, Unpublished.
13. *Proposed* ANSI/ASME HST-2M, Performance Standard for Hand-Chain Manually-Operated Chain Hoists, American Society of Mechanical Engineers, 345 East 47th Street, New York, NY 10017, Unpublished.
14. *Proposed* ANSI/ASME HST-4M, Performance Standard for Electric Wire Rope Hoists, American Society of Mechanical Engineers, 345 East 47th Street, New York, NY 10017, Unpublished.

15. *Proposed* ANSI/ASME HST-5M, Performance Standard for Air Chain Hoists, American Society of Mechanical Engineers, 345 East 47th Street, New York, NY 10017, Unpublished.
16. *Proposed* ANSI/ASME HST-6M, Performance Standard for Air Wire Rope Hoists, American Society of Mechanical Engineers, 345 East 47th Street, New York, NY 10017, Unpublished.
17. Reisinger, Robert R., Work-Rated Hoist Selection, ASME Paper 77-RC-8, American Society of Mechanical Engineers, 345 East 47th Street, New York, NY 10017, 1977.
18. ANSI/NFPA 70–1981, National Electrical Code, National Fire Protection Association, 470 Atlantic Avenue, Boston, MA 02210, 1980.

10.4 BELOW-THE-HOOK LIFTING DEVICES

J. D. Mitchell

10.4.1 Mechanical Lifting Devices

Lifting Slings

The most common below-the-hook lifting devices are the various types of slings. This discussion focuses on those types of lifting slings most often used to connect a load to materials handling equipment. These types are generally categorized as follows: alloy steel chain slings, wire rope slings, metal mesh slings, fiber rope slings, and synthetic webbing slings.

In the case of lifting slings, probably more than other engineered lifting devices, safety considerations are paramount. This is because of the likelihood of misuse in service by operating personnel. When misused, slings are often subjected to overloading, severe wear, kinking, crushing, and impact loading. Without constant diligence on the part of all those involved, the tendency is to ignore the ubiquitous lifting sling until it suddenly fails. For these reasons the engineer must carefully consider the expected field conditions whenever specifying or designing a sling system.

Alloy Steel Chain Slings. Early sling chains were made of wrought iron. General purpose chains made from low-carbon steel are sometimes found being used as lifting slings. However, only alloy steel chains are recommended for overhead lifting. Wrought iron, general purpose (BBB and Proof Coil), and high-test chain are not recommended for overhead lifting and should, therefore, be avoided for use as a supporting member in any below-the-hook lifting device.

Current American codes (ANSI B30.9) specify chain properties, proof loads, working load limits, and inspection requirements for alloy steel chain. Manufacturers are required to affix a durable identification tag to each sling which states size, grade, rated capacity and manufacturer's name. Refer to Table 10.4.1 for strength information and other typical specifications for alloy steel chain.

The advantages of alloy steel chain slings over wire rope, synthetic, and fiber rope slings are that they are better suited for rough loads, abrasive conditions, and high temperatures. However, alloy steel chain must not be heated above 1000°F after being received from the manufacturer. Chain subjected to temperatures in excess of 600°F must have their working load limits reduced.

Table 10.4.1 Typical Specifications for Alloy Steel Chain

Nominal Size of Chain [in. (mm)]	Working Load Limit (lb)	Minimum Proof-Test Load (lb)	Minimum Break-Test Load (lb)	Maximum Length, 100 Links [in. (m)]	Maximum Weight per Unit Length (lb/ft)
$\frac{1}{4}$ (6)	3,250	6,500	10,000	98 (2.49)	0.84
$\frac{3}{8}$ (10)	6,600	13,200	19,000	134 (3.40)	1.75
$\frac{1}{2}$ (13)	11,250	22,500	32,500	156 (3.96)	2.88
$\frac{5}{8}$ (16)	16,500	33,000	50,000	182 (4.62)	4.53
$\frac{3}{4}$ (20)	23,000	46,000	69,000	208 (5.28)	6.55
$\frac{7}{8}$ (22)	28,750	57,500	93,500	234 (5.94)	9.10
1 (25)	38,750	77,500	122,000	277 (7.04)	11.7
$1\frac{1}{8}$ (29)	44,500	89,000	143,000	332 (8.43)	14.3
$1\frac{1}{4}$ (32)	57,500	115,000	180,000	371 (9.42)	17.7
$1\frac{3}{8}$ (35)	67,000	134,000	207,000	391 (9.93)	20.1
$1\frac{1}{2}$ (38)	80,000	160,000	244,000	432 (10.97)	21.9
$1\frac{3}{4}$ (44)	100,000	200,000	325,000	503 (12.78)	30.2

When determining the type of sling to use on an application, the engineer must keep in mind the proverbial saying that a chain is only as strong as its weakest link, and the fact that chain slings will ultimately fail with less warning than wire rope slings.

Figure 10.4.1 may be helpful to illustrate the dramatic effect that the angle of loading can have on the load capacity of all types of lifting slings. Whereas the sling stress is increased only moderately at angles up to 30°, the increase in stress is tremendous beginning at 60°. This elementary engineering fact must not be overlooked when designing lifting systems. Notice that the angle expressed in Fig. 10.4.1 is the *vertical* angle.

Chain manufacturers publish working load ratings for single-sling chains in straight tension, and for double, triple, and quad slings when used at a variety of angles. Quad (4 equal legs) slings are rated at the same capacity as triple slings. This is due to the difficulty in actual practice of achieving equal distribution of load among more than three sling branches.

The American National Standards Institute (ANSI) has established maximum allowable wear at any point of any alloy steel chain link. Table 10.4.2 shows the maximum allowable wear at any point of a link for some common size chains. However, in determining the safety of a used chain sling the inspector must look for nicks and gouges, stretch, and localized bending and shearing, as well as wear. Figure 10.4.2 illustrates the pattern of tensile and compressive stress in a link under load. From this it can be seen that some portions of the link are more sensitive to nicks, gouges, and notches than are other portions. Fortunately, the geometry of a chain link tends to protect tensile stress areas against damage from external causes. That is, the link ends (tensile stress area) are somewhat shielded against damage by interconnected links, and the insides of the barrels (tensile stress area) are sheltered due to their location.

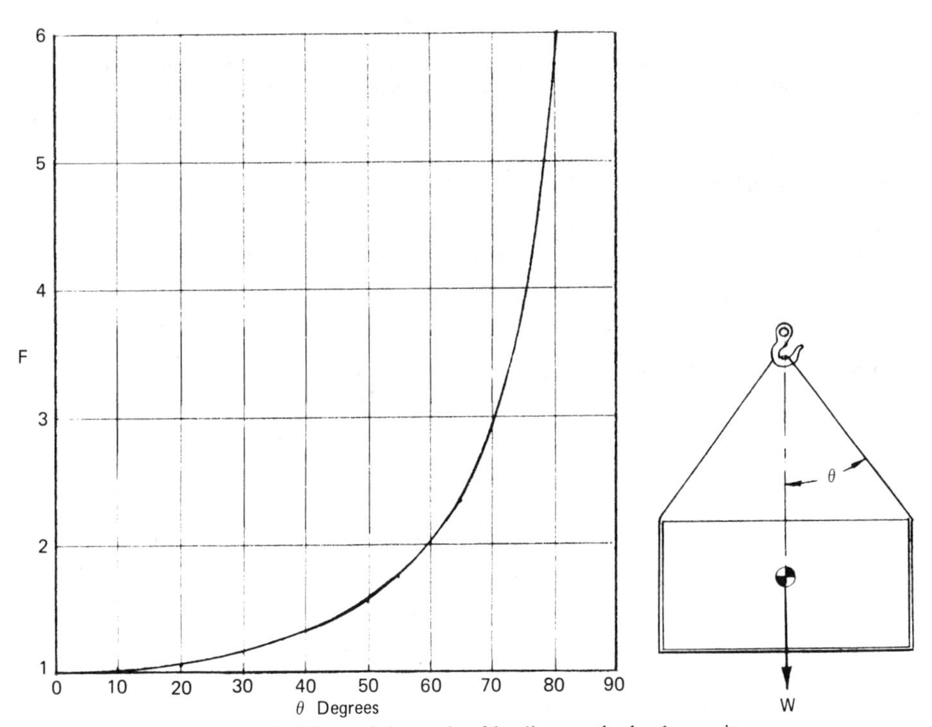

Fig. 10.4.1 Effect of the angle of loading on the load capacity.

$$F = \frac{1}{\cos \theta}, \quad S = \frac{F \times W}{N}$$

where W = load, S = sling tensile stress, θ = vertical angle, N = number of sling legs, f = sling stress factor.

For example: (1) If W = 100 lb and $\theta = 0°$ (vertical hitch), then $S = (1.0 \times 100)/1.0 = 100$ lb; (2) If W = 100 lbs and θ = 45° (basket hitch), then $S = (1.414 \times 100)/2.0 = 70.7$ lb per leg.

Table 10.4.2 Maximum Allowable Chain Link Wear

Chain Size [in. (mm)]	Maximum Allowable Wear [in. (mm)]	Minimum Allowable Diameter at Worn Portion [in. (mm)]
$\frac{1}{4}$ (6)	$\frac{3}{64}$ (1.2)	0.203 (4.8)
$\frac{3}{8}$ (10)	$\frac{5}{64}$ (2.0)	0.297 (8.0)
$\frac{1}{2}$ (13)	$\frac{7}{64}$ (2.8)	0.609 (10.2)
$\frac{5}{8}$ (16)	$\frac{9}{64}$ (3.6)	0.484 (12.4)
$\frac{3}{4}$ (20)	$\frac{5}{32}$ (4.0)	0.594 (16.0)
$\frac{7}{8}$ (22)	$\frac{11}{64}$ (4.4)	0.703 (17.6)
1 (25)	$\frac{3}{16}$ (4.8)	0.813 (20.2)
$1\frac{1}{8}$ (29)	$\frac{7}{32}$ (5.6)	0.906 (23.4)
$1\frac{1}{4}$ (32)	$\frac{1}{4}$ (6.4)	1.00 (25.6)
$1\frac{3}{8}$ (35)	$\frac{9}{32}$ (7.1)	1.09 (27.9)
$1\frac{1}{2}$ (38)	$\frac{5}{16}$ (7.9)	1.19 (30.1)
$1\frac{3}{4}$ (44)	$\frac{11}{32}$ (8.7)	1.41 (35.3)

When loaded to destruction, alloy steel chains usually fail along a plane in the long axis of the link. This is not the case for chains made of softer materials (under 400 Brinell). Failure in these softer steel chains is typically due to shear, and the failure location is approximately 45° away from the long axis on a radial line through the center of curvature of the link end.

In most applications industrial chains are equipped with some type of end fittings. The simplest end attachments are the ring, the oblong link, the pear-shaped link, and various types of slip or grab hooks. These attachments are readily available from chain manufacturers and are usually forged from heat-treated carbon or alloy steel. Obviously, the end attachments must be compatible in strength to the chain that they are used with. Hooks and end links are attached to chain slings by either welded or mechanical coupling links available from the chain manufacturer. Factory-furnished sling assemblies are normally welded, whereas mechanical coupling is the usual method for field assembly. Figure 10.4.3 illustrates some of the common attachments available from established manufacturers. These attachments are designed using curved-beam theory as outlined in most strength of materials textbooks.

Wire Rope Slings. Wire rope slings have largely superseded other sling materials in industrial hoisting operations. This is because of the high strength-to-weight ratio, great flexibility, high reserve strength, and ease of inspection of wire rope. With wire rope slings, warning occurs before failure with the breaking of outer wires. Unlike chain, failure of one of the parts of a wire rope will not

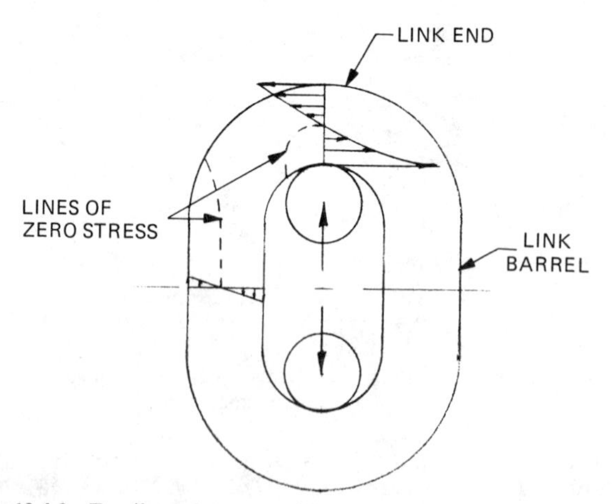

Fig. 10.4.2 Tensile and compressive stress in a chain link under load.

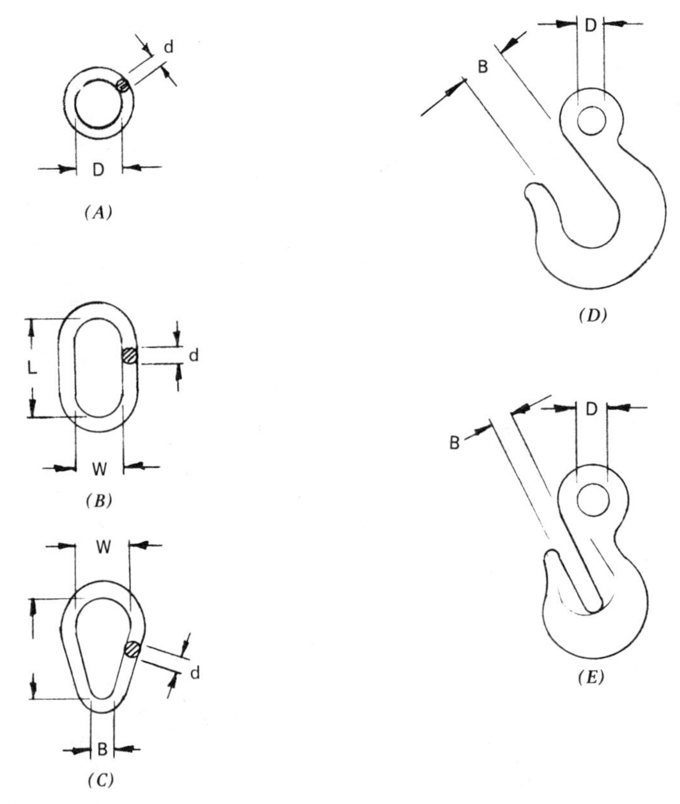

Fig. 10.4.3 Common chain sling attachments (alloy steel). (*A*), Ring, capacities to 135,000 lb (61,400 kg), $D = 3\text{–}10$ in. (76–254 mm), $d = \frac{1}{2} - 2\frac{1}{2}$ in. (13–64 mm). (*B*), Oblong link, capacities to 210,000 lb (95,500 kg), $L = 3\text{–}16$ in. (76–406 mm), $W = 1\frac{1}{2}\text{–}9$ in. (38–229 mm), $d = \frac{13}{32}\text{–}2\frac{3}{4}$ in. (10–70 mm). (*C*), Pear link, capacities to 140,000 lb (63,600 kg), $L = 5\text{–}16$ in. (127–406 mm), $W = 2\frac{1}{2}\text{–}8$ in. (64–203 mm), $B = 1\frac{1}{4}\text{–}4$ in. (32–102 mm), $d = \frac{1}{2}\text{–}2\frac{1}{4}$ in. (13–57 mm). (*D*), Eye slip hook, capacities to 80,000 lb (36,400 kg), $D = \frac{1}{2}\text{–}3$ in. (13–76 mm), $B = \frac{15}{16}\text{–}4$ in. (24–102 mm). (*E*), Eye grab hook, capacities to 80,000 lb (36,400 kg), $D = \frac{1}{2}\text{–}2\frac{1}{4}$ in. (13–57 mm), $B = \frac{3}{8}\text{–}1\frac{1}{2}$ in. (10–38 mm).

result in failure of the entire lifting device. For example, in a 6 × 19 construction wire rope sling, one broken wire will decrease the strength of the sling by less than 1%.

Wire rope slings must not be allowed to become kinked as this causes unequal distribution of the load and severe bending stress in the outer strands. After a wire rope sling is kinked the wire deformations are usually permanent. This causes many sling failures in service. Whenever a wire rope sling comes into contact against a hard, sharp surface every effort must be made to protect the sling with softer materials, such as wood.

Due to the fact that wire rope slings will sustain some compression force, they are more likely than a chain sling to be pushed out of a supporting open hook. For this reason safety considerations dictate that wire rope slings must be positively held in the supporting hook, usually with a safety latch of some sort. This is not to say that chain slings never need a safety latch—just that chain slings are less likely than wire rope slings to be forced out of the hook when the load is slackened.

Wire rope is made up of strands of wires laid together. In wire rope slings the number of wires commonly being used are 19 and 37 and the number of strands 6 or 7. The wire rope strands are laid around a fiber core (FC) or an independent wire rope core (IWRC). The core provides some additional strength, but its main purposes are to hold the shape of the wire rope and help lubricate the wires.

An IWRC wire rope is stronger and less susceptible to crushing than an FC wire rope, but an IWRC rope is less flexible than an FC rope. The strand construction of the wire rope sling also

Table 10.4.3 Specifications for 6 × 19 and 6 × 37 Classification Improved Plow Steel Grade Rope Slings with Fiber Core

Diameter of Rope [in. (mm)]	Rope Construction, FC	Breaking Strength (lb)	Approximate Weight per Unit Length (lb/ft)	Rated Capacities[a] (lb)	
				Vertical	Choker
$\frac{1}{4}$ (6)	6 × 19	5,480	0.105	1,020	760
$\frac{3}{8}$ (10)	6 × 19	12,200	0.236	2,200	1,700
$\frac{1}{2}$ (13)	6 × 19	21,400	0.42	4,000	3,000
$\frac{5}{8}$ (16)	6 × 19	33,400	0.66	6,200	4,600
$\frac{3}{4}$ (20)	6 × 19	47,600	0.95	8,800	6,600
$\frac{7}{8}$ (22)	6 × 19	64,400	1.29	11,800	9,000
1 (25)	6 × 19	83,600	1.68	15,400	11,600
$1\frac{1}{8}$ (29)	6 × 19	105,000	2.13	19,000	14,200
$1\frac{1}{4}$ (32)	6 × 37	123,000	2.63	22,000	16,600
$1\frac{3}{8}$ (35)	6 × 37	148,000	3.18	26,000	20,000
$1\frac{1}{2}$ (38)	6 × 37	176,000	3.78	32,000	24,000
$1\frac{5}{8}$ (41)	6 × 37	206,000	4.44	36,000	28,000
$1\frac{3}{4}$ (44)	6 × 37	238,000	5.15	42,000	32,000
2 (51)	6 × 37	308,000	6.77	56,000	42,000

[a] Rated capacities are for mechanical splices. Hand-tucked splices and hidden-tuck splices would have *lower* rated capacities. Swaged or zinc poured socket end fittings would have *higher* rated capacities. Proof load for single-leg slings and endless slings is to be two times the vertical rated capacity.

affects its flexibility and abrasion resistance, that is, a 6 × 19 construction is less flexible but more abrasion resistant than 6 × 37 construction. For these reasons, 6 × 19 wire rope is normally used in smaller-diameter [through $1\frac{1}{8}$ in. (29 mm)] wire rope slings and 6 × 37 construction is recommended in larger-diameter wire rope slings.

Tables 10.4.3 and 10.4.4 show typical specifications for single-leg wire rope slings made from improved plow steel. The factor of safety for wire rope slings is to be a minimum of 5 over the nominal breaking strength of the wire rope used. ANSI B30.9 specifies that the efficiency of any splicing or end attachment, the number of parts of rope in the sling, the type of hitch used, the angle of loading, and the diameter of curvature around which the sling is bent must also be considered when determining

Table 10.4.4 Specifications for 6 × 19 and 6 × 37 Classification Improved Plow Steel Grade Rope Slings with Independent Wire Rope Core

Diameter of Rope [in. (mm)]	Rope Construction, IWRC	Breaking Strength (lb)	Approximate Weight per Unit Length (lb/ft)	Rated Capacities[a] (lb)	
				Vertical	Choker
$\frac{1}{4}$ (6)	6 × 19	5,880	0.116	1,120	840
$\frac{3}{8}$ (10)	6 × 19	13,100	0.260	2,400	1,860
$\frac{1}{2}$ (13)	6 × 19	23,000	0.46	4,400	3,200
$\frac{5}{8}$ (16)	6 × 19	35,800	0.72	6,800	5,000
$\frac{3}{4}$ (20)	6 × 19	51,200	1.04	9,800	7,200
$\frac{7}{8}$ (22)	6 × 19	69,200	1.42	13,200	9,800
1 (25)	6 × 19	89,800	1.85	17,000	12,800
$1\frac{1}{8}$ (29)	6 × 19	113,000	2.34	20,000	15,600
$1\frac{1}{4}$ (32)	6 × 37	132,000	2.89	24,000	18,400
$1\frac{3}{8}$ (35)	6 × 37	159,000	3.50	30,000	22,000
$1\frac{1}{2}$ (38)	6 × 37	189,000	4.16	34,000	26,000
$1\frac{5}{8}$ (41)	6 × 37	222,000	4.88	40,000	30,000
$1\frac{3}{4}$ (44)	6 × 37	256,000	5.67	48,000	36,000
2 (51)	6 × 37	330,000	7.39	60,000	46,000

[a] Rated capacities are for mechanical splices. Hand-tucked splices and hidden-tuck splices would have *lower* rated capacities. Swaged or zinc poured socket end fittings would have *higher* rated capacities.

the factor of safety. The rated capacity (or working load limit) of a wire rope sling is then equal to the nominal breaking strength divided by the determined factor of safety. The diameter of a wire rope is the diameter of a circle which will enclose it. Therefore, one must be careful when using calipers to measure wire rope to avoid measuring across the adjacent strands.

Wire rope slings with a fiber core should be permanently removed from service if exposed to temperatures in excess of 200°F. Wire rope slings with IWRC should not be used at temperatures above 400°F or below −60°F.

Figure 10.4.4 illustrates the different types of sling hitches used with wire rope slings. The vertical hitch is the simplest sling, consisting of just a single vertical leg. Application of the vertical hitch is straightforward because the only considerations in determining rated capacity are the efficiency of the end attachments and the strength of the wire rope. However, the rated capacity of choker hitches and basket hitches is reduced relative to the vertical hitch because of the angle of loading, the diameter of curvature of the load, and/or concentrated wear at the point of choke. In Tables 10.4.3 and 10.4.4 it can be seen that the rated capacity of a vertical hitch is considerably greater than for the choker hitch.

In a bridle hitch, two, three, or four single hitches are used together to hoist objects that have lifting lugs or other attachments. Ideally the center of gravity of the object being lifted is directly below the crane hook. In practice this is not always possible or practical. For this reason bridle hitches may need to have unequal length legs or unequal strength legs in order to equalize stresses in the slings. If the crane hook is too far to one side of the center of gravity, the object will tilt which can result in an unsafe situation which will need to be immediately corrected. The location of the center of gravity of any load should be approximated before lifting slings are chosen, and definitely before any large lift is attempted.

As with any lifting device, the wire rope sling is not complete and functional until it has been fitted with proper attachments and correctly and safely applied to the job at hand. Wire rope slings must have some sort of an eye connection on each end in order to make attachment to the various

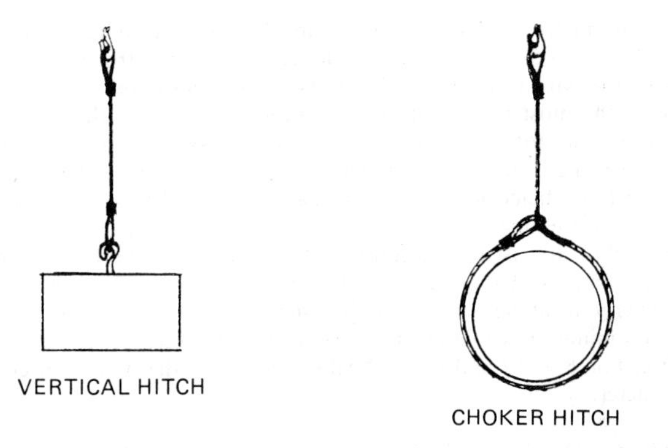

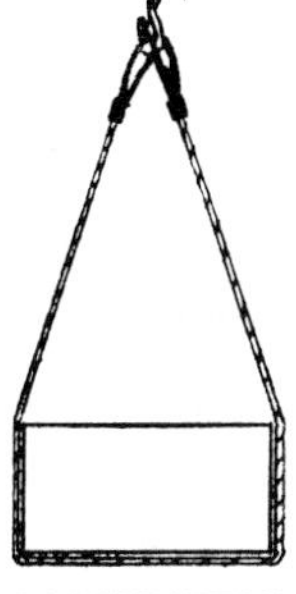

Fig. 10.4.4 Sling hitches used with wire rope slings.

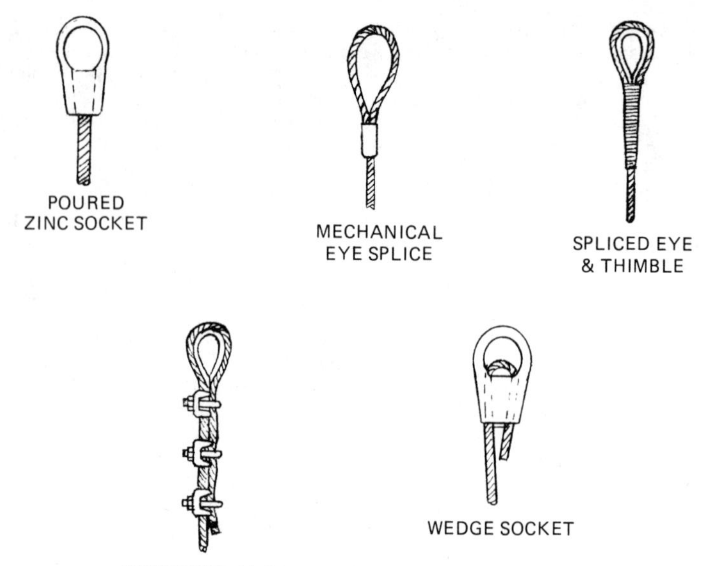

Fig. 10.4.5 Eyes for wire rope slings.

hooks and shackles normally used in lifting a load. The most common methods for making these eyes are shown in Fig. 10.4.5. Some of these connections are as strong as the wire rope and some are not as strong as the wire rope. Since all the types of connections are able to develop the rated capacity of the rope, the most important considerations in choosing the type of connection are the ease of application and the total cost. With the advent of swaging machines being located at wire rope distributors all over the United States in recent years, the swaged mechanical eye splice is becoming very common. Connection efficiencies are listed in Table 10.4.5. The use of wire rope thimbles increases the life of the sling by decreasing wear and distributing bearing loads.

When wire rope clips are used, care must be taken to apply the base of the clip to the live or long end of the eye, and the U-bolt against the dead or short end of the eye. Table 10.4.6 gives information for properly installing wire rope clips when forming sling eyes with this type of connection. ANSI B30.9 specifies minimum cable length (clear length) between splices, sleeves, or end fittings for any wire rope sling. For 6 × 19 and 6 × 37 cable laid slings the minimum clear length is to be 10 times the rope diameter.

Metal Mesh Slings. Metal mesh slings are designed for use on high-temperature or abrasive applications. They are widely used in the lifting of metal bars, plates, and similar shapes that have a tendency to cut synthetic web slings or wire rope. Metal mesh slings manufactured from carbon or stainless steel can withstand operating temperatures in the range from −29°F to +550°F without decreasing the working load limit. Metal mesh slings used to handle materials that could damage the wire mesh, or used to handle loads with soft finishes, are often impregnated by the manufacturer with neoprene or poly vinyl chloride (PVC). These impregnated or coated slings are limited to an operating temperature range of 0–200°F.

Table 10.4.5 Wire Rope Connection Efficiencies

Type of Connection	Efficiency
Poured zinc socket	100%
Mechanical eye splice	100%
Spliced eye and thimble	80–90%
Wire rope clips and thimble	80%
Wedge socket	70%
Wire rope	100%

Table 10.4.6 Installation Details for Wire Rope Clips

Diameter of Rope [in. (mm)]	Minimum Number of Clips Required	Clip to Center Center Spacing [in. (mm)]	Total Length of Rope Turned Back [in. (mm)]
1/4 (6)	2	1 1/2 (38)	8 (203)
3/8 (10)	2	2 1/4 (57)	10 (254)
1/2 (13)	3	3 (76)	12 (305)
5/8 (16)	3	3 3/4 (95)	16 (406)
3/4 (20)	4	4 1/2 (114)	20 (508)
7/8 (22)	4	5 1/4 (133)	24 (610)
1 (25)	4	6 (152)	30 (762)
1 1/8 (29)	5	6 3/4 (171)	36 (914)
1 1/4 (32)	5	7 1/2 (191)	40 (1016)

Figure 10.4.6 illustrates the construction of a metal mesh sling and Fig. 10.4.7 shows how the mesh is made. Standard metal mesh slings are manufactured in three different construction duty ratings: heavy duty, medium duty, and light duty. Refer to Table 10.4.7 for specifications regarding these fabric duty ratings. By slipping the male handle through the female handle of a metal mesh sling a choker hitch is formed which grips the load extremely tight. This characteristic allows a single metal mesh sling to be used for some applications that normally would require two or more chain or wire rope slings. The factor of safety for metal mesh slings is required by ANSI to be a minimum of 5 and the proof load is to be a minimum of 1 1/2 times the rated capacity (see Table 10.4.8).

Fiber Rope Slings. For thousands of years mankind's only material for making slings was various cordage made from animal hairs and fibers of perennial plants such as banana. Abaca or manila is the predominate fiber used for heavy rope and slings, although sisal, hemp, cotton, and jute have been used also.

With the advent of modern chemistry, synthetic fibers were developed that have replaced the natural fibers to a great degree. Nylon, polyester, and polypropylene ropes have the advantages of higher strength, resistance to rot and mildew, and resistance to bacteriological damage when compared to natural rope such as manila. As a rule of thumb, a nylon rope of a given diameter has about 3 times the breaking strength of a manila rope. Likewise a polyester rope has about 2 1/2 times the strength of manila.

Fiber ropes are made with two common types of construction today. These constructions are twisted

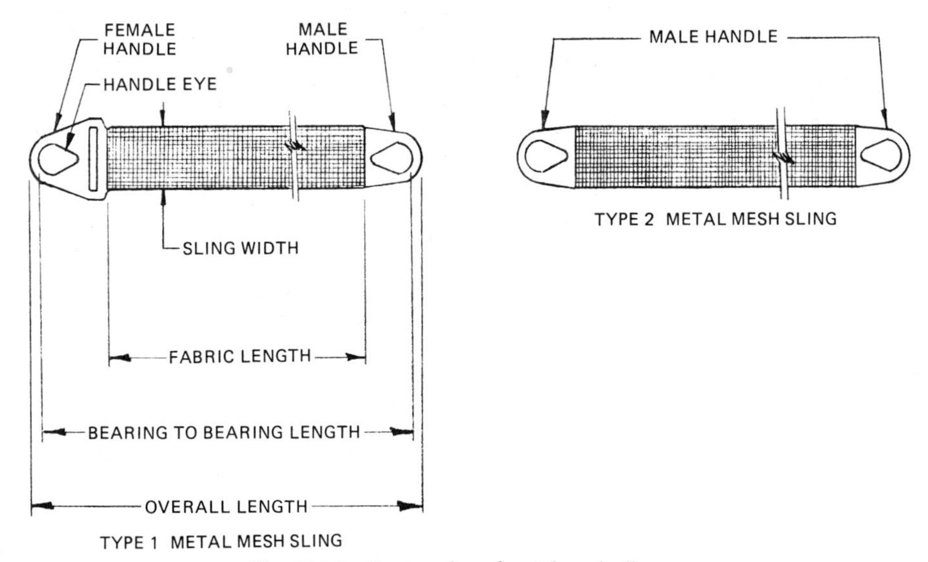

Fig. 10.4.6 Construction of metal mesh sling.

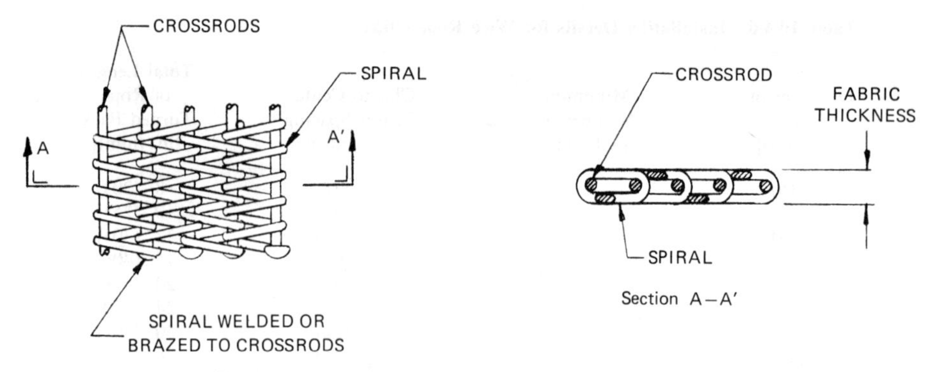

Fig. 10.4.7 Details of construction of metal mesh sling.

rope and braided rope. Since this discussion is addressing slings, only conventional three-strand twisted rope is dealt with here. Twisted rope is made by twisting several threads into a yarn, several yarns into a strand, and then three strands into a rope. Braided rope is used mostly for small-diameter lines such as clothesline and sash and would not be likely used as a lifting sling.

ANSI B30.9 contains specifications for what are called the four "basic" fiber types: manila, nylon, polyester, and polypropylene. The three synthetic fiber ropes gain their strength advantage over manila due, in large part, to the fact that the individual fibers making up the rope run the entire length of the rope. These fibers are technically called continuous filaments or, in the case of some polypropylene ropes, continuous film.

An interesting convention regarding fiber rope is that the rope size is often determined by diameter up to 1 in. (25 mm) and by circumference if larger than 1 in. (25 mm) in diameter. In addition, the fishing and shipbuilding industries use different terminology regarding rope construction than do industrial buyers. In the tables of this section, fiber rope size is specified by diameter only. Refer to ANSI B30.9 for more detailed information regarding natural and synthetic fiber rope sling configurations, safe operating practices, and inspection. Tables 10.4.9, 10.4.10, 10.4.11, and 10.4.12 illustrate rated capacity for manila, nylon, polyester, and polypropylene rope slings. Note that the different materials have different factors of safety over the breaking strength. That is, manila rope slings have a safety factor of 5, nylon a safety factor of 9, polyester a safety factor of 9, and polypropylene a safety factor of 6.

The rated load capacities tabulated for these fiber rope slings apply in the temperature range from −20°F to 180°F, except for wet rope at temperatures below freezing. The rope manufacturer should be consulted for application recommendations for any service temperatures outside the above. Nylon slings are highly resistant to alkalies but should not be exposed to acids. Polyester slings are more resistant to acids and are sometimes used for dipping materials into acids and pickling solutions. Polypropylene resists acids and alkalies, but is softened by industrial solvents. Polypropylene rope is not as strong as nylon or polyester but its light weight allows it to float in water.

Synthetic Webbing Slings. Synthetic webbing slings are woven from nylon, polyester, or polypropylene. Because they are soft and flexible, they are used extensively in industry to handle delicate or polished surfaces that would be marred by alloy chain, wire rope, or metal mesh slings (Fig. 10.4.8). Other advantages of this type of sling are that they are nonsparking, are unaffected by mildew, rot,

Table 10.4.7 Metal Mesh Slings

Parameter	Heavy Duty	Medium Duty	Light Duty
Nominal spiral turns per foot of sling width	35	43	59
Spiral wire size			
USSWG	10 Ga	12 Ga	14 Ga
Diameter	0.135 in. (3.4 mm)	0.105 in. (2.7 mm)	0.080 in. (2 mm)
Nominal crossrods per foot of fabric length	21	30	38
Size of crossrods USSWG	0.162 in. (4.1 mm)	0.135 in. (3.4 mm)	0.080 in. (2 mm)
Nominal fabric thickness	$\frac{1}{2}$ in. (13 mm)	$\frac{3}{8}$ in. (10 mm)	$\frac{5}{16}$ in. (8 mm)

Table 10.4.8 Rated Capacities for Metal Mesh Slings

Sling Width [in. (mm)]	Wire Gauge	Rated Capacities (lb) Vertical or Choker	Vertical Basket
2 (51)	10	1,500	3,000
3 (76)	10	2,700	5,400
4 (102)	10	4,000	8,000
6 (152)	10	6,000	12,000
8 (203)	10	8,000	16,000
10 (254)	10	10,000	20,000
12 (305)	10	12,000	24,000
14 (356)	10	14,000	28,000
16 (406)	10	16,000	32,000
18 (457)	10	18,000	36,000
20 (508)	10	20,000	40,000
2 (51)	12	1,350	2,700
3 (76)	12	2,000	4,000
4 (102)	12	2,700	5,400
6 (152)	12	4,500	9,000
8 (203)	12	6,000	12,000
10 (254)	12	7,500	15,000
12 (305)	12	9,000	18,000
14 (356)	12	10,500	21,000
16 (406)	12	12,000	24,000
18 (457)	12	13,500	27,000
20 (508)	12	15,000	30,000
2 (51)	14	900	1,800
3 (76)	14	1,400	2,800
4 (102)	14	2,000	4,000
6 (152)	14	3,000	6,000
8 (203)	14	4,000	8,000
10 (254)	14	5,000	10,000
10 (254)	14	5,000	10,000
12 (305)	14	6,000	12,000
14 (356)	14	7,000	14,000
16 (406)	14	8,000	16,000
18 (457)	14	9,000	18,000
20 (508)	14	10,000	20,000

Table 10.4.9 Manila Rope Slings, Safety Factor = 5

Diameter of Rope [in. (mm)]	Breaking Strength Minimum (lb)	Approximate Weight per Unit Length (lb/ft)	Rated Capacities (lb) Eye and Eye Sling[a] Vertical	Choker
$\frac{1}{2}$ (13)	2,650	0.075	550	250
$\frac{5}{8}$ (16)	4,400	0.133	900	450
$\frac{3}{4}$ (20)	5,400	0.167	1,100	550
$\frac{7}{8}$ (22)	7,700	0.225	1,500	750
1 (25)	9,000	0.270	1,800	900
$1\frac{1}{8}$ (29)	12,000	0.360	2,400	1,200
$1\frac{1}{4}$ (32)	13,500	0.417	2,700	1,400
$1\frac{1}{2}$ (38)	18,500	0.599	3,700	1,850
$1\frac{5}{8}$ (41)	22,500	0.746	4,500	2,300
$1\frac{3}{4}$ (44)	26,500	0.893	5,300	2,700
2 (51)	31,000	1.08	6,200	3,100
$2\frac{1}{4}$ (57)	41,000	1.46	8,200	4,100
$2\frac{1}{2}$ (64)	46,500	1.67	9,300	4,700

[a] See ANSI B30.9 for rated capacities for endless slings and for basket hitches with various leg-to-vertical angles. The contact surfaces of the hitch eyes must have a diameter of curvature at least double the diameter of the rope from which the sling is made.

Table 10.4.10 Nylon Rope Slings, Safety Factor = 9

Diameter of Rope [in. (mm)]	Breaking Strength Minimum (lb)	Approximate Weight per Unit Length (lb/ft)	Rated Capacities (lb) Eye and Eye Sling[a]	
			Vertical	Choker
$\frac{1}{2}$ (13)	6,080	0.065	700	350
$\frac{5}{8}$ (16)	9,880	0.105	1,100	550
$\frac{3}{4}$ (20)	13,490	0.145	1,500	750
$\frac{7}{8}$ (22)	19,000	0.200	2,100	1,100
1 (25)	23,750	0.260	2,600	1,300
$1\frac{1}{8}$ (29)	31,350	0.340	3,500	1,700
$1\frac{1}{4}$ (32)	35,625	0.400	4,000	2,000
$1\frac{1}{2}$ (38)	50,350	0.550	5,600	2,800
$1\frac{5}{8}$ (41)	61,750	0.680	6,900	3,400
$1\frac{3}{4}$ (44)	74,100	0.830	8,200	4,100
2 (51)	87,400	0.950	9,700	4,900
$2\frac{1}{4}$ (57)	118,750	1.29	13,000	6,600
$2\frac{1}{2}$ (64)	133,000	1.49	15,000	7,400

[a] See ANSI B30.9 for rated capacities for endless slings and for basket hitches with various leg-to-vertical angles. The contact surfaces of the hitch eyes must have a diameter of curvature at least double the diameter of the rope from which the sling is made.

or bacteria, and have elongation characteristics that minimize the effects of shock loading. The light-weight characteristic of synthetic web slings makes them advantageous for applications that require a lot of handling by personnel.

Nylon is the most popular material used for synthetic webbing slings. Nylon has good chemical resistance to many chemicals (see Table 10.4.13) but is not to be used with acids or bleaching agents. Nylon webbing slings should not be used at temperatures exceeding 180°F. Stretch at rated capacity is at about 10% according to most manufacturers.

Polyester webbing slings do not stretch as much as nylon or polypropylene slings (about 3% at rated capacity), and so are recommended for use whenever a minimum of stretch is required and a synthetic webbing sling is needed. Common acids do not affect polyester slings and so they are used for pickling and acid dipping. Do not use them with concentrated sulphuric acid and alkaline solutions,

Table 10.4.11 Polyester Rope Slings, Safety Factor = 9

Diameter of Rope [in. (mm)]	Breaking Strength Minimum (lb)	Approximate Weight per Unit Length (lb/ft)	Rated Capacities (lb) Eye and Eye Sling[a]	
			Vertical	Choker
$\frac{1}{2}$ (13)	6,080	0.080	700	350
$\frac{5}{8}$ (16)	9,500	0.130	1,100	550
$\frac{3}{4}$ (20)	11,875	0.175	1,300	650
$\frac{7}{8}$ (22)	17,100	0.250	1,900	950
1 (25)	20,900	0.305	2,300	1,200
$1\frac{1}{8}$ (29)	28,025	0.400	3,100	1,600
$1\frac{1}{4}$ (32)	31,540	0.463	3,500	1,800
$1\frac{1}{2}$ (38)	44,460	0.668	4,900	2,500
$1\frac{5}{8}$ (41)	54,150	0.820	6,000	3,000
$1\frac{3}{4}$ (44)	64,410	0.980	7,200	3,600
2 (51)	76,000	1.18	8,400	4,200
$2\frac{1}{4}$ (57)	101,650	1.57	11,500	5,700
$2\frac{1}{2}$ (64)	115,900	1.81	13,000	6,400

[a] See ANSI B30.9 for rated capacities for endless slings and for basket hitches with various leg-to-vertical angles. The contact surfaces of the hitch eyes must have a diameter of curvature at least double the diameter of the rope from which the sling is made.

Table 10.4.12 Polypropylene Rope Slings, Safety Factor = 6

Diameter of Rope [in. (mm)]	Breaking Strength Minimum (lb)	Approximate Weight per Unit Length (lb/ft)	Rated Capacities (lb) Eye and Eye Sling[a]	
			Vertical	Choker
$\frac{1}{2}$ (13)	3,990	0.047	650	350
$\frac{5}{8}$ (16)	5,890	0.075	1,000	500
$\frac{3}{4}$ (20)	8,075	0.107	1,300	700
$\frac{7}{8}$ (22)	10,925	0.150	1,800	900
1 (25)	13,300	0.180	2,200	1,100
$1\frac{1}{8}$ (29)	17,385	0.237	2,900	1,500
$1\frac{1}{4}$ (32)	19,950	0.270	3,300	1,700
$1\frac{1}{2}$ (38)	28,215	0.385	4,700	2,400
$1\frac{5}{8}$ (41)	34,200	0.475	5,700	2,900
$1\frac{3}{4}$ (44)	40,850	0.570	6,800	3,400
2 (51)	49,400	0.690	8,200	4,100
$2\frac{1}{4}$ (57)	65,550	0.920	11,000	5,500
$2\frac{1}{2}$ (64)	76,000	1.07	12,500	6,300

[a] See ANSI B30.9 for rated capacities for endless slings and for basket hitches with various leg-to-vertical angles. The contact surfaces of the hitch eyes must have a diameter of curvature at least double the diameter of the rope from which the sling is made.

however. As with nylon, polyester webbing slings are not suitable for use at temperatures exceeding 180°F.

Polypropylene webbing slings are not as strong as nylon or polyester but can be used whenever acid or alkalies are present. These slings should not be used at temperatures exceeding 200°F and the amount of stretch at rated capacity is about 10%.

Synthetic webbing slings are available with many different coatings (latex, neoprene, polyurethane) and treatments in order to improve their abrasion resistance and life. Nylon and polypropylene are adversely affected by ultraviolet light exposure, such as from sunlight or arc welding. Coating of the slings by the manufacturer can improve its resistance to ultraviolet rays. However, the nature of the slings is such that the user should carefully inspect any and all synthetic webbing slings on a regular basis. Sharp or abrasive edges on the lifted materials can cut the outer threads of the slings and seriously reduce the safe lifting capacity of the slings. Any sling that has visible acid burns, melting or charring, snags, punctures, tears or cuts, broken stitches, distortion of end fittings, or other apparent defects must be immediately removed from service until a qualified person can make the determination of whether or not the sling is safe.

The synthetic webbing sling manufacturers publish the rated capacities of the different types of slings they have available with or without fittings, and in vertical, choker, and basket hitches. ANSI B30.9 requires that all synthetic web slings have a factor of safety of at least 5. Since there are different manufacturing methods used for making the slings, a rated capacity chart for synthetic web slings is not included here. Refer to your sling distributor or manufacturer for rated lifting capacities and detailed specifications on synthetic webbing slings that are available.

The most common types of sling configurations available from manufacturers are depicted in Fig. 10.4.9. Fittings used on the slings must be designed to sustain twice the sling rated capacity without

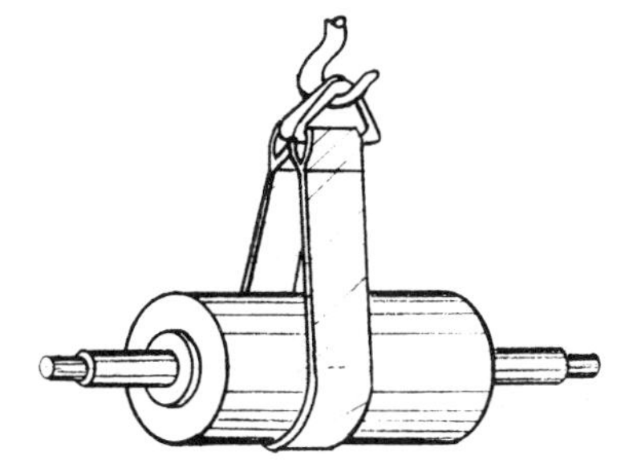

Fig. 10.4.8. Synthetic webbing sling (basket hitch).

Table 10.4.13 Synthetic Web Sling Use

Agent	Nylon	Polyester	Polypropylene
Acids, common	*No*	OK	OK
Acids, sulphuric	*No*	*No*	OK
Alcohol	OK	OK	OK
Aldehydes	OK	*No*	OK
Alkalies, weak	OK	OK	OK
Alkalies, strong	OK	*No*	OK
Bleaching agents	*No*	OK	OK
Dry cleaning solvents	OK	OK	OK
Ethers	OK	*No*	OK
Halogenated hydrocarbons	OK	OK	OK
Hydrocarbons	OK	OK	OK
Ketones	OK	OK	OK
Oil, crude	OK	OK	OK
Oil, lubricating	OK	OK	OK
Soaps, detergents	OK	OK	OK
Water, seawater	OK	OK	OK

permanent deformation. In addition, fittings are to have no sharp edges that would damage the webbing, and shall have a minimum breaking strength equal to that of the sling.

Grabs

Over the years, since the advent of the industrial revolution, a great number and variety of grabs have been invented in order to facilitate the handling of loads by cranes. Many of these devices have not proven their usefulness and are seldom seen in use anymore. However, other standard grabs continue to be used in industry on a regular basis. In addition, improvements in grab design and brand new grasping and holding inventions bless the materials handling industry each and every year.

In fact, there are many companies in business in different countries all around the world whose sole product line consists of these below-the-hook lifting devices called grabs. This huge diversity of firms and products plus the many decades of grab device history has led to confusion regarding the

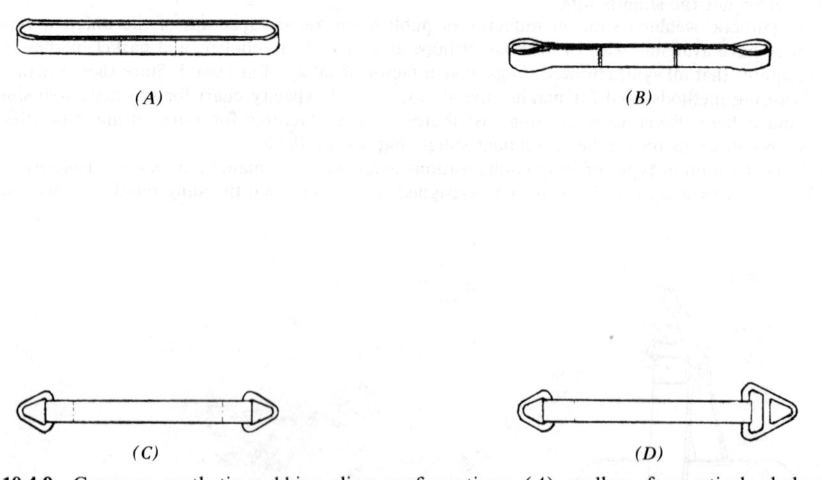

Fig. 10.4.9 Common synthetic webbing sling configurations. (*A*), endless, for vertical, choker, or basket hitches; (*B*), eye & eye, for vertical, choker, or basket hitches; (*C*), triangle/triangle, for vertical and basket hitches only; (*D*), triangle/choker, for vertical, choker, and basket hitches.

classification of and terminology for grabs. In general usage the words grab, grapple, tong, hook, bucket, lifter, clamshell, and clamp are often used synonymously. The lower parts of these devices are also called innumerable names, for example, jaws, legs, tines, claws, grips, teeth, blades, fingers, and scoops. It is true that in some industries the confusion in grab terminology is not great, however, on the whole, no standard terminology has been used. The following is an attempt to straighten out some of this confusion as well as to be helpful in the selection, sizing, and application of grabs.

Classification of Grabs. Grabs used with materials handling cranes can be self-contained hook-on, externally actuated hook-on, or permanently attached to the lifting mechanism. The self-contained hook-on class of grab can be attached to the regular hook of the hoist or crane and serve its purpose without any other connection to power or control. Examples of this type of grab are C-hooks, pallet lifters, rail tongs, lifting clamps, and overbalance hooks. These grabs may or may not require manual assistance by personnel to hook up, but there is no connection to an auxiliary electric, mechanical, hydraulic, or other source. Slings could also fall into this general class of below-the-hook lifting device even though they are not usually considered a grab as such. Figure 10.4.10 illustrates some of the self-contained hook-on type grabs. Notice that these grabs can be placed onto or removed from the crane hook relatively easily.

The externally actuated hook-on type of grab is physically suspended from the crane hook, but must be actuated from a source other than the crane hook. Some examples of this type of grab are self-powered buckets, chain release hooks, motorized lifters, and motorized hook blocks. With these grabs the actuating means is located either on the grab, on the hoist mechanism, or some other place. In each case, however, the load cannot be handled without a signal from the operator to the grab. Figure 10.4.11 shows two of this class of grab. These grabs can usually be removed from the crane hook with just a little more effort than the self-contained grab.

With the permanently attached device, the grab is connected to the crane by reeving, pinning or

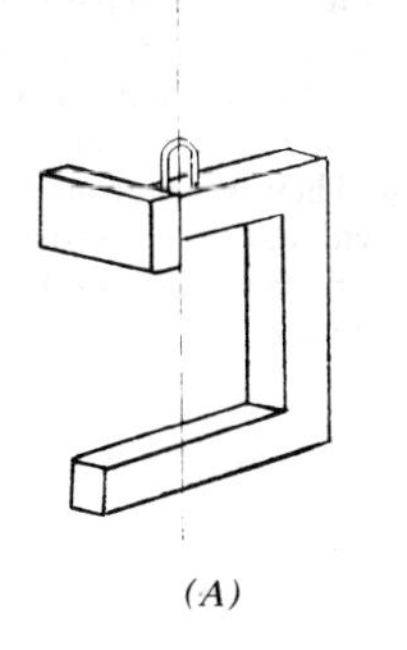
(*A*)

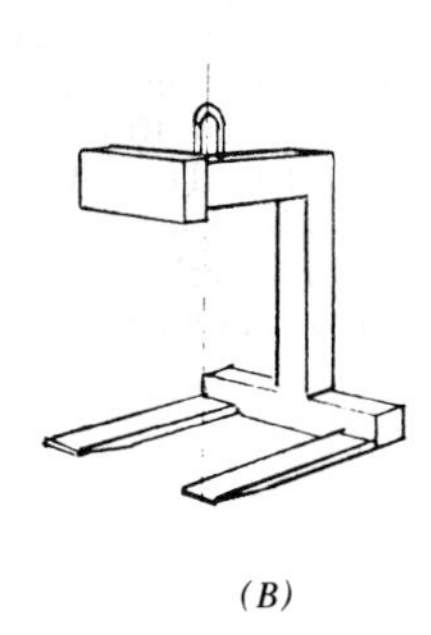
(*B*)

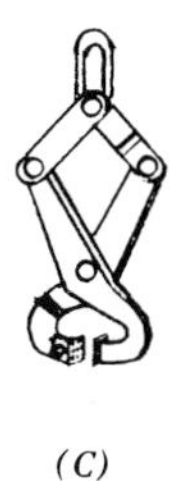
(*C*)

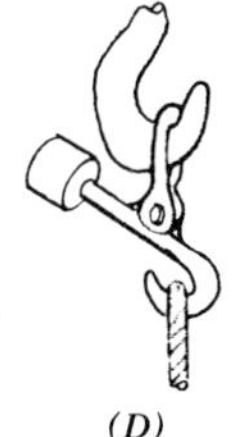
(*D*)

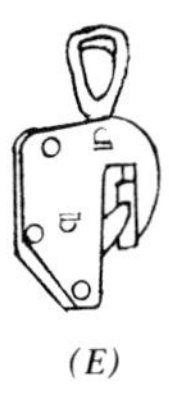
(*E*)

Fig. 10.4.10 Self-contained hook-on grabs. (*A*), C-hook; (*B*), pallet lifter; (*C*), rail tongs; (*D*), overbalance hook; (*E*), lifting clamp.

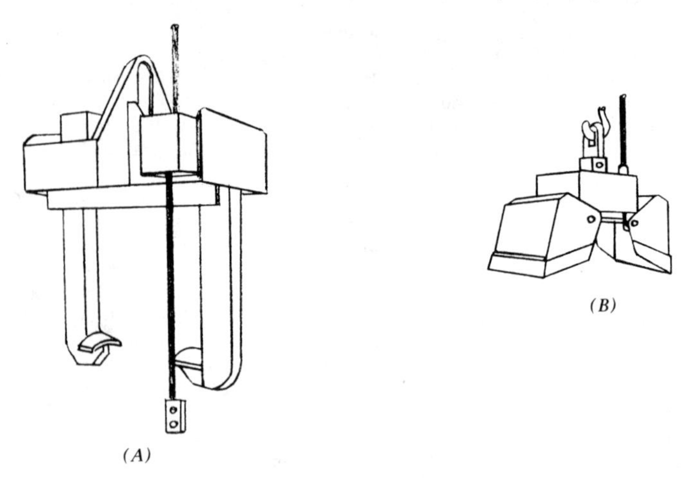

Fig. 10.4.11 Externally actuated hook-on grabs. (*A*), motor-driven coil lifter; (*B*), self-powered bucket.

some other nontemporary means. Most grapples and buckets are in this class of grab and are not designed to be disconnected from the crane except for repair or replacement. See Fig. 10.4.12 for illustrations of some of the permanently attached class of grabs.

In addition to the three general classifications of grabs just mentioned, there are several other methods that are used in practice to describe or classify grabs. Grabs can be separated as to source of energy required to cause them to operate. They can be described according to their type of construction, that is, what they look like. Grabs can also be described by function, that is, what the grab is designed to do or handle. Figure 10.4.13 is intended to display these methods for classifying grabs. By definition, all grabs connect to the load being handled in a mechanical way. They wrap around, pinch into or hook onto the load. Vacuum lifting devices and magnetic lifting devices are not considered to connect to the load with a mechanical means and so are not classified as grabs. See Sections 10.4.2 and 10.4.3 for discussion of and information on vacuum and magnetic lifting devices.

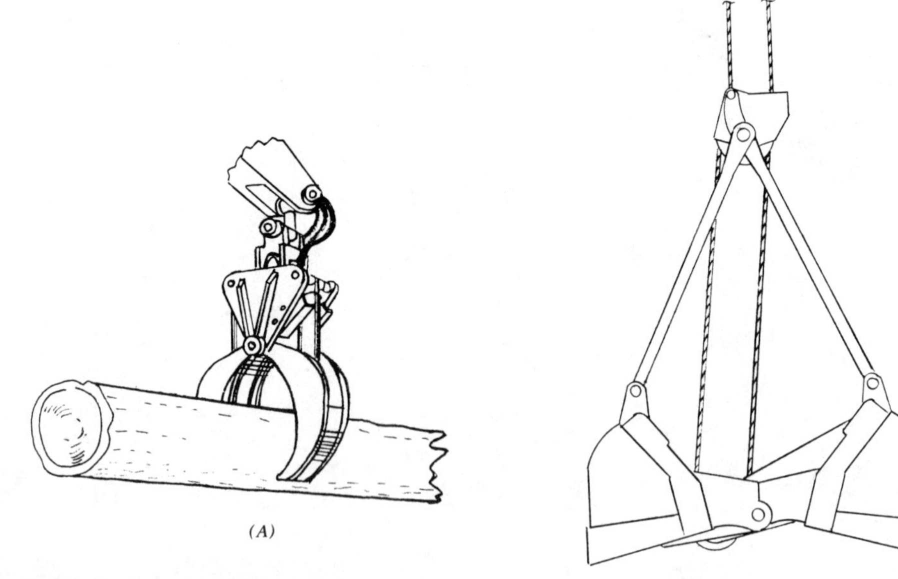

Fig. 10.4.12 Permanently attached grabs. (*A*) hydraulic log grapple; (*B*) bucket reeved to crane-hoist.

By Source of Energy

A. Gravity powered
   1. Manually applied and removed
   2. Manually applied, mechanically released
   3. Automatic

B. Non-gravity powered
   1. Electrical
   2. Hydraulic
   3. Pneumatic

By Type of Construction

A. Curved or straight beam
   1. C-hooks
   2. Overbalance hooks
   3. Lifting beams

B. Simple link
   1. Tongs
   2. Lifting clamps

C. Complex mechanical
   1. Grapples
      a. Bypass
      b. Orange peel
   2. Buckets
      a. 2-blade
      b. Orange peel
   3. Lifters

By Function

A. Buckets: Cleanup, digging, rehandling, rock, ore, slag
B. Grapples: Log, pulpwood, scrap, rock, debris
C. Hooks: Overbalance, chain release solenoid, automatic releasing, trip
D. Lifters: C-hook, pallet, sheet, slab, coil, lifting beam, spreader bar
E. Lifting clamps: Sheet, plate, weldment, beam, rail, structural
F. Tongs: Rail, ingot, timber, pipe, carton, bale, beam, crate, slab

Note: These examples are representative and not exhaustive. There are other grabs made which are not included in this listing.

Fig. 10.4.13 Classification of grabs.

Grab Glossary. The following terms are peculiar to the classification and description of the mechanical below-the-hook lifting devices known as grabs. The understanding of these terms should facilitate the design and application of these devices when used in conjunction with the other information given in this section.

Arm. An upper link member used to develop leverage in a grapple or bucket.

Blade. A lower portion of a bucket or grapple generally shaped like a leaf. See Jaw.

Bucket. A grab that has jaws that when closed form a container resembling a bucket or scoop.

Bypass Grapple. A grapple that has two or more jaws whose points pass by each other when closed.

Clamp, Lifting. A grab using screws, cams, and/or links and which holds a load by binding it tightly between two opposed parts.

Clamshell. A bucket.

Claw. The lower portion of a grapple or bucket that has slender fingers designed to dig or clutch. The end portion of a grapple or bucket jaw.

Finger. A claw.

Glommer. A lifting clamp.

Grab. Any mechanical device for clutching, holding, or lifting objects.

Grapple. A grab with two or more opposed jaws that pivot in order to open and close and to hold a load.

Grip. A lifting clamp.

Hook. A grab consisting of a curved or bent beam designed to hold a load by suspension.

Interlocking Grapple. A bypass grapple.

Jaw. A lower portion of a grab. The part that actually contacts and holds the load.

Leg. The jaw linkage of a tong.

Lifter. Any grab that cannot be classified as a bucket, grapple, clamp, hook, or tong.

Lifting Beam. A lifter consisting of one or more straight beams, designed to be attached to a load at two or more points.

Lip. The cutting edges of a blade or claw.

Orange Peel. A type of bucket or grapple consisting of three or more jaws which open and close in a manner similar to the peeling of the skin off an orange.

Point. The end or tip of a jaw.

Reeve. To pass a rope through the sheaves of a block or hoist.

Scoop. The jaw of a bucket.

Spreader Beam or Spreader Bar. A lifting beam.

Tine. A claw.

Tooth. A point or claw that is often designed to be replaced when worn by the abrasive actions of digging or holding the load.

Tong. A grab consisting of two legs and several arm linkages, and which lifts a load by simple mechanical leverage. Tongs are dependent upon friction and/or the weight of the load for developing the forces necessary to grip and lift the load.

Capacity Rating of Grabs. There are three ways that grabs can be rated for capacity: (1) by the weight they will safely handle, (2) by the maximum opening that they have, (3) by the volume of material they will hold. In many cases a grab is rated by two of these methods, and in some cases by all three. However, usually grabs are rated as follows:

1. By safe working load (weight or force): lifters, grapples, tongs, hooks, lifting clamps
2. By maximum opening (length): lifters, grapples, tongs, lifting clamps
3. By amount of materials per bite (volume): buckets

Most grabs are readily available up through 20,000 lb (9,100 kg) capacity. Many standard grapples and buckets can be procured with openings to 8 ft (2.4 m) and volume capacities of up to 6 yd^3 (5 m^3). For the most part, tongs are not stocked with openings larger than about 4 ft (1.2 m). Due to the fact that the design of a grab can be quite complex, the plant engineer would be well advised to search for an existing design or existing product to do the job before designing a grab.

Grab Selection. Once it has been determined that a grab may be needed for a particular application, then the necessary steps can be taken to decide the best type and size of grab that will be required. A possible series of steps to take in the selection of a grab are as follows:

1. Study the characteristics of the material to be handled.
2. Consider the type of crane to be used.
3. Determine the environmental and safety conditions.
4. Notice clearances and required lift.
5. Correlate grab duty cycle with rest of operation.
6. Estimate grab weight and consider the resulting payload.
7. Research and study standard grabs available to do the task.
8. Calculate cost and probable utility of custom-made grab (if necessary).
9. Using value analysis choose the best grab for the job.

When selecting grabs by using this nine-step method, some of the pertinent questions that may need to be asked are:

1. *Characteristics.* What is the size of the load? What is its density? Will a tong, grapple, or bucket handle the load? Is the material poisonous or radioactive? Can personnel be in the vicinity of the load for hooking or unhooking? Does the load consist of single pieces or multiple pieces? Is the load unit or bulk?

2. *Crane.* Is the crane already existing or will it need to be obtained also? What is the total crane capacity? Is the crane electric, hydraulic, mobile, overhead? What are the other limitations of the crane? What type of crane is normally used for this application?

3. *Environment.* What are the maximum and minimum expected operating temperatures? Are there dusts or other contaminants? Will the grab be subjected to or dipped into water or other corrosive solutions? Can personnel be allowed in the area to hook or unhook the material? How convenient will it be to maintain the grab? Are there other safety considerations?

4. *Clearances.* How high must the loads be lifted? What are the clearances between the grab and loads and/or obstructions?

5. *Duty Cycle.* (An important consideration is the fact that the grab will often be subjected to much greater wear and abuse than the rest of the machine.) What are the duty cycle requirements? What is the expected life of the crane? Will it be more economical to choose a grab that will last indefinitely or to choose one to be partially or totally replaced periodically?

6. *Weight.* What is the estimated weight of the tentatively chosen grab? Will the resulting payload be satisfactory? Can the crane handle both the grab weight and the required payload? What is the best way to connect the grab to the crane?

7. *Standard Grabs.* Are pre-designed grabs available from existing manufacturers? Will they meet the specifications developed in the earlier steps? Are there several types of grabs that will do the job? Which type(s) of grab is usually used for this application? How do any available standard grabs compare with each other so far as specifications and costs?

8. *Custom Grabs.* What factors necessitate designing a custom-made grab? What can be learned from similar standard grab design that will help in the design of the custom grab? What is the estimated cost of the design, manufacture, and maintenance of the custom-made grab?

9. *Final Selection.* Can the cost of the grab be economically justified? Will the grab chosen require unusual research and development or testing? Is the grab safe for personnel and equipment? If there is any doubt about the successful application of the grab to the job, have experts been consulted?

Grab Design. The design of most grabs entails the careful application of the sciences of mechanics and strength of materials as well as considerable laboratory and field testing in order to assure reliability and safety. Beam-type grabs (e.g., counterbalanced lifters and lifting beams) are often simple enough that useful devices can be designed without extensive engineering effort. Therefore, the following information is offered in order to aid in the design of this type of grab for a special application, especially for the cases where a manufactured product is not available.

Counterbalanced Lifters. Counterbalanced lifters are classified by type of construction as curved beam grabs. They are designed to allow the lifting of awkward or bulky items, such as coils and pallets, by an overhead crane. Normally this type of lifter is designed with a counterweight located so as to keep the device level when unloaded. This feature results in the advantages of quicker attachment to the load and better control of the grab compared to one without counterweight. These lifters are designed so that the centers of gravity of the lifter and of the load are in alignment with the hoisting lines.

Curved-beam theory is helpful in determining stress and deflection of the counterbalanced lifter. It is important when designing this type of lifter to minimize the effect of stress risers that can occur at the inside corners of the lifter. The overbalance hook (Fig. 10.4.10) is a special case lifter where the counterweight is greater than the amount needed to balance. This allows the hook to self-release a sling whenever the load is lowered enough to put slack in the lifting sling.

Lifting Beams. When lifting nonsymmetrical loads it is difficult to attach slings so that the center of gravity of the load is directly below the hook. One way to solve this problem is to use a lifting beam where the hook attaches closer to one end of the beam than the other, thereby lining up the center of gravity of the load with the hook (Fig. 10.4.14).

Lifting beams are a necessity for applications where a long load must be lifted by a single hook crane. In these situations, a load handled with bridle slings alone may have a tendency to slip and allow the load to slide out the end of the slings. A beam with two attachment points with sufficient spread allows level and safe handling (Fig. 10.4.15).

There are two other major reasons to use lifting beams:

1. To gain headroom
2. To reduce end damage to loads

Figure 10.4.16 illustrates a lifting beam designed to improve headroom. The lifting beam in Fig. 10.4.17 has been designed to eliminate edge damage on some delicate rolls. Occasionally lifting beams

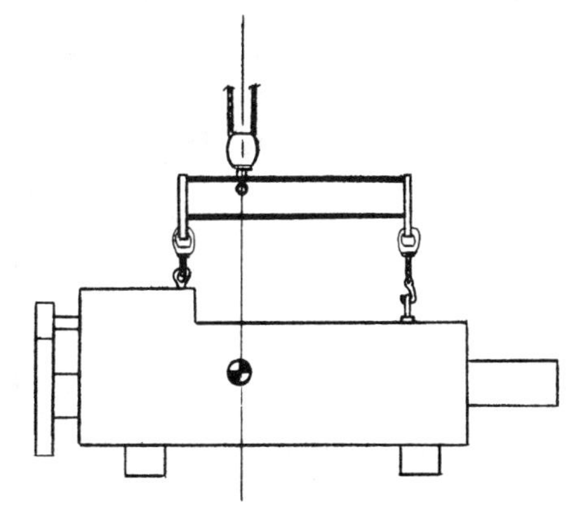

Fig. 10.4.14 Lifting beam for nonsymmetrical load.

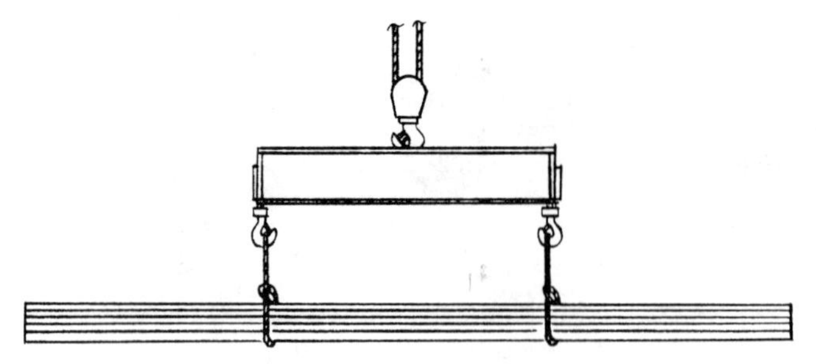

Fig. 10.4.15 Lifting beam for long load.

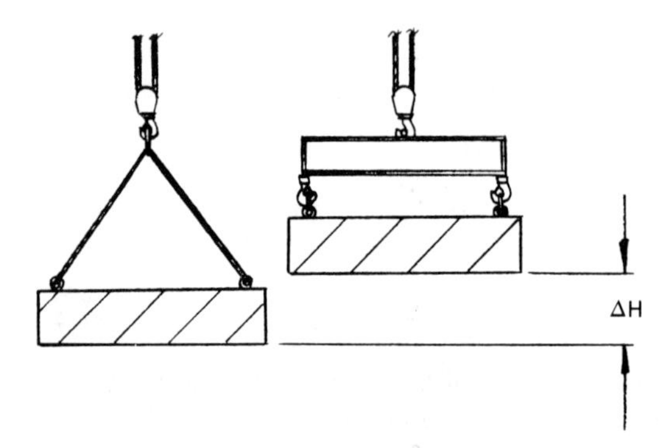

Fig. 10.4.16 Lifting beam used on right to gain headroom (ΔH).

are designed to allow a heavy single point lift to be done by two or more smaller-capacity cranes. This technique can be economical for situations where the heavy lift is required infrequently (Fig. 10.4.18). Another type of crane uses two hoists that are permanently reeved to a spreader beam which has sheaves in each end (Fig. 10.4.19).

Even though a variety of standard design spreader beams are available from manufacturers through 20-ton (18,000-kg) capacity, there are applications where a custom spreader beam must be designed. Spreader beams can be fabricated from American Standard W, S, and C shapes, or they can be welded up from bar or plate stock. Extremely long or high-capacity spreader beams are made of a truss design so as to provide adequate strength with minimum weight. Very efficient small- to medium-size spreader beams can be designed from readily available steel rectangular tubing. On simple spreader beams, bending stress is the parameter that normally determines the size of the section needed. A factor of safety of 5 or more over the published ultimate tensile strength of the material is common practice. On long spreader beams the deflection must be carefully calculated in order to prevent undue bending and whipping. Undoubtedly the most common cause of beam failure is web buckling which occurs due to inadequate lateral stability. For this reason all spreader beams must be designed with sufficient lateral section modulus to assure no web buckling when lifting its rated load. Proof load for a spreader beam is twice its rated capacity.

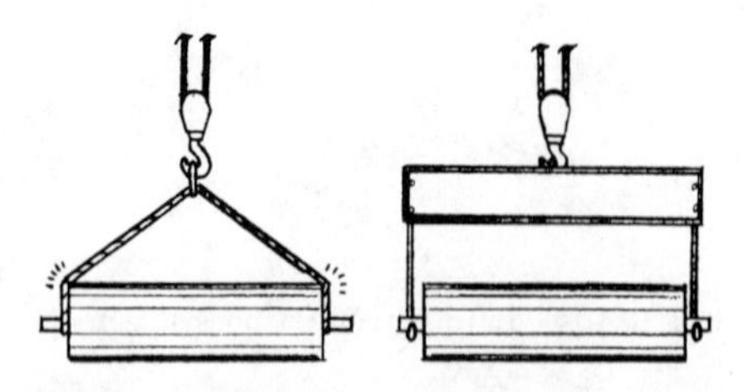

Fig. 10.4.17 Lifting beam used on right to eliminate end damage.

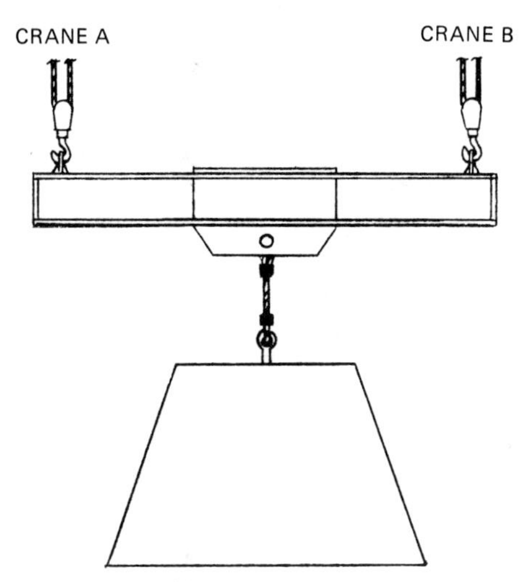

Fig. 10.4.18 Lifting beam for two separate cranes.

10.4.2 Magnetic Lifting Devices

Ferromagnetic iron and steel products can often be efficiently handled using magnets. For most applications DC-powered electromagnets are used, but battery-operated and permanent magnets are also available for some specific situations. For some applications significant savings in time can be realized using magnets instead of handling with slings. Magnets can be used to advantage on ferromagnetic materials that cannot tolerate nicks and gouges caused by slings or grabs. Most magnets can be easily fitted for remote control in order to more safely stack materials and to increase the speed of hooking and unhooking. The lifting capability of any lifting magnet is affected by the load surface condition, load length, load thickness, and load material alloy. Lifting magnets lose their magnetism at very high temperatures, so duty cycle is a very important consideration, especially for electromagnets.

Electromagnets

Higher-capacity lifting magnets are usually electromagnets. These magnets are designed to be suspended from a crane hook to handle large volumes of pig iron, scrap, turnings, punchings, steel slabs, steel plate, and so on.

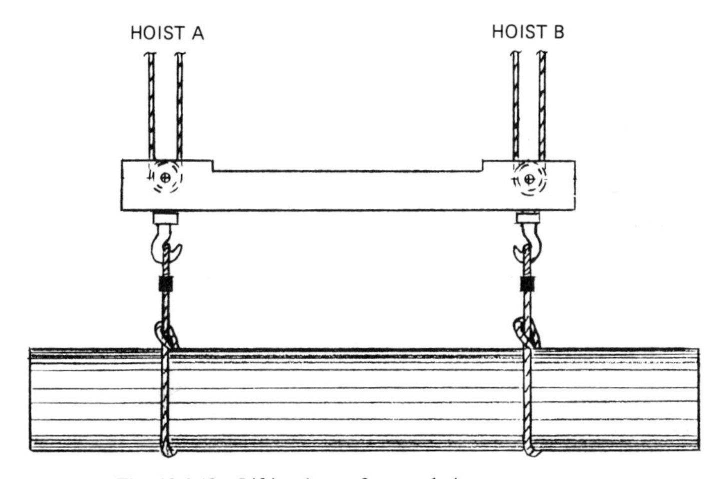

Fig. 10.4.19 Lifting beam for two hoists on one crane.

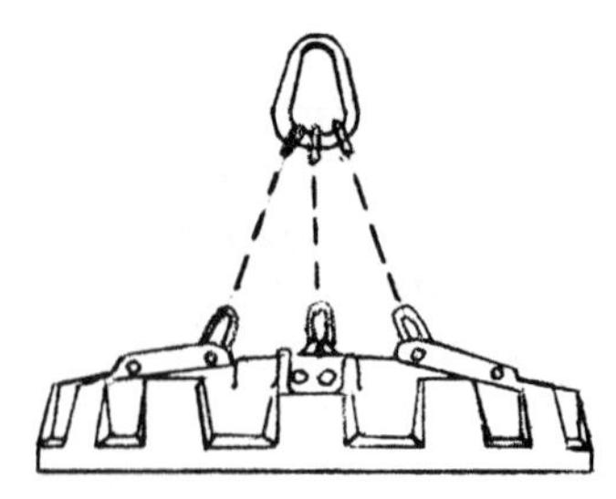

Fig. 10.4.20 Scrap-handling magnet.

Figure 10.4.20 depicts a typical circular scrap-handling magnet. Construction of these magnets consists of three legs of alloy steel chain connected to a pear link on their crane hook end and to the magnet case on their other end. Windings of either strap aluminum or copper are located inside the case which consists of a center pole and an outer pole. The insulated strap windings are used rather than insulated wire because a greater number of turns can be made in the same space. The bottom plate is made of a durable material such as manganese steel to stand up to the wear and abuse caused by contact with the load. A terminal box connects the leads from the crane DC power source to the magnet.

The most popular voltage used with large electromagnets is 230 v DC. A motor-generator set or a rectifier is mounted on the crane, and power is conducted from the crane by means of a cable reel. The cable reel pays a multiconductor cable in and out to the magnet as the crane hook is lifted or lowered. Even though steps can be taken to minimize the possibility of power failure to the magnet, LOADS MUST NEVER BE CARRIED OVER ANYONE. This rule is generally true for any hoisted load held by any means, but must be especially emphasized for electromagnets because the load will drop if magnet power ever fails.

The operator of a circular electromagnet has a control means at his disposal in the crane cab so he can turn the magnet on and off. The experienced operator knows to leave the power off until the magnet settles on the load in order to help keep the magnet cool and to increase the load that the magnet will pick up. The magnet should not be used as a battering ram and should be kept as dry as possible in order to keep maintenance to a minimum. Typical scrap-handling circular lifting magnet specifications and lifting capacities are shown in Tables 10.4.14 and 10.4.15.

Electromagnets are also produced that are powered by self-contained batteries instead of direct current from a motor-generator or a rectifier. The capacity of this type of electromagnet is limited by the power and weight of the battery which subtracts from the payload. Battery magnets have the advantage of no cords or wires from the crane and can, of course, be used on applications where there is no DC power available. Battery lifting magnets are available with built-in battery chargers.

Whereas circular-shaped electromagnets are popular for handling scrap, rectangular magnets are often more practical for handling materials such as beams, plates, rails, and pipe. On thin materials the limiting factor is often the deflection (rather than the weight) of the material which causes the load to "peel" away from the magnet. In these cases the solution may be to provide several magnets suspended from a spreader beam attached to the crane hook.

In order to improve the lifting power of electromagnets, bipolar rectangular magnets are made. These magnets have two widely spaced poles which cause the magnetic field to reach out farther. Bipolar magnets are used when only partial contact can be made with the load or when the magnet

Table 10.4.14 Scrap Handling Magnet Specifications

Magnet Diameter [in. (mm)]	Net Weight (lb)	Magnet Headroom [in. (mm)]	Cold Current Amps @ 230 v	Minimum Cable Size
30 (762)	1,100	30 (762)	17	#8
40 (1,020)	1,800	38 (965)	30	#8
48 (1,220)	2,900	39 (991)	55	#8
57 (1,450)	4,400	51 (1,300)	74	#6
66 (1,680)	6,150	52 (1,320)	89	#6
72 (1,830)	8,300	53 (1,350)	114	#4
87 (2,210)	12,500	60 (1,520)	158	#2
92 (2,340)	15,400	60 (1,520)	184	#2

Table 10.4.15 Scrap Magnet Average All-Day Lifting Capacity[a] (Not Maximum Load)

Magnet Diameter [in. (mm)]	Plate Punchings (lb)	#2 Heavy Melting (lb)	Pipe Scrap (lb)	Steel Turnings (lb)
30 (762)	650	550	300	175
40 (1,020)	1,100	900	650	400
48 (1,220)	2,200	1,750	1,000	600
57 (1,450)	3,100	2,700	1,600	850
66 (1,680)	5,200	4,100	2,400	1,300
72 (1,830)	6,250	4,700	3,300	1,500
87 (2,210)	9,250	6,800	5,400	2,600
92 (2,340)	11,350	8,400	6,600	3,000

[a] Lifting capacities are based on optimum conditions. Variables in the magnetic system or materials will affect performance.

must operate through an air gap due to scale, rough surface, or nonferrous material. A battery-powered bipolar electromagnet is illustrated in Fig. 10.4.21.

Permanent Magnets

Permanent lifting magnets consist of ceramic magnets enclosed in a protective steel housing. Standard units are available with capacities through 2000 lb (900 kg). Larger capacities can be procured on special order. With permanent magnets, the magnetic field is at full strength indefinitely and so is not dependent upon any external power source. This means that permanent lifting magnets, such as shown in Fig. 10.4.22, are ready to be used at any time and are very portable. As with battery-powered electromagnets, there are no electric cords to get in the way of the lifting operation.

Lifting devices made with permanent ceramic magnets are disengaged from the load by mechanical or electric means. The mechanically disengaged type of magnet usually requires an operator at the point of hook-up in order to turn the magnet "off." The electric method turns the magnet off by transferring the path of the permanent magnetic flux from the load to a movable internal bar which is activated by a control coil. The control coil is powered by a self-contained battery that can be operated with a manual switch or by remote push button. A great advantage of the permanent lifting magnets is that they do not lose their magnetism when there is a power failure to the crane. They are not susceptible to dropping the load due to any break in a magnet control wire or other electrical device and so do have an inherent safety advantage over other types of magnets used for lifting.

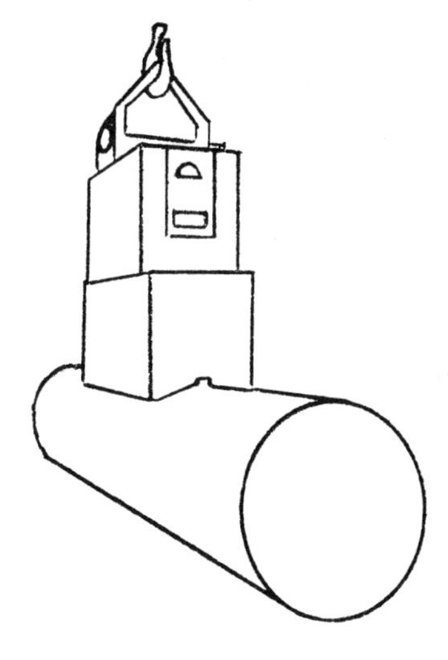

Fig. 10.4.21 Battery-powered lifting magnet.

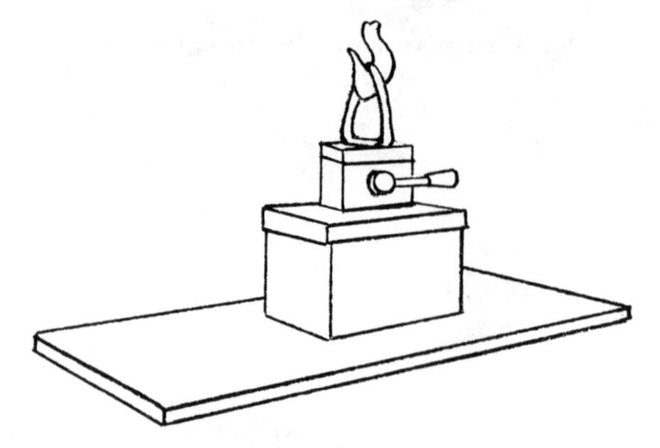

Fig. 10.4.22 Permanent lifting magnet.

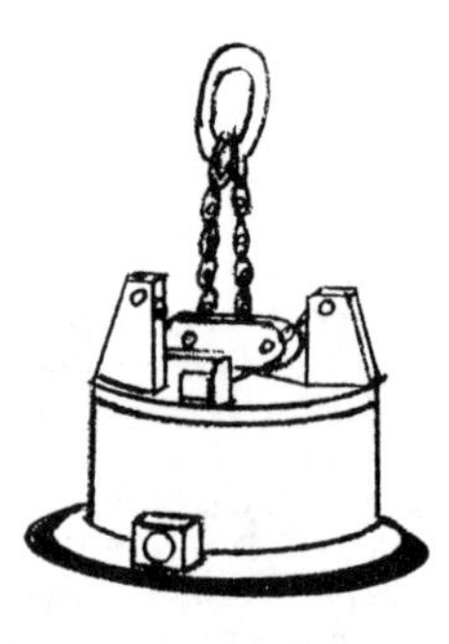

Fig. 10.4.23 Mechanical vacuum lifter.

10.4.3 Vacuum Lifting Devices

Materials such as sheet steel, aluminum, glass, plastic, and wood products can be effectively handled with vacuum lifting devices. Most standard vacuum lifters available are designed to work only on nonporous materials. These devices use the forces of atmospheric pressure to lift a load by evacuating a space (chamber) inside a vacuum pad located between the hook and the load. The vacuum pad or pads are commonly evacuated by means of a mechanical piston mechanism, a motorized electric vacuum pump, or a battery-operated vacuum pump.

The mechanical vacuum lifter in Fig. 10.4.23 is designed to be directly attached to the crane hook by the lifting bale. A flexible sealing ring, constructed of a material such as neoprene, seals out air from the vacuum chamber. These lifters are available with rated capacities up to 5000 lb (2300 kg) in a single unit. Larger loads can be handled by connecting multiple pads (which are evacuated

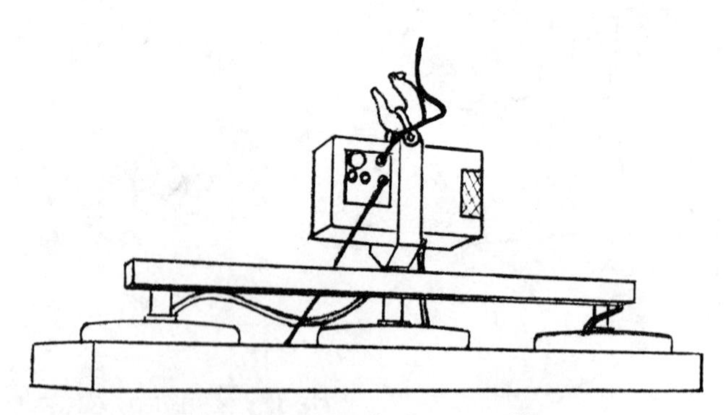

Fig. 10.4.24 Motorized vacuum lifter.

Table 10.4.16 Typical Vacuum Lifter Pad Configurations[a] (For Lifting Sheet and Plate)

Pad Configuration	Maximum Capacity (lb)	Maximum Plate Size [ft (m)]
Two pads in-line	4,800	6 × 24 (1.8 × 7.3)
Three pads in-line	3,000	6 × 30 (1.8 × 9.1)
Four pads in-line	8,000	6 × 40 (1.8 × 12.2)
Four pads on two crossarms	8,000	12 × 30 (3.6 × 9.1)
Eight pads on four crossarms	16,000	12 × 40 (3.6 × 12.2)

[a] Configurations, capacities, and plate sizes are representative only. Consult with the vacuum lifter supplier for recommendations for each individual application.

by a common single vacuum piston) to a spreader beam or frame. Some sort of a vacuum gauge is incorporated into the lifter to inform the operator whether the holding vacuum is sufficient to lift the load.

Motorized and battery-operated vacuum lifters work similarly to the mechanical lifters except that the vacuum chambers are evacuated by a motor and pump. Figure 10.4.24 illustrates a motorized lifter that has three vacuum pads in order to handle relatively long plates. By incorporating a vacuum reservoir into the power package of the lifter the lifting device can be made safer against pump or power failure. Battery-powered units have no electrical cords to the crane which gives them the same flexibility that a mechanical vacuum lifter offers.

The rigidity of the load is an important consideration when handling materials with vacuum lifters. Flexible sheet materials have the tendency to "peel" away from the vacuum pads, like they do with a magnet lifter. A sufficient number of vacuum pads, adequately spaced, will allow fairly large sheets to be safely and efficiently handled (see Table 10.4.16). As with any below-the-hook lifting device, safety considerations dictate that loads held by vacuum lifters must *never* be conveyed over personnel.

CHAPTER 11
BASIC STORAGE EQUIPMENT AND METHODS

HERBERT H. KLEIN

Unarco Materials Storage Division
Unarco Industries, Inc.
Chicago, Illinois

W. SCOTT FOWLER

Lyon Metal Products, Inc.
Aurora, Illinois

WILLIAM DEVANEY

Stanley-Vidmar
Allentown, Pennsylvania

ROBERT B. FOOTLIK

Footlik & Associates
Evanston, Illinois

TERRY STROMBECK

Spacesaver Corporation
Ft. Atkinson, Wisconsin

11.1 STORAGE RACKS, FLOW RACKS, AND STACKING FRAMES

Herbert H. Klein

11.1.1 Introduction

Industrial steel storage racks as they are known today did not exist 35 years ago. Railroad cars filled with material in cardboard cartons were hand unloaded onto four-wheeled platform trucks that were pushed to the elevator of a multistoried warehouse and the cartons hand-unloaded into stacks on floors with 10-ft-high ceilings. People used the same platform trucks to assemble orders from the various stacks, pushing the loaded truck to the elevator and then to the shipping dock where they were hand-unloaded into vehicles for store delivery. Such rudimentary distribution has been radically changed by some relatively simple innovations:

Forklift trucks have replaced man-powered vehicles and loading methods. They have been on the materials handling stage since the early 1900s and received their biggest impetus during World War II when forklifts were the logistical workhorses of the army and navy.

Unit loads of 2000–4000 lb that are easily picked up and transported by forklift trucks make highway trailer loading and unloading a matter of minutes compared to the many hours for the hand-unload method. The same unit load is moved within the warehouse by the forklift vehicles.

Industrial warehouse buildings with one-story ceiling heights of 30 ft and wide column bay spacings replaced the multistoried warehouses and are ideally suited for use with the unit load and forklift concept.

Industrial steel storage racks proved to be the fourth necessary materials handling device to maximize the efficiency of the previous three, because of "selectivity" and "carton crush-out." If one unit load is tiered on top of another up to the 30-ft ceiling height, the bottom load cannot be shipped unless the top ones are removed. If the unit loads are all different (colors, sizes, labeling, etc.) each is best removed without disturbing the others. Likewise if the material or containers of the assembled unit loads cannot be piled very high (or "crushes out") tiering to the ceiling is either not possible or dangerous. Rack structures provide the unit load separation that overcomes these two problems. The cost of racks can be as much as one-half of the money invested in the materials handling equipment, so their choice and application should be handled intelligently.

Computer technology has recently been added to all the foregoing to optimize their use and operation. Today's modern computers tell the distribution manager where to store the incoming unit loads, when and where to move to the picking face, the sequence in which to pick, and what and when to reorder.

All these components together have provided an efficient system that distributes the food, shelter, and clothing of our consumer-oriented society.

11.1.2 Unit Load Considerations

A unit load is a quantity of small items tiered on a platform resulting in a larger and heavier load. Unit loads are moved about on countless different types of platforms constructed of wood, steel, cardboard, plastic, paper, and combinations thereof. Sometimes a side-clamping device is used in place of forks on lift trucks so that no platform at all is required. However, the most common situation is on a wood pallet, as shown in Fig. 11.1.1. 252,000,000 such wood pallets were constructed by pallet manufacturers in the United States in 1981, and they consumed 14% of the nation's lumber supply for the year. About half of these were "expendable" or one-way shipping devices, the other half are substantially constructed to be reused many times and are the most used basic platform for unit loads entering and leaving the warehouse and shipping industries.

Because pallet load sizes determine the rack dimensions and member capacities, they are extremely important to rack suppliers. Table 11.1.1 shows pallet sizes and quantities manufactured over one year: note the first size indicating the vast predominance of the grocery marketplaces' 48 × 40 standard pallet in the industry. Referring to Fig. 11.1.1, note the following:

1. The *stringer* or stored depth must always be specified first. (A rack system designed for a 48 × 40 pallet as shown will not accommodate a load turned the other way.)
2. The pallet must be substantial enough to carry the load across the supporting structure. (An expendable shipping pallet not meant for pallet rack use may collapse on a conventional two-beam shelf.)
3. Pallet load must be secure. Odd-shaped bags or cartons or nonstaggered tier patterns may require gluing, wrapping, tying, or banding.

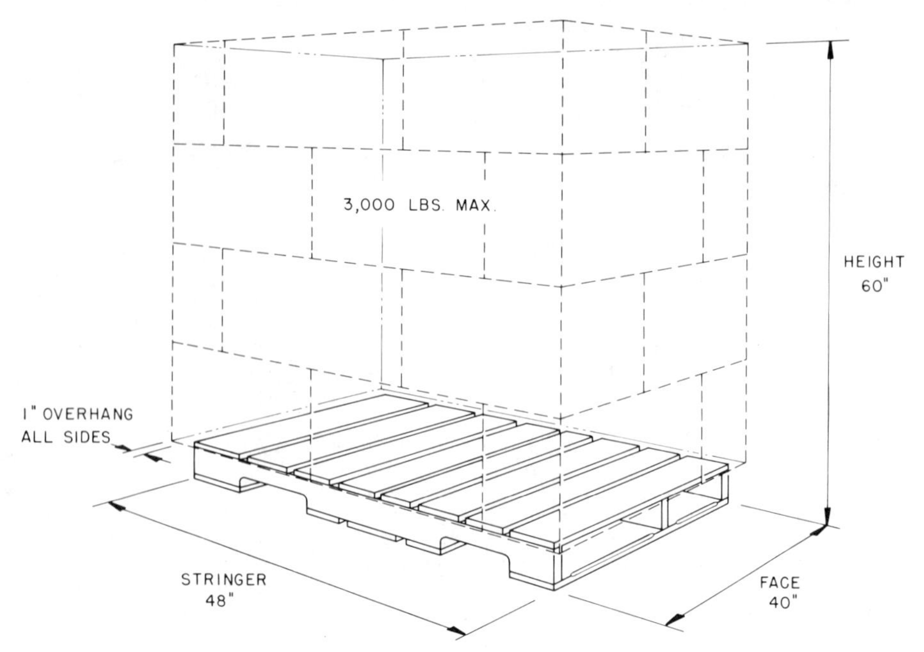

Fig. 11.1.1 Typical pallet load.

4. *Load overhangs* (if any) must be indicated; show maximums.
5. *Overall height* to include the pallet.
6. Load must be factual and include weight of pallet. Maximum loads are usually indicated, but if it is possible to segregate different weight classifications by location, product line, and so on, and this separation can be maintained, by all means do so.

11.1.3 Types of Racks

Selective Pallet Rack

The oldest and still most popular type, the selective pallet rack system utilizes horizontal beams connected to prefabricated upright frames to provide independent, multiple-level storage on either side of an aisle, with rows 15–30 bays long and connected to cross traffic aisles (Fig. 11.1.2). The selective pallet

Table 11.1.1 Pallet Sizes Produced in United States, 1977[a]

Pallet Size (in.)	Quantity Produced (Millions)	Percentage of Total Production
48 × 40	57	27.4%
48 × 48	10	4.8%
40 × 48	10	4.7%
42 × 42	6.5	3.2%
48 × 42	6	3.0%
36 × 36	3	1.7%
36 × 48	3	1.8%
40 × 32	3	1.4%
40 × 40	2	1.2%
44 × 56	2	1.0%
All other sizes		49.8%

[a] 236 million total pallets produced in 1977, 208 million accounted for in this study.

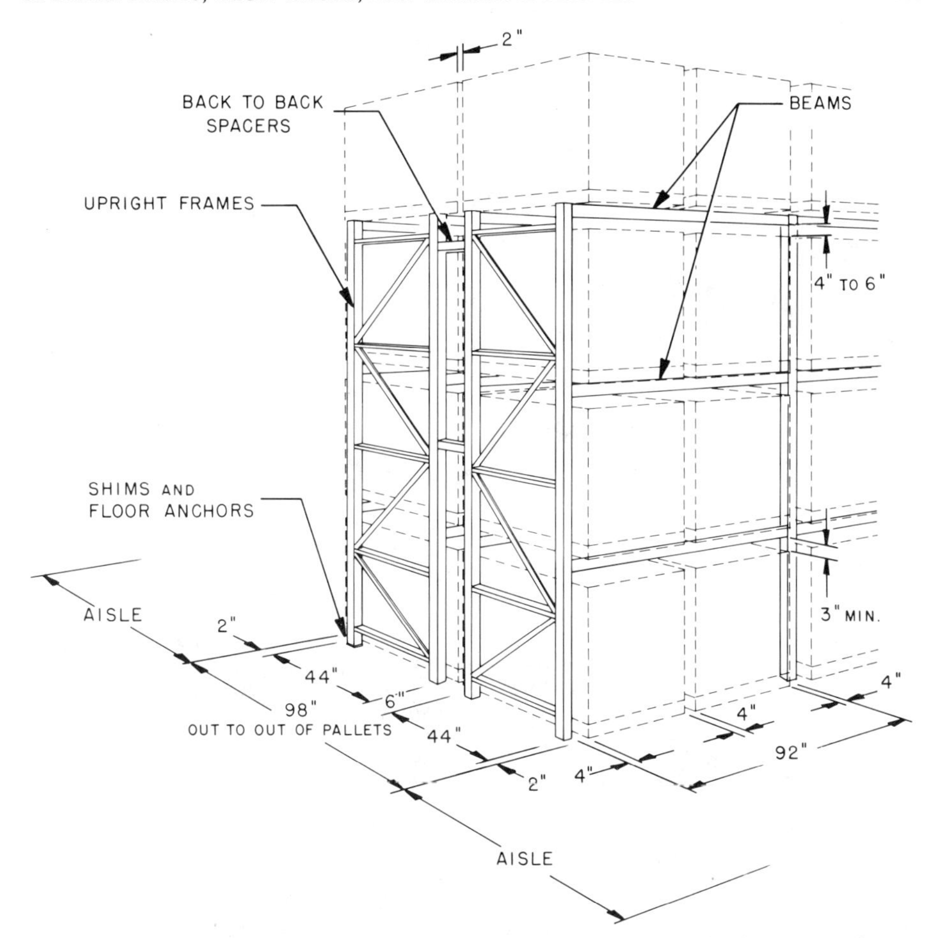

Fig. 11.1.2 Selective pallet rack. Dimensions shown are for 48 in. × 40 in. pallet loads, no overhang.

rack provides 100% access to every load stored, but is comparatively wasteful because of the high percentage of space devoted to aisles. Double rows are connected by back-to-back spacers (starter and ending single rows are similarly fastened to walls with wall ties), and shims and floor anchors are added to the upright frame posts where required. Aisle dimensions between rack rows are wide enough to accommodate right-angle turns by forklift trucks which deposit or remove pallets on two beams which constitute a shelf. Today's advanced forklift designs often rotate the load to either side of the aisle without turning, thereby minimizing aisle widths.

Beam length (bay dimension) normally accommodates two pallet loads with minimum 3-in. side clearances. Rapid operation or high rack structures increase this to 4 in. or even more. For the normal 40 in. wide pallet (no overhang), two pallets plus 3 in. side clearance spaces equals 89 in. Normal industry practice is to use 92 in. most often, but 90 in. and 94 in. are also used. Two pallets add up to 6000 lb total shelf load. This span length and shelf load are the variables entered onto load charts available from many rack manufacturers for fast selection. If pallet loads are lightweight and/or smaller dimensions, three pallets per bay may be a more economic solution. Conversely, large, heavy loads may best be stored in single-pallet bays.

Upright frame depths are usually a function of the stored pallet depth, allowing for a minimal 2 in. pallet extension front and rear as shown in Fig. 11.1.3. Upright height is totaled from loaded pallet heights and clearances and beam depths. Note that higher levels require larger vertical clearances to permit lift truck operators to lift pallets on and off the rack, to avoid dragging the pallet in and out. The required frame capacity is the total of the shelf capacities in one bay. Once again, manufacturers' load charts offer easy component selection based on the height and depth dimensions, the required capacity, and distance from the floor to the first shelf. Although end uprights receive only a half bay load, they are sized identical to interiors to provide interchangeability and reserve strength because of greater exposure to forklift impact.

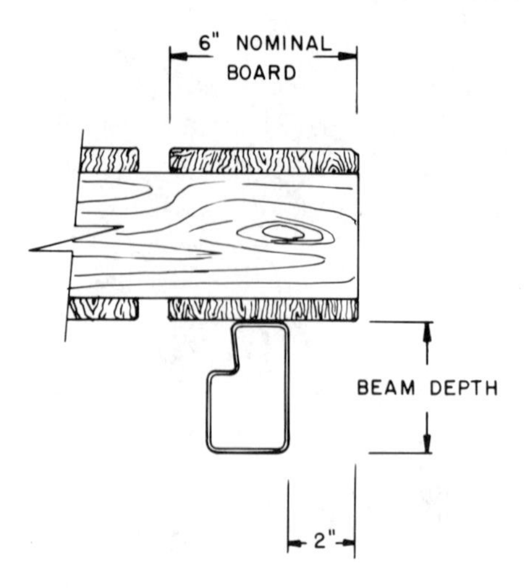

Fig. 11.1.3 Minimum pallet overhang.

Shelf heights are also a function of the manufacturers' adjustability increment. Some have bolted connections with hole patterns punched in the upright posts on 2-in., 3-in., or 4-in. centers. Others have various proprietary boltless connections that are adjustable on the same increments. This quick adjustment feature is used by some merchandisers to accommodate changes in inventory, product design, and so on. One auto parts distributor was able to add 20% more pallet positions to his system by rearranging his height clearances to accommodate his present product mix.

Back-to-back spacers should be used to help maintain rack alignment and increase rack stability. A minimum 2-in. space between back-to-back loads is recommended to keep loads from interlocking and allow latitude in placement (Fig. 11.1.3). If top of storage is more than 25 ft high, water sprinkler specification NFPA 231C requires a 6-in. flue between back-to-back loads. Building columns are often placed in this flue space, and a 12 in. building column dictates a 16 in. back-to-back space.

If in-rack sprinklers are necessary, upright posts must be anchored. Anchoring also helps maintain rack alignment and proper aisle width. (Much of the reluctance to have anchor holes remain in the floor when racks are removed should be overcome by the newer type of anchors that have hole size drilling requirements identical with their diameter. These small holes are easily patched with epoxy.) Floor unevenness or rack out-of-plumb conditions may require shims. If some posts of unloaded racks are free of the floor at time of installation, shims should be inserted under the upright bearing plate.

Front-to-rear supports under pallets (Fig. 11.1.4) must be used in shelves that span traffic aisles and that people pass under, and when pallets extend less than the 2 in. minimum shown in Fig. 11.1.3, or where shorter pallets are intermixed. They may also be used for additional safety on the upper levels of today's higher rack structures.

Checklist for Selective Pallet Rack

1. Aisle width is between pallets, not racks.
2. Aisle width should be sufficient for smooth, rapid operation.

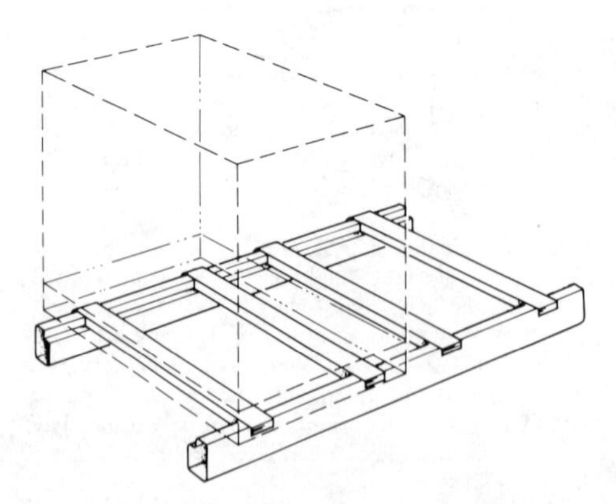

Fig. 11.1.4 Front-to-rear supports.

3. Try to bury building columns in back-to-back flue spacers, or in a pallet tier. Building column centers are often based on an optimum rack layout. Some suppliers have computer printouts that can supply this trial-and-error optimization.
4. Consult with your fire insurance carrier regarding ceiling sprinkler requirements, in-rack sprinklers, and the trade-offs involved.
5. 18-in. minimum clearances between ceiling sprinklers and top of top load.
6. If you have an unguided forklift operation, think about probable vehicle impact problems. Rated rack capacities from some manufacturers are based on static loading with no additional abuse factor other than the required structural design factors. If racks are subject to rough use, upright frames and perhaps beams should have additional reserve strength. The increased expenditure for this may be less in the long run than repairs or replacement.
7. With unguided vehicles, uprights should be protected at aisle corner intersections. Heavy pipe core drilled into the floor provides such protection. Most rack manufacturers also have heavy-duty wraparound anchored impact protectors that may suffice. Roadway guard rail type barriers at the cross aisle ends have been used effectively.
8. If outrigger style forklifts are used, uprights should have impact protection at the bottom front post, anchored to the floor. Various types of other impact protection devices should be considered, depending on type of vehicles used.
9. If pedestrian traffic enters the rack area, vertical and/or horizontal wire mesh barriers should be considered.
10. Make sure forklift can extend high enough for pallet to clear top shelf.

Two-deep Rack

Another variation of the standard selective rack called two-deep rack (Fig. 11.1.5) has become quite popular in the past decade. By moving two rows of double-row rack together and servicing this two-

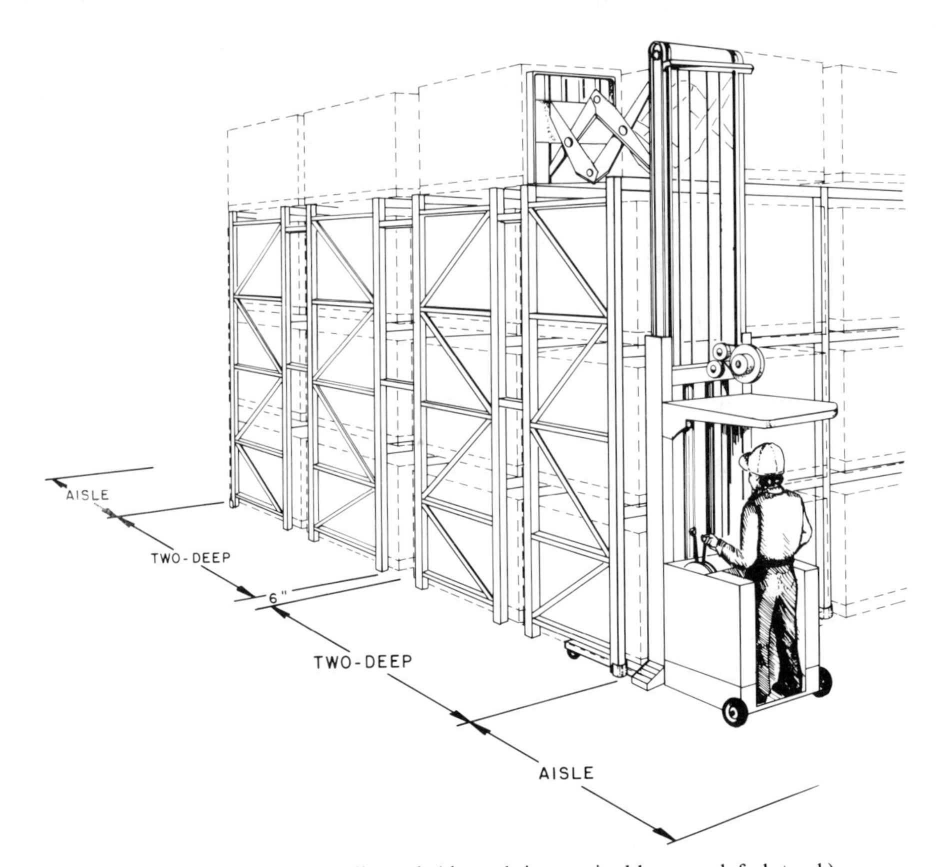

Fig. 11.1.5 Two-deep pallet rack (shown being serviced by a reach-fork truck).

deep system with a pantagraph reach–forklift truck, the number of aisles in a distribution center are cut in half. Aisle size can be further reduced by guided aisle lift equipment with a movable mast that eliminates the need for right-angle turns. The trade-off of this greater load density is a decrease in load selectivity; the rear tier of pallet loads is shaded by the pallets in front, and a front load must be removed to allow access to the rear pallet. This type of storage has gained wide acceptance, and a partial industry user list would include paper products, book printing, auto parts distribution, plastic molding, and public warehousing.

Components and design for two-deep rack can be identical with the selective system referenced previously, with one addition. In installations where the top shelf height is greater than 16 ft, it is advisable to have a continuous front-to-back support two pallets deep (two under the pallets, as in Fig. 11.1.4). This prevents a pallet getting hung up on the shelf beams when engaging the rear pallet, which the operator cannot visually observe at the higher levels. Often a height-reader system working off the mast is used to help solve this problem. Uprights are all interconnected by back-to-back spacers; the longer ones are used at the center to provide the 6-in. flue space usually required for sprinkler protection.

Selective Pallet Rack Variations

Selective pallet rack is adapted to other types of storage by using a variety of front-to-rear support members. Wood skids, steel skid boxes, or wire containers (Fig. 11.1.6) are accommodated by using front-to-rear skid channels.

Some industries store product on a piece of plywood (also called a slave pallet) which necessitates raising it above the shelf beams to allow fork entry, hence the name fork entry supports (Fig. 11.1.7).

Steel coils weighing 20,000 lb each are stored in heavy-duty structural steel pallet rack by placing a drop-in coil support on each shelf (Fig. 11.1.8*A*). Much lighter 55-gallon drums weighing but a fraction of this are placed four or more per shelf, each in its lighter-constructed coil support (Fig. 11.1.8*B*).

Carpet storage (Fig. 11.1.9) is yet another variation that permits individual selectivity of 12-ft-long carpet rolls weighing up to 1500 lb. Two 48-in.-deep rows of selective rack are separated by a 48-in. back-to-back spacer, and a four-beam shelf is decked over with wood or steel. A special long, tapered pole replaces the forks on the lift truck, and this "spears" the cardboard cores of the rolls and makes right-angle turns in 25-ft aisles to deposit rolls in carpet racks. This large aisle size has been reduced by skewing the racks 45° to the aisle, so now only 45° turns (and thus narrower aisles) are required. The resultant space losses at row ends and the one-way traffic pattern tend to cancel out the aisle width economies.

If the shelves in single rows of standard selective pallet rack are decked, they become ideal storage shelves for hand-loaded cartoned merchandise as used in the stay-pressed clothing industry, shoe manufacturing, and others (Fig. 11.1.10). Because a person places the cartons on and/or off the shelves, such systems are generally called hand stack storage. A guided aisle man-aboard platform picking

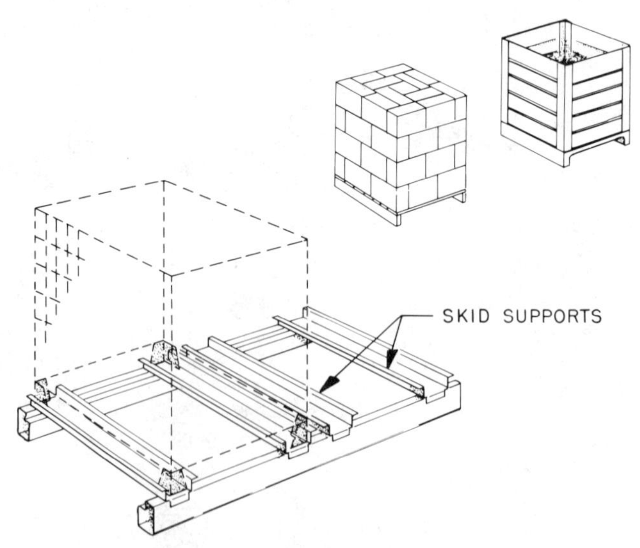

Fig. 11.1.6 Skid supports (no pallet overhang).

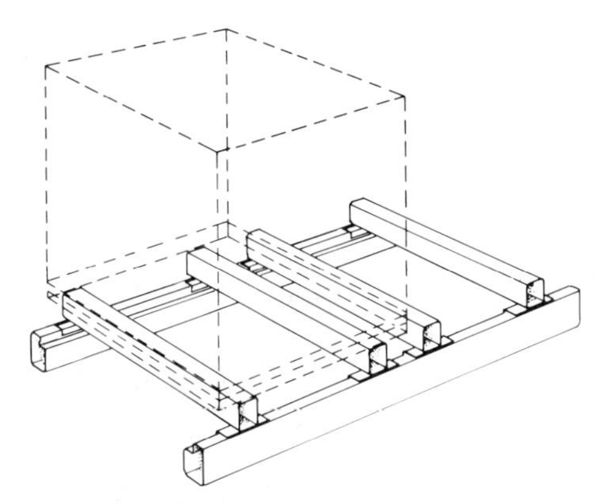

Fig. 11.1.7 Fork entry supports (no pallet overhang).

truck capable of servicing rack heights to 30 ft is often microprocessor controlled to automatically transport the order selector to the proper opening and even instruct the selector how many of what to select for whom. Steel decking can be corrugated roof type or slotted-steel deck planking which permits water penetration. Particle board with front-to-back supports underneath is the most common wood type of deck, with plywood also being used. Welded wire panels with integral front-to-back supports also provide water and light penetration, allow air movement, and prevent dust accumulation. All of these decks can be flush fitting with the beams, because of the step-ledge beam configuration as supplied by most manufacturers. These are available usually to fit $1\frac{1}{2}$ in. deep decking (and sometimes $\frac{3}{4}$ in. deep decking), as shown in Fig. 11.1.11.

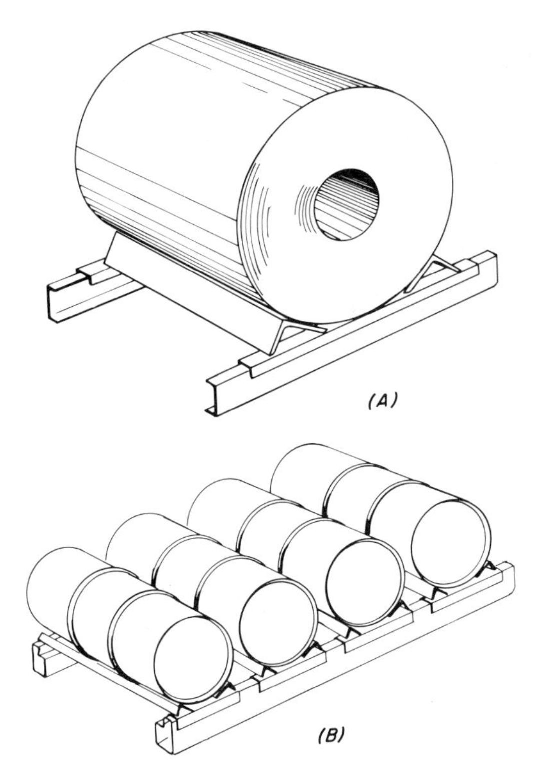

Fig. 11.1.8 (A), heavy-duty coil supports. (B), light-duty coil supports.

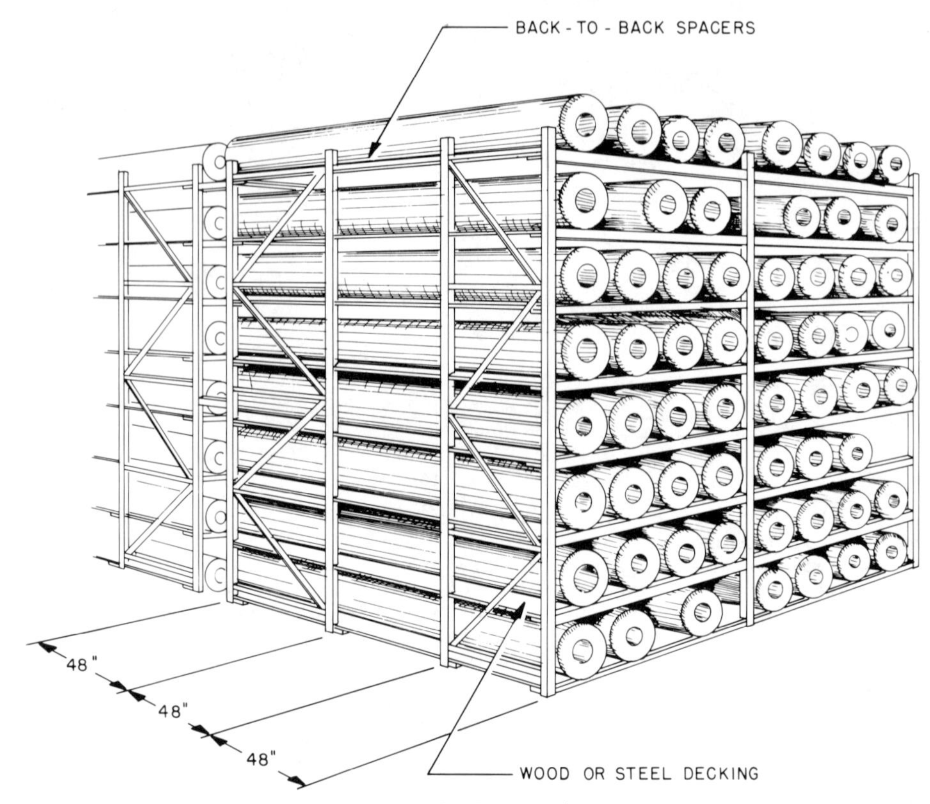

Fig. 11.1.9 Carpet rack.

A current requirement for relatively low (7–10 ft high) hand stack selective storage racks is for a new concept of discount food supermarkets generally called box stores because the shopper selects directly out of the opened carton. Pallet reserve storage can be on the top shelf, above the hand stack shelves (Fig. 11.1.12). Depending on the amenities involved (individual price marking, bring your own sacks, etc.), savings to the shopper can be substantial.

Drive-in, Drive-through Rack

Where palletized or unitized storage of medium or large quantities of like product is required, drive-in or drive-through rack systems should be considered. In this type of storage, when the forklift and

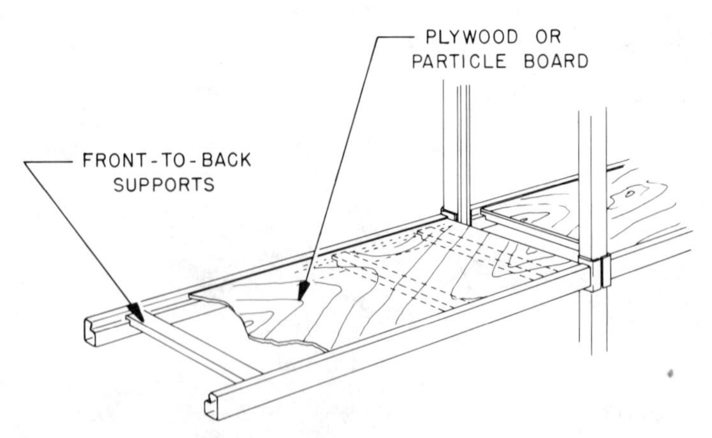

Fig. 11.1.10 Hand stack storage.

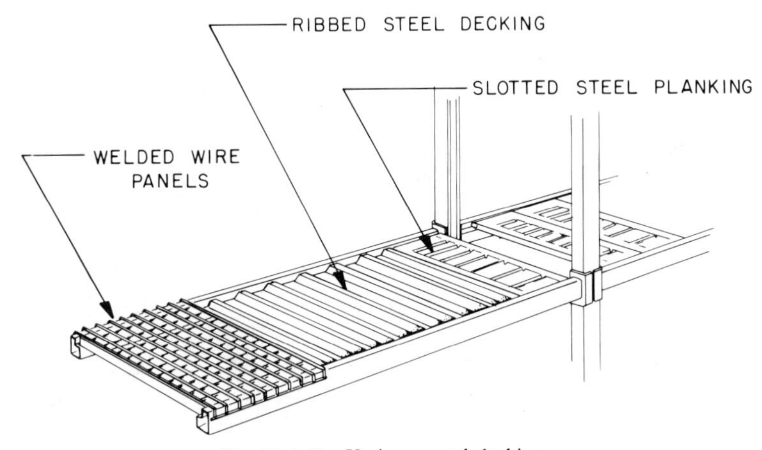

Fig. 11.1.11 Various metal decking.

loaded pallet makes its final right-angle turn it addresses a tunnel opening into which it can drive, with the pallet already elevated to the level it will be stored. The vehicle proceeds into the tunnel until it can travel no further or reaches the end of the tunnel, deposits the loaded pallet on rails that extend out from the sides of the tunnel, and backs out. The rails cantilever out on arms from upright frames that form the sides of the tunnel and are connected across the top by top ties. Spacers may be used between the uprights, and the uprights must be anchored to the floor. Some type of top fix must usually be supplied for drive-throughs (tie to wall, ceiling, building columns, adjacent rack structure, etc.). When the tunnels are unloaded from the opposite end of the tunnel (accomplishing FIFO inventory rotation) the rack system is a drive-through (Fig. 11.1.13).

When the rack system is against a wall or has *beams* across at the rail levels at the rearmost pallet positions, the system is a drive-in (Fig. 11.1.14). With the drive-in configuration, anchor section frames do not have to be overhead tied, because the beams across at the rail levels provide the necessary stability. If the rails, beams, and anchor section uprights are all securely fastened together, the top ties on a three-deep drive-in may be omitted (see Fig. 11.1.14). Beyond three-deep, additional drive-in depth is usually treated like drive-through construction.

Drive-ins do not permit FIFO inventory rotation, because the tunnel is open at one end only, and thus would seem to be less desirable when compared to drive-throughs. Practice has shown otherwise, mainly because in the everyday use of a drive-through rack system, if tunnels are completely emptied

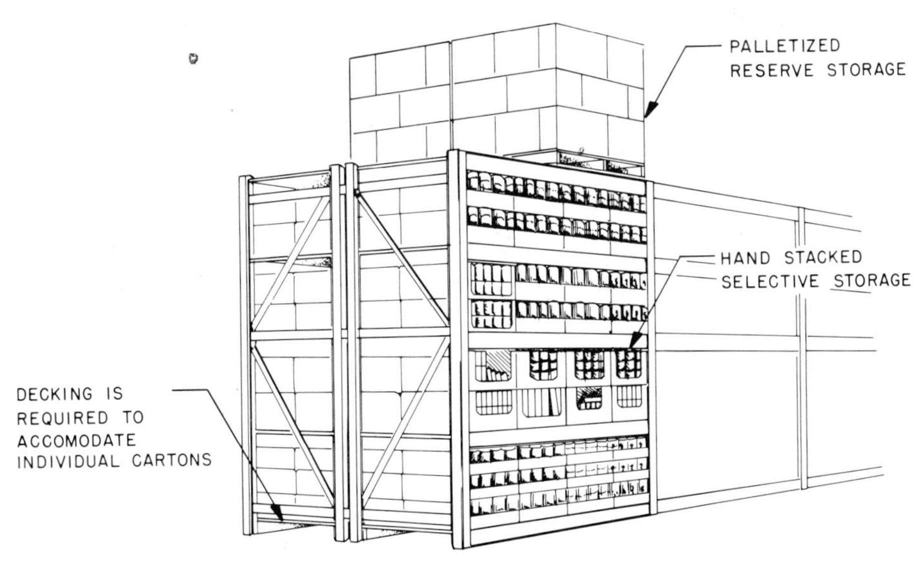

Fig. 11.1.12 Selective storage rack for retail application.

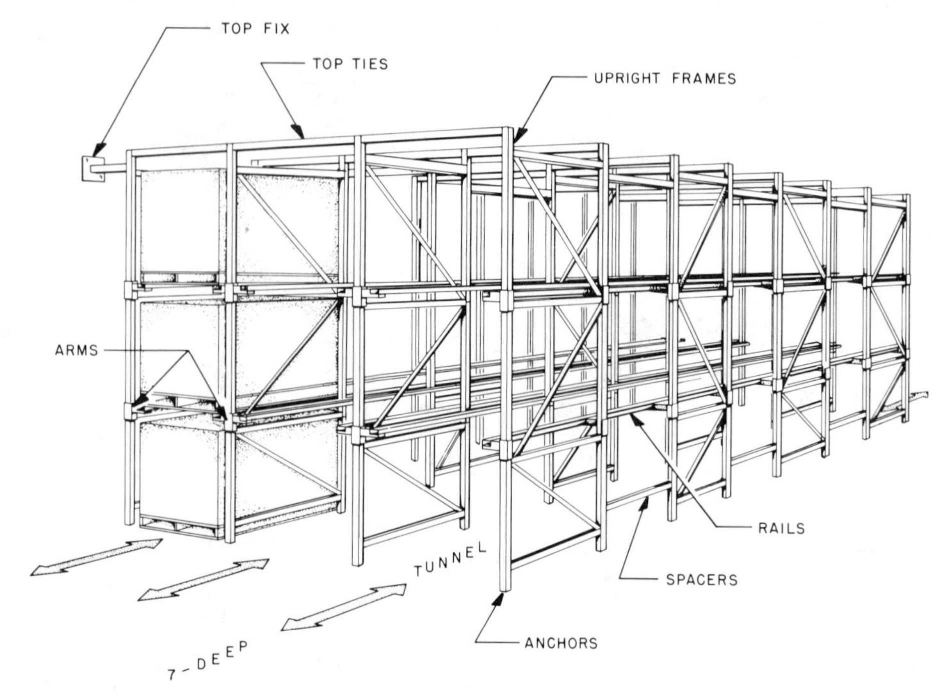

Fig. 11.1.13 Drive-through rack (seven-deep shown).

prior to restocking, the system is, on the average, half full (Fig. 11.1.15*A*) which in most cases is too great a penalty to pay for 100% inventory rotation. (Where this 100% FIFO is important, as with perishable grocery produce like lettuce, drive-throughs are used.) A more economical, stable, and utilitarian system can be supplied with shallow drive-ins (Fig. 11.1.15*B*). Where quantities are in the medium range, space savings important, and inventory rotation given middle priority, drive-ins two- to seven-deep are used. Some examples are bagged pet food, prepared meats, frozen foods,

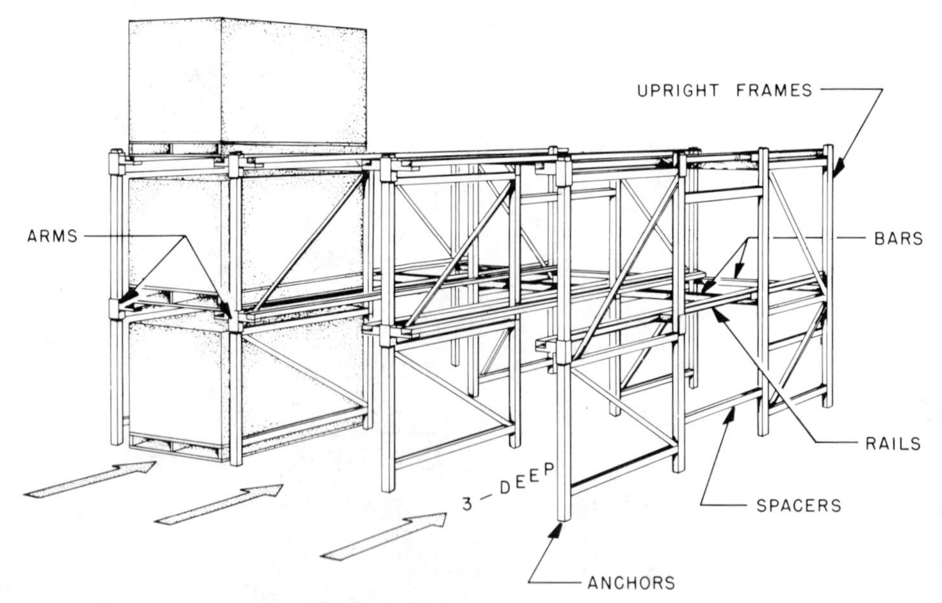

Fig. 11.1.14 Drive-in rack (three-deep shown).

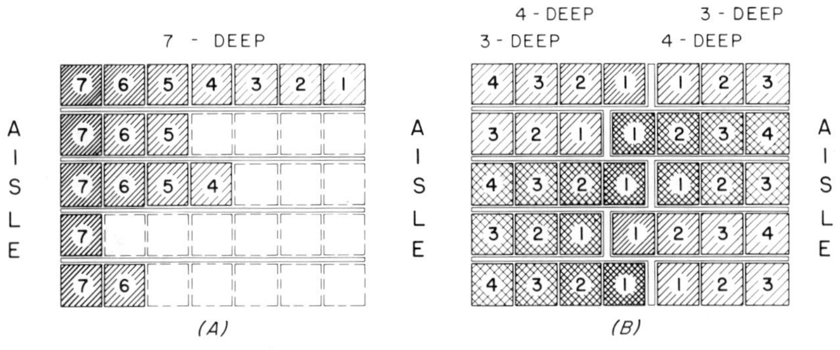

Fig. 11.1.15 (A), half-full drive-throughs. (B), shallow drive-ins.

light bulbs, soap products, paper products, in-process items, and electrical products. Much deeper drive-ins are used when in and out movement is slow, perhaps as long as once per year. Some examples would be apples, which fill drive-ins in cooler buildings 10–20 pallets in depth when near ripe, and are slowly emptied over the remainder of the year. Palletized candy and frozen turkeys are the opposite—deep drive-ins are slowly filled all year long and quickly emptied once or twice per year. U.S. government dairy subsidy policies have filled deep drive-in racked freezers with cheese and butter. Palletized tobacco is cured and stored in 13-deep drive-ins. TV picture tubes are stored in deep drive-ins, allowing an inventory accumulation to permit economical runs when switching sizes on a production line.

Because of the additional steel and greater complexity in construction and fabrication, drive-in and drive-through systems cost approximately 40–60% more than standard selective pallet rack, based on a cost per pallet stored. They also require more forklift time per load moved, because of the slow travel speed when within the rack tunnels. The space savings because of aisle elimination overcomes these drawbacks, and when the quantities are sufficient, drive-in presents an economic solution to materials handling as witnessed by their steady growth pattern over recent years.

Rack design is best left to manufacturers' representatives who should be supplied with sufficient information. This would include loaded pallet data, area available, forklift dimensions, depth of tunnels, and any other pertinent information. One test that the customer can perform is to determine if the loaded pallets can span the load rails without too much sag or deflection. Fig. 11.1.16 shows wood blocking resting on the floor, simulating the position of the rails. A typical maximum loaded pallet resting in its rack position overnight can easily verify the pallets' ability to span the rails.

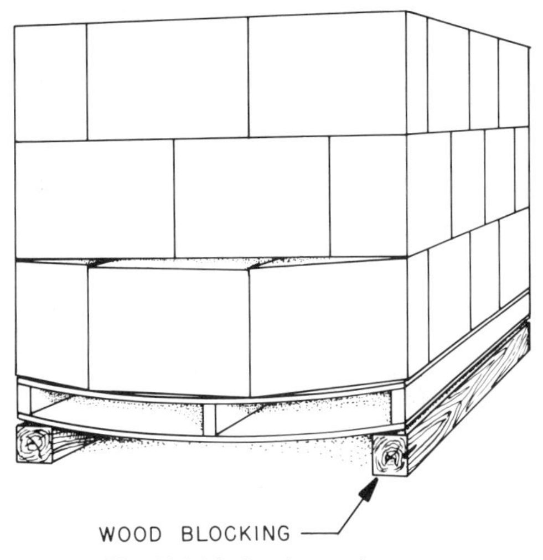

Fig. 11.1.16 Load spanning test.

Cantilever Rack

This is a relatively special purpose rack usually reserved for long, narrow items not easily stored on the post-and-beam type of construction used in selective pallet rack. To store 3000-lb pallet loads on a cantilever system would approximately double the amount of steel required, hence cantilever is not used for consistent, cube-sized pallet loads. On the other hand, steel service centers have hundreds of trays containing various grades, sizes, and lengths of steel bars that must be readily available (100% selectivity) and cantilever rack has proved to be the most practical method of storage.

Cantilever rack components are quite simple (Fig. 11.1.17). All the vertical posts (four, in the case of double-row selective rack) become one central column, moved to the center flue space. From this, a cantilever arm extends out to each side to receive the load, either directly if the columns are spaced close enough (as is the case for bar stock trays) or indirectly from a load shelf, as in the case of furniture storage. Columns are tied together with bracing.

Most cantilever racks are freestanding, and the column is supported by a heavy duty base. Occasionally in lighter-duty cantilever systems it is advantageous to omit the bases and stabilize the columns by cross-aisle ties and floor anchoring (this frees up floor storage space—sometimes good for furniture storage). It is also possible to use a combination of bases and cross-aisle ties. Infrequently both bases and cross-aisle ties are omitted and the columns are fixed at the top to the building roof steel (with the building owner's permission).

Rack manufacturers have devised various types of adjustable connections, some similar to those used in selective pallet rack, particularly in the lighter-duty varieties. Heavier-duty models depend

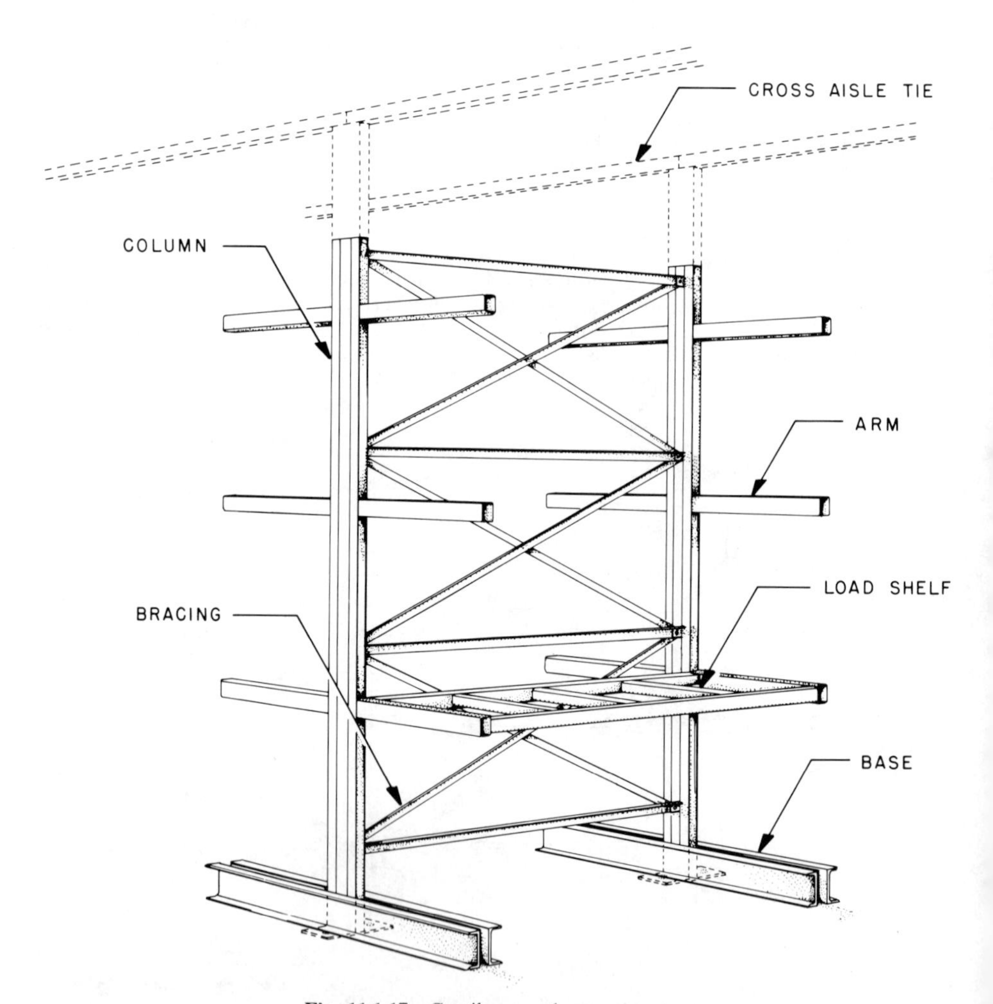

Fig. 11.1.17 Cantilever rack components.

on bolts, pins, or clamp fits. Sometimes when adjustability is not important, arms are welded directly to the columns.

Cantilever rack varieties can be broken down as follows:

1. *Light Duty.* For box goods, shirts, blue jeans, and so on, with a load capacity of 20–30 lb per square foot. Arms are usually 2 ft long and columns are 4 or 5 ft on centers. Particle board spans the arms, and long unobstructed shelves result which are ideal for a person to walk by and select from. With overhead cross-aisle ties and intermediate walkway levels, this configuration makes an excellent two- or three-level mezzanine system. A variation omits the particle board and substitutes long runs of pipe fastened to the arms permitting hanging goods (men's suits, jackets, coats, etc.) to be stored. One such installation has 14 mi of pipe for such storage. The light loading and short arms for this type of decked and hanging goods cantilever system is competitive to hand stack rack discussed previously.

2. *Medium Duty.* Covers a range of 30–50 lb per square foot and has seen wide use in furniture warehouse–showrooms. The long unobstructed decked shelves provide ideal storage for sofas and other pieces of varying lengths. Decking material is normally particle board with plywood and occasionally steel deck planks also being used. The space between bases is also decked 10–12 in. from the floor to accommodate the platform height of the man-aboard guided aisle lift truck from which a person moves furniture on or off the rack shelves (the person wears a tether harness to prevent any fall mishaps). Decking is supported by load shelves, and columns are on 8–12-ft centers (columns may or may not be overhead tied). One advantage to overhead ties is limited column deflection, which, without these ties, can be several inches into the aisle on each side at the 25-ft-height level to the top shelf. (Design practice dictates this critical one-side loading condition, for capacity as well as deflection.) Some manufacturers fabricate tapered columns which minimize this deflection problem.

The furniture warehouse–showroom concept has the prospective customer enter the complex by walking through the central cross aisle of the warehouse area, between many rows of furniture resting on 25-ft top shelf cantilever racks. He then enters the showroom area, where 100 room settings tastefully display items all the way down to the pictures on the wall. In the time required to have a salesman write the order and arrange for payment, a stockman is able to retrieve and set at the loading dock the items purchased, including that sofa for the customer to take home, or they can be delivered at a nominal charge. This replaces the old way of choosing a furniture piece, color, and fabric and waiting 10 weeks for delivery from the factory.

3. *Heavy-Duty Cantilever.* Covers a range from 50 lb to 150 lb per square foot, and includes steel bar stock, lumber storage of bundles of plywood, particle board, and other wood products. Once again, varying long, narrow loads are not adaptable to selective pallet racks and bare-arm cantilever provides the most advantageous method of storage. Forklift equipment can be a four-directional vehicle that operates in narrow aisles usually with a guidance system fastened to the rack bases.

4. *Very-heavy Duty.* A few rack manufacturers service a limited need for very heavy loads such as steel plate storage on 48-in. or 60-in.-long arms with capacities to 6000 lb per arm.

5. A specialty item is two cantilever upright standards with adjustable 12-in. or 18-in.-long arms to store miscellaneous tubing, pipe, rods, and so on.

Some manufacturers supply load tables facilitating cantilever component selection. Others advise representative contact, particularly when used in conjunction with mezzanines and picking systems.

Portable Racks

Portable racks enable the user to stack material, usually in pallet-size loads, on top of one another thus making more efficient use of floor and air space during production, warehousing, and shipping. A wide variety of portable rack designs are available because each manufacturer tends to have his own design. The two general categories are pallet stacking frames and unitized portable racks.

Pallet Stacking Frames. These consist of four posts and top horizontal framework constructed of steel which attaches to a wood pallet permitting multiple stacking of pallet loads one on top of the other (Fig. 11.1.18). When not in use, pallet stacking frames can be quickly disassembled and stored in a minimum of space. The automotive tire industry has huge warehouses that depend on these devices. The U.S. Post Office uses them, and large, light, bulky items such as rolls of insulation, steel tubing frames for shopping carts, and even nested aluminum boats (stored outdoors) are some typical uses. Design and stacking tests for determining capacity are available from the Rack Manufacturers Institute (RMI).

Unitized Portable Racks. These serve the same function as pallet stacking frames but have a steel base framework (sometimes decked with wood or steel) and may or may not have removable posts (Fig. 11.1.19). The steel base provides greater rigidity in stacking, compared to the wood pallet

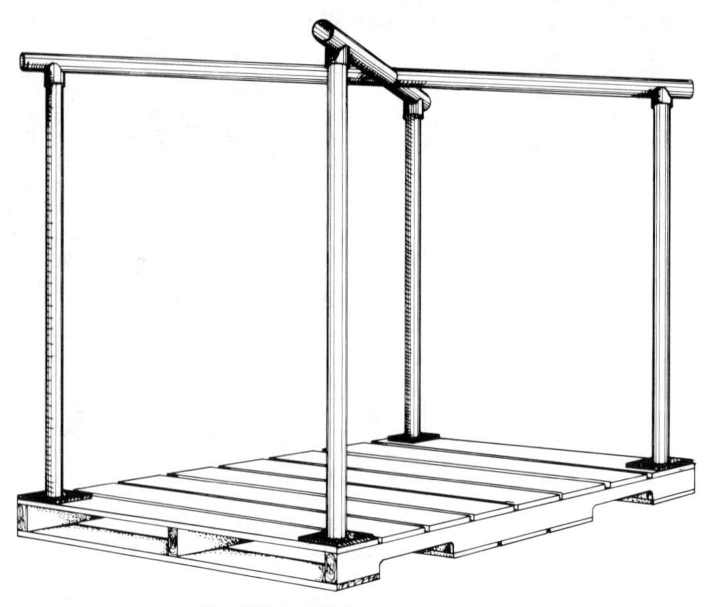

Fig. 11.1.18 Pallet stacking frames.

used in the pallet stacking frame concept, plus the ability for units to interlock when stacking one on top of the other. They are often used as intraplant shipping containers for large steel stampings (barbecue kettles, car fenders) and can have added partitions to accommodate odd shapes. A large retail catalog merchandiser uses them as a transportation and storage aid that is loaded with cartons at the receiving dock, placed on a towveyor cart, and moved several blocks to a storage area spur where they are forklift stacked four high. Once again, the RMI has a specification that covers load capacity testing and load posting. It also cautions not to transport more than one unit load at a time with fork trucks.

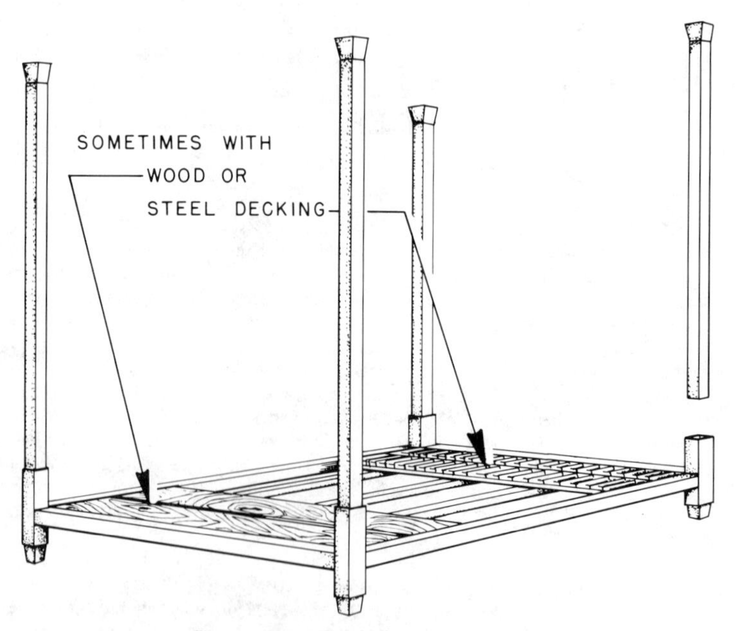

Fig. 11.1.19 Unitized portable racks.

Flow Racks

This series of rack structures accomplishes first in, first out inventory rotation (FIFO), minimal space and labor requirements, easy access to product and convenient restocking, and a host of other advantages. The two major categories are based on size: pallet flow racks operate with large unit loads in the 1000–3000-lb range and are handled by forklift equipment; package flow racks operate with hand-loaded cartons or packages of about 25 lb.

Pallet flow racks have two basic divisions depending on the number of pallets in depth. The shallow type usually consists of a picking module about 30 bays long with a cross section as shown in Fig. 11.1.20. The rack structure is very similar to the two-deep rack described in the selective pallet rack section, but in place of continuous front-to-back pallet supports, two or three flow track wheel conveyor sections are placed under each lane of two pallets (Fig. 11.1.21) and mounted at a slight incline to provide a gravity assist when the rear pallet is moved to the forward picking position. Loads are arranged two levels high at each picking level, and mezzanine supports are added to take advantage of the warehouse ceiling height, resulting in three walkway levels on each side of the take-away belt conveyors mounted in the center of the picking aisle. An order picker walking up and down the aisle next to the belt conveyor selects cartons off the front-loaded pallets and places them on the moving belt. The order picker's efficiency is greatly improved because of a much shorter walking distance, quick take-away, and constant reserve replenishment. This can be further improved by batch picking a number of orders at one time as generated by a computer printout, including a stick-on label for each carton.

When the front pallet load is emptied, the pallet is removed by hand and carried to a return pallet lane where it is stacked and then returned to the forklift aisle. The full pallet is then indexed forward (with human assistance if the gravity pitch is not sufficient to start the pallet load forward) and the forklift operator can replace the indexed pallet when it fits into his schedule.

Modern grocery distribution centers can have anywhere from three to nine of such modules, and the conveyor belts are merged above the picking modules, the batch picked cartons sorted automatically and sent to the proper truck dock via conveyor. Drug store chains use this concept, as well as a fast-growing merchandising house we see in many shopping centers.

The deep type of gravity pallet flow rack is usually used for full pallet shipments, and can be a solid block of storage perhaps 12 pallets wide by 4 pallets high by 5 to 12 pallets deep. To assure

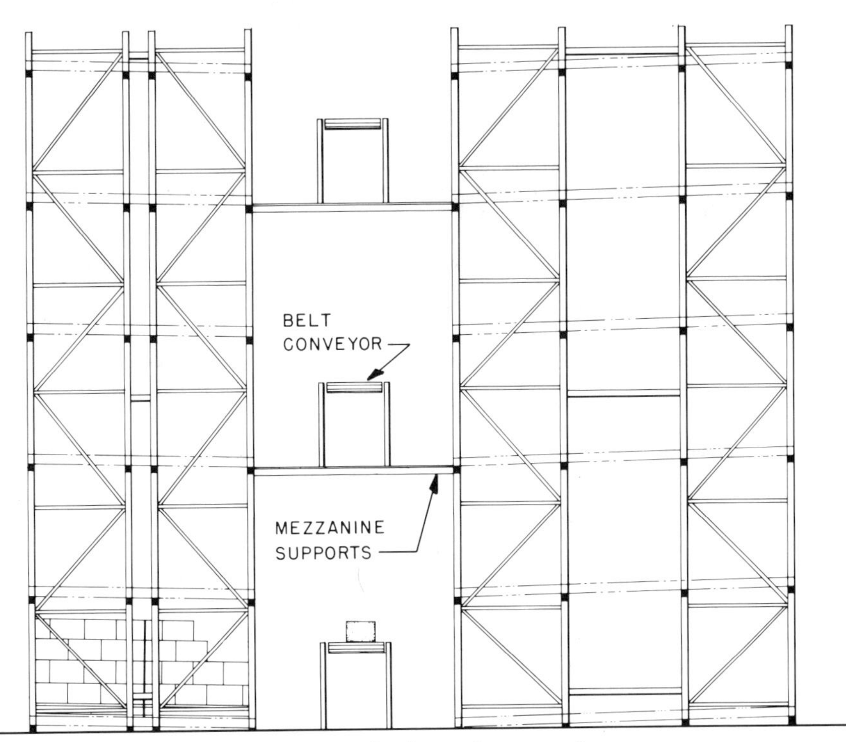

Fig. 11.1.20 Pallet flow rack cross section.

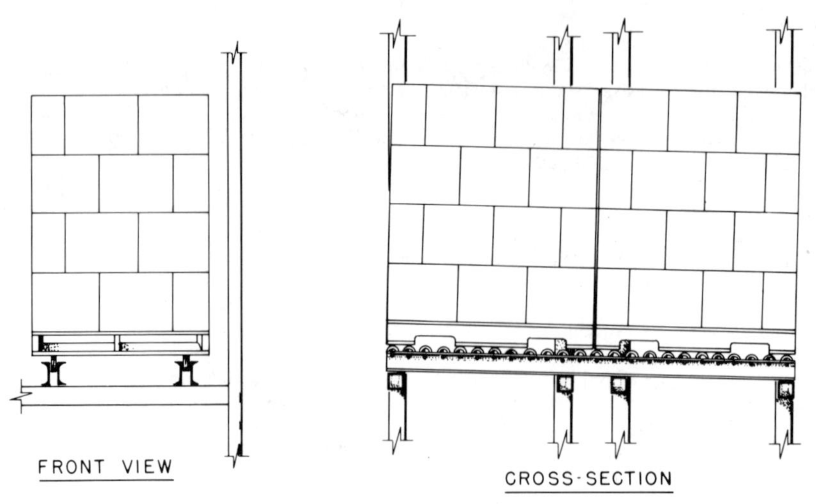

Fig. 11.1.21 Flow track section.

that no pallet load gets hung up in such a gravity flow system, various accessories are added, such as speed retarder devices, pallet side guides, pallet entry devices, pallet exit separators, and perhaps slave pallets with tracking lips. Flow racks of this configuration depend on large volume, some applications being book and catalog signatures, frozen prepackaged potatoes, and frozen hamburgers for the fast-food industry.

One manufacturer has a special soft plastic-tired wheel in his flow track conveyors that controls the gravity speed, thereby eliminating some mechanical devices. Another has a soft air-inflated hose pulsate up and down, thereby engaging and disengaging the flow track wheels to the pallet and minimizing any speed build-up problems.

A concept that serves the same function but eliminates the flow track (and gravity pitch) has a shallow-wheeled shuttle cart travel out on channel-shaped pallet supports from a pick machine situated in a central aisle, elevate slightly, and bring full pallets back to the central aisle. The shuttle also indexes loads toward the central aisle so new stock can be placed in the opposite end. The electrically operated shuttle cart can be captive to the lane or can accompany the picking machine, and can be battery powered and recharged when within the picking machine. It is also possible to work out of a central aisle only, providing lanes are not too deep and FIFO is not a major concern. A European system operates similarly except each load has a wheeled cart, and lane rails are on a very slight incline.

All these deep pallet systems are supported with rack structures similar to those described in standard selective rack, with various accessories where required. Quite naturally, costs increase drastically when adding mechanical equipment; this can be as high as a factor of 10 times the cost per pallet of standard selective pallet rack when speed control and pallet separation devices are added to pallet flow. The various economies such as space saved, fewer people, fewer forklift trucks, and 100% inventory rotation, must be weighed against this.

Package flow racks (Fig. 11.1.22) function in much the same manner and have the same advantages as pallet racks only they are hand loaded with individual cartons rather than unit loads which are forklift loaded. Picking can be either for full case lots (usually deep lanes) or broken case quantities, as illustrated. This type of selection is best suited for medium movement for full case picking. (Fast movers would remain on the pallets and slow moving items would be selected off of hand stack racks.) Figure 11.1.22 shows a two-level mezzanine which would be installed in a distribution center. The outside single rows of pallet racks are serviced by forklift trucks which provide pallet loads for restocking the flow lanes. The narrow restock aisles are between the pallet rack and the package flow rack. The rack itself consists of roller track and lane dividers which support, convey, and guide the cartons. The track and guides engage or snap into horizontal shelf frames which are supported on vertical frames. The vertical frames also provide attachment for the mezzanine supports. The picker has a very concentrated display of products from which to select which maximizes his efficiency.

The picker assembles orders in a cardboard or plastic container which is pushed along the roller conveyor. In between the roller conveyors in the center of the system is a powered belt conveyor on which is placed the container when the order is completed. Empty cartons are disposed of on the overhead trash conveyor.

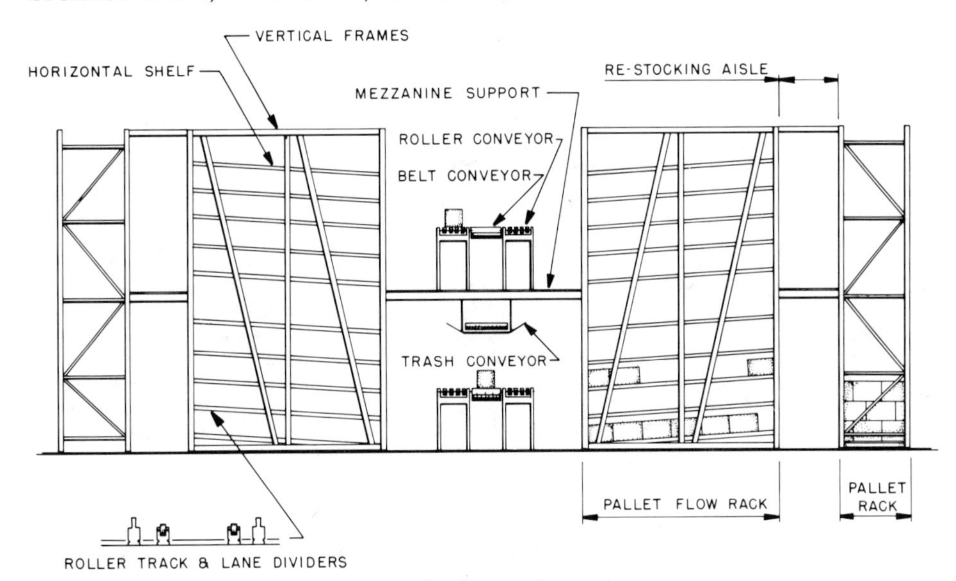

Fig. 11.1.22 Package flow racks.

Many installations do not have enough volume for this elaborate mezzanine structure and one quadrant of the system shown is sufficient to select items most suited to this type of storage. Typical merchandise selected by this method includes cosmetics, patent medicines, toiletries, cigarettes, electrical parts, hardware, art supplies, shoes, and toys.

Manufacturers have devised ingenious track designs, some incorporating small plastic rollers fitting in special light-gauge steel roll-formed track that snap, slide in, or are clamped to the shelf frames. Frames have adjustable connections that readily allow vertical shelf heights to be varied on small increments ($\frac{1}{2}$ in. or $\frac{3}{4}$ in.); one has an infinite adjustment arrangement.

Miscellaneous

1. Movable shelf rack consists of a steel framework shelf (sometimes with a sheet steel deck) which, with the load on it, is removed and replaced in the rack by a forklift truck or other mechanical lifting device. The four corners usually have engaging loops that fit into serrations on the upright frame posts. This system works well for aircraft aluminum sheet storage, where a bundle rests on a shelf and is taken out when several pieces are required at a machine point and then returned. Punch press dies are stored on smaller shelves of this type of system, the in–out being handled by a small hand-pushed stacker crane that is rack supported.
2. Slotted angle rack is a roll-formed steel angle-shape that has a continuous hole pattern that facilitates bolting up into any rack configuration desirable. This material, sold in bundles of ten pieces 10 ft long with bolts and nuts, is best applied to various hand storage applications like in back rooms of stores or stand-up molding storage in lumber yards. Manufacturers' load-capacity charts are available for post and beam rack configurations.
3. 55-gallon drums are sometimes stored horizontally on double cradles that facilitate stacking two at a time by forklifts.
4. Bar stock racks are long cradles that stack one on top of another and are serviced with an overhead crane with a special grab attachment.
5. Bar stock separator racks consist of small U-shaped modules that stack one on top of the other, 90° to the bar stock. Two or more such stacks accommodate one tier of bar stock, and a number of tiers placed next to one another make a neat pigeon hole arrangement for bar stock storage.
6. Coil racks are four posts attached to a V-trough which permits a number of wire coils to be contained for transport and storage.
7. Special user-designed racks can take most any configuration to suit the individual company's needs. Small specialty steel fabricating shops are locally available in most areas to satisfy this portion of the rack industry.

11.1.4 Floor Levelness

Latest developments in floor-running forklift equipment feature machines capable of placing pallet loads on shelves at heights as great as 55 ft from the floor, and able to leave the aisle they service and transfer to adjacent aisles. They are generally classified as "mid-rise" storage vehicles to distinguish from "high-rise" AS/RS storage which go to 100-ft heights and conventional "low-rise" storage with forklift trucks making right-angle turns and using rack structures usually less then 25 ft in height. These mid-rise machines are always aisle guided, have turret or shuttle arrangements that place the pallet load on the shelf without making right-angle turns, and sometimes take the machine operator up to the shelf level where the pallet is. Because these vehicles run directly on the warehouse concrete floor, they follow the contours that result from floor unevenness when the floor was poured. With low-rise equipment, this unevenness does not usually adversely affect the operation, but with mid-rise heights an entirely different set of circumstances arises. Figure 11.1.23 shows an aisle cross section where the floor takes a $\frac{1}{4}$-in. dip over a 4-ft dimension. If the truck has a 4-ft wheelbase and a 40-ft lift, the floor unevenness is multiplied by 10. With aisle guidance, side clearances between the rack structure and loaded vehicles are sometimes as close as 3 in. (assuming perfectly level floors), hence this $\frac{1}{4}$ in. unevenness causes a collision situation. (Actual situations like this have necessitated chipping out the top section of new floors and replacing with a level epoxy mixture.)

Floor Specification

The American Concrete Institute's "Guide for Concrete Floor and Slab Construction" (ACI 302.1R-80) lists four classes of surface tolerances and implies a method of measuring surface tolerance (or unevenness).

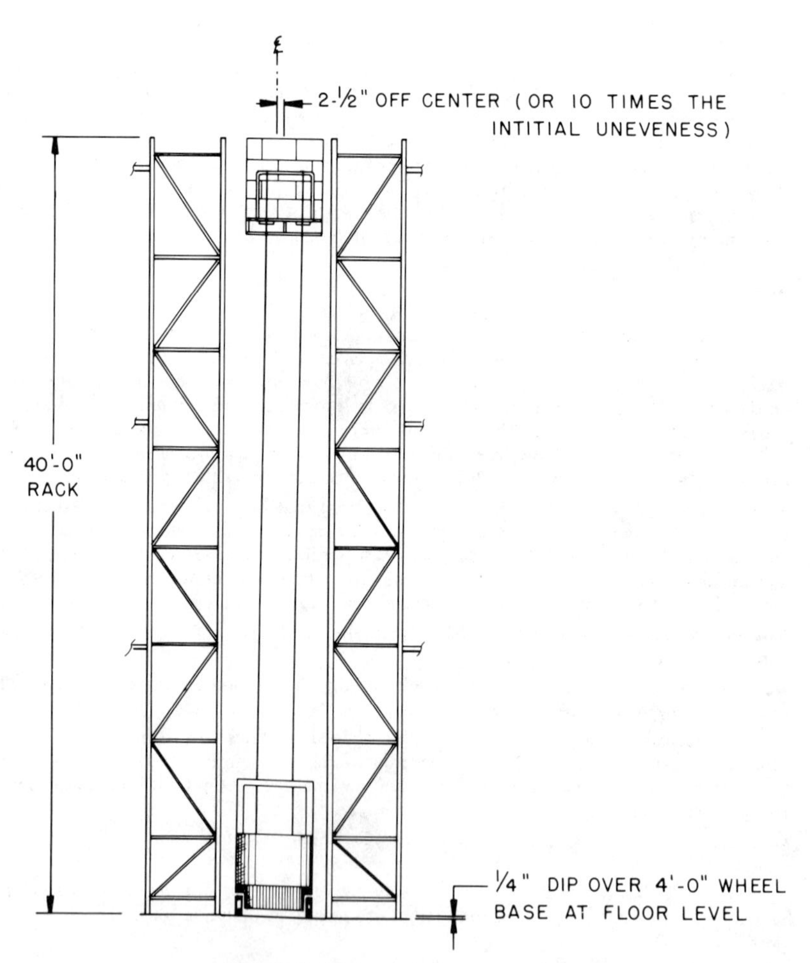

Fig. 11.1.23 Results of an uneven floor.

For class A surface finish tolerance, depressions in floors between high spots shall not be greater than $\frac{1}{8}$ in. below a 10-ft straightedge.

ACI 302 lists three additional classes, AX, BX, and CX, with tolerances of $\frac{3}{16}$ in., $\frac{5}{16}$ in., and $\frac{1}{2}$ in., respectively. There is little or no guidance for the specifier on what tolerance is needed for any given use. The current situation may be characterized as follows:

1. Surface tolerance specifications for flatwork are often unrealistic and ambiguous.
2. Specifiers seldom have a meaningful basis for knowing what surface tolerances are needed and how changes in surface tolerances can affect costs of floors.
3. Surface deviations are seldom measured.
4. There is no standard method for measuring surface deviations or variations.
5. Contractors seldom know what degree of surface accuracy they are providing.

Obviously, the concrete floor contractors are not delivering the floor evenness they say or think they are, and forklift owners (or manufacturers) are asking for floors much better than what is actually required.

It is possible to set steel concrete screed forms level by a laser beam machine to a very close tolerance and finish the concrete to be flush with the screed tops. The closer the screeds are to each other, the less variation the floor will have, and inversely, the more labor intensive and expensive it will be.

Conclusion

A practical method for measuring a finished floor for unevenness is illustrated in Fig. 11.1.24. Besides checking with the forklift manufacturer for his recommendations on unevenness, it might be worthwhile to check the tolerances on a floor where similar equipment is operating satisfactorily. In addition to the local tolerance discussed, an overall floor unevenness anywhere in the building (or room) must be specified to within $\pm\frac{1}{2}$ in. (or $\pm\frac{3}{4}$ in.). (This is a total variation of 1 in. or $1\frac{1}{2}$ in.). At no place is the local 10-ft unevenness tolerance discussed earlier to be exceeded.

11.1.5 Plumbing of Racks

Rack plumbness refers to upright frames being perpendicular to the plane of the floor. Besides floor unevenness, manufacturing and erection inconsistencies can collectively contribute to an out-of-plumb condition. Up to 1 in. deviation in 10 ft is allowed in most rack specifications, but mid-rise requires a closer tolerance for operation aspects, and the forklift supplier should be consulted for recommendations. A vertical plumbness of $\frac{3}{4}$ in. in 40 ft of height is reasonably obtainable. Closer tolerance than this requires additional installation time and possible over-aisle ties, both of which can prove to be costly.

11.1.6 Deflection Limitations

When a horizontal beam has a load placed on it, it "sags" or deflects. Most rack specifications set a limit on this deflection defined as $L/180$ or the span length divided by 180. For a 106-in.-long beam this becomes 0.59 in.

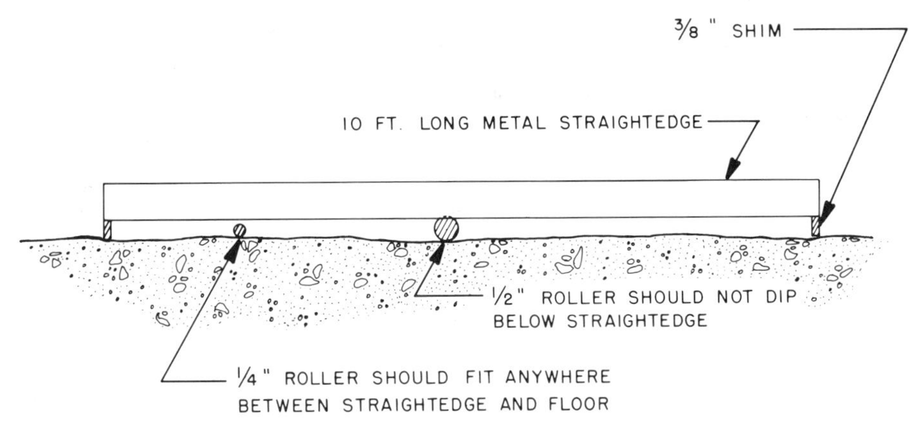

Fig. 11.1.24 Method for measuring floor level tolerances.

Some mid-rise systems specify this deflection be held to $\frac{1}{4}$ in. to provide the forklift vehicle with an ample target to enter the forks in the pallet, and this restriction can add substantially to the cost of the rack system. (One mid-rise job had an add-on cost of $70,000 for this deflection limitation.)

One possible solution is a height selection system on the forklift which reads a piece of reflective tape on the beam, thereby eliminating some of the inaccuracies and accommodating $L/180$ deflections. Another is to have the operator ride up with the load, where he can easily observe the forks engaging the pallet.

11.1.7 Aisle Guidance

The close aisle clearances on mid-rise systems mandate an aisle guidance system when the vehicle is within the picking aisle. The following two variations have evolved:

Wire-guided. These systems depend on a wire buried in a center groove cut in the aisle floor and filled with an epoxy cement. The forklift vehicle has an electronic reader device mounted on the centerline of the truck and when it follows the wire signal, the truck is automatically steered to follow the wire. This system permits the first level of loads to rest directly on the floor.

Roller-guided. Rollers are mounted outboard at the four corners of the vehicle, near the floor, and are guided by heavy steel rails mounted to the floor, which positively positions the vehicle as it proceeds down the aisle. If it is necessary or desirable to keep the first pallet level off the floor, the guides can also be used for pallet support beams. Size and spacing of floor anchors must be sufficient for the loads involved, and flared entranceways are often concrete filled to withstand truck impact.

11.1.8 Fire Protection

As clear available storage heights increase within distribution centers, fire protection problems increase. Racked pallet storage provides the worst possible configuration for fire flue spaces within combustibles (cardboard cartons stored on wood pallets), and several distribution center fires have pointed up the need for adequate research on the problem.

In 1967 the Rack Storage Fire Protection Committee was formed. The membership was composed of fire insurance carriers, users, and rack and warehouse equipment manufacturers who cooperatively developed and financially sponsored a program of full-scale fire testing of combustible unit loads on racks. Close to one million dollars have been spent on these tests, the most comprehensive in the world to date. The major combatant to fire utilized by the test committee has been water sprinklers; and the size, type, quantity, and location under various conditions were carefully controlled over many full-size tests. Fifty-foot-high loaded rack structures were fully instrumented, with thermocouples in rack members, building columns, bar joists, and at the ceiling. The time of the sprinklers opening discharge was noted, as well as the rate of fire travel. After each test, the extent and percentage of damage was charted, as well as maximum temperatures and temperature fluctuations. Tests have shown that a repeat of protection patterns can control fires in rack of any height, and the National Fire Protection Association's publication NFPA 231c, "Rack Storage of Materials—1980," specifies the protection required for any height of storage. Fire insurance carriers depend on NFPA 231c for guidance when outlining a customer's requirement, and should be consulted during the planning stages of a materials storage system.

11.1.9 Rack Design Specifications

Industrial steel storage-rack construction requires proper structural design under the supervision of qualified engineering personnel, guided by applicable specifications. The two publications that apply to all steel building construction, the American Institute of Steel Construction (AISC) "Specification for the Design, Fabrication and Erection of Structural Steel for Buildings" and the American Iron and Steel Institute (AISI) "Specification for the Design of Cold-formed Steel Structural Members" are the basic design criteria. Because of peculiarities and uniqueness of rack structures that are beyond the scope of these two specifications, the indusrty has developed a series of publications to provide additional guidance to produce safe, utilitarian, and economical rack structures. Historically these developed as follows:

1. The 1964 Rack Manufacturers Institute's (RMI) rack design specification addressed the state of the art as of 1964, and many rack manufacturers developed load and application charts in their published brochures facilitating customer selection of rack components. The Rack Manufacturers Institute, a trade association established in 1958, has grown to a current membership of 38 companies who fabricate and market at least 70% of all rack structures in the United States. The institute has been the main source of continued funding at the technical university level to produce up-to-date data for rack design.

2. The 1972 RMI "Interim Specification for the Design, Testing and Utilization of Industrial Steel Storage Racks" and the 1974 "American National Standard Specification for the Design, Testing, and Utilization of Industrial Steel Storage Racks" (ANSI MH16.1–1974) addressed many developing areas such as load limit posting, forklift impact protection, horizontal loading criteria, vertical impact shelf loads, effective post lengths, and others. These two practically identical specifications are much used on large installations, sometimes with customer modifications.

3. The 1979 RMI "Specification for the Design, Testing and Utilization of Industrial Storage Racks" provides additional refinement to the foregoing, and will also be subject to fine-tuning based on extensive testing and research presently being carried on.

Besides the warehouse building, rack structures are often the major portion of the dollar investment on a materials storage project. A potential customer should be satisfied that he or she is spending wisely, and the preceding data are available to assist in this endeavor.

Building Codes

Up until 10 years ago, racks were considered plant equipment and as such were not subject to local municipal building code jurisdiction. Many communities, particularly those in active earthquake zones, now incorporate rack installations in their ordinances, and design criteria for the particular installation is reviewed by the municipal building department and a building permit issued. It is wise to check with the municipality where a potential system will be installed to ascertain if their participation will be necessary.

Earthquake Design

The federal government is concerned with proper building seismic design in all of the United States. Insofar as rack structures resemble such steel-framed configurations, seismic rack design is covered in the 1972 RMI and later rack design codes.

Full-scale loaded-rack testing has been done under the auspices of the National Science Foundation on an earthquake simulator to confirm current state-of-the-art design criteria. Although not entirely conclusive and suggesting further study in some areas, this testing did prove that pallets do not fly off the loaded rack configurations during an earthquake and the present degree of safety provided under existing codes is fairly reasonable. Depending on the seismic zone involved, earthquake considerations can add measurably to the cost of a rack installation and the inclusion of such is up to the customer.

11.1.10 Conclusion

Potential customers have several sources of assistance in gaining further information on storage racks, and should be aware of the following:

1. The Material Handling Equipment Distributor network is ready to supply all the equipment (including racks) required in a warehouse. Some of these organizations have qualified engineering personnel on their staffs who can assist the user in their racking needs.
2. All rack manufacturers who are RMI members have engineering assistance available to potential customers. They are aware of present code design requirements and can keep a user abreast of changes.
3. Consultants provide many levels of services, and can be active with a user from the very beginnings of a materials storage project or on specific, technical needs.

It is important to remember that the materials storage business is dynamic and constantly changing, and professional advice from all sources should be considered.

BIBLIOGRAPHY

Industrial Steel Storage Racks Manual.

Manual of Safety Practices.

Minimum Design Standards for Pallet Stacking Frames.

Rack Storage of Materials 1980 NFPA 231c, National Fire Protection Association, Batterymarch Park, MA 02269.

Specification for the Design and Utilization of Industrial Steel Portable Storage Racks.

Specification for the Design, Testing, and Utilization of Industrial Storage Racks, ANSI MH16.1–1974.

Speers, Ralph E., Concrete Floors on Ground E-B75, Portland Cement Association, 5420 Old Orchard Road, Skokie, IL 60077.

Standard Nomenclature for Industrial Steel Storage Racks.

The preceding six publications are available from the Rack Manufacturers Institute, 1326 Freeport Road, Pittsburgh, PA 15238.

Wood Pallet Information, National Wooden Pallet and Container Association, 1619 Massachusetts Avenue N.W., Washington, D.C. 20036.

11.2 SHELVING AND BINS

W. Scott Fowler

The Industrial Revolution rapidly produced improvements in the production of parts, assemblies, and products, and the volume of support materials increased to meet the new demands. The by-product of this growing volume of materials was the need for increased and more efficient storage in production and distribution, including the capability to store light, hand-loaded materials. The storage media used in this period of time was custom built from wood, and was typically the last area of consideration in the production plan.

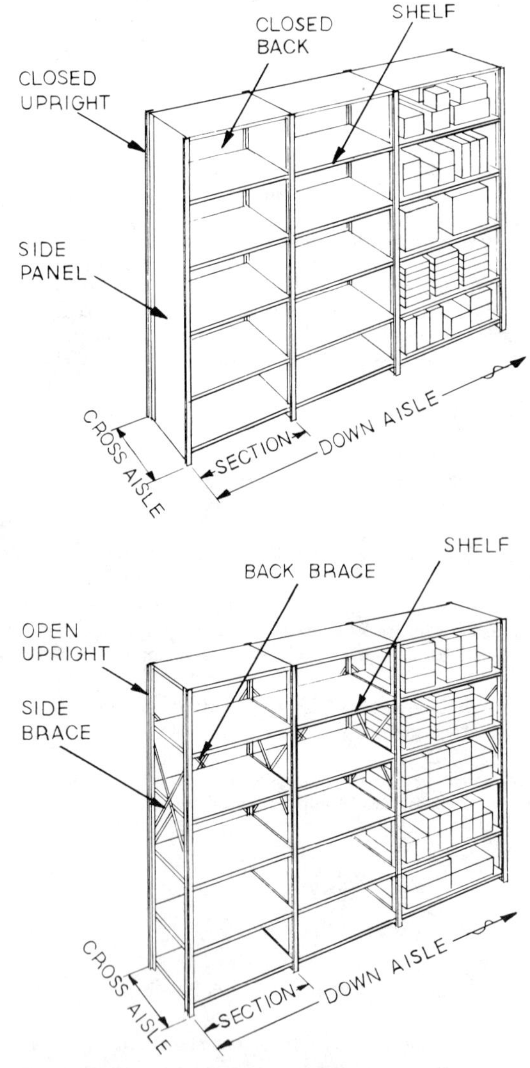

Fig. 11.2.1 Closed and open types of shelving sections.

The use of steel for fabrication of shelves, bins, and drawers emerged shortly after the turn of the century as the need for standardization of the product became necessary to enable companies to adapt to rapidly changing storage requirements. Adjustable shelves, dividers, and drawers became standard as part of the transition from wood to steel because additional holes or slots could easily be added as part of the fabrication process. Steel construction also eliminated hazards unique to wood shelving and bins—such as fire, breakage, rottage, sanitation, bending, chipping, or warping.

From these humble beginnings, we can now look at the products available today for storage of small, hand-loaded items, and how to select, size, and apply the product to the items stored as they relate to the human limitations of the operator.

11.2.1 Shelving

The Shelving Manufacturer's Association (SMA), a product section of the Material Handling Institute, has published a document, "Nomenclature for Steel Shelving," to aid the end user and the supplier in the identification of the product and its components. This publication uses commonly accepted terms and definitions in the shelving industry.

A shelving section is nominally 36 in. (91 cm), 42 in. (107 cm), or 48 in. (122 cm) wide in the down-aisle direction; 12 in. (30 cm), 18 in. (46 cm), or 24 in. (61 cm) deep in the cross-aisle direction; 84 in. (213 cm), 96 in. (244 cm), or 120 in. (305 cm) high. The shelving section is of either open or closed construction, as shown in Fig. 11.2.1.

The correct nomenclature and description for a shelving section, single-row/single-face and double-row/double-face shelving are as follows:

Section. A single structure comprising upright posts and horizontal shelves. This is the basic unit and does not exceed in width or depth the nominal maximum dimensions of one shelf.

Single Row/Single Face. One (generally continuous) row of sections, joined together and side to side, to be served form one service aisle.

Double Row/Double Face. One (generally continuous) row of units, joined together, back to back and side to side, to be served from two service aisles.

A variety of accessories is available to complement the storage of materials and parts as defined below and shown in Fig. 11.2.2:

Bin fronts. Used to retain loose parts stored on a shelf.

Shelf boxes. Containers with dividers for storage of small parts.

Shelf dividers. Front-to-back partitions for dividing shelves in the down-aisle direction.

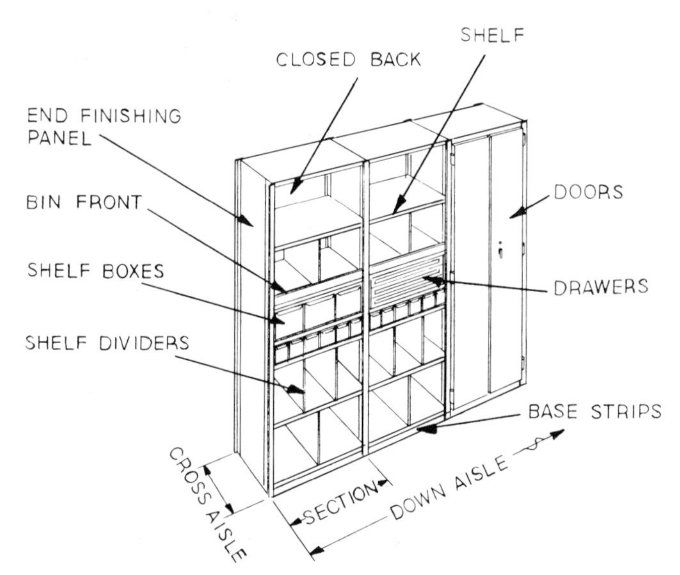

Fig. 11.2.2 Typical shelving components.

Drawers. Roll-out drawers with partitions, dividers, and trays for storage of tools and small parts.

Base strips. A finishing strip to close off the opening between the bottom shelf and the floor.

End finishing panel. A cover for end uprights to conceal fasteners and close off the opening between the side sheet and floor, or bottom shelf and the floor.

Doors. Swinging or sliding doors can be added to secure a section (or sections) of shelving.

A technical document, "Specification for the Design, Testing, Utilization, and Application of Industrial Grade Steel Shelving," has been published by the Shelving Manufacturer's Association (SMA) with the intent of offering information to the users and suppliers of industrial grade steel shelving. The document provides definitions and methods of calculating and testing key elements of a shelving section, such as shelving section load, shelf capacities, and shelf connectors.

Shelving section load capacities are determined by calculation in accordance with the American Iron and Steel Institute (AISI), "The Design of Cold-formed Steel Structure Members," or the "Specification for the Design, Fabrication of Structural Steel Building" (AISC). An alternate method of determining section load capacities is by testing the product in accordance with SMA procedures.

The procedure for testing section loads requires that a two-wide section be erected in a plumb and square condition, with the axial load applied to simulate a field-fixed condition. Figures 11.2.3 and 11.2.4 illustrate methods of applying axial loads in a test environment with simulated shelf loading.

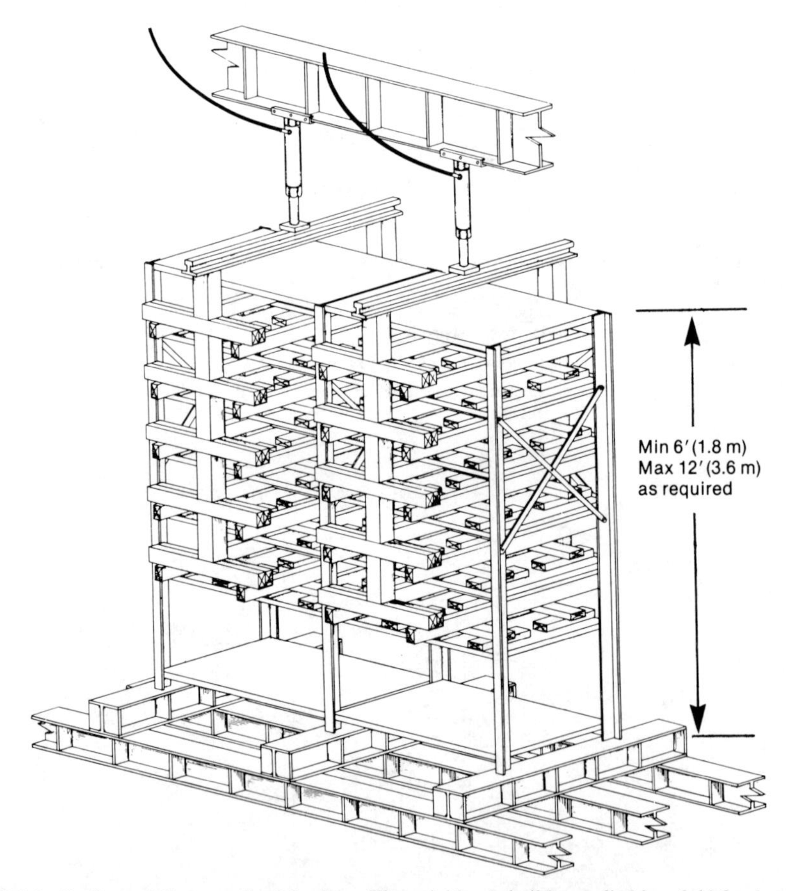

Fig. 11.2.3 Perforated Compression Members. The axial load shall be applied in a field fixed condition on a two-wide section on a unit that is plumb and square. The space between loaded shelves should be spaced as far apart as possible (without exceeding the bottom shelf space) to allow members to react in a normal manner; each shelf level shall have an equally distributed load applied simultaneously and adequate protection shall be used to prevent any individual in the test area from injury.

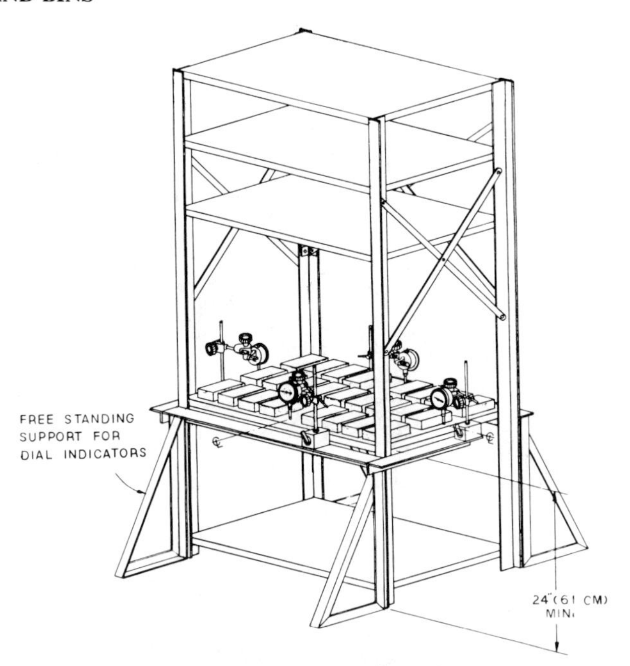

Fig. 11.2.4 Equally Distributed Shelf Loading. Weight of (not to exceed) 25 lb (11.3 kg) increments is equally spaced on test shelf and deflection readings taken at appropriate increments at the center of each edge of the shelf. The individual weights must be spaced with a minimum gap of $\frac{1}{4}$ in. (6.35 mm) for adjacent weights to prevent a bridging effect. When required, weights of equal size can be stacked on top of each other maintaining the minimum gap of $\frac{1}{4}$ in. (6.35 mm).

The rated load capacity of a section of shelving is based on an average of three tests conducted under identical conditions. The capacity is based on the mean ultimate load falling within ±10% of the three test results. A safety factor of 1.92 is applied to the mean ultimate load to determine the shelving section load. The rated capacity is associated to the larger, unsupported shelf located near the bottom of the test units.

Shelf tests must be conducted in a section of shelving with the test shelf located at a minimum distance of 24 in. (61 cm) from the floor.

The loads are applied as equally as possible by hand, in increments of 25 lb or less, or by mechanical or hydraulic force as shown in Figs. 11.2.4 and 11.2.5.

The allowable rated uniform load is based on the smallest of the following test results:

1. The load at which the shelf has deflected to span/140 (span is width or depth) from the center of the edge of the width or depth of the shelf.
2. The yield point load (load at which the deflection curve becomes nonlinear) factored by a safety factor of 1.25.
3. The ultimate load (load at which failure occurs) factored by a safety factor of 1.5.
4. Four times the rated capacity of one shelf connection (a shelf clip is tested to ultimate load and factored by a safety factor of 2).

The rated load capacity of a shelf or connector is determined by an avarage of three tests based on the mean ultimate load value falling within ±10% of the three test results.

A concentrated line load normally supports one-half of the allowable rated uniform load and should be taken into consideration by the user when evaluating loading conditions and selecting the product.

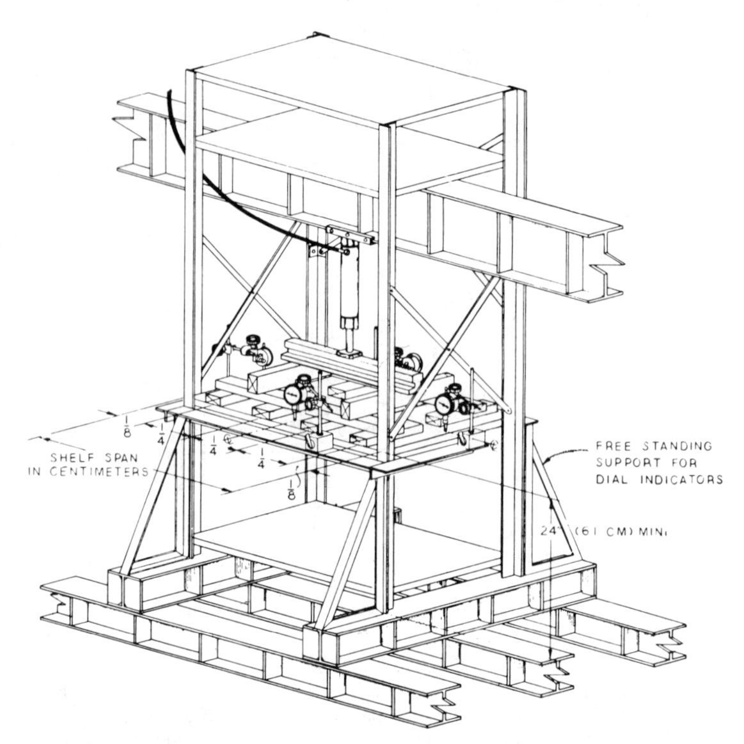

Fig. 11.2.5 Equally Distributed Shelf Loading. Mechanical or hydraulic force is applied to four-point loading of test shelf and deflection readings taken at appropriate increments at the center of each edge of the shelf. The four-point loading is placed $\frac{1}{8}-\frac{1}{4}-\frac{1}{4}-\frac{1}{4}-\frac{1}{8}$ of this test shelf span up to and including 60 in. (152 cm) widths and $\frac{1}{16}-\frac{1}{8}-\frac{1}{8}-\frac{1}{8}-\frac{1}{8}-\frac{1}{8}-\frac{1}{8}-\frac{1}{8}-\frac{1}{16}$ from 60 in. (152 cm) to 96 in. (254 cm) widths. There is a gap of $\frac{1}{4}$ in. (6.35 mm) between the front and rear loading members to prevent bridging within the loading members.

11.2.2 Bin Units

Bins are used for storage of small bulk, light items, and packaged merchandise in tool rooms, repair shops, and light-assembly areas. Bins are available as individual steel units with optional steel, fiber, or plastic boxes, and steel bin racks with optional steel, fiber, or plastic boxes.

Individual steel bin units are nominally 36 in. (91 cm) wide in the down-aisle direction, 12 in. (30 cm) or 18 in. (46 cm) deep in the cross-aisle direction, 39 in. or 66–75 in. (99 cm or 168–191 cm) high, and furnished as a closed unit, assembled with nuts and bolts, and shipped set up. Figure 11.2.6 illustrates a variety of options available, such as labelholders, dividers, boxes with dividers, and closed bases.

Load capacities are based on 15–20 lb per box, or approximately 90–200 lb per shelf, depending on arrangement of dividers or other optional equipment. Load capacities can be determined by using SMA procedures as outlined earlier, but generally are used for storage of light products.

Steel bin racks are nominally 36 in. (91 cm) wide in the down-aisle direction, 12 in. (30 cm) deep in the cross-aisle direction, and 60–75 in. (152–191 cm) high. Each unit has a base, two upright assemblies with sloping shelves supporting fiberboard boxes. Units using plastic boxes are supported from steel members and are removable, as shown in Fig. 11.2.7.

11.2.3 Selection of the Storage Location

A planned approach to the storing of parts in shelving, bins, and drawers is required in order to achieve the maximum use of the available cube and to improve the effectiveness of the personnel picking and storing inventories. A composite installation is shown in Fig. 11.2.8.

The physical properties of each given part will aid in determining the most suitable location for picking. A study of each part, based on size and weight, is made to determine the location in a

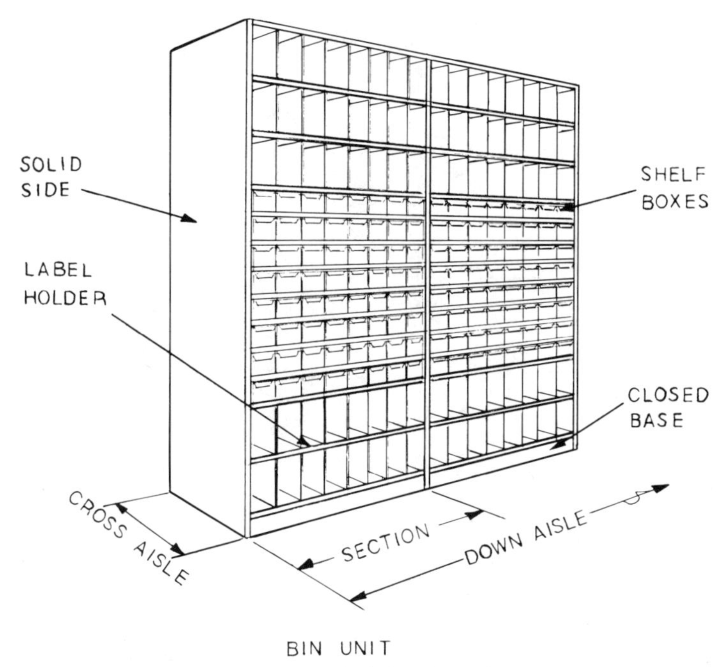

Fig. 11.2.6 Typical bin-unit options.

given bin, drawer, or section of shelving. The parts should be divided into groups of small, medium, and large, as well as light, average, and heavy. This information is used to place the parts in a given elevation and to aid in selecting the depth of the storage units.

The operator who picks and stocks the storage units can best function by picking large, light items stored above the shoulder, because large items are easily identified and light items can be handled from this position. Small items of average weight are best stored between the shoulder and hip so that the operator can see into small openings or down into extended boxes or drawers and yet be able to physically lift and place them. Larger sized, heavy items are best stored below the hip location so that the operator can use leg muscles for lifting, as illustrated in Fig. 11.2.9.

The depth of the storage unit should be determined by the ability of the operator to see and effectively reach the parts being stored. That is, small items in a given section of shelving, bins, or drawers should be stored in shallow units (12–15 in.; 30–38 cm deep) to reduce reaching, stooping, or overextending by the operator in his effort to reach small items located in the back of a given location.

As the items stored increase in physical length, so should the storage unit (18–30 in.; 46–76 cm deep), thus allowing for optimum efficient use of the human operator.

Optimizing the Location with Activity

The activity level (frequency of use for each part) can also be used to determine the optimum location for picking. The use of historical data will usually show that 80% of the activity encompasses only 20% of the inventory. The parts have been divided into three areas, based on size and weight—and now these areas can be divided by part number into areas of high low activity.

The object of this review is to place high-activity items as close to the shoulder–hip zone as possible, thus minimizing operator movement and reducing fatigue (see Fig. 11.2.10).

The placement of high-activity parts in the same section of storage can also be used to help lay out the location of given sections of storage in the floor plan.

Selecting the Product to House the Part

One of the most important rules to follow in storage of parts is that each part must have a separate location unto itself. This rule of thumb is the most misused rule, and poor inventory control generally leads back to poor application of this rule. The separate physical location provides a space that can

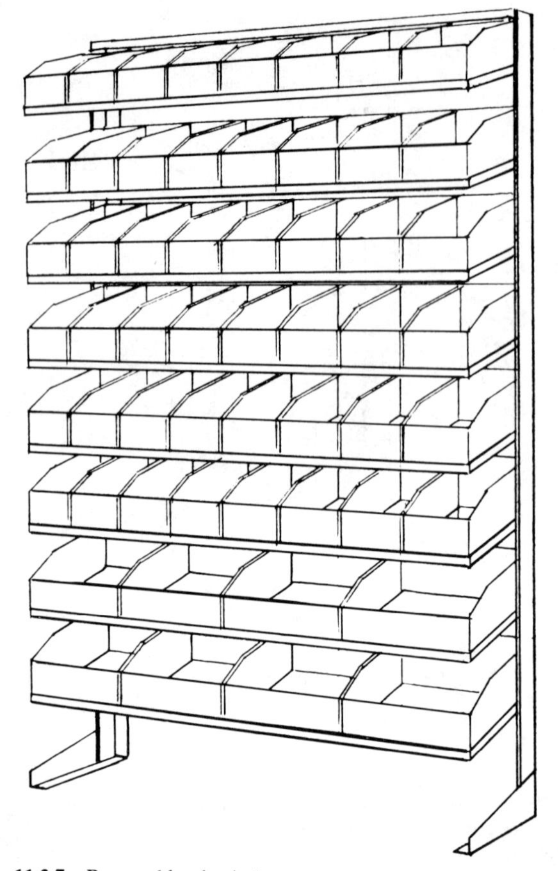

Fig. 11.2.7 Removable plastic boxes are supported on steel members.

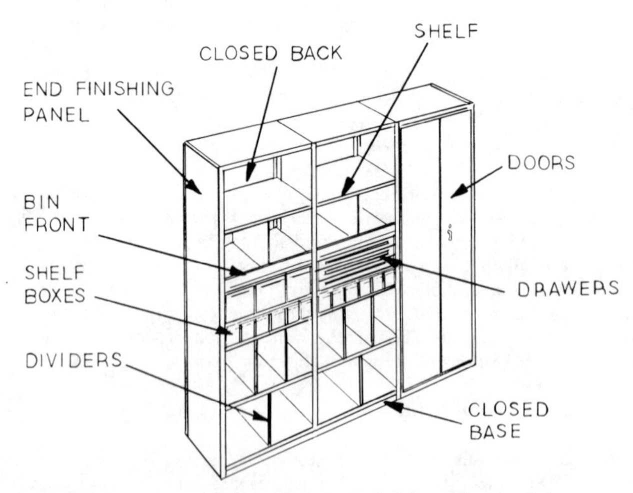

Fig. 11.2.8 Composite installation of shelving, bins, and drawers.

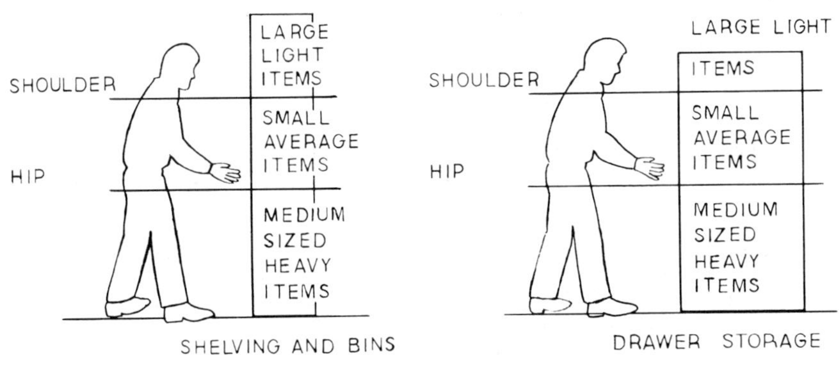

Fig. 11.2.9 For ease of manual handling, items should be segregated as shown.

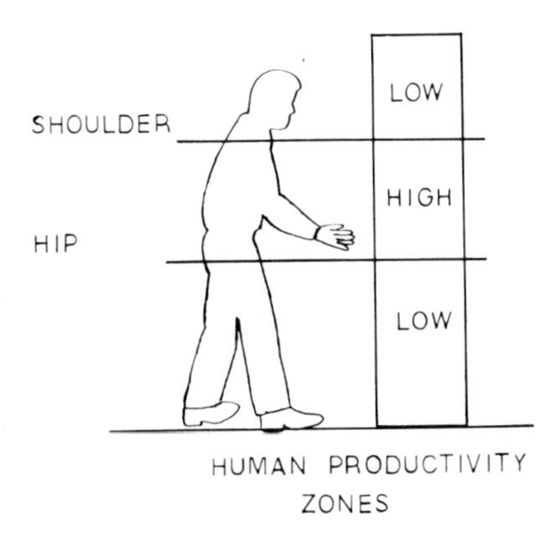

Fig. 11.2.10 High activity items should be located so as to minimize operator movement and fatigue.

be selected to fit maximum inventory quantities required and provide a fixed location that can be identified with coordinates and cross-indexed with the part number.

The selection of each separate location can be made from a variety of storage products, such as boxes, drawers, divided openings, and complete shelf locations. Generally, small loose items and small packaged items are stored in divided boxes, drawers, or bin boxes that, when extended, allow the operator to see into the opening and retrieve or store parts by hand (see Fig. 11.2.11).

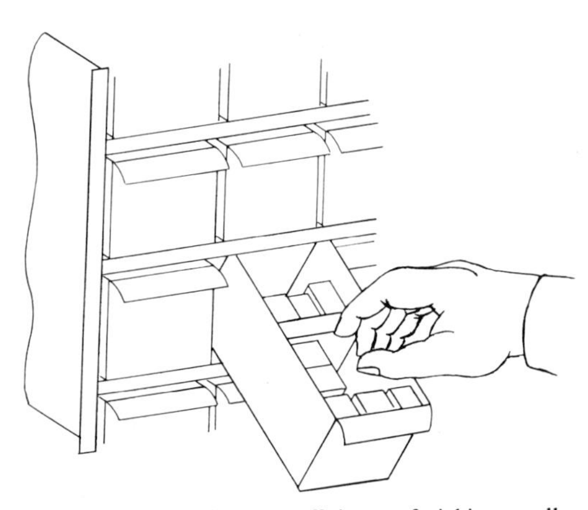

Fig. 11.2.11 Dividers improve efficiency of picking small parts.

Parts that are average in size are typically stored in large boxes, deep drawers, or divided shelf openings, allowing the human hand to grasp the part. Obviously, the human hand must physically fit into this storage media; typically, the opening is not smaller than $4\frac{1}{2}$ in. wide by $4\frac{1}{2}$ in. high (11 × 11 cm). Larger items are stored in large, divided shelf spaces or on complete shelf locations, allowing the operator maximum access for both hands when picking or stocking the parts.

Floor Plan Layout

The floor layout of shelving, bins, and drawers must consider the flow of materials into and out of the physical location. The identification of these key points will be used to determine the direction of rows and the placement of products.

The rows of shelving, bins, or drawers are typically layed out with the beginning of the rows facing the input/output area of activity. This provides a flow of traffic into lanes of storage media in an organized fashion. Pass-through aisles are used to gain access from aisle to aisle. Common depth, width, and height storage units are used to build each row; that is, 12-in. (30 cm) deep, 36-in. (91 cm) wide, 84-in. (213 cm) high shelving sections are used to complete rows of shelving as shown in Fig. 11.2.12.

The placement of high-activity items in given sections or units of storage can play a key role in the floor plan layout. The placement of these sections or units near the input/output area of activity will help to reduce the operator picking and stocking time, as illustrated in Fig. 11.2.12. This type of loading also aids in the supervision of the operators, since they will be visible most of the time.

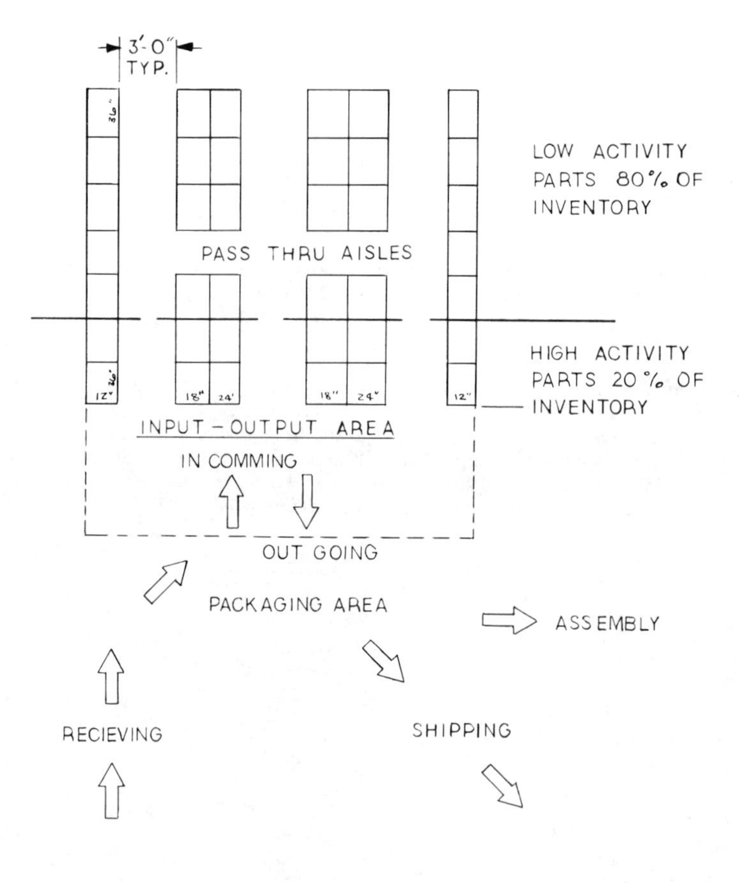

Fig. 11.2.12 High-activity items should be placed near input/output area to reduce operator picking and stocking time.

Maintaining Inventory Storage Location

The success of any storage application begins with the selection of the appropriate storage media for the respective parts.

The continuing success of any storage application relies on the maintaining of inventory control and the storage locations. The quantity of parts and the number of items stored change from year to year, and the user must evaluate these changes and periodically adjust the storage media and locations for the products to be stored. This is the key to the long-term success of an existing installation, as well as the justification for new applications.

11.3 MODULAR STORAGE DRAWERS

William Devaney

Modular storage drawer systems are very efficient for storing small nonbulk items such as parts, tools, and maintenance items. In general, drawers are most efficient when the individual items being stored have a volume of less than 1 ft^3, and the aggregate weight does not exceed 400 lb per drawer.

Both the storage drawers, and the cabinets that support them, are furnished in uniform modules having the same width and front-to-back depth, but of varying heights. Therefore, to achieve maximum storage density, the planning engineer must first select a combination of drawer heights that will more closely receive the inventory to be stored. Then he or she chooses a range of cabinet heights that will not only accommodate the required number of drawers, but also provide for the issue counters or work surfaces needed for necessary retrieval and record keeping.

11.3.1 Typical Storage Drawer Systems

Unlike most shelving drawers, which are simply shelves fitted with perimeter walls to form a box, modular storage drawers should have the following characteristics:

1. Be two feet square and in a configuration that permits access from the front as well as the sides of the drawer. (The average extended human arm length is 25 in.)
2. Be constructed to carry 400 lb per drawer.
3. Be secured within a cabinet housing by a suspension that will prevent drawers from sagging when fully open and fully loaded.
4. Be mounted on bearings that permit the drawer to glide smoothly in and out of the housing when fully loaded, so that the back of the drawer extends beyond the face of the housing when fully opened.
5. Contents at the rear of the drawer should be as readily accessible as contents in the front.
6. Be capable of subdivision into compartments as required by the shapes and sizes of the contents being stored.
7. Permit labeling of individual drawers and compartments.
8. Permit interchangeability of different drawer sizes within the same cabinet housing or between different cabinet housings (see Fig. 11.3.1).

To simplify matching modular drawers to modular cabinets, manufacturers have standardized on drawers that measure 25 in. on each side, and have developed a simple "point" system for drawer

Fig. 11.3.1 Different drawer sizes are used within the same cabinet housing.

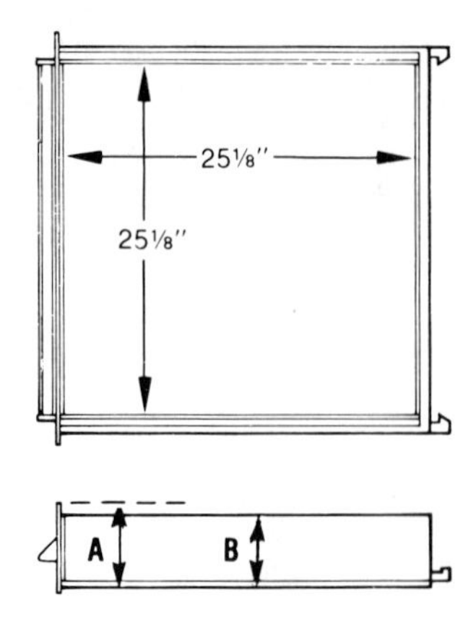

Drawer Model No.	Usable Height "A" (inches) (Note 1)	Body Height "B" (inches)
20	2¼	2
25	3	2
30	3⅞	3⅛
35	4⅝	3⅛
40	5⅜	4¾
45	6¼	4¾
50	7	6⅜
60	8½	6⅜
70	10⅛	6⅜
80	11¾	6⅜
90	13¼	6⅜

Notes:
1. Usable height is measured from bottom drawer to top of drawer front.
2. the suspension system is included with each drawer.
3. Drawer is slotted so that it can be divided into 32 equal spaces left to right and front to back. Spacing is approximately ¾ inch.
4. Drawer sizes other than those listed are available on special request at extra co

Fig. 11.3.2 Using model numbers to determine cabinet capacity.

and cabinet height selection. Typically, five points represent 2 cm (0.7874 in.) of height in a drawer or cabinet housing.

The points assigned to each drawer correspond to its *usable height,* and are expressed as a model number. Likewise, a cabinet's total point capacity for receiving drawers is expressed by its model number. The designer need only determine the number of drawers required in any cabinet, add up their model numbers, and choose a cabinet housing whose model number most closely matches the total (see Fig. 11.3.2).

For example, as shown in Fig. 11.3.2, one manufacturer has 11 standard drawers, ranging from 8 cm to 36 cm ($2\frac{1}{4}$–$13\frac{1}{4}$ in.) in usable height, with model numbers from 20 through 90, inclusive. The company's five standard cabinet housings, as shown in Fig. 11.3.3, vary from desk height to eye-level height, and have model numbers from 135 to 340, inclusive.

The combination of drawers that fit into a cabinet is easily determined: The total of drawer model numbers must equal the model number of the cabinet that will house them. Thus, a combination of four Model 40 drawers and four Model 45 drawers will exactly fill a Model 340 cabinet (8 drawers = 340). Likewise, one Model 45 drawer and four Model 50 drawers will exactly fill a Model 245 cabinet (5 drawers = 245).

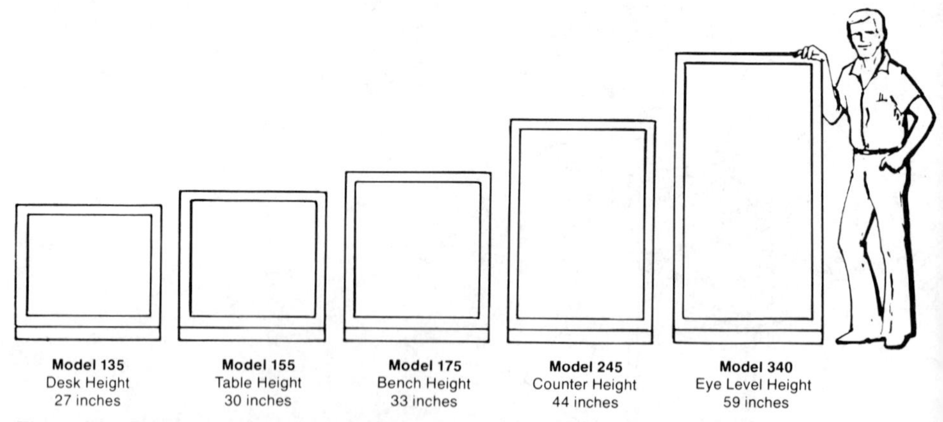

Fig. 11.3.3 Standard cabinet housings. Model numbers are presented here as an example. Each manufacturer uses different numbering system, but the basic modular concept is the same.

11.3.2 How to Determine Drawer Requirements

To achieve maximum storage density, determine how many drawers will be required by taking the following steps:

1. Analyze the existing inventory and determine exactly what articles will be stored in drawers. For example, you may decide that individual items larger than 1 ft^3 should not be stored in drawers.
2. Mentally or physically segregate the articles to be stored in drawers by height groupings (see Fig. 11.3.4).
3. Calculate the total surface area in square feet that each height group will occupy at its average inventory level (Fig. 11.3.5) when closely assembled.
4. The surface storage area of a standard storage drawer is 4 ft^2. Divide the total square feet of inventory to be stored in drawers by 4, to determine the total number of drawers that will be required.
5. Find the correct drawer height, by model number, for each height grouping. Use Fig. 11.3.6 for this calculation.

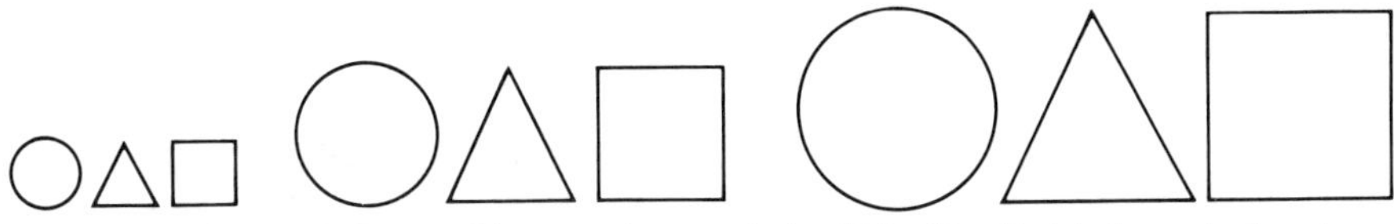

Fig. 11.3.4 Sizes and shapes of items to be stored should be identified before selecting drawer and cabinet configuration.

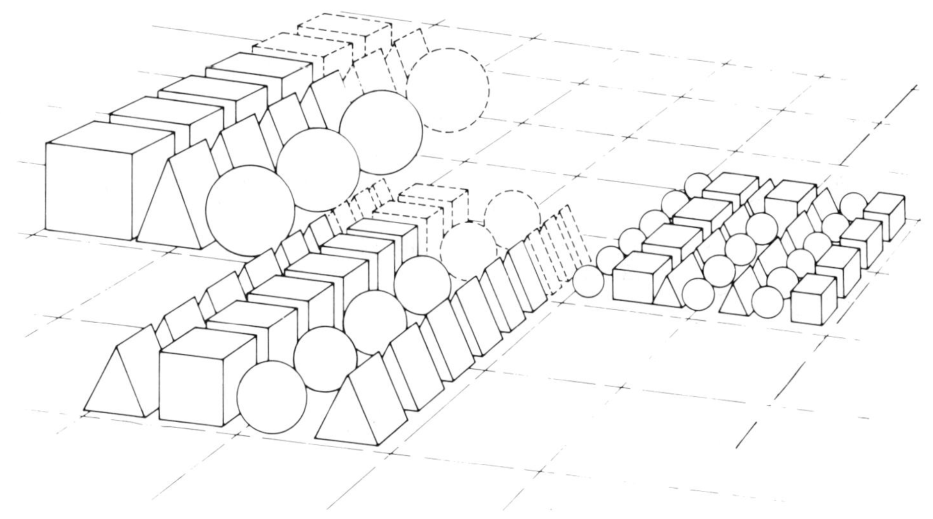

Fig. 11.3.5 Part of planning process is determining total surface area that each height group will occupy at its average inventory level.

Group No.	Description	Height Range	Recommended Drawer Model
1	Very small items	up to $2\frac{1}{4}$ in.	20
2	Small items	2–$3\frac{3}{4}$ in.	30
3	Medium items	$3\frac{1}{2}$–$5\frac{1}{4}$ in.	40
4	Large items	5–7 in.	50
5	Very large items	7–12 in.	60,70,80,90

Fig. 11.3.6 Data for determining correct drawer height for each item height grouping.

6. To allow for expansion, add additional drawers at this time. A 10–25% expansion factor is usually recommended for future growth. Example: If a 25% increase is anticipated, multiply the number of drawers in a height grouping by 1.25 and round off to the next full drawer 34 drawers required for a particular height grouping × 1.25 = 42.5, or 43 drawers required.

11.3.3 Choosing Cabinet Housings

Deciding on which housing, or combination of housings, is most appropriate for your application and involves several factors. If, for example, counter-height cabinets are required for issue areas, Model 245 cabinet housings should be selected. On the other hand, if issue-counter housings are not necessary, choosing only Model 340 housings will provide maximum storage density for the available floor space. In most applications, both sizes will be required.

To complete the cabinet housing selection, simply add up the drawer model numbers previously determined, and select the housing model number that most closely matches this total. (Minor adjustments in drawer model numbers can be made at this time to exactly match the cabinet housing model number.) An example:

Two Model 20 drawers = 40 points of storage
Three Model 30 drawers = 90 points of storage
Four Model 40 drawers = 160 points of storage
One Model 50 drawer = 50 points of storage
One Model 340 housing = 340 points of storage

If, for example, 10 ft of issue-counter work surface is needed, specify four Model 245 cabinets, since each standard cabinet model is 30 ft wide. Again, select drawer heights by model numbers, so that the total points of drawer storage add up to 245.

When fitting even-numbered drawers into odd-numbered cabinets, round out the total of storage points by adding five points to drawer, in order to exactly fill a housing.

Cabinet housings have built-in pallet bases for simple mobility—either within the existing facility or to other facilities. Fully loaded cabinets can be moved easily and will fit three abreast on a tractor trailer. Cover plates supplied with each housing conceal the fork entry guides, and prevent dirt buildup.

11.3.4 Compartmentalization

The most simple and accurate method of compartmentalizing storage drawers is with partitions and dividers supplied with the system. Their use permits the formation of compartments that match the shapes and sizes of the articles being stored, yet they may be adjusted to meet changing requirements.

Drawer partitions are slotted and always span an entire drawer, either front to back or left to right. Prepunched holes in the partitions and in the drawer bottoms line up for easy installation.

Dividers, used in conjunction with and between partitions, further subdivide drawers into smaller compartments. Dividers fit into the slots of the partitions or drawer sidewalls, and can be readily moved and rearranged, as necessary, between the partitions. Layouts of drawer interiors are usually referred to as loading diagrams, and carry the prefix LD (see Fig. 11.3.7).

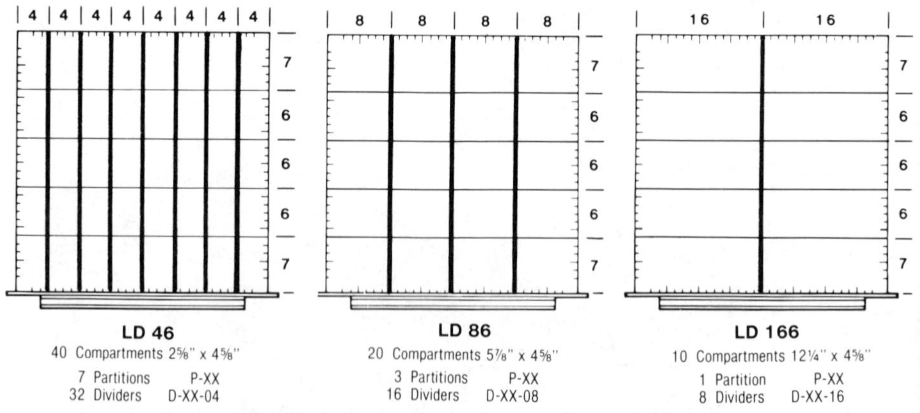

Fig. 11.3.7 Loading diagrams show layouts of drawer interiors. Standardized layouts of the various interiors are usually referred to as drawer loading diagrams. A few examples are shown in this figure. (See manufacturer's Loading Diagram Book for complete selection.)

Among the accessories available from manufacturers of storage drawer systems are label holders for each compartment, plastic bins, grooved trays, and drawer bottom posts.

11.3.5 Storage Drawer Locator System

There are three basic ways to organize and store items in modular storage drawer systems:

1. *Random Storage.* Items of the same height grouping (with no regard as to use) are stored together in order to achieve maximum storage density and the highest degree of storage flexibility.
2. *Family Storage.* Families of similar items are stored together by function (hardware, electrical, etc.), by end use (tune-up kits, repair parts, etc.), or by subassembly (motors, pumps, etc.).
3. *Part Number Sequence.* Storing items in sequence by part numbers is generally regarded as the least efficient method of organizing inventory. Storage density is lost because varying heights must be stored side by side. This method is also inflexible when a change is made in the size of an item being stored.

The most efficient method of storage and retrieval is achieved when items are grouped by height and stored at random, and each random location in the system is assigned an alphanumeric or numeric address. With random storage, a specific article may be stored anywhere in the system. Likewise, incoming inventory is placed anywhere in the system that a drawer compartment (address) of the correct size is available. Random storage is also called "floating slot" storage, since the location of any particular stockkeeping unit can vary, or float, over a period of time, due to quantity changes.

Regardless of which method of storage is used, the following elements are basic to any drawer cabinet location system:

1. Identify a zone or row of cabinets (horizontal).
2. Identify a particular cabinet within the row.
3. Identify a drawer within the cabinet (vertical).
4. Identify a row within a drawer (left to right).
5. Identify a specific compartment within the row (front to back).

After the cabinets are in place, parts are assigned and located in drawers having compartmentation appropriate to the item's size and quantity. Parts of similar size, having similar pick rates, are categorized and assigned to drawer locations throughout the cabinets. As a rule, cabinets that contain mostly high-use items are located near the issue area. For coding of part location, see Fig. 11.3.8.

11.3.6 Inventory Control

When a computer is not to be used with the storage system, manufacturers can provide inventory card tray drawers in the system for locating parts. Inventory cards are normally arranged in part number sequence and are operated similar to a library card system.

When utilizing a computer, all retrievals and restocking can be printed out in optimum "pick-path" sequence based on storage locations, which minimizes travel time and maximizes operating efficiency. In addition, inventory records can be constantly examined to optimize inventory levels and space allocation. Random storage and the ability to rearrange drawers within the cabinets make this possible.

The use of the computer provides tight control over storage and retrieval operations. A terminal, when equipped with high-speed printer and a customized software package, enables the operator to perform a wide range of time-saving tasks.

Computer systems are available that offer many detailed capabilities, including the ability to interface with larger host computers. Furthermore, systems are available that can simultaneously support a number of operators, each performing different functions, while at the same time keeping track of thousands of line items.

The computer may be programmed to keep current real-time records of quantities-on-hand at each storeroom or location, print reorder requisitions when inventory drops to a prescribed minimum level, and, at the same time, track the status of outstanding purchase orders. In addition, inventory reports are available that give specific cost distribution by department or by job number, and when desired, a software program can be written for direct interfacing with in-house CAD/CAM and production control systems.

11.3.7 Cabinet Storage Modes

Unlimited applications of cabinet storage can be configured in building-block progression, ranging from low-profile, "walk-around" systems to high-rise automated systems, all with equal storage density.

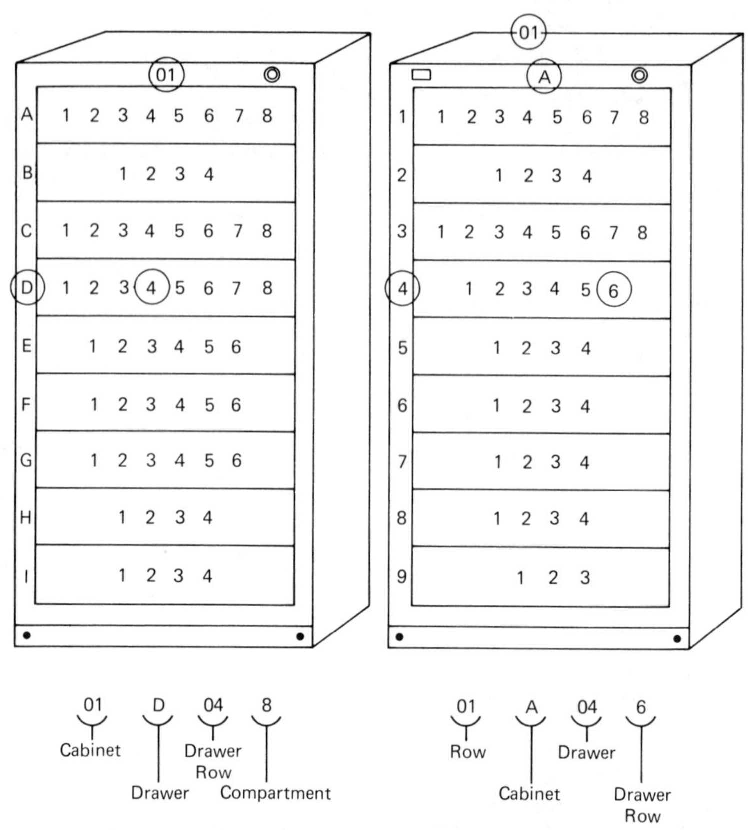

Fig. 11.3.8 The last digit in any random access locater system identifies the location of a part within a drawer. When you get to this level, the part number is identified by a label on the divider of bin within the drawer.

A low-profile, "walk-around" modular storage cabinet installation saves space and manpower, and provides a high degree of organization, security, and control. The inherent nature of a modular storage drawer installation mandates that the user be more efficient.

When inventory and throughput requirements change and space becomes more critical, the mobility and modularity of a storage drawer system allows the designer to make efficient use of available vertical air space. A platform mezzanine mode installation permits stacking of cabinets on cabinets, with the use of access stairs and safety railings. This is efficient platform storage, whereby a mezzanine is actually formed on top of a row of existing cabinets. This platform mode permits expansion of any drawer storage system with a minimum expenditure for additional cabinets, but requires no expenditure for additional new building construction.

As new products are added in the manufacturing process, the number of line items required for adequate inventory levels and storage increases. When this happens and additional storage capacity is required, air space can be utilized in the existing system by stacking cabinets two high (generally 10 ft) and back to back. Access is provided by a rolling ladder to the topmost drawers.

Stacking cabinets up to 40 ft high and using a free-path, or noncaptive aisle man-to-part picking machine for stocking and retrieving inventory increases utilization of existing storage space and drives throughput to the maximum. A high-rise noncaptive aisle system should be considered when line item inventory is high, storage activity is increasing, and throughput of the existing system does not satisfy original design requirements.

Figure 11.3.9 depicts the high-rise mode, using stacked modular storage drawer cabinets for maximum storage density, and the highest attainable levels of throughput with a captive aisle man-to-part picker. Captive aisle machines, which are rail guided top and bottom, can achieve speeds of over 350 ft/min, traveling in the X/Y axis. Unusually high throughput rates of over 120 picks per hour per machine are possible with this system. This advanced high-rise storage system can be designed and made operational in incremental phases, by adding modular storage drawer cabinets, according to

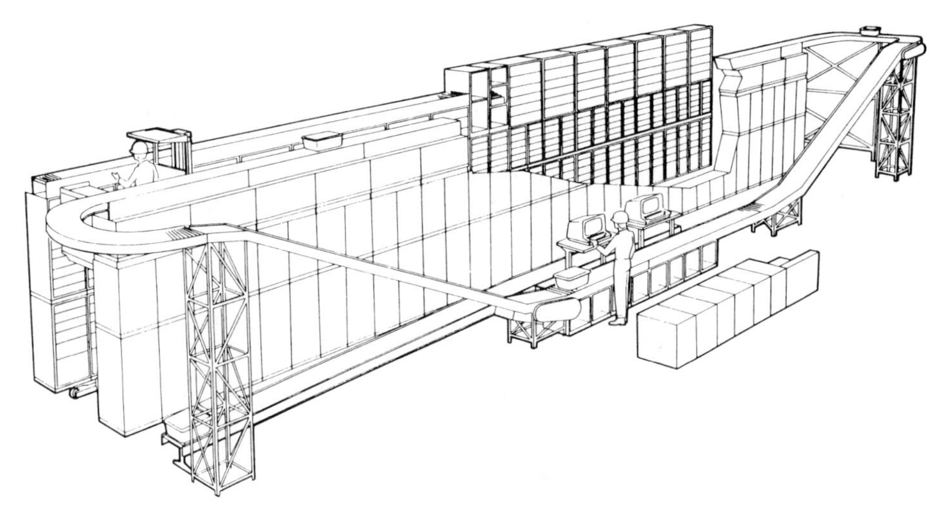

Fig. 11.3.9 High-rise modular storage drawer system.

inventory growth, stacking them, adding automated picking machines, conveyors, and computer controls as storage requirements dictate.

11.3.8 Benefits

One of the most obvious advantages offered by modular storage drawers over conventional storage is space savings. Savings in floor space frequently exceeds 50% in walk-around systems. Savings of 75–80% is not uncommon in high-rise situations.

The reduction in space required for drawer storage is a direct result of more efficient utilization of the storage cube (no vacant space above the items being stored—you're not storing air) and the high density of the stored product within each drawer.

Time, owing to faster retrieval, stocking, and inventory-taking, is another obvious benefit. Storage drawer systems reduce retrieval time by at least one-third, and generally allow inventory to be taken four to six times faster. These important savings are a direct result of better organization of inventory, shorter walking distances, 100% visibility within the drawer, and immediate access to contents within the drawer.

Less obvious, but equally important, are the benefits realized from reduced manpower required to operate a more efficient system and the consequent savings in labor costs, faster and more accurate inventory-taking, the elimination of dust and dirt because all stored items are completely enclosed, less damage, and improved security owing to well-designed cabinet locking systems.

Modular storage drawer cabinets in various configurations are often incorporated into workbenches, desks, and mobile cabinets. The unique flexibility of the drawer is the key.

11.3.9 Useful Empirical Data

In many years of experience with modular storage drawer systems, the following data have been found to be axiomatic:

1. A 59-in. high, eye-level cabinet contains an average of 9.5 drawers. A 44-in. high, counter-height cabinet contains an average of 7 drawers. These are the two most popular cabinet models used in storage drawer systems.

2. One cabinet 59 in. high × 30 in. wide × $27\frac{3}{4}$ in. deep will usually replace 2.5–4 sections of conventional shelving measuring 7 ft 3 in. high, 36 in. wide, and 18 in. deep, with seven shelves per section. (If pull boxes are stored on the shelves, each tier of boxes should be counted as a shelf.)

3. A standard storage drawer provides an average of 4 ft² of usable storage area, when compartmented.

4. When the total surface square footage of items to be stored in drawers has been determined, the number of drawers and housings that will be needed can be calculated by using the 4–7–9.5 rule of thumb. (4 represents the usable square footage of the average drawer, 7 is the average number of

drawers in a Model 245 counter-height cabinet, and 9.5 is the average number of drawers contained in a Model 340 eye-level cabinet.)

STEP 1. Divide the total surface square footage of the inventory by 4. The result is the total number of drawers required to store the inventory.

STEP 2. If counter space is to be provided, decide how many linear feet will be needed. Divide that number by 2.5 (the width, in feet, of a counter-height cabinet). The result is the number of Model 245 cabinet housings required for the counter.

STEP 3. Multiply the number of Model 245 housings obtained in Step 2 by 7. This is the number of drawers that will eventually be used for the issue counter.

STEP 4. Subtract the number of drawers that are used for the issue counter from the total obtained in Step 1. The remainder is the number of drawers that will be required for the Model 340 eye-level cabinets.

STEP 5. Now divide that remainder by 9.5. The answer is the number of Model 340 cabinets that will be required for the installation.

The number of drawers and cabinets calculated with the 4–7–9.5 rule can be used to estimate the cost of modular storage drawer equipment.

11.4 STORAGE MEZZANINES

Robert B. Footlik

A company is a candidate for a storage mezzanine if it needs more floor space, has at least a 14-ft-high ceiling, has material that it does not necessarily need immediately all the time, and if an individual can stand on a ladder and see clear space above existing stock. Any one of these criteria would indicate a need to think more in terms of cubic content than in terms of floor area. Even the most cluttered operation will have areas that are relatively low and could benefit from the use of a mezzanine above them. Natural locations for mezzanines are typically above shelving, packing operations, staging of incoming or outbound goods, washrooms, offices, manufacturing areas, and so on. In fact, any low-height operation that is presently located in a high-ceiling area could probably benefit from a mezzanine.

Typical advantages of mezzanines are that they offer twice the floor area in a given space, providing twice the number of "picking fronts," as well as twice the cube utilization. Mezzanines are an excellent alternative to building new space. The typical cost ratio between mezzanine floor area and a new building has been approximately three to one. In other words, it is impossible to build 9000 ft² or more of mezzanine for the price of 3000 ft² of new buildings. A mezzanine offers additional cost advantages by providing investment tax credits or accelerated depreciation, when compared with new construction, and if you have no land available for additional building, the only way to expand is internally through the use of mezzanine storage space.

Mezzanines are generally more secure than main floor space and provide greater control over an inventory compared with placing stock on the main level. In addition, it may also be possible to use a mezzanine to tie together various parts of an operation on a second level, where it would be impossible to tie them together on the main floor. The first place to start determining what type of a mezzanine you need is to develop a set of purposes or goals and compare these with the basic types of mezzanines available.

Mezzanines can be designed on a modular basis, piece-built above existing or proposed pallet racks or shelving, or formed in some combination of these two basic patterns (see Fig. 11.4.1).

11.4.1 Types of Mezzanines

Almost all freestanding mezzanines are modular in concept (see Fig. 11.4.2) with a basic module size of perhaps 10 ft by 10 ft (3 × 3 m) expandable in four directions. Freestanding mezzanines have posts that are formed from either cold-rolled light-gauge sheet metal, hot-rolled structural shapes, or pipe. Sheet formed uprights offer low initial cost, but have limited impact resistance from materials handling equipment. Since most of these columns taper, they require fairly large projected area on the floor. Freestanding mezzanines utilizing pipe columns and/or structural steel components are generally used for long spans and higher load capacities.

Piece-built freestanding mezzanines are commonly supported on standard steel shelving, pallet or carton flow racks, pallet racks, or drive-in racks. The choice of which type of storage equipment for the supporting level is based on the characteristics of the products to be handled. If both the main floor and mezzanine are to be used for small items, then a shelving supported mezzanine will be the most practical. If, however, these small items are handled in cartons and move out in larger

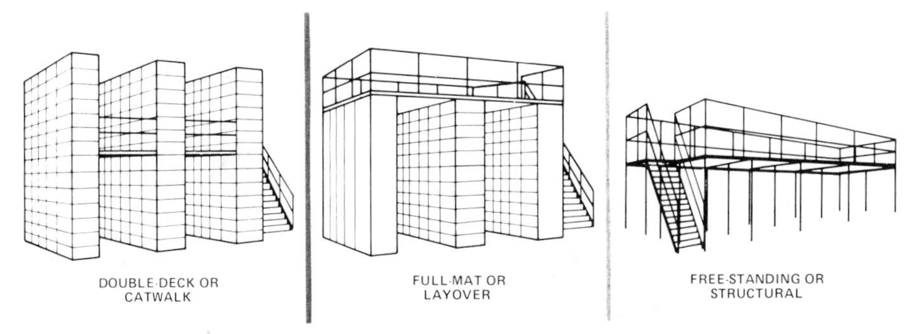

Fig. 11.4.1 Typical mezzanine structures.

quantities, then a flow rack support would be indicated; or alternatively, if the goods will be handled in and out as palletized loads, then a pallet or drive-in rack would be indicated.

Standard selective pallet rack makes an ideal mezzanine system because of the many vertical posts, and short beam spans result in comparatively light loads being transferred to the concrete floor, particularly when the racks are furnished with extra large bearing plates, say 6 in. × 6 in. × $\frac{3}{8}$ in. or larger if required. Floor capacity can be checked by design charts available in the Portland Cement Association's booklet, *Concrete Floors on Ground,* by Ralph E. Spears. (It may be necessary to core drill older floors if the concrete thickness and type of material under the floor cannot be determined from the building blueprints.) Besides the rack shelf loads, the mezzanine loading (usually in the 75–150 lb/ft^2 range) must be added to determine the total post landing. This mezzanine capacity should be factual. Contact should be made with the local building department to determine fire exit restrictions and permit requirements.

Figure 11.4.3 illustrates how easily standard selective pallet rack (hand stack configuration) can be divided vertically into three walkway levels by adding catwalk supports 3 or 4 ft long. Large

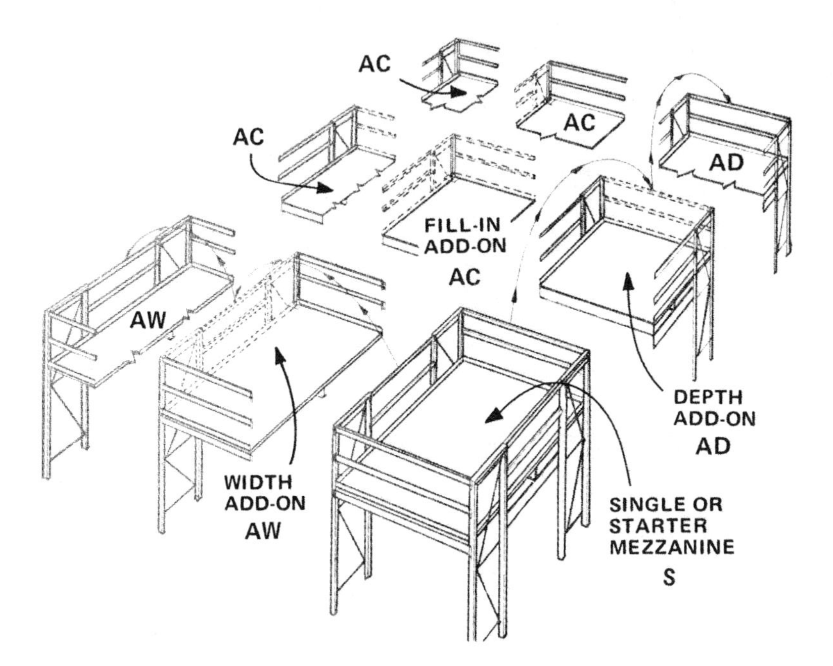

Fig. 11.4.2 Typical free-standing modular mezzanine. A starter size module can be expanded with add-on width (AW) sections, add-on depth (AD) sections, or add-on fill-in (AC) sections to finish out the row or corner. (Courtesy of Equipto.)

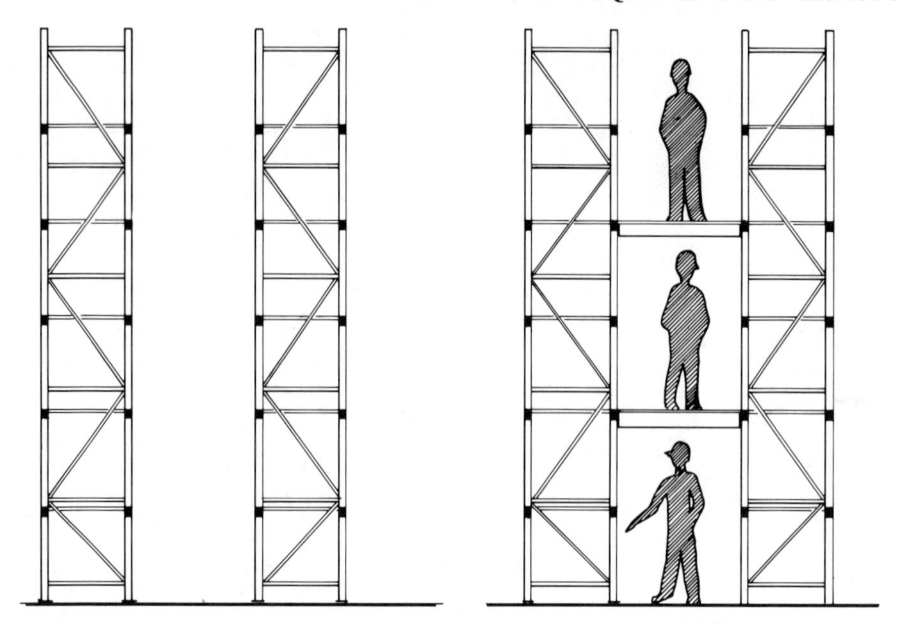

Fig. 11.4.3 Hand-pick mezzanines.

mail-order selection systems have been installed using this basic idea, plus a hand-pushed cart to carry the merchandise. A similar cantilever shelf and hanging goods mezzanine was described earlier. This type of storage is also adaptable to smaller storage requirements such as wood pattern storage for a malleable iron foundry, airplane wing template storage, cardboard records box storage, and the common tool crib in industrial plants.

Figure 11.4.4 shows standard selective pallet rack 10–20 ft high with a mezzanine deck placed on top of the rack. This top area has been utilized in many ways, some being a one-level-high pallet flow selection system for potato chips, miscellaneous light pallet load storage with hand-powered or motorized pallet moving equipment, cardboard carton storage, and 7-ft high shelving units or hand stack rack for order selection.

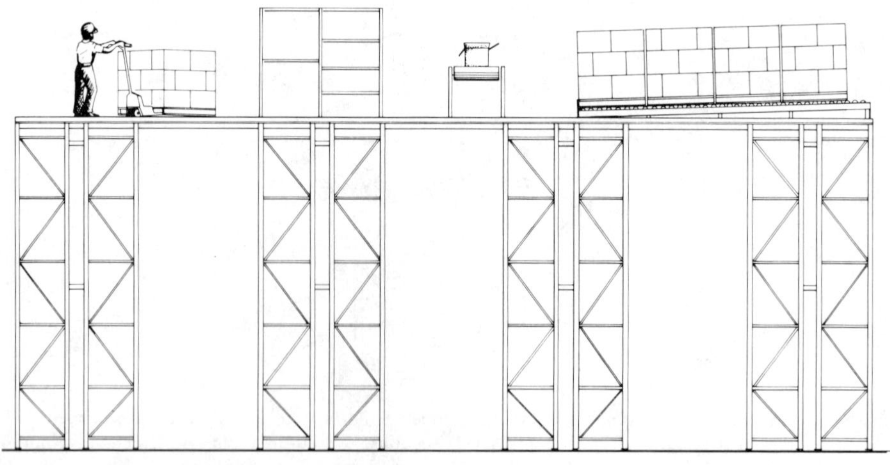

Fig. 11.4.4 Mezzanine deck system.

Many additional large mezzanine configurations have been devised, two of which are described in the flow rack section. Besides conveyors, towveyors (both overhead and in-floor) have been used to provide merchandise movement on carts.

In its simplest form, the form of storage on all mezzanine levels is identical, with the lower supports extended straight up to serve the upper levels, and a floor is utilized only in the aisles. If, however, the storage method on the mezzanine will not be the same as the storage method below, a different type of mezzanine floor, or deck, will be required.

11.4.2 Decking

When the lower and upper storage are identical, the aisles are generally decked with open grating of either the bar or plank type. Bar grating (also known as subway grating) or riveted grating offers high strength and excellent rigidity with a low-profile cross section. Balanced against this is a greater initial cost, and greater resistance to cart and pedestrian traffic. Plank-type formed steel or mesh grating is excellent for low-traffic areas and shorter spans. Its advantage of lower initial cost is offset, however, by a need to increase the supporting substructure, and the problems posed by uneven loading, which will cause uneven deflection of individual planks. Crimping or welding planks together, especially in main aisles, will form it into one unit for more uniform movement. However, the use of plank grating or mesh is not recommended in cart traffic areas because the point loading of cart wheels will cause premature fracture of individual bars.

If the type of storage on the upper level is not the same as the storage or function that will be utilized on the lower level, a full mat type of deck or flooring will be most suitable. This decking should cover the entire area (not just the aisles) and may be formed of open steel grating or solid decking.

Open grating facilitates ventilation and may eliminate the need for heating or additional lighting. In a sprinklered building it generally will not be necessary to provide sprinkler heads underneath an open grating mezzanine, unless goods will be stocked over the entire area.

When the entire floor is going to be covered with stored materials, or if people will be working at fixed locations under the mezzanine, a solid deck is recommended. With solid flooring, dual-level utilities such as sprinklers, lighting, heating, ventilating, and air conditioning will be required. When planning a rack-supported mezzanine utilizing a solid deck more than 8 ft deep, consult with your insurance company to determine fire protection requirements. Rack-supported structures in a sprinklered building must conform to National Fire Protection Association Code 231c. Different types of decking are shown in Fig. 11.4.5.

11.4.3 Layout Considerations

In addition to studying the physical characteristics and methods of storage, it is vital that the space be carefully examined to insure that it meets proper criteria to accept a mezzanine. A 14 ft 0 in. minimum ceiling height (although 16 ft 0 in. plus is preferred), floor loads in excess of 150 lb/ft^2, a minimum of obstructions in all three dimensions, and a relatively long-term lease or commitment to ownership of the facilities are of prime importance. Other areas of consideration include the necessity of dual-level sprinklers (and consequent water supply availability), lighting above and below the mezzanine, problems of ventilation and heat buildup on and under the mezzanine, product characteristics, seismic considerations, labor acceptance, and initial cost. In addition, examine the character of the materials usage above and below the mezzanine. It may not be economically feasible or rational to split product lines between an upstairs and downstairs area to fractionalize supervision and personnel with a mezzanine. A common pitfall of neophyte materials handling engineers is to place the fast moving items on the first level and the slow moving items on the second level, without giving due consideration to the fact that many of these slow moving items are frequently picked in conjunction with the fast moving ones. What is most desirable is to have a fast-moving *family* or product grouping on the main level and the slower-moving families on the secondary level, with these splits based on frequency of movement—by request, not pieces or dollars.

Mezzanines are not necessarily limited to two levels. Excellent results have been achieved with three-high, and even four-high, mezzanines in high bay areas. Many rack-supported mezzanines currently in use have second-level elevations in excess of 15 ft. This is most advantageous with palletized storage on the first level, serviced by lift trucks, and "broken case" and small item picking on the upper area. The only major limitation to doing this is how well the mezzanine can be serviced to get products to and from it.

Virtually, every form of materials handling device can interface successfully with a mezzanine. Obviously, lift trucks, whether counterbalanced or reach type, can be utilized to place goods on the mezzanine. An order-picker or stock-selector type of lift truck will not only allow goods to be placed on the mezzanine, but can also serve as an elevating device for personnel.

Automated storage and retrieval devices are being used increasingly in conjunction with storage mezzanines, especially where this mezzanine can serve also as a bridge to an upper story of an existing

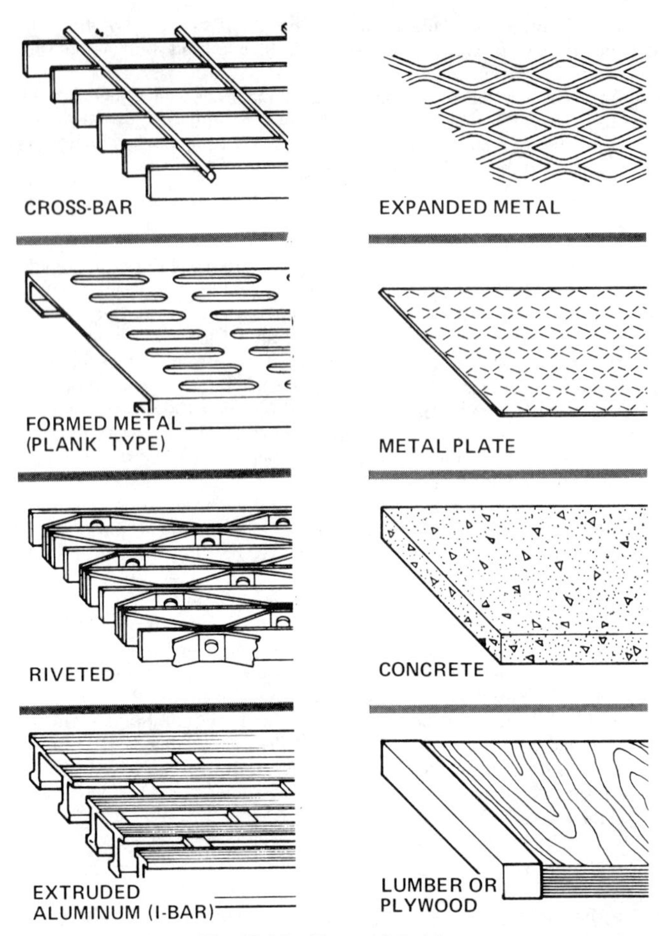

Fig. 11.4.5 Types of decking.

multistory building. By locating the mezzanine at a height half way up the height of the AS/R system, the vertical travel distance of the machine can be effectively cut in half.

Conveyors of all types can be used to service a mezzanine, and installations range from belt conveyor to power-and-free overhead systems and even the use of ramps for towline conveyors and carts. The chief advantage of a conveyor is its low initial cost. Its main disadvantage, however, is the handling of a relatively small quantity of material at a time and the necessity of having someone to put the goods onto the conveyor and someone to take the materials off. Inclined elevators and other lifting devices can be used in cases of intermittent movement, where the mobility of a lift truck is unnecessary. If an elevator or lift truck moves the goods up to the mezzanine, it may still be wise to consider the use of a slide or chute to bring the goods back down.

Any storage mezzanine that will be utilized by personnel must have at least one staircase or ladder and railings. The Occupational Health and Safety Administration (OSHA) has provided specific guidelines for how to design stairs and protective railings for mezzanines. See Table 11.4.1, adapted from ANSI standards.

11.4.4 Standards

In addition to the OSHA codes, sprinkler suppliers and insurance carriers should be consulted to make sure that National Fire Protection Association Codes 231 or 231c are being complied with. Generally, mezzanines are not fastened to the building structure, and do not require building permits for construction. If, however, the mezzanine is tied into the structure of the building, then a building permit is required and all local codes and regulations must be met. This is especially true in seismic areas such as California and Alaska. Design information for seismic zones should be obtained from

Table 11.4.1 Standard for Mezzanine Design

	Standard Number	
Subject	OSHA[a]	Other
General information	1910.21	—
	1910.22	
Stairs, fixed	1910.24	ANSI[b] A64.1–1968
Railings	1910.23	ANSI A12.1–1973
Edge protection	1910.23	ANSI A12.1–1973
Means of egress	1910.36	
	1910.37	NFPA[c] 101–1970
Ladders, fixed	1910.27	ANSI A14.3–1974
Ladders, portable wood	1910.25	ANSI A14.1–1975
Ladders, portable metal	1910.26	ANSI A14.2–1972
Metal bar grating	—	ANSI/NAAMM May 1974
Sprinklers	1910.159	NFPA 231
		NFPA 231c
Hazard marking	1910.144	ANSI 253.1–1967
Definitions	1910.21	—

[a] OSHA = U.S. Department of Labor,
Occupational Safety and Health Administration,
General Industry Standards,
Superintendent of documents,
U.S. Government Printing Office,
Washington, D.C. 20402.
[b] ANSI = American National Standards Institute,
1430 Broadway,
New York, NY 10018.
[c] NFPA = National Fire Protection Association,
60 Batterymarch Street,
Boston, MA 02110.

the specific mezzanine manufacturer. In some municipalities mezzanines should be referred to as "free-standing multilevel shelving" to avoid code language aimed at structural mezzanines and balconies. Some major standards pertaining to mezzanines are summarized in Table 11.4.1.

Installation of a mezzanine provides an excellent opportunity to improve the layout of the area beneath the mezzanine. In designing a layout, strive to get double duty from all aisles. This means that aisles that are necessary for passageway to a means of egress should also be serving for picking or restocking merchandise. All aisles must be as straight as possible and a simple, regular layout will minimize engineering costs and cut down on the number of different components required. Avoid dead-end aisles on the mezzanine, for both efficiency and safety. If a lift truck will be operating under the mezzanine, be sure there is sufficient clear height to clear the mast and overhead guard of the vehicle. In addition, check the freelift on the lift truck and determine at what height the mast of the truck begins to rise with the carriage. Little or no freelift will preclude using this vehicle under a mezzanine. In a rack-supported mezzanine, the bottom posts should be protected from lift truck abuse, particularly at cross aisles. Rack front posts can be reinforced and painted a bright color for increased visibility. All obstructions that can injure an individual or cause damage to the mezzanine or equipment should be painted orange or yellow. Similarly, unit heaters, power boxes, sprinkler risers, and other obstructions should be protected by guards or clearly delineated to minimize damage and interference with the operation. Whenever possible, horizontal runs of conduit, sprinkler pipes, or other main lines should be located with a minimum clear height of 7 ft 0 in., or alternatively, route these utilities above areas where people will not be walking.

Working on the mezzanine is just as safe as working on the main floor, provided common sense safety rules are followed. These range from the obvious, such as no smoking or spitting on an open grate mezzanine, to wearing crepe- or flat-sole shoes, which will minimize tripping. Employees should be prohibited from jumping or climbing down on the merchandise, utilizing racks as ladders, or in any other manner endangering themselves and other employees. Good-natured horseplay that is fun on a main floor can be disastrous on a mezzanine. It is best to continually remind personnel that they must think of their own safety, as well as the safety of other employees and the products being handled. Dropping goods over the side of the mezzanine may be expedient, but it creates hazards for everyone and everything.

A storage mezzanine may be the best means of expanding a building. It is the fifth direction of expansion, once you have exhausted north, south, east, and west. The only limitation to utilization of it is capital and imagination.

11.5 OUTDOOR STORAGE LAYOUT AND HANDLING

Robert B. Footlik

Because of the high cost of new construction, many companies are considering outdoor storage of materials whenever possible. Properly protected, almost any type of item can be stored in an outdoor yard for an indefinite period of time. Within climatic and operational constraints, the overall efficiency of a well-planned yard can be as great or greater than that of an indoor facility.

The potential for use of outdoor storage is limited by several factors, some within the control of a materials handling engineer, others outside his jurisdiction. Weather conditions cannot be controlled to any appreciable extent. Zoning requirements may be negotiable, and layout and product protection can be easily changed. Just as it is important to analyze needs for an indoor facility, it is even more important to analyze them for outdoor storage. Furthermore, it is absolutely imperative that the yard be designed properly, or a good deal of time and effort will be wasted.

11.5.1 Utilizing a Yard

When can a yard be utilized? Probably the most obvious factor affecting yard utilization is climate. In dry, arid regions, outdoor storage is common, because accessibility is excellent for a high percentage of the year. But outdoor storage is not precluded in other regions. For example, if a company's products are utilized mainly during the spring, summer, and fall (such as cast-iron soil pipe used for drainage systems), then yard storage makes sense, because the seasonal requirements of a company's customers are being followed. Climate is also irrelevant if the utilization of materials stored is relatively low, and the parts being stored are very large.

Unless products are extremely lightweight, or are subject to freezing or moisture penetration, the climate is not as relevant as it may first appear. Therefore, rather than starting with climate considerations, it is best to first look at the products. If they are large, bulky, and cumbersome, such as pipe, structural steel, or large weldments, then the wide-open spaces of a storage yard will be perfect for their storage.

Some materials are stored in cardboard cartons, but do not have any physical characteristics that would be changed by outdoor storage. These items can be protected by wrapping them with shrink or stretch film or other plastic films, or they can be simply taken out of their cartons and stored loose.

Sometimes physical climatic problems can be overcome through astute packaging of the materials. On the other hand, some items, as, for example, large fiberglass bathtub and shower modules, are frequently stored outside without any protection at all, since they are generally installed in a dirty environment until after construction crews have finished with the building. However, such units must be protected from high winds, either by sheltering barriers or, alternatively, by tying them down.

Slow-turnover items must be protected from dirt and from ultraviolet deterioration due to sunlight. Each product must be analyzed for its individual characteristics and the type of physical deterioration that will result from the specific climatic conditions likely to be encountered. Most suppliers can provide data on such factors as resistance to sunlight fading, embrittlement from repeated hot–cold thermal cycling, or damage to finishes.

Yard storage is often resorted to because of indoor space limitations. However, it is important that every effort be made to fully utilize indoor space before going out into the yard. The amount of travel distance involved in stocking and retrieving materials outdoors must be considered. Assume, for example, that outdoor storage places materials an extra 100 ft away from where they are needed. This means an extra 200 ft of walking for each retrieval. Multiplying this figure by 20 trips per day, 5 days a week, 52 weeks a year, shows that personnel will have walked an extra 197 miles in the course of a year. Thus, every yard should be treated in the same manner as the interior layout of a building. Space utilization should be maximized, and goods should not be spread over the entire area merely because the space is available.

Security is of utmost importance in planning a storage yard. It is not just a protection against theft, but is also a deterrent to vandalism. A security program should be designed to keep out juvenile offenders, as well as deterring adult professionals.

Safety is another important consideration. All too often, there is a tendency to short-cut or bypass safety procedures when working in the yard. For example, lifting equipment may be operated beyond its rated capacities, or under unsafe or even impossible conditions. And loads that are safe to handle indoors may present special hazards when equipment must operate in high winds or on slippery surfaces.

11.5.2 Site Conditions

The physical shape of the yard generally does not represent the major layout factor. Even the most impossible-appearing shapes can be utilized to some extent. However, yard shapes may be determined not only by obvious boundaries such as fences or survey lines. Other factors come into play, such as easements, railroad sidings, rights of access, set-backs, and special zoning considerations. All these factors will modify the apparent physical shape of a yard.

A good way to begin evaluating a yard site is to list its assets on one sheet of paper, and compare them with disadvantages or problems on a second sheet. Frequently some factors will appear on both pages. For example, a 75-kVA power line separating two adjacent areas, and requiring an easement, may at first appear to be a disadvantage. However, in actuality it is frequently possible to rent the space directly under such a cable for a very nominal fee, and the space can be utilized for additional yard storage without any major expenditure of capital for land acquisition. A good yard layout capitalizes on such advantages, but until they are identified they cannot be utilized. Other potential assets include drive-through traffic flow to minimize backing and turning, proximity to buildings and shed areas, potential for low-cost expansion through leasing of other property, and utilization of common driveways.

Along with the physical horizontal shape of the yard, vertical conditions must also be examined. A small change in grade may be highly desirable for drainage, but if the yard drains in the wrong direction, an undesirable lake may materialize. Therefore, the site survey should include not only the boundaries and shapes, but also the topography and contours of the site and its surrounding environment. For example, a flat area may look ideal for a site at first glance, but if it is at the bottom of a "bowl," with all other land in the area draining into it, it may not be suitable for storage. Items might float away with the first major rainfall. Large yard areas with great changes in elevation may require extensive contouring and terracing to become usable. At the same time, the cost of water removal and drainage may be prohibitive. No yard layout can be made without first examining these topographical features.

Almost any problem of topography can be overcome, but economic feasibility must be considered. If fill is plentiful and inexpensive, it may be possible to build up the site. However, if the cost of fill per cubic yard is extremely high, it may not be economically feasible to raise the yard above the flood plain.

Sites located in a flood plain are not precluded from development, provided the items stored will not be damaged by water or, alternatively, if they can be placed on piers above the flood level. In some cases, it may be acceptable to build up only a specific portion of the area above flood level.

In addition to looking at vertical dimensions and topography, the engineer should also check soil conditions. Wet, spongy soil that is incapable of withstanding the forces of the handling equipment can result in having to tow or dig out equipment or make major expenditures to develop a system of driveways or storage areas capable of sustaining loads from materials and vehicular traffic. Soil tests taken by competent specialists will point out specific problems and indicate recommended courses of action.

Typical driveways for truck traffic should be at least 8-in.-thick concrete over a 6-in. crushed stone base, or 3-in. asphaltic paving over 8 in. of base stone. Forces imposed directly under the tires of a slowly moving highway truck are considerably higher than for a truck moving at a steady 30 mi/hr due to the slip angles of the tires and the dynamics of a relatively static load versus a rolling load. Although 2-in. thick paving may be adequate for a side street or automobile traffic, it will quickly be ground away by backing and maneuvering of semi-trailer trucks.

Paving the entire yard area facilitates lift truck traffic as well as highway vehicle traffic. When doing so, it is best to keep the paving as uniform as possible, and thus provide for greater future flexibility in the use of aisles and storage areas. Using 8 in. of concrete for driveways and 6 in. for storage areas may save some yards of concrete, but as time passes people will begin to forget which are the aisles and which the storage areas. The increased maintenance costs that may result will offset any upfront savings. Concrete and macadam are priced as materials only. All finishing operations take place on the surface. Therefore, the extra labor involved in placing the concrete is minimal, and the incremental cost of using one uniform surface is usually cost effective.

Zoning is another consideration. Most municipalities have specific rules regarding fencing and other factors that may affect feasibility of yard storage. Inquiries should be made with the local zoning board or building commissioner before planning for outdoor storage is started. In many areas storage may not be allowed unless it is screened. Screening can involve anything from a few bushes, to metal or plastic slats placed in a chain link fence, or even to a brick wall around the yard. Although screening may be desirable from an esthetic point of view, it can pose a security threat to the yard owner because police or security service personnel will be unable to look directly into the yard when driving past it.

Just as it is necessary to check with the local zoning body, it is also wise to check with the Environmental Protection Agency if the products involved might pose any hazard to the environment. It is also desirable to alert the local fire department or safety agencies regarding specific product hazards. Doing this up front will generally save problems in the future. At the same time, inputs from the local municipality might be quite valuable for planning purposes.

11.5.3 Principles of Yard Layout

Most of the rules that apply for indoor storage are equally valid outdoors. Straight line flow, clean, simple driveway layouts, and uniform, orderly arrangement are, of course, of primary importance (see Fig. 11.5.1). If there is any single principle to watch for particularly, it is that goods should not be spread out merely because the space is available. As has been pointed out, a relatively modest spread of materials can become quite costly in terms of time and effort involved in retrieval.

As a general rule, stock should be stored together and in a uniform manner. The same storage units that are used indoors may also be obtained with heavy paint jobs or galvanized finishes for use outdoors. Stacking frames, pallet racks, cantilever racks, and other storage devices allow better utilization of available height so that goods are not spread horizontally.

Flexibility is important for indoor storage, but even more so outdoors. Generally most storage yards accommodate a wide variety of goods having unusual sizes or shapes. The very nature of a yard is that it be flexible, and any layout that is developed should enhance this flexibility, not stifle it. For this reason, all driveways should be maintained in easily accessible straight-through configurations. Minimum driveway width for one-way trucks is 12 ft, with access to the drive being provided from a straight-through gate, or alternatively from a wider main aisle. Preference should be given to 15 ft wide aisles for ease of maneuverability.

Main aisles for two-way truck traffic should be at least 24 ft wide, with greater space provided if necessary. If a pedestrian walkway is provided through the yard, it should be at least 4 ft wide, located adjacent to driveways and clearly marked. If highway trailers must be unloaded from the side instead of from the top or the end, driveways must be widened to accommodate this procedure. They should also be sized to accommodate the largest loads traversing them. It is difficult and unsafe, for example, to take 20-ft long loads down a 12-ft wide aisle, because the loads must of necessity be lifted over stored items.

Adequate areas must also be set aside for high-hazard storage, parking, and staging of inbound or outbound materials. In northern climates, easily accessible locations for the piling of snow will make snow removal more efficient, and greatly reduce the effects of bad weather conditions. Whenever possible, all fixed equipment such as gas pumps, jib cranes, production equipment, or fixed craneways should be located out of the path of main expansion of either the building or yard. In addition, such equipment should be protected against possible damage by highway trucks or materials handling equipment.

Typically, posts and bumpers are used to completely surround an item to be protected. All bumpers should be painted orange or yellow for high visibility under overcast and night working conditions. Unpainted steel bumpers may be very visible during the day under 5000 foot-candles of light, but virtually invisible at night under 3 foot-candles.

Orderly storage is absolutely imperative. All aisle markings should be applied to paving or stakes in a highly visible manner. Unless aisle boundaries are clearly delineated, throughput and efficiency in the yard will be greatly diminished. Similarly, any yard that will be used under adverse conditions, such as cloudy overcast days, night, or fog, should be well illuminated. Preferably high-pressure sodium fixtures should be used to delineate aisles, and enable personnel to see materials without glare or dark spots. Alternating light and dark areas should be avoided to maintain optium visibility. It generally takes 3–8 min for a person's vision to adjust completely to night conditions, but only a matter of seconds to lose this night vision when walking into an over-illuminated area.

Location of gates is especially important in obtaining maximum yard efficiency. A single gate offers excellent control with a single point to be watched and a single key to use when it is locked. However, if there is only one gate, trucks must be turned around within the yard, or backed out through the same gate. This procedure is satisfactory in large yards. However, small yards may benefit from having two gates to provide a drive-through path. Control of the gates is then solely reliant on the honesty and integrity of personnel; multiple security devices or guard houses may be necessary.

If a gate is part of a fence at a railroad siding, the railroad must be given a key or other means of entering the yard to deliver cars as necessary. Security may thus be compromised, but the only alternative is to dictate to the railroad that cars can only be spotted during normal business hours, or else an elaborate system of inner and outer fences may be necessary to maintain control.

The type of gate used depends on its function. If the purpose is to simply keep out casual vehicles and truck traffic, then a chain across the fence opening, or around the entire yard, may be sufficient. However, if higher security is desired, then a sliding or rolling gate may be the answer for a large opening. Small openings can use hinged gates, provided that sufficient space is available to allow the swing to take place unobstructed. Security gates that will withstand the impact of a fully loaded semi-trailer truck also are available, but at a high cost. Ideally, all gates should be painted a high-visibility color to minimize the risk of damage by vehicles.

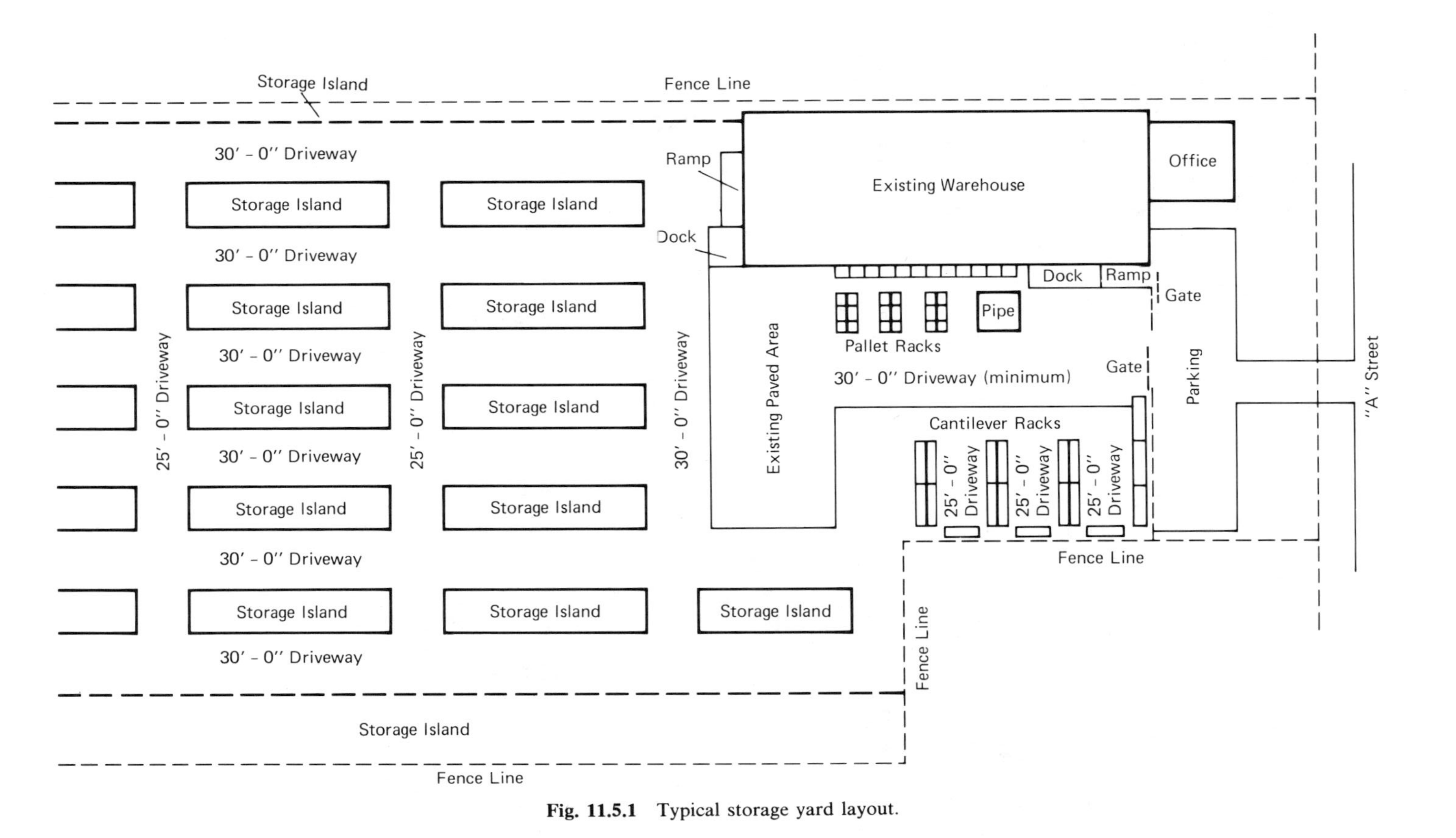

Fig. 11.5.1 Typical storage yard layout.

11.5.4 Materials Handling Equipment Selection

Equipment must be selected on the basis of the rugged handling and abuse it will receive under field conditions, now and in the future. The working environment definitely shapes the type of equipment needed. Working in an unpaved, muddy field, for example, requires far different type of equipment than that for working on a paved, smooth, clean surface. Similarly, four-wheel-drive or hydrostatic-drive equipment may be required for operation during inclement weather. Any proposals on materials handling equipment must spell out in detail all field conditions that are likely to be encountered. Equipment must be specified that will be up to the tasks.

For example, diesel-powered lift trucks that are stored outdoors, at sub-zero temperatures, will be much more difficult to start and maintain reliably in winter than at other times of the year. And, depending on the engine configuration and the type of vehicle, as much as 3–5 min may be required for warmup time before the equipment will operate reliably. Storing the truck indoors, or adding crankcase, head-bolt, battery, and radiator heaters will significantly improve its utilization. However, this type of accessory equipment should be specified in the original proposal, and not added later. Provisions also must be made for plugging in these accessories in the field.

In terms of flexibility, the same general rules apply outdoors as indoors. A wide variety of lift truck configurations is available, as well as a full range of accessories to make working outdoors safer and easier. Virtually every manufacturer of forklift trucks for outdoor use offers an all-weather cab, windshield wipers, headlights, special tires, and other equipment to adapt a forklift to a wide variety of operating conditions.

Somewhat less flexible, and suited primarily for outdoor use, are mobile cranes. This family of equipment offers flexibility by being able to reach over some materials to get to other items. Such cranes are generally more expensive than forklift trucks. However, they can offer savings in such factors as paving costs, particularly if driveways are fully paved but storage areas are unimproved. With a mobile crane, it is possible to go over difficult terrain and still achieve high storage densities.

A disadvantage of such a crane, however, is the need for putting a hook or sling on the load. This operation may require a two-person crew to perform a job that can be achieved with one person and a lift truck. Various attachments and lifting devices can be used to get around this problem, however, providing their cost can be justified. One example is a magnet used for handling ferromagnetic materials without a hook or sling.

All things being equal, the mobile crane is the single most popular device for handling very large or highly irregular objects over irregular terrain under adverse conditions. Also, the mobile crane can be adapted to highway conditions, and taken to the point of delivery, to ease placement of materials at a factory site.

Less flexible and more specialized types of equipment are fixed cranes. These units may be gantries, jibs, monorails, or underhung or top-running cranes. They are all fixed-path devices and confined to a specific yard area. Depending on the size of a yard and its shape, the fixed-path configuration may be an asset or a liability.

A typical application could include the extension of an indoor crane through a special door or hole in the wall so that the unit can work outdoors as well. In this manner, large weldments or similar objects can be moved from the outside directly back into the indoor facility, with a minimum of rehandling.

Horizontal movement of materials can be accomplished with trucks, burden carriers, driverless tractors, carts, dollies, or other wheel or track equipment. Storage yards that already have rail facilities may be able to operate very efficiently with rail-mounted equipment, such as cranes, gondolas, or flat cars, without requiring total paving of the yard. Generally, this type of equipment is used in conjunction with bulk handling, rather than individual piece handling.

Straddle carriers of various forms can also be used successfully for yard storage. Long, thin items, such as pipe and structural steel, can be safely transported at relatively high speed, and stored in a very dense array by utilizing a straddle carrier or mobile straddle hoisting device for both lifting and transportation. With a linear layout, a travel path as narrow as 2 ft can be left between rows to accommodate the wheels of this vehicle, with the loads passing over other stored materials, so that the aisles are "in the sky" instead of on the ground.

All this equipment can be tied together with radio communications or a public address (PA) system. Generally, outdoor speakers have proven to be the least expensive route, except in very large yards. However, they offer only general communication, and broadcast their message to everyone, including neighbors in the area.

Two-way radio communications over FM frequencies provide an extremely reliable means of reaching individuals and giving them separate assignments. Systems for on-line, direct computer access, as well as voice communication, are available from several manufacturers. Use of this type of equipment in conjunction with light pens, which can read uniform product or bar codes, can provide a high degree of reliability in maintaining yard inventories on a computer. In addition, these radio links can also keep track of personnel and assigned tasks, and maintain time standards and facilitate management of a large yard. Efficiency and communication go hand in hand. A yard layout should not be developed without an accompanying communications and control plan.

11.5.5 Stock Arrangement

One weak area in many outdoor storage operations lies in the orderly arrangement and maintenance of inventory control of the stock. All too often, materials are stored haphazardly by a "memory locator system." Before any materials are stored outside, a uniform, consistent method of marking the materials should be established. Various metal and plastic tags can be used for marking items. Generally, they are quite durable. Use of such tags also permits color coding to indicate periodic changes related to product, season, or year of storage, providing a control for product rotation and identification.

Items placed outdoors should be properly protected from the elements. Stretch or shrink wrapping are alternatives for protecting materials that can be harmed by rain or snow. Other types of protective coatings can be used over individual castings, weldments, or other fabricated pieces. Plastic wrapping is not necessarily a universal panacea. In some cases it can permit buildup of condensation due to temperature variations. Such condensation can be corrosive to some products. Thus, selection of any protective coating should be tailored to the physical characteristics of the particular material.

All storage locations should be labeled with some form of easy reference guide to address specific positions. One frequent approach is the "post office system," with odd numbers on one side of a driveway, and even numbers on the other side. Individual drives can be labeled with street names or alphabetical codes. Individual location addresses should be tied in with a system of move tickets or computer addresses to keep track of materials.

A simple move ticket, with one part staying on the item, and a second part coming into the office to be stored in a tub file or computer file, will significantly enhance control of materials stored outdoors. In northern climates, where materials may be covered with snow, the addresses should be prominently displayed on sign posts or suspended ropes. If inventory is taken in winter, having part of the move ticket on the item will make the job much easier, as will a computer printout on a location-by-location basis.

11.5.6 Physical Considerations

All operations should take into account the extra travel distances that are necessitated by the large areas that outdoor storage frequently covers. Movement of personnel and equipment must be kept to a minimum. Therefore, personnel should be trained to try to take something out to storage when something else is being retrieved. In addition, if parts are of a size that will allow for multiple picking and multiple stocking, then all work should be performed in batches. Many facilities have operated quite successfully by utilizing trailers towed by lift trucks so that several storage and retrieval operations can be accomplished on each trip into the yard.

Another excellent time saver is handling all paperwork indoors, so that switch time is minimized. This approach is especially valuable during inclement weather, or where a multitude of very similar looking parts will be placed into storage. Naturally, on-line direct communication, either by personnel or computer, can significantly improve overall efficiency and productivity of personnel.

Cleanliness and safety are operational necessities in the yard. Materials should never be dropped in aisles, where they will impede flow of traffic or create a safety hazard. A continual training program is highly recommended to ensure that personnel understand proper and safe methods for handling materials under all conditions.

All dunnage, strapping, packing materials, and other items that can cause damage to equipment, or present a hazard to people, should be eliminated as soon as possible. Part of the personnel safety training program should stress the importance of not only spotting dangerous situations, but also correcting them immediately. Hazards such as loose and unstable loads, large splinters or nails at eye level, pot holes, and damaged equipment should be reported by personnel so that appropriate action can be taken early.

Additional instruction should include a thorough working knowledge of the equipment that will be used, including information on how to turn off power sources or disconnect the equipment should any potential hazards arise. On a very practical level, the training should also consider times when a person would be expected to run for his life, versus the times when he should stay and combat any particular hazard.

Operational procedures should also cover all aspects of security, including the proper methods of working with outside truckers, creation of bills of lading, proper receiving techniques, handling recovery of damaged materials, treatment of outsiders visiting the yard, and so on. Employees should in general be discouraged from having access to the storage area with their private vehicles. Similarly, control of all vehicular traffic should be maintained at the entry and exit of the storage area.

Detailed treatment of every aspect of outdoor yard storage is virtually impossible. Products, local conditions, management philosophies, zoning, topography, building, soil conditions, and a host of other items all dictate the final size, shape, and utilization of the outdoor area. The rules for indoor storage apply doubly for outside, simply because of the number of factors outside the company's control. One does not normally expect high winds, or damaging hail inside a structure, but these are common occurrences with outdoor storage. When all these factors are taken into consideration, it can be truly said that the sky is the limit when it comes to what can be stored outside successfully and economically.

11.6 MOBILE STORAGE SYSTEMS

Terry Strombeck

The economy of industrial mobile storage systems results from the efficient utilization of the nonproductive space whereby storage units replace the many access aisles required by nonmobile stationary storage racks. This high-density capability of the mobile concept increases storage capacity by 100% or more, as compared to fixed rack or shelf systems requiring the same space, but still maintains 100% selectivity.

For instance, in the same given area with 100% selectivity, a mobile system has a capacity for 1000 pallets, as compared to only 500 pallets in a static (stationary) system with multiple aisles. In effect, mobile systems can double the storage capacity in a given area without doubling the floor space, or reduce up to 50% the space required for current capacity.

11.6.1 Mobile System Operation

In actual operation, the shelves, racks, or other storage components are installed on wheeled carriages that are mounted on rails; the system is compact and requires only one access aisle per bay of carriages. Operated by push-button controls mounted on the face of each carriage, or remotely controlled from forklifts via radio transmitters, one or more carriages can be moved left or right automatically and simultaneously to open up the single "transposable" aisle in the position desired within the module system (see Fig. 11.6.1).

11.6.2 Mobile System Flexibility

The inherent flexibility of the mobile concept permits custom designing of each system to carry almost any material in any configuration and in any plant location. Column obstructions, architectural constraints, and other building barriers require engineering and tailoring a mobile system to the special needs of the user.

A variety of mobile storage units is available for use with wide-span shelving, pallet rack, cantilever rack, industrial containers, special-purpose racking, and other types of various storage equipment available today.

Fig. 11.6.1 Mobile pallet rack system requires only a single "transposable" aisle for entry. Carriages are automatically moved by remote radio control or push buttons on carriages. Mobile carriages can be moved one at a time or all simultaneously, across rails embedded flush with concrete floor.

100% Selectivity. Mobile systems offer 100% rack or shelf selectivity at all times. Some systems can provide storage density but at the expense of selectivity and/or access. The mobile concept delivers maximum density with full 100% selectivity.

Hand or Lift-Truck Loading. The flexibility of mobile high-density storage systems accommodates either order-picking or lift-truck loaded installations—and the systems will handle any material regardless of bulk, size, weight, or configuration. These systems can range from perishable storage to steel warehousing, and they can handle any payload a materials handling and storage operation might require. There are no restrictions to type of rack or shelving. The carriages can be designed to carry any material up to 250 tons per individual mobile carriage, extending to 200 ft in length and 40 ft high.

Fast and Easy Access. Mobile systems can be tailored to the activity level of any type of operation. All that needs to be determined is how often access is needed to a storage position. And regardless of whether it is rack storage or unit storage on shelves, quicker and easier order-filling is possible because it is in a smaller space and therefore requires less travel time for an operator.

CVU Factor. It is commonly known that the average industrial storage floor space consists of 40% storage and 60% access aisles. The elimination of nonproductive access aisles maximizes the available footage of present building space for more material storage (see Fig. 11.6.2). Further, the dollar savings include not only cubic volume utilization (CVU), but attendant energy costs for heating,

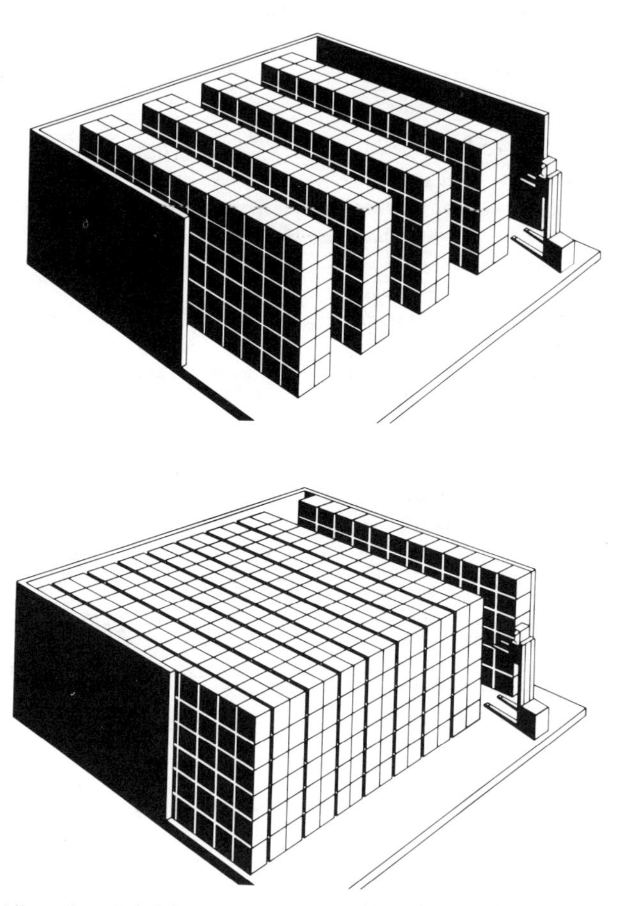

Fig. 11.6.2 Mobile rack and shelving systems save considerable space in comparison to conventional systems because only one aisle is needed. Racks, shelves, and modular drawers can be installed on the track-mounted, motorized carriages, which can be shuttled side-to-side to provide access where needed.

lighting, and air conditioning wasted aisle spaces. In frozen food storage this cost can be exorbitant. A complete CVU analysis of existing storage facilities is essential, whereby recommendations include layouts for the most efficient use of present storage areas and also details for planning future growth and expansion utilizing mobile systems.

Hand-loaded Storage. Case-type, wide-span and other types of storage equipment can be tailored to accommodate any hand-loaded goods stored such as parts, case goods order-picking, supplies, tools, records, automotive TBA parts, electrical parts, and paper goods. Interfacing and mixing of material storage within high-density mobile storage modules can be achieved because of the inherent flexibility of the mobile concept.

Material Control and Security. Spacesaver systems minimize material damage because the concept compacts the material stored into a protected and controlled "cube of storage." Therefore, only a small percentage of storage positions or goods are vulnerable to damage at any one time. Modules, when compacted and locked, reduce the opportunity of pilferage.

Justification for Mobile Storage Systems

1. Cost of floor space.
2. Cost of land.
3. Availability of floor space.
4. Availability of land.
5. Value of revenue-producing space versus storage space.
6. Reduced energy costs (less space to heat, cool, and light).
7. Tax savings over construction—10% investment tax credit. Excellent depreciation. Less real estate tax.
8. Lower insurance cost.
9. Lower maintenance cost (janitorial).
10. Security justification.
11. Expanding capacity in some area reduces transfer and transportation cost.
12. Larger capacity may allow for large quantity purchases, reducing unit cost of item stored.
13. Large stock capacity of finished goods may allow larger production runs, reducing unit production cost.
14. Large capacity in area may allow for better stock control or categorization, reducing stocking and picking time.
15. Reduced fire exposure because of increased density in the cube.
16. Labor reduction, by material placement.

11.6.3 Selecting and Specifying

When selecting, specifying, and installing an industrial mobile storage/materials handling system, factors such as floor loading (live load weights and capacity), turnaround selectivity, and levelness of the installation must be evaluated. The design and engineering of a mobile system depends on the structural configuration of the frame members relative to the dimensions of the materials to be stored and the type of shelving, rack, or other storage facility.

Mainly, it is important to have available trained installation engineers, especially since these large industrial-type mobile systems are tailored and custom designed for each specific application. Space analysis consultation and recommended facilities layouts are available from engineering departments, along with locally available service and maintenance personnel.

Briefly, the following are some specification of a mobile system to be considered:

Wheels. Storage loads are distributed directly to the wheel assemblies. There are three types of load wheels utilized: driven, nondriven, and guide. The load capacity range of each wheel is from 3000 to 30,000 lb. Each wheel carries the payload directly from the upright framing member (see Fig. 11.6.3). Carriage wheels are maintenance free with lifetime lubrication and friction-free heavy-duty flange-mounted bearings.

Rails. Rails are permanently installed by setting into concrete on leveled support pads spaced by load requirements. The top surface of each rail is set flush with the existing or new floor surface. The spacing between rails in equal to the dimension of the rack bay width. Rails are located directly under each rack upright member. Because of the set-flush installation technique, rails present no obstruction to forklift truck travel.

Fig. 11.6.3 Heavy-duty wheel assemblies on the mobile system have load capacities from 6000 to 30,000 lb. Maintenance-free wheels are available, with lifetime lubrication and friction-free, heavy-duty, flange-mounted bearings.

Drive Systems. Each mobile carriage in the system contains its own individual drive system. At the heart of each drive system is a reversible electric motor with integral gearing. The low frictional resistance of tracks and wheels allows the use of either $\frac{1}{4}$, $\frac{1}{2}$, or 1 Hp motors depending on the load requirement of each carriage.

Control Capabilities. Mobile pallet rack systems incorporate the latest state-of-the-art electronics for total control and safety. Push-button controls are installed on each carriage face. Additional control devices are available such as remote fork truck command, central command control unit, key-operated circuits, night circuits, and emergency overriding control circuits.

Height Limitations. The height restriction of the system is that of standard pallet rack systems, that is, 6 times the centerline dimension of the load wheels to the top of the top load. For applications requiring heights above the normal limitations and for seismic zoned areas, a special antitip mechanism is designed into the system.

Carriage Movement. Each carriage contains its own drive and control system. When more than one carriage is to be moved to create an alternate new access aisle, the command is passed via control logic to the individual carriages. Sequentially, the carriages clear each other by approximately 1 in. and maintain this dimension until they reach the desired location.

Safety Systems. There are three standard safety systems on mobile storage systems:

1. *Safety Sweep.* Safety sweep is located at foot level along both sides the entire length of the carriage (see Fig. 11.6.4). When contacted by an obstruction in the aisle during movement, the carriage stops instantly and requires the reactivation of the module by the controls located outside of the access aisle.
2. *Central Safety and Power Control Unit.* Each module of carriages in a system is provided with a central safety and power control unit, with a lockable master switch, a lockable key operated switch, "working and malfunctioning" signal lamps and a 220/24 v, 60 Hz control transformer with an isolated winding.
3. *Warning Horn and Lights.* The system includes warning horns and flashing lights which operate in advance of and during the system carriage movement.

Seismic Criteria. Carriages, racks, wheels, tracks, and footings are beefed up for not only the weight loads (such as for steel), but also to meet the uniform building codes and earthquake seismic Zone IV requirements, where mandated.

Fig. 11.6.4 Each carriage has a full-length, safety floor sweep on both sides of all carriages. Pressure against the sweep causing movement of $\frac{1}{16}$–$\frac{1}{8}$ in. deactivates carriage movement. All safety sweeps are finished with visible warning striping.

11.6.4 Product Application

Regarding product application the markets are virtually limitless, but some markets have greater potential for mobile material handling systems. The largest single potential is manufacturing facilities. A few areas within a large manufacturing operation are:

In-Process Material. This is possibly the single largest storage problem area, as all other areas of a plant could be affected by the handling and storage of this phase of the operation. In addition, it generally occupies the most costly floor area because of its need to be located as close to production as possible.

Purchase Parts. This is material purchased from outside vendors and used in the finished product. For economic reasons, large quantities are purchased and must be stored until needed.

Finished Goods. Ideally, a manufacturer would like to eliminate the need for storing finished products, shipping directly to the customer. For obvious reasons, this is not possible; as a result, the manufacturer is open to any solution to this major overhead cost.

Distribution. Many times the distribution area is located in a leased facility some distance from the plant. For this reason, it becomes a high-cost area. Mobile storage systems could free up space located at the plant site, thus justifying the system.

Parts Room. Always a high-cost area in a manufacturing facility because of its prime location close to production lines. Ideal application for a mobile system.

Tool Room. Prime area because of the low activity. Generally takes up a large area of the plant because of necessity to store fixtures and tooling for production line changes.

Record Storage. All major manufacturing facilities have a records storage area for production and tooling drawings, quality control production records, and inventory control records. Again a slow-through put section utilizing valuable floor area.

Although industrial materials handling mobile systems are a relatively recent development in the United States, they have opened up a new concept in space conservation, and have been accepted as an integral part of materials handling and parts storage and retrieval. These mobile systems are providing solutions to costly space and storage problems.

CHAPTER 12
STORAGE SYSTEMS

CHARLES E. MANLEY

The Raymond Corporation
Greene, New York

DAVID L. FOWLSTON

The Raymond Corporation
Greene, New York

ALEX J. NAGY

Materials Handling and Warehousing Consultant
Farmington, Michigan

BRIAN W. SANFORD

Litton Unit Handling Systems
Florence, Kentucky

W. R. MIDGLEY

Midgley, Clauer & Associates, Inc.
Youngstown, Ohio

DONALD J. WEISS

White Storage & Retrieval Systems, Inc.
Kenilworth, New Jersey

JOHN CASTALDI

Supreme Equipment & Systems Corporation
Brooklyn, New York

12.1 MAN-ABOARD SYSTEMS

Charles E. Manley and David L. Fowlston

12.1.1 Scope

This section is limited to the discussion of floor-supported vehicles and does not cover man-aboard equipment supported only from overhead.

The applications that are discussed include evaluating an existing system, expanding existing storage systems, planning new storage systems, and equipment.

The two primary warehouse functions which apply in this instance include:

Order picking—the fulfillment of customer orders in less than unit or pallet loads; and

Pallet handling—the handling of full pallet loads of materials.

Man-aboard vehicles can fulfill both of these functions.

12.1.2 Evaluating an Existing Warehouse System

Operational data is essential when evaluating the performance of an existing storage system. Without this information, any solution for improved performance will be a "band-aid" fix, addressing the symptom rather than the cause. Operational data will not only supply information as to the efficiency of the physical storage, but also provide a means to evaluate the system financially. A new or improved storage system can be easily justified if it results in a large savings in operational costs.

Four areas must be considered when collecting operational data: physical storage operation, operating costs, activity, and personnel.

Physical Storage Operation

The most obvious starting point to obtain operational data is to look at the warehouse equipment. This includes the load or material handled, storage aids, and storage vehicles. Considerations to identify include:

Number of stock-keeping units (SKU) to be stored.

Type of storage pallet, skid, or container used. A pallet can be constructed as a two-way, four-way, single face, double face, flush, or wing type.

Dimension of the pallets, skids, or containers (length, width, and height).

Type of storage aids, or racks, used to support the materials (height, distance between, length of openings, and material thickness).

Orientation of the materials within the storage systems; that is, how the inventory pattern is arranged (unit loads, less-than-unit load, location due to activity, and selectivity).

Number of inventory "turns" annually.

Activity or frequency of movement in and out of the storage system (the number of loads or orders per shift).

Type of vehicles used to move the material.

Number of people required to operate present system.

Type of system used for location control (how the material locations are assigned, stored, and issued for retrieval).

Type of communication (formal and informal).

Type of paperwork issued and received.

Operating Costs

Operational costs are more difficult to obtain; however, the input provided from this information is critical for any assessment of an existing system. They are usually assigned a dollar-per-square-foot value for total costs of operations.

These continuing costs of operating include:

Building costs/land costs (amortization or depreciation)
Leasing costs
Taxes
Insurance
Fire protection services
Utilities (electric, gas, water, sewage, oil, phone, etc.)
Maintenance (building and grounds)
Security
Services required (snow, garbage, and scrap material removal)
Employee salaries

Operational costs can also be converted to their "cost of storage." The total cost can be divided by the number of locations to give the "cost per location." The type of financial justification will suggest the use of either dollars per square foot or cost per location.

Activity

To determine the nature and volume of activity, several steps must be taken:

Classify the product to determine how to handle it, and how the related storage aid is to be used. All products will fit into one of three categories: small parts, case/carton, and irregular shape.

Quantify product activity, both in unit-load and order-picking modes. Total inbound and outbound quantities per shift are required.

Determine the number of stock-keeping units (SKU) being stored. Usually one slot or pick face is required for each unit, especially if total selectivity is needed. The 80/20 rule always applies here (i.e., 80% of the activity will be on 20% of the SKUs).

Classify SKUs. Two methods are normally used: by activity for each SKU (turnover) and by dollar value of each SKU.

Choose one of two location methods: fixed slots where each SKU has a permanently assigned slot, or random slots where each SKU occupies any available empty slot. A mixture of the two systems can also be utilized. Fixed storage systems usually cause more honeycombing (wasted space).

Personnel

Quantify all personnel involved in the current warehouse operation. Include all personnel from the receiving operation through the shipping department. Examples are management, receivers, order pickers, and stock clerks.

12.1.3 Expanding Existing Storage Systems

Expansion or modification of an existing system is a cost-effective way to improve the materials handling operation. Existing storage aids and associated equipment can be utilized and much of the modification work can be performed by the maintenance department. New materials handling equipment may be required if it is determined that the number of storage slots must be increased or the storage area must be compressed into a smaller area. Certain types of materials handling vehicles require less work area to operate, which means more area available for storage. Some examples are:

Type Vehicle	Approximate Operating Space Saved
Counterbalance	1%
Reach	18%
Straddle	20%
Side reach	25%
Turret	27%
Double reach	35%

The operating requirements are usually supplied by the vendor and should be followed. For example, the normal operating aisle for right-angle stacking vehicles is the minimum recommended aisle plus 12 in. The absolute minimum operating aisle should never be used for right-angle stacking because it usually results in one or all of the following:

Product damage
Rack damage
Equipment damage
Decreased productivity

Examination of the existing materials handling system could indicate that the current material flow may have to be changed. The inbound point, storage, and outbound point may have to be relocated. The new system should minimize crossing flows of materials and duplication of handling. Some systems lend themselves to a straight-through flow, that is, receiving located on one side, shipping on the other, and storage in between. The majority of systems have material flows starting and ending on the same dock. To insure the best system is employed, a complete understanding of material flow is necessary. There are materials handling representatives that have access to computer programs that will simulate storage requirements, vehicle requirements, material flow, and so on, helping customers to understand the impact before any changes take place. These can be very instrumental when trying to determine if a change will be profitable.

The storage aids must be positioned correctly to assist the material flow. It is possible to increase storage density by changing the rack position. The normal position for storage racks is in the long direction of the building or storage area. This usually results in fewer aisles and more storage locations. Racking also provides improved utilization of space. For example, bulk or block storage only allows two or three levels of storage. Storage racks will allow storage between this level and the overhead space that is normally wasted in this type of system (see Fig. 12.1.1).

In some systems it is necessary to combine various styles of storage (e.g., single or double pallet racks, cantilever rack, A-frame racks, drive-in racks, flow through, sliding rack, block storage, shelves, drawers, carousels, and bins) to accommodate many types of inventory and handling requirements.

The location of the material within the storage aids must also be addressed. The 80/20 rule usually applies: 80% of the activity will result from 20% of the inventory. Therefore, it is important to have these materials located in a position that is the fastest and easiest to get to.

Subsystems

Man-aboard equipment is most effectively used when it functions in the storage area where it was intended to function. It is not intended to be a transport vehicle. This then means that subsystems must be developed to feed and take away from the storage system. Many methods and equipment are available that interface with man-aboard equipment. Pickup and delivery (P & D) stations should provide for this equipment for inbound and outbound materials and they should be located such that travel of man-aboard equipment is minimized. P & D stations are usually located at the end of the

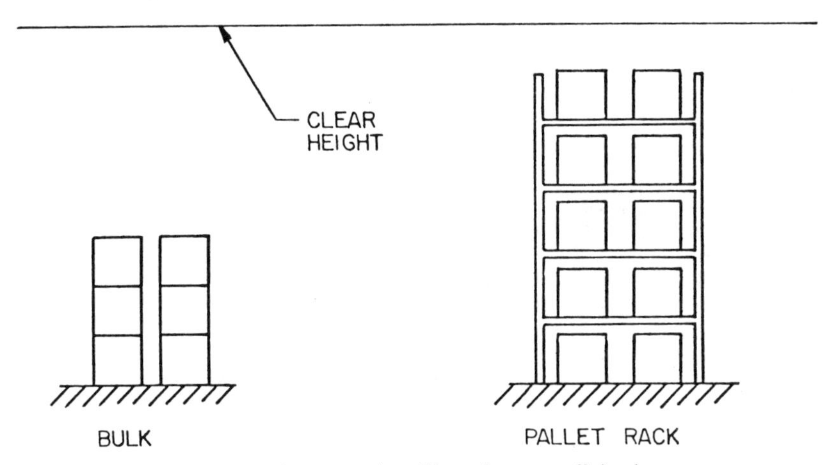

Fig. 12.1.1 Storage racks utilize cube space efficiently.

working aisles. From this point materials are moved forward in the system by transport vehicles, conveyor, AGVS vehicles, and so on.

Location Control Systems

Materials stored in the storage system must be able to be located upon demand. Each storage slot within the system must be identified by a location number. Location numbers locate the slot by rack row, horizontal position down the row, and vertical position. These numbers may be alphabetic, numeric, or alphanumeric. Alternating alphanumeric numbers appears to yield fewer errors due to operator's transposing or misreading location numbers.

Once the location numbering system to be utilized has been established, some method must be used to record and store this data so that each stored SKU can be relocated. This becomes the "Item Location Files." These files also control the balance of each SKU left in inventory. All "In" and "Out" transactions must be recorded by one of two methods:

1. The older method of doing this is by simple 3 in. × 5 in. index cards which record SKU number, location, SKU description, transaction date (In and Out) and current balance.
2. With the recent and continuing advancement in technology, computers become the most cost-effective and least time-consuming method of item location files. Much the same data is filed or stored.

Location Identification

Preassigned. Much time can be saved by the man-up vehicles if location slots are preassigned in receiving the product to the warehouse. To do this, receiving personnel must know if the particular SKU is a large or small user. If warehouse setup is to make high-activity users the most accessible, large movers will be the most available to the activity area. This eliminates the problem of man-up vehicles traveling the warehouse looking for empty slots to store the incoming SKU. It also aids in maintaining control of the warehouse.

Assigned in Stock Room. Location slots can also be assigned as they are received to the stock room. This handling operation could be eliminated if numbers are preassigned.

Product Identification

It is important to identify the product being handled. Storage (storage aid—rack, bins, etc.) handling and equipment specifications are dependent on the size, weight, and shape of the product. Most products fall into one of three categories: case/carton, small parts, and irregular shapes. Each category can demand different warehousing methods and equipment. The following factors must be considered:

Size. What is the physical size and the percentage of each size? Which size is the most active? How should it be stored? Pallet rack? Can bins be used and can they hold the necessary quantity?

Weight. What is the weight of the products and the percentage in each category? Can they be handled by hand or is additional equipment required? How many can be loaded on a pallet? What is total weight of pallet?

Shape. Is it an irregular shape? Does it require a special storage holder to retain it?

Durability. Is it subject to damage easily? Does it require special protective devices? Product damage can be costly if handled improperly.

12.1.4 Planning a New Storage System

The primary consideration when planning a new man-aboard storage system should be *efficiency.* The system must be able to efficiently move the material that is unique to the individual company, whether it is manufacturing, distribution, or warehousing. The system must be designed to meet present needs as well as future requirements. Ideally, system planning should include a five-year forecast. This is normally in the neighborhood of 20–25% of additional storage space to accommodate growth. Another consideration is seasonal or peak storage requirements.

It would be an easy process if there were a completely automated system that could be fitted to each company's materials handling needs. In fact, the totally automated systems apply only to select groups. The majority of companies find the answer somewhere in the area of semiautomation. Total automation is very expensive, and in some cases can never be cost-justified. It is very difficult for smaller companies to install a system that provides no return on investment.

Determining the type of storage system requires a complete understanding of the present system including the type of material, manner of handling the intended flow of material, quantities, personnel, paperwork, and equipment. Equally important are the reasons for considering a new storage system:

The present storage system is too small.
Material handling cannot meet productivity requirements.
Expansion to auxiliary storage areas.
Manufacturing expanding into material storage areas.
Type of material to be stored has changed.

All of these factors are part of the decision-making process required in planning the type of new storage system.

Generally, solutions to consider are:

Building a new facility.
Adding on to present facility.
Leasing or renting (short or long term).
Warehousing in a public facility.
Improving utilization of existing warehouse space.

Once the decision is reached as to the type of expansion or change to be made, the actual storage system must be developed. A good starting point is to visit existing operations to evaluate their systems that are handling similar material. Many ideas can be gathered, and the best features incorporated.

Developing a New Warehouse

The advantage of building a new storage facility is that it can be constructed to fit the product. Material flow can be directed to create the best condition from inbound points and storage, to outbound points. Crossing material flow and double handling can be minimized, and control or discipline for the system can be established at an earlier point. (Modification of existing warehousing sometimes carries over old work habits that are hard to break.)

Layout Considerations. Warehouse layout is dependent on the size of the load to be handled and the type of equipment used. The layout must begin with the greatest dimension of the load and the recommended operating clearance for the equipment vehicle. This combination determines the working aisle. The working aisle is the distance between the loads (if the load overhangs the pallet), pallets (if pallet is larger than load), and racks (if loads are less than slot depth) (see Fig. 12.1.2).

Once the working aisle has been established, the storage aids (racks) can be positioned. The support columns should be positioned to best fit the rack arrangement. The best results are obtained by placing the columns in the back-to-back flue spacing which eliminates any storage loss due to column interference (Fig. 12.1.3).

The least amount of interference is obtained by arranging the work aisles perpendicular to the

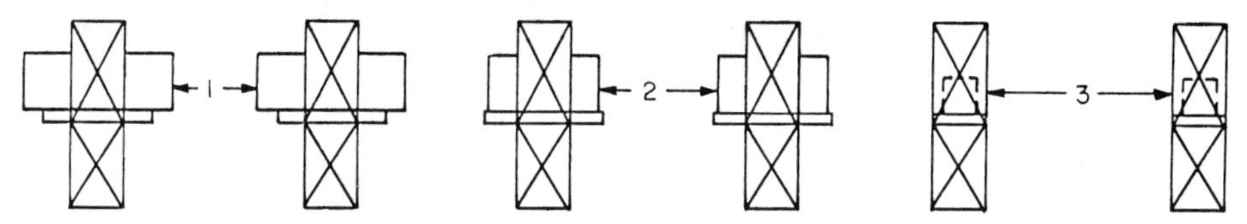

Fig. 12.1.2 Working aisle is distance between, left ro right, loads, pallets, or racks, depending on which is the widest.

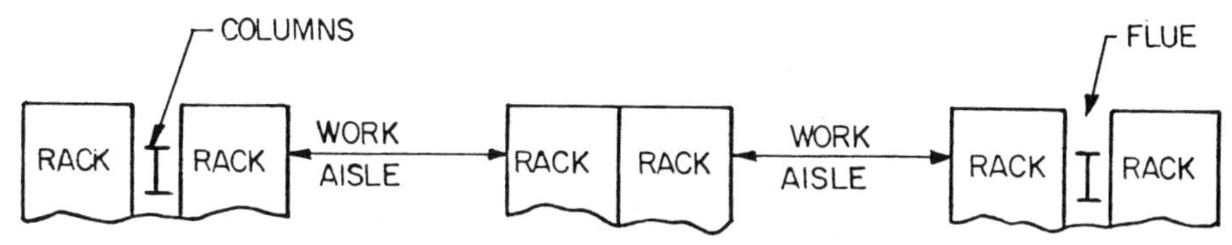

Fig. 12.1.3 Support columns should be positioned to best fit rack arrangement.

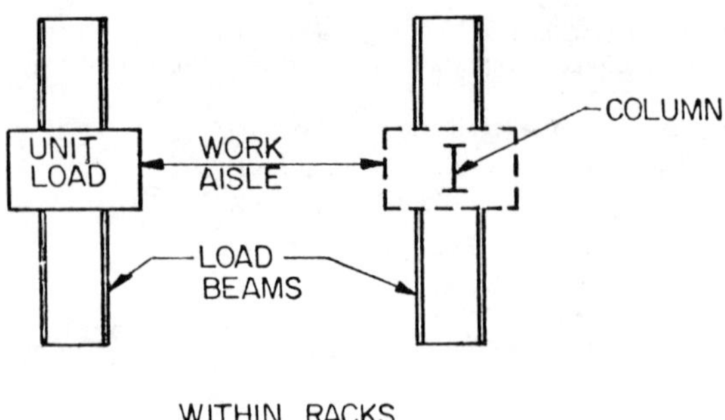

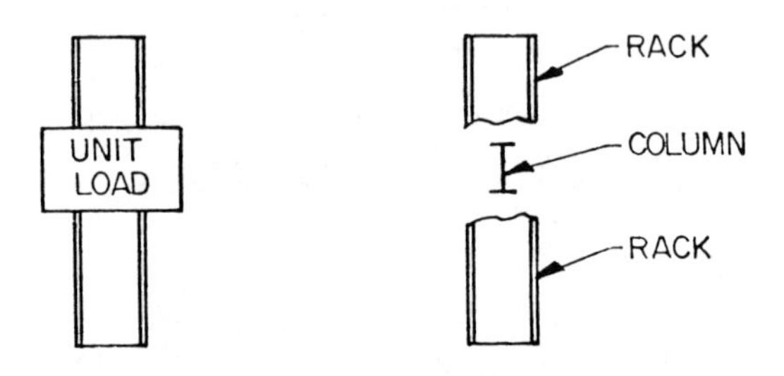

Fig. 12.1.4 Placement of columns to minimize loss of storage slot space.

widest column grid spacing. If it is impractical for the columns to fall in the flue spacing, they should fall within the storage aids, which results in a minor loss of storage slots (Fig. 12.1.4).

The back-to-back dimension or flue spacing is the clearance between the unit load within the storage system. Sufficient clearance should be maintained to satisfy fire insurance underwriters' requirements, prevent unit loads from interlocking, provide room for any in-rack springler system, and create room for air circulation in controlled-atmosphere conditions, such as freezers or coolers. The minimum recommended dimension is 6 in. (Fig. 12.1.5).

Horizontal placement of the unit loads on the load beams affects the productivity of the storage system. The minimum spacing between the unit loads is 4 in. for counterbalance or reach-type operations. The spacing for a straddle or deep-reach vehicle system is dependent on the size of the base legs of the vehicle. If this spacing is allowed to decrease, it becomes more difficult to store or remove the unit loads. Any increase in storage or retrieval time reduces the system's productivity. See Fig. 12.1.6.

The optimum storage condition can be obtained between the column grid and the storage/aisle system by choosing the correct sequence:

Begin and end with storage rack.

Begin and end with work aisle.

Begin with storage rack, end with work aisle.

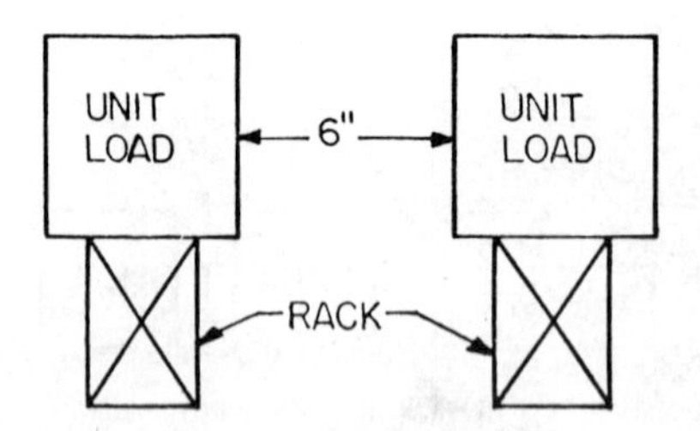

Fig. 12.1.5 Recommended back-to-back or flue spacing dimension.

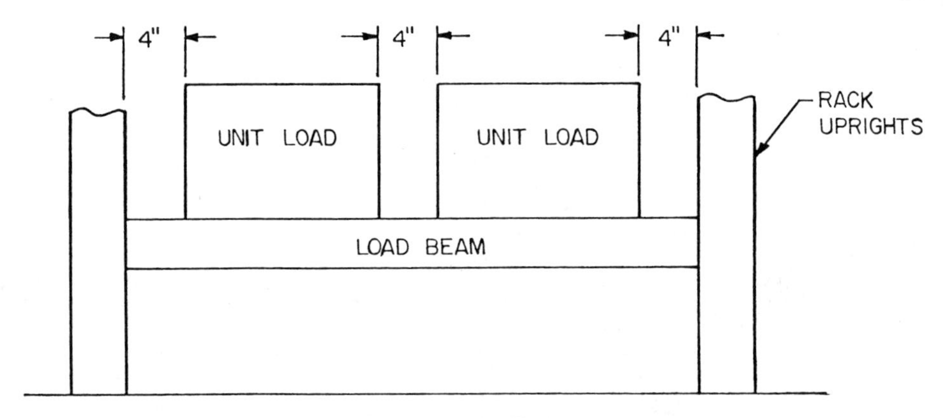

Fig. 12.1.6 Minimum spacing between unit loads is 4 in. for counterbalance or reach-type vehicle operations.

The storage/aisle sequence can be simulated to ensure the best fit with a minimum of storage slot loss. Slight adjustments in work aisle or flue space dimensions across the layout can allow some freedom for the storage pattern.

Cross aisles and main aisles allow access to the storage system. For maximum productivity, a main or cross aisle should be designed into the system every 150–200 ft. Main and cross aisles should be designed to allow two-way travel and are normally 13–18 ft in width. Main aisles are advisable at the end of the storage racks. This avoids "dead-ending" the work aisles against walls, which would create extra travel time by forcing the vehicle to retrace its path when changing to another work aisle. Dead-ending also eliminates an escape route for the warehouse people in case of emergency. Some state laws enforce an OSHA requirement for a minimum width escape route.

Careful consideration must be given to the vertical clearance or building height. The amount of land or space available will have an impact on the building height. If there is not enough room for a 25-ft building to contain the required storage slots, the alternative is to increase the working height of the building. Building costs increase proportionately with building heights, but, in this case, the cost can be justified through the increased number of storage slots. This can also be compared to the costs of other alternatives (e.g., renting, public warehousing, or purchasing a larger lot). The equipment needed to service a higher vertical warehouse is another consideration. This equipment is more complicated and higher priced. Vertical height considerations not only include room for the unit load, but also clearances for:

Roof structures
Sprinklers
Lights
Heaters
Ventilators and ducts
Piping
Power lines
Miscellaneous equipment

Many of these clearances are established by insurance underwriters and should be investigated before the final building height is established. For example, some underwriters require as much as 36 in. between the top unit load and a sprinkler head. Control of these dimensions is critical for layouts. Care must also be taken to position these overhead obstacles away from the working areas. Productivity will be decreased if the vehicles must work around obstructions.

Some other building requirements that should be considered are:

Local building construction codes
Type of flow and construction
Subfloor conditions and requirements
Drainage, industrial waste

Dynamic floor loading
Shipping and receiving docks
Staging and marshalling areas
Parking lots
Access from main highway system
Maintenance areas
Building type (conventional or rack supported)

Floor Requirements. Floor requirements are important when using man-up high-lift vehicles. Consideration should be given to construction methods for geographic location. Wheel loadings and pressures are a factor for the equipment selected and can be supplied by the equipment manufacturer. The floor must be able to withstand these forces over a period of time.

Floor flatness has become a major concern. There is an increasing awareness that a rough, uneven, and low-grade industrial floor can have a major impact on material handling equipment. Improperly designed and/or installed industrial floor can:

Increase vehicle maintenance.
Increase vehicle downtime.
Cause poor steering and guidance problems.
Make storage and retrieval difficult.
Cause excessive shiming of storage aid (rack).
Decrease operator comfort.
Decrease productivity.

Any vehicle can be made to work on any floor but it may *not* provide its optimum performance. Any "hard axle" vehicle (without shocks, springs, etc.) will reflect floor variations as undesirable mast motions. Floor tolerances are keyed to the maximum elevated height used in the particular system (see Fig. 12.1.7). On a given floor, static horizontal load displacement varies directly with the elevated height:

$$\text{H.D.} = \frac{\text{E.H.}}{\text{OAW LW}} \times \text{Floor variation}$$

where: H.D. = horizontal displacements
E.H. = elevated height
OAW LW = overall width load wheels

Acceptable floor variation recommendations can be supplied by the individual equipment manufacturers.

Floor flatness becomes more critical and demanding as the elevated height to be utilized in the system increases. Generally, dollars initially invested in a good floor will be returned many times over.

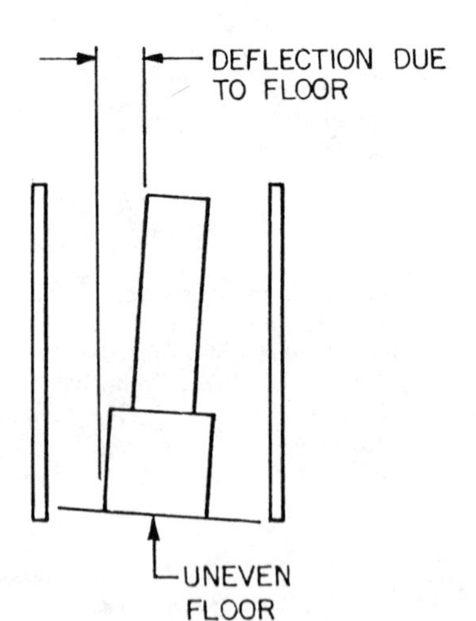

Fig. 12.1.7 Static horizontal load displacement varies directly with elevated height.

12.1.5 Equipment

There are many materials handling applications that require the operator to be elevated with the load or forks. One application is *high-level order picking.* High-level order picking is used when it is necessary to have maximum space utilization in a limited area. It allows order picking to 30 ft. Man-aboard unit-load (pallet) handlers are used when occasional order picking is required. These units have a 30–60-ft working height. The man-aboard unit is also used to eliminate the need for vertical positioning devices at these heights. When the operator is elevated to each storage slot, less space is required for clearance and very expensive unit load can be positioned with confidence. Man-aboard unit load handlers can be divided into two groups: floor supported man-aboard and floor/ceiling-supported man-aboard. Each has particular advantages for individual material handling systems.

Floor-supported Man-aboard Vehicles

Floor-supported man-aboard vehicles (Fig. 12.1.8) can operate to working heights of 40 ft. These units are very mobile and can travel from one working aisle to the next very easily. They are designed to work within the unit storage system and are normally supported with some type of subsystem that supplies and removes the unit loads as required. The design of the load handling device allows these vehicles to operate in very narrow aisles, 18–24 in. wider than the unit load.

These vehicles are also referred to as turret trucks. The turret allows access to both sides of the working aisle without turning the vehicle. Instead, the load handler rotates 180°, from one side to the other. To maintain productivity, it is necessary to have rail or wire guidance in the work aisle.

Rail guidance consists of rollers mounted on the vehicle that guide along steel angles mounted on the floor. This keeps the vehicle mechanically centered in the work aisle. The wire-guided system has a wire buried in the floor that carries a low-frequency signal. The vehicle has a sensor to detect this signal and relays information to the steering system which keeps the vehicle centered over the wire. Without these types of steering systems, the vehicles have to be manually steered which is slower and requires greater clearances.

Consideration must also be given to rack and floor tolerance specification. Deviations in floor levelness are greatly magnified at higher elevations. These deviations, along with mast and fork deflection, can create serious interference problems. Rack variations from the vertical also add to this interference. Equipment manufacturers can supply the recommended tolerances for clearances, racks, and floors. These recommendations should be followed to obtain the maximum productivity.

Floor-supported vehicles are down-rated in capacity at higher elevations. This is required to meet OSHA safety requirements for stability. Static and dynamic tests determine the maximum capacity that can be safely handled at various working heights. This downrating may require a zoning type of storage, that is, storing the heavier loads at the lower elevations to meet OSHA safety standards. Floor-supported units can be used with standard pallet rack. The narrow work aisles allow maximum storage density to be obtained. This can be applied to new storage facilities or conversion of existing systems to take advantage of standard storage aids (racking).

Floor/Ceiling-supported Man-aboard Vehicles

The *floor/ceiling-supported, man-aboard vehicles* (Fig. 12.1.9) are normally used in a working range of 40–60 ft. These vehicles are constructed to operate with an upright support that is physically connected

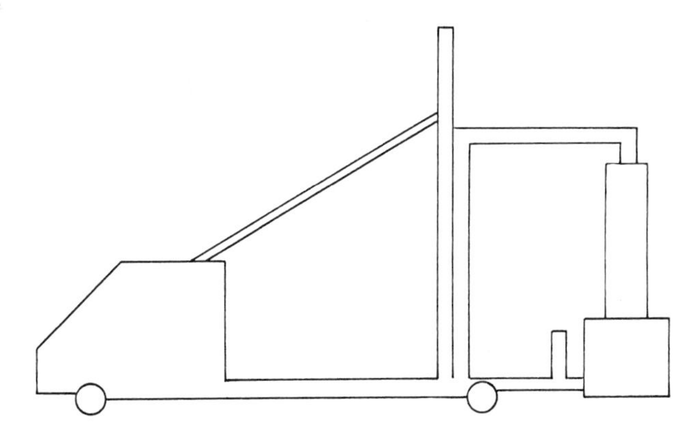

Fig. 12.1.8 Floor-supported man-aboard vehicle.

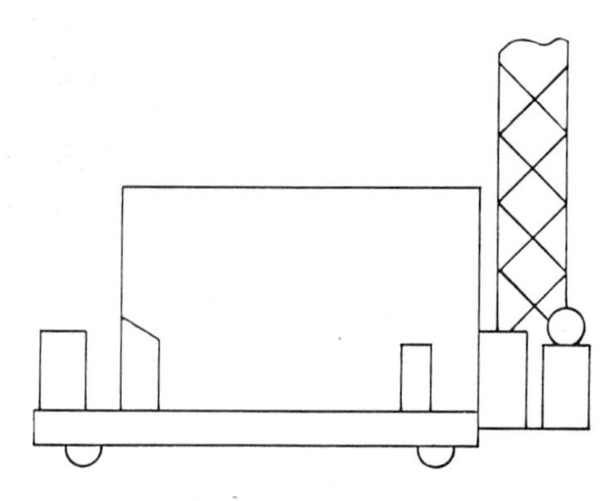

Fig. 12.1.9 Floor/ceiling supported, man-aboard vehicle.

to the ceiling and travels in a steel channel or on a steel rail anchored to the floor. This configuration allows operation with maximum capacity and maximum speed at the vehicle's fully elevated height. Alternating current (AC) power is supplied from electrical wires in the ceiling that are connected whenever the vehicle is in the work aisles. These man-aboard vehicles are designed to operate in work aisles 6 in. wider than the unit loads.

Maximum productivity is obtained when the vehicle is dedicated to one aisle. However, to avoid the expense of having a vehicle in each work aisle, there are configurations that allow travel between aisles. One vehicle design can move from one aisle to the next using direct current (DC) battery power source. Other designs must include a transfer station located across the ends of the work aisles. The vehicle is driven out of one work aisle onto the transfer station and the station is moved to align with the desired aisle. The man-aboard vehicle is then driven off the transfer station into the new aisle.

The unit load handler is normally a shuttle attachment (see Fig. 12.1.10). The shuttle attachment requires a storage aid (racking) that supports the unit load only on the sides. This allows the shuttle room to get under the load when picking it up or setting it down (Fig. 12.1.11).

The floor/ceiling system allows very dense storage because of the extremely narrow work aisles and the high working height. This system can be installed in a conventional or a rack-supported building. Installation costs are normally higher than floor-supported man-aboard vehicle installations.

The man-aboard systems are semiautomated. Some of the positioning functions are automated, however the operator is still needed to control the remaining functions.

The completely automated systems (AS/RS), on the other hand, do not require any operators. These systems demand uniform loads, exact clearances, complete computer control, and automated subsystems for support. The AS/RS is used in special applications. These systems are very difficult to cost-justify. The application requirements are usually more important than the costs. It is necessary to select the proper system for each material handling requirement. All companies require some automation to remain productive and competitive. There are many levels of automation. The challenge is to select the level of integration that best suits the individual needs.

Man-up Vehicle—Order-picker Type

Order picking is the materials handling method employed where product selectivity is required in less than pallet loads in the filling of customer orders. The picking generally involves thousands of

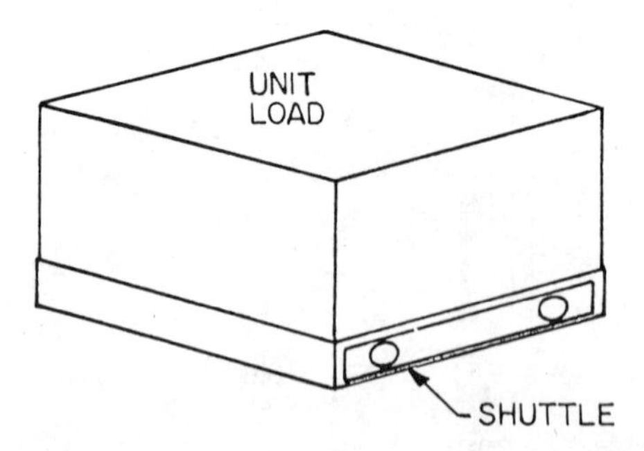

Fig. 12.1.10 Shuttle attachment is normally used for handling unit loads.

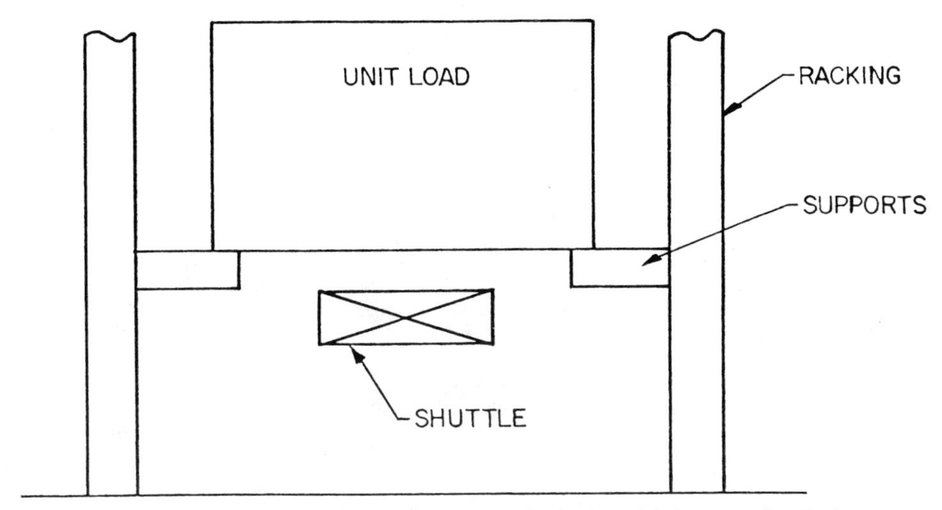

Fig. 12.1.11 Provision made for shuttle to get under load to pick it up or place it down.

repetitive actions, day after day, month after month—performed by a combination of vehicles and people. The order-picking operation alone contributes 25% of the cost of a product. Materials handling accounts for 30–60% of the cost of a product (Fig. 12.1.12). Notice all the handling cycles that occur are in relation to the production cycles. Many methods of order picking exist and attention to selection is important. If pallets are used in the system, they will be the factor that determines the working aisle widths throughout the system.

The following items are some key factors that must be considered in setting up or improving a system utilizing man-up (high-lift) order pickers:

Pick Aisle. The vehicle-working aisle from which product is to be picked. If it is the same aisle where stock is replenished, vehicle contention may be a problem between the pallet handler and order picker. Control will be required.

FIFO. First in–first out. If the product has a shelf life, FIFO may be required.

80/20 Rule. It is important that the product is located. The product location mode (banding, zoning, etc.) will depend on how the product is to be picked: one order per run or batch picking. The most active SKU's (20%) should be the most accessible as these are the pick faces visited most often. The location mode generally dictates whether fixed slot storage or random storage system is used. Fixed slot storage generally results in more honeycombing (wasted space).

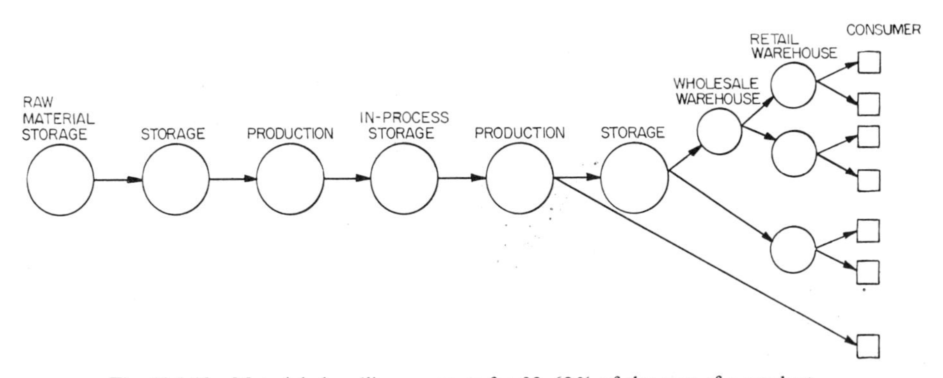

Fig. 12.1.12 Materials handling accounts for 30–60% of the cost of a product.

Elevated Height. Man-up order pickers are most productive up to 22 ft in elevated height. Vehicles are available up to 30 ft but, normally, slow-moving SKUs or back-up stock are located above the 22-ft level.

Physical Specs.

Man-aboard High-lift Order Pickers. These order pickers lift the operator with the load (Fig. 12.1.13). This type of vehicle is provided by many manufacturers. The vehicles are generally powered by an electrical power section (which remains at floor level) in either resistor or SCR control. SCR controls are usually preferred because they result in smoother operation and yield less battery draw. A hydraulically operated telescoping mast section can generally provide access to elevations up to 30 ft. This height is measured at the floor level of the operator's compartment or at the top of the load-carrying forks. Care must be taken that the protective guard for the operator clears the lowest overhead obstruction in the warehouse as the complete compartment rises. The result is that more pick fronts are available in the system. The amount of pick fronts required is dictated by the amount of different SKUs to be picked. In some conditions, it is possible (space permitting) to put more than one SKU in each pick front. Government specifications demand a minimum height of 74 in. in the operator's compartment from floor to underside of protective guard. Operator platform lengths are variable. Operator platform widths are varied to fit the width of the working clear aisle. Picked product is stored on pallets, modules, and so on, located on the load-carrying forks. When full, these are removed and a new load-carrying device is picked up to return to order picking.

Clearances. The most important clearance is that on each side of the operator platform and the working or picking aisle, see Fig. 12.1.14. The minimum clearance per side recommended is 6 in. at low-elevated heights (15 ft). As the height increases, this clearance should increase approximately 1 in. for every 5 ft in height.

The other clearance concern is that the top of the protective overhead guard clears the lowest warehouse obstruction when the operator's compartment is fully elevated.

Aisles. The clear picking aisle width is established by the load length utilized in the warehouse system and thus the unit load handling vehicle used. An exception is when the man-up order picker is used exclusively in the picking aisle for replenishment as well as picking. The load width or vehicle

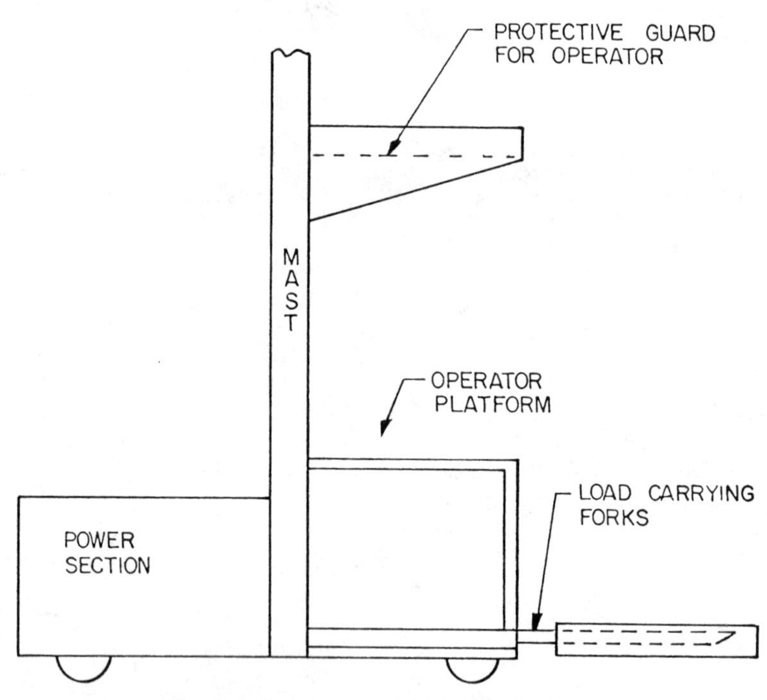

Fig. 12.1.13 Man-aboard, high-lift orderpicker.

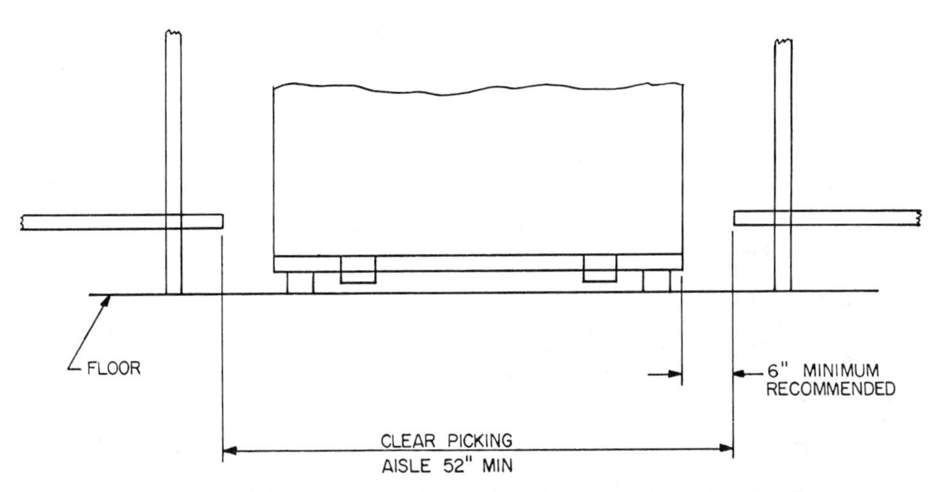

Fig. 12.1.14 Minimum clearance for operator platform is 6 in. at low elevations.

width then becomes the determining factor, whichever is greater. The main or cross aisles (perpendicular to the pick aisles) should be wide enough to allow vehicles to swing into the pick aisles easily and productively. Usually if rail or wire guidance is used, a minimum of 13 ft should be allowed.

Guidance. These vehicles can be equipped to function with both rail and wire-guided systems. Because guidance means the operator does not have to steer the vehicle in the picking aisles, he can travel faster, thus be more productive. Table 12.1.1 indicates the percent increase in productivity using guidance.

Cost of wire versus rail guidance varies depending on the size of the system and the quantity of vehicles required. Wire guidance allows for more flexibility should there be a need to change or expand warehouse space.

Elevated Heights. Man-up order pickers are usually limited to a maximum of 30 ft in elevated height. Arrangement of product in relation to picking height is important to how productive the system will be. Horizontal travel is faster than lift speeds and requires less battery power. Maximum horizontal travel speed is also not as important in high-lift order pickers as is acceleration and deceleration. This is important because the operator is constantly starting and stopping. Arrange product to minimize the vehicle lifting and lowering.

Extended Height. Extended height of most vehicles is the elevated height plus 80–90 in. The extended height is the greatest height on the vehicle when fully elevated. Make sure this dimension will clear the lowest overhead obstruction in the warehouse.

Table 12.1.1 Increase in Productivity Using Rail or Wire Guidance

Number of Picks in Aisle	Percent Increase in Productivity Using Guidance for Given Pick Aisle Length, in Feet								
	50	60	70	80	90	100	110	120	130
1	8.2	8.7	10.9	11.9	13	13.6	14.8	15.5	16.4
2	3.9	4.9	5.8	6.6	7.5	8.2	8.5	9.7	10.3
3	2.2	3	3.8	4.3	5.0	5.8	6.1	6.8	7.2
4	1.3	2.5	2.7	3	3.5	4.0	4.4	5.1	5.4
5	0.7	1.3	1.8	2.1	3.0	3.1	3.3	3.7	4.2
6	0.2	0.7	1.0	1.4	2.5	2.7	2.8	3.1	3.3
7		0.3	0.7	1.0	1.6	1.8	2.0	2.4	2.6
8			0.3	0.7	1.2	1.3	1.7	2.0	2.1

12.2 AUTOMATED HIGH-RISE RACK SYSTEMS

Alex J. Nagy

The impact and widespread use of high-rise systems first began in Europe in the early 1950s. The high cost of real estate and the land-locked constraints placed on plant expansions in France, England, and Germany focused the attention of planners on the better use of "air-rights." Utilization of the vertical cube became the prominent design parameter that led to the 100-ft-plus storage systems that marked the period.

Storing higher meant problems of access for storage and retrieval. The loads had to be compartmentalized and space provided vertically between loads to allow the forks of materials handling equipment to enter without disturbing loads above and below. This required the design of special racks. The limitations of lift and the wide aisle requirements of conventional forklift equipment available at the time necessitated the development of an entirely new concept in materials storage handling and retrieval hardware. Thus was born the stacker crane.

It can be said that the original stacker crane evolved from a conventional overhead traveling crane. The entire crane assembly was tipped on end 90°.

The horizontally traveling gantry beams became the vertical masts of the new stacker: floor-supported on a rail at one end and ceiling-guided in channels at the top end. The former hoist assembly became an elevating carriage. The electric hoist was replaced by a shuttle mechanism on this carriage now moving loads in and out of a compartment instead of traveling up and down.

The evolution progressed from there. The operator in a control cab was eventually replaced by automatic controls. The access aisles were narrowed to a matter of a few inches wider than the length of the load and all of the operational aspects including inventory control became real-time, on-line functions through ancillary microprocessors and minicomputers. A typical storage and retrieval (S/R) machine is shown in Fig. 12.2.1.

There are hundreds of automated high-rise stacker systems in use in the United States today with scores more being designed and implemented every year. It is unquestionably the most dramatic growth segment in the materials handling industry today. As more diversified industries and more and more smaller companies are being attracted by the savings and productivity gains available through such systems, we must be aware of the factors that determine financial, operational, and logistical practicality.

This section outlines the basic steps that should be followed in planning an automatic storage and retrieval system (AS/RS).

The first step in planning an AS/RS is to carefully define what the system is expected to do. What are the specific needs meriting considerations leading to a new system? Some objectives are:

1. Release space now being used for storage of raw stock, work-in-process or finished goods, making it available for valuable manufacturing space.
2. Gain needed floor space without adding brick and mortar.
3. Increase warehouse capacity to meet long-range plans.
4. Eliminate repetitive and unnecessary handling.
5. Minimize damage to product and equipment.
6. Improve stock rotation.
7. Control and reduce inventory.
8. Comply with governmental regulatory edicts.
9. Reduce operating costs.
10. Reduce pilferage.
11. Increase labor productivity.
12. Create a safer and more pleasant working environment.
13. Improve customer service.

Once the goals and objectives are defined, the next step is to organize a task team or project group to conduct an initial feasibility study for management decision-making purposes. It will delineate the scope and purpose of the project and gather the pertinent data required to support costs, savings, and payback. This phase is frequently contracted out to a consulting firm or to a system supplier having expertise in designing, building, and installing systems on a "turn-key" basis; that is, someone having sole responsibility from original concept to the final acceptance tests of a completed system. The study is followed by a firm bid with timetables, implementation schedules, equipment specifications, manning tables, and so on, which form the basis for a fixed contract.

If the task team is to be made up of personnel in-house, the organization should include representatives from several departments. The capital outlay for a project of the magnitude of an AS/RS requires that a project manager be designated from either the industrial or facilities engineering departments to join operations, plant engineering, and business system delegates. Figure 12.2.2 shows a typical

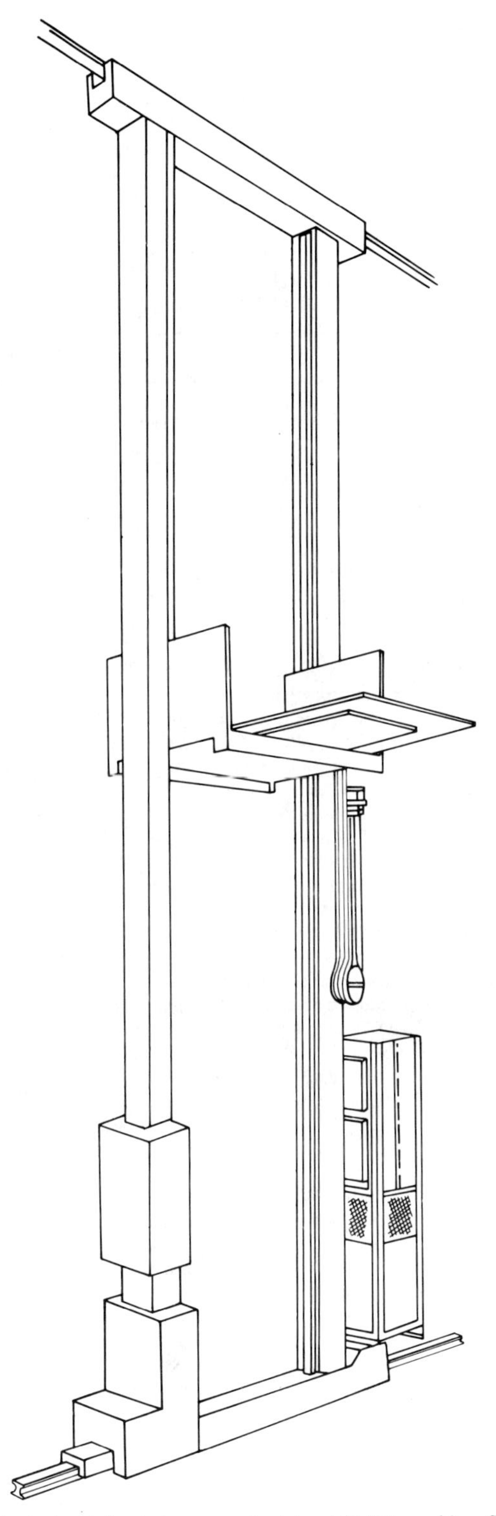

Fig. 12.2.1 Typical unit load storage and retrieval (S/R) machine. Source: Ref. 1.

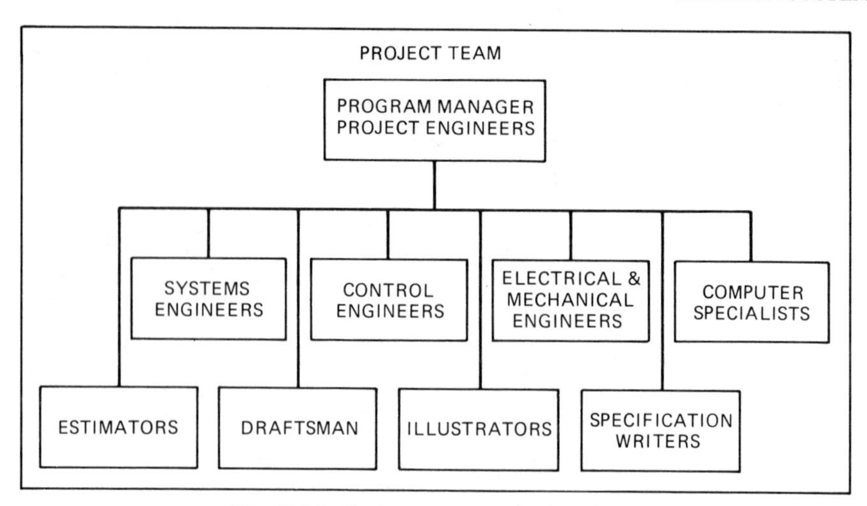

Fig. 12.2.2 Project team organization chart.

project team organization chart. Finance, purchasing, warehousing, materials control, and materials handling representatives are assigned early in the planning stages to make them more effective as the project approaches the bidding stage.

The size and makeup of the project team can vary. Not all members contribute during the entire planning period. The gathering of the required data from the operating departments may be a short-term or intermittent task. If an outside consulting or engineering firm is retained, a spokesman from each discipline should be named to coordinate and verify the information being gathered.

Historical information is not always available and provisions must be made to begin the accumulation of comparative items such as movement, storage volumes, order-picking rates and labor costs—all on an hourly, daily, weekly, or annual basis.

The accumulated data should reflect lows, peaks, and averages over a given period and should be projected to anticipated future levels set by marketing, sales, and production management. The system size and capacity is determined to handle the established limits. Any requirements beyond that level must be considered as future expansion needs. Generally, a 20% allowance for expansion is considered a reasonable rule of thumb for growth during the useful life of the equipment.

An important part of such a study is a review of the material flow, present and proposed. How will the material move? What will be the input and output per hour between receiving, inspection, storage, production, and shipping? This key element, throughput, will affect the selection of the transportation mode (forklift, conveyor, wire-guided vehicle, trailer trains, etc.). It will determine the cycle times, aisle lengths, and the number of stacker cranes that will be required. Can they be captive to an aisle or can one stacker service multiple aisles? Will there be an order-picking function carried out from loads brought from storage and will left-over stock be restored? Will output units be full loads only? What percentages of each? The number and location of the system's ancillary supporting equipment such as pickup and delivery stations (P & D's), weighing scales, and profile checks can then be estimated.

The end result of the feasibility study will be a preliminary set of operational and functional specifications that can provide system suppliers with the design parameters with which to estimate costs (Fig. 12.2.3). All system purveyors can provide standardized worksheets and blank outline forms detailing the elements that they need in order to provide you with a budgetary proposal. To facilitate uniform nomenclature and description during the proposal preparation stage and bid evaluation, a glossary of standard terms as defined and compiled by the Association of AS/RS Manufacturers of the Material Handling Institute is included at the end of this section.

The components of an automated high-rise system with storage and retrieval machines can be categorized into five major elements: storage modules (loads), storage structures (racks), stacker cranes (AS/RS), pickup and delivery (P & D) stations, and controls.

12.2.1 Storage Modules

The most significant factor in the design and layout of an AS/RS is the load/size/weight configuration. The storage module contains and supports material to be stored and handled. It presents material in a configuration that is transferable by AS/RS, and conveyable and storable in high-rise storage racks.

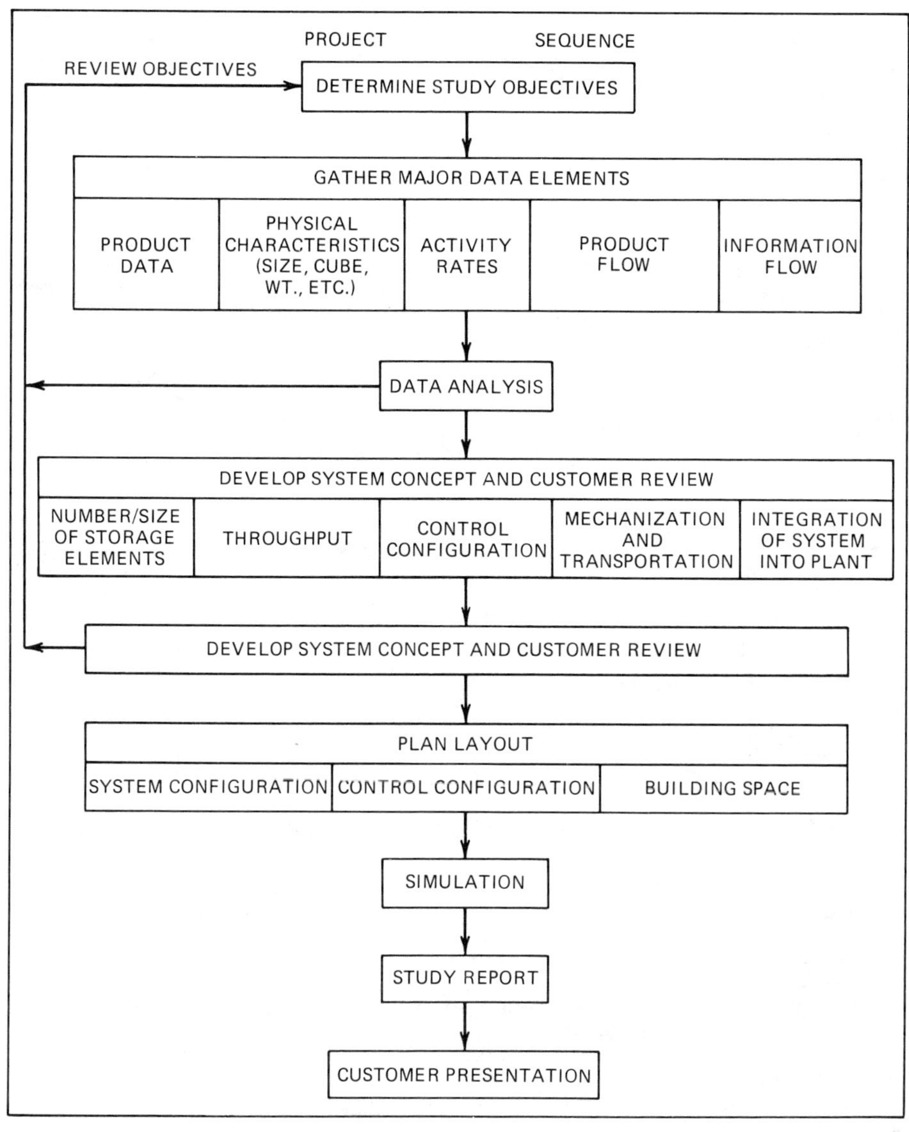

Fig. 12.2.3 Project sequence chart.

This storage module determines the dimension of the storage rack space required for a given inventory.

Modules are generally divided into pallets, slave pallets, metal containers, storage bins and totes, and carts with wheels. Self-contained loads such as boxes, barrels, drums, and steel coils where no further support is needed, are themselves self-contained modules. The base or footprint of the module must be evaluated for compatibility in moving in and out of the system. If conveyors are used, the load base should have a bottom surface which allows bilateral movement. Turntables, transfers, and lift devices affect the allowable shape and construction. Forklift entry slots, nesting pockets, stacking caps, crane hooks, and runners for underclearance are details warranting design considerations.

An AS/RS should have provision for monitoring and rejecting any storage modules that exceed predetermined dimensions. Automatic weighing and profile checks for sizing are safeguards that prevent improper loads from getting into the system.

Common forms of storage modules are shown in Fig. 12.2.4. These units are controlled by stackers in an AS/RS and they must be compatible with industrial trucks, conveyors, and automated guided vehicle systems (AGVS).

Fig. 12.2.4 Common storage modules. Source: Ref. 1, Fig. 5.

Irregularly shaped loads or loads with shorter length and narrower width than the standard load module can be handled on slave pallets that provide a standard load base. The structural shortcomings of vendor pallets can be overcome by placing them directly on such slave pallets for storage (Fig. 12.2.5). Slave pallets are generally plywood sheets thick enough to provide support with minimal deflection. Inbound material is placed on such carriers for storage and removed as the material leaves the AS/RS. The slave pallet remains captive to the system.

Loads with greater length and width than the load module and which overhang the base should be avoided if possible. Such a condition (Fig. 12.2.6) places added structural demands on the support members (load arms) that hold the module in the rack. They must be extended as the load overhang increases.

In summary, define the physical dimensions and gross weights of all the loads to be handled in the system. Determine the number and container type of each to be stored. Of course, system performance is improved and implementation simplified if the loads to be handled are as uniform as practical.

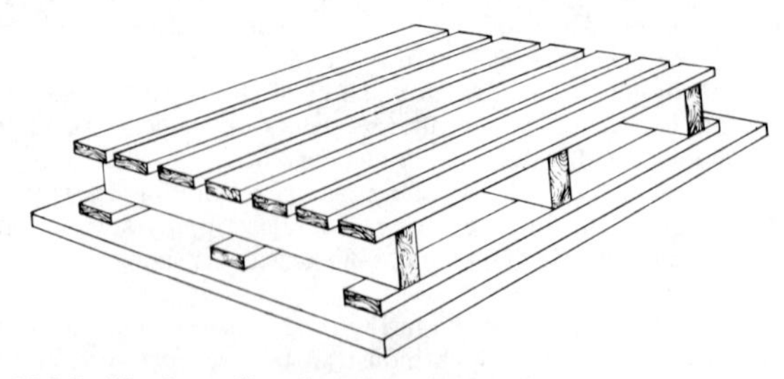

Fig. 12.2.5 Vendor pallet placed on slave pallet. Source: Ref. 1, Fig. 5.

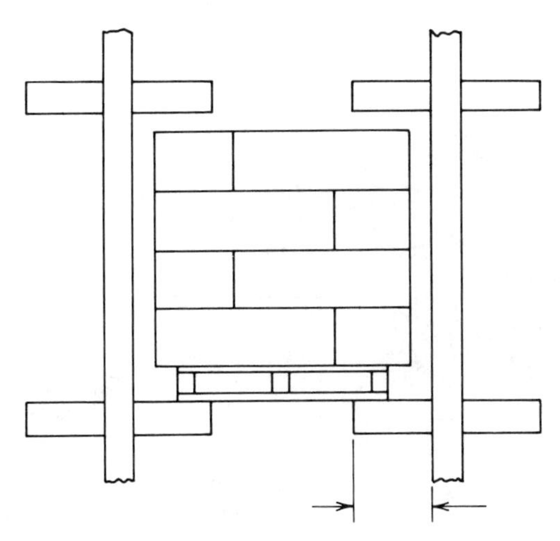

Fig. 12.2.6 Load overhang condition. Length of load support arm increases with load overhang.

12.2.2 Storage Racks

The function of the rack or storage structure in an AS/RS is to support multiple levels of loads that can be individually selected. The rack design is an application of basic structural steel analysis. Factors covered include the size and weight of the loads, the total number of load openings required, and whether the structure will be freestanding within an existing building (Fig. 12.2.7), or used as the support for a new building roof and skin (Fig. 12.2.8). Normal design considerations such as soil-bearing characteristics, slab construction, snow and wind loads, and seismic considerations are important criteria.

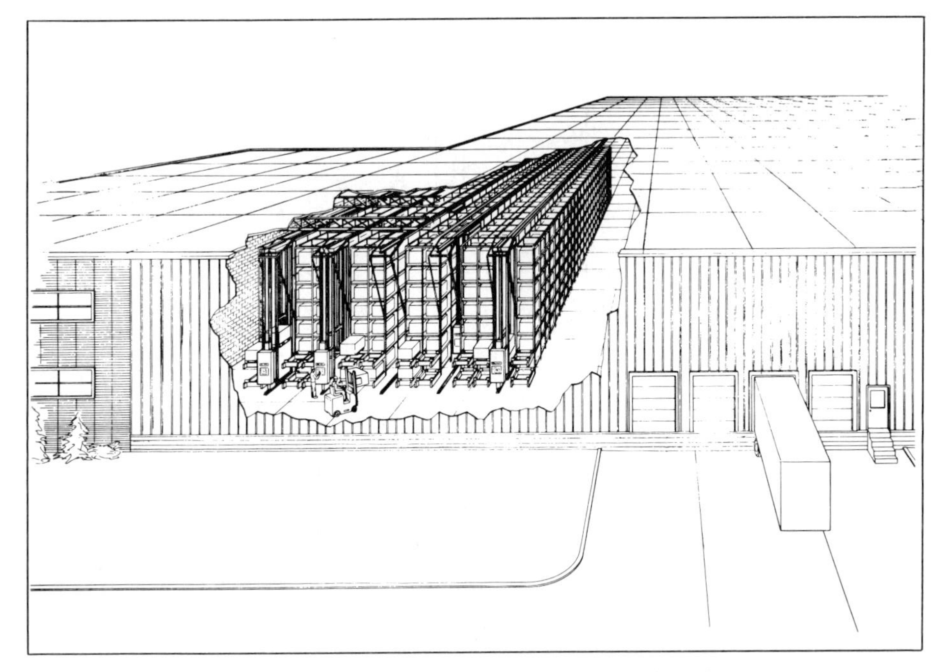

Fig. 12.2.7 Free-standing rack system. Source: Ref. 1, Fig. 5.2*A*.

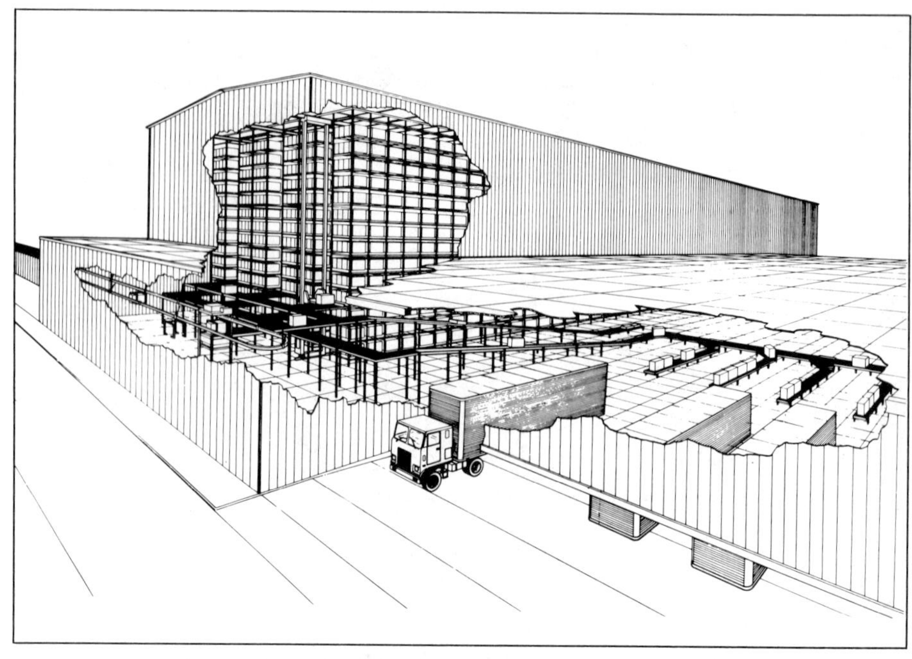

Fig. 12.2.8 Rack-supported building. Source: Ref. 1, Fig. 5.2*B*.

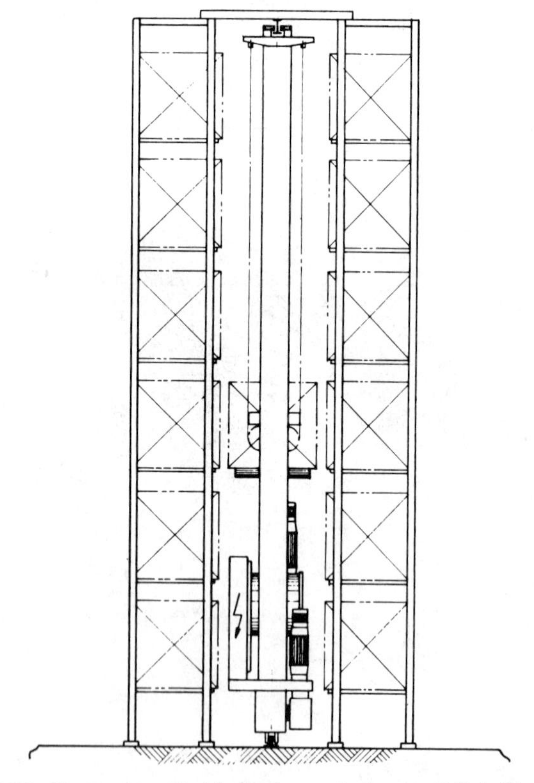

Fig. 12.2.9 Single-wide aisle/single-deep rack. Source: Ref. 1, Fig. 3.1.

The final physical configuration of the structural envelope results from the determination of:

1. The number of loads horizontally in a row
2. The number of loads stored vertically in a bay
3. The number of stackers/aisles and if the loads are stored one deep, or two deep on each side of the aisle (Figs. 12.2.9 and 12.2.10) or two deep in a double-wide aisle (Fig. 12.2.11).

The length of the rack results from the number of loads in a row plus the horizontal clearances. These operating clearances between the loads and the vertical uprights of the supporting rack can be from $1\frac{1}{2}$ to $2\frac{1}{2}$ in. (Fig. 12.2.12). This space allows for variations in rack plumb, load alignment, and the positioning accuracy of the stacker crane. Suppliers can provide specific tolerances for their equipment in a given system. To this overall rack length (number of roads + clearances + structural members) the supplier will add the distance required for crane runout at the rear end of the row and the required space for pickup and deposit stations at the front end.

The vertical number of loads or height of the building envelope is determined by the clearance height under the building trusses (in a freestanding system) and the operating clearances needed for the shuttle mechanism of the stacker. The distance from the top of the top load to the ceiling and the distance from the floor to the bottom of the lowest load again is a function of the stacker crane dimensions and operating requirements (Fig. 12.2.13).

The total width of a system is calculated by adding the number and width of the rack rows to the width and number of stacker aisles to be used. The type of load and load support in a rack dictates the amount of allowable overhang into a stacker aisle. Generally, the load overhangs the front and rear of the storage rack by 2–3 in. so that the width of a rack holding a 48-in. long pallet, for example, would be 44–42 in. wide. Back-to-back, single-row racks are generally spaced 12–6 in. apart to allow a vertical fire tunnel between the overhanging pallet loads.

To determine the number of AS/RS units and aisles, first the throughput must be calculated. That is, how many loads in and how many loads out per hour must be stored and retrieved? This is directly related to production and shipping activity. Use anticipated maximum or peak hourly rates,

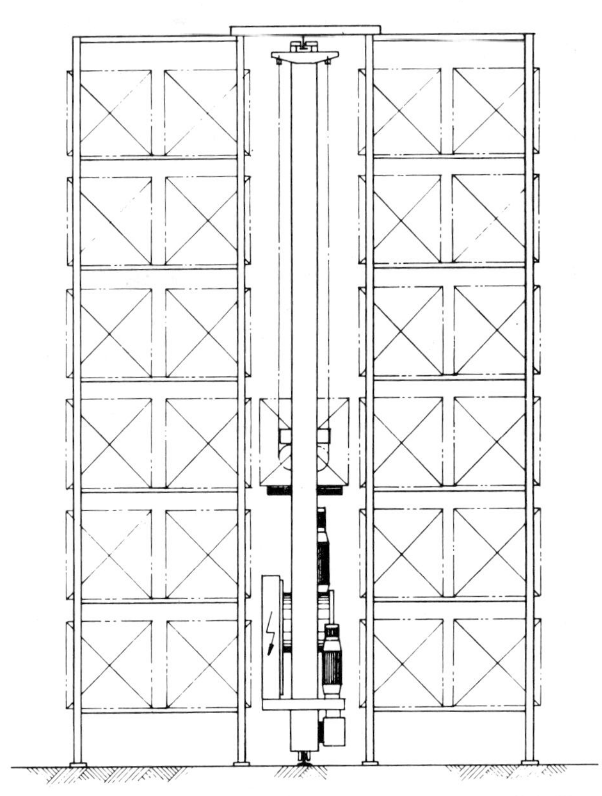

Fig. 12.2.10 Single-wide aisle/double-deep rack. Source: Ref. 1, Fig. 3.2.

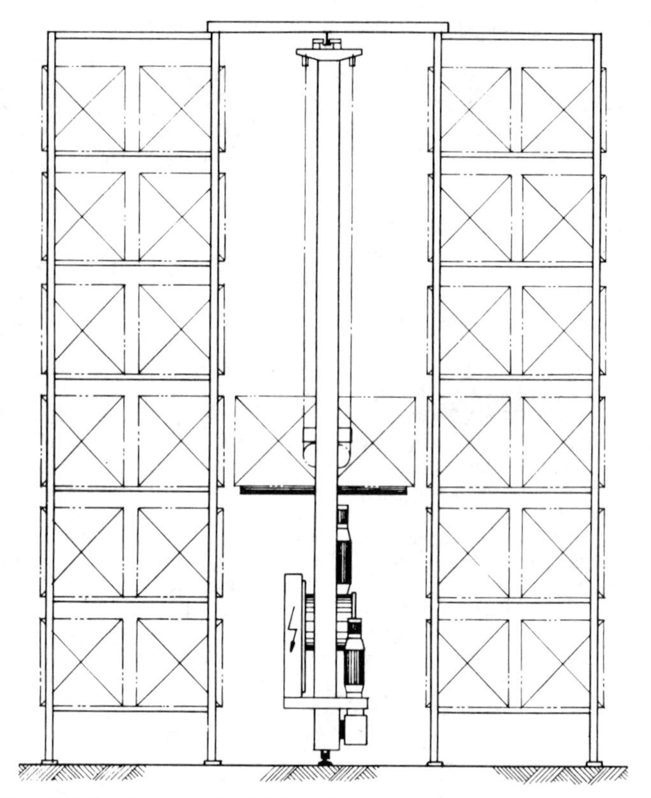

Fig. 12.2.11 Double-wide aisle/double-deep rack. Source: Ref. 1, Fig. 3.3.

not averages per shift, day, or week. This throughput rate, stores and retrieves per hour, will indicate how many cranes will be needed in terms of single and dual cycles. The AS/RS industry has agreed on a basic formula by which cycle time can be measured.

A single-cycle operation means that a crane starting from the pickup station executes either a store or retrieve command and awaits the next command. A dual-cycle operation means that a crane starting from the pickup station executes consecutively a store and retrieve command.

Method of Computing Cycle Time for One-Command Operation (*Fig. 12.2.14*)

STEP 1. Compute all times between operator actuation of S/R machine and completion of the preceding cycle to a point where S/R machine is ready to start a second cycle.

STEP 2. Retrieve a load at home station.

STEP 3. Store the load at a storage location $\frac{1}{2}$ the number of storage addresses along the aisle, and up $\frac{1}{2}$ the number of vertical storage locations in that aisle. (In cases of simultaneous travel, use the longer of the travel or the lift times computed individually.)

STEP 4. Return to home position empty and stop.

Method of Computing Cycle Time for Dual-Command Operation (*Fig. 12.2.15*)

STEP 1. Compute all times between operator actuation of S/R machine and completion of the preceding cycle to a point where S/R machine is ready to start a second cycle.

STEP 2. Pick up a load at home station.

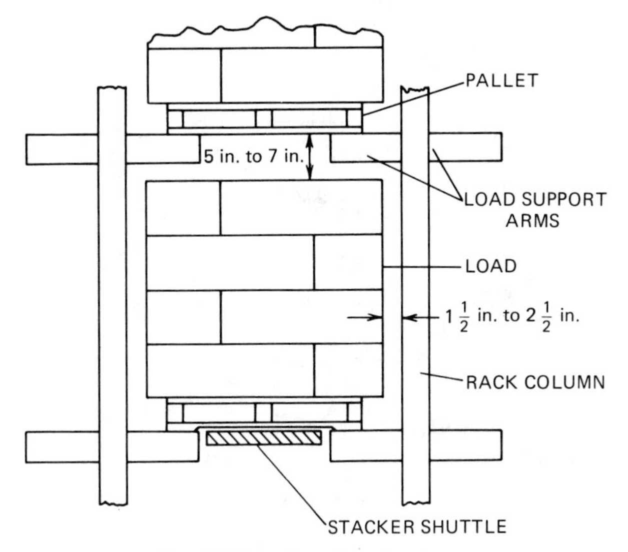

Fig. 12.2.12 Operating clearances.

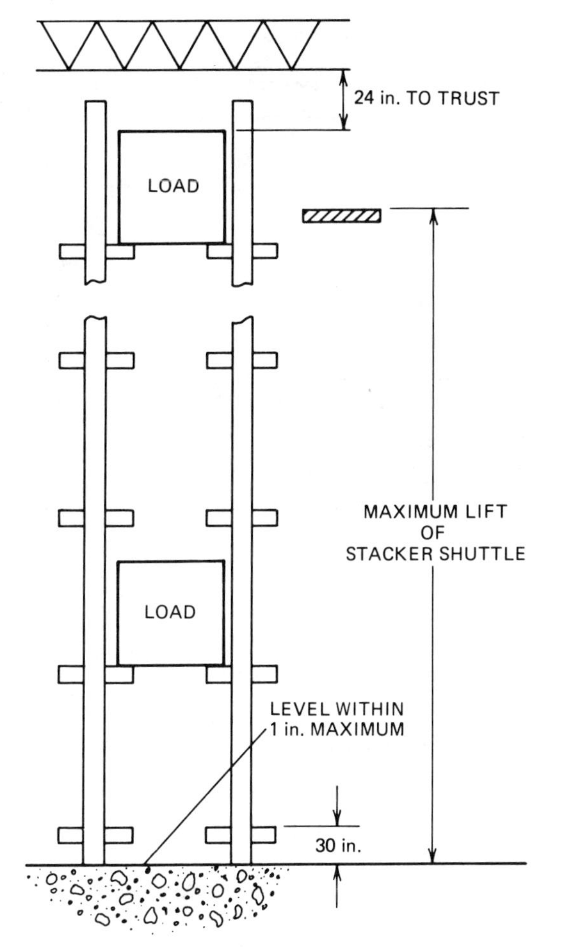

Fig. 12.2.13 Clearances between ceiling and floor.

START AT "HOME" WITH PICKUP.
RETURN TO "HOME" EMPTY.

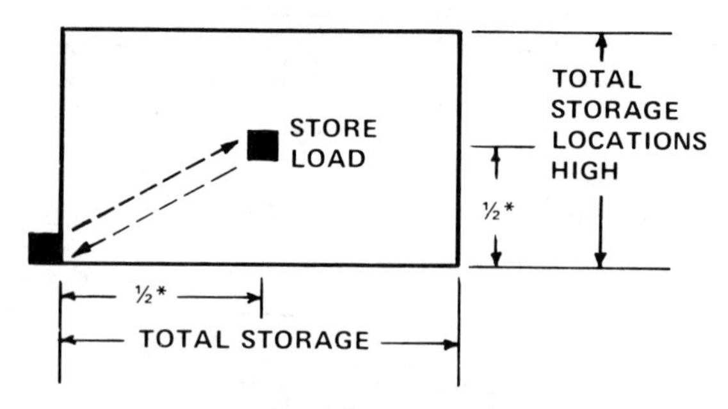

"HOME" AT END OF SYSTEM AND
AT LEVEL OF 1ST STORAGE LOCATION.

Fig. 12.2.14 Method of computing cycle time for one-command operation.

START AT "HOME" WITH PICKUP.
RETURN TO HOME FOR DEPOSIT.

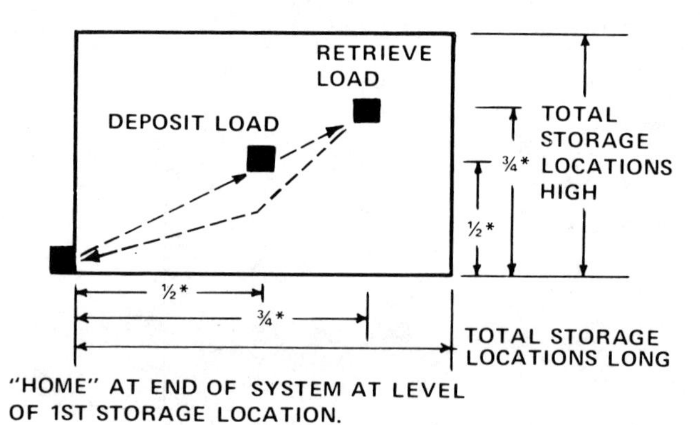

Fig. 12.2.15 Method of computing cycle time for dual-command operation.

STEP 3. Store the load at a storage location $\frac{1}{2}$ the number of storage addresses along the aisle, and up $\frac{1}{2}$ the number of vertical storage locations in that aisle. (In cases of simultaneous travel, use the longer of the travel or the lift times computed individually.)

STEP 4. Retrieve a load at a storage location $\frac{3}{4}$ the number of vertical storage locations in that aisle. (In cases of simultaneous travel, use the longer of the travel or the lift times computed individually.)

STEP 5. Deposit load at home position.

Method of Computing Cycle Time for Double Deep, Single Cycle

STEP 1. Retrieve a load at pickup station (single deep stroke cycle).

STEP 2. Travel with the load to storage location $\frac{1}{2}$ the number of storage addresses along the aisle, and up $\frac{1}{2}$ the number of vertical storage locations in the aisle.

STEP 3. Deposit a load in storage location (average time between single and double deep stroke cycle).

STEP 4. Return to delivery position.

STEP 5. Deposit load.

Dual Cycle

STEP 1. Pick up a load at pickup station (single deep stroke cycle).

STEP 2. Travel with the load to storage location $\frac{1}{2}$ the number of storage addresses along the aisle, and up $\frac{1}{2}$ the number of vertical storage locations in the aisle.

STEP 3. Deposit load in storage location (average time between single and double deep stroke cycle).

STEP 4. Travel without a load to a storage location $\frac{3}{4}$ the number of storage locations in that aisle and up $\frac{3}{4}$ the number of vertical storage locations in that aisle.

STEP 5.* Pick up a load in storage location (average time between single and double deep stroke cycle).

STEP 6. Return to delivery position with the load.

STEP 7. Deposit load (single deep stroke cycle).

STEP 8. Carriage vertical travel to pickup position (applicable only on multilevel P & D stations).

**Shuffle Cycle*

The shuffle cycle (when required) will replace step 5 of the dual-command cycle.

STEP 1. Pick up a load in storage location (single deep stroke cycle).

STEP 2. Travel with the load three storage locations (three bays) horizontally and two storage locations (two levels) vertically.

STEP 3. Deposit load in storage location (double deep stroke cycle).

STEP 4. Return travel from the position outlined in Step 2 to position in Step 1.

STEP 5. Pick up a load in storage location (double deep stroke cycle). For comparison purposes only: Three of the total cycles per hour for an S/R machine shall include shuffle cycles, in order to indicate the throughput degradation compared with dual cycles without shuffles. Actual system operation may require a varying number of SKUs in storage, and the system operation algorithms.

In cases where $\frac{1}{2}$ or $\frac{3}{4}$ of the storage locations (horizontal, vertical) equals a fractional number, round off to the next higher (longer) storage location.

Most stacker crane manufacturers and system purveyors can provide computer simulation programs that take into consideration this throughput and cycle requirement for a given rack row length. The horizontal acceleration, top speed and deceleration, the vertical elevating velocity, and the time for a shuttle cycle in and out all give input that result in determining the number of cranes required.

If the throughput is low it will not warrant a crane being assigned to each aisle, especially if the number of rack openings is large with a big area configuration. In this case you may want to consider a stacker crane transfer car (Fig. 12.2.16). A stacker crane transfer car may be either floor supported or bridge supported. This allows a crane to service multiple aisles. Then add to the total cycle time the time required to move the crane into the transfer car, to move the car to another aisle, and to unload the crane from the car. Generally, a ratio of no less than three aisles per crane is economically justifiable compared to the captive crane per aisle approach. Also, more space is required at the end of the aisle. This runout space, if a transfer car is used, can be used as a convenient maintenance station.

12.2.3 Storage/Retrieval Machines

AS/RS units are categorized into three main types: single masted (Fig. 12.2.17), double masted (Fig. 12.2.18), and man-aboard (Fig. 12.2.19). All are generally floor supported on a track and ceiling

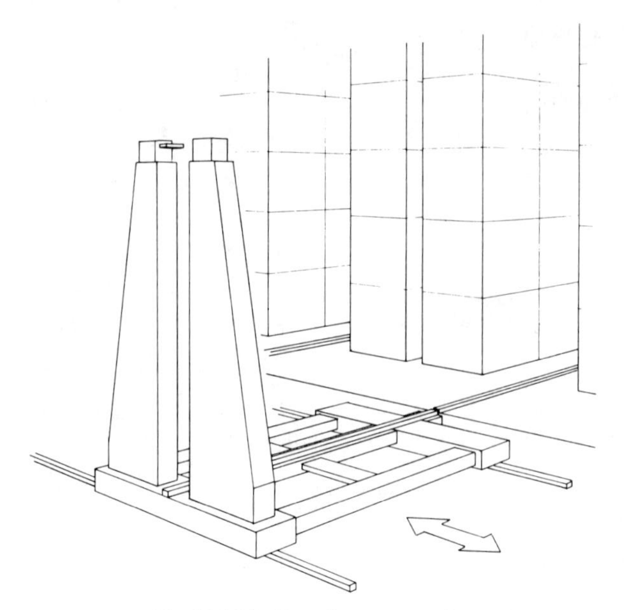

Fig. 12.2.16 Transfer car operation.

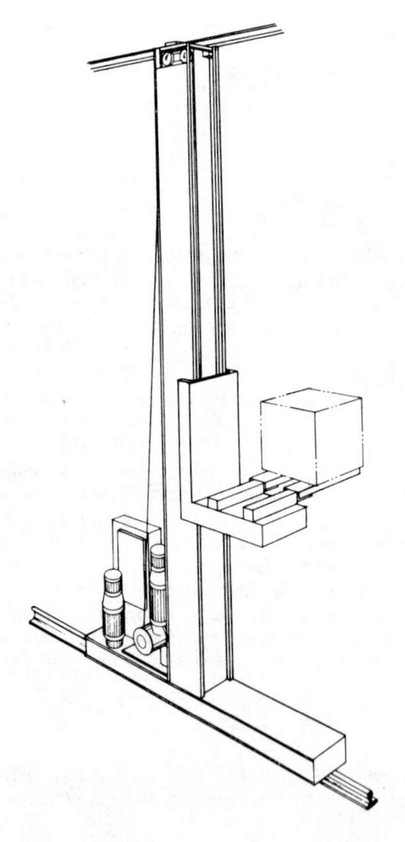

Fig. 12.2.17 Single-mast machine.

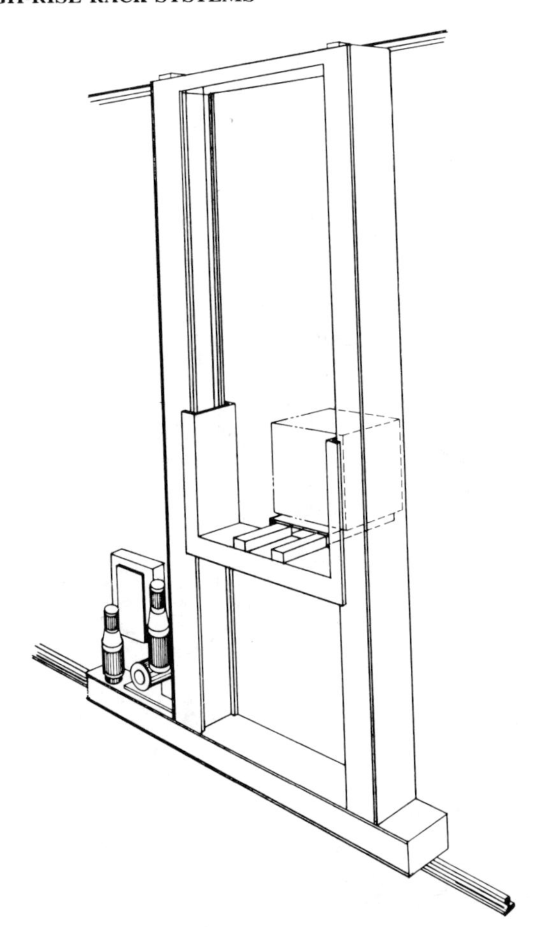

Fig. 12.2.18 Double-mast machine.

guided at the top by guide rails or channels to insure accurate vertical alignment. Some single-masted and man-aboard models are supported from the ceiling to hang in a pendulum mode.

The vertical mast of a typical S/R machine guides and supports a carriage on which unit loads are carried. One or more lateral shuttles or telescoping extraction devices attached to the carriage inject and retrieve loads, one or two loads deep in and out of the storage rack. Storage depths greater than two loads deep on either side of the aisle are classified as deep-lane storage systems.

Such systems use roll-through storage racks with multiple gravity lanes. Loads are placed in at one end and gravity fed to a take-out position at the opposite end (Fig. 12.2.20).

If the loads consist of small binnable items that can be placed either on shelves or directly into containers, the design of the AS/RS crane is modified. A man-aboard style is equipped with a picking platform (Fig. 12.2.21), or the container is brought to the order picker at a work counter (Fig. 12.2.22). Such types of cranes are called ministackers.

12.2.4 Pickup and Delivery Stations

To provide a means for accomplishing throughput to and from the AS/RS and the supporting transportation system, stations are provided to precisely position inbound and outbound loads for pickup and delivery by the crane.

A pickup and delivery (P & D) station can be a simple elevated structural pedestal to accommodate the configuration of the load. They may be designed to convey, rotate, or elevate loads if required. The interface to these stations can be by forklift truck, powered conveyor, in-floor towline, or automatic guided vehicle system (AGVS). These can be integrated into a total system to provide ancillary functions such as counting, automatic weighing, size and profile checks, identification, and labeling. Typical

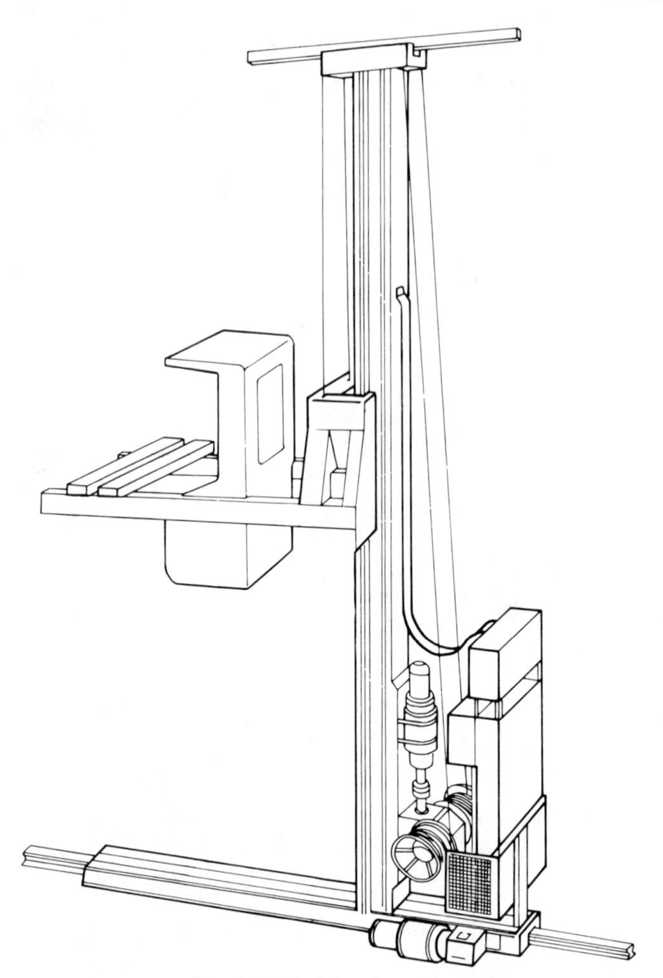

Fig. 12.2.19 Man-aboard machine.

load and unload configurations are shown in Fig. 12.2.23. Depending on the system, load and unload stations can be at different levels or elevations such as a mezzanine, or at opposite ends of the stacker aisle.

12.2.5 Controls

All stacker cranes are equipped with manual controls situated in a suitable cab or platform location on the crane. This permits operation of all crane functions by an onboard operator.

If the system is computer controlled, a switch to this manual mode allows for unlocking the automatic controls for maintenance or emergency purposes using the manual controls.

In the simplest automatic operation, a manual end-of-aisle input station at each crane accepts prepunched function cards or push-button input to command the machine through programmable controllers on each stacker. Inventory control and status of empty/full locations are kept separately by the operator.

In more sophisticated automatic operations, the optimum level of computer control can encompass:

1. Monitoring and controlling the complete transportation system moving loads into and out of the AS/RS equipment, including operation of conveyors, transfers, pallet accumulators/dispensers, weigh scales, elevators, turntables, and profile checks (length, width, height); and rejecting those loads outside of allowable tolerances.
2. Optically identifying loads for quantity, part number, and manufacturer and updating the inventory plus other pertinent readable data.

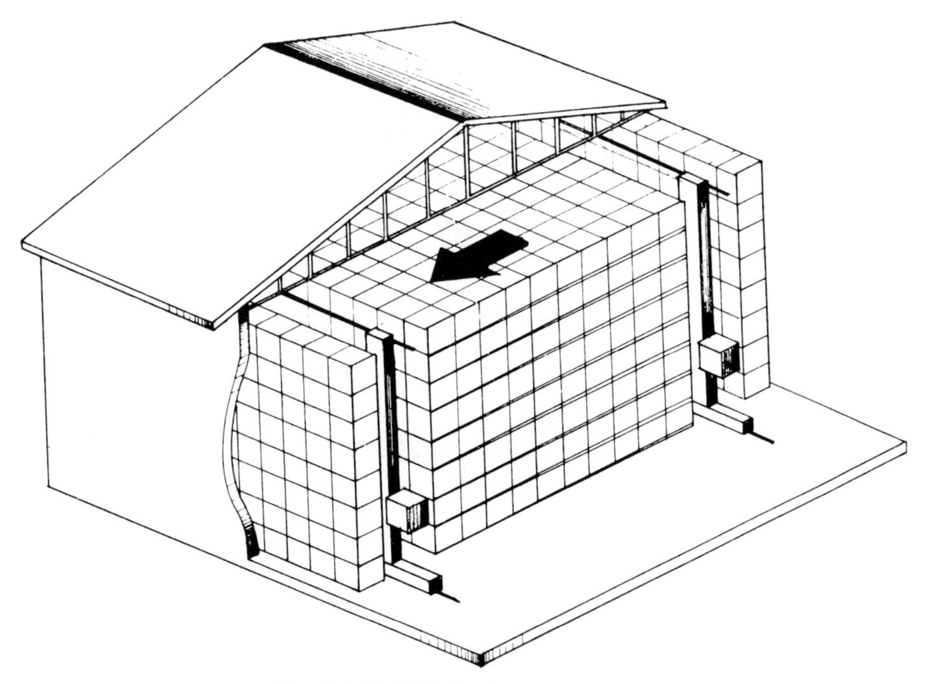

Fig. 12.2.20 High-density storage system.

3. Assigning specific openings (addresses) for each individual load.
4. Allocating material to specific aisles according to family group, popularity, or physical characteristics.
5. Retrieving loads on an FIFO or LIFO basis.
6. Communicating on a real-time basis with other host or hierarchic computers miles away.
7. Scheduling shift output hourly or daily as well as updating purchasing, production, shipping, and accounting procedures.
8. Reporting via computer printouts all of the foregoing information.

The selection and evaluation of those functions that are mandatory or desirable must come from a complete system analysis. The study takes into consideration all the requirements such as inventory, throughput, manpower, and economic payback.

12.2.6 System Justification

An analysis of economic payback and financial justification is the measurement by which management accepts or rejects a proposed AS/R system. The competition for available capital funds in areas of tooling, production, improved product design, distribution, and so forth, makes it mandatory that the bottom line for a decision be based on the most favorable rate of return on investment (ROI). The considerations involved include land investment costs, building and grounds, machinery and equipment, and operating costs.

Land Investment Costs

A high-rise AS/RS 60 ft high requires 60–70% less floor space than a conventional warehouse with 18–20-ft ceilings. The use of "air-rights" provides savings in land costs when expansion is required. The reduction of travel distances to store and retrieve material can produce additional savings in more efficient material flow.

Building and Grounds

Construction economics through the use of a rack-supported structure and the elimination of interior building columns is a most significant factor in determining ROI. Even with a freestanding rack storage

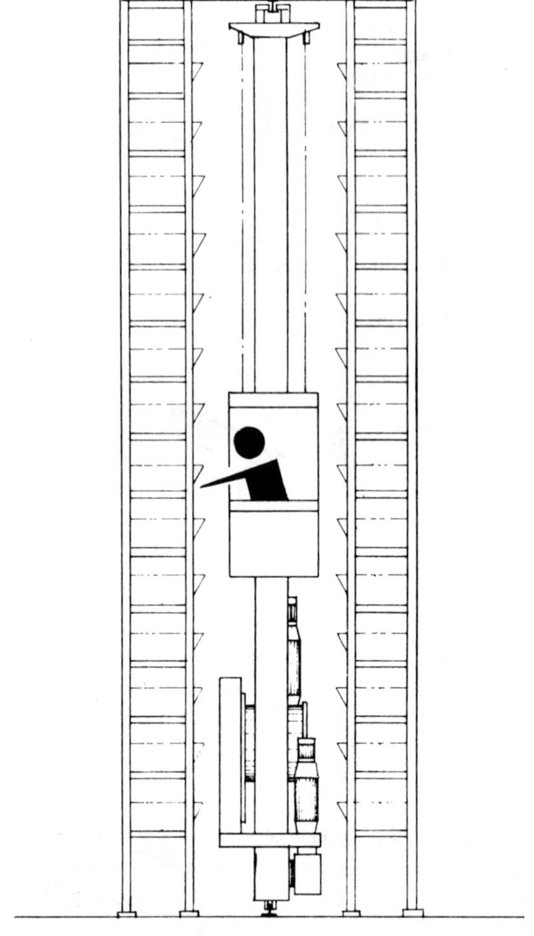

Fig. 12.2.21 Man-to-part concept.

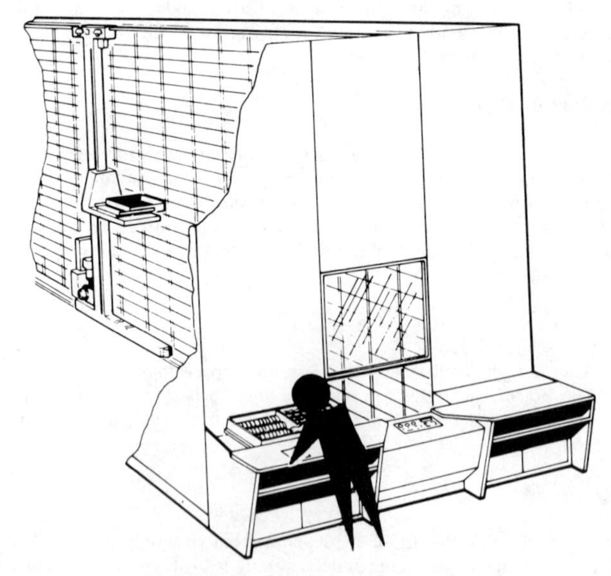

Fig. 12.2.22 Part-to-man approach.

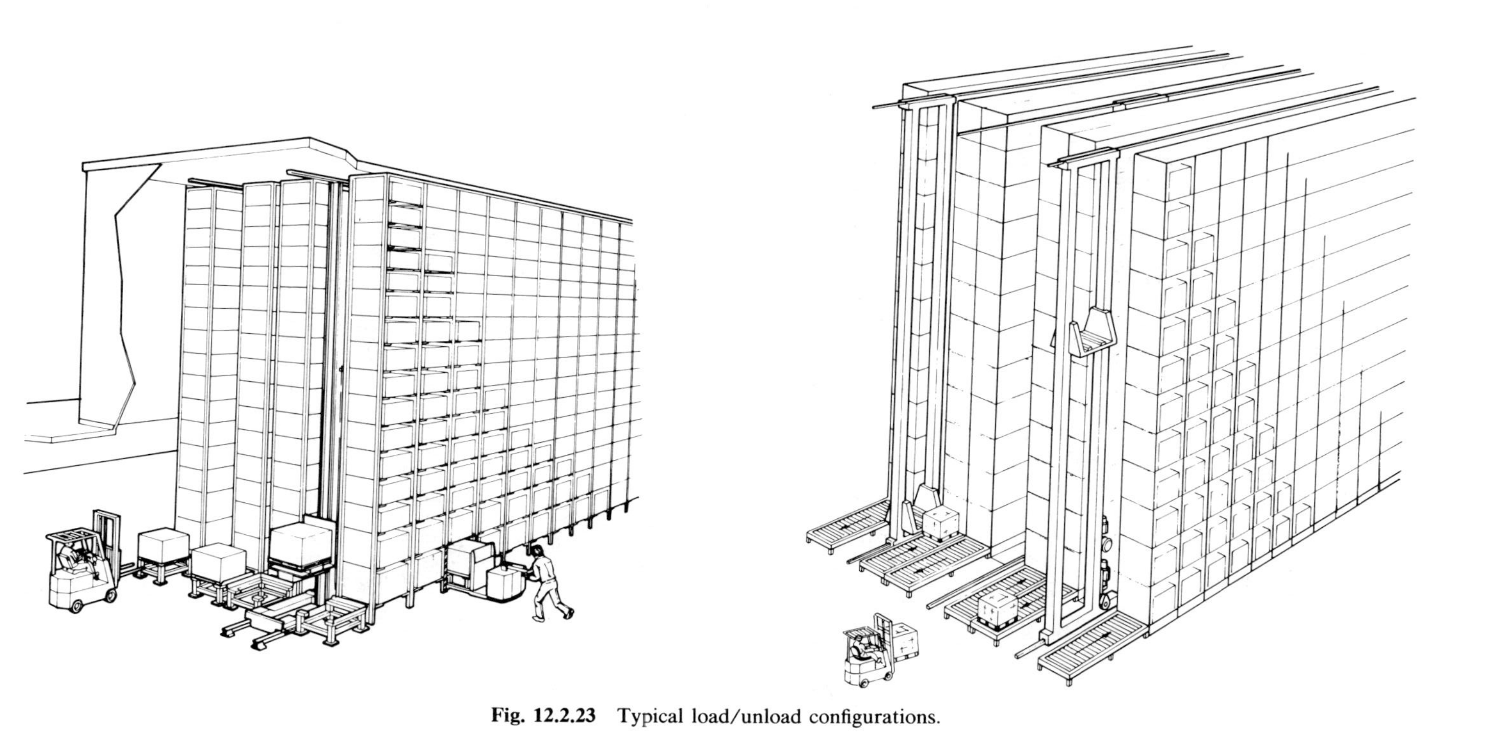

Fig. 12.2.23 Typical load/unload configurations.

structure, the savings in site preparation, foundation costs, utilities, HVAC, fire protection, and security costs may be magnified.

If off-site facilities are being leased or rented, these costs, if eliminated, are direct savings in an investment analysis. By improving the density of inventory, the increased value of sales per square foot of warehouse occupied is higher. Not having to budget for more capacity by adding brick and mortar can be a cost avoidance factor in ROI.

Machinery and Equipment

The investment costs of stacker cranes versus conventional forklift trucks to handle a given input and output volume is higher in an AS/RS. The manning savings and operational costs must be balanced against this differential. Rack costs are generally higher than in a conventional warehouse. Ancillary equipment such as powered conveyor interface, pickup and delivery, weighing, and sizing stations must be compared for labor cost and operating savings.

The obsolescence factor is more favorable in an AS/RS. Upgrading and optimizing the system on a scheduled basis extends the useful system life to 12–15 years in comparison to the 5–7-year depreciated schedule for conventional materials handling equipment and the 30–40 years for a conventional warehouse.

The justification process should include the benefits of applicable investment tax credits and an analysis of the eventual salvage value at the termination of the useful life of the equipment and hardware.

Even at this point, a study should be made of how much it would cost to update and refit present equipment, machinery, and controls to the current state of the art. Usually the modernization costs are much more than acquiring new equipment and contemporary technology.

Additional tax benefits are being allowed in many instances where the close dependency between building and integral equipment allows for the depreciation of the building as machinery. In a rack-supported building, the roof, wall panels, and side framing may qualify for tax incentives.

Operating Costs

Direct and Indirect Labor. Automating the functions of receiving, identification, inspection, dispatching, storage, retrieval, order picking, and shipping provides potential for direct labor savings of 50% or more, especially when fringe costs of 30–40% are included. This degree of direct labor reductions has a heavy impact on indirect and clerical labor savings as well, especially if the system includes a real-time, on-line computer program. The need for support personnel for inventory control, scheduling, and procurement can be substantially reduced.

Inventory Carrying Costs. The average costs of carrying inventory is estimated to be 30–45% of the value of the inventory on an annual basis. Reducing inventory levels as little as 10% on a million-dollar inventory could bring savings of $100,000 which could help justify an AS/RS.

The elimination of paperwork can expedite throughput and accurate scheduling. Cushioning inventory levels to compensate for shrinkage is no longer necessary in a real-time environment. With automatic inventory control, for example, inventory known to be in-house but unlocatable is pinpointed and available. A large dollar value change can also be affected by improving record accuracy and status reporting of in-process material.

Maintenance. In spite of more mechanical complexity and operational sophistication, AS/RS equipment is less costly to maintain and repair. Plug-in modules, built-in diagnostic routines and computer-scheduled preventative upkeep are designed into the system. Downtime is minimized by quick, temporary manual interface. System designers and suppliers usually guarantee a level of operational efficiency as high as 96% before system acceptance.

Energy. The modern AS/RS warehouse can be considered a "black box," with no light, no heat, no people; material simply going in at one end and coming out at the other. Less heat, light, and environmental controls are required than in a conventional warehouse. Unmanned and fully automated systems with remote control insure substantial savings in utility costs.

12.2.7 Cost Comparisons

The following example illustrates a typical cost comparison between a conventional warehouse and a proposed AS/R system to meet the same requirements. Based on a hypothetical model developed from the parameters in the previous five sections, the size, loads, throughput, and controls are to handle a system requiring:

8500 openings

Four stacker cranes in four aisles

Category	Conventional 30-ft Warehouse	AS/RS 65-ft Hi-rise
Land	130,680 ft² (3 acres) @ $26,000/acre = $78,000	32,670 ft² (0.75 acres) @ $26,000/acre = $19,500
Site preparation	$50,000	$13,000
Building	129,000 ft² @ $30/ft² = $3,870,000	32,000 ft² @ $55/ft² = $1,760,000
AS/RS (Freestanding) with computer control		$3,245,000*
Conventional racks	8500 openings @ $35/opening = $297,500*	
Fork trucks	10 trucks @ $14,000 = $140,000*	3 trucks @ $14,000 = $52,000*
Fire protection	$60,000	$120,000
TOTAL	$4,495,500	$5,209,500
* (Less investment tax credit)	$437,500 @ 10% = ($43,750)	$3,297,000 @ 10% = ($329,700)
Total investment conventional	$4,451,750	
Total investment AS/RS		$4,879,800
Additional investment for AS/RS		$428,000

Fig. 12.2.24 Calculation of initial warehouse investment.

Category	Conventional 30-ft Warehouse	AS/RS 65-ft Hi-rise
Manpower		
Direct	34 people @ $20,000 = $680,000	17 people @ $20,000 = $340,000
Indirect	26 people @ $20,000 = $520,000	16 people @ $20,000 = $320,000
Depreciation		
Building, 40 years	Annual $96,750	Annual $44,000
Equipment, 7 years	Annual $36,500	Annual $275,000
Inventory carry charge @ 30% per year	$3,000,000 inventory @ 30% per year = $900,000	$2,700,000 inventory @ 30% per year = $810,000
Interest on facility	$4,451,750 @ 14% per year = $623,245	$4,879,800 @ 14% per year = $683,172
Maintenance	$70,000	$50,000
Utilities	$100,000	$60,000
Product damage	$50,000	$10,000
Pilferage	$80,000	$10,000
Insurance	$40,000	$70,000
Housekeeping	$20,000	$10,000
Total Operating Costs	$3,216,495	$2,682,172
Annual Cost Savings	$534,323	

Fig. 12.2.25 Calculation of annual warehouse operating costs.

Freestanding rack structure

Computer-controlled, automatic inventory

65-ft high-rise building versus a conventional 30-ft high building

Relative costs can be calculated per procedures in Figs. 12.2.24 and 12.2.25.

Additional operating costs not shown on Fig. 12.2.25 could include hidden and intangible factors such as:

Poor customer service resulting from uncertain or inaccurate stock conditions

Lost sales due to errors

Periodic full physical inventory costs

12.2.8 Conclusion

The final measure of the acceptability of an AS/RS is financial favorability. Management must judge the factors of capital outlay, payback, and the rate of return on investment against the gains in productivity, efficiency, and profit. This evaluation must be continually ongoing to insure that the competition does not pass you by.

AS/RS is a relative newcomer that has proven itself again and again in hundreds of industries. Through computer interfaces it has made possible the integration of materials handling and storage with the total manufacturing and distribution process.

Keep adding to your general store of knowledge and continue increasing your confidence level in systems by regularly touring several of the nation's outstanding installations. Your local AS/RS vendor or system supplier will be glad to arrange a visit.

12.2.9 Glossary

Aisle. Space between storage compartments in which the S/R machine operates.

Aisle Hardware. The support devices in the S/R machine aisle. These include: floor rail, top guide rail, end stops, electrification, actuators marking end of aisle, and code plates or markers for bay location.

Aisle Transfer Car. A machine or vehicle for transferring an S/R machine from aisle to aisle, which also normally runs on a rail or rails.

Automated Storage/Retrieval System (AS/RS). A combination of equipment and controls which handles, stores, and retrieves materials with precision accuracy and speed under a defined degree of automation. Systems vary from relatively simple, manually controlled order-picking machines operating in small storage structures to giant, computer-controlled storage and retrieval systems totally integrated into the manufacturing and distribution process.

Automatic Guided Vehicle System (AGVS). A vehicle equipped with automatic guidance equipment, either electromagnetic or optical. Such a vehicle is capable of following prescribed guide paths and may be equipped for vehicle programming and stop selection, blocking, and any other special functions required by the system.

Bay. One series of vertical storage locations when considered as a set.

Baffle. A barrier installed in a horizontal (or vertical) plane within the storage racks at sprinkler levels to concentrate heat on sprinkler heads.

Binnable Materials. Those small items stored either on shelves or directly in containers.

Cab. The operator's compartment of an S/R machine.

Carriage. That part of an S/R machine by which a load is moved in the vertical direction.

Command Cycle. See Dual Command Cycle and Single Command Cycle.

Conveyor(s). Fixed-path handling systems that carry, queue, and position loads.

Deep Lane Storage. Storage depth greater than two loads deep on one or both sides of the aisle.

Double Deep Storage. Loads that are stored two deep on each side of the aisle.

Dual Command Cycle. The time between actuation of the S/R machine and completion of a cycle in which one load is stored and another load is retrieved and the S/R machine is ready to start a new cycle.

Free standing Rack Structure. Installed inside a building of conventional construction, supported only by the floor and not supported or attached to any building structure.

Load Configuration.

Load Length. Maximum overall dimension of a pallet or load module and load in the direction perpendicular to the length of the aisle.

Load Width. Maximum overall dimension of pallet or load module and load in the direction parallel to the aisle.

Load Height. Maximum overall dimension from bottom of pallet or load module to top of load.

Load Overhang. Amount of load projection past base of pallet or load module in any direction.

Man-aboard. A storage and retrieval concept whereby materials are accessed by taking the operator to the materials on board an S/R machine.

Miniload. A storage and retrieval concept whereby materials are accessed by bringing the container to the operator. The term is typically used in small parts applications and/or where the weight of the container does not exceed 750 lb.

Order Picking. The selection of less-than-unit-load quantities of material for individual orders.

P & D Stations. A location at which a load entering or leaving storage is supported in a manner suitable for handling by the S/R machine. (Prior usage has also called this the Transfer Station, I/O, Pickup and Delivery Station, Feed/Discharge Station, etc.)

Queue. A line formed by loads or items while waiting for processing.

Rack-Supported Building Structure. A complete and independent load storage system in which the storage rack is the basic structural system.

S/R Machine. A machine operating on floor or other mounted rail(s) used for transferring a load from a storage compartment to a P & D station and from a P & D station to a storage compartment. The S/R machine is capable of moving a load both vertically and parallel with the aisle and laterally placing the load in a storage location. Common types of S/R machines are: miniload S/R machines, unit load S/R machines, and man-aboard S/R machines.

Shuttle. The load-supporting mechanism on the carriage which provides for movement of loads into or out of storage compartments and P & D stations.

Single Command Cycle. The time between actuation of the S/R machine and completion of a cycle in which one load is stored or retrieved and the S/R machine is ready to start a new cycle.

Single Deep Storage. Loads stored one deep on each side of the aisle.

Slave Pallet. A handling base or container on which a unit load is supported and which normally is captive to a system.

Storage Module. Those items such as pallets, containers, and boxes, containing, holding, or constituting the unit load.

Storage Structure. A system of storage locations or compartments plus any other members that may be required to support and/or guide an S/R machine.

Tier. A set of storage locations having a common elevation.

Unit Load. Any load configuration handled as a single item.

REFERENCES

1. *Considerations for Planning and Installing an Automated Storage/Retrieval System.* AS/RS Association of The Material Handling Institute, Inc., 1326 Hartford Road, Pittsburgh, Pa. 15238. 1st Ed. c. 1977, 2nd Ed. c. 1982.

BIBLIOGRAPHY

Collins, Chas. H., *Basic Information AS/RS.* Detroit Chapter of the International Materials Management Society, 1978.

Nine Simple Steps to Determine the Layout, Design and Estimated Cost of an Automated Storage/ Retrieval System. P & H Harnischefeger Corp., Milwaukee, Wis. 53201, 1977.

Load Design Considerations for Automatic Storage Systems. Clark Equipment Co., Battle Creek, Mich. 49016, 1978.

12.3 HIGH-DENSITY SYSTEMS

Brian W. Sanford

12.3.1 Deep-Lane Concepts

A high-rise unit-load system represents a "random access" warehouse: each unit load is stored in an *X-Y* rack address along an aisle served by a storage/retrieval machine.

In a high-density warehouse the *X-Y* addresses are deep enough to permit storage of two or more unit loads, one behind the other, in a single *X-Y* location (see Fig. 12.3.1). Double-deep racks permit two unit loads per aisle address (*X-Y* location). Deep lane racks permit storage of two or more (usually more than two) unit loads per aisle address (*X-Y* location).

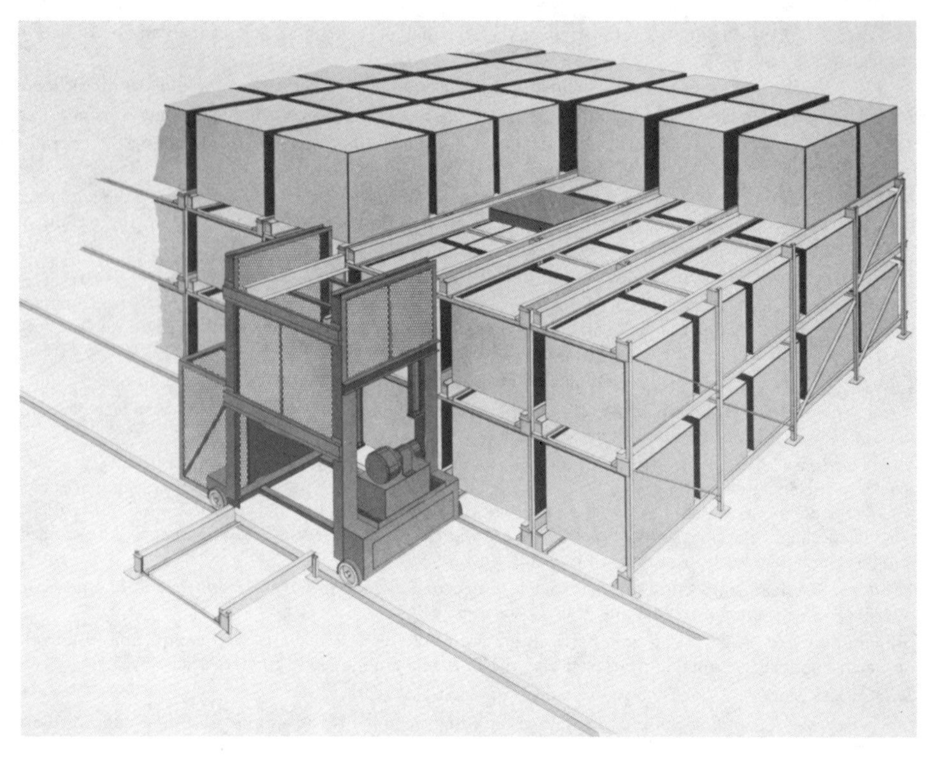

Fig. 12.3.1 High-density storage system.

Double-Deep Systems

A storage/retrieval (S/R) machine normally has a five-element (three racks, two pinions) shuttle table for single-deep storage. This can be replaced with a nine-element (five racks, four pinions) shuttle capable of extending into either the front or rear storage position in a double-deep rack. Because the shuttle table is cantilevered from the S/R lift carriage and does not touch the storage rack, the only rack changes required are the deeper storage addresses, additional rack uprights, and an extra 75–100 mm of clearance at each storage level to accommodate the thicker shuttle.

Deep-Lane Systems

For storage of three or more unit loads deep, the shuttle table is replaced by a deep-lane shuttle car (DLC) which rides the S/R lift carriage to the *X-Y* location. When the S/R machine is positioned, the DLC uses powered wheels to drive from the S/R lift carriage into the rack address, using the storage rails as a track means. The DLC stores (or retrieves) a unit load and then returns onto the lift carriage of the S/R machine.

Deep-Lane Storage Rack

The DLC has about the same vertical thickness as the double deep shuttle. However, deep lane storage rails have a modified shape to support the DLC in the lane. The shape is either a C or Z section (depending on DLC design). The top flange supports the unit loads. The bottom flange provides a riding surface for the DLC. Separation is 165–200 mm.

Deep-Lane Shuttle Car (DLC)

The DLC is a single-purpose vehicle designed to transfer a unit load onto the S/R machine, and transfer it into the storage lane (and retrieve each load in turn). The DLC has a drive to power the wheels. A second drive elevates the load carrier approximately 40 mm, allowing the unit load to be transported 20 mm above the storage rails.

S/R Machine Options

S/R machines for deep-lane service are as varied as the S/R field itself. The purpose of the S/R machine is to transport the DLC and its live load to the selected *X-Y* address. Configuration of the aisle equipment is determined by the dimensions of the building and the function of the system:

Two-wheel (single rail, top supported) S/R machines have been used for service in tall warehouses (over 10 m).

Four-wheel (double rail, freestanding) machines have been used in spaces of less than 10 m clear height.

Four-wheel S/R transfer cars have been used for work-in-process systems and on a machine-per-level basis in multilevel, single-aisle racks (used with high-speed elevators in the conveyor system).

All three types of S/R aisles have been furnished in man-aboard units and in automated systems.

DLC Options

Deep-lane shuttle cars (DLC) have been furnished in both AC and DC power versions. AC-powered vehicles have a cable reel with the cable connected to the S/R machine. DC power is provided by onboard batteries which are recharged each time the DLC returns to the S/R machine. The trend is toward DC power, but AC is still used for heavy-duty, high-cycle-rate applications.

DLC control options also vary by manufacturer and application. Man-aboard S/R machines usually employ some sort of operator control over the DLC, using radio control, infrared, or laser communications. Automated S/R–DLC systems have automated control and feedback to the warehouse computer via the AS/R computer link.

12.3.2 Applications

Warehousing

High-density storage is found primarily at the manufacturing level, where a large quantity of an item is manufactured in a production run, stored in deep lanes, and retrieved at an average daily sales rate, usually on an FIFO basis.

Order Picking Systems

Reserve stock is stored in deep lanes and retrieved on an FIFO basis for delivery to the pick face (which is often a pick tunnel supplied by two S/R machines).

Marshaling Operation

Truck load quantities of unit loads are stored by destination in deep lanes. When the trailer is available for loading, the assembled order is moved to the loading point for that truck door.

Work in Process

Work processed in batch quantities can be held in deep-lane storage between operations. Deep lanes of varying depths are also used to deliver unit loads and fixture tables to process machinery arranged along the material handling aisle.

12.3.3 Design Considerations

High-density systems are more flexible than random access systems in fitting into existing warehouses. Low-activity systems can be configured with a reduced number of aisle machines. Certain design constraints must be considered, however.

AS/R Machine Cycle Time

Double-deep shuttles and DLC vehicles have longer average store–retrieve cycle times than single-deep shuttles. Although the S/R machine time for *X-Y* movements is identical, this *Z*-axis time factor must be considered in determining throughput rate.

Stock-Keeping-Unit (SKU) Segregation

Inventory can be mixed in double-deep or deep lanes. However, retrieval of a unit stored behind other units requires additional S/R machine operations, slowing system throughput. In most deep-lane systems, all units in a lane are homogeneous; the unit closest to the aisle is acceptable for shipping compared to any other unit in that lane.

The owner needs to determine what constitutes homogeneity: same product code, same date code, same order, date code range, production lot code, and so on.

Rack Utilization

SKU lot size is the major consideration in rack lane depth. If inventory is segregated by SKU there will be partially filled lanes (even double-deep lanes). This includes active output lanes and active input lanes for which there are no additional incoming products. The rack utilization due to deep-lane storage is calculated by dividing the number of units in storage by the number of unit load positions in the lanes in use (not empty).

Rack Operating Space

As a practical matter virtually every AS/R system is operated slightly less than full. This permits incoming volume to exceed shipping volume for a limited period of time. The percentage of slack allowance is based on operating experience. Balanced operations usually plan to fill within 95–96% of capacity. Imbalanced facilities (input on three shifts, ship on one shift) might work 90% or even less, to permit back-shift surge. In high-density systems, excess storage positions must be planned to allow for both rack utilization and operating slack.

12.3.4 High-density Warehouse Configuration

The final configuration of a high-density warehouse is a compromise of site availability, material flow, system activity, SKU lot size, operating balance, seasonality, and control requirements. The number of aisles, lane depth, and materials handling and control systems all work together to provide the most useful system for planned future operations.

12.4 RACK-SUPPORTED BUILDINGS

W. R. Midgley

12.4.1 Preliminary Structural Drawings

The preliminary structural drawings furnished the structural designer are usually storage rack plans, elevations, and cross sections. They are normally furnished by a materials handling consultant who has developed them while working with the owner. The cross sections are typically in the longitudinal or down-aisle direction as well as in the transverse or upright direction. These drawings enable the structural analyst to make the computer models necessary to size the structural members for the loads that the rack-supported building will be subjected to.

The size and spacing of the storage rack members shown on the drawings are determined by the size and weight of the product stored. The length of time that a particular product may be stored in the rack also helps determine the location of that product in the rack system. The products that will be stored in the system in the future should also be considered. The probability that a product stored today will also be stored over the life of the rack system is also often considered. The numbers of sizes of each stored product determine the numbers and sizes of the openings.

It is necessary for the materials handling consultant to estimate the structural member sizes before making preliminary drawings. This is necessary in order to lay out the plan view, elevations, and cross sections so that the plan size and height of building are approximately known. There are often plan and/or height restrictions on the building for one reason or another. One such restriction might be the plot size available for the building; another might be the cost of high-masted stackers compared to adding another aisle or stacker, thereby keeping lower-masted stackers but increasing the width of the building.

The materials handling consultant can estimate the sizes of the members by previous experience and by knowing the pallet loads and approximate height of the building (i.e., the number of openings high). For example, for a one- or two-deep pallet-stored stacker-rack system, the post can be sized closely enough by knowing the total load on the post between the floor and the first longitudinal brace. If it is assumed that a 3 in. × 3 in. × 12 gauge cold-rolled C-section post would be satisfactory and later it is found unsatisfactory from the computer results (see Section 12.4.3), the gauge of the post could be increased to 11 gauge or 10 gauge without going to a different overall building plan size. If the 10-gauge post were not sufficient, then the unsupported length of the post could be reduced by adding another one or two longitudinal braces (down-aisle ties); or the post could be reinforced by doubling up the post for some distance by welding another 3 in. × 3 in. C-section to the originally specified post.

12.4.2 Building Codes, Specifications, and Loads

The local government building agency usually specifies the controlling building code, and that code is normally the latest edition of the building code for the state in which the rack building is to be located. The state code might specify that the latest edition of the *Basic Building Code* (BOCA) [1], the *Standard Building Code* (SBCC) [2], or the *Uniform Building Code* (UBC) [3] be used.

From the building code is obtained the wind load, the snow load, and the seismic load criteria. Often the seismic load criteria will be that specified by UBC [3]. The live load or pallet load is determined by the owner and the materials handling consultant. The impact load caused by lowering the pallet load onto a shelf beam and in turn onto the shelf arm, if there is any, is 25% of one unit load as specified in paragraph 7.2 by the *Rack Manufacturers Institute Specification* (RMI) [4]. For a man-aboard two-pallet-deep stacker-rack system, and for a single-reach stacker (i.e., the stacker carries one pallet), the impact is 25% of one pallet load. For a man-aboard two-pallet-deep stacker-rack system, and for a double-reach stacker (i.e., the stacker carries two pallets), the impact is 25% of two pallet loads. However, for an automatic stacker, the moving equipment manufacturer is responsible for supplying to the rack manufacturer, information on maximum vertical static and dynamic loads for the design of racks (see paragraph 7.2 of the RMI specification [4] and paragraph 7.2 of the appendix to the RMI specification [4]).

It is almost always necessary for the structural designer to determine what lateral load is most critical, and then superimpose that lateral load on the critical gravity loading to determine the most critical loading condition for the structure. In addition to the wind load and seismic load, the horizontal load of $1\frac{1}{2}$% specified by paragraph 7.3.1 of the RMI specification [4] must also be considered.

The lateral load caused by the stacker on the top-rail stacker support (the cross-aisle tie) should also be considered. This load is caused by the eccentricity between the load and the mast, and it is transformed to the rack by the top rail support. The cause of the force is schematically depicted in Fig. 12.4.1.

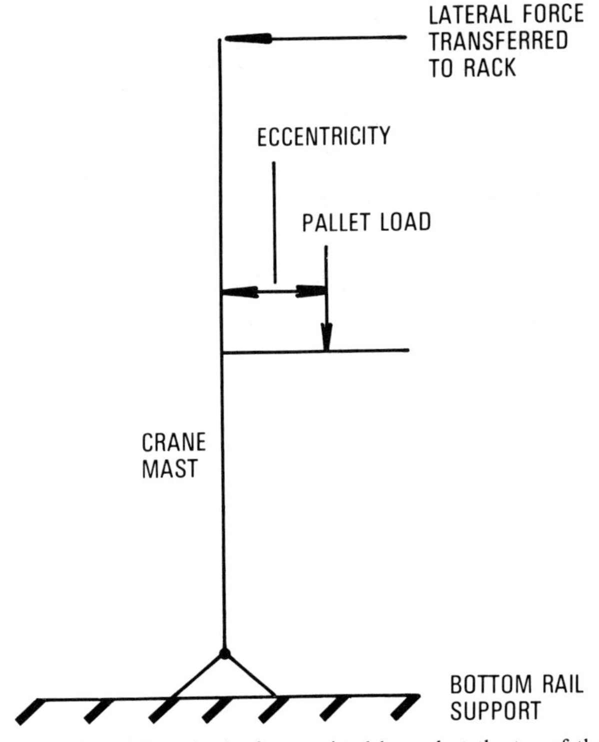

Fig. 12.4.1 Stacker-rack lateral force is the force resisted by and at the top of the rack caused by the eccentricity of the load to the mast of the stacker crane.

The impact load transferred to the rack system and caused by a runaway stacker ramming the endstops at top and bottom must also be investigated. The stacker manufacturer should be consulted for this happening and his opinion obtained as to what that force should be. The ANSI standard for *Controlled Mechanical Storage Cranes* [5] states that the bumpers shall be capable of resisting a force equal to the rated load plus the weight of the stacker traveling at a speed of at least 100% of rated load speed. The total force applied to the stop may be computed from

$$F = \frac{Wv^2}{gs}$$

where: F = total longitudinal inertia force acting at the line of the resultant force of the stacker bumper, in pounds (N)
W = total weight of the pallet load plus the hoist and forks plus a proportionate weight of the mast, pounds (kg)
v = stacker velocity at 100% of full rated speed, ft/sec (m/sec)
s = stroke of the bumper spring at point where the stacker stopping energy is fully absorbed, ft (m)
g = acceleration of gravity, 32.2 ft/sec² (9.81 m/sec²)

For the case where hydraulic bumpers may be used, see AISE Technical Report No. 6 [6] for the force to be resisted by the stop.

The primary seismic code used in the United States is the UBC [3]. The UBC permits either a static analysis or a dynamic analysis. If a static analysis is used, then the total shear and the overturning moment acting at the base should be computed by the three different methods permitted by section 2312 to determine the smallest base shear and base moment. The first method according to 2312 (d) is

$$V = ZIKCSW \tag{12.4.1}$$

where: V = the total lateral force at the base, pounds (N).
Z = numerical coefficient dependent upon the building zone as determined from the Seismic Zone Map of the United States, Figs. 1, 2, and 3 in section 2312. For locations in Zone 1, $Z = \frac{3}{16}$. For locations in Zone 2, $Z = \frac{3}{8}$. For locations in Zone 3, $Z = \frac{3}{4}$. For locations in Zone 4, $Z = 1$.
I = Occupancy Importance Factor as set forth in Table 23-K in section 2312. It can be assumed to almost always be 1.0 for rack-supported buildings.
K = numerical coefficient as set forth in Table 23-I in section 2312.
C = numerical coefficient as specified in section 2312 (d).
S = numerical coefficient for site–structure resonance, and specified in section 2312 (d).
W = the total dead load plus 50% of the rack rated capacity, and the product $CS = 0.2$, pounds (kg). (The value for W is specified in footnote 2 below Table 23-J in Ref. 3.)

For W equal to the total dead load plus 50% of the rack rated capacity, the storage rack units must be interconnected so that there are a minimum of four vertical elements in each direction to resist horizontal forces. This is usually the case for rack buildings since there would be four rack posts in each direction.

The second method for the computation of the base shear via section 2312 (g) [3] is

$$F_p = ZIC_pW_p \tag{12.4.2}$$

where: F_p = lateral forces on a part of the structure and in the direction under consideration, pounds (N). Z and I are the same as before.
C_p = numerical coefficient as specified in section 2312 (g) and is given in Table 23-J of Ref. 3. C_p for racks more than two storage levels high shall be 0.24 for the levels below the top two levels; for the top two levels, C_p = 0.30. It should be pointed out that, via footnote 1 of Table 23-J for elements laterally self-supported only at the ground level, the values for C_p can be $\frac{2}{3}$ of the value given in the table (i.e., $\frac{2}{3}$ of 0.24 and $\frac{2}{3}$ of 0.30).
W_p = the weight of the racks plus contents, pounds (kg). W_p may be less than the rated capacity of the racks and the value should be approved by the owner and engineer. Section 7.5.1 of the RMI specification [4] states that the live load is 75% of the maximum design load, unless specific average load information is available.

The third method for the computation for the base shear is via UBC Standard No. 27-11 [7]. Equation 12.4.1 is used to compute the base shear. The definitions for the terms used in the equation are the same as specified previously; however, the values for K, C, S, and W are as follows:

K = 1.33 for racks or portions thereof where lateral stability is dependent on diagonal or X bracing. Connections for the bracing members shall be capable of developing the required strength of the member.

$K = 1.00$ for racks where lateral stability is wholly dependent on moment-resistant frame action.

$K = 0.67$ for racks where lateral stability is wholly dependent on ductile moment-resistant frame action. For a frame to qualify as a ductile moment-resistant frame, it must have a ductility ratio (maximum stable displacement divided by displacement at initiation of inelastic action) of at least 3.

$K = 0.80$ for racks where lateral stability is dependent upon a combination of bracing and ductile moment-resistant frame action, with the qualifications as specified previously. The bracing system acting independently of the frame shall be designed to resist the total required lateral forces. The ductile moment-resistant frame shall have the capacity to resist not less than 25% of the required lateral force.

$C = \frac{1}{15} T^{1/2} \leqq 0.12$ for racks installed on the ground level. T is the fundamental period of vibration of the rack, in seconds, in the direction under consideration. Properly substantiated calculations or test data for establishing the period T shall be submitted. The product of CS need not exceed 0.14.

S = numerical coefficient for site-structure resonance.

W = weight of rack structure plus contents, pounds (kg).

If a dynamic analysis is used, via UBC section 2312 (i), then properly substantiated technical data must be used to establish the lateral forces and the dynamic characteristics of the structure must be considered. The applied lateral forces are usually obtained from a response spectrum which is defined as the maximum responses (i.e., displacement, stress, acceleration, pseudorelative velocity, etc.) of all possible linear one-degree systems due to a given ground or support motion. The response spectrum is normally determined by a geotechnical consultant who has taken into account the effects of local soil and geologic conditions, as well as statistical results of several studies of site-dependent spectra developed from actual time-histories recorded by actual motion instruments located in various parts of the world. The dynamic characteristic of the racks is considered and a modal analysis of the rack in both the longitudinal and transverse directions is normally performed. The analysis can be elastic or elastoplastic, whichever is specified by the owner.

As mentioned previously in this section, and for a static analysis, it is necessary for the structural designer to determine what lateral load is critical, and then superimpose that lateral load on the critical gravity loading to determine the most critical loading condition for the structure. For a standard pallet rack, the critical loading condition is likely to be the full gravity or pallet load plus an allowance for dead load, or every other beam loaded plus an allowance for dead load. For a stacker-rack system, the critical gravity loading condition for a post is likely to be some combination of alternate loading below the post–span between longitudinal bracing under investigation, with full gravity load above the post–span under investigation. An allowance for dead load should also be included. The allowance for dead load should always include the dead load of sprinkler feeder lines.

12.4.3 The Concrete Slab

The concrete slab is normally designed by a foundations consultant. The slabs are normally on ground and designed for the allowable bearing pressure for the particular soil at the site. Preliminary maximum conservative axial compression and tension forces for the rack posts are often determined from preliminary analyses in order to design the slab and have it poured and cured to receive the racks. The slab is then designed for the greatest compression and tension forces resulting from overturning due to wind load or seismic forces. Normally, overturning does not rule, and the slab is designed for the compression load.

There is usually an edge beam around the perimeter of the slab. The purpose of the beam is first to carry the post loads for the uprights adjacent to the sidewalls, and second to have the edge of the slab extend below the frost line to protect the slab so that the soil below the slab is not subjected to alternate freezing and thawing. Often the post loads are greater near the sidewalls that require a deeper slab or edge beam. In warmer climates, it is common to increase the slab depth starting at some distance in from the edge and tapering it to the required depth at the edge. The slab usually has steel in both directions and should be designed according to *ACI Building Code Requirements for Reinforced Concrete* [8].

12.4.4 Computer Models

The purpose of the models is to load a model of the rack system with the critical loading conditions for the rack system, and from the computer solution, obtain the forces and moments on each of the members of the model. The computer models are determined by taking cross sections through the rack system in the two main directions [i.e., the longitudinal or down-aisle direction and the transverse or upright (truss) direction]. The models should include the main vertical bracing system in both directions. Figure 12.4.2 shows a simplified but typical plan view of a rack building (usually a rack building will have many more uprights than depicted in Fig. 12.4.1). Figure 12.4.3 shows a longitudinal model and Fig. 12.4.4 a transverse model.

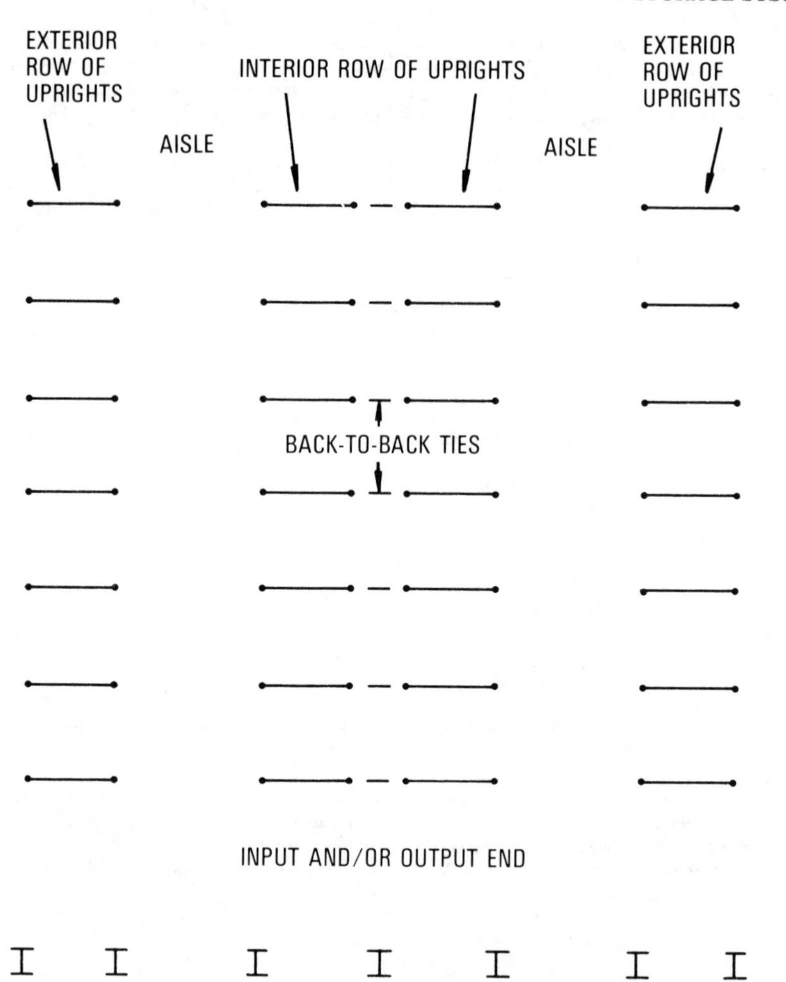

I I I I I I I

Fig. 12.4.2 Schematic plan view of typical rack building depicting four rows of uprights and two aisles as well as an input–output end at the south end.

The longitudinal model shown in Fig. 12.4.3 depicts five bays. A model of five bays is picked since it is assumed that the vertical X bracing ① in the back plane occurs every five bays. The lateral load path for seismic load then would be, for the case in Fig. 12.4.3, bending of the front post in the lateral direction indicated on the figure by A for the shelf located between the longitudinal bracing; then the load is transferred to the front longitudinal bracing indicated at B; then the load is transferred to the back longitudinal bracing at C by means of the X bracing ② located between the longitudinal bracing; and then the load is transferred to the X bracing ① and in turn transferred to the slab.

Figure 12.4.3 depicts the location of the vertical X bracing in the longitudinal direction in the back planes of the interior uprights. Also there is usually vertical X bracing in the longitudinal direction located in the back planes (against the outside longitudinal walls) of the exterior uprights. The location of the load openings are normally opposite each other in the interior and exterior rows as in the longitudinal bracing. Also horizontal plane bracing exists between the longitudinal bracing in the exterior rows so that a similar load path exists for the lateral seismic load in the longitudinal direction for the exterior rows.

The lateral load path for seismic load acting on the transverse model is from the shelf beams (sometimes by friction and sometimes by load stops), to the shelf arms, to the upright, and to the slab. The load path for wind load on the transverse model is from the girts to the exterior uprights to the slab, and across the aisle by means of the roof truss to the interior uprights and then to the slab (see Fig. 12.4.4).

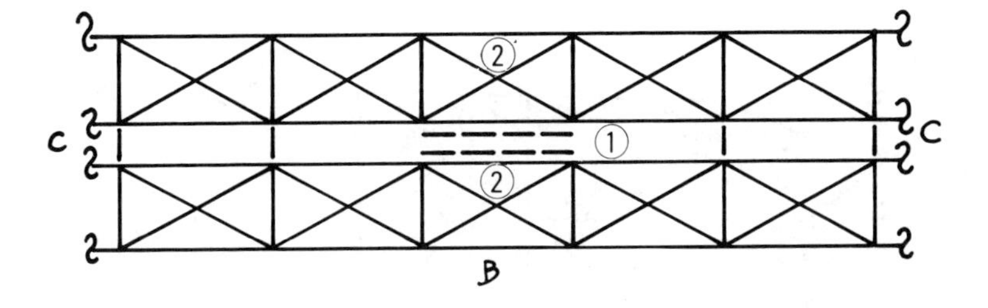

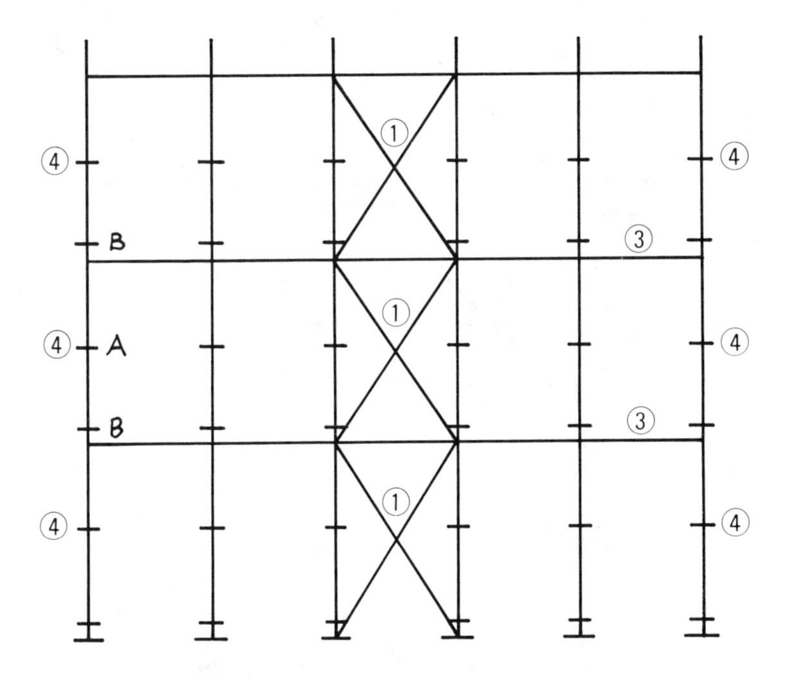

(1) "X" BRACING IN BACK PLANE OF BACK-TO-BACK UPRIGHT TRUSSES

(2) "X" BRACING IN EACH PLANE OF LONGITUDINAL BRACING

(3) LONGITUDINAL BRACING

(4) LOAD ARMS

Fig. 12.4.3 Longitudinal computer model of five bays with vertical X bracing in the center bay.

For the longitudinal model, the longitudinal bracing connections are usually designed to take moment, and the X bracing connections are pinned. The base plates may or may not be designed to resist moment in the longitudinal direction. Thus for the computer longitudinal model, the longitudinal bracing connections are normally fixed, the back plane vertical X bracing connections are pinned, and the post or column bases may or may not be fixed. For the case of standard pallet racks, the shelf beam to post connections may provide enough moment-resisting capacity such that horizontal plan bracing between shelf beams and vertical X bracing in the back plane of the rows of uprights may not be required.

For the transverse computer model, the upright roof-truss and base-plate connections are normally considered pinned. Thus all connections for the transverse model are normally released.

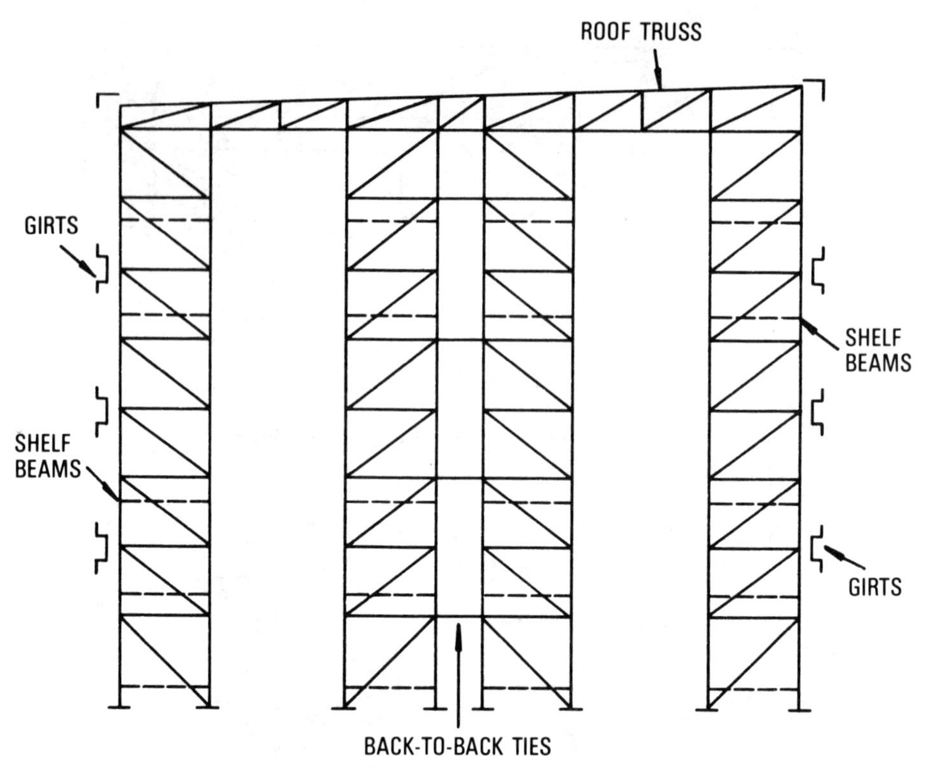

Fig. 12.4.4 Transverse computer model showing four uprights, two aisles, and the roof truss.

A computer program often used for a static analysis of the models is STRESS. If a dynamic analysis is used for seismic force, the SAP IV program may be used. There are many programs available for static or dynamic analysis. A list of "Earthquake Engineering Computer Programs" can be obtained from the National Information Service, see [9].

The end-wall wind loads (i.e., the wind load acting on the north and south walls of the plan view of Fig. 12.4.2) are, for the north wall, taken by girts into the rack posts, then to the longitudinal bracing in both the front and back planes, then to the X bracing between longitudinal bracing for the front plane only, and then to the vertical X bracing in the back plane of posts, and then to the slab. The end-wall wind loads on the south end wall shown in Fig. 12.4.2 are usually taken by girts to the wide flange end wall columns and then by horizontal struts to the vertical X bracing in the rack at the south end of the racks and then to the slab. If possible, the designer will place vertical X bracing in the end bays of the racks specifically to take the end-wall wind loads immediately to the slab, without transferring them through longitudinal bracing of two or three bays.

Furthermore, vertical X bracing in the end walls in the east–west direction (south wall of Fig. 12.4.2) is usually provided to prevent relative displacement between the racks and the end wall, and to prevent too much end-wall distortion.

12.4.5 Sizing Structural Members

The sizing of cold-rolled or cold-formed steel structural members follows the basic specifications, namely *Specification for the Design of Cold-Formed Steel Structural Members,* American Iron and Steel Institute (AISI) [10] and *Specification for the Design, Testing, and Utilization of Industrial Steel Storage Racks,* Rack Manufacturer's Institute (RMI) [4]. The sizing of hot-rolled structural members follows the familiar *Specification for the Design, Fabrication and Erection of Structural Steel for Buildings,* American Institute of Steel Construction, Inc. (AISC) [11].

The shelf beams and shelf arms for a one- or two-pallet-deep stacker-rack system are designed or sized in the normal manner. For a stacker, the impact factor of 1.25 as mentioned in Section 12.4.2 must be considered. The beams in a standard pallet rack must also be investigated for an additional vertical impact load of 25% of one unit load, placed in the most unfavorable position for the particular determination (moment or shear) as stated by paragraph 7.2 of RMI specification [4]. For a standard pallet rack subjected to lateral load, the moments and forces on the beams would be obtained from the computer results, and the beams, of course, would act as beam columns.

The columns in either a pallet rack or a stacker rack would also be designed as beam columns. The moments and forces on each column between beams for pallet racks or longitudinal braces for stacker racks would be obtained from the computer results.

The interaction equation would of course be used to determine if the assumed section of beam or column was overstressed. The allowable column stresses and bending stresses would be determined in the normal manner. Some computer programs, STRUDL for example, have been modified by the users of cold-formed steel members to size the members directly after the program determines the moments and forces on the members. Other users or manufacturers of cold-formed steel members use their own developed programs to size the members and some size the members by hand.

It should be pointed out that for a static seismic analysis and according to Section 2313 (j) G or UBC [3] for seismic zones 3 and 4, all members in braced frames must be designed for 1.25 times the force determined in accordance with Section 2312 (d). Connections shall be designed to develop the full capacity of the members or shall be based on the foregoing forces without the one-third increase usually permitted for stresses resulting from earthquake forces.

The selection of correct column K values for members should be mentioned. The RMI specification [4] permits the use of specified K values or it states that K values may be determined by rational analysis. Paragraph 5.3.1.1 in [3] states that the effective length factor K is 1.7 for pallet racks for the portion of the column between the floor and the first beam above the floor as well as between beams for flexural buckling in the direction perpendicular to the upright frames for racks not braced against sidesway. For racks braced against sidesway, paragraph 5.3.1.2 of the RMI specification states that the effective length factor for pallet racks and stacker racks is $K = 1$, provided that such racks have diagonal bracing in the vertical back plane and horizontal diagonal bracing in the plane of the shelves for pallet racks; and it is implied that horizontal diagonal bracing be in the plane of the longitudinal bracing for stacker racks. This is required so that the front post is laterally supported between longitudinal bracing.

For flexural buckling of the post in the plane of the upright frame, paragraph 5.3.2 of the RMI specification states that $K = 1$ for the portion of the column between braced points provided that the maximum value of the ratio of L short to L long does not exceed 0.15 for uprights with diagonals and horizontals intersecting the posts. For uprights having diagonal braces that intersect the horizontal braces, $K = 1$ providing the ratio of L short to L long does not exceed 0.12. The Commentary to the RMI specification [12] adequately describes the L short and L long dimensions for the uprights.

If one prefers to use rational analysis to determine the K values for buckling of the post perpendicular to the upright, then paragraph 5.3.1 of the Commentary to the RMI specification [12] describes a method that may be used. This method takes into account the semirigid connection of the columns to the floor, and the semirigid connection of the beams to the columns. It also utilizes the nomograph of Fig. C 1.8.2 of the Commentary to the AISC specification [11], which permits one to determine the K value by utilizing the stiffness of the members, beams, and columns, at the ends of the column under consideration. To take into account the semirigid beam-to-column connection, one must determine the spring constant for the connection by performing the portal frame test descibed by the RMI specification and Commentary. Also, it should be mentioned that the RMI Commentary points out in its Table 5.3.1 that the value of $K = 1.7$ may be conservative or unconservative.

It should be pointed out that according to J. A. Yura in *The Effective Length of Columns in Unbraced Frames* [13], the nomograph in Section 1.8 of the Commentary to the AISC specification [11] is based on two principal assumptions; first, elastic action, and second, all columns in a story buckle simultaneously. There are cases for racks where the post will buckle inelastically and the K value obtained from the nomograph will be conservative. However, for the general rack system where a large number of posts in the longitudinal direction exists, one would not take advantage of the second assumption to the AISC nomograph that all columns in a story buckle simultaneously. That is, one would not depend on the two end columns carrying half the gravity load for shear resistance to prevent the large number of columns between the end columns from buckling.

For those cases of inelastic post buckling of racks where Kl/r is less than C_c ($C_c = 109$ for $F_y = 50$ ksi and $Q = 1$), one might consider reducing the K value further. According to AISC, inelastic action is assumed to begin at an average stress of 0.5 F_y; the value of Kl/r corresponding to this stress is the value C_c. For a value of $Kl/r < C_c$, buckling is inelastic and the effective length factor K depends on the tangent modulus and not Young's modulus. Yura suggests an iterative procedure for reducing K in the inelastic range, but a later paper by R. O. Disque, *Inelastic K-Factor for Column Design [14]*, shows that the iterative procedure is not necessary. Yura shows that the factor G used to enter the nomograph Fig. C 1.8.2, section 8, Commentary to AISC, may be modified by multiplying the G elastic by the ratio F_a/F_e^1 to get G inelastic. Disque shows that if f_a is substituted for F_a, that is, if G elastic is multiplied by the ratio f_a/F_e^1, one gets G inelastic and no iterative procedure is required.

The terms are defined as follows:

Kl/r = effective slenderness ratio
K = effective length factor
l = unsupported length, inches (m)
r = least radius of gyration of the cross section, inches (m)
$C_c = (2\pi^2 E/QF_y)^{1/2}$
E = Young's modulus of elasticity of the material, ksi (Pa).
Q = factor that takes into account the effect of post perforations and local buckling on column buckling. Q is to be determined by stub column tests described in paragraph 9.2 of the RMI specification
F_y = yield strength of the material, ksi (Pa)
G elastic $= \Sigma(I_c/L_c)/\Sigma(I_g/L_g)$
I_c = moment of inertia of the column, inches^4 (m^4)
I_g = moment of inertia of the beam, inches^4 (m^4)
L_c = length of column, inches (m)
L_g = length of beam, inches (m)
F_a = allowable column stress, in ksi (Pa)
F_e^1 = Euler buckling equation, in ksi (Pa)
G inelastic $= G$ elastic (F_a/F_e^1)

The value obtained for K from the nomograph in AISC by entering it with G inelastic may then be used to determine the allowable column stress. However, to reduce $K = 1.7$ further may or may not be conservative as pointed out in Table 5.3.1 of the RMI Commentary. To be correct, one should determine the spring constant for their post-to-beam connection by the portal test and in turn determine the actual K value to be reduced.

The purlins, girts, and connections are designed in the normal manner. A separate roof truss may be designed and attached to the tops of the uprights. Another solution is to attach the purlins directly to the tops of the uprights. The posts for the last upright extension to which the purlins would be attached would be cut off at different elevations to match the desired roof slope. For this case a top tie member would usually be provided to tie the tops of the uprights together and also to provide a lateral load path across the top of the uprights.

12.4.6 Special Construction Considerations

The most important construction consideration is probably the erection of the uprights so that they are within tolerance of being plumb. The stacker crane supplier should be consulted to determine what tolerance is acceptable so that the stacker crane will function properly. The post is manufactured to be straight within a maximum of about $\frac{1}{8}$ in. (0.318 cm) in 10 ft (3.05 m). They are erected to about the same tolerance. Paragraph 1.4.10 of the RMI specification states that the maximum tolerance from the vertical is 1 in. (2.54 cm) in 10 ft (3.05 m). Most rack installations are much better than the RMI specified tolerance. The post tops are usually plumb to about $\pm\frac{3}{8}$ in. (0.952 cm). The stacker-rack shelf elevations are normally within $\pm\frac{1}{8}$ in. (0.318 cm).

The rack base plates are shimmed or having leveling nuts below the plates such that all base plates are at the same elevation. The floor or slab is normally surveyed to determine the elevation of the slab at each base plate. The slab is made level to about $\pm\frac{1}{4}$ in. (0.635 cm). The highest point of the slab is located and all other plates are shimmed to that elevation. Sometimes the owner will permit the high points of the slab to be chipped or ground to a lower elevation.

It is very important that the uprights be braced in the longitudinal direction during erection. They are normally braced in the upright direction when they are anchored to the floor. There have been a number of installations that have come down during severe wind storms because they were not adequately braced and anchored to the slab. To be safe, the system should be tied to the slab for the most severe wind load for the site area. One or more cables may be required at each end of a number of uprights in a row depending on the height of the uprights, the number of uprights, and the wind load. Cables may be required at the ends of a number of posts in a row if the horizontal in plane bracing between the longitudinal bracing members is not installed (see Fig. 12.4.3 for the location of the horizontal in-plane bracing ②). When the permanent longitudinal, horizontal, and vertical bracing for a section is installed (Fig. 12.4.3), the temporary bracing may be removed providing the permanent bracing is adequate to take the wind load on all the members in the portion of the rack system that is to braced.

The installation and erection of the uprights is easier for a rack building than for a warehouse. There is usually ample room to move the uprights around and to position them while attached to cables and to a crane hook. About the only problem with handling the upright is that it must be lifted with cables attached at about the third points of the upright so that the posts do not buckle while handling and moving them about.

The end-wall hot-rolled columns located in the input–output end of the rack system normally require separate column footers. These columns may or may not require temporary wind bracing during erection. If the columns and footers are capable of taking the wind load on all the members they support during the erection process and before the permanent end-wall wind bracing is installed, then perhaps temporary bracing is not required.

The heating and ventilating equipment may or may not be supported by the racks. The equipment is designed by a consultant and the supporting structure is provided to position the equipment to his satisfaction. The conveyor system, if provided, is also designed by a consultant and is usually required in the input or output end of the building. This conveyor equipment may or may not be supported by the racks.

12.4.7 Fire Protection

Fire protection is normally provided by ceiling sprinklers and/or sprinklers installed within the racks. The size, number, and location of rack sprinklers are usually specified by the material handling consultant. The National Fire Protection Association Number 231c, *Standard for Rack Storage of Materials 1975* [15], is used as a specification to determine the size, number, and location of ceiling sprinklers and in-rack sprinklers as a function of the commodity stored. A fire barrier or baffle may also be required within the rack to prevent the vertical spread of a fire. The barrier is usually a steel sheet completely covering the stored material on a shelf. The intent of the barrier is to interrupt the vertical flues to prevent the spread of the fire vertically. The barrier is placed just above the sprinklers at a shelf and it covers the full length of the shelves at that level.

The vertical supply pipelines feeding the sprinklers from the top of the racks are located at the back of a single row of uprights or in the vertical space between rows in a double row of uprights. The pipelines are attached to and are supported by the uprights. The horizontal supply pipelines in the racks are located just below a shelf but above the stored material of the shelf below. The main horizontal feeder pipelines are usually at the top of the racks and feed each vertical riser from the top. The main vertical risers feeding the main horizontal feeder lines at the top of the racks are usually located at and attached to the end uprights of a row of uprights.

12.4.8 Tax Benefits

The tax advantage for rack buildings as compared to conventional storage racks in a warehouse is that racks are considered as equipment and therefore have a faster write-off period than buildings.

According to Section 201, Accelerated Cost Recovery System of the Economic Recovery Tax Act of 1981, equipment and thus storage rack capital costs may be written off at 15% the first year, 22% the second year, 21% the third year, 21% the fourth year, and 21% the fifth year. In addition, in 1982 and 1983, a maximum of $5000.00 can be directly expensed; in 1984 and 1985, a maximum of $7500.00 can be directly expensed; and in 1986 and each year thereafter, a maximum of $10,000.00 can be directly expensed.

A building and thus a warehouse may be written off over 15 years based on 175% declining balance depreciation method, and the write-off method changes to a straight-line depreciation method in later years. With reference to a rack building, this would mean that the siding, girts, roofing, purlins, roof beams or roof trusses, and floor slab would be considered building components and would be in the 175% declining balance depreciation category.

It is possible that a portion of the racks in a rack building would fall in the 175% declining balance category since in addition to carrying the pallet loads, they also act as building columns to resist the applied snow, wind, and seismic live loads. The proportion of the racks that would be considered building components would certainly be optional because the amount of time that the racks perform as equipment, that is, support pallet loads, is much greater than the time required to support building live loads. Furthermore, the size and thus the material weight or cost of the racks in a rack building would be greater than the cost of racks required to carry pallet loads only. To determine the ratio of rack-building costs to normal storage-rack costs carrying only pallet loads would require a separate design analysis for each. To determine what portion of the racks in a rack building can be written off as equipment and what portion must be written off as a building, an accountant should be consulted.

12.4.9 Summary of Terms

Symbol	Definition
C	Numerical coefficient
C_c	$\sqrt{\dfrac{2\pi^2 E}{QF_y}}$
C_p	Numerical coefficient
E	Young's modulus of elasticity of the material, ksi (Pa)

Symbol	Definition
F	Total longitudinal inertia force acting at the line of the resultant force of the stacker-bumper, in pounds (N)
F_a	Allowable column stress, in ksi (Pa)
F_e^1	$\dfrac{12\ \pi^2 E}{23\left(\dfrac{KL}{r}\right)^2}$, ksi (Pa)
F_p	Lateral forces on a part of the structure and in the direction under consideration, pounds (N)
F_y	Yield point, ksi (Pa)
G elastic, or G	$\Sigma(I_c/L_c)/\Sigma(I_g/L_g)$
G inelastic	G elastic multiplied by f_a/F_e^1
I	Occupancy importance factor
I_c	Moment of inertia of the column, inches^4 (m^4)
K	Numerical coefficient
K	Effective length factor
L_c	Unsupported length of column, inches (m)
L long, L short	Dimensions of uprights, inches (m)
Q	Stress and/or area factor to modify allowable axial stress
S	Numerical coefficient for site–structure resonance
V	Total lateral force at the base, pounds (N)
W	Total weight of the pallet load plus the hoist and forks plus a proportionate weight of the mast, pounds (kg)
W	Total dead load plus 50% of the rack rated capacity, pounds (kg)
W_p	Weight of the racks plus contents, pounds (kg)
Z	Numerical coefficient dependent on the building zone as determined from the Seismic Zone Map of the United States
f_a	Axial stress, ksi (Pa)
g	Acceleration of gravity, 32.2 ft/sec^2 (9.81 m/sec^2)
l	Unbraced length of member
r	Radius of gyration
s	Stroke of the bumper spring at point where the stacker stopping energy is fully absorbed, ft (m)
v	Stacker velocity at 100% of full rated speed, ft/sec (m/sec)

REFERENCES

1. Building Officials and Code Administrators International, Inc., *The BOCA Basic Building Code,* 17926 South Halsted Street, Homewood, Illinois 60430, 1981 Ed.
2. Southern Building Code Congress International, Inc., *The Standard Building Code,* 900 Montclair Road, Birmingham, Alabama 35213, 1979 Ed.
3. International Conference of Building Officials, *The Uniform Building Code,* 5360 South Workman Mill Road, Whittier, California 90601, 1979 Ed.

4. Rack Manufacturers Institute, Product Section of the Material Handling Institute, *Specification for the Design, Testing, and Utilization of Industrial Steel Storage Racks,* 1326 Freeport Road, Pittsburgh, Pennsylvania 15238, November 1979.
5. American National Standards Institute, Inc., *Controlled Mechanical Storage Cranes,* American National Standard, 1430 Broadway, New York, New York 10018, ANSI B30.13-1977.
6. Association of Iron and Steel Engineers, *Specifications for Electric Overhead Travelling Cranes for Steel Mill Service,* AISE Technical Report No. 6, Suite 2350, 3 Gateway Center, Pittsburgh, Pennsylvania 15222, March 1, 1979.
7. International Conference of Building Officials, *Uniform Building Code Standards,* 5360 South Workman Mill Road, Whittier, California 90601, 1979 Ed.
8. American Concrete Institute, *Building Code Requirements for Reinforced Concrete (ACI 318-77),* Box 19150, Redford Station, Detroit, Michigan 48219.
9. National Information Service, Earthquake Engineering, Computer Programs, Davis Hall, University of California, Berkeley, California 94720.
10. American Iron and Steel Institute, Committee of Sheet Steel Producers, *Specification for the Design of Cold-Formed Steel Structural Members,* 1000 16th Street, N.W., Washington, D.C. 20036, September 1980.
11. American Institute of Steel Construction, Inc., *Specification for the Design, Fabrication and Erection of Structural Steel for Buildings,* 400 North Michigan Avenue, Chicago, Illinois 60611, November 1, 1978.
12. Rack Manufacturers Institute, Product Section of the Material Handling Institute, *Commentary and Supplementary Information—Specification for the Design, Testing and Utilization of Industrial Steel Storage Racks,* 1326 Freeport Road, Pittsburgh, Pennsylvania 15238, 1979 Ed.
13. J. A. Yura, The Effective Length of Columns in Unbraced Frames, *AISC Engineering J.* Vol. 8, No. 2, University of Texas, Austin, Texas 78712, April 1971.
14. R. O. Disque, Inelastic K-Factor in Design, *AISC Engineering J.,* American Institute of Steel Construction, 400 North Michigan Avenue, Chicago, Illinois 60611, Second Quarter, 1973.
15. National Fire Protection Association, *Standard for Rack Storage of Materials,* Number 231c, 470 Atlantic Avenue, Boston, Massachusetts 02210, 1975.

12.5 CAROUSEL SYSTEMS

Donald J. Weiss

12.5.1 The Concept

A carousel is a series of linked bin sections mounted on an oval horizontal track. When activated, the bins revolve, bringing the desired bin to the operator.

12.5.2 History

Industrial carousel systems evolved from light-duty overhead conveyor systems for sorting garments on hangers. The first industrial carousels made in the 1960s were overhead garment conveyors with suspended wire baskets and shelves. In the 1970s, heavier-duty bottom-driven versions were introduced that were capable of carrying weights in excess of 1000 lb per bin. In the 1980s, sophisticated computer and microprocessor controls for carousels were developed and integrated into major automated storage and retrieval systems.

12.5.3 Applications

There are many varied applications of standard carousels which can be classified into the following broad categories:

Order Picking. For less than full case lots of small items for customer or dealer orders in medium to high activity applications.

Production Storage. For small parts or subassemblies required to be stored on the shop floor or in stockrooms.

Toolroom Storage. For moderate activity of tools, maintenance parts or other items that require limited access or security.

Kit Storage. For work-in-process storage of tote trays in assembly operations.

Media Storage. For storage of documents, tapes, films, manuals, blueprints, samples, and so on.

Staging and Accumulation. For temporary storage and transport between multiple access points with individual operator control at each point.

Drawer or Cabinet Storage. For mounting modular drawer or cabinet systems on a rotating pallet base.

Pallet Storage. For use in storing large heavy pallets automatically while saving aisle space—particularly in freezers, ovens, or other environmental enclosures.

Progressive Assembly. Continually revolving carousels with variable speed control that enable work-in-process to be stored and distributed to operator work stations along the carousel.

Burn-In. For electrical testing of products by mounting a continuous oval electrified track on top of the carousel with moving trolleys that enable power to be brought to each individual shelf.

Sorting or Consolidation. By assigning each shelf for a specific order, job, or department, items may be placed into the carousel and sorted into it.

12.5.4 Types of Carousels

All carousels are either top or bottom driven and may be either intermittent or continuous run.

Top Driven. Top-drive units consist of a tubular track supported by poles or stanchions every few feet on which the conveyor chain is mounted. Wire bins are hung from the chain and loosely guided below on a lower track. Wheels are generally lifetime sealed bearings mounted at a 45° angle which support a hinge pin and connecting chain links (see Fig. 12.5.1). The loads to be carried determine the quantity and size of the wheels used. Although larger diameter tracks, wheels, yokes, and hinge pins may be used to carry heavier loads, top-driven units are generally used for lighter loads of continuous-run slower-speed applications. Drives on top-driven units may be mounted above the upper track or recessed below.

Bottom Driven. Bottom-drive units use the same tubular track but use a larger diameter concave steel wheel with axial rollers mounted under a steel bottom plate (see Fig. 12.5.2). The upper guide track is mounted on stanchions and traps the upper guide wheels. Drives are floor mounted and accessible by removable bin backs. Bottom-drive units are more popular because of their ability to carry heavier weights, because they are more dependable, and because they eliminate the potential of product contamination from wheel or track wear, require less overall height, are more adaptable to double tiering, they spread floor load more evenly, and are more rigid from the seismic point of view.

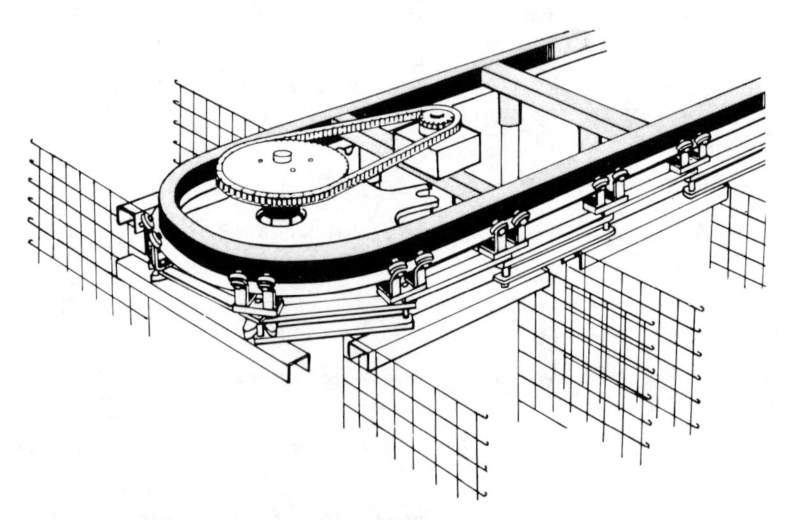

Fig. 12.5.1 Top-driven system.

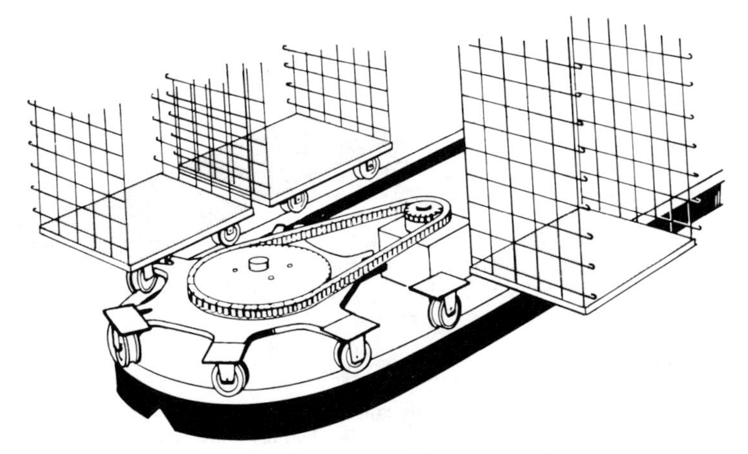

Fig. 12.5.2 Bottom-driven system.

Intermittent Run. Most carousels are used to store and retrieve items and are activated by manual control, microprocessor, or computer when an item is to be loaded or unloaded. When activated, the carousel revolves and the desired bin is delivered to the operator control point by the shortest route.

Continuous Run. Continuous-run carousels are generally unidirectional and often have manually or electrically variable speed drives. They are generally used in assembly operations where work is stored and transported on the carousel while operators are stationed on one or both sides of the unit. Inventory is removed from the moving carousel by any operator and placed back into the carousel also while moving, generally at slow speeds. Continuous-run units are sometimes used for high-volume sorting or timed cycle burn-in applications.

12.5.5 Features and Options

Carousel drives are typically located at either or both ends of the unit. Most drives consist of individual motor, speed reducer, roller chain, pulleys, and sprockets. Dual drives are required for heavier loads and longer units. Motors may be either AC or DC.

AC Drives. Generally used for continuous-run applications or lower-activity applications. AC motors require a clutch to protect the drive and reduce jerking on startup. The range of clutches include simple mechanical devices, air clutches, electric clutches, or electronic soft-start units. While simple mechanical clutches are not adjustable and must be sized differently for the motor horsepower being used, air, electric, and soft-start units are adjustable. A clutch is required for each motor with the exception of an electronic soft-start device which can control two motors.

DC Drives. While somewhat more expensive than AC drives, DC units do not require clutches. DC drives should be used wherever a smooth start and stop is required, where stopping accuracy of less than ±1 in. is desired, or where the loads on the unit vary. Because of the ability to program the maximum speed, acceleration, and deceleration characteristics of the DC motor controller in the field, it is far more flexible than AC drives. Wherever variable speed or the ultimate use of robot picking heads is required, DC drives should be used.

Structure. The supporting structure of top- and bottom-driven carousels consists largely of welded angle iron and channel iron. The overall concept is one of standard drive and idler end sections and standard center sections of different lengths fit together to form any length unit. Expansion of the unit is possible at any of the splice joints.

Bin Design. Standard bins and shelves are generally fabricated of wire. Wire is less expensive than sheet metal, lighter in weight, and dust free. Where heavier weights are involved, standard wire bins must be strengthened by using solid sheet metal supporting bin backs. The solid backs give the bin rigidity, prevent small items from falling through and reduce vertigo effect, but most importantly, support the major portion of the weight of each individual shelf. Shelf load is therefore distributed down through the solid back and both wire sides. Standard bin widths are 21, 24, and 30 in. Other widths can be made but are expensive because they generally require a different chain pitch. Bin heights may be anywhere from 2 ft to 10 ft, but are generally 6, 7, or 8 ft. Standard shelf depths are 14, 18, and 22 in., but other depths are available. Standard bins offer vertical shelf adjustability on 3-in. or 6-in. centers, but other centers are also offered. Special sheet metal or angle iron cantilevered bin designs are quite common.

Shelves. Shelves are adjustable without the use of hardware and each shelf forms a rigid box structure by joining the two sides and back of the bin. Where loads in excess of 100 lb per shelf are required, reinforced shelves should be used. Shelves are generally pitched 7° to the rear.

Clearance. Top-driven units generally require 7 in. below the bin and 9 in. above the bin for the structure. Bottom drive units require 8 in. below the bin and 6 in. above the bin for drive structure. A minimum of 6 in. clearance should be allowed around all sides of a carousel to prevent overhung items from causing a jam up.

12.5.6 Controls

Carousels can be effectively controlled by anything from a simple foot switch to a powerful main frame computer. Although most carousels are controlled by microprocessors of various types, selection of the appropriate control should always take into account both present and future needs. All controls should be plug compatible for interchangeability and future upgrading. The present range of controls include:

Foot Control. A momentary contact bidirectional-type switch usually with OSHA guard.

Hand Control. A maintained switch generally mounts on an arm extending from the carousel.

Keyboard Control. A digital selector that delivers the desired bin via the shortest route contains bidirectional jogging capability, lighted display, and emergency stop.

Floor Mats. Various size electrified mats may be used to activate or deactivate carousels at multiple locations.

Safety Lockouts. For systems with multiple control points, one operator is able to retain control while locking out others.

Carousel Interface Box. The basic building block for any computer-controlled system. This is an intelligent microprocessor that communicates through a single network cable to other carousels, terminals, and computers using RS232C or RS422 protocol.

Microprocessor with Memory. This type of control can monitor from one to six carousels and has the ability to queue and sort up to 500 requests. Bin location data is stored in its memory and a bank of carousels can be activated intelligently. Ideal for users who are not ready for full computer control but want the ability to upgrade by simply unplugging a microprocessor and plugging in a computer at a later date.

Computer Control. Carousels may be controlled by almost any kind of computer ranging from small personal or home computers to large mainframes. It is generally recommended that dedicated host (usually a mini) control the carousels because of real-time constraints. Any computer will communicate with the basic carousel interface box described previously.

12.5.7 Sizes and Layout

Although carousels may be most any length, from a minimum of 10 ft to over 100 ft, the majority are in the 30–50-ft range because of retrieval times and the desire for accessibility. Many carousel installations are multitiered to take advantage of available cubic capacity. The use of robots enables multitiered storage to be effective in high-ceiling areas.

The determination of size is dependent on many factors including shape of the area involved, whether parts are to be stored in random fashion or by fixed locations, the picking frequency, and the size, shape, and weight of the items.

The major space saving benefit of carousels occurs when they are placed side by side or are multitiered. Units can be placed only a few inches apart if no access is desired along the sides and if the items to be stored are not likely to protrude. In many installations, an emergency service aisle is placed between carousels. For this very limited purpose, a maximum of 18 in. is usually sufficient. On the other hand, if a working service aisle is desired, 2 ft or more may be left between units.

When multitiered units are used to take advantage of available ceiling height, mezzanine design

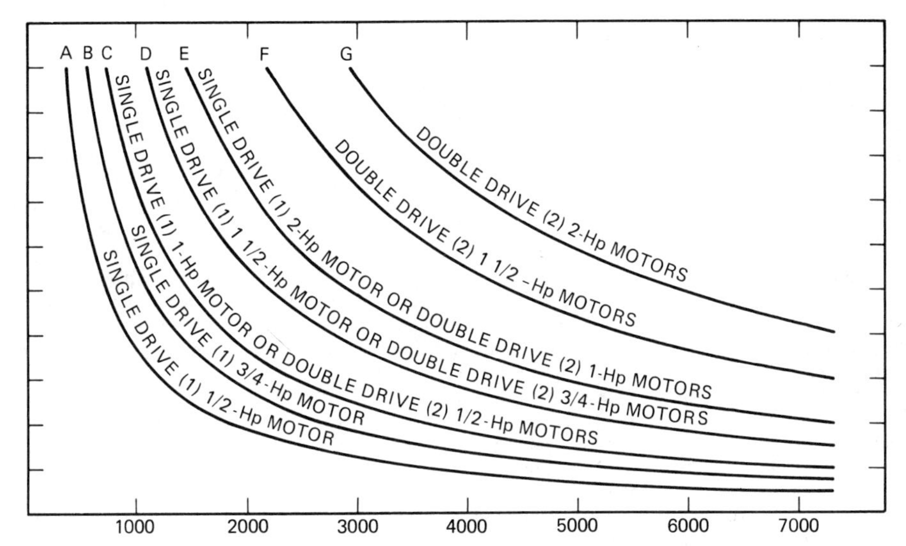

Fig. 12.5.3 Horsepower calculations. For example: conveyor, 40 baskets; live load, 300 lb; speed 60 ft/min; weight of basket and shelves, 100 lb. Total weight = weight of baskets + live load = (40 × 100) + (40 × 300) = 4,000 + 12,000 = 16,000 lb. Answer: use two drive units, $\frac{3}{4}$-Hp motor each. Note: If total horsepower required is 1 Hp (curve C), up to 30 baskets requires one drive, more than 30 baskets requires a double drive.

becomes an important consideration. Mezzanines may cover the entire area or may be used just in the actual picking area across the front of the carousels. With this arrangement there is generally a savings in lighting and sprinkling costs but a loss of accessibility.

12.5.8 Drive Calculation

The horsepower required to drive a carousel can be determined by the following formula:

$$\mathrm{HP} = \frac{L \times N \times F \times S}{33{,}000 \times E}$$

where: S = speed of carousel in ft/min
L = average live load per bin
N = number of bins
F = coefficient of friction (usually 0.03)
E = system efficiency (usually 70%)

This equation tells us the total horsepower required. Whenever the unit is over $1\frac{1}{2}$ Hp or greater than 30 bins in length, it is desirable to use dual drives to provide the total horsepower requirement. A chart plotting typical horsepower calculation is shown in Fig. 12.5.3.

12.6 MINILOAD SYSTEMS
John Castaldi

Miniload storage systems are materials handling systems that store and retrieve bins containing parts totalling approximately 200–750 lb per bin. Bins are automatically selected, extracted from specified storage spaces, brought to a designated delivery point for stocking, cycle counting, bagging, or kitting of their contents by one or more operators, and then automatically returned to the assigned storage space. Miniload storage systems differ from pallet load systems not only in the weights handled and the storage techniques employed, but in the priorities that are assigned to the evaluation of cost effectiveness. Whereas space utilization is the primary determinant in the specifying and planning of pallet load systems, throughput considerations tend to take prime consideration for miniload storage systems.

12.6.1 Typical Miniload System

In a miniload storage system, parts are housed in bins that are automatically delivered to operators at well-defined work stations. As a result, the operators do not have to walk or ride into storage aisles in search of items to be retrieved. At the work stations, bins are presented to the operators at heights that provide maximum convenience for the tasks to be performed and with all bin contents visible. As different bins are delivered to the work stations, operators can make pick after pick without fatigue. Restocking of bins is accomplished with similar ease. When picking and/or restocking of the bin is complete, the bin is automatically removed from the work station in order to be stored. The automatic storage and retrieval system (AS/RS) that makes these operator actions possible is shown in typical form in Fig. 12.2.22.

Bins containing parts to be processed are housed in storage racks that are arranged on both sides of one or more aisles. A microprocessor-controlled stacker, which consists of a column, platform, and extraction mechanism, is located in each aisle. The column of the stacker moves the stacker horizontally from the front to the rear and from the rear to the front of the aisle; the platform moves vertically in up and down directions on the column; and the extraction mechanism moves from side to side on the platform. Under the control of the microprocessor, the stacker extracts and stores bins in the storage racks, and transports the bins to and from work stations or conveyors at the front, rear, or (in some arrangements) the sides of the storage area.

The column and platform on each stacker move simultaneously and independently. When moving from a rest position in front of a storage space, work station, or (when applicable) conveyors, the column and platform individually accelerate to high speed, move at high speed until they approach their destination, and then decelerate to a creep speed. They then move at creep speed in order to stop, precisely aligned, in front of the storage space, work station, or conveyor location, whichever is applicable. The extraction mechanism thereupon moves on the platform at a relatively constant speed long enough to pull a bin onto the platform or to push a bin off the platform.

The design of the AS/RS machine thus begins to hinge on such factors as the size and quantity of bins being stored and retrieved, on the type and quantity of stackers used, and the amount of available storage space. All of these factors affect and are affected by the throughput of the machine.

12.6.2 Justification

The temptation often arises to assign considerations of building cost savings as the major element for cost justification of AS/RS specification. Doing so was a valid procedure when operating within the framework of pallet-load systems. Automatic storage and retrieval systems of pallets were used when storage heights could be increased beyond the 20–25-ft heights that were manageable by manned fork trucks. Increases in building heights to accommodate AS/RS during planning and alteration stages could be achieved economically. Such was the case because building costs do not increase as much as storage volume does when storage area heights are raised. In fact, building height increases of 120% produce increased costs of less than 50%.

However, building cost factors must be examined in a different light when the size and weight of parts permit the use of a miniload storage system. Miniload storage systems have consistently shown 400% increases in floor space utilization as a result of their use of existing building space. Floor-to-ceiling storage spaces are fully utilized by miniload storage systems and the amount of aisle space that is required is held to a minimum. Significantly, however, miniload systems provide cost justifications beyond those realized from space utilization. Manpower reductions and the centralization of personnel activities, which are intrinsic to miniload systems, are the prime considerations. Economies that result from manpower reductions and from the ability to make task assignments that are based on work centralization are most immediately effected by throughput of the AS/RS machine.

12.6.3 Making Throughput Calculations

System throughput in a miniload storage system is most dependent on the speed of machine operations in the storage and retrieval of bins. Machine operating speed is evaluated by calculating machine cycle time. Machine cycle time calculations are of two types for miniload AS/RS: single-command cycles and dual-command cycles. In a single-command cycle (see Fig. 12.2.14) computations are made of the time required to accomplish the following:

Pick up a bin at the work station or conveyor (point 1).

Bring the bin to a storage location (point 2) that is half the distance from the pickup point to the furthest horizontal and vertical reaches of the storage area.

Return the column and platform to the original bin pickup point (point 1).

In the dual-command cycle (see Fig. 12.2.15) computations are made of the time required to accomplish the following operations:

Pick up the bin at the work station or conveyor (point 1).

Store the bin at a location (point 2) that is half the maximum horizontal distance and half the total height of the storage area.

Move to and extract a bin that is at a location (point 3) which is $\frac{3}{4}$ of the maximum distance in the horizontal and vertical directions.

Bring the conveyer to the work station or conveyor (point 1).

In the calculation of the single- and dual-command cycle times, the following variables are elements in the computation:

Hv = the horizontal velocity of the column, in ft/sec (m/sec)
Vv = the vertical velocity of the platform, in ft/sec (m/sec)
Ha = the horizontal acceleration or deceleration to and from high and creep speeds of the column, in ft/sec^2 (m/sec^2)
Va = the vertical acceleration or deceleration to and from the high and creep speeds of the platform, in ft/sec^2 (m/sec^2)
Zv = the velocity of the extraction mechanism, in in./sec (cm/sec)
Hp = the distance between horizontal storage space openings (horizontal pitch), in ft (m)
Vp = the distance between vertical storage space openings (vertical pitch), in ft (m). In certain installations there may be more than one value for Vp, in which case, vertical pitches would be expressed as Vp_1, Vp_2, and so on
$\frac{Cv}{d}$ = the time required to travel at creep speed (creep velocity divided by distance traveled), in sec
CL = the length of the storage bin, in in. (cm)
Kz = a fixed time constant that is part of every extraction cycle performed by the extraction mechanism, in sec
X = a quantity of horizontal coordinates to be traversed by the column when moving from point to point. When movement is made between more than one set of two points, the quantity is expressed as X_1 and X_2, respectively
Y = a quantity of vertical coordinates to be traversed by the platform when moving from point to point. When movement is made between more than one set of two points, the quantity is expressed as Y_1 and Y_2, respectively

These variables are computed in the relationships shown in Fig. 12.6.1 for single-command cycle computations and in Fig. 12.6.2 for dual-command cycle computations. The machine cycle time in each case is equal to the sum of the solutions computed for the relationships expressed. The quantities represented by X, X_1, X_2, Y, Y_1, and Y_2 in the equations of Figs. 12.6.1 and 12.6.2 become small factors in overall machine cycle time, even though they represent the size of the storage space. At that, these functions of horizontal and vertical distance have much smaller values in miniload AS/RS than in pallet load systems. Size of storage space to be traveled by the machine, so important to calculations of pallet load system machine cycle time computations, loses its meaning in miniload AS/RS computation. Operating time for the extraction mechanism is the prime consideration in the miniload system, confirming the need to calculate throughput.

Quick estimates of dual-command machine cycle time can be made using the nomograph of Fig. 12.6.3. To use the nomograph, proceed as follows:

STEP 1. Determine the length and the height of the active storage space in the AS/RS unit.

STEP 2. Locate the length of the active storage space on the Active Length (ft) scale of the nomograph.

STEP 3. Extend a straight edge horizontally through the point located in step 2 so that it intersects the Active Height (ft) scale. If the value read from the Active Height scale is equal to or greater than the active height determined in step 1, the value at the point of intersection of the straight edge and the MHI Dual-cycle Time (sec) scale is the machine cycle time. If the value read from the Active Height scale is less than the value determined in step 1, proceed to step 4.

STEP 4. Locate the height of the active storage space of the AS/RS unit on the Active Height scale of the nomograph.

STEP 5. Extend the straight edge horizontally through the point located in step 4 so that it intersects the MHI Dual-cycle Time scale. The value at the point of intersection is the machine cycle time.

As an example of the first case described by the preceding steps, consider the situation in which the calculation is being performed for an AS/RS machine in which the active storage length is 50 ft and the active storage height is 25 ft. Extending the straight edge horizontally through the 50 increment on the Active Length scale produces an intersection on the Active Height scale with the 20 increment. Since the actual height is greater than the value read at the point of straight-edge intersection, the

OPERATION	REQUIRED TIME (in seconds)		
Pick up bin at point 1		$\frac{CL}{Zv} + Kz$	
Accelerate to high speed	$\frac{Hv}{Ha}$	or	$\frac{Vv}{Va}$
Travel X storage bays/ Y tiers to point 2	$\frac{*X(Hp) - \frac{Hv^2}{Ha}}{Hv}$	(whichever is greater)	$\frac{Y(Vp) - \frac{Vv^2}{Va}}{Vv}$
Decelerate to creep speed	$\frac{Hv}{Ha}$		$\frac{Vv}{Va}$
Creep time travel		$\frac{Cv}{d}$	
Deposit bin in storage space at point 2		$\frac{CL}{Zv} + Kz$	
Accelerate to high speed	$\frac{Hv}{Ha}$	or	$\frac{Vv}{Va}$
Travel X storage bays/ Y tiers to point 1	$\frac{X(Hp) - \frac{Hv^2}{Ha}}{Hv}$	(whichever is greater)	$\frac{Y(Vp) - \frac{Vv^2}{Va}}{Vv}$
Decelerate to creep speed	$\frac{Hv}{Ha}$		$\frac{Vv}{Va}$
Creep time travel		$\frac{Cv}{d}$	
Single-command machine cycle time		(Sum of the above)	

* Equation applies only for computed values greater than or equal to 0: that is

$$X(Hp) \geq \frac{Hv^2}{Ha}$$

Fig. 12.6.1 Single-command machine time calculation. Point 1 = work station or conveyer; point 2 = storage space at horizontal coordinate X, vertical coordinate Y with X and Y equal to one-half total horizontal and vertical storage area distances.

dual-cycle time is read at the point of intersection of the straight edge and the MHI Dual-cycle Time scale; that is, 55.0 sec.

Were the active height to be a value of 17 ft when the active length is 50 ft, as in the first example, the second case described by the preceding steps is illustrated. Extending the straight edge through the 50-ft increment on the Active Length scale reveals that the actual height is less than the height increment at the straight-edge intersection on the Active Height scale. The straight edge is therefore moved to the 17-ft increment on the Active Height scale. The machine cycle time is then read from the point of intersection of the straight edge and the MHI Dual-cycle Time scale; that is, at the value of 52.0 sec.

Were the active height of an AS/RS unit to be the value that is on the same horizontal as the active length intersection on the Active Length scale, the value on the MHI Dual-cycle Time scale is read directly.

12.6.4 System Interfaces

Once computation of machine cycle time indicates that a selected AS/RS stacker and storage arrangement are the means of achieving optimum throughput for the application being considered, it is necessary to take proper advantage of that throughput by establishing the most effective machine-to-man interface for the system. The efficiency and cost effectiveness of the machine-to-man interface is determined by the type of configuration that is selected for the delivery of a retrieved bin to an operator and the return of that bin to storage from the operator. Configurations that provide the interface include:

End-of-aisle delivery
Side shuttle work station delivery

OPERATION	REQUIRED TIME (in seconds)		
Pick up bin at point 1		$\frac{CL}{Zv} + Kz$	
Accelerate to high speed	$\frac{Hv}{Ha}$	or	$\frac{Vv}{Va}$
Travel X_1 storage bays/ Y_1 tiers to point 2	$\frac{X_1 \times Hp - \frac{Hv^2}{Ha}}{Hv}$	(whichever is greater)	$\frac{Y_1 \times Vp - \frac{Vv^2}{Va}}{Vv}$
Decelerate to creep speed	$\frac{Hv}{Ha}$		$\frac{Vv}{Va}$
Creep time travel		$\frac{Cv}{d}$	
Deposit bin in storage space at point 2		$\frac{CL}{Zv} + Kz$	
Accelerate to high speed	$\frac{Hv}{Ha}$	or	$\frac{Vv}{Va}$
Travel X_2 storage bays/ Y_2 tiers to point 3	$X_2 \times Hp - \frac{Hv^2}{Ha}$	(whichever is greater)	$Y_2 \times Vp - \frac{Vv^2}{Va}$
Decelerate to creep speed	$\frac{Hv}{Ha}$		$\frac{Vv}{Va}$
Creep time travel		$\frac{Cv}{d}$	
Pick up bin at point 3		$\frac{CL}{Zv} + Kz$	
Accelerate to high speed	$\frac{Hv}{Ha}$	or	$\frac{Vv}{Va}$
Travel $X_1 + X_2$ storage bays/ $Y_1 + Y_2$ tiers to point 1	$\frac{(X_1 + X_2) \times Hp - \frac{Hv^2}{Ha}}{Hv}$	(whichever is greater)	$\frac{(Y_1 + Y_2) \times Vp - \frac{Vv^2}{Va}}{Vv}$
Decelerate to creep speed	$\frac{Hv}{Ha}$		$\frac{Vv}{Va}$
Creep time travel		$\frac{Cv}{d}$	
Deposit bin at point 1		$\frac{CL}{Zv} + Kz$	
Dual-command machine cycle time		(Sum of the above)	

Fig. 12.6.2 Dual-command machine time calculation. Point 1 = work station or conveyor; point 2 = storage space at horizontal coordinate X_1, vertical coordinate Y_1 with X_1 and Y_1 equal to one-half total horizontal and vertical storage area distance; point 3 = storage space at horizontal coordinate X_2, vertical coordinate Y_2 with X_2 and Y_2 equal to three-quarters total horizontal and vertical storage area distance.

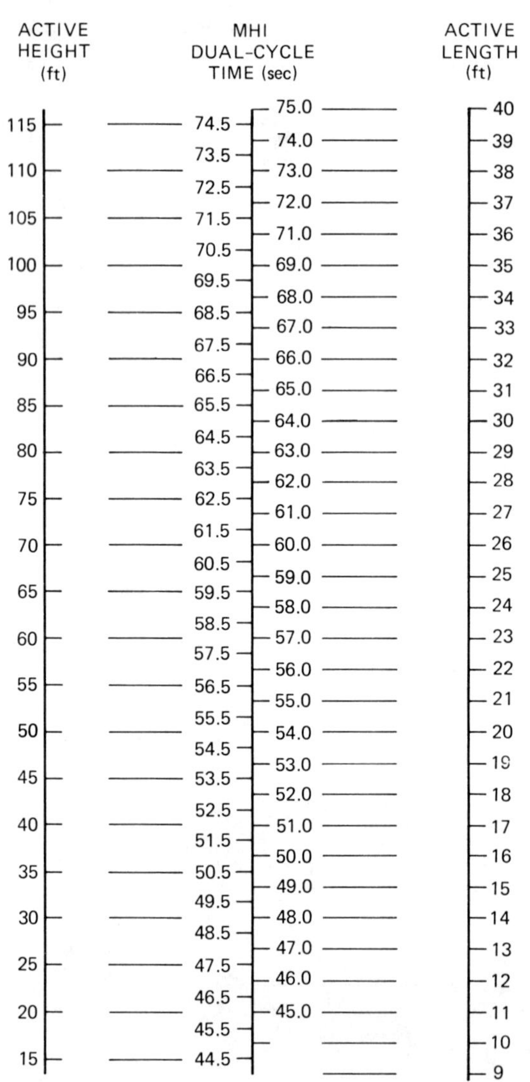

Fig. 12.6.3 Machine cycle-time computation nomograph. MHI dual-cycle times were obtained from HP41C program, DCTA dated 16 APR 82, with the following constants: Container length = 48 in.; Horizontal velocity = 5.0 ft/sec; Horizontal acceleration/deceleration = 1.5 ft/sec^2; Vertical velocity = 1.5 ft/sec; Vertical acceleration/deceleration = 1.0 ft/sec^2.

Dual point delivery
Horseshoe conveyor
Loop conveyor

The choice of one of these configurations as the AS/RS interface in a given installation is a function of the tasks to be accomplished by the system and the cost of accomplishment.

End-of-Aisle Delivery Configuration

An end-of-aisle delivery configuration is shown in the plan view of Fig. 12.6.4. In systems using this configuration, the stacker brings retrieved bins to the end of the aisle and then moves additionally forward through a window to extend the platform beyond the storage rack area and into a cutout in the counter of a work station. The operator picks and stocks bin contents, with the bin remaining in

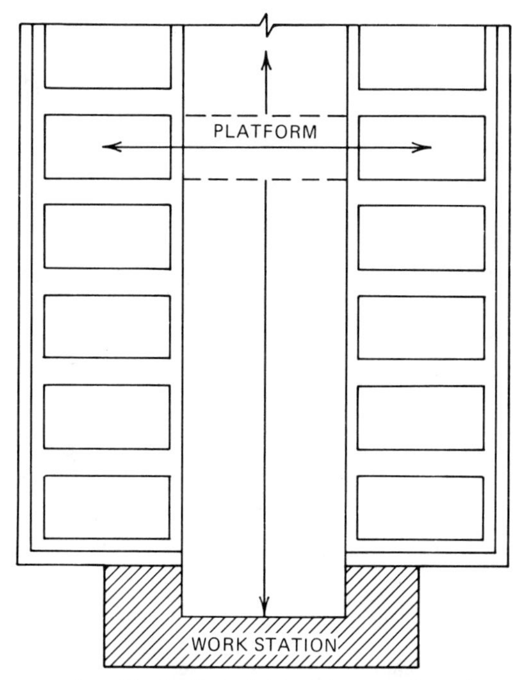

Fig. 12.6.4 End-of-aisle delivery configuration.

place on the platform. This arrangement can be accomplished at the least expense. However, since the operator works on the bin while the bin is on the platform, the stacker is immobilized until released by the operator. High throughput rates can be dissipated in this configuration since operator work time exceeds the time for the machine to store and retrieve bins.

Side Shuttle Work Station Delivery Configuration

Installation of a side shuttle work station at the end of the storage aisle permits, at the least, a partial release of the AS/RS stacker for more active duty. In a side shuttle work station delivery configuration, shown in Fig. 12.6.5, the platform is extended beyond the storage area limits as in the end-of-aisle delivery configuration. Moreover, extraction cycles push and pull bins to and from side shuttles on each side of the platform. The operator then works on a bin without tying up the stacker. The side shuttle work station delivery configuration functions at its best in situations that require uniform pick and restock operation times. For optimum use, the operator's bin work time should equal the single-command machine cycle time. However, machine time calculations require the addition of the times required to move the bin on and off the platform. It must also be remembered that whenever work on a bin at either side shuttle work station exceeds the adjusted machine cycle time, throughput is reduced to one-half of its value. Fundamentally, side shuttle work station delivery configurations are most effective when operations on bins are to be performed sequentially or serially, rather than when concurrent operations on two bins are required.

Dual-Point Conveyor Configuration

A dual-point conveyer configuration permits operators to perform concurrent bin operations. In configurations of this type, shown in Fig. 12.6.6, the stacker brings bins to and takes them from conveyors on each side of an extended platform. Bins delivered to a conveyer form a queue on that conveyor. Operators at work stations fed by the conveyors work on bins on a one-at-a-time basis as in the configurations described previously, but with the advantage that the conveyor provides a time and space buffer between the stacker and the operator. When an operator is finished working on a bin, the bin is pulled from the conveyor by the stacker. Two operators are thereby enabled to work concurrently and independently. In addition, bins can be worked on at a third location in an operation

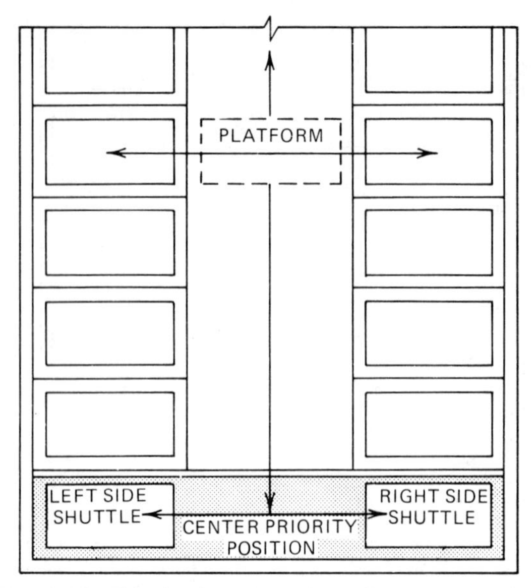

Fig. 12.6.5 Side shuttle work station delivery configuration.

called a window pull. In window pulls, the platform is extended through the storage rack window, but the bin is not pushed onto the conveyor. The third operator can then pick or stock the bin on the platform as though working on an end-of-aisle delivery system.

The major disadvantage of the dual-point conveyor system is that a bin must be taken from a work station as soon as an operator is finished working on its contents to provide him with access to the next bin to be processed. If a stacker is occupied with storage or retrieval operations for bins associated with the work of the operator at one work station, the operator at the other station must wait until the stacker can accommodate his needs. Similarly, if both operators require the attentions of the stacker at the same time, only one can be accommodated, while the other is forced to wait his turn. The waiting time in each case signals reduced manpower utilization.

Horseshoe Conveyor Configuration

A horseshoe conveyor configuration, shown in Fig. 12.6.7, provides isolation of the operators from the stacker so that they need not rely on the stacker for immediate access to or release from bins. The conveyor in this configuration is arranged in a horseshoe shape in front of the storage area window.

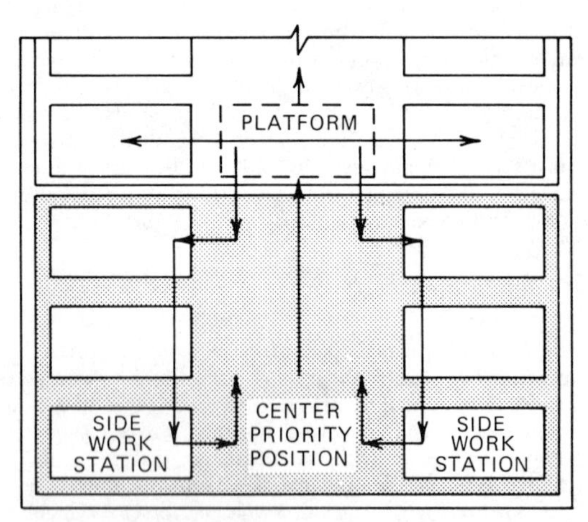

Fig. 12.6.6 Dual-point conveyor configuration.

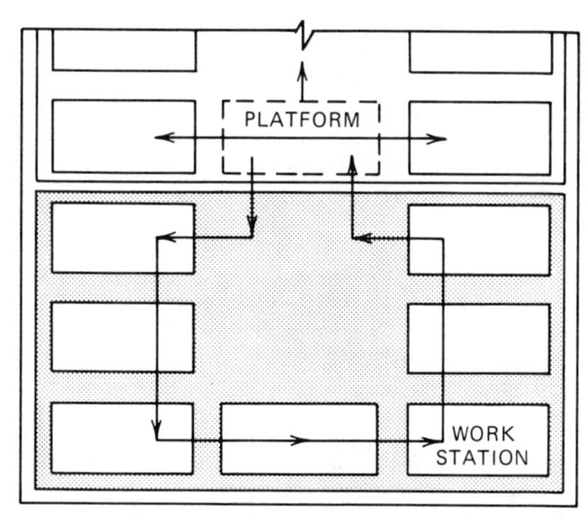

Fig. 12.6.7 Horseshoe conveyor configuration.

Conveyors provide buffers in the bin flow to and from the work station. When a multiple-aisle system is used, pairs of aisles can be terminated with horseshoe conveyor configurations in an adjoining arrangement. If the directions of conveyor flow are opposite for members of a pair, one operator can process bins from two aisles, thereby spelling his neighbor on a temporary basis or, when appropriate, for a full duty shift during hours of reduced work activity. Horseshoe conveyor configurations are ideal for quick pull operation. When the time to pull or restock grows longer, all aisle operations can be held up until an operator releases a bin. Other drawbacks are that it is difficult for two people to work the same aisle and window pulls are inconvenient or impossible.

Loop Conveyor Configuration

The configurations described to this point are all intended to function best when work is performed on one storage bin at a time. With the exception of the horseshoe conveyor configuration, all provide for single bin pick or stock operations in a window pull action. When kits are being made up by the picking actions, however, a significant increase in operating time is experienced. Operating time increases are also experienced when a small work force is used to handle a multiple aisle system, that is, when operators move from work station to work station or from window to window in order to make their picks. These problems are compensated for by a loop conveyor configuration. In a loop conveyor configuration, stackers bring bins to and remove them from conveyors that form loops which pass the aisle windows and the work stations. In such arrangements, each operator has access to bins in all aisle locations and all aisle locations are accessible to all operators.

When a loop conveyor is used, some of the work stations can be assigned on a dedicated basis; that is, some can be assigned to picking operations only, and others to filling operations only. The remaining work stations are used for both picking and filling.

The loop conveyor is particularly valuable for kitting of parts. Two types of kitting sequences are applicable. One sequence, batch pulling, involves pulling a number (batch) of parts from each bin, when the bin is accessed, to permit a quantity of kits to be formed at the same time. In the other sequence, parts are pulled from a number of bins in sequence in order to form a single kit. The loop conveyor is most effective when batch pulling is employed. The different bins from which batches are to be pulled are queued on the conveyor so that the stackers and the operators are both buffered.

Loop conveyors present a limitation in that, as their size increases, the travel path for the bins gets longer, increasing the time required to make emergency pulls. To compensate for this, a delivery and pickup level can be arranged at the aisle window that is at a different height than the conveyor level. Window pulls can then be made at the added level. Alternately, additional work stations can be located off the conveyor loop, for example, on the exterior side of the outermost aisles. These work stations provide access to bins in the associated aisle only. Aisles served by these work stations should therefore be used to store materials that are involved in extended time activities such as bin content cycle counting.

12.6.5 Operator Assist Devices

Productivity, manpower reductions, and additional cost savings can be further improved by judicious selection of equipment designed to provide further assists to the operator. Some examples are described.

Ticket Printer

Ticket printers can be installed at work stations and operate under the control of the microprocessor directing stacker activities. They provide printed records of the transactions performed at the work station, permitting items picked from bins to be identified by such information as part number, quantity, next higher assembly, date of transaction, and kit number.

Card Reader

Card readers can be installed at work stations to direct operations performed therein. Punched cards identify the bin to be retrieved, the work station that is to receive the bin, and, when applicable, the data to be printed by the ticket printer. The card readers operate under the control of stacker microprocessors to perform operations otherwise the responsibility of the operator.

Computer Control

A minicomputer, added to the microprocessors in the stackers, can coordinate the activities of a number of stackers by determining the sequence in which bins are to be called, the sequence of arrival of bins at individual work stations, the sequence of pickup from work stations, and the sequence of bin restorage. With a command display device fed by the minicomputer, digital readouts describing such information as the part to be pulled from the bin, quantities to be pulled, where parts are located in the bin, and so on, can be provided to operators at the work stations. Minicomputers can also be used to provide inventory control and to control displays on cathode ray tube (CRT) terminals. These displays can provide work station operators and remotely located personnel such as loading dock people and supervisors with data that describes the status of parts requests, the location of parts, bookkeeping considerations, backorder and reorder requirements, and so forth.

A larger, "data-crunching" minicomputer can provide master inventory control, cross-indexes of parts-to-storage locations, parts-per-kit lists, part life tracking cycles, and so on.

12.6.6 Summary

The specification and design of miniload storage systems should begin with the calculation of machine throughput required and machine throughput available. Thereafter, advantage should be taken of the throughput achieved by selection of the best machine-to-man interface for the intended system operation. Finally, consideration should be given to the selection of the operator assist devices that increase production, reduce manpower loadings, and establish the lowest overall cost for AS/RS operations.

CHAPTER 13
WAREHOUSING

JAMES A. TOMPKINS

JERRY D. SMITH

Tompkins Associates, Inc.
Raleigh, North Carolina

JOHN R. HUFFMAN

Semco, Sweet & Mayers, Inc.
Los Angeles, California

KENNETH B. ACKERMAN

The K. B. Ackerman Company
Columbus, Ohio

13.1 FACILITY DESIGN

James A. Tompkins

13.1.1 Functions and Objectives of Warehousing

Warehousing is simply defined as the holding of goods until they are required. This definition can be extended by considering the basic functions or activities of warehousing. Warehousing consists of *receiving* goods from a supplier, *storing* the goods until requested by a user, *picking* the goods from storage when required, and *shipping* the goods to the user.

From the standpoint of how these basic functions are accomplished, every warehouse is unique. The specific methodologies used to receive, store, pick, and ship goods in a given warehouse are naturally dependent on the environment in which the warehouse exists. From the standpoint of the types of activities performed within a warehouse, however, warehouses are not unique. Every warehouse is concerned with the same basic types of activities: receiving, storing, picking, and shipping. Consequently, all warehouses face the same types of problems in specifying the most efficient and effective methodologies for receiving, storing, picking, and shipping goods. The ultimate answers to these similar problems will be unique since every warehouse operates within its own unique environment. The approach to addressing and solving these problems, however, is exactly the same in every warehouse. This problem-solving approach is best defined in light of the general objectives of a warehouse.

The primary objective of a warehouse is to maximize the effective use of resources while satisfying customer requirements. All warehouses have three scarce resources: (1) space, (2) equipment, and (3) people. The customers of a warehouse have two basic demands: (1) that the right product be available at the right place at the right time, and (2) that the product be received in a condition usable by the customer. Based on these resources and customer requirements, the primary objectives of a warehouse are more clearly defined to be:

1. Maximize the effective use of warehouse space.
2. Maximize the effective use of warehouse equipment.
3. Maximize the effective use of warehouse labor.
4. Maximize the accessibility of all goods.
5. Maximize the protection of all items.

The general warehouse planning methodology is directed at properly planning the resources of the warehouse to meet these objectives.

13.1.2 Space Planning

Warehouse space planning is concerned with quantitatively documenting the space requirements of the warehouse over some specified planning horizon. The general methodology for warehouse space planning is:

1. Determine what is to be accomplished.
2. Determine how to accomplish it.
3. Document the space requirements of each element required to accomplish the activity.
4. Determine the total space requirements.

Two major activities require space planning within a warehouse. The first activity deals with the transition of goods into and out of the warehouse, namely the receiving and shipping activities. The second activity deals with the space required to store the planned inventory requirements of the warehouse.

Receiving and Shipping Space Planning

As the names imply, the receiving and shipping functions of the warehouse are concerned with getting materials into and out of the warehouse. The efficiency and effectiveness of the warehouse as a whole is typically directly related to the efficiency and effectiveness of the receiving and shipping activities. Unfortunately, the most often neglected areas of a warehouse are the receiving and shipping areas.

This neglect typically manifests itself in the failure to properly define the space requirements of the receiving and shipping functions.

The steps necessary to properly define the space requirements of the receiving and shipping areas include:

1. Define the materials to be received and shipped.
2. Determine the receiving and shipping dock bay requirements.
3. Determine the vehicle interface requirements.
4. Determine the allowances for maneuvering inside the warehouse.
5. Determine the buffer and staging area requirements.
6. Determine space requirements for dock related activities.

Define Materials to Be Received and Shipped. The first step in space planning for receiving and shipping activities is to define the materials to be received and shipped. A useful tool in accomplishing this is a shipping and receiving analysis chart (SRAC), as shown in Fig. 13.1.1. The first five columns of the SRAC define the physical characteristics of the items to be received and shipped. The sixth column defines how many of any one item will be received or shipped at any one time. The seventh column defines when the receipts and shipments can be expected to occur. For an existing warehouse, the best source of the information required for the first seven columns of the SRAC are past receiving reports and shipping releases. These documents will define what has actually been received and shipped in the past. This historical information should then be adjusted as required based on future receiving and shipping plans. For a new warehouse, the information requirements of the first seven columns of the SRAC must be totally based on the anticipated inventory and throughput projections for the new facility.

Column eight of the SRAC documents the nature of the carrier to be used to transport goods to and from the warehouse. This information is necessary to ensure that proper dock facilities are provided to accommodate the range of carriers to be encountered. The types of specifications of interest include the length, width, and height of a truck, and the height of the truck bed above the roadway. For an existing warehouse, this information is best obtained by contacting freight carriers used in the past to determine the types of equipment the carrier expects to use over the next 5–10 years. Likewise, for new facilities, prospective freight carriers should be queried concerning their projected equipment plans.

Columns 9 and 10 of the SRAC document the planned materials handling methods and the projected time required to accomplish the receiving and/or shipping of a given type of material. The time required to load or unload carriers should ideally be based on calculated labor standards derived from time

Company ____________ Date ______ ______ Raw Materials ____________ Finished Goods

Prepared by ____________ Sheet ____________ of __________ ____________ Plant Supplies

Description	Unit Loads				Size of Shipment (unit loads)	Frequency of Shipment	Transportation		Materials Handling	
	Type	Capacity	Size	Weight			Mode	Specifications	Method	Time

Fig. 13.1.1 Shipping and receiving analysis chart.

study, work sampling, or predetermined time standards. Historical data reflecting the time required for carrier loading and unloading in the past can be utilized so long as it reflects future labor requirements.

The proper philosophy for documenting the goods to be received or shipped using a tool such as the SRAC in Fig. 13.1.1 is to classify the materials into generic categories and then complete the SRAC for each generic category. Two items classified within the same generic category should have similar materials handling and storage characteristics. In other words, two items within the same category will likely be stored and handled using similar storage and materials handling methods. The advantages of documenting the items to be received and shipped through the use of generic categories of goods are twofold. First of all, the magnitude of the effort is greatly diminished. Typically, a warehouse with several thousand different individual and distinct items, or stockkeeping units, can be defined based on less than 10 to 20 generic categories of items. Secondly, the classification of the goods into a relatively small number of categories of items greatly increases the value of the information. Since much of the information of interest will be obtained from forecasts of future expectations, describing large, generic categories of materials is typically much more accurate than similar forecasts for individual stockkeeping units.

Receiving and Shipping Dock Bay Requirements. A dock bay is a position at which a carrier is spotted at the warehouse for loading or unloading. The number of dock bays needed is a function of the number of carriers requiring service during a given time frame, the interarrival rates of the carriers, and the time required to service the carriers at the dock. The two most valuable methods for quantitatively determining the number of dock bays needed are waiting-line analysis and simulation. Waiting-line analysis is most effectively utilized when the arrivals of carriers at the warehouse and the time required to service the carriers at the dock occur in a completely random manner [1]. Where the interarrival and service rates of the carriers follow some predictable pattern, simulation will be a more effective technique for quantifying dock bay requirements [1, 2].

Once the number of dock bays has been determined, the proper configuration of the dock bays must be specified. Two basic types of truck dock configurations exist: 90° docks and finger docks. With a 90° dock, shown in Fig. 13.1.2A, the truck is positioned at a 90° angle with the horizontal

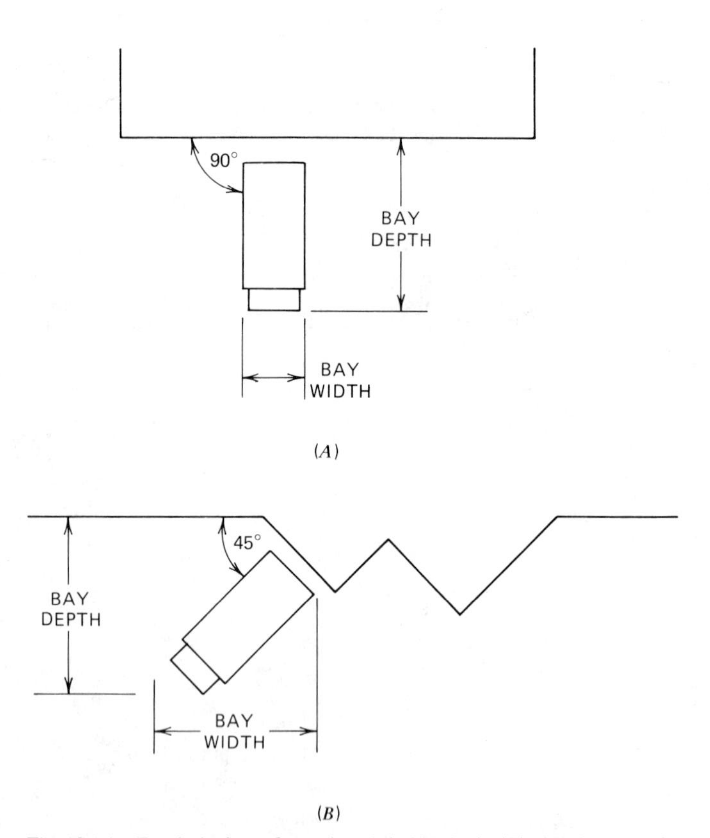

Fig. 13.1.2 Truck dock configuration. (*A*), 90° dock; (*B*), 45° finger dock.

Table 13.1.1 Dock Configuration Factors for 90° Docks

Truck Length ft (m)	Dock Berth Width ft (m)	Apron Depth ft (m)
40	10 (3.0)	46 (14.0)
(12.2)	12 (3.7)	43 (13.1)
	14 (4.3)	39 (11.9)
45	10 (3.0)	52 (15.8)
(13.7)	12 (3.7)	49 (14.9)
	14 (4.3)	46 (14.0)
50	10 (3.0)	60 (18.3)
(15.2)	12 (3.7)	57 (17.4)
	14 (4.3)	54 (16.5)
55	10 (3.0)	65 (19.8)
(16.8)	12 (3.7)	63 (19.2)
	14 (4.3)	58 (17.7)
60	10 (3.0)	72 (21.9)
(18.3)	12 (3.7)	63 (19.2)
	14 (4.3)	60 (18.3)

face of the dock. At a finger dock, the angle between the truck and the dock is less than 90°. For example, Fig. 13.1.2B illustrates a 45° finger dock. The differences between 90° docks and finger docks lie in the space requirements of the dock inside and outside the warehouse. As shown in Fig. 13.1.2, a 90° dock requires less bay width and more bay depth than a finger dock. Consequently, a 90° dock requires less space inside the warehouse building dedicated to a given dock bay, and more space outside the warehouse building for truck maneuvering. The amount of inside space required by finger docks decreases as the angle of the finger dock increases, but it is always greater than the space required by a 90° dock. The amount of outside space required by finger docks increases as the angle of the finger dock increases, but will always be less than the space required by a 90° dock.

Since inside warehouse space is typically more expensive than outside warehouse space, 90° docks are more popular than finger docks. Where outside space is at a premium, however, because of the shape of the warehouse site or the need for future expansion, finger docks are often more attractive than 90° docks.

Another dock configuration issue is the berth width of the dock bays. The most common berth widths are 10 and 12 ft (3.0 and 3.7 m). Again, the issue becomes one of inside versus outside space requirements. The greater the berth width, the greater the inside warehouse space dedicated to each dock bay. On the other hand, as berth width increases, the apron bay depth requirements outside the warehouse decrease. Tables 13.1.1 and 13.1.2 illustrate the relationships between the two types of docks, varying berth widths, and outside apron depth requirements. Note that the apron depths given in these tables are for an unobstructed dock. Where trucks must back up to the dock alongside other trucks, the apron depths must be increased by the length of the truck.

Vehicle Interface Requirements. Vehicle interface requirements deal with the space provisions for allowing carriers access to and maneuverability at the dock area. One important vehicle interface

Table 13.1.2 Dock Configuration Factors for Finger Docks for 65-ft Tractor Trailer

Dock Berth Width ft (m)	Finger Angle	Apron Depth ft (m)	Bay Width ft (m)
10 (3.0)	10°	50 (15.2)	65 (19.8)
12 (3.7)	10°	49 (14.9)	66 (20.1)
14 (4.3)	10°	47 (14.3)	67 (20.4)
10 (3.0)	30°	76 (23.2)	61 (18.6)
12 (3.7)	30°	74 (22.6)	62 (18.9)
14 (4.3)	30°	70 (21.3)	64 (19.5)
10 (3.0)	45°	95 (29.0)	53 (16.2)
12 (3.7)	45°	92 (28.0)	54 (16.5)
14 (4.3)	45°	87 (26.5)	56 (17.1)

is the traffic flow pattern into the dock area. The direction in which vehicles travel about the warehouse has a significant impact on the efficiency of carrier spotting and the space requirements of that activity.

Trucks should enter a receiving or shipping dock area in a counterclockwise direction of travel to allow the truck to back into the dock bay in a clockwise direction of travel. As shown in Fig. 13.1.3A, clockwise backing enables the truck driver to clearly see the rear of the truck as it turns into the dock bay. On the other hand, when backing counterclockwise, the truck driver must rely on a rearview mirror for visibility or on human guidance to properly position the vehicle. Counterclockwise backing requires that the apron bay depth be approximately 20 ft (6.1 m) greater than that required for clockwise backing. Where counterclockwise travel about the warehouse is not possible due to site restrictions, either a truck turnaround area should be provided to allow clockwise backing, or the additional 20 ft (6.1 m) of apron depth should be provided.

The dimensions of the service roads leading into and around the warehouse depend on whether one- or two-directional travel is desired. One-directional service roads should be 12 ft (3.7 m) wide; two-directional service roads should be 24 ft (7.3 m) wide. Gates through which carriers travel should be 20 ft (6.1 m) wide for one-directional travel and 30 ft (9.1 m) wide for two-directional travel. Where pedestrian travel will also pass through the gate, the gate width should be increased by 6 ft

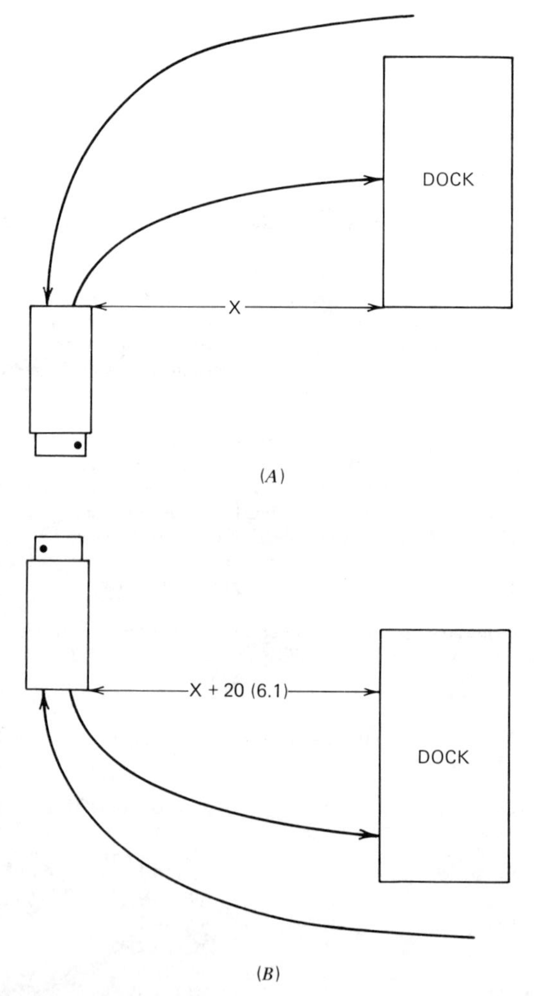

Fig. 13.1.3 Counterclockwise versus clockwise direction of travel. (*A*), counterclockwise direction of travel allows clockwise backing in X ft (m); (*B*), clockwise direction of travel allows counterclockwise backing in X + 20 ft (X + 6.1 m).

(1.8 m). Intersections of the service roads about the facility should be recessed or Y-shaped to allow easier and safer carrier turns. The minimum truck-turning radius for service road curves and intersections is 35 ft (10.7 m).

Maneuvering Allowances Inside the Receiving and Shipping Areas. Maneuvering space on a receiving or shipping dock consists of two components: (1) space for a dock-leveling device to allow access into and out of the carrier, and (2) a maneuvering aisle to allow access between the carrier and the receiving and shipping buffer areas (Fig. 13.1.4).

A dock-leveling device is an apparatus used to adjust the height of the dock to the height of the bed of the carrier to be loaded or unloaded. The amount of space required for the dock leveler depends on the type of dock leveler used and the height differential between the dock and the vehicle bed. Generally, temporary portable dock levelers occupy 3–7 ft (0.9–2.1 m) of space measured from the face of the dock. Permanent dock levelers occupy 4–10 ft (1.2–3.0 m) of space inside the building. Other types of dock levelers are installed outside the warehouse or on the carrier itself and require no inside warehouse space.

The width of the maneuvering aisle is dependent on the type of materials handling equipment used to load or unload the carrier. Manually operated hand trucks and pallet jacks require a maneuvering aisle width of approximately 6–8 ft (1.8–2.4 m). Powered forklift trucks require a maneuvering aisle approximately 8–12 ft (2.4–3.7 m) wide. The maneuvering aisle should not be used as a main warehouse aisle; it should be restricted primarily to materials handling traffic related to loading and unloading carriers. Pedestrian traffic in this area should be minimized to reduce the potential for accidents. The maneuvering aisle should begin immediately behind the area allocated for the dock leveler.

Receiving and Shipping Buffer Area Requirements. A buffer area in a receiving or shipping department acts as a holding area for materials during transfer of control of the goods between the warehouse and the carrier. In a receiving department, the buffer area allows for faster receiving throughput and minimizes demurrage and detention changes by allowing dock personnel to concentrate first on unloading the goods into the warehouse to maximize carrier turnaround. Once the carrier is unloaded and released, a more thorough check-in and inspection of the goods can take place within the buffer area as required. In a shipping department, the buffer area allows the accumulation of outbound freight in one place to ensure that the proper material will be shipped on the right carrier to the right customer.

Sufficient buffer space is critical to the existence of efficient and effective receiving and shipping operations. The amount of buffer space required is dependent on the physical characteristics of the unit loads to be received or shipped, the amount of material on a given inbound or outbound shipment, and the degree of control that can be exercised over when shipments arrive at or depart from the warehouse.

Two extreme degrees of control over receiving and shipping activities exist. The first extreme implies

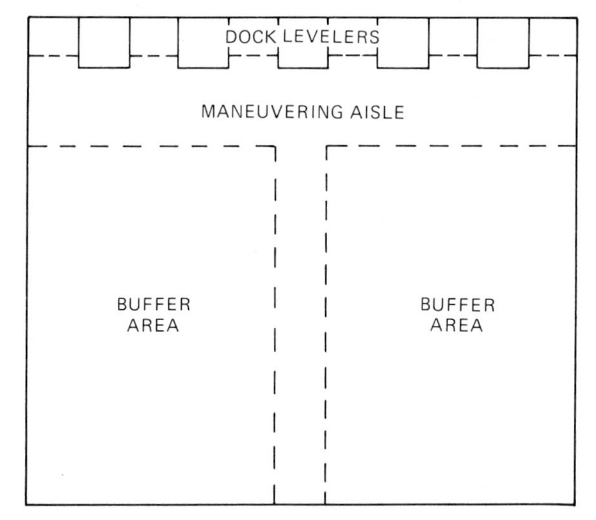

Fig. 13.1.4 Receiving and shipping area configuration.

that the arrivals of carriers at the receiving or shipping dock are largely controlled by warehouse management. In this case, the amount of buffer space required typically is at a minimum. For example, if the arrivals of incoming shipments can be scheduled throughout the day, then the same buffer space can be reused over and over again to minimize the total buffer space requirements. The material on a given shipment will be unloaded into the buffer area, the material will be checked in as required in the buffer area, and the material will be removed from the buffer to storage before the scheduled arrival of the next shipment to use the same buffer space. As the degree of control over the arrivals of carriers increases, the space requirements for buffer areas decreases.

The second extreme implies that the arrivals of carriers at the warehouse are outside the control of warehouse management. In this case, it is not unusual to have two or more carriers arrive simultaneously and compete for the same buffer area. To accommodate this situation, the amount of buffer space must be determined based on the peak demand or workload. In other words, the buffer area must be sufficiently large to allow the peak demand of shipments to be handled so that carrier waiting times, and thus potential detention and demurrage charges, are minimized.

Requirements for buffer areas in existing warehouses should be determined by analyzing historical receiving and shipping patterns to identify the surge periods and volumes. For new facilities, the anticipated throughput of the warehouse should be forecasted to project anticipated surges in activity.

Space Requirements for Dock-related Activities. Several dock-related activities require space within the receiving and shipping areas. Among the more important and common dock-related areas are office space, quality control hold areas, trash disposal, empty pallet storage, and the truckers' lounge.

Office space must be provided for receiving and shipping supervisory and clerical activities. Approximately 125 ft^2 (11.6 m^2) of space should be provided for each dock employee who regularly requires office workspace. Consideration should be given to proper cubic space utilization by possibly locating receiving and shipping offices on mezzanines so that the floor level can be used as a buffer or hold area; or, two-floor office areas should be considered to minimize the floor space occupied by offices.

A quality control hold area is essential for accumulating received material that has been rejected during a receiving or quality control inspection and is awaiting disposition. A separate and distinct hold area must be allocated. The amount of space required is a function of the type of material likely to be rejected, the specific inspection process followed and the timeliness of disposition of the rejected material.

A means of disposal must be provided for the large volume of trash generated in receiving and shipping areas. Failure to allocate space to trash disposal will result in the use of other valuable dock space not originally intended for that function. The result will be poor housekeeping, congestion, unsafe working conditions, and a loss of productivity.

If it is a practice to palletize materials when received or to unpalletize materials before shipping, then space must be allocated to accommodate the flow of empty pallets through the receiving and shipping areas. In addition, other unitizing techniques, such as shrink and stretch wrapping, require specialized equipment which must be apportioned space on the dock.

A truckers' lounge is an area where the truck driver is required to be if not exercising the right to be at the truck when the truck is being loaded or unloaded. The purpose of a truckers' lounge is to control the trucker's activities while on warehouse property. A truckers' lounge is a proven approach to minimizing potential safety, pilferage, and labor relations problems caused by having truck drivers on warehouse premises. The truckers' lounge should include private toilet facilities and a pay phone or a telephone routed through the warehouse switchboard for the truckers' use. At least 150 ft^2 (13.9 m^2) should be provided for a truckers' lounge.

Storage Space Planning

Storage space typically comprises the great majority of the total space contained within a warehouse. A critical element of the warehouse planning process is the quantification of the storage space requirements of the various commodities to be stored within the warehouse. The steps necessary to properly define the storage space requirements of a warehouse include:

1. Define the materials to be stored.
2. Determine the proper storage philosophy.
3. Determine the space requirements for alternative storage methods.

Define the Materials to Be Stored. The first step in determining storage space requirements is to define the materials to be stored. A useful tool in facilitating this process is a storage analysis chart (SAC), similar to that shown in Fig. 13.1.5. The first five columns of the SAC define the physical characteristics of the materials to be stored. Columns 6 through 8 define how much of each type of material is to be stored. Columns 9 through 12 of the SAC specify the methods used to store the materials. The same generic category approach should be exercised in defining the materials to be stored as was described in the discussion of receiving and shipping space analysis.

Company		Date		Raw Materials		In-Process Goods						
Prepared by		Sheet ___ of ___		Plant Supplies		Finished Goods						
Description	Unit Loads				Quantity of Unit Loads Stored			Storage Space				
	Type	Capacity	Size	Weight	Maximum	Average	Planned	Method	Specifications	Area (sq. ft.)	Ceiling Hgt. Req'd.	

Fig. 13.1.5 Storage analysis chart.

Storage Philosophies. Columns 6 and 7 of the SAC shown in Fig. 13.1.5 define the maximum and average inventories to be expected for each category of materials stored in the warehouse. Column 9 specifies the inventory level upon which the storage space requirements will be based. The relationship between the planned inventory level and the maximum and average inventory levels is dependent on the philosophy used in allocating material to storage space. Three basic material storage philosophies exist: (1) random location storage, (2) assigned location storage, and (3) combination location storage.

With random location storage, any stockkeeping unit (SKU) may be allocated to any available storage location. During one time period, a given SKU may be stored in one location, while at some later date, the same SKU may be allocated a different storage location. With assigned storage, each SKU is assigned to a fixed location within the warehouse. A given SKU will only be stored in its assigned location, and no other SKU will be stored in that location. As the name implies, combination location storage involves the combination of the random and assigned location systems for each SKU. A common combination system will include the storage of the overstock of a commodity in a random location and storage of a smaller quantity of the commodity in an assigned location area to facilitate order-picking productivity.

The amount of space planned for a stockkeeping unit is directly related to the method of assigning space. If assigned location storage is used, then the space requirements should be based on the maximum inventory level. If random location storage is used, then the space requirements should be based on the average inventory level. Finally, if combination location storage is used, then the planned inventory level should fall at a value between the maximum and average inventory levels.

The decision to use random location storage, assigned location storage, or combination location storage should be made in light of the specific circumstances of each individual warehouse. Each philosophy has its applications, advantages, and disadvantages. Choosing one storage philosophy over another can only be accomplished by carefully evaluating the trade-offs inherent in the alternative approaches. Table 13.1.3 summarizes a qualitative comparison of the three options on three important attributes: space utilization, accessibility to material, and materials handling efficiency.

Table 13.1.3 Comparison of Storage Philosophies

Attribute	Assigned Location Storage	Random Location Storage	Combination Location Storage
Space utilization	Poor	Excellent	Good
Accessibility	Excellent	Good, with a good stock location system Otherwise, poor	Good
Materials handling efficiency	Good	Good	Poor

Space utilization in an assigned location system is poor due to the need to allocate space based on the maximum expected inventory level, when in fact, actual inventory levels normally approach the average inventory level. Consequently, a great deal of empty space is common in assigned location storage. Random location storage is extremely space efficient since the space allocated is based on the average expected inventory, which corresponds to the level typically on-hand. Space utilization efficiency for combination location storage is dependent on the degree to which random storage is approximated. The more material stored randomly, the more space efficient is combination storage.

Assigned location storage provides excellent accessibility to the materials in storage since each SKU has a unique, fixed location which does not change on a regular basis. Consequently, the whereabouts of each item in storage is guaranteed by the storage philosophy. Random location storage provides good material accessibility so long as there exists a good stock location system to track the current whereabouts of each SKU. If, however, a formal stock location system does not exist, or is poorly designed or maintained, then accessibility with random location storage will be extremely poor. Combination location systems typically have good material accessibility so long as a good stock location system exists to track the randomly stored goods, or the percentage of goods stored randomly is very small.

Assigned location storage and random location storage score equally well on the materials handling attribute. In each philosophy, material is received, stored, retrieved from storage, and shipped to a user. Materials flow is straightforward and economical. Materials handling efficiency is not as good, however, with some combination location storage systems. Oftentimes material is received, stored in the random location storage area, retrieved, stored in the assigned location storage area, and retrieved and shipped to a user. The result is several extra handling steps not required with either strictly assigned or random stock location systems.

Determining Space Requirements for Alternative Storage Methods. The storage space requirements of a warehouse are dependent on two primary factors:

1. The nature of the product to be stored
2. The nature of the storage and handling methods used

The first factor, the nature of the product, concerns defining the physical characteristics of the products to be stored and the planned inventory levels to be carried in the warehouse. This information can be adequately defined by using the storage analysis chart previously given in Fig. 13.1.5. The second factor, the nature of the storage and handling methods, can be defined by documenting three use-of-space, or accessibility, factors: (1) aisle allowances, (2) honeycombing allowances, and (3) ingress/egress allowances.

Aisle allowance is the percentage of space occupied by aisles within a storage area. The amount of aisle allowance depends on the storage method, which dictates the number of aisles required, and on the materials handling methods, which dictate the width of the aisles. Aisle allowances can be determined for any given storage and materials handling alternative under consideration by simply sketching a reasonable facsimile of the anticipated layout and calculating the percentage of space dedicated to aisles.

Honeycombing allowances are the percentage of storage space lost because of ineffective use of the capacity of a storage area. As illustrated in Fig. 13.1.6, honeycombing occurs whenever a storage location is only partially filled with materials. To place materials into the vacant spaces shown in Fig. 13.1.6 before the other materials occupying those storage locations are removed would result in blocked stock and/or poor stock rotation. Consequently, the vacant spaces represent honeycombing losses to be incurred until each storage location is completely vacated.

Ingress and egress allowances refer to the space requirements necessary to allow materials to be easily placed into and removed from storage. For example, in a typical pallet rack application, the size of the pallet rack opening is specified such that approximately 4–6 in. (10–15 cm) of space exists between the rack uprights and the pallet loads resting on the rack crossarms. In addition, approximately 6–8 in. (15–20 cm) of space is provided between the tops of the loads resting on the crossarms and the bottom of the next set of crossarms immediately above the pallet loads. This "clearance" space is required to allow for the insertion of pallets into and removal of pallets from the rack opening in an efficient and effective manner.

Every storage and handling method has some degree of aisle allowance, honeycombing allowance, and ingress/egress allowance. To calculate the space requirements of a given inventory of a given product requires that the space to be occupied by the unit loads of material being stored be adjusted by the anticipated allowances for aisles, honeycombing, and ingress/egress. Table 13.1.4 presents several guidelines for different storage methods which can be used to facilitate the determination of storage space requirements.

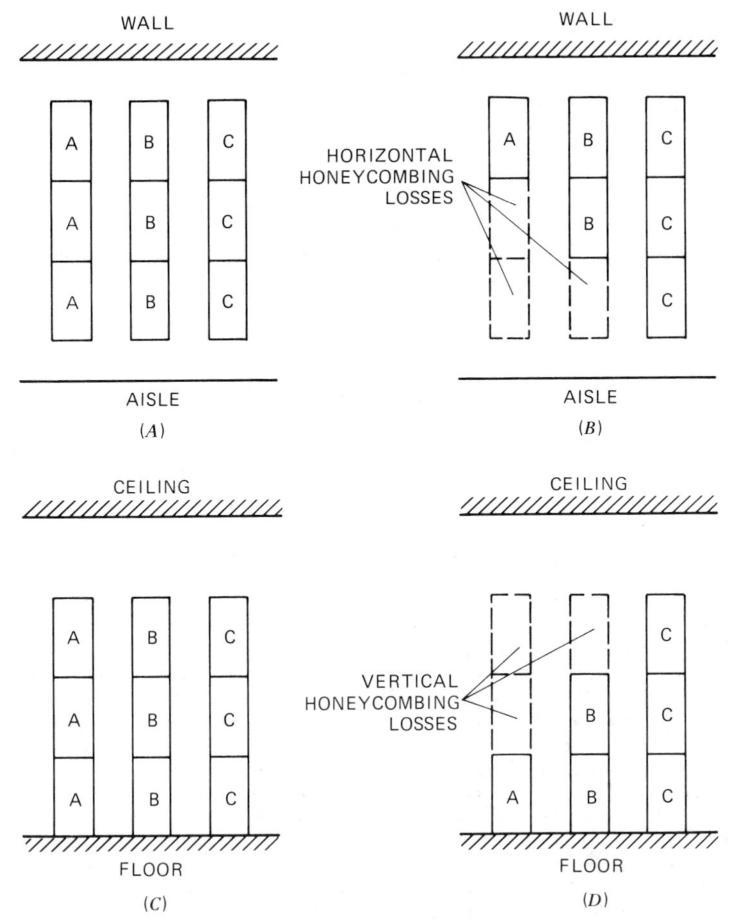

Fig. 13.1.6 Honeycombing allowances in bulk storage. (*A*), top view, no honeycombing; (*B*), top view, horizontal honeycombing; (*C*), elevation view, no honeycombing; (*D*), elevation view, vertical honeycombing.

Table 13.1.4 Characteristics of Pallet Storing Methods

Factor	Bulk Storage	Portable Racks	Drive-in Racks	Flow-through Racks	Sliding Racks	Pallet Racks
Aisle allowances[a]	10%	10%	45%	30%	30%	65%
Honeycombing allowances	25%	25%	25%	30%	0%	0%
Accessibility of unit load	10%	10%	30%	30%	100%	100%
Load crushing	Bad	Nil	Nil	Some	Nil	Nil
Stability of load	Poor	Fair	Good	Fair	Good	Good
Ease of relocation	N.A.	N.A.	Fair	Difficult	Difficult	Good
Speed of installation	N.A.	N.A.	Good	Fair	Slowest	Fastest
Rotation of stock	Poor	Poor	Poor	Excellent	Good	Good

[a] Including access and working aisles.

Personnel Services Space Requirements

The personnel-related areas and services of interest in a warehouse are:

Warehouse offices
Food services
Lavatories
Locker rooms

Office space must be provided for all warehouse employees engaged in clerical and administrative tasks. Private space should be allocated based on the following scale:

Position	Square Feet	Square Meters
Senior executive	300–400	27.9–37.2
Junior executive	200–250	18.6–23.2
Supervisors	100–175	9.3–16.3
Staff personnel	100–150	9.3–13.9

General office space should consist of approximately 125 ft^2 (11.6 m^2) for each clerical employee to work in the area. Aisles and corridors should be 4 ft (1.2 m) wide in the private office area of the warehouse and 6 ft (1.8 m) wide around the general office area. To allow ease of ingress and egress, desks in a general office area should not be closer together than 2 ft (0.6 m) and desk chairs should be at least 3 ft (0.9 m) from surrounding fixed objects (except, of course, their corresponding desks) and 4 ft (1.0 m) from surrounding movable objects.

The need for food services within a warehouse can become a critical factor if this is not recognized before the warehouse is built. The Occupational Safety and Health Administration (OSHA) has ruled that food may not be eaten nor may any beverage be consumed in any place where toxic substances are present. The OSHA definition of toxic substance is extremely liberal and encompasses almost any material found in a warehouse. Therefore, the picnic tables on the receiving docks of many warehouses are illegal. The alternatives are either to require employees to eat off the warehouse premises or to provide an area, specifically designated for food services, that meets OSHA regulations. Because most warehouse management does not want employees leaving the premises for lunch, the second alternative is usually chosen. Warehouse food service areas generally take the form of a segregated lunch area with vending machines from which food may be purchased. The required size of the lunch area depends on the number of employees who will use the area at any given time. The area required for a vending machine food service should conform to the following scale:

Number of Persons Using Area at Any Given Time	Square Feet of Area Required per Person (m^2)
25 or less	13 (1.2)
26–74	12 (1.1)
75–149	11 (1.0)

Subpart J, paragraph 1910, 141(c–f) of the Occupational Safety and Health Act defines very clearly the lavatory requirements in a warehouse. From a warehouse design viewpoint, the important initial considerations are:

1. Separate facilities shall be provided for each sex, not to be farther than 200 ft from the location where workers are regularly employed.
2. The number of water closets for each sex shall be determined by the following table:

Number of Persons	Minimum Number of Water Closets
1–9	1
10–24	2
25–49	3
50–74	4
75–100	5

One additional facility is required for each additional 30 persons.

3. Where more than 10 men are employed, the number of water closets given in the preceding table may be reduced by one for each urinal provided. The number of water closets must remain at least two-thirds of the number specified in the table.
4. At least one sink with adequate hot and cold water shall be provided for every 10 employees, or portion thereof, up to 100 persons; for over 100 persons, one sink for each additional 15 persons must be provided. A sink will be at least 24 in. (6.0 cm) wide and shall have an individual faucet.
5. Where 10 or more women are employed, at least one bed for resting is required. Two beds are required if between 100 and 250 women are employed, and one additional bed is required for each additional 250 women employees.

For space planning purposes, 12 ft^2 (1.1 m^2) should be allowed for each water closet, 5 ft^2 (0.5 m^2) for each urinal, and 6 ft^2 (0.6 m^2) for each sink. Entrance doorways should be designed so that the interior of the lavatory is not visible from outside the room when the door is open. An allowance of 14 ft^2 (1.3 m^2) should be made for this entrance. For each bed required in the women's lavatory, 60 ft^2 (5.6 m^2) should be allocated.

OSHA requires that separate locker rooms be provided for each sex whenever it is the practice to change from street clothes to working clothes. If locker rooms are not provided, OSHA requires that facilities be provided for hanging outer garments. For planning purposes, 6 ft^2 (0.6 m^2) should be allocated for each person using a locker room. Often, it is convenient to combine a lavatory and locker room. Although it is desirable to locate locker rooms for the convenience of the employees, they should, if possible, be away from the primary flow of materials and in a location that provides good ventilation. Mezzanines or locations along an outside wall are, therefore, often good places for lavatories and locker rooms.

13.1.3 Warehouse Layout Factors

The achievement of an efficient and effective warehouse operation is largely dependent on the existence of a good warehouse layout. The warehouse layout typically dictates the degree of material accessibility, material flow patterns, and the locations of traffic bottlenecks, labor efficiencies, personnel safety, and warehouse security. The objectives of a warehouse layout should be:

1. To ensure maximum space utilization
2. To provide for the most efficient materials handling
3. To provide the most economical storage in relation to the costs of equipment, space, material damage, and warehouse labor
4. To provide maximum flexibility to meet changing storage and handling requirements
5. To make the warehouse a model of good housekeeping

The general methodology for designing a warehouse layout consists of five steps:

1. Define the location of all obstacles.
2. Locate the receiving and shipping areas.
3. Locate the primary and secondary order-picking and storage areas.
4. Assign stock to storage locations.
5. Evaluate the alternative warehouse layouts.

Step 1 of this methodology involved identifying and locating the physical characteristics of the warehouse facility that are beyond the control of the warehouse planner. Common examples of such obstacles are building support columns, emergency exit doors, utilities, stairwells, elevator shafts, and fire protection equipment. Such obstacles should be identified and located on the warehouse layout plan before more flexible objects and activities are included in the layout. Otherwise, problems will arise with conflicts between desired layout plans and what can actually be accomplished in the actual facility.

Oftentimes, the location of the receiving and shipping areas is dictated by the site plan on which the warehouse is or will be located. If this is not true, then the receiving and shipping areas should be located in the warehouse layout in a manner which maximizes the efficiency of these operations. Factors to be considered include the location of highways, service roads, railways, the desired traffic pattern around the facility, desired traffic patterns within the warehouse, and labor productivity in accomplishing the receiving and shipping activities. Energy efficiency factors should also be considered.

Where possible, the receiving and shipping dock bays should typically face in a direction away from the prevailing winds around the warehouse site to reduce heat loss by blocking the wind from entering the warehouse through open dock doors.

Once the receiving and shipping areas have been located in the layout, the material storage areas should be addressed. The types of storage areas and equipment to be used will dictate to some extent the configuration of the storage layout and the necessary aisle requirements. Primary storage and order-picking areas may be differentiated from secondary areas to provide maximum space, materials handling, and labor efficiency for the most important materials to be stored.

The assignment of stock to storage locations within the layout alternative ensures that space allowances have been made for all the items to be stored. In addition, it allows the layout planner to mentally simulate the performance of the alternative layout for the day-to-day activities expected in the warehouse.

Finally, each alternative warehouse layout must be evaluated to determine the extent to which it achieves the desired objectives. Each warehouse layout should be evaluated based on the degree to which several basic warehouse layout philosophies are followed. Among these layout philosophies are the following:

1. Popularity
   a. Store goods having the greatest turnover as near as possible to the point of use.
   b. Store goods having the greatest turnover in as deep a space block as possible.
2. Similarity
   a. Items received and shipped together should be stored and inventoried together.
   b. Items that have strong correlation with regard to type of item should be stored together.
3. Size
   a. Store heavy, bulky, hard-to-handle goods close to their point of use.
   b. Provide a variety of storage locations and sizes.
   c. Heavy items should be stored in low-ceiling areas and lightweight, easy-to-handle items should be stored in high-ceiling areas.
   d. Look not only at the size of individual items, but also at the size of the total stock of an item.
4. Characteristics of materials
   a. Design the layout to properly accommodate perishable items.
   b. Provide effective layout and storage techniques to maximize space utilization for oddly shaped and crushable items.
   c. Plan for the protection of hazardous materials from causes of accidents and protect other materials from hazardous materials in the event of an accident.
   d. Design the layout around the compatibility of items stored within close proximity of each other.
   e. Design the layout to maximize the protection of security items by location.
5. Space utilization
   a. Conserve the use of space by maximizing the concentration of goods in storage, maximizing cubic space utilization and minimizing honeycombing losses.
   b. Design the layout around obstacles and other limitations on space use.
   c. Aisles should be straight and lead to doors.
   d. Aisles should be wide enough to permit efficient operation, but not wasteful of space.
   e. Every face of an island of storage should have access from an aisle.
   f. Blocked stock should be avoided.
   g. Stacks of material should be uniform, straight, stable, and easily accessible.
   h. Aisle markings should be used to maintain the integrity of the aisles.
   i. Void spaces within the storage areas should be avoided.
   j. Stock location records should be maintained.

REFERENCES

1. Tompkins, James A. and John White, *Facilities Planning,* Wiley, New York, 1983.
2. Tompkins, James A. and Jerry D. Smith, *How to Plan and Manage an Efficient Warehouse,* American Management Association, Boston, MA, 1982.

13.2 STORAGE APPROACHES

James A. Tompkins

13.2.1 Equipment Considerations

The types and styles of warehouse storage equipment are almost unlimited. There exists a storage equipment variation ideal for almost any given warehouse situation. The major types of warehouse storage equipment, however, can be broadly classified into ten categories:

1. Pallet rack
2. Cantilever rack
3. Drive-in/drive-through rack
4. Flow-through rack
5. Mobile rack
6. Portable rack
7. Shelving
8. Modular drawer units
9. Carousels
10. Automated storage and retrieval systems

These equipment and system items are covered in Chapters 11 and 12.

13.2.2 Stock Location Systems

The objective of a stock location system is to track the whereabouts of each item throughout its stay in the warehouse. Stock location systems fall into three basic categories: (1) memory systems, (2) assigned location systems, and (3) random location systems.

Memory location systems depend on the recall of the individuals responsible for placing material into and retrieving material from storage. Memory systems work well if the following restrictions are met:

1. Only one person works with a given storage area.
2. The number of different SKUs stored in that area is relatively small.
3. The number of different storage locations in that area is relatively small.

Otherwise, memory location systems can be disastrous. Given the unlikelihood of all three of these restrictions being met, memory location systems are generally ineffective and inefficient for most warehouse operations.

As discussed previously, in an assigned location system, each SKU is assigned to a specific and unique location in storage. A particular item will always be stored in its assigned location, and no other item may be stored in that location, even though that location may currently be empty. In an assigned location system, the location of every item is always known.

However, even assigned location systems require some method of keeping track of which items are assigned to which locations. One common method is to rely on the memories of the people working the storage area. Because items are assigned to fixed locations, one individual need not be concerned with the actions taken by another individual. In addition, labor absenteeism and turnover are less of a problem in an assigned location system than in a strictly memory system because a permanent record of material location assignments can be made to train replacements. Still, the limited capacity of the human memory limits the effectiveness and efficiency of this type of assigned location system. Alternatives are (1) the assignment of items to locations in item-number sequence and (2) the preparation of permanent records to cross-reference materials with their locations. Storing materials in item-number sequence eliminates the need to memorize stock locations and does not incur the expense necessary to produce permanent records. A disadvantage of storing materials in item-number sequence is the inability to deliberately place high-throughput items closest to the point of use and low-throughput items farthest from the point of use for the purpose of improving materials handling efficiency. To eliminate this shortcoming, assigned location systems can be designed to rely on permanent records to cross-reference items and their locations. The process of assigning items to assigned location systems does not have to follow any pattern. High-throughput items can be assigned to locations close to the point of use, and low-throughput items can be assigned to locations more remote from the point of use. Human memory does not have to recall the location of an item; instead, each storage location is given a name or address that uniquely identifies it. A permanent cross-reference record can be queried to identify the address of each item in storage. The permanent cross-reference record is usually in the form of a printed directory, a catalog, or a computer record accessible by some input/output device such as a teletype or cathode ray tube (CRT).

Random location systems allow material to be stored in any location currently available in the storage area. As in assigned location systems that use cross-reference location records, each storage location is given a unique name or address. When an item is randomly placed in the storage area, the address of the specific location into which it is put is recorded for future reference. Maintenance of random location system records may be either manual or computerized. Random location systems generally function in the following manner:

1. When an item is received, the stock location records (either manual or computerized) are checked to determine if any of the items is currently on hand and, if one is, to find its location.
2. If a location contains that item and if sufficient space remains, the newly received material is stored in that location and the stock location record for that item is updated to reflect the new quantity stored in the location.
3. If no location contains the item or if there is insufficient space in the current location, the newly received material is placed in any available storage location.
4. The stock location record for that item is updated by recording the address of the storage location in which the material is stored and the quantity of material stored in that location.
5. When retrieval of the item is desired, the stock location record is queried to determine the address of the storage location in which the item is stored.
6. The material is located, the quantity desired is removed from the location, and the stock location record is updated by adjusting the recorded inventory level in the storage location.

A number of guidelines can be followed as a major step toward maintaining an effective stock location system:

1. Keep the stock location system simple. The system must be completely understood by everyone involved. Do not try to do too much with the stock location system; remember that the primary objective is to track the location of materials in storage. Prolonged difficulty in locating material, reporting locations, and updating records is an indication that the stock location system is too complex.
2. Require valid written authorization for the removal of merchandise from a storage location. Moving materials into and out of storage locations, without updating the location records, is a primary cause of failure of stock location systems.
3. Keep the stock location records current. Received materials and shipped materials must be posted to the stock location records in a timely manner to ensure good service to the warehouse's customers and to maintain accurate location records.
4. Report and rectify discrepancies in the stock location system records. Human errors are inevitable. Warehouse employees should be encouraged to report errors when identified; otherwise, the integrity of the stock location system will be compromised, and the system will fail.
5. Store similar items together as often as possible. Doing so makes it easier to locate materials which have become "lost" because the location of the missing material can be narrowed to a specific area of the warehouse where that type of item is generally stored.

13.3 ORDER-PICKING METHODS

James A. Tompkins

The effectiveness and efficiency of the picking function are directly related to the effectiveness and efficiency of the storage function. If we do not know where we have stored an item, then the task of removing that item from storage is more difficult. Consequently, the first step in developing a sound picking function is to develop a sound storage function, complete with a good stock location system. A good stock location system depends on a good information system, which tells the order picker what items are to be picked, where the items are located, how much of each item is to be picked, and in what order the items are to be picked. The output of the information system that contains these instructions is the picking document.

There are two ways to pick material to fill customers orders. One way is to have one order picker—human or mechanical—pick all the material on a given customer order. In this case, the picking document is in the form of a standard customer order, a prerouted customer order, or a prerouted and preposted customer order.

The second way to pick is to have a given order picker pick only a portion of a given customer order. This is called zone picking, and the picking document is generally a prerouted customer order for the items to be picked by each order picker.

The use of a standard customer order as a picking document requires that the order picker locate and pick the requested items in the sequence in which they are listed on the customer order. The sequence of items on the standard customer order usually does not reflect the location of the material in the warehouse. Instead, standard customer orders often list items sequentially by customer item number. Consequently, a standard customer order used as the picking document usually creates an inefficient picking system. The ratio of time spent traveling to time spent actually picking material is extremely high because there is no logical sequence to the order in which items are picked. A tremendous amount of backtracking is required to pick an item that was already passed in the process of following the sequence in a standard customer order.

A prerouted customer order is simply a standard customer order that has been altered to sequence the items to be picked according to their location in the warehouse. With a prerouted customer order, backtracking of order pickers is eliminated because each item is picked the first time it is passed. The total distance traveled by the order picker is reduced. If the distance traveled per customer order is reduced, the productivity of each order picker is increased. As a result, fewer order pickers are needed for a prerouted customer order-picking system.

The efficiency of a prerouted customer order as a picking document can be increased by also preposting the customer order. Preposting is the act of verifying the on-hand status and updating the inventory records of the requested items of an order before the picking document is produced. If an item is not on hand, then it is omitted from the picking document. If the quantity on hand is lower than the quantity ordered, then the picking document reflects the quantity the order picker can expect to find. Preposting eliminates the picking inefficiencies created when the order picker wastes time searching for items called for by the picking document which are not currently in inventory.

The benefit of zone picking is the minimization of order-picker travel per item picked through the maximization of the number of picks along the path traveled by the order picker. In a system in which one order picker picks an entire order, the order picker typically spends a high percentage of time traveling from one storage location to another. This is true even when the picking document is prerouted because the items on a given order may be scattered throughout the warehouse. In a zone-picking system, however, an order picker picks all the items for each customer order that are stored within the zone of the warehouse assigned to that order picker. Consequently, the order picker spends a much larger percentage of time actually picking items.

To illustrate this concept, consider a warehouse in which 100 min are required to pick the average prerouted customer order. Of this 100 min, approximately 70 min are consumed traveling between storage locations, and 30 min are spent actually picking material. If an order picker is required to pick an entire customer order, then the time required to pick two prerouted customer orders is 200 min. If zone picking is employed, where one order picker will pick half of a customer order and another order picker the other half, then the total time required to pick two prerouted customer orders is computed as follows:

		Order Picker 1	Order Picker 2
Walk time:	Orders 1 and 2	35 minutes	35 minutes
Pick time:	Order 1	15 minutes	15 minutes
	Order 2	15 minutes	15 minutes
Total per order picker		65 minutes	65 minutes
Grand total		130 minutes	

Therefore, the time required for zone picking two prerouted customer orders is 70 min less than that required for having each order picker pick entire customer orders. The reason for this time reduction is the fact that with zone picking, the walk time for the two orders is combined; the total distance traveled is half that required to individually pick the two orders.

In zone picking, however, time and space must be allocated for the consolidation of the individual portions of a customer order into a complete order. In the example just given, the amount of picking time saved by zone picking is actually less than 70 min. The time required to consolidate the individual portions of a customer order can be minimized by having the order pickers segregate the items by customer order as they pick them. This will eliminate the need to sort through a large group of items to identify the components of each customer order. When order consolidation is relatively easy and quick, zone picking will likely be much more productive than picking of entire customer orders. Therefore, the alternative of zone picking should always be carefully evaluated before a decision is made to have each order picker pick entire customer orders.

13.4 COMPUTERS IN THE WAREHOUSE

John R. Huffman

13.4.1 Bases and Justifications for Computer Applications

Each transaction processed by the software and hardware required by manual warehouse computer applications reflects a material movement. Each movement:

1. Is triggered by information.
2. Requires instructions defining what is to be moved, how much is to be moved, the locations from which and to which it must be moved and when.
3. Affects the inventory of an SKU in two locations.

For example, the information on a customer's order triggers the preparation of picking instructions in the form of picking documents. Information that picking has been completed results in the creation of invoices and the transfer of inventory from the picking position and the warehouse to accounts receivable. Information that the inventory in a picking position is low initiates an instruction to replenish it from reserve stocks; executing that instruction increases one inventory and decreases the other.

In addition to preparing movement instructions and updating inventories, the computer can estimate the labor and equipment hours each move requires and can accumulate totals. It can also record the difference between the expected and actual results. Because a manual warehouse computer system utilizes these capabilities, it can assist warehouse management:

1. To provide clear specifications and schedules for material movements (or operations).
2. To estimate labor requirements and schedule manpower assignments.
3. To operate the warehouse as planned and, therefore, to realize the planned labor productivity.
4. To maintain accurate, current inventories that can be verified readily.

Computer Applications

Uses of the computer in a manual warehouse can be classified as:

1. Picking and shipping applications.
2. Receiving and stocking applications.
3. Inventory applications.

Picking and shipping applications consist of:

1. Preparing picking documents and invoices.
2. Planning delivery operations and vehicle loads and preparing route manifests.
3. Planning the daily operations schedule.
4. Estimating the manpower and equipment hours required by each operation during each schedule period and for the entire shift.
5. Assigning picking positions randomly.
6. Predetermining the merchandise repacked in each tote or carton.

Receiving and stocking applications consist of preparing:

1. A receiving document or documents.
2. Instructions as to the area in which a receipt should be stored.
3. Instructions for palletizing or otherwise preparing each receipt for storage.
4. Documents for moving the merchandise to a specific storage location.
5. Documents for verifying that received merchandise was stored in the location specified by the computer.
6. Instructions for replenishing picking locations.

Inventory applications consist of:

1. Maintaining inventory records in total and by warehouse location for each SKU.
2. Providing periodic inventory/purchasing reports.
3. Triggering inventory verifications and providing documents for that purpose.
4. Preparing documents for verifying the location of and/or for counting inventory rapidly and efficiently.

The first four picking and shipping applications, the first three receiving and stocking applications, a single inventory per SKU, and inventory/purchasing reports can be installed independently of one another and without major effects on the remaining manual systems. Inventory records are usually installed first because they are required for accounting purposes.

Savings and Added Costs Resulting from Computer Applications

Computer applications reduce warehouse labor costs because they:

1. Provide factual information that enables management to plan more efficient operations.
2. Provide instructions that maximize the productivity of labor.

Computer applications reduce delivery costs because they provide factual information for planning delivery routes and rating common carrier shipments and, finally, applications provide customer services which will maintain or increase sales and profits.

Computer applications increase warehousing costs because:

1. An initial investment is required for programming or adapting packaged programs and for training warehouse, data processing, and clerical personnel.
2. Additional operating personnel are required to maintain agreement between the physical inventories and the computer records of those inventories.
3. Additional data processing equipment and personnel may be necessary.

A computer application should be installed only if it is justified by the anticipated net savings. These savings will be realized if warehouse and data processing personnel: (1) conform to any new operating methods, systems, and procedures; (2) understand the effect of each material movement on the computer inventory records; and (3) recognize the physical significance of inventory adjustments.

Picking and shipping applications are usually installed immediately after computer records of item inventories because these operations require most of the warehouse labor and are, therefore, the easiest to cost justify and because applications in this area will improve customer service.

Prerequisites for Computer Applications

A computer system in a manual warehouse requires as a minimum:

1. To input customers' requirements—an order entry system.
2. For access to computer files—item and customer numbers.
3. To identify receipts and shipments—purchase order and invoice numbers.
4. To guide pickers and stockmen—warehouse location numbers.

Advanced applications of the computer to warehouse operations require:

1. To plan deliveries—route and stop numbers and, as applications become more sophisticated, item weight, cube, freight class, and DOT classification.
2. To plan and schedule operations—production standards based on experience or industrial engineering techniques.
3. To reduce picking costs:
   a. Item descriptions that correspond to the product description on the piece, inner pack, or case to be picked.
   b. Picking documents that specify only merchandise on hand according to the computer inventory records.
4. Clearly defined and accurate repack picking units and vendor case packs.
5. Accurate, current inventories.

Accurate, current inventories are essential when operations in a warehouse require multiple computer inventories per SKU. If this requirement is not met, it will not be possible to execute consistently the move instructions issued by the computer. The resulting confusion will result in unacceptably high labor costs and "markouts."

Limitations on Computer Applications in a Warehouse

Potential applications of the computer in a warehouse are limited by operating policies with respect to picking and shipping orders and the order entry cut-off time.

There are two policies with respect to picking and shipping orders:

1. Maintain order integrity during picking and shipping operations.
2. Destroy order integrity during picking operations and restore it on the dock prior to shipment.

When order integrity is maintained, there is only one warehouse inventory and one picking position per SKU in the computer. The only material movements that can be planned and/or recorded by the computer are those to and from this inventory. Computer applications dealing with material movements in such a warehouse are minimal; computer applications for planning operations are possible but require interpretation by experienced management.

Order integrity is destroyed and restored to obtain the labor savings that result when:

1. Broken case quantities are picked by order from one set of picking locations.
2. Individual full cases to be shipped or to replenish broken-case picking positions and nonconveyable items are bulk picked from another set of picking positions.

Additional labor savings may justify the maintenance of computer records of randomly located reserve inventories and computer direction of the movements from receiving to these locations and from them to picking positions. For this reason many more warehouse computer applications are possible when order integrity is destroyed and restored than when it is maintained.

There may be one or more order entry cut-off times. The only one or the final one may occur hours prior to the start of the picking shift or during the shift. If a cut-off time occurs during the picking shift, computer assistance when shipments and operations are planned is limited but other applications are not hampered. They are carried out on what is, or approximates, a real-time basis.

The ultimate applications of the computer to a manual warehouse in which order integrity is destroyed and restored create a system that is very comparable to any that supports an automated storage and retrieval system; the major difference is that the instructions are executed by personnel instead of equipment.

Specific Applications

This chapter covers two sets of common computer applications in a manual warehouse when (1) order integrity is destroyed and restored and (2) computer assistance in planning is possible because the entry of orders is cut off sufficiently in advance of the start of the picking shift. The second set of applications provides more inventories per SKU and therefore includes more computer applications than the first set. The following section also includes additional computer applications that improve customer service and a description of the system modifications necessary when order entry cut-off occurs during the picking shift. The section concludes with examples of an emerging group of computer applications that depend on microcomputers and bar-coded labels; it shows how these applications can be adapted to a warehouse in which order integrity is destroyed and restored.

13.4.2 Computer Applications When Order Integrity Is Destroyed and Restored

Typical Warehouse Layout and Merchandise Flows

When operations destroy and restore order integrity, repack, full case, nonconveyable, and any other categories of merchandise are located in separate areas of the warehouse. Reserve inventories are separated from picking inventories. Pallet loads and small lots are stored in random reserve locations. Repack and nonconveyable picking locations are always fixed because the size of the location depends on the item; locations from which vendor-pack full cases are picked may be fixed or assigned randomly. Multiple inventories per SKU are customary. All categories of merchandise are manually picked simultaneously by order or in bulk, assembled by order on the dock for piece count verification, loaded and shipped. Repack merchandise is checked as necessary and packed in the picking area. The replenishment of picking positions may occur during the picking shift or a separate shift.

Figure 13.4.1 is a schematic layout showing the warehouse areas and the movements between these areas when order integrity is destroyed and restored. These movements are classified as picking (P); sorting, order assembly, and shipping (S); replenishment (R); and receiving and storage (B).

The sorter is a key element of this layout because it segregates—by customer or by delivery route for subsequent manual sortation by customer—the stream of full cases and repack cartons or totes conveyed to it in random sequence so far as customer is concerned. Typical sorting systems consist of conveyor spurs and:

1. A powered conveyor loop from which cases and totes are manually pushed onto the spurs.
2. A powered conveyor, key-coding station, and air-operated pushers.
3. A slat sorter, tray sorter, or high-speed sorting conveyor and a scanner backed up by a key-coding station.

Merchandise movements, all by conveyor and via the sorter, are:

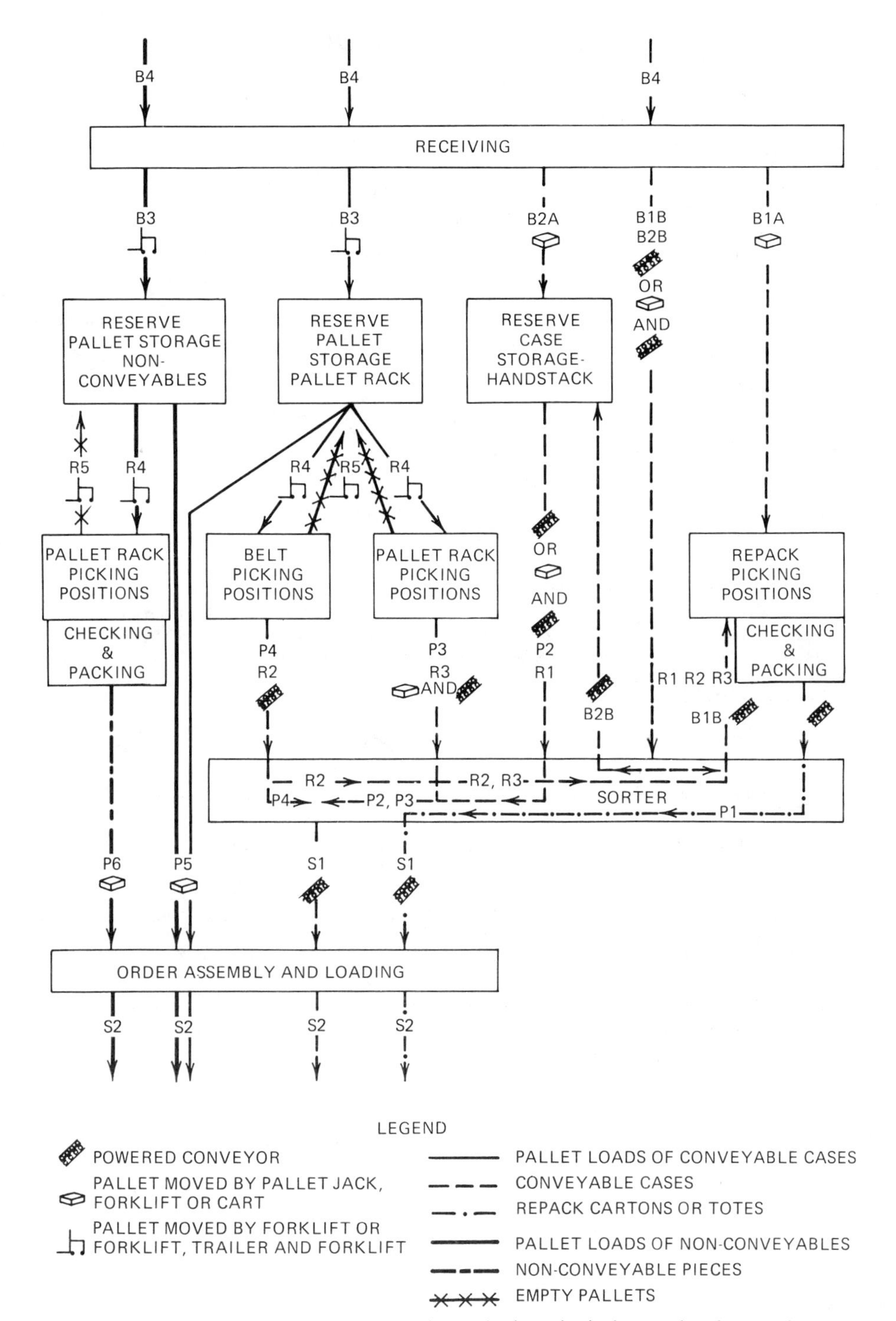

Fig. 13.4.1 Merchandise movements when order integrity is destroyed and restored.

1. Receipts consisting of a few cases each from receiving to repack picking positions or hand stack storage (B1B, B2B).*
2. Totes or cartons of loose merchandise from receiving to repack picking positions (B1B).
3. Totes or cartons of broken case merchandise picked by order (P1) and conveyable vendor pack cases bulk picked from pallet positions at a picking belt (P4, R2), from pallet positions to a pallet (P3, R3), and from hand stack racks (P2, R1) to spurs (R1, R2, R3) which accumulate cases to replenish repack picking positions or to shipping spurs (S1), each of which is assigned to receive merchandise for one stop or one route.

   Merchandise for one stop is unloaded from the conveyor into the truck. Merchandise for one route is manually sorted from the spurs to one or more pallets for each stop prior to truck loading and delivery (S2).

All other merchandise moves by pallet (B3, R4, R5, P5, P6). In some warehouses, receipts move from the dock to the repack picking positions and hand stack racks by pallet (B1A, B2A) instead of via conveyor (B1B, B2B).

Summary of Computer Applications

Three sets of computer applications, or systems, are possible in this warehouse. Two require documents prepared by a mini or larger computer to specify and/or to report all movements. They are:

A picking-oriented system

A total system

The third set of applications is emerging. It utilizes the mini or large computer as a host for microcomputers and screens at the points of operation and the transmission of information and instructions from computer to computer rather than by documents; it relies on bar coding to minimize identification errors. These applications eliminate documents and manual information processing required by the other systems.

All of these systems utilize the mini or large computer:

1. To plan deliveries by captive trucks (S2).
2. To schedule warehouse operations.
3. To project warehouse manpower requirements by operation during short interval schedule periods and for the entire shift.
4. To prepare the identical documents required to implement these plans.
5. To determine repack picking position replenishment quantities.

In the two systems that require documents for all movements the mini or large computer prepares:

1. A location-sequenced document for picking repack (P1) and a label for identifying the repacked carton or tote.
2. Labels for picking and identifying for shipment vendor-pack full cases, nonconveyable merchandise, and pallet loads (P2–P6).
3. Labels for picking replenishment cases (R1–R3).

Repack picking locations are fixed; reserve pallet and hand-stack picking/reserve locations are randomly assigned when either system is in use.

The picking-oriented system provides guidance for the processing of receipts (B4) and minimal instructions for the movement of receipts to repack picking locations (B1A, B1B) and hand-stack storage (B2A, B2B). It provides no instructions for the movement of pallets (B3, R4, R5); the documents that control these movements are prepared manually.

When a picking-oriented system is used, the computer maintains for each SKU:

1. A total inventory.
2. A repack picking position inventory.
3. An inventory for each random hand-stack location.

It assumes that all the palletized inventory of an SKU is located in its fixed full-case picking position whether it is at a belt or in the pallet racks. This position is replenished by a manual system.

---

* References in parentheses are to movements shown in Fig. 13.4.1.

The total system provides detailed instructions for processing receipts (B4) and the movement of receipts to repack picking positions (B1A, B1B) or to random hand-stack storage (B2A, B2B). It provides documentation for the movement of pallets from receiving to reserve storage (B3), the replenishment of positions where cases are picked from pallets (B4), and the removal of empty pallets from these positions (B5). This system introduces the random assignment of positions for pallets from which cases are picked.

It maintains the inventory in every storage and picking location in the warehouse. This requires, in addition to the inventories maintained by a picking-oriented system, the inventory and vendor case pack of the SKU on the pallet:

1. In each position where cases are picked from pallets.
2. In each reserve pallet location.

Typical documents required by each system to perform each move in Fig. 13.4.1 and to verify inventories are indicated in Fig. 13.4.2.

Emerging computer applications publicized to date have been limited to the storage and picking of pallet loads (B3, P5) and repack (B1A, P1) and the sortation of repack (S1). They can, however, be expanded to eliminate many of the documents required by a total system.

Prerequisites to All Three Systems

All prerequisites (except item weight and cube) must be satisfied before any one of the three systems can be implemented. Order entry must be cut off before the picking shift to implement the shipping and operations planning and manpower estimating applications of the computer that are common to all three systems.

Planning Shipping Operations: All Systems

After order entry cut-off time, each line on all orders to be filled and shipped during the next working shift is processed past the item file to:

1. Eliminate items not stocked.
2. Determine the type of pick (repack, full case, or nonconveyable) required.
3. Determine the number of picking units ordered and the number that can be picked.
4. Calculate the weight and cube of each picked line if that data is available by picking unit.
5. Determine the number of totes (or cartons) of repack from the weight and cube data or an average number of lines per tote or repacked carton.
6. Determine the freight and DOT classification of each piece to be shipped.
7. Store the information necessary to print all documents or transmit all instructions.

The computer then prints a route summary (Fig. 13.4.3) and a route analysis (Fig. 13.4.4) for each route.

The "Stops," "Lbs," and "Cube" columns of the route summary enable warehouse personnel to determine which captive delivery vehicles are overloaded and which can transport more merchandise. Stops are transferred from one route to another and additional routes are created as necessary to comply with the weight, cube, and stop capacity of each vehicle; stop numbers are revised to minimize vehicle travel while complying with customer delivery time restrictions. All changes are recorded on the route analysis.

The changes are input; the computer prints a revised route summary and route analysis for review. This interactive process is repeated until all routes are acceptable.

Planning and Scheduling Warehouse Operations and Manpower: All Systems

Stops are then assigned to short-interval schedule periods of "waves" manually or by an algorithm. The objectives of either assignment method are:

1. To restore order integrity on the shipping dock with a minimum of personnel and errors.
2. To minimize the picking operation delays that occur when the dock falls behind schedule.

Attaining these objectives depends on the method of assigning stops to waves, the number of spurs per shipping door, and the length of these spurs. Typically a number of trucks are loaded simultaneously, one per shipping door. The initial stop or group of stops to be loaded in each truck is assigned to the first wave, the next stop or group of stops is assigned to the second wave, and so

Movement	Order Integrity Destroyed and Restored: Picking Oriented System	Order Integrity Destroyed and Restored: Total System	Order Integrity Maintained
	PLANNING SHIPPING AND WAREHOUSE OPERATIONS		
S1, S2	Route Summary (Fig. 13.4.3) and Route Analysis (Fig. 13.4.4)		
P1–P6	Wave Analysis and Wave Summary		
	PICKING REPACK FOR SHIPMENT		
P1	Pick List plus Man-Readable (MR) Label (Fig. 13.4.7)	Pick List plus MR Label (Fig. 13.4.7) or Man-Readable and Scannable (MR&S) Label (Fig. 13.4.14)	Pick List, Sometimes with MR Label; Otherwise Stencil
	PICKING NONCONVEYABLES AND PALLET LOADS FOR SHIPMENT		
P5, P6	—MR Labels in Bulk or Order Pick Sequence Followed by Trailer Labels— (Fig. 13.4.8)	(Fig. 13.4.15)	Pick List
	PICKING CONVEYABLE FULL CASES FOR SHIPMENT AND REPACK REPLENISHMENT		
P2, P3, P4, R1, R2, R3	MR Labels in Bulk Pick Sequence Followed By Trailer Labels (Fig. 13.4.8)	MR or MR&S Labels in Bulk Pick Sequence Followed by Trailer Labels (Fig. 13.4.15)	Pick List
	Trailer Label Control Sheet (Fig. 13.4.16)		No Document
	ORDER ASSEMBLY AND SHIPPING		
S1	Labels for Merchandise Movements P1 through P4		Route Manifest; Manual Bill of Lading
S2	Route Manifest (Fig. 13.4.5), Nonconveyable and Full Case Manifest (Fig. 13.4.6), Labels for Merchandise Movements P1 through P6; Bill of Lading		

		REPLENISHMENT OF FULL CASE AND NONCONVEYABLE MERCHANDISE PICKING POSITIONS WITH PALLETS	
R4	Pallet Tag (Fig. 13.4.9)	Let Down List (Fig. 13.4.12)	No Document
R5	No Document	Empty Pallet Pick-up List (Fig. 13.4.13)	No Document
		RECEIVING	
B4	Purchase Order Copy Showing Fixed Picking Locations; Misc. Tie/High etc. Listings	Receiving Report (Fig. 13.4.10)	Purchase Order Copy Showing Fixed Picking Locations
		STORING RECEIPTS	
B1A	Manual or Computer Prepared Labels (Page 26)	Receiving Report (Fig. 13.4.10)	Manual or Printed Label
B2A	Pallet Tag (Fig. 13.4.9)	Manual Hand Stack Storage Tag	
B1B	Sorting Spur and Repack Picking Position Written on Label Affixed to Each Case or Tote	MR & MR&S Label Showing Sorting Spur Only on Each Case; Receiving Report (Fig. 13.4.10)	Receipts Move to Stock via Mobile Equipment
B2B	Pallet Tag on One Case; Spur Number Written on Label Affixed to Each Case	Reserve Storage Tag (Fig. 13.4.11) on One Case; MR or MR&S Label Showing Sorting Spur Only Affixed to Each Case	
B3	Manually Prepared Pallet Tag (Fig. 13.4.9)	Reserve Storage Tag (Fig. 13.4.11)	No Document
		INVENTORY APPLICATIONS	
		Reports Adapted to Each Warehouse	
		Receipt Placement Verification/ Walk-Off List (Fig. 13.4.17)	
		Inventory Verification/Cycle Count/ Periodic Inventory List (Fig. 13.4.18)	

Fig. 13.4.2 Documents provided to systems that rely on printed documents.

PICK DATE 8/20/82 ROUTE SUMMARY 12:59:30 PAGE 1

ROUTE	STOPS	LBS	CUBE
201	10	18,332	902
202	17	34,271	1,275
203	12	27,510	.
204	17	.	.
.	.	.	.
.	.	.	.
.	.	.	.
.	.	.	.
209	.	.	.
TOTAL	127	376,271	12,327

Fig. 13.4.3 Route summary.

on. There are usually two spurs (A and B) per door. Conveyable full cases and totes or cartons of repack picked during wave 1 are sorted to spur A. They are loaded in the truck or sorted from the spur to separate pallets for multiple stops, checked and loaded. While these operations are underway, pieces generated during wave 2 accumulate on spur B. While the pieces on spur B are being processed, the pieces from wave 3 accumulate on the empty spur A. Nonconveyable pieces and pallet loads picked and delivered to the dock, usually a wave early, are merged with the conveyable pieces.

Ideally each spur would be of sufficient length to hold all of the conveyable pieces sorted to it during one wave. That length is, however, usually excessive because the number of conveyable cases for some stops is very large. To make reasonable length spurs possible the computer system must be able to assign the cases for one stop to successive waves.

It is possible to sort conveyable pieces to pallets for stops when there is only one spur per door. Pallets are provided initially at each spur for the stops comprising waves 1 and 2. When all pieces for stops assigned to wave 1 have been transferred from the spur to the appropriate pallets, they are removed and replaced by pallets for wave 3 stops. This operating method should be used only when the number of stops per wave is two or three, because errors when sorting pieces from the conveyor to the pallets will increase as the number of stops increases.

Increasing the number of waves per day will reduce the number of stops per spur per wave and, consequently, the number of sorting errors. If carried too far, this practice may reduce the number of cases bulk picked per wave so much that case picking productivity decreases substantially.

When it is necessary to assign a route or routes (each comprised of a number of stops) to each wave, the stops on the route or routes that comprise wave 1 are assigned to the spurs at successive doors. The stops on routes assigned to wave 2 are allocated to the same spurs in the same manner. The number of stops assigned to spur during a wave should recognize the sorting problems described previously. Pallets of merchandise ready for loading by stop are moved to the door at which the truck being loaded is located.

However they are developed, the stop assignments to waves are input to the computer. At this time route assignments to spurs are also input if they have not been designated permanently. The computer prints:

1. A work period or "wave" analysis identical in format to the route analysis but summarizing the same information by work period and showing the spur to which each stop assigned to each wave will be sorted.
2. A work period or "wave" summary that:
   a. Lists for each wave, by operation, the work load and the labor-hours required to complete it.
   b. Totals these figures for the shift.

The computer also calculates and prints a wave schedule in which working hours per wave are proportional to the repack lines per wave. Such a schedule maintains a uniform repack picking work force because that operation usually requires the largest component of the warehouse work force and is most likely to fall behind schedule.

When the number of conveyable totes or repack cartons and cases to be shipped per day approaches the capacity of the conveying and sorting system, wave schedules must be based on conveyable pieces rather than repack lines.

In any case, adjusting the assignment of stops to work periods, inputting these adjustments, and reprinting both documents may be necessary to achieve an operating plan that will maximize productivity. As soon as it has been established, the computer:

PICK DATE 8/20/82 — ROUTE ANALYSIS, ROUTE 201 — 13:01:50 — PAGE 1

ROUTE /STOP	SPUR NO.	INVOICE NUMBER	CUSTOMER NAME AND ADDRESS	RPK LINES	NON CONV LABELS	FULL CASE LABELS TO BELT	FULL CASE LABELS TO CART	FULL CASE LABELS FROM HAND STCK	FULL CASE LABELS TOTAL	FORK LIFT PICKS	R/S TOTALS LINES	R/S TOTALS LBS	R/S TOTALS CUBE
201–39		1739		119	21	33	9	40	82	0	191	4,175	270
		1792		30	0	1	0	3	4	0	34	100	15
201–27		1810		18	10	5	1	9	15	0	34	3,913	170
201–24		1547		83	19	14	7	10	31	0	126	2,566	155
201–17		1654		13	3	2	0	5	7	0	21	299	25
201–13		1712		267	8	18	9	17	44	0	315	1,305	83
201–09		1779		127	8	6	5	4	15	0	140	793	55
201–07		1740		0	0	1	0	53	54	7	40	1,575	39
201–04		1727		0	0	6	1	17	24	0	18	912	32
201–03		1607		0	0	5	0	7	12	0	10	320	37
201–02		1699		0	0	2	0	11	13	0	9	473	21
ROUTE TOTALS				657	69	93	32	176	301	7	838	18,332	902
ROUTE MAN-HOURS				13.14	2.35	1.56	0.54	2.99	5.09	0.47		TOTAL	21.05

Fig. 13.4.4 Route analysis.

1. Assigns a wave number and a revised route and stop number—if one was input—to each document lying in memory.
2. Prints, for each captive truck route, each common carrier, UPS, and so on, a route manifest (Fig. 13.4.5) and a nonconveyable and full case route manifest (Fig. 13.4.6) (or one document combining the information on both) for the picking-oriented and total systems.
3. Prints, as necessary, individual shipment and truckload bills of lading which indicate pieces and weight shipped by freight and DOT classifications.

At this point the computer has provided:

1. A schedule for operations and an estimate of the manpower by wave required to complete it. If this information is based on reasonably accurate productivity standards for each operation, it provides a basis for controlling labor costs and for measuring one aspect of each supervisor's managerial competence.
2. Documents or information for checking merchandise before loading, for obtaining missing full cases and nonconveyables and for delivering the picked merchandise one wave at a time.

The next step is printing the warehouse operating documents required by either system one wave at a time.

Picking-Oriented Systems

Repack Picking Document. Figure 13.4.7 shows a two-copy, post-billed, location-sequenced, broken-case picking document (P1) like those found in warehouses using picking-oriented systems. This document maximizes picker productivity because:

1. Order header information cannot be confused with picking instructions.
2. The instructions for picking each SKU are provided on one line in the sequence required for efficient order picking.
3. The warehouse item number, which does not appear on much of the merchandise in the warehouse, need not be read by the picker but is adjacent to any recorded picking exceptions to expedite their input.
4. Cost and retail price information used by the customer hampers neither the picker nor the control clerk responsible for entering exceptions.
5. There is adequate space between lines.

The document also shows the wave during which the order should be picked. This information is essential for efficient warehouse operation.

The document in Fig. 13.4.7 lists the merchandise that should fill two totes, based on a predetermined average number of lines per tote, and provides a pressure sensitive label for each one. The information on each label also prints on the backing which is perforated to provide two additional, glueable labels when necessary. Each label includes the picking document page number. Labels may also be printed separately from picking documents but this practice decreases productivity.

One copy of this document is a packing slip; it is sent to the shipping dock. The other copy is returned to the control clerk who retains both copies of additional pages on which appear lines specifying full cases, nonconveyable, and pallet load picks. The clerk revises them as necessary from returned picking labels. On the last page of the document appear all items ordered but not listed for picking and why they could not be picked. The packing list copy of each of these pages is also sent to the dock. Thus the complete packing list provides a disposition for every item ordered.

The line below the picking instructions (Fig. 13.4.7) provides the information which enables the control clerk to verify that all pages of each order have been processed for picking adjustments.

Picking Labels. Figure 13.4.8 shows full-case and nonconveyable picking labels of the type used in a picking-oriented system when sortation is manual or key coded. They are used:

1. To pick, and to identify for sorting, conveyable cases for shipment to customers (P2–P4) or for the replenishment of repack picking positions (R1–R3).
2. To pick pallet loads (P5) and nonconveyable items (P6).
3. As shipping labels.

Labels are printed in location sequence within work period for bulk picking. Labels for picking cases and nonconveyables from different warehouse areas (P2–P4, P6, R1–R3) are printed in separate groups each followed by a trailer label (Fig. 13.4.8C). A label for picking a full pallet (P5) specifies

ROUTE MANIFEST 01/28/82 ROUTE 000

DATE PAGE 1

DRIVER	TRUCK	ASSEMBLER
DEPART TIME	RETURN TIME	LOADERS
MILEAGE OUT	MILEAGE IN	% OF LOAD

WAVE	STOP	CUST#	CUSTOMER	CITY & STATE	PALLETS	REPACK BOXES	TOTES	CASES	TOTAL BOXES, TOTES CASES	NON-CONV. LBLS	TOTAL
0001	001	093302	WIGHTMAN CRANE STUART	SAN FRANCISCO, CA.	1/10		5	50	55	24	89
0001	002	024141	D & D SPORTING GOODS & HDWE	OAKDALE CA.			1		1		1
0001	003	086561	UMPQUA BLDG HDWE	REEDSPORT ORE / ONC, /.			2	2	4	3	7
0001	004	025148	EASTSIDE HDWE DBA MEYERS HDWE	MARYSVILLE, CA.			1		1		1
0001	005	093666	WIMER CITY HALL FEED & HDWE	ROGUE RIVER, OR.			1		1		1

Fig. 13.4.5 Route manifest (Courtesy of California Hardware Company).

10/06/82 08:56:58			CALIFORNIA HARDWARE COMPANY NON-CONVEYABLE AND FULL CASE MANIFEST REPORT ROUTE 002						PAGE: 2
WAVE	STOP	ORDER #	CUSTOMER	TYPE	PRODUCT DESCRIPTION	ITEM #	CASES OR BUNDLES	UNITS	LOCATION
0004	02A	100501290	TAYLOR LUMBER & HDWE	CS	41276 40A IF G.E LAMP 24PKG.		4	96	M-32-97-6-1
0004	02B	100656740	TAYLOR LUMBER & HDWE		1/2 ALUM RW FLEX CONDUIT		4	400	S-40-79-1-3
0004	05A	100501780	IRVINE CO AGRICULTURAL DIV	CS	731 EVEREADY LANTERN BATT 6V		1	6	K-37-64-2-1
0004	06A	100657690	CLARK DYE HARDWARE	CS	DA3000 MAKITA 3/8"ANGLE DRILL		3	3	N-23-90-1-3
0004	06A	100657690	CLARK DYE HARDWARE	CS	2400B MAKITA 10"MITRE SAW		1	1	A-80-04-2-0
0004	06A	100657690	CLARK DYE HARDWARE	CS	2401B MAKITA 10"MITRE SAW		3	3	N-25-73-2-2
0004	08A	091756561	MORROW MEADOWS	CS	D8340 1/4X4 RH. HD. TOGGLE 50/BX		1	8	L-39-61-3-1
0004	09A	100656650	RICOH ELECTRONICS INC	CS	40015(40-16) 12 OZ WD-40 PREV		1	12	D-42-91-2-3
0004	09A	100656650	RICOH ELECTRONICS INC	CS	40015(40-16) 12 OZ WD-40 PREV		3	36	E-42-75-3-3
0004	10A	100500830	LOS ANGELES TIMES		B-616-36 SLEDGE HANDLE-36"		1	2	S-67-31-1-1
0004	12A	092301123	ABC LUMBER CO		322 SP57/5C R5 WTR HTR BL H92		1	4	S-31-32-1-1
0004	12B	100655080	ABC LUMBER CO		371 5/8X18 CARPNTR WRCKING BAR		1	3	S-61-39-1-2
0004	12B	100655080	ABC LUMBER CO	CS	CW-50 OXYGEN CYLINDER		2	12	L-33-62-3-2
0004	12B	100655080	ABC LUMBER CO		SC311-US3 1 1/2X30 CONT HINGE		1	5	S-64-43-2-2
0004	12B	100655080	ABC LUMBER CO	CS	R7-22D RUBBER STRAP 12/DISPLAY		1	1	K-30-65-2-2
0004	12B	100655080	ABC LUMBER CO	CS	R7-31D RUBBER STRAP 12/DISPLAY		1	1	K-35-83-2-1
0004	12B	100655080	ABC LUMBER CO	CS	R7-9D RUBBER STRAP 12 PER DISP		1	1	K-33-62-1-3
0004	12B	100655080	ABC LUMBER CO	CS	R7-15D RUBBER STRAP 12/DISPLAY		1	1	L-37-66-4-2
0004	12B	100655080	ABC LUMBER CO	CS	9512 B&D CAR LITE 12VOLT		2	2	N-23-72-1-2
0004	12B	100655080	ABC LUMBER CO		28 HANDLE FOR #26 SANDER 48"		1	2	S-58-42-1-1
0004	12B	100655080	ABC LUMBER CO		7360 HH 12 LB SLEDGE HAMMER		1	2	S-65-35-1-3
0004	12B	100655080	ABC LUMBER CO		JTM 60" LUMA THREAD HANDLE		1	6	S-67-15-2-1
0004	12B	100655080	ABC LUMBER CO		JT 1 1/8X60 PS-THR HARDWD HDLE		1	8	S-67-17-2-1
0004	12B	100655080	ABC LUMBER CO		B-8-28 BOYS SNGLBIT AXE HDLE		1	4	S-67-37-1-3
0004	12B	100655080	ABC LUMBER CO		1/4-20X36"ZP REDI ROD TH 10/PK		1	1	S-67-44-2-2
0004	12B	100655080	ABC LUMBER CO	CS	684433 14ZP SI JACK CH 200'RL		1	1	L-30-80-3-3
0004	12B	100655080	ABC LUMBER CO	CS	684434 12ZP SI JACK CH 200'RL		1	1	A-72-03-1-0
0004	12B	100655080	ABC LUMBER CO	CS	574 SKIL 7 1/4"CIRCULAR SAW		6	6	K-35-63-1-2
0004	12B	100655080	ABC LUMBER CO	CS	MW-210 MICROWAVE STARTER SET		1	1	N-19-77-2-2
0004	12C	100655090	ABC LUMBER CO	CS	65-6' MASTER SELF-COIL CABLE		1	12	L-32-74-4-3
0004	12C	100655090	ABC LUMBER CO		694-36IN HALL MACK TOWEL BAR		2	6	S-32-23-2-3
0004	12C	100655090	ABC LUMBER CO	CS	MN1300B2 D BATTERY <12>		1	2	L-40-76-4-3
0004	12C	100655090	ABC LUMBER CO	CS	MN2400B2 DURACELL BATTERY<12BX		1	2	K-32-97-1-2
0004	12C	100655090	ABC LUMBER CO		10472 F40CW GE LAMP 6PKG		8	48	S-62-42-1-3
0004	13A	100657570	DE NAULTS HDWE -HOME CTR #2	CS	90550 50PC SKOKIE GRINDING WHL		1	1	L-32-81-3-3
0004	13B	100657580	DE NAULTS HDWE -HOME CTR #2		60" FIBREGLASS SCREEN 18X16		1	1	S-60-43-1-3
0004	13C	100657590	DE NAULTS HDWE -HOME CTR #2	CS	FC4101 3M SCOTCHGARD 16 OZ		1	12	E-42-85-4-3
0004	13C	100657590	DE NAULTS HDWE -HOME CTR #2	CS	HDW-12 PUMICE SCOUR BAR		1	12	K-39-71-1-2
0004	13C	100657590	DE NAULTS HDWE -HOME CTR #2	CS	8OZ SILVER CLEANER EZ EST DIP		1	12	L-29-77-4-1
0004	13C	100657590	DE NAULTS HDWE -HOME CTR #2		8280 CAMPER/TRAILER BROOM 6 PK		1	6	S-31-41-2-2
0004	13C	100657590	DE NAULTS HDWE -HOME CTR #2		109 QUICK CHANGE MOP HANDLE HD		1	6	S-55-15-1-1
0004	15A	892800250	BECKMAN INSTRUMENTS	CS	155A FARBER 55 SUP S S URN		1	1	M-32-74-6-3

Fig. 13.4.6 Nonconveyable and full case manifest (Courtesy of California Hardware Company).

```
WAVE: 02                       *** TREE OF LIFE, INC. ***                        DATE: 08/15/82

ROUTE: 02 KNOXVILLE TRUCK           P A C K I N G   S L I P
 STOP: 450                          *****FLOW RACK*****          ORDER#      PAGE:    3

CUST: G640            SHIP TO: GOOD FOOD STORE FRONT             001354
                               403 EAST FLEMING DR.
                               MORGANTON       ,NC  28655

****************************************************************************************************
 LOCATION        DESCRIPTION                    SIZE  PK US  ORD PCKR  SHPD   ITEM#        COST RETAIL
----------- ----------------------------- --- --  --- ----- ----- ------       ------- ------

A  05-09-3-5 A.H.BONUS SUP.AC.PL.500MG300'S  12 EA    1 (   ) (   )   4216-8     9.210  15.35

A  05-18-4-1 CASHEW PIECES LARGE FANCY5 LB.   6 01    1 (   ) (   )  16856-7     9.150   1.83

A  05-22-1-4 TOL CARETN.CAROB DROPS    5LB.   6 01    1 (   ) (   )  15556-4    13.650   2.73

A  05-24-2-3 D.S.WH.WHT.PASTRY FLOUR  2 LB.  12 EA    1 (   ) (   )   7850-1     0.913   1.37

A  05-24-4-3 TOL CARAFFECTION RAISIN   5LB.   6 01    1 (   ) (   )  15596-0    14.150   2.83

A  05-25-3-1 TOL EAST COAST APPLE JCE. GAL.   4 EA    2 (   ) (   ) 232982-9     3.460   5.19

    *** LAST PAGE OF ZONE "A" (*****FLOW RACK***** ) OF "ABCDEFG   J "     LINES=  22 ***


  0   2   0   2   4   5   0           0   2   0   2   4   5   0

SHIP TO:       08/15/82        WAVE ORDER#   SHIP TO:        08/15/82        WAVE ORDER#
G640          REPACK "A"        02  001354   G640           REPACK "A"        02  001354
GOOD FOOD STORE FRONT                        GOOD FOOD STORE FRONT
403 EAST FLEMING DR.           PAGE LABEL    403 EAST FLEMING DR.            PAGE LABEL
MORGANTON       ,NC  28655       3    1      MORGANTON       ,NC  28655        3    2
```

Fig. 13.4.7 Repack picking document with man-readable tote or carton labels (Courtesy of Tree of Life Inc.).

the full-case picking position; the location of the pallet to be picked is obtained from a manual file of reserve pallet position inventories. Each label for a case to be shipped shows, in matrix print, the shipping spur, route, and stop number.

The issue of labels for an area to a picker is recorded on a label control sheet similar to the example in Fig. 13.4.16. As soon as picking is completed, that individual returns the trailer label and unpicked labels to a control clerk who:

1. Records their return on the label control sheet.
2. Uses the unpicked labels as a source document:
   a. For revising route manifests, and for correcting the quantities picked on packing list pages listing full cases only.
   b. Inputting picking corrections to revise post-billed invoices.

Repack Picking Position Replenishment. Typically, 75–80% of the repack picking positions are replenished by receipts. There is no reserve inventory of these items; the picking position houses the entire warehouse inventory. For the other 20–25% of the SKUs picked as repack, the computer maintains not only a warehouse inventory for the SKU but also a subinventory for its repack picking position. A reorder point triggers replenishment of the latter; the inventory capacity of the position determines the maximum number of vendor pack cases comprising the desired replenishment. The actual replenishment quantity may be limited by the reserve inventory on hand.

The computer stores the item identification, number of replenishment cases, case pack, and both picking locations as a message. It sorts accumulated messages by case picking location, and prints each one on a replenishment case picking label (Fig. 13.4.8). Instead of the shipping spur, route, and stop number, this label shows the replenishment spur (which indirectly identifies the aisle location of the pick position), the shelving section level, and position on the level of the repack picking position being replenished.

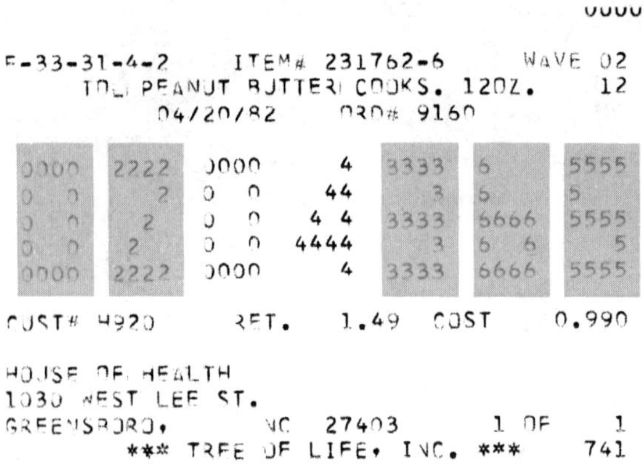

```
                                                         UUUUL
F-33-31-4-2       ITEM# 231762-6          WAVE 02
     TOL PEANUT BUTTER COOKS. 12OZ.          12
          04/20/82       ORD# 9160

0000  2222  0000      4  3333  6     5555
0  0     2  0  0     44     3  6     5
0  0    2   0  0    4 4  3333  6666  5555
0  0   2    0  0   4444     3  6  6     5
0000  2222  0000      4  3333  6666  5555

CUST# H920       RET.    1.49   COST     0.990

HOUSE OF HEALTH
1030 WEST LEE ST.
GREENSBORO,         NC   27403         1 OF    1
       *** TREE OF LIFE, INC. ***        741
```

(A)

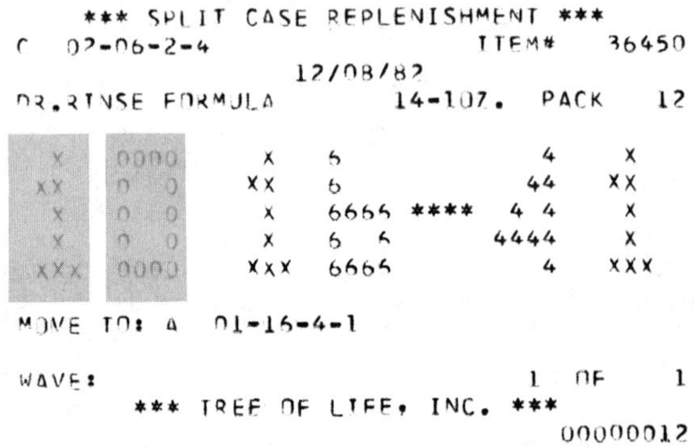

```
      *** SPLIT CASE REPLENISHMENT ***
C   02-06-2-4                   ITEM#    36450
                 12/08/82
DR.RINSE FORMULA           14-10Z.    PACK    12

  X   0000      X   6               4     X
 XX   0  0     XX   6              44    XX
  X   0  0      X   6666  ****    4 4     X
  X   0  0      X   6  6         4444     X
 XXX  0000     XXX  6666            4   XXX

MOVE TO: A   01-16-4-1

WAVE:                              1  OF     1
       *** TREE OF LIFE, INC. ***
                                        00000012
```

(B)

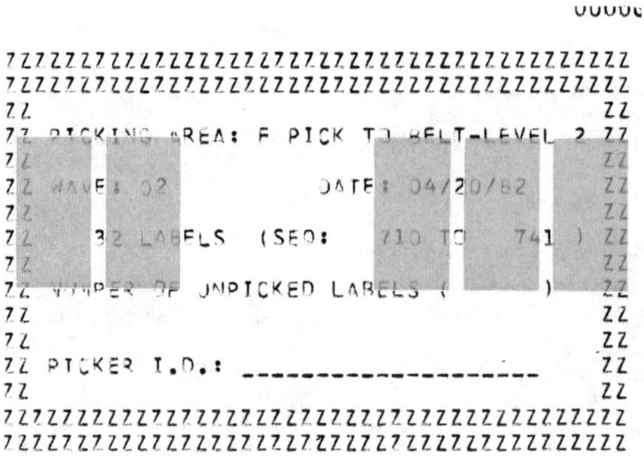

```
                                                  UUUUL
ZZZZZZZZZZZZZZZZZZZZZZZZZZZZZZZZZZZZZZZZZZZZZZ
ZZZZZZZZZZZZZZZZZZZZZZZZZZZZZZZZZZZZZZZZZZZZZZ
ZZ                                          ZZ
ZZ PICKING AREA: F PICK TO BELT-LEVEL 2     ZZ
ZZ                                          ZZ
ZZ WAVE: 02               DATE: 04/20/82    ZZ
ZZ                                          ZZ
ZZ    32 LABELS    (SEQ:   710 TO   741 )   ZZ
ZZ                                          ZZ
ZZ NUMBER OF UNPICKED LABELS (          )   ZZ
ZZ                                          ZZ
ZZ                                          ZZ
ZZ PICKER I.D.:  ____________________       ZZ
ZZ                                          ZZ
ZZZZZZZZZZZZZZZZZZZZZZZZZZZZZZZZZZZZZZZZZZZZZZ
ZZZZZZZZZZZZZZZZZZZZZZZZZZZZZZZZZZZZZZZZZZZZZZ
```

(C)

Fig. 13.4.8 Man-readable full-case and nonconveyable picking labels (Courtesy of Tree of Life Inc.). (*A*), conveyable full case and nonconveyable picking label; merchandise to be shipped (Picking position F-33-31-4-2; shipping dock spur 02; Route 04; Stop 365). (*B*), conveyable full case picking label; repack replenishment (case picking position C-02-06-2-4; repack replenishment spur 10; repack picking position A-01-16-4-1). (*C*), trailer label.

The case picker picks replenishment cases in the same manner as cases to be shipped; the repack stockman accumulates cases on the spur by repack picking location and stocks each one.

The computer can schedule the picking in two ways:

1. All cases to replenish merchandise picked on Monday during a separate wave prior to the start of the Tuesday picking shift or prior to the Monday shift if necessary to provide inventory for Monday picking.
2. One or more waves prior to the one during which the cases can be stored in the repack position. This schedule requires a replenishment picking wave before the picking shift begins.

When the first schedule is followed, the initial processing of order lines described earlier is the only one necessary. The second schedule reduces the inventory in the repack picking positions for fast moving items but at the cost of processing each order line twice. The first processing is as described earlier except that the storing of information to print all documents is omitted. The second processing occurs after a satisfactory wave schedule has been established. At this time, the processing is repeated and all document-printing information is stored complete with the correct wave numbers.

To guide the resizing of repack picking positions, the computer prints, on call, a repack replenishment frequency report which lists each SKU replenished since the last report was printed, the frequency of replenishment, the number of cases supplied, and the date of the previous report. The information is listed in declining replenishment frequency.

Receiving and Stocking Information. The computer assistance to receiving and stocking consists of a receiving department copy of the purchase order which shows the permanent full-case picking position if there is one and the repack picking position if no full-case picking position exists. This document is supplemented by periodic computer reports which list, for each SKU, these fixed picking positions, random hand-stack locations, inventories, case packs, pallet tie and high, storage area restrictions, and vendor case dimensions, cube, and weight. Pallet ties and highs and storage area limitations are often left to warehouse personnel rather than printed.

Receiving and Storing Pallet Loads. The storage of pallet loads of conveyable vendor pack cases and the replenishment of picking positions where cases are picked from pallets utilize a two-part pallet tag (Fig. 13.4.9). It is completed, except for the reserve storage location, by receiving department personnel who obtain all necessary information including the picking location from the receiving report and computer listings. A forklift driver records on the tag the random reserve pallet location in which he stores the load (B3), leaves one copy on the load, and places the other copy behind the other tags for previously received merchandise in a sleeve adjacent to the picking position.

Replenishing Fixed Full-Case Picking Positions Housing Pallets. When the inventory on a pallet in a picking position is reduced to two or three cases or nonconveyable pieces, a picker transfers them from the pallet to the floor. A forklift driver, noting the empty pallet, removes the oldest tag from the sleeve, picks up the empty pallet, and delivers it to receiving (R5). He then transfers the pallet load of merchandise in the reserve location shown on the tag to the pick position (R4). The possibility that the replenishment will be too late to provide needed cases is minimized by assigning two adjacent pick positions to one SKU and treating them as one location.

Receiving and Stocking Repack in Fixed Picking Positions. Receipts of conveyable full cases or less-than-case lots in a tote to be stored in a repack picking position are identified in the receiving area:

1. For movement (B1A) via mobile equipment by affixing a manually or computer-prepared label showing the item number, description (if computer prepared), number of cases or pieces, and location (a) to one case of the receipt or (b) to the bundled loose pieces before placing them in a tote designated for loose receipts to be stored in a range of pick locations.
2. For movement (B1B) via conveyor and sorter by affixing to each case or tote an additional written or computer-prepared label bearing the replenishment spur number to which it must be sorted.

Receiving and Stocking Full Cases in Random Hand-stack Locations. Random storage of the hand-stack inventory of conveyable cases reduces stocking costs, insures FIFO movement of reserve inventory, utilizes hand-stack space more fully and familiarizes warehouse personnel with the concepts and requirements of random reserve inventory storage.

The receiver uses a rule, such as "20 cases or one-third of a pallet maximum," to determine if the receipt of an SKU should be stored in the hand-stack racks. To store a receipt there randomly (B2A, B2B) he completes a pallet tag similar to the one in Fig. 13.4.9 (but omits the picking position),

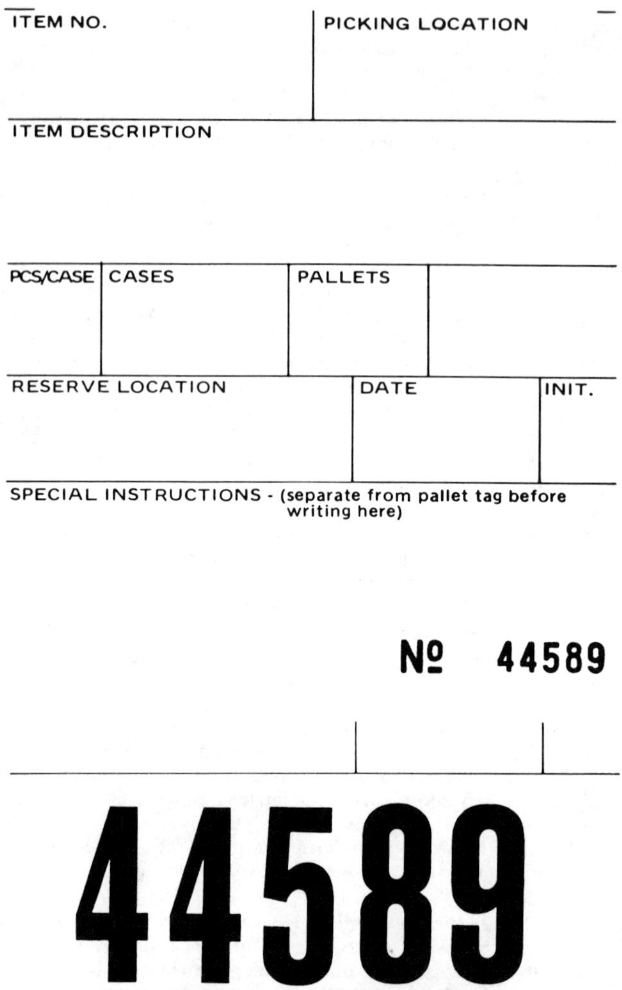

ITEM NO.			PICKING LOCATION	
ITEM DESCRIPTION				
PCS/CASE	CASES	PALLETS		
RESERVE LOCATION		DATE	INIT.	
SPECIAL INSTRUCTIONS - (separate from pallet tag before writing here)				

№ 44589

44589

Fig. 13.4.9 Manually prepared pallet tag.

writes "Hand Stack" on it, and attaches it to one of the cases each of which he also identifies—for movement—by item number or item number and spur number.

The stockman stores the merchandise delivered to him in any available location of sufficient size to hold the receipt (storage locations may be restricted to zones based on merchandise characteristics), records the location on the tag, leaves one copy on the merchandise, and returns the other to the receiving department. Receiving personnel record the receipt of the tag to verify that merchandise received was stored and located before forwarding the tag to data processing or using it themselves to update the inventory of the SKU by location and in total.

The computer files are augmented by a record for each random hand-stack location. Each of these records contains, as a minimum, the storage location number, the item number, the number of cases stored in the position, the vendor case pack, and storage date. The record may contain additional information such as the expiration date of salability and stockman identification. Successive records are chained to insure that the computer accesses them in FIFO order. If the merchandise is stored and also picked from a pallet, any random hand-stack locations are in addition to the permanent location of the pallet from which cases are picked.

To provide space for the storage of future receipts in hand-stack rack, the computer may assign the picking of cases to hand-stack locations first. As the inventory in one location is exhausted, the computer assigns picks to the next one. When all cases in hand-stack locations have been designated for picking, the computer assigns those remaining to be picked to the pallet position. As a result of this logic, cases for one customer may be picked from several locations.

Since the record of each random hand-stack location contains the vendor case pack, the computer can be programmed to adjust customer orders or replenishment requirements (and the repack picking position inventory replenishment) to the case pack stored—but only for cases picked from hand-stack locations. All cases picked from a pallet are assumed to be packed as shown in the item master file.

Because random hand-stack storage requires synchronization of the computer and warehouse inventories of an SKU and because it requires the reporting of the results of storage operations to the computer by warehouse personnel, random storage of merchandise in hand-stock locations should be limited initially to a few SKUs. As inventory verifications indicate satisfactory agreement between the computer and warehouse inventories, random storage may be expanded gradually to all items stored in the hand-stack racks.

Inventory Applications. When the computer system is picking oriented, the computer prints periodic reports showing for each SKU:

1. The total inventory of the SKU.
2. The inventory in each repack picking position and random hand-stack reserve storage/picking location.
3. The location where merchandise is picked from a pallet and the inventory there and in all reserve pallet storage positions.

The computer can provide only this information for any inventory verification triggered by a picking discrepancy, for any cycle counts, or for periodic inventory of the entire warehouse.

Total System

Operating Practices and Additional Computer Applications. Warehouse locations housing pallets holding the picking inventories of conveyable cases are randomly assigned in a warehouse using a total system. All other operating practices are identical with those in a warehouse using a picking-oriented system.

A total system maintains the total inventory of an SKU, its inventory in repack and random hand-stack locations, and its inventory in every reserve and picking location housing a pallet. The vendor case pack is shown for the merchandise in every full-case location. The pallet location inventories permit computer control of the pallet moves which are directed by the pallet tag and manually in a picking-oriented system. The case packs make it possible to consider all case pack changes when processing order lines or repack picking position replenishments.

Receiving and Stocking Information. When a total system is installed in a warehouse, the computer provides:

1. A receiving report that minimizes the necessity for reference to supplementary listings.
2. A storage tag that specifies the storage location for each pallet load.

The receiving report (Fig. 13.4.10) includes not only the item number, description (and quantity ordered if receiving is not "blind"), and fixed repack picking location but also as much of the following information as possible:

1. Specification of whether the order quantity of a vendor-pack full-case item will be stored on a pallet or in the hand-stack racks.
2. Palletizing instructions when necessary.
3. Any restrictions as to the reserve storage zone in which a pallet may be stored.
4. The numbers of the accumulation spurs on which repack and hand-stack receipts should accumulate for stocking.
5. Instructions to store a palletized receipt of vendor-pack full cases of nonconveyables in a picking position rather than a reserve position.
6. A request for weight and cube information and space to supply it when necessary.

The receiving report may list receipts to be stored in repack picking positions by spur so the merchandise can be received as it is being stocked. Thus this document can include complete instructions for receiving and stocking each SKU received.

Stocking Merchandise. To store pallet loads of vendor-pack full cases or nonconveyable merchandise (B3) receiving personnel select a two-part reserve storage tag (Fig. 13.4.11) imprinted with a location number from a file of such tags for the storage area specified on the receiving report. After

**** RECEIVING COPY ****

P.O.# 21840 REC.NO.

PO DATE: 31DEC82 DATE DUE: 14JAN83

BUYER: 04
VENDOR: AUDIO SPECIALISTS INC.
003848 2134 TRANS CANADA HWY. S.
MONTREAL PQ
H9P 2N4

IN044 PAGE 1

PRINTED
DATE: 06JAN83
TIME: 15:58:25

FREIGHT- PREPAID
TERMS- NET 30M

COMMENTS SUB TO CONFIRMA TION

ITEM#	DESCRIPTION	SIZE LINE	QTY ORDER	BACK ORDER	QTY REC	PACK WHS VNR	R F	C NC	SPUR	REPACK LOCATION	QTY	LANDED COST	CUBE - CM L * W * D	MIN/ MAX
503800-5	TDK AD-C60 CASS. TAPE	001	600				R	Y	4	1-02-03-2-5		[illegible]	18 12 8	50 250
503805-4	TDK AD C-90 CASS. TAPE	002	500				R	Y	4	1-02-03-2-2		[illegible]	18 12 8	50 250
503810-4	TDK D C-60 CASS. TAPE	003	400				R	Y	4	1-02-03-2-6		[illegible]	18 12 8	50 250
503815-3	TDK D C-90 CASS. TAPE	004	300				R	Y	4	1-02-03-2-3		[illegible]	18 12 8	200 400
503820-3	TDK SA-X C-60 HI-BIAS CAS S. TAPE	005	200				R	Y	4	1-02-03-2-4		[illegible]	18 12 8	50 250
503825-2	TDK SA-X C-90 HI-BIAS CAS S. TAPE	006	400				R	Y	4	1-02-03-2-1		[illegible]	18 12 8	50 250
503830-2	TDK SA-C90 HI BIAS	007	2,500				R	Y	4	1-02-02-1-5		[illegible]	18 12 8	50 250
503835-1	TDK SA-C60 HI BIAS TAPE	008	400				R	Y	4	1-02-03-1-3		[illegible]	19 12 8	50 250

Annotations: Case Pack; Units to Repack Location; If R, Picked as Repack or Repack and Full Case / If F, Picked as Full Case Only; If Y, Vendor Case is Conveyable / If N, Vendor Case is Non-conveyable

************ TOTAL ORDER ************

WEIGHT	CASES	VOLUME	LANDED COST
0	1,040	1.80864	[illegible]

RECEIVING COMPLETE
~~RECEIVING INCOMPLETE~~
ITEMS NOT ON P.O.

WRONG GOODS SHIPPED
~~OVERSHIPPED~~
P.O. # DIFFERENT

NO PACKING SLIP

SIGNATURE

MADE IN CANADA

Fig. 13.4.10 Receiving report (Courtesy of London Drugs Limited). Case Pack; Units to Repack Location; If R, Picked as Repack or

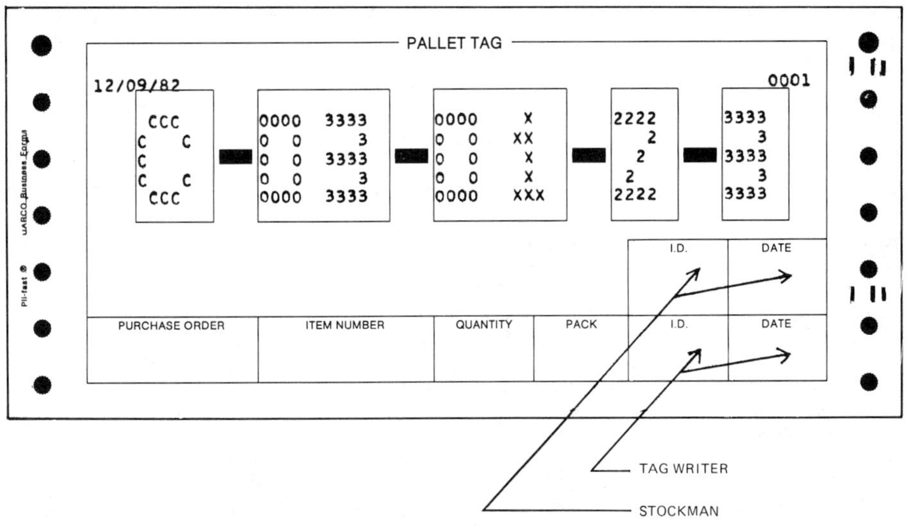

Fig. 13.4.11 Reserve storage tag (Courtesy of Tree of Life Inc.). Tag Writer; Stockman.

completing the tag, the individual affixes it to a pallet load and notes each location or the number of tags issued for pallet loads of any one SKU on the receiving report. A forklift driver stores the merchandise in the location shown on the tag and verifies that he has chosen the correct location by comparing the location number on the tag with that on the bar or floor below. He leaves one copy of the tag attached to the pallet, removes the other, initials it, and returns it to receiving. There it is the basis for input to the computer of the same inventory by location information maintained for hand-stack locations. A record, on the receiving report, of tags or locations issued permits remedial action if all tags issued are not returned.

Reserve storage tags are usually printed in batches on demand. Tag control is possible if the computer maintains a record of all locations for which tags were printed and records each tag as storage of merchandise in the location is input or when unused tags are returned to initiate the printing of a new batch.

Conveyable vendor-pack cases are randomly stored in hand-stack rack as described for the picking-oriented system except that:

1. Blank reserve storage tags are used instead of pallet tags.
2. When cases are conveyed to the hand-stack area, a man-readable or man-readable and scannable label indicating the sorting spur number is affixed to each case.

Recording, on the receiving report, the fact that a hand-stack tag was issued provides the information necessary to verify that a copy of each tag is returned.

When the appropriate sorting spur appears on the receiving report, a computer-prepared label showing the appropriate sorting spur can be affixed to each case or tote of loose merchandise intended for those locations. Repack can be simultaneously counted and stocked in the picking positions.

An on-line, real-time system for receiving and storage offers some important advantages over a batch system. When the quantity of an item received is input, the computer can:

1. Print the required number of tags each complete with the information that is handwritten when the computer system operates in the batch mode.
2. Account for all tags issued and prepare lists of tags issued but not confirmed as stored.
3. Provide up-to-date repack pick locations and sorting spurs.

When reserve storage tags are prepared on-line, the time previously required to prepare them can be devoted to checking the quantity of merchandise to be stored. The processing capacity of the computer and the programming must be such that receiving personnel are not delayed by slow computer response.

Computer Applications in Picking and Shipping. The computer applications of the picking-oriented system for picking and shipping are augmented by these additional applications in the total system:

1. The preparation of man-readable *or* man-readable and scannable labels.
2. The random slotting© or restricted random slotting© of pallet positions from which full cases are picked.†
3. The replenishment of these positions from reserve pallet inventories (R4).
4. The removal of empty pallets from these positions (R5).

These additional uses of the computer are possible because the number of cases picked justifies a scanner to control sortation and requires many pallet movements.

Full-Case Picking Position Replenishment. Since the inventory in cases on every pallet is known by the computer, the start-of-shift inventory in any pallet position from which vendor-pack conveyable cases or nonconveyable merchandise are picked is known. An additional pallet is transferred to another random pick position (B4) only when all the cases on the pallet already in the picking position have been reserved for picking *and* an additional case or cases is required. If the inventory on the pallet in a case picking position at the end of the picking shift will be zero, no pallet is moved to a picking position. Multiple pallet loads will be moved to picking positions when necessary.

The computer selects the pallet to be transferred from reserve storage to a picking position on an FIFO basis. It assigns a random picking location to this pallet from a list of open positions (random slotting©); it will also assign pick locations to pallets from one of the series of such lists for different picking position areas (restricted random slotting©) on the basis of limitations such as pallet size and load height, and merchandise flammability. When two pallet positions, one above another, are accessible from the floor or a walkway for picking cases, large or light cases may be assigned to upper-level positions and small or heavy cases may be assigned to lower-level positions if the item records in the computer are so coded. In a warehouse where multiple streams of cases, totes, and/or cartons from separate but identical case picking areas are merged into a single stream for key coding or scanning, it is possible to devise algorithms that consider the flow rate of each stream when assigning picking locations.

The let-down list (Fig. 13.4.12) is the document used to transfer pallets from reserve storage to picking locations (R4). It provides space for inputting corrections of pallet load inventories.

The computer usually schedules the let-down of pallets to pick positions just prior to the start of the picking shift. If a picking location is occupied or the merchandise on the pallet in the reserve location is not that specified on the let-down list, the forklift driver disregards the move instruction and reports the facts to his supervisor for remedial action.

Empty Pallet Pick-up. A forklift driver removes empty pallets from locations shown on the empty pallet pick-up list (Fig. 13.4.13) and transfers them to the receiving department (R5). High production rates are possible because the list is printed in location sequence and because the forklift operator picks up the first pallet, moves to the second location, drops the empty on the pallet there, picks up both pallets and repeats the process until he has a load of pallets limited by the height of the pick position or the stability of the load. If there is inventory on any pallet, the forklift driver records it on the document and delivers the merchandise to a specified location for subsequent processing. Any unexpected inventory such as this or any inability to let down a full pallet as scheduled triggers a check of the inventory of the SKU in all locations.

When pallets are random slotted in pick positions, the warehouse layout should provide pick positions in each area for the maximum projected number of SKU's picked in the area plus the projected maximum number of pallets to be let down into that area on any day. This number of positions is usually adequate because there is no warehouse inventory of some SKU's and no pallets of others in picking locations. At intervals the activity of each SKU in a picking position should be reviewed; pallets of out-of-season merchandise or merchandise not picked during 30 or 60 days should be returned to reserve storage.

If the number of picking positions is still inadequate, pallet pick-up and let-down can be scheduled by wave. When scheduling let-downs in this manner, the computer must recognize any repack replenishment case picking and allow a sufficient number of waves for picking and stocking the replenishment cases; such scheduling of replenishments requires that each order line be processed twice.

Picking Documents. A typical total system repack picking document complete with man-readable and scannable tote or carton label is shown in Fig. 13.4.14. If the anticipated contents of the tote or carton requires two document pages, the label on the second page is defaced during printing.

---

† "Random slotting" and "restricted random slotting" are terms copyrighted by Semco, Sweet & Mayers, Inc.

11/23/82
03:39:36

CALIFORNIA HARDWARE COMPANY
CONVEYABLE LET-DOWN REPORT
PRIORITY

PAGE 3

FROM LOCATION	TO LOCATION	DESCRIPTON	ITEM #	UNIT	QUANTITY	PACK	# OF CASES	PALLET SIZE
V-59-40-5-2	L-29-77-4-1	20X20X2 FURNACE FILTER	17335167	EACH	48	12	4	SM
V-60-20-4-2	K-39-69-2-1	24" FIBREGLASS SCREEN 18X16	22435036	CTN	20	1	20	SM
V-62-19-6-3	K-39-75-1-1	P-41-OC CORNING PETITE COVERS	30023006	BOX	100	4	25	SM
V-62-28-5-1	L-29-75-3-1	0686-CTG ACRYLIC LATEX-DRK BRN	37002656	EACH	300	12	25	SM
V-62-40-4-2	K-29-69-2-3	220-2"ALLPURPOSE GLUE 5CD/BOX	37025707	BOX	110	10	11	SM
V-63-23-6-3	K-33-83-1-3	1/4-20 PS HEX NUTS 100/BOX	19440908	EACH	57600	4800	12	SM
V-64-15-7-1	K-29-69-2-1	4MILX16' CLEAR POLY FILM	22430284	ROLL	35	1	35	SM
V-67-41-3-2	K-39-63-2-2	DCM15 DRIPCOFFEEMKR W/STARTER	29311255	EACH	10	1	10	SM
V-67-43-4-1	K-39-65-1-1	440-49 4PC MIX BOWL SET	30500359	SET	36	4	9	SM

Fig. 13.4.12 Let-down list (Courtesy of California Hardware Company).

11/23/82
04:48:28

CALIFORNIA HARDWARE COMPANY
EMPTY PALLET-PICKUP REPORT

PAGE 1

LOCATION	DESCRIPTION	ITEM #	UNIT	WAVE
D-42-73-1-3	1601 KRYLON	35803394	EACH	0004
D-42-79-2-1	204 1 QRT PLASTIC ROOF CEMENT	35518083	EACH	0004
D-42-83-2-3	32-QT CRYSTAL CLR FINISH RESIN	37023447	EACH	0004
D-42-93-2-3	750 SENTRY PRO-FUEL CYLINDER	26625137	EACH	0001
D-43-76-1-3	7710 13 OZ SPRAY RUSTOLEUM 910	35800143	EACH	0001
D-43-82-2-1	101 1 GAL ECONOMY ROOF COAT	35518059	EACH	0004
D-43-86-1-2	7776 QTS RUSTOLEUM OLD 412	35500024	EACH	0004
E-42-89-4-1	208 1 GAL WET SURF ROOF CEMENT	35518141	EACH	0004
E-42-91-3-3	7777 13 OZ SPRAY RUSTOLEUM7278	35800432	EACH	0004
E-42-93-3-1	7775 1/2 PTS RUSTOLEUM OLD 977	35500941	EACH	0004
E-43-80-3-3	NO 5 RED WD 1 QT BEHR PAINT	35719475	EACH	0004
E-43-82-4-3	10-QT POLY GLS ENAMEL WHITE	35550771	EACH	0004
E-43-90-3-3	7779 13 OZ SPRAY RUSTOLEUM 634	35800036	EACH	0001
F-42-73-5-2	8060 CLEAR SPRAY SHELLAC	35802941	EACH	0001
F-42-73-6-3	0387-SPRAY SPEED-E-NAML GLSBLK	35800903	EACH	0002
F-42-83-5-2	93 PT EXT SATIN VARATHANE	35714195	EACH	0004
F-42-89-6-1	GAL JASCO 103 PAINT REMOVER	36211050	EACH	0002
F-42-91-5-3	440 11 OZ. COVE BASE ADHESIVE	37050200	EACH	0004
F-43-78-6-1	PTS JASCO 101 PAINT REMOVER	36211035	EACH	0003
F-43-78-6-2	8074 SPRAY PRIMER RED OXIDE	35802776	EACH	0004
G-41-76-2-1	303 GAL ASPH DRVWY CRACK FILLR	36100055	EACH	0003
G-41-78-1-1	203 GAL COLD ROOF CEMENT	35510072	EACH	0004
G-41-79-2-3	7011 SPRY NAT LBR GRY MET PRMR	35809854	EACH	0004
G-41-83-1-3	208 1 GAL WET SURF ROOF CEMENT	35518141	EACH	0002
G-41-91-2-3	2081 13 OZ SPRAY RUSTOLEUM	35800267	EACH	0004
G-41-92-1-1	9612 FABSPRAY VINYL WHITE	35802859	EACH	0004
G-41-95-2-2	90 LG/SC CLEAR GLOSS VARATHANE	35714062	EACH	0004
G-42-74-1-3	0537 SPRAY EPOXY CLEAR	35802057	EACH	0001
G-42-86-2-1	40011(40-12) 9 OZ WD-40 PREV	20225710	EACH	0004
G-42-96-1-3	7434 13 OZ SPRAY RUSTOLEUM H3	35800333	EACH	0001
J-02-00-0-0	14X20X1 AGII FILTER	17335209	EACH	0003
K-29-83-1-1	1181 ALMOND RM DRAINER TRAY	33023276	EACH	0004
K-30-86-1-3	4CH10 4M10X25 CLEAR POLY FILM	22400055	ROLL	0003
K-30-87-2-3	N-2-1/2 RANGETOP SAUCEP CORNFL	29900990	EACH	0004
K-30-91-2-2	19001 ARMOR-ALL COUNTER DISPLY	20220067	DISP	0003
K-30-96-2-2	R7-31D RUBBER STRAP 12/DISPLAY	19925312	DISP	0002
K-32-66-1-1	7112 WHITE R-MAID SHPWER MAT	31702145	EACH	0004
K-32-71-1-2	A-2-8 SPICEOLIFE SAUCEPAN	29903861	EACH	0004

Fig. 13.4.13 Empty pallet pick-up list (Courtesy of California Hardware Company).

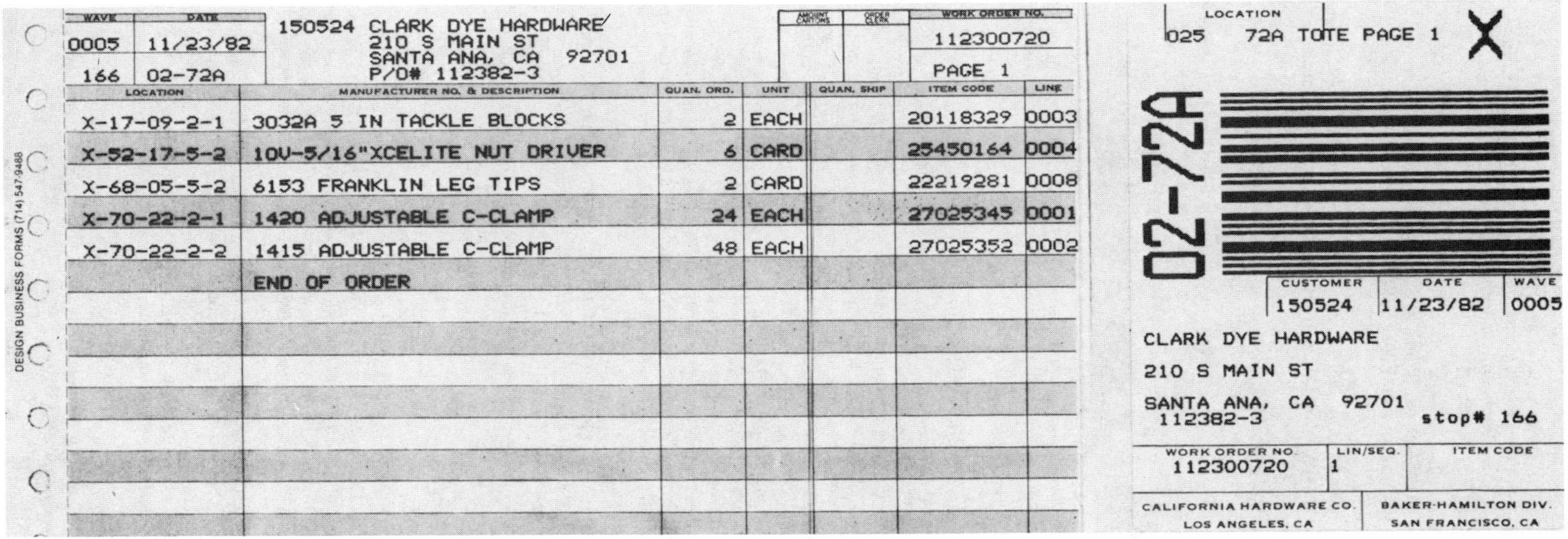

WAVE	DATE	
0005	11/23/82	150524 CLARK DYE HARDWARE 210 S MAIN ST SANTA ANA, CA 92701 P/O# 112382-3
166	02-72A	

AMOUNT CARTONS	ORDER CLERK	WORK ORDER NO.
		112300720
		PAGE 1

LOCATION	MANUFACTURER NO. & DESCRIPTION	QUAN. ORD.	UNIT	QUAN. SHIP	ITEM CODE	LINE
X-17-09-2-1	3032A 5 IN TACKLE BLOCKS	2	EACH		20118329	0003
X-52-17-5-2	10V-5/16"XCELITE NUT DRIVER	6	CARD		25450164	0004
X-68-05-5-2	6153 FRANKLIN LEG TIPS	2	CARD		22219281	0008
X-70-22-2-1	1420 ADJUSTABLE C-CLAMP	24	EACH		27025345	0001
X-70-22-2-2	1415 ADJUSTABLE C-CLAMP	48	EACH		27025352	0002
	END OF ORDER					

DESIGN BUSINESS FORMS (714) 547-9488

LOCATION

025 72A TOTE PAGE 1 X

02-72A

CUSTOMER	DATE	WAVE
150524	11/23/82	0005

CLARK DYE HARDWARE

210 S MAIN ST

SANTA ANA, CA 92701
112382-3 stop# 166

WORK ORDER NO.	LIN/SEQ.	ITEM CODE
112300720	1	

CALIFORNIA HARDWARE CO. LOS ANGELES, CA — BAKER-HAMILTON DIV. SAN FRANCISCO, CA

Fig. 13.4.14 Repack picking document with man-readable and scannable label for Stop 72A on Route 2 (Courtesy of California Hardware Company).

Man-readable and scannable labels (Fig. 13.4.15) are used to pick conveyable vendor-pack full cases for shipment or repack picking position replenishment; man-readable labels (Fig. 13.4.15) are used to pick pallet loads from reserve locations and nonconveyables. Labels are printed as in a picking oriented system. They are accompanied by trailer labels, the issue and return of which are monitored by a label control sheet (Fig. 13.4.16).

Computer Inventory Applications. Since the computer maintains a record of the inventory in every location in the warehouse, it can prepare:

1. A receipt verification list (Fig. 13.4.17) which includes each hand-stack or reserve pallet location in which a vendor-pack case SKU was stored during the previous shift and complete identification of the item.
2. A walk-off list (Fig. 13.4.17) which shows in location sequence for a specified range of locations only the location, item description, and item number.
3. An inventory verification request (Fig. 13.4.18) which lists by location each inventory for all SKUs to be checked.
4. A cycle count and annual inventory list (Fig. 13.4.18) which lists in numerical sequence every warehouse location to be counted and the merchandise, if any, occupying it.

Items different from those listed are recorded in the Exception column of the first two lists. Inventories are recorded on the second two lists. Corrections and additions to the items listed on them are entered along with the inventory counts. Using such documents can reduce the time required to the inventory by 50%.

Order Entry Cut Off During the Picking Shift

When order entry is cut off during the picking shift, there is typically a cut-off time for each route—usually one or two hours prior to its scheduled departure. Order takers typically call customers according to a schedule designed to obtain orders from the customers on routes in their scheduled departure sequence; the orders are commonly entered into the computer via a CRT as the order is being taken. The order cut-off, picking completion, order assembly, loading completion, and truck departure times for each route are scheduled.

When cut-off times occur during the picking shift, it is not possible to prepare the route analysis, route summary, wave analysis, and wave summary. All of the other documents necessary to operate the warehouse, however, can be prepared.

Typically, orders to be shipped on routes with cut-off times sufficiently far in advance of the start of the picking shift are divided into one or more batches. The orders in each batch are processed past the item master and inventory records. During this processing the computer prepares:

1. Let-down and empty pallet pick-up lists if inventories of merchandise on pallets are maintained.
2. Prebilled repack picking documents/packing lists with pages on which vendor-pack full cases, nonconveyables, and unpicked merchandise are listed.
3. Picking labels for merchandise to be shipped or to replenish repack picking positions and trailer labels.
4. Route manifests.

Orders taken during the picking shift are accumulated and processed in the same manner at 20–30-min intervals which terminate at the same time as one or more cut-offs. Although orders for routes for which order taking has been cut off comprise the bulk of the work load in each batch, there are usually some orders for routes with future cut-off times. Warehouse supervision sequences the repack picking documents for these orders in accordance with the scheduled time for the completion of picking. The routes for which repack may be picked at any one time are specified by signs placed where they can be seen by all picking and checking personnel. The sign for a route is removed as soon as all orders on the route manifest have been checked. Let-downs are completed immediately after printing to provide the inventory required to fill the vendor-pack full-case and nonconveyable merchandise picking labels. Labels for routes not scheduled for picking when a batch is processed may be held by the computer and printed when the route is scheduled for picking.

The control clerk or clerks process documents in the sequence specified by the picking completion schedule.

When a warehouse operates in this fashion there must be one shipping spur for each route scheduled for picking plus additional spurs for a number of the routes to be shipped later. Two or more routes are usually assigned to each spur each day.

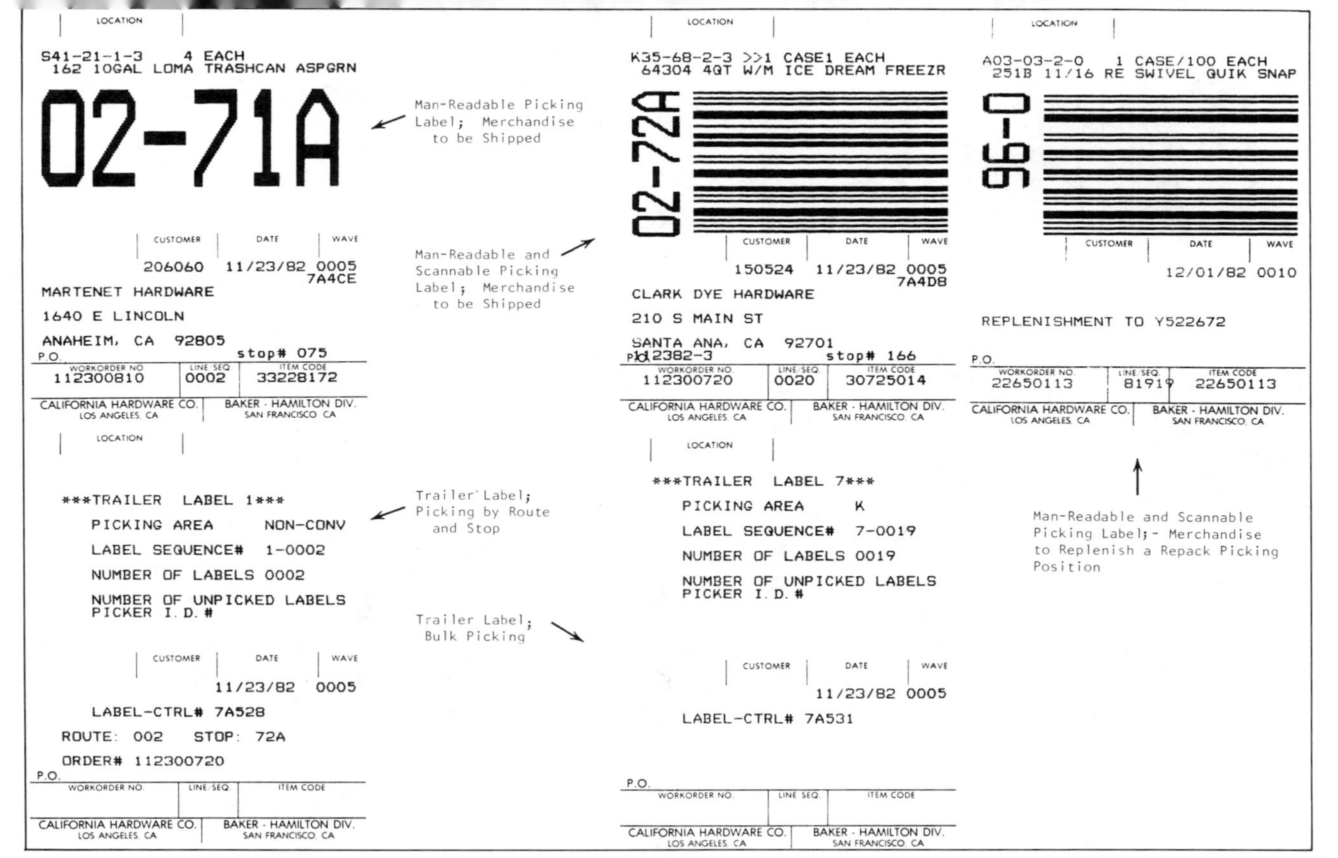

Fig. 13.4.15 Man-readable and scannable full case and nonconveyable merchandise picking labels and trailer labels (Courtesy of California Hardware Company). Man-Readable Picking Label—Merchandise to be Shipped; Man-Readable and Scannable Picking Label—Merchandise to be Shipped; Trailer Label—Picking by Route and Stop; Trailer Label—Bulk Picking; Man-Readable and Scannable Picking Label; Merchandise to Replenish a Repack Picking Position.

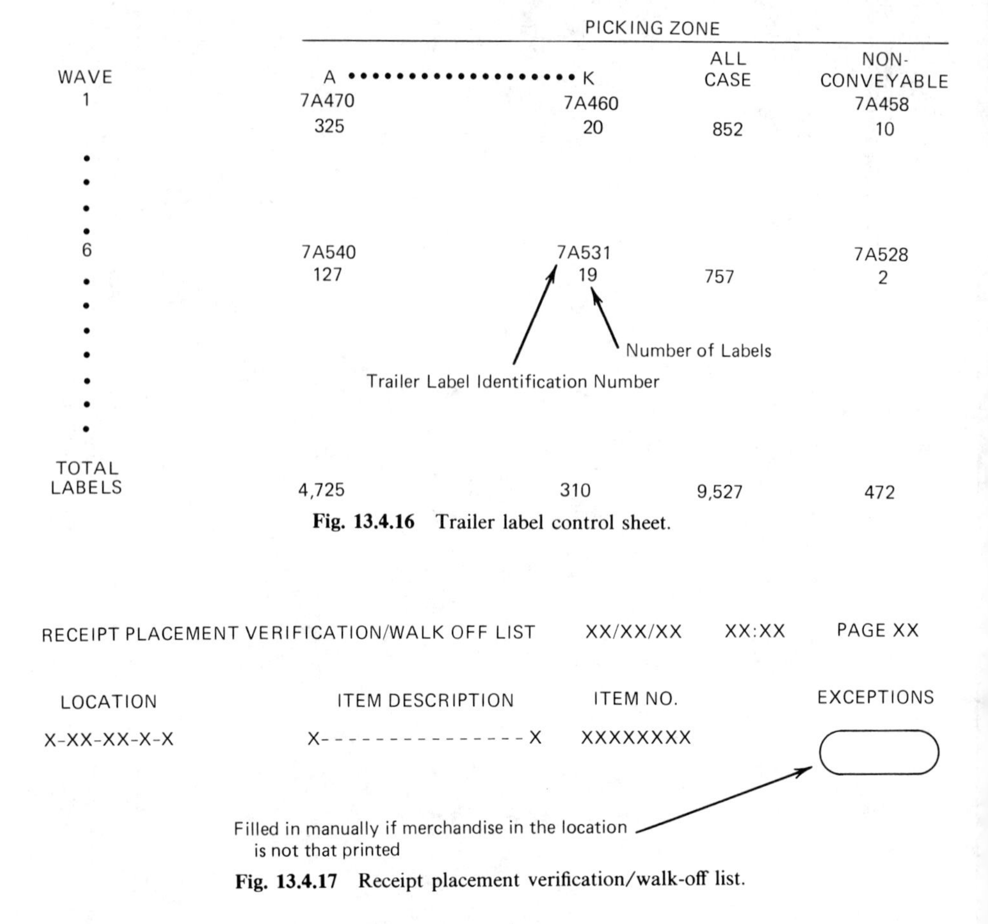

Fig. 13.4.16 Trailer label control sheet.

Fig. 13.4.17 Receipt placement verification/walk-off list.

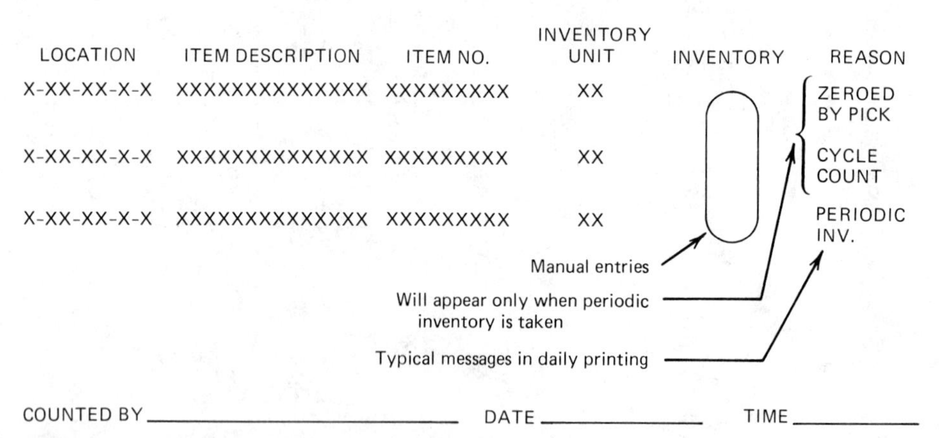

Fig. 13.4.18 Inventory verification/cycle count/periodic inventory list. Manual Entries; Will Appear Only When Periodic Inventory Is Taken; Typical Messages in Daily Printing.

Computer Applications to Improve Customer Service

The preparation of price tickets for application by the customer is a common customer service provided by the computer; the tickets are part of the full-case or nonconveyable merchandise picking label. Repack picking documents are printed on sheets of price stickers. Document header information is printed on the first row of stickers. The picking location and all other picking information are printed on the stickers in each row at the left of the sheet. The balance of the stickers in a row are price tickets, each of which carries the document number so tickets returned to the control clerk indicate merchandise not picked. The balance of the price tickets are placed in the tote or carton with the merchandise.

When repack merchandise is simultaneously picked and ticketed, the picking document consists of:

1. A header printed on the first row of stickers.
2. Row after row of price tickets bearing the picking location and printed in location sequence across the sheet.

Pickers are assigned to zones.

Order entry is cut off far enough in advance of the picking shift that there is time for the computer to adjust the number of items in each zone to provide a full day's work for the picker assigned to it.

If the repack merchandise stocked in one area of the customer's store is shipped in repacked cartons or totes, each identified by area, the customer's labor to check the shipment and stock the merchandise will be reduced. To provide this saving, the repack merchandise is located in the picking area by department. Merchandise is assigned to cartons or totes on the basis of weight or cube. The computer identifies the contents of each one by printing the invoice line or page numbers of the contents on the label—if the merchandise shipped is also listed on the invoice by department. If it is not, customer departments are identified by different stop numbers and separate invoices. Overflows are placed in totes or cartons identified by a handwritten label bearing the same information.

When the repacking picking document is printed on stickers, tote or carton labels are printed separately and attached manually to the document.

Emerging Systems

Emerging systems rely heavily on microcomputers, CRT screens, and bar-coded labels at the location where the operation is performed. The microcomputer converts information transmitted from a host via radio or loaded into its memory from a data port into instructions displayed to the warehouseman [1]. It records the results of executing these instructions for transmission to the host via radio or data port. Wanding bar-coded labels replaces manual recording; it provides error-free input with respect to the identity of the warehouseman, merchandise, and order-processed and warehouse locations. This input can also trigger display of the instructions for processing the merchandise or the location of the next operation.

Emerging systems have been installed primarily in warehouses that ship full or broken case merchandise but not both. In one warehouse [2], individual pairs of boots in bar-coded, chipboard boxes are picked from all levels of pallet rack with order-picker trucks. Operations for each picking shift are planned much as described previously. Bulk picking instructions are stored by the host computer and input to the memory of a microcomputer onboard each truck via data port. An onboard screen displays picking instructions; the microcomputer positions the vehicle at the picking location. The picker wands the first box picked; if a "beep" is heard, he or she is picking the correct merchandise. The picker fills the order, placing the units in a special cart, and inputs the number of units picked via a keyboard. The latter input initiates movement of the truck to the next picking position and display of the next set of picking instructions. Picked cartons are mechanically transferred from the special cart to a sortation system controlled by a scanner. It assembles all the cartons for one stop on a spur where packers check them against a computer-prepared packing list and strap them for shrink wrapping and identification by computer-prepared shipping label.

To store receipts, a driver loads the microcomputer memory with empty warehouse locations via a data port. The computer displays these locations; inputting one moves the truck to the location. Wanding the bar code on the cartons to be stored establishes the identity of the merchandise; inputting the number of units stored completes the transaction. Inputting the next unoccupied location starts the process over again. After all of the merchandise has been stored, the memory of the onboard computer is transmitted to the host via a data port for the updating of inventory files.

In another warehouse [3], merchandise is stored and picked in pallet loads by swing-reach trucks equipped with an onboard display, keyboard, printer, and FM transmitter–receiver. Inserting a key-punched item card and inputting the gross weight of the load as it comes from manufacturing triggers the preparation and application of a label bearing a man-readable rack storage location and a man-readable and scannable load identification number on each pallet load of finished product. A conventional forklift moves the pallet to the storage area. As soon as the pallet has been stored by the swing-reach truck and identified by wanding the label, the latter information is transmitted by radio to a host computer which supplies a picking instruction whenever possible. This procedure minimizes the time that the truck travels empty and maximizes its productivity. Data processing is also used extensively in this warehouse for planning.

Hand-held units complete with radio have been used in a similar fashion [4].

Repack has been picked from flow racks equipped with a signal light in each bay to indicate it houses merchandise to be picked and a digital display at each pick position which indicates the number of units to be picked [5].

Potential Applications When Order Integrity Is Destroyed and Restored. The storage of palletized receipts and the replenishment of positions where cases are picked from pallets by computer-, display-, and keyboard-equipped trucks is an obvious possibility in a warehouse that destroys and restores order integrity. If the host computer is on-line, it could, when information about a pallet load receipt and its location in the receiving department is input:

1. Print a pallet label bearing this information in man-readable and scannable form.
2. Store the same information for transfer to a forklift via data port.

The forklift driver could proceed to the receiving dock location shown on the screen, wand the label on the pallet, and store it if the display shows he has found the correct pallet. Wanding the bar code on the rack location after storing the pallet would verify placement of the item in the specified location. Transmitting the results of the wanding would confirm placement and up-date the computer inventory. If, at any point, the forklift driver could not follow instructions, he could call up a trouble screen and report the information it specifies.

The let-down list and empty pallet pick-up list could be stored and transferred to the onboard computer memory via data port. Wanding the bar code on the label of a pallet load and on both rack location labels would make let-down foolproof and eliminate the document. The empty pallet pick-up list could be eliminated by similar practices. Finding a carton on a supposedly empty pallet, when input, would trigger the storage of a transaction that could not be completed until the carton was identified and this identification was input along with any necessary inventory adjustments. Wands, hand-held microcomputers, and the bar coding of locations offer the possibility of storing cases identified by bar codes in hand-stack rack and shelving without a document.

Picking full cases for shipment and replenishment and sorting them could be accomplished without labels in a warehouse sorting with a scanner to one stop per spur if the inventory consisted of merchandise bearing bar codes. Picking instructions could be printed on a list or stored in a hand-held unit. While the Department of Defense requirement that cases delivered to its facilities be bar coded will result in a high percentage of all cases being bar coded, there will undoubtedly be enough uncoded cases to require the continued use of picking labels in warehouses distributing the products of many manufacturers. Picking labels will probably be necessary, since they serve as shipping labels, even if all merchandise is bar coded, when shipments via LTL or common carrier are common.

These emerging systems will reduce labor costs and improve inventory accuracy. The savings created must be balanced against the initial cost of the equipment and programming, the variety of bar codes now in use [6], the care with which they must be printed, and the fact that malfunctions, programming errors, and computer downtime may close the warehouse.

13.4.3 Computer Applications When Order Integrity Is Maintained

Typical Warehouse Layout and Merchandise Flows

When order integrity is maintained in a warehouse that ships broken and full-case merchandise, a well-planned layout usually includes one area devoted to shelving and—possibly—flow rack, an adjacent area or areas of hand-stack rack and pallet rack, and floor stack areas. There may be separate areas for flammables and high-value items.

Orders for merchandise shipped in broken-case or broken and vendor-pack full-case quantities are filled from fixed picking locations in the shelving and flow racks. To minimize the need for picking position replenishment, the reserve inventories of SKUs located in the shelving are combined with the picking inventory or stored on the dust cover whenever possible. The reserve inventories of other items are stored in fixed hand-stack reserve locations close to the picking positions or in pallet rack. Nonconveyable SKUs and those shipped only as full cases are picked from fixed picking positions in the lower levels of the pallet rack; their reserve inventories are stored above the picking positions. When receipts of these items are less than pallet loads, they may be picked from fixed positions in the hand-stack rack. The computer maintains one inventory and one picking position per SKU.

In many such warehouses each order is picked to one or more pallets or carts which are delivered to checker, packers, and the shipping dock as an entity.

Merchandise is typically picked with a location-sequenced document that lists every SKU ordered in two groups:

1. Items picked in broken and vendor-packed full-case quantities.
2. Items picked only in vendor-pack full-case quantities and nonconveyable SKUs.

If an item is picked in broken-case or vendor-pack full-case quantities, the quantity to be picked is shown on the picking document in broken-case picking units. If the item is picked as vendor-pack full cases only, the picking unit is a case; the nonconveyable unit is usually each but may be case.

The picker may recognize that a pick quantity specified in broken-case units is a full case. If he does and if he knows where the reserve inventory of the item is located, he may pick a full case; he may pick the merchandise from the repack position. Repack picking positions are replenished when that is necessary by the picker who picks the last unit from the location or at the close of the day.

The few pallet-load picks are identified by the picker or a leadman, either of whom arranges for the delivery of such a pick to the dock. The only order assembly operation consists of "marrying" these picks and the balance of the merchandise required to fill the order.

Figure 13.4.19 is a schematic layout showing the warehouse areas and the movements between these areas when order integrity is maintained. The movements (B1A, B2A, B3, B4, etc.) are the same as those shown in Fig. 13.4.1.

Computer Applications

The computer controls or records the results of these movements, which are shaded in Fig. 13.4.17:

1. The receipt of merchandise.
2. The movement of merchandise from picking positions through checking to order assembly and shipping (P1, P3, P6).
3. The shipment of merchandise to the customer (S2).

The computer is aware of the receiving function because it maintains purchase order records; it reflects receipts (B4) as they are input to increase the one inventory maintained per SKU. The computer is not aware of the multiple locations for the inventory of one SKU and does not reflect the other merchandise movements shown in Fig. 13.4.17. They are initiated and performed by warehouse personnel responding to verbal or visual information as they were trained.

The computer prepares picking documents and invoices and maintains the single inventory per SKU. It provides minimal information for receiving and stocking merchandise and for replenishing picking positions. It also provides inventory and purchasing reports.

It is important to note that expanding computer applications beyond the location-sequenced picking document enables management to operate the warehouse as effectively as possible with the limited information available to the computer. If order entry is cut off sufficiently far in advance, the computer can:

1. Plan deliveries by captive trucks.
2. Project total warehouse manpower requirements.
3. Prepare route manifests.

It can also provide documents which will reduce the costs of stocking merchandise and replenishing repack picking positions.

Prerequisites for Computer Applications

The only prerequisites for these applications are:

1. An order entry system.
2. Item, customer, and warehouse location numbers.
3. Route and stop numbers and possibly item weight.
4. An overall estimate of the productivity of warehouse or picking and shipping personnel in lines per labor-hour.
5. Case packs, not necessarily accurate, for merchandise picked from repack picking positions if the computer assists the replenishment of these positions.

Accurate, current inventories are not necessary if the picking document includes every item ordered by the customer.

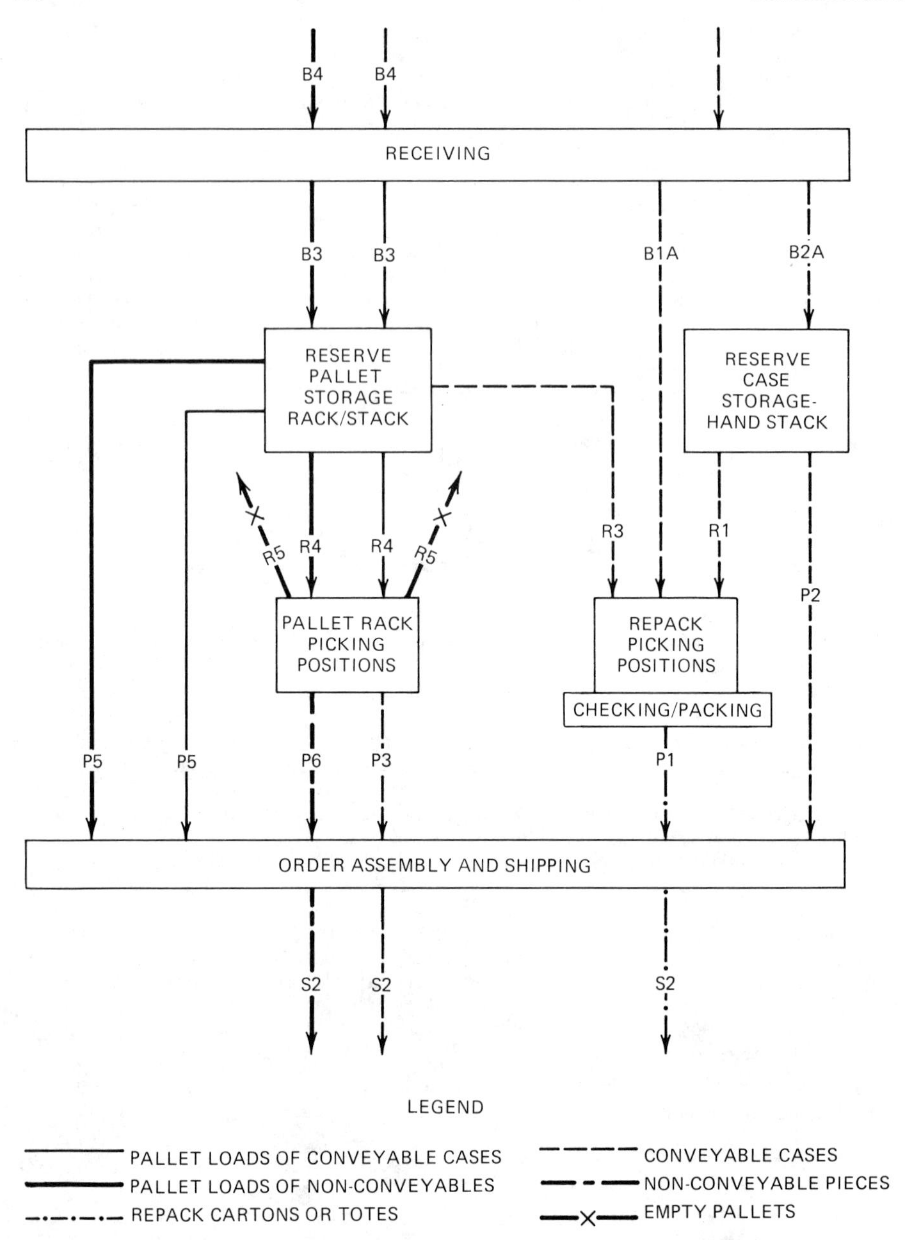

Fig. 13.4.19 Merchandise movements when order integrity is maintained.

Planning Shipping Operations

The computer processes orders past the item file and prepares a route summary (like that shown in Fig. 13.4.3) except that the information provided for each route consists of lines, pounds, or lines and pounds. It also prepares route analyses (like the example in Fig. 13.4.4) which include this information and total warehouse labor-hours or picking and shipping labor-hours based on historical records of lines per labor-hour. Using its empirical knowledge of the capacity of trucks in lines or pounds, warehouse supervision plans routes as described previously. The computer then prepares a route manifest like that shown in Fig. 13.4.5. It may show no quantities to be shipped or only the number of vendor-pack full cases for SKUs picked as full cases only. The missing information is added manually as the orders are checked prior to loading. A nonconveyable and full-case manifest (see Fig. 13.4.6) is useless since some lines may be filled with full cases despite the fact that the pick unit is each or inner packs.

Planning and Scheduling Warehouse Operations

In a typical small warehouse, picking documents are printed before route manifests. Scheduling operations consists of manually sequencing these picking documents in filling order.

If picking documents are not printed prior to route manifests, the stops shown on the route analysis can be assigned to waves manually with the objective of equalizing lines per wave. These assignments can be input to the computer which prepares wave analyses, a wave summary, and a wave schedule and prints picking documents by wave.

Planning Warehouse Manpower

The computer estimate of the total personnel or picking and shipping personnel required by the orders to be filled during one shift will assist supervision planning the assignment of individuals to operations. If the manpower required is less than that available, picking and shipping personnel will be assigned to other operations. If the requirement exceeds the available manpower, receiving and stocking personnel will be transferred to order filling and shipping operations or overtime will be scheduled. The extent of these reassignments is determined by experience.

Picking Document

The picking document in a warehouse that maintains order integrity should be the same as the one shown in Fig. 13.4.7 except that it will not include the pressure-sensitive labels. The document may be one or two parts. It may incorporate one or more gummed, tear-off labels showing the route and stop number, customer address, invoice number, and so on. These are applied by the packer, who prepares any necessary additional labels with a stencil. Manual procedures for processing the picking document after checking are essentially the same as described previously.

Repack Picking Position Replenishment

The computer can prepare a list of the cases that must be picked to replenish repack picking positions by accumulating the broken case units picked until they equal at least one full case. The replenishments required are then printed in a report which lists the repack picking location, item description, item number, and the estimated numbers of replenishment cases and broken case units required. Pickers or stockmen utilize their knowledge of the location of reserve inventories to pick the cases for stocking at the close of the picking shift.

This computer application is effective only if the accumulated quantity picked is adjusted when merchandise is marked out and if vendor case packs are reasonably accurate. The computer can prepare a repack replenishment frequency report.

Receiving and Stocking Applications

The computer lists the picking location of each SKU ordered on the receiving department copy of the purchase order; receiving department personnel write this location number on a pressure-sensitive label and affix it to the merchandise. Stockmen store the merchandise in the picking location, the reserve location, or both. They may create a reserve location for an SKU or consolidate the receipt with the existing reserve inventory after moving it to a larger-capacity reserve location. No information about these stocking operations is input to the computer.

Manually prepared labels can be replaced by computer-prepared labels showing the item number, item description, quantity received in broken and full case units, and the picking location. These labels are prepared when the purchase order is printed and filed with the receiving department copy.

13.4.4 Programming and Implementing Applications

Programming Specifications

Programming specifications for computer applications in a manual warehouse should:

1. Be based on and include a merchandise flow chart (Figs. 13.4.1 and 13.4.19) and a warehouse layout showing all location numbers.
2. Specify the information that must appear on the instructions for each move.

3. Classify each warehouse location by use (repack pick, floating slot, palletized reserve, etc.) and specify the inventory information to be maintained for each classification.
4. Specify sufficient information so that detection of any differences between the information in the computer files and the corresponding merchandise, inventory, or location in the warehouse is easy.
5. Provide manual and data processing procedures for handling such disagreements.
6. Include document designs.
7. Include a brief description of operations and their relationships to each document and the manual and data processing procedures.
8. Include a schedule for cut-off times, order processing, delivery and operations planning, document printing, and warehouse operations.
9. Include customer requirements with respect to labels, packing lists, invoices, and identification of merchandise at the time of delivery.
10. Include specifications for, interfaces with, or required changes in, data processing systems for order entry, selecting orders to be shipped, maintaining inventory, invoicing, and so on. In a large warehouse some of these systems may be processed on a host computer.
11. Include the projected numbers of SKUs, inventory locations, lines per day, documents per day, and other operating parameters necessary to define requirements.

To minimize faulty communication with programming personnel and revisions, specifications should be prepared after an operations manual has been written or outlined in detail.

Programming

The picking position replenishment schedule determines whether each order line must be processed once or twice if routes or stops cannot be preassigned to the waves of a warehouse that destroys and restores order integrity.

Programs for printing labels should be planned carefully because this operation is slow, especially when labels are bar coded or include expanded numbers (Fig. 13.4.5). Labels can be printed two, three, or four across. They should, however, be separable into vertical strips of single labels, 25 or 50 per strip, in picking-location sequence because this practice maximizes picker productivity.

Operating experience normally results in program revisions. For this reason, each computer application should be program module whenever possible. Modifications affecting only one module should be possible without reprogramming orders.

Package Programs

Software houses and materials handling equipment manufacturers provide package programs that include some or all of the computer applications described. Complete specifications facilitate selection of the appropriate package and determination of the cost of any necessary modifications.

Implementing Applications

Computer applications in the warehouse will be successful upon implementation only if they are *thoroughly* tested prior to that time. Test data should represent operations performed as anticipated *and* as many deviations from normal operations and errors as can be devised. When possible, applications should be installed one or more at a time rather than all at once; each application should be debugged before installing the next one or next group.

Discrepancies between warehouse movements and inventories and the corresponding computer records are the major implementation problem. To avoid them:

1. Key warehouse personnel should be cross-trained in data processing operations and vice versa.
2. Responsibility for input quality and warehouse data processing should be assigned to a supervisor in each area before implementation begins.

Computer Processing Time

A significant amount of time is required to perform the processing required by the picking, shipping, and picking position replenishment applications of the computer in a warehouse that destroys and restores order integrity. In one warehouse using a total system with predetermined tote contents and replenishment by wave, 6–8 hr of IBM 4331 computer time are required. During this period the computer processes 200 orders and 16,000–20,000 lines. Approximately half of this time is required to print 14,000–16,000 lines on repack picking documents and 6000–8000 man-readable labels like those in Fig. 13.4.8 two across. In another warehouse utilizing the same computer applications, 1500 picking labels with expanded print are printed four across and 9000 repack lines are printed on pick lists. Three to four hours of PDP-1145 time are required.

REFERENCES

1. Warehouse Customer Service: Tomorrow's Levels Today!, *Modern Materials Handling,* November, 1977, pp. 73–75.
2. Computer Integrated Warehousing—Advanced Controls Boost Productivity 40%, *Modern Materials Handling,* February 5, 1982, pp. 48–53.
3. Real-Time Control of a Yard Operation, *Modern Materials Handling,* November 19, 1982, pp. 39–42.
4. Rubber Queen's Paperless Warehouse, *Production and Inventory Management Review,* May, 1982, pp. 28, 29.
5. Computer-Aided System Gives 52% More Picks per Worker, *Modern Materials Handling,* November 5, 1982, pp. 50–51.
6. Automatic Identification—Now It's Really Taking Off, *Modern Materials Handling,* September 21, 1982, pp. 32–43.

BIBLIOGRAPHY

The ABC's of Warehousing. Market Publications, Inc., Washington, D.C. 1978.

Ackerman, K. B., *Warehousing: A Guide for Both Users and Operators.* The Traffic Service Corporation, Chicago. 1977.

Ackerman, K. B., R. W. Gardner, and L. P. Thomas, *Understanding Today's Distribution Center.* The Traffic Service Corporation, Chicago. 1972.

Blanding, W. and K. E. Way, *100 Ways to Improve Warehouse Operations.* Market Publications, Inc. Washington, D.C. 1978.

Briggs, A. J., *Warehouse Operations: Planning and Management.* Wiley, New York. 1960.

Falconer, P. and J. Drury, *Industrial Storage and Distribution.* Wiley, New York. 1975.

General Services Administration. *Warehouse Operations.* U.S. Government Printing Office, Washington, D.C. 1969.

Jenkins, C. H., *Modern Warehouse Management.* McGraw-Hill, New York. 1972.

Material Handling: Warehousing. *Plant Engineering,* 1977.

Rack Storage of Materials, 1975, 231c. National Fire Protection Association, Boston. 1975.

Warehouse Modernization and Layout Planning Guide. Department of the Navy, Washington, D.C. 1978.

13.5 WAREHOUSE AUTOMATION

Kenneth B. Ackerman

Planning for warehouse automation is no different from planning for any other kind of capital improvement in the warehouse or in the production plant. Basically, the decision involves a weighing of costs and benefits, as well as a consideration of the risk of failure versus the gain if the program is a success.

There are three reasons why an executive might be afraid to consider automated handling:

The capital investment involved is substantial—often more than many companies would risk if there were a chance of failure.

Even if the system works, savings might not meet expectations.

The system may be too complex to run smoothly or be managed effectively.

In essence, mechanization is the substitution of machinery for work formerly done by hand.

In considering the cost/benefit relationships, it is well to look at decisions reached by others, particularly in the early days of developing automated warehouse systems. One of the first such systems in the United States was installed at the finished goods warehouse of a manufacturer of frozen bakery items. In that situation, the ultimate justification for the equipment was not based on cost savings, but rather on human factors. Because of the extremely hostile environment of the freezer, the manufacturer had difficulty in recruiting workers, and the cost of mechanizing the function was justified because it allowed the manufacturer to get a job done which people did not want to perform. Other early applications of automation in warehousing took place in situations where expansion of a conventional warehouse was impossible. At the first International Conference on Automation in Warehousing, held at Nottingham, England, in 1975, a Czech professor reported that in his country industrial plants cannot expand into agricultural land without paying the state a fee equal to the estimated value of 200 years of crops. Faced with this enormous cost penalty, some warehouse operators chose a high-rise automated system simply because conventional expansion cost was greatly inflated by this government regulation. In other cases, tax policy has been the benefit that caused the investment to be made. The most common example is the rack-supported building, in which depreciation regulations for taxation in the United States frequently allow the useful life of most of the building to be calculated as the relatively short life of storage rack. Without question, some early decisions for automation were motivated primarily by desire for prestige, somewhat like the early computer installations. Many automated installations contain observation balconies for plant visitors, a clear sign that management considers the installation to be a showcase as well as a warehousing tool.

13.5.1 Strategic Planning

Before considering specific automated order-picking applications, develop a long-range strategic plan for materials handling within the company. Such a plan can provide the framework for future implementation of systems that will assure compatibility of components and handling concepts over at least a 10-year planning horizon. This type of planning is even more important for companies with multiple divisions and manufacturing/distribution locations, since the proper application of modular handling concepts can provide a great deal of compatibility and flexibility in meeting long-range handling and storage requirements [1].

The strategic plan must consider the shelf life of the products that will be warehoused in the automated system; the product life will in effect govern the life of the system itself. If the size and shape of the product have changed radically over the past 10 years, what is the likelihood that similar changes will take place over the next 10 years? Will the automated system have the flexibility to cope with this major change?

13.5.2 Partial Mechanization

It is a mistake to mechanize every item that can be mechanized. True, this will save the most labor, but for many items the incremental mechanization investment is not justified by the labor savings. Typically, some items should be handled conventionally, some should be mechanized with some equipment, and some items should be fully mechanized.

For example, there could be a situation where mechanizing every item in the warehouse will give an average 5-year payback, which seems acceptable on the surface. However, many items may have had an individual payback on their incremental investment of 1 year or less, whereas other items had a poor payback, such as 10 years or more. In fact, only those items with a good enough payback should be mechanized. As a result, one might mechanize one-half to two-thirds of the items with an average payback of 3–4 years, and no item with more than a 5–6-year payback.

Therefore, in order to make the proper decision, each item must be individually analyzed to determine what mechanization, if any, is justified. To justify, calculate the dollar labor savings with mechanization and compare this to the incremental dollar investment to mechanize. If the payback (or return-on-investment) meets company standards, then mechanization is justified.

Payback of no more than 6 years in labor savings is a reasonable company standard to justify mechanization. Then, if sufficient items are justified for mechanization to permit a reasonably sized system, these items are grouped together and mechanized. The remaining items are handled conventionally.

If mechanization is justified for only a few items, then it may not be worth mechanizing. The decision actually rests on the total dollar savings expected [2].

13.5.3 Mechanization for Individual Items

In general, more mechanization is justified for faster-moving items because there are greater labor savings to pay back the fixed investment. For example, use of a conveyor for high cube movement items may be justified because the conveyor saves much travel to and from the dock over nonmechanized selection.

Specifically, the decision for each item is made by calculating the labor savings dollars per year, the investment dollars, and then the payback, by this ratio:

$$\text{Payback years} = \frac{\text{Investment \$}}{\text{Labor savings \$ per year}}$$

Thus, if the item's payback time is short enough, mechanization is justified.

Before completing the ratio, it is advisable to construct detailed models of the labor needed in the nonmechanized and mechanized systems. The difference between these is the labor savings achieved by mechanizing. For example, a set of labor models can be developed to express the labor for each case handled with added labor for larger cases and heavier cases [2].

13.5.4 Examples of Cost and Savings

Using a nonmechanized system as an example, representative savings are

3 min per pallet with pallet flow rack, due to easier replenishment

2 sec per shipped unit with case flow rack, due to easier selection and reduced "walk by"

6 sec per case with conveyor, due primarily to reduced travel

The incremental mechanization investment is determined for each item. Representative costs are

$100 per lane for two-deep pallet flow rack

$25 per lane for a case flow rack

$100 per item for conveyor in a typical two-high selection situation ($200 in a one-high selection situation)

Figure 13.5.1 shows how to determine the economics of a typical hand-stacked item moving 18 cases per week, with $\frac{1}{2}$ ft^3 per case. Figure 13.5.2 shows how to determine the economics for a typical pallet item moving 50 cases per week, with 1 ft^3 per case. The concepts applied here can be used in many different situations.

13.5.5 Separations of Product

In considering a mechanized system, products in the warehouse should be divided into four categories: fast movers, slow movers, items requiring repack, and items that cannot be handled in the mechanized system. The last category would be products that have unusual packaging or some characteristic that makes them unsafe for use in the system. Clearly, the mechanized system can best be used for the fast-moving items. Either a different system or no system at all will be considered for the other three categories of product.

Rack in: Taken Away by:	Conventional Conventional	Case Flow Conventional	Case Flow Conveyor
Cases per hour			
Selection	160	174	199
Savings			
Seconds per case	Base system	1.8	4.4
$ per year	↓	4.68	11.44
Investment $	↓	25	40
Payback years	↓	5.3	3.5

Fig. 13.5.1 Economics for a typical hand-stacked item moving 18 cases per week, or 936 per year; $\frac{1}{2}$ ft^3 per case. Labor is assumed at $10 per hour.

Rack in: Taken Away by:	Conventional Conventional	Pallet Flow Conventional	Pallet Flow Conveyor
Cases per hour			
Replenishment	700	4000	4000
Selection	160	160	194
Savings			
Seconds per case	Base system	4.2	8.1
$ per year	↓	30.33	58.50
Investment $	↓	100	200
Payback years	↓	3.3	3.4

Fig. 13.5.2 Economics for a typical pallet item moving 50 cases per week, or 2600 per year; 1 ft^3 per case (assume 40 cases per pallet and $10 per hour labor).

13.5.6 Flexibility

Perhaps the toughest question to be answered in evaluating mechanized systems is whether the system can cope with anticipated change. Shocking examples of such failure are found in many warehouses. One highly mechanized system for picking and shipping of drug and pharmaceutical products was installed in an older building originally designed for tire storage. Dock doors in the building were rebuilt to fit the loading of low-bed retail delivery trucks. Conveying systems and aisle layouts were especially designed for the drug operation. After a relatively short period of use, the wholesale drug company had labor and financial problems which caused it to go out of business. At this point, the automated equipment was worth more than the value of the building. Furthermore, the high cost of removing this equipment made relocation an uneconomical alternative. Neither the dock doors nor the storage layout were adaptable to general warehousing purposes. As a result, a high percentage of this investment in automation was lost.

What might have been done differently? In this situation, specialized equipment was used when more standardized systems were available. Systems with wide application in grocery distribution were rejected in favor of equipment especially designed for drug products. No consideration for future relocation of the equipment was made, and therefore it was impossible to move the system economically.

13.5.7 The Learning Curve

One automated picking system was justified primarily because it offers ease of training order-picking employees. In this installation, illuminated signboards show the order picker the number of pieces and the location of the picking lane from which goods should be pulled. Through the use of this system, the picking operator can function without any papers, and the job of order picking can be learned in a few minutes by someone who has absolutely no experience with the product line.

13.5.8 Alternatives for Mechanization

There are three popular alternatives for the picking of smaller cases or parts:

Move the person to the merchandise to be picked

Move the merchandise to the person

Move the merchandise to a processing area

Examples of the first include man-aboard unit-load machines as well as order-picking trucks. These systems are particularly effective when the picking route can be designed to include many picks in each trip down the aisle.

The most common example of the merchandise-to-person concept is the carousel system. Carousels are particularly useful when existing clear building height is limited, or in cases where volume and throughput are not high enough to justify other alternatives.

The merchandise-to-process application usually involves the delivery of full bin quantities or tote trays to a point of use for picking, processing, or assembly [1].

13.5.9 The Future of Automated Systems

Investment in an automated system is a decision that balances the return on investment against the risks involved. The savings from an automated system will frequently not arise from materials handling operations within the warehouse, but rather from improved control over materials handling operations, and therefore a better ability to react to increased service levels and sales. Automated systems probably will not be adopted by any company that does not forecast long-term market growth. One of the major risks, that of technical unreliability, has been considerably reduced as the construction of automated systems has matured.

The circumstances in which the risks are minimal and the benefits at their greatest can be summarized as follows:

1. An expanding market.
2. An expanding market serviced from specific sites as a result of rationalization.
3. The need to improve and maintain competitive service to retain or increase market share.
4. High retained profits.
5. When experience can be transferred; that is, favoring multiple installations with considerably reduced risks after the first successful installation. This suggests a large multiple-plant organization as a prerequisite.

6. Where logistics cost is identifiably high. Again, this favors multilocation component manufacturing and/or high throughput assembly from a large range of component parts.
7. A top management and financial management structure that understands and accepts the risks of technical innovation and has the ability to motivate its management and staff to welcome innovation. This is difficult to achieve unless it is against a background of expanding sales which guarantees job security.
8. An understanding that the adoption of advanced technology is in many cases a function of the market share achieved. That market share is a function of price and service which is a function of the technology of the production and distribution system. We require less departmentalized decision making and more integration if the true potential of technology is to be achieved [3].

13.5.10 Conclusions

The enormous risk and desire for prestige which marked early applications of warehouse automation have been tempered by improving reliability, lower costs, and greater acceptance. Yet automated systems, particularly those involving high-rise installations, are most popular in parts of the world where the price of land is extremely inflated and therefore the ability to expand the warehouse laterally is limited. A decision to invest in automation involves strategic planning considerations, and no automation should be made without extensive planning. There are steps toward partial automation or mechanization that should be considered before a plan to automate is completed. The payback can be calculated on the basis of individual items. A likely result of such analysis is to automate part of the warehouse and leave the rest of it on a manual system. Such separation should be done on the basis of speed of movement, whether or not items require repack, and segregation of items that won't fit in the system. Automated systems are most likely to be adopted by relatively large organizations that have an expanding market.

REFERENCES

1. Glude, Terry, in *Warehousing and Physical Distribution Productivity Report,* Vol. 15, No. 11. Marketing Publications Inc., Silver Spring, MD. November 1980.
2. Phipps, J. R. and J. T. Brown, Jr., in *Warehousing and Physical Distribution Productivity Report,* Vol. 15, No. 6. Marketing Publications Inc., Silver Spring, MD. June 1980.
3. Williams, John, "Automated Storage and Retrieval Systems." National Materials Handling Centre, Cranfield, England. September 1982.

13.6 METRICATION IN THE WAREHOUSE

Kenneth B. Ackerman

Throughout the decade of the 1970s, there was increasing government and public relations rhetoric about the need for the United States to accelerate a conversion to the metric system. Perhaps the high-water mark of this movement was the Metric Conversion Act of 1975.

Table 13.6.1 Imperial and Metric Pallet Dimensions

Imperial (rounded inches)	Hard Metric (rounded metrics)	Hard Metric (not-rounded inches)	Soft Metric (not-rounded metrics)
24 × 32 in.	600 × 800 mm	23.64 × 31.52 in.	609 × 812 mm
32 × 40	800 × 1000	31.52 × 39.40	812 × 1016
32 × 48	800 × 1200	31.52 × 47.28	812 × 1219
36 × 42	900 × 1060	35.46 × 41.75	914 × 1066
36 × 48	1060 × 1200	35.46 × 47.28	1066 × 1219
40 × 48	1000 × 1200	39.40 × 47.28	1016 × 1219
42 × 54	1060 × 1370	41.75 × 53.96	1066 × 1371
48 × 60	1200 × 1500	47.28 × 59.10	1219 × 1523
48 × 72	1200 × 1800	47.28 × 70.90	1219 × 1828
36 × 36	900 × 900	35.46 × 35.46	914 × 914
42 × 42	1060 × 1060	41.75 × 41.75	1066 × 1066
48 × 48	1200 × 1200	47.28 × 47.28	1219 × 1219

Source: This table was prepared by Bob Promisel, General Manager, Industrial Engineering, G. C. Murphey Co. for an article prepared for *Warehousing and Physical Distribution Productivity Report,* Vol. 18, No. 1.

In materials handling, the primary opportunity to coordinate with the metric system is in the sizing of the pallets. The 48 in. × 40 in. four-way pallet is the most common size used for warehousing and unitizing of cased merchandise in this country. The 48 in. × 40 in. pallet is virtually the same size as the pallet measuring 120 cm × 100 cm. In fact, the area of the two pallets differs by only 3%, and the lateral measurements are insignificantly different. It is possible that one reason why the 48 × 40 was adopted in the United States was its ease of conversion to metrics. The sizes are so close that it is not likely that a 120 × 100 pallet would be detected as anything significantly different if it were substituted for a 48 × 40 pallet.

It is equally fortunate that the dozen most-common pallet sizes used in the United States also readily convert to metric measurement. As can be seen from Table 13.6.1 each pallet has an equivalent in reasonably rounded metric measurements. While none of the pallets is exactly the same size in imperial and metric measurements, the size variations in each case are almost insignificant.

CHAPTER 14
CONTAINERS

FRED ZACHARIAS

Buckhorn, Inc.
Cincinnati, Ohio

A container simply defined is "a thing in which material is held or carried." Yet as simple as it is, the container is probably the most overlooked part of a materials handling or storage system. Many companies could have saved themselves thousands of dollars had they considered the container as an initial part of their system, rather than an afterthought.

Choosing the right container can be one of the most important steps in designing a materials handling system. For a system to function properly, the container must properly function within it.

14.1 SELECTION FACTORS

But how do you choose the right container?

First, consider how and where the container will be used. How much weight will be put in it, and how it will be handled. Also, will it be used for storage, distribution, or both?

The environment in which the container will function will be a determining factor in the type of material selected. High humidity or exposure to moisture can rust steel. These conditions can also cause corrugated containers to break down and untreated wood to rot and split. Extreme temperatures are another factor, especially when considering plastic, paper, or wood. High temperatures can cause most plastics to melt or lose strength. With paper or wood, there is the possibility of fire. Low temperatures can cause some materials to become brittle and easily break upon impact.

Chemicals are another consideration. Many plastics are impervious to most chemicals, but not all. Metal containers can be affected by acids and caustic solutions.

The elements can also affect the container. Weather conditions, such as rain and temperature, can have effects as previously mentioned. Ultraviolet light from the sun can break down the structural integrity of plastic, if it is not properly compounded.

Containers used in the assembly or storage of sensitive electronic parts should be made of antistatic or conductive materials. These type materials can help reduce damage due to static electricity.

The item(s) handled can also have an effect on the container, its performance, and the material from which it is made.

Ideally, the ratio of the container weight to the weight of the container *plus* contents should be as low as possible. This is especially true for a container which will be handled manually. A lighter container can help minimize worker fatigue and increase efficiency. Light containers, used for distribution and shipping can help reduce freight costs. They can also have an effect on operating costs by putting less wear on equipment.

In some situations, containers must comply with certain government regulations, especially containers used to handle food, where there is direct contact with the product. These containers must be manufactured from approved materials; the approval of which is based on their level of carcinogenic or toxic compounds. Most often approval of containers for use in the food, chemical, or pharmaceutical industries is granted by the Food and Drug Administration (FDA). There are, however, other agencies that regulate and approve containers, their use, and materials. They are: Department of Transportation (DOT), U.S. Department of Agriculture (USDA), and the Nuclear Regulatory Commission (NRC).

Consideration must also be given to the configuration of the container and how it will actually perform (i.e., stack only, nest only, stack and nest, etc.). Although containers can be found in almost any shape, the most widely used is rectangular. The rectangular-shaped container offers the greatest ease of handling, stability, conveyability, and adaptability to available space. It can also be cross-linked on pallets, providing a stable load.

14.2 TYPES OF CONTAINERS

Many containers are categorized by the method in which they perform. These are as follows:

14.2.1 Stacking-only Containers

These are containers that, whether used or stored, are always in the stacking position (Fig. 14.2.1). Typically, they have relatively straight or vertical sides to maximize the cube of the container and are normally used only in static storage situations. Their straight walls make them good for use on conveyors, because they do not ride up as tapered-end boxes can do if line pressure becomes too great. However, some caution is needed here, as some stacking-only containers have a very large top lip for strength. This lip can cause the container to act like tapered-wall containers if line pressure is excessive.

Fig. 14.2.1 Stacking containers, modular type.

14.2.2 Nesting-only Containers

These containers have tapered sides which allow one container to fit inside the other without preorientation of one container to another (Fig. 14.2.2). Nesting-only containers cannot be stacked unless they are used with lids. They are best suited to applications where space occupied by unused containers is of importance (e.g., truck backhaul space, back room storage, or valuable manufacturing space).

The nesting of containers is normally expressed in a "nesting ratio." This can range from 2:1 to 5:1, depending on the container design. A nesting ratio of 4 to 1, for example, means that each additional container added to a stack of nested containers will add $\frac{1}{4}$ of the height of one container to the stack. Obviously, the higher the nesting ratio, the greater the space conserved.

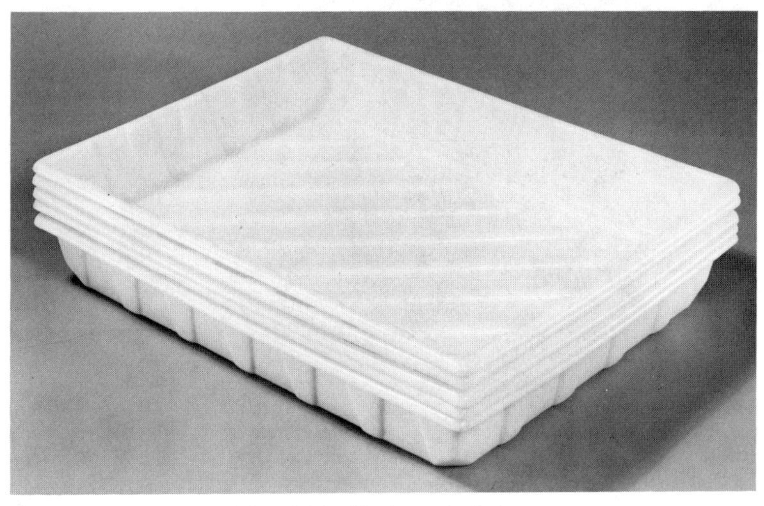

Fig. 14.2.2 Nesting-only design.

14.2.3 Stacking and Nesting Containers

These containers stack and nest without the use of a lid; they do, however, require orientation or the use of support mechanisms such as bails. In the case of the latter, either a flipping or sliding pair of bails takes the place of a lid upon which the container is stacked (Fig. 14.2.3).

Two popular styles that do not involve mechanical mechanisms for stacking are 90° and 180° stacking-and-nesting designs. Ninety percent stacking-and-nesting containers work on two principles. The first is the cross-stack principle. The container has tapered sides similar to a nesting container; stacking is accomplished by turning a container 90° to the container upon which it is stacked (Fig. 14.2.4). These containers are always rectangular (not square), and they usually have grooves in the bottom which engage the top edge or lip of the container beneath. This may also be achieved by small protrusions in the top of the lower container fitting into detents in the bottom of the upper container. In all other aspects, these containers act the same as nesting containers.

Fig. 14.2.3 Stack/nest containers with bails.

Fig. 14.2.4 90° cross-stack and nest type.

Fig. 14.2.5 180° stack and nest type.

The second form of a 90° stack-and-nest container involves the length, width, and height of two opposite side walls in relation to the adjacent end walls. When turned 90°, the container actually sits one inside the other. However this approach offers no more than a 2 to 1 nesting ratio.

The 180° stacking-and-nesting method requires 180° rotation from either the nested or stacked position to achieve positive stacking or nesting (Fig. 14.2.5). The end or side walls have alternating support indentations which allow nesting in one direction and stacking in the other.

Combinations of the foregoing approaches have also been used to obtain containers with multiple stacking levels.

Normally, stacking-and-nesting containers are utilized only in static applications. In dynamic applications, such as use as a distribution container, the disengagement of the stacking interference can cause one container to drop into the next which can result in product damage. Caution! When using stack-and-nest containers in a dynamic situation, the depth of engagement and the contact surface area should be taken into consideration.

14.2.4 Collapsible Containers

There are a number of types of collapsible containers, from die-cut corrugated sheets which can be folded, taped, glued, or stapled, either manually or mechanically, to rigid collapsible containers made from wood, steel, and so on. The more rigid types are all characterized by sides, bottoms, and sometimes tops with hinged members. Some types collapse completely flat, similar to a die-cut corrugated box, taking up no more space than the bottom of the container (Fig. 14.2.6). Typically, containers of this design offer a collapsed stacking increment (similar to nesting ratio) of 3:1 to 5:1.

Their advantages include vertical or straight side walls for maximum cube utilization. A disadvantage, however, is the time and labor required for assembly and disassembly and the wear and tear of the hardware required for their setup and disassembly.

14.3 CONTAINER SIZE

Another important aspect of container selection is size and how to determine the best size for the application.

Obviously, the size of the item(s) to be handled will have a great deal to do with the size of container selected.

Consider, too, the weight of the container, empty and full, and the method(s) by which the container will be handled. Will much of the handling be done manually, by forklifts, on conveyors, or on pallets?

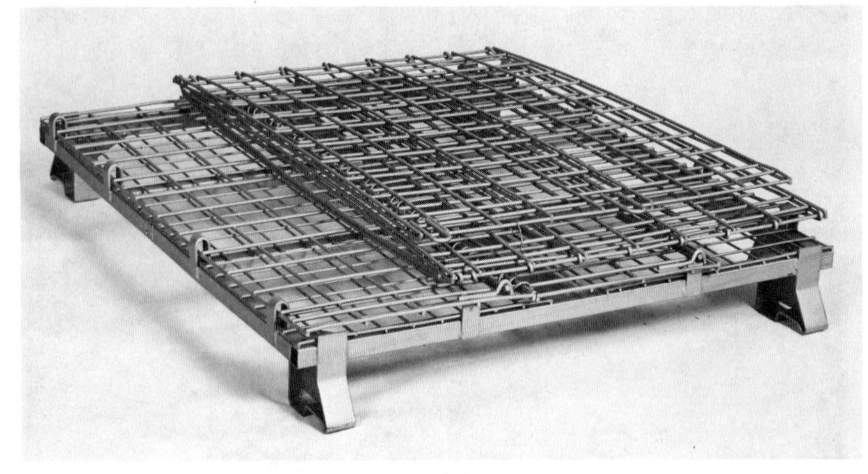

Fig. 14.2.6 Collapsible wire container.

When the containers are filled, will they be stacked? If so, how high and by what method (i.e., manually, forklift, etc.)? Will the container be required to support the load, the contents, or both?

Is the container to be handled on conveyors? If so, the size and capacity of the conveyor plus any openings in walls through which it passes must be considered. Also, is there adequate room to clear pipes, vents, conveyor guards, and so forth? (This is an important reason why the container should be considered in the initial planning of a system.)

How much space will be required for the storage of empty containers? This also can be an influence on the size container selected.

The ideal size container for any system should:

1. Fully utilize the cube of the mode of transportation and/or storage module, that is, internal dimensions of vans, trailers, rail cars, carts, shelves, pallet rack, and so on.
2. Hold enough product to be economically sound, yet when loaded be compatible with the weight limitations of each handling step.
3. Be compatible with the dimensional limitations of all equipment, as well as any physical characteristics of the environment in which it operates.
4. Maximize the amount or quantity of product held while satisfying all of the preceding criteria.

When the container is often the last item to be considered, the ideal container size is usually sacrificed because too much time and money has been spent to select, purchase, and often install the equipment preventing selection of the ideal size container. In most cases, had the container been considered *first,* the incremental cost difference to purchase compatible equipment with the ideal container would have been immaterial to the project.

Other points worthy of some discussion are security and labeling and identification.

14.4 SECURITY

As long as people are a part of the handling system, product losses will occur. Some losses will occur from damage, due to improper handling or a container that does not meet the physical or environmental characteristics of the handling cycle. Often overlooked, however, is the problem created by pilferage and carelessness. The value and attractiveness of the contents are normally the key to this problem. Very few containers are pilferproof; however, several designs provide resistance or deterrent to the problem by using a lid or cover and some method of sealing it.

The means of sealing takes many forms such as wire or plastic tie wraps, rivets, banding, or even tape (Fig. 14.2.7). They can be applied manually or automatically. The number of containers involved, and the cost of labor and equipment usually dictate the method. The additional cost of these deterrent accessories must be added to the total cost justification.

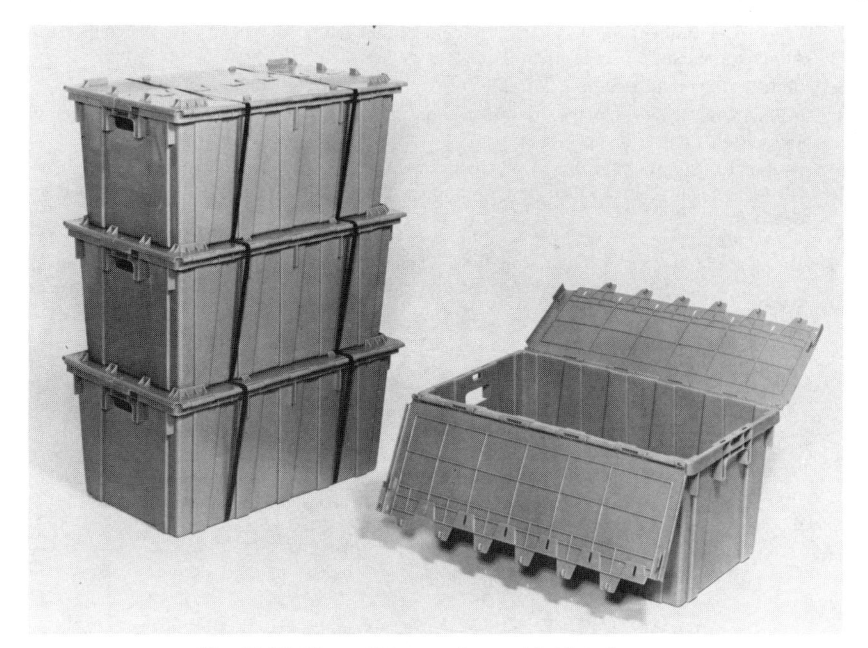

Fig. 14.2.7 Strappable containers with hinged cover.

14.5 LABELING AND IDENTIFICATION

Labeling and identification of containers can range from simple pressure-sensitive labels to more permanent types of identification such as stenciling, hot stamping, color coding, bar coding, and printing. Other methods include identification plates and ticket or label holders that are an integral part of the container or attached as a secondary operation. Caution! Attachments to a container can sometimes interfere with the container's function, such as nesting. These added-on items may also cause interference with other handling equipment.

Labels or other forms of identification can be used to identify for a variety of purposes: content, routing information, storage position, tare weights, ownership, and even warnings. An often overlooked benefit of using the company name and/or logo in some permanent fashion is that it can serve to support the company identity program, build pride, and, for reusable containers, aid in the proper return of the container.

14.6 COST FACTORS

One last, but important point in selecting a container—cost, or more appropriately, return on investment and payback. Cost elements that should be considered include:

1. Initial purchase price each.
2. Initial purchase freight (i.e., landed cost).
3. Disposable versus returnable or reusable. If reusable, the number of trips or uses must be estimated to develop a *cost per use* figure for comparison versus disposable containers.
4. Capital expenditure versus expense expenditure. Most companies have policies that dictate how the purchase is handled. These policies can have a significant impact on a company's return on investment and cash flow.
5. Labor costs:
   a. *Setup and breakdown* of corrugated or collapsible containers.
   b. *Repair* of broken slats, wire welds, torn sides, or bottoms for wood and metal containers.
   c. *Cleaning,* if required, of metal, wire, or plastic containers.
6. Space required to store containers not in use, whether it be reserve stock or the normal required quantity. In the case of reusables, space for the returns. In seasonable businesses, nonpeak usage storage must be considered also.

7. Auxiliary equipment, such as carton folders and assembly equipment, staplers, tape machines, gluing equipment, and washers.
8. Control systems to minimize losses of reusables and insure balance and flow.
9. Product loss due to damage, as well as shrink due to pilferage. Deterrents to pilferage can be used but their cost must be added.
10. Insurance costs can vary, based on the design and material of the container. Obviously, steel will not typically burn. Open design containers will allow a sprinkler system to work more effectively.

A materials handling system will function most efficiently and be of lowest cost over time if the container to be used in the system is considered first rather than last during the planning process.

CHAPTER 15
PACKAGING AND MATERIALS HANDLING

STERLING ANTHONY, JR.

Sterling Anthony, Inc.
Detroit, Michigan

The presence of packaging is pervasive in an industrialized society. That statement is easily supported when one considers that an industrialized society is characterized by the mass production, mass marketing, and mass distribution of a vast and ever-increasing variety of products. Studies have repeatedly shown that the degree of industrialization and the quality of life in a society bear a direct correlation to that society's per capita consumption of packaging materials. The major portion of an industrialized society's output must be packaged, to some degree, in order to be handled, stored, transported, and distributed.

This chapter is devoted to an overview of packaging as an industry function, with emphasis placed on its relationship to materials handling. Packaging and materials handling are inseparable. That is suggested by the fact that there exists a Society of Packaging and Handling Engineers. But more so, the two functions are inseparable because of their complete interdependency. The interdependency can be summed up this way:

> The manner in which goods are packaged affects the ease, cost, efficiency, and safety of handling them. Restated, the packaging influences the handling method.
>
> The manner in which goods are handled affects the materials, shape, dimensions, and weight of the packaging. Restated, the handling influences the packaging method.

Both present and evolving conditions place increased importance on the effective interactions between the packaging and materials handling functions. Among the most important are:

1. *Productivity.* Packaging and materials handling are components of the system of flows, whereby material moves into a production facility, is converted into a finished product, and that finished product distributed for consumption. As vital parts of that sourcing–production–distribution continuum, packaging and materials handling are expected to contribute toward the improved productivity of the overall system.

2. *Costs.* The materials, components, assemblies, and finished products that fill the supply and distribution pipelines of industry are most often packaged, and are frequently handled a number of times before arrival at final destination. Packaging and materials handling costs, combined, can account for a large portion of a product's selling price. Although indispensable functions, packaging and materials handing are purely expense items. Therefore, systematically managing the amount of packaging and handling performed on a product is a reliable means of decreasing product cost and increasing profit margins.

3. *Customer Service.* Increasingly, companies compete, not so much strictly on price, as on the concept of customer service. In other words, business volume can be won and maintained, based on the quality and consistency by which customer orders are received, processed, shipped, and delivered. Packaging and materials handling are customer service components that influence product assortment, speed of order picking, and delivery cost, among other important variables.

4. *Automation.* The three previous discussion points are influenced by automation. Productivity can be increased, costs decreased, and customer service made more effective—all with the proper type and amount of automation. As industry pursues the concept of the "automated factory" certain consequences loom for packaging and materials handling. The movement and handling of materials within a facility will be mechanized, and performed at higher speeds, and with higher degrees of accuracy. Such automated handling systems, in order to perform optimally, will require packaging of a higher standardization, tighter tolerances, and possibly greater strength.

5. *Imports/Exports.* Industry is unmistakably moving toward the global sourcing and distribution of products. Over time, the probability will increase that a given product contains imported components, and our given products will be exported. Packaging will truly have to be international in terms of its handling characteristics. A move in that direction is the international standardization of pallet sizes and intermodal container dimensions.

All of the foregoing paragraphs were meant to introductorily present packaging as an activity of large scope, complexity, and potential. However, in order to best utilize packaging, the materials handling professional must have an understanding of the basic nature of packaging, in terms of what packaging is, its various functions, and the different types of packaging materials, as well as the process by which a package is designed and developed.

15.1 DEFINITION OF PACKAGING

Undoubtedly, materials handling has been expertly defined a number of times within this handbook. But, what is packaging? That seemingly simple inquiry is likely to receive a multitude of different answers. For instance, to the marketing person, packaging is a means for presenting the product in a sales-generating light. To the distribution person, packaging is a means for protecting the product during handling, storage, and transportation. And, to the retail consumer, packaging is a means for deriving product usage satisfaction. While all such definitions are accurate, they are limited in scope, reflecting a particular user's interface with the packaging.

The best definition of packaging is that of a system. The system concept of packaging explains it as, not merely the physical container, but an interrelated set of activity components consisting of:

Basic raw materials (i.e., wood, sand, ore, and chemicals)

Converting operations that form packaging materials and containers

Production operations whereby the package is filled, closed, sealed, and quality checked

Unitizing or other preparations for distribution

Distribution through channels, involving storage, handling, and transportation

Emptying of packaging through product usage

Disposal, reuse, or recycling of the packaging

Three important points concerning systems packaging are worth bearing in mind:

1. As with all systems, the individual components are linked. The decisions made and implemented at one component level will carry ramifications at other levels.
2. The parties to the system are numerous and diverse, including: raw material suppliers, converters, package machinery suppliers, product manufacturers (package users), services (e.g., of designers and consultants), distribution channel intermediaries (e.g., carriers, warehousers, wholesalers, and retailers), and consumers of packaged products.
3. The packaging system itself is a component of the larger system of sourcing, manufacturing, and distributing products.

15.2 FUNCTIONS OF PACKAGING

Categorically, there are only four functions that packaging can perform. The four functions of packaging are: (1) containment, (2) protection, (3) communication, and (4) utility. Those functions should be found in all types of packaging, regardless of the product or industry involved.

Containment refers to the package's ability to serve as a receptacle: to hold its contents. When product spills, leaks, or otherwise escapes from its packaging, the containment function has been compromised. The containment function engineered into the packaging should reflect product characteristics, economics, and the recognized consequences of the product escaping from its packaging. For example, the manufacturer of a hazardous material would seek 100% product containment in the packaging. By contrast, the manufacturer of rock salt could choose packaging (e.g., shipping bags) which would occasionally result in product leakage. Because of product value and the innocuous effects of spilled salt, this approach is economically more justifiable than packaging having 100% integrity.

Protection is the function that enables the packaging to shield its contents from various hazards imposed by handling, transportation, storage, and atmospheric conditions. Closely associated with the containment function, the protection function, too, can be 100% or less. Generally, the more expensive, or the more critical in importance a product is, the greater the justification for engineering packaging that will provide the highest level of protection.

Communication is the function called upon to convey information and messages through the use of shape, size, color, graphics, symbols, and the printed word. Communications, such as display of brand name and product claims, are commonly-thought-of examples. However, the wire-mesh container that allows the observer to view the contents, and the corrugated shipping container with a pallet pattern printed on its top panel, are also examples of "communicating" packaging.

Lastly, utility is the function that facilitates the interaction between the package and those who come into contact with it. Too often, the utility function is associated strictly with the packaging of retail products. Features such as easy opening, reclosability, and convenient dispensing of the contents are examples. But, when the package can be easily handled through a facility such as a production plant or warehouse, that too is part of the utility function.

The main points to be remembered about the functions of packaging are:

1. Containment, protection, communication, and utility are the *only* functions of packaging. Never assign packaging a task that cannot be categorized under one of those four major headings.

2. A package should display a given balance and emphasis on particular functions based on product and market considerations. For instance, a military package might show strong emphasis on the containment and protection functions, with the communication and utility functions being addressed strictly from a fundamental standpoint. Contrastingly, a retail consumer package may exhibit some compromise of the protection and containment functions, while showing strong aesthetic emphasis in the communication and utility functions.
3. Any package can display a function ranging from a low emphasis to a high emphasis. The accepted generalization is that the ultimate cost of the packaging goes up in accordance with how high a degree each function is engineered into it. So the challenge becomes the development of packaging that displays a cost-effective balance of the functions, based on known conditions.

15.3 PACKAGING MATERIALS, CONTAINERS, EQUIPMENT, MACHINERY

Diversity is perhaps the best description of packaging materials, containers, equipment, and machinery. The industries are constantly challenging the packaging decision maker with new alternatives from which to meet his packaging objectives.

The major groups of packaging materials are:

Paper and paperboard
Metal
Glass
Plastics
Wood
Fabric

The major container categories are:

Fiberboard boxes
Folding and rigid cartons
Cans, drums, and pails
Bottles and carboys
Bags, sacks, and pouches
Wooden crates and boxes
Racks, bins, and other durable containers

Categorizing packaging systems and equipment by function, one can list:

Forming and assembly
Filling and loading
Weighing and counting
Overwrapping
Closing and sealing
Unitizing
Labeling and coding
Coating and laminating

When commenting on the size and makeup of the packaging industry, one has to define what is included. Most quoted figures reflect only the value of materials and containers purchased by the end-user (product manufacturer). Under those conditions, packaging has for decades been the largest consumer of paper and glass, one of the largest users of steel, and an increasing user of plastics. Actual statistics on the size and makeup of the various packaging categories are compiled by a number of sources, including:

The U.S. Department of Commerce
The Packaging Institute, U.S.A.
The Society of Packaging and Handling Engineers
The Fiber Box Association
The Packaging Machinery Manufacturers Institute

Some concluding comments about packaging industries follow:

1. The materials and containers industries are highly competitive. Companies compete against companies within the same industry, and against companies in different industries. For example, the plastics industry has taken certain volume from the glass and metal industries.
2. The user of packaging components should formulate procurement strategies for the choice and evaluation of packaging vendors. This is extremely important if the annual expenditures for packaging are a large part of the product's cost, as is often the case in the food, drug, and cosmetic industries.

15.4 PACKAGE DEVELOPMENT

Package development is the procedural framework through which a package comes into existence. It is the "birth process" for a package. Some companies perform package development in a hit-or-miss fashion, with predictably, the results being hit or miss. Others approach package development as a systematic and disciplined process, and experience a higher incidence of success and satisfaction with the resulting package.

Acknowledging the incredible variety of goods that are packaged, the reader might reasonably wonder if there can be a package development process that is universally applicable. The answer is yes.

The package development process that is advocated consists of five basic steps: (1) gather information, (2) formulate requirements, (3) design, (4) test, and (5) improve the packaging.

STEP 1. Gather Information. Successful package development is absolutely dependent on accurate and relevant information. If this initial step is not performed correctly, the project is slated for failure at the onset, or, at best, will not achieve the level of success possible.

It is imperative to recognize that packaging is an interdisciplinary activity, and that correctly performed package development cuts across functional lines. Every major function within the company—marketing, distribution, finance, manufacturing, engineering, and so on—has requirements and expectations of packaging. Package development should start by receiving the inputs of all concerned and affected parties.

Here are some suggestions for better managing the information gathering stage:

1. Formally establish the lines of communication between the packaging function and the other functions in the company that are served by packaging. This involves agreeing upon responsibilities and expectations, and reducing the same to policy statements. Such a statement, for instance, might read, "the physical distribution (or materials management, or logistics) department is responsible for providing the information necessary to develop packaging that is compatible and efficient with the company's materials handling, transportation, and warehousing methods."
2. Develop comprehensive checklists to lend structure and organization to information gathering. This minimizes the chances for oversight. A checklist for information related to materials handling should detail receiving facilities and equipment for incoming packaging supplies, storage conditions of those supplies, how packaging supplies are moved to the production line, how the packages are filled, closed, and unitized, in-plant storage of packaged product, number and type of handlings encountered during distribution, and the handling methods of the receiver of the packaged product.
3. Start the information gathering process early. It is always advisable to pair package development with product development. Such an approach permits changes to be made in the product that makes the product easier or less expensive to package, without impairing the product's basic function and character. Also, the more time allowed for development, the less likelihood of getting backed into a rush job situation, along with mistakes that so often accompany eleventh-hour decisions.

STEP 2. Formulate Requirements. The information gathered in step 1 is used to formulate packaging requirements. Requirements state what the packaging must do. As explained earlier, packaging can perform only four functions: containment, protection, communication, and utility. Packaging requirements that are formulated around the four functions eliminate the possibility of assigning packaging a task that it is not equipped to perform. Consider these suggestions:

1. Express packaging requirements in the form of written specifications and standards. Keep the specifications and standards revised and updated.
2. State packaging requirements in measurable terms, whenever possible. For instance, rather than state that the package must protect its contents throughout the distribution system, state protection requirements in terms of established test methods that simulate the expected distribution conditions.

3. To the extent practical, allow the providers of the information on which the requirements are based to approve those requirements. Sometimes, the package developer misinterprets the supplied information or fails to understand it fully. Faulty requirements will result unless a mechanism exists for detecting such miscommunications.

STEP 3. Design. At this juncture, information has been gathered and used to develop packaging requirements. The choice of materials and the design of packaging to meet those requirements is the next logical step.

Some coverage was given to the different types of materials and containers earlier in this writing. However, it is worth commenting that the degrees of complexity associated with this step run the gamut. At times, the choice of material or container is automatic, perhaps as a result of regulation, tradition, or economics. More often, the package developer must choose among competing material and container alternatives.

How well this step is performed is inextricably tied to the awareness of available or possible materials and containers. Attempting to stay abreast of the constant parade of new materials, designs, and processes in packaging can be a full-time undertaking in itself. Yet, unless the choice of material, container, and related systems is an informed choice, this step will be underperformed.

Remember these points:

1. The final package is a compromise of many different (and often conflicting) requirements. You can't please everyone, or, more correctly, you can't please everyone equally. Choose the package that best meets a set of prioritized requirements.
2. Design for simplicity. When the materials or design of a package are needlessly complex, the package costs extra time, effort, and money throughout its life. Supplier competition is reduced, quality control is more difficult, setup, filling, and closing operations are slower, and so on.
3. Be aware of the possible pitfalls of innovation. Base packaging innovation on specific criteria, rather than on creative whims. If a product has a history of acceptance in a certain type of container, switching to a radically different container may cause problems in product recognition and product acceptance. It would be advisable, under such circumstances, to stay within the traditional container category, but distinguish the product through innovative graphics, or utility features, for instance.

STEP 4. Test the Packaging. How does the package developer know if the chosen package meets the requirements set by the gathered information? By testing. The purpose of testing is to determine packaging performance under specified conditions. These conditions should bear strong correlation to actual "field" conditions.

Testing has gained importance as a result of increased government regulations, consumer concerns for product safety, and the costs associated with product loss and damage due to inadequate packaging. But package testing is far from an exact science—a truth that has caused frustration and worse to those who have made decisions without questioning packaging test results. Test data always must be viewed through the analytical visors of common sense and experience.

The following points should be considered when designing tests for packaging:

1. The most important requirements should always be tested. Laboratory tests can determine package/product interaction and how well the package guards against moisture, gases, and so on. Consumer perception tests lend insight into the package as a marketing tool. Distribution tests measure the package's protective capabilities under transportation, storage, and handling conditions.
2. Tests should be performed according to written procedure. To have true validity, test results should be reproducible. This is possible only if testing is repeatedly performed under the same procedure.
3. Abandon or revise a test procedure as soon as it is discovered that the test is not predictive.
4. Stay sensitive to costs. Testing should be less expensive, in time and money, than acquiring the information by other means. Also, investigate the opportunities of substituting outside testing services and vendor certification and testing in lieu of in-house testing.

STEP 5. Improve the Packaging. The last step is to improve the package in response to changes in information, requirements, materials, processes, regulations, consumer preferences, distribution channels, or a host of other factors.

The life blood of any company is, in large part, its ability to successfully develop and market new products and to extend the profitable lives of existing products. Packaging has remarkable value in new product introduction, boosting sales of current products, decreasing costs and increasing profit margins, and expanding distribution boundaries.

Keep these suggestions in mind:

1. Make packaging improvement an ongoing and organized effort. The more successful marketers of packaged goods develop a budget and specific objectives for package improvement.
2. Do not think of package improvement as being synonymous with reducing the cost of packaging. A more attractive, stronger, or otherwise improved package can pay for the increase in package cost through benefits and savings in other areas.
3. Devise an auditing process to uncover areas for package improvement. Make sure the audit covers major areas of package performance, such as distribution, marketing, and manufacturing. The purpose of the audit, of course, is to uncover opportunities to make packaging more efficient and responsive to needs and conditions of the company.

15.5 THE CONCEPT OF STANDARDIZATION

Standardization is a highly important determinant of an efficient interface between packaging and materials handling. Standardization denotes an attempt to limit the amount of variability of the packaging. As stated earlier, the method of packaging influences the method of handling. Therefore, the more standardized (less variable) the packaging, the more standardized (less variable) can be the handling method. Lower investment in handling equipment, as well as higher utilization of handling equipment are direct consequences of standardized packaging.

When speaking of standardized packaging, one is almost always referring to the physical characteristics of dimensions, shape, and weight, rather than the packaging material. The reason is that the physical characteristics of packaging influence the type and capacity of the materials handling equipment, more so than does the material. For example, a dump truck can handle corrugated fiberboard boxes, corrugated plastic boxes, or paper-wrapped rolls—different materials—as long as their physical characteristics are within the capacity constraints of the equipment.

Packaging standardization, however, does not only permit standardization of materials handling equipment. Facilities and methods are also influenced. In other words, time-consuming and costly revisions of shipping and receiving areas, storage areas, line feeding procedures, and packaging filling operations, for instance, can be held to a minimum. Furthermore, packaging standardization carries the potential for lower packaging material costs through volume related purchase discounts, reduced packaging inventories, and reduced warehousing space requirements.

One major caution is in order. The main obstacle to standardized packaging is nonstandardized products. When one attempts to minimize the different types of packaging used for a variety of different products, penalties can result in the form of dunnage requirements and overall decreased package density.

Packaging standardization takes two major forms: (1) standardization of auxiliary handling devices (i.e., pallets, skids, slipsheets, and durable containers) and (2) standardization of the individual package (i.e., boxes and barrels) that are placed on or in the auxiliary handling devices.

In standardizing packaged loads for handling purposes, the recommended sequence is to first standardize the auxiliary device, and then standardize the individual packaging. For example, in the case of palletized handling, one would standardize the pallet dimensions first, and then standardize the dimensions of the packaging, let's say cartons, so as to make the best utilization of the surface area of the pallet.

It is imperative that packaging standardization be performed with a systematic approach. When seeking to standardize packaging to achieve better handling, consideration should be given to the entire range of activities and conditions existing between initial receipt of the packaging to final disposal of it. The standardized packaging should represent the best compromise obtainable, having factored in such variables as: purchasing price and arrangements, inbound transportation, in-plant receiving, storage and handling, in-plant manufacturing procedures, distribution channels, and disposal or recycle requirements. A package with physical characteristics that best combines with a company's handling equipment and methods may not be the best package in terms of utilizing storage and transportation spaces, and vice versa. Therefore, in order to manage package standardization systematically, competing factors should be prioritized, and the final package reflect consideration for the factors in order of their prioritized importance.

An evolving trend is the computerized determination of standardized packaging. Companies exist that provide the service of calculating by computer, packaging dimensions that best suit a particular set of handling, transportation, and storage conditions. Such services determine the dimensions of the primary package units, shipping cartons or other secondary packaging, and the unitized loads. Also, an increasing number of manufacturers have developed in-house capabilities for this use of the computer.

Relatedly, the American National Standards Institute published the "American National Standard for Unit-Load and Transport-Package Sizes" (ANSI MH10.1M-1980). The standard sets forth a number of unit-load and transport package dimensions (expressed in metrics) that efficiently utilizes the cubic areas of the international general-purpose freight containers (containers, trailer trucks, and railroad boxcars). Additionally, the standard presents guidelines for determining the linear dimensions of the filled transport packages so as to make the most utilization of the areas of the unit-loads specified in the standard. While ANSI MH10.1M-1980 treats mainly the transportation element of packaging standardization, it can be a useful reference document, if the user recognizes its limited scope and makes the necessary analytical allowances.

15.6 DURABLE CONTAINERS

Durable containers, as their name implies, are packages of long service lives. Unlike expendable packaging, durable packaging is reused repeatedly, commonly over a period of years. Examples of durable containers include racks, bins, and welded wire-mesh containers.

Durable containers are an important facet of the packaging/handling interface in a large number of industries. Durable containers are most compatible with the following conditions:

1. High-volume products
2. Frequent movement or shipments between points in a flow system

The preceding section on standardization is particularly relevant to durable containers since they represent a capital investment. Therefore, a large number of different types of durable containers can be very much more costly than standardizing on fewer types.

It is generally agreed that the design of durable containers should provide efficiencies in the handling, storage, and transportation areas. So, product densities, transport vehicle dimensions, storage area characteristics, and materials handling equipment capacities are of primary concerns.

Three principles should govern the design of durable containers: durability, stackability, and flexibility. Durability, of course, is required as a function of the long-term usage of the container. For that reason, high-strength steel is the most common material for constructing durable containers. Containers should be capable of being stacked, in order to better utilize storage space. It should be remembered that the stacking requirement also influences the durability requirements of the structural members that must bear the stacking stresses. Lastly, flexibility relates to making the container useful for different products. Grouping products by density, adjustable height extenders, and effectively designed dunnage are the best means to achieve flexibility.

The major advantages of durable containers are:

Favorable economics over expendable packaging, under high volume, frequent shipments between set points

Reduces solid waste disposal costs and problems associated with expendables

Reduces handling costs and allows for greater standardization of equipment and procedures

Can be stacked higher than unsupported expendable packaging

Reduces fire hazards, since returnable containers have few flammable components

Improves inventory control when counts within a container are standardized

When properly sized and dunnaged, provides better product protection from expendables

Possibility of negotiating with the carrier for rate-free return of containers

Possibility for increased systems efficiency, especially between supplier and manufacturer, between manufacturer and customer, and between sites of a multifacility manufacturer

The disadvantages of durable containers include:

The requirement of an upfront initial capital investment

Maintenance and repairs of containers

Administration and control of containers

High volume and shipping requirements necessary to economically justify use of durables

The major issues that must be addressed in getting a durable container program started are:

Numbers and types of containers needed

The allocation of the containers among facility sites

Ownership (whether centrally owned, or owned by one division, for instance)

Accounting for the location of the containers (best done by computer, for large volumes)

Method for charging for repairs and maintenance

Incentives and penalties that promote compliance of the parties to the rules of the durable container program

15.7 INTERMODAL CONTAINERS

Intermodal containers are enclosures of standardized dimensions (e.g., $8 \times 8 \times 20$ ft and $8 \times 8 \times 40$ ft), transportable by various modes (truck, rail, ship, and plane). They find their greatest usage in international shipments and domestic shipments of great distances, such as coast to coast.

Intermodal containers are discussed here because they have a direct impact on packaging and materials handling efficiency. Specifically, they:

Reduce the degree of product handling, since the entire container is handled as a unit.

Reduce the required amount of packaging (possibly) since the product is handled less.

Reduce the incidence of loss and stolen products, since the container can be locked and sealed until final arrival.

Increase utilization of the benefits of the various transportation modes since the containers can be transported by different conveyances.

15.8 MAJOR COMPONENTS OF THE PACKAGING AND MATERIALS HANDLING INTERFACE

With bulk handling being the primary exception, the vast majority of industry's output must be packaged in order to be economically and safely handled, stored, and transported. The packaging and materials handling interface, then, is a vital component in a logistics/material flow system (see Fig. 15.8.1).

In attempting to manage the packaging and materials handling interface, the decision maker will invariably grapple with three interrelated concerns:

How to package the product for best handling

How to handle the product for least damage, and best utilization of labor, equipment, space, and capital

How to package and handle the product for best interaction with other systems' functions (i.e., transportation, warehousing, manufacturing, etc.) for best overall systems performance

Imaginative, aggressive, and effective responses to the foregoing concerns are a sure path to reduced costs and greater productivity. This chapter concludes by discussing the major components of the packaging and materials handling interface: labor, equipment, space, technology, and management.

15.8.1 Labor

Even as advances in automation become commonplace, and the prospects for the automated factory loom a distinct possibility, the task of moving material throughout the flow cycles is still rather labor intensive. Labor, an expensive element of flow costs, can be reduced or made more efficient when:

1. Packaging allows the movement of a given quantity of material through a minimum number of individual moves. In other words, it is more efficient to move case loads and unit loads than individual packages.
2. Packaging allows for fewer labor-related mistakes and facilitized performance in inspection, order picking, and inventory control activities. When the packaging is communicative and functional, there is less opportunity for mistaken identity, the wrong goods shipped, inventory miscounted, and so on.
3. Packaging allows for the safe interaction with the human element. Every company wants to minimize the chances of injured people, lawsuits, and negative publicity stemming from packages that topple, strain the workers' physical strength limits, or fail to protect the worker from exposure to harmful contents.

15.8.2 Equipment

Often mistakenly thought of as the opposite of labor, equipment is an adjunct to labor, its purpose being to make labor more productive. The manner in which equipment is chosen and acquired can affect a company's liquidity, profitability, and competitive position for a long time. The equipment issue, therefore, deserves careful analysis of many factors, not the least of which is packaging.

Let's examine some basic relationships between packaging and the three major equipment categories: trucks, conveyors, and cranes and hoists.

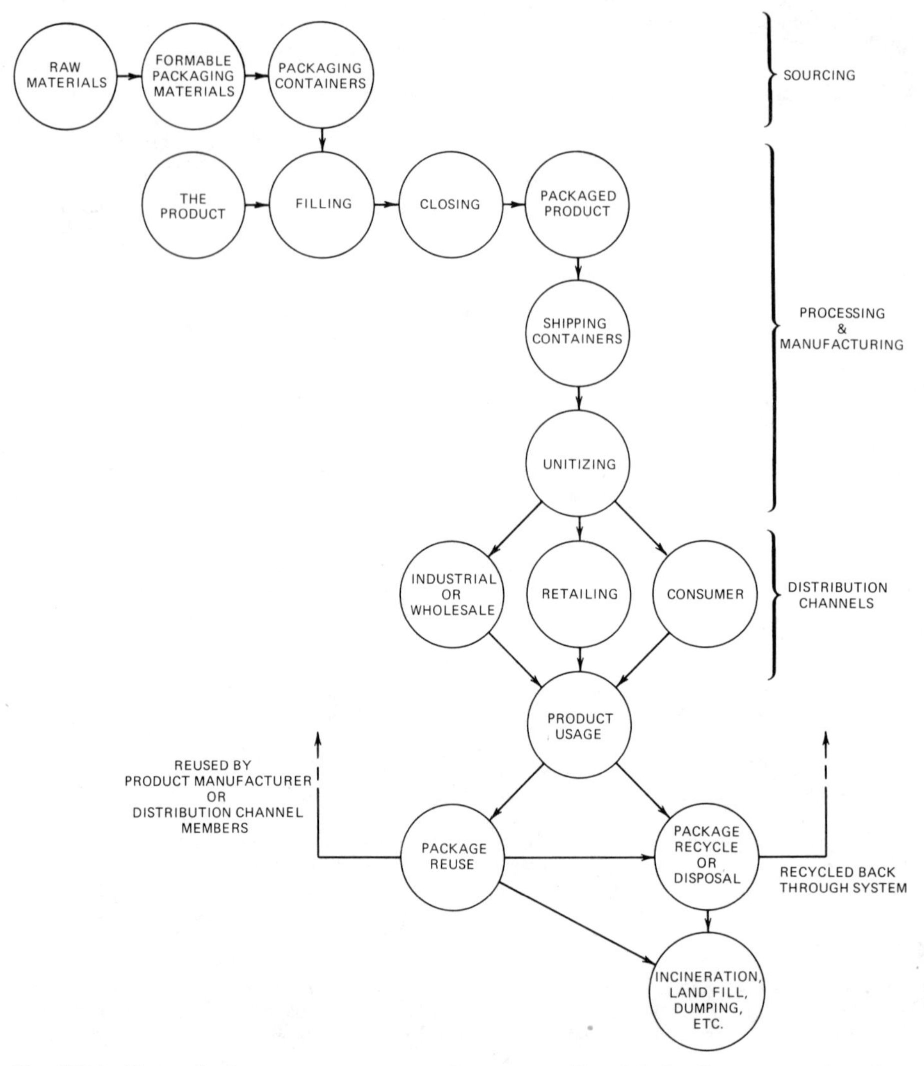

Fig. 15.8.1 The packaging systems concept: a flow system. Materials handling occurs throughout the system, from start to finish.

Trucks

The fork truck, the workhorse of industry, is inseparably associated with the pallet. Is a pallet a packaging or materials handling item? An argument can be made for either case. But such an argument would be pointless if it fails to realize that the pallet's identity crisis underscores the close interface between packaging and materials handling.

There are many opportunities to improve the utilization of trucks with packaging and unitizing that better challenges the lift capacity of the trucks.

Furthermore, attachments such as clamp and push–pull are directly related to the type of packaged loads handled, and they affect equipment utilization, operator skill requirements, and the costs for equipment acquisition, operation, and maintenance.

Conveyors

Infeed, outfeed, integrating, and linking of operations; these are the major functions of conveyors. Conveyors carry individual packages up through complete unit loads. Packaging, in order to run efficiently on conveyors, must be conscientiously designed to do so. A package too small or too large for the conveyor on which it runs spells trouble. Likewise does a package too light to take the inclines and declines that a conveyor route may take through a facility, or one too heavy for the drive mechanism that powers the conveyor.

And, of course, conveyorized movement imparts shocks, vibrations, and impacts to the packaging; hazards which should be considered in determining the packaging's strength requirements.

Cranes and Hoists

Packaging handled by cranes and hoists must be suitable by providing the means for attaching the slings, for instance.

Just as basic is the recognition that acceleration, deceleration, and gravitational forces act on a package as it is lifted, held suspended, and lowered. Careful analysis is called for here, for if a load becomes unhitched, or the bottom of, say, a crate falls out, product damage is a certainty, and personnel injury is a real possibility.

15.8.3 Space

Packages and unit loads occupy space—in storage and in transportation vessels. That truth should be reflected in the determination of what the packaging should be in terms of shape, size, and weight.

Simply put, packaging and unitizing should be designed for the efficient use of cube. It is difficult to overstress that need in these times of transportation deregulation, and the trend toward consolidation of warehousing facilities.

15.8.4 Technology

Some of the newer technologies upon which industry is spearheading its productivity drive are directly related to the packaging and materials handling interface. Specifically: automated storage and retrieval, optical scanners, automatic guided vehicle systems, and robotics.

As an example, unit loads, in order to be compatible with automated storage and retrieval systems, must conform to the weight and cubic limitations of the storage addresses.

Optical scanners, in addition to relying on the package to carry the machine-decipherable code, are influenced by the printing quality and contrast of the code, as well as the orientation of the package as it passes the code reader.

Automatic guided vehicle systems transport unitized loads. Those loads should be secure, stable, and within stated dimensional limitations so that they do not tilt over, come apart, or otherwise prove incompatible with the method of handling.

And robots can be programmed to palletize, load and unload shipping containers, and even perform package testing.

This generalization tends to hold true: Automated systems of the type just described are most productive under repetitious and standardized conditions. Translated: Automation is basically inflexible. Standardized or modular packaging can permit a greater degree of automation than is possible under conditions of greater variability of packaging.

An additional challenge to packaging takes the form of the need to be more exacting in holding quality and dimensional tolerances so that the packaging will be more compatible with automation.

15.8.5 Management

The packaging field is, in a word, dynamic. New materials, containers, and technology are the rule, not the exception. For instance, there have been quite recent innovations in all the major means of unitizing—pallets, slipsheets, durable containers, stretch wrapping, shrink wrapping, strapping, and adhesives. The result is that it is virutally impossible to stumble on an efficient packaging system that integrates well with materials handling and other material flow functions. On the contrary, the path toward such packaging must be charted with systematic analysis and decision making.

Here are some checklist items by which to evaluate the management of the packaging and materials handling interface:

To what degree does the packaging for raw material received from your suppliers facilitate your operations of receiving, unloading, inspection, and movement to storage or production?

To what degree do captive handling units (i.e., pallets, bins, trays, racks, and hoppers) facilitate in-process storage, inventory control, and production line feeding and out-feeding?

To what degree does the packaging of the finished product facilitate labor, machinery, materials, and space used in assembling the packaging?

To what degree does the packaging of the finished product facilitate storage, order picking, loading into transport carrier, and low transportation damage rates?

To what degree does the packaging of the finished product facilitate your customers' receiving, unloading, inspection, storage, and production operations?

To what degree does the packaging facilitate disposal, reuse, recycling, and so on?

The flow of raw materials from supply sources into the manufacturing location, the flow of finished goods from the manufacturing location to the customer, and finally, the flow of refuse, reclaimable materials, and recalls back through the system—all can be made more efficient and productive by better integration and management of the packaging and materials handling interface. As industry plans layouts for new facilities and flow systems, and revises existing facilities and flow systems in its pursuit of greater productivity, reduced costs, better customer service, and improved sales and profits, it should look to the packaging and materials handling interface as an indispensable building block.

BIBLIOGRAPHY

Apple, James M., *Material Handling Systems Design.* Ronald Press, New York 10016. 1972.

Brody, Aaron and Jack Milgrom, *Packaging in Perspective.* Arthur D. Little, Inc., Cambridge, MA 02140.

Friedman, Walter F. and Jerome J. Kipnees, *Distribution Packaging.* Robert E. Krieger Publishing Co., Huntington, NY 11743. 1977.

Hanlon, Joseph F., *Handbook of Package Engineering.* McGraw-Hill, New York. 1971.

CHAPTER 16
FREIGHT ELEVATORS

WILLIAM S. LEWIS

Jaros, Baum & Bolles
Consulting Engineers
New York, New York

16.1 THE FREIGHT ELEVATOR AS A MATERIALS MOVER

16.1.1 Freight Elevators as General Conveyances

Vertical conveyances have existed since early Roman times as a direct result of the application of the wheel to either a pulley or windless driven by man or animals. These vertical conveying devices were all basically batch conveyors more often used for materials than for people due to the relatively unsafe arrangements "invented" by their owners or local craftsmen. The early conveyances using animal power were replaced by the Industrial Revolution in the early part of the nineteenth century with steam-driven engines as multiple-story factories became an important evolution of the cottage industries. The relatively primitive horizontal transportation systems dictated an early concentration of the cottage industries into multiple-story complexes that required the movement of raw materials to the work place and the movement of finished goods to the horizontal transportation interface. The equipment put together by local foundries and machine shops often resulted in major accidents due to the failure of ropes or chains, since there is no way in which to defeat gravity when the vital supporting link fails.

In the true sense of the word, the elevator did not become a mature concept until 1853 when Elisha Graves Otis invented the "safety" elevator that retained its integrity when the hoisting ropes parted. At that time guided platforms with wooden rails were common and Otis equipped the platform with a wagon spring that, when the tension was removed from the ropes, engaged a cog design guide rail at each side of the platform (see Fig. 16.1.1).

Elisha Otis' safety brake in 1853 made the freight elevator as safe a means of vertical conveyance as George Westinghouse's air brake made the railroad car a safe means of horizontal conveyance. The parallel development of the elevator with the railroad soon brought about the dispersing of manufacturing facilities from the centralized concept around waterways and permitted the migration of manufacturing facilities to areas where land was more plentiful and factory construction costs could be reduced by single-story concepts. In these, the horizontal material movers were most cost effective and less labor intensive than the vertical factory with its labor-intensive vertical transport.

The freight elevator as a materials mover must respond to the needs of the manufacturing process to transport raw, semifinished, or finished materials, as a batch conveyor to what is essentially a continuous flow process. In such a context, the freight elevator became a batch mover that had to respond to the conveyor mover because of its larger size and handling capacity on its vertical round trip. In this context, whenever the single elevator, limited in size by the current technology, could not handle the throughput required by the manufacturing or warehousing process, a multiplicity of elevators were installed, each operated by an attendant who responded to known schedules or voice commands shouted into the hoistway. This manual system was labor intensive and expensive to heavy manufacturing, such that the migration to less expensive land and horizontal factories became a major development as the railroad provided ubiquitous and reliable horizontal transportation to user or water-shipping points.

The current computer, minicomputer, and microprocessor technologies have permitted the automation of freight elevators so that they can become a cost-effective slave to a major materials handling system operating on multiple levels of a multistory facility. The multistory factory, processing plant, or warehouse, located in a major activity center, can now be retrofitted with an automation system for vertical materials movement as effectively as the high-rise office building has been automated for pedestrian movement. Most often, this has taken the form of retrofitting existing facilities with automated conveyors that deliver pallets, carts, tote boxes, or other containers to multiple levels for processing, warehousing, receiving, and shipping. The future may hold that the multistory factory will return to the major activity centers as efficiencies can be demonstrated. This will be accelerated whenever the cost of the horizontal commutation of factory workers increases substantially.

Freight Elevator Defined

Elevator Definition. The freight elevator is defined in the A17.1 Safety Code for Elevators as follows:

Elevator—A hoisting and lowering mechanism, designed to carry passengers or authorized personnel, equipped with a car (elevator car) which moves in fixed guides and serves two or more fixed landings.

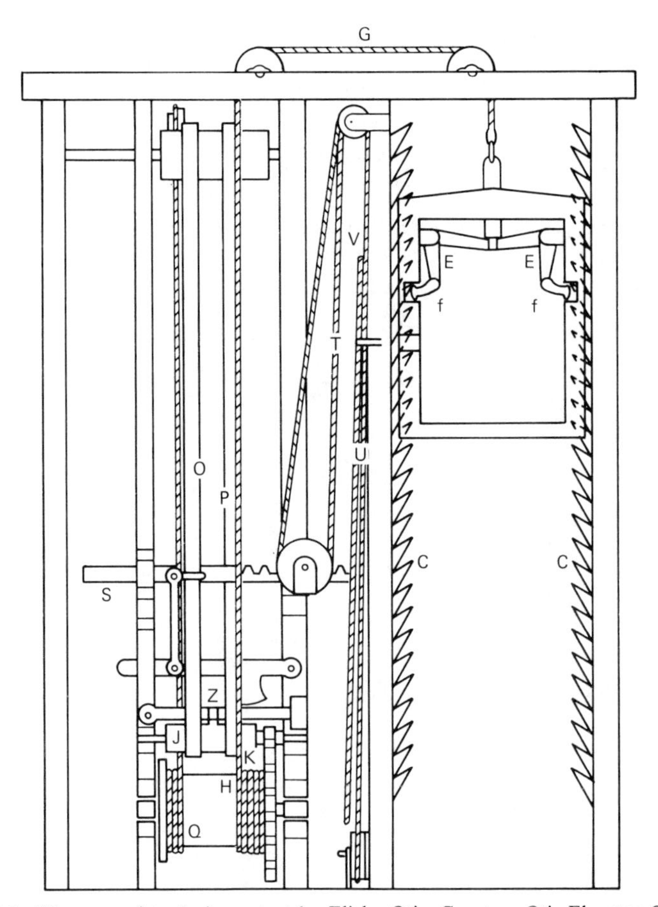

Fig. 16.1.1 Elevator safety device patent by Elisha Otis. Courtesy Otis Elevator Company.

Freight Elevator:

Elevator, Freight—An elevator primarily used for carrying freight and on which only the operator and the persons necessary for unloading and loading the freight are permitted to ride. Its use is subject to the modifications specified in Section 207.

Rule 207.1:. . . Passenger elevators and freight elevators permitted by Rule 207.4 to carry employees shall conform to the requirements of Rule 207.8.

Rule 207.4—Carrying of Passengers on Freight Elevators: Freight elevators shall not be permitted to carry passengers.

Rule 207.8—Additional Requirements for Passenger Overload: Passenger elevators and freight elevators permitted by Rule 207.4 to carry employees shall be designed and installed to safely lower, stop and hold a car with an additional load of up to 25% in excess of the rated load.

These quotations of applicable rules for freight elevators are taken from the 1981 edition of the A17.1 Safety Code for Elevators and should be used as a reference only since this code is annually revised. The current edition should always be referred to for specific projects along with the annually published interpretations.

It should be noted that this code recognizes only two categories of elevators, passenger and freight. For practical purposes, if passengers are to be carried on a freight elevator with any consistent traffic pattern the elevator should be designed in accordance with passenger elevator capacity design relating to the inside net area, and equipped with appropriate passenger elevator doors. This elevator more often is shaped narrow and deep as a common freight elevator, as opposed to the usually configured passenger elevator which is wide and shallow for ease of passenger transfer.

The following equations are used in determining the minimum rated load W for passenger elevators or freight elevators consistently carrying passengers. The reference is Rule 1300.1 of A17.1 Code.

For an elevator having an inside net platform area of not more than 50 ft² (4.65 m²):

$$W = 0.667\,A^2 + 66.7\,A$$

For an elevator having an inside net platform area of more than 50 ft² (4.65 m²):

$$W = 0.0467\,A^2 + 125\,A - 1367$$

where A = inside net platform area, in ft² (m²), and W = minimum rated capacity load, in lb (kg). Figure 16.1.2 is a graphical representation of this equation.

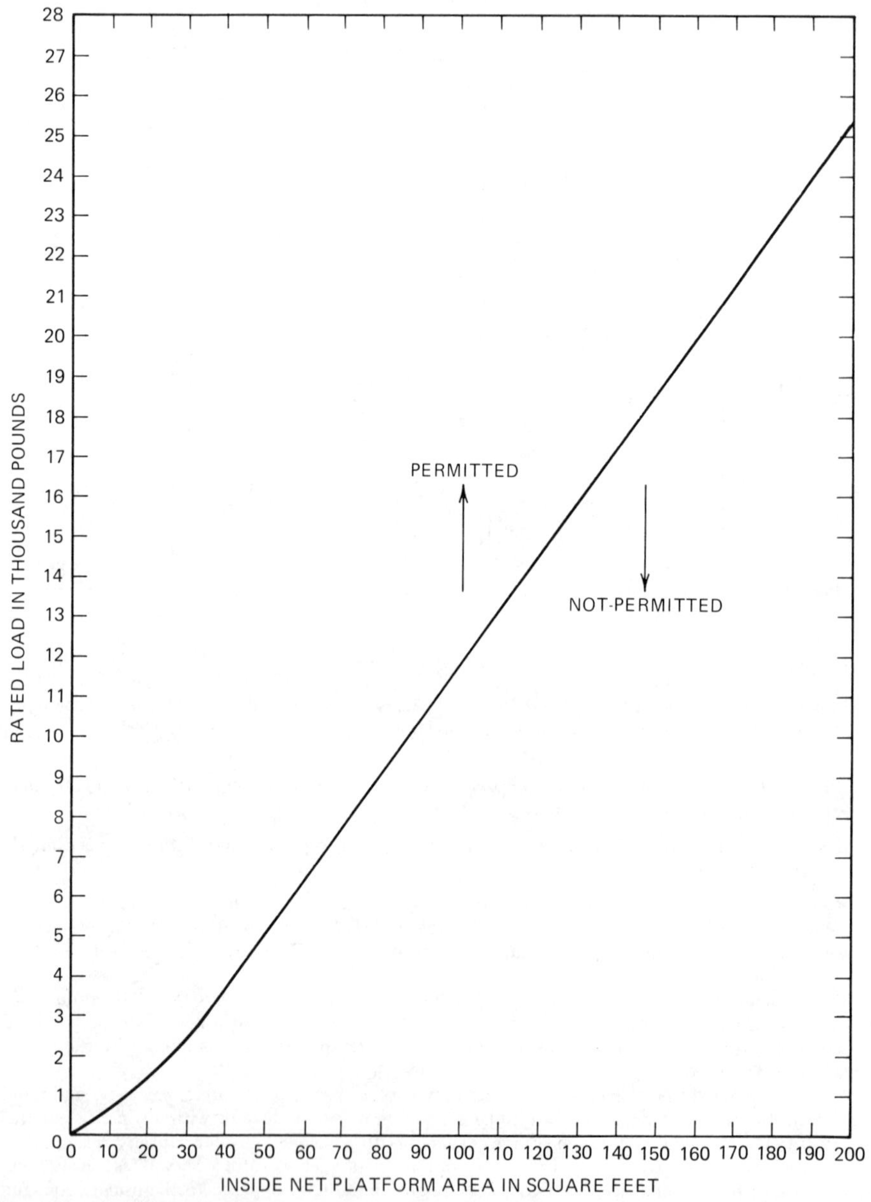

Fig. 16.1.2 Minimum rated load for passenger elevators by A17.1 code. Courtesy of Jaros, Baum & Bolles.

16.1.2 Conflict in Codes

"Paragraph Deleted"

16.1.3 Hydraulic Elevators

The least expensive and slowest-speed elevator is the oil hydraulic elevator utilizing a direct-acting hydraulic piston, with oil pumped by a dedicated hydraulic system to raise the elevator (see Fig. 16.1.3). The elevator travels downward by gravity alone. The speed in both directions and the leveling of the platform are controlled by a complex hydraulic oil-control valve which gradually accelerates an elevator car, travels it at a constant speed in either direction and decelerates it when approaching a landing for an accurate stop. Hydraulic elevators have the advantage of low overhead heights and shallow pit depths as well as remote machine rooms where permitted in some local jurisdictions. The code permits hydraulic elevators to be installed without car safeties, which are required for electric traction elevators. This deletion concentrates the hydraulic elevator structural reaction in the elevator pit under all operating conditions. Hydraulic elevators can operate at UP speeds up to 200 ft/min (1.0 m/sec), but most often operate at speeds of 150 ft/min (0.75 m/sec). The maximum vertical rise is limited to approximately 70 ft (20 m) for the smaller size freight elevators.

16.1.4 Geared Elevators

Geared elevators are used for intermediate speeds and for higher travels than permitted by hydraulic elevators (see Fig. 16.1.4). While a winding-drum geared elevator is permitted by the code, under Rule 208.1, its use is restricted to elevators that do not travel in excess of 50 ft/min (0.25 m/sec) and whose travel does not exceed 40 ft (12.2 m). For practical purposes, this elevator is of such limited design as to not be a current product from most manufacturers. An elevator with a counterweight is called a traction elevator and, in the geared version, has a current upper speed limitation of 450 ft/min (2.3 m/sec) with a maximum travel of approximately 20 stories depending on the size of the freight elevator and the necessity of special machine designs for handling capacity loads. The geared elevator is a cost-effective design which uses a high-speed, low-torque electric motor with a worm-to-worm gear reduction between the motor shaft and the driving sheave. It is usually installed with a maximum vertical acceleration and deceleration of 3.0–3.5 ft/sec² (0.9–1.1 m/sec²). In small freight elevators it is usually roped 1:1 between the car and the counterweight, whereas in larger elevators, to reduce the loading on the traction drive sheave shaft, it is usually roped 2:1, so that one-half of the structural supporting reaction can occur on dead-end hitches in the machine room for both the car and the counterweight.

16.1.5 Gearless Elevators

A gearless traction elevator has a low-speed, high-torque electric motor with a drive sheave mounted directly on the motor armature shaft. It is installed at speeds from 300 ft/min to 1000 ft/min (1.5–5 m/sec) for large freight elevators. Vertical rise is virtually unlimited; however, it is most effectively installed with a roping of 2:1 to the car and to the counterweight, which probably would limit its travel to 500 ft (150 m) (see Fig. 16.1.5). It is usually installed with a maximum vertical acceleration and deceleration of from 4.0 to 4.5 ft/sec² (1.2–1.4 m/sec²) when roped 1:1.

16.1.6 Freight Elevators as Material Movers

The freight elevator as a material mover has been the elevator's traditional heritage for all materials and equipment heavier or bulkier than a person could handle while traveling between floors on a stairway or a ramp. Material movers in activity centers have developed an industry devoted to what has become known as materials handling systems. This industry has as many concepts in vertical material movement as there are specialized problems requiring solutions. However, the manual freight elevator in industry has declined in importance in recent years as manual and even fork truck loading and unloading have become labor intensive in those manufacturing and storage facilities that still justify vertical orientation.

The automation of horizontal pallets and cart conveyors has only recently been extended into vertical movements with automated elevators, conveyors, or lifts. The use of the word *lifts* in this chapter refers to those conveyances falling under the scope of Part XIV—Material Lifts and Dumbwaiters

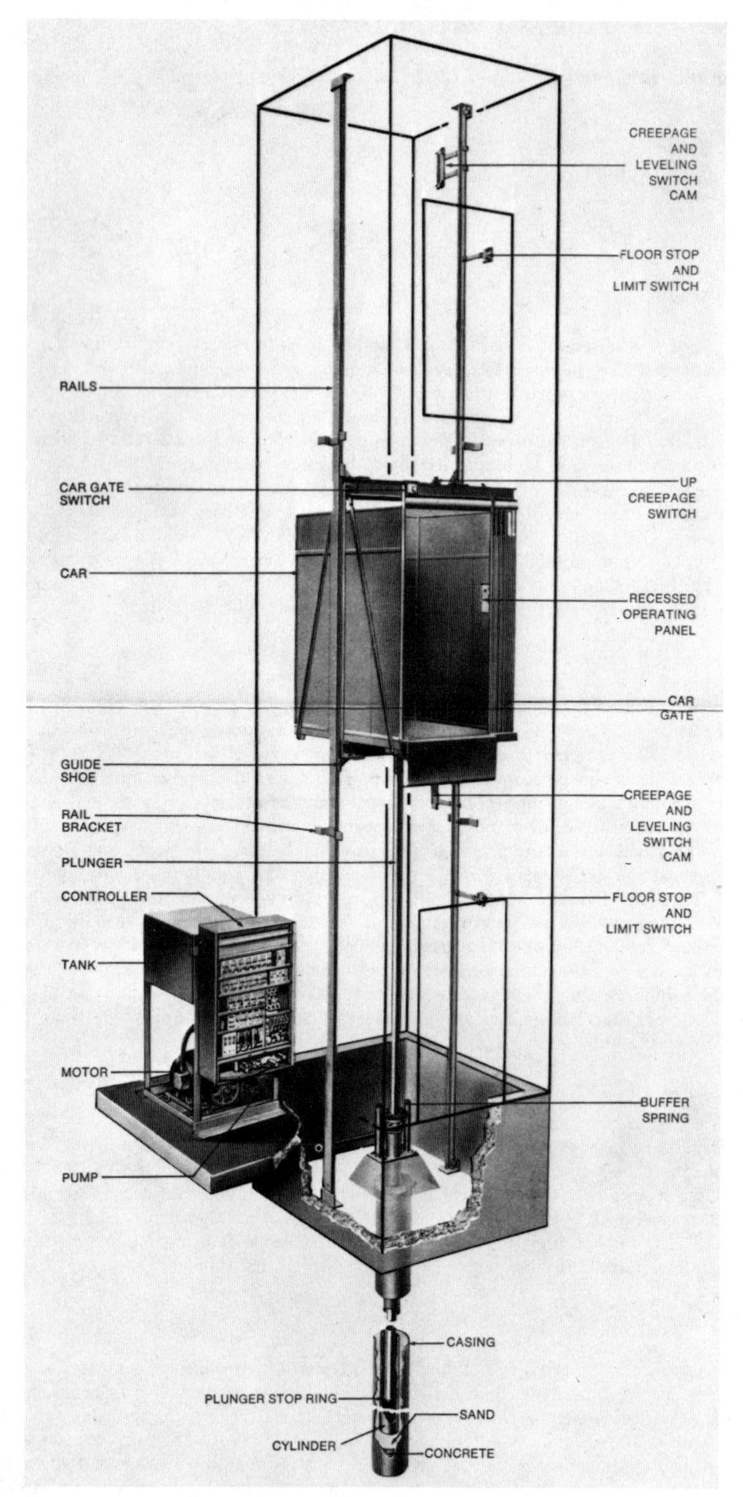

Fig. 16.1.3 Hydraulic freight elevator with direct-acting plunger. Courtesy of Otis Elevator Company.

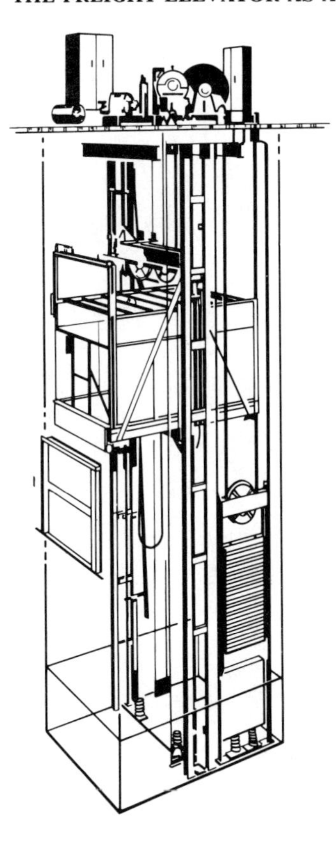

Fig. 16.1.4 Geared electric freight elevator with counter-balanced bi-parting landing doors.

With Automatic Transfer Devices as used in the A17.1 Code. In these installations, the lifts become slaves to addressed containers, pallets, or carts. These containers, pallets, or carts have readable addresses so they can be automatically delivered to the destination floor. In multiple floor installations, the computer, minicomputer, or microprocessor are programmable so that they can select priority origins and destinations, and automatically deliver the containers, carts, or pallets for maximum cost-effective throughput capability. Such equipment has been installed in automated supply systems for both the processing industries and automated storage and retrieval systems for warehousing functions.

The basic uses and concepts of freight elevators have not changed substantially since the Elisha Otis invention of the safety brake in 1853. Essentially, the freight elevator is either a manually loaded and manually unloaded, as well as manually operated, freight elevator or it is an automated or semiautomated lift which is automatically loaded and unloaded and may be completely unattended depending on the supervisory systems.

16.1.7 Standard Freight Elevator

The standard freight elevator has been traditionally loaded and unloaded by a hand truck for loads which are multiple rather than single unit loads. The extension of the hand truck to a battery-powered walkie truck allows for manual loading and unloading of loads heavier than that carried on hand trucks or dollies. The riding fork truck provides the most efficient loading and unloading process with the maximum loads of 3000–5000 lb (1400–2300 kg) on a pallet. However, the fork truck imposes substantial loading impacts as it crosses from the landing sill to the elevator platform as well as reactions in the guide rails when it applies its brake suddenly with the gross weight of the truck and pallet which may be over 10,000 lb (4500 kg). However, the standard freight elevator is still the most commonly installed freight elevator in a range of sizes from 2000 to 8000 lb (900–3500 kg).

16.1.8 Freight Elevators as Material Lifts

Freight elevators as material lifts that are now classified under Part XIV of the A17.1 Code involve automated freight elevators to handle containerized loads with automated transfer devices for carts,

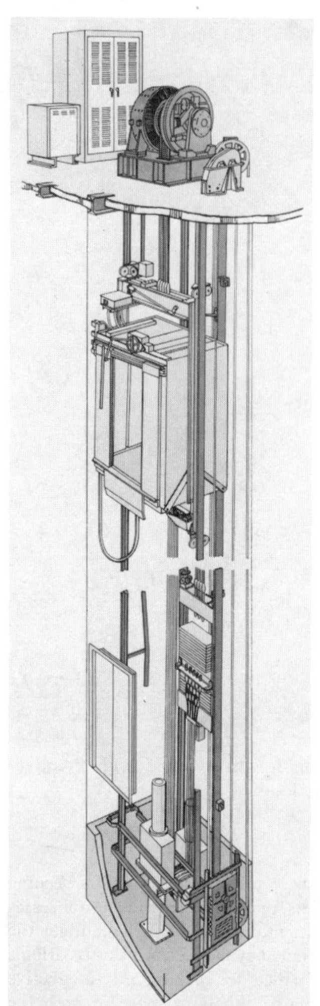

Fig. 16.1.5 Gearless electric elevator for high-speed or high-rise passenger and freight service. Courtesy of Otis Elevator Company.

pallets, or other containers. This usually takes the form of in-floor towlines for carts (Fig. 16.1.6), overhead power-and-free conveyors for carts (Fig. 16.1.7) and for powered rollered conveyors for pallets (Fig. 16.1.8), and other nonvehicular containers. Regardless of the number of unit loads carried by the lift, the lift is essentially still a batch system operating in a fashion to approximate a continuous flow requirement.

Freight elevators can also be configured to provide the vertical transportation of robot vehicles (Fig. 16.1.9), which position themselves automatically by following an on-floor guidance path to the elevator entrance, calling the elevator, and self-propelling itself into the elevator. It then transmits its destination address code to the elevator logic which takes the elevator to the proper floor and permits the robot vehicle to exit, preferably at the opposite end from its entrance. Such a robot vehicle usually follows a guide wire buried in the floor or some other path that is sensed by measuring changes in an electric field, a magnetic field, or in reflected light conditions.

16.1.9 Dumbwaiters as Material Movers

The dumbwaiter is a "junior" freight elevator with a limitation on its floor area and its car height to discourage and hopefully prevent personnel from riding on it. The limitations imposed by the A17.1 Safety Code address this concept although it cannot prohibit personnel from riding on it if they are intent on doing so.

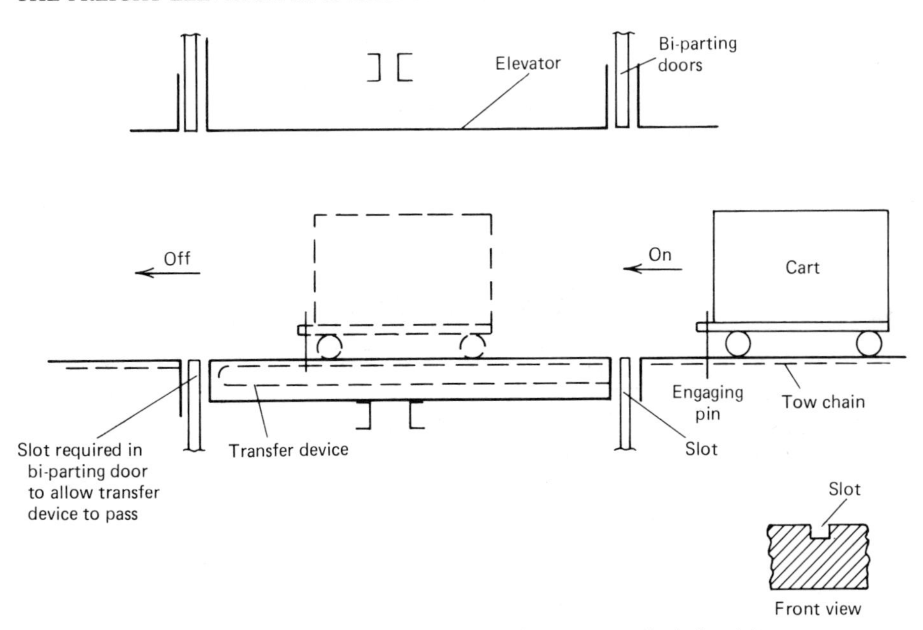

Fig. 16.1.6 Freight elevator with in-floor towline conveyor for industrial carts.

The dumbwaiter is defined in the A17.1 Code as follows:

Dumbwaiter—A hoisting and lowering mechanism with a car of limited capacity and size which moves in guides in a substantially vertical direction and is used exclusively for carrying material.

The dumbwaiter is limited to a net inside platform area of 9 ft² (0.84 m²) and to a total inside height of the car and doors of 4 ft (1.22 m). The dumbwaiter is governed by Part VII of the A17.1 Code by inserting special provisions and referencing provisions of the elevator parts that still apply to dumbwaiters.

16.1.10 Standard Manual Dumbwaiter

The standard dumbwaiter in the manual configuration is arranged to stop either at counter height or at floor level, which changes the need for pit and overhead requirements. The counter-height stopping arrangement is usually configured to have the car platform and door sill at 2 ft 6 in. (760 m) above

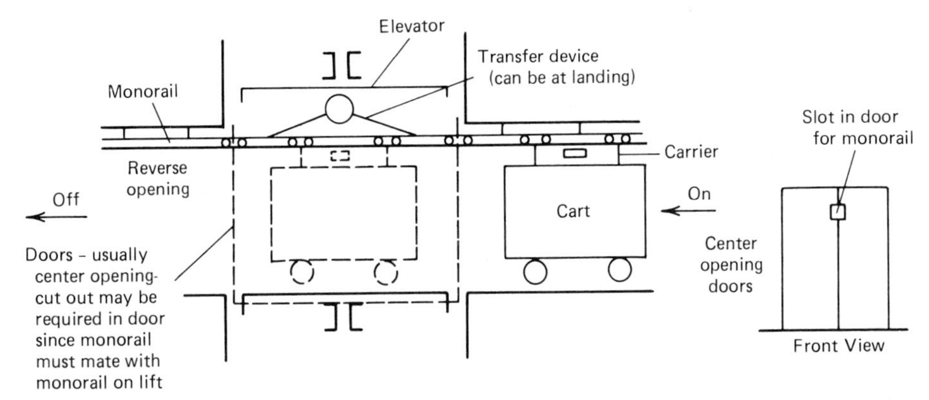

Fig. 16.1.7 Freight elevator with overhead powered-and-free track conveyor for industrial and institutional carts.

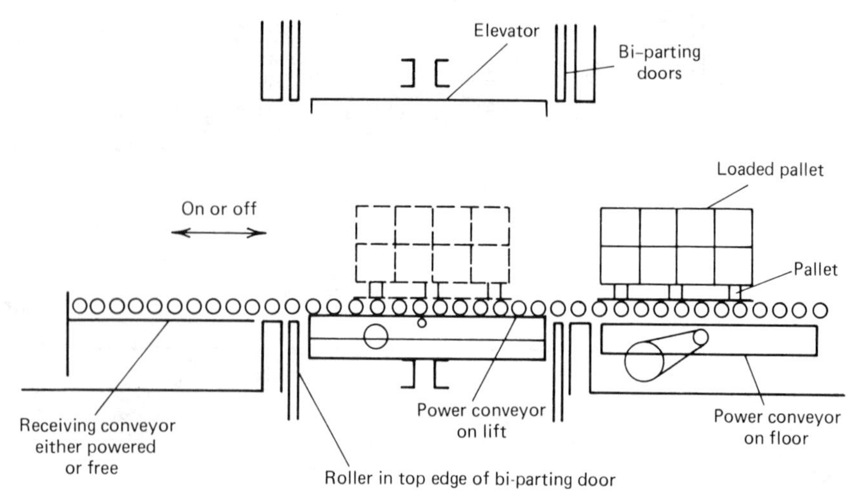

Fig. 16.1.8 Freight elevator with powered roller conveyor for pallets.

the finished floor to facilitate manual loading and unloading of general merchandise or boxes, roll drawings, or bulk paper products. The floor stopping arrangement permits the use of carts for handling mail or other general purposes which relate to their secondary movement on a floor before or after its vertical movement.

16.1.11 Automated Dumbwaiters with Transfer Devices

A special section of the A17.1 Code (Part XIV) governs the use of either the counter-stopping or floor-stopping dumbwaiter when it has an automatic transfer device incorporated into its car or platform construction. It also governs the door operation, special door controls, and door sequencing when such equipment is to be used unattended and with limited access to authorized personnel.

A counter-stopping dumbwaiter with a manual transfer device is shown in Fig. 16.1.10 to handle tote boxes. Such equipment can handle other containerized materials provided the container is of uniform size in its three major dimensions. It may or may not have a loading table on the out-feed conveyor. If the conveyor on the landing is a passive roller or skatewheel conveyor, it is arranged for a decline so that the boxes position themselves at the end of the gravity conveyor in the order of arrival. Such a conveyor obviously will not permit self-loading at typical floors but only automatic gravity unloading.

With the addition of a reversible loading table or conveyor, the lift becomes an automatic loading and unloading tote box lift usually described as an inject–eject device.

A floor-stopping arrangement is shown in Fig. 16.1.11 where a transfer device extends from the dumbwaiter and picks up the cart by engaging an on-cart coupler and drawing it into the lift. The

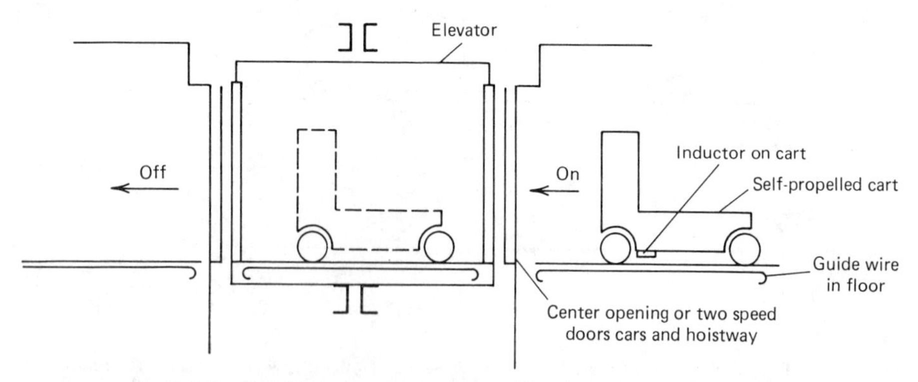

Fig. 16.1.9 Freight elevator or passenger with robot self-propelled vehicle.

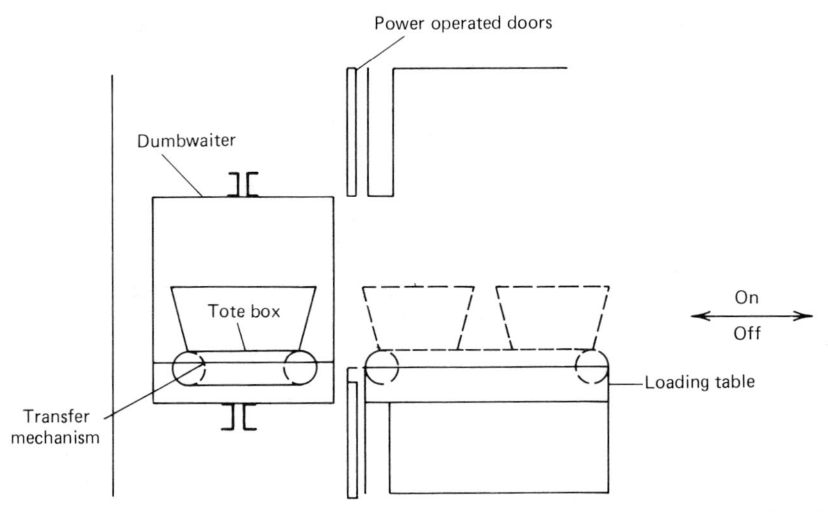

Fig. 16.1.10 Dumbwaiter with powered belt conveyor for automatic loading and unloading of tote boxes or trays.

reverse operation occurs on the destination floor with the transfer device depositing the cart onto the floor, releasing itself, and withdrawing the transfer device into the car for redirection to another pickup. The cart lift of this arrangement, without an in-floor conveyor, will inject only one manually positioned cart while it will eject automatically a multiplicity of carts at a single destination without attention until the carts meet obstruction.

16.2 STANDARD FREIGHT ELEVATOR DESIGN

16.2.1 General Design Considerations

The major consideration for design of a standard freight elevator is to assess the maximum requirements of throughput in both the up (UP) and down (DN) directions. The method of loading and unloading

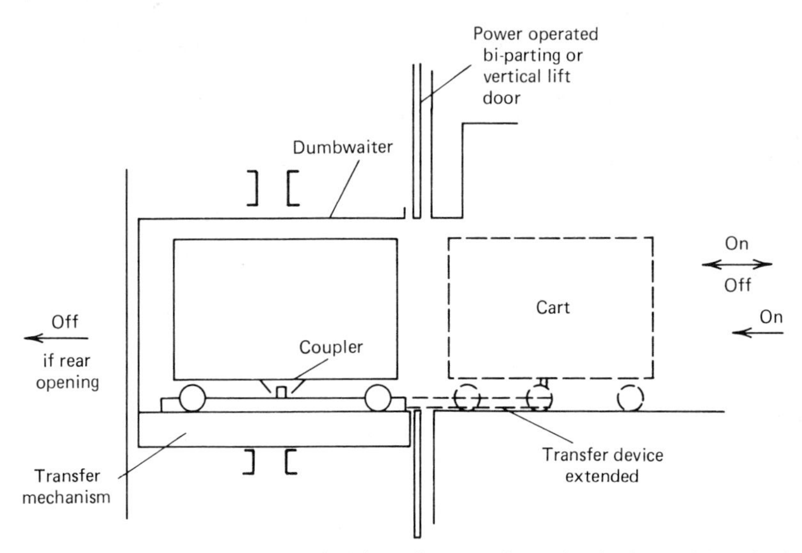

Fig. 16.1.11 Dumbwaiter with automatic inject–eject transfer device for industrial or institutional carts.

the freight elevator will determine the classification with respect to the requirements of design of car platform, car frame, and the truckable sill on the car platform and on the landing door.

Leveling accuracy is an important consideration with respect to the linear error permitted between the landing door truckable sill and the car sill. As a part of this accuracy specification, it will be necessary to assess the suitability of the wheeled vehicles to handle this error. In conjunction with the running clearance gap required by code, this requirement is stated in Section 108—Horizontal Car and Counterweight Clearances. Rule 108.1D specifically requires the clearance as follows:

108.1D—Between Cars and Landing Sills: The clearance between the car platform sill and the hoistway edge of any landing sill, or the hoistway side of any vertically sliding counterweighted or counterbalanced hoistway door or of any vertically sliding counterbalanced biparting hoistway door, shall be not less than $\frac{1}{2}$ in. (13 mm) where side guides are used not less than $\frac{3}{4}$ in. (19 mm) where corner guides are used. The maximum clearance shall be not more than $1\frac{1}{2}$ in. (38 mm).

With the new technology of pulse counting for leveling logic, the accuracy of leveling can now provide a tolerance of $\pm\frac{1}{8}$ in. ($\pm$3 mm) in short rises and a tolerance of $\pm\frac{1}{4}$ in. ($\pm$7 mm) in higher rises. The change in accuracy is a function of cumulative errors that occur in the suspension and measurement system.

The control system relates to the speed control of the elevator and determines its leveling accuracy, its acceleration, deceleration, and contract or rated speed. Usually, the electric freight elevator system uses either a two-speed alternating current motor or a generator field controlled direct current motor for its speed control between the leveling speed and the contract running speed, depending on the speed ratio. The maximum ratio for a two-speed alternating motor is 6:1 with leveling speeds required in the vicinity of 25 ft/min (0.125 m/sec). All equipment should provide for a releveling capability so that as the elevator is loaded or unloaded, the vertical relationship between the sill on the car and the sill on the door panel (and landing sill) will be maintained within an acceptable tolerance for continued loading and unloading operations.

The operation of the elevator is the logic input to the system that relates to its landing destinations as registered from the landing to call an elevator, or as registered in the car to dispatch an elevator to the desired loading or unloading landing. This usually is done on a per load basis since an operator is required for each elevator. However, multiple unit installations, when justified, may be tied together in duplex or triplex configurations so that only one elevator will answer the call registered at a landing rather than registering calls at a landing for each elevator.

16.2.2 Capacity Specification and Loading Classification

The rated load capacity of a freight elevator should, in all cases, be conservatively based on its anticipated maximum use so that unsafe conditions will be avoided in either the operation of or the equipment change in a manufacturing facility. The equipping or reequipping operation may, in many cases, determine the capacity and size of a freight elevator over the normal operation of material movement. This also relates to the proper determination of the car width, door width, and door height for the minimum single unit size of equipment that the elevator must accept.

The classes and loading of freight elevators is detailed in Section 207—Capacity and Loading. Rule 207.2b specifies, in detail, the appropriate minimum capacity and the loading classification which determines the structural design of the platform, the car frame, car main guide rails, rail brackets, and the landing door truckable sill. The variations in this rule should be carefully studied and be given in detail to the elevator supplier so that the equipment can be appropriately designed for its maximum use and abuse. Rule 207.2b is quoted in its entirety as follows:

207.2b—Classes of Loading and Design Requirements: Freight elevators shall be designed for one of the following classes of loading:

(1) Class A: General Freight Loading. Where the load is distributed, the weight of any single piece of freight or of any single hand truck and its load is not more than $\frac{1}{4}$ the rated load of the elevator, and the load is handled on and off the car platform manually or by means of hand trucks.

For this class of loading, the rated load shall be based on not less than 50 lb/ft² (2.39 kPa) of inside net platform area.

(2) Class B: Motor-Vehicle Loading. Where the elevator is used solely to carry automobile trucks or passenger automobiles up to the rated capacity of the elevator.

For this class of loading, the rated load shall be based on not less than 30 lb/ft² (1.43kPa) of inside net platform area.

(3) Class C: There are three types of Class C loading as follows:

Class C1: Industrial Truck Loading. Where truck is carried by the elevator.

Class C2: Industrial Truck Loading. Where truck is not usually carried by the elevator but used only for loading and unloading.

Class C3: Other Loading with Heavy Concentrations. Where truck is not usually used.

These loadings apply where the weight of the concentrated load, including an industrial power or hand truck, if used, is more than $\frac{1}{4}$ the rated load and where the load to be carried does not exceed the rated load. (For concentrated loads exceeding the rated load, see Rule 207.6.)

(a) The following requirements shall apply to Class C1, Class C2 and Class C3 loadings:

The rated load of the elevator shall be not less than the load (including any truck) to be carried, and shall, in no case, be less than load based on 50 lb/ft² (2.39 kPa) of inside net platform area;

The elevator shall be provided with a two-way automatic leveling device. (See definition.)

(b) For Class C1 and Class C2 loadings, the following additional requirements shall apply:

For elevators with rated loads of 20,000 lb (9070 kg) or less, the car platform shall be designed for a loaded truck of weight equal to the rated load or for the actual weight of the loaded truck to be used, whichever is greater. For elevators with rated loads exceeding 20,000 lb (9070 kg), the car platform shall be designed for a loaded truck weighing 20,000 lb (9070 kg), or for the actual weight of the loaded truck to be used, whichever is greater.

(c) For Class C2 loading the following additional requirements shall apply:

The maximum load on the car platform during loading or unloading shall not exceed 150% of rated load.

For any load in excess of rated load on elevators with a rated load of 20,000 lb (9070 kg) or less, the driving machine motor, brake and traction relation shall be adequate to sustain and level the full 150% of rated load.

For any load in excess of the rated load on elevators with a rated load exceeding 20,000 lb (9070 kg), the driving motor, brake, and traction relation shall be adequate to sustain, and level the rated load plus either 10,000 lb (4540 kg), or the weight of the unloaded truck to be used, whichever is greater.

NOTES [*Rule 207.2b(3)*]:

(1) When the entire rated load is loaded or unloaded by an industrial truck in increments, the load imposed on the car platform, while the last increment is being loaded or the first increment unloaded, will exceed the rated load by part of the weight of the empty industrial truck.

(2) This Rule does not prohibit the carrying of an industrial truck on a freight elevator of Class C2 or C3 loading, providing that the total weight on the elevator does not exceed the rated load of the elevator and the elevator is designed to meet the requirements of Section 1301 or 1303, as appropriate, for the load involved.

Figure 16.2.1 diagramatically shows the loading with the general restrictions.

16.2.3 Rated Speed Specification

The rated speed specification should be related to the throughput requirements as discussed previously, which will set the type of elevator equipment provided. As indicated previously, the maximum speed for a hydraulic elevator is currently limited to 200 ft/min (1 m/sec) but more often is installed with a practical limit of 150 ft/min (0.75 m/sec).

The geared elevator has two drive versions. The two-speed AC drive is limited to a speed of 150 ft/min (0.75 m/sec) because of the specification of leveling speed and cost-effective speed ratio between the two synchronous windings on the electric motor. The generator field control drive has a limitation of approximately 400 ft/min (2 m/sec) limit in the lower capacity range and approximately 200 ft/min (1 m/sec) in the upper capacity range. The lower capacity range is usually defined as capacities within the range between 2000 and 8000 lb (900 and 3500 kg) capacity and the higher capacity range between 10,000 and 20,000 lb (4500 and 9000 kg). This is an arbitrary separation and borderline cases should be referred to prospective elevator suppliers. It should be noted at this point that the two-speed AC drive will probably produce a leveling accuracy of $\pm\frac{3}{8}$ in. (±11 mm) with medium loads and under heavy loads, may produce a leveling accuracy of $\pm\frac{1}{2}$ in. (±15 mm). If such a drive is used for the high-capacity freight elevator and industrial truck loading is required, zoned inching control should be provided to permit releveling on a manual basis. This will also assist the loading and unloading process where rope stretch may be a significant factor.

The speeds available from gearless traction equipment are limited only by the practical rope speed

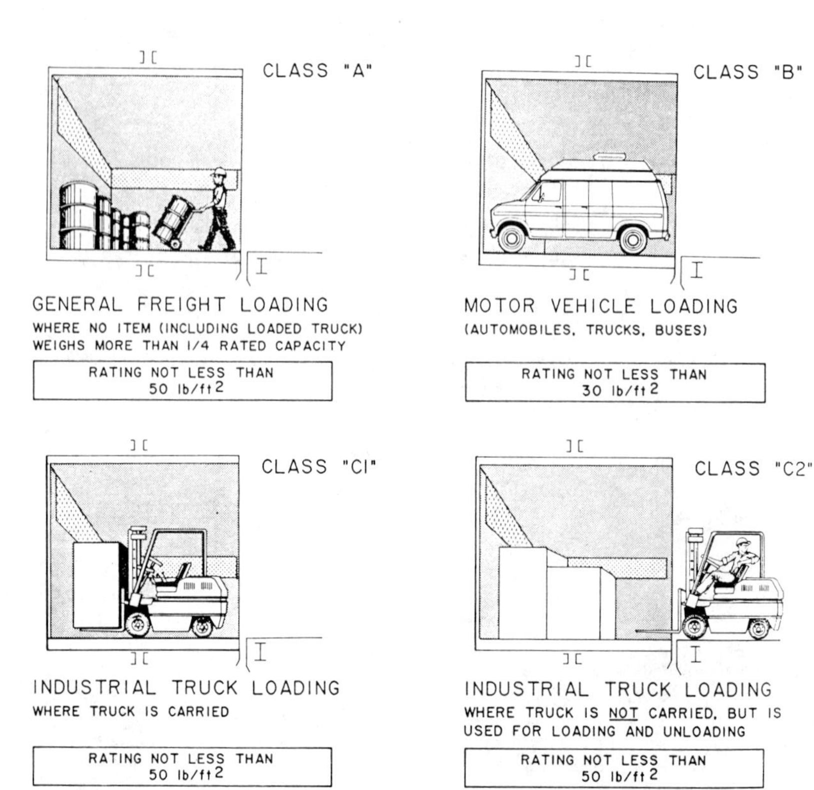

Fig. 16.2.1 Types of freight elevator loading to be used in design specification. Courtesy of National Elevator Industry Inc.

VERTICAL TRANSPORTATION STANDARDS

GENERAL INFORMATION FOR FREIGHT ELEVATORS WITH
CLASS "A", "B" & "C" LOADING
(REFER TO ANSI/ASME A17.1, RULE 207.2b)

NOTE

THE PICTORIAL FREIGHT LOADING SHOWN FOR THE DIFFERENT CLASSES OF LOADING IS INTENDED FOR BOTH ELECTRIC AND HYDRAULIC ELEVATORS. SOME DIFFERENCES OCCUR IN RAIL FORCES.

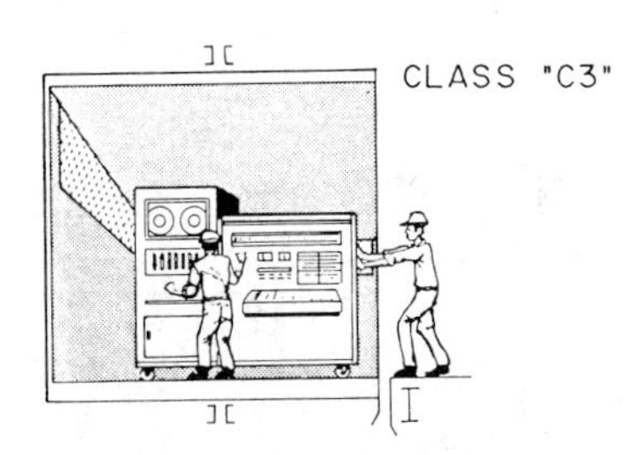

CONCENTRATED LOADING

(NO TRUCK USED) BUT LOAD INCREMENTS ARE MORE THAN 1/4 RATED CAPACITY. CARRIED LOAD MUST NOT EXCEED RATED CAPACITY

RATING NOT LESS THAN 50 lb/ft^2

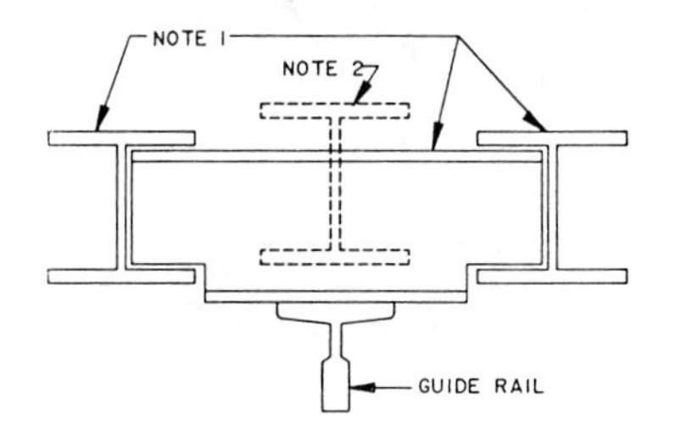

NOTE 1 VERTICAL RAIL COLUMN SUPPORTS AND CROSS TIE MEMBERS ARE REQUIRED AND PROVIDED BY OTHER THAN THE ELEVATOR SUPPLIER WHEN RATED LOAD EXCEEDS 8000 lb. THE SIZE OF THE RAIL COLUMNS ARE DETERMINED BY OTHERS FROM RAIL FORCES FURNISHED BY THE ELEVATOR SUPPLIER.

NOTE 2 ALTERNATE METHOD OF RAIL COLUMN SUPPORT, WHEN RATED LOAD IS 8000 lb OR LESS THE SIZE OF THE COLUMNS ARE DETERMINED BY THE OTHERS FROM RAIL FORCES FURNISHED BY THE ELEVATOR SUPPLIER.

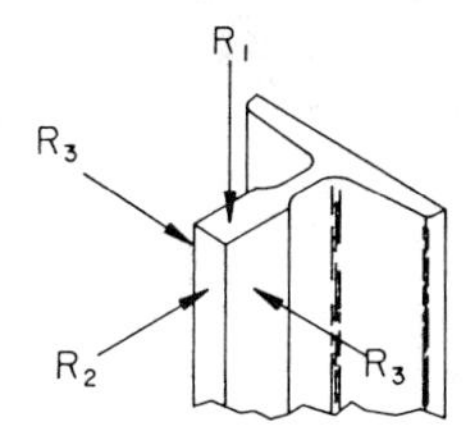

GUIDE RAIL FORCES (FOR A SINGLE RAIL)	ELECTRIC ELEVATOR	HYDRAULIC ELEVATOR
	lb	lb
R_1	FOR THESE FORCES, CONSULT ELEVATOR SUPPLIER	
R_2		
R_3		

Fig. 16.2.1 *(continued)*

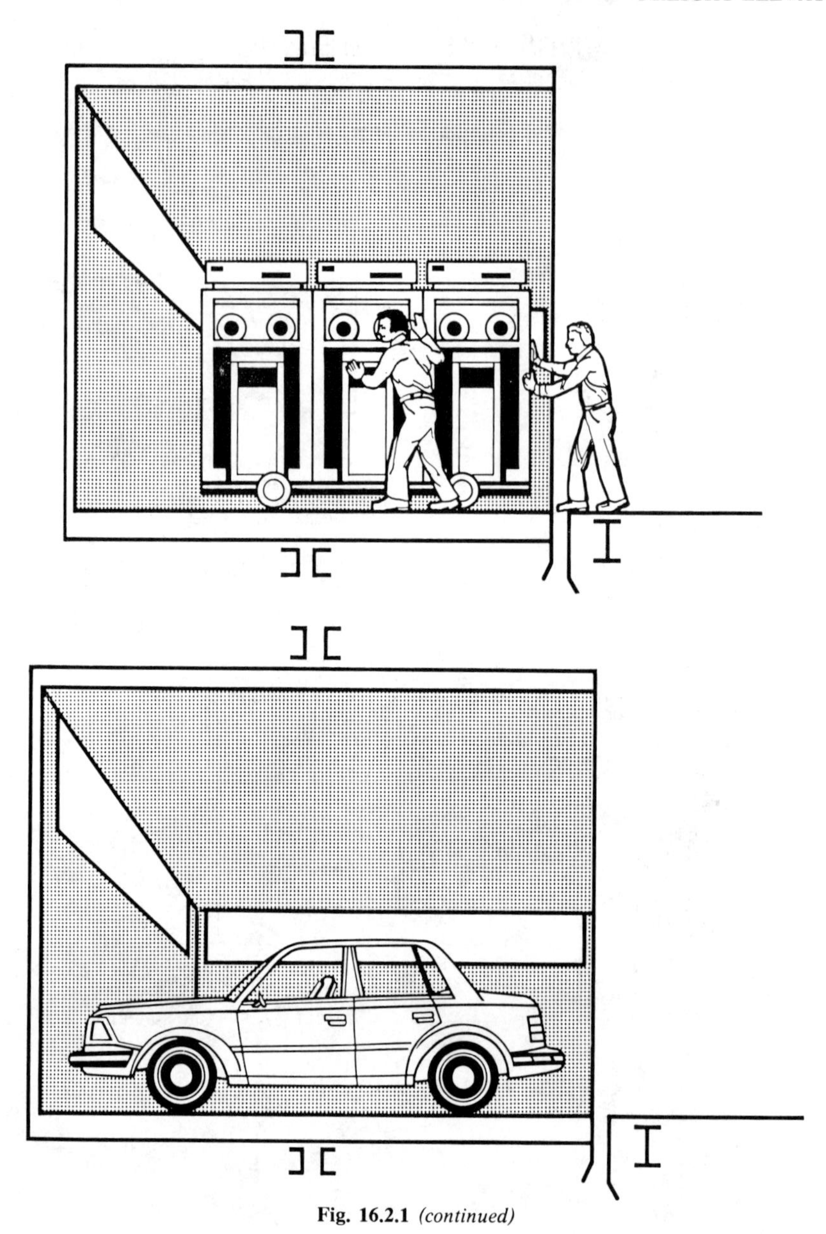

Fig. 16.2.1 *(continued)*

passing over the driving sheave and the 2:1 roping relationship of the elevator car and counterweight. Currently, this limitation appears to be 1400 ft/min (7.0 m/sec) relating to an elevator speed of 700 ft/min (3.5 m/sec). The lower speed of a gearless freight elevator probably is set at the lower limit of 300 ft/min (1.5 m/sec) for cost purposes. It may be indicated in special circumstances that the gearless elevator should be provided instead of a geared elevator since the design of the gearless machine permits substantially larger shaft loadings than the geared elevator. This condition would exist when there is a large elevator car with a high rated load capacity. Since the dead weight of the car is repeated at 100% on the counterweight and the rated load of the car is repeated at 40–50% on the counterweight, it is apparent that load and physical size may dictate the gearless drive over the geared drive under certain circumstances.

16.2.4 Car Platform Design

The car platform design is the most important initial design parameter after this specification of rated speed, rated capacity, and loading classification. The platform should be designed for either single-ended or double-ended loading and unloading, depending on its position in the manufacturing or warehousing facility and the resulting traffic patterns generated by its location. The single-ended arrangement provides the most efficient loading and unloading configuration for riding fork trucks since it permits loading from only one end on a last in–first out (LIFO) basis. The problem with double-ended fork truck loading occurs in the abuse of the car gate at each opening of the elevator car and the potential problems with loading on a first in–first out (FIFO) basis. The single-ended platform also permits substantial reinforcement of the rear of the car enclosure so that the unskilled fork truck operator cannot damage the car gate and disable the elevator for a substantial period of time.

The hand truck and the battery-powered walkie truck for pallet loading can conveniently use either the single- or the double-ended platform design as well as the standard car design. The abuse of the car enclosure by this method of loading and unloading is usually insignificant and does not seriously affect the elevator's downtime, except in extreme cases. However, the use of a riding fork truck, with its protruding forks, can cause serious damage to even a standard freight elevator design that is single-ended. Serious damage can easily be incurred with double-ended elevators since the car gate at the nonloading end will usually be in the closed position. Its construction usually is of a light material, designed to prevent personnel accidents and not to absorb the impact of a fork truck with its load. It is suggested that the platform be supplemented with one or two perimeter channels mounted on edge to prevent the forks or the pallets from penetrating the car enclosure. For medium duty, it is suggested that a 12-in. (300-mm) high channel be mounted on edge to absorb this abuse. In more heavily loaded elevators, two channels can be mounted to provide maximum protection. In addition, the car enclosure, which is made of a much lighter material, should be protected immediately above the rubstrakes by a 2 in. (50 mm) wide by 12 in. (300 mm) high oak rubstrake. The rubstrakes should be made of oak, maple, or other similar material and fastened in such a manner as to permit replacement as they become damaged. They absorb the impact of off-center and overhanging loads. Additional rubstrakes may be considered if high loads continuously are overhanging at levels which are above the primary rubstrake (see Fig. 16.2.2).

The floor of the platform should be a wear-resistant material. Traditionally, it was a double hardwood flooring with the top flooring of maple, birch, or oak. However, the heavy fork truck usage has indicated that this is a difficult material to maintain and replace. Current practice is to provide a plywood subflooring, hardwood top flooring onto which is placed aluminum floor plates removable in manageable sections from inside the car and covered with a flowed-on mastic compound containing an abrasive surface. This eliminates the exposed diamond plate design, which is extremely noisy and objectionable for cart and fork truck loading, and eliminates the slipping and skidding that occurs when the plate is smooth. The mastic compound with abrasive material is easily replaced on a maintenance basis rather than a repair shutdown. The floor plates can be replaced easily since they are applied to the top flooring, not a part of the car platform construction.

16.2.5 Car Sling Design

The car sling is the structural support of the platform that surrounds the car enclosure and interfaces with the car guide rails to transmit the loading and unloading impact forces to the building structure and to restrain the elevator during its travel in what most often will be an unbalanced loading condition.

For slow-speed elevators with relatively light capacities, roller guides may be provided that accommodate the loading up to approximately 15,000 lb (7000 kg). Higher capacities will require sliding guides that present a large surface area to the rail face so that the pressures generated during loading and unloading will be at an acceptable level. The roller guides provide an operation that does not require lubrication of the guide rails and hence, for maximum fire safety or when other hazardous conditions prevail, are preferred. Sliding guides on elevators above 15,000 lb (7000 kg) should consider the use of sliding guides with an intensive preventive maintenance program to assure that the rails are not overlubricated and generating a fire hazard, particularly with the debris that accumulates in the elevator pit.

A current design practice has been to include in the car sling design provision for the concept of using a chain fall or chain hoist inside the elevator car to lift flexible loads for ease of loading and unloading, particularly large rolls of carpet. This hoist beam or hoist eye must be accommodated from the crosshead member at the top of the car sling and supported in such a fashion that the lower flange and most of the web protrude into the car enclosure for the installation of a beam clamp or other device. It is recommended that such hoisting devices be temporarily located in the elevator and used when only necessary. Such a design of a hoist beam and the car enclosure and car crosshead member of the sling is shown in Fig. 16.2.3.

Fig. 16.2.2 Freight elevator with rolled section steel rubstrakes and hoist beam. Courtesy of Jaros, Baum & Bolles.

16.2.6 Car Enclosure Design

The car enclosure preferably should be totally enclosed on the two sides, rear and top, so that a protection is provided against falling material from above. The side and end car panels for the special design using perimeter steel channels should be set back from the face of the channel so that the car enclosure will receive the most protection from overhanging loads.

The car operating panel and the associated intercommunication or telephone equipment should be recessed into the side panels so that overhanging loads will not damage this operating equipment and cause extended outages.

The illumination provided should be flush-mounted to the top of the car enclosure with the lamps easily replaceable from inside the car or from the car top, as appropriate. A suitable nonshattering glass or plastic lens should be provided with guard so that loads will not cause breakage. Adequate illumination will help alleviate unnecessary impact with the rear panels or the rear gate.

16.2.7 Freight Elevator Landing Doors

The standard arrangement of landing doors on freight elevators is the counterbalanced biparting variety that provides a bridging sill between the landing sill and the platform sill. This truckable sill transfers the impact of the truck wheels as it crosses the gap from the landing sill directly back into the building structure, either at the landing sill or through other supporting members. This introduces two gaps as the wheels cross from the landing sill to the platform sill. Truckable sills are designed to relate to the capacity of the freight elevator and to the class of loading.

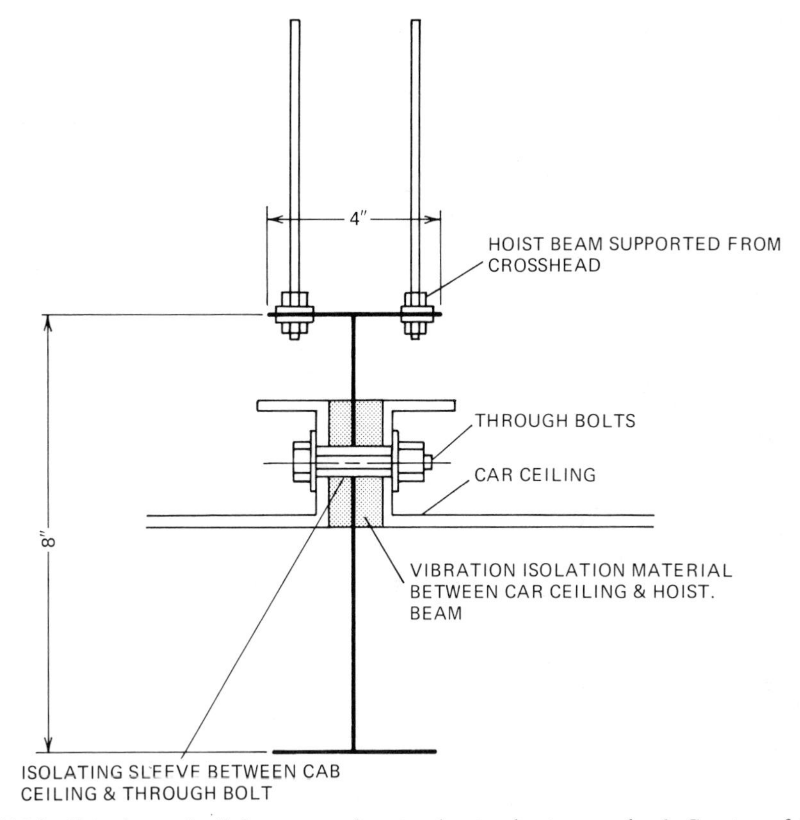

Fig. 16.2.3 Hoist beam detail for structural connection to elevator crosshead. Courtesy of Jaros, Baum & Bolles.

The counterbalance biparting door has two vertically sliding panels; the bottom panel slides down and the top panel slides up. The height of the door, plus $\frac{1}{2}$ of the door height, oftentimes exceeds the floor height of the facility. In this case, it is necessary for the lower door panel on one floor to pass the upper door panel on the lower floor on parallel tracks. This arrangement permits the bottom door panel to slide down in front of the top panel of the floor below. This arrangement is called a pass-type door as opposed to the regular door. In this case, the top panel is offset into the hoistway approximately the thickness of the bottom door panel above plus the running clearances. This also requires that the truckable sill be extended into the hoistway so that the top panel can bear directly on it with its resilient safety astragal when the door panels are in the closed position.

In addition to the regular and pass-type arrangement, there are other compound arrangements for unusually short floor heights and unusually high door heights. Special considerations should be given in designing a facility that demands their installation. (See Fig. 16.2.4 for typical pass-type entrance section.)

When passengers are permitted to ride on the freight elevator by the local authority having jurisdiction, they must be equipped with sequenced power-operated landing doors and car gates so that, in opening, the hoistway door shall be opened at least two-thirds of its travel before the car gate can start to open; and, in closing, the car gate shall be closed at least two-thirds of its travel before the hoistway door can start to close. The gate should also be equipped with a reversing edge so that, if it meets an obstruction, it will reverse to the open position to be reinitiated by the momentary pressure of the DOOR CLOSE button located in the car and at each landing. Section 112, Power Operation, Power Opening and Power Closing of Hoistway Doors and Car Doors or Gates of the code covers this arrangement under Rules 112.5 and 112.6.

The car gate design for the single-ended car enclosure can be of a standard variety, usually of an open expanded wire mesh or perforated sheet steel, so that it provides vision through the gate but rejects a 2 in. (50 mm) ball for safety reasons. The double-ended gate should be designed for abuse by truck traffic so that, when the car is being loaded or unloaded, the forks of the fork truck and the pallet will not damage it and shut down the elevator for extensive repairs. It is suggested that

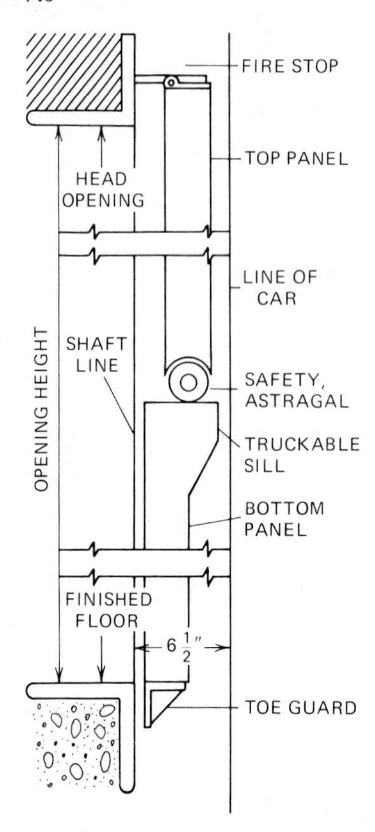

Fig. 16.2.4 Typical pass-type counterbalanced bi-parting freight elevator doors.

the lower 2 ft (600 mm) be reinforced and rigidized to accept this abuse without disengaging the gate track or becoming deformed.

16.2.8 Leveling Accuracy

The leveling accuracy of an elevator is basic to its function of speed control as outlined previously. The most accurate leveling will be accomplished by the latest technology of pulse counting from the machine or from the governor so as to precisely locate the elevator in its deceleration, leveling, and final stopping modes. This current technology with generator field control or solid-state motor drive, depending upon the conversion technology from AC to DC, will provide a leveling accuracy approaching $\pm\frac{1}{16}$ in. ($\pm$1.6 mm). This has been demonstrated on passenger elevators and it can reasonably be inferred that $\pm\frac{1}{8}$ in. ($\pm$3 mm) can be expected in quality freight elevator installations.

The traditional leveling of an elevator from a selector located in the machine room provided poor leveling of $\pm\frac{3}{8}$ in. ($\pm$9 mm) which often was too inaccurate for power truck loading when the load was half removed. Then the leveling misalignment could be exacerbated by the rope stretch reduction. The reverse condition would exist in loading an empty elevator. To correct this, the feature of "zoned inching" was developed which permitted the elevator to be manually releveled during the loading and unloading process if the misalignment became so great that it caused severe impact in the fork truck transfer across the opening to the car sill.

Subsequent techniques were developed which brought the elevator into the floor approximately with the machine room selector, but the final leveling was determined by permanent magnet switches on the car and a series of vanes permanently mounted in the hoistway that related to the level condition at each landing. This created a consistent leveling of $\pm\frac{1}{4}$ in. ($\pm$7 mm). This leveling accuracy was the industry standard until the pulse-counting techniques were recently applied to the elevator industry with the adaption of minicomputer and microprocessor technology. It is recommended that all Class C electric elevators be equipped with the latest leveling technology to minimize the abuse in the loading and unloading of the elevator by fork truck.

16.2.9 Loading Dock Design

The design of a freight elevator to work properly with a loading dock requires an understanding of the arrival rate of raw materials entering the facility and the departure rate of finished materials with the appropriate time-slotting. It also should be understood that it is not practical to weatherstrip freight elevator doors. As a result, the total design of the loading dock and truck parking facilities should take into consideration the air transfer possible around the running clearances of the door panels. Experimentation with dense nylon brushes around the perimeter of door openings on passenger elevators has recently indicated that the application may be appropriate on freight elevators in cases where the temperature conditions are extreme and the dock is continuously open to ambient temperature and wind velocities.

It is axiomatic that there must be receiving and shipping space on the dock between the elevator and the truck itself so that the elevator can work appropriately, handling two-way traffic in such a way as to not tie up either the elevator or the truck. It is suggested that the minimum dock area for temporary storage of materials should be 300–400 ft² (28–38 m²) to accommodate the full load of a semitrailer truck with maneuvering room for the fork truck in its loading or unloading operation.

All intensely used loading docks should be provided with either a truck leveler or a dock leveler so that various heights of semitrailer trucks can be accommodated easily with fork truck loading or unloading. The lack of this feature can provide unsafe conditions in the use of manually placed ramps to accommodate the normal misalignment.

A part of the loading dock design should be a ramp between the dock and grade that will permit the use of manual carts or dollies with vans and pickup trucks. The ramp should be the basic design that is supplemented by steps, but steps should never replace a ramp. Loading docks equipped with only steps have been the source of justifiable criticism.

16.2.10 Elevator Time–Distance Parameters

The time–distance parameters for a standard freight elevator are shown in Fig. 16.2.5 as a family of curves relating to the rated speed of the elevator and the type of elevator provided. These curves show the approximate time to make an express run of the distances indicated, including the acceleration up to full speed running and deceleration times. It is assumed that the subsequent door operation will be sequential to that shown for travel time.

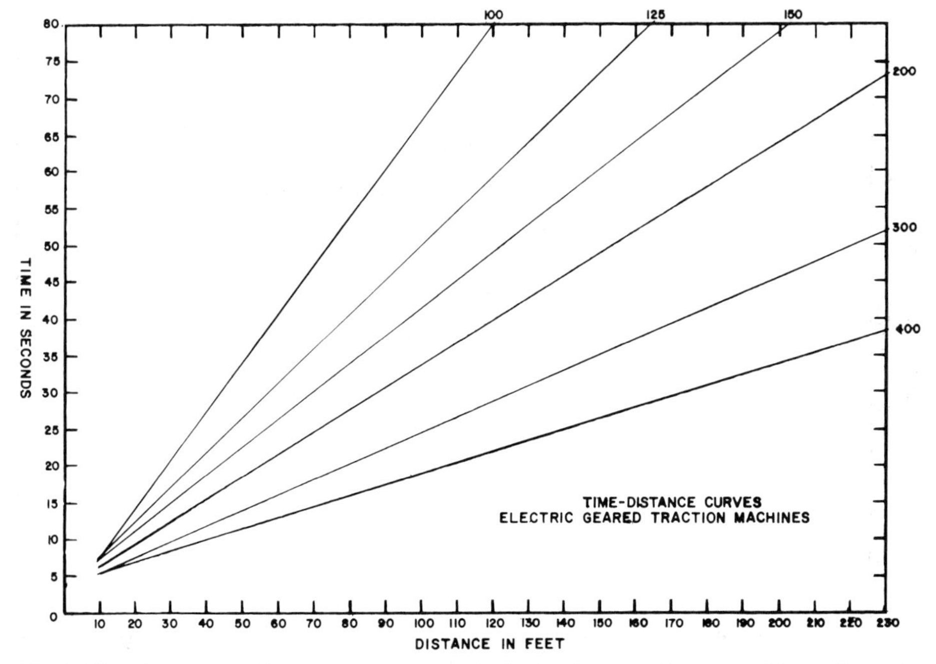

Fig. 16.2.5 Time versus distance chart for electric freight elevators. Courtesy of Jaros, Baum & Bolles.

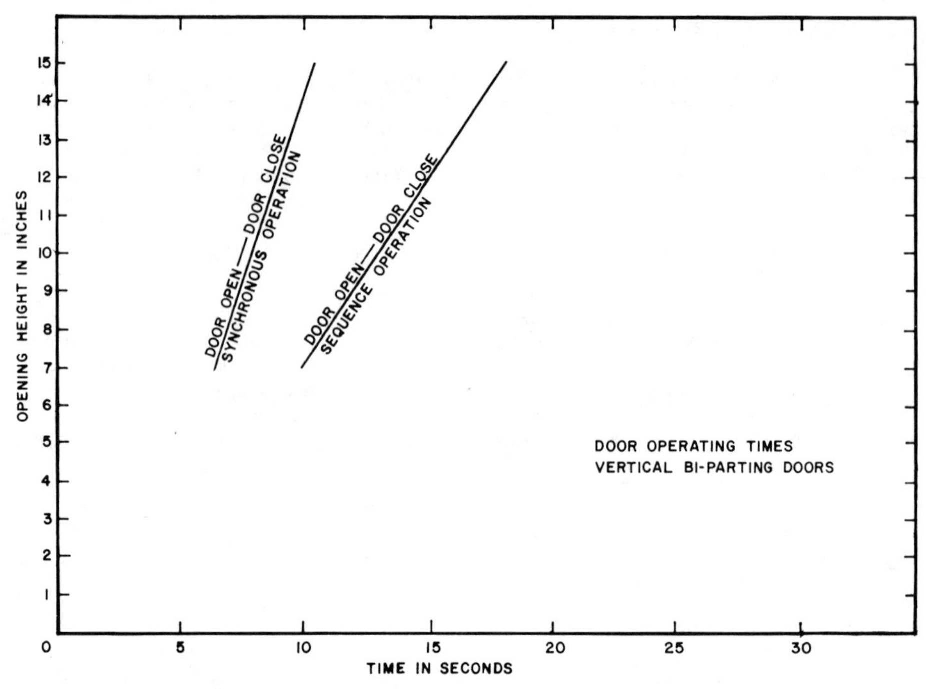

Fig. 16.2.6 Door operating time versus height for vertical bi-parting freight elevator doors. Courtesy of Jaros, Baum & Bolles.

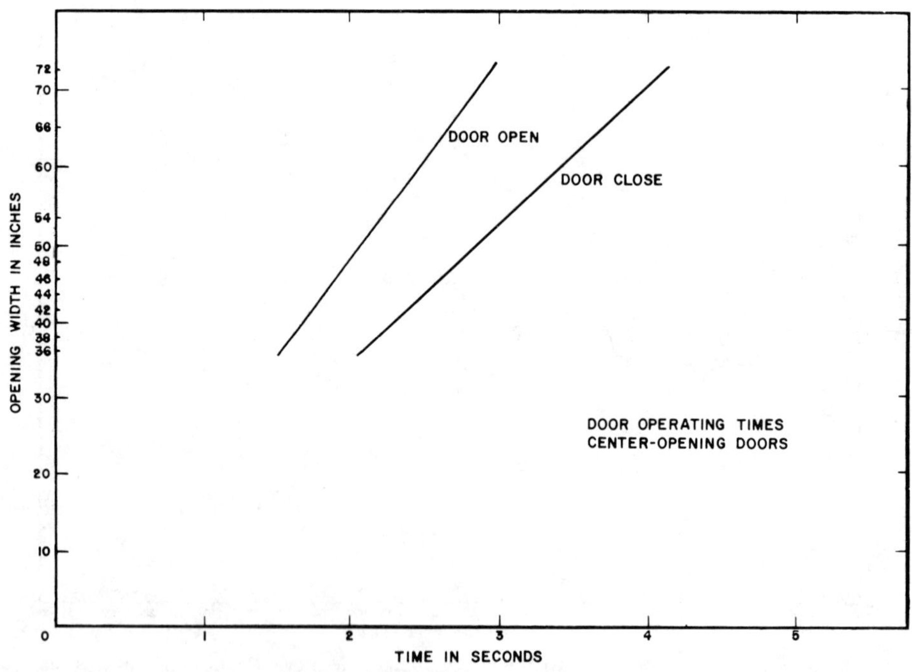

Fig. 16.2.7 Door operating time versus width for horizontal center-opening passenger and freight elevator doors. Courtesy of Jaros, Baum & Bolles.

If speeds in excess of 400 ft/min (2.0 m/sec) or heights in excess of 200 ft (60 m) are encountered, the extrapolation of the curves will give results precise enough for calculating round-trip times and throughput capacities.

The maximum speed of 700 ft/min (3.5 m/sec) is related to an optimum rope speed of 1400 ft/min (7.0 m/sec) which has been accepted in the industry for usual installations. This relates to the fact that an electric freight elevator is best arranged for a 2:1 roping arrangement so that the shaft loading can be reduced to commercial machine sizes and that the structure will directly receive one-half of the load at the dead-end hitches at the top of the hoistway for both the car and the counterweight.

16.2.11 Door Operation

The time parameters for counterbalanced biparting doors versus their height is shown in Fig. 16.2.6 related to the height of the door based on the total mass involved. In special applications where passengers are involved or where overhead tracks may be desired, it is appropriate to consider the installation of a center-opening door instead of a vertically sliding door similar to those provided for passenger elevators. It is appropriate to consult with the elevator supplier and the regulatory authorities to consider the appropriate penetration of the door panels in a nonstandard configuration. For reference purposes, the door opening time and the door closing time for representative center-opening doors are shown in Fig. 16.2.7.

16.3 SPECIAL FREIGHT ELEVATOR DESIGN

Special freight elevators or material lifts usually involve the automation of the elevator to be a slave to a materials handling system which dictates each origin-to-destination trip or trips, depending on the multiplicity of containers carried on each elevator at a given time that are addressed for multiple destinations. To respond to this category and to an activity that started in the 1950s, Part XIV of the A17.1 Code was eventually incorporated to assure that the equipment and its installation was as safe as possible under the circumstances of operation of this type equipment. As outlined earlier, it is appropriate, prior to installing special vertical transportation equipment, to consult with the local regulatory authorities as to whether the equipment under consideration falls under the scope of Part XIV Material Lifts of the A17.1 Safety Code for Elevators or under the B20.1 Safety Code for Conveyors which recognizes reciprocating vertical lift conveyors which perform similar functions.

The general considerations of capacity, speed, operation and acceleration–deceleration are identical with those for standard freight elevators. In the consideration of leveling accuracy, it may be important under circumstances of these material lifts to consider not the linear misalignment but the angular misalignment of leveling accuracy between the on-floor conveyor and the on-car conveyor or transfer device. Under certain circumstances, this consideration will be more appropriate than the linear accuracy due to the dynamics of transfer. The transfer of the container must usually bridge the car sill-to-landing sill dimension taken up by the counterbalanced biparting or center-opening doors and appropriate car gate. In the case of double-ended lifts, the dedication of in-feed conveyors at one side and out-feed conveyors at another can permit the misalignment to be predicted so that the container is always traversing from a higher to a lower conveyor transfer condition in its movement.

The class of loading of the elevator should be appropriate to the impact encountered as discussed previously.

16.3.1 Car Platform Design

The car platform should be designed to receive the transfer device and any vehicle load. The access to the transfer device, if located on or within the platform, should be by removable panels with sufficient access space and convenient power receptacles so that repairs may be effected safely without exposing maintenance personnel to potentially moving elevator parts. Space should be provided on the car platform to permit maintenance and repair of all on-car equipment.

16.3.2 Car Enclosure Design

The car enclosure design should, in general, conform to that discussed previously with emphasis on panels and guards to protect maintenance personnel from moving elevator parts in the maintenance of transfer devices located on the elevator car. It should be noted that there are special requirements for in-car controls required under Part XIV of the A17.1 Code.

16.3.3 Door Selection

The selection of special freight elevator or material lift doors is similar to those of the standard freight elevator. The only difference would be in the use of center-opening doors if a power-and-free or in-floor tow chain conveyor is used. The use of an overhead power-and-free conveyor in the materials handling system usually requires a transfer device on the car that engages the in-floor or overhead power-and-free conveyor system. Such a transfer device may demand a door penetration most easily accommodated by center-opening doors or their sills as opposed to vertically opening doors. In such instances, the Underwriters Laboratories have made specific exemptions to fire test requirements and have provided equivalency opinions for equipment that is specially designed if in their judgment the equipment does not incur an exposure different from the standard design.

16.3.4 Material Lift Transfer Devices

The transfer devices in the various materials handling systems between the floor-related and car-related components are varied and many units are specifically designed to provide a one-of-a-kind solution. However, most fall into the following general categories for essentially automated operation with respect to freight elevator or material lift equipment.

In-floor Tow Chain Conveyor

The in-floor tow chain conveyor is used in conjunction with a material lift transports cart (Fig. 16.3.1) and on the lift (Fig. 16.3.2) by means of positioning a cart in front of the lift entrance, releasing it

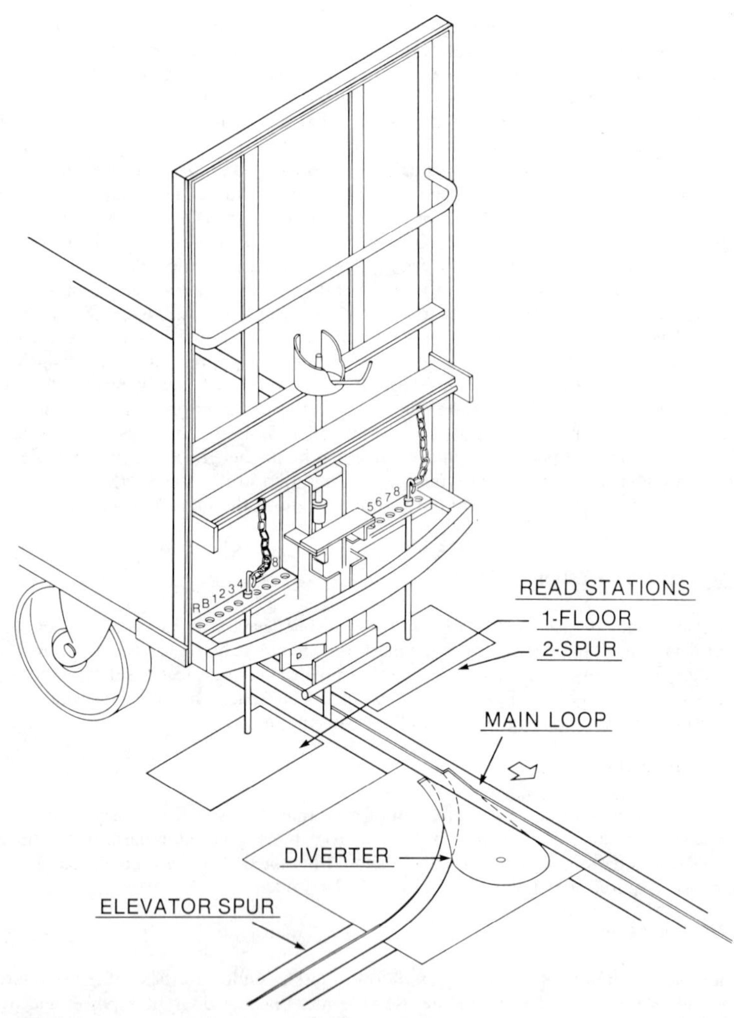

Fig. 16.3.1 In-floor towline cart entering elevator spur. Courtesy of Otis Elevator Company.

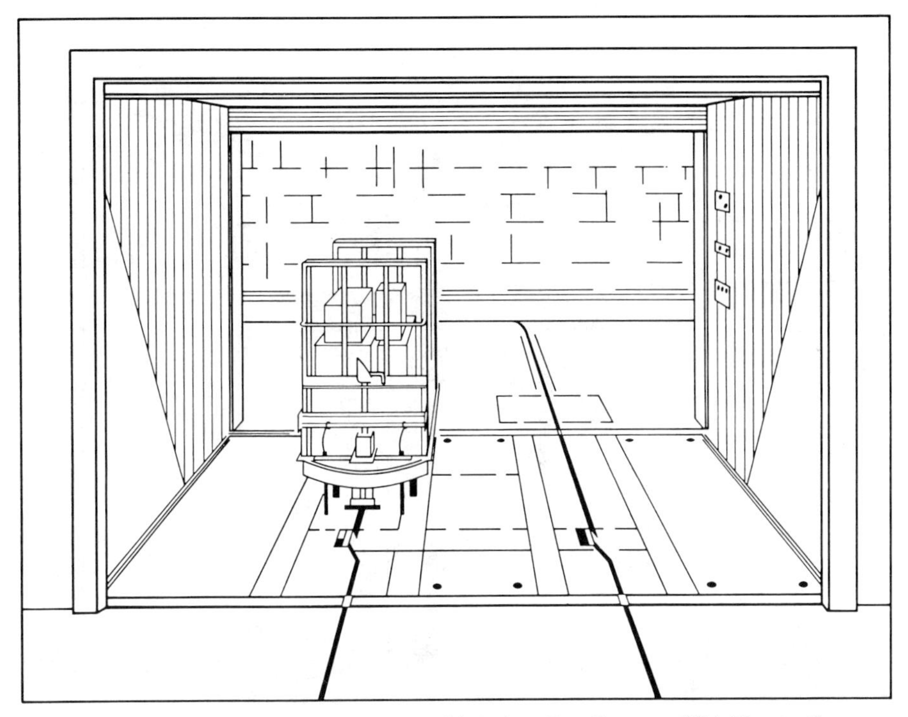

Fig. 16.3.2 Loaded cart entering freight elevator with dual towline. Courtesy of Otis Elevator Company.

from the tow chain and reading the destination address of the pin positions at the front of the cart. The lift is called to the landing, the doors open, and the transfer mechanism from the lift reaches out and injects the cart into the lift and onto its dedicated tow chain. The lift shown in Fig. 16.3.2 is a dual lift that is double-ended, which permits the pickup and delivery simultaneously of two carts in the up and down directions of travel. The reverse action takes place at the destination floor with the delivery of the cart and the pickup of another. This type of system has been used most extensively in general merchandise warehousing where stocking and shipping cannot be accommodated by standardized containers or pallets but must be transported on general purpose carts. These units have been installed in capacities from 3000 to 6000 lb (1350–2800 kg) at speeds of from 100 to 600 ft/min (0.5–3.0 m/sec).

Overhead Power-and-Free Conveyor

The overhead power-and-free conveyor transports carts on a monorail utilizing carriers which carry an escort destination (see Fig. 16.3.3). These carts are injected into the lift, usually a single cart per lift, and transported to the destination floor where they are ejected at the opposite end by the on-car transfer device. This conveyor system requires essentially continuous free track and hence the doors may require a penetration which dictates that they normally are center-opening rather than vertically opening. These units are most often used in health-care facilities for supply, processing, and distribution of food and supplies. In larger installations for industrial use, they would become part of a multifloor manufacturing facility with higher capacities. These units range in capacity from 600 to 5000 lb (270–2300 kg) operating at a speed of from 50 to 500 ft/min (0.25–2.5 m/sec).

Powered Roller Conveyor

The powered roller conveyor accepts pallets from a fork truck on the loading conveyor station section. The operator designates the floor destination and the pallet is moved automatically on the in-feed conveyor, unit-section by unit-section, until it forms a queue at the lift doors. When the lift arrives, the doors open and the pallet is accepted into the lift with as many other pallets as the lift has capacity for. The lift travels to the destination floor and ejects the pallet onto an out-feed conveyor at the opposite end of the lift on a FIFO basis. A more sophisticated version would have the pallet

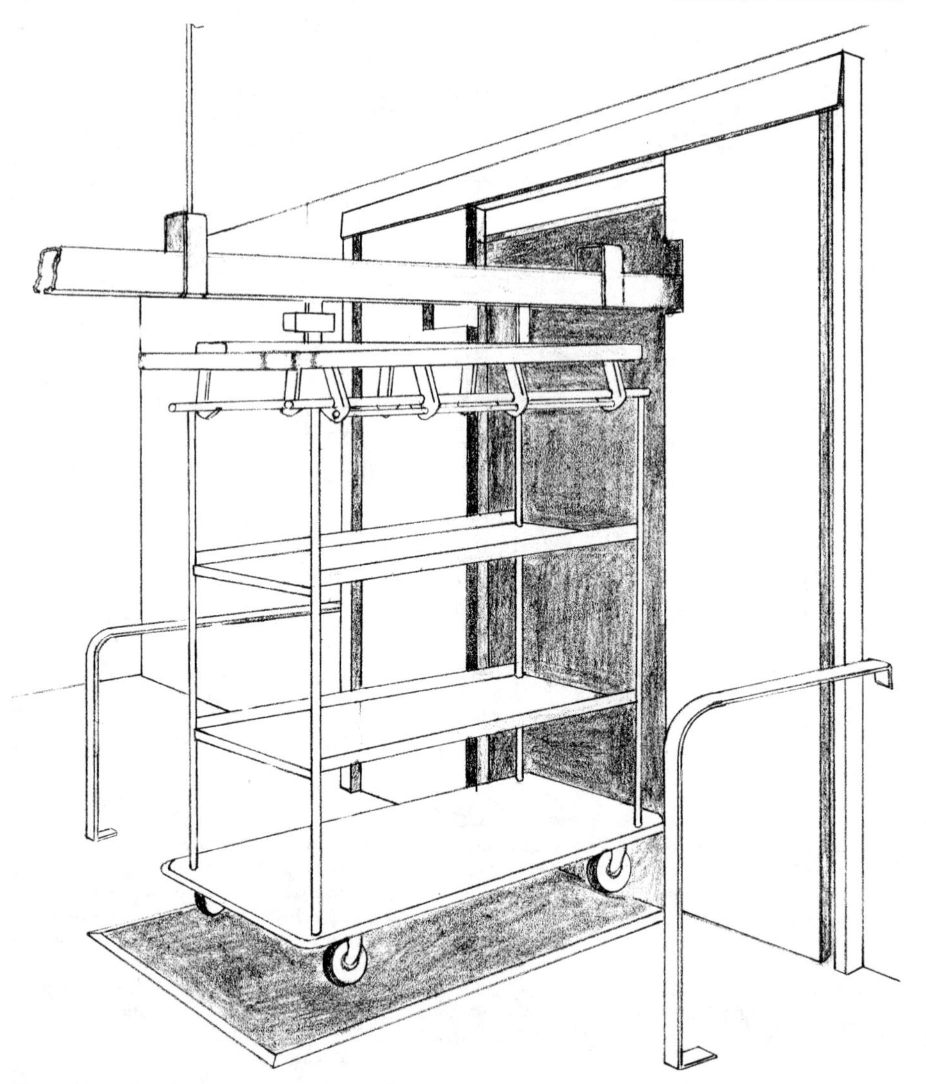

Fig. 16.3.3 Overhead power-and-free conveyor institutional cart entering freight elevator (material lift). Courtesy of Jaros, Baum & Bolles.

contain an escort address which is read as it enters the lift, storing the address in the memory of the minicomputer for destination determination. The use of a multiple-unit capacity lift and in-feed and out-feed conveyors at each landing permits management-established priorities to control the movement of pallets among the various functions served, such as incoming raw material storage, materials in process, and shipping functions according to time of day, length of waiting queues, and day of week concepts. The use of the minicomputer technologies permits reprogramming as conditions warrant to provide the optimum in flexibility. This concept is shown in its simplest form in Fig. 16.3.4.

Inject–Eject Cart Device

The inject–eject cart transfer device is an above-floor transfer unit that reaches out from the lift car, engages a cart coupler on the underside of the cart, and injects it into the cart lift. At the destination floor it ejects it at front or rear opening according to the configuration and material being transported. The cart is positioned in front of the lift with the address of its destination either registered in normal elevator fashion or escort on the cart so that the lift can read its destination once it is inside the car enclosure. The cart lift system automatically picks up the cart and delivers it to the floor destination

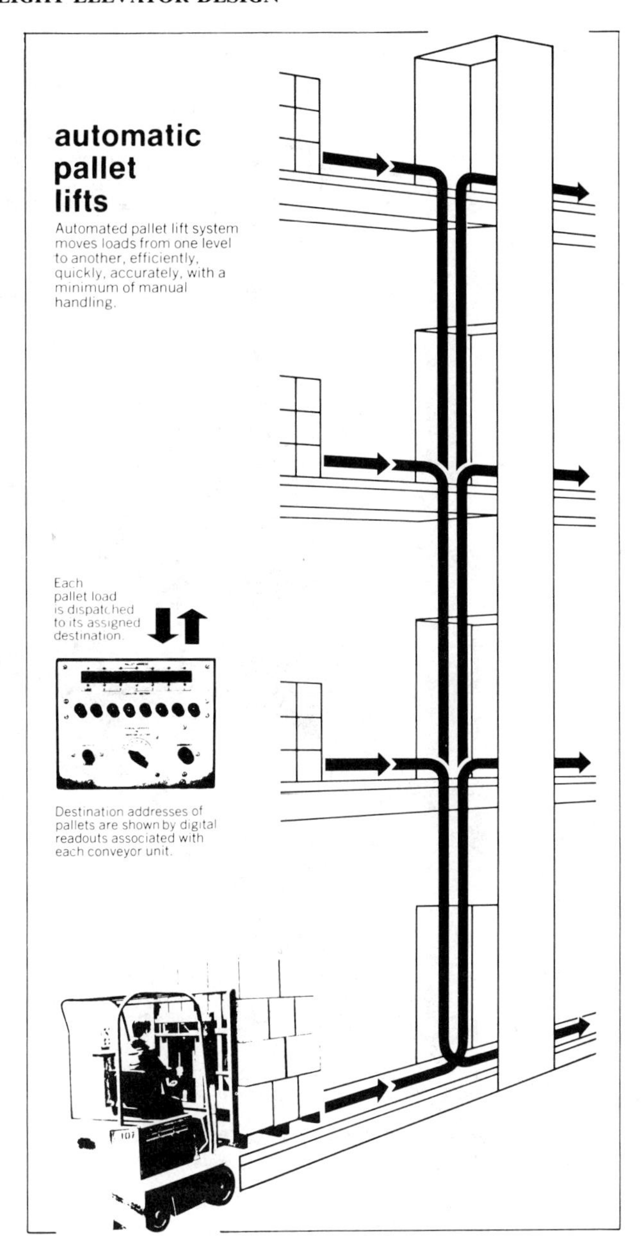

Fig. 16.3.4 Automated pallet lift (freight elevator) with automated in-feed and out-feed conveyors. Courtesy of Otis Elevator Company.

on a one-at-a-time basis. The cart lift shown in Fig. 16.3.5 utilizes in-floor wheel tracks to guide the casters in an eject transfer so that the cart tracks in a straight line. This is most effective in overcoming poor floor leveling and poor maintenance conditions of swivel casters in commercial, institutional, and industrial applications. The lifts usually are rated from 800 to 1000 lb (350–450 kg) and operate at speeds from 200 to 1000 ft/min (1.0–5.0 m/sec).

A variant of the transfer device with coupler configuration is the cart lift that ejects a sliding bed which lifts the dead weight of the cart from its casters and draws it into the lift. Similar devices could be configured that use forks and similarly transport pallets as in the automated storage and retrieval systems.

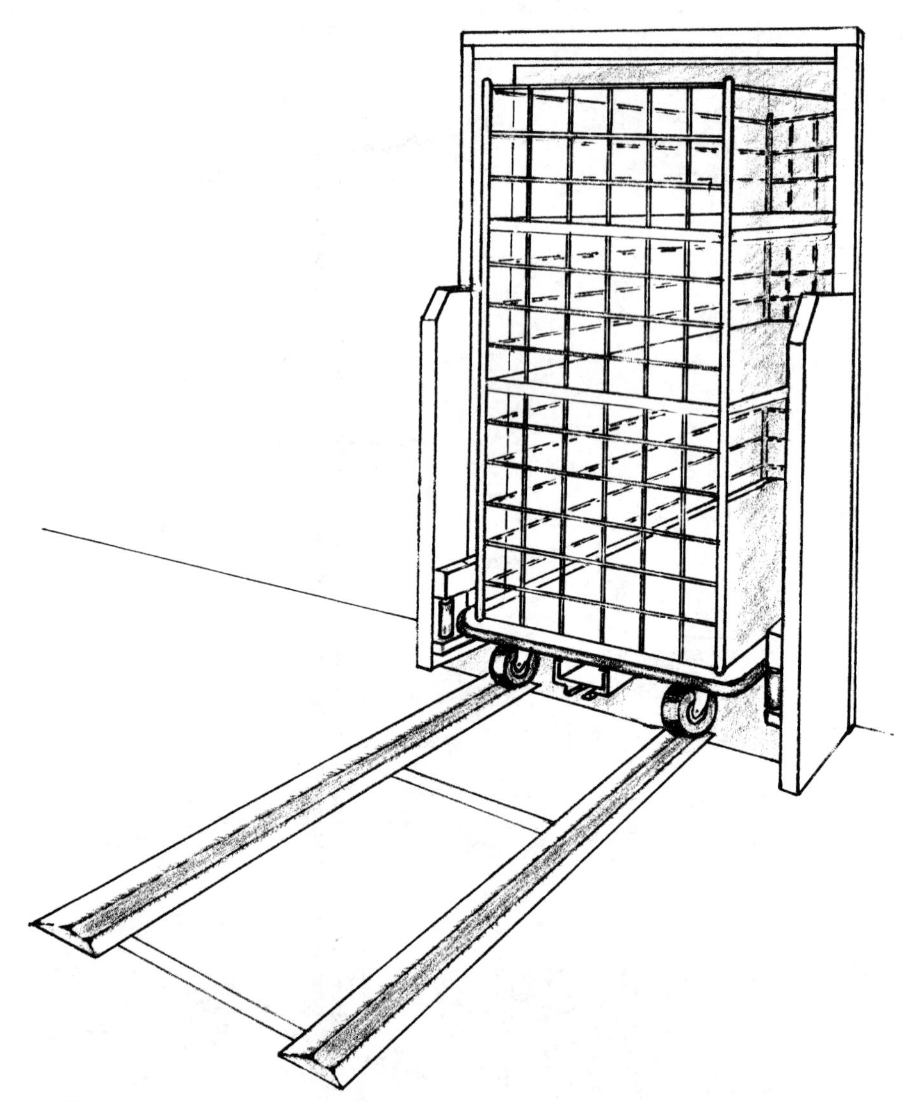

Fig. 16.3.5 Inject–eject cart lift with in-floor tracks. Courtesy of Jaros, Baum & Bolles.

16.3.5 Container Designs

In each of the foregoing transfer device concepts, it is essential that they be designed for standardized containers, even to the extent that they are special for a specific installation and dedicated to the in-house system. It is an essential ingredient in the design of such total vertical and horizontal materials handling systems to evaluate first the on-floor distribution system and its container so that the lift can be appropriately designed and the transfer device selected that most cost-effectively accommodates the throughput function. The most common of these containers are carts, either passive or active, pallets, or tote boxes.

The in-floor tow chain conveyor transports carts in the industrial and institutional fields. The overhead power-and-free conveyor transports carts or tote boxes in the industrial and institutional fields. The roller conveyor transports pallets or tote boxes primarily in the industrial area. Inject–eject carts transport carts primarily in the institutional and commercial applications.

16.3.6 Addressing Systems

The addressing systems for material lifts and their transfer devices relate to either a manual input address for each individual container, cart, or pallet or an escort address reading system which reads the encoded destination address on each container. Whenever the batch lift accepts more than one container per trip, it is necessary to estabish a means of remembering the first-in pallet and delivering it to its destination prior to traveling to the destination of the second-in pallet for its delivery. Such a complexity dictates the necessity of stepping along in the computer memory the address of each pallet so that it is constantly monitored in its travel vertically and horizontally through the system.

16.3.7 Material Lift Operation

The operation of the material lift is a slave to the management of the materials handling system that operates the entire facility. Whenever such an installation is contemplated, it is important to provide sufficient memory in the system so that reprogrammability is accommodated with ease and without extensive delays. The need for the constant changing of priorities may require a powerful central processing unit (CPU) to handle complicated priorities by a relatively elementary vertical lift system.

16.4 STANDARD AND SPECIAL DUMBWAITER DESIGN

Dumbwaiters like standard and special freight elevators and material lifts can be configured to accommodate special cart or tote box sizes that perform specific functions. It often is cost effective to design special containers to perform specific functions rather than using standard containers and attempting to adapt them to specialized tasks for which they may not be suited.

Dumbwaiters are limited to: (1) an area of 9 ft^2 (0.84 m^2) net inside, (2) a car and door height limited to 4 ft (1.2 m), and (3) a capacity limited to a rated load not exceeding 500 lb (2.27 kg). Smaller areas are rated appropriately in Section 702—Capacity and Loading of the A17.1 Code governing the structural capacity and to the rated capacity with respect to net platform area. These capacities and limiting dimensions are net, including the transfer device.

For practical commercial purposes, dumbwaiters are separated into two categories depending on their stopped position with respect to the finished floor level. The counter-height dumbwaiter stops usually 2 ft 6 in. (630 mm) above the floor level. This height can vary according to specific purposes and the ease with which the interior can be accessed under manual loading conditions or when equipped with a transfer device, the in-feed and out-feed conveying devices which may be part of an on-floor manual system. The counter-height-stopping dumbwaiter with transfer device handles a container usually categorized as a tote box due to its ease of transport. The floor-stopping dumbwaiter must handle a cart, either manually or with a transfer device, for convenience of loading and unloading. The cart can take any configuration appropriate to its superstructure again related to its on-floor transportation parameters.

The justification of automation for dumbwaiters as well as elevators lies in the efficiencies achieved by moving materials between floors without manual escort and the use of general use elevators. In vertical buildings, the employee away from the "home floor" becomes practically unsupervised, and hence productivity is substantially reduced. Automated and semiautomated materials handling systems have now become a major factor in maintaining a high level of productivity while reducing the overall labor content of inter- as well as intrabuilding movements. This is true regardless of whether the building is industrial, institutional, or commercial. These equipments originated in the industrial and health-care sectors and now have become prevalent in the commercial sector. It is possible to configure combination functions to provide special tasks. One health-care facility found it appropriate to "double-deck" the concept with a cart on the lower deck with an inject–eject transfer device and with a tote box on the upper deck with a belt transfer device.

16.4.1 Dumbwaiter with Tote Box Device

The tote box transfer device shown in Fig. 6.1.10 can be configured in two forms based on the on-floor personnel available. If the delivery system is essentially a one-way system, the transfer device can feed the tote box to a passive out-feed gravity conveyor which queues the tote boxes awaiting transfer to carts or other on-floor means of distribution. The reverse direction requires individual manual loading which may be inconvenient and time consuming depending on the weight of the containers.

A dumbwaiter equipped with a reversible transfer device, with a similar transfer capability on the on-floor conveyor, permits unattended operation depending on the master priority programming of the dumbwaiter. Such a conveyor configuration permits the unattended pickup and delivery of containers according to the central dispatching and receiving mode selected. Obviously, a double-ended arrangement will provide the greatest throughput.

16.4.2 Tote Box Designs

Tote boxes are available in many designs to accommodate special tasks for general or special contents. The usual tote boxes used in industrial, institutional, and commercial applications are:

A Box—approximately 10 in. wide × 14 in. long × 10 in. high (250 mm × 360 mm × 250 mm)
B Box—approximately 14 in. wide × 18 in. long × 10 in. high (350 mm × 450 mm × 250 mm)
C Box—approximately 22 in. long × 16 in. wide × 8 in. high (560 mm × 410 mm × 200 mm)

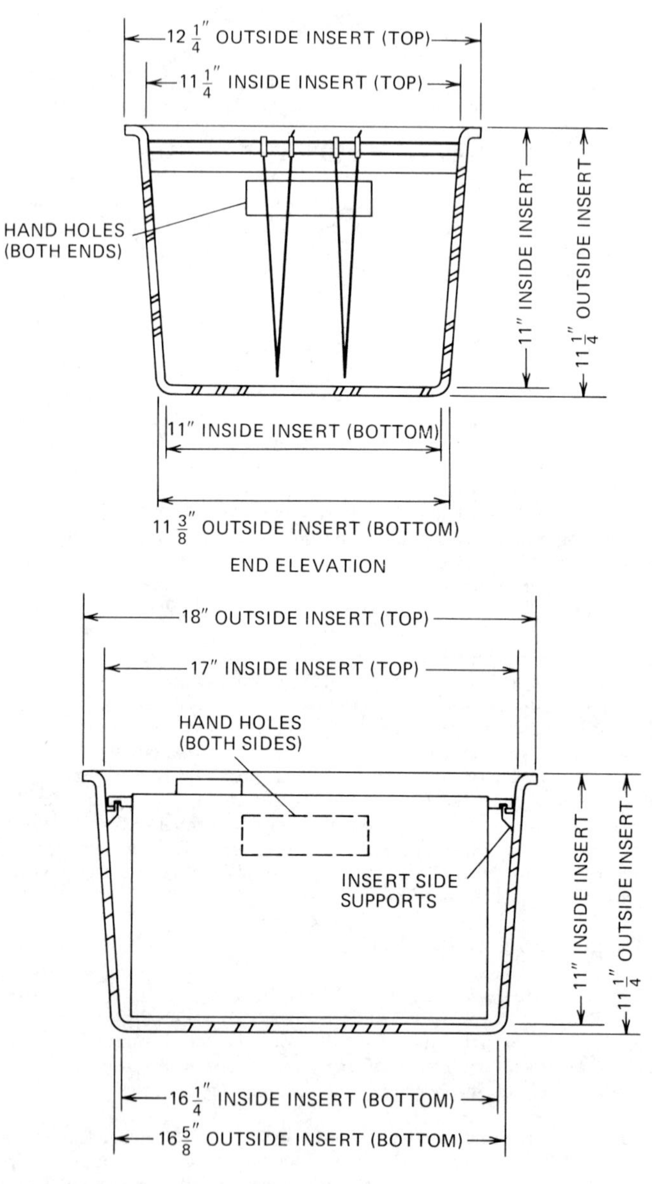

Fig. 16.4.1 Special tote box for mail delivery by vertical conveyor for transfer to on-floor delivery by manual or robot cart system. Courtesy of Jaros, Baum & Bolles.

These units can be equipped with lids in a variety of configurations, including lockable. The nominal capacity of these tote boxes are A Box, approximately 20 lb; B Box, approximately 40 lb; and C Box, approximately 60 lb.

Special function boxes can be obtained provided a significant quantity justifies the special mold costs. An example of such a special design is the tote box shown in Fig. 16.4.1 which permits the presorting of mail into hanging folders dedicated to the tote box for ease of vertical transport as well as horizontal distribution in a manual cart dedicated to the floor. An on-floor mail cart for tote boxes is shown in Fig. 16.4.2.

16.4.3 Dumbwaiter with Cart Transfer Device

The dumbwaiter with a cart transfer device stopping at floor level usually is an automated inject–eject device connecting with a coupler on the underside of each cart so that it picks up one cart

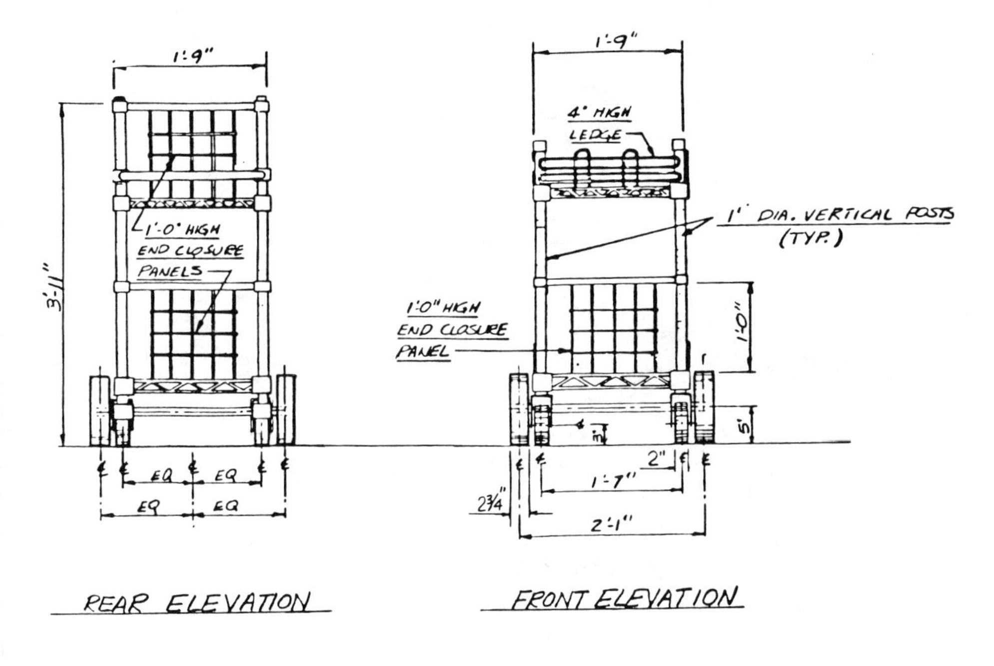

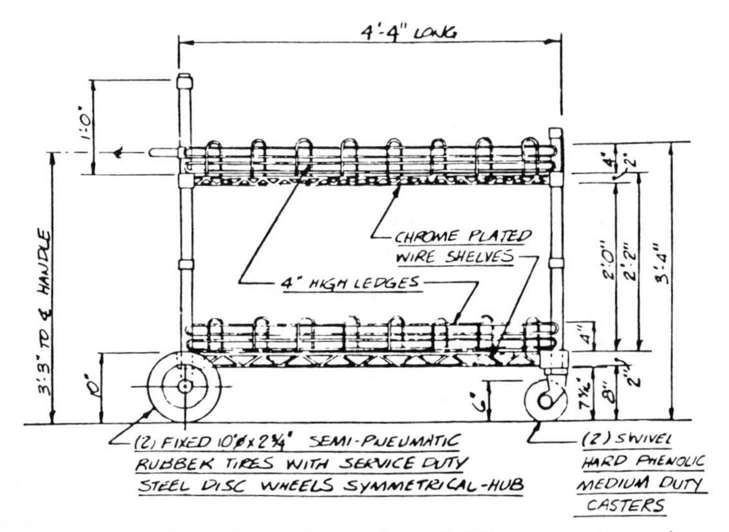

Fig. 16.4.2 On-floor mail cart for tote box regular mail delivery or random package or bulk delivery.

automatically, injects it into the dumbwaiter and ejects it at its destination floor. This configuration is shown in Fig. 16.4.3. Such equipment can be configured depending on space availability, traffic patterns, and intensity of single direction traffic during peak dispatch and pickup periods with single- or double-ended operations at each floor. The double-ended operation permits dedicated out-feed sides and dedicated in-feed sides so that the two functions may be conducted simultaneously in anticipation of the priority dispatch or priority pickup mode. This configuration is essential in a health-care facility which permits one side to be dedicated to clean delivery and the other side to be dedicated to soiled return.

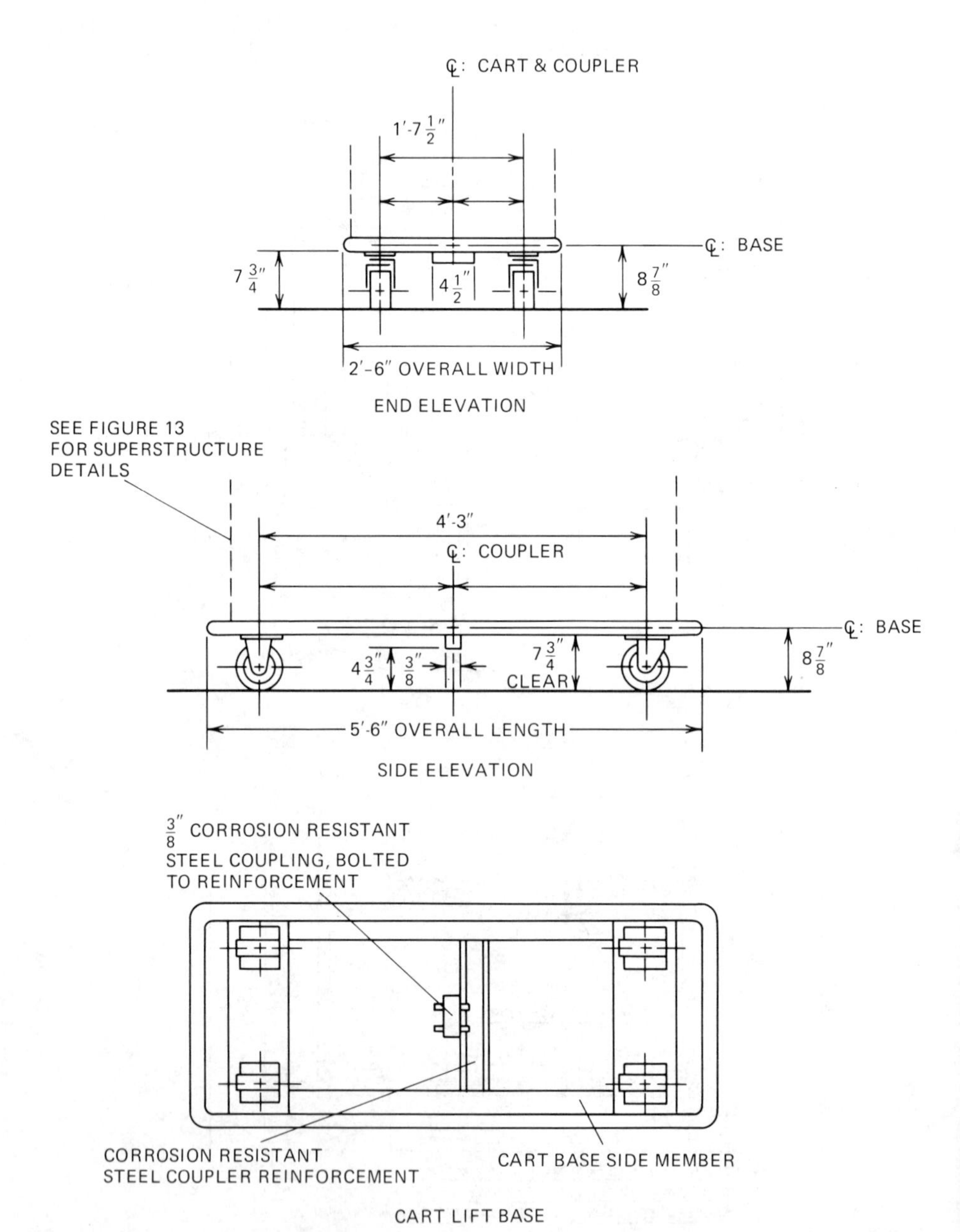

Fig. 16.4.3 Inject–eject cart base for use with multiple superstructures in institutional or industrial applications.

16.4.4 Cart Designs

The cart design is more easily adapted to special functions than the tote box, since the dumbwaiter cart transfer device recognizes any uniform cart base with identical caster and coupler locations. The superstructure above the uniform base can be configured at any height permitted by the door opening and the perimeter dimensions of the base. It is appropriate to point out that each installation can configure its own "system-standard base" dimensions appropriately for the major function of the predominant superstructure. Once the system-standard base has been selected, other superstructures may be configured to handle different unit volumes or special larger loads.

The institutional cart design for health-care facilities usually uses the maximum interior dimensions of the dumbwaiter and transports the materials in an enclosed or covered cart for sepsis control. The commercial equivalent for mail and bulk package handling is a similar open cart for the loose accommodation of packages or in conjunction with tote boxes for the presorted delivery of mail simultaneous with the bulk delivery of supplies. Such combinations are shown in Figs. 16.4.4 and 16.4.5.

16.4.5 Performance Parameters

The door-closing and door-opening speed for both the counter-stopping and floor-stopping dumbwaiters is 1 ft/sec (300 m/sec) for each sliding door panel. It should be recognized that both types of dumbwaiters use counterbalanced biparting doors that generate a maximum door opening or door closing time of 2 sec since only one-half the door height will be traveled by a door panel.

The transfer device will usually be powered by a torque motor which can be stalled for an extended period of time before its overload trips out. This is due to the necessity of accommodating stalled tote boxes or carts with queues excessive for the available out-feed dimensions. The transfer device operates usually at speeds approximately 30 ft/min (0.15 m/sec). Door-open and door-close dwell times are inserted in the sequence of operations to assure stability of the injected or ejected load. A dwell time of 3–5 sec is a standard arrangement.

The safety features of the door operation consist either of a deflecting leading edge of the top door panel that reverses the door or a stall feature which, if an obstruction is met, the door reverses for its door-open dwell time and attempts to close again.

16.4.6 Intercommunication Capability

In all major installations, it usually is essential to provide a dedicated intercommunication system among the typical floors with the main dispatching and receiving floor so that the efficiency of the system can be optimized for maximum throughput. Since the dumbwaiter is a batch system, its throughput capability must not be a function of unsupervised on-floor personnel. The intercommunication facility, preferably an intercom as opposed to a telephone, provides dedicated single- or multiple-station communication without interference.

16.5 UNIQUE FREIGHT ELEVATOR APPLICATIONS

Freight elevators for handling materials or unit loads have been used in special application in both fixed towers and moving towers that respond to a specific function and environment. The fixed tower is essentially a large elevator in a tower that serves a multiplicity of stalls for parking motor vehicles or racks with "pigeon holes" for pallet storage. The moving tower elevator expands the vertical capacity of the elevator into the horizontal. These arrangements can be configured for either manual or automated operation, usually in a short-term storage facility so that the content of specific stalls or pigeon holes can be retrieved under programmed instructions.

16.5.1 Fixed Tower Elevator—Moving Carriage

The fixed tower–moving carriage concept is a large elevator with a laterally moving carriage that positions a transfer dolly to pick up a single or series of pallets or an automobile to draw the load onto the elevator and reposition it at another landing's pigeon hole or stall. The parking garage version is shown diagrammatically in Fig. 16.5.1.

The transfer dolly can be replaced by a series of combing fingers which lifts the load and deposits it on matching combing fingers in each stall or pigeon hole. This concept raises the combing fingers and lowers the load onto the stall fingers for storage with a reverse action when the load is retrieved. The parking garage version of this concept is shown diagrammatically in Fig. 16.5.2.

16.5.2 Moving Tower Elevator

The moving tower elevator is an extension of the fixed elevator utilizing a horizontal tower technique that horizontally levels the tower to accommodate the vertical level of the elevator so that a substantially

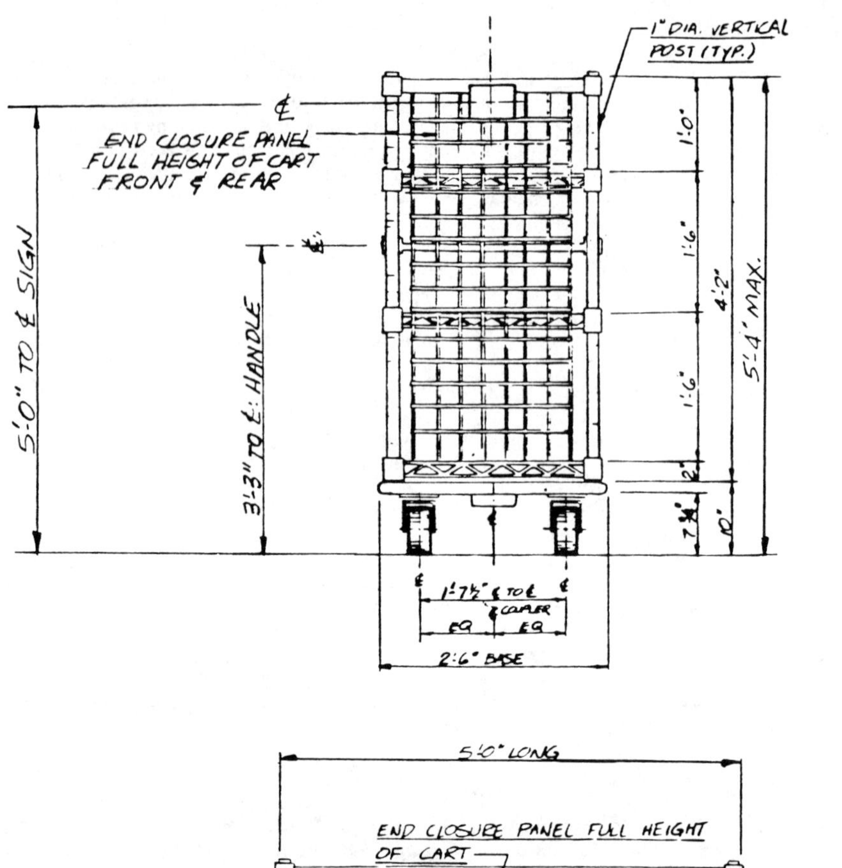

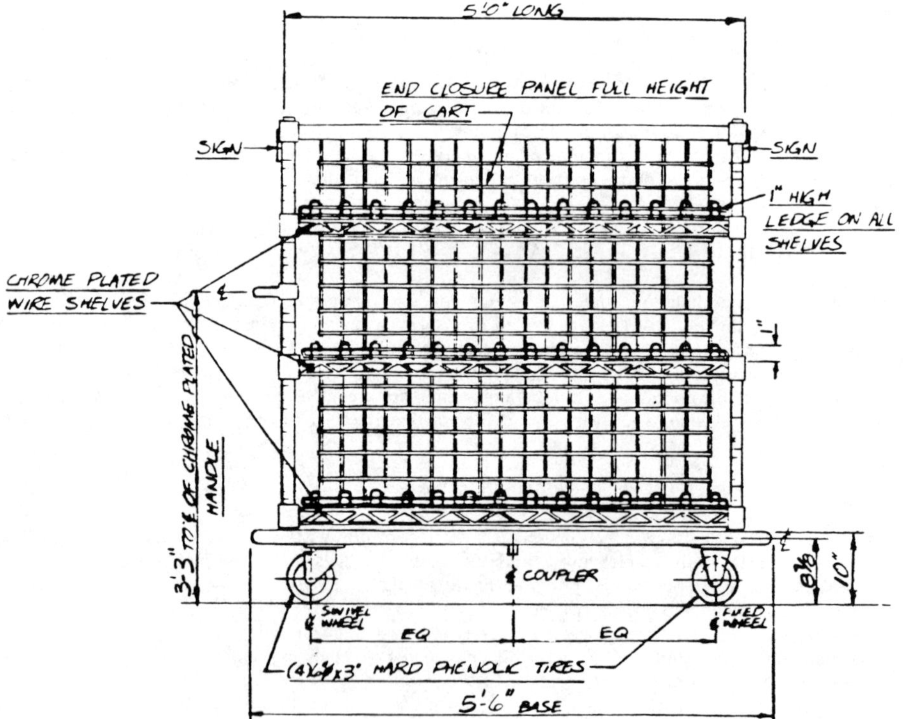

Fig. 16.4.4 Inject–eject laundry or exchange cart for health-care facilities.

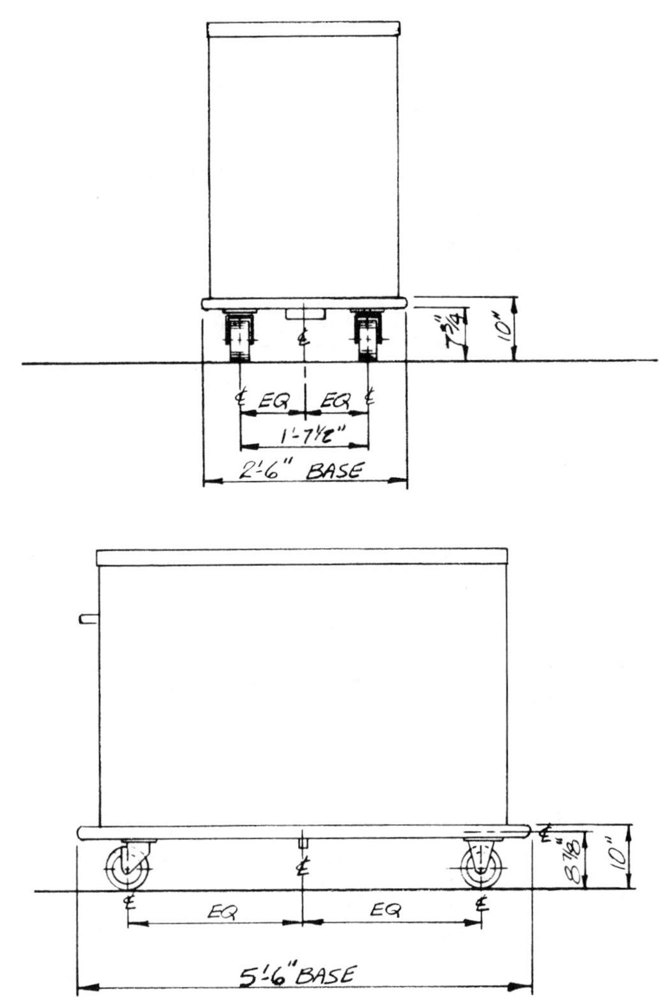

Fig. 16.4.5 Inject–eject surgical case cart for health-care facilities.

greater number of pigeon holes or stalls can be accessed. The concept of using a transfer dolly or combing fingers is identical to that of the fixed tower elevator.

As in the fixed tower concept, the moving tower concept permits either manual or automated operation to store and retrieve. In these combinations, it is a larger version of the now ubiquitous stacker crane used in automated storage and retrieval warehousing. The stacker crane is now a mature discipline of its own.

16.6 BUILDING INTERFACES

The freight elevator system interacts with its associated building in a manner that must accommodate its peculiar requirements and simultaneously comply with the building codes affecting fire and smoke control as well as life safety. These interface requirements may be significantly different from those required by other conveying systems. This conflict was discussed earlier; however, these requirements essentially fall into the same categories for evaluation purposes as freight elevators. These interface categories are as follows:

1. *Structural.* The forces exerted on the building structure by the elevator equipment during loading, unloading, running, acceleration, deceleration, and emergency stopping.
2. *Electrical.* The energy source of all elevators to power and illuminate the elevator.

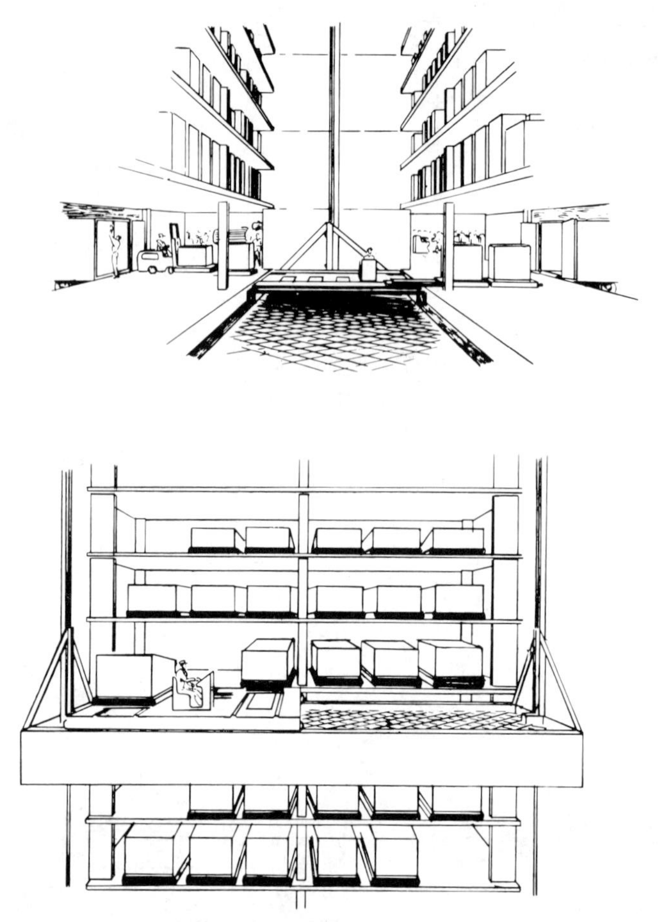

Fig. 16.5.1 Fixed-tower freight elevator with moving carriage and transfer dolly.

3. *Mechanical.* The ventilation and/or air conditioning equipment necessary to accommodate the heat release of the equipment primarily located in the machine room. The requirements for fire protection systems as required by local codes.
4. *Architectural.* The treatment of the hoistway walls and other enclosures required by code that surrounds the elevator equipment and its machinery as well as the appropriate access for maintenance and repair.
5. *Environmental.* The environmental conditions peculiar to some installations requiring special consideration of atmospheric contaminants.

16.6.1 Structural

Structural support of a freight elevator or material lift during loading, unloading, and running requires guide rails for the elevator car and its counterweight running in a substantially vertical and parallel relationship and enclosed by a hoistway. This support is obtained at every floor interval or more often. Intermediate supports may require special structural members acting as columns to provide the intermediate support when the reactions exceed the level of stress permitted with floor-related bracket intervals. These conditions exist for both electric traction elevators as well as hydraulic elevators. The necessity of transferring these loads to the building structure may be more acute with hydraulic elevators due to the alignment sensitivity of the ram with respect to its packing.

The machine room of hydraulic elevators may be located adjacent to the hoistway or located remotely. Some jurisdictions, however, require that the hydraulic elevator machine room be located contiguous to the shaftway and vented directly to it for fire safety considerations. The electric traction elevator machine room usually is located overhead and imposes on the structure the total support

Fig. 16.5.2 Fixed vertical conveyor with moving horizontal conveyor acting as a freight elevator. Courtesy of Jaros, Baum & Bolles.

requirements of the dead weight of the car, the counterweight, and the live load. The machine room may be located adjacent and contiguous to the hoistway for smaller freight elevators. However, the large-capacity freight elevators require overhead machine rooms that are extended above the top of the hoistway.

The pit of each elevator absorbs any vertical reactions that may occur due to deceleration of an overtraveling elevator car. The hydraulic elevator is supported entirely from the pit through the hydraulic cylinder which imposes its structural reaction on the pit floor. The electric traction elevator imposes a reaction on the pit floor only when the car or the counterweight overtravels the bottom terminal landing and engages a buffer to retard its movement. These buffer impacts, although they may occur only under test conditions, must be accommodated appropriately.

The identification of structural reactions for the hoistway, machine room, and pit accommodations are the responsibility of the supplier of the freight elevator, cart lift, or dumbwaiter so that the structural engineer may appropriately design the supporting members at the interface with the building.

Hoistway Requirements for Rail Bracketing

The elevator car guide-rail configuration is a relatively weak member for horizontal loads and hence must be bracketed to transfer the loading and unloading forces to the building structure in a manner that does not exceed the stress limitations (see Fig. 16.6.1). Most elevators, from 2000 to 8000 lb (900–3600 kg), will be accommodated by a standard floor height interval not exceeding 14 ft (4.3 m).

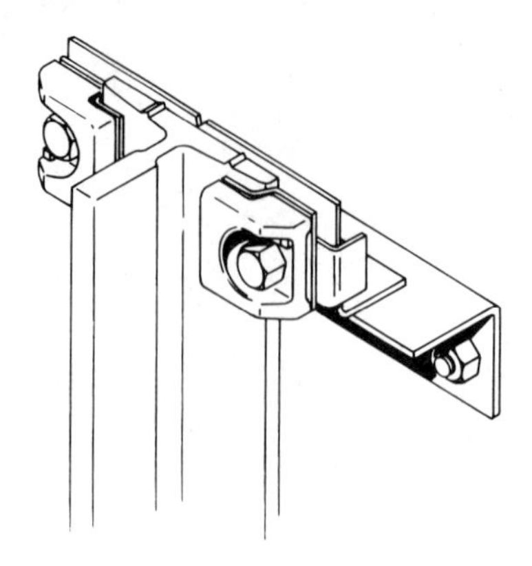

Fig. 16.6.1 Standard light-duty freight elevator rail clip detail fastening the rail to the bracket. Courtesy of Otis Elevator Company.

Freight elevators with a capacity of 10,000–20,000 lb (4500–9000 kg) will require column backing behind each main car guide rail so that the loading and unloading forces may be transferred through the rail immediately to a supporting structure. From 20,000 to 100,000 lb (9,000–45,000 kg) the requirement may be such as to require two guide rails on each side of the car with more extensive column rail backing.

In all cases, freight elevator installations require the supplier to provide the structural engineer with the required location of bracket supports. Figure 16.6.2 diagrammatically represents the rail arrangement for the car and counterweight in a standard floor-to-floor bracket interval. Figure 16.6.3 indicates the column backing for a single main car rail. Both figures indicate the type of information that must be supplied to accommodate all of the static loading and unloading as well as the dynamic running forces imposed. It should be noted that all electric traction elevators impose a safety reaction in which the elevator car guide rail acts as a column to retard an overspeeding elevator. This reaction usually is a major consideration in supporting the guide rails, with respect to deflection and with respect to pit reactions. This may also occur on a counterweight guide rail if a counterweight safety is required.

Machine Room Requirements

The elevator machine rooms of an electric traction elevator impose structural reactions as a function of the location of the machine and the dead-end hitches when the elevator is roped 2:1. In the case of the overhead machine room, these reactions are acting downward on the supporting structure. The A17.1 Code specifies the fiber stress and deflection requirements for the supporting members directly related to the elevator machine beams provided by the elevator supplier.

The adjacent contiguous electric traction elevator machine room configuration imposes an uplift from the machine due to the ropes traveling from the machine room up to the top of the hoistway and over supporting sheaves to directly support the car and counterweight in the same manner as if the machine room were located overhead. The machine room reactions from these arrangements usually impose both a vertical uplift as well as a lesser horizontal component due to deflecting sheaves or other unusual machine arrangements in the confined spaces available.

There are no structural reactions from a hydraulic elevator in the machine room except for the location of the pump unit and controller. However, it is suggested that, in major hydraulic installations where a significant amount of hydraulic fluid will be stored, the machine room floor be depressed with a concrete revetment that will contain the oil from other parts of the building if the flexible connections or any hydraulic equipment fails and releases the full charge of hydraulic fluid.

Pit Requirements

The pit requirements for a hydraulic elevator essentially relate to spreading the load from the hydraulic ram across the pit floor for the support of the dead and live loads. The pit must also provide a waterproof connection to the hydraulic cylinder or its enclosing casing if the water table reaches the level of the bottom of the hydraulic cylinder.

The pit requirements for an electric traction elevator must accommodate the car and counterweight buffer reactions that occur when the car or counterweight overtravels the lower terminal landing as well as the car or counterweight safety reactions in the guide rails to retard an overspeeding car or counterweight. A counterweight safety, however, is required only when there is a space beneath the

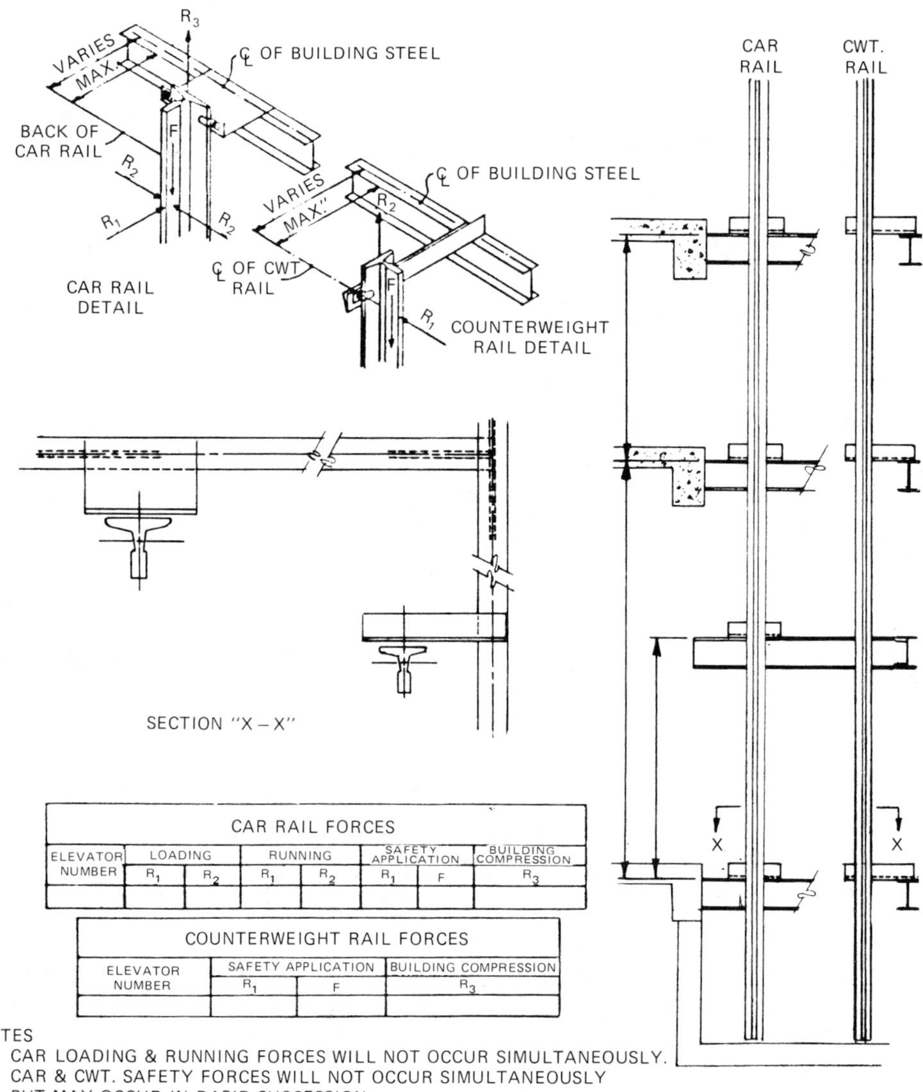

CAR RAIL FORCES							
ELEVATOR NUMBER	LOADING		RUNNING		SAFETY APPLICATION		BUILDING COMPRESSION
	R_1	R_2	R_1	R_2	R_1	F	R_3

COUNTERWEIGHT RAIL FORCES			
ELEVATOR NUMBER	SAFETY APPLICATION		BUILDING COMPRESSION
	R_1	F	R_3

NOTES
1– CAR LOADING & RUNNING FORCES WILL NOT OCCUR SIMULTANEOUSLY.
2– CAR & CWT. SAFETY FORCES WILL NOT OCCUR SIMULTANEOUSLY BUT MAY OCCUR IN RAPID SUCCESSION.
3– FORCE R_3 WILL OCCUR DUE TO BUILDING COMPRESSION.
4– CAR & CWT. BUILDING COMPRESSION FORCES (R_3) WILL OCCUR SIMULTANEOUSLY. FORCES EXPECTED TO BE LIMITED BY SLIDING RAIL CLIPS DURING & AFTER CONSTRUCTION
5– MAXIMUM ALLOWABLE DEFLECTION 1/8 IN.

Fig. 16.6.2 Freight elevator main car and counterweight rail bracket design and rail force data (standard duty). Courtesy of Jaros, Baum & Bolles.

elevator pit that may be or become occupied. These reactions also are a part of the required information to be shown on the elevator layout drawings for design purposes.

16.6.2 Electrical Requirements

All elevators require electric power for propulsion as well as for lighting of the car enclosure and for door operation. Under certain circumstances, it may be appropriate to install a telephone or intercommunication system between the elevator car enclosure, elevator machine room, and other remotely located management functions in a facility.

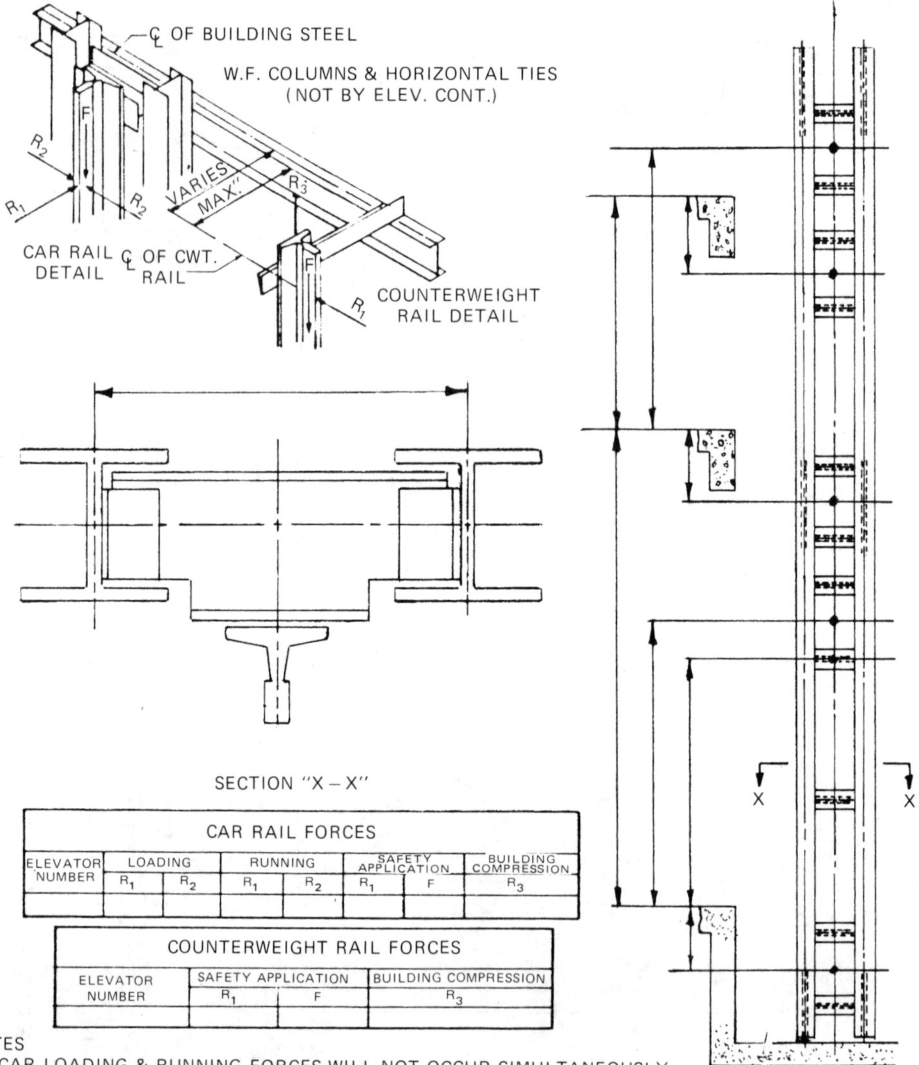

ELEVATOR NUMBER	CAR RAIL FORCES: LOADING		RUNNING		SAFETY APPLICATION		BUILDING COMPRESSION
	R_1	R_2	R_1	R_2	R_1	F	R_3

ELEVATOR NUMBER	COUNTERWEIGHT RAIL FORCES: SAFETY APPLICATION		BUILDING COMPRESSION
	R_1	F	R_3

NOTES
1– CAR LOADING & RUNNING FORCES WILL NOT OCCUR SIMULTANEOUSLY.
2– CAR & CWT. SAFETY FORCES WILL NOT OCCUR SIMULTANEOUSLY BUT MAY OCCUR IN RAPID SUCCESSION.
3– FORCE R_3 WILL OCCUR DUE TO BUILDING COMPRESSION.
4– CWT. BUILDING COMPRESSION FORCES EXPECTED TO BE LIMITED BY SLIDING RAIL CLIPS.
5– RAIL BACKING COLUMN CONNECTIONS TO BUILDING STRUCTURE MUST ACCOMODATE BUILDING COMPRESSION.

Fig. 16.6.3 Freight elevator main car and counterweight rail bracket design and rail force data (heavy duty). Courtesy of Jaros, Baum & Bolles.

The power, lighting, and telephone interface should occur in the machine room so that troubleshooting and repair can be conducted from the one location by elevator service personnel. The previous practice of supplying the lighting circuit and the telephone circuit at the halfway point of the hoistway is not considered a desirable interface location.

16.6.3 Mechanical Requirements

The mechanical interface with elevator equipment involves the plumbing (sprinklers) and heating, ventilating, and air conditioning requirements, with respect to the heating and heat release of the electromechanical equipment and with the fire- and life-safety considerations. Additionally, the building

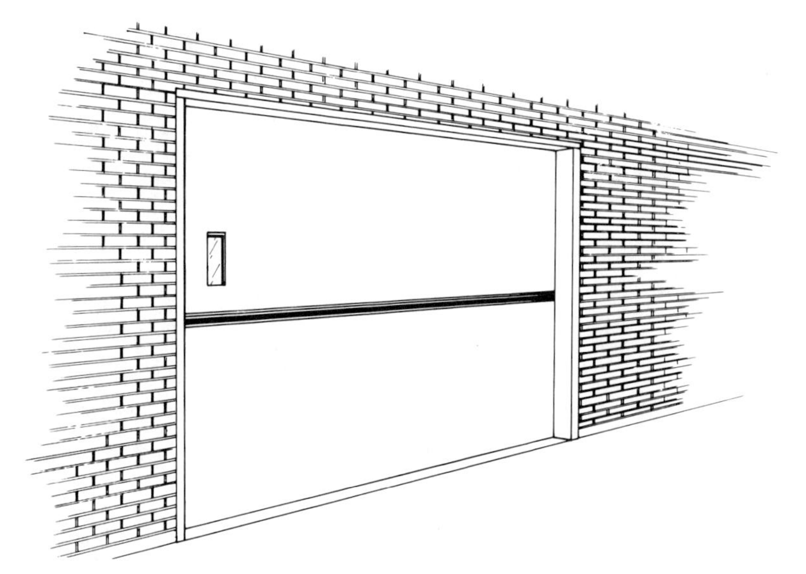

Fig. 16.6.4 Channel head and side jambs in masonry wall. Courtesy of Otis Elevator Company.

code may require that the hoistway must be vented to relieve smoke and hot gases in a manner prescribed as a function of the building height and the location of the machine room.

The machine room should be designed so that it will be ventilated to maintain a maximum of 95°F (35°C) exit air temperature. This air temperature may be increased to 104°F (40°C) if the control equipment is entirely conventional relay logic. The minimum ambient air temperature should not be lower than 50°F (10°C).

The governing codes stipulate the fire-safety requirements for sprinklers and/or smoke detectors. The machine room and top of the hoistway should always avoid sprinklers unless absolutely necessary from the standpoint of water damage due to spurious discharge.

The pit seldom requires sprinklers but the governing code will stipulate them if required.

16.6.4 Architectural Requirements

The architectural requirements relate primarily with the enclosure that surrounds the hoistway, pit, and machine rooms and the fire rating required by the building codes. Specific requirements for freight elevators are as follows:

1. Channel frames—to mount door tracks
2. Angle sill—to support lower door panel during loading and unloading
3. Access—to machinery spaces

The channel frames for the side and head jamb of the vertically sliding counterbalanced-biparting doors should extend the full height between floors to accommodate the mounting of the door tracks on the inside of the hoistway and to provide protection of the hoistway wall against abuse during loading and unloading. Figure 16.6.4 shows the arrangement desired for a concrete masonry unit (CMU) wall which is preferred for the front hoistway wall due to use and abuse considerations.

The sill angle, located in the floor between the channel side jambs, should be of sufficient strength to support the loading and unloading impact from a loaded fork truck and must also be capable of absorbing the impact load as the truck passes over the lower door panel which contains its own structural sill.

Access to the machinery spaces that contain machines and controllers will require a 60° ladder if the space is superelevated above the roof level by more than 3 ft (900 mm).

16.7 CONCLUSION

Additional descriptive and technical information concerning elevators and elevator systems and their application can be found in *Vertical Transportation: Elevators and Escalators,* 2nd Ed., 1983, by G. R. Strakosch, and in Chapter SC-4 Vertical and Horizontal Transportation in the *Monograph on the Planning and Design of Tall Buildings,* 1980, by The Council on Tall Buildings and Urban Habitat.

CHAPTER 17
POSITIONING EQUIPMENT

DANIEL J. QUINN

Southworth, Inc.
Portland, Maine

GUY A. CASTLEBERRY

Conco-Tellus, Inc.
Mendota, Illinois

ROBERT DeCRANE

K. L. Cook & Associates, Inc.
Sugarcreek, Ohio

SAUL B. GREEN

Green Associates, Inc.
Pittsburgh, PA

BEN BAYER

Aero-Go, Inc.
Seattle, Washington

WILLIAM R. TANNER

Productivity Systems, Inc.
Farmington, Michigan

17.1 LIFT TABLES

Daniel J. Quinn

17.1.1 Definitions and Operating Description

This section is about lift tables: what they are, how they work, and how to select one for a particular application. Put simply, a lift table is a mechanical scissorlike device employed to make work easier and safer, while increasing productivity.

Lift tables are used in schools, factories, grocery stores, and warehouses, to name a few applications. They are used for stacking, unstacking, loading, and unloading as well as tilting, lowering, work positioning, and assembly. Many more applications are possible, accentuating one of a lift table's best features—its versatility.

The mechanical construction of a lift table consists of structural steel base frame made of angles, flat bars, or wide flange beams (Fig. 17.1.1). The tabletop is constructed of steel plate and various angle and channel beam members. Legs of a lift table may be made from steel plate or structural tubes with bearing surfaces and bases welded. The legs are tied together with angle, tubular, or pipe cross members. The pivot pins and shafts are made of high-strength steel, polished at points where bushings operate. Bearing points have teflon-lined lifetime-lubricated dry bushings. Rollers are either cold-rolled steel (CRS) with oil-impregnated porous bushings or needle-bearing cam followers.

Most lift table cylinders are of the single-acting type (pressure on one side only). Cylinders are made from honed tubing of either steel or stainless steel. Various types of seals are used; the V-cup is most common due to its lower leakage. Other types of seals used are cast iron rings, O-rings with back-up washers and cup packings. Some lift tables use displacement-type rams, which have vee packing rings and take-up nut.

Since all cylinders will leak a small amount and more after wear, it is desirable to provide a return line from the opposite end of the cylinder to the top of the tank. This will return small leakage to the tank and also prevent the cylinder from taking in dirt or dusty air when the lift table is lowered.

Lift tables are generally designed with a 3-to-1 safety factor. This is normal for general machinery. Economy models may have a smaller safety factor and lifts for severe and mill duty or personnel lifts usually have larger safety factors.

Motors

The four types of motors that are used to power hydraulic lift tables are open drippable, totally enclosed fan cooled, totally enclosed nonventilated, and explosion proof.

Open Drippable. These motors are vented with the windings exposed to air. They are the least expensive but do not offer the safety aspects of the others.

Totally Enclosed Fan Cooled. These motors are completely enclosed within a metal case and have a fan attached to the rear of the motor which is used for cooling. Though more expensive than the open drippable type, these motors are more advantageous because they filter out contaminants in the atmosphere such as oil, dust, and dirt.

Totally Enclosed Nonventilated. These motors are the same basic construction as totally enclosed fan-cooled motors but do not have the fan mounted on the rear of the motor. The biggest advantage of these motors is that they are particularly suited for intermittent duty and need not rely on the fan to keep the unit cool.

Explosion Proof. These motors are of totally enclosed nonventilated construction. Also the wire connections are potted to assure a hermetical seal.

Hydraulics

Lift table hydraulics are one of the simplest types of hydraulic systems. In most cases the motor is run causing the pump to force oil into the cylinders to raise the table (see Fig. 17.1.2). To lower the table the hydraulic oil is drained out of the cylinder, through a solenoid-operated valve, back to the tank. The system components consist of a hydraulic tank, suction filter, pump, check valve, relief valve, solenoid lowering valve, flow control, and hydraulic cylinder. The tank consists of a reservoir

Key #	Description	Qty.	Key #	Description	Qty.
1	Motor (Electric)	1	13	Leg Roller Assembly	4
1A	Capacitor (Single Phase Only)	1	15	Center Scissor Pin	1
2	Gasket for Pump	1	15A	Scissor Pin Bushing	2
3	Coupling for Pump and Motor	1	16	Lower Hinge Pin Bushing	2
4	Pump w/relief & Check Valve	1	17	Upper Hinge Pin Bushing	2
5	Hydraulic Cylinder Assembly	1, 2, or 3	*18	Up-Limit Switch	
6	Flow Control	1	23	Cylinder Pin	1, 2, or 3
7	Down Valve w/Solenoid	1	24	Cylinder Pin Bushing	2, 4, or 6
8A	Filter Cap & Screen Assembly	1			

* Optional

Fig. 17.1.1 Typical lift table construction.

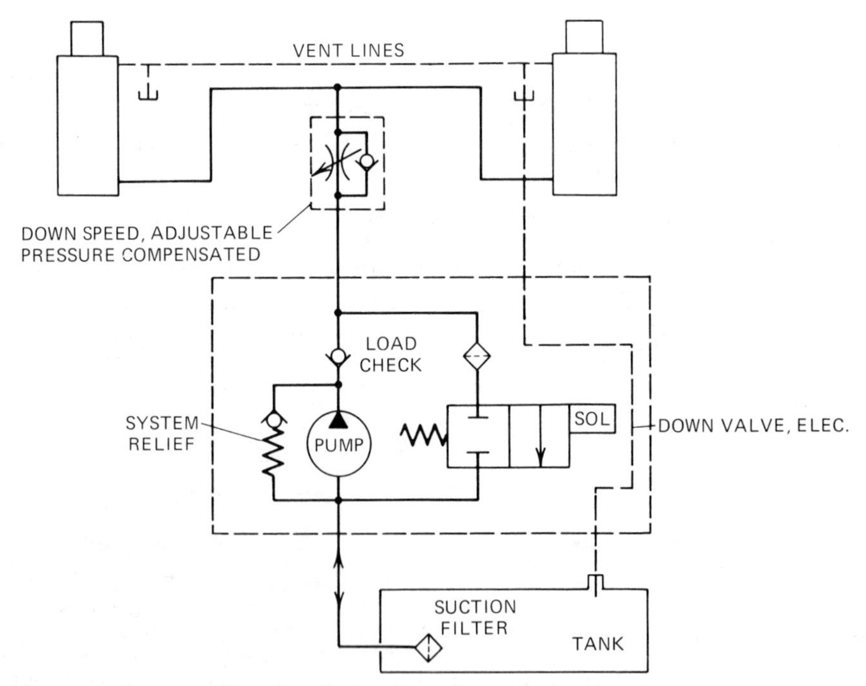

Fig. 17.1.2 Typical hydraulic circuit for lift table.

of sufficient size to hold enough oil to fill the cylinders as required to raise the lift—about 20% extra capacity is desirable. Due to intermittent use, a larger tank is not required. Most large tank systems are required to keep the hydraulic oil cool for continuous operation.

A suction filter is a screen-type filter mounted in the pump suction line either at the tank or between tank and pump. Suction hose size and length is important to be sure adequate oil can flow to the pump at all times. The pumps used for lift tables are of the fixed displacement gear type. They are the simplest type pump and, at a given speed, pump a nearly constant volume of oil. The pump volume is measured in gallons per minute. A check valve is used after the pump so when the lift table is loaded, the oil in the cylinders cannot flow back through the pump.

A relief valve is used to limit the maximum pressure at which the oil can be pumped. When the relief valve setting is exceeded, the valve opens, diverting the oil from the lift. This prevents overloading the lift table; safe pressure should not be exceeded.

Required pressure depends on the type of lift, manufacturer, and components in the hydraulic system. The down valve used is normally of closed poppet type. This is used for low leakage and usually is rated to leak a maximum of five drops per minute at 3000 psi. The poppet valve is used so the lift table will leak a minimum amount when holding a load in the raised position. There are usually only three places a lift table can leak: the check valve, down valve, and the cylinder packing.

The flow control valve is provided to control the lowering speed of the lift. This usually can be adjusted as desired and usually is pressure compensated so the lift will come down at about the same speed with full load as with a light load. Placing the flow control as close to the cylinders as possible also provides a safety feature. If a line between the flow control and power unit bursts or leaks, the machine will come down no faster than normal. The pressure used in hydraulic lift tables is usually between 1000 psi and 2500 psi and varies from one manufacturer to another.

Most lift tables use what is termed a standard general purpose hydraulic oil. These oils are fortified with rust and oxidation inhibitors and antiwear properties. They have a viscosity of 150 Saybolt Seconds Universal (SSU) at 100° F. It is possible to use SAE 10 nondetergent motor oil and/or automatic transmission fluid in lift table systems. For outdoor use, an aircraft grade oil can be used for the lower temperatures. Water-base and fire-resistant fluids can be used but usually require special seals and hose throughout the system.

In conclusion, the ideal motor to use in hydraulic scissor lifts is a totally enclosed, nonventilated type because of its suitable design for intermittent duty.

Voltages and Amp Draw

Table 17.1.1 shows various horsepowers and the current they draw at various voltages.

Controls

There are many devices used to control table position, rise height, and so forth.

Up–Down Push Button. This is the basic hand-held unit supplied with many lift tables. In single-phase units, the up button operates the motor. In three-phase units, the up button operates the coil of a starter, which in turn operates the motor. In all cases the down button operates the down solenoid.

Foot Switch. The foot switch operates the same way as the push button except the switch is actuated by a person's foot instead of a finger. The foot switch must be supplied with a guard to prevent accidental actuation of the switch.

Table 17.1.1 Current Draw for Different Horsepower and Voltage Levels[a]

	Current Draw (A) for Given Voltage/Phase/Frequency Conditions				
Horsepower	120/1/60	240/1/60	240/3/60	480/3/60	575/3/60
$\frac{1}{2}$	8				
1	16	7.6	3.6	1.8	1.4
2			6.8	3.4	2.7
3			9.6	4.8	3.9
5			15.2	7.6	6.1

[a] Values are approximate and may vary according to manufacturer.

Linestarter. A linestarter is used when the supply voltage is three phase or when the single-phase current of the motor is greater than the rating of the push button or foot switch. A linestarter also provides overload protection for the motor by means of a device that disconnects the motor if it draws more current than it should.

Transformer. A transformer is used to reduce the operating voltage from a high level such as 240 or 480 V to a safer level of 120 or 24 V. Push buttons and foot switches are less apt to cause severe electrical shock at 120 or 24 V than they are at the higher voltages of 240 or 480 V.

Solenoid. The solenoid operates the down valve of the lift table. The coil voltage is always the same as the transformer.

Limit Switches. Limit switches are used to control the raised and lowered height to specific heights. The up limit controls the motor and the down limit controls the down solenoid.

Wiring Techniques

All basic wiring should be done to satisfy the basic requirements of the National Electric Code. However, if there are further requirements, the following standards should apply.

JIC Wiring Practices and Standards. The Joint Industrial Council (JIC) standards take off where the National Electric Code ends. JIC requires such things as:

1. All wires should be terminated at terminal strips.
2. All enclosures are to be NEMA 12 coil and dust tight.
3. All wires are to be run in conduit.
4. All fittings are to be gasketed.

They are the type of standards which add a considerable amount of quality and cost to the machine.

NEMA 12. NEMA 12 (National Electrical Manufacturers Association) is a standard that states that all electrical enclosures used on a machine will be oil and dust tight. These enclosures are gasketed and all seams are welded.

NEMA 7 and 9. NEMA 7 and 9 are standards that state that all enclosures will be explosion proof. These enclosures have machine covered surfaces, $\frac{1}{4}$-in.-thick plates, and welded seams.

17.1.2 Applications

Stacking and Unstacking

Lift tables have been used in stacking and unstacking operations on a multitude of products that have one thing in common. The height at which the work takes place is fixed, and the product received in a pile has to be elevated to this fixed height.

A pile with one product per layer is the easiest to stack and unstack. More than one per layer, but all oriented in the same way, is the next easiest. Products that are in a pinwheel pattern are the most difficult to handle. Piles that have alternative layers also create problems.

The most common type of stacking and unstacking piles is the simple lift that brings each layer up to the work height and the product is then pushed or pulled off the pile by hand. Normally the unstacking is done more easily if the bottom of the layer to be unstacked is slightly above the working height. This type of unstacking and stacking is easy to justify by the increased productivity that it gives the operator. This simple type of system can also be justified on intermittent base operation if the number of people needed to do the job is limited. Also a justification can be made regarding the safety aspects to avoid back injuries.

Stacking is the exact opposite of the unstacking where the operator pulls or pushes the product over the pile. The operation is usually easier if the pile is slightly below the working height.

Piles that have alternate layers and some with pinwheel patterns sometimes require a turntable to be mounted on top of the lift table. The following are factors for both stacking and unstacking turntables:

1. *Product Weight.* If the weight is too great, for safety reasons, the operator should not have to lean over the pile to push or pull the product on the back side of the pile. This is particularly true if only one operator is used.
2. *Product Bulk.* If the product is too bulky it may make it difficult for the same reasons as for product weight.

3. *Product Orientation.* The orientation of the product to the machine or to the work being done should be correct. It may be inconvenient to have to turn the product after it is unstacked or difficult to reorient once on the pile.

Before going into more complex stackers and unstackers, there is one stacking system often overlooked. It is used primarily on large sheet products or on long products that have individual layers. Lumber is an example. This is the tilt unit with the lift table in the bottom of the tilt frame.

The unit is usually equipped with a backstop to prevent the product from unstacking until it is at the proper height. The unit looks like a capital L with a lift table in the short base and the backstop in the vertical line. The pile is first placed on the unit and the unit is tilted back to a predetermined angle. The lift is then raised until the top layer is free to slide off and over the backstop.

In actual use, lumber usually slides off, but large sheets like corrugated paper require an operator to assist in the unstacking. On these large sheets, the operator only has to shake the top sheet to break the friction bond and the sheet will slide off the pile.

Complex stackers and unstackers are usually designed by companies who specialize in building machinery for specific industries such as bottling equipment, food processing, cement, and bag handling companies.

Some advantages for using a lift table in a complex type stacker are:

1. The company can purchase a catalog item and avoid the engineering and development charge of building a lifting device.
2. The lift table is self-contained and self-supporting from the floor, thus freeing all sides and top for other equipment.

When designing one of these complex stackers or unstackers, the original equipment manufacturer should supply the lift table manufacturer with as much technical information as possible, including duty cycle, description of operation of weight, height, and space limitations, and special environmental problems. Most manufacturers of lift tables have engineered packages to add to their lift tables to handle these problems. Lift tables can be designed to inch up in very accurate increments or to come down in virtual free fall without damaging the product or machine.

The stacking and unstacking problems that plague most industries today are largely ones of oversight. For example, an operation that runs once or less per week but has caused two back injuries over the past three years should be looked at closely.

An operation that gets good productivity and has two or more operators in a time study might show that a small portion of the operator's time is spent doing the actual work with most of the time used to hand stack and/or unstack.

Indications where a lift table is needed in any plant are as follows:

1. Use of two or more skids to bring product up to the correct height to be transferred to the work area.
2. Use of a fork truck when a much less expensive lift table could be used.
3. Two or three people lifting a product up to a machine.

Lift and Tilt

When applying a tilt function in conjunction with lift table applications, a multitude of combinations may be utilized to accomplish an end result that will suit individual applications.

The basic applications for lift and tilt, tilt and lift are discussed next.

Lift and Tilt Applications. Primarily, the lift and tilt function is used for dumping or feeding small, individual materials in bulk or in mass (see Fig. 17.1.3 and Fig. 17.1.4). This function can also be used to feed manually for in-line production process such as:

Bin feed or dump
Gondola feed or dump
Hopper feed or dump

Lift and tilt equipment can be applied in metal working and chemical and food processing.

Metal Working. Casting or stamping small parts in volume in which small parts go into hopper or bins, then when full, the hopper or bin is placed on lift and tilt equipment to be dumped or fed to further production steps of washing, drilling, machining, or other in-line production process.

Fig. 17.1.3 Lift and tilt function.

Chemical and Food. Mass product dumping or feeding of product or material for mixing or filling for in-line production process.

Lift and tilt equipment can also be utilized on in-line production process stations where accumulation of materials is dictated by time work in process. For example, sometimes, in production line process, production is faster from one station to another. The result is accumulation of materials. If this occurs, materials can be placed in hoppers or bins and then on lift and tilt equipment. From there, materials can be fed or dumped to next station in production process.

Tilt and Lift

The primary function of the equipment is dumping or feeding in a controlled mode. This is done so that materials in bulk or piece can be "shingled off" in a careful manner, with little risk of damage to the materials. For example:

Bin feed or dump

Fig. 17.1.4 Full tilting unit.

Gondola feed or dump

Hopper feed or dump

Tilt and lift equipment can be applied in the packaging industry and the chemical and food industry.

Packaging Industry. Finished package product is accumulated in bin or gondola, then placed on tilt and lift equipment to be fed onto conveyor for shipment.

Chemical and Food Industry. Tilt and lift equipment can be utilized in control mode to tilt and lift bulk materials for mixing products in large vats or mixers.

Tilt and lift equipment is designed to control the flow of material based on the process and the requirement. Most tilt and lift equipment can be interfaced with conveyors, mixers, vats, bins, hoppers, and washers. Lift and tilt equipment can be designed to perform numerous material handling tasks in a controlled mode from 500 to 50,000 lb capacities.

Work Positioning

Almost all applications where lift tables are employed can be described in terms of positioning. If the application is to load or unload production machines, the elevating and lowering action of the lift tables is simply a positioning function. This creates the ideal height interface between the work pieces and the feed side of the machine (elevating) and a stack of finished pieces with the off-load side of the machine (lowering). If the application is to elevate personnel to stock high shelves or to work (paint, weld, etc.) on large products, the actual function again is to position the worker in relation to the job. The function of lift tables is more accurately understood when they are perceived as positioning tables.

This perception is more important to understanding this section than any other typical lift table application. Here, the up/down characteristics of the lift table are merely available directions used to attain the most efficient location for the workpiece during assembly. Although this is one of the most basic applications for lift tables, it is perhaps the least understood.

A better understanding can be gained by viewing positioning tables in terms of the study of ergonomics. Ergonomics is the science of maximizing a person's working efficiency through more efficient work station design. It involves the incorporation of individual physical factors such as the operator's height, reach, strength, and skill into the design or modification of the work station. The fruits of the ergonomist's efforts are the dual benefits of increased productivity and improved worker morale which is generally associated with noteworthy quality improvements.

Another direct benefit or by-product that evolves by using positioning tables in an ergonomic effort is the enhancement of safety in the production environment. The positioning table, in its various modifications, offers many different means of gaining these benefits by allowing more effective utilization of the worker's effort.

Many attempts have been made to design the ideal work station. The common goal of these efforts is to reduce a worker's mental and physical fatigue by eliminating the repetitive movements of bending, stretching, stooping, squatting, and so on.

In order to accomplish this, each of these designs has attempted to position the workpiece in an ideal relationship to the average worker. The most repetitive pitfall of these schemes has been the lack of planning concerning the average worker. Specifically, the 34-in. work station may be ideal for the tall operator on the first shift but uncomfortable for the short operator on the second shift. Or it may be ideal for one operator to work on the top of the workpiece but may force him to squat to finish the bottom and vice versa.

Other problems result from repetitive functions at the ideal work height; they include strain, tendonitis, and fatigue, among others. They are caused by forcing the worker to constantly repeat the same actions from a fixed position.

The solution that satisfies the goals of ergonomics and the realities of production is to reposition the table. The key, then, is variable height. Remembering that there is not an average worker in terms of physical characteristics or work style preference, that the workpiece has a top-to-bottom range, and that fixed-position actions present safety and production problems, varying the height of the work station becomes a viable answer.

Versatility of the positioning table makes practically any work station requirement a potential application. Within standard design, there are tables that lower to 6 in. of overall height and tables that can elevate to 168 in. When there is a recessed floor, the collapsed height can be flush with the floor or lower, depending on the requirement. This range of travel allows for infinite, incremental adjustments to gain the ideal position. The tops of these tables can be as small as 12 × 24 in. or as large as any product may require. The tops can be flat or shaped. They can be mounted with clamps, rollers, turntables, tilting devices, ball transfers, or practically any other feature that might aid in the particular production process. The tables have capacities to position products that are weighed in ounces to products far exceeding shipping limitations.

In summary, whether the orientation be production, safety, quality, or economics, the positioning table provides many solutions to the work station in a universal design.

First, by increasing the efficiency of operator movements and reducing fatigue through ideal positioning, production requests are answered. Second, by eliminating awkward, uncomfortable movements and lessening repetitive fixed-position actions, safety requests are answered. Third, the combination of factors resulting in production and safety benefits has a direct effect on improving worker morale—and high morale has a corresponding relationship with high quality; therefore the quality requests are answered.

Finally, the variable height of the work station offers flexibility making a standard design suitable to virtually any operator. By consequently eliminating the costly need for customizing, the economic requests are answered.

Upenders and Downenders

An upender or downender (the two are usually identical, but used for different purposes) is basically two platforms attached at a right angle somewhat resembling a large L. The primary purpose of the upender/downender is to change the vertical or horizontal axis of a product or process up to and including 90° by vertical rotation. Thus the upender/downender can be described as a turntable standing on its side.

Some upender/downenders are designed to change this axis up to 180°, although this type of unit is usually referred to as an inverter. The L design of the upender/downender can be pivoted at the point of junction of the L, or can be supported by arms attached between the supporting base and each leg of the L. Each method has its purpose. An upender/downender with the pivot point located at the junction of the L not only changes the axis of the product, but also shifts the position of the product by the distance relationship of each platform of the L. This occurs because actually the L is tipping over and laying now on what was previously the upright leg of the L.

This type of application is desirable when the upender/downender is contained within a conveyor or other transport system, and it is required that the product be received by the upender/downender from one conveyor or machine, upended or downended, and transferred to another conveyor or machine.

This transfer can be made in a straight-line or 90° direction as desired. This type of machine can also be utilized at the end of a conveyor line to rotate the product for correct orientation for pickup by a fork or clamp truck.

The upender/downender with each leg of the L supported by links to the supporting base rotates about a point that is centered within the area described by the L. This allows the product to have its axis rotated, yet remain in the same position as it was prior to rotation. Maintaining the original position of the product while changing its axis may be desirable in certain applications such as in-line installation in a conveyor or other transport system.

Beyond this difference in pivot methods, upender/downender units are conceptually similar. Two designs are available for handling a wide range of products from very small to very large. The method of powering upender/downender units usually depends on the application. Units can be powered hydraulically through one or more hydraulic cylinders and a self-contained, remote, or master hydraulic power unit; an electric motor gear reducer with chain and sprocket or cam/arm drive; cylinders operated by an air-over-oil system; and straight air-powered cylinders with proper restrictor valving.

Usually it is desirable to have positive control over the operation, so the latter method is rarely used. The upender/downender unit can be designed so that conveyor or other transfer components can be mounted on either or both legs of the L. In some cases the platform(s) of the L are depressed in a V-groove to contain a cylindrical product during the upending/downending process. In others, the platform(s) may have special fittings or attachments to allow the product to be precisely positioned on the platform for critical alignment during assembly or other process.

Upender/downender units are used in a variety of applications. In papermaking, large rolls of paper are placed on the unit either manually by overhead crane, clamp or fork truck, or by conveyor for upending to a vertical position for palletizing or stacking.

During the manufacture of large electric motors, the motor body is to be positioned vertically to receive the rotor, placed in the motor by overhead crane. A manufacturer of computer disk drives utilized a small upender to position a 150-lb drive 90° from its normal horizontal position to install the drive belt and adjust the braking mechanism. In this installation, the upender is built into, and is an integral part of, the production line conveyor system.

Upender/downenders can be integrated with other production equipment such as turntables and lifts for an almost limitless variety of production requirements. They can be used in a wide variety of industries and for all types of products and processes.

Dock Lifts

The efficiency and safety of most loading areas can be greatly improved by adding a modified lift table commonly referred to as a dock lift. Regardless of the physical description of the shipping/receiving area (ground level, low dock, high dock, etc.), there are usually material flow bottlenecks and safety hazards caused by vehicles that cannot vertically align with the existing facility. These inefficiencies and risks can be eliminated by equalizing the height differences with a dock lift.

Essentially, a dock lift is a loading platform that handles any cargo on a level plane in an infinitely adjustable vertical range from 0 to 59 in. It provides maximum efficient load handling while eliminating costly concrete platforms and ramps, dangerous inclines, and expensive truckwells. Its versatility not only guarantees that any vehicle can be accommodated but when interfaced with an existing dock, it provides ground access. This multifunctional dock offers safety enhancements in terms of injury reduction as well as protection of unstable loads.

Dock lifts are generally categorized as medium duty or heavy duty. The medium-duty dock lift is designed to handle loads up to 6000 lb and is generally used in conjunction with carts, hand trucks, and pallet jacks. The platform is a diamond plate surface and is normally 72 in. wide and 96 in. long. The axle roll-on capacity of this series is 50% of the overall capacity. Thus, load dispersement over each axle of the transport equipment must be considered when sizing the lift, and many never exceed half the rated capacity.

Heavy-duty dock lifts can handle loads up to 20,000 lb and are commonly used in conjunction with powered transport equipment such as straddle stackers and fork trucks. The standard platforms for this series are the same as the medium-duty lift except they are of a heavier construction and 144 in. long. The axle roll-on capacity is again 50% of the rated capacity on the sides but 75% on the ends. This is a feature designed to accommodate the typical transport equipment (fork trucks, etc.) which generally carry 80% of their load on the front wheels.

Both series of dock lifts offer a wide variety of optional platform sizes and features but always include the following as standard equipment:

Bevel toeguard painted safety yellow.

Removable pipe rails 42 in. high with mid-rail, 4-in. kick plate, and snap chain opposite throw-over plate.

Diamond-tread deck surface.

Hinged split diamond-tread throw-over plate. Travels from 15% below horizontal to 15% beyond vertical for total swing of 120°.

Weatherproof up/down constant-pressure push button on 20 in. Koil Kord.

Linestarter-transformer for 24-V controls—not included on single phase machines.

Hydraulic velocity fuses at cylinders.

Standard voltage: 208/3/60, 230/3/60, or 460/3/60.

10,000-lb minimum roll-over capacity equal to lift capacity on heavy-duty models.

Adjustable lowering time in pressure-compensated flow control valve.

Removable lifting eyes for ease of installation.

Complete installation package including 20 ft of Koil Kord on control power unit reservoir filled with hydraulic oil 5 ft of hydraulic hose with power unit.

17.1.3 Sizing Guidelines

The selection of the proper lift for an application is very important. A hasty choice based on a few major parameters can cause downtime and frequent parts replacement as well as shorten the life of the machine.

The listed capacity of any lift is for uniform centric loads. If the load is placed on the lift from overhead, stacked on, or put on while the lift is lowered, there is no need to increase capacity. If the load extends over the end or side of the lift while it is raised, closer examination is warranted.

One rule that applies here is that a point load of X extending over the side or end requires a lift 2X capacity. Given that a point load is not possible, it then depends on the physical size of the item coming onto the table. A lift with a standard size top (one where the top is the same size as the base frame) is the best selection here. A load should not be brought in from the side or end of a lift with a top that hangs over the base frame. If this is done, the load is then cantilevered off the end or side causing a lever action, promoting tipping and causing strain on the lift components.

Capacity

First, determine what capacity lift is required. Just because a load is 2000 lb doesn't mean a 2000-lb capacity lift is required. Consider the following:

1. How much weight the lift table will be handling including the load, pallet, and any other platform-mounted accessories (conveyor, turntable, etc.)
2. How the lift is being loaded—from overhead, over the sides, or over the end
3. Whether the lift will be raised or lowered when being loaded
4. How the lift is being unloaded
5. Whether the lift will be raised or lowered when being unloaded
6. The physical size of the load
7. Whether the load is uniform and centric or eccentric

Some manufacturers will modify a standard lift to make the base frame the same width as the platform; this is desirable for side loading and nonperformance loads.

Power Unit

Generally, lift tables are powered by either an electrohydraulic or an all-hydraulic power unit.

The size of the power unit determines the use time of the lift table. A higher-horsepower motor will turn the pump faster and force fluid into the cylinders faster.

Most manufacturers offer their standard lifts with an internal intermittent-duty electrohydraulic power unit (rated for 120 motor starts per hour max.). Other power unit options available are:

1. External power unit.
2. Larger horsepower power units for faster rise times (these usually have to be external due to their size).
3. Air/hydraulic power units—these require 80 psi and 80 ft^3/min air supply and cannot be internally mounted.
4. Power units with continuous-duty motor rated for 160 motor starts per hour.
5. Power units with extra directional valves—used for double-acting cylinders and other powered options.

6. Full JIC continuous-running power units—these have many extra features including oversize tanks, extra gauges and filters, tank clean-outs, mill, chemical-duty motors, and high service factors. They are designed for heavy duty industrial use.
7. Other special power units: NEMA-rated power units (explosion proof), temperature-rated power units, and other configurations.

Many manufacturers supply lifts without power units and supply the displacement required so an existing hydraulic system (if pressures, etc., are compatible) can be used.

Vertical Travel

The vertical travel of a lift is the difference between the fully elevated height and the fully collapsed height. Determine the highest maximum height that the lift must go. All lifts can be controlled to go any intermediate height within its travel.

If considering a pit-mounted lift, the maximum height achieved from the floor must be equal to the lift's vertical travel (assuming the lift platform is flush with floor while in the pit). The maximum vertical travel of a lift cannot be altered in the field for the following reasons:

1. The rods in the cylinders will not be long enough.
2. The angle the leg forms with the table top will become too large for the lift to maintain its rated capacity.

Tabletop

Consider what the largest item going onto the lift will be, then figure on extra space (for a person, a pallet truck, tools, or other equipment).

Usually, the higher the vertical travel of the lift, the longer the table must be. This is because the legs become longer as a lift's vertical travel increases. There is a solution to this, but it is expensive to go to a double or triple scissor lift. For example, one manufacturer uses a single scissor lift with 60 in. of vertical travel and has an 86-in. long min. top. That same manufacturer makes a double scissor lift with 60 in. of vertical travel with a 53-in long min. top. Tabletop length is a function of vertical travel.

If a load is particularly long (or wide), most manufacturers offer tandem lifts—two regular lifts attached end to end (or side by side). There are limits as to how long a standard lift can be; this is usually determined by how long a relatively flat tabletop the manufacturer can get.

On standard straight-sided tables, one can get a platform that is the same size as the base frame. Pit-mounted machines that require a beveled toeguard platform must have a tabletop slightly larger than the base frame (usually 4–6 in. all the way around).

For lifts being unloaded over the sides or ends while the lift is raised the same comments apply.

To size lifts carrying nonuniform eccentric loads, the same general sizing applies. The center of gravity (c.g.) should be the same as the base frame of the lift. Again if the c.g. is right on the side or end, the capacity should be doubled.

Controls

Manufacturers usually offer ranges of controls for lifts. Consider the following controls for starting and stopping of lifts:

1. Controls for actuation of lift
   a. Constant pressure monetary (requires holding) push-button or foot-switch controls.
   b. Maintained-contact controls (depress button or pedal and lift automatically rises or lowers).
   c. Key-operated controls.
   d. High NEMA rating.
2. Controls for intermediate stopping of lift
   a. Upper travel limit switch
   b. Lower travel limit switch
3. Controls for leveling—usually for stacking or unstacking applications such that top of load stays at a given height as lift raises or lowers.
   a. Photoelectric package—includes photo cell and reflector such that lift raises or lowers if beam is broken.
   b. Feeler switch—includes limit switch mounted on a stanchion such that lift will shut off when top of load hits and trips limit switch.

4. Miscellaneous controls—most manufacturers supply a device to step primary voltage (220/440/3/60) down to control voltage (115/1/60 for inside use, 24 V for outside use). This unit usually houses the required motor starters.

Tables can be designed for a variety of safety interlocks so a lift will not raise or lower until something else has been done. Examples of this would be an electrical interlock in a gate on a personnel lift so the lift will not raise unless the gate is fully closed, or using an interlock on a lift with a manual turntable so a lift cannot be lowered unless the turntable is in the proper orientation. Tables can be modified to run in highly automated conditions requiring computerized controls.

Options

Many different options are available to suit particular requirements. The following are some of the most popular:

1. Mobility options
   a. Two swivel casters and two rigid casters with foot-operated floor lock. These can be mounted either under the left base frame or to the sides of the frame.
   b. Flanged or V-groove wheels mounted to base frame.
   c. Semi-live package includes two fixed steel rollers, tank, and skid spotter. This is best where mobility is required on an infrequent basis.
2. Platform options
   a. Gravity roller or skatewheel conveyor.
   b. Powered roller conveyor.
   c. Manual turntable with spring-loaded detents every 90°.
   d. Powered turntable.
   e. Handrails.
   f. Diamond plate, aluminum, or stainless steel top.
3. Safety options
   a. Bellows guard—pleated neoprene accordian skirting.
   b. Beveled toeguard for pit-mounted lifts. Side of tabletop angles in, forcing one's foot from under lift during lowering.
   c. Pressure-sensitive tapeswitches on side of tabletop. These are wired so that any exertion on the side of the tabletop (e.g., by a person) will actuate the tapeswitch and cause the lift to shut off.
4. Powered options
   a. Double-acting cylinders for power up /power down (standard lifts are gravity down). This option is beneficial when a fast lowering time is required.
   b. Lifts with powered traverse—gear motor drive to rear wheels for powered mobility usually requires drive housing adjacent to base frame of lift.

High-Cycle Lifts

Lifts that cycle generally more than 30 times per hour should have this package. It usually includes heavy-duty bronze, greasable bushings at all pivot points and needle-bearing camfollower leg rollers. It is designed to double the expected life of a lift table.

Finally, lifts can be painted with many different finishes to suit requirements. Most manufacturers will also export boxed lifts for overseas shipping.

Thus, we have covered what a lift table is, its possible applications, and how to select the proper one. Most importantly, lift tables are versatile tools used to reduce time, money, and waste, while increasing production, safety, and worker morale. For any workpiece or product requiring positioning for assembly, testing, painting, or any other manufacturing process, there is a lift table that can handle it.

17.2 INDUSTRIAL MANIPULATORS

Guy A. Castleberry

17.2.1 Operating Description

Frequently, remotely operated manipulators become necessary because of the remoteness of a function or environments that are hazardous or otherwise inaccessible to humans. Applications for these manipulators are important in industry, space, and undersea projects involving materials handling. Although manipulators have many applications, this section deals specifically with industrial materials handling manipulators.

Manipulators and robots often are put in the same category. This has occurred because most work on robotics to date, both in research and application, has been in the context of industrial manipulator requirements.

The technical sophistications and differences between manipulators and robots are indicated mainly by the robots' ability to make their own decisions during operations. Manipulators can be called robots only if they have programmability.

Robots have three basic elements: the manipulator, the power supply, and the controller. The manipulator is the mechanical unit that does the work. It is often referred to as the "arm." A power supply provides the strength. There are three types of power: hydraulic, electric, and pneumatic. A controller is the brain that remembers the tasks and controls the motions. These range in complexity from simple air-logic to a minicomputer.

The manipulator, power supply, and controller are supplied as integral parts of a robot. The manipulator is referred to as the arm because it is no more than that; no hand is involved. A hand is added later. It is designed for a specific task and is commonly referred to as a grabbing device.

Industrial manipulators included in this section consist of elements similar to those involved in a robot: a manipulator (arm), power supply, and a grab (hand); however, it is controlled by a human instead of a controller. This explanation should help to provide the analogy between a robot and an industrial manipulator.

A manipulator (sometimes referred to as jib crane or articulated jib crane) is often combined with a lifting device for handling materials. These lifting devices include air lifts, hoists, balancing hoists, and balancing cylinders.

The installation of industrial manipulators generally requires more of a system than a unit approach. The following steps typically are involved in installing an industrial manipulator.

1. Define the required task as a foundation for the industrial manipulator.
2. Establish the acceptable range of operation parameters for equipment and processes; provide opportunities for possible changes to optimize the manipulator installation.
3. Design the system using a systematic approach for integrating workers, machines, materials, and the manipulators.
4. Prepare required documentation for the project.
5. Monitor the installed manipulator.

Industrial manipulators can have a direct effect on improving a company's profits, but they cannot do it alone. They must be used in conjunction with people, raw materials, the facility, and other equipment. All of these factors must be integrated.

17.2.2 Justification for Use

There are many justifications for using industrial manipulators. These vary from system to system. The most common are discussed here.

Increased Productivity

Cost justification for CNC (computer numerically controlled) lathes or turning systems generally is based on productive running time. An "ideal" operation (from the standpoint of cost) might utilize these expensive machine tools almost continuously to produce pieces requiring long cycle times.

Even multishift operations are concerned with the cycle time for each part—because loading and unloading time, considered nonproductive, can be significant. Machines making pieces requiring *short* cycle times have to be loaded and unloaded more frequently, hence their actual running time is reduced even further. Pieces that are heavy take even longer to load or unload manually, further reducing productivity of the expensive machines.

Industrial manipulators used to load and unload CNC lathes are reported to have paid for themselves in a few months due to increased productivity; additionally they reduce handling damage to finished parts.

Operator Fatigue

Industrial manipulators provide the operator with increased strength and reach capability. These increased capabilities not only improve the operator's utilization but also reduce operator fatigue.

There are two forms of operator fatigue: physiological fatigue, where the operator's muscles are overstressed; and psychological fatigue, which may be the result of too many sequences of operations, complexity of operations, and/or high accuracy demands.

A properly designed manipulator, by doing virtually all of the work, eliminates physiological fatigue. Manipulators also can be programmed to control a sequence of operations—utilizing push buttons—and thereby contribute significantly toward reducing psychological fatigue.

Dependable Product Handling

When workpieces are heavy or awkward, they are typically difficult and time consuming to move. The ease and speed of handling these parts can be critical. Industrial manipulators designed specifically for certain type products and handling requirements not only can make the job easier and faster, but also reduce damage to parts and other equipment, such as machine tools, fixtures, and cutting tools.

Maintenance of Quality Standards

Certain foundry operations must be performed almost immediately after a product leaves a furnace. For example, the enameling operation on a cast iron bathtub is performed almost immediately after the tub leaves the furnace, and the finish may be compromised. Yet, when the furnace door is opened, temperatures exceeding 1700° F come forth—and the tub may weigh 300–400 lb. An industrial manipulator equipped with special grab forks enables one operator to quickly move a tub from furnace to enameling table because the manipulator arm permits the operator to function at a comfortable distance from the oven; the mechanics of the manipulator minimize the weight of the tub as it is moved, and the specially engineered grab fork permits quick, precise loading/unloading and secure handling of the tub. But besides making it possible to achieve a high standard of quality, the industrial manipulator increases the efficiency of the operation and frequently improves worker safety.

Cost Considerations

Cost considerations are important for determining justification for manipulator use. Approximate cost comparisons are as follows:

Equipment	Cost Factor
Jib crane with lift device	1
Articulated jib with lift device	1.4–2.6
Balancer with hook	1 –2.9
Balancer with grab	1.4–6.4
Low-profile unit with hook	2.2–3.9
Low-profile unit with grab	4.4–9.8
Robot	12.0–123.4

These cost factors apply only to equipment, not installation or area preparation. Area preparation involves a much greater variance than the equipment itself. For example, on a jib crane installation, the jib crane equipment represents approximately 60 to 80% of the total cost of the system. On a robot installation, the robot represents approximately 20 to 40% of the total cost of the system. Since the equipment listed here is not directly comparable, these cost factors should be used only as a rough guide.

17.2.3 Types of Manipulators

Within the manipulator family a wide range of choices is available, depending on the object to be handled and the work to be performed. Some units can pick up and move objects vertically or horizontally and provide an interface with other handling equipment. Other manipulators have the ability to perform functions such as rollover, tilting, pouring, and dumping—which might be required during production, machine tool loading, or inspection. The most common manipulators are described next.

Jib Cranes

The jib crane is the simplest manipulator. It has a rotating boom attached to a column that is held in the vertical position by floor, ceiling, or wall bracket mounting. Jib cranes generally have pneumatic or electric hoists for lifting and lowering a load.

Jib cranes are the least expensive way to handle loads. They can be placed in production areas for assembling heavy parts, loading machine tools, forging presses, and so on. However, this type manipulator is very clumsy, requiring the operator to pull and push the load into position; excessive jogging of the hoist is required to obtain the correct elevation of a load. Thus the jib crane is not recommended when positioning of a load is paramount.

Articulated Jib Cranes

Greater versatility is obtained by mounting the hoist on an articulating jib crane (see Fig. 17.2.1). This type differs from standard jib cranes in that it has a secondary 360° pivoting arm attached to the primary arm. The combination of the two pivots allows the jib to reach where standard cranes cannot—around building supports, columns, or machines.

Articulating jibs have a working area of up to 40 ft in diameter plus extremely free horizontal movement. They may be ceiling or floor mounted, even mounted overhead on a trolley, depending on the application. Although either an electric or pneumatic hoist can be used, the articulating jib's internal piping makes it ideal for pneumatic hoists. The articulating jib is especially practical where overhead space is limited, where loads are awkward or heavy, or where smooth, lateral movement along a conveyor is essential. If overhang presents a problem, the articulated boom can reach in and under an obstruction.

Balancers

Balancers provide free triaxial movement—vertical, horizontal, and rotational—in a near-weightless condition within the work area of the machine (see Fig. 17.2.2). True balancers incorporate arms in a parallelogram; these are proportional so that vertical movement of a load is accomplished without any horizontal displacement, and horizontal movement is made without vertical displacement. Balancers utilize the natural movements of the human body and carry the load with little or no restriction.

The word *balancer* is used because these manipulators literally balance a load. The load is supported and controlled at all times. This type of manipulator is self-compensating in supporting the load in all positions. Thus, with the load totally balanced, the operator has complete control and manual flexibility in guiding it into machinery, racks, furnaces, or other equipment; the operator's efforts are magnified without affecting his precision. The operator's function is merely to guide the load. Minimal force is used to overcome the inertia of the load, because natural body motions are used. A recent advance in this type of manipulator has been the addition of parallel linkage arms besides the balancing arms. This feature extends reach capability and allows more precise load positioning, while maintaining the object's physical orientation and path of motion in an exact plane.

In considering a balancer in an industrial application, there are several limitations which must be addressed; these are:

1. *Headroom Requirements.* Overhead clearance must be considered; the minimal requirement is approximately 12 ft.

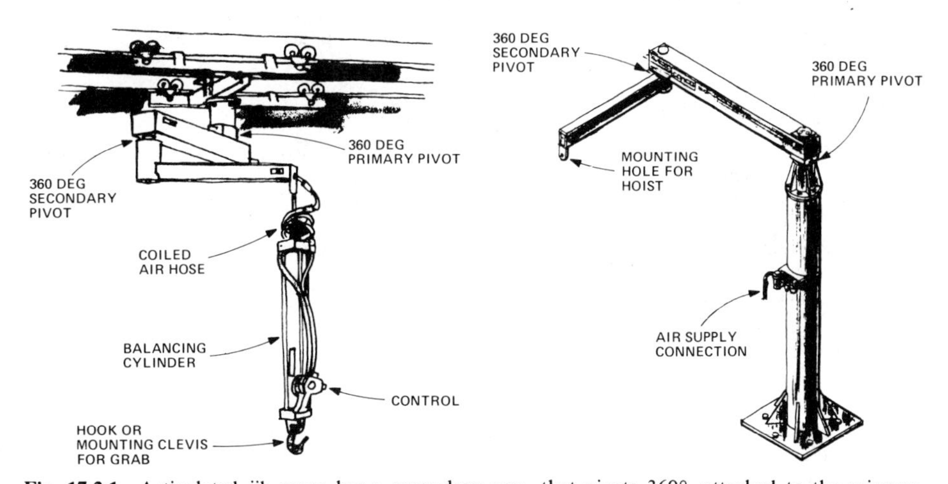

Fig. 17.2.1 Articulated jib crane has a secondary arm, that pivots 360°, attached to the primary arm. The unit thus has a considerable range of accessibility and reach. It can be mounted from an overhead trolley (left) or on a floor pedestal.

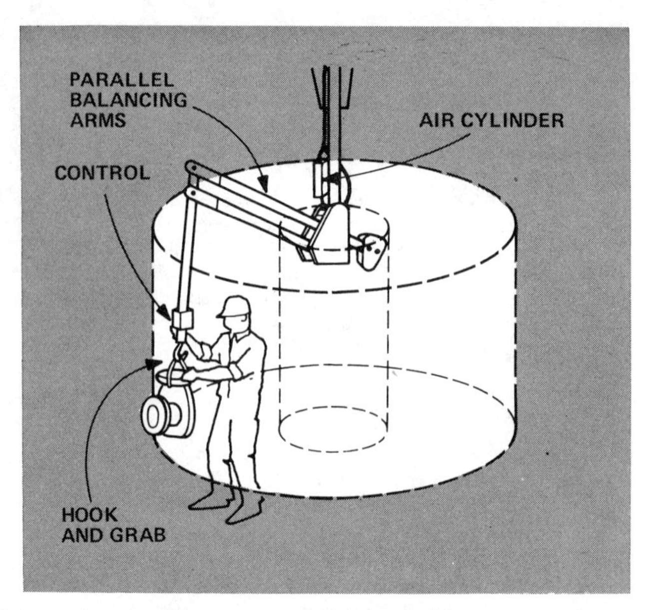

Fig. 17.2.2 Balancer has parallelogram arms and free triaxial movement. Operator merely guides the load. Large working envelope is provided.

2. *Lift.* Each series balancer has a specific vertical lift. The lift must be evaluated for each individual job; and the correct height of the pedestal or ceiling spacer must be determined, based on the lift requirement and series of balancer selected.
3. *Reach and Capacity.* The total horizontal travel of the balancer is defined as the reach; the reach capability of each unit is a function of the lifting capacity of the balancer. The maximum lifting capacity of these units is 1700 lb.
4. *Reach-In.* The reach-in capability of the balancer is the ability of a unit to be extended beyond the normal work envelope for special applications. This is accomplished by using the parallel linkage arms in conjunction with the balancing arms.
5. *Cycle Time.* Cycle time limitations are dependent mainly upon what is to be accomplished with the balancer and an operator's dexterity.
6. *Side Loading.* The side-loading capability of the balancer has a finite limit; applications requiring side loading must be checked with the particular unit under consideration.

Low-Profile Manipulator

Low-profile manipulators (Figs. 17.2.3 and 17.2.4) are generally used where overhead obstructions or low ceilings exist. This type of manipulator incorporates parallelogram linkage and multiaxis rotation to provide triaxial movement. In standard low-profile manipulators, the operator manually activates horizontal sluing around the machine pedestal, elbow, and wrist, plus vertical rotation at the wrist. Generally, hydraulic rotary actuators can replace manual action, if desired.

The low-profile work envelope is provided by the manipulator extending out from the horizontal plane and grabbing the load in that plane. By using this concept, headroom requirements are reduced significantly.

Limitations to be considered for low-profile manipulators are:

1. *Reach Capacity.* The reach capability of these units is a function of lifting capacity, and varies depending on the particular manufacturer. The maximum lifting capacity can vary from 150 to 1000 lb.
2. *Speed.* Most low-profile manipulators use a hydraulic power source which generally functions slower. The key advantage in using hydraulics is higher working pressure, hence smaller linear actuators (cylinders).

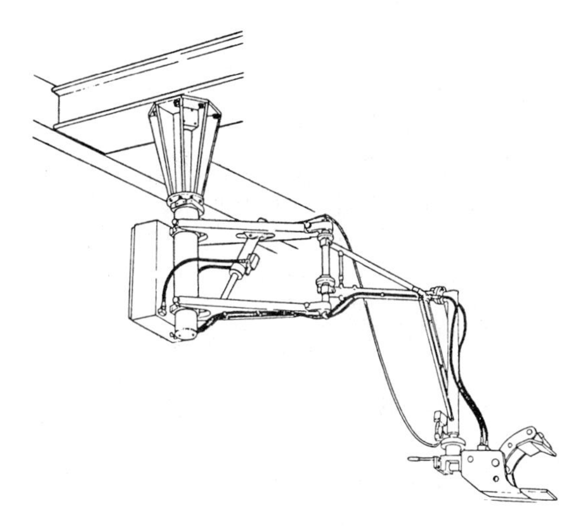

Fig. 17.2.3 Low-profile manipulator is mounted overhead. Courtesy Positech Corp.

3. *Space Requirements.* If hydraulic power sources are used, additional floor space is required for the reservoir and pumps.
4. *Side Loading.* The capability of the unit has a finite limit; applications requiring side loading must be checked with the individual unit under consideration.
5. *Operation Position.* The operator must stand on the side of a unit instead of directly behind it. This could present positioning problems on certain types of reach-in applications.
6. *Work Envelope.* The effective work volume of this type manipulator is greatly reduced in comparison with other types of manipulators. This is because the structural space requirement of the arms limits the effective volume in which the manipulator can work. The work envelope is not rectangular; therefore, special consideration must be given to the layout of the manipulator. Because of the geometry of low-headroom manipulators, it is difficult to maintain true vertical and/or horizontal movement.
7. *Lift.* The vertical lift must be evaluated for each individual job. The correct height of the pedestal or ceiling spacer is determined by the lift requirement and the series of low-profile manipulator selected.

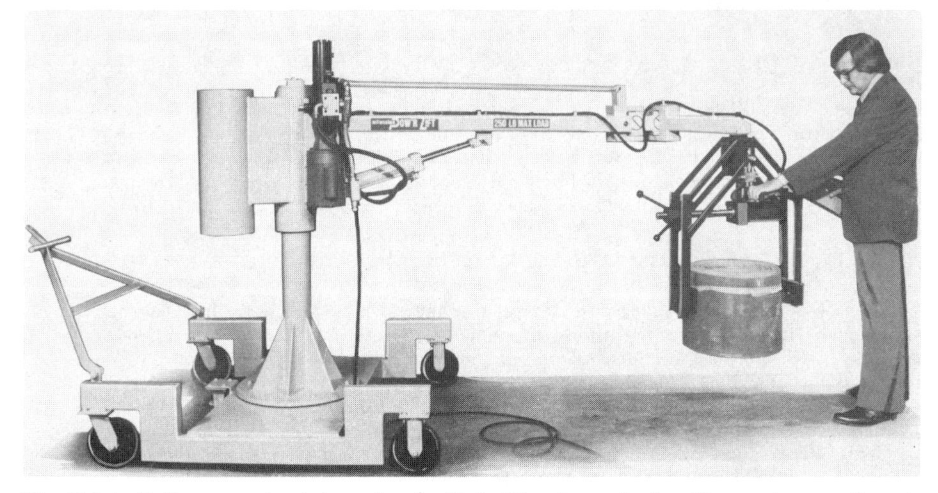

Fig. 17.2.4 Dolly-mounted unit is equipped with holding fixture for handling sandpaper rolls and other cylindrical objects. Courtesy Milwaukee Cylinder.

Manipulator Adjuncts

Used in conjunction with manipulators—and often an integral part of the manipulator—are devices called hoists, balancing hoists, and air lifts.

Hoists. The simple industrial manipulators will always need a lifting and lowering device. The hoist is one of the most practical methods of accomplishing this task. Many variations of hoists are available, including wire rope, link chain, and roller chain hoists. Hoists can be either electric or air powered, and they are available in a number of lift speeds and mounting.

The primary consideration in selecting a hoist for a manipulator application is positioning. Hoists have limitations in control; thus jogging is required for final positioning. One of the main differences between a hoist and a balancing hoist or balancing cylinder is that the balancing hoist allows the load to "float." On a hoist, the load is positioned by utilizing a switch for "powering" the load.

Balancing Hoists. Air-powered balancing hoists can perform three basic functions. One is to balance the tool and fixture suspension (where the tool remains on the hook at all times). The second has a pistol-grip metering valve with up and down controls. It is best suited for handling varying weight loads in more traditional hoisting-type applications. The third type utilizes a balancing circuit having unload, low-load, and high-load modes, controlled via a selector handle. This type is best suited to constant weight-handling applications. The second and third types are mated with a custom grab.

All three types of balancing hoists have a floating or cushion action which enables easy and precise load positioning by an operator. The units usually are mounted on a simple bridge crane which, in turn, is mounted on precision track runways to give far-reaching, versatile overhead positioning. These units can handle loads weighing up to 1400 lb.

Limitations that must be considered for balancing hoists are:

1. Loads over 500 lb generally do not have full float capabilities with pistol-grip models.
2. Heavier loads do not have as much overhead bridge ease of movement.
3. When the balancing hoist is significantly off center of the bridge crane, overhead movement may be somewhat restricted.
4. Side loading or "yarding" makes movement much more difficult and causes excess wear on wire rope and drum surfaces.
5. Strict attention must be paid to (a) securely gripping load; (b) having unit in proper up/down mode; and (c) having load in place when starting up cycle (to avoid a "flying hook").
6. To realize float action and flexible positioning benefits, the operator generally must keep the load reasonably well centered under bridge crane and in relation to work station.

Balancing Cylinder. These devices (see Fig. 17.2.1) are used when the load to be handled does not exceed 800 lb and the vertical distance the load is to be lifted does not exceed 3 ft. This type of lift reduces drag or overtravel problems that can occur with hoists. It gives the operator a better feel for the load. This is achieved through the use of the pneumatic power source and the metering valve.

Size and Capacity

Manipulators are available in a variety of models; specifications are determined by the weight of the object to be handled, the horizontal and vertical reach needed, and the type of mounting required. Light-duty units typically handle loads weighing 90–200 lb. The work area varies from approximately 8 to 20 ft in diameter. Medium- and heavy-duty models usually can handle loads weighing from approximately 250 to 1700 lb within the same work area.

17.2.4 Types of Manipulator Mounting

A prime consideration in selecting an industrial manipulator should be versatility. This applies not only to functional capabilities but also to the mounting of the unit. Therefore, most manipulators have been designed to be either mounted on the floor, trolley mounted, dolly mounted, or suspended from the ceiling; due to heavy loading imposed on the support, manipulators seldom are wall mounted. The most common method of mounting a manipulator is on the floor.

Manipulators that are floor mounted on a pedestal also can be supplied with a castered platform, commonly referred to as a dolly. This dolly mounting greatly enhances a manipulator's versatility by allowing the unit to be moved from one work area to another. It also can eliminate the inconvenience of reserving floor space for a permanently mounted pedestal. A disadvantage of the dolly mount is the fact that the unit must be leveled at each setup to insure proper operation. If the manipulator is not properly leveled, the arms will drift inward, outward, or rotate about the pedestal. This motion will make the manipulator uncontrollable and difficult to gracefully handle materials. When floor mountings are not desirable, or when floor space is not available, the manipulator can be supplied with a ceiling mounting.

Ceiling mounting offers some obvious advantages compared with floor mounting. It provides a clear floor area with few restrictions on equipment layout and operator movement; however, it does require more customer preparation prior to installation of the unit. The overhead mounting structure must be level, structurally adequate to support the manipulator and its rated load, and have sufficient rigidity to resist twisting when the load is at maximum reach in any possible position. The structure also must be designed at the correct height to orientate the manipulator in the desired work area. Spacers are sometimes used to help locate the unit at the proper elevation.

An excellent addition to the ceiling-mounted manipulator is a manipulator mounted on a trolley. As with the dolly option, the trolley allows a single manipulator to function in more than one work area. The trolley is supported by two parallel I-beams which are suspended from an overhead structure in the building. These trolleys either can be power driven or pushed manually. It should be noted that to move a push trolley down the runway, the operator must be parallel with the runway. Otherwise, when the operator tries to move the trolley, he will rotate the manipulator around the trolley until parallel to the runway; only then will the trolley move down the runway.

The trolley's function is to transfer the manipulator to a new location so it can handle the material at that location—not to transfer the load from one location to another. When it is necessary to transfer a load, the following guidelines must be followed closely. (A rough guideline for using a power driven unit and a push unit is dependent on total weight of manipulator and load.) If a push trolley is to be used for continuous use with long runways, the total weight should not exceed 1600 lb. This would require the operator to exert approximately 30 lb of force to move the trolley. For short runways, with intermittent use, the total weight of the unit should not exceed 3200 lb. This would require the operator to exert approximately 60 lb of force to move the trolley. Power-driven trolleys are used when these guidelines are exceeded or because of increased speed, dirty environment, frequent starting and stopping, and so on. Power-driven trolleys, in general, increase the efficiency of a system by enabling the operator to do more work in less time, with greater accuracy and less fatigue. Power-driven trolleys usually are single-speed units using squirrel cage motors or air motors providing one speed (50 ft/min) in each direction. When a trolley is used, the customer generally assumes more responsibility to assure that the trolley runway is parallel and properly supported; he also is responsible for the trolley's power supply.

The work area of a manipulator in the plan view is shaped like a doughnut. The manipulator should be mounted in a location which best utilizes the work area. In most instances the manipulator centerline should be alongside of assembly lines and machine tools, not directly above. This part is illustrated in Fig. 17.2.5, showing good mounting and poor mounting techniques for a manipulator servicing a conveyor. If in doubt about proper location, the manipulator manufacturer should be able to provide assistance.

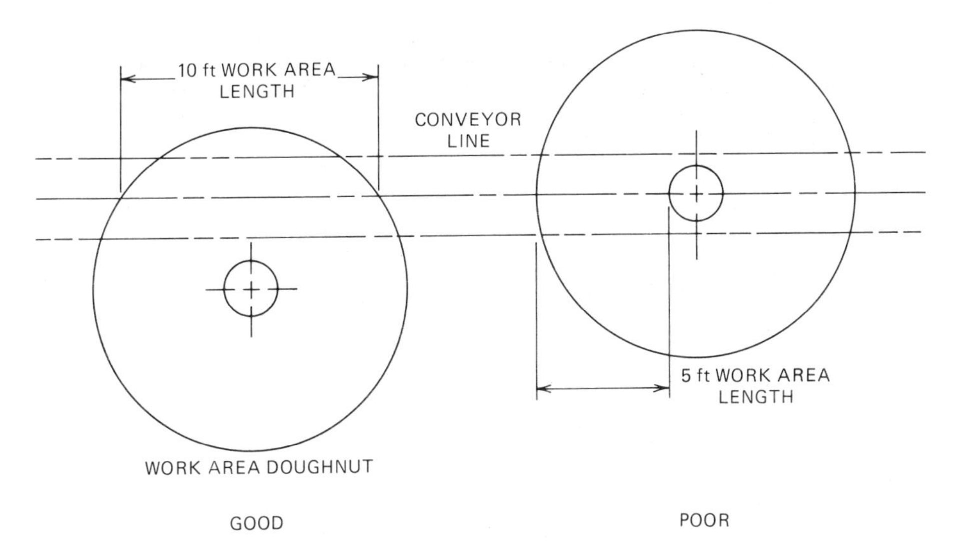

Fig. 17.2.5 Comparison of manipulator mountings.

Manipulator Mounting Requirements

A manipulator unit must be mounted level for proper operation. Out-of-level mounting may cause drifting of the unit or abnormal wear on pivot points. Lack of rigidity in the support will cause the unit to be out of level. (Out of level should not exceed $\frac{1}{8}$ in. in 3 ft.)

Manipulators normally operate smoothly, and it is not necessary to design supports for an impact factor.

Overhead mounting must be made to a rigid support. When the support is building framing, the effect of other loading on the support must be considered. Deflection or twisting due to other loads may result in misalignment of the manipulator.

Overhead mounting, particularly in high-bay buildings, may require extensive support framing to get the support point down to the proper mounting height. An inverted pedestal may be used to support the unit but requires a very rigid support, since any rotation will tend to amplify through the pedestal.

Base-mounted units should be provided with a concrete foundation pad; concrete floors are generally inadequate. Reinforced concrete floors 8 in. and thicker may be adequate for light-capacity units.

Foundation pads are to be designed so that the allowable soil bearing is not exceeded and there is no uplift under the pad. Where plastic soils are encountered, the allowable bearing should be reduced to minimize differential settlement (which could tilt the pad). Soil with a bearing value less than 1500 lb/ft^2 should be compacted to increase the bearing before a pad is installed.

Concrete in the pad should have minimum compressive strength of 2500 psi. The top of the foundation pad should be finished smoothly for good contact with the unit pad. Normally, reinforcing is not required in the foundation pad. Pads larger than 6-ft square should have reinforcing to minimize cracking.

High-capacity manipulators that would require heavy bases and anchors are frequently embedded in the foundation pads. The mast is embedded a sufficient length to prevent crushing of either the mast or the concrete. Reinforcing ties should be placed in the top portion of the pad to assist in holding the concrete together. This type foundation is generally deeper to accommodate the embedded length of the mast.

Base-mounted units are most commonly connected to concrete with anchor bolts. This permits the unit to be relocated with minimum alterations. Anchor bolts are normally cast in the concrete and must have sufficient embedment to develop their full strength. Expansion type anchors may be used but must be checked carefully (they are not as strong as cast-in-place anchors).

Dolly mounting requires the user to counterweight the dolly to prevent the unit from tipping over during operations. Dollies are designed to provide adequate space for counterweighting these units.

The operating envelope of the manipulator must be checked for interference with other equipment or building components. It may be necessary to provide stops or guards to restrict the operational area of a unit.

17.2.5 Trolley Design

Overhead trolleys used in conjunction with industrial manipulators significantly increase the versatility of the manipulators. The user who writes the specifications and requirements for the trolley generally writes a rather loose spec compared with the manipulator specification. This generally is due to a lack of understanding for the importance of this portion of the manipulator system. Yet, frequently the success of the manipulator system is dependent on trolley selection and design. In this section, trolley design is discussed as it relates to wheel selection and runway design. There are two types of trolleys that can be used with industrial manipulators. These are underhung and top-running trolleys. A top-running trolley runs on top of the runway. These are used on heavy-duty manipulator applications where the bottom flange of the runway (I-beams) cannot adequately support the trolley, manipulator, and rated load. However, the most common type of trolley for manipulator use is the underhung trolley. It is used primarily because most manipulator applications are relatively light in weight compared with an industrial crane application. The underhung trolley rides hanging from the bottom flange of the runway I-beams system.

Wheel Selection

The wheels on industrial manipulator trolleys can be either flanged or plain with guide rollers. The flanges or guide rollers are required to transfer any side thrust or skewing forces into the runway structure.

Since the trolley wheels will ride on a tapered surface (the I-beam flanges are tapered), the wheels will be either machined to match the rolling surface of the lower flange, or will have a convex surface so it can easily rotate on the tapered I-beam flange. Both wheel designs should have approximately

Table 17.2.1 Maximum Wheel Loads for Under-Running Trolleys

Wheel Diameter (in.)	Wheel Load (lb) for Given Flange Width					
	Contour Tread[a,b]			Convex Tread[c]		
	$\frac{1}{2}$ in.	1 in.	$1\frac{1}{2}$ in.	$\frac{1}{2}$ in.	1 in.	$1\frac{1}{2}$ in.
4	2,000	4,000	6,000	1,200	2,400	3,600
5	2,500	5,000	7,500	1,500	3,000	4,500
6	3,000	6,000	9,000	1,800	3,600	5,400
7	3,500	7,000	10,500	2,100	4,200	6,300

[a] Wheel load $P = 1000\ WD$ (in lb), where W = width of tread exclusive of flanges (in in.), and D = diameter of wheel (in in.).
[b] Contour tread matches the rolling surface of the lower flange of the track beam.
[c] Wheel load $P = 600\ WD$ (in lb), where W and D are as in footnote *a*.

the same rolling resistance for industrial manipulator applications; therefore, neither offers any advantage in the effort required to power the trolley. The advantage in the different wheel designs involves their load-carrying capabilities. The contour tread designs have approximately a 65% increase load-carrying capacity over the convex tread design. Also, larger wheels on the same runway have greater load-carrying ability than smaller wheels.

Drive wheels should be matched pairs and have the same diameter within a tolerance of 0.010 in.

The guidelines for wheel capacity are found in Table 17.2.1.

The wheel should be designed for the *maximum* wheel load under normal conditions without undue wear. The maximum wheel load is that load produced with the trolley handling the rated load in the position that produces the maximum reaction at the wheel. Impact loading need not be considered in the recommended formula.

The wheels should be made from either steel or cast iron; steel wheels generally are preferred. The steel should be rolled or forged from an open hearth, basic oxygen, or electric furnace process. Unhardened wheels have tread hardness of approximately 260 BHN and are sufficient for most industrial manipulator applications. If cast iron wheels are used, an acceptable class of iron is required; consideration also should be given to brittleness and impact strength of cast iron.

One of the common oversights in designing the runway for the trolley is the proper sizes of the I-beams. In sizing these beams, not only must they be designed for gross bending between their supports, but also for localized bending of bottom flanges. The localized bending stresses must then be combined with the gross bending stress to determine the maximum principal stress and the maximum tensile stress. These maximum stresses must then be compared with the allowable stresses for the material being used.

The localized bending stresses can be analyzed using the equation $S = K_m\ (6P/t^2)$, in which P is the wheel load and t is the flange thickness. In this equation, K_m is a dimensionless coefficient that is dependent on the location of the load and the part under consideration. Values for K_m can be found in Table 17.2.2. The wheel loading should be considered as a concentrated load where the wheel contacts the flange of the beam. An impact factor of 1.15 should be used in conjunction with this equation.

Table 17.2.2 K_m Factors for Localized Bending Stress Equation[a]

	d/a						
c/a	0	0.25	0.50	1.0	1.5	2	∞
1.0	0.509	0.474	0.390	0.205	0.091	0.037	0
0.75	0.428	0.387	0.284	0.140	0.059	0.023	0
0.50	0.370	0.302	0.196	0.076	0.029	0.011	0
0.25	0.332	0.172	0.073	0.022	0.007	0.003	

[a] In this table a = flange width, c = distance from web to centerline of wheel, and d = distance from centerline of wheel to stress application point.

After the localized bending stresses are calculated, they must be combined with the gross bending stresses. These stresses are combined using the Mohr's circle theory. The principal stresses shall not exceed an allowable stress of 0.60 F_y for tension and compression; and the maximum shear stress should not exceed 0.40 F_y, where F_y = yield strength of the material.

17.2.6 Power Sources

Most industrial manipulators have a pneumatic source powering some function of the manipulator. Pneumatic components can be built up into a variety of systems offering a high degree of dependability and safety with relatively low cost. A well-planned and developed system can provide excellent trouble-free service, plus flexibility for future expansion. Pneumatic power source is available in most manufacturing facilities. This fact alone makes using pneumatics an attractive choice.

The maximum system pressure at a pneumatic power source is controlled by the compressor control system, which generally maintains a pressure range. The compressor may start automatically when the system pressure drops to 120 psi, and automatically stop again when the pressure in the receiver reaches 140 psi. Because of pressure drops in the pneumatic lines, the minimum line pressure to the manipulator can be a minimum of 90 psi; industrial manipulators are generally designed to operate at this pressure. Industrial uses of high pressure pneumatic systems have been limited because of the cost of producing air pressure above 150 psi in large quantities. Operating pressure limits the load capabilities unless large cylinders are used.

In order to maintain a continuous pressure at the manipulator and eliminate compressor fluctuations (or to utilize pressures lower than compressor output), a pressure regulator is used on the manipulator. In some manipulator applications, several pressure regulators are used to control pressure for the grabbing devices.

The need for dry, clean compressed air is paramount to successful manipulator operation. Unless injurious elements are removed thoroughly, compressed air is highly destructive. It will corrode, rust, and ruin cylinders, valves, gauges, instruments, and brakes. In fact, wet air will damage the air system itself.

Generally the moisture in the air is removed by an air dryer located near the compressor. The air generally is filtered at the compressor, but a final filtering still must occur at the manipulator. The filter at the manipulator removes contamination from the air before it reaches the directional valve and actuating cylinder or air motor. Air line filters are generally fitted with a filter element that removes contaminants of around 5 μm.

Pneumatic-powered manipulators offer advantages and disadvantages compared with electrical and hydraulic-powered manipulators.

1. Pneumatic controls do not have the self-destructive characteristics of systems that are subject to resistance and heat. Specifically, air-piloted valves can operate indefinitely without malfunction compared with solenoids—which have a finite life expectancy and can burn out if overheated.

2. Because of heat generated by hydraulic and electrical systems, cooling air must be present to remove excess heat. If cooling air must pass through the electrical components, the air must be clean and dry. Hydraulic components produce heat but they can be cooled by passing the fluid through heat exchangers. *Pneumatic systems do not require cooling.*

3. Through the use of thermoplastic tubing and small components, air controls can be installed easily in compact metal boxes (similar to electrical switches, relays, etc.).

4. Hydraulic fluids have a higher density than dry air. The higher inertia of the hydraulic fluid is of concern when sudden acceleration and deceleration is experienced because of the shock loads. Also, hydraulic systems require floor space for locating reservoirs.

5. Pneumatic control component sizes are much smaller, compared with electrical or hydraulic components.

6. Hydraulic actuators offer the highest torque (or force)-to-inertia ratio in comparison with most mechanical, pneumatic, and electrical systems.

7. With hydraulic components, it is inherently difficult to prevent leaks during normal usage; thus they have earned a reputation for messiness. Any leaks in pneumatic components will not present this problem.

8. Hydraulic systems can require an initial running time to allow the hydraulic fluid to warm up to a normal operating temperature. On pneumatic and electrical systems, no initial warm-up is required.

9. One main difference between applying pneumatic and hydraulic-powered systems is the compressibility factor of the two media. The pneumatics can offer some advantages if the manipulator is designed to float the load. This is accomplished by the compressibility of the pneumatic media, which makes precise positioning of the load easy to accomplish.

10. If hydraulics are used, an additional electrical power supply is required to run the pump. This requires additional floor space and also additional servicing and troubleshooting if problems arise. When pneumatics are used to power the industrial manipulator, the main factory compressor will require no additional power supply, floor space, or servicing.

11. Pneumatic systems are explosion proof.

12. For electrical systems, generally skilled maintenance and service people are available; not necessarily so for hydraulic and pneumatic systems.

13. Hydraulic systems are noisy in comparison to electrical and pneumatic systems.

17.2.7 Grabs and Manipulators

Industrial manipulator manufacturers and users share the goal of designing and manufacturing a safe product. To achieve this objective, there are several design guidelines which, when followed, insure a safe and reliable manipulator design. These guidelines are:

1. The grab and manipulator structural parts should have a 3:1 safety factor on the yield strength, or a 5:1 safety factor on the ultimate strength of the material being used. This is based on the maximum stress that will occur in the parts during the most severe operation for which the grab or manipulator was intended.

2. Grabs relying on the friction force developing by clamping to hold a load must use a *fail-safe design.* That is, the clamping force must be maintained even if the activating force is lost. The friction force developed must be a minimum of two times the maximum load to be handled, based on the best available data regarding the coefficient of friction between the load and clamping material. If data regarding the two materials are not available, appropriate tests must be made to determine the coefficient of friction.

3. Grabs relying on vacuum to hold the load must generate a sufficient vacuum to hold a minimum of three times the maximum load to be handled. There must be a valve located as close as possible to the vacuum heads in order to prevent loss of vacuum at the support points, should the source of the vacuum be lost. Vacuum grabs are to be used only for lifting nonporous material having smooth, flat surfaces. When the vacuum source is a vacuum pump remote from the grab, there must be provisions to alert the operator to drops in vacuum to one-half of the design value. The manipulator control should prevent an operator from lifting the load until proper vacuum is reached. When the vacuum is turned off, the manipulator control goes to a no-load condition (balance control) or the vacuum's force must be maintained (metering control). On multiple-vacuum-cup systems, where a loss of vacuum at one vacuum cup could occur, the vacuum circuit must be designed so that the loss of vacuum at one or more cups will not cause a vacuum loss at the remaining cups or a loss in the manipulator control.

4. If a magnet is used on a grab, it must have the capacity to lift three times the maximum load to be handled. All electromagnets must have a backup power source and proper crossover switching to hold the load in the event of a power failure. When batteries are used for the power source, there must be provisions for a warning when the batteries are discharged to a point where the lifting capacity falls below the 3:1 safety factor.

5. All grabs are to be identified with the maximum rated load capacity.

6. The grab must be designed with enough clearances or guards to eliminate, during normal use and foreseeable misuse, "pinch points" and other hazards (a) within the grab mechanism, (b) between the grab and the load, and (c) between the grab and the machinery being loaded.

7. Incorporated into the manipulator and grab designs must be considerations for the strength of the operator. Some general guidelines for consideration of the human strength for various operating functions and positions are listed below; these figures are nominal values for adult females. (For males, these figures could be increased by 40%.)

   Finger press: 5 lb

   Finger pull: 5 lb

   Forearm lift: 25 lb

   Two-hand scissor action at elbow height with hands about 8 in. apart, standing position: 18 lb

   One-hand lever-type control, standing position, push: 75 lb; pull: 70 lb

   Hand cranking at elbow height with a 4–10-in. diameter crank: 30 in.-lb

8. In manipulator and grab designs, there are some common situations that should be avoided. Many of these situations occur because the designer has failed to appreciate the importance of the manipulator in terms of its compatibility with human capabilities and limitations. The following are situations which *should be avoided:*

Systems that require operators to apply maximum strength for long periods of time. (Powered controls should be provided in situations that require excessive operator strength.)

Systems that require continuous, rapid movements for long periods of time, such as push–pull cycling and cranking.

Systems that force operators to maintain a tight grip or to hold some device in a fixed position for long periods of time without intermittent rest periods.

Systems that require operators to hold their arms above their heads for long periods of time.

Systems that impair an operator's view of the work area or operation.

9. In designing grabs, it is important to consider the grab forces applied to the product being handled. These grab forces can sometimes be high enough to impair the product being handled. For example, a common method of lifting sheet metal parts utilizes a vacuum grab. In this application, vacuum grabs may lift the sheet metal product but the product itself sometimes is incapable of supporting its own weight; product damage results. Clamping grabs also can produce forces large enough to crush the product being handled.

10. Special consideration must be given to the design of manipulators and grabs on high-temperature applications. On such applications, dimension changes occur due to expansion and contraction of the material. While the effects of natural expansion and contraction cannot be totally controlled, they can be minimized by allowing the various components to expand more or less freely without geometric interference. As temperatures increase, the fatigue strength of the materials decreases rapidly. Special alloy steels must be used in the design. When high temperatures are encountered, the guidelines listed in item 1 do not apply; a conventional fatigue analysis must be performed on all components subject to high temperatures.

Many times hot parts or grabs will be cooled by quenching. Some metals will crack as a result of this thermal shock. If the change is not too severe, repetitions of temperature and stress gradients in metals may be enough to cause eventual failure. This is referred to as thermal fatigue. Because of thermal fatigue, grabs are normally designed for a specific production cycle and have a finite life.

High-temperature environments can limit the type of bearing and cylinders used because of the limitation on seals.

11. The manipulator and grabbing device can provide many degrees of freedom; common motions are:

Manipulator motions:

- Radical movement about the pedestal
- Vertical movement (lift) measured from the mounting surface
- Horizontal movement (reach) measured from the pedestal

Grab motions:

- Tilt motion from the horizontal plane to the vertical plane
- Swivel motion providing radial movement about the horizontal plane, parallel to the grab mounting surface
- Wrist motion providing radial movement about the vertical plane, perpendicular to the grab mounting surface

17.2.8 Safety Considerations

During the operation of an industrial manipulator the safety of the operator and any surrounding equipment is paramount. Accordingly, special consideration must be given to the design of the manipulator, the manipulator controls, and any grabbing device that may be supplied with the industrial manipulator.

Hazardous conditions can occur in the operation of a manipulator and result in the operator losing control of the load and/or the manipulator. If these conditions occur, the manipulator can injure the operator or damage surrounding equipment. Consequently, three potentially hazardous conditions require safety systems:

1. The interruption of the power supply due to failure or pinching of the hoses. Loss of power would result in the manipulator arms dropping rapidly.
2. Loss of load due to the grab losing grip on the product. If the manipulator is powered by a pneumatic cylinder, loss of the load will result in rapid expansion of air in the lifting cylinder. The manipulator arms would then rise at a high velocity.
3. Accidental release of the grab by the operator. This would allow the product to be dropped.

Any of these situations could result in the operator losing control of the load and/or the manipulator. To minimize the risk in operating an industrial manipulator, a number of hazard control devices are included in the manipulator design. The control devices should be incorporated in any manipulator—regardless of the type of power supply. A pneumatic power supply was considered for the following devices:

1. *Up-Speed Safety Control.* This limits the upward velocity of the manipulator arms. It is achieved by filling the low-pressure side of the lifting air cylinder with a relatively incompressible fluid such as oil. During operation of the manipulator, the displaced fluid flows to and from a fluid reservoir via a flow control valve. The flow control restricts the flow of the fluid so that the velocity of the cylinder rod is controlled to a safe rate. This feature provides protection in the event that the load is lost.
2. *Down-Speed Safety Control.* A down-speed safety control is provided by restricting the velocity of air exhausting from the lift cylinder when the manipulator arms are being lowered. The restriction is provided by a pneumatic flow control valve. The flow control valve is designed to provide a safe rate of down speed in the event of loss of air supply.
3. *Interlock.* An interlocking pneumatic circuit is provided to integrate the control of the pneumatic clamping grab and the main lifting circuitry. The interlock will not allow the operator to release the grab from the product until the load has been set down. The interlock is provided by sensing air pressure in the lifting cylinder and using the signal to operate a four-way control valve which, in turn, operates the grab clamping cylinder.
4. *Grab—Lost-Load Control.* This safety device should be incorporated in all grab controls. It prevents a load from being released from the grab due to loss of air pressure. (This situation occurs when a power supply line to the grab actuators is cut or ruptured.)

A final comment on safety: Absolute safety cannot be built into a manipulator. Safe manipulator operation is based on a combination of manipulator safety devices and, most important, a competent operator. All operators of industrial manipulators must be trained. This training should cover how the manipulator is to be used, all the safety features that have been built into the manipulator, and the limitations of the manipulator.

17.2.9 Types of Grabs

In some manipulator applications, the difference between a successful and an unsuccessful installation can be the grabbing device. The manipulator operator should understand, appreciate, and enjoy using the total system and particularly the grab. Therefore, the grab must be easy to operate and require minimum strength and effort.

The strength and effort required are keys to selecting the grabbing device and determining whether the device will be manually or power actuated. Other key considerations are how the product is to be handled and any interference present. Possible grab interferences can include interference between the grab and machine tool, jigs, fixtures, and so on.

There are many types of grabs that have been developed for industrial use. In most industrial applications a grab must be specifically built for the application because of the wide range of objects, shapes, and sizes found in various applications. The cost of the grab typically will be equal to or greater than the manipulator itself.

Grabbing devices can be classified into four broad categories: (1) mechanical grabs, (2) vacuum grabs, (3) powered external clamp grabs, and (4) powered internal clamp grabs. All but the vacuum grab can be designed to support the weight of an object mechanically or through friction. The advantage of supporting the object mechanically is that the weight, size, and cost of the grab is lower than for other conventional grabbing methods.

Mechanical Grabs

A mechanical grab is a device that requires no power source to activate the grab. These grabs are activated either by the operator through a mechanical linkage (Fig. 17.2.6), self-activated by the weight of the material being handled (Fig. 17.2.7), or the grab is designed to cradle the material between a mechanical finger and a stop (Figs. 17.2.8 and 17.2.9). This type of grab is the most economical but is limited in the functions it can successfully perform. Generally, with this type grab the material being handled can be transferred in only one plane.

Vacuum Grabs

The vacuum grabs can be used to pick up a variety of parts. Generally the objects handled have flat surfaces or surfaces with gradual contour. Examples are: glass sheets, sheet metal, rubber sheets, office

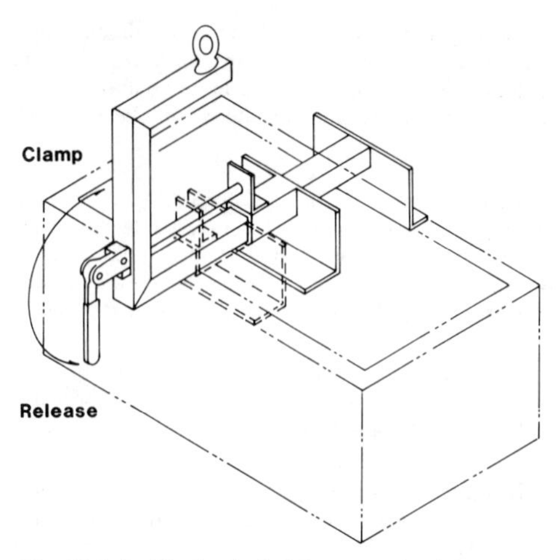

Fig. 17.2.6 Mechanical C-frame grab with manually expandable internal jaws. Part is supported by internal edge.

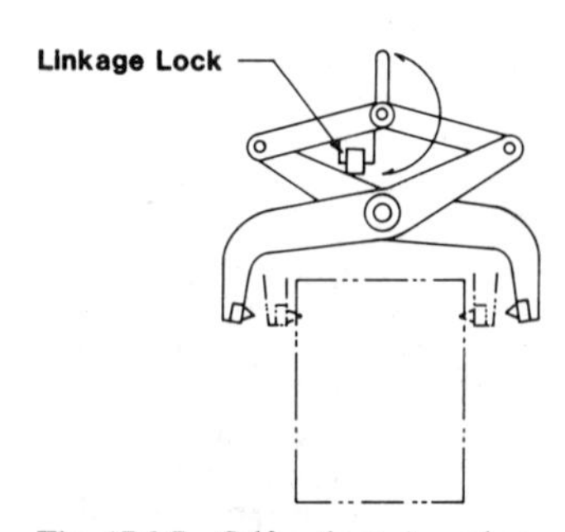

Fig. 17.2.7 Self-activated mechanical tong grab. The clamping force is proportional to the weight of the product being lifted.

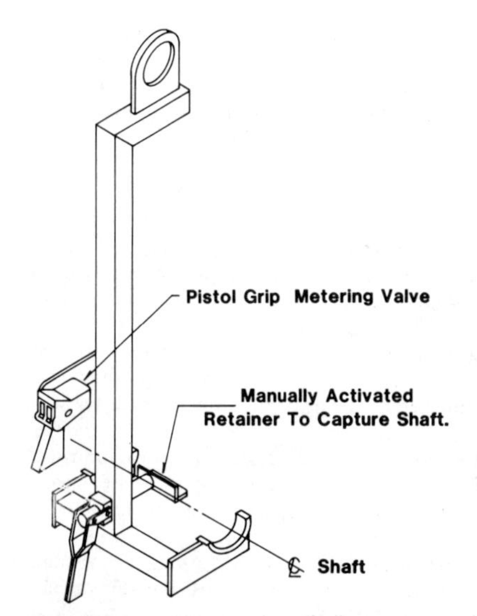

Fig. 17.2.8 Mechanical grab used for horizontal transferring of large shafts. Grab cradles the shafts and the retainer is activated to capture the shaft in position.

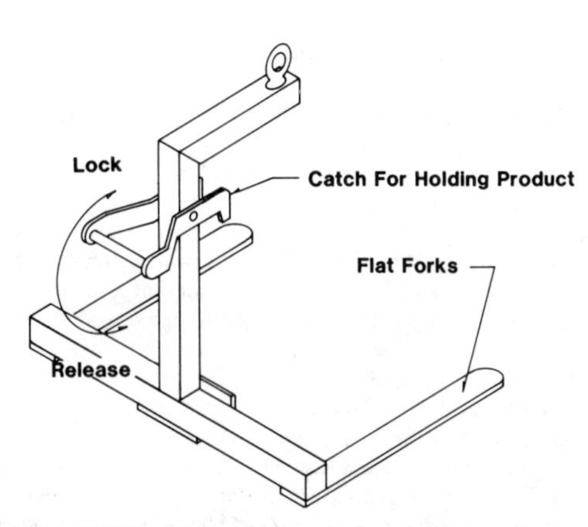

Fig. 17.2.9 Mechanical grab for picking up boxes and transferring them horizontally; grab hooks are inserted in hand holes of box while forks support the boxes in the vertical plane.

furniture, television picture tubes, household appliances, and cartons. Products handled vary in size and weight. Upper limits are 8 ft × 10 ft sheets and 1000 lb weight.

Many functions can be incorporated into the vacuum grab. There are a variety of vacuum cups in many capacities, sizes, and shapes to help accommodate the product being handled. These cups can be arranged geometrically to support the product. Typical vacuum grabs can have one or several vacuum cups, adjustable cup spacing to handle several sheet sizes, and powered or manual tilt to transfer product from the horizontal plane to the vertical plane. Typical grabs are shown in Figs. 17.2.10 and 17.2.11.

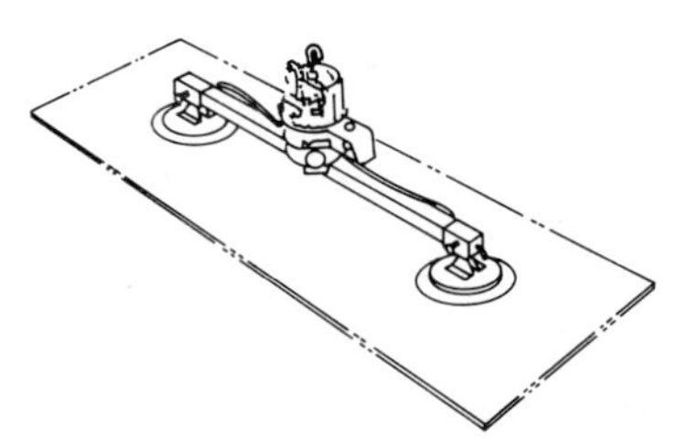

Fig. 17.2.10 Two-pad vacuum grab for the horizontal transfer of sheet products.

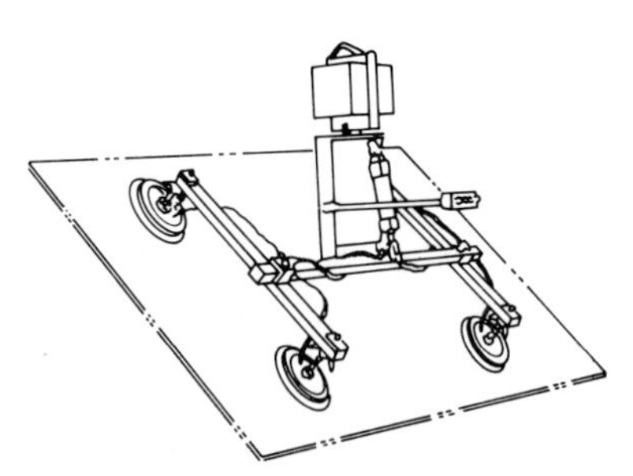

Fig. 17.2.11 Four-pad power tilting vacuum grab. Product can be tilted 90° for transferring sheet from horizontal to vertical plane.

Powered External Clamping Grabs

These are simply mechanical grabs with either hydraulic or pneumatic powered means for activating the clamps. With this type of grab more functional flexibility can be built in. Typically, because of the power source and a linkage external to the product, greater clamping force can be obtained, which allows heavier loads to be handled. Typical grabs are shown in Figs. 17.2.12, 17.2.13, and 17.2.14.

Powered Internal Clamping Grabs

Powered internal clamping grabs provide a particular flexibility not available with other types of grabs. The greatest advantage provided by this type grab is that it frees the outside surface of the object being handled and allows this surface to be used for other purposes. A limitation for this grab is that the linkage used to develop the clamping force is restricted in size and cannot develop tremendous clamping force. Also, this type of grab *generally* relies on friction to hold the product. Accordingly, objects handled with these grabs are generally smaller in size and weight than those handled with external clamping grabs. A typical grab is shown in Fig. 17.2.15.

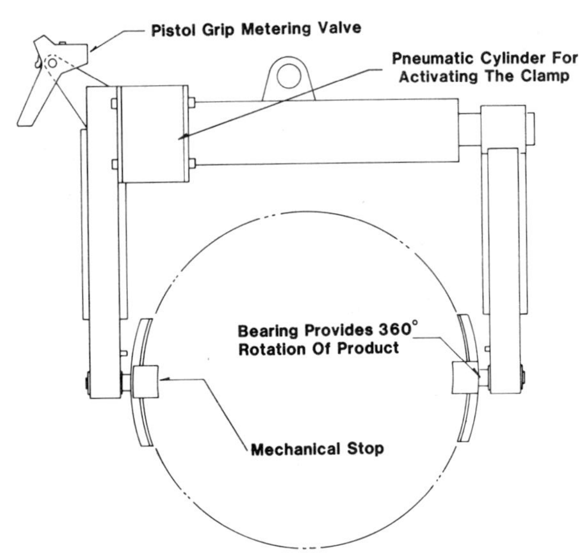

Fig. 17.2.12 External clamping grab for picking up large-diameter parts in the horizontal or vertical plane and providing 360° part rotation.

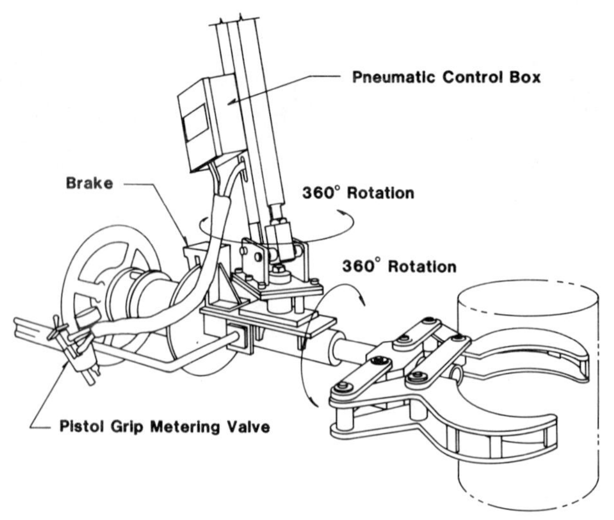

Fig. 17.2.13 External clamp for small-diameter parts with rotation and air control holding brake. Brake is released by tilting hand wheel.

One clamping principle used for this type of grab is the expanding mandrel. There are several types of expanding mandrel grabs but the two most frequently used are:

1. An elastomer mandrel expanded by contracting a cylinder (shown in Fig. 17.2.16).
2. An expanding hose attached to a probe (shown in Fig. 17.2.17).

In operation, the mandrel is inserted in the bore and compressed air expands the compressible material against the inside bore. For delicate workpieces, different types of materials can be attached to the mandrel and the grab operates at a lower pressure to prevent damage to the parts. Both of these types of grabs have been used successfully to handle spools of wire and yarn.

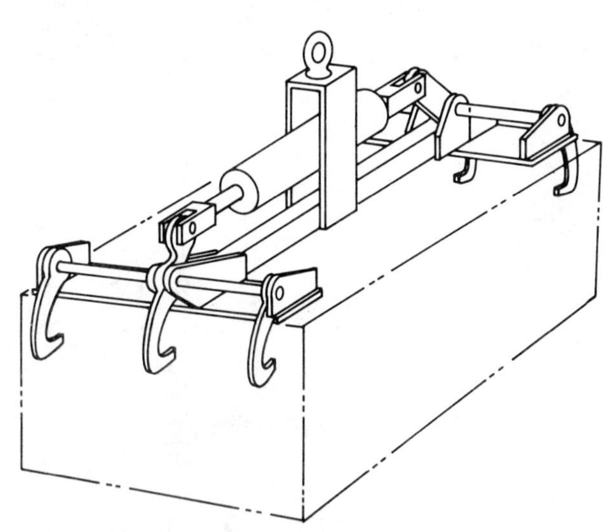

Fig. 17.2.14 External hook grab. Powered hooks pierce the material for lifting and transferring the product in the horizontal plane.

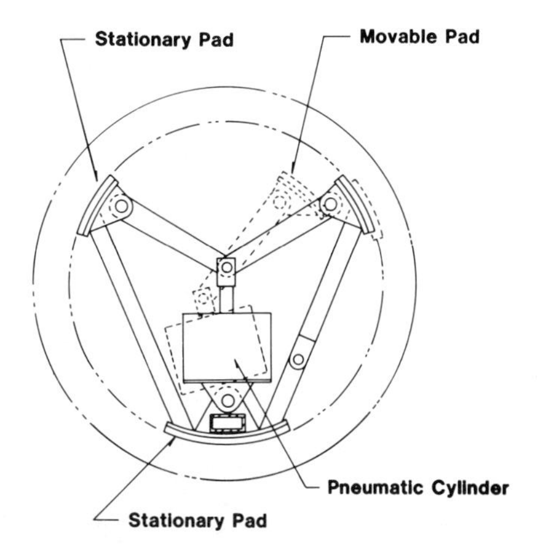

Fig. 17.2.15 Internal clamping grab used for horizontal transfer of large-diameter parts. Parts can be stacked on one another using this type of grab.

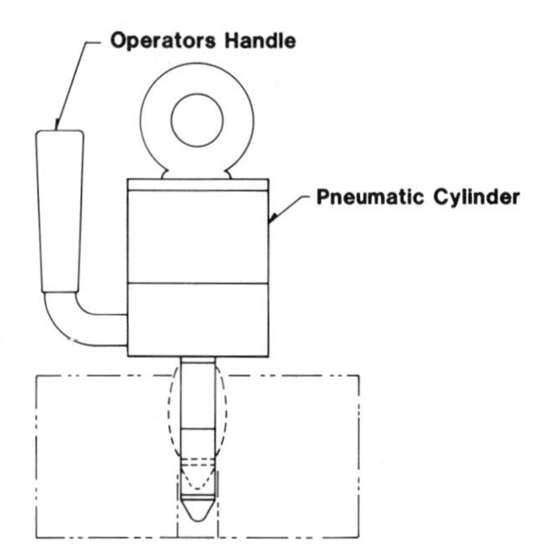

Fig. 17.2.16 Internal clamping grab. Air cylinder compresses elastomer spring axially, causing it to expand and hold product being handled. This type grab is used for transferring the product in the horizontal plane.

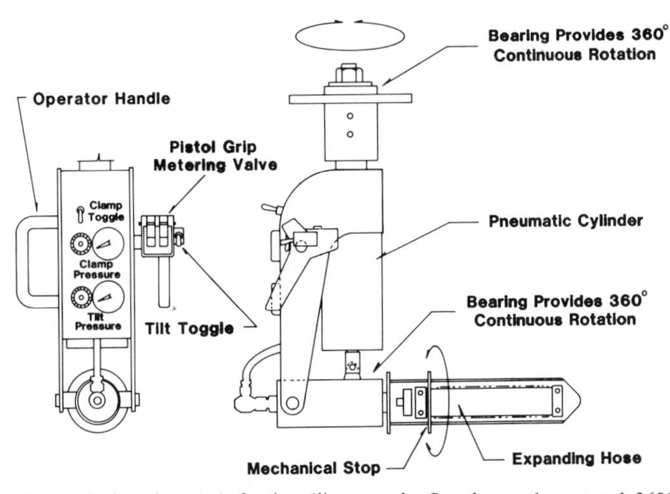

Fig. 17.2.17 Internal clamping grab for handling spools. Spools can be rotated 360° continuously about two planes, and spools can be tilted 90°. This grab is used to transfer spools from a pin rack to a packing carton.

17.3 DUMPERS

Robert DeCrane

17.3.1 System Components

Uniform bulk containers for parts and materials have become an important cost-saving development in plant modernization. Generally plants have converted from small bins, tote boxes, cans, bags, and other small unit loads to larger uniform-size containers to reduce handling costs. The size, shape, and capacity of the in-plant container is determined by the physical characteristics of the material it will hold, how the container can be integrated into the manufacturing process, and if it can be handled economically and safely at every point along the line.

One of the associated problems being solved by materials handling engineers is that of discharging the contents of these bulk containers, whether it be to supply parts to a machine operator at the work position or adding materials in a mixing or batching operation. In either application, the discharge or flow of the materials from these heavy bulk containers MUST BE CONTROLLED. It is our contention after working with these problems for 30 years that positive control is best achieved with hydraulic dumpers.

A number of factors determine the design of the dumper that will perform the unloading operation to job requirements.

Material

The first factor to consider is the material to be handled. Every material has an angle above horizontal at which it will move. The term *dumping angle* is used to define the required position of the container at which the material will move and completely empty the container.

Containers of metal parts empty themselves at a 45° angle above horizontal. The primary control objective with this material is to provide a gentle sliding of parts into the work tray to avoid damage and spillage.

There are, however, many materials whose pouring characteristics must be known so that controlled dumping can be assured. This should be determined by the end user.

Container

The second factor to be considered is the container. This is the starting point in design. In fact, the dumper's frame size and power requirements are usually determined by the size, shape, composition, and load capacity of the container to be emptied. Will the shape of the container affect the flow of the material? For example, it must be known if the container opening will restrict the flow of material, or perhaps the shape of the container might be such that there would be internal places where the material would hang up. And another important factor is whether or not the material has a tendency to glomerate, such as pumice and pigments if they are being dumped from shipping containers in which the material has packed down in transit. Dumpers designed for containers of this type can be equipped with vibrators to overcome the bridging characteristics of the material within the container.

Dumpers made for bulk powered chemical containers are designed with a hood and chute which snugly fits the top edge of the box to prevent a fall back of material when partial dumps are made. If required, further control for the dumping of chemicals and pigments can be provided by an iris valve at the discharge end of the chute. The hood and valve also act to control the generation of dust.

Receptacle

The receptacle or receiving place of the discharged load is the next factor to be considered. This might be a table, tray, conveyor, hopper, mixing machine, and so on. The elevation of the receptacle and the container's size will determine the overall height and other space requirements of the dumper.

Space

The materials handling engineer must take into consideration the always critical space factor for a dumper installation. Figure 17.3.1 shows the necessary dimensional data that affect dumper design.

Means of Loading

How will the loaded containers be brought to the dumper? What kind of equipment will be used? Will it be a fork truck, a pallet mover, a dolly, a hand truck, or conveyor? Will the dumper be loaded at floor level or should the receiving carriage be elevated?

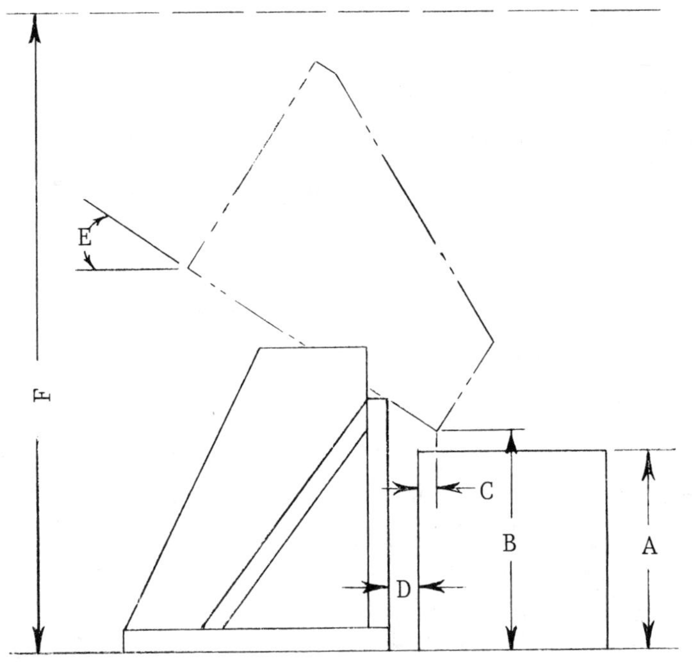

Fig. 17.3.1 Typical dumper dimensional data. A = height of table, hopper, conveyor, etc.; B = A, plus desired height above receptacle or opening the discharge edge or dumper should stop, at the extreme dumping angle; C = minimum allowable distance the discharge edge of dumper at extreme dumping angle should extend over near side of receptacle; D = minimum distance the dumper frame can be placed from receptacle. Is it necessary to span any obstruction between floor and discharge height? E = dumping angle, if other than 45°. (45° above horizontal is most common); F = overhead clearance.

Special Requirements

Product contamination is sometimes a factor. If the material being dumped is susceptible, it might be necessary that the dumper frame and parts be enameled, galvanized, or plated as required, and that areas of direct product contact be stainless steel or aluminum.

There are two other questions that the materials handling engineer must consider in providing controlled dumping: (1) should the dumper installation be stationary or portable, and (2) are there special electrical requirements? The proper electrical controls in most instances eliminate needless labor. Timing devices can automatically control the dumper operation throughout the cycle unattended. Remote controls permit the operator to control a discharge from any vantage point. A basic specification sheet for dumpers is presented in Fig. 17.3.2.

17.3.2 Twin-Cylinder Container Dumper

What Is It?

The twin-cylinder container dumper is a dumper with hopper attached to the frame by pivot shaft (see Fig. 17.3.3). It uses one hydraulic cylinder on each side of hopper for rotating the load around a fixed pivot point to obtain desired dump height and dump angle. The dumper can be anchored to the floor, mounted on a stand, made portable on wheels, or made portable by fork trucks.

What Can It Do?

The dumper can dump maximum load of 10,000 lb up to 60 in. dump height (does not include stand, if any) at 45° dump angle above horizontal. Dumpers are available with maximum 60° dump angle.

MATERIAL HANDLED

Name of material:________________________________

Indicate: Solid________ Liquid________ Free Flowing________________

Operating conditions: Dry________ Moist________ Corrosive________________

Is dumper washed down after use?

CONTAINER TO BE DUMPED

Indicate by check:

___Vat___Barrel___Drum___Box___Other (Submit sketch & dimensions)

Container is: ___Steel___Stainless___Aluminum___Wood___Fiber

___Other

___Round Containers ___Square, Rectangular, Other

Height:________________________ Height:________________

Dia. at opening________________________ Dumping side:________________

Dia. at widest point________________________ End:________________

Note: Allow for container bulge.

WEIGHT

Weight of container:________________ Weight of material:________________

Total Weight:________________________________

If container is special size or shape it is advisable that a sample container be shipped to manufacturer for testing with dumper.

RECEIVING POINT OF DUMPED LOAD (DUMPING HEIGHT)

Height of receiving level from floor________________________

Dumping height required if different from receiving level________________

Identify receptacle (i.e., table, tray, conveyor, hopper, opening, etc.)________________

If pouring into opening: Length____ Width____ Dia.________________

SPACE AVAILABLE TO INSTALL DUMPER

Ceiling height__________ Floor area length________________________

Width:________________________________

CHUTE

Material for chute construction to be:

Mild Steel____________ Stainless (Indicate Type)________________

Is sound deadening required?____ Is flow control required?________________

Is dust control required?________________________

DUMPER CONSTRUCTION

Dumper to be portable____________ Stationary________________________

Wheels to be steel____________ Floor Saver________________________

Special finish required Gal.________ Epoxy________ Other________________

POWER

Voltage__________ Phase__________ Hz________________________

Motor: TENV(Standard)__________ Explosion Proof________________

NEMA__________ Class__________ Division________________

Waterproof____________ Weather Proof________________________

Control Circuit: Volts____________ Hz________________________

Special Specifications—Automotive, USDA, etc.—Send Copy with Request

LOADING METHOD

Type of Equipment that will be used to load dumper:

Fork Truck________ Pallet Truck________ Other________________________

Can dumper be loaded above floor level?________________________

Fig. 17.3.2 Dumper specification sheet.

Table 17.3.1 provides dimension ranges. It can handle all kinds of containers—steel or wooden boxes, cardboard containers, carts, part bins, and in-process containers.

How Does It Work?

The dumping is done by two cylinders. Cylinder size depends on load capacity, load center, and maximum allowable hydraulic pressure (not more than 1750 psi). The hydraulic fluid is supplied to the cylinders by a hydraulic pump driven by electric motor. Push-button control starts or stops hydraulic power at any point in dumping cycle.

Cylinder sizes vary from 2-in. bore to 4-in. bore. Hydraulic pumps vary from 2.4 to 7 gal/min. Electrical motors vary from $1\frac{1}{2}$ to 10 Hp.

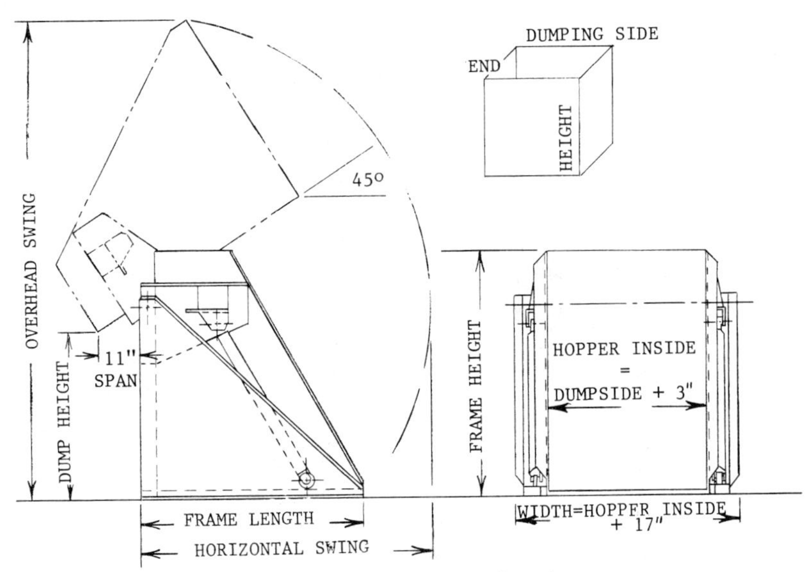

Fig. 17.3.3 Twin-cylinder container dumper.

17.3.3 Low-Level Drum Dumper

What Is It?

The low-level drum dumper is a dumper with hopper attached to the frame by pivot shaft. The unit uses only one hydraulic cylinder in back of the hopper for rotating the load around a fixed pivot point to obtain desired dump height and dump angle. The dumper can be anchored to the floor or made portable on wheels.

What Can It Do?

It can dump maximum load of 1500 lb up to 60 in. dump height at 45° dump angle. (Dumpers available with maximum 60° dump angle.) It can handle drums up to 55-gal capacity. Details are provided in Fig. 17.3.4 and Table 17.3.2.

Table 17.3.1 Dimensions (in.) for Twin-Cylinder Container Dumper

Container						
End Width	Height	Dumping Height	Overhead Swing	Frame Height	Horizontal Swing	Frame Length
48	36	36	115	63	70	55
48	42	42	126	69	75	55
48	48	48	136	75	79	55
48	54	54	147	81	84	55
48	60	60	157	87	88	55
42	36	36	111	63	66	49
42	42	42	122	69	71	49
42	48	48	132	75	76	49
42	54	54	143	81	81	49
42	60	60	153	87	85	49
36	36	36	107	63	62	43
36	42	42	118	69	67	43
36	48	48	128	75	72	43
36	54	54	139	81	77	43
36	60	60	149	87	82	43

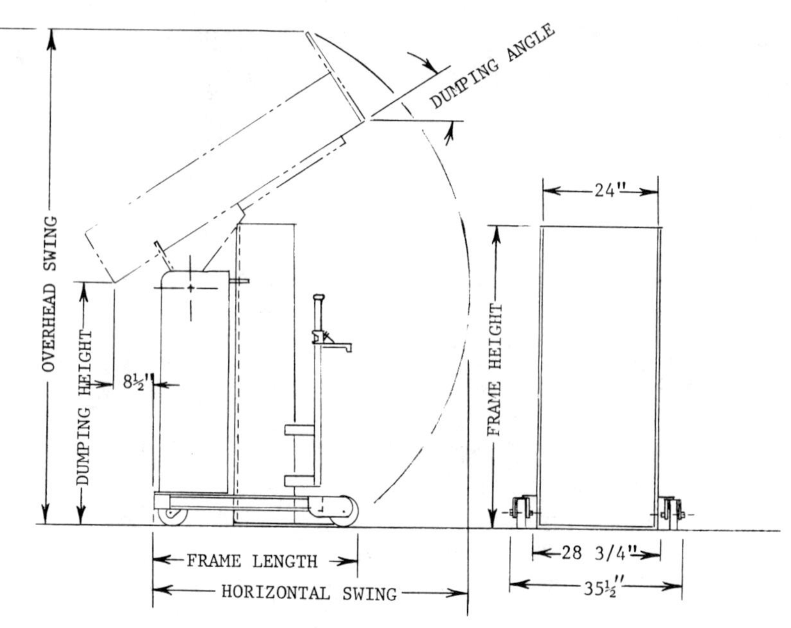

Fig. 17.3.4 Low-level drum dumper.

How Does It Work?

The dumping is done by one cylinder. The cylinder size depends upon load capacity. The hydraulic fluid is supplied to the cylinder by hydraulic pump driven by electric motor. The dumper is equipped with locking clamps (one on each side of hopper) for keeping drums secure during dumping.

Dumpers up to 750 lb capacity have a 2-in.-bore cylinder, 1.0 gal/min pump, and $\frac{1}{2}$ Hp motor. 1000-lb-capacity dumpers have a $2\frac{1}{2}$-in.-bore cylinder, 1.4 gal/min pump, and $\frac{3}{4}$ Hp motor. 1500-lb-capacity dumpers have a 3-in.-bore cylinder, 1.4 gal/min pump, and $\frac{3}{4}$ Hp motor.

17.3.4 High-Level Drum Dumper

What Is It?

The high-level drum dumper uses a single cylinder in back of the hopper to raise it vertically to dump height and then pivot it for dumping. The cylinder has chain-over-sprocket design to allow two inches of hopper travel for every inch of cylinder stroke. This dumper can be anchored to the floor or made portable by wheels.

What Can It Do?

This dumper provides complete control dumping into high vats, bins, hoppers, and mills. It can dump up to 1000 lb load at higher than 6 ft dump height and at maximum 45° dump angle. Data are presented in Fig. 17.3.5 and Table 17.3.3.

Table 17.3.2 Dimensions (in.) for Low-Level Drum Dumper with Dump Angle of 45° above Horizontal

Dumping Height	Horizontal Swing	Overhead Swing	Frame Length	Frame Height
36	64	100	51	66
42	64	100	51	$57\frac{1}{2}$
48	69	111	56	$63\frac{1}{2}$
54	74	121	61	$69\frac{1}{2}$
60	79	132	67	$75\frac{1}{2}$

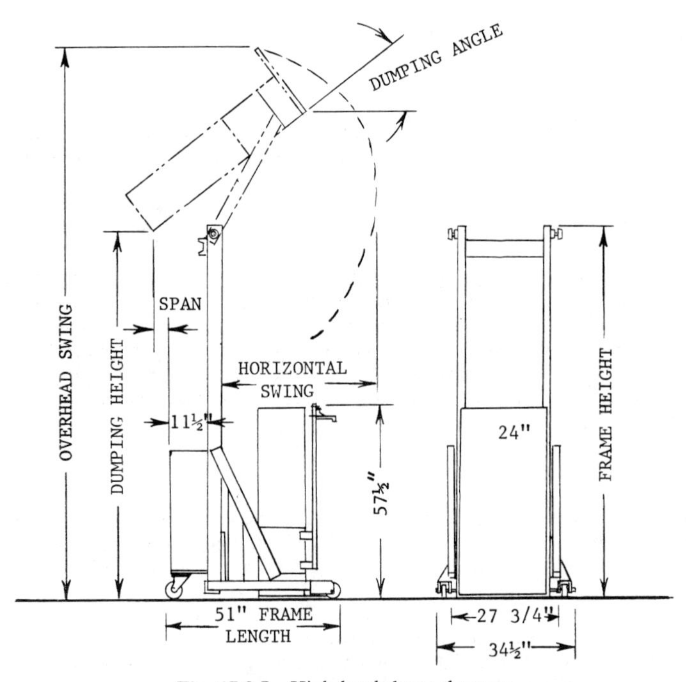

Fig. 17.3.5 High-level drum dumper.

How Does It Work?

The lifting is done by single cylinder. After the hopper is raised vertically to dump height, pivot channels, mounted one on each side of hopper, engage the pivot roll bolted on the frame. This action forces pivoting of the hopper to obtain maximum dump angle.

A 750-lb-capacity dumper uses a 3-in.-bore cylinder, and roller chain; a 1000-lb-capacity dumper uses a 4-in.-bore cylinder, and leaf chain. Units with 126 in. and over dump height have guide channels to support the cylinder rod for protection against failure in buckling.

Table 17.3.3 Dimensions (in.) for High-Level Drum Dumper with Dump Angle of 45° above Horizontal

Dumping Height	Horizontal Swing	Chute Span	Overhead Swing	Frame Height
72	42	7	129	84
78	42	7	135	90
84	42	7	141	96
90	42	7	147	102
96	42	7	153	108
102	42	7	159	114
108	42	7	165	120
114	42	7	171	126
120	42	7	177	132
126	42	7	183	138
132	42	7	189	144
138	42	7	195	150
144	42	7	201	156
150	42	7	207	162
156	42	7	213	168
162	42	7	219	174
168	42	7	225	180
174	42	7	231	186
180	42	7	237	192

17.3.5 Center-Pivoting Drum Dumper

What Is It?

The center-pivoting drum dumper grabs the drum in the center and pivots it at the center of the grab to pour the material out of the drum in front of the dumper. This unit comes in only portable type.

What Can It Do?

The dumper provides fast, clean pouring of liquids or solids. Capacities up to 1000 lb make this unit ideal for mixing, batching, and feeding. It takes all the manual work out of handling, lifting, and dumping steel or fiber drums. This dumper can handle 16–24-in. diameter drums. The dumper has the unique feature of providing variable dump height because its lift cylinder can be stopped at any point of travel and the drum can be rotated at that point. The dumper picks up the drum in front and dumps it in front also. If pouring liquid, the unit can dump from the open-top drum of liquid. Data are provided in Fig. 17.3.6.

How Does It Work?

The center-pivoting drum dumper uses one double-end cylinder for grabbing, one single-acting cylinder for lifting, and two single-acting cylinders for dumping. Hydraulic fluid is supplied to these cylinders by hydraulic pump driven by DC motor. Manual control valves are located in back within easy reach of operator.

This unit is available with a telescopic mast for use where overhead clearance is a factor and high elevation is needed. The dumper comes in two versions: (1) 24-V battery system with battery charger; fully powered for grabbing, lifting, dumping, and travel; travel speed is 1.4 mi/hr (max.). (2) 12-V battery system with built-in battery charger; powered for grabbing, lifting, and dumping.

17.3.6 Twin-Cylinder, High-Dump Units

Description: Twin-cylinder, high-dump (TCHD) units combine the heavy capacity (over 1000 lb) of twin-cylinder dumpers and the high dump heights (over 6 ft) of high-dump units (see Fig. 17.3.7).

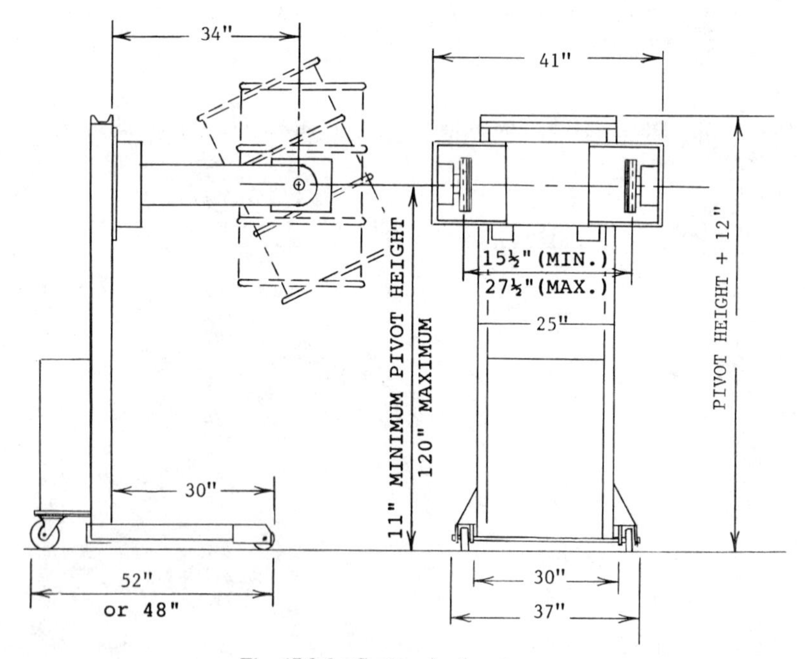

Fig. 17.3.6 Center-pivoting dumper.

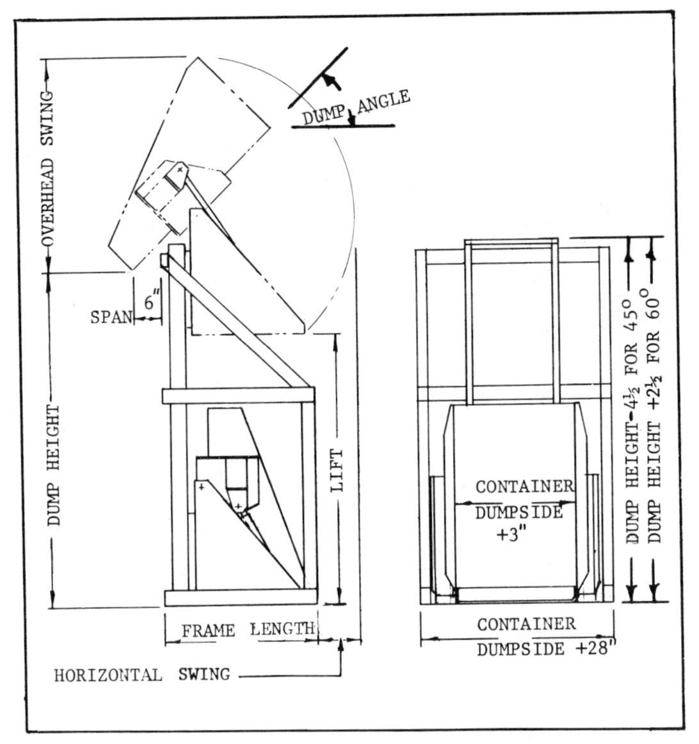

Fig. 17.3.7 Twin-cylinder, high-dump unit.

The standard twin-cylinder dumper frame is modified to accommodate rollers, chains, and hoses so that it can be lifted up in the high-dump frame. High-dump frame structure is designed to lift the TC dumper. It is recommended that this frame be tied into a nearby existing structure for extra rigidity. Load lifting is done by a single cylinder. The chain-over-sheave design gives 2 in. of lift for every 1 in. of lift cylinder travel. This concept reduces the travel time. Once the load is lifted to the proper height, the lift cylinder stops extending and two dump cylinders start rotating the load around a fixed pivot point. The two dump cylinders extend simultaneously until load is rotated to the proper dump angle above horizontal.

The tolerance on the dump angle is $\pm 2°$. The tolerance on the dump height is $\pm 1°$. The span is the clear distance between the back of the lift frame and the discharge point of chute in dump position, usually 6 in.

17.3.7 Sealed Lifting Dumper

Description: The sealed lifting dumper provides an almost perfect seal around the lip of the container to prevent powder or dust leak in the atmosphere while dumping. This helps keep the environment clean and eliminates nuisance to the operator.

Before dumping, the container is first lifted straight up about 8–10 in. into a band. The band is a truncated cone 4 in. high. Top dimension is $1\text{–}1\frac{1}{2}$ in. less than container top, and bottom dimension is $1\text{–}1\frac{1}{2}$ in. more than container top. This allows easy entry of container into sealing area. The effectiveness of the seal also depends on the condition of container top edge. A battered and uneven edge will allow dust to leak.

The band is welded to the hood. The discharge end of the hood can be square, rectangular, or round. It can be furnished with an iris, butterfly, or slide type control valve. The control valve is available with manual, air, hydraulic, or electric actuators. For a restricted opening in the receiving hopper, a spout can be furnished at the end of the discharge valve.

Figure 17.3.8 shows a drum dumping application, and Fig. 17.3.9 illustrates container dumping.

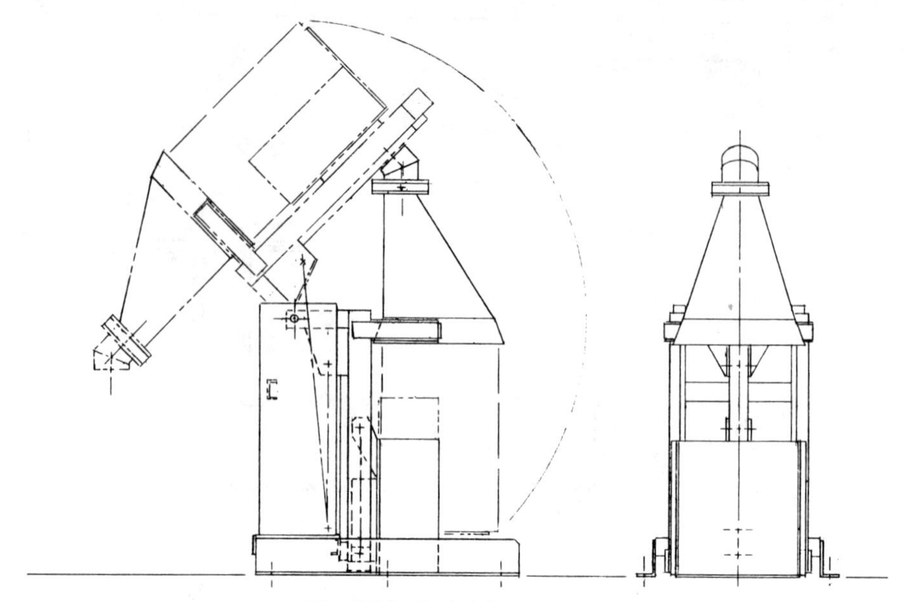

Fig. 17.3.8 Sealed drum dumper.

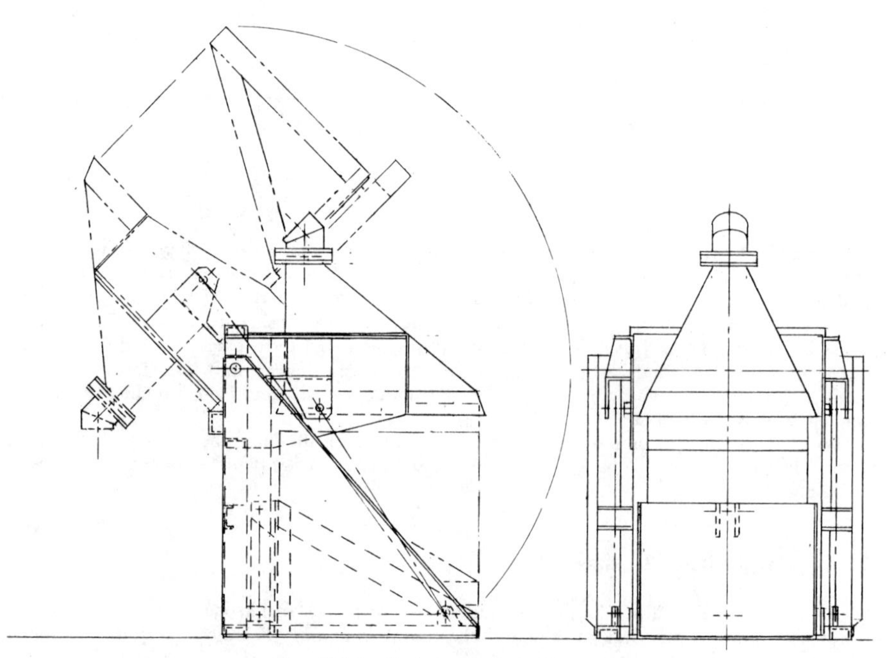

Fig. 17.3.9 Sealed container dumper.

17.4 DIE HANDLERS

Saul B. Green

17.4.1 Introduction: The Die Storage Problem

One of the most difficult industrial engineering concepts is the handling and storage of dies, jigs, fixtures, and molds. This section suggests some approaches for the designer in using current ideas to handle and store this type of item.

The complication in designing an effective storage system is caused by the wide variety of sizes of dies, the variation of their weights, and the low throughput or minimal retrieval requirements.

Another reason for complication is the need to have the dies close to the machine area, with the ability to locate, transport, and load the machinery quickly and effectively.

Some older methods of storage used marked-off areas on the floor for the placement of larger molds and dies. Another was the use also of light-gauge shelving and wood structures. Hydraulic and hand-cranked lift tables have been used to move dies from the machines and also narrow aisle trucks that were pushed manually have been used to handle and move the dies and fixtures.

Other and newer methods include crane loading racks with outrigger shelves that pull out; heavy-duty racks with bridge stringers and outriggers for crane loading; ministackers; modular drawers; and special purpose shelving for the smaller dies, fixtures, and die components.

This section outlines some of these approaches and looks at specific design examples.

17.4.2 Die Handling Designs

The designer planning practical die storage systems has to analyze die handling and usage in manufacturing. Each machine used in processing has as part of its system a die, fixture, or jig. The die is a basic part of the machining process. The die must be changed when the part that the machine is making changes. At other times the die needs maintenance or refurbishing. The speed with which the die is changed and the availability of the new die is considered in the design of the storage device. The number of times dies have to be changed and the quantities in the system affect the design of the storage system.

When the die needs to be changed, then space around the machine and between machines for movement with speed and efficiency become important. Most plants do not allow gas vehicles into the machining area for safety and other reasons. Handling is done by hoists and chain drops that are connected to jibs and booms. The die can also be moved onto a table with wheels (sometimes called a die table) or onto dollies or buggies. The die is then pushed by hand either to a staging area or to a maintenance or storage area. Sometimes the die is met by a forklift and then transported in this fashion to the storage area.

The die system has a number of components, including clamping devices and hardware. When the die is moved from the machine, all of the components usually stay with that die.

In designing a basic die handling system all of the constraints must be considered, including the process the die system will follow, the size and mass of each unit, the speed with which the die should be available, identification in storage, and control of a complete assembly.

17.4.3 Dies, the Storage Unit

For simplification, we will consider *all of the various areas covered in this section as die storage.* Die storage will include dies of all types such as blanking, piercing, progressive, compound, broaching, forming, and trimming, as well as drawing and swagging dies. Die storage will also refer to jigs and fixtures and includes clamping and indexing devices, drilling and tapping jigs of all kinds, milling, broaching, and grinding fixtures, as well as mechanical and air-operated die devices. Die storage also includes specialized tooling, press dies, and molds.

The term *die storage* also includes all variability in size and weight. The term die storage could refer to a substantial number of the same item (in various configurations) or it could mean a broad range of dies and fixtures or jigs.

With the development of hardened steels and carbides, large progressive dies such as those found in container manufacturing are simply called dies. These dies, because of detail and cost, require careful handling and specialized protected storage.

17.4.4 Die Storage Equipment

The equipment described and illustrated covers a range of storage devices for die storage. The storage devices are divided into three sections based on size. Small stamping and punching dies of up to $\frac{1}{2}$ ft^3 are considered one section; the intermediate die which runs up to about 2 ft^3 is another; and the large die and mold that is 3500 lb is also considered as a division.

Storage for Smaller Dies

Shelf Units. Utilizing heavier-gauge shelves and with the use of dividers on the shelf as clip-on labeling devices, the standard shelf (with open or closed back) can be a good storage device for small dies. The standard shelving unit depends on placing the dies on the shelf by hand. The shelf can be designed with oil drainage (should it be needed) and various materials glued to the shelf such as plastic or masonite to keep the die stationary and preventing it from slipping around when it is in storage.

Modular Drawer Systems. Another method for storing small dies is in heavy-duty modular drawer systems. A heavy-duty modular drawer has a capacity of 400–600 lb per drawer. The drawers are 4–7 ft^2 (in bottom drawer surface). One advantage of drawers is the variability in storage location. Each drawer has combs welded around the inner perimeter of the drawer. These combs divide the drawer into grid patterns using partitions and dividers. Drawer units usually have variability in height so that the designer will have a total choice in storage. Drawer units most often pull out 100% on carriages, which permits access to large numbers of small dies and die components. Modular drawers offer a system of easily identified addressable locations so that the die and its components can be stored and located quickly. Drawer units offer the options of having a closed and secure storage device close to the machine so that small components can be available as necessary. Some devices are designed to stand next to the machine and have shelves that either swing out or pull out to hold tools and dies and supplementary parts. These cabinets are designed in various sizes for easy access to the machine itself. Modular drawer systems are described and illustrated in detail in Chapter 11.

Pull-Out Shelf Racks. Pull-out shelf racks (Fig. 17.4.1) are another method of storage for small dies that are too large to be put on a standard shelving unit or into a drawer. Pull-out shelving units usually have different size shelves that can be pulled out partially in the front or rear directions, front only, or even cranked out for heavy-duty size shelves. A number of dies can be placed on the surface of each pull-out shelf, which allows full access to these dies. The pull-out shelf unit is usually under 10 ft and operates with some type of chain drop, hoist, or grappling device. Dies that are 100 lb or 200 lb can be moved on and off the shelf simply and efficiently. Various materials such as fiberboard or masonite can be attached to these shelves to assist in eliminating skidding because of oil-slick surfaces. Pull-out shelf racks sometimes have rims around the perimeter to hold small dies from toppling over the edge of the shelf.

The Intermediate Sizes

Standard Beam. Storage of medium-size dies is commonly managed by using select storage or beam racks. Beams are attached between a pair of uprights (with occasional front-to-back connectors) and wooden pallets are used to hold either an individual die or a number of dies. The loaded pallet is moved to and from the machines and other areas with a fork truck or some device.

Movable Shelf Rack. The movable shelf rack (Fig. 17.4.2) or shelf pallet rack (Fig. 17.4.3) utilizes a set of uprights, a fixed top shelf, together with a back brace, and is secured to the floor by lagging.

Fig. 17.4.1 Example of pull-out shelf rack. Courtesy Rack Engineering Co.

Fig. 17.4.2 Movable pull-out shelf rack. Courtesy of Rack Engineering Co.

Fig. 17.4.3 Movable pallet shelf rack. Courtesy of Rack Engineering Co.

The shelf connects to the four posts of the uprights by various connecting methods. The shelf is the pallet and is moved, together with its load, from one place to another. Movable shelves can accommodate a number of dies on a shelf or a single die on a shelf. The movable shelf rack system can utilize a die table that holds the whole movable shelf with loads up to 2500 lb and can use a fork system similar to those used in small electric trucks or hand-crank trucks. The movable shelf rack offers variations in shelf design with lips on the perimeter or grid configurations designed as part of the shelf. Shelves in this system have pallet legs and/or stirrups for stability. Movable shelf racks also work with ledges on the uprights. The ledges support the pallet and are adjustable.

Heavy-Duty Pull-Out Shelf Units. Pull-out or crank-out shelf systems are designed for the medium-size dies. These racks have capacities of approximately 1000 or 2000 lb per shelf for manual operation and up to 5000 lb for hand cranking. Chain drops or hoists are designed to move the die off the shelf as opposed to mechanical equipment such as fork trucks.

Ministacker Systems. The ministacker system (Fig. 17.4.4) is another form of the movable shelf rack concept. Instead of a single line of movable shelf sections, the stacker system has two rows of sections facing each other. On top of the row is a light-gauge rail. A bridge system runs on the rails. Inside the bridge is a trolly and usually the trolly has a large thrust bearing design to hold a mast. The bearing permits the mast to rotate a full 360°. The mast carries an apron that runs up and down the mast which is connected to a chain hoist device. Hanging on the apron are a set of forks. The ministacker system allows complete access with a narrow aisle that permits access to all of the shelves within the system. The whole stacker system can be placed near the machine area where the

Fig. 17.4.4 Ministacker system. Courtesy of Rack Engineering Co.

dies are used. The operator will have full and speedy access to the dies on an either random-addressing or fixed-position basis in the system. Ministackers handle dies ranging between 1000 and 3000 lb. Where there are large numbers of dies there are long rows of sections of racks. Double-aisle systems are available with the bridge powered as well as the trolly powered. Ministacker systems have a person walking along holding the mast and pacing the system on his walk-along.

Storage and Retrieval Systems. As the size of the storage problem becomes larger in numbers of units, there is a need for a storage system that is larger than a ministacker system. Most ministacker systems are only 10 ft high because of the problem of not being able to see pallets over that height. When there are a great number of pallet loads and utilization of the height is required, then a man rides aboard a cabin attached to the mast and travels with the forks (Fig. 17.4.5). There is no limit to the height of the system and the density that pallet loads can be stored. Systems that are designed with automatic stopping levels or bar code readers are automated and can be used to bring down a pallet load of material (in this case dies) without an operator.

Large and Heavy-Duty Die Storage

Movable Shelf Racks for Storage Capacities Up to 5 Tons. Movable shelf racks have a variety of capacities. Extremely large dies and molds can be stored on movable shelf racks based by sliding the dies and molds into the rack utilizing standard fork truck equipment. The only modification in movable shelf rack is the addition of front-to-back runners on the shelf to allow sliding the die onto the shelf for some handling device.

Fig. 17.4.5 Example of man-aboard stacker. Courtesy of Rack Engineering Co.

Dolly or Wheeled Shelves. A practical solution to the heavier die system storage is the use of a wheeled shelf or dolly rack (Fig. 17.4.6). The dolly rack utilizes a bin or tub device that is set on four V-grooved wheels. These V-grooved wheels run on a set of runners from the front to the back of the rack section. The dolly itself has capacities of up to 8000 lb. A row of heavy-duty sections can face a row of heavy-duty sections. A bridge stringer is set between the two rows of rack. This bridge stringer acts as a runner for the dolly rack and the dolly is pulled out by hand to the center of the aisle from the shelves. The dolly itself has four lugs on the corners that permit pickup of the dolly by a chain drop (hoist) or a crane. The die is moved to the machining area. Many times the dolly itself can be placed on the floor and moved. In situations where the storage requirement requires only a single row of racks, outriggers hung on the rack face can be used to pull the dolly onto it and then pick it off with cranes. These designs are for the heaviest-duty molds and dies.

Heavy-Duty Cantilever Rack. It can be built with shelves and used for storage of heavy dies and molds. The advantage is that there is a continuous running shelf surface in the rack. A set of arms is designed and then a shelf is secured to the arms which permits a number of dies or molds to be placed on the continuous shelf.

Heavy-Duty Bolted or Welded Shelves. For the ultra heavy dies, such as the progressive dies over 5 tons, a heavy-duty, welded or bolted rack structure is designed utilizing high-beam or heavy-duty channel front-to-back runners so that the die can be placed on the system. Normally, these are no more than two high utilizing the floor for the first level.

Die Movement and Handling depends on the size of the installation. Dies and components are generally code marked and sets maintained as part of an assembly. These assemblies are moved from location to location through maintenance, through staging, from storage, to the machine tool in sets, and identified by some identification device. The sets or components develop an addressable location in the storage system designated whether it is modular drawers, heavy-duty shelving, pull-out shelving, ministackers, or AS/RS.

A computer program can be designed to solve random storage (i.e., first in/first out) or specific storage (a specific home location) for each tool and die system.

Software, computer programming, and signaling devices are available for location of die storage components.

Fig. 17.4.6 Heavy-duty dolly rack. Courtesy of Rack Engineering Co.

17.5 AIR FILM DEVICES

Ben Bayer

17.5.1 Principles of Air Film Devices

Evolution

At the outset, a distinction should be made between air film devices and hovercraft or surface effect vehicles. Although both technologies are based on $F = P \times A$ (Force = Pressure × Area), commercial hovercraft operate at relatively low pressures, 1 psig or less (6.9 kPa), and high volume, 10^4 ft^3/min (10^4 liter/sec). Air film devices operate with pressures as high as 50 psig (345 kPa) but at 1/1000 the volume. The operating characteristics, power requirements, and surface conditions are widely different for the two technologies.

Air film technology had its genesis in the late 1950s as General Motors Corporation sought to advance the well known theory of the air bearing. For years heavy loads had been moved by forcing air between two machined surfaces, notably in the die-making industry. This method of movement had two significant prerequisites: both operating surfaces required a very smooth finish and both had to be very flat (Fig. 17.5.1). Two of the goals that General Motors had were:

1. Develop a unitized product on which a load could be placed, moved, and then unloaded.
2. Develop a product that could be utilized on surfaces like factory floors rather than on a machined surface.

The results of the General Motors efforts have come to be known as today's air bearing technology, although it should properly be called compliant air bearing technology.

Within the last 10 years, one manufacturer has advanced the state of the art to utilize water and machine cutting fluids as the film medium. In the movement of heavy loads (thus far up to 4500 tons) the cost of pumping water is about $\frac{1}{8}$ the cost of compressing air. Machine base plates, designed as mobile fixtures, utilize the cutting fluids for flushing the machining chips as well as for operating the bearing. This innovation has led to yet another, more definitive name: compliant fluid film technology.

For the balance of this section air is to be considered as the fluid used. Although the principles apply to other fluids as well, incompressibility becomes a major factor in design and control and the manufacturer should be consulted for planning assistance.

Description

The analogy of a short-stroke air cylinder serves in describing the operation of the air bearing. The compliant diaphragm forms the cylinder wall. When pressurized air is introduced, the diaphragm, once sealed against the operating surface, inflates and lifts the load just as if a rigid cylinder were beginning to stroke. As the internal pressure increases, the air bearing continues to inflate and lift. At some time, the increasing air pressure reaches the point where the pressure times the effective lift area equals the downward force of the load ($F = P \times A$). At this point, the lower surface of the diaphragm breaks contact with the operating surface and a thin escape path is formed around the entire periphery. The exhaust of the air through this thin path creates a dynamic "film," and this film virtually eliminates friction.

Now the cylinder is no longer a simple closed system. It is a dynamic system: an open-ended cylinder with a bubble of compressed air captured within a compliant diaphragm and with a controlled exhaust. So long as air is supplied to the system at or above the operating pressure and at a sufficient volume to maintain the flow through the thin pathway, the load will float free of the friction we

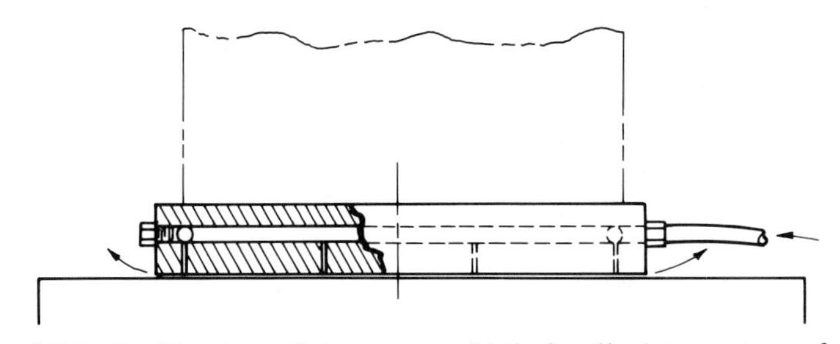

Fig. 17.5.1 Traditional use of air to create a frictionless film between two surfaces.

normally expect. Omnidirectional movement with significantly lower forces, often manpower alone, is now possible so long as the dynamic bubble is maintained.

In actual application, the friction factor is reduced to as low as 0.001 (dimensionless).

Construction

The design of the General Motors compliant air bearing consists of a molded elastomer for the flexible diaphragm which is attached around the circumference and in the center to a top plate. Air is introduced through the top plate into the interior of the diaphragm. "Communicating" holes are spaced around a smaller diameter, through the diaphragm material. A cross section of this design in operation is shown in Fig. 17.5.2.

In the 1960s, the Boeing Company undertook a development program to improve upon the GM design. Among the objectives were:

1. Manufacturing methods that would permit the use of more rugged materials.
2. An increase in effective lift height to provide better compliance to irregular surfaces.
3. A design that would allow higher operating pressures for greater capacities in limited spaces.
4. Dynamic damping provisions to provide greater stability in the case of air pressure or load changes.

The advanced design, patented by Boeing, is called an Aero-Caster (Fig. 17.5.3). It eliminates the necessity of molding the diaphragm shape. Two pieces from flat stock of high-strength, fabric-reinforced materials are vulcanized into a full torus and then vulcanized to a backing plate. In this design, air is introduced through the plate but then divides, feeding both the flexible torus and the plenum or bubble. The principles of operation are the same for both designs, but the Aero-Caster's inflated height, or lift, is twice that of the GM design. Because the ability to conform to undulating surfaces is a function of the lift, this design is more compatible with industrial floor surfaces.

The materials and fabrication methods permit operational ratings as high as 50 psig (345 kPa). The torus, designed with internal orifices, provides automatic damping features without the necessity for external chambers.

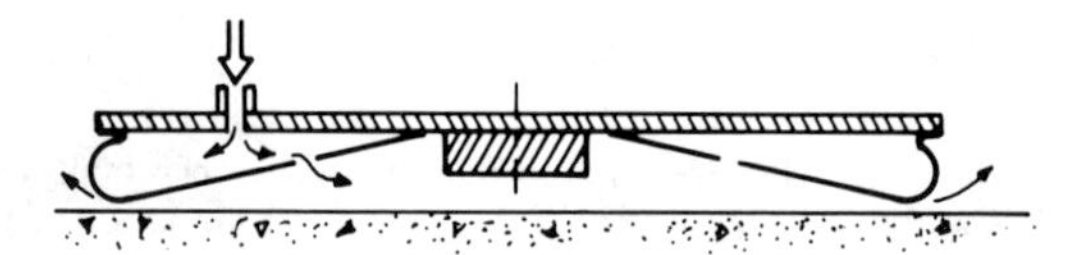

Fig. 17.5.2 Floating air bearing design.

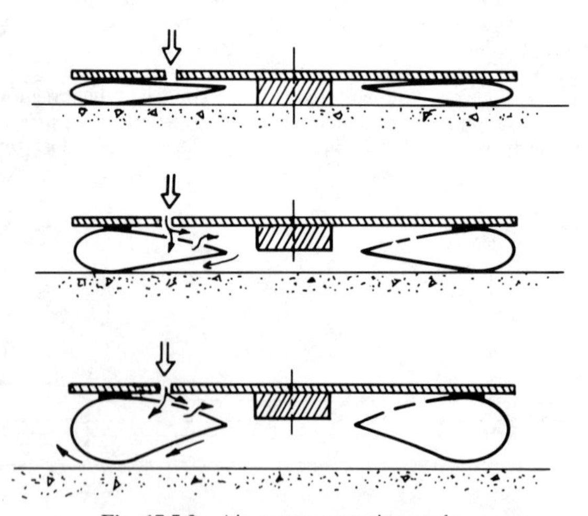

Fig. 17.5.3 Air caster operating cycle.

Configuration

No load is too heavy to move on air bearings provided enough support area is available. Quantities of more than 100 of the 48-in. bearings have been used at one time in the shipbuilding industry. However, there are a minimum number that can be used from a stability consideration.

Placing a load on top of a single air bearing is similar to trying to balance on top of a basketball. Unless other devices are used, a minimum of three air bearings are required and in most industrial applications four or more are used. Because of differing lift heights and for the convenience of interchangeability, the air bearings should be of the same size.

For successful operation, the center of gravity (c.g.) of the load must be within the envelope of lines connecting the centers of the air bearings (Fig. 17.5.4*A*). With proper air control the c.g. does not have to be at the exact center of the envelope but care must be taken to avoid overloading individual air bearings (Fig. 17.5.4*B*). Loads with the c.g. greatly off center can be handled by assymetrical location of the air bearings (Fig. 17.5.4*C*).

Lifting capacity is directly dependent on the area, and air requirements are more dependent on the associated total perimeter. For this reason, as well as for simplicity of controls, lower operating pressures, and conformity to the surface, it is best to use the fewest number of larger air bearings than to use many small ones.

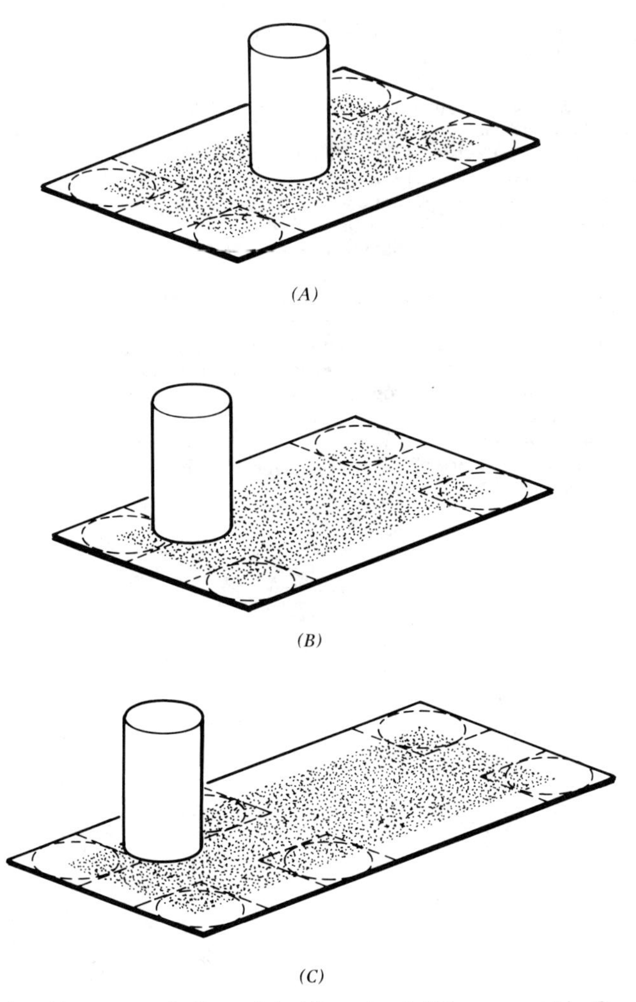

Fig. 17.5.4 Load positions on an air film pallet. (*A*), centered. (*B*), unacceptable. One air bearing is overloaded. (*C*), assymetrical location of air bearings to compensate for eccentric loading.

17.5.2 Air Supply

The required air supply is determined by the size and number of air bearings, the weight of the load, and the type and condition of the operating surface. Since all commercially available air bearings are designed to operate at 50 psig (345 kPa) or less, plant air supply is frequently used. In the event that existing operations require near-capacity output of the plant supply, stationary receiver tanks can be installed which are charged to the volume desired during slack periods and then used only when actually moving a load on air bearings.

When determining the capacity of a plant compressor, the manufacturer's nameplate should contain the volume rating in "CFM" or "l/s." If this information is missing, a rule of thumb is that SCFM equals 4 to 5 times the rated horsepower. (S l/s are approximately 2 times rated horsepower).

Nonrecurring moves, such as encountered in rigging and maintenance, often are accomplished with the use of rented, portable compressors.

Facilities without a permanent air supply installation, such as theaters, school gymnasiums, or warehouses, make use of lower-pressure sources such as rotating-lobe or vane-type blowers. These devices furnish a good volume of air but are limited to a maximum of 10–15 psig (69–103 kPa) operating pressure. These units can be electrically powered or gas or LP engine driven.

Movement of relatively light loads over short distances can be accomplished easily using compressed air or gas tanks available from commercial welding distributers.

Distribution

The supply of a proper volume is of more concern than any specific pressure. If, for some reason, the pressure is limited, larger air bearings can be used, thus increasing the effective lift area. But no compensation can be made for a lack of adequate volume. Therefore, a well-designed air distribution system will have large diameter, smooth inside surface pipe, and/or hose with no restrictive fittings.

Most plant installations have main headers in the nominal 4–6 in. range (11.43–16.83 cm O.D.). These are usually more than adequate for an air bearing system. However, in providing for the use of air tools, most feed lines or air drops are $\frac{3}{8}$–$\frac{1}{2}$ in. (1.72–2.13 cm O.D.) and quite often incorporate a filter and an oiler. If this is true, separate air drops should be provided for air bearing operations. The size depends on the system requirements, with 1 in. (2.67 cm O.D.) or larger not uncommon. Filters and oilers should be omitted. A shutoff valve at the end of the drop provides safety and operating convenience.

Shutoff valves should be of the ball or gate type (Fig. 17.5.5). Globe valves, commonly used for water, should be avoided as they are severely restrictive.

All fittings and accessories should be the same size as the pipe or hose and be of the flow-through type. Quick disconnects are often used and, when safety regulations specify the automatic shutoff type, they must be oversized to allow proper volume flow.

The final link to the air bearing system is the supply hose. It should be of a length that permits the move to be operationally practical yet not so long as to be inconvenient to handle. If adequate air drops are provided at proper spacing, a long move can be made with a minimum number of interruptions to reconnect the supply hose.

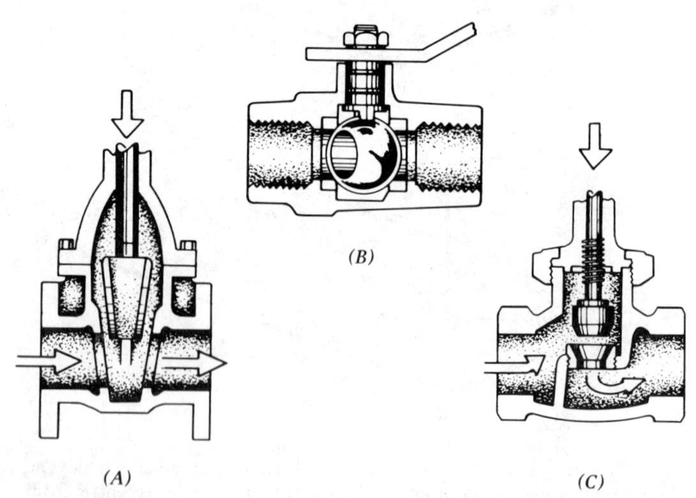

Fig. 17.5.5 Types of valves. (*A*), gate; (*B*), ball; (*C*), globe (unacceptable).

Self-rewinding hose reels can be ceiling or column mounted or can be an integral part of the air bearing system.

Overhead, tracked distribution systems are available but are generally acceptable only for small (low air requirement) systems. Care must be taken to insure the availability of sufficient volume.

Regulation

Since it is rare that the weight distribution of a load will be equally shared by each air bearing, some method must be provided to adjust the air pressure/flow to the individual units. Referring to Fig. 17.5.4*B*, if no provision for regulation were made, the air would escape through the two air bearings on the right because they are so lightly loaded that the exhaust point would be reached before the upper left one had even lifted.

In addition, the conditions of the operating surface may change and require compensation by an adjustment in the air supplied to the air bearing.

The most versatile system is one with positive regulation to each air bearing. This allows "tuning" the system so that the pressure to each air bearing is in direct relation to the portion of the load that the individual air bearing "sees" at any given time. It also allows adjustments as any one (or more) air bearing(s) encounters a change in the operating surface conditions.

Incorporation of a pressure regulator for each air bearing within the distribution system provides this versatility. Regulators with good flow characteristics over a wide range of pressure settings should be used. A gauge at each regulator will provide monitoring capability.

Quite often, fixed path movement of known loads can utilize flow controls rather than pressure regulators. For instance, on an assembly line, the use of automatic flow controls can provide the regulation needed for changes in weight or c.g. without any manual adjustments. These devices have a flexible orifice which responds to a pressure drop and when properly selected, provide automatic compensation for changing demands over a wide range of the air bearing's capacity.

Since the operation of automatic flow control devices is based on the pressure drop across an orifice, they are ineffective with low-pressure sources and should not be used in such applications.

Adjustable flow controls allow for the adjustment of the diameter of a fixed orifice, thus providing variable flow control.

Needle valves are also used, permitting very fine control of air flow within the range of their fixed, nonadjustable flow capacity.

Use of flow controls is not recommended for rigging or maintenance type moves. With uncontrolled operating conditions and the wide variety of sizes, weights, and c.g. locations, positive pressure control systems are best. Since flow controls depend on a restrictive orifice their use could be detrimental.

Also, when using flow controls, a single pressure regulator should be included in the supply line in order to set the overall system pressure.

Consumption

The volume of air required to operate a compliant air bearing is dependent on its effective perimeter, the load it is carrying, and, to the greatest extent, the surface on which it is operating.

On a perfect surface (flat, smooth, and nonporous) the minimum amount of air is required for a given load on a given size air bearing. This is a function of the perimeter of the effective lift area and size gap through which the air film flows (approximately 0.003–0.005 in., or 0.08–0.13 mm).

As the load increases, both the pressure and flow increase until the maximum rating of the air bearing is reached. Forcing additional air into the air bearing will not improve performance or enlarge the exhaust gap but can introduce a "hop" or "bounce" to the system. The most productive pressure and flow values are those necessary for flotation.

Volume is rated in SCFM (Standard cubic feet per minute)* or S l/s (Standard liter per second)† which is the standard usually used in compressed air technology.

Required volumes for air bearings have been converted to this standard and are therefore compatible with the compressor industry.

17.5.3 Surface Characteristics

Roughness and Porosity

The quality and condition of the operating surface will have the greatest effect on air consumption. Since the thickness of the air film is only 0.003–0.005 in. (0.08–0.13 mm), any operating surface

---

* SCFM is the volume at 14.696 psi absolute, 60°F and dry air.

† S l/s is the volume at 101.33 kPa absolute, 15.6°C and dry air.

Table 17.5.1 Operating Surface Index Numbers

Surface Description	Index Number
Polished plate glass	1
Smooth, steel-troweled, sealed concrete	2
Smooth Epoxy coating	1–2
Vinyl tile (well laid)	1.5–2
Galvanized sheet	1.5–2
Hot-rolled sheet or plate	1–2
Varnished hardwood	1–2
Hand-troweled concrete with visible troweling marks	3–5
Smooth troweled concrete (not sealed)	2–4
Worn concrete, smooth surface (not recommended)	3–4
Brush-finished concrete (sidewalk) (unacceptable)	7–10
Asphalt (unacceptable)	10–15

must be smooth for proper operation. Within the compliant fluid film industry, polished plate glass is used as the standard reference and is designated as Index No. 1. This is the surface that requires the least amount of air for flotation of a given load. All other surfaces are referenced to this standard (see Table 17.5.1). The index number is determined by the additional air required for the same load with the same friction to float as it would on an Index No. 1 surface.

Thus, smooth, steel-troweled, sealed concrete is rated Index No. 2 because it requires twice the volume of air to float a load as it does on Index No. 1.

The index number reflects the quality of both the surface smoothness and its porosity. In cross section, a sidewalk looks like mountains and valleys when viewed on the same scale as the air film. Abrasion is high, not only from the peaks but also because the film of air is imperfect and there is physical contact between the air bearing and the surface. Grinding or polishing improves the texture.

Concrete is porous, thus permitting air to flow through it rather than establishing the air film. This loss must be made up by increasing the volume. In some cases, the loss is so great that sufficient air cannot be supplied. All concrete surfaces should be treated with a penetrating sealer. A top coat of epoxy-type material will both smooth and seal a concrete surface.

Planarity

The development of the compliant air bearing was aimed toward a product that could be used on factory surfaces that were installed under existing specifications of the construction industry. In actual application, requirements for surface finish and planarity are within the normal construction specifications covered by the American Concrete Institute Standards 301 and 302. (The American Concrete Institute Standards are referenced only for finish quality and planarity. Other considerations such as set rate, slump, shrinkage, etc., are not relevant to an air bearing discussion.)

Licensed contractors routinely comply with these standards which call for a steel-troweled finish with a tolerance of $\frac{1}{4}$ in. in 10 ft (6.35 mm in 3.05 m). This is equivalent to a newly installed warehouse floor.

In actual operation, an air bearing can conform to undulations (convex or concave) of approximately one-fourth of their lift height across the effective lift diameter. For example, a 21-in. air bearing will traverse a crown with a rise of about $\frac{9}{32}$ in. (7.14 mm) within about 19 in. (483 mm).

Levelness is a concern in that with a friction factor of only 0.001, a slope as low as 0.1% will cause drift and may require restraints or guidance.

The orientation of the air bearing has no effect on its operation. Air bearings have been used vertically in guidance systems and inverted for handling steel plate, containers, and so on.

Discontinuities

The operating surface should be continuous, with no steps or other interruptions through which air losses can occur. Hairline cracks cause no problem but larger cracks and joints should be filled smooth and flush with the adjoining surfaces. Two part mixes of urethane or epoxy-based compounds are commercially available and are satisfactory in filling cracks or joints while still providing for expansion and contraction.

Steps or projections should be ground to 1:10 to 1:20 and the ground area sealed. The resultant height differential should not exceed one-fourth of the lift height of the air bearing.

Drains, railroad tracks, utility troughs, and so on, must be avoided or brought up to the smoothness, nonporosity, and continuity necessary for proper operation (see following discussion of overlays).

Changes of elevation serviced by ramps require that the areas of transition leading onto and off of the ramp be rounded to present a more gradual, continuous operating path. Care should be taken to avoid "high centering" a load at the top of a ramp.

Overlays

Quite often part or all of the proposed travel path may be unacceptable for air film operation. Older or deteriorated floors, wood block floors, rough and porous shipping docks, doorway thresholds, or chipped and cracked surfaces are frequently encountered. Upgrading or resurfacing may be impractical or uneconomical, especially if the move is a one-time or infrequent occurrence.

Temporary overlays of a suitable material are used in order to provide a smooth, nonporous, and continuous surface. If the existing floor has the load-bearing strength required, relatively thin material can be used such as 22-gauge sheet metal (0.76 mm), nonembossed vinyl linoleum, or 10-mil (0.25 mm) plastic. The roughness of the surface must be considered because thin overlay materials will "telegraph" the underlying texture.

For rougher surfaces, low load-bearing capacities, or open gaps (drains, tracks, etc.), thicker supporting overlay materials must be used. $\frac{1}{4}$-in. (6.35 mm) steel plate, butted, welded, and ground, has the strength for many of these conditions. Supplementary cribbing may be necessary for proper load distribution.

Gaps at railroad tracks can be filled with plaster of paris, wet sand, or any load-bearing material and then overlayed with thin material.

When using thin metal, shingle the sheets in the direction of the move. Tape the leading edge to facilitate a transition onto the overlay.

Thicker sheets of metal that are overlapped should have the edges beveled to 1:10 to 1:20.

17.5.4 Sizes and Capacities

Tables

Data for a wide range of air bearings are presented in Table 17.5.2 (in English units) and Table 17.5.3 (in metric units).

Table 17.5.2 Engineering Data for Air-Film Equipment (English Units)

Capacity[a] (lb)	Pressure[b] (psig)	Caster Diameter[c] (in.)	Lift Area[d] (in.^{2})	Effective Lift[e] (in.)	Air Flow[f] (ft^3/min)			
					$\frac{1}{4}$ Load	$\frac{1}{2}$	$\frac{3}{4}$	Full Load
500	12.5	8	40	$\frac{3}{8}$	2	3	4	6
2,000	25.0	12	80	$\frac{3}{4}$	8	9	11	12
3,500	25.0	15	140	$\frac{7}{8}$	8	10	12	14
7,000	25.0	21	280	$1\frac{1}{8}$	14	14	15	16
12,000	25.0	27	480	$1\frac{3}{8}$	16	16	17	18
20,000	25.0	36	800	$1\frac{3}{4}$	17	19	22	24
40,000	25.0	48	1600	$2\frac{5}{8}$	18	21	23	26
14,000	50.0	21	280	$1\frac{1}{4}$	18	22	26	30
24,000	50.0	27	480	$1\frac{1}{2}$	20	25	32	40
40,000	50.0	36	800	$1\frac{7}{8}$	22	28	37	45
80,000	50.0	48	1600	3	40	43	46	50

[a] Maximum load at maximum rated pressure. Minimum loads are normally 5% for normal-duty bearings and 10% for heavy-duty bearings. Consult the factory for critical low-load applications.

[b] Maximum rated pressure.

[c] Nominal outside diameter of air caster.

[d] Effective pressurized lift area of air caster.

[e] Nominal effective lift at maximum load and minimum air flow. Lift height measured from bottom of landing pad to floor surface. (Lift increases with increased air flow.)

[f] Volume of air flow required for free-floating performance on Index No. 2 or better surface (smooth, hand- or steel-troweled surface, sealed concrete or equal).

Table 17.5.3 Engineering Data for Air-Film Equipment (Metric Units)

Capacity[a] (kg)	Pressure[b,c] (kPa)	Caster Diameter[d] (cm)	Lift Area[e] (cm^2)	Effective Lift[f] (cm)	Air Flow[g] (liter/sec)			
					$\frac{1}{4}$ Load	$\frac{1}{2}$	$\frac{3}{4}$	Full Load
227	86.3	20.3	258	1.0	0.9	1.4	1.9	2.8
907	172.4	30.5	516	2.0	3.8	4.2	5.2	5.7
1,588	172.4	38.1	903	2.3	3.8	4.7	5.7	6.6
3,175	172.4	53.3	1,805	2.8	6.6	6.6	7.1	7.6
5,443	172.4	68.6	3,094	3.6	7.6	7.6	8.0	8.5
9,072	172.4	91.4	5,157	4.3	8.0	9.0	10.4	11.3
18,144	172.4	121.9	10,314	6.6	8.5	9.9	10.9	12.3
6,350	344.8	53.3	1,805	3.1	8.5	10.4	12.3	14.2
10,886	344.8	68.6	3,094	3.8	9.4	11.8	15.1	18.9
18,144	344.8	91.4	5,157	4.6	10.4	13.2	17.5	21.2
36,288	344.8	121.9	10,314	7.4	18.9	20.3	21.7	23.6

[a] Maximum load at maximum rated pressure. Minimum loads are normally 5% for normal-duty bearings and 10% for heavy-duty bearings. Consult the factory for critical low-load applications.

[b] Maximum rated pressure.

[c] $kPa \times 1.02 \times 10^{-2} = kg/cm^2$.

[d] Nominal outside diameter of air caster.

[e] Effective pressurized lift area of air caster.

[f] Nominal effective lift at maximum load and minimum air flow. Lift height measured from bottom of landing pad to floor surface. (Lift increases with increased air flow.)

[g] Volume of air flow required for free-floating performance on Index No. 2 or better surface (smooth, hand- or steel-troweled surface, sealed concrete or equal).

Use of Tables

Since the compliant air film technology is based on $P \times A = F$, the tables can be entered with the known load and either the operating pressure or the effective lift area. The data listed are for each individual air bearing.

Example:

Given a total load of 20,000 lb and an operating pressure of 20 psig, four air bearings of 250 in.^{2} lift area each will support the load.

$$\frac{20{,}000 \text{ lb}}{20 \text{ lb/in.}^2} = 1000 \text{ in.}^2 \text{ total lift area required}$$

For four air bearings, 250 in.^{2} lift area required for each bearing.

The 7000-lb-capacity unit has 280 in.^{2} lift area, so four of these units will be suitable. The four air bearings will move 20,000 lb with an operating pressure of about 18 psig.

$$280 \text{ in.}^2 \text{ lift area} \times 4 = 1120 \text{ in.}^2 \text{ total lift area}$$

$$\frac{20{,}000 \text{ lb}}{1120 \text{ in.}^2} = 17.86 \text{ psig}$$

This is about $\frac{3}{4}$ load:

$$\frac{18 \text{ psig operating}}{25 \text{ psig capacity}} = 0.72$$

Using the air flow data, 15 ft^3/min will be required for a $\frac{3}{4}$ load for each air bearing.

$$15 \text{ ft}^3\text{/min} \times 4 \text{ air bearings} = 60 \text{ ft}^3\text{/min required}$$

As noted in the table footnote, this volume is needed for operation on an Index No. 2 surface. Referring to Table 17.5.1, operation on an epoxy surface could require half as much, whereas an unsealed, smooth concrete surface could require twice as much.

17.5.5 Selection Charts

Many of the factors to be considered in selecting an air bearing system are the same as those considered for wheeled vehicles. Weight and size of the load, surface quality, maneuverability, overhead and side clearances, floor loading limitations, frequency of movement, and the operating environment all must be reviewed.

The starting point is the weight of the load. Figure 17.5.6 is an example of a preliminary selection chart incorporating the parameters for using *four* air bearings. Entering from the bottom with the total load weight, a vertical line intersects one or more of the dashed curves representing the capacity of an individual air bearing. A horizontal line drawn from this intersection to the right will give an indication of the operating pressure. A horizontal line to the left, drawn from the intersection of the load line and the heavy solid line, will give an indication of the volume requirement for four air bearings on an Index No. 2 surface.

Thus, a 24,000-lb load could be handled by three different size air bearings (Fig. 17.5.6*A*):

Four 36N using approximately 7.5 psig
Four 27N using approximately 12.5 psig
Four 21N using approximately 22 psig

Any of these systems will use about 65 ft^3/min on an Index No. 2 surface.

More precise figures will be obtained from Tables 17.5.2 or 17.5.3.

The choice of which size to use is determined by the physical dimensions of the load, anticipated future moves of heavier equipment and the desired operating pressure. Surface conditions will affect the operating requirements so that the larger-size air bearing should be selected if it is known that higher than the volume shown on the chart will be needed for unsealed, rough, or unusual surfaces.

Form or Arrangement

Once the size of the air bearing has been determined, the configuration of the system can be considered. As mentioned earlier, individual air bearings with separate air controls provide the greatest versatility in a system. Manufactured products that vary in one dimension can be moved by use of planks which can be located at any distance apart, relative to each other. If successive loads are basically similar, within a defined size and weight range, a pallet may offer significant operating advantages. A single-

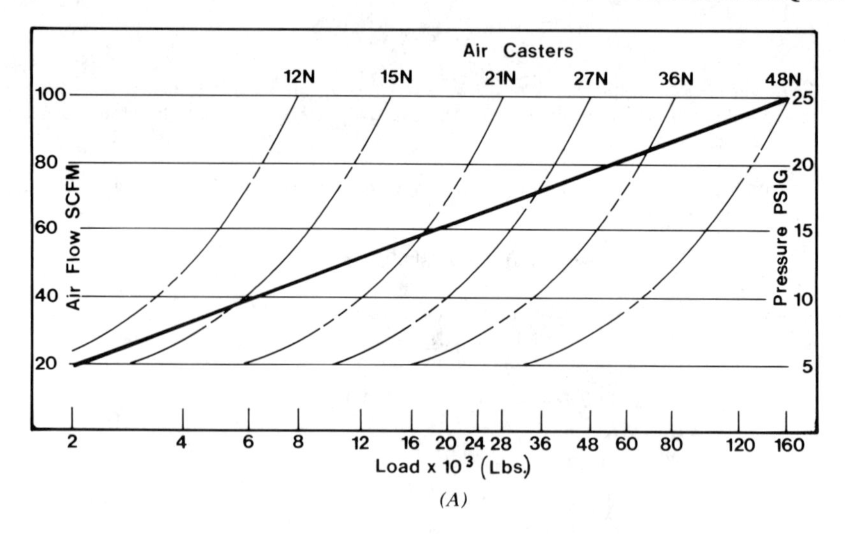

(A)

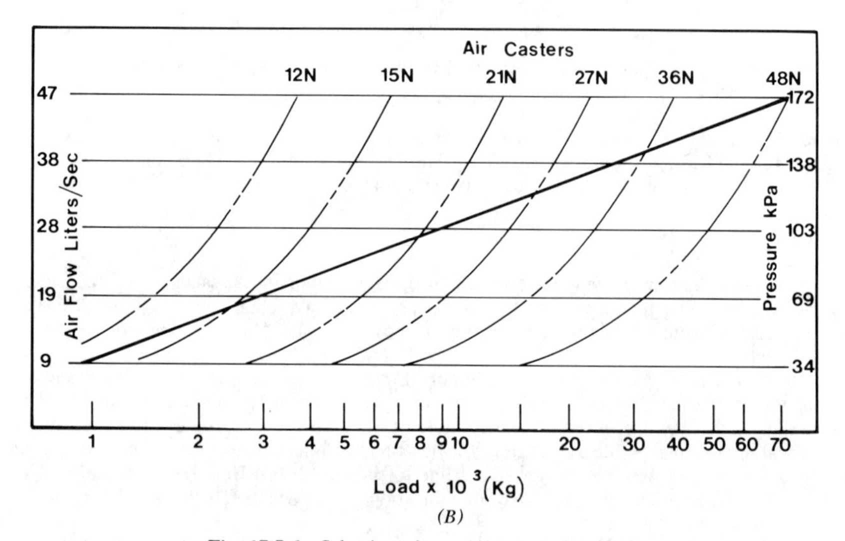

(B)

Fig. 17.5.6 Selection chart. (*A*), English; (*B*), metric.

piece pallet, with integral automatic or adjustable air controls, will replace the several air bearings, hoses, and regulation system. In many cases, the pallet is designed for the load to remain on top permanently as in the case of a portable scissor-lift system. Utilizing the high lift of the air caster, skids can be designed that allow clearance for the air pallet to be inserted, moved on air, and set back down on the skid runners when the air is turned off.

Any air bearing apparatus can be transformed into a turntable by establishing a fixed pivot point. An air-loaded, spring-return pressure foot is one method of converting a pallet into a turntable for rotation around a fixed point. In the released position, the pallet is once again free floating in any direction.

The air bearing system can incorporate a wide variety of accessories: power drives, guide wheels, safety devices, locating or guidance mechanisms, custom fixtures, pilot-operated remote controls, automatic sequencing, and interlocks can be provided.

Figure 17.5.7 shows some of the systems and accessories that are available. These should not be considered as the only possibilities. The fluid film technology lends itself to intermarriage with many other technologies and methods.

The separate bearings and planks, shown in the upper row, depend on the structural rigidity of

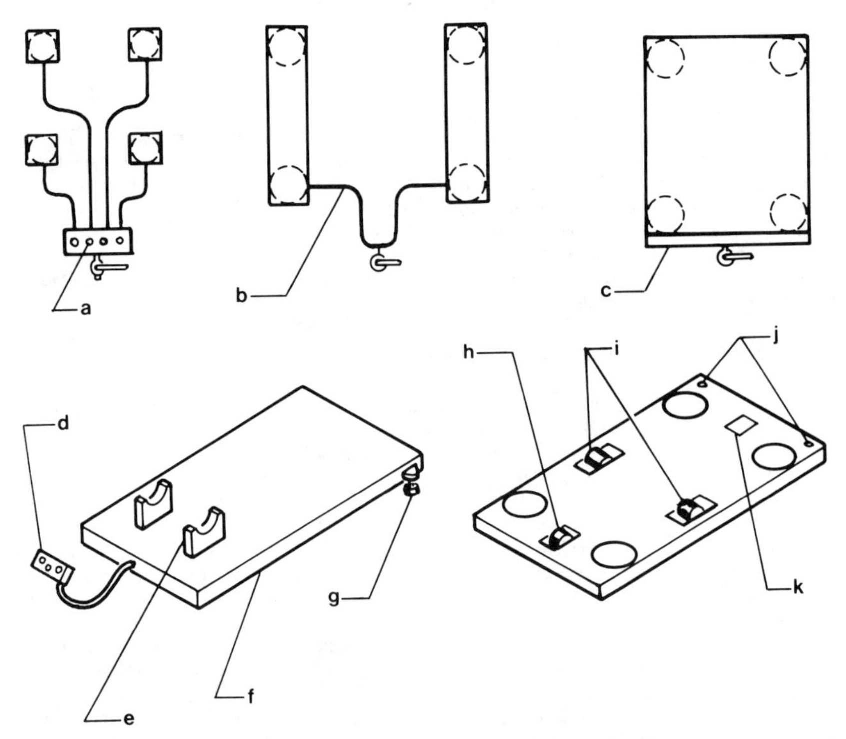

Fig. 17.5.7 Variety of fluid film devices. a, separate air bearings with individual regulation. b, planks. Flexible hose allows variable spacing. c, air pallet with automatic regulation integral with the structure. d, remote, pilot-operated controls. e, special fixtures. Can be removable. f, structural steel pallet. g, roller guide. h, air-operated drive wheel. i, air-operated guide wheels. j, locating shot pins. k, wire guide sensor for automatic steering.

the load for proper operation. The stiffness of the pallets will insure that the bearings remain parallel to the operating surface.

17.5.6 Applications

As knowledge and experience with fluid film increase, the application of the technology to manufacturing, commercial, entertainment, construction, research, and transportation industries has proven that it is a manpower- and cost-effective tool for any materials handling situation.

Initially, very heavy and cumbersome loads presented the opportunity for productive use of fluid film. After a decade and a half of application history, the predominant value of the technology appears to be in the broad area of location and alignment.

The ease of movement provides the capability of accurate, minute positioning, with manual adjustment in the X-Y plane to an infinite degree, of the vast majority of commonly moved weights (500–10,000 lb, or about 200–4500 kg).

The maneuverability and freedom from a fixed path system allows greater latitude in assembly-line layout. Revisions to material flow paths can be made quickly and at no cost. The low floor loading and convenience of almost friction-free movement means better use of floor space and utilization of areas not serviced by overhead or wheeled equipment. Productivity is increased when the waiting time for materials to be positioned or taken away is reduced. Initial building construction costs can be lower since floor thickness might be reduced, added support structure for overhead machinery may not be required and rail or support-leg anchors might be eliminated. Subassemblies can be larger and can be done off-line.

The uses are limitless, depending only on an air supply and a proper operating surface.

BIBLIOGRAPHY

Loomis, A. W., Ed. *Compressed Air and Gas Data,* 3rd Ed., Ingersoll Rand, Washington, NJ 07882.

17.6 INDUSTRIAL ROBOTS

William R. Tanner

In many plants, materials handling is the last bastion of nonautomated work. Sophisticated automatic equipment operates at manufacturing work stations, but between work stations, workers lug baskets of parts by hand or with forklifts. Somehow, the least sophisticated and challenging tasks have been preserved, while those requiring the most skill have been automated. Robotics offers the materials handling engineer the opportunity to change much of this.

When materials handling engineers consider applying robots, they may find themselves suggesting radical and fundamental changes in their plants. Using robots to move parts entails detailed knowledge of individual manufacturing processes; manufacturing and materials handling engineers must work closely to install most robots.

17.6.1 An Exemplary Application of Robotics

A Watertown, New York, plant produces more mercury clinical thermometers than any other plant in the world. At one time, workers transferred boxes of thermometers from one process to another. The thermometers moved through a succession of hot and cold baths, vibrators, and centrifuges. Employees stood for entire shifts, tediously lifting boxes from one work station to another, getting sore feet and raw skin. Needless to say, many thermometers never reached the hospitals and clinics, since many were dropped by tired, bored, or careless workers.

Such unfavorable working conditions prompted the manufacturer, Cheseborough-Pond, to seek an industrial robot to perform these tasks [1]. Engineers designed a work station in which one robot moved boxes of thermometers from hot water to chilled water tanks, to a vibrator and a centrifuge, a surge table, and finally to conveyors. One employee now runs the entire system. Since the 1976 installation, the company saved $900,000. Employees who formerly worked at these undesirable jobs now work at more-desirable and better-paying jobs. The robot produces thermometers of a more uniform high quality and it rarely drops a box.

This industrial robot application illustrates many aspects of modern robotics. Materials handling can be particularly ill-suited for workers. By its nature, it is often repetitive and unchallenging. Robots ignore such problems, and indeed are especially good at them. Materials handling often presents hazardous situations to workers; again robots feel no effects from these situations. They perform well in handling heavy loads, reaching overhead, handling hot or toxic parts or materials, and working in noxious environments.

Robots also work with incredible accuracy, timing, and pace. As for their scale of operations, they can handle such heavy parts as automobile motor blocks or such fragile items as fluorescent light bulb tubes.

Because robots, by definition, can be programmed and reprogrammed, they can perform different tasks in series, as did the thermometer-handler discussed previously. This flexibility, combined with contact sensors in the end-of-arm tooling, allows them to handle complex situations. In palletizing and depalletizing, for instance, robots can adjust for progressive loading position changes. They can synchronize their movements on several axes in order to work with moving conveyors. For all these reasons, robots are fast becoming a popular tool for solving materials handling problems.

Robots essentially transport, manipulate, and sense objects in the industrial setting. These capabilities, in materials handling, translate into reduced costs, elimination of unpleasant and hazardous work for humans, improved product quality, and increased productivity.

17.6.2 What Is a Robot?

We all know the differences between a person and a robot. The differences between robots and other forms of automation are not quite so clear, however.

The Robot Institute of America defines a robot as a "reprogrammable multi-functional manipulator designed to move material, parts, tools, or specialized devices through variable programmed motions for the performance of a variety of tasks" [2].

This definition excludes automation devices that perform sequences of tasks that are fixed or preset, or that repeat fixed instructions. The key concept in robots is flexibility.

These flexible devices today grind castings in foundries, load and unload stamping presses, weld, inspect parts, and paint car bodies. Others assemble components, load bricks into refractory ovens, or pack air conditioners. Because operators can reprogram robots, it is easy to adapt to changing shapes of parts, or of altered timing elsewhere on the line, or many other changes.

Hard automation devices can perform only one task. A robot, by contrast, can run through a whole series of tasks, such as one robot that removes dust from a harrow disk, lifts it to a registration table, regrips it in a centered position, loads it in two successive grinders, and delivers it to an output table. Thus, robots must be viewed as general purpose devices, rather than dedicated machines.

Another key difference between hard automation and robots is in the extent of manipulation functions. Robots can move objects through space while at the same time reorienting their positions. This capacity for manipulation creates robots' uncanny resemblance to human hands and arms. And because these movements possess such high degrees of accuracy and repeatability, and such a range of strength, they can often surpass human performance.

More and more robots are bringing sensory capabilities to their industrial settings. Although these capabilities still represent the leading edge of the technology, they promise much, especially to materials handling functions. Tactile sensing, pattern recognition, three-dimensional vision, collision avoidance, and infrared sensing will be applied to many industrial tasks. Among other possibilities, safety problems may be eased, parts orientation made more flexible, and quality control inspection may be combined with materials handling.

General Characteristics of Robots

The industrial robot basically manipulates. It is a mechanical device with as many as seven discrete, synchronized axes of motion. Its control system and memory enable it to perform sequences of motion and function without outside direction. An operator can program or "teach" the robot to perform a different job at any time.

Every robot consists of three major components: the manipulator which performs the required task, the controller which stores data and directs the movement of the manipulator, and the power supply which provides energy to the manipulator. The controller and manipulator are linked by feedback devices which sense and signal the positions of the various axes or joints of the manipulator.

Robot Manipulators

Manipulators, the action part of a robot, come in one of four configurations or a combination (see Figs. 17.6.1 through 17.6.4). These configurations—cylindrical, spherical, jointed arm or anthropomorphic, and cartesian or rectangular—provide robots with their major degrees of freedom. A "wrist" at the end of the robot arm may provide as many as three more degrees of freedom. Wrist axes include roll (rotation perpendicular to end of arm), pitch (rotation vertical to the arm), and yaw (rotation horizontal to arm).

Robot Controllers

Controllers run the gamut from simple step sequencers through pneumatic logic systems, diode matrix boards, electronic sequencers, and microprocessors to minicomputers. Controllers may be integrated into the body of the manipulator or housed separately.

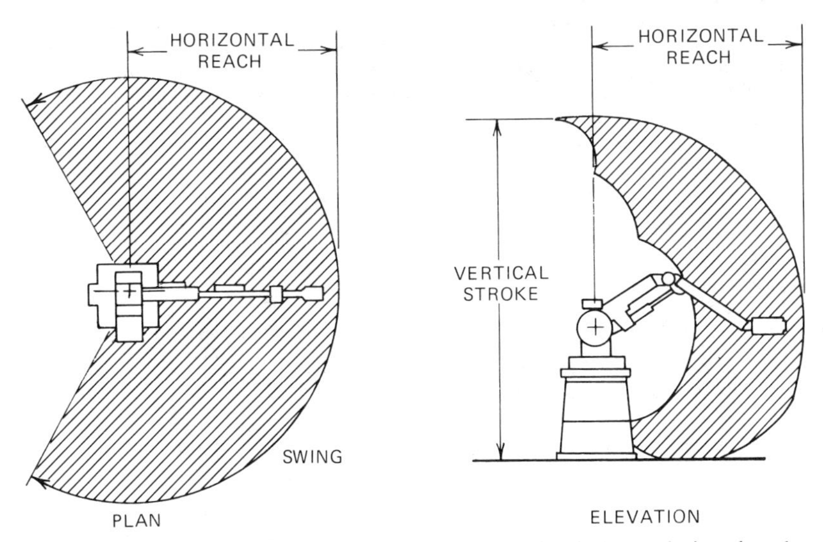

Fig. 17.6.1 Basic jointed arm (anthropomorphic) robot. Cross-hatched area depicts the robot work envelopes—the volume of points in space that can be reached by end-of-arm tooling. All work stations that interact with the robot must be located within the work envelope. Similar cross-hatched areas are indicated in Figs. 17.6.2 to 17.6.4.

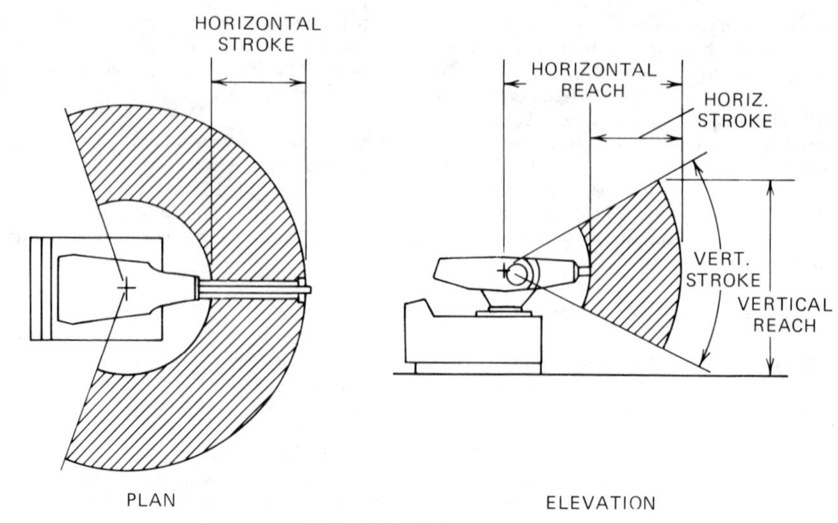

Fig. 17.6.2 Spherical robot.

Controllers function in three ways. First, they initiate and terminate motions of the manipulator in a desired sequence and at desired points. Second, they store position and sequence data in memory. Third, they interface with the "outside world," namely the equipment around them.

The complexity of the controller depends upon the nature of the robot's capabilities. Simple nonservo devices usually employ some form of step sequencer. More complex controllers can employ virtually any technology available in the computer or industrial control fields.

Robot Power Supplies

Robot actuators receive energy from electrical, pneumatic, or hydraulic systems. Pneumatically powered robots usually receive power from a remote compressor. Hydraulically actuated robots normally include their power supply as an integral part of the manipulator, or as a separate unit. They can be operated on petroleum-based or fire-retardant fluids.

Servo versus Nonservo Robots

Nonservo robots often earn the names "end point," "pick and place," "bang-bang," or "limited sequence" robots. These terms imply limited capability and restricted application, which is not necessarily the case. The term "nonservo" provides a more descriptive and less restrictive meaning.

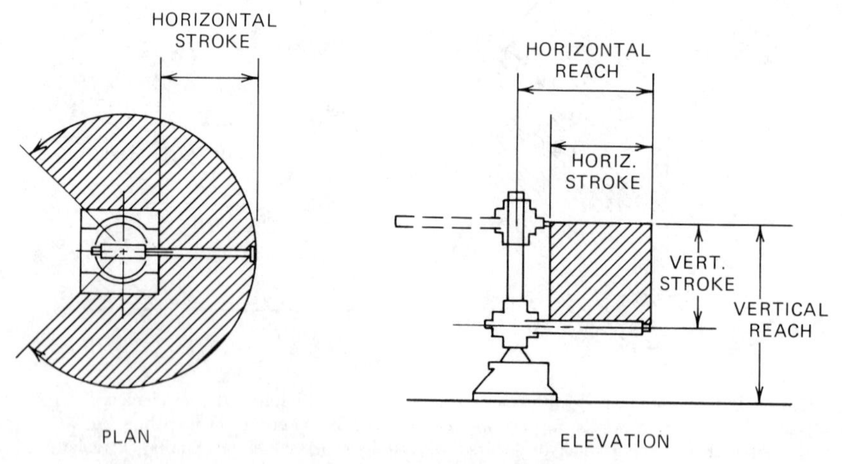

Fig. 17.6.3 Cylindrical robot.

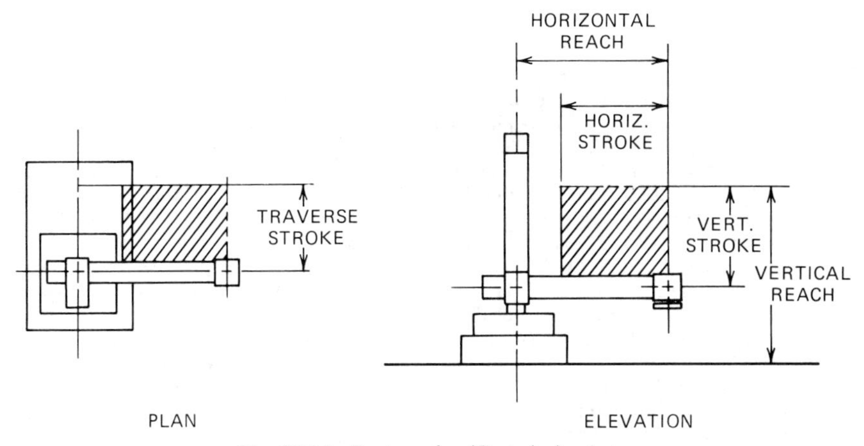

Fig. 17.6.4 Rectangular (Cartesian) robot.

A typical hydraulic or pneumatic nonservo robot operates relatively simply. The sequencer/controller signals the valves to open, admitting air or oil to the actuators. The members move until they are physically restrained by contact with an end stop. Switches then signal the end of travel to the controller, which then commands the control valves to close. The sequencer then indexes to the next step and the controller sends the next signals. This process repeats until the entire sequence of steps has been executed.

In a nonservo robot, there are usually only two positions for each axis to assume. The sequencer provides the capability for many motions in a program, but only to the end points of each axis. It is possible to decelerate at the approach to the stops by valving or shock absorbers, and a limited number of intermediate stops can be installed on some axes.

Nonservo robots can perform well in high-speed operations, since the manipulator is generally smaller in size and there is full flow of air or oil through the control valves. Repeatability to within 0.010 in. (0.25 mm) is possible on larger units, and to within 0.001 in. (0.025 mm) on some smaller units. The nature of nonservo programming limits their flexibility and positioning capability. However, they are relatively low in cost, simple to operate, and highly reliable.

Servo-controlled robots provide greater performance possibilities. A typical hydraulic servo-control sequence begins with the controller addressing the memory location of the first command position and reading the actual position of the various axes (by the position feedback system). The differences, or "errors," between these two sets of data are compared. These error signals are amplified and transmitted as command signals to servo valves for the actuator of each axis. The servo valves, operating at constant pressure, control flow to the manipulator's actuators, which move in response. Feedback devices send new position data back to the controller. There, new error signals are generated, amplified, and sent as commands. This process continues until the error signals are eliminated, whereupon the servo valves reach null, flow to the actuators is blocked, and the axes come to rest at the desired position. The controller then addresses the next memory location and responds appropriately to the data stored there. This may be another positioning sequence for the manipulator, or a signal to an external device. The process repeats sequentially until the entire set of data, or program, has been executed.

On a servo-controlled robot, the manipulator's members can be commanded to move and stop anywhere within their limits of travel, rather than just at their extremes. Since the servo valves modulate flow, it is feasible to control the velocity, acceleration, and deceleration of various axes.

Servo robots display smooth, controlled, flexible motions. Their controllers and memory systems permit storage and execution of more than one program, with selection of the programs from memory in any order often possible via externally generated signals. Larger servo-controlled robots can achieve repeatability of ± 0.060 in. (± 1.5 mm) and the smaller servo-controlled robots may have repeatability within ± 0.004 in. (± 0.1 mm). Because these robots are more complex, they are more expensive and more difficult to maintain than nonservo robots. Reliability also suffers somewhat due to their complexity.

At the upper end of the scale, in terms of load capacity and working range, point-to-point servo-controlled robots perform a wide variety of parts- and tool-handling tasks. These robots employ a "record playback" method of teaching and operation. The operator "walks through" the motions required with the robot, and the positions are thus programmed. This initial programming is usually not difficult, although often it is time-consuming. Servo robots provide great flexibility to users, and most often employ hydraulic or electric drives.

17.6.3 Robot Usage

Although estimates vary, about 6000 robots presently operate in the United States, about the same in Europe, and more than 14,000 in Japan. Certainly the United States potential for robots far exceeds actual installations. One recent analysis [3] estimates that robots have displaced one out of every 1500 of the estimated seven million production workers in jobs suitable for robots. In a survey of 300–400 manufacturing plants, the same authors found that materials handling occupied a quarter of the robot installations. Spot welding of automobile bodies represented the most important application, with 40% of all U.S. robots so used.

In most plants, robots have been applied to jobs that are unpleasant or hazardous for workers. This is really just the first generation of robot applications; they perform less than 2% of all operations in most of the plants where they are installed. Future generations of application will undoubtedly see robots furthering overall automation of manufacturing processes.

Most U.S. industrial robots inhabit large, sophisticated operations. The report cited [3] found that most robots had been placed by automobile, aerospace, and other large equipment manufacturers. Over 80% of the installations are less than 5 years old. And companies that use robots also seem to use sophisticated systems such as numerical control, computer-aided design, and computer-aided testing. It is predicted that robot use will make a transition in the coming years. Rather than primarily performing undesirable jobs for workers, they will help manufacturers boost productivity, reduce operating costs, and improve product quality.

Robot Performance

Manufacturers often quote design lives of 40,000 performance hours for their products. In actual experience, robots can function well for about 10 years, provided that they are well maintained. Most seem to require a general rehabilitative overhaul after about 10,000–15,000 hr of operation.

Manufacturers also claim a downtime of only 2%. That is close; users can probably count on downtime of 2–4%, with proper maintenance.

17.6.4 Robots as Materials Handlers

Materials handling was the first area of robot application in the United States. Today, the majority of nonwelding robots perform materials handling tasks. The most common application would be a nonservo (pick-and-place) robot essentially transferring parts from one point to another. Current materials handling applications include:

Loading bottles that contain poisonous liquids
Transferring auto parts from machine to overhead conveyor
Depalletizing wheel spindles onto conveyors
Packaging air conditioners
Palletizing cereal boxes
Loading transmission cases from roller conveyor to monorail
Loading refractory bricks into kiln cars
Palletizing glass tubes

Advantages of Robots for Materials Handling

Generally speaking, robots measure up well as materials and parts handlers. They offer some distinct advantages over both manual parts handling and special-purpose automation. Cost effectiveness stands out notably. A robot's operating costs fall significantly below wages or piece rates for manual labor in most industries. The thermometer-moving robot, for instance, generated $900,000 of direct labor savings in its first 3 years (16,000 hr) of operation. Its $90,000 installation cost seems trivial by comparison.

Special-purpose automation often costs more than a comparable robot installation. Since the robot's flexibility avoids obsolescence which could result from product or process changes, the additional expense of hard automation looks even less attractive. Plant engineers and even work station operators can easily reprogram and retool many robots to accommodate such changes. They can even assign the robot to an entirely new location if they wish.

As a materials handler, a robot may be faster, more accurate, and stronger than many workers at a comparable job. The more complex robots do not, however, outpace a worker, on the average. It is their steady, unvarying pace which actually tends to increase productivity.

Robots can have greater reach and capacity than workers. They are largely unaffected by heat, noise, dust, fumes, toxic substances, and other circumstances which would be hazardous or undesirable to a worker. Robots are neither distracted nor bored by tedium and repetition of tasks. They can, with proper end-of-arm tooling, also handle several parts at a time.

Some modern plants are being built with plant floor computer communication systems in mind. Robots, being computer controlled, will fit easily in these advanced plant designs.

Robot Limitations

Like workers, robots bring their own set of problems to the workplace. In some cases, these limitations make workers more appropriate for the job than robots. In others, hard automation does the job better. Some large new plants have chosen not to use robots at all. Plant engineers and managers need to make careful decisions in deciding how to apply robots to their workplace. They should be aware that the sophistication and cultural mythology of robots are highly seductive; it sometimes takes a great deal of objectivity to overcome these forces.

With few exceptions, robots are deaf, dumb, and blind. They require disciplined, orderly working environments. In this respect, robots are at variance with typical materials handling and packaging methods.

Most robots, for instance, reach for a part at one particular location, in one particular orientation. This means that parts delivered to the robot must align carefully with these requirements. On the other end of the process, the robot delivers with equal consistency. Pallets and containers have to be carefully positioned for access. If dunnage is required for retention of a parts orientation and positioning, its handling must also be considered.

This limitation of robots requires that the workplace be redesigned around robot dimensions and capacities. Where a worker once operated a work station, he or she could simply reach into a basket or tub of jumbled parts. This will not do for a robot however. The parts must be delivered by parts feeders, magazines, accumulating conveyors, storage racks, or some method of palletizing. Perhaps a worker will need to initially move parts from bulk containers to feeders or conveyors.

Robots cannot, for the most part, apply judgment in performance of tasks. Although advanced robots can be outfitted with sensors and inspection devices, most cannot distinguish between "good" and "bad" or "right" and "wrong" parts. The vast majority of robots on the market require the manufacturing process to make these decisions before the parts reach the robot. Thus, presorting or in-line inspection may be needed upstream from the robot.

Tactile and visual sensory capabilities come on a few robots today. These devices, however, are expensive and relatively rare in the market. In general, they need further development before a wide range will work well in real time. Meanwhile, most factory managers will probably have to work around the less intelligent robots' limitations.

Of course, many robot capabilities make up for their shortcomings. Engineers can largely overcome these problems by following three simple ground rules in designing the robot process. Throughout the task(s) it is best to:

Define the part orientation at the point where the robot picks it up.

Preserve that orientation throughout the process.

Never let the robot drop, but rather, carefully, release a part.

These rules keep the process relatively simple. In the example of the thermometer manufacture, a single robot arm moves the boxes through seven locations. At each step it controls the orientation of the boxes so that they can be retrieved from each step in a predictable fashion. It can, for instance, set the boxes into the hot water bath by reaching right in—something a human worker would not do. In this way, it preserves part orientation.

Robots perform best in a particular industrial niche. Compared to workers or hard automation, robots seem best suited for the principle of "moderation in all things." Hard automation pays off for long production runs with high volume. A three-shift operation usually calls for hard automation, especially where the tasks can be broken down into simple steps.

For very low-volume, or small-batch operations, workers often perform best. This becomes especially apparent in operations involving one or two shifts, and highly complex tasks.

Robots rank somewhere between workers and hard automation. They usually require a two-shift operation to pay off and handle tasks of medium complexity and speed.

Robots present many users with programming problems. Unfortunately, for most robots, no program language exists. Teaching usually is conducted at the work site with a push-button programming device. Sometimes teaching turns out to be more difficult than expected. It would be helpful to have an off-line programming language for teaching, and to have the language compatible with computer-aided manufacturing (CAM) systems and communication networks.

Another limitation shows up for materials handling of delicate or intricate parts. The market offers few small, light robots. Typically, they come large and heavy, capable of lifting only 10% of their weight. In light industry, a small robot with a relatively greater load capacity would be more suitable.

17.6.5 The Application Decision

Because robots are glamorous in industry these days, many people are eager to install them in their plants. While this enthusiasm indicates innovative management, any company should be careful to justify such an installation. An installation that fails will sour both management and workers and may delay or prevent future installations which might work very well. Failed installations are also expensive mistakes.

Perhaps the most important suggestion is to consider a first robot installation as a sort of trial run. Look at a fairly simple task, one which assures success almost completely. Do not try to install a Cadillac model on your first installation, one which does half a dozen tasks in the twinkling of an eye. The possibility of this device causing production interruptions and general frustration is high. A simpler robot is more likely to smooth the way for more sophisticated devices later on.

A few rules of thumb can be a guide in deciding about robots. First, it can generally be assumed that the installed price of a robot will amount to double the purchase price of the machine. The additional costs include applications engineering, installation costs, training, and so forth.

Second, a robot will usually not generate savings unless you can eliminate at least two worker positions per day by its installation.

Third, rule two may be irrelevant if you want to solve some noneconomic problems. These factors may be very important in justifying a robot. Important noneconomic factors include:

1. *Increased Productivity.* This results from the constant pace of robot performance.

2. *Quality.* Manufactured goods may be of high, more consistent quality than worker-produced goods. Robots also reduce scrap, breakage, and so on.

3. *Undesirable Tasks.* Robots work well in the presence of noise, dust, fumes, heat, dirt, heavy loads, fast pace, or monotony. Workers often respond to such conditions with high turnover, work stoppages, poor workmanship, absenteeism, grievances, and sabotage. Robots may eliminate some of these problems, especially if management introduces the robots in full consultation with workers. Robots can also work well in clean-room conditions.

4. *Advancement of Technology.* Installation of a robot may familiarize management and workers with a new technology which may be applied more fully later on.

5. *Competitive Position.* Robots may provide a competitive advantage by producing goods at less cost than competitors. Robots also give users inherent flexibility to meet shifts in market demands by increasing or decreasing production rates, or to alter the production method. These changes could be incorporated without having to retool, as with hard automation.

6. *Hazardous Operations.* Some tasks endanger workers or do not comply with safety regulations. In such cases, robots may be the most cost-effective and humane solution.

7. *Other Potential Savings.* Robots may reduce costs of protective clothing, safety equipment, lighting and ventilation levels, parking, dining, washroom and locker room facilities, supervisory work loads, and others. Although single items here may be minor, the total may add up to a solid justification for a robot installation.

Fourth, if the batch size is too small, or the set-up process too time-consuming, robots may not be appropriate. Actual application experience indicates that if set-up processes (reprogramming, part locating, end-of-arm tooling changes) consume more than 5–10% of production time, they are simply too expensive.

17.6.6 Adapting to Robots

Successful implementation of a robot operation requires close cooperation among managers, engineers, and workers. This is particularly true with materials handling robots, since they must interface closely with individual manufacturing processes.

Resistance to robots comes as much from mid-management personnel as from floor workers. Without going into all the reasons for this, suffice it to say that management tends to be conservative—often for good reason. Do not be content with upper management utterances about commitment, productivity, and progress. Make sure they are involved.

To obtain management support and commitment, the person trying to get the robot installed should remember some basic principles of group process. First, the people concerned should be involved in the actual choosing of the device. Once people are involved in evaluating various options, some of the threatening aspects of robots disappear. Second, be sure that upper management supports the innovation. If this support is well communicated, then the inevitable difficulties of the transition period will fell less risky to mid-management people. Shop rearrangement, debugging, and various temporary problems will occur. Mid-managers are oriented by production goals, time horizons, and reliance on formalized rules and communication channels. Top management support will alleviate mid-management stress during transition.

One writer [4] has suggested the following approach to bringing floor workers in on the robot installation:

First, workers must enter the planning process at the earliest possible stage. Although this is unorthodox, it is absolutely critical, regardless of the type of shop involved. Ideally, workers should be assured that the robot will not make them jobless, but that they will take different jobs in the plant. This process involves discussing plans with the workers and their union representatives.

Once workers are reassured that they will not be made jobless, clarify the benefits of robot installation. Frequently, health and safety benefits will be important, as well as the economic health of the company.

As soon as planning for the robot begins, planning for employment of displaced workers should begin. If they are left in the dark about their futures, needless personnel problems result. Some of the skilled workers displaced may be retrained to maintain the robot itself.

Finally, it helps to conduct a public relations effort with the nearby general public during the planning process and after installation. News releases along the way keep everyone informed, and an "open house" for media reporters after installation generates excitement. Provide plenty of illustrations.

Generally speaking, installing a robot takes as much time solving the human and organizational problems as it does to redesign the workplace and deal with engineering issues. It pays to take the organizational work seriously before installation, or you could wind up with management and workers hindering the fine engineering work you do. It is better to invest this time and effort before the installation than afterward; afterward is usually too late to solve the problems.

17.6.7 Types of Applications

To provide a more concrete idea of materials handling capabilities of robots, we will describe some actual installations. Each example illustrates a different robot capability. Most of the examples come from the automobile industry; this reflects the fact that auto makers have been the first to make widespread use of robots in American industry, but does not imply that it is the only industry where robots apply.

Accurate Positioning

In an operation that assembles automobile speedometer housings, a robot transfers and mates components. This simple, pneumatic nonservo robot picks up a subassembly from a feed chute and transfers it to a fixture pallet on an indexing conveyor of an assembly machine. On the pallet, a mating part has been previously installed. The robot inserts the subassembly into the mate. Positioning accuracy required for this operation is ±0.010 in. (25 mm). The robot repeats this task every 6 sec.

Similar operations have nonservo robots placing parts with accuracies of 0.0005 in. (0.01 mm), and repeat cycles of less than 2 sec.

Destacking with Sensory Feedback

In another plant, a nonservo robot helps transfer glass display panels for computer terminals. Three innovations increased the robot's capabilities beyond pick-and-place functions.

First, the robot was placed on an indexing mechanism. This provided lateral positioning in front of any of three rows of glass on the rack. Second, the robot was fitted with a tactile feedback sensor on its end-of-arm tooling. This controlled its longitudinal stroke. The arm is programmed to extend to a position beyond the last piece of glass in a row. Upon extension of the arm, the sensor, a pneumatic limit switch, interrupts movement at contact with a piece of glass. The robot then lifts the glass out of the rack, reorients it to a horizontal position, and transfers it to a belt conveyor. The third innovation, a timer, controls the robot's cycle. The transfer cycle time varies with the distance the robot has to travel to reach each piece of glass. The timer inhibits the robot's release of glass until the belt has traveled the required distance.

Other servo and nonservo robots employ sensors to load metal blanks into forming presses, transfer parts from process racks to conveyors, load and unload machine tools, and remove flat parts from shipping containers.

Sorting

A nonservo robot inspects complex sheet metal stampings and segregates good from bad parts. The robot's laser/photo detector inspection device determines the presence of various holes and notches. The plates arrive at the work station stacked in a neat pile, but in random orientation. They are not symmetrical.

The robot picks up a plate from the stack and positions it over the inspection array. If all holes and notches are present, all the photo detectors will be activated and a signal instructs the robot to dispose of the part on the "good" stack. Then it picks up the next part.

If, however, all holes and notches are not detected, the part is either defective or upside down. To find out which is true, the robot flips the part and presents it for another inspection. If all holes and notches are now detected the part goes to the "good" stack. Otherwise, the robot releases it to the "reject" stack.

With fiber-optic arrays, parts with as many as 100 small holes of various sizes and locations can be reliably sorted at rates of up to 750 per hour. Other robots can sort parts according to size or weight.

Handling Heavy Parts

A servo robot moves parts weighing up to 80 lb in a plant producing truck rear axle assemblies. This robot handles ring gear blanks through three successive machining operations. It takes raw forgings of several weights and diameters through a vertical lathe, a broach, and a multiple-spindle drill, at a rate of one per minute. In two of the operations, the parts must be inverted during transfer.

The robot's end-of-arm tooling automatically compensates for the different part diameters. It also incorporates details to permit grasping parts on either the inside or outside diameter, as required by the machining operation.

To integrate these operations, the workplace required modifications. Two machine tools were relocated, and two gravity-roll conveyors and a part stand were installed.

Previously, in these operations, a hoist was used to lift the heavy parts at each machine tool, and workers occasionally suffered back and hand injuries.

Other robots have been installed on floor or overhead traverse devices. They can load and unload as many as 12 machine tools, and transfer parts through successive machining operations. Some, equipped with multiple-gripper end-of-arm tooling, load and unload more than one part at a time or reduce the load/unload cycle time.

Handling Hot Parts

A large servo robot transfers hot steel parts in a plant producing high-pressure industrial gas containers. The 2300°F parts are moved from a forging press to a draw bench. They range in weight from 60 to 300 lb. The robot can reach over 9 ft and rotate more than 180° in a horizontal plane. Because it easily reaches from press to draw bench, no equipment relocation was required.

The robot removes parts from the press, reorients them, and places them on the draw bench in about 35 sec. After each cycle, it cools its end-of-arm tooling in a tank of water. The operation runs 18 hr a day, producing from 1000 to 1400 parts daily.

The same task formerly required four operators per day, working with a hoist and transfer rail. The hot, heavy work created a serious worker turnover problem. The robot relieved these workers, providing a 2-year payback in labor savings alone.

Other hot-part handling applications have robots transferring parts from furnaces to induction heaters to die-forging machines, from forging presses to trim presses, and from pouring lines to shakeout tables in foundries. Others handle parts in heat-treating operations. These applications reduce labor costs, yield higher production rates, and remove workers from hot, noisy, dirty, and dangerous environments.

Palletizing

A servo-controlled robot integrates two powered-roll indexing conveyors into a system that palletizes automobile catalytic converters in standard wire shipping containers. The 45-lb parts arrive, correctly oriented, on one conveyor. Shipping containers are transported to and from the load station on the second conveyor. At the load station, the containers are tipped about 15° from vertical; this provides easier access for the robot and prevents parts from falling over after they are released.

The robot stacks 58 parts in each container, two rows across and two rows high. It handles 600 parts per hour, three shifts per day. The system operates completely unattended—the robot controls automatic delivery of parts and delivery and removal of containers. Because this robot was designed into a new facility rather than an existing operation, no workers were displaced. It is estimated, however, that more than 12 people per day would have been needed for the same level of production.

Palletizing is a common application for robots. In other plants, robots handle fragile powdered metal preforms, glass tubes and sheets, heavy machined parts, unfired bricks, synthetic fiber bobbins, and injection-molded plastic parts. Often the robots handle several parts simultaneously. In addition to labor savings and worker relief from monotonous tasks, palletizing robots improve packing density, reduce damage, and attain higher handling rates.

Working with Moving Conveyors

A servo robot transfers automobile windshields from a process rack to a continuously moving overhead conveyor. It removes the parts from an autoclave after laminating and hangs them on the conveyor for transport through a washer. The windshields stand vertically in the racks, and being covered with oil, are handled by the edges.

A limit switch sensor on the robot's end-of-arm tooling "finds" each windshield for pickup. The robot synchronizes its movements through a switch and resolver actuated by the conveyor. Windshields can thus be deposited in the carriers, regardless of conveyor speed.

Two workers formerly loaded the 35-lb windshields. The company thus saved the cost of two people per shift on a multishift operation. In addition, fewer of the slippery parts are broken during handling.

Both servo and nonservo robots can work on line-tracking operations. In other applications they load machines from moving conveyors, transfer sheet metal components from moving assembly fixtures to conveyors, handle large vacuum-formed plastic parts on a moving conveyor, and transfer foundry cores from a core-making machine through a core dip operation and onto a conveyor.

Packaging

A robot that packs air conditioners in shipping containers combines several interesting features, including interaction with a worker. The servo robot picks up 35-lb units from a moving conveyor. The pickup is signalled when the unit breaks a light beam from a photoelectric control mounted alongside the conveyor. After picking up the unit, the robot moves it to a shipping container, the container moves to a second position. Here a worker adds spacers between the units and a sheet of dunnage for the second layer. The robot then packs in another six units.

This robot is equipped with a safety post which prevents the arm from striking the worker, should a malfunction occur. Another safety switch shuts down the air conditioner delivery conveyor if a unit has not been picked up within the 19-sec transfer cycle.

Other packaging applications include cereal box packing, health and beauty products, photochemical developing kits, and office copier machine packaging.

REFERENCES

1. Becker, Richard, "Experiences in Applying Robots in Light Industry and Future Needs," *Industrial Robots, Vol. 2, Applications,* 2nd ed., William Tanner, Ed., Society of Manufacturing Engineers, Dearborn, MI, 1981.
2. Tanner, William, *Industrial Robots, Vol. 1, Fundamentals,* p. 2, Society of Manufacturing Engineers, Dearborn, MI, 1981.
3. Sanderson, R., J. Campbell, and J. Meyer, *Industrial Robots: A Summary and Forecast for Manufacturing Managers,* p. 48, Tech Tran Corporation, 1982.
4. Skole, Robert, "An Industrial Public Relations Checklist," *Industrial Robots, Vol. 1, Fundamentals,* William Tanner, Ed., Society of Manufacturing Engineers, Dearborn, MI, 1981.

CHAPTER 18
DOCK OPERATIONS AND EQUIPMENT

E. RALPH SIMS, JR.

The Sims Consulting Group, Inc.
Lancaster, Ohio

The American Trucking Association has a motto that says "If You Got It, It Came By Truck." With rare exceptions this is true. The intermodal structure of the transport system increasingly interlocks the highway truck with other modes. As a result, almost every manufacturing and distribution facility requires a truck dock capability to deal with a variety of highway trailers, straight trucks, and special vehicles in a safe and efficient manner. Rail docks, although not available in every location, must also be designed to accommodate a variety of vehicle designs.

18.1 THE HIGHWAY VEHICLE

Figure 18.1.1 shows the dimensions of a typical highway trailer. This is the most commonplace shape and dimensional pattern in modern over-the-road truck transport. Trailers vary in length but most of them have the same height and width dimensions. The tailgate height, the over-the-road height, and the width of the vehicle are generally the limiting dimensions in the design of the truck dock. Almost all states, and Canada, limit over-the-road height to 13 ft 6 in. The 96-in. load width is universal except that Canada, Connecticut, and Hawaii allow 102-in. widths over-the-road. There is also a proposal in the United States Congress to allow a 108-in. maximum width. The majority of highway vans have tailgate heights of between 48 and 56 in., the mode of this dimensional pattern being a height of 51 in. Thus, there is a degree of trailer standardization that can be accommodated with proper dock design. Conversely, straight trucks vary in height, and delivery vehicles are usually lower. Furniture vans are also lower because of smaller wheels. All of these vehicles must be accommodated at the truck dock.

Figure 18.1.2 shows the concept of the International Standards Organization (ISO) standard highway/intermodal container. The chassis on which the container rides is usually specially built for containers. In some cases, a flat-bed vehicle is also used. When a flat bed is used, complications arise in dealing with the dock because of the extra height of the load and the usual gap between the end of the tailgate of the flat bed and the doors of the container. In most cases, these containers are either 20 or 40 ft in length, and they are 8 ft × 8 ft in section. Some 10-ft long containers are also in use. Some of the container lines are still using nonstandard containers. In situations where open-top or flat-bed vehicles are commonly used for shipping or delivery of heavy machinery, steel, lumber, pipe, and so on, side loading with forklifts or overhead loading with cranes can replace truck dock operations. Side loading is also common in European trucking operations.

18.2 LOADING/UNLOADING OPERATIONS

One of the features of truck dock operation is the option of driver load versus shipper load. In many instances, trailers are dropped at the docks, and the shipper loads the trailer and has an on-site tractor for shifting and spotting trailers. This tractor may be supplied by the carrier or by the shipper. Conversely, in a carrier load operation the highway van backs up to the dock and the shipper either drives the load on with a forklift truck or tailgates the load for the driver to stack in his van. In such instances, the usual procedure is to leave the tractor attached to the trailer. Figure 18.2.1 shows a switch tractor with an elevating fifth wheel unit. These are quite common in marine terminal operations for shifting chassis-mounted containers into and out of the loading areas. They are also quite applicable to transfer and spotting operations in a truck yard.

Safety is an absolute essential in the design of truck docks. Government highway regulations and OSHA have also placed safety demands on the docks and vehicle designers. The most important truck dock safety issue is the relationship between the forklift driver, the forklift truck, and the highway trailer. Figure 18.2.2 demonstrates the dangers involved in driving a forklift truck onto a highway trailer. These dangers are accentuated by the movable tandem on the trailer as shown in the drawing. When the tandem wheels are in the forward position, the moment arm at the tailgate is such that the entry of a forklift truck can easily tip the trailer up on its rear and dump the driver and the forklift truck into the gap between the trailer and the dock. This can be fatal! This is a particularly dangerous situation when bridge plates without T bars are used. Such bridge plates should never be used in loading operations where power vehicles are present. The power vehicle has a tendency to "squirt" the plate out from under its wheels and, thus, drop the forklift truck or other power vehicle into the gap between the truck and the dock. The proper bridge plate design with a safe T bar is shown in Fig. 18.2.3. Highway vans should also be tied down at the nose and jacked at the rear to prevent this type of accident. In the case of rail dock operations, a similar but modified version of the bridge plate should be used as shown in Fig. 18.2.4.

Figure 18.2.5 shows a safely rigged highway trailer being loaded by a fork truck using a dock leveler in place of a bridge plate. The highway trailer is chocked, jacked, and tied down. These operations

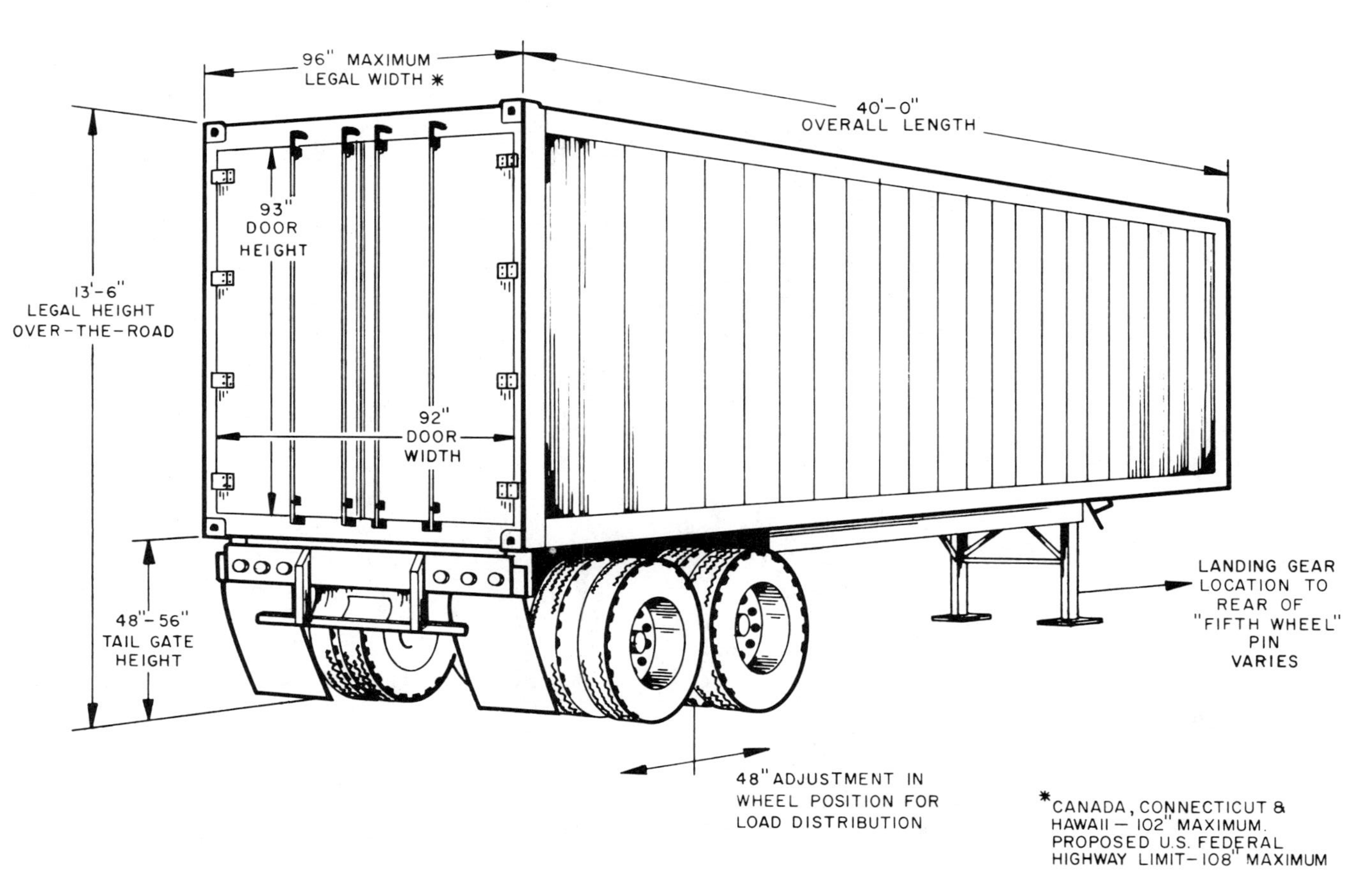

Fig. 18.1.1 Typical highway trailer dimensions.

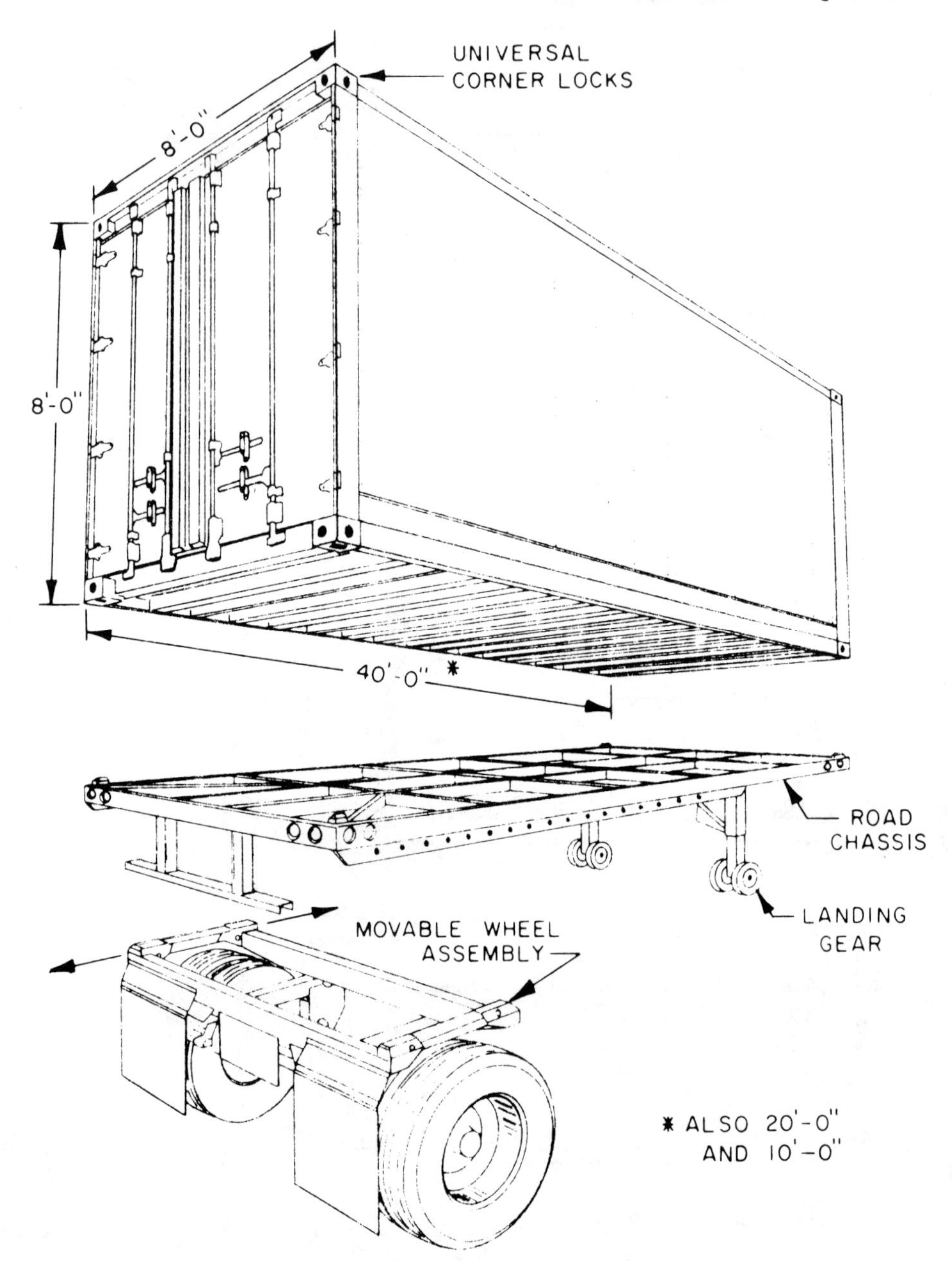

Fig. 18.1.2 Typical ISO container system.

should be accomplished by the driver as a part of his service to the customer. The jacks, chocks, and "come-along" tie-down units should be chained to the apron pavement so that they cannot be stolen. The tie down procedure can be enforced by telling the carrier operators that they will not be loaded or unloaded with forklift trucks until the safety equipment is in place. In a shipper load operation the shipper personnel will, of course, have to do this work.

A recent addition to the list of available dock safety equipment is the trailer lock shown in Fig. 18.2.6. This device locks the ICC bumper bar on the rear of the trailer to the docks. It is intended to do the same job as chocks in preventing the trailer from rolling away from the dock face.

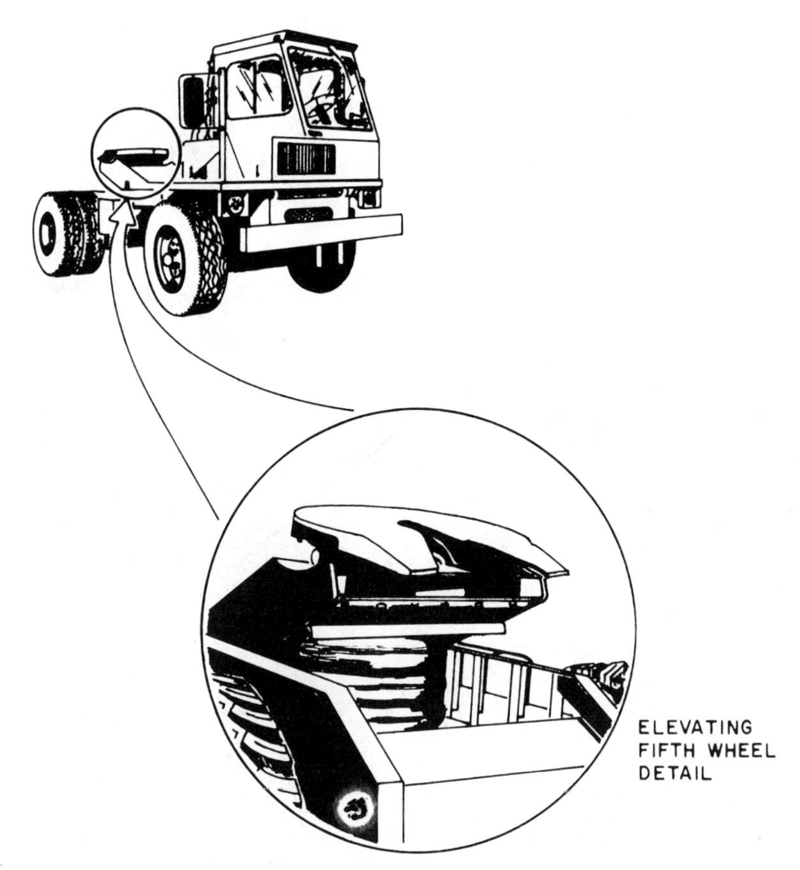

Fig. 18.2.1 Internal transfer truck (fifth wheel latching and elevating, controlled from cab).

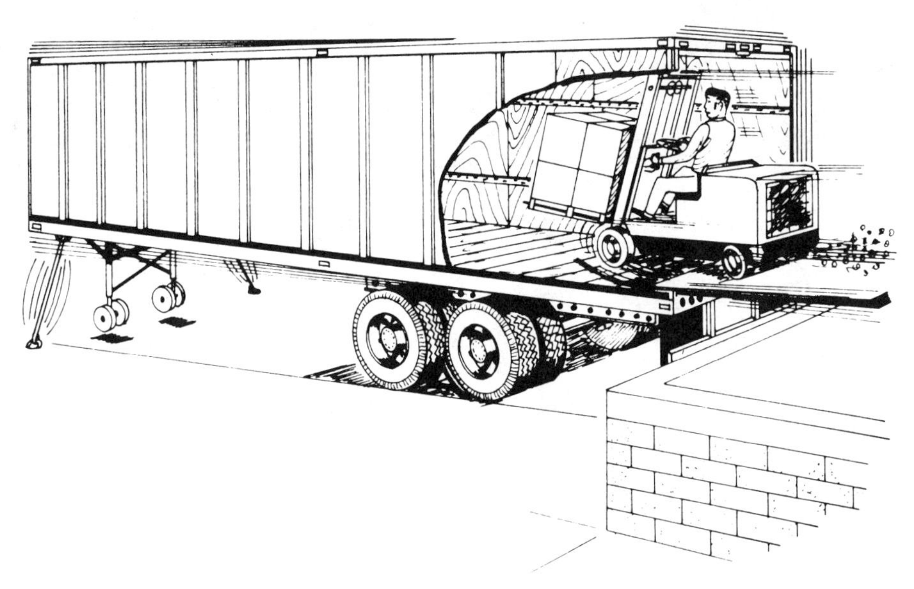

Fig. 18.2.2 Empty trailer tilt dangers.

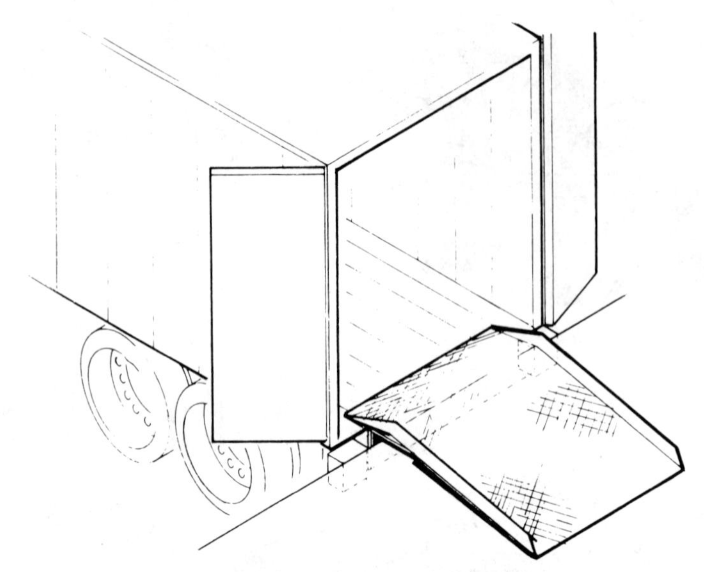

Fig. 18.2.3 Typical truck dock board.

Fig. 18.2.4 Typical rail dock board.

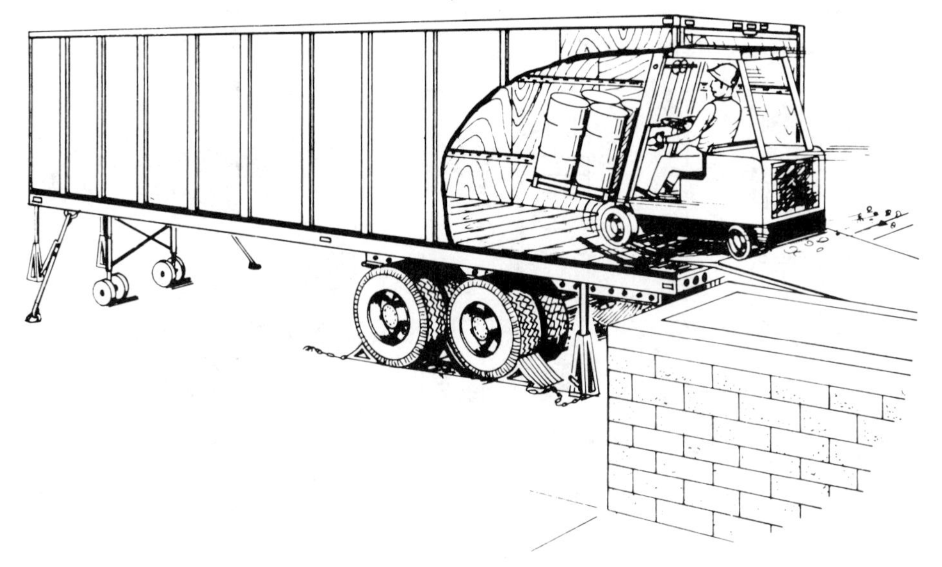

Fig. 18.2.5 Trailer loader safety equipment.

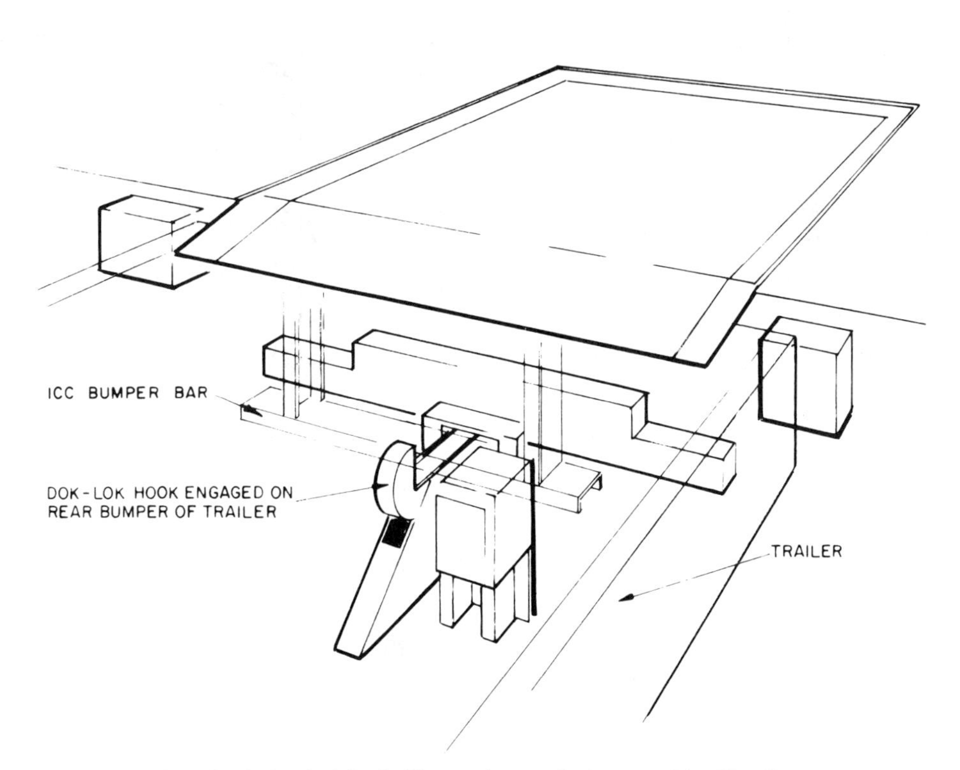

Fig. 18.2.6 Method for locking truck to dock. Courtesy Rite-Hite Corp.

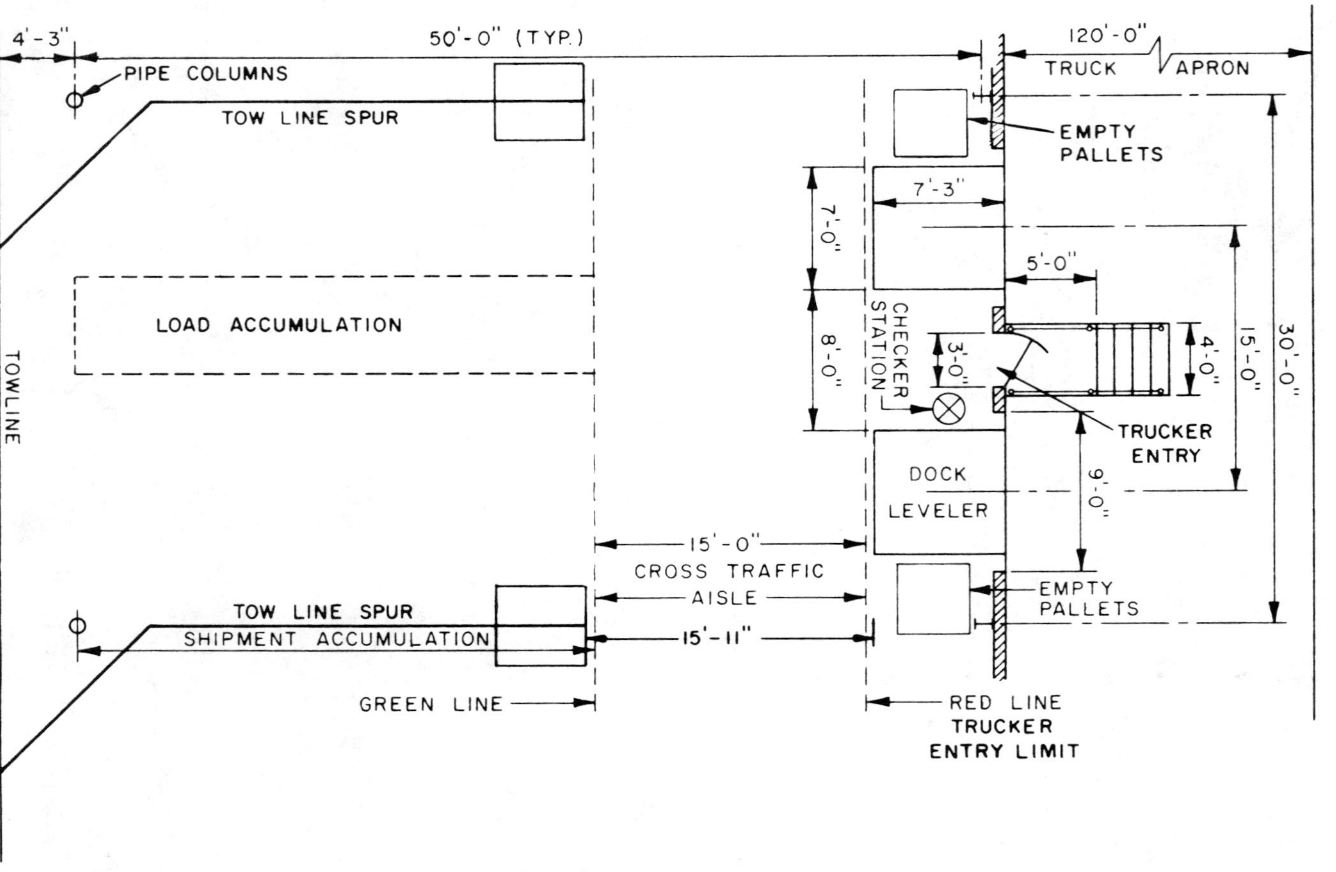

Fig. 18.3.1 Typical dock layout.

18.3 DOCK LAYOUT

A shipping or receiving dock is a very active area. Limitation of the space behind the truck door or dock face often inhibits the effectiveness of the dock operation. In addition, poor layout can permit pilferage and other misdemeanors to occur in the dock area. The work space behind the truck dock should be designed with several basic criteria in mind. It should:

1. Permit efficient loading and unloading of highway trucks with forklift trucks, hand jacks, conveyors, or manual handling methods.
2. Provide safe working conditions and sufficient room for people to work without fear of being run down by a forklift truck.
3. Have sufficient space for order and goods accumulation or receiving checking and proper scheduling in the dock area without blocking traffic or creating congestion at the rear end of the truck.
4. Be secure and so arranged that drivers and other external personnel do not have direct access to documents or merchandise that are not destined for their vehicle.
5. Be so arranged that merchandise moving into or out of the dock area can be handled automatically or mechanically, in support to the loading or unloading operation.

Figure 18.3.1 shows a basic dock layout. In this instance, the recommended dimensions are presented as minimum standards. It should be noted that the dock spacing is 15 ft on centers. This permits the opening of truck doors, the maneuvering of vehicles, and the movement of people between trucks without any interference. However, a more important reason for this spacing is to allow sufficient room around the tailgate of the truck and the dock leveler for piling empty pallets, movement of forklift trucks, and positioning of load checkers without mutual interference.

This arrangement also provides a measure of security. A 15-ft cross aisle is inserted between the tailgates of the highway vehicles and the shipment accumulation area. By using a red line to mark the aisle on the truck side, the driver can be limited to the area between the red line and the cab of his vehicle. The green line on the warehouse side of the aisle emphasizes that this area is open for transverse movement of in-house personnel. This 15-ft aisle permits maneuvering of forklift trucks in lateral service traffic patterns.

The choice of the 28-ft towline spur in the illustration is purely arbitrary. However, in most instances, this spur length would safely accommodate four or five towline carts with one pallet of merchandise on each. In a nontowline operation, this area might be equipped with pallet racks to permit cube utilization in the accumulation of shipments. In a nonpalletized function there might also be storage furniture and pallet racks to utilize cube. In any case, the shipment would be accumulated across the aisle from the vehicles by carrier load at the dock. Forklift trucks or dock hands would move the merchandise to the tailgate of the truck or into the vehicle and the driver, or shipper personnel, would load the van.

The bay spacing of approximately 30 ft × 50 ft is typical of modern warehouse design. The 30-ft spacing is compatible with the 15-ft dock-to-dock centerline dimension and the arrangement of pallets and pallet racks in the warehouse layout. The 50-ft spacing works very well with pallet spacing and rack arrangement and keeps columns out of the dock area. This pattern is usually quite economical in a good warehouse design. The dimensions can vary with the particular facility, usually plus or minus 1–2 ft in each direction, depending on the pallet size, the rack pattern, the size of the aisles, the type of equipment used in the warehouse, and other features of the storage operation. Pipe columns are recommended for the interior structure of the warehouse to minimize corners, vermin traps, and other undesirable structural features. They also provide a strong, clean design and a good appearance.

18.4 DOCK OFFICES

One of the major problems in designing shipping and receiving operations is the need to isolate external personnel from both the internal clerical people and the merchandise. Figure 18.4.1 shows a typical design for a shipping and receiving office in a warehouse. A separate truckers' room is provided with a talk window access to the shipping office. This prevents the truckers from getting into the office where they can become involved with the in-house personnel and paperwork. They are also limited in their access to the dock and controlled by the shipper. The truckers' lobby should be equipped with a coin telephone and coffee and other vending machines for the convenience of the drivers. By using a separate toilet, they are further isolated from in-house personnel. The talk window between the dock side of the shipping office and the interior of the truck dock permits access to the shipping office from the dock as well as from the truckers' lobby, by both warehouse personnel and drivers, without permitting the truckers to wander about the warehouse. Another security feature is the prevention of outside truckers from having contact with the people who write the shipping documents. Forged

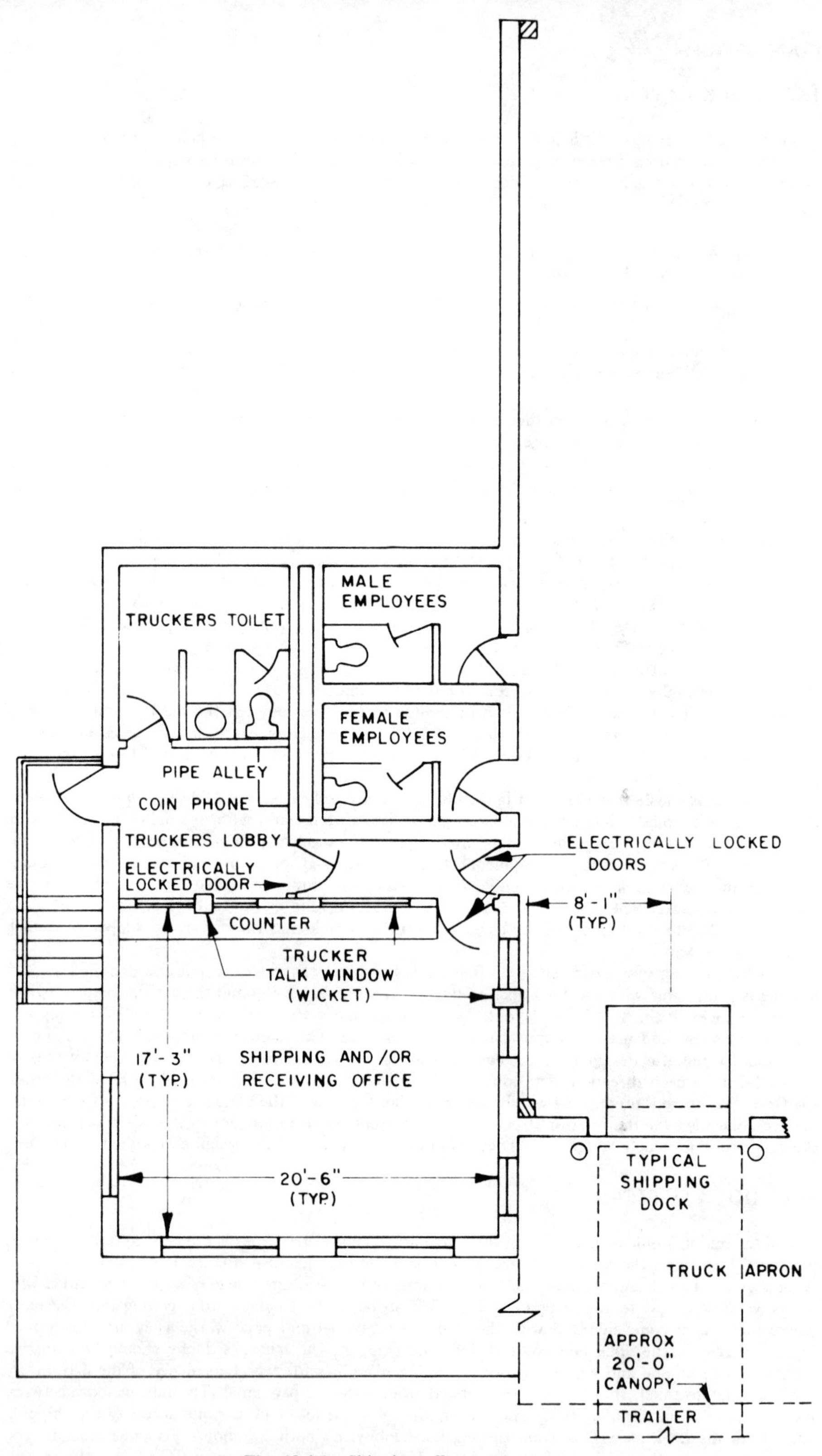

Fig. 18.4.1 Shipping office arrangement.

documents are a very simple means of burglarizing a warehouse. Collusion is essential to a warehouse robbery.

Shipping and receiving operations should be conducted at separated docks for smooth uncongested flow and good security. Separate and mirror image offices at opposite ends of the truck dock are a common layout pattern.

18.5 DOCK AND DOOR DIMENSIONS

The dimensioning of the dock and the ramp is critical to the design of a good truck operation. As mentioned earlier, it is appropriate to think in terms of 51 in. from the truck dock to the bottom tangent line of the wheels. All of the designs presented in this section are based on that height, which accommodates the full range of trucks. The use of a dock leveler that will elevate 12 in. and drop 9 in. will usually accommodate all truck bed heights for conventional highway vans. In the case of "lowboys," furniture vans, and other special vehicles, external ramps might be necessary to bring truck beds up to dock height. Extra-long dock levelers can also sometimes be used to deal with low vehicles.

The cross section shown in Fig. 18.5.1 has some specific features that are worthy of discussion. It should be noted that the apron slopes down 6 in. to a point 20 ft from the dock face, and the canopy extends to the same point over a drain at that location. This arrangement assures the ability to keep the trailer's rear tandem area clear of debris, ice, and snow. It can be flushed out with a hot-water fire hose, or it can be equipped with under-the-concrete heating units. This arrangement also assures a level and safe rear tandem position. The apron slopes upward from the drain to the end of the 65-ft truck apron. This apron area should be concrete or at least have concrete wheel pads to avoid pitting of the ramp area by vehicle wheels. The overall apron width from the dock face to the outside edge of the pavement should be at least 120 ft. The 120-ft dimension from the dock face to the nearest obstruction allows for cross traffic, turnout, and parking as shown in Fig. 18.5.2. As a general rule, the apron space should be designed to allow 65 ft from the dock to the clear area for vehicle spotting at a covered dock. Most highway vehicles are between 50 and 55 ft in length. However, they should have extra space for maneuvering, opening and closing doors, and other safe operations. In addition, the design of the facility should anticipate the lengthening of highway vehicles in the future.

Figure 18.5.3 shows the application of these same design features to an inside truck dock with a mezzanine office above the trucks. It should be noted that sprinklers are required inside the truck area below the mezzanine floor. It is also recommended that a signal light be installed to assure the driver's knowledge of the door position before moving out of the dock area. In some instances, a fire door is required between the enclosed truck dock and the interior of the plant. This should be a guillotine door wherever possible. A 15-ft clearance is required for the truck doors which should not be less than 12 ft wide on 15 ft centers. The same dock section and apron dimensions apply in this situation as in the exterior dock.

Figure 18.5.4 shows the dimensioning of a typical building face door. Figure 18.5.5 shows the arrangement of a weather seal and bump posts on such a door.

Figure 18.5.6 shows the application of an air curtain to a truck dock door. This method or the use of a blanketing heater as shown in Fig. 18.4.1 can conserve energy and prevent both open-door heat loss and dock area discomfort. It is generally recommended when servicing conventional highway vans that the door opening be 9 ft × 9 ft with appropriate bumpers and safety devices. The installed dock leveler should be on a 7 ft × 7 ft design with a 51-in. service height from the bottom level tangent line of the truck wheels and a rise of 12 in. and a drop of 9 in. from level. Figure 18.5.7 shows typical dock leveler pit designs. In all cases, the bottom of the pit should slope outward to permit drainage and flushing of the pit with water to clean it out. The pits should be rimmed with curb angles and constructed of poured-in-place concrete. The dock leveler should be a spring or counter-balanced type with safety devices to prevent it from dropping when the truck moves away, cross-travel locks to permit forklift truck cross travel, and an extendable lip to assure proper contact with the vehicle. In most cases, a mechanically counterbalanced dock leveler is quite appropriate. The type that can be released to rise and then placed by walking onto the leveler is usually adequate. Hydraulic-powered levelers are also available.

There are several additional features that should be mentioned in this connection. Figure 18.5.8 shows a Z section track guard on the overhead door. The use of the Z sections protects the door tracks from impact by fork trucks, pallets, and other devices. It also serves as an additional security measure. In order to break into a warehouse through a truck dock door, a burglar generally tries to pry the door track loose from the wall with a pry bar to slip in along the side of the door. Using the Z section track guard over the guide rail helps to prevent prying of the rail from the interior of the wall. Placement of a door on the outside of the warehouse is a clear invitation to burglary and weather leakage. The isolation of the shipping and receiving offices and the wide cross aisle in the dock area also help to limit access to the interior of the warehouse.

Figure 18.5.9 shows a typical dock construction of poured concrete. The floor deck should extend

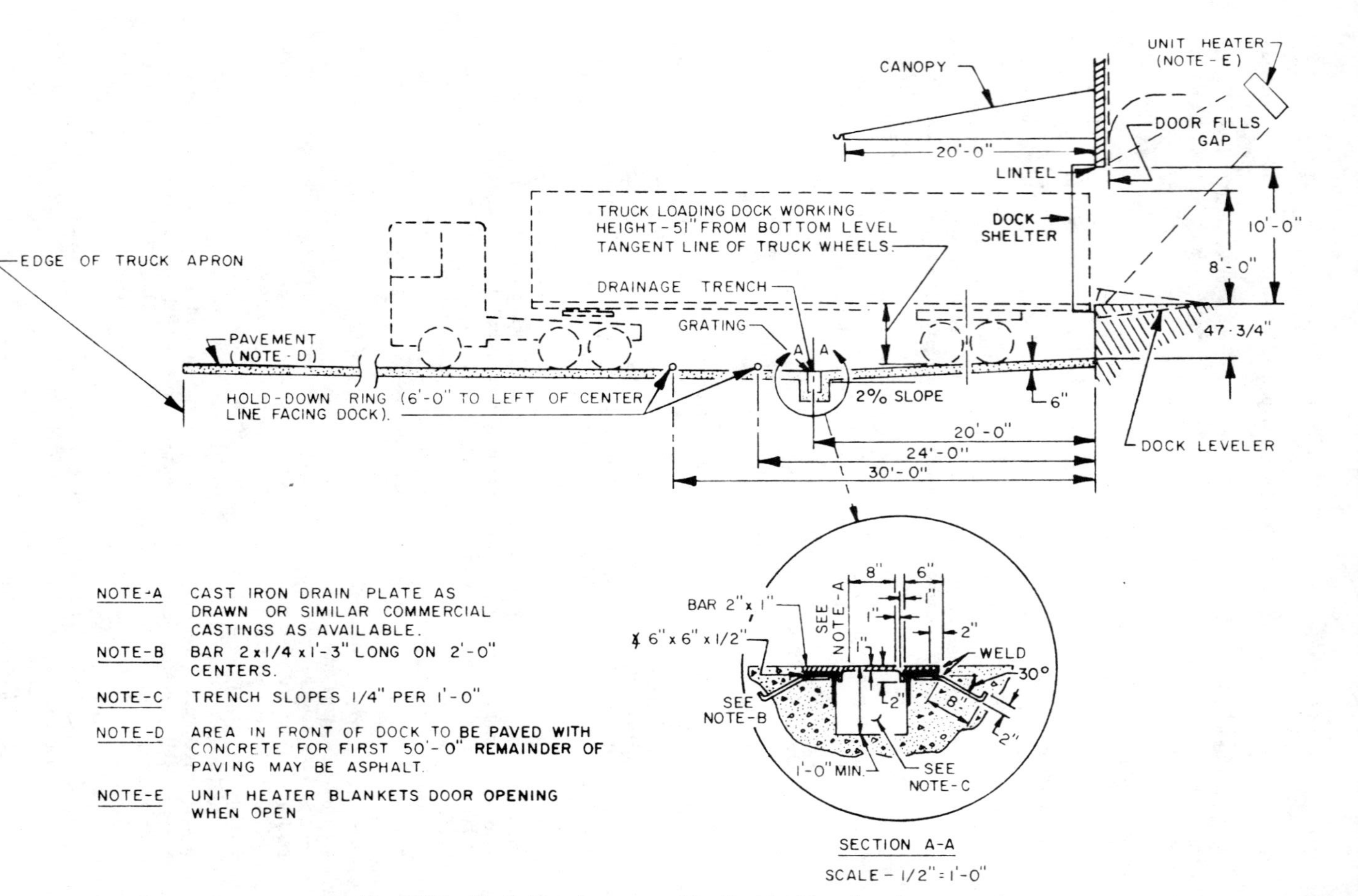

Fig. 18.5.1 Typical truck apron and loading dock, section view.

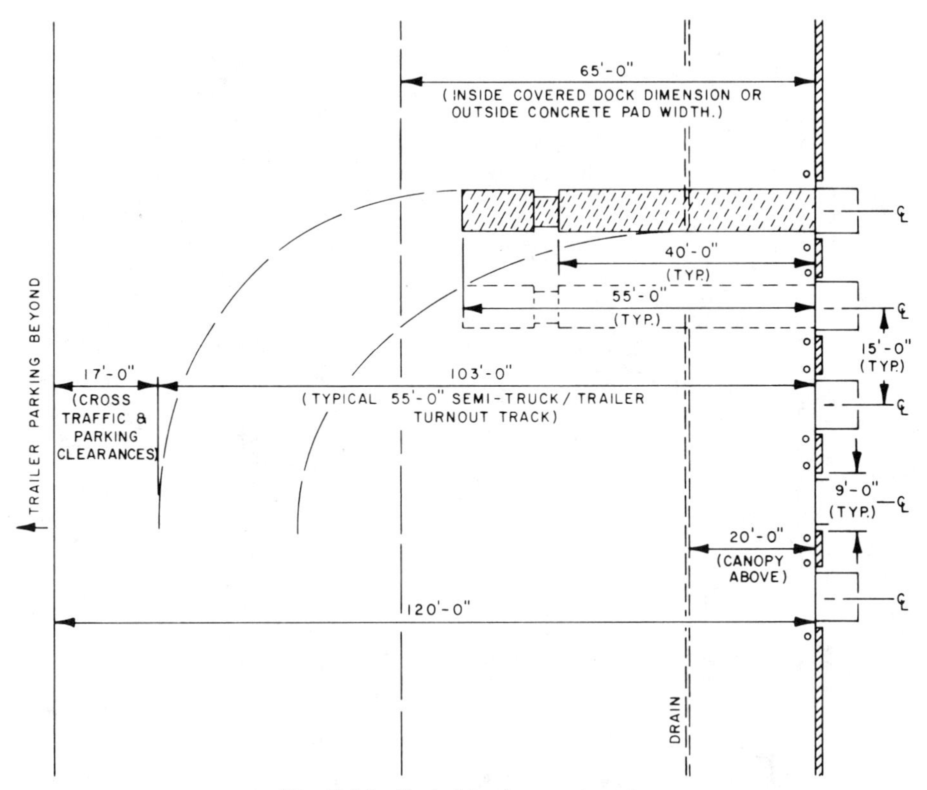

Fig. 18.5.2 Typical truck apron layout.

beyond the soil beam or concrete wall, and the wall should be poured concrete with steel edging along the face and a timber face bumper. Proper construction of a truck dock is essential, since it takes a heavy beating from the impact of vehicles. Damage to a truck dock can occur repeatedly and at any time.

Figure 18.5.8 also shows some of the building protection techniques which can be used around truck docks. In this instance, it is suggested that placement of 12-in.-diameter concrete-filled pipes adjacent to the dock can limit the ability of a highway truck to hit the building. Also, the rubber bumpers and timbers on the dock face absorb the shock when it does hit. It is frequently desirable to put heavy concrete-filled pipe bumpers on the building corners to protect the structure when these corners are adjacent to truck maneuvering space.

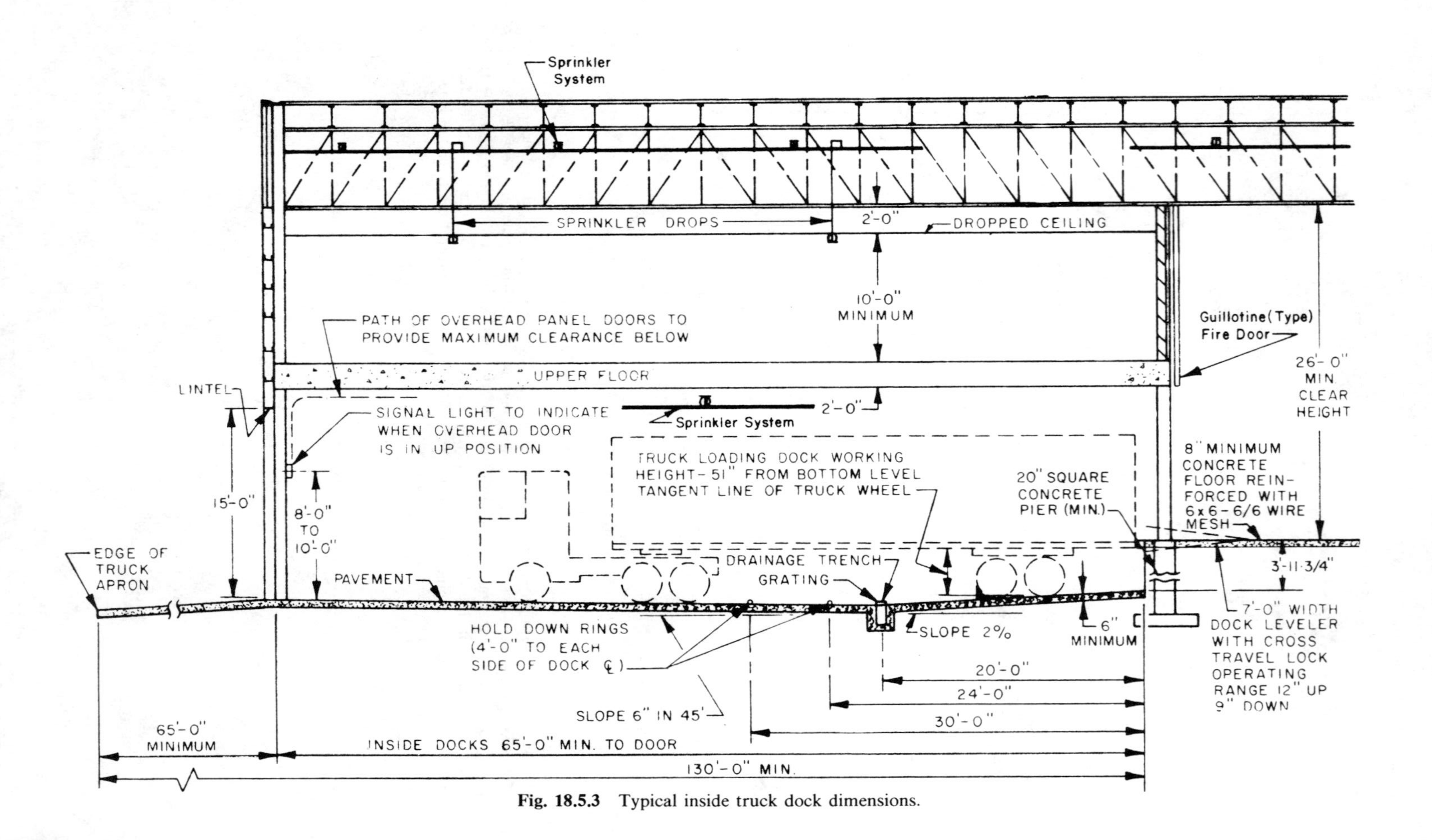

Fig. 18.5.3 Typical inside truck dock dimensions.

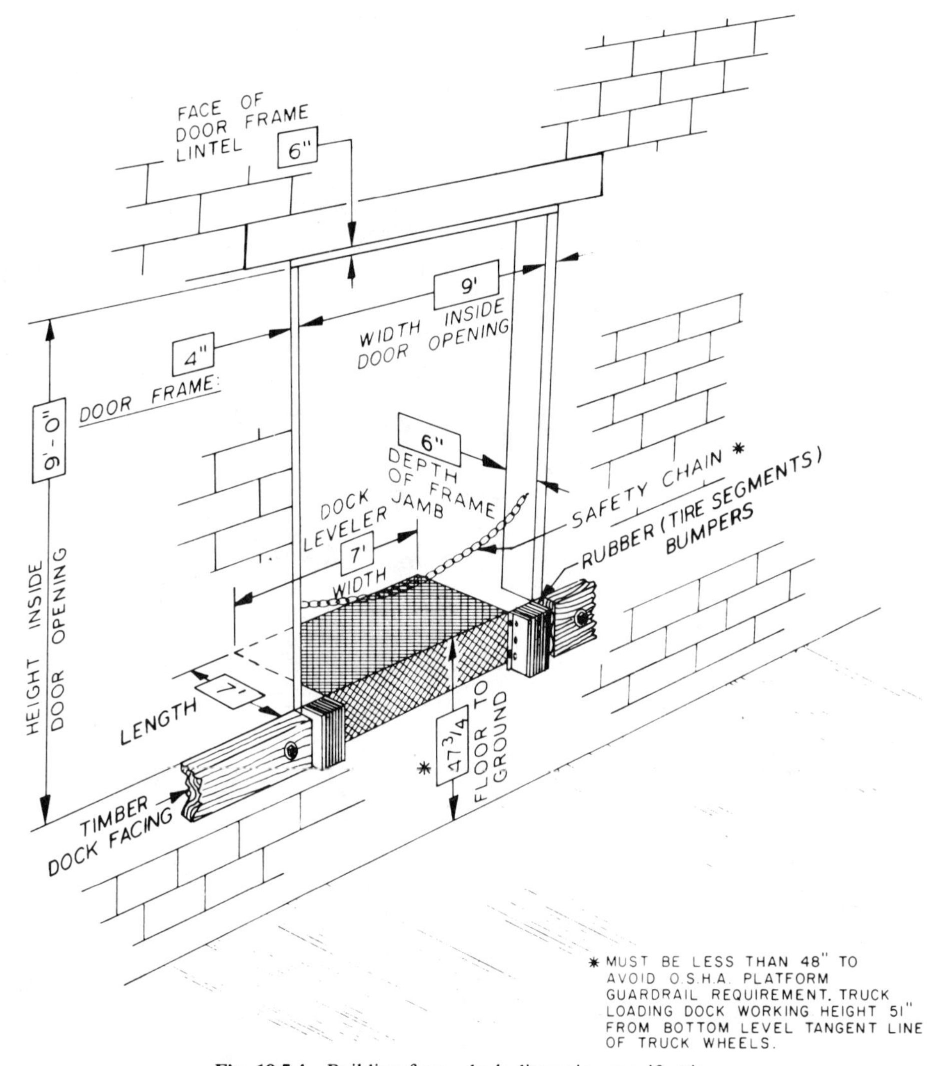

Fig. 18.5.4 Building face—dock dimension specifications.

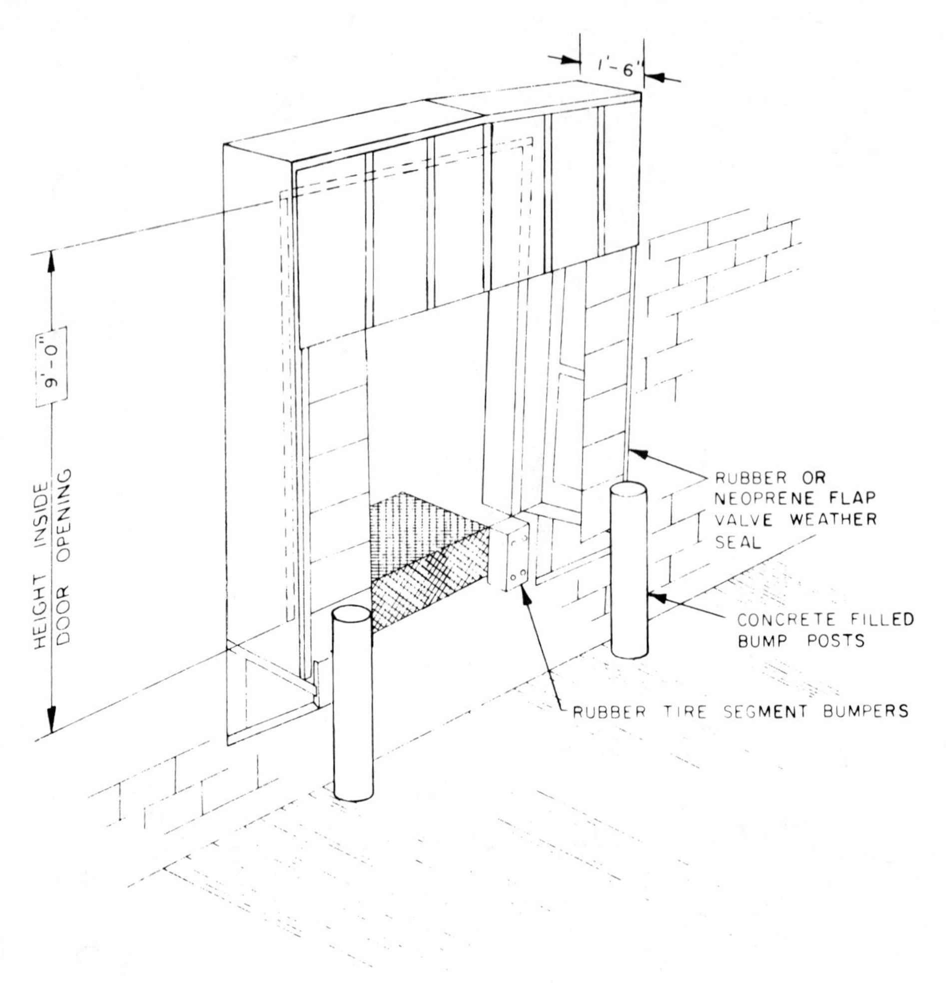

Fig. 18.5.5 Typical dock weather seal.

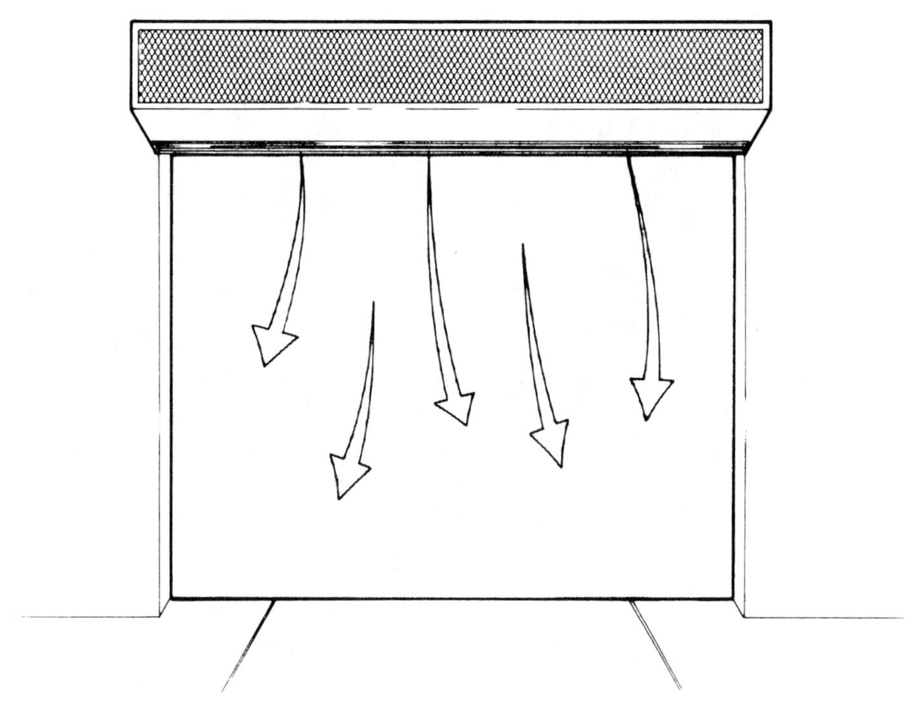

Fig. 18.5.6 Truck dock air curtain

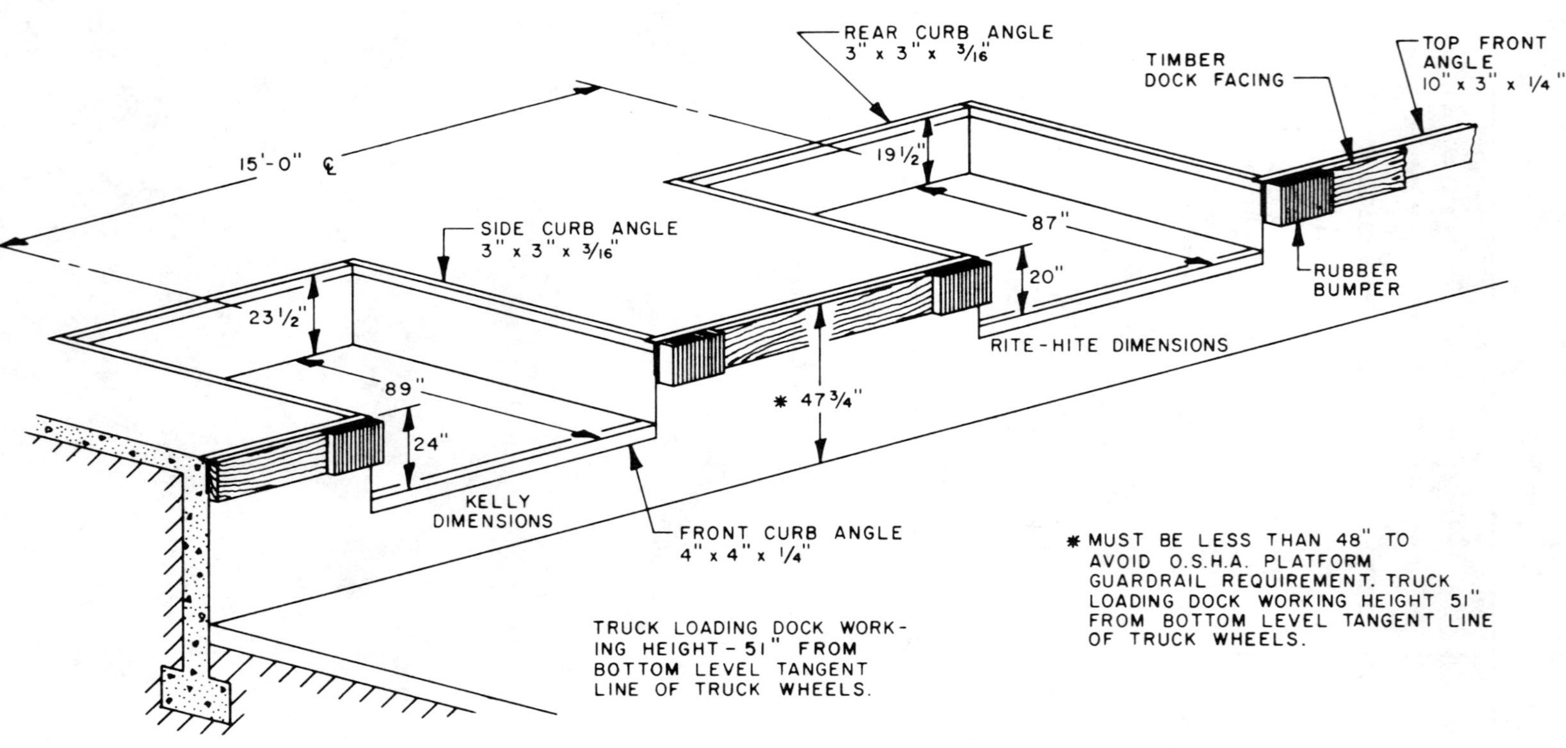

Fig. 18.5.7 Typical dock board pit, nominal 8 ft x 7 ft dock leveler.

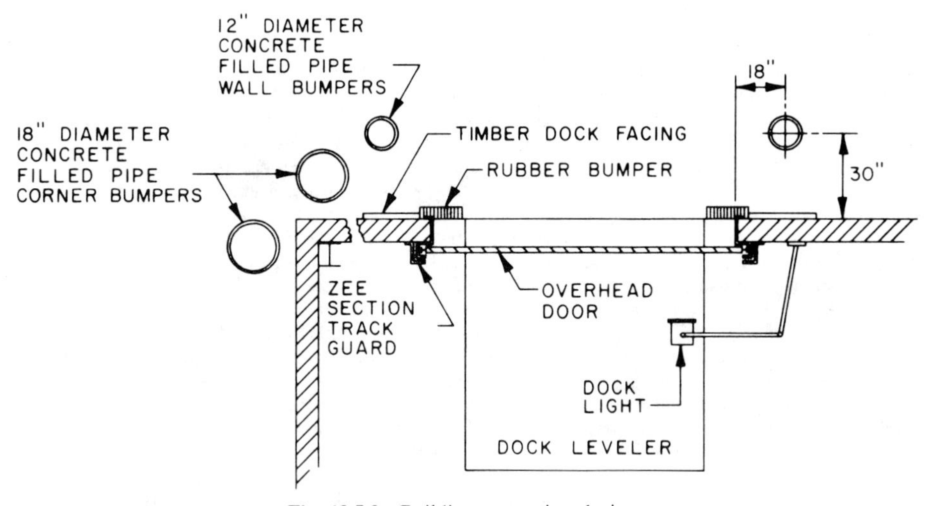

Fig. 18.5.8 Building protection devices.

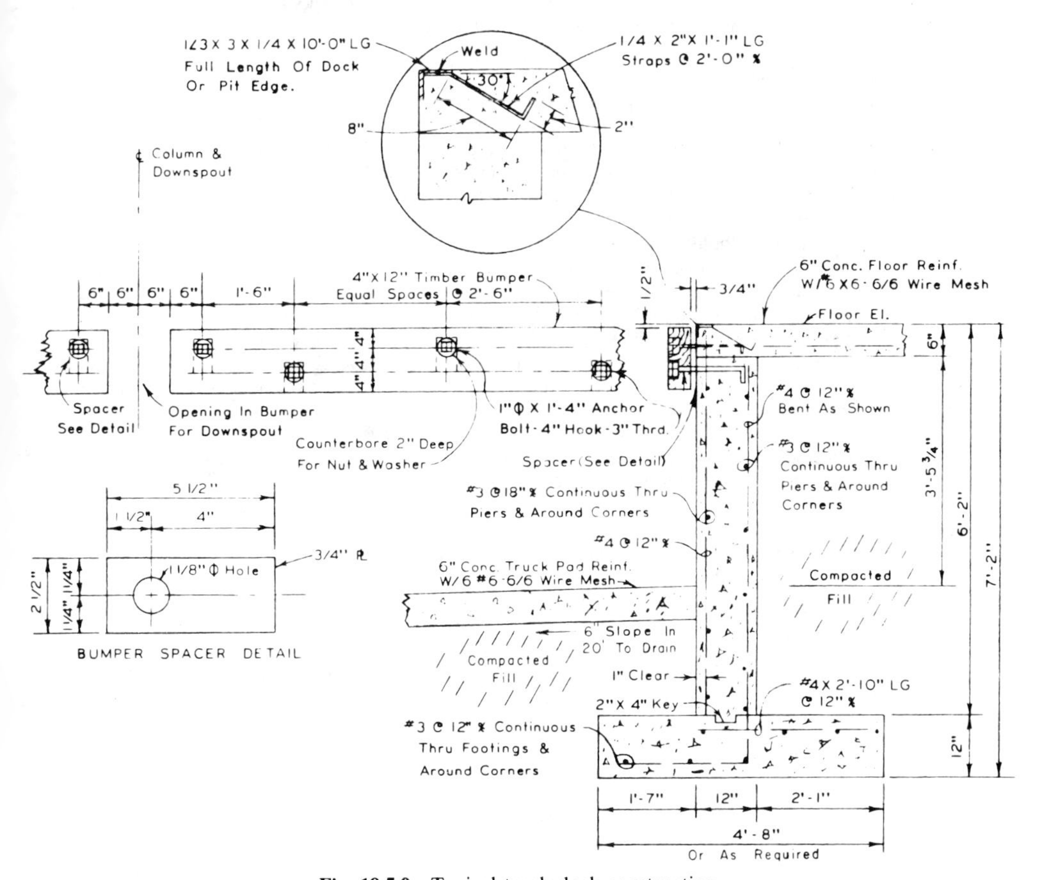

Fig. 18.5.9 Typical truck dock construction.

18.6 NUMBER OF DOCKS

There are many variables in the determination of the required number of docks. The basic element in the analysis is the loading/unloading time required for each vehicle. This varies with the type of cargo, the condition of the load (palletized, loose packages, drums, pallet boxes, bulky goods like pipe, furniture, machinery, etc.) and the type of vehicle (box van, flat bed, open-top van, straight truck, pickup, low boy, etc.). In developing the required dwell time factors, the loading and unloading times must be analyzed and timed or the time estimated by synthetic means (MTM, work sampling, experience, history, etc.). Truck dock turnaround between loads must also be estimated.

With these figures in hand, the analyst must then sample or predict the number of trucks per shift and define the peak arrival and departure times. By synthesizing or simulation of the vehicle activity pattern, the analyst can define the net required truck dock occupancy hours. In general, a 70–80% occupancy at peak hours can be used as a base requirement. Business activity projections should then be applied and the projected number of docks should be rounded upward to the next highest amount at the 5-year future activity level. In most instances a minimum of two docks should be built to provide a safety factor for growth, surges, and broken-down vehicles blocking operations. Although there are many mechanized techniques available, no one has as yet devised a good way to mechanically load miscellaneous freight. All of the mechanical methods require custom application designs and high-volume operations for their economical use.

The design of the truck dock in a manufacturing or distribution facility is not just a matter of cutting a hole in the wall and putting a door in it. The truck dock must be designed in such a manner that the operations can be handled safely and efficiently, security can be maintained, and the dock can be weatherproof. The correct number of docks must be built and suitable dock equipment should be applied. It is essential to recognize that the truck dock is the interface between the internal controlled structure of a business organization and the external and out-of-control transportation element of the company's business. The dock should be designed to provide adequate facilities for the trucker, to provide good shipping and receiving service, and to prevent the truckers from interfering with operations within the warehouse or factory. The whole structure of a dock system should be built around the objectives of efficiency, safety, security, and customer service.

18.7 RAIL DOCKS

The interior work area layout and building and door considerations of a rail dock operation are similar to those required for trucks. However, because of the varying rail car lengths and configurations, and the longer loading and unloading cycles, outside rail docks are losing favor and are being replaced by inside rail wells. The inside operation eliminates weather and security problems and permits more flexible work scheduling.

The inside rail dock should always be placed along a wall to avoid creation of a cross-traffic barrier. It should also be laid out at right angles (90°) with the truck docks to permit independent expansion of capacity and joint use of equipment and personnel. Figure 18.7.1 shows a section through a typical inside rail dock. Figure 18.7.2 shows the standard rail clearance dimensions.

Whenever possible, rail sidings should permit bypass traffic by laying an extra siding outside the building. This can permit on-site car spotting and reduce demurrage charges. A private car mover (Fig. 18.7.3) can be a good investment in large rail operations. It is also useful to have a private piggy-back ramp when heavy "pig" traffic is a factor.

Rail dock boards are similar to those used for trucks and dock levelers are not usually appropriate. However, dock-mounted, swing-up plates (Fig. 18.7.4) are often used in place of dock boards. They reduce accident risk, cannot be misplaced, store safely, and accommodate various car deck conditions (box cars, "reefers," flat cars, etc.) and cushioned cars with safety. Rail dock edges should also have steel angle curbing to avoid sprawling and chipping.

The same conveyor loading techniques as in truck operations can be applied to rail docks. However, trucks, piggy-back, and container freight seem to be taking packaged cargo from rail operations. The most common industrial rail operations involve bulk or tank cars and docks are not required.

In heavy operations it is often useful to place a bridge or gantry crane over the rail dock (Fig. 18.7.5) to unload large or heavy items. The bridge should straddle the track and part of the dock area to permit the unloading of heavy items, pipe, steel, machinery, and so on, from flat or gondola cars.

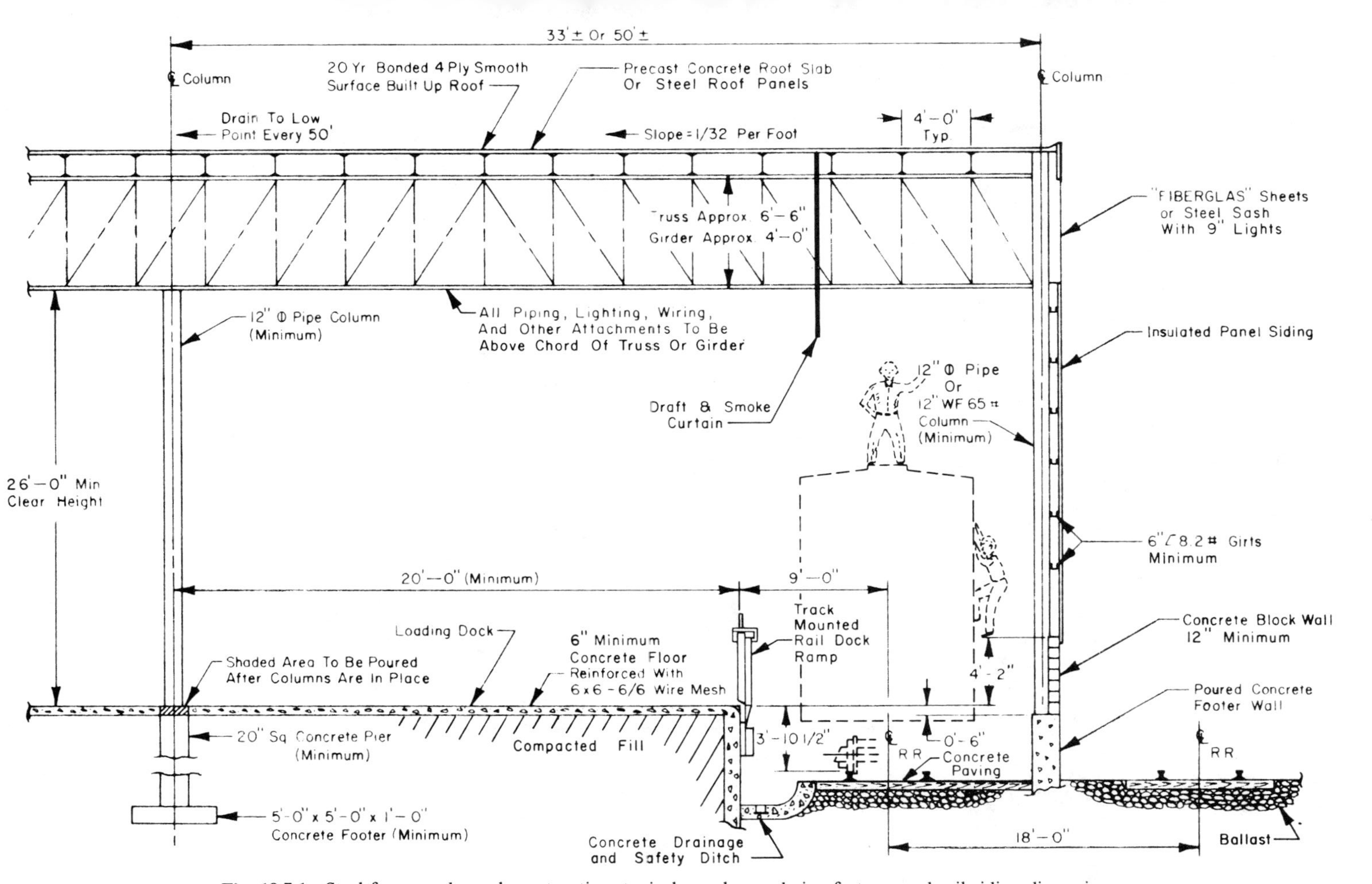

Fig. 18.7.1 Steel frame and panel construction, typical warehouse design features and rail siding dimensions.

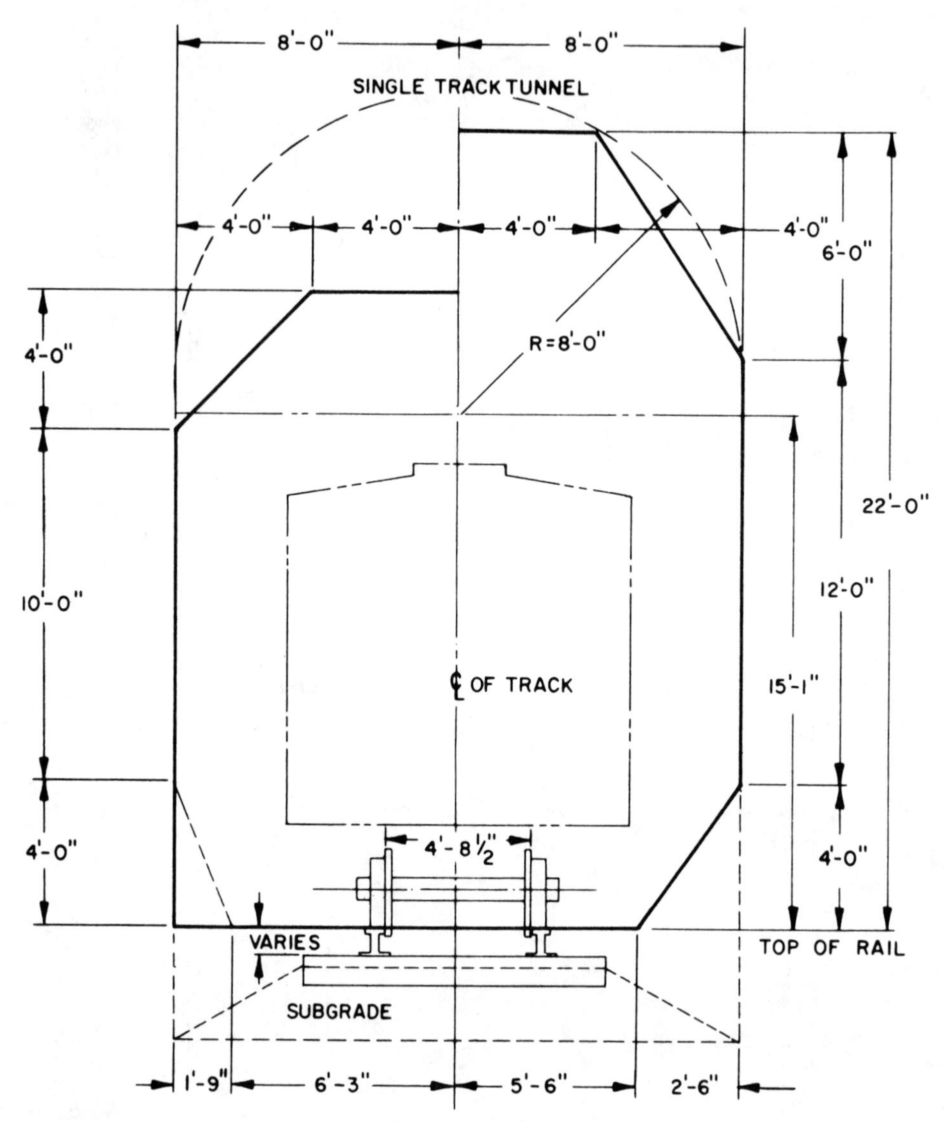

Fig. 18.7.2 Railway clearances (single track).

Fig. 18.7.3 Rail car mover.

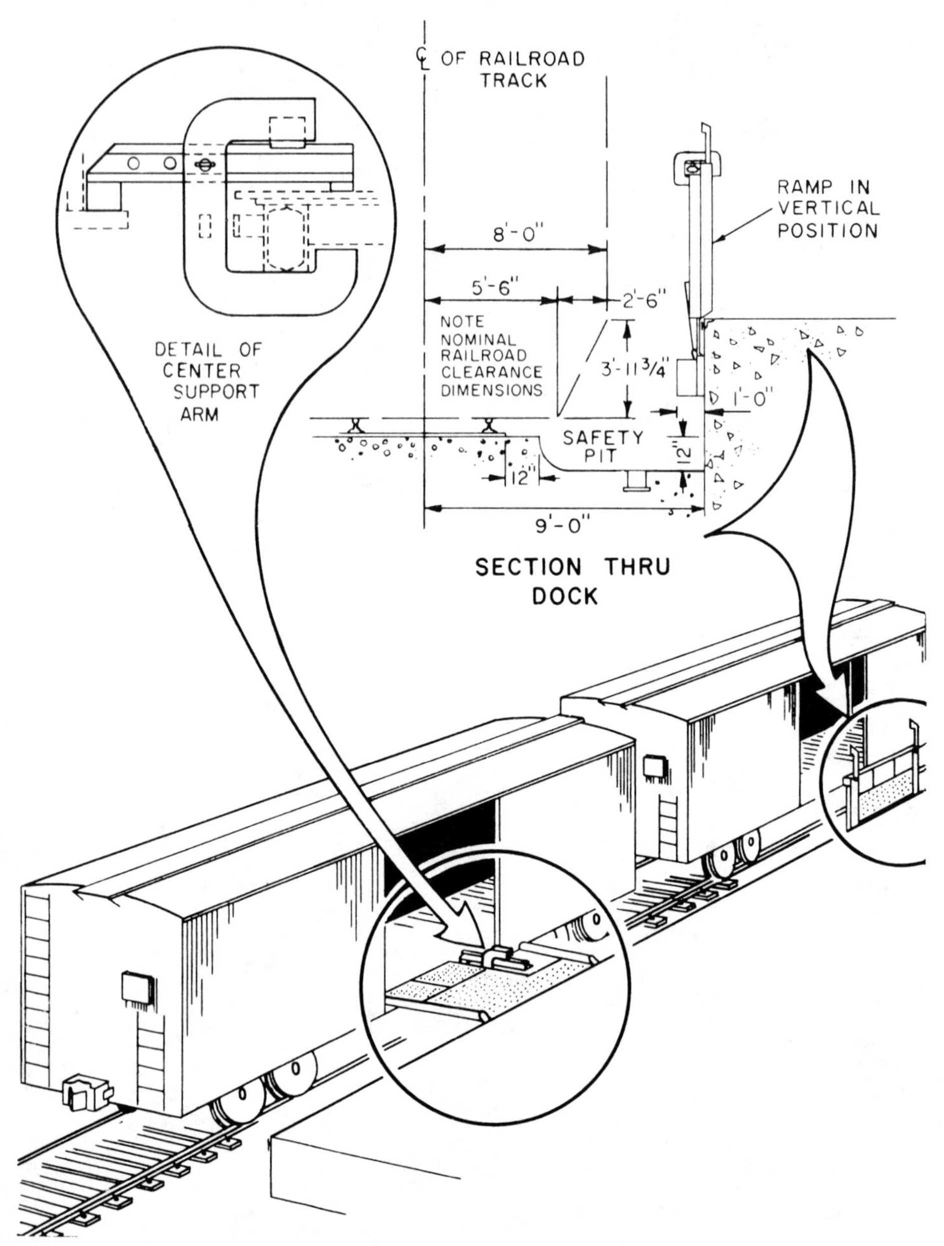

Fig. 18.7.4 Dock-mounted, swing-up plates for rail cars.

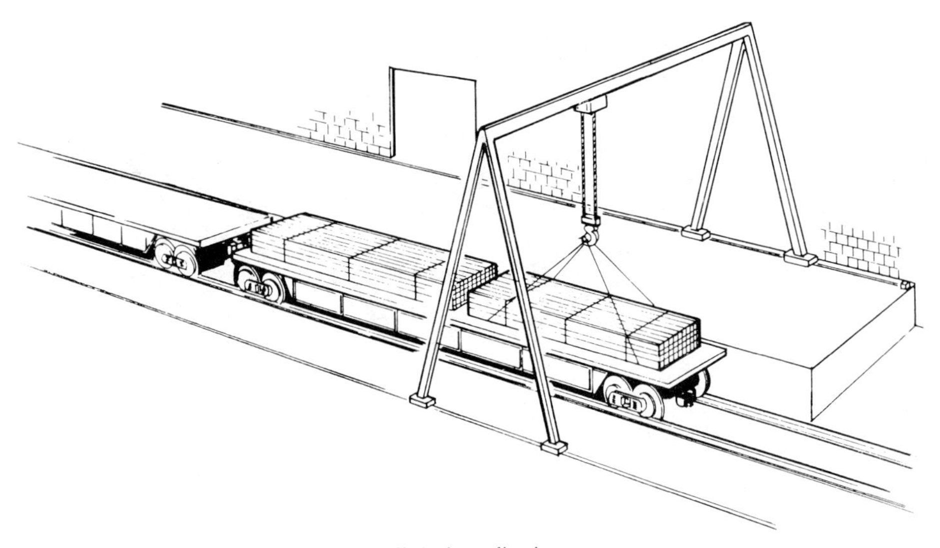

Fig. 18.7.5 Rail dock application, gantry crane.

18.8 MECHANIZED LOADING

In many instances, high-volume loading operations can be designed to use conveyor-supported dock facilities. When uniform or similar packages or bagged goods are to be loaded, extendable conveyors can often be used to deliver product from picking lines or production directly into the vehicle. In such instances, it is often desirable to also provide for palletized bypass delivery of goods into the van (Fig. 18.8.1).

When palletized goods are to be unit-load shipped, it is often desirable to retrieve the pallet and load the goods with clamp trucks or on expendable slipsheets. Figure 18.8.2 shows a pallet retrieval system for a high-volume drum and bag shipping facility. In this case, the drums are loaded by a clamp-equipped forklift and the slipsheeted bags are removed from the pallets for loading.

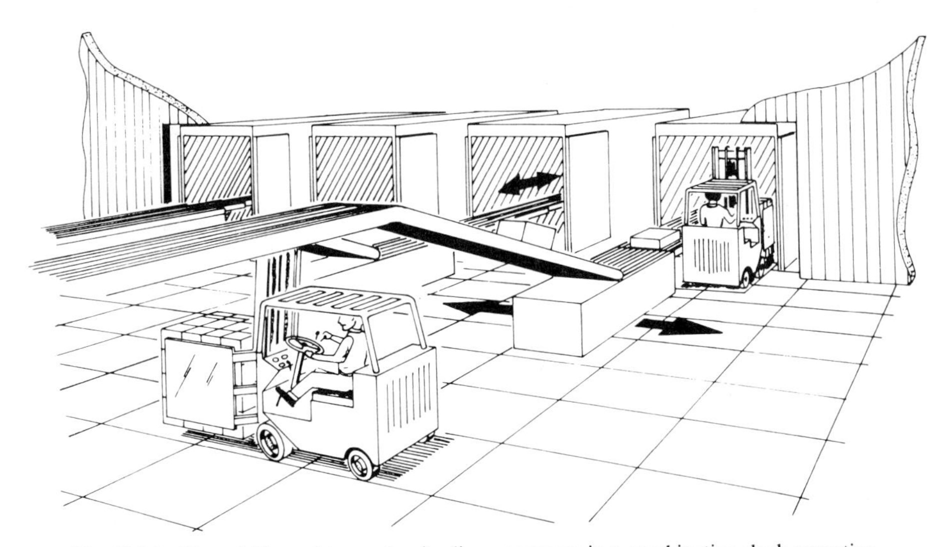

Fig. 18.8.1 Extendable and traversing loading conveyors in a combination dock operation.

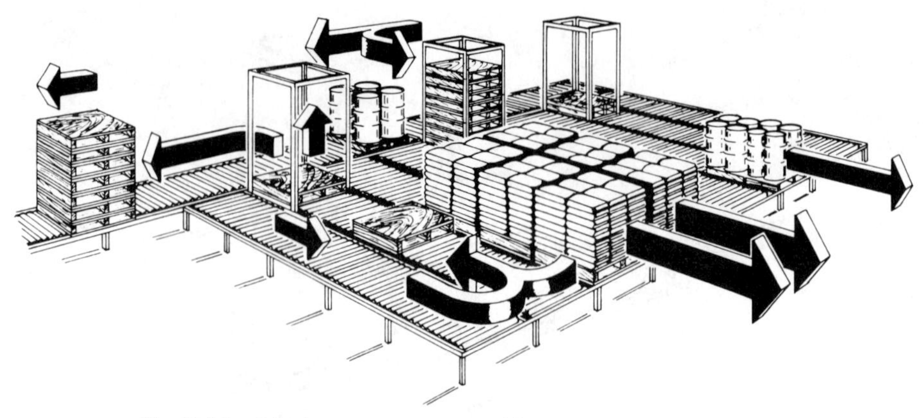

Fig. 18.8.2 Shipping conveyor system with automated pallet retrieval.

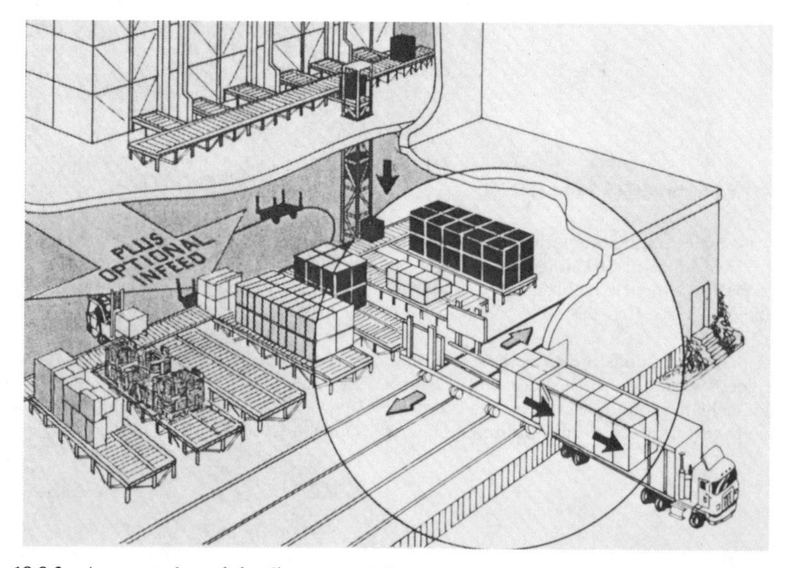

Fig. 18.8.3 Automated truck loading system. Courtesy Automatic Truckloading Systems, Inc.

Mass loading of trucks is also possible by prearranging loads on conveyors or slider docks and mechanically pushing the whole load into the vehicle. This accelerates dock turnaround and reduces labor cost. Figure 18.8.3 shows one such system.

The most common mechanized loading procedure uses forklift trucks and pallets. In this case, the pallet size is critical to achieve a snug and safe load fit and the forklift must have high freelift and a low-headroom mast and overhead guard.

PART III
BULK MATERIALS HANDLING

CHAPTER 19
INTRODUCTION

JAMES NOLAN

Stubbs, Overbeck and Associates, Inc.
Houston, Texas

Bulk materials have varied and sometimes strange personalities. The Conveyor Equipment Manufacturers Association (CEMA) has specified 37 characteristics that can be identified. These can change with temperature, humidity, time, and so on. This chapter discusses the classification of materials, methods of handling, storage, retrieval, and equipment used. The engineering of equipment and systems for bulk material handling requires an understanding of the characteristics, knowledge of the factors that can change these characteristics, consideration of the effect of the characteristics on the systems, and the ability and ingenuity to design and select components to cope with the problems.

This chapter serves only to introduce data and material presented in detail in subsequent chapters. Original data and data common to more than one chapter are presented.

19.1 MATERIAL CLASSIFICATION

An understanding of the material to be handled is essential in the design and selection of any bulk handling system. ANSI/CEMA publication No. 550–1980, *Classifications and Definitions of Bulk Materials,* is excellent for reference. (ANSI = American National Standards Institute.) The engineer must consider the characteristics of the material in all of the conditions that may be encountered, including atmospheric conditions. These conditions change from season to season and from location to location. This is especially true for materials that are transported over long distances. The effect of the transportation must be considered. Factors such as vibration, exposure to climatic changes, and changes caused by artificial atmospheres (both intentional and accidental) may be important. It is usually valuable to see the material in the form, quantity, and surrounding conditions where the system will be used.

A microphotograph showing the shape, size, and interrelationship of material particles helps the engineer visualize how the material will perform during handling, storage, and retrieval (see Fig. 19.1.1).

Special attention must be paid to the safety aspects of the system. Consider the possibilities of reactions due to time, atmospheric changes, stray materials coming in contact with the materials being handled, fire, or accidental spills. It is important that bulk materials characteristics be verified for each situation. Generalizations can cause the engineer to overlook characteristics that could seriously affect the system.

The classification factors identified in ANSI/CEMA 550 show a material code as follows:

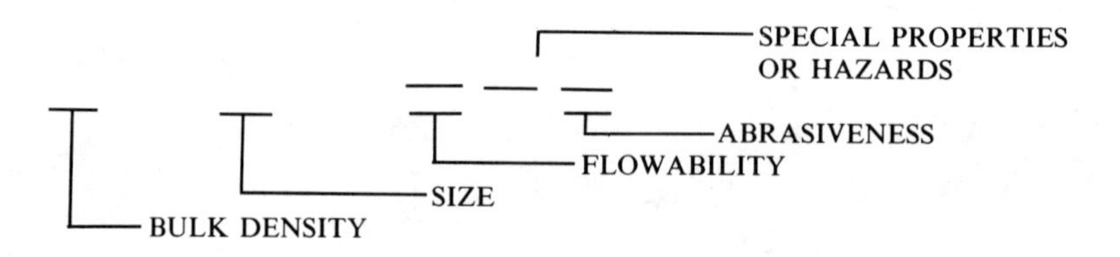

Materials classifications are shown in Table 19.1.1.

Vibrating screens are frequently used for determining classification, particle size, and distribution. Laboratory models can select screen sizes based on visual observation of the material to be tested. Dry granular material is easily checked. Particle sizes may show a typical distribution as shown in Table 19.1.2.

Normally slight losses occur because some material sticks to screens. A skillful technician starts each run with clean screens. Normally a balance scale that can weigh to 1 mg is adequate. At least three runs should be made and averaged. Maximum and minimum limits should be noted. This data may be of help in determining the type and size equipment required.

Coarse materials (over $\frac{1}{2}$ in., or 12.7 mm) would normally require larger screens and larger samples. References for testing screens are ANSI/CEMA 550–1980, Z23.1–1939, and ASTM E11–39. See Table 19.1.3 for ASTM standard screen sizes.

Air classification can be used. However, it is not as reliable because exact size separation is difficult to achieve. Very fine materials can be photographed with aid of an electronic microscope. Enlargements where one micron is equal to 1 mm are used as illustrated in Fig. 19.1.1.

The code table (Table 19.1.1) does not cover all situations. Figure 19.1.2 lists additional information to help classify material characteristics. This chart is useful as a check list to insure that all factors are considered in classifying materials. Using this chart will point out trouble areas that may require additional testing before bulk handling equipment is specified or designed.

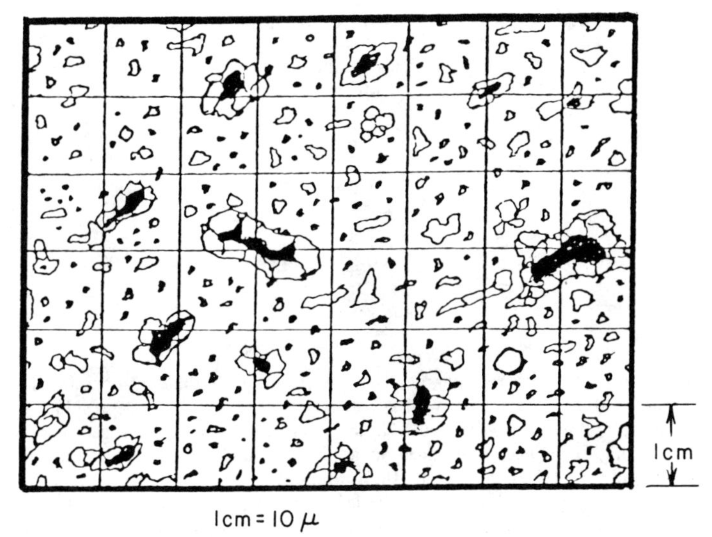

Fig. 19.1.1 Copy of microphotograph of dust particles.

19.1.1 Similarity Is Not Sameness

A perfect example is feathers and goosedown. A problem existed at one time when feathers were used for stuffing pockets in sleeping bags. Preparation included washing and drying. The actual finished specifications required goosedown. A system that was designed on the testing of feathers worked beautifully. When the goosedown arrived and the production line was ready to go, the goosedown did not behave in the same way as the feathers. After it was washed and dried, the goosedown floated in the air like cigarette smoke. It was very difficult to control and measure the correct quantity to put into one pocket of a sleeping bag.

Table 19.1.1 Materials Classification Codes

Parameter	Material Characteristics	Class
Size	Very fine; 100 mesh and under	A
	Fine; $\frac{1}{8}$ in. mesh and under	B
	Granular; $\frac{1}{2}$ in. and under	C
	Lumpy; containing lumps over $\frac{1}{2}$ in.	D
	Irregular; being fibrous, stringy, or the like	H
Flowability	Very free flowing; angle of repose up to 30°	1
	Free flowing; angle of repose 30–45°	2
	Sluggish; angle of repose 45° and up	3
Abrasiveness	Nonabrasive	6
	Mildly abrasive	7
	Very abrasive	8
Other characteristics	Contaminable, affecting use or saleability	K
	Hygroscopic	L
	Highly corrosive	N
	Mildly corrosive	P
	Gives off dust or fumes harmful to life	R
	Contains explosive dust	S
	Degradable, affecting use or saleability	T
	Very light and fluffy	W
	Interlocks or mats to resist digging	X
	Aerates and becomes fluid	Y
	Packs under pressure	Z

Table 19.1.2 Typical Particle Size Distribution

Mesh	Particles Retained on Mesh	% of Sample
10	9 g	4.5
20	42 g	21.0
30	121 g	60.5
50	25 g	12.5
Pan	3 g	1.5
Total	200 g	100.0

Another product has very different characteristics under slightly different conditions—asphalt. Asphalt is a familiar product used in highway construction; it is a low grade residual by-product of petroleum processing by refineries. Mixed with sand and fine gravel, it makes a very good road surface. The same material, highly refined, is used as an additive in drilling fluids for controlling viscosity. This is an entirely different situation. A very slight amount of moisture affects the way it handles in grinding, storage, mixing, bagging, and use in the field when added to drilling mud.

Another product that is very difficult to handle is fertilizer. It can be ammonium phosphate, ammonium nitrate, or urea. Very slight changes in atmospheric conditions, such as humidity, can change a free-flowing material to a sticky mess. In a warehouse storing ammonium phosphate there will seem to be dust on the structure. But it is evident that it is not dust when it is touched. It is wet and sticky due to the hygroscopic quality of the ammonium phosphate. Most chemical fertilizers exhibit the same characteristic of being very hygroscopic. They attract moisture and change their flow characteristics.

Coal is another product that can vary drastically with a slight change in moisture content. Fines will build up on chutes and transfer points on belt conveyors, bucket elevators, and other equipment when a slight amount of moisture is present. Very fine dry coal will be no problem. It will flow freely, will fluidize easily, and can be moved by pneumatic or belt conveyors.

There is a vast amount of knowledge about many materials already known by most engineers. These materials are easily handled with the existing knowledge. The state of the art is such that there are very seldom any problems handling known materials. The problems occur with materials whose characteristics are still unknown. These must be ascertained accurately before any design work is started on the materials handling system.

Table 19.1.3 Partial List Testing Screen Sizes, ASTM Standard

Nominal Opening[a]	Mesh Size	Aperture in. (mm)	Wire Diameter in. (mm)
Inches			
1		1.000 (25.4)	0.156 (3.96)
$\frac{1}{2}$		0.500 (12.7)	0.108 (2.74)
$\frac{1}{4}$	3	0.250 (6.35)	0.073 (1.85)
$\frac{3}{16}$	4	0.187 (4.76)	0.055 (1.41)
$\frac{5}{32}$	5	0.157 (4.00)	0.048 (1.23)
$\frac{1}{8}$	6	0.132 (3.36)	0.043 (1.09)
Microns			
2000	10	0.079 (2.00)	0.033 (0.84)
1000	18	0.039 (1.00)	0.021 (0.52)
500	35	0.020 (0.50)	0.012 (0.32)
250	60	0.010 (0.25)	0.007 (0.19)
149	100	0.006 (0.15)	0.0044 (0.111)
105	140	0.004 (0.11)	0.0030 (0.075)
74	200	0.003 (0.074)	0.0021 (0.053)
44	325	0.0017 (0.044)	0.0014 (0.036)

[a] $\frac{1}{2}$ in. and up considered coarse by ANSI 550; $\frac{1}{2}$ in. down to 140 mesh considered granular; > 140 mesh considered powder.

MATERIAL: ______________________

SIGNIFICANT PROPERTIES

	SUB MICRON	POWDER 200 MESH	GRANULAR -10 MESH	-1" -3"
PARTICLE SIZE	-12"			-3"
ANGLE OF REPOSE	0° 70°			35° 35°
BULK DENSITY	5#/C.F. 400#/C.F.			100#/C.F. 100#/C.F.
SHAPE	FIBRILLAR			SPHERICAL
UNIFORMITY COEF.	50			2
FLOWABILITY	0			100
FLOODABILITY	100			0
HAZARDOUS	DEADLY			INERT
CORROSIVE	SEVERE			INERT
ABRASIVE INDEX (CEMA 550 A-1)	416			1
HYGROSCOPIC	10% CRIT.HUMID.			100% CRIT.HUMID
DEGRADABILITY	VERY			NOT LIKELY
VALUE	$10.00/#			$0.10/#
TEMPERATURE	100°C -18°C			20°C 20°C
FLAMABILITY	VERY			INERT
STATIC ELECTRICITY	HIGH POTENTIAL			NIL
DECOMPOSABLE	LIKELY			NEGLIGIBLE
PURITY	99%			80%
COHESIVENESS	SEE TABLE I			
DUST PRODUCING (SEE CEMA B-8)	5%			0%
	HARMFUL			HELPFUL
SPECIAL PROPERTIES (DESCRIBE BELOW)	VERY SPECIAL CONSIDERATION	SPECIAL CONSIDERATION	CAUTION IN THIS AREA	ROUTINE MATERIAL

STUBBS OVERBECK & ASSOCIATES, INC.

SPECIAL PROPERTIES: ______________________

Fig. 19.1.2 Bulk materials evaluation chart. Courtesy Stubbs Overbeck & Associates, Inc.

19.1.2 Testing Materials

There is a vast amount of data on material characteristics available to the engineer. When conditions vary from previous experience it is necessary to investigate the particular material. Certain characteristics are determined by vision, feel, laboratory tests, pilot plant, and microscopic examination. It is easy to determine the angle of repose, the angle of slide, and the horizontal angle of shear of most materials with a simple laboratory examination. When material characteristics are unknown, testing should be done under the appropriate conditions. All of these procedures are discussed in later chapters in detail.

19.1.3 Hazardous and Toxic Materials

Definition of hazardous materials depends on whether the substance is pure when shipped for sale. This category is under the auspices of the Department of Transportation and is listed in federal code regulations CFBR 49, parts 100 through 199. In brief, any material that is flammable, toxic, corrosive, or explosive is designated as a hazardous material. Other materials such as waste and garbage are a nuisance although not necessarily hazardous. They do require special consideration. Handling of nuclear materials is an entirely different matter controlled by national and international regulations.

Environmental requirements for safety should be followed by those who generate, store, dispose, or transport hazardous materials from one place to another. These requirements are controlled by local, state, provincial, or national governments.

Many compounds have been declared hazardous in the national emissions standards for hazardous air pollution, and very strict exposure limits to these compounds are specified in the regulations.

Safety requirements on the transportation and handling of hazardous materials of the various international agencies center around labeling, recording, and manifesting. The efforts are directed toward any spill of hazardous materials. Proper labeling of shipment of hazardous materials lets those at a spill site know precisely what is in the shipment. The manifest spells out the contents of the truck, tank car, barge, ship, or pipeline. With this information, people who are involved in protection of the public against spills, fire, and/or explosion will know how to handle the specific situation.

Obviously, for health reasons, a person should be protected from contact with hazardous materials and their vapors. People handling or working around hazardous materials should have proper safety

Table 19.1.4 List of Standards

Standard Number[a]	
ANSI B20.1	Safety Standards for Conveyors and Related Equipment
ANSI Z35.1	Accident Prevention Signs, Specifications for
ANSI C136–71	Sieve or Screen Analysis of Fine and Coarse Aggregates, Method of Test for
ANSI C85.1	Automatic Control, Terminology for
ANSI B30.5	Crawler, Locomotive and Truck Cranes, Safety Code for
ANSI PTC9	Displacement Compressors, Vacuum Pumps and Blowers
ANSI Z94.6	Facility Planning
ANSI MH113.1	Metal Belts, Glossary of Terms and Definitions, Standards and Specifications
ANSI S5.1	Measurement of Sound from Pneumatic Equipment, Test Code for
ANSI A12.1	Floor and Wall Openings, Railings, and Toeboards, Safety Requirements for
ANSI B15.1	Mechanical Power Transmission Apparatus, Safety Standards for
ANSI A121.1	Vibrating Screens—Terms and Definitions
ANSI Z210.1	American National Standards, Metric Practice
ASTM A454–72	Steel Conveyor Chain, Specifications for
ASTM E105–58	Probability Sampling of Materials, Practice for
ASTM D518–61	Resistance to Surface Cracking of Stretch Rubber Compounds, Method of Test for
ASTM D430–59	Ply Separation and Cracking of Rubber Products, Methods of Dynamic Testing for
ASTM E11–70	Wire-Cloth Sieves for Testing Purposes, Specifications for
CEMA 102	Terms and Conveyor Definitions
CEMA 350	Screw Conveyors
CEMA 300	Screw Conveyor Components
CEMA	*Belt Conveyors for Bulk Materials,* CBI Publishing Co., Inc., Boston, MA, 1979
ANSI/CEMA 550	Bulk Materials, Classification and Definitions of
NFPA 61D	Agricultural Commodities for Human Consumption, Prevention of Fire and Dust Explosion in the Milling of
NFPA 653	Coal Preparation Plants, Prevention of Dust Explosions in
NFPA 63	Prevention of Dust Explosions in Industrial Plants, Fundamental Principles for the
NFPA 13	Installation of Sprinkler Systems
NFPA T3.24.69.5	Mechanical Vacuum Pumps and Blowers, Method of Rating for
UL 877	Circuit Breakers and Enclosures for Use in Hazardous Locations, Class I, Groups A, B, C and D, and Class II, Groups E, F and G, Safety Standards for
UL 894	Switches for Use in Hazardous Locations, Class I, Groups A, B, C and D, and Class II, Groups E, F and G, Safety Standards for

[a] ANSI = American National Standards Institute; ASTM = American Society for Testing and Materials; CEMA = Conveyor Equipment Manufacturers Association; NFPA = National Fire Protection Association; UL = Underwriters Laboratory.

clothing such as rubber suits, rubber gloves, eye shields, and other devices available at all times. The main consideration here is that approved threshold limits with these hazardous compounds are not exceeded. The National Fire Protection Association (NFPA) has several standards applicable to these hazardous substances. These citations are in the 1981 edition for the National Fire Code. See the list of standards in Table 19.1.4 for regulations and recommendations.

19.2 BULK SOLIDS STORAGE AND RETRIEVAL

The storage of large quantities of material which may amount to millions of tons requires the most careful study of all facts. A proper choice can be made in each case for the method and kind of storage best suited to the purpose and to the material. All things considered, how the material is received, maintained in storage, and retrieved are very important. Solutions to the problems are not always simple. They start with the origin of the materials which may be a mine, a beneficiation plant for ore, or a port receiving bulk products by ship, barge, truck, or railroad for export or import. Every means must be investigated to determine the best method of handling the specific incoming material before placing it into storage. The design of storage facilities requires consideration of the various functions and the effect of material characteristics on each function. These functions include receiving, handling, storage, protection, and reclaiming for either reshipping or use.

As discussed in Section 19.1, knowledge of material classification and characteristics is an absolute must before determining methods of materials handling systems and equipment. Several points to be considered are: Is the material biodegradable, hazardous, toxic, a producer of dust, chemically reactive to the atmosphere, or inert? We present on the following pages several materials that need to have protected storage. Protected storage includes silos with some nitrogen blanketing or other inert gas, buildings for protection from weather, and roofed areas. Storage in the open is suitable for some materials.

Grain can be stored in silos. In the open, a certain percentage of deteriorated material will be on the surface of the pile.

Some materials are inert no matter what you do to them. Barite, limestone, granite, and sand and gravel can be piled outdoors. Piles can be retrieved by belt in tunnels, front-end loaders, portable cranes, or special retrieval machines mounted on gantries traveling on rails. When excess moisture is present materials may not flow from hoppers easily. This is typical of most granular materials when moist.

Some materials such as fertilizers are hygroscopic and need to have special protection in storage. Some materials such as polyethylene, polystyrene, or polypropylene powders have residual gases remaining in the product. When these are stored in a hopper or bin the emitted gases are subject to explosion if there is a source of ignition such as static electricity and if oxygen is within combustible limits. All these things must be considered in storage and retrieval of many bulk materials. Subsequent chapters deal with various problems in bulk materials handling, such as bin flow, belt conveyors, bucket elevators, screw conveyors, and equipment necessary to handle these materials. Above all, we must remember that material characteristics are very important.

Shown in Fig. 19.2.1 is a modern cement plant illustrating several types of bulk materials handling equipment.

Figure 19.2.2 shows the results of testing inclined screw conveyors. These curves are averages and should not be used with materials of unknown characteristics.

Figure 19.2.3 shows the composite results in testing screw elevator capacities. Abrasive materials can be handled, but at the expenses of very high maintenance and downtime (not recommended except for infrequent and short-life uses).

Figure 19.2.4 shows the results of testing various free-flowing material. The two curves are not absolute limits; however, with known material characteristics, staying within these limits should not present problems. These data are not to be used for metering without some means of adjusting the outlet area.

19.2.1 From Source to Destination

Many different types of equipment are used to transfer material from source to destination. Truck, rail, barge, ship, and conveyors are used for distance transportation of material. Front-end loaders and bulldozers handle relatively short distances.

Millions of tons of iron ore are shipped on the Great Lakes. At port facilities this ore is transferred to railroad cars for delivery to steel mills. On the return trip these same cars carry coal from Appalachia to the Great Lake ports. Very elaborate systems have been set up to handle tremendous volumes of coal and iron ore at the shipping points around the Great Lakes. It is possible to load a ship in Duluth, Minnesota, with iron ore in just a few minutes. That same ship travels across Lake Superior through the locks to Detroit, Toledo, Cleveland, Pittsburgh, Erie, or Buffalo. There it can be unloaded with large clamshell-type unloading machines operating on gantries up and down the dock. The ship can be turned around in just a few hours. There is some spillage taken care of by reclaim hoppers and belt conveyors. To fill the large hoppers which serve the ship for loading, rail systems have been developed which move the rail car up to a point where it is turned over and dumped in a continuous motion as the train moves through. Some installations use bottom dumps with specially designed cars.

At one time, only canoes could go through the rapids between Lake Superior and Lake Huron. As a result of the solid rock configuration of the Sault Ste. Marie locks, ore and coal ships have been developed that are long, slender, and of shallow draft. They barely squeeze through the locks, and travel slowly. However, once they are in the open waters, the long slender design permits very good speed with relatively low horsepower.

A large quantity of materials now is transferred by truck and rail from the point of origin to a consolidation point. Grain is an example of this situation. It is picked up by truck from the farms where it is produced, transferred to a grain elevator near a railroad site, transferred to the rail car and shipped a considerable distance either to a port on the west coast, gulf coast, or east coast for export. It is shipped also to consuming facilities throughout the United States.

One of the things that needs to be considered in storing bulk materials in open stockpiles is the method of reclaiming. This can be with large expensive reclaim machines mounted on gantries and traveling on rails. These are justified only when there are several thousand tons of material to be reclaimed per day. Examples: feeding a large power plant or a large ship-loading installation for export. Some of these methods are very good.

Fig. 19.2.1 Modern cement plant. Limestone is received by barge and unloaded, stored, and retrieved by bridge crane. A belt conveyor system unloads railroad cars and feeds grinding mills and kilns. Inside the plant are bucket elevators, bagging machines, and palletizers. A pneumatic conveyor system is

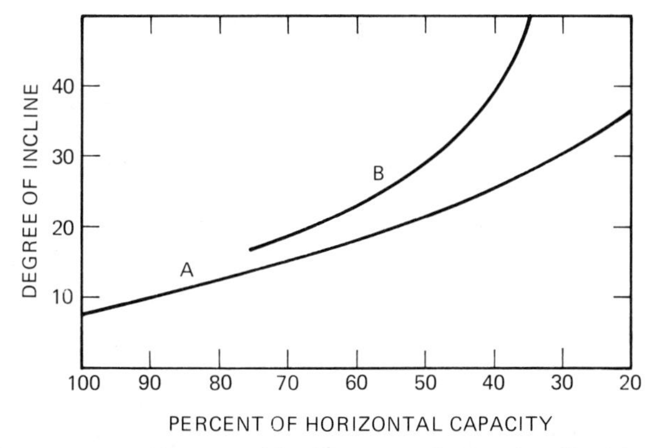

Fig. 19.2.2 Inclined screw conveyor capacities. Use screw elevator data for sizing screws over 45°. A = standard pitch in U trough; B = close fit in pipe casing. Standard horizontal screw, approximately 85% volumetric efficiency. 90% efficiency with close-fitting housing.

A smaller machine typically called a front-end loader can be used for some reclamation operations. Front-end loaders or bulldozers can reclaim material from dead storage to live storage in a very short period of time. Three large front-end loaders handling wood chips can move more than 1000 tons an hour from a stockpile to a conveyor loading point. Handling coal and crushed rock, one front-end loader with smaller buckets can transport several hundred tons an hour.

Normally front-end loaders are used for reclamation of materials where operation is intermittent, as in storage of fertilizer in a warehouse. The shipping requirements dictate that large storage warehouses are necessary. The production of fertilizer is a constant tonnage per hour. It requires large storage capacity accumulation. The fertilizer will be reclaimed, loaded, and shipped in a short period of time.

Even though the production capacity of the plant is scheduled for 10–12 tons an hour, 24 hours a day, seven days a week; usually only one shift is required every two or three days to ship out material.

The same facility may be loading trucks for bulk shipments. The large front-end loader may also be used to feed hoppers which in turn supply bagging machines for bagged materials. Most bags of fertilizer today are 50 lb. These are filled with automatic bagging machines at the rate of as many as

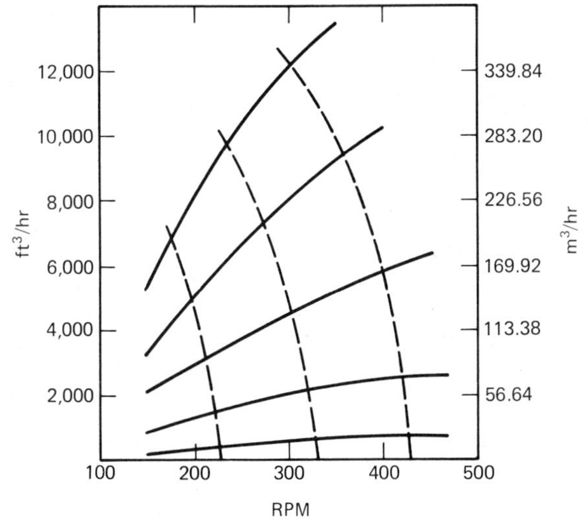

Fig. 19.2.3 Vertical screw elevator capacity.

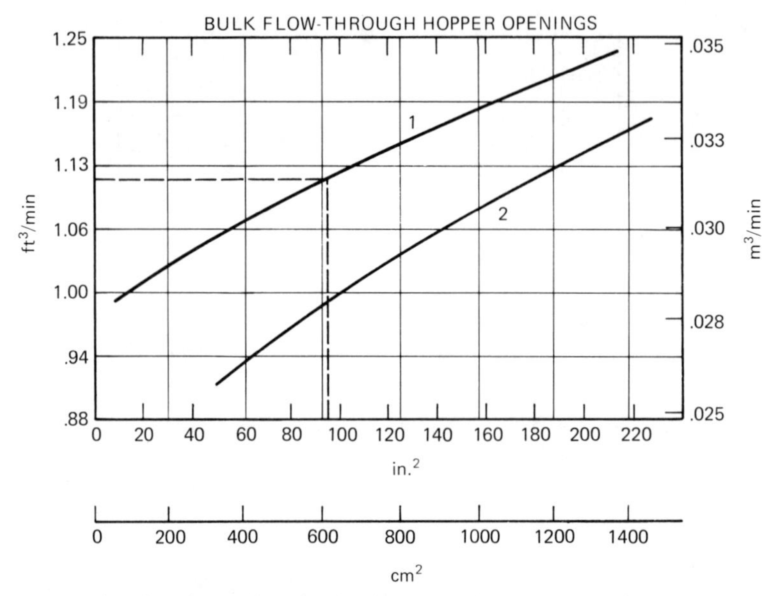

Fig. 19.2.4 Results of testing various free-flowing materials. 1, fine, dry, free-flowing materials; 2, coarse with fines and/or some moisture. Opening must be more than three times size of largest lumps to prevent clogging.

30 bags per minute, 1500 lb/min. The same front-end loaders are easily adapted to feed such a system. Normally a system of conveyors would be used to move the bags from the feeding machine to either trucks, rail cars, or interim storage.

The rate at which materials must be recovered from a stockpile will influence the choice of the method used to reclaim. Obviously a reclaim rate of 50,000 tons in 24 hours will justify some rather expensive and elaborate equipment but it must be reliable. Large power plants can justify this in coal pile reclaiming. Some cement plants justify this kind of equipment. However, their tonnage requirement is not nearly as great as power plants. Export facilities loading ships and barges can usually justify elaborate expensive equipment for loading from stockpiles. However, there has to be a very accurate economic study made of all these systems to determine which is the most cost effective over a long period of time. Amortization of capital must be considered.

One of the methods frequently found in cement plants utilizes bridge cranes with clamshell buckets servicing the storage area. Raw materials are frequently stored under a shed roof when there is no concern about atmospheric humidity, only about rain. Some incoming raw material may be stored in the open in these same plants. Crawler cranes with clamshell buckets or dragline buckets are frequently used for stockpiling and reclaiming in small installations such as sand and gravel pits or ready-mix concrete plants. This type of crane is versatile and portable and can move from one pile to another in order to feed a belt conveyor which feeds a process system. These cranes can be either wheel mounted or crawler mounted. They can transfer material directly from rail cars, barges, open stockpiles, or to and from surge piles, to a dump truck, a rail car, or a large stockpile. They are very mobile. Cranes work well in restricted places and can be used to stockpile and reclaim in odd-shaped areas which are poorly adapted to the use of more elaborate fixed equipment. When the tonnages are low the crane can move from one stockpile to another in a matter of few minutes.

A dump truck can be loaded by a crane or front-end loader with 20 tons from one stockpile. The next dump truck may come from a sand and gravel plant needing 20 tons from a different pile. When the operation is complete, the equipment moves to another pile and continues other operations. This could involve feeding sand and gravel aggregate to a ready-mix concrete plant.

Bulldozers, particularly crawler mounted, are readily adaptable for transferring material from a dump point to a stockpile and reclaiming back to a loading point. They are capable of moving a tremendous amount of material. They have a low footprint area so they can climb up on top of the stockpile and spread it out and make the pile bigger. This is very advantageous to some installations.

Scrapers, sometimes called earth movers, are also used to build stockpiles, distributing the material evenly over a long distance. Scrapers are self-loading vehicles and are thus able to pick up loads from small initial piles formed especially for this purpose. In modern coal handling, rock mining, and sand and gravel operations they are used to move material to a common stockpile. In cut and fill, they take the material from a hill and deposit it in a valley. They do not compact material. Other means must be used for fill to be compacted.

When fines are present in coal stockpiles there is a potential source of fire from spontaneous combustion. For fire prevention these stockpiles must be sprinkled and maintained in a wet condition. Degradation is often a serious problem except when storing coal which is to be crushed and pulverized for use in power plants for boiler fuel or for carbonizing in coking ovens.

Advantages are obtained using the scraper laying method. Scrapers can stockpile using large rubber tires. Under the weight of the heavily loaded scraper, the hauling and spreading will compact the coal by as much as 25% of its original loose or broken volume. This makes it possible to store greater quantities in smaller space. However, once this coal is compacted in these stockpiles, it is difficult to retrieve with underground tunnels, even with special types of feeders. More significantly, the laying and spreading largely overcomes segregation and eliminates voids in the mass. This, combined with the high degree of compaction, diminishes oxidation to a point where the hazards of spontaneous combustion are vastly reduced. The danger of spontaneous combustion in coal storage piles cannot be ignored. Tests and experiments lead to the conclusion that the hazard of self-generated fires can be lessened in two ways. First, the coal can be compacted in the pile, thus reducing the voids and squeezing out the air. Second, the coal can be sprinkled with water to keep it wet and cool.

19.2.2 Storage Methods

Many bins and hoppers are not well designed because of the space limitations in a plant or facility. As a result, the bin's actual capacity is merely a small rat hole down the center with the sides and the corners holding material possibly even for years. The actual capacity of the bins is maybe 10% of its intended design. This is a situation that can be alleviated by knowing the characteristics of the material to be stored. Do not compromise the bin design and the material flow. The best results can be achieved by understanding the basic classification of material and flow characteristics in and out of a bin or hopper. All these things are discussed in detail later in Chapter 21.

Other types of bins (sometimes called bunkers) are used in power plants and coke feeding ovens to provide a means of storage. These bins or bunkers are constantly supplied with coal or lignite from an outdoor storage pile. Coal is fed into grinders and thus to the fire boxes of the boilers. Usually there is a traveling tripper belt conveyor system that distributes the coal uniformly along this long row of bunkers which feed several pulverizers and boilers.

Most coal power plants have very elaborate conveying systems to receive coal either by barge, truck, or rail car. Coal is transferred to stockpiles and then to reclamation systems that convey it to the storage bunkers near the boilers. Some systems use mixing bins to blend various grades of coal from different sources into the feed system. This ensures that the boilers of the power plant will have a constant Btu fuel content. This is desirable from an economic standpoint.

Some coals are very high grade in Btu content and others are very low. Some are highly contaminated with sulfur, others with a lot of ash content. Lignite is very high in combined water. As a result the Btu content is very low. If transportation costs are low it would be a good source of fuel. It cannot be transported over a long distance and be financially competitive with bituminous coal. Surface moisture on lump or granular coal has very little effect on its Btu content as far as feeding a boiler is concerned.

Important storage equipment used for storing grain, coal, and iron ore are silos. Silos are tall vertical structures usually formed of reinforced concrete, welded steel, and concrete staves with hoops or riveted plates. Large clusters are usually made with slip-form concrete construction methods which are most economical in large-capacity storage systems. Typical of a slip-formed concrete silo structure is a grain elevator. It may hold as much as 10 million bushels of grain. These are always served by railroads, trucks, barges, or ships. Unloading facilities accumulate grain from the immediate neighborhood, then store the grain until it is ready to be loaded into cars. These facilities are also usually equipped with the cleaning facilities and grain fumigating equipment. Grain silo installations have an elaborate system of bucket elevators, vibrating screens, dust collecting equipment, and belt conveyors.

One of the many hazards in a grain elevator is dust and the danger of an explosion at any time. Grain dust, with the proper air/dust ratio and a source of ignition, will create a considerable amount of damage when it explodes. Over the years, there have been many grain elevator explosions. All precautions should be taken to eliminate generation and accumulation of dust. Adequate ventilation is very desirable in head houses and screening towers. Bucket elevators, belt conveyors, screens, dedusting, and other types of equipment in grain elevators should always use static-conducting belts. All equipment should be grounded. Motors should be explosionproof. All gear reducers should be grounded and should not have contact with metal which could possibly create a spark. All power transmission equipment used in these elevators should have adequate protection and proper grounds.

Other means of storage of bulk materials are closed or open sheds which are primarily to protect the material from the weather. Protection is needed from rain, cold, and wind. A shed open at one end will not do much good for some materials. When wet stockpiles are frozen, other solutions must be found to feed the material into a process system. In inclement weather, particularly when it is cold, materials need to be pushed into the outlet for underground tunnel storage retrieval systems with a bulldozer or front-end loader. It does not make much difference what the material is; it will not flow when it is wet and frozen.

In this introductory chapter, all we need to say is that free-flowing materials without any tendency to compact have very few problems in discharging from bins. However, most of the materials used in modern industry do have problems with compaction and consolidation. This makes it difficult to discharge material in bins. There are many devices on the market that will assist in the flow. None of them is perfect and none of them will work for all materials. Some of these devices work very well for certain materials under certain conditions. These problems have to be understood by the engineer designing the hopper and the materials handling storage and retrieval system. When there are questions about the characteristics of the materials and whether or not they will permit flow, it is obvious the material needs to be tested.

There are many ways to fill a bin or hopper. The most common is the dragline or clamshell bucket for sand plants or ready-mix concrete plants. The next is probably a conveyor or a bucket elevator delivering material to the top. The most popular modern system for dry bulk materials in a powdery form is a pneumatic conveyor. All of these systems work under planned conditions if they are properly applied. Placing the material in the bin is relatively easy. Getting it out is another situation. When material tends to attract moisture, packs with pressure, or is crystalline in shape and is cohesive, then additional problems exist. The cohesiveness of a powder is the tendency of each crystal to intermesh with an adjacent crystal. When these particles are irregular in shape they compact in the hopper. This compaction is caused by pressure, vibration, and static electricity. All these factors can contribute to the propensity of this particular type of material to form a solid mass in the hopper. This also can occur with powders in the plastic industry where temperature has an effect, making particles cling together and form one solid mass.

19.2.3 Discharge Methods

As mentioned previously there are many types of devices on the market to assist in delivering material from a bin. Various feeders have been designed for particular materials handling problems. The apron feeder is one that is commonly used with coarse, abrasive, large, or lumpy materials. The belt feeder serves many uses but is better for the less severe conditions that generally involve powdered or granular materials. Both kinds provide a constant volume at the feed. Volume at the feed is emphasized, not weight. Where there is no variation in moisture content or change in the size of the particles, most data are relative to a volumetric-type feed. If the bulk density varies, then the feed rate will vary. This is true of almost all volumetric feeders.

There are some feeders on the market used when discharging from bins. These feeders are augmented by a load cell system of scales. These scales can be electronic devices which feed back to the flow control mechanism and regulate the flow based on weight.

Vibrating feeders are very good with foundry sands prepared and ready to go into the mold. Other materials fed with vibrating feeders can be sticky, granular, or very coarse materials up to large lumps. Reciprocating feeders (which are similar to vibrating feeders) are generally operated at a reduced frequency with larger amplitudes. They are also very good for handling materials such as crushed rock in a variety of sizes.

Screw feeders are good for handling powders, granules, and small lumps. If the powder is free flowing and aerated these screw feeders can be lengthened in order to feed the material accurately and at a constant rate. There have to be very close-fitting tolerances between the screw and the housing and close control over the rpm. Some type of electronic feedback system from the weight control device is needed to adjust the speed of the screw feeder.

A horizontal rotary table feeder is available which works well for granular and lumpy materials but is not the answer to all feeding applications. As long as the bulk density is constant, rotary table feeders are accurate.

A study of the conditions relative to the material classification will influence the choice of equipment and the procedure to be followed. The engineer who has experience and knowledge dealing with material in a particular field is best equipped to make proper decisions. There are constantly new problems with new materials. It is absolutely necessary that the new materials be investigated and tested. Then apply this basic information to solve handling the problem.

19.2.4 Conclusion

Many factors must be considered for the solutions to the problems of storage, retrieval, and shipping of bulk materials. No rule of thumb suggested may be practical. A similar material that was stored under apparently identical conditions may behave differently because the proposed site and climate situations are not exactly the same.

Heavy materials sink out of sight in a storage pile. Providing a concrete reinforced slab with pilings below would be a prohibitive cost in a majority of applications. However, in some cases it is justified. The nature of the material, its transportability, the time requirements for loading and unloading from the stockpile, and the cost of the reinforced area required are factors.

Large stockpiles of bulk materials that are held in reserve for 2, 3, or 5 years do not present many problems because some degradation of the product is anticipated and storage and retrieval methods are not necessarily adaptable to high reclaim rates. These are strictly reserve piles. Storage piles used and/or frequently turned around, like coal supplying systems to a power plant, may normally have on hand a 90-day supply of fuel. They are constantly adding to this reserve and constantly reclaiming a portion.

The provision of the proper storage and retrieval equipment and the operation of this system requires a considerable amount of study. Management should have knowledge of the materials to be stored, the rates for retrieval, and the amounts of material received in a given period of time.

A train arrives with a hundred cars or 10,000 tons per day. There are many problems involved with selecting equipment required to unload the train. Are the cars to be rollover or bottom dumps? Shall they be stored in the yards and unloaded in the next day or two? All these things have to be considered before a decision can be made as to what type of unloading and storage and retrieval equipment is required.

The retrieval method has to be at a constant rate for a boiler-fired steam-generating plant. The boiler retrieval system has to be reliable and tuned to the requirements of the plant. There obviously has to be a surge capacity for breakdown and routine maintenance requirements. The durability and reliability of equipment used in these types of facilities have to be without question the very best. A power plant cannot afford to be shut down because of equipment failure. A gear motor failing on a drive at a transfer point could knock out a power system. The planning required for handling these kinds of tonnages and the maintenance and operation techniques must be well planned and engineered ahead of time to make sure that system reliability is as close to 100% as possible.

CHAPTER 20
PROPERTIES OF BULK SOLIDS

RICHARD C. WAHL

Vibra Screw, Inc.
Totowa, New Jersey

20.1 INTRODUCTION

Solids are generally the most difficult state of matter to handle. All types of liquids and gases are readily stored in tanks or vessels and accurately dispensed with the ease of turning a valve. Bulk solids are seldom that easy. What is more, they present a vast range of difficulties between types and even between grades of one type.

If a material has been handled by a company before, its properties will be familiar and any areas of difficulty in flow or processing will already be known. If not, further investigation may be required. Even a well-known product may be affected by a change in handling, as from mechanical to pneumatic conveying or from surge to bulk storage hoppers.

Several systems are used today for classifying materials according to the degree of difficulty in handling, or the kind of difficulty. The properties of materials that cause them to be classified as difficult may affect one or more of the stages of processing. Failure of a material to flow from storage by gravity, for example, is a frequent problem resulting from a combination of properties such as density and compressibility. A useful grouping of materials into four classes has been devised to aid in the selection of storage hoppers, and to determine what kind of flow aid devices and their sizes will be needed (Table 20.1.1).

How Properties Affect Design of Storage Hoppers

Once the material classification has been established for a hard-to-handle material (Classes I through IV), the following questions must be answered in order to assure flow from storage:

1. What is the optimum storage bin size? Determined by bulk density, the capacity of the bin, retention time, and method of filling the bin.
2. What slope is needed for the cone section? The slide and flow characteristics of the material in question will dictate slope.
3. What size bin discharger is required? For example: $\frac{2}{3}$ the diameter of the bin for Class III materials such as powdered lime, which is sluggish, fluidizing.
4. What size outlet will be needed? For example: a 7:1 ratio for a bin discharger outlet area to feeder screw area for Class III materials.

This frequently occurring subject is covered in detail by Winters [1] and it can be approximated by use of a simple calculator.

20.2 REVIEW OF PROPERTIES

A traditional view of materials divides them into two groups: crystalline and amorphous. A more practical division in modern industry is two quite different categories: easy to handle and hard to handle. "Hard to handle" is a simple term for a complex phenomenon. Materials handling engineers look beyond the mechnical, electrical, and optical properties of solids to examine the effect their properties have on flow.

Any thorough consideration of a process should take into account the following properties:

Abrasion

Knowing a material's abrasiveness is important to proper design of equipment to protect against wear. Such materials as coke and foundry sand will wear hoppers, chutes, screw feeders and conveyors, and pneumatic handling systems. Hardened steels, wear-resistant liners, and high-density plastics must be considered for contact materials.

Adhesion

The distinction between adhesion and cohesion is sometimes blurred; it may be worthwhile, therefore, to remark the difference applicable here: simply stated, cohesion is internal, adhesion is external. Adhesion is the sticking together or adhering of substances in contact with each other (Fig. 20.2.1). An extreme example of this is kaolin clay, which is so tacky that it will stick to a wall when thrown against it. This can create unusual problems in moving this material from storage; adhesive materials tend to bridge in storage and thus require external assistance.

Table 20.1.1 Material Classification Table

Material Class	Description	Average Flow Rates in ft^3/hr under Free-Fall Conditions for Given Outlet Size[a]				
		8 in.	10 in.	12 in.	20 in.	30 in.
I	Material is granular and free flowing. Would normally flow unassisted but temperature and moisture changes may cause it to bridge occasionally. Example: Granular salt, sugar, plastic pellets. Slide angle 30°. Bin activator should be $\frac{1}{4}$–$\frac{1}{3}$ the diameter of the bin.	2,600	4,000	6,000	16,000	36,000
II	Material is a sluggish powder 100–300 mesh. Would not normally flow by gravity alone. Example: Flour, starch. Slide angle 35–55°. Bin activator should be $\frac{1}{3}$–$\frac{1}{2}$ the diameter of the bin.	1,250	2,000	3,000	8,000	18,000
III	Material is a powder that tends to be readily adhesive or becomes easily fluidized (−325 mesh). Example: Adhesive materials (TiO_2, and pigments in general); fluidizing materials (hydrated lime, cement, talcum powder, confectionery sugar). Bin activator should be $\frac{1}{2}$–$\frac{2}{3}$ bin diameter.	1,000	1,700	2,500	7,000	15,000
IV	Material is fibrous or flaky with a relatively low bulk density of 3–20 lb/ft^3. Particle sizes are from $\frac{1}{8}$ in. strands to 1 in. or larger chips. Has tendency to interlock and absorb vibration. Example: Woodchips, slivers, shavings, asbestos fibers, flaked grain. Bin activator should correspond to bin diameter.	330	500	750	2,000	4,500

[a] Always use largest possible outlet diameter compatible with downstream equipment.

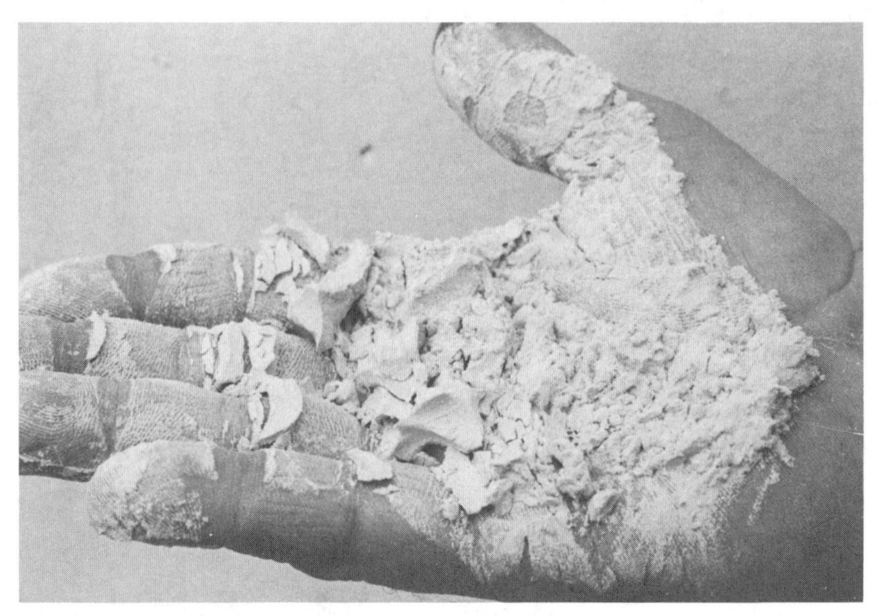

Fig. 20.2.1 Adhesion is the sticking together of substances in contact with each other.

Angle of Repose

This measurement gives a direct indication of how free flowing a material will be. It provides an indirect indication of other properties affecting flow: particle size and shape, porosity, cohesion, fluidity, surface area, and bulk.

Angle of repose is defined as the included angle formed between the edge of a cone-shaped pile formed by falling material and the horizontal. A low angle of repose indicates that the material flows readily; a high one, that it does not (Fig. 20.2.2).

Angle of Fall

After measuring the angle of repose, the cone-shaped pile of material is jarred as by dropping a weight near it. The pile will "fall" resulting in a new, shallower angle with the horizontal—the angle of fall. The greater the angle of difference (between angles of repose and fall) the more free flowing is the material. The way the pile falls is of special interest. If particles fall and spread out along the

Fig. 20.2.2 Angle of repose is the angle formed between the edge of a cone-shaped pile formed by falling material and the horizontal. A low angle of repose (right) indicates that the material flows readily; a high one (left), that it does not.

slope of the pile, only the degree of flowability in indicated. If the entire pile collapses, it indicates that the material contained entrained air and is prone to flushing.

Angle of Slide

This is the angle at which a material will slide down a flat surface due to its own weight. Angle of slide provides an indication of the material's flowability and is particularly useful in hopper and chute design. A simple test device is shown in Fig. 20.2.3.

Fig. 20.2.3 Simple test for angle of slide—the angle at which a material will slide down a flat surface due to its own weight.

Angle of Spatula

To measure a material's internal friction, a spatula is covered to a maximum with material and lifted out of the heap. A measure is made of the angle of the side of the material pile. After jarring the pile with a falling weight, the angle is remeasured. The average of the two angles is the angle of spatula. The higher the angle, the greater the flow resistance.

Cohesion

Cohesion is defined as the molecular attraction by which particles of a body or material are united or held together. Its importance lies in the fact that material with a high cohesion factor does not flow readily and particular care is therefore required in designing a storage bin, hopper, or feeder using that material.

A direct measurement of cohesion is difficult and requires sophisticated equipment such as a Cohetester. An alternative measurement can be made of the material's uniformity coefficient. This is a numerical value obtained by dividing two factors: the width of a sieve opening that will pass 60% of the sample, divided by the width of the opening that will pass 10% of the sample. The flowability of a material is directly affected by its uniformity in both size and shape. Obviously the greater the diversity of the material in size and shape, the less readily it will flow. In addition to indicating particle size and shape, the uniformity coefficient of a material also provides an indication of its compressibility. The more compressible a material is, the more trouble it will be to move it through the various stages of a process. Generally, if a material is not compressible, it will flow readily.

Compressibility

Compressibility is defined as a function of packed bulk density and aerated bulk density expressed and calculated by the equation [2]:

$$C\ (\%) = \frac{100\,(P - A)}{P}$$

where: C = compressibility
P = packed density
A = aerated density

The dividing line between free flowing (granular) and non-free flowing (powder) is about 20–21% compressibility. A higher percentage indicates a powder that is not free flowing, and will be likely to bridge in a hopper. In this event external aid of some kind is required, preferably a bin discharger.

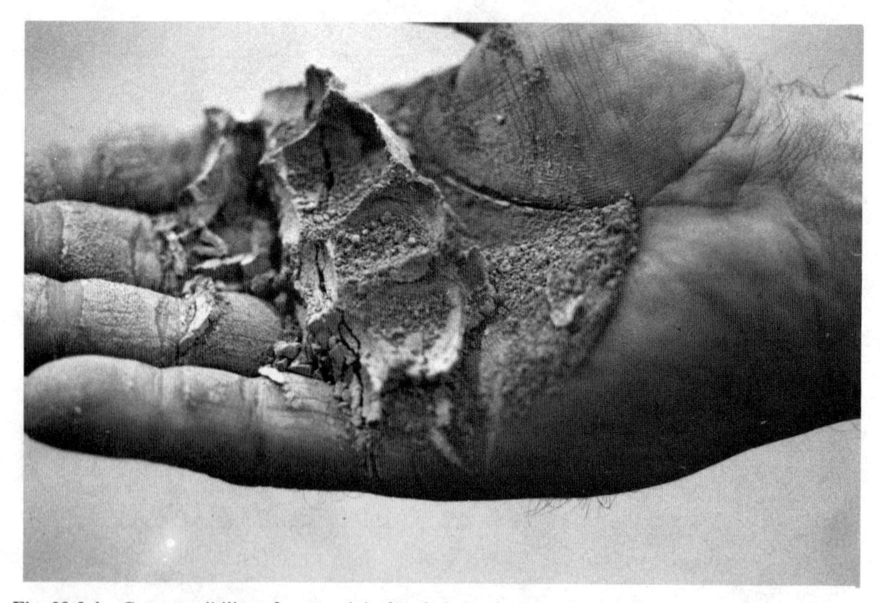

Fig. 20.2.4 Compressibility of a material often helps to indicate its cohesion and moisture content.

The compressibility of a material often helps to indicate uniformity in size and shape of the material, its deformability, surface area, cohesion, and moisture content (Fig. 20.2.4).

Corrosion (Chemical, Rust, etc.)

When corrosive materials are processed, they must be handled in equipment with contact surfaces of alloy steel, special plastics, or coated with corrosion-resistant paint.

Density

Among the commonly handled materials in industry, there is an enormous range of densities (expressed in pounds per cubic foot) of the magnitude of at least 100:1. Carbon black powder with a bulk density of from 4–7 lb/ft^3, offers a decided contrast to powdered nickel, at 150–200 lb/ft^3. Bulk density (Table 20.2.1) is important for:

Feed range computations
Calculating hopper or bin capacity
Determining compressive or compacting strength of a material that can occur in a hopper or bin

There are three kinds of bulk density that apply to materials handling calculations:

Aerated
Packed
Dynamic or working

The figure most often used in computations is dynamic or working bulk density, and it is a function of aerated and packed density. It is computed by the following equation [2]:

$$W = (P - A)\,C + A$$

where: W = dynamic working density
P = packed density
A = aerated density
C = compressibility

Aerated density is obtained by loosely filling a container of known size and weighing. Packed density is obtained by filling a container of known size, tamping, refilling to top off, and weighing.

Dispersibility

This is the basic property that causes a material to flood or to produce dustiness in the surroundings. Dispersible materials are generally of low bulk density and fine particle size, which causes them to behave more like a gas or a liquid than a solid. Materials with a dispersibility rating of more than 50% are very floodable, and are likely to flush from a storage bin unless measures are taken to prevent this occurrence.

Friability (Degradation)

This quality can usually be determined by a simple squeeze test, similar to that used in checking compressibility.

If it is undesirable to have any breakdown of the product in process, as in food products such as cereals or vegetables, it is mandatory to use equipment whose design or performance will prevent such breakage.

Moisture Content and Hygroscopicity

Materials that have an inherently high percentage of moisture, such as wood chips (which when air dry contain from 6 to 40% by weight of moisture) may pose a handling problem, but not because of the moisture content in itself. Free, surface, or combined moisture as in sludges or wet ores is what causes the classic problems of sticking and poor flow. Generally, free moisture over 5–10% is considered risky. Likewise, those materials that are hygroscopic or absorb moisture are almost certainly a problem. When storage bins are not covered and are therefore exposed to weather, materials such as salt will cake and refuse to flow. Others such as sugar and lime will cake and flow erratically or not at all in conditions of high humidity. If it is the only factor, this condition is amenable to several remedies: keeping the storage bin covered, keeping the system enclosed, air conditioning, or moving the plant to a dry climate.

Table 20.2.1 Material Densities

Material Description	Bulk Density (lb/ft³)	Average Particle Size (U.S. Std. Sieve, Unless Noted)
Adipic acid	45	100
Almonds, broken	28–30	$\frac{1}{2}$–$\frac{1}{4}$ in.
Alum, fine	45–50	5–10
Alumina	50–65	200–350
Aluminum chips, dry	7–15	Usually long, irregular
Ammonium chloride	45–52	100
Ammonium nitrate	45–50	20–30
Ammonium perchlorate	50–70	100–200
Asbestos, fine	15–20	$\frac{1}{8}$ in.
Baking powder	40–55	100
Baking soda (sodium bicarbonate)	40–55	100
Barite, powder	120–180	100
Barium carbonate	72	100
Barley, whole	36–48	$\frac{1}{2}$–$\frac{1}{4}$ in.
Bauxite, dry, ground	68	5–10
Bisphenol—A	40–45	50–100
Bread crumbs	20–25	$\frac{1}{4}$ in.
Brewer's grain, spent, wet	55–60	$\frac{1}{2}$ in. and smaller
Calcium carbide	70–80	50–200
Calcium stearate	10–20	300–350
Carbon, activated, dry, fine	8–20	$\frac{1}{8}$ in.
Carbon black, powder	4–7	100–200
Caustic soda	88	$\frac{1}{8}$ in. and smaller
Cement	60–70	100–150
Chalk, pulverized	65–75	100
Charcoal, ground	18–28	100
Citric acid	50–60	200
Coal, anthracite, sized—$\frac{1}{2}$ in.	55–60	$1\frac{1}{2}$ in.
Coal, bituminous	45–55	2 in.
Coal, lignite	40–45	$\frac{1}{4}$ in. and smaller
Cocoa, powdered	30–35	100
Coconut, shredded	20–22	1–2-in. strands
Coffee, ground, dry	25	40–50
Coke, breeze	25–35	$\frac{1}{4}$ in. and smaller
Coke, petroleum	35–45	3–6 in.
Copper ore, ground	120–150	2 in. and smaller
Cork, granulated	12–15	$\frac{1}{2}$ in.
Corn cobs, ground	17	$\frac{1}{2}$ in.
Corn flour	30–40	100–200
Cornmeal	38–40	5–10
Corn starch	30–50	200–300
Cottonseed meal	35–40	5–10
Diatomaceous earth	11–14	40–50
Dolomite, crushed	80–100	$\frac{1}{2}$–$\frac{1}{4}$ in.
Flour	35–40	40–50
Fly ash	30–45	100–300
Glass batch	80–100	$\frac{1}{2}$–$\frac{1}{8}$ in.
Gypsum, calcined, powdered	60–80	100
Ice, flaked	40–45	$\frac{1}{2}$ in.
Iron oxide	25	100–200
Kaolin clay	42–56	50–100
Lead oxide	30–180	200–300
Lime, hydrated	32–40	50–100
Lime, pebble	53–56	$\frac{1}{2}$ in.
Magnesium chloride (magnesite)	33	$\frac{1}{2}$ in. and smaller
Magnesium dioxide	70–85	100–200
Magnesium oxide	20–50	300–325
Meat meal	40–50	10–300
Milk, powdered	20–45	10–50
Molding plaster	30–40	200

Table 20.2.1 (*Continued*)

Material Description	Bulk Density (lb/ft³)	Average Particle Size (U.S. Std. Sieve, Unless Noted)
Mushrooms	40	$\frac{1}{2}$–$\frac{1}{4}$ in.
Nickel powder	170–200	200–300
Nylon, flakes, pellets	30–40	$\frac{1}{2}$ in. and smaller
Oats	20–30	$\frac{1}{4}$–$\frac{1}{16}$ in.
Oxalic acid	60–70	200–300
Peanuts	40	$\frac{1}{2}$ in. and smaller
Phosphate rock, ground	90–110	100–200
Polymer powder	20–40	200–325
Potato, cut, flakes	15–40	$2\frac{1}{8}$ in.
Quartz dust	70–80	100–150
Rice	55	$\frac{1}{2}$–$\frac{1}{4}$ in.
Rouge	20–30	200–300
Rubber, granular	40	$\frac{1}{2}$ in.
Salt, fine	70–80	5–10
Sand, foundry, prepared	65–75	5–10
Sand, foundry, shake out	90–100	5–10 lumps
Sand, olivene	100	100
Sawdust, dry	10–13	5–20
Sea coal	65	5–10
Soap, detergent	15–50	5–10
Soda ash	55–65	5–10
Soy bean meal	40	5–10
Starch	25–50	50–100
Sugar, refined, granulated, dry	50–55	5–10
Sulfur, powdered	50–60	50–100
Talcum powder	50–60	200–300
Titanium dioxide	50–80	300–350
Urea prills	35–45	5–10
Wheat	45–48	$\frac{1}{2}$ in.
Wood chips, hogged fuel	15–25	Long and stringy
Wood chips, screened	10–30	1–3 in.
Zinc dust	150–200	50–100
Zinc oxide	30–35	100–200

Particle Size and Shape

In some industries such as plastics manufacture, the particle size and shape of a material may more strongly influence its handling difficulty than any other factor. Initially, this may be predicted by visual inspection. For materials handling purposes, materials such as coal or plastic powders and pellets are classified by size, whereas fibrous materials which may mat or interlock are classified by shape. These materials include wood chips, chopped fiberglass, film scrap, and flake polyester.

In most cases, powders range from coarse (24 mesh or larger) to very fine (350 mesh or smaller). Most materials are classified by use of six testing sieves (with a range of 24, 40, 60, 100, 200, and 350 mesh) (Fig. 20.2.5). The sieve that passes through most of the powder when it is shaken or tapped is the one that establishes the rating for the material. When more comprehensive, subsieve analysis is required for materials whose fineness is in the micron range, sedimentation methods are required.

Static Charges

Some materials are subject to a buildup of static electricity which can cause sticking and clogging in storage and processing equipment. High-volume production, requiring high feed rates, can generate a static buildup of sufficient magnitude to require some means of circumventing the problem. The usual remedy is a static inhibiter coating or grounding strap. In plastics products, static problems occur with foam or film scrap during pneumatic conveying. Similar steps must be taken to prevent blockages of flow.

Fig. 20.2.5 Testing sieves (range of 24, 40, 60, 100, 200, and 350 mesh) used to classify materials.

Temperature Limits

In the plastics industry, some materials are stored when they are hot and then allowed to cool. As the hot material cools, the particles may adhere and form agglomerates, with the result that flow characteristics are adversely affected. Materials with low melting points are most affected.

Similarly, in grinding plastics, temperature limits must be taken into account. Care must be taken with all elastomers in this respect.

Other materials such as dry cement and gypsum become extremely fluidizable when hot, radically altering their low-temperature behavior.

20.3 MEASUREMENT OF PROPERTIES

Traditionally, simple tests and procedures were all that was needed to measure many of the previously mentioned properties. Heft the material for weight. Squeeze a handful of the material to determine if it is compressible or adhesive. Spill material onto a table so it builds a cone to determine angle of repose. These simple tests help keep a focus on efforts expended where they will be most productive.

In response to the increasing need for more accurate evaluation of a greater number of material characteristics, especially as they pertain to flow in bins, hoppers, feeders, and conveyors, Hosokawa Micromeritics Laboratory of Osaka, Japan, has designed the Powder Characteristics Tester (Fig. 20.2.6) based on the works of Ralph L. Carr, Jr. [3]. It has already achieved wide acceptance and is used by major oil companies, chemical, food, and pharmaceutical companies, and various private and government research departments.

The powder tester is a device that permits seven separate measurements to be made with the use of one instrument: angle of repose, angle of fall, angle of spatula, cohesiveness, compressibility, dispersibility, and density. By means of these seven measurements, a material may be classified into one of

Fig. 20.2.6 Powder characteristics tester.

seven categories of flowability ratings, or one of five categories of floodability ratings (see Tables 20.2.2 and 20.2.3). The flowability ratings provide indices for determining when bridge-breaking measures, using a vibrator or bin discharger, are required. The floodability ratings provide indices for determining the necessity of rotary seals or valves to prevent flushing (flooding).

Another recent instrument development is a cohesion tester [4]. The unit measures cohesive force and displacement of powder. Tensile strength is detected by a strain gauge, and displacement by a linear voltage differential transformer (see Fig. 20.2.7).

20.4 PROBLEM MATERIALS

The basis of all bulk material handling is the precise definition and accurate classification of materials according to their individual handling characteristics, which consist of a specific combination of particle size and shape, physical factors such as cohesiveness, bulk density, and others. From a handling point of view, solids are generally more complex and more difficult to handle than liquids or gases.

Of the top 50 chemicals handled by industry, 30% are solids, the rest are liquids or gases [6]. Virtually every one of these solids can be classified as a problem material. These and other common problem materials are listed here, together with rankings in the top 50 and with suggested solutions of handling difficulties which have been proved out by successful installations in industry. This information provides details that may be immediately helpful or may indicate the need for further research or laboratory testing before undertaking volume production or processing.

Adipic Acid. White crystalline solid. Bulk density 55 lb/ft^3. Average flowability, mildly abrasive. Used in the manufacture of nylon and polyurethane foams, food additives, adhesives. Rank in volume of chemicals produced in U.S., no. 50.

Alumina (Aluminum Oxide) Al_2O_3. Bulk density 16–80 lb/ft^3; varies from fines to large chunks. Material is abrasive, and when finely ground, it is fluidizable. When conveying the fines, a screw should be used; for the coarse product, a belt or vibrating pan feeder can be used.

Aluminum Sulfate (Alum) $Al_2(SO_4)_3$. Fine powder or white crystals, mildly abrasive. Bulk density 60–70 lb/ft^3. Used in sizing paper; pH control in paper industry; catalyst, waterproofing agent for concrete. Hygroscopic, average flowability. Number 37 in volume of production.

Table 20.2.2 Flowability Ratings

Degree of Flowability	Flowability Index	Necessity of Bridge-Breaking Measures	Angle of Repose		Compressibility		Angle of Spatula		Uniformity[a]		Cohesion[b]	
			Degree	Index	%	Index	Degree	Index	No.	Index	%	Index
			<25	25	<5	25	<25	25	1	25		
Very good	90–100	Not required	26–29	24	6–9	23	26–30	24	2–4	23		
			30	22.5	10	22.5	31	22.5	5	22.5		
			31	22	11	22	32	22	6	22		
Fairly good	80–89	Not required	32–34	21	12–14	21	33–37	21	7	21		
			35	20	15	20	38	20	8	20		
		Sometimes vibra-	36	19.5	16	19.5	39	19.5	9	19		
Good	70–79	tor is required	37–39	18	17–19	18	40–44	18	10–11	18		
			40	17.5	20	17.5	45	17.5	12	17.5		
		Bridging will take	41	17	21	17	46	17	13	17		
Normal	60–69	place at the	42–44	16	22–24	16	47–59	16	14–16	16		
		marginal point	45	15	25	15	60	15	17	15	<6	15
			46	14.5	26	14.5	61	14.5	18	14.5	6–9	14.5
Not good	40–59	Required	47–54	12	27–30	12	62–74	12	19–21	12	10–29	12
			55	10	31	10	75	10	22	10	30	10
		Powerful mea-	56	9.5	32	9.5	76	9.5	23	9.5	31	9.5
Bad	20–39	sures should be	57–64	7	33–36	7	77–89	7	24–26	7	32–54	7
		provided	65	5	37	5	90	5	27	5	55	5
		Special apparatus	66	45	38	4.5	91	4.5	28	4.5	56	4.5
Very bad	0–19	and techniques	67–89	2	39–45	2	92–99	2	29–35	2	57–79	2
		are required	90	0	>45	0	>99	0	>35	0	>79	0

Source: Reprinted from *Chemical Engineering,* pp. 166 and 167 of January 18, 1965 issue, with approval of Mr. Ralph Carr, Jr., and the copyright owner, McGraw-Hill Inc., New York, NY 10036, USA, Copyright 1965.

[a] Use these figures for granules or granular powder with which the uniformity can be measured.

[b] Apply these figures for fine and cohesive powders with which the cohesion can be measured.

Table 20.2.3 Floodability Ratings

			Flowability		Angle of Fall		Angle of Difference		Dispersibility	
Degree of Floodability	Floodability Index	Measures for Flushing Prevention	Index from Table 20.2.2	Index	Degree	Index	Degree	Index	%	Index
			>60	25	<10	25	>30	25	>50	25
			59–56	24	11–19	24	29–28	24	49–44	24
Very high	80–100	Rotary seal must be used	55	22.5	20	22.5	27	22.5	43	22.5
			54	22	21	22	26	22	42	22
			53–50	21	22–24	21	25	21	41–36	21
			49	20	25	20	24	20	35	20
			48	19.5	26	19.5	23	19.5	34	19.5
			47–45	18	27–29	18	22–20	18	33–29	18
Fairly high	60–79	Rotary seal is required	44	17.5	30	17.5	19	17.5	28	17.5
			43	17	31	17	18	17	27	17
			42–40	16	32–39	16	17–16	16	26–21	16
			39	15	40	15	15	15	20	15
			38	14.5	41	14.5	14	14.5	19	14.5
Tends to flush	40–59	Sometimes rotary seal is required	37–34	12	42–49	12	13–11	12	18–11	12
			33	10	50	10	10	10	10	10
			32	9.5	51	9.5	9	9.5	9	9.5
May flush	25–39	Rotary seal is necessary depending on flow speed and feeding conditions	31–29	8	52–56	8	8	8	8	8
			28	6.25	57	6.25	7	6.25	7	6.25
			27	6	58	6	6	6	6	6
Won't flush	0–24	Not required	26–23	3	59–64	3	5–1	3	5–1	3
			23	0	>64	0	0	0	0	0

Source: Reprinted from *Chemical Engineering,* pp. 166 and 167 of January 18, 1965 issue, with approval of Mr. Ralph Carr, Jr., and the copyright owner, McGraw-Hill Inc., New York, NY 10036, USA, Copyright 1965.

Fig. 20.2.7 Cohesion tester.

Ammonium Nitrate (Norway Saltpeter) NH_4NO_3. Colorless crystals. Bulk density 40–50 lb/ft^3. FGAN (fertilizer-grade, prilled and coated with kieselguhr). Used in fertilizers, explosives, as an oxidizer (solid rocket propellants). Number 10 in volume production. Hazardous. Hygroscopic material bridges severely in storage, flows spasmodically or not at all. Bin discharger required.

Ammonium Sulfate $(NH_4)_2SO_4$. Granular crystals, 65–80 lb/ft^3 bulk density. Mildly abrasive. Used in fertilizers, water treatment, fermentation, viscose rayon. Ranks no. 29 in volume of production. Sticky, corrosive. Hygroscopic and builds up and hardens in storage. Needs bin discharger.

Animal Feed. A Class IV material. Contains 18–40% moisture and has a bulk density of about 25 lb/ft^3. Feed hangs up in storage bin and plugs the outlet. Use of flow aid needed.

Asbestos (Shredded). This fibrous material has an average bulk density of 15–45 lb/ft^3. Toxic by inhalation. Used in fireproofing and in brake linings. Tends to mat, interlock, and pack, with resultant poor flow characteristics. Assistance is required to maintain flow and control. Since material is light and fluffy, a screw feeder should be used to control the flow rate and dust. Bridging is most common problem.

Barium Carbonate $BaCO_3$. Toxic. A fine powder with an average bulk density of 55–150 lb/ft^3. Material is fairly sluggish; assistance is required to maintain flow and control. Slightly abrasive. Screw feeders are normally recommended for handling. Uses: as a rodent killer, in barium salts, ceramic flux, optical glass.

Bauxite. Basically an aggregate of aluminum-bearing minerals. Most important ore of aluminum. Mohs hardness 1–3. Used in abrasives. Varies from fines to large chunks with bulk densities of 30–75 lb/ft^3. When dry, this material flows fairly well, but as the moisture content increases, its flowability decreases. Material has a tendency to build up and harden as it dries out. Also slightly abrasive. For the fines, a screw feeder should be used; for the large chunks, a belt or pan feeder can be used. To obtain a continuous flow from storage and to maintain control, assistance is required.

Bentonite. A thixotropic gel derived from clay (becomes fluid when shaken). Bulk density 30–65 lb/ft^3. Light and fluffy, it aerates; it also packs under pressure. Used as a bonding agent for sand in foundry industry, and is usually applied by means of vibrating screw feeders.

Calcium Chloride $CaCl_2$. White, deliquescent crystals, granules, lumps, or flakes. Bulk density around 50 lb/ft^3. Used in deicing roads; thawing coke, coal, sand, stone, ore; drilling muds; pulp and paper industry; pharmaceuticals; refrigerating brines; as drying agent. Rank in volume, no. 40. Hygroscopic, corrosive. Appropriate precautions must be taken.

Calcium Sulfate (Gypsum) $CaSO_4$ or $CaSO_4 \cdot 2H_2O$. In hydrated form, known as plaster of paris. Sets when water added. Flows erratically from hoppers, due to bridging, jamming, or ratholing. Is also difficult to meter with precision. A common solution is a live bottom bin (a static bin and bin discharger) and a weigh belt feeder.

Carbon Black. Bulk density ranges from 5 to 45 lb/ft^3. Fine powder subject to dusting because of light weight. When pelletized, product agglomerates and packs under pressure. Vibrating bin discharger removes entrained air, conditions material before discharge from storage. Metering is accomplished with vibrating screw feeder. Rank no. 34 in volume production. Used as reinforcement for rubber, as pigment in carbon paper, typewriter ribbons.

Cement. See Portland Cement.

Clay (see also Kaolin) Hydrate Aluminum Silicate. Types include bentonite, kaolinite, and attapulgite. Absorbs water to produce a plastic, cohesive mass. Very sticky. Material hangs up and bridges in storage. Use of bin discharger usually required to prevent wedging of material.

Coal. Varies from fines to large chunks with bulk densities of 30–70 lb/ft^3. Flows fairly well when dry, but flowability decreases as moisture content increases; this is a particular problem when in open pile storage. Generally abrasive, and requires assistance to maintain flow from storage and control. Pulverized coal requires a positive means of control of its fluidity. When moist, it tends to build up and harden. A screw feeder should be used for pulverized coal; for large chunks, a belt or pan feeder can be used. High-sulfur coal (above 4%) can corrode mild steels; stainless steels preferred.

Diatomaceous Earth (also called Kieselguhr and Diatomite). Soft and bulky (88% silica). Bulk density from 7–45 lb/ft^3. Used in filtration, as an absorbent, extender, mild abrasive, paper coating. Sluggish flow. Tends to pack into lumps, particularly when moist. When dry, may fluidize. Bin discharger used for positive flow, screw feeder for metering.

Dolomite (see Limestone) $CaMg(CO_3)_2$. A carbonate of calcium and magnesium. Bulk density, from fines to large chunks, of 60–120 lb/ft^3. Mohs hardness 3.5–4.
Used as a refractory, in fertilizers, in removal of SO_2 from stack gases. Assistance is required to maintain continuous flow and control. Screw feeders are recommended for handling the fines while belt or pan feeders handle the larger sizes.

Elastomers. See Polymers.

Fertilizer.

1. *Peat Moss.* Semicarbonized residue of plants formed in bogs and marshes. Water content about 85%; when field dried, 30–40%. Very fibrous. Product cakes and bridges. Bulk density is 30–35 lb/ft^3. Bin discharger eliminates bridging which occurs with unaided flow.
2. *Phosphates.* See Phosphate Rock.
3. *Urea.* See Urea.

Fiberglass, Chopped. Agglomerates. Matting and clustering of material is a problem in feeding. Requires vibrating screw feeder, bin discharger.

Flour, Wheat. Fine powder, very dusty. Bulk density is 35–40 lb/ft^3. Subject to contamination. Flour is prone to bridging, flooding, and ratholing from supply hoppers. Assistance of bin discharger needed for steady flow.

Fly Ash. Class III material, with a bulk density of 37 lb/ft^3. Average flowability, moderate abrasiveness. Aerates, and has tendency to fluidize. Fluidization due to high velocity may be eliminated by increasing hopper outlet and feeder inlet diameters.

Food. Sanitary, food-grade equipment required. Important to check out particulars with USDA, FDA on Food, Drug & Cosmetic Act, and Environmental Protection Agency (EPA).

Foundry. Sand is the workhorse of foundries. In constant use, it tends to lose its natural fluidity and become sticky and pack, bridge, or rathole. High clay-content foundry sand, with 0.5–8% moisture content, and bulk densities from 70–110 lb/ft^3, is often stored in tanks of 500-ton capacity. The prepared sand in some casting operations is so tacky that it will stick to a vertical wall. Hot return sand in a storage bin will often form massive ratholes, and the hot sand segregates, emptying the middle first and the sides last, with uneven cooling of the sand causing casting defects.

Various devices have been tried to keep sticky sand flowing, including sledge hammers, aeration pads, and pneumatic rappers, but their usefulness has been limited. Troublesome foundry sands are generally handled by bin dischargers. Properly installed and of proper design, they eliminate packing, bridging, and ratholing; by ending segregation they provide first-in, first-out flow. Result is improved casting quality and reduction of rejects.

Graphite (Black Lead, Plumbago). Powder or flakes, with bulk densities from 28 to 50 lb/ft^3. Powder dusty, and material becomes fluid and flushes from storage. Use of bin discharger recommended, to vibrate material to constant density by eliminating entrained air. Enclosed system recommended to contain dust problem.

Gypsum. See Calcium Sulfate.

Ilmenite. Iron ore, Class III. Sluggish, fluidizing. Dust problem with belt feeder; use screw feeder.

Iron Oxide. Fine powder, with bulk densities from 80 to 165 lb/ft^3. Used extensively as pigment and in foundry industry. Has a tendency to bridge in storage and to pick up moisture, becoming lumpy, creating flow problems in automatic feed equipment. Material is also dusty, creating both handling and environmental problems. Assistance from a bin discharger and a vibrating screw feeder helps maintain flow and control.

Kaolin Clay. Most refractory of all clays. Bulk densities from 13 to 100 lb/ft^3. Varies from pulverized to large chunks. Average flowability, but assistance is required to maintain continuous flow from storage and control while feeding.

When pulverized, material is fluidizable. A screw feeder should be used with pulverized kaolin to control flow and dusting. With the larger particles, a belt or pan feeder can be used. A bin discharger is helpful in removing material from storage.

Lime. Ranks no. 3 in chemical volume. *Quicklimes* (pebble lime), CaO, bulk density 55–60 lb/ft^3, particle size 2–$\frac{1}{4}$ in. *Hydrates,* $Ca(OH)_2$ (caustic, slaked), bulk density range from 25 to 40 lb/ft^3, particles mainly 200 mesh. *Limestones,* $CaCO_3$, bulk density 87–95 lb/ft^3 (avg.); in designing silo or bin, common practice is to use 30 lb/ft^3 for hydrated lime, and 55 lb/ft^3 for quicklime.

Lime, Hydrated. Class III material. When air conveyed to bin, contents may become highly aerated, causing erratic discharge. Problem solved by use of bin discharger and vibrating screw feeder or rotary valve.

Limestones may be divided into three classes based on degree of reactivity, from most to least:

High-calcium, containing at least 95% $CaCO_3$.

High-magnesium, which is still mainly $CaCO_3$, but containing more than 5% $MgCO_3$.

Dolomitic, containing equal ratio of calcium and magnesium.

In scrubber reactions for removal of SO_2, the presence of $MgCO_3$ tends to make the reactivity of the limestone impractical.

Since limestones must be pulverized to react effectively, they are not likely to possess acceptable uniformity of particle size. Material exhibits average flowability but becomes progressively more floodable the finer it is ground.

Magnesite (Magnesium Oxide) MgO. Bulk density is 13–26 lb/ft^3. Mohs hardness 3.5–4.5. Material is sluggish, abrasive.

Metals. Most are crystalline, conductors of electricity. Many are quite hard.

Copper Powder. Bulk densities from 60 to 150 lb/ft^3. Average flowability.

Nickel Powder. Bulk densities from 170 to 200 lb/ft^3, particle size from 3 to 5 microns. Average flowability.

Silver. See Silver.

Zinc. See Zinc Powder.

Mica. A crystalline silicate with bulk density of 30–40 lb/ft^3. Cleaves into thin sheets which are flexible and elastic. Material varies from fine flakes to a pulverized product. Flowability is sluggish, and assistance is required to maintain continuous flow and control. Since mica is light and fluffy, fluidization may occur, requiring control. First choice of metering equipment is a screw feeder, but if set up properly, a belt feeder can be used.

Paper. Papermaking begins with chipping of wood. Handling is covered under Wood Chips.

Peat Moss. See under Fertilizer.

Perlite, Expanded and Powder. This material varies from fine powder to $\frac{1}{2}$-in. particles, and bulk densities range from 3 to 60 lb/ft^3. To maintain flow and control of this material, assistance is required. When the proper prefeed equipment is used, either a screw feeder or a belt feeder can be used to meter this material.

Phenol C_6H_5OH. White crystalline mass. Hygroscopic. Bulk density 50 lb/ft^3. Used in manufacture of adipic acid, salicylic acid, and as disinfectant. Rank in volume, no. 36.

Phosphate Rock. Broken to pulverized. Material varies from large chunks to pulverized particles; bulk densities range from 60 to 100 lb/ft^3. Assistance is required to maintain flow and control. Care should be taken since material is slightly abrasive and corrosive. For the pulverized material, a screw feeder should be used; for the large chunks, either a belt feeder or a pan feeder is recommended.

Plastics. See Polymer Powders.

Polymer Powders. Polymers are macromolecules formed by joining together five or more monomers. Compressible, sluggish. Need assistance from storage. Metering, conveying not a particular problem.

Portland Cement. A hydraulic cement made by burning and grinding a mixture of pure limestone and clay. Bulk density is 40–100 lb/ft^3. Material builds up and hardens in storage. It is subject to deterioration with age, and it aerates. Assistance from a bin discharger provides reliable hoppering.

Potassium Sulfate K_2SO_4. Colorless or white, hard crystals or powder with bulk densities from 40 to 90 lb/ft^3. Used as reagent, fertilizer, in alum manufacture, glass manufacture, and as a food additive. A sluggish material which requires assistance to maintain flow and control. Either a screw feeder or a belt feeder can be used to meter this material if set up properly.

PVC (Polyvinyl Chloride) Powder and Pellets. A synthetic thermoplastic polymer. Presently under restriction by FDA in rigid and semirigid food containers. Varies in shape and size of pellets. Bulk density ranges from 15 to 35 lb/ft^3. Pellets are free flowing, requiring no special equipment. However, material is slightly abrasive and does build up a static charge. This flow problem can be eliminated by grounding entire system. Powder is generally compressible and requires assistance to maintain flow and control. Screw feeders, belt feeders, or pan feeders can be used to meter these products.

Salt NaCl. Crystalline, somewhat hygroscopic and mildly corrosive; bulk density ranges from 30 to 90 lb/ft^3. Freshly mined rock salt in fine crystal granular powder flows by gravity, but when stored for prolonged periods, it tends to compact and resist flow. In high-temperature or humid areas, salt can set up hard and cause flow problems. To insure continuous flow and control, assistance is desirable. Most metering equipment can be used to handle this material.

Salt Cake (Sodium Sulfate) Na_2SO_4. White crystals or powder, bulk density of 70–95 lb/ft^3. Used in kraft paper, paperboard, glass, ceramic glazes, processing textile fibers. Subject to caking, and is preferably not stored in bulk for prolonged periods, unless bin discharger is used.

Sand. See under Foundry.

Sawdust. Material is generally very fine, mixed with some coarse slivers. Bulk density ranges from 6 to 40 lb/ft^3. Has strong tendency to absorb moisture, which makes it difficult to handle. To insure continuous flow and control, assistance is required. Most metering devices can be used to feed this material.

Silver Ag. In powder form, silver has a bulk density of about 125 lb/ft^3; the granular crystals are about 165 lb/ft^3. Can fluidize; bin discharger recommended.

Soda Ash (Sodium Carbonate) Na_2SO_3. Overall bulk density from 6 to 100 lb/ft^3. Powder, 55–65 lb/ft^3; light, 25–30 lb/ft^3. Small white crystals or powder. Some uses: glass manufacture, chemicals, pulp and paper, soaps and detergents, water treatment.

Rank no. 12 in volume production. Material has tendency to hang up in storage bin. Bin discharger eliminates erratic flow and flow stoppages caused by bridging and compaction. For accurate feeding, vibrating screw feeder needed.

Sodium Silicate (Water Glass). Made by fusing sand with soda ash. Bulk density from 58 to 90 lb/ft^3. White powders to $\frac{3}{4}$-in. particles. Forms gel with acids. Used in soaps and detergents; water treatment, bleaching agent, binder, waterproofing mortars and cements. Rank no. 45 in volume production. Average flowability, hygroscopic, and can cake.

Sodium Tripolyphosphate (STPP) $Na_5P_3O_{10}$. White powder or granules. Bulk density 40–50 lb/ft^3. Used in water softening, as a food additive. Rank no. 47 in volume production. Average flowability.

Soy Flour (Meal). Fine-ground powder, 100 mesh or less, of average flowability. 5% moisture. Bulk density 20–40 lb/ft^3. Powder aerates and becomes fluid. Has explosive potential. Flakes used mainly for animal feeds. Freshly made meal is difficult to store because of its tendency to set during storage. Stored meal bridges and ratholes while being emptied from bin. Massive bridges known to form and release causing structural problems. Consistent and safe discharge requires assistance of bin discharger.

Starch. A carbohydrate polymer, with a bulk density range from 15 to 50 lb/ft^3. White, amorphous powder or granules. Used in adhesives, paper, textiles (filler and sizing). Aerates and becomes fluid. Starches can also be sluggish and somewhat pressure sensitive. Assistance is required to assure flow from storage.

Sugar. Refined sugar is hygroscopic, and the pulverized powder is extremely tacky; it will stick to the walls of storage bins. Bulk density ranges from 5 to 60 lb/ft^3. It is free flowing when dry, but frequently bridges in storage. It packs under pressure, and care must be taken against contamination. Steady, uniform flow can be provided with bin dischargers.

Talc (Talcum, Soapstone, Steatite). A natural anhydrous magnesium silicate. Material varies from very fine powder to $\frac{1}{2}$-in. particles with bulk densities of 20–60 lb/ft^3. Foliated varieties are called talc. Mohs hardness 1–1.5. An absorbent. Material aerates and becomes fluid. To insure positive flow and control, assistance is required. Because of its fluidity when pulverized, a screw feeder is recommended to control discharge. A belt feeder can also be used if the proper prefeed is used.

Terephthalic Acid (TPA). White crystals or powder. Bulk density from 35 to 70 lb/ft^3. Used in production of polyester resins, fibers, and films. Rank no. 34 in volume production. Average flowability.

Titanium Dioxide TiO_2. White powder in two crystalline forms. Bulk density ranges from 20 to 100 lb/ft^3. Used in pigments and paints. Notably difficult to handle. It tends to agglomerate and pack in storage. It cannot be fed by gravity. For feeding, open screw required to avoid packing and binding. Rank no. 48 in volume production.

Toner (Plastic and Carbon Black). Raw material has a bulk density of about 40 lb/ft^3 and a particle size of about 12 microns. It is used with copier machines and computers. It is extremely difficult to handle: entrained air will cause it to flood; when it is deaerated, because of its small particle size and low melting point, it picks up heat rapidly and may fuse into lumps, causing erratic flow. Assistance is required to assure both flow and control.

Tripolyphosphate. See Sodium Tripolyphosphate.

Urea (Carbamide) $CO(NH_2)_2$. White crystals or powder. Usually handled in form of prills. Bulk density of 35–50 lb/ft^3. Hygroscopic; absorption of moisture causes caking. Material tends to agglomerate, lump, and bridge in storage. Can be moved successfully with bin discharger. Used in fertilizer, plastics, adhesives, pharmaceuticals, cosmetics, dentifrices. Rank no. 13 in volume production.

Vermiculite. The ore varies considerably in moisture content—from 5 to 15%—and in winter it is often partly frozen. Bulk density is about 110 lb/ft^3. Vermiculite, expanded, consists of fluffy $\frac{1}{2}$-in. particles, with a bulk density of 10–15 lb/ft^3. Because of its lightness, it may become windswept, and proper precautions must be taken to avoid material spillage and losses. Assistance for both heavy ore and light expanded form needed from storage.

Water Treatment. See Lime.

Wollastonite. A natural calcium silicate used in ceramics, paint, rubber, paper, welding, and in the construction industry. Material is brittle and fibrous. In dry powder form, its bulk density is from 30 to 50 lb/ft^3. In fluffy mass, when it is principally minus 200 mesh, it tends to interlock and bridge. Flow is aided by use of a bin discharger.

Wood Chips. Wood is reduced to chips to allow penetration and diffusion of the chemicals needed for digestion into pulp. Since wood is hygroscopic, moisture content varies, depending on changes in humidity and temperature, as much as 50%. Further, wood is anisotropic, exhibiting unequal changes in its three major directional axes. Even though they are irregular, chips are relatively free flowing when freshly chipped, but in storage, they are among the most intractable materials processed in industry.

The ever-increasing size of silos is a major factor. Modern silos are often 30 and 40 ft in diameter, and when they are full they may contain more than a million pounds of chips, all bearing down on the bottom cone where compaction occurs.

Flow interruptions are frequent because of a continuing tendency of the chips to interlock. When this occurs, the chips form a stubborn arch suspended above the outlet. The problem is aggravated in winter weather. Another common problem is ratholing, sometimes described as "funnel flow." Bin dischargers have successfully solved most handling problems.

Zinc Powder. Bulk density 200 lb/ft^3. Average flowability, but sticky and tends to agglomerate. Use of a vibrating screw feeder recommended for controlled rate of feed.

20.5 EFFECT ON MATERIALS SELECTION AND EQUIPMENT

Two strategic ways to use a knowledge of material properties exist: (1) in the selection of a type or grade of material or if necessary, a substitute and (2) in the selection of the equipment to store, meter, and convey.

The first alternative may be a matter of economics, as in the choice of using bag or bulk lime, or in using nearby limestone or transporting lime.

The second alternative may be more complicated, and may often require laboratory or field testing. Before this need be undertaken, some first steps described in the preceding pages may be taken, beginning with a visual inspection, and squeezing a handful of material to check on its compressibility and adhesiveness. Unless a material is hygroscopic, it is no trouble to discover that it is free flowing.

It has been estimated that 90% of all handling problems originate at the cone-shaped outlet or discharge end of the storage bin. Depending on the properties of the material, there are ways to determine approximately how large a bin discharger may be needed without the need for running a material test [5].

Peak performance of feeders, conveyors, blenders, and packaging devices is totally dependent on the system used to discharge bulk materials from storage. The flow of material from the bin must be continuous and on demand. And the material itself must be of uniform density and unsegregated. Segregation can occur both in the size of the particles and the duration of time in storage (first in, first out). When there are changes in the speed of the process line, the system must be able to compensate for them automatically, and it must be able to override variations in grade or other changes in material characteristics, or changes in the environment such as variations in humidity. For most processors, the system should be able to handle a variety of different materials without major changes or prolonged downtime.

Abrasive materials such as large particle coal may tear conveyor belts; fluidizable materials may spill.

In such cases, it may be necessary to control the flow of any similar materials by feeding them onto the belt with a bin discharger or prefeed the materials with a screw feeder. Ceramic linings may be desirable on bin dischargers that handle abrasive materials such as coke. Kraus has comprehensively reviewed one aspect of this subject in his book, *Pneumatic Conveying of Bulk Solids*.

Compressible materials should not be used in screw feeders except under known favorable circumstances. Often a material such as titanium dioxide requires a special screw design. Instead of the conventional solid flight screw, an open wire screw should be used to provide freeness for the material so that it can flow without packing, which can break the screw.

Hygroscopic materials such as salt and lime are a particular problem in humid climates. If the systems cannot be enclosed, sometimes the only remedy is air conditioning.

When friability must be contended with, it is better to use a belt feeder rather than a screw feeder.

When a choice can be made of two different equipment alternatives, be sure to keep in mind the downstream effect of the alternatives. A tissue mill faced the choice of drying paper sludge either in a filter press or a vacuum dryer. The filter press produces long, flat particles which tend to mat and interlock; the vacuum dryer, in which the sludge is not pressed, produces smaller particles which are much easier to handle.

Fibrous materials are invariably difficult to handle because they mat and interlock in storage. This is especially true of wood chips and fiberglass. Wood chips are relatively light in weight, but since they are used in great volumes, there may be more than a million pounds of material pressing down on the narrow discharge cone at any given moment. Use of a properly designed bin discharger is essential.

References are supplied for specialized materials requiring more detailed information, such as materials used in environmental control, asbestos, fly ash, explosives, and certain highly corrosive materials.

Role of the Test Laboratory

A frustrating and expensive situation may occur when a smooth-running pilot plant is expanded into a full-scale production process which then unexpectedly fails. Whenever problem materials are involved, it is risky to assume that a process will succeed even on the basis of a successful pilot operation. It is even riskier to do so on the basis of superficial information or testing. The relatively small cost of test laboratory studies or other consultation is well worthwhile when the alternative of guesswork and possible failure is considered. The best downstream equipment in the world is unequal to the challenge of handling a material which stubbornly refuses to budge from its storage bin. The old saying, "experience is the best teacher" has never been more apt than in the capricious world of bulk solids, with its infinite variety and its inevitably costly surprises.

REFERENCES

1. Winters, R. J., "Improving Flow from Bins," *Chemical Engineering Progress,* September, 1977.
2. Powder Characteristics Tester Brochure, Operating Instructions, Hosokawa Micron Corp., Osaka, Japan.
3. Carr, R. C., Jr., "Evaluating Flow Properties of Solids," *Chemical Engineering,* January 18, 1965.
4. Cohetester Brochure, Hosokawa Micron Corp., Osaka, Japan.
5. Wahl, R. C., "The Design and Operation of Bin Activators," *Bulk Solids Handling,* W. Germany, February, 1981.
6. "50 Chemicals," *Chemical and Engineering News,* April 4, 1981.

BIBLIOGRAPHY

Classification and Definitions of Bulk Materials, Conveyor Equipment Manufacturers Association, Washington, DC 20005.

Hawley, G. G., Ed., *Condensed Chemical Dictionary,* Van Nostrand Reinhold Co., New York, 1981.

Kirk-Othmer, *Encyclopedia of Chemical Technology,* John Wiley, New York.

Kraus, M. N., *Pneumatic Conveying of Bulk Materials,* McGraw-Hill, New York. *Chemical Engineering* 1980.

Lime, Handling, Application and Storage, National Lime Association, Washington, DC, 1976.

"Materials," *Scientific American,* September, 1967.

Modern Plastics Encyclopedia, McGraw-Hill, New York, 1981–82.

Panshin and DeZeeuw, *Textbook of Wood Technology,* McGraw-Hill, New York, 1977.

Stacking, Blending, Reclaiming of Bulk Materials, Trans Tech Publications, Clausthal, West Germany, 1975–77.

Steam, Its Generation and Use, The Babcock & Wilcox Co., New York, 1975.

Technical Association of the Pulp & Paper Industry, New York.

Vibra Screw Test Laboratory. Files on more than 14,000 materials handling applications in industry.

Wahl, R. C., "Handling Bulk Materials," *Plastics World,* November, 1978.

CHAPTER 21
DESIGN OF BINS AND HOPPERS

JOHN W. CARSON

JERRY R. JOHANSON

Jenike and Johanson
No. Billerica, Massachusetts

21.1 INTRODUCTION

21.1.1 Importance of Proper Design

A properly designed bin, hopper, and feeder can produce reliable flow of solids on demand at controllable rates without flooding or flushing. Unfortunately, costly production delays are often traced to improper working of bins and hoppers. Reduction in live storage capacity to 10% of the total volume is not uncommon and often causes the bin to act as a chute with no effective retention time to deaerate powders or to provide surge capacity for process control.

Improperly designed bins and hoppers cause safety problems. Manual prodding of bin hangups often results in injury to personnel when the obstruction suddenly collapses. The shock load from a suddenly released hangup may cause structural damage. Shifts in solids flow patterns or pressure concentrations from flowing solids in bins often cause structural failure.

This chapter deals with the many considerations necessary to design a bin and hopper to handle a bulk solid with special emphasis on the importance of selecting design parameters consistent with the flow properties of the solid.

21.1.2 Definitions

Bin. A container for bulk solids with one or more outlets for withdrawal either by gravity alone or by gravity assisted by flow-promoting devices.

Included within this definition are silos, vessels, grain elevators, coal bunkers, and other specialized terms used in various industries. They are all "bins."

A bin can be divided into two main sections: cylinder and hopper. The cylinder section has a constant cross-sectional area and is usually square, circular, or rectangular in plan view. In the hopper, the cross-sectional area changes from top to bottom—usually decreasing ("converging hopper"). At the outlet, a feeder, discharger, or gate is often used to stop and start the flow and, in many cases, to control the discharge rate.

Bulk Solid. A material that is handled in bulk form and consists of discrete solid particles ranging in size from submicron to several inches.

Using this definition, materials as diverse as fumed silica, plastic pellets, coal, and mine-run ores all are bulk solids.

21.1.3 Typical Flow Problems

There are several common problems that can develop when handling bulk solids in bins and hoppers [1, 2].

No flow

This condition, by far the most serious, describes what happens when a gate is opened or a feeder started and nothing comes from the bin. "No flow" can be due either to the formation of a stable obstruction at the outlet, commonly referred to as an arch or bridge, or due to the formation of a stable hole or pipe directly above the outlet extending to the top of the bin. This latter problem is usually referred to as ratholing or piping.

Erratic Flow

This problem often occurs because of a switch from one no-flow problem to another. For example, consider the situation when a rathole has developed and the bin is hit by a sledge hammer or vibrated by a railcar passing close by. If the vibration is sufficient to collapse the rathole, material falls to the outlet region and some is discharged. However, the free fall velocity is always greater than the rate at which material can exit from a bin so material is decelerated at the outlet region. This deceleration compacts the material and may give it sufficient strength to support a stable arch at the outlet. Hence the flow goes from (1) no flow, due to ratholing, to (2) some flow as the rathole collapses, to (3) no flow, due to arching. The reverse can also happen as a bin is vibrated and a rathole starts to form.

Flushing

If the erratic flow sequence just described occurs with a fine powder, the problem is compounded. When a material falls through air, the particles become separated. If this material consists of fine particles, the separation is slight so the air becomes trapped and cannot easily escape, particularly once the material stops falling and is collected in a pile. As the air attempts to escape, the powder takes on the characteristics of a fluid.

Hence when a rathole collapses with fine material, the region above the outlet quickly becomes filled with a fluidized powder which then usually floods uncontrollably through the feeder. Such floods may rapidly empty the entire contents of the bin. For example, in one fly ash bin, the bin emptied through 40 ft of horizontal screw conveyor under it.

Limited Capacity

If a stable rathole develops in a bin, the live or usable capacity is only the volume of the rathole which may be as low as 10–20% of the total bin volume.

Segregation

Many bulk solids segregate when they are handled. The filling of a bin is no exception. If the bulk solid consists of a range of particle sizes, fines usually concentrate under the fill point and coarse particles concentrate at the periphery. With a center-loaded and center-discharged bin, this region of fines concentration is located directly above the outlet. If, in addition, the flow pattern upon discharge of the bin is such that the material above the outlet is withdrawn before that at the periphery of the bin, the discharge consists mostly of fines if the bin has just been filled, or if the charge rate is greater than that of withdrawal. Later when filling has stopped, either a stable rathole forms which prevents further discharge or, if the material has insufficient strength to support a rathole, the larger particles from the side walls slough off into the top of the center flowing channel. Then the discharge consists of a high percentage of coarse particles. A third condition exists when the rate of charge and withdrawal are equal. Here a stable flow channel forms and the discharge consists of roughly the same percentage of fines and coarse as the charge.

Degradation

Degradation in the form of spoilage of food products, oxidation leading to spontaneous combustion of coal, and caking of chemical powders is usually caused by a first-in, last-out sequence of flow from the bin. Degradation due to particle breakage is caused by impact during fall into the bin or by excessive pressures combined with excessive interparticle motion during flow in the bin and hopper.

Level Control

The contents of a bin can be accurately determined by using load cells but this is often impractical with large bins. Hence numerous level control devices are on the market that attempt to measure the volume of the bin's contents. Many of these devices do not work reliably when the contour of the top surface has an extreme variation—such as when a rathole develops and most of the solid remains on the bin walls.

21.2 FLOW PATTERNS

Most of the flow problems noted thus far are related to the flow pattern, that is, the manner in which material flows as it discharges from a bin [3, 4]. Basically there are two major types of flow patterns: funnel flow and mass flow. A third type, expanded flow, is a combination of the first two.

Table 21.2.1 shows the relation between typical flow problems and the type of flow pattern.

21.2.1 Funnel Flow

This flow pattern, also known as core flow, describes a condition in which some material in the bin is in motion while some is stationary. Funnel flow bins have shallow hopper walls or even a flat bottom; consequently a relatively large volume of storage can be achieved with a minimum of headroom. Unfortunately, this benefit is often offset by the first-in, last-out sequence of flow. Stable ratholes develop if the material has sufficient cohesive strength causing severe loss of live capacity. Pseudostable ratholes may develop causing erratic flow. Most fine powders exhibit flushing if placed in a funnel flow bin because, once they become deaerated outside of the flow channel, they are capable of supporting a stable rathole. Finally, funnel flow bins exacerbate the problem of segregation.

Table 21.2.1 Relationship between Flow Problems and Flow Patterns

Problem	Funnel Flow	Mass Flow
No flow		
Arch	√	√
Rathole	√	
Erratic flow	√	
Flushing	√	
Lack of design capacity	√	
Segregation	MAX.	MIN.
Degradation of product	√	
Level control	√	

Funnel flow bins are only suitable for coarse, free-flowing, nondegrading solids when segregation is unimportant.

Two common types of funnel flow bins are shown in Fig. 21.2.1. A flat-bottom bin with a small single outlet or multiple outlets always gives a funnel flow pattern under gravity-flow conditions since there is material on the sides of each outlet. Most pyramidal hoppers also give a funnel flow pattern because the inflowing valleys tend to trap material. Similarly, shallow conical hoppers cause funnel flow.

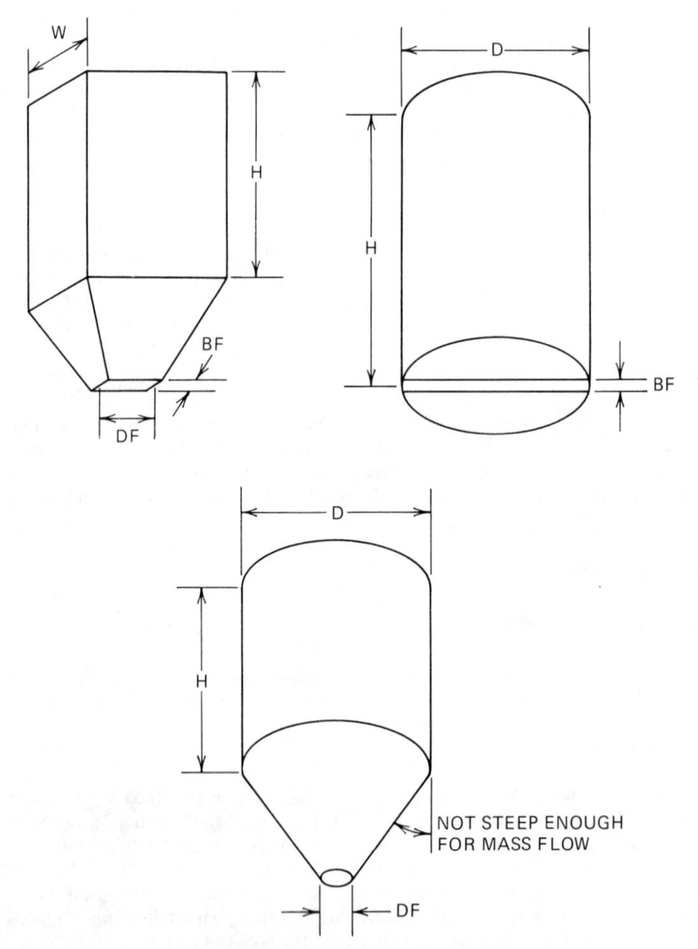

Fig. 21.2.1 Typical funnel flow bins.

Note two important observations. First, the flow pattern is set by the hopper not the cylinder. If no hopper is present (as in the case of the flat-bottom bin) or if the hopper walls have a shallow slope or are rough, a funnel flow pattern develops. The roughness and shape of the cylinder walls have little effect on this.

Second, note that the volumes given in Fig. 21.2.1 are the total volumes for these geometries. They are not necessarily the live or usable volumes, particularly if a rathole develops.

21.2.2 Mass Flow

This flow pattern describes a condition in which all the material in a bin is in motion whenever any is withdrawn. It is not necessary that the velocity across the cross section be constant, only that all of the material be in motion.

Mass flow bins require more headroom than funnel flow bins because the hopper walls must be smooth and steep. In return for this expense, the flow pattern is one of first-in, first-out. Ratholes do not develop, so fine powders have time to deaerate after being charged into a bin. Material bulk density at the outlet is relatively constant and segregation is minimized because particles at the center and side walls of the bin are discharged simultaneously. Mass flow bins are suitable for cohesive solids (including most fine powders), solids that degrade with time, and when segregation needs to be minimized.

Single Stationary Hoppers

Various types of mass flow bins are shown in Fig. 21.2.2. The conical hopper requires an angle (measured from the vertical) no greater than θ_C; otherwise material will cease flowing at the walls and result in a funnel flow pattern. Provided the length to width ratio of a transition hopper outlet is at least 3:1, the side walls can be designed for θ_P which is usually some 10–12° less steep than θ_C. Note that the end walls of a transition hopper are limited to θ_C. The same 3:1 minimum outlet ratio and θ_P side wall angle apply to the chisel hopper geometry.

A wedge hopper with vertical end walls can be designed with a side wall angle up to θ_P and a minimum length-to-width ratio at the outlet of 2:1. Pyramidal hoppers give a mass flow pattern if the valley angle (see Fig. 21.2.1) is no greater than θ_C (from the vertical).

Insert Hoppers

The insert hopper is a special type of mass flow bin that consists of a central hopper which has the required minimum slope to achieve mass flow, that is, θ_C for a cone or θ_P for a wedge. Another hopper is placed outside of the first one, such that material can flow through both the inner hopper

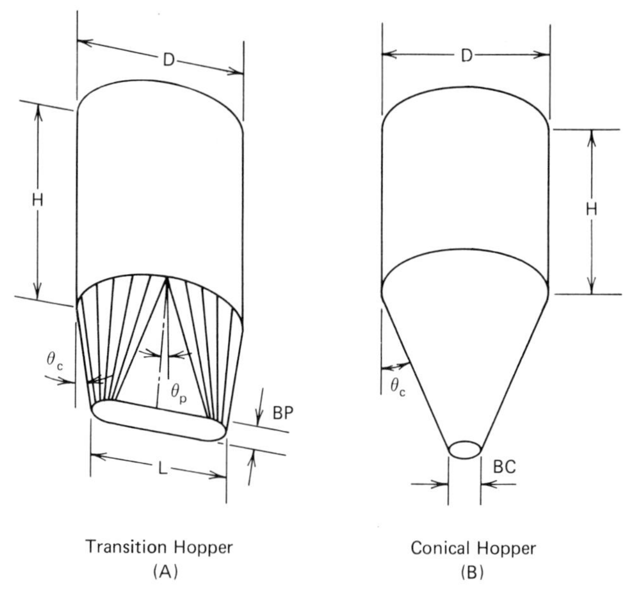

Fig. 21.2.2 Typical mass flow bins. *A,* transition hopper. *B,* conical hopper.

and the annulus between the inner and outer hoppers. It has been found that the outer hopper angle can be twice the angle from the vertical as the inner one and still achieve mass flow. See Section 21.6.4 for a more complete description of this device.

Vibrated Conical Hopper

Several vibrated hoppers are on the market. These devices impart a lateral motion to a conical hopper wall which mobilizes the wall friction component perpendicular to the direction of solids flow (downward). As a result, the downward component of friction is greatly reduced along with the effective coefficient of friction for flow in the hopper. The extent of this reduction depends on the ratio of lateral to vertical velocities of the solids. If the vertical component down the wall is zero, the effective coefficient of friction with lateral vibration is zero. This allows the solids to flow in mass flow on a very shallow cone. Thus it is not uncommon for a vibrated cone with θ_C equal to 45° to exhibit mass-flow behavior.

Since the effective friction coefficient depends on the downward velocity, it is likely that the downward velocity for mass flow will be significantly limited. This is especially true near the outlet of the hopper where the small cross-sectional area requires the highest downward velocity and the lateral motion (because of rotation about the hopper centerline) is the smallest. When this occurs, the flow generally pulsates between funnel flow and mass flow oftentimes with lack of control of the outfeed, especially with fine powders.

Flat-Bottom Bins and Mass Flow

The beneficial effects of hopper vibration are limited to a flow channel whose diameter approximates the top diameter of the vibrated hopper. If it is attached to a static upper section whose walls are not steep and/or smooth enough for mass flow, ratholing will still occur if the material is cohesive enough. If no static hopper section is used, material pressure from the cylinder may accentuate nonuniform flow conditions, again resulting in ratholes.

A flat-bottom bin can provide essentially a mass flow condition except for a small portion of the solids at the bottom of the bin. The basic requirement is that the outlets be arranged so that the flow pattern above each outlet intersects another and intersects the vertical cylinder walls. This is guaranteed by the configuration in Fig. 21.2.3. The bin on the right shows a slot across the entire

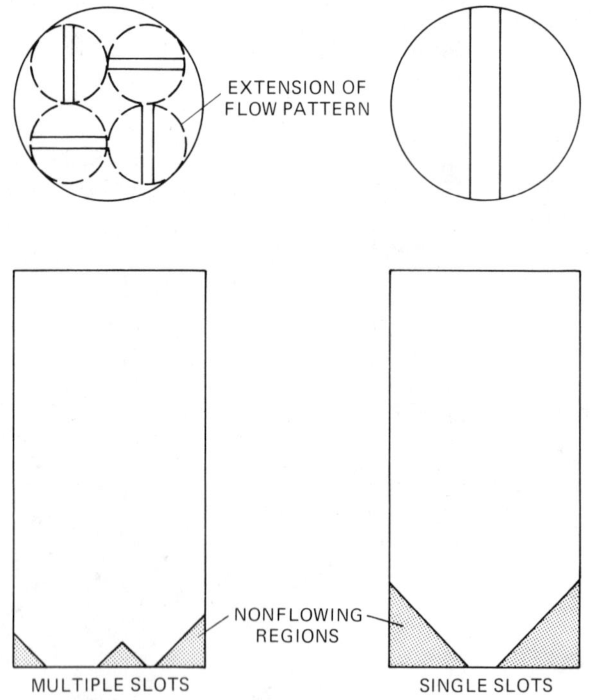

Fig. 21.2.3 Flat-bottom bins that create mass flow.

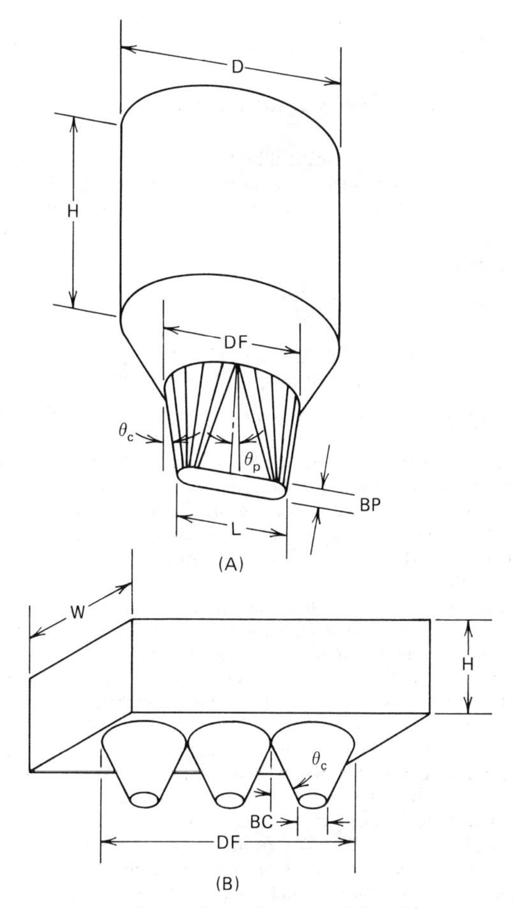

Fig. 21.2.4 Typical expanded-flow bins.

diameter of the bin which virtually guarantees flow on the vertical cylinder walls when the solids ıevel is high enough (about one bin diameter) to develop solids contact pressures compatable with the expanded flow channel indicated. The bin on the left has four slot outlets arranged so that the circle circumscribed around the diagonal of each slot touches an adjacent circle and the vertical bin walls. The head of material required to force flow along the cylinder walls in this case is about half that of the bin on the right.

These pseudo-mass flow bins are useful for extremely abrasive solids that might cause excessive wear along hopper walls.

21.2.3 Expanded Flow

This flow pattern develops when a mass flow hopper section is located below a funnel flow section. The mass flow hopper section just above the outlet ensures a more uniform bulk density upon discharge than is possible with a complete funnel flow bin. In addition, by expanding the flow channel out to the size of the top of the mass flow hopper section, ratholing can often be eliminated in the funnel flow hopper section. Thus the total bin height is often significantly less than if a full mass flow pattern had developed, while retaining many of the benefits of mass flow.

Two types of expanded flow bins are shown in Fig. 21.2.4. The transition hopper has the required slope and smoothness to provide mass flow and therefore expands the flow channel to a diameter that is large enough to make the rathole unstable. Above this a cone section is used which does not have the required slope or smoothness for mass flow, yet provides clean-out in a funnel flow pattern.

By using multiple mass flow hoppers whose sides touch each other, it is possible to expand the flow channel to a diameter equal to the sum of the diameters of the individual hoppers as shown in Fig. 21.2.3.

Expanded flow bins are useful for retrofitting existing silos to reduce hangups or to reduce the cost of new installations. They are most practical when the cylinder diameter exceeds 30 ft.

21.3 MEASUREMENT OF FLOW PROPERTIES

The flow of bulk solids has received significant attention over the past 30 years [5]. Seven primary variables have been identified which describe flow properties [6]. They are tabulated in Table 21.3.1 along with their typical variations with moisture content and solids pressure level.

21.3.1 Wall Friction Angle

The wall friction angle ϕ' is defined as follows:

$$\tan \phi' = \frac{\text{Frictional resistance to sliding of a bulk solid across a sample of bin wall material}}{\text{Solids pressure acting normal to wall sample}}$$

Another way of expressing this frictional resistance is through the coefficient of sliding friction μ which is equal to $\tan \phi'$.

This frictional resistance is often measured in a shear cell as shown in Fig. 21.3.1. The sample of bin wall material is fixed in position and a circular ring is placed on top of it. This ring is then filled with the bulk solid and a special cover with shearing bracket placed on the top. The ring is rotated slightly to raise it just off the plate so that the only contact is between the bulk solid and the plate. A series of weights is placed on the top cover and the force required to push the shear cell across the plate is measured for each weight.

A typical result is shown is Fig. 21.3.2. Note that the frictional resistance does not necessarily go to zero at zero solids pressure. This effect is called adhesion. Also note that the general shape of the curve is either a straight line or one which is concave downward.

Using the definition of wall friction angle ϕ' given previously, the general variation of ϕ' with pressure is as given in Fig. 21.3.3. Note that as the solids pressure decreases, the relative resistance to material slide (as denoted by ϕ') increases.

Besides pressure, several other variables affect ϕ' such as temperature, surface finish, chemical reaction between bulk solid and wall, temperature, moisture film at the wall, storage time at rest, particle size and shape, and relative hardness of particle and wall. As examples: Stainless steel is sometimes better than carbon steel since the latter can corrode with time. Often 2B finish stainless sheet has a lower frictional resistance than #1 (mill) finish stainless plate. However, very smooth walls such as glass liners often have high values of ϕ' because of adhesion due to the larger surface area in contact with the bulk solid. Plastics such as polyethylene or Teflon may have low friction with soft particles such as food products, but have high friction with hard minerals. A slight increase in moisture at the wall may have a tendency to lubricate coal sliding on stainless steel, but cause severe sticking of a clay-type ore.

21.3.2 Flowfunction

This material property is defined as follows [7]:

> *Flowfunction:* the relationship between cohesive strength of a bulk solid and consolidating pressure.

It is often measured in a direct shear cell as shown in Fig. 21.3.4. The base and ring are of the same size and usually circular in cross section. Material is placed in them and the cover placed on the top surface. Weights are then added to compress the sample and a shearing force applied to the ring/cover combination while the base is fixed in position.

A typical result is shown at the lowest part of Fig. 21.3.3. Note that as the consolidating pressure increases, so does the material's cohesive strength.

Primary variables that affect the flowfunction are:

Temperature. Temperatures other than ambient and temperature changes sometimes result in material having more cohesive strength than at ambient temperature.

Time of storage at rest. Strength often increases with increasing time.

Moisture. Strength increases up to 85–90% of saturation above which the voids between particles become essentially full and strength decreases due to decrease in surface tension between particles.

Particle size. Cohesive strength increases as particle size decreases.

Table 21.3.1 Basic Bulk-Flow Properties of Solids

Property Name	Symbol and Units	Description	Change with Increased Moisture	Change with Increased Consolidating Pressure, σ
Effective angle of internal friction	δ, deg	The kinematic friction condition during steady flow	Usually increases	Decreases significantly at low consolidating pressures
Angle of internal friction	ϕ, deg	The friction condition as a bulk solid starts to slide on itself at the onset of flow	Usually decreases	Usually increases
Kinematic angle of surface friction	ϕ', deg	Tan ϕ' is the coefficient of kinematic friction between the bulk solid and a wall surface	Sometimes increases significantly; sometimes decreases	Usually decreases slightly
Bulk density	γ, lb/ft^3	The unit weight of the bulk powder	Usually decreases at low consolidating pressures	Increases
Unconfined yield strength	f, lb/ft^2	The measure of cohesion and agglomerating tendency of the powder, usually expressed as a function of consolidating pressure	Increases significantly as long as saturation is not approached	Increases significantly
Compressibility factor	β	The slope of log γ vs. log σ	Increases	Little change at usual pressure levels, but tends to zero at both very high and very low pressures
Permeability	K, ft/sec	The superficial flow velocity of air through a solid, with a gas pressure gradient equal to γ	Usually increases as long as saturation is not approached	Decreases significantly

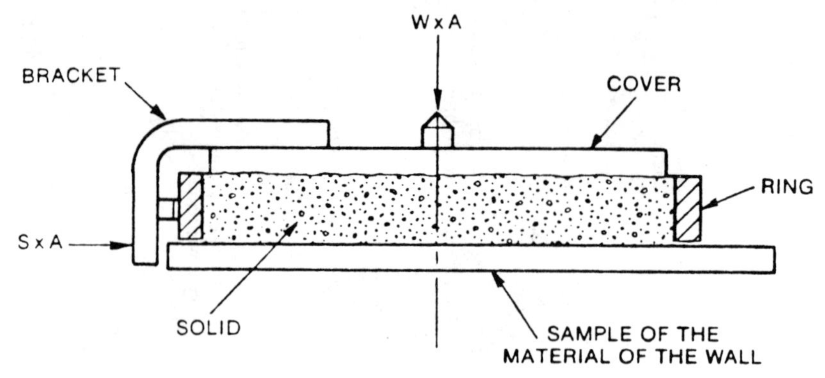

Fig. 21.3.1 Device for measuring wall friction.

21.3.3 Effective Angle of Internal Friction

The effective angle of internal friction δ [7] is defined as the interparticle kinematic friction angle that exists during steady flow.

A more rigorous definition of δ is

$$\sin \delta = \frac{\sigma_1 - \sigma_2}{\sigma_1 + \sigma_2} \tag{21.3.1}$$

where: σ_1 = major principal stress during steady flow
σ_2 = minor principal stress during steady flow

δ is measured in the shear cell at the same time as the flowfunction. A typical result is shown in Fig. 21.3.3. Note that δ typically decreases with increasing pressure, particularly at low pressures.

δ is affected by the same variables as the flowfunction with the exception of time at storage at rest. The latter is not a consideration since δ relates only to steady flow conditions. The direction of change with the other three variables is similar to the direction of change of the flowfunction with these variables.

21.3.4 Angle of Internal Friction

The angle of internal friction ϕ is the interparticle friction angle as a bulk solid starts to slide on itself at the onset of flow.

It is measured with the shear cell at the same time as the flowfunction and δ. A typical result is shown in Fig. 21.3.3. Increasing pressure usually causes an increased value of ϕ but this is not always true. Other variables which affect ϕ are temperature, time of storage at rest, moisture, and particle size.

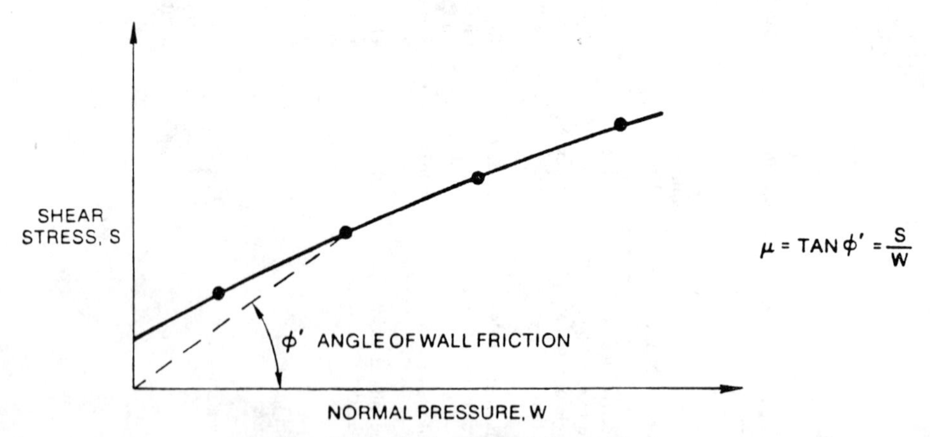

Fig. 21.3.2 Typical wall friction results.

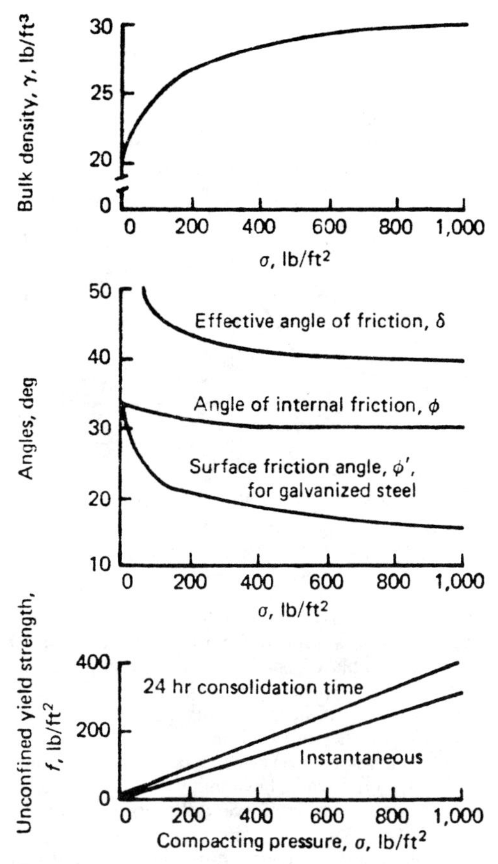

Fig. 21.3.3 Typical variation of material properties.

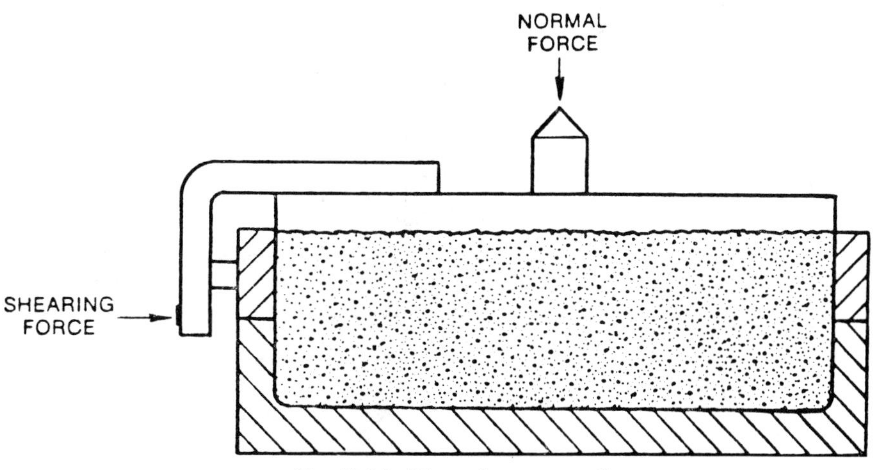

Fig. 21.3.4 Direct shear tester cell.

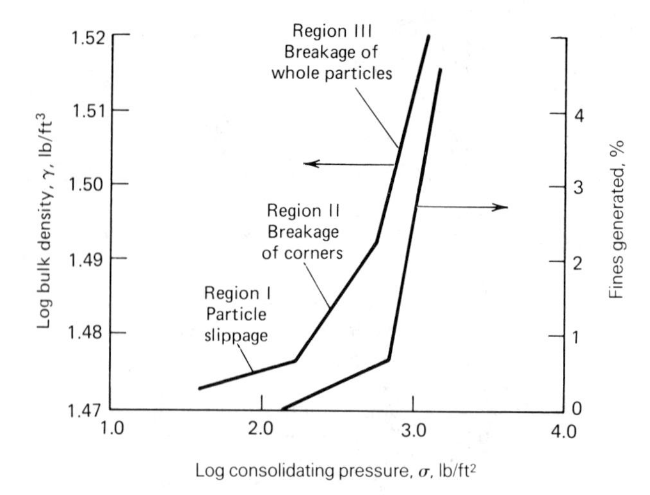

Fig. 21.3.5 Typical bulk density variation when particle breakage occurs.

21.3.5 Bulk Density

Bulk density γ is the weight per unit volume of bulk solid.

It is measured with a cup of known cross-sectional area which is filled with the bulk solid and a top placed on it. A series of weights is placed on the top cover and a dial gauge measures the depth of solid as a function of these weights.

A typical result is shown in Fig. 21.3.3. If the data points (bulk density as a function of consolidating pressure) are plotted on a log-log chart, they often fall on a straight line, the slope of which is denoted β. Exceptions to this occur at very low and very high pressures where the bulk densities level off (i.e., β goes to zero). If particle attrition, such as breaking of corners or whole particles, occurs within the range of pressures tested, a series of linear ranges with increasing slope will result as shown in Fig. 21.3.5.

Other variables that affect bulk density are temperature, moisture, and particle size.

21.3.6 Permeability

The permeability coefficient K is defined [6] as the superficial velocity of gas through a bulk solid associated with an upward gas pressure gradient equal to the solid's bulk density.

A device for measuring this is shown in Fig. 21.3.6. It consists of a cylinder, at the bottom of

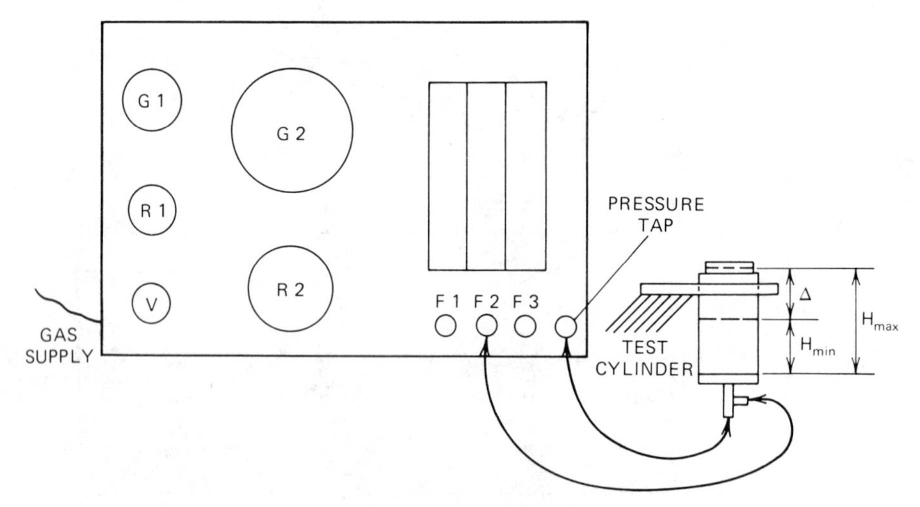

Fig. 21.3.6 Permeability tester.

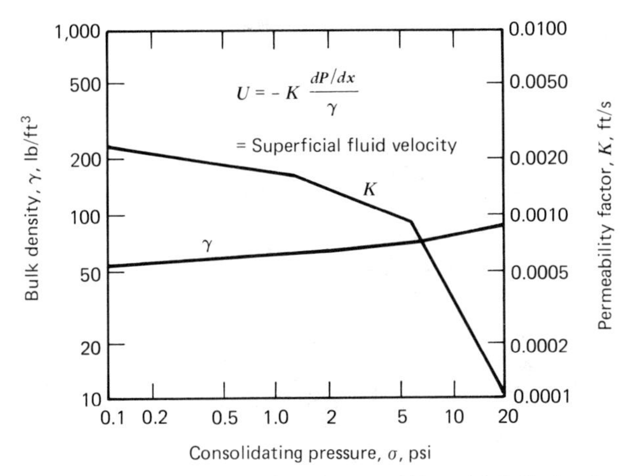

Fig. 21.3.7 Typical variation of permeability with bulk density.

which is a porous membrane attached to an air or gas supply. The cylinder is filled with the bulk solid and the gas supply increased until the pressure tap at the bottom indicates a value such that the difference in pressure between the bottom and the top, divided by the height of the cylinder, is equal to the solid's bulk density.

The solid is compacted in a series of steps, and the bulk density and gas flow measured at each step. The results are as shown in Fig. 21.3.7. Note that if the data points are plotted on a log-log plot of permeability coefficient K versus bulk density γ, they fall approximately on a straight line.

Higher moisture generally results in higher permeability because agglomerates form. The major effect of temperature is usually only on air or gas viscosity for which the results can easily be corrected. Particle size has a strong effect on permeability in that as the particle size decreases, so does the permeability.

21.4 DESIGN METHODS

The following methods of calculation for wall angles and hopper outlet sizes generally follow the methods of Jenike [4, 7, 8]; however, they have been simplified here for ease of understanding.

21.4.1 Wall Angle for Mass Flow

The required wall angle for mass flow is a function of the frictional resistance to material sliding against the wall and the hopper geometry. The higher the wall frictional resistance, the steeper the wall angle must be in order to achieve mass flow. This frictional resistance is denoted by the wall friction angle ϕ' as described in Section 21.3.1. As noted, this angle varies with solids pressure acting normal (perpendicular) to the wall.

Three typical distributions of this normal pressure are shown in Fig. 21.4.1. While no values are stated, they show in general how the pressure varies as the level of material drops. Note two important conclusions:

In the lower half of a mass flow hopper section, the normal pressure at a given point is practically independent of the amount of material in the cylinder. Extending this further, the height and diameter of the cylinder have little effect on these pressures.

In the lower half of the hopper section, the pressure decreases as the outlet is approached. Theoretically this pressure goes to zero at the apex and varies with the span of the hopper as follows:

$$\sigma_1 = \frac{ff\, B\, \gamma}{H(\theta)} \tag{21.4.1}$$

where: ff = flowfactor of hopper (described below)

B = hopper span expressed either as the diameter of a conical hopper or the width of a wedge-shaped hopper

γ = bulk density of solid

$H(\theta)$ = a function that depends on the hopper type and angle as shown in Fig. 21.4.2.

This equation defines the value of major consolidating pressure (principal stress) σ_1. The pressure acting normal to the wall is proportional to this value.

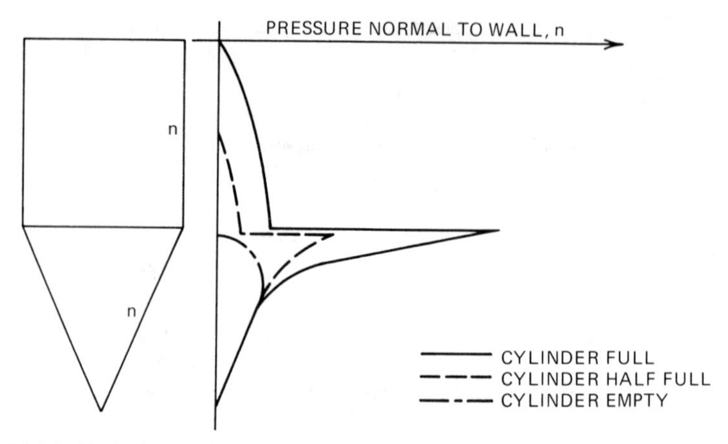

Fig. 21.4.1 Typical mass-flow wall pressure distribution with varying head of material.

Once σ_1 has been calculated at a certain point in the hopper, a plot such as shown in Fig. 21.3.6 can be used to find the wall friction angle ϕ' at this point.

Limiting mass flow conical hopper angles computed by Jenike [7] for various values of effective angle of internal friction δ are given in Fig. 21.4.3. For wedge-type hoppers, the limiting angle is not clearly defined. As a general rule, an angle θ_P 10° or less greater (i.e., less steep) than that for the conical hopper is a conservative design value.

The procedure used with these figures is as follows:

1. Select the type of hopper that will be studied—either cone or wedge.
2. Choose a hopper span B. It is conservative to take this value as the outlet size but often at this step in the process, the outlet size has not been determined. Therefore, select a series of B values.
3. From (21.4.1), compute σ_1, the major consolidating pressure for each value of B.
4. Go to a plot such as shown in Fig. 21.3.2 and, for each value of σ_1, determine the value of wall friction angle ϕ'.
5. Using the values of ϕ', go to the design chart (Fig. 21.4.3) and determine the maximum conical hopper angles (measured from the vertical) for mass flow. For example, if $\phi' = 20°$ and $\delta = 50°$, the maximum angle for a conical hopper θ_C is 23°. Using this angle plus 10° for a wedge hopper, θ_P is 33°.
6. Repeat this process for various hopper spans and for both hopper geometries. Because the wall friction angle ϕ' often decreases with increasing solids wall pressure, significant headroom can sometimes be saved by making the hopper wall progressively less steep in going from the

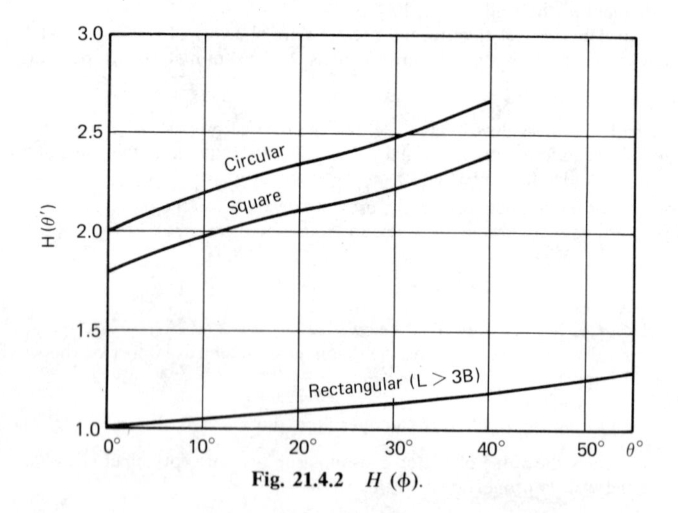

Fig. 21.4.2 $H(\phi)$.

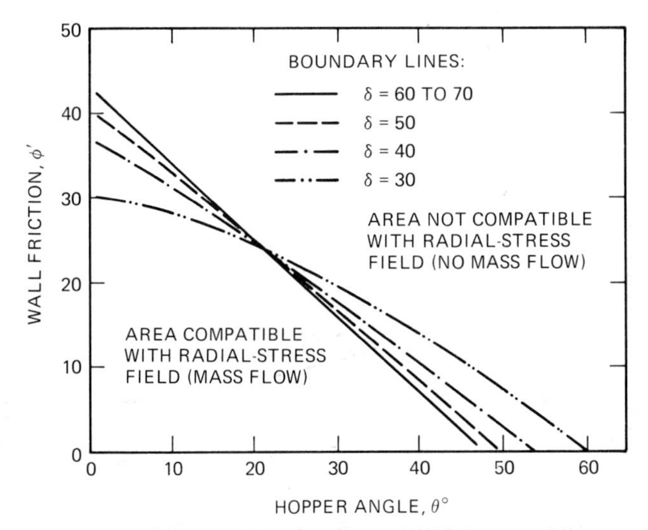

Fig. 21.4.3 Q' ϕ chart for various δ (cones only).

outlet region to the cylinder. It is usually easier to do this as a series of step changes rather than a continuous change in slope.

Note that wedge hoppers require significantly less steep hopper walls for mass flow and these wall angles are less sensitive to slight differences in wall friction angle ϕ' than is the case in a conical hopper.

21.4.2 Outlet Size

To ensure flow when an outlet gate is opened or feeder is started, it is necessary that the outlet size be large enough to prevent arching (bridging, doming) if a mass flow hopper is used, or ratholing (piping) if a funnel flow hopper is used.

Arching in Mass Flow Bins

This can be due to the interlocking of a few particles that are large with respect to the outlet or due to the formation of a cohesive arch.

To overcome the first type of arching, it is sufficient that the outlet size be at least five to six times the largest particle size if a circular outlet is used, or three to four times (i.e., minimum width) if a rectangular outlet is used.

The procedure used to ensure that a cohesive arch will not form is somewhat more complicated. First, it is necessary to compute the flowfactor of the hopper, defined as:

Flowfactor (*ff*) = (major consolidating pressure at a point in the hopper)/(minimum stress in the abutment of an arch at this point)

Various flowfactor values are given in Fig. 21.4.4 for limiting mass flow hopper wall angles. By selecting the type of hopper geometry (cone or wedge) and effective angle of internal friction δ, a value of flowfactor *ff* can be obtained.

Next, go to the flowfunction such as shown in Fig. 21.3.2 and draw a line starting at the origin which has an inverse slope of *ff*. There are now several possibilities:

1. The flowfunction lies entirely below the flowfactor. This indicates a material that is easy flowing. Consideration of particle interlocking and flow rate dictate the minimum outlet size, not cohesive arches.

2. The flowfunction intersects the flowfactor. Calling the value at the point of intersection $\overline{\sigma}_1$, the minimum outlet size B required to prevent cohesive arches is:

$$B = \frac{\overline{\sigma}_1 H(\theta)}{\gamma} \tag{21.4.2}$$

where: $B = BC$ for circular outlet, BP for rectangular (see Fig. 21.1.2)
$H(\theta)$ = function as shown in Fig. 21.4.2
γ = bulk density of solid

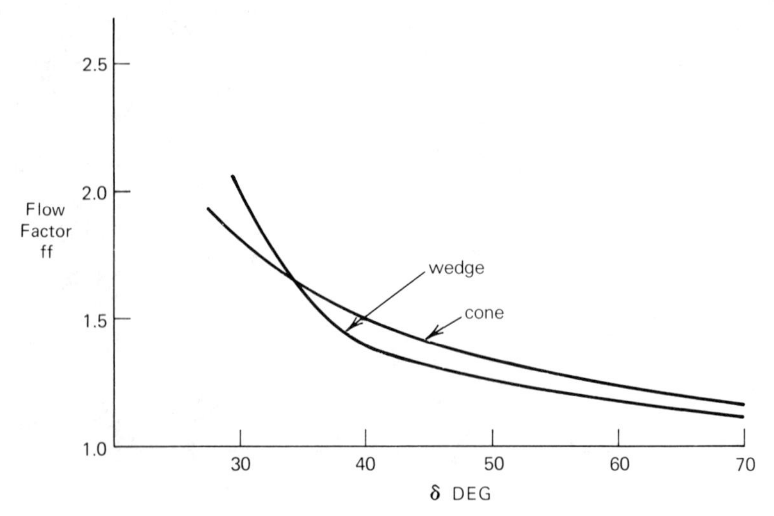

Fig. 21.4.4 Hopper flowfactors (ff) as a function of δ for cone and wedge limiting conditions for mass flow.

Since the flowfactors for conical and wedge hoppers are essentially the same for the same value of δ, the fact that $H(\theta)$ for a circular outlet is approximately double that of a rectangular one suggests that the minimum outlet size of a conical hopper is approximately twice the minimum width of a wedge hopper. This indicates another significant advantage of the transition, chisel, or wedge hopper: not only can the hopper slope be less steep than that of a cone and still achieve mass flow, but also the outlet size can be considerably smaller.

Note also that (21.4.2) expresses the minimum outlet size in a dimensional value. Thus no matter what size bin is used, unless the outlet is at least this size or larger, material will not flow.

3. The flowfunction lies entirely above the flowfactor. This indicates that gravity alone is not sufficient to overcome cohesive arching. Flow aids such as vibrators or air blasters must be used, or the material's flowfunction must be changed by, for example, lowering its moisture content or temperature or adding a free flow additive. See Section 21.6.3.

Arching and Ratholing in Funnel Flow Bins

In order to assure satisfactory flow in a funnel flow bin, it is necessary that the outlet be large enough so that ratholing and arching do not occur. Flow should not rely on flow-promoting devices. Typical shapes of outlets are shown in fig. 21.2.1. Denoting the major dimension of each outlet as D, the flow channel which tends to form will be circular and have this diameter. To make this channel unstable, D must be selected at least equal to the critical rathole diameter DF computed in the following way:

Using the height h and hydraulic radius R, where R = (cross-sectional area)/(length of perimeter of the bin), estimate the effective consolidating head EH:

$$EH = \frac{[1 - \exp(-\mu\, K\, h/R)]R}{\mu K} \tag{21.4.3}$$

Compute major consolidating pressure σ_1 as follows:

$$\sigma_1 = EH\, \gamma \tag{21.4.4}$$

Using this value of σ_1, find $\bar{\sigma}_1$ from the time flowfunction and the angle of internal friction ϕ.
Compute the critical rathole diameter DF

$$DF = \frac{\bar{\sigma}_1\, G(\phi)}{\gamma} \tag{21.4.5}$$

Function $G(\phi)$ is plotted in Fig. 21.4.5.

In a square or circular outlet computed for no-ratholing, arching cannot occur. In a rectangular outlet of diagonal D, the minor dimension of the outlet (width BF shown in Fig. 21.2.1) must also be computed so that arching does not occur. A flowfactor ff = 1.7 is used here.

This assures a flow channel angle not less than 30° from the vertical at the sides of the rectangle. BF is calculated as follows:

Draw the flowfactor ff = 1.7 over the time flowfunction of the solid and determine the value of $\bar{\sigma}_1$ at the point of intersection (Fig. 21.3.3).

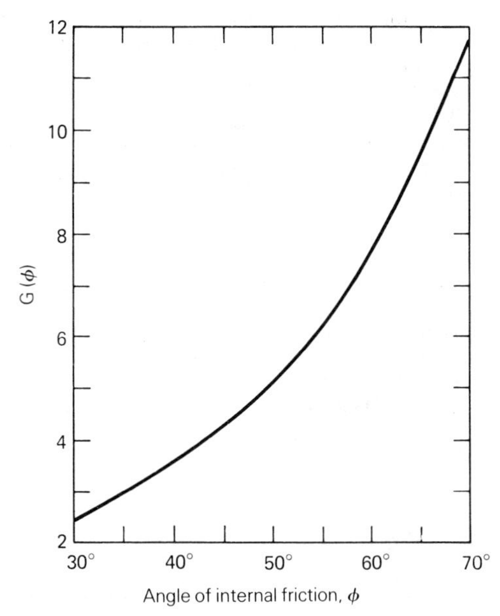

Fig. 21.4.5 Function G(ϕ).

Compute BF from (21.4.2). For $H(30°) = 1.15$, this equation becomes

$$BF = \frac{1.15\overline{\sigma}_1}{\gamma} \tag{21.4.6}$$

Flow Rate

If a material is coarse and free flowing, the maximum rate of discharge Q through an outlet is given as follows by Johanson [9]:

$$Q = \gamma A \sqrt{\frac{(B\,g)}{2(1 + m)\tan\theta}} \tag{21.4.7}$$

where: γ = bulk density of solid
A = cross-sectional area of outlet
B = diameter of circular outlet or width of rectangular outlet
g = acceleration due to gravity
θ = hopper angle measured from the vertical
m = 1 for conical hopper, 0 for wedge hopper

One extreme to a coarse, free-flowing material is a fine powder. Such a material's flow rate is often several orders of magnitude less than that of a coarse solid having the same bulk density.

The flow of fine powders can take three steady-state flow-rate-dependent modes. The first occurs at flow rates below a limiting condition which can be calculated based on the permeability and compressibility of the powder. This type of flow is characterized by the uniform gravity flow of fully deaerated material. The limiting condition occurs when compaction in the cylinder forces too much gas out through the top surface of the bin. This causes a slight vacuum to form as the material expands while flowing through the converging portion of the bin. The result is a gas counterflow through the hopper outlet which limits the solids flow rate.

The second mode of flow occurs at flow rates greater than the limiting rate, and can be achieved only by the use of a gas permeation system at an intermediate point in the bin to replace the lost gas.

The third mode occurs when the flow rate is too high to allow the material to fully deaerate. In the extreme the material is completely fluidized and floods or flushes through the outlet.

A complete understanding of these rate phenomena requires the calculation of the two-phase flow system associated with the gas entrainment in solids [21]. However, the necessity of such detailed calculation can be established by considering the permeability and density change of the solid as follows:

$$CF = [(\gamma_1/\gamma_{\min}) - 1]\frac{H}{K} \tag{21.4.8}$$

where: CF = correlation factor proportional to the time required for gas to escape from a bulk solid
γ_1 = bulk density at a consolidating pressure head equal to half the bin diameter
γ_{min} = minimum bulk density as solids enter the bin. Note: if the solid is conveyed pneumatically to the bin, this will be the fluidized density.
H = total bin height
K = the permeability of the solid at a density about halfway between γ_1 and γ_{min}

Table 21.5.1 provides an indication of the types of problems likely to correlate with *CF*.

21.4.3 Bin Loads

Bin failures are always expensive in terms of lost production, equipment repairs and replacements, and downtime.

The main causes of bin failures are uneven foundation settlement, faulty construction (e.g., missing rebars), explosions, and the designer not anticipating loading conditions. The last two causes are often related, since bin failures can initiate explosions, such as at a grain elevator.

Failures due to unanticipated loading conditions in bins fall into one of six categories [10–16]:

Formation of large voids with subsequent dynamic loads imposed on the structure by the collapse of hung-up solid.

Denting of circular walls caused by off-center channeling adjacent to a wall.

Overpressures at the point where the flow channel intersects the cylinder wall in funnel flow and expanded flow.

Development of mass flow in bins designed for funnel flow, causing overpressures.

Asymmetric pressures caused by inserts, such as beams across the cylinder of a bin, or by nonuniform withdrawal of multiple-outlet bins.

Drastic means of flow promotion, for example, explosives, excessive vibrations, or air injection.

Each of these conditions is related to the flow pattern of solids in the bin and can be eliminated or designed for when that pattern is predictable and consistent.

Currently available bin design codes use either the Janssen or the Reimbert methods to determine the wall loads. Using these methods, attempts are made to cover contingencies by magnifying factors based on the bin size and the generic name of the stored solid without explicit consideration of its flow properties and the flow pattern.

Some attempts have also been made to allow for off-center unloading and the resultant eccentric load distribution but, in the main, the codes recommend axisymmetric loads on the cylindrical part of the bin.

Symmetric Pressures in Cylinder Section

Janssen's equation is developed by consideration of equilibrium of a horizontal layer of solid in a vertical cylinder and is expressed as follows:

$$q = \frac{\gamma R}{\mu K}\,[1 - \exp(-\mu\, K\, Z/R)] \tag{21.4.9}$$

where: q = average vertical pressure at depth Z of the solid
γ = bulk density of the solid (assumed constant in derivation)
μ = coefficient of friction between the solid and cylinder
R = hydraulic radius of cylinder, that is, the ratio of cross-sectional area to length of perimeter
K = ratio of horizontal pressure acting on the wall of the cylinder to average vertical pressure at the same elevation, assumed constant.

The coefficient of friction μ can vary between a low value obtained during sliding and a high value obtained after the solid has been at rest. If the solid flows in a channel within stagnant material, the lower value is equal to sin δ where δ is the effective angle of internal friction of a solid (see Section 21.3.3).

In order to allow for unavoidable deviation of the tested samples from field conditions, Jenike [17] recommends that the lower bound of μ be decreased by 0.05 and the upper bound increased by the same magnitude.

If the flow channel coincides with the cylinder walls, consideration must be given to the shape of those walls. Unfortunately, the cylinder section of a bin used to store a bulk solid is never perfect. The walls diverge and converge due to fabrication errors, distortion due to weld shrinkage, and other causes. To cover these undefined variations, the ratio K in (21.4.9) needs to vary and the bin loads need to be computed on the basis of bounds of K that maximize the loads. Jenike [16] recommends the following bounds where the variations and imperfections in the cylinder are minimal:

$$KL = 0.25 \text{ (lower bound)} \qquad KH = 0.6 \text{ (upper bound)} \tag{21.4.10}$$

The highest pressures P acting normal to the wall (horizontal) are obtained by using the low bound on μ and the high bound on K (KH). This yields the following equation for P:

$$P = \frac{\gamma R}{\mu[1 - \exp(-0.6\,\mu\, Z/R)]} \tag{21.4.11}$$

The maximum vertical force per unit length of circumference imposed by shear on the wall acts when the vertical pressure q is at a minimum. This is computed by taking upper bounds of both μ and K and results in the following equation:

$$V = \gamma R \frac{Z - R[1 - \exp(-0.6\,\mu\, Z/R)]}{0.6\,\mu} \tag{21.4.12}$$

Mass Flow Hopper Pressures

The pressures acting normal (perpendicular) to the converging hopper walls, P, are computed from the Janssen equation modified for the change of section which results in the following equation for axial symmetry [17]:

$$P = \gamma \text{ rad}K \left[\frac{h - Z}{N} + \left(\frac{q}{\gamma} - \frac{h}{N}\right)\left(\frac{1 - Z}{h}\right)^{N+1}\right] \tag{21.4.13}$$

where: $\text{rad}K = 6 \tan \theta_C\, SA/[4\, SA(\tan \theta_C + \tan \phi') - 1]$
$SA = \sigma'/\gamma B$
$N = 2 \text{ rad}K\ (1 + \tan \phi'/\tan \theta) - 3$

Variables h, Z, and θ are defined in Fig. 21.4.6. The parameter m is equal to 1 for conical hoppers and 0 for the side walls of wedge hoppers. ϕ' is the wall friction angle. $\sigma'/\gamma B$ is given by Jenike [4]. For mass flow hoppers near the limiting angle for flow at the walls, this parameter has a narrow range of 0.4–0.65 for conical hoppers and 0.8–1.3 for wedge-shaped hoppers. Using the larger value in the upper part of the hopper produces a conservative result. The value of q is determined from (21.4.9) using the dimension of the vertical cylinder at the top of the hopper.

Parameters rad$K$ and N are determined and then P is computed from (21.4.13).

The foregoing approach gives maximum values in the upper part of the hopper. A similar calculation should be run using rad$K$ = 1.0 to calculate an upper bound to the pressures in the lower part of the hopper.

Flow Irregularities

With very cohesive solids, the critical diameter to prevent arching can approach or even exceed the diameter of the cylinder section. If this occurs, the solid is capable of arching throughout the bin. Various flow aid devices such as vibrators, aerators, or air cannons are often used to initiate flow

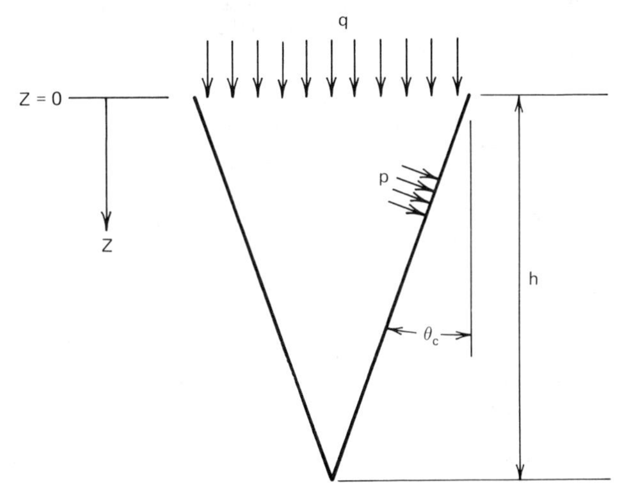

Fig. 21.4.6 Definition of Z, h, θ_C, p, q.

from the hopper (see Section 21.6.3). Using such devices, the hopper may become empty while the cylinder remains full of material. If and when this material drops, it may tear the hopper from the cylinder section or break through the hopper wall. Under these conditions the head of the solid in the cylinder should not exceed one diameter and the hopper should have the strength to withstand the impact of a falling mass.

Another type of flow irregularity that has structural implications occurs when the bin vibrates as material is discharged. Bin vibrations can range from those that are hardly detectable and a mere annoyance to those that pose a serious structural problem. The more severe vibrations, prevalent in funnel flow bins and silos containing such materials as coal and cement clinker, are usually low in frequency, high in amplitude, and irregular. Yet by understanding the causes of vibrations, it is often possible to eliminate or at least minimize them.

Some vibrations are caused by outside influences, such as feeders or conveying systems, and transmitted to the bin and its contents. Others are caused by material flow, and it is these which are the hardest to analyze and correct. The basic phenomenon of such bin vibration is as follows: a portion of the material expands as it flows in a steady state primary flow channel, causing regions of low density or voids to form. This region of expansion or void formation enlarges, thus reducing support for the surrounding dead material which can eventually become unstable and suddenly begin to flow in a secondary unstable flow channel. The result is that this action fills the void or compacts the loose material which abruptly stops and again supports the surrounding material. The process then starts over as the material expands again.

There are basically two types of instability required to trigger this mechanism:

1. Instabilities related to changes in physical properties of the material, such as: differences in kinematic and static coefficients of friction of the material at the walls or on itself (slip stick); an increase in cohesion with time causing bridges and voids to form at the transition between the cylinder and the hopper in a storage bin; or an increase in the wall friction angles with time, causing severe changes in the flow pattern.
2. Instabilities related to basic flow pattern limits of steady flow, such as limiting wall friction conditions or the limited central flow channel of funnel flow bins.

In each of these cases it is possible to have unstable incipient flow stress conditions in the nonflowing regions.

Asymmetric Bins

Pressures that act on the bin walls at a given elevation can vary around the circumference due to eccentric flow, inserts, and multiple outlets. Eccentric flow can be due to off-center withdrawal or improper design of a feeder. This causes the flow channel to intersect the cylinder wall at a lower elevation on one side. Above this point the wall pressures from the flowing solid are smaller than from the stagnant solid. When the cylinder is circular, these nonuniformities cause horizontal bending of the walls leading to vertical cracks in concrete silos and inward denting of steel bins on the side adjacent to the flow channel.

An insert placed within the cylinder of a bin causes a transition from an overpressure exerted on the insert and reflected upward onto the wall. A single beam placed across a cylinder projects overpressures that cause an elliptic pressure distribution on the wall at the level of the reflected overpressures. Such a cylinder should be adequately reinforced to absorb the resultant bending moments.

The effect of an insert placed within the hopper of a bin is less drastic because pressures are generally lower.

Multiple-outlet bins usually must be designed for funnel flow since it is not possible to ensure discharge through all of the outlets simultaneously and uniformly at all times. The most severe condition occurs when only a single off-center outlet near the wall is active and material forms a stable rathole above it. The pressures exerted by the stagnant solid above the other outlets remain unchanged while the pressure on the wall portion adjacent to the rathole is zero. This leads to exceedingly high bending moments within the cylinder section. It is important that all of the possible single and multiple flow channels be investigated to determine the critical design condition.

Multiple-outlet bins may also experience arches at the top of one or more of the hoppers used to store a cohesive solid when operated as follows: normal operation for the bin is to have all outlets active at about the same flow rate. This causes a large vertical pressure, hence large bulk solid strength, to occur at the top of the hopper. Under these conditions, arches do not form at the top of the hopper because the large vertical pressure provides the arch-breaking force necessary to maintain flow. At some later time most of the outlets become inactive with only one or two still active when the large vertical force is no longer available to break the arch at the top of the active hoppers.

The configuration of the solid at the top of the silo also needs to be considered. Off-center charging into the bin may cause significant asymmetric loading as well as the formation of off-center flow channels.

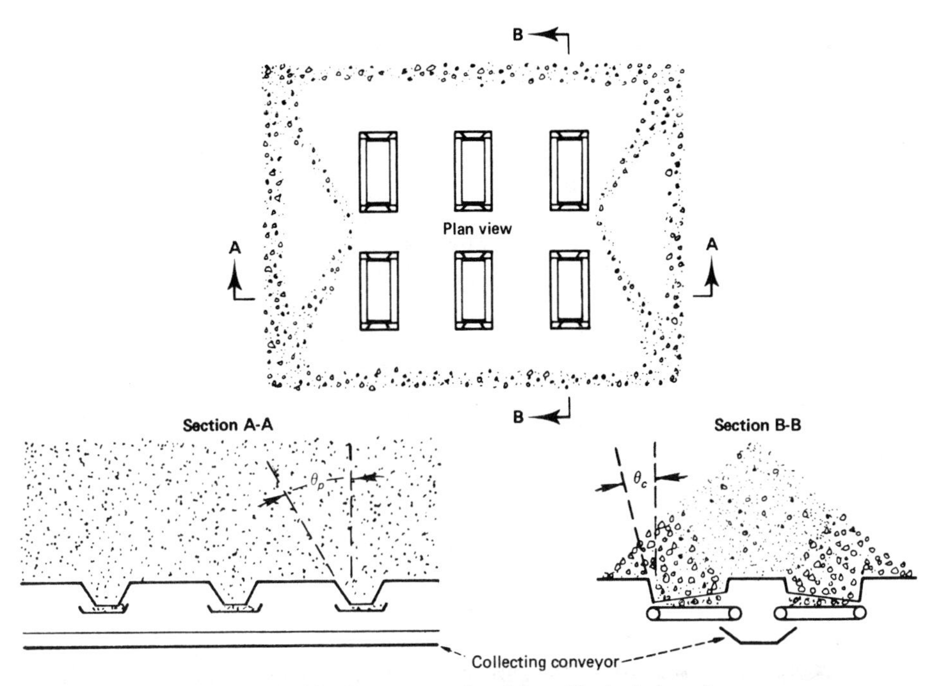

Fig. 21.4.7 Bottom-tunnel reclaim with single tunnel.

21.4.4 Bottom Reclaim Stockpiles

Stacked on a concrete or earthen pad, stockpiles may be covered by a roof structure or left uncovered, depending on whether there is a need to keep the material dry. Stockpiles are formed using various overhead conveyors (such as a tripper conveyor) or mobile equipment (such as front-end loaders and scrapers). Stacking with conveyors often produces segregation in stockpiles, but this problem is less pronounced with mobile equipment. Segregation can never be eliminated, and excessive compaction of the solids may occur with either method.

One common method of material recovery is bottom-tunnel reclaim [18]. Figure 21.4.7 shows the layout for a single-tunnel system, which provides a maximum of capacity for the lowest feeder investment. Minimum hopper dimensions are determined from the flow properties of the material. Since mass flow is essential, the hopper walls must be inclined at the angles θ_P and θ_C (as defined in Fig. 21.4.7) for the particular material. The outlet just above the belt must be at least as large as the minimum width necessary to prevent arching, *BP*, for a long slotted outlet. As indicated in the figure, the top dimensions of the short, mass flow hopper should be designed for funnel flow; the width should exceed the minimum needed to prevent arching in a funnel flow slot, *BF*, whereas the length must exceed the critical rathole diameter, *DF*.

The layout shown in Fig. 21.4.7 provides the smallest mass flow hoppers and feeder lengths allowable for a given material. In some cases, it may be more economical to provide fewer, but longer, slots. If a single slot can be used, the collecting conveyor can be eliminated. The dimensions of the slot are determined by the feeder's ability to pass material uniformly along its full length. The maximum allowable length for uniform withdrawal as a function of feeder width is about 10:1.

Unfortunately, as a pile is drawn down, the design shown in the figure tends to feed fines first, then coarse material. If segregation hinders subsequent processing, this design should not be used.

Figure 21.4.8 shows a more expensive bottom-tunnel reclaim system in which the slots feed perpendicular to a common collecting conveyor. This arrangement promotes remixing of fines and coarse particles. The hopper design criteria are similar to those described previously. However, if the spacing of the slots is equal to or less than the diagonal of the top of the hopper, the flow channels from adjacent slots will join and the length of the slot need not satisfy the rathole criterion. This design is effective for solids with extremely large critical rathole diameters.

Bottom-tunnel reclaim lends itself to automation. For materials that dust and must be covered, bottom-tunnel reclaim ensures dust-free operation. Other methods, because of the lifting action of the reclaimer, tend to aggravate this difficulty. However, bottom-tunnel reclaim is not suited for handling extremely cohesive materials. For such solids, top reclaim by bucket wheel, scraper truck, or front-end loader should be considered.

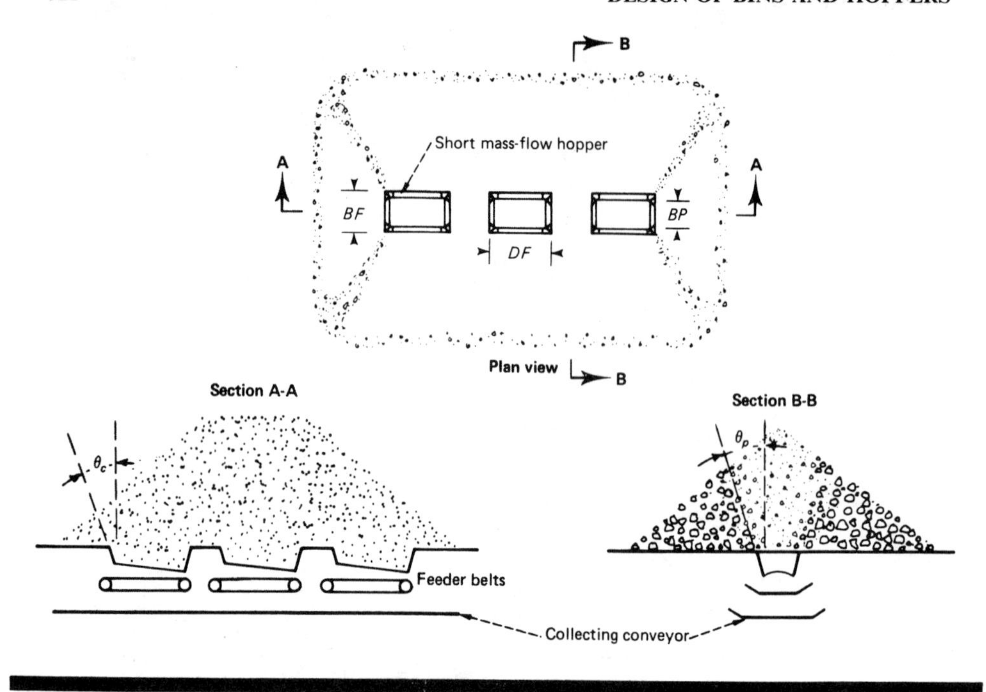

Fig. 21.4.8 Bottom-tunnel reclaim with slots aligned perpendicular to conveyor.

21.5 FEEDERS, DISCHARGERS, AND GATES

It is important to differentiate between a feeder and a discharger. A feeder is a device used for controlling the rate of withdrawal of a bulk solid from a bin. A discharger, on the other hand, is used to enhance material flow from a bin but, by itself, is not capable of controlling the rate of withdrawal.

With either feeders, dischargers, or gates, it is important that the full area of the outlet be "live." This can be accomplished through proper design of the feeder or discharger and through proper operation of the gate. That is, the gate must be operated either fully open or fully closed, and not used to control the rate of discharge unless the bulk solid is easy flowing.

21.5.1 Feeders

Criteria for Feeder Selection

Often, personnel at a plant may have personal preference for a certain type of feeder because of past experience, availability of spare parts, or to maintain a uniformity for maintenance throughout the plant. These personal preferences can usually be accommodated.

In general, several types of feeders can be used in a given application provided they are designed to provide uniform flow across the entire outlet. Most of the bad experience associated with a particular type of feeder comes from improper design of the bin and feeder combination.

For example, if a constant-pitch screw that is several pitches long is placed under a slot outlet, feed will occur only at the far end of the screw resulting in a funnel flow pattern even if the bin is designed for mass flow. With a cohesive solid, arching will likely occur along the length of the screw and a stable rathole will form at the far end. With a more free-flowing solid, the torque on the screw will be excessive as the material is prevented from feeding along the length of the screw because there is no volume available in the constant-pitch screw. Similar results occur in other types of feeders when they are improperly designed.

Table 21.5.1 lists some of the major considerations in deciding on the type of feeder to use. As should be apparent from some of the items, material properties and bin design significantly affect the type and size of feeder, so an iterative procedure is often needed to select the best feeder for a given application.

Table 21.5.1 Behavior of Dry Material in Bin

Correlation Factor (CF)[a]	Material Behavior
1	No fluidization; Eq. 21.4.8 predicts flow rate very well.
10	Significant variation in limiting rate as predicted by Eq. 21.4.8. No serious flushing problems.
100	Very likely to need an air permeation system to achieve the desired flow rate. Very likely to experience flooding and flushing problems in funnel flow bins.
1000	Likely to remain fluidized in mass flow bins unless sufficient retention time is provided. If bin size is below the critical, then it will be very difficult to handle the material in other than a pneumatic conveying system.

[a] Proportional to time required for gas to escape from bulk solid.

Screw Feeders

Screw feeders come in a variety of types, the most common of which is a single helicoidal or sectional flight screw with end discharge. Other types include multiple screws with end discharge and single and multiple screws with center discharge. Since the single screw with end discharge is the most common, we concentrate on its design.

The key to proper screw feeder design is to provide an increase in capacity in the feed direction. Unfortunately the normal tolerance of fabrication is such that extending the length under the hopper to greater than six to eight times the screw diameter often results in a poorly performing screw.

A design that has been widely used with good success in mass flow bins is shown in Fig. 21.5.1.

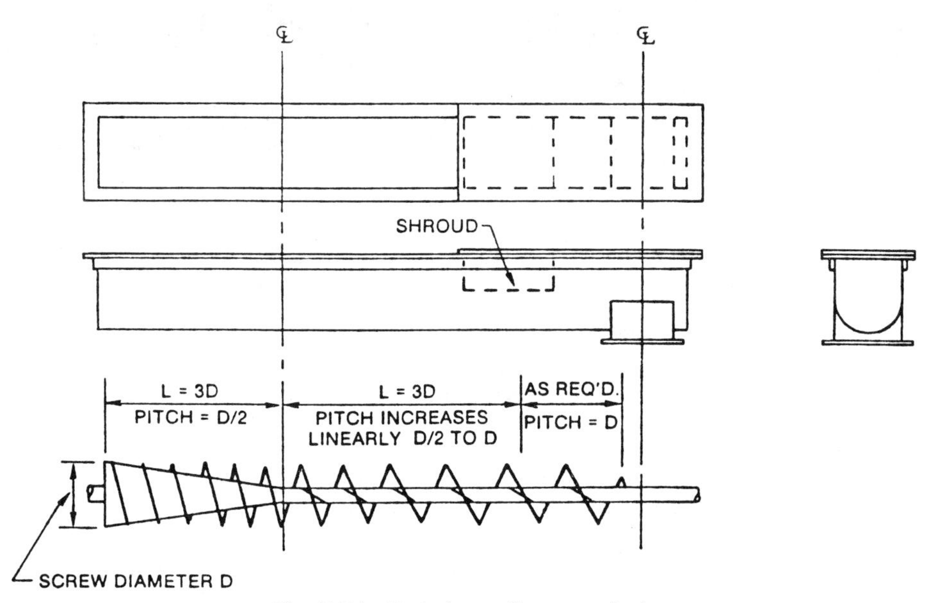

Fig. 21.5.1 Typical mass flow screw feeder.

Table 21.5.2 Capacities of Various Types of Feeders

	Screw										
Variable	Single	Multiple	Center Discharge	Belt	Apron	Rotary Valve	Siletta	Acrison	Table	Rotary Plow	Vibrating Pan
Temperature	——Up to 1000°F——			Up to 450°F (neoprene)							
Flow rate	Low	Low	Low	High	High	Very low	Very low	Very low	Low	High	Low
Maximum ton/hr ($\gamma = 100$ lb/ft^3)	600	1000		5000	5000						
Feeder speed	——2–40 rpm—— (80 intermittent)			5–300 ft/min (500 at steady state)	10–50 ft/min				2–10 rpm		0–80 ft/min (sloping down), 50 ft/min (horiz.)
Impact (direct)	Fair	Fair	Fair	Poor	Good	Poor	Poor	Poor	Good		Good
Discharge point	End	End	Center	End	End	Center	Center	Center or end	Side	Side	Near center (end or side)
Hopper outlet configuration	Rectangular	Rectangular	Rectangular	Rectangular	Rectangular	Square, round, or rectangular (star)	Square, round, or rectangular	Square, round, or rectangular	Round	Round	Square or round
Max. slot length : width ratio	6		15	Dictated by bed depth		4	2	4	—	Unlimited	—
Gravimetric or volumetric[a]	Generally volumetric	Volumetric	Volumetric	Either	Volumetric	Volumetric	Generally volumetric	Generally volumetric	Volumetric	Volumetric	Volumetric

Ability to seal against gas pressure	Depends on material properties (0.2 psi/ft free flowing, 5 psi/ft cohesive)	Poor	Poor	Poor	Poor		30 in. H_2O	Poor	Poor	Poor	Poor
Turndown	10:1	10:1	10:1	10:1	10:1	10:1	10:1	10:1	3:1	3:1	10:1 or more (?)
Return spillage	No problem	No problem	No problem	Bad	Bad	No problem	No problem	No problem	Some problem going to conveyor	Some problem	No problem
Material:											
Arch dim	$BP \leq 2$ ft		$BP \leq 2$ ft	$BP \leq 3$ ft	$BP \leq 3$ ft	$BC \leq 2$ ft	$BC \leq 5$ ft	$BC \leq 8$ ft	$BC \leq 6$ ft	$BP \leq 2$ ft	$BC \leq 2$ ft
Sensitive to overpressure		——be careful——						Lock intrometer and feed screw			
Max. particle size	$\simeq\frac{1}{2}$ in. (larger with special screw)			$\simeq 2$ in.	+12 in.	$\simeq\frac{1}{2}$ in. (normal) $\simeq 1$ in. (special) (striker plate or star feeder)	$\simeq\frac{1}{2}$ in.	$\simeq\frac{1}{2}$ in.	$\simeq 2$ in.	$\simeq 2$ in.	+12 in.
Dry powder	Good	Good	Poor	Fair	Poor	Good	Good	Good	Poor	Poor	Poor (OK with valve)
Abrasive				Poor	Good	Poor		Poor	Good		Poor
For:											
Dust control enclosure	Good	Good	Good	Poor	Poor	Good	Good	Good	Poor	Poor	Good (if enclosed)
Cleanout				Good							Good
Degradable material (food grade)	Poor	Poor	Poor	Good	Poor	Fair	Good	Poor	Poor	Poor	Good
Convey beyond bin	Fair	Poor	Poor	Good	Poor	Poor	Poor	Poor	Poor	Poor	Fair

It consists of a cone section with constant half pitch and an increasing pitch section followed by a constant pitch conveying section. To avoid hanger bearings, the total length usually has to be limited to about 12–14 ft. U-troughs are preferable to V-troughs and, in order to prevent a lip at the hopper outlet, the hopper outlet width should be equal to the screw diameter. Since standard U-troughs have an inside dimension 1 in. larger than the screw diameter, this ensures no lip.

Approximate capacities of various size screws are given in Table 21.5.2. It is wise to limit the rotational speed to no greater than 40 rpm in order to prevent significant wear or power requirements. Low rpm's, say two or less, often require large, expensive reducers. Nonuniform discharge may also be a problem at low screw speeds, particularly with cohesive solids. To minimize this problem, a double or triple flight section is often used at the discharge end of the screw. It is important that the material loading in the screw be sufficiently reduced before this section to ensure that it will not be 100% loaded before discharge.

In order to ensure increasing capacity in the feed direction, the following maximum tolerances and lengths should be used:

		Maximum Length-to-Diameter Ratio	
Screw Diameter	Assumed Pitch Tolerance	Cone Section	Increasing Pitch Section
6 in.	$\pm\frac{1}{4}$ in.	2.2	3.8
9 and 10 in.	$\pm\frac{1}{4}$ in.	2.9	4.5
12 and 14 in.	$\pm\frac{3}{8}$ in.	3.0	4.5
$>$ 14 in.	$\pm\frac{1}{2}$ in.	3.0	4.5

The length of the hopper outlet should generally be no greater than the cone section ratio given (multiplied, of course, by the screw diameter) plus three diameters of the screw. The remainder of the increasing pitch section is used to decrease the material loading in the conveying portion.

The screw flights should be terminated no more than 1 in. past the discharge opening and a short section of reverse flight used beyond the opening to prevent packing of cohesive solids near the end bearing.

Belt and Apron Feeders

As with screw feeders, the key to proper belt or apron feeder design is to provide increasing capacity (draw) along the length of the bin outlet. Without this, material channels at one end of the hopper and disrupts mass flow.

An effective way to increase capacity is to cut a converging wedge in such a way that it is closer to the feeder at the back of the outlet than at the front. This provides expansion in both plan and elevation (see Fig. 21.5.2).

With small belts (say, 12 in. or less), flat idlers can be used. However, with larger belts, sag between the idlers forces a rhythmic movement of material up into the hopper as it passes over each idler. This increases power and belt wear and may cause particle attrition. It is therefore better to use troughing idlers for these larger sizes if possible. Generally this is no problem unless the belt is a weigh feeder in which case a flat belt may be required.

There are three types of troughing idlers—equal length, unequal length, and picking—and at least three standard idler angles (from the horizontal)—20, 35, and 45°. Of these, 35° equal-length troughing idlers are the most common although unequal-length and picking idlers often allow the use of narrower belts for the same capacity. A typical belt/hopper interface that accomplishes the above objectives is shown in Fig. 21.5.2. It is important that the bed depth of material at the front of the outlet be at least 1.5 to 2 times the largest particle size to prevent blockage.

Through proper design of the interface, the loads—both shear and normal—imposed on the feeder section under the outlet are relatively low. Factors that increase these loads are as follows:

1. Belt sag if idlers are not closely spaced under the bin outlet.
2. Differential deflection between the feeder and hopper if the two are independently supported.
3. Refilling of the bin from empty while the feeder remains stopped.

These forces can be minimized in the following ways:

1. Mount the feeder on elastic supports designed to maintain the feeder in its proper position while running, and deflect sufficiently when an additional load is applied.
2. Support the feeder from the hopper and then run it at a slow rate while first filling the bin from an empty condition.

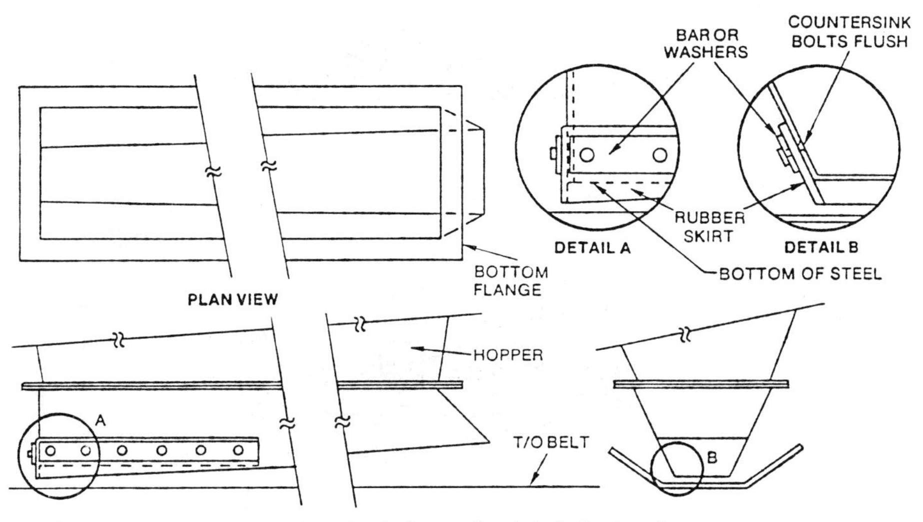

Fig. 21.5.2 Typical mass flow belt feeder interface.

After this initial procedure, the feeder loads will be low as long as a minimum head of material (about two outlet widths) is maintained within the hopper.

Rotary Valves

Two common problems that are often experienced when using rotary valves are:

1. Material tends to flow faster from the hopper on the upside of the rotary valve. This tendency can be reduced or eliminated by adding a short vertical section having a height of about 1.5 outlet diameters between the hopper outlet and rotary valve inlet or by off-setting the valve.
2. Gas backflow from a downstream higher-pressure environment. For example, if material is being fed into a positive-pressure pneumatic conveying line, a nonvented rotary valve acts as an effective pump on the upside. This gas backflow into the hopper often slows down the material discharge rate and can even cause arching. This is effectively overcome by venting the valve on the return side. Such venting may sometimes be necessary even with rotary valves exposed to atmospheric conditions if the required rates are high or the material impermeable to air.

Table Feeders

These feeders should be designed using the same principles as outlined for screw and belt feeders. The skirt is raised above the table in a spiral fashion to provide an increasing capacity in the direction of rotation as shown in Fig. 21.5.3. The plow is located outside the bin and removes only the material that flows from under the skirts.

Vibratory Feeders

Vibratory feeders can also provide uniform flow along a slot opening of limited length, as shown in Fig. 21.5.4. The distance between the feeder pan and the hopper should be increased in the feed direction.

Slot length is limited by the motion of the feeder. In a long slot, the upward component of motion will not be relieved by the front opening. As a result the solid will pack, possibly causing flow problems with sticky solids and large power requirements with free-flowing materials. To get around this problem, vibratory feeders and reciprocating plate feeders have been designed to feed across the slot. This kind of feeder may require several drives to accommodate extreme width, although the drives will be small because of the feeder's short length.

Rotary Plow Feeders

Because of lower capital and operating costs, as well as ease of maintenance, rotary plow feeders are often used under large stockpiles instead of belt feeders or a series of small outlets. This system can

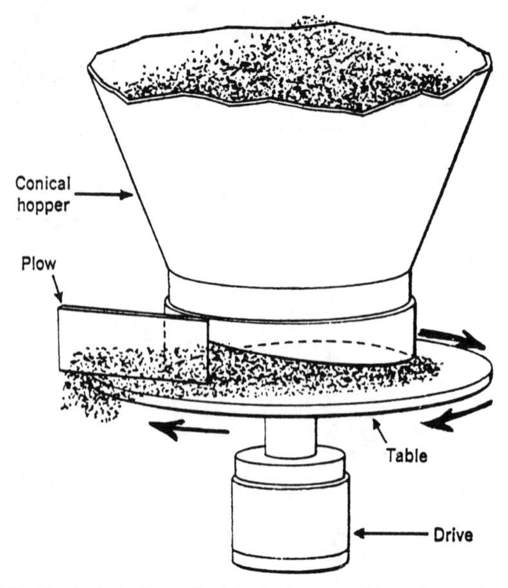

Fig. 21.5.3 Typical design of table feeder to achieve increasing capacity.

be used to move minerals—ranging from coal to iron ore—stored at mine sites, processing facilities, and power plants.

The mechanism by which a rotary plow moves material is as follows: When a rotary plow begins operating, it loosens material in a narrow vertical channel above it. If twin plows are used and both are operating, two channels will form independent of each other. The pressures exerted on the adjacent material are generally low and proportional to the size of the flow channel.

If a plow is stationary, and the material has sufficient cohesive strength, the channel eventually empties out, forming a rathole.

As a plow traverses the stockpile, a narrow flow channel lengthens. Material on either side of it either remains stationary or slides, depending on the wall friction angle along the sloping wall, the wall angle, and the head of material.

If the material slides, it does so only for a small distance since, as the material in the flow channel is compressed, the pressure it exerts on the adjacent material increases, resulting in a stable mass.

If the side material does not slide, the level of material in the flow channel drops and material sloughs off the top surface.

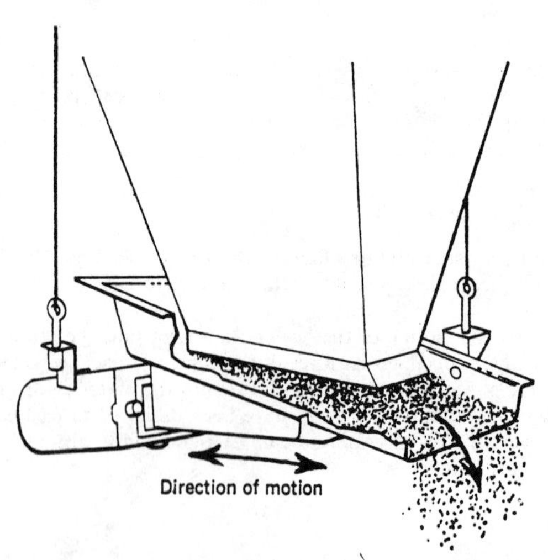

Fig. 21.5.4 Typical vibratory feeder to achieve increasing capacity.

Even if the side material does not start sliding immediately, it may start when the level reaches a certain point because of the smaller support offered from the flow channel. Likewise, the side material may stop sliding at a lower elevation if the relative support offered to it by the walls becomes greater.

These effects need to be analyzed as a function of plow size, material flow properties, wall friction angle between solid and stockpile walls, and wall slopes. This allows determination of proper sizing of plows, geometry of stockpiles, and the necessity of liners, as well as minimum plow traversing speed to avoid product degradation such as spontaneous combustion of coal.

21.5.2 Gates

Various types of gates can be used such as clamshell, slide, or spile bar. If the bin is designed for mass flow, it is vitally important that the gate be operated only in a full-open or full-closed position. A partially opened gate sets up stagnant regions above it resulting in a funnel flow pattern.

In general, the simpler the gate arrangement, the better. For example, a motorized slide gate above a screw feeder provides little benefit over a series of flat plates which are driven in between the opening when the screw trough is dropped $\frac{1}{2}$ in., but it is considerably more expensive in both capital equipment and maintenance. In addition, a motorized gate provides an easy means for an operator at some remote location to partially close the gate and thereby upset the flow pattern in the bin without knowing what has been done.

21.6 SPECIAL CONSIDERATIONS

21.6.1 Limited Flow Rate of Fine Powders

As noted earlier, the critical flow rate of a fine material is significantly less than that of a coarse material and this may impose restrictions on the operation. These restrictions are imposed by the interaction of the interstitial gas and solids [19–24]. One way to increase the flow rate is through the use of an air permeation system, a typical application of which is shown in Fig. 21.6.1. The system is composed of an air supply, pressure regulator, gauge, rotometer, and piping. The rotometer is adjusted for a maximum air rate with the pressure regulator set for a certain reading on the gauge and the bin empty. This pressure regulator setting is for the maximum feed rate of the powder; some field adjustment in pressure is sometimes necessary if fluctuations in flow rate are observed.

This system has been used successfully when the proper pressures and air flow rates have been determined by two-phase flow analysis. However, it is easy to put too much air into the system, and cause extreme flooding conditions. Air permeation systems should never be used without proper controls on pressure and air rates.

If a screw feeder is used, it is desirable to place a pressure tap at the top of the screw trough

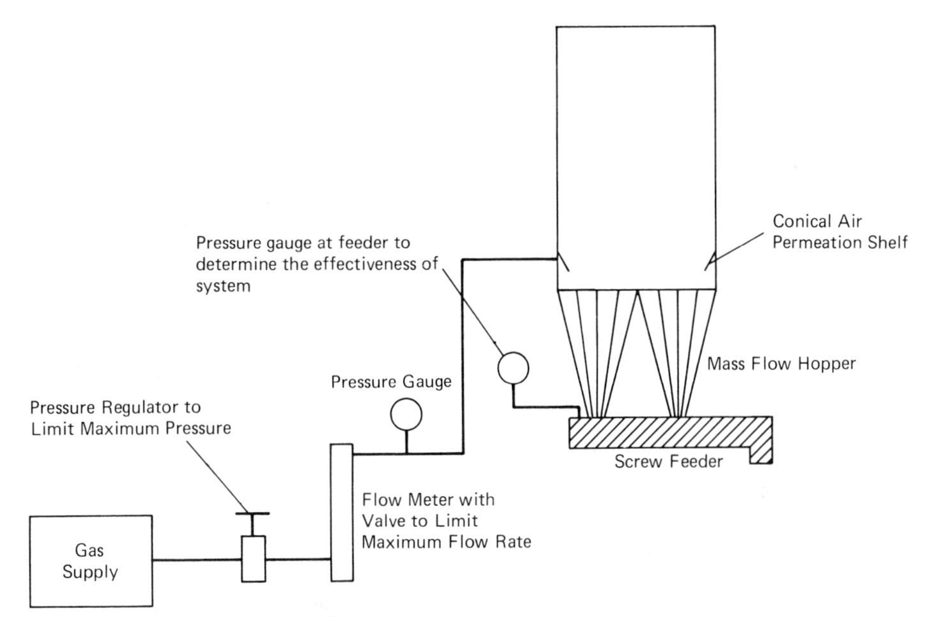

Fig. 21.6.1 Typical air permeation system.

near the bin/feeder interface. For the best control of the feed rate, the pressure regulator should be adjusted so that this pressure tap indicates atmospheric pressure during steady-state flow. As the feed rate varies, however, the regulator pressure to achieve this will vary. In practice, the reading on the pressure tap can be allowed to vary by several inches of water (either positive or negative) and still maintain good accuracy on the volumetric feed. Hopefully, only one regulator setting will be required for the required range of feed rates. However, wide variation in material properties may require field adjustment of the pressure and maximum flow rate.

During the operation of the bin, the pressure regulator holds a constant pressure. The air flow rate then adjusts automatically between zero and the maximum set by the rotometer. When operating with a low head of solid, the air flow rate will be consistently at the maximum allowed by the rotometer and the air pressure will adjust to accommodate this flow rate between the maximum pressure set on the regulator and zero psig. When this pressure drop occurs, the maximum allowable solids rate may decrease.

If the solids flow is to be stopped for less than a day, the air permeation should be maintained to ensure immediate flow of solids at the required rate when the feeder is started. If solids flow is to be stopped for several days, the air may be turned off but should be turned on again several hours before solids flow restarts.

21.6.2 Standpipes

Vertical cylinders filled with moving solids are often used to take a gas or liquid pressure drop, either increasing pressure from top to bottom or decreasing pressure. These cylinders, known as standpipes, are finding extensive use in many new oil shale processing facilities and are commonly found in coal-fired power plants using pressurized pulverizers.

In the latter instance, hot gases from the pulverizer cannot be allowed to enter a bunker, since they may cause a fire. To prevent this, cool inert gas at a slightly higher pressure is often added to the housing of the feeder which is between the bunker and pulverizer.

Without a standpipe, the gas pressure at the bunker outlet would be greater than atmospheric and the resulting upward gas pressure gradient and counterflow could cause erratic coal flow and even flow stoppage, or arching.

Unfortunately many of the methods for choosing the proper diameter and height of standpipes are somewhat arbitrary. Part of the problem is that the mechanism that allows a standpipe to be effective is not well understood.

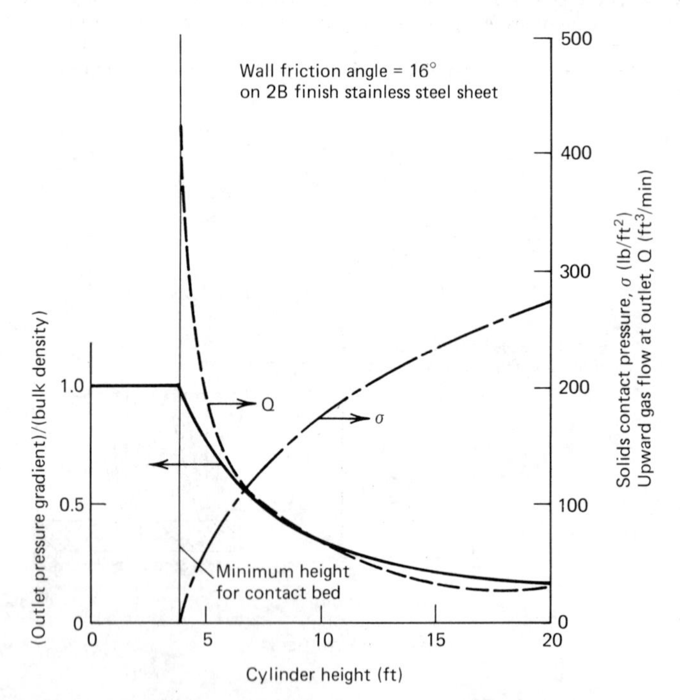

Fig. 21.6.2 Calculated effect of cylinder height for 3-ft-diameter cylinder using coal at 10.9% moisture sealing against a pressure of 30 in. w.g.

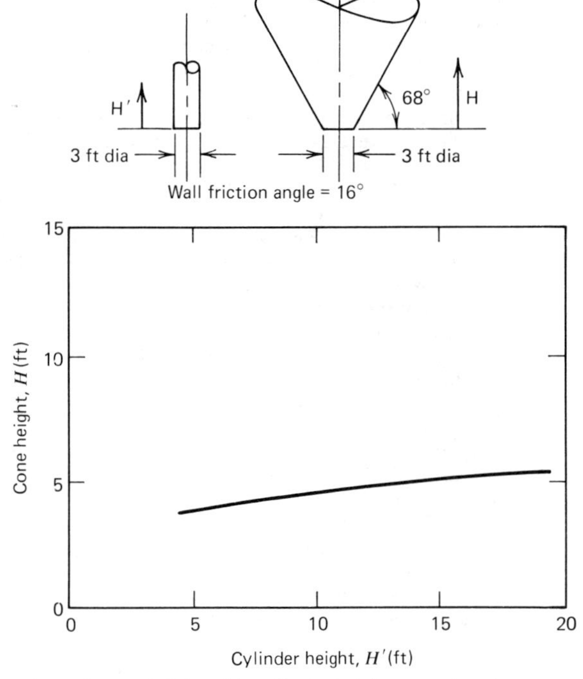

Fig. 21.6.3 Comparison of cone height with cylinder height to seal against a pressure of 30 in. w.g. using coal with 10.9% moisture.

The most important variable to consider is the upward gas pressure gradient which acts at the outlet of a bunker or the outlet of a standpipe. This gradient is a body force similar to, but acting in the opposite direction of, gravity. It not only retards flow but can increase the critical arching dimension of a cohesive solid to the point that it will not flow.

A two-phase flow computer program has been developed by Jenike & Johanson, Inc. to analyze the effects of solids and gases flowing together. Figures 21.6.2 and 21.6.3 show the results of a computer analysis on 11% moisture coal using a 68° cone of varying height with a 3-ft-diameter outlet [24]. Figure 21.6.2 shows a 3-ft-diameter cylinder of varying height. In each case, 30 in. water pressure was assumed at the bottom with a 16° wall friction angle on the cone or cylinder wall. In each figure is plotted: the pressure gradient at the outlet divided by the bulk density of the material at that point; the vertical solids contact pressure at the outlet, σ; and the upward gas flow at the outlet, Q. It is interesting to compare the height of a cone to that of an equivalent cylinder. To do so requires matching one of the three variables—pressure gradient, solids pressure, or gas flow rate. Fortunately the results are practically the same no matter which variable is chosen. As shown in Fig. 21.6.3, a cone is not as effective as a cylinder in sealing against gas counterflow. For example, a 10-ft-tall cone is equivalent to a 4.5-ft-tall cylinder. Doubling the cone height to 20 ft results in its effectiveness increasing only about 20%.

In general, a cylindrical standpipe is recommended when feeding to a pressurized environment. It should be sized so that the upward pressure gradient is not sufficient to support an arch and the gas counterflow is limited to the design value. It is not uncommon to find this upward pressure gradient high enough to double the outlet size required to prevent arching. Erratic flow is likely if this occurs and the outlet is designed based only on gravity flow without a pressure gradient.

21.6.3 Flow Aids

Air Blasters

The rapid release of air from a blow tank into a mass of bulk material has been used for a number of years in underground blasting of coal and, more recently, in the dislodging of bulk solids in bins and hoppers so as to promote solids flow. The system consists of a tank of pressurized air, a quick-release valve, and a nozzle that is built into the bin or hopper to allow the air to enter. As the quick-release valve is opened, the gas velocity is limited by the sonic velocity; however, when the air encounters the solids the permeability of the porous solids provides a limit that quickly decreases the velocity below sonic. It is also important to note the significantly higher pressure gradient at the bottom of the standpipe compared with the top. This is typical in a compressible, reasonably impermeable solid.

The air spreads out in essentially a hemispherical fashion around the point of entry, and causes flow pressure gradients that eventually reach the surface of the arched material. These gradients add to the force of gravity, and if the combined effect is large enough, the arch will break and solids flow will be established. In general, air blasters are most effective when used to break arches in materials that have an increase of cohesive strength with time at rest.

The calculated pressure distribution, as a function of time, from an air-blaster discharge nozzle for an oil sand is shown in Fig. 21.6.4 [21]. This air blaster has a volume of 24.5 ft³ and an initial blast pressure of 90 psig. The entire process occurs rapidly (less than a second) as indicated by the time sequence associated with each of the pressure distributions.

The effectiveness of an air blaster in breaking bridges depends on the distance between the air injection nozzle and the bridge. The arch reduction factor for this example is shown in Fig. 21.6.5 as a function of distance from the blaster inlet. The use of the graph can be demonstrated as follows:

Suppose the outlet of the bin has a 3-ft diameter and, because of additional gain in strength with time, the material is capable of arching over a 6-ft diameter. The required arch reduction factor to cause flow is the ratio $\frac{6}{3} = 2$. Figure 21.6.5 indicates that this occurs when the air blaster is not more than 8 ft away from the arch. This suggests that for this blaster to be effective in breaking the arch, it would have to be placed within 8 ft of where the arch would occur.

To complete the blaster analysis, we must consider the possibility of the material forming an arch above the point of application of the air blaster. Here the arch has to collapse by gravity alone. If the critical arching dimension were 6 ft, the blaster would have to be placed above the level in the bin where the diameter was 6 ft. If both of these conditions could not be satisfied simultaneously, two or more blaster injection levels would have to be considered.

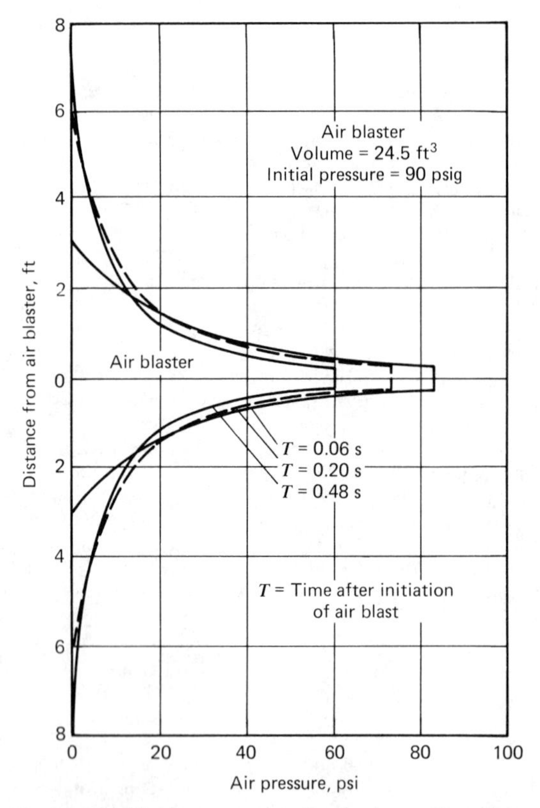

Fig. 21.6.4 Air blaster pressure distribution on bin walls.

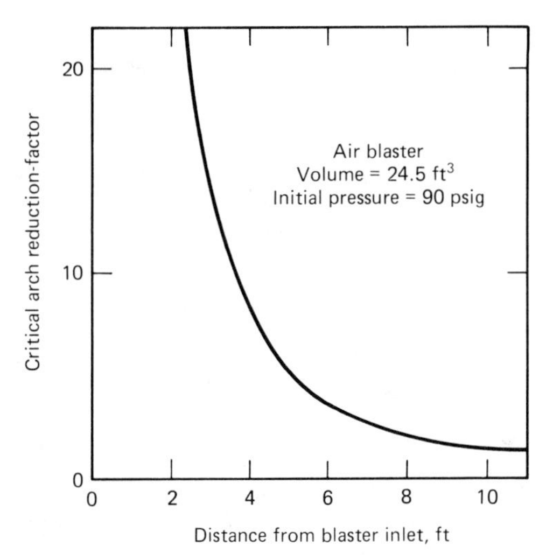

Fig. 21.6.5 Air blaster effects on critical arching dimensions.

External Vibrators

Such devices are often effective in getting material to start flowing after having set for some time at rest. An external vibrator should not be used continuously to ensure flow, nor should they be used with solids that are pressure sensitive. Generally, it has been found that low-frequency, high-amplitude vibrators are more effective for this application than those which have a high frequency and a low amplitude.

Free Flow Additives

Various materials can be mixed with a bulk solid to decrease its strength. Among these are chemical sprays, fumed silica, and fly ash added to coal.

Usually the costs of the chemical and application system are high so this becomes quite an expensive alternative if applied continuously. However, in the case of consumer products such as flour and salt, there may not be any alternative.

Freeze Conditioning Agents

Freeze conditioning agents are used in many industries where large quantities of wet or moist bulk solids need to be handled or stored reliably under subfreezing conditions. The major industries involve the transportation and storage of various metallic ores and coal. With a cost of approximately $1 per ton of bulk solid, it is often not possible to justify freeze conditioning agents addition unless the consequences of unreliable solids flow are significant; for example, coal flow in a power plant is far more critical than at a mine load-out silo or at a steel plant coke battery.

Moist or wet bulk materials tend to freeze to equipment as well as to form large strong lumps that are not easily broken or handled. Often a product that solves one problem (e.g., lump formation) has little or no effect on the other problem (wall adhesion). Indeed they go by different names. The former are called freezing conditioning agents whereas the latter are called side release agents.

Various types of freeze conditioning agents and side release agents have been developed to reduce the handling problems associated with freezing. The principal types are:

1. *Oil Based.* This type usually has organic salt modifiers to reduce the strength of frozen moist material. It is advantageous under conditions where freeze conditioning agents retention and frozen adhesion to equipment are problems. However, it is usually limited to temperatures not much less than about 20°F. Often it becomes quite viscous at low temperatures making spray application difficult.

2. *Glycol Base.* This type reduces frozen moisture/material strength, freezing temperature, and adhesion. It usually has modifiers to accentuate these characteristics and can be effective over a wide range of subfreezing temperatures. Disadvantages may include toxicity and water solubility, which allows dilution and leaching under open storage conditions.

3. *Calcium Chloride.* This is used in either a solid form with materials of high moisture content or in a 25–35% solution. This freeze conditioning agent primarily melts ice and lowers the moisture freezing point which reduces the material's strength and adhesion. The main disadvantages are its water solubility with consequent dilution and reduction in effectiveness, and corrosion of steel equipment when present in large concentrations. The latter problem is a significant consideration in most applications; however, this type of freeze conditioning agent can be manufactured in a chromated or "pacified" condition at greater cost.

4. *Wall Liners.* Various types of nonmetallic liners are on the market which purport to reduce the adhesion of a bulk solid to the wall.

Factors influencing the effectiveness of freeze conditioning agents and side release agents are:

1. *Particle Size.* In general, a fine bulk material has a greater surface area per unit volume than a coarse material. Consequently, the finer material requires a greater application rate to achieve a satisfactory concentration at the points of contact where primary particle freezing occurs.

2. *Particle Size Distribution.* An optimum particle size distribution exists where the area of interparticle contact per unit volume is at a maximum. Approaching this condition increases a material's cohesive and frozen strength because of the increasing bonding area.

3. *Moisture Content.* Increasing the moisture content up to 80–90% of saturation usually increases both the material cohesion and the frozen strength. This is most severe with materials that are fine and/or optimally packed.

4. *Clay Content.* Clay is cohesive and adhesive by nature of its moisture retention and extremely fine particle size, making handling difficult even at room temperature. Because of these properties, its frozen strength is extreme.

5. *Temperature.* In general, the strength of ice (pure and solutions thereof) increases as the temperature decreases. This is caused by crystal growth and solution phase changes.

6. *Freeze Conditioning Agents Distribution.* Because freeze conditioning agents or side release agents are usually applied in relatively small quantities, uniformity of distribution is critical. Regions of insufficient application will show little or no handling improvement while an overapplication may not provide a corresponding improvement in handling.

One way to evaluate effectiveness of a freeze conditioning agent is by direct uniaxial compression to failure of a large number of samples. However, this procedure only works well with strong samples, and often ignores the effect of consolidating pressure on material strength. Another method is to determine the stress required to fail samples in bending. Again, this method is not satisfactory at low sample strengths, is often not utilized as a function of consolidating pressure, and appears to yield a greater deviation in generated data. A third method involves freezing trays of bulk solid, dropping the contents onto a grid, and measuring the amount that breaks up enough to drop through the grid. Many variables such as particle size, grid size and shape, fall height, and stress concentrations at impact, are often not considered with this approach.

A fourth method consists of direct shear tests at 28–30°F on the −6 mesh fraction of a sample with a controlled amount of freeze conditioning agent added with an atomizer. This method is applied over a range of consolidating pressures and is particularly effective at lower sample strengths where a uniaxial compression or bending specimen is not reliable; however, it is not reliable at higher strengths because of test equipment limitation and difficulty in attaining adequate shear planes. Consequently, at higher sample strengths, the samples of "hockey pucks" are sheared across a plane coincident with the sample's cylindrical axis.

This testing method provides a good representation of conditions within a bin and can be used to test either a freeze conditioning agent or side release agent. Unfortunately it poorly represents the discharge and subsequent material breakup from a rotary dump rail car.

In all determinations of the effectiveness of freeze conditioning agents and side release agents, it is imperative that the test conditions represent the actual storage and handling process. It is not always easy to determine or control the effect on material strength due to variables such as freeze conditioning agent or side release agent distribution and absorption, bulk solid variability, and particle size distribution.

21.6.4 Inserts

A properly designed bin and hopper results in savings through better flow-rate control, larger live-storage capacity, and improved uniformity of withdrawn materials. Unfortunately, most bins are not designed for proper flow. Changes in the basic design of existing bins are difficult to justify economically. Using flow-corrective inserts can sometimes cause proper flow without costly basic design changes.

Inserts are static devices placed within a bin—usually in the hopper section—whose purpose is usually to expand the flow channel of a funnel flow bin to approach mass flow. Two of the more common types of inserts are inverted cones, or pyramids, and the insert hopper cone-in-cone.

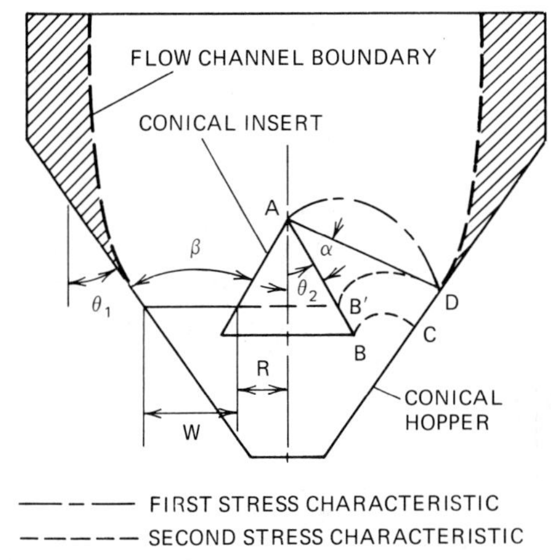

Fig. 21.6.6 Region of influence of conical inserts.

Inverted Cone, or Pyramid

Flow-corrective inserts sometimes can solve flow problems in existing bins. However, the placement and size of such inserts is critical. If they are too small, the flow pattern will not be changed; if they are too large, flow may stop completely. The usual procedure for finding the proper size and placement is by trial and error. Using the results presented by Johanson [25], it is possible to predict the proper size and placement of the insert based on the properties of the granular material to be stored in the bin.

In an application in which the granular solid presents no arching or ratholing problems, segregation often occurs during the loading of the bin. When material is withdrawn, the fine central core is removed first, leaving the coarse material to come out last. Such an undesirable withdrawal can be eliminated by causing the entire mass of material to move uniformly, at least in the vertical section of the bin. A steep mass-flow hopper will accomplish this. However, the proper insert in the existing hopper will accomplish the same thing.

Figure 21.6.6 shows the proper placement for an insert. The following procedure will provide the selection of the insert of the smallest diameter with a preselected θ_1 value that will cause the desired flow pattern. Larger inserts will work equally well but are not necessary.

Procedure for creating mass flow in bins where arching and ratholing are not problems follows:

1. Select the slope angle θ_2 of the insert. In cases where cleanout is not required, a flat plate will be the least expensive and $\theta_2 = \pi/4 + \delta/2$. When cleanout is necessary, $\theta_2 = 30°$ is sufficient for almost all materials. Some advantage may be gained with some materials by using a cylindrical insert.
2. Look up the critical W/R and α values in Figs. 21.6.7 and 21.6.8. For this step the frictional properties ϕ' and δ as well as the hopper angle must be known.
3. Make an outline drawing of the existing bin.
4. Starting at point A, draw in line AB with slope angle $(\pi/2 - \alpha - \theta_2)$ from the horizontal.
5. Draw line CD inclined at angle α from the vertical where $\tan \alpha = (\tan \theta_1)/(1 + W/R)$. Points on this line represent values of critical W/R.
6. Draw line BE at slope angle θ_2. Point E, the intersection of DE and BE, locates the bottom of the insert. The size and placement of the insert are now determined.

It may be desirable to add a factor of safety to this procedure by reducing the critical W/R as obtained from Fig. 21.6.7. Usually a 10% reduction is sufficient.

The foregoing procedure is completely valid only for materials that do not arch and/or rathole. With a less free-flowing material, there are additional problems to consider such as:

1. Calculate the minimum opening required to prevent arching over the annular ring-shaped opening between the insert and the wall. This can be accomplished by using the procedure described by Johanson [25] for calculating the critical diameter to prevent arching over a circular opening in a flat-bottom bin. The critical width W is about three-fourths of this critical diameter. Information regarding material flow in the existing bin can give a clue to this critical dimension.

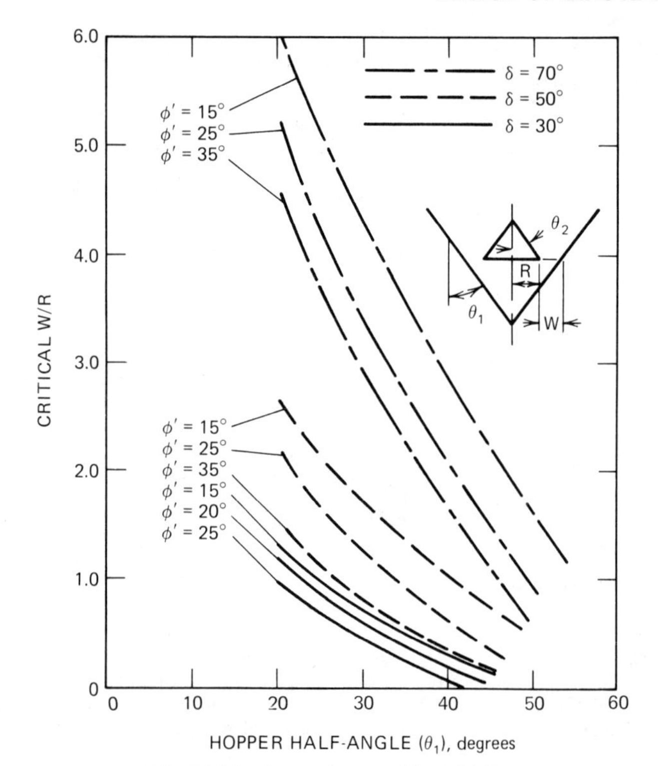

Fig. 21.6.7 Approximate critical W/R.

If flow occurs through the outlet in the existing bin without arching problems, the critical width W is less than three-fourths of the diameter of the outlet.

2. Locate the position F, in Fig. 21.6.9, at which the horizontal distance W between the wall and line CD is equal to the critical width for arching. No insert may be placed below F without causing arching problems.
3. Check to be sure that point E is above point F. If it is not, the insert will have to be raised and increased in size. If this is impossible to achieve, the bin problem cannot be solved by static inserts. However, in most cases, point F will be very low in the bin.
4. Check to see if a stable rathole will form below the upper insert. If material ratholed in the existing bin without any insert, a rathole may form below the insert if the natural angle of repose of the material allows a depth H of material above the hopper opening in excess of the opening diameter d. When this occurs, a lower insert such as that shown in Fig. 21.6.9 may be used to prevent this ratholing. With some materials the lower insert may be required below point F. In this case the opening in the hopper will have to be enlarged so that the lower insert can be moved above point F or eliminated.

When the only purpose of the insert is to prevent ratholing in the bin, it is not necessary to design the insert for flow along the vertical bin walls. The design (for the prevention of ratholing) can be based on a vertical channel equal to the critical diameter required to make the rathole unstable. This critical diameter may be known through experience, or it can be calculated when the flow properties of the material are known.

Insert Hopper

A new concept has been developed for controlling and expanding flow patterns in bins by using a mass flow hopper-within-a-hopper [26]. The insert hopper is designed to provide mass flow through the central channel and also through the annular space between the insert and the outer hopper. The concept is of considerable use in eliminating particle segregation, saving headroom for mass flow bin designs, eliminating ratholing problems in existing funnel flow bins by creating mass flow, and blending solids by either a single pass or recycling the solids through the bin.

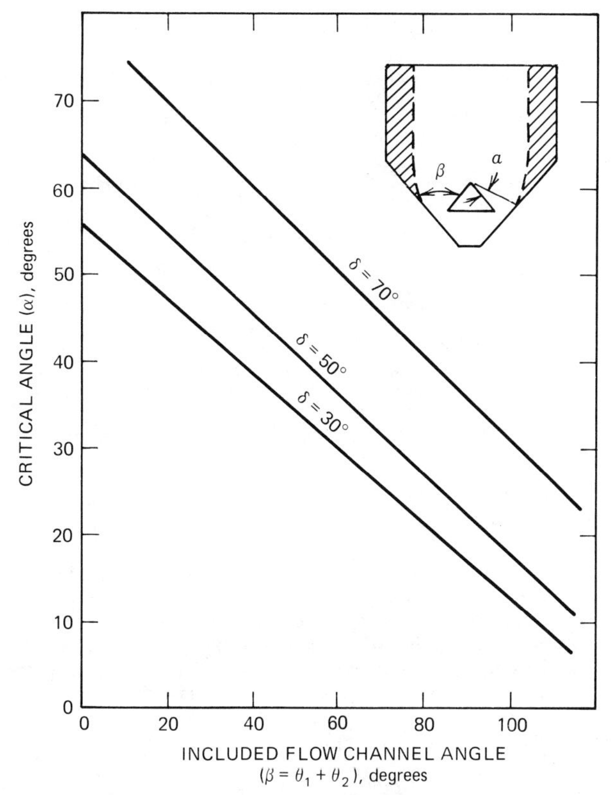

Fig. 21.6.8 Approximate angle α to determine the limit of flow along the hopper walls.

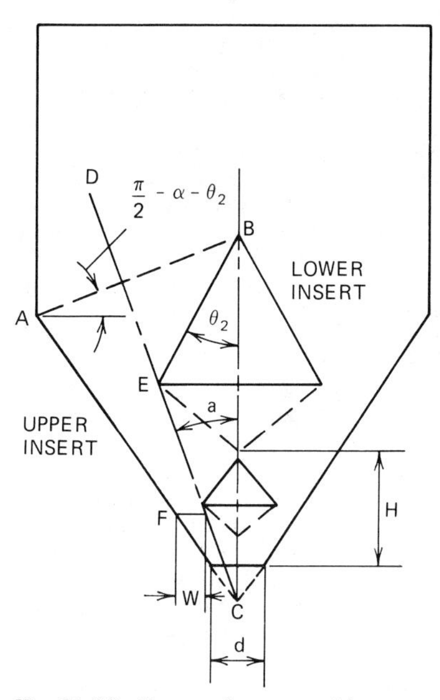

Fig. 21.6.9 Proper placement of inserts.

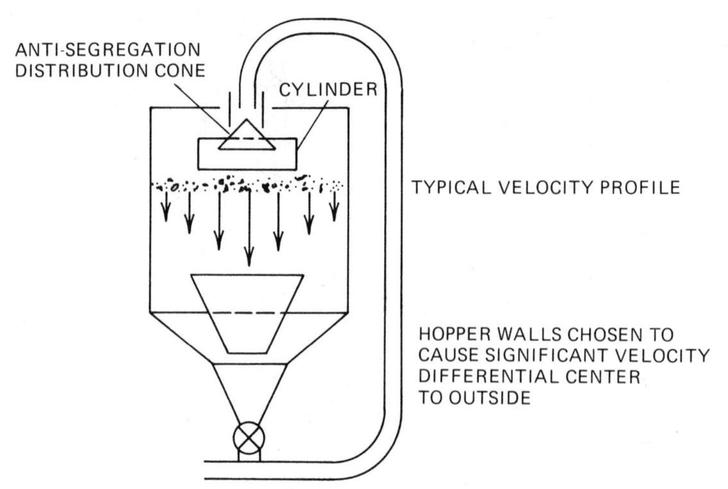

Fig. 21.6.10 Recirculating blender design for badly segregating material.

The unit, shown in Fig. 21.6.10, consists of a central hopper and an additional hopper outside the first which allows material to flow through both the inner hopper and the annulus formed between the two hoppers. The inner hopper has the required minimum slope to achieve mass flow along the walls and the outer hopper angle can be twice the angle (from the vertical) of the inner hopper and still achieve mass flow. This concept applies equally well to wedge-shaped hoppers with vertical end walls. By varying the geometry in the discharge region of the bin, the solids velocity profile can be changed at will to provide the desired results.

The hopper is particularly well-suited to applications in which segregation must be eliminated or in-bin blending is desired.

Whenever a solid with a variety of particle sizes is charged into a bin or process vessel, segregation of coarse and fine material generally occurs. If this segregation is the result of the sifting mechanism [27], the fines accumulate at the point of charging while the coarse material concentrates at the periphery of the container. If the bulk solid contains extremely fine material, the fines are likely to be airborne and hence concentrated at the periphery of the container. When a mass flow hopper without a sufficient vertical bin section or a funnel flow hopper is discharged, the central core of material tends to discharge first, followed by the outer region. This type of flow produces segregation that often leads to difficulty in further processing of the material.

By proper use of the hopper shown in Fig. 21.6.10, the flow pattern can be controlled so that even when the level of the material drops into the conical hopper, the velocity in the inside and outside are essentially the same. Thus, even though segregation occurs during charging of the container, the material is remixed upon discharging through the outlet. An important feature of an antisegregation insert hopper design (in addition to a properly designed insert) is the vertical section just below the bottom of the insert. This vertical section with a mass flow outlet hopper forces a uniform flow pattern at the outlet of the insert. This uniform flow pattern is then enforced in the remainder of the vessel above by the walls of the insert.

Recent tests of the hopper for a pharmaceutical application have indicated that the unit was effective in eliminating segregation problems in 98% of the material. Only the last 2% of the discharge from the container did not meet specifications. This was particularly encouraging since prior to use of the hopper entire batches were often rejected.

Other successful antisegregation applications of the hopper have been for instant coffee, in which an existing bin was modified to provide a uniform flow pattern, and oil shale, where a free-flowing shale was extremely prone to segregation.

Another approach to reducing segregation is to prevent it from occurring when the bin is charged. Hopper patent [28] also incorporates a device to prevent this type of segregation, as shown in Fig. 21.6.10. It consists of a cylinder to feed material onto a diverging conical chute with a deflector cylinder at the bottom of the chute. Since coarse particles tend to have a lower coefficient of friction than fine particles, they move at a higher velocity down the conical chute and impact the cylinder deflecting them back into the stream of fines which do not impact the cylinder. This causes a mixed delivery of coarse and fines to the surface of the bin. In actual applications this device has proven extremely effective in preventing segregation.

The insert hopper concept is also ideally suited for in-bin blending because it can provide a means of controlling the flow pattern throughout the bin and hence the effectiveness of the blending.

In-bin blending is accomplished by providing a significant velocity gradient across the bin [29] so that some of the first material charged into the bin is delayed and mixed with some of the last material to be charged. This is often done with a series of tubes inserted in the bin drawing material from various levels and recirculating the entire mass or with various conveying devices (e.g., air injection or a vertical lift screw) within the bin taking material near the bottom and conveying it up to the top. The screw-type blender often degrades the material, air injection may require significant quantities of compressed air, and the tube blender is limited to free-flowing materials. The hopper blending concept has no moving parts, can be designed for cohesive as well as free-flowing materials, and derives its blending action entirely from the geometry of the bin and insert. The insert hopper design criteria for a cohesive solid are identical to that for any mass flow hopper and are easily determined from the flow properties of the bulk solid.

In the blending configuration of the insert hopper, the bottom portion of the bin is designed as a mass flow cone. The flow pattern achieved in this mass flow cone (namely a high velocity in the center and low velocity at the outside) propagates up through the insert hopper portion of the bin.

REFERENCES

1. Johanson, J. R., "Feeding," *Chemical Engineering,* Deskbook Issue, Oct. 13, 1969, pp. 75–83.
2. Jenike, A. W. and J. R. Johanson, "Solids Flow in Bins and Moving Beds," *Chemical Engineering Progress,* Vol. 66, No. 6, June 1970, pp. 31–34.
3. Johanson, J. R. and H. Colijn, "New Design Criteria for Hoppers and Bins," *Iron and Steel Engineer,* Oct. 1964, pp. 85–104.
4. Jenike, A. W., "Storage and Flow of Solids," University of Utah, Engineering Experiment Station, Bulletin No. 123, Nov. 1964.
5. Jenike, A. W., P. J. Elsey, and R. H. Woolley, "Flow Properties of Bulk Solids," ASTM, *Proceedings,* Vol. 60, 1960, pp. 1168–1181.
6. Johanson, J. R., "Know Your Material—How to Predict and Use the Properties of Bulk Solids," *Chemical Engineering,* Deskbook Issue, Oct. 30, 1978, pp. 9–17.
7. Jenike, A. W., "Gravity Flow of Bulk Solids," University of Utah, Engineering Experiment Station, Bulletin No. 108, Oct. 1961.
8. Jenike, A. W., "Why Bins Don't Flow," *Mechanical Engineering,* Vol. 86, No. 5, May 1964, pp. 40–43.
9. Johanson, J. R., "Methods of Calculating Rate of Discharge from Hoppers and Bins," SME of AIME, *Transactions,* Vol. 232, Mar. 1965, pp. 69–80.
10. Jenike, A. W., "Denting of Circular Bins with Eccentric Drawpoints," ASCE, *Journal,* St. 1, Feb. 1967, pp. 27–35.
11. Jenike, A. W. and J. R. Johanson, "Bin Loads," ASCE, *Journal,* St. 4, Apr. 1968, pp. 1011–1041.
12. Jenike, A. W. and J. R. Johanson, "On the Theory of Bin Loads," ASME, *Journal of Engineering for Industry,* Vol. 91, Ser. B, No. 2, May 1969, pp. 339–344.
13. Jenike, A. W., J. R. Johanson, and J. W. Carson, "Bin Loads—Part 2: Concepts," ASME, *Journal of Engineering for Industry,* Vol. 95, Ser. B, No. 1, Feb. 1973, pp. 1–5.
14. Jenike, A. W., J. R. Johanson, and J. W. Carson, "Bin Loads—Part 3: Mass-Flow Bins," ASME, *Journal of Engineering for Industry,* Vol. 95, Ser. B, No. 1, Feb. 1973, pp. 6–12.
15. Jenike, A. W., J. R. Johanson, and J. W. Carson, "Bin Loads—Part 4: Funnel-Flow Bins," ASME, *Journal of Engineering for Industry,* Vol. 95, Ser. B, No. 1, Feb. 1973, pp. 13–16.
16. Jenike, A. W., "Load Assumptions and Distributions in Silo Design," Norwegian Society of Chartered Engineers, Conference on Construction of Concrete Silos, 1977.
17. Jenike, A. W., "Effect of Solids Flow Properties and Hopper Configuration on Silo Loads," ASME Century 2, *Emerging Technology Conferences*—1980, Unit and Bulk Materials Handling, p. 97.
18. Johanson, J. R., "Design for Flexibility in Storage and Reclaim," *Chemical Engineering,* Deskbook Issue, Oct. 30, 1978, pp. 19–26.
19. Bruff, W. and A. W. Jenike, "A Silo for Ground Anthracite," *Powder Technology,* Vol. 1, 1967/68, pp. 252–256.
20. Johanson, J. R. and A. W. Jenike, "Settlement of Powders in Vertical Channels Caused by Gas Escape," ASME, *Journal of Applied Mechanics,* Vol. 39, Ser. E, No. 4, Dec. 1972, pp. 863–868.
21. Johanson, J. R., "Two-Phase-Flow Effects in Solids Processing and Handling," *Chemical Engineering,* Jan. 1, 1979, pp. 77–86.

22. Reed, G. B. and J. R. Johanson, "Feeding Calcine Dust with a Belt Feeder at Falconbridge," ASME, *Journal of Engineering for Industry,* Vol. 95, Ser. B, No. 1, Feb. 1973, pp. 72–74.
23. Turco, M., C. Gaffney, and J. R. Johanson, "Feeding Dry Fly Ash without Flooding and Flushing," *Proceedings,* Powder and Bulk Solids Conference, Philadelphia, 1979.
24. Carson, J. W. and J. A. Marinelli, "Establishing Reliable Coal Flow in Power Plants," *Power Engineering,* Nov. 1981, pp. 90–93.
25. Johanson, J. R., "The Use of Flow-Corrective Inserts in Bins," ASME, *Journal of Engineering for Industry,* Vol. 88, Ser. B, No. 2, May 1966, pp. 224–230.
26. Johanson, J. R., "Controlling Flow Patterns in Bins by Use of an Insert," *Bulk Solids Handling,* Vol. 2, No. 3, Sept. 1982, pp. 495–498.
27. Johanson, J. R., "Particle Segregation and What to Do About It," *Chemical Engineering,* May 8, 1978, pp. 183–188.
28. Johanson, J. R., "Blending Apparatus for Bulk Solids," United States Patent No. 4,286,883, Sept. 1, 1981.
29. Johanson, J. R., "In-Bin Blending," *Chemical Engineering Progress,* Vol. 66, June 1970, pp. 50–55.

CHAPTER 22
FEEDERS

GEORGE A. SCHULTZ

Epstein Process Engineers
Chicago, Illinois

22.1 FEEDERS IN BULK SOLIDS HANDLING SYSTEMS

To assure a uniform flow of a solid between storage, handling, and processing equipment requires careful application of a bulk feeder. Feeders must not only meter material at the exact rate the receiving unit is designed for but also eliminate surges to assure a constant flow.

Tasks that feeders can be called upon to perform include transfer of raw materials from rail or truck dump hoppers to conveyors, raw or semiprocessed material from bins to process units such as crushers, dryers, grinders, blenders, mixers, and finished products into final packing operations. Consideration must be given to the changing characteristics of a solid material as it passes from raw through semiprocessed to finished product state. Also, the behavior of the material as it flows into and out of bins or hoppers, rests under its own weight, and moves into and out of a piece of processing equipment all must be carefully analyzed.

Consideration of these factors is essential when attempting to answer questions such as:

What size and shape of opening in bin or hopper will ensure flow through the opening into the feeder? Based on volumetric capacity (ft^3/hr), often the inlet size of a feeder selected from the manufacturer's literature is smaller than the most desirable bin opening. Too small an opening will result in erratic material flow. On the other hand selecting the feeder based on the most desirable bin opening could necessitate an oversized feeder, requiring it be run at a slower speed.

Will flow rate be hampered by material restrictions, as for example lump size? Lump size is one of the more critical factors in feeder selection and bin opening size. Figure 22.1.1 presents a rule-of-thumb method for determination of a feeder opening based on lump size.

Will the feeder be required to prevent an uncontrolled release of material (flooding) through the opening? Handling a fine powdered material which deaerates when compacted but fluidizes with freefall will result in the solid acting differently at the bin opening. The bin opening, feeder, and most likely a flow-aid device could be required in combination to assure a continuous flow when handling this type of material.

Are flow characteristics of the material such that an extraction type feeder is required? Many stringy, flaky, or extremely cohesive materials must be agitated to remain in a flowable condition to enable passage through the bin feeder opening. Extraction-type feeders can be used to provide both agitation and feeding action.

What are the dust containment or spillage requirements? Dusty or explosive materials may require totally enclosed or inert-gas blanked feeders requiring specially designed bin discharge openings.

Traditionally, feeders are designated by their conveying element or flow actuating mechanisms. In addition to conveying and metering functions, feeders can be combined with a process such as mixing, screening, and removal of metal particles. Two or more feeders can be interlocked with controls to provide a mixed blend of two or more solids. However, for purposes of selection as a flow regulator, feeders are classified into volumetric or gravimetric types.

Volumetric feeders deliver a specific volume per unit of time. Multiplying volumetric flow rate by bulk density of the material will result in the rate of flow. Generally, volumetric feeders have a $\pm 5\%$ variation because feed rates are affected by variations in bulk density of the solid.

Where the feed rate must be controlled within a $\pm 1\%$ range, a gravimetric feeder is required. Gravimetric or weigh feeder operation is based on a direct rate of flow reading and are therefore not affected by variations in bulk density. Many types of volumetric feeders can be adapted to gravimetric operation by application of feedback controls.

22.2 TYPES OF FEEDERS

Frequently, several types of feeders are suited to meet the requirements of a given application making selection of the right type of feeder a difficult task. The dominant factor in selection is the experience and preference of individual engineers and operators. Often an earlier misapplication of a feeder will eliminate it from future consideration in a particular plant or even a total industry. Following is a listing of generally available types of feeders which will be described:

Apron
Belt
Chain curtain
Extraction type
En masse
Flight

- Gravimetric
- Pneumatic and fluidizing
- Portable
- Reciprocating plate
- Rotary table
- Roll
- Rotary vane
- Rotary plow
- Scalper
- Screw
- Vibrating

22.2.1 Apron

Apron or pan feeders are made up of overlapping flights called aprons or pans which are connected to, and supported on, steel chains or bars forming a continuous steel belt. Pans can be constructed of high-carbon steel, manganese steel, or nickel chrome–molybdenum steel. Sprockets mounted in head and tail terminals of the feeder engage the chain or bars to drive the unit. Steel rollers, mounted to the pans, rods, chain, or fixed to the frame between the terminals, are provided to support the flights (Fig. 22.2.1).

Apron feeders are well suited for impact loading conditions and for handling coarse, hot, abrasive materials. Materials typically handled on apron feeders include ores, minerals, cement, clinker, slag, and metal scrap. Major applications include the feeding of solids from hoppers into crushers (Fig. 22.2.2) and the handling of discharge from bins or surge piles to other types of conveyors. Their positive chain drive enables these units to be started under full load and to be "jogged" to feed large lump so as not to overload a crusher. Apron feeders are not suited for handling fine abrasive material which can get into the joints and cause wear.

Apron feeder capacities are based on feeding a volumetric quantity of material. The quantity is controlled by the cross-sectional area of the bed of material on the feeder and the speed at which the feeder operates. In general practice, apron feeder operation is limited to an incline of 10° with special pan designs required for greater inclines. Apron feeder speeds range from 5 ft/min (0.02 m/sec) to 75 ft/min (0.38 m/sec) dependent on application; a good target speed is 30 ft/min (0.15 m/sec). The cross-sectional area is limited to a width equal to the normal pan width minus 4–6 in. (101.6–152.4 mm) and depth approximately 75% of the skirt width. Ordinarily, it is better practice to increase the depth of load and operate pans at a slower speed for high capacities.

Standard units are available in widths up to 96 in. (2438.4 mm) and lengths to 100 ft (30.48 m). In standard practice however, length of apron feeders are generally not much over 30 ft (9.144 m) with the final length of the feeder dependent on the type of material handled and the method of loading.

Various types of drives are available. The apron feeder can be driven by an electric motor or can be powered by a combination of chain and/or gear trains from the equipment being fed. Motor–reducer–chain drive combinations are common and it is not unusual for countershafts to be included in the drive train.

22.2.2 Belt

Belt feeders are available in two types—roller and slider bed construction. The more common roller belt feeder employs flat or picking idler rollers to support the belt (Fig. 22.2.3). The slider bed belt feeder utilizes a flat steel plate in place of these rollers.

The belt feeder is basically a short belt conveyor and consists of a belt, drive pulley, take-up pulley, drive, skirts, and gate. This feeder operates by having a surcharge of material over a belt

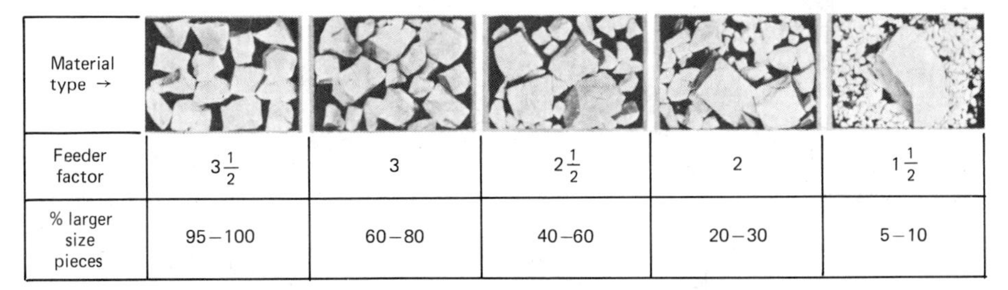

Material type →					
Feeder factor	$3\frac{1}{2}$	3	$2\frac{1}{2}$	2	$1\frac{1}{2}$
% larger size pieces	95–100	60–80	40–60	20–30	5–10

Fig. 22.1.1 Rule-of-thumb method for determining feeder throat opening: (1) select picture of type of material to be fed, (2) multiple feeder factor by largest dimension of biggest piece to determine hopper throat opening to minimize bridging.

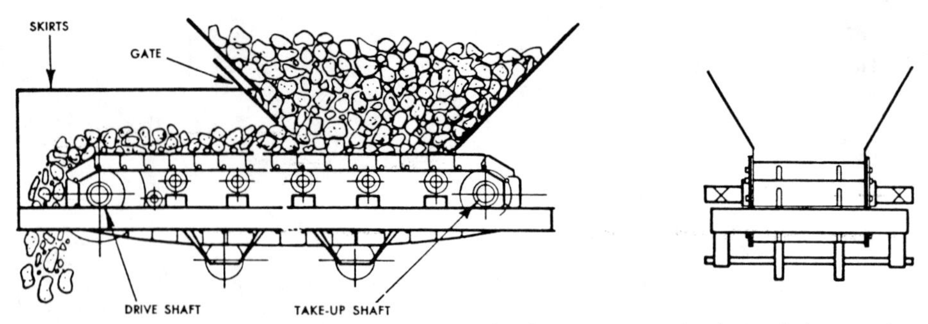

Fig. 22.2.1 Apron feeder. Materials are carried on overlapping pans, mounted on heavy chains operating on rails or rollers. Courtesy of Barber-Greene Co.

which is operating at a constant speed. A cut-off gate assures that a constant volume of material is discharged. The rate of feed can be regulated by changing the belt speed or the gate opening.

Because they are self-cleaning, belt feeders can be used for applications requiring the handling of different materials without contamination. With the adjustable gate and variable speed drive, accurate control of material can be maintained for precise blending of materials. It is in the handling of relatively fine materials that belt feeders have found their most widespread acceptance. A belt feeder is not recommended for an application where the material is too sticky to discharge properly from the belt, where the material is too hot to handle on a belt, or where large lumps of material with sharp edges could tear the belt or cause excessive wear.

Belt feeders are usually not used for 6-in. or larger lumps (152.4 mm) and the design should be such that even smaller lump material does not impact directly on the belt. Belt feeders and for that matter apron feeders, when handling certain materials, produce dead spots (Fig. 22.2.4). A combination belt feeder and tapered hopper can be designed to handle the material by feeding from a long slotted outlet (Fig. 22.2.5).

Standard belt feeders are available in widths up to 72 in. (1828.8 mm). Operation at a slight upward or downward incline is practical but horizontal operation is preferred. Speed is generally limited to 50 ft/min (0.254 m/sec) or less. The selection of a properly constructed feeder belt for wear is essential for continuous operation.

Spillage and dust is a problem and special provisions must be made. Spillage can be eliminated under the skirts by means of adjustable seal strips and carry back on the return run and by a properly

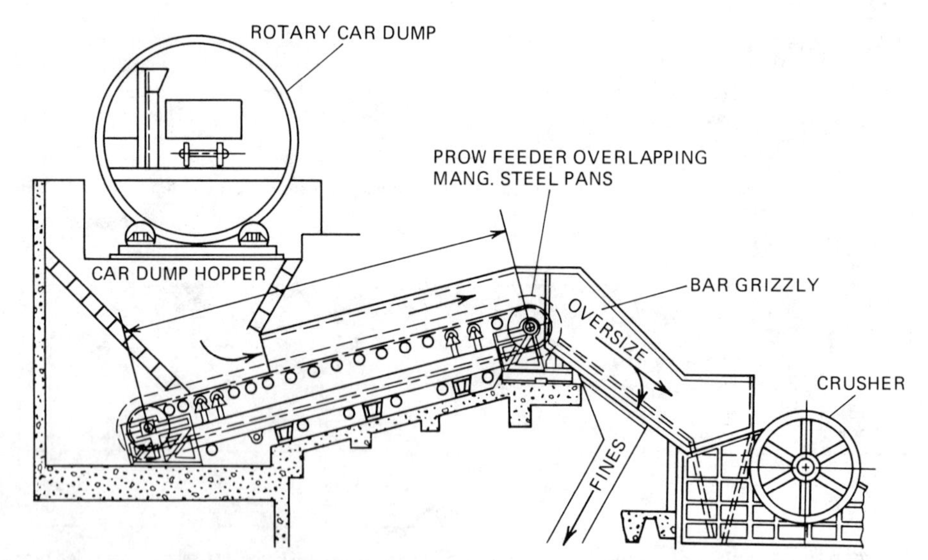

Fig. 22.2.2 Typical apron feeder application. Extra-heavy-duty apron feeder feeding crusher. Courtesy of Stephens-Adamson Inc.

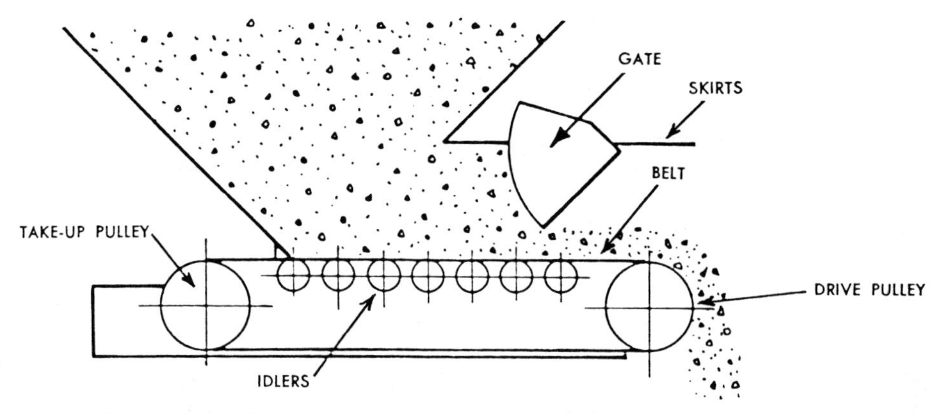

Fig. 22.2.3 Belt feeder. Short endless belt operating over idler rolls. Courtesy of Barber-Greene Co.

mounted belt cleaner at the head shaft. Dust control can be accomplished by total enclosure of chutes or application of dust collection equipment.

Drives are generally electric motors operated in conjunction with V-belts or chain drives with speed reducers.

22.2.3 Chain Curtain

The chain curtain feeder is a power-operated curtain of endless lengths of chain, resting on and retarding the flow of bulk materials on an inclined chute (Fig. 22.2.6).

This device is primarily used for heavy-duty rock handling applications used to control feed to crushers or screens in mining operations. The unit is sized based on empirical data provided by the manufacturers. It can be designed to serve openings from 10 in. (254 mm) wide × 22 in. (558.8 mm) high to 60 in. (1524 mm) wide × 92 in. (2336.8 mm) high. Maximum desirable rpm of the head drum is based on experience. Horsepower is based on the rpm of the drum times the feed width times empirical factors provided by the manufacturer.

22.2.4 Extraction Type

Due to existing bin conditions or building limitations, a free-flowing bin or hopper design often cannot be provided. Sometimes when handling materials that are stringy or sticky the ideal bin or hopper design that would result in a free flow is not practical. For these conditions there are available a wide variety of extraction-type feeders which combine agitation with metering. These units usually

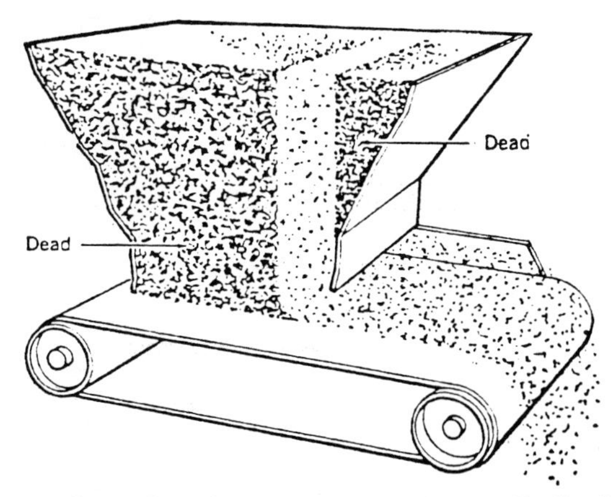

Fig. 22.2.4 Belt or apron feeder. Operation can produce dead spots. "Feeding," Johanson, J., *Chem. Engr.,* Oct. 1969.

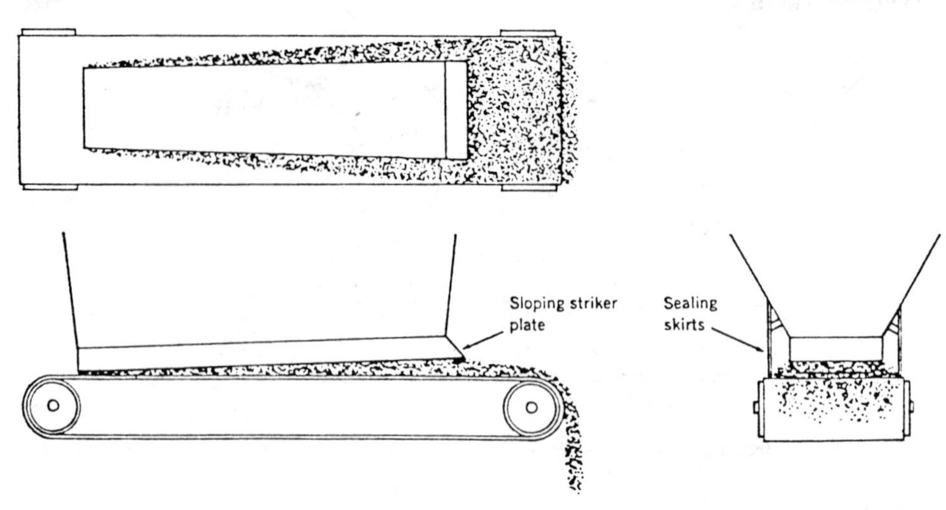

Fig. 22.2.5 Slotted opening can help produce a more uniform flow. "Feeding," Johanson, J., *Chem. Engr.*, Oct. 1969.

involve a combination of a flow-actuating mechanism with a control-feeding element. An extraction feeder should not be confused with a flow aid device which promotes solids flow but does not control it.

Most extraction feeders are proprietary items and as such are covered by the designer's or manufacturer's patent and typically have a registered trade name.

These units generally involve a mechanical or vibratory action with conventional feed mechanism. An arm that extends up into the material in the cone of a cylindrical bin and is attached to the center of a rotary plow feeder is one of the more typical mechanical types. Another employs a vibratory hopper and screw feeder arrangement which can be attached to the bottom of a bin.

It is recommended that if an extraction-type feeder appears to be required that the user work closely with several manufacturers and set up field tests to make their final selection to assure proper operation.

22.2.5 En Masse

The en-masse feeder consists of a series of skeleton or solid flights mounted on endless chain or other linkage which can operate over a horizontal or slightly inclined path. The chain and flights operate within a closely fitted casing for the carrying run with the bulk material carried in a substantially continuous stream within the full cross sections of the casing, hence the name en masse. This unit is seldom applied as an individual feeder and is used primarily as part of a total en-masse conveyor system which is covered in another section of this text.

22.2.6 Flight

The flight feeder consists of a flat pan over which material is moved by a series of bar flights extending the width of the feeder pan and mounted at each end on a chain. The movement of flights over the

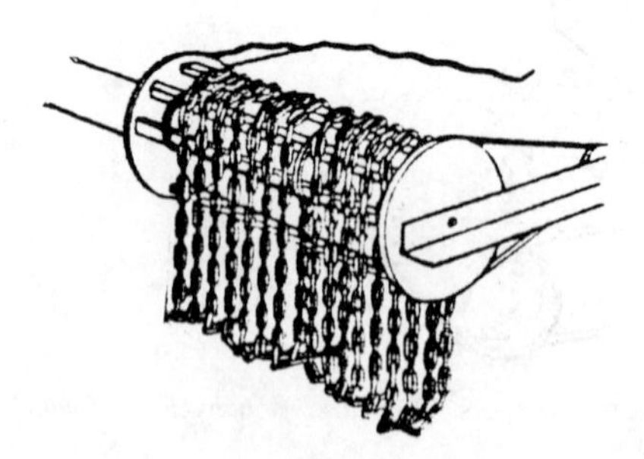

Fig. 22.2.6 Chain curtain feeder. Courtesy of Rexnord Inc.

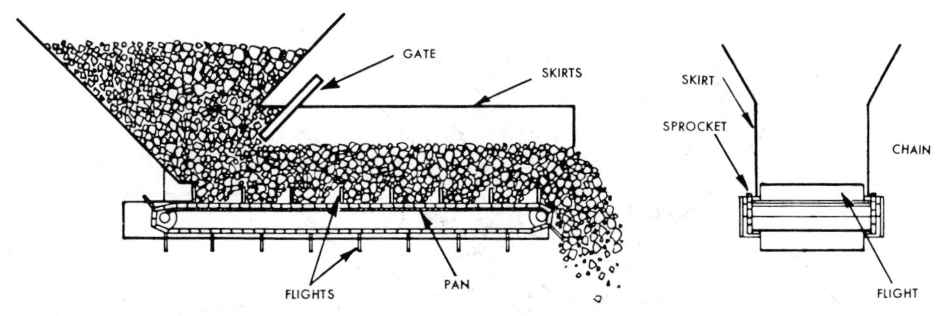

Fig. 22.2.7 Flight feeder consists of bars or flights attached to two strands of chain which slide along a flat plate. Courtesy of Barber-Greene Co.

pan results in a positive feeding action by which the material is dragged out of the hopper or bin under which the feeder is mounted (Fig. 22.2.7).

The flow of material is controlled by a cut-off gate which regulates the depth of material discharged by the feeder. Capacity is regulated by changing the speed of the flights or the depth of the gate. Flight feeders can be installed horizontally or set at an incline. Flight feeders are generally used for lightweight, nonabrasive materials such as coal or wood chips. Materials with large lumps and/or those consisting entirely of fines are best handled by other types of feeders.

Standard units are available to handle capacities up to 300 ton/hr of 50 lb/ft^3 (800.95 kg/m^3) material. Widths range from 12 in. (304.8 mm) to 36 in. (914.4 mm) and from 5 ft (1.524 m) to 12 ft (3.66 m) in length. Speed ranges from 20 to 50 ft/min (0.102–0.254 m/sec). Larger sizes can be provided along with special abrasion-resistant pans and flights.

22.2.7 Gravimetric

A type of belt feeder manufactured in combination with a constant-weight measuring device is the basis of the design of many of the gravimetric feeders available today. Initially, these continuous weight measuring devices used primarily mechanical-actuated linkages, as many still do today. However, in recent years gravimetric feeders using electronic and pneumatic closed-loop controls have been introduced and have gained acceptance.

These units are the proprietary designs of individual manufacturers, identified by trade names and covered by patents.

22.2.8 Pneumatic and Fluidizing Feeders

Pneumatic pumping and air-activated fluidizing units are available for feeding bulk materials. These units are used primarily for handling dry pulverized and granular materials such as cement, fly ash, lime, fine coal, and pulverized rocks. Additional data regarding these units will be found under the subject of pneumatic conveying systems elsewhere in this text.

22.2.9 Portable

Often when handling bulk solids there is a need to feed material to or from a moving target. Examples include discharging a bottom-dump rail car that cannot be accommodated at a fixed track location, and feeding bags or drums of bulk material that must be fed into several production machines located within a factory. There are available for these and similar applications a variety of standard feeders adapted for portability and to these special handling needs.

Portable inclined and horizontal screws, belts, and flight feeders are some of the more common types available.

22.2.10 Reciprocating Plate

The reciprocating feeder consists of a flat pan resting on rollers. The pan is actuated by a connecting rod fastened to an eccentric or crank arm. On the forward stroke the material is carried forward, but on the return stroke the pan slides from under the material allowing a portion of material to drop off. Capacity is regulated by a gate, the length of stroke, and/or the speed of the crankshaft (Fig. 22.2.8).

This type of unit is suitable for feeding nonabrasive lumpy materials such as coal. The feed is

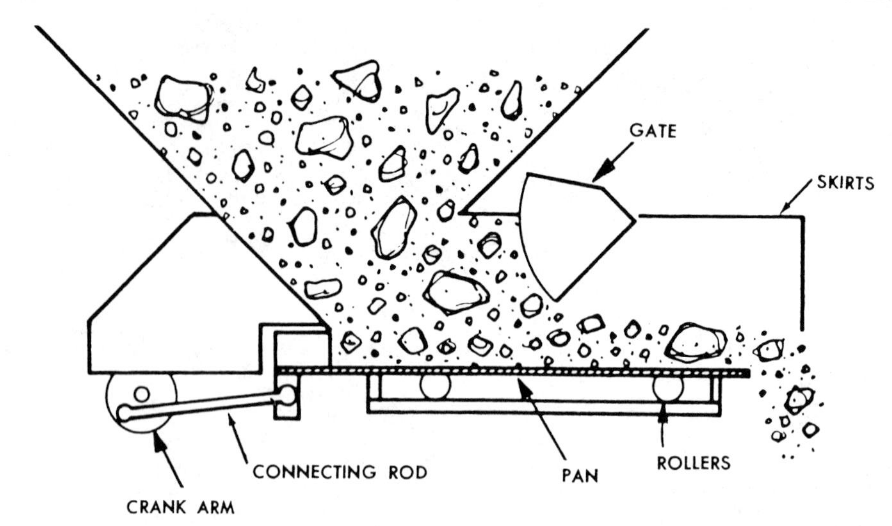

Fig. 22.2.8 Reciprocating feeder consists of reciprocally driven plates or pans impacting or pulsating forward product flows. Courtesy of Barber-Greene Co.

intermittent and volume can easily be regulated by varying the speed and stroke of the reciprocating plate. Sometimes these units are furnished with double plates for discharging from double hoppers for a more continuous flow.

The principal advantages of reciprocating feeders are their low cost and their ability to handle a wide range of miscellaneous materials including lumps. These feeders, however, are not self-cleaning. When the hopper is empty the last plateful must be removed manually if the feeder is to be used for handling a different material and contamination is undesirable. These feeders are not recommended for highly abrasive materials due to the sliding action of material on the pan, unless the unit's pan is made of abrasive resistant material.

Reciprocating feeders are available for handling capacities to 1000 ton/hr of material weighing 100 lb/ft^3 (1601.9 kg/m^3). Standard units range from 18–60 in. (457.2–1524 mm) wide and up to 8 ft (2.44 m) in length.

These units can be driven by means of chain drives from the machines they feed or by a separate gear motor and linkage.

22.2.11 Rotary Table

A rotary table feeder is a circular disc that rotates horizontally beneath an open-bottom bin. The table is larger in diameter than the opening of the bin. The space between the bin opening and the table is small enough to prevent the material from flowing out over the edge of the table when it is not moving. As the table rotates, the material feeds out under an adjustable collar onto the edge of the table. A fixed plow scrapes the material, at the controlled rate, from the table onto a conveyor or to an item of processing equipment (Fig. 22.2.9).

Rotary table feeders are used in a variety of industries including foundries, sintering plants, and pulp and paper mills. They make it possible to feed just the right amount of material for mixing with other materials. They assure a constant flow, eliminating starving and flooding.

These units can be provided in diameters from $1\frac{1}{2}$ to 20 ft (0.46–6.1 m). Volume of discharge is regulated by raising or lowering the collar, by the positioning of the plow, and/or by varying the rpm of the table. The larger-diameter units are used for handling chips and lighter materials, and the smaller-diameter units are used for heavier materials.

22.2.12 Roll

The roll or rotary drum feeder consists of a cylindrical roll similar to a pulley. It is located in a chute just below the discharge opening of a hopper (see Fig. 22.2.10).

A roll feeder is dependent on the friction between the roll surface and the material, so its feed is not positive. The rolls are sometimes made of cast iron with flanged ends. The face of the roll can be grooved to aid in carrying the material up and over the drum face for discharge into the chute below.

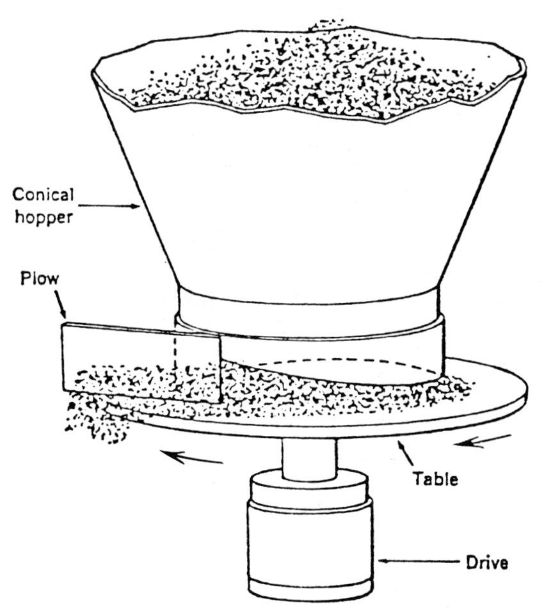

Fig. 22.2.9 Rotary table feeder. The plow located outside the bin plows material that flows under the skirts. "Feeding," Johanson, J., *Chem. Eng.,* 1969.

The rotary drum unit can also be used as a rotary air lock on the outlets of dust collectors and similar points where material must be discharged from a container under pressure but slightly above or below atmospheric.

22.2.13 Rotary Vane

Rotary vane feeders are manufactured in two basic forms, open and enclosed. The open unit is best suited for handling material containing small lumps; the enclosed unit, made with smaller vanes and designed to act as an air lock in handling the material, prevents fines from flowing through the feeder if the material becomes aerated.

Rotary vane feeders are often connected to the bottoms of bins and hoppers, and the pockets formed by the vanes offer a positive discharge action. The combination of the rpm at which the vanes rotate and the cubic capacity of the individual pockets determines the volume of material discharged. Except for very special designs, wet or sticky material should not be handled because the operation of the devices relies on gravity.

There are many types of rotors available and some of those more commonly available are discussed in Fig. 22.2.11. Rotary vane feeders are provided in a variety of special designs including blow-through, side entry, vertical construction, and punged seals, and can be built in a wide variety of materials.

Normally V-belts or chains from motor reducer combinations provide the drive power. A variable-speed drive makes these units especially adaptable to meet blending requirements.

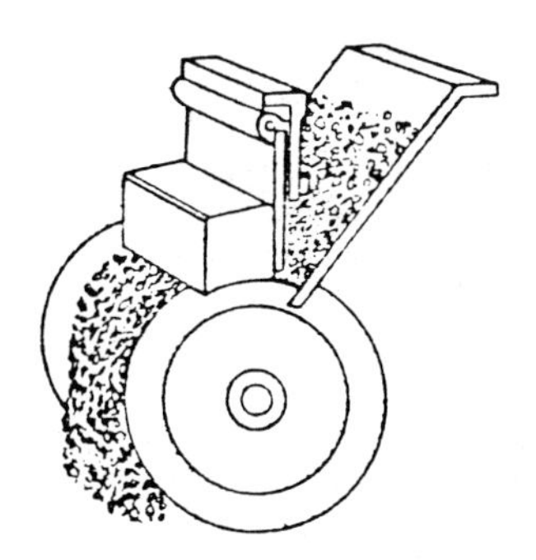

Fig. 22.2.10 Roll feeder is a smooth drum that rotates. Courtesy of Rexnord Inc.

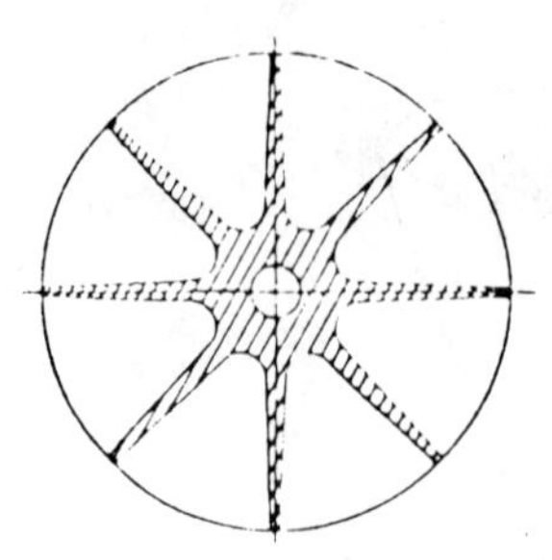

ROTOR TYPE A

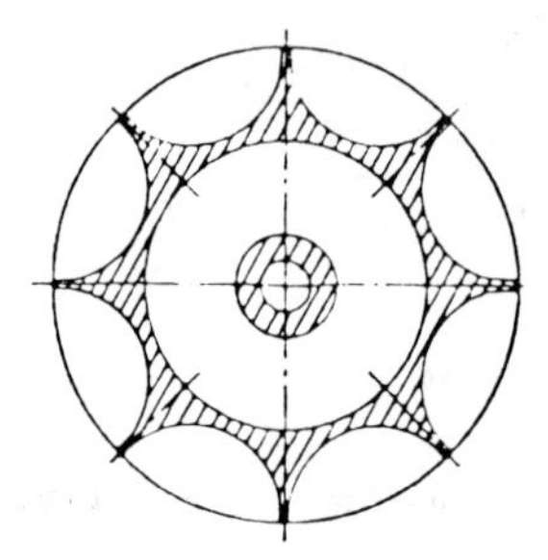

ROTOR TYPE B

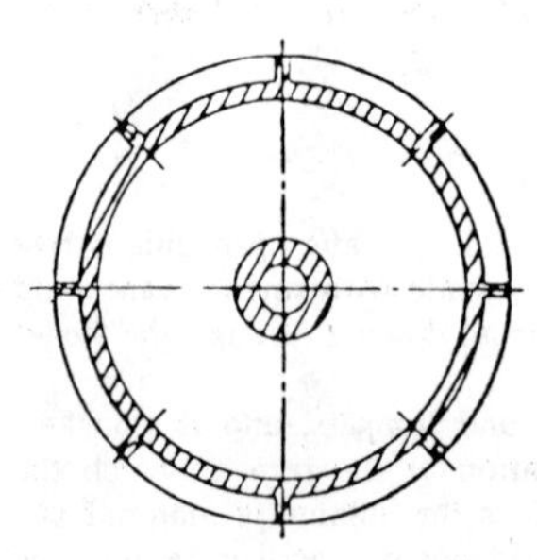

ROTOR TYPE C

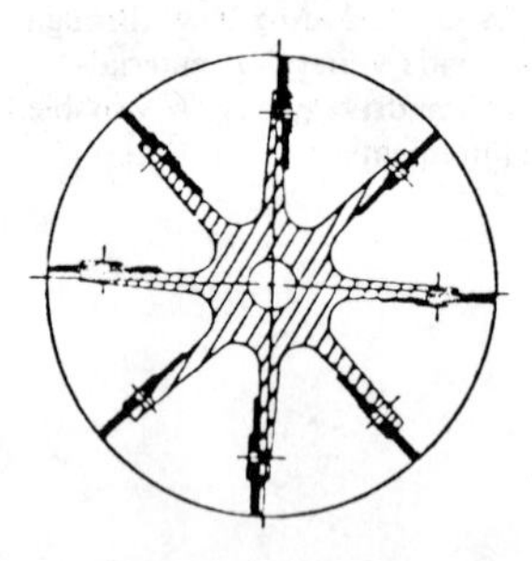

ROTOR TYPE D

Fig. 22.2.11 Rotary vane feeder. Typical rotors Type A, B, and C are used primarily where the toray vane feeder must function as a metering device or to hold a low pressure differential. Type D is provided with adjustable tips for ware. Feeder reprint, *Rock Products,* Aug. 1964, p. 82.

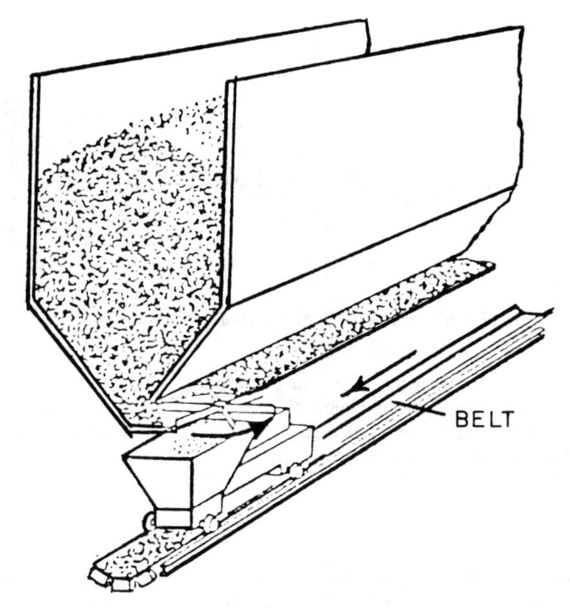

Fig. 22.2.12 Rotary plow feeder consists of a traveling carriage having a bladed rotor for plowing bulk materials from a continuous shelf to a collecting conveyor. "Feeding," Johanson, J., *Chem. Eng.*, Oct. 1969.

22.2.14 Rotary Plow

The rotary plow feeder is a series of plows mounted on a vertical shaft so that the plows rotate in a horizontal path with the plows and its driving mechanisms mounted on a self-propelled traveling carriage. Material flows by gravity from a bin or series of bins onto a shelf from which the plows scrape the material for discharge onto a horizontal belt conveyor (Fig. 22.2.12).

The rotary plow feeder is suited for handling ores, coal, and material with fine or small lumps. It has been used successfully for reclaiming from stockpiles where large volumes of materials must be stored and reclaimed. The carriage drive is reversible and provisions can be made to adjust the traversing speed *and* plow speed during operation. Limit switches can be provided to automatically reverse the carriage travel for continuous operation over a predetermined travel distance.

22.2.15 Scalper

Vibrating and reciprocating feeders can, by means of a grizzly section located at their discharge, act both as a feeder and coarse-product sizing device. However, the material characteristics or the process needs can require a special combination of feeding and scalping actions, and special feeders are available.

There are several types of these specialty feeders consisting of a surface of reciprocating steel bars or rows of specially designed rotating rolls (Fig. 22.2.13). Quarry-run or mine material can be

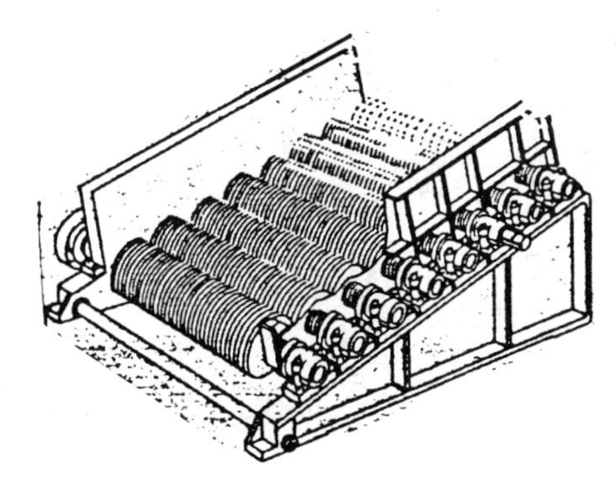

Fig. 22.2.13 Scapler feeder of the live roll grizzly type consisting of a series of spaced, rotating, parallel rolls with fixed size openings. Courtesy of Rexnord Inc.

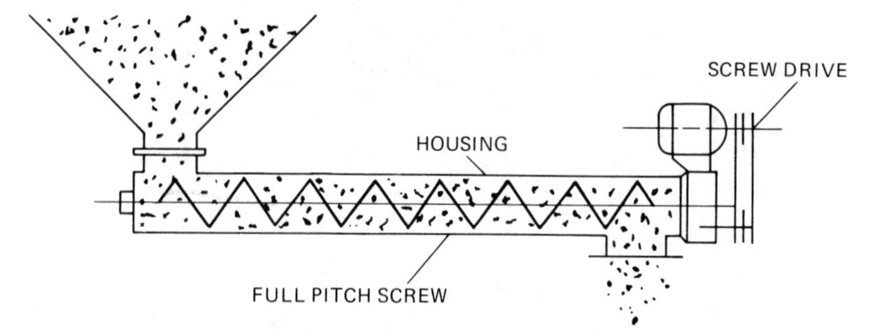

Fig. 22.2.14 Screw feeders are short screw conveyors.

dumped directly on these feeders by truck or shovel. The load is forced into contact with bars by gravity and the fines pass through the unit with the "oversize" material passing over. These special feeder-scalpers can handle virtually any bulk material and are especially useful when handling damp or sticky materials. However, due to their more complicated construction and drive-mechanism requirements their initial capital cost is high.

22.2.16 Screw

A screw or spiral feeder consists of metal flighting mounted on a shaft. The mounted flighting is suspended within a U-shaped, round, or flared trough driven by an electric motor through a speed reducer or chain drive. The flighting can be full pitch or short pitch and is available in uniform and tapered diameters (Fig. 22.2.14).

Dimensions and, as a consequence, replacement parts are standardized throughout the industry. Special custom designs have been made for feeding material against pressures up to 15 psig and over. Screw feeders have an advantage over belt and apron feeders in that there is no return element to spill solids. Also, the screw can be completely enclosed for dust control.

Screw feeder can handle a wide range of materials from powders to lumps. Lump size is generally limited to lumps smaller than the smallest pitch and to a bin opening less than 6 times the screw diameter. Special flighting may be required for handling sticky or cohesive materials. Extremely free-flowing fine powders will flood through a screw feeder resulting in an uncontrolled flow. The normal $\frac{1}{2}$ in. (12.7 mm) clearance between the flighting and trough can result in contamination and can be reduced when handling multiple products.

Various designs of screw feeders are available and are fully covered in manufacturers' catalogs. Standard full pitch is suitable for handling fine free-flowing materials where the feed opening is limited to 1–1$\frac{1}{2}$ pitches. If the diameter of the screw is uniform the feed of the material will come from only a portion of a slotted inlet and not across the entire length (Fig. 22.2.15).

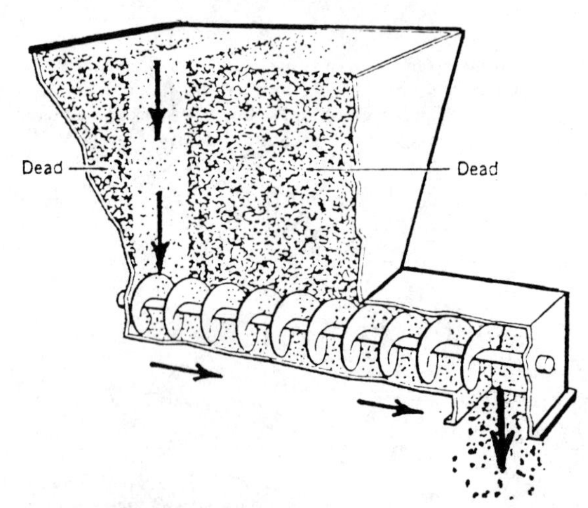

Fig. 22.2.15 Screw feeders with full pitch screw leave dead spots in the bin. "Feeding," Johanson, J., *Chem. Eng.*, Oct. 1969.

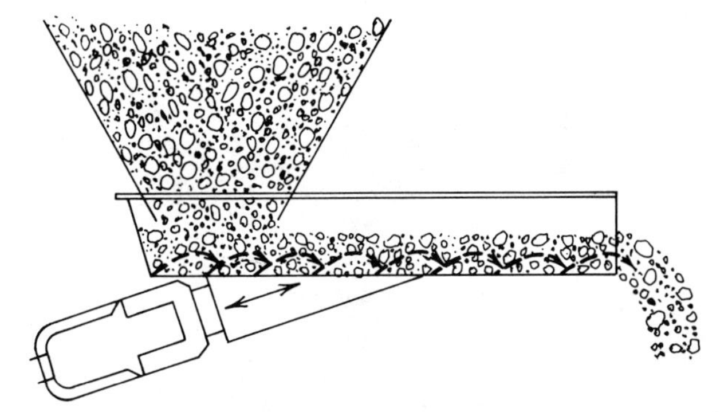

Fig. 22.2.16 Vibrating feeder indicating particle movement on vibrating pan.

Screw feeder conveyor pitch can be varied increasing in the direction of flow to provide a more uniform discharge across a slotted bin opening. Another variation to accomplish the same effect is to utilize a tapered flight diameter. However, in order to assure the best uniform bin discharge across a slotted bin opening requires a combination of tapered flight diameter and increasing pitch.

The capacity and horsepower of a screw conveyor is in part based on experience and judgment. Capacity tables and horsepower factors are given in all manufacturers' catalogs, but with some variation.

22.2.17 Vibrating

A vibrating feeder consists of a pan mounted on springs actuated by a mechanism that causes the material to move by throwing it upward and forward on the pan (Fig. 22.2.16). There are two general types of vibrating feeders available.

1. *Direct Force.* Where the force is produced by rotating counterweights and applied directly to the pan. This design is essentially constant-rate and does not provide precise feed control due to changes in rate when loaded. Low in cost, they handle a wide range of solids and can operate in poor environments (Fig. 22.2.17).
2. *Indirect Force.* Where the vibrating force is produced by an exciter mass which is amplified by a spring suspension. This design is the most commonly used, requiring less horsepower and maintenance than the direct-force type. Two excitation systems are in common usage, electromagnetic (Fig. 22.2.18) and electromechanical (Fig. 22.2.19).

Electromechanical feeders are competitively priced against electromagnetic at rates over 5 ton/hr. The electromechanical unit requires less specialized maintenance and can be easily built in explosion-proof construction. They can also handle a wider range of materials than their electromagnetic counterpart.

Feed rate is a function of bulk density, width and depth of loading on the pan, and linear speed along the trough. Declining the pan will increase conveying speed. A critical factor in the proper application of a vibrating feeder is the geometry of the hopper, hopper opening, and feeder pan and/or skirts. Manufacturers' catalogs not only include information on handling rates, equipment dimensions,

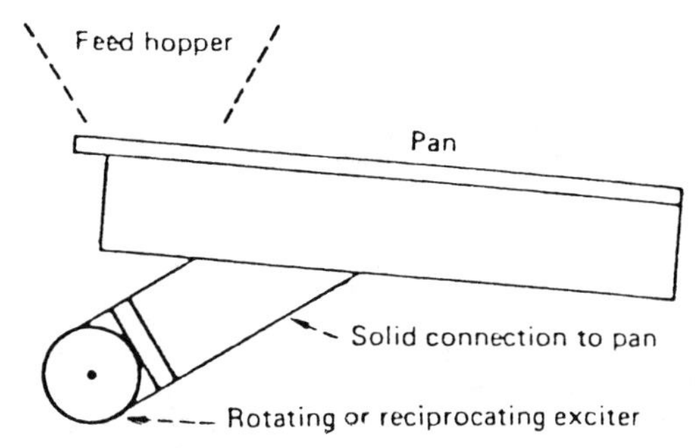

Fig. 22.2.17 Vibrating feeder with direct-force single mass action. "Solids Handling," *Chem. Engr.*, McGraw-Hill, 1981.

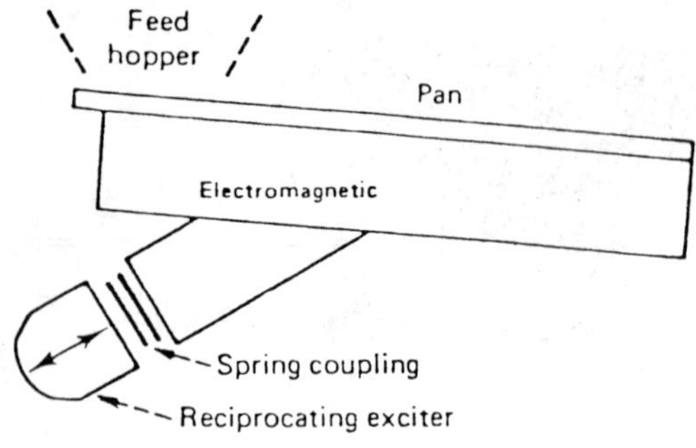

Fig. 22.2.18 Vibrating feeder with an indirect force resulting from an alternating direct current vibrator coupled to the pan by metal or fiberglass leaf springs. "Solids Handling," *Chem. Engr.*, McGraw-Hill, 1981.

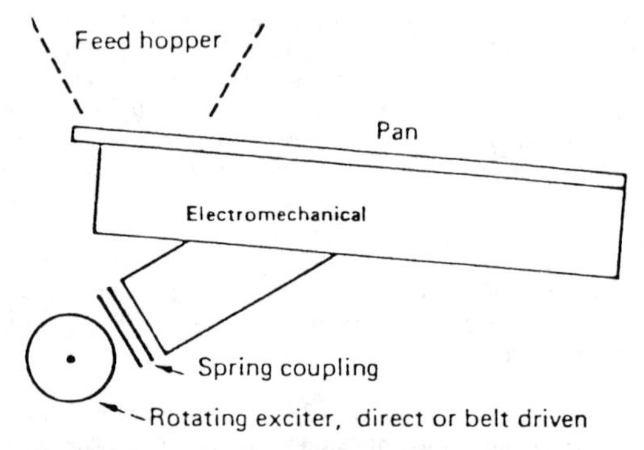

Fig. 22.2.19 Vibrating feeder with an indirect force resulting from an eccentric weight driven by an electric motor. "Solids Handling," *Chem. Engr.*, McGraw-Hill, Feeders, p. 119.

and power requirements but also include recommended geometry for installation. Performance of vibrating feeders is more sensitive to particle size than other types of feeder. When handling very fine powders or cohesive solids expert knowledge is required to minimize unexpected results.

Sizes of available units range from electromagnetic units a few inches wide to almost a 1 ft wide. (0.305 m) long handling a few pounds per hour, to electromechanical units 76 in. (1930.4 mm) wide × 20 ft (6.10 m) long, handling over 1000 ton/hr. Special designs can be provided for units 100 in. (2540 mm) wide × 35 ft (10.7 m) long handling 10,000 ton/hr.

22.3 FEEDER SELECTION AND SIZING

Selection of feeders for handling solids is influenced by hopper/bin geometry, dimensions and locations of discharge openings, material characteristics, configuration of equipment involved, or space restrictions and personal preference. Working with a proper data base, the best means to size a feeder is to use manufacturers' catalogs which outline capacities in ft^3/hr, dimensions of available units, and their drive/horsepower requirements.

When obtaining a recommendation and cost for a given type of bulk feeder, users often encounter differences in unit size and horsepower as quoted by various bidders. Even when a particular type, size, and horsepower feeder is specified by an experienced bulk engineer, suppliers will often propose an alternative or take an exception. These variations result from the fact that the determination of the capacity and horsepower of bulk feeders is based on many factors, some which require testing and others which are set empirically.

In general, the capacity of a bulk feeder is based on a volume of bulk material with a given bulk density passing through a controlled space at a fixed speed. If no other factors played a role in bulk feeder capacity determination, this would be a simple calculation. For example, with reference to Fig. 22.3.1, assume a material with a bulk density of 50 lb/ft^3 (800.9 kg/m^3) being handled at a conveying speed of 100 ft/min (0.508 m/sec) passing through a controlled space with a fixed height

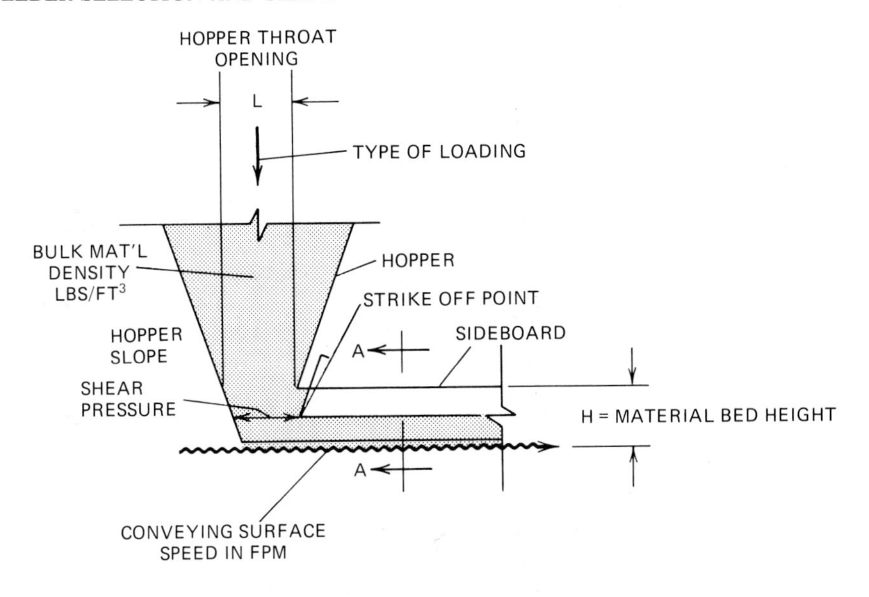

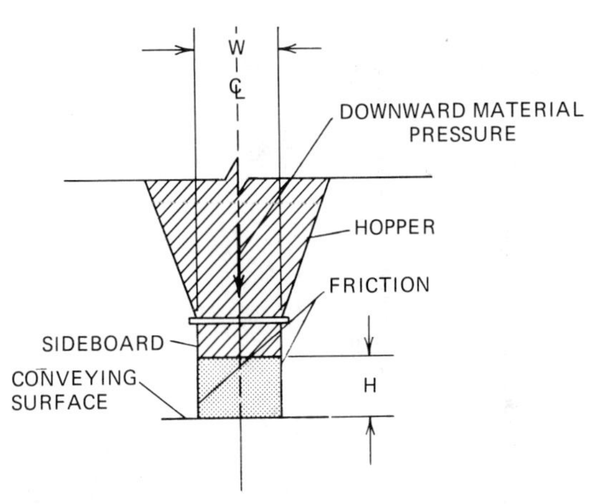

Fig. 22.3.1 Bulk feeder forces necessary for consideration in capacity and horsepower determination.

of 7.2 in. (182.9 mm) and width of 20 in. (508 mm). This bulk feeder would discharge the following theoretical capacity:

$$\text{Bulk density} \times \quad \text{Volume} \quad \times \quad \text{Speed} \quad = \quad \text{Capacity}$$

$$50\ \text{lb/ft}^3 \quad \times \frac{7.6\ \text{in.} \times 20\ \text{in.}}{144.\ \text{in.}^2/\text{ft}^2} \times 100\ \text{ft/min} = 5000\ \text{lb/min}$$

This capacity of 5000 lb/min multipled by 60 min/hr and then divided by 2000 lb/ton equals the units of theoretical bulk handling capacity of 150 ton/hr.

$$\frac{5000\ \text{lb/min} \times 60\ \text{min/hr}}{2000\ \text{lb/ton}} = 150\ \text{ton/hr}$$

Starting with this simplified calculation it is then necessary to modify this value for any or all of the following factors to determine the actual capacity of the unit.

Is hopper slope sufficient to assure the material *will not* rathole or arch and starve the feeder?

Is the hopper throat opening L and W sufficient to pass lumps assuring that the opening will not plug, interrupting material passage?

Table 22.3.1 Feeder Selection Data

Parameter	Type of Feeder							
	Apron	Belt	Flight	Reciprocating Plate	Rotary Table	Rotary Vane	Screw	Vibrating
				Material Characteristics				
Size								
Ex. coarse (48″ × 0″)	*							
Coarse (12″ × 0″)	*			*				
Medium (3″ × 0″)	*	*	*	*	*		*	*
Fine ($\frac{3}{4}$″ × 0″)	*	*	*	*	*	*	*	*
Very fine (10 M × 0″)					*	*	*	*
Flowability								
Free	*	*	*	*	*	*	*	*
Flooding						*	*	
Sticky	*	*			*			

Other								
High temperature	Good	Poor	Good	Good	Good	Good	Good	Good
Abrasive	Good	Poor	Poor	Avg.	Poor	Poor	Poor	Good
				Feeder Characteristics				
Feed Requirement								
Uniform discharge	Good	Good	Avg.	Avg.	Good	Good	Avg.	Good
Accuracy of weight	Avg.	Avg.	Poor	Avg.	Avg.	Avg.	Avg.	Avg.
Feed regulation range	Good	Good	Poor	Avg.	Good	Avg.	Avg.	Good
Cost								
Initial capital	High	Avg.	Avg.	Low	High	Low	Avg.	High
Maintenance	High	Low	High	Low	Low	Avg.	High	Low
Power	High	Low	High	Avg.	Avg.	Low	Avg.	Low

Are the characteristics of the materials being handled such that as the material passes under the "strike-off" point it will remain at a uniform 7.2 in. (H) depth and not "swell up"?

What is the frictional efficiency of the feeder carrying surface?

What happens to capacity if the conveying surface is sloped upward or downward?

What effect, if any, will the frictional drag of the skirtboards, strike-off point, and so on, have on capacity?

In what manner is the material fed to the feeder?

Is it free fall, cushioned fall, or mass release?

Is the feeder steadily being filled while it is being loaded or is unloading taking place from a static mass?

Some engineers and manufacturers run tests on the material or with their specific feeder to determine flow factors, opening sizing, operating characteristics, and so on. Others rely solely on past experience, while still others base their selection on their own methodology.

Determination of bulk feeder horsepower, although not as empirical as capacity selection, is also not a clear-cut technical calculation. Again, with reference to Fig. 22.3.1, the power necessary to drive a bulk feeder in general consists of the following factors:

The horsepower necessary to move the feeder mechanism in its empty state

plus

The power necessary to move the material including lifting, if any,

plus

The power necessary to overcome material friction along the feeder sides (skirts)

plus

The power necessary to shear the material from the hopper opening.

Very often within these horsepower calculations is included a special material factor, which has never been uniformly established by the conveyor industry. Also, startup horsepower, which is greater than the running horsepower, is often arbitrarily factored.

Although reputable manufacturers more than likely will quote their best unit for any given application, due to the variables involved there will always be an honest difference in price between manufacturers. Consequently, any capacity and horsepower formulas given in manufacturers' catalogs or in other texts must be carefully analyzed by the owner's engineers for the particular application.

Feeder selection is further complicated by the fact that individual types are available in a variety of constructions and careful consideration is necessary to assure long-term reliable service. For example, the choice of a heavy-duty carbon-steel apron feeder over a heavy-duty manganese apron feeder to reduce initial capital cost, could often result in increased downtime and maintenance costs.

With so many variables involved in feeder selection, it is possible to establish only general guidelines. In Table 22.3.1 some of the more pertinent factors that affect selection of some of the more common types of solid feeders have been summarized. An asterisk (*) indicates an ideal application; this does not mean, however, that an individual feeder cannot be adapted to work satisfactorily under other conditions.

CHAPTER 23
BELT CONVEYORS

GEORGE A. SCHULTZ

Epstein Process Engineers
Chicago, Illinois

23.1 INTRODUCTION

A belt conveyor is one of the most versatile types of bulk handling equipment available. It is suited for handling a variety of bulk materials (Table 23.1.1) over a wide range of capacities (Table 23.1.2). It provides an economical and practical means for transporting bulk materials over long distances and over terrains requiring a wide range of paths of travel (Fig. 23.1.1).

Belt conveyor systems can be arranged to convey bulk materials as well as to weigh, sort (magnetically), sample, batch, and blend when necessary. Properly designed, belt conveyors are noted for their long-term, dependable, and low-cost operation.

23.2 BASIC DESIGN ELEMENTS

Belt conveyors belong to a class of non-self-contained equipment. Unlike self-contained units such as pumps and compressors the efficiency or effectiveness of a belt conveyor depends on a skillful choice of five essential elements that make up a particular conveyor, coupled with the unit's proper integration into a system. The essential elements of typical belt conveyors are (Fig. 23.2.1):

1. The *belt,* which forms the moving and supporting surface on which the conveyed material rides.
2. The *idlers,* which form the supports for the troughed carrying strand of the belt and the flat return strand.
3. The *pulleys,* which support and direct the belt and control its tensions.
4. The *drive,* which imparts power through one or more pulleys to move the belt and its load.
5. The *structure,* which supports and maintains alignment of idlers, pulleys, and drive.

Assuming proper selection of these five elements, an engineer must then devise the proper integration of the belt conveyor within a system. How the conveyor is best fed and discharged; how it is supported, either permanently or as a portable unit; accessability for operation and maintenance; electrical starting and stopping needs; and many other factors must be studied and carefully coordinated.

Although the design and application of a simple belt conveyor is no mystery, considerable experience is necessary to design the wider, longer, high-capacity units—complex conveyor systems or belt conveyors that must operate under unusual conditions. Expert assistance in these areas is available from a variety of consultants, engineering contractors, and manufacturers located throughout the world.

This section on belt conveyors contains design information for average operating conditions. The design information has been arranged within a framework normally employed when designing a belt conveyor. As in most technical fields there are some differences of opinion regarding calculations of belt tensions, horsepower requirements, and friction factors. Although these differences have little effect on conveyors under 500 ft in length they can have a considerable effect on a longer and more complex unit.

Table 23.1.1 Typical Materials Handled by Belt Conveyors

Material Characteristics	Example
Maximum size lumps, sized or unsized	
Mildly abrasive	Coal, earth
Very abrasive, not sharp	Bank gravel
Very abrasive, sharp and jagged	Stone, ore
Half max. lumps, sized or unsized	
Mildly abrasive	Coal, earth
Very abrasive	Slag, coke, ore, stone, cullet
Flakes	Wood chips, bark, pulp
Granular, $\frac{1}{8}$–$\frac{1}{2}$ in. lumps	Grain, coal, cottonseed, sand
Fines	
Light, fluffy, dry, dusty	Soda ash, pulverized coal
Heavy	Cement, flue dust
Fragile, where degradation is harmful	Coke, coal
	Soap chips

Table 23.1.2 **Belt Conveyor Capacity Ranges**

Belt Width (in.)	Weight per Cu. Ft of Material	Capacity[a] (ton/hr) for Given Belt Speed (ft/min)											Cross Section of Load (ft²)	Cubic Feet per Hour @ 100 ft/min	Cubic Yards per Hour @ 100 ft/min	Bushels per Hour[b] @ 100 ft/min
		100	150	200	250	300	350	400	450	500	550	600				
14	30	8	12	16	20	24	----	----	----	----	----	----				
	50	14	21	27	35	41	----	----	----	----	----	----				
	75	20	30	40	50	60	----	----	----	----	----	----	0.090	541	20	265
	100	27	41	54	68	71	----	----	----	----	----	----				
	125	34	51	68	85	102	----	----	----	----	----	----				
	150	40	60	81	100	112	----	----	----	----	----	----				
16	30	11	17	23	29	34	----	----	----	----	----	----				
	50	19	29	38	48	57	----	----	----	----	----	----				
	75	29	43	57	72	86	----	----	----	----	----	----	0.126	758	28	379
	100	38	57	76	95	114	----	----	----	----	----	----				
	125	48	72	95	122	148	----	----	----	----	----	----				
	150	57	86	114	143	171	----	----	----	----	----	----				
18	30	15	23	30	38	45	53	60	----	----	----	----				
	50	25	38	50	63	75	88	99	----	----	----	----				
	75	38	57	75	94	112	131	149	----	----	----	----	0.165	990	37	425
	100	50	75	99	124	149	174	198	----	----	----	----				
	125	63	94	124	156	187	218	248	----	----	----	----				
	150	75	113	149	187	224	262	297	----	----	----	----				
20	30	19	28	38	49	59	67	75	----	----	----	----				
	50	32	48	63	79	94	110	125	----	----	----	----				
	75	48	71	94	120	146	167	188	----	----	----	----	0.208	1,250	46	622
	100	63	94	125	162	198	224	250	----	----	----	----				
	125	79	118	157	201	245	279	313	----	----	----	----				
	150	95	142	188	240	292	333	375	----	----	----	----				
24	30	29	43	58	73	87	102	116	130	145	----	----				
	50	48	73	97	121	145	169	193	217	241	----	----				
	75	72	109	145	182	218	254	290	326	362	----	----	0.321	1,930	72	980
	100	97	145	193	242	290	338	386	435	483	----	----				
	125	121	186	242	302	363	423	483	543	603	----	----				
	150	145	218	290	363	435	507	579	652	724	----	----				
30	30	48	72	95	119	143	167	191	215	239	----	----				
	50	79	119	159	199	239	279	318	358	398	----	----				
	75	119	179	238	298	358	418	477	537	596	----	----	0.530	3,180	118	1620
	100	159	239	318	398	477	557	636	716	795	----	----				
	125	198	298	397	497	596	696	795	895	994	----	----				
	150	238	358	477	597	716	836	954	1074	1193	----	----				

Table 23.1.2 *(Continued)*

Belt Width (in.)	Weight per Cu. Ft of Material	Capacity[a] (ton/hr) for Given Belt Speed (ft/min)											Cross Section of Load (ft^2)	Cubic Feet per Hour @ 100 ft/min	Cubic Yards per Hour @ 100 ft/min	Bushels per Hour[b] @ 100 ft/min
		100	150	200	250	300	350	400	450	500	550	600				
36	30	71	107	142	178	213	248	284	319	355	390	426				
	50	119	178	236	296	355	414	473	532	591	650	709				
	75	178	266	354	444	532	621	710	798	886	975	1064	0.788	4,730	175	2430
	100	237	355	473	592	710	828	946	1064	1182	1301	1419				
	125	296	444	591	740	888	1035	1183	1330	1477	1626	1773				
	150	356	533	709	888	1065	1242	1419	1596	1773	1951	2128				
42	30	99	148	198	247	297	346	395	445	494	543	593				
	50	165	247	329	412	495	577	659	742	824	906	989				
	75	247	371	494	618	742	865	988	1112	1236	1359	1483	1.098	6,590	244	3400
	100	330	494	659	824	989	1153	1318	1483	1648	1812	1977				
	125	412	617	823	1030	1236	1441	1648	1853	2060	2265	2472				
	150	495	741	988	1236	1484	1730	1977	2224	2472	2719	2966				
48	30	131	197	263	329	394	460	526	591	657	723	788				
	50	219	329	438	547	657	766	876	985	1095	1204	1314				
	75	328	493	657	821	985	1150	1314	1478	1642	1806	1971	1.460	8,760	324	4520
	100	438	657	876	1095	1314	1533	1752	1971	2190	2409	2628				
	125	547	822	1095	1368	1642	1916	2190	2463	2737	3010	3285				
	150	657	986	1314	1642	1971	2299	2628	2956	3285	3613	3942				
54	30	168	252	336	420	504	588	672	756	840	924	1008				
	50	280	420	560	700	840	980	1120	1260	1400	1540	1680				
	75	420	630	840	1050	1260	1470	1680	1890	2100	2310	2520	1.868	11,200	415	5810
	100	560	840	1120	1400	1680	1960	2240	2520	2800	3080	3360				
	125	700	1050	1400	1750	2100	2450	2800	3150	3500	3850	4200				
	150	840	1260	1680	2100	2520	2940	3360	3780	4200	4620	5040				
60	30	210	315	420	525	630	735	840	945	1050	1155	1260				
	50	350	525	700	875	1050	1225	1400	1575	1750	1925	2100				
	75	525	787	1050	1312	1575	1837	2100	2362	2625	2887	3150	2.333	14,000	518	7250
	100	700	1050	1400	1750	2100	2450	2800	3150	3500	3850	4200				
	125	875	1312	1750	2187	2625	3062	3500	3937	4375	4812	5250				
	150	1050	1575	2100	2625	3150	3675	4200	4725	5250	5775	6300				

[a] Capacities given are for horizontal conveyors having a uniform feed and load. Inclined conveyors, as a rule, will not handle as many tons per hour and this should be taken into consideration. If there are peak loads, belts of sufficient capacity to handle material at maximum rate should be used. Where tripper is to be used belt speed generally should not be less than 300 ft/min.

[b] Loading for grain, etc., on 20° standard troughing idlers.

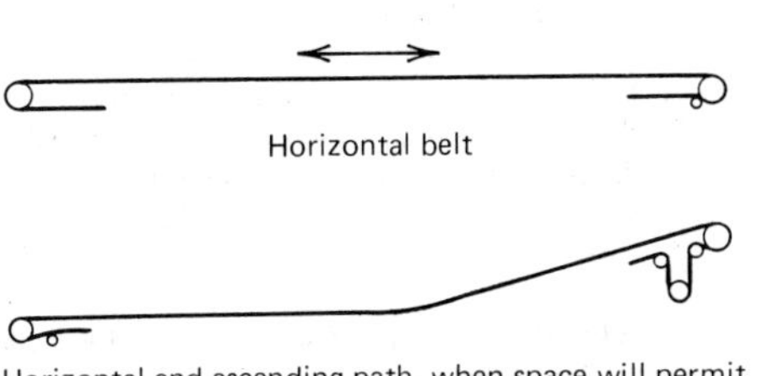

Horizontal belt

Horizontal and ascending path, when space will permit vertical curve and belt strength will permit one belt.

Ascending and horizontal path, when belt tensions will permit one belt and space will permit vertical curve.

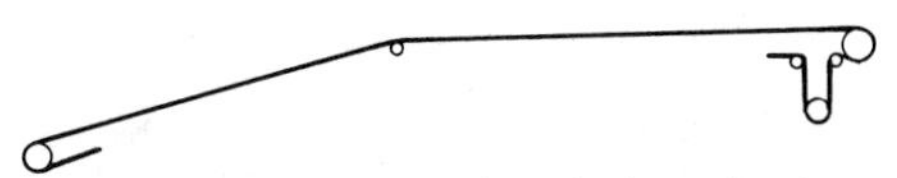

Ascending and horizontal path; or horizontal and descending path, when space will not permit vertical curve but one belt can be used.

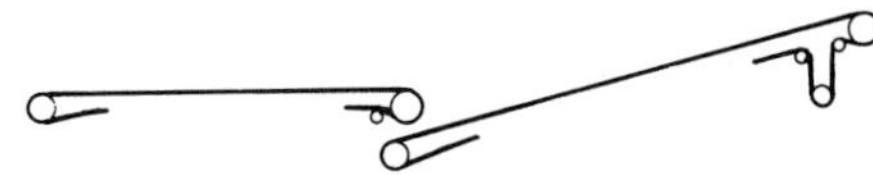

Possible horizontal and ascending path, when space will not permit a vertical curve or when the conveyor belt strength requires two belts.

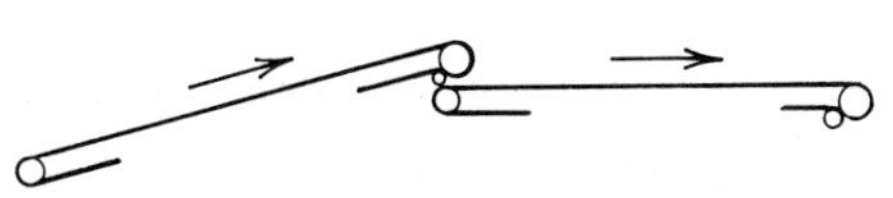

Ascending and horizontal path, when advisable to use two conveyor belts.

Possible horizontal and ascending path, when space will not permit vertical curve belt strength will permit one belt.

Compound path with declines, horizontal portions, vertical curves, and incline.

Loading can be accomplished, as shown, on minor inclines or declines.

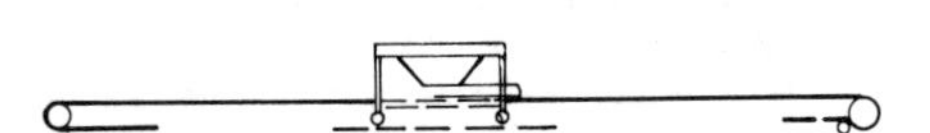

Traveling loading chute to receive materials as a number of points along conveyor.

Fig. 23.1.1 Typical belt conveyor profiles.

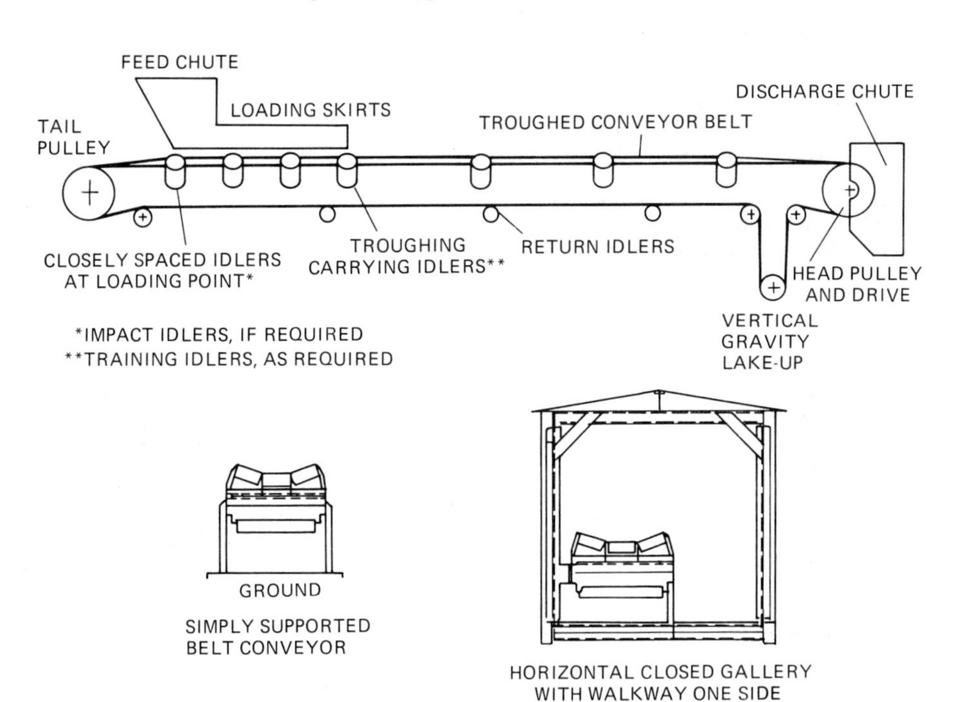

Fig. 23.2.1 Typical belt conveyor arrangement.

Table 23.2.1 Worksheet: Information Necessary to Design or Select a Belt Conveyor

Questions to Be Answered	To Determine These Facts
1. What is the horizontal distance over which the material is to be conveyed?	1. Will a belt conveyor serve the purpose?
2. What is the vertical height that the material is to be lifted or lowered?	2. What width of conveyor is required to handle the capacity and maximum size of lump?
3. What kind of material is to be handled and what is its weight per cubic foot?	3. At what speed shall we run the conveyor?
4. What is the average required capacity in tons per hour?	4. How much power is required to drive the conveyor?
5. What is the maximum required capacity in tons per hour?	5. What arrangement of pulleys should we use to drive this conveyor?
6. How will the flow of material be controlled?	6. What is the maximum belt tension?
7. What are the dimensions of the largest lumps?	7. What belt shall we select?
8. What percentage of the total volume to be handled will consist of this maximum size lump?	8. What diameters of pulleys are required for this belt?
9. Is the material hot, cold, wet, dry, sticky, oily, abrasive, or corrosive? To what degree?	9. What shaft diameters are required for these pulleys?
10. How many loading points are there and where are they located?	10. What diameter, style, and spacing of idlers shall be used?
11. How is the material to be discharged from the conveyor: over the head pulley or through a tripper?	11. What type of take-up shall we use and where will it be located?
12. How many discharge points are there and where are they located?	12. What type of drive equipment shall be used?
13. What would be the most convenient location for the drive?	13. What accessories shall be provided?
14. What is the prime mover, an electric motor or an internal combustion engine?	
15. If an electric motor, what are the current characteristics?	
16. What is its output speed, and what is the size and keyseat of the output shaft of the prime mover?	
17. Is the material to be weighed in transit on the belt?	
18. Do you wish to remove tramp iron from the material as it passes over discharge pulley?	
19. Do you wish to sample the material as it is being conveyed or discharged?	
20. What safety devices should be applied to the conveyor?	

Throughout the world, various universities and companies have developed equations for computation of belt tension as a function of load, speed, length, and friction factors. Some of these have been standardized by the DIN# 22100 in Germany and CEMA (Conveyor Equipment Manufacturers Association) in the United States. The factors given herein are conservative and final design will necessarily depend on the individual engineer's preferences, or specific standards.

The common objective is to design a belt conveyor that will deliver maximum performance at minimum cost per ton of material handled. To do so, it is necessary to obtain answers to key questions and to develop the facts necessary to determine if a belt conveyor will suit the purpose for which it is to be designed and arranged. A summary of key questions and their purpose is outlined in Table 23.2.1.

23.2.1 Applicability of Belt Conveyors

The potential range of sizes, speeds, and capacities of belt conveyors is so great that these factors seldom determine whether a belt conveyor will serve the purpose.

Table 23.2.2 Maximum Conveying Incline Angles

Material Carried	Maximum Angle of Incline[a] (degrees)	Material Carried	Maximum Angle of Incline[a] (degrees)
Alumina, dry, free flowing	18	Ore (see stone)	15–20
Beans, whole	8	Packages	15–25
Coal, anthracite	16	Pellets, depending on size, bed of material and concentricity (taconite, fertilizer, etc.)	5–15
Coal, bituminous, sized, lumps over 4 in.	15	Rock (see stone)	15–20
Coal, bituminous, sized, lumps 4 in. and under	16	Sand, very free flowing[d]	15
Coal, bituminous, unsized	18	Sand, sluggish (moist)[c]	20
Coal, bituminous, fines, free flowing[b]	20	Sand, tempered foundry	24
Coal, bituminous, fines, sluggish[c]	22	Stone, sized, lumps over 4 in.	15
Coke, sized	17	Stone, sized, lumps 4 in. and under, over $\frac{3}{8}$ in.	16
Coke, unsized	18	Stone, unsized, lumps over 4 in.	16
Coke, fines and breeze	20	Stone, unsized, lumps 4 in. and under, over $\frac{3}{8}$ in.	18
Earth, free flowing[b]	20	Stone, fines $\frac{3}{8}$ in. and under	20
Earth, sluggish[c]	22	Wood chips	27
Gravel, sized, washed	12		
Gravel, sized, unwashed	15		
Gravel, unsized	18		
Grain	15		

[a] For ascending conveyors when uniformly loaded and with constant feed.
[b] Angle of repose 30–45°.
[c] Angle of repose over 45°.
[d] Very wet or very dry, with angle of repose less than 30°.

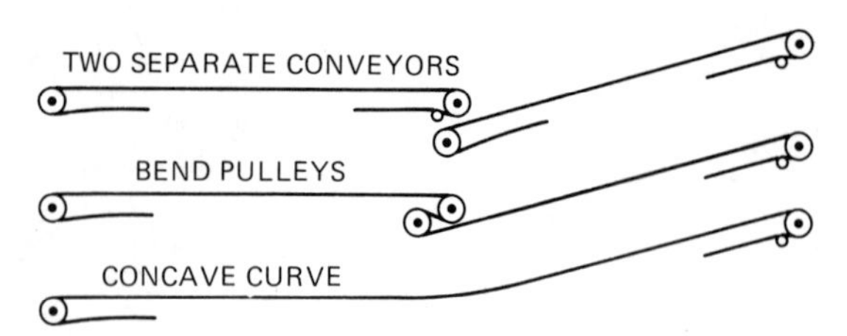

Fig. 23.2.2 Concave change in vertical paths of conveyor.

Of all the main physical characteristics of the plant which will determine whether a belt conveyor can be used, the most important is the relation between the horizontal distance and the vertical distance the material is to be lifted and the flowability of the material to be transported. These factors determine the angle of inclination of the conveyor.

If this angle of inclination is so great that the material will roll back on the belt, it may be necessary to combine a conveyor with a specially designed sidewall and cleated belt or elevator for lifting the material. Table 23.2.2 gives the maximum safe inclinations of troughed belt conveyors for handling various bulk materials.

We may find that the horizontal distance available is more than sufficient to accommodate the height the material is to be lifted. In this case, it may be more economical, rather than using a shallow angle of inclination for the entire conveyor length, to run the conveyor horizontally for a certain distance and then incline it at its discharge end. When a transition is made between a horizontal run of conveyor and an inclined run, or between an inclined run and a steeper incline, it is necessary to have a concave change in the travel of the material. Concave changes in direction can be made by using one of the methods indicated in Fig. 23.2.2.

The concave curve must be designed to insure that the belt will remain in contact with, and not lift off, the troughing idlers at startup. Since the lifting and spilling of a material at concave curves is a common occurrence, it behooves the owner/engineer to double-check designs. Although these calculations are complex and have even been computerized by some designers and manufacturers, Fig. 23.2.3 can be used for a quick approximation of minimum recommended radii. The values given in Table 23.2.3 can be used for plotting the coordinates of a concave curve. In some cases it is necessary to change direction of the belt from an incline path to horizontal. This is accomplished by means of either a bend pulley or closely spaced idlers as illustrated in Fig. 23.2.4. Although the design of this convex bend is not nearly as complex as the concave change, care must be taken so that, as the belt flattens out, spillage is held to a minimum or not allowed to occur.

23.2.2 Determining Conveyor Width

The capacity of a belt conveyor depends on its belt width and speed and the cross-sectional area of material on the belt.

Lump size has also an effect on belt width, as shown in Fig. 23.2.5. Quite often it is desirable to crush the material to be conveyed before it is loaded onto a belt conveyor.

The carrying capacity of a belt conveyor in tons per hour is determined as follows:

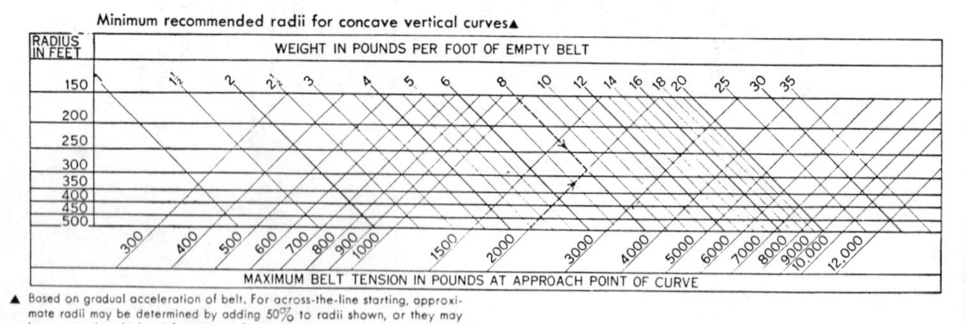

▲ Based on gradual acceleration of belt. For across-the-line starting, approximate radii may be determined by adding 50% to radii shown, or they may be accurately calculated from inertia forces.

Fig. 23.2.3 Minimum recommended radii for concave vertical curves.

Table 23.2.3 Data for Plotting a Concave Curve

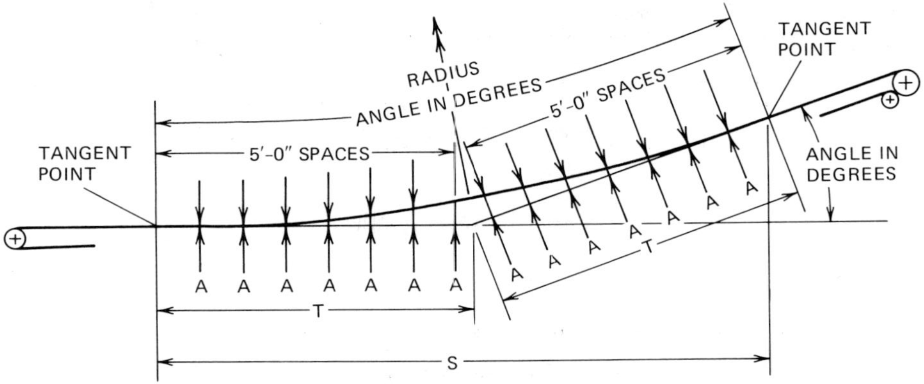

Radius of Concave Curve (ft)	Angle (deg) 5	6	7	8	9	10	11	12	13	14	15	16	18	20
	S, Base Length of Curve (ft and in.)													
150	13– 0$\frac{7}{8}$	15– 8$\frac{1}{8}$	18– 3$\frac{3}{8}$	20–10$\frac{1}{2}$	23– 5$\frac{9}{16}$	26–0$\frac{9}{16}$	28– 7$\frac{7}{16}$	31– 2$\frac{1}{4}$	33– 8$\frac{15}{16}$	36– 3$\frac{1}{2}$	38–9$\frac{7}{8}$	41– 4$\frac{1}{8}$	46–4$\frac{1}{4}$	51– 3$\frac{5}{8}$
200	17– 5$\frac{3}{16}$	20–10$\frac{7}{8}$	24– 4$\frac{1}{2}$	27–10	31– 3$\frac{7}{16}$	34–8$\frac{3}{4}$	38– 1$\frac{15}{16}$	41– 7	44–11$\frac{7}{8}$	48– 4$\frac{5}{8}$	51–9$\frac{3}{16}$	55– 1$\frac{9}{16}$	61–9$\frac{5}{8}$	68– 4$\frac{7}{8}$
250	21– 9$\frac{1}{2}$	26– 1$\frac{9}{16}$	30– 5$\frac{5}{8}$	34– 9$\frac{1}{2}$	39– 1$\frac{5}{16}$	43–4$\frac{15}{16}$	47– 8$\frac{7}{16}$	51–11$\frac{3}{4}$	56– 2$\frac{7}{8}$	60– 5$\frac{3}{4}$	64–8$\frac{1}{2}$	68–10$\frac{15}{16}$	77–3$\frac{1}{16}$	85– 6$\frac{1}{16}$
300	26– 1$\frac{3}{4}$	31– 4$\frac{5}{16}$	36– 6$\frac{3}{4}$	41– 9	46–11$\frac{1}{8}$	52–1$\frac{1}{8}$	57– 2$\frac{15}{16}$	62– 4$\frac{1}{2}$	67– 5$\frac{13}{16}$	72– 6$\frac{15}{16}$	77–7$\frac{3}{4}$	82– 8$\frac{3}{16}$	92–8$\frac{1}{2}$	102– 7$\frac{1}{4}$
350	30– 6$\frac{1}{16}$	36– 7	42– 7$\frac{7}{8}$	48– 8$\frac{1}{2}$	54– 9	60–9$\frac{5}{16}$	66– 9$\frac{3}{8}$	72– 9$\frac{1}{4}$	78– 8$\frac{13}{16}$	84– 8$\frac{1}{16}$	90–7$\frac{1}{16}$	96– 5$\frac{11}{16}$	108–1$\frac{7}{8}$	119– 8$\frac{1}{2}$
400	34–10$\frac{3}{8}$	41– 9$\frac{3}{4}$	48– 9	55– 8	62– 6$\frac{7}{8}$	69–5$\frac{1}{2}$	76– 3$\frac{7}{8}$	83– 2	89–11$\frac{3}{4}$	96– 9$\frac{3}{16}$	103–6$\frac{5}{16}$	110– 3$\frac{1}{16}$	123–7$\frac{5}{16}$	136– 9$\frac{11}{16}$
450	39– 2$\frac{11}{16}$	47– 0$\frac{7}{16}$	54–10$\frac{1}{8}$	62– 7$\frac{1}{2}$	70– 4$\frac{3}{4}$	78–1$\frac{11}{16}$	85–10$\frac{3}{8}$	93– 6$\frac{11}{16}$	101– 2$\frac{3}{4}$	108–10$\frac{3}{8}$	116–5$\frac{5}{8}$	124– 0$\frac{7}{16}$	139–0$\frac{11}{16}$	153–10$\frac{15}{16}$
500	43– 6$\frac{15}{16}$	52– 3$\frac{3}{16}$	60–11$\frac{1}{4}$	69– 7	78– 2$\frac{9}{16}$	86–9$\frac{7}{8}$	95– 4$\frac{7}{8}$	103–11$\frac{1}{2}$	112– 5$\frac{11}{16}$	120–11$\frac{1}{2}$	129–4$\frac{15}{16}$	137– 9$\frac{13}{16}$	154–6$\frac{1}{8}$	171– 0$\frac{1}{8}$

Table 23.2.3 *(continued)*

T, Distance from Tangent Point to Intersection (ft and in.)

Radius of Concave Curve (ft)	Angle (deg) 5	6	7	8	9	10	11	12	13	14	15	16	18	20
150	6– 6 9/16	7–10 3/8	9–2 1/16	10– 5 7/8	11–9 11/16	13– 1 1/2	14– 5 5/16	15–9 3/16	17– 1 1/16	18– 5	19– 9	21–1	23–9 1/16	26– 5 3/8
200	8– 8 3/4	10– 5 3/4	12–2 3/4	13–11 13/16	15–8 7/8	17– 6	19– 3 1/8	21–0 1/4	22– 9 7/16	24– 6 11/16	26– 3 15/16	28–1 5/16	31–8 1/8	35– 3 3/16
250	10–11	13– 1 1/4	15–3 1/2	17– 5 3/4	19–8 1/8	21–10 1/2	24– 0 7/8	26–3 5/16	28– 5 13/16	30– 8 5/16	32–10 15/16	35–1 5/8	39–7 1/8	44– 1
300	13– 1 3/16	15– 8 11/16	18–4 3/16	20–11 3/4	23–7 5/16	26– 2 15/16	28–10 5/8	31–6 3/8	34– 2 1/8	36–10	39– 5 15/16	42–1 15/16	47–6 3/16	52–10 3/4
350	15– 3 3/8	18– 4 1/8	21–4 7/8	24– 5 11/16	27–6 9/16	30– 7 7/16	33– 8 7/16	36–9 7/16	39–10 1/2	42–11 11/16	46– 0 15/16	49–2 1/4	55–5 3/16	61– 8 9/16
400	17– 5 9/16	20–11 9/16	24–5 9/16	27–11 11/16	31–5 3/4	34–11 15/16	38– 6 3/16	42–0 1/2	45– 6 7/8	49– 1 3/8	52– 7 15/16	56–2 9/16	63–4 1/4	70– 6 3/8
450	19– 7 3/4	23– 7	27–6 1/4	31– 5 5/8	35–5	39– 4 7/16	43– 3 15/16	47–3 9/16	51– 3 1/4	55– 3	59– 2 15/16	63–2 15/16	71–3 1/4	79– 4 3/16
500	21–10	26– 2 1/2	30–6 15/16	34–11 9/16	39–4 3/16	43– 8 15/16	48– 1 3/4	52–6 5/8	56–11 9/16	61– 4 11/16	65– 9 15/16	70–3 1/4	79–2 1/4	88– 2

A, Length of Ordinate (ft and in.)

Radius of Concave Curve (ft)	Distance from Tangent Point (ft) 5	10	15	20	25	30	35	40	45	50	55	60	65	70	75	80	85	90
150	0–1	0–4	0–9	1–4 1/16	2– 1 3/16	3– 0 3/8	4–1 11/16	5–5 3/16										
200	0–0 3/4	0–3	0–6 3/4	1–0 1/16	1– 6 13/16	2– 3 1/8	3–1	4–0 1/2	5– 1 9/16	6–4 3/16								
250	0–0 5/8	0–2 7/16	0–5 7/16	0–9 5/8	1– 3 1/16	1– 9 11/16	2–5 9/16	3–2 11/16	4– 1 1/16	5–0 3/8	6–2	7–3 11/16	8–7 9/16					
300	0–0 1/2	0–2	0–4 9/16	0–8	1– 0 1/2	1– 6 1/16	2–0 1/2	2–8 3/16	3– 4 11/16	4–2 3/8	5–1	6–0 3/4	7–1 1/2	8– 3 3/8	9–6 5/8	10–10 3/8		
350	0–0 7/16	0–1 11/16	0–3 7/8	0–6 7/8	0–10 3/8	1– 3 7/16	1–9 1/16	2–3 1/2	2–10 11/16	3–7 1/16	4–4 3/16	5–2 5/16	6–0 15/16	7– 0 7/8	8–1 9/16	9–3 3/16	10–5 13/16	11–9 1/4
400	0–0 7/16	0–1 1/2	0–3 7/16	0–6	0– 9 7/16	1– 1 1/2	1–6 7/16	2–0 1/16	2– 5 3/16	3–1 5/8	3–9 9/16	4–6 5/16	5–3 3/4	6– 2 1/16	7–1 1/4	8–1	9–1 9/16	10–3 1/16
450	0–0 3/8	0–1 3/8	0–3	0–5 3/8	0– 8 5/16	1– 0 1/16	1–4 1/4	1–9 7/16	2– 2 15/16	2–9 7/16	3–4 7/16	4–0 1/4	4–8 3/4	5– 5 3/4	6–3 5/8	7–2 1/8	8–1 1/16	9–1 1/16
500	0–0 5/16	0–1 1/4	0–2 11/16	0–4 13/16	0– 7 1/2	0–10 3/4	1–2 3/4	1–7 3/16	2– 0 7/16	2–6 1/8	3–0 3/8	3–7 3/8	4–3 5/16	4–11 3/16	5–7 7/8	6–5 3/16	7–3 5/16	8–2 1/8

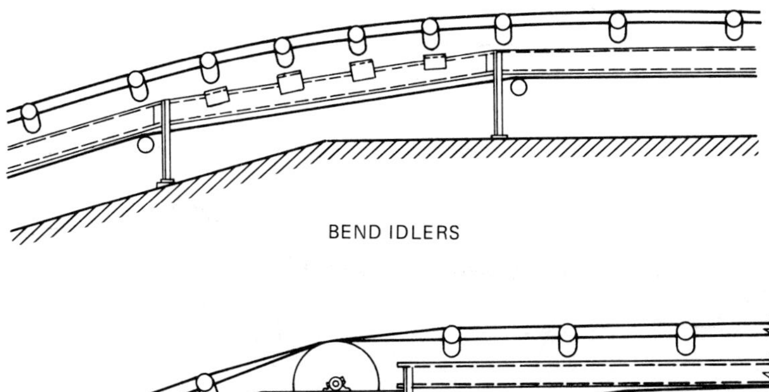

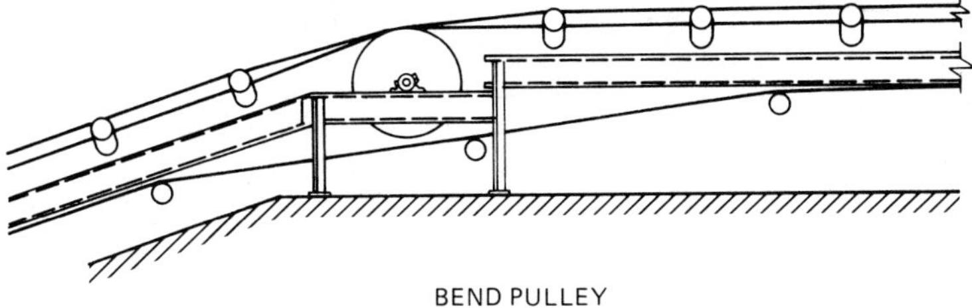

Fig. 23.2.4 Convex bends.

$$\text{TPH} = \frac{A \times BD \times S}{4800}$$

where: A = cross-sectional area, in in.^{2}
BD = bulk density, in lb/ft^3
S = belt speed, in ft/min

Figure 23.2.6 shows a typical cross section of a troughed belt conveyor. In North America, the standard troughing angles are 0°, 20°, 35°, and 45°.

The angle of surcharge is often called the dynamic angle of repose. As the conveyor belt passes successively over each carrying idler, the material is correspondingly agitated. This agitation tends to work the larger pieces to the surface of the load and tends to flatten the surface slope. This explains why the angle of surcharge is less than the static angle of repose.

Table 23.2.4 shows a general relationship between flowability, angle of surcharge, and angle of repose.

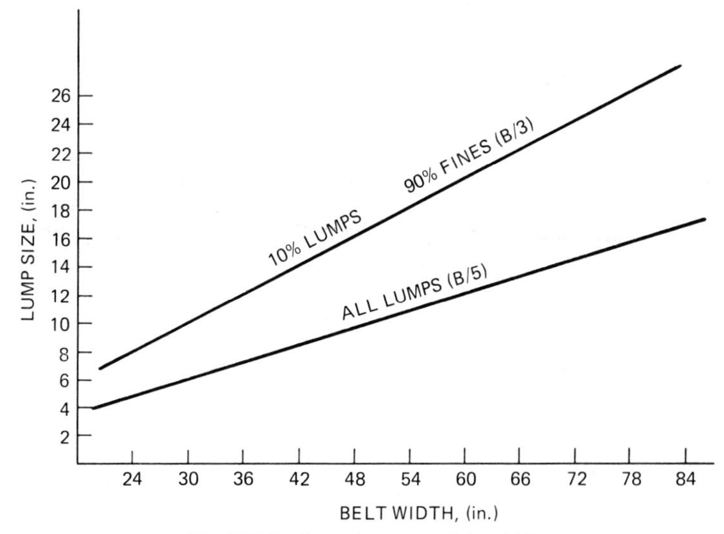

Fig. 23.2.5 Lum size versus belt width.

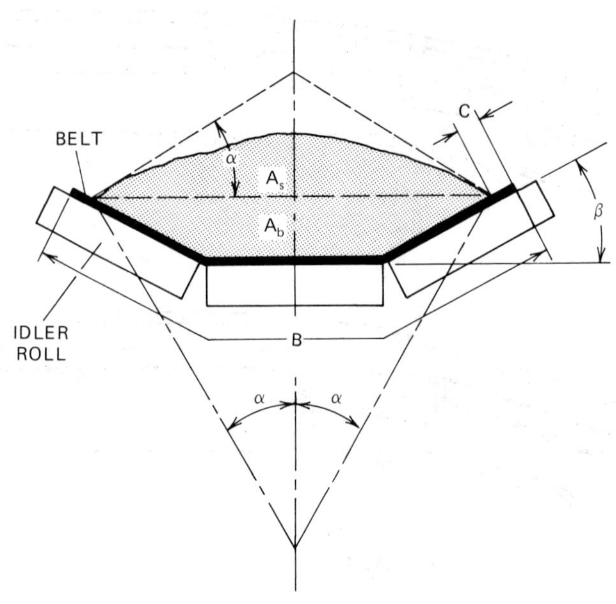

Fig. 23.2.6 Cross-sectional area of conveyor belt. B = belt width; A_s = area of surcharge; A_b = area of base trapezoid; α = angle of surcharge; β = troughing angle; C = free edge distance (= $0.055B + 0.9$ in.).

Table 23.2.4 Relationship between Flowability, Angle of Surcharge, and Angle of Repose

Very Free Flowing, 5° Angle of Surcharge	Free Flowing, 10° Angle of Surcharge	Average Flowing		Sluggish, 30° Angle of Surcharge
		20° Angle of Surcharge	25° Angle of Surcharge	
5°	10°	20°	25°	30°
0–19° Angle of Repose	20–29° Angle of Repose	30–34° Angle of Repose	35–39° Angle of Repose	40 and Up Angle of Repose
		Material Characteristics		
Uniform size, very small rounded particle, either very wet or very dry, such as dry silica sand, cement, wet concrete.	Rounded, dry polished particles, of medium weight, such as whole grain and beans.	Irregular, granular, or lumpy materials of medium weight, such as anthracite coal, cottonseed meal, clay.	Typical common materials such as bituminous coal, stone, most ores.	Irregular, stringy, fibrous, interlocking material, such as wood chips, bagasse, tempered foundry sand.

Table 23.2.5 Maximum Conveyor Belt Speeds

Material Being Conveyed	Belt Speeds (ft/min)	Belt Width (in.)
Grain or other free-flowing, nonabrasive material	500	18
	700	24–30
	800	36–42
	1000	48–96
Coal, damp clay, soft ores, overburden and earth, fine-crushed stone	400	18
	600	24–36
	800	42–60
	1000	72–96
Heavy, hard, sharp-edged ore, coarse-crushed stone	350	18
	500	24–36
	600	Over 36
Foundry sand, prepared or damp; shakeout sand with small cores, with or without small castings (not hot enough to harm belting)	350	Any width
Prepared foundry sand and similar damp (or dry abrasive) materials discharged from belt by rubber-edged plows	200	Any width
Nonabrasive materials discharged from belt by means of plows	200, except for wood pulp, where 300–400 is preferable	Any width
Feeder belts, flat or troughed, for feeding fine, nonabrasive, or mildly abrasive materials from hoppers and bins	50–100	Any width

23.2.3 Determining Conveyor Speed

Certain material such as coal, coke, and other friable substances, when transferred from one conveyor to another, suffer breakage and deterioration of the lumps. Other materials such as large lump ore, rock, slag, or other abrasive substances, when transferred from one conveyor to another, may cause severe damage to the belt. Consequently, certain maximum speeds are recommended by the industry for the handling of various types of bulk materials.

Table 23.2.5 gives the maximum speeds that are considered good practice with various widths of belts handling several kinds of materials.

23.2.4 Determining Conveyor Power

Any belt conveyor horsepower equation is a summation of four power components:

1. Power to move the belt system empty (i.e., overcome empty friction)
2. Power to move the loaded belt (i.e., overcome loaded friction)
3. Power to raise or lower the load
4. Friction from ancillary equipment, such as skirts, scrapers, trippers

All these equations include friction factors for idler rotation and belt/load flexing. There is some difference of opinion as to what these values should be, whether they are constant or a function of belt speed and/or belt loading and/or belt length, and/or belt sag.

CEMA has published a detailed design manual for belt conveyors, *Belt Conveyors for Bulk Materials,* and the publication is recommended for use by everybody involved in belt conveyor design. Since power requirements for belt conveyors are dependent on many variables related to conveyor profile, type of drive pulley arrangements, belt tensions and belt speed, type of idlers, and idler spacing, any detailed discussion of this subject is referred to the CEMA publication.

The CEMA power equation is as follows:

$$\text{Belt Hp} = L\,K_t(K_x + K_y W_b + 0.015\,W_b) + K_y L W_m + H W_m \frac{S}{33{,}000}$$

where: L = length, in ft
W_b = weight of belt, in pounds per foot of belt length
W_m = weight of material carried, in pounds per foot of belt
H = elevation, in ft
S = belt speed, in ft/min
K_t = temperature factor (dimensionless)
K_x = resistance factor for rotating idlers (lb/ft)
K_y = resistance factor for moving belt and load (dimensionless)

Various charts and tables have been published by CEMA, in *Belt Conveyors for Bulk Materials*, for values of the factors K_t, K_x, and K_y.

Because it is often desirable for a field engineer to make some "slide-rule" calculations for quick approximation of belt tensions, the Goodyear equations for computing belt tensions are commonly used:

$$\text{Horsepower} = \frac{T_E \times S}{33{,}000} + \text{Accessories}$$

$$T_E = C(L + L_0)\left(Q + \frac{100T}{3S}\right) \pm \frac{100T}{3S}$$

where: T_E = effective tension or belt pull at drive pulley, in lb
C = composite friction factor (see Table 23.2.6)
L = belt conveyor length, in ft (projected length between pulley counters)
L_0 = equivalent length, in ft (see Table 23.2.6)
Q = weight factor in pounds per linear foot of conveyor, representing the weight of moving parts of belt conveyor (see Table 23.2.7)
T = capacity, in tons/hr (2000 lb/hr)
S = belt speed, in ft/min
H = elevation, in ft
$\frac{100T}{3S}$ = weight of material carried on the belt, in lb/ft

In metric: T in tonne/hr, Q in kg/m, H and L in m, S in m/sec, and T_E in kg. The Goodyear equation becomes:

$$T_E = C(L + L_0)\left(Q + \frac{T}{3.6S}\right) \pm \frac{T}{3.6S} H$$

$$\text{N (Hp)} = \frac{T_E \times S}{75} + \text{Accessories}$$

Table 23.2.6 Friction and Length Factors[a]

Class of Conveyor	Friction Factor, C	Length Factor, L_0 (ft)
For conveyors with permanent or other well-aligned structures and with normal maintenance.	0.022	200
For temporary, portable, or poorly aligned conveyors. Also for conveyors in extreme cold weather that are either subject to frequent stops and starts or are operating for extended periods at −40°F or below.	0.03	150
For conveyors requiring restraint of the belt when loaded.	0.012	475

[a] The C and L_0 factors have proven to be satisfactory for the great majority of conveyor belt tension and horsepower calculations. However, when long, relatively level, heavily loaded conveyors are encountered where power requirements are large and made up primarily of friction, it is recommended that Goodyear (Akron) be consulted for additional engineering assistance in selecting these factors.

Table 23.2.7 Weight Factor for Moving Parts

Width (in.)	Light-Service Material (to 50 lb/ft³)		Medium-Service Material (50–100 lb/ft³)		Heavy-Service Material (over 100 lb/ft³)	
	B_w[a]	Q[a]	B_w	Q	B_w	Q
14	1	7	2	13	3	19
16	2	8	3	14	4	21
18	3	9	4	16	5	23
20	4	10	5	18	6	25
24	5	14	6	21	7	29
30	6	19	7	28	8	38
36	7	26	9	38	11	52
42	9	33	11	50	14	66
48	12	40	15	60	18	82
54	14	50	18	71	22	97
60	17	62	21	85	27	115
66	20	75	24	103	32	135
72	22	88	28	121	36	155

[a] B_w = weight of belting; Q = weight factor.

23.2.5 Conveyor Pulley Arrangements

In any belt drive, whether it is for a transmission, or a conveyor, or an elevator, there exists a difference of tension in the belt on the two sides of the drive pulley. The larger tension is called the "tight side" (T_1) and the smaller is called the "slack side" tension (T_2). Without slack side tension to prevent slipping, the belt cannot be driven. The difference between the tight side and slack side tension is known as the effective tension (T_E); this is the tension which actually does the work (see Fig. 23.2.7).

Equations have been given to calculate the total belt horsepower. From this value the effective tension may be calculated as follows:

$$T_E = \frac{\text{Total belt Hp} \times 33{,}000}{\text{Belt speed}}$$

Similarly, the effective tension may be calculated from watt-meter power readings of actual installations. In this case motor and drive losses must be deducted from the electrical input to the motor, leaving the power absorbed by the belt. Substituting this latter value in the foregoing equation, with the proper value of belt speed S, also provides the effective tension. It should be noted that drive and motor losses must be added to the electrical output of the motor (generator) to obtain the belt horsepower in the case of decline belts requiring restraint.

Usually the slack side tension is obtained by a counterweight (Fig. 23.2.8) or by a screw-type (Fig. 23.2.9) take-up. The former is preferable, because it maintains a constant tension automatically, and may be set at the lowest tension at which the conveyor can be driven. This type maintains a constant tension under all conditions of load, starting, stretch, and so on.

The amount of slack side tension necessary is determined by multiplying the effective tension by the drive factor C_w. Values of C_w depend on the arc of contact between belt and drive pulley or pulleys, type of take-up, and whether drive pulleys are bare or lagged.

For a certain type of belt drive, the ratio T_1/T_2 is constant and is governed by the friction coefficient existing between belt and the drive pulley and the arc of contact.

$$\frac{T_1}{T_2} = e^{f\theta} = \frac{1 + C_w}{C_w}$$

where: e = natural logarithm
f = coefficient of friction
θ = angle of wrap around pulley, in radians (1° = 0.0174 radian)
C_w = wrap factor = $1/(e^{f\theta - 1})$

Table 23.2.8 shows some of the f values for bare and lagged pulleys. Also:

$$T_1 - T_2 = T_E$$

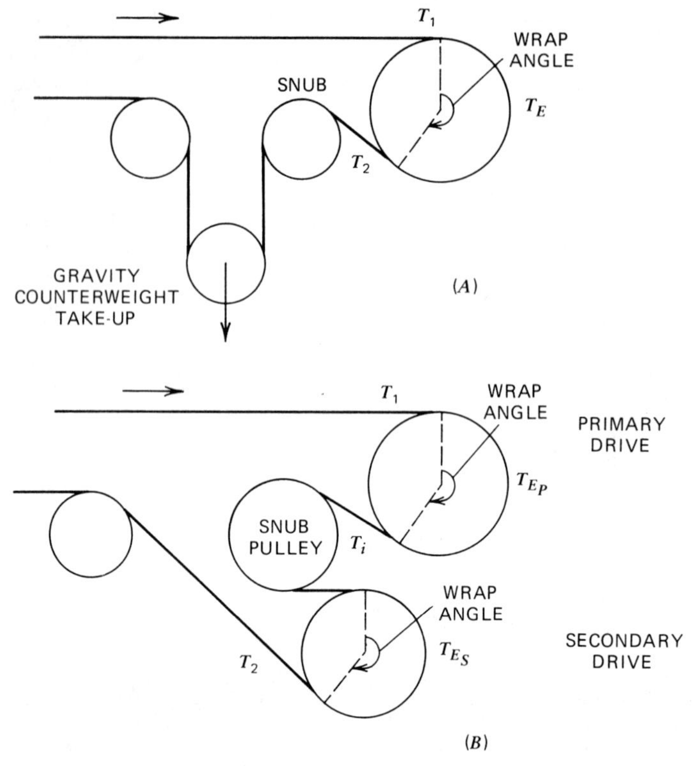

Fig. 23.2.7 Drive pulley arrangements. (*A*), single-head drive; (*B*), dual-head drive.

Therefore $T_2/T_E = C_w$, or slack side tension $T_2 = C_w T_E$. Table 23.2.9 shows values for C_w, using $f = 0.25$ for bare pulleys, and $f = 0.35$ for lagged pulleys.

When a screw take-up is used, the wrap factor is usually increased in order to guarantee sufficient slack side tension T_2, even after the belt stretches. A gravity take-up compensates automatically for belt stretch.

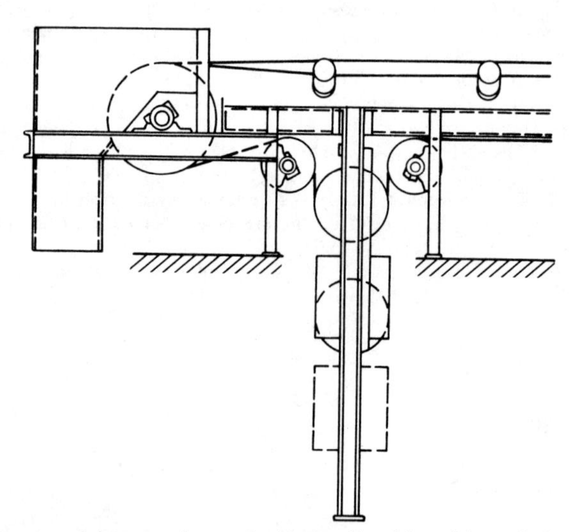

Fig. 23.2.8 Counterweight-type take-up. Snubbed pulley drive with vertical gravity take-up.

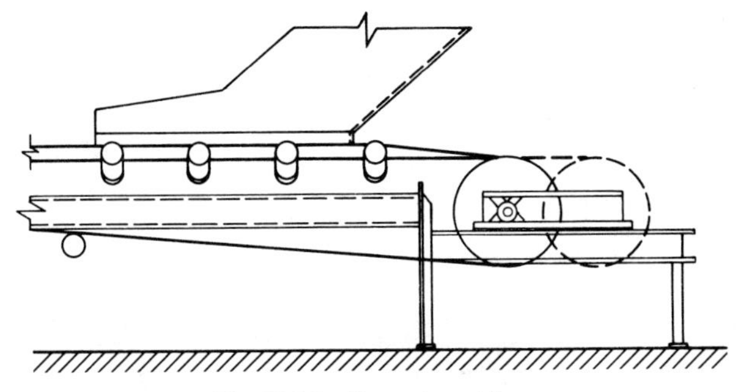

Fig. 23.2.9 Screw-type take-up.

Table 23.2.8 Friction Factors for Bare and Lagged Pulleys

	Friction Coefficient, f	
Belt Condition	Bare Pulley	Lagged Pulley
Dry	0.5	0.6
Lightly wet	0.2	0.4
Wet	0.1	0.4
Wet and dirty	0.05	0.2

23.2.6 Conveyor Belt Tension

Operating Maximum Belt Tension

The operating maximum belt tension is defined as the maximum belt tension occurring when the belt is conveying the design load from the loading point continuously to the point of discharge. Operating maximum tension usually occurs at the discharge point on horizontal or inclined conveyors and at the loading point on regenerative declined conveyors. On compound conveyors, the operating maximum belt tension frequently occurs elsewhere. Because the operating maximum belt tension must be known to select a belt, its location and magnitude must be determined.

Conveyors having horizontal and lowering, or horizontal and elevating, sections can have maximum tensions at points other than a terminal pulley. In this case, belt tensions can be calculated by considering the horizontal and sloping sections as separate conveyors.

Table 23.2.9 Wrap Factors for Bare and Lagged Pulleys

		Wrap Factor, C_w			
		Gravity Take-Up		Screw Take-Up	
Type of Drive	Arc of Contact Wrap, θ	Bare Pulley	Lagged Pulley	Bare Pulley	Lagged Pulley
Plain	180°	0.84	0.50	1.2	0.8
Snubbed	200°	0.72	0.42	1.0	0.7
	210°	0.66	0.38	1.0	0.7
	220°	0.62	0.35	0.9	0.6
	240°	0.54	0.30	0.8	0.6
Dual or tandem	380°	0.23	0.11	0.5	0.3
	420°	0.18	0.08	—	—

Temporary Operating Maximum Belt Tension

A temporary operating maximum belt tension is that maximum tension that occurs only for short periods. For example, a conveyor with a profile that contains an incline, a decline, and then another incline, may generate a higher operating tension when only the inclines are loaded and the decline is empty. These temporary operating maximum belt tensions should be considered in the selection of the belt and the conveyor machinery.

Starting and Stopping Maximum Tension

The starting torque of an electric motor may be more than $2\frac{1}{2}$ times the motor full-load rating. Such a torque transmitted to a conveyor belt could result in starting tensions many times more than the chosen operating tension. To prevent progressive weakening of splices and subsequent failure, such starting maximum tensions should be avoided. Likewise, if the belt is brought to rest very rapidly, especially on decline conveyors, the inertia of the loaded belt may produce high tensions.

The generally recommended maximum for starting belt tension is 150% of the allowable belt working tension. On conveyors with tensions under 75 lb per ply inch or the equivalent, the maximum can be increased to as high as 180%. For final design allowances, conveyor equipment or rubber belt manufacturers should be consulted.

23.2.7 Belt Selection

Practically all belt conveyors for bulk materials use rubber-covered conveyor belts, made of a woven carcass (Fig. 23.2.10) having strength enough to pull and support the load, and protected from damage by rubber covers which vary in thickness for different applications:

1. *Conventional Belting.* This belting has plies of fabric made of cotton, cotton-nylon, rayon, rayon-nylon, and others. The plies are impregnated with rubber and are separated by a skim coat of rubber for added flex life. Tension ratings vary from 140 to 500 pounds per inch of belt width (piw) for cotton-nylon combinations and up to 1500 piw for rayon-nylon combinations.
2. *Steel Cable Belts.* This belting, made up of spaced steel cables suspended in rubber and wrapped in a fabric envelope, is used where very high strengths and minimum stretch are required. Ratings up to 6000 piw are available.
3. *Heat Service Belts.* Special belts are available for jobs where hot materials must be handled. They must retain their physical properties at temperatures up to 250°F and resist abrasion by the conveyed material. These belts utilize carcasses of nylon, polyester, cotton, nylon, or glass. Covers are usually butyl, chloro-butyl, or EPDM (ethylene-propylene-dipolymer).

Applications for belt conveyors range from a few pounds per minute to thousands of tons per hour, and a great variety of materials can be handled.

Actual belt selection is dependent on an analysis of a variety of factors:

Required belt tension requirements
Length and speed of conveyor
Abrasiveness of material handled
Size of lumps and their tendency to cut or tear the cover
Characteristic of material being handled (i.e., not oily, acid, wet)
Method of loading conveyor
Type of take-up

The ability of the loaded belt to trough properly on the carrying idlers is called troughability. It is important that conveyor belting not be too stiff or too thin. Consequently, belting width, thickness, and construction must be carefully analyzed as per Fig. 23.2.11 to enable the belt to trough properly.

Other factors that must not be overlooked in the selection of conveyor belting are the amount and control of stretch. Change in belt length due to climatic conditions, especially under severe conditions, and stretch due to tensions, especially on long heavily loaded conveyors, must be carefully analyzed because it can influence drive and take-up selection and location. Another important factor to be considered is belt splice. It must always be remembered that the strength of the splice, which holds the belt together to make it endless, determines its strength.

Conveyor belts can be fastened together by means of metal splices or can be field vulcanized. The latter is always preferable but cost and accessibility often limit justification of its use. Therefore, on shorter, low-tension conveyor belts metal fasteners are commonly used.

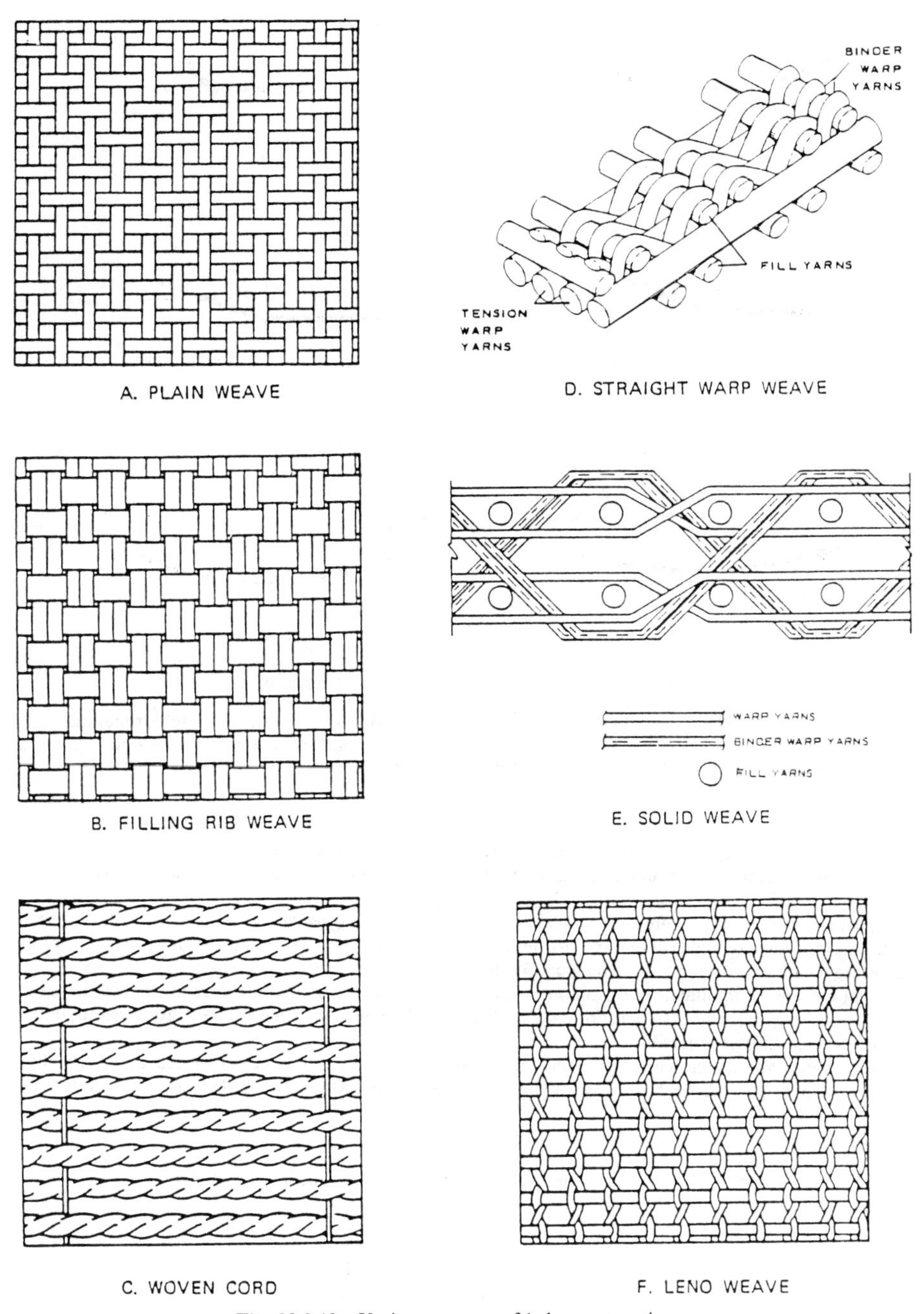

Fig. 23.2.10 Various weaves of belt construction.

Due to the variety of factors involved in conveyor belting and the broad range of belt types available, it is recommended that various manufacturers' literature be carefully studied before selection.

23.2.8 Pulley Diameter

The selection of the proper diameter of pulley is necessary to prevent separation of the belt plies and/or excessive stress on the felt fabric plies as they flex around the pulley. Many factors are involved

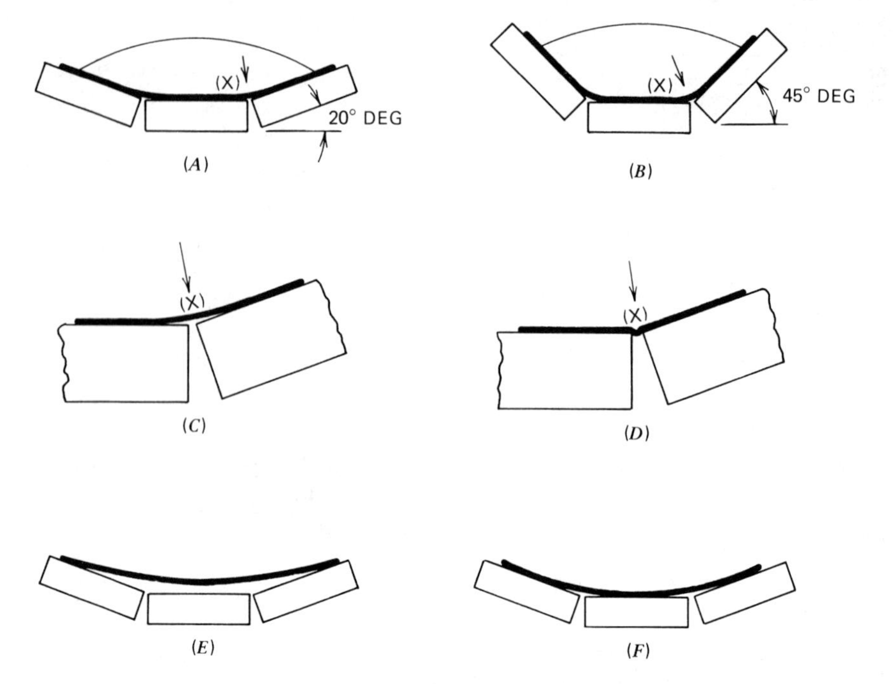

Fig. 23.2.11 Belt conveyor troughability. (*A*), idler troughing angle 20°; (*B*), idler troughing angle 45°; (*C*), belt design unaffected by load weight; (*D*), belt design affected by load weight; (*E*), stiff belt, improper troughing; (*F*), flexible belt, proper troughing.

in the selection of the proper pulley diameter such as the amount of wrap, belt tension at the pulley, space, characteristic of the material handled, belt life expectancy, shaft and bearing size, and size and ratio of reducer.

Large pulleys require more space, greater torque, and larger speed reducers. However, the conveyor belting constitutes a high percentage of the cost of a belt conveyor. Prolonging the belt's life by use of a larger-diameter pulley whenever possible more often than not is a wise choice.

Belt conveyor manufacturers' handbooks indicate recommendations for minimum pulley diameters as, for example, in Table 23.2.10. For each application, the data provided by the belting manufacturer should be carefully examined.

The pulley should be wider than the belting. Widths of standard pulleys exceed belt width by 2 in. for belting up to 42 in. However, for belting width above 60 in., the pulleys are 4 in. wider than

Table 23.2.10 Minimum Pulley Diameters for Reduced-Ply Belts[a]

	Minimum Pulley Diameters (in.)		
Maximum Belt Tension	80–100% Tension	60–80% Tension	40–60% Tension
To 100 piw	14	12	12
To 150 piw	16	14	12
To 200 piw	18	16	14
To 300 piw	24	20	18
To 400 piw	30	24	20
To 500 piw	36	30	24
To 700 piw	42	36	30

[a] For multi-ply or steel cable belts refer to CEMA. In all cases obtain manufacturer's recommendations.

the belting. Special conditions, such as very long conveyors, complex terminals, or when handling sticky material, can dictate even greater dimensional differences.

Crowned or curved-face pulleys have a definite and desirable centering and grinding effect on the belt. However, there are conditions where their use may not be desirable, for example, under conditions of high stress or when there are severe reverse bends found within complex belt terminals. Snub pulleys are used to assure proper wrap around the drive pulley and to relieve return idlers of excessive loading.

Pulleys are often rubber lagged to aid in transmitting horsepower and to protect the surface of the belt as it passes around the pulley. Although lagging with certain materials can aid in keeping the pulley free of material buildup, there are specially designed self-cleaning pulleys made for this purpose.

23.2.9 Shaft Diameter

Shafting to be used with belt conveyor pulleys cannot be selected independently from the pulley since the shaft and pulley must be treated as a single structure.

The resultant load on a pulley is the vector sum of belt tensions, pulley weights, and weight of shaft. This is graphically illustrated in Fig. 23.2.12 for a simple drive shaft arrangement subject to a combination of bending and torque.

Pulley design is generally based on the use of any commercial shafting material, such as AISI C1018 steel. Ratings are not increased when higher-strength shafting is used. High-strength shafting can be of advantage when the end of the shafts are turned down to a smaller diameter or to withstand added torsional stresses. ALTHOUGH THE USE OF HIGH-STRENGTH STEEL INCREASES THE STRENGTH OF THE STEEL, IT DOES NOT DECREASE DEFLECTION.

It is essential that the diameter of the shaft, shaft material, and bearing centers be known for design. Tentative shaft diameters can be determined from application of approximate factors from Fig. 23.2.13.

Extreme caution must be exercised to avoid excessive shaft deflection which can increase the stress and deflection in pulley end discs of the standard welded steel pulleys. A pulley made with very thick end discs has a minimum shaft deflection and therefore only the manufacturer can determine

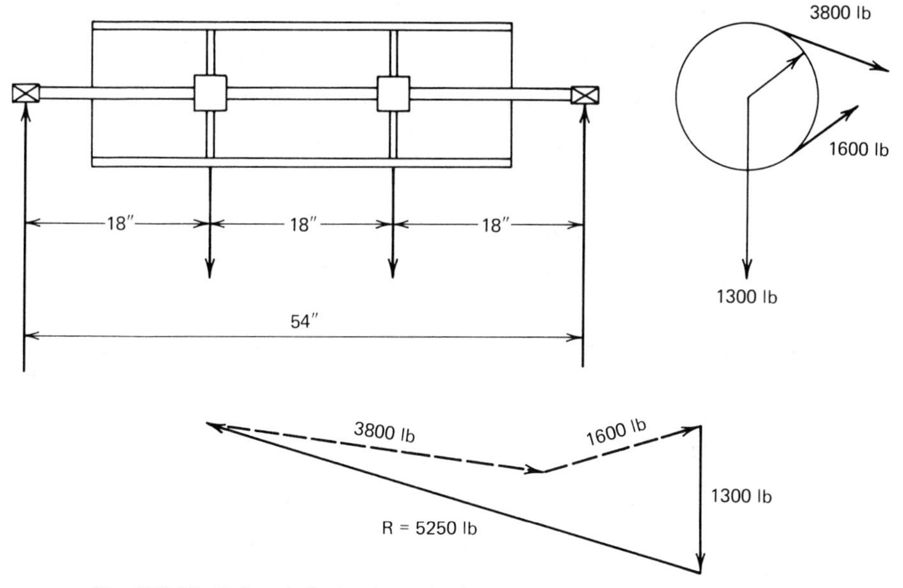

Fig. 23.2.12 Pulley shaft size determination.

$$\frac{5250}{2} = 2625$$

$$\text{Maximum Bending } M_B = 18'' \times 2625 = 47{,}250 \text{ in.-lb}$$

$$M_T = 12(3800 - 1600) = 26{,}400 \text{ in.-lb}$$

Assume 6000# stress with shaft with keyseat

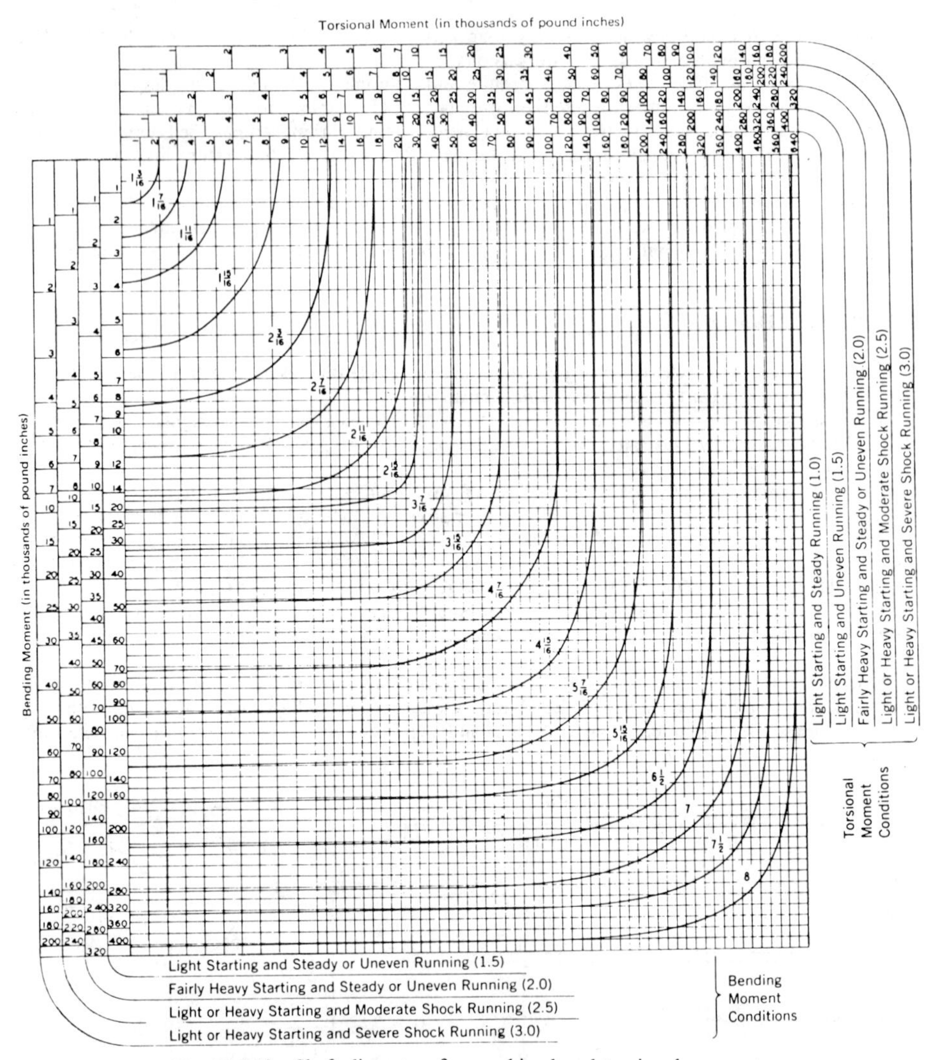

Fig. 23.2.13 Shaft diameters for combined and torsional moments.

the actual deflector. For preliminary design it is recommended that shaft deflection be limited to 0.01 in./ft of bearing centers. It is recommended that complete pulley loading information be included with all pulley orders or requests for quotations.

23.2.10 Idler Sizing

Idlers must be selected to properly protect and support the belt and load to be carried. A wide variety of available belt conveyor idlers are illustrated in Fig. 23.2.14. They are designed to incorporate various roll diameters, fitted with antifriction bearings and seals mounted on shafts (see Fig. 23.2.15).

Roll diameter and bearings and seal requirements constitute major components affecting frictional resistance which influences belt tension and horsepower requirements. Selection of the proper roll diameters and size of bearing and shaft is based on type of service, operating conditions, load carried, and belt speed. For aid in idler selection the various idler designs have been grouped into classifications (see Table 23.2.11).

Factors to consider when selecting idler spacing are belt weight, material weight, idler rating, sag, idler life, belt rating, and belt tension.

If too much sag of a loaded troughed belt is permitted between the troughing idlers, the material may spill over the edges of the belt. For the best design, and especially on long-center belt conveyors,

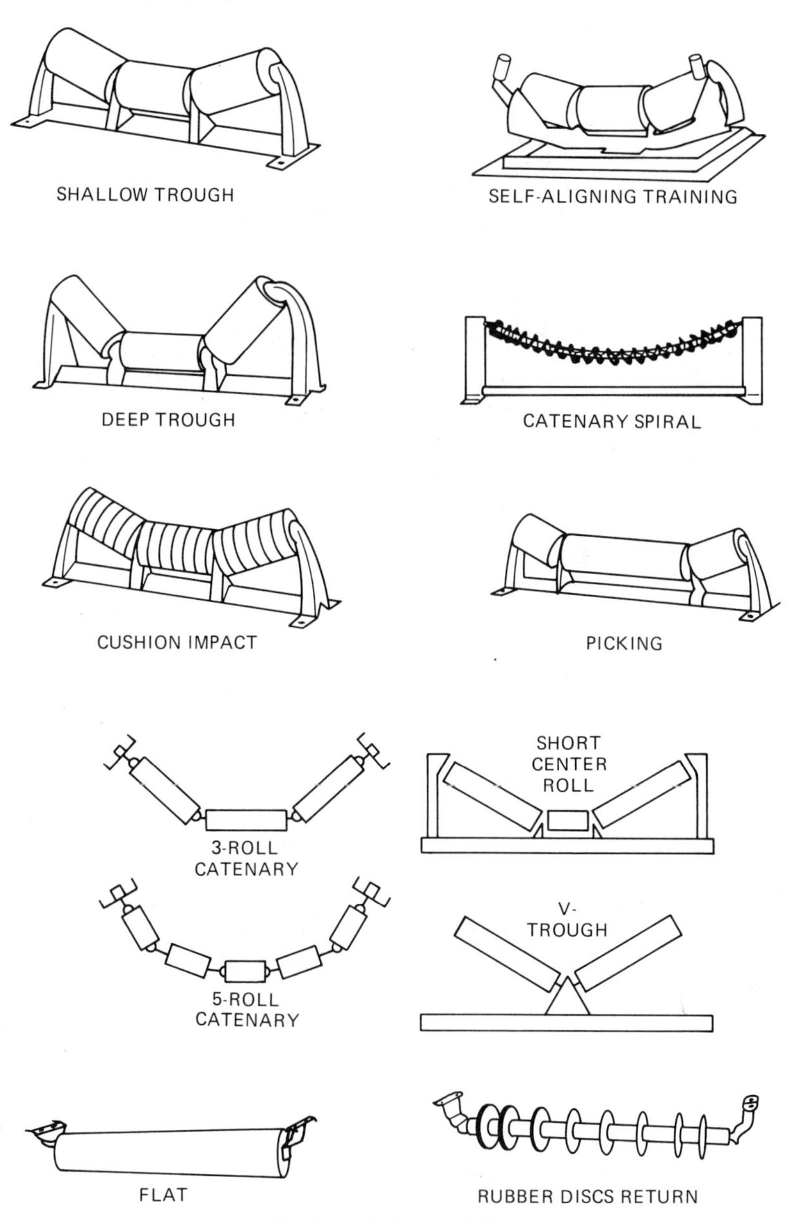

Fig. 23.2.14 Types of idlers.

the sag between idlers must be limited. Table 23.2.12 lists suggested normal troughing idler spacing for use in general engineering practice, when the amount of belt sag is not specifically limited.

Conveyor systems have been designed successfully utilizing extended idler spacing and/or graduated idler spacing. Extended idler spacing is greater-than-normal spacing and is sometimes applied where belt tension, sag, belting strength, and idler ratings permit. Advantages are lower idler cost (fewer used) and better belt training.

Graduated idler spacing is greater-than-normal spacing at high-tension portions of the belt. As the tension along the belt increases, the idler spacing is increased. Usually this type of spacing occurs toward and near the discharge end.

Extended and graduated spacing are not commonly used but if either is employed, care should be taken not to exceed idler rating and sag limits during starting and stopping.

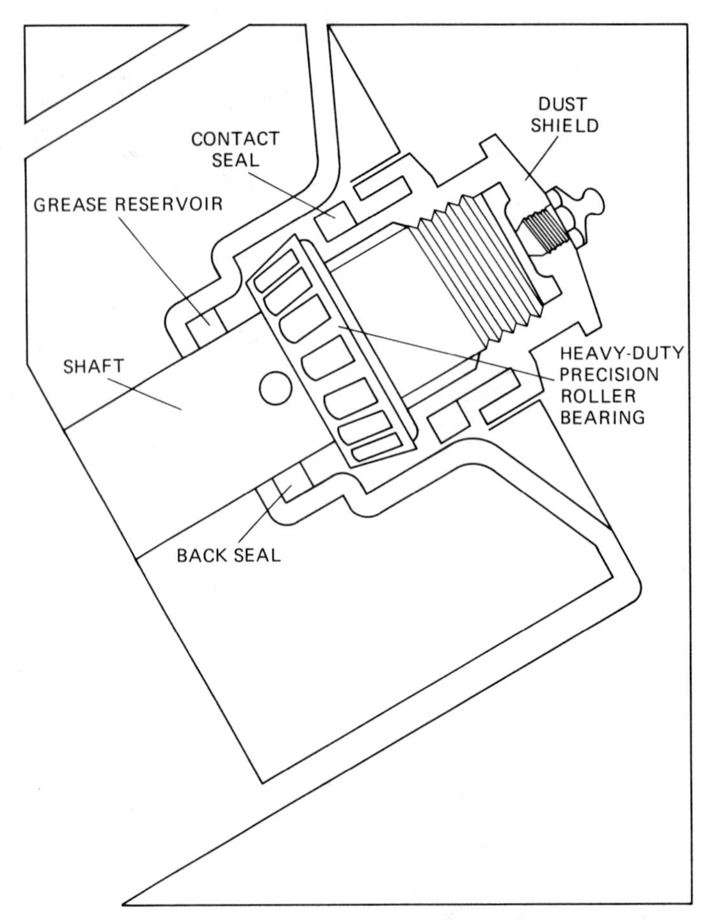

Fig. 23.2.15 Typical idler bearing assembly.

Table 23.2.11 Idler Classification

Classification	Former Series No.	Roll Diameter (in.)	Description
A4	I	4	Light duty
A5	I	5	Light duty
B4	II	4	Light duty
B5	II	5	Light duty
C4	III	4	Medium duty
C5	III	5	Medium duty
C6	IV	6	Medium duty
D5	NA	5	Medium duty
D6	NA	6	Medium duty
E6	V	6	Heavy duty
E7	VI	7	Heavy duty

Table 23.2.12 Idler Spacing[a]

Belt Width (in.)	Spacing of Troughing Idlers (ft) for Given Weight of Material Handled (lb/ft³)						Spacing of Return Idlers (ft)
	30	50	75	100	150	200	
18	5.5	5.0	5.0	5.0	4.5	4.5	10.0
24	5.0	4.5	4.5	4.0	4.0	4.0	10.0
30	5.0	4.5	4.5	4.0	4.0	4.0	10.0
36	5.0	4.5	4.0	4.0	3.5	3.5	10.0
42	4.5	4.5	4.0	3.5	3.0	3.0	10.0
48	4.5	4.0	4.0	3.5	3.0	3.0	10.0
54	4.5	4.0	3.5	3.5	3.0	3.0	10.0
60	4.0	4.0	3.5	3.0	3.0	3.0	10.0
72	4.0	3.5	3.5	3.0	2.5	2.5	8.0
84	3.5	3.5	3.0	2.5	2.5	2.0	8.0
96	3.5	3.5	3.0	2.5	2.0	2.0	8.0

[a] Spacing may be limited by load rating of idler.

At loading points, the carrying idlers should be spaced to keep the belt steady and to hold the belt in contact with the rubber edging of the loading skirts.

23.2.11 Take-Ups

Every belt conveyor should be equipped with a take-up to:

1. Allow for stretch and shrinkage of the belt.
2. Insure that the minimum tension in the belt is sufficient to prevent undue sag between idlers.
3. Insure that the tension in the belt in back of the drive pulley (or pulleys) is sufficient to permit such pulley(s) to transmit the load.

Belt stretch varies with temperature, atmospheric conditions, and tension. It is customary for the general run of conveyors to allow 1 ft of take-up for every 100 ft of conveyor length. However, where space permits, 1 ft 6 in. per 100 ft is preferable since most belts undergo an initial stretch when operated under load. In the case of belts that are spliced with metal lacing, it is a simple matter to cut out a piece after the belt has been run in and then resplice.

When one or more splices in long belts are to be vulcanized, it is well to "run in" the belt with one splice of metallic lacing until the initial stretch has occurred, the lacing may be removed and the belt stepped down for a vulcanized splice. Sufficient take-up must be allowed to hold the amount of belt ultimately required for vulcanizing, plus 1.5% for belt movement after vulcanization.

As previously mentioned, there are two types of take-ups in general use, screw type and gravity type. Screw take-ups are limited in the amount of their adjustment; the maximum in the larger stock sizes is about 36 in. They should not be used on conveyors in excess of 250 ft long. However, in special cases, screw take-ups larger than 30 in. may be obtained. The advantages of screw take-ups are cheapness and compactness. Their disadvantage is that they leave the adjustment in the hands of the operator. The operator may, through carelessness or ineptitude, neglect to tighten the take-up to prevent undue sag of the belt between idlers and possibly cause slipping of the drive pulley; or he may tighten it too much and produce unnecessary and excessive tension in the belt. A second disadvantage is that screw take-ups require manual adjustment to provide for belt stretch or shrinkage, whereas the gravity type is fully automatic. In spite of these objections, screw take-ups are generally used on short conveyors.

Screw take-ups are located at the tail end of conveyors which are driven at their head end as in Fig. 23.2.16*A*, or at the head end of conveyors with tail drive such as boom conveyors (Fig. 23.2.16*B*). Occasionally on conveyors with internal drives (Fig. 23.2.16*C*), they are placed directly behind the drive.

Gravity take-ups may be horizontal, inclined, or vertical. The amount of take-up can be set anywhere from 3 ft to 40 ft or more. With proper counterweighting, the belt tension can be regulated throughout the entire conveyor so the minimum tension is always available to prevent undue sag and the slack tension in back of the drive is always enough for the drive pulley to transmit its load.

The choice of take-up may be determined by the availability of space, operating conditions, or point of minimum belt tension. On horizontal conveyors with head drives, the minimum belt tension

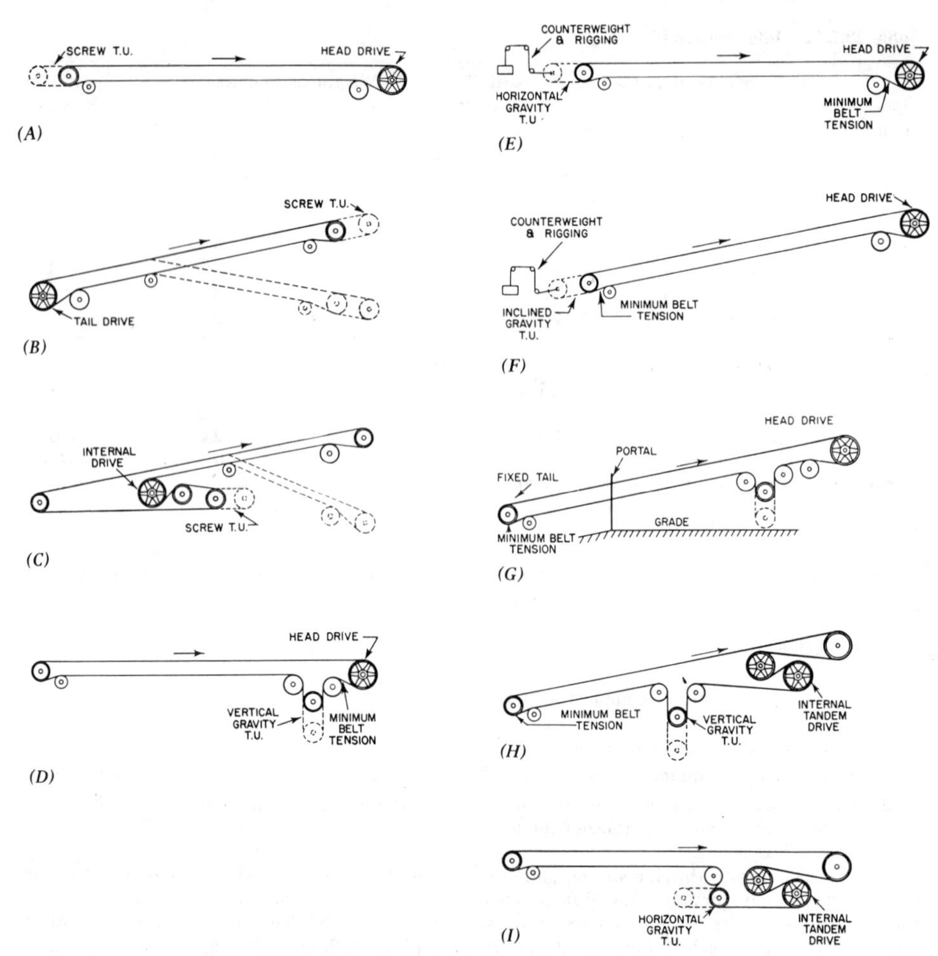

Fig. 23.2.16 Typical selections and location of take-ups.

occurs directly behind the drive, and a vertical gravity take-up may be used, as in Fig. 23.2.16*D*. However, its travel may be such that a deep pit is required which, in addition to its expense, presents drainage and cleanout problems and makes the take-up bearings inaccessible. Or it may use space in a building that could be used to advantage for other equipment. In this case, a horizontal gravity take-up at the tail end may be desirable even though more counterweight is required (Fig. 23.2.16*E;* note that this arrangement is less expensive because three fewer pulleys are required). It is also preferable from the standpoint of belt life because, as the illustration shows, 360° of belt wrap has been eliminated.

On inclined conveyors, the minimum belt tension usually occurs at the tail end and an inclined gravity tail take-up may be desirable (Fig. 23.2.16*F*). However, on some mine slope conveyors, the tail end is loaded by shuttle cars or by a mine conveyor and a tail take-up may be in the way. Also, the amount of travel required may be such that space is not available for the counterweight travel. In this case, a vertical gravity take-up behind the drive above ground may be called for (Fig. 23.2.16*G*).

Figures 23.2.16*H* and 23.2.16*I* illustrate two examples of the use of gravity take-ups used with internal tandem drives. In either of these cases, a tail take-up could be used if conditions warranted.

23.2.12 Conveyor Drives

All belt conveyor installations involve the proper application of conveyor drive equipment including speed reduction, electric motors and controls, and safety devices. The preferred drive location for a belt conveyor is that which results in the least maximum belt tension. For simple horizontal and inclined conveyors this is usually at the discharge end. For decline conveyors the preferred location is usually at the load end. Special conditions and requirements can require that the drive be located

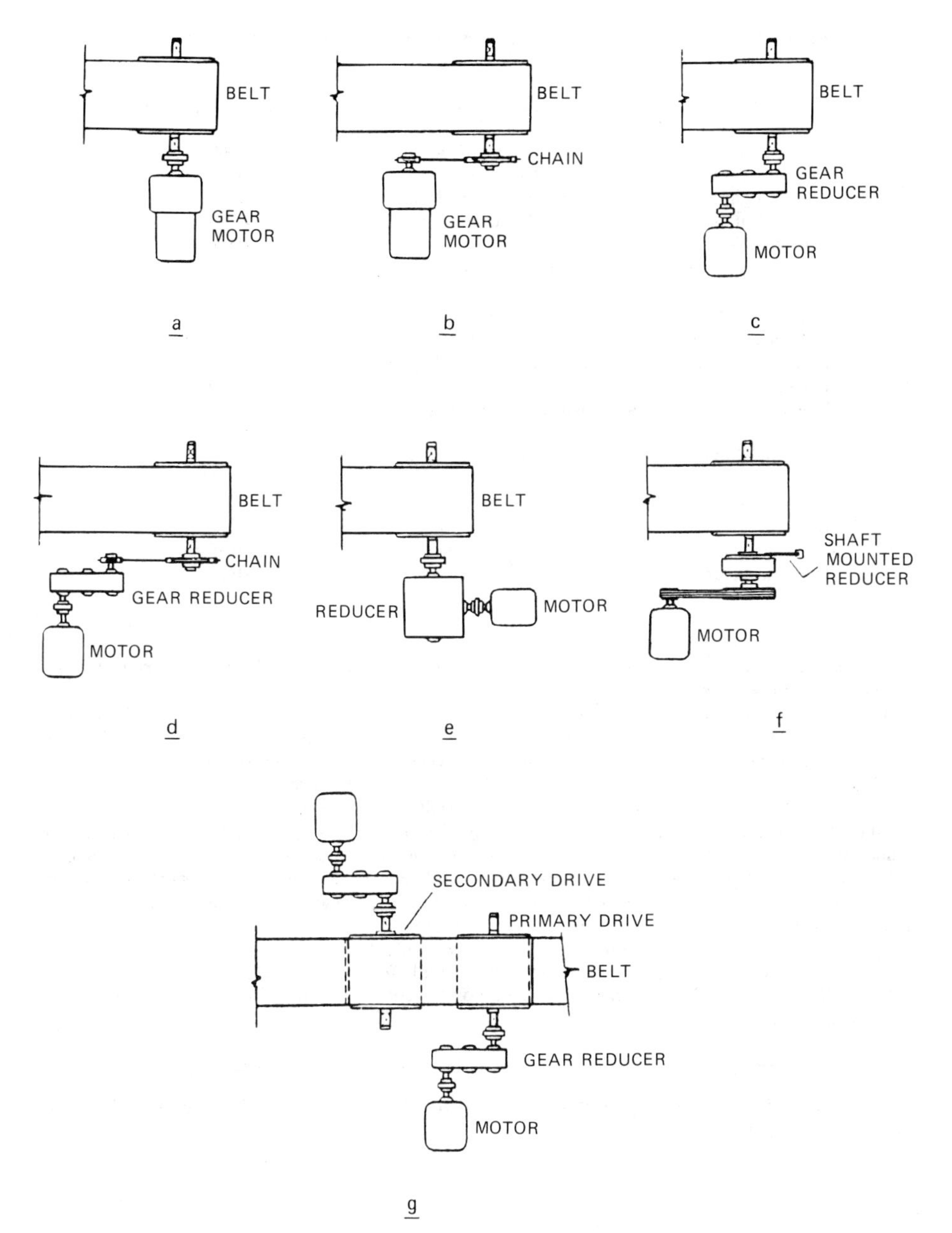

Fig. 23.2.17 Typical drive arrangements. (*A*), gearmotor directly connected by flexible coupling to drive shaft, a simple, reliable, and economical drive. (*B*), gearmotor combined with chain drive to drive shaft, one of the lowest-cost flexible arrangements and substantially reliable. (*C*), parallel-shaft speed reducer directly coupled to the motor and to drive shaft, versatile, reliable, and generally heavier in construction and easy to maintain. (*D*), parallel-shaft speed reducer coupled to motor, and with chain drive, to drive shaft, provides flexibility of location and also is suitable for the higher horsepower requirements. (*E*), spiral-bevel helical speed reducer, or worm-gear speed reducer, directly coupled to motor and to drive shaft, often desirable for space saving and simplicity of supports. The spiral-bevel reducer costs substantially more than the worm-gear speed reducer but is considerably more efficient. (*F*), drive-shaft-mounted speed reducer with V-belt reduction from motor, provides low initial cost, flexibility of location, and the possibility of some speed variation and space savings where large speed reduction ratios are not required and where horsepower requirements are not too large. (*G*), dual-pulley drive is used where power requirements are very large, and use of heavy drive equipment may be economical by reducing belt tensions.

Table 23.2.13 Efficiency Factors for Individual Drive Components

Type of Speed-Reduction Mechanism	Approximate Mechanical Efficiency
V-belts and sheaves	0.94
Roller chain and cut sprockets, open guard	0.93
Roller chain and cut sprockets, oil-tight enclosure	0.95
Single-reduction helical or herringbone gear-speed reducer or gearmotor	0.95
Double-reduction helical or herringbone gear-speed reducer or gearmotor	0.94
Triple-reduction helical or herringbone gear-speed reducer or gearmotor	0.93
Double-reduction helical gear, shaft-mounted speed reducers	0.94
Low ratio (up to 20:1 range) worm-gear speed reducers	0.90
Medium ratio (20:1 to 60:1 range) worm-gear speed reducers	0.70
High ratio (60:1 to 100:1 range) worm-gear speed reducers	0.50
Cut spur gears	0.90
Cast spur gears	0.85

elsewhere. Often internal drives are utilized on longer conveyors and inclined boom conveyors for reasons of economy, accessibility, or maintenance.

Belt conveyor drive equipment normally consists of a motor, speed reducer, drive shaft, and necessary machinery to transmit power from one item to another; the simplest arrangement using the least number of components is the best. Often however, special-purpose components must be provided to modify starting and stopping, provide for a hold-back, or vary belt speed.

Figure 23.2.17 illustrates some of the more commonly used drive equipment assemblies.

The final selection of the speed-reduction mechanism is based on preference, capital cost, power limitations, speed reduction characteristics, space, and/or drive locations. There is also available a motorized head pulley which combines the motor and reducer within the framework of a pulley. These units must be carefully selected initially because once a choice is made as to horsepower and speed any change could require an entire unit replacement.

The drive motor horsepower must reflect the division of the horsepower at the drive shaft by the overall efficiency of the reduction machinery. The overall efficiency of the drive train is the product of the multiplication of the efficiencies of all its components as given in Table 23.2.13. Drive train components are illustrated in Fig. 23.2.17*F*.

V-belt and sheaves (0.94) × shaft-mounted reducer (0.94) = 0.884 efficiency.

Therefore, if the motor horsepower calculated at the drive shaft was 20.5 horsepower, the motor horsepower required is:

$$\frac{20.5}{0.884} = 23.2$$

Thus, a 25-Hp motor would be required to drive the conveyor.

The general-purpose squirrel cage motor will fulfill the requirements of most belt conveyor drives. Open motors are objectionable in dusty or exposed locations. Splashproof or totally enclosed fan-cooled motors should be used whenever drives are not housed.

The starting torque of the normal-torque motor is usually sufficient for operation of normal conveyors. However, when a large amount of the total power is required to overcome friction, a high-torque motor may be needed. As a rule of thumb, if twice the friction load plus the lift load is greater than the starting torque of a normal-torque motor, use a high-torque motor. On the larger drives (50 Hp+), resistance starting may be desirable. The power otherwise may be wasted in rapid acceleration resulting in higher belt and shaft stresses. Smooth starting may be accomplished by use of torque couplings. Smooth starting is essential on conveyors with drives furnished for future extensions with vertical curves or with belt trippers.

Conveyor drives with large-horsepower motors requiring several steps of resistance starting may be more economically served by wound rotor motors.

Variable-speed drives and creeper drives applied to enhance maintenance or endure cold weather climates are special forms of belt conveyor drives.

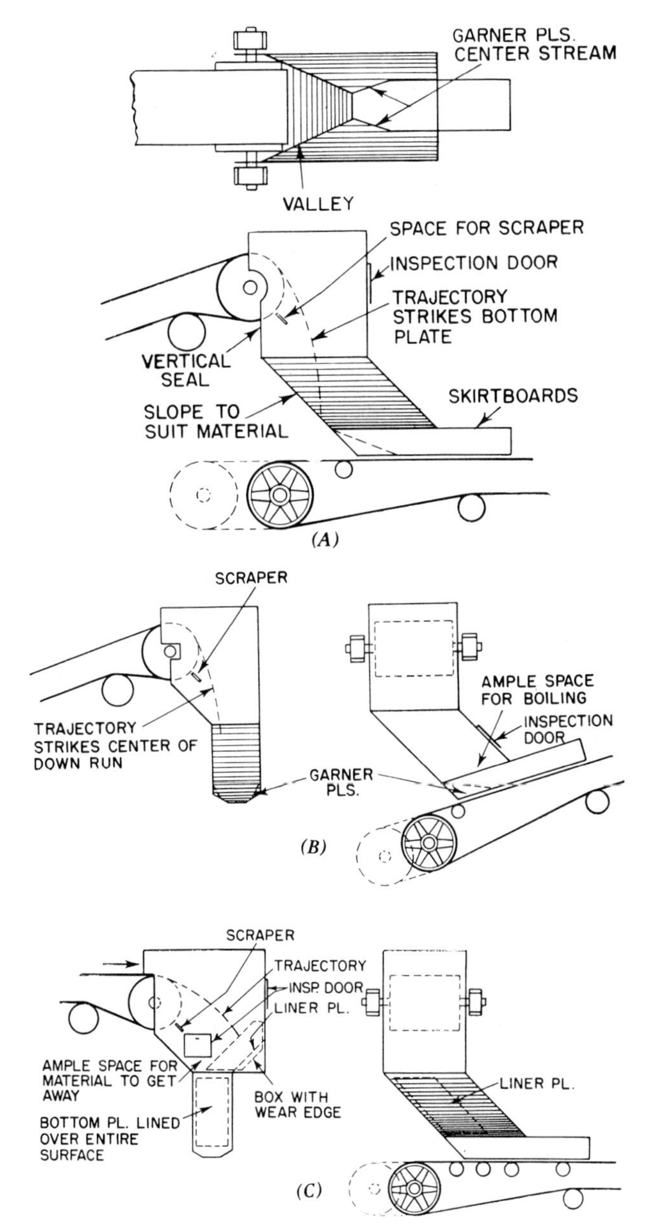

Fig. 23.2.18 Typical belt conveyor chute arrangements. (*A*), with in-line conveyor transfer; (*B*), with right-angle conveyor transfer; (*C*), with "stone box" to minimize wear when handling abrasive material.

A loaded inclined belt with a steep slope may move backward when its forward motion is stopped for any reason. This action can cause material to collect at the tail, damaging the belt, and could even cause a safety hazard. To prevent this motion a backstop should be used when the force to lift the load vertically is greater than one-half the force required to move the belt and load horizontally.

23.2.13 Accessories

Probably the most important belt conveyor accessory is the chute used to load or discharge material onto or off of the belt (see Fig. 23.2.18). Application of the correct method and equipment for loading the belt increases belt life, reduces spillage, and aids in keeping the belt trained. The design of chutes

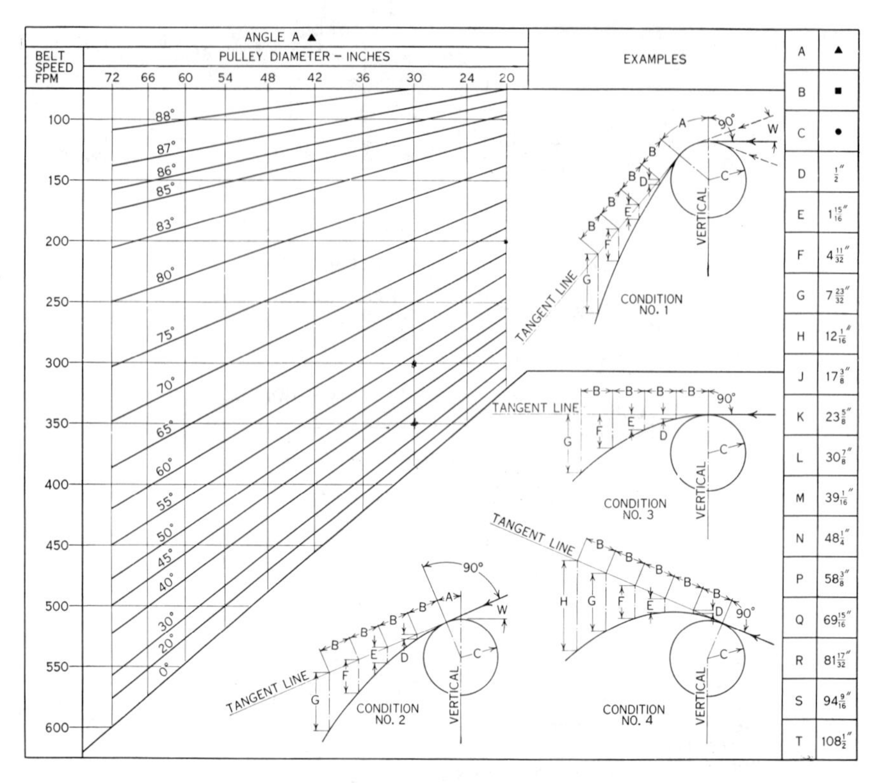

Chart F may be used to determine the trajectory of material from discharge pulley for the conditions illustrated above and as described below:

Condition 1 applies to horizontal and ascending belts when angle A exceeds 0°, and for descending belts when angle A exceeds angle W.

Condition 2 applies to descending belts when angle A is less than angle W.

Condition 3 applies to horizontal belts when angle A = 0°

Condition 4 applies to inclined belts when angle A = 0°

▲ A = Angle from vertical at which material will leave belt as it travels over discharge pulley. It is found at the point on chart where a line projected to the right from the belt speed intersects a line projected downward from the pulley diameter.

■ B = 1" per 100 feet per minute speed of belt (4" for 400 FPM, 2½" for 250 FPM, etc.) measured along tangent line at same scale used for indicating dimension C.

● C = Radius of discharge pulley in inches plus 1" (19" for 36" diameter pulley, 25" for 48" pulley, etc.). This 1" added to the pulley radius is intended to represent the approximate thickness of the belt and therefore the lower particles of the material. The chart is based on this value but if it should be desired to find A where C is different, then,

$$\cos A = \frac{V^2}{G\,C}$$

where V = belt speed in feet per second
G = acceleration of gravity = 32.16
C = distance in feet from center of pulley to tangent line desired.

Fig. 23.2.19 Trajectory of material over discharge pulley.

and other loading devices is influenced by capacity, material characteristics, and whether the belt is loaded at single or multiple positions. The principal requirements for a chute to properly load a belt are:

1. To load the material on belt at a uniform rate
2. To center load on the belt
3. To reduce impact of material falling on the belt
4. To deliver material in the direction of belt travel
5. To deliver material to belt at a velocity as near the speed of the belt as possible
6. To maintain a minimum angle of inclination of belt at loading point

In order to design chutes to meet these factors to properly load a belt, the trajectory of material over the discharge pulley must be calculated. The data given in Fig. 23.2.19 can be used to determine the path the material will follow so chute clearances, wear points, and proper feed to the take-away belt can be determined.

Materials can be discharged from belt conveyors to meet a wide variety of requirements. Discharge can be confined to single or multiple points, or the material may be along the entire length of the conveyor. Some of the methods of discharging from belt conveyors include:

1. Discharging over an end pulley
2. Discharging over one or more fixed trippers
3. Discharge over movable trippers
4. Plowing material from one or both sides of a belt by fixed or traveling plow

Figure 23.2.1 indicates typical arrangements of belt conveyors.

23.3 SYSTEM CONSIDERATIONS

23.3.1 Cleaning the Belt, Pulleys, and Idlers

The wide diversity of materials and their characteristics handled on belt conveyors has resulted in the development of a wide variety of means of cleaning belts on applications where the material tends to adhere.

Rubber- or metal-bladed wipers located on or near the head pulley of the conveyor provide adequate cleaning and economical construction and maintenance for a large number of applications. Note scraper locations given in Fig. 23.2.18. Generally, the blades are mounted in a pivoted frame and are held in contact with the belt by means of springs or counterweights.

When the conveyor is handling material that cannot be satisfactorily cleaned from the belt by means of such rubber- or metal-bladed cleaners, it may be necessary to consider the use of water spray, compressed air, or fixed or power-driven revolving brushes. Sometimes it is necessary to use two or more of these devices in combination.

On applications where it is anticipated that material will tend to build up on the faces of conveyor pulleys, it is desirable to have the pulleys lagged with rubber. Proper selection of the grade of rubber and, if necessary, the application of grooving to the rubber, will generally avert the problem. In the most difficult cases it may be necessary to apply pulley scrapers, arranged so that the scrapings are deflected from the path of the belt.

Consideration of return belt rubber tread idlers should be given on applications where the material may tend to build up to an undesirable degree on the return idlers.

23.3.2 Weighing

When it is necessary to weigh materials in transit on belt conveyors and record the amount delivered to certain points of a processing system, automatic recording scales are used. These scales can be either mechanical, electronic, or air operated. Impulses from electronic scales can be used to control the feeders delivering materials to the belt.

The scales can be furnished for standard-width conveyors and are accurate, compact, and do not disrupt the continuous flow of material on the belt.

There are also available batch feeders that consist of belt feeders, weighing or measuring material as it is discharged from hoppers.

23.3.3 Magnetic Separation

Tramp iron can be removed from materials carried on belt conveyors by either permanent or electromagnetic pulleys.

The pieces of tramp iron are drawn to the belt surface as they pass over the magnetic pulley. The pieces then fall free as the belt leaves the pulley, falling into a chute or bin.

Other types of separators and metal detectors are available that are suspended over the stream of material on the conveyor. Metal detectors can indicate the presence of both magnetic or nonmagnetic metals.

23.3.4 Sampling

There are sampling systems available that take a representative sample of the material as it passes over the conveyor discharge. Samples may be taken for various reasons and can be collected on a continuous or intermittent basis. The sampling system crushes, sizes, and prepares the sample for laboratory analysis.

23.3.5 Dust Control

Dust control and protection of personnel can be accomplished by enclosures. Where required, the entire belt conveyor and its terminals can be totally enclosed and the dust exhausted to dust collecting systems. Many applications require no more than enclosures at transfer points, with or without dust collecting systems.

23.3.6 Supports and Galleries

Belt conveyor supports are simple and are easily designed to suit a wide range of conditions.

Galleries and housings are used to enclose belt conveyors where the conveyors are carried across open spaces. They can be incorporated in a bridge structure and can be designed for convenient access to the conveyor. A wide selection of modern materials such as roofing, siding, and window and door framing often makes it possible to design conveyor galleries uniformly blending with the architecture of adjoining buildings and structures.

Fig. 23.2.20 Typical galleries and housings.

Housings prevent ice and wind from causing a belt to run off-center and an empty belt from being blown off the idlers. They also decrease deterioration of the belt by protecting it from the sun.

Typical galleries and housings are illustrated in Fig. 23.2.20.

23.3.7 Safety Devices

A wide selection of safety devices is available to be applied for varying arrangements of conveyors and conditions surrounding their operation. Safety pull cords can be strung the length of the belt conveyors. Pulling on the safety cord at any point immediately shuts off the power.

Terminals and drive machinery can be protected by guards as necessary depending on exposure to personnel. For the highest degree of protection, expanded metal guards can completely enclose all moving parts. Generally, guarding of high-speed rotating parts and pulleys is adequate.

Backstops can play an important part in safety to personnel as well as protection of the conveyor equipment.

Automatic take-up machinery should be completely enclosed with expanded metal guards or the like. In addition, a counterweighted take-up can be supplied with a means to avert its free fall in the case of accidental parting of the belt.

23.4 GENERAL SYSTEM DESIGN CRITERIA

Following the preceding outline will result in the selection and design of a belt conveyor but it will not ensure that the conveyor will properly function within a given bulk materials handling system.

In the study of a materials handling system involving belt conveyors, the number of conveyor transfers should be kept at a minimum in order to reduce material degradation, dust production, and cost. All belt conveyors should be elevated a few feet above ground or otherwise made accessible to facilitate inspection, maintenance, and clearance. Railroad, plant roadway clearances, cranes, and other mobile equipment must be considered. Working points of belt conveyor at transfer points and between process equipment, bins, hoppers, feeders, steel supports, and building structures must be carefully determined and established.

Materials carried by a belt conveyor can be discharged from the belt in different ways to effect certain desired results. Although the simplest discharge from a conveyor belt is to allow the material to pass over a terminal pulley, the use of shuttles, trippers, plows, stackers, traveling hoppers, and other specialized equipment designed around belt conveyors for unloading, storing, stockpiling, reclaiming, and loading must be understood as part of a total materials handling system.

ACKNOWLEDGMENT

Development of a design criteria for belt conveyors has been an on-going effort since before the turn of the century. Much of this initial work was carried out by the manufacturers of bulk material belt handling conveyors and components. Firms such as Link Belt Company, Hewitt-Robins, Rex Chain Belt, Jeffey Manufacturing, Continental Gin, Stephens Adamson, Goodyear, Goodrich, and many others played an important and essential role. The data, equations, tables, and figures still used today stem from their pioneering work.

The author acknowledges the importance of the industries' contribution to the contents of this chapter, with special thanks extended to H. Colijn for generous aid and assistance in its preparation.

CHAPTER 24
CHAIN CONVEYORS: APRON, PAN, AND FLIGHT

GEORGE A. SCHULTZ

Epstein Process Engineers
Chicago, Illinois

24.1 INTRODUCTION

A well-designed chain conveyor made up of high-quality material is an excellent means of conveying abrasive and high-temperature materials, or withstanding the effects of impact when handling large lumps.

Chain conveyors employ single or double strands of continuous chains wrapped around head and tail end sprockets. The units are generally operated by motor drives attached to the head/drive shaft. Material can be carried directly on aprons or pans or pushed in a trough by flights attached to the chain(s). The chain conveyor derives its name from the type of attachment, that is, apron (Fig. 24.1.1), pan (Fig. 24.1.2), or flight (Fig. 24.1.3).

There are four types of chain conveyors based on whether the chain slides or rolls and whether the material is pushed or carried (Fig. 24.1.4). Units can be arranged for operation horizontally, inclined, or in combination. With proper component selection, chain conveyors can be designed to operate at

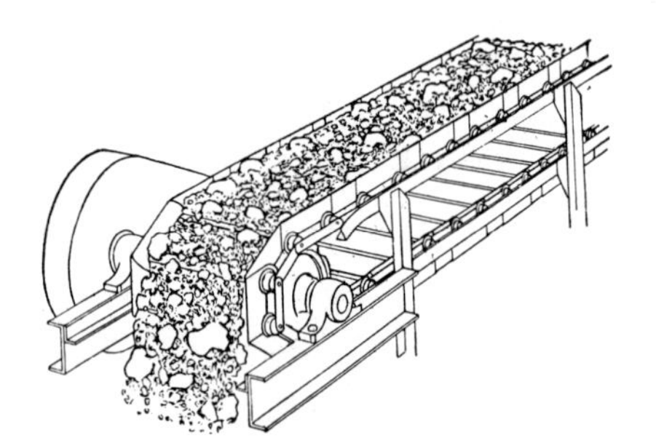

Fig. 24.1.1 Apron conveyor—a conveyor in which overlapping horizontal plates (pans) are attached to twin chains to form a bed for carrying bulk materials.

Fig. 24.1.2 The pan conveyor is comprised of one or more endless chains with interlocked metal belt arranged to form a carrying surface.

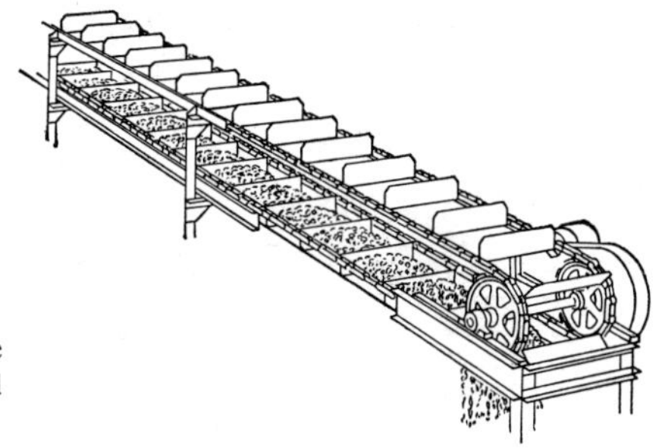

Fig. 24.1.3 The flight conveyor is comprised of one or more endless chains to which flights are attached which push material though a trough.

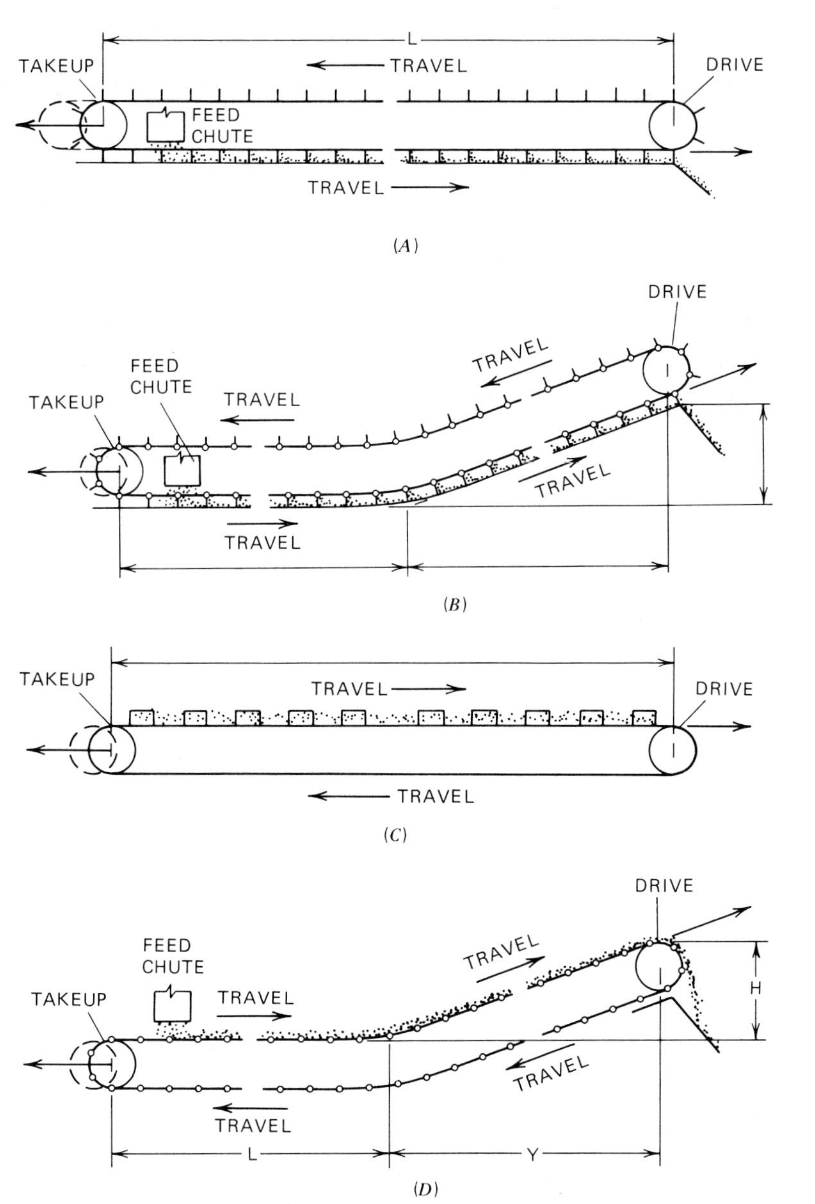

Fig. 24.1.4 Types of chain conveyors. (*A*), chain and material sliding; (*B*), chain rolling and material sliding; (*C*), chain sliding and material carried; and (*D*), chain rolling and material carried.

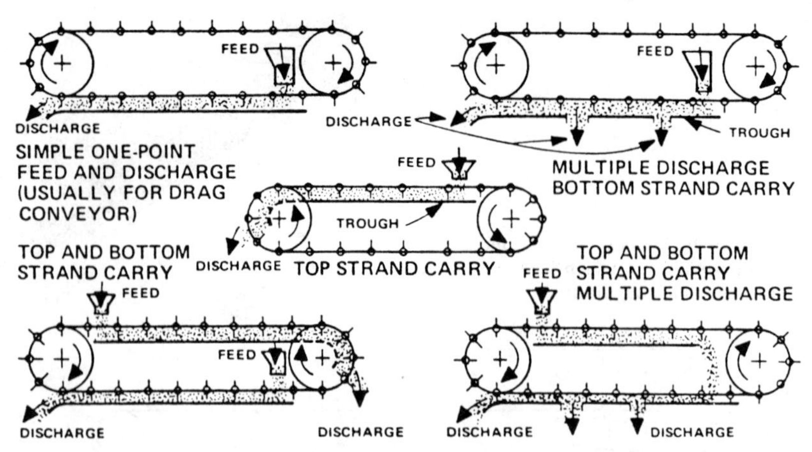

Fig. 24.1.5 Alternate feed and discharge arrangements for flight conveyor.

inclines up to 45°. Flight conveyors can be easily enclosed for dust containment and arranged to serve multiple filling and discharge points (Fig. 24.1.5).

A critical component in the design of a chain conveyor is the "engineered chain" used for performing the function of conveying as compared to a chain used to transfer power from one point to another. Engineered chains are available in a variety of material including malleable iron, Z-metal, chilled iron, manganese steel, cast steel, welded steel, and others. Selection is based on required strength, expected operating speed, abrasiveness of material handled, and type of attachment required. Working strengths of engineered chains are accurately given in manufacturers' literature. When severe abrasion conditions are encountered special surface hardening is often required. Figure 24.1.6 is a comparative

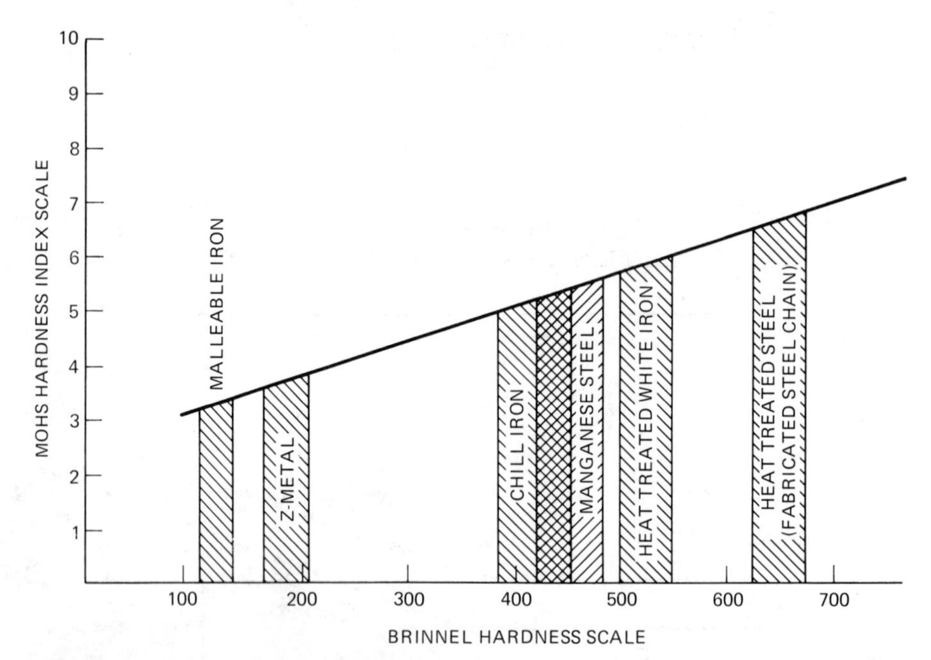

Fig. 24.1.6 Mineral hardness and chain joint hardness. This chart, developed by Rexnord Engineers, can be used to determine the hardness of various materials and for selecting chains of proper hardness to handle the material. Mohs scale of hardness: 1, talc; 2, gypsum; 3, limestone, triple super phosphate, bauxite; 4, cement rock, taconite pellets, fluorite; 5, apatite, mill scale; 6, cement clinker, rutile sand, feldspar, orthoclase; 7, quartz, iron ore sinter, silica sand; 8, topaz; 9, aluminum oxide, sapphire, corundum; 10, diamond.

Table 24.1.1 Coefficient of Friction for Various Materials on Steel Plate

Material	Coefficient
Anthracite coal	0.33
Bituminous coal	0.59
Cement	0.93
Clay	0.60–0.70
Coke	0.36
Grains	0.30–0.40
Hog fuel (dry)	0.60
Hydrated lime	0.65
Pulverized limestone	0.58
Sand and damp ashes	0.68
Soda ash	0.65
Starch	0.78
Sugar, fine, granulated	0.67
Wood chips	0.35

hardness guide for abrasion defined in terms of a Mohs index number. Temperature and corrosiveness must also be given consideration when selecting which chain material to specify.

The capacity of a chain conveyor is the product of the unit's available cross-sectional area multiplied by the chain speed. The speed of the chain conveyor is based on the material to be conveyed and should always be kept as slow as possible. For example, very abrasive material should be conveyed at speeds of less than 10–20 ft/min (0.051–0.10 m/sec), while mildly abrasive materials can be moved at speeds of 100 ft/min (0.51 m/sec) or more.

Drive requirements for chain conveyors are in part determined by the frictional properties of the material being conveyed. The chain pull on which the drive motor is based must be sufficient to carry or push the conveyor load but must overcome *all* sliding material and chain frictional forces. Table 24.1.1 is a list of typical coefficients of friction for various material sliding or steel plate.

With apron or pan conveyors the entire load is supported by the chains and one of the forces needed to be considered is the friction load required to move the loaded chain. When stationary sides are used to confine the material as it moves along in apron or pan conveyors, the rubbing of the material against the sides causes an additional functional load.

Special consideration must be given to the starting effort, especially under load. The starting effort for an apron or pan conveyor may be 2 to 4 times that required to keep these units in motion. With a horizontal conveyor the frictional forces of the return chain strands must also be considered. An inclined apron or pan conveyor's frictional resistance of the return strand can have a positive or negative effect on chain pull dependent upon the result of the force of gravity.

Although the equations for calculating apron and pan conveyor chain pull are basic, many of the friction factors used in them are empirical. For example, friction coefficients can be based on average axle diameters or on average operating conditions such as handling slightly moist coal. Friction coefficient can be arbitrarily increased when handling gritty material or, on the other hand, working under extremely good, well-lubricated conditions, frictional factors can be arbitrarily reduced. The same is true of the equations used to determine sliding friction. Once a chain pull is calculated it can be modified by judgment factors covering the amounts of terminal and drive friction necessary to determine final drive horsepower. When purchasing a chain conveyor drive horsepower and terminal component, selection should be carefully analyzed by the buyer's engineers.

A chain conveyor requires extra care in erection if it is to function properly. Assuming properly designed and selected components, it is essential that all components, and especially the head and tail sprockets, be in line. Twin chain units should be checked for matching both at startup and while running. Components should be carefully checked for loose or missing bolts, cotter pins, and so forth.

Proper and continued adjustment checks are essential for long-term, maintenance-free operation. Lubrication of chains should be carried out per manufacturer's recommendations contingent upon usage. Lubrication of roller chain when handling certain types of abrasive dusty material may not be advisable.

24.2 APRON CONVEYORS

Apron conveyors are commonly used to handle bulk materials such as ores, stone, sand and gravel, coal, cullet, foundry, refuse, and similar materials. They excel in the following applications:

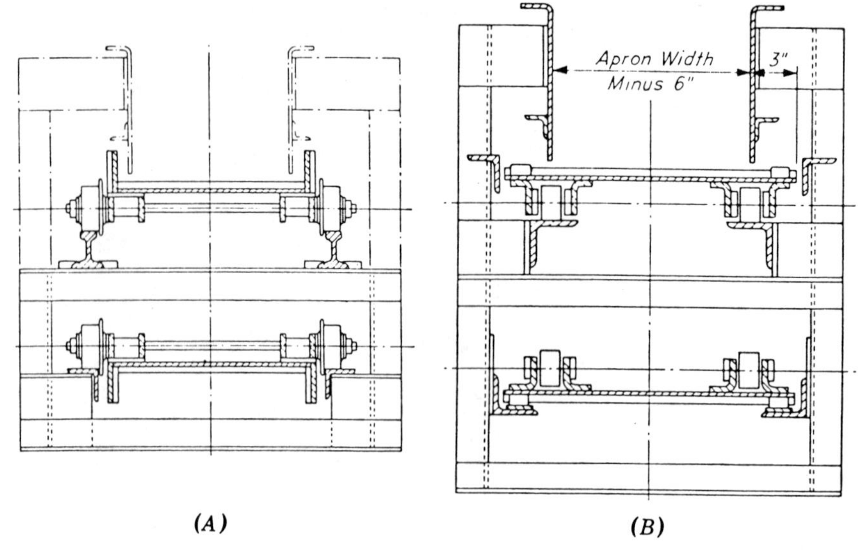

Fig. 24.2.1 Apron pans. (*A*), style A pan, leak-proof type with wings; (*B*), style A pan mounted on K-type attachments, *no* wings.

Table 24.2.1 Common Types of Apron Pans

Manganese Steel Pans

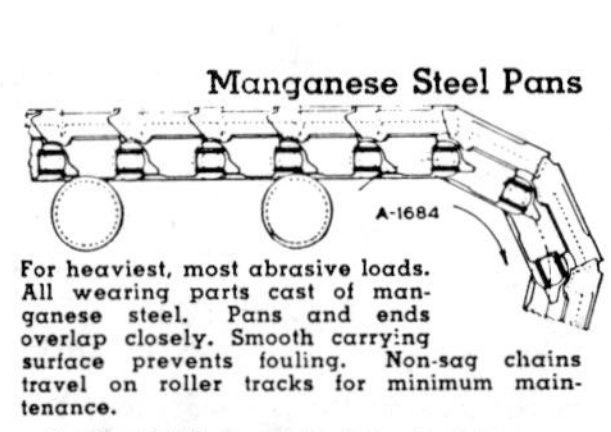

For heaviest, most abrasive loads. All wearing parts cast of manganese steel. Pans and ends overlap closely. Smooth carrying surface prevents fouling. Non-sag chains travel on roller tracks for minimum maintenance.

Style "A" Double Beaded Pans

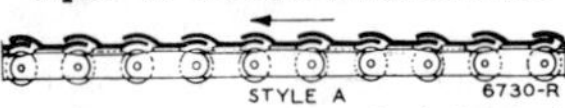

A rugged, shallow type pan for feeders horizontal or inclined to 20°. Mounted on G-4 unbushed, steel roller chain. Usually operated with skirt plates to prevent spillage of load.

Style "B" Single Crimped Pans

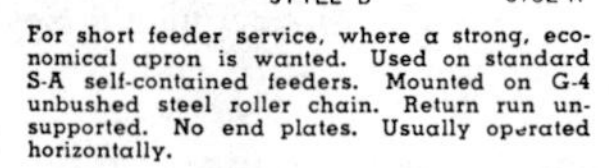

For short feeder service, where a strong, economical apron is wanted. Used on standard S-A self-contained feeders. Mounted on G-4 unbushed steel roller chain. Return run unsupported. No end plates. Usually operated horizontally.

Style "C" Double Beaded Pans

Similar to style "A" except that they are mounted on malleable iron roller chains with side attachment links, which permit return run to be suported and longer units possible. Usually furnished with end plates. Suitable for inclines to 25°.

Style "D" Single Beaded Pans

Deeply beaded section suitable for wide conveyors. Will convey on inclines up to 30°, and discharge gently with minimum drop. Mounted between long pitch, steel bushed roller chain with deep inside links.

Style "PT" Pans

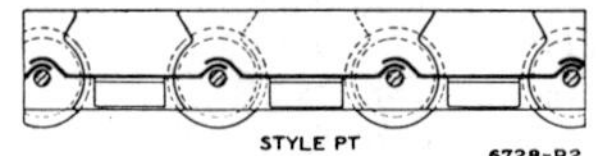

Shallow type, double beaded pans with comparatively long pitch which permits material to spread evenly over the carrying surface for picking or inspection. Mounted between long pitch, steel bushed roller chains with wide inner bars for sides.

Style "E" Deep Pans

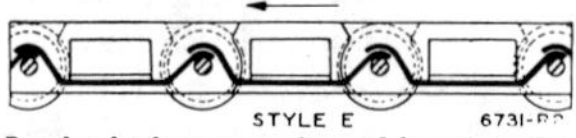

Popular for large capacity and heavy service. Rigid enough for wide pans, and can be furnished with armored filler blocks. Mounted on long-pitch, steel-bushed, roller chain with cross rods to insure alignment. Deep inside chain links increase carrying capacity.

Style "F" Deep Pans

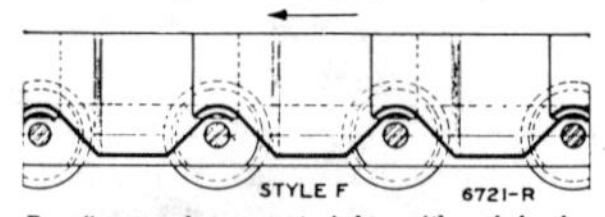

For fine or lump materials, with reinforcing end plates separate from chain links. Built in larger sizes than Style "E" pans and suitable for greater capacities. For use horizontally or on inclines to 30°.

Style "G" Conveyor Buckets

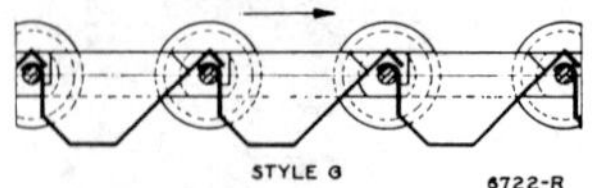

Heavy buckets for fine or lump material. Have high capacities operating up inclines to 50°. Mounted between strands of long pitch, steel bushed roller chain with cross shafts to insure alignment. Formed end plates.

Style "H" Deep Pans

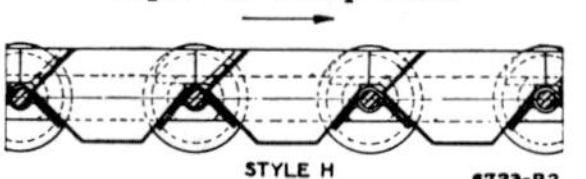

Rugged, bucket shaped pans for long conveyors to handle large volumes up inclines as steep as 50°. Pans hinged on cross rods, the ends of which form chain pins. Separate end plates to reinforce pans.

Style "J" Heavy Buckets

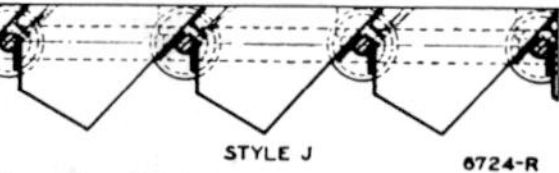

Deep type buckets for angles of 30 to 70°. Buckets are hinged upon cross rods, the ends of which serve as chain pins. Steel angles reinforce lips of buckets and prevent spillage.

Style "M" Heavy Buckets

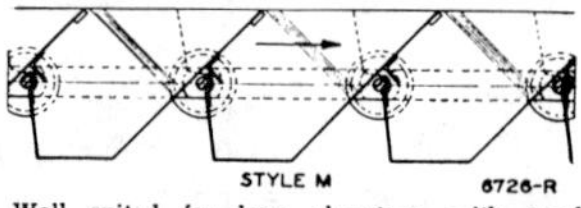

Well suited for long elevators, with good capacity for inclines of 40 to 70°. When furnished without upper half of lip — "M" buckets will convey large volumes horizontally.

Style "R" Heavy Buckets

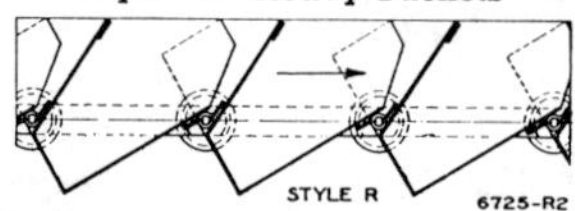

For largest and heaviest elevator service, for inclines of 50 to 70°. Buckets are hinged upon the cross rods which serve as chain pins. Built in standard widths up to 96 inches.

Style "W" Cast Pans

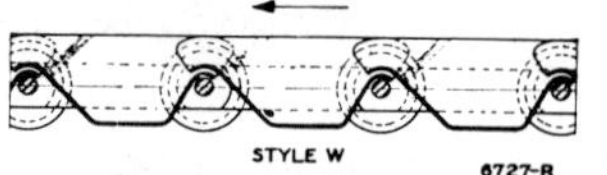

Overlapping, malleable or cast iron pans for handling hot, abrasive or corrosive materials. Cast sides and deep section permit large capacities and inclines up to 25°. Mounted between long pitch, steel bushed roller chains with cross rods.

Carrying material horizontally, up and down inclines, or in combination.
Conveying hot or abrasive materials.
Handling large lumps.
Picking conveyor applications.

The component parts that make up a complete apron conveyor should be selected in the following order:

Type of apron (pan)
Thickness of pan
Type of chain and attachment
Sprockets
Skirtboard requirements
Support frame construction

Apron pans can be flat or equipped with side plates to increase capacity and reduce spillage (Fig. 24.2.1). Some of the more commonly used types of pans are described in Table 24.2.1. The thickness of the apron pan is dependent upon the weight to be supported on each apron, impact of falling lump, and abrasiveness or corrosiveness of the material. Table 24.2.2 is a general guide to apron pan selection.

Table 24.2.2 Recommended Thickness of Apron Pans

Recommended Apron Pan Thickness (in.)	Application
$\frac{3}{16}$	Light, mildly abrasive materials.
$\frac{1}{4}$–$\frac{5}{16}$	Medium-weight material having some corrosive and abrasive properties. Moderate-impact duty.
$\frac{3}{8}$ or greater	Heavy, abrasive, corrosive materials; high-impact duty.

The more commonly used chain for apron conveyor applications is a steel, bushed roller chain with pans bolted to the chain by Types A or K attachments (Fig. 24.2.2). For nonabrasive materials weighing 50 lb/ft³ (801 kg/m³) or less, with a minimum lump size, a 4-in. (101.6 mm) pitch chain is adequate. For most other materials a 6-in. (152.4 mm) pitch and larger chains are used. Manufacturers' literature should be referred to for specific application information.

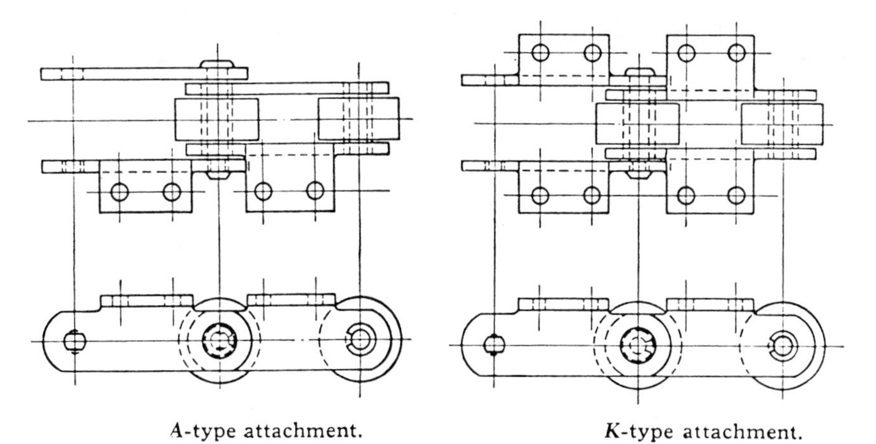

A-type attachment. *K*-type attachment.

Fig. 24.2.2 Chain attachments, links having suitable projections with holes to which apron pans can be attached.

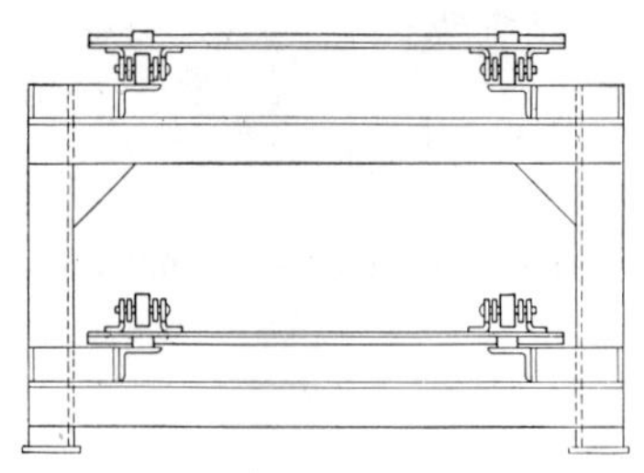
ARRANGEMENT A—Apron without apron pan ends, angle track both runs

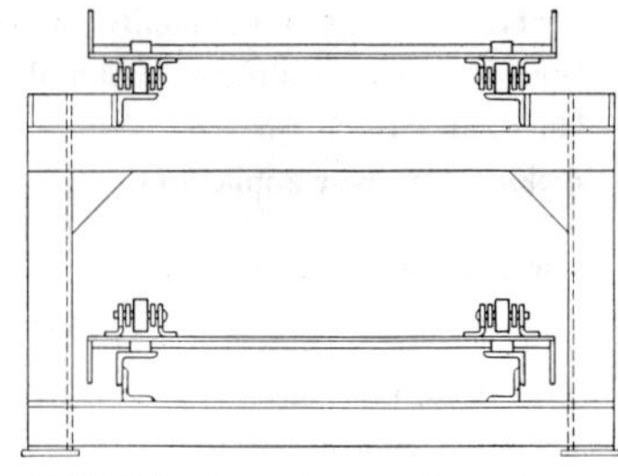
ARRANGEMENT B—Apron with apron pan ends, angle track both runs

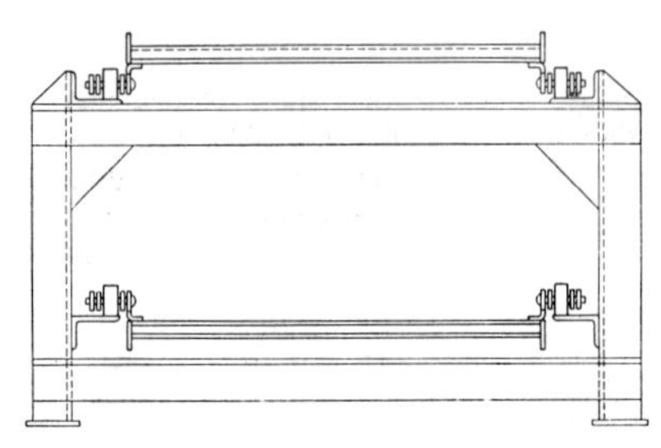
ARRANGEMENT C—Apron with apron pan ends, angle track both runs

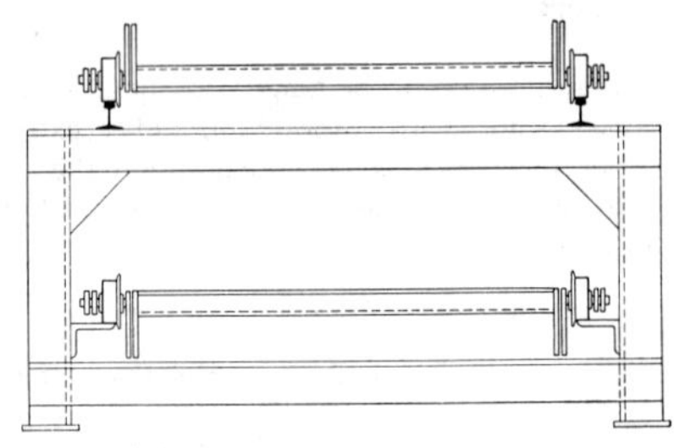
ARRANGEMENT D—Apron with apron pan ends, rail track carrying run, angle track return run

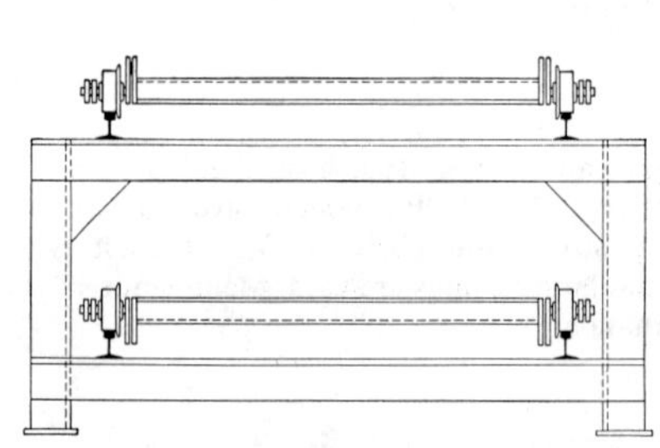
ARRANGEMENT E—Apron with apron pan ends, rail track both runs

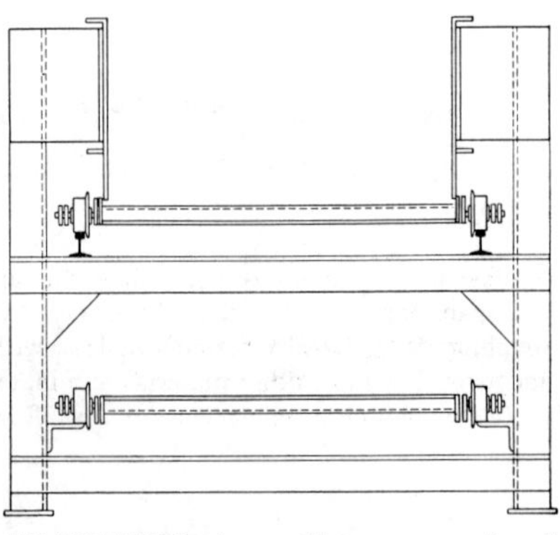
ARRANGEMENT F—Apron with apron pan ends, rail track carrying run, angle track return run

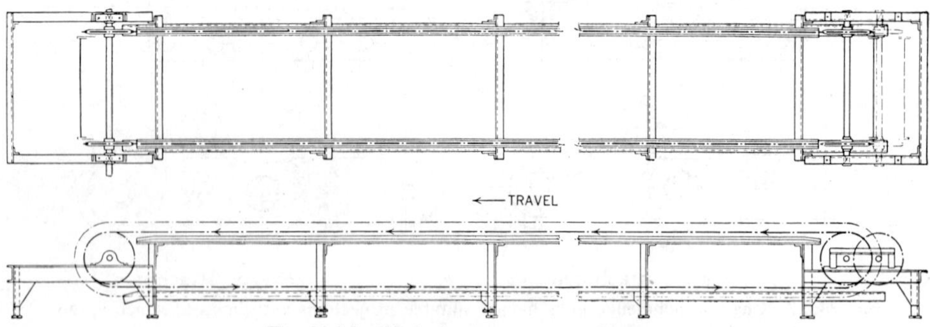

Fig. 24.2.4 Typical apron conveyor frame.

The pitch diameter and number of teeth of the drive and tail sprockets are very important factors. If the sprockets are too small in diameter or there are too few teeth, the unit may not operate smoothly or cause the pans to properly overlap as they pass around the terminal sprockets.

Stationary skirtboards can be used to increase the depth of material carried. When skirtboards are used, it is generally recommended that the depth of material carried not exceed $\frac{2}{3}$ the width of the conveyor. The frictional drag resulting from the material rubbing against the skirtboards must also be taken into consideration when drive horsepower calculations are being made.

Standard available frame cross sections are illustrated in Fig. 24.2.3. Their purpose is to provide support for running tracks, as well as to support stationary skirtboards when necessary. Terminal sprockets, shafts, and bearing supports must be extremely rigid and designed for easy adjustment for alignment as indicated in the typical apron conveyor frame in Fig. 24.2.4.

The volumetric capacity of an apron conveyor is dependent on the width, depth, and speed of the carried stream of material. Width and depth of the material stream is in part determined by the lump size of the material to be handled. Table 24.2.3 serves as a general guide for this selection. The slope of the carried stream of material can reach a critical angle where the load will begin to slip or roll back. This angle is steeper for angular material or when handling fines with embedded lumps as compared to rounded lumps. Even at a critical angle range material can still be carried at a limited depth above the top of the apron (Fig. 24.2.5). Use of special deep pans or flights welded to the apron pans can increase the inclined conveying capacity of a particular unit.

24.3 PAN CONVEYORS

Pan conveyors are a form of apron conveyor designed to handle sharp, highly abrasive, hot materials, machine parts, and similar products. In addition they can be designed to handle scrap, castings, heat-treated parts from quench tanks, chips, stampings, and other materials including food products and garbage.

A major component of the pan conveyor is the carrying surface commonly referred to as steel belting, piano-hinged belt, or hinged steel belting. This steel belting can be provided in a variety of widths and constructions. The pans are made from heavy-gauge steel formed to interlock on a belt pin mounted between two standard steel roller side chains.

Typical types of steel belting are illustrated in Fig. 24.3.1. The dimpled construction is used to assure drop-off of small flat light gauge parts; perforated construction enables coolants to drain from products handled or, as with the multiwiring perforated construction, allows air passage for cooling. Cleats can be welded to any of these designs to eliminate fall-back on inclines.

Standard chains (Fig. 24.3.2) are provided in $2\frac{1}{2}$ in., 4 in., 6 in., or 9 in. (63.5, 101.6, 152.4, 228.6 mm) pitch construction. Normally fabricated from high-carbon steel, they can be modified to suit individual needs. Sprockets are made of cast iron with chilled rims. Stock sizes include $2\frac{1}{2}$ in. (635 mm) P—5, 6, 8, 10, and 12 teeth; 4 in. (101.6 mm) P—6, 8, and 10 teeth; 6 in. (152.4 mm) P, and 9 in. (228.6 mm) P—6, 8 teeth. Hardened steel sprockets are also available.

Preengineered head, tail, and intermediate support sections are readily available from a variety of manufacturers. Standard sections can be arranged to suit horizontal construction as well as an inclined application. A critical point in proper design is the skirt design. In order to handle most products, the pan is manufactured with wings arranged in a standard frame (Fig. 24.3.3). When the material tends to be bulky it is often necessary to extend the standard frame by adding skirts (Fig. 24.3.4). With this construction, due to the nature of the types of materials handled (paper scrap, metal turning,

Table 24.2.3 Capacities of Horizontal Apron Conveyors[a]

Width of Apron between Skirts (ft)	Maximum Lump Size (in.)	Depth of Material on Apron (in.)	Capacity (ton/hr) at 20 ft/min
2	6	12	45
2.5	8	12	56
2.5	10	15	70
3	12	12	68
3	14	18	102
4.5	14	12	79
4.5	18	24	158
4	18	12	90
4	24	24	180
5	18	12	113
5	28	24	225

[a] For material with density of 50 lb/ft^3.

VOLUME OF CONVEYORS OPERATING UP INCLINES

Allowable slope is steeper for material with angular than with rounded lumps and is also increased by the presence of fines enough to embed and hold the lumps. For any given material there is a range of slopes, where slippage appears, but where material can still be carried at a limited depth above the top of the pan. In this range the use of flights or the presence of large lumps embedded in fines increases the depth. The following table and diagram show what can be expected from material such as coal or stone. For gravel or material having rounded lumps the slopes must be 2° to 4° flatter. "M" is the size of average lumps in inches.

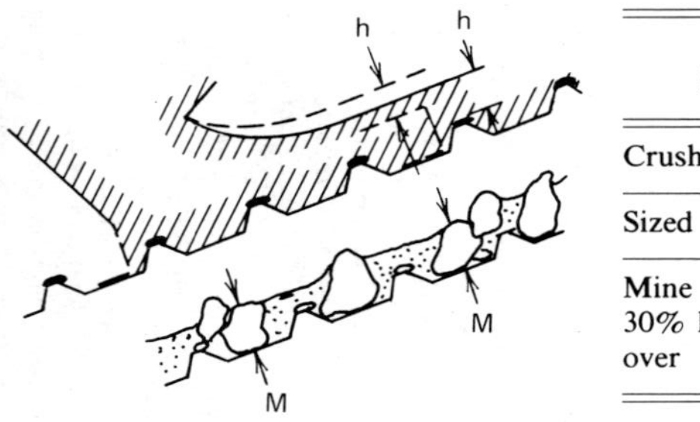

Size of Material	Slopes in Degrees, from Horizontal	Maximum Depth over Pan or Flight
Crushed to $1\frac{1}{4}''$ maximum	25° to 30°	4″ to 8″
Sized lump	22° to 27°	3″ to size of lumps
Mine run, containing 20 to 30% lumps "M" inches or over	25° to 30°	"M" inches minus depth of pan or flight

Fig. 24.2.5 Volume of apron conveyor operating up inclines.

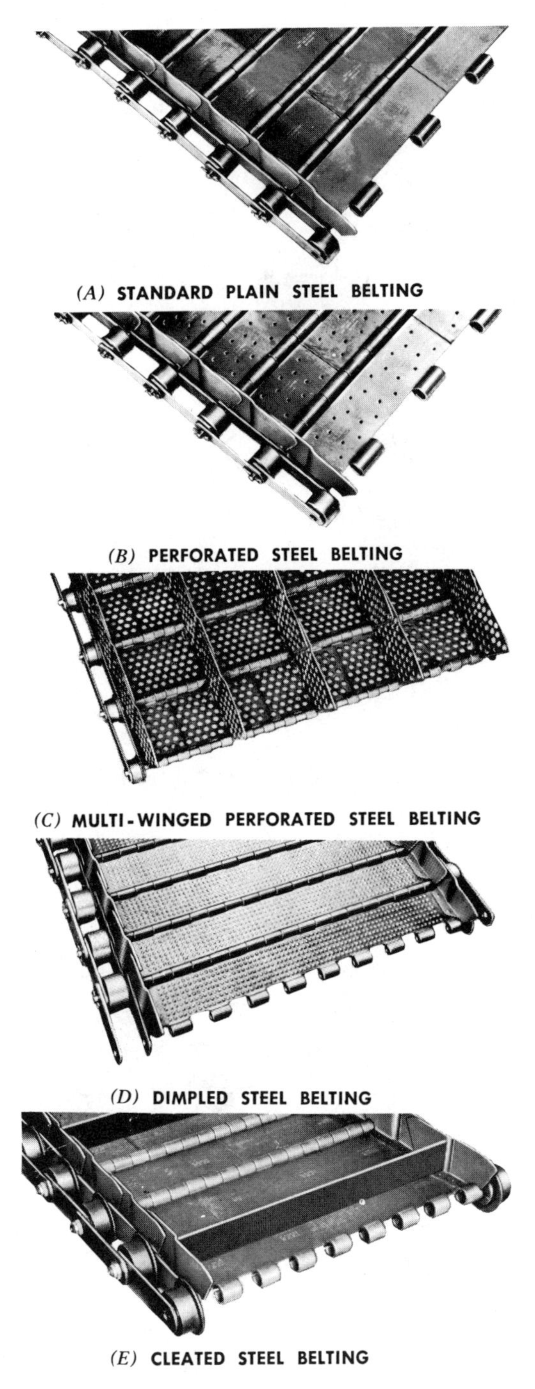

Fig. 24.3.1 Typical types of hinged steel belting. (*A*), closely fitted to eliminate fall-through; (*B*), wings provide compartmentalized effect and large perforations provide cooling for hot forgings, castings, and so on; (*C*), facilitates coolant drain from products handled; (*D*), assures drop-off of small, flat, light-gauge parts; (*E*), eliminates fall-back on sharp inclines, can be welded to standard, dimpled or perforated belting.

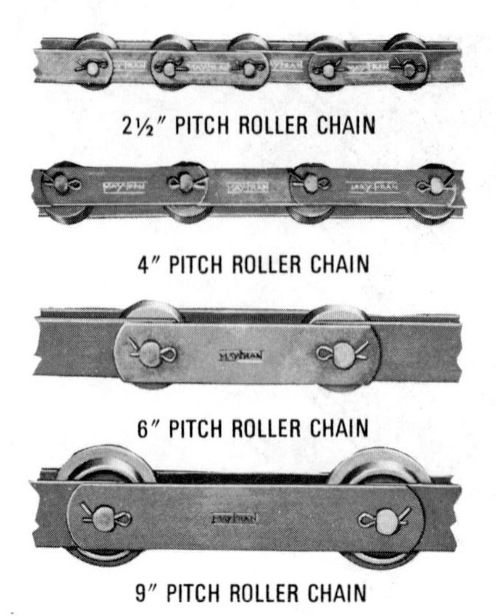

Fig. 24.3.2 Typical pan conveyor side chains.

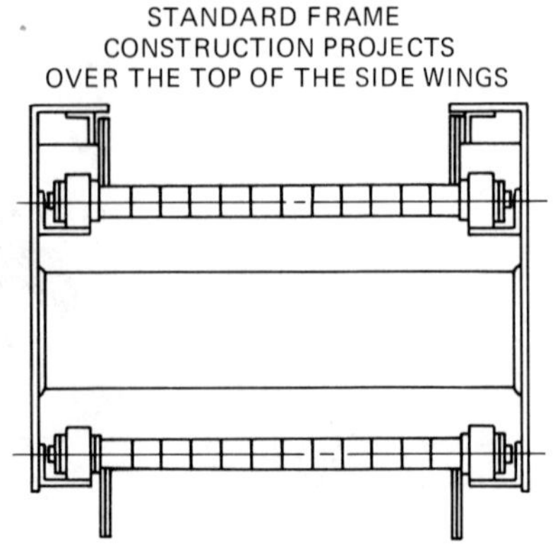

Fig. 24.3.3 Skirt as an integral part of the basic frame construction. It is recommended for handling individual parts or pieces rather than bulk items. However, if the conveyor system is relatively straight and horizontal, considerable bulk can be accommodated.

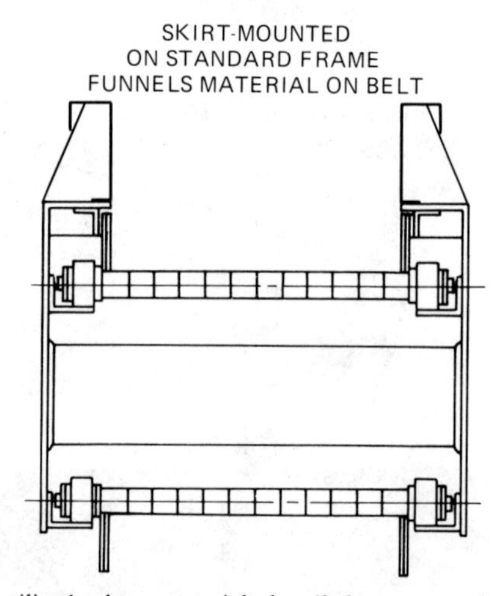

Fig. 24.3.4 Skirt is now utilized where materials handled run toward volume rather than weight. Conveyors provided with this type of skirt do an excellent job of handling light-gauge sheared scrap and trim as well as bulky formed metal parts.

garbage, etc.), it can become entangled in the wings of the metal belt and a guard must be added for additional protection (Fig. 24.3.5).

Pan conveyor construction has been standardized for use as either fixed and portable units for handling small metal scrap, chips, and small parts (Fig. 24.3.6). This metal belt construction has also been modified into a flat construction for use on units adapted to floor installation as assembly conveyors. The current interest in solid waste handling and recovery has led to development of special heavy-duty pan conveyor designs for use in these applications. Pans, frames, and drives must be specially designed to meet these heavy-duty requirements.

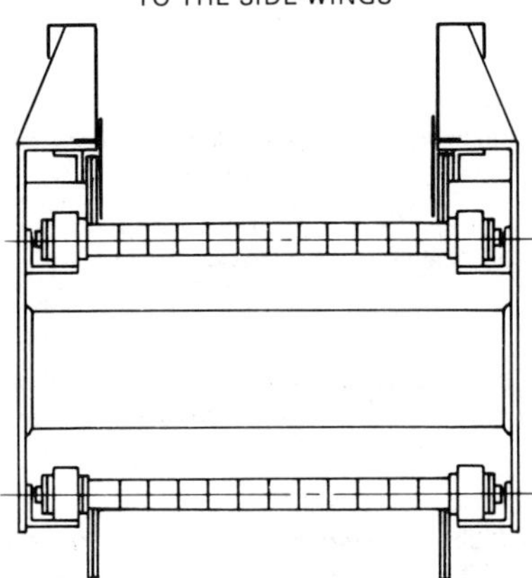

Fig. 24.3.5 Skirt is a further development of Fig. 24.2.8 skirt and provides a guard that projects down toward the belt surface and overlaps the side wings which prevents materials from being entangled with the side wings.

Fig. 24.3.6 Portable pan conveyor.

The capacity of a pan conveyor, like the apron conveyor, is based on volume, dependent on the width, depth, and speed of the carried stream of material.

Selection of the most advantageous width and depth of material for a conveyor is a more complex process than for the apron conveyor. For example, the pan conveyor used for a waste handling application must handle material with bulk densities as light as 3–5 lb/ft^3 (48.1–80.1 kg/m^3) which can "ball-up" into an object several feet (meters) in diameter.

Pan conveyors are generally powered by a motor operating via a reducer and chain drive. The reducer and chain drive should be carefully designed and selected for the variable and shock loading typically encountered. Often the standard pan conveyor drive may not have sufficiently heavy-duty construction and should be up-graded or provided with safety clutch to withstand potential overloads.

24.4 FLIGHT CONVEYORS

Flight conveyors have been used to convey bulk materials such as coal, wood chips, hog fuel, lump lime, crushed ice, foundry sand, sludge, and certain fruits and vegetables. Although these units are within limits suitable for handling granular or lump materials, they are not suitable for conveying sticky, sluggish, very abrasive, or corrosive materials.

The following factors are essential for selecting a flight conveyor:

Type of material and its characteristics

Capacity (maximum in tons per hour)

Maximum size of lumps *and* percentage of maximum lump in total volume

Length and incline of unit

Service requirements

Generally available types of flight conveyors are presented in Fig. 24.4.1. Sliding chain conveyors are simpler, with fewer moving parts than a roller chain type but horsepower requirements are higher. A roller chain unit operates with less pulsation than a sliding chain unit. The lower friction of the roller chain units permits design of longer units, with lower horsepower and reduced operating costs, but may be susceptible to jam-ups.

Flight, normally manufactured from steel, have been made of wood, malleable iron, and other materials spaced at distances from 12 in. (304.8 mm) to 36 in. (914.4 mm). Flight spacing is varied to suit the size of lumps, required capacity, and slope of unit. Table 24.4.1 is a brief resume of size and capacity of available units, and additional data can be found in manufacturers' catalogs. If a flight conveyor is operated at an incline, the capacities given in Table 24.4.1 should be multiplied by the following factors.

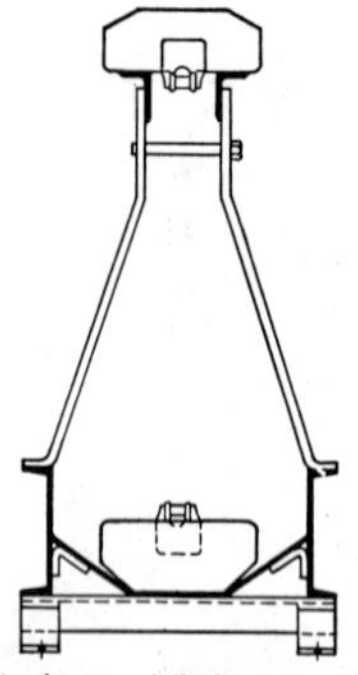

Single strand flight conveyor with scraper flights

Single strand flight conveyors with scraper flights consist of malleable iron flights attached to a single strand of chain and are designed for sliding directly on a steel trough. This type of conveyor is suitable for handling free-flowing materials with lumps no larger than 4 inches. Heavy flights operating in heavy troughs can be made to suit requirements.

Single strand flight conveyors with shoe-suspended flights consist of steel flights attached to a single strand of chain suspended from malleable iron sliding shoes and operating over a trough within limited clearances. Sliding shoes are attached to the sides of the flights near the upper edge and slide on flat, renewable steel bars.

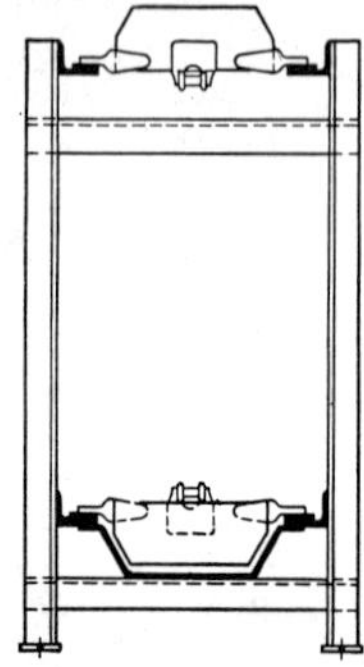

Single strand flight conveyor with shoe-suspended flights

Single strand flight conveyors with roller-suspended flights are of the same general construction as the shoe-suspended type but have the advantage of decreased chain pull resulting in lower power requirements.

Double strand flight conveyors with roller-suspended flights consist of steel flights attached to double strands of chain supported by the chain rollers. The chains operate on flat, renewable steel bars attached to the top of the channel sides of the trough. These conveyors are used for larger capacities and longer paths than the single strand flight conveyors and can handle material containing lumps up to 16 inches in size.

Double strand flight conveyors with sliding chain-suspended flights consist of steel flights attached to double strands of rivetless chain. The chains operate on flat renewable steel bars attached to the trough. These conveyors are used for larger capacities and longer paths than single strand flight conveyors and can be made to convey on both runs.

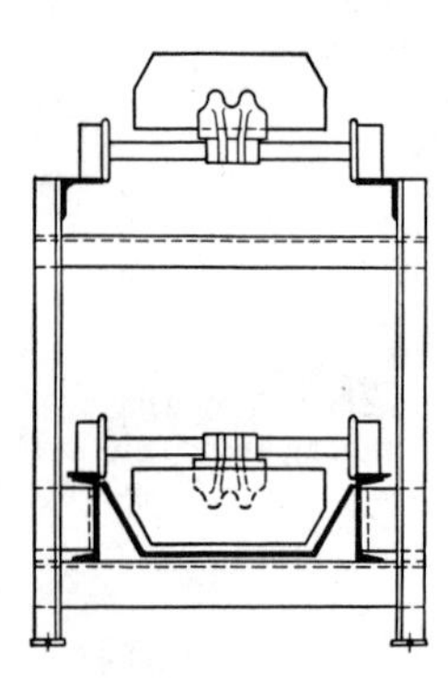

Single strand flight conveyor with roller-suspended flights

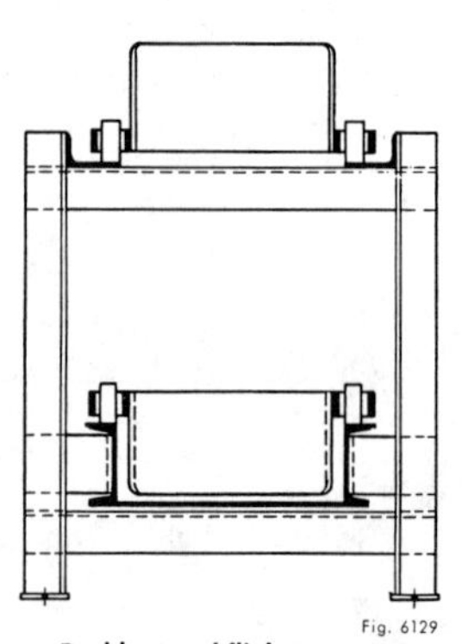

Double strand flight conveyors with roller-suspended flights

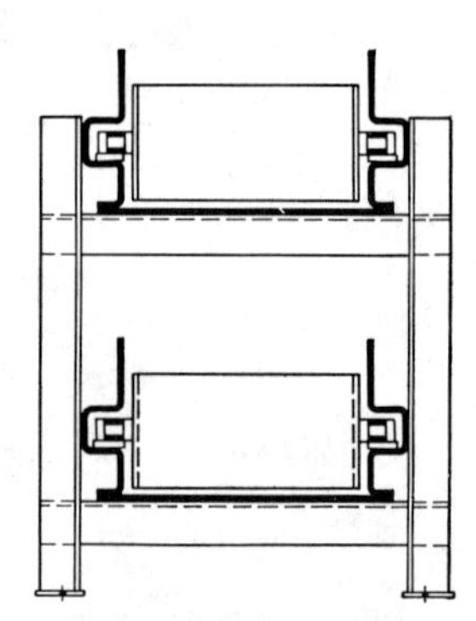

Double strand flight conveyor with sliding chain-suspended flights

Fig. 24.4.1 Types of flight conveyors.

Incline	Multiplication Factor
Horiz. to 20°	1.00
To 25°	0.98–0.90
To 30°	0.90–0.80
To 35°	0.85–0.70
Above 35°	Special considerations

One of the advantages of flight conveyors is their adaptability to a variety of feed and discharge arrangements as shown in Fig. 24.4.2. Also a flight conveyor can be designed so that it is self-feeding; it does not require a separate feeder if arranged as in Fig. 24.4.3.

Table 24.4.1 Capacity of Flight Conveyors[a]

Flight Dimensions (in.) (Width × Depth)	Lump Size (in.)[b] Single Strand	Lump Size (in.)[b] Double Strand	Approximate Capacity (ton/hr) at 100 ft/min
12 × 6	$3\frac{1}{2}$	4	60
15 × 6	$4\frac{1}{2}$	5	73
18 × 6	5	6	84
24 × 8	—	10	174
30 × 10	—	14	240
36 × 12	—	16	360

[a] For materials with density of 50 lb/ft³.
[b] Lumps not to exceed 10% of total volume.

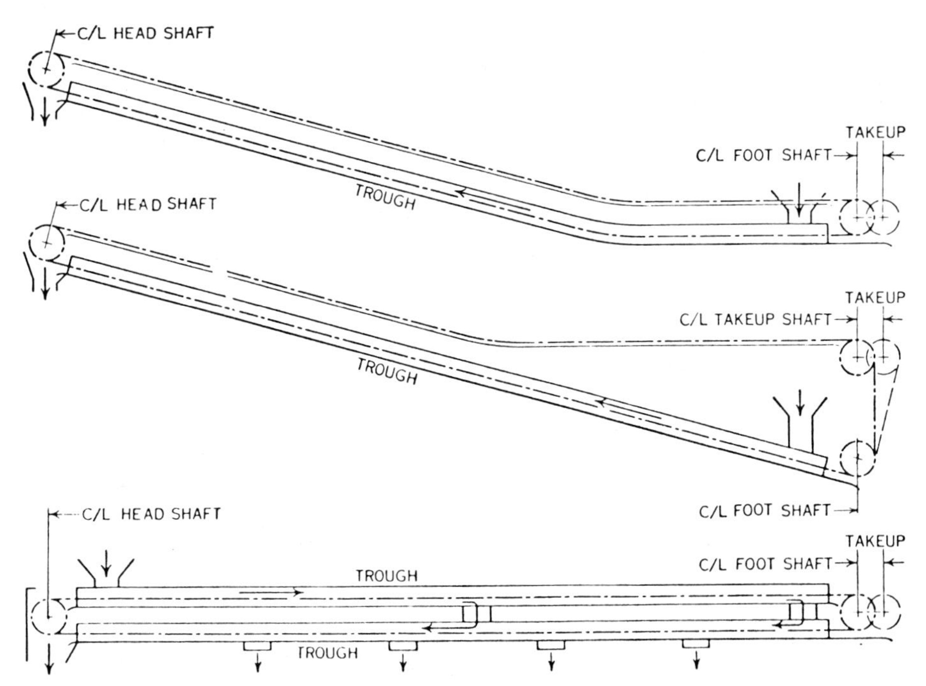

Fig. 24.4.2 Typical arrangement of flight conveyors.

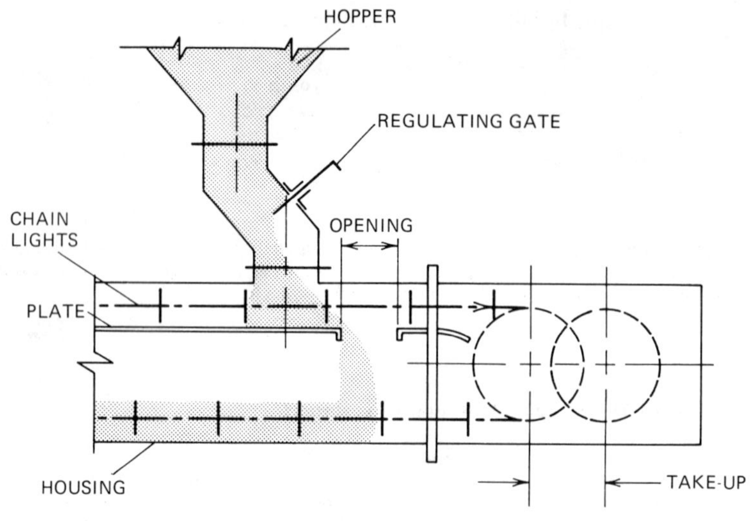

Fig. 24.4.3 Flight conveyor arranged for self-feeding. Limited to fine and granular materials.

Flight conveyor horsepower required consists of three factors:

I The horsepower required to run empty conveyor

plus

II The horsepower required to carry load over horizontal distance

plus

III The horsepower required to light load

This third component for lift can be disregarded in figuring a horizontal conveyor. These values can be expressed by the formula:

$$\text{Hp} = \overset{\text{(I)}}{\frac{0.06\ W_c L S F_c}{1000\ F_D}} + \overset{\text{(II)}}{\frac{T L_L F_L}{1000\ F_D}} + \overset{\text{(III)}}{\frac{TH}{1000\ F_D}}$$

Table 24.4.2 Chain Friction Factors F_C[a]

	Method of Mounting Flights			
Type of Chain Used	Flights Sliding on Bottom of Trough	Flights Sliding on Shoes on Side Guides	Flights Sliding on 2 Strands of Chain on Guides	Flights Rolling on 2 Strands of Roller Chain
Rivetless, detachable or similar (drag) chains	0.5	0.4	0.4	
No. 1130 roller				0.262
Steel bushed roller				0.150

[a] If return run can be carried on large diameter rollers spaced 10–12-ft centers, F_C for the run can be reduced to approximately 0.1.

where: S = speed of conveyor in ft/min
T = tons (2000 lb) of material handled per hour
W_C = total weight of single run of chains and flights per foot
L = horizontal length of loaded run of conveyor, in feet
L_L = horizontal length of loaded run of conveyor, in feet (total length of loaded runs of conveyor carrying on both runs)
H = lift in feet, vertical projector
F_C = friction factor for chain and flights (Table 24.4.2)
F_L = friction factor for load (Table 24.4.3)
F_D = friction factor for drive loss (Table 24.4.4)

Table 24.4.3 Load Friction Factors F_L

Material	Factor F_L	Material	Factor F_L	Material	Factor F_L
Coal slack, dry	0.65	Coal sized, wet	0.55	Wood chips, dry	0.27
Coal slack, wet	0.93	Fly ash	0.93	Grain, clean	0.48
Coal sized, dry	0.50	Sewage sludge, dried	1.1	Bicarbonate of soda, dry pulv.	1.00

Table 24.4.4 Drive Loss Factor F_D[a]

For Each Reduction (Sprockets or Sheaves)		Reduction for Each Helical Gear Speed Reducer		
Steel Roller Chain	V-Belt	Single	Double	Triple
0.90	0.93	0.95	0.93	0.91

[a] Average rating may vary with type and condition of unit.

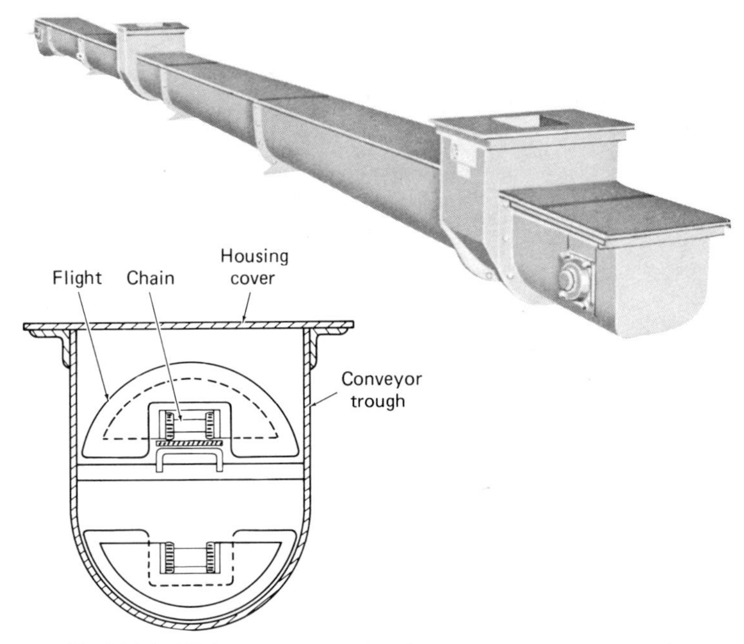

Fig. 24.4.4 Flight conveyor mounted in screw conveyor type trough.

Again, these friction factors F_C, F_L, and F_D have been set by, and must be modified by, individual experience to suit the individual application.

Although this discussion deals primarily with flight conveyors in their true definition (i.e., flights pushing material in a trough) there are other special versions of the chain/flight conveyor. One version (Fig. 24.4.4) utilizes chain and specially formed flight which travels through an enclosed, modified screw-conveyor-like trough. Special head/drive and tail/take-up terminals are required. The advantages of these units are:

Standardized parts are readily available.

They are self-cleaning.

They are gentle handling.

These units have been used extensively in the feed, grain, and fertilizer industries with great success. The unit pictured comes in sizes for 6–24 in. designed to handle 800–10,300 ft^3 at a chain speed ranging from 100 to 150 ft/min. There are larger versions of this unit used exclusively for grain. A 42-in.-wide unit of this type, for example, running at 250 ft/min will handle 79,000 bu/hr of wheat at 60 lb/bu. These units are available from a variety of manufacturers, most of whom also manufacture screw conveyors.

There are still other modifications of the chain/flight conveyor that utilize various combinations of chain and bar-type flights for conveying materials in enclosed troughs. These units are usually proprietary design of a given manufacturer. These types of units are also close cousins to the drag and en-masse conveyor, covered elsewhere in this handbook.

CHAPTER 25
METAL SCRAP AND CHIP HANDLING CONVEYORS

DANIEL T. FITZPATRICK

Prab Conveyors, Inc.
Kalamazoo, Michigan

25.1 SOLVING A NEED

Every firm involved in metal turning, boring, or milling has as one of its products a certain amount of scrap. That scrap presents problems. It must be removed from the production area; it is usually difficult to handle; it creates housekeeping problems because it is dirty, and in its raw state it has little or no value to offset its handling cost.

State and federal regulations involving hazardous wastes and ground water pollution also make some form of scrap processing or containment imperative. The plant manager who continues to leach cutting oils into the ground is simply borrowing time against the inevitable. At the same time, rocketing costs make some form of cutting-oil recovery economically attractive.

There is no simple—or single—answer to these problems. The type of work being done, production volume, the physical characteristics of the plant and its equipment, along with other specifics, all play a part in what can and should be done. This calls for an assessment of the particular situation, an understanding of the equipment and processes available, and, finally, an economic analysis of any proposed solution.

This section attempts to make understanding, choosing, and operating scrap handling equipment a little easier.

25.2 TYPES OF METAL SCRAP

Bushy Steel Scrap

Bushy steel scrap (machine shop turnings and borings) will include some fine particles but mostly it consists of spirals of various lengths and cross sections generated by turning and boring. A typical machine department generates spirals ranging from very fine hairlike pieces up to heavy brittle helixes. Unless the material is conveyed away from the machine as generated, it develops into large wads, bundles, or "Brillo" pads in containers. Cutting lubricants and coolants commonly make up a considerable amount of the weight of this material—up to 30%. Bulk density ranges from 5 to 20 lb/ft^3.

Broken Steel Chips

Bushy chips that have been passed through a continuous crusher are usually classified as shoveling chips, with a particle size no more than 2 or 3 in. in the major dimension. Milling machines, broaches, or lathes fitted with chip breakers can also generate this type of scrap. Due to the irregular nature and large surface area, a great deal of coolant or lubricant can be included. Bulk density ranges from 20 to 100 lb/ft^3.

Bushy Aluminum Chips

Generated in the same manner as bushy steel chips, this material is very light and fluffy with an extreme tendency to tangle on itself. Density range can be as low as 2 lb/ft^3 up to 10 lb/ft^3.

Broken Aluminum Chips

Generated in the same way as steel shoveling chips, this material is not necessarily any higher in density than the bushy aluminum chips. Cornflake-type particles are extremely difficult to move by gravity.

Brass Scrap

Most brass machining operations generate broken chips with a high percentage of small particles and fines. Some operations produce needlelike chips, and thread cutting will generate a bushy scrap; however, it is generally mixed in with broken chips and is a low percentage of the overall press scrap output in a given plant. Bulk density of brass scrap ranges from 40 to 100 lb/ft^3.

Cast Iron Scrap

Machining operations generate particles of various sizes ranging from a maximum of around $\frac{1}{8}$ in. down to fine dust. A considerable amount of coolant—generally water soluble—can be carried. Bulk density ranges from 60 to 100 lb/ft³. Certain types of cast iron scrap are cementlike and density may be from 100 to 130 lb/ft³.

Stamping Scrap

Particle size can range from very small slugs, which are generated by perforating operations, up to fairly large, heavy-gauge, irregular pieces, which are the result of trimming automotive parts. Electric motor lamination scrap is a particularly difficult material due to the thin gauges and tiny scrap particles. Lubricant content ranges from very small amounts on stamping operations to very large amounts used on deep-draw forming operations. Bulk density is from 50 to 150 lb/ft³.

Die Casting Scrap

This material consists of sprues, gates, risers, and so on, trimmed from die castings after they are removed from the die casting machines. It also includes scrap parts. A small amount of die lubricant usually adheres to the material, although not in sufficient quantities to severely limit remelt operations. Particle size may range from small chunks to long awkward pieces of irregular shapes. Die casting scrap has a tendency to catch on conveyors and tangle. This applies to aluminum, zinc, brass, and magnesium scrap.

25.3 TYPES OF METAL SCRAP AND CHIP HANDLING CONVEYORS

When choosing any conveyor, whether it is a single unit or a complete system, there are always several basic goals to be achieved:

1. An effective solution to the specific conveying need
2. The greatest economy, but balanced with reliable operation
3. Low operating cost and minimum maintenance
4. Flexibility to meet future needs

The first step is to clearly identify the needs and relate them to the particular advantages or disadvantages of the many types of conveyors available.

In this section is a brief description of the available types of metal scrap handling conveyors. There is no attempt to provide an exhaustive study because the subject is too broad. Individual situations almost always involve special characteristics so that judgment, generally based on experience, usually plays an important role.

The following scrap handling conveyors are illustrated in Fig. 25.3.1. It is important to remember that most of these units are manufactured in a great range of sizes and operating capacities. As with most mechanical equipment, they also have optimum performance characteristics within their total range, characteristics which are dependent on the type of scrap being handled.

Single Chain Drag Conveyer

The single chain drag conveyor handles curly turnings, tangled bundles, large or small stamping scrap, including motor lamination, die castings or die casting scrap, or any combination of these materials whether they are wet, dry, or flooded with coolant. Its flights are hinged to ride over jams, skimming part of the material. Successive flights reduce the jam and gradually eliminate it. For short or long runs, straight or inclined up to 45°.

Ram Conveyor

A liquid-tight trough is fitted with barbs to prevent backward movement of material in this type of conveyor. A track or wear plate supports the moving ram. The ram assembly has equally spaced pusher devices, and an actuating device—usually a long-stroke hydraulic cylinder—that provides a reciprocating slow-speed movement of approximately 5 ft in each direction.

This type of conveyor is usually mounted in or below the floor and covered by a series of floor plates except at input points. It is suitable for horizontal movements only and can be made in extremely long lengths with a single drive—upward of 500 ft. Common sizes range from 12 in. × 12 in. up to 36 in. × 36 in. depending on capacity required. It can be installed on a grade either with or against the flow of material to provide coolant drainage. Where coolant flow is extremely high, side baffles can be provided to protect against possible flooding of coolant onto the factory floor, or starving the coolant return system.

This type of conveyor is especially well suited to handling bushy turnings and mixtures of fine and bushy chips. It can be fitted with pressure relief valves, timers, and so on, which signal jam-ups

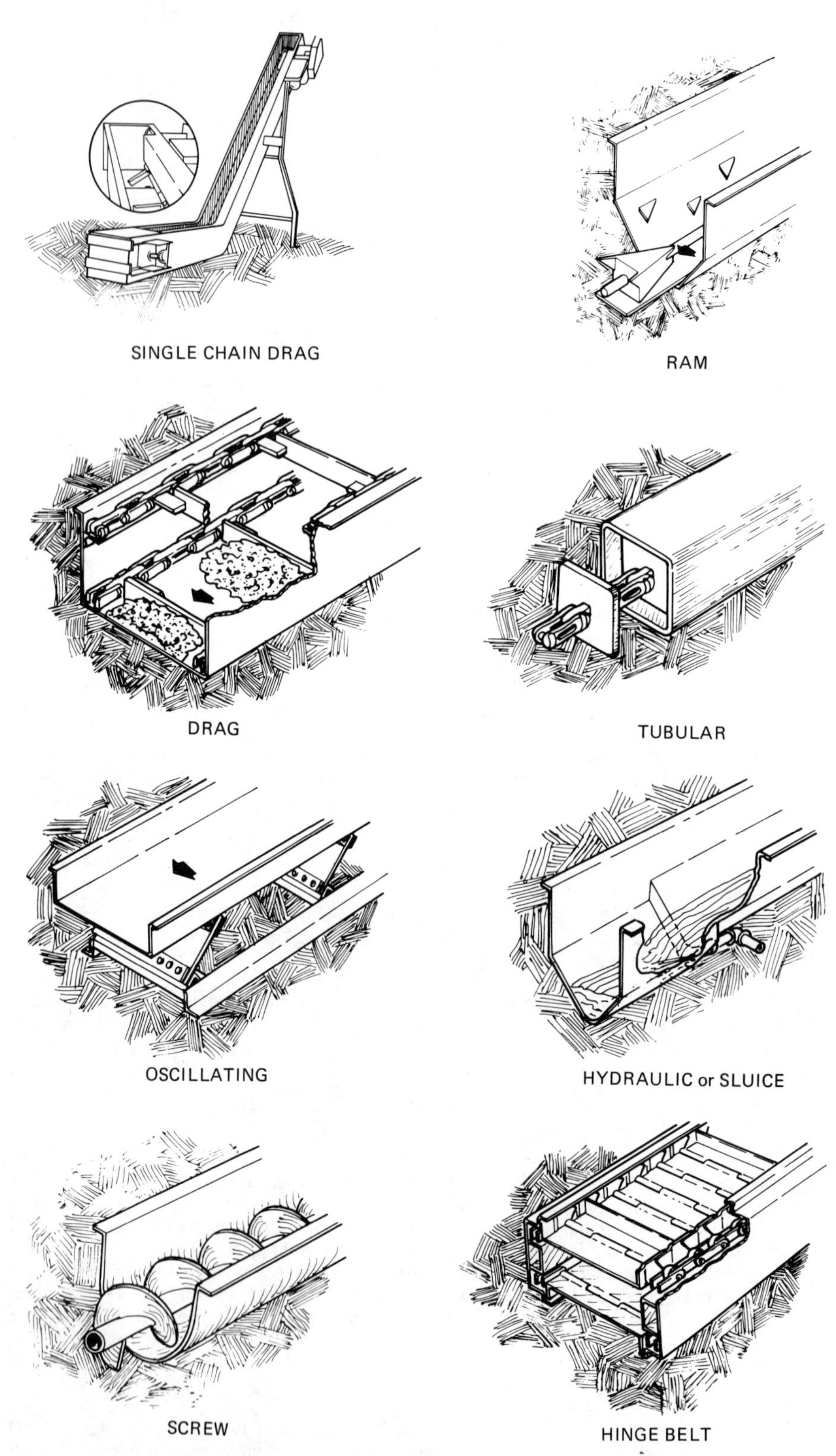

Fig. 25.3.1 Basic types of conveyors for scrap and chip handling.

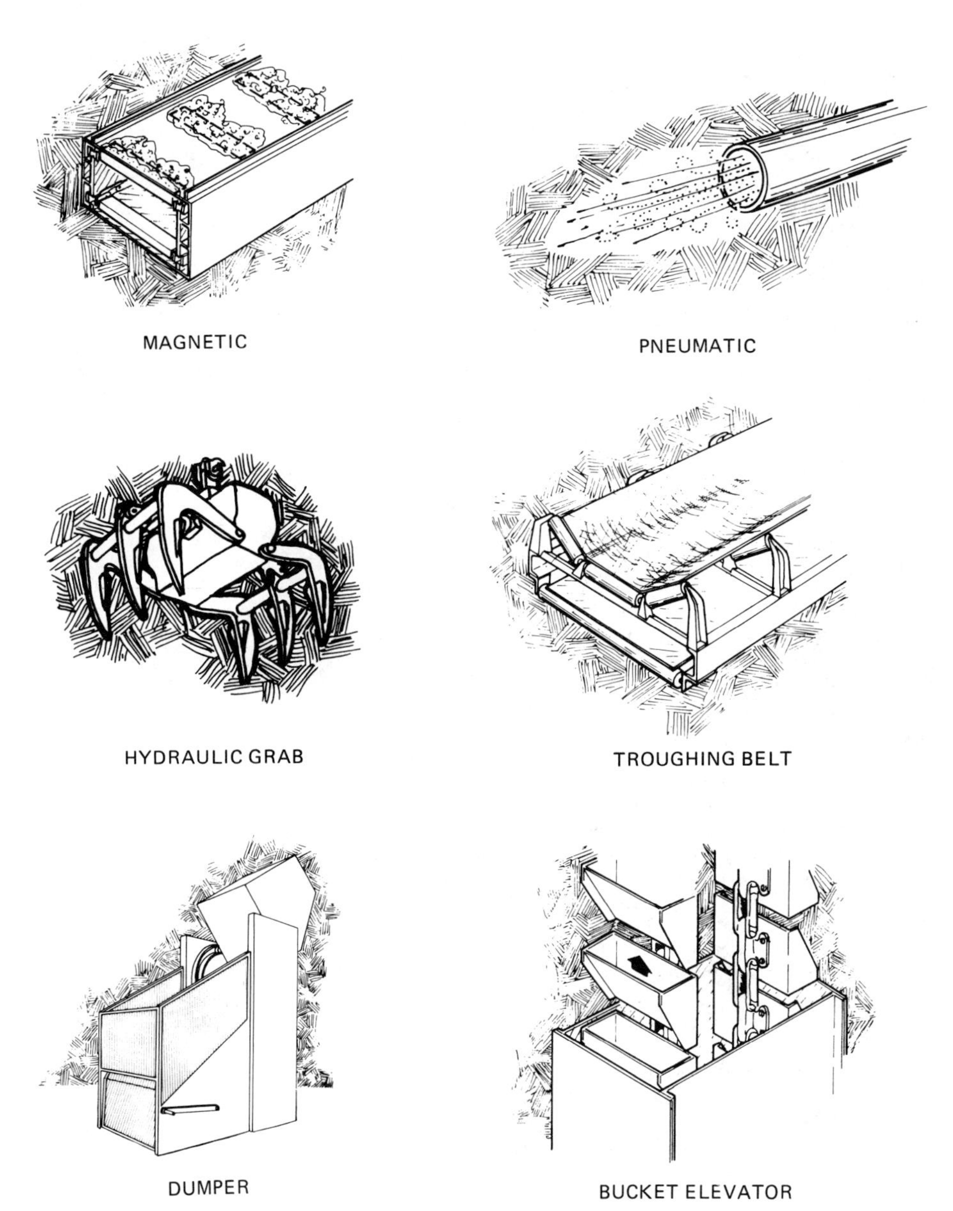

Fig. 25.3.1 (*Continued*)

that might be caused by bar ends or foreign material, preventing damage to the conveyor itself. This conveyor is low in cost and has extremely low maintenance and operating costs. It is available in both held-down and non-held-down models. The latter is best suited for most applications.

Drag Conveyor

The usual form of the drag conveyor consists of a rectangular trough 12 in. or more in width fitted with wear bars and support angles to accommodate parallel strands of chain in the upper and lower corners of the trough. Drag plates or flights are attached to the chains at a spacing commonly ranging from 12 to 24 in. depending on the application. Drive is through a headshaft fitted with suitable sprockets and driven at a very low speed, usually to provide a flight speed of around 10 ft/min.

This type of conveyor is well suited to handling finely divided materials with or without coolant flow in either direction. It is also capable of conveying on an incline. Material is carried on the bottom run of chain. The return portion of the chain runs back over the top of the carrying area. For this reason any material that clings to the chain or flights is automatically recirculated and given another chance to discharge. Thus the drag conveyor compares with other single-path conveyors in that carry-over problems are minimized. Maintenance on this type of conveyor is easy, since all components are visible when the cover plate is removed.

Multiple discharge points are easily provided on drag conveyors, since the carry run is on the bottom side and simple discharge gates are easily incorporated in the design.

Tubular Conveyors

Similar in function to the double-chain drag conveyor, the tubular (en masse or tubular drag) conveyor is designed with a single strand of chain, usually of the universal jointed type. This construction permits changes in direction along both horizontal and vertical planes. Extra fluid capacity can be obtained by proper sizing of the conveyor casing.

Tubular conveyors have several major application advantages. First of all, a single conveyor with a single drive can serve several different locations and alignments. Because the movement of the chain is in a loop configuration, it is possible to use all of the conveyor effectively rather than only one-half of it as is the case in over-and-under conveyor types such as belt conveyors and double-chain drag conveyors.

Providing the material to be conveyed is reasonably dense, a high-capacity conveyor can be installed in a relatively small and easily accessible trench. Narrow floor plates can be used to cover the trench, giving convenient access to the conveyor at any point on its length.

An elevating section can usually be provided as part of the collection conveyor, and this feature incidentally provides easy access to the drive.

This conveyor should not be used for masses of curly, stringy turnings, stamping scrap, or die casting scrap. It should only be used for handling finely divided free-flowing materials.

Oscillating Conveyors

A horizontal trough of any desired shape is reciprocated at a small angle to the horizontal with a frequency ranging from 400 to 600 cycles/min. This action provides steady movement of material at speeds of approximately 15 ft/min, depending on the nature of the material being conveyed. Ample capacity within a relatively small space is provided by suitable proportioning of the trough. Sticky fluid may impede material movement, since flow is not positive. A rigidized metal line often alleviates this problem. This type of conveyor is low in initial cost and has very low maintenance cost due to the small number of moving parts. Oscillating conveyors must be solidly anchored to the floor, otherwise material flow is reduced and progressive destruction of the conveyor and its mountings will occur. Only horizontal or downward sloping material movement should be planned with oscillating conveyors.

Hydraulic or Sluice Conveyors

Crushed or finely divided scrap can be conveyed along sluicing trenches provided ample coolant is available and the trenches are carefully planned to avoid hang-up points. Velocity nozzles are arranged at strategic points along the trench to maintain material flow. This type of conveyor is simple and quite maintenance free. However, it is only feasible in connection with large amounts of coolant and considerable energy output is required in order to keep the material moving.

Screw Conveyor

Heavy-duty screw conveyors, operating at relatively low speeds, have proven to be very effective in the straight-line conveying of crushed or flowable metal scrap. It should be emphasized that light-duty agricultural-type screw conveyors cannot be used in this application. Moderate upward slopes can be negotiated. This conveyor is particularly well suited for multiple discharge or continuous distribution.

Hinge Belt Conveyor

More versatile than any other type of metal scrap conveyor, the hinge belt conveyor can be used to handle any type of metal scrap from bushy material to fines, wet or dry, in any volume, and in a wide variety of conveyor paths. It is capable of combining horizontal and elevating movements. Counterbalancing these benefits is the fact that the hinge belt conveyor has a great many moving parts and a tendency to jam up unless it is very carefully applied. It serves best as an elevating conveyor on wet or dry materials that do not contain a large percentage of fines.

Magnetic Conveyor

A magnetic chip conveyor has a sealed frame with a stainless steel top surface which is a fixed part of the frame.

There are no external moving parts except for the headshaft and sprocket. The magnet assemblies, moving inside the frame below the stainless steel carrying surface, attract the particles of metal and move them toward the discharge area.

This type of conveyor is particularly suitable for applications involving very small particles, metallic sludges, or submersion in coolant tanks. Because very fine particles can be handled, action of the conveyor has a filtering effect on the coolant.

Sheet metal scrap in particular may contain very fine pieces with sharp corners which tend to hang up in the moving parts and cracks to be found on hinge belt conveyors. Although much costlier than a hinge belt conveyor, the magnetic conveyor may be less expensive in the long run when such fine particles predominate.

Pneumatic Conveyors

Crushed steel, crushed aluminum, brass, cast iron, and so forth, can be conveyed successfully in high-velocity penumatic conveyors. Pressure-type conveyors are ordinarily used to move fairly large volumes of chips over long distances, for example, crushed and dried chips conveyed from the chip house to overhead railroad storage hoppers. Specially reinforced elbows are a must due to the abrasive action. This equipment is very cost effective on long runs at high capacities.

Hydraulic Grab

Originally developed for log handling, hydraulic grabs have been used successfully in loading tangled material into crushers. They also have been used for loading sheet metal scrap into rail cars. Powered by a self-contained electric-motor-driven power unit, the grabs are relatively inexpensive and very rugged. Individual operators are required.

Troughing Belt

Most kinds of metal scrap can be handled on heavy-duty low-speed troughing belt conveyors, but they are generally limited to high-volume applications. They are suitable for long runs and gradual elevations not exceeding 20°. Normal applications for this conveyor are outdoors in scrap yards.

This type of conveyor has a tendency to spill material and a certain amount of carry-over has to be expected due to sharp particles sticking to the relatively soft belt surface. It is not suitable for impact of heavy objects or large bundles of tangled turnings.

Dumper/Lifter

The dumper/lifter is commonly employed to elevate tote bins of metal scrap for dumping into processing operation such as crushing or wringing. It is low in cost and requires little floor space. Effective use requires a standardized type and size of tote bin.

Bucket Elevator

When vertical movement of material is a requirement of a system layout, a heavy-duty chain-type bucket elevator can be used. A single or double strand of high-strength chain is fitted with closely spaced hardened-steel buckets.

Material to be conveyed is received in the boot section, and the buckets literally dig the material out of the boot section, then elevate and discharge with a centrifugal action.

High chain speeds are possible in this type of conveyor resulting in high capacity for the amount of cost involved. However, it should be emphasized that light-duty agricultural-type bucket elevators cannot be used for this application. Further, if material tends to pack, jamming in the boot section becomes a problem.

25.4 METAL SCRAP CONVEYOR SYSTEMS

25.4.1 Conveyor Selection

Table 25.4.1 relates various types of ferrous and nonferrous metal scrap to the metal scrap handling conveyors just described. It indicates the suitability or unsuitability of a particular conveyor to the materials and conditions noted.

Table 25.4.1 Metal Scrap Conveyor Selection[a]

		Straight-Line Primarily Horizontal				Combination Straight and Incline					Multidirection Path				Vertical Path	
Scrap	Coolant Volume	Oscil-lator	Screw	Ram	Trough Belt	Hinge Belt	Compac-veyor	Drag	Single Chain Drag	Magnetic	Grab	Tubular/ En Masse	Pneumatic	Hydraulic	Dumper/ Lifter	Bucket Elevator
Die casting	—	X				X			X						X	
Bushy steel	Low			X		X	X		X		X				X	
	High			X		X	X		X						X	
Broken steel,	Low	X	X		X	X	X	X	X	X		X	X	X	X	X
3 in. maximum	High					X		X	X	X		X		X	X	
Bushy	Low			X		X	X		X		X				X	
aluminum	High			X		X	X		X							
Broken	Low	X	X		X	X	X	X	X			X	X	X	X	X
aluminum	High					X		X	X			X		X	X	
Brass	Low	X	X	X	X		X	X	X			X	X	X	X	X
	High							X	X			X		X	X	
Cast iron	Low	X	X		X		X	X	X	X		X		X	X	X
	High							X	X	X		X		X	X	
Stamping	Dry	X			X	X			X	X					X	
	Sticky	X				X			X	X					X	

[a] X indicates a conveyor suitable for the material being conveyed.

It is evident that among the more than 100 situations covered in the table, more than one type of conveyor can be used in most cases. In practice, when the solution needed involves more than a short conveyor run, two or more conveyors used in combination often provide the best results.

The best solution will be the one that avoids the greatest number of potential problems. All scrap conveyor types have been developed to handle specific problems, but unfortunately they are commonly misapplied when chosen on the basis of minimum cost and inexperience. Failure to study the type of scrap to be moved is a prominent cause of misapplication. Scrap is normally not a homogenous material with reliable characteristics. The worst possible combination of coolant flow or lack of it, fine and coarse materials, and large and small foreign matter must be taken into consideration before final product selections are made.

Unless the scrap drops directly from the production machine into the conveyor, the conveyability of the scrap may be changed by crowding or packing it into containers. If something bad can happen, it will happen, and suitable provision to handle it in the conveyors is a must.

25.4.2 Typical Scrap Handling Conveyor Systems

In order to illustrate some of the ways in which basic conveyor types can be applied—either singly or in combination—to create systems, the following pages describe some examples of metal scrap handling systems now in operation. Among them are large and small systems, combinations of equipment, and methods of handling various types of commonly encountered metal scrap. Again, the units shown are not intended to be exhaustive of the possibilities. They are simply an indication of the potential for moving metal scrap more efficiently in modern plants.

Steel Stamping Press Scrap System

This system (Fig. 25.4.1) is one of three similar systems installed in a roller bearing plant. The systems are designed to move stamping scrap resulting from bearing retainer production. The scrap is relatively fine in nature and rather oily in some parts of the plant. While oscillating conveyors are not suitable for extremely sticky materials, this oily scrap is handled successfully in an oscillator/steel belt system.

Purpose. To collect scrap from production areas and discharge it into an open-top trailer outside the plant. The discharge conveyor is designed to distribute the load as it accumulates, avoiding any concentrated pile-up which might have to be shifted manually.

Design. Three types of conveyors have been interconnected to produce a system that fulfills the purpose while overcoming some particular operating hurdles.

The main run consists of a series of oscillating conveyors which were chosen because they are low in first cost, low in maintenance costs, able to handle surges in material flow, and are relatively quiet when solidly anchored as they must be.

Since the length of oscillating conveyors is limited to approximately 80 ft to avoid heavy drive loads, this system uses several sections. The height and width of the conveyor is altered at each inter-

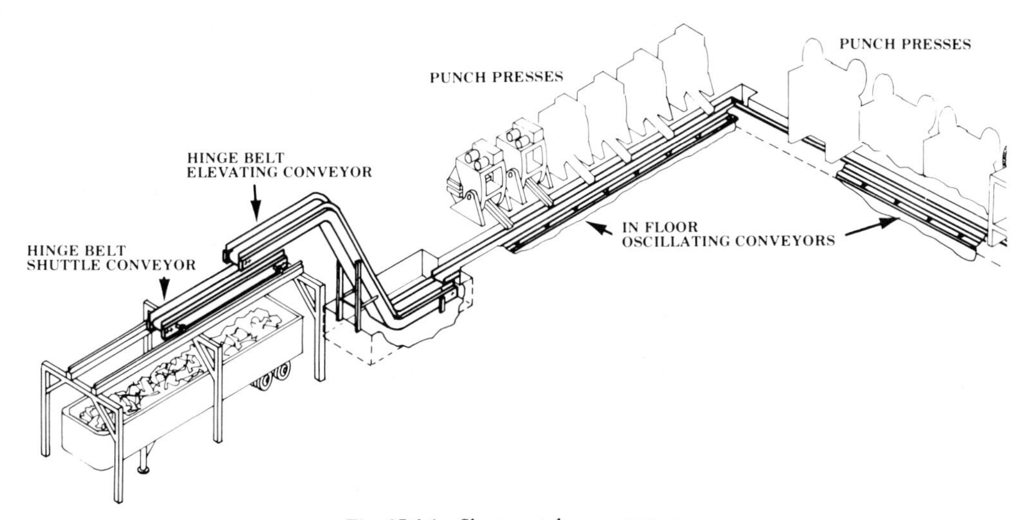

Fig. 25.4.1 Sheet metal scrap system.

change to avoid pinch points in the system which might create a resistance. This is important in oscillating conveyors because the flow of material is not positive and pinch points create jams.

Hinged steel belt conveyors are located outside the plant wall to elevate the discharge from the oscillating conveyor to the receiving trailer height in a very short horizontal distance.

Because of the oily nature of this scrap, there is a tendency for the scrap to carry beyond the discharge point, which can create two problems: (1) it can accumulate on the underside of the steel belt to create damaging jam-ups or, (2) it may discharge by gravity at random points where it is not wanted.

The solution was twofold. The top, horizontal run of the hinged steel belt conveyor was designed to be longer than usual, giving overrunning scrap more room to discharge by gravity before it might cause problems.

A hinged steel belt shuttle conveyor was added. Moving along the length of the receiving trailer, it distributes the load evenly. It also catches overrun scrap from the horizontal run of the elevating conveyor.

There is much discussion about the sound levels of oscillating conveyors that calls for clarification. In themselves, most oscillating conveyors are quiet as a result of design innovations. Sound can be generated from the material being conveyed, through improper anchoring or as a result of being run too fast. Both anchoring and operating speed are solved by proper system design and, where the material being handled creates a problem, trench installations lower the sound below any objectionable level in the working environment.

Oily Brass Turning System

This drag conveyor system was designed and successfully put into use conveying oily brass scrap with occasional bar ends (Fig. 25.4.2).

Purpose. To provide an in-floor system to collect oily brass scrap from two lines of automatic screw machines, convey to bar end separator, from bar end separator to continuous centrifuge, then elevate for discharge.

Design. All conveyors in this system are heavy-duty drag conveyors built around double strands of rivetless forged-steel chain fitted with floating drag flights. This type of conveyor was adaptable to

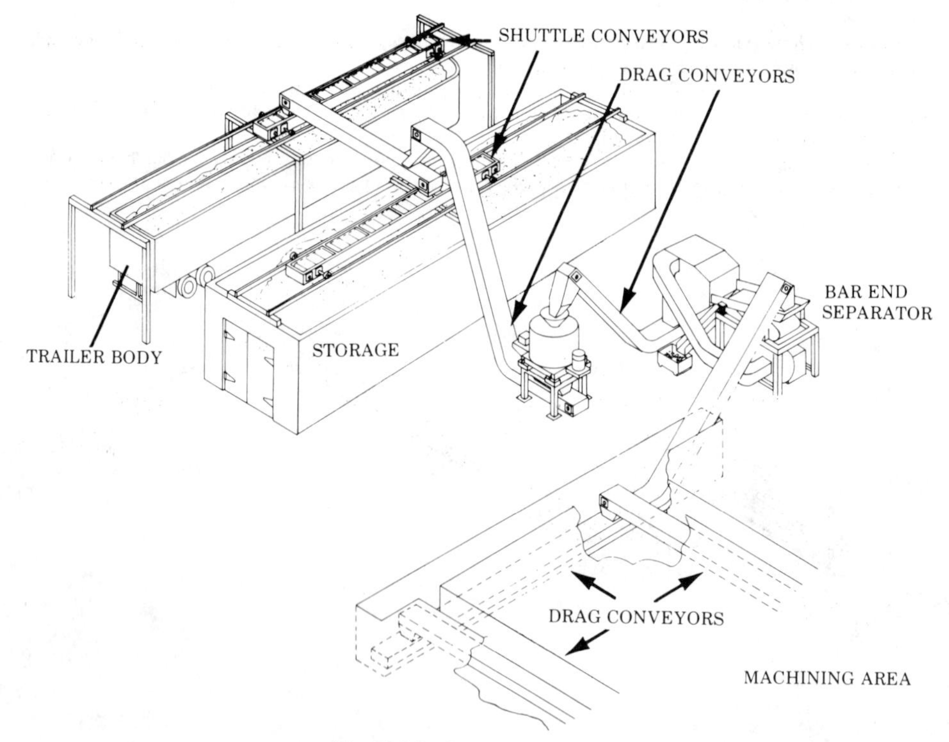

Fig. 25.4.2 Brass scrap system.

all runs so a high degree of part interchangeability was possible. (Note: ram conveyors work equally as well on brass scrap and should be considered for in-floor scrap collection.)

Brass scrap is a high-value product, and it was considered necessary to provide a secure storage area for the material to accommodate the production flow when it was not possible to load directly into a trailer truck.

The dry chips from the centrifuge are elevated to a horizontal conveyor which discharges into either one of two shuttle discharge conveyors. The shuttle conveyors are so arranged as to move fore and aft over a rectangular area to achieve uniform loading over the entire length of the area. The shuttle conveyor above the loading dock distributes the scrap into an open-top trailer body. The shuttle conveyor in the storage area is similarly arranged to distribute the material in the area for maximum use of the space.

When no truck is available for loading at the dock, scrap is loaded in the storage area and then later transferred to a truck by means of a front-end loader.

System for Chips of All Types

This system is unusual in that it is a multipurpose system and involves five different types of chip conveyors, each one selected to best perform its assigned function (Fig. 25.4.3).

Purpose. To collect steel chips of three different kinds from three different areas, separate the two kinds of coolant oil from the chips and channel them to separate filtration units, then feed all the chips through processing equipment and elevate the crushed chips to a storage hopper above a railroad track. Total capacity 300 lb/hr.

Design. Three machining areas are involved. Area 1 consists of bar machines which generate bushy steel turnings coated with heavy coolant oil. Area 2 consists of 90 gear hobbers and shapers which generate fine chips with heavy coolant oil. Area 3 consists of chuckers and bullards which generate bushy steel chips and fine steel chips coated with water-soluble coolant.

Conveyor A is a ram conveyor, 100 ft long, selected because it is the most trouble-free method of conveying bushy steel chips combined with coolant. This unit feeds into Conveyor B, and at this point it is desired to have the chips go in one direction, toward the heavy oil side.

Consequently, Conveyor B is a hinge belt conveyor with a sloping configuration so that the oil on the chips will drain through the perforated belt to the bottom cover of the conveyor. From there it drains back into Conveyor E. The bushy chips proceed to Conveyor C.

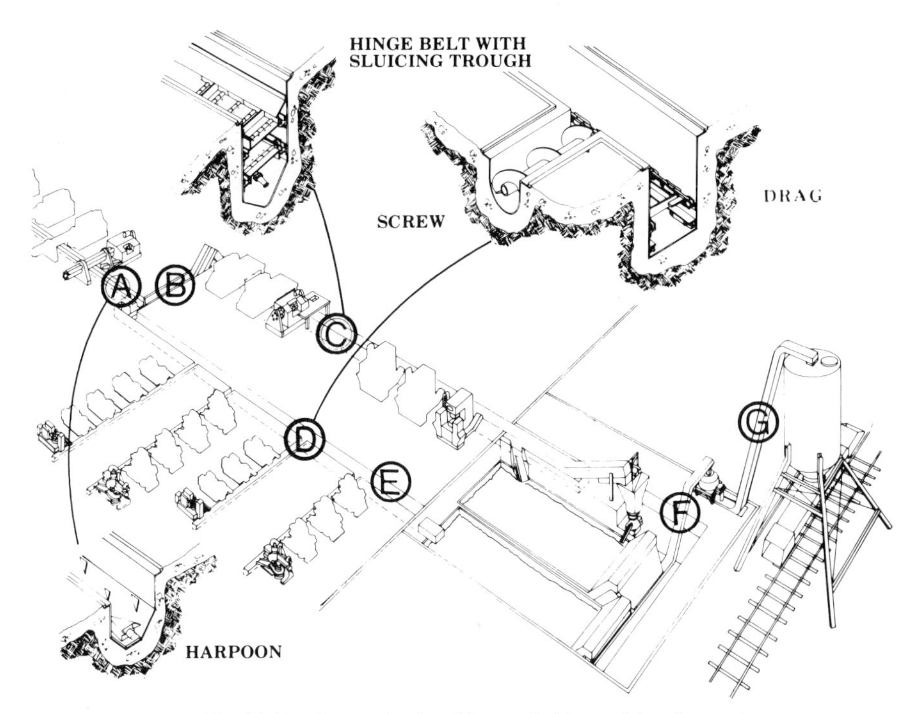

Fig. 25.4.3 System for handling steel chips and turnings.

Conveyor C is also a hinge belt conveyor, 230 ft long, but it is combined with a hydraulic/sluice conveyor running beneath it.

Large quantities of very flowable water-soluble coolant oil wash through the perforated hinged belt, taking with them a sizable quantity of fines. The combination of water-soluble oil and fines is propelled to the filter unit by means of high-velocity nozzles strategically placed along the length of the sluice. These nozzles receive fluid from the filter unit.

Conveyor C has an inclined section at the discharge end. It feeds a crusher which then directs the crushed bushy chips into the filtration unit.

Conveyors D are screw conveyors, selected as the most economical means to handle combination of fine steel chips and heavy oil. These screw conveyors vary in length from 80 to 120 ft, and serve 90 machines.

The fine oily chips are fed into Conveyor E which is a double-chain drag conveyor, 200 ft long, providing ample capacity for the total quantity of heavy oil generated by the system as well as all of the fine chips. This conveyor then discharges its entire load into the heavy-oil filtration unit.

Conveyor F is a double-chain drag conveyor, selected for its capability to elevate the combination of crushed bushy chips and fine chips into centrifuges.

Conveyor G is a hinge belt conveyor which elevates the dried chips to a storage hopper.

This example points up the fact that scrap handling systems usually require several different types of conveyors in order to obtain the most practical design. This system represents the thoughtful consideration of both functional and cost factors in order to achieve system objectives at minimum cost with minimum operating headaches.

CHAPTER 26

SCREW, VIBRATORY, AND EN MASSE CONVEYORS

B. J. HINTERLONG

Continental Screw Conveyor Corporation
St. Joseph, Missouri

CONVEYOR EQUIPMENT MANUFACTURERS ASSOCIATION

Vibrating Equipment Section
Washington, D.C.

A. D. SINDEN

Stephens-Adamson Manufacturing Co.
Aurora, Illinois

26.1 SCREW CONVEYORS

B. J. Hinterlong

26.1.1 Introduction

The first screw conveyor as such was designed by Archimedes for removing water from the hold of a ship built for King Hiero of Syracuse. The first Archimedian conveyor was of an internal helix design mounted at an angle with the lower end in the water and the upper end arranged to discharge. These early designs were used primarily to pump water. The American inventor John Fitch prepared drawings a little before 1790 showing a section of screw conveyor flighting used to propel his steamboat. This flighting was very similar to today's design of flighting.

However, the big growth in screw conveyors came about 1783 when a man named Oliver Evans laid out the first mechanized flour mill which incorporated screw conveyors. These units were powered by a system of wooden-tooth gears, wooden pulleys, and leather belts which were driven by a single waterwheel. The first flour mill built by Evans in 1745 incorporated the screw conveyor with a round wooden core on which a series of helical forms or wooden plows were mounted. The whole assembly revolved in a wooden trough or box as it was then called. Later Evans improved the design by mounting helical sheet metal sections on the wooden core. These metal sections were originally of the sectional flight variety and were later mounted to a section of pipe and butt-welded together.

The next technological advance occurred when Frank C. Caldwell patented the continuous one-piece screw flight formed by rolling a continuous strip through two cones. This today is what we call helicoid flighting.

The original screw conveyors used wooden bearings which are still used by many in the grain industry today. Cast iron bearings originated about the same time the steel flight and trough evolved. Today there are a great variety of nylons and other synthetic materials that are used for bearings.

Enclosed drive speed reduction gears replaced the wooden gears and leather belts allowing the screw conveyor to be driven by various means of power (hydraulic, electrical, etc.), which gives the designer a great versatility in applying the screw conveyor. The screw conveyor is capable of handling a great variety of products; many products that are toxic to human beings or toxic at certain stages of the processing can easily be handled. A screw conveyor is a totally enclosed piece of equipment capable of containing dust, gas, or fumes, depending on the type of seals that have been designed into the equipment.

Screw conveyors are not limited to conveying horizontally. With a specialized design the unit may operate at a slope or in the vertical position. In general the screw conveyor is the oldest form of conveyor known today and is still one of the most useful mechanical handling devices used by industry.

26.1.2 Applications

A screw conveyor is a bulk material handling device capable of handling a great variety of materials that have good flowability. The organization Conveyor Equipment Manufacturers Association (CEMA) has defined in their Publication No. 550, *Classification and Definitions of Bulk Materials,* a system of denoting the various degrees of flowability in materials. However, some other materials may have differing characteristics based on the temperature and moisture content of the given material at the time being conveyed.

One of the primary advantages of the screw conveyor is the number of feed inlets and discharge openings that can be provided for. This allows the screw conveyor to receive and distribute material for in-plant storage to a number of locations. Typical applications include grain storage, feed mills, cereal processing, and chemical plants.

Conveyors can also be adapted for volume control. When utilized in this form they are called screw feeders and are attached to the bottom of bins, hoppers, bag dumps, storage piles, and so on. These units are designed to regulate the flow of material to the downstream equipment.

Screw conveyors also allow the heating and cooling of material while in transit. This can be accomplished by jacketing the trough for circulating cooling or heating medium through the trough jacket. Similarly, this can be accomplished by circulating fluids through the helix itself.

Toxic materials can easily be handled with a screw conveyor because the enclosed trough can be built tight enough to contain the toxic dusts or vapors and reduce personnel hazards. On the other hand, materials that must be free from contaminants may also be handled because the enclosed screw conveyor trough protects the materials from outside influences.

26.1.3 Design Considerations

The key to any successful screw conveyor design is, first, a thorough understanding of the characteristics of the material to be handled. Table 26.1.3 gives general information on a variety of materials that can be handled. Second, the action of a screw conveyor is important to understand. In conveying, material tends to tumble and shear; this action may or may not have detrimental effects on the final output. Finally, it is most important to have a thorough knowledge and understanding of the way material flows and the effects of variations in the flow. Capacities of bulk material handling equipment are usually given in tons per hour, pounds per hour, or pounds per minute. Maximum capacity is not usually the average daily or hourly output. Also, the apparent density in the material could vary. The conveyor size and speed must be based on the maximum volume and the apparent as-conveyed density of the material. To determine the correct volume entry conveyor capacity, maximum poundage should be divided by the least apparent density expected in the material.

Surge loads are sometimes experienced in most conveyor systems. The surge load is dependent on what type of device is used to initiate the flow. Even with the more sophisticated feed regulations, surges do occur and care should be exercised to assure that the conveyors have the capacity to handle these surge volumes.

Material Characteristics

Many studies have been made to define the characteristics of bulk materials. Table 26.1.4 gives general descriptions of the various types of characteristics readily recognizable. The CEMA Publication No. 350, *Screw Conveyors,* offers a more detailed in-depth analysis of the various materials that may be handled by screw conveyors.

It should be kept in mind that the action of a screw conveyor could possibly change the characteristics of the material in transit. It should be checked closely.

If possible all materials should be subject to a screen analysis test. This will determine the percentages of the different size material within a given test. The fine dusting materials may require gaskets or seals while others of a larger particle size may not.

Material with lumps should also be defined closely. Table 26.1.5 indicates the maximum lump size that can be handled in a given size conveyor.

Irregular and interlocking materials may cling together and require special consideration. They may wrap around the pipe or the hanger resulting in a jammed condition. Materials that pack under pressure also need special consideration and should be studied in detail in respect to their action within the conveyor.

In determining the flowability of the material two considerations must be studied. One is the angle of slide, and the other is the internal angle of friction of the material. The angle of slide may be determined by tilting the plane carrying a quantity of material. By measuring the angle at which the material slides, a sliding friction factor can be determined. The internal angle of friction can be determined through the shear cell test. Any changes in moisture, temperature, particle size, or corrosion characteristics of the material will affect the flowability which will in turn affect horsepower and other design principles. The abrasiveness of materials is also a subjective quantity and is not easily defined. Tables 26.1.1 and 26.1.2 contain data on bearing, shaft style, and thickness for various abrasive groupings. Table 26.1.3 indicates which grouping is suggested. A selection of components for handling abrasive

Table 26.1.1 Selection of Component Group[a]

	Components		
Component Group	Hanger Bearing	Couplings	Weight of Flights and Trough[b]
A	Babbitted or nylon	Cold-finished steel	Standard duty
B	Oil-impregnated wood or nylon	Cold-finished steel	Standard duty
C	Self-lubricating bronze	Cold-finished steel	Standard duty
D	Hard iron	Hardened steel	Medium duty
E	Hard iron	Hardened steel	Heavy duty
F	Babbitted or nylon	Cold-finished steel	Heavy duty
G	Oil-impregnated wood or nylon	Cold-finished steel	Medium duty
H	Babbitted or nylon	Cold-finished steel	Medium duty
J	Self-lubricating	Cold-finished steel	Medium duty

[a] Refer to Table 26.1.3 for component group used based on the material to be handled in the conveyor.
[b] See Table 26.1.2.

Table 26.1.2 Specifications of Flights and Trough

Component Groups A, B, and C[a]

Diam. of Conv.....		4″	6″	9″		10″		12″			14″		16″
Coupling Size.....		1	$1\frac{1}{2}$	$1\frac{1}{2}$	2	$1\frac{1}{2}$	2	2	$2\frac{7}{16}$	3	$2\frac{7}{16}$	3	3
Conv. No.	Helicoid	4H204	6H304	9H306	9H406	10H306	10H412	12H408	12H508	12H614	14H508	14H614	16H610
	Sectional	.....	6S307	9S307	9S407	.....	.....	12S409	12S509	12S612	14S509	14S612	16S609
Gauge	Trough	16	16	14	14	14	14	12	12	12	12	12	12
	Cover	16	16	16	16	16	16	14	14	14	14	14	14

Component Group D[a]

Conv. No.	Helicoid	4H206	6H308	9H312	9H412	10H306	10H412	12H412	12H512	12H614	14H508	14H614	16H614
	Sectional	.....	6S309	9S312	9S412	10S309	10S412	12S412	12S512	12S616	14S512	14S616	16S616
Gauge	Trough	14	12	12	12	12	12	10	10	10	$\frac{3}{16}$	$\frac{3}{16}$	$\frac{3}{16}$
	Cover	16	14	14	14	14	14	14	14	14	14	14	14

Component Groups E and F[a]

Conv. No.	Helicoid	4H206	6H312	9H312	9H316	.....	.....	.....	.....	12H614	.....	14H614	16H614
	Sectional	.....	6S312	9S312	9S416	10S309	10S416	12S416	12S516	12S616	14S512	14S616	16S616
Gauge	Trough	12	$\frac{3}{16}$	$\frac{3}{16}$	$\frac{3}{16}$	$\frac{3}{16}$	$\frac{3}{16}$	$\frac{3}{16}$	$\frac{3}{16}$	$\frac{3}{16}$	$\frac{3}{16}$	$\frac{3}{16}$	$\frac{3}{16}$
	Cover	16	14	14	14	14	14	14	14	14	14	14	14

Component Groups G, H, and J[a]

Conv. No.	Helicoid	4H206	6H308	9H312	9H412	10S306	10H412	12H412	12H512	12H614	14H508	14H614	16H614
	Sectional	.....	6S309	9S312	9S412	10S309	10S412	12S412	12S512	12S616	14S512	14S616	16S616
Gauge	Trough	14	12	12	12	12	12	10	10	10	$\frac{3}{16}$	$\frac{3}{16}$	$\frac{3}{16}$
	Cover	16	14	14	14	14	14	14	14	14	14	14	14

[a] See Table 26.1.1.

Table 26.1.3 Materials

Material	Average Weight per Cu. Ft	Loading Classification[a]	Type Conveyor to Use[b]	Horsepower Factor, F_m	Symbol[c]
Alfalfa meal	17	II	A,B,C	0.5	f
Almonds, broken or whole	30	IIA	E	0.9	d
Alum, lumpy	50–60	II	A,B,C	1.0	v
Alum, pulverized, fine	45–50	I	A,B,C	0.6	v
Alumina (aluminum oxide)	75–100	III	E	1.7	v
Ammonium chloride, crystalline	45–50	II	A,B,C	0.4	—
Asbestos, shred	20–25	IIA	A,B,C	1.0	f,m,p
Ashes, dry	40	IIA	E	2.0	c
Asphalt, crushed	45	II	A,B,C	1.0	—
Bakelite and similar plastics, powdered	35–45	II	G	1.4	v
Baking powder	40–55	I	B	0.6	v
Barley	38	I	A,B,C	0.4	e
Barytes, powdered	120–140	IIA	E	1.0	v
Bauxite, crushed, dry	75–80	III	E	1.8	v
Beans, castor	36	I	A,B,C	0.4	—
Beans, navy	48	I	A,B,C	0.4	—
Bentonite, minus 100 mesh	50–60	IIA	E	0.9	a,v
Bicarbonate of soda	45–55	II	A,B,C	0.4	—
Bone black, pulverized	20–25	IIA	E	0.9	v
Bone meal	55–60	II	E	1.2	v
Bones, crushed	35–50	IIA	E	1.0	v
Borax, powdered, fine	50–60	II	A,B,C	0.6	v
Boron	75	II	A,B,C	1.0	—
Bran	16	II	A,B,C	0.4	e,f
Brewer's grain, spent, dry	28	II	A,B,C	0.4	v
Brewer's grain, spent, wet	58	II	H	0.6	v,c
Buckwheat	42	I	A,B,C	0.4	e
Calcium carbide, crushed	70–80	IIA	E	1.0	—
Carbon, activated, dry, fine	8–20	IIA	E	1.2	d,v
Carbon black, pellets	25	II	C	1.6	v
Carbon black, powder	4–6	II	C	0.4	v
Cement clinker	75–80	III	E	1.8	v
Cement, portland, aerated	65	IIA	E	0.6	v
Chalk, crushed	88–95	IIA	E	1.0	p
Chalk, pulverized	70–75	IIA	E	0.7	a,p
Charcoal	15–34	IIA	E	1.2	d
Cinders, coal	40	III	E	1.8	v
Clay, brick or tile, dry ground	65–80	IIA	E	1.5	v
Coal, anthracite, sized	55–60	II	F,G,H	0.9	d,h,v,c
Coal, fines or slack	50	I	H	0.7	c
Coal, pulverized	35	I	A,B,C	0.6	c
Coal, sized	50	II	H	0.6	c
Cocoa, powdered	30–35	II	B	0.6	p

Table 26.1.3 Materials (*continued*)

Material	Average Weight per Cu. Ft.	Loading Classification[a]	Type Conveyor to use[b]	Horsepower Factor, F_m	Symbol[c]
Cocoa beans	35	IIA	E	0.9	d
Coffee, ground	25	II	B	0.6	d
Coffee beans, green	25–30	II	A,B,C	0.4	d
Coffee beans, roasted	25–30	I	A,B,C	0.4	v
Coke, loose	22–30	III	E	2.0	d,m
Coke, petroleum	35–45	IIA	E	1.0	m
Copper sulphate	60–70	II	F,G,H	0.5	v
Cork, ground	8–10	II	A,B,C	0.5	a,f
Corn, shelled	45	I	A,B,C	0.4	e
Corn grits	40–45	II	A,B,C	0.4	v
Cornmeal	40	II	A,B,C	0.4	—
Cottonseed, dry	20–25	II	A,B,C	0.4	v
Cottonseed, cake, cracked	43	II	A,B,C	0.5	—
Cottonseed hulls	12	II	A,B,C	0.5	f
Cottonseed meal	38	I	A,B,C	0.4	—
Cottonseed meats	40	I	A,B,C	0.4	—
Cullet (broken glass)	80–120	III	E	1.8	v
Diatomaceous earth (diatomite), dry	12–17	III	E	1.6	v
Dicalcium phosphate (calcium phosphate)	40–50	II	A,B,C	0.5	—
Dolomite, crushed	90–100	IIA	E	1.0	v
Epsom salts (mag. sulphate)	40–60	II	A,B,C	0.4	v
Feldspar, ground	65–80	IIA	E	1.5	v
Feldspar, lumps	90–100	IIA	E	1.5	v
Ferrous sulphate	60–75	IIA	E	0.9	v
Flaxseed	45	I	A,B,C	0.4	e
Flaxseed cake	50	II	A,B,C	0.6	—
Flaxseed meal	25	II	A,B,C	0.4	—
Flour, wheat	35	II	B	0.6	d,v
Flourspar	95–105	IIA	E	1.0	v
Flue dust, blast furnace	35–45	III	E	1.6	a,v
Fly ash, boiler house	35–40	IIA	E	1.0	v
Foundry sand, dry	90–95	III	E	1.7	—
Fullers earth, raw	38	III	E	1.0	—
Fullers earth, spent, 35% oil	63	III	E	0.9	—
Gelatin, granulated	32	II	A,B,C	0.6	—
Glass batch	90–100	III	E	1.8	v
Graphite, flake	40	II	A,B,C	0.4	—
Graphite, flour	28	I	A,B,C	0.4	a
Graphite, ore	65–75	IIA	F	0.4	v
Grass seed	10–12	I	A,B,C	0.4	e,f
Gypsum, calcined	58	IIA	E	0.9	v
Gypsum, crushed	75–80	IIA	E	1.0	v
Hops (humulus), spent, dry	35	II	F	0.5	—
Hops (humulus), spent, wet	55	II	F	0.5	c,p
Ice, crushed	40	II	F	0.4	—
Ilmenite ore	140–160	III	E	1.7	v
Lead ores (cerussite), crushed	180–230	IIA	E	1.0	v

Lignite (brown coal)	45–50	II	F	0.5	v
Limanite (brown ore)	120	III	E	1.7	—
Lime, ground (unslaked)	60–65	II	D	0.6	p,v
Lime, hydrated (cal. hydroxide)	20–35	II	F	0.8	a,p,v
Lime, hydrated, pulverized and air separated, 200 mesh	32–40	II	D	0.6	a,p,v
Lime, pebble	56	IIA	E	2.0	w
Limestone, crushed	85–95	IIA	E	1.6	v
Limestone dust	70–80	IIA	E	1.0	v
Limestone screenings	88	IIA	E	1.6	—
Linseed cake, pea size	50	II	A,B,C	0.4	—
Litharge, pulverized (lead oxide)	200–250	IIA	E	1.0	v
Lithopone	120–140	II	A,B,C	1.0	v
Malt, dry, crushed	20–30	I	A,B,C	0.4	e,f,v
Malt, dry, whole	20–30	II	A,B,C	0.4	e,v
Malt, wet or green	60–65	II	H	0.4	v
Malt meal	38	II	B	0.4	—
Manganese ore (crushed)	150–200	III	E	2.0	v
Marble, crushed	80–95	III	E	2.0	v
Marl (clay)	80	IIA	E	1.5	—
Mica, flake, pulv. or ground	20–30	II	D	0.9	v
Milk, dried, flake	5–6	II	B	0.4	d,v,h
Milk, malted	27	II	B	0.5	d,p
Milo maize	56	II	A,B,C	0.4	e
Oats	26	I	A,B,C	0.4	e
Oyster shells, ground, under $\frac{1}{2}$ in.	50–60	IIA	E	0.9	v
Paper pulp stock	—	IIA	G	1.0	m,v
Paraffine cake, broken	30–45	II	H	0.5	v

Peanuts, shelled	35–45	II	H	0.4	d,v
Peanuts, unshelled	15–20	II	H	0.6	v
Phosphate, rock granular	90–100	IIA	E	1.7	v
Pumice, ground	40–45	III	E	1.6	v
Quartz, pulverized or granular	85	III	E	1.8	—
Rice, clean	45	I	A,B,C	0.4	—
Rice, rough	36	II	A,B,C	0.4	e
Rice bran	22	I	A,B,C	0.4	—
Rice grits	45	I	G	0.4	—
Rice hulls	16	I	A,B,C	0.6	—
Rye	44	I	A,B,C	0.4	e
Salt, coarse	45–50	IIA	E	1.0	c,v
Salt, dry fines	70–80	IIA	E	1.7	c
Salt cake, dry, broken	65–85	II	F	1.0	c
Sand, dry	100	III	E	1.7	—
Sand, silica, dry	90–100	III	E	2.0	v
Sawdust	10–12	I	A,B,C	0.5	v
Shale, crushed	85–90	IIA	E	1.0	v
Shavings, wood	8–15	II	A,B,C	0.5	f,m,v
Silica gel	45	III	E	1.7	h
Slate, crushed	80–90	II	F	1.0	v
Sludge, sewage, dry	45–55	IIA	E	0.8	v,c
Soap, powdered or chips	15–25	II	A,B,C	0.4	d,v
Soda ash, dense (heavy)	60–65	IIA	D	0.8	v
Soda ash, light	20–30	IIA	D	0.8	f
Sodium phosphate	50–65	IIA	A,B,C	0.9	v
Sodium sulphate	65–85	II	F	0.5	v
Soybeans, cracked	32–36	II	A,B,C	0.5	e
Soybeans, whole	45–50	III	E	1.6	e,v

Table 26.1.3 (*continued*)

Material	Average Weight per Cu. Ft	Loading Classification[a]	Type Conveyor to Use[b]	Horsepower Factor, F_m	Symbol[c]	Material	Average Weight per Cu. Ft	Loading Classification[a]	Type Conveyor to Use[b]	Horsepower Factor, F_m	Symbol[c]
Soybean meal	40	II	A,B,C	0.5	—	Talc, pulverized	80–90	II	B,C	0.6	a,v
Starch	45–50	II	A,B,C	1.0	v	Tobacco scraps	15–25	II	G,H	0.5	f,v
Steel chips, crushed	25–85	IIA	E	1.6	v	Vermiculite, expanded	16	II	A,B,C	0.5	—
Sugar, raw	55–60	II	G,H	1.0	p,r,v	Vermiculite ore	70–80	II	G,H	0.5	v
Sugar, refined	50–55	II	G,H	0.5	d,v	Wheat, cracked	48	II	A,B,C	0.4	e
Sugar beet pulp, dry	10–15	II	A,B,C	0.8	v	White lead, dry	75–100	II	E	1.0	v
Sugar beet pulp, wet	25–45	II	H	0.9	v	Wood shavings	8–15	II	E	0.5	f,m,v
Sulphur, lumpy	70–75	II	G,H	0.6	e,v	Zinc oxide, heavy	30–35	II	A,B,C	1.0	p,v
Sulphur, powdered	60–65	II	A,B,C	0.6	e,v	Zinc oxide, light	10–15	II	A,B,C	2.0	f,p,v

[a] See Table 26.1.5.
[b] See Tables 26.1.1 and 26.1.2.
[c] See Table 26.1.4.

Table 26.1.4 Characteristics of Materials

Symbol	Characteristics
a	Aerates and seeks its own level.
c	Corrosive.
d	Degradable or contaminable, affecting its use. Where contamination of the product must be avoided, use oil-impregnated wood bearings.
e	Explosive dust possible.
f	Fluffy and light.
g	Contains harmful dust or fumes which should be confined.
h	Hygroscopic, absorbs and retains moisture.
m	Mats or interlocks (fibrous).
p	Packs under pressure.
r	Tends to stick to flights and pipe; ribbon conveyor recommended.
v	Weight varies considerably. Wide variations in weights occur in some materials due to conditions of material as produced for specific purposes, or to moisture content. Check the weight of product whenever possible.
w	May require special-diameter flights or trough due to the size of the lumps causing a wedging action between flights and trough.

material should be reviewed in regard to the service to which a conveyor will be subject. Twenty-four-hour-a-day operations will cause more wear than a few hours per day. A more detailed explanation on all types of material characteristics is provided in CEMA Publication No. 550, *Classification and Definitions of Bulk Material.* Chapter 2 of that publication fully explains size classification and coding, flowability coding, and abrasive coding.

The information offered in Table 26.1.1 is only a guide. The material code and material factor are based on our experience. In all cases a sample of the material should be obtained and the various design parameters should be developed for each individual sample.

Conveyor Size and Speed

To determine the conveyor size and speed it is first necessary to turn to Table 26.1.3 and establish the loading classification for the given material. After the loading classification has been determined, refer to Table 26.1.5 for a cross-sectional loading area. Proper performance of a screw conveyor assumes the operation is controlled with volumetric feeders and the material is uniformly fed into the conveyor housing at all times. Be sure to check lump size limitations after selecting the conveyor diameter.

For further determining capacity in nonstandard units, the following equation may be used.

$$\frac{C}{\text{rpm}} = \frac{0.7854\,(D_s^2 - D_p^2)\,P\,K\,60}{1728}$$

where: C = capacity, in ft^3/hr
rpm = revolutions of screw per minute
D_s = diameter of screw, in in.
D_p = outside diameter of pipe or tube, in in.
P = pitch of screw, in in.
K = percentage of trough loading (0.33, 0.45, etc.)

After determining the loading characteristics and the lump size, the speed of the conveyor can readily be determined by dividing the maximum cubic feet per hour handled by the rate per cubic feet per revolution shown in Table 26.1.5. In calculating capacity and horsepower, efficiency factors must be considered for: conveyors with regular flights and mixing paddles, cut flights, and cut flights with mixing paddles. The delivered capacity cubic feet per hour will be the rate of capacity given in Table 26.1.5 multiplied by the factor e given in Table 26.1.6 under loading classifications and opposite the type of conveyor. To obtain the proper size to use, divide the required capacity of cubic feet per hour by the factor e in Table 26.1.6. This will give the rated capacity in cubic feet per hour for the proper conveyor size, which then can be selected from Table 26.1.5. For other variations in screw design, consult a screw conveyor manufacturer for effects on capacity. The horsepower required must then be figured on the basis of rated capacity using equations given later.

The size of a screw conveyor depends not only on a capacity required but on the size and proportion of the lumps in the material to be handled. Also, the characteristic of a lump needs to be considered and whether or not it will break up in transit as it is being conveyed. The allowable lump size in the screw conveyor is a function of the radial clearance between the outside diameter of the central pipe

Table 26.1.5

LOADING CLASSIFICATION

CLASS I—(45% FULL) Pulverized, small size, friable non-abrasive and free flowing materials. Also medium weight, non-abrasive granular or small lump material mixed with fines.

CLASS II—(30% FULL) Non-abrasive materials consisting of fines, granular, or medium lumps mixed with fines.

CONVEYOR				CLASS I—(45% FULL)							CLASS II—(30% FULL)						
SIZE		RATING		Maximum Lump Size			Capacity†				Maximum Lump Size			Capacity†			
							T.P.H. For 100 Lb. Mat'l		Cubic Feet Per Hour					T.P.H. For 100 Lb. Mat'l		Cubic Feet Per Hour	
Diameter	Coupling size	Max. H.P. at 100 RPM	Max. In-Lbs. Torque	Lumps 25% of total	All Lumps	Max. RPM*	at Max. RPM	at 1 RPM	at Max. RPM	at one RPM	Lumps 25% of total	All Lumps	Max. RPM*	at Max. RPM	at one RPM	at Max. RPM	at one RPM
4	1	1.5	950	½	¼	175	5.6	.032	112	.64	¾	⅜	130	2.86	.022	57	.44
6	1½	5.0	3200	¾	½	165	18.8	.114	376	2.28	1	⅝	120	9.00	.075	180	1.5
9	1½ 2	5.0 12.0	3200 7560	1½	¾	150	60.0	.40	1200	8.00	1½	¾	105	28.3	.27	565	5.4
10	1½ 2	5.0 12.0	3200 7560	1½	¾	145	80.0	.55	1600	11.00	1¾	⅞	95	36.1	.38	725	7.6
12	2 2 7/16 3	12.0 15.0 25.0	7560 9500 16000	2	1	140	135.0	.96	2700	19.3	2	1¼	90	58.5	.65	1175	13.0
14	2 7/16 3	15.0 25.0	9500 16000	2½	1¼	130	200.0	1.54	4000	30.8	2½	1½	85	89.3	1.05	1790	21.0
16	3	25.0	16000	3	1½	120	285.0	2.36	5700	47.3	3	2	80	125.5	1.57	2510	31.4
18	3	25.0	16000	3	2	115	390.0	3.4	7800	68.0	3½	2	75	171.0	2.27	3420	45.5
20	3 3 7/16	25.0 40.0	16000 25000	3½	2	105	490.0	4.6	9800	93.0	4	2½	70	217.0	3.1	4350	62.0
24	3 7/16	40.0	25000	4	2¼	100	810.0	8.1	16200	162.0	4½	2¾	65	352.0	5.4	7030	108.0

LOADING CLASSIFICATION

CLASS IIA—(30% FULL) Moderately abrasive materials consisting of fines, granular, or medium lumps mixed with fines.

CLASS III—(15% FULL) Highly abrasive lumpy or stringy material which must be carried at a low level in trough to avoid contact with hanger bearings or interference with hanger frames.

FEEDERS—(95% FULL)

	CLASS IIA—(30% FULL)							CLASS III—(15% FULL)							FEEDERS—(95% FULL)		
	Maximum Lump Size			Capacity†				Maximum Lump Size			Capacity†					Capacity	
				T.P.H. For 100 Lb. Mat'l.		Cubic Feet Per Hour					T.P.H. For 100 Lb. Mat'l.		Cubic Feet Per Hour				
Conveyor Diameter	Lumps 25% of Total	All Lumps	Max. RPM*	at max. RPM	at one RPM	at max. RPM	at one RPM	Lumps 25% of Total	All Lumps	Max. RPM*	at Max. RPM	at One RPM	at Max. RPM	at One RPM	Max. Size Lumps 25% of Total	TPH For 100 Lb. Mat'l at one RPM	Cu. Ft. Per hour at one RPM
4	¾	⅜	65	1.43	.022	29	.44								½	.068	1.38
6	1	⅝	60	4.5	.075	90	1.5	1¼	¾	60	2.28	.038	45	.75	¾	.237	4.75
9	1½	¾	50	13.5	.27	270	5.4	2	1½	50	6.75	.135	135	2.7	1½	.84	16.8
10	1¾	⅞	50	19.0	.38	380	7.6	2	1½	50	9.5	.190	190	3.8	1½	1.19	23.8
12	2	1¼	50	32.5	.65	650	13.0	2½	2	50	16.3	.325	325	6.5	2	2.04	40.8
14	2½	1½	45	47.3	1.05	945	21.0	3	2½	45	23.6	.525	473	10.5	2½	3.26	65.2
16	3	2	45	70.6	1.57	1430	31.4	3½	3	45	35.3	.785	708	15.7	3	5.0	100.
18	3½	2	40	91.0	2.27	1820	45.5	4	3	40	45.7	1.14	915	22.8	3	7.2	144
20	4	2½	40	124.0	3.1	2480	62.0	4½	3½	40	62.0	1.55	1240	31.0	3½	9.8	195
24	4½	2¾	40	216.0	5.4	4320	108.0	4½	4	40	108.0	2.70	2160	54.0	4	17.0	340

*Maximum R.P.M. for conveyors using Hard Iron Bearings $= \dfrac{190}{\text{Diam. of Coupling}}$ However, the lower speed of that obtained from the formula and that given in the table above should be used.

and the inside radius of the screw conveyor trough, as well as the portion of lumps in any given mix. In any case where a large percentage of lumps are expected, consult a screw conveyor manufacturer for recommendations.

Screw Conveyor Components

Selection of bearing material for intermediate hangers is based primarily on experience and the conveyed material characteristics. Table 26.1.1 offers general guidelines and a selection of intermediate bearing material. In addition to these bearing materials, there are several new man-made products available on the market today which could be utilized to satisfy many of the special requirements.

Coupling shafts are fabricated from an AISI C1018 standard cold-rolled steel or equivalent. For

Table 26.1.6

Type of Screw	Class of Conveyor Loading			
	I (45% Full)	II & IIA (30% Full)	III (15% Full)	FEEDERS (100% Full)
Regular Flight with Mixing Paddles	.90	.88	.87	.90
Cut Flight	.72	.66	.62	.72
Cut Flight with Mixing Paddles	.59	.53	.48	.59

hard iron bearings the shaft is surface hardened to approximately 58–60 Rockwell C. The following are guidelines used with the different types of bearings.

1. Babbit bearings must be lubricated frequently and have a maximum operating temperature of approximately 130°F. Bronze bearings may be operated in temperatures up to 225°F. Care should be exercised in utilizing these bearings since bearing material may contaminate the product being used.
2. Oil-impregnated wood bearings are manufactured of hard maple to have a maximum operating temperature of approximately 160°F. Graphite-impregnated bronze has a maximum operating temperature of approximately 200°F. Plastic and reinforced fibers require no grease or lubrication and are usually run dry and are best suited for material being handled in a wetted form. Maximum operating temperature varies greatly and the bearing manufacturer should be consulted.
3. Commercial-grade carboñ bearings may be operated in temperatures up to 750°F.
4. Ball bearings are preferred when handling granular or pellet-type material not containing any fine powder. Maximum operating temperature is roughly 225°F.
5. Hard iron, white iron, or Ni-hard bearings are always used with hardened coupling shafts and abrasive material. Hardened steel or stellite inserts may be used with hard iron bearings. Maximum operating temperature is approximately 500°F. Caution should be exercised when using hard iron bearings. The maximum rpm for conveyors using hard iron bearings is equal to 190 divided by the diameter of the coupling.

Horsepower

To operate a horizontal screw conveyor, the horsepower must be based on proper installation, alignment, and uniformity in feed rate to the conveyor. The following equations may be used for horsepower requirements:

$$\mathrm{Hp}_f = \frac{L\,N\,F_d\,F_b}{1 \times 10^6}$$

$$\mathrm{Hm}_m = \frac{C\,L\,W\,F_f\,F_m\,F_p}{1 \times 10^6}$$

$$\mathrm{Hp}_{\mathrm{total}} = \frac{(\mathrm{Hp}_f + \mathrm{Hp}_m)\,F_o}{e}$$

where: C = capacity, in ft³/hr
e = drive efficiency, Table 26.1.16
F_b = hanger bearing factor, Table 26.1.7
F_d = conveyor diameter, Table 26.1.8
F_f = flight factor, Table 26.1.9
F_m = material factor, Table 26.1.3
F_o = overload factor, Table 26.1.10
F_p = paddle factor, Table 26.1.11
L = total length of conveyor, in ft
N = operating speed in rpm
W = density of material, Table 26.1.3

Table 26.1.7 Hanger Bearing Factor (F_b)

Bearing Type	F_b
Ball	1.0
Babbitt	1.7
Bronze	1.7
Oil-impregnated wood	1.7
Plastic	2.0
Nylon	2.0
Hard iron	4.4

Table 26.1.8 Diameter Factor (F_d)

Screw Diameter (in.)	F_d
6	18
9	31
12	55
14	78
16	106
18	135
20	165
24	235

Table 26.1.9 Flight Factor (F_d)

	Conveyor Loading			
Flight Type	15%	30%	45%	95%
Standard	1	1	1	1
Cut	1.1	1.15	1.2	1.3
Cut and folded	Not recommended	1.5	1.7	2.2
Ribbon	1.05	1.15	1.2	—

Table 26.1.10

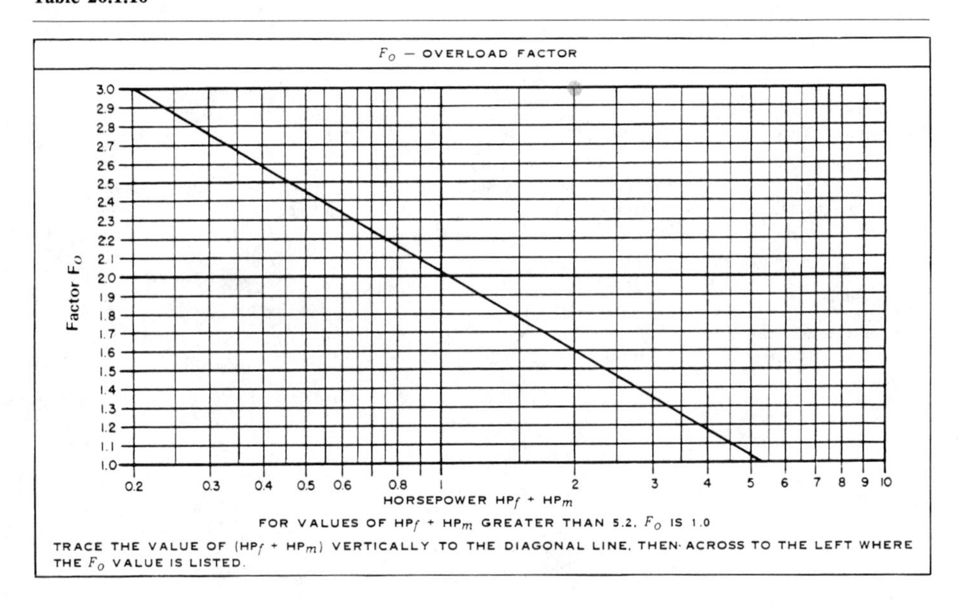

Table 26.1.11 Paddle Factor (F_p)

	Paddles per Pitch Set at 45° Reverse Pitch				
	None	1	2	3	4
F_p	1.0	1.3	1.6	1.9	2.2

The derivation of these equations is given in CEMA Publication No. 350. The factors F_m, F_d, and F_b are important factors in determining capacity, conveyor size, and speed. The factor F_b is related to the friction in the hanger bearings due to rubbing of the shaft in the bearing journal and is empirically derived. Factor F_d is the average weight per foot of the heaviest rotating parts as related to the coupling shaft diameter. F_m is dependent on the characteristics of the material and is derived primarily through experience. An overload factor F_o is used to correct calculated horsepowers of less than 5. This correction factor is utilized to overcome the lower torque ranges in small motors.

Factors F_f and F_p are additional correction factors for the various types of screw flight forms. These, again, are empirically derived and are related to the net effective area of the screw flight.

Consideration should be given in sizing horsepower to how the screw conveyor will be expected to operate. Good operating procedure is to run the conveyor until it is empty; however, there may be situations where this is not practical and the conveyor will be required to start under load. The material needs to be studied closely in these cases to determine if there are any packing characteristics present.

All power transmitting components should be sized to safely handle the rated horsepower. If 7.3 Hp is required by the horsepower equation and a 10-Hp motor is used, all components should be sized to handle the starting load of the 10 Hp. In Table 26.1.5, a maximum inch-pounds of torque rating is given for each size shaft. This number represents the minimum torque rating of either the coupling bolt or the pipe in bearing. This number can be calculated with the equation torque $T =$ 63,025 times horsepower (Hp) divided by rpm. Screw conveyors are limited in length by the amount of torque that can be transmitted through the pipe, bolts, and couplings. For torsional ratings higher than those shown in the table, please refer the problem to the manufacturer. For problems of this nature, several options are available, for example, dual drives, higher-grade bolts, or schedule 80 pipe.

Deflection

If screw conveyors of standard lengths are used, deflection is usually no problem. However, when longer than standard lengths are utilized without intermediate hangers, care should be taken to assure that the screw does not deflect excessively and contact the trough, causing excessive wear. As a rule of thumb, deflections greater than $\frac{1}{4}$ in. should be referred to the manufacturer. For calculating deflection at the mid-span, the following equation may be used:

$$A = \frac{5\ WL^3}{384\ EI}$$

where: A = deflection at mid-span, in in.
W = total weight of screw, in lb
L = length of screw between bearings, in in.
E = modulus of elasticity for steel
I = moment of inertia

The nomograph shown in Fig. 26.1.1 may be used for simpler approximation of deflection.

26.1.4 Screw Conveyor Components

The following section describes the uses of the various components of a screw conveyor that facilitate the appropriate selection for solving a particular screw conveyor problem and reveals the great flexibility that a screw conveyor can offer.

Conveyor Screw Flighting

Screw conveyor flighting is manufactured in one of two ways, helicoid or sectional. Helicoid flights are formed by rolling a flat bar or strip through two cones into a continuous helix. The flighting manufactured in this manner has a thinner outer edge than inner edge. Sectional flights are formed from a flat disc and the thickness is uniform throughout. A continuous helix is then manufactured by butt-welding the flights together on a piece of pipe. Standard screw specifications are shown in Table 26.1.12. The screw conveyor can be formed in either a right- or a left-hand design. The hand may clearly and easily be ascertained by looking at the end of the screw. Figure 26.1.2 shows the means of telling the different hands. The hand of the screw is a very important consideration in ordering a screw conveyor. Once built and installed, a certain hand and the direction of the rotation is fixed for the desired direction of material transportation. A replacement must be of the same hand to avoid disastrous results.

Screw Flight Mounting

The helix of the screw is mounted on either a hollow tube or solid shaft. Normally, schedule 40 carbon steel pipe is used, but, depending on the design parameter, schedule 80 and or mechanically

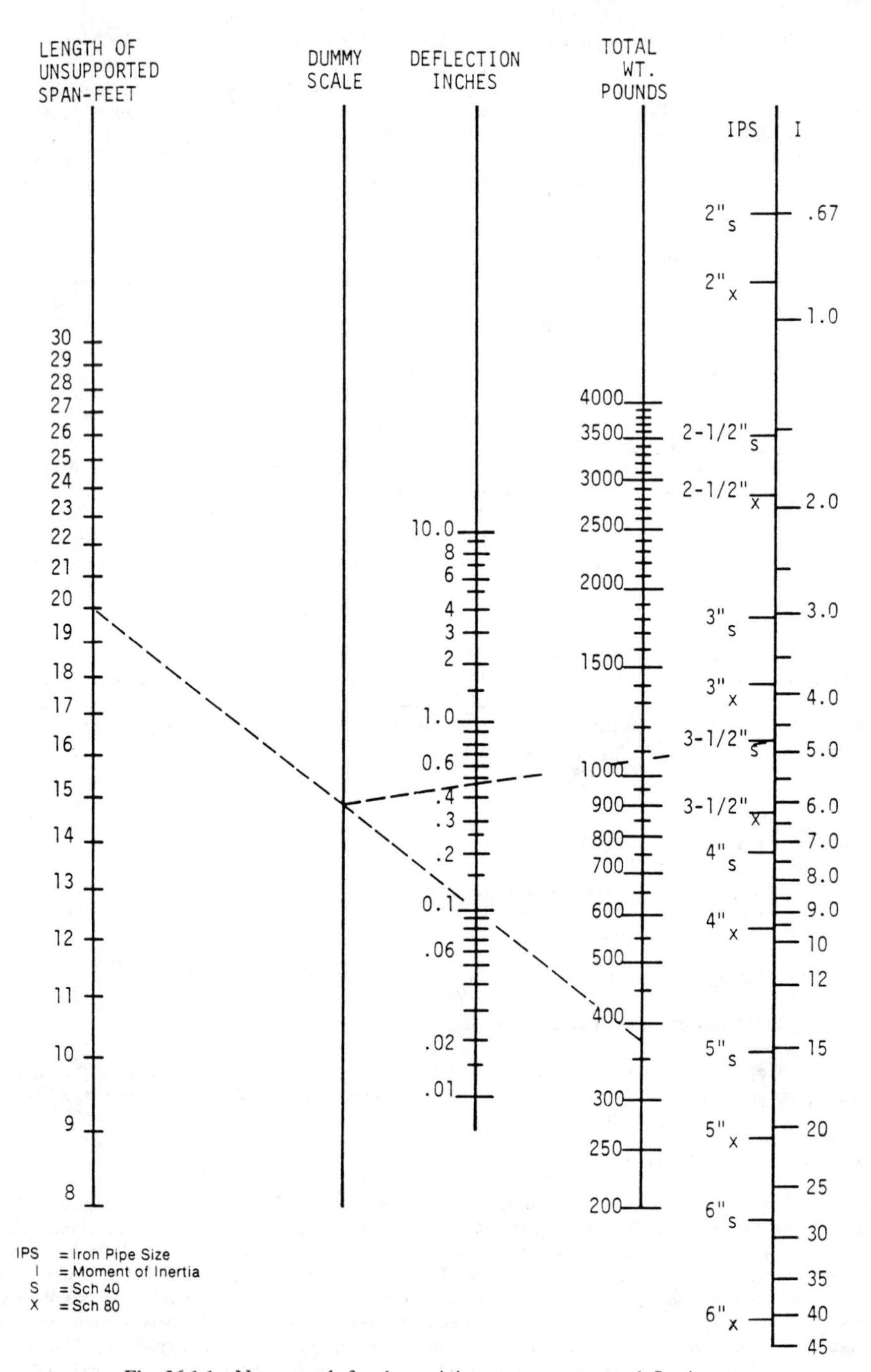

Fig. 26.1.1 Nomograph for determining screw-conveyor deflection.

Table 26.1.12.*A* Helicoid Conveyor

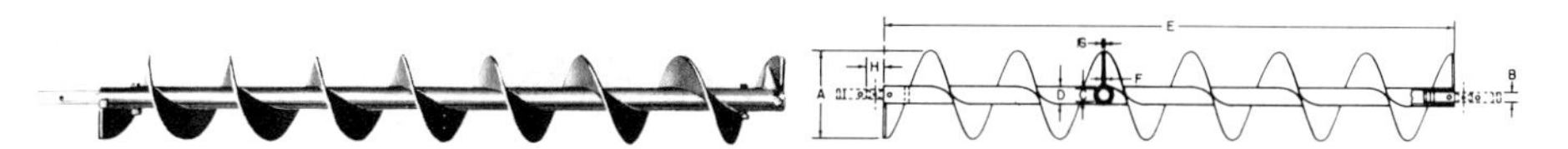

STANDARD LENGTH SECTIONS

Dimensions in Inches, Average Weights in Pounds and Horsepower Ratings

Conveyor Screw Size Designation***	Part No.	Diam. of Coupling	Pipe Sizes: Nominal Inside Dia.	Pipe Sizes: Approx. Outside Dia.	Thickness of Flights: Next to Pipe	Thickness of Flights: Outer Edge	Length of Standard Section Center to Center of Hangers in Feet	Length of Space for Hanger Bearing	* Exact Net Length of Std. Section Ft. & In.	Weight Per Standard Section: Mounted Conveyor	Weight Per Standard Section: Flighting Only	+ Max. Horse-power at 100 R.P.M.
	Part No.	B	C	D	F	G		H	E			
4H204	220000	1	1 1/4	1 5/8	1/8	1/16	8′	1 1/2	7′-10 1/2″	25	7.2	1 1/2
4H206	220002	1	1 1/4	1 5/8	3/16	3/32	8′	1 1/2	7′-10 1/2″	32	12.8	1 1/2
6H304	220003	1 1/2	2	2 3/8	1/8	1/16	10′	2	9′-10″	52	14.0	5
6H308	220005	1 1/2	2	2 3/8	1/4	1/8	10′	2	9′-10″	62	28.0	5
9H306	220007	1 1/2	2	2 3/8	3/16	3/32	10′	2	9′-10″	70	31.0	5
9H406	220008	2	2 1/2	2 7/8	3/16	3/32	10′	2	9′-10″	91	30.0	10
9H312	220010	1 1/2	2	2 3/8	3/8	3/16	10′	2	9′-10″	101	65.0	5
9H412	220011	2	2 1/2	2 7/8	3/8	3/16	10′	2	9′-10″	121	60.0	10
9H414	220012	2	2 1/2	2 7/8	7/16	7/32	10′	2	9′-10″	140	85.0	10
10H306	220013	1 1/2	2	2 3/8	3/16	3/32	10′	2	9′-10″	81	48.0	5
10H412	220014	2	2 1/2	2 7/8	3/8	3/16	10′	2	9′-10″	130	76.0	10
12H408	220015	2	2 1/2	2 7/8	1/4	1/8	12′	2	11′-10″	140	67.0	10
12H508	220016	2 7/16	3	3 1/2	1/4	1/8	12′	3	11′-9″	168	64.0	15
12H412	220018	2	2 1/2	2 7/8	3/8	3/16	12′	2	11′-10″	180	102.0	10
12H512	220019	2 7/16	3	3 1/2	3/8	3/16	12′	3	11′-9″	198	96.0	15
12H614	220020	3	3 1/2	4	7/16	7/32	12′	3	11′-9″	228	120.0	25
14H508	220022	2 7/16	3	3 1/2	1/4	1/8	12′	3	11′-9″	170	84.0	15
14H614	220023	3	3 1/2	4	7/16	7/32	12′	3	11′-9″	254	132.0	25
16H610	220024	3	3 1/2	4	5/16	5/32	12′	3	11′-9″	228	120.0	25
16H614**	220025	3	4	4 1/2	7/16	7/32	12′	3	11′-9″	324	180.0	25
18H610	220027	3	3 1/2	4	5/16	5/32	12′	3	11′-9″	251	137.0	25
18H614**	220028	3	4	4 1/2	7/16	7/32	12′	3	11′-9″	332	190.0	25
20H610	220029	3	3 1/2	4	5/16	5/32	12′	3	11′-9″	266	151.0	25
20H614**	220030	3	4	4 1/2	7/16	7/32	12′	3	11′-9″	354	211.0	25

*The exact overall length of a standard section of conveyor is the distance from end to end of pipe. This is the length that will be furnished when a "standard section" or "regular length" is ordered.

**Can be furnished with 3-7/16 dia. coupling.

+Maximum Horsepower that can be safely applied with standard construction. Horsepower at other speeds in direct ratio.

***Most sizes can be formed in stainless steel.

drawn steel tubing may be used. The pipe sections are bushed at each end and holes are drilled for the coupling bolt connections; this is standard practice. The drive coupling and tail shafts are manufactured of the same diameter and drilled to match the drill holes in the ends of the pipe. Most screw pipe shafts are bushed so the shafting slips in easily with a tolerance of ±0.015. The coupling shafts may be hardened for abrasive material. The coupling bolts are a special hexagon head bolt with a short length of threads so that the threaded portion does not come into the load area where the screw pipe and shaft contact. Coupling bolts are standardized for a given size pipe.

Hangers and Hanger Bearings

To support the screw, hangers and hanger bearings are utilized when one or more sections of conveyor are required. The industry has developed a great number of alternate designs for hangers and bearings to accommodate just about every application. Figure 26.1.3 shows the great variety and gives a brief explanation as to their application.

Trough Ends

Fabricated plate is used to enclose the end of the screw conveyor trough and support the bearing that holds the tail shaft or drive shaft. Depending on the design requirements, standard flange-mounted bearings can be mounted to the flat plate or pillow blocks may be used whereby a pedestal-type trough end is required. Trough end bearings may be manufactured of babbitt, bronze, ball, or roller type, and in some cases special bushing materials. Roller and ball bearings have the capability of handling nominal thrust loads. If sleeve bearings are utilized, special thrust washers or alternate design

Table 26.1.12.*B* Sectional Flight Conveyor

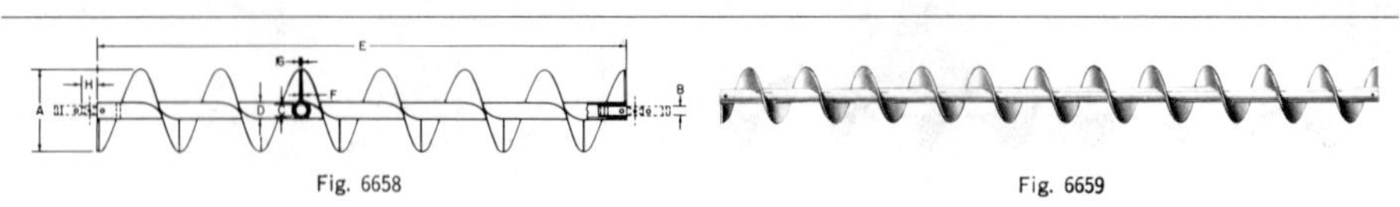

Fig. 6658 Fig. 6659

STANDARD LENGTH SECTIONS

Dimensions in Inches, Average Weights in Pounds and Horsepower Ratings

Conveyor Screw Size Designation	Part Number	Diameter of Coupling	Pipe Size		Thickness of End Flights	Length of Standard Section Center to Center of Hangers in Feet	Length of Space for Hanger Bearing	* Exact Net Length of Std. Section Ft. and In.	Weight Per Std. Section	† Maximum Horse-power at 100 R.P.M.
			Nominal Inside Diameter	Approx. Outside Diameter						
		B	C	D	F		H	E		
6S304	220032	1 1/2	2	2 3/8	10	10′	2	9′-10″	55	5
6S309	220033	1 1/2	2	2 3/8	10	10′	2	9′-10″	65	5
6S312	220034	1 1/2	2	2 3/8	3/16	10′	2	9′-10″	75	5
6S316	220035	1 1/2	2	2 3/8	1/4	10′	2	9′-10″	90	5
9S305	220036	1 1/2	2	2 3/8	10	10′	2	9′-10″	65	5
9S309	220037	1 1/2	2	2 3/8	10	10′	2	9′-10″	80	5
9S312	220038	1 1/2	2	2 3/8	3/16	10′	2	9′-10″	95	5
9S405	220039	2	2 1/2	2 7/8	10	10′	2	9′-10″	85	10
9S409	220040	2	2 1/2	2 7/8	10	10′	2	9′-10″	100	10
9S412	220041	2	2 1/2	2 7/8	3/16	10′	2	9′-10″	115	10
9S416	220042	2	2 1/2	2 7/8	1/4	10′	2	9′-10″	130	10
9S424	220043	2	2 1/2	2 7/8	3/8	10′	2	9′-10″	160	10
10S305	220044	1 1/2	2	2 3/8	10	10′	2	9′-10″	70	5
10S309	220045	1 1/2	2	2 3/8	10	10′	2	9′-10″	85	5
10S405	220046	2	2 1/2	2 7/8	10	10′	2	9′-10″	90	10
10S412	220048	2	2 1/2	2 7/8	3/16	10′	2	9′-10″	120	10
10S416	220049	2	2 1/2	2 7/8	1/4	10′	2	9′-10″	135	10
12S407	220050	2	2 1/2	2 7/8	3/16	12′	2	11′-10″	125	10
12S412	220051	2	2 1/2	2 7/8	3/16	12′	2	11′-10″	156	10
12S416	220052	2	2 1/2	2 7/8	1/4	12′	2	11′-10″	205	10
12S507	220053	2 7/16	3	3 1/2	3/16	12′	3	11′-9″	148	15
12S512	220054	2 7/16	3	3 1/2	3/16	12′	3	11′-9″	180	15
12S612	220055	3	3 1/2	4	3/16	12′	3	11′-9″	200	25
12S616	220056	3	3 1/2	4	1/4	12′	3	11′-9″	215	25
12S624	220057	3	3 1/2	4	3/8	12′	3	11′-9″	280	25

For additional listings of standard sections—see next page. Sectional flight screw conveyor notes on next page.

*The exact overall length of a standard section of conveyor is the distance from end to end of pipe; thus, a standard 10 ft. section of a 9″ x 2″ conveyor will have a pipe 9′-10″ long, 2″ being deducted for the applicable hanger.

†Maximum horsepower that can be safely applied with standard construction. Horsepower at other speeds in direct ratio.

must be used to absorb the thrust. Trough end seals are used to minimize the leakage of material around the drive and tail shaft. A variety of seal designs is available depending on the abrasive or corrosiveness of the material. Figure 26.1.4 shows several different styles with a brief explanation of their application.

Troughs

A trough for a standard screw conveyor is normally U-shaped and is radiused at the bottom with standard heights and lengths designed to make a convenient and rigid enclosure for the conveyor as well as a supporting structure. Basically there are four types of trough: the angle flanged, single flanged, double flanged, and flared trough. Figure 26.1.12 shows these four designs. Covers for the trough, primary function to protect personnel from the moving parts of the conveyor and to control the dust and material being conveyed within the conveyor housing. Under no circumstances should conveyors be used as a step or walkway. Care should always be taken to assure that the power is turned off whenever the covers are removed. A lock-out should be provided in such a manner that when the covers are removed all power is shut off. Most screw conveyor manufacturers provide safety stickers and safety hazard warning labels at no cost. Covers are fastened to the trough by a wide variety of fasteners; typical are bolts, spring clamps, or C-clamps. Discharge spouts and gates are provided to direct and control the flow of material. Occasionally, an end discharge is used, which is simply a modified trough end. To support the trough, feet or saddles are utilized. Feet are bolted to the trough flanges, saddles may be welded at any point between the flanges. Supporting distances are dictated by the conveyor weight and the material being conveyed. In all cases, supporting structure should be designed at the recommendation of the manufacturer.

All components described in the foregoing have been standardized by the screw conveyor industry, and the specifications for these components can be found in CEMA Standard No. 300.

Table 26.1.12.*B* (*Continued*)

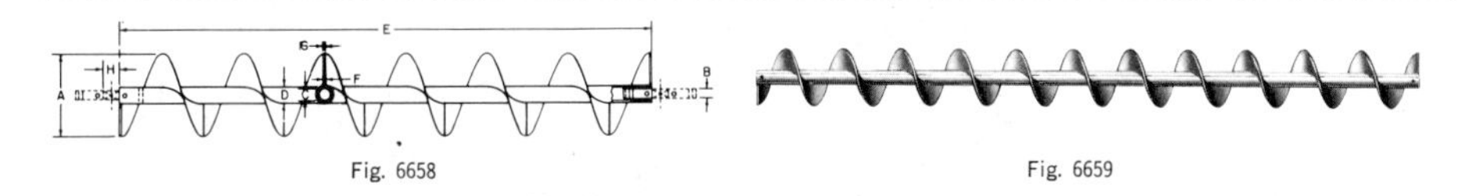

Fig. 6658 Fig. 6659

STANDARD LENGTH SECTIONS—(Continued)

Dimensions in Inches, Average Weights in Pounds and Horsepower Ratings

Conveyor Screw Size Designation	Part Number	Diameter of Coupling	Pipe Size: Nominal Inside Diameter	Pipe Size: Approx. Outside Diameter	Thickness of End Flights	Length of Standard Section Center to Center of Hangers in Feet	Length of Space for Hanger Bearing	* Exact Net Length of Std. Section Ft. and In.	Weight Per Std. Section	† Maximum Horse-power at 100 R.P.M.
		B	C	D	F		H	E		
14S507	220058	2 7/16	3	3½	3/16	12′	3	11′-9″	172	15
14S512	220059	2 7/16	3	3½	3/16	12′	3	11′-9″	215	15
14S612	220060	3	3½	4	3/16	12′	3	11′-9″	222	25
14S616	220061	3	3½	4	¼	12′	3	11′-9″	245	25
14S624	220062	3	3½	4	⅜	12′	3	11′-9″	342	25
16S609	220063	3	3½	4	3/16	12′	3	11′-9″	210	25
16S612	220064	3	3½	4	3/16	12′	3	11′-9″	235	25
16S616	220065	3	3½	4	¼	12′	3	11′-9″	285	25
16S624	220066	3	3½	4	⅜	12′	3	11′-9″	365	25
18S609	220067	3	3½	4	3/16	12′	3	11′-9″	215	25
18S612	220068	3	3½	4	3/16	12′	3	11′-9″	245	25
18S616	220069	3	3½	4	¼	12′	3	11′-9″	295	25
18S624	220070	3	3½	4	⅜	12′	3	11′-9″	425	25
20S612	220071	3	3½	4	3/16	12′	3	11′-9″	300	25
20S616	220072	3	3½	4	¼	12′	3	11′-9″	360	25
20S724	220073	3 7/16	4	4½	⅜	12′	4	11′-8″	475	41
24S712	220074	3 7/16	4	4½	3/16	12′	4	11′-8″	440	41
24S716	220075	3 7/16	4	4½	¼	12′	4	11′-8″	510	41
24S724	220076	3 7/16	4	4½	⅜	12′	4	11′-8″	595	41

*The exact overall length of a standard section of conveyor is the distance from end to end of pipe; thus, a standard 10 ft. section of a 9″ x 2″ conveyor will have a pipe 9′-10″ long, 2″ being deducted for the applicable hanger.

†Maximum horsepower that can be safely applied with standard construction. Horsepower at other speeds in direct ratio.

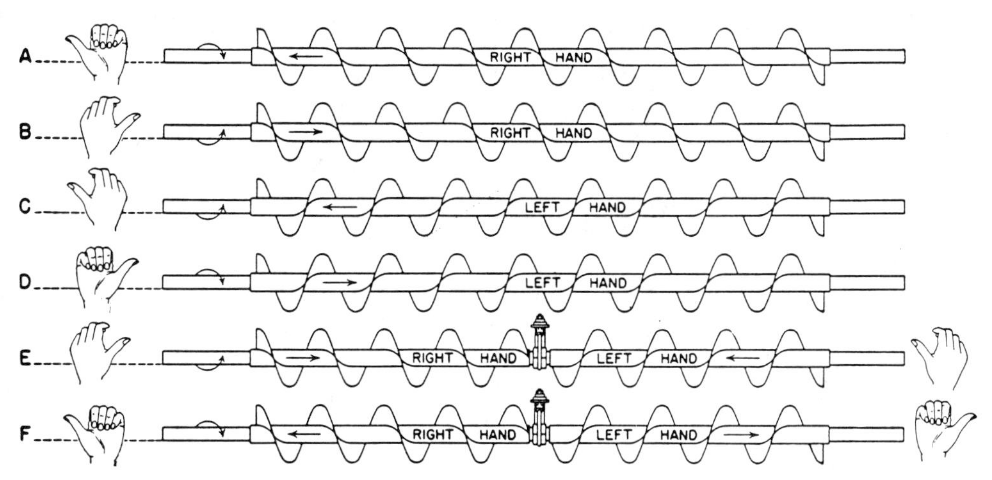

Fig. 26.1.2 A simple method of selecting the "hand" of a conveyor is to extend the thumb of your hand to indicate the direction of the flow of material, with the extended fingers in the direction of rotation. The "hand" of the conveyor will be opposite that used; that is, you use your left hand to analyze a right-hand conveyor. The above may seem complicated, but by checking with it a few times you will soon appreciate its value.

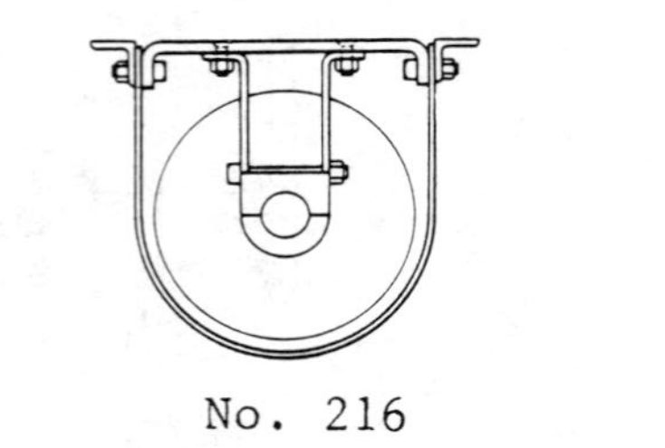

No. 216

Rigid steel frame hanger, used generally with chilled or white iron bearings for handling heavy abrasive materials like cement. Also used with wood or babbitted bearings.

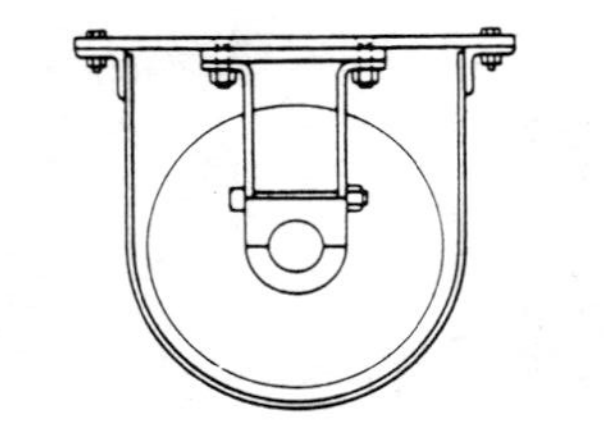

No. 116

Rigid hanger same as 216 except for top mounting in lieu of side mounting.

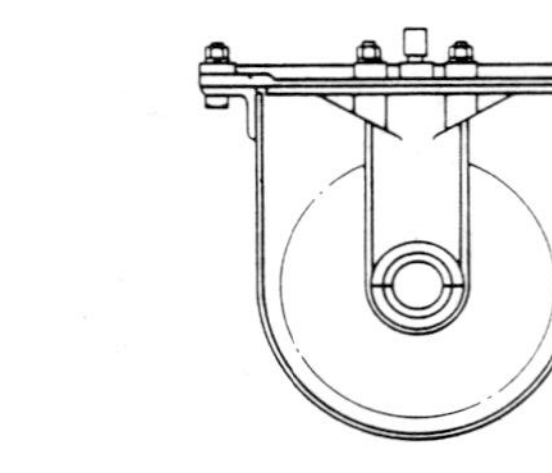

No. 19-B

Cast frame hanger, "U" bolts and babbitted caps available on repair order basis only.

No. 226

Steel frame hanger using babbitted or white iron bearings. Presents a minimum obstruction to the flow of material. Used principally in conveyors carrying cement, chemicals, etc.

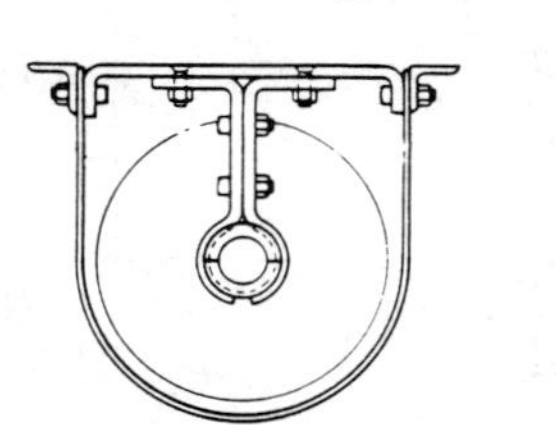

No. 226

Steel frame hanger similar to No. 220 except more rugged. Hanger bolts below the trough cover. Fitted with babbitt, wood, nylon or hard iron bearings.

No. 326

Steel frame hanger similar to No. 226 but providing for longitudinal expansion of the conveyor. Used generally where hot chemicals, gypsum, etc. are conveyed.

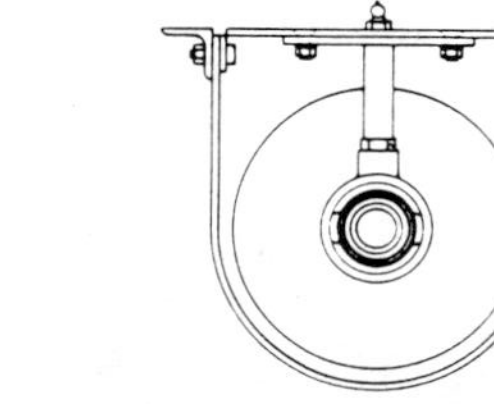

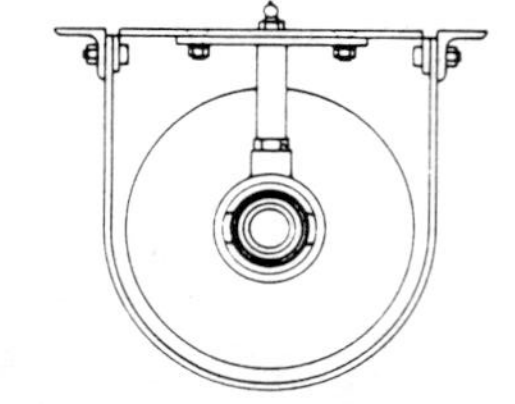

No. 270-B

Steel frame hanger, adjustable, with ball bearings. Presents a minimum obstruction to flow of materials. Designed for dusty installations. Interchangeable with No 226 hanger.

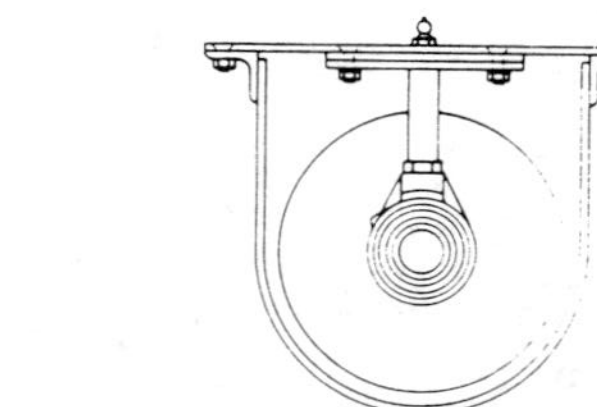

No. 290-B

Steel frame hanger, adjustable, with ball bearings. Presents a minimum of obstruction to flow of material. Designed for dusty installations. Interchangeable with No. 116 and 220 hanger.

Fig. 26.1.3 Types of frame hangers.

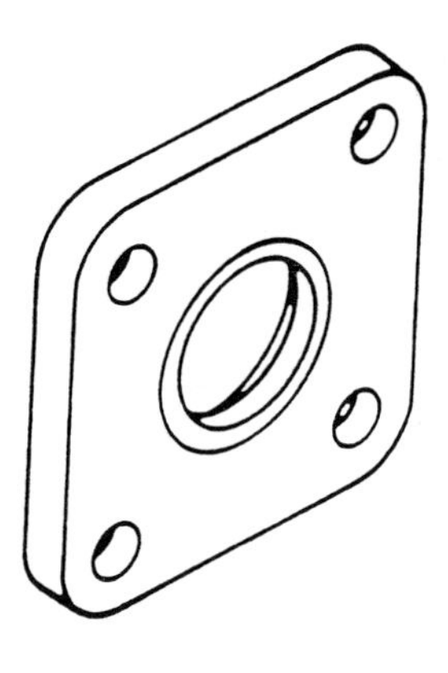

Plate Seal

The plate seal is an economical, effective sealing device, designed for exterior mounting between the end bearing and the trough end. Standard units employ lip type seals to contact the shaft but other types of commercial seal cartridges also may be used. The seal plate and the end bearing are bolted to the trough end by one set of bolts.

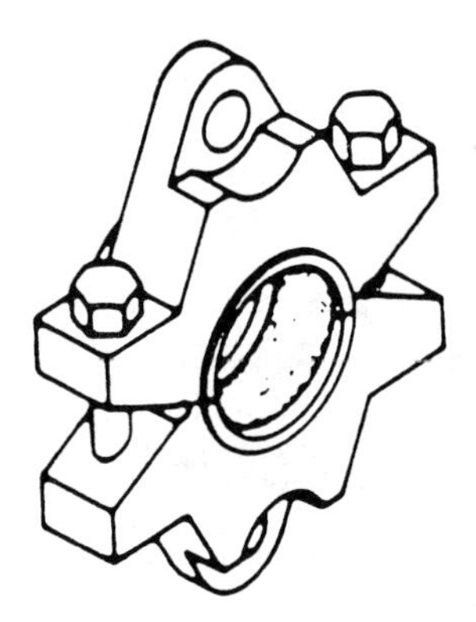

Split Gland Seal

Split gland seals are designed for interior or exterior mounting. They provide a seal which is effective for many applications.

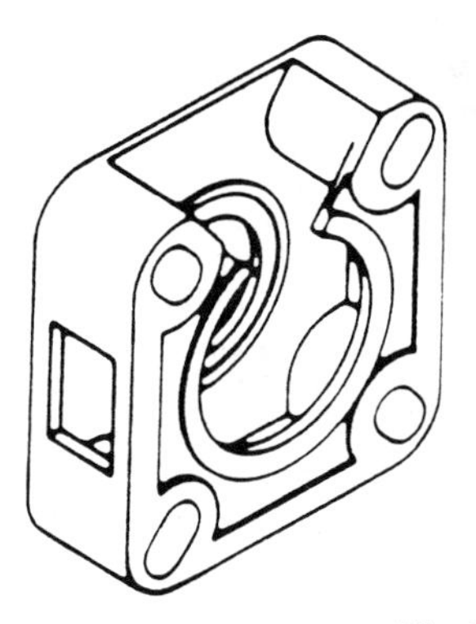

Waste Packing Seal

This universal type seal is arranged for use with waste packing or with cartridge type lip or felt seals. An opening at the top of the seal housing facilitates waste repacking, and exposes the waste for oiling.

The packing seal housing is mounted outside the trough end between it and the end bearing.

Fig. 26.1.4 Various types of trough end seals.

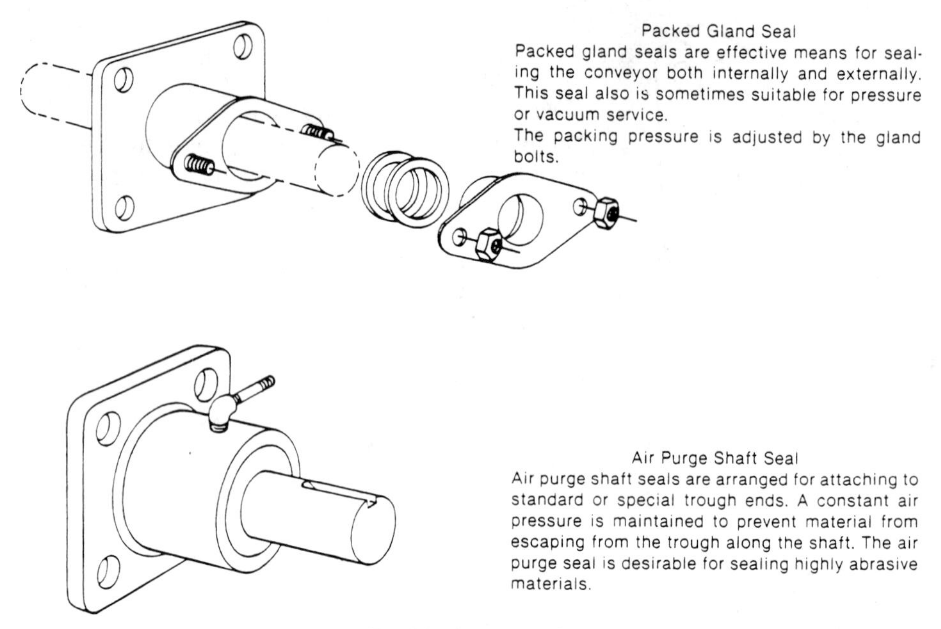

Fig. 26.1.4 (*continued*)

26.1.5 Materials of Construction

Standard screw conveyor flighting, pipe, shafts, troughs, covers, and so forth, are usually manufactured of a low-carbon hot-rolled steel in the form of sheets, plates, bars, strips, angles, and so on. Standard shafting is normally of a cold-rolled, cold-finished mild steel and occasionally these shafts are surface hard depending on the conveyors. Flanged ball bearings, roller bearings, or sleeve bearings are used for mounting on the trough ends.

Hanger bearing material varies with the substance the units will be subjected to. Examples of typical hanger bearing material are babbitt, bronze, oil-impregnated hardwood, plastic, or nylon. White cast iron and ball bearing cartridges are also utilized. Table 26.1.1 offers a guide for construction utilizing standard components. However, common sense should be used in determining construction materials.

Classes of Enclosure

The industry for many years has used such terms as dust-tight, commercially dust-tight, weather-type, and waterproof in relation to the types of enclosures required. These terms can be very ambiguous and broadly interpreted by different individuals. To help overcome this problem, CEMA has established recommendations for classes of conveyor enclosures. The following generalizes the CEMA enclosure classifications:

Class IE. Class IE enclosures are those provided primarily for the protection of operating personnel or equipment, or where the enclosure forms an integral or functional part of the conveyor or structure. They are generally used where dust control is not a factor or where protection for, or against, the material being handled is not necessary—although a certain amount of protection is afforded.

Class IIE. Class IIE enclosures employ constructions that provide some measure of protection against dust or for, or against, the material being handled.

Class IIIE. Class IIIE enclosures employ constructions that provide a higher degree of protection in these classes against dust, and for or against the material being handled.

Class IVE. Class IVE enclosures are for outdoor applications and under normal circumstances provide for the exclusion of water from the inside of the casing. They are not to be construed as being water-tight, as this may not always be the case.

Table 26.1.13 shows the method of construction to obtain the various enclosures.

Table 26.1.13 Enclosure Construction

Component Classification	Enclosure Classifications			
	IE	IIE	IIIE	IVE
A. Trough Construction				
Formed and angle top flange				
1. Plate type end flange				
a. Continuous arc weld	X	X	X	X
b. Continuous arc weld on top end flange and trough top rail	X	X	X	X
2. Trough top rail angles (Angle top trough only)				
a. Staggered intermittent arc and spot weld	X			
b. Continuous arc weld on top leg of angle on inside of trough and intermittent arc weld on lower leg of angle to outside of trough		X	X	X
c. Staggered intermittent arc weld on top leg of angle on inside of trough and intermittent arc weld on lower leg of angle to outside of trough, or spot weld when mastic is used between leg of angle and trough sheet		X	X	X
B. Cover Construction				
1. Plain flat				
a. Only butted when hanger is at cover joint	X			
b. Lapped when hanger is not at cover joint	X			
2. Semiflanged				
a. Only butted when hanger is at cover joint	X	X	X	X
b. Lapped when hanger is not at cover joint	X			
c. With buttstrap when hanger is not at cover joint		X	X	X
3. Flanged				
a. Only butted when hanger is at cover joint		X	X	X
b. Buttstrap when hanger is not at cover joint		X	X	X
4. Hip roof				
a. Ends with a buttstrap connection				X
C. Cover Fasteners for Standard Gauge Covers				
1. Spring screw, or toggle clamp fasteners or bolted construction*				
a. Max. spacing plain flat covers	60 in.			
b. Max. spacing semiflanged covers	60 in.	30 in.	18 in.	18 in.
c. Max. spacing flanged and hip-roof covers		40 in.	24 in.	24 in.
* For bolted construction use:				
$\frac{1}{4}$" bolts, 4"-10" dia. screws (min. dia.)				
$\frac{5}{16}$" bolts, large dia. screws (min. dia.)				
D. Gaskets				
1. Covers				
a. Red rubber or felt up to 230°F		X	X	
b. Neoprene rubber, when contamination is a problem		X	X	X
c. Closed-cell foam-type elastic material to suit temperature rating of gasket		X	X	X
2. Trough end flanges				
a. Mastic type compounds		X	X	X
b. Red rubber up to 230°F		X	X	X
c. Neoprene rubber, when contamination is a problem		X	X	X
d. Closed-cell foam-type elastic material to suit temperature rating of gasket		X	X	X
E. Trough End Shaft Seals*				
1. When handling nonabrasive materials			X	X
2. When handling abrasive materials	X	X	X	X
* Lip-type seals for nonabrasive materials				
Felt type for mildly abrasive materials				
Waste type for highly abrasive materials				
F. Dust Collecting Systems				
1. Provisions should be made for connecting to external dust collecting systems			X	

Weld Finishes

Various types of application require that the preparation of welds require certain degrees of smoothness. The term "grind smooth" is very general and offers various interpretations. To help establish criteria, CEMA has offered recommended classes of finishes:

Class I finish has weld spatter and slag removed but no grinding is done on the welds.

Class II finish is a refinement of the as-welded condition with the welds rough-ground to remove heavy ripple or unusual roughness.

Class III finish has the welds medium-ground with some pits and crevices permissible. This finish is recommended for materials that do not tend to contaminate or hang up in the pits or crevices.

Class IV and V finishes have the welds ground fine with no pits or crevices. The only difference between the two finishes is the degree of polish. These finishes are recommended for sanitary regulations.

26.1.6 Special Features of a Screw Conveyor

Figure 26.1.5 illustrates some of the more commonly used special features of a screw conveyor. These features greatly expand the range of the usefulness of screw conveyors.

26.1.7 Installation, Operation, and Maintenance

Because of variations in length and installation conditions, screw conveyors are usually shipped as subassemblies. Most components are manufactured to the standards of the Conveyor Equipment Manufacturers Association (CEMA). Manufacturers will design and manufacture special components for unusual requirements. Conveyors can be ordered as complete units, shop assembled, and match marked before shipping, or as individual components to be aligned and assembled in the field. When the manufacturer engineers the conveyor, complete specification drawings are generally furnished. Manufacturers' instructions should be followed.

26.1.8 Safety

Conveyor assemblies or components must be installed, maintained, and operated in such a manner as to comply with the Occupational Safety and Health Act, all state and local regulations, and the American National Standard Institute safety code. Taking into consideration all the physical aspects of the installation, the following safeguards may be required to protect the operations and those working in the immediate area of the conveyor.

1. *Covers and gratings.* Use rugged gratings in all loading areas and solid covers in other areas. Covers, guards, and gratings at inlet points must be such that personnel cannot be injured by the screw.

2. *Guards.* For protection of the operator and other persons in the working area, purchaser should provide guards for all exposed equipment such as drives, gears, shafts, and couplings. *NOTE:* DO NOT STEP OR WALK ON CONVEYOR COVERS OR GRATING OF POWER-TRANSMISSION GUARDS. A warning label to this effect is shown in Fig. 26.1.6.

Precautions for Hazardous Operations

Standard screw conveyors are not equipped to operate under conditions that may be hazardous, nor with hazardous materials. The manufacturer should be consulted if there is any indication that a hazardous condition or material is involved. Several situations may create these conditions. A few of the more common follow:

1. Where the product area is under pressure or vacuum, or the trough is provided with jackets for heating or cooling, special precautions are required. Standard components are not designed for this service.
2. Materials may be explosive, flammable, toxic, noxious, or so forth. Special provisions for safety are required. Do not use standard components.
3. Conveyors handling foodstuffs are subject to special codes for materials, construction, location, and accessibility; investigate before ordering standard components. Food conveyors often require hinged access doors for inspection or drop-bottom troughs for cleaning. Special precautions should be taken for protection of personnel against contact with the screw. Extensive use of padlocks, with keys in the hands of management personnel, is one means frequently used.

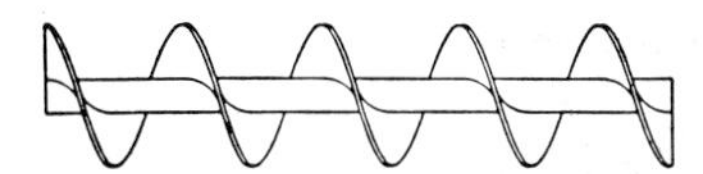

—Standard Pitch Conveyor—Standard assembly for conveying material horizontally or on inclines not to exceed 20° except in special cases.

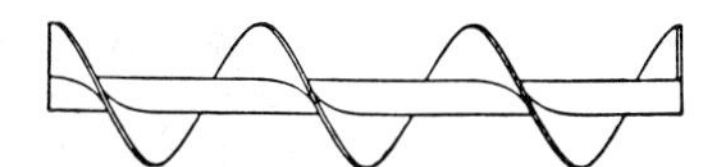

—Long Pitch Conveyor—Standard assembly for high capacity, free flowing material.

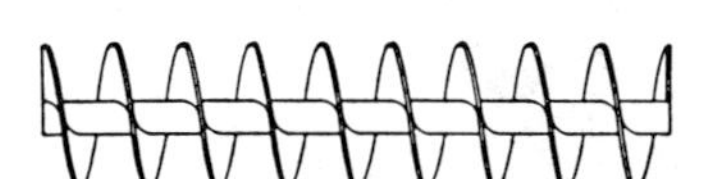

—Short Pitch Conveyor—used generally in feeders or where the material is to be conveyed slowly for cooling, heating, drying or cooking.

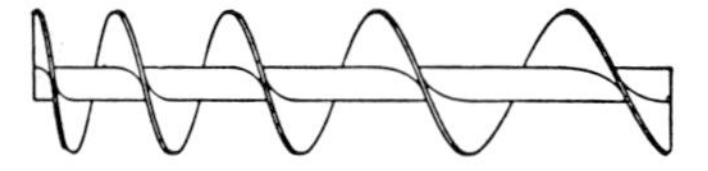

—Variable Pitch Conveyor—for use in special feeders.

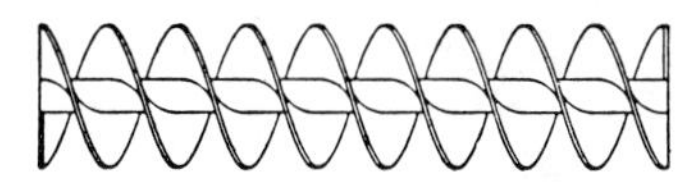

—Double Flight Standard Pitch Conveyor —used largely in feeders giving slightly higher capacity and a more even flow and discharge of material.

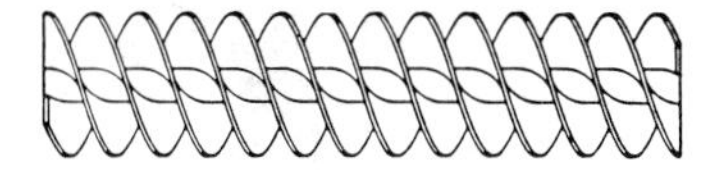

—Triple Flight Standard Pitch Conveyor —used principally in feeders for free flowing material.

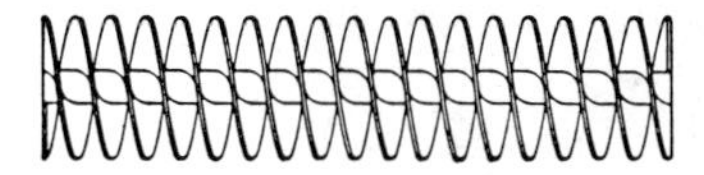

—Double Flight Short Pitch Conveyor—used primarily in feeders for smooth discharge of slow moving material or to prevent flushing of free flowing material under bins.

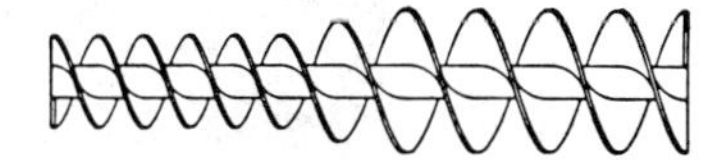

—Tapered Double Flight Conveyor—used in special feeders.

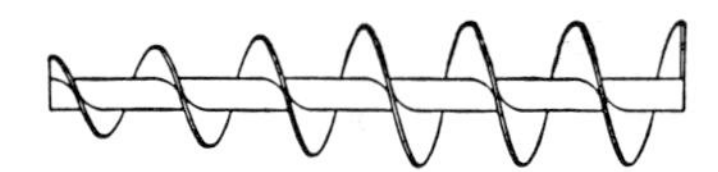

—Tapered Standard Flight Conveyor—for use in special feeders.

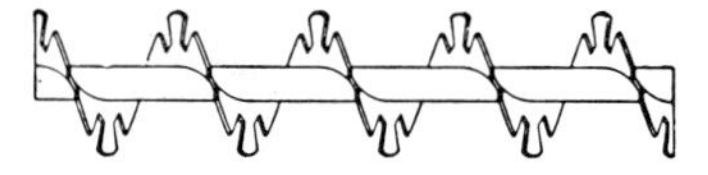

—Cut Flight Conveyor—generally used for mixing or retarding material.

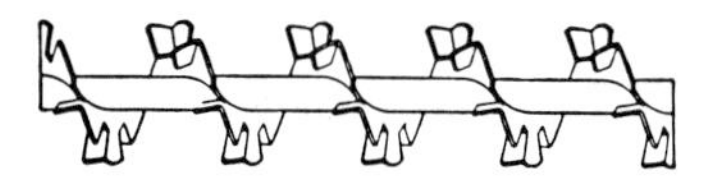

—Cut and Folded Flight Conveyor—for mixing and retarding material.

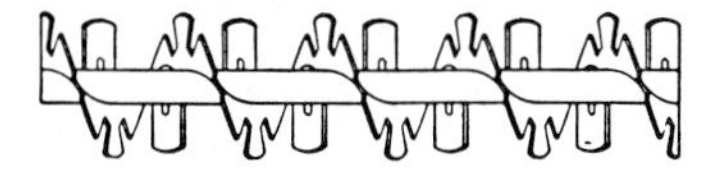

—Cut Flight Conveyor with Paddles—for mixing and retarding material.

Fig. 26.1.5 Special screw conveyor features.

Fig. 26.1.6 Warning label for electrical lock-out.

Electrical

Conveyor component manufacturers generally do not provide electrical equipment to control the conveyors. In selecting electrical control equipment to be used with any conveyor installation, the purchaser must use equipment conforming to the National Electrical Code, the National Electrical Safety Code, and other local or national codes. Consideration should be given to some or all of the following devices and to others that may be appropriate.

1. Overload protection. Devices such as shear pins, torque limiters, and so forth, to shut off power whenever operation of conveyor is stopped as a result of excessive material, foreign objects, excessively large lumps, and so on.
2. No-speed protection. Devices such as zero speed switches to shut off power in the event of any incident that might cause conveyor to cease operating.
3. Safety shut-off switch with power lock-out provision at conveyor drive.
4. Emergency stop switches readily accessible whenever required.
5. Electrical interlocking to shut down feeding conveyors whenever a receiving conveyor stops.
6. Signal devices to warn personnel of imminent startup of conveyor, especially if started from a remote location.
7. Special enclosures for motors and controls for hazardous atmospheric conditions.

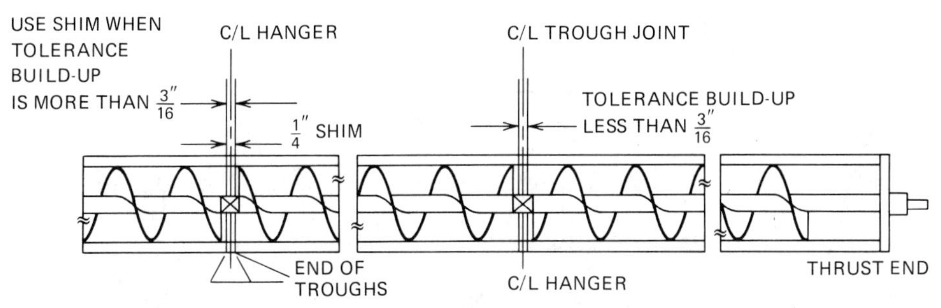

Fig. 26.1.7 Aligning trough joints and couplings with shims.

26.1.9 Installation Procedures

Check all assemblies and parts against shipping papers, and inspect for damage on arrival. Look for dented or bent trough and bent flanges, flighting, pipe, or hangers. Minor damage incurred in shipping can be readily repaired in the field.

For severely damaged parts, file an immediate claim with the carrier. Before proceeding with erection, make sure that all supplementary instructions are included. If anything is missing, consult the supplier.

Screw conveyor troughs must be assembled straight and true with no distortion. If anchor bolts are not in line, either move them or slot the conveyor feet or saddle holes. Use shims under feet as required to achieve correct alignment (Fig. 26.1.7). Do not proceed with installation of shafts and screw until trough has been completely aligned and bolted down.

Conventional Conveyor Screws

1. When shipped as loose parts, assemble bearings to trough end plates.
2. If trough ends are factory assembled with trough, check bearings, and seals for possible misalignment which may have occurred during shipment. Realign if necessary.
3. Place troughs and trough ends in proper sequence with discharge spouts properly located. Connect the joints loosely. Do not tighten the bolts. Align trough bottom and centerline perfectly using piano wire, as in Fig. 26.1.8. Then tighten joint bolts and all anchor bolts.
4. Begin assembly of screw sections, working from the thrust end. (Drive shaft and thrust bearings are normally at the discharge end to place the conveyor screw in tension.)
5. Place the first screw section in the trough, fitting it onto the end shaft. Install coupling bolts. If reinforcing lugs are on ends of flighting, install screw so lugs are opposite the carrying side of the flight.
6. Insert coupling shaft into opposite end of conveyor pipe; install coupling bolts.

"A" Dim. should be equal for full length of trough, bottom to be smooth thru joint.

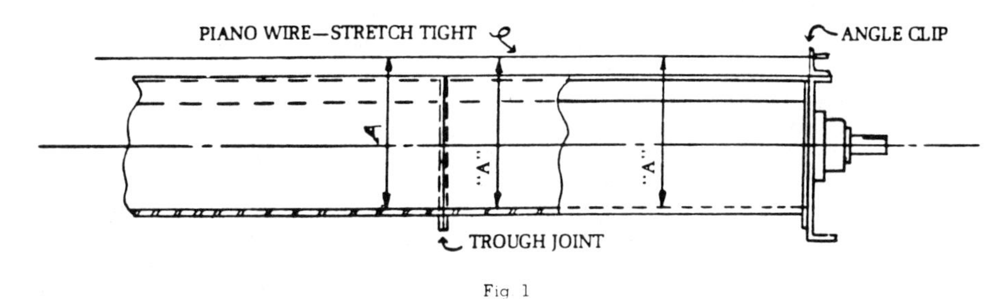

Fig 1

Assembly of conveyor screws should always begin at the thrust end. If the unit does not require a thrust unit, assembly should begin at the drive end. If a thrust end is designated, assemble trough end and thrust bearing. Insert the end, or drive shaft, in the end bearing. Do not tighten set screws until conveyor assembly is completed.

Fig. 26.1.8 Aligning trough bottom and centerline with piano wire.

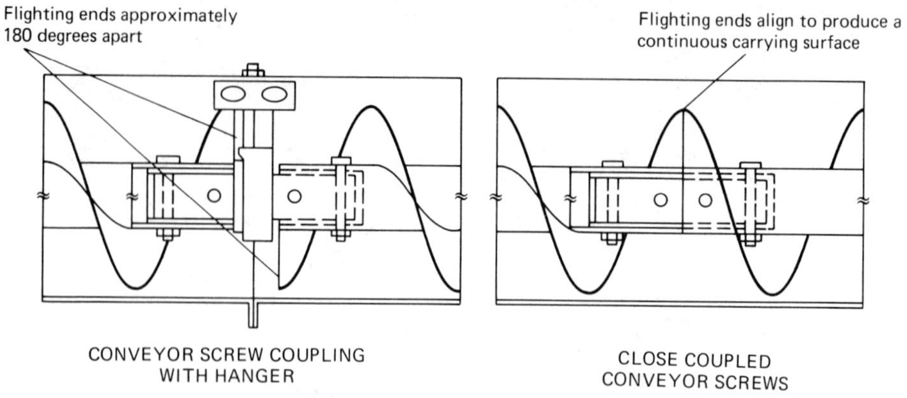

Fig. 26.1.9 Installing screw section in trough and fitting it onto coupling.

7. If screws are not close coupled, slide hanger over coupling and bolt to trough.
8. Pull conveyor screw away from discharge end of conveyor to seal the thrust connection and remove any play in coupling bolts.
9. Place next screw section in trough and fit onto coupling so that flighting end is about 180° from end of flighting of first section (see Fig. 26.1.9). Install coupling bolts. For close-coupled conveyors without hangers: Assemble screws so that flighting at adjoining ends of screw sections aligns to provide a continuous surface. In the case of material supplied on orders for components only, the coupling bolt holes are drilled in only one end of the coupling shafts and it will be necessary to mark and drill the other end in field. Remove shaft from screw before drilling: DO NOT USE SCREW PIPE AS DRILL JIG.
10. Insert coupling shaft into opposite end of pipe; install coupling bolts. Install hanger and pull out on pipe to remove any play (see Step 8).
11. Go back to hanger installed previously; center the bearing between ends of pipes, and tighten hanger mounting bolts. Revolve screw to check alignment. If screw does not turn freely, adjust hanger mountings until it does. Then proceed with installation of next screw section.
12. Alternately assembly screw sections, couplings, and hangers as in Steps 9 through 11 until all screw sections except the last one have been installed. Remove trough end to install last section.
13. Install tail shaft trough-end bearing and fasten into last screw section with coupling bolts. Check freedom of rotation of entire screw.
14. When trough end seals are used, be sure shafts are centered in seal openings.
15. Tighten collar set screws in any antifriction bearings in trough ends and hangers. Check and tighten all hanger assembly and mounting bolts.
16. Tighten packing gland-type seals only enough to prevent leakage; if tightened excessively they may impose a drag on the conveyor and wear rapidly.
17. Fill waste-packed-type seals with waste packing loosely, but sufficiently to encircle the shaft and fill the corners, to prevent packing from rotating with the shaft.
18. Remove all debris from trough (bolts, nuts, shipping materials, etc.). Install covers in proper sequence to locate inlet openings. Handle covers with care to avoid warping and bending, and attach them with fasteners provided. Do not tighten excessively, especially when using gaskets, because leaks may occur when covers are permanently kinked.
19. Install drive at proper location in accordance with separate instructions provided. After electrical connections have been made and before handling any material, check screw rotation for proper direction of travel. Incorrect screw rotation can result in serious damage to the conveyor and to related feeding, conveying, and drive equipment. If rotation is incorrect, have electrician reverse motor rotation.
20. Lubricate drive and all bearings in accordance with manufacturer's instructions. DRIVES ARE GENERALLY SHIPPED WITHOUT OIL.
21. MAKE SURE HAZARD LABELS AFFIXED TO TROUGHS AND/OR COVERS ARE IN PLACE AND NOT OBSCURED.

26.1.10 Operation Methods

Only persons completely familiar with the following precautions should be permitted to operate the conveyor. The operator should thoroughly understand these instructions before attempting to use the conveyor. Failure to follow these precautions may result in serious personal injury or damage to equipment.

1. ALWAYS operate conveyor in accordance with these instructions and those on hazard label (Figs. 26.1.6 and 26.1.10).
2. DO NOT place hands or feet in conveyor opening.
3. NEVER walk on conveyor covers or gratings.
4. DO NOT put conveyor to any other use than that for which it was designed.
5. AVOID poking or prodding material in conveyor with bar or stick inserted through openings.
6. ALWAYS have a clear view of conveyor loading and unloading points and all safety devices.
7. Keep area around conveyor, drive, and control station free of debris and obstacles.
8. NEVER operate conveyor without covers, grating, guards, and other safety devices in position.

Initial Startup (Without Material)

1. REMEMBER—screw conveyor drive is generally shipped WITHOUT oil. Add oil to drive in accordance with manufacturer's instructions.
2. MAKE SURE before initial startup that conveyor is empty, that end bearings and hangers are lubricated, and that all covers, guards, and safety equipment are properly installed.
3. If conveyor is part of a materials handling system, make certain that conveyor controls are interlocked electrically with those for other units in system.
4. Check direction of conveyor rotation in each unit to assure correct flow of material.
5. Operate conveyor empty for several hours, making a continuous check for heating of bearings, misalignment of drive, and noisy operation. If any of these occur, proceed as follows:
   a. If antifriction bearings are used, check supply of lubricant. Either too little or too much lubricant can cause high operating temperatures.
   b. Lock out power supply and check for misalignment in trough ends, screws, and hangers. Loosen, and readjust or shim as necessary. If unable to eliminate misalignment, check parts for possible damage during shipment.
   c. Check assembly and mounting bolts.

Caution

Guards, access doors, and covers must be securely fastened before operating this equipment.

Lock out power before removing guards, access doors, and covers.

Failure to follow these instructions, may result in personal injury or property damage.

Fig. 26.1.10 Hazard warning label used with conveyor.

Initial Startup (With Material)

1. CHECK that the conveyor discharge is clear before feeding material.
2. Increase feed rate gradually until rated capacity is reached.
3. Stop and start conveyor several times, and allow to operate for several hours.
4. Shut off conveyor and lock out power supply. Remove covers and check coupling bolts for tightness. Check hanger bearings, realign if necessary, and retighten mounting bolts.
5. Replace covers.

Extended Shutdown

If conveyor is to be inoperative for a long period of time, it is advisable to permit it to operate for a period of time after the feed has been cut off in order to discharge as much material as possible from the trough. However, there is a nominal clearance of $\frac{1}{2}$ in. between the screw and the trough and this procedure will allow a small amount of material to remain in the trough. Therefore, if the material is corrosive or hygroscopic or has a tendency to harden or set up, the trough should be cleaned completely after the conveyor is shut down and power locked out.

26.1.11 Maintenance Programs

Establish routine periodic inspection of the entire conveyor to insure continuous maximum operating performance. Practice good housekeeping. Keep the area around the conveyor and drive clean and free of obstacles to provide easy access and to avoid interference with the function of the conveyor or drive.

1. Lock out power to motor before doing any maintenance work—preferably with a padlock on control.
2. Do not remove padlock from control, nor operate conveyor, until covers and guards are securely in place.

Servicing of Conveyor Components

In most cases this involves removing an unserviceable part and installing a replacement. The installation procedures are outlined in the previous section. Specific instructions for the removal of conveyor components follows:

Conventional Conveyor Screws. To remove a section or sections of conventional conveyor screw, proceed from end opposite the drive.

Remove trough end conveyor screw sections, coupling shafts, and hangers until all screw sections have been removed, or until damaged or worn section is removed.

To reassemble, follow above steps in reverse order or see assembly instructions in Section 26.1.9.

Sections of conventional conveyor screw equipped with split flight couplings may be removed individually with a minimum of disturbance or adjacent sections.

Couplings and Hangers. Replace couplings and hanger bearings when wear in either part exceeds $\frac{1}{4}$ in. Replace coupling bolts when excessive wear causes play.

Lubrication

Frequency of lubrication depends on factors such as the nature of the application, bearing materials, and operating conditions. Weekly inspection and lubrication is advisable until sufficient information permits establishment of a longer interval.

Drive. Lubricate the drive following manufacturer's instructions provided for the speed reducer and the other drive components requiring lubrication. Speed reducers are generally shipped WITHOUT oil.

Ball or Roller Bearings. Ball and roller bearings may be furnished in trough ends or hangers. Lubricate in accordance with manufacturer's instructions provided.

Babbitted or Bronze-Bushed Bearings. Babbitted or bronze-bushed bearings may be furnished in trough ends or hangers. Lubricate in accordance with manufacturer's instructions.

Other Bearings. For oilless or graphite bronze, hard or chilled iron, oil-impregnated wood, or plastic-laminate hanger bearings, no lubrication is required.

26.1.12 Handling of Hot Materials

When handling hot materials in a screw conveyor, it is necessary to recognize the length change in a conveyor as the hot material is being conveyed. As a general practice, the supports of the trough should allow movement of the trough feet during the trough expansion and contraction. The drive end of the conveyor should be fixed and expansion allowed to take place from that point.

Additionally, screw conveyors may expand or contract in length at different rates through the trough; expansion hangers are generally recommended to compensate for this change. The change in screw conveyor lengths can be determined by the following equation.

$$\Delta L = L(T_1 - T_2)c$$

where: ΔL = increment of change in length, in in.
L = overall length of conveyor, in in.
T_1 = upper limits of temperature, °F
T_2 = lower limits of temperature, °F
c = coefficient of linear expansion, in./°F

Coefficients for various metals are as follows:

Metal	Coefficient
Hot-rolled carbon steel	6.5×10^{-6}
Stainless steel	9.9×10^{-6}
Aluminum	12.8×10^{-6}

26.1.13 Screw Feeders

Screw feeders are totally enclosed conveyors whose function is to control or regulate the flow or feed of a predetermined volume of material at a uniform rate from a bin or hopper. There are four types: full pitch, regular and half or short pitch, regular full pitch tapered, and half pitch tapered. Full-pitch feeders are usually used for handling free-flowing materials where it is not objectionable for the feed to be from the rear of the hopper or bin instead of being fed uniformly across the length of the opening. It also can be used more economically and very satisfactorily where the length of the opening is not over twice the pitch of the conveyor. Half-pitch regular feeders are usually used where the materials are of a nature that they may flood and overload the conveyor being fed. Screw feeders with taper flights are generally used to handle material that may contain a considerable amount of lumps. They are also used extensively when it is necessary or desirable to draw the material from the bin or hopper uniformly across the length of the feed opening rather than from the rear of the bin or hopper. Uniformity in drawdown eliminates dead areas of nonmovement of material in the bin or hopper. Using a tapered-flight conveyor instead of a regular-flight conveyor will, in most cases (especially where the feed opening is long), consume much less horsepower. An alternate to tapered flights is a design utilizing a tapered pipe section. This design further eliminates dead areas of material in the trough.

Feeders with conveyor extensions are necessary when the material must be conveyed a distance that would require the use of intermediate hangers. In this case, a larger-diameter conveyor and a standard trough are used in combination with the feeder's spiral, unless the combination is given in Table 26.1.14. Calculation for horsepower on screw feeders is very similar to that for a screw conveyor. Basically, it is divided into the energy required to run the feeder empty and the energy to move and overcome the material friction.

Horsepower for single screw feeder:

$$\text{Hp} = \frac{(\text{Hp}_a + \text{Hp}_b)F_o}{e}$$

Horsepower for single screw feeder with extension conveyor:

$$\text{Hp} = \frac{(\text{Hp}_a + \text{Hp}_b + \text{Hp}_f + \text{Hp}_m)F_o}{e}$$

Table 26.1.14

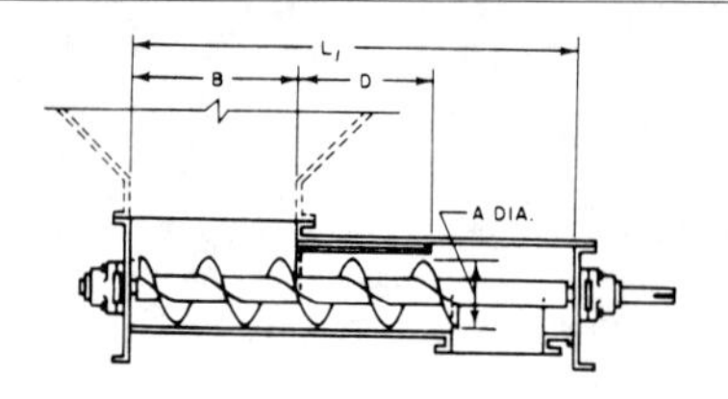

STANDARD SCREW FEEDER

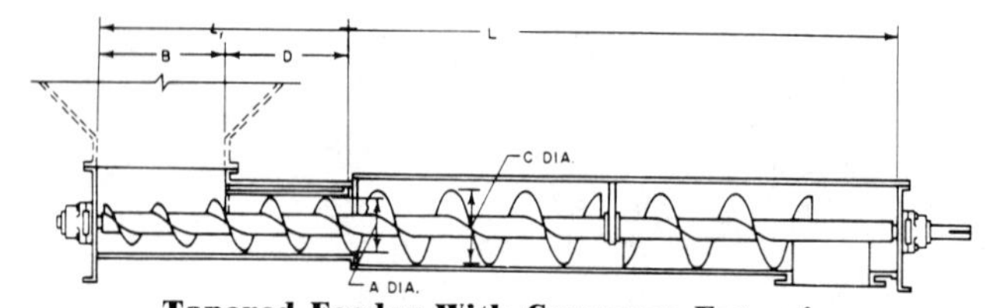

Tapered Feeder With Conveyor Extension

FEEDER (95% Full)						SCREW CONVEYOR EXTENSION									D	
Screw		Recommended		Capacity		Class I Materials			Class II & IIA Materials			Class III Materials				
						Conveyor			Conveyor			Conveyor				
Diam. A	Pitch	Small dia. of Taper	Max. length of Feed Opening B	T.P.H. for 100 lb. Mat'l at 1 RPM	Cu. ft. per hour at 1 RPM	Diameter C	Pitch	% Full	Diameter C	Pitch	% Full	Diameter C	Pitch	% Full	Max. *	Min. †
6	6	4	36	.237	4.75	9	6	43%	9	9	28%	10	10	20%	12	9
‡6	4	4	36	.158	3.17	....	....	....	9	6	28%	10	10	14%	8	8
9	9	5¾	60	.84	16.8	12	12	41%	14	14	26%	16	16	17%	18	14
‡9	6	5¾	60	.56	11.2	10	10	47%	12	12	27%	14	14	17%	12	12
10	10	6¼	70	1.19	23.8	14	14	36%	14	14	36%	18	18	16%	20	15
‡10	6⅝	6¼	70	.80	15.9	12	12	39%	14	14	25%	18	12	16%	14	14
12	12	7½	80	2.04	40.8	16	16	41%	18	18	28%	20	20	21%	24	18
‡12	8	7½	80	1.36	27.3	14	14	42%	16	16	27%	18	18	19%	16	16
14	14	8¾	95	3.26	65.2	18	18	45%	20	20	33%	24	24	19%	28	21
‡14	9⅜	8¾	95	2.18	43.5	16	16	44%	18	18	30%	24	24	13%	19	19
16	16	10	110	5.0	100.0	20	20	51%	24	24	29%	....	....	....	32	24
‡16	10⅝	10	110	3.34	66.7	18	18	46%	20	20	34%	24	24	19%	21	21

* For fine or pulverized materials
† For lumps 1/4" and under (When the material has a majority of 1/4" lumps or over use a cutoff plate instead of a shroud)
‡ Short pitch is recommended for pulverized materials which tend to flood (Materials that become fluid cannot be regulated)

where empty feeder friction power is:

$$\mathrm{Hp}_a = \frac{L_1 N F_d F_b}{1.0 \times 10^6}$$

Feeder material friction power:

$$\mathrm{Hp}_b = \frac{\mathrm{C\,W\,L_1}\, F_m}{1.0 \times 10^6}$$

Empty extension conveyor friction power:

$$\mathrm{Hp}_1 = \frac{L N F_d F_b}{1.0 \times 10^6}$$

Extension conveyor material friction power:

$$\mathrm{Hp}_m = \frac{C W L F_f F_m F_p}{1.0 \times 10^6}$$

where: C = capacity, in ft^3/hr
W = apparent density of material as conveyed, in lb/ft^3
L = length of extension conveyor, in ft
L_f = equivalent length of feeder, in ft. See Table 26.1.15 for method of arriving at values of L_f for various types of screw flighting.
L_1 = length of feeder, in ft, as shown in Table 26.1.14
N = speed of screw conveyor, in rpm
F_b = hanger bearing factor, Table 26.1.7
F_d = conveyor diameter factor, Table 26.1.8
F_m = material factor, Table 26.1.3
F_o = overload factor, Table 26.1.10
e = efficiency of the drive selected, see Table 26.1.16

Where screw feeders are mounted at the bottom of bins and hoppers the screw must perform under heavy loads of material. Under certain conditions with certain materials, the startup torque can be very high, resulting in larger than normal drives and heavier components.

26.1.14 Multiple Screw Feeders

Multiple screw feeders are made up of several horizontal screws arranged closely side by side to completely cover the area of flat bottom bins. They are used for materials that have a tendency to

Table 26.1.15 Equivalent Length of Feeder (L_f)

Maximum Particle Size[a] (in.)	Flight Type under Inlet	Value of L_f (ft)
$\frac{1}{8}$	Standard pitch, uniform diameter.	$L_1+\frac{B}{6}+\frac{D}{12}$
	Short pitch, uniform diameter.	B and D from Table 26.1.14
	Standard pitch, tapered diameter.[b]	$L_1+\frac{B}{12}+\frac{D}{12}$
	Short pitch, tapered diameter.[b]	B and D from Table 26.1.14

[a] For larger size, consult conveyor manufacturer.
[b] Variable pitch of constant diameter may be used in place of tapered diameter and constant pitch flighting.

Table 26.1.16 Mechanical Efficiencies of Speed Reduction Mechanisms

Type of Speed Reduction Mechanism	Approximate Efficiencies
V-belts and sheaves	0.94
Precision roller chain on cut-tooth sprockets, open guard	0.93
Precision roller chain on cut-tooth sprockets, oil-tight casing	0.94
Single-reduction helical or herringbone enclosed gear reducer or gear motor	0.95
Double-reduction helical or herringbone enclosed gear reducer or gear motor	0.94
Triple-reduction helical or herringbone enclosed gear reducer or gear motor	0.93
Single-reduction helical gear, enclosed shaft-mounted speed reducers and screw conveyor drives	0.95
Double-reduction helical gear, enclosed shaft-mounted speed reducers and screw conveyor drives	0.94
Low ratio (up to 20:1) enclosed worm gear speed reducers	0.90
Medium ratio (20:1 to 60:1) enclosed worm gear speed reducers	0.70
High ratio (60:1 to 100:1) enclosed worm gear speed reducers	0.50
Cut-tooth, miter or bevel gear, enclosed countershaft box ends	0.93
Cut-tooth spur gears, enclosed, for each reduction	0.93
Cut-tooth miter or bevel gear open-type countershaft box ends	0.90
Cut-tooth spur gears, open for each reduction	0.90
Cast-tooth spur gears, open for each reduction	0.85

pack or bridge under pressure. A collecting screw conveyor that runs at right angles to the bin is used to discharge material to another conveyor or process equipment. Depending on the number of screws involved, they may have single or multiple drives. Capacities required, operating horsepower, and so on, can be calculated by the same method used for single feeders, making the calculation for one screw and then multiplying by the number of screws used. Due to factors affecting the operation of multiple screws, such as loads and the weight of material in the bin, any multiple screw feeder having an inlet area of more than 12 ft² should be referred to a screw conveyor manufacturer.

26.1.15 Inclined Units

There are many occasions where a screw conveyor can be utilized on an inclined path to solve many conveying problems with a minimum amount of equipment and space required. Besides the obvious advantages, there are certain disadvantages that must be recognized in designing an inclined screw. The primary disadvantage is that the capacity of a given screw conveyor decreases with the increase of the incline and, too, the horsepower per unit of capacity changes. The reasons for these effects are numerous.

As the angle of inclination increases, there is a reduction in the effective angle of flight as it pushes against the material. Depending on the pitch and the angle of incline, a certain portion of the helical flight is on a horizontal plane and does not urge the material forward. The reduction of the ability of a flighting to move the material forward causes material turbulence and tumbling. The turbulence and tumbling of the material requires more horsepower than the power normally required to convey the material. Also, the cross-sectional loading is increased, so intermediate hangers present an obstacle for the material to flow past. The U shape of the conveyor trough is such that the material is allowed to fall backward over the top of the rotating screw on the inclined conveyor. Again, increasing the turbulence and cross sectional loading. Figure 26.1.11 shows the effect on capacity as the inclination is increased. A number of things can be done to overcome the problems associated with inclined screws. They are: (1) limit the use of standard screw conveyors to inclines of less than 20°, (2) use close tolerances between the trough and the screw, (3) increase the speed over the horizontal screw application, (4) use short pitch such as $\frac{2}{3}$ or $\frac{1}{2}$, (5) eliminate hanger bearings, and (6) use tubular trough with a minimum clearance between the trough and screw. See Fig. 26.1.12 for trough designs.

Horsepower

The horsepower of an inclined screw may be approximated by using the following calculations:

1. Calculate the horsepower of the screw conveyor as though it were a horizontal screw conveyor, using the horsepower equations shown earlier.

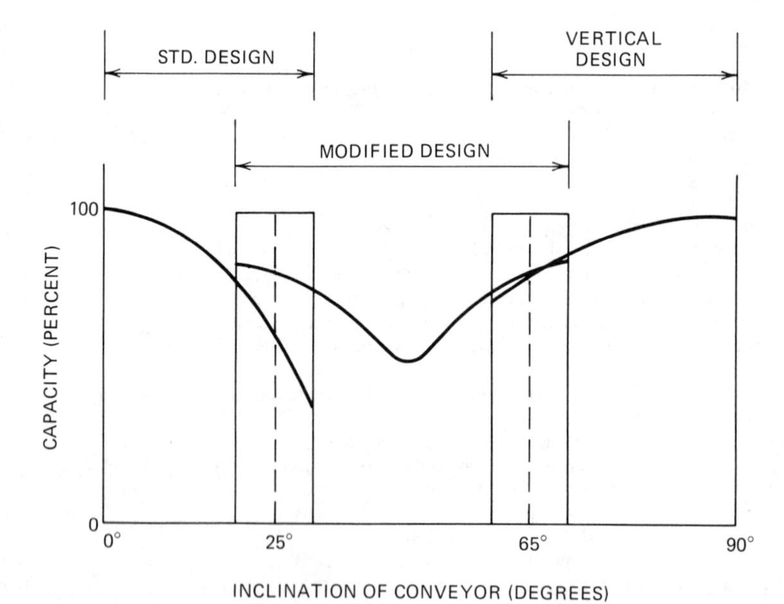

Fig. 26.1.11 Effect of degree of incline on conveyor capacity.

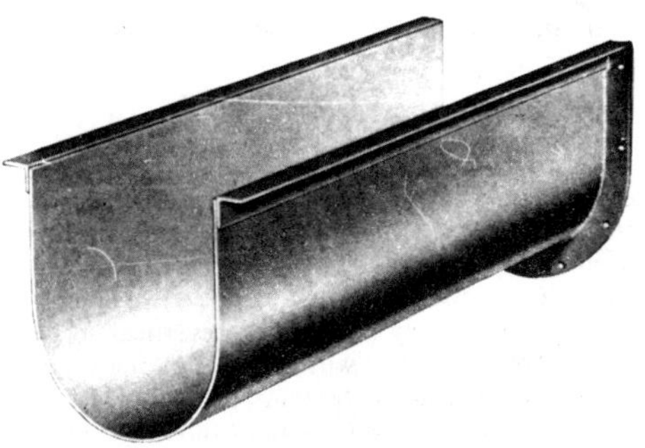

Angle Type Trough
Fig. 4.15A
The angle type trough here illustrated is typical. Dimensional data is published in C.E.M.A. Screw Conveyor Standard 300-1971.

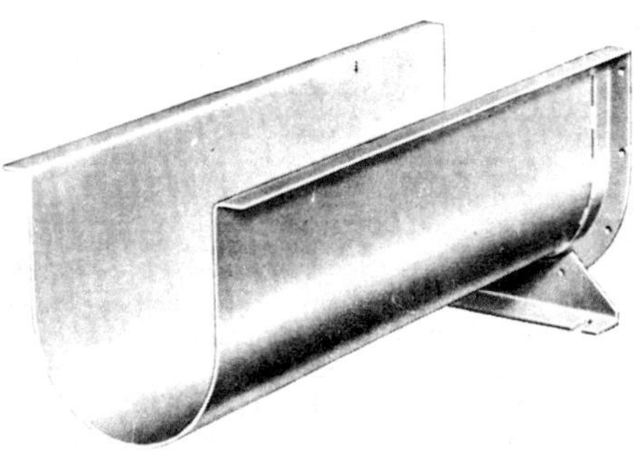

Single Flanged Trough
FIGURE 4.15B
The single flanged trough is generally an alternate construction when the heavier gauges of steel are required.

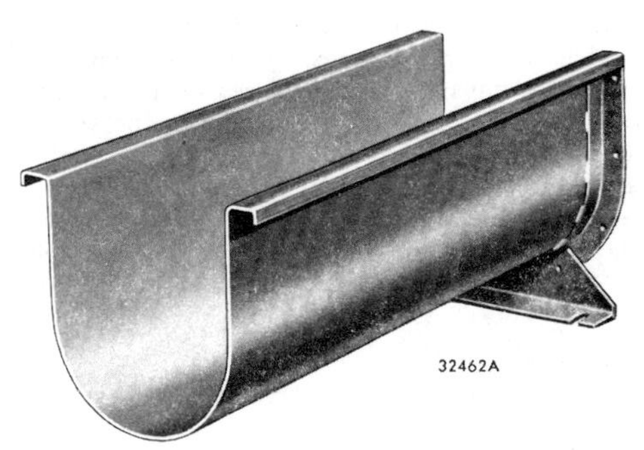

Double Flanged Trough
FIGURE 4.15C
The double flanged trough conforms to the same general dimensions as the angle type and single flanged type. It is an alternate construction when the lighter gauges of trough are required.

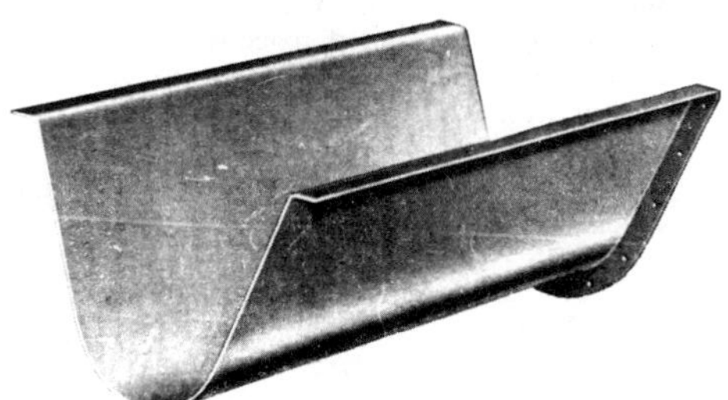

Flared Trough
Fig. 4.15D

Fig. 26.1.12 Basic trough designs.

2. Calculate the actual horsepower to lift the material the total height of the incline. This may be done as follows:

$$\frac{(\text{lb/min})\ (\text{Height of lift, ft})}{33{,}000} = \text{Hp of lift}$$

If the rate of conveyance of the material to be handled C is in cubic feet per hour, and the apparent density of the material W is in pounds per cubic foot, then:

$$\text{lb/min} = \frac{CW}{60}$$

3. Estimate the horsepower required to overcome the decrease in efficiency due to the extra agitation and tumbling of the material. Obviously, this factor will vary with each application. It is wise to consult a screw conveyor manufacturer for the benefit of his experience.
4. Add the horsepowers calculated in 1, 2, and 3. This will be the approximate total horsepower to operate the loaded inclined screw conveyor, not considering the efficiency of the drive.
5. Divide by the drive efficiency to select the motor Hp.

In any case the drive should be placed at the top or discharge end of the screw conveyor. Note that if the speed reduction unit is tilted at the angle of the conveyor, care should be taken to see that the oil level is not jeopardized, that the shaft seals will be satisfactory to retain the lubricant, that the oil level gauge will indicate the necessary level, and that the oil filling and drain plugs are accessible and operative at the tilted position of the reducer.

26.1.16 Vertical Units

A vertical screw elevator, often called a lift or an elevator, is a unit that conveys material vertically. Vertical screw elevators can satisfy many conveying problems and have the further advantage of being compact, requiring less space than other forms of elevating equipment. Basically, the vertical screw elevator can handle many of the materials shown in Table 26.1.3. Exceptions to this rule are materials containing large lumps, and materials that are very dense or extremely abrasive. The vertical screw elevator consists of a tubular trough enclosing a rotating screw conveyor with an inlet at the lower end and a discharge at the top end. The drive may be placed either at the bottom or at the top. The top bearings must absorb the thrust because the rotating element is hanging. The method of feeding a vertical screw elevator is most important. It is recommended that vertical screw elevators be fed by another screw feeder and not be gravity fed. Material can be fed to the vertical unit by a straight or offset intake horizontal feeder. A straight intake unit is simple and effective for those materials that will not become damaged by jamming or forcing. The offset intake is very often used, especially for more fragile materials. Standard practice in design varies between manufacturers and it is important to consult them for their recommendations. Vertical screw elevator speeds must be adequate to convey and to overcome a fallback of material in the distance between the housing and the screw. The material in a vertical screw moves en masse through the tube. Typical capacities of standard size vertical screws are shown in Table 26.1.17. Vertical screw elevators are fabricated of materials similar to those for horizontal screws; however, tolerances in some units are smaller, based on the material that the unit is required to convey. Intermediate hangers or stabilizers are used with the vertical screw where extended lengths are utilized. This eliminates deflection or "whipping" of the rotating element within the casing. Long vertical screws should not be run empty and should not be run at less than 30% of design capacity; doing so will cause excessive damage to the unit. The horsepower for a vertical screw may be approximated by using the following equation. Because of the many variables associated with a screw elevator installation, it is recommended that you consult with the manufacturer on horsepower.

$$\mathrm{Hp} = \frac{(\mathrm{Hp}_f + \mathrm{Hp}_v)}{0.90}$$

where: Hp_f is the horsepower to drive the empty conveyor
Hp_v is the horsepower to convey the material vertically

$$\mathrm{Hp}_f = \frac{L_1 N F_d F_b}{1{,}000{,}000}$$

where: L_1 = total length of the vertical screw conveyor, in ft
N = speed of vertical conveyor screw, in rpm
F_d = conveyor diameter factor from Table 26.1.8
F_b = hanger bearing factor from Table 26.1.7

$$\mathrm{Hp}_v = \frac{C L W F_v}{1{,}000{,}000}$$

where: L = total lift height, in ft, measured from the centerline of the opening to the bottom of the discharge opening
C = capacity, in $\mathrm{ft}^3/\mathrm{hr}$
W = apparent density of the conveyed material, in $\mathrm{lb}/\mathrm{ft}^3$
F_v = manufacturer's empirical factor

Table 26.1.17 Capacities for Vertical Screws

Rotor Lift Size	Cubic Feet per Hour
6 in. diameter	400
9 in. diameter	1300
12 in. diameter	3000
16 in. diameter	6000

26.1.17 Drives

The drive of a screw conveyor normally includes an electric motor, a speed-reduction gear box, a drive shaft, and a means to transmit power from one unit to the other. The simplest drive using a minimum number of units is by far the best. Physical limitations or other reasons may often dictate a more exotic drive arrangement.

A self-contained screw conveyor drive is most commonly used for horizontal screw conveyors. It consists of a speed reducer, either single or double reduction with a special low-speed shaft and mounting adaptor for the trough end. A motor mount is usually furnished with this drive and the motor and sheaves are connected with V-belts and V-belt sheaves. The drive shaft of a screw conveyor drive is designed so as to extend through the trough end and bolt to the standard connection of the screw. The shaft transmits all radial and thrust loads directly into the speed reducer which is designed to receive these loads. The flange mounting adaptor incorporates seals to prevent leakage of material into the reducer and also retains a lubricant of the reducer to prevent it from leaking into the trough. Since the motor is connected to the input shaft of the reducer by means of a V-belt, a wide range of screw conveyor speeds can be obtained by the use of standard speed-reducer ratios and standard V-belt drive ratios.

Another simple drive includes a standard shaft-mounted reducer mounted directly on the drive shaft of the screw conveyor. The motor is connected to the speed reducer by V-belts and the motor can be mounted above or to the side. An adjustable tie rod or torque arm prevents the rotation of the speed reducer and affords a simple means of putting tension in the V-belts.

A third means of driving a screw conveyor is with a gear motor. This unit is connected to the drive shaft of the screw conveyor by means of a precision roller chain drive. With this design you have great flexibility in locating the gear reducer and motor. The in-line gear motor provides a wider range of power capacities than the standard shaft-mounted reducers and screw conveyor drives. Speeds can be changed easily by selection of the appropriate sprockets. The chain drive is usually mounted in an oil-tight casing. Other types of drive arrangements include parallel shaft reducers and worm gear reducers. These units are usually used where large horsepowers are required or a larger reduction in speed is necessary. In utilizing drive mechanisms other than a screw conveyor drive, care needs to be exercised in the design of the drive shaft. It is necessary that the thrust load be isolated independently from the speed reducer. This is usually accomplished by means of thrust rings or thrust collars mounted directly to the drive shaft. Screw feeders are ideal for use in process control. Variable-speed drives allow changes in speed as required for the process. Screw feeders have a constant-torque variable horsepower speed characteristic. The simplest means of varying the speeds is mechanically through the use of a variable-pitch motor sheave. Other ways include DC motors, AC variable frequency, and eddy current couplings. Hydraulic motors may also be used for speed control and torque control.

In general, inclined screws utilize the same type of drives horizontal screws use. However, care should be taken to assure that at the incline, proper lubrication is afforded the reducer. Gear reducer manufacturers can assist on this problem. Vertical screw conveyors have a specially designed right-angle drive designed specifically for the vertical application. These drives can usually mount at the top or the bottom and accommodate thrust and radial loading.

In selecting the proper horsepower motor the drive efficiency must be taken into consideration. Table 26.1.16 shows the efficiency for various types of mechanical speed-reduction mechanisms. These efficiencies represent conservative figures for the components of the drive train, taking into account possible slight misalignments, uncertain maintenance, and the effects of temperature change. Appropriate service factors for various power transmission components should be determined from the manufacturer's catalog taking into account the intended service, hours of operation, and the type of operating condition. The American Gear Manufacturers Association (AGMA) designates different service factors for different types of speed reducers. No single system of service factors is available for gear speed reduction mechanism. The Mechanical Power Transmission Association's "Standard Specifications for Drives Using Narrow Multiple V-belts," dated 1977, and "Specifications for Drives Using Classical Multiple V-belts," dated 1977, tabulate recommended service factors for V-belt drives.

BIBLIOGRAPHY

Classification and Definitions of Bulk Materials, Publication No. 550, Conveyor Equipment Manufacturers Association, 1974.

Conveyor Terms and Definitions, Standard No. 102, Conveyor Equipment Manufacturers Association.

Jenike, A. W., *Flow of Solids in Bulk Handling Systems,* ASME Paper 54-SA-34, June 1954.

Jenike, A. W., *Storage and Flow of Solids,* Bull. No. 123, Utah Engineering Experiment Station, Eighth Printing (Revised), April 1980.

Johanson, J. R., "Feeding," *Chemical Engineering,* Deskbook Issue, October 13, 1969, Jenike & Johanson, Inc.

Safety Standard for Conveyors and Related Equipment, American National Standard B20.1–1976.

Screw Conveyor Dimensions, Standard No. 300, Conveyor Equipment Manufacturers Association.

Screw Conveyors, Publication No. 350, 2nd Ed., Conveyor Equipment Manufacturers Association, 1980.

26.2 VIBRATORY FEEDERS AND CONVEYORS

Conveyor Equipment Manufacturers Association
Vibrating Equipment Section

26.2.1 Introduction

Traditionally engineers have been interested in eliminating vibration. Fortunately a few individuals saw that vibration was useful as a means of accomplishing the movement of bulk granular materials. Nearly all production and design engineers are aware of the existence of vibratory conveyors and feeders, although many are not familiar with their fundamental characteristics. Demands for greater productivity, improved performance, cost savings, and better space utilization require a thorough understanding of what vibratory equipment is, what makes it work, and, consequently, how to select the right piece of equipment for a particular job.

Vibratory conveying has been used in the United States for more than a century; however, it has been only within the past few decades that the concepts of vibratory conveying and feeding have attained general acceptance in industry, resulting in a rapid growth in the number and types of vibratory equipment being manufactured. There is also an ever increasing demand for this type of equipment and, consequently, an increase in the number of applications in all types of industry.

Much of the work done in the United States on the analysis of vibratory conveying has been kept confidential and used by the manufacturers to design various types of proprietary vibratory materials handling equipment. Through the combination of theoretical knowledge, development efforts, and practical knowledge obtained from numerous applications, designers have been able to provide vibratory conveyors and feeders that have proven to be some of the most reliable pieces of equipment in modern plants.

26.2.2 Glossary

The movement of material by vibration is a relative newcomer to the bulk handling field and as a result much confusion exists as to design, correct application, and even terminology.

To remove the confusion involving the terminology used to describe this relative new type of materials handling equipment, the Conveyor Equipment Manufacturers Association (CEMA) has developed a Glossary of Terms and Definitions to serve as the industry standard.

Adjustable rate vibrating feeder. A vibrating feeder in which the material flow rate within a specific range can be changed while the feeder is operating (see Fixed-Rate Vibrating Feeder).

Amplitude. A value equal to one-half the stroke.

Angle of attack. See Stroke Angle.

Back end. See Feed End.

Balanced and isolated vibrating conveyor. A balanced vibrating conveyor that is mounted on springs to further reduce the transmitted dynamic forces.

Balanced vibrating conveyor. Any vibrating conveyor designed to reduce the transmitted dynamic forces to the supporting structure.

Brute force drive. One in which the only forces applied to the vibrating members are generated by a directly (rigidly) connected vibrating mechanism.

Deck. See Trough.

Depth of bed. Thickness of the layer of material traveling a conveyor or feeder surface.

Discharge end. The end of a conveyor or feeder most downstream, usually where the material is fed off the unit.

Dynamic balancer. A device that reacts to the reversing forces of a vibrating conveyor to achieve the design of a balanced vibrating conveyor.

Dynamic load. The reversing forces transmitted to the supporting structure by the vibrating equipment when operating. Peak dynamic loads may occur at other than the running frequency.

Eccentric weight. A weight that is attached to a shaft or fly wheel mechanism to produce an unbalanced moment.

Electromagnetic vibrating equipment. A type of vibrating device that uses an electromagnet to develop the driving force.

Electromechanical vibrating equipment. A type of vibrating device that uses a motor along with eccentric crank or eccentric weights to develop the driving force.

Explosionproof. Equipment designed in accordance with existing codes and standards such that it will operate in a specified hazardous environment without causing an explosion.

Feed end. The end of a conveyor or feeder most upstream, usually where the material is fed onto the unit.

Fixed-rate vibrating feeder. A feeder where the rate of material flow is constant during operation. The material flow rate may be changed with a mechanical adjustment while the feeder is turned off (see Adjustable Rate Vibrating Feeder).

Grizzly. A heavy-duty screening surface consisting of a series of space bar, rail, or pipe members running in the direction of material flow used for the rough sizing of bulk materials.

Hand. A designation of right or left used to indicate a specific side of a conveyor or feeder. It is determined when facing in the direction of material flow, as it moves away from the viewer.

Head load. The weight of the material in a hopper or bin imposed on a vibrating feeder.

Leg angle. The angle between the centerline of the supporting legs and a line normal to the deck surface or base.

Line of action. The line that defines the machine's movement or stroke.

Liner. A member or material added to the trough or chute designed for a specific purpose such as to resist wear, heat or cool the deck, reduce noise, reduce friction, or prevent material buildup.

Material depth (mat. depth). See Depth of Bed.

Natural frequency conveyor or feeder. A machine that has its operating frequency close to or at the natural frequency of the internal spring mass system.

Oscillating conveyor. See Vibrating Conveyor.

Pan. See Trough.

Scalping. The process of removing oversize lumps on a continuous basis from a stream of bulk material.

Screen. A perforated or meshed surface used to separate coarse from fine particles.

Spring rate. Force per unit length of deflection of a spring usually expressed in lb/in. or kg/cm.

Springs, coupling. See Springs, Reactor.

Springs, isolation. The springs or isolation devices used to support a vibrating feeder or conveyor and to isolate vibrations from supporting structure.

Springs, reactor. The primary springs in a vibrating system which alternately store and release energy.

Springs, suspension. See Springs, Isolation.

Springs, tuning. Weights and/or springs added to or subtracted from an electromagnetic feeder in order to tune the feeder to the desired natural frequency.

Static load. Force on the supporting structure resulting from the weight of equipment plus material when not operating.

Stroke. The total peak-to-peak displacement occurring each operating cycle of a vibrating conveyor or feeder.

Stroke angle. The angle of the machine line of action with respect to the conveying surface.

Timing the drive. Twin shaft: Adjusting the phase relationship between shafts that carry eccentric weight to produce the desired vibratory motion. Conveyor: Adjusting the eccentric drive so that its neutral position is at the mid-point of the stroke.

Tray. See Trough.

Trim bar. See Spring, Tuning.

Trough. A channel generally longer than its width, open at the top or fitted with a cover, which contains the material being conveyed. The shape of the cross section depends on the type of conveyor or feeder involved.

Tuning weight. Weight added to or subtracted from an electromechanical feeder in order to tune the feeder to the desired natural frequency.

Variable rate vibrating feeder. See Adjustable Rate Vibrating Feeder.

Vibrating absorber. See Dynamic Balancer.

Vibrating conveyor. A machine that transports material using an oscillating or vibrating motion. The machine may be designed with a wide range of frequencies and strokes.

Vibrating feeder. A machine incorporating a trough and a vibrating drive designed to control the rate of delivery of material at a controlled weight from storage bins, silos, hoppers, and so on.

26.2.3 General Principles of Operation

Vibrating feeders and conveyors consist of a material-transporting trough driven by a controlled vibratory force system which imparts a tossing, hopping, or sliding-type action to the material.

Basically, a vibrating feeder is applied where it is necessary to meter or control the flow of material from a hopper or bin, whereas a vibrating conveyor is normally applied to move a material from point A to point B. The versatility of these units allows material to be processed or manipulated while in motion.

Material Conveyance

The first theoretical investigations of the movement of particles by vibration was done by C. Schenck in Germany in the first part of this century. Numerous other investigations have been made until correlations are now available combining theoretical analysis and practical results. Material property variations are such that there is no exact solution explaining why, even today, many manufacturers rely on experimental results for accurately determining material travel rates. The motion of particles resting on a trough is influenced by stroke angle, frequency of operation, length of stroke, and material properties such as the coefficient of friction and cohesion. The motion can be broken down into horizontal and vertical components. Although, in reality, resultant motion may be somewhat elliptical for many classes of vibratory equipment, sufficient accuracy of analysis may be obtained by considering this motion to be essentially straight line. Straight-line motion means that the horizontal and vertical components are in phase; it is a basic assumption in virtually all work relating to feed rate theory.

The theory of particle movement on both vibratory conveyors and feeders is the same. It is the design and construction of the equipment itself that makes these two types of equipment suitable for entirely different applications. The vibration generator or exciter for a conveyor or feeder may be electromagnetic, electromechanical (eccentric weight), pneumatic, or hydraulic producing a trough vibration of nearly sinusoidal motion at some angle to the trough. When the unit is operating, the trough is vibrating along a path with a controlled amplitude and direction. The angle between the directed vibration and the trough bottom is called the stroke angle. The result of this directed linear vibration is a repetitive series of throws and catches that moves the material on the trough (Fig. 26.2.1). The particle is in contact with the trough for a portion of the drive cycle as shown in point A to point B. When the particle leaves the trough it travels with a uniform horizontal speed but the vertical speed gradually decreases due to gravity. At some later point the trough again contacts the particle and the process is repeated. This process conveys material along the trough from 0 to 120 ft/min or more, depending on the combination of drive frequency, amplitude, stroke angle, and material properties.

A vibratory motion that results in a vertical acceleration component less than that of gravity (32 ft/sec^2) will transport materials with a gentle "shuffling" manner without impact or noise generation between the material and the conveyor trough. Actually, the material never leaves the trough surface, but moves ahead with a gentle sliding-type action when the pressure between it and the trough is at a minimum. This mode of operation, especially with a long stroke applied at a low frequency, conforms with most noise level restrictions.

In many applications, it is desirable to actually have the material leave the trough and then impact upon it again, for example, shaking sand from castings or removing sawdust from a sawmill waste. This mode of operation increases the sound pressure level but special trough designs are available to maintain an acceptable noise level.

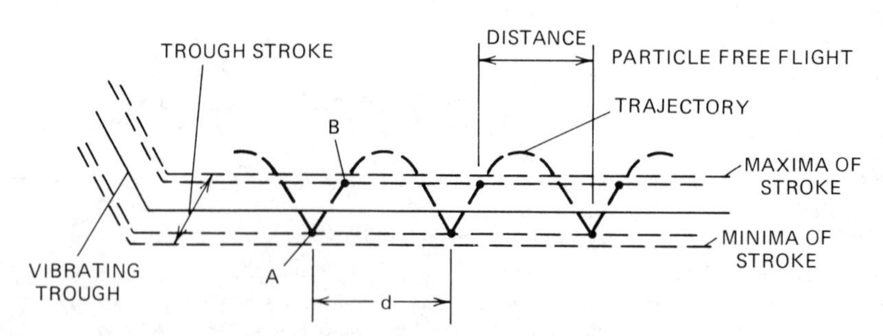

Fig. 26.2.1 Simplified illustration of particle motion on a vibrating trough. Particle contacts trough at its position labeled A and rides with it until position B, where the vertical acceleration reaches 1g and the particle leaves the trough in free flight, landing again at trough position A but displaced by horizontal distance *d*.

With proper flow depths and trough widths, material can be delivered at rates of several hundred tons per hour on vibratory conveyors, and several thousands tons per hour on vibratory feeders.

26.2.4 Vibrating Conveyors

For more than a century American industry has been aware of the practical use of vibration to move materials; however, it has been only in the last few decades that design engineers have been able to intelligently apply the principles of vibration to production equipment. Although conventional materials handling equipment such as belts, aprons, screws, and drags are widely used throughout industry, vibrating equipment has been more extensively accepted because of its unique capabilities, versatility, relatively low maintenance characteristics, and ease of installation.

General Applications

One major factor that differentiates a vibrating conveyor from conventional materials handling equipment is the fact that the material is "live" and moves independently of the conveying medium; whereas, on a conventional belt conveyor, for example, the material is static and the conveying medium is moving. This important characteristic, plus a number of other unique advantages, offers the design engineer solutions to many difficult materials handling problems.

In the majority of conveyor applications, the only purpose of the unit is to convey material from one point to another; however, a vibrating conveyor offers advantages since the material can be processed while in transit. Some of the features offered by vibrating conveyors are:

Scalping and screening can be performed while conveying.
Material may be cooled or dried while being conveyed.
Extremely hot materials can be handled.
Equipment can withstand heavy impact loading.
Highly abrasive materials can be handled.
Different products can be handled on a single unit by use of divided troughs.
Material can be distributed through a number of discharge points.
Material can be de-watered.
Hot material can be water quenched.
Various sized materials can be easily oriented.
Vibrating conveyors are inherently self-cleaning and leakproof. No return run eliminates spillage.
Conveyors can be designed to meet stringent sanitation requirements.
Leaching can be performed with a liquid flowing counter to the material movement.
Conveyor troughs can be easily made dust-tight or designed for gas-tight operation.
There are no moving parts, except the trough, in contact with the material being conveyed.
Friable or easily degradable materials can be handled safely.
Individual unit lengths of 300 ft are possible.
Foundation or support vibration can be eliminated by balanced designs.

Vibrating Conveyor Motion

The general design of a vibrating conveyor consists of a vibrating work member, driven by a controlled vibratory force system. Vibrating conveyor operating frequencies normally range from 200 to 3600 vibrations per minute with an amplitude or stroke range from 0.03 to 1.5 in. total movement.

The controlled stroke is applied to the trough in an angular relationship, line-of-action, which results in both horizontal and vertical force components. The line of action of the applied force is varied for specific applications. An important consideration is that the noise level generated by a vibrating conveyor, as well as the degree of vibration transmitted to the supporting foundation, is a function of the design of the unit.

Basic Designs

There are a variety of vibrating conveyor designs available which differ in detail since most manufacturers have patented and proprietary designs unique to their equipment; however, vibrating conveyors generally have similar basic elements (Fig. 26.2.2):

A trough in which the material is conveyed.
A base which mounts the conveyor in place and ties all of the other elements together.

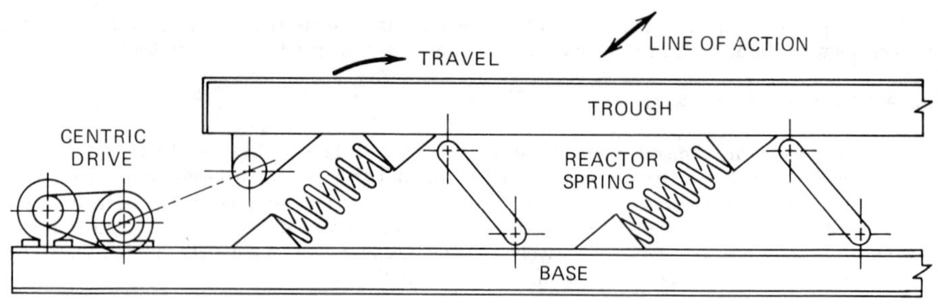

Fig. 26.2.2 Basic elements of vibrating conveyors.

A trough supporting system to direct the motion of the trough.

An eccentric drive assembly which is the source of the controlled vibrating motion applied to the conveyor.

Many designs also include:

A reactor spring system which alternately stores and releases energy at each end of the trough stroke.

Reviewing these elements:

1. The trough is the only portion of the vibrating conveyor that comes in contact with the material being conveyed. It can be fabricated in a variety of materials in almost any shape and size and can be adapted to perform various processes while the material is in motion.

2. The base is primarily a means of mounting the conveyor and is usually of a simple design incorporating structural steel members. It can be designed as an elaborate trusslike structure or can be simplified so all corners are eliminated to meet sanitation specifications for the food, chemical, and other related industries.

3. The trough supporting system's primary function is to control and direct the motion of the trough. This system can assume a variety of shapes and may be cast or fabricated assemblies incorporating maintenance-free flexible connections at each end or a simple flexible slat.

4. The drive is the prime element in a vibrating conveyor because it is the source of the controlled vibration. The drive may be in the form of a positive direct-connected linkage, a positive flexible-connected linkage, or a nonpositive motorized counterweight assembly. These latter two are found primarily on conveyors that take advantage of the natural frequency phenomena whereas the first is generally used on brute-force units.

5. The reactor spring system can assume many forms including steel coil springs, flexible steel or glass slats, rubber blocks, circular rubber toroids, and torsion bars. The particular application involved often makes one type more advantageous than another.

Drive Designs

There are several approaches to the design of vibrating conveyors, but the factors given primary consideration are brute-force or natural frequency. This latter design is also occasionally described as a resonant frequency design. Both designs have their place in the materials handling field and among the factors that dictate the correct selection are: the application involved, the width and length of the trough, the total vibrating mass or weight, the environment into which the unit is to be installed, and economic considerations. See Fig. 26.2.3.

Electromagnetic Drive. In some applications multiple electromagnetic feeder-type drive units (Fig. 26.2.3*D*) are located along the length of a single trough, especially in those applications requiring the ability to obtain finite adjustment of the speed of material movement.

Brute-Force Drive. A brute-force conveyor design (Fig. 26.2.3*A*) is one in which all the force necessary to vibrate the trough is derived from the drive mechanism.

Natural Frequency Drive. A conveyor utilizing natural frequency design (Fig. 26.2.3*B–E*) supplements the drive mechanism with a reactor spring system designed to reduce the vibrating drive force.

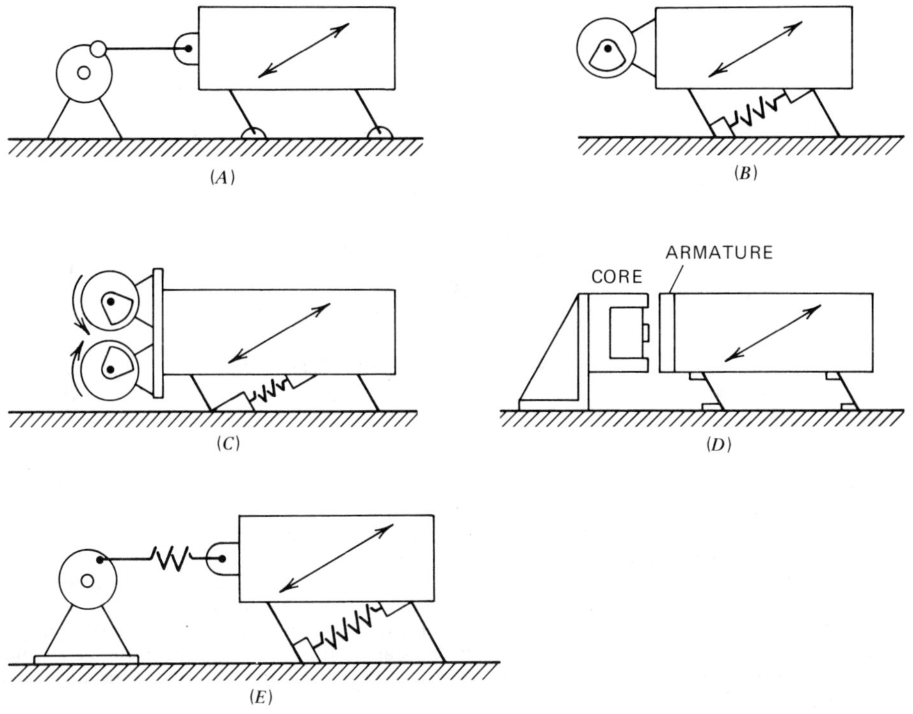

Fig. 26.2.3 Typical conveyor drives. (*a*), positive direct-connected brute-force eccentric crank; (*b*), single-rotating eccentric weight; (*c*), double-rotating eccentric weight; (*d*), electromagnetic; (*e*), positive flexible-connected.

The basic design consideration is that the natural frequency of the vibrating weight and its reactor spring system complement the operating speed of the conveyor drive to take full advantage of the sub-resonant natural frequency phenomena in which the stroke of the conveyor and deflection of the reactor springs are able to respond freely to varying conditions of material loading, for example, the stroke increasing slightly under a surge load condition when additional input force is required. The drive is sized to provide the power required to overcome drive mechanism friction and damping friction of the material movement. The vibratory drive mechanism on a natural frequency conveyor allows the trough and individual reactor springs to vibrate at their natural frequency, alternately storing and releasing energy at each end of the trough stroke. The reactor springs distribute the drive forces uniformly along the length of the conveyor, minimizing operating stresses, and distribute the reactive forces to the base and supporting structure.

In some designs a flexible drive-to-trough connection is designed to transmit and cushion the starting and stopping sequences, generally the period of highest force requirement since the mass must be accelerated from a static position or decelerated to a stop, but remain essentially rigid while the unit is operating. However, it must retain a degree of flexibility to allow the trough mass and reactor spring system to alter their stroke in response to material loading.

The advantages of this design are low internal stresses, low horsepower requirements, the ability to vibrate large masses, and the capability to design conveyors extending 300 ft or more in length with a single drive.

Foundation Reactions

All vibrating conveyors, regardless of their individual design details, are subject to the same basic physical laws and impart both a static and dynamic loading or reaction to the foundation or supporting structure to which they are mounted. Both values must be considered to adequately install a vibrating conveyor.

Unbalanced Conveyors. The static loading is the entire weight of the unit, including the base and all machinery components, plus the weight of the maximum anticipated material load. This loading

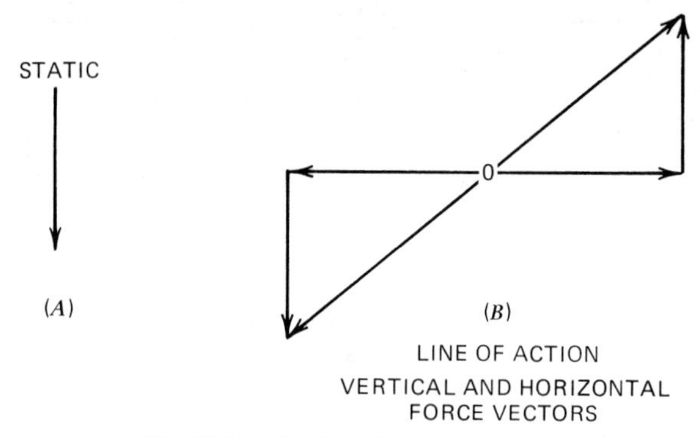

Fig. 26.2.4 Static and dynamic loading.

is a downward acting force comparable with that of any other piece of production machinery (Fig. 26.2.4*A*).

The dynamic loading of a vibrating conveyor must be given careful consideration, as it is the result of a mass that is accelerated and decelerated at a specific frequency, thus subjecting the foundation or supports to a reversing load condition.

The dynamic reaction is actually the resultant force produced by the deflection of each of the springs in the reactor system. Generally, this force is developed along the line of action of the conveyor and therefore can be resolved into component forces acting in both vertical and horizontal planes. Since this is a vibrating conveyor which is moving back and forth along a specific line of action, the resolved forces result in both an upward and downward vertical vector and a back and forth horizontal vector (Fig. 26.2.4*B*).

A review of the vertical vectors shows that the downward-acting force attempts to push the unit into the foundation, whereas the upward force attempts to lift the unit off the foundation or separate it from the supporting structure. This requires that the conveyor be suitably welded to embedded steel or supports or otherwise held in position by an adequate anchor bolting system.

The horizontal vectors are applied in a manner that applies a shearing action to the anchor bolts or welds holding the unit in place. These horizontal forces are generally of greater magnitude than the vertical and must be given full consideration when rigidly mounting a vibrating conveyor.

An interesting aspect of the dynamic reaction force is the fact that, because the reactor system springs are uniformly distributed along the length of the unit, the total resultant dynamic reaction force on a natural frequency conveyor can be considered as uniformly distributed. Conversely, a brute-force conveyor may have an additional concentrated force in the drive area, especially when starting and stopping.

Vibrating conveyor foundations and supports must be designed to withstand the dynamic and static load reactions of the conveyor without causing objectionable vibrations or deflections. The allowable deflection in supports subjected to vibrating forces is considerably less than that for structures involving only static loading conditions. In addition to this deflection limitation, the supporting structure must be of sufficient rigidity that its natural frequency exceeds the operating frequency of the vibrating conveyor. This will prevent even a small vibrating force from being magnified and causing sympathetic excitation elsewhere in the structure. Engineering guidance should be requested from the vibrating conveyor manufacturer or other qualified sources.

Balanced Conveyors. When it is necessary to install a vibrating conveyor from overhead or in an area involving a questionable supporting structure or soil conditions, a variety of balanced conveyor designs are available for these applications. Also, modern vibrating conveyors sometimes approach a size and mass where it is impractical to consider installing them without some type of balancing in order to keep foundation and supports economically feasible.

A balanced vibrating conveyor is designed to reduce the unbalanced reaction force transmitted to the foundation or supporting structure. The degree of balancing will determine the resultant force transmitted to the supporting structure or foundation. The designer must give primary consideration to the degree of balancing as well as the static weight of the unit when designing the supporting structure. Generally, following manufacturers' suggested recommendations will eliminate potential problems involving objectionable secondary transmitted vibration.

In general, a balanced vibrating conveyor is one in which a secondary mass vibrates 180° out of phase with the vibrating trough, to reduce the inertia forces at each end of the trough stroke.

An inexpensive balancing system is one where a counterbalance mass is added to the conveyor base and the entire unit is then mounted on isolation mountings which absorb the residual movement. This balancing method can be designed to reduce transmitted forces by approximately 90–95%.

Other balancing systems are available that will absorb 98% or more of the reaction forces. These designs normally involve mounting the secondary balancing mass on a reactor spring system similar to that used for the conveying trough. A single eccentric drive assembly may be arranged to positively drive the two masses out of phase with each other.

A variety of balanced designs are available from various manufacturers that make it possible to eliminate virtually 100% of the reaction force. One design uses a free-floating balancer which sympathetically responds to any change in trough stroke or base movement to compensate for changing load conditions. This design must be mounted on flexible isolation mountings to assure proper response.

Selection of Capacities

A vibrating conveyor is essentially a volumetric conveying device, the capacity of which is readily determined by consideration of the cross-sectional area of the material bed and the material travel rate. The capacity can range from a few tons per hour to as much as 1000 ton/hr.

Conveyor widths can be as narrow as 6 in. or as wide as 12 ft, depending on the application and material size. Very often the width of the conveyor trough is dictated not by capacity requirements, but by such factors as material size or space limitations.

Most granular, free-flowing bulk materials typically convey with a velocity of 50–60 ft/min; however, some material characteristics will sometimes alter this rate. Extremely fine materials that tend to aerate, materials with high moisture content or with high interparticle slippage generally convey at a slower speed, whereas coarse materials that have a tendency to mat together, can be moved at speeds exceeding 100 ft/min. Generally, fine mesh or slow-moving materials convey best in a shallow bed (2–3 in.) whereas average materials can be moved in depths ranging from 6 to 12 in. It may be desirable to design for a slow travel rate to prevent degradation of friable material, to aid in cooling or other processing functions, or to keep noise to a minimum.

Conveyor manufacturers should be contacted for their recommendation on handling materials for each specific application.

Special Designs

Recent advances in the design of vibrating conveyors now allow units to be inclined in excess of the conventional 5° limitation always considered as maximum. A variety of bulk materials can now be conveyed at inclines approaching 30°.

Special troughs designed to accommodate thermal expansion can be provided for handling hot material such as 1800°F castings.

Troughs can be lined with a variety of materials, including rubber, nonstick plastics, ceramic bricks, or alloy steel, depending on the application requirements.

Units may be provided with perforated opening or bar-type grizzly sections for sizing or separating functions.

Automatic remotely controlled or manual discharge gates or plows may be located as required for intermediate discharge.

Troughs may be any shape, rectangular, tubular, made completely dust- and gas-tight, and include inspection or cleanout ports.

Conveyors can be designed in many configurations: horizontal, horizontal with an inclined portion built into the same unit, or arranged as a true vertical spiral unit.

With today's technology, the only limitation placed on the application of vibrating conveyors is the extent of the design engineer's imagination.

26.2.5 Vibratory Feeders

Vibratory feeders are designed to feed granular solids at a controlled rate. They normally are located beneath a storage silo, reclaim tunnel, or surge bin. The units are designed to withstand head load of the material in the hopper and can be equipped with rate controls to vary the output from minimum to 100% with reasonable linearity. The units are normally self-contained with respect to drive components and reaction forces, and the vibration is isolated from the supporting structure through suitable isolation springs.

Design Principles

It is important to realize there are various designs of vibratory feeders, those commonly known as electromagnetic and those known as electromechanical, depending on the type of excitation.

Important also is the classification based on the presence of two or more spring-connected moving masses, as opposed to a single-mass design. The drive systems for vibratory feeders can be categorized as either brute-force or natural-frequency (tuned) systems depending on how the force is applied to the trough. Considerations in choosing the right design for a particular application include trough weight and dimensions, environment, the type of material being handled, and of course the feed requirements for the specific process.

Brute-Force Feeders. Brute-force feeders are called single-mass systems because the vibratory excitation or drive means is connected directly to the pan or trough of the feeder. Although some brute-force feeders utilize electromagnetic drives, the electromechanical drive is most commonly used.

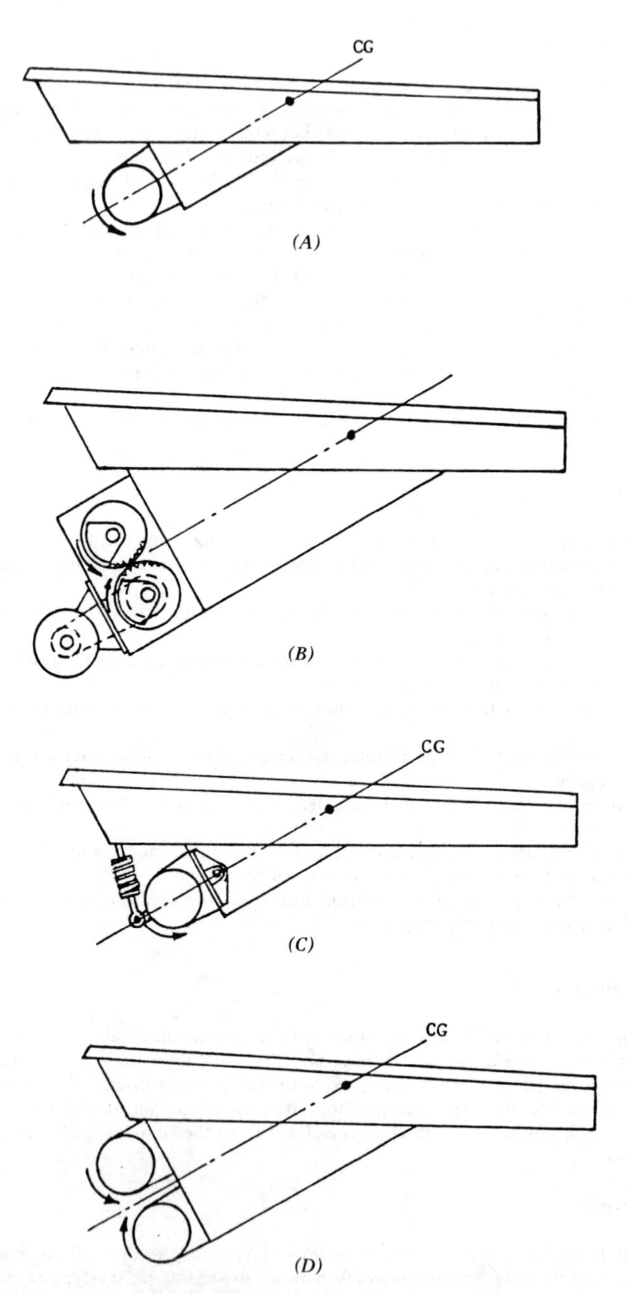

Fig. 26.2.5 Brute force designs. (*a*), single rotary; (*b*), geared counter-rotating eccentrics; (*c*), dual rotary; (*d*), single rotary with pivot mount.

These electromechanical feeders derive 100% of their vibratory drive force from heavy centrifugal counterweights (Fig. 26.2.5*A*). The forces developed are transmitted directly to the trough through heavy-duty bearings. Linear motion is generated by the use of counter-rotating shafts with timing gears operating in an oil bath housing and driven through a V-belt drive to the motor (Fig. 26.2.5*B*). Other designs utilize two synchronizing vibratory motors with counterweights mounted on the motor shafts (Fig. 26.2.5*C*) or they will have a single motor, pivoted to allow transmittal of the generated force essentially in a predetermined direction and allowing adjustment of the resulting stroke angle (Fig. 26.2.5*D*). Brute-force feeders are usually applied as constant rate devices. Their feed rate can be adjusted by changing the slope of the trough, the size of hopper opening, or the amount of counterweight and stroke. In some cases, mechanical or electrical variable-speed drives are applied to vary the frequency and feed rate. The regulation and control range, however, is somewhat limited.

The trough stroke obtained by the brute-force machine is dependent on a simple relationship between the rotating eccentric weight of the exciter and the total weight of the machine, such that it is entirely independent of the operating speed. Therefore, an advantage of the brute-force system is that tuning need not be considered. Any drive frequency is acceptable that is sufficiently higher than the natural frequency of the isolation system to maintain stable operation of the equipment. More exacting design information is available in brief feeder application manuals published by the feeder manufacturers.

Natural-Frequency Feeder. Natural-frequency feeders, often referred to as tuned or resonant feeders, employ two or more spring-connected moving masses. The most common tuned feeder is a two-mass, spring-connected vibratory system. One of the masses is the trough; the other mass, the reaction or excitation mass. Because of the selection of spring constants, a relatively small exciting force is amplified to generate a vibratory motion. In essence a tuned feeder takes advantage of the natural magnification of vibrational amplitude that occurs when a vibrating system is operated near its natural frequency or resonance condition. The excitation or drive force may be supplied by either electromagnets or rotating eccentric weights.

Figure 26.2.6 shows a two-mass feeder and its spring arrangement, schematically represented by the two spring-coupled masses M_1 and M_2, the trough and exciter mass, respectively. The spring rate of the spring system and the weight of the feeder trough and exciter determines the natural frequency of vibration. In order for the two-mass system to take advantage of the natural magnification factor, the system must have its natural vibration frequency close to the operation frequency. For the most advantageous results, the operating or driving frequency should be below the natural frequency of the loaded feeder. This is referred to as subresonant tuning. Nearly all feeder manufacturers design their vibratory feeders to have subresonant tuning. A subresonant system can operate favorably under high head loads which may be caused by large hopper openings, wide heavy troughs, or the necessity to convey the material a considerable distance from the hopper to the end of the trough.

It is not the weight of the head load on feeder operation, but rather the damping capacity of the bulk material being handled that must be considered in feeder design and selection. The damping effect of the material is a direct measure of the energy that is absorbed by the material in moving from the hopper and along the vibrating pan.

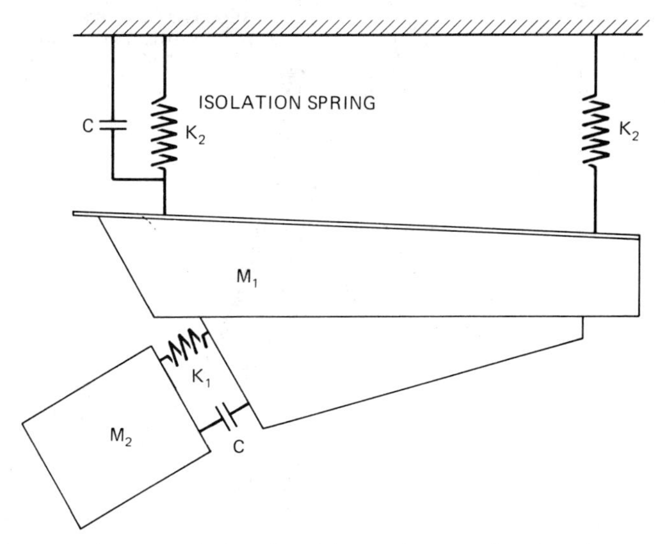

Fig. 26.2.6 Example of two-mass feeder. C = inherent damper; M_1 = trough mass; M_2 = exciter mass; K_1 = reactor springs; K_2 = isolation springs.

The operation of a two-mass system is explained graphically by the magnification factor curves of Fig. 26.2.7. These curves plot the dimensionless amplitude as a function of the tuning ratio for various degrees of damping. A tuning ratio of 1.0 is resonance where operating frequency and natural frequency are the same.

Points to the left of the curve peak determine subresonant operation, operating frequency is less than natural frequency. In an unloaded condition a subresonant tuned feeder might be represented by point A on the curve.

If weight were added to the trough side of the feeder, such as a pure head load imposed by granular material, the natural frequency of the machine would tend to be reduced, causing the point of operation to be closer to resonance, point B on the curve. The result would be an increase in the trough stroke, consequently an increase in the feeder output.

All granular materials, however, have internal damping; that is, they dissipate energy within themselves when vibrated. This damping is comprised of interparticle friction and the energy is absorbed when the particles are deformed because they are not purely elastic. Damping absorbs the energy of vibration and attenuates the amplitude of the vibratory feeder. The amplitude of the feeder under a real material is influenced by a combination of head load and damping and will actually lie somewhere on a vertical line between points B and C depending on the amount of damping that the material exhibits.

Electromechanical Feeders

Tuned feeders that use rotating eccentric weights are referred to as electromechanical feeders. The usual operating frequencies for these feeders are 720–1800 rpm. Spring systems may consist of steel coil springs, rubber in compression, rubber in shear, steel leaf springs, pneumatic springs, or any combination of these.

Drive Designs. Drive excitation may be accomplished by a single motor with a double extended shaft mounted with eccentric weights as shown in Fig. 26.2.8*A*. Dual counter-rotating motors may

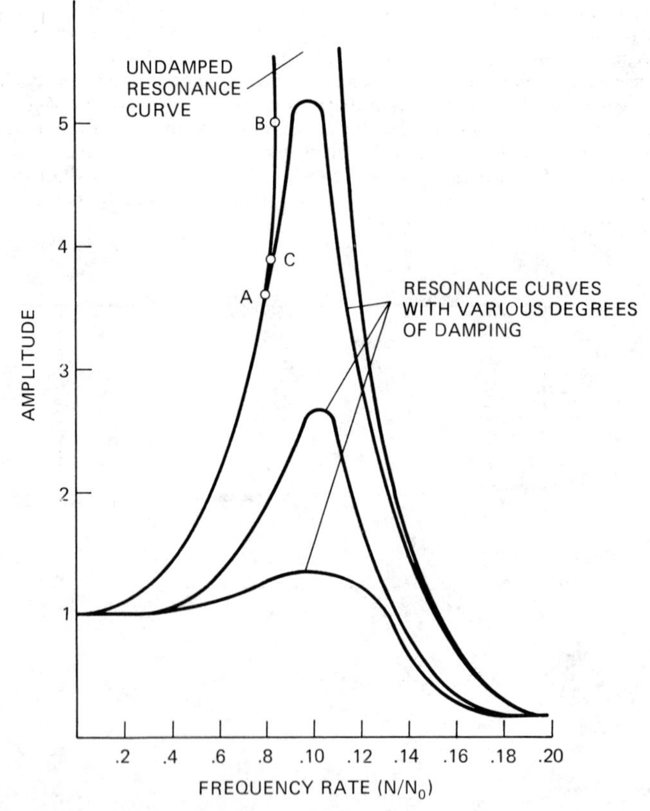

Fig. 26.2.7 Typical magnification factor curves for a tuned two-mass system.

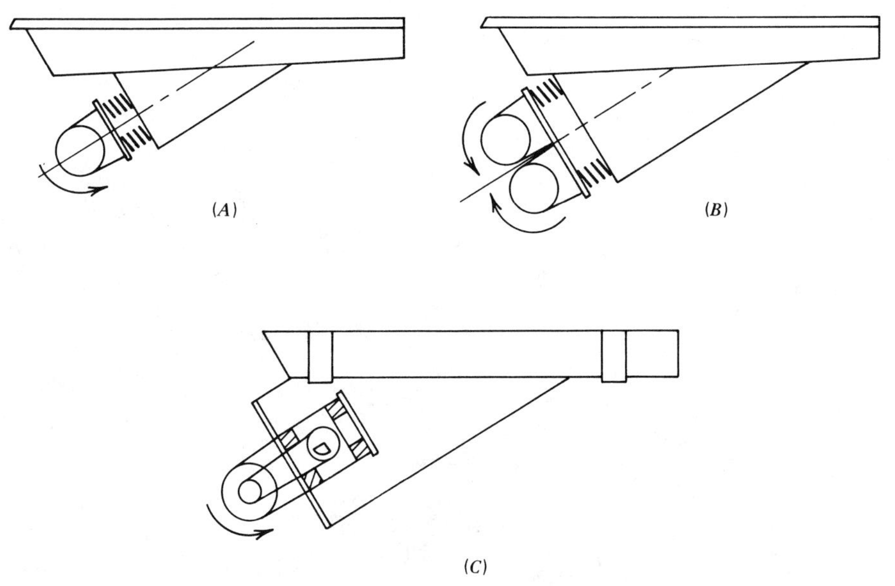

Fig. 26.2.8 Examples of tuned two-mass electromechanical feeder designs. (*a*), electromechanical single motor; (*b*), electromechanical dual motor; (*c*), electromechanical with belt-driven exciter.

be used (Fig. 26.2.8*B*) without mechanically connecting their shafts, provided they are on a common rigid frame, in which case they will synchronize. The same effect may be obtained by using a single motor driving two sets of gear-connected shafts by means of a V-belt. Each of the shafts have equal eccentric weight. Other systems use eccentric weights on a shaft driven by a single squirrel cage AC motor through a V-belt (Fig. 26.2.8*C*).

Tuning or adjustment of the natural frequency of an electromechanical feeder is accomplished by several means. These include adjusting the maximum operating speed, changing spring rate by adding or deleting the number of springs used or providing in the design to add or delete mass (weight) on the trough or exciter structures.

Feed Rate Controls. Variable feed rate of subresonant vibrating feeders is accomplished by varying either the operating speed of the AC motor or the magnitude of the force generated by the rotating eccentric weights. Controls that vary the voltage or adjust frequency of current to the motor are used to obtain variation in speed of AC motors.

Two methods of varying voltage are the auto transformer and SCR controller. The auto transformer requires manual adjustment while the SCR control is capable of accepting remote or process electrical signals to vary motor speed.

Varying the frequency of the supply current will also change motor speed and is accomplished by frequency controllers.

Varying the force of eccentric weights involves using a variable pressure signal which changes the position of eccentric weights on the rotating shaft of the drive. The use of transducers allows this control to also accept a remote or process electrical signal.

Electromagnetic Feeders

Electromagnetic vibratory feeders are dynamically balanced two-mass vibrating systems consisting of the trough and trough connecting bracket coupled to the electromagnetic drive by means of leaf springs. This mechanical system is caused to vibrate by magnetic force impulses supplied by an electromagnet. The frequency of vibration is equal to, or a derivative of, the frequency of the AC voltage input to the electromagnet. Figure 26.2.9 illustrates diagramatically the component parts of a typical electromagnetic feeder.

Drive Design Considerations. This vibrating system is designed and manufactured to have a natural frequency with no load of material approximately 8–15% higher than the frequency of the alternating current supplied to the electromagnetic drive. Natural frequency in this case can be defined as that frequency at which the system will vibrate freely when power is disconnected from the electromagnet.

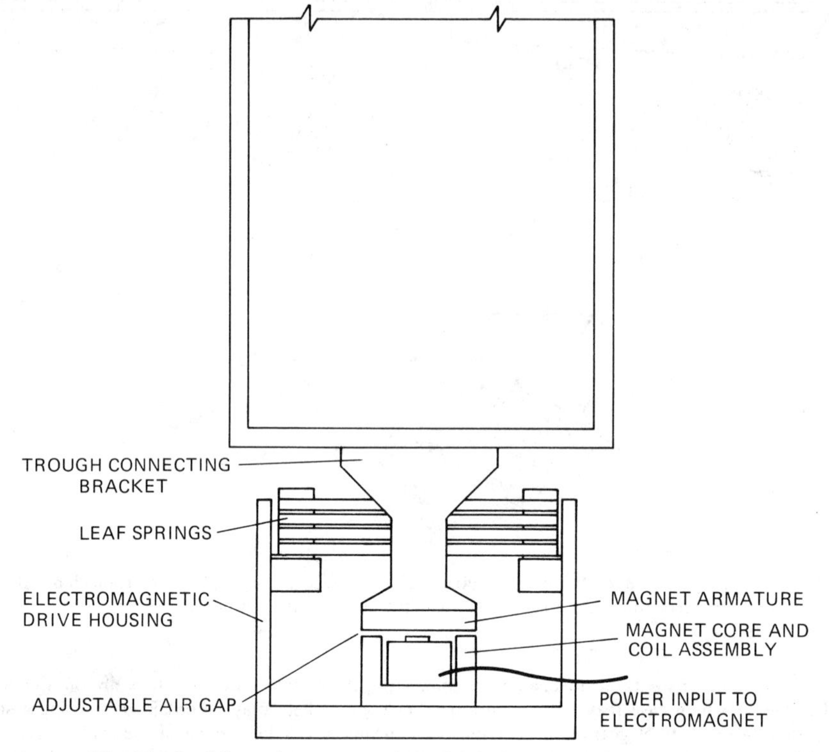

Fig. 26.2.9 Schematic representation of a typical electromagnetic feeder.

The feeder is tuned to the desired natural frequency by adjusting the number and size of the leaf springs that couple the trough and the trough connecting bracket to the electromagnetic drive. The natural frequency is a function of the total spring rate of the leaf springs and the masses of the vibrating system.

Drive units can be positioned either above or below a trough (Fig. 26.2.10). A below-deck drive is most commonly used, but above-deck drives can be supplied for installations where there is insufficient space below the trough. However, an above-deck drive may slightly reduce feeder capacity.

Figure 26.2.11 illustrates typical arrangements of electromagnetic drives connected to troughs. Combining a number of electromagnetic drive units creates feeders ideally suited for special applications.

Multiple-magnet drive units, positioned one behind the other, may be used in a long, vibrating conveyor. When an especially wide material layer is desired, multiple-magnet drives can be placed side by side on extra-wide feeder troughs. The number of driving magnets required for the trough is determined by its width. Dual twin-magnet drives—two sets of twin magnets, one set placed behind the other—will provide both increased capacity and the ability to handle exceptionally heavy loads.

Feed Rate Controls. The electromagnetic feeder's output is varied by adjusting the power of the electromagnet, the level of applied voltage, and thus the current flow to the coil of the magnet. Several control schemes, depending on the manufacturer of the specific feeder, may be used. These include the use of solid state control components such as SCRs or Triacs, variable auto transformers, or rheostats in series with the electromagnet coil. Some controls take advantage of the tuning of the feeder to vary feed rate, these controls are used with feeders tuned at half the frequency of the supply voltage. By allowing more or less of the full supply frequency to be applied to the feeder, its output will vary as the higher applied frequency component affects the natural frequency of the feeder and thus its vibrating stroke. Solid state diodes and SCRs are rectifiers, allowing current to flow in only one direction. When they are used in series with the electromagnetic coil, the feeder operates at half the frequency of the supply voltage. Some manufacturers use a permanent magnet in the design of the electromagnetic drive. The principal effect of the permanent magnet is analogous to the rectifier, in that its magnetic field opposes the electromagnetic field set up by the applied current in one direction. The feeder, therefore, operates at half the supply frequency.

Electromagnetic vibratory feeders generally operate from single-phase 60 Hz (or 50 Hz) power

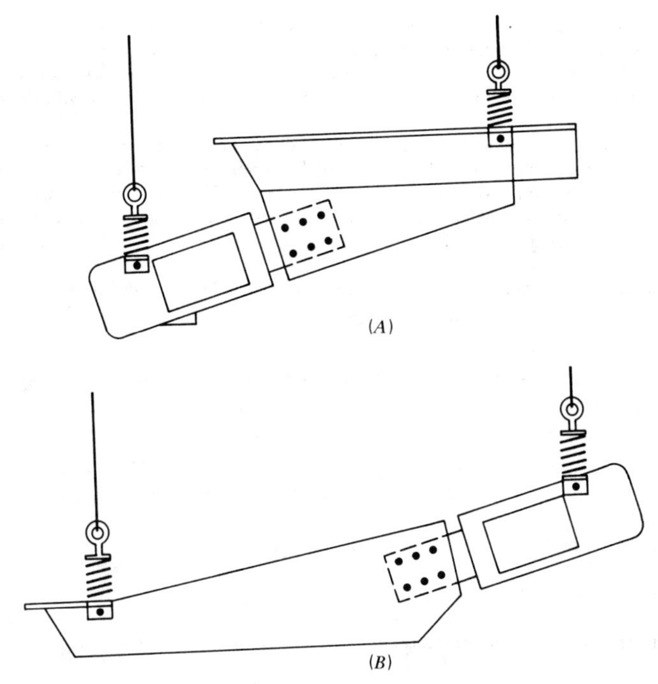

Fig. 26.2.10 Illustrations of feeder installations. (*a*), suspension mounting; (*b*), suspension mounting overhead magnet.

source and, consequently, their operating frequency is 3600 vibrations per minute (or 3600 vib/min). Some electromagnetic feeders operate at submultiples of the power source frequency utilizing special "split-wave"control circuitry. In such cases, the feeders may operate at 1800 or 1200 vib/min from 60 Hz (1500 and 1000 vib/min from 50 Hz) power sources.

Application

Vibratory feeders have been servicing industry for more than 50 years. They have traditionally ranked with equipment requiring minimum maintenance and operating expense. By proper initial selection

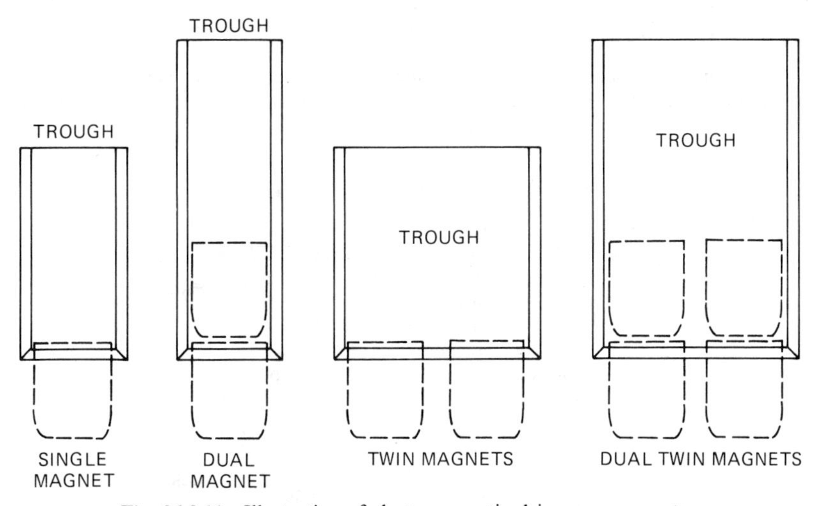

Fig. 26.2.11 Illustration of electromagnetic drive arrangements.

of the equipment and its associated control system, both the first cost and maintenance costs can be minimized. As the cost of the installation and maintenance labor and power continues to spiral upward, it becomes essential to consider equipment capabilities in the early stages of plant design or expansion. Table 26.2.1 summarizes the advantages that may be gained through the use of vibratory feeders.

Industries. Vibratory equipment is used in virtually all industries requiring the transport and controlled metering of bulk solid materials of all kinds from mineral ores to breakfast cereals. Included in the industry users list would be hard rock mining, coal mining and processing, iron and steel mills, foundries, industrial chemical processing and packaging, food processing and packaging, and drug and cosmetic processing and packaging. An applications summary for this equipment includes:

Feeding mineral ores, or quarried stone, to a primary, secondary, or tertiary crusher.

Metering of ores and coal from outside stockpiles to conveyor belts.

Feeding scrap iron and steel.

Glass batch feeding to weighing and conveying systems.

Feeder/grizzly for conveyor belt padding and crusher load control.

Feeder/conveyor for furnace charging, ash and clinker removal.

Feeders for packaging, blending, batching, mixing, flaking, freezing, and drying operations.

Metering of all types of bulk solids from storage bins and hoppers to vibrating conveyors, belt conveyors, or other apparatus.

Spreading and table feeders for inspection of produce or other goods.

Material Classifications. Generally speaking, the industrial materials encountered in applying vibratory equipment may be broken down into several classifications. The five categories listed give some insight into the various material characteristics and the problems that might be associated with handling them.

CLASS 1. Powder; sluggish. These materials are generally dry powders, such as flour, starch, and other materials in the size range of 100–300 mesh, with a repose angle of 40–55° and slide angles of 35–50°.

CLASS 2. Powder; sluggish, adhesive. These include materials that adhere readily to metal surfaces, such as titanium oxides and color pigments. It also includes materials of Class 1, when moisture is present.

CLASS 3. Powder; sluggish, fluidizes. These materials are in the size range of 325 mesh to a few microns, such as cement, hydrated lime, fly ash, and talcum powder. These materials can be readily overcompacted by excessive vibrations and yet when arching or bridging is broken, they will just as easily fluidize.

CLASS 4. Granular; free flowing. These materials have granular particles larger than 100 mesh. Densities are usually above 25 lb/ft^3. They flow readily under gravity and have relatively low repose and slide angles (10–35°). Minimal vibrations are necessary to move this material.

CLASS 5. Fibrous; flaky, flocculent. These are materials such as asbestos, fibers, shredded tobacco, fiberglass, wood chips, and many materials in the mill feed industries. They cover a wide range of

Table 26.2.1 Advantages of Vibratory Equipment

Virtually no degradation of conveyed material
Scalping and screening or picking can be done
Hot material can be handled
Abrasive materials can be handled
One unit can have a divided flow stream
One unit can have multiple discharge points
Cooling or drying can be done while conveying
Dewatering can be performed
Units are self-cleaning
Units are able to meet sanitation standards
Units can be covered, made dust-tight

bulk densities, particle sizes, have different angles of repose and slide characteristics. These materials absorb the vibration energy readily. They usually require long strokes at lower frequencies to feed at competitive conveying capacities.

Feeder Size Selection. Feeder capacities are dependent on the width of the trough, depth of material flow, and the conveyability of the material. Maximum flow rates of 20–80 ft/min can be obtained from vibratory feeders. The flow rate is a function of the material characteristics and downslope of the feeder trough.

Capacity in tons per hour is expressed by the simple relationship:

$$C = \frac{Wd\gamma R}{4800}$$

where: W = trough width, in in.
d = depth of material, in in.
γ = bulk density, in lb/ft^3
R = linear feed rate, in ft/min; a function of material properties

These factors are influenced by many other factors including the hopper design.

The feed rate is influenced by stroke, frequency, feed angle, material of trough, density, particle size, internal damping, interparticle friction, cohesiveness, adhesiveness, hopper design, and moisture content.

The feed depth is influenced by particle size, density, and moisture content.

The hopper design is influenced by wall slopes, throat opening, gate height, width, amount of divergence, and skirt divergence in vertical and horizontal planes.

Several of the parameters are fixed by the manufacturer as a good compromise for most materials including feeder stroke, feeder operating frequency, and drive angle. One other parameter, the trough material, may be selected on the basis of the material being handled.

From the capacity equation it is noted that for a specific installation and material the width of the trough will determine the maximum capacity of the feeder.

The capacity chart shown in Fig. 26.2.12 offers a guide for selecting the size feeder required for a particular application.

Feeder Trough Design. Manufacturers offer several different trough options for special applications. Troughs for vibratory feeders can be furnished in the following designs:

Flat bottom
Radius bottom
V shape
Half round
Tubular
Covered with complete dust sealing
Belt-centering discharge
Diagonal discharge
Screen decks
Grizzly section
Water-jacketed troughs

Feeder troughs are available in mild steel, abrasion-resistant steel, stainless steel, or special alloys to meet various applications. Replaceable liners are furnished in the foregoing materials as well as rubber, plastic, or ceramics. Liner materials should be selected with consideration of materials being handled as well as economics.

A very common material used as a liner is Type 304 stainless steel. This is particularly adaptable to materials that have a corrosive effect as well as abrasiveness. The stainless steel material is often used for this application. For abrasive applications, alloy materials such as T-1 and Jalloy are often used.

Hopper Design Considerations. The basic material properties cannot be altered, therefore the hopper must be well designed to assure good flow by taking into account some basic characteristics.

The prime requisite of feeding bulk materials is that the material first must be delivered to the feeder trough. This requires proper design of the hopper and sizing of the opening. Improperly designed transition sections often reduce feeder capacities by 20–30% and, in extreme cases, reduce the flow to zero. It is important that proper consideration be given to the transition section between hopper

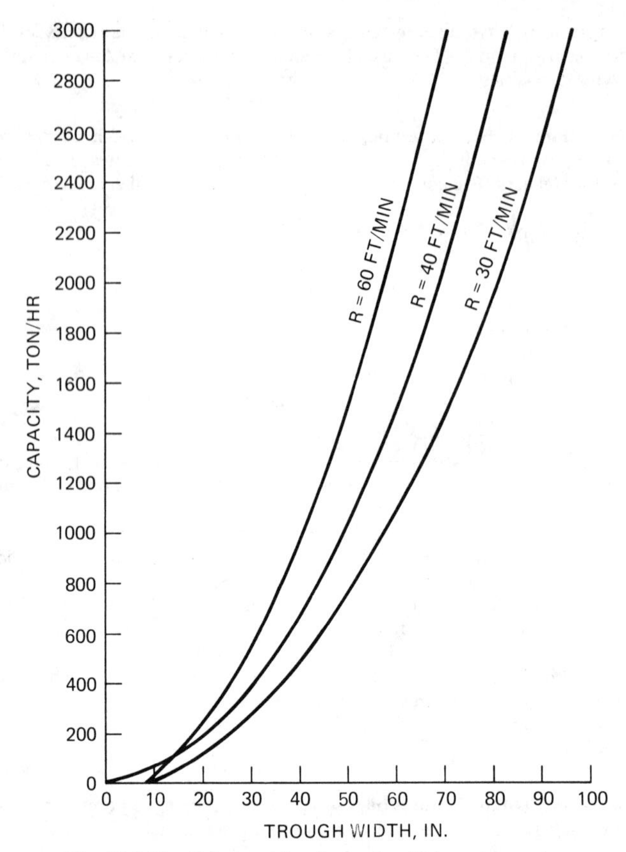

Fig. 26.2.12 Relationship of trough width and capacity.

or bin and the feeder. The feeder manufacturer should be contacted if a unique problem is present to assure proper operation of the system.

A properly designed hopper permits near-uniform material flow. Too large a hopper opening may result in stagnant material in the rear portion of the feeder, while too small a hopper opening may result in flow blockage or arching.

It is suggested that generally accepted hopper design parameters be employed [1, 2, 3], based on measured flow properties. In addition, special consideration should be given to the outlet onto the feeder. Typically, a hopper outlet as in Fig. 26.2.13 will result in the smallest and lowest-priced vibratory feeder for a given capacity. As the hopper throat is increased, feeder size, and hence cost, increase. This increase may be unavoidable if arching, due to lump size or material cohesiveness, predicates a large opening. In the extreme case, however, the increase in hopper opening is self-defeating as a stagnant wedge of material will form on the rear of the feeder, absorbing energy.

With proper hopper design, the effective throat opening in Fig. 26.2.13 can be just large enough to overcome arching. Then the width of the hopper outlet forms a rectangular opening which is virtually always large enough for gravity flow of the required capacity. Ratholes should be eliminated by proper design of the transition hopper, sufficient to provide flow along walls.

Where possible, the suggested hopper design of Fig. 26.2.13 should be used. Care must be taken to assure the hopper is large enough to pass the maximum flow required.

There is a natural tendency for vibratory feeders to withdraw from the front portion of the hopper opening. Proper design will permit material to be placed on the feeder at the rear of the trough.

In such a design, consideration should be given to the following points as outlined in Fig. 26.2.13:

1. The slope of the angle of the rear wall (A) should be steep enough to permit material flow along the rear wall (60° or more).

2. The slope of the front wall (B) should be just enough to permit material flow. Too shallow a slope will result in the buildup of material above the gate opening, and too steep a slope may disturb

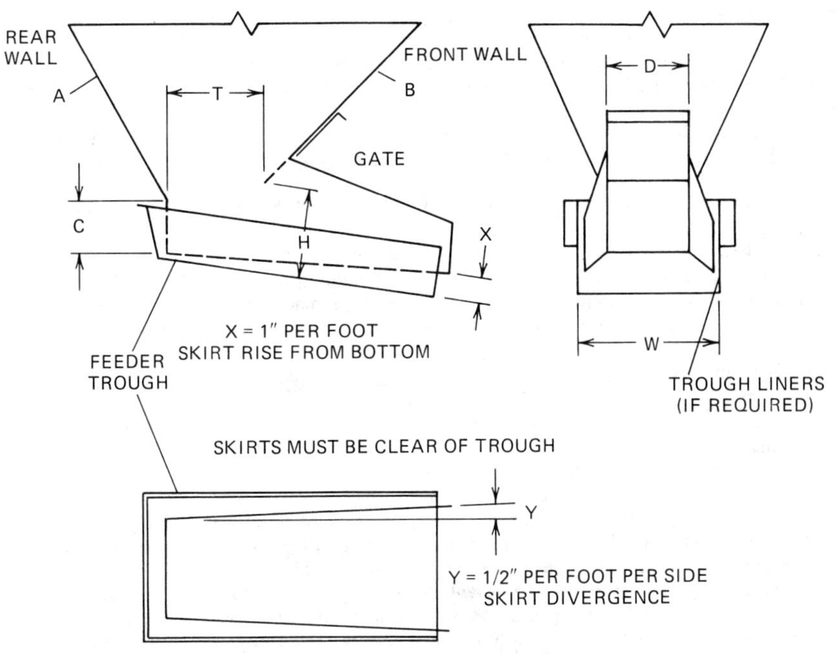

Fig. 26.2.13 Vital points of proper hopper design.

flow patterns within the hopper. Generally, an acceptable slope for front wall (B) would be 5° less than rear wall (A).

3. A short vertical section (C) should be provided just above the rear of the trough. The height of this section (C) must be at least the height of the trough. A preferred height of 1 ft is recommended in high-tonnage applications.

4. The throat dimension (T), for randomly sized particles, should be at least $2\frac{1}{2}$ times the diameter of the largest particle of material. On applications where particles are nearly the same size, (T) dimension should be 5 times the size of these particles. This will prevent material from interlocking and lodging in the throat opening.

5. The gate opening (H) should be at least twice the size of the largest particle of material. The gate height should increase proportionally to the capacity required and the particle size. For best flow patterns within the hopper, throat dimension (T) should be equal to or slightly larger than one-half the gate opening (H). If (T) dimension is greater than $1\frac{1}{2}$ times the (H) dimension, a disturbance may be caused in the flow pattern due to greater velocities at the front of the hopper. The dimension (H) should be measured from the bottom of the trough with the trough at some downslope, preferably 8°.

6. The width of opening (D), for randomly sized particles, should be at least $2\frac{1}{2}$ times the diameter of the largest particle of material. On applications where particles are nearly the same size, (D) dimension should be 5 times the size of those particles. This opening must be consistent with the capacity requirements.

7. Skirt boards should diverge at a rate of approximately $\frac{1}{2}$ in./ft of length so that the opening at the front of the feeder trough is greater than at the hopper opening. The skirt boards should also rise slightly away from the trough bottom at a rate of approximately $\frac{1}{2}$ in./ft of length from hopper to front of trough. This is to prevent material from becoming blocked between the skirts and the trough bottom.

Where any installation may be questionable the feeder manufacturer or representative should be contacted.

Installation. Basic to the proper installation of any vibratory feeder is properly designed supporting structure to eliminate possible resonant vibration. Although most feeders are supplied with isolation systems that are more than 90% effective, any supporting member in resonance with the operating frequency of the feeder will tend to vibrate, taking energy from the feeder's ability to transport material.

Normally bins, silos, and hoppers are sufficiently rigid due to their own inherent loadings. A good rule of thumb is to have the structural supporting members at four times the operating frequency.

Vibratory feeders are usually installed by suspension mounting by four flexible wire cables. They can also be supplied for floor mounting on a solid base or with a combination of floor mounting and suspension mounting. Clearance must be maintained between the bin or skirt boards and the feeder trough. Trough vibration must not be impeded by rigid attachments to adjacent objects. Any connections (such as dust seals) between the trough and adjacent objects must be flexible, preferably cloth or rubber.

Effective isolation from the supporting structure is accomplished by the use of steel coil springs, elastomer springs, or pneumatic springs.

For tuned two-mass feeders it is important to note that no modifications to the trough or exciter should be made without consulting the manufacturer. Welding to or cutting from the trough or base structure could affect the tuning of the unit causing serious machine damage or improper operation. Such field modifications could also weaken the structure causing failure by fatigue stress.

26.2.6 Limitations of Equipment

As seen from the foregoing, vibratory feeders and conveyors are used to meter and transport a wide variety of materials over a large application base. There are few limitations, with today's technologies, to the successful application of this class of equipment. Some materials that tend to fluidize, or adhere to the feeder trough may be more efficiently handled by other means. Also a dusting problem may occur at the outlet of a feeder handling some dry, powdery materials, and require special handling. It is shown that feed extraction from hoppers is not fully positive, therefore, in applications requiring extreme accuracies, a weighing system of some type may be recommended. The structural integrity of vibratory equipment is extremely important. This equipment is subject to high accelerations and high reversing stress cycles; consequently, to achieve acceptable industrial equipment fatigue life, the structures must be designed with stress limits much below that considered good design practice for structures not subjected to vibration. Special attention must also be paid to the design of weldments and welding techniques.

REFERENCES

1. Jenike, A. W., "Storage and Flow of Solids," Bulletin 123, University of Utah, Salt Lake City, November 1964.
2. Johanson, J. R., "Method of Calculating Rate of Discharge from Hoppers and Bins," *Transactions AIME* 69, March 1965, p. 232.
3. Johanson, J. R. and H. Colijn, *Iron & Steel Engineering,* October 1964, p. 36.

26.3 EN MASSE CONVEYORS

A. D. Sinden

26.3.1 Characteristics of En Masse Conveyors

En masse conveyors are so called because they convey materials by causing them to flow in a compact and unbroken stream through a conduit. The conveyor consists of a stationary conduit or casing and a moving, articulated element which is pulled through it. The moving element, comprising a series of flights, occupies only a fraction of the space inside the conduit; the remainder of the space is filled with the material that is being conveyed. Assuming correct design, a flowable material like grain or dry cement will move with the moving element rather than remaining stationary and allowing the flights to pass through it.

Advantages of En Masse Conveyors

The chief advantages of en masse conveyors are:

1. They are inherently enclosed.
2. They are versatile as to layout.
3. They are relatively small in cross section.
4. They are self-feeding.

Wherever a conveying problem calls for tightness against leakage of dust, liquid, or gas, en masse conveyors should be considered. Wherever arrangement is difficult or space is limited, en masse conveyors often provide the best, and sometimes the only feasible, solution. En masse conveyors do not compete with belt conveyors or bucket elevators in the fields of handling coarse mineral products or large tonnages over long distances.

26.3.2 Types of Flights

The flights on en masse conveyors may be vanes that approximately cover the cross section of the casing or they may be of skeleton type, consisting of relatively narrow bars shaped to sweep a majority of the perimeter. In either case the spacing, or pitch, of flights must be kept within a certain limit; for best results, pitch should not exceed average cross-sectional width of casing. Skeleton flights were used in the prototype of the en masse conveyor (known as the Redler conveyor, named after the inventor) and have continued to be a popular design in the conveyors which have been marketed under this classification. The cutaway view of a Redler conveyor in Fig. 26.3.1 shows the skeleton-type flights. An example of rigid vane-type flights is shown in the cutaway view of a Bulk-Flo conveyor (Fig. 26.3.2). Still another type uses flat vanes pivotally mounted on a chain and a device to tilt the flights when passing a discharge outlet. Figure 26.3.3 shows a portion of the moving element from an elevator of the latter type.

The casing may be rectangular or circular in cross section. Manufacturers generally prefer the rectangular shape because of the greater ease in fabrication for bend sections, for casings with renewable liners where access to the interior is necessary.

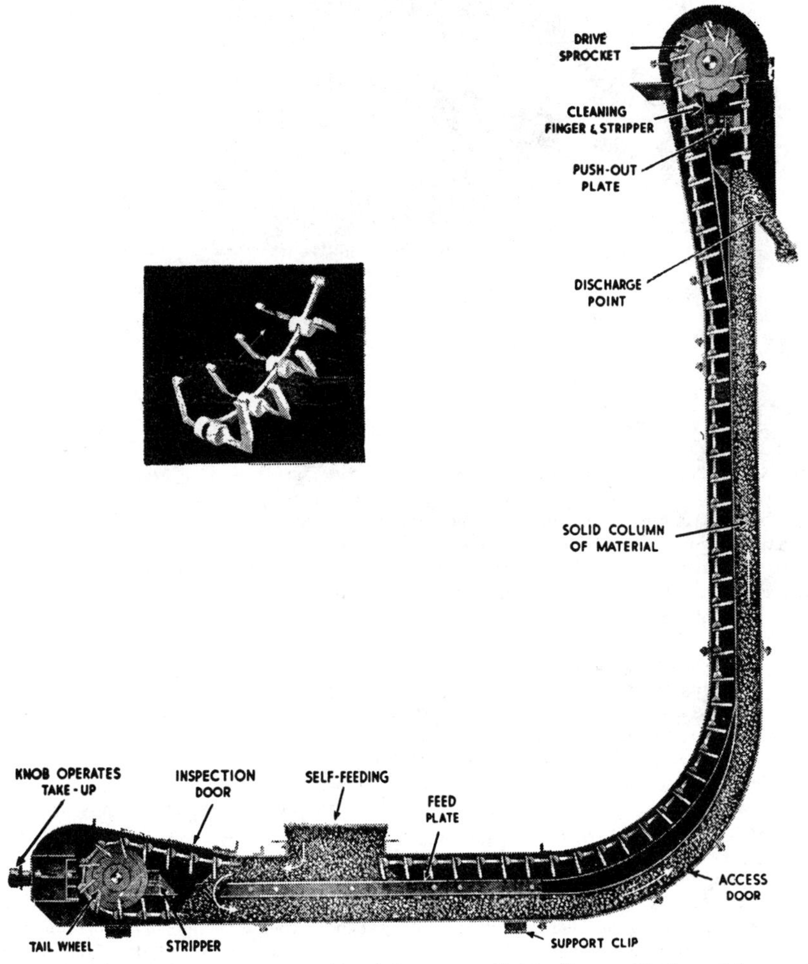

Fig. 26.3.1 Redler en masse conveyor using skeleton-type flights. Source: Stephens-Adamson Mfg. Co.

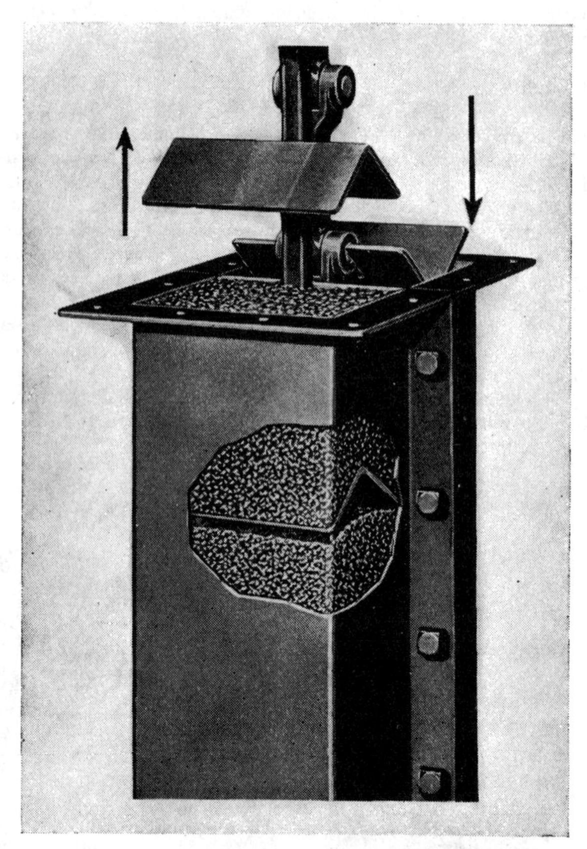

Fig. 26.3.2 Section of Bulk-Flo elevator using double-sloped vanes for discharging. Source: Link-Belt Co.

26.3.3 Conveyor Shapes

A wide variance is possible in the path and shape of the conduit—one of the greatest assets of the en masse conveyor. However, a basic limitation exists wherever the path of the material-carrying portion of the casing slopes steeper than the angle of slide for the material. The shape must then be such that material enters where the slope of the casing is well below the angle of slide. While a path in more than a single plane is possible, such a conveyor requires a universally flexible or double-articulated moving element. Conveyors of this kind have generally been found troublesome and are shunned by most manufacturers.

In Fig. 26.3.4 (illustrations *a* to *g* inclusive) are shown a variety of common shapes. The shapes marked *a, b, c, d,* and *e* have the return run of the chain parallel to the carrying run, while shapes *f* and *g* are circuit types. Parallel return-run casings are generally integral with carrying-run casings. The two runs are necessarily separated by a partition in portions that slope above the angle of repose. In horizontal or gradually sloping portions the runs may be partitioned off, but more often they have the return-run flights carried on tracks with center open, as in Fig. 26.3.5. Even where the runs are parallel, they also may be in separate casings, as shown in Fig. 26.3.6.

Shape *c* (Fig. 26.3.4) is referred to as a loop-boot elevator, shape *d* as an L-type elevator or conveyor-elevator, shape *e* as a Z-type, *f* as a vertical closed circuit or a run-around, and *g* as a horizontal closed circuit. The loop of shape *c* may have an alternate shape in which there is a short horizontal portion at the base, as shown by dotted lines.

In shapes such as *e* and *f,* where the carrying run makes a bend at the top of a vertical or steeply sloping portion, it is usual to provide a "drum-bend corner." The moving element and material bear against the rim of a rotatable drum, thus reducing sliding friction. Both of the circuit types have the ability to recirculate any material that is not discharged—a feature of particular importance in the horizontal closed-circuit elevator-conveyor.

Fig. 26.3.3 Tilting-vane-type flights. One flight shown titled for discharging. Source: Chain Belt Co.

26.3.4 Horizontal Closed-Circuit Conveyors

Figure 26.3.4, previously referred to, shows one of the possible shapes of an en masse conveyor of the horizontal closed-circuit type. A typical cross section through one run of such a conveyor is shown in Fig. 26.3.7.

These conveyors have the unique ability to receive material from multiple open inlets, taking no more than enough from each to acquire a normal load, and to discharge through multiple open outlets, giving out just as much to each as is needed to keep each receiving chute filled. Such characteristics are exceedingly useful and the applications are many. An example is the drawing of coal from any of several bunkers and delivering it to a number of stoker hoppers, automatically keeping them supplied under varying demands. Another common application is the supplying of a line of packaging machines from one or more supply bins. A principal advantage attending this operation is the ability of the conveyor to return undelivered material to the feed point, where only enough new material is added to complete a full normal load.

26.3.5 Operating Speed

Rate of conveying in terms of volume per unit of time depends on cross section of the casing, on speed of the moving element, and on a capacity factor Q, which expresses the ratio of average speed of the stream of material to that of the flights.

Standard casing sizes, as listed by manufacturers, range from about 0.07 ft^2 of cross section to about 1.7 ft^2, with some eight or nine sizes in this range. Since designation of sizes is not systematized among manufacturers, it is not feasible to list the various sizes accurately. Table 26.3.1 shows approximate sizes in terms of casing widths. These sizes correspond roughly to commercially available conveyors. Depth of casing runs from 85% of width in the smaller sizes to 60% in the larger sizes.

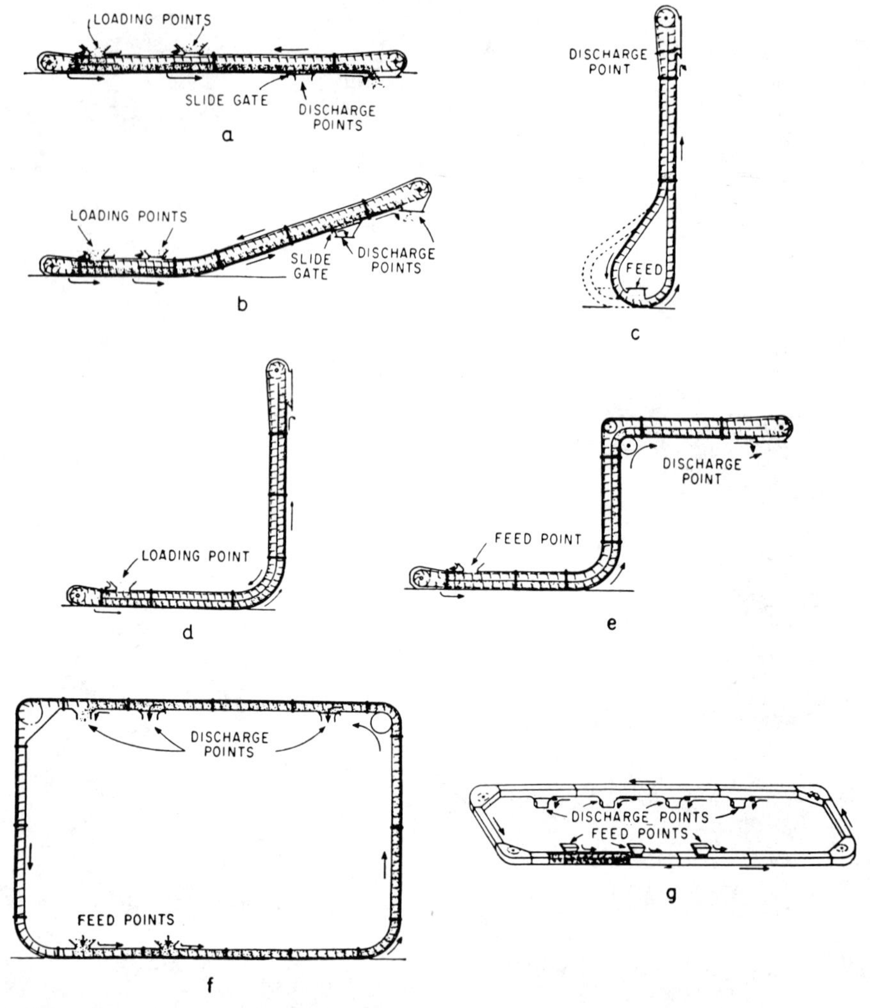

Fig. 26.3.4 Typical common shapes of en masse conveyors and elevators. Source: Stephens-Adamson Mfg. Co.

Chain speeds are limited by a rather complex set of factors. These factors include:

1. Running qualities of chain on sprocket
2. Ability of the particular material to enter and leave the conveyor at the conveying speed
3. Tendency of the material to fluidize with too rapid motion
4. Considerations of wear and maintenance

As a general guide to maximum allowable speeds, free-flowing, nonabrasive materials such as grain can be handled with chain speeds up to 100 ft/min. For aeratable materials like cement or pulverized starch, chain speeds should be kept below 50 ft/min. Stringy, flaky, sticky, or sluggish materials require handling at slower speeds according to the nature and extent of their unfavorable properties, generally keeping well under 50 ft/min. For limiting rate of wear and reducing maintenance requirements, chain speed is only one of a large array of considerations. Where the material handled is abrasive, such as coke, chain speed is seldom over 30 ft/min.

26.3.6 Capacity Factor

The capacity factor Q, defined previously, is usually 1.0 for horizontal conveyors. For elevating, Q varies from 0.85 for granular materials loaded close to the bend, to 0.50 for loose powders. Values

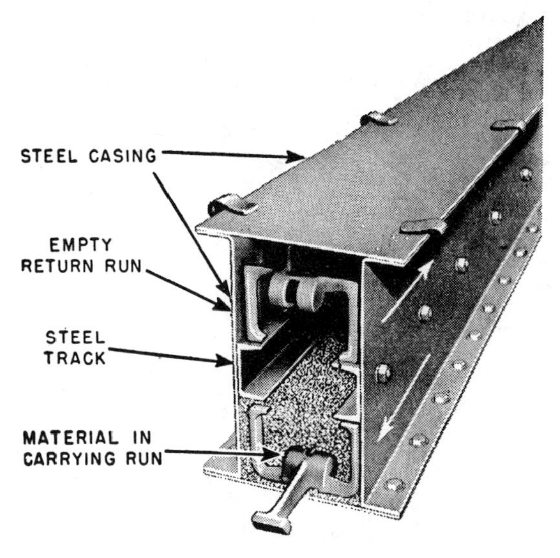

Fig. 26.3.5 Double-run casing with open return-run tracks. Source: Stephens-Adamson Mfg. Co.

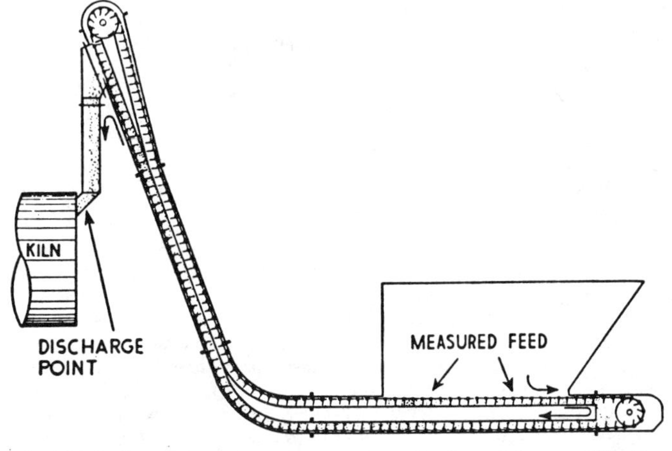

Fig. 26.3.6 Parallel runs in separate casings. Stephens-Adamson Mfg. Co.

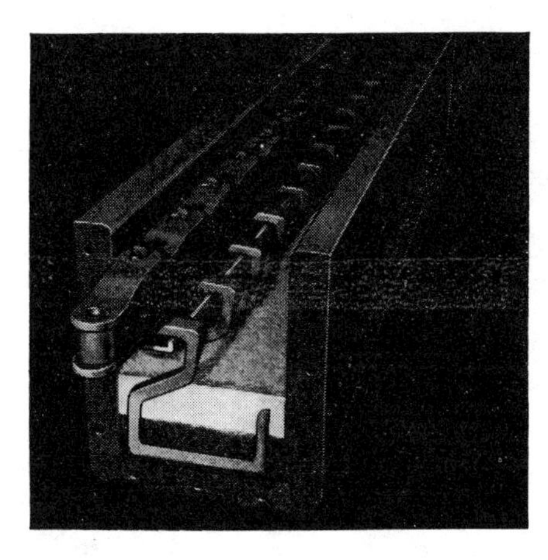

Fig. 26.3.7 Typical cross section through a horizontal closed-circuit conveyor. Source: Stephens-Adamson Mfg. Co.

Table 26.3.1 Casing Widths and Cross-Sectional Areas

Casing Width (in.)	Cross-Sectional Area (ft^2)
4	0.07
6	0.15
8	0.3
10	0.4
12	0.6
14	0.8
16	1.0
18	1.3
20	1.7

are higher as the horizontal casing length from loading point to bend is greater. If conditions are such that Q tends toward the smaller values, it will also tend to be variable and uncertain.

Volumetric capacity may be determined from the equation

$$C = A \times Q \times S$$

where: C = capacity, in ft^3/min
A = cross-sectional area, in ft^3 (see Table 26.3.1)
Q = capacity factor
S = chain speed, in ft/min

Capacity by weight, or tonnage, may be determined from the equation

$$T = \frac{W \times A \times Q \times S}{33}$$

where: T = capacity, in ton/hr
W = weight of material, in lb/ft^3

Surges

The fact that en masse conveyors may be self-feeding has led to an erroneous idea that they are basically incapable of handling surges. Actually, where they are expected to handle a variable rate of flow from another conveyor or from a processing machine, they can be designed to convey at a rate in excess of any expected surge. They will then handle any lesser amount also, just as is the case with belt conveyors or bucket elevators. Size and chain speed should be selected to allow not only for surges but also for uncertainties in the capacity of factor Q and weight per cubic foot. Since en masse conveyors, unlike centrifugal-discharge bucket elevators, have a wide latitude in workable speeds, it is quite easy to make corrections in capacity after installation. To this end, speeds should be assumed below the maximum that would be allowable, and drive equipment should be designed for easy change of ratio.

26.3.7 Power Consumption

Although the power needed to operate en masse conveyors is greater than that for the belt conveyor or bucket elevator, it is not generally large enough to be a serious deterrent to their use. The accurate determination of power requirement, especially for shapes having bends, must take into account the characteristics of the material handled. Since the properties which affect performance of en masse conveyors are often of an intangible nature and are recognizable only as a result of experience, it is well to have horsepowers checked by manufacturers in any case where power requirement might be a critical factor.

Power consumption may include two or more of five main components:

1. Sliding friction of material on casing walls
2. Sliding friction of conveying element in return run
3. Overcoming of gravity in case of lift
4. Friction of moving element on curve plates of bends
5. Internal friction in material at bends

The more common basic shapes of en masse conveyors are shown in Fig. 26.3.4. The following approximate equations give motor horsepowers for the various shapes.

Table 26.3.2 lists weights per cubic foot for numerous materials handled by en masse conveyors and numerical factors to be used in the equations for calculating horsepowers for the various shapes of these kinds of conveyors.

$$\text{For shape } a,\ \text{Hp} = \frac{ELT}{1000}$$

$$\text{For shape } b,\ \text{Hp} = \frac{ELT + HT}{1000}$$

$$\text{For shape } c,\ \text{Hp} = \frac{GT(H + C/2)}{1000}$$

$$\text{For shapes } d, e, \text{ and } f,\ \text{Hp} = \frac{(FL + GH + K)T}{1000}$$

$$\text{For shape } g,\ \text{Hp} = \frac{JDT(1 + 0.07N)}{1000}$$

where: C = casing width, in in.
D = distance around horizontal closed circuit, in ft
H = height of loaded run, vertical or steep slope, in ft
L = length of loaded run, horizontal or gradual slope, in ft
N = number of corners in horizontal closed-circuit conveyor
T = tons per hour
E, F, G, J, K = factors from Table 26.3.2

Numerical factors for various materials handled by en masse conveyors are listed in Table 26.3.2. A comprehensive list of such materials would be impractical; therefore, only familiar, representative kinds are included. Each material listed has a well-defined and distinctive set of properties. Factors for other materials may be assumed to be the same as for the materials in the table having similar properties.

The equations are based on "fed loads." That is, they assume the material drops through the inlet at a controlled rate such that it is never allowed to back up into the supply chute. Many en masse conveyors, however, are self-feeding. The inlet is then "choked" and the moving stream of material in the casing has to slide out from under the stationary mass of material in the supply chute, hopper, or bin. Obviously, more power is needed for a choked feed than for a fed load. To allow for the additional horsepower, it is necessary to increase the value of L by four times the length of the choked inlet. If the inlet is in the return run, as in Fig. 26.3.6, five times the length of the choked inlet should be added to the loaded length of the carrying run.

Table 26.3.2 Factors for Various Representative Materials[a]

Material	Weight (lb/ft³)	*E* (for units 5 in. wide)	*F*	*G*	*J*	*K*
Beans, dry	54	1.3	2.5	3.9	1.0	100
Bran	26	3.3	6.6	3.4	2.3	0
Cement, dry Portland	85	2.5	6.0	5.2	2.0	0
Coal, dry slack	50	2.1	3.8	3.6	1.6	40
Coal, wet slack	55	2.8	4.7	4.4	2.0	40
Coffee, ground	28	1.9	3.9	2.9	1.4	20
Corn flakes	12	2.9	6.0	2.1	1.8	0
Flour, wheat	35	2.6	5.6	3.2	1.8	0
Lime, dry burned, small lumps and dust	50	2.8	4.2	6.1	1.8	200
Salt, dry granulated	80	1.7	3.3	5.0	1.4	80
Starch, lump	30	1.7	3.2	3.4	1.2	90
Sugar, dry granulated	50	2.3	4.8	7.9	1.8	160
Wheat, dry, fairly clean	48	1.5	2.7	4.5	1.1	40
Wood chips	25	2.7	5.1	2.5	1.7	40

[a] For use in equations for calculating horsepowers for the various conveyor shapes.

26.3.8 Limitations of Length and Height

The limiting factor in length and/or height of en masse conveyors is the working tensile strength of the moving element. A standard design is offered by each manufacturer. This design is a compromise between excessive cost for short units and undue limitation of lengths. Some manufacturers also offer an extra strong chain at a higher cost. This chain is specified when the combination of horsepower pull and height-weight factor is too great for the standard element. While the cost of a conveyor on a per-foot basis is less as the length increases within the limit of a standard chain, the cost per foot will suddenly rise as the length goes into the class requiring high-strength chain. As length is further increased, a point is reached where an adequate chain would be impractical in size and prohibitive in cost.

For most shapes of conveyors the length limitation refers to the carrying run only. However, since it is assumed that a horizontal-circuit conveyor should be capable of carrying a full load completely around the circuit, the limit for this type refers to total chain length. An offsetting factor is the possibility of dual drives for horizontal circuits, which double the length allowable with one drive. Table 26.3.3 is a guide to maximum conveyor lengths that are feasible.

Allowable lengths as given in Table 26.3.3 may be increased in the case of the solid-flight type by regulating the relation of feeding rate to chain speed so as to reduce the cross section of load. In the case of the skeleton flight this can be accomplished only by going to a smaller size unit. However, if conveyor lengths are thus carried far beyond the normal limits, the economic setup becomes questionable. Attempts have been made to produce a moving element having an extreme tensile capacity by use of high-strength alloys. Generally, experience has shown that the metal with a lower yield point and greater ductility is more reliable and more resistant to fatigue.

26.3.9 Feed Inlets

The feed inlet consists of an opening in the upper wall of a portion of casing that is horizontal or sloping at an angle less than the angle of slide of the material. Where rate of feed is controlled by means independent of the conveyor, as for material flowing from a continuous drier, there is usually no special problem concerning the inlet. However, most installations take advantage of the unique ability of en masse conveyors to feed themselves at just the correct rate without any extra equipment. The choked feed inlet, although basically sound, has certain limitations.

Since material above the inlet is relatively compact and quiescent, some types of products, like soap chips, may arch across the opening. Therefore, since the effective width of inlet cannot be greater than the width of the moving element, the selection of conveyor size may depend on necessary width of inlet to prevent arching, rather than on tonnage requirement.

Often an inlet is made as long as the length of a bin or hopper, thus saving important space. In operation under a full bin, the material below the inlet is under much more than normal pressure. Not only is wall friction thus increased, but also the moving stream must be sheared from the stationary material above it. The chain load per foot of choked inlet is approximately five times as great as it is per foot of covered casing.

With shapes having double-run casings it is customary to feed into the upper, or return, run, as illustrated in Fig. 26.3.4. This simple construction is not suitable for all materials. In materials containing lumps, some of the lumps will tend to catch between the edges of flights and the end of the inlet, thus causing excessive chain loads and possible bending or breaking of parts. This difficulty can be avoided by feeding the material into the carrying run where flights are of the skeleton type. With construction as shown in Fig. 26.3.8, it is evident that lumps cannot become jammed between flights and inlet.

Table 26.3.3 Allowable Lengths of Chains for En Masse Conveyors

Shape of Unit	Maximum Length for Standard Chain (ft)		Maximum Length for High-Strength Chain (ft)	
	Wheat	Cement	Wheat	Cement
Straight horizontal	300	110	450	170
Straight vertical	110	80	180	130
Horizontal followed by equal length of vertical	90	45	160	75
Horizontal closed circuit with one drive	320	120	—	—

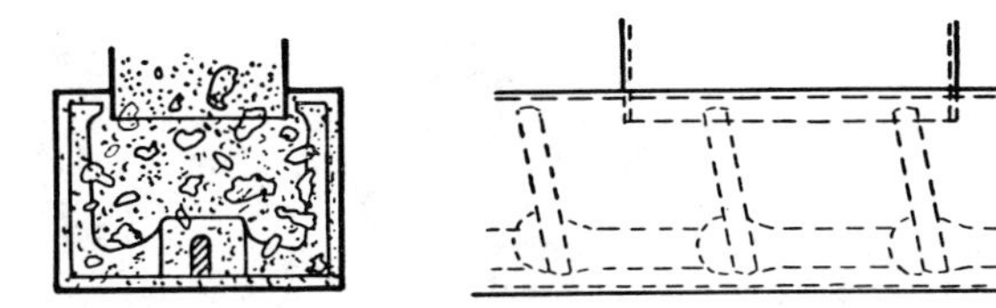

Fig. 26.3.8 Construction of inlet to avoid jamming of lumps. Source: Stephens-Adamson Mfg. Co.

Handling Finely Divided Dry Materials

Many finely divided dry materials may become aerated in handling and will then behave much like liquids. Typical examples occur in the case of cement and pulverized starch. Such materials do not require any special treatment in horizontal conveying, but in any kind of uphill path, where gravity opposes motion of the moving element, a fluid material will tend to stand still or run back in a conveyor of standard design. By certain refinements in construction, however, this difficulty can be overcome to such an extent that fluidity in any degree is no bar to successful handling in any direction.

The only special layout requirement is that the material must enter in a horizontal or downhill section and the casing must be straight for several chain pitches on each side of the inlet. The flights must form complete barriers covering the cross section of the casing. The clearance spaces between periphery of flights and casing walls must be so small that the material in its most fluid condition cannot flow back through them at a rate anything like that at which the flights are moving it forward. Preventing this backflow may increase manufacturing costs considerably. In conveyors using the skeleton type of moving element, special flights called web flights are inserted at intervals to provide barriers.

26.3.10 Frictional Resistance at Bends

The great versatility in layout of en masse conveyors is due largely to their ability to carry materials around bends. It is always desirable, however, to avoid bends if practical, the alternative usually being to use two separate units instead of one. In each case it is necessary to decide whether the advantage of a single unit will offset the limitations imposed by a bend.

Additional friction is caused in bends by:

1. Pressure of the flights against the inner wall resulting from component forces of chain tension
2. Internal shifting of material

The friction resulting from rubbing of flights on a bend is greater in proportion to chain tension at the point where the bend occurs. Thus a bend near the feed point may be of little consequence, while the same bend located far beyond the inlet may be quite serious. As the flights advance around a bend, the chain tension builds up cumulatively because of friction. Therefore the total increase of tension is greatly dependent on the angle, or degree, of a bend, but is not affected by the radius of curvature. While the increase in chain tension is reflected in higher power requirement and need for a stronger chain, the bend friction also may create a wear problem. Whether or not the problem is so serious that a contemplated conveyor layout becomes impractical depends on numerous factors, as discussed in Section 26.3.13. No hard and fast rules can be stated.

Internal friction in material results from the shifting of particles from their normal positions when the stream changes from straight to curved and then back to straight. Some materials, such as flour, offer little resistance to internal movement, while others, such as granulated sugar, offer so much resistance that a bend may become a major factor in determining power and chain tension. Unlike friction of flights on the casing, internal friction in material is independent of chain tension and also of bend angle beyond 20 or 30°, but it does vary inversely with the radius of curvature. Bend material friction is greater in a bend followed by a vertical run than in a bend from vertical to horizontal.

For bends from vertical or steeply sloping runs to horizontal or gradual slopes it is practical to substitute a rotating drum for the stationary inside curve plate. Thus most of the friction of flights on the casing is eliminated. The saving is of considerable importance because the chain tension tends to be high at the top of a vertical run. Dribble of material through operating clearances around the drum is disposed of by running it into the return run. Since this method is not usually feasible in a bend at the base of a vertical run, drum-bend corners are generally specified only in Z-type or run-around shapes. (See types *e* and *f* illustrated in Fig. 26.3.4.)

26.3.11 Discharge of Materials

Discharge is effected through an outlet in the bottom of a horizontal or sloping portion of casing or in any wall of a vertical casing. Free-flowing nonsticky materials are discharged by gravity. Although such materials will tend to flow out of the spaces between flights through an opening in a vertical wall, a portion resting on any surface of the moving element that slopes less than the angle of slide will be carried beyond. Various devices are employed to prevent flights from carrying material past an outlet, or at least from carrying over the head sprocket and into the return run. Bulk-Flo flights are sloped two ways, as shown in Fig. 26.3.2. In some other makes, flights are pivoted and are caused to tilt by means of a cam opposite the outlet. Skeleton flights do not have flat surfaces large enough to carry appreciable quantities of material, but web flights hold considerable amounts which are poured off as the flights are tilted in passing around the head sprocket. Whatever system is used, little difficulty is experienced with free-running materials.

Special problems of discharging arise with materials that are sticky, stringy, or sluggish. Outlets comprising two, three, or four walls of the casing and lengths greater than standard may be a solution where these properties exist in moderation. If skeleton-type flights are used, a push-out plate may be installed to force material out of the area inside the flights. Thus sticky materials can be handled successfully that could not be discharged by gravity alone. A particularly favorable arrangement for feeding and discharging difficult materials is the vertical circuit (Fig. 26.3.4*f*) equipped with skeleton flights. The material can enter and leave the casing through the open side of the flights, thus encountering no obstruction to its free passage.

An important advantage of en masse conveyors is their adaptability to multiple feed and discharge points. In general, only a simple slide gate is needed to start or stop the process of feeding or discharging at any location along the path of a conveyor. For discharging, the gate may be used to reduce the discharge to a very low rate. A partial discharge, adjustable as to amount, may take place at several points simultaneously. These methods are feasible because the flights continually pass over the outlets and prevent arching over the narrowest openings.

26.3.12 Cleanout

In a few conveyor applications it is essential that the conveyor be able to clear itself of material. Some types, such as belt conveyors, naturally clear themselves completely by running for a short time after the feed is stopped. Others, like screw conveyors and bucket elevators, clear themselves except for a residue in the trough or boot. Since the latter types of conveyors will not eliminate the residue by any amount of running, they may be unsatisfactory for some kinds of service. An example is the handling of a material that would be subject to spoiling or infestation if allowed to remain in the conveyor during a shutdown period.

Among the conveyors that are completely enclosed, the en masse types have excellent self-cleaning ability. Where they are horizontal, or on slopes below the angle of repose of the material, the cleanout action is inherent even with standard designs. Since the flights tend to bear directly against the casing bottom as the bulk of the material is swept out, the removal of remnants is nearly complete. The addition of wiper strips along the edges of several flights will insure cleanout of the last vestiges.

Conveyors having portions of the carrying run vertical or steeply sloping present a special problem regarding cleanout. When the feed is stopped and the casing below the bend has been emptied, normal conveying above the bend ceases. Fine materials start to drop back through the clearance spaces between flights and casing walls. If the rate of backward flow is less than the conveying rate, a portion of the vertical casing will be cleared. However, any material leaking past the flights will remain in the conveyor unless it is removed by some means other than the normal en masse action. Minimum clearances that are practical with ordinary manufacturing procedures are bound to permit appreciable leakage with most materials, even where wiper flights are used in conveyors having vertical or nearly vertical runs.

In certain makes of en masse conveyors all or part of the flights may be made of a type that will elevate and discharge a quantity of material without the benefit of a supporting column of material underneath. A special construction is also necessary in the head section in cases where it discharges on the vertical. This provision is made to insure that the carrying flights will so discharge their load that none of it falls back into the up-run casing or is carried over into the down-run casing. When fed at a normal rate the material is propelled in a solid stream. As soon as the feed is stopped, the cleanout action commences and continues until practically all material has been discharged from the casing. The time required for cleanout varies from 1 min to 20 or 30 min according to the height and speed of elevator and character of the material.

Complete cleanout assumes nonsticky material that does not cling to the chain or flights. Although many sticky materials are successfully handled in en masse conveyors, complete self-cleaning in such cases is out of the question.

26.3.13 Problems of Wear

With most en masse conveyors wear is no problem. Even when handling abrasive materials, en masse conveyors can show good economy if properly specified and designed. The places where wear may occur, the conditions which cause or increase wear, and the steps that may be taken to offset wear must all be recognized.

Even under adverse conditions, covers and return-run casings, where there is no positive pressure of the flights against the surfaces, are not subject to wear. Carrying-run surfaces not under positive pressure from the flights are only slightly affected if the material, even though abrasive, is fine grained or pulverized. For example, in handling fine sand, there would be practically no wear on any of four sides of a vertical carrying run or on the sides or top of a sloping or horizontal carrying run.

Surfaces that have rubbing contact from the flights under positive pressure are subject to wear, even in return runs and in handling relatively nonabrasive materials, but the tendency increases with the abrasiveness of the material. If preventive measures are not taken, wear may occur, for example, in the return-run tracks of horizontal conveyors, in the partition plate supporting the return chain of a sloping elevator, in the bottom of the carrying run of a horizontal or sloping unit, and in the bottom of the horizontal carrying run of an L-type elevator. To a still greater degree, wear may occur in the inside curve plate of a loop boot, in any of the curve plates of a bend section (although primarily in the inside curve plate of the carrying run), and in the curve plate at the base of an elevator head section. The rate of wear on curve plates decreases as the radius of curvature of the plate increases and as the tension in the chain at the curve is less. Therefore, the curve-plate wear of an L-type elevator is less as the length of horizontal run decreases.

Wear on the casing walls of carrying runs is greatly increased when large, hard lumps become wedged in the clearance spaces between flights and casing. This wear may occur not only in all the regular wearing areas but also in the four walls of a vertical carrying run and the sides of a sloping or horizontal carrying run. The rate of wear will increase in proportion to the hardness, roughness, size, and number of the lumps in the material being conveyed.

In general, casing wear due to wedging of lumps is not accompanied by corresponding wear in the flights. However, casing wear due to rubbing contact from the flights, as on curve plates and return tracks, is also accompanied by a tendency to wear the flight surfaces that make the rubbing contact. For example, the tops of flights are subject to wear on the curve of an L-type elevator, while both the tops and bottoms of flights are subject to wear on horizontal conveyors in which part of the carrying run may occasionally operate without load.

Methods for Increasing Wear Life

To combat wear in en masse conveyors, the following methods may be used:

1. The chain speed may be reduced by using a larger conveyor.
2. The wearing parts may be made thicker.
3. The wearing parts may be made of harder materials.
4. The parts subject to wear may be made easily renewable.

Although each of these measures will add to the first cost of the equipment, each will increase the wear life per dollar of initial cost.

26.3.14 High Temperature Effects

En masse conveyors are well suited to handling hot materials or to operating in hot surroundings. Their heat-resisting ability particularly surpasses that of conveyors that have vital parts made of rubber, such as belt conveyors, belt elevators, and certain types of vibrating conveyors. Up to about 300°F no special problems are encountered. Higher temperatures may bring about any or all of the following effects unless provided against in the design:

1. Breakage of casing or fastenings caused by expansion
2. Buckling or overstressing of chain induced by unequal expansion between chain and casing
3. Permanent weakening of chain resulting from changes in grain structure of metal
4. Rapid corrosion
5. Chain breakage or stretching because of lowered strength at the high temperature

Expansion of casings may be allowed for by use of telescoping sections, which allow terminals and chute connections to remain fixed, or by sliding supports, which require that at least one terminal shall move. Unequal expansion between chain and casing may be compensated for by gravity or spring take-ups, which will add 10 to 20% to normal chain load.

Danger of embrittlement of the chain from heat begins at about 500°F for malleable iron, at 800°F for heat-treated malleable iron, and at 900°F for the lower-steel alloys. Higher alloys are obtainable which are stable up to the highest practicable operating temperatures.

Temporary loss of strength from excessive heat for all metals begins at about 700°, with a reduction of 50% at 900°. Above 900°, only metals designed for elevated temperatures should be considered. Many of these are available, each suitable for a given set of conditions and all being much more expensive than the common steel alloys. With some metals, temperatures as high as 1200°F are possible.

26.3.15 Protection against Explosions

For safety in the bulk handling of materials containing flammable vapors or dust, a primary requisite is that the conveyor be vapor- or dust-tight. Assuming total enclosure, certain other characteristics are also needed:

1. Inability to produce sparks
2. Minimum enclosed space that could contain an explosive mixture
3. Means for relieving pressure before a destructive bursting can take place

En masse conveyors are adapted to the prevention of explosions in all of the ways noted.

1. Sparks may be caused by accumulation of static electricity or by friction. If all parts of an en masse conveyor are made of metal, which is usual, and the casing is grounded, the possibility of static sparks may be ruled out. Friction sparks are unlikely in the normal operation of any type of conveyor, but are always possible as the result of an accident. In the case of a sudden jamming, sparks are less likely to occur as the speed of the moving element is slower, and as the materials from which the conveyors are constructed contain less iron or steel. The relatively low speeds of en masse conveyors favor safety, and the speeds may be kept even lower than normal by using conveyors of larger sizes for a given capacity. Nonsparking metals such as manganese bronze are sometimes used for chains and sprockets. Some kind of device—such as shear pins, for quick relief of load if jamming occurs—is essential if there is a chance of friction sparks causing an explosion.

2. A design in which there was no space at all that could contain an explosive mixture would be impractical in any type of conveyor. Although en masse conveyors necessarily contain vacant spaces in the return runs and terminals, the volume relative to conveying capacity is less than in other types such as bucket elevators and screw conveyors. If, in spite of small vacant space and measures to prevent sparking, explosions present a great hazard, it is advisable to fill the casing with an inert gas such as carbon dioxide.

3. A simple way to prevent serious damage in the event of an explosion occurring in the casing of an en masse conveyor is the inclusion of pressure relief panels. These panels are located at convenient points in the walls of the return run or terminals. They may consist of flanged openings covered with a durable and easily replaceable material, such as rubberized fabric, which will burst and relieve the pressure long before any deformation would occur in the body of the casing. In considering the advisability of relief panels it should be noted that an explosion may originate outside a conveyor and travel into it through the connecting chutes.

26.3.16 Noise Reduction

Under most circumstances en masse conveyors run quietly, sometimes to the point where no sound will reveal whether they are running or stopped. Under certain other circumstances they may be noisy to an objectionable degree. The noise, usually an uneven squeal, is nearly always produced by rubbing of moving parts against the casing in the presence of a conveyed material having certain noise-potential properties. The vibration may be transmitted through the entire casing so that its source is difficult to locate.

In general, the property that promotes noise may be described as "chalkiness." It is found to high degree in talc and soda ash, and to a moderate degree in cellulose acetate. Materials such as cereals, especially if oily, like soybean meal, do not produce any noise.

If noise-producing materials are to be handled in a location where noise would constitute a serious disturbance, it may be the determining factor in the design and selection of conveyors. Shapes should be chosen which cause only a minimum of rubbing between moving element and casing. A loop elevator (Fig. 26.3.4*c*) would be much more suitable for installation that an *L*-shaped conveyor (Fig. 26.3.4*d*).

Noise may be eliminated, or satisfactorily reduced, by use of various nonmetallic surfaces at the points of rubbing on the casing. Manufacturers of en masse conveying equipment are in a position to furnish data on effectiveness, wear resistance, and cost of noise-prevention materials to suit any given set of circumstances. The main objective is not to overlook the question of noise if it appears to be possible and objectionable.

26.3.17 Combating Corrosion

Attention should be given to any possible corrosive effect between the conveyed material and the surfaces with which it comes in contact. The effect of corrosion is generally more pronounced wherever there is rubbing that would tend to scour off the corroded surfaces and continually expose new metal to the corrosive action. The alternate rusting and scouring away of the rust is especially severe in the handling of wet coal for boiler-plant use. Not only does the coal from open cars or outside storage average high in moisture, but also it is common practice to run coal conveyors intermittently and allow them to remain full of wet material while stopped. Under these conditions a casing made of mild steel of ordinary thickness may have a life much too short for good economy.

Two methods are available for combating corrosion: increasing thickness of parts, and making them of metals or materials more resistant to the particular type of corrosion. In specifying greater thickness, it is unnecessary to use the heavier material for covers, return runs, or portions of terminal sections not coming in contact with the moving stream of material. But the affected parts may well be several times normal thickness, since cost in this case increases but little in proportion to weight.

Although there is a wide selection of materials of construction that resist corrosion, the list is greatly narrowed when cost and economy are taken into account. For example, to install stainless steel or brass equipment for the handling of boiler-house coal would increase the cost of equipment out of proportion to the benefit gained. Galvanizing is of little advantage for coal-handling equipment since the surfaces subject to rubbing are soon denuded. Steel alloys, however, which provide much greater overall economy than ordinary steel, are available for construction of conduits. They contain small percentages of alloying elements such as chromium, nickel, and copper, which increase conveyor cost less than 7% and produce metals that last approximately twice as long as mild steel of equal thickness.

Corrosion from conveyed materials or from gases and/or elevated temperatures involves a wide variety of chemical reactions which may be controlled or minimized by use of suitable materials of construction. Materials used for casings may include stainless steel of various types, stainless-clad steel, brass, cast iron, bakelite, numerous kinds of plastics, and ordinary steel galvanized, metalized, or coated with vitrified enamel. For moving elements, stainless steels of SAE types 302, 304, and 316 are commonly used, as are manganese bronze and Monel metal, as well as malleable iron or steel, galvanized or metalized. If possible, specifications for corrosion-resisting construction should be based on previous experience by the prospective user, because manufacturers of equipment do not always have a knowledge of corrosive action equal to that of producers or users of the particular material to be handled.

26.3.18 Protection against Overloads

Accidental shutdowns or damage to equipment are generally caused by overloads, foreign materials, or choked outlets. Overfeeding is normally impossible in en masse conveyors, but excessive chain pulls may result from a change in the character of material, for instance, an increase in moisture content. Such overloads usually build up gradually, and an ordinary thermal relay will generally give adequate protection to the motor. The sudden jamming that would occur, however, if foreign material—say a steel bar—should drop between a moving flight and the edge of the feed inlet would not be adequately relieved by an overload relay. The reason is that the inertia of the high-speed elements of the drive would be more than the chain could absorb under a sudden stop. It is essential, therefore, to have some device, such as shear pins, located in the low-speed portion of the drive to give instantaneous relief. Fluid or friction couplings are not suitable because they are too expensive if large enough to serve at low speed, and fail to eliminate the effects of inertia when installed at the high-speed end of the drive.

En masse conveyors, in common with nearly all other types, cannot withstand choking of the final outlet. While it is true that material backing up into the head section will ultimately cause a shutdown through the motor overload relay or the shear-pin device, the conveyor may be seriously damaged before it stops. The reason is that when the chain running onto the head sprocket is buried in material, it may "climb the teeth." When this happens the entire chain—return run as well as carrying run—may be put under tension enough to start cracks or cause permanent distortion before the drive torque builds up sufficiently to cause the shear pins to act. Therefore, devices to prevent backing up of material into the drive terminal are essential for trouble-free operation.

26.3.19 Representative Installation

A typical layout of a storage and reclaiming system using en masse conveyors is shown in Fig. 26.3.9. The truck delivery point is shown in the right-hand view, and the general layout of the storage bins

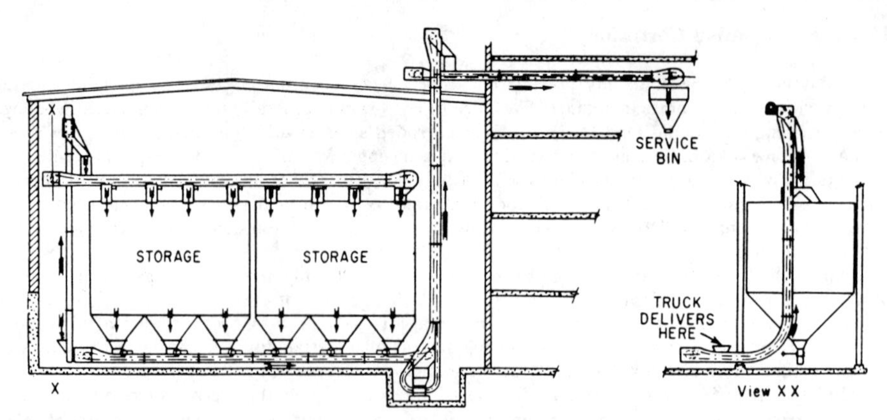

Fig. 26.3.9 Layout typical of storage and reclaiming system using en masse conveyors.

and the conveyor system is diagrammed in the left-hand view. Incoming bulk material is distributed overhead to the various storage bins and reclaimed from them through the lower conveyor line. It is then fed into an elevating conveyor and delivered by a horizontal conveyor to the service bin.

CHAPTER 27
BUCKET ELEVATORS

TED P. SMYRE

Jeffrey Manufacturing Div.
Dresser Industries, Inc.
Woodruff, South Carolina

27.1 GENERAL INFORMATION

When bulk raw materials and grains must be elevated while maintaining a continuous material flow, bucket elevators should be considered. These vertical conveyors are, in general, especially useful where moderate to high capacities of bulk materials are handled and where floor plan or site space is limited. With proper selection of bucket elevator type and proper materials handling system design, most free-flowing bulk raw materials and grains can be handled. Normal lifts range up to 100 ft, but by special construction, lifts up to 300 ft are not uncommon. Above this height, loads developed in both conveyor and structure are generally beyond the economical design range of this equipment, at this time.

Bulk materials handling elevators have two general classifications, spaced-bucket elevators and continuous-bucket elevators. The two general classifications can be subdivided into particular types.

Three common types of spaced-bucket elevators are:

- Centrifugal-discharge elevators
- High-speed elevators for handling grain, woodchips, and similar nonabrasive materials
- Marine-leg elevators

In continuous-bucket elevators, the three most common types of elevators are:

- Standard continuous-bucket elevators
- Super-capacity elevators
- High-speed continuous elevators for handling bulk raw and process materials

The latter type can be further subdivided according to conveyor chain/belt and bucket combination.

There is a wide range of standard and special engineered units from which the bucket elevator type and size best suited for the particular service can be selected.

Most free-flowing materials can be handled. It is extremely important that feed be controlled to the elevator in a continuous and uniform rate and be directed so that the buckets are evenly loaded across their width. For most bulk material applications, feeds with uniform particle size of 1/2 in. and maximum oversize of 4 in., not exceeding 10% of the whole, will handle well.

When handling fine and ultrafine materials that aerate through transfer but can compact to twice the aerated density (such as cement), special sizing, application, and design techniques must be observed.

Standard high-speed grain-type bucket elevators have been developed for handling light free-flowing grains.

For handling large-lump bulk materials up to 12 in., standard low-speed continuous-bucket, super-capacity elevators are used.

Although bucket elevators are generally applied to free-flowing materials, with careful application some sticky materials can be successfully handled. Special care must be taken in design of the loading zone and selection of buckets. Since these materials tend to pack and to build up in buckets, extra-capacity and free-discharge buckets must be used.

Numerous special bucket elevators have been built to handle ultrahot materials in controlled atmosphere, explosive materials, fluids, shredded municipal solid waste, muck, and many other materials. However, these designs should not be attempted by the uninitiated!

27.1.1 Controlled Feed and Open Discharge

Operating speeds of various types of elevators vary from 50 to 800 ft/min. This fact requires different loading and discharge designs among the various elevator types.

However, because each bucket elevator has a constant volumetric capacity, to deliver the rated capacity and give satisfactory performance, they must be fed at a uniform metered rate. Grain and similar materials flow freely and evenly such that the volume of flow can be regulated by a gate or valve. Most other materials are more difficult to handle and are subject to varying degrees of sluggishness and flooding. These materials must be fed to the elevator boot under definite control, preferably by means of pan, belt, vibrating, or vane feeder, or similar positive volumetric feeder.

The feeder with its bunker or surge bin can be remote from the elevator, provided the connecting conveyor is constant speed and constant volume. Air slides and conveyors subject to similar surging can result in unsatisfactory loading.

The discharge chute from the elevator must be of sufficient size to accommodate the volume of discharge and multiples of largest piece size. If the chute is inclined, it must be steep enough to assure flow. If material backs up into the elevator, buckets may be stripped off in the head and/or boot section. Almost certainly, unsatisfactory operation will result.

Flow of material feed to the bucket elevator should be in line with buckets, not at an angle (as viewed in plan). If a change in direction is required at the elevator, it should be accomplished with the orientation and design of the discharge chute.

27.1.2 Capacity

The capacity equation for either spaced- or continuous-bucket elevators is as follows:

$$T = \frac{VMS}{6400s} \text{ with buckets 75\% full}$$

$$T = \frac{VMS}{8000s} \text{ with buckets 60\% full}$$

where: T = peak capacity, in ton/hr
S = speed, in ft/min
M = weight of material, in lb/ft^3
s = distance from center to center of buckets, in in.
V = struck volume of each bucket, in in.^{3}

Generally, bulk materials aerate through transfer chutes and in the elevator loading zone. Therefore, bucket conveying capacity for most materials must be derated. This is normally done based on the at-rest bulk density of material being conveyed as illustrated in the preceding equations. Further derating for some materials may be required.

27.1.3 Bucket Elevator Components

The principal elements of a bucket elevator are:

1. Head shaft with pulley for belting or sprockets for chain. This assembly is normally mounted in antifriction bearing pillow blocks supported by the housing head section.

2. The drive, gear reducer, and motor drives, which may be V-belt (in light-duty elevators), ANSI drive chain and sprockets, or direct-coupled with flexible coupling or shaft-mounted reducer.

The most common gear reducers utilized are shaft mounted, concentric shaft, and, for heavy-duty applications, parallel shaft.

Mechanical holdbacks are necessary to prevent reversal of the elevator in case of power failure. Dependent on elevator size and application, the holdback may be integral to the reducer or mounted separately on the head shaft.

3. Foot shaft with pulley or sprockets. Normally the foot shaft assembly is mounted in bearings or guides for take-up adjustment of the belt/chain and bucket assembly.

In some special designs and when handling material that tends to pack, the foot shaft is fixed with take-up at the head shaft.

4. Elevator buckets mounted on belting or chain. This assembly defines the elevator and must be matched to the application.

5. The elevator enclosure houses the bucket and belting or chain assembly and generally provides mounting and enclosure for the rotating machinery. Loading and discharge chutes are integral to the enclosure.

Bucket elevator housings are comprised of boot section, intermediate casing, and head section. They are generally fabricated of mild structural steel shapes and plates. Except in large high-lift elevators, the casing supports the live load and the machinery load and accepts the drive reactions. Bucket elevators, except in special designs, must be laterally restrained.

6. Platforms, ladders, and hoist beams are frequently mounted on elevator housings for maintenance access.

27.2 SELECTION OF BELTING OR CHAIN

After the correct selection of elevator style and buckets has been made, the selection of belting or chain is of prime importance. Chain can be selected knowing tight- (ascending) side tension only; however, for belt selection, tight-side and slack- (descending) side tension must be known. In either case, both tensions must eventually be known to size head shaft machinery. Thus a universal approach in calculating tensions for component selection may be taken.

$$T_1' = (W_C + W_B + LL)\,\mathrm{H}$$

where: T_1' = tight-side tension, in lb
W_C = weight of belt or chain with fasteners and attachments, in lb/ft
W_B = weight buckets, in lb/ft
LL = weight of live load, in lb/ft
H = elevator shaft centers, in ft

$$T_2 = (W_C + W_B)\,H$$

Tight-side tension T_1 must have an additional factor added to compensate for digging in the boot. Elevator manufacturers have developed standards for this factor and should be consulted. In tall elevators and continuous elevators, digging forces are of less concern than in short or spaced elevators. This is due to the compensating effect of materials design safety factors; therefore,

$$T_1 = T_1' + T_D$$

where: T_1 = tight-side tension with digging factor
T_D = additional loading for digging in boot

Manufacturers' ratings for elevator chains should be used and a selection made where the working load is within the theoretical calculations obtained from the preceding equations. This working load is normally based on a factor of at least six to one in relation to ultimate chain strength. Judgment is required in the selection of the proper chain. If the selection is good, the chain installed should provide several years of operation without replacement or unusual maintenance.

Combination-type chains, steel knuckle, welded steel, and roller-type chains are three types of chain with wide ranges of usefulness in elevator service.

27.2.1 Selection of Belting

Correct selection of belting requires accurate evaluation of several design features. Some of these are:

1. Adequate thickness, plies, or special fasteners to prevent bucket bolts from pulling through the belting
2. Proper combination of pulley and belt to assure friction drive between them
3. Adequate slack-side tension to prevent slippage at head pulley
4. Sufficient tension rating
5. Proper covers and carcass for the application

Other special design features must often be considered. Belting manufacturers have developed straightforward procedures to select belting for common industrial applications and more detailed procedures to select belting for engineered class elevators.

With the variety of standard and special belting available and the general accessibility to manufacturers' data and selection procedures, their current data and procedures should be used in belting selection.

27.2.2 Horsepower for Driving Elevators

The horsepower required to drive bucket elevators can be estimated, in most cases, from the following equations:

$$\mathrm{Hp} = \frac{H \times T}{500} \text{ for spaced-bucket elevators with digging boot}$$

$$\mathrm{Hp} = \frac{H \times T}{550} \text{ for continuous-bucket elevator with loading leg}$$

where: T = tons per hour
H = vertical lift in ft

Obviously these two equations become progressively conservative for elevators with high lifts, as there is no difference in drive efficiency in high- or low-lift elevators. Hence the greater reserve capacity of the drive is available for loading conditions at the boot and for starting the elevator from rest with a full load in the buckets.

However, in high-capacity engineered elevators, thorough mechanical calculations based on the difference between tight- and slack-side tension plus machinery friction, digging and power transmission losses are required.

27.3 CENTRIFUGAL-DISCHARGE BUCKET ELEVATORS

Centrifugal-discharge bucket elevators have buckets mounted at spaced intervals on engineered class chain or bucket elevator type belting. This type elevator is used to handle bulk materials that can be picked up by spaced buckets as they pass under the foot wheel and discharged by centrifugal force as the buckets pass over the head wheel (Fig. 27.3.1).

Spaced-bucket centrifugal-discharge elevators are most frequently used for conveying low to medium capacities of sized and free-flowing bulk materials. Typical materials conveyed are coal, borax, salt, grain, cement, and aggregate. Material particle size should range from minimums of 10 mesh to $\frac{3}{4}$ in. with maximums of 2–4-in. cubes, with the latter constituting no more than 10% of the whole. For average industrial work, the capacity of centrifugal-discharge elevators is seldom over 150 tons/hr. This type elevator offers an inexpensive, compact apparatus for raising bulk materials in mining and process operations.

General design features include rigid, seamless, smooth buckets of such shape as to assure proper material pickup in the boot and clean discharge. Buckets are bolted to alloy steel engineered class chains that are heat treated for strength, toughness, and hardness or to elevator belting specially selected for adequate tension, resistance to bucket bolt pull-out, and other considerations particular to bucket elevator application.

There is a definite ratio required between bucket speed and head-wheel diameters. This ratio is based on achieving a combination of good bucket loading conditions and good centrifugal discharge. Table 27.3.1 shows recommended head-wheel diameters for different bucket speeds. Depending on the nature of material handled, the speed and head-wheel sizes may be varied after checking the discharge trajectory.

Caution must be used if bucket speeds are increased over those listed in Table 27.3.1. At higher speeds "fanning" action may develop because of centrifugal force of buckets passing around the foot wheel. This action results in underfilling the buckets. Excessive speed at the head shaft will also cause the material to leave the buckets earlier than it should for correct discharge, and will allow part of the material to fall back down the elevator casing instead of being thrown out through the discharge opening.

Several types of standard buckets are commonly combined with either chain or belting, the choice being based primarily on the type of material being handled (Table 27.3.2).

The most commonly used bucket for centrifugal elevators is Style AA. These are round bottom buckets with a reinforced digging lip. A similar bucket is AARB which has additional reinforcement on front and back to provide additional wear life and more substantial attachment to chain or belt.

Style B buckets have a lower front than Style AA buckets and are used to better advantage in handling coarsely broken or more abrasive material such as coke, ore, stone, and similar materials.

Style C buckets have flat fronts and are used to handle materials that tend to stick or pack in other type buckets.

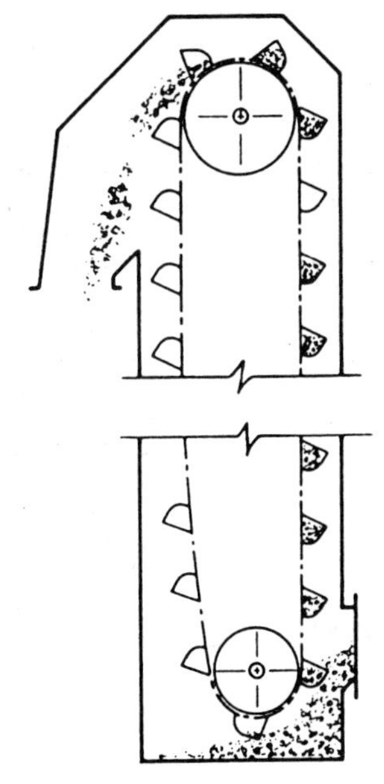

Fig. 27.3.1 Centrifugal-discharge bucket elevator.

Table 27.3.1 Recommended Head-Wheel Diameters for Various Bucket Speeds

Head Wheel Diameter (in.)	For High-Speed Grain Elevators		For General Industrial Purposes and Low-Speed Grain Elevators	
	rpm of Head Wheel	ft/min of Belt	rpm of Head Wheel	ft/min of Belt or Chain
12			55	180
15			50	200
18			47	230
24	69	427	42	270
30	61	479	38	305
36	56	527	35	335
42	52	573	32	355
48	49	615	29	370
54	46	650	28	400
60	43	675	27	425
72	40	754	25	480
84	37	813		
96	34	855		

Table 27.3.2 Elevator Selection Table

Material	Class Material	Average Weight[a] (lb/ft³)	Elevator Type	Elevator Class
Acid phosphate (pulv.)	A	60	C, S	Chain
Alum (lumpy)	NA	55	C, S	Chain or belt
Aluminum chips	A	15	C	Chain or belt
Aluminum oxide	A	120	C	Chain or belt
Ashes	A	40	S[b]	Chain or belt
Asphalt (crushed)	A	45	C, S	Chain or belt
Bakelite (powdered)	NA	35	C	Chain or belt
Baking powder	A	55	C, S	Chain or belt
Barley[c]	NA	38	S	Chain or belt
Bauxite (aluminum ore)	A	80	S[b], C, SC	Chain or belt
Beans[c]	NA	50	S	Chain or belt
Bones (crushed)	NA	40	S[b]	Chain or belt
Borax (powdered)	A	53	S	Chain or belt
Bran[c]	NA	16	S	Chain or belt
Brewer's grain (dry)[d]	NA	28	S[d]	Chain or belt
Brewer's grain (wet)	NA	55	S	Chain or belt
Carbon black (pelletized)	NA	25	C, S[b]	Chain or belt
Cement clinker	A	75–80	C, SC	Chain or belt
Cement (fluffed)	A	65–70	C, S[b]	Chain or belt
Chalk (crushed)	A	90	C, S	Chain or belt
Charcoal	A	15–34	C	Chain or belt
Cinders, blast furnace	A	57	C, S[b]	Chain or belt
Clay (dry)	A	63	C, S	Chain or belt
Coal	NA	50	C, S, SC	Chain or belt
Coffee beans	NA	25–30	C, S	Chain or belt
Coke	A	25–40	C, S	Chain or belt
Cork	NA	12–15	C	Chain or belt
Corn (shelled)[c]	NA	45	S	Chain or belt
Cornmeal[c]	NA	40	S	Chain or belt
Cottonseed (dry)	NA	25	S	Chain or belt
Cottonseed meal	NA	38	S	Chain or belt
Dolomite (crushed)	A	95	C, S, SC	Chain
Feldspar (ground)	A	75	C, S[b]	Chain or belt
Flaxseed[c]	NA	45	S	Belt
Flour[c]	NA	35	S	Belt
Fuller's earth (raw)	NA	40	S	Belt

Table 27.3.2 (*continued*)

Material	Class Material	Average Weight[a] (lb/ft³)	Elevator Type	Elevator Class
Fuller's earth (burnt)	A	40	C, S	Belt
Fuller's earth (oily)[d]	NA	60	S[b]	Belt
Glass batch	A	90	C, S	Belt
Grain	NA	30–40	S	Chain or belt
Gravel	A	100	C, S, SC	Chain or belt
Gypsum (crushed)	A	100	C, S[b], SC	Chain or belt
Hops, spent (dry)	NA	35	S	Chain or belt
Hops (wet)	NA	50	S	Chain or belt
Ice (crushed)	NA	35–40	C, S[b]	Chain
Iron ore (crushed)	A	110–150	C, SC	Chain or belt
Lime (ground)	A	64	C, S, SC	Chain or belt
Lime (hydrated)	NA	40	C	Chain
Limestone (pulverized)	A	70	C, S, SC	Chain or belt
Limestone (crushed)	A	96	C, SC	Chain or belt
Malt, whole (dry)	NA	20–30	S	Chain or belt
Malt (wet)[d]	NA	60–65	S	Chain or belt
Oats[c]	NA	26	S	Belt
Phosphate rock	A	75–85	C, SC	Chain or belt
Plastics	NA	—	C, S	Chain or belt
Quartz (crushed)	A	100	C, SC	Chain or belt
Rice (hulled)[c]	NA	45	S	Chain or belt
Rice (rough)	A	36	S	Chain or belt
Rubber (ground)	NA	23	C, S	Chain or belt
Rye[c]	NA	45	C, S	Chain or belt
Salt (coarse)[d]	NA	45	S[b]	Chain or belt
Salt (cake)[d]	NA	85	S[b]	Chain or belt
Salt (fine)[d]	NA	80	S[b]	Chain or belt
Sand (dry)	A	100	C, S	Chain or belt
Sand (damp)	A	120	C, S	Chain or belt
Sand (foundry)	A	100	S[b]	Belt
Sand (silica)	A	100	C, S	Belt
Sawdust	NA	12	S	Chain or belt
Shale	A	92	C, S	Chain or belt
Slag, furnace	A	70	C, SC	Chain or belt
Soda ash (heavy)	A	60	C, S	Chain or belt
Soda ash (fluffy)	A	30	P	Chain
Soybeans (cracked)[e]	A	32–36	S	Chain or belt
Soybeans (whole)[e]	A	45–50	S	Chain or belt
Soybean (cake)	A	42	S	Chain or belt
Soybean flour	NA	27	S	Chain or belt
Soybean meal	NA	40	S	Chain or belt
Starch[c]	NA	45	S	Chain or belt
Stone (crushed)[f]	A	100	C, SC	Chain or belt
Sugar (raw)	NA	60	S[g]	Chain
Sugar (refined)	NA	55	S[b]	Chain or belt
Sulphur	NA	90–125	S[b]	Chain or belt
Tanbark (ground)[d]	NA	55	S[b]	Chain
Wheat[c]	NA	48	S	Chain or belt
Wood chips	NA	10–30	C, S[b]	Belt

[a] The weight per cubic foot of material varies widely, depending upon the size of material and condition (moisture content, etc.); therefore the weight per cubic foot should be determined as accurately as possible. The weights listed here are average weights most frequently encountered.

[b] A fixed-bearing boot is recommended where material is lumpy or tends to pack, cake, or build up in the boot.

[c] Light-gauge steel Salem elevator buckets may be substituted for the malleable iron buckets and the number of plies in the belt reduced when handling materials of this nature.

[d] When handling foods or corrosive materials, chain and buckets, either specially plated or made from special materials, as well as specially prepared belts, may be required to protect chain and buckets and to prevent contamination of food.

[e] Soybean hulls are extremely abrasive.

[f] Use class S elevator for crushed stone only when the material contains lumps not over $\frac{1}{2}$ in.

[g] Use Style C malleable iron buckets.

Table 27.3.3 Vertical Spaced-Bucket Elevators, Chain-Mounted Style AA Buckets, Centrifugal Discharge

Maximum Size Pieces		Capacity[a] (ton/hr) for Given Material Weight (lb/ft³)				Buckets Style AA M. Iron		
Uniform Size	10% of Whole[b]	25	50	75	100	Size (in.)	Spacing (in.)	Chain Speed (ft/min)
$\frac{3}{4}$	3	7	14	21	28	8 × 5	16	250
1	$3\frac{1}{2}$	12	25	37	50	10 × 6	16	250
$1\frac{1}{4}$	4	20	40	60	80	12 × 7	16	250
$1\frac{1}{2}$	$4\frac{1}{2}$	28	55	80	110	14 × 8	18	250
$1\frac{3}{4}$	$4\frac{1}{2}$	31	60	95	125	16 × 8	18	250

[a] Buckets filled to approximately 75% of catalog rating. Horsepower at head shaft based on buckets 100% full.
[b] Mixed with fines.

Table 27.3.4 Vertical Spaced-Bucket Elevators, Belt-Mounted, Style AA Buckets, Centrifugal Discharge

Maximum Size Pieces		Capacity[a] (ton/hr) for Given Material Weight (lb/ft³)				Buckets Style AA M. Iron		Belt	
Uniform Size	10% of Whole[b]	25	50	75	100	Size (in.)	Spacing (in.)	Synthetic Carcass Width (in.)	Speed (ft/min)
$\frac{3}{4}$	3	9	18	28	35	8 × 5	13	9	260
1	$3\frac{1}{2}$	13	26	40	50	10 × 6	15.5	11	260
$1\frac{1}{4}$	4	22	44	65	90	12 × 7	15.5	14	265
$1\frac{1}{2}$	$4\frac{1}{2}$	30	60	90	120	14 × 8	18	16	265
$1\frac{3}{4}$	$4\frac{1}{2}$	34	68	100	135	16 × 8	18	18	265

[a] Buckets filled to approximately 75% of catalog rating. Horsepower at head shaft based on buckets 100% full.
[b] Mixed with fines.

All of these buckets are commonly available in cast malleable iron and plastic. They are also available in fabricated steel. The Style AA when fabricated from light-gauge steel is commonly referred to as "Salem Steel" bucket.

Tables 27.3.3 and 27.3.4 contain typical data for centrifugal-discharge elevators, chain and belt types, when equipped with Style AA buckets. Larger elevators of this type are utilized; however, analysis usually indicates a different style elevator to be more economical.

27.4 CONTINUOUS-BUCKET ELEVATORS—STANDARD TYPE

Continuous-bucket elevators have buckets mounted in a continuous manner on engineered class chain or bucket-elevator-type belting (Fig. 27.4.1). Continuous-bucket elevators are used where judgment indicates slow-moving operation or higher capacities than available with standard centrifugal-discharge elevators are required. Although speeds are usually about 50% of centrifugal elevators, closer spacing of higher-capacity buckets can give greater capacity than a centrifugal elevator of the same relative overall size. The material to be conveyed is directed to the buckets through an internal loading leg and is discharged over the face of the preceding bucket while passing around the head wheel.

Continuous elevators may be used to handle the same kinds of material as the centrifugal type; however, they are especially recommended for handling lumpy, abrasive, and friable materials, or materials that would be difficult to pick up in the boot.

General design features include steel fabricated buckets with flat sloping fronts. Since the buckets do not load by digging, reinforced lips and fronts are not required for most buckets handling the majority of materials. For extra-wide buckets and buckets handling highly abrasive materials, reinforcements and braces are frequently used.

In this type of elevator, the steel buckets, almost touching one another, are bolted on belt or chain. Normally when mounting buckets more than 16 in. wide, two strands of chain are used. As each bucket discharges onto the back of the preceding bucket, the material slides away and thus does not depend on centrifugal action for discharging. The holes for attaching the buckets to the chain or belt are placed in the bottom half of the bucket, thus allowing the lip of the bucket to rise as it goes over the head pulley or sprocket. This action delays the discharge slightly and allows the preceding bucket to come into position for correctly unloading the elevator.

The integral loading leg directs the material into the buckets after they have passed around the foot wheel and are on the ascending side. Thus the buckets are not required to dig the material out of the boot. Some material, especially fines, can be expected to spill into the boot bottom and a sweeping action by the buckets does occur.

Special attention must be given to assure that the height of the loading point is sufficient to allow an adequate number of buckets to be exposed to the loading leg. If the loading zone is not properly designed, the material handling characteristics of the elevator will be lost.

Tables 27.4.1 and 27.4.2 contain typical data for continuous bucket elevators, chain and belt types,

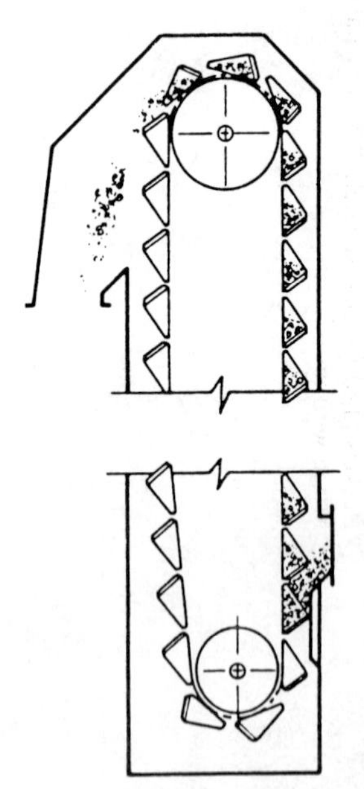

Fig. 27.4.1 Continuous-bucket elevator.

Table 27.4.1 Vertical Continuous-Bucket Elevators, Chain-Mounted, Style D Buckets

Maximum Size Pieces		Capacity[a] (ton/hr) for Given Material Weight (lb/ft^3)				Buckets Style D 10 Ga.		
Uniform Size	10% of Whole[b]	25	50	75	100	Size (in.)	Spacing (in.)	Chain Speed (ft/min)
3/4	2 1/2	10	20	30	40	8 × 5 × 7 3/4	8	160
1	3	15	30	45	60	12 × 6 × 11 3/4	12	160
1 1/2	3	25	50	80	105	12 × 8 × 11 3/4	12	160
1 3/4	4 1/2	30	60	92	125	14 × 8 × 11 3/4	12	160
2	5	35	70	105	140	16 × 8 × 11 3/4	12	160

[a] Buckets filled to approximately 75% of catalog rating. Horsepower at head shaft based on buckets 100% full.
[b] Mixed with fines.

Table 27.4.2 Vertical Continuous-Bucket Elevators, Belt-Mounted, Style D Buckets

Maximum Size Pieces		Capacity[a] (ton/hr) for Given Material Weight (lb/ft^3)				Buckets Style D 10 Ga.		Belt	
Uniform Size	10% of Whole[b]	25	50	75	100	Size (in.)	Spacing (in.)	Synthetic Carcass Width (in.)	Speed (ft/min)
3/4	2 1/2	15	30	45	60	8 × 5 × 7 3/4	8	9	240
1	3	22	45	65	90	12 × 6 × 11 3/4	12	14	240
1 1/2	4	40	80	120	160	12 × 8 × 11 3/4	12	14	240
1 3/4	4 1/2	45	90	140	185	14 × 8 × 11 3/4	12	16	240
2	5	55	110	160	210	16 × 8 × 11 3/4	12	18	240

[a]Buckets filled to approximately 75% of catalog rating. Horsepower at head shaft based on buckets 100% full.
[b]Mixed with fines.

when equipped with standard D style buckets. Bucket widths of 24 and 26 in. are commonly used; however, analysis may indicate a different style elevator to be more economical.

27.5 SUPER-CAPACITY BUCKET ELEVATORS

Super-capacity elevators are a special type of continuous-bucket elevator. Instead of mounting the buckets on chain or belting through bolts in the back of the buckets, they are end-mounted between two strands of chain (Fig. 27.5.1). Usually medium- to heavy-gauge fabricated steel buckets and high-strength, high-grade roller or alloy steel knuckle chain is required for this kind of elevator. By bolting the chain to the ends of the bucket, the super-capacity bucket can be designed to project on both sides of the chain centerline (Fig. 27.5.2). This feature substantially increases the capacity per foot of bucket–chain assembly over the standard continuous elevator. It also precludes utilizing belting in the conveying assembly. Usually, a more functional high-strength elevator can be designed than would be achieved by long projection or extra-wide continuous style buckets, and where correctly applied, a smaller overall machine will be achieved.

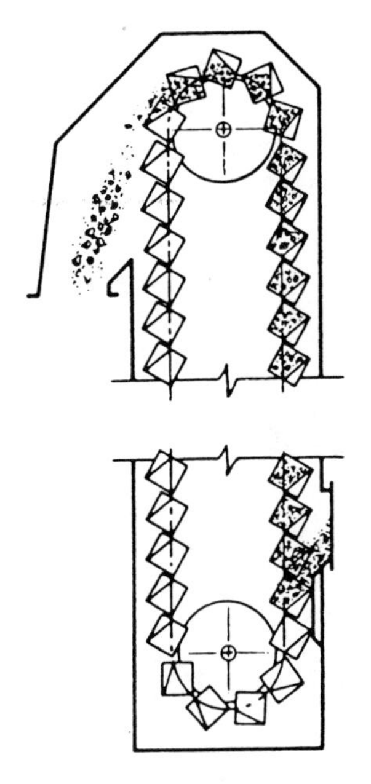

Fig. 27.5.1 Super-capacity bucket elevator.

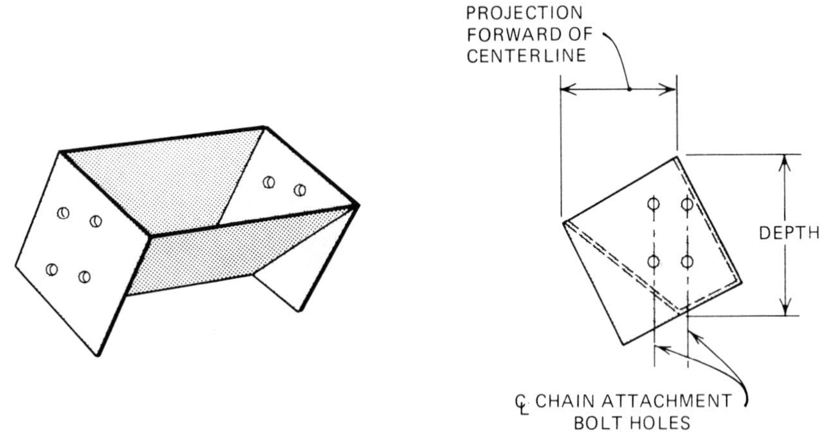

Fig. 27.5.2 Super-capacity bucket. Bucket projects on both sides of chain centerline. An individual strand of chain bolts to each end of bucket.

Table 27.5.1 **Super-Capacity Bucket Elevators, Two Chains, End-Mounted Buckets, Steel Knuckle Chains**

Maximum Size Pieces		Capacity[a] (ton/hr) for Given Material Weight (lb/ft^3)		Buckets Steel Super Capacity		
Uniform Size	10% of Whole[b]	50	100	Size (in.)	Spacing (in.)	Chain Speed (ft/min)
$2\frac{1}{2}$	6		150	$16 \times 8\frac{3}{4} \times 11\frac{5}{8} \times 10$ ga.	12	100
$2\frac{1}{2}$	6	75		$16 \times 8\frac{3}{4} \times 11\frac{5}{8} \times 10$ ga.	12	100
$2\frac{1}{2}$	6		200	$20 \times 8\frac{3}{4} \times 11\frac{5}{8} \times 10$ ga.	12	100
$2\frac{1}{2}$	6	100		$20 \times 8\frac{3}{4} \times 11\frac{5}{8} \times 10$ ga.	12	100
$3\frac{1}{2}$	8		280	$16 \times 12\frac{3}{4} \times 17\frac{5}{8} \times \frac{3}{16}$	18	125
$3\frac{1}{2}$	8	140		$16 \times 12\frac{3}{4} \times 17\frac{5}{8} \times \frac{3}{16}$	18	125
$3\frac{1}{2}$	8		350	$20 \times 12\frac{3}{4} \times 17\frac{5}{8} \times \frac{3}{16}$	18	125
$3\frac{1}{2}$	8	175		$20 \times 12\frac{3}{4} \times 17\frac{5}{8} \times \frac{3}{16}$	18	125
$3\frac{1}{2}$	8		400	$24 \times 12\frac{3}{4} \times 17\frac{5}{8} \times \frac{3}{16}$	18	125
$3\frac{1}{2}$	8	200		$24 \times 12\frac{3}{4} \times 17\frac{5}{8} \times \frac{3}{16}$	18	125
$3\frac{1}{2}$	8		520	$30 \times 12\frac{3}{4} \times 17\frac{5}{8} \times \frac{3}{16}$	18	125
$3\frac{1}{2}$	8	260		$30 \times 12\frac{3}{4} \times 17\frac{5}{8} \times \frac{3}{16}$	18	125
$3\frac{1}{2}$	8		600	$36 \times 12\frac{3}{4} \times 17\frac{5}{8} \times \frac{3}{16}$	18	125
$3\frac{1}{2}$	8	300		$36 \times 12\frac{3}{4} \times 17\frac{5}{8} \times \frac{3}{16}$	18	125

[a]Buckets filled to approximately 75% of catalog rating. Horsepower at head shaft based on buckets 100% full.
[b]Mixed with fines.

Super-capacity elevators are used to handle a wide range of friable, heavy, or abrasive materials ranging from lumps to fines, or where high conveying capacity is required. Common applications include fertilizer prills, which require gentle handling at high capacities, to bulky materials of 10–12 in. They have been built for capacities of 1000 ton/hr. This special type continuous elevator operates at conveying speeds from 150 ft/min in standard sizes to 50 ft/min in the very large sizes or when special handling conditions require low speed. These elevators have been built with bathtub-size buckets.

Table 27.5.1 contains typical data for a representative range of common super-capacity elevators. This class elevator is commonly sized and engineered for particular applications and capacities. Therefore, the data in the table should not be considered as limiting parameters. And manufacturers should be consulted for application.

27.6 SPECIAL HIGH-SPEED CONTINUOUS ELEVATORS

Numerous engineered class high-speed continuous elevators have been built for special applications. Several designs have been standardized by manufacturers. The units considered here are primarily applicable to handling bulk raw materials and should not be considered in the same category as high-speed grain elevators. It should be noted that high-speed grain elevators have been utilized, sometimes with modifications, in conveying specific bulk raw materials.

One of the more common high-speed continuous elevators is referred to by many manufacturers as the cement mill elevator (Fig. 27.6.1). Cement mill elevators incorporate features from both centrifugal-discharge and continuous-type elevators. This type elevator is used to handle readily free-flowing materials, but where gentle handling of materials is not required. Typical materials conveyed are raw mix, cement, and coal.

Cement mill elevators are furnished with style AC buckets, usually mounted on a single strand of engineered class chain. Style AC buckets are designed to load either by digging in the boot or continuous loading via a loading leg. Discharge is by centrifugal force as the buckets pass over the head wheel. The buckets are usually cast ductile iron, but can be cast aluminum, plastic, or fabricated steel. They are generally heavy duty with reinforced lip. Conveying chain is attached by bolting through the back of the bucket.

Due to the long bucket projection, high bolt pull-out forces generally preclude use of belting.

When the buckets are closely spaced, material is directed to the buckets through a loading leg. If the buckets are more widely spaced, loading is accomplished through a combination of scooping material from the boot and loading through the loading leg.

Table 27.6.1 lists typical data for cement mill elevators. When application parameters include lower capacities, usually a standard centrifugal-discharge elevator is used. When higher capacities are required, a special type of the cement mill elevator, the saddlebag elevator, may be used.

By projecting the saddlebag bucket to both sides of the chain centerline, capacity of the AC bucket, for a given width, is increased by the bucket volume projected behind the chain, in a manner similar to the capacity increase from continuous to super-capacity buckets.

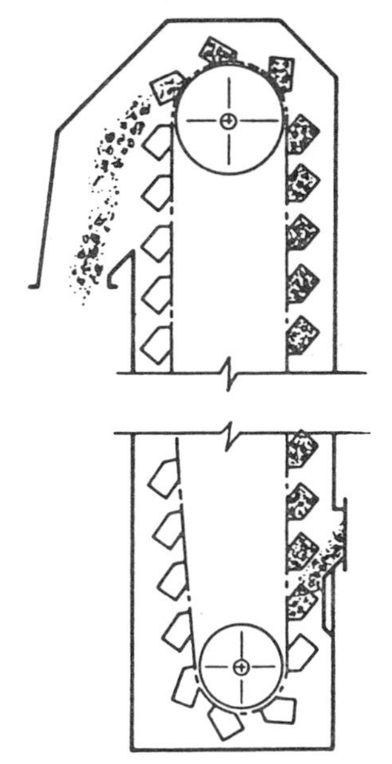

Fig. 27.6.1 Cement mill bucket elevator.

Table 27.6.1 Cement Mill Elevators, Chain-Mounted, Style AC Buckets

Maximum Size Pieces		Capacity[a] (ton/hr) for Given Material Weight (lb/ft^3)		Buckets Style AC M. Iron		
Uniform Size	10% of Whole[b]	50	100	Size (in.)	Spacing (in.)	Chain Speed (ft/min)
$1\frac{1}{2}$	$4\frac{1}{2}$	55	110	12 × 8	18	265
$1\frac{1}{2}$	$4\frac{1}{2}$	80	160	12 × 8	12	265
$1\frac{3}{4}$	$4\frac{1}{2}$	75	150	16 × 8	18	265
$1\frac{3}{4}$	$4\frac{1}{2}$	110	220	16 × 8	12	265
2	5	120	240	18 × 10	18	265
2	5	180	360	18 × 10	12	265

[a] Buckets filled to approximately 75% of catalog rating. Horsepower at head shaft based on buckets 100% full.
[b] Mixed with fines.

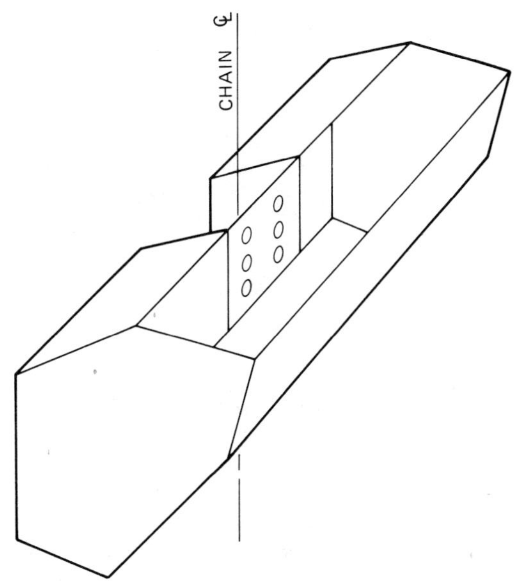

Fig. 27.6.2 High-speed ACS saddlebag bucket. End view shows internal material pockets and recess for mounting chain.

In the saddlebag bucket, the chain, or chains, are bolted to the back of the bucket, but assembles in a recess formed by projecting pockets of the bucket (Fig. 27.6.2). The buckets are mounted continuously. Loading is through loading leg on the upward strand and discharge is by centrifugal force.

These units are generally used for extra-high lifts at high-capacity requirements. For functional design in high-lift, high-capacity requirements, relatively high chain conveying speeds of 250–350 ft/min are frequently used. At these high speeds, special design features must be introduced at both feed and discharge.

The most recent bucket elevator introduced for elevating bulk raw materials at high speed and extra-high lifts has been the high-capacity, high-speed continuous belt-type elevator.

Buckets are generally designed for continuous loading through a loading leg and for centrifugal discharge. Special attention must be given to design of boot and associated area for proper material transfer.

Lifts in excess of 250 ft and capacities of more than 15,000 ft^3/hr of roller mill product, consisting primarily of fine-mesh limestone, have been constructed.

The crucial component of the elevator is the steel-cable-reinforced belting. With this belt, greater lifting strengths and higher speed combinations can be achieved than would be practical with chain. The belting must be specially designed for the application with particular attention to rubber compounds, method of bucket mounting, and belt splicing. Key advantages of steel-cable-reinforced belting over ply types are increased load rating, the relatively small amount of stretch in high-lift heavy-load applications, and better strength and aging characteristics.

27.7 MARINE-LEG BUCKET ELEVATORS

The marine-leg type of centrifugal discharge is used for unloading bulk cargoes from ships and barges. It consists of a vertical elevator leg that digs into the bulk cargo and usually discharges onto a horizontal conveyor. The conveyor takes the material to the dock for warehousing or processing. The elevators are supported on a pivoting boom frame so that the elevator can be raised or lowered as it digs into the bulk cargo at the varying positions of the cargo in the hold and under varying tide conditions. The elevators are usually arranged so that they can be moved thwartwise across the hatch openings. Their usefulness depends on their being able to dig at the full rating of the elevator, and this requirement means that the material under the decks has to be brought to a position where the elevator can reach it.

Power shovels or bulldozers working in the ship continually crowd the material to the boot of the elevator. Drag scrapers operating between the elevator boot and the skin of the ship have been designed with remote-control points located on deck. These remotely controlled drag scrapers are frequently operated by compressed air, by means of electrically actuated air valves. This arrangement gives excellent flexibility in the control of the scraper buckets, which operate between blocks placed on pad eyes on the frames of the ship and blocks mounted on the lower part of the elevator casing. The elevator buckets are of the high-speed design. They are usually mounted on a heavy precision elevator chain, and the drive is arranged for an operating speed of 350–400 ft/min.

For complete flexibility, the marine-leg elevators have been mounted on traveling gantry frames so that they can dig from the several hatches without having to move the ship. Complicated structural steel design problems must be solved to provide stability for the elevator and structure in all operating positions, and to withstand high winds when the hinged boom is in the stowed position. The boom is usually balanced by a counterweight system designed to pick up the correct amount of counterweight for all the operating positions of the boom and the correct amount of counterweight for all the operating positions of the boom and elevator.

A moderate-size marine-leg elevator has a capacity of somewhat less than 250 ton/hr of free-digging grain or raw sugar.

27.8 GRAIN ELEVATORS

The most common elevator used at this time to handle small grains such as corn and wheat has spaced buckets mounted on belting. There are two subdivisions in this type of elevator, high speed and low speed.

High-speed grain elevators (Fig. 27.8.1) are available in standard designs up to 12,500 ft^3/hr, 10,000 bushels/hr. Special designs can handle up to 40,000 bushels/hr and operate at up to 750 ft/min. Since operating speeds are generally quite high, loading is generally on the returning or downside of the elevator. This method of loading reduces degradation and generally results in a bucket, or as more commonly referred to, cup, loading of approximately 75–90% available volume. With a properly designed boot, the free-flowing grain is readily carried around the boot pulley and up the ascending leg. By feeding on the downside, less degradation of grain occurs at high operating speeds than would be found if feed were introduced to the rapidly ascending cups.

Discharge is by centrifugal force. As speeds increase, special care must be taken in design of the head section. At higher speeds the retaining capability of the cup is less effective because the cup begins its arc around the head pulley. Therefore, the head casing must provide an open path for the grain from an early discharge through the head section discharge flange.

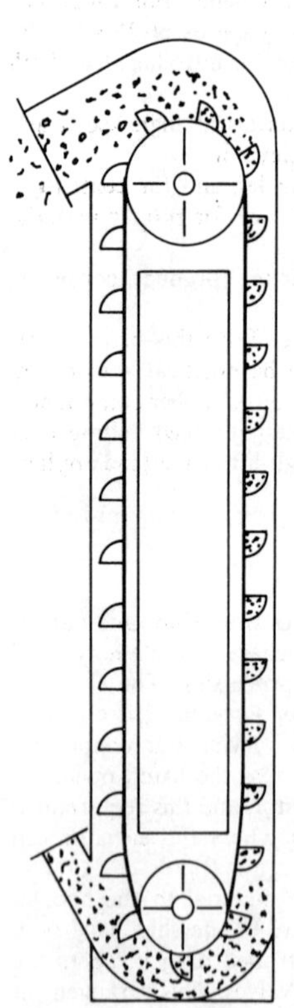

Fig. 27.8.1 High-speed grain bucket elevator.

Commonly the ascending, elevating, and return run of belting/cup assembly is enclosed in individual intermediate casings or legs. These casings are generally self-supporting, but must be laterally restrained. Lifts approaching 300 ft are not uncommon.

To achieve higher capacities at reduced grain degradation, low-speed, high-volume elevators are available. By utilizing multiple rows of large steel buckets on a wide belt, some of the operating problems associated with high-speed elevators, such as belt whip and excessive dusting, are reduced.

Low-speed elevators generally operate at 350 ft/min. At this speed, loading is still on the down-side and discharge is centrifugal. With a double row of 36 in. × 21 in. buckets, capacities of up to 60,000 bushels/hr are achieved. However, higher live loads and machinery, belting, and bucket loads set a practical lift height at this time of less than 200 ft.

Either type of elevator will service a typical grain silo of 120–170 ft height.

In the design and application of grain elevators, special care must be taken to exclude foreign materials from the equipment, and to control dusting.

If considered applicable for relief of pressure buildup leading to potential explosions, venting should be provided. Other safety devices and operating procedures should also be included as required.

CHAPTER 28
PNEUMATIC CONVEYORS

FREDERICK A. ZENZ

Frederick A. Zenz, Inc.
Garrison, New York

IVAN STANKOVICH

Ivan Stankovich, Inc.
Piedmont, California

FRANK GERCHOW

Koppers Co. Inc.
Sprout, Waldron Div.
Muncy, Pennsylvania

M. R. CARSTENS

Atlanta, Georgia

28.1 FLUIDIZATION ENGINEERING FUNDAMENTALS

Frederick A. Zenz

Pneumatic handling presents a safe, reliable, economical, and flexible way of handling materials, and one that is harmless to the environment. One of the main functions of pneumatic handling is transfer of material from one location to another. Other functions are: heat exchange, mixing, drying, particle growth, adsorption, synthesis reaction, cracking and reforming of hydrocarbons, carbonization and gasification, calcining and clinkering, and gas–solid reactions.

The idea of doing useful work with air is very old. The first pneumatic conveying systems appeared at the end of the nineteenth century. But only since 1945 have pneumatic handling systems received proper attention.

Condensed within these pages is an enormous background of literature (select examples are cited in the text) with which the reader is urged to familiarize himself or herself in the event questions arise as to the interpretation of any of the equations or graphs. The material is presented solely from the viewpoint of providing the basic tools for design calculations.

28.1.1 Flow Characteristics of Bed Solids

Angular Properties

The two principal characteristic angles exhibited by bulk solids, pertinent to their behavior in fluidized processes, are the angle of repose and the angle of internal friction. The former represents an equilibrium between stationary bulk solids and the surrounding medium, usually recorded as the slope of a poured pile of the solids relative to the horizon. The latter has also been referred to as the angle of shear, the angle of draw, the wedge of maximum thrust, and so on, among the various engineering disciplines. It represents an equilibrium between moving bulk solids and stationary bulk solids [1]. Neither of these angles are calculable; they should be measured experimentally for any specific material or, at best, estimated from data for other similar solids [1]. In general, angles of repose fall in the range of 35–40° and angles of internal friction predominantly in the range of 65–75°.

Gravity Flow

The gravity flow of bulk solids into an orifice or slot has been demonstrated [2] to follow the identical relationships developed for liquids and gases represented by the dimensionless relationship:

$$W^2 = \frac{g \rho_B h \Delta\rho}{\tan \alpha} \tag{28.1.1}$$

where: W = weight of solids per unit time times net unit area
g = gravitational acceleration
ρ_B = solids bulk density, in weight/unit volume
h = narrowest dimension of slot or diameter of orifice or port (minus 1.5 times mean particle diameter)
$\Delta\rho$ = Difference between solids bulk density and density of surrounding gaseous or liquid medium, in weight/unit volume
α = Bulk solids angle of internal friction, in degrees

In lieu of experimental data, α may be taken as 70°.

Pressured Flow

If pressure is exerted on bulk solids above a port, by a superimposed cocurrent flow of fluid through their interstices, the solids flow is increased, above that given by (28.1.1), in proportion to this relative fluid pressure above and below the port. This has been correlated empirically [3] as shown in Fig. 28.1.1 in terms of the multiple of the rate calculable from (28.1.1).

A more proper correlation taking into account true interstitial fluid velocity profiles, as opposed to the simplistic plug flow differential in the abscissa of Fig. 28.1.1, has never been reported. Such sophistication is generally found unnecessary in view of the substantial enhancement of flow resulting from relatively small pressure differentials.

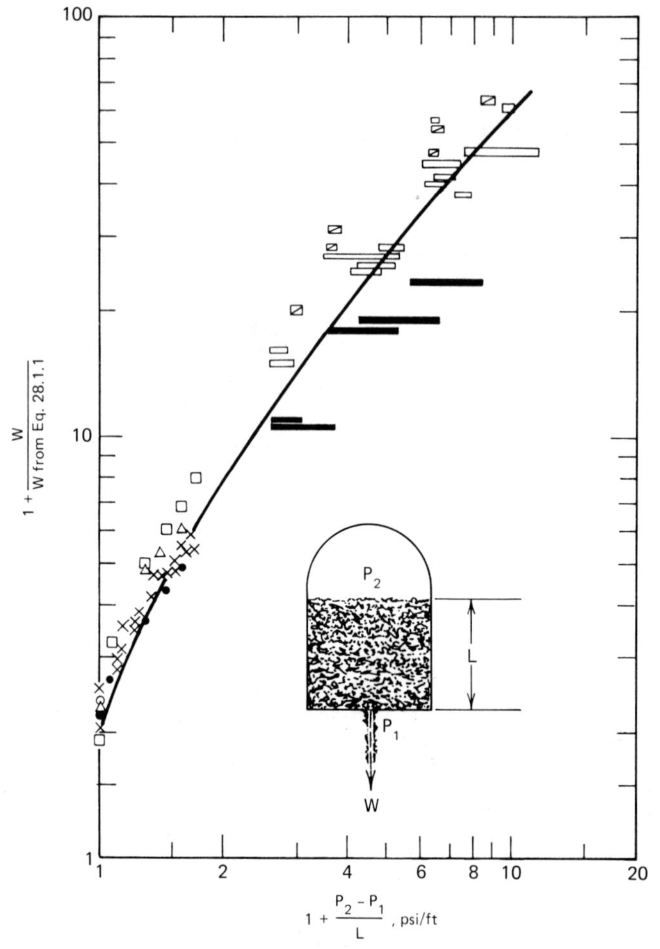

Fig. 28.1.1 Effect of pressure differential on bulk solids flow rate.

Fluidized Flow

The flow of fluidized solids through ports has been shown [4] to obey the conventional orifice equation with a coefficient of 0.445 so that if $V = C_0(2g\Delta H)^{1/2}$ then:

$$W = 0.445\rho_B\sqrt{2gH} \tag{28.1.2}$$

where H = depth or height of column of fluidized solids above the port.

28.1.2 Bed Fluidizing Gas Flow

Incipient Fluidization Velocity

From a hydrodynamic or aerodynamic point of view, incipient fluidization has been defined as the minimum superficial fluid velocity through a bed of solids at which its frictional resistance in passing through the interstices equals the bulk solids weight. Depending on the solids dilatancy (the ability of the particles to reorient into looser yet stable configurations) any additional gas will rise through the bed in the form of bubbles. In the event that the solids exhibit a dilatant characteristic, the higher superficial velocity through the loosest packing at which bubbles first form is referred to as the incipient bubbling velocity. This latter velocity is of principal significance in reaction kinetics; the former should more properly be referred to as the velocity at incipient buoyancy. The prediction of either one requires knowledge of the void fraction ϵ in the beds which in turn relates to particle and bulk densities since

$$\rho_B = \rho_P(1-\epsilon) \tag{28.1.3}$$

where ρ_p = apparent individual particle density (including internal pores, if any).

The superficial fluid velocity corresponding to any bed void fraction at which the pressure drop equals the bed weight has been related to the fluid and particle properties in the generalized correlation [5] of Fig. 28.1.2, which can be expressed analytically for purposes of computer programming in the form:

$$\begin{aligned}\epsilon = 1.4673081 &+ 0.0147726\ (\text{ordinate}) \\ &- 0.26864208\ (\ln \text{abscissa}) \\ &+ 0.18601248\ (\ln \text{ordinate}) \\ &+ 0.00814086\ (\ln \text{ordinate})^2\end{aligned} \tag{28.1.4}$$

Grid Design

The principal criterion for the design of multiport grids is the provision of sufficient pressure drop to assure a near-equal flow of fluidizing medium through each and every port. This has been explored experimentally and demonstrated industrially over perforated plates of all dimensions, to be a minimum

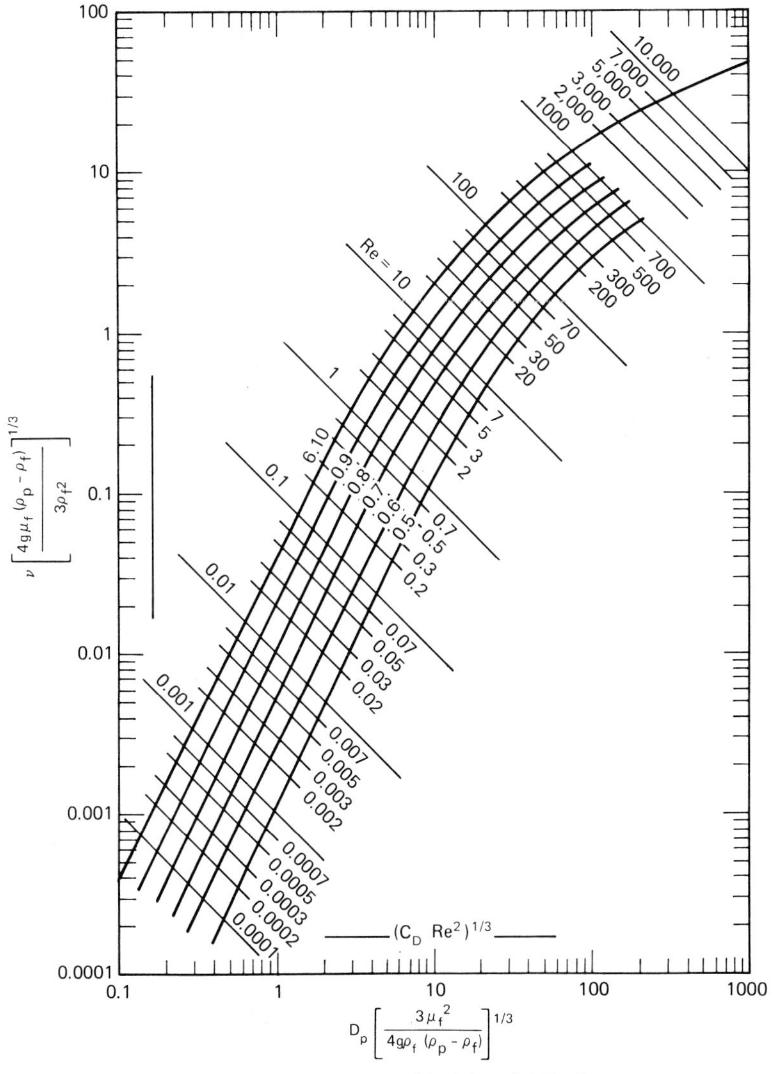

Fig. 28.1.2 Correlation of incipient fluidization.

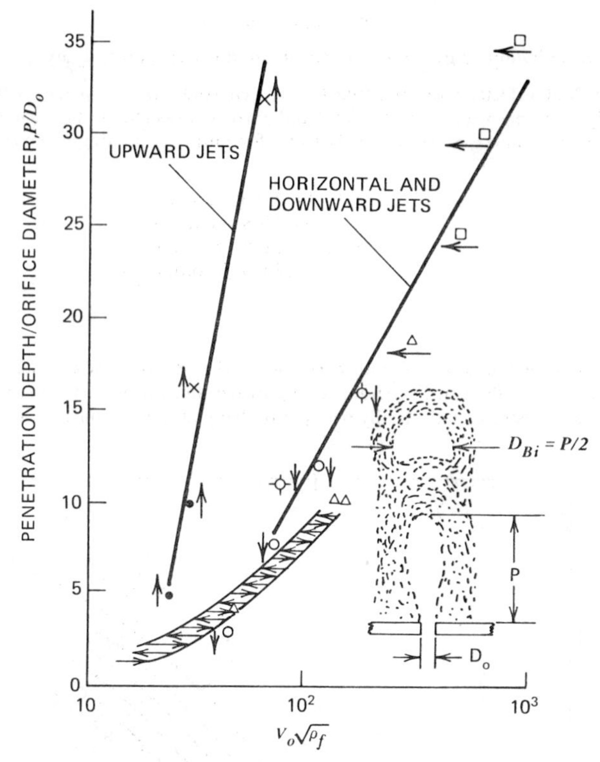

Fig. 28.1.3 Bubble formation at grid ports.

of 30% of the pressure drop through the bed above [6]. Again, the dimensionless orifice equation suffices with the appropriate 0.8 coefficient, because of the bubbling [7] into the downstream side of the port:

$$V_0 = 0.8\sqrt{\frac{2g(0.3L\rho)}{\rho_f}} \tag{28.1.5}$$

where: V_0 = velocity of fluidizing medium through port, distance/unit time
L = depth of bed above grid plate, length
ρ = density of bed above grid plate, weight/unit volume
ρ_f = density of fluidizing medium, weight/unit volume

The total volumetric flow of fluidizing medium divided by the required port velocity from (28.1.5) establishes the maximum hole area. This can then be divided into as many holes (with appropriate diameters) as desired. In general it is desirable to design for the maximum number of the smallest reasonable-size holes, which on plate steel might be limited by the cost of punching, to holes with an aspect ratio no greater than unity.

Bubble Formation

Fluidizing media enter the bed through the grid ports forming prolate spheroid-shaped jet penetrations [8] through whose surfaces the fluid passes at the incipient bubbling velocity [9]. The penetrating length of these jets has been correlated [8] as shown in Fig. 28.1.3. The bubble that results when the jet is pinched off by inflowing bed solids has a diameter equal to half the penetration depth.

Velocity of Bubble Rise

The velocity of rise of gas bubbles in fluidized beds of solids follows the same correlations [10] developed for the rise of gas bubbles in liquids, where, over the range of interest:

$$C_D = \frac{4gD_B(\rho - \rho_f)}{3\rho V_B^2} = 2.67 \tag{28.1.6}$$

where: C_D = drag coefficient, dimensionless
D_B = bubble diameter
V_B = bubble velocity

so that solving (28.1.6) for bubble velocity (because $\rho >>> \rho_f$):

$$V_B = 0.7067\sqrt{gD_B} \tag{28.1.7}$$

Mechanism of Bubble Rise

It has been experimentally demonstrated that bubbles in beds of solids simply represent holes into which the solids at the walls can slide by gravity if their angle of internal friction is less than 90°. The thickness of the down-flowing shell of solids [11] surrounding a rising bubble has also been demonstrated to be a quarter of the bubble's diameter. The bed of solids beyond a radial distance of 0.75 bubble diameters from its center are presumably unaffected by its upward passage.

Bubble Growth

When two bubbles attain so close a proximity that their down-flowing shells of solids touch, they are inevitably drawn into coalescence, forming a common bubble of somewhat larger diameter. Since this bubble will then rise faster, as prescribed by (28.1.7), it will overtake others and continue to grow and accelerate by this mechanism, which is quantitatively definable by the relationship:

$$\frac{D_B}{D_{Bi}} = 0.15\frac{L}{D_{Bi}} + 0.85 \tag{28.1.8}$$

where: D_B = bubble diameter at bed depth L
D_{Bi} = bubble diameter generated at a grid port

Equation 28.1.8 has been shown to be in excellent agreement with experimental data as well as with the more empirical correlations of several investigators [9].

Bubble Stability

The upper surface of a gas bubble in a fluidized bed consists of layers of particles bound by the equivalent of a surface tension force resulting from the flow of the fluidizing medium through its interstices at the incipient fluidizing velocity that is supportive of the bed weight. As evident from Fig. 28.1.2, particularly at Reynolds numbers less than about 5, the curves are essentially parallel so that

$$\nu = \frac{gD_p(\rho_p - \rho_f)C}{18\mu_f} \tag{28.1.9}$$

where: D_p = geometric weight mean particle diameter of bed
C = a fractional constant relating the ordinate of Fig. 28.1.2 at ϵ to the ordinate at $\epsilon = 1$, at any value of the abscissa
μ_f = the viscosity of the fluidizing medium

Equation 28.1.9 can be rearranged to:

$$\frac{18\nu\mu_f}{Cg} = D_p^2(\rho_p - \rho_f) \tag{28.1.10}$$

either side of which bears the units of surface tension. The classical experiments by Rowe and the visible demonstration by Danckwerts [10], in combination with the surface tension of (28.1.10), have well established the stability of the bubble roof; particles do not rain from the roof into the rising bubble.

Bubble Effects on Reaction Kinetics

In order that the gases fed to a fluidized bed undergo reaction they must be purged from the bubbles into the interstices of the surrounding catalyst bed before the bubbles reach and burst through the bed surface. The product of the volume fraction of the feed gases achieving interstitial contact and the catalyst surface area per unit of bed volume, forms a measure of the contact effectiveness from which the relative effects of bubble size, bubble velocity, bed depth and all system variables can be evaluated [12]. Analytically this product takes the dimensionless form:

$$C_{\text{eff}} = aL\left[\frac{3L\nu}{V_B D_B}\left(1 - \frac{\nu}{V}\right) + \frac{\nu}{V}\right] \tag{28.1.11}$$

where: C_{eff} = relative contact effectiveness factor
a = effective catalyst surface area/unit volume of bulk solids
V = total fluidizing gas flow, volume/unit time × unit area

Equation 28.1.11, evaluated at a fixed V and L, when plotted against ν for conventional values of the product $V_B D_B$ based on Fig. 28.1.3 and (28.1.7) and (28.1.8), suggests that the optimum mean particle diameter for a high-conversion gas catalytic reaction is one having an incipient bubbling velocity in the range of 0.1–0.3 ft/sec (3–9 cm/sec) under reaction conditions.

Equation 28.1.11 serves as the basis for scaling up slugging pilot plant reactors to the bed depths required in large freely bubbling reactors to achieve equal contact with the identical catalyst at identical superficial velocities and operating conditions. Such equal values of (28.1.11) under slugging versus freely bubbling conditions lead to a first approximation [13] for sacle-up given by the dimensionless relationship:

$$\frac{L_{ind}}{L_{pp}} = 1.68 \left(\frac{D_B}{D_T}\right)^{0.75} \tag{28.1.12}$$

where: L_{ind} = bed depth required in freely bubbling industrial reactor
L_{pp} = bed depth in slugging pilot plant reactor
D_B = mean diameter of bubbles rising in the freely bubbling reactor
D_T = inside diameter of the pilot plant reactor

28.1.3 Particle Entrainment

Analogy to Vapor–Liquid Equilibria

In a bed of wide particle-size distribution, the smallest particles having the lowest terminal velocity will be preferentially entrained or elutriated. It therefore becomes obvious that a bubbling fluidized bed of mixed particle sizes behaves like a boiling mixture of liquids of different molecular weights [14]. As illustrated in Fig. 28.1.4 there is direct similarity between temperature and velocity, boiling point and terminal velocity, molecular weight and particle size, condenser and cyclone, and splash height and transport disengaging height (TDH). The prediction of entrainment therefore requires first the determination of the equilibrium distribution of the entrained particles in the exiting gas stream and then the absolute amount that will saturate this gas stream. This again suggests analogy to relative volatility and to vapor pressure.

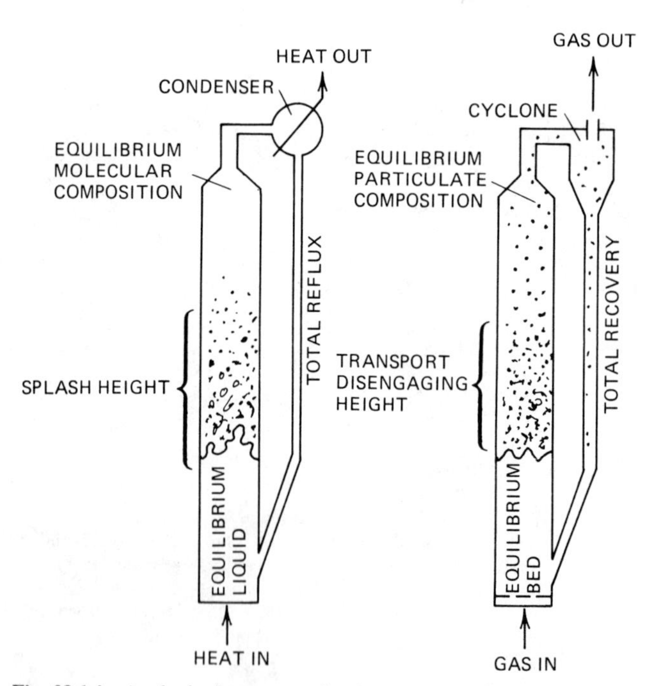

Fig. 28.1.4 Analogies between molecular and particulate equilibria.

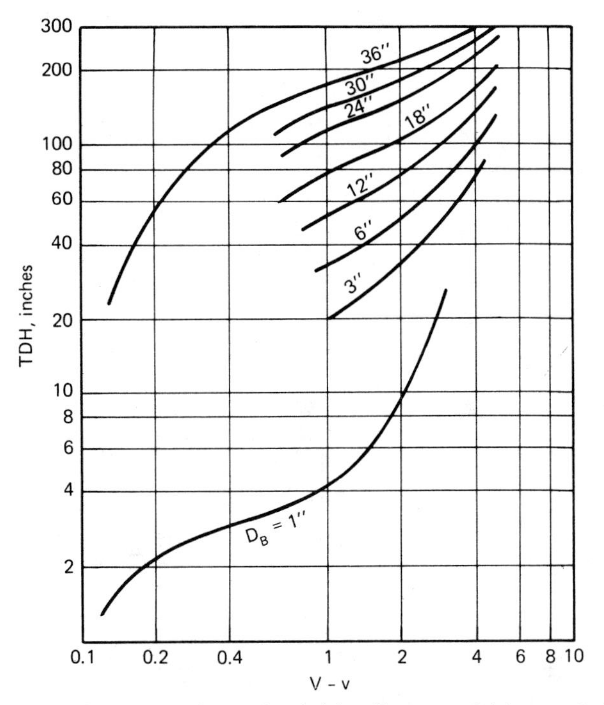

Fig. 28.1.5 Correlation of transport disengaging height. V = superficial gas velocity (ft³/sec) per free area (ftª), in ft/sec; ν = superficial bed surface velocity at onset of bubbling, in ft/sec; $V - \nu$ = ft³/sec of bubble gas per ftª of free cross section; D_B = diameter of bubble bursting at bed surface, in in.

Transport Disengaging Height

As in the case of distillation trays where entrainment of liquid droplets may limit tray spacing, so in the case of fluidized beds an equally empirical correlation determines the level above which particle agglomerates will all have fallen back, leaving solely an individual particle-saturated stream pneumatically conveying the solids in a dilute suspension [15]. The height TDH above the bed surface beyond which entrainment is constant, is a function of the size and frequency of the bubbles bursting at the surface. This is represented in the empirical correlation of Fig. 28.1.5.

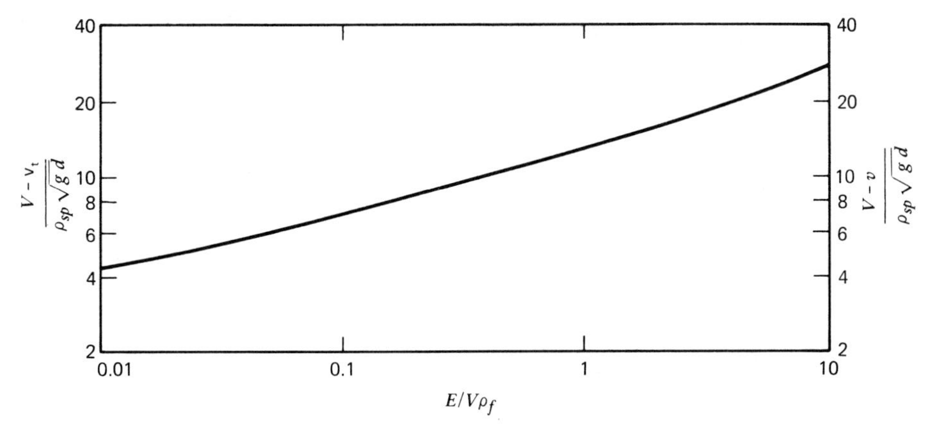

Fig. 28.1.6 Correlation of entrainment rate and composition.

Table 28.1.1 Calculation of the Composition of the Entrainment from a Bed Fluidized at 2 ft/sec Superficial Ambient Air Velocity[a]

Bed Composition		Velocity (ft/sec)		Figure 28.1.6[b]			Entrainment Composition	
Weight Fraction, w	Average Diameter, d (μm)	v_t	$V - v_t$	Ordinate $\frac{V - v_t}{\rho_{sp}\sqrt{gd}}$	Abscissa $\frac{E}{V\rho_f}$	$\frac{wE}{V\rho_f}$	Weight Fraction	Cumulative Weight Fraction Smaller than d
0.02	21	0.1104	1.8896	16.69	2.25	0.0450	0.28518	
0.03	29	0.205	1.795	13.49	1.14	0.0342	0.21673	0.2852
0.05	35	0.292	1.708	11.68	0.71	0.0355	0.22497	0.5019
0.10	45	0.465	1.535	9.26	0.30	0.0300	0.19012	0.7269
0.10	55	0.667	1.333	7.27	0.11	0.0110	0.06971	0.9170
0.10	69	0.990	1.010	4.92	0.0196	0.00196	0.01242	0.9867
0.10	85	1.402	0.598	2.63	0.00137	0.000137	0.00087	0.9992
0.10	102	1.875	0.125	0.50	nil	nil	nil	1.0000
						$\Sigma = 0.157797$	$\Sigma = 1.00000$	
	↑		Entrainment composition above TDH				↑	↑

[a] Particle apparent specific gravity = 2.4.

[b] At values of the ordinate less than 5: Abscissa = 0.0000225 $(\text{Ordinate})^{4.25}$.

Entrainment Composition

The size distribution of the particles entrained from a fluidized bed above TDH is calculable from the bed composition and the left-hand ordinate of Fig. 28.1.6 which in conjunction with the abscissa may be regarded as a relative volatility curve. The calculation procedure is illustrated in Table 28.1.1 carried out in English units; the abscissa and ordinates of Fig. 28.1.6 are, however, dimensionless. In the example of Table 28.1.1 only 60 weight % of the bed consisted of particle subject to entrainment above TDH since the remainder had terminal velocities greater than the superficial bed fluidizing rate. Though the bed contained only 10 weight % smaller than about 29 microns, the entrainment contains about 28 weight % smaller than 29 microns because of the relative elutriability or volatility of the finer sizes.

Entrainment Rate

The fluidized bed can be regarded as a saturation feeder able to supply from its inventory more than the gas can carry, without the particles getting within each other's wakes and condensing out as fall-back agglomerates. The particle saturation of the exiting gas stream at the superficial fluidizing velocity (analogous in an ideal vapor–liquid system to a partial pressure equaling the product of vapor pressure and concentration in the liquid) can be determined from the dimensionless correlation represented by the right-hand ordinate of Fig. 28.1.6. For the example in Table 28.1.1, the abscissa of Fig. 28.1.2 is 4.39 which, at a minimum bubbling voidage of 0.5, yields an abscissa of 0.026 and hence a minimum superficial bubbling velocity of 0.06 ft/sec. From the last column in Table 28.1.1, the geometric weight mean particle diameter of the entrained particles is about 35 μm so that the right-hand ordinate of Fig. 28.1.6 is equal to 13.3, giving an abscissa of 1.1 and hence a total entrainment rate of $0.6 \times 0.0765 \times 1.1 = 0.051$ lb of particles per cubic foot of gas leaving the bed above TDH. The figure 0.0765 represents the gas density, and 0.6 the weight fraction of entrainable particles in the bed.

28.1.4 Particle Separation from Effluent Gases

Alternative Methods and Equipment

Entrained particles must generally be separated from the effluent gases simply to be returned to the bed to maintain a catalyst inventory. This is usually accomplished via cyclones located within the

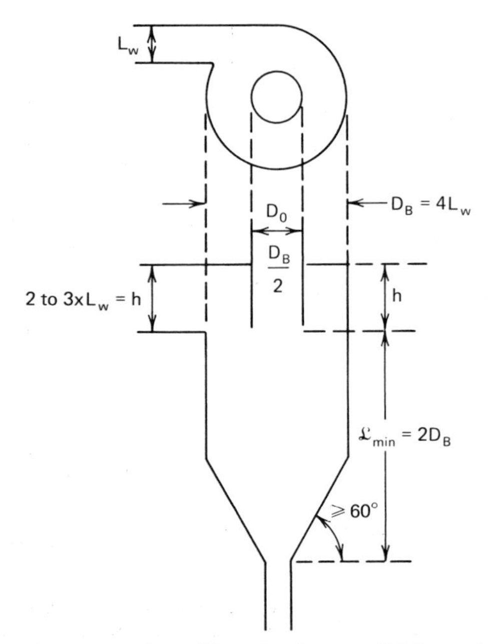

Fig. 28.1.7 Optimum cyclone proportions. V = effective superficial entraining velocity, in ft/sec; v = minimum bed bubbling velocity, in ft/sec; v_t = terminal velocity of particle of diameter d, ft/sec; ρ_{sp} = particle apparent specific gravity; g = 32.2 ft/sec²; d = particle diameter, in ft; D = Geometric weight mean particle diameter of entrained material, ft; ρ_t = entraining gas density, lb/ft³; E = maximum lb of solids entrainable/sec × ft².

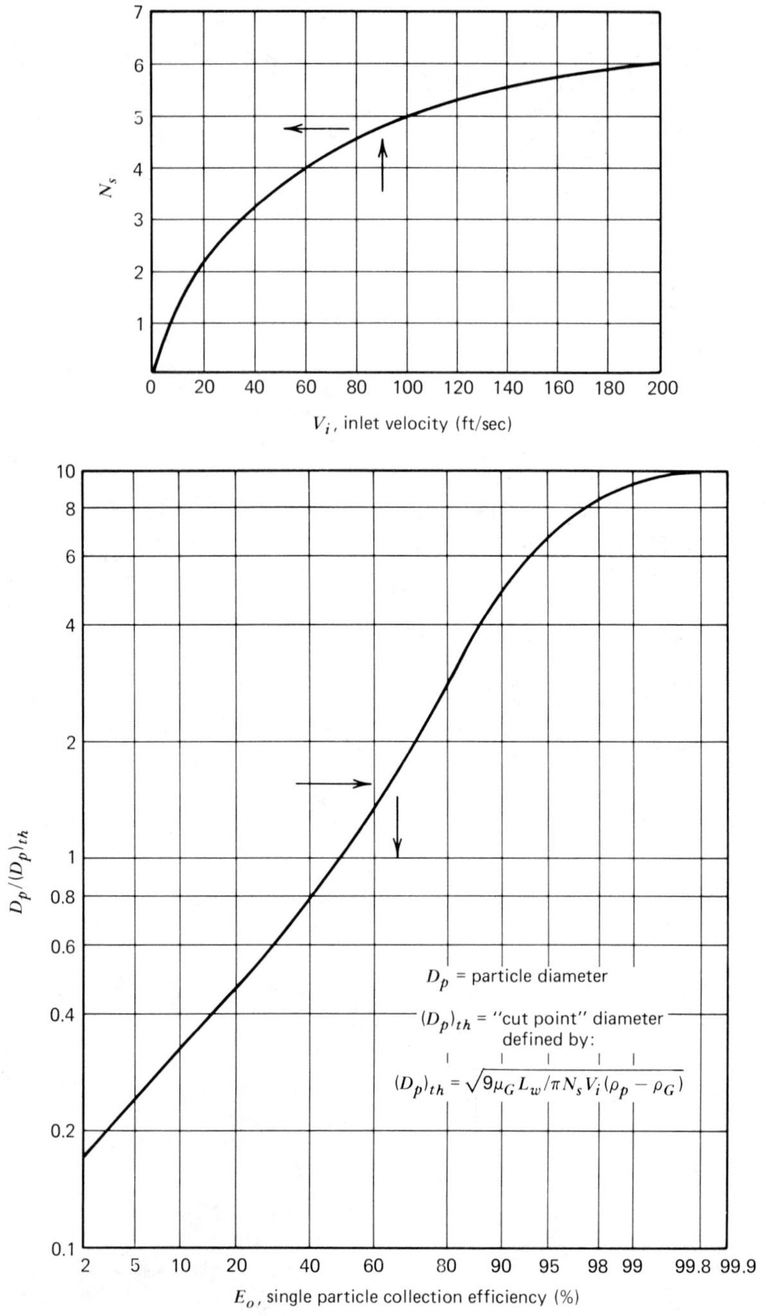

Fig. 28.1.8 Correlation of fractional efficiency of well-proportioned tangential inlet cyclones (see Fig. 28.1.7).

reactor or externally [16]. If one stage of cyclones does not have the economically necessary efficiency to avoid loss of costly catalyst or to maintain the necessary catalyst surface area inherent in the finer bed particles, then additional stages are provided in series with the first.

A secondary need for particle separation lies in governmental regulations restricting effluents, and a third, in the possible use of the effluent gas to drive a power-recovery turbine whose economical blade life varies inversely with size and concentration of impinging particles. Where power recovery

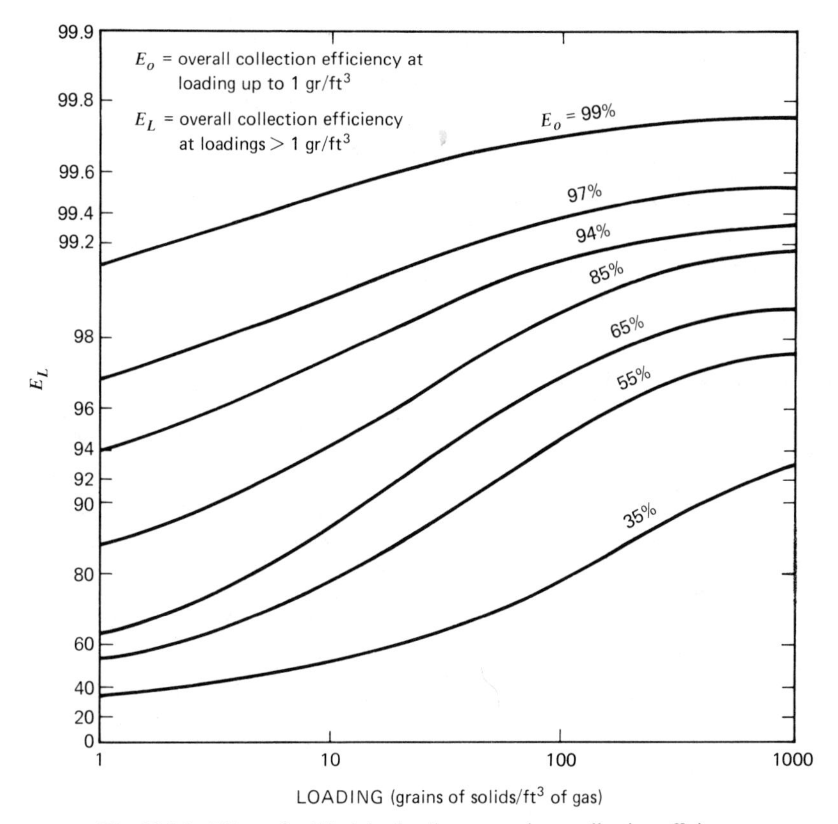

Fig. 28.1.9 Effect of solids inlet loading on cyclone collection efficiency.

is attractive, the feed gases are generally at a high temperature so that cyclones, granular bed filters, and electrostatic precipitators represent the only feasible dry recovery equipment. To date, granular bed filters are as yet in developmental stages though proven in principle to have exceptional efficiencies [17]; electrostatic precipitators are very costly and require substantial maintenance and operating costs. Cyclone development has therefore proceeded to the point where satisfactory levels of particle separation can generally be assured [16].

Imposed constraints on the maximum allowable particle concentrations in gases exhausted to the atmosphere can be met more economically with tray, spray, or venturi-type wet collectors and with cloth bag filters where temperatures permit.

With the exception of pilot plants using shorter-life, low-capacity, sintered metal filters, and phthalic anhydride reactors for which special fiberglass filters have been developed, it is the common practice to recover entrained solids with high-capacity, low-pressure-drop parallel trains of low-cost cyclones.

Cyclone Separators

The prediction of cyclone catch and loss particle size distributions when the inlet loading exceeds 1 grain/ft^3, when the gas outlet tube is necked down to promote saltation in the barrel, or when cyclones are arranged in series with tangential transitions, is an extremely sophisticated process [17], minor but significant portions of which have been studied in detail and still remain proprietary to certain specialized research laboratories. Optimum cyclone dimensions are, however, well documented, as represented in Fig. 28.1.7. Tangential inlet units exhibit higher collection efficiency than axial flow designs or those with so-called volute, scroll, or wrap-around inlets. The basic correlations for predicting the particle collection efficiency of a tangential inlet cyclone conforming to the optimum dimensions of Fig. 28.1.7 are given in Figs. 28.1.8 and 28.1.9.

The pressure drop across a conventional optimum dimensioned cyclone located within the particle disengaging height above a fluidized bed is composed of the sum of an inlet contraction loss, a particle acceleration loss, a barrel friction loss, a flow reversal loss, and an exit contraction loss, all of which can be totalled to yield to a first approximation the relationship:

$$\Delta P \approx V_i^2\left[0.0113\ \rho_f + \frac{Ldg}{167} + 0.000188\ \rho_f N_s\left(\frac{D'}{d'}\right)\right] \tag{28.1.13}$$

where: ΔP = cyclone pressure drop, in in. of water
ρ_f = inlet gas density, in lb/ft³
Ldg = inlet solids loading, in lb of solids/ft³ of gas
N_s = number of spiral traverses within the barrel
$\frac{D'}{d'}$ = ratio of barrel inside diameter to hydraulic diameter of the rectangular inlet

REFERENCES

1. Zenz, F. A. and D. F. Othmer, *Fluidization and Fluid–Particle Systems,* Chapter 2, Reinhold, New York, 1960.
2. Zenz, F. E. and F. A. Zenz, *Industrial & Engineering Chemistry Fundamentals,* Vol. 18, No. 4, pp. 345–348, 1979.
3. Bulsara, P., R. Eckert, and F. A. Zenz, I. & E.C. Proc. Des. & Dev., Vol. 3, No. 4, pp. 348–355, 1964.
4. Zenz, F. A., *Petroleum Refiner,* Vol. 41, No. 2, pp. 159–168, 1962.
5. Zenz, F. A., *Petroleum Refiner,* Vol. 36, No. 8, pp. 147–155, 1957.
6. Nat'l. Petr. Ref's. Assoc., Question and Answer Session on Refining Technology 1970, p. 86; Zenz, F. A., and D. F. Othmer, *Fluidization and Fluid–Particle Systems,* Reinhold, New York, 1960, p. 171.
7. Zenz, F. A., *Petroleum Refiner,* Vol. 33, No. 2, pp. 99–102, 1954.
8. Zenz, F. A., Proceedings of Tripartite Chemical Engineers Conference, Montreal, Sept. 25, 1968; Institution of Chemical Engineers, (London) Symposium Series No. 30, pp. 136–139.
9. Zenz, F. A., *The Fibonacci Quarterly,* Vol. 16, No. 2, pp. 171–183, April 1978.
10. Van Krewelen, D. W. and P. J. Hoftijzer, *Chemical Engineering Progress,* Vol. 44., p. 529, 1948.
11. Zenz, F. A., *Hydrocarbon Processing,* Vol. 46, No. 4, pp. 171–175, April 1967.
12. Zenz, F. A., *Petroleum Refiner,* Vol. 36, No. 11, pp. 321–328, 1957.
13. Zenz, F. A., *Hydrocarbon Processing,* pp. 155–156, January 1982.
14. Gugnoni, R. J. and F. A. Zenz, *Fluidization,* J. Grace and J. Matsen, Eds., Plenum Press, Henniker N.H. Conference, August 1980.
15. Zenz, F. A. and N. A. Weil, A.I.C.H.E. *Journal,* Vol. 4, No. 4, pp. 472–479, 1958.
16. American Petroleum Institute, Publication No. 931, Chapter 11, May 1975.
17. Ciliberti, D. F., D. L. Keairns, and D. H. Archer, Fifth International Conference on Fluidized Bed Combustion, Dec. 13, 1977; see also DOE Contract No. DE-AC21-80 MC 14141 final report on Phase I, Task 3, 1981.

28.2 CLASSIFICATION OF PNEUMATIC HANDLING

Ivan Stankovich

The classification of pneumatic handling may be based on different functional and technical characteristics such as: (1) direction of transportation (horizontal, vertical, inclined); (2) degree of concentration of the solid phase (dense, medium-dense, dilute); (3) size of transported particles (fines, coarse); (4) continuity of transportation (batch, continuous); (5) magnitude of pressure in the conveying pipelines (high, medium, low); (6) sign of pressure in conveying pipelines (positive, negative, negative–positive); (7) function (freight pipelines, miscellaneous devices, such as feeders, dischargers, bin aerators); and (8) design and physical characteristics (capsule pipelines, air lifts, pneumatic and air-activated conveyors).

The classification (Fig. 28.2.1) is based on different characteristics and represents a practical approach to ease communication.

Pneumatic freight pipelines, also known as solid, solid freight, commodity, and cargo pneumatic pipelines, are pipelines used for the transport of different solid materials including manufactured products. The conveying medium is gas, or more often, atmospheric air.

There are at least three basic variations of pneumatic freight pipelines: pneumatic capsule pipelines, pneumatic pipelines, and air-activated gravity conveyors.

Pneumatic pipelines are freight pipelines, carrying fluidized bulk material or sometimes solids, such as aluminum foil, paper, and waste from one point to another.

Pneumatic pipelines may be divided into air lifts and pneumatic conveyors.

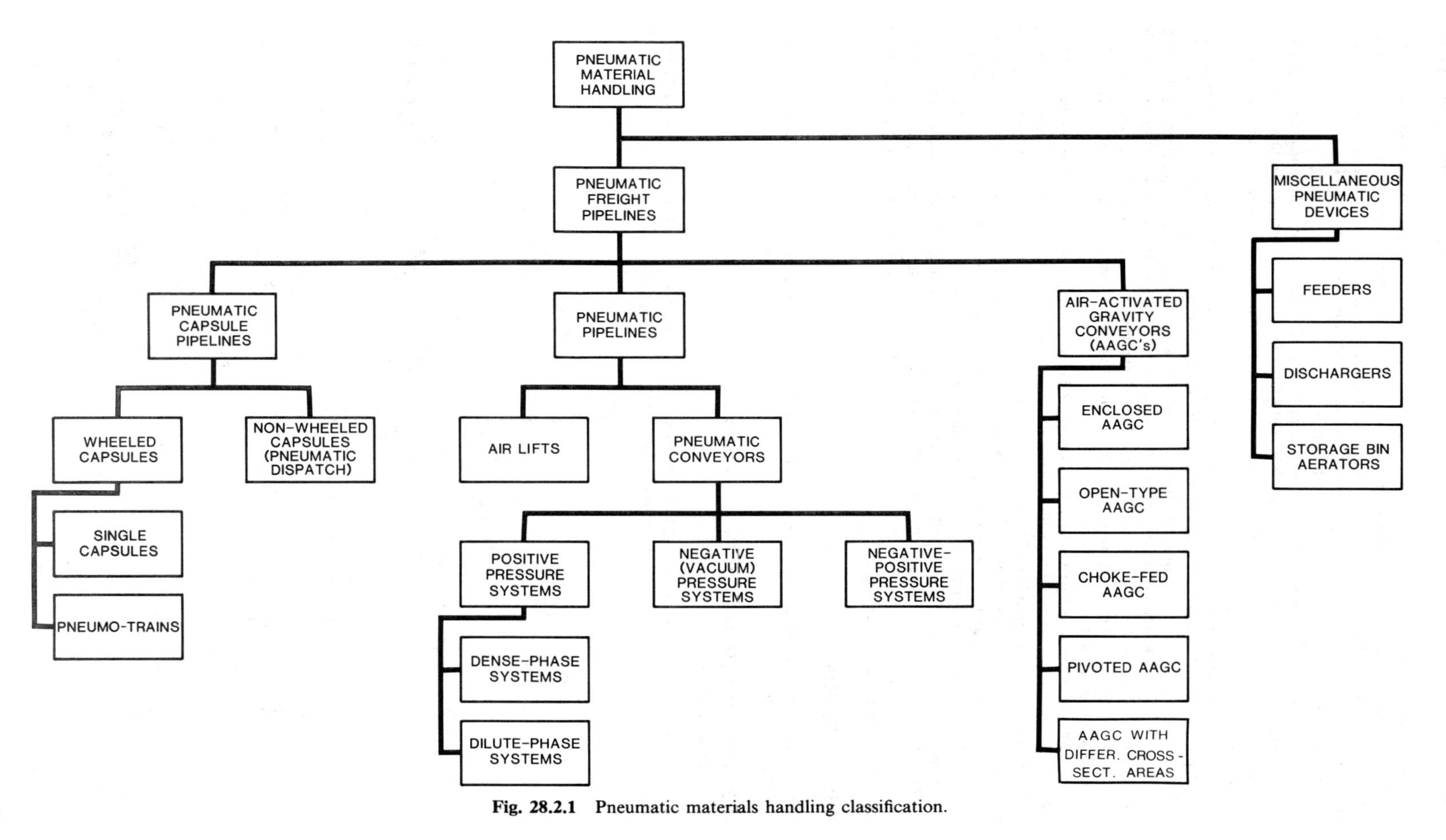

Fig. 28.2.1 Pneumatic materials handling classification.

Pneumatic conveyors (Section 28.4) are pipelines used to transport different bulk materials or other solids to a distance up to 1.5 mi (2.4 km), practically in any direction—horizontally and vertically. The feeding of material into the conveyor is either continuous or batch type. Depending on the pressure in the line, the conveyor may be of positive, negative, or negative-positive type. The material in the pipeline may be conveyed in a dense or dilute phase.

Air lifts (Section 28.6), with capacities up to 200 ton/hr (181,437 kg/hr), are pneumatic pipelines used on relatively short vertical runs only. Bulk material is fed into the vertical pipeline continuously by a dense-phase fluid bed feeder. At the highest point, the fluidized material is discharged into a disengaging vessel or storage bin (Fig. 28.6.1).

Pneumatic capsule pipelines (Section 28.7), with capacities up to 200 ton/hr (181,437 kg/hr), are freight pipelines (with a cross section slightly smaller than the inside of the pipeline), carrying bulky and/or commodity solids to a distance up to 15 mi (24 km). Systems with capacities up to 2500 ton/hr (2.27×10^6 kg/hr) and transporting distances up to 120 mi (193 km) are studied. Capsule pipelines may be of the wheeled or nonwheeled type. Sometimes the wheeled capsules are carried as single units; sometimes they are connected in a stirring of capsules. Such a variety of capsule pipeline is known as pneumotrain.

Air-activated gravity conveyors (AAGC) (Section 28.5), with capacities up to 2500 ton/hr (2.27×10^6 kg/hr), are pneumatic freight pipelines carrying bulk material in a dense-phase state usually to a relatively short distance up to 0.3 mi (0.48 km) (see Fig. 28.5.1). Two basic means are used to transport the material: (1) fluidization, that is, friction between particles and between particles and pipe walls is decreased essentially; and (2) gravity force, by declining the conveyor downstream. AAGC may be of the enclosed, open, or pivoted type. Recently, choke-fed air-activated gravity conveyors have gained a reputation of being a very useful, self-controlling, transporting, feeding, and discharging device.

Miscellaneous pneumatic devices (Section 27.8), such as different feeders, dischargers, storage bin aerators, and air locks are extensively used in the industry. The number, type, and application of such devices is increasing steadily.

BIBLIOGRAPHY

Stankovich, I. D., "Pneumatic Handling Classification," BHRA, Pneumotransport–4 Conference, Carmel, CA, 1978.

28.3 PNEUMATIC HANDLING SYSTEMS

Ivan Stankovich

Large and sophisticated pneumatic systems are used in different branches of industry presently (Fig. 28.3.1). A *system approach* means to compile and apply simultaneously, different considerations, such

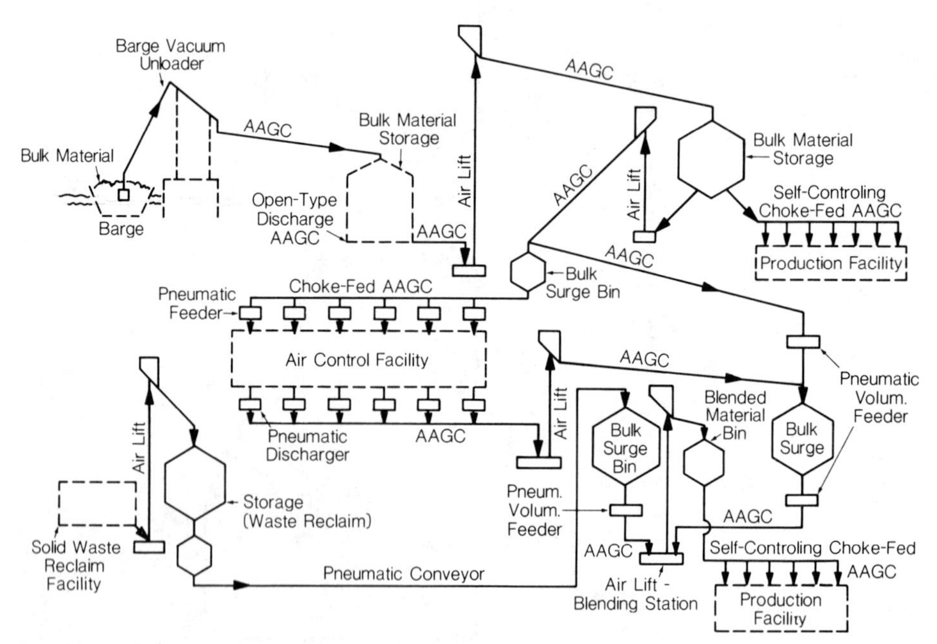

Fig. 28.3.1 Contemporary pneumatic handling system (schematic). AAGC = air-activated gravity conveyor.

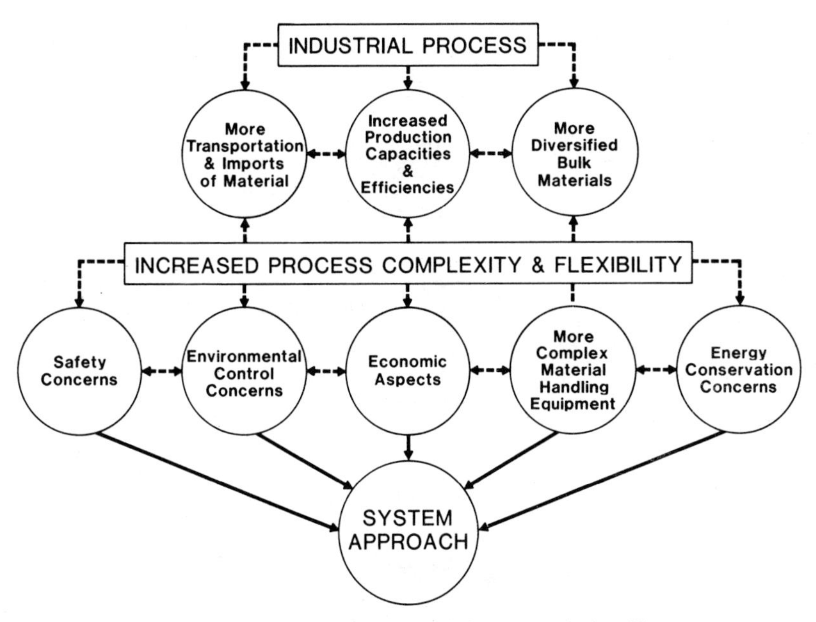

Fig. 28.3.2 The system approach in pneumatic handling.

as pneumatic handling functions, equipment design, maintenance and operation, automation, and economics, taking into account such requirements as safety, environmental control, and energy conservation, as well as the possible feedbacks and interconnections (Fig. 28.3.2).

28.3.1 System Elements

After the overall objectives of a materials handling project are known and decision has been made to use a complex pneumatic system, single elements, such as unloading and storing (Table 28.3.1) have to be selected in a most convenient and rational manner. Process, environmental, economic, and other considerations have to be taken into account at this stage of design. Selecting system elements is the conceptual system design.

28.3.2 Equipment

Pneumatic handling equipment is arbitrarily divided into functional equipment and auxiliary equipment (Table 28.3.2). Functional equipment is performing a basic function in the system (conveying, unloading,

Table 28.3.1 Bulk Materials Handling Elements

Type	System Element
Functional	Unload
	Convey
	Storage
	Discharge
	Feed
	Distribute
	Blend
	Mix
	Classify
Auxiliary	Collect dust
	Weigh
	Meter
	Rate
	Remove oversize particles

Table 28.3.2 Pneumatic Materials Handling Equipment Type and Function

Type	Function	Note
A. Functional equipment		
1. Pneumatic conveyor (all types)	Convey material in all directions	Section 28.4
2. Air lift	Convey material vertically Blend material	Section 28.6
3. AAGC	Convey material Blend material	Section 28.5
a. Open-type	Discharge material Convey material Feed material	Storage facilities
b. Choke-fed	Distribute material Convey material Feed material	
4. Unloader		
a. Vacuum	Unload material Convey material	Rail cars, ships, barges, trucks
b. Positive pressure	Unload material Convey material	
5. Miscellaneous equipment		
a. Air-activated feeder	Feed material	Section 28.8
b. Air activated	Discharge material	
c. Air-seal	Feed material Seal compartments Convey material vertically	
d. Bin aerators	Facilitate discharge	
e. Air-activated mixer	Mix material Blend material	
f. Air-activated distributor	Distribute material	
B. Auxiliary equipment		
1. Dust collectors	Collect particles	Section 28.9
2. Material metering device	Meter material flow	
3. Air-activated particle trap	Remove oversized particles	

feeding, etc.). Auxiliary equipment, such as dust collectors, supports the function of basic equipment. An important phenomenon is the development of miscellaneous pneumatic equipment, such as feeders, dischargers, distributors, and air seals (Section 28.8).

28.3.3 Equipment Interface

Great attention and care should be devoted to the selection and design of equipment and subsystem equipment interface. Some typical examples are given in Table 28.3.3.

28.3.4 System Design

The system design is a simultaneous, complex approach to all elements, aspects, and feedbacks involved in the project. None of them should be resolved separately, outside of the system.

An example of sequences in a system approach is given in Fig. 28.3.3. The main steps are as follows:

Careful study of given objectives

Selection of system elements (Section 28.3.1)

Appraisal (testing) of handled material

Selection of equipment (Section 28.3.2)

Design of equipment interface (Table 28.3.3)

28.3.5 System Reliability

System reliability may be an extremely important factor in system design. The following example illustrates that importance.

Table 28.3.3 Some Examples of Equipment Interface[a]

No.	Interface Combination	Requirements, Conditions, and Recommendations
		A. Between pneumatic handling elements
1	AAGC–air lift	1. Equipment shall be sized: $CAP_{al}^{des} > CAP_{aagc}^{des}$
		2. AAGC and air-lift feeding column shall be properly ventilated (Section 28.9)
		3. Adjustments shall be provided for: AAGC capacity in a certain range; Ventilation; Air-lift feeding column fluidizing air
2	Air lift–AAGC	4. Equipment shall be sized: $CAP_{aagc} > CAP_{al}^{des}$
		5. Deaeration of conveyed material in air lift disengaging vessel shall be adequate (Section 28.6)
		6. Ventilation adjustments shall be provided
3	AAGC–pneumatic conveyor	7. Equipment shall be sized: $CAP_{aagc}^{adj} > CAP_{pc}^{des}$
		8. Choke-fed AAGC (Section 28.5) is often a proper selection
		9. Pneumatic conveyor fluidizing vessel and AAGC shall be ventilated adequately (Section 28.9)
4	Pneumatic conveyor–AAGC	10. Equipment shall be sized: $CAP_{aagc} > CAP_{pc}^{des}$
		11. Special requirements: Adequate material deaeration at pneumatic conveyor terminal; An air seal to separate pneumatic conveyor from AAGC shall be provided
		B. Between pneumatic and nonpneumatic handling elements
5	AAGC–belt conveyor (or screw conveyor, vibrating feeder, and other)	1. Equipment shall be sized: $CAP_{bc} > CAP_{aagc}^{adj}$, or $CAP_{sc} > CAP_{aagc}^{adj}$, or $CAP_{vf} > CAP_{aagc}^{adj}$, etc.
		2. Bulk material, before transferred from AAGC to other mechanical equipment, shall be deaerated adequately
		3. Transfer points shall be ventilated and designed adequately (Section 28.9)

[a] AAGC = air-activated gravity conveyor; CAP = capacity; CAP^{adj} = capacity, adjustable; CAP^{des} = capacity, design; CAP_{aagc} = capacity, AAGC; CAP_{al} = capacity, air lift; CAP_{pc} = capacity, pneumatic conveyor; CAP_{bc} = capacity, belt conveyor; CAP_{sc} = capacity, screw conveyor; CAP_{vf} = capacity, vibrating feeder.

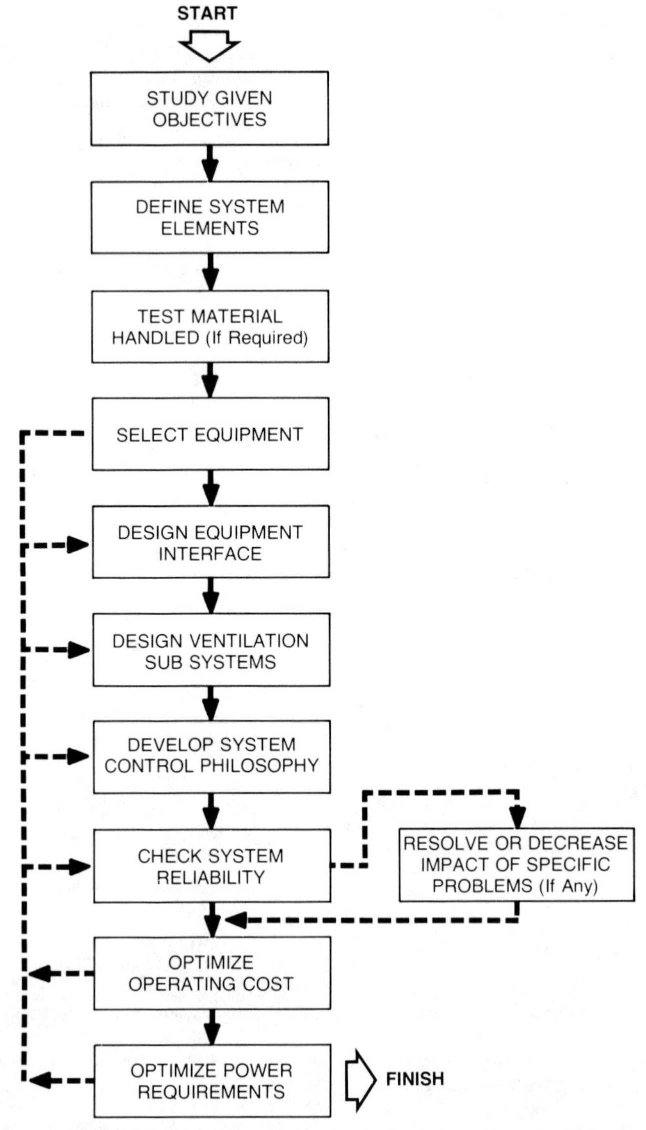

Fig. 28.3.3 Sequence of pneumatic handling system design.

A pneumatic materials handling system is illustrated in Fig. 28.3.4. The reliability of system components (Table 28.3.4) can be determined from the equipment service record or by experience from similar applications.

The reliability of the system R_s, as shown in Fig. 28.3.4 and using durations of component continuous service—Case A (Table 28.3.4), is:

$$R_s = R_1 \times R_2 \times R_3 \times \cdots \times R_{17}$$
$$R_s = 0.41$$

Usually, this is unsatisfactory for a contemporary materials handling transfer system.

The system reliability can be improved (R_s^{im}) by using a surge bin as a by-pass for material transfer (Fig. 28.3.4):

$$R_s^{im} = R_1 \times R_2 \times R_3 \times \cdots \times R_{10} \times [1 - (1 - R_{11})(1 - R_{12}) \cdots (1 - R_{17})]$$
$$R_s^{im} = 0.64$$

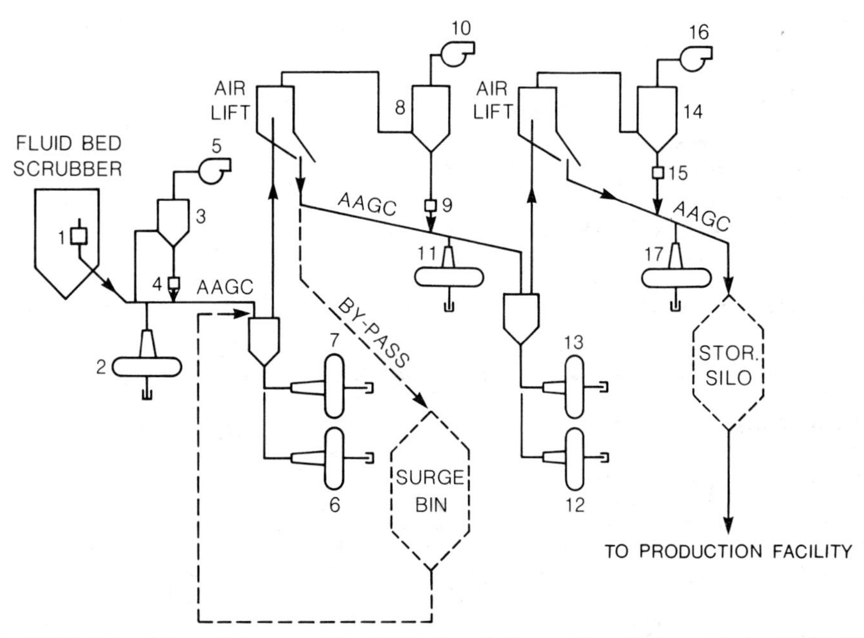

Fig. 28.3.4 Material transfer system (simplified scheme). 1—scrubber discharge device; 2, 11, 17—AAGC blower; 3, 8, 14—nuisance dust collector; 4, 9, 15—discharge valve; 5, 10, 16—exhaust fan; 6, 7, 12, 13—air lift blower; AAGC—air-activated gravity conveyor.

This relatively low, but costly system reliability was considered also as an unsatisfactory solution.

The system reliability can be improved by improving reliabilities of single most-important and most-vulnerable components. For instance, dust collector reliability can be increased to 0.99 by increasing the duration of dust collector continuous service from 2 weeks to 1 year. Similarly, fluid bed scrubber discharge and air-lift operations can be elevated to 0.99 (Table 28.3.4). The system reliability, using durations of component continuous service—Case B, C, and D, will be 0.78, 0.80, and 0.83, respectively. These results often may be considered as satisfactory.

Finally, the system reliability (R_s^{im2}) can be improved by using simultaneously both methods with higher achieved component continuous service (Table 28.3.4):

$$R_s^{im2} = 0.90$$

Obviously, in this case the highest reliability will be achieved at highest cost.

To achieve reasonable system reliability: (1) use minimum possible system components, (2) assure

Table 28.3.4 Reliability of System Components

System Component	Case	Duration of Component Continuous Service, H (hr)	Average Duration of Outage, H_1 (hr)	Component Reliability $R = 1 - H_1/H_1 + H$
Fluid bed Scrubber discharge operation	A	712 (or 1 mo)	72 hr (or 3 days)	0.91
	B	8544 (or 12 mo)		0.99
AAGC blower		8544 (or 12 mo)		0.99
Nuisance dust collector	A	336 (or 2 wk)		0.82
	B	2136 (or 3 mo)		0.97
	C	4272 (or 6 mo)		0.98
	D	8544 (or 12 mo)		0.99
Exhaust fan		8544 (or 12 mo)		0.99
Discharge valve		8544 (or 12 mo)		0.99
Air-lift blower	A	2136 (or 3 mo)		0.97
	B	8544 (or 12 mo)		0.99

highest component reliability, and (3) use material transfer bypass. The last two conditions should be considered in conjunction with economical conditions.

BIBLIOGRAPHY

Stankovich, I. D., "A System Approach to Pneumatic Handling," ASME Century 2—Emerging Technology Conferences, San Francisco, CA, 1980.

28.4 PNEUMATIC CONVEYORS

Frank Gerchow

28.4.1 Types of Systems

This section describes various types of pneumatic systems and their advantages and disadvantages which must be understood before one can select the type of system to be used on a given application.

Vacuum System

The first is a straight, simple negative-pressure or vacuum system (Fig. 28.4.1). This consists of the piping and the pickup manifold, which can be a Y-branch, where the material drops into the air system and is pulled by the vacuum, thereby eliminating dust at pickup source.

This type of system can be used in those operations where products must be fed into a dump hopper. The conveying air can be used to control the dust that is generated from the dumping operation. An air intake scoop or filter is required depending on the products to be handled. The receiver can be a cyclone connected with a centrifugal fan that will tolerate small amounts of product and dust passing through it. However, if a rotary positive-displacement blower is the vacuum source, the receiver must be a filter type to prevent any product passing through the blower because of its close tolerances (Fig. 28.4.2).

These systems are particularly suited to moving material from multiple pickup points to a single location, the reason being that the bulk of the system's expense is in the terminal end where the receiver, rotary valves, and vacuum source are located. The pickup points are Y-branches or manifolds with cutoff valves that can be manually or automatically operated.

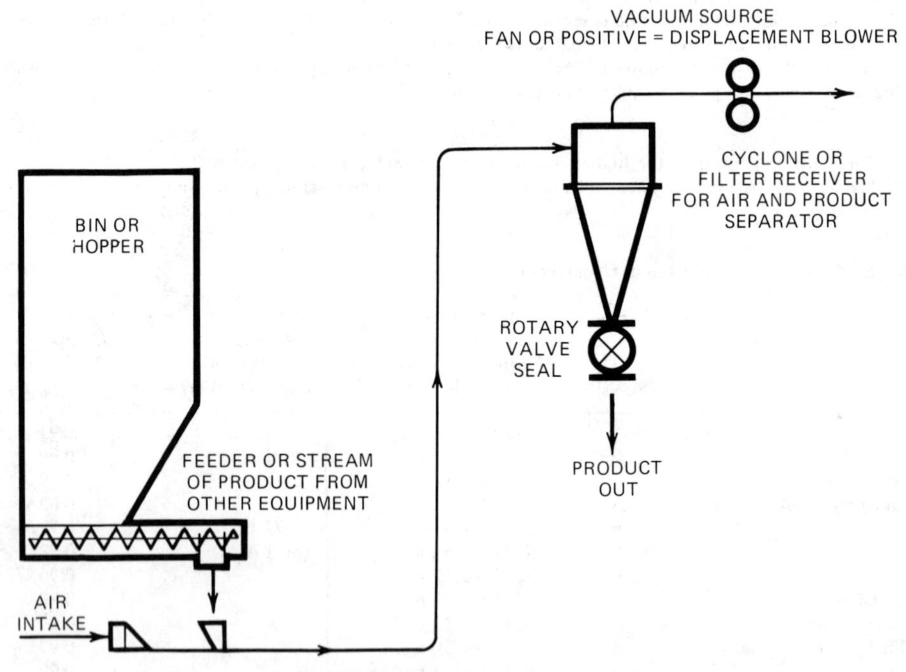

Fig. 28.4.1 Simple vacuum system.

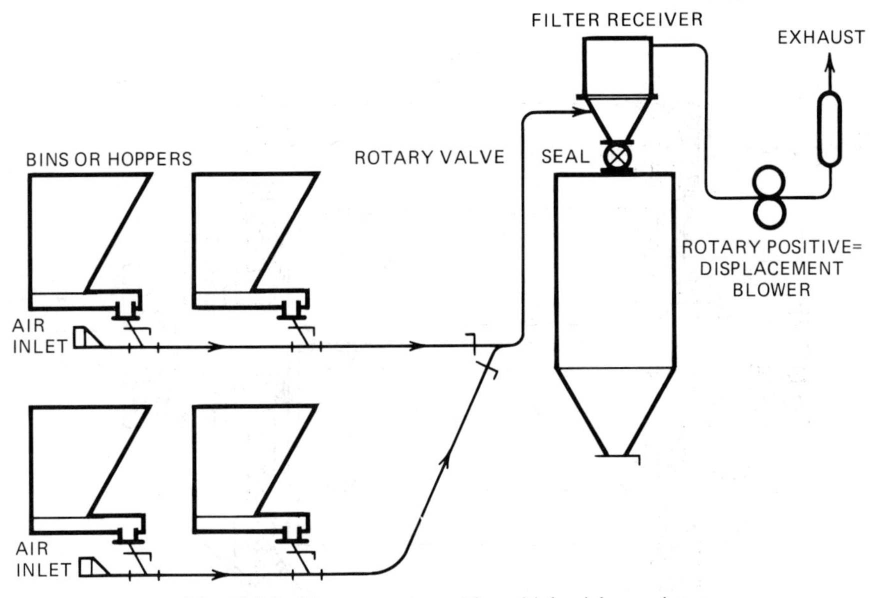

Fig. 28.4.2 Vacuum system with multiple pickup points.

Pickup from more than one location simultaneously, requires an airlock seal on the point farther downstream so that conveying air does not short circuit and sufficient velocity is maintained to convey from the upstream to downstream points.

If pickup is at one point at a time, only a cutoff valve is required and as many pickup points as necessary may be included. Concern about blowback air is eliminated since all leakage is inward. The diverter valves can be the Y-branch type with more knife cutoff gates so that suction can be provided to the leg from which product will be conveyed.

These systems are particularly suited for unloading railroad cars from above the rail. Most hopper cars do not have sufficient clearances to get an air seal between the hopper outlet and the top of the rail. Therefore it is impossible to convey product by positive pressure. The Air Slide car is an exception. It has about 12 in. of clearance beneath the hopper outlet and the top of the rail so it is possible to convey by means of an air seal device under the car or to convey into a vacuum system by using a paddle-type feeder designed for feeding into a negative system.

These systems are also used for picking up from boxcars or flat storage. Many times companies prefer to store the products flat in an area inside of a warehouse or building rather than use storage tanks. Typical is an installation handling English walnuts with negative-type pickup nozzles (similar to a drum nozzle) that come down from the ceiling or roof and reach into the various flat storage areas to pick up the walnuts.

Some mechanized units on the market can be used inside a boxcar to loosen and pick up material that has packed and hardened in transit, and then feed it into the inlet of a vacuum system.

Another advantage of the vacuum arrangement: it provides air to purge grinding equipment, pulls the product into the grinder, and dissipates the heat of grinding. Any moisture that is driven off the product from the heat of grinding is also dissipated with one of these vacuum systems.

Because all leaks are inward, no sophisticated seals are required on any of the diverter valves or even on the rotary valves under the filter receiver. The life span of a rotary valve can be increased and the buildup of product between the closed ends of the rotor eliminated by a natural purging action.

By eliminating stuffing boxes and putting vent holes in the end, one can break the vacuum in the valve and get a natural air purge over the shroud area (Fig. 28.4.3). The shroud area is the clearance between the shroud and the bore of the rotor housing, and that clearance is about the same as that on the rotor tips. This natural air purging can improve the life of a valve and virtually eliminate the wear caused by product getting into the shroud area.

If the seal is unbroken, air can leak from below, carrying the product through the area. This creates a sandblasting effect. Due to leakage, a tight vessel underneath will soon be under the same vacuum as that in the filter receiver or cyclone collector. It is necessary, then, to break the vacuum on the vessel be it a screw conveyor or other type. In some cases, covers on screw conveyors are sucked in because of leaking and unvented valves. It is a decided advantage to vent a rotary valve on a vacuum system in this manner, because the packing maintenance is eliminated.

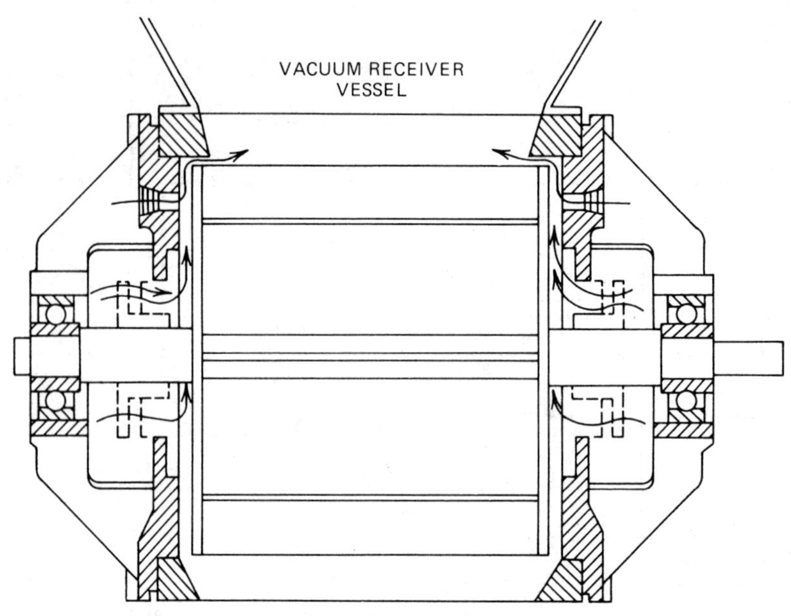

Fig. 28.4.3 Air-purge system.

These systems are excellent for handling toxic materials or corrosive materials because all leakage is inward so there is less danger of any product escaping into the atmosphere. Vacuum systems have been used extensively for flash drying and cooling. For example: wheat flour can be dried from 14% to about 10% moisture. When the product is conveyed 100 to 200 or 250 ft at 5000 ft/min, it is not in the system very long. The air is heated to about 300°F; that heat dries the flour. The flour comes out of the system at approximately 120°F without any damage. The same thing can be done in the cooling operation, depending on the particle size of the product. If the particle size is quite large, it will take awhile for the heat in that particle to move to the surface so the air can remove it from the particle. Rapid cooling has to be done strictly on a finely divided particle.

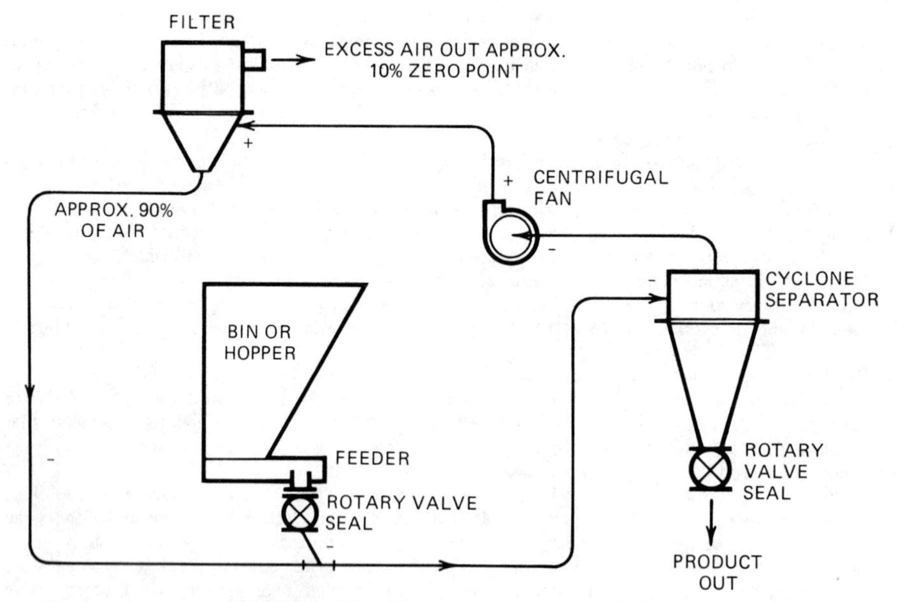

Fig. 28.4.4 Closed-loop vacuum system with fan.

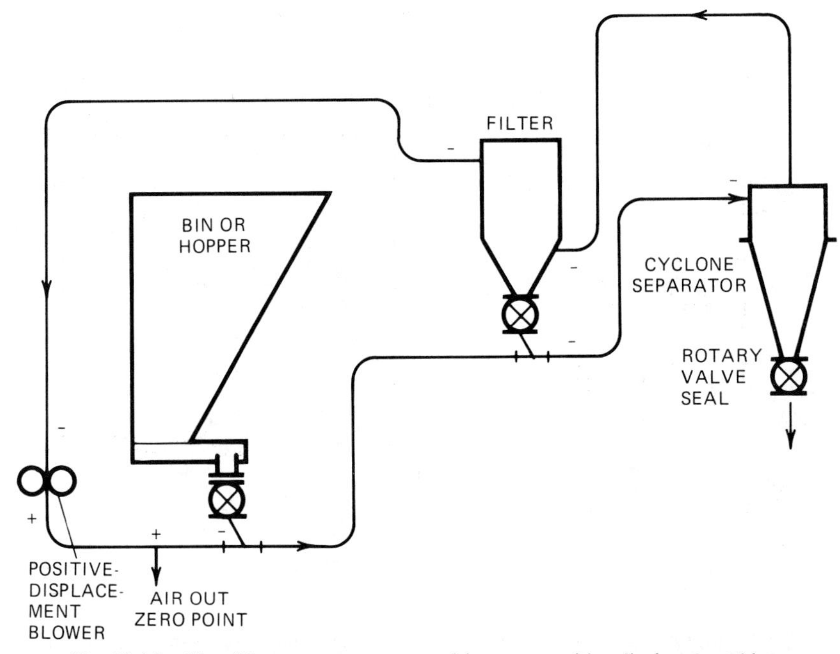

Fig. 28.4.5 Closed-loop vacuum system with rotary positive-displacement blower.

These systems can be arranged in a closed loop to assure that no product gets out into the atmosphere or that a minimum of product gets into the atmosphere should a bag break (Fig. 28.4.4). It also helps to economize on the size of the filter in this particular arrangement because the filter is only handling about 10% of the conveying air. The primary cyclone collector has a carryover of 1% or less, passing through the fan and being blown into the filter. The filter is vented to the atmosphere so that only the excess air actually goes through the filter bag and exits to the atmosphere. The other 90% of the air continues on down the outlet of the hopper and goes back around the horn again so that the dust in the air eventually gets separated. The dust that is separated was missed in the cyclone the first time and is knocked down by the material stream on the second time around. The dust escaping out of the cyclone is 1% or less, depending on the product and its separating ability in a cyclone collector. This gives a rather economical closed loop, but rotary valve or other seal must be at all of the inlet and exit points of the system. No seals are needed at the filter outlet because the return line is connected to close the loop. This is called a series hookup for the filter.

The same thing can be accomplished, making it a closed-loop vacuum system, utilizing a rotary positive-displacement blower (Fig. 28.4.5). But in this case, the filter must handle 100% of the air because the blower cannot tolerate any product passing through it. The zero point is on the discharge of that blower (as the system is entirely vacuum) as indicated in the figure by the negative signs all the way to the inlet of the blower.

The only pressure in the system is from the blower discharge to the zero point, which need be nothing more than a tee open to the atmosphere. Since this is a vacuum system, air will leak in through the rotary valves at the pickup and terminal points of the system. This filtered leakage air is exhausted at the zero point. The dust that is collected by the filter is set up to drop right back into the conveying line, eliminating the need for rehandling the dust.

Pressure System

The next pneumatic system is the straight pressure type, which is ideally suited for conveying from one pickup location to many discharge locations (Fig. 28.4.6). The greatest cost in this system is at the pickup location where the rotary valve and the blower are located. At the terminal end, the system is blowing into some bins and utilizes some diverter valves or a manual switch station. If the product is the same in all bins, just one filter is necessary by interventing the bins. Generally, this type of system is more economical when going from one point to several. The one drawback of this type of system is that the air leakage at the pickup point must be handled. Because of the clearances in the rotary valve and the displacement of the rotor, there is a constant flow of air passing out of the system at the pickup location. This has to be vented in some manner, either by a simple sock

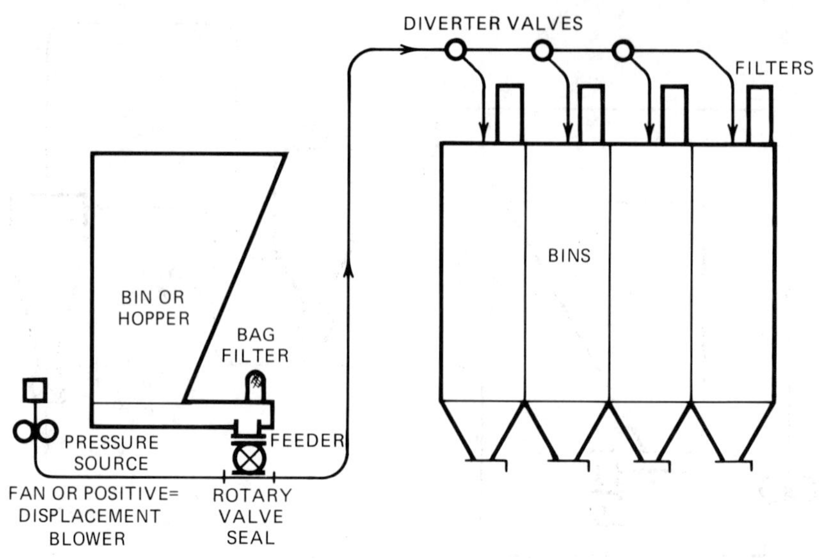

Fig. 28.4.6 Pressure system.

(depending on the product) or an elaborate and expensive secondary dust-collecting method which may make the system more expensive in the final analysis than a vacuum type.

These systems operate up to a maximum of 15 psig, normally the pressure of rotary positive-displacement blowers that are commercially available. Unless a high-pressure tank system is available, rotary valves generally cannot do an effective job of sealing when pressure exceeds 15 psig. Fifteen psig, as compared to the vacuum system described earlier, using the same rotary positive-displacement blower, can provide only about 16 in. of mercury, or roughly 8 psig. Thus a larger conveying job can be done with a smaller pipe in a pressure system than in a vacuum system, but the power consumption should be the same because the same amount of work is being done. A vacuum system with a bigger pipe is not going to take more power to do the same work; the same amount of product is being moved over the same distance. Power consumption should be the same, provided that the same vacuum or pressure source is being used. If one compares a fan on a vacuum system to a rotary positive-displacement blower on a pressure system, then a difference may be seen in the power requirement because the fan is not as efficient as the blower. The power consumption should be very close to the same except for any difference in the efficiency of the air prime movers. The advantage is that it enables a smaller line to be used and increases the density of product loading in the conveying air. Degradation of the product is reduced because the product does not rattle around as much in the conveying pipe. A pressure system of this type generally conveys with a product-to-air ratio of about 20 lb of material per pound of air, or approximately $1\frac{1}{2}$ lb of material/ft^3 of air (or 20 ft^3 of air/ft^3 of product).

A vacuum system utilizing a rotary positive-displacement blower at about 16 in. of mercury conveys about half of these ratios. In other words: 10 lb of material per pound of air, $\frac{3}{4}$ lb of material per cubic foot of air, or 40 ft^3 of air/ft^3 of product.

Because the air volume in a pressure system is considerably reduced, heating and cooling applications usually require a jacketed conveying line using water or some other medium in the jacket.

A closed-loop pressure system is illustrated in Fig. 28.4.7. The zero point has been moved from the discharge of the blower to the inlet of the blower. This is the only difference between the pressure and vacuum closed-loop arrangement. The zero point has been moved from the discharge to the inlet. This assures that there is pressure in the system everywhere except the pipe between the zero point and the blower inlet. If this were a pure air system the zero point would be nothing more than a tee with an intake filter to provide makeup air to offset that leaking out at the pickup, discharge, and other points. If this same system is used on inert gas, the gas is usually introduced at some pressure slightly in excess of standard atmospheric pressure as a means of preventing infiltration of air or oxygen.

If a product is particularly hard to filter for one reason or another then all filters in the system can be eliminated (Fig. 28.4.8). There is a cyclone separator and centrifugal fan but no zero point. The zero point will roam, depending on the leakage into and out of the system at the fan shaft and other points. The pressure against which the fan is operating is going to have an influence on the leakage, so this will affect the zero point location. If manometers were put on the system to analyze it, a check one time might show a certain pressure at some point; a check later may show it has

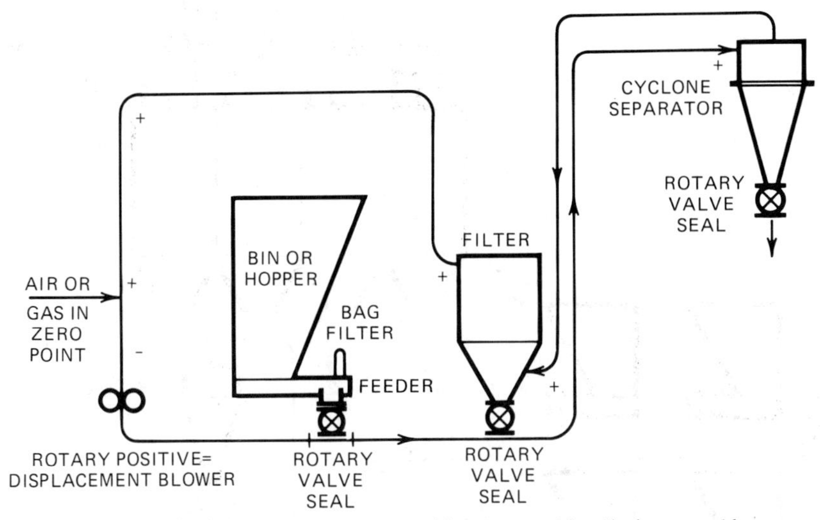

Fig. 28.4.7 Closed-loop pressure system with rotary positive-displacement blower.

changed to a vacuum. Since it is not known whether there will be a vacuum or pressure at the pickup point, a pressure-type pickup tee or manifold must be used at the pickup point. A pressure-type manifold will convey under both pressure and vacuum, but the negative or simple Y-branch type may not, depending on the product being conveyed. The air leakage up through the rotary valve tends to suspend product in the vertical part of the Y-branch and prevent it from falling fast enough to get into the air stream, so a positive-pressure pickup tee is generally used at that point. This is a very economical system because it eliminates filters (although it does add rotary valves in order to seal the pickup and discharge locations).

Vacuum–Pressure System

A combination of the vacuum and pressure system has obvious advantages. Such a system can pick up from a railroad car and convey by pressure to storage tanks, or pick up from any other device that requires a vacuum-type intake to handle particularly fine, dusty material that is difficult to introduce into a pressure system because of the blowback. In going from many pickup points to many discharge

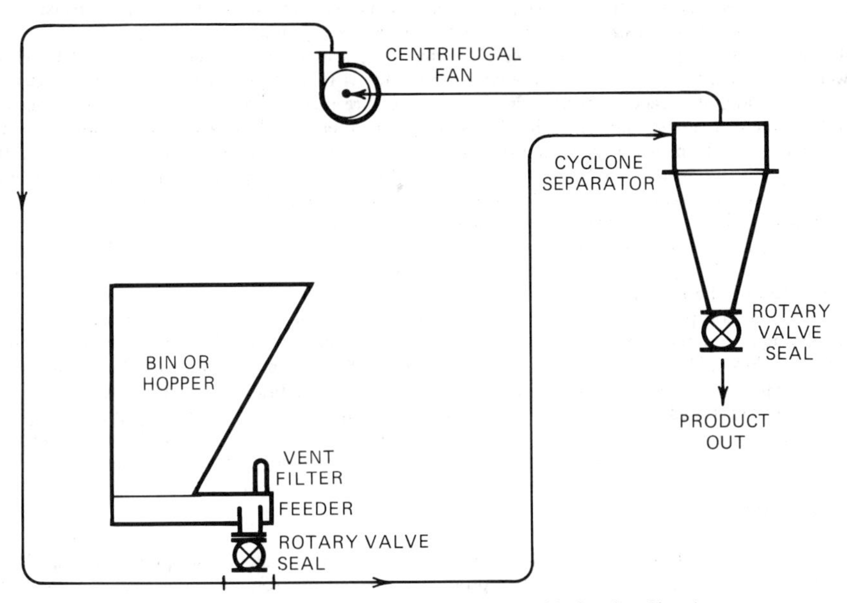

Fig. 28.4.8 Closed-loop pressure system with fan (no filters).

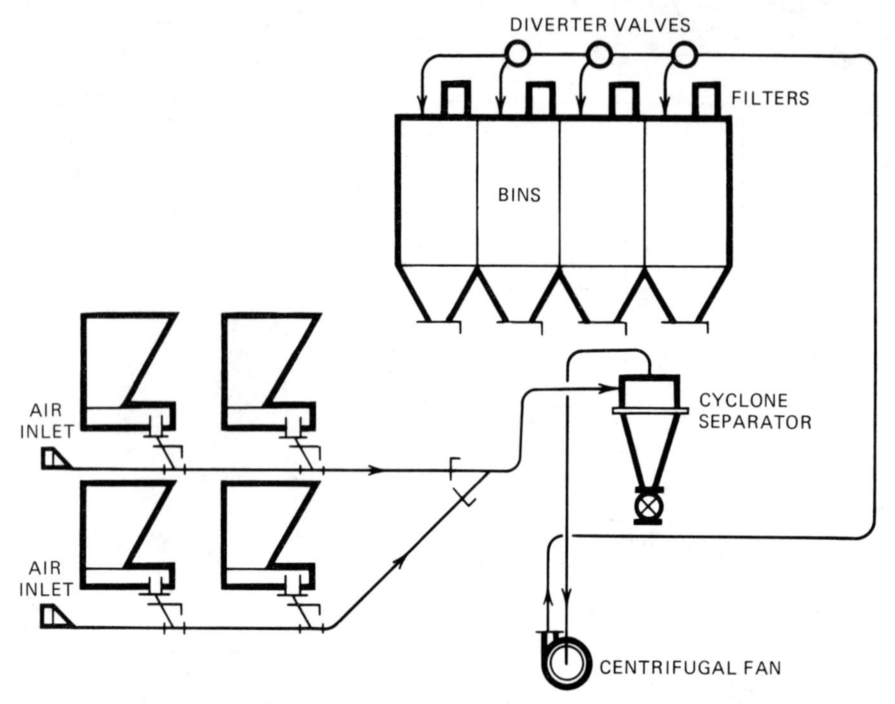

Fig. 28.4.9 Combination vacuum–pressure system.

points, it is usually more economical because the meat of the system—a filter receiver discharging to the pressure side—is in the middle.

One type of vacuum–pressure system utilizes a centrifugal fan with a cyclone separator. This uses the vacuum side of the fan to pull the product into the cylone and the pressure side of the fan to convey away from the cyclone to the discharge points (Fig. 28.4.9). The fan can tolerate the product from the cyclone passing through it without any difficulty. Such a system does have some drawbacks because it is limited to the total pressure drop that can be overcome with a centrifugal fan. Single-stage fans normally develop about 65 in. of water. A multistage fan may get as high as 120 in. of water. This makes the size of the conveying line larger than it would be with rotary positive-displacement blowers. This increases the size of the filters at the discharge location.

For best results, a pull–push system should utilize a separate blower on the vacuum and a separate blower for the pressure side. The main advantage of this is that the vacuum side can be sized for a full 16 in. of mercury vacuum while allowing the pressure side to be sized for a full 15 psig. This helps to maintain a smaller pipe on each side and keeps the filters, diverter valves, and blowers considerably smaller. It also provides good control over the volumes on each leg. When both legs are tied together to a common fan or blower and velocity is too low on the pressure side (which is usually the case because of compression ratio), the blower must be sped up to increase the velocity on the pressure side. Velocity on the negative side then goes out of sight and product degradation or abrasion increases. Separate blowers for the vacuum leg and the pressure leg can be fine-tuned independently of each other. Velocity can be increased on the vacuum side by changing the drive on the blower, without affecting the pressure side or vice versa.

In a closed-loop pull–push system the vacuum and pressure legs are closed independently of each other. Trying to combine these becomes difficult because the volumes are quite different due to compression ratios. These pull–push systems are ideally suited for portable use to unload railroad cars. Many freight terminals buy portable units to unload material from the railroad car and put it into trucks so it can be distributed to their customers. Portable systems have been used for picking up spillage from train wrecks with good success in salvage operations. Pull–push systems lend themselves to portable systems since any type of hand nozzle can be used for picking up the product.

Product System

One of the simplest and oldest pneumatic conveyors is the product system. An upright vacuum cleaner is an example of the product system. The product is sucked up from the floor, passes through the fan and is blown into the bag. It is an economical system consisting of only pipe, fan, cyclone, and no rotary valve. It has the advantages of the pull–push system, providing negative suction at the pickup and pressure at the discharge point. This simple system is often overlooked because the fans that are used can only generate about 45 in. of water pressure. This means the pipe is somewhat larger. If the product is not dusty, it can be blown into a cyclone collector.

These systems are used extensively for picking up from grinding equipment because they offer the advantage of pulling the air down through the grinder, dissipating heat and moisture. In some cases, the fan is mounted on the shaft of a hammer mill. The product is pulled from under the mill and turns 180° into the eye of the fan and is blown to a cyclone collector. This is not the best arrangement because the power of the fan takes away from the power available for grinding. The two elbows at the pickup point reduce capacity. It works better independently of grinding equipment. The product can then go straight into the eye of the fan. The fan becomes the bottom 90° elbow. There is then only one elbow to come from the vertical pipe into the horizontal pipe going to the cyclone collector at the terminal end of the system (Fig. 28.4.10). This gives the advantage of the fan throwing the product up the vertical conveying pipe. With such a system, it is usually desirable to have a good straight run on the discharge of the fan to take advantage of the inertia that the fan gives the product.

One consideration in this type of system is whether the product is going to be degraded by passing through the fan. The product should also not be exceptionally abrasive because this will shorten the fan life.

Gravity-Flow Systems

A vacuum system, eliminating rotary valves, but still taking advantage of the vacuum provided by a rotary positive-displacement blower, consists of a receiver that can take a vacuum of 16 in. mercury. Mounted on top of it is a filter also capable of taking vacuum. Between the blower and the filter is a three-way valve which consists of two butterfly valves on a slave linkage so that when one is open, the other is shut. The product is pulled into the receiver. When picking up from a hand nozzle or a hopper car manifold where the flow of product cannot be stopped, a vacuum breaker valve pops open and short-circuits the air at that point, eliminating the suction at the nozzle and stopping the flow of product. After a time delay, allowing the conveying line to clean itself, the three-way valve diverts so that the blower is now open to atmospheric air, eliminating the vacuum on the filter and the receiving device. This allows the counterbalanced hinged door to swing open from the weight of the product and lack of vacuum. After the timer is timed out, indicating that the unit is empty, the diverter valve switches back, placing the filter-receiver under vacuum. This causes the hinged door to be sucked shut again and the vacuum breaker again closes and the process is repeated. This is a

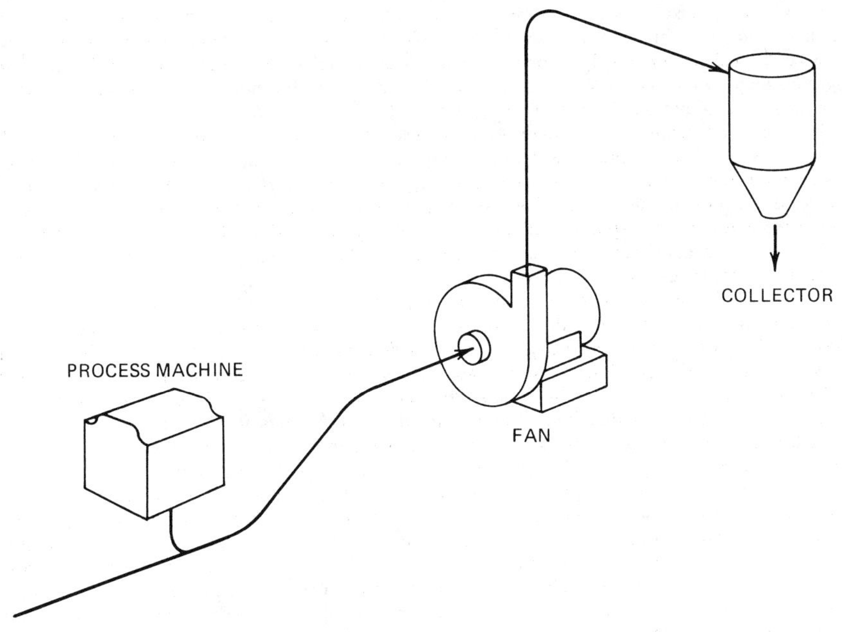

Fig. 28.4.10 Product system.

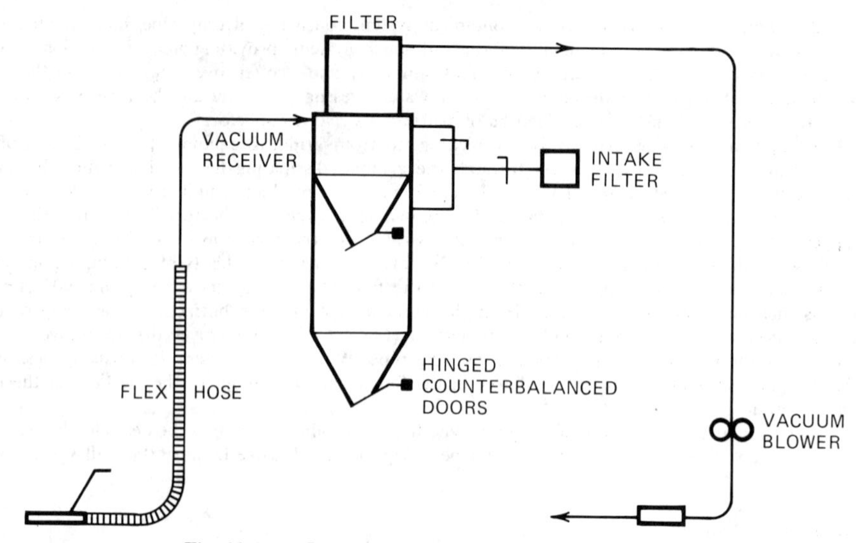

Fig. 28.4.11 Vacuum gravity-flow system (continuous).

rather simple vacuum system where rotary valves are undesired. If the product is excessively abrasive, for example, rotary valves have a relatively short life span and must be avoided. If the product is extremely friable, it may be desirable to eliminate rotary valves so that there is no shearing action on the product. The hinged doors are easily repaired in abrasive service, because they generally have just a piece of rubber as a seal, and natural gum rubber is a good abrasive-resistant material that is rather easily repaired or replaced.

The continuous vacuum arrangement accomplishes the same thing on a continuous basis without stopping the flow of product from the pickup nozzle. It utilizes two hoppers connected by a breather line (Fig. 28.4.11). In such a system the upper hopper is always under vacuum. The lower hopper alternates between zero pressure, or atmospheric, and vacuum, and is controlled by the valve arrangement on the side of the vessel. When the valve going to the intake filter is closed and the upper valve is open, the vessels are under the same vacuum, such that the intermediate counterbalanced door swings free, so that the product that is conveyed is collected in the lower hopper. After a time delay or a level device is tripped, the upper valve closes and the lower valve (going to the intake filter) opens. The lower hopper is then opened to the atmosphere so there is no differential pressure across the lower counterbalanced door. The upper door has been sucked shut so the product is collected in the upper chamber while the lower chamber empties. The cycle continues in this manner so there is a continuous flow of product without interruption at the pickup source.

The pressure version of this system is a blow pot. This system has a three-way valve on the blower discharge rather than on the blower intake (Fig. 28.4.12). A balance line goes into the top of the vessel with a manual balance valve assuring enough air pressure at the top of the vessel to force the product down into the air stream. The metering valve on the bottom is closed, the three-way valve is discharged to the atmosphere and the upper valve is open while the pressure pot is filled. The displaced air propagates down through the vent line to the conveying line and is vented out through the filters at the terminal end of the system.

When it is full as indicated by level controls, weigh cells, or a timing device, then the upper valve shuts, the three-way valve diverts the blower air to the conveying line to the discharge point. Then the metering gate opens to a preset point which gives the proper flow of product out of the hopper into the conveying line. In some cases for non-free-flowing product it may be necessary to fluidize the bottom of the pot or install a simple screw conveyor or other device to help the product to flow out the bottom of that hopper and meter it into the conveying line. But if it is free-flowing product, usually all that is needed is a simple metering gate. A continuous version of this consists of more than one pressure pot. While one is being filled, the other is being discharged.

Combining the vacuum and pressure provides a continuous vacuum–pressure setup without rotary valves (Fig. 28.4.13). Three hoppers are used, but still only two have counterbalanced gates. The upper hopper is always under vacuum, the lower hopper is always under pressure, and the intermediate hopper varies between pressure and vacuum. Normally this operates with the upper valve open, the lower valve closed. The lower hopper is then under pressure, the intermediate hopper is under the vacuum, as is the upper vessel. The upper valve is hanging open so the product is conveyed into the intermediate hopper. Then, again through a timing device or level controls, the upper valve closes,

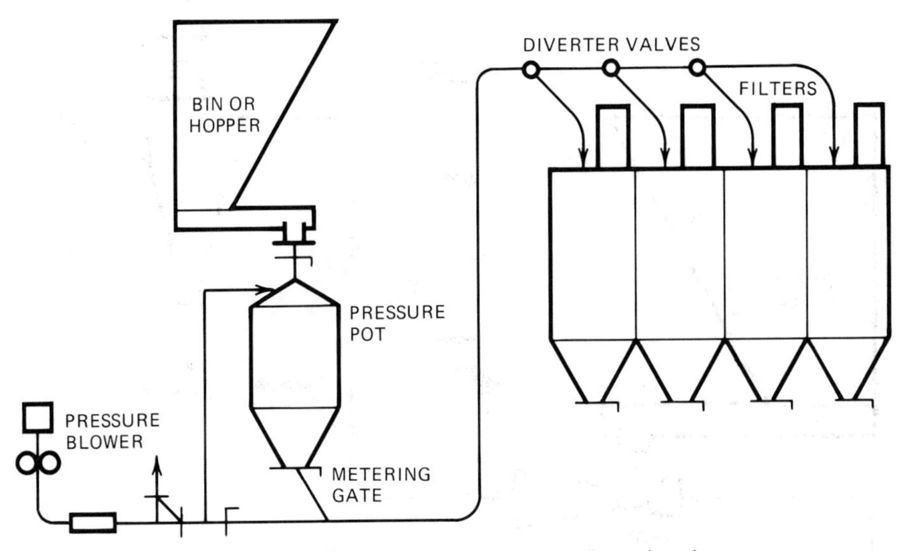

Fig. 28.4.12 Positive gravity-flow system (intermittent).

the lower valve opens, and the intermediate hopper now is under the same pressure as the lower hopper, thereby allowing the lower valve to swing open and the upper valve to swing shut so the product can pass from the intermediate hopper into the lower hopper.

Again after a time delay, the lower valve closes and the upper one again opens, repeating the cycle so that you have a continuous flow of product down through the hoppers from the negative side to the pressure side, without any rotary valves being utilized. These systems can be closed loop also, if needed, as indicated before.

Venturi System

The last type of dilute phase system is the venturi system (Fig. 28.4.14). In this system one should not exceed 1 psig or approximately 25 in. of water. At pressures higher than 1 psig the throat of the venturi becomes so small to overcome the pressure, that it is impractical to get any quantity of product

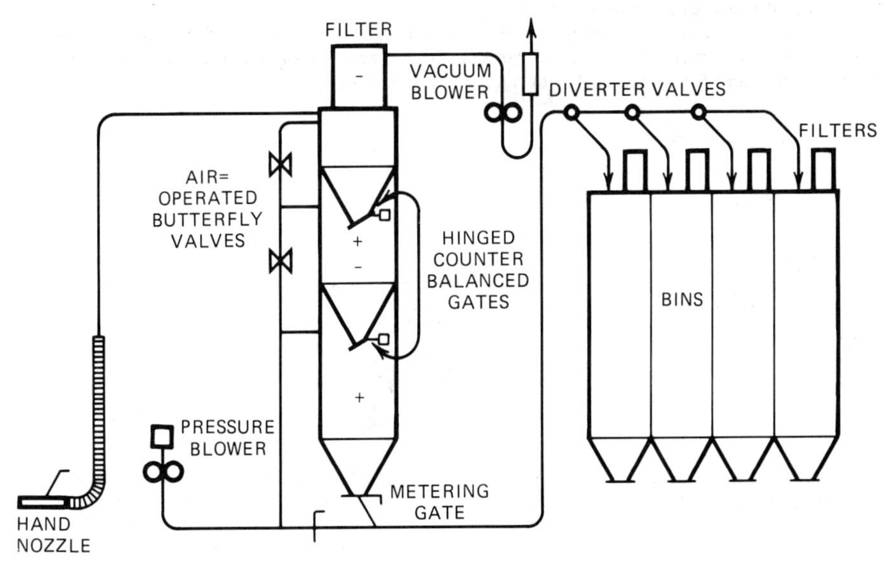

Fig. 28.4.13 Vacuum positive gravity-flow system (continuous).

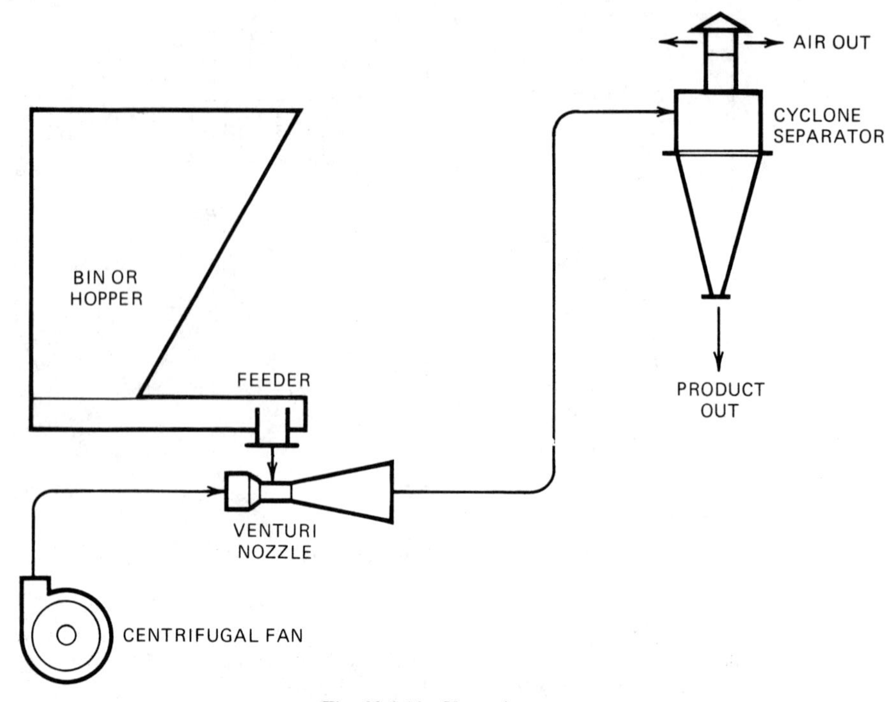

Fig. 28.4.14 Venturi system.

through. The pressure drop through the venturi also becomes excessive. There are venturi nozzles on the market that operate with the conveying system pressure around 4 or 5 psig. The upstream pressure ahead of the venturi, to get 4 or 5 psig downstream from the venturi, is about 11 or 12 psig. There might be a pressure drop in the venturi of 7 or 8 lb which is burned up in wasted horsepower. With this low-pressure type of venturi there is a similar loss, but with 25 in. downstream pressure, the upstream pressure would be about 40 in. of water, depending on the design of the venturi. The bigger the throat area, the less pressure drop from the intake side of the venturi to the discharge side. The velocity through the throat of the venturi to overcome 25 in. of water is 20,000 ft/min when the conveying velocity is 5000 ft/min, so it has one-fourth the area of the conveying line. Even at 25 in. of water the throat area is rather small.

This is a simple system consisting of a fan, venturi nozzle, a cyclone, and pipe and elbows. This type of system is particularly suited for handling highly abrasive or friable material. It can handle elbow macaroni and frozen vegetables, or aggregates, for example.

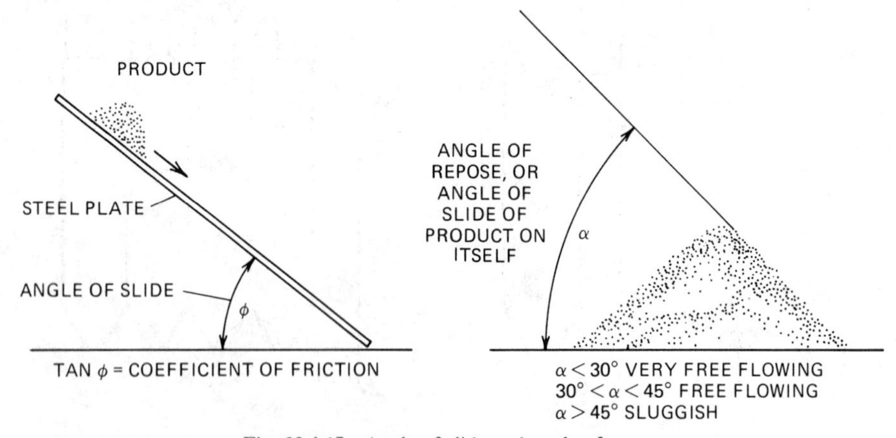

Fig. 28.4.15 Angle of slide and angle of repose.

28.4.2 Design Parameters

Before designing or selecting a pneumatic system, some factors must be known. Most important is the capacity to be conveyed, the distance over which the product will be moved, and the number of elbows involved. Elbows create large pressure drops in an air system; therefore, it is highly desirable to minimize the number of elbows. The number and type of pickup locations must be known, as well as any air requirements for them. If, for example, a piece of size-reduction machinery requires a certain amount of air purge for proper function, this information should be available; it, rather than the pressure drop, may dictate the size of the system to be selected. If any heating, cooling, or drying is to be accomplished while the product is being conveyed, these factors, too, could be the major sizing factor. The required velocity must be determined.

Is the conveying medium going to be air or gas? Is the product explosive or does it require inert gas blankets? A closed-loop system to condition the air for product preservation may be necessary. Should the system be open or closed? Are rotary valves permissible? Is the product nonabrasive so that rotary valves will work satisfactorily, or will it be degraded when passing through a rotary valve?

A pneumatic system has to be fed evenly to prevent plugging of the system. Is the system to be fed evenly by existing equipment or must the feeding device be included with the system?

28.4.3 Product Characteristics

The product characteristics are used to determine the velocities needed and the type of feeding devices to be utilized. The first item is bulk density of the product, an important factor in sizing volumetric devices or items such as rotary valves because the air system itself is not a volumetric machine. The particle size also has an influence on the conveying velocity needed, in addition to the bulk density.

The angle of slide on a steel plate is very important (Fig. 28.4.15). The tangent of the angle of slide is the coefficient of friction of the product on a steel plate. This is used in calculating pressure drop to determine the pipe size needed to convey the product over the distance and number of elbows. The angle of repose is that formed when a sample of the product is poured onto a flat surface. It is used to determine the type of metering and feeding devices coming out of bins or hoppers. Is the product free flowing or is it sluggish? A rule of thumb is that any angle of repose of less than 30° indicates the product to be very free flowing. Greater than 30° and less than 45° tells us that it is considered to be free flowing, and above 45° it is sluggish, which may necessitate a live-bottom construction or aeration of the product.

Can the product be fluidized and, if so, what is its apparent bulk density afterwards? Is the product friable? Is it abrasive? All these things can have an influence on the size and type of the elbows needed. If the product is extremely friable, elbows with a greater radius than normal may be required. Generally, elbows in a conveying line are 4 ft radius. Something extremely friable or abrasive might require an increase in radius to 6 or 8 ft to reduce the centrifugal force when the product goes around the elbow, to reduce abrasion on elbow or product.

Is the material hygroscopic, requiring conditioned air or a closed-loop system? Is it explosive? If so, a closed-loop inert gas system or explosion venting may be necessary. If the product is toxic or corrosive, a negative system may be called for so that there is not a possibility of leaking to the atmosphere.

Is the product coming into the system at an elevated temperature? If a closed-loop or a vacuum system is used, it may be necessary to temper the air or cool it down before it hits the filter or the intake of the rotary positive-displacement blower. Generally, most blower manufacturers do not like to exceed an inlet temperature of approximately 100°F. This varies and depends on the pressure differential across the blower, so it is a good idea to check with the manufacturer before selecting a blower and determining whether a heat exchanger ahead of the unit is necessary. Note that some suppliers call heat exchangers "aftercoolers" and normally install them on the discharge; therefore, they are unfamiliar with intake applications.

Another point of concern is *heat of fusion* of the product. If the product is one for which the heat of compression from the blower may cause problems with fusion, an aftercooler on the blower discharge should be considered. Generally, if the product is coming in at ambient conditions and the blower is operating at approximately 10 lb pressure, the discharge temperature may be as high as 200°F from the blower. If the product is coming in at 70°, the mixture temperature of the two—the air and the product—is only 5 or 10° above the original product temperature. A heat balance should be made to be sure.

Static Properties

If the product exhibits a tendency toward static electricity, it may be necessary to install static eliminators in the filter receivers, cyclones, or elsewhere to keep the product from building up on the filter bags or walls of the filter–receiver. It may also be necessary to add moisture to the conveying air to reduce the static properties.

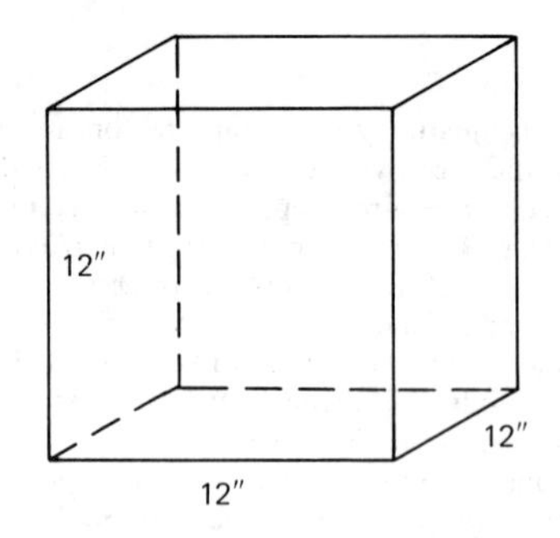

$$1\ FT^3\ H_2O = 62.4\ LB$$

$$\frac{62.4\ LB}{144\ IN.^2} = 0.4333\ LB/IN.^2 = 12\ IN.\ H_2O$$

$$\text{THEREFORE } 1\ LB/IN.^2 = 12\ IN.\left(\frac{1}{0.4333}\right) = 27.69\ IN.\ H_2O$$

Fig. 28.4.16 Basic pressure relationships.

Moisture Content of the Product

If the product has a high moisture content, a point of concern is whether any moisture will be lost to the conveying air and, if so, will it condense out somewhere. Possibly in a filter? If it does, it will form mud in the product around the filter bags. This will blind off the filter. Closed-loop systems are especially susceptible to this. Air coolers and water separators should be used or proper heat and/or insulation used to prevent condensation.

Fat and/or Oil Content

Fats and oils in the product may build up in the elbows, cyclone collectors, or wherever there are centrifugal forces that separate the fats and oils from the product. Sometimes adding heat to the elbow or the cyclone will prevent this buildup. Increasing the radius of the elbow is one possible answer; flexible hose can sometimes be used, and an occasional rap will clear it.

If the product has an objectionable odor, a closed-loop system may be called for or deodorizing of the exhaust air. The amount and type of impurities in the product are important considerations as they can completely alter the product's characteristics, such as abrasion.

28.4.4 Sizing the System

A few basic facts are needed to determine system size. One is that 1 psig is equal to 27.7 in. of water pressure or 27.7 in. of water in a column (Fig. 28.4.16). Next, the type of system must be selected: venturi, a fan, or blower system. The available pressure dictates the size of the conveying line. Figure 28.4.17 shows typical pressures available. One can get 45 in. of water out of a product

INDUSTRIAL FANS (AIR AND PRODUCT)	$= -45$ IN. H_2O $+ 50$ IN. H_2O
PRESSURE FANS (SINGLE STAGE)	$= -65$ IN. H_2O $+ 77$ IN. H_2O
MULTISTAGE OR SPECIAL LARGE DIAMETER	$= -120$ IN. H_2O $+ 170$ IN. H_2O
ROTARY POSITIVE-DISPLACEMENT BLOWERS	$= -218$ IN. H_2O (16 IN. HG) $+ 415$ IN. H_2O(15 PSIG)
VENTURI	$= +25$ IN. H_2O

Fig. 28.4.17 Typical pressures available.

$$KE = \frac{1}{2} MV^2 = \frac{WV^2}{2G}$$

$$\frac{1000 \text{ LB/HR}}{60 \text{ MIN/HR}} = 16.7 \text{ LB/MIN}$$

$$\text{IF } V = \text{FT/MIN}$$

$$\frac{(16.7 \text{ LB/MIN}) \left(\dfrac{V \text{ FT/MIN}}{60 \text{ SEC/MIN}} \right)^2}{2 \times 32.2 \text{ FT/SEC}^2} = 0.000072 \ V^2$$

$$@ \ 5000 \text{ FT/MIN} = 1800 \text{ FT-LB/MIN}$$

Fig. 28.4.18 Acceleration loss.

fan on a vacuum application. On a pressure application that same fan will develop 50 in. of water, because the fan's pressure is dependent on the density of the air at the inlet of the fan. On a pressure application the air is more dense at atmospheric pressure than when it is on a vacuum application. A pressure fan, not designed to tolerate any substantial amount of product passing through it, is a more sophisticated fan designed more efficiently with larger-diameter impellers and more blades. It generally develops approximately minus 65 in. of water in a single-stage unit and 77 in. on a straight pressure application.

The same fan in a two-stage series hookup develops about minus 120 in. of water vacuum or plus 170 in. of water in a pressure application. The rotary positive-displacement blowers produce a vacuum of 16 in. of mercury, which is about 218 in. of water or roughly 8 psig. Some blowers on the market can produce 22 in. of mercury vacuum but the compression ratio then begins to get too high. The terminal velocity gets high, but the pickup velocity can be quite low as the compression ratio increases. Fifteen inches of mercury has a 2-to-1 compression ratio and $7\frac{1}{2}$ psi differential pressure. The same blower that produces 15 psig, or roughly twice the pressure of 15 in. of mercury, also has a 2-to-1 compression ratio. Beyond this, the pickup velocity becomes critical because at 2-to-1 compression ratio, the pickup velocity is $\frac{1}{2}$ the terminal velocity.

The venturi system is limited to a maximum of 25 in. of water. Sixteen inches of water is better: at 16 in., a velocity of 16,000 ft/min through the throat of the venturi is needed, instead of 20,000 ft/min. This means that roughly one-third of the area of the pipe, rather than one-fourth, is needed, so the lower the venturi system pressure drop, the larger the throat area. The available pressures and selection of system type set the parameters for sizing the system. The available pressure drop cannot be exceeded when calculations are made. Different pipe sizes must be tried until the pressure drop calculation is within the limits of the pressure available. The calculations are the same regardless of the system type.

To get into the actual losses due to the product in the system three things must be considered. First is the acceleration loss (Fig. 28.4.18). When the product is introduced into the conveying line it must be started moving down the pipe, giving rise to the acceleration loss, equal to $0.000072 \ V^2$. The velocity is in feet per minute. At 5000 ft/min the acceleration loss for every 1000 lb/hr of product that is to be moved is equal to 1800 ft-lb/min.

The next loss to be considered is the product loss going around the elbows, which is the highest single loss (Fig. 28.4.19). This is $0.000226 \ V^2$ times the coefficient of friction as the foot-pounds per minute loss for each and every 90° elbow. If the velocity is 5000 ft/min and the coefficient of friction = 1.0, the elbow loss is 5650 ft-lb/min. Compared to the acceleration loss discussed earlier, only 1800 ft-lb/min for every 1000 lb/hr, the elbow loss is over three times greater than the acceleration loss for just one 90° elbow.

Horizontal and Vertical Losses. The loss is equal to the force that is the weight of the object times the coefficient of friction times the distance moved along the horizontal pipe (Fig. 28.4.20). If we elevate that product then we have the weight of that product times the distance we are going to elevate. So the losses are the same, with the exception that the horizontal would have a coefficient of friction in the equation and the vertical loss would not make the force, then, equal to the weight.

If we eliminate the negligible coefficient from the horizontal loss, and take the 16.7 lb/min times the distance moved when the distance is 60 ft, the result is 1000 ft-lb/min. It therefore takes 1000 ft-lb/min to move 1000 lb an hour over a 60-ft distance horizontally or vertically, neglecting the coefficient of friction in the horizontal loss calculation. The coefficient of friction (C.F.) should be used when long horizontal distances are encountered.

The loss around the elbow with a C.F. of 1 is 5.65 times as great as the horizontal and vertical distance loss for 60 ft. That would be equal to 339 ft, equivalent feet of pipe. So one 90° elbow is

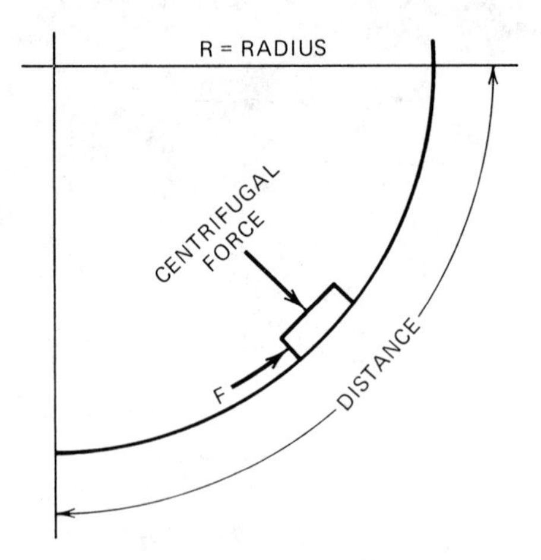

$$F = \text{CENTRIFUGAL FORCE} \times \text{COEFFICIENT OF FRICTION} = \frac{WV^2}{GR} \times \text{C.F}$$

$$\text{DISTANCE} = \text{ARC LENGTH} = \frac{2\pi R}{4} \text{ FOR } 90^\circ \text{ ELBOW}$$

$$\text{TOTAL LOSS} = \text{C. FORCE} \times \text{C.F.} \times \text{ARC LENGTH}$$

$$= \frac{WV^2}{GR} \times \text{C.F.} \times \frac{2\pi R}{4}$$

$$= \frac{WV^2}{G} \times \text{C.F.} \times \frac{\pi}{2}$$

$$= \frac{W\pi}{2G} V^2 \times \text{C.F.} \quad (V = \text{FT/MIN})$$

$$\frac{(16.7 \text{ LB/MIN})(\pi)}{2(32.2 \text{ FT/SEC}^2)} \left(\frac{V \text{ FT/MIN}}{60 \text{ SEC/MIN}}\right)^2 \text{C.F.} = \frac{16.7(3.14)}{2(32.2)} \left(\frac{V}{60}\right)^2 \text{C.F.} = \frac{16.7(3.14)}{64.4(3600)} V^2 \text{ C.F.}$$

$$= 0.000226 \; V^2 \text{ C.F. FT-LB/MIN}$$

@ 5000 FT/MIN, AND C.F. = 1 = 5650 FT-LB/MIN

Fig. 28.4.19 Elbow Losses.

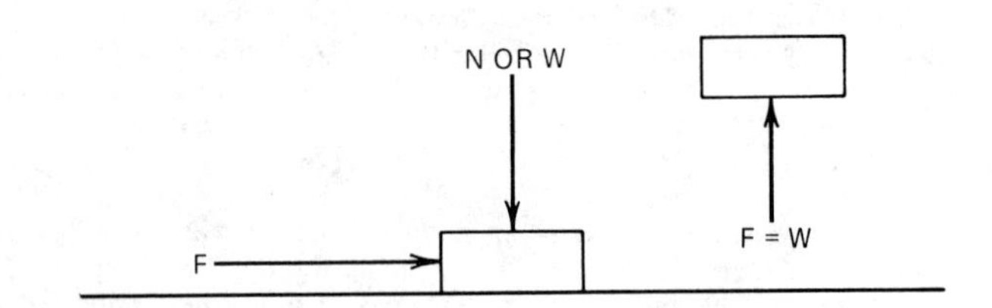

$$\text{WORK} = \text{FORCE} \times \text{DISTANCE} = (\text{LB/MIN})(\text{FEET}) = \text{FT-LB/MIN}$$

$$\frac{1000 \text{ LB/HR}}{60 \text{ MIN/HR}} = 16.7 \text{ LB/MIN}$$

$$16.7 \text{ LB/MIN} \times 60 \text{ FT} = 1000 \text{ FT-LB/MIN}$$

Fig. 28.4.20 Horizontal and Vertical Losses.

$$W = PV$$

$$\text{IF } P = 1 \text{ IN. } H_2O$$

$$V = 1 \text{ FT}^3\text{/MIN}$$

THEN:

$$W = \frac{1 \text{ IN.}}{27.7 \text{ IN./LB/IN.}^2}(144 \text{ IN.}^2\text{/FT}^2)(1 \text{ FT}^3\text{/MIN})$$

$$= \frac{1 \text{ IN.}}{27.7}\left(\frac{\text{LB/IN.}^2}{\text{IN.}}\right)(144 \text{ IN.}^2\text{/FT}^2)(1 \text{ FT}^3\text{/MIN})$$

$$= \frac{144}{27.7} \text{ FT-LB/MIN} = 5.19 \text{ FT-LB/MIN}$$

TO PUSH 1 FT^3/MIN AGAINST 1 IN. H_2O

Fig. 28.4.21 Moving against static pressure.

equivalent to moving that product 339 ft either vertically or horizontally, with a C.F. of 1. It is, therefore, very important to minimize the number of elbows in a dilute-phase pneumatic system.

It takes 5.19 ft-lb/min to move 1 ft^3/min against 1 in. static pressure (Fig. 28.4.21). Converting these figures to pressure, we have the acceleration at 1800 ft-lb/min, the elbow loss for one elbow is 5650 ft-lb/min, 60 ft of pipe is 1000 ft-lb/min, so the total is 8450 ft-lb/min to move 1000 lb/hr over 60 ft through one elbow. If the amount being conveyed is 2000 lb/hr, we then have two times 8450, which gives a total pressure drop for 2000 lb divided by 5.19 or 5.2 ft-lb/min to move 1 ft^3/min against 1 in. of water, and also divided by the system ft^3/min. For a 6-in. system handling 1000 ft^3/min, the result is equal to 3.256 in. of water pressure drop for the material losses alone.

The next topic is air friction losses. To determine the system size, all the pressure drops must be known. From published information the entry loss to bring the air into a well-designed air inlet is 1.24 Hv. The Hv is velocity pressure in inches of water. The branch loss (the loss any time there is a Y-branch in the conveying line, a pickup tee, or manifold, etc.) is equal to 0.2 Hv. A high-efficiency cyclone collector normally has a pressure drop of 2 Hv. Consult the manufacturer's tables and charts to find out what the pressure drop of a particular collector may be, since there are low-efficiency cyclones on the market that have a relatively low pressure drop and other high-efficiency collectors that may have a higher pressure drop. Filters usually, if sized and operated properly, operate in the range of 4–6 in. of water. Hv for 5000 ft/min is equal to 1.56 in. of water. The entry loss is then 1.24 times 1.56, or 1.934. Side branches are 0.312 H_2O. Cyclones are 3.12. To determine the loss due to the friction of air moving through the pipe, we refer to a pressure drop chart and determine the pressure drop, given in inches of water per 100 feet of pipe as in Fig. 28.4.22.

Let's take a look at the sample calculations (Fig. 28.4.23). This system is going to convey plastic pellets that have a 0.5 coefficient of friction. First calculate the material losses, because generally the material losses are higher than the air losses, but this is not always true. It depends on the capacity, distances, and so on, therefore, calculate the material losses first to obtain a preliminary pipe diameter, then calculate the air losses.

Total the two to see if they are within range of the pressure source that has been selected for the air system. The material losses, at 5000 ft/min, are acceleration 1800 ft-lb/min. Then add two elbows with a coefficient of 0.5, which equals 5650 ft-lb. The line loss between is 120 ft or 2000 ft-lb, giving a total of 9450 ft-lb/min per 1000 lb/hr. Since the capacity is 30,000 lb/hr, multiply 9450 times 30 and then divide by 5.19. Then try various pipe sizes. The first one tried was 4-in. Sch. 10 pipe which, to give 5000 ft/min, requires 495 ft^3/min. Using that figure gives a drop for the product of 110 in. of water. Since a rotary positive-displacement blower is being used here, the available pressure will be 10 or 12 lb/in.^{2}. Ten pounds is equivalent to 277 in. of water, so 110 in. looks like it is a little bit on the low side. Therefore try 3 in., which is 290 ft^3/min. The material losses are 187.99 in., which begins to look a little more in range. Then just to be sure try a 2-in. line, which requires 127 ft^3/min. The pressure drop for the material alone is 429.28 in. of water, which is well in excess of 10 lb, so that pipe is too small. Go back to the 3 in., which is the next line size that appears to be in the ball park and calculate the air loss. This gives an entry of 1.9 side branch or pickup tee 0.3. For the line loss of 120 ft multiply 1.2 times the pressure drop in inches of water per 100 feet of pipe at that velocity in a 3-in. Sch. 10 pipe. This is read off the chart to be 14 in., so the total friction loss for the air through that 120 ft of pipe is 16.8 in. of water. For the cyclone add 3 in., so that the total air friction loss moving through that pipe arrangement is 22 in. Add that to 187.99 in. of material loss giving a total loss in that system for product and air of 209.99 in. of water. Divide that by 27.7 to get it into psig to select the blower. This gives 7.58 psig.

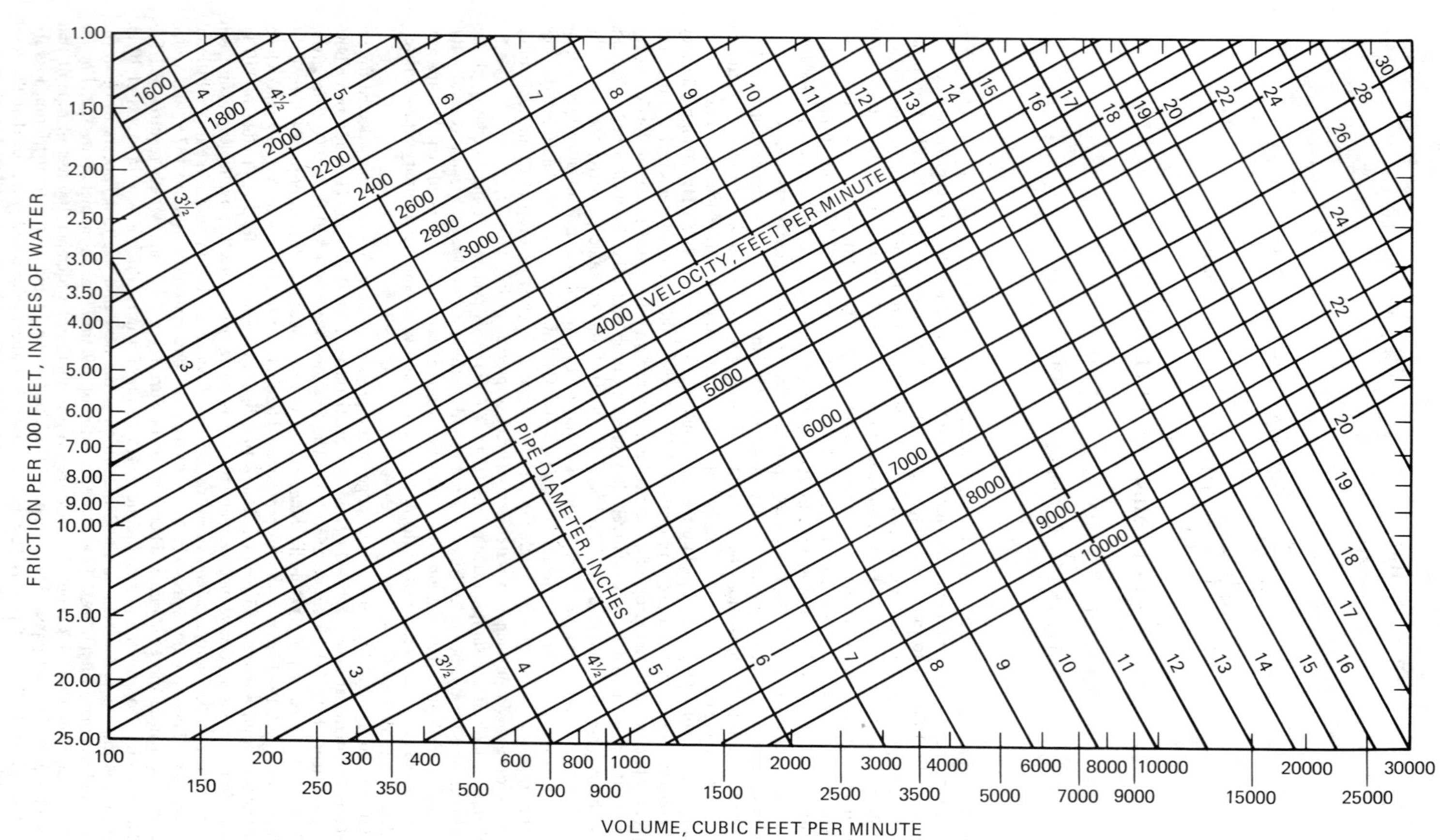

Fig. 28.4.22 Friction loss in round pipes.

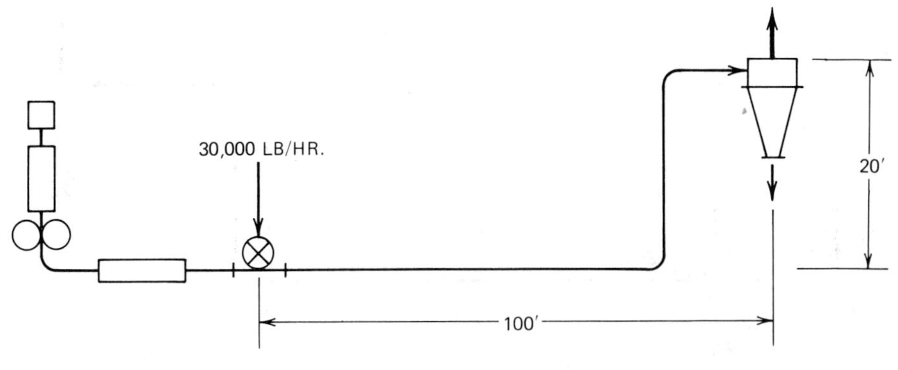

30 LB/FT³ PLASTIC PELLETS
5000 FT/MIN C.F. = 0.5

MATERIAL LOSSES	AIR LOSSES
ACCELERATION = 1800	ENTRY = 1.9
ELBOW = 5650 × 0.5 × 2 = 5650	SIDE BR. = 0.3
LINE LOSS = 16.7 × 120 = 2000	LINE LOSS = 1.2 × 14 = 16.8
$\frac{9450 \times 30}{5.2 \times 495} = 110.14$	CYCLONE = 3.0
	22.0
290 = 187.99	187.99
127 = 429.28	209.99 IN. H_2O
	$\frac{209.99}{27.7} = 7.58$ PSIG
	USE 9 PSIG, 3-IN., SCH. 10 PIPE

Fig. 28.4.23 Sample loss calculations.

For a little bit of reserve, select the blower for $8\frac{1}{2}$–9 psig depending on how much reserve is desired. This usually affects only the size of the motor when selecting the blower size. Therefore, 290 ft³/min in a 3-in. Sch. 10 pipe at 8.5 or 9 psig will do that job.

Velocity. Velocity considerations must be made next (Fig. 28.4.24). In a pressure system much like the system just calculated, the air coming in is the blower inlet SCFM or standard ft³/min (standard air at standard conditions). CFM at the pickup point is CFM_1 and CFM at the terminal point of the system is CFM_2. The CFM coming into the blower inlet and the CFM going out the end of the system have to be the same with the exception of any losses that have occurred in the system. The velocity is equal to the CFM divided by the cross-sectional area of the pipe, so V_1 (pickup velocity) and V_2 (the terminal velocity) show the same relationship as that between CFM_1 and CFM_2. The CFM_1 is equal to the air coming into the blower times the compression ratio, and if this system is operating at 10 psig, the compression ratio is 14.7 over 24.7, or 0.595. Substituting, V_1 is equal to 0.595 V_2. The average velocity in the system is V_1 plus V_2 divided by two; substituting this relationship of V_1 and V_2 into that equation, the average velocity is equal to 0.7975 V_2 at 10 psig. If this material is to be conveyed at 5000 ft/min, V_2 is equal to 6270 ft/min. This can be converted to CFM by multiplying it by the cross-sectional area of the pipe in square feet. V_1 is 3730 ft/min.

Another velocity consideration has to do with the distances to be conveyed (Fig. 28.4.25). Generally, for 200 ft total conveying distance, one can convey with an average velocity of 4000 ft/min. This is equivalent to a terminal velocity of 5000 ft/min at 10 psig. From 200 to 500 ft one would have to go to an average velocity of 5000 ft/min with that same product, and between 500 and 1000 ft, to 6000 ft/min.

These figures are modified depending on the product. If one has not conveyed the material and has no experience, it might be a good idea to run a conveying test to be sure that the material is going to behave as expected. The bulk density of the product is also considered to determine the necessary velocity. If the product is a $\frac{1}{4}$ in. or smaller particle, and weighs up to 55 lb/ft³, use the velocities as previously given. If, however, it is greater than 55 lb/ft³ but less than 85 lb/ft³, increase these by 1000 ft/min. If it is greater than 85 lb/ft³ but less than 115 lb/ft³, increase the velocities by 2000 ft/min. For particles larger than $\frac{1}{4}$ in. and above 115 lb/ft³, check the minimum velocity based on the particle size or the density of an individual particle of the product (Fig. 28.4.26). Assume a 1-in. cube is to be conveyed and weighs 0.144 lb, and the area is 1 in.². This means the pressure to

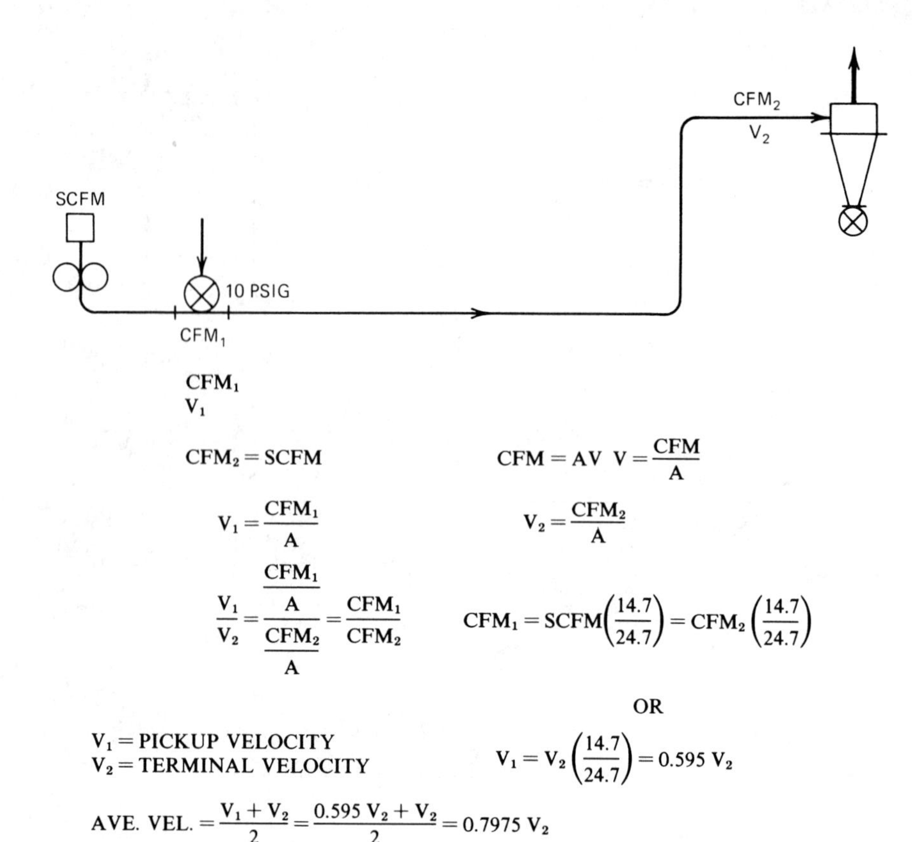

CFM_1
V_1

$$CFM_2 = SCFM \qquad CFM = AV \quad V = \frac{CFM}{A}$$

$$V_1 = \frac{CFM_1}{A} \qquad V_2 = \frac{CFM_2}{A}$$

$$\frac{V_1}{V_2} = \frac{\frac{CFM_1}{A}}{\frac{CFM_2}{A}} = \frac{CFM_1}{CFM_2} \qquad CFM_1 = SCFM\left(\frac{14.7}{24.7}\right) = CFM_2\left(\frac{14.7}{24.7}\right)$$

OR

V_1 = PICKUP VELOCITY
V_2 = TERMINAL VELOCITY

$$V_1 = V_2\left(\frac{14.7}{24.7}\right) = 0.595\ V_2$$

$$\text{AVE. VEL.} = \frac{V_1 + V_2}{2} = \frac{0.595\ V_2 + V_2}{2} = 0.7975\ V_2$$

If 5000 FT/MIN AVE. VEL. USED,

THEN:

$$0.7975\ V_2 = 5000\ \text{FT/MIN} \qquad V_2 = \frac{5000}{0.7975} = 6270\ \text{FT/MIN}$$

$$V_1 = 3730\ \text{FT/MIN}$$

Fig. 28.4.24 Velocity considerations.

DISTANCE CONSIDERATION

TOTAL DISTANCE	AVERAGE VELOCITY
200 FT	4000 FT/MIN
500 FT	5000 FT/MIN
1000 FT	6000 FT/MIN

BULK DENSITY CONSIDERATIONS

BULK DENSITY	AVERAGE VELOCITY
UP TO 55 LB/FT^3	USE ABOVE
$>55\ LB/FT^3$, $<85\ LB/FT^3$	INCREASE ABOVE BY 1000 FT/MIN
$>85\ LB/FT^3$, $<115\ LB/FT^3$	INCREASE ABOVE BY 2000 FT/MIN
$>115\ LB/FT^3$	CHECK MINIMUM VELOCITY FOR PARTICLE SIZE (SEE FIG. 28.4.26)

Fig. 28.4.25 Distance and bulk density considerations.

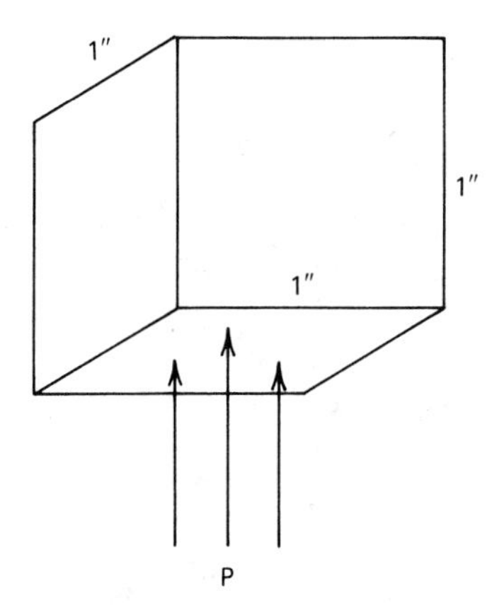

$$P = \frac{\text{WEIGHT}}{\text{AREA}} = \frac{0.144\ \text{LB}}{1\ \text{IN.}^2} = 0.144\ \text{PSI} = 4\ \text{IN. H}_2\text{O}$$

PSI = VELOCITY PRESSURE

$$V = 4000\sqrt{\text{Vp}}$$

$$V = 4000\sqrt{4} = 8000\ \text{FT/MIN}$$

Fig. 28.4.26 Minimum velocity for large particles.

support that particle is 0.144 psi, which is equivalent to 4 in. of water. This dictates a minimum velocity of 8000 ft/min to convey that product.

28.4.5 Sizing and Selecting Components

The third part of this outline deals with sizing of the components for a pneumatic system. These components are primarily the rotary valve, cyclone collectors, and filters. The prime movers of the air are the fans and/or blowers.

Rotary Valve

The one item in a pneumatic system that is hardest to properly size is the airlock or the rotary valve (Fig. 28.4.27). These rotary vane valves can be airlocks, or feeder valves, or a combination of the two.

One type is the gravity feeder. It is usually underneath a container, bin, or hopper where it measures out an amount of product into another piece of equipment without any differential pressure across the valve. There is no pressure differential to consider, so air flow and leakage are not concerns. To size such valves, take the displacement figure of the rotor in cubic feet per revolution and figure out the speed the valve must run to deliver and multiply by cubic feet per hour.

Another typical gravity feeder application is the same unit feeding out of a container into a negative system at the pickup point, such that air leakage will be down through the rotary valve in the same direction of flow as the product. This indicates that there is no reverse flow of air or blowback to consider. This valve is sized the same as the previous one because it, too, serves as a metering device with no regard for any problems with reverse air flow.

An airlock serving as an air seal only, is used when introducing product from another feeding device into a positive-pressure pneumatic system. This valve's primary function is to minimize the amount of air going back out of the valve, to insure that the air in the conveying line is going to be adequate to maintain conveying velocity.

Another application of a pure airlock is at the outlet of a vacuum system. This is similar to the inlet of a pressure system. There is a pressure differential across the valve and the leakage air is going to move from atmosphere into the vacuum receiving vessel in a reverse direction to the product passing through the valve. The product coming into this vacuum receiver is fed into the vacuum system by another means, so that it is on stream with no head of product above the valve. This is the way to differentiate between a feeder and an airlock. If there is a head of material above the valve, it is considered a feeder. If it is sealing a pressure differential at the same time, it is an airlock

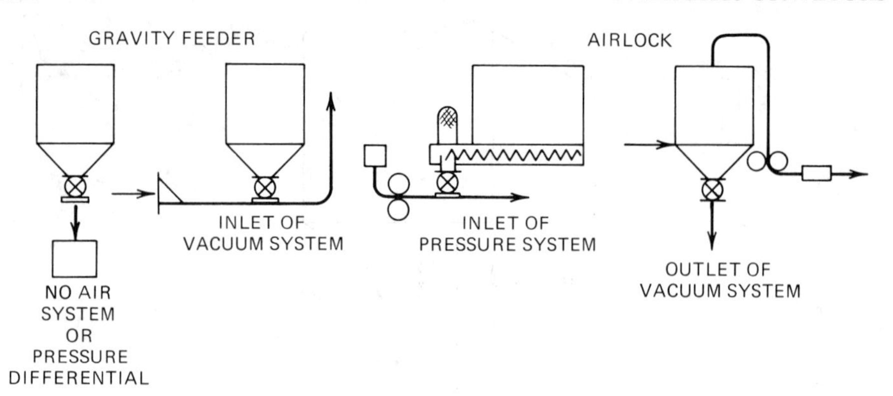

SELECTION TABLE

SIZE	FT³/REV.	NORMAL RPM FOR AIRLOCK	TOTAL DISP. (FT³/HR)	50% TOTAL DISP. (FT³/HR)	GRAVITY FEEDER, 100% CAPACITY AT 25 RPM
4 × 3	0.0119	45	32	16	18
6 × 4	0.03	45	82	41	45
8 × 6	0.116	45	314	157	171
10 × 8	0.24	45	648	324	361
14 × 10	0.68	45	1,836	918	1,021
16 × 14	1.2	45	3,306	1,653	1,800
20 × 18	2.5	35	5,250	2,625	3,750
24 × 22	4.5	31	8,370	4,185	6,760
30 × 26	8.3	30	14,940	7,470	12,450

EXAMPLE:

REQ'D	AIRLOCK SIZE AND RPM		AIRLOCK FEEDER SIZE AND RPM		GRAVITY FEEDER SIZE AND RPM	
500 FT³/HR	14 × 10	45	14 × 10	13	14 × 10	13
1000 FT³/HR	16 × 14	45	16 × 14	14	14 × 10	25
3000 FT³/HR	24 × 22	31	24 × 22	11	20 × 18	20

Fig. 28.4.27 Sizing rotary valves.

feeder. If it is just a feeder, it would have a head of material with no pressure differential. If it is just an airlock, it would have no head, but does have pressure differential.

The table immediately below the illustrations in Fig. 28.4.27 shows the criteria for picking a size of valve for a particular application. The first column gives the size of the valve. The diameter of the rotor is first, and the length second. A 4 × 3 valve has a 4-in. diameter rotor, 3 in. in length. The next column is the cubic foot displacement of the rotor every time it makes a revolution.

The third column is the normal speed for that valve if it is acting as an airlock at the outlet of the vacuum system or at the inlet of a pressure system. These speeds have been determined from experimental work; above them centrifugal forces do not allow the product to enter the rotor properly, and the rotor pocket is not exposed to the product long enough to allow proper filling of the pocket.

The next column gives the total displacement of the rotor in cubic feet per hour at the recommended maximum airlock speed. The next column is exactly 50% of the total displacement to allow for fluidization or aeration of the product that takes place in an airlock situation. The air that is blowing out through the pocket tends to suspend the product if it is a light and fluffy material or aerates readily. The 50% figure is a safety factor to allow a reserve for inefficiency of pocket filling and aeration.

The last column shows the displacement of the rotor at 25 rpm in cubic feet per hour. Twenty-five rpm is the maximum optimum speed when the valve is operating under a head of material serving as a feeder rather than as an airlock. This is a safe speed to insure proper filling of the pockets of the rotor to give maximum efficiency. If, for example, 500 ft³/hr is needed as an airlock, the 50% column shows that a 14 × 10 rotary valve displaces 918 ft³/hr. It would operate at 45 rpm.

Suppose that same valve is going to be also a feeder. It is going to be not only in a position as an airlock, but it is also going to be called upon to meter that product into or out of that system.

Since it is an airlock feeder, the valve is selected from the next to last column indicating that it is a 14 × 10 valve. The speed of the valve to feed 500 ft³/hr is determined by dividing the cubic foot per revolution into the required displacement per minute. This indicates 13 rpm of a 14 × 10 valve will give approximately 500 ft³/hr at 100% efficiency. The drive for that valve is selected to give 13 rpm. Chances are that, in reality, that valve will never do that job at 13 rpm because of the aeration and so on, so the valve would have to run faster. It is a simple matter to change sprockets or sheaves to increase the speed. It is always a good idea to select the drive for the slowest speed because it can be increased rather easily.

The last column is strictly a gravity feeder. For 500 ft³/hr, a 14 × 10 valve, operating at 13 rpm is needed, so, in that case, both valves are the same.

For 1000 ft³/hr, the next to the last column indicates a 16 × 14 valve operating at its standard airlock speed of 45 rpm is needed. If it were to be an airlock feeder, that valve would run at 14 rpm to accomplish the same job. A pure feeder for this application from last column is a 14 × 10 valve at approximately 25 rpm. A valve for gravity feeding is a little bit smaller than when it is called upon to be both the feeder and airlock or simply an airlock.

Three thousand cubic feet an hour as an airlock would require a 24 × 22 valve operating at 31 rpm; as an airlock feeder (because of its displacement of $4\frac{1}{2}$ ft³/rev) it would operate at 11 rpm. If it were a pure gravity feeder, it would use a 20 × 18 valve at 20 rpm.

One of the problems with rotary valves, especially when they are operating under a head of material going into a pressure-type pneumatic system (Fig. 28.4.28), is the venting of the valve to take care of the blowback air or the leakage air that is going to be passing through the valve because of the clearances and the displacement of the rotor as it comes back up empty. In such a situation, a vent hopper with an internal baffle is needed to form an air chamber so that the air can get out of the pockets of the rotor as they come around and the material can then enter the pockets of the rotor. If material is being metered into the valve, so the valve is strictly an airlock, then an internal baffle is not required. In fact, an internal baffle can become a detriment if it is not needed because it then creates a high-velocity area on the vent side of the baffle and can throw more product and material up the vent line. Since the valve is being fed to, it would have a stream of material coming in; thus the entire vent hopper is a settling chamber.

A side inlet valve takes advantage of the angle of repose of the product so that it is impossible to get 100% pocket filling. With the rotor turning counterclockwise in this case (Fig. 28.4.29), the material then falls back into the rotor pocket, eliminating any possibility of shearing the product between the rotor and casing bore.

These valves generally have a built-in vent that can then be vented to a dust system, vent sock, or perhaps back into the top of the storage bin from which you are conveying. If a side inlet valve is used at the outlet of a vacuum system, for example, the blowback air must be vented (Fig. 28.4.30).

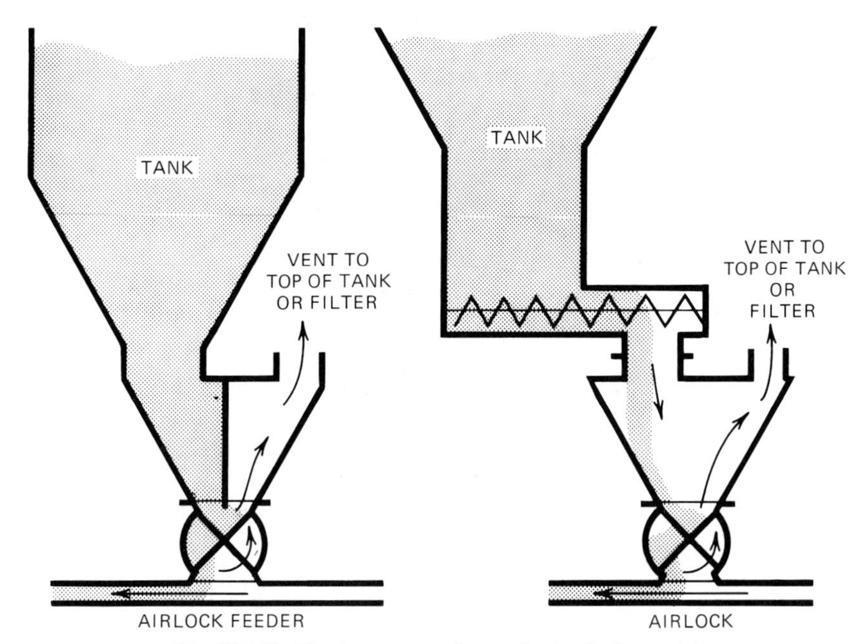

Fig. 28.4.28 Venting rotary valves under head of material.

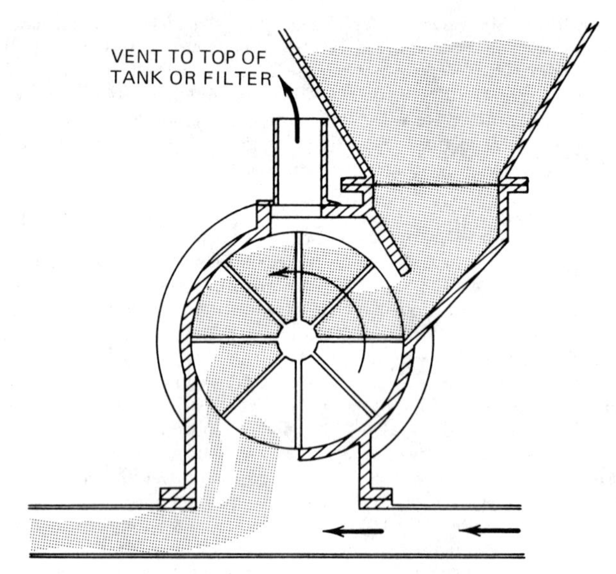

Fig. 28.4.29 Side inlet air lock feeder valve.

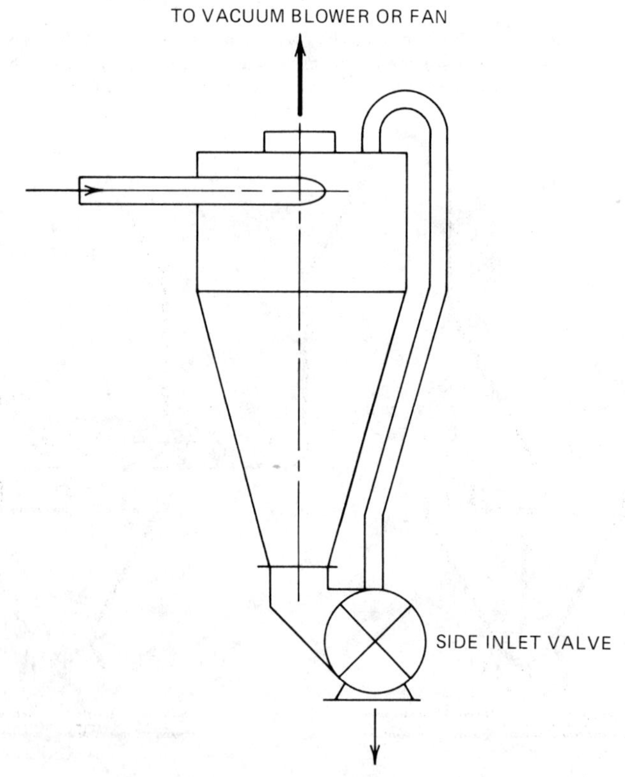

Fig. 28.4.30 Cyclone under vacuum.

If not, the air leakage through the valve would pass up through the same throat area where the product is trying to enter the rotor, and this would reduce the efficiency of the valve substantially.

Cyclone Collector

Manufacturers' charts are available and should be referred to when selecting a cyclone for pneumatic systems (Fig. 28.4.31). The charts usually give the diameter in the first column, the CFM they are rated for plus the major dimensions. A cyclone collector is designed with the inlet, the exhaust stack on top, and the diameter of the cylinder selected for the volume of air or gas that is being handled. The only thing on a cyclone that is sized for the product being handled is the conical outlet. This is the only place where the bulk of solid material is going to be moving. The amount of air or gas to be handled is what dictates the size of the collector. If a rotary valve should be installed underneath the cyclone, then chances are the conical outlets will be sized to meet the inlet of the rotary valve. Generally, the pressure drop is two velocity heads, but the manufacturers' tables must be used to determine the drop for a particular unit. Normally, the smaller the cyclone, the more efficient the cyclone and, of course, the higher the pressure drop.

Filter

Sizing of the filter–receiver or secondary filters on pneumatic systems involves two types. One is the intermittent shake-type unit which has the product dust collected on the inside of its filter bags. In order to clean such a unit, the air flow to the filter bag must be stopped while it is being shaken. Otherwise, the material is re-entrained into the filter bags. Two or more of these can be compartmented so that while one compartment is down and shaking, the others are in service to obtain continuous operation.

One of the drawbacks with this type of a filter is that during shutdown and bag shaking, there is a sudden slug of product landing in the filter hopper, where it can then bridge and create a problem in emptying the hopper.

In a continuous automatic reverse air filter, the dust is collected on the outside of the bags and a header with compressed air blows back through the bags to force the dust off the outside of the bag. These are used widely in pneumatic systems; one of the advantages is the bag is continuously cleaned so there is no sudden slug of material into the hopper, as with the intermittent shake type. This usually means one can get by with just an airlock seal at the bottom of the filter hopper without any danger of bridging. The intermittent shake units are generally rated at 3–4 ft^3/min per square foot of cloth area and can be considerably less, depending on the product being handled. The continuous reverse air filter can operate at a ratio of 6 or 12 ft^3/min to 1 ft^2, depending on the product.

On extremely hard-to-filter materials this might be as low as 2 or 4 to 1. One consideration in selecting filters (particularly if there is a light, fluffy material with a large surface area) is the can velocity. Most products use a can velocity of 3 ft/sec, but on extremely hard-to-filter products it might be as low as 2 ft/sec.

Rotary positive-displacement blowers are selected from the manufacturer's tables or curves for the ft^3/min and pressure or vacuum needed. However, the blower should be selected to have at least 25% reserve capacity in ft^3/min. This is to allow the blower to be operated at a lower speed in order to reduce noise. Also, this allows fine tuning of the system after it starts up. The velocity can be increased if necessary.

Fans

In selecting fans, the manufacturer's charts should be consulted. The fan tends to run wild, making motor selection difficult.

A typical fan curve is shown in Fig. 28.4.32 to help illustrate the run-wild situation. In this case, if the air loss is 30 in. H_2O at the design conditions and the material losses are 20 in., then the total drop is 50 in. H_2O. The air losses are plotted on the curve along with the air and material losses.

Where the air and material loss curve crosses the static pressure curve of the fan, is where the fan will operate while conveying. Hence the fan will consume 21 BHp. When the material is no longer being fed to the system, then the fan will operate where the air loss curve crosses the static pressure curve. The fan will now consume about 26 BHp, requiring a 30-Hp motor to allow for this.

This run-wild horsepower can be calculated by taking the horsepower at design conditions times the square root of the total pressure over the air loss. In this case it would be 21 Hp times the square root of 50/30, which equals 27.11; therefore, a 30-Hp motor is necessary.

If an automatic air flow control valve were used, the valve would add resistance equal to the material losses so that the ft^3/min or velocity would remain constant when the material feed was stopped. This would allow a 25-Hp motor to be used.

The air control valve operates from relays that sense the fan motor amperage to adjust the control valve so the amperage or horsepower remains constant at all operating conditions.

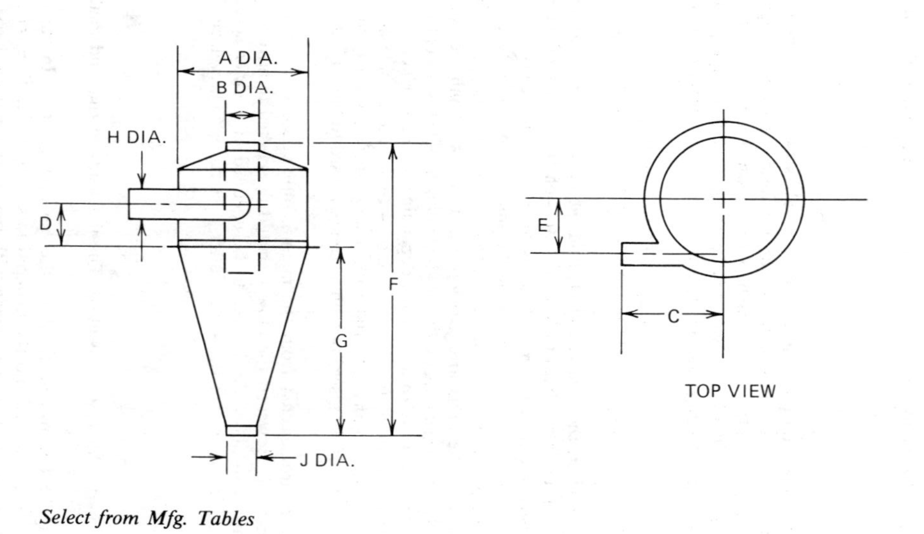

Select from Mfg. Tables

A	CFM	B	C	D	E	F	G	H	J
10″	100–155	4″	8″	$6\frac{1}{4}$″	$4\frac{1}{8}$″	2′–11″	2′–2″	$1\frac{1}{2}$″	3″
12″	155–225	5″	10″	$6\frac{1}{4}$″	$4\frac{7}{8}$″	3′–$4\frac{3}{4}$″	2′–$6\frac{1}{2}$″	2″	4″
14″	225–315	6″	10″	7″	$5\frac{3}{8}$″	3′–11″	2′–$11\frac{1}{2}$″	3″	4″
16″	315–425	7″	—	—	—	4′–$4\frac{3}{4}$″	3′–4″	—	6″
20″	425–630	8″	12″	$10\frac{1}{2}$″	$7\frac{7}{8}$″	5′–$4\frac{1}{2}$″	4′–1″	4″	6″
24″	630–910	10″	15″	$12\frac{1}{2}$″	$9\frac{3}{8}$″	6′–$6\frac{1}{2}$″	4′–11″	5″	6″
28″	910–1235	11″	18″	$14\frac{1}{4}$″	$10\frac{7}{8}$″	7′–$6\frac{3}{4}$″	5′–$8\frac{3}{4}$″	6″	8″

Sized for CFM @ 2 Hv
1. Inlet
2. Diameter
3. Exhaust stack

Outlet sized to fit airlock if used.

Pressure drop increases by square of CFM. Theoretically more eff. @ larger volumes. Oversized collectors desirable for low-density products with large surface areas.

Fig. 28.4.31 Chart for selecting cyclone for pneumatic system.

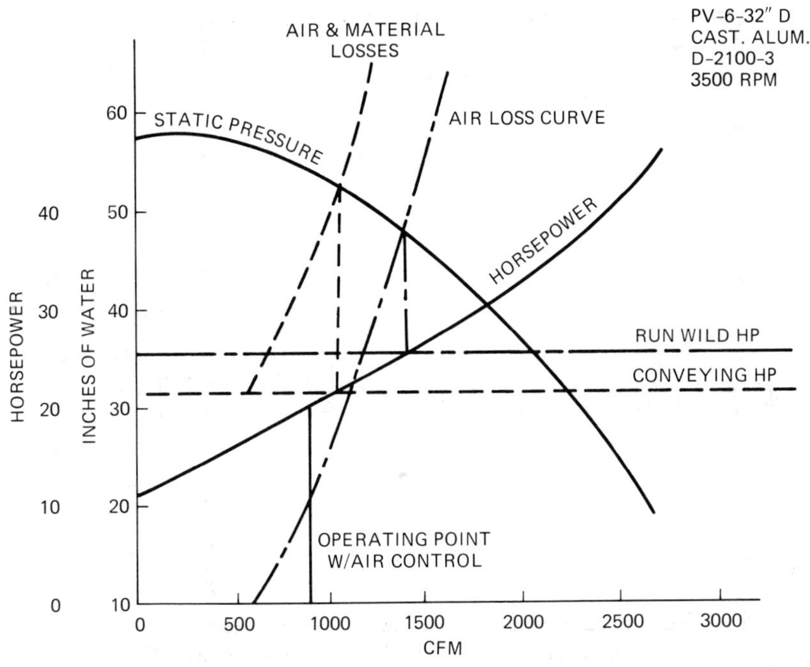

Fig. 28.4.32 Run-wild situation depicted in typical fan curve.

28.5 AIR-ACTIVATED GRAVITY CONVEYORS

Ivan Stankovich

The air-activated gravity conveyor (AAGC), also known under different trade names, such as Airslide (Fuller Co.), Air-Float (Kennedy Van Saun Corp.), Air Trough (Halliburton Co.), Flo-Tray Gravity Conveyor (Ducon Co.), and others, is a device for transporting "fluidizable" (Section 28.10) dry bulk materials from a higher to a lower point, using the gravity force and low-pressure air to fluidize the conveyed material (Fig. 28.5.1).

The work of AAGC is based on the principle that fluidized material can flow in a certain direction continuously. This is achieved by: (1) fluidization of the bulk material (by reduction of the angle of friction both between the conveyed particles and between the conveyor walls and the particles) and (2) by declining the conveyor.

28.5.1 Capacity of AAGC

The capacity of the AAGC depends on: (1) the physical characteristics of the conveyed material, (2) the size of the AAGC, and (3) the angle of declination of the AAGC. However, in practice, one of the advantages of AAGC over the gravity chutes is the possibility of near horizontal conveying. Therefore, the angle of declination is usually held to a minimum for a certain type of conveyed material and level of fluidization. Approximate AAGC capacities are given in Table 28.5.1.

28.5.2 Porous Media

The porous or fluidizing media (Fig. 28.5.1) is the most important part of the AAGC. The purpose of the porous media is to evenly distribute the fluidizing air over the entire conveying surface. An ideal membrane is usually described as having low permeability, expressed as cubic feet per minute of air per square feet of media surface, per inch of water gauge pressure. For example, a typical requirement for a woven fabric porous media is: permeability = 2–5 ft^3/min at 2 in. of water gauge differential pressure (Fig. 28.5.2).

Following are some requirements for the porous media: (1) pore size shall be smaller than the particles of the conveyed material, to prevent blinding or leakage through the pores; (2) $\Delta P_{pm} \geqslant \Delta P_{cm}$, where ΔP_{pm} = pressure drop through porous media, and ΔP_{cm} = pressure drop through fluidized conveyed material; (3) pressure drop through the porous media should be even, otherwise conveying will be sporadic.

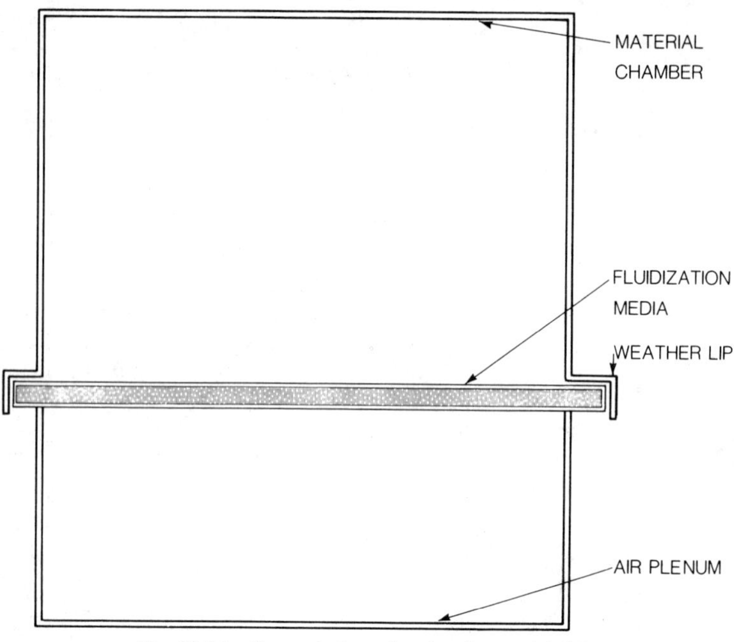

Fig. 28.5.1 Cross section of enclosed-type AAGC.

Different porous fluidizing media are used for AAGC (Table 28.5.2). The selection will depend on the temperature, physical and chemical characteristics of conveyed material, and cost of porous media.

28.5.3 Angle of Declination

The minimum angle of declination (slope) of an AAGC depends basically on the physical characteristics of the conveyed material and shall be determined experimentally for each type of bulk material (Section 28.10). Often the same materials, but with different size distributions, will require different minimum slopes. Basically, porous media has a minor effect on the minimum required conveyor slope. For most bulk materials the minimum required AAGC slope is between 1 and 8°.

Table 28.5.1 Approximate Capacities of Air-Activated Gravity Conveyors[a]

Nominal Width of Air Trough		Approximate Capacity of Air Trough	
(In.)	(mm)	(ft^3/hr)	(m^3/sec)
6	152.4	1,000	0.472
8	203.2	2,000	0.944
10	254.0	3,000	1.416
12	304.8	6,000	2.831
14	355.6	8,000	3.775
16	406.4	10,000	4.719
20	508.0	14,000	6.607
24	609.6	18,000	8.494

Source: Courtesy of Halliburton Co.
[a] For a certain type of conveyed material and angle of declination.

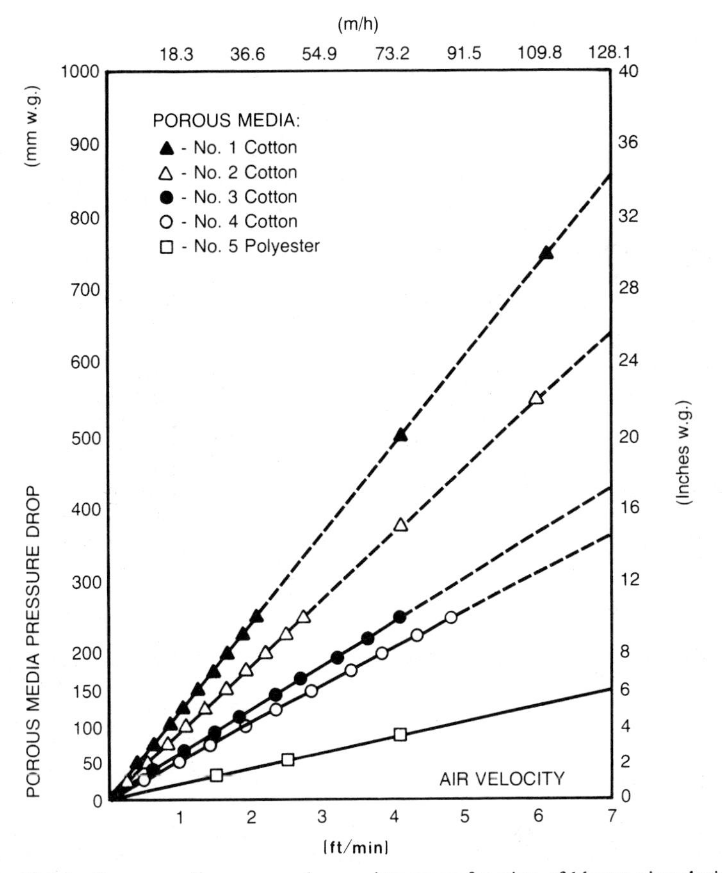

Fig. 28.5.2 Porous media pressure drop resistance as function of blower air velocity.

28.5.4 Air Requirements and Blower Piping

Low-pressure blower air is used to fluidize the bulk material and activate the conveyor. The ideal quantity (or velocity) of air passing through the porous media for most conveyed materials is just the velocity of air forming the bed at *minimum fluidization*. Such a bed, considered to be a dense phase, has a clearly defined upper surface. Excessive air velocities (significantly above the minimum fluidization level), causing instabilities with bubbling and channeling of air, should be avoided.

The required quantity of air depends primarily on the physical characteristics of the conveyed material and shall be determined in each case experimentally. However, for most applications it varies from 4 to 12 ft^3/min (6.8–20.4 m^3/hr).

The air pressure must be adequate to overcome the resistance in blower piping, air plenum, porous media (Fig. 28.5.2), and fluidized material in the AAGC. Conditions are considered ideal if the pressure at the top of the material chamber is close to atmospheric. High pressure may cause an excessive carryover of fine particles into the ventilation system. In practice, fluidizing air is supplied from a centrifugal or positive-displacement blower. The blower pressure depends on the design of the piping system, type of porous media, and conveyed material. However, for most enclosed-type AAGC (Fig. 28.5.1) the blower pressure is 12–16 oz/$in.^2$ (3.66–4.88 kg/m^2). For open-type AAGC (Fig. 28.5.3), the blower pressure, depending on the head of material over the porous media, varies from 16 to 96 oz/$in.^2$ (4.88–29.28 kg/m^2).

The blower piping should be designed to provide approximately equal pressure at each AAGC air plenum inlet. Usually, the total pressure drop in the blower piping system does not exceed 6 oz/$in.^2$ (1.83 kg/m^2). Such a system, made of light-gauge or plastic pipes, often includes also a filter-silencer, modulating valves, air bleed-out points, pressure gauges, and so on.

Table 28.5.2 Porous Fluidizing Media

Type of Fluidizing Media	Temperature: Recommended (°F)	Recommended (°C)	Maximum (Not to Exceed) (°F)	Maximum (Not to Exceed) (°C)	Note
Woven fabric					
Cotton, mildew resistant	200	93	200	93	Generally for low- and mildly abrasive and low-corrosive materials
Cotton, fire retardant	250	121	275	135	
Polyester	275	135	350	177	
Asbestos	600	316	700	371	
Plate					
Plastic	180	82	200	93	Relatively low cost, suitable for food stuffs, some chemicals, and plastics
Ceramic	450	232	1000	538	For corrosive material and impact points
Metal (stainless steel)	2000	1093	>2000	>1093	For corrosive material and/or high and extreme temperature
Sintered metallic (carbides)	2000	1093	2500	1374	For high and extreme temperature, suitable for discharges from furnaces
Aluminum oxide blocks	700	371	1000	538	For corrosive, very-abrasive materials and/or impact points
Metallic wire mesh	1200	649	1300	704	For corrosive materials and/or high temperature

28.5.5 Type of AAGC

Different types of AAGC are shown in Fig. 28.2.1. The two basic types are enclosed (Fig. 28.5.1) and open (Fig. 28.5.3) AAGC. The enclosed AAGC consists of a trough-shaped lower section (air plenum) and an inverted trough-shaped upper section (material chamber). The two sections are separated by the porous (fluidization) media (Fig. 28.5.2). The open-type AAGC, used mainly to discharge bulk material from the bottom of storages, is the same enclosed-type AAGC, except it is not fitted with the product chamber. The porous media must be sealed (silicon caulking, special high-temperature putty, or other) to prevent leakage of the fluidizing air. The AAGC are equipped with air inlets for fluidizing air supply and ventilation hoods (Section 28.9) for exhaust of fluidizing air (Fig. 28.9.2).

The *choke-fed* AAGC is physically the same enclosed AAGC, except the fluidized material fills the material chamber of the conveyor completely, similarly as any fluid fills an enclosed vessel. Therefore,

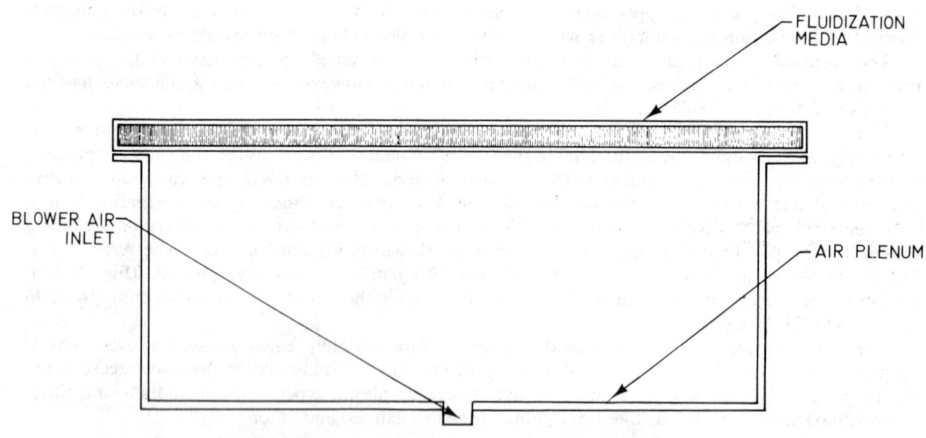

Fig. 28.5.3 Cross section of open-type AAGC.

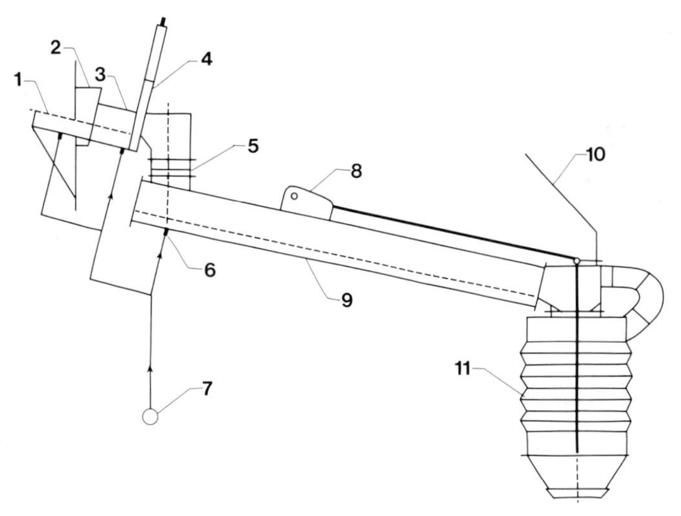

Fig. 28.5.4 Loading system; silo to rail cars. 1—open air trough; 2—silo adaptor; 3—enclosed air trough; 4—air-operated control valve; 5—swivel; 6—swivel air connection; 7—blower; 8—electric winch; 9—enclosed air trough; 10—supporting cable; 11—retractable loading spout. Courtesy of Halliburton Co.

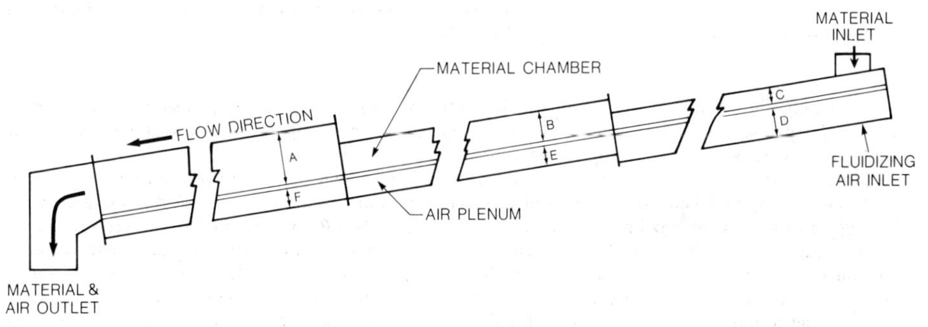

Fig. 28.5.5 AAGC with different cross-sectional areas. A > B > C, F < E < D.

transported material may be discharged at any time to practically unlimited number of points, or not discharged at all (Fig. 28.3.1).

The pivoted AAGC is an AAGC rotating around a fixed axis. Usually a pivoted AAGC is an important part of a loading or unloading arrangement (Fig. 28.5.4).

Often it is inconvenient or even impossible to install a blower air piping and ventilation exhaust system. In this case an *AAGC with different cross-sectional areas* can be used (Fig. 28.5.5). The material chamber is increasing and the fluidizing air plenum is decreasing in incremental steps. The total amount of conveyed material and fluidizing air is discharged from the AAGC usually into a storage facility, where the deaeration process takes place.

BIBLIOGRAPHY

Leitzel, R. E., "Air-Float Conveyors," BHRA, Pneumotransport-4 Conference, Carmel, CA, 1978.

Razumov, I. M., "Pnevmo i gidrotransport v khimicheskoi promishlenosti," *Khimiya,* Moscow, 1979. (Pneumo and Hydrotransport in the Chemical Industry.)

28.6 AIR LIFTS

Ivan Stankovich

An air lift is a vertical pneumatic pipeline (Fig. 28.2.1), transferring bulk material continuously in a fluidized state. Air lifts are similar to pneumatic conveyors, but there are differences in physical principles, design, and application.

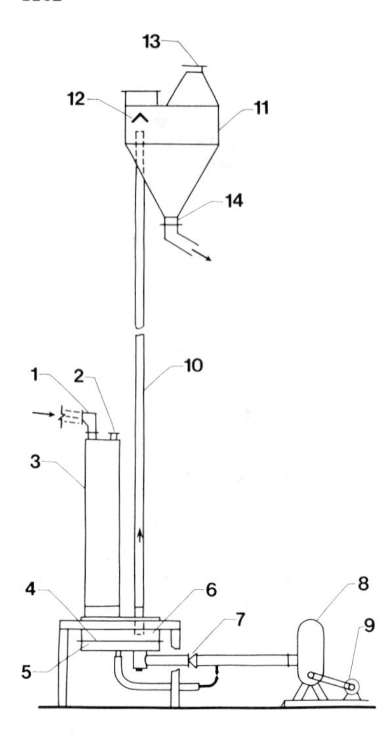

Fig. 28.6.1 Air lift. 1—material inlet; 2—vent; 3—feed column; 4—porous media; 5—air plenum; 6—fluidizing box; 7—check valve; 8—air blower; 9—electrical motor; 10—vertical pipeline; 11—disengaging bin; 12—baffle plate; 13—vent; 14—material discharge. Courtesy of Halliburton Co.

An air lift, shown in Fig. 28.6.1, consists basically of a dense-phase fluid bed feeding compartment, a vertical pipeline, conveying dilute-phase bulk material, a disengaging bin for material deaeration, fluidizing and transporting air supply, and ventilation points.

Material can be fed into the air lift practically by any known bulk material transfer device, such as AAGC, screw and belt conveyors, and even gravity chute. The necessary conditions of material feeding into the vertical pipeline can be described by $\Delta P_F \geqslant \Delta P_T$, where ΔP_F is the pressure in the fluidized bed contained in the feed column and ΔP_T is the pressure in the vertical pipeline. The foregoing relation is the basis for understanding both the physical conditions and the design of an air lift. If fluidization pressure ΔP_F is absent, there will be no vertical transfer of material at all. The greater the ΔP_F, the proportionally more material will be conveyed through the same pipeline cross-sectional area.

To design an air lift, first, the vertical conveying pipeline must be selected for the given capacity and bulk material, similarly as in design of pneumatic conveyors (Section 28.4). Second, based on the calculated transporting pressure for the vertical pipeline ΔP_T, the feed column is designed. The last one is based either on testing results or on principles of dense-phase fluid bed engineering. Special attention should be given to design of the ventilation system (Section 28.9). For more details on air lift design, see Stankovich and Woolever [1].

Air lifts are widespread in the cement and aluminum industries; however, any material conveyed in an AAGC can be transferred successfully in an air lift. In practice, the capacity of air lifts has been limited to 200 ton/hr (181,437 kg/hr) and the conveying height, up to 300 ft (91.44 m). Theoretically and practically, larger capacities and greater lifts are possible. As an example, a 200 ton/hr (181,437 kg/hr) air lift conveying alumina 120 ft (36.58 m) high, requires a 20-in. (508 mm) diameter pipe and approximately 7000 ft³/min (3.30 m³/sec) at 6 lb-ft/in.² (41,368.5 Pa) of blower air. In practice, low- and medium-pressure positive-displacement blowers are used as a source for fluidizing and transferring bulk material in air lifts.

REFERENCE

1. Stankovich, I. D. and K. Woolever, "A Method of Air Lift Design," BHRA, Pneumotransport–4 Conference, Carmel, CA, 1978.

28.7 PNEUMATIC CAPSULE PIPELINES

M. R. Carstens

Pneumatic capsule pipelines are a transport mode that has been in existence for over a century and a half but which has been commercially limited to small-diameter pipelines for transport of documents and other small items around sprawling complexes like hospitals and air terminals. Three developments in the latter half of the twentieth century have expanded the potential usage: (1) development of the technology for cold bending, welding, and laying large-diameter steel pipelines, (2) development of solid-state electronic controllers, and (3) development of electronic digital computers.

28.7.1 Industrial Applications

With the three developments listed, pneumatic capsule pipelines can be designed for periodic releases of capsules (vehicles) into a pipeline through which many capsules are simultaneously rolling, being pushed along by flowing air in the pipe. Under these operating conditions, the performance characteristics of pneumatic capsule pipelines are similar to other types of pipelines, particularly like coarse-slurry pipelines. Because air is compressible and because Roots-type blowers are employed as the power source, booster pump stations are required about every 8 km or so. Although this short distance between pump stations does not preclude consideration of long-haul capsule pipelines, this feature is an incentive to consider short-haul capsule pipelines where the alternative modes (trucks, conveyor belts, and coarse-slurry pipelines) have disadvantages that can be overcome by pipeline transport. For example, transport of coal from a barge terminal for an electricity-generating plant by truck or by conveyor belt may be politically blocked by environmental concerns, whereas movement through a buried capsule pipeline is environmentally innocuous. In any event, pneumatic capsule pipelines are best suited for continuous operation, carrying ore, coal, or similar cargo at large mass-flow rates, say 1 million tons per year or greater.

28.7.2 Principles of Design

The design process can be described in three stages.

First, the most economical pipeline diameter and speed, the power required, the required number of capsules, and a preliminary cost estimate are calculated based on the design mass rate-of-flow and on the assumptions (1) that the pipeline is uniformly sloped from station to station, (2) that the air expands isothermally in the pipeline, and (3) that the rolling resistance of each capsule is the product of coefficient of rolling resistance and the component of weight normal to the assumed pipeline axis. The algorithm employed to calculate the most economical combination by means of an electronic digital computer involves summing the pressure, air density changes, and unit lengths (capsule plus air pocket) between capsules from the downstream end of the pipeline where the pressure is assumed to be atmospheric.

If the first stage on preliminary design appears favorable for a pneumatic capsule pipeline, the second stage is to simulate operation of the system. In addition to pipe diameter and capsule release period obtained from the preliminary design (first stage), pipeline profile is input to the simulation program. Simulation is a physics program in which the dynamic response of each capsule is accounted for as capsules progress through the pipeline. There are two main reasons for simulating the operation even though the algorithm is more complex because computer memory requirements are large, and many computations are required because new solutions are obtained at successive time intervals. One reason for simulating the operation is to determine whether a resonant condition will develop. The movement of wheeled capsules (lightly damped masses) separated by air pockets (springs) can develop a resonant oscillation as the capsules roll uphill and downhill. Changes in pipeline profile can be incorporated in order to eliminate resonance. The second reason for simulation is to select specific blowers for the project.

The third stage is to complete the design including hardware components and project costs. Standardized component designs have been developed, so hardware components can be fabricated without further engineering-design effort being required.

28.7.3 Operating and Maintenance Problems

Operation is automatic, utilizing feedback control in the loading and unloading stations. The results of the first-stage analysis are indicative that the most economical operation is achieved with capsule-release periods in the range of 4 to 7 sec and is achieved with air velocities ranging from 6 to 9 m/sec in a 400-mm-diameter pipeline and from 10 to 14 m/sec in a 900-mm-diameter pipeline.

All of the hardware components have been designed for reliability and ease of maintenance with many parts being common to many units. Reliability is enhanced by means of solid-state controllers with dual sensors at each location and by means of Roots-type blowers at the compressor stations inasmuch as this type of blower has an excellent record for being reliable and for requiring little maintenance. All maintenance work is concentrated at the terminals. Properly laid pipelines between stations are maintenance-free for decades.

Capsule maintenance is to be accomplished at repair shops at the terminals. A scheme for identifying capsules with excess rolling resistance is used to identify those capsules that are to be diverted for repair. The capsules are four-wheeled vehicles built in three subassemblies. The wheel assemblies are bolted onto the ends of the cylindrical cargo container which is open along the top quadrant for

Table 28.7.1 Capsule Geometry

Inside Diameter of Pipe (mm)	Geometric Ratios			
	Length[a]	Volume[b]	Weight[c]	Wheel Diameter[d]
397	5.12	0.317	0.200	0.384
499	5.83	0.367	0.154	0.409
597	6.09	0.372	0.147	0.426
746	6.03	0.396	0.133	0.408
897	6.03	0.414	0.124	0.396

[a] Overall length divided by inside diameter of pipe.
[b] Cargo volume divided by inside volume of pipe occupied by capsule.
[c] Capsule weight divided by weight of water that would occupy a one-capsule length of pipe.
[d] Wheel diameter divided by inside diameter of pipe.

ease in loading and unloading. Capsules are self-righting. Capsule geometries listed in Table 28.7.1 are expressed in dimensionless ratios. Capsules are designed to flow through minimum radius bends which can be achieved by cold bending, that is, radius of curvature approximately 40 pipe diameters. The expectation is that wheel-bearing replacement and wheel-alignment will be the principal maintenance tasks. On the other hand, long wheel-bearing life is anticipated because the wheels are large (Table 28.7.1) and the capsule speeds are low.

28.7.4 Advantages and Disadvantages

Advantages

Environmental. Buried pipelines through which capsules roll are environmentally innocuous. Except for the terminals and booster pump stations, pneumatic capsule pipelines are invisible. The earth over and around the pipeline will attenuate sound with the result that noise pollution is nonexistent. Since atmospheric air is the propellant being injected into and discharged from the pipeline at the terminals and booster pumps, air-pollution control is minor, involving only dust control at the loading and unloading stations.

Safety. Inasmuch as the capsules move in an enclosed guideway, the pipeline, there is no traffic interference with other transport modes, including pedestrians. Since the maximum air pressure is less than two atmospheres absolute, a construction accident on an adjacent construction site which breaks the pipe is not dangerous. Being in an enclosed guideway, cargo is protected from weather, from vandals, and from thieves.

Performance. Pipelines can be placed along routes that are difficult for trucks and conveyor belts. For example, underwater river crossings are common in pipeline technology. Because pneumatic capsule pipelines can be operated at slopes of 25% or more and because pipeline cold bends are quite sharp, pipeline routes can be shorter than truck or train routes in hilly terrain. Capsules rolling in a downhill section of the pipeline act as piston pumps thereby recovering potential energy in contrast to waste of potential energy as trucks move downhill.

Because of automatic feedback control, no operating personnel are required for a pneumatic capsule pipeline. All maintenance work is done at the terminals.

Disadvantages

The principal disadvantage of penumatic capsule pipelines is the lack of growth flexibility, that is, in order to design for future growth, an under-utilized system would have to be installed. In this sense pneumatic capsule pipelines and conveyor belts are alike, in contrast to trucks which can be added as demand increases.

As of 1982, operating experience has been very limited. Pneumatic capsule pipelines which require a large capital outlay to install will be underexploited until performance and reliability can be proved with operating systems.

28.7.5 Cost Comparisons

One system went through the three stages of design in order to develop a cost comparison with a covered conveyor belt. The 1.15-km-long system was designed to transport waste rock at a rate of

91 kg/sec. The cost of installing the pneumatic capsule pipeline was about the same or slightly less than the covered conveyor belt.

28.7.6 Experience with Performed Projects

Other than a 500-m-long, 400-mm-diameter pneumatic capsule pipeline located at Houston, Texas, operating systems had not been installed in the United States as of 1982. The Texas system was installed as a demonstration facility and as a hardware-development facility. A pneumatic capsule pipeline for hauling limestone has been installed at a plant in northern Japan. Several systems are in operation in the USSR.

BIBLIOGRAPHY

Carstens, M. R., "Analysis of a Low-Speed Capsule-Transport Pipeline," *Proceedings,* First International Conference on the Hydraulic Transport of Solids in Pipes, sponsored by the British Hydromechanics Research Association, Coventry, England, September 1970, pp. C4-73 through C4-88.

Carstens, M. R. and B. E. Freeze, "Pneumatic Capsule Pipeline," *Proceedings,* Fifteenth Annual Meeting, Transportation Research Forum, Vol. XV, No. 1, 1974, pp. 206–213.

Carstens, M. R. and D. W. Leva, "Design of a Capsule Pipeline," American Society of Mechanical Engineers, Paper No. 76-pet-34, pp. 1–5. Presented at the Petroleum Mechanical Engineering and Pressure Vessels and Piping Conference, Mexico City, September 1976.

Leva, D. W. and B. E. Freeze, "A Maintenance Program for a Pneumo-Capsule Pipeline," *Journal of Pipelines,* Vol. 1, (1981), pp. 225–232. Published by Elsevier Scientific Publishing Co., Amsterdam.

28.8 MISCELLANEOUS PNEUMATIC DEVICES

Ivan Stankovich

The number of different devices used in pneumatic handling is increasing constantly. These devices, performing different functions, do not have moving parts. They all use air as a primary mover and are fairly simple. A few typical examples are given in the following discussion.

The *air-activated feeder discharger* (Fig. 28.8.1) is a device to feed or discharge bulk material from one point to another. Often such a device can be used as an air lock to separate two pneumatic installations with different internal pressure. Material (1), discharged by gravity into the compartment (2), self-chokes. Timer (11) activates valve (8). As a result, air enters into the plenum (6), fluidizes material in compartment (2). Fluidized bulk material overflows partition (5) and settles in compartment (3). In accordance with the program, timer (11) closes valve (8) and opens valve (9). Air enters into the plenum (7) and the fluidized material in compartment (3) is discharged (12) from the device. The cycle can be repeated with a desired frequency.

The purpose of the *self-activating pneumatic feeder,* the cross section of which is shown in Fig. 28.8.2, is to constantly keep the storage hopper (1) filled up with material. This is achieved by balancing the fluidizing pressure P_1 with the weight of the solids above the porous media P_2 and the resistance

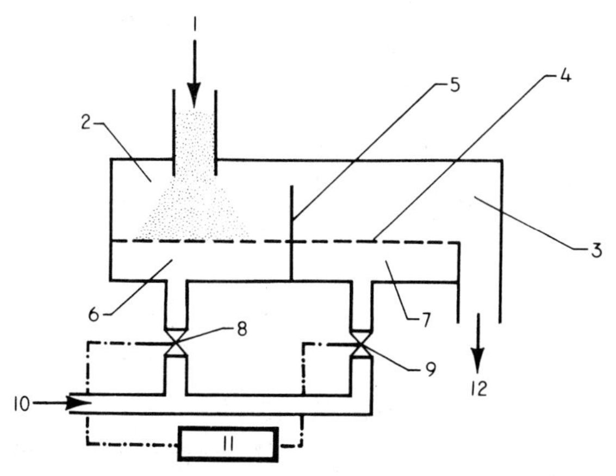

Fig. 28.8.1 Air-activated feeder-discharger: 1—material feed; 2—material compartment; 3—material compartment; 4—porous media; 5—partition; 6—fluidizing air plenum; 7—fluidizing air plenum; 8—valve; 9—valve; 10—fluidizing air source; 11—timer; 12—material discharge.

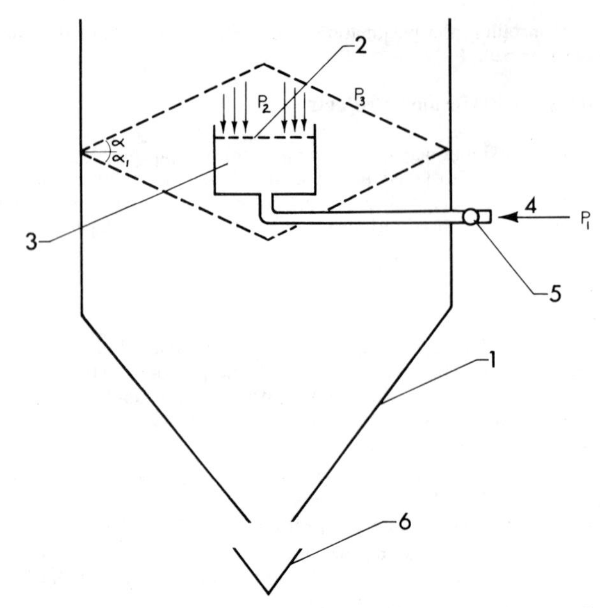

Fig. 28.8.2 Self-activating pneumatic feeder, U.S. Patent No. 4,016,053: 1—bulk material hopper; 2—porous media; 3—fluidizing air plenum; 4—fluidizing air source; 5—valve; 6—hopper discharge valve; P_1—fluidizing air pressure, P_2—weight of solids above porous media, P_3—pipe and porous media pressure drop; α—material angle of repose at system balance; α_1—material angle of repose immediately after material is discharged from the hopper.

to the air flow through the pipe and porous media P_3, that is, $P_1 = P_2 + P_3$. When system is balanced, the angle of repose of the material is α. After a certain amount of material is discharged through the valve (6), the angle of repose of the material will be α_1. The balance will be upset. As a result, the open-type AAGC (2 and 3) is activated by itself, transfers new material into the hopper (1) until the balance is restored and material takes the angle of repose α. The feeder cycle follows every opening of the discharge valve (6).

The *pneumatic air lock* (Fig. 28.8.3) transfers material from the discharge compartment (1) with internal pressure P_1 to the transfer conveyor (6) with internal pressure P_2. $P_1 \neq P_2$. This is achieved

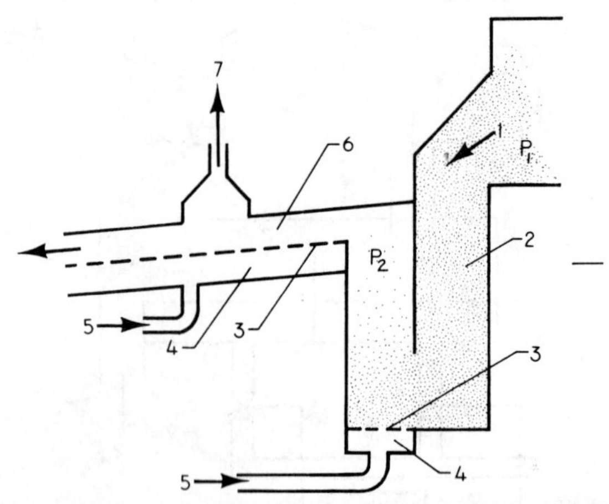

Fig. 28.8.3 Pneumatic air lock. 1—material discharge compartment; 2—material seal column; 3—porous media; 4—fluidizing air plenum; 5—blower air source; 6—transfer conveyor; 7—ventilation; P_1—pressure in the discharge compartment; P_2—pressure in the transfer conveyor.

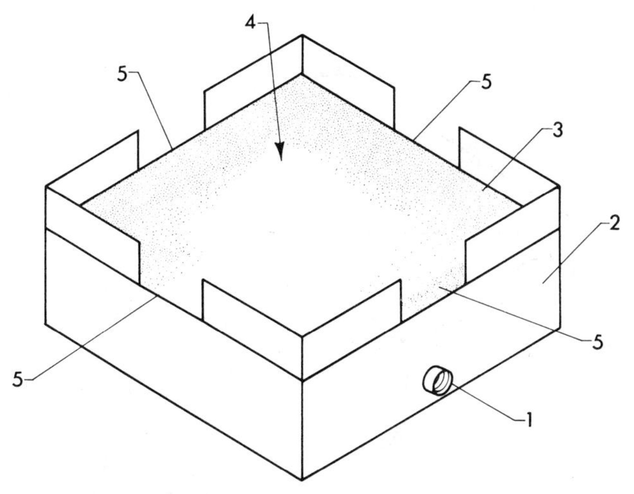

Fig. 28.8.4 Air-activated material distributor. 1—fluidizing air inlet; 2—fluidizing air plenum; 3—porous media; 4—material feed; 5—material distribution points.

by using a material seal column (2). The bulk material is transferred continuously from (2) to (6) by supplying blower air (5) into the plenum (4).

The purpose of the *air-activated material distribution* (Fig. 28.8.4) is to divide a flow of material into a few separate streams. This is achieved by leveling the device horizontally and sizing the distribution points (5). If equal distribution of material is desired, all the openings (5) should have equal dimensions. If the distribution should be not equal, the openings (5) will be sized accordingly.

Aeration pads (Fig. 28.8.5) are used to facilitate the discharge of bulk material from the hopper. They consist of air inlet (1) and porous media (3), mounted inside the hopper wall (2). The number and size of aeration pads will depend on the design and size of the hopper and the physical characteristics of the bulk material.

The *air-activated material trap* (Fig. 28.8.6) is used to separate oversized particles from the main flow of material. Bulk material fed into the trap (2) through inlet (1), is fluidized over the porous media (5). Normal size material, overflowing the skirt (3), is discharged through the outlet (9). Lumps, debris, heavy, and oversize particles (4) are concentrating in the lower corner of the trap, from where they are removed periodically by vacuuming (6).

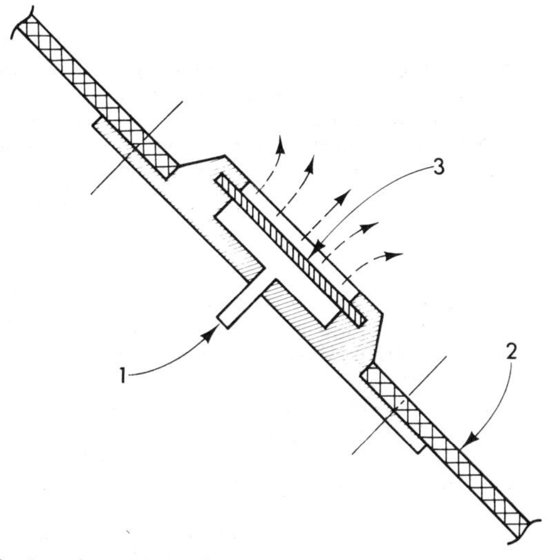

Fig. 28.8.5 Aeration pad. 1—fluidizing air inlet; 2—hopper wall; 3—porous media.

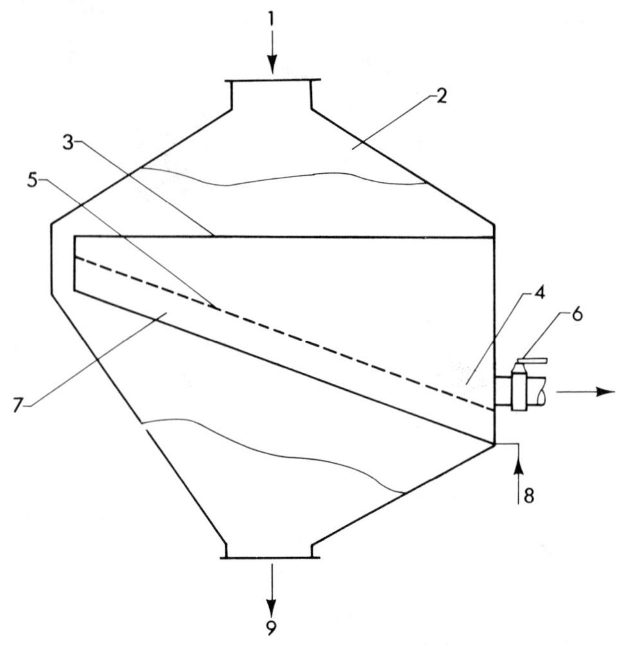

Fig. 28.8.6 Air-activated material trap. 1—material inlet; 2—trap body; 3—skirt; 4—oversize particles and debris; 5—porous media; 6—outlet to vacuum cleaning system; 7—air plenum; 8—fluidizing air inlet; 9—material outlet.

BIBLIOGRAPHY

Colijn, H., "Designing Bulk Solids Handling Systems. IV. Mechanical and Air-Induced Bin Dischargers," *Plant Engineering,* December 10, 1981, pp. 85–88.

Stankovich, I. D. et al., "Feeding Particulate Material," U.S. Patent 4,016,053, 1977.

Stankovich, I. E., "Pneumatic Handling Classification," BHRA, Pneumotransport-4 Conference, Carmel, CA, 1978.

28.9 PNEUMATIC HANDLING VENTILATION

Ivan Stankovich

Ventilation is a very important part of pneumatic handling design. Transfer and interface points are especially dependent on ventilation. Often the performance of an interface point relies on the degree of deaeration, that is, on separating the fluidized and entrained air from solid particles. Ventilation subsystems should be designed carefully and incorporated into the pneumatic handling system, as shown in Fig. 28.9.1. Computerized calculations are used to determine the system pressure drop, and to select the ducts and the fan, as shown in Table 28.9.1 (see also [1]).

28.9.1 Hood Design

A method of efficient hood design in conjunction with pneumatic handling of abrasive bulk material, proven in practice, is shown in Fig. 28.9.2. Such arrangements in comparison with conventional arrangements [1] have demonstrated much higher abrasive resistance. This was achieved by designing the hood (2) for face velocities not exceeding 100 ft/min (0.51 m/sec). The hood extends in a large-diameter vertical duct designed for velocities 150–200 ft/min (0.76–1.02 m/sec). The relative coarse particles, entrained into the duct (2) strike the top of the duct and fall back into the conveying system. The finer particles are entrained into duct (5) and further exhausted to the dust collector through the main duct (6). Minimum possible negative pressure shall be maintained by adjusting the valve (4) and measuring the static pressure at (7).

Such hood arrangement could be used to ventilate AAGCs, air-lift disengaging vessels (Fig. 28.6.1), and other pneumatic handling devices.

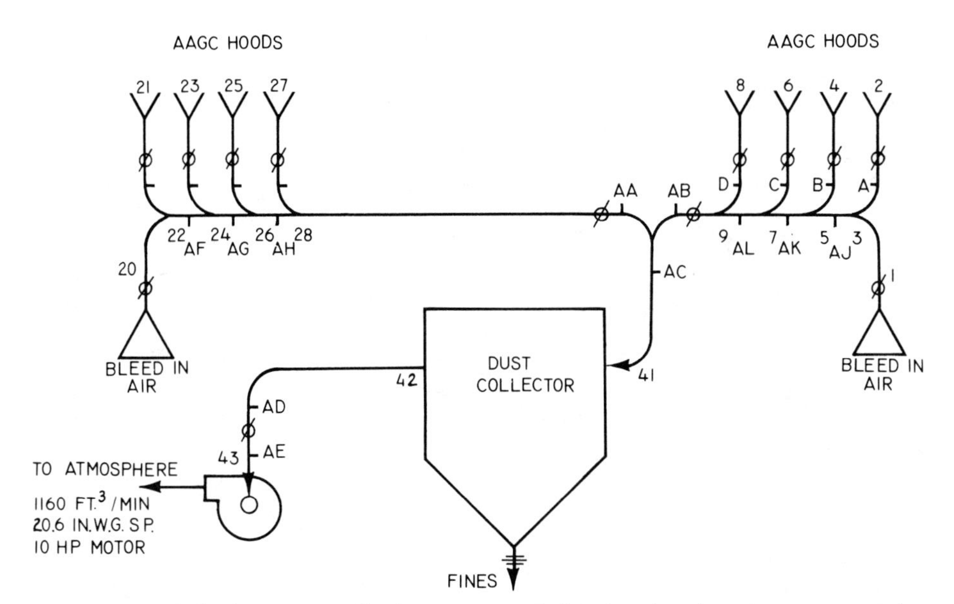

Fig. 28.9.1 Typical nuisance dust collection system ventilating air-activated gravity conveyors (schematic).

Table 28.9.1 Nuisance Dust Collection System, Computerized Calculations[a]

Branch	Duct Diameter (in.)	ft³/min	Velocity (ft/min)	Hood Static Pressure (in.)	Branch Static Pressure (in.)	Test Point	Velocity Pressure (in.)
1–3	3.5	160	2395	0.50	0.98	—	0.33
2–3	2.5	80	2347	0.80	0.98	A	0.32
3–5	3.5	240	3592	—	3.41	AJ	0.75
4–5	2.5	80	2347	0.80	3.41	B	0.32
5–7	4.0	320	3667	—	5.37	AK	0.78
6–7	2.5	80	2347	–.80	5.37	C	0.33
7–9	4.5	400	3622	—	7.70	AL	0.32
8–9	3.0	180	3667	1.96	7.70	D	0.78
9–40	5.5	580	3515	—	10.64	AB	0.72
20–22	3.5	160	2395	0.50	0.98	—	0.33
21–22	2.5	80	2347	0.80	0.98	J	0.32
22–24	3.5	240	3592	—	3.41	AF	0.75
23–24	2.5	80	2347	0.80	3.41	K	0.32
24–26	4.0	320	3667	—	5.37	AG	0.78
25–26	2.5	80	2347	0.80	5.37	L	0.32
26–28	4.5	400	3622	—	7.70	AH	0.76
27–28	3.0	180	3667	1.96	7.70	M	0.78
28–40	5.5	580	3515	—	10.64	AA	0.72
40–41	7.5	1160	3781	—	10.86	AC	0.83
41–42	See Dust Collector						
42–43	12.0	1160	1450	—	17.86	AD	
					18.16	AE	
43–44	See Fan						
44–45	Match fan exhaust	1160	2127	—	—		

[a] See Fig. 28.9.1 for example.

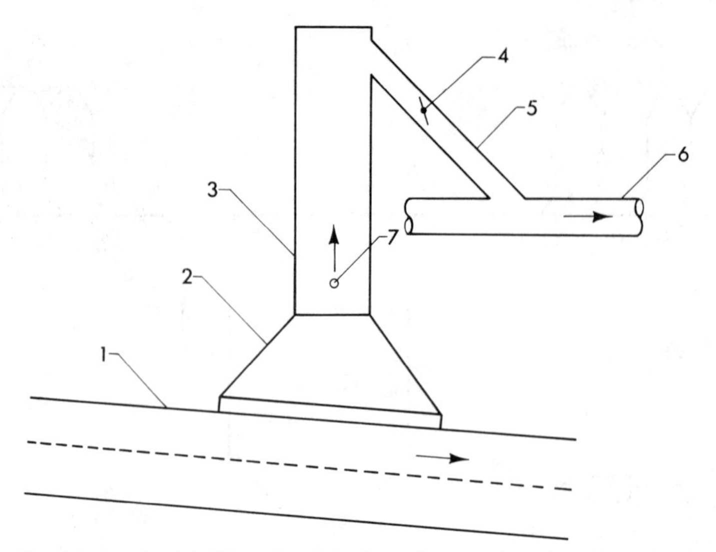

Fig. 28.9.2 Hood design. 1—AAGC; 2—hood; 3—large diameter duct; 4—adjustable valve; 5—small-diameter duct; 6—main exhaust duct; 7—pressure test point.

28.9.2 Cartridge Filters

One of the problems of the conventional pneumatic handling ventilation systems is the erosion of exhaust ducts and the complexity of the ventilation system, requiring relatively large systems of ducts, adjustable valves, baghouses, exhaust fans, and dust discharge valves.

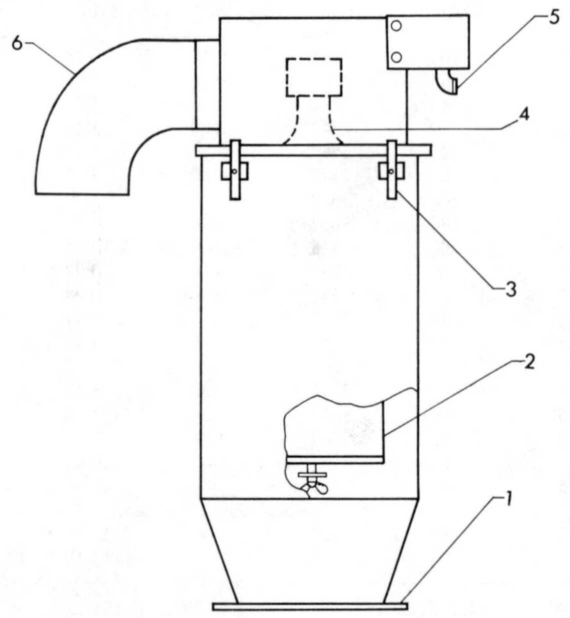

Fig. 28.9.3 Cartridge dust collection unit. 1—flange; 2—TD cartridge filter; 3—strap latch; 4—venturi; 5—pulse air inlet; 6—exhaust duct. Courtesy of Torit Division, Donaldson Co., Inc.

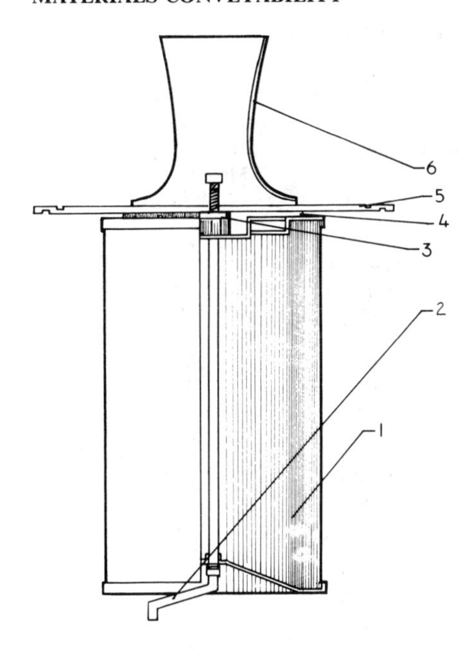

Fig. 28.9.4 TD cartridge filter. 1—filter media; 2—hand crank; 3—hanger pivot; 4—hanger bracket; 5—tube plate; 6—venturi. Courtesy of Torit Division, Donaldson Co., Inc.

This has been overcome in certain applications (AAGC, air lifts, etc.) by using cartridge filter dust collector units (Fig. 28.9.3). Such a dust collecting device can be mounted directly on top of an AAGC, operated under low positive pressure. The cartridge filter (2) is periodically cleaned by pulsing air (5). The clean air is exhausted through the duct (6). No ducts, adjustable air flow valves, baghouses, exhaust fans, and dust discharge valves are required for such an application. The cartridge filter (Fig. 28.9.4) is easily replaceable.

REFERENCE

1. *Industrial Ventilation, A Manual of Recommended Practice,* Committee of Industrial Ventilation, American Conference of Governmental Industrial Hygienists, 15th Ed., Lansing, MI, 1978.

BIBLIOGRAPHY

Stankovich, I. D., "Problems in Control of Particulates in Bulk Material Handling," ASME Winter Annual Meeting, Atlanta, GA, 1977.

28.10 MATERIALS CONVEYABILITY

Ivan Stankovich

The design of a pneumatic handling system must begin with an accurate appraisal of the characteristics to be handled by a particular pneumatic device system. The basic question is, can a material be conveyed pneumatically at all.

Some AAGC manufacturers use the following approximate criteria to determine the material conveyability. Particle size distribution: 100% of material shall pass through the 20-mesh screen. From this amount 10–15% shall pass through the 200-mesh screen. Free moisture content should be below 1%. However, coarser materials with larger amount of fine material and materials with higher moisture have been conveyed. Similar approximate criteria are used for other pneumatic handling devices.

To properly select the equipment and design the pneumatic handling systems, often additional material and material–system characteristics, such as material-to-air ratios, conveying velocities, abrasiveness, attrition, buildup tendencies, dust collection requirements, minimum slopes for AAGC, and conveying rates should be carefully investigated. A list of some materials successfully handled by AAGC, air lifts, and pneumatic conveyors is presented in Table 28.10.1.

Table 28.10.1 Materials Handled by AAGC, Air Lifts, and Pneumatic Conveyors[a]

Material	Bulk Density lb/ft³	Bulk Density kg/m³	Dense-phase Pneumatic Conveying; Material/Air lb/lb (g/g)
Alumina, precipitator dust	54	865	83
Alumina, reacted	58–65	929–1042	68
Alumina, sandy	55–60	881–961	109
Aluminum chloride	92	1474	115
Aluminum fluoride	50	801	45
Barium sulfate	115	1842	201
Bath, crushed	100	1602	30
Bentonite	50	801	155
Calcium carbide	62	993	153
Carbon, black	42	673	134
Carbon dust	38	609	69
Cement, portland	94	1506	180
Cement dust	50	801	134
Clay, kaolin	32	513	102
Clays	42	673	103
Coal, pulverized	52	833	58
Diatomaceous earth	16	256	46
Dolomite	39	625	45
Feldspar	62	993	40
Flour, wheat	30	481	82
Fluorspar	113	1810	80
Fly ash	40	641	129
Gypsum	54	865	55
Iron powder	150	2403	21
Lime, hydrated	37	593	82
Limestone, ground	59	945	46
Magnesite	40	641	39
Perlite dust	60	961	112
Petroleum coke	55	881	40
Phosphate rock dust	70	1121	80
PVC resin	32	513	52
Sand, foundry	104	1666	30
Soda ash	54	865	48
Sodium bicarbonate	30	481	76
Starch	33	529	137
Talc	25	400	241
Zinc oxide	32	513	37

[a] Data applicable to dense-phase pneumatic conveying. (Courtesy of Consolidated Fluidflo Co.)

BIBLIOGRAPHY

Zenz, A. F., "Conveyability of Materials of Mixed Particle Size," *Industrial & Engineering Chemistry Fundamentals,* Vol. 3, p. 65, 1964.

28.11 ABRASION IN PNEUMATIC PIPELINES

Ivan Stankovich

Reasons for rapid wear of pneumatic pipelines are: high abrasiveness of conveying material, inadequate selection of conveying equipment (system), poor design of the pipeline, excessive conveying velocities, and inadequate installation of the pipeline. Usually the abrasion of a pipeline is a result of a few combined factors.

To decrease the degree of abrasion in a pipeline, the following is recommended:

1. Use dense-phase pneumatic conveyors where possible and applicable.
2. Use lowest possible conveying velocities; abrasion W is $W = kv^3$ for straight pipelines and $W = k_1 v^4$ for fittings, where k and k_1 are coefficients related to material conveyed and pipe characteristics (material, hardness) and v is the conveying velocity.

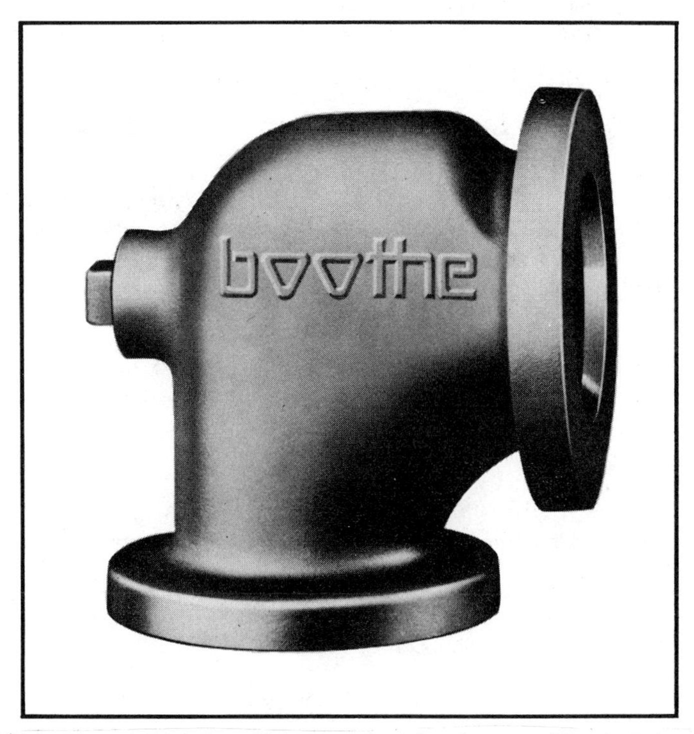

Fig. 28.11.1 Flanged 90° fitting with booster pipe inlet. Courtesy of Boothe Industries, Inc.

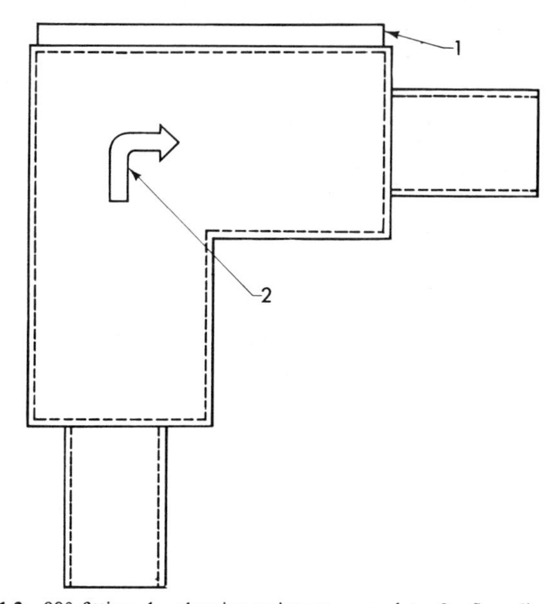

Fig. 28.11.2 90° fitting. 1—abrasive-resistant wear plate; 2—flow direction arrow.

3. Use continuous conveying (e.g., double blow tanks) whenever possible and practical, to eliminate the filling and pipeline purging periods of the conveying cycle, which causes most of the pipe abrasion.

4. Use specially designed abrasive-resistant pipes and fittings (Figs. 28.11.1 and 28.11.2) made of abrasive-resistant iron and steel alloys, ceramic, with wear-backs, replaceable backs, with longer ends to remove the joints from wear impact zones as much as possible and practical.

5. Use larger size pipe and fittings at the pipeline terminal than at the main pipeline section, to minimize the impact of the terminal velocities caused by air expansion.

6. Whenever possible and practical, use smaller pipe size to minimize the impact of material pulsing.

7. Properly support all pipe joints and fittings to avoid vibrations and pipe misalignment.

28.12 ATTRITION OF CONVEYING MATERIAL

Ivan Stankovich

Particles are subject to attrition whenever they are in motion and especially during the start of the motion, and when the motion is abruptly halted or interrupted.

The reasons for material attrition are collisions between the solid particles and collisions of solid particles with pipe walls.

From observations, records, test results, and publications, the following general conclusions can be made:

1. Attrition increases if particles are subjected to shocks.
2. Attrition is maximum at particle-to-surface shock angles of 30–50° from the plane of the surface.
3. Attrition A is basically proportional to the cube of particle transporting velocity v:

$$A = kv^3$$

where k is a characteristic of conveyed material.

4. Attrition occurs mainly from breakdown of the coarser particles. Attrition increased gradually when particle size becomes larger than 0.05 in.

5. Fines or very small particles do not get finer at any significant rate. Attrition decreases rapidly when particle size becomes smaller than 0.03 in.

6. Fine production by attrition occurs mostly at the first movement, which leads to a rounding of edges.

7. The rate of attrition for spherical particles R_{as} is approximately one-half of the rate for nonspherical particles R_{ans}, that is:

$$R_{as} = 0.5\, R_{ans}$$

8. Generally, the attrition A is inversely proportional to the breaking material tensile stress, perpendicular to the crack P, that is:

$$A = \frac{1}{p}$$

Table 28.12.1 Installation with Circulating Solids

	Installation			
Parameter	Granular Filtration	Gel Adsorption	Pebble Heating	Hydrocarbon Processing
Material handled	Gravel	Silica granular gel	Alumina pebbles	Sand
Material size (in.)	$\frac{1}{8}-\frac{1}{4}$	$\frac{1}{32}-\frac{1}{8}$	$\frac{3}{8}-\frac{1}{2}$	$\frac{1}{32}$
Density (lb/ft³)	100	40	55	95
Relative hardness	Hard	Very hard	Very hard	Very hard
Temperature range (°F)	300–600	325–350	1600–1900	300
Average attrition rate (as % of the handled material)	0.0300	0.0018	0.0080	0.2000

and

$$P = \sqrt{\frac{2E\sigma}{l}}$$

where: E = material Young's modulus
σ = specific surface energy of the material
l = length of the initial crack

The rate of particle attrition or fine production, where fines are described as attrition to −325 mesh, is the weight ratio of the newly produced fines to the total weight of the handled material during a certain period of time.

There are many theories relating material physical characteristics to attrition, but there are no methods of calculating attrition rates yet.

Results of attrition tests of different materials for different process installations are presented in Table 28.12.1. Each installation did include an air lift.

BIBLIOGRAPHY

Zenz, A. F., "Find Attrition in Fluid Bed," *Hydrocarbon Processing,* pp. 103–105, Feb. 1971.

Zenz, A. F., Studies of Attrition in Fluid-Particle Systems, Particulate Solid Research, Inc., Riverdale, Bronx, NY 10471.

CHAPTER 29
SLURRY PIPELINES

THOMAS C. AUDE
TERRY L. THOMPSON

Pipeline Systems Incorporated
Orinda, California

29.1 INTRODUCTION

29.1.1 Definition and System Elements

Slurry pipelines are defined as: Two-phase flow pipelines which transport a solid material dispersed in a liquid carrying vehicle.

Slurry pipelines may be classified in several different manners. The primary classifications relate to (1) the product pumped and its physical state and (2) the type of flow (i.e., homogeneous, heterogeneous, etc.). Figure 29.1.1 illustrates the various types of pipelines. Slurry pipelines provide a reliable method of long-distance high-tonnage transport of bulk freight materials. In this discussion slurry pipelines refer to slurry pipeline *systems,* the complete facilities necessary to provide the bulk freight transportation link. Typical slurry systems are shown in Fig. 29.1.2. As shown, the boundaries of the slurry system depend on the material transported. For minerals, such as iron, copper, and limestone, the slurry pipeline is a "long spot" in the process, since the beneficiation process usually includes crushing, grinding, slurry preparation, and some sort of dewatering step after concentration. The slurry pipeline simply separates the steps in the process. For minerals transportation the "pipeline system" refers only to the pipeline, pump stations, and tankage at the terminal. For coal pipelines, on the other hand, slurry preparation and dewatering are done only for purposes of transportation. Therefore, "coal slurry pipeline systems" include slurry preparation and dewatering.

29.1.2 Applications of Slurry Pipelines

Slurry pipelines have been utilized for disposal of mineral wastes (i.e., tailings) for more than 100 years. Since the 1950s, with the development of slurry pipeline technology, useful products such as coal, copper concentrate, iron concentrate, limestone, phosphate, and other minerals have also been successfully transported by pipeline. Nearly 30 such "freight pipelines" are in existence transporting more than 60 million tons of materials per year.

A summary of typical materials amenable to slurry transport, showing specific gravity of solids, maximum particle size, and concentration, is given in Table 29.1.1.

Coal Pipelines

One of the major uses of slurry pipelines is the transportation of steam coal from the mines to the power plant locations. The 108-mi Consolidation Coal Pipeline in Ohio and the 273-mi Black Mesa Pipeline in Arizona have demonstrated the reliability and economics of coal pipeline transport. The

PRODUCT TRANSPORTED:	NATURAL GAS	GAS PRODUCTS	GAS WASTES	WATER	OIL	LIQUID PRODUCTS	LIQUID WASTES	COAL	MINERALS	SOLID WASTES
STATE OF PRODUCT:	GAS			LIQUID				SOLID		
STATE OF VEHICLE:	GAS		LIQUID	GAS	LIQUID			GAS		LIQUID
STATE OF TRANSPORTED PRODUCT:	GAS	GAS-GAS	GAS-LIQUID	LIQUID-GAS	LIQUID-LIQUID	LIQUID		SOLID-GAS		SOLID-LIQUID
								PNEUMATIC	CAPSULE	SLURRY

Fig. 29.1.1 Types of pipelines.

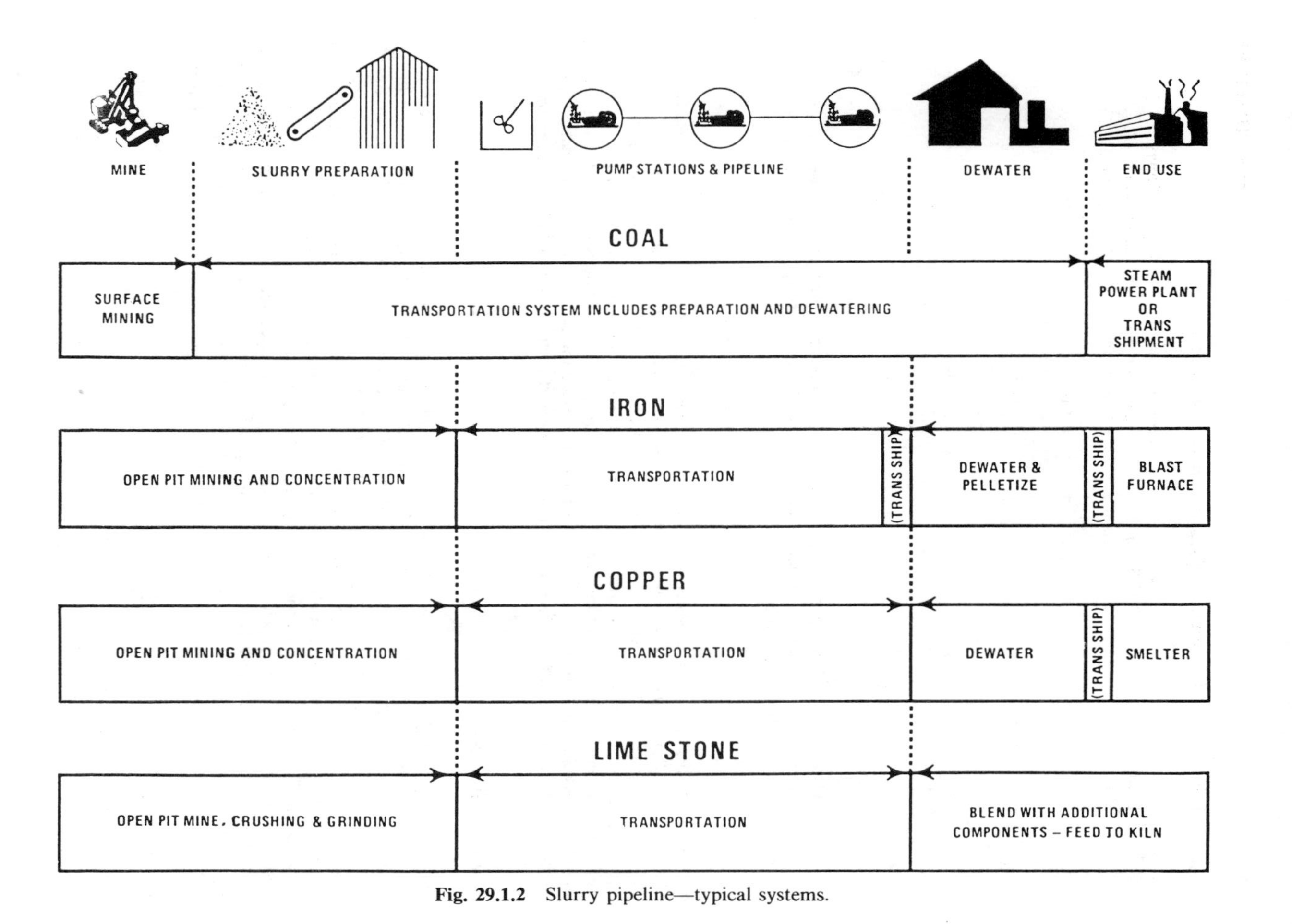

Fig. 29.1.2 Slurry pipeline—typical systems.

Table 29.1.1 Properties of Typical Transported Minerals[a]

Transported Mineral	Solids, Sp. Gr.	Approximate Maximum Particle Size	Typical Slurry Concentration, % Solids by Weight
Coal	1.4	8 Mesh (2.380 mm)	50
Limestone	2.7	48 Mesh (0.297 mm)	70
Phosphate concentrate	3.2	65 Mesh (0.210 mm)	60
Copper concentrate	4.3	65 Mesh (0.210 mm)	55
Iron concentrate	5.0	100 Mesh (0.149 mm)	60

[a] Data are for long distance pipelines using established fine-product pseudo-homogeneous technology. Coarser products can be successfully transported, but for much shorter distances.

long distances between coal deposits and markets in countries such as the United States, Canada, China, and the USSR are creating an increasing demand for coal pipeline transport. In order to transport the coal long distances by pipeline, it must be crushed and ground to about 14 mesh (1.19 mm) by zero mesh. Because coal for pulverized coal boilers must be ground even finer (i.e., 80% passing 200 mesh, or 0.074 mm), this does not impose a penalty on the process. Adding water for slurry transport, however, does add a cost to the system in the form of dewatering costs and some thermal penalty in the boiler. These costs must be offset by savings in transportation costs, reliability, and resistance to inflation.

Minerals Pipelines

The major use to date for "freight pipelines" has been in the transport of minerals. The biggest use has been in iron ore concentrate transport. Ten such systems have been installed with more in the planning stages. The largest system is the 245-mi, 20-in. diameter SAMARCO pipeline in Brazil. This system is designed to transport up to 12 million tons of iron concentrate per year. Since mineral concentrates often go through a slurry stage in their processing, pipeline transportation does not add dewatering requirements over and above those of the normal process. In addition, most metallurgical processes that produce mineral concentrates require fine grinding to liberate the minerals. It is for these reasons that the slurry pipeline concept has been readily accepted in the minerals industry. It is a natural extension of minerals processing. Slurry pipelines have been used for copper concentrates, phosphate concentrate, limestone, and other minerals. In general, almost any finely ground mineral that is not harmed by the addition of a carrying vehicle such as water can be transported by pipeline.

Tailings Pipelines

Tailings refers to the waste products from minerals processing. The objective is to dispose of the material. Therefore, there is an incentive to make the transportation as inexpensive as possible. Historically, tailings lines have been relatively short, above ground, short-life systems. Design has been, to a large extent, based on experience and rules of thumb, rather than theory and detailed design. With development of more remote ore bodies and with increasing environmental awareness, the length of tailings pipelines has been steadily increasing. At the same time, slurry pipeline technology has been advancing as a result of applications in transport of coal and minerals. These two factors have resulted in more attention being paid to design of the tailings pipelines.

Since, by definition, tailings are waste products, there is no incentive to prepare the tailings (i.e., crush or grind) for transport. Rather, the tailings pipeline is called on to transport whatever size or concentration exists in the tailings thickener or other source. Therefore, tailings lines are often heterogeneous flow systems transporting fairly coarse material. As such, velocities must be higher and wear becomes a factor. (This is not always true; some tailings are fine and of a high enough concentration to be handled in much the same manner as mineral concentrates.) In addition, tailings volumes are usually much higher and variable than the mineral concentrates. Slurry pipeline technology is finding an increasing role in tailings applications. Reliability can be improved and system life extended.

29.1.3 Factors in Selecting Slurry Pipelines

The economics of slurry pipelines are site-sensitive. Therefore, it is difficult to make generalizations about slurry transport economics.

One major variable in comparing transport alternatives is system length. Since pipelines can take a fairly direct route, they are often significantly shorter than rail or truck routes, which have more severe grade and construction restrictions, or barge routes, which have obvious length and location restraints. In addition, generalized comparisons are difficult where existing alternate transport modes are available and pricing strategies involve "sunk" versus "incremental" costs. A specific detailed cost comparison of a coal pipeline from Wyoming to Arkansas was presented by Aude, Thompson, and Wasp [1]. The reader is referred to this paper for a methodology of making specific capital, operating, and annual cost comparisons, including the difficult but critical consideration of the effects of inflation.

The following general observations can be made regarding conditions under which long-distance slurry pipelines are used.

Terrain

Slurry pipelines are often selected where terrain is difficult and the location is remote. Pipelines are easier to construct in remote areas than are railroads or truck haul roads, since pipelines have a less restrictive grade requirement and can be installed at several-kilometer-per-day rates by conventional long-distance pipeline construction techniques. In addition, because the pipeline is buried and pump stations can usually be spaced 50–100 mi apart, remote operations and maintenance are relatively simple.

Design of Slurry

It is true that virtually any combination of size consist and solids concentration can be pumped as long as it flows. However, in order to design a system that can be shut down and restarted, that will not wear out the pipe, and that can be operated under predictable and suitable flow conditions, it is necessary to put fairly strict limits on size consist and solids concentration. Figure 29.1.3 shows slurry flow regimes as a function of size consist, concentration, and solids specific gravity. The identification of the flow regime areas in the exhibit is a generalization based on experience with commercial pipeline slurries at normal pipeline velocities of approximately 4–7 ft/sec [2]. Most long-distance commercial slurry pipelines are either in the pseudohomogeneous or complex flow regimes. Tailings pipelines, as previously noted, often are in the heterogeneous flow regime simply because it is not practical or economical to perform any processing to enhance their transportation characteristics.

For useful solid materials, on the other hand, it is often economical to grind or thicken the product to improve its flow characteristics. This approach has been the cornerstone of the development of slurry technology to date. Wasp and his co-workers at Consolidation Coal in the 1950s held the philosophy that they would tailor the slurry to be compatible with existing pipe materials, slurry pumps, and long-distance oil and gas pipeline construction techniques rather than develop new hardware or exotic materials to be compatible with the slurry [3]. They felt that a fundamental understanding of the slurry flow behavior in a stable and controlled environment was superior to developing new hardware or materials compatible with a wider range of size consists and concentrations, even though this approach limits the commercial applications. This approach, which has proven valid over the past quarter century, has given the solids pipeline industry a technologically sound beginning and has provided a firm basis from which excursions into advanced technology can be made.

Throughput

A throughput of one million tons per year or more is usually needed to make transport by a new slurry pipeline competitive with other transport modes. Slurry lines enjoy significant economy of scale benefits similar to oil and gas pipelines. For certain applications, however, especially higher-value, remotely located materials such as copper, annual throughputs of only a few hundred thousand tons per year are commercially viable. This is about the lower limit because pressure drops at such small diameters become excessive.

Distance

For slurry pipelines that require slurry preparation and separation facilities, a minimum distance of 50–100 mi is generally necessary to spread the cost of the end-point facilities. For mineral pipelines

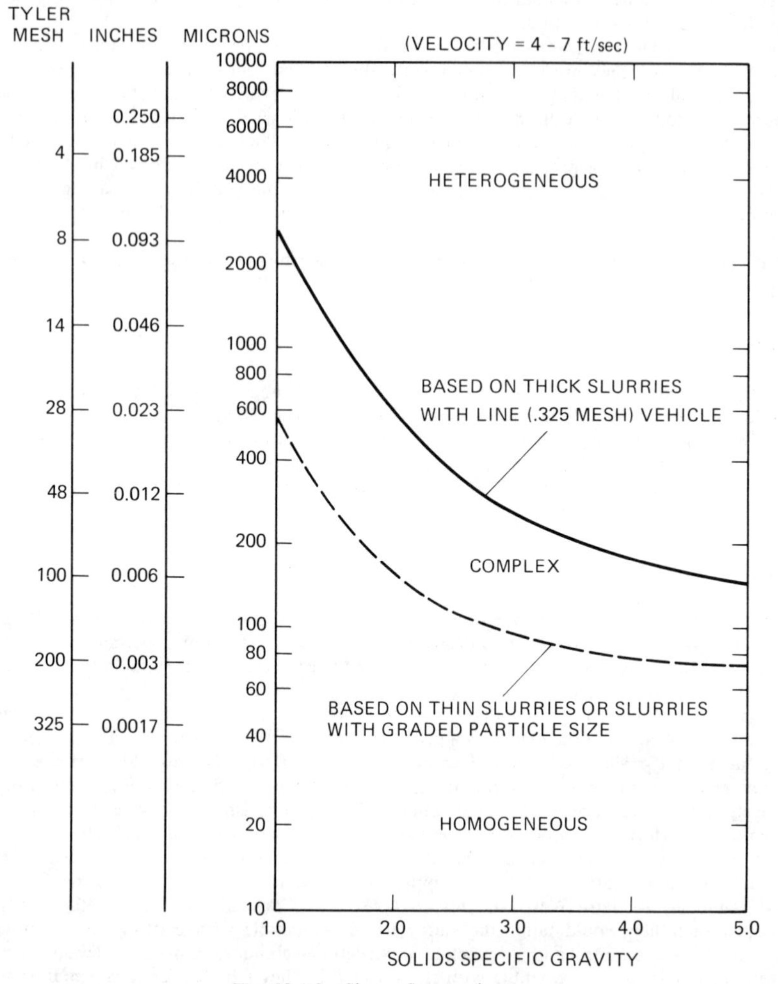

Fig. 29.1.3 Slurry flow regimes.

in which the pipeline is simply a "long spot" in the process (i.e., no additional process facility investment required due to pipeline transport), even pipelines as short as 10–20 mi have been commercially viable. Tailings pipelines, of course, can be as short as a few hundred yards.

29.1.4 Design Procedures

Today, the practicing engineer has the tools available to confidently design a long-life slurry pipeline *if* he or she has control of the design of the slurry (i.e., specification of material size consist and solids concentration). For long-distance long-life slurry pipelines it is imperative that the slurry engineer select and specify these parameters. In most commercial applications the utilization process imposes certain size consist and concentration restrictions which limit the engineer's choices and provide a starting place for evaluations and trade-off studies.

For cases where it is not possible or economical to design the slurry, such as with tailings, the systems can still be designed and engineered but certain penalties in energy costs, pipe wear, flexibility, and shutdown/restart characteristics have to be accepted. For the more general case in which the slurry is specified, the design steps are shown in Table 29.1.2.

Table 29.1.2 Slurry Pipeline Design Steps[a]

I. Process Design
  A. Hydraulics
    1. Select carrier fluid.
    2. Obtain or prepare sample(s) of representative slurry and fluid (if available).
    3. Characterize slurry in laboratory for:
      a. Size consist
      b. Solids specific gravity
      c. Particle shape factor
      d. Particle settling velocity
      e. Slurry pH
      f. Rheology as a function of concentration
    4. Select appropriate rheological model.
    5. Calculate friction losses and spatial distribution of particles (i.e., C/CA) as a function of pipe diameter, velocity, and concentration.
    6. Repeat steps 2 through 5 with alternative size consists (if commercially practical). Select design size consist and range.
    7. Determine deposition velocity and/or laminar–turbulent transition velocity as a function of pipe diameter and concentration.
    8. Identify feasible diameter, concentration, and velocity combinations. Calculate required pumping power using route elevation profile.
    9. Select diameter and operating concentration range based on economic analyses.
    10. Calculate surge pressure "envelopes."
  B. Corrosion–Erosion
    1. Establish desired pipeline life (e.g., 20–50 years).
    2. Test slurry corrosivity in laboratory:
      a. Uninhibited
      b. With various corrosion inhibitors, pH control, and/or oxygen scavengers
    3. Select inhibitor systems for both slurry and carrier fluid.
    4. Estimate wear rate (i.e., metal loss) as function of velocity (for high-velocity or heterogeneous systems only—no wear expected in conventional designs).
    5. Calculate extra pipe wall thickness for corrosion–erosion allowance.
    6. Evaluate system hydraulics for slack flow areas, especially during batching. Avoid slack flow, if possible.
  C. Abrasivity
    1. Measure slurry abrasivity in laboratory, with and without inhibitors (for selection of pump type and estimate of pump parts lives).
  D. Operability
    1. Establish shutdown/restart requirements.
    2. Evaluate shutdown/restart characteristics of selected slurry, using laboratory tests and models.
    3. Select maximum allowable pipeline slope (after consideration of economic impact on slope criteria).
    4. Select route in field, considering balance of operability, minimum length, construction costs, environmental impacts, pumping power, and avoidance of slack flow.

II. Mechanical Design
  A. Pump Stations
    1. Select pump type based on slurry abrasivity and particle size, total head, and economic evaluation.
    2. Select number of stations and location.
    3. Select drivers (e.g., electric motor, diesel, etc.).
    4. Establish speed control requirements and select speed control device.
    5. Specify maximum allowable pulsations and select dampeners.
  B. Automation—Control
    1. Select remote supervisory, local supervisory, or manual control.
    2. Select specific control system.
    3. Select communications carrier compatible with selected control system.
  C. Pipeline
    1. Select pipe grade based on economics, schedule, and quality.
    2. Establish degree of tapering and select wall thickness change points after considering surge pressures and maximum pressures due to batching.
    3. Specify external corrosion protection (i.e., coating and cathodic protection).
    4. Select buried or above-ground installation. Provide freeze protection (if required) and establish burial depth.
    5. Provide for energy dissipation, if required (e.g., choke stations).
  D. Storage
    1. Evaluate system storage requirements under varying operating conditions.
    2. Provide tank or pond storage and calculate required sizes.

[a] From Thompson and Aude [4].

29.2 SLURRY PREPARATION

29.2.1 Introduction

Slurry preparation, as used in the context of slurry pipelines, is the process of combining in specified proportions appropriately sized solids with the carrier liquid to form a slurry with the characteristics required for the particular pipeline transportation application. In many applications one or both of the process factors, percent solids and particle size distribution (PSD), are constrained by other process steps. Iron pellet plant feed, copper liberation from the ore matrix, cement kiln feed, and phosphate concentrator liberation requirements usually set a PSD that is satisfactory for pipeline transportation. In limestone slurry pipelines, for wet-process cement making, a very high percent solids is desired for the kiln feed. This often justifies the use of thinning agents in the slurry to achieve satisfactory rheological properties. Conventional coal slurry preparation is the current exception. In that case, the slurry preparation is dictated by the requirements of the transportation system. The slurry specification represents a balance between the requirements of the slurry preparation, pipeline, and dewatering elements of the system.

Some proposed slurry pipeline applications would introduce unique slurry preparation requirements. The proposed slurries are:

Coal/methanol
Coal/CO_2
Coarse coal
Coking coal
Coal/water fuel
Stabilized coal
Oil agglomeration
Deep or chemical cleaning

Each of these proposed slurries has its particular preparation specifications which are established by transportation and separation or other elements of the use chain. Most of the process steps utilize commercially available equipment adapted to the particular requirements. In some cases new equipment has been developed to meet certain process requirements.

The following discussion deals with the basic slurry preparation process of particle size reduction, percent solids control, and slurry storage.

29.2.2 Comminution

Comminution is reduction to a smaller size. It may be accomplished by crushing of larger-size solids in the dry state or by grinding finer solids, usually in a slurry. In the context of cross-country slurry pipelines, the need for comminution is restricted to coal transportation applications which is the subject of the following discussion.

Crushing

Coal is commonly crushed to smaller than 2-in. size before shipment from the mine. It may or may not have been washed. The first step in a coal slurry preparation process is to further crush the coal before grinding.

No generally accepted theories are available for sizing crushing equipment. Care must be taken, therefore, in extrapolating experience, particularly where the coal is significantly different. It is generally necessary to perform pilot or vendor tests before final selection of crushers can be made. The reduction ratio for this stage of grinding is about 20. For instance, the feed would be about 80% passing 1 in., and the product, 80% passing 14 mesh.

Cage mills seem to have the unique characteristic that product PSD is not greatly affected by throughput or cage wear. It has been observed that horsepower draw increases over time as the strikers wear and become less efficient. Striker replacement and cage renovation is necessary at regular intervals after approximately 150,000–200,000 tons of coal throughput.

Grinding

The grinding step can accomplish two objectives: reducing the coal to the specified particle size distribution, and thoroughly mixing the coal with the liquid carrier. The grinding step has been accomplished at the Ohio and Black Mesa pipelines using rod mills.

A preliminary sizing of rod mills can be made using the Bond [5] methodology, where:

$$W = \frac{10\ Wi}{\sqrt{P}} - \frac{10\ Wi}{\sqrt{F}}$$

where: W = required work input, in kWh/short ton
Wi = work index
P = product size at 80% passing, in microns
F = feed size at 80% passing, in microns

Work indices are not generally available for coal; however, they can be estimated using another Bond equation as follows:

$$Wi = \frac{435}{(Hd)^{0.91}}$$

where Hd = Hardgrove grindability index (ASTM D409-71).

The Hardgrove index may understate grinding requirements for low-rank western coals [6].

The reduction ratio in the Black Mesa rod mills is on the order of 3.0, far less than 12–20 range felt to be most efficient for rod mills [5].

Metal loss due to abrasion of rods and liners can be a significant factor in the rod mill operation; however, according to Rexnord [7], the rod wear at Black Mesa is 0.08 lb/ton of coal processed and liners are still good after 11 years of operation. Metal wear is a function of the amount and characteristics of the ash associated with the coal and whether the coal has been washed.

29.2.3 Slurrification

Slurrification means mixing the solid and liquid phase in the specified proportions. This may be accomplished in two ways:

1. Proportioning water into a mixing vessel based on the weight of solids indicated by a belt scale
2. Controlling slurry density to a preset value using radiation density meter-controlled addition of dilution water

In many applications, the first method is used to establish the basic proportions whereas the second establishes the final percent solids before the slurry enters the pipeline.

In most applications, mixing the solid and liquid phases to form the slurry is not a specific process step. The mixing often is accomplished during the wet-milling step or in a sump or tank upstream of the slurry pumps. Under some circumstances, where wetting the solids is difficult, special effort may be required to accomplish slurrification.

29.2.4 Product Quality Control

The quality parameters of concern in slurry preparation are:

Particle size distribution
Oversize particle control
Percent solids
Slurry consistency

Achieving the specified particle size distribution (PSD) is a fundamental activity of the slurry preparation process; however, means of continuous monitoring of PSD are not well proven. It is necessary, therefore, to confirm the product PSD by regular, typically hourly, analysis of slurry samples.

The preferred method of oversize particle control is the use of screens in the product stream. These screens may also serve to close the grinding circuit. Alternatively, they may simply act as a safety screen which passes all the product, except tramp oversize and trash, when the process is functioning correctly, but as a blind when the product becomes coarse.

The percent solids of the slurry is controlled by continuously monitoring instrumentation as discussed previously. Checks normally are made at regular intervals by drying samples in the laboratory.

Slurry consistency can be affected by PSD, percent solids, solids surface chemistry, and additives. The first two variables were discussed earlier. The solids may interact with the liquid phase differently depending on the location in the mine being worked and the degree of weathering of the solids. For instance, Black Mesa Pipeline notices changes in several slurry properties when near-outcrop (i.e., coal near the earth surface) coal is being shipped.

In some cases, notably limestone slurry, thinning agents are added to reduce slurry yield stress and achieve lower friction losses in the pipeline. Proper addition can be confirmed by in-line viscosity measurements or laboratory checks.

In many systems the final confirmation of slurry quality is a hydraulic test section upstream of the mainline pumps in the first pump station. This safety loop is instrumented to provide readings of unit friction losses, and slurry density. An alarm indicates if pressure losses are outside predicted limits. Since it takes hours for the complete pipeline to be filled with the off-specification material, the operators have ample time to react to the alarm.

29.2.5 Slurry Storage

Slurry pumping applications invariably require surge storage at some location in the system. Most frequently storage is provided just upstream of the first pump station to isolate the pipeline from upsets in the slurry preparation process. This storage may range in size from an unagitated sump to several mechanically agitated tanks. The capacity of the storage vessels may range from a few minutes to several hours of pipeline flow. Slurry storage may also be provided as follows:

For batches of slurry that will be kept separate in the system, for instance, different grades of coal or caps and tails

At pump stations to collect slurry flushed from pumps and piping or to smooth out concentration waves

At downstream terminals as surge storage between the pipeline and downstream processes

At junctions in the pipeline system where feeder lines join the mainline or the mainline branches to distribution lines

Downstream of the pipeline where the process may have extended shutdowns while the pipeline continues to operate or requires a continued supply of material when feed to the pipeline is interrupted

The storage may be active or inactive. Active storage uses agitation to keep the solids in suspension while the solids are allowed to settle out in inactive storage.

Active Slurry Storage

Agitated storage tanks are provided in slurry pipeline systems where a requirement exists for immediately retrievable slurry volume, that is, "live" or "active" storage or "surge" capacity. The criteria for the storage volume required depends on the specific location of the requirement in the system; however, the volume ranges from a few minutes to 24 hours or more of system flow. An agitated tank is used when the maximum volume of unagitated, tapered bottom sumps is not sufficient as limited by slurry consistency requirements. The upper limit of tank capacity is set by agitator technology.

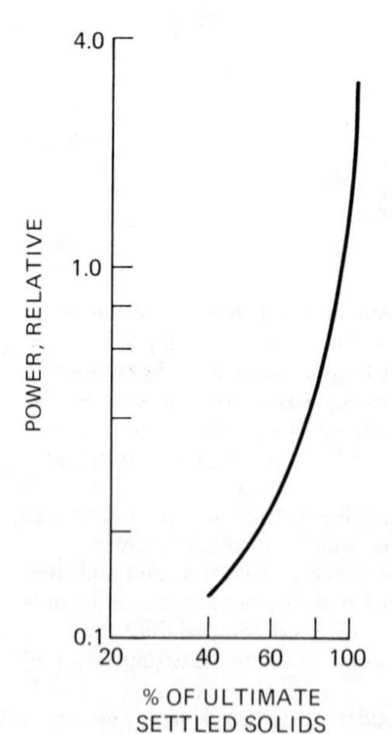

Fig. 29.2.1 Increase in power as weight percent solid concentration approaches the ultimate settled solids condition.

Agitated slurry storage tanks are generally of welded steel construction, cylindrical, and flat-bottomed with diameter more or less equal to height. API Standard 12D "Specification for Field Welded Tanks for Storage of Production Liquids" and API Standard 650 "Welded Steel Tanks for Oil Storage" provide guidance for tank designs which must be modified as necessary to support agitators.

Tank Sizing. The gross tank capacity will not be available for operating use. The active capacity of a slurry tank will be less than the volume due to the following factors: agitator blade elevation, withdrawal pump NPSH requirements, and freeboard requirements.

The minimum operating level in the tank should be about a foot above the top of the (lower) agitator blade. This limit is set for two reasons:

1. Solids suspension cannot be maintained if the agitator blade is partially (or wholly) uncovered.
2. Some agitators can develop violent mechanical instability when the blade is partially uncovered.

The lower operating level of the tank must provide enough NPSH at required flow to the withdrawal pump. Normally, the withdrawal pump and piping should be designed so they are not the limiting factor on tank drawdown.

Freeboard must be allowed at the top of the tank for the overflow weir and normal surface waves and swirl. The freeboard required by the overflow depends on a weir sized to take the maximum tank inflow volume. Add to the overflow freeboard some operating room, say 1 ft, to establish the upper operating level.

These considerations have the following effect on the operating volume of the 2-hr-capacity Black Mesa coal slurry tanks:

Gross volume	640,000 gal
Tank size (49 ft dia.)	45 ft high
Less:	
Agitator blade plus 1 ft	− 7 ft of height
Freeboard	− 1 ft of height
	− 8 ft of height
Operating height	37 ft
Operating volume	525,000 gal

Agitators. Final selection of the agitator motor size for slurry storage tanks generally involves, as a minimum, laboratory analysis of representative samples of the slurry to be agitated. Even a complete description of solids and slurry properties is not sufficient to predict the mixing requirements for thick slurries. The physical and chemical properties of the particle surfaces play an important role in the rheology of fine, dense slurries. Agitator suppliers have bench-scale test procedures that allow agitator size selection based on a small slurry sample.

Pipeline slurry tank agitators generally use an axial flow impeller with a pitch angle between 20° and 45°. Optimization of the blade angle may be done with pilot tests or in the field after installation. The ratio of the impeller diameter to the tank diameter (D/T) plays an important part in mixing power requirements. The trade-off in costs between tall tanks with low mixing requirements and short tanks with low wall pressures (less steel plate) has resulted in the approximately unity tank-height-to-diameter ratio mentioned previously. Overall height will be limited by local soil bearing characteristics.

A typical curve, Fig. 29.2.1, of horsepower versus percent solids shows the effect that percent by weight solids have on the power required of the system. Actually, weight percent solids is not the only variable, but the particle size determines the viscosity at a given weight percent solids. Viscosity is as important a correlating parameter as percent solids. However, viscosity measurements are not yet standardized in the field, and weight percent solids for a particular ore is used, along with pilot testing of the proposed ore. An economic balance is needed to see what the capital and operating cost requirements are for various percentage of solids, as well as the cost of the entire tank and installation to derive the economic optimum for the system.

Stationary fillets in flat-bottom tanks should be considered in storage tanks as they do provide the possibility of marked reduction in horsepower with little change in the effective volume of the system.

Materials of Construction. Slurry tanks are commonly constructed of carbon steel with an allowance for internal corrosion depending on the slurry properties. Corrosion inhibitors used in the pipeline may be added to the slurry tank with tank protection resulting. Agitator shafts are also generally carbon steel, whereas impellers are often rubber covered. Adjustable and replaceable impeller blades provide flexibility for operation and maintenance.

Typical slurry storage tank features are shown on Fig. 29.2.2 and discussed below.

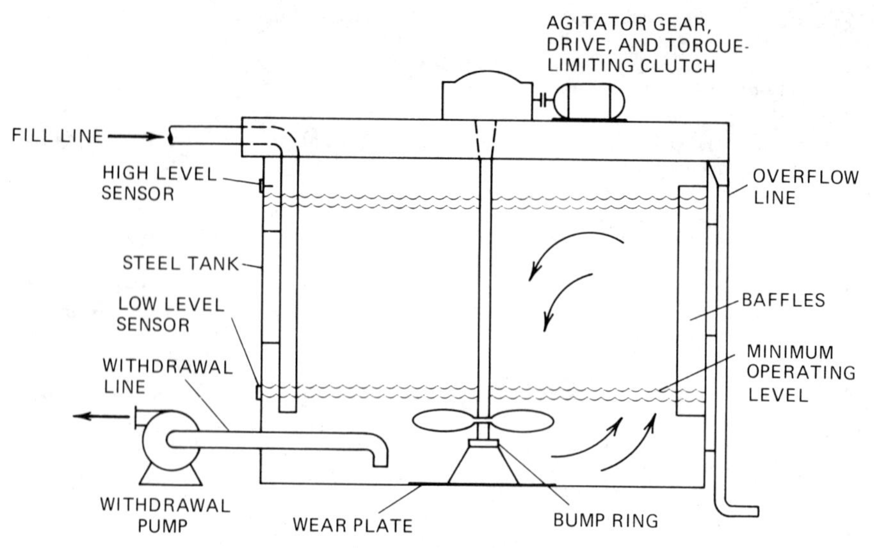

Fig. 29.2.2 Typical slurry storage tank.

1. *Agitator.* The several manufacturers have widely differing mechanical designs for their agitators; however, the following should be considered in specifying an agitator.

Protection of the agitator from damage during startup of a static tank, for instance, after a power outage. A torque-limiting clutch is desirable.

Most agitators are fitted with a shaft guide mounted on the tank bottom or above the blade; however, bottom bearings should be avoided as they are a bothersome maintenance problem.

A motor ammeter and a trip alarm are very useful in operation of storage tanks. The alarm is to alert the operator if the agitator unexpectedly shuts down so he can take immediate action to avoid "sanding out" the agitator.

2. *Tank.* The tank should have the following features:

Access doors. Access doors should be big enough so that a wheelbarrow can be rolled into the bottom of the tank for periodic removal of trash material.

Drains. Drains should be at the very bottom of the tank, below the normal suction, to permit complete drainage for tank inspection and repair.

Lighting. Adequate lighting is needed at the top of the tank platform to show the slurry level.

Ladder. A ladder with rings spaced at 1-ft intervals is handy not only for access, but also for visual observation of tank level.

Overflow. A must, it should be sized to take the maximum fill rate.

Air and water connections. Suitable air and water connections at the top of the tank for maintenance and cleanup purposes are desirable. These connections can also be used after a power failure for lancing to get a stalled agitator restarted.

Level probes. As a minimum, probes with alarms should be installed at the maximum and minimum working levels.

3. *Fill Line.* The pipe feeding the tank is, where possible, brought over the top rim of the tank and sloped to be self-draining into the tank. It should discharge below the minimum operating level to eliminate splash and air entrainment, which can cause a housekeeping problem and raise dissolved oxygen levels (contributing to corrosion).

4. *Withdrawal Line.* The withdrawal line should be extended through the tank wall into the most active zone of agitation; it should be fitted with a 90° elbow, looking down to guard against sanding of the line when no withdrawals are being made. A valve should be fitted outside the tank with water piped for clearing before startup and for flushing after shutdown.

Storage Ponds

The basic distinction between tanks and ponds is that ponds are static storage; that is, no energy is put in to maintain suspension of the solids so they settle to the bottom of the ponds. Ponds, therefore, lend themselves to intermediate to long-term, high-volume storage requirements.

Slurry storage ponds can be classified into two basic types; semiactive and dead storage.

Semiactive ponds are equipped for recovery and reslurrying on short notice. Depending on design, it may be necessary to remove all the stored material before refilling the pond.

Dead storage ponds require considerable equipment and effort to recover the solids. The range in sophistication varies—from the impoundment type used as dump ponds to lined basins with french drains.

The major consideration in pond design is recovery of a uniform solids size consist. Segregation of the solids by size can occur during the filling or recovery operation. The resulting coarse and fine slurry slugs will be difficult or impossible to handle in the downstream processes. The difficulty and expense of recovery must be balanced against the expense of built-in recovery equipment, and shaping and lining the pond. Basic recovery methods include mechanical and dredge. They are described below.

Mechanical recovery utilizes conventional earth-moving equipment to remove the settled solids. Generally these ponds would have the lowest first investment. Prior to the start of recovery operations, the bulk of the water must be removed from the pond either by natural drainage, well points and pumps, or an underdrain system. This type of recovery would be used for dead storage ponds. Such ponds were installed at Mohave for Black Mesa coal slurry.

Dredge recovery utilizes conventional dredging practice to recover the solids. In this case, a water layer or working pool of water must be present to float the dredge. The recovery dredge may be maintained on site or moved in for the recovery operation depending on a requirement for semiactive or dead storage. This type of pond is used at the downstream end of the Samarco iron concentrate pipeline.

29.2.6 Typical Preparation Systems

Commercial coal slurry preparation plant experience is available from only two applications, the Consolidation Coal plant at Cadiz, Ohio, and the Black Mesa plant in northeast Arizona.

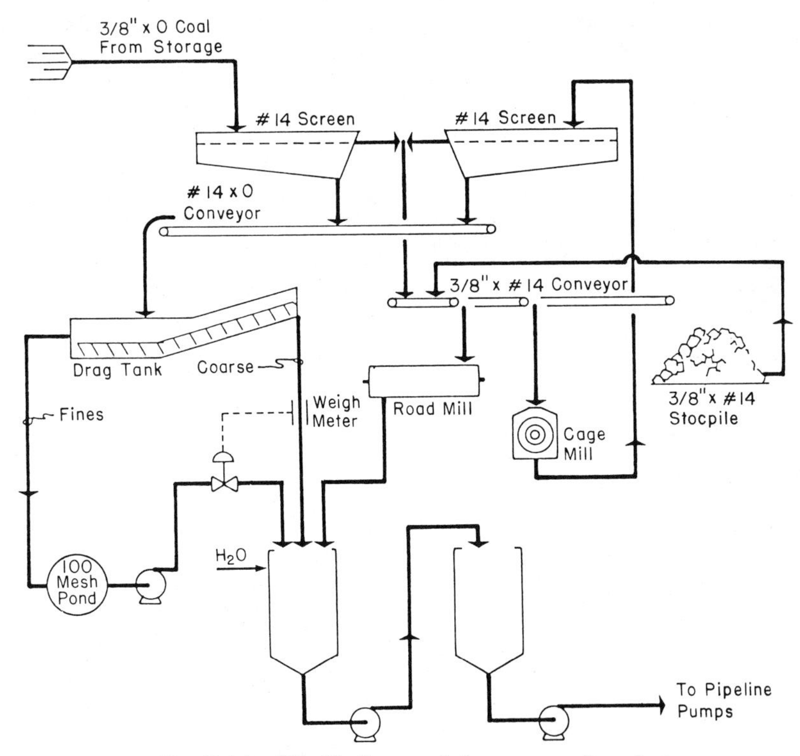

Fig. 29.2.3 Ohio Pipeline—coal slurry preparation plant.

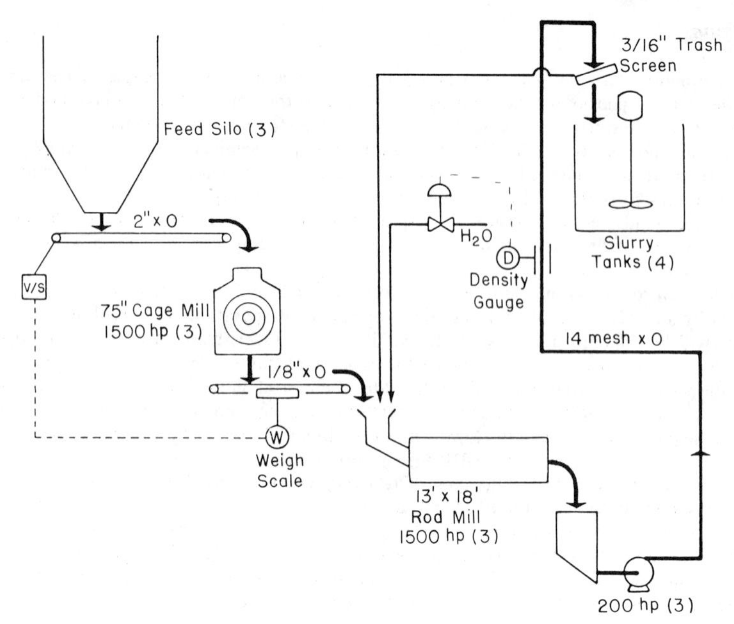

Fig. 29.2.4 Black Mesa Pipeline—slurry preparation plant.

Ohio Pipeline

The Ohio coal pipeline slurry preparation plant is located at Consolidated Coal Company's Georgetown preparation plant near Cadiz, Ohio. The slurry preparation plant operated for 6 years, between 1957 and 1963. A simplified flow sheet of the plant in its final configuration is shown in Fig. 29.2.3. Feed to the slurry preparation plant was minus $\frac{3}{8}$ in. product from the adjacent coal cleaning plant. At one stage in its operation, coal recovered from the black water pond by flotation supplemented the minus 100 mesh fraction.

In the early operation, roll mills in closed circuit were used to crush the $\frac{3}{8}$-in. × 14-mesh fraction; however, these were later replaced by cage mills which required less maintenance [8]. The rod mills finally used to produce the fine component were the peripheral discharge type.

Black Mesa

The Black Mesa coal slurry preparation plant has three mill lines like those shown in Fig. 29.2.4. Each mill line has a capacity of 290 ton/hr when new wearing parts are installed in the cage mills.

The initial Black Mesa configuration used hammer mills in parallel with a $\frac{3}{8}$-in. scalping screen. This resulted in too coarse a product from the plant. At one stage, rods were replaced by balls in the wet mills and the circuit was closed with a cyclone. Oversize could not be controlled due to the necessity of producing a thick, difficult-to-classify product, so the experiment was abandoned in favor of the cage mill/rod mill in open circuit [9].

29.3 SLURRY PIPELINE TRANSPORTATION

29.3.1 Process Factors

Flow of mixtures of solids and liquids (i.e., slurries) in pipes differs from flow of homogeneous liquids in several important ways [2, 10]. With liquids, the complete range of velocities is possible, and the nature of the flow (laminar, transitional, or turbulent) is defined by the physical properties of the fluid and system. With slurries, additional distinct flow regimes and several more physical properties are superimposed on the liquid system. The two major regimes of slurry flow are:

1. *Homogeneous Slurries.* Here, the solid particles are homogeneously distributed in the liquid media, and the slurries are characterized by high solids concentrations, and fine particle sizes. Such slurries often exhibit non-Newtonian rheology (i.e., the effective viscosity is not constant, but varies with the applied rate of shearing strain). Some examples are sewage sludge, clay slurries, and cement–kiln–feed slurry.

2. *Heterogeneous Slurries.* Here, concentration gradients exist along the vertical axis of a horizontal pipe even at high flow rates (i.e., the fluid phase and the solid phase retain their separate identities). Heterogeneous slurries tend to be of lower solids concentration and have larger particle sizes than homogeneous slurries. Florida phosphate rock is a good example. Coarse coal slurry is another example (i.e., 2-in. by 0).

Many slurries encountered commercially are of mixed character; the finer particle-size fractions join with the liquid media to form a homogeneous vehicle, while the coarser sizes act heterogeneously. Pipeline coal slurry is a prime example of the mixed characteristic.

In order to evaluate a slurry system the designer must determine if the dominant characteristic is homogeneous or heterogeneous. Figure 29.1.3 can be used as an aid in determining the dominant characteristic (at velocities of 4–7 ft/sec). Considering a slurry either as completely homogeneous or completely heterogeneous may be used as a first approach. Many slurries, however, exhibit compound, or homo–heterogeneous flow, and as such, require more sophisticated analysis.

Velocity

The two types of slurries, homogeneous and heterogeneous, have entirely different critical-velocity characteristics, as seen in Fig. 29.3.1. Curve A illustrates behavior typical of heterogeneous slurries. The characteristic hook in the friction-loss-versus-velocity curve results from deposition of solids on the bottom of a horizontal pipe, as shown in the figure. (In vertical pipes, the solids that in a horizontal pipe would be deposited, are easily transported, because their settling velocity is usually much lower than normal flow velocities.)

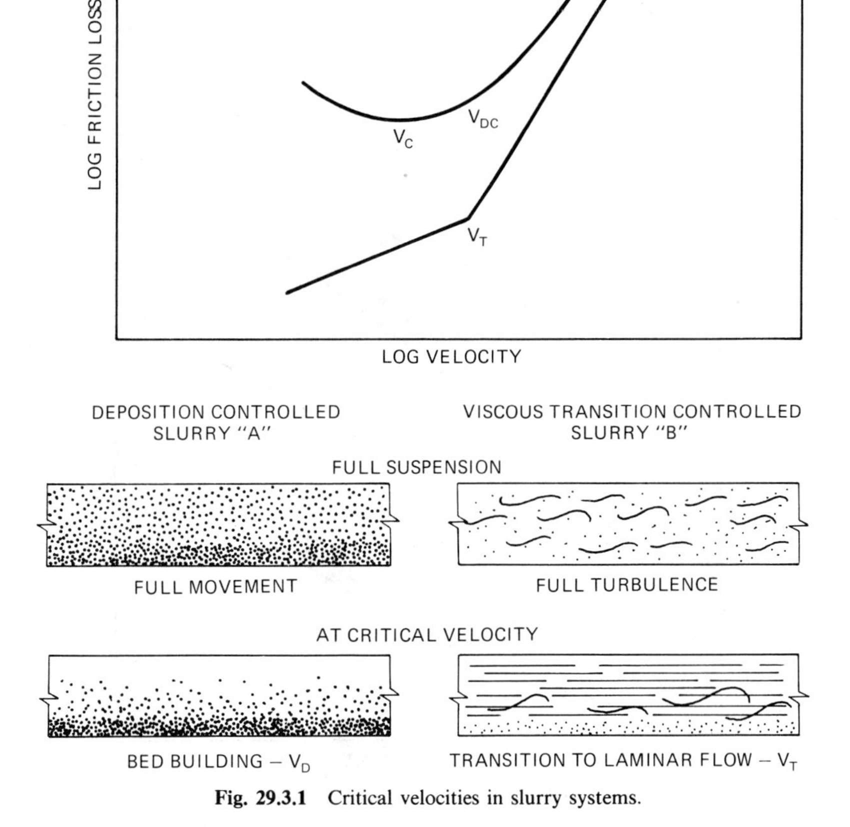

Fig. 29.3.1 Critical velocities in slurry systems.

The deposition velocity is directly related to the settling velocity of the coarser particles in a heterogeneous slurry and the degree of turbulence in the pipe; it therefore increases with increasing particle size or specific gravity. Deposition velocity generally exhibits an increase proportional to pipe diameter to the one-third power or less.

Curve B shows viscous-transition critical velocity, which is characteristic of homogeneous slurries. Although design of a system for operation below the transition critical velocity is theoretically acceptable for truly homogeneous slurries, no turbulent forces exist to suspend even trace amounts of heterogeneous particles. Nearly all long-distance pipelines are designed to give turbulent flow.

The transition velocity and laminar friction losses are very sensitive to the rheological properties of a homogeneous slurry. Transition velocity tends to increase with viscosity and therefore increases with solids (volume) concentration, greater quantity of fines, and lower solids gravity. Transition velocity for slurries with a yield stress (Bingham plastic) is very little affected by pipe diameter, whereas it is directly proportional to diameter for slurries with Newtonian rheological properties.

Concentration and Rheology

The concentration of solids in a slurry is often controlled by the upstream or downstream process. In this case, the dependent variables of slurry viscosity (or rheology) and specific gravity are also fixed. When addition of liquid before transportation is required, the amount of dilution may be an important factor in the downstream process, as well as in the slurry transportation.

Solids concentrations up to about 40% by volume are readily handled when the solids surface area is low (coarse slurries); however, fine materials such as clay or sewage sludge may not be fluid at even a 10% solids concentration. As a general rule, solids that are less than 25% finer than 325 mesh (0.44 microns) will be quite fluid when slurried to a 40% solids volume. For thicker and finer slurries, a knowledge of the relationship between slurry rheology and concentration may be required to select a pumpable concentration.

A good estimate of a comfortable pumping concentration can be made by reducing the static settled-slurry concentration 10–15 percentage points in volume concentration. Slurries approaching static settled concentration can be pumped, providing the high friction losses are tolerable (as, for instance, in short systems). The static settled concentration is determined by allowing a slurry of

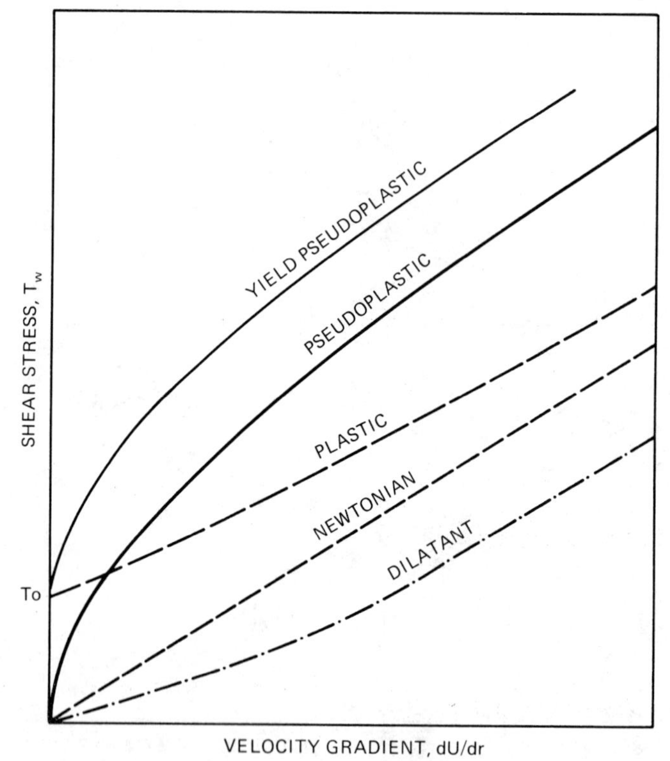

Fig. 29.3.2 Fluid classification of slurries.

Table 29.3.1 Rheological Models of Time-Independent Fluids[a]

Type of Fluid	Industrial Examples	Shear Rate–Shear Stress Relationship[b]
Newtonian	Water, oil, air	$T_w = \frac{\mu}{g_c}\left(\frac{du}{dr}\right)$ Newtonian Model
Plastic	Drilling muds, thick limestone slurries, sewage sludge, thick mineral slurries	$T_w = T_0 + \frac{\eta}{g_c}\left(\frac{du}{dr}\right)$ Bingham Plastic Model
Pseudoplastic	Napalm, cellulose acetate in acetone, paper and pulp suspensions, mayonnaise, some paints and lacquers, detergent slurries, some mineral slurries	$T_w = \frac{K}{g_c}\left(\frac{du}{dr}\right)^n$ Power Law Model
Yield pseudoplastic	Clay water suspensions, polymer solutions	$T_w = T_0 + \frac{K}{g_c}\left(\frac{du}{dr}\right)^n$ Yield Power Law Model
Dilatant	Starch suspensions, quicksand (rarely encountered in practice)	$T_w = \frac{K}{g_c}\left(\frac{du}{dr}\right)^n$ Power Law Model

[a] Adapted from Zandi [12].
[b] Where: T_w = shear stress at pipe wall (lb force/ft^2); T_0 = yield stress (lb force/ft^2); μ = viscosity (lb mass/ft sec); η = coefficient of rigidity (lb mass/ft sec); g_c = acceleration of gravity (ft-lb mass/lb force sec^2); K = consistency index (dimensionless); $\frac{du}{dr}$ = slope of velocity profile (sec^{-1}); n = flow behavior index (dimensionless).

known concentration to settle under static conditions in a graduated cylinder until the slurry/water interface reaches equilibrium. The settled concentration is then calculated from the starting concentration and the beginning and ending volume. If the slurry does not settle, it is already too thick.

Much of the foregoing discussion of solids concentration is based on consideration of the slurry rheology (or viscosity). The rheological properties of a slurry determine the "viscosity" used for friction loss calculation and for the transition critical velocity of fine, thick slurries.

Determining the slurry rheological characteristics is of basic importance. Despite years of extensive work by many researchers, there is still not a general equation of state relating the rate of shearing strain to shear stress. Nevertheless, several simplified flow models are available for solution of practical problems. The coefficients and exponents of these models are determined by laboratory measurement. The reader is referred to Govier and Aziz [11] for a detailed discussion of the subject. Figure 29.3.2 shows various types of flow behavior (shear stress versus shear rate) encountered in slurry systems. Newtonian fluids can be totally described by viscosity, which is the slope of the shear stress versus shear rate relationship. Bingham plastic, pseudoplastic, dilatant, as well as time-dependent fluids, require more complex formulas or parameters to describe them. A summary of the most commonly used rheological models is shown in Table 29.3.1 [12]. Of these, the Bingham plastic model has the most applicability. At high rates of shear it can be used to describe pseudoplastic as well as Bingham plastic materials. Bingham plastic materials may be described by two parameters, the coefficient of rigidity, η, which is the slope of the shear stress velocity gradient plot, and by the yield stress, T_0, which is the shear stress intercept (i.e., the shear stress at the point of zero shear rate). These values can be measured in the laboratory using bench-scale rotational or small tube viscometers.

Preliminary Sizing Procedures

For a preliminary analysis, the following basic design-calculation steps may be used:

1. Classify the slurry as being either homogeneous or heterogeneous.
2. Select the slurry concentration (if a variable).
3. Select a trial pipe size, based on the system's throughput requirement.
4. Calculate the critical velocity (transition or deposition).
5. Check that the design velocity is at least 1 ft/sec above, but not excessively above, the critical velocity. It may be necessary to select another trial pipe size and repeat the calculations until an acceptable relationship between critical velocity and design velocity is achieved.

6. Calculate the design friction losses (distinguishing between horizontal and vertical pipe for heterogeneous slurries).
7. Calculate the system pressure gradient and pump discharge pressure.

29.3.2 Pipeline Control Systems

Most long-distance slurry pipelines utilize some sort of central supervisory control system. This is due to the fact that facilities can be spread out over hundreds of miles. Controls are required for starting and stopping pumps, regulating flow, opening and closing valves, and so on. In addition, various operating parameters also need to be measured and reported, including pump speeds, valve status, pressures, flow rates, density of slurry, and temperatures. The supervisory control signals and data are transmitted over a communications system, such as microwave, leased telephone, cable, or VHF radio. The supervisory control system, although not the most costly part of the pipeline system, is critical to its success. Both oversophistication and oversimplification can greatly damage the system's efficiency and ease of operation.

The recent major advances in solid state technology and programmable controllers (PCs) have greatly improved and simplified supervisory control systems. PCs can take the place of field remote terminal units (RTUs). Relays and field wiring are reduced, as are costs. For a complete discussion of pipeline supervisory control systems, the reader is referred to References 13 and 14.

29.4 SLURRY DEWATERING

At the terminal end of a slurry pipeline, there is usually some form of dewatering, that is, separation of the product solids from the carrier vehicle. Recovery of the solids is the principal aim. In most cases, however, the carrier vehicle is water. which can be reused for make-up cooling water or some other beneficial application.

The process into which the product feeds determines the dewatering flow sheet and equipment selection. Wet-process cement making, for example, only requires thickening, because slurry feed to the kiln is a part of the normal process. For coal feed to conventional coal-fired power plants, on the other hand, fairly extensive dewatering is required, including filtration, centrifugation or, in some cases, thermal drying in order to attain a coal moisture that will allow the coal to flow in bunkers, yet still not be so dry as to be dusty. This moisture range is usually between 7 and 11% surface moisture.

29.4.1 Process Factors

In any dewatering process, the objective is to attain the desired dryness at minimum cost. Since different pieces of equipment are more effective for specific particle sizes, most flow sheets have split streams in which the various particle sizes are dewatered in machines most effective for that particular particle size. Most pipeline slurries have a broad size consist range rather than a "graded" or narrow size consist.

The most important variable in determining dewatering characteristics is the surface area of the particles. The finer the particles, the greater is the surface area and the greater the attraction to moisture. Therefore, the percent of minus 325 mesh (44 microns) particles in the slurry is a useful indicator of difficulty of dewatering. Important process variables are:

Particle size (i.e., surface area)
Liquid viscosity
Feed concentration

Dewatering can be improved by increasing particle size. This can be accomplished, in some cases, by using flocculating agents. Lowering the viscosity of the liquid also improves both centrifugation and filtration performance. A common approach is to heat the slurry before dewatering, to reduce viscosity. (The viscosity of pure water is 1.005 cp at 20°C and only 0.284 cp at 100°C.) Increasing the feed concentration to centrifuges will also increase the rate of solids processed.

For preliminary screening of dewatering equipment, the following information is needed:

1. Particle size distribution by sieve sizes (for very-fine slurries a subsieve analysis is desirable).
2. Particle (i.e., solids) density.
3. Feed concentration range.
4. Temperature of incoming slurry.
5. Rate of slurry delivery.
6. Dewatering rate requirements.

7. Desired product surface moisture range.
8. Equilibrium moisture of solids.

As noted earlier, the finest particles usually exert the greatest influence on dewatering. In most flow schemes the fines are treated separately from the coarser particles.

Bench-scale testing is also quite helpful in determining the difficulty of dewatering a slurry. Leaf filter tests and laboratory basket centrifuge tests are useful in projecting plant cake moistures. In addition, small-scale centrifuge tests performed by manufacturers can be helpful in determining the practicality of dewatering a slurry by centrifugation.

REFERENCES

1. Aude, T. C. et al., "Coal Transportation Costs/Inflation," Proceedings of 4th International Technical Conference on Slurry Transportation, Slurry Transport Association, Las Vegas, Nevada, March 28–30, 1979, pp. 122–129. (Proceedings available from Slurry Transport Association, 490 L'Enfant Plaza East, S.W., Suite 3210, Washington, DC 20024.)
2. Aude, T. C. et al., "Slurry Piping Systems: Trends, Design Methods, Guidelines," *Chemical Engineering,* June 28, 1971, p. 78.
3. Wasp, E. J. et al., "Solid–Liquid Flow Slurry Pipeline Transportation," *Trans Tech Publications,* Clausthal, Germany, 1977, preface.
4. Thompson T. L. and T. C. Aude, "Slurry Pipeline Design and Operation—Pitfalls to Avoid," ASME paper presented at Joint Petroleum Mechanical Engineering and Pressure Vessels and Piping Conference, Mexico City, Mexico, Sept. 20, 1976, pp. 4–5.
5. Bond, Fred C., "Crushing and Grinding Calculations," *Chemical Engineering,* January 1961.
6. Vecci, S. J. and G. F. Moore, "Determine Coal Grindability," *Power,* March 1978.
7. Wunderlin, Mark, Product Sales Manager—Grinding Mills, Rexnord Process Machinery Division, Milwaukee, private communication, July 14, 1982.
8. Snoek, P. E. and F. B. Raymer, "Preparation of Coal for Pipeline Transportation," *Mining Congress Journal,* March 1982, pp. 27–30.
9. Wasp, E. J. et al., "Terminal Facilities for Western Coal Slurry Pipelines," 1976 AIME Annual Meeting, Las Vegas, Nevada, Feb. 1976.
10. Thompson, T. L. and T. C. Aude, "Slurry Pipelines, Design, Research and Operation," *Journal of Pipelines,* Vol. 1, 1981, p. 33.
11. Govier, G. W. and K. Aziz, *The Flow of Complex Mixtures in Pipes,* Van Nostrand Reinhold, New York, 1972.
12. Zandi, I., *Advances in Solid–Liquid Flow in Pipes and Its Application,* Pergamon Press, Oxford, England, 1971, p. 3.
13. Chapman, J. P., "Programmable Controller Based Control Systems in the Materials Handling Industry," Mining Eng. and AIME Transactions, 1983.
14. Pitts, J. D. and J. P. Chapman, in A. Weiss (Ed.), *Computer Methods for the '80s,* Society of Mining Engineers, New York, 1979, pp. 816–817.

CHAPTER 30
SAMPLING AND WEIGHING

S. H. RASKIN

Tri-Cell Company
Rockwall, Texas

CHARLES ROSE

Coal Technology Consultants, Inc.
Birmingham, Alabama

30.1 BULK WEIGHING SYSTEMS

S. H. Raskin

Modern electromechanical systems for measuring the weight of bulk material are the result of recent progress in a number of technical disciplines. Handling equipment is moving increasingly large masses of materials at higher speeds. Basic electronic weighing instruments include microprocessors offering programmable options for tolerating a variety of field conditions. Entire operating facilities are being centrally controlled, each by a single minicomputer, with dozens of selectable flow paths instantly available at the push of a button. Such progress offers unprecedented weighing opportunities.

Whether a particular weighing system lives up to its potential depends on the scope of expertise and degree of cooperation that are applied to decisions affecting design, operation, test, and maintenance. The data here presented are directed toward the decisions and alternatives frequently encountered in satisfying needs for weighing systems.

30.1.1 System Scope

Weighing of bulk material includes utilization of equipment, personnel, and methods to obtain accurate weight at a reasonable cost, while causing minimal interference to operations.

Cost considerations include investigation of possible trade-offs among different types of costs, covering investment in equipment, personnel training expense, labor for inspection, test and operation, anticipated operating downtime, and anticipated material shrinkage based on some level of weighing error. The cost investigation continues through a post-installation review.

As to weighing system compatibility with operations, the ideal weighing system will utilize 100% of production material as the actual measurement test sample and will control weighing calibration without delay to the production flow.

30.1.2 Weight Measurement

The measurement of weight of a product is the act of determining the gravitational force acting on the product mass. Gross weight is the weight force acting on both the material and its container or

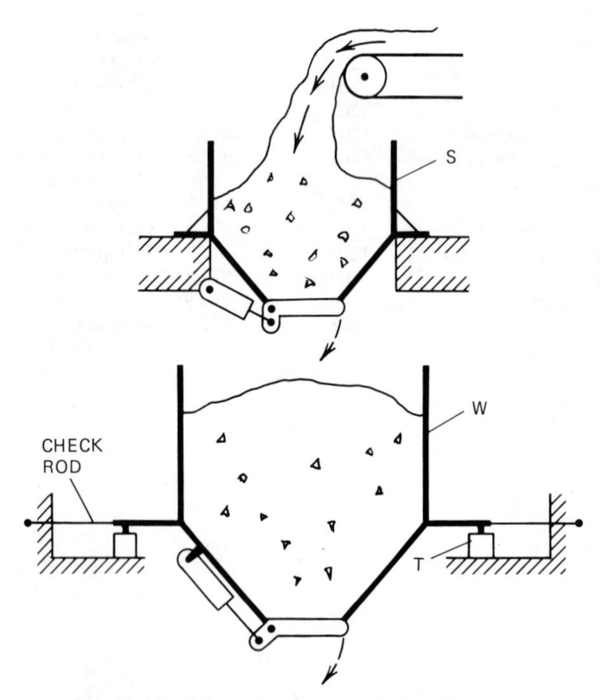

Fig. 30.1.1 Schematic of automatic batching scale.

support means. *Tare* is the weight of an empty container which, when subtracted from the gross weight, leaves a *net weight* value for the contained material.

In the case of belt scales and certain fixed-container scales, such as hopper and tank scales, the tare may be referred to as the *dead-load* of hopper or belt, plus other structure which supports the *applied load* of material. The dead load is eliminated from weight readings by rebalancing the empty measurement system to an indicated zero, by means of a manual rebalance adjustment, a manually activated zero balance push button, or an automatic zero function triggered by the system itself. After the dead load is balanced out, the applied load of material is directly indicated as the net weight.

Static weight readings are obtained when the material is at rest. Dynamic or motion weighing involves measurements taken when the material and/or container are in motion or are being subjected to significant vibratory forces while remaining in place.

30.1.3 Batch Weighing

The advantage of batch weighing is that a quantity of material is statically weighed while isolated from the flow. In its broadest context, batch weighing includes measurement of the weight of any quantity of material that has been collected in a container. If a movable container has been filled and sealed, it may be weighed as a discrete object within the scope of general weighing practice and will not be dealt with here. When a movable container is on a scale while being filled by weight on the scale, however, the weighing process serves as a filling control when applied to small bottles or cans, and as a load-out control when applied to trucks, railway cars, and other large vehicles.

Automatic batching occurs when a fixed hopper, bin, or tank W is mounted on weight-sensitive transducers T, as shown schematically in Fig. 30.1.1, and is furnished with instruments and controls to interrupt the flow of material at a preselected level. Flowing material is temporarily stored in surge bin S while material in weigh bin W is statically weighed and then discharged. The system performs static weighing to produce precise weight data. Control of individual batch mass is less precise, because the material cutoff is a function of dynamic weighing.

The system shown in Fig. 30.1.2 contains features of both the automatic batching and filling control systems, being variously known as trim-filling and batch-trimming systems. Two weigh bins W_1 and W_2 are used for coarse filling of one batch while the other batch is being trimmed to precise size.

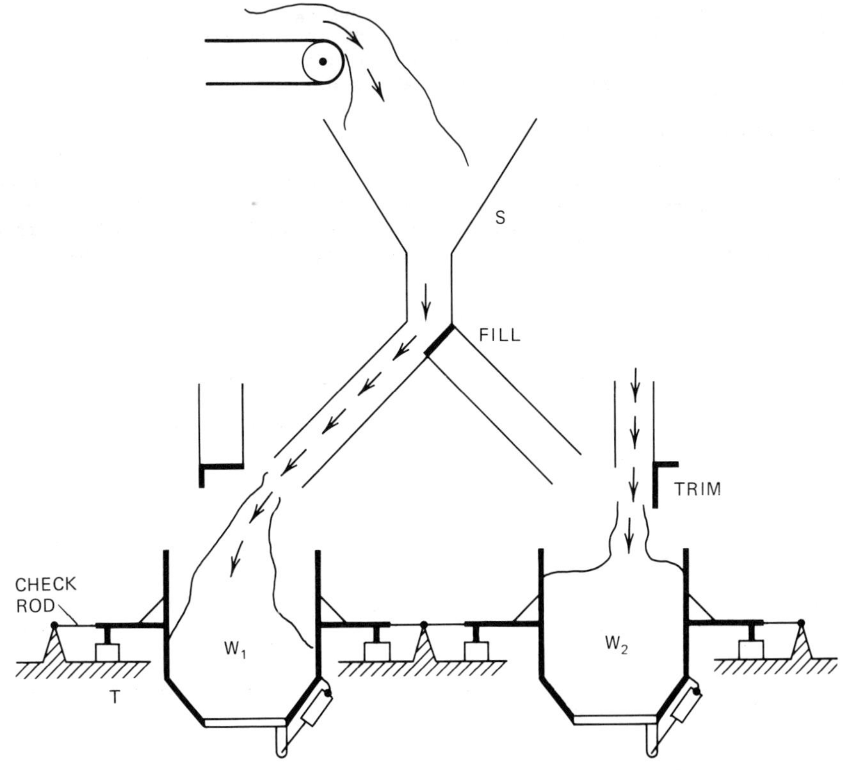

Fig. 30.1.2 Schematic of a dual batch-trimming system.

The coarse filling cutoff is set below the desired weight level, so that the trimming process is in alternate steps of adding small trim increments of material, then statically weighing the result, and repeating until the weight is precisely within a desired range.

The surge bins S of Figs. 30.1.1 and 30.1.2 must be dimensioned according to each system's time requirement for an interruption of material flow, so that the bin scales can complete the required static weighing cycle.

Automatic electronic instruments for controlling small trim-filling systems, designed for the pharmaceutical and food processing industries, can sometimes be applied as controls for heavy-duty bulk weighing systems.

30.1.4 Continuous Weighing

The most common example of continuous weighing is a belt scale, as shown schematically in Fig. 30.1.3. The example shown in the figure is a four-idler scale, because the four weight-sensitive idler rolls R support a weighing portion of the belt. The weighing portion of the belt is deemed to have a weight-bearing length L, where $L = 4A$, and both L and A are expressed in feet.

The applied load of material sensed by the scale will include any material that lies on the belt anywhere within length L. The device of Fig. 30.1.3 also is known as a weighbridge type of scale, because the weight-sensitive rolls are supported by a weighbridge B that rests on transducers T. The cumulative value of the maximum applied load and dead load of the weighbridge must be known for selection of transducer capacity.

In use, however, the working value for the applied load is expressed as a load factor in pounds per foot.

Speed pickup sensor S is shown in Fig. 30.1.3 for generating an electrical signal proportional to belt speed. The transducer signal, which is proportional to weight, and the speed signal both are transmitted to an instrument. Weight of all applied load that passes over the scale is incrementally measured and integrated. If, for example, the load factor on the belt is 200 lb/ft, one ton would pass with each 10 ft of belt. If the speed is sensed to be 600 ft/min, 10 lineal feet of applied load will pass in one second, and the belt scale totalizer will register a one-ton flow of material.

A single-idler pivot-bridge belt scale is shown in Fig. 30.1.4. The fixed pivot P acts to restrain the weighbridge against horizontal drag from the belt friction, and check rods are not used. The applied load acting on transducer T will span a length $L' = A$.

Recent developments in single-idler weighing are continuing to increase its useful ranges of belt speeds and load factors. Another technological development involves nuclear instruments for measuring the density of a material flow. If the volumetric flow can be measured, as in the case of slurry in a pipeline, the combined measurement will be the product of density × volume = mass flow.

30.1.5 Supports and Restraints

Some amount of displacement or deflection will occur in the weight-sensitive transducers and structures which support the applied load mass. The amount of transducer deflection is fixed at time of selecting the transducer. Deflection of containers and supports is controlled by the materials handling machinery designer. The two deflections each act as stiff springs in spring-mass systems and influence vibration characteristics.

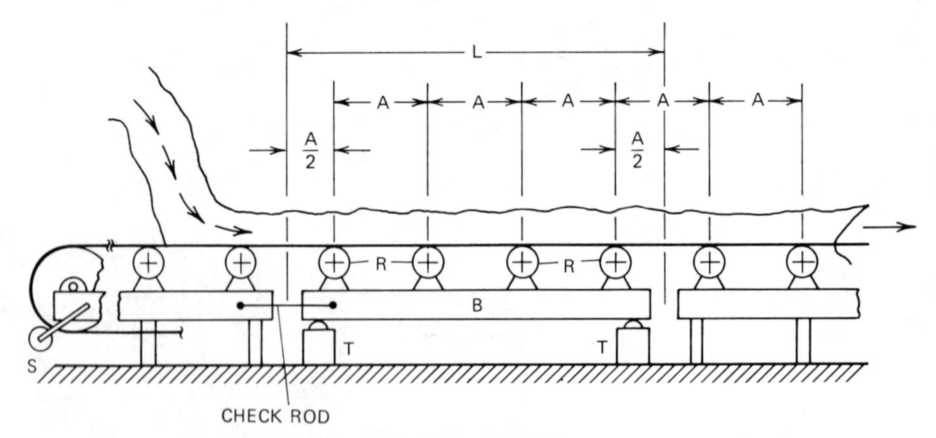

Fig. 30.1.3 Schematic of four-idler, weighbridge-type conveyor belt scale.

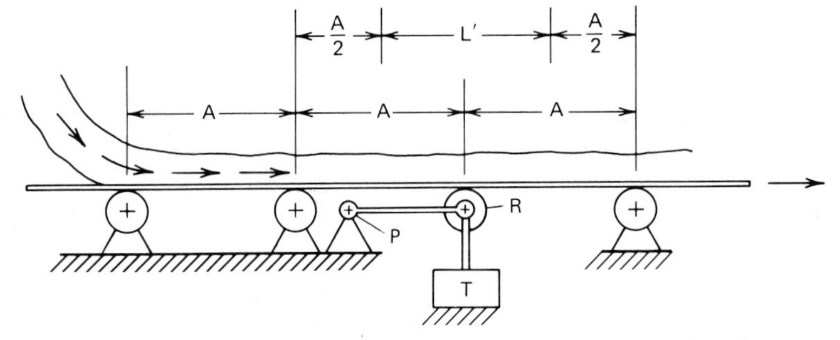

Fig. 30.1.4 Schematic of single-idler pivot-bridge conveyor belt scale.

Meanwhile, the electronic instrument is equipped with some means of filtering or averaging to eliminate unwanted electrical noise and anticipated dynamic forces. The instrument rejection of erroneous dynamic signal must handle the natural frequencies of oscillation of the transducer, the structure, and any harmonics of the two, which otherwise would cause intermittent errors and are difficult to troubleshoot.

Instability in supports under load also can lead to static weighing errors. Allowing the transducer to move out of a plumb condition, which is oriented to vertical weight force, creates an angular instability which can cause check rods to go out of static adjustment. A single adjustment of horizontal restraint must serve under both no-load and applied load conditions.

Misalignment of dead idlers and weight-sensitive idlers on some belt scales will have a direct effect on weighing calibration. Poor alignment not only can cause nonlinear errors of indicated weight, but the errors are difficult to evaluate under most methods of scale testing.

30.1.6 Weight-Sensitive Transducers

Excepting nuclear devices, the role of a transducer is to sense the vertical weight force exerted by gravity on the mass of the applied load, and to generate a signal proportional to that load.

Almost any technological approach to transducers may be used—including optical, pneumatic, and hydraulic—but material handling needs are usually served by two types of transducers: variable transformers and strain-gauge load cells.

Variable transformers, which offer no reactive force to support the load, measure displacement under load in a spring-mass arrangement. Movement of a support under load causes a movable core, which is attached to the support, to move within a stationary electrical coil. The resulting signal is proportional to a relatively large structural displacement of, say, 0.1 in.

Strain-gauge load cells act as a direct, stiff-spring support of load. Internal deformation of the load cell structure causes a change in electrical resistance which can be measured by appropriate instruments. Load cells are used in the vast majority of applications, and they have a deflection under maximum load on the order of 0.001–0.020 in., depending on cell design.

Horizontal side loads must be prevented from erroneously affecting variable transformers and some types of load cells. Self-checking load cells may be protected against damage from side loads, but side loads entering such cells during the weighing cycle may cause excessive weighing errors. Side-load-compensating load cells are a recent development, where the side loads are isolated by structure internal to the cell so that factory calibration includes calibration of horizontal restraints.

30.1.7 Instruments

Advances in some functions of electronic processing and readout instruments have outpaced the quality of available transducer signals. False hopes sometimes arise from considering instrument specifications of, perhaps, readout resolution to one part in 50,000, when the transducer is reliable to, for example, only 0.1% of rated capacity, or one part in 10,000.

Instrument design in other cases has enabled system capability to exceed transducer capability as, for example, where a microprocessor program is used to correct for known nonlinearity of a transducer. The system designer is faced with such an assortment of available instrument options that the selection process contains risk of applying the wrong option to a given need.

Instrument manufacturers sometimes include sophisticated interlocks between optional programs. In some cases, a casual and minor human error in field use of the instrument can set up an interlock that can only be unlocked by turning off system power. Installation and startup problems can be minimized by considering the instrument's optional features and interlocks during preliminary design.

30.1.8 Controls

Operating controls are best kept separate from weighing controls during early design of a weighing and handling system. Then, as individual features of the system uncover commonalities, individual functions can be combined. Ideally, test controls should be under the jurisdiction of the production operator, but with limited access to calibration and program adjustment available only to designated maintenance personnel.

30.1.9 Electrical Protection

Weight measurement systems need to be protected from lightning strikes either on the power line or on the structure in the vicinity of transducers. The transducers must be protected by ground straps of heavy braided cable, well tied to structure on either side of the transducer to provide a bypass to ground. A single-point ground rod to serve the entire system is used to minimize errors caused by RF noise and ground loops. Weighing instruments must be protected from voltage surges through the power line by means of an isolation/regulation transformer. The transformer is used in addition to any voltage regulation within the instrument, and the transformer may be furnished with automatic or manual reset means.

30.1.10 Environmental Protection

Use of the best junction boxes and sealants may serve to seal in moisture, if cables are unprotected and moisture enters exposed ends of cables awaiting installation. Instruments headed for cushioned mounts in an air-conditioned humidity-controlled environment may travel a disastrous path before reaching the intended destination.

Environmental protection starts with shipping container specifications to manufacturers and storage specifications to installers, and ends with operating specifications for later protection against weather, side load protection from wind on outdoor structures, and antispill guards.

30.1.11 Specifications and Tolerances

The design, operation, and approval of weight measurement systems should be performed according to three separate sets of specifications and tolerances. Specifications define the requirements, and tolerances are a form of specification defining allowable error in deviating from a reference or standard.

The design specifications must control selection and fit of system components and interface requirements. Next, the operating specifications are intended to define the methods of operation and test that will assure proper performance and indicate when maintenance or repair is needed. Finally, approval specifications set forth who will conduct acceptance tests and periodic service tests, what references or standards will be used for comparison, and how the tests will be conducted.

Too often, tolerances are stated in terms of percent accuracy in measurement operations, without a definition of test method for determining deviations; such a tolerance is incomplete.

Care also must be exercised to not let critical aspects of component tolerances affect the validity of system tolerances as a result of tradition and habit. An example of that risk can be seen by referring to Table 30.1.1, which is excerpted from United States Air Force PRAM Report 77-1, published in 1978. The table presents composite specifications as a representative sample of responses from 165 vendors of load cells contacted in a 1977 survey of the state of the load cell art. Of significance here is the fact that the first tolerance, rated output, is traditionally stated as a percentage tolerance relative only to a rated capacity load. The next eight tolerances are either directly related to rated capacity load or indirectly related through reference to rated output.

In selecting load cells, care must be exercised to not pass that type of component tolerance through to become a system tolerance. In the example of Fig. 30.1.3, assume a load factor equal to 200 lb/ft over an effective belt length of 10 ft. The applied load at rated capacity RC is: $RC = 2000$ lb. If the dead weight of the belt, idlers, and weighbridge is 2100 lb, the maximum weight to be carried by four load cells is 4100 lb. Therefore, selection is made to use a readily available set of four 1500-lb-rated-capacity load cells.

The combined rated capacity of the four load cells is 6000 lb. According to Table 30.1.1, the linearity tolerance of the load cell is 0.05% of rated capacity, permitting an error equal to 3 lb under any condition of applied load. Since the minimum applied load most frequently specified by regulatory agencies for acceptance test is $\frac{1}{2}$ of the rated load factor, the total applied load during test may be 1000 lb, not the 2000-lb design load. A test deviation of 3 lb relative to an applied load of 1000 lb is equal to 0.30% of applied load. Unless the system described in this example is provided with an instrumentation capability to reduce that one deviation, the system will have a hard time meeting acceptance tolerance of 0.25% of applied load as required by some agencies.

The last tolerance in the list of Table 30.1.1 relates to side loads and was not answered by any respondent with respect to measurement accuracy. Subsequent to the study, a new type of triple-

Table 30.1.1 Composite Specifications and Tolerances for Strain-Gauge Load Cells[a]

Rated output:	3 millivolt/volt (at rated capacity load)
Linearity:	0.05% of rated output
Repeatability:	0.02% of rated output
Hysteresis:	0.03% of rated output
Creep:	0.03% of rated output (in 20 min)
Temperature effect:[b]	
On rated output:	0.0008%/°F of rated output
On zero balance:	0.0013%/°F of rated output
Overload rating:	
Safe:	150% of rated capacity
Ultimate:	400% of rated capacity
Side load rating:	(See text)

[a] As published in manufacturers' literature.
[b] Over a range of 15–115°F (−9.4–46.1°C)

beam load cell has been marketed with specifications and tolerances for side load compensation during the weighing cycle.

30.1.12 Types of Tests

A material test is based on a sample of actual product that has been statically weighed on a calibrated scale and then fed into the weighing system being tested. Indicated weights of the two scales for the same material are compared.

A simulated load test is one in which an object other than actual product is weighed statically on a reference scale and then weighed by the scale in question. Test weights for simulated load tests may be solid mass, dead weights, or, in the case of belt scales, may be test chains. Simulated loads are useful in establishing a calibration factor for maintenance, but they do not reproduce real world conditions.

30.1.13 Rules and Regulations

A scale that is used for measuring weight of a commodity in a commercial transaction is subject to regulation as a commercial device. In the United States, the 50 states have their own weights and measures laws and agencies to enforce the laws. Most states have adopted the specifications and tolerances of National Bureau of Standards Handbook 44; other states have laws that generally conform to the handbook.

If commercial transactions are connected with railroad operations, they may be subject to regulation under a scale handbook prepared by the American Railway Engineering Associated (AREA) and used by member roads of the Association of American Railroads (AAR), even though the scales may be of a type other than railway track scales. Individual states usually will accept a test conducted according to AREA/AAR, because railroad tolerances are sometimes as small as one-half of Handbook 44 tolerances.

In the special case of grain weighing, static weighing is required according to requirements published by the Federal Grain Inspection Service of the United States Department of Agriculture.

State inspectors are stationed in many cities and may be a source of advice as to location of possible reference scales for comparative testing, even when legal tests are not required.

30.1.14 Standards

The National Bureau of Standards maintains standards in the form of test weights whose calibrated mass is traceable back to the world standard "Paris kilogram" stored in a laboratory in France. Thus, calibration of a weighing system using standards that are traceable to the bureau establishes a validity in international transactions.

Each state furnishes standard weights for regulatory tests it conducts, but the states cannot leave their standards on site. Whenever possible, each weighing system should be provided with on-site standards, calibrated relative to traceable state standards and kept available for maintenance tests on as needed.

30.1.15 Capability and Performance

Regulatory tests permit recalibration of a weighing system as part of preparations for testing the system, in the event that the first trial shows the system to not be in proper calibration. Few, if any, states require that the condition "as found" be reported, only that the "as calibrated" condition be within tolerance. Demonstration of a scale's capability for recalibration satisfies the law, but does not serve the user in evaluating day-to-day performance.

In 1981 the performance test methods and tolerances shown in Table 30.1.2 were specified by International Marine Terminals (IMT), covering four belt scales to be used in loading and unloading of barges and ships. Acceptance test for each scale was specified according to the usual "as calibrated" condition. Warranty, however, is based on "as found" conditions from the last 10 periodic tests in any sequence of maintenance tests.

30.1.16 Applications and Trade-Offs

The four belt scales shown in Table 30.1.2 were accepted in service in 1982. In addition, two bin (hopper) scales and one platform scale were installed as shown in the simplified schematic of Fig. 30.1.5.

The belt scales C, D, E, and F could have been furnished with test chains for simulated load tests. Instead, test chains were eliminated in a trade-off to obtain small, more economical dead weights. While dead weights do not enable test of belt tension, they are more readily usable than are chains for simulated load testing. In addition, the trade-off included fitting two surge bins A and B with four load cells each. The load cells, shown as T in Fig. 30.1.5, are each of 200,000-lb capacity, so that each surge bin can accommodate an applied load of 600,000-lb capacity as a hopper scale. Thus,

Table 30.1.2 Tests of Belt-Scale Performance in Lieu of Capability[a]

1. Simulated Load (Dead Weight) Test
   Method:
   Warm up empty belt as directed.
   Read and record change in weight for 10 laps of belt at 100% speed.
   Add two dead weights to weighbridge; measure and record belt speed.
   Read and record change in indicated weight for 10 laps at 100%; repeat at 50% speed (78% of rated capacity).
   Change belt speed to 50% and repeat 10-lap recording (39% of rated capacity).
   Operate with only one dead weight at 100% belt speed to repeat 10-lap test (39% of rated capacity).
   Report "as found" readings before any recalibration.

   Tolerances: (Referenced to computed value of dead weights.)
   Of the last 10 simulate load tests recorded:
   7 or more deviations shall not exceed 0.25%
   9 or more deviations shall not exceed 0.5%
   None of the 10 deviations shall exceed 1.0%

2. Material Test
   Method:
   Warm up belt as directed, record 10-lap reading for empty belt.
   Balance weigh bin(s) to zero readout.
   Read and record totalizer for each belt scale.
   Load bin(s), read and record static weight of material.
   Feed material from bin to belt at required flow rate and belt speed for the belt scale(s) involved.
   Read and record empty bin reading.
   Read and record final totalizer indication for belt scale(s).
   Report "as found" readings before any recalibration.

   Tolerances: (Referenced to actual static weight of material sample.)
   Of the last 10 material tests recorded:
   9 or more deviations shall not exceed 0.5%[b]
   None of the 10 deviations shall exceed 1.0%[b]

[a] Frequency of tests and allowable elapsed time since last calibration are established by manufacturer's warranty and owner's maintenance policy.

[b] Any deviation in excess of 0.25% shall require that the belt scale be recalibrated to an acceptance deviation of 0.25% or less by means of an additional material test.

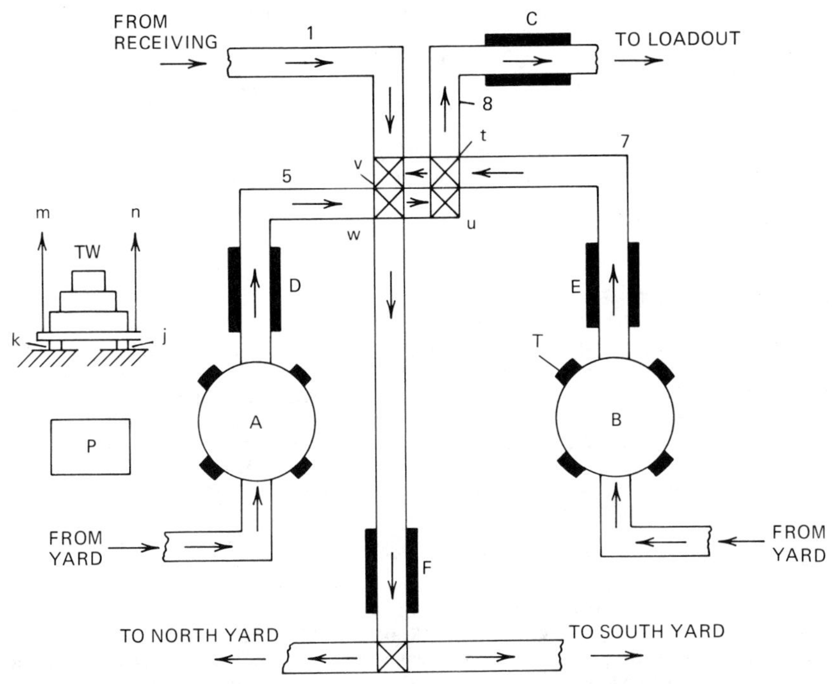

Fig. 30.1.5 Simplified schematic of a coal-transfer terminal designed for material test during loadout.

600 tons of statically weighed product are available to each of belt scales C and F, and 300 tons are available to each of belt scales D and E.

In normal operations, bins A and B serve as surge bins. During operation, the ability to dynamically weigh product in the bin enables use of remote transmission of percent bin load for minimizing over- and underloading of bins. Set-points on the bin scales offer automatic interlock features to stop loading of a bin that is already full, and to prevent feeding into an empty bin before the bottom gate is closed to protect the lower conveyor belt.

If scales were not a consideration, bin A would have fed material to belt 5, and then only to belt 8 at transfer point t. Similarly, bin B would have fed only belt 7 and into belt 8 at transfer point u. Addition of transfer points v and w, however, enable bin A and/or bin B to feed material from the yard back to the yard over belt scale F.

Compatibility of weighing and handling operations is achieved, where bin scale A can be feeding preweighed material to calibrate belt scales D and F, while bin B is serving as a surge bin for loadout. The roles can be reversed to loadout from bin A while bin B is used for calibrating belt scales E and/or F.

A complete test cycle for calibration of belt scale D or E at half of the rated 3600 ton/hr requires a 10-min material run at 1800 ton/hr, or 300 tons. Charging a bin to 300 tons at full speed requires 5 min. This time, added to the 10-min material test for scales D, E, and F, totals approximately 15 min.

Belt scale C has a rated flow of 7000 ton/hr. Bins A and B are both required for delivery of a minimal test load of approximately 580 tons in 10 min. Again, total time for charging the bins and performing the test is approximately 15 min.

Static calibration of the bins is performed by use of on-site standards, shown in side elevation view as weights TW in Fig. 30.1.5. Attachment of test weights to the bin is by means of cables m and n. When not in use, weights TW rest on blocks k and j, ready for immediate use. In practice, a dedicated set of weights is kept under each bin, ready for use with hydraulic jacks.

Traceability of calibration for weights TW is accomplished using the static platform scale P, which is certified by the state and traceable to the National Bureau of Standards.

Scales A through F are all calibrated and checked against each other prior to start of load out to a ship. Confidence in calibration can be maintained during loadout operations by test frequency as desired. Once each hour, the flow of material can be interrupted for 30 sec at bins A and B. When the interruption causes empty portions of belt to reach scales D and E, the totalizer reading for each scale is recorded. The same type of reading of empty belt totalizer condition is recorded for belt scale C shortly thereafter.

At the end of 1 hr the interruption to flow is repeated. Tonnage that passed scale C in that hour is compared to the total tonnage over scales D and E.

In this manner, 100% of operating material flow can be monitored by comparison test, where the actual product forms the material test sample. In the event that intolerable deviations occur, necessary steps can be taken. Meanwhile, monitoring of handling can take place at a cost of less than 1% of operating time.

The example of Fig. 30.1.5 also offers a method for performing legally acceptable material tests in a bulk-handling facility without truck scales or rail scales for statically weighing a material sample.

30.2 SAMPLING SYSTEMS

Charles Rose

30.2.1 Basic Concepts

The market value of a given quantity of bulk solid is a function of the mean quality of the material in respect to the characteristics of interest to the purchaser and user. Differences in the reported mean quality of a consignment as measured by the sampling and analysis process are often of equal or greater importance than equivalent differences in reported weight. For example, a 1% difference in the mean moisture content of a consignment of an ore may have approximately the same effect as a 1% difference in weight because moisture represents useless material. However, depending on contractual specifications, relatively small differences in the reported mean moisture (or any other characteristic) may determine the acceptance or nonacceptance of the consignment. In such cases, the accuracy of the sampling and analysis process demands even greater attention than accuracy in the determination of weight.

In general, the true mean quality of a characteristic of a consignment of bulk material will remain unknown. Only the estimator given as a result of the sampling and analysis process is known. The accuracy of this estimator may be conceptually defined as the degree of agreement between such estimates of the mean quality and the true and unknown mean quality, and depends on both the precision of the sampling and analysis process and the degree of systematic error, or bias, generic to the process [1].

Precision is usually stated in terms of an index of precision, given in many bulk sampling standards as $\pm 2\,\sigma$, where σ is the standard deviation of the statistical distribution of the final analysis results [2,3]. It is generally assumed that statistical distributions resulting from repeated sampling and analysis of bulk solids are normal, or nearly so. Thus, with repeated sampling of an individual consignment using the identical probability sampling plan (assuming there is no true change in the mean quality of the consignment between the taking of the samples), only about 5% of the measurements would be expected to differ from their average by more than $2\,\sigma$, and only about 5% of any two samples should differ by more than $2(2)^{1/2}\sigma$.

Bias is defined as the difference between the expected value of the result of the sampling and analysis process and the true mean quality of the characteristic of the consignment being measured. Bias and precision are illustrated in Fig. 30.2.1.

Unlike the taking of census, or the sampling of individual units produced by a manufacturing process, the basic sampling unit is generally not defined *a priori* in the sampling of bulk solids. In bulk solids, a single segment of the consignment taken as a sample is known as an *increment* and the size of the increment is given in terms of weight [4]. Because of the heterogeneity of bulk solids

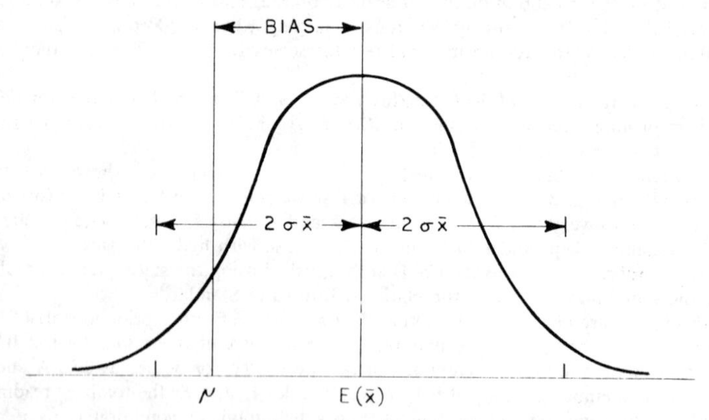

Fig. 30.2.1 Bias and precision.

for a given weight of gross sample, precision may be improved by increasing the number of increments collected at the expense of individual increment weight. However, increment weights must not be so small as to introduce bias by selective rejection of particle size or post-collection changes in the quality of increments such as loss of moisture by drying.

An entire consignment of a bulk material may be thought of as consisting of a numbered set of sampling units, with each sampling unit of the size of the increments to be collected. This numbered set is called a *frame* in statistical terminology. In order that statistical inference be properly utilized and confidence statements be made concerning the results of the sampling process, sampling units to be collected as a sample should be selected by a random method, if possible. Otherwise, neither expected values of the estimates of the mean quality nor variances of the estimates have valid and unambiguous meanings [5].

30.2.2 Minimum Dimensions of Sample Collecting Devices

If one screens a specimen of a bulk solid and separately analyzes the portion of small and large particles, one will usually find the portion of small particles to contain a higher percentage of surface moisture than the large particles. This is because the portion of small particles has a higher percentage of surface area per unit weight. Likewise, one will often find other characteristics to differ considerably between portions of small and large particles. It may be generally stated that characteristics of a bulk solid are not uniformly distributed with respect to particle size.

A device for collecting increments that selectively rejects particles by size will, therefore, cause the sample to be biased. In order that the larger pieces not be rejected, it is important that any sampling device, whether it be a hand scoop, stopped-belt divider, auger, or mechanical cutter of moving streams, has minimum linear dimensions at the opening of 2.5 to 3.0 times the nominal top size of the bulk material, where nominal top size is defined as the size of screen opening through which 95% of the material by weight will pass.

Frequently this requirement will determine the minimum size of increment that may be collected.

30.2.3 Sampling of Stationary Stockpiles

Stockpiles of materials such as coals and ores with a fairly wide size distribution are highly segregated with respect to particle size. During the building of a stockpile, larger pieces tend to roll down the sides of the pile and to the outside, leaving the finer material toward the center. Thus increments collected from only the outside of the pile do not fairly represent the entire stockpile. Moreover, the concept of probability sampling requires random selection from the sampling frame, and some units for sampling will surely be selected from locations other than at the outside or top of the pile.

It is not economically practical to collect increments by hand from locations other than within a meter or so from the surface of a stockpile. Even with the aid of mechanical augers or probes, increments cannot be routinely and economically collected from the center of relatively large stockpiles.

It is for these reasons that the sampling of stationary stockpiles is not recommended as a routine procedure. In the event that judgment of the quality of a stockpile is necessary and no movement of the pile is feasible, the analysis result of the mean quality of increments collected should be treated as from a specimen, and not a true sample. No valid statistical inference is possible.

30.2.4 Advantages of Sampling from Flow Streams

Almost without exception, bulk solids are heterogeneous and anisotropic with respect to the characteristics to be measured. The mean qualities of segments of the material are positively correlated with respect to their location within the consignment, and the degree of correlation varies with spacial direction [6]. For a quantity of material to be homogeneous, complete mixing of the entire quantity of material is necessary, and no such mixing is possible with the normal quantities dealt with as a consignment. Heterogeneity begins at the point the material is mined or manufactured, and is often increased by the common methods of handling and storing, which cause segregation by size.

Considering heterogeneity, it is evident that one must have full access to the entire consignment for the collection of increments in order to obtain a valid sample and use the methods of probability sampling, even if the mathematics of probability sampling are used only to provide an approximation of sampling precision.

Practical access to the entire consignment is only possible when the material is in transit on conveyors or in chutes or similar devices, and with lesser ease of access, in trucks, rail cars, or other such modes of conveyance.

30.2.5 Stopped-Belt Sampling

One may collect increments by stopping a conveyor belt at selected locations and completely removing the increment of material within dividers placed on the conveyor, as shown in Fig. 30.2.2. The minimum

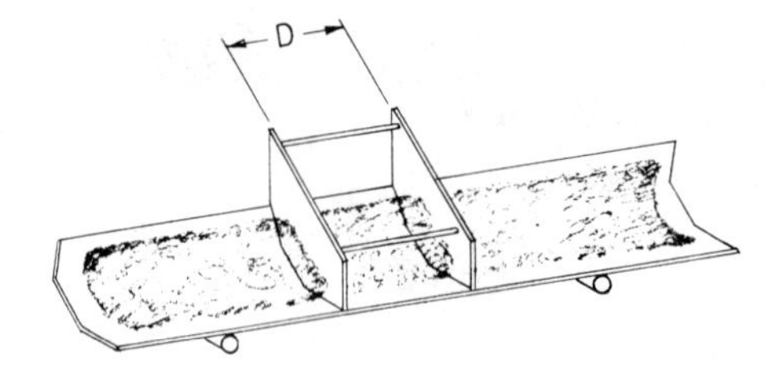

Fig. 30.2.2 Stopped-belt increment collection.

distance between the two dividers should be no less than three times the nominal top size of the material in order that there is no selective rejection of larger pieces of material. This method of physical collection of increments is accepted within the industry as being unbiased. One should note, however, that this does not imply the sampling procedure is necessarily unbiased.

Stopped-belt sampling is not ordinarily used as a routine method due to the necessary interruption of material flow, but is used for special studies and as a referee method for the assessment of other methods of increment collection. Should mechanical sampling be utilized at some point in the materials handling system, conveyor belts upstream of the mechanical sampling system and the first conveyor immediately downstream of the system should be designed for frequent stopping under loaded conditions. Ideally, these conveyors should be capable of stopping at least once per hour during a single 8-hr shift. This is necessary in order that bias tests may be conducted on the mechanical system by comparing mechanically taken increments to stopped-belt increments.

30.2.6 Mechanical Sampling from Moving Streams

A common method of collecting primary increments from a flowing stream or material at a transfer point is shown in Fig. 30.2.3. The primary cutter must move at a constant velocity across the stream; and if the cutter is to reverse direction for collection of the succeeding increment, the normal vector of the cutter opening surface area must be perpendicular to the cutter velocity vector. It is suggested that the cutter opening width be three times the nominal top size of the material. However, in the case of materials of small particle size with surface moisture contents sufficient that bridging could occur across the cutter opening, the cutter width should be no less than 30 mm.

It is important that the cutter has sufficient travel distance that it is completely outside of the material stream when at rest between periods of increment collection.

Cutter lips at the opening are normally constructed of cast 440 stainless steel or other material of high Brinell hardness. Other parts of the cutter body subject to contact by the material stream are

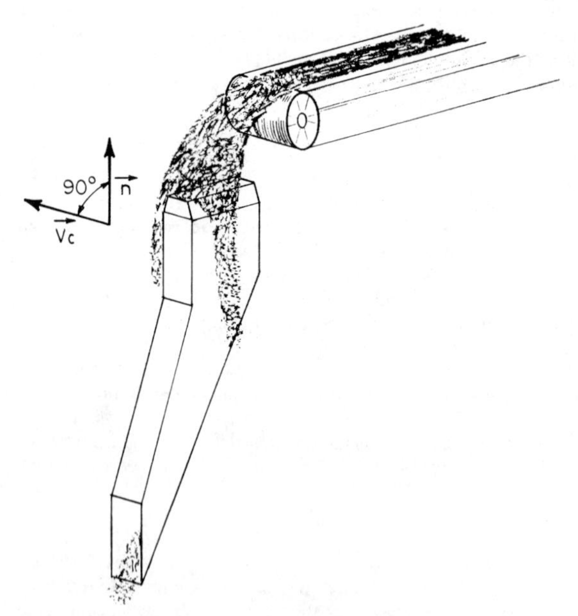

Fig. 30.2.3 Free stream cutter.

often constructed of 304 or 316 stainless steel of sufficient thickness to withstand impact and abrasion. Liners of higher-grade material are sometimes used in high-capacity systems.

There is a maximum velocity that the cutter may move through the free-falling stream without rejection of the larger particles. Beyond this velocity, some of the largest particles impacting the forward edge of the cutter that should fall into the cutter opening will travel over the trailing cutter edge. In effect, they bounce over the cutter rather than into the cutter. The velocity at which this begins to occur depends on a number of factors, including the ratio of the width of the cutter to the largest particle size, the mass of material per unit volume within the flow stream, the velocity of the flow stream, the coefficient of restitution of the material, and to some degree, the shape of the particles of material.

There is no general industry agreement on the maximum safe cutter velocity for a specific material under given conditions. An ASTM coal sampling standard recommends that cutter velocities not exceed 0.46 m/sec, and results of bias tests suggest this to be a safe value for coal under normal conditions [7]. Others feel this value to be too conservative for ideal conditions and have suggested velocities up to 1 m/sec. Unfortunately, little experimental work has been done in this area. It is recommended that until further experimental evidence is available, design velocities of cutters traversing free streams of material should not exceed 0.46 m/sec unless the factors mentioned are thought to be favorable, and in the case of favorable conditions, design velocities should not exceed 0.6 m/sec. Should it be necessary that a cutter velocity exceed 0.6 m/sec, an appropriate test for cutter bias is strongly recommended.

The mass of material collected by a free stream cutter as described above, and where no particle rejection occurs, is given by

$$M_c = \dot{M}\frac{W}{V_c} \tag{30.2.1}$$

where: M_c = mass of increment, in kg
$\dot{M}$ = mass flow rate of the free stream, in kg/sec
W = cutter width, in m
V_c = velocity of the cutter, in m/sec

Another type of cutter called a hammer sampler is shown in Fig. 30.2.4. The hammer sampler collects an increment directly from the moving belt and is used primarily to provide mechanical sampling in existing material handling systems where sufficient headroom is not available for installation of a free stream cutter. It is important that the hammer have sufficient mass to travel through the material without loss of velocity and that it be equipped with appropriate brushes and wipers so as to ensure complete removal of the material from the belt that is between the path of the two hammer plates. Permissible hammer velocities are much higher than for free stream cutters and increment masses are lower. The recommended opening between plates of the hammer is three times the nominal top size of the material.

Fig. 30.2.4 Hammer sampler.

The mass of increment collected by a hammer sampler is given by

$$M_c = \dot{M}\frac{W}{V_b} \quad (30.2.2)$$

where: V_b = velocity of the conveyor belt, in m/sec and other quantities are as previously defined.

30.2.7 Reduction and Compositing of Increments

A primary increment collected by stopping a conveyor belt or by cutting a moving stream will generally have a mass on the order of 10–1000 kg or more. In order to meet sampling precision requirements, as many as 100 or more primary increments may be needed from a single consignment or material. Yet, quantities required for testing and analysis are very small in comparison, in some cases only 1 gram. Thus it is necessary that these large quantities of material be reduced in mass in such a way as to limit bias and variance of reduction to reasonable levels.

In the case of stopped-belt sampling or other manual methods, reduction is often accomplished by the use of laboratory crushers and riffles. This is a labor-intensive procedure.

Increments collected for size analysis are screened without crushing, and usually without division to smaller quantities prior to screening.

Mechanical sampling systems substantially eliminate much of the labor involved in sampling, not only by mechanical collection of primary increments, but also by reducing on line the quantity of gross sample to amounts easily handled and transported to the laboratory. This is accomplished by on-line crushing and division of the collected primary increments. The primary increments are reduced by second- and, if needed, third-stage increment collection. It is necessary that at least one increment be collected at each stage of sampling from each of the increments collected at the previous stage.

Figure 30.2.5 illustrates a typical three-stage sampling system for coal. The gross sample collected as primary increments is reduced in quantity to an amount on the order of 25–50 kg. Note that all increments are composited, thus no information is available for direct estimation of statistical confidence limits. This is a common practice in the bulk solids industry. In such cases, estimates of precision may only be made from prior studies of the material and knowledge of the number of increments collected.

With most bulk solid materials, on-line crushing is necessary in order to reduce the mechanically collected sample to reasonable quantities. If the sample is to be used for a moisture determination, on-line crushing to a nominal size of less than 10 mm is not recommended due to possible loss of

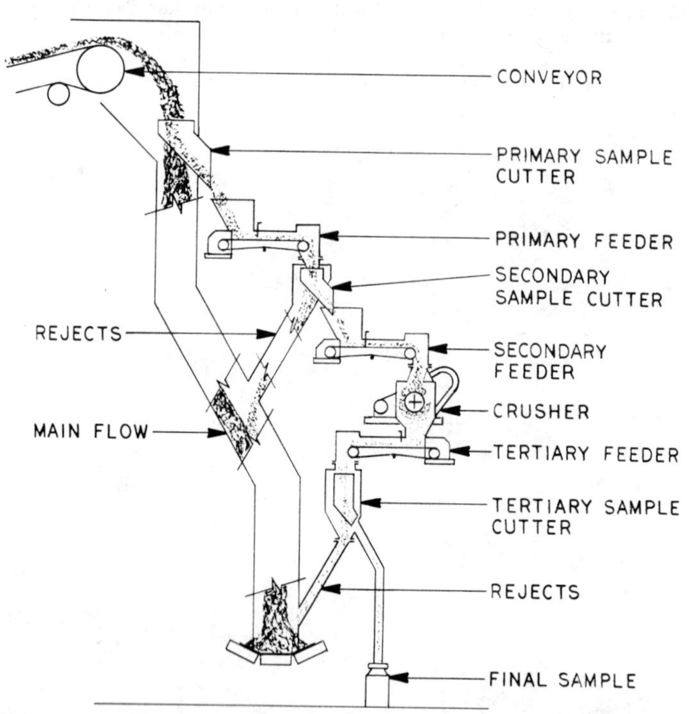

Fig. 30.2.5 Three-stage sampling system for coal. (Courtesy of Ramsey Engineering Company.)

moisture in the sample. For materials with a surface moisture in excess of 5–6%, maintaining an even larger particle size may be necessary. Air drying the material prior to laboratory crushing is recommended unless one can show by tests that no significant loss of moisture will result by immediate crushing to the next particle size dictated by the routine procedure.

The variance of reduction of final quantities collected by a mechanical sampling system, or collected by manual methods, to quantities needed for analysis is often underestimated and should be periodically determined. This may be accomplished by first dividing the collected sample into two equal-sized portions using the routine procedure of division and then continuing the reduction on each of these quantities to the amounts needed for analysis. The analysis of each split will yield two estimates, or one difference. A set of 30 or so of these differences will yield a reasonably good estimate σ_{ra}^{12}, of the variance of reduction and analysis. By using a slightly more complex procedure, the variance of reduction and variance of analysis may be estimated separately [8]. Normally, one finds the variance of analysis to be quite small compared to the variance of reduction.

30.2.8 Mechanical Sampling Methods

Mechanical sampling systems often collect increments systematically at either time or mass intervals. With the time-based systematic method, if n increments are to be collected from a quantity of material conveyed during a total time T, the time interval of increment collections T_I is given by

$$T_I = \frac{T}{n}$$

With mass-based systematic sampling, if n increments are to be collected from a total quantity of mass M_T, the mass interval M_I between increment collection is given by

$$M_I = \frac{M_T}{n}$$

With either the time- or mass-based systematic method, one is, in effect, dividing the entire quantity to be sampled into n strata and selecting one increment from each of these strata. Since increments are normally composited and information on the strata masses and increment masses is lost, it is important that the mass of increment collected be proportional to the strata from which it is drawn. Otherwise, high sampling variances, and possibly bias, may result.

Proportional sampling is accomplished to some degree with time-based systems by maintaining a constant cutter velocity for the collection of all increments. Equations 30.2.1 and 30.2.2 show that the increment mass is proportional to the mass flow rate. However, in sampling streams with varying flow rates, time-based systems do not take precisely proportional increment masses, because changes in flow rates do not coincide exactly with strata boundaries in the stream. It is for this reason that the more costly mass-based systems are preferred for situations where the coefficient of variation of flow rates at the intervals T_I would be more than 20%. The percent coefficient of variation is defined as

$$CV = \frac{\text{Standard deviation}}{\text{Mean value}} \times 100\%$$

A mass-based system is capable of proportional sampling as long as flow rates vary within the range for which the system is designed. This is accomplished by use of a programmable controller which monitors the output of a belt scale or similar weighing device, and signals the primary cutter to operate at the selected mass interval and with a cutter velocity that will result in a given constant mass of increment.

In the case of heterogeneous materials, proportional systematic sampling generally yields a much lower sampling variance than would simple random sampling. However, it is not without drawbacks. Since there is only a single random start (the first increment to be collected) and succeeding increments are collected at intervals of time T_I or intervals of mass M_I, even if the mean quality of each increment were measured, no consistent estimate of the sampling variance could be calculated from the information. In addition, should there be periodicities of quality variation within the stream with wavelengths a multiple of the period of increment collection, the variances of systematic samples could be severely affected.

It is for these reasons that should such periodicities of quality be present or should it not be shown that they do not exist, proportional stratified random sampling is preferred. In proportional stratified random sampling, the total quantity is divided into n strata as with systematic sampling. However, increments are selected at random intervals within the strata and thus periodicities will have no adverse effect.

Proportional stratified random sampling may be accomplished with mass-based systems by programming the controller to signal the primary cutter for operation at random intervals within the strata, and by providing sufficient surge capacity within the system for handling back-to-back collection of primary increments, which will occasionally occur.

30.2.9 Sampling Precision

The total variance σ^2 of sampling, reduction, and analysis consists of the three components

$$\sigma^2 = \sigma_s^2 + \sigma_r^2 + \sigma_a^2 \tag{30.2.3}$$

With simple random sampling of an entire consignment, the variance of the mean sampling result σ_s^2 is given by

$$\sigma_s^2 = \frac{\sigma_I^2}{n} \tag{30.2.4}$$

where σ_I^2 is the variance of sampling units over the entire consignment and n is the number of increments collected randomly.

If estimates $\sigma_I'^2$, $\sigma_r'^2$, and $\sigma_a'^2$ of the respective variances are available, the number of increments needed to meet a given index of precision $\beta = 2\sigma$ is estimated by

$$n = \sigma_I'^2 \left[\left(\frac{\beta}{2} \right)^2 - \sigma_r'^2 - \sigma_a'^2 \right]^{-1} \tag{30.2.5}$$

However, sampling efficiency can usually be significantly improved and sampling costs reduced by subdividing a consignment into several lots (strata) and randomly sampling and analyzing each lot individually. With subdivision of a consignment into L equal-mass lots, the number of increments required per consignment for a given consignment precision β is estimated by

$$n_L = L\bar{n} = \sigma_{I,L}'^2 \left[\left(\frac{\beta}{2} \right)^2 - \frac{\sigma_r'^2 + \sigma_a'^2}{L} \right]^{-1} \tag{30.2.6}$$

where $\bar{n}$ is the number of randomly selected increments per lot and $\sigma_{I,L}'^2$ is the estimate of the average variance of sampling units within lots. One not only gains the benefit of decreasing the effect of the variance of reduction and analysis by L, but since

$$\sigma_{I,L}^2 = \sigma_I^2 - \sigma_L^2$$

where σ_L^2 is the variance of the mean quality of lots within the consignment, then $\sigma_{I,L}^2 < \sigma_I^2$. Thus n_L is always less than n.

As an example, assume

$$L = 4$$
$$\sigma_I'^2 = 2.30$$
$$\sigma_{I,L}'^2 = 2.00$$
$$\sigma_r'^2 + \sigma_a'^2 = 0.01$$
$$\beta = 0.40$$

From (30.2.5) and (30.2.6), we find $n = 77$ and $n_L = 54$. Thus the number of increments required in this case for a consignment index of precision of 0.40 is reduced by about 30% by subdividing the consignment into four equal mass lots.

With mechanical sampling of flow streams of a bulk material, proportional systematic sampling is often assumed to yield approximately the same variances as proportional stratified random sampling. Thus the total variance is written [9]

$$\sigma^2 = \frac{\sigma_{I,S}^2}{n} + \sigma_r^2 + \sigma_a^2 \tag{30.2.7}$$

where $\sigma_{I,S}^2$ is the average variance of sampling units within the n strata, each of from which one increment is drawn.

Assuming an infinite linear population, one mathematical model gives the variance within strata as [6]

$$\sigma_{I,S}^2 = \frac{\sigma_{1,\infty}^2}{m^g} \left(1 - \frac{1}{k^g} \right) \tag{30.2.8}$$

where: $\sigma_{1,\infty}^2$ = the variance of unit masses over the theoretical infinite population
m = the increment mass
k = the number of sampling units per stratum
g = the relative entropy of the infinite population with respect to the characteristic of interest

and $0 < g \leq 1$, where lower values of g indicate greater material heterogeneity.

Note from (30.2.7) and 30.2.8) that with heterogeneous materials ($g < 1$), for a given mass of sample nm, one may improve precision by increasing n and decreasing m.

Estimates of $\sigma^2_{1,\infty}$ and g can be useful in designing sampling systems to meet specific precision requirements.

One method of maintaining a check on the overall precision of sampling and analysis being obtained is to direct and composite every odd-numbered increment into one container, and every even-numbered increment into another container. Two analysis values are then available on each characteristic measured from each consignment or lot sampled. A variance estimate calculated from the two values available from a single consignment or lot is not very useful, because the estimate has only one degree of freedom. However, by maintaining a set of 10 or more such duplicates in continued sampling of consignments or lots by discarding the oldest set each time a new set is available, control charts may be maintained on sampling precision.

An often practical method of estimating maximum sampling equipment and manpower requirements is to determine the number of increments required using (30.2.5) or (30.2.6) with worse-case estimates of the sampling, reduction, and analysis variances, knowing that this will result in an overestimate of the variances that will be obtained in practice with systematic or stratified random sampling. The number of increments actually taken during continued operation may then be adjusted on the basis of the results of control charts prepared from duplicate analyses.

30.2.10 Testing for Bias

No mechanical sampling system should be considered for commercial use until it has been shown by an appropriate bias test to yield samples that are not biased by more than an acceptable value. In the continued routine sampling of bulk materials, bias is of greater importance than precision, since precision of the quality estimates of larger and larger quantities improves as additional consignments are sampled. However, if sampling bias exists, the sampling estimate of the quality of these increasingly larger quantities approaches the biased value, not the true mean value.

As an example of the monetary effects of bias, assume a mechanical sampling system is used to sample coal being received at a utility power generating station which consumes 10,000 metric tonnes of coal daily. At today's fuel prices, even a one-quarter of a percent moisture bias in favor of the coal seller will cost the utility approximately one-half million dollars annually.

Bias is defined as the numerical difference between the expected value of the sample mean and the true mean quality of the quantity being sampled, that is,

$$\text{Bias} = E(\bar{x}) - \mu$$

Since the true mean quality of a lot or consignment of bulk material is not known, it is only possible to test for bias against some reference method of sample collection. The reference method generally used is stopped-belt sampling. One tests for a statistically significant difference between the sample estimates of the quantities $E(\bar{x})$, the expected value of the sample mean of the mechanically collected and processed increments, and $E(\bar{y})$, the expected value of the sample mean of stopped-belt increments. The statistical test used is the paired test for the comparison of two means [9]. Generally, 25 to 35 increment sets, each consisting of a mechanically collected and processed increment and a corresponding stopped-belt increment, are required for an acceptable comparison.

Letting x_i and y_i be the mean qualities of the mechanically collected and stopped-belt increments, respectively, of the ith paired set, the difference d_i is

$$d_i = x_i - y_i$$

and the sample variance of the differences is given by

$$s_d^2 = \frac{1}{n-1} \sum_i^n (d_i - \bar{d})^2$$

where n is the number of sets collected and

$$\bar{d} = \frac{1}{n} \sum_i^n d_i$$

The sample standard deviation of the mean difference is

$$s_{\bar{d}} = \frac{s_d}{\sqrt{n}}$$

Using a table of Student's t, one may make one or more of the following probability statements.

$$Pr\,(\bar{d} - t_{\nu,1-\alpha/2}\, s_{\bar{d}} < B < \bar{d} + t_{\nu,1-\alpha/2}\, s_{\bar{d}}) = 1 - \alpha$$

$$Pr\,(B > \bar{d} - t_{\nu,1-\alpha}\, s_{\bar{d}}) = 1 - \alpha$$

$$Pr\,(B < \bar{d} + t_{\nu,1-\alpha}\, s_{\bar{d}}) = 1 - \alpha$$

where $(1 - \alpha)$ is the probability that the statement is true, $\nu = n - 1$ is the number of degrees of freedom of the variance estimate, and the bias B measured against the reference method of increment collection is

$$B = E(\bar{x}) - E(\bar{y})$$

The confidence limits of the foregoing probability statements are improved by increasing n and/or minimizing s_d. The standard deviation of the differences may be minimized by taking the members of the pair in the closest proximity possible from the material stream. In the case of very heterogeneous materials, the standard deviation of the differences can also be reduced, generally by as much as 30–40%, by first dividing each increment into two equal parts and duplicating reduction and chemical analysis.

In testing multistage mechanical systems, each stage of crushing and division is usually checked for analysis purposes. Design mechanical layouts of sampling equipment should allow for future bias testing by providing bias test chutes or diverters at appropriate locations.

30.2.11 Materials Handling Aspects of Sampling

Too many mechanical sampling systems have been designed and installed where insufficient attention was given to materials handling of the sample and rejected material. The following are a few general rules.

1. Minimize the materials handling path of the sample. The shorter the path from primary increment collection to final sample container storage, the less the moisture bias and potential contamination of the sample.
2. Totally enclose the sample path.
3. Use sufficiently steep chutes, constructed of materials with a low coefficient of friction, for handling the sampled material so that it does not lodge in the system and possibly contaminate future samples.
4. Provide adequate and easily operated access doors for frequent cleanout and maintenance.
5. Do not crush the sample material to a size small enough to cause moisture bias or crusher plugging when higher-moisture materials are encountered.
6. Consider the use of low-speed roll crushers rather than high-speed hammermills.
7. Do not use bucket elevators or augers to elevate the sample material. These devices are not reliably self-cleaning.
8. Consider using belt feeders rather than vibrating feeders for handling sample material.
9. Liberally size conveyors and other devices for handling rejected material.
10. Leave sufficient space around the equipment for easy maintenance.

REFERENCES

1. "Use of the Terms Precision and Accuracy as Applied to Measurement of a Property of a Material," ASTM E 177-71, *Annual Book of ASTM Standards,* Part 41, 1982.
2. "Acceptance of Evidence Based on the Results of Probability Sampling," ASTM E 141-69, *Annual Book of ASTM Standards,* Part 41, 1982.
3. "General Rules for Methods of Sampling of Bulk Materials," JIS M 8100 (1973), Japanese Standards Association, Tokyo.
4. Duncan, A. J., "Bulk Sampling Problems and Lines of Attack," *Technometrics,* Vol. 4, 1962, pp. 319–344.
5. Deming, W. E., *Some Theory of Sampling,* Wiley, New York, 1950, Chap. 3.
6. Rose, C. D., "Variances in Sampling Streams of Coal," *Journal of Testing & Evaluation,* JTEVA, Vol. 11, No. 5, September 1983, pp. 320–326.
7. "Collection of a Gross Sample of Coal," ASTM D 2234-76, *Annual Book of ASTM Standards,* Part 26, 1982.
8. Tomlinson, R. C., "Experiments to Determine the Errors Occurring in the Preparation of Coal Samples for Laboratory Analysis," *Journal of Institute of Fuels,* Vol. 27, 1954, pp. 515–522.
9. "Statistical Interpretation of Data—Comparison of Two Means in the Case of Paired Observations," ISO 3301-1975(E), *ISO Standards Handbook 3,* Geneva, 1979.

CHAPTER 31
ABOVE-GROUND HANDLING AND STORAGE

A. T. YU
DANIEL MAHR

ORBA Corp.
Fairfield, New Jersey

31.1 GROUND STORAGE

Ground storage is the most economical method to stockpile solid bulk material, particularly for large tonnages. It minimizes the investment and maintenance of structures and auxiliary equipment. It is a common practice used in a variety of industries. Ground storage is most applicable to materials that:

1. Are non-hygroscopic
2. Have particle sizes greater than 200 microns
3. Are nonaerating
4. Will not be windswept
5. Retain their properties when exposed
6. Will not be contaminated when exposed
7. Have a relatively moderate cubic valuation

Typical materials are coal, ores, pellets, rock, and sand [1]. Ground storage is not limited, however, to this criteria. It is often used for salt (which is hygroscopic), wood chips (whose fines may be windswept), and other materials. The ultimate use, local climate, and operating conditions are important factors to consider for materials that deviate from the criteria.

Ground support capabilities for the stockpile yard are important considerations in evaluating ground storage. Depending on bulk density, piles higher than 3–6 m (10–20 ft) can exceed the allowable ground pressure. Bulk materials have been stockpiled to 31 m (100 ft) or higher. When in doubt, a geotechnical investigation should be undertaken to determine the height criteria. It is desirable to maximize pile height because a stockpile's volumetric capacity, as formed by a stacker, is a cubic function of height. See Section 31.2 for examples of different pile configurations. In determining ground pressure, an allowance should be made for compaction and partial saturation of the stockpile.

Stockpile design and management techniques categorize ground storage as live, active, and dead storage. Live storage refers to material that is reclaimed by gravity. It applies to tunnels and hoppers that take advantage of the natural flow properties of bulk materials. Live storage applies only to material reclaimed by gravity; above-grade systems require mechanical assistance. Because this chapter is devoted to above-ground handling, live storage is not discussed further here. Active storage is material that is mechanically reclaimed during normal operations. Mechanical reclaim may be via a bucket wheel, bulldozer, portal scraper, or other device. Dead storage describes material that will be stockpiled for a relatively extended period of time. To reduce capital cost and properly prepare the pile, dead storage is usually stockpiled and maintained by mobile equipment. Special preparations are often required to insure that the material's properties are retained.

The amount of ground storage is an important factor to determine before any equipment evaluations are undertaken. The annual usage, allowable pile height, and yard configuration generally determine overall pile dimensions, height × width × length. The proportioning of the stockpile into active and dead storage is influenced by the turnover rate and buffer requirements. The turnover rate is directly proportional to active storage. Buffer requirements are influenced by the normal pattern of receipts, seasonal fluctuations, and potential for strikes and other interruptions. Rules of thumb useful in designing stockpiles are:

1. Stockpile at least 10% of the annual usage or throughput.
2. Allocate at least 10% of the stockpile as active storage.
3. Increase the preceding percentages by 50% when blending materials.
4. Provide active storage for at least 1.5 times the inbound or outbound lot size.
5. Provide storage for seasonal, supply, or other fluctuations.

These rules of thumb are for average conditions. It is common for some industries, like power generation, to increase the foregoing. In the extreme, some power plants maintain a stockpile of 50% of their annual usage if interruptions are imminent. On the other hand, some industries may wish to lower their raw material inventory cost and reduce these percentages.

31.2 STACKERS

There are a variety of stackers in common use. These are often readily adaptable to new design requirements or unique machines can be engineered to suit unusual situations. Stacker selection is an important process since it largely sets yard capital and operating costs.

Table 31.2.1 Stacker Types

Stacker Type	Description	Advantages	Disadvantages	Average Stacking Volume[a] m^3 (ft^3) $\times$ 1000
Fixed	A conveyor that extends over the stockpile. Its single-point discharge forms a conical pile.	Least expensive Least complicated Most labor efficient	Lowest volume	< 15 (500)
Radial	Usually a wheel-mounted conveyor that rotates about a pivot. Its semicircular discharge forms a kidney-shaped windrow pile.	Inexpensive Uncomplicated Labor efficient	Limited volume	3–60 (100–2,000)
Linear	A wheel-mounted conveyor with a linear travel. Its discharge forms a windrow pile on either one or both sides of its travel.	Large volume Labor efficient	Somewhat expensive Somewhat complicated	30–600 (1,000–20,000)
Elevated	Normally a shuttle conveyor, cascading conveyors, or a tripper that is elevated above the stockpile. It forms either multiple conical piles for fixed units or a windrow pile for moving units.	Large volume Land efficient Labor efficient Easily enclosed	Generally higher structural and foundation costs Somewhat complicated	30–600 (1,000–20,000)
Portable	Typically a tire-mounted stacker of any of the above designs. They are most often used to form multiple stockpiles of different materials or grades over a wide area. Mobile equipment is used to move portable stackers from one pile to another.	Inexpensive Flexible	Labor intensive	> 30 (1,000)
Mobile	Often either tire- or crawler-mounted stackers with their own drive train. Extremely large machines use walking mechanisms. Mobile stackers are practical where large areas are very active.	Large volume Flexible	Expensive Complicated Generally requires more maintenance	> 600 (20,000)

[a] Commonly extended by bulldozing.

One useful method of classifying and evaluating different types of stackers is presented in Table 31.2.1. This table classifies stackers by both their physical design and stacking volume capabilities. They are listed according to rank of capability. Features which are often incorporated into stacker designs are listed in Table 31.2.2.

The fixed stacker is the simplest and therefore usually most trouble free. It can be thought of as a conveyor that extends over the yard to stockpile material into a conical pile. It is best applied to facilities that require a limited amount of immediate storage. Because a fixed stacker has just one point of discharge, bulldozing to extend the pile is common. For safety, however, the pile should only be bulldozed on nonstacking hours or when flow is periodically stopped. Because of the fixed discharge height, this stacker is often equipped with either a telescopic chute or lowering well to minimize dusting. These are generally more economical than adding a luffing feature. Screening plants often use multiple fixed stackers to economically stockpile several grades of sized material simultaneously.

The modern radial stacker is designed as a wheel-mounted conveyor slewing about a pivot. This design provides a large radius with a relatively small overhung structure. It facilitates elevating the material to increase pile height. The luffing feature is often optioned to control pile discharge height and reduce degradation and dusting.

The tower radial stacker is another design. Normally a tower radial stacker is elevated and uses a boom conveyor extended over the pile. The tower design is no longer preferred because of its high structural cost compared to its capability, unless it is incorporated as a part of a building.

The linear or traveling stacker is widely used and available in a variety of designs. Its travel is linear and the machine is usually supported on rails at grade level or on slightly elevated berms. It can stack a long, high stockpile. By order of increasing cost and complexity, the linear stacker can be engineered with a:

Single boom—for one side discharge

Twin booms—for two side discharge

Slewing boom—for two-side discharge, flat top piles, and blending

Figure 31.2.1 illustrates a twin-boom, luffing stacker.

Because the stacker track and material delivery system are grade mounted, the linear stacker's travel and the resulting amount of material stockpiled can often be greatly increased at little additional

Table 31.2.2 Stacker Features

Stacker Features	Description
Luffing	Allows the stacker's conveyor boom to be hoisted or lowered. This minimizes the discharge distance to reduce dusting of fines, material degradation, and conveyor horsepower.
Slewing	Permits the stacker's conveyor boom to rotate horizontally. Material can be stockpiled over a wider area or on both sides of a linear stacker travel.
Twin boom	For a linear stacker, an economical method to stockpile material on both sides of travel.
By-pass	Gate used to direct material back onto the yard belt downstream of the stacker and thereby by-pass the stockpile.
Telescopic chute	Discharge chute whose sections nest. It fully encloses material from the discharge point to the pile surface by automatically extending or retracting to suit pile height.
Lowering well	Discharge chute that often doubles as the fixed stacker's head-end support. It partially encloses material from the discharge point to the pile surface. Openings in the lowering well allow material to spill onto the pile. Hinged doors may be provided over the openings to more fully enclose the discharge.
Slinger	A high-speed accelerating device, usually a flat belt, attached to the stacker discharge. Slingers throw and direct material to extend the reach of the stacker. Slingers are uncommon in stacking situations because they increase particulate emissions.
Safety	There are a variety of safety features used in stacker designs to stop, hold, and control the machine in operating and nonoperating conditions. These include tie-downs, brakes, boom rests, locking pins, anemometer high-wind shutdown systems, motion alarms, rail scraper, travel and other motion limits, and bumpers. See Section 31.9 for a more detailed description.

Fig. 31.2.1 Twin-boom, luffing stacker. Photo courtesy Robins Engineers and Constructors.

cost. This is the overwhelming appeal of the linear stacker. As travel distances increase, the linear stacker becomes the attractive alternative.

The elevated stacker can be engineered in a variety of designs. Table 31.2.3 summarizes typical examples. Fixed trippers and cascading conveyors require multiple flow gates and can only form conical piles. Cascading conveyors, in addition, require multiple conveyor drives, one for each conveyor. Fixed trippers and cascading conveyors are therefore equipment intensive and somewhat limited in stockpile capability. The traveling tripper and shuttle conveyor require single conveyor and travel drives. They form continuous windrow piles. The traveling tripper and shuttle conveyor are therefore less equipment intensive and have greater stockpile capability. The supporting structure that allows a continuous discharge increases initial capital cost per length of elevated structure. This is, however, the wrong cost evaluation gauge to use. Considering the cost per volume stockpiled will often highlight that the traveling tripper or shuttle conveyor may be a very cost-effective alternative.

Real estate, especially in highly industrialized areas, is a precious commodity. One factor often neglected in selecting a stacker is the additional land costs that grade-level equipment should be charged. Grade-level equipment by necessity occupies real estate and diminishes that area available for stockpiling. Because grade-level stackers are large, a service roadway is also often required. Considering the stacker track, roadway, and the lost volume due to the slope of the stockpile, a linear stacker will easily occupy an equivalent volume larger than it will stockpile. An elevated structure can, therefore, be more economical. It can also be housed and fitted with telescopic chutes to form a totally enclosed and protected system. Linear stackers, on the other hand, require an open-topped yard conveyor to lift the belt onto the stacker.

Table 31.2.3 Elevated Stackers

Type	Piles	Equipment
Fixed trippers	Multiple conical	Single conveyor drive Multiple flop gates
Cascading conveyors	Multiple conical	Multiple conveyors Multiple flop gates
Traveling tripper	Windrow	Single conveyor drive Single tripper traverse drive
Shuttle conveyor	Windrow	Single reversible conveyor drive Shuttle travel drive

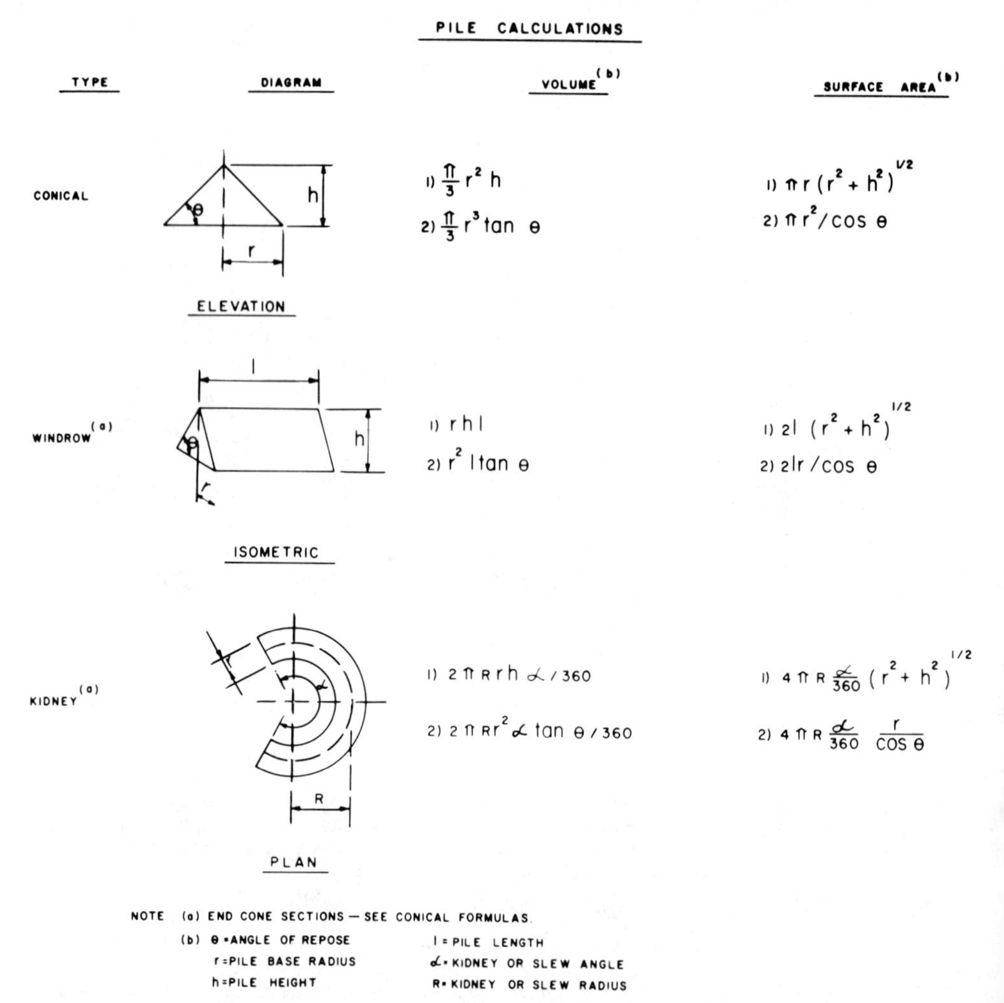

Fig. 31.2.2 Pile calculations.

Portable stackers are typically tire mounted and repositioned by mobile equipment. They are commonly used to stockpile different materials or grades of material over a wide area rather than a restricted location. Portable stackers are extremely flexible and give the operator much latitude in placing and rearranging stockpiles. Because of the number of movements required and the attention that must be devoted to them, portable stackers are labor intensive. They are advantageously used where (1) many materials must be handled, (2) the volume of each individual pile is relatively small, and (3) the mix of materials is unpredictable. Portable stackers are economical but limited in capacity. Design restrictions and operating characteristics usually prevent portable stackers from being applied to major materials handling facilities.

Mobile stackers are tire-, crawler-, or coordinated walker-mounted machines. Tire-mounted mobile stackers are, because of allowable tire pressures, light-duty vehicles. They may more appropriately be considered self-propelled portable stackers. Crawler and especially coordinated walker stackers could be enormous. They and the mobile support system they require are expensive, complicated, and labor intensive. Mobile stackers were developed to meet the vigorous demands of open pit mining. Large volumes, high rates, and vast work areas are all difficult conditions readily accommodated by mobile stackers. They have also been successfully used in land reclamation and large dam construction projects. A mobile stacker and its supply system can take the place of an army of trucks; by comparison then, the mobile stacker can, in these situations, be extremely labor efficient.

There are three basic pile shapes formed by stackers: conical, windrow, and kidney-shaped windrow piles. Other configurations are possible. These are normally a derivative of the basic three and are

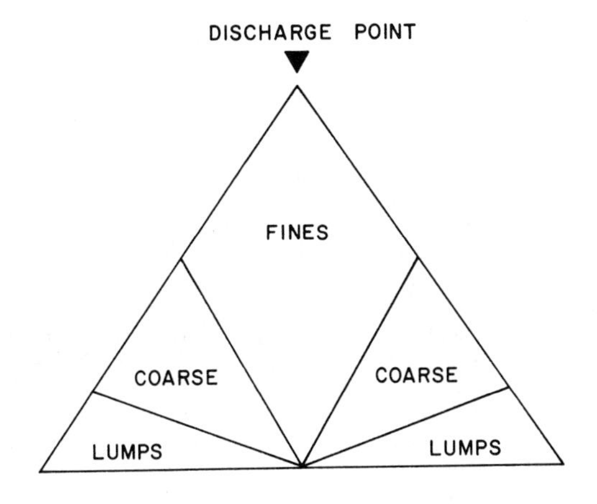

Fig. 31.2.3 Theoretical pile segregation (single discharge).

often formed with flat tops. They can easily be calculated by either adding or subtracting volumes from the basic three. A flat top cone, for instance, can be calculated by subtracting a small cone, the missing top, from a larger cone.

Figure 31.2.2 illustrates the basic piles. Equations [2,3] are given with and without the height variable to simplify computations. The windrow and kidney pile equations are given without the end cone influence. This is often suitable for quick approximations. For exact calculations, include the end cones using the conical pile equations. The volume equation is most useful in evaluating the relative efficiency of stackers; the ratio of cost/cubic is a good parameter to use in comparing alternatives. Surface area equations are given to aid in calculating particulate emissions and surface cover/binder quantities.

Solid bulk material is often a heterogeneous commodity. Its components can vary in their chemistry, hardness, density, and other physical characteristics. These differences are often experienced as a wide range of particle sizes, fines of 200 μm to 100 mm lumps. When stacking a pile, the lumps tend to roll and collect at the toe of the stockpile while the fines accumulate in the center. Figure 31.2.3 illustrates a typical example [4]. Size segregation is associated with chemical segregation and therefore undesirable in many applications. Segregation can be minimized by using slewing and luffing features to form many smaller piles within the large stockpile. Figure 31.2.4 depicts the resulting pile section. Lumps are then distributed throughout the stockpile and the stockpile can be made more homogeneous.

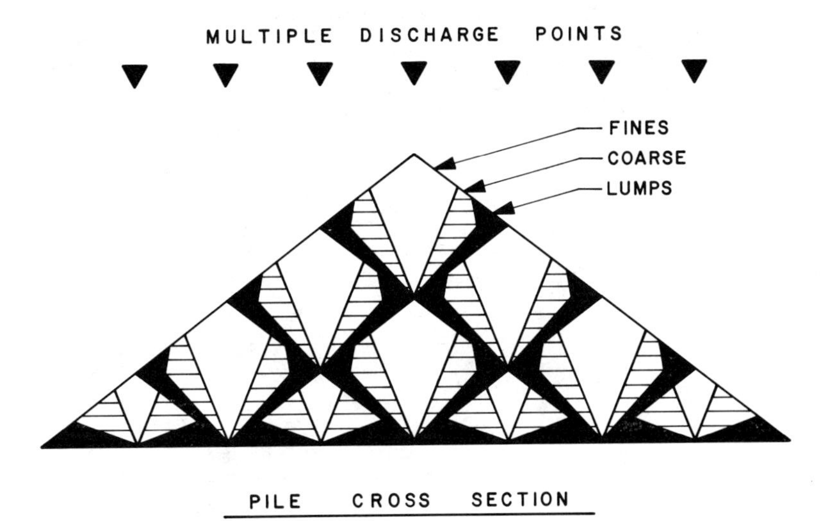

Fig. 31.2.4 Theoretical pile segregation (multiple discharge).

31.3 RECLAIMERS

Above-ground reclaimers are widely used and available in many styles. They can be categorized by their arrangement, structure, and device. These are shown in Table 31.3.1. Many of these arrangement, structure, and device categories can be combined, although some are not practical. Typical handling rates for above-ground reclaimers are summarized in Table 31.3.2.

The functional attributes of each arrangement have been discussed for stackers and are not duplicated here, to avoid repetition. Portable arrangements are not identified for reclaimers although some designs such as dozer traps and movable hoppers may be classified as portable reclaimers. Since these require support equipment like wheel loaders or bulldozers, portable reclaimers are not identified as a separate category.

The structure for an above-ground reclaimer is either a boom, portal, or bridge. A boom is supported at one end of the structure. It is often pinned at that end and guided or reeved to its mid-span. A portal or bridge structure is supported at both ends. The terms *portal* and *bridge* are sometimes used interchangeably. Portal, however, usually refers to a symmetrically inclined structure that straddles the stockpile, whereas bridge normally identifies a horizontal structure either elevated or near grade elevation. The potential stockpile area for boom-type reclaimers is basically unlimited. Bulldozers and other mobile equipment can be used to extend the stockpile and later reclaim material beyond the reach of the boom. The outboard supports for portal and bridge structure reclaimers limit stockpile size to the span of the portal/bridge structure. The entire area in that span is, however, within reach of the machine itself. The portal/bridge structure then is cost effective where the entire stockpile must be within reach of reclaimer as required for precision bed blending, for instance. The boom structure, on the other hand, can have the option for the stockpile to grow beyond the reach of the machine.

Above-ground reclaim devices can be categorized as either continuous or cyclic machines. The bucket wheel, drum, and scraper reclaimers are continuous reclaiming devices. Reclaim rates are largely determined by rotation speed, depth of cut, and advance speed through the pile. The grab bucket, dragline, and shovel are cyclic machines. They reclaim a finite load with a series of motions. This series is repeated as the reclaim cycle. Cyclic reclaimers are labor intensive. The operator normally must precisely coordinate several motions and develop a keen sense of control to maintain reclaim rates. Programmable controllers are being used to relieve these operators of some of their repetitive operations and improve the rates of less experienced or less attentive operators. In either case, however, high operator turnover can have detrimental effects on production rates and equipment maintenance.

Cyclic machines are most advantageous in heavy-duty handling situations. They are typically used in mining applications where lump size, digging forces, outreach distance, and reclaim depth are large [5]. Figure 31.3.1 illustrates a shovel in a quarry application. Hydraulic excavators have become popular in some mine truck loading operations [6]. Grab bucket bridges are common to shipping terminals. Here they can serve several functions including reclaiming, stockpiling, and vessel unloading. To provide even more flexibility, the grab bucket bridge can be fitted with a hook attachment to also handle piece goods and containers. Maintenance cost, however, is high and generally other designs are often optioned. Cyclic machines are best suited to these applications and have become obsolete in typical stockpiling situations.

Continuous machines are applicable to general stockpile reclaiming. They can also be effectively used in mining applications where hard rock is not present such as Germany's open cast brown coal mines [7] and land reclamation projects. Such uses are unique and should be evaluated on a case-by-

Table 31.3.1 Classification of Above-Ground Reclaimers

Arrangement	Structure	Device
		Grab bucket
		Dragline
	Boom	
		Shovel
Linear		
	Portal	
Radial		
Mobile		
	Bridge	
		Scraper
		Bucket chain
		Rotary disc
		Bucket wheel
		Drum

Table 31.3.2 Reclaim Rates

Device	Rates (m^3/hr)[a]	
	Normal	Largest
Grab bucket	40–230	
Dragline	50–270	1,800
Shovel	70–390	
Scraper	180–1800	10,000
Bucket chain	1000–4000	
Rotary disc	1000–1800	
Bucket wheel	200–4000	13,000
Drum	500–1500	

[a] 1 m^3/hr = 1.31 yd^3/hr; 1 m^3/hr = 1.77 ton/hr of 100 lb/ft^3 material.

case basis. For loose material stockpiles, the digging forces are much less than those experienced in banked material. Lump size is usually controlled and well within the reclaimer's capability. One notable exception is exposed stockpiles in regions subject to freezing rain and sudden temperature drops. Here the top few feet of material can become hard frozen and unreclaimable. For power plants and process industries that rely on a dependable supply of material, underground reclaim or covered storage for the active pile should be considered whenever freezing will occur. If these designs are uneconomical, bulldozers can sometimes be judiciously used to push away the crust in slabs and expose the unfrozen inner pile.

The scraper uses a circulating chain with scraper flight attachments. Reclaim is a relatively slow, typically less than 0.6 m/sec, digging and dragging action. Within limits, scrapers are suited to abrasive or some interlocking material but also handle a variety of commodities including [8]:

bauxite	limestone	salt
clay	marl	sand
coal	ores	shale
fertilizers	potash	slag
gypsum	rock	wood chips

Fig. 31.3.1 Cyclic machine. Shovel operating in a granite rock quarry in Arkansas. Photo courtesy of Bucyrus—Erie.

Fig. 31.3.2 Full portal scraper. This portal type scraper is a unique stacker/reclaimer. It operates on a coal stockpile for a cement plant in Texas. Photo courtesy of PHB Weserhutte A G.

For some of these materials, covered storage is necessary for protection. Normal reclaim rates for scrapers vary from 180 to 1800 m^3/hr although larger or dual machines are possible.

Scrapers are available in a number of popular designs. The full portal scraper straddles the stockpile and reclaims from the longitudinal face of the stockpile as seen in Fig. 31.3.2. This particular machine also stacks. The semiportal scraper has one leg elevated on either a wall or roof structure. It also reclaims from the longitudinal stockpile face. The bridge scraper is grade mounted and reclaims by slicing the entire cross section of the stockpile. A harrow or other raking device is used with the bridge scraper; it dislodges material along the cross section and cascades material to the grade level scraper. A boom-mounted scraper can be arranged to reclaim from either the face or cross section of the pile.

Scrapers reclaim to one side of the stockpile. Their resulting thrust in that direction must be considered in designing the track support system. Reclaimed material is discharged either: (1) from a stockpile shelf onto a below-grade conveyor or (2) into a short boot fitted on the machine which allows the scraper to elevate material onto an above-grade belt conveyor.

The rotary disc reclaimer is designed to continuously reclaim material from the entire cross section of the stockpile. It is specifically intended for blending applications where maximum homogenization is required. Radial arms with finger attachments agitate the cross section of the pile. This surface cascades to the outer ring. The ring collects material and discharges it as a low-velocity curtain onto the yard conveyor. The rotary disc is supported by a bridge which spans the stockpile. The disc can be tilted infinitely from 50° in one direction to 50° in the other. It can, therefore, be adjusted to suit the angle of repose of the stockpile and easily change reclaim direction [9,10].

There are several limitations of the rotary disc reclaimer. The diameter of the wheel tends to dictate the maximum pile width. Reclaiming very sticky materials can be a problem. The stockpile base should be an ellipse that matches the wheel at maximum tilt to assist in cleanup. Theoretically the rotary disc is, however, the ultimate in bed blending devices. It does not slice the cross section and therefore its reclaim action can be described as a zero batch operation. For other reclaimers, each slice represents a given number of tons as seen in Fig. 31.3.3.

The bucket chain is similar in concept to the scraper but is different in application. Like the scraper, it uses a circulating chain, however, the attachment is a bucket instead of a scraper flight. The bucket makes the reclaimer more suitable for sticky materials and changes the reclaim direction. The bucket chain rotates with an upward digging action and elevates material onto an overhead bridge conveyor. The bucket chain has been used in applications where material is stockpiled in a pit [11]. This arrangement can be very efficient for enclosed stockpiles because the walls of the pit can become part of the building structure.

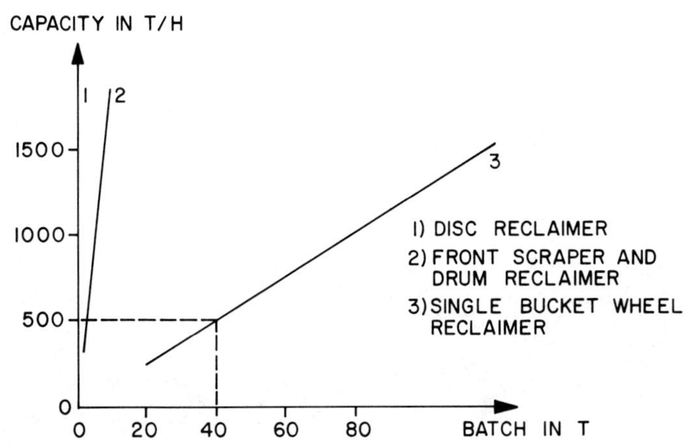

Fig. 31.3.3 Blending equipment comparison. The batch or slice of several different types of reclaimers is compared. Note that the disc reclaimer has a zero batch.

Because of its versatility, the bucket wheel has become a popular above-ground reclaim device. It has been designed as a bridge, boom, or been directly mounted. It can be found on mobile and rail-mounted equipment. The bucket design enables this continuous reclaimer to handle relatively sticky material, free flowing material, and material which will degrade if handled by scrapers or bulldozers. The following materials have been successfully reclaimed by bucket wheels [12]:

bauxite
clay
coal
coke
iron ore
limestone
pellets
sinter

The bridge-mounted bucket wheel is used in blending applications. It has a bucket wheel that traverses the length of the bridge to reclaim from the cross section of the stockpile. The buckets discharge onto a belt conveyor mounted on the bridge. Because of the sloped sides of the stockpile, the bucket wheel traverse speed is programmed to increase as a square function of the distance from the peak to the toe of the pile. As the bucket wheel traverses this cross section, it reclaims different material components at any one time. To improve blending, the bridge-mounted bucket wheel reclaimer is often fitted with multiple bucket wheels to reduce the quality fluctuations of each slice. A harrow or other device to agitate the pile cross section is also often incorporated. The drum reclaimer, Fig. 31.3.4, may be considered the ultimate bridge-mounted bucket wheel reclaimer. The drum spans the entire width of the pile cross section and is therefore better than multiple bucket wheels. A drum reclaimer's slice is usually considered to be one rotation of the drum. See Section 31.6 for a description of how these reclaimers are used for blending.

The boom-mounted bucket wheel is widely used for general above-ground reclaiming. Mobile arrangements are used to reclaim large blocks of material while linear or rail-mounted arrangements reclaim windrow piles. Mobile bucket wheels are used with a variety of support equipment to extend their reach into the pile including: (1) yard conveyor with traveling hopper, (2) reversible linear stacker, (3) mobile conveyors and bandwagons, and (4) shiftable conveyors.

The reclaim action of the boom-mounted bucket wheel is important to stockpile and system design. If the stockpile material is free flowing like taconite pellets, the boom-mounted bucket wheel can reclaim material at the toe of the pile. Material will cascade down the pile to the reclaimer. If the material is non-free flowing, the boom-mounted bucket wheel may have to reclaim the pile in benches. Typically, the height of a bench is half the diameter of the wheel. A 16-m-high pile being reclaimed by an 8-m bucket wheel is reclaimed in 4 benches; each bench is 4 m high. As the boom-mounted bucket wheel slews through the pile, it reclaims a crescent-shaped cut of material (see Fig. 31.3.5). The deep center of the crescent is in the direction of machine travel. The tapered ends of the crescent are at 90° to the direction of travel. To compensate for the varying depth of the crescent, the boom

Fig. 31.3.4 Drum reclaimer. Reclaimer used in a coal liquifying plant. Photo courtesy of PHB Weserhutte A G.

slewing speed should be varied as the secant of the boom slew angle from the direction of travel [13]. This increases the width of the cut to obtain a constant reclaim area of depth × width. Reclaiming the extreme tapered ends of the crescent is inefficient. The depth becomes too shallow and boom slewing must slow, stop, and reverse direction. The total slew angle for mobile boom-mounted bucket wheel reclaimers should be limited to about 160°.

Rail-supported designs require a gap in the center of the crescent for the yard conveyor, machine travel, and access roadway. Unfortunately, this gap removes the widest part of the crescent and limits slewing to approximately a 60° arc on either side of travel. In addition, the material's angle of repose slopes the stockpile from its peak to the toe of the pile. The upper benches, therefore, begin further away from the reclaimer which again reduces the slew angle. The sloped face of the pile reduces the height of the crescent cut so the buckets are only partially filled. As seen in Fig. 31.3.6, the upper benches of a rail-supported, boom-mounted, bucket wheel reclaimer are particularly difficult to reclaim with reduced cut depth, slewing angle, and cut height. Stockpiling a flat-top pile will help to mitigate these disadvantages if the geometry of the reclaimer and slope of the pile are compatible. A steep stockpile angle of repose is best for rail-supported, boom-mounted, bucket wheel reclaimer. If the material's angle of repose is very shallow, a mobile bucket wheel reclaimer may be more efficient.

The mechanical excavator (Fig. 31.3.7) is an example of a mobile, boom-mounted, bucket wheel reclaimer. This particular machine features a unique angled wheel which encircles the boom conveyor instead of positioning it to one side of the boom. The angled wheel reduces both the digging moment and bucket discharge height resulting in a lighter machine. The boom-mounted bucket wheel tends to leave small piles of material unreclaimed. A payloader is often used to scoop up this material and level the yard.

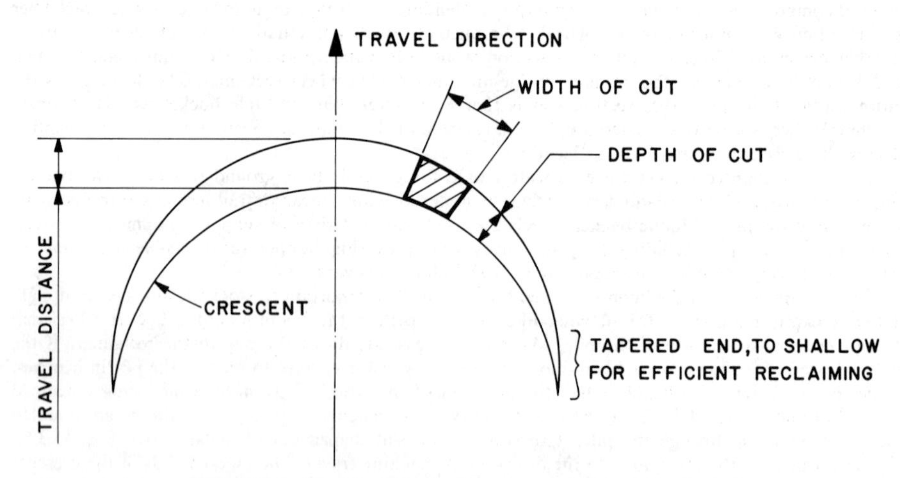

Fig. 31.3.5 Reclaim plan for boom-mounted bucket wheel.

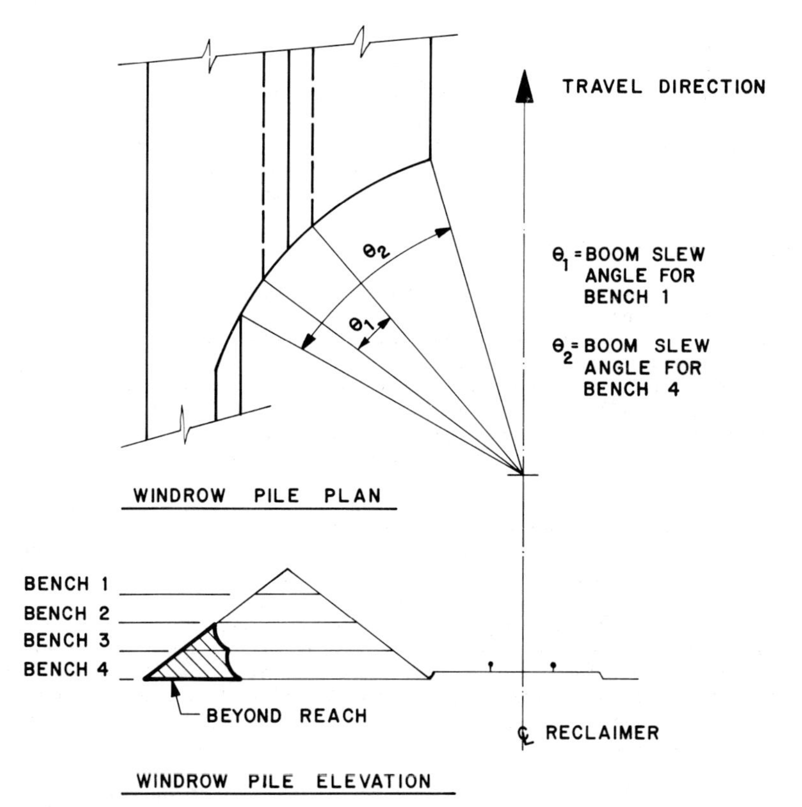

Fig. 31.3.6 Reclaim geometry for rail-supported, boom-mounted bucket wheel.

Fig. 31.3.7 Mobile bucket wheel reclaimer. Photo courtesy of Mechanical Excavators, Co.

Another variation of the mobile bucket wheel reclaimer the wheel mounted directly on the carriage of the reclaimer. Elimination of the boom structure results in a substantially lighter design. This machine reclaims material from the toe of the pile. It can be used with a traveling hopper, reversible stacker, or mobile conveyor depending on the pile width to be reclaimed. It is suitable for free-flowing material that has a shallow angle of repose. Its wide buckets will clean the yard leaving a flat smooth surface. Non-free-flowing material should, however, be avoided. This material reduces reclaim efficiency. In addition, high piles of non-free-flowing material could be a safety hazard if the material will stand up as a wall. Potentially, the wall can cave in and bury the reclaimer or bystanders.

31.4 STACKER/RECLAIMER

The term stacker/reclaimer describes any device that as one unit performs two functions, stacking or reclaiming. It can be a portal scraper with a common boom conveyor, a stacker/rake, or any of several other variations. By far the most common arrangement is the bucket wheel stacker/reclaimer. This is a rail-supported, boom-mounted, bucket wheel reclaimer that has been fitted with the appurtenances to make it a dual-functioning machine. These appurtenances often include a reversing yard conveyor, elevating tripper, and reversing-boom conveyor. The rail-supported, boom-mounted, bucket wheel reclaimer has become so popular that it is synonymous with the term stacker/reclaimer or S/R. To be consistent, the popular terminology will be used here. Other designs will be specifically identified where they are discussed.

The dual functions of the S/R are features that can be either most advantageous or a constant operating impediment [14]. By combining two features into one machine and yard belt, capital cost for the stockpile yard is often most economical for S/R's. The yard must, however, comply to this either/or feature; it cannot perform two functions at once. Too often the initial economic advantage is seized during the planning stages and the operating impacts are totally disregarded. The either/or feature is suitable for operations which are untroubled by the warehousing concerns of FIFO, blending, or demurrage/usage restraints. Often, at least some of these concerns are important. A common method to partly address these concerns is to incorporate an economical fixed or radial stacker into the yard design. This provides temporary storage for the inbound while the S/R is busy reclaiming. The temporary pile is then reclaimed later by mobile equipment. Since the operating cost to do this is high, the temporary storage pile option should be used only in situations where there will be a minimal number of conflicts.

Maintenance on an S/R is another topic affected by the dual nature of this machine. A material handling system can be divided usually into three entities: (1) an inbound or receiving system, (2) the yard or storage system, and (3) the outbound or plant system. The inbound and outbound can have different handling rates. As long as these do not vary too widely, a single S/R should be able to accommodate them. What is often neglected, however, is that the S/R must work with both the inbound and outbound systems. Usually most of the material goes into storage; otherwise, an S/R may not be justified. Because of this double duty, the S/R experiences basically twice the usage of the balance of the system. The S/R then becomes a critical member of both the inbound and outbound systems, works twice as hard as the rest of the system, and is the one element that is most difficult to schedule maintenance for.

The S/R is often required to service both long-term or dead storage as well as active storage. As noted earlier, the dead storage pile is often built by mobile equipment used to extend the pile beyond the reach of the S/R and compact material to minimize weathering. When completing this task, mobile equipment disturbs the active stockpile. As noted before, the S/R has a complicated reclaim methodology. It is critically controlled by pile geometry. If this geometry is disturbed by mobile equipment, reclaim rates will suffer. To mitigate these problems, it is desirable to have one side of the S/R track dedicated to dead storage and the other to active storage. In this way, the normal daily reclaim activity is maintained at maximum rates and dead storage requirements are satisfied. To do this, however, requires the yard to be twice as long since only one side of the track is considered active.

The operation of the S/R during reclaiming is important to the design of its interfacing system and the performance of the outbound or plant system. Successful reclaiming with the S/R requires a skilled operator who is familiar with pile characteristics. The stockpile's susceptibility to sluffing, the reclaim sequence, and pile geometry all affect production rates.

The method used for reclaiming of a typical triangular or windrow pile is the side block bench method as shown in Fig. 31.4.1. The pile is reclaimed in blocks of convenient length. Each block bench is completely reclaimed before proceeding to lower benches. When all benches in a block are reclaimed, the operator proceeds to the next block. Ideally, a complete block is reclaimed during a single reclaim cycle to reduce the exposed pile surface area.

The typical reclaim sequence involves several steps. The boom is first positioned at the proper bench level and located to give a proper cut depth for the bucket wheel. Slewing the boom with a rotating bucket wheel across the face of the pile reclaims one "cut" of material. The S/R is then advanced along the track. The boom is then slewed in the reverse direction to reclaim the next "cut." As long as the S/R is reclaiming on one bench, only the slewing and traveling motions are involved.

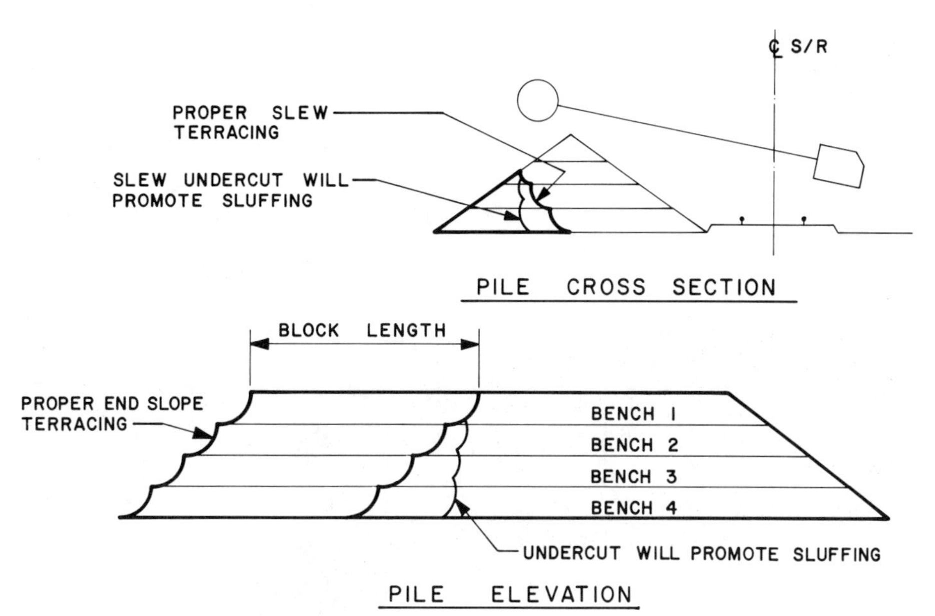

Fig. 31.4.1 Side block bench reclaim method.

Fig. 31.4.2 Boom-mounted bucket wheel stacker/reclaimer. This S/R operates on the coal yard of Pennsylvania Power's Bruce Mansfield Plant. Photo courtesy of Dravo Corporation.

Fig. 31.4.3 Stacker/rake. This stacker/reclaimer uses a combination of a scraper for reclaiming and a belt/scraper flight for stacking. Photo courtesy of Dravo Corporation.

As the S/R reclaims the stockpile, its reclaim rate will constantly vary. The rate will be greatest as the boom swings though the center of the pile for "cream digging." As the boom reaches the sloped sides of the pile, the buckets will be only partially filled and the reclaim rate will rapidly decrease. If the boom slews beyond the sloped end of the pile, it will "dig air" and the production rate will be zero until the boom is reversed and slewed back into the pile. Production will then increase until the bucket wheel is again reclaiming the center of the stockpile. This ever changing rate makes it difficult for an unskilled operator to judge and meet overall production objectives. It also highlights that the designer must consider this varying rate. It can be accommodated by: (1) providing a surge hopper and feeder on the S/R which will balance these fluctuations, (2) increasing the size of part of the handling system to the cream digging rate and then providing the surge hopper, or (3) increasing the size of the entire handling system to the cream digging rate. The best alternative depends upon overall system design requirements. Average reclaim rates will also be affected by the S/R switching from one bench to the next.

During reclaiming, material can sluff or cascade down the reclaimed surfaces of the pile. Three factors that contribute to this condition are:

1. The circular shape of the bucket wheel and the corresponding wall of material can result in steep pile slopes for each bench. These range from horizontal below the wheel to angles approaching 90° at the back side of the wheel.
2. Slewing the boom too far away from the track on lower benches will leave a steep face on the remaining pile.
3. Reclaiming a lower bench too close to an upper bench can result in a steep pile slope of the end of a block.

These factors are controlled by:

1. Limiting the vertical height of each reclaim bench by properly locating bench limits.
2. Reducing the slew angle to terrace lower benches.
3. Terracing the block to control the end slope of the pile.

Examples of these causes of sluffing can be seen in the elevation and cross section of the pile in Fig. 31.4.1.

In spite of controls, material characteristics can vary continually. Sluffing and caving can still occur and bury the wheel or damage the machine. The operator must be vigilant and evade sluffing if possible. One method to do this is to reduce the depth of cut. This, however, decreases the overall reclaim rate.

It is important for the engineer evaluating an S/R to consider the reclaim idiosyncrasies of S/R's. Often literature describing reclaim capabilities states only "cream digging" rates. If these are taken verbatim, production will never meet expectations. Likewise if only average rates are considered, the operator would still have to overload the system during cream digging if a surge hopper is not provided. Considering all aspects of the design is then critical to the successful design of S/R systems.

There are many different arrangements and devices used for S/R's. Figure 31.4.2 depicts a reversing design used at Pennsylvania Power's Bruce Mansfield Plant. Here the stockpile is located off the inbound and outbound systems. Figure 31.4.3 illustrates the stacker/rake.

The stacker/rake incorporates a boom that combines a belt conveyor with a scraper conveyor. Material is loaded onto the top strand for stockpiling. The scraper flights create a pocket to contain material and thereby allow the boom to be hoisted for stacking at an incline much steeper than belt conveyors alone can handle. This gives the stacker/rake a relatively low carriage profile and reduces the trailer length to make more of the yard usable. The bottom strand of the boom is used very much like the standard scraper described in Section 31.3.

31.5 MOBILE EQUIPMENT

Mobile equipment—bulldozers, scrapers, loaders, and trucks—is an important element in above-ground handling and storage. All too often, though, mobile equipment is inadequately investigated during facility planning. The owner is concerned about the initial capital cost and mobile equipment often makes only a minor contribution. The facility engineer is involved with systems evaluations and design. Facility operators, however, are keenly aware of mobile equipment. It makes a significant impact on operating budgets for:

1. Equipment operators
2. Fuel and lubricants
3. Repair
4. Tire replacement
5. Machine replacement

For a given production rate, a facility with a high initial capital cost will often have a low mobile equipment operating cost and vice versa. The most economical yard design is one that minimizes their combined cost. To do this, the application and costs of mobile equipment must be evaluated during facility planning.

The four major types of mobile equipment are the bulldozer, wheel loader, truck, and scraper, as illustrated in Fig. 31.5.1. They serve a variety of functions. Each has its own unique capabilities. For any given situation, the most efficient fleet of equipment is the one that recognizes these different capabilities and uses one type to complement the other.

The bulldozer is commonly used to assist in extending or reclaiming the stockpile. Bulldozing production curves are given in Fig. 31.5.2. This chart must be modified by factors given in Fig. 31.5.3 and Tables 31.5.1 and 31.5.2. For maximum production, proper blade selection is important. This is influenced by material characteristics and bulldozer limitations. Material particles with sharp edges resist the natural rolling action of a dozer blade. These particles require more horsepower to move a given volume than material with rounded edges. Loose material, material that has voids, is easier to move than material that lacks voids. Material lacking voids is either well graded or banked. It is generally heavy and the surface of particles contact one another forming a bond which must be broken. Materials that lack moisture are difficult to move because particle bonds increase. Excess moisture makes the material heavy and also difficult to move. There is an optimum moisture level that reduces dust and offers the best condition for dozing ease and operator comfort. Excess moisture and freezing conditions increase the material's bond strength and decrease dozing rates. The traction and available horsepower of the bulldozer determine its ability to push. Traction is dependent on the weight of the machine, dozing grade, and the material's coefficient of traction. Figure 31.5.3 illustrates the relation-

Fig. 31.5.1 Mobile equipment. Illustrated are common types of mobile equipment, top to bottom, the bulldozer, wheel loader, truck, and scraper. Courtesy of Caterpillar Tractor Co.

ship of grade to the dozing factor. Table 31.5.1 gives the coefficient of traction for different material conditions and machine carriages [15].

Job conditions also affect bulldozer production rates. The operator's skill, weather conditions, work rules, and so on, all are important factors in considering a day's production. Factors that can be applied are given in Table 31.5.2.

As seen from these tables and charts, the bulldozer is most effective for digging and moving material over short distances and on steep grades. The high traction coefficient of track-type bulldozers makes them an ideal choice for handling clay, coal, crushed rock, and shale. Wheel-type dozers are more efficient for slightly longer distances. The wheel-type dozer is also available from some sources with a bucket-type blade that partly carries, partly pushes the load to reduce the rolling action of the blade. Bulldozers are also useful in tight quarters and small piles because of their ability to turn with a zero radius. Excessive maneuvering, however, will increase track carriage wear.

The wheel loader has several common names. It may be referred to as a payloader, front-end loader, or simply a loader. The wheel loader is subject to many of the same production characteristics described for bulldozers. These are not repeated, to avoid repetition. One additional consideration is the bucket fill factor. The bucket can be partially filled, struck or level filled, or heaped. The ability of the wheel loader to lift and carry a load without tipping or losing control should also be considered.

Wheel loaders are basically cyclic machines. Their production can be calculated via a cycle time consisting of: (1) load, dump, and maneuver, which is 0.40 min for an average articulated loader, and (2) travel time for those situations where the wheel loader must carry the load some distance. The nomograph in Figs. 31.5.4 and 31.5.5 provides a simple method to estimate production in a truck loading application. Note that travel time must be added to this.

For long haul distances, trucks are an asset for above-ground handling. Trucks interface with wheel loaders, shovels, mobile reclaimers, or bins. Trucking operations are often compared to other handling methods including railroads, overland conveyors [16], and aerial tramways. The economical

ESTIMATED DOZING PRODUCTION • Universal and Straight Blades • D7 through D10

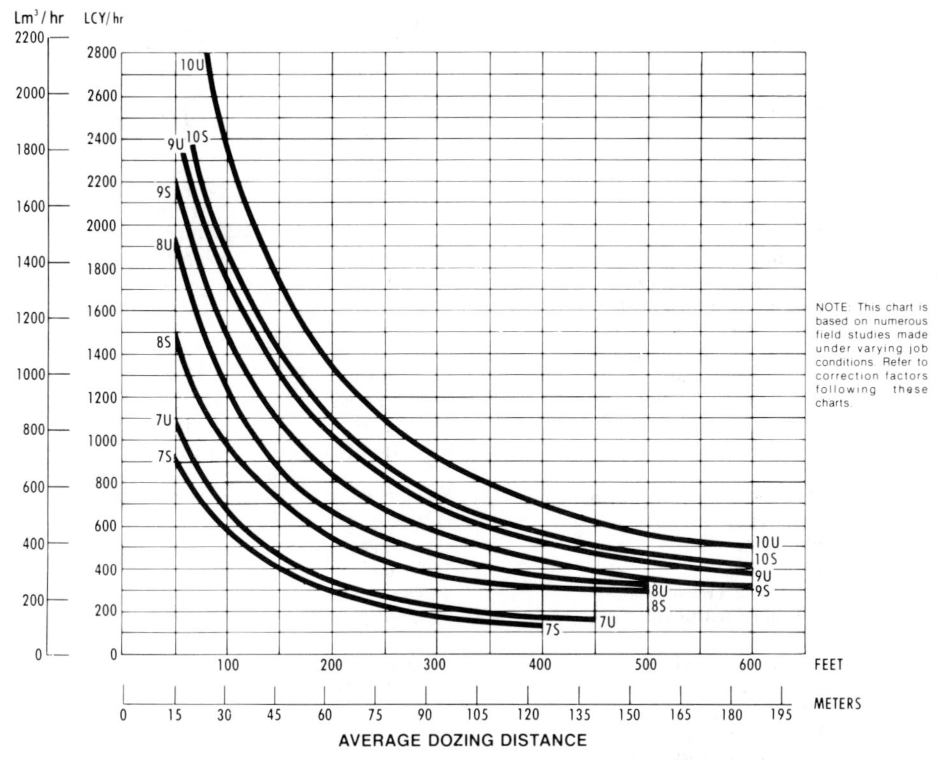

Fig. 31.5.2 Estimated dozing production. Courtesy of Caterpillar Tractor Co.

alternative is determined by comparing the truck operating cost plus the cost of road upkeep with the capital/operating financial evaluation of the other alternatives. Trucks are usually an economic possibility for applications that are low to moderate in tonnage, have haul distances from 10 to 200 km, and have moderate grades to negotiate.

Because of the haul distance, trucks are basically travel vehicles. Their cycle time is largely determined by their average speed. Loading and dumping accounts for less than 5 min normally. Travel time is affected by truck capability, load, grade, rolling resistance, climate, driver skill and attitude, road maintenance, and the speed of the slowest truck in the fleet. For short cycle times less than 30 min, Fig. 31.5.6 depicts the estimated production of various truck capacities.

An evaluation of truck systems should scrutinize the loading and dumping installations. It is easy to improve truck efficiency by increasing the number or capability of loading and dumping stations. This, however, is short sighted since the station waiting time increases. An optimization of the number of trucks and the capability of stations should be developed using queuing theory.

Scrapers are wheel-mounted vehicles that have a greater operating range than bulldozers, but a lower range than trucks. They are ideally suited for travel distances of 300 m to more than a kilometer. Scrapers are self-loading and unloading. The engine and transmission are, however, primarily designed for hauling materials. When tough digging is encountered, a tandem powered scraper or push–pull

Table 31.5.1 Traction Coefficients

Condition	Machine	Coefficient of Traction
Well-compacted coal	Track type	0.75–0.80
	Wheel type	0.40–0.50
Loose coal	Track type	0.60
	Wheel type	0.30–0.40

Table 31.5.2 Job Condition Correction Factors

Job Condition	Track-Type Tractor	Wheel-Type Tractor
Operator		
Excellent	1.00	1.00
Average	0.75	0.60
Poor	0.60	0.50
Material		
Loose stockpile	1.20	1.20
Hard to cut; frozen—		
with tilt cylinder;	0.80	0.75
without tilt cylinder;	0.70	—
cable controlled blade	0.60	—
Hard to drift; "dead" (dry, noncohesive material) or very sticky material	0.80	0.80
Rock, ripped or blasted	0.60–0.80	—
Slot Dozing	1.20	1.20
Side by Side Dozing	1.15–1.25	1.15–1.25
Visibility		
Dust, rain, snow, fog, or darkness	0.80	0.70
Job Efficiency		
50 min/hr	0.84	0.84
40 min/hr	0.67	0.67
Direct Drive Transmission (0.1 min fixed time)	0.80	—
Bulldozer[a]		
Angling (A) blade	0.50–0.75	—
Cushioned (C) blade	0.50–0.75	0.50–0.75
D5 narrow gauge	0.90	—
Light material U-blade (coal)	1.20	1.20
Blade bowl (stockpiles)	1.30	1.30
Grades—See following graph		

[a] Angling blades and cushion blades are not considered production dozing tools. Depending on job conditions, the A-blade and C-blade will average 50–75% of straight-blade production.

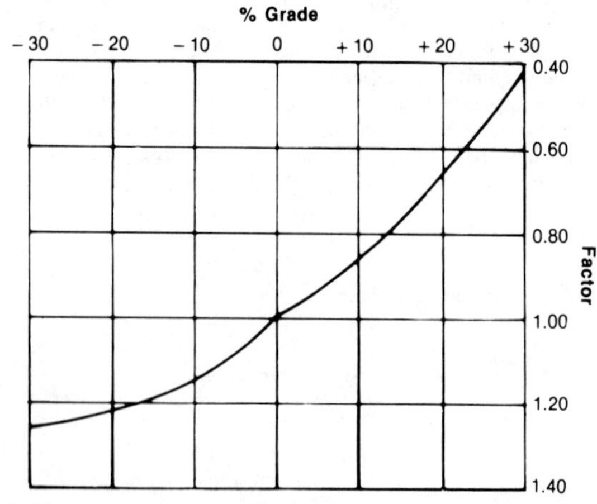

Fig. 31.5.3 Percent grade vs. dozing factor. Courtesy of Caterpillar Tractor Co.

Wheel Loaders | Production and Machine Selection Nomograph
• To find required bucket payload and bucket size

1. Enter required hourly production on Scale B 230 m^3/hr (300 yd^3/hr).
2. Enter cycles per hour on Scale A (60÷.6=100 x .75=75 cycles/hr.)
3. Connect A thru B to C. This shows a required payload of 3 m^3 (4 yd^3) per cycle.
4. Enter estimated bucket fill factor on Scale D (0.90).
5. Connect C thru Scale D to E for required bucket size 3.4 m^3 (4.5 yd^3).
6. Transfer cycles per hour Scale A and required payload Scale C to the following

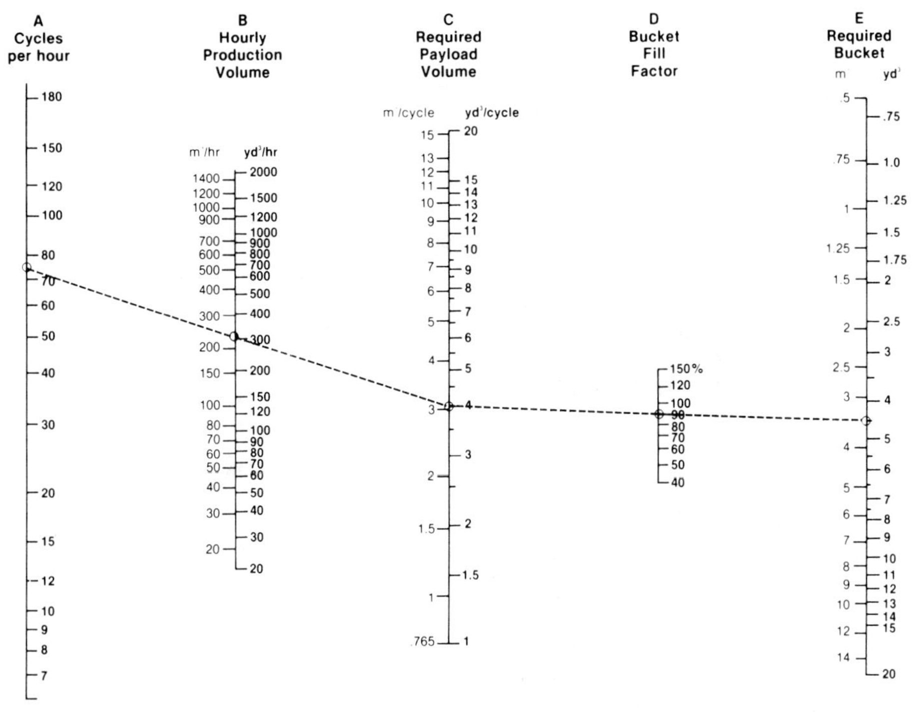

Fig. 31.5.4 Wheel loaders. Production and machine selection nomograph. Courtesy of Caterpillar Tractor Co.

design should be used. An elevating scraper is another design. It incorporates a self-loading elevator or ladder to load material into the bowl. It improves loading capabilities but cannot handle sticky material or large lumps. Because of the self-loading and unloading feature, queuing is often not a major concern. Many of the characteristics discussed for bulldozers and trucks also apply to scrapers. Figure 31.5.7 is an example of a production curve of a large push–pull scraper.

When a dead storage pile is made, it should be properly formed to minimize weathering and material degradation. It is commonly recommended to layer the dead storage pile with each layer being approximately 200–300 mm. Using this method, each layer is well compacted. Table 31.5.3 illustrates typical compaction values for different types of mobile equipment on coal. After the dead storage pile is fully formed, it should be sealed, when practical, with a layer of fine compacted material to prevent intrusion of water. It should then be topped with a layer of coarse material to prevent erosion. In this manner, a well-compacted and protected dead pile is formed. Compaction is particularly critical for coal piles where spontaneous combustion can be a serious hazard.

31.6 BLENDING

Blending has historically been important to many industries that use bulk solid materials. Process facilities, cement plants, steel mills, and so on, have used blending to reduce the variability of their

Production and Machine Selection Nomograph | **Wheel Loaders**

• To find payload weight and tons per hour

7. Enter material density on Scale F 1780 kg/m^3 (3000 lb/yd^3).
8. Connect C thru Scale F to Scale G to give payload weight per cycle 5300 kg (11,500 lb).
9. Compare Scale G quantity 5300 kg (11,500 lb) with full turn operating load on the following page. 966D with 3.5 m^3 (4.5 yd^3) bucket=full turn 6082 kg (13,412 lb).
10. For hourly tonnage, draw a straight line from Scale G thru Scale A to Scale I 400 metric tons (450 U.S. tons).

• • •

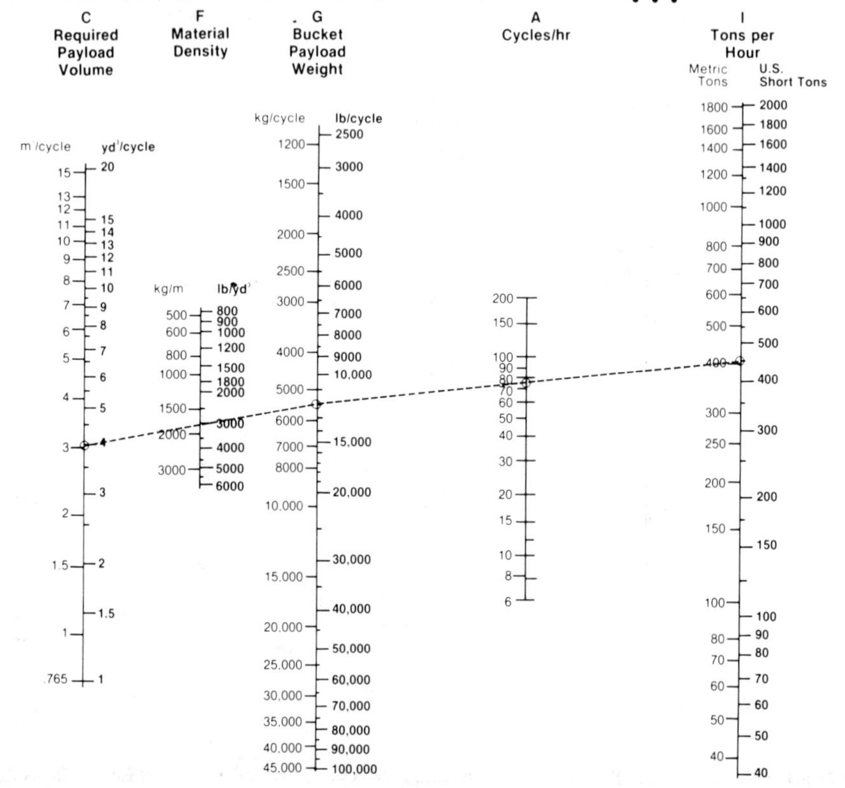

Fig. 31.5.5 Wheel loaders. Production and machine selection nomograph. Courtesy of Caterpillar Tractor Co.

bulk raw materials and improve their products. This is particularly true when there are several sources and mines whose seams vary in quality. Blending is also being used to combine different materials or materials with different properties. The power industry, for instance, combines environmentally attractive low-sulfur coal with high-sulfur coal. This produces the most economical fuel that conforms to emission and boiler requirements [17,18].

Sometimes a distinction is made between blending and mixing. Blending normally refers to the combining of materials or lots of the same material to homogenize characteristics. Mixing usually implies materials are proportioned as ingredients. Regardless of the primary purpose of blending, both homogenization and mixing are often concurrent. The most sophisticated homogenization blending system should, for instance, be able to sweeten the blended pile with a high degree of reliability. Most proportional blending involves some layering simply because lots are stockpiled together. Rather than be incumbered by semantics, all systems shall be described as blending.

Homogenization is one purpose for blending. It reduces the variability of the material. By standard statistical theory, the measure of variability is the standard deviation expressed as [19,20]:

$$\sigma = \sqrt{\frac{\Sigma(\bar{X} - X_i)^2}{N - 1}}$$

Hauling Unit Production Per 60 Minute Hour
- Metric Tons (graph)
- Cubic Meters (table)

Off-Highway Trucks

NOTE: Table below is keyed to grid lines. Read graph first, then follow ton line across to table for conversion to m³ according to material density. E.g., a 45.4 metric ton truck hauling 600 tons an hour would be producing 600, 400, 300, 240, 200 or 171 m³ per hour, depending on material weight. (Interpolate as needed.)

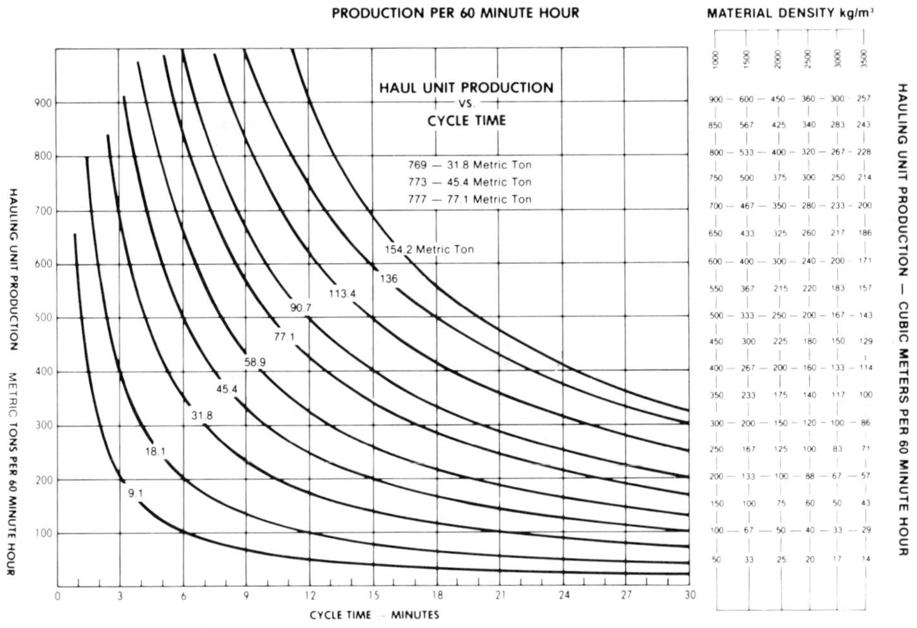

1000	1500	2000	2500	3000	3500
900	600	450	360	300	257
850	567	425	340	283	243
800	533	400	320	267	228
750	500	375	300	250	214
700	467	350	280	233	200
650	433	325	260	217	186
600	400	300	240	200	171
550	367	215	220	183	157
500	333	250	200	167	143
450	300	225	180	150	129
400	267	200	160	133	114
350	233	175	140	117	100
300	200	150	120	100	86
250	167	125	100	83	71
200	133	100	88	67	57
150	100	75	60	50	43
100	67	50	40	33	29
50	33	25	20	17	14

Fig. 31.5.6 Off-highway trucks. Hauling Unit Production. Courtesy of Caterpillar Tractor Co.

where: σ = standard deviation
$\overline{X}$ = the population mean
X_i = the sample mean
N = number of samples taken

The objective of blending, in this instance, is to reduce the standard deviation to an acceptable level. This is done by establishing the permissible variation and the confidence interval. Figure 31.6.1 illustrates a typical example. The normal distribution curve is drawn for a critical component whose mean is 30% of the material. This component has a standard deviation of 4.5%. For a two sigma confidence level, it can be predicted that this component will have a mean value between 21 and 39% of the total material. This wide variation may be unacceptable for the process. If this is so, blending can reduce the standard deviation to 1.5%, for instance. For a two sigma confidence level, this component would then have a mean value between 27 and 33% of the total material. By using different levels of sophistication, other values can be obtained. It is important to note, however, the mean of the population never changes regardless of how much homogenization blending is done. Only by adding a sweetener via proportioning techniques will the blend ever improve. To reduce the standard deviation, the common practice is to increase N. This is typical of bed blending where material is stockpiled in longitudinal layers. The pile is then reclaimed by slicing all layers with each layer representing one sample. In bed blending then, N is the number of layers in the pile.

The second purpose of blending is to combine several different materials into one specific product via recipe blending. The following equation can be used:

$$U_b = \sum_{i=1}^{n} \alpha_i U_i$$

$$= \alpha_1 U_1 + \alpha_2 U_2 + \alpha_3 U_3 + \cdots + \alpha_n U_n$$

Wheel Tractor-Scrapers

657B Push-Pull Bm^3 (BCY)/hr
- Distance vs. Production
- 37.5-39 Tires

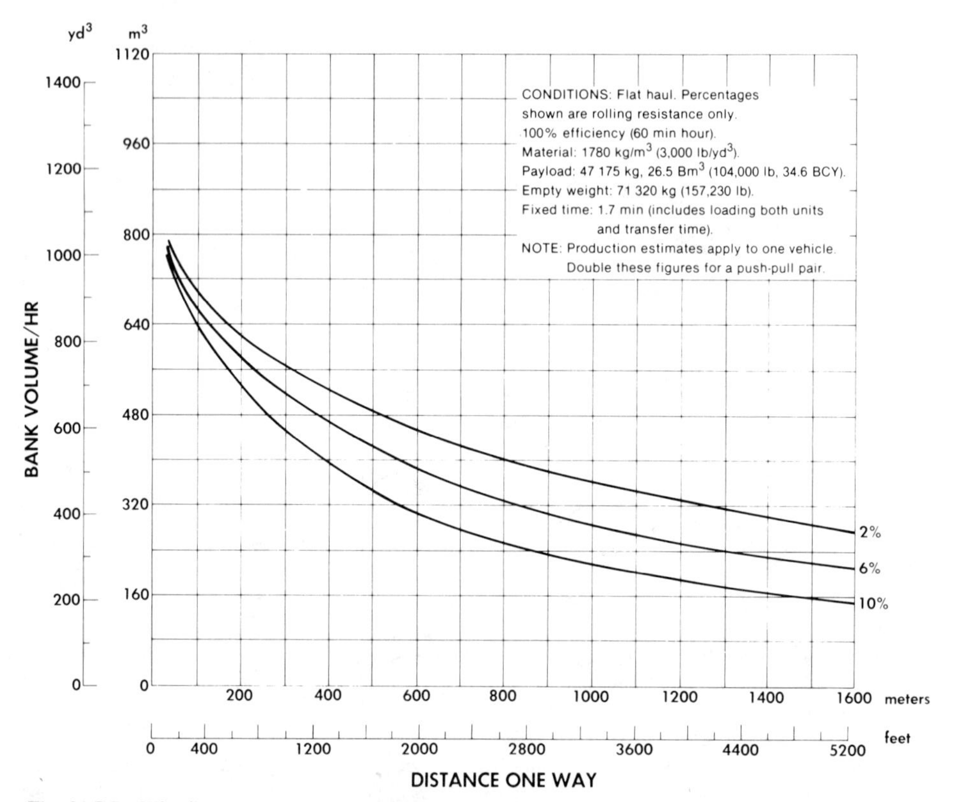

Fig. 31.5.7 Wheel tractor-scrapers. Distance versus production. Courtesy of Caterpillar Tractor Co.

where: U = the value being controlled
α = the proportional amount of each component
i = any material being blended
$1,2,3, \ldots, n$ = various specific materials being blended
b = the blended product

For precise recipe blending applications, this equation assumes that the mean values of each material are known rather than being estimated by statistical methods. Each lot of material is sampled before being blended. For noncritical blending applications or instances where the mean variance is within tolerance, sampling each individual lot is unnecessary.

Table 31.5.3 Typical Compaction Values

	Compaction Value		
Machine	kg/m^3	lb/ft^3	lb/yd^3
Track-type tractors	950–1040	58–65	1550–1750
Wheel dozers	1010–1070	62–67	1700–1800
Wheel loaders	1040–1130	65–70	1750–1900
Wheel tractor–scrapers	1070–1190	67–74	1800–2000

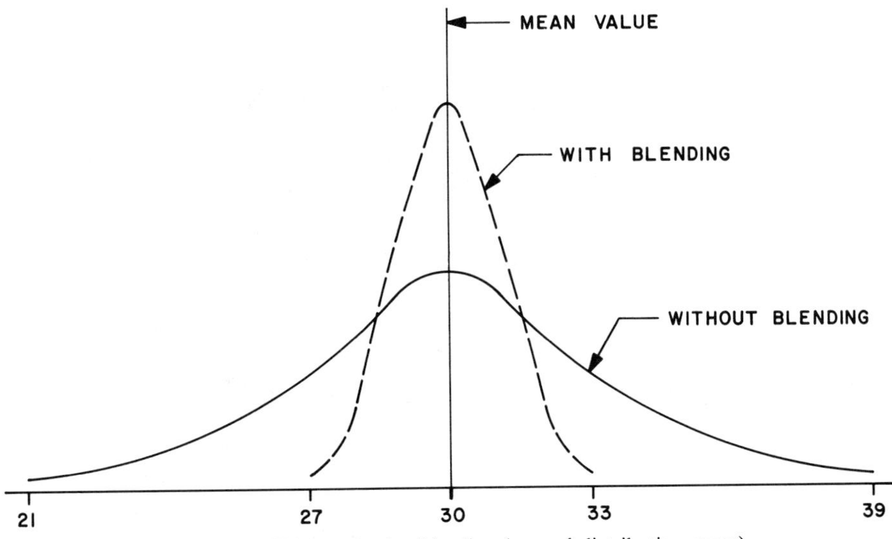

Fig. 31.6.1 Homogenization blending (normal distribution curve).

As illustrated in Fig. 31.6.2, there are several methods to stockpile materials for blending. Each method uses different components and arrangements. They represent different amounts of flexibility, varying ability to homogenize material, and a range of investment requirements.

The chevron method stacks material at the center discharge peak. Each subsequent layer completely covers the previous layer thus forming a chevron cross section. Each successive layer has a larger volume. To maintain a constant layer thickness, the travel speed of the stacker must be variable; it must decrease as the layers build the pile. The chevron method represents a most economical approach to homogenizing material whose lump size is consistent. The single line discharge at the peak of the pile permits use of a short luffing boom stacker. If large lumps are present, however, these will roll down the side of the pile and collect at its base. This segregation would defeat the purpose of blending. Because the pile is symmetrical, each reclaim cut of the cross section represents two slices.

The windrow method stacks material using multiple discharge peaks. Many small piles are formed within the larger pile. Each small pile is a layer. The windrow method minimizes segregation and should be used whenever significant size variability is anticipated. Because of the multiple discharge peaks, the stacker boom must slew and be long enough to reach the peak furthest from the track. This is a more expensive machine than the chevron method stacker. Since the windrow pile is not symmetrical, one reclaim cut represents one slice. Because the piles can all have the same cross-sectional area, a constant travel speed can be used.

The chevron/windrow method is sometimes used. It combines the two methods and enables the operator to stockpile several layers before having to slew the boom. This reduces the amount of slewing movements and the resulting wear. The chevron/windrow method has the attributes of the windrow method. It minimizes segregation but requires a long boom slewing stacker.

Skewed chevron blending is a method devised to bed blend material using a portal scraper. This method is different than other bed blending methods; the portal scraper reclaims from one face of the pile instead of the cross section. As seen in Fig. 31.6.2, material is stockpiled so it cascades down the far side of the pile. Each layer is then exposed on the portal scraper side. The portal scraper travels along the length of the pile lowering its boom for each pass. To form the pile, a short luffing and slewing boom stacker is one arrangement that can be used. A more economical arrangement is to incorporate a luffing, shuttling boom onto the portal scraper structure. This transforms the portal scraper into a type of stacker/reclaimer. It features some of the attributes of stacker/reclaimers, reduced capital cost but operating limitations. Skewed chevron blending also requires one additional reclaim step. The reclaim face of the pile must first be dressed by reclaiming any uneven ripples on the pile face. Dressing the pile does not reclaim all layers which may be objectionable for precise blending situations. If so, the dressing material should be recycled onto another blending pile.

Separate pile blending is often identified as belt blending. Material is blended by reclaiming from several piles at once and layering each pile on the reclaim belt. This method is designed to blend known multiple pile characteristics into the required material. Often any of several recipes can be made by varying the number of piles or proportions reclaimed. In this way, several processes could be fed from one yard or the most economical component materials can be consistently used to minimize

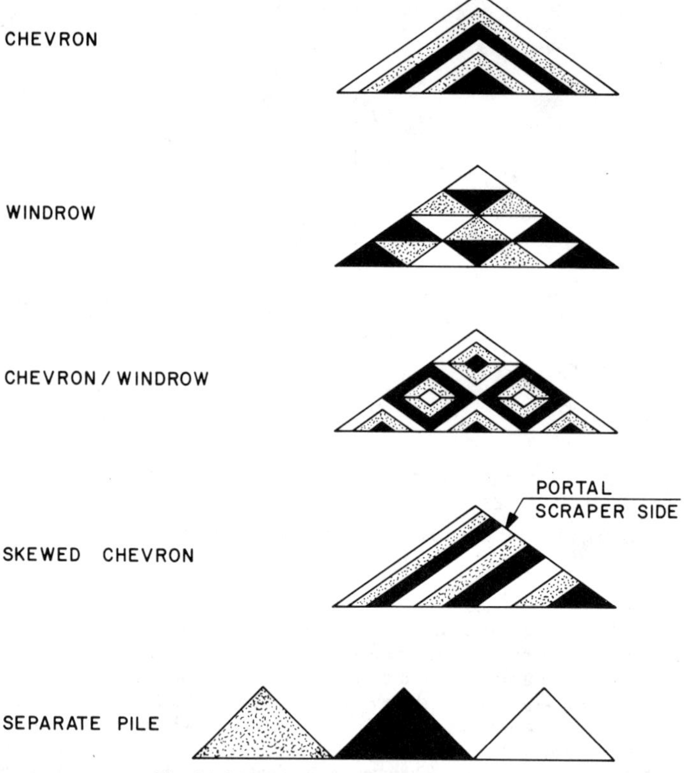

Fig. 31.6.2 Blending stockpiling methods.

cost. Belt blending is, therefore, very flexible. If precise characteristics are desired, it may be possible to increase sampling of the inbound material and adjust pile averages accordingly. To maximize homogenization and precisely control pile characteristics, the use of both bed and belt blending may be necessary.

Bed blending can be applied to both linear and radial yard arrangements. The linear arrangement incorporates rail mount stackers and reclaimers with a linear travel. Typically, rather long narrow piles are formed as described in Section 31.3. Before reclaiming can begin, a pile must be fully formed to provide a complete cross section. While one pile is being built, another is being reclaimed. This doubles the stockpile area required.

The radial arrangement is also known as circular since it is a 360° operation of both the stacker and reclaimer. The material can be stacked using a chevron/cone shell method. This method layers material on the inclined leading section of the pile. The stacker slews clockwise and then counterclockwise through a preselected angle. This angle is determined by the layer length. The angle progresses clockwise with each successive layer, as shown in Fig. 31.6.3. The reclaimer follows at some angle behind the stacker. The radial arrangement's major advantage is its compact continuous stockpile. It follows itself stacking and reclaiming as it goes. Real estate required for the yard is minimized. The diameter and height of the stockpile determine the total capacity of the yard. The maximum practical pile volume is approximately 100,000 m^3 [21]. Multiple radial yards can be used to increase yard capacity and provide redundancy. Because of its set size, blending requirements should be carefully investigated before proceeding with a circular design.

Many types of equipment can be used for bed blending. The portal scraper, bridge scraper, bucket wheel on bridge reclaimer, disc reclaimer, and drum reclaimer are common. A comparison of how some slice a blending pile is illustrated in Fig. 31.3.3. Each slice is termed a batch. The operation of each reclaimer is described in Section 31.3. Mobile equipment can also be used to bed blend material. Bulldozers and scrapers can layer material while wheel loaders reclaim it. This method was used by a major steel mill for many years. The appropriate selection balances accuracy and the capital/operating cost economics.

Belt blending is accomplished with a tunnel or bins that have multiple feeders. These can be belt feeders, vibrating feeders, rotary plows, or just gates, depending on material characteristics and blending accuracy requirements. Typically the feed rate is adjusted on several feeders under different piles. In

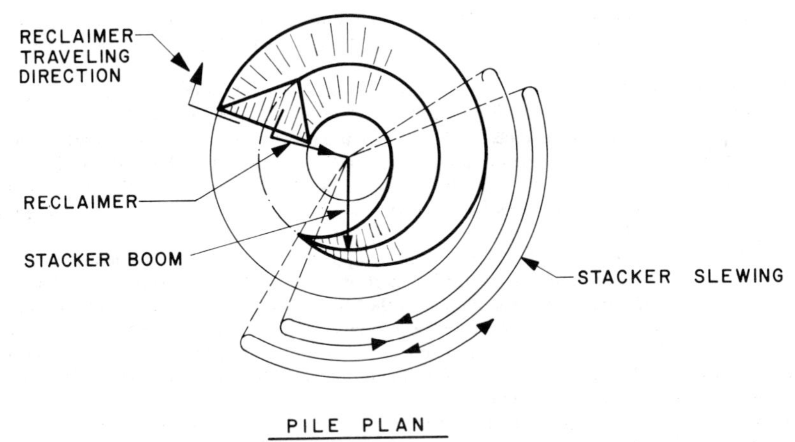

Fig. 31.6.3 Circular pile blending.

this manner, several piles can be reclaimed at once. Custom blends can be obtained by merely changing feeders or rates. Belt scales located between piles are used to monitor and control the feeders.

The Four Corners Power Plant is an example of how bed blending techniques serve the Arizona Public Service Company. The plant is fueled with coal from Utah Construction and Mining Co.'s immense Navajo Mine adjacent to the plant. The mine has up to seven seams, some of which are quite shallow and contain lower-quality coal. While adding significantly to the recoverable reserves, these shallow seams have a much lower heating value and numerous lenses and partings. This makes the coal's ash content quite variable. The Btu content of mined coal has varied from 7,000 to 10,200 Btu/lb. Over a 6-year period, a target average of 9000 Btu/lb was maintained with a variance of 47 Btu or about 0.5%. Because of this uniformity which was anticipated, the boilers were designed with relatively close tolerances and were substantially less expensive than wide-latitude boilers.

The Pride Transloader belt blends coal to meet sulfur dioxide (SO_2) requirements to Georgia Power. This facility is a transshipment terminal in the transportation chain from mines to power plants. Coal is received by barge from a variety of sources. It is classified by sulfur and Btu content and stockpiled by this classification. Classified coal is reclaimed from several piles at once via belt feeders and layered onto tunnel belt conveyors. This belt-blended coal is then loaded into unit trains destined for Georgia Power's plants. Twenty different blend recipes are used to meet SO_2 limits. Recipe selection depends on the current cost, characteristics, and availability of coal. As the unit train is loaded, the blended coal is sampled and the analysis is teletyped to the plant before the unit train arrives. The criteria for blending allows 1% variation. If a 30/70 blend is desired, for instance, the allowable range is 29–31/71–69.

31.7 SYSTEM DESIGN

Any above-ground system must function as a coherent unit of a larger unified entity. It must be designed so the stockpile, yard arrangement, fixed equipment, and mobile equipment are complementary and best meet plant requirements. A convenient method for developing successful system designs is the purpose-oriented approach which is patterned on scientific problem solving. The purpose-oriented approach defines the facility's purpose, evaluates its functions and interactions, balances the capital and operating cost, and incorporates a master plan for expansion.

Four purposes should be examined to define the above-ground system. These are [22]:

Link
Warehouse
Distributor
Processor

The link system literally ties two different operations together. It must meet the normal and peak demands of the two interfaces. A typical example is tying sequential operations. Topics of major concern are hourly handling rates, operating hours, and system availability.

The warehouse system provides storage for fluctuations between operations. These fluctuations may be daily if one interface is continuous and the other intermittent. Seasonal fluctuations are also common. The freezing of the Upper Mississippi and Great Lakes, for instance, halts shipping for several months. Fluctuations may also be somewhat unpredictable, like carrier or supplier strikes. Topics of major concern for warehouse systems are the annual tonnage, usage fluctuations, and the economic consequence of delays.

The distributor system services several interfaces. If interfaces have different material requirements, the distributor system is a convenient focal point for operations. It may be possible to minimize equipment, labor, and real estate, for instance. If all or several interfaces have a common material requirement, however, the distributor system can also provide volume advantages. One large stockpile can be used and bulk material purchases can be centralized. Topics of major concern for distributor systems are the number of flow paths, hourly rates for each path, combination of paths that can be simultaneous, and path priority if conflicts occur.

The processor system performs a unique function to add value to the raw material. For above-ground systems, this is commonly a blending operation. Crushing, sizing, separating, concentrating, weighing, sampling, cleaning, and packaging are other functions often associated with yard and plant operations. In any case, the bulk material has been upgraded and its value increased. Topics of major concern are the techniques used to confirm, control, and verify operations; statistical deviations of the delivered product; allowable tolerances for the operation; availability of critical blending components; and the ability to use or dispose of by-products.

The second step in the purpose-oriented approach is to evaluate the system's functions and interactions. What is the economic consequence of each purpose? Which interfaces are more important? When should one operation be performed instead of another? In belt blending, for instance, the most economical blend is an objective, but should we rely always on a critical component? How important are material characteristics and tolerances? In reality, the economic solution is the one that optimizes all objectives. To do this, the purposes and their major topics of concern should be quantified in economic terms. This provides an absolute gauge for all influences. If this is not possible, they should be ranked by order of importance and a weighted factor applied so at least a relative, rather than an absolute, quantification can be made. Completely subjective evaluations should be avoided.

Capital cost is an important consideration in system design. It is closely scrutinized by investors to determine the viability of new facilities. Considerable engineering is often expended optimizing capital cost. Too often, however, operating cost is only casually investigated. Operating cost is, however, an important cost component. Figure 31.7.1 illustrates this relationship. For any given annual requirement, it is often possible to decrease capital cost by minimizing hourly handling rates. As seen, the capital cost reduction should be somewhat linear, however, the operating cost is not. A facility that is underdesigned could have a lower capital cost but a very high operating cost. Likewise, optimizing operating cost alone is not cost effective. A balance of capital and operating costs must be obtained to minimize the total cost. If there are several different production levels possible, a family of curves similar to those in Fig. 31.7.1 should be evaluated.

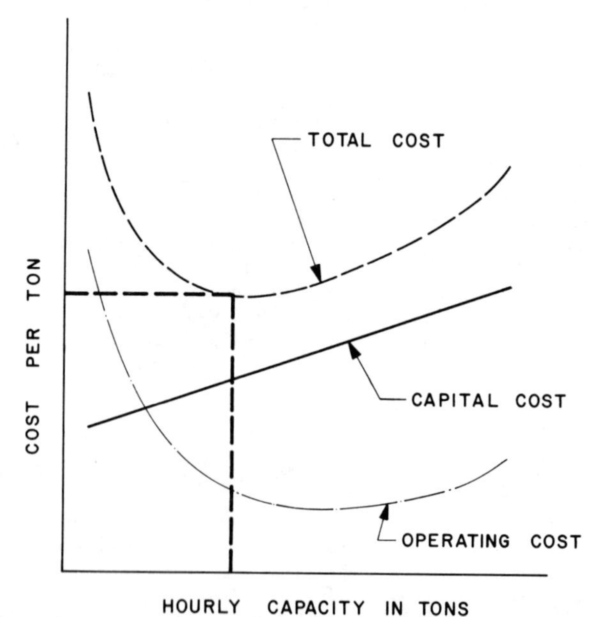

Fig. 31.7.1 Fixed annual tonnage cost.

The final step in the purpose-oriented approach is the development of a master plan. This plan recognizes that over the 20–40-year life of the facility, the original economic conditions may be drastically altered. Technology is rapidly outdating many facilities. Government regulation can strongly influence markets. World events can unexpectedly affect opportunities. The master plan examines the results of several levels of increased and decreased production rates and answers a host of "what if" questions. When necessary, tools such as simulation and computer modeling should be used to obtain expeditious solutions. The master plan reserves property for future expansion so additions can be accommodated in a logical progression. In this way, the facility does not become boxed in and future flow paths and processes do not become unwieldy.

System design and use of the purpose-oriented approach examines the system itself. It defines, evaluates, balances, and plans the system and its future. The system should be designed before equipment selections are made rather than attempting to design a system around equipment selections. A compendium of the purpose-oriented approach that has been a helpful guide is, "the emphasis on material handling equipment obscures the real problem: the handling of material" [23].

31.8 ENVIRONMENTAL DESIGN

Above-ground handling and storage is normally an uncovered operation. It is subject to the elements and becomes part of the landscape. As a result, sound environmental design techniques are necessary. They are required to meet permit requirements and establish the facility as a good neighbor. Before preliminary plans are set, the latest local and federal legislation should be consulted to verify design requirements. Primary controls are often essential in controlling dust emissions, runoff, noise, and esthetics.

Dust emissions for above-ground systems may be classified as either point or fugitive sources. Point sources are transfers. They can be the transfers from one belt to another, from a discharge boom to the pile or, on reclaim, from the bucket wheel to a conveyor. At a transfer, dust can become airborne and thus an environmental problem. Methods of control for point sources include:

Enclosures
Collection
Suppression
Design

Transfer enclosures reduce the amount of dust that becomes airborne since the wind is blocked. They also help to contain the dust. Adding collection to a transfer enclosure produces a negative pressure so that airborne dust is collected. Equations that have been used to size collection rates are [24]:

$$Q_1 = A_1 S_1 - 110 \sqrt[3]{\frac{TH^2 A_3^2}{GL}}$$

$$Q_2 = A_2 S_2 + \frac{T}{G}\left(110 \sqrt[3]{\frac{TH^2 A_3^2}{GL}} - Q_1 + 33.3\right)$$

where: Q_1 = head pulley exhaust rate, in ft³/min
Q_2 = tail pulley exhaust rate, in ft³/min
A_1 = head chute inlet area, in ft²
A_2 = skirtboard outlet and open area, in ft²
A_3 = material stream cross-sectional area in chute, in ft²
S_1 = inlet belt speed, in ft/min
S_2 = outlet belt speed, in ft/min
T = material flow rate, in ton/hr
H = height of material fall, in ft
G = material density, in lb/ft³
L = average material lump size, in in.

If $S_1 < \{110[(TH^2A_3^2)/(GL)]^{1/3}/A_1\}$, then $Q_1 = 0$. If the material hits a stone box or sloped plate in the chute, then $H = H_1 + H_2 \sin \alpha$. Here H_1 is the fall height, H_2 is $H - H_1$, and α is the sloped plate angle from the horizontal plane. Depending on the required efficiency and operating characteristics, baghouses, cyclones, or scrubbers are used to purify the collected air.

Dust suppression is a second means to control dust. This method wets the airborne dust particle and thereby increases its weight and settling rate. Being able to properly coat dust particles and the amount of water that must be added are major concerns. This is affected by particle size and the surface tension of the suppression medium. Some systems use a wetting agent to reduce the surface tension of water and increase suppression effectiveness [25,26]. Other systems introduce air into a water–agent solution to produce foam, which both enhances wetting and reduces the amount of water necessary. Foam systems are ideally suited for applications where increased material moisture is undesirable. The strategic placement of both water and foam sprays is perhaps the most important element [27]. Dust suppression has the advantage of a carry-over effect. Once properly wetted, the dust particle will remain wetted for a period and tend not to become airborne again. Systems with multiple successive transfers could suppress dust at the first few transfers and then take advantage of the carry-over effect for succeeding transfers and the pile itself.

Design is the final method for controlling dust. By minimizing the free fall impact height and properly directing the material stream, the amount of airborne dust can be reduced. Simplifying the system design to eliminate transfers and reducing rehandling will also help.

Fugitive sources of emissions are dirt roads used to service the above-ground system and the pile itself. A factor that has been used to estimate emissions from dirt roads is 0.1 lb/vehicle mile. This is dependent on a variety of factors. Fugitive emissions can be controlled by either suppression or surface treatment. Dirt roads can be periodically wetted with water or compounds which agglomerate or bind the dust. Topping a dirt road with coarse stone will help to control dust but heavy traffic will tend to promote rutting. Paved roads are, of course, the ultimate solution for light-duty vehicle roads. For heavy or track-mounted equipment, this is impractical in many situations. Road systems on large stockpiles are particularly difficult to control. They can resist wetting or be soft so binders quickly break down. Often the best solution is to simply cover the dusty roadway on a pile with a fresh moist layer of material.

Controlling stockpile emissions can be accomplished by suppression or protection. Water is commonly used to wet the pile surface. This can be accomplished by water wagons, agriculture-type irrigation sprays, or a large sophisticated spray system. The Superior Midwest Energy Terminal uses a series of 1200 and 600 gal/min water cannons mounted 36 m (120 ft) above grade with a spray radius of 106 m (350 ft). If the surface of the pile will not be disturbed for a long period, the stockpile should be shaped, compacted, and the use of surface binders considered for protection. For small active stockpiles, an enclosure may be practical.

Uncontrolled dust emissions have been calculated using the following equations [28,29]:

$$E_T = \frac{0.0018 \dfrac{S}{5} \dfrac{U}{5} \dfrac{H}{10}}{\left(\dfrac{M}{2}\right)^2}$$

$$E_P = 0.05 \frac{S}{1.5} \frac{d}{235} \frac{F}{15} \frac{D}{90}$$

where: E_T = transfer source emission, in lb/ton
E_P = pile source emission, in lb/ton
S = silt content, % less than 74 mm
U = mean wind speed, in mi/hr
M = moisture content, in %
d = number of dry days per year
F = time wind speed exceeds 12 mi/hr, in %
D = material storage time, in days
H = height of drop, in ft

As can be seen from these equations, spraying a pile most every dry day can, with a system like that at Superior, drastically reduce emissions.

Control efficiencies for emission sources which have been used [30] are:

Source	Control	Efficiency
Transfer	Enclosed plus baghouse collection	99
Transfer	Telescopic chute	75
Transfer	Wet spray suppression	70
Stockpile	Regular watering	75

To find the controlled emission quantities use the following equation:

$$\text{Controlled emission} = \text{Uncontrolled emission} \times \frac{100 - \text{Efficiency}}{100}$$

An effective environmental dust control system recognizes the hierarchy of control techniques. Prevention always outranks cure, so first make a conscious effort to prevent dust generation. Reduce the number of transfers and reduce transfer heights. Examine collection, suppression, containment, and dilution (ventilation for enclosed areas). Use the most efficient method or combination of methods. Spin-off benefits can include [31]:

1. Improved productivity via a better work environment
2. Reduction in cleaning costs
3. Reduced maintenance particularly if the dust is abrasive or corrosive
4. Reduced hazards if the dust is explosive

Runoff from above-ground storage areas is another important environmental issue. Civil design handbooks are useful in determining yearly precipitation, pond and pan evaporation, and runoff factors for different soil conditions. Treatment is dependent on the characteristics of the runoff. It may be necessary to screen out foreign materials, adjust the pH, settle suspended solids before discharge, or even design a closed-loop system. Studies are being conducted to determine how much of a rainstorm should be treated. The first few hours normally produce the heaviest concentration of pollutants, then the runoff is often relatively clear.

Designing a runoff system should use a water balance. A sample equation is as follows:

Net H_2O = total precipitation + inbound H_2O − discharge − infiltration − pond evaporation − normal pan evaporation − artificial evaporation − shipped H_2O

The net precipitation should include the entire stockpile area from which runoff is collected and the pond area itself.

Noise from above-ground handling systems is an important factor if the operation is close to neighboring industry or the community. Equipment must meet OSHA standards. These are work environment levels and it is desirable in some situations to reduce this to be a good neighbor. Several methods can be helpful in reducing noise levels. Above-ground equipment can be located away from property lines. Distance itself is important in reducing noise levels. Transfers can be sheeted with sound-absorbing materials. Sound barriers like berms, landscaping, walls, or the dead storage pile can be strategically placed. Equipment that may generate less noise can be incorporated into the design. Operating hours can be limited to daylight hours.

The esthetics of a facility should be given careful consideration. Often a minor deviation of design, materials, or even lighting and color can enhance facility appearance. The tools and access to clean and maintain the facility are important. An austere monotonous system can be transformed into a modern and upbeat operation. A good appearance helps community relations. Likewise it improves employee morale. Anything that attracts and interests quality employees will benefit the operation. Forethought, and the tools to clean and maintain the system will not only improve the esthetics of a facility, but ultimately its economic value.

31.9 SAFETY

Safety is important in both the design and operation of above-ground handling and storage systems. OSHA and ASME's Standard B20.1 are excellent sources for design fundamentals. Some of the safety features that should be used are [32,33]:

Mechanical

1. OSHA guards for couplings, pulleys, and other rotating equipment.
2. Guards at floor and grade levels below belt take-up counterweight.
3. Backstops for inclined conveyors to prevent roll back.
4. Failsafe brakes on decline or luffing conveyors to prevent runaway conditions. Brakes should be sized to absorb the heat of at least two full load stops.
5. Hinged skirtboard clean-up doors.
6. Torque overload devices.
7. Rail clamps and locks for large rail-mounted equipment.

Electrical

1. System lock-out switch for maintenance.
2. Emergency stop switches at drive and operator locations.
3. Emergency pullcord switches and extended pull cords along the full walkway length of conveyors.
4. Audible and visual starting alarms.

5. Interlocking devices to automatically stop the system if a conveyor stops, chute or bin plugs, or other condition that blocks material flow.
6. Sequential starts and stops for normal conditions to eliminate plugging between components with different start/stop times.
7. Underspeed switch to warn of an impending overload condition and use in sequential starts.
8. Overspeed switch on decline or luffing conveyors to warn and control an impending runaway condition.
9. Travel and overtravel limit switches for all rail-mounted equipment.
10. Electrical overload protection.
11. Wind speed alarm for rail-mounted or large mobile equipment.

Structural

1. Handrails, ladders, walkways, service platforms, and so on, per OSHA.
2. Provide access to all areas requiring service. Remember that tools and lubricants must often be hand carried.
3. Allow sufficient area at transfer station platforms for clean-up with a long-handle shovel.
4. Crossover stiles for long conveyors.
5. Emergency fixed bumpers for rail-mounted equipment.
6. Wind tie-down or boom rest for large equipment.
7. Provide hooks or trolley beams for maintenance where practical.

Operation

1. Before starting the system or any component, verify that all workmen are clear, sound the warning alarm, and announce starting on the public address system.
2. Do not ride conveyor belts.
3. Cross over or under conveyor belts only where protected access is provided unless your crew properly tags the belt out for maintenance.
4. Repairs and clean-up shall only be done on equipment that is stopped and properly tagged out.
5. Guards must be in place when equipment is in service.
6. Slides of loose material are possible on the pile. When working on a side slope or in valleys or pockets, a ladder shall be used to provide stable footing. A lifeline shall be attached to those on the pile and they shall be attended.
7. Do not go under an overhang; always loosen it from the sides.
8. When operations are finished, secure all equipment to be locked or tied down.
9. Install safety signs to restrict access to hazardous areas. This is especially important for areas that are opened and unguarded for maintenance.
10. Use a tag-out system when any component is serviced.

REFERENCES

1. *Classification and Definition of Bulk Materials, Book No. 550–1970;* CEMA Engineering Conference, Washington, DC, 1970.
2. Matthews, C. W., "Calculating the Size of Stockpiles," *Stockpiling of Materials,* Maclean-Hunter, Chicago, 1970, pp. 100, 101.
3. Selby, Samuel M., *Standard Mathematical Tables,* The Chemical Rubber Co., Cleveland, 1965, p. 495.
4. Yu, A. T., "Bulk Sampling," *Skillings Mining Review,* Vol. 61, No. 7, Feb. 12, 1972, p. 16.
5. Pundari, N. B., "Selecting and Using Large Walking Draglines for Deeper Overburden Stripping," *Mining Engineering,* April 1981, p. 377.
6. "Mass Excavation with Hydraulic Excavators," *Mining Engineering,* March 1982, p. 272.
7. Chugh, Yoginder P., "Opencast Mining of Brown Coal in Eastern Europe," *Mining Engineering,* November 1980, p. 1587.
8. "Reclaiming Scrapers—References," PHB Weserhutte, Bulletin 704.10.
9. Gertel, Von. A. W., "The Homogenization of Bulk Material in Blending Piles," Delft University of Technology, Netherlands, 1980.

10. *Homogenizing Stores, Type CDR,* F. L. Smith, Denmark, 1981.
11. "Reclaimer," Buhler—Miag, Bulletin, 1979.
12. "Bucket Wheel Reclaimers—References," PHB Weserhutte, Bulletin 79.12.10, 1979.
13. Schuster, J. W. and T. D. Wertz, "Operation and Control Philosophy of Bucket Wheel Stacker/Reclaimers," *Unit and Bulk Materials Handling,* ASME—Century 2, San Francisco, 1980, p. 5.
14. Dunn, Michael B., "Stacker/Reclaimers—An Operator's Viewpoint," *Bulk Systems,* June 1979, pp. 8–11.
15. *Caterpillar Performance Handbook,* 11th Ed., Caterpillar Tractor Co., Peoria, 1980.
16. Benavides, F. M. and R. M. Schuster, "Economic Comparison and Evaluation of an Overland Conveyor Versus Alternate Transportation Methods," *Mining Engineering,* February 1982, pp. 176–181.
17. Morgan, James and Daniel Mahr, "Evaluation of Coal Blending Systems," *Combustion Engineering,* April 1981, pp. 28–35.
18. Mahr, Daniel, "Coal Blending at Power Plants," *Power Engineering,* June 1981, pp. 86–89.
19. Yu, A. T., "In-Plant Blending," *Bulk Materials Handling,* Vol. 3, University of Pittsburgh, 1975, p. 84.
20. Yu, A. T., "Bed Blending and Quality Control," *Decision Making in the Mineral Industry,* CIM, Canada, 1972.
21. "Main Data for the Design of Circular Blending Beds with the PEHA-CHEVCON®-System," PHB Weserhutte AG, Bulletin 81.08.40.
22. Yu, A. T. and Daniel Mahr, "A Purpose-Oriented Approach to Transshipment Terminal Development," *Skillings Mining Review,* Duluth, February 2, 1980, pp. 14–19.
23. Yu, A. T., "A Profit-Oriented Approach to Bulk Material Handling," *Engineering and Mining Journal,* March 1973, p. 167.
24. Colijn, H. and P. J. Conners, "Belt Conveyor Transfer Points," *Bulk Materials Handling,* Vol. 2, University of Pittsburgh, 1973, p. 82.
25. Glaess, Harvey and Don Werle, "Scavenging Action of Rain on Air-borne Particulate Matter," *Industrial and Engineering Chemistry,* Vol. 48, No. 9, September 1956, pp. 1512–1516.
26. Glanville, James O. and James P. Wightman, "Wetting of Powdered Coals by Alkanol–Water Solutions and Other Liquids," *Fuel,* Vol. 59, August 1980, pp. 557–561.
27. Siebel, Richard J., "Dust Control at a Transfer Point Using Foam and Water Sprays," *Bureau of Mines Respirable Dust Program,* Technical Progress Report 97, U.S. Department of the Interior, May 1976.
28. Noble, George, "Coal Handling and the Air Pollution Confusion," *Coal Mining and Processing,* July 1981, p. 85.
29. Noble, George, Interview, April 26, 1982.
30. Ibid.
31. Yu, A. T., "The Battle Against Dockside Dust," *Skillings Mining Review,* Duluth, February 17, 1973.
32. Webb, J. C., *American Standard Safety Code for Conveyors, Cableways and Related Equipment—B20.1,* The American Society of Mechanical Engineers, New York, 1958.
33. Mahr, Daniel and W. F. Lawson, *Operating Procedures for the Bruce Mansfield Yard Department,* ORBA Corporation, Fairfield, N.J., October 1981.

CHAPTER 32
EXCAVATORS

CLIFTON H. HUBBELL

Sauerman Bros. Inc.
Bellwood, Illinois

ALEXANDER POMERANTSEV

California State University at Fullerton
Fullerton, California

32.1 DRAG SCRAPERS

Clifton H. Hubbell

Drag scrapers are long-range material handling machines used for moving bulk materials or for excavating and moving materials. They use bottomless scrapers which are pulled by wire ropes actuated by hoists. Industries that use drag scrapers include: cement, grain, fertilizer, sugar, coke, steel, coal, chemical, sand and gravel, mining, construction, power, and shipping. Drag scrapers are generally used to move materials a few hundred feet up to 1000 ft (305 m). They can dig to depths of 100 ft (30 m) or from banks more than 200 ft (61 m) high.

32.1.1 Applications

Moving Bulk Materials

1. Reclaim bulk materials from storage buildings. Materials handled include cement, soybean meal, phosphate rock, citrus pellets, and sugar.
2. Move bulk materials in or out of outdoor storage areas. Materials include clay, coal, petroleum coke, iron ore, oyster shells, potash, carnalite, and sand.
3. Reclaim bulk materials from barges or ships. Materials include cement, coal, sand and gravel, triple super phosphate, and urea.
4. Move bulk materials from inaccessible locations. Applications include removing mill scale from underneath rolls and tables in steel mills, reclaiming corrosive chemicals from storage buildings, cleaning pipes and culverts, cleaning settling ponds or ash and sludge lagoons, and handling clay slurry.

Excavating and Moving Materials

1. Excavate sand and gravel from dry or wet deposits.
2. Dig artificial lakes.
3. Clean out ponds and reservoirs.
4. Reclaim land.

32.1.2 Construction and Operating Principles

A drag scraper machine has a bottomless scraper for handling material. The scraper is pulled by cables actuated by a hoist equipped with clutches and brakes. Blocks are used to support and guide the cables. Additional equipment can be used to change the line of operation or raise the scraper for the return part of the cycle. This additional equipment can include additional drums on the hoist or separate winches and movable towers, trolleys, or bridle frames.

The bottomless scraper is the key element of a drag scraper machine. When properly designed and applied, the scraper will dig and fill in a short distance and move a full load without excessive spillage. The most efficient design is crescent shaped. This type of scraper is strong and light and requires a relatively low line pull to move material. It is shaped to ride over the material when backhauled.

When scrapers are used to excavate hard-packed materials, they are equipped with replaceable digging teeth. When scrapers are used to handle loose materials or are operated on floors of buildings, in holds of ships or barges, in lined troughs, or in culverts, teeth are omitted.

The handling capacity of a drag scraper is greater for short hauls than long hauls. When high capacities are required, often short hauls can be made.

Drag scrapers can be classified in two general types:

1. Straight haulage machines in which the scraper is pulled back and forth on the material. These machines are usually built in sizes from $\frac{1}{2}$ to 5 yd^3 (0.382–3.82 m^3) with lengths of haul up to 500 ft (152 m).
2. Track cable machines in which the loaded scraper is pulled in on the material, but the scraper is returned empty through the air. Sizes of these machines range from 3 to 15 yd^3 (2.29–11.47 m^3) with spans up to 1000 ft (305 m).

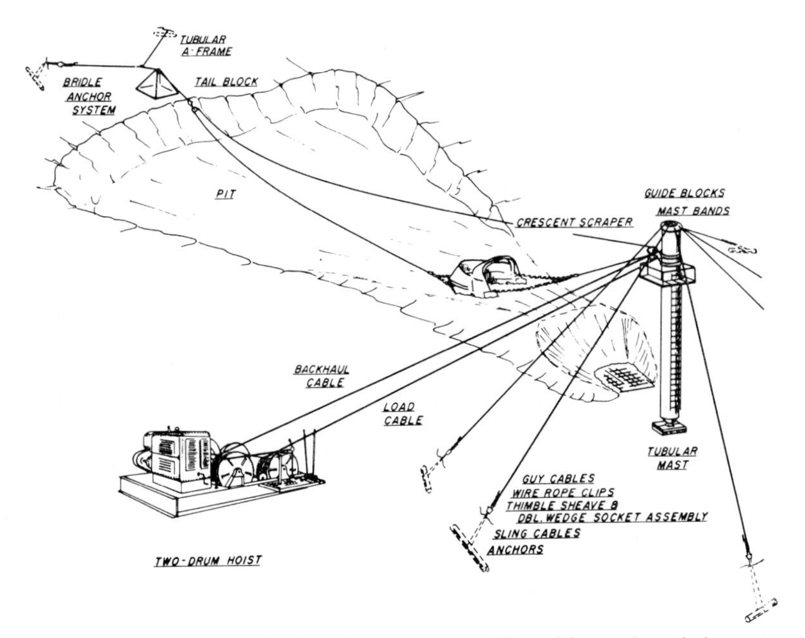

Fig. 32.1.1 Straight haulage drag scraper machine with two-drum hoist.

Straight Haulage Machines

Two-Drum Machines. The simplest type of drag scraper machine is a straight haulage machine using a two-drum hoist (Fig. 32.1.1). One drum of the hoist is used to pull the loaded scraper and the second drum is used to pull the empty scraper back. With this type of machine the scraper operates back and forth essentially in one straight line.

Machines with Provision for Shifting the Line of Operation. A second type of straight haulage machine has provision for changing the line of operation so that the scraper operates over an area instead of in a single line. Changing the line of operation can be accomplished by using the third drum of the hoist to move a bridle frame with a tail block on a cable or a trolley (Fig. 32.1.2). A traveling tower can also be used to move the tail end of the machine.

Overhead trolleys are used in buildings or in ships and barges to make it possible to place the scraper on top of the material for better handling of noncaving materials (Fig. 32.1.3).

When drag scrapers are used to move materials in one direction, the line speed for backhauling the empty scraper is generally faster than the line speed for pulling the loaded scraper, to increase the handling rate and to make use of the full potential of the power unit.

Drag scrapers are also used for moving materials in both directions. An example of this is the use of a drag scraper to put materials into a storage area which are later reclaimed to a hopper by the drag scraper. This is accomplished by reversing the scraper. For this type of installation two drums of the hoist are capable of pulling the loaded scraper and have the same line speed.

Drag scraper hoists usually have adjustable weight-set drag brakes which are interlocked with the clutches to control slack in the cables. Booster brakes are sometimes used to increase the braking action.

Clutches can be manually controlled or pneumatically controlled with straight air valves for ease of operation or electropneumatically operated for remote control.

Generally drag scrapers use an operator (Fig. 32.1.4). The operator can be near the hoist or in a remote operator's cab, which can be located where desired, and multiple operator's cabs can be used to provide control from several points so that the operator can be located for good visibility of the operation.

The operator is able to control the hoist which actuates the scraper from his station and in addition is able to control the movement of bridle frames, trolleys, or self-propelled traveling towers, so that only one operator is required. With this type of operation the need for personnel to enter the work area during the operation is eliminated. This assures safe operation away from free-flowing, hot, dusty,

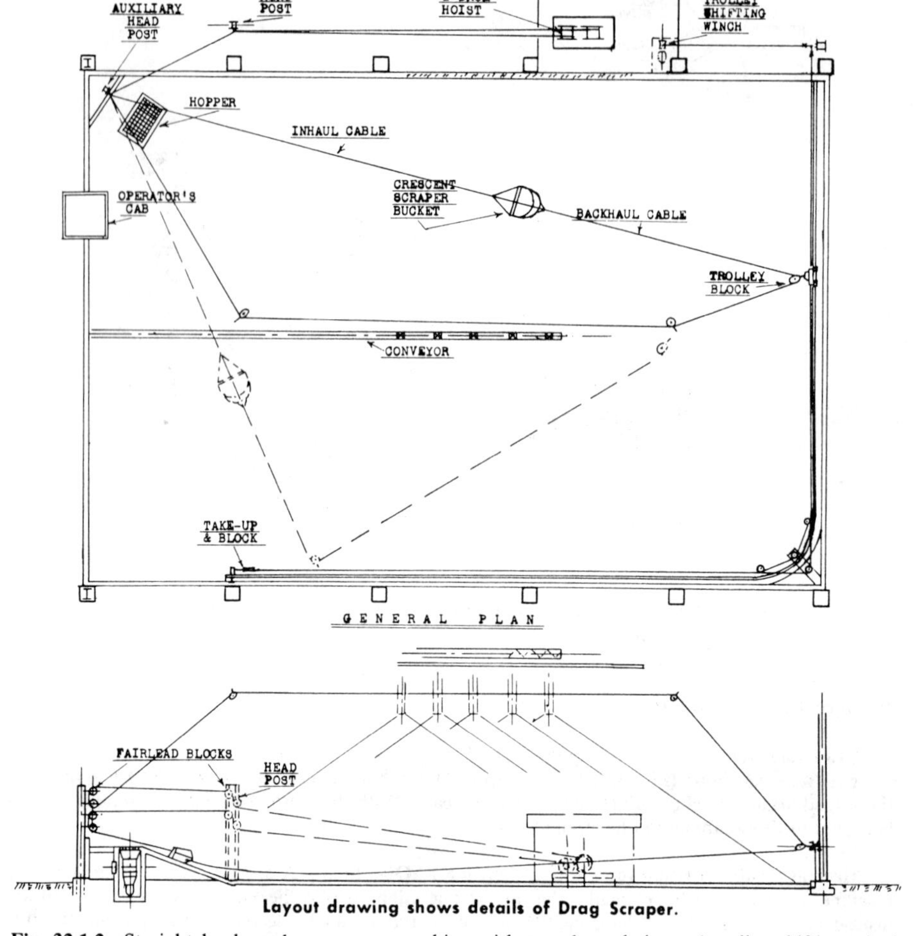

Fig. 32.1.2 Straight haulage drag scraper machine with two-drum hoist and trolley shifting winch.

or hazardous materials which might endanger personnel and equipment, since only the scraper and cables contact the material. The hoist and other machinery can also be located outside the work area.

The ratio of payload to total weight (payload plus dead load) is 70–80% for drag scrapers. For mobile equipment in which the power unit and operator move with the material, this figure is 15–55%.

Some straight-haulage drag scrapers can be automated. The machines that operate in one line lend themselves to relatively simple automation. A typical application is an automated drag scraper machine for handling mill scale below the rolls and tables in a steel mill. The machine generally operates between two fixed points controlled by limit switches, delivering mill scale to a grizzly. Often a timer is used to control the number of trips per hour so that the machine can be regulated to suit material handling requirements.

Other automatic arrangements can be used. A drag scraper can be programmed to increase the length of haul each trip of the scraper by an adjustable amount. Also, the machine can be arranged so that the line of operation can be changed periodically. In addition to limit switches and timers, the drag scraper machine can be controlled by counters, bin-level indicators, and other such devices. These serve to coordinate the drag scraper operation with other operations in a plant.

Table 32.1.1 shows handling rates for straight haulage drag scraper machines from $\frac{1}{2}$ to 5 yd^3 (0.382–3.82 m^3) for various lengths of haul. Allowances should be made for maintenance, downtime, shifting the line of operation, and difficult digging conditions.

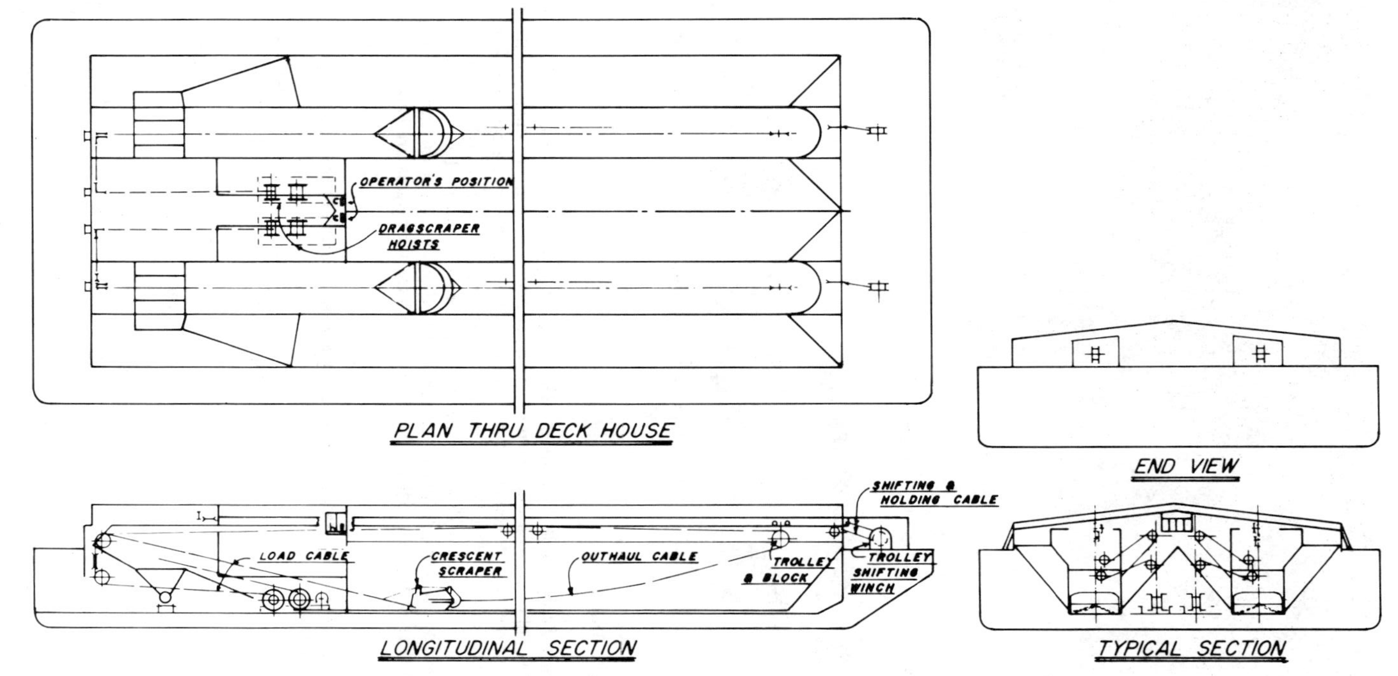

Fig. 32.1.3 Straight haulage drag scraper machine with overhead trolleys in barge.

Fig. 32.1.4 Straight haulage drag scraper machine with trolley handling sugar in storage building.

Track Cable Machines

Two-Drum Track Cable Machines. The simplest type of track cable machine is one in which the scraper is pulled in with one drum of the hoist and is returned through the air by gravity to the digging point. With this arrangement the scraper is suspended from a carrier that rides a track cable on the return trip (Fig. 32.1.5). The track cable is slack when the loaded scraper is being inhauled and tensioned by a second drum of the hoist for gravity return. To provide the proper return by gravity there must be enough heights provided at both ends so that the scraper returns down a slope and is kept clear of the material on the return trip. Greater heights are needed for longer returns of

Table 32.1.1 Rated Handling Capacities of Straight Haulage Drag Scraper[a]

	Drag Scraper Machine Size in yd³							
Length of Haul	$\frac{1}{2}$	$\frac{3}{4}$	1	$1\frac{1}{2}$	2	3	4	5
100 ft yd³/hr	40	60	80	120	160	265	354	442
30 m m³/hr	31	46	62	93	124	205	273	341
200 ft yd³/hr	24	36	48	72	96	163	217	271
60 m m³/hr	19	28	37	56	74	126	168	210
300 ft yd³/hr	17	26	34	51	69	117	156	195
90 m m³/hr	13	20	27	40	53	91	121	151
400 ft yd³/hr	13	20	27	40	53	92	122	153
120 m m³/hr	10	15	21	31	41	71	95	118
500 ft yd³/hr	11	16	22	33	44	75	100	125
150 m m³/hr	8	13	17	25	34	58	78	97

Source: Sauerman Bros. Inc.

[a] Handling rates are based on free-caving material. Machines $\frac{3}{4}$–5 yd³ are diesel powered with torque converter. The $\frac{1}{2}$-yd³ size is a straight diesel unit. For electrical powered machines deduct approximately 10%.

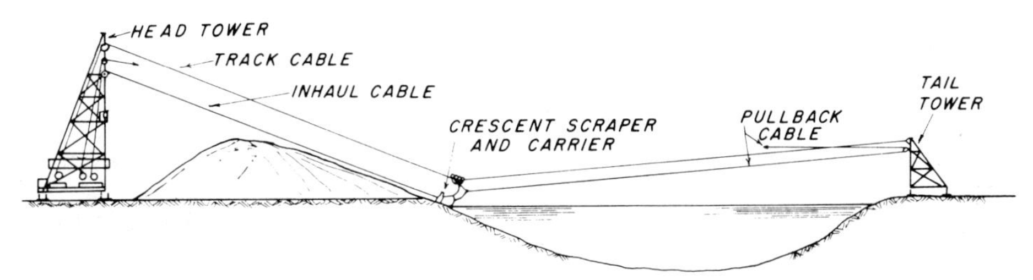

Fig. 32.1.5 Track cable drag scraper machine with pullback and movable head and tail towers.

the scraper. The necessary heights can be provided by stationary guyed masts or towers or by movable towers.

Track Cable Machines with Pullback. For long spans it is often desirable to provide a third drum for pullback of the scraper through the air rather than depending on gravity for return of the scraper. Heights must be provided at both head and tail ends so that the scraper can be conveyed through the air clear of the material to the digging point. The track cable is slack when inhauling and is tensioned for aerial return.

The head end is generally higher than the tail end to provide for stockpiling of material at the head end and to provide for the necessary lead distance between the hoist and blocks.

Pie-shaped areas can be worked with a stationary head end and movable tail end. Rectangular areas can be worked when both head and tail ends are movable. Noncaving materials require more frequent changing of the line of operation than free-caving materials.

Large drag scraper machines are generally of the track cable type. Track cable machines have several advantages over straight haulage machines. With a track cable machine, wear on the scraper is reduced because the scraper is returned through the air rather than being dragged back on the material, higher speeds can be used for aerial return, and more control is provided for the placement of the scraper.

Table 32.1.2 shows the rated handling capacities of track cable drag scraper machines. Allowances should be made for maintenance, downtime, shifting the line of operation, and difficult digging conditions.

Table 32.1.2 Rated Handling Capacities of Track Cable Machines[a]

	Track Cable Machine Size in yd^3							
Length of Haul	3	4	5	6	8	10	12	15
100 ft yd^3/hr	213	284	355	426	568	710	852	1065
30 m m^3/hr	164	219	274	328	438	547	657	821
200 ft yd^3/hr	141	188	235	283	377	471	565	706
60 m m^3/hr	109	146	182	218	291	364	437	546
300 ft yd^3/hr	106	141	176	211	282	352	423	529
90 m m^3/hr	82	109	136	164	218	273	327	409
400 ft yd^3/hr	84	113	141	169	225	282	338	422
120 m m^3/hr	65	87	109	131	174	218	262	327
500 ft yd^3/hr	70	94	117	141	188	234	281	352
150 m m^3/hr	55	73	91	109	145	182	218	272
600 ft yd^3/hr	—	80	100	120	161	201	241	301
180 m m^3/hr	—	62	78	93	125	156	187	233
700 ft yd^3/hr	—	—	88	105	140	176	211	263
210 m m^3/hr	—	—	68	82	109	136	163	204
800 ft yd^3/hr	—	—	—	94	125	156	187	234
240 m m^3/hr	—	—	—	73	97	121	145	182
900 ft yd^3/hr	—	—	—	—	—	140	168	211
270 m m^3/hr	—	—	—	—	—	109	131	163
1000 ft yd^3/hr	—	—	—	—	—	128	153	191
300 m m^3/hr	—	—	—	—	—	99	119	148

Source: Sauerman Bros. Inc.

[a] Handling rates are based on free-caving material. Machines 3–5 yd^3 are diesel powered with torque converter. Machines 6–15 yd^3 are electric powered.

Tautline Scraper Machines. Track cable machines can be arranged with a tautline track cable. The track cable is kept taut and raising and lowering of the scraper is done with a separate hoisting cable suspended from the carrier which is operated by a hoisting winch. With this arrangement the track cable is held well above the material and precise control of the scraper is provided.

Application Data

The following information should be furnished for the proper selection of either a straight haulage or a track cable drag scraper machine:

1. General description of the work to be done, preferably with sketches or drawings.
2. Type of material to be handled, including weight per cubic foot or kilograms per cubic meter, size and grade, moisture content, angle of repose, cohesive nature of the material, compaction effect or chemical reaction while in storage, and characteristics of the material, including whether it is free-caving, noncaving, or corrosive.
3. Desired hourly handling capacity.
4. Number of hours of operation of the machine per day.
5. Type of power to be used for the hoist, such as diesel or electric, and, if electric, the current characteristics including number of phases, Hertz, and voltage.

Scrapers Used with Draglines and Other Excavating Machines

The reach of a conventional dragline excavator can be extended by using a scraper with carrier (Fig. 32.1.6). One drum of the dragline is used for dragging the loaded scraper and the second drum is used for handling a track cable which is anchored several hundred feet away from the dragline. The scraper is returned to the digging point by gravity. To provide stability for the dragline excavator a boom support is generally used. A scraper can usually be at least 50% larger than the dragline bucket used with a dragline, because the scraper is lighter and does not carry the material.

Because dragline hoists have relatively slow speed and limited cable spooling capacity, a separate hoist can be mounted between the dragline and the boom support to provide a higher operating speed, increased spooling capacity, or pullback of the scraper.

A scraper can also be used in place of a dragline bucket in a casting operation. In this operation the scraper can also be larger than the dragline bucket. With this arrangement the scraper can excavate at grade or below grade and also pull down banks or piles. The bottomless scraper in a casting operation can be equipped with a lifting hitch so that a partial load can be lifted. The load that is lifted is about 50% of the rated load of the scraper.

A hydraulic excavator can also be converted to a drag scraper machine by providing a boom and hoist mounted on the hydraulic excavator.

32.2 SLACKLINE CABLEWAYS

Clifton H. Hubbell

32.2.1 Applications

Slackline cableways are used to dig material, lift it free of the excavation, convey it to a discharge point, and dump it automatically. The distances over which the machines operate vary from several hundred feet up to 1000 ft (305 m). Slackline cableways are recommended for deep digging [up to 125 ft (38 m) below water], long hauls, and delivery of material to an elevated point such as a stockpile or hopper. Typical applications for slackline cableways are:

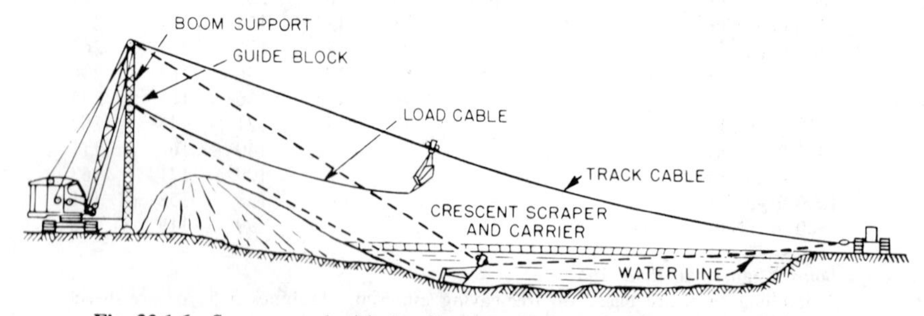

Fig. 32.1.6 Scraper used with dragline to extend reach of the boom machine.

Excavation of sand and gravel
Cleaning out of settling ponds
Deepening of rivers and channels
Reclaiming industrial wastes

32.2.2 Construction and Operating Principles

Construction

A slackline cableway consists of a bucket and carrier assembly, a track cable with a multiple-part tensioning cable, a load cable, blocks to guide the cables, and a guyed mast to provide height at the dump end, a two-drum hoist with one drum for the tensioning cable and one drum for the load cable, and an anchorage at the tail end for the track cable (see Fig. 32.2.1). The hoist is usually designed for providing a slow digging speed with a faster speed to convey the loaded bucket through the air to the dump point. For most applications the bucket returns to the digging point by gravity and it is necessary that the mast or head end be high enough in relationship to the tail end for gravity return (Fig. 32.2.2). The operator is generally located near the hoist and often in an elevated position for better visibility of the work area.

Operating Principles

In operating the slackline cableway the operator tensions up on the track cable by taking in the tensioning cable. This raises the bucket suspended from a carrier which travels on the track cable. The bucket is then allowed to travel down the track cable to the digging point by gravity. The tensioning cable is then slackened, which in turn slackens the track cable and lowers the bucket into the excavation. The bucket is pulled forward with the load cable to dig a load. When the bucket has a load, the tensioning cable is then spooled in, which tightens the track cable and raises the bucket clear of the material. The load cable is then used to pull the loaded bucket through the air. The bucket is conveyed through the air until the traveler block on the chain assembly contacts a button located on the track cable. When the traveler block is in contact with the button and the load cable is pulled, the bucket is dumped (Fig. 32.2.3). The bucket dumps automatically at the place determined by the position of the button on the track cable. The button can be moved to a new position on the track cable to change the dumping point. After the bucket is dumped it is again returned down the track cable by gravity to the digging point and the cycle is repeated.

Slackline cableways work on a straight line of operation. This line of operation can be changed

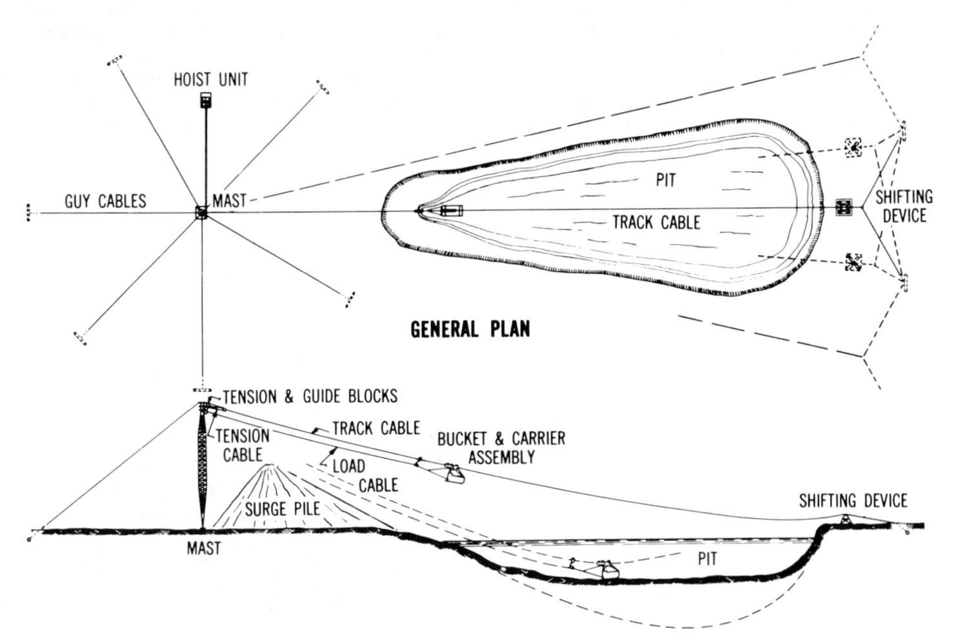

Fig. 32.2.1 Slackline cableway layout.

Fig. 32.2.2 View of slackline cableway looking toward mast.

by moving the tail end so that the machine will work over an area. The typical tail end uses a tower to provide some elevation of the track cable, which is anchored to a bridle cable. The line of operation is shifted by moving the tower and its track cable connection on the bridle cable by other equipment.

Another type of tail end is a rapid-shifting system with two towers with a cable system between them. A bridle frame can be shifted with a motor-powered winch controlled by the operator. Movable towers can also be used to anchor the track cable. Noncaving materials require more frequent movement of the tail end than free-caving materials.

Since the bucket returns to the digging point by gravity, it is necessary that the head end be high enough in relation to the tail end. If the bucket is to discharge into an elevated hopper, care must be taken to allow sufficient dumping clearance for the bucket.

The machines are generally made in sizes $\frac{1}{2}$–$3\frac{1}{2}$ yd^3 (0.382–2.68 m^3). Hoists with diesels with torque converters are usually used because they provide heavy line pulls for digging and also high speeds for conveying the loaded bucket through the air to the dumping point. Torque converters also provide a smooth transition between the slow digging speed and the high conveying speed.

Hoists can also be equipped with electric motors rather than diesel units. Special arrangements such as both high- and low-speed gearing for the load drum are used to provide a low digging speed and a high conveying speed. Hoists powered with diesel units with torque converters are generally preferred because they are smoother, operate at higher speeds, and have higher handling capacities. Table 32.2.1 shows handling capacities for various sizes of slackline cableways over various haul distances. The handling capacities shown in Table 32.2.1 are for the handling of free-caving materials

Fig. 32.2.3 Slackline cableway dumping on storage pile.

Table 32.2.1 Rated Handling Capacities of Slackline Cableways[a]

	Slackline Cableway Machine Size, in yd^3						
Length of Haul	$\frac{1}{2}$	$\frac{3}{4}$	1	$1\frac{1}{2}$	2	$2\frac{1}{2}$	$3\frac{1}{2}$
Lgth of span, ft:	400	500	600	700	800	900	1000
Lgth of span, m:	120	150	180	210	240	270	300
150 ft yd^3/hr	25	57	83	—	—	—	—
45 m m^3/hr	19	44	63	—	—	—	—
200 ft yd^3/hr	22	54	80	97	—	—	—
60 m m^3/hr	17	41	61	74	—	—	—
250 ft yd^3/hr	20	51	70	91	114	150	—
75 m m^3/hr	15	39	54	70	87	115	—
300 ft yd^3/hr	17	46	65	84	110	147	182
90 m m^3/hr	13	35	50	64	84	112	139
400 ft yd^3/hr	—	—	55	72	98	125	161
120 m m^3/hr	—	—	42	55	75	96	123
500 ft yd^3/hr	—	—	—	61	84	107	143
150 m m^3/hr	—	—	—	47	64	82	109
600 ft yd^3/hr	—	—	—	—	76	95	129
180 m m^3/hr	—	—	—	—	58	73	99
700 ft yd^3/hr	—	—	—	—	—	82	115
210 m m^3/hr	—	—	—	—	—	63	88

[a] Handling capacities are when digging 30 ft (9 m) below the mast. Machines are diesel powered with torque converter. Handling capacities are for operations in free-caving materials such as ordinary sand and gravel.

with hoists powered with diesels with torque converters. The handling capacities for electric-powered machines are usually less than for machines powered with diesel torque converters.

Allowances should be made for maintenance, downtime, shifting the line of operation, and difficult digging conditions. Digging at greater depths than shown will also reduce the handling capacities.

Application Data

For the proper selection of a slackline cableway the following information should be furnished:

1. General description of the work to be done, preferably with sketches or drawings showing the shape and dimensions of the area to be worked.
2. The type of material to be handled, including the weight per cubic foot or kilograms per cubic meter, size, and characteristics such as whether the material is free-caving or noncaving.
3. Elevation of the delivery point in relation to the digging point and the location of the tail end or anchor of the track cable.
4. Desired maximum depth of excavation.
5. Delivery point of the material such as stockpile or hopper.
6. Desired hourly handling capacity.
7. Number of hours of operation of the machine per day and length of the operating season.
8. Type of power for the hoist such as diesel or electric, and if electric, the current characteristics including number of phases, Hertz, and voltage.

32.3 BUCKET WHEEL EQUIPMENT

Alexander Pomerantsev

Considerable increase in the efficiency of the modern open cast mining and stockyard operation technique is due to the use of continuously operating machines. Being the first unit in the production line, bucket wheel excavators (Fig. 32.3.1) are the leading element. The main features of these machines are continuous excavation of material and uninterrupted discharge onto any loading system. When certain parameters, such as digging height and capacity of machines, are suitably matched, the efficiency of bucket wheel excavators is extremely good. Their application ranges from the lightest to the heaviest of soils and practically to any hard materials (even frozen material is handled successfully in Canada

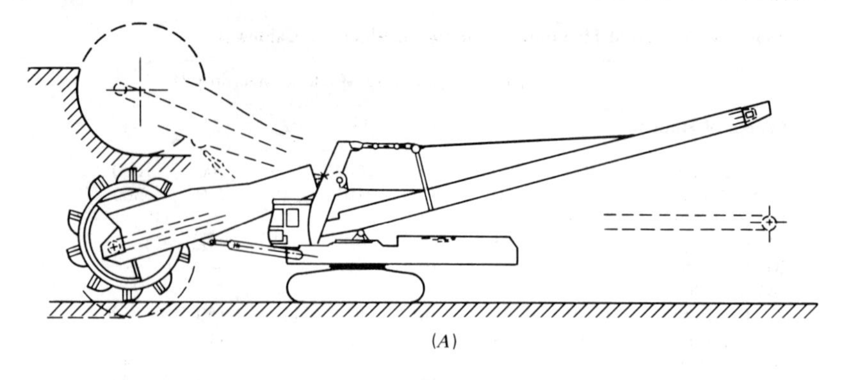

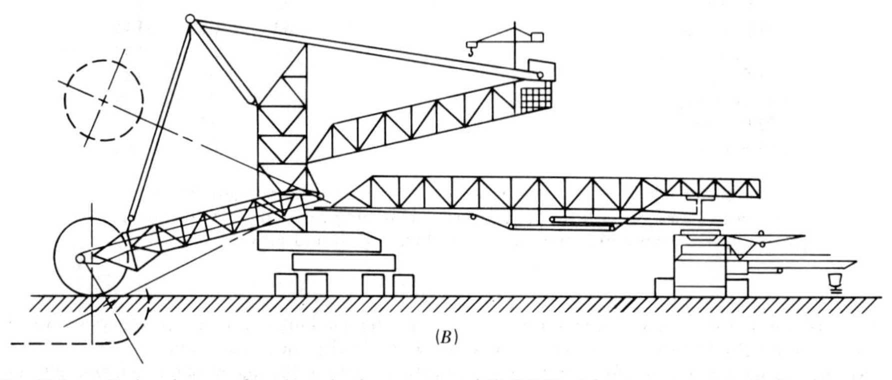

Fig. 32.3.1 Basic schemes of bucket wheel excavators. (*A*), BWE without appropriate discharge system; (*B*), BWE with appropriate discharge system.

and the USSR), and a big advantage of bucket wheel excavators is clean selective mining of various superimposed material layers.

Bucket wheel machines, particularly reclaimers, were covered to a certain extent in Chapter 30. This section provides more detailed, technical information.

Bucket wheel excavators (BWEs) are principally used in full block operation for excavating uniform material and in the so-called side block operation where there are layers of different types of materials. With independent mobile shiftable conveyors or movable conveyor bridges (belt wagons) with mobile feeders, not only the block width can be considerably raised, but also the digging height can be essentially increased when operating in steps (Fig. 32.3.1*B*).

The range of application of bucket wheel excavators is not limited only to open cast mining. These machines are of outstanding suitability everywhere where material is to be excavated, conveyed, and handled: earth-moving jobs, canal and trench excavation, dam and highway construction, and aggregate production.

As for bucket wheel reclaimers (BWRs) (see Fig. 32.3.2), they have found wide application for operation above bunker trenches and bunkers as well as for handling bulk material, particularly to reclaim the raw material at the mine triangular or flat-top stockpiles, harbors, steel mill stockyards, and so on. The demands for the stockyards are excellently met by the crawler-mounted or rail-mounted bucket wheel reclaimers. The deciding factors are very good adaptivity to the shape of the stockpile, digging height, and material to be handled.

The essential characteristic difference between bucket wheel excavators and reclaimers is the digging force at cutting lips of the bucket wheel. Contrary to solid materials, only low digging forces are required for loose bulk materials, which are in close relation to the loading and dimensioning of the machines. Bucket wheel reclaimers, especially those designed for use in stockyards, permit high outputs and thus a high degree of economy for a relatively low design weight. There are various possibilities for adapting the design of the bucket wheel reclaimer to particular application such as using the same machine for stacking as well as for reclaiming materials.

BWEs and BWRs dig with a number of evenly spaced buckets on a wheel periphery, the wheel being mounted on the extreme end of the slewing, liftable and lowerable noncrowd or crowd-type

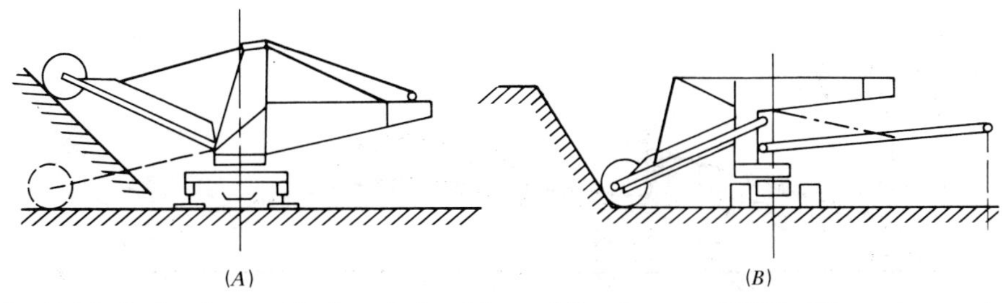

Fig. 32.3.2 Basic schemes of bucket wheel reclaimers. (*A*), rail-mounted BWR; (*B*), crawler-mounted BWR.

wheel boom carrying a take-away belt conveyor. Excavated material flows in a continuous stream via the transfer points to the belt conveyor carried by the discharge boom (BWEs and crawler-mounted BWRs) and further to the belt conveyor discharge system. Booms can move simultaneously in the horizontal and vertical planes (see Fig. 32.3.3).

Usually BWEs and BWRs have a machinery deck and superstructure mounted on a 360° swing circle.

Patented first in the United States in 1881, BWEs and BWRs are manufactured now mostly in West and East Germany (Krupp, Orenstein, and Köppel, Buckau-Wolf, Lauchhammers), in the United States (Mechanical Excavators, Bucyrus-Erie, Barber-Green (right-side-mounted double wheels BWE), in the USSR, Japan, and Great Britain.

32.3.1 Classification of Equipment

The following parameters are used by designers and manufacturers to characterize and describe the capability and features of the BWEs and BWRs:

1. Maximum theoretical output of loose material excavated per hour by machine. Machines are subdivided by output as follows:

Small output, up to 500 m^3/hr
Medium output, up to 2500 m^3/hr
High output, up to 5000 m^3/hr
Superhigh output, above 5000 m^3/hr

2. Digging force per unit area or per unit length (BWEs only) of machines:

Normal digging force, up to 7 kg/cm^2
Increased digging force, above 7 kg/cm^2*

3. Mine layouts. BWEs with multibench operation:

High BWEs, when the depth of digging is not more than a half-diameter of the wheel (see Fig. 32.3.4)
High–deep BWEs, when the depth of digging is more than a half-diameter of the wheel (see Fig. 32.3.5)
BWRs are chosen for:
Rectangular storage yard with a modular or bench type of operation
Circular storage yard (special design of BWRs)
Storage trench

4. Method of discharge. Excavated material could be loaded into trucks, a train, shiftable belt conveyor, belt wagon, or dumped directly onto the piles.

5. Method of thrusting:

Advancing the crawlers or rail wheels

* Conversion (transition) from digging force per unit area to the digging force per unit length will be shown below.

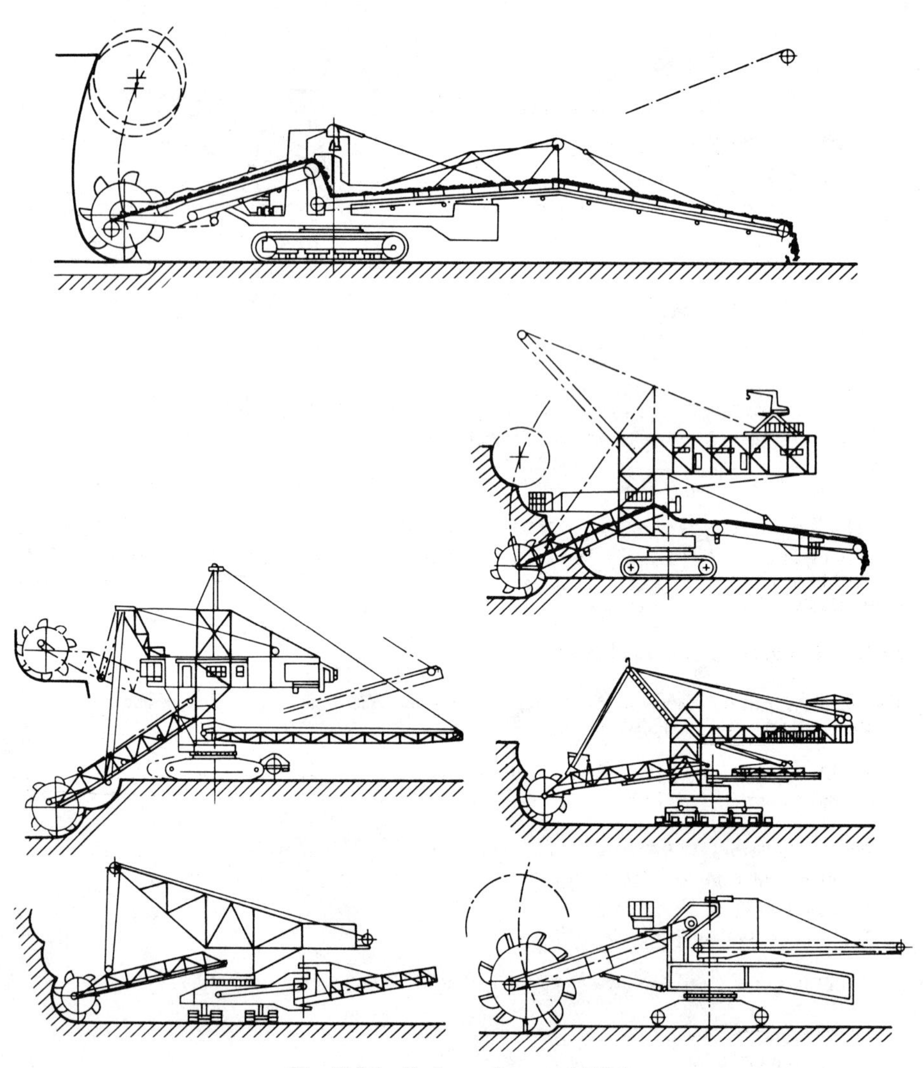

Fig. 32.3.3 Various schemes of BWEs.

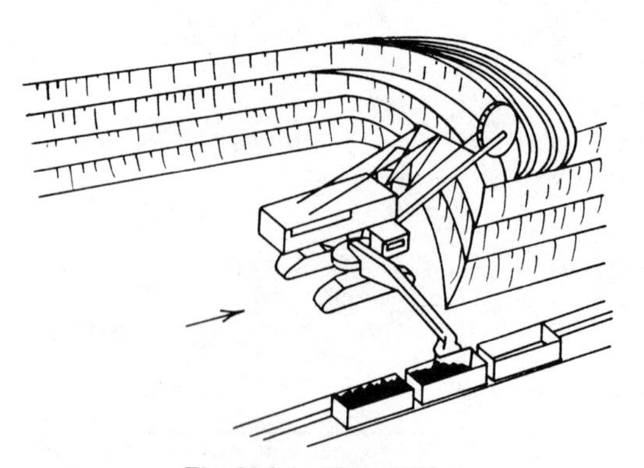

Fig. 32.3.4 High BWE.

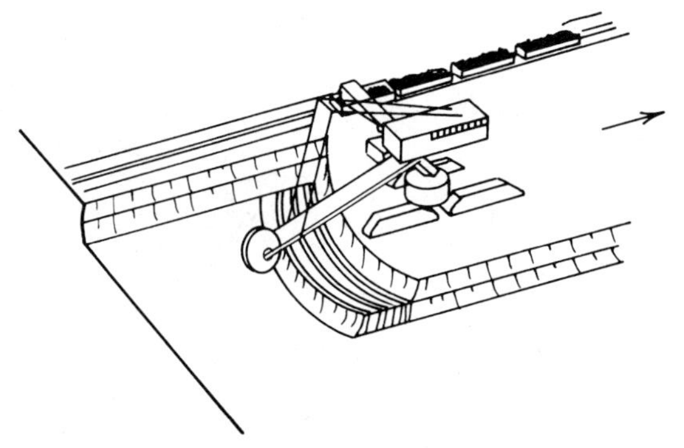

Fig. 32.3.5 High–deep BWE.

Thrusting the boom forward (see Fig. 32.3.6) (The latter method is very favorable for selective mining.)

6. Type of propelling unit. Machines are subdivided as follows:

Crawler-mounted machines
Rail-mounted BWRs
Rail-walking mounted or rail-crawler mounted BWEs

32.3.2 Working Processes

Wheel excavation can be fulfilled either by a terrace or dropping cut (see Fig. 32.3.7). In this figure the parameters of cut sickles are presented as follows:

c = current slice depth, in m or cm
c_{max} = maximum slice depth, in m or cm
b = slice width, in m or cm
h_c = slice height, in m or cm

Figure 32.3.7 also shows the difference between the bench profiles for different methods of thrusting. It is necessary to point out that the advantage of the drop cut is in the possibility to control lump size of excavating material and in the improvement of the efficiency of the machine in hard or frozen material digging, but the usage of the drop cut always needs substantial (up to 25%) additional power consumption. Depending on requirements, the propel unit and the bucket wheel, BWEs and BWRs can be used in three different types of operation:

Face of front operation: machine transfers along the working face.

Full-block operation: a face block is excavated in several bench cuts by raising, lowering, and continuously slewing of the bucket wheel boom (see Fig. 32.3.8).

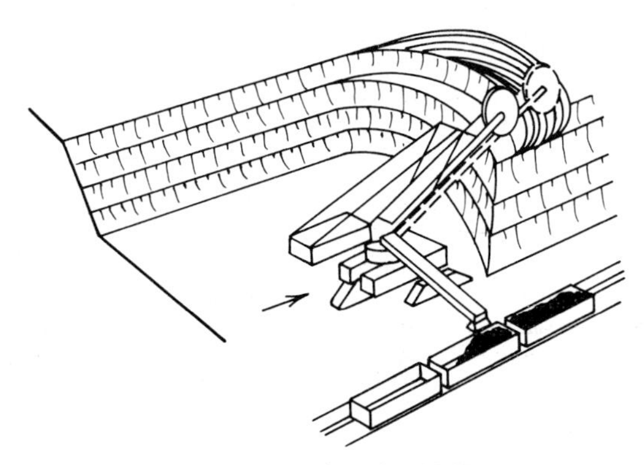

Fig. 32.3.6 Thrusting BWE.

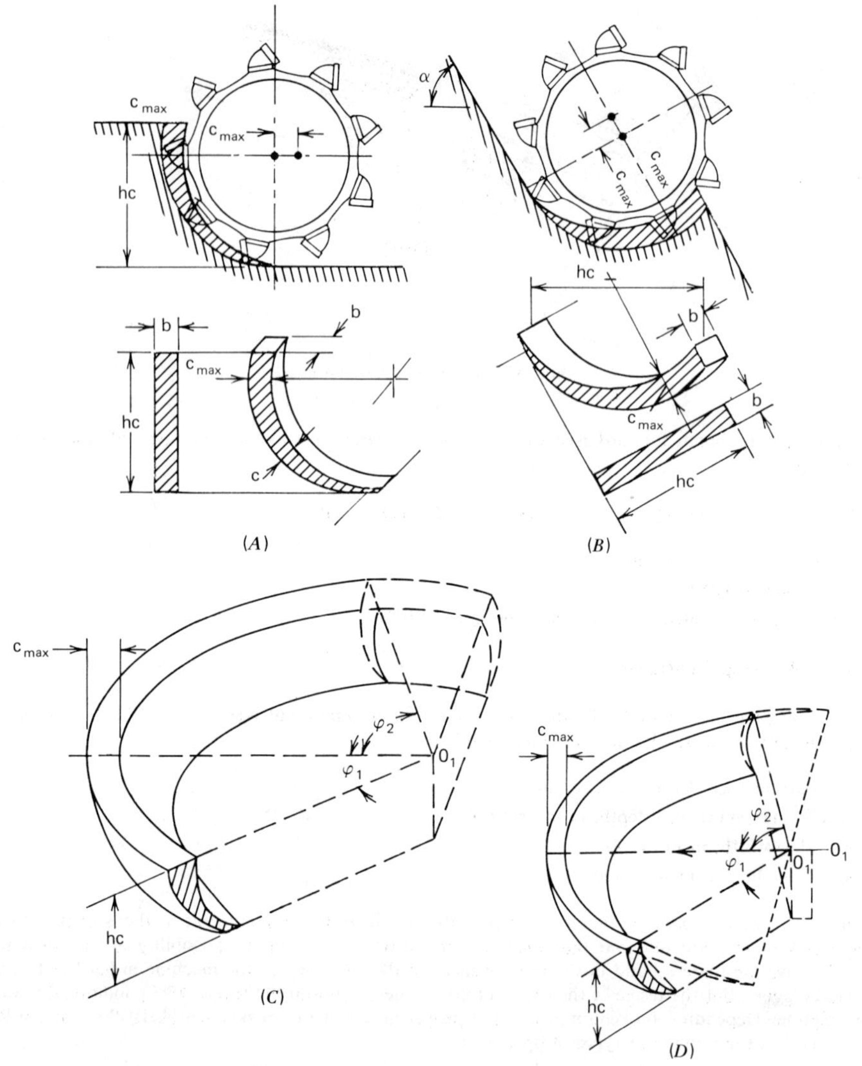

Fig. 32.3.7 Terrace, dropping cuts, and types of benches. (*A*), terrace cut; (*B*), dropping cut; (*C*), bench developed by thrusting BWE; (*D*), bench developed by thrustless BWE.

Half-block operation: the traveling lengths are greater than those necessary for full block operation, but much shorter than for front face operation.

Face of front operation for BWRs is shown in Fig. 32.3.9.

32.3.3 Machine Elements, Their Parameters and Selection

Bucket Wheel

Taking into consideration the type of the material to be excavated and conveyed, engineering mining plan, and required output, the selection of the machine elements has to be started from the bucket wheel. The bucket wheel can have gravitation or inertial (very rare application) method of discharge; each of these methods can be fulfilled with the forced discharge of the buckets. The gravitation discharge wheels can have three different types of construction: cell-less (Fig. 32.3.10), celled (Fig. 32.3.11), and combined (short cell). Most of the recent units have cell-less (reversible) or combined types of bucket wheels, because these types of wheels with a greater speed have a discharge capacity about

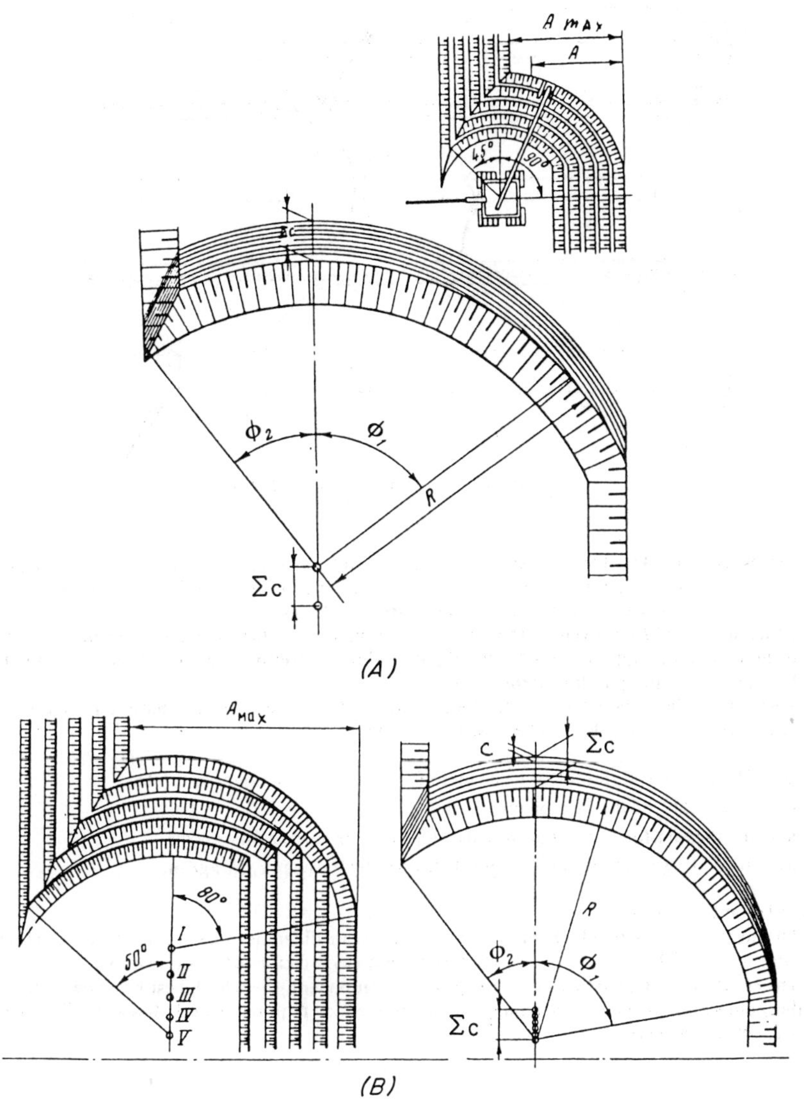

Fig. 32.3.8 Mine layout for BWEs, (*A*), thrusting BWE; (*B*), thrustless BWE.

double that of similar-sized celled units. After each bucket passes the ring space cover plate, the cell-less wheels empty continuously either onto slope chute (*A*) or to conical rotary ring (*B*), roller table (*C*), rotary drum (*D*), disc feeder (*E*), belt feeder (*F*), two belt feeders (*G*), scrab drum feeder (*I*) (see Fig. 32.3.12), or directly to the belt (*H*).

It is possible to discharge material without changing flow direction; for this purpose the bucket wheel should have horizontal and vertical tilts, as shown in Figs. 32.3.12 and 32.3.13 where also the basic mutual arrangements of the bucket wheel and the booms are indicated.

The bucket consists of a tough wear-resistant cutting blade (lip) with or without teeth, a strong load-bearing, but ductile and resilient support, and the body, which can be solid or flexible (chain body). The profile of the lips could be different, as is shown in Fig. 32.3.14. It is also necessary to point out that the lips can be vertical or slanted (dotted line in Fig. 32.3.20). To provide for satisfactory cutting geometry and lump size, the bucket can be equipped with precutters and the bucket wheel can have additional cutting lips without a bucket body.

The parts of the bucket wheel drive (such as a reduction gear box and overload clutch) can be mounted on the side of the bucket wheel, inside the bucket wheel, on the conveyor belt side, or on

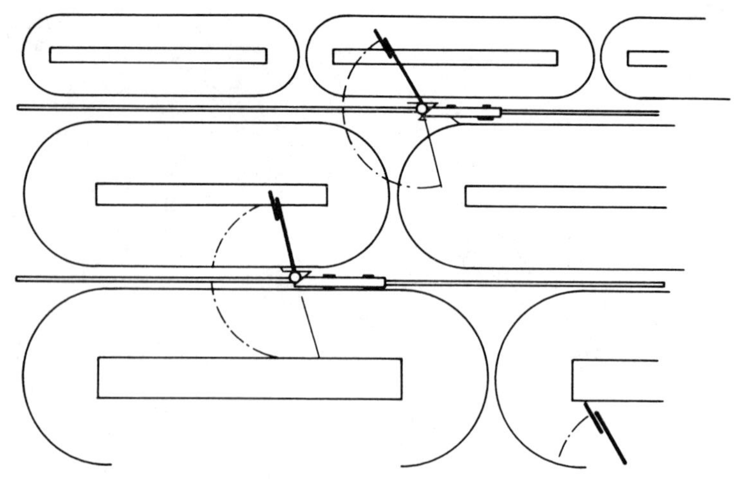

Fig. 32.3.9 Mine layout for BWRs.

both sides (see Fig. 32.3.15). The angles α_1, α_2, β_1, and β_2 represent very important parameters that obviously have to be taken into consideration in designing and operating BWEs and BWRs. Locations of these components can be quite different in hydraulic drives.

Preference should be given to the outside location with statically determined three-point support or to the two-point support (flexible) for the disc clutch. Overload clutches could be a notch clutch, fluid coupling, or multiple-disc friction clutch.

Selection of the parameters of the BWEs and BWRs is based on the basic characteristics of the digging force and the machine. These basic characteristics are:

Q_{th}—the theoretical output of loose material excavated per hour, in m^3/hr

H_m—the height of the face or pile, in m

K_1—the digging force per unit area developed by excavator, in kg/cm^2

K_l—the cutting force per unit length developed by excavator, in kg/cm

The relation between K_l and K_1 is shown in Table 32.3.1.

The digging force with which it is necessary to excavate different types of soils (materials) is presented in Table 32.3.2. Table 32.3.2 also shows density and swell factor for those materials.

The selection of the parameters of the primary component, the wheel, best begins with the selection of the number of buckets. Then one can determine the required wheel diameter D and the total capacity of the buckets.

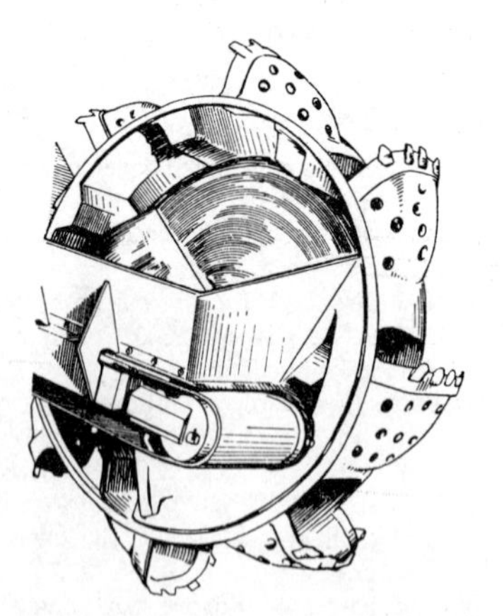

Fig. 32.3.10 Cell-less type of bucket wheel.

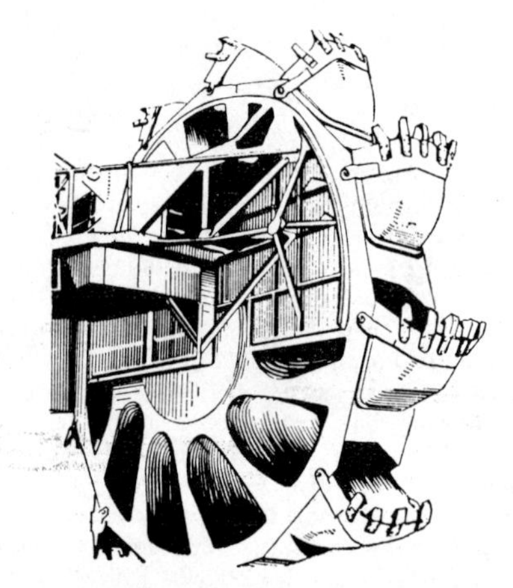

Fig. 32.3.11 Celled type of bucket wheel.

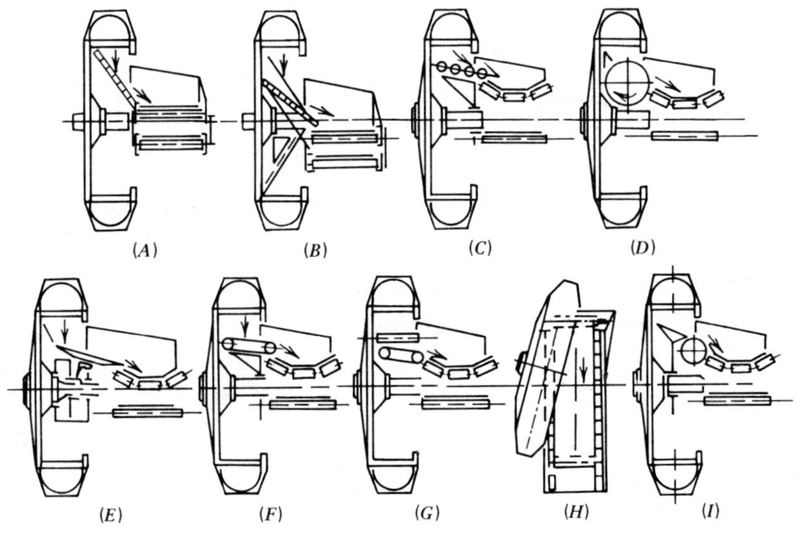

Fig. 32.3.12 Various types of transfer points for material from bucket wheel to the conveyor belt.

For the purpose of selecting the number of buckets mounted on the digging wheel, one needs to examine the effect that this number has on the digging process. It is instructive to consider the torque M_1 (kg × cm) produced by the tangential component (P_{01}) of the digging force about the wheel axis. This torque is given by

$$M_1 = (P_{01} + P_{011} + \cdots + P_n) \times \frac{D}{2} \tag{32.3.1}$$

Equation 32.3.1 is the sum of torques of tangent components of digging forces of all the buckets that are in contact with the material (see Fig. 32.3.16).

The values of the forces P_{01}, P_{011}, P_{012} are functions of the number of buckets installed on the wheel, configuration of buckets, number of buckets digging simultaneously and cutting parameters.

So, (23.3.1) can be rewritten as follows:

$$M_1 = K_1 \times \frac{V_t \times \pi \times D^2 \times \cos^2 \epsilon \times C \times \sin(\alpha + K\,\phi)}{2 \times V_w \times Z}, \text{ kg, cm} \tag{32.3.2}$$

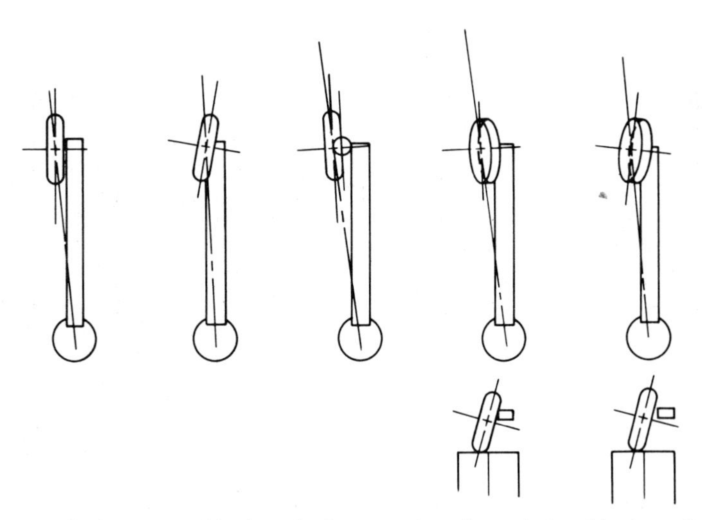

Fig. 32.3.13 Various types of bucket wheel arrangements in vertical and horizontal planes.

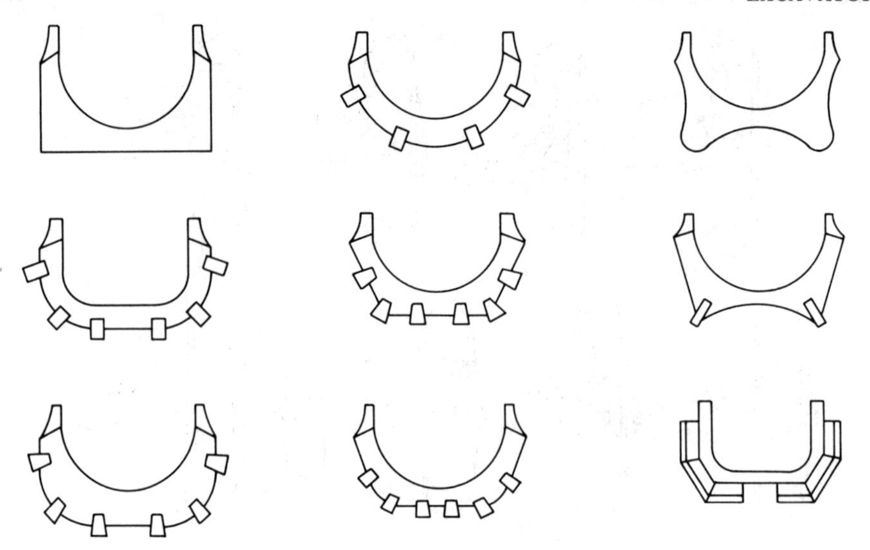

Fig. 32.3.14 Various types of cutting lips of buckets.

where: D = wheel diameter, in cm
Z = number of buckets
V_t = tangential speed of bucket due to rotation of machine on its axis, wheel slewing speed, in m/min
V_w = tangential speed of bucket due to rotation of wheel, in m/min
ϵ = angle between V_t and V_w, $\epsilon = \text{arc tan}\,(V_t/V_w)$
c = current slice depth, in cm
α = angle between vertical and cutting edges of first bucket making contact with face
π = 3.1416

The values of the velocity of the bucket due to rotation of the machine on its axis and the rpm of the wheel as a function of the wheel diameter for different values of K_v are shown in Fig. 32.3.17.

ϕ = pitch angle of bucket
k = from 0 to $(Z_e - 1)$

The symbol Z_e represents the maximum number of buckets that dig simultaneously (see Table 32.3.3).
The variation of M_1 as a function of the number of buckets on the wheel varying from 6 to 15

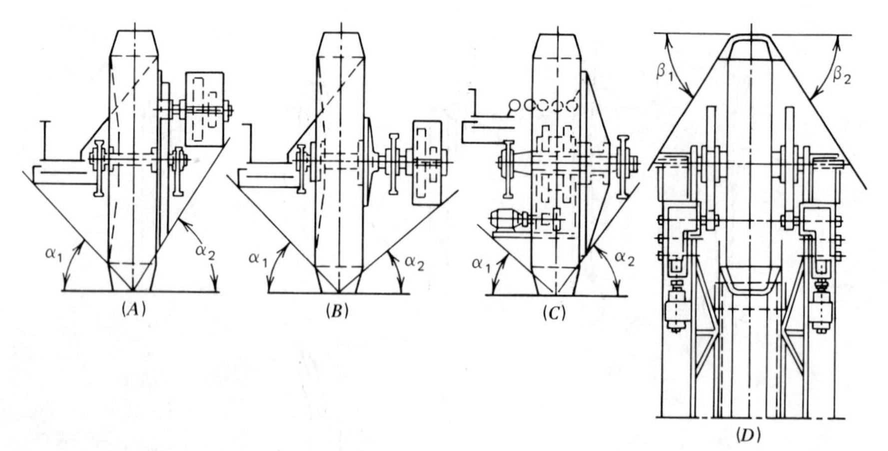

Fig. 32.3.15 Various bucket wheel drive arrangements. (*A*), (*B*), on the bucket wheel side; (*C*), inside the bucket wheel; (*D*), both sides of the bucket wheel.

Table 32.3.1 Relationship between Digging (K_1) and Cutting (K_l) Forces

Capacity of Bucket (m^3)	0.1	0.4	1	1.5	2	2.5	4
K_l/K_1	8.7	12.2	16.5	18.4	19.7	20.6	22.3

for the height of the bench $H = 0.5D$ (sawtooth curve) is shown in Fig. 32.3.18. Torque M_1 for $Z = 6$ is assumed to be an arbitrary unit.

The range from 0.5 to 1.5 denotes the break of one of the buckets from the contact with the material. Examination of Fig. 32.3.18 and Table 32.3.3 shows that the maximum torque decreases rapidly as the total number of buckets on the wheel increases, as long as an increase in the total number of buckets does not lead to an increase in the number of simultaneously excavating buckets.

Table 32.3.2 Characteristics of Materials

No.	Material	K_1 for Bucket Wheel Excavator[a] (kg/cm²)	Density (ton/m³)	Swell Factor (K_M)
	Easily Excavated Material			
1	Sand, sandy loam, loam (soft, moist, and loose)	0.4–1.3	1.2–1.5	1.05–1.16
2	Loam, fine pit gravel, clay (soft and moist	1.2–2.5	1.4–1.9	1.14–1.26
	Average Material			
3	Loam (strong), pit gravel (strong), very soft claystone and siltstone, very soft coal	2.0–3.8	1.6–2.0	1.24–1.30
4	Very strong loam with broken stone, strong clay, and soft shale	3.0–5.5	1.9–2.2	1.28–1.37
5	Coal, manganese and phosphate ore, sandstone (middle strength)	5.2–7.6	2.1–2.5	1.35–1.42
	Difficultly Excavated Material			
6	Chalk, gypsum (middle strength), strong coal, soft frozen soil	7–12	2.3–2.6	1.4–1.45
7	Limestone, strong chalk, marl, strong gypsum, strong sandstone, and frozen soil (middle strength)	12–30	2.3–2.8	1.42–1.48

[a] The digging force per unit area reflects the digging process for BWEs and BWRs more correctly than the cutting force per unit length; that is usually used for shovel and other earth-cutting machine designs. The relation between these forces can be established by the equation:

$$K_l = \frac{K_1}{(0.015 - 0.02)\,D}$$

The highest value of the coefficient in this equation shall be used for BWEs and BWRs, with the capacity of the bucket more than 0.45 m³.

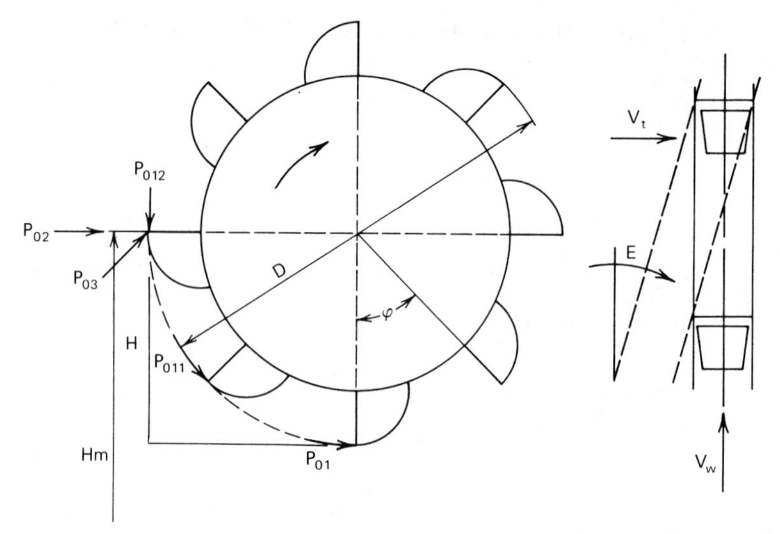

Fig. 32.3.16 Digging scheme.

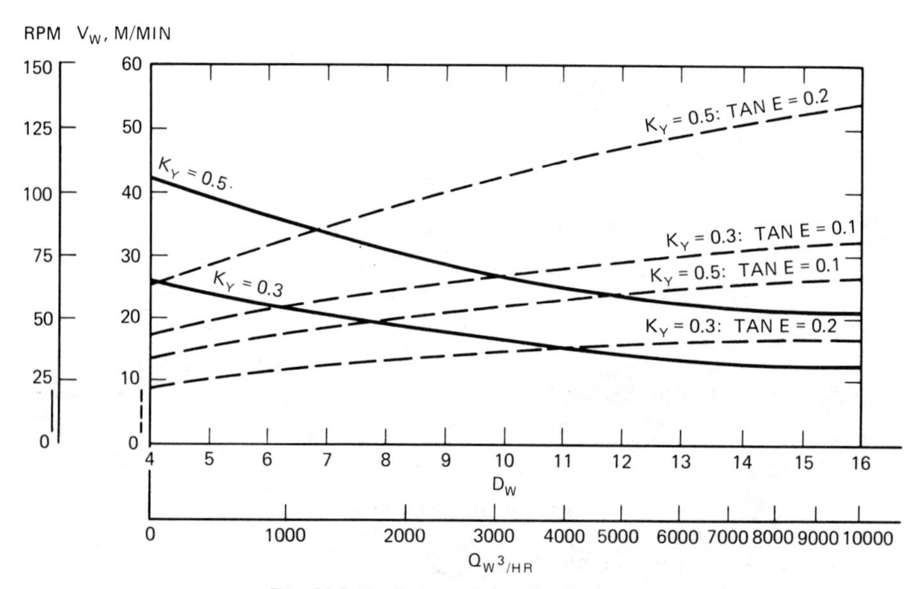

Fig. 32.3.17 Rpm and the wheel velocity.

Table 32.3.3 Total Number of Buckets Mounted on a Wheel Versus Number of Simultaneously Digging Buckets

Z Number of Buckets on Wheel	Z_e Maximum Number of Simultaneously Digging Buckets
6, 7	1 or 2
8, 9	2
10, 11	2 or 3
12, 13	3
14, 15	3 or 4

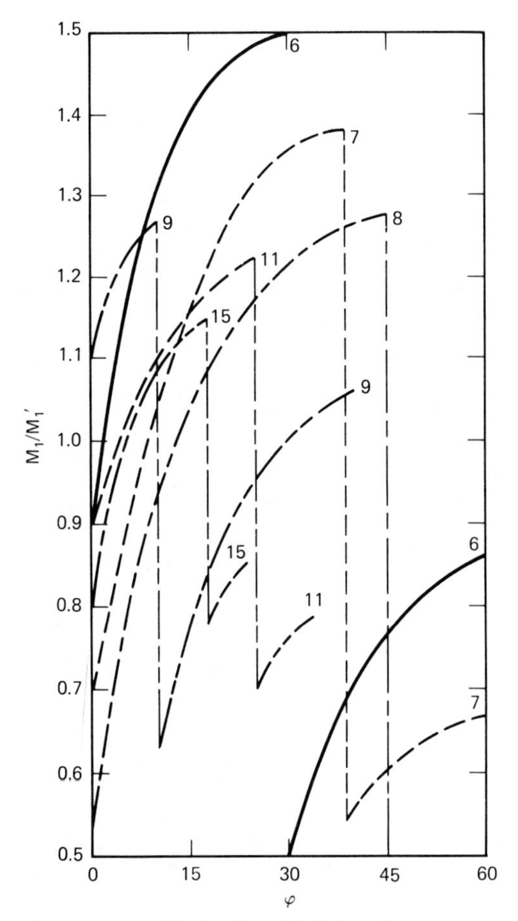

Fig. 32.3.18 Variation of the moment M_1 for $H = 0.5\ D$ (numbers represent the number of buckets on the wheel).

One may also note that no significant decrease in the maximum torque results from increasing the number of buckets beyond 10.

Figure 32.3.19, unlike Fig. 32.3.18, shows the variation of M_1, but for various H (from 0.4 D to 0.85 D). The torque value for $Z = 6$ and $H = 0.5\ D$ is an arbitrary unit. One may note that the maxima and minima converge as the number of buckets increases and the height of the cut increases. Thus, use of more buckets and large H leads to less variation in the torque, and, therefore, to smoother operation of the machine.

Numerous experiments have shown that use of a greater number of buckets has a favorable effect on the way they discharge their load, while requiring practically no increase in the power needed to drive the wheel. Those experiments also allow us to establish the number of buckets that should be mounted on a wheel depending on the digging forces (as is shown in Table 32.3.4), that are necessary for excavation of the material.

Selection of other parameters of the wheel is based on the relation between the output rate and the parameters that describe the local conditions.

Theoretical output of the loose material of the wheel machinery is

$$Q_{th} = 60 \times Z \times q \times n,\ \text{m}^3/\text{hr} \tag{32.3.3}$$

where: n = revolution per minute of the wheel
q = capacity of the bucket

On the other hand, effective output of the material is

$$Q_{eff} = 60 \times V_t \times c \times H,\ \text{m}^3/\text{hr} \tag{32.3.4}$$

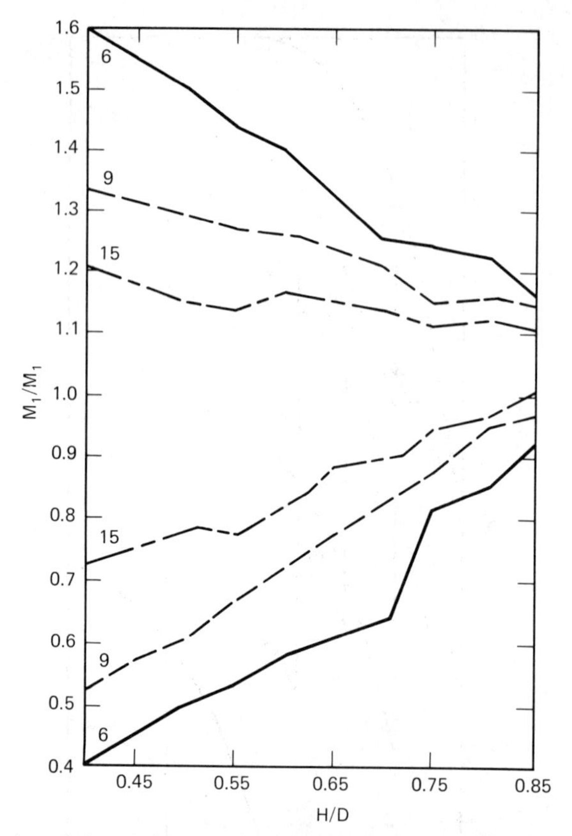

Fig. 32.3.19 Variation of the maximum and minimum of the moment M_1 at $H = (0.4\text{–}0.85)\ D$ (numbers represent the number of buckets on the wheel).

So,

$$Q_{eff} = \frac{\eta_e \times Q_{th}}{K_M}$$

where: η_e = degree of bucket fill
K_M = swell factor

All the components can be expressed in terms of D, and therefore, diameter of the wheel is related to the theoretical output Q_{th} for $\eta_e = 1$:

$$D = 0.273 \sqrt[5]{\frac{K_q^2 \times Q_{th}^2}{K_v^2 \times K_b^4 \times K_H^4 \times K_c^4 \times K_M^4 \times \tan^4 \epsilon}} \qquad (32.3.5)$$

where: K_q = ratio of bucket capacity to cube of its height
K_v = ratio of tangential velocity of the wheel to tangential velocity at which centrifugal acceleration is equal to gravitational acceleration
K_b = ratio of distance between cutting edge lips of adjacent buckets to the length of the bucket
K_H = ratio of height of cut to the wheel diameter
K_c = ratio of thickness of cut to the height of the bucket

Table 32.3.4 Choosing Number of Buckets

K_1	3	5	7	8	9
Z	7–8	8–9	9–10	12–14	14–16

$\tan \epsilon$ = ratio of tangential velocity of bucket due to rotation of wheel, to the tangential velocity of bucket due to rotation of machine on its axis

Using the most unfavorable combination of the coefficients of (32.3.5), it has been found that the foregoing complicated expression may reasonably be approximated by:

$$D = (0.3 - 0.4) \times Q_{th}^{0.4}, \text{ m} \tag{32.3.6}$$

Once the diameter of the wheel has been selected, one may determine the total capacity of the buckets from

$$\Sigma_q = \frac{\pi \times K_b \times K_c^2 \times K_H^2 \times K_M^2 \times \tan^2 \epsilon \times D^3}{K_q}, \text{ m}^3 \tag{32.3.7}$$

or approximately, from (for cell-less type wheel)

$$\Sigma_q = 0.01 \times D^3, \text{ m}^3 \tag{32.3.8}$$

The components of the digging force P_{01}, P_{02}, and P_{03} can be found as follows:

$$P_{01} = 40 \times K_1 \times D^2, \text{ ton or kg} \tag{32.3.9}$$

$$P_{02} = (0.25 - 0.4)\, P_{01} \tag{32.3.10}$$

$$P_{03} = \frac{V_t}{V_w} P_{01} \tag{32.3.11}$$

The rpm of the bucket wheel is equal to

$$n = \frac{21}{\sqrt{D}} \tag{32.3.12}$$

and the tangential speed of the bucket is

$$V_w = 1.11 \sqrt{D}, \text{ m/sec} \tag{32.3.13}$$

Once one has established the total capacity of the buckets, it is very easy to establish the capacity of a single bucket by dividing Σ_q by Z. But it is necessary to determine the parameters that define any individual bucket. Two bucket configurations are used: those with vertical lips (line BC of Fig. 32.3.20), and those with slanted lips (line AB of Fig. 32.3.20). Taking into consideration the capacities of various subparts of the buckets, the capacity of a bucket with vertical lips can be expressed by

$$q_r = 2 \times f \times h^3 \times (1 - 2 \tan \epsilon), \text{ m}^3 \tag{32.3.14}$$

and the one with the slanted lips is given by

$$q_c = \left\{(2f - \tan \beta) \times \left[f - \tan \epsilon \times (2f - \tan \beta) - \frac{\tan \delta \times \tan \beta}{4 \cos \beta \times f}\right] + \tan \epsilon \times \left(f - \frac{\tan \delta}{2 \times \cos \beta}\right)\right\} \times h^3, \text{ m}^3 \tag{32.3.15}$$

where the symbols represent the dimensions indicated in Fig. 32.3.20, and f denotes the coefficient of soil resistance to shift. Table 32.3.5 illustrates the capacities of buckets with lips slanted at $\beta = 15°$.

But, for practical need, the capacity of the bucket can be estimated from the following equations:

$$\text{Easily excavated material, } q = 0.8\, h^3 \tag{32.3.16}$$

$$\text{Average material, } q = 0.95\, h^3 \tag{32.3.17}$$

$$\text{Difficultly excavated material, } q = 1.1\, h^3 \tag{32.3.18}$$

The width of the bucket should be equal to its length and the length shall be equal to 1.5–2.0 × height of the bucket.

Wheel Boom

The designers and manufacturers provide basically two types of the boom structure: truss or truss-less with different cross sections, rectangular or circular.

The slope of a conveyor is the boundary condition for establishing the length of the wheel boom. The length of noncrowd wheel boom of the BWE can be equal to 1.3 of the height of the face (H_m), and the length of the crowd wheel boom can be equal to 1.55 H_m, and 1.7 and 2.2 for BWR, respectively. We should also say that the height of the face (H_m) excavated by the BWE and the height of the pile reclaimed by the BWE can be equal to 2.5–3.0 times the wheel diameter.

The conveyor belt system in BWEs and BWRs, usually rated 50% more than Q_{th}, has two important transfer points: from the bucket wheel to the bucket wheel conveyor belt (the feature of this transfer point has already been discussed) and from the bucket wheel conveyor belt to the discharge boom. The complexity of the second transfer point consists of the following: the speed of the discharge conveyor belt can differ from the speed of the bucket wheel conveyor belt, the dropping height can be sufficient, and besides that, the discharge belt can be slewing up to ± 100° from the longitudinal axis of the bucket wheel conveyor belt.

Special attention should be paid to belt cleaning.

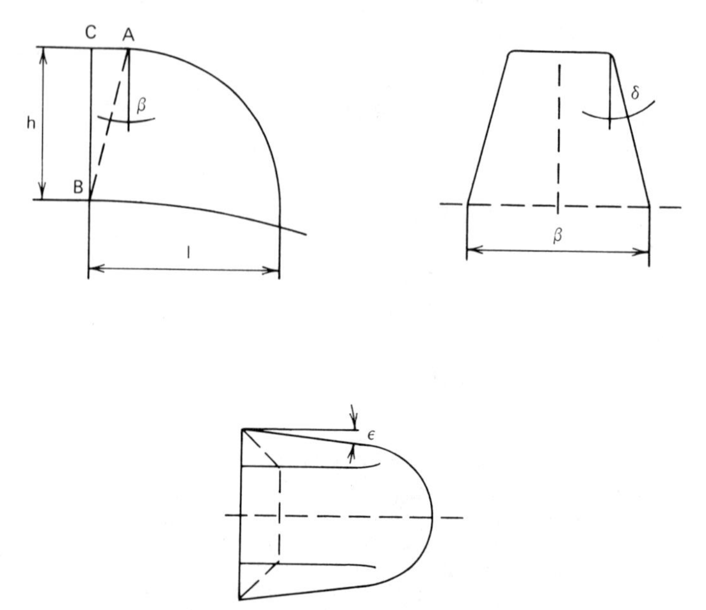

Fig. 32.3.20 Main dimensions of the bucket.

Table 32.3.5 Capacity of the Bucket with Slanted Lips[a]

δ \ ε	6°	8°	12°
10°	0.920 h^3	0.906 h^3	0.756 h^3
20°	0.874 h^3	0.821 h^3	0.710 h^3
30°	0.819 h^3	0.766 h^3	0.685 h^3

[a] $2f = 1.5$.

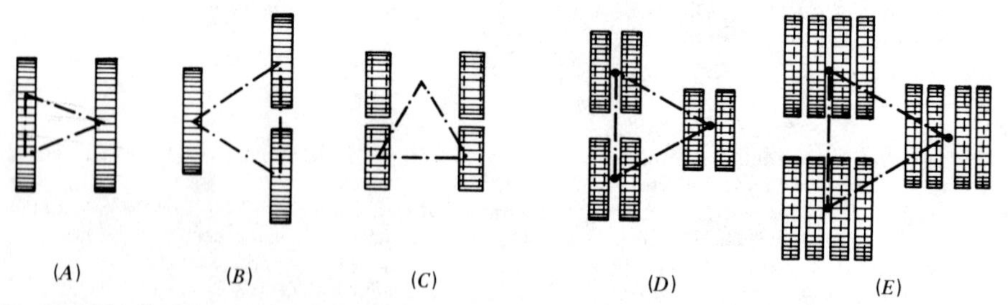

Fig. 32.3.21 Various crawler assemblies. (*A*), two-crawler unit; (*B*), three-crawler unit; (*C*), four-crawler unit; (*D*), six-double crawler unit; (*E*), twelve-double crawler unit.

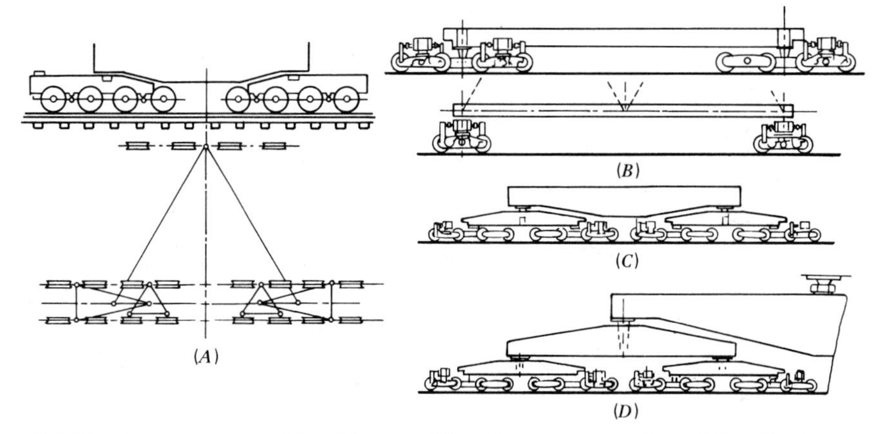

Fig. 32.3.22 Various rail assemblies. (*A*), and (*B*), with two-stage equilizer; (*C*), with three-stage equilizer; (*D*), with four-stage equilizer.

Table 32.3.6 Calculating Moments and Forces

Acting Moments and Forces	Equation
Maximum bending moment (ton × m) for:	
Bucket wheel boom	$0.07\ K_1 \times D^3(1 + 0.158\ D)$
Discharge boom	$0.00274\ K_1 \times D^4$
Counterweight boom	$0.00586\ K_1 \times D^4$
Superstructure	$0.056\ K_1 \times D^3(1 + 0.594\ D)$
Maximum forces (ton) acting in turntable:	
Vertical	$0.00061\ K_1 \times D^2(1 + 1.43\ D)$
Horizontal	$0.00035\ K_1 \times D^2(1 + 0.5\ D)$
Maximum vertical forces (ton) transferred to:	
Ball race	$0.00061\ K_1 \times D^2(1 + 1.98\ D)$
Propelling unit	$0.00061\ K_1 \times D^2(1 + 3.38\ D)$

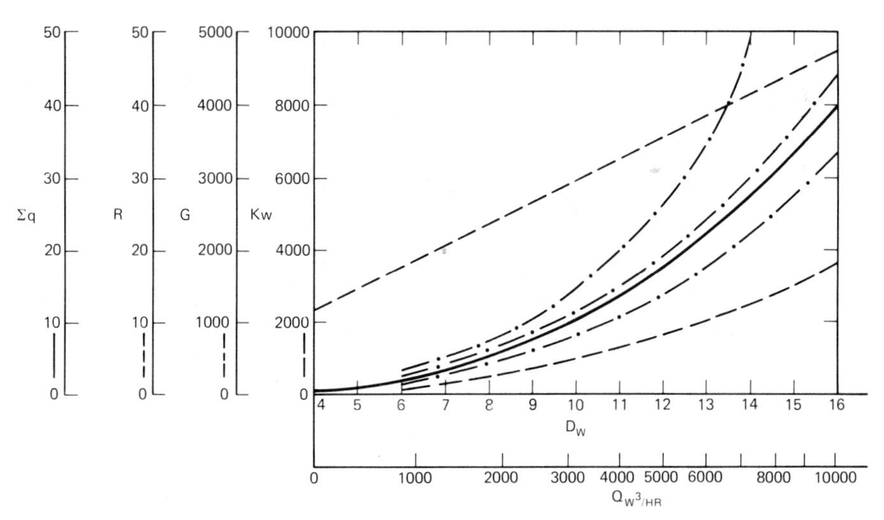

Fig. 32.3.23 General chart for preliminary estimation of the basic parameters of BWEs and BWRs.

Table 32.3.7 Characteristics of BWE Excavators

Parameters	$40\frac{13^a}{0}$	$70\frac{6.5}{0.5}$	$100\frac{14}{0}$	$120\frac{12}{0.8}$	$175\frac{10}{0.5}$	$250\frac{12}{5}$	$400\frac{17}{1.5}$	$400\frac{10}{0.7}$	$550\frac{12.8}{5}$	$1200\frac{22}{2}$	$1500\frac{30.5}{5}$
1. Theoretical output of loose material excavated, Q_{th} (m³/hr)	216	325	453	825	630	1125	1370	1540	1260	3450	5860
2. Digging force per unit area, kg/cm²	7.9	2.9	4.4	10.1	4.05	6	6.5	6.5	4.9	6.4	8.7
3. Maximum radius of digging, m	29.6	9.0	22.9	15.75	10.75	15.5	21	10.8	19.6	29.5	5.2
4. Maximum radius of discharge, m	—	17	20	15.3	20.0	20	25	21.4	20	58	104
5. Total power installed, kW	95	80	130	250	280	430	580	380	645	2060	5080
6. Service weight, G (ton)	100	64	133	80	206	312	547	270	476	1285	3450
7. Average ground pressure, kg/cm²	—	0.91	1.0	0.8	1.0	0.9	1.0	1.0	1.1	1.17	1.0
8. Wheel diameter, m	3.1	2.8	3.7	4.0	1.8	5.2	6.45	5.8	6.2	8.2	11.5
9. Number of buckets	8	6	8	12	7	7	9	7	8	8	10
10. Bucket discharge per minute	90	78	90	115	60	75	57	64	60	48	65
11. Bucket wheel belt width, m	0.65	1.0	0.8	1.0	0.9	1.2	1.2	1.2	1.4	1.6	2.0
12. Bucket wheel belt speed, m/sec	1.68	2.5	2.1	2.2	3.5	2.8	4.6	3.5	2.2	3.8	4.0
13. Discharge belt width, m	0.68	1.0	0.8	1.0	0.9	1.2	1.2	1.2	1.4	1.6	2.0
14. Discharge belt speed, m/sec	3.0	2.5	2.1	2.2	3.5	3.0	4.8	3.5	2.5	4.0	4.5
15. $G/Q_{th} \times 1000$ m³/hr	410	197	290	200	327	279	400	175	510	370	580
16. kW/Q_{th}	0.58	0.25	0.29	0.62	0.44	0.38	0.42	0.25	0.52	0.6	0.88
17. kW/G	0.95	1.25	0.98	3.13	1.36	1.37	1.06	1.41	1.35	1.6	1.47

Table 32.3.7 (Continued)

Parameters	$1600\frac{40}{7}$	$1900\frac{30}{5}$	$2800\frac{50}{20}$	$3800\frac{52}{25}$	$6320\frac{50}{18}$	$630\frac{20}{3}9^{b}$	$765\frac{10}{3}14$	$850\frac{26}{3}11$	$1200\frac{24}{4}20$	$1600\frac{40}{10}31$	$2400\frac{40}{4}25$
1. Theoretical output of loose material excavated, Q_{th} (m^3/hr)	5000	5950	10,200	8700	15,900	1700	4200	2450	3450	4500	7200
2. Digging force per unit area, kg/cm^2	11.4	5.1	4.85	6.5	7.0	6.5	6.5	6.6	3.8	7.7	5.5
3. Maximum radius of digging, m	74	50	95	79	70.5	21	38	23.5	44.7	66	68
4. Maximum radius of discharge, m	59	89	42	102	125	16	77	27	58	35	112
5. Total power installed, kW	—	3145	5,820	8400	15,000	990	2070	1490	1815	3940	5650
6. Service weight, G (ton)	4245	3900	5,000	7400	12,900	705	1510	1700	1438	3300	4585
7. Average ground pressure, kg/cm^2	1.2	1.1	1.1	1.4	1.5	1.4	1.4	1.2	1.31	0.9	1.26
8. Wheel diameter, m	16.3	13	14	17.5	21.6	7	7.3	8.2	8.2	11.5	12.5
9. Number of buckets	10	10	10	10	18	8	9	10	8	10	10
10. Bucket discharge per minute	52	52	60	38	42	45	45	45	48	50	50
11. Bucket wheel belt width, m	2.0	2.2	2.4	2.6	3.2	1.4	1.4	1.7	1.6	1.8	2.2
12. Bucket wheel belt speed, m/sec	4.0	3.5	4.0	3.8	5.2	2.8	4.6	4.0	3.9	3.3	3.6
13. Discharge belt width, m	2.0	2.2	2.4	2.6	3.2	1.4	1.4	1.7	1.6	1.8	2.2
14. Discharge belt speed, m/sec	4.0	3.5	4.0	4.0	5.2	3.2	5.1	4.5	3.9	3.3	4.2
15. $G/Q_{th} \times 1000$ m^3/hr	850	660	490	850	811	415	360	690	415	735	635
16. kW/Q_{th}	—	0.58	0.58	0.96	0.94	0.58	0.49	0.61	0.56	0.88	0.78
17. kW/G	—	1.14	1.16	1.4	1.16	1.4	1.35	0.87	1.26	1.19	1.24

[a] $q\frac{H_m}{h_m}$ = typical designation of thrustless BWEs. q = capacity of bucket, in liters; H_m = height of face, m; h_m = depth of face, m.

[b] $q\frac{H_m}{h_m} L_t$ – typical designation of thrusting BWEs. L_t = length of bucket wheel boom thrusting, m.

Table 32.3.8 Bucket Wheel Excavator Installations

Location	Material	Capacity (m^3 per Bucket)	No. of Buckets	Type of Wheel	Wheel Diameter (m)	Cutting Speed (m/sec)	Theoretical Output Loose Q_{th} (m^3/hr)	Installed Power (kW)	Machine Weight (tons)	$\frac{G}{Q_{th}} \times 1000$	$\frac{kW}{Q_{th}}$	$\frac{kW}{G}$
Sumatra	Chalk/lignite	0.25	7	Cell-less	5.2	2.1	500	430	283	570	0.86	1.52
Yugoslavia	Clay/lignite	0.35	8	Cell-less	6.2	2.4	1260	715	418	330	0.57	1.7
India	Sandstone	0.7	9	Cell-less	8.0	3.35	2500	1873	1135	450	0.75	1.65
Morocco	Phosphate	0.1	8	Cell-less	3.5	2.06	540	130	121	220	0.24	1.07
California	Dredge tailings (earth) and cobbles to 0.3 m	1.4	8	Cell-less	9.0	2.35	3460	1050	551	160	0.35	1.9
Mauritania	Broken iron ore	0.65	8	Celled	6.6	1.5	1380	663	513	400	0.48	1.29
California	Loose, semiloose, and rock overburden	2.0	10	Cell-less	9.1	2.83	6880	1490	726	105	0.22	2.05
S. Africa	Overburden in copper mine	0.35	8	Cell-less	6.2	2.4	1260	—	418	330	0.57	1.7

The Structure of the BWEs and BWRs

The structure of BWEs and BWRs consists of two parts:

1. The substructure carrying the main slewing mechanism and supported by crawler or rail groups.
2. The revolving upper superstructure with supporting frames for bucket wheel, discharge booms, permanent or movable counterweight (rocker type for BWRs) and machine desk. The slewing mechanism consists of slewing gear and the turntable transmitting the loads. Some machines may have two turntables: one for the bucket wheel boom rotation, and the other for the discharge boom. The turntable can have cylindrical or tapered rollers on circular tracks welded to both the super- and the substructures, or it can be constructed with ball races or a ball-bearing slewing ring. There are three- or four-point supports of the superstructure and different types of the crawler and rail group arrangements (see Figs. 32.3.21 and 32.3.22).

Crawler-mounted BWEs and BWRs cost less, can cover a wider range of the works, can work with shiftable conveyors, and their changing of location is virtually unlimited, but auxiliary equipment such as dozers or graders are sometimes necessary for landscaping. Rail-mounted BWRs can provide a higher degree of automation and high travel speed.

The structures of BWEs and BWRs should be in the form of three-dimensional frameworks, because the resultant force does not lie in a single plane of action. So, the structure usually consists of numerous special frameworks, which permit constant free movements relative to each other in one to three planes.

The construction material should be of fine-grained steel and high tensile bolts; bolts should be preloaded. The center of gravity of the BWEs and BWRs must always remain within the ball race core circle diameter under all operation load conditions.

Table 32.3.9 BWR Installation

Machine Designation		Location	Material	Theoretical Output Q_{th} (m^3/hr)	Machine Weight (tons)	$\frac{G}{Q_{th}} \times 1000$
	$\frac{100 \cdot 800}{25}$ [a]	Angola	Iron ore	900	180	200
C	$\frac{340 \cdot 1000}{28}$	Italy	Coal	1000	225	225
C	$\frac{550 \cdot 1200}{34.6}$ [b]	USA	Coal	1500	390	260
	$\frac{670 \cdot 1400}{40}$	Spain	Phosphate	2800	380	136
	$\frac{900 \cdot 1400}{50}$	Morocco	Phosphate	3600	466	129
C	$\frac{290 \cdot 1200}{30.5}$	Australia	Iron ore	4000	505	126
C	$\frac{750 \cdot 1400}{37}$	Netherlands	Iron ore	5000	740	148
C	$\frac{1500 \cdot 1800}{45}$	Netherlands	Iron ore and coal	6000	750	125
R	$\frac{360 \cdot 1200}{19 + 26}$ [c]	Liberia	Iron ore and coal	2750	260	96
R	$\frac{220 \cdot 1200}{13 + 20}$	Argentina	Iron ore and pellets	1250	145	116

[a] $\frac{q \cdot W}{L_t}$ = typical designation of rail-mounted reclaimer.

[b] $C_m \frac{q \cdot W}{L}$ = typical designation of rail-mounted double-action reclaimer (stacker/reclaimer).

[c] $R \frac{q \cdot W}{L + L_1}$ = typical designation of crawler-mounted reclaimer.

Variables: q = capacity of buckets, liters; W = width of the conveyor belt, mm; L = length of the bucket wheel boom, m; L_1 = length of the discharge boom, m; R = crawler-mounted reclaimer; C_m = double-action reclaimer.

Assuming the foregoing, bending moments and acting forces in basic elements of BWEs and BWRs could be calculated by using the equations that are shown in Table 32.3.6.

32.3.4 Total Weight of Equipment

Preliminary estimation of the weight G of the BWEs and BWRs may be made from the following expressions. For a crawler-mounted machine with a noncrowd digging boom,

$$G^n = 0.22 \times K_1 \times D^3 \times P, \text{ tons} \tag{32.3.19}$$

where P = permissible specific ground pressure exerted by the machine (usually 1 kg/cm²). The weight of the excavator with crowd digging boom is:

$$G^m = 0.28 \times K_1 \times D^3 \times P, \text{ tons} \tag{32.3.20}$$

The rail-mounted machines are 30% heavier.

The total weight of the BWEs and crawler-mounted BWRs could be distributed between separate units in the following way: wheel boom with a wheel and a wheel drive, 12%; counterweight, 18%; discharge boom, 3.5%; overstructure, 9%; slewing frame and equipment, 17.5%; and crawler unit and equipment, 40%.

The total weight of the rail-mounted BWRs could be distributed in the following way: wheel boom with the wheel and drive, 11%; counterweight, 22%; discharge boom, 10%; slewing frame and equipment, 26%; and rail group assemblies, 31% (41% if the machine does not have the discharge boom).

32.3.5 Electrical Instrumentation Equipment

Machines with medium, high, and superhigh output can be equipped with DC motors (single or group drives) with adjustable voltage control, particularly for slewing mechanism; machines with a small and medium output can be equipped with hydraulic drives; the machines with a small output are usually equipped with wound rotor motors.

The total power (kW) required may be estimated from

$$\text{kW} = 0.026 \times Q_{th} \times (K_1 + 0.52 \times H_m + 0.2 \times K_1 \times \sqrt{D^{2.5}}) \tag{32.3.21}$$

This estimate includes all power requirements for driving the wheel, for the conveyors of the digging and discharge booms, for lifting these booms, and for slewing and moving the machine.

Power supply for the machine takes place by trailing cables from the fixed point. Cable reels are mounted on the machines or on separate vehicles, which accompany them.

The machine with high output shall have special equipment for communications and all machines must have special devices to prevent damage from the placement of the bucket wheel on the ground when hoisting cable failure takes place.

As for automation purposes, the following devices are useful for operation of the machines:

Programmed automatic adjustment of slewing speed
Automatic initiation and programmed distance of advance for setting the slice depth
Automatic limiting of the slewing angles and automatic reversal of the slewing direction

32.3.6 Operator's Cab

The operator's cab can be in permanent position, or it can be raised and lowered together with the bucket wheel boom (special suspension system) or separately.

32.3.7 Lubrication Technology

In BWEs and BWRs the following types of lubrication systems are usually used:

Force feed
Splash
Stand-oil
Pressurized grease

32.3.8 Specific Equipment Characteristics

Three characteristics are usually taken to evaluate the capability and efficiency of using BWEs and BWRs:

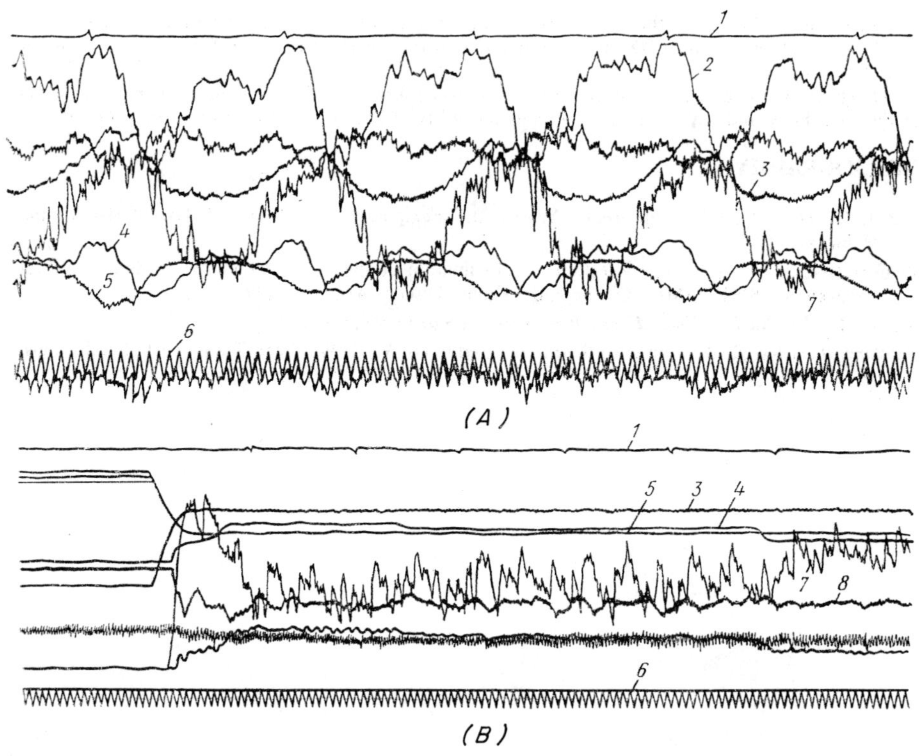

Fig. 32.3.24 Typical oscillograms of BWE when the clay (a) and sandy loam (b) are developed. 1—rpm recorder; 2—tangential speed of the bucket due to rotation of the wheel; 4 and 5—voltages of slewing and bucket wheel motors; 6—time recorder; 7 and 8—the currents of the bucket wheel and slewing motors.

1. Ratio of the weight of the service machine to its 1000 m³/hr output; for well-designed excavators this ratio should be equal to 350–400; for BWRs, 150–200.
2. Ratio of the total power installed to the output of the machine. This ratio should be equal to 0.35–0.45; for reclaimers, 0.2–0.3.
3. Ratio of the total power installed to the weight of the machine. This ratio should be equal to 1.05–1.3 depending on the digging forces developed by the excavator; for reclaimers, 0.8–1.0.

32.3.9 Estimating Equipment Parameters

The earlier equations for preliminary estimation of basic parameters of bucket wheel excavators such as diameter of the wheel, the total capacity of the buckets, the length of the wheel boom, the total weight of the machine and the total power required as a function of the capacity are given in a diagram form in Fig. 32.3.23. The weight of the machine and the total power required are shown for two values of the digging forces per unit area: 2 kg/cm² and 5 kg/cm².

The technical data of some existing BWEs are shown in Table 32.3.7 and some BWE and BWR installations in Tables 32.3.8 and 32.3.9.

32.3.10 Equipment Testing

The basic method of testing of the principal elements of these machines is the strain gauge measurements. Figure 32.3.24 shows the typical oscillograms taped when the strong clay and sandy loam were developed by BWE.

32.3.11 Design Trends

It is not surprising that industrialized countries have recognized the advantages of the BWEs and BWRs and are trying to develop and modernize the existing methods of operation, design, and construction of those machines. At the present time, manufacturers can offer a wide range of BWEs and BWRs from 100 m^3/hr to 20,000 m^3/hr outputs, with the length of the bucket wheel boom from 10 to 100 m.

Designers and manufacturers now are working in the direction of unification of the basic elements of these machines and make the economy and versatility of bucket wheel equipment even better.

BIBLIOGRAPHY

Aiken, G., *Open Pit and Strip-Mining System and Equipment, Continuous Methods,* SME Mining Handbook, 1977.

Pomerantsev, A., "Preliminary Selection and Estimation of the Basic Parameters of Bucket Wheel Excavators," SME-AIME Meeting and Exhibit, Tuscon, Arizona, 1979.

Rasper, L., *The Bucket Wheel Excavators,* Trans. Tech. Publications, 1975.

Stacking, Blending, Reclaiming of Bulk Materials, edited by R. H. Wöhlbier, Trans. Tech. Publications, 1977.

PART IV

TRANSPORTATION INTERFACE

CHAPTER 33
RAIL AND WATER TRANSPORTATION

WILLIAM E. SABINA

W.E. Sabina, Inc.
Materials Handling Consultant
Denver, CO

JOHN F. OYLER

Dravo Engineers and Constructors
Pittsburgh, Pennsylvania

33.1 RAIL INTERFACE

William E. Sabina

The discussion in this chapter on the rail and water transportation interface is intended as a guide in selecting and sizing equipment which provides the handling transition between the two modes of transport and their associated storage systems.

In the case of water transportation the effect of the inland waterway's locks and dams is discussed as it relates to barge and towboat fleet sizes.

The interface between rail transport of bulk materials and their loading and unloading facilities requires the comparison of numerous variables in train loading, unloading, and positioning equipment available to comprise a terminal system. The major factors to consider in the design of the interfaces are:

1. Annual tonnage to be handled
2. Climate
3. Soil properties and groundwater considerations
4. Handling properties of the bulk material
5. Moisture content of the bulk material
6. Ownership of the rail cars
7. Demurrage charges and free unloading time

If the annual tonnage approaches one million tons, railroads will usually give substantial decreases in hauling rates if the "unit train" concept is employed. This concept basically involves a train set or number of train sets which are dedicated to one service from one load point to one user.

However, rail loading and unloading terminals vary widely in configuration due to the large number of variables affecting each installation.

This text addresses itself to the variables and the important aspects of their comparison.

33.1.1 Rail Car Loading

Design of rail car loading systems is heavily influenced by the annual tonnage to be loaded during a given time period and the free loading time permitted by the railroad serving the facility.

A loading system can be as simple as a front-end loader scooping material from a surge pile adjacent to the track and discharging it into the car or cars on a spur or siding track. A second approach would be to have a storage hopper positioned over the track whose capacity is determined by the number and size of cars to be loaded and the capacity of the system which replenishes the hopper.

As an example of the second approach, assume that ten 100-ton (90.7-tonne) cars were to be loaded in 2 hr. A simple approach would be to provide a 1000-ton (907-tonne) hopper over the track, which would then load all 10 cars from one charge of the hopper. However, since a hopper recharging system is required in any case, a smaller hopper could be provided to load the cars in the same time period provided that the recharge rate is established accordingly.

A simple equation will provide the hopper size and recharge rate required for a given load-out quantity and desired loading time. The equation would be as follows:

$$H = nC - R_1 t_1 \qquad (33.1.1)$$

where: n = number of cars
C = capacity per car, in ton/car
H = hopper capacity, in tons
R_1 = hopper recharge rate, in ton/hr
t_1 = allowed load-out time, in hr

Using the equation, it is quite simple to vary the hopper size and calculate the recharge rate required to load a given consignment of cars in the allowed time. Hopper size and recharge rate can be varied in a given circumstance to find the most economical combination of equipment for the installation.

It is necessary of course to examine the system from a base point of the lowest recharge rate which is based on the minimum time between trains in which case the hopper size can be represented by the following expression:

$$R_1(\text{min.}) = \frac{nC}{t_1 + t_2} \tag{33 1.2}$$

where t_2 = time between trains, in hr.

In high-capacity loading terminals where unit trains of 100 cars or more containing 100 tons (90.7 tonne) each are to be loaded in 2 hr or less, the storage hopper capacity increases to be equal to the train capacity, or more if consecutive trains are to be loaded head to tail. Such hoppers take the form of large slip-form concrete silos, 70 ft (21.3 m) or more in diameter, with heights approaching 200 ft (61 m). Silos with dimensions such as this are normally found at large western United States coal mines which require the capacity to load five or six 10,000-ton (9070-tonne) trains per day from a single silo. In extreme cases such as this, the silo recharge rate can reach as high as 10,000 ton/hr (9070 tonne/hr).

The fact that train arrivals can be extremely erratic makes precise analysis of a loading terminal impossible. Therefore, judgment must be used to maximize return on investment for a given facility, because it is not economical to design a facility for "worst case" train arrivals and not logical to design for the ideal situation of trains arriving evenly spaced in time.

Types of Loaders

Bulk loading of rail cars can be accomplished by the following methods:

1. Front-end loader
2. Mobile elevating conveyor
3. Fixed conveyor from storage
4. Hopper mounted above the car which is fed by fixed conveyors
5. Compound storage hopper with batch weigh hoppers beneath
6. Storage silo over the track with direct flood loading of trains
7. Conical storage pile with underpass track for flood loading of trains

Modern designs are being strongly influenced by the railroad companies which are requiring that the 263,000 lb (119,296 kg) gross car weight limit not be exceeded. Therefore, in the design of any new loading system, consideration must be given to some manner of weighing function. Modern, "pitless" track scales can be provided at relatively low cost with sufficient accuracy even for commercial purposes.

The first three basic loading configurations are for relatively low annual throughput installations and can be equipped with a low-cost track scale to control the gross load. Item three (3) could be equipped with a very low-cost belt scale to measure the weight of material loaded, which can easily be added to the tare weight of the car, shown on the car body, to limit the gross load of the car. One problem with this method is that it does not account for material left in the car due, for example, to freezing problems at the unloading point. For absolute accuracy a track scale would have to be added to obtain the actual weight of the car entering the load zone.

The next level of sophistication deals with somewhat higher annual through-load terminals which load in fairly large numbers of cars, say 30 to 40, in each train. As discussed earlier, instead of providing a 4000-ton (3628-tonne) hopper to load the 40 100-ton (90.7-tonne) cars in say 2 hr, better economics might dictate a 2000-ton (1814-tonne) hopper with a recharge rate of 1000 ton/hr (907 tonne/hr). However, if we were to examine minimum recharge rate requirements per (33.1.2) on the basis of, for example, t_2 equal to 24 hr between trains, the minimum recharge rate of 154 ton/hr (140 tonne/hr) with a hopper size of 3692 tons (3349 tonne) could be the obvious choice for that situation. In any case, the equations permit examining each installation and quickly narrowing down the choices.

Items 5, 6, and 7 are all variations of approach to minimum total costs for unit train loading applications. The goal is to load a unit train in the amount of time allowed by the railroad company without exceeding their gross load limitations. The main issue to be dealt with here is economics. There is no point in providing loading rates that exceed the rate beyond which no economic gain is realized. If the railroad company allows 4 hr free time for loading, then the system should be designed to use most of that time with perhaps 30 min saved for positioning the train, clearing the train after loading, and minor mishaps which might delay loading. If however, a terminal has a large number of trains to load per day it might be well to reduce the loading time well below the allotted time to minimize the problem of trains being backed up causing congestion at the loading site; but again, only if the terminal is contractually penalized for the delays or benefitted for preventing them. The flood load chutes and virtually instantaneously acting shut-off gates make practical rates up to 10,000 ton/hr (9070 tonne/hr) permitting a 100-car, 100 tons (90.7 tonne) per car, unit train to be loaded in 1 hr. It is also difficult for the terminal to save labor costs since round-the-clock loading capability will be insisted upon in any case by the railroad company.

Weighing the unit trains accurately as they are flood loaded in motion is a rather difficult task requiring as many as three track scales end to end properly interconnected with control logic to determine the car loading. It is not simply a matter of determining the final gross load, but a problem of loading the car as close to the limit allowed by the railroad company without exceeding it. Flood loading accuracies have been controlled within ± 1% of gross weight, which provides a satisfactory basis for system design.

Another flood loading approach is the weigh bin wherein one weigh bin is loaded from a hopper above while the second is discharging its entire predetermined contents into the moving car below. The weigh hopper's contents can be controlled within ± 100 lb (45 kg) by mounting them on load cells which dictate the closing of the gate from the hopper above. By utilizing an in-motion track scale ahead of the load zone, accurate tare weights can be obtained from each car permitting the appropriate weigh hopper to be filled with the exact quantity of material allowed for each car to be loaded precisely to its gross load limit.

The incentive for such great maximum load accuracy is the penalty imposed by the railroad companies for removing and unloading overweight cars before permitting them on their track system. Since commercially accurate scales are required for the basis of payment for the commodity in any case, the addition of maximum load control is not of great consequence.

While many modern terminals accommodate trains whose cars are all the same dimensions, the flood loading in motion principle is also applicable to random trains that have cars of varying sizes. The loading rates for such trains do not reach the 10,000 ton/hr (9070 tonne/hr) of unit trains but can be designed for rates up to 5000 ton/hr (4535 tonne/hr) or more depending upon the magnitude of the variations in car dimensions.

Train Positioning

Most modern train positioning is provided by locomotives, especially at high-capacity installations. Loading in motion using locomotives requires a special speed control device to obtain speed stability which main-line locomotives do not normally have at speeds below 1 mi/hr. However, the speed control is available and the railroad company serving the facility can utilize the special speed controls and obtain the desired speed.

Car positioning at small-capacity loading terminals can also be accomplished by locomotives but other devices are available for the purpose. Alternate devices for consideration are as follows:

1. Trackmobile
2. Rope-operated train positioner
3. Car puller winches

These car positioning devices are available in a variety of sizes and prices and their selection should be the result of a comparative cost study for machines that fit the application. Prices and capacities for such equipment are available from the various manufacturers.

33.1.2 Rail Car Unloading

Unloading bulk materials from rail cars obviously is influenced by the type of car chosen for the situation. Open-top cars can be inverted by a variety of car dumpers whose selection will be discussed later.

The most commonly available car supplied by the railroads is the old-style hopper bottom which is available in a range of capacities up to 100 net tons (90.7 tonne) and can be discharged through the bottom doors or unloaded by inverting the car if the material is not free flowing. A variation of the hopper-bottom car has been developed in recent years and is referred to as the "rapid-discharge" car. This car has doors in the bottom which provide a much larger opening of the bottom than the conventional hopper bottom car along with steeper slope sheets in the car ends. The total design provides a car which, under ideal conditions, can discharge the contents of the car in as little as 12 sec.

Another unloading method, recently developed, is the side-dump car. This style of car is equipped with large air cylinders that tilt the car bottom sideways while simultaneously lowering the side of the car on the discharge side. The final tilted position provides a single smooth sliding surface comprised of the bottom and side of the car.

An old style of car unloading, the tilt table unloader, is used to discharge grain out the side door of a boxcar. The tilt table tilts endwise and sidewise in a pattern that permits discharging the grain through the side door to a hopper below. More modern practice in grain handling is to utilize covered hopper-bottom cars which are loaded through openings in the roof of the car and unloaded through the bottom, a method which is more economical than the tilt table.

Car Dumpers

Figure 33.1.1 depicts, in simplified form, the various styles of car dumpers available when unloading by inverting the car.

Of the five methods shown, random rotary (Figs. 33.1.1*A* and 33.1.2) are the most economical device where random or variable size cars are to be unloaded. The center of rotation of this style of dumper is, ideally, a few feet above the floor of the car which requires the least rotating horsepower, the smallest ring diameter, and the shallowest hopper pit.

The second style of rotary dumper, Fig. 33.1.1*B*, is utilized in unit train applications wherein the cars are unloaded without being uncoupled from adjacent cars. This style of dumper requires that the center of rotation be on the centerline of the couplers. Each car is equipped with a rigid (nonrotating) coupling on one end and a swivel coupling on the other. As can be seen, the ring diameter for this style of rotary dumping is larger than for the random style. Also, because the live load is more eccentric to the center of rotation, the rotating horsepower is greater. Finally, the larger end rings require a deeper, and therefore more expensive, hopper pit. The economic advantages of the unit train dumper lie with the reduced freight rates offered by most railroads for this type of dedicated train service and the reduced manpower requirements versus any unloading method that requires uncoupling the cars to dump them. The unit train installation, properly designed, requires one person to unload the train, whereas any uncoupling–recoupling operation requires a minimum of two people, and the probability of three or more, depending on ancillary equipment and total system design. In any case, a careful analysis of manpower requirements is mandatory in arriving at the most cost effective system design. The cost analysis should be done on a "present worth" or equivalent basis since it is otherwise impossible to compare systems where one might be capital intensive while the other is labor intensive.

The tippler style of dumper, Fig. 33.1.1*C*, is most often selected where site groundwater is shallow, making pit excavation disproportionately expensive. Simple observation reveals that the material being unloaded is simultaneously lifted and tilted and therefore requires a more massive dumper design and higher horsepower. A variation of the tippler dumper is currently under development which involves the use of a rotary dumper style of structure but has the centerline of the car offset laterally from the center of rotation of the dumper. Since this design is only a variation of structural configuration, its advantages will have to result from economics of manufacture.

The rollover dumper, Fig. 33.1.1*E*, is a seldom-used variation of rotary dumping. The design of this style of dumper can vary widely depending on the diameter of the end rings and the slope of the ramp. With small-diameter rings and a relatively steep ramp the receiving hopper could be kept small and close to the surface approximating the advantages of the tippler dumper.

Another somewhat outmoded style of dumper is shown called the high-lift dumper (Fig. 33.1.1*D*). Again, its variations are virtually endless, limited only by economics. As can be seen, the receiving hopper can be placed entirely above ground level, greatly minimizing civil costs from that aspect. However, the dumper itself is massive and expensive due to lifting the entire loaded car, due also to large counterweights which are employed to minimize dumping horsepower, and, finally, due to the great weight of the pan which must, at its receiving end, be the full length of the car. So in total, the high-lift dumper is heavy, expensive, and quite complex, requiring considerable attention to maintenance. An advantage of the high-lift dumper that merits attention is that it can minimize degradation of the product by limiting the number of times the material is dropped in the handling system. As an example, these dumpers (Fig. 33.1.3) have been utilized to discharge coal directly from cars into the holds of a ship. Many examples can be seen at Great Lakes or Eastern Seaboard ports. In general, however, it must be concluded that the high-lift style of dumper has been outmoded by lighter, more economical dumper systems.

Receiving Hoppers and Product Removal

In most every case, a car unloading system requires a receiving hopper to allow for the surge requirements of relatively instantaneous unloading of large volumes of product. This is true with the exception of some "rapid discharge" bottom dump installations, which drop the coal into a storage yard area beneath a trestle, or the side dump applications, which basically unload over a length of track to permit broadcasting solid wastes at a disposal site.

The term *hopper,* as used in the case of car unloading, is a misnomer since the word hopper implies storage and a car dump hopper should not be sized on that premise. Rather, a car dump hopper is, in reality, a catch bin whose length is defined by the car length, whose width is determined by the width of the falling stream of material and whose depth is a function of the valley angle between the sloping sides appropriate for successful bulk flow out of the container for the material(s) in question. The only other volumetric consideration to be applied to the container is that it supply

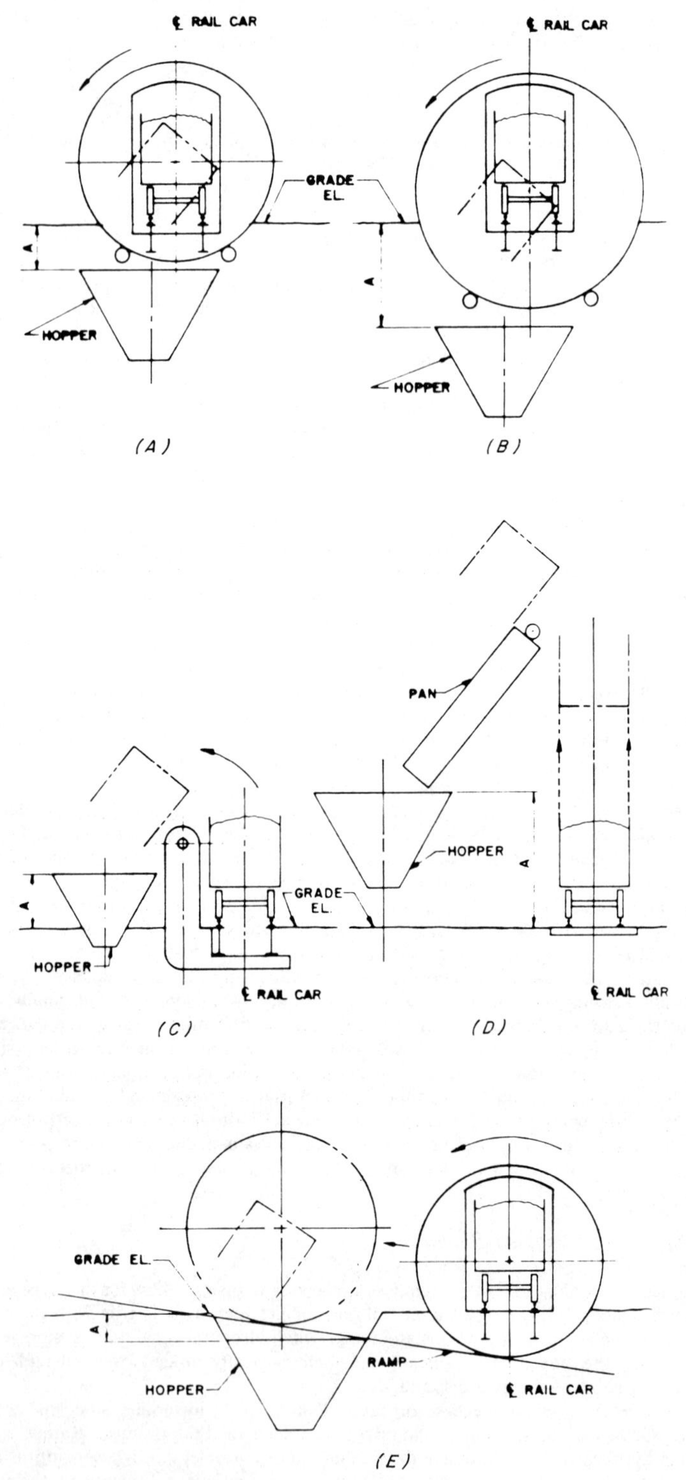

Fig. 33.1.1 Schematic illustration of car dumper styles showing their relative hopper elevation with respect to grade elevation. (*A*), random rotary; (*B*), unit train rotary; (*C*), tippler; (*D*), high lift; (*E*), rollover.

Fig. 33.1.2 Typical rotary dumper with car inverted near final position.

Fig. 33.1.3 High-lift dumper discharging directly into ship holds to minimize degradation.

sufficient surge capacity between the intermittent discharges of the cars and the continuous withdrawal from the hopper by the feeder/conveyor system. A capacity of approximately $1\frac{1}{2}$ times the largest car volume normally satisfies the function of this surge hopper, provided the feeder/conveyor takeaway rate matches the car dumping rate. To size a dumper-receiving hopper for significantly more capacity than this is patently incorrect since material storage capacity is always provided elsewhere at the unloading site in a much more economical manner than underground "storage."

As previously stated, the subject of hopper sizing is appropriate to all car unloading installations with the exception of the rapid-discharge and side-dump applications.

Rapid discharge, or, as it is sometimes referred to, unloading-in-motion, is a particular style of bottom dumping from a continuously moving train passing across a trestle that is 30–40 ft above ground level. The doors of the cars are triggered by a low-voltage electrical signal from a segmented "hot" rail along the trestle which is contacted by a pickup shoe on each car. The electrical impulse triggers a solenoid-operated air valve which permits air to flow to the pneumatic door opening system. Sections of the "hot" rail are selectively energized establishing the zone in which the cars are discharged. This zone dumping permits discharging down the face of the initial pile, thus minimizing the distance the material must fall and, therefore, the amount of dust generated. The length of the trestle is of course a function of the total train capacity or the capacity of two or three trainloads if such number can arrive within a short period. Dust generation from this trestle style of dumping is a major factor to consider when evaluating the unloading-in-motion system. Deluge dust suppression has been employed in an attempt to control the fugitive dust, and some proposals have been made to enclose the trestle below the rails for the same purpose. Since this style of dumping is unique it deserves consideration due to its potential benefits. Those benefits are, minimum turnaround time for the train, which can reduce freight rates, and minimum operating personnel, provided the reclaim system is designed to eliminate the need for mobile equipment.

Conventional bottom-dump hoppers can be designed in the normal manner, bearing in mind that the normal discharge rate can be quite slow due to the 45° slope sheets and the small doors that must be opened manually. Complete discharge can take 2 or 3 minutes and even then with the help of a device to shake the cars. Side dump cars could obviously be discharged into a conventional hopper. However, until the present their main application has been to discharge solid wastes over a long stretch of slightly elevated tracks at waste disposal sites. The dumped material is then removed and spread over the disposal site by mobile equipment.

Train Positioning

Each car-unloading installation must be provided with a method of moving the empty cars away from the dumping point and the loaded cars into the dumping position. Again, annual throughput will guide this analysis. The car or train positioning method is largely dictated by the number of cars that must be moved simultaneously. Methods that require consideration are:

1. Gravity
2. Car-puller winches and ropes
3. Trackmobiles
4. Shifter engines
5. Automatic train positioners
6. Hyraulic cylinder-actuated train positioner

Gravity feed and removal systems are used in conjunction with a variety of ancillary equipment to control the velocity of the cars at various locations along the track system. Of course, the natural terrain must favor gravity feed since to create the necessary slopes artifically would be prohibitively expensive. However, since the slope needs to be only slightly greater than the rolling resistance of the cars, or in the range of $\frac{1}{4}$ to $\frac{1}{2}$%, gravity feed should be given consideration where such slopes are naturally available.

Small electrically powered rope winches with rope reeving and connecting tackle can be used to position a number of loaded cars in addition to pushing the empty cars out of the dump zone. The number of cars handled will obviously vary depending on the pulling power of the winch, slopes and curves of the track, and size of the cars; but this style of car positioning would normally not exceed the handling of 10 to 15 cars.

Another device for installations handling a small number of cars is the trackmobile. This device is a small diesel or gasoline engine-driven cart which is normally used to feed rail cars into the car dumper in situations where feed rates of relatively few cars per hour are required. It is a uniquely versatile machine which has two sets of wheels, one set enabling it to run on the rails, the other off the rails. In low-capacity operations, the operator of the trackmobile can insert loaded cars into the dumper, dump the load, eject the empty car from the dumper coupling it to previously processed cars and return alongside the track rails to a position behind the next loaded car to begin the next cycle. The process or cycle can be improved by the addition of a separate dumper operator who can provide various other manual functions such as uncoupling the loaded cars.

Higher-capacity installations require the use of a locomotive which can be either a main-line locomotive or a smaller shifter engine as the need dictates. A discussion will be required with the railroad serving the facility in question to determine whether or not a locomotive needs to be purchased for the operation. In any case, this feed device working in harmony with the dumper operator plus yard crew provides a car feeding and removal system that can achieve the next plateau of feed rates. This operation generally involves four workers; the locomotive operator, the dumper operator, and one person at each of the entry and exit ends of the dumper to pull coupling pins, align couplers after dumping, signal the locomotive operator, and so on. The locomotive operator can be eliminated by using a radio-controlled locomotive operated by the dumper operator. This method is normally restricted to dumping coupled cars with swivel couplings. The upper limit feed rate of such an operation is greatly influenced by the number of operations required of the locomotive which is encumbered by a group or "cut" of cars consisting of 15 to 20 cars weighing as much as 2600 gross tons. Burdened with such a load, the locomotive must:

1. Accelerate the "cut" of cars
2. Push the empty car out of, and clear of, the dumper
3. Decelerate the cut
4. Accelerate the cut in the reverse direction
5. Decelerate the cut, spotting the next loaded car on the dumper
6. Accelerate the cut again in reverse to clear the dumper

The locomotive function can be improved by the addition of a device to insert the loaded car, another to eject the empty car, and a car retarder in the dump zone to stop the loaded car. This combination frees the locomotive to simply spot the next loaded car adjacent to the dump zone during the unloading cycle. The locomotive in this case becomes simply a start/stop indexer and the dumping cycle is reduced.

One of the most modern train positioning devices, used exclusively for unit trains, is the automatic train positioner (Fig. 33.1.4). The use of this type feed device should be limited to the handling of swivel-coupled unit trains. Such cars are, in a given circumstance, identical to each other and therefore permit fully automatic positioning and dumping without uncoupling the cars.

As with any device, a variety of forms are available. Functionally, however, the device utilizes a mechanically propelled, carriage-mounted arm that travels back and forth alongside the rail car tracks on a parallel track system. The arm is automatically entered between cars either gripping the engaged coupling or entering between a coupling half and the car striker. After such engagement, the arm

Fig. 33.1.4 Automatic trail positioner mechanism with propulsion arm disengaged.

and carriage is propelled forward until the next loaded car, or group of cars, is indexed into the dumper, the entire train being thus indexed one or more car lengths. At this time, a train-holding device is engaged to hold the train in position while the pusher arm exits from between cars, returns one car length and reengages in preparation for the next cycle. During the arm disengage, return, and reengage period, the dumper rotates, emptying the car on the dumper, and returns to its seated position in preparation for the next train-indexing stroke.

The subject of positioner location does not appear overly important and, in fact, both entry- and exit-end positioner locations have been employed and are operating. However, a close examination will reveal some important differences that must be considered.

The three areas to be considered are:

1. Operator location
2. Equipment operation
3. Costs

Operator location is extremely important to this equipment in order to promote efficient train handling and to minimize the possibilities of accidents. In addition are the elements which an operator should see with an attempt to place these elements in the order of their importance:

1. Rotary dump car clamps
2. Train wheel locks
3. Car paint strips (indicating swivel- or rigid-coupling half)
4. Positioner carriage
5. Dump hoppers

If the operator has as good a view as possible of all these elements, operator location will be optimized.

Items 1 and 5 can be viewed equally well from either end of the dumper. Items 2 and 3 can only be viewed from an entry end operation position. Item 4 is best viewed by locating the operator at whichever end the positioner is located. Reexamination of this list makes it obvious that, from a standpoint of operator location, the train positioner should be located at the entry end of the dumper.

Equipment operation, the second area of consideration, needs to be reviewed for:

1. Complexity of equipment design
2. Train turnaround time (assuming equal dump-position cycle time)
3. Availability of caboose

There are a few differences in design consideration between entry and exit end location of the positioner. The exit end location requires a photoelectric (or other) device to sense location of the first loaded car in front of the dumper in order to obtain precise stopping location control of the positioner arm at variable locations. The variable stopping location of the arm is necessitated by the presence of seven or eight draft gears (springs) between the positioner arm stopped locations and the wheel locks at the far end of the dumper. The entry-end positioner location avoids that problem since there is not more than one draft gear between the positioner arm and the wheel locks when the arm reaches its stopped position.

Regarding train turnaround time, again, the entry-end positioner location is preferable provided the train is equipped with a caboose for positioning the last loaded car on the dumper. Assuming the caboose is available, the entry location positioner requires initial manual engagement of the positioner arm with the train plus manual spotting of the last car using the positioner. The exit-end positioner requires locomotive positioning of the first two cars at the head end of train followed by initial manual engagement of the positioner arm with the train. The net time difference for the entry-end positioner versus the exit type is small, but still favors the entry-end location.

The final equipment operation category, availability of caboose, is perhaps the most critical. The entry-end positioner location requires a caboose for the positioner to spot the last loaded car on the dumper. The caboose need not be the same length as the unit train cars, but can be any standard caboose. Positioning the last car by use of the locomotives is virtually out of the question because the distance between the locomotives and the last car all but prevents the necessary spotting accuracy. The exit-end positioner, on the other hand, spots the last car the same as all other cars and does not require a caboose. So the exit-end positioner has the advantage in spotting the last car for dumping provided that the train has no caboose.

Costs, the last item of consideration, will favor the entry-end positioner. Although the actual positioner and dumper costs will be equal for either entry or exit design, the total installation will be higher for the exit design. This is mainly due to configuration of the operator's cab or control room.

With the positioner at the exit end, there is no location where the operator can see all the functions he should see without the assistance of closed-circuit television or an assistant operator. The operator's cab or control room in this case is usually much larger, hence more expensive, than that required for an entry-end positioner.

The final conclusion must be drawn that entry-end location of an automatic train positioner is superior in virtually every way to exit-end location. The only situation that is in favor of exit-end location is that where either the train does not have a caboose or the caboose design precludes its use in positioning the last car.

Positioners have been installed that are equipped to handle uncoupled random cars and swivel-coupled unit-train cars at the same dumper installation. A careful analysis of random car handling, using a positioner, reveals a very time-consuming labor-intensive operation. In such unusual cases where both types of cars are actually involved, economics will favor the use of a locomotive to process the random cars.

The train-holding device, previously mentioned, merits a separate discussion. To date, this device consists of chocks or wedges which are mechanically entered in front of and behind the wheels of one axle or one truck assembly of a loaded car on the entry side of the dumper. Due to locomotive clearance requirements, chock or wedge height is restricted. This restriction directly affects the holding power of this device. Since the holding power is limited, it is necessary to pay particular attention to track slopes and positioner carriage deceleration rate. The force on the holding device is in general the sum of (1) the gravity component of the train weight, which is in direct relation to track slope, and (2) dynamic forces induced by deceleration of loaded cars which are still in motion after the pusher arm has disengaged for its return stroke.

A current development in train-holding devices involves the use of a fixed-location holding arm engaging the coupler or striker. The holding arm is nearly identical to the carriage-mounted arm which pushes the train. Where anticipated holding forces are high (above approximately 40,000 lb; 18,144 kg), the only practical means of gripping the car is at its location of greatest strength: the coupler or striker. One area of caution needs mentioning; an inrun track slope in excess of 0.25% can cause serious design and/or operating problems.

Dust Control

Environmental regulations require that careful attention be given to fugitive dust emissions when designing an unloading system.

The random and unit train rotary dumper can be covered, except for entry and exit doors, when the product handled presents dust problems. In addition to the building, a wet or dry dust control system can be applied; the resulting design can be made to meet fugitive dust regulations. The dry collection system will require approximately 120,000 ft^3/min (56 m^3/sec) air flow and correct sizing and location of pickup hoods for successful operation. The wet-type system differs from conventional wet suppression in that it is designed to provide a moisture cloak over the rising dust cloud, wetting the particles and causing them to fall back into the dump hopper. The dry system has become more popular due to broader application without regard to freezing climate and also to better total economics. Wet systems require constant use of a commercial wetting agent for successful operation and its costs make the present worth cost of the dry system less than the wet system for installations with long life expectancy.

Bottom dumping into a track hopper produces less dust per ton of coal unloaded and the hopper dimensions are smaller permitting more effective use of a dry or wet collection system. Rapid discharge or unloading-in-motion systems discharging from an elevated trestle are all but uncontrollable if the product is fine and/or dry. Future systems of this type should be designed to discharge into a long, covered trench.

The tippler dumper can be provided with a unique building whose only opening is virtually plugged by the body of the car during the period of maximum dust generation. In addition, the building is equipped with a dry dust collection system which helps to provide a highly efficient dust control system for this style of dumping.

Cold Weather Unloading

Unloading rail cars containing a frozen product is, at best, a serious problem. If the bulk material has high surface moisture, below-freezing conditions can greatly reduce, and sometimes stop, the unloading operation.

Assuming that it is not economical to remove the surface moisture by drying the product, the most efficient means of dealing with freezing conditions are suggested as follows:

1. Selection of inverted dumping as opposed to bottom dumping.
2. Car shakers or vibrators to aid flow.

3. Infrared heaters to free the product from the car body.
4. Addition of a freeze inhibitor to the product.
5. Lump breakers to reduce frozen lumps to acceptable size for handling.

Item 1 represents the most critical and difficult decision and ideally requires a thorough knowledge of the material to be handled. Item 2, car shakers, are not strongly recommended but should be thought of as a remedial solution which is only occasionally required. Item 3, infrared heaters, are commonly applied where long periods of below-freezing conditions exist regularly. They usually do not provide a total solution since it is only economical to provide sufficient heat to free the material from the car body and not enough to completely thaw the frozen material. Freeze inhibitors, item 4, come in basically two forms: chemical additives to reduce the strength of the ice which forms or, as in the handling of coal, application of oil which serves a similar function to the chemicals. In any case, the additives must not detrimentally affect the end use of the product being handled. Lump breakers (Fig. 33.1.5), item 5, should be considered wherever frozen lumps are anticipated. Large lumps can plug hopper throats and shut down an entire unloading system. Therefore, if a lump breaker is required it must be located above a grizzley which will pass normal size lumps and support oversized ones on top until they are reduced in size by the lump breaking device.

Ancillary Equipment

The function of indexing and unloading cars is enhanced in most cases by ancillary equipment which helps to provide the total unloading interface function. Most of these items have already been discussed, but a review list is offered for use as a guide in system concept design.

1. Trackmobile
2. Empty-car ejector
3. Loaded-car inserter
4. Car retarders
5. Thawing equipment
6. Car vibrators
7. Frozen lump breakers
8. Weigh scales
9. Sampling equipment

Fig. 33.1.5 Lump breaker on grizzley beneath rotary car dumper.

10. Dust suppression or collection
11. Damage protection devices
    a. Dump and seated position buffers (for invert-type dumping)
    b. Controlled torque coupling at drive motor (for invert-type dumping)
    c. Car position sensors
12. Wheel chocks or holding arms

Since most of the items listed are provided in a wide variety of forms it is advisable to obtain descriptions of each item from the manufacturers when evaluating the quality of each.

Some dumpers are offered with a weigh scale as part of the car platen which can obtain gross and tare weights in the dump position without delaying the unloading operation. The weighing devices are capable of obtaining sufficient accuracy for commercial purposes.

On-board primary cutters are available on some dumper designs which can save valuable height at a later transfer point. It is necessary to state that such devices do not, as yet, meet the ASTM sampling guidelines but have been bias tested and proven to be acceptable for commercial purposes by agreement between the buyer and seller of the bulk commodity in question.

The damage protection category is important. Almost all invert-type dumpers are heavily counterweighted to minimize dumping horsepower. In certain cases, due to inattentive maintenance, dumpers have lost braking ability and progress violently to either the dump or seated position with potential for great damage. An energy-absorption system at each location can minimize or eliminate damage due to the counterweights. The drive motor slip coupling serves to isolate the energy of the motor armature which, due to extremely high gear ratios, could otherwise cause significant damage to the drive train. Car position sensors simply prevent activating a dumping motion with a car or locomotive in a hazardous position. Photoelectric cells are available for this type of industrial application and are widely used for this function.

Locomotive and Caboose Influence

In a large percentage of cases, dumper system designs require that clearance be provided for a locomotive or caboose to pass. Clearance diagrams can be obtained from the railroad company serving the facility for both. It should be clarified here that it is not necessary to provide the large over-the-road clearances for the locomotive and caboose but the smaller in-plant clearance dimensions.* It is significant to note that if a caboose is involved, it will often establish the maximum vertical clearance requirements, whereas the locomotive more often establishes the horizontal or width clearance criteria.

Dump System Economics

Determination of the most cost-effective system for unloading a given quantity of rail cars is profoundly complex and requires absolute objectivity in comparing viable alternatives.

The first determination that must be made is whether or not the material in question can be bottom dumped or if the cars must be inverted for regular successful discharge of the material, for it is easily provable that bottom dumping is, in all ways, the most economical unloading basis. This will not hold true, however, if freezing or poor flow conditions add large unexpected labor and material costs to the unloading operation. If bottom dumping the material is recommended then it must be decided whether to use railroad-supplied hopper-bottom cars or privately owned rapid-discharge cars. The rapid-discharge cars would have to be purchased for the application since railroads do not as yet supply such cars.

The subject of car ownership is not a topic relevant to this chapter but bears mentioning since it can strongly influence the selection of the unloading equipment. For example, if the railroad company will supply only standard hopper-bottom cars, an invert type of unloading system may be required although bottom-dumping or rapid-discharge cars were otherwise a viable system.

If bottom dumping is not deemed advisable and an invert dumping method is required, then the particular invert style must be decided upon.

If the installation is served by unit train, the unit train style of dumper should receive first consideration. To utilize any other style of dumper requires uncoupling and recoupling the cars in addition to reconnecting air hoses. Labor costs plus demurrage charges from the railroad company will certainly offset any apparent savings in initial capital investment.

If the installation is not served by unit trains then the random rotary and the tippler dumper should be considered. The random rotary style equipment will be cheaper but the associated civil work could offset the savings. Comparative concept designs and pricing, including civil work, must be studied to make the final determination.

---

* *The Carbuilders Encyclopedia* contains a great deal of useful information of use to a dump system designer.

BIBLIOGRAPHY

Bulk Solids Handling—The International Journal of Storing and Handling Bulk Materials, Trans Tech Publications (Federal Republic of Germany).

Copeland, C. T., "Operational Evaluation of Freeze Conditioning Agents," *Coal Technology '80 Proceedings,* Vol. 3, Coal Technology, Nov. 1980.

Sabina, W. E., "Rotary Car Dumper Systems," *Journal of Engineering for Industry,* ASME Publications, Feb. 1979.

Ziegelmiller, W., "Cold Weather Unloading," *Coal Technology '80 Proceedings,* Vol. 3, Coal Technology, Nov. 1980.

33.2 WATER INTERFACE

John F. Oyler

33.2.1 General Categories

The transportation of bulk solids on water in North America can be subdivided into four general categories—inland waterways, the Great Lakes, coastal shipping, and transoceanic. Although the same general types of equipment and material handling technology apply to all four categories, it is constructive to discuss each one separately and to define the characteristics unique to each.

33.2.2 Inland Waterways

The canalization of the Mississippi River and its major tributaries has produced one of the most significant bulk materials transportation systems in the world utilizing push towing. In reality, the upper portion of the system is a series of essentially quiescent pools, bounded by large dams and joined by modern locks. The lower portion of the Mississippi is open river. The dimensions of the locks dictate the design of the fleet of towboats and barges which are the backbone of the system. Table 33.2.1 lists the major locks in the Mississippi system and provides sufficient dimensional information to understand the sizing of barge tows.

For an example of the interdependence between lock size and tow size, consider the standard "large" lock on the Ohio River, which is nominally 110 ft wide by 1200 ft long, like the one shown in Fig. 33.2.1. Such a lock is capable of transferring a large towboat pushing a tow of 15 standard hopper barges from one pool to another without "breaking up" the tow. The standard 1500 net ton capacity barge is nominally 35 ft wide by 195 ft long. A tow of 15 barges lashed together—three abreast and five long—plus a 200-ft-long towboat practically fills the 110 ft by 1200 ft lock.

The economics of barge transportation depends largely on the time required to make a complete round trip. This involves travel upstream as well as downstream, locking time, and the time required for loading and unloading. Travel time depends on the horsepower of the towboat and the size of the tow. It may range from 5 mi/hr to 15 mi/hr. A 4200-Hp towboat with a tow of fifteen 1500-net-ton-capacity barges will travel 11 mi/hr in still water and can be "locked through" a large lock in 2 hr. Loading and unloading times are a function of the equipment used and will be discussed separately.

33.2.3 Barges

In order to understand the types and sizes of river barges normally in use, it is necessary that a distinction be made between dedicated and common-carrier tow service. In situations where there is a long-term commitment to haul a specific cargo in large quantities between two specific points, economics favors dedicating a fleet of towboats and barges to this service and custom designing them to take full advantage of the unique characteristics of the particular service. The net result is the dedicated (unit) tow which normally consists of a group of barges that always are lashed together in the same orientation. In contrast with this is the common-carrier service in which each barge is an independent unit which may be located at any position in a randomly positioned tow. This discussion begins with the so-called standard barges used in common-carrier service and then treats custom-designed barges for dedicated unit tows.

Through the first half of the twentieth century the standard hopper barge was 175 ft long by 26 ft wide and about 10 ft deep. It was designed for compatibility with the locks on the Allegheny and Monongahela Rivers, carried about 900 net tons of cargo, and had two rake (curved hull) ends. The early wooden barges were replaced by riveted and finally welded steel barges.

Since the early 1950s, a number of variations have been designed into the barge fleet, many of which must now be considered as standard construction. One variation has been the development of barges with vertical hopper sides, for the coal hauling service, to maximize automated unloading and cleanup. Wider (35 ft) and longer (195 ft and 200 ft) barges have been designed for use on waterways

Table 33.2.1 Ohio River System Locks

Mile	Designation	Dimensions (ft) Large Lock	Small Lock
	Ohio River		
6.2	Emsworth	110 × 600	56 × 360
13.3	Dashields	110 × 600	56 × 360
31.7	Montgomery	110 × 600	56 × 360
54.4	New Cumberland	110 × 1200	110 × 600
84.2	Pike Island	110 × 1200	110 × 600
126.4	Hannibal	110 × 1200	110 × 600
161.7	Willow Island	110 × 1200	110 × 600
201.0	Belleville	110 × 1200	110 × 600
237.5	Racine	110 × 1200	110 × 600
279.2	Gallipolis	110 × 600	110 × 360
341.0	Greenup	110 × 1200	110 × 600
436.2	Meldahl	110 × 1200	110 × 600
531.5	Markland	110 × 1200	110 × 600
604.4	McAlpine	110 × 1200	110 × 600
720.7	Cannelton	110 × 1200	110 × 600
776.1	Newburgh	110 × 1200	110 × 600
846.0	Uniontown	110 × 1200	110 × 600
918.5	Smithland (completed in 1980)	110 × 1200	110 × 1200
938.9	Lock 52	110 × 1200	110 × 600
962.6	Lock 53	110 × 1200	110 × 600
	Monongahela River		
11.0	Lock #2	110 × 720	56 × 360
23.1	Lock #3	56 × 720	56 × 360
42.3	Lock #4	56 × 720	56 × 360
61.7	Maxwell	84 × 720	84 × 720
84.1	Lock #7	56 × 360	None
91.3	Point Marion	56 × 360	None
102.1	Morgantown	84 × 600	None
108.3	Hildebrand	84 × 600	None
115.4	Opekiska	84 × 600	None
	Allegheny River		
7.2	Lock #2	56 × 360	None
15.3	Lock #3	56 × 360	None
24.4	Lock #4	56 × 360	None
30.1	Lock #5	56 × 360	None
36.3	Lock #6	56 × 360	None
46.7	Lock #7	56 × 360	None
53.2	Lock #8	56 × 360	None
62.4	Lock #9	56 × 360	None

Fig. 33.2.1 Typical Ohio River lock.

with larger, modern locks. Barges with one (or both) square ends, instead of rake ends, were developed to maximize payload for a given length barge located within the body of a tow. Rolling and lift-off covers have been provided for service in which weather protection is important. Table 33.2.2 lists the major types of standard barge currently available and their significant characteristics.

Barge capacity, for a specific set of physical dimensions, depends on the unit weight of the cargo being transported and the draft (water depth) available. Table 33.2.3 is provided to illustrate this relationship and the variety of size variations available in one barge type (195 ft by 35 ft, single rake end).

Because of the availability of different lengths and end configurations, it is possible to combine "standard" barges into a very efficient integrated tow arrangement. The advantage of this approach is significant. For constant towing power and water depth, an increased capacity of 4.8% and towing speed of 10.0% is achievable when a fully integrated (one rake end only on each leading and trailing barge) combination is used for a tow of 15 195 ft by 35 ft barges as contrasted with 15 double-rake-end barges of the same size.

Custom fleet design is also cost effective for the case of the "dedicated" tow when all of the operating constraints are well defined and reasonably limited. An example of this is an integrated tow of tank barges for hauling liquid products from the Gulf Intracoastal Waterway west of New

Table 33.2.2 Barge Data

Barge Type	Cargo	Width (ft)	Length (ft)	Depth (ft)	Rake Ends	Covers	Capacity (tons)
Standard	Coal	26	175	11	2	No	
Rake Stumbo	Coal	26	195	11	1	No	1157
Double Rake							
Stumbo	Coal	26	195	11	2	No	1110
Jumbo	Coal	35	195	12	1	No	1536
Box	Coal	35	200	12	0	No	1676
Jumbo	Bulk	35	195	12/13/14		1	Varies
Box	Bulk	35	200	12/13/14		0	Varies

Table 33.2.3 Additional Barge Data

Hull Depth	12 ft	13 ft	14 ft
Hopper Volume	71,050 ft^3	75,940 ft^3	80,700 ft^3
Draft		Tons Capacity	
8′ 0″	1314	1307	1284
9′ 0″	1522	1515	1491
10′ 0″	1731	1724	1698
11′ 0″	—	1935	1907
12′ 0″	—	—	2117

Orleans throughout the Mississippi River system. The Gulf Intracoastal Waterway locks are 75 ft wide; those on the Mississippi system, 110 ft wide. Optimum barge width consequently is 54 ft, to permit "double string" locking in the Mississippi system and single string locking in the Gulf Intracoastal Waterway. Length of the barges is influenced by three criteria: overall tow length of 1180 ft (including towboat) on the Gulf Intracoastal Waterway, a maximum (American Bureau of Shipping) length-to-depth ratio of about 25, and a practical half tow length of 600 ft to permit double locking in the shorter Mississippi system locks. The result is a tow consisting of two locking modules. The first has a pair of lead barges with single rake ends, each 295 ft long, followed by a pair of long box barges, each 295 ft long. The second locking module has a pair of short box barges 145 ft long, a pair of trailing barges with single rake ends, each 295 ft long, and a 150-ft-long towboat. Figure 33.2.2 is a photograph of such a tow.

33.2.4 Towboats

Variation in the size of towboats in use on the inland waterways is directly related to horsepower required to propel (or restrain against the current) a tow of barges. As the size of tow increases, so does the power required to push it. River current is, of course, the other governing design parameter. As tow size grew between 1950 and 1970, available towboat size grew accordingly and reached an economical plateau at about 7000 horsepower. Although channel and lock size restrictions placed an upper economic limit on tow size, a number of U.S. Army Corps of Engineer improvements intended to maintain or increase channel depths resulted in significantly higher current velocities in the lower Mississippi. This resulted in the development of the 10,500-Hp towboat which presently is the largest size in common use. Although most towboats are custom designed to meet the specific needs of the owner, the availability of standard diesel engines forces them to naturally fall into standard size classifications. Table 33.2.4 lists a number of these classifications currently available, and their significant characteristics.

Fig. 33.2.2 Fully-integrated unit tow.

Table 33.2.4 Towboat Data

Horsepower	Length (ft)	Width (ft)	Draft
2,800	110	34	7 ft 9 in.
4,200	140	42	8 ft 6 in.
5,600	140	42	8 ft 6 in.
6,400	168	45	9 ft 0 in.
10,500	190	54	9 ft 0 in.

33.2.5 Loaders

In general, coarse bulk solids that are too dense to be handled pneumatically are loaded into hopper barges utilizing a positioned loader which essentially consists of a high-speed (1200 ft/min) belt conveyor mounted on a luffing (capable of being raised or lowered) boom. In some applications the boom may also have a shuttle motion, to provide the capability of loading the barge at several transverse points. Figure 33.2.3 shows a typical barge loader.

Normally, economic considerations dictate that loading and unloading equipment for river barges be located in a fixed position, with the barges being moved during the loading or unloading operation. Total elapsed time for loading or unloading a single barge is short enough that frequent (several times an hour) moves are required to replace one barge with another.

33.2.6 Barge Haul Systems

A variety of barge haul systems have been developed to move barges during these operations. Basically, they consist of winches and wire rope cables to haul the barges, plus some provision to breast (restrain laterally) the barges. Breasting can be accomplished in several ways: against stationary cell structures, using mechanical arms attached to stationary structures, or using breasting hulls free to move vertically as the water level changes.

The hauling motion may be incremental, as required by a grab bucket unloading operation, where the barge is repositioned after the unloader has emptied all it can reach, or continuous, as required

Fig. 33.2.3 Typical barge loader.

by a bucket ladder unloading operation. The hauling applications are similar mechanically but differ dramatically in electrical control and duty. Power requirements are dictated by river current, size of barge (or string of barges) being hauled, and speed and acceleration of the hauling system.

In most high-speed operations, the time spent removing one barge that has been loaded (or unloaded) and replacing it with another is critical. To minimize this lost time a "shuttle barge" system is frequently used. This system consists of a small barge, permanently attached to the barge haul system, and equipped with winches and lines to lash a standard barge to each of its square ends, moving both the shuttle and the barge being unloaded. While this is occurring, a work boat removes the empty barge on the other side of the shuttle and replaces it with the next barge to be unloaded. Like most barge haul systems, this variation incorporates a pair of opposed winches with one hauling and one providing a moderate countertorque to minimize sag of the slack line.

33.2.7 Barge Loader Systems

Barge loading, like many other materials handling applications, provides an opportunity for optimization of operations by careful selection of system constituents and consideration of their relationship to each other and to the system as a whole. Figure 33.2.4 shows a coal-handling facility that exemplifies this principle. This terminal is required to load 22,500 tons of coal into 15 hopper barges in a manner that will minimize turnaround time for the tow of barges. To accomplish this, a tow of 15 empty barges, three abreast and five long, is moved upstream of the fixed loader. The transverse fleet lashing is removed, converting the tow into three strings of five barges. The string nearest the shore is attached to the barge haul system and is moved into position to begin loading. Once loading begins, the string of barges is hauled slowly (about 10 ft/min) past the loader. The loading spout is bifurcated and equipped with a flop gate so loading can be transferred from the aft end of one barge to the fore end of the next without interruption of the flow. When the first string of barges is fully loaded, the loading operation is interrupted briefly while the filled string is hauled forward far enough to permit the next string to be moved transversely into position to be hauled forward so loading can begin again.

33.2.8 Unloaders

There are two common methods for unloading bulk solids from river barges—clamshell grab bucket unloaders and continuous-bucket ladder unloaders. The clamshell unloader is the descendant of a traditional approach utilizing grab buckets mounted on cranes. Early versions (whirler cranes) were mounted on cells in the river and possessed both slewing and luffing motions. Such rigs are still in

Fig. 33.2.4 Coal transfer terminal.

use where unloading requirements are too small to justify the capital expenditure for larger capacity units. For high-tonnage requirements, stationary in-line unloaders are applicable. These units employ rope-operated systems in which a fairly light trolley is pulled inward and outward on a stationary boom. The trolley supports sheaves from which wire ropes lead down to a grab bucket.

The most commonly used grab bucket in North America is the bight-of-line vertically reeved clamshell type as shown in Fig. 33.2.5. It is supported in the bight of two separate lines, the "hold" line, which is reeved through a sheave on the main frame of the bucket, and the "close" line, which is reeved through a pair of sheaves, one on the main frame of the bucket and the other on its hinged jaws. When the "close" line is pulled, the jaws are moved together and the bucket closed. Once the bucket is closed, the "close" line supplements the "hold" line and helps hoist the bucket and its load. Figure 33.2.6 illustrates a different style of grab bucket, the scissor arm, direct-reeved clamshell bucket. This style is used widely on large ship unloaders in Europe. It has an advantage in payload-to-bucket weight ratio because of the digging leverage provided by its reeving system, with the accompanying disadvantage of a much more complicated wire rope cable system.

The wire rope system used for hoisting and closing grab buckets is a direct function of the reeving system and the weight of the grab bucket and its payload. Available capacities are directly related to the sizes (and strengths) of wire rope which are commercially available. Table 33.2.5 illustrates a series of hoists currently in common use for grab bucket unloaders with direct-reeved systems. The significance of the two-digit series number is the design lifted load, in kips. Note the decrease in design load for line speeds in excess of 760 ft/min; this is illustrated in Fig. 33.2.7.

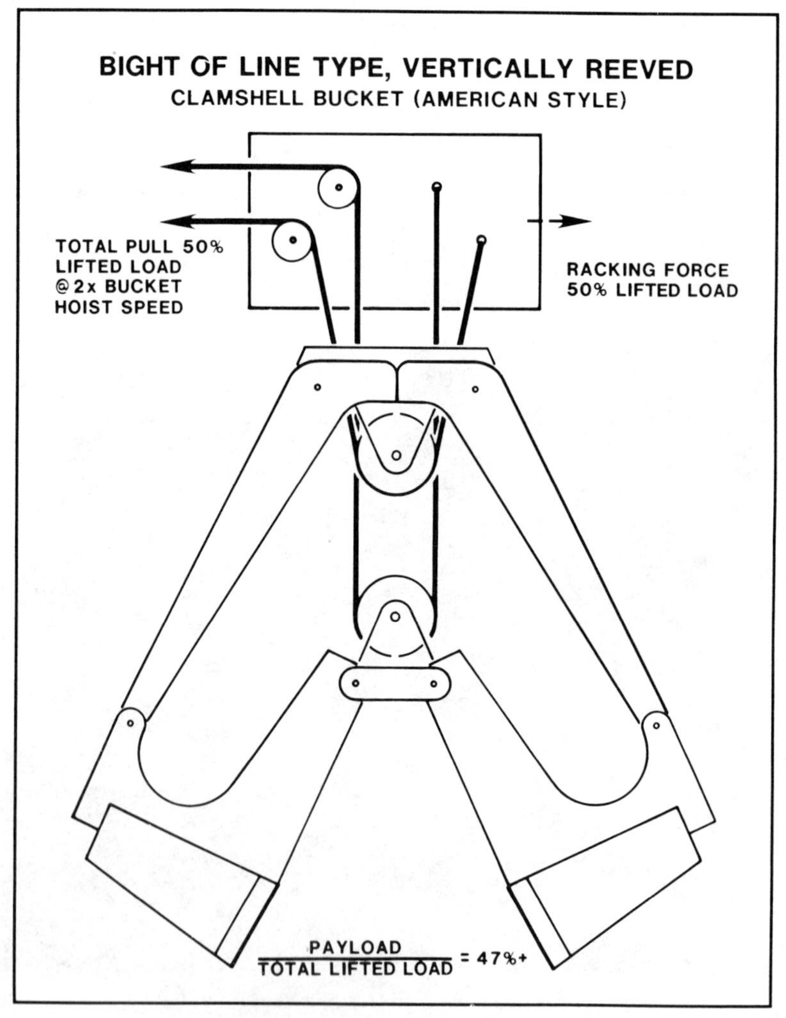

Fig. 33.2.5 Bright-of-line clamshell bucket.

Table 33.2.5 Hoist Data

Hoist type	CD40	CD65	CD80	CD110	CD140
Line pull (kips) at 760 ft/min	10.0	15.0	20.0	27.5	35.0
Line pull (kips) at 1200 ft/min	5.6	9.5	12.5	17.4	24.4
Wire rope diameter (in.)	$\frac{7}{8}$	$1\frac{1}{8}$	$1\frac{1}{4}$	$1\frac{1}{2}$	$1\frac{5}{8}$
Horsepower	250	375	500	675	900

Unloading a barge using a straight line, rope-operated clamshell grab bucket unloader is accomplished by a series of incremental moves of the barge relative to the unloader. At each barge position the bucket is cast into the barge at a variety of positions transversely across the barge until the area which can be reached from that position is essentially emptied.

33.2.9 Definition of Capacity

One of the biggest sources of confusion in the materials handling field is the definition of capacity, in tons per hour. Nowhere is this more important than in the evaluation of systems for loading and

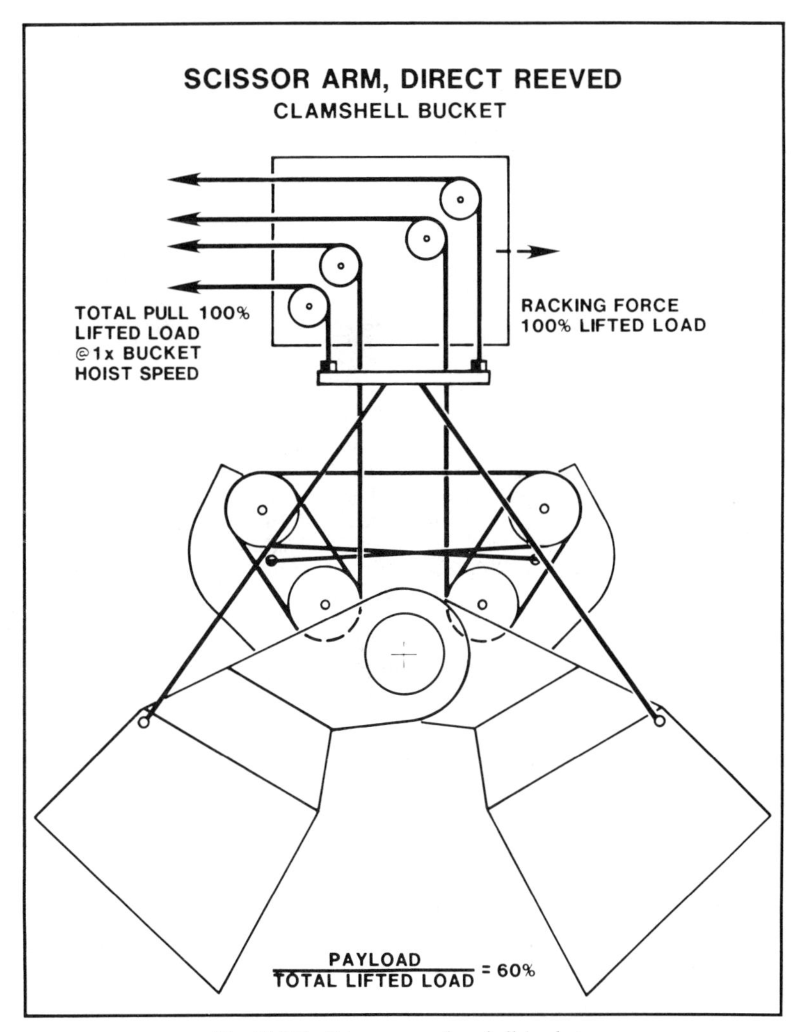

Fig. 33.2.6 Scissor arm clamshell bucket.

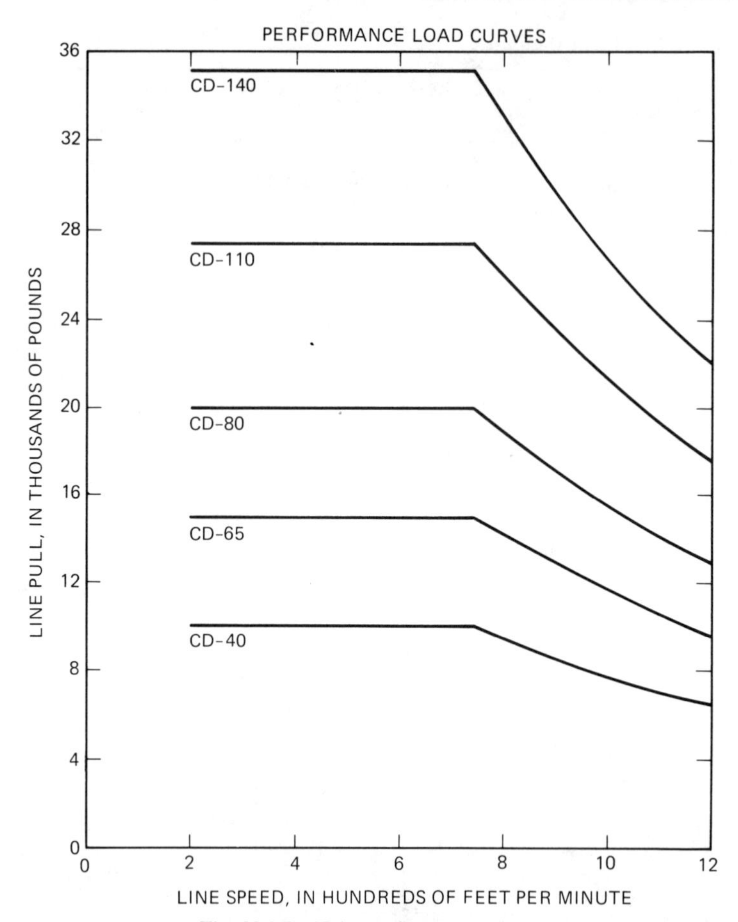

Fig. 33.2.7 Hoist performance curves.

unloading bulk materials into and from vessels. The materials handling engineer must realize that this responsibility lies with the designer of the overall system and that the equipment supplier can only be held responsible for some type of calibration rate for sustained operation of his equipment.

To illustrate this principle, it is constructive to consider an actual example. Figure 33.2.8 is a photograph of a pair of traveling clamshell unloaders, unloading iron ore pellets from a typical Great Lakes ore boat. From the standpoint of the operator the important parameter for sizing or evaluating the performance of the unloading facility is the "time in port" required to unload the vessel. This time includes the time required to remove hatch covers, the time required for the unloaders to travel from hatch to hatch (probably in accordance with a prescribed sequence which optimizes vessel list and trim, rather than unloading efficiency), the time required for the unloader to travel between alternate digging points in one hatch, the time required for lowering and hoisting a bulldozer into the hatch for cleanup, and the time required to complete cleanup using the unloader—all in addition to the time required to perform the actual unloading. Each of these operations affects the time in port and is independent of the unloader's actual design capability of performing its principal function.

In addition, it is not easy to evaluate the actual unloading operation for a variety of situations. The distance the grab bucket must travel in one complete cycle from digging point to discharge hopper and back to digging point varies considerably even in the unloading of a single vessel. The width of the vessel, the vessel's draft (its vertical position in the water), and the vertical location of the material being unloaded all influence the location of the digging point. Also, the efficient operation of this kind of equipment is extremely operator-sensitive. Consequently, it is difficult for the supplier of an unloader to guarantee an average unloading rate for all applications.

It is, however, practical and usual for the equipment supplier to guarantee a calibration rate for the unloader. This rate, commonly called the "free digging rate" (in tons per hour) is the product of the bucket payload and the number of cycles per unit time the unloader can perform using a predeter-

Fig. 33.2.8 Two traveling clamshell unloaders.

mined digging point. Normally, this digging point is at the mean waterline and at the transverse centerline of some prescribed vessel. This rate should be guaranteed and verified by actual performance test operation of the unloader. To estimate an average unloading rate for a complete vessel, it is necessary to begin with the calculated free digging rate, to exercise some judgment regarding operator skill and motivation, and to evaluate the time required for all of the operations not directly related to unloading. The average unloading rate will be much lower than (often less than 50% of) the free digging rate. To improve the average digging rate it is prudent for the materials handling engineer to provide enough surplus capacity (typically 10%) in the conveying system that accepts the discharge from the unloader, to enable the operator to take full advantage of the cream digging situations where the digging point gives a shorter duty cycle.

33.2.10 Continuous Unloaders

The requirement for unloading rates significantly higher than those practically achievable by grab bucket unloaders led to the development of continuous-bucket ladder unloaders. There currently are two distinct types of continuous-bucket ladder unloaders in common usage—the hinged single ladder unloader and the hoisted twin-ladder unloader. Both types utilize a pair of heavy-duty roller chains to move a series of wide (8–12 ft) buckets through a digging area, to hoist the loaded buckets to a discharge position, and to return them to the digging area.

The hinged concept relies on rotational positioning of the digging head relative to a hinge near its top, to accommodate vertical variations in digging point due to a barge draft and changing water level. The digging head itself (the lowermost part of the digging arm) includes a tensioning sprocket on a pivoted arm which ensures a constant orientation of the bottom of the chain of buckets regardless of the orientation of the arm itself. The arm and its hinged support are mounted on a trolley that can move across the width of the barge. The normal sequence of operation is for the digging head to make one pass down the center of the barge, "hogging out" a major portion of its contents. The pass is made by hauling the barge relative to the digging head with the head stationary at the centerline of the barge. A second, cleanup pass is made with the digging head being moved back and forth transversely ("sashaying") as the barge is hauled past it.

Where significant (greater than 27 ft) differences in water level are experienced, the hoisted twin-head continuous-bucket ladder unloader is used. In this design the entire runway on which the digging head trolleys travel is hoisted or lowered to maintain a constant position of the digging head relative to the barge regardless of water level or draft. Neither the hinged top support nor the tensioning lower sprocket is required. This design typically has two digging heads although a single head is occasionally used (similarly the hinged approach can also be used with two digging heads). The operating mode incorporates three passes, the first a "hogging" pass with both digging heads at the barge centerline, the second with each head as close to the hopper sides as possible, and the third, a cleanup pass with both heads back at barge centerline.

The same concerns regarding average capacity for a complete barge that were discussed in regard to grab bucket unloaders apply to continuous ladder unloaders. The unloader supplier must be responsible for well-defined free digging rate. The barge haul supplier must be responsible for hauling speeds at prescribed loads. The system designer must consider all other factors and operations when determining the average rate.

Fig. 33.2.9 Transfer terminal.

Fig. 33.2.10 Ocean-going barge.

Table 33.2.6 World Bulk Carriers

Vessel Size (dwt)	Number of Vessels
0– 50,000	3622
50,000–100,000	614
Over 100,000	331
Total	4567

33.2.11 Transfer Terminals

The interface between inland waterway transportation and oceangoing vessels consists of a number of interesting transfer terminals. Typically, such a terminal consists of equipment to load and unload river barges and oceangoing vessels, and frequently the capability of stockpiling and reclaiming bulk materials. Figure 33.2.9 is a photograph of a typical land-based terminal. A recent development has been the use of mid-stream transfers. This concept involves a large hull on which is mounted a continuous-bucket ladder unloader for unloading bulk material from river barges and a conveying and loading system to load it into oceangoing vessels.

33.2.12 Coastal Shipping

Coastal shipping in North America includes the two major intracoastal waterways (Gulf and Atlantic) as well as open sea movements close to the shore. The Gulf Intracoastal Waterway has been mentioned previously; it is more important as a commercial transportation system than is the Atlantic Intracoastal Waterway. A significant segment of the bulk materials transportation depends on the use of large (18,000–30,000 tons) seagoing barges like the one shown in Fig. 33.2.10. These barges operate quite economically along the coast and in the Gulf of Mexico. They are pushed by towboats dedicated to this service. This approach permits the motive power (the towboats) to be kept in nearly continuous operation, while the barges are being loaded and unloaded.

Shown in the photograph of the seagoing barge is a hoisted twin continuous-bucket ladder unloader of the type previously described, with one significant exception. In this application the unloader travels on trucks on a runway, while the vessel is moored in a stationary position. The economic transition point between moving the vessel and moving the loader or unloader occurs at vessels of about 15,000 tons capacity. It is also interesting to note that this is the only continuous unloader currently unloading seagoing vessels in North America. The inherent advantage of continuous loading versus grab bucket unloading has not yet been confirmed for oceangoing vessels, although several continuous unloaders are currently being planned or constructed.

33.2.13 Great Lakes Shipping

Transportation of bulk materials on the Great Lakes has been significant for most of the twentieth century. The construction of the Eisenhower locks opened this area to smaller oceangoing vessels in recent years. The laker trade was developed to haul iron ore from Minnesota to the steel mills of Indiana, Ohio, and Pennsylvania. This was principally done using ore boats of the type previously shown in Fig. 33.2.8. These vessels have a number of individual hatches and are unloaded by grab bucket unloaders or by Hulett unloaders. The Huletts require the movement of a large mechanical arm in each cycle, but nonetheless have proven to be dependable unloaders for many years.

At this time the fleet of bulk carriers plying the Great Lakes numbers over 250, ranging from a

Table 33.2.7 Typical Bulk Carriers

Vessel	Dwt	Type	Draft (ft)
Marcona	270,000	Ore/oil	68
Sysla	223,000	Ore/oil	65
Cetra Centaurus	170,370	Ore/oil	60
Tibetan	149,950	Ore/bulk/oil	55
Robina	113,000	Bulk/oil	50
Aegir	82,400	Bulk	45
Aquabelle	46,700	Bulk	40
Atlantic Splendour	19,900	Bulk	32

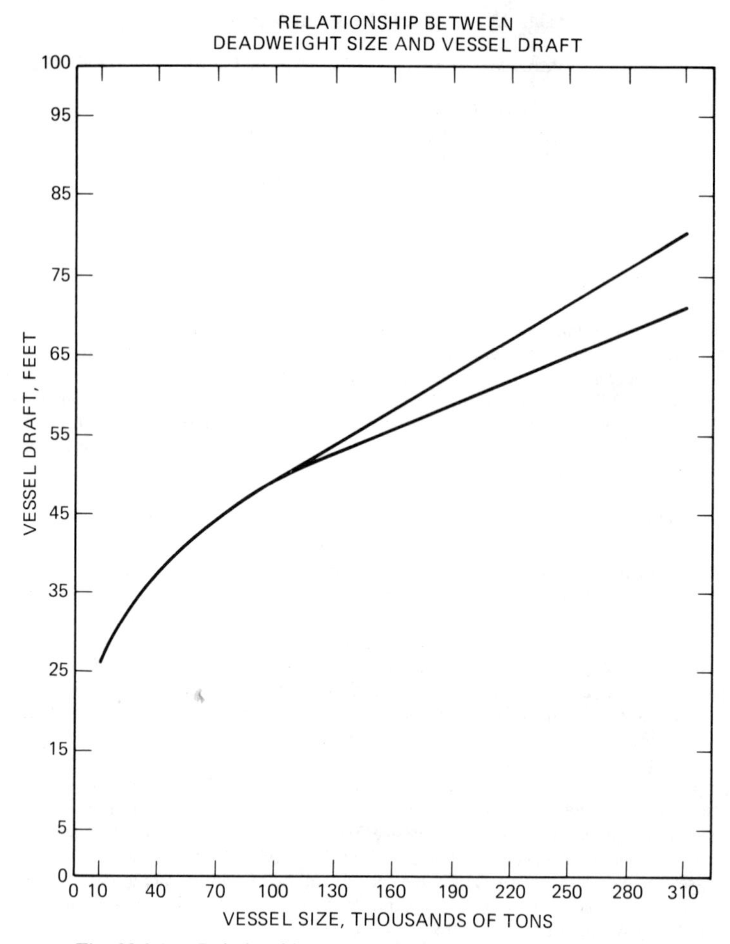

Fig. 33.2.11 Relationship between deadweight size and draft.

minimum size of 259 ft by 43 ft to the current maximum of 1004 ft by 105 ft. The largest bulk carriers can transport about 30,000 tons of bulk cargo and draw 28 ft of draft. About 40% of this fleet are self-unloading vessels.

Self-unloading vessels have been increasing in numbers for many years on the Great Lakes, and several large high-capacity units have been added in recent years. The economic considerations influencing the decision to incorporate self-unloading capability into a vessel merit discussion. Because the self-unloading system in a vessel is custom designed for a specific vessel and cargo, it normally can achieve a much higher unloading rate than a land-based unloading system which must handle a variety of cargoes and vessels. The rapid unloading and reduced turnaround can be quite beneficial economically, particularly for large, expensive vessels. Obviously, this advantage decreases with longer hauls since the time in port becomes a smaller percentage of the total trip time, while the effect of the loss of cargo capacity displaced by the weight of the unloading system remains constant. Under normal conditions a trip time of about 3 days is about the maximum for economic justification of self-unloading.

33.2.14 Transoceanic Shipping

The fleet of bulk carriers that serve the transoceanic trade changed dramatically in the 1960s with the advent of the very large cargo carrier. In 1955, the largest bulk carriers afloat were in the 35,000 dwt class. By 1975, there were several 225,000 dwt vessels in active service. Table 33.2.6 is a current listing of dry-bulk carriers in operation. It is interesting to note that the largest 20% of the vessels represent well over 50% of the fleet capacity. Table 33.2.7 lists characteristics of typical vessels. Most of the very large bulk carriers are engaged in the transport of iron ore. This transition was the direct result of an effort to duplicate the economics already being realized by the use of very large tankers in the oil trade. Universal use of the very large carriers has been hindered by the lack of ports with sufficient draft available. Figure 33.2.11 relates required draft to vessel tonnage.

As has been mentioned earlier, practical considerations demand that even the smallest oceangoing vessels be moored during loading and unloading, and that the materials handling equipment servicing them travel along the length of the vessel during the loading or unloading operation. The addition of the traveling capability and the much higher operating rates are the principal characteristics that distinguish ship loaders and unloaders from barge loaders and unloaders. At this time, a loading rate of 12,000 ton/hr appears to be a practical maximum for a single unit. The maximum unloading rate today features a bucket with a payload of 50 tons; it can achieve a free digging rate of over 3,500 ton/hr. This probably is a practical maximum for a single grab bucket unloader; to achieve higher rates a continuous unloader must be considered.

BIBLIOGRAPHY

Collins, Richard, Michael J. Gawinski, and Louis H. Meece, "The Vital Interface of Ocean Shipping and Inland Waterways," Third International Symposium on Transport and Handling of Minerals, Vancouver, British Columbia, 1979.

Gawinski, Michael J., "Solving the Energy Crisis—World Coal Trade 1980–2000," Bulk Handling and Transport Conference, Amsterdam, 1981.

Mandella, Bruce F. and James N. White, "The Hall Street Coal Transfer Terminal in St. Louis," AIME Annual Meeting, New Orleans, Louisiana, 1979.

Meyer, Russell N., "Facilities for the Loading/Unloading of River Barges," University of Pittsburgh Seminar on Transportation Systems for Bulk Solids, Pittsburgh, Pennsylvania, 1972.

Oyler, John F., "Handling of Bulk Solids at Ocean Ports," pp. 567–597 in *Stacking, Blending, Reclaiming of Bulk Materials,* R. H. Wohlbier, Ed., Trans Tech Publications, Clausthal, Germany, 1977.

Price, W. L., "Coal, U.S.A.—Mine to Market via Inland Rivers System," Coal Handling and Storage Symposium, Chicago, 1980.

Sjogren, Jack G., "New Coal Transfer Facilities on the Inland Waterways," Dravo Corporation, Pittsburgh, Pennsylvania, 1980.

Terry, Dennis W., "Coal Barge Unloaders," Coal Technology '80 Conference, Houston, Texas, November, 1980.

CHAPTER 34
TRUCK AND AIR TRANSPORTATION

FRANK J. MEINERS

Carlisle Engineering Management, Inc.
Carlisle, Massachusetts

ROBERT PROMISEL

The Austin Co.
Cleveland, Ohio

34.1 TRUCK INTERFACE

Frank J. Meiners

34.1.1 Preparation of Documents and Materials

Few of us think of ourselves as being in the shipping business. After all, we're high tech! . . . or volume producers . . . or astute brokers, or whatever. Yet each of us relies on shipping to finish the job. "Cash and carry" works at retail, but seldom elsewhere. So we've got to be shippers; and we have to know how to do it. Otherwise, we can suffer losses in a lot of ways:

Outright theft; more common than we like.

Accidental loss in transit; but we usually say it was stolen.

Discovered damage, which is a loss to shipper, receiver, and carrier.

Hidden damage, which may ultimately cause product failure.

Undershipment, which causes administration costs.

Overshipments, a hidden loss, seldom recovered.

Shipment delay, a "hidden loss" and an administration cost.

Extra operating cost, the result of ineffective actions taken to avoid the other losses.

Effective, efficient, and organized methods are needed to plan and control the shipping operation. This chapter recognizes the total problem, but it specifically addresses the problem from an operational point of view.

Minimize Losses

Management always gets the blame, and usually deserves it. This is true in shipping especially. Management action can directly influence the level of losses in shipping:

As part of product design, management can produce packages for products that are popularly sized and modular to each other.

Written procedures, methods, and standardized controls are vital to the shipping function.

Efficient, orderly picking methods used in a controlled environment will reduce losses.

Packaging unit load design, protection, and handling methods and controls all contribute to loss prevention.

Dockside staging, checking, other dock methods, and controls produce good results.

Finally, truck loading, sealing, and dispatching methods and controls all contribute to damage and loss prevention.

The common theme throughout this list is "methods and controls." So it is up to management to give its supervisors all the training and supervisory tools needed to get the job done. A reverse flow of control information should assure management that the job was done.

Start with Product Design

Our interest in product design is limited to effects on order picking and shipping. Simplification, modularization, standardization, and self-explanation are the management tools employed in product design:

Products should be packaged in numbers that are popular with users. This avoids order picking, split cases, and their resulting counting errors.

The primary numbers which customers like should be combined into larger units without division, and in modular fashion: 4 six-packs = 1 case.

Larger case-units should likewise modularly combine into still larger units which palletize in a regular pattern. Hard to do, but the try can pay dividends: 8 six-packs = 2 cases = 1 jumbo case.

Pallet-loads produced should be easily combined into truck-load shipments: pallet-load weight should approximate its share of the truck-load cube. Then the truck can be loaded without voids or dunnage.

Order forms or price lists should make it easy to order in the primary quantities and desired multiples.

The case should be marked with content and count information that is identical to shipping document information.

The handling unit also should be the counting unit: if you handle cases, count cases; if you handle pallets, count pallets.

Material identifiers should be their common names: "Storm windows" has meaning: "Glazed Infiltration Controls" does not.

Each product that is different in any way from all others must be identified with a single unique stock number. No more than one number should be applied to each product.

As you can see, there are conflicts between these and other requirements which must be reconciled. Case marking can be less rigorous where products are not easily confused; if industry practice is to order in "squares," roof shingles may not be salable as "bundles," and so on. Anyway, we should know what we would like the system to be, and be ready to change it when the opportunity arises.

Written Procedures

The overall control mechanism employed in dock operations is a set of well-organized procedures. The procedures normally will be included as part of an operations manual, along with other material covering personnel, equipment, and safety. A "procedure" for a physical operation includes a step-by-step job guide, the materials used, and equipment required. For a control operation, the procedure might cover a paper documentation method or computerization. Paper documentation is still the primary control mechanism, but computerized systems are gaining on them.

Paper Documentation Procedures Are "Standard"

These include the form format, its "form number," and any permanent copy included on it. Usually, a sample copy of the form provides all of this. The procedure also includes the document's and user's usage, responsibility, content, timing, location, rationale, and interfacing methods for each step in its life cycle:

Input information for preparation.

Document preparation and distribution.

Document control on the dock.

Information inputs to the document before loading (other document numbers, changes, carrier, etc.).

Reporting information entered on the document during the operation.

Retention or review of the document in the operating department prior to return.

Return routing for information and retention.

The system outlined is a manually generated printed form implementation. Today, there are few documents not using printed forms or books. There are fewer still where complete computer control is in use. A blend of computer-generated printed forms used with some manual records (Truck Logs) and some computerized on-line systems (Inventory, Order Picking) probably is most common.

Computer Systems Are Replacing Paper Documents

Until recently, written documents alone controlled the shipping operation; and gaps in written documentation usually were looked upon as lacks in control. To some extent, this viewpoint is still valid, but a major exception must be made due to the introduction of on-line computer control. Careful design of a computer control system usually upgrades a manual system because of the rigorous analysis required. This is particularly true in the area of written procedures, which are a necessary part of computer program documentation. In general, internal functions (inventory, pick lists, logging, etc.) are more easily converted to computerized on-line control. Legal and other considerations hinder the acceptance of nontraditional methods for operations and documents which concern outside organizations (Bills of Lading, Load Sheets, etc.).

Computer System Procedures

On-line computer system procedures are designed to meet the needs of the physical operation as well as the data processing operation. The principal operating requirements are the same as those outlined for paper systems, but special care is needed to ensure that lines of responsibility and authority are not blurred:

The operation of the on-line system with the ongoing dock operation must be understood and accepted by all levels of responsibility.

Prior to installation, the program must be tested and validated for operational needs.

Program safeguards are needed to limit access to responsible persons only.

Input information must be validated prior to entry, and the program should validate it for format.

Routine dockside interactions with the system must be formally defined and included in the written procedure.

Exception and error handling must be provided for by the system in a straightforward way.

Inputs and outputs must be monitored periodically, by management or staff to ensure continued effectiveness. Operation of the system should be continuously reviewed to ensure that it continues to help supervision and management to optimize the operation.

Because of its gargantuan appetite for work, and for speed and reliability, the computer can become a crutch. This should not be allowed to happen; market changes and new handling methods ultimately will require new procedures and methods. Critical supervisors and a nosey management must constantly supervise the computer as an operational tool. In no way does a computer relieve supervision from its responsibility for all aspects of running an efficient operation, or management's duty to provide the tools and information to run the job.

34.1.2 Paper Documents

The several documents and records used in shipping are described later, but they have common characteristics that enable their use together as a system:

Pick Lists and Load Sheets usually apply to more than one order, whereas Shipping Orders and Bills of Lading apply to single orders.

Measurement units are carefully selected to promote accuracy. Pick List quantities must translate into Shipping Order quantities, then into Load Sheet quantities, and onto the Bill of Lading. This translation must be done using standardized conversion factors, which reflect, for example, units per case, cases per unit load, and so on. Documentation must be designed to clearly establish responsibility for counts and the resultant unit conversions. With this system, each quantity is counted at a given level only once (see Fig. 34.1.1).

Each unique product has its own number and description which are used unchanged on all documents. The product number format and content are designed to promote readability. The description should describe the product as it is identifiable in the warehousing and in the shipping cycle.

Shipping Order

The shipping order is the driving document of the loading environment. It is the source document upon which others are based and functions as a "Shop-Order" for the Shipping Department. Normally it is launched by a sales office, after coordination with credit checks, inventory status, and other factors. The information contained on the order includes:

Date of scheduled shipment

Order date

Customer or destination identification

Separate mailing address

Quantity per item

Other measurement units for the item (optional)

Item code and description

Item storage location for picking (optional)

Load assignment for LTL (optional)

Total weight (optional)

Special instructions for preparation, shipping, or billing

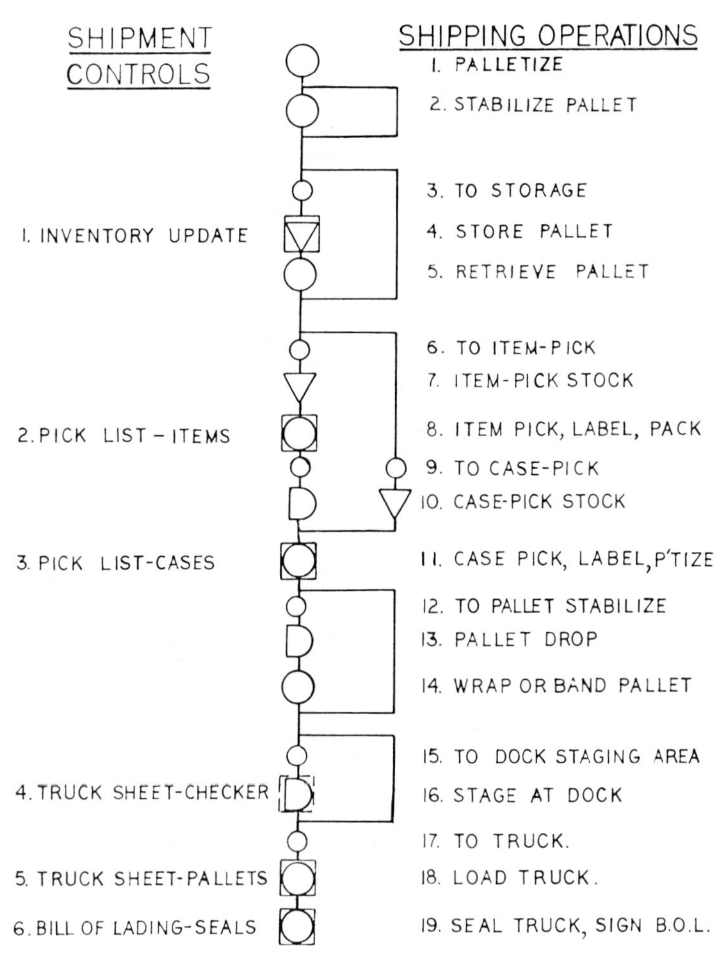

Fig. 34.1.1 Shipping flow versus controls.

Bill of Lading

The Bill of Lading is the primary legal instrument of the shipment. Its format and content are subject to the requirements of the law; for trucks, this is usually an Interstate Commerce Commission jurisdiction. Information requirements are:

Bill number (Serial number)
Seal numbers, for seals affixed by shipper
Date of shipment
Name, address of shipper
Name, address of receiver
Quantity, description, and weight of each item
Total weight

The Bill of Lading establishes the transfer of ownership of merchandise when it is loaded on the truck.

Truck Sheet

When more than one shipment is moved on a truck, the information on all the Bills of Lading may be combined into a single Truck Sheet, or Manifest. The Truck Sheet does not replace the Bills of

Lading; each still must be prepared, signed, and handled to comply with legal requirements. Truck Sheets are used to simplify loading and checking of combined shipments.

Truck Sheets can cover various load types, generally dividing into classes of loads where the goods are aggregated and physically handled as though they were going to a single addressee, and into classes of loads where individual shipments must be kept segregated through loading and subsequent transport. Of course, there are also combinations of the two types used. In either case, the Truck Sheet may contain these items, depending on loading and checking needs:

Serial Load Sheet numbers

Truck load identification (if used)

Date of shipment

Carrier routing, or other identifiers for private or leased carriage

For each load increment (unit load, line-item, or other identifiable lot), a visual-check quantity. This can be number of cartons or bags per item, number of unit loads, and so on.

Totals for the above increments

Weights by increments and total load weight

A load diagram to guide dock personnel

Destination information

Seal numbers, for seals affixed by shipper

Pick Lists

Pick Lists are used to improve efficiency in load gathering. The order information contained in Shipping Orders and Bills of Lading usually is presented to comply with commercial and legal requirements. To maximize order-picking effectiveness, the same information is repeated on the Pick List, but operational factors are stressed:

The location of each item in the picking area is indicated in a simple, unambiguous format.

The product number and name of each item also must be clear.

The quantity of each item must be in measurement units that correspond to the containers picked (i.e., "2 bags" instead of "70 lb," "10 cases" instead of "120 doz.," etc.).

Space on the Pick List must be clearly defined for reporting of quantity picked, shorted items, and other variances.

A simple listing of information does not exhaust requirements for a good Pick List. The document, like a Load Sheet, serves as a step-by-step guide, leading the picker in a logical way through the task. Order pickers do not work at well-lighted desks. This places an extra burden on management to provide a picking document that, in the clearest, simplest way, tells the picker exactly what he or she must do. Its color, type style, and size, even the paper itself, are all important.

Package Labels

These are widely used instead of pick lists. A label is prepared for each package in the order. As the order is picked, labels are applied to each piece by the order picker. If each piece picked is labeled and no labels are left over, the shipment quantity is correct.

Pallet Slip

Pallet slips may be used in two general ways:

To identify pallet loads of inventory. In this case, the slip information would be limited to the pallet serial number, product number, product quantity, date code, and quality tracing information.

Slips also identify pallet loads that have been order picked. In this case, they should identify the order number, shipment identification, pallet number, and date. A serial number system for the pallet load may also be employed.

The objective in using pallet slips on the shipping dock is to enable the loader to confine his attention to handling and counting pallets only. The slip should symbolize that order picking is completed, and that the pallet load contains a definite part of the shipment which is identifiable on the pick list or bill of lading.

The slip is glued or fastened on the side of the load that faces an approaching fork truck. Very-low-height loads may allow placement atop the load and still be visible. The pallet should always be positioned so that the slip is properly positioned for the next fork truck. On flow racks, this requires a 180° pallet "spin" before entering the rack.

Pallet slips may be stick-on labels, multipart forms, punch cards, other cards, reusable, and so on. The primary physical requirement is that plant conditions should not separate the slip from the pallet. Color coding is often useful, in conjunction with large print. Remember, the fork truck driver must read the slip under poor conditions, as are usually found in a warehouse environment.

Truck Logs

As a backup to the other documents, a log should be kept of all vehicles serviced at the dock. This log should include:

Vehicle identification (usually the name of the carrier)
Driver name
Time in
Time out
Bills of Lading numbers loaded or unloaded
Seal numbers applied or removed

This document helps control operations, particularly in a longer-term sense. It furnishes valuable operating information for scheduling. It also confirms the validity of more formal documentation if the latter is challenged. To ensure validity of the log, every vehicle should be entered, including nonfreight types such as delivery vans, parts suppliers, and private cars.

Truck Seals

Although not a "document" in form, the door seals used on trucks serve much the same purpose. They are a symbolic representation by management that unauthorized entry to the truck is prohibited. Also, breaking a sealed doorway in Interstate Commerce is unlawful. Therefore, doorway seals must be handled with the same precision as a Bill of Lading:

Seals should be consecutively numbered, and this number should be entered on the Bill of Lading.

Seals should reject attempts to open them without breaking, or to reseal them unnoticed.

Seal numbers should be recorded on each individual Bill of Lading or Truck Sheet and issued with them.

Extra seals should not be allowed on the dock.

Where trucks are received, the seals should first be examined to make sure they are unbroken.

Finally, the seals should be applied and removed only by reliable dock personnel. This can vary from place to place, but this should never be the truck's driver. To a driver, a seal in the hand can be a temptation to steal undetected. The seal can serve its purpose of reinforcing honesty only if it is used properly.

34.1.3 Order Picking and Load Assembly

These two functions are different only in their scale of operation; a truckload is only a giant-size carton into which pallet-load items are "picked" by fork trucks. We will look at the operations from the small end, order picking, first.

The first thing to say about order picking is to avoid it. We already have mentioned the interaction of package sizing and order quantities with this factor. If customers can be encouraged to order in quantities of full cases or full pallets, the picking cost is saved. However, you cannot expect a customer to order an inconvenient quantity, so the containers and cartons must be sized to match the customer's needs. Also, you cannot expect him to guess what you want, so price lists and order blanks must steer him to the desired quantities.

Item Picking for Less-Than-Case Lots

Item picking should be done directly into the shipping container to avoid extra handling. A range of container sizes can be used to match popular shipment quantities fairly closely.

But if the picked material closely matches the case size, it must be made easy to pack. This means we must pick larger-sized items first because they normally would be packed first. To do this, we need a picking layout where large items are at the beginning of the pickers' travel path, with a progression to smaller sizes. The pick list also must be designed with large items at the top, reflecting the physical layout of order picking.

If cases cannot be used for order picking, goods may be placed into larger tote boxes, or picked directly onto conveyors. Picking to a conveyor is done the same as if the goods were going directly into a carton. That is usually what happens at the far end of the conveyor. If picked into totes, reverse the picking order from the "direct-to-case" method because the items that are picked first will be packed last.

Product labels or other identification should be visible in the tote box or unsealed carton. This allows ready checking without further handling. Cartons may even be left unsealed with labels faced out on opposite sides so that two-high loads can be checked. In any event, only one person should be responsible for the contents of a case. That person should give the last check before closing it, and it should be identified by a label or marker by order number and case number.

The key elements of successful item picking are an orderly environment, coherent information provided to the picker, and each picker being responsible for the contents, labeling, and close-up of his own output. Following close-up, the case is dispatched to the dock, directly or via a downstream case-picking operation.

Case Picking to Pallets or Belts

Full-case picking shares many requirements with item picking;it is needed to provide an orderly operation, coherent information, and fixed responsibility:

Case sizes should be designed as modules that combine into good pallet loads.

The case-picking layout should promote a good pallet load. Large, heavy cases should be picked first and form the bottom of the load.

If possible, case picking should be done directly onto the pallet to avoid intermediate handling and blurred lines of responsibility.

When palletized, case labels and marks should face out to allow easy identification.

In this way, the picker has total responsibility for the pallet load.

Bulk Picking Saves Labor

A second alternative exists: bulk picking. In this system, a group of orders is combined and "pre-picked" to a belt or pallet as a first operation. A second operation sorts the cases' output into individual orders. This greatly reduces walking time for small-lot picking. The small lots are easily split up later, and responsibility is fixed at the final separation point.

Large-Case Picking Systems

A third alternative, picking to a belt, exists for case picking. This alternative is most commonly used in very large operations where belt transport and multiple picking levels make sense. Fixing responsibility for the pallet load is difficult with this system because high-speed handling denies the palletizer an opportunity to realistically perform the checking function. Accuracy using this system often is acceptable despite this difficulty; the high organization and information levels needed to run these systems offset the bad effects of mixed responsibility. High-volume picking often can terminate in floor-stacking directly in the truck. In this case, products are checked "on the fly," which humans often fail to do reliably. Bar-code readers and computers are a ready solution to this problem and are accurate and reliable.

Pallet Identification

Order-picked cases usually are staged in designated locations at the dock. If the operation has been completed in good order, no further checking is required, but the pallet must be identified for the truck loaders. The pallet slip used to identify each pallet should be fastened to the face of the load which is visible to the approaching fork truck driver. The dock should be organized so that pallets are worked on one side only so identification is never obscured. This requirement is lost with flow racks, and this loss is their major disadvantage.

The pallet slip is a symbolic statement by the order picker that the pallet load is ready to ship. The picker also initials the marked pick list to signify that the pallet load is ready. However, sometimes the pallet load is not ready; back orders, damaged goods, missing stock, and other disturbances occur. When this happens, the pick list must be clearly marked to indicate what actually has been done. The pallet slip should also clearly indicate that the pallet is not ready. Supervision and management can then handle the situation and it is their responsibility to clear the load for shipment.

Pallet Picking Is for High Volumes

Full pallet loads are also "order picked," often to a dock staging area, other times for case picking and item picking stock. Where layout, equipment, and order characteristics allow it, the pallet load is best handled directly from storage onto the truck. Volume handling operations always consider this a first choice because of low cost, low space usage, easiest control, and so on. In these operations, the Bill of Lading is the pick list also. The fork truck driver performs the entire operation, including checking the load. When properly done, with safeguards for orderly operations and coherent information, the driver-checker is as accurate as any other manual system.

Modified Pallet Picking

The full-pallet-only system can be extended to include previously prepared pallet loads of mixed cases and item-picked cases. The loader should not be asked to perform a case-by-case check of these pallet loads; checking should be completed before the pallet is staged. Pallet slips should indicate each pallet's identification number ("No. 1 of 6—Order 12345"). This identification should be echoed to the truck loader on a Truck Sheet; the Bill of Lading seldom will be found to have enough flexibility to routinely handle these system needs.

Checking Pros and Cons

"Checkers" often are employed on loading docks to personally check and label cases in the staging area. When used in this way, the checker bears responsibility for the load. There are good reasons for not using checkers:

They complicate the loading operation; every complication can be a source of errors.

Checkers slow down the loading process, limiting response time.

Checkers represent extra cost.

Checkers give a false sense of security, not always justified.

The checking function requires extra space.

Despite the listed drawbacks, there are often very good reasons to use checkers and load staging:

Fork trucks used in storage racks may not be suitable for loading trucks, so staging is needed.

Pallet loads that are staged from varying areas require a single final check of the pallets at loading to confirm that the entire load is shipped.

If loading docks are used intensively, it may be necessary to stage the load so that rapid shipment is possible.

If load wrapping or banding is needed, staging is usually also needed.

Scheduling needs may require a staging buffer between order makeup and loading.

Inventory shortages may result in occasional staging. Constant shortages legitimize a staging operation.

Complexity of shipments contributes to staging requirements.

One Viewpoint

The use of a checker in a volume-type full-pallet handling environment is counterproductive. Load sheets and pallet slips can be combined into a simple tally-counting operation by the fork truck loader. This is usually the most accurate check.

If the case-picking operation works smoothly, pallet-load output from it will be reliable and properly labeled. Under these circumstances, these pallet loads can be loaded from a staging area by the fork truck loader without further checking.

Where large, conveyorized systems are used, automatic checking with bar codes and computer-input scanners should be used. Few manual checkers can reliably perform under these work loads so automation is a must.

34.1.4 Pallet Stabilization

Loads may need to be stabilized if the material shape, weight, packaging, strength, compressibility, and other factors require it. Other factors that combine with these are truck and trip characteristics, pallet load type, load exposure in transit, carrier regulations, and economy considerations. Usually, the need for stabilization arises from a combination of factors, but it often can be avoided. Sometimes, however, stabilization can be used to enhance the benefits of the total system:

The consideration of economy may make column-stacked cases on a pallet best for delivery purposes. This method requires stabilization for some cases.

Stabilization deters breaking down, or breaking into, pallet loads. This benefits security and reduces unnecessary partial pallets in stock.

Square cases do not interlock in a pallet load; they column-stack only. This may be avoided if the product allows a change to an oblong case which does not need stabilization.

Case materials are trending toward slicker finishes. This promotes instability. Broad, printed panels are slicker than plain kraft, so a design change may avoid a stability problem. In addition, printing reduces case strength.

Special antislide finishes are available from box manufacturers. Coatings can be applied also as part of packaging and palletizing operations.

There are differences in plastic film—some have low friction coefficients. This should be considered to reduce sliding tendencies.

If multitudes of case sizes are used, they will not form good unit loads. A modular case-size system avoids this problem.

Bags tend to roll when forced sideways. This tendency is reduced with glued-end bags which form more square, flatter shapes. These bags are also more economical; they are used for automatic filling.

Truck loading methods can reduce stabilization requirements by producing tight loads.

Truck suspension characteristics should be examined for vibration problems. Partial load conditions may allow excessive vibration of the load.

Road conditions probably always will be poor in the given political and economic climate; smoother roads would help avoid stabilization needs.

Stabilization Methods

Package modifications, materials, and coatings are "lost causes" for dock supervisors, but management should pursue them to avoid handling and transit damage.

Where wrapping or banding is used for all pallets, it is best applied during or right after palletization. These operations are conducted as a last step in the production process, mechanized, and using manpower under scheduling controls. However, some pallet loads may be wrapped or banded at the dock.

Antislide Methods

If packages tend to slide off top tiers, but the load is stable below, we can try local cures (Fig. 34.1.2).

1. The simplest method of preventing top-case slides is to tie them all together. One or two rows can be tied. With very light loads, such as cases of empty cans or bottles, this tie can be made with twine; often light string will do. Then the whole top tier will move as a unit and move much less than before. This type of binding can be applied mechanically with ease, with the pallet elevated on the forks of a lift truck right after palletization. Plastic bands and narrow film wraps also are used in this way, and these almost always are applied mechanically. Steel strapping has been used for this also.

2. Corrugated or paper sheets can be inserted between ties of a pallet load. They can be used at several heights through the load. The sheets raise the friction coefficient and "de-couple" sliding cases from those below. This method is best used during palletization; it is easy to include in manual operations. Sheet feeders are common options for palletizers.

Banding Methods

These methods are used for column-stacked loads, or those others with little interlocking. Banding can be used also with corrugated wrapping and pallet boxes to provide a "gathering" function, or where packaging strength is needed. Banding is typically applied vertically; methods vary in the way its compressive force is applied (Fig. 34.1.3).

Bare banding often is done with large, rigid loads that cannot be damaged or deformed by the banding. Plywood or steel sheets are examples.

If the banding can cause damage, corner protectors protect the products. These can be made from folded, corrugated kraft. They also can be bought "ready made" from steel, plastic, or dense chipboard. The palletized products must be large enough to unitize with two or three bands per side. A balance between tension requirements and damage susceptibility must be struck for this method.

Top sheets are used for small, irregularly shaped cases on pallets. An extreme example of this is

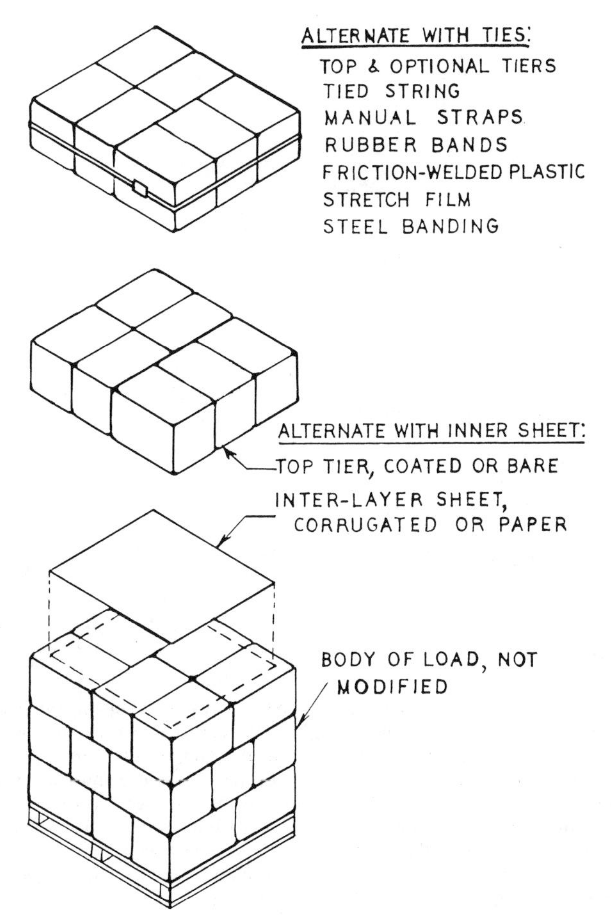

Fig. 34.1.2 Top-tier sliding safeguards.

bulk pallets of cans; they have chipboard sheets between each layer, and a plywood sheet on top. Heavy banding pressure holds it all together.

Top caps are used instead of sheets for looser loads, those that cannot resist heavy pressure, and for loads with more irregularity. The depth of caps is varied to provide as much "gathering" as needed. Usually, they are fabricated of corrugated, chipboard, or other paper material.

Wrapping Methods

Film wrapping and paper wrapping have differing uses. Paper wrapping is used to keep products clean and to provide a "gathering" quality to loads which are otherwise tied or banded. Paper seldom is used for strength.

Film, on the other hand, provides all of paper's advantages, and adds its own strength (Fig. 34.1.4). It almost always is used alone to bind and protect loads. Two film types are in common use: shrink film which is prestretched and springs back when heated, and stretch wrap which is simply stretched around the load. Film also may be plastic netting for products that need air circulation. Film can be opaque if security warrants.

Film application methods are well developed for production application at the palletizers. There are good reasons for applying film by machine: tension control, lower film usage, controlled coverage, and lower operation cost. When film is applied outside the production area, manual methods are common, but stretch machines are widely used also. These have a turntable on which the pallet turns while film is "bandaged" over the load by a stationary head. These machines are widely used, but the track record of shrink wrap machines has not been as good. The tendency of early shrink wrap ovens to break down spurred development of handguns to replace them. As a result, dockside shrink wrap ovens have diminished in use. Common manual methods are shown in Fig. 34.1.4.

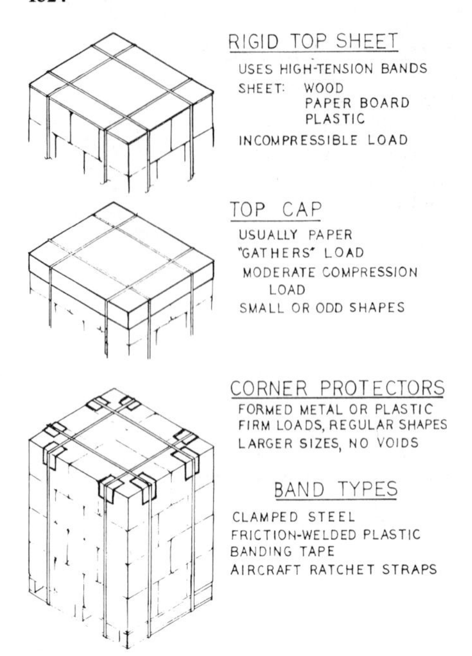

Fig. 34.1.3 Load-banding methods.

Manual shrink wrapping requires bags large enough to allow manual placement. The shrink percentage of the film must be large because of the loose fit of the bag. The manual gun used in this method burns propane, ignited electrically. Heat alone shrinks the film readily, but too much heat can be generated easily, which blows holes in the hot film. Placement of the bag and manual heating consume 3 min or less. This is as short as the cycle time for one-man use of the bag and fixed oven. If a larger operation requires speed, the common method is to limit order pickers to dropping the load onto a conveyor serving the oven. Others then operate the shrink wrap and move it to staging. The loss of one-man responsibility probably is partly responsible for the popularity of gun-shrinkage.

Dockside stretch wrapping often is a simple manual operation. However, cart-mounted roll mounts

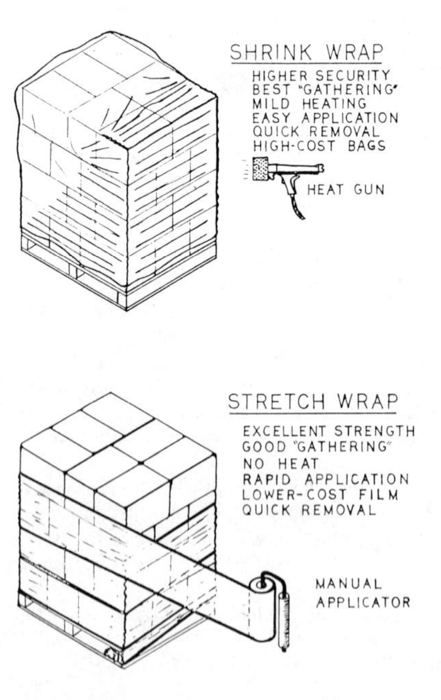

Fig. 34.1.4 Film application—manual methods.

also are available to do the same thing. The carts are available in manual types, with powered travel, and with automatic controls and steering. The operation of all is the same: First a leader of film is wadded up and jammed into the pallet at a corner. Second, the film is stretched around the load in the designed pattern, and third, the film is cut off and tucked under a previous fold to hold it. Stretch wrapping takes less time than shrink wrapping and does not require heat.

34.1.5 Truck Load Damage Control

A close fit of the load to the truck betters the chances for a damage-free trip. Light, oblong packages that fit the truck's interior well have the best chances. The tendency today is toward building higher pallet loads to save in-plant handling costs: more on the pallet per trip; but this results in fewer pallets per truck due to weight restrictions, so more voids show up. In addition, current packaging tends to produce denser products, which adds to the problem. To combat these effects, various stabilization methods are employed (Fig. 34.1.5).

Rear Braces

Front-to-rear motions result from engine acceleration, which is mild, and braking action, which is up to four times more powerful. All trucks have some longitudinal voids, combated by various means:

The load can be stair-stepped at the rear to lessen package fallover from the rear of the load.

If the load leaves a small gap at the rear, dunnage can be used to fill it. Dunnage can be solid pads of multi-ply corrugated kraft up to 3 in. thick, or it can be the inflatable bag type. The inflatable type expands to fill up voids of the load, but is not as resistant to compression as the solid type.

Larger gaps require transverse barriers which may be friction load bars, a built-in load bar system, or load doors. Load doors are similar to bars; they lock in place and provide a large bearing surface to the load. The same effect can be had with plywood sheets and load bars.

Side-Load Braces

Gaps at the sides occur within truck loads because unit loads are a poor fit, or because they are single-filed in the truck to meet load restrictions. Several methods of combating gaps have been commonly used:

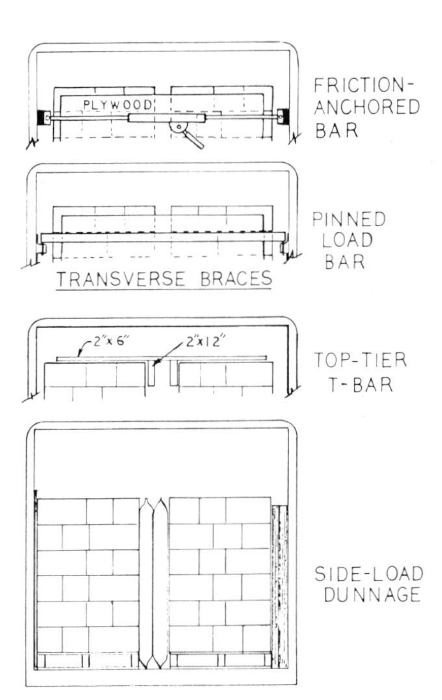

Fig. 34.1.5 Load-side aids.

Inflatable or multi-ply dunnage is most commonly used. Deflation of the air bags leaves a gap beside the pallet load to allow easier unloading. Multi-ply dunnage also helps. Multi-ply dunnage, or even thin corrugated kraft sheets, often is used beside pallet loads, solely to provide smooth, sliding surfaces for unloading.

Top-mounted load bars with short, vertical dividers are used as well. They are held by gravity and the short dividers prevent top tier slides. They are cheap and easy to handle. This is sufficient where top sliding is the only problem.

34.1.6 Truck Loading Operations

Dock and Apron Control

Trucks and drivers can be integrated into a dock operation if the results are beneficial to them as well as yourself. The benefits to the truckers are simple:

Minimum delays

No favoritism

Simple procedures

Delays can be minimized by scheduling trucks. This can be done for definite hours, or by 4-hr time slots. Within a time slot, trucks should be serviced on a first-come, first-served basis. Unscheduled arrivals should not be serviced before previous arrivals.

When trucks arrive, drivers should immediately log in on the Truck Log. A perfect system would immediately tell them the location and time at which they will be loaded or unloaded. Most systems are not perfect and drivers must wait until called.

A waiting room for drivers saves your time as well as theirs. With a toilet, coffee or vending machine, good seats, and suitably aged magazines, it is reasonable to expect drivers to stay put until called. Lacking this, drivers should stay with their trucks. Drivers should not be allowed elsewhere while waiting for services. If waiting in the yard, drivers can be called with a bullhorn, on an electrical callboard, CB radio, or other means.

When called, drivers move their vehicles to the proper location. If the truck is sealed, the dock foreman, or his or her representative, should break the seals, save them, and enter seal numbers on the Receiving Sheet or Bill of Lading. The driver places the wheel chocks, opens his own doors, removes bars, dunnage, and so on, and then remains to check the load on or off the truck. Plant personnel should place dock boards and position loading lights and should check wheel chocks and other safety equipment.

Safety Requires Attention

When the truck is at the dock with its doors open, it becomes an extension of the dock's interior, but it does not have sprinklers or other protective equipment. The driver must remain with the truck at the dock, ready to move it if needed. If trailers are dropped for service, a yard jack should be available for emergency moves.

The Driver Should Be Responsible

Local work rules determine who does what during truck loading, but the principle that each person should be responsible for his or her own output should be preserved. Though rules may dictate that plant personnel do the loading, drivers should be encouraged to ensure that loads are properly positioned, with dunnage and load bars properly placed. They also should be expected to take an active part in load checking and be provided with their copies of Load Sheets or Bills of Lading for this purpose.

Truck Loading with Pallets

The Load Sheet should indicate, by sketch or diagram, the position of each pallet load in the truck. As each load is placed, the Load Sheet should be tally-marked by the loader. The loaders should indicate, in the sketch, the position of dunnage, load bars, and other parts in the load.

When loads are placed, they should butt at least three directions—front, rear, and one side. The load interference with these surfaces helps to keep it intact. If remaining gaps are larger than package sizes, they must be filled with dunnage. Loads should be compressed to the front as allowed by the product; loose loads can tighten through vibration, leaving gaps behind.

Truck Loading with Conveyors

Alternate loading methods using conveyors are widely used with automated order picking. Conveyors can either be nonpowered and manually placed, or be the telescopic, powered type. Automatic system controls are the primary load verification. When checkers are used, they generally are responsible for label application as well. With conveyors, cases are most often floor-stacked for multiple destinations.

Because manual stacking allows much tighter fitting of the load to the truck, fewer load-stabilization devices are used. Usually, stair-stepping provides protection for partial loads between stops. Load bars are used also.

34.1.7 Completing the Load

Dock supervision should ensure that the load is complete and secure before accepting it from the loader. Then Truck Seals are recorded on the Truck Sheet or on the Bill of Lading. On trucks that unload at several points, seals may be issued for each leg of the trip. However, seal numbers are important only for the first leg. One cannot be sure what number was used after that. It would be well to use seals with numbers not in the regular sequence, or unnumbered seals for remaining legs of the trip. Best practice would be for the carrier to provide seals for secondary trip legs.

Completing the Documentation

Final documentation is then completed by the truck driver and appropriate party, who may be either the supervisor or a shipping clerk. Either way, the Bill(s) of Lading and Truck Sheets are signed, with seal numbers and load quantity corrections entered by both the driver and appropriate plant personnel. The last dock transaction is to enter truck time of completion and departure on the Truck Log.

Out the Gate

In larger facilities, the truck must pass a guard gate to leave. At this point, the guard should double-check the truck seals against the numbers on the Bill of Lading. It also is not unusual to log the time of departure, seals, and numbers as an independent check on the truck's movements within the facility.

34.2 AIR INTERFACE

Robert Promisel

34.2.1 Introduction

In this chapter, the various ways packages may be shipped to points anywhere in the United States, as well as internationally, will be covered.

Although the air freight field has grown in leaps and bounds over the years, the general information contained herein may change while this chapter is being written. For information on specific methods for shipping various products the air carrier should be contacted.

Right now out of Newark Airport, LaGuardia Airport, and Kennedy Airport (they make up the New Jersey/New York air cargo center), more than 1800 cargo carrying flights leave each day. This is an average of one every 48 seconds.

Airborne Foreign Commerce Statistics

Figures from the United States Census Bureau show that airborne foreign commerce in the United States was close to 2 million tons in calendar year 1981. Estimated value of these imports and exports was about $77 billion. Volume was up 120,000 tons, and value increased by $3 million from the previous year.

New York's JFK Airport was far and away the export–import leader, accounting for one-third of all tonnage and close to one-half of all dollar value. Los Angeles International Airport was second in both categories. The top ten airports are show below:

Airport	Volume (tons)	Value (billion $)
1. New York (JFK)	721,118	37.50
2. Los Angeles	209,009	10.00

	Airport	Volume (tons)	Value (billion $)
3.	Chicago (O'Hare)	159,977	5.90
4.	Miami	117,597	4.00
5.	San Francisco	92,769	6.80
6.	Boston (Logan)	50,389	2.38
7.	Atlanta	21,443	0.65
8.	San Juan	20,052	0.40
9.	Seattle–Tacoma	18,932	0.75
10.	Honolulu	16,174	0.70

Rates

Rates are based on the zone charts supplied by the individual air carriers. Rates vary from carrier to carrier and vary with time. Thus they will not be covered here. Individual organizations considering shipping by air should consult the air carriers individually, and get confirmed rates from them.

34.2.2 Methods of Shipment

The Air Transportation Network

An air carrier will pick up and deliver to as many as 480 airports and approximately 12,000 delivery points nationwide.

Forwarder Pickup and Delivery. A service where the air carrier may use a freight forwarder to pick up from source point and deliver from arrival point to customer. In this instance, the forwarder is not part of the air carrier's organization.

Air/Truck Service. Shipments can be interchanged with air carriers, air freight forwarders, or air service of your choice.

Door-to-Door Service. Pick up at your door, transfer to an air carrier, and, once in the destination city, its counterpart will make the delivery, providing door-to-door service on a single air bill. Pickups of air freight may be in trucks that range from 10-ft (3.05 m) econolines to 45-ft (13.7 m) tractor trailers, as well as flat-bed trailers.

Some offer next day service before 10:30 A.M., others by noon, and still others 24 hr from the time they pick up the packages.

Package Classifications. Another major air carrier has several package methods of delivering material quickly:

Priority. This is a door-to-door overnight service of up to 70 lb that insures delivery up to noon the next day.

Overnight Letter. Up to 2 oz delivered overnight.

Courier Pak—Overnight Envelope. Up to 2 lb door-to-door delivery.

Courier Pak—Overnight Box. Up to 5 lb delivered overnight door-to-door.

Courier Pak—Overnight Tube. Up to 5 lb delivered in a tube.

Express Mail Next Day Service. Up to 2 lb can be delivered on overnight service to more than 1200 cities where there are over 3000 post offices in the United States. Currently over 70,000 packages are delivered this way by the United States Postal Service.

Small Package Dispatch. This method guarantees cross-country service on a one-day or a same-day delivery basis, usually predicated by a flat charge, airport to airport. There is also a flat charge pickup delivery service by city zone. The restrictions on this type of package are 50 lb (22.7 kg) maximum, 90 in. (228.6 cm) maximum total dimension. Other air carriers sometimes increase this level to 70 lb (31.8 kg) with some 90 in. (228.6 cm) dimensions. It is usually picked up within $\frac{1}{2}$–3 hr of the call to an 800 number provided by shipper. It is usually delivered with a prepaid delivery receipt within $\frac{1}{2}$–3 hr at the destination point.

First Freight. Guaranteed on a specified flight with a general commodity rate plus 30% more than regular freight.

Restrictions. No limit to pieces or weight, individual pieces or boxes. Pickup and delivery may be by shipper's truck or by air carrier truck. Each piece must be labeled and segregated in freight terminal. Each shipment is computer-monitored by air freight information systems.

Regular Air Freight. Usually goes on passenger as well as jet freighters. It is not a priority since it does not require same day service.

Dispatch. Air freight organizations provide a variety of services that may include 24-hr-a-day dispatch, 7 days per week. This type of service is used primarily for: (1) critical repair parts, (2) documents, and (3) blood.

Distant Services

Alaska. From continental United States to Anchorage, next business-day delivery; from Anchorage to continental United States, second business-day delivery.

Hawaii. From continental United States to Hawaii, next business-day delivery; from Hawaii to continental United States, second business-day delivery.

Puerto Rico. From continental United States to Puerto Rico, next business-day delivery; from Puerto Rico to continental United States, by noon next business day.

Assembly or Distribution Services

1. Assembly service will be performed by the carrier, subject to the following:
   a. The carrier will accept two or more parts of a shipment from one or more shippers at point of origin and will assemble such parts, at one location at point of origin, into one shipment for transportation for one airport of origin to one destination airport for delivery to one consignee at one destination address if, no later than a time of receipt by the carrier of the first of the parts assembled, the carrier receives written instruction to provide assembly service for those parts from the shipper (or, from the consignee, if there be more than one shipper); provided, however, that all parts of the shipment, other than those mislayed, shall be delivered to the consignee at one time.
   b. All charges applicable to shipments receiving assembly service shall be paid by the consignee.
   c. The shipper indicates in writing with each part tended, "ASSEMBLY SERVICE REQUESTED."
   d. The shipments moving under carriers' customs bond and assembly service will be accepted only when all parts receiving such service are under such customs bond.
   e. All parts of a shipment to be assembled must be tendered to the carrier within a 24-hr period ending at 2:00 A.M. daily. Any parts received after 2:00 A.M. will be considered as separate shipments at the rate or charge applicable thereto, or if requested, as part of the next day's assembly.
   f. The carrier will not perform assembly service in connection with any shipment that is a quartered distribution service.
   g. Each part of an assembly shipment shall be subject to charges for cubic dimensional weight as provided in specific rules, and to charge for declared value as provided in specific rules.
   h. The service charge for assembling parts of a shipment varies from carrier to carrier, subject to a minimum charge dependent upon the carrier's rules for shipment.
   i. The carrier will not perform assembly service in connection with any shipment that is a quartered containment loading service.
2. Distribution service will be performed by the carrier, subject to the following:
   a. Upon receipt of written instructions to provide distribution service and when a manifest is given the proper breakdown of a shipment an individual manifest listing the goods to be delivered to each address, is received by the carrier from the shipper or consignee (if there be more than one consignee, only from the shipper) no later than the time of receipt by the carrier of a shipment the carrier will accept a shipment from one shipper at one time at one address, receipted in one lot for transportation from one airport of origin and will segregate the parts of the shipment at destination where the carrier will deliver such parts to the consignee or consignees, provided, however, that if the parts of the shipment are to be delivered to more than one consignee the shipment must be prepaid. Distribution service is not available for shipments from United States points to Canada points.

b. The carrier will not perform distribution service and connection with any shipment which is accorded assembled service.

c. The service charge for distributing parts of a shipment varies from carrier to carrier, subject to a minimum charge dependent upon the carrier's shipment rules.

d. Each part of a distribution shipment shall be subject to charges for declared value as provided for in a specific rule.

3. When the carrier has received written instructions to provide assembly service, and only one part of a shipment is tendered within a 24-hr period ending at 2:00 A.M., the service charge named herein shall be applied to such part.

4. When either pickup or delivery service is requested for parts of a shipment, such services will be provided, subject to applicable rates and charges applied individually to each part.

5. Notwithstanding paragraph 1a, one or more parts of a shipment that receives assembly service may be released by the carrier to the consignee upon the demand of the consignee; provided that there is a charge assessed of a figure to be stipulated by the specific carrier, subject to a minimum charge dependent on a carrier for one or more parts released at one time; provided that transportation charges are paid on such part or parts at the rate which would have been applied had such a part or parts been a separate shipment; and provided the charges for the remaining part or parts of the assembly shipment are recomputed to reflect the exclusion of the part or parts released. Nothing shall affect the duty of the carrier to collect its full tariff charges for providing assembly service on the entire shipment including those assembly charges applicable to the part or parts released.

34.2.3 Aircraft Cargo Capacity and Specifications

The aircraft shown in Fig. 34.2.1 are the most commonly used for transporting air freight anywhere in the continental United States. The 747 is used primarily for overseas shipments. The capacities on these aircraft can range from 12,830 lb (5,733.5 kg) to as much as 91,000 lb (40,950 kg).

34.2.4 Air Freight Containers and Specifications

Figure 34.2.2 provides examples of air freight containers that may be used. Four of the containers are supplied by air carriers and four of the containers are usually provided by the shippers. Table 34.2.1 also provides container data.

Air freight is usually picked up and assembled in an air cargo container to be shipped to a consignee in another part of the country.

There are a variety of different size containers available to the shipper including an E container that is stocked for sale to the shipper.

Containerization protects shipments from loss or damage and has the advantage of favorable rates. In addition, the contents of the container can be distributed to any number of consignees on request. Container capacities can vary anywhere from 500 lb (226.8 kg) and 16 ft^3 to as much as 12,500 lb (5670 kg) and 460 ft^3. With certain air carriers there is no rental charge if the loaded container is returned within 48 hr.

Various types of material handling equipment are used to move these containers around:

For the type A container, a dolly transporter is used.

For the type D container, a standard lift truck is required.

For the L3 container, a dolly transporter is used.

For the type LN container, a standard lift truck is used.

For the type L7, a dolly transporter is used.

For the type L11, a dolly transporter is used.

Container Type	Tare Allowance per Container
A	Actual weight
D	63 lb (28.6 kg)
E, E2	18 lb (8.2 kg)
EH	9 lb (4.1 kg)
FTC	Actual weight
L3, L5, L7	Actual weight
LN	100 lb (45.4 kg)
M1, M2	Actual weight
Q	13 lb (5.9 kg)

Other container types have no tare allowance.

Container Specifications

1. Type A Container
   a. A pallet, with load properly restrained and contoured within the maximum dimensions of a Type A container, and not exceeding the maximum gross weight for such container type, shall be considered a Type A container.
   b. Type A containers owned by the shipper or consignee must be certified by the Federal Aviation Administration as being air worthy, and a copy of such certification must be supplied to the carrier by the shipper. Tendering such containers, the shipper affirms that the container has been maintained in conformity with the Federal Aviation Administration.
   c. Empty weight shall be marked on at least one outside vertical face of the container.
   d. Furnished by carrier, shipper, or consignee.
2. Type D Containers
   a. A pallet with a restrain net, straps, or banding to restrain the load shall not be considered a container. EXCEPTION: a pallet with load properly restrained and contoured within the maximum dimensions, weight limitations, and top loading capability of a Type D container shall be considered a Type D container. Dimensions, gross weight, and top loading capability must be shown on 2 sides of the pallet-supported shipment.
   b. Provisions for mechanical handling are required. Fork tine entries shall provide a minimum of $2\frac{1}{2}$ in. clear height and 8 in. width for each tine and be located for standard fork tine application.
   c. Container characteristics outlined in Fig. 34.2.2 include maximum external dimensions, actual container dimensions, actual empty weight, and minimum height and width.
3. Types E and EH
   a. A pallet with load properly restrained and contoured within the maximum dimensions, weight limitations, and top loading capability of a Type E or EH container, shall be considered a Type E or EH container. Dimensions, gross weight, and top loading capability must be shown on 2 sides of the pallet-supported shipment.
   b. Container type, maximum external dimensions and inches, actual container dimensions and inches, actual empty weight and pounds, and minimum top loading capability must be marked on two outside vertical faces of the container in letters and numbers and not less than $\frac{3}{4}$ in. (1.9 cm) high and $\frac{1}{2}$ in. (1.3 cm) wide.
   c. Furnished by shipper or consignee.
4. Type F Containers
   a. Containers must be rectangular with each of the adjoining sides perpendicular (at right angles) to one another.
   b. Maximum external dimensions in inches must be marked on two outside vertical faces of the containers in numbers and letters and not less than $\frac{3}{4}$ in. (1.9 cm) high and $\frac{1}{2}$ in. (1.3 cm) wide.
   c. Furnished by shipper or consignee.
5. Type L Containers
   a. A pallet with a restraint net, straps, or banding to restrain the load shall not be considered a container, except:
      Pallet with load properly restrained and contoured within the maximum dimension and weight limitations with Type L7 containers shall be considered a Type L7 container.
      A pallet properly restrained with overall dimensions of 60 in. (152.4 cm) in width, 125 in. (317.5 cm) in length, and not exceeding 64 in. (162.6 cm) in height and which does not exceed the maximum weight limitation for Type L5 containers shall be considered a Type L5 container.
   b. Furnished by carrier or shipper.
6. Type LN Containers
   a. A pallet with a restraint net, straps, or banding to restrain the load shall not be considered a container, except:
      A pallet with load properly restrained and contoured within the maximum dimensions of a Type LN container and not exceeding the maximum gross weight for such container type, shall be considered a Type LN container.
   b. Provisions for mechanical handling are required. Fork tine entries shall provide a minimum of $2\frac{1}{2}$ in. (6.4 cm) clear height and 8 in. (20.3 cm) width for each tine and be located for standard fork tine application.

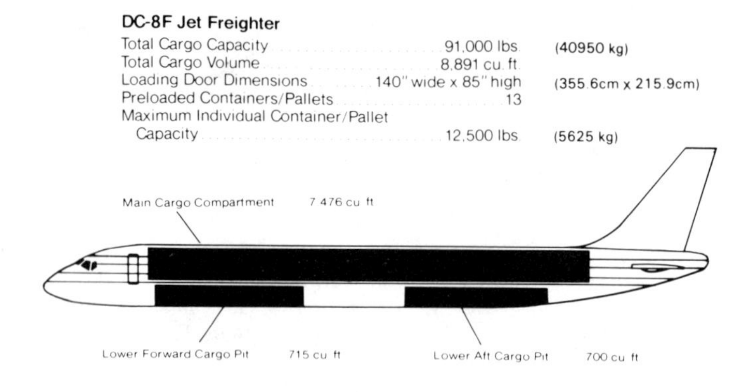

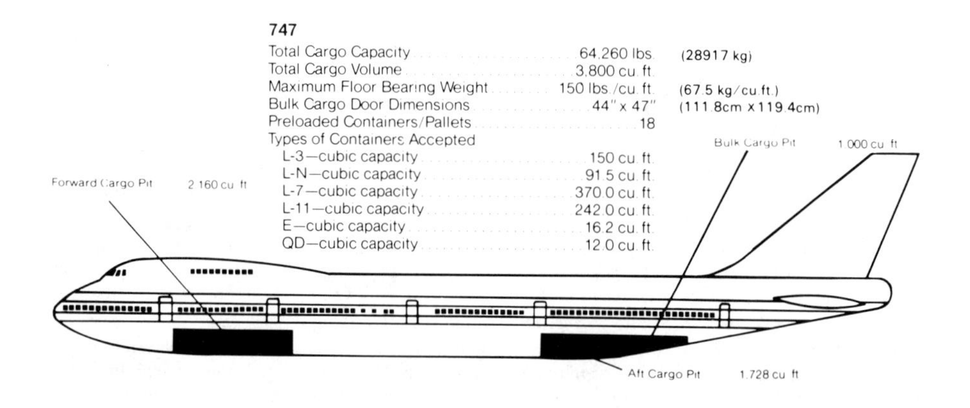

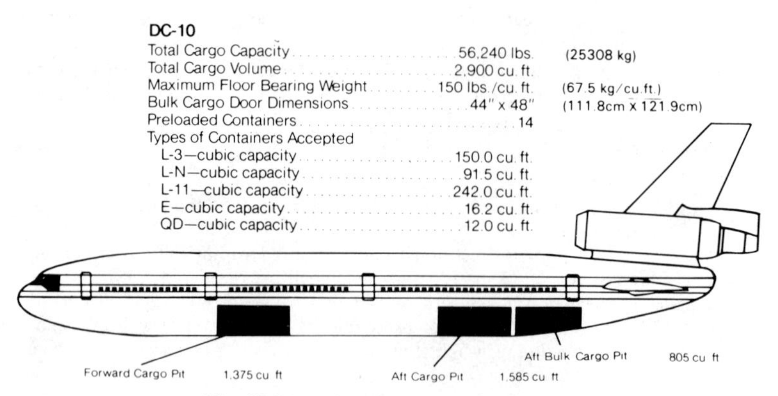

Fig. 34.2.1 Aircraft cargo capacity.

DC-8/Super DC-8

	DC-8	Super DC-8	Metric DC-8	Metric Super DC-8
Total Cargo Capacity	20,850 lbs.	23,615 lbs.	9382.5 kg	10626.8 kg
Total Cargo Volume	1,390 cu. ft.	1,868 cu. ft.		
Maximum Floor Bearing Weight	120 lbs./sq. ft.	24,225 lbs.	54 kg/.09m²	10901.3 kg
Bulk Cargo Door Dimensions	36" x 44"	36" x 44"	91.4cm x 111.8cm	91.4cm x 111.8cm

Types of Containers Accepted
E—cubic capacity . . . . . 16.2 cu. ft.
QD—cubic capacity . . . . . 12.0 cu. ft.

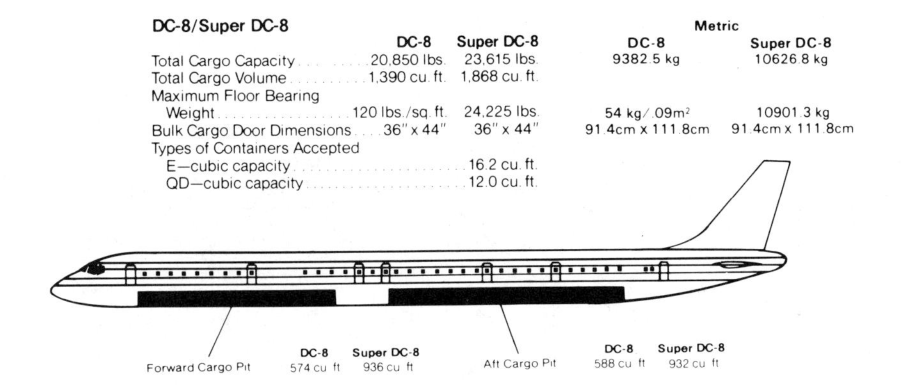

727/Super 727

	727	Super 727	Metric 727	Metric Super 727
Total Cargo Capacity	12,830 lbs.	21,000 lbs.	5773.5 kg	9450 kg
Total Cargo Volume	910 cu. ft.	1,454 cu. ft.		
Maximum Floor Bearing Weight	150 lbs./cu. ft.	150lbs./cu. ft.	67.5 kg/cu.ft.	67.5 kg/cu.ft.
Cargo Door Dimensions	48" x 45"	42" x 55"	121.9cm x 114.3cm	106.7cm x 139.7cm

Types of Containers Accepted
E—cubic capacity . . . . . 16.2 cu. ft.
QD—cubic capacity . . . . . 12.0 cu. ft.

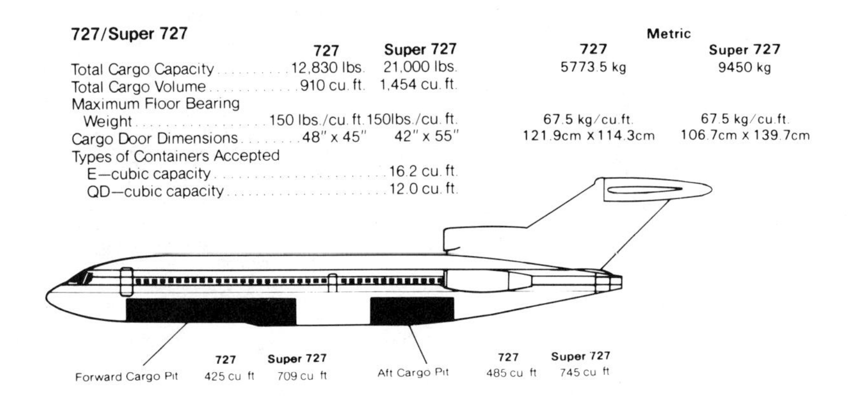

737

Total Cargo Capacity	12,985 lbs.	(5843.3 kg)
Total Cargo Volume	875 cu. ft.	
Maximum Floor Bearing Weight	150 lbs./cu. ft.	(67.5 kg/cu.ft)
Cargo Door Dimensions	FWD 48" x 35"	(121.9cm x 88.9cm)
	AFT 48" x 33"	(121.9cm x 83.8cm)

Types of Containers Accepted
E—cubic capacity . . . . . 16.2 cu. ft.
QD—cubic capacity . . . . . 12.0 cu. ft.

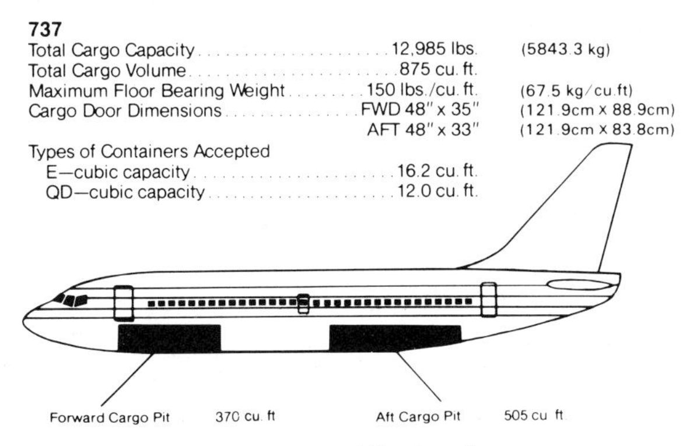

Fig. 34.2.1 (*Continued*)

Container General Commodity

Air Freight Containers

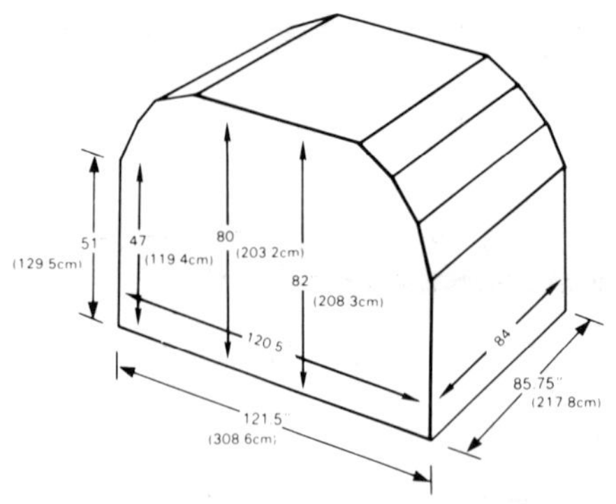

TYPE A

(EQUIVALENT TO IATA TYPE 3)
CONTOURED FOR THE DC-8F
PROVIDED BY AIR CARRIER

External Displacement 460 Cubic Ft.
Internal Capacity 440 Cubic Ft.
Allowable Tare Weight Actual Weight
Maximum Gross Weight 12,500 Lbs. (5625 kg)
Maximum Floor Bearing Weight 200 Lbs. per sq. ft. (90 kg per .09m²)

Dolly Transporters Available.

A pallet with load properly restrained and contoured within the maximum dimensions of a Type A container will be considered a container.

45"
(114.3cm)
42" (106.7cm)
58"
(147.3cm)

TYPE D

FOR THE DC-8F
PROVIDED BY THE SHIPPER

Maximum External Dimensions L-42", W-58", H-45" (106.7cm-147.3cm-114.3cm)
External Displacement 57.1 Cubic Ft.
Internal Cubic Capacity Varies (Container may be any size up to maximum dimensions)
Allowable Tare Weight 63 Lbs. (28.4 kg)
Maximum Gross Weight 2,000 Lbs. (900 kg)
Maximum Floor Bearing Weight 200 Lbs. per sq. ft. (90 kg per .09m²)
Top Loading Capacity 1,200 Lbs. (540 kg)

Provisions for standard fork lift handling are required.

Container type, actual dimensions, actual empty weight and top loading capability must be marked legibly on two sides of the container.

A pallet with load properly restrained and contoured within the maximum dimensions of a Type D container, will be considered a container.

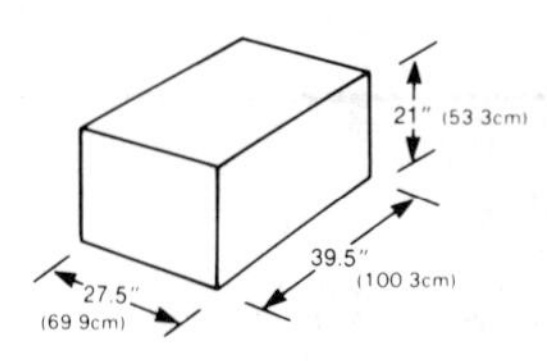

TYPE QD

FOR NARROW BODY AIRCRAFT
PROVIDED BY THE SHIPPER

Maximum External Dimensions L-39.5", W-27.5", W-21" (100.3cm-69.9cm-53.3cm)
External Displacement 12 Cubic Ft.
Internal Cubic Capacity Varies
Allowable Tare Weight 13 Lbs. (5.9 kg)
Maximum Gross Weight 400 Lbs. (180 kg)
Maximum Floor Bearing Weight 200 Lbs. per sq. ft. (90 kg per .09m²)

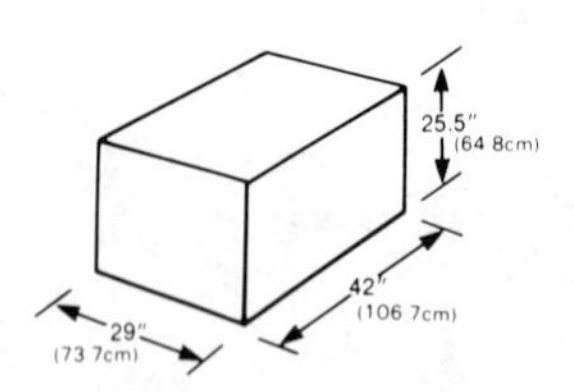

TYPE E

FOR NARROW BODY AIRCRAFT
PROVIDED BY THE SHIPPER

Maximum External Dimensions L-42", W-29", H-25.5" (106.7cm-73.7cm-64.8cm)
External Displacement 16.2 Cubic Ft.
Internal Cubic Capacity Varies
Allowable Tare Weight 18 Lbs. (8.1 kg)
Maximum Gross Weight 500 Lbs. (225 kg)
Maximum Floor Bearing Weight 200 Lbs. per sq. ft. (90 kg per .09m²)

Fig. 34.2.2 Examples of air freight containers.

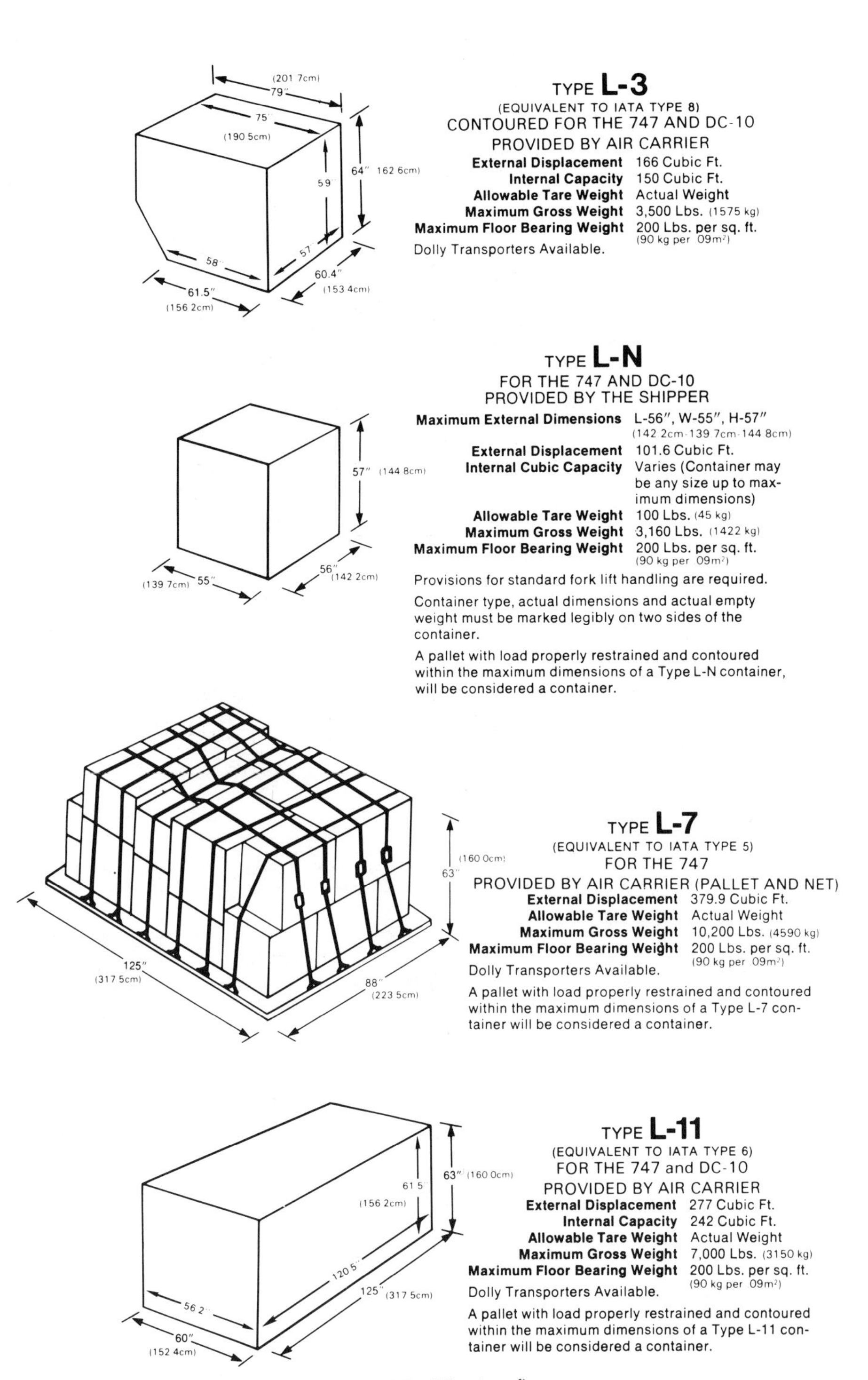

Fig. 34.2.2 *(Continued)*

Table 34.2.1 Air Freight Containers[a]

Container Type	Owner	Capacity (ft^3)	External Dimensions (in.)	External Dimensions (cm)	Displacement (ft^3)	Minimum Chargeable Weight (lb)	Minimum Chargeable Weight (kg)	Maximum Gross Weight (lb)	Maximum Gross Weight (kg)	Handling Features for Shippers[b]
M1	Air carrier	572	L: 125 W: 96 H: 96	318 244 244	666	4,400	1,980	15,000	6,750	A
M2	Air carrier	1077	L: 240 W: 96 H: 96	610 244 244	1,280	12,363	5,563	25,000	11,250	A
L6	Air carrier	310	L: 160 W: 60.4 H: 64	406 153 163	358	2,800	1,260	7,000	3,150	B
LD7	Air carrier	355	L: 125 W: 88 H: 63	318 224 160	401	2,800	1,260	10,400	4,680	B
L10	Air carrier	Varies	L: 125 W: 60.4 H: 63	318 153 160	275	1,694	762	6,500	2,925	B
A1	Air carrier	393	L: 88 W: 125 H: 87	224 318 221	554	3,000	1,350	13,300	5,985	B

A2	Air carrier	440	L: 88	224	554	3,200	1,440	12,500	5,625	B
A3	Air carrier	440	W: 125	318						B
			H: 87	221						
B	Shipper	Varies	L: 84	213	214	1,800	810	5,000	2,250	C
			W: 58	147						
			H: 76	193						
LW	Air carrier	Varies	L: 98	249	100	500	225	1,200	540	B
			W: 42.2	107						
			H: 41.6	106						
B2	Shipper	Varies	L: 42	107	107	900	405	2,500	1,125	C
			W: 58	147						
			H: 76	193						
EH	Shipper	9	L: 35.4	90	9	100	45	250	113	D
			W: 21	53						
			H: 21	53						
Q	Shipper	Varies	L: 39.5	100	13	100	45	400	180	E
			W: 27.5	70						
			H: 21	53						

[a] Dimensions and weights are approximate and may vary slightly between carriers.
[b] A, picked up or delivered on conventional truck trailer chassis; B, dolly transporters available; C, lift trucks; D, consult air carrier; E, side handles recommended.

   c. Container type, maximum external dimensions in inches, actual container dimensions in inches, actual empty weight in pounds, and minimum top loading capability must be marked on two outside vertical faces of the container in letters and numbers not less than $\frac{3}{4}$ in. (1.9 cm) high and $\frac{1}{2}$ in. (1.3 cm) wide.
   d. Furnished by shipper or consignee.
7. Type M Containers
   a. A pallet or marine container, with load properly restrained and contoured within the maximum dimensions of a Type M1, M2 container and not exceeding the maximum gross weights for such container types, shall be considered a Type M1 container or a Type M2 container.
   b. Furnished by carrier or shipper.
8. Type Q Containers
   a. A pallet with a restraint net, straps, or banding to restrain the load shall not be considered a container.
   b. Container type, maximum external dimensions in inches, actual container dimension in inches, actual empty weight in pounds, and minimum top loading capability must be marked on two outside vertical faces of the container in letters and numbers not less than $\frac{3}{4}$ in. (1.9 cm) high and $\frac{1}{2}$ in. (1.3 cm) wide.
   c. Furnished by shipper or consignee.
9. Type FTC Container.
   a. Actual empty weight shall be marked on at least one outside vertical face of the container.
   b. Furnished by shipper.

Additional Container Data

The containers are rented based on the distance that they will be shipped (e.g., the rate from Akron, Ohio, to Chicago, Illinois, is lower than the rate from Akron, Ohio, to Denver, Colorado).

Containers are made primarily of metal (returnable to air carriers) or combinations of corrugated with and without wood base. These containers may be in two individual categories: disposable and reusable.

34.2.5 Materials Handling Equipment

Typical methods of handling are as follows:

1. By lift trucks onto a flat-bed truck.
2. Intermodal container hooked up to a truck chassis which moves right to the air freighter, bypassing the freight terminal completely.
3. Break bulk system to deliver packages to individual consignees in individual cities.
4. ISO (International Standards Organization) containers being used on 747 freighters overseas.

One such small package carrier operates approximately 100 aircraft to a 500,000 ft^2 hub in the southeast. Sorts approximately 400,000 packages in a 2-hr span at night with about 1000 people on a floating shift. These packages are then placed back on their original aircraft and the packages are returned to the destination point and distributed.

34.2.6 Procedure for Preparing an Airbill

Code numbers, with explanations, will help familiarize the individual with the easy steps for filling out this airbill (see Fig. 34.2.3):

1. Enter the origin station code (e.g., SFO, LAX, EWK) (see Table 34.2.2).
2. Declared Value—the value declared for carriage when specified by the shipper. When no value is declared the appreciation NVD should be entered in this box.
3. Routing—shipper may specify the desired routing of the shipment.
4. Airport of Destination—enter the name of the city or airport of destination.
5. Prepaid/Collect—check appropriate box to show who will pay the charges.
6. Consignee—the name, address, city, state, and zip code of the firm or individual who will receive the shipment. The words "to order of" or any similar phrase should not be used. If additional information is needed for delivery, enter in section 10 (instruction to carrier).
7. Code Number—when applicable, the trucker number should be entered in this section.

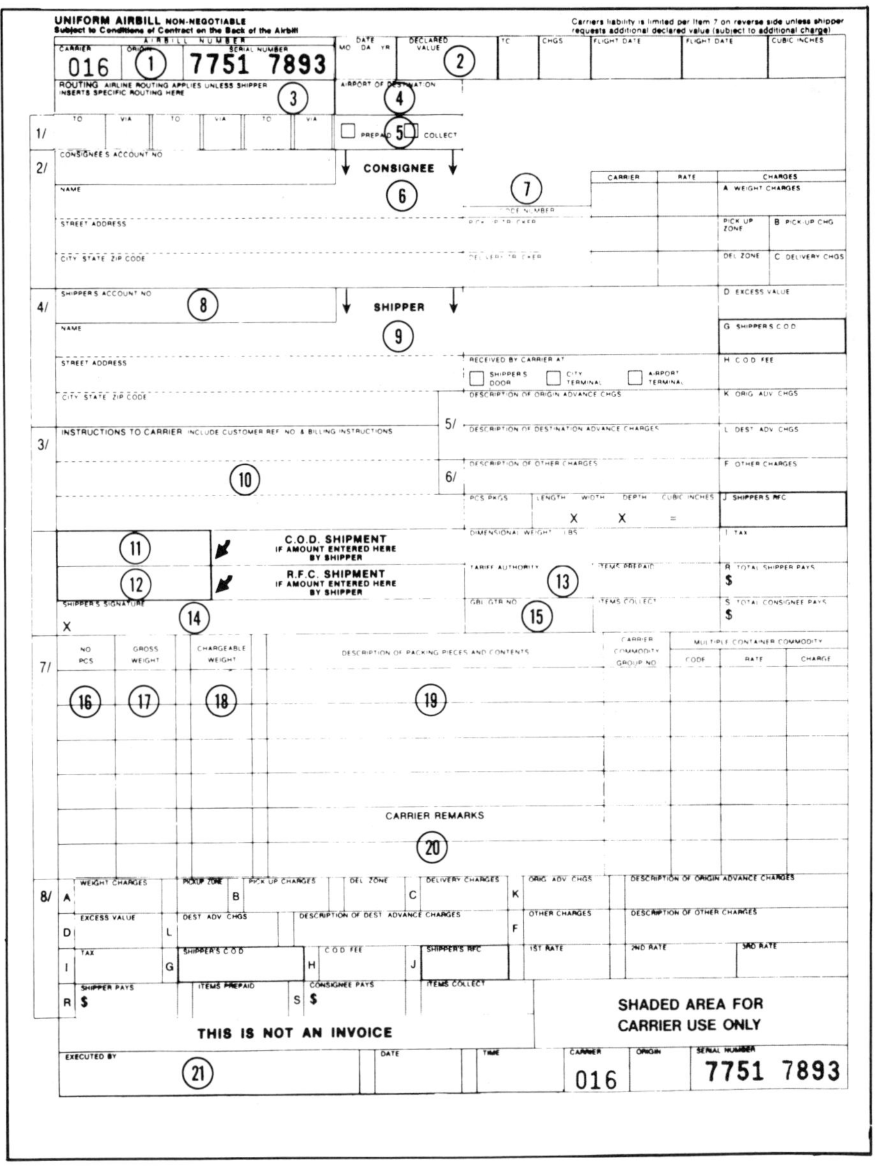
UNIFORM AIRBILL NON-NEGOTIABLE
Subject to Conditions of Contract on the Back of the Airbill
Carriers liability is limited per Item 7 on reverse side unless shipper requests additional declared value (subject to additional charge)
AIRBILL NUMBER
CARRIER 016 | ORIGIN (1) | SERIAL NUMBER 7751 7893
DATE MO DA YR | DECLARED VALUE (2) | TC | CHGS | FLIGHT DATE | FLIGHT DATE | CUBIC INCHES
ROUTING AIRLINE ROUTING APPLIES UNLESS SHIPPER INSERTS SPECIFIC ROUTING HERE (3)
AIRPORT OF DESTINATION (4)
1/ TO | VIA | TO | VIA | TO | VIA
PREPAID (5) COLLECT
2/ CONSIGNEE'S ACCOUNT NO
CONSIGNEE (6)
NAME
STREET ADDRESS
CITY STATE ZIP CODE
(7)
CARRIER | RATE | CHARGES
A WEIGHT CHARGES
PICK UP ZONE | B PICK-UP CHG
DEL ZONE | C DELIVERY CHGS
D EXCESS VALUE
4/ SHIPPER'S ACCOUNT NO (8)
SHIPPER (9)
NAME
STREET ADDRESS
CITY STATE ZIP CODE
G SHIPPER'S C.O.D.
RECEIVED BY CARRIER AT
SHIPPER'S DOOR | CITY TERMINAL | AIRPORT TERMINAL
H C.O.D. FEE
DESCRIPTION OF ORIGIN ADVANCE CHGS | K ORIG ADV CHGS
3/ INSTRUCTIONS TO CARRIER INCLUDE CUSTOMER REF NO & BILLING INSTRUCTIONS
(10)
5/ DESCRIPTION OF DESTINATION ADVANCE CHARGES | L DEST ADV CHGS
6/ DESCRIPTION OF OTHER CHARGES | F OTHER CHARGES
PCS PKGS | LENGTH X WIDTH X DEPTH = CUBIC INCHES | J SHIPPER'S RFC
DIMENSIONAL WEIGHT LBS | I TAX
(11) C.O.D. SHIPMENT IF AMOUNT ENTERED HERE BY SHIPPER
(12) R.F.C. SHIPMENT IF AMOUNT ENTERED HERE BY SHIPPER
TARIFF AUTHORITY (13) | ITEMS PREPAID | R TOTAL SHIPPER PAYS $
SHIPPER'S SIGNATURE X (14)
GBL GTR NO (15) | ITEMS COLLECT | S TOTAL CONSIGNEE PAYS $
7/ NO PCS (16) | GROSS WEIGHT (17) | CHARGEABLE WEIGHT (18) | DESCRIPTION OF PACKING PIECES AND CONTENTS (19) | CARRIER COMMODITY GROUP NO | MULTIPLE CONTAINER COMMODITY: CODE | RATE | CHARGE
CARRIER REMARKS (20)
8/ A WEIGHT CHARGES | PICKUP ZONE | B PICK UP CHARGES | DEL ZONE | C DELIVERY CHARGES | K ORIG ADV CHGS | DESCRIPTION OF ORIGIN ADVANCE CHARGES
D EXCESS VALUE | L DEST ADV CHGS | DESCRIPTION OF DEST ADVANCE CHARGES | F OTHER CHARGES | DESCRIPTION OF OTHER CHARGES
I TAX | G SHIPPER'S C.O.D. | H C.O.D. FEE | J SHIPPER'S RFC | 1ST RATE | 2ND RATE | 3RD RATE
R SHIPPER PAYS $ | ITEMS PREPAID | S CONSIGNEE PAYS $ | ITEMS COLLECT
THIS IS NOT AN INVOICE
SHADED AREA FOR CARRIER USE ONLY
EXECUTED BY (21) | DATE | TIME | CARRIER 016 | ORIGIN | SERIAL NUMBER 7751 7893

Fig. 34.2.3 Elements of an airbill.

8. Shipper's Account Number—number assigned by air carrier to identify the customer should be entered here.
9. Shipper—full name, address, city, state, and zip code of the firm and individual making the shipment.
10. Instructions to Carrier—enter shipper instructions such as:

   New line
   Special handling requirements
   Specific information for onward carriage or delivery
   Consignee telephone number
   Reference to related air waybills or shipping documents
   Billing instructions

Table 34.2.2 Airport Codes for U.S. Airports

Code	Location	Code	Location
ABO	Albuquerque, New Mexico	MCI	Kansas City, Missouri
ALB	Albany, New York	MCO	Orlando, Florida
AMA	Amarillo, Texas	MEM	Memphis, Tennessee
AUS	Austin, Texas	MIA	Miami, Florida
BDL	Hartford, Connecticut	MOB	Mobile, Alabama
BHM	Birmingham, Alabama	MSP	Minneapolis/St. Paul, Minnesota
BNA	Nashville, Tennessee	MSY	New Orleans, Louisiana
BOS	Boston, Massachusetts	OAK	Oakland, California
BTR	Baton Rouge, Louisiana	OKC	Oklahoma City, Oklahoma
BUF	Buffalo, New York	ONT	Ontario, California
BWI	Baltimore, Maryland	ORD	Chicago, Illinois
CLE	Cleveland, Ohio	ORF	Norfolk/Va. Beach, Virginia
CMH	Columbus, Ohio	PDX	Portland, Oregon
CRP	Corpus Christi, Texas	PHL	Philadelphia, Pennsylvania
CVG	Cincinnati, Ohio	PHX	Phoenix, Arizona
DAY	Dayton, Ohio	PIT	Pittsburgh, Pennsylvania
DEN	Denver, Colorado	PSP	Palm Springs, California
DFW	Dallas/Fort Worth, Texas	RNO	Reno, Nevada
DTW	Detroit, Michigan	ROC	Rochester, New York
ELP	El Paso, Texas	SAN	San Diego, California
EWK	Newark, New Jersey	SAT	San Antonio, Texas
HNL	Honolulu, Hawaii	SDF	Louisville, Kentucky
HRL	Harlingen, Texas	SEA	Seattle, Washington
HSV	Huntsville, Alabama	SFO	San Francisco, California
IAH	Houston, Texas	SHV	Shreveport, Louisiana
ICT	Wichita, Kansas	SJC	San Jose, California
IND	Indianapolis, Indiana	SJU	San Juan, Puerto Rico
JAN	Jackson, Mississippi	SLC	Salt Lake City, Utah
JFK	New York, New York	SMF	Sacramento, California
LAS	Las Vegas, Nevada	STL	St. Louis, Missouri
LAX	Los Angeles, California	SYR	Syracuse, New York
LBB	Lubbock, Texas	TPA	Tampa, Florida
LGA	New York, New York	TUL	Tulsa, Oklahoma
LIT	Little Rock, Arkansas	TUS	Tucson, Arizona
MAF	Midland/Odessa, Texas	WAS	Washington, D.C.

11. C.O.D. Shipment—enter any amount to be collected from the consignee on behalf of the shipper (does not include the C.O.D. fee).
12. R.F.C. Shipment—not applicable, leave blank.
13. Tariff Authority—enter any reference, required to obtain rate classification.
14. Shipper's Signature—legible signature of shipper or agent of shipper.
15. GBL/GTR Number—the complete government bill of lading for government transportation request numbers when applicable.
16. Number of Pieces—the exact number of pieces being shipped. When there is more than one entry, the number(s) should be totaled at the bottom.
17. Gross Weight—the actual weight of the pieces should be inserted on the same line as the respective number of pieces. When there is more than one entry, the weights should be totaled at the bottom.
18. Chargeable Weight—enter chargeable weight of the shipment, dimensional weight when applicable. Gross weight minus tare weight for containers.
19. Description of Packing Pieces and Contents—enter the accurate commodity description of the goods comprising the shipment and the type of packing used (carton, bundle, etc.). NOTE: Entries for restricted articles must include: proper shipping name; classification; type of label required (if any); and cargo aircraft only (if applicable).
20. Carrier Remarks—this box may be used to enter any additional information related to the shipment.
21. Executed By—obtain signature of carrier representative.

Commercial Invoice

The commercial invoice provides the details necessary for the proper completion of the export documents.

The key to export documentation is the commercial invoice. All other documents are prepared from it. It is used to verify the shipment's value in the event of an insurance claim and is used by foreign customs to assess import duties. The proper number of copies may be supplied by the shipper or will be prepared by the international air carrier.

The commercial invoice must contain the following information:

Name and address of shipper.

Name and address of consignee.

Name and address of intermediate consignee if available. This can be a bank, a customs broker, an agent, and so on.

Country of origin of the merchandise.

Terms of sale (FOB-CIF, etc.)

Discounts.

Complete description of the merchandise, unit price, and extensions.

Certified true and correct and originally signed.

While the international air carrier is shipping the material overseas, the drafts are prepared, and consular documents, certificates of origin, or any other documents required to successfully accomplish an international transaction are being done.

While the shipment is airborne, the carrier is sending the details of the shipment and the arrival information to their associates at the destination airport. On request, the carrier will give the same information that is sent by the shipper to the customer overseas. On arrival the overseas associate will notify the customer and final clearance and delivery will be accomplished.

34.2.7 Special Procedures for High-Value Shipments/Shipments Not Acceptable

High-Value Shipments

Shipments that meet any two of the following three criteria must meet special handling requirements:

1. Declared or stated value is $5000 or more.
2. Declared or stated value is more than $100 per pound.
3. Contents are "articles of extraordinary value" as follows:

    Artwork
    Bills of exchange
    Bonds
    Bullion or precious metals
    Currency
    Deeds
    Dore bullion
    Evidences of debt
    Furs, fur clothing, fur trim clothing
    Gems, cut or uncut
    Gold bullion, coined, uncoined, cyanides, dusts, or sulphides
    Jewelry (other than costume)
    Money
    Pearls
    Platinum
    Promissory notes
    Securities, negotiable
    Silver bullion, coined, uncoined, concentrates, cyanides, precipitates, or sulphides
    Stamps, postage or revenue
    Stock certificates
    Watches

Restrictions/Shipments Not Acceptable

1. Articles that are listed as not acceptable for transportation by air carriers under the terms of D.O.T. (U.S. Dept. of Transportation) hazardous materials regulations.
2. Restricted articles Tariff No. 6D,CAB82.
3. Live animals.
4. Corpses.
5. Cremated or disinterred remains.
6. Shipments that require the carrier to obtain a federal, state, or local license for the transportation.
7. Shipments that would be likely to cause damage or delay to equipment, personnel, or to other shipments.
8. Shipments prohibited by law.
9. Firearms.

Restricted Articles: Special Service for Special Materials

As part of the total service capability most air carriers will handle restricted article shipments provided they meet all applicable federal and tariff regulations. These regulations are detailed in Title 49, Code of Federal Regulations (CFR), and in the official air transportation restricted articles, tariff and circular number 6-D. The tariff and circular are maintained in each air carrier freight office and any air carrier staff will provide any assistance required for every special requirement.

34.2.8 Other Shipment Information

Disposition of Shipments

Air carriers will usually hold a shipment at destination without charge for 24 hr, beginning at 6:00 P.M. (local time) after arrival.

Following the expiration of such free storage time, the air carrier will:

1. Continue to hold the air shipment for the consignee at a service charge of $.75 per day.
2. Notify the shipper by mail at the address shown on the airbill, and dispose of the shipment as directed by the shipper at the shipper's expense.
3. If no instructions are received from the shipper within 30 days after the day of mailing such notice and the consignee has failed to pick up the shipment, the carrier will dispose of it at public or private sale.

The shipper and consignee remain liable for any deficiency in the amount due the carrier resulting from such sale.

International Air Freight Shipments

What gets shipped overseas is identical to what gets shipped in the domestic United States, Hawaii, Alaska, and Puerto Rico. However, the complication begins to set in when keeping up with the day-to-day changes for documentation requirements, export limitations, flight schedules, and the complexities of foreign trade.

A shipper's letter of instructions has to be completed. The shipper's letter of instructions (SLI) is considered as the authorization for the air carrier to prepare documents and arrange the transportation to the airport of destination.

The United States government requires a shipper's export declaration (SED) for:

1. All international shipments valued at $500 or more.
2. Shipment of commodities which require a validated export license.

International air carriers will usually complete the SED free of charge after the shipper has filled in the shipper's letter of instruction (SLI). As of August 1, 1979, all shippers are required to list their internal revenue identification number on a specific line of the SED immediately following the shipper's name.

SED Information

1. Exporter named on the validated license or company entitled to export under applicable general license.

2. Internal Revenue Service (IRS) identification number; if none, use Social Security number.
3. Company name who is ultimate consignee on validated export license. Buyer and/or user at country of destination.
4. Foreign bank, if required by letter of credit (LC) consignee, customs broker, or agent.
5. Insurance should be requested when invoice value exceeds $9.00 per pound.
6. A validated export license is required by the United States Department of Commerce for certain types of commodities. When required, the air carrier will assist the shipper in obtaining the export license.
7. If shipper prefers to negotiate the letter of credit (LC), it is desirable that a copy be attached to SLI to insure proper completion of documents.
8. Information containing the name and telephone number of the consignee's broker will expedite customs clearance at destination point.

Claim Procedure

All claims must be made in writing to the originating or delivering carrier within a period of 270 days after the date of acceptance of the shipment by the originating carrier. In computing the time period under this paragraph, the first day of the period shall be the day after acceptance of the shipment by the originating carrier.

Damage and/or loss discovered by the consignee after delivery and after a clear receipt has been given to the carrier must be reported in writing to the delivering carrier at destination, within 15 days after delivery of the shipment, with privilege to the carrier to make inspection of the shipment and container(s) within 15 days after receipt of such notice. If more than 15 days elapsed between date of the delivery of a shipment by the carrier and notice of loss or damage by the consignee, the consignee shall show good cause why the loss or damage has not been discovered earlier and timely notification given.

Receipt of the shipment by the consignee without complaint shall be *prima facie* evidence that the shipment has been delivered in good condition and in accordance with the airbill.

Consignee acceptance of a sealed container shall be *prima facie* evidence that the seal(s) and a container are intact and that no loss of the contents occurred while the shipment was in carrier's possession.

The airbill shall be *prima facie* evidence of the conclusion of the contract, of the receipt of the shipment, and of the conditions of transportation.

The statements and the airbill relating to the weight, dimensions, and packing of the shipment, as well as those relating to the number of packages, shall be *prima facie* evidence of the facts stated; that is, those relating to the quantity, volume, and the condition of the shipment do not constitute evidence against the carrier except so far as they have been, and are, stated on the airbill to have been checked by the carrier in the presence of the shipper, or relate to the apparent conditions of the shipment.

No claim with respect to a shipment, any part of which is received by the consignee, will be entertained until all transportation charges have been paid. The amount of the claim may not be deducted from the transportation charges. When the consignee does not receive any part of a shipment, a claim with respect to such shipment will be entertained, even though transportation charges thereon are unpaid.

Package Computer Tracking

Almost all air carriers have one computer system or another for tracking each individual package or container that is shipped via their system. Real language printouts are available and a computer interface is also available through the various air carriers. Various types of office terminals and computer tracking and tracing methods and systems are used. Some airlines use bar codes on airbills for tracking air traffic packages.

Insurance

A small package shipment shall have a declared value of $50.00 unless the shipper declares a higher value on the airbill (not to exceed $500.00).

Postal Zip Codes

Postal zip codes are used for package delivery all over the United States. The zip codes are usually placed on the airbill and when the packages get to the destination city they are sorted by the zip codes for delivery to the consignee.

34.2.9 Glossary

Chargeable Weight. The gross weight less the tare allowance.

Container. A container that complies with the specifications and requirements in Rule No. C10.

DOT Hazardous Materials Regulations. Hazardous materials regulations issued by the Materials Transportation Bureau of the Department of Transportation entitled 49 of the Code of Federal Regulations, Parts 171 through 177 (49 CFR171-177).

Gross Weight. Total actual weight of the container and the contents of such container.

IATA. International Air Transportation Association.

Net Weight. The same as chargeable weight (gross weight less tare allowance).

Pallet Supporter. A portable conveyor base placed under a container for the purpose for positioning such container for loading and unloading while in the possession of the consignor or consignee.

Tare Allowance. The following number of pounds per container.

ULD. Unit load devices.

CHAPTER 35

RADIO CONTROLS AND AUTOMATIC IDENTIFICATION SYSTEMS

S. GENE BALABAN

Dynascan Corporation
Telemotive Product Group
Chicago, Illinois

EDMUND P. ANDERSSON

Computer Identics Corporation
Canton, Massachusetts

35.1 REMOTE RADIO CONTROLS

S. Gene Balaban

35.1.1 Introduction

To place the concept of remote radio controls in perspective, it would be appropriate to reflect on controls in general and their industrial evolution which inexorably created the demand for radio remote applications. From the outset, the invention and application of the electric motor and related electromagnetic powering devices required some means to practically and safely start them, control the speed, and stop them. These needs applied to a myriad of devices in addition to electric motors. Examples are the solenoids to control valves, actuators as applied to brakes, and mechanical linkages utilized for linear or rotary motion and for a host of machine tools.

The earliest forms of controls are represented by knife switches which directly interrupted the electric current needed to operate these electromagnetic devices and systems. Later, with the introduction of relays, it became possible to operate the motors and other devices remotely. That is to say, two benefits were achieved; first, it became possible to interrupt the power to the prime mover via a pilot device through a hardwired control circuit, and second, operation could be via a conveniently mounted switch or push button which permitted an operator to control from distances of a few feet or hundreds of feet.

The advantage, of course, was that it permitted the person to be remote from the machine, pump, valve, or mill, particularly if the work environment were not safe or comfortable for humans. On this basis, from the earliest days of electrical control until the present time, hardwire has been the predominant method of controlling and/or carrying messages, from the simplest light switch to complex data transmissions. Further, the techniques of remote control were, and are, necessary to implement modern technological growth, because without it we could not have centralized systems: the closed loop and/or computerized control.

Examples are the extensive continuous process complexes in the petrochemical industry, the steel mill centralized control for rolling mills and basic oxygen furnaces, the remote control and alarm systems in utilities and atomic energy facilities, and, of course, the type of hardwired remote links where conductor bar or festooned cables are used in our contemporary materials handling automatic storage and retrieval systems.

On the other hand, radio, while venerable in communications since the Marconi era, is relatively recent as a control means. It gradually came into increased usage because it does things that can be done no other way in some materials handling applications. In others, it provides a more efficient, safer, and economical method of moving material.

However, it was not until we put a man on the moon by radio and tracked his heartbeat, temperature, and other gastrointestinal functions that the people in industry were sufficiently comfortable with the concept of applying it to materials handling.

35.1.2 Applications in Materials Handling

While long used in military and aerospace applications, remote radio control has been a recognized industrial tool only since the early 1960s. Acceptance in industry has grown steadily since then. To date, most industrial applications have involved control of materials handling equipment, particularly electric traveling bridge cranes. The following are some of the most common uses of radio remote controls:

Electric traveling bridge cranes
Transfer cars
Yard locomotives
Freight car haulers
Oil tanker loading arms
Automatic storage and retrieval systems
Computerized weigh scale, scrap batching
Computer-controlled automatic-guided vehicles
Order selection vehicles
Mining machines

Front-end loaders
Furnace and soaking pit doors

Because the predominant use of radio remote control for materials handling is for overhead bridge cranes, this section highlights this application in greater detail. Later we discuss the advantages of applying radio to automatic storage and retrieval systems and other applications including distance-detection anticollision systems.

The extent to which radio is applied to cranes can be determined by noting below the accessory items that are also controlled by radio in addition to movement of the bridge, trolley(s) and hoist(s):

Grabbers
Hook rotators
Magnets
Sheet lifters
Audible warning devices
Crane lights
Limit and bypass switches
Reset functions
Weigh scales
Readouts

Other types of cranes to which radio has been successfully applied are gantry cranes, monorails, and split cranes with traveling trolleys.

As noted, radio remote control of overhead traveling cranes represents the greatest usage of this technology in materials handling. Why radio control? Let's review some of the reasons.

35.1.3 Justification Criteria

Operational Benefits

Essentially, control by radio eliminates the need for a cab operator on an overhead crane working in conjunction with a floor man, sometimes known as a hooker. The floor man hand signals the cab operator the direction in which he wants him to move the bridge or trolley, and when to raise or lower the hoist. He also hooks the load, adjusts the grabs, and so on, as may be required.

With radio, one man can control the crane from the floor and can reduce damage and improve load positioning while saving the costs of one man per shift. When the crane is controlled by a cab operator, the crane movements are facilitated by the operation of various levers located in the cab. These levers start and stop and control the speed of the various crane motors via crane controls. In older systems, these controls were of the drum type, and, as will be discussed later, eventually were supplanted by magnetic (contactor type) and ultimately stepless solid-state controls.

Control by radio is made possible by permitting the operator to control the crane from the floor by operating miniature motion switches located on a portable transmitter, cigar box size, worn on the belt as shown in Fig. 35.1.1. Signals are transmitted to a crane-mounted receiver and then, via an intermediate relay system, to operate the crane controls that move the crane.

Of course, we must recognize the very widespread one-man method of controlling a crane by pendant. Pendant control, which was made possible by the conversion of manual drum-controlled cranes to magnetic contactor controlled techniques, is a good example of the transition from local manual control to hardwired remote control mentioned earlier. A pendant, as shown in Fig. 35.1.2, consists of a cable of wires connected to a hanging box containing push-button type controls as well as switches whose operation permits electric control of the contactors located on the crane. Although pendant control increases productivity by having one operator do the job of two, it does have its disadvantages as compared to radio. Eliminating the pendant prevents walking-the-load-type accidents (Fig. 35.1.2). With radio the floor man does not have to walk a pendant path and can position himself in a safe location to hook the load and move it.

Further, the advent of radio control permitted the possibility of a pitch and catch operation. This is commonly called two-box control where two people can control one crane on a first-come-first-served basis (Fig. 35.1.3). Other variations are possible only with radio remote control. Examples are a multiple box as well as multiple crane operation (Fig. 35.1.4). In the lower figure, the first operator can pick up a load at the siding and send the crane down to the second operator on the right by transmitting the necessary signals. He or she may then shift the transmitter key and transfer control to the second operator who can then continue to move the crane and the load to the storage position. In the upper diagram is shown a multiple-control operation in a major auto manufacturing plant where an operator can control any one of six cranes by dial selecting from a single radio transmitter.

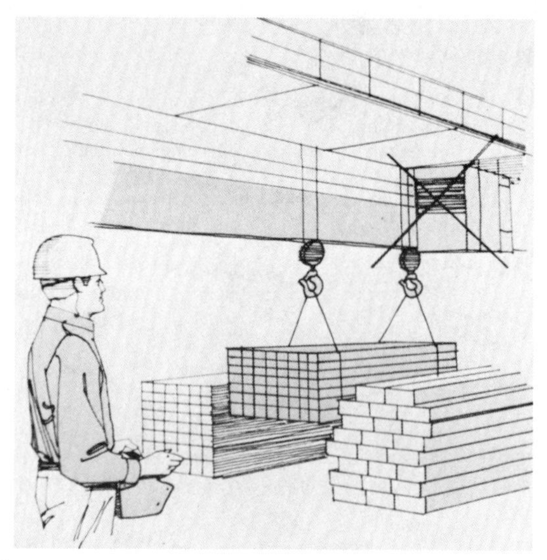

Fig. 35.1.1 Elimination of cab operator reduces damage and improves load positioning.

Thus, the operator no longer is required to climb a ladder to each crane whenever he has to make an infrequent or alternate lift.

An example of an integrated operation in a foundry application is shown in Fig. 35.1.5. Here each operator, A and B, has an identical transmitter capable of controlling each of the two monorail trolleys on a first-come-first-served basis. In this manner, operator A, for example, can send the first trolley of the monorail from the melt position to the pour position controlled by operator B.

35.1.4 Application Specifications

General

What should be considered before applying radio to control a crane? Although the manufacturers of radio crane control will generally assist the user in selecting the optimum system, some of the most frequently asked and important questions to review are:

What radio frequency should be used?

Will one-transmitter or two-transmitter operation be optimum?

Shall the system be controlled range?

What type of backup controls are necessary, if any?

What are the industrial relations considerations? Union problems?

What are the environmental factors?

What type of system would best suit the application as well as plant maintenance capabilities (i.e., analog, tone modulated, digital, or other)?

What are the economics? Is radio remote control justified for this application?

Shall the transmitter batteries be disposable or rechargeable?

Is this for a new crane or an installed crane; and is the crane, if installed, controlled by drum, magnetic controls, or stepless solid-state controls?

When considering the application of radio control to materials handling equipment, it is recommended that some type of an application data form be completed. This will assist the purchaser and the supplier of the radio controls to determine the optimum system to suit the materials handling equipment. A typical application data form generally provided by the supplier of the radio controls is shown in Fig. 35.1.6.

While some of the things that must be considered to properly apply radio to a crane application are self-evident, the cost effectiveness, and accordingly the payback, will be affected by the type of crane and more particularly the type of crane motor controls on the crane in question. The type of controls furnished with the crane, as well as the selection of which frequency to use, are perhaps

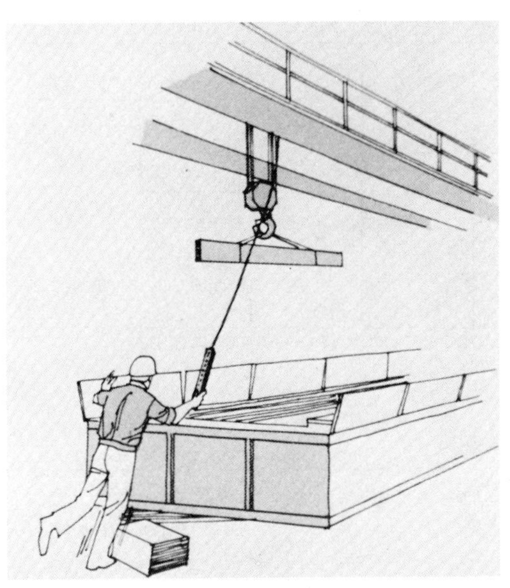

Fig. 35.1.2 Eliminating the pendant prevents "walking-the-load" accidents.

two of the most important factors affecting both costs and performance. We first review the types of controls and how they affect the interfacing with remote radio control.

Crane Controls

First it must be observed that remote radio control can be retrofitted to installed cranes as well as new cranes. As a matter of fact, most of the earliest installations of crane controls were as retrofits to existing cranes, some as old as 75 years. Although there are no industry statistics available, it would appear that in recent years the number of installed cranes converted to radio as compared to new cranes purchased varies from as low as 40% to as high as 60%, depending on the state of the economy.

This being the case, radio must be capable of interfacing with some of the oldest types of controls as well as with the most recent developments.

Fig. 35.1.3 "Pitch and catch" operation of a radio-controlled crane.

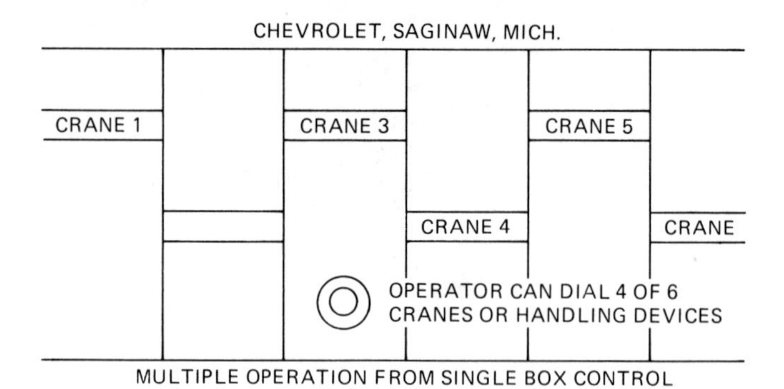

Fig. 35.1.4 Multiple box and multiple crane operation.

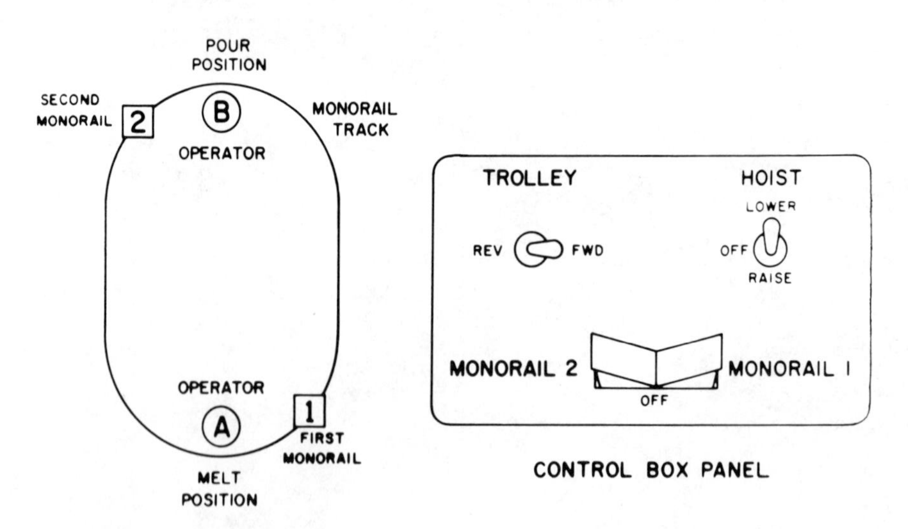

Fig. 35.1.5 Integrated foundry operation.

Drum

These early types of motor controls, perhaps deriving their name from the shape of the drum-shaped rheostat box, were primarily used for DC series motors. These were manual controls requiring a cab operator to rotate a lever that varied the resistance to DC motors and thus controlled the current and subsequently the speed of the motors. This is similar to the types of controls used on the obsolete traction vehicles or street cars and subsequently applied to cranes. It is not possible to readily operate this type of crane by radio since there would be no way to operate the control lever on the manual rheostat-drum from a remote point without the use of a servomotor which could be unsafe and costly. It is unlikely that today any new crane will come equipped with drum controls in the United States. Therefore, application of radio to cranes equipped with drum controls can only be accomplished if the drum controls are converted to the more modern types developed later.

Contactor (Magnetic Controls)

The motors used in hoisting and crane service vary in size from as small as a $\frac{1}{2}$ Hp for light-duty monorails to 150 Hp depending on the type of loads lifted.

As mentioned at the outset of this chapter, the introduction of relay-contactor controls eventually made obsolete the earlier drum controls required for the speed steps. The number of speed steps used in these cranes and the number of motors affect the cost and complexity of a radio system. For many years direct current controls were favored for heavy-duty cranes for precision load spotting. But in more recent years, alternating current controls have been improved and have become more popular. The usefulness of alternating current controls improved with the capability of providing better speed control than was possible previously. Although AC controls in the past 30 years have been very acceptable for light-duty crane applications, most mill duty requirements demand direct current controls because the power is more readily available and widely used for furnaces and other machinery.

Radio is readily adapted to these types of controls by the use of interfaces located in the same area as the crane controls which are usually mounted on the bridge of the crane.

Solid State

In the past 10 years, the most recent advance in control technology has been the addition of solid-state components, supplanting the stepped-switch contactors. The usual approach to static AC crane control is through the use of saturable core reactors or saturable transformers.

By using a stepless master switch and SCR (silicon controlled rectifier) controls, infinitely variable DC current can be supplied to the reactor, which varies the AC output to the motor. The advantage of the solid-state controls is primarily the elimination of the contactors. Removing the active contactors and replacing them with static components eliminates contact wear and reduces maintenance.

Radio control can be readily applied to solid-state stepless crane controls, and in the past 5 years, the rate of these conversions has accelerated.

Frequency Selection

The choice of a radio frequency for a radio remote control application involves the consideration of the following factors:

1. The frequencies that are made available by the licensing bodies—the FCC (Federal Communications Commission) and IRAC (Interdepartmental Radio Advisory Committee).
2. The type of user.
3. The location of the user.
4. The existence of other authorized or licensed radio signals in the area.
5. The extent of radio and/or electrical interference (noise) in the area.
6. The maximum operating range desired for the system.
7. Specific technical performance requirements for the radio system.
8. Operational safety considerations.
9. The capability of the radio equipment supplier to provide equipment that will operate at the desired radio frequency.

The federal government has been given the authority by Congress to control the use of the radio frequencies. The operation of any radio frequency device is controlled by the federal government under guidelines that have been adopted by international treaties. The federal government has divided users of radio frequencies into the following groups for licensing purposes:

APPLICATION DATA—RADIO CONTROL

Installation Location: Area: ______________________
(Describe) (Plant Sect., Bldg. No., Bay No., etc.)
Company: ______________________
Street: ______________________
City: ______________ State: __________ Zip Code: __________

Application Type: Crane: ______________ Other: ______________
(Mfg. Serial No. and Capacity) (Conveyor, Lift Truck, Furnace Doors, Soaking Pits, Transfer Cars, Ingot Buggy, etc.)

If other, describe: ______________________

A. Crane Controls:
Controllers are: Stepless ______ Magnetic ______ Drum ______ Manufacturer ______
Crane is now controlled by: Cab: ______ Pendant ______ Radio ______
Is manual control to remain after conversion to radio? ______

B. Motion Functions

	Motion Functions	Horse-Power	RPM	NEMA Size of Contractors	Number of Speed Steps on Present Switches	Minimum Required Radio Speed Steps
					(In each direction)	
(1)						
(2)						
(3)						
(4)						
(5)						

C. Bridge Brake
Hydraulic (Make) ______ (Type) ______
Electrical (Shunt) ______ (Series) ______ (Other) ______ (None) ______

D. Electrically Operated Power Disconnect Type
Manual/Magnetic ______ Protective Panel ______ Main Line Contactor ______ Other ______

E. Hoist Limit Switch
Power ______ Control ______ Other ______

F. Crane Ht. ______ ft Span ______ ft Runway Length ______ ft

G. Would control shared by two operators (one at a time) increase efficiency? Yes ______ No ______

H. Moving or stationary cabs? ______

I. Can antenna be mounted under moving cab? Yes ______ No ______

Auxiliary Equipment:

A. Is warning device to be radio controlled? Yes ________ No________

Type: Bell ______ Horn ______ Siren ______ Other ________________

B. Are any Limit RESET switches required to be operated from the radio control? Yes ______ No ______

If Yes, please describe: ________________________________

____

C. What other auxiliary equipment is to be operated from the radio control? Include magnets, motorized grabbers, etc. ________________

Number of commands required ________________________________

D. Does crane now have anticollision protection? Yes ______ No ______

Is automatic anticollision protection required? Yes ______ No ______

Power Provided:

A. Give voltage and frequency of power source and magnetic controls of moving machine ________________

____

Environment:

A. Is this an indoor or an outdoor installation? ________________

B. Ambient temperatures at vehicle ranges from ____________ to ____________ degrees (Centigrade)

C. Ambient temperature at floor ranges from ____________ to ____________ degrees (Centigrade)

D. Describe atmosphere in which moving device operates. Is it corrosive? Heavy dust condition? What kind?

____

____

E. Will standard NEMA-12 enclosures be satisfactory? ________________

Range Requirements:

A. The usually recommended controlled range of 75 to 100 feet operating distance? Yes ________ No ________

B. If extended range is required (in excess of 200 feet), do you have an existing FCC license for 72–76 MHz devices? Yes ________ No ________ 450–470 MHz? Yes ________ No ________

NOTE! Range can be selected by function.

EXAMPLE! Bridge may be commanded from extended range but hoist and trolley could be operated only within a "SAFETY CIRCLE" ″ of 100 feet.

Installation: Are plant maintenance personnel available to do installation? Allow about four man days. Yes ______ No ______

Drawings:

A. Can you provide us with electrical schematics and wiring diagrams of the existing or proposed power and control circuits of equipment to be controlled? Yes ______ No ______

B. Are they enclosed with this form? Yes ______ No ______

Delivery: What is a realistic delivery requirement? (6–8 wks. normal) ________________

Special Requirements: Use separate sheet to describe any pertinent special conditions (such as space limitations, etc.) and attach.

Completed by: ________________________ Date: ________________

Fig. 35.1.6 Application data form for radio controls.

1. Federal government users.
2. Non-federal government users.

Federal government users must obtain permission to operate radio frequency services from IRAC (Interdepartmental Radio Advisory Committee). Non-federal government users must obtain permission to operate radio frequency services from the FCC (Federal Communications Commission).

Table 35.1.1 lists the frequencies in the very low frequency (VLF), low frequency (LF), medium frequency (MF), high frequency (HF), very high frequency (VHF), and ultra high frequency (UHF) bands. The FCC and IRAC restrictions have made these suitable for consideration as frequencies for radio remote control operations.

Table 35.1.2 lists some of the general advantages and disadvantages of operation in each of the available frequency bands.

In regulating its frequencies, the FCC has chosen to divide its users into groups of "radio services." The technical restrictions and the frequencies made available under the FCC rules governing each of these services varies.

Table 35.1.3 lists the radio services covered under the FCC rules that are most likely to be encountered in radio remote control applications. The class of users must apply for a license in accordance with the services listed unless they elect to use low frequency, described later.

Transmitters used for radio remote control applications are usually mobile and the operating distance from the unit being controlled is relatively short. For radio remote control applications, a frequency that has been limited by the FCC or IRAC to a low transmitted signal power is important because it reduces the portable transmitter battery current drain.

One of the variables also to be considered is: Shall the system be controlled range? The concept of controlled range was first developed in the early days of radio control in 1961. The initial objections to radio control of cranes came from the operating and safety people who were concerned that, with a small transmitter buckled to his waist, an operator might attempt to operate a crane from a long,

Table 35.1.1 Radio Frequencies

Band Designation	FCC Frequencies	FCC Rules Reference	IRAC Frequencies	IRAC Rules Reference
VLF LF MF Up to 3 MHz	10–490 kHZ 510 kHz to 1.6 MHz (Any number of channels) No license required.	15.111	10–490 kHz 510 kHz to 1.6 MHz (Any number of channels) No authorization required.	7.9.2
HF 3–30 MHz	26.995–27.255 MHz (6 channels) Radio control Radio service License required	95.216	26.970–27.260 MHz (Any number of channels) No authorization required.	7.9.2
VHF 30–300 MHz	72.02–76.00 MHz (30 channels) Manufacturers Radio Service and other industrial radio services License required	90.79	29.89–50.00 MHz (357 channels) Authorization required. 162–174 MHz (477 channels) Authorization required.	4.3.6 4.3.7
UHF 300 MHz to 3 GHz	457–470 MHz (8 channels for telecommand) (377 half channels) Business radio service License required	90.75	406.1–420 MHz (235 channels) Authorization required.	4.3.9

Table 35.1.2 Frequency Band Properties

Band	Advantages	Disadvantages
VLF LF MF Up to 3 MHz	1. No user license necessary. 2. Lack of multipath dropout effects. 3. Easily controlled (limited) range. 4. Minimum technical restrictions (except for radiated field strength). 5. Large number of available frequencies.	1. Many natural sources of interference. 2. Uncontrolled co-users (but major users readily identifiable as listed).
HF 3–30 MHz	1. Minimum geographical restrictions.	1. License required. 2. High usage of frequencies (personal—citizens band). 3. Multipath dropout effects. 4. Interference from long distance.
VHF 30–300 MHz	1. Relatively easy to meet frequency stability requirements.	1. License required. 2. Limited number of frequencies available. 3. High usage of frequencies. 4. Multipath dropout effects. 5. Moderate cost.
UHF 300 MHz–3 GHz	1. Large number of available frequencies.	1. License required. 2. Higher cost. 3. Relatively hard to achieve frequency stability. 4. Multipath dropout effects.

Table 35.1.3 FCC Licensed Radio Services for Radio Remote Control

Service	
I. Personal Radio Service	FCC Rules Part 95
A. Radio Control Radio Service (Citizens Band)	(Subpart C)
II. Private Land Mobile Radio Service	FCC Rules Part 90
A. Industrial Radio Services	(Subpart D)
1. Business Radio Service	(90.75)
2. Manufacturers Radio Service	(90.79)
3. Power Radio Service	(90.63)
4. Forest Products Radio Service	(90.67)
5. Petroleum Radio Service	(90.65)
6. Motion Picture Radio Service	(90.69)
7. Special Industrial Radio Service	(90.73)
B. Land Transportation Radio Services	(Subpart E)
1. Railroad Radio Service	(90.91)
C. Public Safety Radio Services	(Subpart B)

unsafe distance, say 300–1000 ft away where he could not be sufficiently close to the load to view his lifts and move the crane safely. The concept of controlled range was mandated, and the most popular technique to accomplish this is by utilizing very low frequency type system operating between 200 and 400 kHz. The very low frequency band facilitates controlled range operation because of the presence of near field effect radiation. In the near field range, the transmitter and receiver antennas are inductively coupled within the first tenth of a wave length. Although not requiring an FCC license by the user, the radio manufacturer must provide a system which is certified in accordance with FCC Rules, Part 15, Subpart E.

Various manufacturers offer controlled range operation utilizing VHF or UHF transmission reported to be accomplished via reduction of receiver sensitivity. It is not established whether this type of controlled range is possible repetitively. An accepted technique for those desiring VHF transmission is to utilize a very low frequency channel to enable system operation up to the extent of the range limitations.

In most controlled range applications, however, the user will require that the range be limited to between 75 and 150 ft distance from a plumb line dropped from the hoist to prevent the operator from moving beyond that distance. Radio systems are, therefore, designed to cause all crane motion to cease automatically if operation beyond the prescribed limitation is attempted (Fig. 35.1.7).

Governmental Standards

In the United States, except for FCC regulations, there are very few governmental or institutional standards governing the operation of radio remote control of materials handling equipment. Although OSHA standards have been revised considerably since their inception, the OSHA standard which relates to this type of operation suggests that the requirements for safety will be met provided that loss of transmitted signal will result in cessation of operation of the crane. Interestingly enough, in the United Kingdom governmental requirements mandate that all overhead crane remote operating systems limit the range.

35.1.5 Typical Industrial Installations

Overhead Cranes

Over the years, cab and pendant operated cranes, both AC and DC, have been converted to radio operation. Although hard statistics are not available, it is estimated that about 75% of the radio operated cranes in the United States were previously operated by cab or pendant and were converted after the crane had been installed.

The first crane application in a steel mill was started up in September of 1961. The crane was built by Morgan Engineering Company in 1907 and amply illustrates that new technology need not wait for new cranes for its best application. This is a four-motor, DC crane.

In this installation, the auxiliary hoist lifts a $1\frac{1}{2}$ ton valve seat of the type used on gas engine generating sets. The safety department of this steel works insisted that the operating range be reduced to a maximum distance of 70 ft from the transmitter. In this way the operator has good visibility for running the crane over gas engines or other high obstructions.

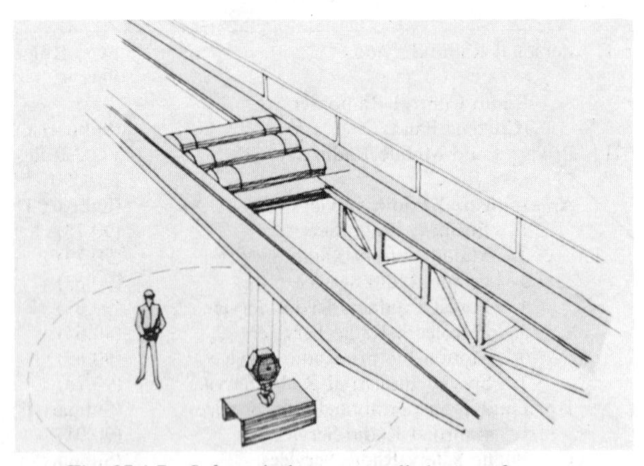

Fig. 35.1.7 Safety circle or controlled range feature.

Although many of the initial applications for radio control were in the metals industry, the auto industry was next to utilize radio to increase productivity. In one installation, two four-motion cranes on one runway handle steel stock three shifts a day. This was the first system to provide direct radio control of open and close motions of coil grab rotators.

In a major auto stamping plant where 22 cranes are operated by radio control under one roof, overhead cranes handle multi-ton stamping dies and sheet coils. One of the features of these systems is that they all utilize low-frequency transmission from 200 to 400 kHz, making it possible to operate them in close proximity with each other without duplicating frequencies.

The use of radio for controlling traveling trolleys on split (interconnecting) bridges is an example of another industrial use for radio control.

In this aircraft facility where the fuselage and wing skins for heavy jets are produced, more than 50 cranes with traveling trolleys on interconnecting bridges (formerly controlled by multiple manned pendants) were converted to radio to improve productivity. Six-foot-wide, 105-ft-long aircraft skins weighing up to 7300 lb are handled by 165-ft-long cranes.

Another example of split cranes helps to remove the bottlenecks in a major steel warehouse (Fig. 35.1.8). In this type of operation, one man operates half of a split crane for hoisting a short load and controls both sections for lifting long loads by using the radio control to interlock the dual segments of the split crane to bridge the 95-ft-wide bay.

Radio control can be utilized in hot metal handling operations controlling pouring in the refining plant of a major zinc corporation (Fig. 35.1.9).

Miscellaneous Materials Handling Applications

Up to this point, application and justification criteria have been described for utilizing radio remote controls for various types of overhead materials handling cranes, which are examples of the most predominant industrial use. The selection and application guidelines for use in other applications to be described now are similar to those outlined previously. This is particularly true of frequency selection although the radio hardware and principles of operation, to be described later, sometimes vary according

Fig. 35.1.8 Split crane operated by remote radio control.

Fig. 35.1.9 Hot-metal pouring is controlled by radio.

to the application and installation environment. The typical installations described were selected to highlight the flexibility of radio remote control when used to augment safety and improve productivity.

Industrial Doors

A far cry from the garage door opener is the system employed at a major specialty metals company to control 14 gas-fired furnace doors. The charger, a specially designed, heavy-duty, diesel-powered truck removes ingots with its hydraulic pinchers through the door.

Tanker Loading Arms

One of the problems in oil tanker loading and unloading is the length of time it takes to couple and uncouple loading arms used to pipe petroleum and other products to and from the giant tankers.

The operator has full control of the movements of the loading arms by manipulating the transmitter switches and sending signals to a receiver located on the dock. The transmitter is intrinsically safe for use in explosive hazardous areas. Prior to conversion to radio, it was necessary to haul a very-heavy explosion-proof electrical pendant aboard the tanker from the dock in order to control the pumps and to manually couple and uncouple the arms. Use of radio increases productivity by reducing the tanker's costly and wasteful time at the unloading pier.

Yard Locomotives and Car Haulers

Our railroad industry and freight yard operations are making increased use of radio remote control (Fig. 35.1.10). One-man-operated switch engines and rail car spotters can now accurately and speedily

Fig. 35.1.10 Walking alongside the train while it is backing up, the engineer equipped with remote radio can best observe and react to track problems.

move railroad cars with appreciable savings over conventional cab-operated methods. Productivity and safety are also greatly increased because the remote operator can spot cars from track level without the need to rely on an additional brakeman. Rail car movers are operated remotely by radio when using the rail wheel mode of operation and pulling the cars.

Scrap Weighing and Handling System

This application demonstrates the flexibility of radio for transmission of data and control. A typical system is utilized in a steel mill scrap yard where a proper mixture of scrap by weight must be accumulated and eventually transferred to a furnace by a furnace charging container.

In this operation the cab operators control transmitters to perform the necessary functions. The scrap handling system employs two bridge cranes located over scrap cars, each spanning 100 ft and containing a cab with an operator. A third transverse crane functions as a transfer crane. The operation is as follows:

1. Each top-running crane equipped with electromagnets unloads scrap metal from the various incoming gondola cars.
2. On each transfer car a large digital display board, readable by the crane operator, shows the actual buildup in weight of a given type of scrap metal.
3. At an appropriate time, the crane operator sends a radio signal which triggers a transmitter on the transfer car which then sends the data to a central computer indicating the car number, type of scrap, and the weight of the scrap in a particular transfer car.
4. The operator also controls the diesel electric transfer car by radio and, if required, moves it down to the second crane operator to add additional quantities of scrap.
5. Finally, the transfer car is released again and the operator of a transfer crane at the end of the yard can take over control by radio. The container can then be lifted off the transfer car and delivered to the furnace area.

Industrial Trucks—Grocery Warehouse

Turning to grocery warehousing, radio control of industrial trucks facilitates order selection of dry groceries. Truck movement is controlled by an order picker who can perform tasks without repetitive boarding of the truck to move it as required.

Tractor Train—Furniture Warehouse

Radio control is also used in connection with a computer-controlled automatic guided-vehicle system.

The order selector in a furniture warehouse requests by radio for a tractor train at home base to be dispatched to his area. The call is received by the system computer and a train is dispatched to his location. The computer talks to the operator by means of a voice synthesizer and provides information on vehicle status. When the tractor train arrives at the operator station, the belt-mounted transceiver is used to move the train during order selection similar to the grocery warehouse operation previously described.

Automatic Storage and Retrieval Systems

More recently, radio remote control has been used in connection with automatic storage and retrieval system (AS/RS) machines. The justification relates to reliability and space economics. Signals transmitted by radio can be more reliable than hardwire in certain situations. Some of the maintenance problems involved are eliminated. The application of radio telemetry to AS/RS comes to the fore where the automation or logic control has been removed from the AS/RS machine and placed in a protected location.

In this case, communication from the control point to the moving machine using conductor bar systems or festooned cables is at a disadvantage. The advantages of the radio data links are as follows:

1. Festooned mechanical systems utilizing cables and trolleys and requiring I-beam support systems are subject to wear and require periodic replacement of the cables or trolleys.
2. The space required for cable hanging and the collection area for accumulated cable loops can be eliminated. These present a difficult mechanical problem in automatic bridge-type AS/RS.
3. Collector bar systems providing both full and half duplex communication also are limited because of the requirement that the control be mounted on the AS/RS machine. Where logic control must be removed from the machine, conductor bars sometimes cannot be used at all (noise, etc.).
4. Although faulty cables or worn collectors may require a complete shutdown of the S/R machine, the radio system downtime should not usually tie up the machine for the same length of time.
5. Not all AS/RS applications lend themselves to radio remote data links, but candidates would include outdoor systems, overhead stacker cranes, S/R machines in food freezers, hammerhead cranes used in underwater conditions, and applications in hot or corrosive environments.

In these instances, cables, conductors, and collector replacement would occur more frequently than usual and industrial radio remote systems designed for the life of the crane appear to be more economical. It should be noted that the types of systems used in view of the information requirement, to our knowledge, have been digital radio links as compared to analog or other methods of transmission.

35.1.6 Radio Equipment Selection

The choice of radio remote control equipment is dependent on a number of criteria, some of which have previously been outlined. To review what has been discussed, the selection of the appropriate radio remote equipment should be preceded by an evaluation of the following:

1. The type of materials handling equipment to be controlled (i.e., overhead traveling crane, yard locomotive, AS/RS, etc.).
2. The number of motor functions (motions) or type of data required.
3. The number of speed steps and method of acceleration (i.e., stepped or stepless).
4. The type and design of interface required. These could vary from electrical or solid state relays, to electropneumatic or electrohydraulic depending on the application (i.e., overhead cranes, yard locomotives, oil tanker loading arms, etc.).
5. Radio transmission frequency, which is governed by the FCC or IRAC; thus the selection of frequency band depends on user classification, existing radio frequency usage, distance requirements, and so on.

35.1.7 System Components

As previously discussed, the basic radio remote control system components usually consist of a transmitter, a receiver, and an interface. As noted previously, there is a wide variety of installations. Thus the number of transmitters and receivers and type of interfaces necessary to perform the material handling functions will vary as required. Further, the specific design of each of these depends on the manufacturer's radio design criteria as adapted to the user's specifications. Recognizing these adaptation variations, we discuss the basic system components as applied to materials handling equipment.

Transmitter

Most transmitters used in materials handling are portable and worn around the waist. There are some fixed transmitters for special applications. The portable transmitters must be relatively lightweight, from 6 to 9 lb, and, of course, battery powered. In addition to containing the electronics and battery, the face plate of the transmitter supports the various motion levers and switches needed to operate the crane or other materials handling equipment.

The transmitter usually contains a built-in antenna of the wound ferrite rod or slotted printed

Fig. 35.1.11 Typical transmitter.

circuit board type, but some transmitter antennas are of the external "rubber duck" or hanging-wire design.

A typical transmitter is shown in Fig. 35.1.11. Note the key switch on the side of the transmitter for application of the power; also note the "feel-coded" knobs of various shapes that permit the operator to recognize by the shape of the knob the particular function, bridge, trolley, hoist, and so on. Auxiliary functions, such as brakes, alarm signals, and grabs, are controlled by various switches and buttons in the front section. Most transmitter manufacturers will design the operating switch configuration (face plate) in accordance with the user's specification.

Also note the safety handle which also supports the operator's palm and is recommended to prevent accidental movement of the switches. All motion switches are dead-man type for safety purposes. Although there are no government or industry standards, most manufacturers will provide safety-start interlocks and secure radio signals in accordance with the requirements. The transmitter shown in Fig. 35.1.11 is a good example of the flexibility of radio crane control because it is designed to control either of two cranes as selected by the floor operator.

The type of battery selected can be of paramount importance to the performance and productivity of the system. Unlike consumer-type transistor radios which are usually operated by a 9-V battery or 4 cells developing 6 V, the transmitters for industrial applications generally have specially designed battery packs utilizing the type and quantity of cells necessary to satisfy the design of the transmitter. For example, some of the types of batteries used are mercury, lithium, nickel–cadmium, and lead–acid. Although the mercury and lithium batteries are nonrechargeable, the other types mentioned are capable of being recharged, with the frequency of recharge requirements varying widely with the system selection, transmitter power, and so forth. The mercury throw-away battery is very popular because of its life, which varies depending on operating conditions, but can in some cases last a few months. On the other hand, some rechargeable battery–transmitter combinations may require recharging every shift, which can sometimes be neglected and can cause downtime. Longer times between recharging are claimed to be available from some manufacturers. The user is advised to carefully evaluate the supplier's recommendation (and reputation!!) in comparison with the performance and utilization factors.

Receiver

The receiver and interfacing equipment are usually mounted in a convenient location on the crane or locomotive or other materials handling device to be controlled by radio. When mounted on an older crane, the receiver can usually be mounted somewhere in the cab or on a catwalk adjacent to the crane controls. As previously mentioned, the signals from the transmitter are received by an antenna which must be mounted on the materials handling equipment, usually visible from the floor, to assure a clear path for signal transmission. Receiving antennas can be of various sizes and shapes depending on the transmission frequency and the manufacturer's choice. On some of the newer cranes, built without cabs, the receiver is sometimes welded or bolted to a girder.

Figure 35.1.12 shows two enclosures mounted on the crane just above the hoist in an outrigger

Fig. 35.1.12 Two receiver enclosures mounted on crane just above the hoist in an outrigger position.

position. The receiver is on the left and is the smaller of the two enclosures, whereas the intermediate relay enclosure (interface) is adjacent on the right. What may be located within the receiver enclosure will come in various shapes and sizes depending on the manufacturer as well as the type of system.

Interface

As previously indicated, the signals received by a radio receiver for materials handling equipment are generally low powered and the output from a multichannel or digital receiver is too small to power the heavy-duty relays or contactors found on cranes, AS/RS equipment, locomotives, and so on. The output signals must, therefore, be suitably amplified or conditioned before used with relay or solid-state motor controls.

Although most radio receivers are modular in construction, with size varying with the number of functions to be controlled, the variety of interface equipment is almost limitless. Generally, the purchaser of these systems must depend on the expertise, experience, and reliability of the radio control manufacturer to properly design an interface system to suit the materials handling equipment.

System Design

The first variable to be considered in system design relates to frequency selection. This has been covered in detail earlier in this chapter particularly with respect to the user's involvement in frequency selection. (Refer to Tables 35.1.1, 35.1.2, and 35.1.3.)

Probably the second most frequently encountered variable is whether the system is *analog* or *digital.* Analog design has been in use since the beginning of control schemes and probably represents the greatest portion of all installed systems. For example, the low-frequency systems are all analog and utilize what is commonly called a frequency multiplex scheme for coding purposes. In these systems, a multiplicity of narrow-band crystal-controlled unmodulated carriers are transmitted, each of which is designed to control a separate motion or auxiliary function. The security of such a system, that is, its immunity from interference from other signals or plant electrical noise, is greatly dependent on various features such as amplitude matching, fault detection circuits, narrow-band filtering, and multiple channel code requirements. These types of systems are generally the easiest to maintain and understand by the purchaser. The disadvantages of such systems are limited distance where longer range is desired and the fact that they do not readily lend themselves to stepless or proportional control or data transmission.

Moving higher in the frequency bands, it should be observed that both analog and digital systems are provided in either VHF or UHF bands. In these cases, the analog systems utilize tone modulation for coding and signal selection to identify the various functions required. However, regardless of which frequency is used, only a single channel is required for a multiplicity of functions as previously described.

On the other hand, there are numerous modulation schemes employed in communicating digital data over a radio frequency link. These modulation schemes can be placed into one of four categories:

1. Amplitude modulation (AM)
2. Frequency modulation (FM)
3. Phase modulation (PM)
4. Amplitude–phase modulation (APM)

The purchaser will have to depend on the expertise of the manufacturer or his or her own staff to properly evaluate the optimum system for the requirements. As with instrumentation and general control loops, the trend is toward digitally designed systems. The advantages are the ability to readily provide the capacity for data transmission as well as proportional control. Another advantage is the ability to interface with computer and microprocessing equipment. There are those who suggest that the disadvantages of digital systems are that they are more difficult to maintain and sometimes more susceptible to interference.

One of the considerations that may influence the choice of one type of system architecture over another may be the type of radio remote control systems installed if one contemplates additions to an existing plant. Compatibility and experience of maintenance and service personnel can, in many cases, justify the continued use of a less sophisticated system, whereas a green field installation may dictate alternate choices. Above all, the purchaser must, with these types of systems as with other materials handling equipment, carefully evaluate the experience, capabilities, and flexibility of the supplier to fulfill present as well as future requirements.

35.1.8 Anticollision Systems

Introduction

Radio remote control techniques are also applied as a distance-detection method when used for automatic anticollision protection of materials handling equipment. Similar distance-detection techniques are used to maintain safe prescribed operating distances between cranes in multiple installations to prevent rail overloading (see Fig. 35.1.13). There have been efforts to raise the state of the art relative to automatic anticollision systems for aircraft and land vehicles (i.e., automobiles, trucks, etc.). There has been very little prior art in the materials handling industry with certain exceptions. These are the relatively short-range anticollision devices commonly found in the front of automatic guided vehicles and various types of user-designed photo cell or mechanical limit switch devices to prevent overhead traveling crane collisions.

Prior to the advent of OSHA, the type of anticollision systems attempted for cranes consisted of limit switches on the cranes operated by stops on the rails which would remove the power to the bridge and apply the brakes when a crane trespassed beyond its zone of operation. The disadvantage is that cranes are thus confined to specific zones of operation. Another type of anticollision system consisted of wands or bars attached to cranes which operated switches on opposing cranes when one crane came within, say, 10–15 ft of another. Photo cell and infrared detectors apparently had met with minimal success. Ultrasonics, previously short-range devices, are sometimes adversely affected by temperature variations, and other sonics have been attempted, but no successful installations have been reported.

Current methods are reported to employ a combination of electronic generators and signal monitors

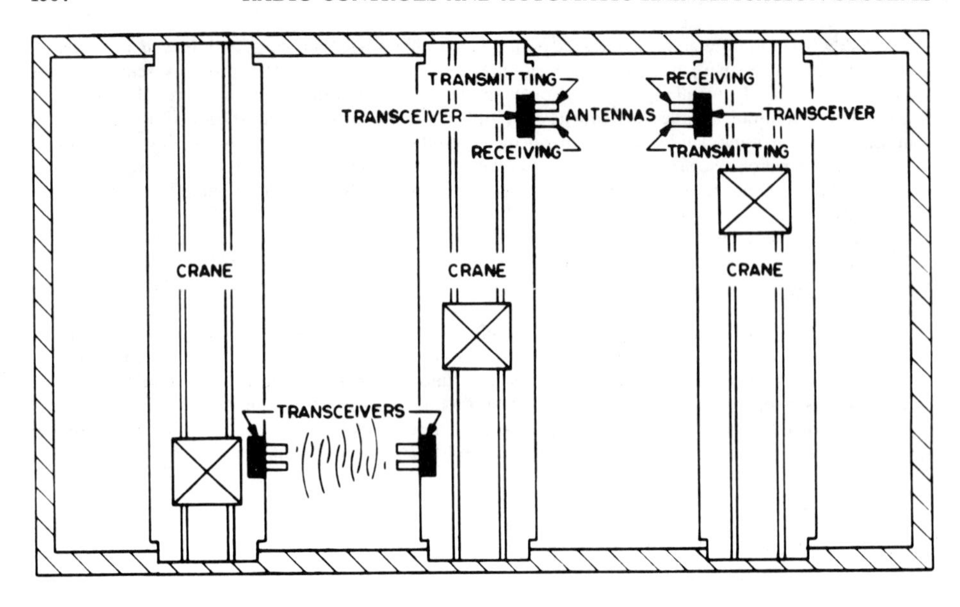

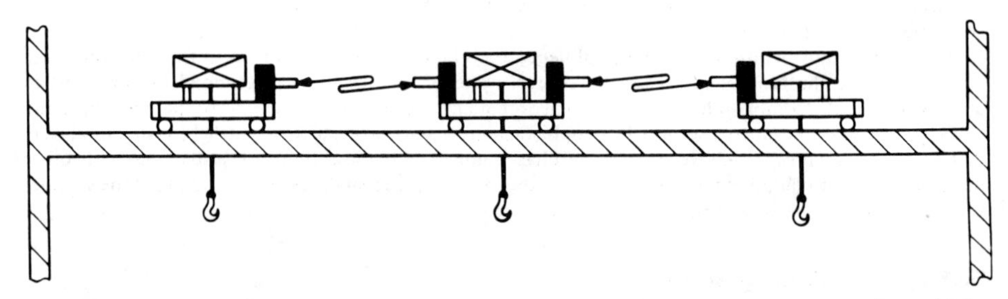

Fig. 35.1.13 Anticollision detectors maintain safest operating distances between cranes in multiple installation.

mounted on cranes whose outputs are connected to each crane by a pair of wires or conductor bars strung parallel to the crane runways. A suitable bridge-type comparator is used to judge distance. This system is in limited use today, probably because the cost of running parallel wires along runways and the use of collectors involves substantial installation and, possibly, maintenance expense. However, such systems are reported to have an advantage of being immune to interference.

It was not until the publication of OSHA's Rule 1910.179 regarding bumper requirements that stronger motivation for development of a more practical anticollision system was present. It was recognized that a bumper was a shock-absorbing device, whereas the industry needed an anticollision system as a practical means of automatically and repeatedly preventing collisions from occurring in the first place.

Although plant safety personnel and OSHA no doubt provided the initial stimulus for movements in this direction, anticollision systems are being used today additionally as a loss prevention means to avoid the disastrous effects of product liability lawsuits.

The system to be described later has been reported to be the type in most widespread use in this country and in Canada. Usage has also spread to the Common Market countries as well as Latin America and Japan. It is categorized as a radio remote control system because, as a matter of fact, it does use low-frequency radio waves as its transmission and detection means. The system has been

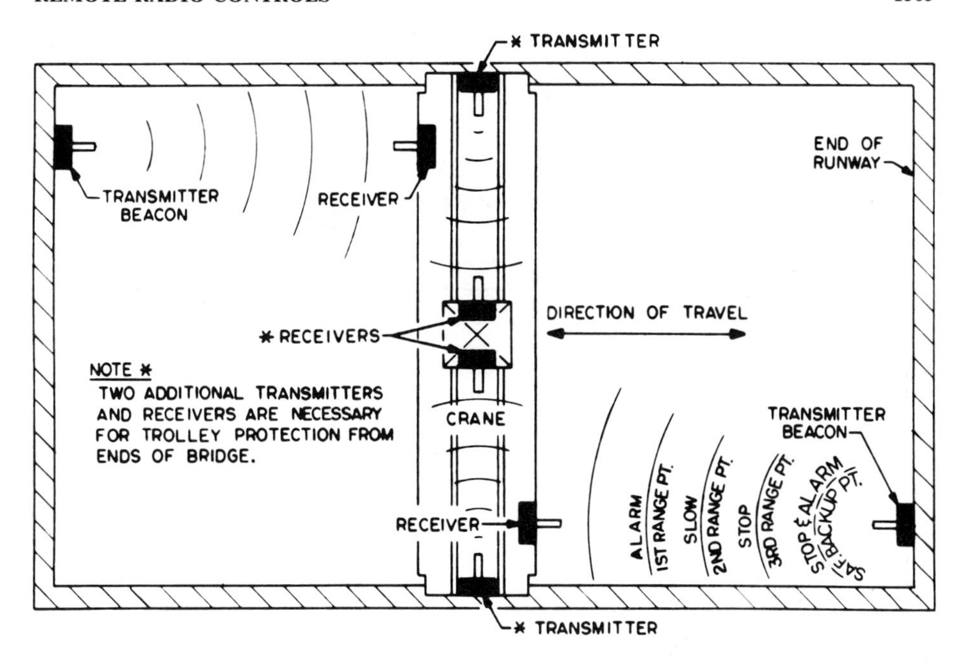

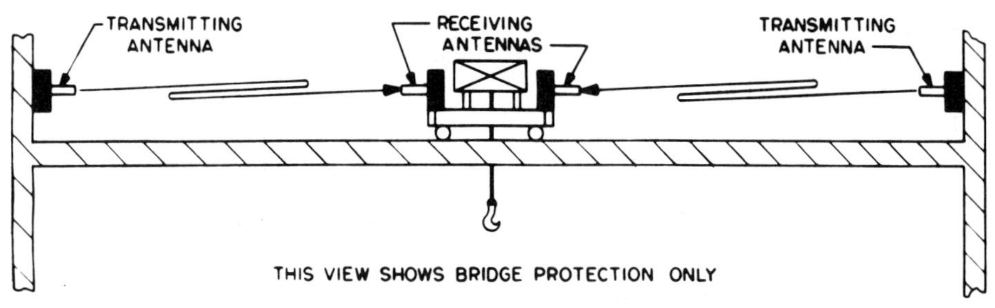

Fig. 35.1.14 Anticollision system for single crane in a bay.

applied to overhead traveling cranes, automatic stacker cranes, and various industrial materials handling vehicles.

Tracked Equipment

A schematic drawing of a single crane in a crane bay illustrates how the system works (Fig. 35.1.14). In these views, the bridge of the crane moves to the left and to the right along the rails as indicated. A low-frequency beacon transmitter and antenna are mounted at either end wall of the bay to provide a signal against which can be measured the limits of travel of the crane within the bay.

The receivers are mounted on the crane with each respective receiving antenna oriented toward the opposing transmitting antenna. Each receiver's sensitivity is adjusted so that it is effective in the near field electromagnetic region and acts as an override to the normal method of control. The system detects the position of the crane by measuring the strength of the transmitted signal and permits it to decelerate gradually until the stopping point to prevent dangerous load swing upon sudden stop.

Traveling in the direction indicated, the system can be adjusted so that the first position reached by the crane (note "1st Range Pt.") will trigger an alarm and slow the crane at a predetermined distance from the transmitter. The second position of the system may be adjusted to reduce the speed

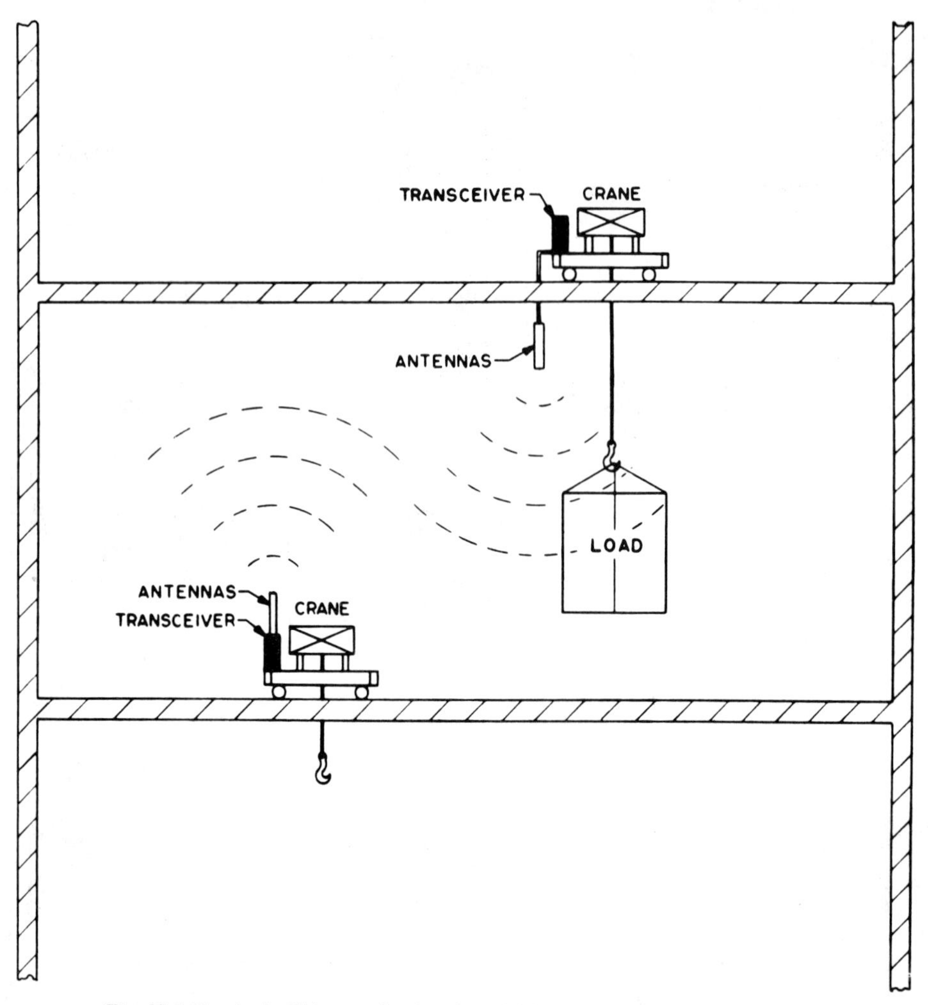

Fig. 35.1.15 Anticollision application for vehicles operating at different heights.

of the crane even further when it reaches a preset position, whereas the third and stronger signal detection point of the system may be adjusted to bring the crane to a stop.

The technique of control to prevent collision in a multiple-crane installation is similar to that previously described, except that each crane bridge is provided with one or more transceivers (Fig. 35.1.13). Each transceiver is capable of generating a signal and also capable of receiving a signal from its opposing transceiver. For example, the transmitting antenna on left crane transmits its signal to the receiving antenna on center crane while center crane transmits its signal back to left crane.

Another unique but important application of a distance-detection system is to automatically prevent the collision of two-tiered cranes operating at different heights within a bay (Fig. 35.1.15). Collision is prevented between the moving hoisted load of an upper crane and the bridge or trolley of a lower-level crane, or between a crane and a tracked vehicle passing under the crane's load.

One transceiver is required for each crane or vehicle. As shown in the diagram, the antenna of the upper crane is mounted below the crane, directed at the vertically mounted antenna atop the lower crane. The upper crane's transceiver is only turned on by a limit switch when the hoist is in a lower "collision" position.

Trackless Equipment

The predominant applications of automatic anticollision systems have been for tracked materials handling equipment such as overhead bridge cranes, gantry cranes, automatic stacker cranes, as well as various

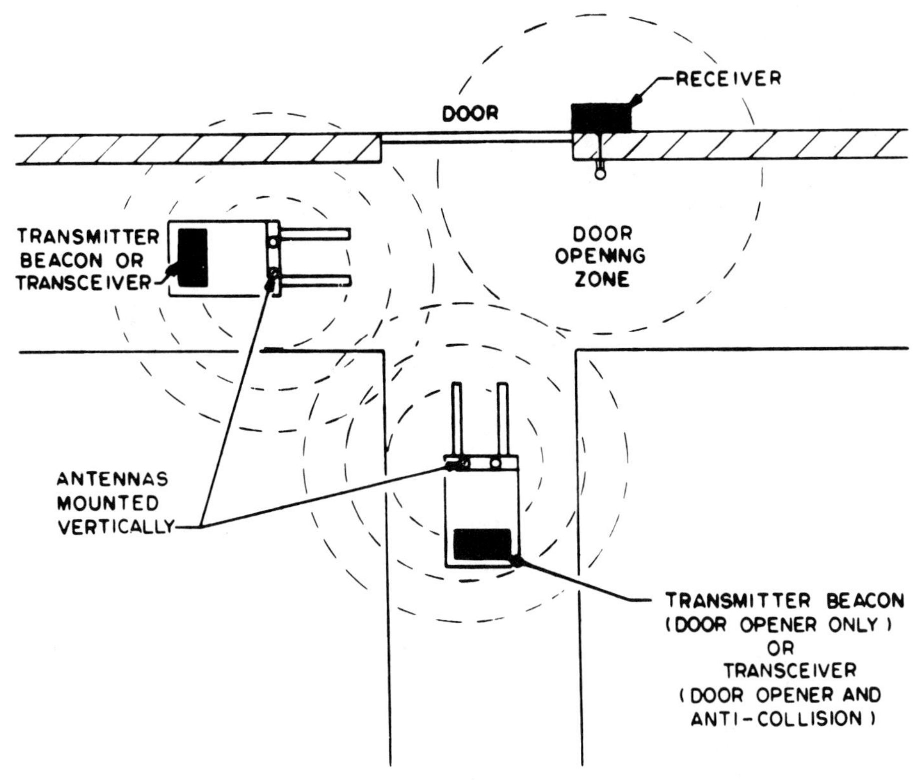

Fig. 35.1.16 Door opener or door opener and anticollision application.

types of transfer cranes. More recently, there has been an increased interest in the application of this high-technology equipment to trackless vehicles such as lift trucks and large haulage trucks.

Lift truck applications can be varied and serve to prevent lift truck collisions around blind corners. Distance-detecting anticollision equipment can also serve as door openers. Satisfactory operation in either case, as shown in Figs. 35.1.16 and 35.1.17, is dependent on the vertical mounting of antennas which provide omnidirectional patterns, since the trucks do not follow unidirectional tracked paths. In general, the principles of operation are similar to the tracked materials handling equipment, and although the electronic portions can be identical, the interface requirements may vary considerably depending on the controls utilized on the trucks.

Drivers of large haulage trucks for surface mines have blind spots encompassing 65% of their field of vision because of the enormous dimensions of their vehicle. Blind spots immediately in front, around the front right side, and to the rear have resulted in haulage truck operators driving forward or backward over nearby smaller vehicles and/or individuals. Truck haulage accidents have become the greatest single cause of accidents in open-pit mines.

Over the years, various types of systems have been introduced to improve the field of vision in these haulage trucks. Many techniques have been used, including advanced mirror systems with special lenses and closed-circuit TV. These quickly got dirty, out of focus, and broke easily. And all of these visibility aids require a driver to concentrate on several places at the same time.

Similar to the foregoing examples, the same principles of low-frequency radio electromagnetic signaling and detection have been applied. The significant differences are to prevent the large haulage truck from backing over a small service truck by alerting the driver who will then manually stop the large truck. Figure 35.1.18 indicates the idealized radiation patterns and equipment location. It is important to note that beacon-type transmitters are provided for the smaller service vehicles which send continuous signals that are detected within, say, 20 or 30 ft of the haulage truck. The haulage truck utilizes front and back receiving antennas connected to receivers which measure the strength of the transmitting antennas' signal. At specific preset distances, an annunciator is alarmed within the haulage truck cab alerting the driver to the proximity of the smaller vehicle.

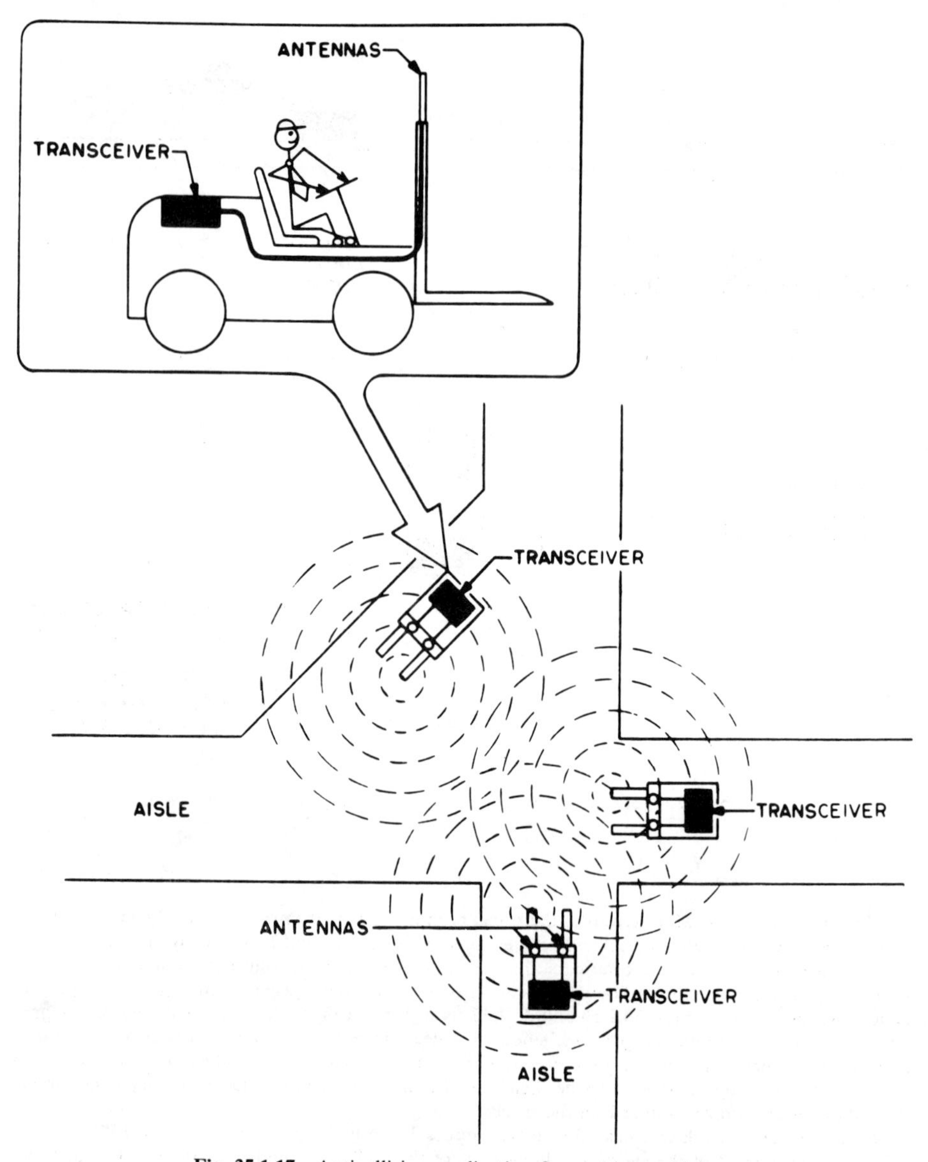

Fig. 35.1.17 Anticollision application for multiple vehicles.

BIBLIOGRAPHY

Federal Register II, Dept. of Labor, OSHA, Vol. 36, No. 105, May 27, 1971, Revised Aug. 13, 1971.

Illig, F. H., "Eliminating Pendant Control Confusion," National Safety Congress, 1976.

Iron and Steel Engineer, Feb 1974, p. 41.

Material Handling Engineering Magazine, Penton/PC Inc., Cleveland.

Modern Material Handling Magazine, Cahners Publishing Company, Boston.

1980 Forum Proceedings, Material Handling Institute, Inc., Pittsburgh.

Plant Engineering Magazine, Technical Publishing Company, Barrington, Ill.

"Report on Survey of Incidents of Personal Injury and Property Damage Accidents Involved in the Operation of Overhead Traveling Cranes," National Safety Council, Chicago, 1975.

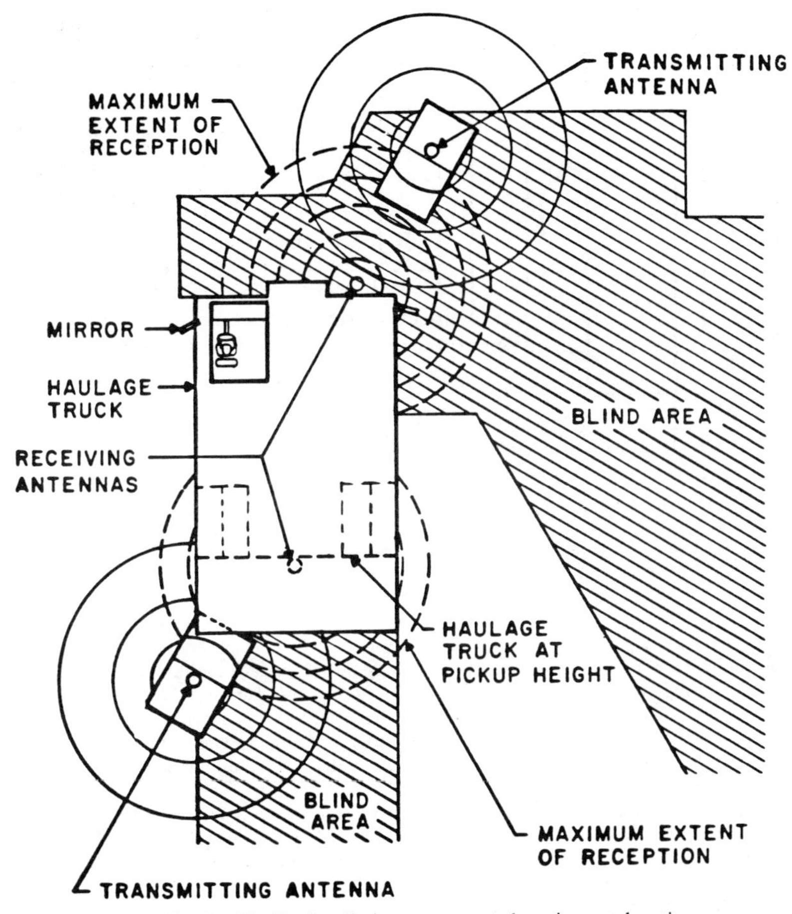

Fig. 35.1.18 Idealized radiation patterns and equipment location.

35.2 AUTOMATIC IDENTIFICATION SYSTEMS

Edmund P. Andersson

35.2.1 Introduction

Automatic identification in the materials handling industry is usually synonymous with bar code scanning and the scanning of marks by fixed-beam reading devices.

Let's begin by examining some milestones. Figure 35.2.1 shows the most important milestones for industrial scanning. The so-called fixed-beam readers were first introduced in the 1960s. These were the first step up the ladder. The Sylvania KARTRAK system followed, demonstrating the feasibility of scanning a machine-readable code several times as the code passed by the scanning device. This was the beginning of the moving-beam approach to scanning. Until 1971, however, scanners were always looking at machine-readable codes comprised of retro-reflective materials. These were expensive and impractical for most industrial applications. The first big breakthrough occurred with the introduction of the laser scanner and its capacity for reading simple black and white printed marks—those things we now call bar codes.

But, how does one create the bar code? Bar code printing devices came along soon after the laser scanners, and things began to happen. The UPC (Universal Product Code) was adopted by the food industry in 1973, giving the concept of bar codes a tremendous lift. Bar codes literally became a household item. This created a credibility for scanning that obscure industrial applications could not. But, the latter started to grow rapidly because of the publicity associated with retail point-of-sale.

The microprocessor came along and scanning equipment and printing devices were redesigned to

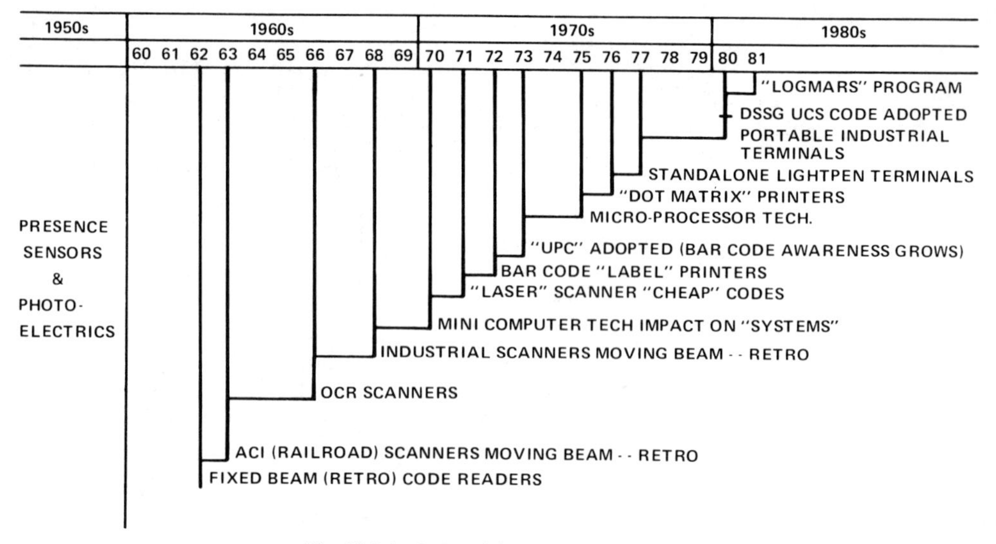

Fig. 35.2.1 Industrial scanning milestones.

take advantage of its capabilities. Product prices came down, new products were created, opening more opportunities for integrating scanning technology into materials handling functions. Controllers followed, accelerating the integration of scanners in system configurations. In 1976, the micro was used in a new type of printing device—the dot matrix printer—which made many system designers reevaluate business forms and documents used in order processing. This printer has revolutionized forms printing by treating them as an output of a computer, including bar code symbols, sequential numbering, and so on.

Stand-alone bar code data entry terminals, which use a lightpen for data entry, came along in 1977. This gave scanning a new dimension in materials handling and manufacturing. Portable versions were introduced in 1980, a milestone that closed the loop for scanning. With the arrival of the portable bar code terminal, the systems designer had all the tools necessary to identify and track products no matter where they were, moving on a conveyor, sitting on a shipping platform, or stored in inventory. And, all scanning could be done using the same product label—the bar code.

In 1980, the Distribution Symbology Study Group (DSSG) announced the selection of standard bar code symbols for printing on corrugated cartons. This milestone may have proven to be the single most important event of this decade. It does for distribution and manufacturing what UPC does for retail stores or what ACI could have done for the railroads. Learn about the UCS, sometimes called the "UPC Case Symbol," and its implications as soon as you can.

The final milestone is the U.S. Department of Defense LOGMARS program. If a company sells to the federal government, it will be encoding products with bar codes prepared to government specifications. They will use the bar codes to improve the procurement and logistics system. The various milestones are illustrated in Fig. 35.2.1.

35.2.2 Materials Handling Opportunities for Automatic Identification

Conveyors

The most common types of conveyors in materials handling are:

- Belt and roller
- Tilt tray
- Power and free
- In-floor towline

In all these applications, the common denominator is *movement,* sometimes at very slow speeds (e.g., power and free, towline), sometimes at extremely fast speeds (e.g., tilt tray). The conveyor merely conveys the product from one place to another. The control system needs input to determine when and where to activate diverters. The method of input is too frequently human, via a control console. In nearly all such situations, this is a poor way to utilize expensive labor. In high-speed conveyor operations, the hardware is underutilized because it does not operate at design speed because the humans simply cannot keep up. Some high-speed tilt-tray conveyors operate with only every fourth or fifth tray loaded. The line was flying but was clearly underutilized because the induction system was under human input.

Another problem is human error. What does it cost? What problems does it create? Even on slow conveyors, scanners are probably justified just to eliminate human data entry and associated errors. Besides, humans can press buttons on a control console, but they forget it immediately. A scanning system does not; the information is passed along and has utility elsewhere in the system.

Automated Guided Vehicle Systems (AGVS)

Guided vehicles have grown in popularity, especially in the aftermath of all the publicity on factory automation. AGVs have removed the driver from the vehicle, placing their operation and guidance under computerized control. But AGVs require some type of input before they can perform. By adding scanning devices at key pickup and delivery stations, and at data input/output terminals, the AGVS can be upgraded and made even more productive. Some of the benefits of scanning include:

Auto load ID (product or destination)
Auto vehicle monitoring and control
Auto inputs to AS/RS for load ID
Confirmation of transaction completion
Improved productivity and equipment utilization

Robots

Robotics is a new factor in materials handling and production. As with everything else, robots attempt to remove the human from certain functions. Robots are most often used at work stations and their *most frequent* functions are:

Handling parts
Working with manufacturing or process equipment, machines, and conveyors

Robots are *infrequently* used for:

Assembly
Inspection

Most robots are deaf, dumb, blind, incapable of applying judgment, and unable to discern between good and bad, correct and wrong. In the context of this section, therefore, we can look to robotics for materials handling and in-process functions. How does automatic identification technology integrate here? Some of the benefits of scanning are:

It tells robot *when* unit may be handled.
It gives robot *identification* of unit being handled.
It enables robot to *activate program* for that unit.
Based on ID, it provides *handling instructions* when robot's functions are completed.
It communicates *production data* to information system.

Automatic Storage/Retrieval Systems (AS/RS)

Load identification is the primary role of automatic identification in AS/R systems. The scanners are located at the induction or transfer location, to scan a product identification code. The data are sent to the AS/RS computer, which, upon receipt of load identification, assigns and directs the load to the storage location.

Working this sequence in reverse can effectively update inventory files based on transaction confirmation.

Scanners also play an important role in integrating AS/RS, AGVS, conveyors, and robotics in the automated factory by providing discrete load or product information to the appropriate controllers/computers as transfers occur.

35.2.3 Automatic Identification Tools

The systems designer has a formidable arsenal of automatic identification products—or tools—at his or her disposal. There is literally a product for every type of application that the manufacturers know about. The following sections briefly present these products and their intended uses.

Scanning Moving Products

Dynamic scanners are used for collecting information from moving products. There are several versions of these devices. Their most noteworthy attributes are:

They read passing bar code symbols several times, even at very fast speeds (e.g., up to 500 ft/min).

They tolerate wide variations in bar code location and orientation.

Several different types are available to fit specific applications.

Figure 35.2.2 illustrates a typical scanning curtain for a dynamic scanner. These scanners, as we will see in later sections, can be and are mounted in various positions, sometimes beside a conveyor, sometimes above, rarely underneath. They are very flexible and even come in small packages for installation in tight places.

Photocells report the presence of passing objects to the scanner's decoding units, which are programmed to look for a bar code message soon after. If none is received and the photocells report the presence of another passing item, the decoder will record a "no read." This is an important aspect in any system design because no-reads (which may be unlabeled products) must be accommodated in some manner.

Dynamic scanners range in cost from less than $5,000 to as much as $70,000. At the low end is a scanner with an integrated decoding unit capable of general-purpose scanning at most throughput speeds. For about $9,000–$10,000, there is a more sophisticated scanner capable of a wide range of very difficult scanning functions (e.g., these offer the best optical performance over the widest operating range). For $50,000–$70,000, you can acquire a scanner that can read a bar code over a 360° frontal radius to the scanner. These devices are normally mounted above product flow and require that the bar codes be face-up and remain at approximately a 90° angle to the scanner. The choice of scanners is usually dictated by economics and/or application specifications.

If bar code orientation is a problem in a system, there are some choices. They are:

1. Use an omnidirectional bar code, if space permits, and a $5,000–$10,000 scanner.
2. Use an array of $5,000–$10,000 scanners with different scanning angles (e.g., 2 to 5 are less expensive than a single $50,000–$70,000 unit).
3. Use a combination of 1 and 2.
4. Select a $50,000–$70,000 scanner.

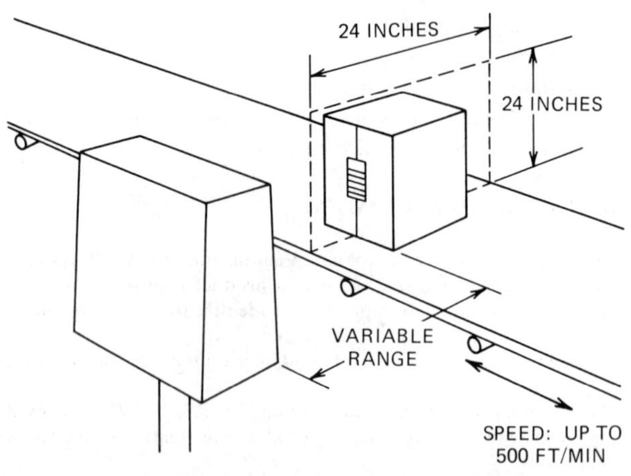

Fig. 35.2.2 Typical scanning curtain for a dynamic scanner.

Scanning Stationary Products

Beginning with the hand-held lightpen and continuing to computers and controllers, we have seen a variety of new products to improve data collection. Bar codes still dominate the field in product types, but OCR users should not be discouraged. Some new bar code symbols include both bar code and OCR characters in the same symbol, with the human readable in OCR. If someone has an investment in OCR and wants to take advantage of bar codes (let's say in product sortation, or receipts processing), there is no reason not to.

In the context of materials handling and production, these scanning products fall into three categories:

1. Stand-alone terminals
2. Portable terminals
3. Mobile scanners

Figure 35.2.3 presents five types of scanning devices and terminals that could be used in these applications. They are:

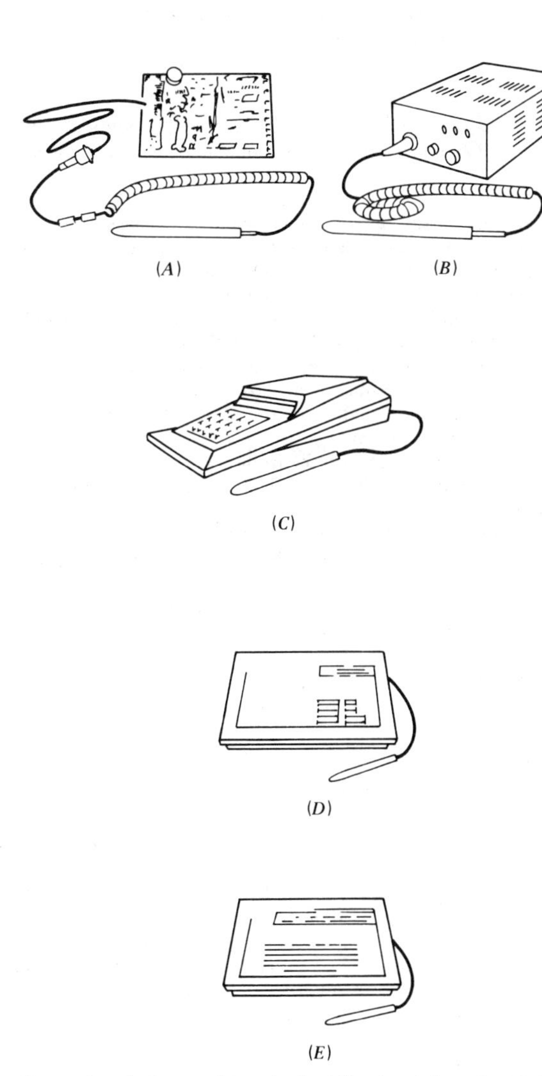

Fig. 35.2.3 Types of scanning devices and terminals. (*A*), circuit board with lightpen; (*B*), interface unit with lightpen; (*C*), light-duty terminals with lightpens; (*D*), heavy-duty terminals with lightpens (numeric); (*E*), heavy-duty terminals with lightpens (alphanumeric).

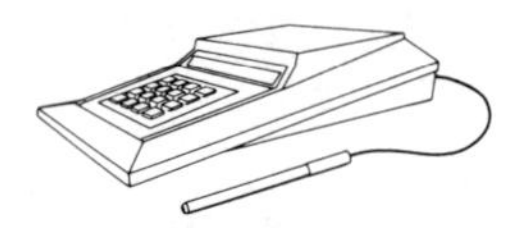

Fig. 35.2.4 Portable terminals.

Circuit board and lightpen—designed for integration within a CRT or similar terminal. Upgrades dumb terminal to accept bar code data entry.

Lightpen with decoder and interface—used externally to CRT or as stand-alone requiring no keyboard entry or display.

Light-duty terminal with lightpen—has keyboard and display, requires no supporting terminal.

Heavy-duty terminal with lightpen—rugged cast aluminum housing, full keyboard, and display.

Figure 35.2.4 presents a portable bar code data collection terminal. This one has a full keyboard and alphanumeric display. Another version uses bar code symbols in place of a keyboard. These devices differ from the units found in retail applications in that they are more rugged and their operation is geared to industrial applications and hourly labor. Communication is normally from the terminal, through a communications interface, and then to a computer or controller via communications lines. These devices also read the popular bar code symbols used in materials handling and production. Custom programming and downloading is possible at additional cost.

It is also feasible to install scanning devices and terminals aboard vehicles. For example, a picking car could integrate a lightpen and decoding unit aboard the vehicle and scan each product being picked. A dynamic scanner could be mounted on an AS/R vehicle or AGV to scan products that are stationary while the vehicle passes them.

The products we have just seen range in price from $800 to $2500 for the stand-alone terminals and $1800 to roughly $2200 for the portable units. These prices include the lightpen and a choice of bar code symbol decoding.

All these products should be viewed in a systems context. Look beyond the immediate benefit of data collection speed and efficiency and examine such things as data base integrity, reduction of errors, and associated delays.

Part of the integration process includes the important supporting products and services necessary for automatic identification technology. These include:

Bar code and OCR *printing devices*

Film masters needed to prepare artwork for bar code symbol printing on cartons, packaging, and so on.

Automatic *label applicators*

Verifiers, to check quality and readability of bar codes being produced in volume

Microcontrollers and *communications devices*

Systems services, representing unique experience in real-time, automatic data collection/control

All these products, services, and expertise are available from member companies of the Material Handling Institute's (MHI) *Automatic Identification Manufacturers* (AIM) Product Section. AIM is also an excellent source of information on bar code symbology, a subject in itself. The most common symbols used in material handling are: Interleaved 2-of-5 (AIM USD-1), and Code 3 of 9 (USD 2 and 3). These and other symbols are available for a small fee. AIM also offers a paper on, "New Techniques in Bar Code Printing Technology."

Other Code-Reading Techniques

In a number of applications, the environmental constraints or absence of line-of-sight access to the code preclude the use of optical identification techniques (e.g., item or carrier identification into or through a paint-spraying, baking, or machining operation). Here the need for a machine-readable tag that will provide reliable feedback in spite of extremes of temperature and contamination by paint, coolants, and so on can be critical to overall system integrity. For such applications, microwave or RF (radio frequency) identification systems should be considered. These systems are characterized by:

An active (battery-operated) or passive (external-signal-activated) chip mounted on a circuit board housed in an industrial package suitable for mounting to high-value products (e.g., automobiles, appliances, and aircraft components) or product carriers (totes, pallets, lift trucks, AGVs, etc.).

A reader that transmits microwave or radio signals at a unique level and that receives and decodes return signals from the encoded tags for transmission to the handling system controller.

Depending on the application, reader ranges vary from one foot or less to several feet.

Nonoptical Identification Systems: Radio and Microwave

For years a variety of companies have worked on the development of identification systems employing RF (radio frequency) or microwave technology for vehicle identification in the transportation market (e.g., rail cars, buses, subway cars, etc.). In the early 1980s, new developments in the basic technology shifted the focus to potential applications in industrial process control and materials handling. Each system contains three primary components:

1. The transponder or tag which is typically a printed circuit card incorporating receiving and transmitting capability and housed in an industrial package, some of which are capable of handling extremes of temperature ranging from −40°F to 400°F. Two basic types of tags are available:
   a. A passive tag which responds to external RF signal input with a unique 4- to 14-digit number. This tag is preprogrammed at the factory to the user's specification. A field-programmable tag is also available.
   b. A battery-powered tag is available that can receive, store, and transmit data. In other words, it can actually carry instructions to the point or points at which the material it identifies is to be processed, and can receive and store data along the way, thereby creating an audit trail that can be readily checked at quality control or prior to release for shipment.
2. An antenna which is installed adjacent to the line of tagged item movement and transmits signals and receives data from the tags for processing by the system "reader."
3. The reader or interrogator which controls one or more antenna(s), decodes and processes tag data content for transmission to data logging devices, programmable controllers, or mini- or microcomputers.

Radio and microwave systems typically use tags that are permanently attached to high-value products (e.g., automobiles, large appliances, farm equipment chassis, etc.) or such reusable product carriers as tote boxes, bins, containers, pallets, and overhead power-and-free conveyor trolleys. They are particularly useful for product identification tasks in environments that preclude the use of optical alternatives, for example, tracking products/carriers that move through paint spraying, coolants, ovens, and so on. Tag costs are higher than for bar code labels and comparable to the permanent code plates used in fixed beam code reading. Reader costs vary with the application.

PART 5

SAFETY, ENVIRONMENT, AND HUMAN FACTORS

CHAPTER 36
SAFETY, ENVIRONMENT, AND HUMAN FACTORS

GARY E. LOVESTED

Deere and Company
Moline, Illinois

JOHN W. RUSSELL

Liberty Mutual Insurance Co.
Glastonbury, Connecticut

ROBERT KOLATAC

West Milford, New Jersey

LYLE F. YERGES

Yerges Consulting Engineers
Downers Grove, Illinois

36.1 MATERIALS HANDLING SAFETY IN INDUSTRY

Gary E. Lovested

Of all accidents occurring in the workplace, materials handling accidents are the most common. The National Safety Council estimates that 20–25% of all occupational injuries are the result of handling materials either manually or mechanically [1]. This is probably a very conservative figure. Materials handling injuries are pervasive throughout the workplace regardless of the industry and operation.

It is reported that on the average, industry moves about 50 tons of materials for each ton of finished product produced [2]. Some industries move 180 tons for each finished ton of product. With this amount of materials handling, accidents will occur. However, many of them are preventable through common sense and through proper job design.

Tremendous strides have been made in automated materials handling systems such that little, if any, manual materials handling is performed by workers. Yet much manual materials handling remains in industry. Workers in most operations still manually handle materials to and from their machines, warehouse workers still manually pick parts, and so forth. All of this involves the manual handling of parts, whether it is for inspection purposes or for production purposes.

Although much has been done to improve the materials handling activities within industry, the fact remains that employees are still being injured in materials handling accidents—either by manual or mechanical materials handling.

With the increase of working women in plant environments, designers must now consider the needs—both physical and mental—of a greater range of workers. This can be a real challenge!

Humans have both physical and mental limitations. Not considering them in materials handling work design may result in worker stress that exceeds their physical and mental capabilities resulting in injury and work inefficiency.

Tremendous amounts of money are expended each year on new processes, methods, and equipment in this country without enough thought being given to the human–system interface. In materials handling applications, such questions that might be asked are: "Can all workers lift the loads expected of them with reasonable safety?", "Does the cockpit (work station) seat of a forklift truck provide lower back support when operating the controls and traveling?", and so on. The answer to these and many other questions will help in the planning of materials handling activities which will reduce accidents and at the same time increase work efficiency.

36.1.1 Safety Considerations

The most common type of manual materials handling injuries are sprains and strains to the back and other body areas, and fractures and contusions to the body extremities. Many are the result of unsafe work practices by workers such as improper lifting, carrying too heavy of a load, incorrect gripping of objects, failing to wear personal protective equipment, and so forth.

However, not all of the unsafe work practices attributable to workers may be their fault. For example, if warehouse parts pickers must lift parts out of pallet boxes located in such a position that workers are forced to bend over to lift parts, then so-called "proper" lifting techniques cannot be used. Workers should not be faulted for something they have no control over—these are situations created by management and are under the control of management.

Definition of an Accident

An accident is an undesirable event that may or may not result in physical harm to people and/or property.

Several thoughts are contained in this definition. An accident may or may not result in an injury (e.g., close call). A "close call" or "near-miss" accident is an accident—only no one is injured.

For example, suppose a worker is picking parts from an elevated picking truck 18 ft from the floor when the truck's hydraulic system fails and the picking platform free-falls to the floor. Suppose also the worker was able to grab onto a rack and prevent his fall. That was a near-miss accident!

Later, the worker uses the same truck without it having been repaired and is picking parts from the same level when the same accident occurs again. This time the worker was not fortunate enough to grab onto a rack and the worker falls and is killed. There were two accidents—one was a near miss and the other resulted in a fatality. Both accidents occurred from the same cause—a faulty hydraulic system. If steps had been taken to determine the cause of the first accident and the cause corrected, then the second accident would not have occurred.

Another part of the accident definition is that it is an undesirable event. This implies that no one would intentionally design into a job things that would surely result in accidents. Yet many times, upon investigation of an accident, poor job design was determined to have caused or contributed to the accident. Good intentions are not enough—knowledge of safe job design is required.

Accidents are symptomatic that something is wrong in the work environment and they also indicate a less-than-efficient production environment. All accidents are caused! Identifying the cause and correcting it will ensure that the same accident will not be repeated. Safety, efficient production, and high quality go hand in hand. One does not exist without the other.

Materials Handling Injuries

Materials handling injuries can be divided into two main groups: trauma and nontrauma. Trauma injuries are mostly sudden injuries resulting in a wound to living tissue, caused by an exterior force (e.g., lacerated finger from sharp sheet metal striking it). Nontrauma injuries, on the other hand, are injuries from voluntary or involuntary motions in which the body is stressed beyond its physical capacity (e.g., muscle strain).

Trauma Injuries. Trauma injuries in materials handling results from many sources such as handling sharp materials like sheet metal, dropping parts on the feet, slipping and falling from oil spilled on the floor, and collisions of powered industrial trucks.

The National Safety Council reports that the dropping of parts (e.g., parts slipping from the hands) accounted for 7% of all work injuries and 5% of the total workers' compensation paid [3].

A major portion of manual materials handling injuries occurs to the legs and feet, with the greater percentage occurring to the feet. Foot protection, although not *preventing* the accident, may prevent many of these types of trauma injuries. Foot protection should include metatarsal (instep) protection which affords better protection than toe protection.

Manual materials handling hand and finger injuries, such as those from handling sharp material, can be prevented or reduced by gloves or other hand protection. However, workers should not wear gloves around rotating machinery. Extensions of the hand, such as tongs for feeding materials into metal forming presses and handles or holders attached to objects, are examples of tools that can effectively reduce hand injuries.

The eyes, trunk, and other body parts can also be injured during manual materials handling. For example, the cutting of wire- or metal-strap-bound bundles or containers could result in a whip-back of the cut ends, striking the body. Eye protection and gloves will help to prevent injury.

Anytime materials are moved or handled manually, there exists the possibility for injury, which adds to the ultimate cost of the product. To reduce this exposure and increase work efficiency, manual materials movements should be minimized and mechanical materials movement maximized.

Nontrauma Injuries. Nontrauma materials handling injuries may result from lifting objects, from repetitively bending and twisting, from pushing or pulling or carrying objects, and so on. All of these activities may result in injury to the musculoskeletal system given enough time and stress. In fact, a major difference between trauma and nontrauma injuries is the time of occurrence.

Back injuries account for approximately 20% of the total number of disabling or lost-time work injuries in the United States [4]. These back injuries represented approximately 30% of the total U.S. workers' compensation costs paid.

But these are just average costs! Some industries, because of the high amount of manual materials handling performed have even higher costs. It has been reported that the average cost of a workers' compensation case involving the back is over $6000. Back and other types of overexertion injuries are increasing and are increasingly costly.

Back injuries are the number one problem facing industry today. With the great strides many companies have made in reducing trauma-type (contact) injuries over the years, back injuries now loom as the single largest problem both in frequency and cost. Yet the knowledge exists to reduce both frequency and severity of back injuries by applying established human factors.

36.1.2 Ergonomic Considerations

In designing any materials handling systems that will involve workers, the designer must consider both the physical and mental capabilities of the worker(s) who must interface with this system. Applying these human characteristics to design is called human factors engineering. In Europe, where it began, applying human factors is known as ergonomics. The purpose of human factors engineering or ergonomics is to apply this knowledge in the design of the work environment, which will increase work efficiency and thereby decrease work injuries.

Workers come in all sizes and shapes, strengths, and ages. In the workforce, there are the young and old, male and female, skilled and unskilled, and strong and weak. Some women are stronger than men, although the "average" male is stronger than the "average" female. Each of these factors can have a strong bearing on how the job should and can be performed.

Some fields of human factors engineering are:

Anthropometry—study of body dimensions and mobility.

Biomechanics—deals with mechanical forces and their effect on the musculoskeletal system. Combines engineering with anatomy and physiology.

Work physiology—concerned with metabolic and circulatory responses of individuals performing work tasks.

Psychology is another field that is also involved in human factors as it pertains to the behavior of individuals and their attitude toward work. A brief review of the three main fields pertinent to the discussions in this chapter follows.

Anthropometry

Anthropometry is the practice of measuring various human physical characteristics such as size, strength, and mobility [5]. Engineering anthropometry is the application of such data to workplace design, equipment design, and so on, in order to maximize safety, efficiency, and operator comfort.

Body dimensions of the user population should be considered by every engineer in the design of work layouts and equipment. For example, body dimensions of people change with age. The largest dimension change is during the growing years to the late teens or early 20s. The next largest change is later in age when body weight increases as well as body circumference. Designers need to know this if, for example, they are designing seats. Strength changes with age. Other variables include sex, ethnic groups, and nationality, and they can have a profound effect on design.

Thus, in the workplace, engineers should consider these physical variables in the design of equipment and work stations in order to accommodate the greatest range possible of worker population variables.

An unfortunate fallacy applied by some is to design equipment or work stations to fit the "average" man or woman. The problem with this is that there is no such person and this can lead to serious user problems.

Designing to the "average," or 50th percentile, means that half of the population conceivably could be excluded. For example, if a doorway was designed for the height of an average male, approximately half of the male population would have to stoop to go through it. If storage racks were set at a height where the average worker (50th percentile) could reach, half of the workers would not be able to reach that height.

Another fallacy of the "average person" concept is the assumption that average-sized individuals are essentially average in all dimensions. This is not so, because studies indicate that body proportions are not more or less constant between individuals [6].

Neither does the "average person" concept apply to strength, mobility, and other biomechanical data. If factory lifting requirements were set to the lifting strength capabilities of the average of 50th percentile male worker, then half of the male workers would not be able to perform these lifts. If women are in the same factory workforce, most of them would not be able to meet the 50th percentile male lifting requirements.

Thus, a capable engineer will look beyond using averages and instead attempt to incorporate the largest population possible into the design. Ideally, the engineer should try to accommodate 100% of the user population in the design, but this is seldom practical. Generally the range between the 5th (weakest, shortest, etc.) and 95th (strongest, tallest, etc.) percentile is used, where possible.

The key to accommodating the workplace population anthropometric ranges is adjustability [7]. Equipment or work stations built to any one set of dimensions will seldom accommodate everybody in the user population. Thus, if the equipment or work station does not "fit," the results might be work inefficiency, poor work quality, and unnecessary physical strain, fatigue, and accidents.

Biomechanics

Biomechanics is the study of mechanical forces in movement or rest on the musculoskeletal system of the human body. While physiological functions are also involved, discussion here involves only the mechanical aspects. Biomechanical study is concerned with two basic areas: statics and dynamics and the forces that act upon them at rest or in motion.

One of the basic concepts in mechanics is force—either internal or external—usually defined as a push or pull. External forces are called loads, such as in lifting an object. Internal forces are called stresses, which are the internal resistances of a material(s) to an external force.

Force is further characterized by: (1) magnitude, (2) action line of the force, (3) direction of force, and (4) point of application where the force is applied.

Application of biomechanics in analyzing lift, lower, push, pull, and carry tasks are useful for identifying the physical forces to workers in specific manual materials handling applications.

Work Physiology

Oxygen consumption, metabolic energy expenditure rate, and heart rate are the most common responses used to determine maximum work intensity. This is useful in determining at what level an individual can work without incurring excessive fatigue [8].

Two types of muscular activity exist: dynamic and isometric. Dynamic activity causes the muscles to shorten resulting in movement of the bones around skeleton joints. Activities such as walking and bicycling exemplify this activity. In isometric activity, the muscles do not shorten or result in bone movement; holding an object in a fixed position is an isometric activity. Sustained contractions in isometric activity result in localized muscle fatigue.

Understanding physiological responses to work and applying them to specific work tasks helps in reducing worker fatigue. For example, workers who squat to perform a two-hand squat lift expend more energy than those who bend at the waist and lift. Thus, although the two-hand squat lift method may be a good method of lifting under some circumstances, it is not a good method when it must be performed repeatedly, because of the energy expenditure.

General

In general, the design of materials handling systems—whether they be manual or automated—must include consideration of human physical characteristics and capabilities for optimum results. Although there is always the chance for injury or property damage when workers are physically involved in a materials handling system, the opportunity for worker error should be minimized through design.

In a materials handling system that is well planned with shortened material flow routes, ample aisles, and so on, materials handling equipment is required that oftentimes supplements manual materials handling. Human factors should also play a major role in the selection and purchase of this equipment. Each piece of equipment should be purchased based on how it will operate individually and as part of an operating system. Equipment design should be state of the art and reflect human factor and safety considerations.

The work stations where workers operate or use the equipment (e.g., cockpit of a forklift truck) should be designed to fit the workers' movements and strengths. All work surface heights (including seats) should be adaptable to the individual worker, work stations should be uncluttered and provide ample work area so as not to cramp body movements, and machine controls including emergency buttons should be within easy functional reach for the workers' hands or feet.

Additionally, workplace design should avoid unnecessary strain on workers' muscles, joints, ligaments, and respiratory and circulatory systems. To do this requires elementary knowledge of human physical characteristics and what workers are physically capable of performing without injury to themselves.

General questions that should be considered in reducing back and other manual materials handling injuries are [9]:

1. Can the job be engineered to eliminate the materials handling activity?
2. Can the job be engineered to eliminate manual materials handling by moving or conveying material mechanically?
3. Can the material itself cause injury (e.g., rough, sharp, etc.)?
4. Can handling aids be provided workers that will make their jobs easier and safer?
5. Will the use of personal protective equipment or clothing help prevent or minimize injuries should an accident occur?

Human factors can provide solutions to many different materials handling problems. Lack of human factor considerations may result in inefficient plant layouts, improper use of equipment, and poor systems planning—all of which may result in accidents and costly materials handling inefficiencies.

Readers are encouraged to refer to the reference sources at the end of this chapter for further information on anthropometry, biomechanics, and work physiology.

36.1.3 Back Injuries

Characteristics of Back Injuries

It is oftentimes difficult to establish the cause(s) of back pain or attribute it to a single cause, because so many factors may be involved. Contributing factors listed in the National Institute for Occupational

Safety and Health's (NIOSH) *Work Practices Guide for Manual Lifting* [10] that may be involved in back pain are:

1. Fatigue
2. Postural stress
3. Trauma
4. Socioeconomic stress
5. Personality
6. Degenerative changes
7. Congenital defects
8. Reduction in the size and shape of the spinal canal and intervertebral foramina
9. Genetic factors
10. Stretching, angulation, compression, or adhesion of nerve roots
11. Neurological dysfunction
12. The duration of symptoms
13. Physical fitness
14. Body-awareness

Further compounding of the problem is that more than one of these factors may be present in an incident of back pain. At present, the state of the art is such that not much is known about the interaction of these factors nor the extent to which they are contributive causes.

Although back pain may come from several sources—not all of them work related—this section deals with the work-incurred back injury and some of the traditional approaches used to deal with it.

Back overexertions from lifting heavy objects can result in acute or chronic injury to the back muscles, lumbar vertebrae, or the intervertebral disks. Additionally, various types of abdominal hernias can result from heavy lifting. If excessive strain is combined with degenerative changes in the back, the risk of injury is high.

In a major U.S. insurance company study, 191 compensable back cases from 32 states were investigated [11]. The study noted that 70% of the back injuries involved manual tasks such as lifting, lowering, pushing, pulling, or carrying. Lifting accounted for 49% of these back injuries. Manual handling tasks such as lifting, lowering, pushing, pulling, and carrying were involved in 70% of the back injuries.

Further analysis of the study data indicated that low lifting tasks (0–30 in., or 76 cm, from the floor) were particularly dangerous. A warehouse study by the author confirms the foregoing and found that 69% of the warehouse overexertion injuries occurred in the floor-to-knuckle area [12].

Besides location of lift, repeated lifting for long periods of time can cause excessive fatigue. Fatigue can affect workers' performance when it becomes excessive resulting in muscle impairment. Many back injuries do not occur from a one-time situation such as lifting but are the result of gradual abuse to the back over a period of time.

Thus, judgments should not be summarily made after a back incident that the incident could not have occurred because the worker was not lifting anything heavy. Repeated lifting over time may be the cause and the weights being lifted may be of secondary importance.

It is apparent then that many variables are involved in back injuries. In lifting tasks, the main variables may be summarized in three categories: task, environmental, and human.

Lifting task variables include weight and size of object, the height from which the object is lifted and the distance lifted, the rate or number of times the object is lifted during a time period, and the horizontal distance between the object and the back.

Environmental variables include temperature, humidity, and air flow in which the lifting task is being performed, as well as the presence of any air contaminants or oxygen deficiencies.

Human variables include age, body physique, physical fitness, sex, and work position.

Traditional Approaches to Reducing Back Injuries

Approaches over the years to reduce back injuries have not been particularly effective, as indicated by the increasing number of back injuries occurring in industry. Two traditional approaches commonly used in industry to curb back injuries are worker preplacement medical examinations, and workers' training programs in lifting.

The following was found in a major insurance company study concerning preplacement medical exams and lifting training programs [13].

Preplacement Medical Examinations. The common medical-based selection techniques used by companies in the study were found not to be effective in controlling or reducing low-back injuries. The selection techniques included medical histories, medical examinations, and low-back x-rays. The study revealed that just as many low-back injuries occurred in companies using these selection techniques as by companies that did not use them.

The use of back x-rays are predicated on the hypothesis that back abnormalities predispose individuals to a greater incidence of low-back injury. Substantial evidence does not support this.

Lifting Training Programs. Lifting training programs were found in the study not to be effective in controlling or reducing low-back injuries. The study did not attempt to evaluate the quality of the training programs provided, only whether it was or was not presented. Just as many back injuries were experienced by companies who provided workers' training in lifting as those companies that did no lifting training.

The study found that the only effective control for low-back injuries is to design the job to fit the capabilities of workers through use of human factor approaches. However, even this approach will only be partially effective. If manual materials handling tasks can be raised so that 75% or more of the worker population can perform them without overexertion, then perhaps 67% of the back injuries could be reduced. The other 33% of the back injuries will occur anyway no matter what the job.

These sobering study conclusions attack traditional approaches used in industry today for curbing low-back injuries. The only hope given is for adopting human factors/ergonomics measures. This is not to say that preplacement examinations or training programs should be dropped. But it is saying that if the entire back injury prevention program is based solely on these two activities, this study and others indicate it probably will not be effective.

36.1.4 Developing a Back Injury Prevention Program

Because manual materials handling back injuries are a significant problem for most companies, this section discusses the rather broad aspect of what a back injury prevention program may consist of and how to implement it. Because most work-incurred back injuries result from manual lifting tasks, the main emphasis of this section is oriented toward this task.

The following major program activities which will be discussed are:

1. Employee preplacement activity
2. Training and awareness activity
3. Job physical stress study
4. Human factors engineering control
5. Measuring and monitoring program results

The discussion on job physical stress study also includes how to analyze lifting tasks and determines their acceptability using the NIOSH two-hand lift formula.

Each of these activities is important by itself but combined together, they present an over-all program which should be effective in reducing back injuries. Because of the necessary limited coverage of these topics, readers should refer to the reference sources at the end of this chapter for additional in-depth information.

Program effectiveness will require the cooperation of the safety, engineering, and medical functions as well as that of the workers. The extent to which this cooperation is obtained will determine the program's overall effectiveness.

Employee Preplacement Activity

The purpose of this activity is to determine whether job applicants or transferred workers are medically suited to the job that will be assigned and whether they can work safely and productively.

Several influences today affect the hiring of workers which employers may not have faced in the past [14]. More women are working in industry today than ever before, the workforce is getting older, the mandatory retirement age for workers has been increased, and handicapped individuals can no longer be excluded from the workplace without justification.

The Equal Employment Opportunity Act (EEO) now influences employers' decisions on who they can hire. The EEO mandates that employers must justify their hiring decisions by meeting the guidelines established by the Equal Employment Opportunity Commission.

A screening program using personal interviews, medical examinations, and medical histories attempts to weed out prospective workers who do not "fit"—both psychologically as well as medically—the employer's employment criteria.

Typical medical screening approaches include such activities as clinical examinations, developing medical histories, and perhaps taking spinal or low-back x-rays.

As previously discussed, studies have shown that these medical screening activities have not been particularly effective in identifying individuals who are susceptible to future back problems. This is not to imply that back medical screening should be discontinued. However, careful consideration should be given before disqualifying job applicants based solely on this information.

Preemployment Strength Testing. A program that holds promise for employers is the preemployment strength testing program. This program is based on studies that show that when job lifting requirements approach or exceed workers' strength capability, as demonstrated by an isometric job test simulation, the likelihood of sustaining a back or musculoskeletal injury increases [15].

If the job strength demands and an individual's strength capabilities are known, then rational decisions can be made on matching workers to jobs.

Biomechanical job evaluations are performed on all or selected jobs where such data as weight lifted, horizontal distance of object to ankle, vertical distance of object from floor, object size, lift frequency, and hand location are recorded. This is to establish the physical strength required on the job.

Secondly, three isometric tests which reasonably simulate the strength required on a job are run on individuals. The three tests include testing:

1. Leg-lifting strength, to simulate using a leg/squat-lift technique to lift compact objects from near the floor area.
2. Torso-lifting strength, to simulate using a bent-back lift technique to lift bulky objects from the floor.
3. Arm-lifting strength, to simulate using an arm lift to lift an object from a table or bench.

The testing equipment simply consists of static handles attached to a load cell which measures the forces exerted that can be recorded. This test equipment is currently commercially available from several sources.

Next, a strength comparison is made of strength required on the job to that of the individual. If the required job lifting strength approaches or exceeds that of an individual, there is strong likelihood of back or musculoskeletal injury.

This leaves two options open should the required job strength exceed those of an individual—disqualify the individual from the job and hire superstrong individuals, or redesign the job so that worker strength capabilities are not exceeded. Actually, the last option is the most realistic one. Due to federal workplace regulations on hiring, companies would not be prudent in pursuing exclusionary policies in hiring based on strengths of a few.

The preemployment strength testing program holds promise as a technique to reduce back and musculoskeletal injuries by matching the physical strengths to those required on the job. Readers interested in pursuing this technique further are encouraged to read the National Institute for Occupational Safety and Health's book, *Pre-employment Strength Testing in Selecting Workers for Material Handling Jobs* [15]. The publication contains much useful information for implementing the program.

Training and Awareness Activity

As was previously indicated, worker training programs on lifting were found not to be particularly effective in reducing back injuries. This is not to say that training should be discontinued! But, if back injury prevention efforts are oriented mainly toward training workers on lifting and little else is done, there probably will be few results forthcoming.

The object of training is to increase knowledge, skills, and awareness. Obviously, if workers already have these attributes, no training is needed. Operating on the assumption that training is the cure-all to safety problems and that "too much training cannot hurt," results in waste and cost inefficiencies, not to mention a loss of credibility with the workers.

Lifting training programs should be implemented only after an investigation and analysis has revealed that a lack of knowledge or skills in lifting resulted or contributed to an injury.

Workers should undergo lifting training through new-worker orientation programs, job safety analysis programs, personal or group safety contact programs, and so on. Workers transferred to other production areas requiring different or additional lifting should also receive training.

The need for training should always be preceded by problem identification which identifies a lack of knowledge or skills as being the real cause of a problem. If improper lifting techniques by workers have been identified as the main cause of back injuries, then the next step is to determine whether the improper lifting techniques used reflect poor workplace design or methods.

For example, if workers are required to lift parts out of deep metal tubs, they may be forced into using improper lifting techniques such as stooping over and lifting heavy parts. Therefore, workers' training programs on two-hand squat lifting would be ineffective. In this example, the container should be redesigned or raised in order for workers to assume proper lifting positions.

If the use of improper lifting techniques under the control of workers are found to be the real cause, then lifting training programs would be beneficial (if they are followed).

A training program should consider more than the actual training phase itself. If back problems from using improper lifting techniques that are under the control of workers are a problem, then this need should be communicated to workers via plant posters, bulletin board notices, and so on. The object is to create an awareness among workers of a problem and the need to do something about it.

Operating management and supervision should be made aware of the problem through injury reports, management and supervisory meetings, and so on. Since it is their workers who are being injured, they should be a part of the program.

Either line supervision should attend worker training sessions on lifting or special training programs should be held for them. It is important that they know what workers are being taught in order to answer worker questions, to understand the lifting techniques themselves as well as for enforcement purposes.

If classroom training is to be provided, class size should be kept small so that information exchanges between workers and the instructor(s) can take place informally. Some of these information exchanges can be valuable to management for identifying substandard work conditions or methods that may be unknown to them. This provides management the opportunity to correct a condition or method before an accident or injury occurs.

All lifting training should include practice training. Unless this practice training is provided, workers may not know how to correctly perform the lift even though they may be able to explain how and why it should be done. This also offers the instructor the opportunity to correct any poor lifting techniques before they become habits.

What should be taught in the lifting training programs may vary. Making the worker aware of the dire consequences improper lifting can have to his or her body, showing them how to avoid unnecessary stress in lifting, and teaching them to become aware of their physical limits are general items which may be taught.

NIOSH's *Work Practices Guide for Manual Lifting* [16] recommends worker lifting training include such topics as the anatomy of the spine and trunk, the effects of lifting on the body, safe lifting postures, and several other aspects.

It is important that workers not be burdened with a lot of unnecessary anatomical descriptions or biomechanical detail that would detract from the main message. The information provided should be relevant and address worker needs by showing them how they can perform a skillful lift and the reasons for doing so.

Two-Hand Squat/Lift Method. Generally in industry, one technique is taught for lifting—the two-hand squat/lift method. This traditional lifting method has been taught (not necessarily used) for many years. The object of which is to lift a load from near the floor level by bending the knees and squatting and straddling the object to lift it.

This is a good lift method when it can be or should be performed. Unfortunately, it has its limitations as a general overall lifting technique [16].

Limitations of this method are:

1. When a load cannot be positioned between the knees and must be lifted from in front of the knees, large lifting moments are created on the lower back resulting in high muscle forces and spinal compressive forces.
2. Many Workers do not have the quadricep (front thigh) muscle strength to raise the body much less the load (i.e., "lift with the legs") when they are in the squat position.
3. Repeated lifting using the squat/lift method is not feasible because of the extra energy required to lift the upper torso. This is similar to performing deep-knee bends all day long.

With these limitations, it is obvious that objects requiring manual lifting should not be generally stored on or near the floor level.

If squat-lifting is required—subject to the limitations previously noted—the following is a description of this method [17].

Before a worker attempts to lift a load and carry it to another location, he or she should inspect the path of travel to see that no obstructions exist which could cause the worker to slip, trip, or fall. Oil and water on the floor results in many back injuries when its presence is not known and the worker slips.

The load should be inspected first for sharp or rough edges and hand protection worn when necessary. If the load is oily, greasy, or wet, it should be wiped off first before attempting to lift.

The load should be sized up before being lifted to determine whether the weight (and bulk) is within the lifter's capability. If it is not, help should be sought.

According to the National Safety Council, the squat/lift method consists of six basic steps.

1. Correct feet positioning. Straddle the load with one foot alongside of the load in the direction of movement and the other behind the other side of the load. The rear foot will provide thrust to the body in the lift. If the feet are too close together, a loss of balance may occur, resulting in back injury.
2. Keep the back straight and bend the knees. A straight back (not necessarily vertical) keeps the spine fairly rigid thus evenly distributing the pressure on the lumbar intervertebral disks. A bent or arched back causes unnecessary stress on the back muscles and uneven pressure on the disks.
3. Use a good grip. Using a full-palm grip instead of the finger tips to lift a load will reduce stress to the arm muscles. This also provides more arm strength and reduces the possibility of a load slipping out of the hands.
4. Pull load as close to the body as possible. Tuck elbows and arms in close to the body for optimum strength. Keep arms straight where possible. Avoid arm or shoulder flexing which imposes unnecessary stress to the upper arm and chest muscles.
5. Tuck the chin in. This straightens the spine and raises the chest and shoulders for more efficient arm action.
6. Keep the body weight over the feet. This aids in body balance.

Figure 36.1.1 illustrates the proper lift position prior to lifting. Lift primarily by straightening the legs. As the lift proceeds, the straight but inclined back will return to a vertical position. After completing the lift, if the worker wants to turn, he or she should turn the whole body instead of twisting the upper torso. This can be accomplished by stepping out with the foot toward the direction of travel. Twisting the body should be avoided as this produces unequal stress on the lumbar intervertebral disks.

Another thing to avoid in lifting is attempting an awkward lift. Also, avoid lifting at arm's length as it dramatically increases the stress to the lower back. Avoid continuing to lift if the load is too heavy—set it down immediately and get help.

The two-person lift can be tricky where substantial weights or bulkiness are involved and should be discouraged in favor of finding other methods to efficiently and safely move these materials. The two-person lift requires good timing and communication between the lifters. Many back overexertions or contact injuries have occurred because one end of a load slipped from a worker's grasp. Additionally, different size workers lifting a common load can result in unequal load-weight distributions. Depending on hand grasp point, the smaller worker may bear the brunt of the weight.

Balanced One-Hand Lift Method. Oftentimes in lifting objects, the two-hand squat/lift method cannot or should not be performed by workers on their jobs because of lifting location, part size, or lifting frequency.

In many industries such as manufacturing and warehousing, lifting parts out of containers is a common practice [18]. However, many of these containers present lifting problems to workers when the part level is half full or less. Most back injuries occur while lifting from the floor-to-knuckle area (30 in.; 76 cm) which is the same location as when part levels in most containers are low. This

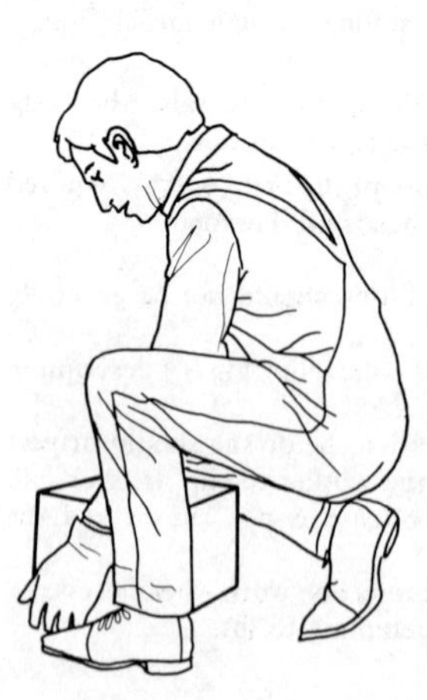

Fig. 36.1.1 Example of two-hand squat-lift method. Note straight back, spread feet, and the load is pulled between knees.

low-lifting situation requires workers to bend over into a container to lift parts and in so doing increase the stress to their backs.

Normally, workers lifting parts out of containers assume their lift position by leaning their thighs or waist against a container, their knees locked straight and the back bent forward in order to reach the parts with both hands. As the part level in the container gets lower, the more horizontal the back becomes. This creates high lifting moments to the back that may result in a back injury.

The traditional rules of two-hand lifting do not apply when lifting parts out of containers because of the impossibility of bending the knees. Yet, in many industries, perhaps half of the parts and materials being processed are moved around in containers.

An alternative method to the two-hand squat/lift method is the balanced one-hand lift method [19]. This method is a natural lift and is easily taught to workers. It is a method that can be used in lifting parts out of containers with reasonable safety.

The balanced one-hand lift method consists of lifting a part out of a container with one hand while the other hand is pushing down on the top edge of a container. The lifting stress is shared with the developed arm and shoulder muscles instead of solely with the back muscles. The free hand should continue to push down until the back is in a vertical position.

With the free hand pushing down on a container top while the other is grabbing a part, the back is better supported and the larger muscles in the shoulder take the stress. Figure 36.1.2 illustrates the proper one-hand balanced lift method. Limitations to this lift method are part weight, size, and grasp area.

Although not much research has been performed to determine the weight that can be lifted with one hand, the writer has noted in factory observations that weights between 15 and 20 lb (7–9 kg) appear to be reasonable. An analysis of part weights by one large manufacturer indicated that the majority of parts in containers, as they were moving through various work processes, could be lifted with reasonable safety using the balanced one-hand lift method.

The balanced one-hand lift method can also be used in situations where no containers are involved. The free hand can be placed above the knee and used as an anchor point while the other hand reaches for an object on the floor or low area. Pushing down on the anchor point acts the same as pushing down on the top edge of a container.

This lift method, when it can be used, should be taught in worker lifting training programs and made part of the work methods.

Physical Fitness and Diets. Programs to upgrade the physical fitness or dietary habits or workers are beneficial, if followed.

A weak and therefore vulnerable back is usually characterized by poor posture, abdominal obesity, and poor trunk muscle strength.

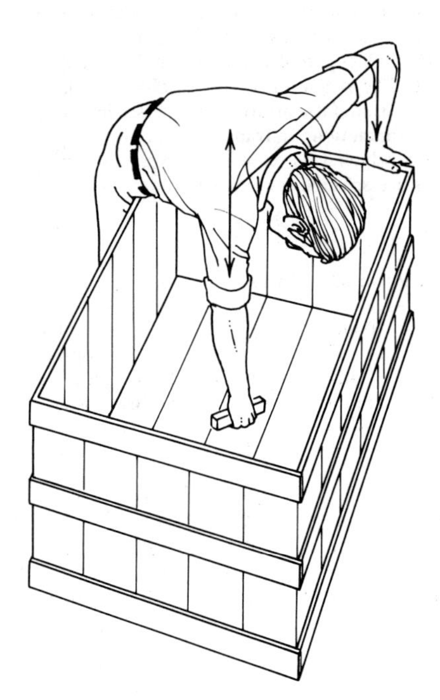

Fig. 36.1.2 Example of balanced one-hand lift method. Lift is made by pushing down on top of container which shares stress across shoulders instead of the back.

Most experts agree that flabby abdominal muscles are a major factor in back pain. The abdominal muscles are very important in lifting because they support the lower back from the inside through the creation of intra-abdominal pressure. Dr. Laurens Rowe of Kodak found that workers who lost time from work because of low back pain were four times more likely to have weak abdominal muscles than workers without back pain.

This indicates that exercise to strengthen abdominal and back muscles will have a positive effect on reducing back injuries, whether at work or at home.

Nutrition and dietary control is also necessary to reduce obesity. Obese persons tend to have posture problems such as swayback which puts more stress on the lower back. The weak back then is at risk in performing bending and lifting tasks that might not otherwise affect other more trim workers.

Many programs and materials are currently available which provide help in exercising and dietary control. The problem is motivating workers to follow them. Since a strong back is important to any workplace human factor effort, consideration should be given to implementing worker exercise programs. Five to ten minutes of exercising prior to beginning work would be beneficial to most workers.

Job Physical Stress Study

The purpose of a job study is to systematically collect pertinent facts on manual materials handling tasks which when later analyzed can be used to make rational decisions. Some of the major elements making up this activity are:

1. Problem identification. Identifying and prioritizing back injury experience by job classification or work process will help to narrow down the field to those selected areas which are incurring the majority of back problems.
2. Performing the job study. This step includes the actual job observation and the measuring and recording of various manual materials handling tasks and task variables.
3. Analyzing the data. After the job has been studied and various task parameters recorded, the data is analyzed to determine if a problem exists and the magnitude of the problem.

Problem Identification. Identifying problem areas and setting job study priorities are accomplished by identifying those jobs or work processes incurring a high proportion of the low-back injuries. Basically, this can be accomplished by establishing injury incidence rates (and possibly severity rates) for each job or work process incurring low-back and musculoskeletal injuries.

For definitional purposes in this discussion, musculoskeletal injuries are sprains and strains to other areas of the body besides the back, such as the shoulder or upper arm. Nonback musculoskeletal injuries are recorded separately from injuries to the back because the causes may be different. Sprains refer to joint involvement and the stretching of ligaments; strains refer to muscle stretching.

Incidence rate information is normally secured from the plant safety department. In the absence of available incident rate information, an analysis of the plant's accident data should be made.

In sizeable work populations, an analysis of the OSHA (Occupational Safety and Health Act) recordable injuries occurring annually should be sufficient. The definition of an OSHA recordable case is those occupational incidents resulting in death or illness or nonfatal injuries, or illnesses involving medical treatment, loss of consciousness, restriction of work or motion, or transfer to another job.

From this information, a priorities list is established from which jobs or work processes incurring the majority of low-back and other musculoskeletal injuries are selected for further study.

Table 36.1.1 Sprain/Strain by Job Classification, OSHA Recordable Cases

Occupation	Number of Workers	No. Hours Worked	Number of Back Injuries	Incident Rate[a]	Number of Other Sprains	Incident Rate[a]	Total Sprains	Total Sprain Incident Rate[a]
Assembler	928	1,759,488	18	2.0	14	1.6	32	3.6
Swing frame grinder	25	47,400	5	21.1	5	21.1	10	42.2
Foundry laborer	123	233,208	6	5.1	4	3.4	10	8.5
Material handler	148	280,608	2	1.4	1	0.7	3	2.1
Maintenance helper	10	18,960	0	0.0	1	10.5	1	10.5

[a] Incident rates per 200,000 hr of exposure.

Table 36.1.1 illustrates one type of format that can be used in determining job study priorities. In this example, swing frame grinders with an overall incidence rate of 42.2 would have a higher job study priority than for materials handlers.

After the job classifications or work processes have been selected for study, a "job injury analysis" form may be developed that will provide specific injury experience for each job selected. Table 36.1.2 shows a format that may be used.

This form is divided into three general diagnosis classifications: contact injury, musculoskeletal, and back sprains/strains.

The contact injury category includes all trauma cases (i.e., cuts, lacerations, bruises, fractures, etc.) which that job classification experienced. Although contact injuries are not normally associated with back or other musculoskeletal injuries, a University of Michigan study noted that as back and other musculoskeletal injuries decreased on a job, so did the contact injuries [20]. The reason for this is not fully understood but it may be associated with fatigue.

Musculoskeletal injuries, as indicated earlier, are separated from the back injuries because this may be a large category by itself. This category may also suggest additional sources of physical stress on the job besides lifting.

Across the top of the form is a general breakdown of the body part involved and also an incident rate column. At the bottom is a section for background data such as: analysis period, number of workers in job classification, total number of production workers, and estimated hours worked.

The use of a form like this presents data in a convenient form that can be used as a job injury history. Later, after remedial measures have been implemented, this form can be used to monitor and measure performance improvements.

Performing the Job Study. After the job classifications or work processes have been identified for study, a job study is then performed. This is simply observing a job or work process and recording the various task variables. The person performing the study should be familiar with the work activity.

Equipment required to make measurements need not be extensive. In addition to wearing personal protective equipment that might be required in the work area, observers should have a tape measure for measuring distances, a watch to measure approximate job cycle times or task frequency, a calibrated spring scale to determine material weights or forces, and a clipboard and paper to record observations.

Next, determine which specific workers within the job classification or work process to study. Select good workers with good work records who are performing the job according to established job methods.

The number of job observations to be made will depend on the uniformity of job tasks within a job classification or work process. If everyone within a job classification performs exactly the same tasks using the same equipment, perhaps only a few job studies need to be made to verify the uniformity of the tasks.

If, for example, assemblers are assembling various products, or there are various progressive stages of assembly, then many job studies within that job classification or work process will have to be made.

Table 36.1.2 Job Injury Analysis[a]

	Number of OSHA Recordable Cases					
	Body Part Injured				Incident Rate[b]	
Diagnosis Classification	Upper Extremities	Shoulder and Back	Lower Extremities	Head, Neck, Abdomen	For This Job Classification	For Total Unit
	2[c]		1[c]		12.6[c]	6.9[c]
Contact injury	14	0	4	5	97.0	15.8
Musculoskeletal	5	0	0	0	21.1	4.7
Back sprain	0	5	0	0	21.1	3.4

[a] Background Data.
Analysis Period: FY 1982
No. of Workers in Job Class: 25
Total Number of Productive Workers: 943
Hours Worked:
Job Class: 47,400 (% of total = 2.6%)
Total Productive Hours: 1,787,928

[b] Per 200,000 hr of work exposure.

[c] Lost-time injuries.

In job classifications where variable operations are performed under one job classification title, it saves considerable study time if on the accident investigation report the specific operation being performed is noted. This permits prioritizing operations within a job classification for study.

There are certain task variables that are important to note in any manual materials handling situation. In lifting, there are at least six task variables to note.

1. Object weight
2. Horizontal location of object
3. Vertical location of hands at start of lift
4. Vertical travel distance of lift
5. Frequency of lifting
6. Duration or period of lifting

Each of these variables interacts with each other. For example, object weight as force on the lower back is dependent on the vertical and horizontal location of the weight, how frequently it is lifted, and how far. Thus, these variables have to be considered collectively in analyzing lifting situations.

Reliance on only a couple of criteria is liable to produce a false reading as to the real lifting stresses on a job. For example, a person is normally strongest when lifting with the legs and back such as in the squat-lift position. If strength were the only criterion used, objects would be left on the floor to be lifted. Yet studies indicate that low-back disc compression and cardiovascular stress may be high when lifting objects off the floor level.

Additionally, there exists a large individual variability in injury risk and lifting performance today in industry. Some lifting situations may be so hazardous that only a few can perform them with reasonable safety. For example, with the increasing number of women in the workforce, traditional male jobs requiring heavy lifting may be beyond the capability of female workers. These tasks, when discovered, such as via a job study, should be modified through job redesign to reduce the lifting stresses to accommodate a wide range of workers.

To record the various task variables in a job study, a form such as illustrated in Fig. 36.1.3 can be used. The top part of the form is used to record background information of the job as well as on the individual being studied. Below this are task descriptions where the various task variables are recorded. Since there may be several lifting (or pushing, pulling, lowering, and carrying) tasks on a job, the parameters for each one must be recorded. This necessitates repetitive task identifications on the report.

All measurements must be accurate and reproducible. Measurements for distances and weights can be rounded to a whole number. In measuring and recording lifting task variables, the following measurement methods should be observed:

1. *Object Weight.* Weigh object. If object weight varies during an operation, record average and maximum weights.
2. *Horizontal Distance.* Measure horizontally from midpoint of ankle bone to midpoint of handgrasp. If feet are spread apart, measure from midpoint between ankle bones to midpoint of handgrasp. This distance is normally between 6 and 32 in. (15–80 cm). Generally objects cannot be brought closer to the spine than 6 in. (15 cm) because of body interference, and most individuals cannot reach objects more than 32 in. (80 cm) away.
3. *Vertical Distance at Start of Lift.* Measure vertical distance from floor or standing surface to handgrasp at origination of lift. Vertical distance is normally between 0 and 70 in. (175 cm) which is within the reach of most people.
4. *Vertical Travel Distance.* Vertical distance lifted between origin and destination measured at handgrasps.
5. *Frequency of Lift.* Record the average number of lifts per minute.
6. *Duration.* Note how long task is performed during the work shift (i.e., performed one or two hours, or is it continuous throughout the shift).

In pushing, pulling, or carrying tasks, additional, nonlifting task variables need to be recorded.

On pushing and pulling tasks, the initial and sustained forces needed to start and maintain load movement should be recorded, as well as the distances pushed or pulled. A spring scale can be used to determine these forces.

Carrying tasks require distances carried as well as vertical height which load is carried.

Analyzing the Data. After all of the measurements have been recorded, the next step is to analyze the data to determine if manual materials handling problems exist and where. Various reference sources exist which will help the reader in this effort. A couple are discussed here.

One of the big problems in industry is to determine what is too heavy to lift. What is heavy for

UNIT________ STUDY DATE________

ANALYST________ JOB PHYSICAL STRESS EVALUATION Page ____ of ____

1. DEPT. NO.	2. JOB CLASSIFICATION	3. OCC. CODE	4. PAY CATEGORY A. INCENTIVE B. HOURLY

5. JOB DESCRIPTION

6. PART NAME (if app'l)

7. PART NUMBER	8. OPER. NO. (If app'l)	9. STD. HRS. PER 100 PCS (if app'l)	10. MACHINE NO. OR DESCRIPTION

EMPLOYEE DATA	SEX: A. MALE B. FEMALE	12. BODY TYPE: A. ROTUND B. MUSCULAR C. THIN	13. AGE	14. TIME ON JOB MOS	15. DID EMPLOYEE RECEIVE TRAINING ON LIFTING? A. YES B. NO

16. IF 'YES', HOW LONG AGO GIVEN? MOS.	17. FOOTING A. WET B. DRY C. SLIPPERY	18. WORKSPACE: A. ADEQUATE B. CROWDED

NOTES: (Cond. of stock etc.)

TASK 1.	A. DESCRIPTION OF MATERIAL/EQUIPMENT HANDLED:

B. SIZE (inches)	C. WEIGHT (lbs.) Max. Avg.	D. FREQUENCY OF HANDLING–TIMES PER: Minute Hour	E. DURATION 1 Hour continous

F. TASK DESCRIPTION

G. 1. LIFTING 2. LOWERING 3. PUSHING 4. PULLING 5. I-HAND or NON-SYMMETRICAL LIFT 6. CARRYING	H. (For F1, 2, 5 only) 1. FLOOR TO KNUCKLE 2. KNUCKLE TO SHOULDER 3. SHOULDER TO ARM REACH	I. DISTANCES (in) Origin Destination 1. HORIZONTAL ____ ____ 2. VERTICAL ____ ____	J. EFFORT (Push-Pull) (lbs.) 1. INITIAL ____ 2. SUSTAINED ____

K. POSTURE 1. STAND 2. SIT 3. SQUAT 4. DEEP SQUAT 5. STOOP 6. LEAN 7. SPLIT

TASK 2.	A. DESCRIPTION OF MATERIAL/EQUIPMENT HANDLED:

B. SIZE (inches)	C. WEIGHT (lbs.) Max. Avg.	D. FREQUENCY OF HANDLING–TIMES PER: Minute Hour	E. DURATION 1 Hour Continuous

F. TASK DESCRIPTION

G. 1. LIFTING 3. PUSHI...

Fig. 36.1.3 Example of format that may be used to record various task variables during a job study.

one worker may not be for another. Additionally, it is not feasible to establish and administer individual work standards when there is a large work population whose individual strengths and other physiological capabilities range substantially.

To feasibly administer such a program requires using lifting standards on population ranges which will be within the capability of the majority of workers.

This need has been partially met recently by the National Institute for Occupational Safety and Health (NIOSH) in their publication, *Work Practices Guide for Manual Lifting* [21]. This publication provides guidelines for determining what is an acceptable lift. In the following discussions for analyzing lifting tasks, this guide is used.

This guide is limited to two-hand smooth, symmetrical lifting tasks where the hands are not spread over 30 in. (75 cm) apart and a favorable ambient environment exists. The lifting posture is unrestricted and good coupling (between handgrasp, shoes, floor surface) is provided. It is also assumed that other tasks such as pushing and pulling are minimal.

Two criteria have been established—the Action Limit and the Maximum Permissible Limit. These denote lower and upper limits whereby lifting tasks falling into these areas may require administrative and/or engineering controls.

The rationale for the establishment of these two limits based on epidemiological, biomechanical, physiological, psychophysical criteria are:

1. *Maximum Permissible Limit* (*MPL*). Studies have shown that back and other musculoskeletal incidence rates increase significantly when work is performed above the MPL. The biomechanical compression forces on the L5/S1 disc when working above the MPL would not be tolerable for most workers. The metabolic rates for most workers would exceed 5.0 kcal per minute. Only around 25% of the males and less than 1% of the females have the muscle strengths to work above this level.
2. *Action Limit* (*AL*). Studies have shown only a moderate increase in musculoskeletal injury rates at this level with the compression forces on the L5/S1 disc being tolerable to most young,

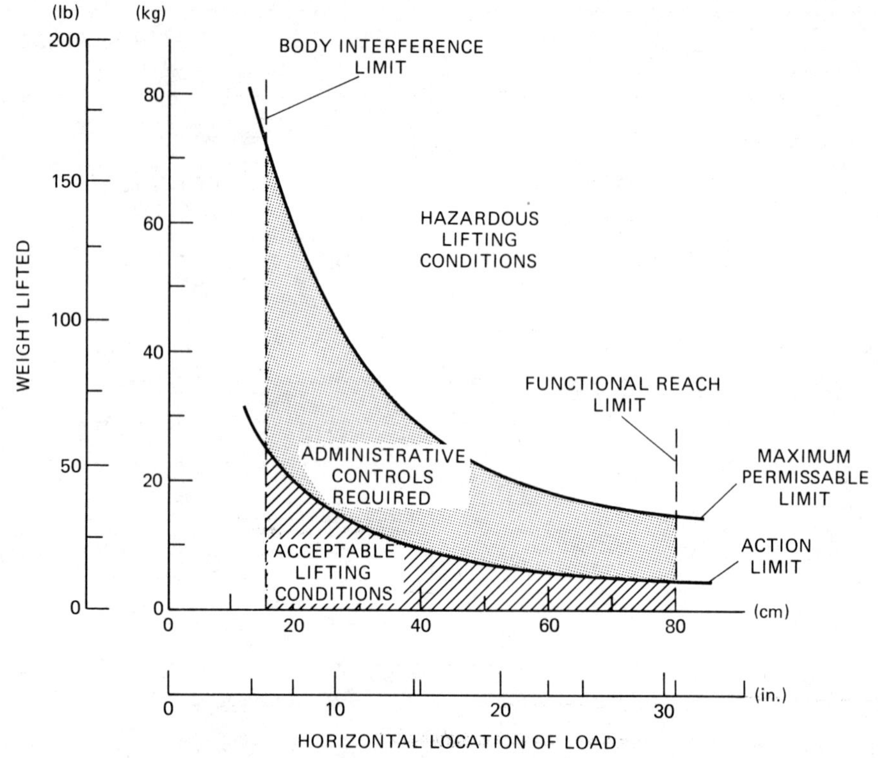

Fig. 36.1.4 Three regions of lifting showing maximum weight by horizontal location for infrequent lifts from floor-to-knuckle area [8]. From *Work Practices Guide for Manual Lifting,* National Institute for Occupational Safety and Health, DHHS (NIOSH) Publication No. 81-122, U.S. Government Printing Office, Washington, D.C., 1981.

healthy workers. Metabolic rates would be in the acceptable range and over 99% of the males and over 75% of the females would have muscle strengths to lift loads described by the AL.

Figure 36.1.4 illustrates the AL and MPL for the three regions of lifting using a particular lift frequency and location. Thus, all lifts, after being analyzed, will fall into one of three areas:

1. Lifts below the AL which represent nominal lifting risks to most workers.
2. Lifts between the AL and MPL are considered unacceptable and will require administrative and/or engineering controls.
3. Lifts above the MPL are considered unacceptable and will require engineering controls.

With the information recorded during the job study, it is now possible to determine whether a lifting problem exists by combining the task variables into an equation:

$$\text{AL (lb)} = 90\left(\frac{6}{H}\right)(1 - 0.01|V - 30|)\left(0.7 + \frac{3}{D}\right)\left(1 - \frac{F}{F_{\max}}\right)$$

$$\text{AL (kg)} = 40\left(\frac{15}{H}\right)(1 - 0.004|V - 75|)\left(0.7 + \frac{7.5}{D}\right)\left(1 - \frac{F}{F_{\max}}\right)$$

$$\text{MPL} = 3\ (\text{AL})$$

where: H = horizontal distance; cannot be less than 6 in. (15 cm) because of body interference
V = vertical distance at start of lift
D = vertical travel distance; distance lifted between origin and destination of lift. If travel is less than 10 in. (25 cm), set $D = 10$
F = frequency of lift (average number of lifts per minute); assumed to be between one lift every 5 (0.2) min and $F_{\max}$. Set $F = 0$ if lift is less frequent than once per 5 min
$F_{\max}$ = maximum lift frequency per minute; use value from Fig. 36.1.5

	Average Vertical Location at Origin of Lift, in.	
Lifting Duration	Standing $V > 30$ in. (75 cm)	Stooped $V \leq 30$ in. (75 cm)
1 hr	18	15
8 hr	15	12

Fig. 36.1.5 Values based on time and vertical distance at origination of lift which are used in NIOSH formula. Adapted from: *Work Practices Manual for Manual Lifting,* National Institute for Occupational Safety and Health, DHHS (NIOSH) Publication No. 81–122, U.S. Government Printing Office, Washington, D.C., 1981.

Applying the NIOSH Equation. An example will be used to illustrate how to use the NIOSH equation and how to interpret the results. From a job previously studied and recorded, the following lifting situation was found to exist. A worker was continuously, for 8 hr, lifting a 55-lb (25-kg) part with two hands from a 6-in. (15-cm)-high pallet to his 38-in. (96-cm)-high workbench. The horizontal distance was determined to be 16 in. (41 cm). He made approximately three lifts per minute. The equation with the variables inserted would be:

$$\text{AL (lb)} = 90\left(\frac{6}{16}\right)(1 - 0.01|6 - 30|)\left(0.7 + \frac{3}{32}\right)\left(1 - \frac{3}{12}\right)$$

$$= 90\ (0.38)(0.76)(0.79)(0.75)$$

$$\text{Al} = 15 \text{ lb (7 kg), rounded}$$

$$\text{MPL} = 45 \text{ lb (20 kg), rounded}$$

The 55-lb (25-kg) part is above the Maximum Permissible Limit of 45 lb (20 kg) and therefore requires engineering controls. The majority of workers will not have the strength or physical capacity to perform work safely at this level. Worker training programs on lifting will not help in this case. The job should be redesigned to reduce the lifting stresses.

Horizontal distance H had the most impact in the equation with a value of 0.38. Reducing horizontal distance would improve this factor.

By way of a corrective example, let us assume that the same job was redesigned by having the worker lift the 55-lb (25-kg) part off of a 30-in. (75-cm)-high platform instead of near the floor. He lifted them about 10 in. (25 cm). Because the worker can now pull the part close to him without bending prior to lifting, the horizontal distance is reduced to 10 in. (25 cm).

Inserting these task variable changes into the equation, the following results:

$$\text{AL (lb)} = 90\left(\frac{6}{10}\right)(1 - 0.01|30 - 30|)\left(0.7 + \frac{3}{10}\right)\left(1 - \frac{3}{12}\right)$$

$$= 90\ (0.60)(1)(1)(0.75)$$

$$\text{AL} = 40 \text{ lb (18 kg), rounded}$$

$$\text{MPL} = 120 \text{ lb (54 kg), rounded}$$

The 55-lb (25 kg) part came close to the Action Limit area of 40 lb (18 kg), requiring administrative controls. By raising the height of a load, a better lift position could be assumed by the worker, which decreased the horizontal distance. By reducing the horizontal distance, two other equation factors were improved. The majority of workers should be able to lift this load with reasonable safety providing administrative controls are applied.

Administrative controls, for the purposes of the NIOSH guide, include worker selection and training programs. Since both areas have been already covered, they will not be discussed further.

Engineering controls include job redesign in order to reduce the task variables adversely affecting a lift. Some examples of this will be provided later.

Pushing, Pulling, and Other Tasks. If manual tasks involve pushing, pulling, or carrying, the various task variables recorded in the job study can be used in evaluating what percentage of the work population can be expected to perform a task without unreasonable risk in overexertion or excessive fatigue.

While various printed references on task strengths are currently available, the one that will be discussed was developed by Snook [22]. A set of task tables was developed by Snook et al. after years of psychophysical study on numerous industrial workers. The tables show the acceptable weights by population percentages that can be handled with relatively safety by males and females for pushing, pulling, carrying, lowering, and lifting tasks, by various frequencies and force locations.

A set of these tables also appears in *Material Handling: Loss Control Through Ergonomics* [23]. This book discusses how to apply the tables as well as other human factor/ergonomic criteria in manual materials handling systems.

By way of an example, suppose a job required pushing a picking cart about 25 ft (7.6 m) once a minute. The cart handle was located a little below waist height. During the job study, it was recorded that 100 lb (45 kg) initial force was required to start the cart moving and 60 lb (27 kg) sustained force was required to keep it moving. We want to determine whether this job is reasonably safe for most workers.

Reference to the "push" table indicates that less than half of the male workers and even fewer female workers would be able to perform this push task with relative safety all day long. The table shows that 75% of the male workers and around 50% of the female workers would be able to exert an initial force of 62 lb (28 kg) and sustained force of 35 lb (16 kg). Further, 90% of the male workers and about 75% of the female workers would be able to exert an initial force of around 46 lb (21 kg) and sustained force of 24 lb (11 kg).

It is obvious in this example that the forces required to push this cart should be drastically reduced by design for safety and productivity reasons.

Similar table comparisons can be made for pulling, carrying, and lowering tasks. One note of caution: the values in the tables just discussed and those determined from the NIOSH two-hand lifting equation are not absolutes. They are relative values and should be used as guidelines.

Human Factor Engineering Controls

Human factor engineering controls means using engineering approaches to design out or reduce the physical stresses in the work environment.

A program to engineer out these stresses and physical hazards will provide long-term results and more results over any other single thing done.

In applying the NIOSH formula or other tables, if the strengths required on the job are higher than most workers can handle safely, there are three options available:

1. Hire only superhuman people with high physical strengths to do these tough jobs.
2. Establish worker training programs on lifting.
3. Engineer out the physical stresses on the job.

Of the three options, the last is the only option that will give positive results. Under the NIOSH equation, any lifting that exceeds the Maximum Permissible Limit is unacceptable and requires ergonomic or engineering controls. Actually, this should be the first choice at any time.

Many times in lifting situations, engineering out is thought of as putting a hoist on the job. While this may be considered as a viable alternative, the job method should be evaluated first. Maybe the only thing that needs to be done is to bring the parts in on a platform so the worker does not have to squat to lift them. This reduces the vertical location factor in the equation and may bring the lift up to where it is in the acceptable lift category.

If the job requires frequent bending to lift lightweight parts one at a time, the solution may be to have the worker lift multiple parts at one time, thereby reducing the frequency of the lift. There are many options open to choose once the specific problem is identified.

Anytime a lift or other manual effort can be eliminated or reduced by installing more conveyors, lift platforms, and so on, less chance will exist that an overexertion injury will occur.

Some general tips to apply in reducing back injuries:

1. Reduce the weight of the load lifted. (Exception: very light loads may be increased to reduce frequency.)
2. Reduce the horizontal distance between the load and the body. (The closer the load is to the body, the less it weighs to the back in terms of reduced stress.)
3. Avoid designing a work station requiring workers to twist while handling materials.
4. Raise loads off the floor. (High physical energy is expended to squat-lift—some workers cannot)
5. Reduce frequency of lift. (Guide: 1 lift every 5 min for heavy loads; 1 lift per minute for lighter loads.)
6. Minimize vertical and horizontal travel distances (saves physical effort and energy).
7. Design in good "couplings." (Floor–shoe–handgrasp relationships; they should all be compatible.)
8. Design out housekeeping problems. (Provide storage, unobstructed and clean floors, waste containers, etc.)

9. Provide work stations with sufficient and uncluttered work space for unrestricted body movement.
10. Provide materials handling equipment, where possible (tilt platforms, conveyors, personnel hoists, powered industrial trucks, etc.).

Measuring and Monitoring Program Results

Any program worth doing is worth managing—effectively. Any program implemented in an effort to reduce back or other musculoskeletal injuries (or any type of injury) should be periodically monitored to determine its progress. Monitoring efforts usually incorporate measuring systems of some type.

During the problem identification phase, a "sprain/strain by job classification" form and a "job injury analysis" form were used to determine job study priorities. Both of these forms recorded the current injury history before any corrective measures were taken. To monitor program results, the original injury histories can be used as a basis for updating and comparing current program results—this provides a "before" and "after" comparison.

These and other reports can provide a very useful management tool for cost justifying other similar human factor programs in other areas. Periodic monitoring of overall experience and specific program steps will uncover any changes or modifications that need to be made.

36.1.5 Storage and Movement of Materials

This section discusses the major safety and human factor aspects of materials storage and movement. Because full coverage of this subject is not possible here, only major items which are commonly involved in injury are discussed.

The reader, not finding information on a particular item, is encouraged to consult the references at the end of this chapter for further help.

Rack and Bin Storage

Permanent or temporary materials storage should be neat and orderly. This promotes efficiency as well as safety. Materials strewn around the floor or haphazardly piled increase the potential for accidents as well as damage to materials.

All storage racks or bins should be secured to the floor, wall, and to each other to prevent them from falling over. The corners of racks or bins which are exposed to forklift truck abuse should be protected to prevent the rack or bin legs from being damaged and collapsing. Storage capacities should not be exceeded.

Many accidents to workers occur because of improper storage practices:

1. Parts fall off of overloaded containers in racks.
2. Sheet metal parts slide off rack or bin shelves.
3. Stored parts project into aisles and workers walk into them.
4. Stacked cardboard parts cartons break down, because of humidity, and fall.
5. Unstable parts or cartons stored on tops of bins fall.

Continuous monitoring of parts storage is required to control these conditions.

Today, high rack storage is used by many companies [24]. Materials handling equipment used in these areas are specially designed and may be manually or computer controlled. Care must be taken when workers are in the area whether they work there or are walking through the area. Worker safety should be considered during the design concept including maintenance trouble-shooting procedures. Applicable safety standards for this type of equipment include the American National Standards Institute (ANSI) B56 series: "Powered Industrial Trucks"; and the crawler crane section of ANSI B30.5, "Safety Code for Crawler, Locomotive and Truck Cranes."

Aisles

Aisles should be free of obstacles and trash. Trash containers should be liberally provided and workers encouraged to use them. Aisle lines should be marked and all containers, stock, and storage should be in back of the aisle lines. If forklift trucks are used to move and store materials across aisle lines, the turning radius of the truck should be considered in determining aisle width. Aisles should be wide enough so workers can move about freely (e.g., walking, picking stock from bins, etc.) and also leave a safe amount of room for powered industrial trucks with loads to pass. Floors should be kept in good condition and level where materials are being stored to prevent them from toppling over.

At inside blind intersections, overhead mirrors can be installed to prevent collision with other vehicles or pedestrians. Warning signs and signals at such locations can serve as reminders to workers of a potentially bad intersection.

Picking Ladders

In sequential picking where individual stock or parts are manually picked, part location should be considered for picking efficiency and safety. Where stock or parts are stored in racks or bins requiring climbing, heavier stock requiring two hands to lift and carry should be stored in lower levels where workers do not have to climb.

Several injuries including fatalities have occurred to workers who fall while descending a ladder trying to carry parts with both hands. The climbing of stock, racks, bins, or even containers should not be permitted because of the hazard potential of falling.

If climbing is required to relatively high storage areas, heavy-duty materials handling platform ladders with steps should be used. These ladders are usually on rollers and have a braking mechanism with rubber feet that is actuated when weight is imposed upon it, such as when a worker is on the ladder.

At the top of a platform ladder is a working platform enclosed with standard guardrails to protect the worker from falling. Many platform ladders also come equipped with standard railings on the sides of the steps depending on its length. When picking parts, one hand should always be free to use the railing when descending.

Alternately, if climbing a short distance, warehouse shelf ladders may be used. Generally this type of ladder has steps instead of rungs and has a horizontal extension on top that hooks or rests on rack or bin shelves to provide stability. Normally, they are used to climb only a few steps to pick stock.

If rows of bins are used and their height is around 7 ft, requiring climbing to pick the upper levels, specially designed bin-steps which bridge between the bin rows may be used. The worker places the bin-step into position and steps up to pick stock with both hands from the upper bin levels. This is safer than stepping up on a lower bin shelf and holding onto the bin with one hand while the other picks stock. Not only may this practice damage the bin shelving or fronts, it presents a slipping hazard.

Ordinary stepladders, rung ladders, or 2- or 3-step stools are not recommended for climbing because they provide little stability to the worker reaching and picking a part. In fact, fall fatalities have been recorded from as little as 2 to 3 steps from the floor level using this type of equipment. This has occurred because workers, while carrying stock, trip on a step and had no railing to grab to arrest their fall.

For efficient use of time and safety, enough climbing equipment should be provided and conveniently placed so that workers will use it. This might include providing lightweight warehouse shelf ladders or bin-steps at the ends of each rack or bin row. Experience indicates that if the equipment is not readily available when workers want to use it, it will not be used.

Cutting Metal and Plastic Strapping

Many items in plant and warehouses are bound with metal or plastic strapping. Improper cutting of the strapping has resulted in many serious injuries or fatalities when the strapping under pressure is suddenly released. The strap ends can fly back and strike workers or the bound load when released rolls onto workers.

To reduce this accident potential, workers should be trained in the proper procedures for cutting strapping. Minimally, eye and hand protection as well as substantial clothing should be worn. The load should be sized up first before cutting to determine what effect will result when the load is cut.

A special "safety" band cutter should be used to cut strapping. Safety band cutters hold both ends of a band after it has been cut preventing a sudden whiplash of the ends. Wire cutters or metal shears are not recommended. Steel strap should never be broken by applying leverage with a claw hammer, crowbar, or other leveraging tools.

Before cutting strapping, workers should make sure fellow workers are not standing in the area where they could be injured. To cut banding, the worker should place one gloved hand on the portion of the strap near him. Thus, when the strap is cut, the end nearest the worker is held in place and the other end flies away from the worker's face. The worker's head should be away from the direct line of the strap. Because of the potential energy release the cut strap represents, eye protection should be worn.

Strap should never be cut at an angle. They should be cut square since angled cuts increase the sharpness of the ends. Trash containers should be available in the area to dispose of the strapping.

Dockboards

The potential for serious injury or fatality always exists around loading or unloading docks. A portable dockboard may move or a truck trailer may walk away from the dock resulting in a forklift truck and operator falling between the truck trailer and dock. Because of this serious potential, the dock area should be constantly monitored to correct unsafe worker practices and to inspect for unsafe conditions.

Dockboards used to load or unload truck trailers or boxcars should be designed to carry four times the heaviest load expected to be carried over them [25]. They also should be wide enough to permit easy maneuvering of materials handling equipment. Flat and unsecured pieces of boiler plate should not be used as dockboards.

Powered dockboard such as automatic dock levelers and fixed hydraulic dockboards should have all pinch points guarded such as on the side edges where pinch points are formed when the dockboard descends flush with the dock surface. It is recommended that the sides of the movable sections raised above the dock surface be painted yellow to denote a potential tripping hazard. All dock levelers and hydraulic dockboards will require regularly scheduled maintenance. According to OSHA [Subpart D, 1910.30 (a)(iii)], all powered dockboards must be designed and constructed in accordance with Commercial Standard CS202-56 (1961), "Industrial Lifts and Hinged Loading Ramps," published by the U.S. Department of Commerce.

Portable dockboards must be secured into position to prevent slipping before they are used. The sides should be turned up at right angles, or curbed, to prevent materials handling equipment from running over the edges. They should be kept free from oil, grease, water, snow, or ice and have a nonskid surface to prevent slipping.

Trailer trucks being boarded by powered industrial trucks must have their wheels—on both sides—chocked or blocked (or equivalent protection provided) to prevent trailer movement. The truck brakes should be set if the truck is still attached to the trailer.

When the trailer is not connected to a truck, jacks should be placed under the front end of the trailer to prevent upending. Two jacks are recommended because one jack placed under a trailer front may not be sufficient to prevent a heavily loaded trailer from tipping over. On extended bed trucks, jacks should be placed in the rear.

Railroad cars, like truck trailers, require positive protection from movement when they are being loaded or unloaded by materials handling equipment. Special railroad chocks are available for this purpose. To warn train crews not to move a railroad car while it is still being loaded or unloaded, standard blue flags should be mounted in the daytime and blue lights used at night.

Additional information concerning OSHA requirements is provided in Subpart D, 1910.30(a), Subpart N, 1910.178(k) and in other areas of the "Occupational Safety and Health Standards for General Industry."

Containers

Two types of containers are discussed in this section: small containers, which are lifted and carried, and large containers from which materials are manually lifted out and which normally require materials handling equipment to move.

Small Containers. In some industries, small containers such as tote boxes are lifted and carried quite frequently. Injuries commonly associated with this activity are back overexertions or dropping the container or its contents onto the legs and feet.

At least three criteria should be considered in small container design: weight, size, and grip area. Each criterion could be a limiting factor or interact with the others. Obviously, if a loaded container is too heavy and a poor grip is used, the container and its contents could slip from the grasp.

If a container is to be lifted, a model using the NIOSH two-hand lifting equation previously discussed can be developed to determine what weights can be lifted with reasonable safety. This will be helpful in determining what the total design weight limitations—load plus container weight—should be.

The second criterion is container size. Containers that are long reduce the lifting capability of the worker. Generally, they should not be longer than 30 in. (76 cm) unless extenuating circumstances exist and not much weight is involved. Container width should be as narrow as feasibly possible to reduce horizontal distance and keep the load center of gravity as close to the body as possible.

Third is the grip area on the container. On most containers, there are generally two types of grips used: hook grip and power grip. Many types of tote boxes use the hook grip where the fingers are flexed around a handle or hooked under the rolled top edge of the container and the thumb is not used in lifting or carrying. This is the least desirable grip and may be the most commonly used method. Many containers that require this type of grip have sharp edges, too small of a handle diameter, or insufficient hand clearance between the handle and the container.

The power grip in which the container rests on the palms of the hands (palms and fingers horizontal, with the thumbs vertical) provides a good gripping force. With this method, care has to be taken on where the container is going to be placed after being lifted. Injuries to the fingers and knuckles can occur if the container is set down on the fingers or if the fingers strike against the edge of a bench prior to setting the container down. The container should first be set on the edge then pushed onto the bench.

Another consideration is floor surface–worker shoe sole friction. If there is low friction, workers could slip, twisting the back or falling. Workers should not be expected to climb or step up onto other surfaces when carrying containers.

Large Containers. Many industries use various types of large containers to carry materials through various processing areas. These containers are often referred to as tubs, skids, pallet boxes, or other names.

Many of these containers present problems to workers attempting to lift parts or materials out of them day after day. Back overexertions oftentimes result from lifting materials out of containers when the part or material level is less than half full. Bending over to lift materials out of the bottom of a deep container results in substantial stress to the lower back due to the high lifting moments.

In designing containers, or investigating worker complaints or back injuries involving containers, it will be helpful to use the NIOSH two-hand lifting equation in determining the material weight limits that should be stored in a container.

Figure 36.1.6 illustrates a model that can be developed to determine various weight profiles using the NIOSH two-hand lifting equation. In this example, the model is intended to represent a wood pallet box. Each set of figures represents the Action Limit/Maximum Permissible Limit which is calculated for that area of the container.

The horizontal distance assumes that the worker's feet are close in and partially under the front side of the pallet box. Normally, lifting parts out of containers when they are full or near full does not present as much of a problem as when the part level gets lower. Note in Fig. 36.1.6 that as the horizontal distance increases or the part level decreases, the less weight should be lifted.

The design of a container oftentimes dictates the posture that workers use to lift parts and materials. When a container is full, workers can reach a part without much bending of the back. As the part level gets lower, the farther over (horizontally) the back must bend to reach and lift the part. The two-hand squat-lift method normally cannot be used as a lifting method. Repeated bending of the back to lift materials may, after a period of time, result in back injury.

Sometimes in warehouse activities pallet boxes are stored under rack shelves. The rack horizontal beam may be 15–20 in. (38–51 cm) above the pallet box, forcing the worker to bend into this limited space to reach a part. In situations like this, the rack beam should be raised and/or the front top half of the pallet box removed.

If the front of containers (pallet boxes, skids, tubs, etc.) were designed so the top half could be

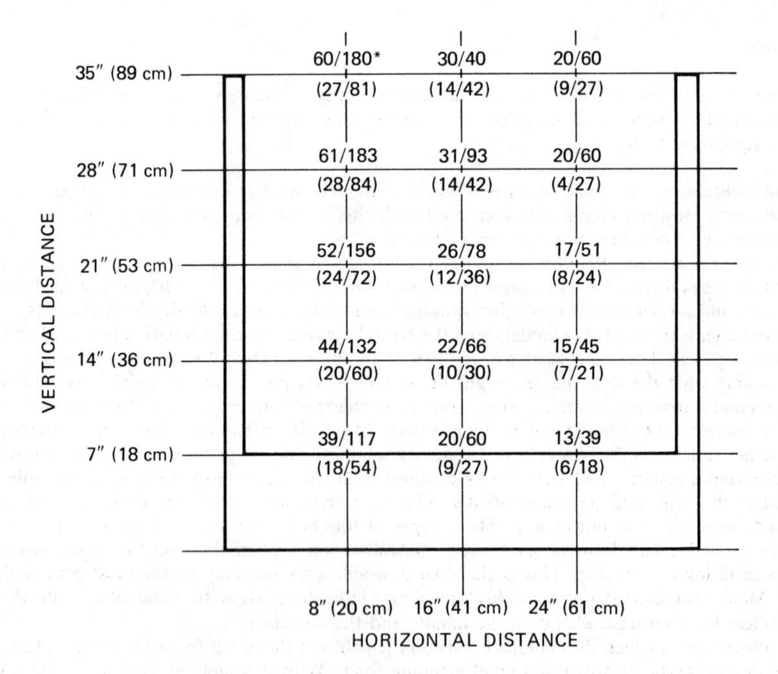

Fig. 36.1.6 Various weights that can be lifted (AL/MPL) from various container locations using the NIOSH lifting equation. Calculations based on a lift frequency of once per minute for 8 hr and a vertical travel distance to the top of the container.

removed or dropped when the part level was half full, workers could assume a better lifting position. Workers could squat, pull the parts close to them and use two-hand lifting methods to lift parts (depending on frequency). Depending on the part weight, size, and grasp area, the part could also be lifted using the balanced one-hand lift method.

Alternatively, instead of using one container of normal height, consider using two shorter ones that can be stacked on top of each other. The top container can be picked out of without much bending. When it is empty, it can be removed and the lower one can be picked using a good lifting position.

Container design should always include considering how the container will be used. Some are designed for maximum volume without any consideration being given to workers who will use them. Long and wide containers with deep bottoms present reach problems to many workers. In these types of containers, workers may not even be able to reach materials near the bottom or on the far side.

Containers should have open outside bases so workers can stand with their feet under the containers. This helps in reducing the worker horizontal distance to the part by getting closer to the load. The inside bottoms of containers should be at least 12–14 in. (30–36 cm), or higher, off of the floor or raised to the equivalent height.

The following approaches should be considered in presenting parts/materials to workers so they can be lifted with relative safety. Provide:

1. Drop or cut sides on containers so workers can assume better lifting positions when the material level is half full or less.
2. Tilt dollies so containers can be raised and tilted to provide workers with a better lift position.
3. Hydraulic or scissor lifts to set containers or materials on, which can be raised or lowered to waist height.
4. Platforms/rails on which containers can be set to raise the inside bottom off of the floor.
5. Hoists to lift the materials out of containers. This option may be more costly when compared to other options.

If workers have to lift repeatedly out of containers (picking parts, lifting parts to their machines, etc.), the six task variables that were previously discussed in lifting should be considered in container design and/or placement. Each factor's effect should be minimized.

Figure 36.1.7 illustrates the effect repositioning a container from the floor area onto a tilt dolly has on the weights that can be reasonably lifted. The upper illustration shows a part being lifted out of a container at 12 in. (30 cm) from the floor area. The horizontal distance is 17 in. (43 cm) because the back is near horizontal in order to reach and lift the part. The frequency of lift is once per minute for 8 hr.

Inserting the task variables in the NIOSH lifting equation, we find:

$$\text{AL (lb)} = 90\left(\frac{6}{17}\right)(1 - 0.01|12 - 30|)\left(0.7 + \frac{3}{20}\right)\left(1 - \frac{1}{12}\right)$$

$$= 90\,(0.35)\,(0.82)\,(0.85)\,(0.92)$$

$$\text{AL} = 20 \text{ lb (9 kg)}$$

$$\text{MPL} = 3(20) = 60 \text{ lb (27 kg)}$$

The Action Limit of 20 lb (9 kg) indicates that this type of container should be limited to lightweight parts if workers are expected to lift out of it repeatedly.

The biggest penalty factor in the equation is the horizontal distance of 0.35. If the horizontal distance were decreased, the factor would increase.

The second highest penalty factor is the vertical distance at the beginning of the lift which is 0.82. If the vertical distance were increased, the factor would increase.

The lower illustration in Fig. 36.1.7 shows the effect on the AL and MPL if the container were raised onto a tilt dolly. The horizontal distance has been reduced to 8 in. (20 cm) and the vertical distance at the start of the lift was increased to 30 in. (76 cm).

In applying the NIOSH lifting equation, the Action Limit has now increased to 63 lb (29 kg), which indicates that heavier parts can be lifted with reasonable safety from this position.

Both illustrations merely depict the relative differences that vertical location has on material weights that can be lifted. It is always preferable to keep part weights as close as possible to or below the Action Limit in order that the majority of workers can lift with reasonable safety.

Forklift Trucks

Equipment used to move materials must be adapted to the physical capabilities and limitations of the workers operating them, both for safety and efficiency. Also, they must meet the requirements of the Occupational Safety and Health Act (Subpart N, 1910.178, Powered Industrial Trucks).

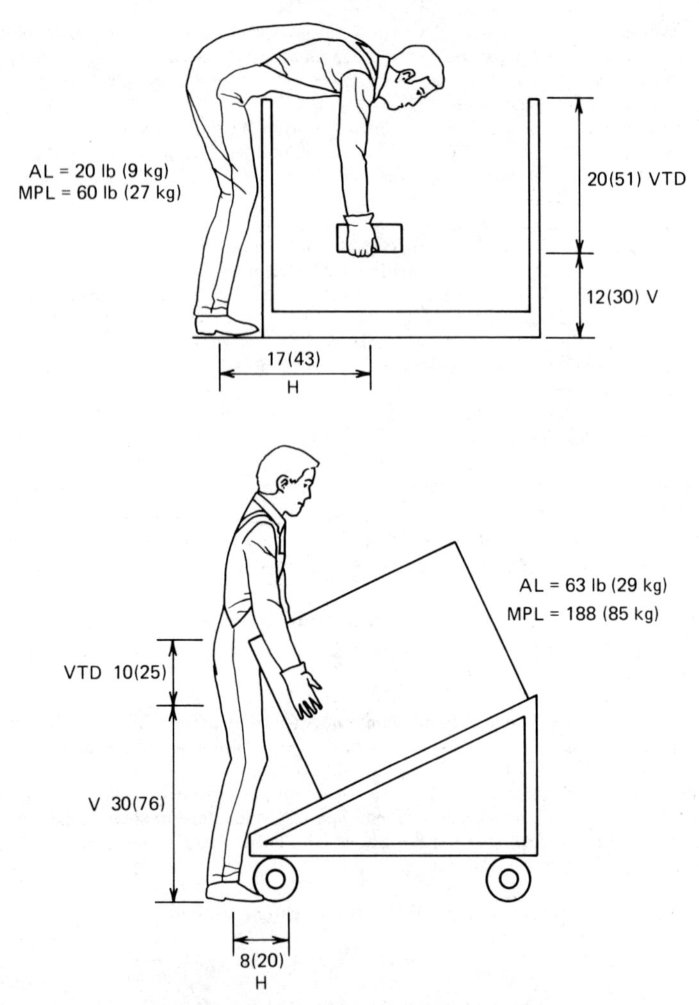

Fig. 36.1.7 Effect that raising a container has on the AL and MPL using the NIOSH lifting formula. Based on lift frequency of once per minute for 8 hr. Distance measurements are in in. (cm).

Occupational Safety and Health Act (OSHA) requirements for powered industrial trucks include design criteria, operation, and operator training. OSHA makes no distinction in its training requirements as to types of powered industrial trucks used. It merely states that "Only trained and authorized operators shall be permitted to operate a powered industrial truck."

There is good reason for this because forklift trucks are involved in many types of accidents involving fatalities, injuries, property damage, and near misses.

Two high-risk types of accidents inherent with forklift truck operations are collision and overturn. Collision with other forklift trucks, vehicles, plant pedestrians, or building parts has resulted in serious injuries and fatalities to forklift truck operators and others. Overturning a forklift truck while speeding around a corner or down a ramp, or falling into a dockwell (i.e., between loading dock and truck trailer) has resulted in severe injuries or fatalities to forklift truck operators.

Human engineering principles are often violated in the design of forklift trucks, which contributes to accidents [26]. In a study on materials handling equipment, it was noted that there are many human factors/safety design weaknesses on powered forklift trucks such as missing or inadequate safety devices, poor visibility, lack of standardization in braking, cramped driver compartments, poor seat design to cushion the transmission of impacts and vibrations, and insufficient body support.

The physical environment in which the forklift truck operates is important. Outside the plant, pedestrians and vehicles are separated by sidewalks and streets. In the plant, moving vehicles and pedestrian traffic is mixed together offering a potential hazardous situation if not controlled.

Although the speeds of vehicles in the plants are less than those in the streets, the problem of stopping in time or avoiding a developing situation still exists. An average forklift truck, fully loaded and traveling 10 mi/hr (16 km/hr) can stop in approximately 2.2 sec, after traveling approximately 22 ft (6.6 m). This may not allow enough time to stop should a worker step out into its path.

Forklift truck speeds should be limited to around 3 mi/hr (4.8 km/hr) when operating around workers. This is about average walking speed and leaves the forklift operator a reasonable amount of stopping room, providing the operator is alert. Top speed in plant areas where no workers are present should be around 8 mi/hr (12.8 km/hr). Governors or specially designed forklift trucks may be used to control top-end speeds.

An analysis of the top 10 forklift truck accidents occurring at a major U.S. manufacturer indicated the following types [27]:

1. *Worker Struck by Forklift Truck.* This was the most common type of accident and is potentially fatal. Examples of this are: workers being struck when they stepped out in the path of a moving forklift truck, or the leading rear steering wheels of a forklift truck run over a worker's foot as the operator turned a corner.

2. *Worker Struck by Object.* Typically, this type of accident occurs when a forklift truck shoves a container or large part into a worker who is standing between the forklift truck and a fixed object.

3. *Operator Struck by Falling Object.* The forklift truck operator is usually off of his/her vehicle and is manually maneuvering a load or part on the forks or floor when load falls on the legs and feet.

4. *Other Worker Struck by Object.* Overloaded containers or unsecured parts may fall from a forklift truck onto fellow workers. This is particularly apt to occur when the forklift truck is traveling around corners.

5. *Mounting or Dismounting Forklift Truck.* The act of getting on or off forklift trucks has resulted in many sprained knees or ankles from stepping on objects on the floor or from the sudden transfer of body weight onto one foot.

6. *Forklift Truck Overturns.* Speed going around corners or down ramps results in potentially serious overturn hazards.

7. *Collision with Other Vehicles.* Running into other vehicles, particularly at intersections or blind aisles, can be hazardous—particularly if the other vehicle struck is a smaller vehicle.

8. *Forklift Falling off Dock.* Serious injuries or fatalities occur when forklifts back or run off the edge of a loading dock. Many times this occurs during loading/unloading of a truck trailer that "walks" away from the dock because the wheels on both sides of the truck trailer were not blocked or chocked.

9. *Body Part Struck by Object.* If part of the operator's body is protruding outside the running lines of the forklift truck, it is liable to be caught between some part on the forklift truck and a fixed object.

10. *Parts Falling Back onto the Operator.* All forklift trucks should have overhead canopy guards when any loads are lifted or carried above the forklift truck's mast. This is required by OSHA. However, there is usually an exposed area between a forklift truck's mast and the overhead canopy guard where parts can fall onto the operator's lap.

Three main operator causes of forklift truck accidents are: lack of knowledge or skill; operator not attentive to job at hand; and operator taking chances. Many of these types of accidents can be avoided by: training the new operator thoroughly; monitoring the operator's performance; and enforcing safe practices by positive (reward) and negative (punitive) control measures.

Back problems are a common ailment among many forklift truck operators because of daily forklift truck vibration and jostling. To minimize this, care should be taken in the selection and purchasing of forklift trucks so that the design chosen is safe and has had human factor engineering. The cockpit of the forklift truck is the operator's work station. It should be designed to accommodate him or her—not the so-called "average" person.

Some of the specific design criteria to consider in the selection of forklift trucks are:

1. Operating controls should not require the operator's arm(s) to be extended and unsupported for long periods of time. This may result in fatigue and stress to the arms, shoulders, and neck. Each control lever should be easy to operate and identified by shape, location, or label as to its function. Lever position movements should not operate differently than what the operator expects (i.e., raise mast, raise lever; lower mast, lower lever, etc.). Clutch and/or brake pedals should not protrude too far out into the cockpit. The distance between the face of the clutch and/or brake pedals and the accelerator pedal should not be excessive. Manual gear shift controls should not require the operator to lean forward to operate.

2. Gauges on the forklift truck should be visible, functional, and simple to read. Gauge placement

should be based on what the operator needs to know most often. Gauges should inform operators what they need to know when they need to know it. For example, if operators are not to exceed 3 mi/hr (4.8 km/hr) in the plant, some gauge should be available to inform the operators of their speed.

3. Seats on forklift trucks should be comfortable, firm, and support the lower back. They should not be canted back. The seat bottom should not be too long or raised near the front which will impede blood circulation to the operator's thighs or restrict foot and leg movement for operating the foot controls. Seats should be adjustable both vertically and horizontally to accommodate the various sizes of operators, both male and female.

4. The center portion between the mast uprights should be uncluttered with roller chain or hydraulic systems and provide good forward visibility.

5. Maintenance accessibility should be considered for changing batteries, refueling, changing filters, engine and hydraulic maintenance, and so on. A preventive maintenance program should be established to keep forklift trucks in peak operating and safe condition.

6. Horns or other audible warning devices are necessary to warn others of the forklift truck's presence. This is especially important for electric forklift trucks, which are quieter than other trucks.

Readers are referred to the American National Standard Institute B56 series on powered industrial truck standards as well as the OSHA general industry standards for additional information. Forklift truck manufacturers are also a source of information including films and operator safety training programs.

Walking Electric Transporters

Many industries move a lot of materials and stock efficiently in skids, pallet boxes, tubs, flats, and so on, using walking electric transporters. They are also referred to as walking powered lift trucks.

There are generally four basic types of transporters used in industry: pallet, platform, combination pallet and platform, and the stacker type. Each has its own specialized use in moving containers (or other items such as press dies) depending on the container bottom configurations.

The steering handle incorporates the controls used to stop and reverse the direction of travel. It also serves to actuate the braking control. The transporter is battery powered and has separate controls for raising and lowering the platform and mast (if applicable).

To stop the transporter, the steering handle is raised or lowered to set the brake. The steering handle should automatically return to the braking position if it is released in case of an emergency. Figure 36.1.8 illustrates the suggested braking positions.

Although walking electric transporters are simple to operate, there are some safety precautions to note. When walking with the load behind, the operator should walk to one side of the transporter facing the direction of travel. This prevents the operator from being pinned between the transporter and a fixed object, tripping over obstacles, or running the transporter onto the heel of the foot.

Operators should always face the direction of travel. Reverse travel should be used when a load obscures the operator's vision, when leaving a boxcar or truck trailer, and when carrying a load down

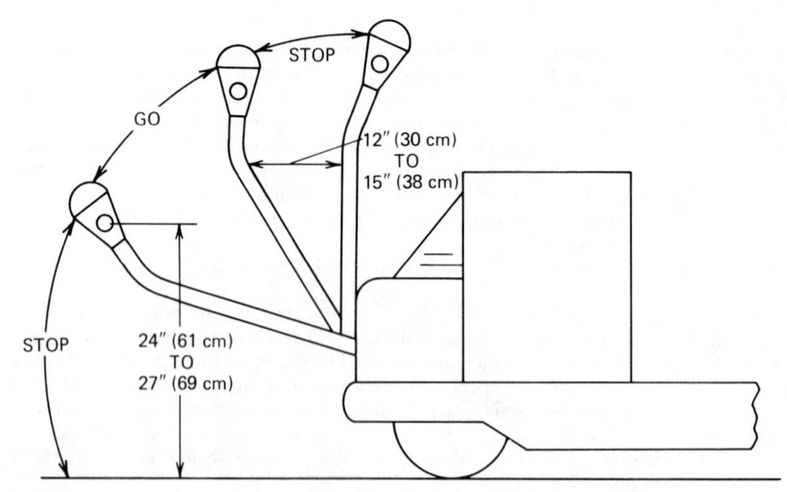

Fig. 36.1.8 Suggested upper and lower handle braking positions on walking electric transporters.

an incline. Loaded transporters should never be turned on a ramp or incline as there is always the possibility of upset. On inclines and ramps, like forklift truck operation, the load should always face the top whether going down or up.

Severe injuries can result from improper operation of the transporter. For example, the push-off foot is often injured after reversing direction and the transporter runs over the foot. Severe internal injuries can result by backing the transporter up and not facing the direction of travel. The operator may back into a fixed object and the transporter handle strikes him in the stomach. Fractured ankles and/or other injuries may result if operators sit on the battery case and operate the controls. Operators should never ride on these transporters unless it has been specially designed for this.

Transporter operators should be trained in the use of this equipment before being allowed to operate them. OSHA requires that operators of powered industrial equipment be trained. For more safety information on this type of equipment, contact your local supplier.

Carts (Push/Pull)

The pushing or pulling of carts is pervasive in some industries. In some types of warehousing, picking carts are used extensively as a mobile container to place parts after they have been manually picked. The extent of back or other body part overexertions from pushing or pulling these carts is not precisely known. Yet, studies show that forces on the back can be excessive depending upon cart design, body position, foot positions, and cart handle height.

Some items to consider in cart design and use are:

1. *Handle Height.* Of critical importance is the height of the handle from the floor. Handle height significantly affects pushing and pulling capabilities. A lower handle height is better for pulling motions than pushing motions. To achieve optimum push or pull forces, cart handles should be adjustable vertically or vertical push/pull bars installed. This permits various sized individuals to assume better biomechanical postures resulting in greater leverage.
2. *Cart Wheels.* The size of cart wheels should be large enough to minimize rolling resistance. The use of small caster wheels on a large heavy cart should be avoided where possible. Consider designing the front wheels of a cart that is pushed so they cannot swivel—let the rear wheels swivel. This aids in steering and turning.
3. *Body Positions.* The body positions used to move carts will depend on whether carts are being pushed or pulled. Generally, better leverage in starting cart movement is applied when the feet are spread apart (one foot in back of the other). In most instances of pushing or pulling, one foot serves to push or pull off. Frictional relationships between the floor surface and workers' shoe soles are important. Obviously, if the floors are slippery, the worker will assume a different position to keep from slipping there by reducing the amount of force which can be applied to move the cart.
4. *Push/Pull Forces.* It is always preferable to push than pull carts because more force can be generated in a push. Two forces are encountered: initial force and sustained force. The initial force, or that force to get the cart started, is higher than the sustained force to keep the cart moving. To determine the relative forces that male and female workers are capable of exerting with reasonable safety, the reader is referred back to the section on "Pushing, Pulling, and Other Tasks" which discusses this.

36.1.6 Materials Handling Equipment Safety Standards

The following materials handling safety standards are available through the American National Standards Institute (ANSI), 1420 Broadway, New York, NY 10018. Those standards with an asterisk at the end of the listing denote that they are also OSHA standards.

Conveyors

Conveyors and Related Equipment, Safety Standards For, ANSI B20.1.*

Industrial Trucks

Electric-Battery-Powered Industrial Trucks, Safety Standards For, ANSI/UL 583 (B56.3).

Internal-Combustion-Engine Powered Industrial Trucks, Safety Standard For, ANSI/UL 558 (B56.4).

Powered Industrial Trucks, ANSI/NFPA 505 (B56.2).*

Powered Industrial Trucks—Low Lift and High Lift Trucks, Safety Standard For, ANSI B56.1.*

Tow Tractors, Electrical Guided Industrial, ANSI B56.5.

Trucks, Powered Industrial—Rough Terrain Fork Lift Trucks, Safety Standard For, ANSI B56.6.

Lifting Devices

Base Mounted Drum Hoists, Safety Code For, ANSI B30.7.

Controlled Mechanical Storage Cranes, ANSI B30.12.

Crawler, Locomotive, and Truck Cranes, Safety Code For, ANSI B30.5.*

Derricks, Safety Code For, ANSI B30.6.*

Handling Loads Suspended From Rotorcraft, Safety Standard For, ANSI B30.12.

Hooks, Safety Standard For, ANSI B30.10.

Mobile Hydraulic Cranes, Safety Standard For, ANSI B30.15.

Monorails and Underhung Cranes, ANSI B30.11.

Overhead and Gantry Cranes (Top-Running Bridge, Multiple Girder), Safety Standard For, ANSI B30.2.0.*

Overhead Hoists, Safety Standard For, ANSI B30.16.

Slings, Safety Standard For, ANSI B30.9.*

REFERENCES

1. *Accident Prevention Manual for Industrial Operations—Engineering and Technology,* National Safety Council, 8th ed., Chicago, Ill., 1980.
2. Kroemer, K. H. E., *Material Handling: Loss Control through Ergonomics,* Alliance of American Insurers, Chicago, Ill., 1979.
3. *Accident Facts—1981 Edition,* National Safety Council, Chicago, Ill., 1981.
4. Antonakes, John A., "Claims Cost of Back Pain," *Best's Review,* September 1981, pp. 36–40.
5. Van Cott, H. P. and R. G. Kincade, Eds., *Human Engineering Guide to Equipment Design,* Rev. ed., U.S. Government Printing Office, Washington, DC, 1972.
6. National Aeronautics and Space Administration, *Anthropometric Source Book, Vol. 1: Anthropometry for Designers,* NASA Reference Publication 1024, U.S. Government Printing Office, Washington, DC, 1978.
7. Kroemer, K. H. E., *Material Handling: Loss Control through Ergonomics,* Alliance of American Insurers, Chicago, Ill., 1979.
8. *Work Practices Guide for Manual Lifting,* DHHS (NIOSH) Publication No. 81-122, National Institute for Occupational Safety and Health, U.S. Government Printing Office, Washington, DC, 1981.
9. *Accident Prevention Manual for Industrial Operations—Engineering and Technology,* National Safety Council, 8th ed., Chicago, Ill., 1980.
10. *Work Practices Guide for Manual Lifting,* DHHS (NIOSH) Publication No. 81-122, National Institute for Occupational Safety and Health, U.S. Government Printing Office, Washington, DC, 1981.
11. Snook, S. H., R. A. Campanelli, and J. W. Hart, "Three Preventive Approaches to Low Back Injury," *Professional Safety,* July 1978, pp. 34–38.
12. Lovested, G. E., "Reducing Warehousing Material Handling Strains," *Human Factor Society Proceedings,* 1980, pp. 653–654.
13. Snook, S. H., R. A. Campanelli, and J. W. Hart, "Three Preventive Approaches to Low Back Injury," *Professional Safety,* July 1978, pp. 34–38.
14. Ayoub, M. A., "Preemployment Screening Programs that Match Job Demands with Worker Abilities," *Industrial Engineering,* March 1982, pp. 41–46.
15. Chaffin, D. B. et al., *Pre-employment Strength Testing in Selecting Workers for Material Handling Jobs,* National Institute for Occupational Safety and Health, U.S. Government Printing Office, Washington, DC, 1977.
16. *Work Practices Guide for Manual Lifting,* DHHS (NIOSH) Publication No. 81-122, National Institute for Occupational Safety and Health, U.S. Government Printing Office, Washington, DC, 1981.
17. *Accident Prevention Manual for Industrial Operations—Engineering and Technology,* National Safety Council, 8th ed., Chicago, Illinois, 1980.
18. Lovested, G. E., "Reducing Warehousing Material Handling Strains," *Human Factors Society Proceedings,* 1980, pp. 653–654.
19. Lovested, G. E., "The One-Hand Lift as a Lifting Method Alternative," *National Safety News,* June 1981, pp. 26–27.

20. Chaffin, D. B. et al., *Pre-employment Strength Testing in Selecting Workers for Material Handling Jobs,* National Institute for Occupational Safety and Health, U.S. Government Printing Office, Washington, DC, 1977.
21. *Work Practices Guide for Manual Lifting,* DHHS (NIOSH) Publication No. 81-122, National Institute for Occupational Safety and Health, U.S. Government Printing Office, Washington, DC, 1981.
22. Snook, S. H., "The Design of Material Handling Tasks," *Ergonomics,* Surrey, England, December 1978, pp. 963–985.
23. Kroemer, K. H. E., *Material Handling: Loss Control through Ergonomics,* Alliance of American Insurers, Chicago, Illinois, 1979.
24. *Accident Prevention Manual for Industrial Operations—Engineering and Technology,* National Safety Council, 8th ed., Chicago, Illinois, 1980.
25. *Accident Prevention Manual for Industrial Operations—Engineering and Technology,* National Safety Council, 8th ed., Chicago, Illinois, 1980.
26. Coleman, P. J. et al., "Human Factors Analysis of Materials Handling Equipment," DHEW (NIOSH) Contract 210-76-0115, National Institute for Occupational Safety and Health, Morgantown, West Virginia, 1978.
27. Lovested, G. E., "Top Ten Forklift Truck Accidents," *National Safety News,* September 1977, pp. 123–127.

36.2 APPLICABLE REGULATIONS

John W. Russell

36.2.1 Introduction

Rules and regulations have a definite effect on manual and mechanical materials handling. The manufacturers and users of materials handling devices must be aware of the rules and regulations that may apply to them.

Almost any material that is manufactured, stored, transported, or disposed of is subject to one or more categories of rules and regulations. The purpose of this section is to identify some of the agencies, organizations, and other sources that can be contacted concerning materials handling rules and regulations. Since these regulations can change significantly from time to time, it is important to stay informed on any new standards and their modifications.

Table 36.2.1 shows some of the public and private agencies and organizations that are involved in materials handling regulations. At the end of this section, there is a list of addresses for the agencies

Table 36.2.1 Most Common Sources of Materials Handling Regulations[a]

Federal, State, and Local Government Sources

Occupational Safety and Health Administration (OSHA), Department of Transportation (DOT), Environmental Protection Agency (EPA), National Institute for Occupational Safety and Health (NIOSH), Mine Safety and Health Administration (MSHA), Nuclear Regulatory Commission (NRC), Federal Aviation Administration (FAA), Federal Railroad Administration (FRA), General Services Administration (GSA), Federal Highway Administration (FHA), U.S. Department of Commerce, Labor Departments, and U.S. Coast Guard.

Private and Other Sources

American National Standards Institute (ANSI), National Fire Protection Association (NFPA), National Safety Council (NSC), Material Handling Institute (MHI), International Material Management Society (IMMS), American Conference of Governmental Industrial Hygienists (ACGIH), American Society for Testing and Materials (ASTM), Underwriters Laboratories, Inc. (UL), manufacturers associations such as the American Trucking Association (ATA), Compressed Gas Association (CGA), Crane Manufacturers Association of America (CMAA), engineering societies (AIIE, ASME, ASSE, SAE, SME, etc.), insurance companies and associations, local accident prevention organizations, and educational institutions.

[a] Refer to listing in the text for addresses.

and organizations in the table. Governmental agencies such as the Occupational Safety and Health Administration (OSHA), Department of Transportation (DOT), and the Environmental Protection Agency (EPA) have standards that apply to materials handling in the workplace, transportation, and environmental areas. There are also rules and regulations that may be instituted by state and local agencies.

Organizations such as the National Fire Protection Association (NFPA), American National Standards Institute (ANSI), and others have established standards and codes that affect the handling of materials. Other sources such as the National Safety Council (NSC), International Material Management Society (IMMS), Material Handling Institute (MHI), American Trucking Association (ATA), manufacturer associations, and engineering societies can provide a wealth of information concerning materials handling rules, regulations, and standards. Also some governmental agencies and insurance companies provide consulting services that pertain to the safe handling of materials. It is important to realize that there are also voluntary standards and good methods of practice that may definitely be advantageous to use.

The first step in complying with the various rules and regulations is to evaluate the operations. An inventory of the different tasks and equipment used by the organization should be established. Once this is completed, it will be somewhat easier to identify which standards may apply to the operation.

Figure 36.2.1 shows some of the areas of materials handling affected by rules and regulations. This is meant only as a guideline. Other specific guidelines may apply to the situation, depending on the specialty area.

36.2.2 Areas Affected by Regulations

The following is a breakdown of some of the areas that may be affected by these rules, regulations, and standards. The appropriate agencies, organizations, and equipment manufacturers—shown in par-

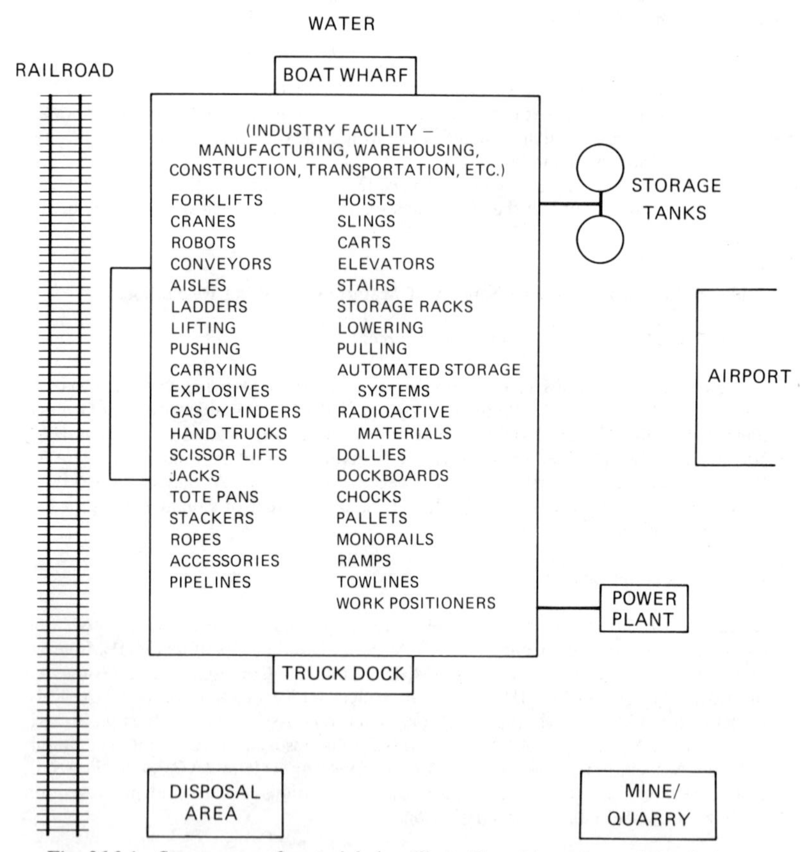

Fig. 36.2.1 Some areas of materials handling affected by rules and regulations.

entheses—and others can be consulted for additional guidelines and answers to specific questions and problems. These are only a few of the items contained in each category. It is impossible to discuss all of the specifics in each one. Typical equipment items are presented in Fig. 36.2.2.

Conveyors (OSHA, ANSI, NFPA, Manufacturers)

If a conveyor is used in a hazardous environment, it should be approved for the installation. Static electricity should be controlled while transporting dusts, liquids, and so on.

Guards should enclose the nip points and/or screw mechanism.

Overhead conveyors should be provided with barriers so that the articles being transported will not fall on someone.

Conveyor shutoffs should be installed in accessible spots. Trip stop wires should be installed if required.

Conveyor crossover stairs for individuals may have to be installed.

Alarms may be necessary to alert individuals to the startup of the conveyor.

Fire protection methods should be reviewed for conveyors going through fire walls or from one floor to another.

Cranes (OSHA, ANSI, NFPA, Manufacturer)

Regular written inspections of the unit, supporting hardwire, and so on, should be made. The manufacturer should be contacted concerning the type and method of inspection.

Daily visual checks of the cranes should be made. A written record of this should be maintained. The manufacturer should be contacted for further information.

Load capacity should be indicated on the unit and this capacity should not be exceeded.

Standardized crane signals should be used by the operators and employees.

Training should be given to the individuals operating the equipment.

Rigging should be inspected before each use and defective equipment should be removed or repaired.

Elevators (ANSI, OSHA, Manufacturer)

Inspections of the unit should be made in accordance with applicable federal, state, and local codes.

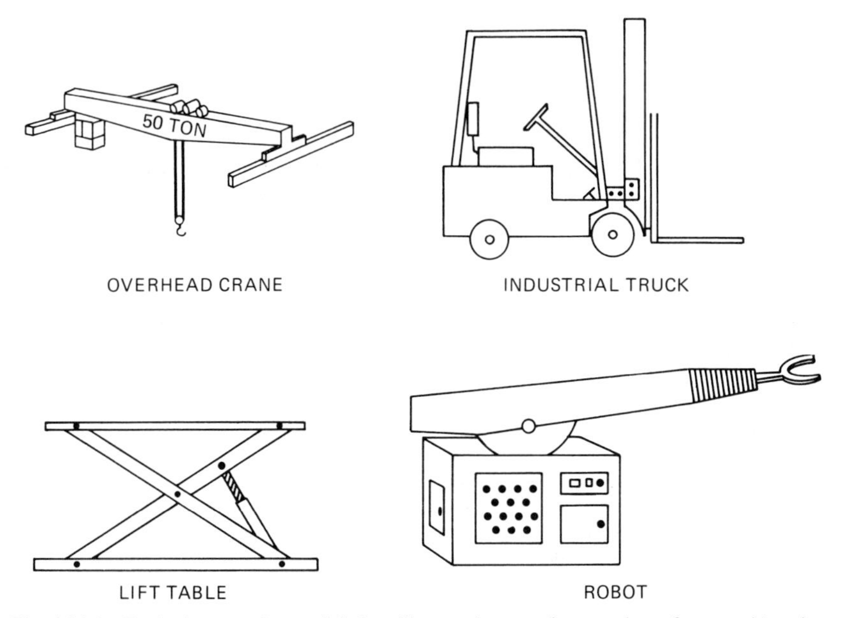

Fig. 36.2.2 Typical types of materials handling equipment that require safety considerations.

The capacity of the elevator should be clearly indicated and it should not be exceeded.

Freight elevators are meant for the transportation of materials, not individuals.

Forklifts (*OSHA, ANSI, NFPA, Manufacturer*)

Inspect for defects before each shift use or on a daily basis (whichever is more frequent) and provide a written defect report.

Use the proper type of vehicle in hazardous environments (See Chapter 7).

Driver training and refresher courses (including vehicle speed, traffic control, etc.) should be provided.

Machine guards should be kept in place, including overhead canopies and vertical back rests.

Proper handling of the fuel, battery recharging, and so on, is important. No smoking should be allowed in these situations.

Work platforms, attachments, and so on, should be used properly.

Other items such as adequate lighting and controlled levels of engine carbon monoxide should be checked.

Hand Trucks, Dollies, and Carts (*OSHA, Manufacturer*)

The units should be maintained in good condition and regular checks should be made. This includes checking for sharp edges, pinch points, and wheel or caster defects.

Hoists (*OSHA, ANSI, NFPA, Manufacturer*)

Regular written inspections of the hoists, overhead rails, and hardware should be made. The manufacturer should be contacted concerning the type and method of inspection.

Below-hook attachments such as magnetic and vacuum lifters should be inspected on a regular basis.

Employees should be trained to use the equipment in a safe manner.

Written sling and wire rope inspections should be made. Alloy steel chain slings should be tagged and certified. Mesh slings also have to be tagged. Only qualified sources such as the manufacturer should repair, modify, or tag these units.

The load capacity should be clearly indicated on the unit and it should not be exceeded.

Manual Handling (*OSHA, NIOSH, ANSI, NFPA, ACGIH*)

The employees should be provided training on how to handle materials including hazardous substances such as asbestos, and carcinogens. To help do this, Safety Data Sheets concerning these items should be requested from the manufacturer.

The employees should not be exposed to higher than permissible levels of vapors, dusts, and so on. Engineering controls are preferred, but the employees should be provided personal protective devices if needed or desired. Depending on the exposure, head, ear, eye, face, hand, foot, and body protection may be required. This includes protection from burns, radiation, noise, vapors, dusts, and so on.

Ladders, stairs, and floor surfaces should be free from defects such as openings, sharp edges, and slip and fall hazards. Adequate lighting should also be provided. Where applicable, floor load limit regulations should be observed.

Training in manual lifting methods in order to minimize back injuries is recommended. Whenever a lifting, lowering, pushing, pulling, or carrying task is judged to be too physically demanding for the employees, consideration should be given to using mechanical handling methods.

Adequate procedures for investigating materials handling accidents and injuries should be established.

Robots (*OSHA, Manufacturer*)

The robots should be isolated from the employees if possible, to prevent someone from being struck by the robot.

If the robot is used in a hazardous environment, the unit should be approved for this use.

Emergency power shutoffs should be provided.

The unit should be locked and tagged out during maintenance.

Storage (OSHA, NFPA, DOT, ANSI, Manufacturer)

Combustible and flammable liquids should be stored in approved and properly labeled containers. Only certain amounts should be stored in specific areas. Storage areas may require diking, explosion-proof wiring, ventilation, fire extinguishing systems, and so on.

Materials should be stored with secure piling methods and certain items should be segregated from each other. For instance, caustics and alkalis may react with each other. Oxygen and acetylene compressed gas cylinders should be stored separately from each other.

Racks should be anchored and should not be overloaded. High-rise automated storage and retrieval systems (AS/RS) may require specialized fire extinguishing systems.

Individuals working above the floor level should be protected from falling when picking orders from shelving units.

Tanks, silos, and other confined spaces may require special ventilation and protection from static electricity, lightning, and so on. Persons working in these confined spaces may require personal protective devices, possibly including supplied air line systems, life lines, and so on.

Transfer lines should be labeled and checked periodically for leaks.

When dispensing liquids, bonding and grounding methods should be used when necessary to minimize static electricity.

Explosives and radioactive items must be stored and handled with special precautions. Federal, state and local laws (regarding these exposures) should be followed.

Transportation (DOT, EPA)

Specific sizes and types of containers should be used for the storage and transportation of gases, liquids, solids, and so on.

The containers should be labeled indicating their contents and other characteristics such as if they are flammable, caustic, radioactive, explosive, and so on.

Transportation vehicles may require placarding indicating what is being carried.

Regulations exist concerning the inspection of the vehicle, maintenance of written records concerning the operation of the vehicle, and so forth.

Loads should be secured and protected from damage.

Waste materials must be disposed of with only certain methods. Specific methods concerning the containers, labeling, transportation, and disposal should be followed. The transportation routes may be restricted, depending on what is being transported.

As previously mentioned, this material provides a starting point for further study to develop specific details. Also, modifications of materials handling equipment should only be made by authorized individuals. Proper maintenance procedures should be established for all of the equipment.

Specialized industries have specific rules and regulations that apply to them. These run the gamut of all industries from A to Z and include agriculture to construction and food handling to longshoring. For instance, the following is a list of a few of the materials handling areas of concern in construction:

Handling of sharp, toxic, and hot materials may require personal protective equipment.

Flammable and combustible liquids and gases should be handled and stored with certain procedures.

Rigging equipment for materials handling has certain inspection, use, and storage requirements.

Handling of waste materials should follow particular guidelines.

Cranes, derricks, hoists, elevators, conveyors, aerial lifts, and so on, have certain inspection, use, and storage requirements. Personnel using these devices should understand how to use them. In some cases, the operators must be licensed to use the equipment.

Explosives are handled in accordance with particular regulations.

Handling, storage, and disposal of construction materials such as steel and lumber should follow certain guidelines.

Longshoring includes, but is not limited to, the following material handling regulations:

Equipment such as cranes, hoists, ropes, and winches, must be inspected and certified on a regular basis.

Personal protective equipment may be needed while handling the cargo.

Equipment operators may require training and licensing to operate certain types of equipment.

Certain procedures regarding the storage and handling of the cargo should be followed. Securing loads and observing deck load capacities require attention.

Spillage of material should be handled with certain methods.

If the organization is medium or large, it is important to have the engineering, manufacturing, purchasing, and safety departments work together. They should be actively involved in adhering to the materials handling rules, regulations, and standards.

For a small company, it may be practical to appoint one individual responsible for providing attention to materials handling requirements.

It is essential to take advantage of the resources of the federal, state, and local agencies, private organizations, manufacturers, and insurance companies. If you do not, there is the risk of not complying with the law. But there is also the possibility of legal fines, cost of corrections, accidents, injuries, and the chance of loss of life.

Organizations that Generate Safety Information

Alliance of American Insurers, 20 North Wacker Drive, Chicago, IL 60606.
American Conference of Governmental Industrial Hygienists, 2205 South Road, Cincinnati, OH 45238.
American Industrial Hygiene Association, 475 Wolf Ledges Parkway, Akron, OH 44311.
American Institute of Industrial Engineers, 345 East 47th Street, New York, NY 10017.
American National Standards Institute, 1430 Broadway, New York, NY 10018.
American Society of Mechanical Engineers, Inc., 345 East 47th Street, New York, NY 10017.
American Society of Safety Engineers, 850 Busse Highway, Park Ridge, IL 60068.
American Society for Testing and Materials, 1916 Race Street, Philadelphia, PA 19103
American Trucking Association, Inc., 1616 P Street, N.W., Washington, DC 20036.
Canadian Standards Association, 178 Rexdale Boulevard, Rexdale, Ontario M9W. 1R3, Canada.
Compressed Gas Association, Inc., 500 Fifth Avenue, New York, NY 10036.
Crane Manufacturers Association of America, 1326 Freeport Road, Pittsburgh, PA 15238.
Department of Transportation
Federal—400 7th Street, S.W., Washington, DC 20590
State—Consult local listing.
Environmental Protection Agency
Federal—Washington, DC 20210
State—Consult local listing.
Federal Aviation Administration, Dept. of Transportation, 800 Independence Avenue, S.W., Washington, DC 20591.
Federal Highway Administration, Dept. of Transportation, 800 Independence Avenue, S.W., Washington, DC 20591.
Federal Railroad Administration, Washington, DC 20590.
General Services Administration, 18th and F Streets, N.W., Washington, DC 20405.
International Material Management Society, 3310 Bardaville Drive, Lansing, MI 48906.
Interstate Commerce Commission, Washington, DC 20423.
Institute of Makers of Explosives, 420 Lexington Avenue, New York, NY 10017.
Material Handling Institute, 1326 Freeport Road, Pittsburgh, PA 15238.
Mine Safety and Health Administration, U.S. Department of Labor, 4015 Wilson Boulevard, Rm. 516, Arlington, VA 22203. Also check for local listings.
National Fire Protection Association, 470 Atlantic Avenue, Boston, MA 02210.
National Institute for Occupational Safety and Health, U.S. Department of HEW, Cincinnati, OH 45226. Also check for local listings.
National Safety Council, 444 North Michigan Avenue, Chicago, IL 60611. Also check listings for local councils.
Nuclear Regulatory Commission, Washington, DC 20555.
Occupational Safety and Health Administration, U.S. Department of Labor, 200 Constitution Avenue, N.W., Washington, DC 20210. Also check local listings.
Society of Automotive Engineers, Inc., 2 Penn Plaza, New York, NY 10001.
Society of Manufacturing Engineers, 20501 Ford Road, Dearborn, MI 48128.
Underwriters Laboratories, Inc., 207 East Ohio Street, Chicago, IL 60611.
U.S. Bureau of Mines, Department of the Interior, 2401 E Street, N.W., Washington, DC 20241.
U.S. Coast Guard, Department of Transportation, 400 Seventh Street, S.W., Washington, DC 20590.
U.S. Department of Commerce, Washington, DC 20423.

BIBLIOGRAPHY

American National Standards Institute, *Powered Industrial Trucks,* ANSI B56.1-1969, ANSI, New York, NY, 1969.

Federal Highway Administration, *Federal Motor Carrier Safety Regulations,* U.S. Department of Transportation, Washington, DC, 1981.

Materials Transportation Bureau, *General Requirements for Shipments and Packaging,* U.S. Department of Transportation, Washington, DC, 1980.

McElroy, Frank E., *Accident Prevention Manual for Industrial Operations,* Vols. 1 and 2, 8th Ed., National Safety Council, Chicago, IL, 1981.

National Fire Protection Association, NFPA No. 505-1978, Boston, MA, 1978.

National Institute for Occupational Safety and Health, *Work Practices Guide for Manual Lifting,* U.S. Department of Health and Human Services, Cincinnati, OH, 1981.

Occupational Safety and Health Administration, 29CFR 1910–*General Industry,* U.S. Department of Labor, Washington, DC, 1981.

Occupational Safety and Health Administration, 29CFR 1918–*Longshoring Industry,* U.S. Department of Labor, Washington, DC, 1975.

Occupational Safety and Health Administration, 29CFR 1926—*Construction Industry,* U.S. Department of Labor, Washington, DC, 1979.

Tanner, William R., *Industrial Robots,* Vols. 1 and 2, Society of Manufacturing Engineers, Detroit, MI, 1981.

36.3 DUST CONTROL

Robert Kolatac

36.3.1 Introduction

The processing industries utilize many different materials in the formulation of end products. These materials may be solids, liquids, or gases. Each state has its own unique handling properties.

The most difficult form to handle is the bulk solid materials. Expertise in bulk solids handling equipment has lagged behind industry needs. This may be due to slower technological developments with solids than that of the practical experience developed for liquids and gases. Standards or universal laws have been formulated, proven, and used for many years in liquid and gas handling. Man was investigating fluids and developing data hundreds of years before the evaluation of the solid state.

Particulate solids do not exhibit uniform handling properties as do liquids and gases. Since basic properties vary significantly, one can theorize that the lack of research in the solids handling field, was attributed to this nonconsistent state.

The need for clean air in living and work areas cannot be understated. Elimination of airborne pollutants can be accomplished using today's technology. There are, however, many factors that must be considered with any pollution abatement system. These range from public health and welfare to pure economic considerations.

All industrial operations must be concerned about, or have a need to control, particulates. Emission of fine particles into the air we breathe can cause reduced visibility, respiratory irritation, and other health problems. The typical urban industrial atmosphere may contain lead compounds, fly ash from the combustion of solid and liquid fuels, particles from metallurgical, pharmaceutical, food processing, mining operations, and others.

The industrial worker is generally more affected than others, since the working environment can become contaminated by a higher concentration of pollutants.

Industry can no longer regard a geographical area as an entity unto itself, but must realize that the earth is a closed-loop system. Pollutants, from whatever source and geographical area, eventually affect everyone.

Dust arising from processing operations cause maintenance problems, especially those dusts considered to be abrasive. Abrasive dusts get into gear drives and bearings and cause a failure. Conveyor and motor drive belts can become coated, and begin to slip.

Compressed air systems, pneumatically actuated devices, lubrication systems, mechanical actuators, and so on, all require contaminant-free operation. This is not possible in a "dirty environment." Equipment service life is greatly reduced, therefore, if dust is not controlled.

When solid materials are handled, particulates may be formed. "Dusting" may result from attrition between larger agglomerates, or be caused by processing equipment. This breakdown may be caused by a physical and/or chemical change. Any operation involving the crushing of large agglomerates into fine particles for refinement generates large amounts of dust. Screening and classifying, metallurgical refining operations, and so forth, all result in dust and fume generation.

Dusts we inhale fall into two groups: soluble or insoluble (in body fluids). Most soluble dusts can be absorbed in the body and pass without problems. Insoluble dusts, however, accumulate on lung tissue or pass directly into the blood stream and are transported throughout the body. Others may chemically react with body fluids to create toxic by-products.

36.3.2 Particulate Classifications

The individual particle is the main topic in fabric filtration. An understanding of the particle is essential for proper evaluation.

In dust control, the size of a particle is measured in microns. A micron is 1/1000 of a millimeter, or 1/25,400 of an inch. Particles larger than 50 microns can be seen by the naked eye. Small particles (down to 0.005 microns) can only be observed by using powerful microscopes.

Particulates are classified according to the methods in which they are performed. These are: dusts, fumes, smoke, mists/fogs, and sprays. These are transported in a gaseous media; the combination is referred to as an aerosol.

Dust

Dusts are small, solid particles that are created by the disintegration of larger particles in processing operations such as grinding, crushing, explosion, screening, packaging, and conveying. Particle sizes range from submicron and up, although they typically tend to be larger than 1 micron.

Fumes

Fumes are fine, solid particles formed by the vapor condensation of solids. They are largely a result of combustion, distillation, or sublimation. Average particle size is less than 1 micron. The composition of the fume particulate can be different from the original material due to oxidation/reduction reactions which may occur.

Smoke

These are fine, solid particles resulting from the incomplete combustion of organic materials. Particle sizes range from 1 micron to 0.01 micron and finer.

Mists, Fogs, and Sprays

These are liquid particulates or droplets formed by condensation and/or mechanical disintegration. Particle sizes range from in excess of 10 microns to 0.01 microns.

Solids Concentration

Solids concentration is a measure of the total mass of all the solids entrained in a given volume of gas. Concentration is most often indicated in grains of solids per cubic foot of gas *or* grams per cubic meter of gas. The relationship of "grains to grams" is 1.0 gr/ft^3 = 2.3 g/m^3.

Control systems are rated according to the solids concentrated in the gas to be cleaned.

Function	Concentration
Air filtration	0.001–5 gr/1000 ft^3
Industrial air filtration	5–100 gr/1000 ft^3
Dust control	0.1–100 gr/ft^3

For a dust control system, dust concentrations are further classified:

Designation	Concentration
Light	0.5–2 gr/ft^3
Medium	2–3 gr/ft^3
Moderate	3–5 gr/ft^3
Heavy	5 and up gr/ft^3

Pressure

Pressure is a force exerted over a specific area. Pressure at sea level is approximately 14.7 lb/in.^{2} and decreases with increased elevation.

A conveying gas in an air cleaning system exerts a pressure on the interior of the system while atmospheric pressure exerts an exterior pressure on the system. Pressure differentials are an important consideration in air filtration systems. Pressure is referred to as either absolute or gauge pressures. Absolute pressure starts from a zero base, where gauge pressure starts from atmospheric pressure. Absolute pressure equals gauge pressure plus atmospheric pressure.

Temperature

The temperature of a gas is a measure of the mean molecular kinetic energy of the gas. Gas temperature must be known in order to determine the amount of heat that may have to be removed from a gas/solid as they flow through a system.

Density

Conveying gas density is a measure of its mass per unit volume. The gas density varies with temperature and pressure changes.

Flow Rate

Flow rate is the volume per unit time of conveying gas flowing through the system. Air moving equipment and collector are sized to meet the volume of contaminated gas that must be moved through the system in a given time.

Pressure Drop or Resistance

All gas cleaning devices offer resistance to the motion of the carrier gas flowing through it. The pressure upstream of the cleaning device must be greater than the downstream pressure in order to overcome system resistance.

Air-to-Cloth Ratios

Air-to-cloth ratio is the relation between the number of square feet of cloth area to the amount of contaminated air passing through the collector. Air-to-cloth ratios vary with material collected as well as the type of cleaning device used.

36.3.3 Particulate Solids

An important item in designing any solids handling or collection system is the material to be handled. The basic characteristics of the material determine, in many cases, collector capacities, configuration, contact materials, metering device or devices, discharging devices, and other process components. An understanding of a material's characteristics and how to apply these characteristics are essential when designing or modifying any solids handling system. The Conveyor Equipment Manufacturers Association (CEMA) has developed a code that classifies a wide variety of particulate solid materials according to their handling characteristics.

Manufacturers and users of dust collection equipment concur that a dust collector can never be too large. Excess cloth area results in fewer maintenance problems, lower pressure drop, additional capacity for future expansion, and so on. Unfortunately, economics require proper, closer sizing of equipment to meet process requirements. The answer to this problem is to select the highest filter rate maintaining good operation. To do this, material characteristics must be closely examined. The precise function of each characteristic may not be easily defined mathematically. Some variables defy logic completely.

An initial review of the material being handled will eliminate future problems.

Particle Size

Material particle size is a derived property pertaining to the entire mass of particles. A screen analysis and visual inspection of material show that particle size and quantity of each specific size and shape of the individual particle varies within a single material type. Particle size distribution plays an important part in the design of the conveying equipment. A screen analysis should always be conducted on the material being conveyed and a particle size distribution curve established. The distribution curve will assist the designer in determining if the material will readily flow out of the collector hopper, and if the material will flow down a chute, duct, or airlock.

For air cleaning applications, particles are classified as being extremely fine, fine, medium, or coarse, according to the size range, in microns, in which 50% or more are of the same relative size.

Particle Size	Micron Range (50% of Population)
Extremely fine	0.5–2
Fine	2–7
Medium	7–15
Coarse	15+

Particle size evaluation should be made in conjunction with specific equipment manufacturers to determine clear-cut understanding of particle size distribution, prior to equipment selection.

Particle size and shape also dictate the amount of exposed surface area a material will have. Fine particles with large surface areas may readily explode in the presence of air.

Temperature

Temperature gradients may affect a material if the temperature involved induces a physical change or initiates a chemical reaction. Temperature changes can result in condensation forming within conveying lines, hoppers, or metering devices. Pelletized material containing latent heat that, when collected, can result in a resolidification of all particles into a single mass or masses of considerably larger sizes that will not flow out of the collector hopper. Heat generated from processing equipment should not initiate endothermic chemical reactions.

Substantial changes can occur in materials due to freezing of moisture within the particulates. An ice bond that develops impairs flowability.

A normally dry mixture of materials that flows readily from a storage vessel may become a pasty mass in the conveying line caused by an increase in air temperature from compressive heat generated in the air supply blower.

Pressure

Pressure differentials must be considered in the design of bulk solids handling systems. Pressure differentials, whether positive or negative, can cause significant changes in the material.

Abrasiveness

Abrasiveness is defined as the material's ability to abrade or grind away the surface of other materials it comes in contact with. A material's abrasiveness determines the type of contact material used in the collector, conveying lines, and discharging devices. Abrasiveness is also a function of particle shape, size, and density. Extremely hard particles may have a perfectly spherical shape (no sharp cutting edges) to gouge into conveying lines, chutes, and so on. Particle shape is related to the manner in which they are formed in process. A free-flowing particulate may go through a reduction/oxidation reaction and form long chains or agglomerates. Photomicrographs of particulate samples are extremely useful for particle shape evaluation.

Extremely abrasive materials will require wear plates or liners of very hard material within the collector hopper. Chutes and elbows must receive special construction considerations.

Hygroscopicity, Deliquescence, and Efflorescence

A material is hygroscopic if it has a tendency to pick up moisture from the surrounding atmosphere. Hygroscopic materials will tend to cake up, causing hard agglomerates. Changes in particle size, due to hygroscopicity, can seriously alter the flow of material from the collector. Some materials physically deteriorate when conveyed. Others give up moisture contained in their molecular structure.

Materials of the this nature require an enclosed system where a dry, inert gas is substituted for ambient air. Condensation within the system, caused by heat differentials, should be eliminated by maintaining system temperature above the dew point.

Chemical Properties

Chemical properties are those specific properties of a particulate that are unique to the material. These include, among others, explosiveness, reactivity/stability, odor, corrosiveness, polymerization activity, and crystal structure. How these factors affect the system operation may have to be determined from full-scale laboratory tests or from actual field installation.

36.3.4 History of Dust Control and Collection

Because reclaiming of process dusts is now a requirement to meet air quality standards, baghouse utilization has expanded at a rapid pace.

The first type of "dust collector" was nothing more than a simple cloth worn over the face to protect desert travelers from the swirling sand. Doctors wore gauze masks to protect themselves and their patients from infection.

The first fabric filters used commercially were applied in the nonferrous smelting and refining industries. These were large cloth bags that would fill with solids, and which would then be manually shaken to remove the dust, thereby cleaning the bags. In many areas of the flour milling industry today, these same large fabric bags are still used for bin venting purposes.

Bag cages or thimble sheets were used in some dust collectors before 1900. Automatic shaking or cleaning devices were quickly incorporated.

Reverse air collectors and collectors with mechanical shakers were perfected in the 1920s and 1930s. Some of the original designs of these early model collectors are still being used today.

In 1957, the pulse jet collector was developed. This device used timed, bursts of compressed air to clean the filter media, thereby eliminating many mechanical components previously used for cleaning.

Baghouses have remained virtually unchanged since that time. Innovations have been made to increase collection efficiency, add maintenance reliability, and improve fabric construction.

The requirement for dust collection devices has kept pace with the times, however. Few solids handling systems are complete without dust collection equipment. Revised environmental standards enforce the importance in controlling emissions from processing equipment.

36.3.5 Materials Handling Applications

Specific areas in materials handling systems generate dust. At these locations, devices for collecting and containing dusts must be provided. For example, belt conveyors, bucket elevators, and screw conveyors all emit dust along pulleys, side boards, shafts, and at transfer points. Also, inlet and discharge (gravity) chutes and storage bin vents are all points of common dust emission.

Capacities

Prior to designing an efficient dust collection system, explicit capacity requirements should be known. Consideration should be given to:

Operating capacity—average capacity of system expected under normal operating conditions.

Peak capacity—maximum capacity ever expected even during short, trouble-free periods. Recycled product loads considered in peak capacity.

Rated capacity—average capacity during ideal operating conditions.

Design capacity—capacity from which structural and mechanical design calculations are made.

Preselection Guidelines

Prior to equipment selection some important considerations should be made.

1. Determine capacity requirements, system distance and path to be followed, and operating environment.
2. Determine the material characteristics and process constraints on equipment selected.
3. Any previous experience with material and equipment selected?
4. Consider the geometry and space requirements available for the equipment being considered.
5. Does the type of equipment selected fit into the total system being considered?
6. Compare the type of equipment selected to other devices equally suited. Consider their operation, initial cost, operating and maintenance costs, and service and parts availability.

Types of Dust Collectors

The type of dust collection equipment selected is based on several factors. These include: material factor, process factor, conveying gas factor, temperature effect, dust loads, and device limitations.

Material Factor. The basic characteristics of the solids being collected. How they react in process and conveying.

Process Factor. This includes conveying gas flow rate, collector efficiency, total allowable pressure drop within the system, and particulate flow rate.

Conveying Gas Factor. The conveying gas should not affect the solids being conveyed. The conveying gas may be explosive, corrosive, or toxic, which will require special considerations.

Temperature Effect. Practical experience has shown that additional filter cloth area is required as conveying gas temperatures increase. Changes in gas viscosity and density cause this phenomenon.

Device Limitations. Determine the physical dimensions and design of the device being considered, with relative space available for the device. Do not overlook the loaded weight and the installation area.

Collection Devices

Dry collection devices include: gravity settling chambers, cyclones, baghouses, and electrostatic precipitators.

Gravity Settling Chambers. The gravity settling chamber is one of the simplest and oldest dust collection devices. This is nothing more than an expansion chamber, where the increase in cross-sectional area of the chamber relative to the cross-sectional area of the inlet causes a decrease in conveying gas and solid velocity. The resultant decrease in velocity, along with gravitational effects, cause the particles to settle out. For optimum results with gravity chambers, solid size is generally larger than 10 microns, with conveying gas velocity less than 60 ft/min. Design equations for gravity chambers are readily available.

Cyclones. A cyclone is a centrifugal collection device. This type of collector may be used in dry or liquid applications. The cyclone is more efficient than the gravity settling chamber, handling particles as fine as 2 microns with 40% efficiency. Contaminated gas enters the cyclone tangentially at the top; the solids swirl around and separate from the lighter conveying gas, dropping to the bottom of the conical shaped body. Clean gas exits at the top. Cyclones are manufactured in different efficiency and capacity ratings. For best results, minimum solid sizes should be between 5 and 10 microns.

Fabric Filters. Fabric filters (or baghouses) are used most extensively for controlling the emissions of solids from conveying gas streams. They are relatively inexpensive and can handle a wide variety of materials, high gas flows, and have good separation efficiencies. New breakthroughs in fabric materials have resulted in improved bag cleaning methods, and better collection efficiencies, with reduced pressure drops across the filter. Solids as fine as 0.1 micron have been collected with excellent efficiency, and conveying gas flows from 50 ft^3/min and greater.

Electrostatic Precipitators. This device uses electric forces to collect suspended particles contained in dirty gas. The particulates are precharged and are collected on collecting electrodes, or plates. Dry, submicron-size solids are collected in these devices.

36.3.6 Filter Selection Requirements

Fabric-type dust collectors used in separating fine solids from process gas streams have some distinct advantages.

1. Many different types of solids may be handled, so long as the conveying gas stream temperature is maintained above the dew point to prevent condensation from forming in the unit.
2. With a good maintenance program, collector efficiency remains constant.
3. Unlike other types of gas cleaning equipment, the efficiency of the filter fabric is consistent and uniform, and is independent of particle size, dust loading characteristics, and variations in the properties of the conveying gas.
4. Performance is predictable, meeting approval by local air pollution control authorities. Operating pressure drop and power requirements are predictable and uniform.
5. Many different designs, capacities, fabric cleaning, and filter media are available for design flexibility.

The major disadvantages of baghouses are the limitations on their application due to temperature that the fabric can withstand, moisture levels in the conveying gas, and some space restrictions.

The first step in designing a filtration system is to clearly define all process variables, material variables, codes or emission standards that must be met, physical installation requirements, and economic considerations.

1. *Gas Flow Rate.* What are the minimum, average, and maximum flow rates? (Remember statements regarding capacities.) What are the gas and material temperature, moisture, dust loading, and other related chemical properties?
2. *Particulate Characteristics.* List all material characteristics as previously defined.
3. *Code and Emission Standards.* What is required to comply with present codes? How about future emissions standards?
4. *System Requirements.* What is unique about the system as a whole? Develop heat and material balances. What could occur during a system upset?
5. *Installation or Space Availability.* What space has been allocated for equipment; is it indoors or outdoors? Can equipment be easily maintained? Are utilities readily available for auxiliary equipment? Does device meet operation requirements (light, medium, or heavy duty)?
6. *Economical Considerations.* Is there a more economical device available to meet all requirements? What are initial costs? What are operational and maintenance costs?

Once a logical assessment has been made of the problem, the solution can then be sought.

36.3.7 Operational Problems

Filters provide relatively few problems if properly maintained. Some typical problems are bag failure, plugging in hoppers and ducts, and blower or fan, belt drive, and valving problems.

Bag Failures

Common causes of fabric failure are excessive dust accumulation and mechanical and design problems.

1. Excessive dust accumulation causes abrasion of fibers, binding, and plugging. Excessive dust deposits caused by condensation cannot be removed by normal cleaning. Also, burn holes are caused by hot particles.
2. Mechanical problems include fabric failure at seams and tears between cloth and support cages.
3. Design problems cause bags to rub together, improper bag tensioning, poor sealing around fabric and support collars, and wear caused by cleaning mechanism.

Plugging in Hoppers and Ducts

Difficulties often arise from dust plugging in the inlet duct due to solids buildup.

Material accumulated in collection hoppers may bridge and resist discharge, requiring auxiliary discharging devices such as bin activators, side vibrators, and air pads.

Material buildup within rotary airlocks and screw conveyors can cause emptying problems. Quick cleaning devices or "poke" and access ways may be required.

Dust can plug in the filter tubes due to an overaccumulation of dust in the collection hoppers. Reliable high-level indicators should be installed in the hoppers to eliminate this problem.

Blowers, Fans, Drives, and Valves

Standard preventative maintenance procedures as recommended by the equipment manufacturers eliminate or drastically reduce problems of this nature.

36.4 NOISE CONTROL

Lyle F. Yerges

36.4.1 Introduction

A wealth of information on the physics of sound and the mechanics of vibrating bodies is available to the engineer; and many excellent books and references on noise and vibration control can be found in almost any library or bookstore. In this section, only the information necessary for any competent engineer to design noise out of equipment or to control the unavoidable noise and vibration generated by such equipment and its operation is discussed.

Sound is defined as a vibration in an elastic medium. *Noise* is simply unwanted sound, whatever its nature. Sound and vibration are generated by almost any moving bodies, particularly those involved in materials handling systems and processes.

Sound originates with a *source*—any rotating, oscillating, vibrating, sliding, reciprocating, or magnetostrictive equipment—or by turbulent flow, combustion, impacts, or similar actions.

Sound travels via *paths,* virtually any solid or fluid medium in contact with the source, including the air surrounding it; and it usually reaches a *receiver*—normally a human is the receiver of interest.

The significant human responses to sound and vibration are hearing and feeling. The effects of these phenomena on humans are so well known and documented that this section deals with them only insofar as criteria and standards for their control dictate.

When the frequency of sound exceeds about 15 Hz (cycles per second), humans begin to hear the sound; but we are almost completely deaf to sound at frequencies higher than about 16–20 kHz (16,000–20,000 cps).

Within normal intensity levels, sound is a vital and necessary part of our lives, conveying much of the information necessary for our activities, including communication and signals. The normal ear begins to respond to sound at 0 dBA (20 μPa); however, long-term exposure to high noise levels—in excess of 85 dBA (just under 6×10^{-5} lb/in.^{2})—can produce harmful effects on humans, particularly irreversible hearing loss. Since literally millions of industrial workers are exposed to levels in excess of 85 dBA, it is apparent that the design engineer must be aware of the noise output of any equipment he or she designs or installs.

Humans sense vibration as "feeling," whether by tactile sensation or by physical responses such as quivering, shaking, pulsations, or other bodily reactions. At some low frequencies (under about 20 Hz), we may experience nystagmus as our eyes resonate to the vibration; dizziness or vertigo, as our inner ear structure vibrates; and muscular vibration and chest pulsations. Long-term exposure to severe vibration can produce degenerative effects such as "white finger" (Raynaud's disease).

Control of noise and vibration to safeguard exposed workers and operators, as well as to prolong the life of equipment, is the subject of this section. Noise can be controlled at the source, anywhere along the transmission path, or at the receiver. Ear protection will not be discussed, except to say that it is normally only a last resort and usually the least-successful option available.

36.4.2 Noise Fundamentals

A brief review of some fundamental aspects of motion will simplify analysis of noise and vibration and should lead the engineer most directly to solutions to the design problems. In general, it is not necessary even to consider the wave equations, but only to look at the basic relationships of motion.

A vibrating or oscillating body varies constantly in its velocity and direction of motion. Hence, the motion involves the acceleration of mass, which involves force:

$$\text{Force} = \text{Mass} \times \text{Acceleration}$$

The moving mass involves energy:

$$\text{Kinetic energy} = \tfrac{1}{2}\,\text{mass} \times \text{Velocity}^2$$

The energy and force involved in most noise and vibration problems are small. Table 36.4.1 compares noise levels with power and pressure for the usual range of interest to the designer. (The reader is

Table 36.4.1 Decibel Notation in Acoustics[a]

Power (Watts)	Sound Pressure Level: dB	Sound Pressure Level: N/m^2 (Pa)	Pressure (lb/in.^{2})	Common Noise Levels
10^8	200	200,000	3×10	
10^6	180	20,000	3	
10^4	160	2,000	3×10^{-1}	Jet engines
10^2	140	200	3×10^{-2}	Pain threshold
1	120	20	3×10^{-3}	
10^{-2}	100	2	3×10^{-4}	OSHA limit
10^{-4}	80	0.2	3×10^{-5}	
10^{-6}	60	0.02	3×10^{-6}	Normal speech
10^{-8}	40	0.002	3×10^{-7}	
10^{-10}	20	0.0002	3×10^{-8}	Whisper
10^{-12}	0	0.00002	3×10^{-9}	Hearing threshold

[a] The "decibel" is always a *ratio,* never a unit. In this table, Power means watts/m^2, and dB is a *level.*

Table 36.4.2 Acoustical Impedance of Various Materials

	Acoustical Impedance	
Material	(lb/in²-sec)	(kg/m²-sec)
Cork	165	116×10^3
Pine	1,900	$1{,}340 \times 10^3$
Water	2,000	$1{,}410 \times 10^3$
Concrete	14,000	$9{,}870 \times 10^3$
Glass	20,000	$13{,}400 \times 10^3$
Lead	20,500	$14{,}450 \times 10^3$
Cast iron	39,000	$27{,}500 \times 10^3$
Copper	45,000	$31{,}725 \times 10^3$
Steel	58,500	$41{,}240 \times 10^3$

referred to any good handbook on acoustics for the nomenclature of noise and vibration, particularly the terms "decibel, dB, and dBA.")

Vibrating sources radiate energy to their surroundings, either through the structures to which they are attached, or by "driving" energy into the air in contact with their vibrating parts (or into any fluids directly associated with or connected to them).

Radiation from the vibrating parts is governed by the *impedance* of the source, the impedance of any transmission paths, and the radiation efficiency of the vibrating parts.

In most materials handling equipment, hard, rigid metal or plastic parts exhibit very high radiation efficiency; some soft, elastomeric parts, on the other hand, have a relatively low efficiency.

Impedance can best be described as the rate at which unit mass accepts energy. Table 36.4.2 lists the acoustical impedance of various materials commonly used in machines and building design.

As in other engineering problems, impedance match (or mismatch) determines the rate at which energy can be transmitted between various media. A good impedance match will encourage energy transmission; a poor impedance match will retard or minimize transmission.

Already it should be apparent that the designer is in control of the noise or vibration generated by the equipment, and that each design decision dictates an acoustical result.

36.4.3 Noise Sources

Typical elements or components of materials handling equipment capable of producing significant noise or vibration include:

Air and gases: blowers, nozzles, valves, exhausts, suction devices, fans, and compressors
Bearings
Belts
Bowl feeders
Buckets
Cables
Chains
Conveyors: bottle, can, parts, and so on
Elevators
Engines
Forklift trucks
Gears
Gravity and free-fall conveyors
Hoppers
Linkages
Motors
Pipes
Plates
Pulleys
Pumps

Rollers
Screws
Shakers
Stops: lugs, cams, gates, and so on
Tubes
Vibratory screens and feeders

The operating parameters of equipment or equipment components significantly affect the noise or vibration generated or transmitted from the source (see Table 36.4.3). Thus, the designer can influence the acoustical characteristics of equipment by choice of operating parameters, as well as by the physical characteristics of the components.

36.4.4 Control Fundamentals

There are several noise control options available to the engineer.

1. *Do not Make the Noise.* Often the simplest and most effective approach to noise control involves simple design changes to avoid impacts (tapping, pounding, impact stops, etc.); that is, reducing velocity, rpm, power, pressure, or otherwise controlling operating parameters (see Table 36.4.3). Even improving balance of rotating parts and proper lubrication of gears and bearings, as well as adequate maintenance (tightening parts and attachments) can often provide significant noise reduction.

2. *Modify or Substitute Processes.* Substituting electric motors for internal combustion engines; belts for gears; belt conveyors for shakers, vibrators, blowers, or impact feeders is frequently the most feasible noise control approach. Avoiding free-fall conveying processes and part-upon-part impact to move parts along a line is often advisable. Optimizing operations, such as using bowl or vibratory feeder equipment for maximum transport efficiency and minimum noise has proved highly effective in many instances.

3. *Use Vibration Isolation, Impedance Mismatch, or Damping.* Cushioning impact points or substituting plastics or tough elastomers for metal contacts is particularly effective in many instances.

Using flexible connectors in pipe lines and resilient support of piping and fluid lines are possible alternatives to rigid piping and attachments. Vibration isolation—spring or elastomer support of rotating, oscillating, or reciprocating components—is a well-established and standard means of avoiding vibration transmission to structure and surroundings. Damping of vibrating panels with applied mastics or sheet materials, selective stiffening of panels to change panel resonances, substituting perforated or expanded metal for solid panels to minimize radiation from their surfaces, or even resilient attachment of panels at strategic points on the structure where movement is at a minimum are useful options in many instances.

Table 36.4.3 Effect, in dB, of Operating Parameters on Machine Noise[a]

Noise Source	Noise Output, dB
Internal combustion engines	$10 \log_{10}$ horsepower ratio
	$30 \log_{10}$ speed (rpm) ratio
Fans	$10 \log_{10}$ horsepower ratio
	$10 \log_{10}$ pressure head ratio
	$50 \log_{10}$ rotational speed (rpm) ratio
Pumps	$17 \log_{10}$ horsepower ratio
	$40 \log_{10}$ speed (rpm) ratio
Gas flow	$80 \log_{10}$ velocity ratio (Mach 1 and higher velocities)
	$60 \log_{10}$ velocity ratio (velocities less than Mach 1)
	$30 \log_{10}$ pressure ratio (at velocities less than Mach 1)
Liquid flow	$60 \log_{10}$ velocity ratio (without cavitation)
	$120 \log_{10}$ velocity ratio (with cavitation)

[a] To determine the effect of changing an operating parameter (such as horsepower, rpm, etc.) on the noise output of a particular piece of equipment:

1. Determine the ratio of its present operating parameter to that of the proposed new parameter; for example, *double* the speed or *halve* the horsepower.
2. Insert this number (2 or $\frac{1}{2}$, for example) into the equation.
3. Add, algebraically, this value to the existing noise level of the machine to determine the new or changed level resulting from the change in operating parameter.

4. *Absorb Acoustic Energy.* The use of acoustical absorbents within housings, enclosures, and buildings is a frequent option in noise control; but it is limited in its applications to many materials handling systems. At best, it can normally provide less than 10 dBA noise reduction, and it is much less effective in attenuating low-frequency noise. In conjunction with other approaches, however, it can be helpful. Mufflers for exhausts, blow-offs, and valve discharges are usually highly effective in eliminating some of the noisier, more annoying noise problems in handling systems.

5. *Shield or Enclose the Noise Source or the Receiver.* A widely used noise control method involves partial or complete enclosure of the noise source. Properly done, this can be very effective, but it can also be a complete failure, or it can even amplify the noise from the source. A tight, complete enclosure of adequately heavy, air-tight panels or septa can provide more than 30 dBA attenuation, *if* it is lined with adequate absorption; even a partial booth or a simple panel shield which cuts off line-of-sight between source and receiver can provide from 5 to 15 dBA reduction, particularly when lined with absorbents. Usually access to the interior of enclosures, for operating adjustments, maintenance, or other purposes, dictates less than complete and somewhat "leaky" designs, seriously compromising the potential value of the enclosure.

Enclosures or shields must normally have a surface density of 2 lb/ft^2 (10 kg/m^2) or more; and they may be of rigid or flexible materials, properly supported, and with reasonably air-tight seams or joints. They should be attached to points of minimum vibration on the structure of the equipment, or on adequate vibration isolation supports. Openings, if any, must be as small as possible. Ventilation or material access should normally be through proper acoustically lined ducts or channels.

Invariably, all but the simplest problems require combination solutions, involving several or all of the alternatives discussed. When the required noise reduction is determined, the lowest cost, simplest method likely to provide the necessary attenuation should be chosen; further steps should follow until the desired improvement is achieved.

36.4.5 Materials and Systems

Noise and vibration control is a process involving various methods and approaches to controlling noise generation and transmission. Materials and systems used in the process (often erroneously called "acoustical materials") are only elements in the process, and must be chosen for their particular contribution to the economical solution to the particular problem. There are no generic acoustical materials, and rarely are there "standard" solutions to noise and vibration control problems.

Acoustical Absorbents

Acoustical absorbents are normally lightweight, porous, usually fibrous or foamed matrices, which provide optimum impedance and carefully controlled air flow resistance within their structure as the oscillating pressure of the impinging sound wave in the air reaches them. They can only absorb sound that reaches them, and they do not attract sound to themselves in their normal usage. The oscillation of the air entrapped within their porous structure is a somewhat viscous, "lossy" movement, converting some of the acoustic energy to heat.

Typical absorbents are glass fiber or mineral wool batts, blankets, or boards, of various densities and thickness; urethane foams, either closed-cell or reticulated; or matrices of porous particles bound together with cements or binders which permit their structures to be readily permeable to air flow.

Their absorptivity normally ranges from about 50% to 90% or more of the sound reaching them, varying with the frequency of the sound and the density and thickness of the absorbent. Absorptivity is rated in *absorption coefficients* (the decimal fraction of the energy absorbed at a particular frequency or frequency range, normally from 125 to 4000 Hz), or in *noise reduction coefficient* (NRC)—the arithmetic average of the absorption coefficients in the four octave bands from 250 Hz (center frequency) to 2000 Hz—the so-called typical noise frequencies. Standardized test procedures in recognized laboratories produce values that are widely published in various handbooks and in the advertising literature of materials manufacturers.

Used to line interior surfaces of enclosures (or rooms), absorbents reduce the energy that would otherwise be reflected back into the enclosure to build up the energy level within the enclosure (often well in excess of the level that might otherwise exist in free, unenclosed space). In typical small enclosures, even very efficient absorbents (NRC of 0.90 or more) can provide no more than about 10 dBA reduction. As barriers, their lightweight, porous structure is particularly ineffective, and sound transmits readily directly through them.

Because they are soft and fragile, and prone to absorb fluids and gases, they are frequently protected by perforated facings or thin, flexible plastic films. Fire resistance, cleanability, maintainability, and durability requirements often dictate the type and location of absorbents.

The use of absorbents in lined ducts or tubes, or within carefully "tuned" cavities to produce mufflers is common practice. Where absorptive material is not feasible, a tuned "reactive" chamber, utilizing the Helmholz resonator principle, is also a practical type of muffler. Air and gas discharges through such mufflers or through porous, sintered metal or ceramic nozzles can provide considerable attenuation. In general, mufflers can provide from 5 to 30 dBA noise reduction if properly chosen and used.

Vibration Isolation and Damping

"Floating" noise or vibration sources on resilient supports provides a significant impedance mismatch, with inefficient energy transport between source and support. An impedance mismatch of more than 100/1 is quite feasible by using rubber, for example, between two steel parts, or between steel and concrete (see Table 36.4.2). Energy is reflected back to the source, or stored in the resilient mount and returned to the source with each vibration. If the source cannot "drive" attached surfaces or structures, it cannot transmit acoustic energy into those elements.

Rubber and dense fiberglass blocks, steel springs, and similar resilient materials or devices are commonly used as vibration isolation mounts to support noisy equipment or to isolate vibration or noise sources. From 10 dBA to more than 40 dBA isolation can be provided by standard, readily available devices; and even higher isolation can be obtained from special devices, such as the "air bag" assemblies used to isolate against very-low frequencies.

Isolation efficiency depends on the static deflection of the mounts under the dead load of the supported equipment, the driving frequency of the supported equipment, and the amount of damping (if any) inherent in the resilient mount or material. Isolation from 70 to 99% of the driving force is feasible, with static deflections normally ranging from about 0.10 in. to as much as 9 in. (0.25–23 cm).

Generally, the natural frequency of the resiliently supported system is determined by the static deflection of the mounts, calculated thus:

$$f = 3.13\sqrt{\frac{1}{d}}$$

where: f = frequency, in Hz (cps)
d = static deflection, in in.

or:

$$f = 5\sqrt{\frac{1}{d}}$$

where d = static deflection, in cm.

The natural frequency of isolation mounting systems is normally designed for about $\frac{1}{5}$ to $\frac{1}{3}$ of the lowest driving frequency of the supported vibration source or the lowest frequency of any moving element in the supported equipment.

Damping is a means of reducing the "ring" and reverberation from vibrating parts that radiate sound to the surroundings. It can be accomplished in several ways; for example, inherently highly damped or "lossy" materials (such as plastics) can be substituted for steel in panels, gears, or rollers. Or vibrating panels can be stiffened to move their vibrational modes to higher or less objectionable frequencies where it is easier to damp the vibration.

Damping compounds—viscoelastic sheet or mastic—are forms of "lossy" compounds with high internal viscosity or friction, which heat up internally when flexed or stretched, converting vibrational energy to heat. They are intimately applied to the areas of vibrating surfaces where the amplitude of vibration is maximum; and normally they are applied in a thickness approximately equal to the thickness of the part being damped. Restrained layer damping—viscoelastic sheet sandwiched between two relatively rigid or stiff sheets, such as steel or aluminum or aluminum foil—is a particularly effective approach.

Damping is relatively effective at high frequencies—above 200 Hz—providing as much as 3–10 dBA reduction in radiated noise; but it is normally of little help at lower frequencies.

Enclosures and Barriers

Interrupting the transmission path of airborne sound by enclosing the source or erecting a shield or barrier between the source and the receiver is often an effective noise control approach. As mentioned earlier, air-tight septa—whether panels or sheets, rigid or flexible—prevent much of the energy impinging upon them from passing through them, reflecting it back to the source. When enclosing the source is not feasible, shielding or enclosing the receiver (operator or worker) in a partial or complete room or booth may be practical.

Barriers are made of plywood, steel, aluminum, plastic, loaded vinyl fabrics, glass, and a host of materials common to building construction. To be effective sound barriers, their surface density must be high enough to reflect *most* of the sound impinging upon them (99.9% or more). Thus porous, lightweight absorbents are quite ineffective, and their use must be restricted to absorbing sound within the enclosure or preventing reflections from the barrier or shield to other critical areas.

Simple, straight barriers—partitions, panels, or curtains—must be at least 8 ft (2.5 m) high and 12 ft (4 m) long to provide 4–7 dBA attenuation between a receiver and a relatively small, concentrated point noise source, and they must cut off line-of-sight between the source and receiver. Folded into a three-sided, partial booth, this same amount of material around either the source or the receiver can provide 5–10 dBA attenuation in most instances. Usually the side of the panels facing the noise source or the inside of a partial booth surrounding the source should be covered with a good absorbent, providing at least 0.65 NRC. In the case of a personnel booth, the absorbent should be inside the booth, where the operator is located.

Where vision through the panel is required, areas of $\frac{1}{4}$-in. laminated glass or acrylic sheets can be used as viewing areas.

Closed boxes, booths, or rooms can provide up to 40 dBA attenuation if relatively air tight and without penetrations. When openings are required for access, entry or exit of materials, or for ventilation, the attenuation may be sharply reduced. Properly designed, acoustically lined "tunnels" or "mazes," where feasible, can minimize sound transmission through such penetrations and restore the effectiveness of the enclosures.

Combination or Integrated Solutions

Rarely is noise control in materials handling systems accomplished by the use of a single material or approach. Usually several materials or combination materials or systems are used; an integrated approach is required. As mentioned earlier, substituting processes or materials should be investigated first to minimize noise or vibration at the source. Then the simplest, most economical means of providing the maximum protection for operators and exposed personnel must be determined, using available materials and products, when acceptable in the particular environment.

Figure 36.4.1 indicates an integrated approach used to solve the problem of excessive noise in a

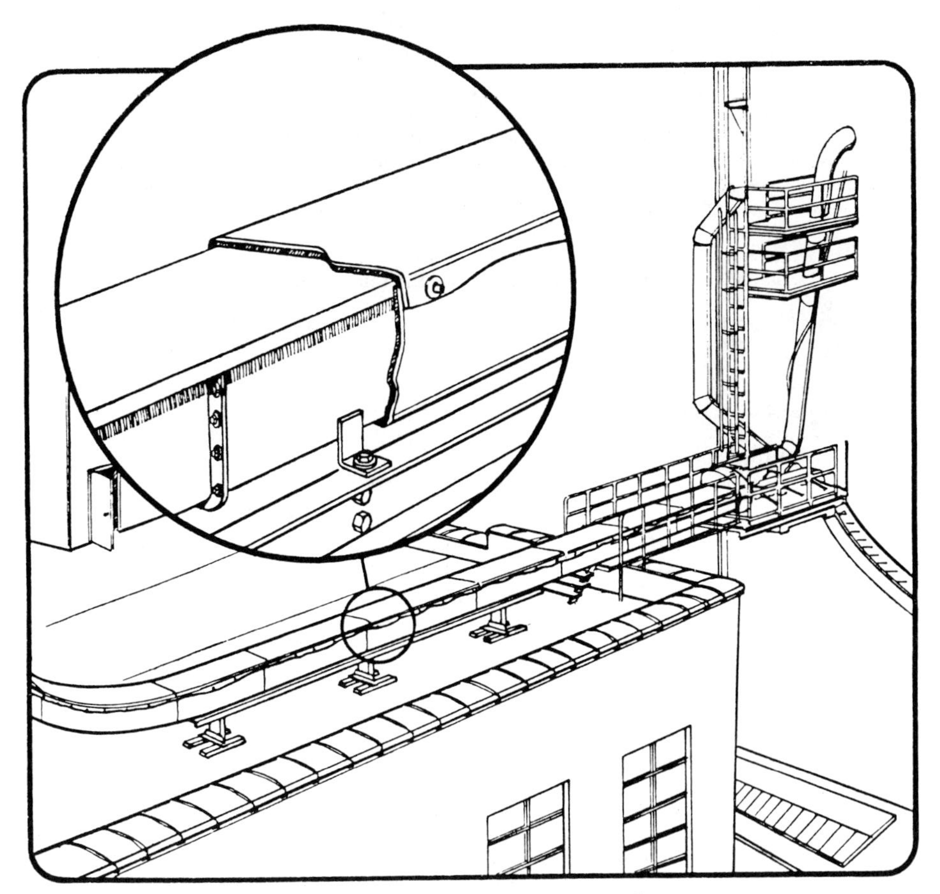

Fig. 36.4.1 Noise control in conveyor lines.

large, high-capacity conveyor system used to move granular material throughout a large chemical plant. Noise levels up to 110 dBA produced by material moved by buckets and screw conveyors through steel pipes and tubes exposed employees in many areas of the plant. Ready access to the conveyors was imperative (to clear jams, sample the stock, and maintain the equipment).

The conveyors were wrapped with a composite of urethane foam and leaded vinyl sheet, with a snap-on fastening arrangement to facilitate access to any point in any line. Noise level reductions of almost 30 dBA were effected by the combination of damping, absorption, and flexible barrier material. As a result, a significant noise problem, affecting many workers, was eliminated.

36.4.6 Sound in Rooms and Enclosures

Since materials handling systems are usually installed in rooms or enclosed plants, the effect of the enclosure may be important. Workers exposed to the direct sound field of the source—usually 3 ft (1 m) to 10 ft (3 m) from the source—are relatively unaffected by the reflections or reverberation of the room (unless they are very near a large, reflective surface, such as a wall or partition).

As might be expected, sound contained within an enclosure is not free to propagate indefinitely and disperse freely. The energy that reflects from enclosing surfaces is added to the sound field of the source and affects it according to this equation:

$$\text{SPL} = \text{PWL} + 10 \log_{10}\left[\frac{Q}{4\pi r^2} + \frac{4}{R}\right] + 10.5 \text{ dB}$$

where: SPL = sound pressure level, in dB
PWL = sound power level, in dB
r = dimensions, in ft

$$R = \frac{\alpha S}{1 - \alpha}$$

S = total area of the room surfaces
α = average absorption coefficient of the surfaces
Q = directionality factor of the sound source; usually 2 for sources on a flat plane

Note: when dimensions are in meters, the last figure (10.5 dB) changes to 0.2 dB All of the terms following PWL in the equation can be lumped into a quantity sometimes called the "room effect" (actually, the distance plus room effects). For large rooms, and reasonable distances from the sound source, this room effect is a negative quantity, and the Sound Pressure Level (SPL) is lower than the Sound Power Level (PWL, usually available from manufacturers of the equipment or obtained by measurement) by several dB. For typical industrial spaces, the room effect can be taken from Table 36.4.4.

Acoustically absorptive surfaces in a room, unless very near the operator or the source, normally can reduce levels only about 0–5 dBA; distance from the source attenuates the sound significantly, as is apparent.

Table 36.4.4 Room Effect, in dB[a]

Distance from Source		Hard, Reverberant Spaces	Soft, Absorptive Spaces
ft	(m)		
10	(3)	12 dB	17 dB
20	(6)	15	22
30	(9)	17	25
40	(12)	18	27
50	(15)	19	29
100	(30)	20	34

[a] Never use this table for calculation of small enclosures (such as machine enclosures), machine housings, fan housings, or similar spaces. It is intended only as a quick, reasonably accurate means of forecasting the SPL in a work space if machine or source Sound Power is known.

Never use this table for calculations where the directivity (Q) of the source is high. A horn or any peculiarly directional source will focus the sound sharply.

In the near field of a very large sound source, there may be little attenuation within the first 10 ft (3 m) from the surface of a source, even in a totally absorbent field.

Small enclosures (as contrasted to rooms or factories) contain the sound within the direct sound field of the source or machine. They are made of tight, dense, heavy panels or surfaces to prevent sound transmission outside the enclosure. The effectiveness of enclosing panels or surfaces in stopping sound transmission through them is called their Sound Transmission Loss—that is, the ratio of the energy impinging on them to the energy radiated from the opposite side, expressed in dB. Since the effectiveness of panels in containing sound varies with the frequency of sound (they are much less effective at low frequencies than high), a sort of empirical average figure is used to describe their performance. The STC (Sound Transmission Class) is approximately the average loss through the panels in the frequency range from 125 to 4000 Hz. For practical purposes, a reasonable rule of thumb for calculating the sound isolation performance of a panel is:

$$\mathrm{SPL}_{\text{quiet side}} = \mathrm{SPL}_{\text{source side}} - \text{STC of panel}$$

expressed in dBA.

However, if a source is completely enclosed within a relatively small, tight enclosure (as machines are frequently enclosed), the radiated sound energy must either escape or build up to where the escaping energy just equals the generated energy. Within a hard, reflective enclosure, even when built of heavy, tight panels or surfaces, the Sound Pressure Level at a machine may be much *higher* than it would be if there were no enclosure. An operator within the enclosure would be exposed to higher levels than if there were no enclosure. Further, however good the sound attenuation (STC or Sound Transmission Loss) of the enclosure, it will accomplish little shielding if there is no acoustical absorption within it.

Applying acoustical absorbents to the *interior* surfaces of the enclosure will cause some of the energy within it to be absorbed, and the noise level within the enclosure will build up only by the amount not absorbed. For example, if the average absorptivity of all surfaces within an enclosure were only 10%, the level inside could build up by 10 dB; if average absorptivity were 50%, the buildup would be only 3 dB; and if 90%, the buildup would be less than 1 dB.

As mentioned earlier, any opening, open joint, or leak is a major sound transmission path. One square *inch* of an opening transmits as much sound as 100 square *feet* of a partition which provides 40 dB attenuation.

36.4.7 Choosing the Noise Control Option

Materials and systems used in noise control cover a wide range of products, devices, and systems, of widely differing performance and effectiveness. Products and data change frequently, and should always be chosen from the most current data available from the manufacturer. The claims and rhetoric of advertisements should always be investigated in detail, and the product(s) or system(s) should be chosen for their applicability to the particular problem and the specific environment of the installation.

It is wise, before attempting to choose a noise control approach, to weigh the potential costs against the possible benefits of the procedure. This usually means asking:

1. How many workers are involved?
2. Are workers concentrated in a small area or widely distributed throughout the plant?
3. Are the noise sources concentrated or distributed?
4. Which approach will protect the most workers?
5. Is it simpler to isolate the noise source or the operators?
6. Which approach will provide the greatest noise reduction at the lowest cost; that is, how many dollars per dB are involved? How many dollars per worker?

Table 36.4.5 Noise Control Cost/Benefit Analysis[a]

Control Option	Attenuation dBA	Cost[b] per sq ft, or each	Productivity Penalty
Absorption	3–5	$.50–2.00	None
Damping	3–10	$.20–4.00	None
Barriers	5–15	$3.00–8.00	Up to 15%
Machine enclosures	5–50	$5.00–25.00	Up to 25%
Worker shelters	5–25	$500–3500 each	None

[a] These data are feasible ranges of performance, typical cost ranges, and productivity penalty data based on actual on-the-job case studies in the United States.

[b] Multiply the cost/ft^2 by 11 to obtain cost/m^2

7. Can the option(s) chosen actually provide the reduction required?
8. What productivity penalty will result?

Table 36.4.5 will provide guidance in answering these questions.

BIBLIOGRAPHY

Beranek, L. L., *Noise and Vibration Control,* McGraw Hill, New York, 1971.

Harris, C. M., *Handbook of Noise Control,* 2nd ed., McGraw Hill, New York, 1979.

Yerges, L. F., *Sound, Noise and Vibration Control,* 2nd ed., Van Nostrand Reinhold, New York, 1978.

APPENDIX A
METRIC CONVERSION

RALPH M. COX

Industrial Handling Engineers
Houston, Texas

A.1 INTRODUCTION

This section covers conversion factors between the English and metric (International System of Units, or SI) systems of measurement. The conversion factors included have been specifically selected for use by engineers and managers involved with materials handling. Please note that English–English and metric–metric conversion factors have not been included. They are readily available in many reference texts.

All entries have been made in their conventional and reciprocal forms. As a further aid to materials handling personnel, factors involving a "double" conversion have been included, as in 1 foot/minute = 0.00508 meters/second or 1 meter/second = 196.8 feet/minute. In addition, in an effort to promote familiarity with both systems, typical quantities and values for the English system are expressed in the metric system, as in 50 pounds = 22.68 kilograms.

A.2 CONVERSIONS

Area

1 square inch = 6.452 square centimeters
1 square inch = 645.2 square millimeters
1 square foot = 0.0929 square meters
i.e., 50 square feet = 4.65 square meters
1,000 square feet = 92.9 square meters
50,000 square feet = 4,645 square meters
100,000 square feet = 9,290 square meters
250,000 square feet = 23,225 square meters
1 acre = 4,047 square meters
1 square mile = 2.590 square kilometers
1 square centimeter = 0.155 square inches
1 square meter = 1,550 square inches
1 square meter = 10.76 square feet
1 square kilometer = 247.1 acres
1 square kilometer = 10,760,000 square feet
1 square kilometer = 0.3861 square miles

Bulk Density

1 pound/cubic foot = 16.02 kilograms/cubic meter
i.e., 10 pounds/cubic foot = 160.2 kilograms/cubic meter
30 pounds/cubic foot = 480.6 kilograms/cubic meter
60 pounds/cubic foot = 961.2 kilograms/cubic meter
90 pounds/cubic foot = 1,441.8 kilograms/cubic meter
120 pounds/cubic foot = 1,922.4 kilograms/cubic meter
1 kilogram/cubic meter = 0.06243 pounds/cubic feet

Dry Volume

1 cubic inch = 16.39 cubic centimeters
1 cubic inch = 0.01639 liters
1 cubic foot = 28.32 liters
1 cubic foot = 0.02832 cubic meters
i.e., 64 cubic feet = 1.81 cubic meters
100 cubic feet = 2.83 cubic meters
1000 cubic feet = 28.3 cubic meters
2560 cubic feet = 72.49 cubic meters
5000 cubic feet = 141.6 cubic meters
1 bushel = 0.03524 cubic meters
1 cubic centimeter = 0.0610 cubic inches
1 liter = 61.02 cubic inches
1 liter = 0.03531 cubic feet
1 cubic meter = 35.31 cubic feet
1 cubic meter = 28.38 bushels

Dry Volumetric Flow

1 cubic foot/minute = 0.472 liters/second
1 cubic foot/minute = 1.6992 cubic meters/hour
i.e., 5 cubic feet/minute = 8.50 cubic meters/hour
25 cubic feet/minute = 42.48 cubic meters/hour
50 cubic feet/minute = 84.96 cubic meters/hour
75 cubic feet/minute = 127.44 cubic meters/hour
100 cubic feet/minute = 169.92 cubic meters/hour
1 liter/second = 2.118 cubic feet/minute
1 cubic meter/hour = 0.5885 cubic feet/minute

Length

1 inch = 25.4 millimeters
i.e., 12 inches = 304.8 millimeters
36 inches = 914.4 millimeters
40 inches = 1016 millimeters
48 inches = 1219.2 millimeters
72 inches = 1828.8 millimeters
1 foot = 0.3048 meters
i.e., 10 feet = 3.05 meters
250 feet = 76.2 meters
500 feet = 152.4 meters
750 feet = 228.6 meters
1000 feet = 304.8 meters
1 mile = 1.609 kilometers
1 millimeter = 0.03937 inches
1 centimeter = 0.3937 inches
1 meter = 3.281 feet
1 kilometer = 0.6214 mile

Liquid Volume

1 gallon (U.S.) = 3.785 liters
1 gallon (U.S.) = 0.003785 cubic meters
1 liter = 0.2642 gallons (U.S.)
1 cubic meter = 264.2 gallons (U.S.)

Power

1 horsepower = 0.7457 kilowatts
i.e., 2 horsepower = 1.49 kilowatts
5 horsepower = 3.73 kilowatts
10 horsepower = 7.46 kilowatts
25 horsepower = 18.64 kilowatts
50 horsepower = 37.29 kilowatts
1 kilowatt = 1.341 horsepower

Pressure

1 pound/square inch = 6.895 kilopascals
i.e., 5 pounds/square inch = 34.48 kilopascals
10 pounds/square inch = 68.95 kilopascals
50 pounds/square inch = 344.8 kilopascals
125 pounds/square inch = 861.88 kilopascals
1000 pounds/square inch = 6895 kilopascals
1 kilopascal = 0.145 pounds/square inch

Temperature

[Temperature (°F) − 32] × $\frac{5}{9}$ = Temperature (°C)

i.e., 0°F = −17.78°C
32°F = 0°C
50°F = 10°C
100°F = 37.78°C
200°F = 93.3°C

[Temperature (°C) + 17.78] × 1.8 = Temperature (°F)

Velocity

1 foot/minute = 0.00508 meters/second
i.e., 20 feet/minute = 0.10 meters/second
75 feet/minute = 0.38 meters/second
125 feet/minute = 0.64 meters/second
750 feet/minute = 3.81 meters/second
3000 feet/minute = 15.24 meters/second
1 mile/hour = 1.6093 kilometers/hour
i.e., 2 miles/hour = 3.22 kilometers/hour
5 miles/hour = 8.05 kilometers/hour
25 miles/hour = 40.23 kilometers/hour
50 miles/hour = 80.45 kilometers/hour
100 miles/hour = 160.9 kilometers/hour
1 foot/minute = 0.01829 kilometers/hour
1 meter/second = 196.8 feet/minute
1 kilometer/hour = 0.6214 miles/hour
1 kilometer/hour = 54.68 feet/minute

Weight

1 ounce = 28.3495 grams
1 pound = 453.5924 grams
1 pound = 0.4536 kilograms

i.e., 10 pounds = 4.54 kilograms
50 pounds = 22.68 kilograms
100 pounds = 45.36 kilograms
1000 pounds = 453.6 kilograms
40,000 pounds = 18,144 kilograms

1 ton (short) = 0.91 metric tonnes
1 ton (short) = 907.18486 kilograms
1 gram = 0.03527 ounces
1 gram = 0.0022 pounds
1 kilogram = 2.2046 pounds
1 kilogram = 0.001102 tons (short)

Weight Flow

1 pound/minute = 0.4536 kilograms/minute
i.e., 5 pounds/minute = 2.27 kilograms/minute
10 pounds/minute = 4.54 kilograms/minute
25 pounds/minute = 11.34 kilograms/minute
50 pounds/minute = 22.68 kilograms/minute
100 pounds/minute = 45.36 kilograms/minute

1 pound/hour = 0.4536 kilograms/hour
i.e., 100 pounds/hour = 45.36 kilograms/hour
2,500 pounds/hour = 1,134 kilograms/hour
10,000 pounds/hour = 4,536 kilograms/hour
50,000 pounds/hour = 2,268 kilograms/hour
100,000 pounds/hour = 44,536 kilograms/hour
1 ton/hour = 907.18486 kilograms/hour
i.e., 5 tons/hour = 4,535.92 kilograms/hour
10 tons/hour = 9,071.85 kilograms/hour
25 tons/hour = 22,679.62 kilograms/hour
50 tons/hour = 45,359.24 kilograms/hour
100 tons/hour = 90,718.5 kilograms/hour
1,000,000 pounds/year = 453,600 kilograms/year
i.e., 10,000,000 pounds/year = 4,536,000 kilograms/year
50,000,000 pounds/year = 22,680,000 kilograms/year
100,000,000 pounds/year = 45,360,000 kilograms/year
500,000,000 pounds/year = 226,800,000 kilograms/year
1,000,000,000 pounds/year = 453,600,000 kilograms/year
1 ton/year = 907.18486 kilograms/year
i.e., 5,000 tons/year = 4,535,924.3 kilograms/year
25,000 tons/year = 22,679,621 kilograms/year
100,000 tons/year = 90,718,486 kilograms/year
250,000 tons/year = 226,796,210 kilograms/year
500,000 tons/year = 453,592,430 kilograms/year
1 kilogram/minute = 2.2046 pounds/minute
1 kilogram/hour = 2.2046 pounds/hour
1 kilogram/hour = 0.001102 tons (short)/hour
1,000,000 kilograms/year = 2,204,600 pounds/year
1 ton (short)/year = 907.18486 kilograms/year

A.3 FURTHER CONSIDERATIONS

Commonly used conversion factors are shown in Table A.1. Each conversion factor is listed as a number greater than one and less than 10, with six or fewer decimal places. This number is followed by the letter E (for exponent), a plus or minus symbol, and two digits indicating the power of 10 by which the number must be multiplied to obtain the correct value.

Thus, to convert from inches to meters, multiply by 2.54×10^{-2}. The asterisk in the table indicates that the conversion factor is exact, and all subsequent digits are zero. Therefore, 1 inch = 0.0254 meters (exactly).

Changing from U.S. to metric units does not just involve applying a conversion factor. In some cases, it represents a change in the magnitudes of numbers in which engineers commonly think. An example is provided in Table A.2, which compares conveyor belt velocities in U.S. and SI terms. Another comparison, involving belt tensions, is shown in Table A.3. Finally, Table A.4 provides a handy summary of common conversion formulas.

Table A.1 Commonly Used Conversion Factors

Quantity	Conversion	Factor[a]
Plane angle	degree *to* rad	1.745 329 E − 02
Length	in *to* m	2.54* E − 02
	ft *to* m	3.048* E − 01
	mile *to* m	1.609 344* E + 03
Area	in^2 *to* m^2	6.451 600* E − 04
	ft^2 *to* m^2	9.290 304* E − 02
Volume	ft^3 *to* m^3	2.831 685 E − 02
	U.S. gallon *to* m^3	3.785 412 E − 03
	in^3 *to* m^3	1.638 706 E − 05
	oz (fluid, U.S.) *to* m^3	2.957 353 E − 05
	liter *to* m^3	1.000 000 E − 03
Velocity	ft/min *to* m/s	5.08* E − 03
	ft/sec *to* m/s	3.048* E − 01
	km/h *to* m/s	2.777 778 E − 01
	mile/h *to* m/s	4.470 4* E − 01
	mile/h *to* km/h	1.609 344* E + 00
Mass	oz (avoir) *to* kg	2.834 952 E − 02
	lb (avoir) *to* kg	4.535 924 E − 01
	slug *to* kg	1.459 390 E + 01
Acceleration	ft/s^2 *to* m/s^2	3.048* E − 01
Force	kgf *to* N	9.806 65* E + 00
	lbf *to* N	4.448 222 E + 00
	poundal *to* N	1.382 550 E − 01
Bending, torque	kgf-m *to* N · m	9.806 65* E + 00
	lbf-in *to* N · m	1.129 848 E − 01
	lbf-ft *to* N · m	1.355 818 E + 00
Pressure, stress	kgf/m^2 *to* Pa	9.806 65* E + 00
	$poundal/ft^2$ *to* Pa	1.488 164 E + 00
	lbf/ft^2 *to* Pa	4.788 026 E + 01
	lbf/in^2 *to* Pa	6.894 757 E + 03
Energy, work	Btu (IT) *to* J	1.055 056 E + 03
	Calorie (IT) *to* J	4.186 8* E + 00
	ft-lbf *to* J	1.355 818 E + 00
Power	hp (550 ft-lbf/s) *to* W	7.456 999 E + 02
Temperature	°C *to* K	$t_K = t_C + 273.15$
	°F *to* K	$t_K = (t_F + 459.67)/1.8$
	°F *to* °C	$t_C = (t_F - 32)/1.8$
Temperature interval	°C *to* K	1.0* E + 00
	°F *to* K *or* °C	5.555 556 E − 01

Source: From ASME Guide SI-1, Orientation and Guide for Use of SI (Metric) Units.
[a] * = exact relationship in terms of base units.

Table A.2 Comparison of Belt Velocities

m/sec	ft/min	m/sec	ft/min	m/sec	ft/min	m/sec	ft/min
1.00	197	2.00	394	3.00	591	4.50	886
1.25	246	2.25	443	3.25	640	5.00	984
1.50	295	2.50	492	3.50	689	5.50	1083
1.75	344	2.75	541	4.00	787	6.00	1181

Table A.3 Comparison of Belt Tensions

N	lbf	N	lbf	N	lbf	N	lbf
1,000	225	10,000	2,248	50,000	11,240	200,000	44,962
2,000	450	15,000	3,372	60,000	13,489	250,000	56,202
3,000	674	20,000	4,496	70,000	15,737	300,000	67,443
4,000	899	25,000	5,620	80,000	17,985	350,000	78,683
5,000	1,124	30,000	6,744	90,000	20,233	400,000	89,924
6,000	1,349	35,000	7,868	100,000	22,481	450,000	101,164
8,000	1,798	40,000	8,992	150,000	33,721	500,000	112,404

Table A.4 Frequently Used Conversion Formulas

Pound-force, lbf	× 4.4482	= N
Mass, lb	× 0.4536	= kg
Length, ft	× 0.3048	= m
Velocity, ft/min	× 0.0051	= m/sec
Mass per length, lb/ft	× 1.4882	= kg/m
Acceleration, ft/sec^2	× 0.3048	= m/sec^2
Area, ft^2	× 0.0929	= m^2
Volume, ft^3	× 0.0283	= m^3
Horsepower, Hp (U.S.)	× 745.7	= W (Watts)

APPENDIX B
SOURCES OF INFORMATION ON MATERIALS HANDLING

Catalogs and Directories

Material Handling Engineering Handbook and Directory, Penton/IPC, Inc., 614 Superior Ave., West, Cleveland, OH 44113.

MHI Literature Catalog, The Material Handling Institute, Inc., 1326 Freeport Rd., Pittsburgh, PA 15238.

Modern Materials Handling Casebook/Directory, Cahners Publishing Co., 221 Columbus Ave., Boston, MA 02116.

Plant Engineering—Directory and Specifications Catalog, Technical Publishing Co., a company of The Dun & Bradstreet Corp., 1301 S. Grove Ave., Barrington, IL 60010.

General Literature

Lesson Guide Outline, College–Industry Council on Material Handling Education, 1326 Freeport Rd., Pittsburgh, PA 15238.

Material Handling Review and Certification Guide, International Material Management Society, 3900 Capital City Blvd., Suite 103, Lansing, MI 48906.

Plant Engineering—Material Handling Library, Technical Publishing, a company of The Dun & Bradstreet Corp., 1301 S. Grove Ave., Barrington, IL 60010.

Business and Trade Magazines

Distribution/Warehouse Cost Digest, Marketing Publications, Inc., 217 National Press Building, Washington, DC 20045.

Engineer's Digest, Walker–Davis Publications, Inc., 2500 Office Center, Willow Grove, PA 19090.

Handling & Shipping, Penton/IPC, Inc., 614 Superior Ave., West, Cleveland, OH 44113.

Industrial Engineering, Institute of Industrial Engineers, 25 Technology Park/Atlanta, Norcross, GA 30071.

Industrial Equipment News, Thomas Publishing Co., One Penn Plaza, 250 W. 34th St., New York, NY 10001.

Industry Week, Penton/IPC, Inc., 614 Superior Ave., West, Cleveland, OH 44113.

Material Handling Engineering, Penton/IPC, Inc., 614 Superior Ave., West, Cleveland, OH 44113.

Material Handling Product News, Gordon Publications, Inc., 20 Community Place, Morristown, NJ 07960.

Materials Management and Distribution, Maclean–Hunter Publishing Co., 481 University Ave., Toronto, M5W 1A7, Ont., Canada.

Modern Materials Handling, Cahners Publishing Co., 221 Columbus Ave., Boston, MA 02116.

New Equipment Digest, Penton/IPC, Inc., 614 Superior Ave., West, Cleveland, OH 44113.

Plant Engineering, Technical Publishing, a company of The Dun & Bradstreet Corporation, 1301 S. Grove Ave., Barrington, IL 60010.

Production Engineering, Penton/IPC, Inc., 614 Superior Ave., West, Cleveland, OH 44113.

Traffic Management, Cahners Publishing Co., 221 Columbus Ave., Boston, MA 02116.

Technical Societies and Industry Associations

American Institute of Plant Engineers, 3975 Erie Ave., Cincinnati, OH 45208.
American Society of Mechanical Engineers, 345 E. 47th St., New York, NY 10017.
American Production and Inventory Control Society, Watergate Office Bldg., Suite 504, 2600 Virginia Ave., N.W., Washington, DC 20037.
Caster and Floor Truck Manufacturers Association, 3525 W. Peterson Ave., Chicago, IL 60645.
Conveyor Equipment Manufacturers Association, 1000 Vermont Ave., N.W., Washington, DC 20005.
The Industrial Truck Association, Suite 210, 1750 K St. NW, Washington, DC 20006.
Institute of Electrical and Electronic Engineers, 345 East 47th St., New York, NY 10017.
Institute of Industrial Engineers, 25 Technology Park/Atlanta, Norcross, GA 30092.
International Material Management Society, 3900 Capitol City Blvd., Suite 103, Lansing, MI 48906
Material Handling Equipment Distributors Association, 201 Rt. 45, Vernon Hills, IL 60061
The Material Handling Institute, Inc., 1326 Freeport Rd., Pittsburgh, PA 15238. Affiliated associations:

Automatic Identification Manufacturers
Crane Manufacturers Association of America, Inc.
Hoist Manufacturers Institute
Monorail Manufacturers Association, Inc.
Rack Manufacturers Institute

National Wooden Pallet and Container Association, 1619 Massachusetts Ave., N.W., Washington, DC 20036.
Packaging Institute, Inc., 342 Madison Ave., New York, NY 10017.
Society of Packaging and Handling Engineers, 14 E. Jackson St., Chicago, IL 60604.
Society of Manufacturing Engineers, 20501 Ford Rd., P.O. Box 930, Dearborn, MI 48128.
Robotic Industries Association, P.O. Box 1366, Dearborn, MI 48121.
Robotics International of SME, One SME Drive, P.O. Box 930, Dearborn, MI 48121.

INDEX